HANDBOOK OF
LASERS

The CRC Press
Laser and Optical Science and Technology Series
Editor-in-Chief: Marvin J. Weber

A.V. Dotsenko, L.B. Glebov, and V.A. Tsekhomsky
Physics and Chemistry of Photochromic Glasses

Andrei M. Efimov
Optical Constants of Inorganic Glasses

Alexander A. Kaminskii
Crystalline Lasers:
Physical Processes and Operating Schemes

Valentina F. Kokorina
Glasses for Infrared Optics

Sergei V. Nemilov
Thermodynamic and Kinetic Aspects
of the Vitreous State

Piotr A. Rodnyi
Physical Processes in Inorganic Scintillators

Michael C. Roggemann and Byron M. Welsh
Imaging Through Turbulence

Shigeo Shionoya and William M. Yen
Phosphor Handbook

Hiroyuki Yokoyama and Kikuo Ujihara
Spontaneous Emission and Laser Oscillation
in Microcavities

Marvin J. Weber, Editor
Handbook of Laser Science and Technology
Volume I: Lasers and Masers
Volume II: Gas Lasers
Volume III: Optical Materials, Part 1
Volume IV: Optical Materials, Part 2
Volume V: Optical Materials, Part 3
Supplement I: Lasers
Supplement II: Optical Materials

Marvin J. Weber, Editor
Handbook of Laser Wavelengths

HANDBOOK OF
LASERS

Marvin J. Weber, Ph.D.

Lawrence Berkeley National Laboratory
University of California
Berkeley, California

CRC Press
Taylor & Francis Group
Boca Raton London New York

CRC Press is an imprint of the
Taylor & Francis Group, an **informa** business

CRC Press
Taylor & Francis Group
6000 Broken Sound Parkway NW, Suite 300
Boca Raton, FL 33487-2742

First issued in paperback 2019

No claim to original U.S. Government works

ISBN 13: 978-0-367-45546-0 (pbk)
ISBN 13: 978-0-8493-3509-9 (hbk)

**Visit the Taylor & Francis Web site at
http://www.taylorandfrancis.com**

**and the CRC Press Web site at
http://www.crcpress.com**
Printed and bound by CPI Group (UK) Ltd, Croydon, CR0 4YY

Library of Congress Cataloging-in-Publication Data

Weber, Marvin J., 1932-
   Handbook of lasers / Marvin J. Weber.
      p.   cm.-- (CRC Press laser and optical science and technology series)
   Includes bibliographical references and index.
   ISBN 0-8493-3509-4 (alk. paper)
      1. Lasers--Handbooks, manuals, etc. I. Title. II. Series.

TA1683 .W44 2000
621.36′6--dc21

00-057894

Library of Congress Card Number 00-057894

Preface

Lasers continue to be an amazingly robust field of activity, one of continually expanding scientific and technological frontiers. Thus today we have lasing without inversion, quantum cascade lasers, lasing in strongly scattering media, lasing in biomaterials, lasing in photonic crystals, a single atom laser, speculation about black hole lasers, femtosecond-duration laser pulses only a few cycles long, lasers with subhertz linewidths, semiconductor lasers with predicted operating lifetimes of more than 100 years, peak powers in the petawatt regime and planned megajoule pulse lasers, sizes ranging from semiconductor lasers with dimensions of a few microns diameter and a few hundred atoms thick to huge glass lasers with hundreds of beams for inertial confinement fusion research, lasers costing from less than one dollar to more than one billion dollars, and a multibillion dollar per year market.

In addition, the nearly ubiquitous presence of lasers in our daily lives attests to the prolific growth of their utilization. The laser is at the heart of the revolution that is marrying photonic and electronic devices. In the past four decades, the laser has become an invaluable tool for mankind encompassing such diverse applications as science, engineering, communications, manufacturing and materials processing, medical therapeutics, entertainment and displays, data storage and processing, environmental sensing, military, energy, and metrology. It is difficult to imagine state-of-the-art research in physics, chemistry, biology, and medicine without the use of radiation from various laser systems.

Laser action occurs in all states of matter—solids, liquids, gases, and plasmas. Within each category of lasing medium there may be differences in the nature of the active lasing ion or center, the composition of the medium, and the excitation and operating techniques. For some lasers, the periodic table has been extensively explored and exploited; for others—solid-state lasers in particular—the compositional regime of hosts continues to expand. In the case of semiconductor lasers the ability to grow special structures one atomic layer at a time by liquid phase epitaxy, molecular beam epitaxy, and metal-organic chemical vapor deposition has led to numerous new structures and operating configurations, such as quantum wells and superlattices, and to a proliferation of new lasing wavelengths. Quantum cascade lasers are examples of laser materials by design.

The number and type of lasers and their wavelength coverage continue to expand. Anyone seeking a photon source is now confronted with an enormous number of possible lasers and laser wavelengths. The spectral output ranges of solid, liquid, and gas lasers are shown in Figure 1 and extend from the soft x-ray and extreme ultraviolet regions to millimeter wavelengths, thus overlapping masers. By using various frequency conversion techniques—harmonic generation, parametric oscillation, sum- and difference-frequency mixing, and Raman shifting—the wavelength of a given laser can be extended to longer and shorter wavelengths, thus enlarging its spectral coverage.

This volume seeks to provide a comprehensive, up-to-date compilation of lasers, their properties, and original references in a readily accessible form for laser scientists and engineers and for those contemplating the use of lasers. The compilation also indicates the state of knowledge and development in the field, provides a rapid means of obtaining reference data, is a pathway to the literature, contains data useful for comparison with predictions and/or to develop models of processes, and may reveal fundamental inconsistencies or conflicts in the data. It serves an archival function and as an indicator of newly emerging trends.

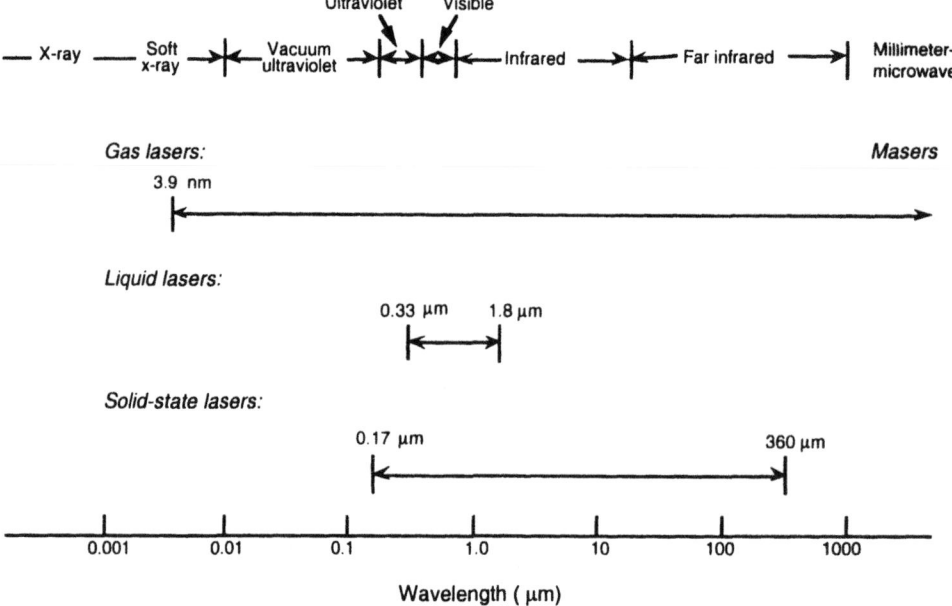

Figure 1 Reported ranges of output wavelengths for various laser media.

In this volume lasers are categorized based on their media—solids, liquids, and gases—with each category further subdivided as appropriate into distinctive laser types. Thus there are sections on crystalline paramagnetic ion lasers, glass lasers, polymer lasers, color center lasers, semiconductor lasers, liquid and solid-state dye lasers, inorganic liquid lasers, and neutral atom, ionized, and molecular gas lasers. A separate section on "other" lasers which have special operating configurations or properties includes x-ray lasers, free electron lasers, nuclear-pumped lasers, lasers in nature, and lasers without inversion. Brief descriptions of each type of laser are given followed by tables listing the lasing element or medium, host, lasing transition and wavelength, operating properties, and primary literature citations. Tuning ranges, when reported, are given for broadband lasers. The references are generally those of the initial report of laser action; no attempt is made to follow the often voluminous subsequent developments. For most types of lasers, lasing—light amplification by stimulated emission of radiation—includes, for completeness, not only operation in a resonant cavity but also single-pass gain or amplified spontaneous emission (ASE). Thus, for example, there is a section on amplification of core-valence luminescence.

Because laser performance is dependent on the operating configurations and experimental conditions used, output data are generally not included. The interested reader is advised to retrieve details of the structures and operating conditions from the original reference (in many cases information about the output and operating configuration is included in the title of the paper that is included in the references). Performance and background information about lasers in general and about specific types of lasers in particular can be obtained from the books and articles listed under Further Reading in each section.

An extended table of contents is provided from which the reader should be able to locate the section containing a laser of interest. Within each subsection, lasers are arranged according to the elements in the periodic table or alphabetically by materials, and may be further separated by operating technique (for example, in the case of semiconductor lasers, injection, optically pumped, or electron beam pumped).

This *Handbook of Lasers* is derived from data evaluated and compiled by the contributors to Volumes I and II and Supplement 1 of the *CRC Handbook Series of Laser Science and Technology* and to the *Handbook of Laser Wavelengths*. These contributors are identified in following pages. In most cases it was possible to update these tabulations to include more recent additions and new categories of lasers. For semiconductor lasers, where the lasing wavelength may not be a fundamental property but the result of material engineering and the operating configuration used, an effort was made to be representative with respect to operating configurations and modes rather than exhaustive in the coverage of the literature. The number of reported gas laser transitions is huge; they constitute nearly 80% of the over 16,000 laser wavelengths in this volume. Laser transitions in gases are well covered through the late 1980s in the above volumes. An electronic database of gas lasers prepared from the tables in Volume II and Supplement 1 by John Broad and Stephen Krog of the Joint Institute of Laboratory Astrophysics was used for this volume, but does not cover all recent developments.

Although there is a tremendous diversity of laser transitions and types, only a few laser systems have gained widespread use and commercial acceptance. In addition, some laser systems that were of substantial commercial interest in past years are becoming obsolete and are likely to be supplanted by other types in the future. Nevertheless, separate subsections on commercially available lasers are included thoroughout the volume to provide a perspective on the current state-of-the-art and performance boundaries.

To cope with the continued proliferation of acronyms, abbreviations, and initialisms which range from the clever and informative to the amusing or annoying, there is an appendix of acronyms, abbreviations, initialisms, and common names for lasers, laser materials, laser structures and operating configurations, and systems involving lasers. Other appendices contain information about laser safety, the ground state electron configurations of neutral atoms, and fundamental physical constants of interest to laser scientists and engineers.

Because lasers now cover such a large wavelength range and because researchers in various fields are accustomed to using different units, there is also a conversion table for spectroscopists (a Rosetta stone) on the inside back cover.

Finally, I wish to acknowledge the valuable assistance of the Advisory Board who reviewed the material, made suggestions regarding the contents and formats, and in several cases contributed material (the Board, however, is not responsible for the accuracy or thoroughness of the tabulations). Others who have been helpful include Guiuseppe Baldacchini, Eric Bründermann, Federico Capasso, Tao-Yuan Chang, Henry Freund, Claire Gmachl, Victor Granatstein, Eugene Haller, John Harreld, Stephen Harris, Thomas Hasenberg, Alan Heeger, Heonsu Jeon, Roger Macfarlane, George Miley, Linn Mollenauer, Michael Mumma, James Murray, Dale Partin, Maria Petra, Richard Powell, David Sliney, Jin-Joo Song, Andrew Stentz, Roger Stolen, and Riccardo Zucca. I am especially grateful to Project Editor Mimi Williams for her skill and help during the preparation of this volume.

Marvin J. Weber
Danville, California

General Reading

Bertolotti, M., *Masers and Lasers: An Historical Approach*, Hilger, Bristol (1983).

Davis, C. C., *Lasers and Electro-Optics: Fundamentals and Engineering*, Cambridge University Press, New York (1996).

Hecht, J., *The Laser Guidebook* (second edition), McGraw-Hill, New York (1992).

Hecht, J., *Understanding Lasers* (second edition), IEEE Press, New York (1994).

Hitz, C. B., Ewing, J. J. and Hecht, J., *Understanding Laser Technology,* IEEE Press, Piscataway, NJ (2000).

Meyers, R. A., Ed., *Encyclopedia of Lasers and Optical Technology*, Academic Press, San Diego (1991).

Milonni, P. W. and Eberly, J. H., *Lasers*, Wiley, New York (1988).

O'Shea, D. C., Callen, W. R. and Rhodes, W. T., *Introduction to Lasers and Their Applications*, Addison Wesley, Reading, MA (1977).

Siegman, A. E., *Lasers*, University Science, Mill Valley, CA (1986).

Silfvast, W. T., Ed., *Selected Papers on Fundamentals of Lasers*, SPIE Milestone Series, Vol. MS 70, SPIE Optical Engineering Press, Bellingham, WA (1993).

Silfvast, W. T., *Laser Fundamentals,* Cambridge University Press, Cambridge (1996).

Svelto, O., *Principles of Lasers*, Plenum, New York (1998).

Townes, C. H., *How the Laser Happened: Adventures of a Scientist*, Oxford University Press, New York (1999).

Verdeyen, J. T., *Laser Electronics*, 2nd edition, Prentice Hall, Englewood Cliffs, NJ (1989).

Yariv, A., *Quantum Electronics*, John Wiley & Sons, New York (1989).

The Author

Marvin John Weber received his education at the University of California, Berkeley, and was awarded the A.B., M.A., and Ph.D. degrees in physics. After graduation, Dr. Weber continued as a postdoctoral Research Associate and then joined the Research Division of the Raytheon Company where he was a Principal Scientist working in the areas of spectroscopy and quantum electronics. As Manager of Solid State Lasers, his group developed many new laser materials including rare-earth-doped yttrium orthoaluminate. While at Raytheon, he also discovered luminescence in bismuth germanate, a scintillator crystal widely used for the detection of high energy particles and radiation.

During 1966 to 1967, Dr. Weber was a Visiting Research Associate with Professor Arthur Schawlow's group in the Department of Physics, Stanford University.

In 1973, Dr. Weber joined the Laser Program at the Lawrence Livermore National Laboratory. As Head of Basic Materials Research and Assistant Program Leader, he was responsible for the physics and characterization of optical materials for high-power laser systems used in inertial confinement fusion research. From 1983 to 1985, he accepted a transfer assignment with the Office of Basic Energy Sciences of the U.S. Department of Energy in Washington, DC, where he was involved with planning for advanced synchrotron radiation facilities and for atomistic computer simulations of materials. Dr. Weber returned to the Chemistry and Materials Science Department at LLNL in 1986 and served as Associate Division Leader for condensed matter research and as spokesperson for the University of California/National Laboratories research facilities at the Stanford Synchrotron Radiation Laboratory. He retired from LLNL in 1993 and is presently a scientist in the Center for Functional Imaging of the Life Sciences Division at the Lawrence Berkeley National Laboratory.

Dr. Weber is Editor-in-Chief of the multi-volume *CRC Handbook Series of Laser Science and Technology*. He has also served as Regional Editor for the *Journal of Non-Crystalline Solids*, as Associate Editor for the *Journal of Luminescence* and the *Journal of Optical Materials*, and as a member of the International Editorial Advisory Boards of the Russian journals *Fizika i Khimiya Stekla* (Glass Physics and Chemistry) and *Kvantovaya Elektronika* (Quantum Electronics).

Among several honors he has received are an Industrial Research IR-100 Award for research and development of fluorophosphate laser glass, the George W. Morey Award of the American Ceramics Society for his basic studies of fluorescence, stimulated emission and the atomic structure of glass, and the International Conference on Luminescence Prize for his research on the dynamic processes affecting luminescence efficiency and the application of this knowledge to laser and scintillator materials.

Dr. Weber is a Fellow of the American Physical Society, the Optical Society of America, and the American Ceramics Society and has been a member of the Materials Research Society and the American Association for Crystal Growth.

Advisory Board

Contributors

William L. Austin
Lite Cycles, Inc.
Tucson, Arizona

Guiuseppe Baldacchini
ENEA - Frascati Research Center
Roma, Italy

Tasoltan T. Basiev
General Physics Institute
Moscow, Russia

William B. Bridges
Electrical Engineering and Applied Physics
California Institute of Technology
Pasadena, California

John T. Broad
Informed Access Systems, Inc.
Boulder, Colorado
(formerly of the Joint Institute of
Laboratory Astrophysics)

Eric Bründermann
Lawrence Berkeley National Laboratory
Berkeley, California

John A. Caird
Laser Program
Lawrence Livermore National Laboratory
Livermore California

Tao-Yuan Chang
AT&T Bell Laboratories
Holmdel, New Jersey

Connie Chang-Hasnain
Electrical Engineering/Computer Sciences
University of California
Berkeley, California

Stephen R. Chinn
Optical Information Systems, Inc.
Elmsford, New York

Paul D. Coleman
Department of Electrical Engineering
University of Illinois
Urbana, Illinois

William B. Colson
Department of Physics
Naval Postgraduate School
Monterey, California

Christopher C. Davis
Department of Electrical Engineering
University of Maryland
College Park, Maryland

Robert S. Davis
Department of Physics
University of Illinois at Chicago Circle
Chicago, Illinois

Bruce Dunn
Materials Science and Engineering
University of California
Los Angeles, California

J. Gary Eden
Department of Electrical Engineering/Physics
University of Illinois
Urbana, Illinois

Raymond C. Elton
Naval Research Laboratory
Washington, DC

Michael Ettenberg
RCA David Sarnoff Research Center
Princeton, New Jersey

Henry Freund
Science Applications International Corp.
McLean, Virginia

Claire Gmachl
Lucent Technologies
Murray Hill, New Jersey

Julius Goldhar
Department of Electrical Engineering
University of Maryland
College Park, Maryland

Victor L. Granatstein
Naval Research Laboratory
Washington, DC

Douglas W. Hall
Corning Inc.
Corning, New York

John Harreld
Materials Science and Engineering
University of California
Los Angeles, California

Thomas C. Hasenberg
University of Iowa
Iowa City, Iowa

Alexander A. Kaminskii
Institute of Crystallography
USSR Academy of Sciences
Moscow, Russia

David A. King
Ginzton Laboratory
Stanford University
Stanford, California

David J. E. Knight
DK Research
Twickenham, Middlesex, England
(formerly of National Physical Laboratory)

Henry Kressel
RCA David Sarnoff Research Center
Princeton, New Jersey

Stephen Krog
Joint Institute of Laboratory Astrophysics
Boulder, Colorado

William F. Krupke
Lawrence Livermore National Laboratory
Livermore, California

Chinlon Lin
AT&T Bell Laboratories and
   Bell Communications Research
Holmdel, New Jersey

Roger M. Macfarlane
IBM Almaden Laboratory
San Jose, California

Brian J. MacGowan
Lawrence Livermore National Laboratory
Livermore, California

Dennis L. Matthews
Lawrence Livermore National Laboratory
Livermore, California

David A. McArthur
Sandia National Laboratory
Albuquerque, New Mexico

George Miley
Department of Nuclear Engineering
University of Illinois
Urbana, Illinois

Linn F. Mollenauer
AT&T Bell Laboratories
Holmdel, New Jersey

James M. Moran
Radio and Geoastronomy Division
Harvard-Smithsonian Center for Astrophysics
Cambridge, Massachusetts

Peter F. Moulton
MIT Lincoln Laboratory
Lexington, Massachusetts

James T. Murray
Lite Cycles, Inc.
Tucson, Arizona

Joseph Nilsen
Lawrence Livermore National Laboratory
Livermore, California

Robert K. Parker
Naval Research Laboratory
Washington, DC

Dale Partin
Department of Physics
General Motors,
Warren, Michigan

Stephen Payne
Lawrence Livermore National Laboratory
Livermore, California

Alan B. Peterson
Spectra Physics, Inc.
Mountain View, California

Maria Petra
Department of Nuclear Engineering
University of Illinois
Urbana, Illinois

Clifford R. Pollock
School of Electrical Engineering
Cornell University
Ithaca, New York

Richard C. Powell
Optical Sciences Center
University of Arizona
Tucson, Arizona

Donald Prosnitz
Laser Program
Lawrence Livermore National Laboratory
Livermore, California

Charles K. Rhodes
Department of Physics
University of Illinois at Chicago Circle
Chicago, Illinois

Harold Samelson
Allied-Signal, Inc.
Morristown, New Jersey

Anthony E. Siegman
Department of Electrical Engineering
Stanford University
Stanford, California

William T. Silfvast
Center for Research and Education in
  Optics and Lasers
University of Central Florida
Orlando, Florida

David H. Sliney
U.S. Army Environmental Hygiene Agency
Aberdeen Proving Ground, Maryland

Jin-Joo Song
Center for Laser Research
Oklahoma State University
Stillwater, Oklahoma

Phillip A. Sprangle
Naval Research Laboratory
Washington, DC

Andrew Stentz
Lucent Technologies
Murray Hill, New Jersey

Richard N. Steppel
Exciton, Inc.
Dayton, Ohio

Stanley E. Stokowski
Lawrence Livermore National Laboratory
Livermore California

Rogers H. Stolen
AT&T Bell Laboratories
Holmdel, New Jersey

Henryk Temkin
AT&T Bell Laboratories
Murray Hill, New Jersey

Anne C. Tropper
Optoelectronic Research Centre
University of Southhampton
Highfield, Southhampton, England

Riccardo Zucca
Rockwell International Science Center
Thousand Oaks, California

Contents of previous volumes on lasers from the
CRC HANDBOOK OF LASER SCIENCE AND TECHNOLOGY

VOLUME I: LASERS AND MASERS

VOLUME II: GAS LASERS

SUPPLEMENT 1: LASERS

HANDBOOK OF LASER WAVELENGTHS

Marvin J. Weber

HANDBOOK OF LASERS

TABLE OF CONTENTS

APPENDICES

Section 1: Solid State Lasers

Section 1
SOLID STATE LASERS

1.0 Introduction

Solid state lasers include lasers based on paramagnetic ions, organic dye molecules, and color centers in crystalline or amorphous hosts. Semiconductor lasers are included in this section because they are a solid state device, although the nature of the active center—recombination of electrons and holes—is different from the dopants or defect centers used in other lasers in this category. Conjugated polymer lasers, solid-state excimer lasers, and fiber Raman, Brillouin, and soliton lasers are also covered in this section.

Reported ranges of output wavelengths for the various types of solid state lasers are shown in Figure 1.1. The differences in the ranges of spectral coverage arise in part from the dependence on host properties, in particular the range of transparency and the rate of non-radiative decay due to multiphonon processes.

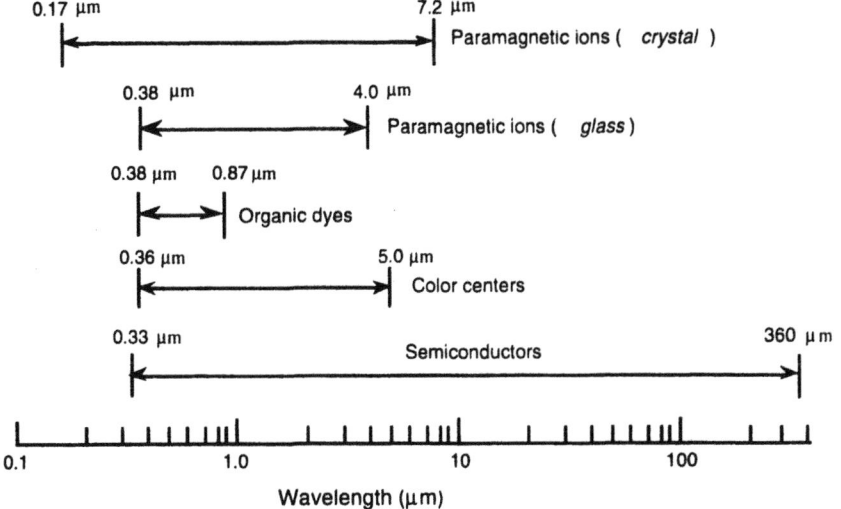

Figure 1.1   Reported ranges of output wavelengths for various types of solid state lasers.

Further Reading

Cheo, P. K., Ed., *Handbook of Solid-State Lasers*, Marcel Dekker Inc., New York (1989).

Koechner, W., *Solid-State Laser Engineering* (fourth edition), Springer Verlag, Berlin (1996).

Powell, R. C., *Physics of Solid State Laser Materials*, Springer-Verlag, Berlin (1997).

Powell, R. C., Ed., *Selected Papers on Solid State Lasers*, SPIE Milestone Series, Vol. MS31, SPIE Optical Engineering Press, Bellingham, WA (1991).

See, also, Tunable Solid-State Lasers, *Selected Topics in Quantum Electronics* 1 (1995), Diode-Pumped Solid-State Lasers, *Selected Topics in Quantum Electronics* 3(1) (February 1997), and the following proceedings of the Advanced Solid State Laser Conferences, all published by the Optical Society of America, Washington, DC:

OSA Trends in Optics and Photonics: Advanced Solid State Lasers, Vol. 26, Fejer, M. M., Injeyan, H. and Keller, Ursula, Eds. (1999).

OSA Trends in Optics and Photonics: Advanced Solid State Lasers, Vol. 19, Bosenberg, W. R. and Fejer, M. M., Eds. (1998).

OSA Trends in Optics and Photonics: DPSS Lasers: Applications and Issues, Vol. 17, Dowley, M. W., Ed. (1998).

OSA Trends in Optics and Photonics: Advanced Solid State Lasers, Vol. 10, Pollack, C. R. and Bosenberg, W. R., Eds. (1997).

OSA Trends in Optics and Photonics: Advanced Solid State Lasers, Vol. 1, Payne, S. A. and Pollack, C. R., Eds. (1996).

Chai, B. H. T. and Payne, S. A., Eds., Proceedings Vol. 24 (1995).

Fan, T. Y. and Chai, B., Eds., Proceedings Vol. 20 (1994).

Pinto, A. A. and Fan, T. Y., Eds., Proceedings Vol. 15 (1993).

Chase, L. L. and Pinto, A. A., Eds., Proceedings Vol. 13 (1992).

Dubé, G. and Chase, L. L, Eds., Proceedings Vol. 10 (1991).

Jenssen, H. P. and Dubé, G., Eds., Proceedings Vol. 6 (1990).

Section 1.1
CRYSTALLINE PARAMAGNETIC ION LASERS

1.1.1 Introduction

The elements that have been reported to exhibit laser action as paramagnetic ions (incompletely filled electron shells) in crystalline hosts are indicated in the periodic table of the elements in Figure 1.1.1. These are mainly transition metal and lanthanide group ions and generally involve intraconfigurational transitions. Typical concentrations of the lasing ion are ≤1%; however, for some hosts and ions concentrations up to 100%, so-called stoichiometric lasers, are possible. Also included in italics in Figure 1.1.1 are several ions for which only gain has been reported (see Section 1.1.8).

Energy level diagrams and lasing transitions for iron group ions are shown in Figures 1.1.2 and 1.1.3, for divalent lanthanide and trivalent actinide ions in Figure 1.1.4, and for trivalent lanthanides in Figures 1.1.5–1.1.9. The properties of lasers comprising these ions are listed in Sections 1.1.4–1.1.6.

The general operating wavelengths of crystalline lanthanide-ion lasers are given in Figure 1.1.10 and range from 0.17 mm for the 5d→4f transition of Nd^{3+} to 7.2 μm for the 4f→4f transition between J states of Pr^{3+}. Whereas f→f transitions of the lanthanide ions have narrow linewidths and discrete wavelengths, d→f transitions of these ions and transitions of many iron group ions have broad emission and gain bandwidths and hence provide a degree of tunability. The tuning ranges of several paramagnetic laser ions in different hosts are shown in Figure 1.1.11; the ranges for explicit host crystals are included in the laser tables. Tunable lasers are based almost exclusively on vibronic transitions of iron transition group elements.

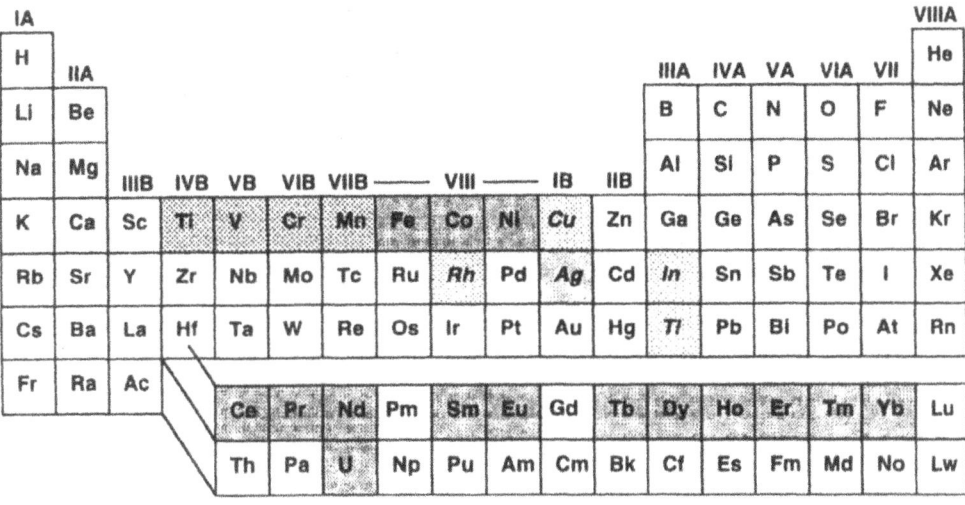

Figure 1.1.1 Periodic table of the elements showing the elements (shaded) that have been reported to exhibit laser action as paramagnetic ions in crystalline hosts. Gain has been reported for elements shown in italics.

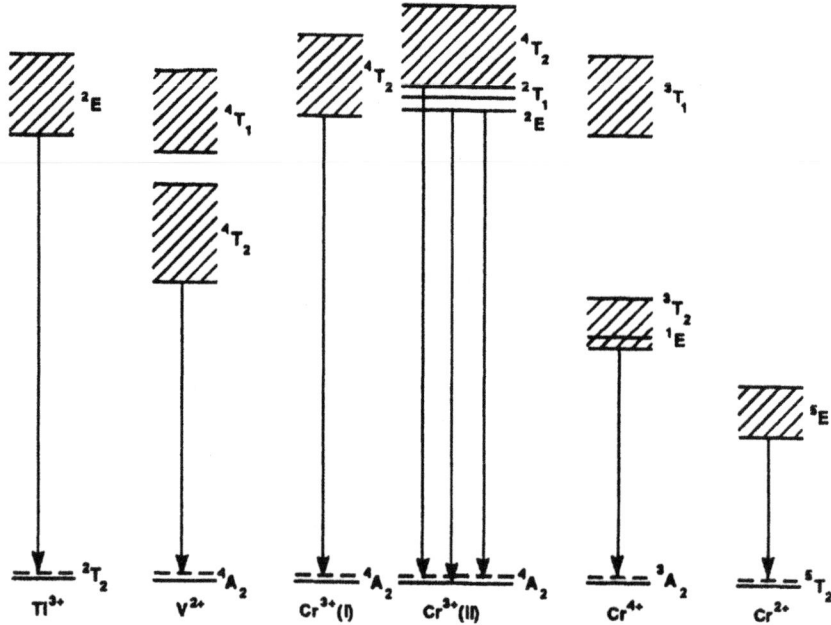

Figure 1.1.2 Energy levels and laser transitions of crystalline titanium, vanadium, and chromium ion lasers. The two energy level schemes for trivalent chromium correspond to chromium ions in different crystal field environments. Dashed levels are associated with laser transitions terminating on vibronic levels.

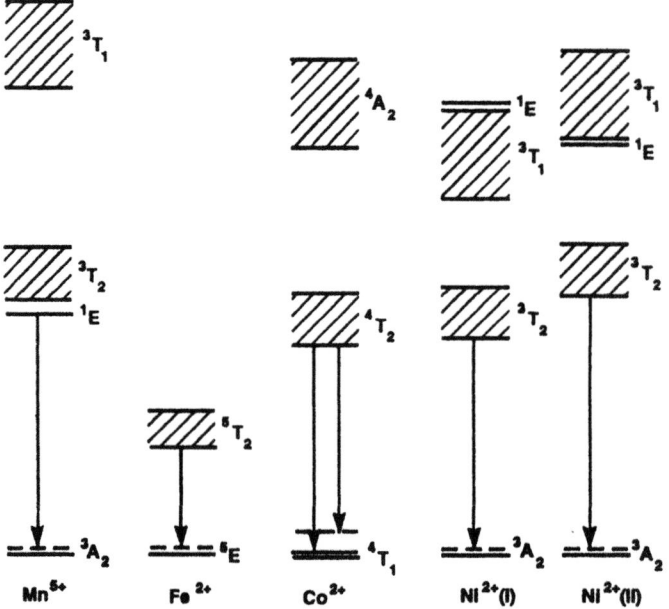

Figure 1.1.3 Energy levels and laser transitions of crystalline manganese, iron, cobalt, and nickel ion lasers. The two energy level schemes for divalent nickel correspond to nickel ions in different crystal field environments. Dashed levels are associated with laser transitions terminating on vibronic levels.

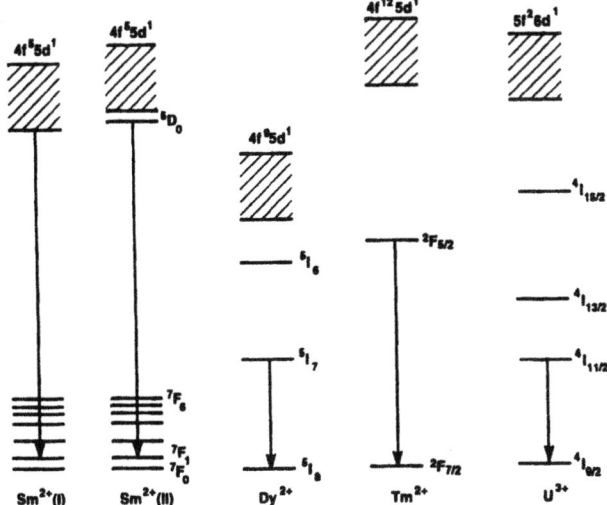

Figure 1.1.4   Energy levels and laser transitions of crystalline divalent lanthanide and actinide ion lasers. The two energy level schemes for divalent samarium correspond to samarium ions in different crystal field environments.

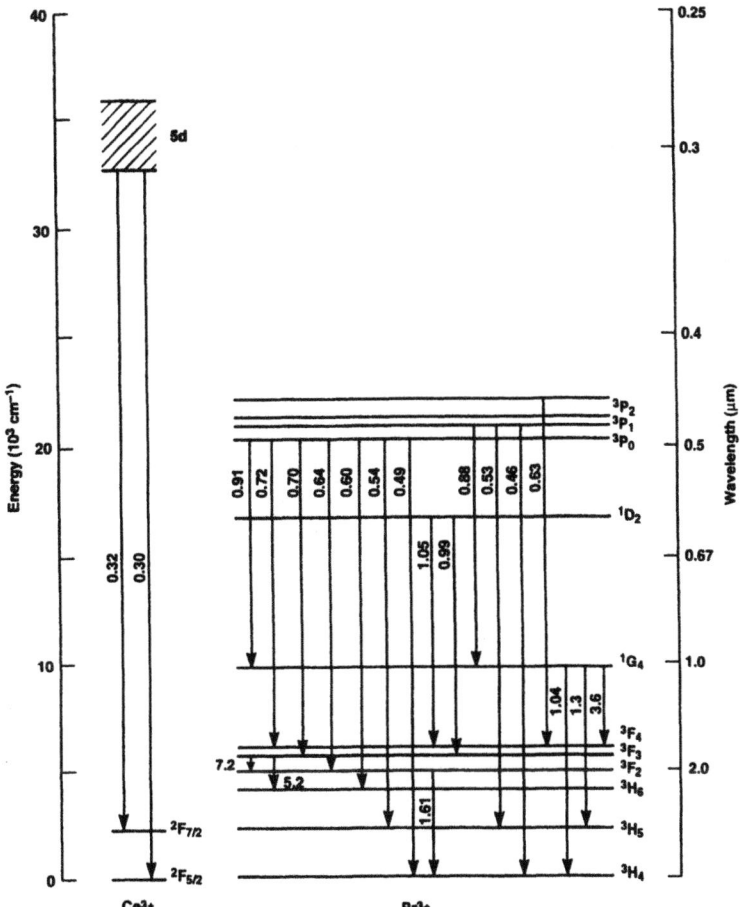

Figure 1.1.5   Energy levels, laser transitions, and wavelengths (microns) of crystalline cerium and praseodymium ion lasers.

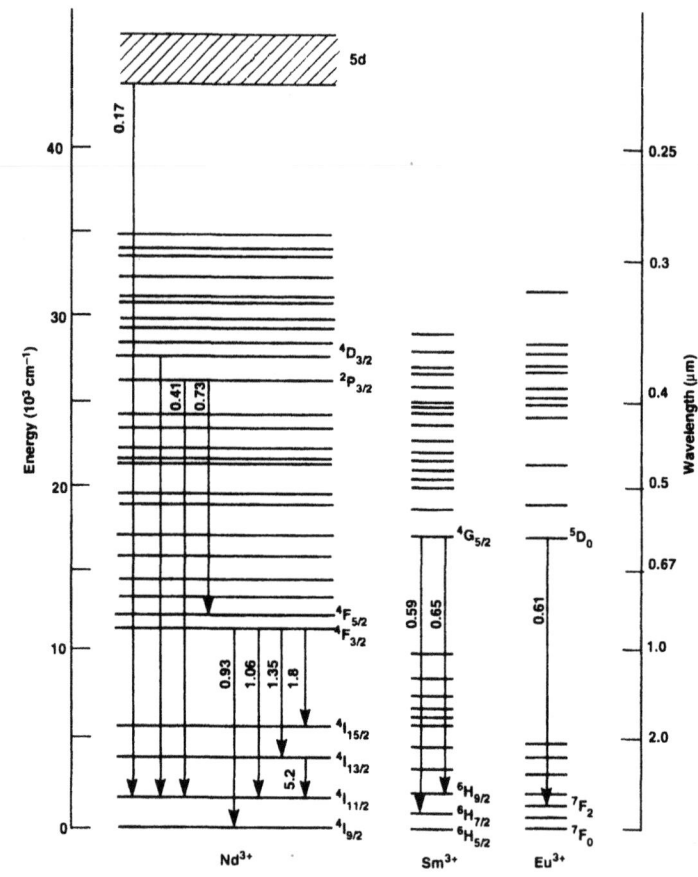

Figure 1.1.6   Energy levels, laser transitions, and approximate wavelengths (microns) of crystalline neodymium, samarium, and europium ion lasers.

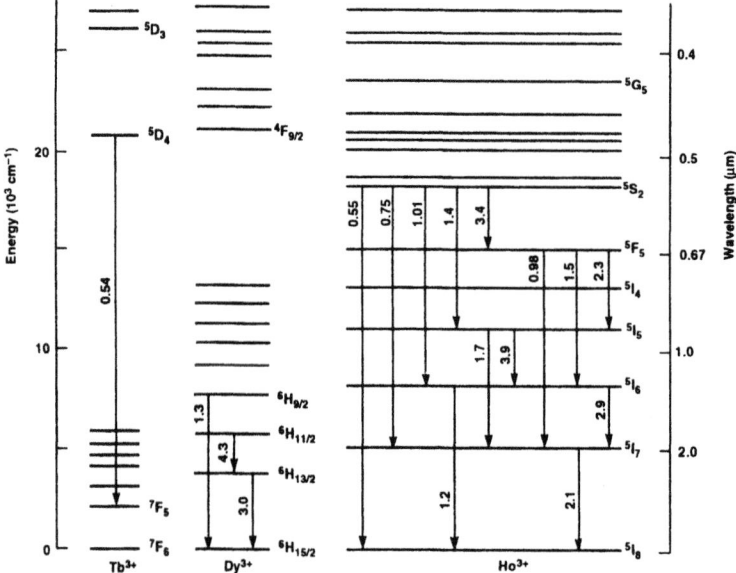

Figure 1.1.7   Energy levels, laser transitions, and approximate wavelengths (microns) of crystalline terbium, dysprosium, and holmium ion lasers.

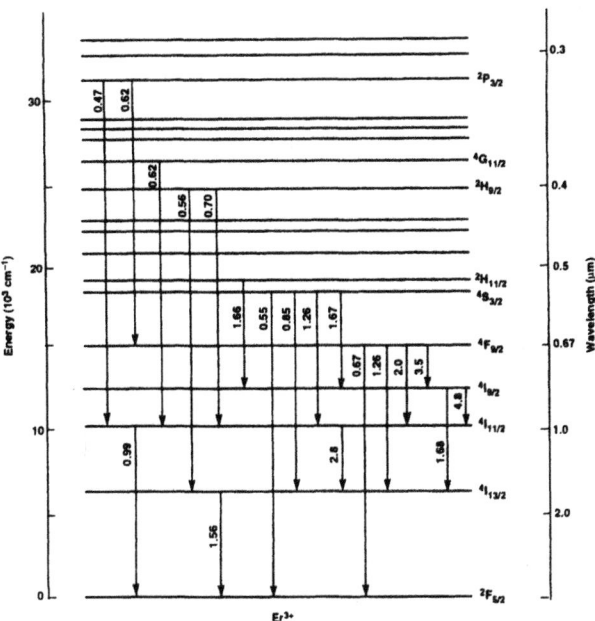

Figure 1.1.8   Energy levels, laser transitions, and approximate wavelengths (microns) of crystalline erbium lasers.

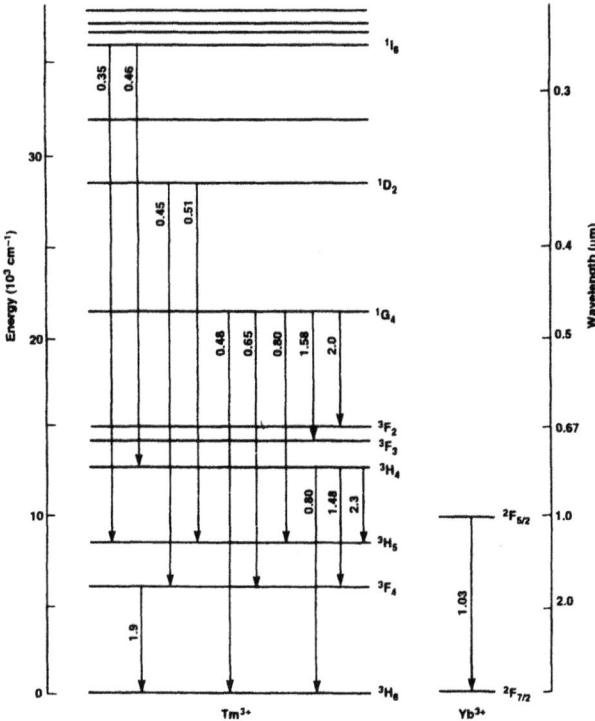

Figure 1.1.9   Energy levels, laser transitions, and approximate wavelengths (microns) of crystalline thulium and ytterbium ion lasers.

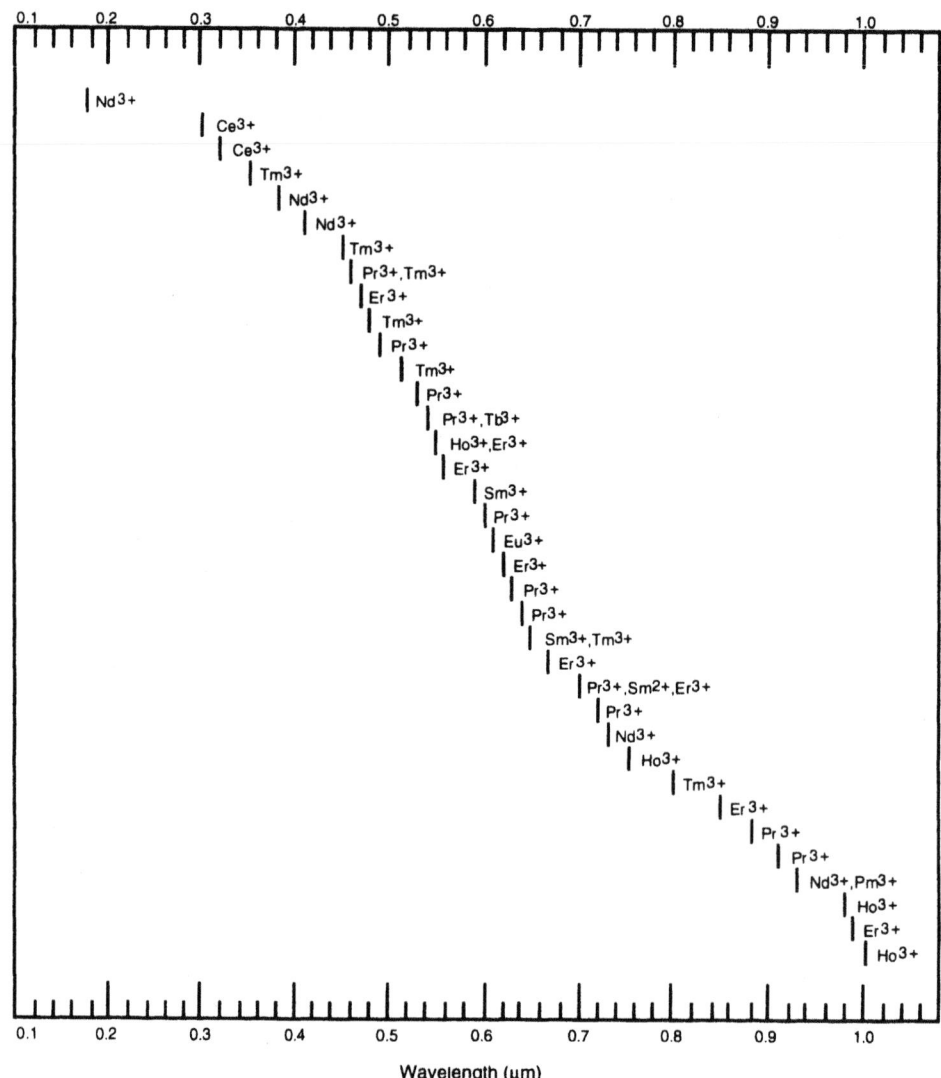

Figure 1.1.10a   Approximate wavelengths of crystalline lanthanide-ion lasers; exact wavelengths are dependent on the host and temperature and the specific Stark levels involved (from the *Handbook of Laser Wavelengths*, CRC Press, Boca Raton, FL, 1998).

Codopant ions have been added to improve the optical pumping efficiency of laser ions via fluorescence sensitization. Lanthanide laser ions and codopant sensitizing ions that have been reported are summarized in Table 1.1.1. Sensitizing ions, if present, are listed with the host crystal in the laser tables in Section 1.1.6.

Codopant ions have also been added to relax the terminal laser level and prevent self-terminated laser action. Laser transitions and codopant deactivating ions are listed in Table 1.1.2.

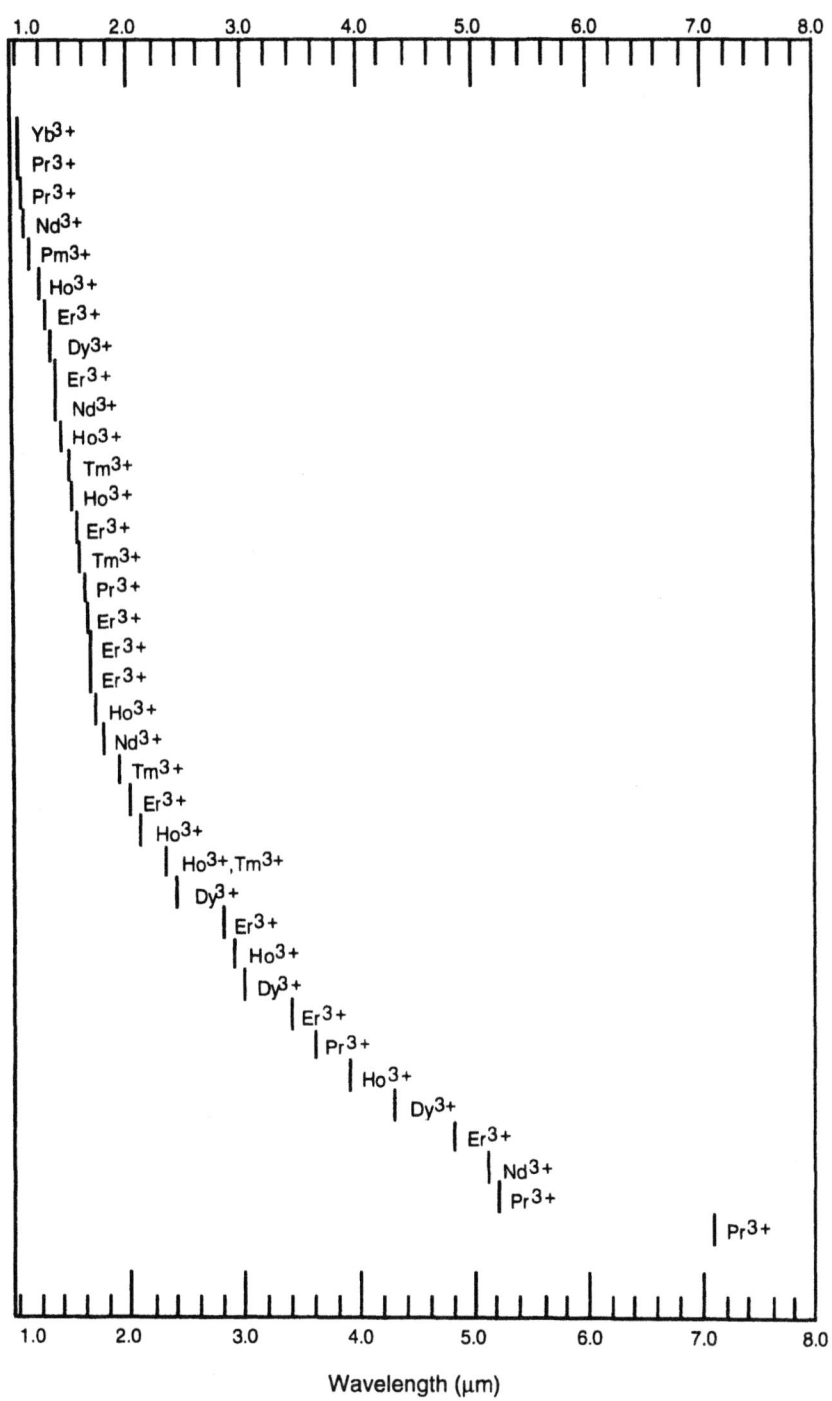

Figure 1.1.10b   Approximate wavelengths of crystalline lanthanide-ion lasers; exact wavelengths are dependent on the host and temperature and the specific Stark levels involved (from the *Handbook of Laser Wavelengths*, CRC Press, Boca Raton, FL, 1998).

Table 1.1.1
Codopant Ions Used to Sensitize Lanthanide Laser Ions

Laser ion	Sensitizing ions	Laser ion	Sensitizing ions
Pr^{3+}	Yb^{3+}	Ho^{3+}	Cr^{3+}
			Cr^{3+},Tm^{3+}
Nd^{3+}	Cr^{3+}		Cr^{3+},Yb^{3+}
	Ce^{3+},Cr^{3+}		Fe^{3+}
Sm^{3+}	Tb^{3+}	Er^{3+}	Ce^{3+}
			Ho^{3+}
Tb^{3+}	Gd^{3+}		Yb^{3+}
			Cr^{3+}
Dy^{3+}	Er^{3+}		Cr^{3+},Yb^{3+}
	Yb^{3+}		
		Tm^{3+}	Er^{3+}
Ho^{3+}	Er^{3+}		Er^{3+},Yb^{3+}
	Tm^{3+}		Cr^{3+}
	Yb^{3+}		Cr^{3+},Er^{3+},Yb^{3+}
	Er^{3+},Tm^{3+}		
	Er^{3+},Yb^{3+}	Yb^{3+}	Nd^{3+}
	Tm^{3+},Yb^{3+}		Cr^{3+},Nd^{3+}
	Er^{3+},Tm^{3+},Yb^{3+}		

Table 1.1.2
Codopant Ions Used to Deactivate the Terminal Laser Level

Laser ion	Lasing transition	Crystalline host	Codopant ion	Ref.
Ho^{3+}	$^5I_6 \rightarrow {}^5I_7$	$Y_3Al_5O_{12}$	Nd^{3+}	673
Er^{3+}	$^4S_{3/2} \rightarrow {}^4I_{13/2}$	$LiYF_4$	Pr^{3+}	224
	$^4I_{11/2} \rightarrow {}^4I_{13/2}$	$LiYF_4$	Pr^{3+}	1070
		$Y_3Al_5O_{12}$	Nd^{3+}	672
		CaF_2-ErF_3	Ho^{3+}	791
		CaF_2-ErF_3	Tm^{3+}	791
		$Er_3Al_5O_{12}$	Tm^{3+}	882
		$LuAlO_3$	Tm^{3+}	1120
		$K(Y,Er)(WO_4)_2$	Ho^{3+}, Tm^{3+}	1119
		$Er_3Al_5O_{12}$	Ho^{3+}, Tm^{3+}	1130
		$Lu_3Al_5O_{12}$	Ho^{3+}, Tm^{3+}	1131
		$Y_3Al_5O_{12}$	Ho^{3+}, Tm^{3+}	1132
		$Lu_3Al_5O_{12}$	$Nd^{3+}, Ho^{3+}, Tm^{3+}$	1131
Tm^{3+}	$^3F_4 \rightarrow {}^3H_4$	$LiYF_4$	Tb^{3+}	700

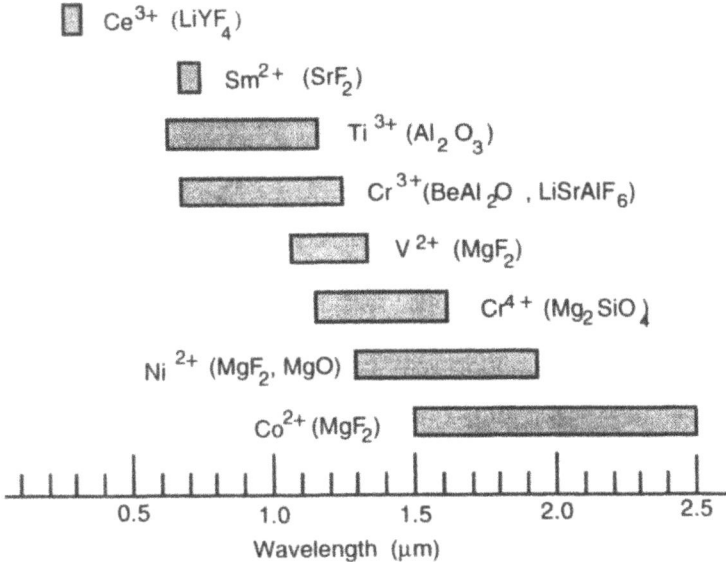

Figure 1.1.11   Reported wavelength ranges of representative tunable crystalline lasers operating at room temperature (from the *Handbook of Laser Wavelengths*, CRC Press, Boca Raton, FL, 1998).

Upconversion processes make possible many additional lasing transitions and excitation schemes. Upconversion excitation techniques include multi-step absorption, ion-ion energy transfer, excited state absorption, and photon avalanche processes. Lasers based on upconversion schemes are noted in the mode column of the laser tables. Transitions involved in upconversion processes are given in Table 1.1.3 and can be identified by reference to the relevant energy level diagrams for the ions in Figures 1.1.4–1.1.8. The success of many of the schemes depends upon the degree of resonance of energy transfer transitions and the rate of nonradiative transitions by multiphonon emission and thus varies with the host crystal.

Cascade and cross-cascade lasing schemes have also been employed; transitions involved in cascade and cross-cascade lasing schemes are summarized in Tables 1.1.4 and 1.1.5. For examples of avalanche-pumped upconversion lasers, see References 18 and 1037.

Table 1.1.3
Multi-step Upconversion Excitation Schemes

$\rightarrow$ optical transition $\Rightarrow$ ion-ion energy transfer transitions $\rightarrowtail$ nonradiative transition

Laser ion	Upper laser level	Codopant ion	Upconversion excitation scheme
Pr^{3+}	$^3P_0$	—	1) $^3H_4 \rightarrow {}^1G_4$ 2) $^1G_4 \rightarrow {}^3P_1 \rightarrowtail {}^3P_0$
		Yb^{3+}	1) $^2F_{7/2} \rightarrow {}^2F_{5/2}$ (Yb^{3+}) 2) $^2F_{5/2} - {}^2F_{7/2}$ $(Yb^{3+}) \Rightarrow {}^3H_4 - {}^1G_4$ (Pr^{3+}) 3) $^1G_4 \rightarrow {}^3P_{1,0}$
Nd^{3+}	$^4D_{3/2}$	—	1) $^4I_{9/2} \rightarrow {}^4F_{5/2} \rightarrowtail {}^4F_{3/2}$ 2) $^4F_{3/2} \rightarrow {}^4D_{3/2}$
			1) $^4I_{9/2} \rightarrow {}^4G_{5/2} \rightarrowtail {}^4F_{3/2}$ 2) $^4F_{3/2} \rightarrow {}^4D_{3/2}$
	$^2P_{3/2}$	—	1) $^4I_{9/2} \rightarrow {}^4G_{5/2} \rightarrowtail {}^4F_{3/2}$ 2) $^4F_{3/2} \rightarrow {}^4D_{3/2} \rightarrowtail {}^2P_{3/2}$
Ho^{3+}	$^5S_2$	Yb^{3+}	1) $^2F_{7/2} \rightarrow {}^2F_{5/2}$ (Yb^{3+}) 2) $^2F_{5/2} - {}^2F_{7/2}$ $(Yb^{3+}) \Rightarrow {}^5I_8 - {}^5I_6$ (Ho^{3+}) 3) $^2F_{7/2} \rightarrow {}^2F_{5/2}$ (Yb^{3+}) 4) $^2F_{5/2} - {}^2F_{7/2}$ $(Yb^{3+}) \Rightarrow {}^5I_6 - {}^5S_2$ (Ho^{3+})
	$^5I_7$	Yb^{3+}	1) $^2F_{7/2} \rightarrow {}^2F_{5/2}$ (Yb^{3+}) 2) $^2F_{5/2} - {}^2F_{7/2}$ $(Yb^{3+}) \Rightarrow {}^5I_8 - {}^5I_6$ $(Ho^{3+}) \rightarrowtail {}^5I_7$
Er^{3+}	$^2P_{3/2}$	—	1) $^4I_{15/2} \rightarrow {}^4I_{11/2}(Er_1^{3+})$ 2) $^4I_{15/2} \rightarrow {}^4I_{11/2}(Er_2^{3+})$ 3) $^4I_{11/2} - {}^4I_{15/2}(Er_1^{3+}) \Rightarrow {}^4I_{11/2} - {}^4F_{7/2} \rightarrowtail {}^4S_{3/2}$ (Er_2^{3+}) 4) $^4S_{3/2} - {}^4I_{15/2}(Er_2^{3+}) \Rightarrow {}^4F_{9/2} - {}^2K_{13/2}$ $(Er_3^{3+}) \rightarrowtail {}^2P_{3/2}$
	$^4G_{11/2}$	—	1) $^4I_{15/2} \rightarrow {}^4I_{13/2}$ (fourfold) $\Rightarrow {}^4G_{11/2}$
	$^2H_{9/2}$	—	1) $^4I_{15/2} \rightarrow {}^4I_{11/2}(Er_1^{3+})$ 2) $^4I_{15/2} \rightarrow {}^4I_{11/2}(Er_2^{3+})$ 3) $^4I_{11/2} - {}^4I_{15/2}(Er_1^{3+}) \Rightarrow {}^4I_{11/2} - {}^4F_{7/2}$ $(Er_2^{3+}) \rightarrowtail {}^4S_{3/2}$ 4) $^4I_{15/2} \rightarrow {}^4I_{11/2} \rightarrowtail {}^4I_{13/2}$ (Er_3^{3+}) 5) $^4S_{3/2} - {}^4I_{15/2}(Er_2^{3+}) \Rightarrow {}^4I_{13/2} - {}^2H_{9/2}(Er_3^{3+})$

Table 1.1.3—*continued*
Multi-step Upconversion Excitation Schemes

Laser ion	Upper laser level	Codopant ion	Upconversion excitation scheme
	$^4S_{3/2}$	—	1) $^4I_{15/2} \rightarrow {}^4I_{9/2} \Rightarrow {}^4I_{11/2}$ 2) $^4I_{11/2} \rightarrow {}^4F_{5/2,7/2} \rightarrow \Rightarrow {}^4S_{3/2}$
		—	1) $^4I_{15/2} \rightarrow {}^4I_{11/2}(Er_1^{3+})$ 2) $^4I_{15/2} \rightarrow {}^4I_{11/2}(Er_2^{3+})$ 3) $^4I_{11/2} - {}^4I_{15/2}(Er_1^{3+}) \Rightarrow {}^4I_{11/2} - {}^4F_{7/2} \Rightarrow {}^4S_{3/2}\ (Er_2^{3+})$
	$^4F_{9/2}$	Yb^{3+}	1) $^2F_{7/2} \rightarrow {}^2F_{5/2}\ (Yb^{3+})$ 2) $^4I_{15/2} \rightarrow {}^4I_{13/2}(Er^{3+})$ 3) $^2F_{5/2} - {}^2F_{7/2}\ (Yb^{3+}) \Rightarrow {}^4I_{13/2} - {}^4F_{9/2}\ (Er^{3+})$
		Yb^{3+}	1) $^2F_{7/2} \rightarrow {}^2F_{5/2}\ (Yb^{3+})$ 2) $^2F_{5/2} - {}^2F_{7/2}\ (Yb^{3+}) \Rightarrow {}^4I_{15/2} - {}^4I_{11/2}\ (Er^{3+})$ 3) $^2F_{7/2} \rightarrow {}^2F_{5/2}\ (Yb^{3+})$ 4) $^2F_{5/2} - {}^2F_{7/2}\ (Yb^{3+}) \Rightarrow {}^4I_{11/2} - {}^4F_{7/2}\ (Er^{3+}) \Rightarrow {}^4F_{9/2}$
	$^4I_{11/2}$	—	1) $^4I_{15/2} \rightarrow {}^4I_{13/2}(Er_1^{3+})$ 2) $^4I_{15/2} \rightarrow {}^4I_{13/2}(Er_2^{3+})$ 3) $^4I_{13/2} - {}^4I_{15/2}\ (Er_1^{3+}) \Rightarrow {}^4I_{13/2} - {}^4I_{9/2} \Rightarrow {}^4I_{11/2}\ (Er_2^{3+})$
Tm^{3+}	$^1I_6$	Yb^{3+}	1) $^2F_{7/2} \rightarrow {}^2F_{5/2}\ (Yb^{3+})$ 2) $^2F_{7/2} - {}^2F_{5/2}\ (Yb^{3+}) \Rightarrow {}^3H_6 - {}^3H_5\ (Tm_1^{3+}) \Rightarrow {}^3F_4$ 3) $^2F_{7/2} \rightarrow {}^2F_{5/2}\ (Yb^{3+})$ 4) $^2F_{5/2}\ {}^2F_{7/2}\ (Yb^{3+}) \Rightarrow {}^3F_4\ {}^3F_3\ (Tm_1^{3+}) \Rightarrow {}^3H_4$ 5) $^3F_3 - {}^3H_6\ (Tm_1^{3+}) \Rightarrow {}^3F_3 - {}^1D_2\ (Tm_2^{3+})$ 6) $^2F_{7/2} \rightarrow {}^2F_{5/2}\ (Yb^{3+})$ 7) $^2F_{5/2}\ {}^2F_{7/2}\ (Yb^{3+}) \Rightarrow {}^1D_2\ {}^3P_J\ (Tm_2^{3+}) \Rightarrow {}^1I_6$
Tm^{3+}	$^1D_2$	—	1) $^3H_6 \rightarrow {}^3H_4$ 2) $^3H_4 \rightarrow {}^1D_2$
			1) $^3H_6 \rightarrow {}^3H_4\ (Tm_1^{3+})$ 2) $^3H_6 \rightarrow {}^3H_4\ (Tm_2^{3+})$ 3) $^3H_4 - {}^3H_6\ (Tm_1^{3+}) \Rightarrow {}^3H_4 - {}^1D_2\ (Tm_2^{3+})$
Tm^{3+}	$^3H_4$	Yb^{3+}	1) $^2F_{7/2} \rightarrow {}^2F_{5/2}\ (Yb^{3+})$ 2) $^3H_6 \rightarrow {}^3H_5 \Rightarrow {}^3F_4\ (Tm^{3+})$ 3) $^2F_{5/2} - {}^2F_{7/2}\ (Yb^{3+}) \Rightarrow {}^3F_4 - {}^3F_2\ (Tm^{3+}) \Rightarrow {}^3H_4$

Table 1.1.3—*continued*
Multi-step Upconversion Excitation Schemes

Laser ion	Upper laser level	Codopant ion	Upconversion excitation scheme
Tm^{3+}	$^1G_4$	Yb^{3+}	1) $^2F_{7/2} \rightarrow {}^2F_{5/2}$ (Yb^{3+})
			2) $^2F_{7/2} - {}^2F_{5/2}$ $(Yb^{3+}) \Rightarrow {}^3H_6 - {}^3H_5$ $(Tm^{3+}) \Longrightarrow {}^3F_4$
			3) $^2F_{7/2} \rightarrow {}^2F_{5/2}$ (Yb^{3+})
			4) $^2F_{5/2}$ ${}^2F_{7/2}$ $(Yb^{3+}) \Rightarrow {}^3F_4$ ${}^3F_2 \Longrightarrow {}^3H_4$ (Tm^{3+})
			5) $^2F_{7/2} \rightarrow {}^2F_{5/2}$ (Yb^{3+})
			6) $^2F_{5/2} - {}^2F_{7/2}$ $(Yb^{3+}) \Rightarrow {}^3H_4 - {}^1G_4$ (Tm^{3+})

Table 1.1.4
Cascade Laser Schemes

$\rightarrow$ lasing transition $\Longrightarrow$ nonradiative transition

Laser ion	Cascade transitions
Pr^{3+}	$^3P_0 \rightarrow {}^1G_4 \rightarrow {}^3F_4$
	$^3P_0 \rightarrow {}^1G_4 \rightarrow {}^3H_5$
Nd^{3+}	$^4F_{3/2} \rightarrow {}^4I_{13/2} \rightarrow {}^4I_{11/2}$
Ho^{3+}	$^5S_2 \rightarrow {}^5I_5 \rightarrow {}^5I_6$
	$^5S_2 \rightarrow {}^5I_5 \rightarrow {}^5I_7$
	$^5S_2 \rightarrow {}^5I_6 \rightarrow {}^5I_8$
	$^5S_2 \rightarrow {}^5I_7 \rightarrow {}^5I_8$
	$^5S_2 \rightarrow {}^5I_5 \Longrightarrow {}^5I_6 \rightarrow {}^5I_7$
	$^5S_2 \rightarrow {}^5I_5 \Longrightarrow {}^5I_6 \rightarrow {}^5I_8$
	$^5S_2 \rightarrow {}^5I_5 \Longrightarrow {}^5I_6 \rightarrow {}^5I_7 \rightarrow {}^5I_8$
	$^5S_2 \rightarrow {}^5F_5 \Longrightarrow {}^5I_4 \Longrightarrow {}^5I_5 \rightarrow {}^5I_6 \rightarrow {}^5I_7$
	$^5I_6 \rightarrow {}^5I_7 \rightarrow {}^5I_8$
Er^{3+}	$^4S_{3/2} \rightarrow {}^4I_{9/2} \rightarrow {}^4I_{11/2}$
	$^4S_{3/2} \rightarrow {}^4I_{9/2} \rightarrow {}^4I_{13/2}$
	$^4S_{3/2} \rightarrow {}^4I_{11/2} \rightarrow {}^4I_{13/2}$
	$^4S_{3/2} \rightarrow {}^4I_{13/2} \rightarrow {}^4I_{15/2}$
	$^4S_{3/2} \rightarrow {}^4I_{9/2} \Longrightarrow {}^4I_{11/2} \rightarrow {}^4I_{13/2}$
	$^4S_{3/2} \rightarrow {}^4I_{9/2} \Longrightarrow {}^4I_{11/2} \rightarrow {}^4I_{13/2} \rightarrow {}^4I_{15/2}$
	$^4F_{9/2} \rightarrow {}^4I_{11/2} \rightarrow {}^4I_{13/2}$
	$^4I_{11/2} \rightarrow {}^4I_{13/2} \rightarrow {}^4I_{15/2}$
Tm^{3+}	$^3F_4 \rightarrow {}^3H_5 \Longrightarrow {}^3H_4 \rightarrow {}^3H_6$

Table 1.1.5
Cross-Cascade Laser Schemes

$\rightarrow$ lasing transition $\Rightarrow$ nonradiative energy transfer transitions

Laser ions	Cross-cascade transitions
$Er^{3+} + Ho^{3+}$	$^4S_{3/2} \rightarrow {}^4I_{13/2} (Er^{3+})$ $^4I_{13/2} - {}^4I_{15/2}(Er^{3+}) \Rightarrow {}^5I_8 - {}^5I_7 (Ho^{3+})$ $^5I_7 \rightarrow {}^5I_8 (Ho^{3+})$ $^4I_{11/2} \rightarrow {}^4I_{13/2} (Er^{3+})$ $^4I_{13/2} - {}^4I_{15/2}(Er^{3+}) \Rightarrow {}^5I_8 - {}^5I_7 (Ho^{3+})$ $^5I_7 \rightarrow {}^5I_8 (Ho^{3+})$
$Er^{3+} + Tm^{3+}$	$^4S_{3/2} \rightarrow {}^4I_{13/2} (Er^{3+}) \Rightarrow$ $^4I_{13/2} - {}^4I_{15/2}(Er^{3+}) \Rightarrow {}^3H_6 - {}^3F_4(Tm^{3+})$ $^3F_4 \rightarrow {}^3H_6(Tm^{3+})$ $^4I_{11/2} \rightarrow {}^4I_{13/2} (Er^{3+})$ $^4I_{13/2} - {}^4I_{15/2}(Er^{3+}) \Rightarrow {}^3H_6 - {}^3F_4(Tm^{3+})$ $^3F_4 \rightarrow {}^3H_6(Tm^{3+})$
$Tm^{3+} + Ho^{3+}$	$^3H_4 \rightarrow {}^3H_5 \Rrightarrow {}^3F_4(Tm^{3+})$ $^3F_4 - {}^3H_6(Tm^{3+}) \Rightarrow {}^5I_8 - {}^5I_7 (Ho^{3+})$ $^{55}I_7 \rightarrow {}^5I_8 (Ho^{3+})$ $^3H_4 \rightarrow {}^3F_4 (Tm^{3+})$ $^3F_4 - {}^3H_6(Tm^{3+}) \Rightarrow {}^5I_8 - {}^5I_7 (Ho^{3+})$ $^{55}I_7 \rightarrow {}^5I_8 (Ho^{3+})$
$Er^{3+} + Tm^{3+} + Ho^{3+}$	$^4I_{11/2} \rightarrow {}^4I_{13/2} (Er^{3+})$ $^4I_{13/2} - {}^4I_{15/2}(Er^{3+}) \Rightarrow {}^3H_6 - {}^3F_4(Tm^{3+})$ $^3F_4 - {}^3H_6(Tm^{3+}) \Rightarrow {}^5I_8 - {}^5I_7 (Ho^{3+})$ $^{55}I_7 \rightarrow {}^5I_8 (Ho^{3+})$

Further Reading

Caird, J. and Payne, S. A., Crystalline Paramagnetic Ion Lasers, in *Handbook of Laser Science and Technology, Suppl. 1: Lasers*, CRC Press, Boca Raton, FL (1991), p. 3.

Hanna, D. C. and Jacquier, B., Eds., Miniature coherent light sources in dielectric media, *Opt. Mater.* 11, Nos. 2/3 (1999).

Kaminskii, A. A., *Crystalline Lasers: Physical Processes and Operating Schemes*, CRC Press, Boca Raton, FL (1996).

Kaminskii, A. A., *Laser Crystals, Their Physics and Properties*, Springer-Verlag, Heidelberg (1990).

Moulton, P., Paramagnetic Ion Lasers, in *Handbook of Laser Science and Technology, Vol. I: Lasers and Masers*, CRC Press, Boca Raton, FL (1995), p. 21

1.1.2 Host Crystals Used for Transition Metal Laser Ions

Table 1.1.6
Host Crystals Used for Transition Metal Laser Ions

Crystal	Ti^{3+}	V^{2+}	Cr^{2+}	Cr^{3+}	Cr^{4+}	Mn^{5+}	Fe^{2+}	Co^{2+}	Ni^{2+}
Oxides									
Al_2O_3	●			●					
$Ba_3(VO_4)_2$						●			
$BeAl_2O_4$	●			●					
$BeAl_6O_{10}$				●					
$Be_3Al_2Si_6O_{18}$				●					
$CaGd_4(SiO_4)_3O$					●				
$CaY_2Mg_2Ge_3O_{12}$									●
Ca_2GeO_4					●				
$Ca_3Ga_2Ge_3O_{12}$					●				
$Ca_3Ga_2Ge_4O_{14}$				●					
$Gd_3Ga_5O_{12}$				●					
$Gd_3Sc_2Al_3O_{12}$				●					
$Gd_3Sc_2Ga_3O_{12}$				●					
$La_3Ga_5GeO_{14}$				●					
$La_3Ga_{5.5}Nb_{0.5}O_{14}$				●					
$La_3Ga_{5.5}Ta_{0.5}O_{14}$				●					
$La_3Ga_5SiO_{14}$				●					
$LiNbGeO_5$				●					
Mg_2SiO_4					●				
MgO									●
$ScBO_3$				●					
$ScBeAlO_4$				●					
$Sr_3Ga_2Ge_4O_{14}$				●					
$SrGd_4(SiO_4)_3O$					●				
$YAlO_3$	●								
Y_2SiO_5					●				
$Y_3Al_5O_{12}$				●	●				
$Y_3Ga_5O_{12}$				●					
$Y_3Sc_2Al_3O_{12}$				●					
$Y_3Sc_2Ga_3O_{12}$				●					
$ZnWO_4$				●					
Halides									
$CsCaF_3$		●							
$KMgF_3$								●	●

Table 1.1.6—*continued*
Host Crystals Used for Transition Metal Laser Ions

Crystal	Ti^{3+}	V^{2+}	Cr^{2+}	Cr^{3+}	Cr^{4+}	Mn^{5+}	Fe^{2+}	Co^{2+}	Ni^{2+}
$KZnF_3$				•				•	
$LiCaAlF_6$				•					
$LiSrAlF_6$				•				•	
$LiSrCrF_6$				•					
$LiSrGaF_6$				•					
MgF_2		•							•
MnF_2									•
$Na_3Ga_3Li_3F_{12}$				•					
$SrAlF_5$				•					
ZnF_2								•	
Chalcogenides									
CdMnTe			•						
ZnS			•						
ZnSe			•				•		
Phosphide									
n-InP								•	

1.1.3 Host Crystals Used for Lanthanide Laser Ions

Table 1.1.7
Host Crystals Used for Divalent Lanthanide Laser Ions

Crystal	Sm^{2+}	Dy^{2+}	Tm^{2+}
Halides			
CaF_2	•	•	•
SrF_2	•	•	

Table 1.1.8
Host Crystals Used for Trivalent Lanthanide Laser Ions

Crystal	Ce^{3+}	Pr^{3+}	Nd^{3+}	Sm^{3+}	Eu^{3+}	Dy^{3+}	Ho^{3+}	Er^{3+}	Tm^{3+}	Yb^{3+}
Oxides										
$Al_2(WO_4)_3$									•	
$Ba_{0.25}Mg_{2.75}$-$Y_2Ge_3O_{12}$			•							
$Ba_2MgGe_2O_7$			•							

Table 1.1.8—*continued*
Host Crystals Used for Trivalent Lanthanide Laser Ions

Crystal	Ce^{3+}	Pr^{3+}	Nd^{3+}	Sm^{3+}	Eu^{3+}	Dy^{3+}	Ho^{3+}	Er^{3+}	Tm^{3+}	Yb^{3+}
Oxides										
$BaGd_2(MoO_4)_4$			•							
$BaLaGa_3O_7$			•							
$Ba_2NaNb_5O_{15}$			•							
$Ba_2ZnGe_2O_7$			•							
$Ba_3LaNb_3O_{12}$			•							
$Bi_4Ge_3O_{12}$			•				•	•		
$Bi_4Si_3O_{12}$			•							
$Bi_4(Si,Ge)_3O_{12}$			•							
$Bi_{12}SiO_{20}$			•							
$Ca_{0.25}Ba_{0.75}$- $(NbO_3)_2$			•							
$CaAl_4O_7$			•					•		
$CaGd_4(SiO_4)_3O$			•							
$CaLa_4(SiO_4)_3O$			•							
$CaMg_2Y_2Ge_3O_{12}$			•							
$CaMoO_4$			•					•	•	
$Ca(NbO_3)_2$		•	•					•	•	
$Ca(NbGa)_2$- Ga_3O_{12}									•	
$CaSc_2O_4$			•							
$CaWO_4$		•	•				•	•	•	
$CaYAlO_4$			•							
$CaY_2Mg_2Ge_3O_{12}$			•						•	
$CaY_4(SiO_4)_3O$			•				•		•	
$Ca_2Al_2SiO_7$			•					•		
$Ca_2Ga_2Ge_4O_{14}$			•							
$Ca_2Ga_2SiO_7$									•	
$Ca_3Ga_2Ge_3O_{12}$			•					•		
$Ca_3Ga_2Ge_4O_{14}$			•							
$Ca_3Ga_2SiO_7$			•							
$Ca_3Ga_4O_9$			•							
$Ca_3(Nb,Ga)_2$- $(Ga_3O_{12}$			•							
$Ca_3(NbLiGa)_5O_{12}$								•		
$Ca_3(VO_4)_2$			•							
$Ca_4GdO(BO_3)_3$			•							
$Ca_4La(PO_4)_3O$			•							

Table 1.1.8—*continued*
Host Crystals Used for Trivalent Lanthanide Laser Ions

Crystal	Ce^{3+}	Pr^{3+}	Nd^{3+}	Sm^{3+}	Eu^{3+}	Dy^{3+}	Ho^{3+}	Er^{3+}	Tm^{3+}	Yb^{3+}
CeP_5O_{14}			•							
$CsLa(WO_4)_2$			•							
$CsNd(MoO_4)_2$			•							
$ErAlO_3$							•	•	•	
$ErVO_4$										
$Er(Y,Gd)AlO_3$							•		•	
Er_2O_3							•			
Er_2SiO_5							•	•	•	
$Er_3Al_5O_{12}$							•			
$Er_3Sc_2Al_3O_{12}$							•			
$Ga_3Al_5O_{12}$			•							
$GdAlO_3$			•				•	•	•	
$GdGaGe_2O_7$			•							
GdP_5O_{14}			•							
$GdScO_3$			•							
$GdVO_4$			•				•			
$Gd_2(MoO_4)_3$			•							
$Gd_2(WO_4)_3$			•							
Gd_2O_3			•							
$Gd_3Al_5O_{12}$			•					•		
$Gd_3Ga_5O_{12}$			•				•	•		•
$Gd_3Sc_2Al_3O_{12}$			•					•		•
$Gd_3Sc_2Ga_3O_{12}$			•				•		•	
$HfO_2\text{-}Y_2O_3$			•							
$Ho_3Al_5O_{12}$							•			
$Ho_3Ga_5O_{12}$							•			
$Ho_3Sc_2Al_3O_{12}$							•			
$KEr(WO_4)_2$								•		
$KGd(WO_4)_2$			•							
$KGd(WO_4)_2$			•				•	•		•
$KLa(MoO_4)_2$			•				•	•		
$KLu(WO_4)_2$			•				•			
$KNdP_4O_{12}$			•							
$KY(MoO_4)_2$			•							
$KY(WO_4)_2$			•							
$K(Y,Er)(WO_4)_2$							•	•		
$K_3(La,Nd)(PO_4)_2$			•							
$K_5Bi(MoO_4)_4$			•							

Table 1.1.8—*continued*
Host Crystals Used for Trivalent Lanthanide Laser Ions

Crystal	Ce^{3+}	Pr^{3+}	Nd^{3+}	Sm^{3+}	Eu^{3+}	Dy^{3+}	Ho^{3+}	Er^{3+}	Tm^{3+}	Yb^{3+}
K$_5$Nd(MoO$_4$)$_4$			•							
LaAlO$_3$			•							
LaAl$_{11}$MgO$_{19}$										
LaBGeO$_5$			•							
LaGaGe$_2$O$_7$			•							
LaMgAl$_{11}$O$_{19}$			•							
LaNbO$_4$			•							
LaP$_5$O$_{14}$			•							
(La,Nd)P$_5$O$_{14}$			•							
(La,Pr)P$_5$O$_{14}$		•								
LaSc$_3$(BO$_3$)$_4$			•							
(La,Sr)(Al,Ta)O$_3$			•							
LaSr$_2$Ga$_{11}$O$_{20}$			•							
La$_2$Be$_2$O$_5$			•							
La$_2$O$_3$			•							
7La$_2$O$_3$-9SiO$_2$			•							
La$_2$Si$_2$O$_7$			•							
La$_3$Ga$_5$GeO$_{14}$			•							
La$_3$Ga$_5$SiO$_{14}$			•							
La$_3$Ga$_{5.5}$Nb$_{0.5}$O$_{14}$			•							
La$_3$Ga$_{5.5}$Ta$_{0.5}$O$_{14}$			•					•		
Li(Bi,Nd)P$_4$O$_{12}$			•							
Li(La,Nd)P$_4$O$_{12}$			•							
Li(Nd,Gd)P$_4$O$_{12}$			•							
LiGd(MoO$_4$)$_2$			•							
LiLa(MoO$_4$)$_2$			•							
LiNbO$_3$			•				•	•	•	•
LiPrP$_4$O$_{14}$		•								
LuAlO$_3$		•	•				•	•		
LuScO$_3$			•							
Lu$_2$SiO$_5$			•							
Lu$_3$Al$_5$O$_{12}$			•				•	•	•	•
(Lu,Er)$_3$Al$_5$O$_{12}$								•		
Lu$_3$Ga$_5$O$_{12}$			•							•
Lu$_3$Sc$_2$Al$_3$O$_{12}$			•							•
β''-Na$_{1+x}$Mg$_x$-Al$_{11-x}$O$_{17}$										
NaBi(WO$_4$)$_2$			•							
NaGaGe$_2$O$_7$			•							

Table 1.1.8—*continued*
Host Crystals Used for Trivalent Lanthanide Laser Ions

Crystal	Ce^{3+}	Pr^{3+}	Nd^{3+}	Sm^{3+}	Eu^{3+}	Dy^{3+}	Ho^{3+}	Er^{3+}	Tm^{3+}	Yb^{3+}
$NaGd(MoO_4)_2$			•							
$NaGd(WO_4)_2$			•							
$NaGdGeO_4$			•							
$NaLa(MoO_4)_2$			•							
$NaLa(WO_4)_2$			•							
$NaLuGeO_4$			•				•			
$NaNdP_4O_{12}$			•							
$NaYGeO_4$			•							
$NaY(MoO_4)_2$			•							
$NaY(WO_4)_2$			•							
$Na(Nd,Gd)(WO_4)_2$			•							
$Na_3Nd(PO_4)_2$			•							
$Na_3(La,Nd)(PO_4)_2$			•							
$Na_5(Nd,La)(MoO_4)_4$			•							
$Na_5(Nd,La)(WO_4)_4$			•							
$NdAl_3(BO_3)_4$			•							
$Nd(Ga,Cr)_3(BO_3)_4$			•							
$NdGaGe_2O_7$			•							
$Nd_3Ga_5O_{12}$			•							
$Nd_3Ga_5GeO_{14}$			•							
$Nd_3Ga_5SiO_{14}$			•							
NdP_5O_{14}			•							
$PbMoO_4$			•							
$PbWO_4$			•							
$Pb_5Ge_3O_{11}$			•							
PrP_5O_{14}		•								
$RbNd(WO_4)_2$			•							
$ScBeAlO_4$			•						•	
Sc_2O_3			•							
Sc_2SiO_5								•		
$SrAl_2O_4$			•				•			
$SrAl_4O_7$			•				•	•	•	•
$SrAl_{12}O_{19}$			•		•				•	
$Sr_xBa_{1-x}(NbO_3)_2$			•							
$SrGdGa_3O_7$			•							•
$SrGd_4(SiO_4)_3O$			•				•			

Table 1.1.8—*continued*
Host Crystals Used for Trivalent Lanthanide Laser Ions

Crystal	Ce^{3+}	Pr^{3+}	Nd^{3+}	Sm^{3+}	Eu^{3+}	Dy^{3+}	Ho^{3+}	Er^{3+}	Tm^{3+}	Yb^{3+}
SrLaGa$_3$O$_7$		•						•		
SrMoO$_4$		•	•				•			
SrWO$_4$			•				•			
SrY$_4$(SiO$_4$)$_3$O							•	•	•	
Sr$_2$Ca$_3$(PO$_4$)$_3$			•							
Sr$_3$Ca$_2$(PO$_4$)$_3$			•							
Sr$_3$Ga$_2$Ge$_4$O$_{14}$			•							
Sr$_3$Ga$_2$GeO$_{14}$									•	
Sr$_4$Ca(PO$_4$)$_3$			•							
Sr$_5$(PO$_4$)$_3$			•							•
Tm$_3$Al$_5$O$_{12}$							•			
YAlO$_3$		•	•				•	•	•	
(Y,Er)AlO$_3$							•	•	•	
(Y,Gd)AlO$_3$								•		
YAl$_3$(BO$_3$)$_4$			•							
YP$_5$O$_{14}$			•							
(Y,Nd)P$_5$O$_{14}$			•							
YScO$_3$			•				•	•		
YVO$_4$		•	•		•		•		•	
Y$_2$O$_3$			•				•	•		•
Y$_2$O$_3$-ThO$_2$			•				•	•		
Y$_2$SiO$_5$			•							
Y$_3$Al$_3$O$_{12}$			•							
Y$_3$Al$_3$O$_{12}$			•							
(Y,Ce)$_3$Al$_5$O$_{12}$			•							
(Y,Lu)$_3$Al$_5$O$_{12}$			•							
Y$_3$Ga$_5$O$_{12}$			•							
Y$_3$Sc$_2$Al$_3$O$_{12}$			•							
Y$_3$Sc$_2$Ga$_3$O$_{12}$			•				•	•		
Yb$_3$Al$_5$O$_{12}$							•	•		
ZrO$_2$-Er$_2$O$_3$							•	•	•	
ZrO$_2$-Y$_2$O$_3$			•							
Halides										
BaF$_2$			•							
BaF$_2$-CeF$_3$			•							
BaF$_2$-GdF$_3$			•							
BaF$_2$-LaF$_3$			•							
BaF$_2$-YF$_3$			•							
BaEr$_2$F$_8$							•	•	•	

Table 1.1.8—*continued*
Host Crystals Used for Trivalent Lanthanide Laser Ions

Crystal	Ce³⁺	Pr³⁺	Nd³⁺	Sm³⁺	Eu³⁺	Dy³⁺	Ho³⁺	Er³⁺	Tm³⁺	Yb³⁺
$Ba(Y,Er)_2F_8$								•		
$Ba(Y,Yb)_2F_8$							•			
$BaYb_2F_8$	•	•				•	•	•	•	
CaF_2			•				•	•	•	•
$CaLu_2F_8$			•							
$Ca_2Y_5F_{19}$			•							
$CaF_2\text{-}CeF_3$			•							
$CaF_2\text{-}ErF_3$							•	•		
$CaF_2\text{-}ErF_3\text{-}TmF_3$								•	•	
$CaF_2\text{-}ErF_3\text{-}TmF_3\text{-}$ YbF₃							•			
$CaF_2\text{-}GdF_3$			•							
$CaF_2\text{-}HoF_3$			•				•	•		
$CaF_2\text{-}HoF_3\text{-}ErF_3$								•		
$CaF_2\text{-}LaF_3$			•							
$CaF_2\text{-}NdF_3$			•							
$CaF_2\text{-}ScF_3$			•							
$CaF_2\text{-}SrF_2$			•							
$CaF_2\text{-}SrF_2\text{-}BaF_2\text{-}$ YF₃-LaF₃			•							
$CaF_2\text{-}YF_3$			•				•	•		
$CaF_2\text{-}YF_3\text{-}NdF_3$			•							
CdF_2			•							
$CdF_2\text{-}CeF_3$			•							
$CdF_2\text{-}GaF_3$			•							
$CdF_2\text{-}GdF_3$			•							
$CdF_2\text{-}LaF_3$			•							
$CdF_2\text{-}LuF_3$			•							
$CdF_2\text{-}ScF_3$			•							
$CdF_2\text{-}YF_3$			•							
$CdF_2\text{-}YF_3\text{-}NdF_3$			•							
$CeCl_3$			•							
CeF_3			•							
$CsGd_2F_7$			•							
CsY_2F_7			•							
$ErF_3\text{-}HoF_3$								•		
$ErLiF_4$								•		
$GdF_3\text{-}CaF_2$			•							
$GdLiF_4$			•							
$HoLiF_4$							•			

Table 1.1.8—*continued*
Host Crystals Used for Trivalent Lanthanide Laser Ions

Crystal	Ce^{3+}	Pr^{3+}	Nd^{3+}	Sm^{3+}	Eu^{3+}	Dy^{3+}	Ho^{3+}	Er^{3+}	Tm^{3+}	Yb^{3+}
KYF_4			•							
KY_3F_{10}			•							
K_7YF_5			•							
$K_5(Nd,Ce)Li_2F_{10}$			•							
$K_5NdLi_2F_{10}$			•							
$LaBr_3$		•								
$LaCl_3$		•								
$(La,Pr)Cl_3$		•							•	
LaF_3	•	•	•				•	•		
$LaF_3\text{-}SrF_2$			•							
$LiCaAlF_6$	•									
$LiErF_4$								•		
$LiGdF_4$			•							
$LiHoF_4$							•			
$LiKYF_5$			•							
$LiLuF_4$	•	•	•					•	•	
$LiSrAlF_6$	•									
$LiYF_4$	•	•	•			•	•	•	•	
$Li(Y,Er)F_4$							•	•		
$LiYbF_4$		•					•	•		
MgF_2									•	
MnF_2								•		
$\alpha\text{-}NaCaCeF_6$			•							
$\alpha\text{-}NaCaErF_6$		•				•	•		•	
$\alpha\text{-}NaCaYF_6$			•							
$5NaF\text{-}9YF_3$			•							
$Na_{0.4}Y_{0.6}F_{2.2}$			•							
$PbCl_2$			•							
$PrBr_3$		•								
$PrCl_3$		•								
PrF_3		•								
SrF_2			•						•	
$SrF_2\text{-}(Y,Er)F_3$							•			
$SrF_2\text{-}CeF_3$			•							
$SrF_2\text{-}CeF_3\text{-}GdF_3$			•							
$SrF_2\text{-}ErF_3$								•		
$SrF_2\text{-}GdF_3$			•							
$SrF_2\text{-}LaF_3$			•							
$SrF_2\text{-}LuF_3$			•							
$SrF_2\text{-}ScF_3$			•							

Table 1.1.8—*continued*
Host Crystals Used for Trivalent Lanthanide Laser Ions

Crystal	Ce^{3+}	Pr^{3+}	Nd^{3+}	Sm^{3+}	Eu^{3+}	Dy^{3+}	Ho^{3+}	Er^{3+}	Tm^{3+}	Yb^{3+}
SrF_2-YF_3			•							
$Sr_2Y_5F_{19}$			•				•	•	•	
TbF_3				•						
YF_3			•							
Oxyhalides										
$BaCaBO_3F$									•	•
$Ba_5(PO_4)_3F$			•							
CaF_2-CeO_2			•							
$Ca_3Sr_2(PO_4)_3F$										•
$Ca_4Sr(PO_4)_3F$										•
$Ca_5(PO_4)_3F$			•						•	•
$Na_2Nd_2Pb_6$-$(PO_4)_6Cl_2$			•							
$Pb_5(PO_4)_3F$			•							
$Sr_5(PO_4)_3F$			•							
$Sr_5(VO_4)_3Cl$			•							
$Sr_5(VO_4)_3F$			•							
Chalcogenides										
La_2O_2S			•							

1.1.4 Tables of Transition Metal Ion Lasers

Table 1.1.9
Transition Metal Ion Lasers

Optical pump

AL	— alexandrite ($BeAl_2O_4$:Cr) laser
ArL	— argon-ion laser
D	— frequency doubled
DL	— dye laser
ErLYF	— Er:$LiYF_4$ (YLF) laser
ErYAG	— Er:$Y_3Al_5O_{12}$ (YAG) laser
Hg	— mercury arc lamp
KrL	— krypton-ion laser
NdGL	— Nd:glass laser
NdL	— neodymium laser
NdYAG	— Nd:$Y_3Al_5O_{12}$ (YAG) laser
NdYLF	— Nd:$LiYF_4$ (YLF) laser
NdYAP	— Nd:$YAlO_3$ (YAP) laser
RL	— ruby (Al_2O_3:Cr) laser
RS	— Raman-shifted
TiS	— Ti:sapphire (Al_2O_3) laser
TmYAP	— Tm:$YAlO_3$ (YAP) laser
TmHoYAG	— Tm,Ho:$Y_3Al_5O_{12}$ (YAG) laser
W	— tungsten arc lamp
Xe	— xenon arc lamp

Mode of operation

AML	— actively mode-locked
cw	— continuous wave
p	— pulsed
qcw	— quasi-continuous wave
qs	— Q-switched
PML	— passively mode-locked
SML	— synchronously mode-locked

Titanium (Ti^{3+}, $3d^1$)

Host crystal	Laser transition	Wavelength (µm)	Temp. (K)	Optical pump	Mode	Ref.
Al_2O_3	$^2E \rightarrow {}^2T_2$	0.66–1.178	300	Ar laser	cw	82–89
			80	Ar laser	cw	83
			300	dye laser	p	83, 90–96
			300	Xe lamp	p	91, 109–112
			300	DNdYAP	p	83, 84, 86, 92, 97–108
			300	Cu laser	p	1039
			510	DNdYAG	p	84
			300	DNdYAG	qs	99
			300	Ar laser	AML	110
			300	DNdYAP	SML	113
		0.700–0.818	300	sun	cw	1155
$BeAl_2O_4$	$^2E \rightarrow {}^2T_2$	0.73–0.95	300	DNdYAG	cw	170
			300	DNdYAG	p	171,172
		0.753–0.946	300	Xe lamp	p	189
$YAlO_3$	$^2E \rightarrow {}^2T_2$	0.6116	300	DNdYAP	p	59

Vanadium (V^{2+}, $3d^3$)

Host crystal	Laser transition	Wavelength (μm)	Temp. (K)	Optical pump	Mode	Ref.
CsCaF$_3$	$^4T_2 \rightarrow {}^4A_2$	1.24–1.33	80	Kr laser	cw	582
MgF$_2$	$^4T_2 \rightarrow {}^4A_2$	1.07–1.16	80	Ar laser	cw	261, 303–305
		1.1213	77	Xe lamp	p	488

Chromium (Cr^{2+}, $3d^4$)

Host crystal	Laser transition	Wavelength (μm)	Temp. (K)	Optical pump	Mode	Ref.
CdMnTe	$^5E \rightarrow {}^5T_2$	2.515	300	RS NdYAG	p	1031
	$^5E \rightarrow {}^5T_2$	2.17–3.01	300	TmHoYAG	p	1157
ZnS	$^5E \rightarrow {}^5T_2$	2.286–2.530	300	Co:MgF$_2$ L	p	914
		~2.35	300	Co:MgF$_2$ L	p	915
		2.134–2.799	300	Co:MgF$_2$ L	p	914
		~2.35	300	Co:MgF$_2$ L	p	915
ZnSe	$^5E \rightarrow {}^5T_2$	2.134–2.799	300	Co:MgF$_2$ L	p	914
		2.138–2.760	300	TmYAP	cw	1124
		~2.35	300	Co:MgF$_2$ L	p	915

Chromium (Cr^{3+}, $3d^3$)

Host crystal	Laser transition	Wavelength (μm)	Temp. (K)	Optical pump	Mode	Ref.
Al$_2$(WO$_4$)$_3$	$^4T_2 \rightarrow {}^4A_2$	0.80	300	Kr laser	cw	210
Al$_2$O$_3$	$^2E \rightarrow {}^4A_2$	0.6929(R$_2$)	300	Xe lamp	p	125
		0.6934	77	Hg lamp	cw	126, 127
		0.6934	77	Ar laser	cw	128
		0.6943(R$_1$)	300	Xe lamp	p	131–2, 138
		0.6943(R$_1$)	300	Hg lamp	cw	133–4, 297
		0.6943(R$_1$)	300	Kr laser	cw	135
		0.7009(N$_2$)	77	Xe lamp	p	153
		0.7041(N$_1$)	77	Xe lamp	p	153
		0.6943–0.6952	300–500	Xe lamp	p	137
		0.7670	300	Xe lamp	p	197
Be$_3$Al$_2$Si$_6$O$_{18}$	$^2E \rightarrow {}^4A_2$	0.685	300	RS-DNdL	p	123
	$^4T_2 \rightarrow {}^4A_2$	0.720–0.842	300	Kr laser	cw	164, 165
		0.720–0.842	300	Xe lamp	p	166
BeAl$_2$O$_4$	$^2E \rightarrow {}^4A_2$	~0.680	77	Xe lamp	p	118
		0.6803	300	Xe lamp	p	120

Chromium (Cr^{3+}, 3d^3)—*continued*

Host crystal	Laser transition	Wavelength (μm)	Temp. (K)	Optical pump	Mode	Ref.
BeAl$_2$O$_4$	$^2E \rightarrow {}^4A_2$	0.6804	300	Xe lamp	p	121,122
	$^4T_2 \rightarrow {}^4A_2$	0.70–0.82	—	Hg lamp	cw	140–142
			300	Kr laser	cw	143
BeAl$_2$O$_4$	$^4T_2 \rightarrow {}^4A_2$	0.70–0.82	—	Xe lamp	cw	141, 142
			300–330	Xe lamp	p	120, 144–6
			330–370	Xe lamp	p	120, 145, 148, 149
			548–583	Xe lamp	p	142, 146
			300–370	Xe lamp	p	141, 142, 144, 147, 148, 150
				Xe lamp	PML	142, 151
				Xe lamp	AML	142
		0.701–0.818	300	Xe lamp	p	121, 154
		0.744–0.788	300	Hg lamp	cw	179
BeAl$_6$O$_{10}$	$^4T_2 \rightarrow {}^4A_2$	0.79–0.87	300	DNdYAG	p	204
Ca$_3$Ga$_2$Ge$_4$O$_{14}$	$^4T_2 \rightarrow {}^4A_2$	0.87–1.21	300	RL, DL	p	241, 1017
(Gd,Ca)$_3$-(Ga,Mn,Zr)$_5$O$_{12}$	$^4T_2 \rightarrow {}^4A_2$	0.774–0.814	300	Xe lamp	p	198
Gd$_3$Ga$_5$O$_{12}$	$^4T_2 \rightarrow {}^4A_2$	0.769	300	Kr laser	cw	174
Gd$_3$Sc$_2$Al$_3$O$_{12}$	$^4T_2 \rightarrow {}^4A_2$	0.75–0.81	300	Xe lamp	p	183–185
				Kr laser	cw	174–5, 182
				Ar laser	cw	174–5, 182
Gd$_3$Sc$_2$Ga$_3$O$_{12}$	$^4T_2 \rightarrow {}^4A_2$	0.742–0.842	300	Xe lamp	p	177, 178
				Kr laser	cw	174–176
				Ar laser	cw	174–176
KZnF$_3$	$^4T_2 \rightarrow {}^4A_2$	0.766–0.865	300	Kr laser	cw	193
				dye laser	p	191, 192
				ruby laser	qcw	194
				Xe lamp	p	195
		0.775–0.816	80	Kr laser	cw	191, 192
		0.790–0.825	200	Kr laser	cw	191, 192
(La,Lu)$_3$(La,Ga)$_2$-Ga$_3$O$_{12}$	$^4T_2 \rightarrow {}^4A_2$	0.83	300	Kr laser	cw	174
La$_3$Ga$_5$GeO$_{14}$	$^4T_2 \rightarrow {}^4A_2$	0.88–1.22	300	ruby laser	p	241, 242, 1017
La$_3$Ga$_{5.5}$Nb$_{0.5}$O$_{14}$	$^4T_2 \rightarrow {}^4A_2$	0.9–1.25	300	ruby laser	p	240, 1017

Chromium (Cr^{3+}, $3d^3$)—*continued*

Host crystal	Laser transition	Wavelength (μm)	Temp. (K)	Optical pump	Mode	Ref.
$La_3Ga_5GeO_{14}$	$^4T_2 \to {}^4A_2$	0.88–1.22	300	ruby laser	p	241, 242, 1017
$La_3Ga_{5.5}Nb_{0.5}O_{14}$	$^4T_2 \to {}^4A_2$	0.9–1.25	300	ruby laser	p	240, 1017
$La_3Ga_5SiO_{14}$	$^4T_2 \to {}^4A_2$	0.815–1.22	300	Kr laser	cw	209, 1017
				ruby laser	p	208
$La_3Ga_{5.5}Ta_{0.5}O_{14}$	$^4T_2 \to {}^4A_2$	0.925–1.24	300	ruby laser	p	240, 241
$LiCaAlF_6$	$^4T_2 \to {}^4A_2$	0.72–0.84	300	Kr laser	cw	162
				Xe lamp	p	163
$LiSr_{0.8}Ca_{0.2}AlF_6$	$^4T_2 \to {}^4A_2$	0.750–0.950	300	Xe lamp	p	186
$LiSrAlF_6$	$^4T_2 \to {}^4A_2$	0.780–1.010	300	Xe lamp	p	201
		0.78–0.92	300	Kr laser	cw	196, 199
		0.809–0.910	300	LD	PML	894
		0.815–0.915	300	NdYLF	p	200
		~0.825–0.875	300	LD	PML	1066
$LiSrCrF_6$	$^4T_2 \to {}^4A_2$	0.890	300	TiS laser	p	243
$LiSrGaF_6$	$^4T_2 \to {}^4A_2$	0.820	300	Kr laser	p	212, 1025
$Na_3Ga_3Li_3F_{12}$	$^4T_2 \to {}^4A_2$	0.748–0.832	300	Kr laser	cw	180
$ScBeAlO_4$	$^4T_2 \to {}^4A_2$	0.792	300	Kr laser	cw	206
$ScBO_3$	$^4T_2 \to {}^4A_2$	0.787–0.892	300	Kr laser	cw	162, 202, 203
$Sr_3Ga_2Ge_4O_{14}$	$^4T_2 \to {}^4A_2$	0.895	300	ruby laser	p	1017
		0.90–1.15	300	ruby laser	p	241, 242
$SrAlF_5$	$^4T_2 \to {}^4A_2$	0.852–1.005	300	Kr laser	cw	227, 228
$Y_3Al_5O_{12}$	$^4T_2 \to {}^4A_2$	0.6874	~77	Xe lamp	p	124
$Y_3Ga_5O_{12}$	$^4T_2 \to {}^4A_2$	0.74	300	Kr laser	cw	173
$Y_3Sc_2Al_3O_{12}$	$^4T_2 \to {}^4A_2$	0.767	300	Kr laser	cw	196
$Y_3Sc_2Ga_3O_{12}$	$^4T_2 \to {}^4A_2$	0.76	300	Kr laser	cw	173
$ZnWO_4$	$^4T_2 \to {}^4A_2$	0.98–1.09	77	Kr laser	cw	271
		0.98–1.09	300	dye laser	p	271

Chromium (Cr^{4+}, $3d^2$)

Host crystal	Laser transition	Wavelength (µm)	Temp. (K)	Optical pump	Mode	Ref.
$CaGd_4(SiO_4)_3O$	$^3T_2 \to {}^3A_2$	1.37	300	NdYAG	p	701
Ca_2GeO_4	$^3T_2 \to {}^3A_2$	1.348–1.482	300	NdYAG	p	629
		1.390–1.475	300	laser diode	cw	1032
$Ca_3Ga_2Ge_3O_{12}$	$^3T_2 \to {}^3A_2$	1.32–1.43	300			433
$LiNbGeO_5$		1.31–1.52	~110	Nd laser	p	1045
		1.38–1.44	~200	Nd laser	p	1045
Mg_2SiO_4	$^3T_2 \to {}^3A_2$	1.221	300	NdYAG	p	708
		1.235	300	NdYAG	p	706, 709
		1.244	300	NdYAG	cw	707
		1.195–1.295	300	NdYAG	SML	507
		1.2–1.32	77	NdYAG	cw	514
		1.204–1.277	300	NdYAG	AML	507
		1.236–1.300	300	laser diode	cw	1032
		1.244	300	NdYAG	cw	552
			300	alex. laser	p	558
$SrGd_4(SiO_4)_3O$	$^3T_2 \to {}^3A_2$	1.44	300	NdYAG	p	701
$Y_3Al_5O_{12}$	$^3T_2 \to {}^3A_2$	1.32–1.53	300	NdYAG	p, cw	585, 589
		1.350–1.560	300	NdYAG	p,c w	647
		1.35–1.5	300			981, 982
		1.37–1.51	300	NdYAG	AML	664
Y_2SiO_5	$^3T_2 \to {}^3A_2$	1.18–1.29	77	NdYAG	p	508
			77	Cr:LiSAF	qs	508
		1.3	300	AL, RL	p	567
$Y_3(Sc_x,Al)_{5-x-}$ O_{12}: 0<x<1.7	$^3T_2 \to {}^3A_2$	1.309–1.628	300	NdYAG	p	572

Manganese (Mn^{5+}, $3d^5$)

Host crystal	Laser transition	Wavelength (µm)	Temp. (K)	Optical pump	Mode	Ref.
$Ba_3(VO_4)_2$	$^1E \to {}^3A_2$	1.1810	300	dye laser	p	509

Iron (Fe^{2+}, $3d^6$)

Host crystal	Laser transition	Wavelength (µm)	Temp. (K)	Optical pump	Mode	Ref.
n-InP	$^5T_2 \to {}^5E$	3.53	2	dye laser	p	748
ZnSe	$^5T_2 \to {}^5E$	3.98–4.54	15–180	ErYAG	p	1158

Cobalt (Co^{2+}, 3d^7)

Host crystal	Laser transition	Wavelength (µm)	Temp. (K)	Optical pump	Mode	Ref.
KMgF$_3$	$^4T_2 \rightarrow {}^4T_1$	1.62–1.90	80	Ar laser	cw	192
		1.821	77	Xe lamp	p	535,1003
KZnF$_3$	$^4T_2 \rightarrow {}^4T_1$	1.65–2.15	27	Ar laser	cw	785
		1.65–2.15	80	Ar laser	cw	192,786, 1003
MgF$_2$	$^4T_2 \rightarrow {}^4T_1$	1.5–2.3	80	NYAG	cw	114,543, 768
			80	NYAP	p	704,762
			225	NYAP	p	766,768
			299	NYAG	p	769
			80	NYAP	p, qs	766,768
			80	NYAG	qs	770
			80	NYAG	AML	771
			80	NYAG	AML/qs	772
		1.63–2.11	80	NYAG	cw	754
		1.750	77	Xe lamp	p	535,801
		1.800–2.450	300	NYAG	p	799,1003
		1.8035	77	Xe lamp	p	535,801
		1.99	77	Xe lamp	p	535,801
		2.05	77	Xe lamp	p	535,801
ZnF$_2$	$^4T_2 \rightarrow {}^4T_1$	2.165	77	Xe lamp	p	535,801

Nickel (Ni^{2+}, 3d^8)

Host crystal	Laser transition	Wavelength (µm)	Temp. (K)	Optical pump	Mode	Ref.
CaY$_2$Mg$_2$Ge$_3$O$_{12}$	$^3T_2 \rightarrow {}^3A_2$	1.46	80	NdYAP	p	261,704
KMgF$_3$	$^3T_2 \rightarrow {}^3A_2$	1.59	80	Xe lamp	p	261,774
MgF$_2$	$^3T_2 \rightarrow {}^3A_2$	1.61–1.74	80	Kr laser	cw	261,775
			80	NdYAG	cw	261,754
			80	NdYAP	p	704
			80	NdYAG	qs	754
			80	NdYAG	AML	771
		1.623	77	Xe lamp	p	535,739
		1.636	77–82	Xe lamp	p	740
		1.674–1.676	82–100	Xe lamp	p	535,739
		1.731–1.756	100–192	Xe lamp	p	535,739
		1.75	200	NdYAG	cw	806
		1.785–1.797	198–240	NdYAG	cw	535,739

Nickel (Ni^{2+}, $3d^8$)—*continued*

Host crystal	Laser transition	Wavelength (μm)	Temp. (K)	Optical pump	Mode	Ref.
MgO	$^3T_2 \rightarrow {}^3A_2$	1.3144	77	Xe lamp	p	535
		1.316	82	NdYAG	cw	662
		1.318	80	NdYAG	cw	261
		1.328	131	NdYAG	cw	662
		1.369	153	NdYAG	cw	662
		1.409	235	NdYAG	cw	662
MnF_2	$^3T_2 \rightarrow {}^3A_2$	1.865	20	Xe lamp	p	535
		1.915	77	Xe lamp	p	535
		1.922	77	Xe lamp	p	535
		1.929	85	Xe lamp	p	535
		1.939	85	Xe lamp	p	535

1.1.5 Tables of Divalent Lanthanide Lasers

Table 1.1.10
Divalent Lanthanide Ion Lasers

Optical pump	Mode of operation
Hg — mercury arc lamp	cw — continuous wave
W — tungsten arc lamp	p — pulsed
Xe — Xe arc lamp	

Samarium (Sm^{2+}, $4f^6$)

Host crystal	Laser transition	Wavelength (μm)	Temp. (K)	Optical pump	Mode	Ref.
CaF_2	$5d \rightarrow {}^7F_1$	0.7083	65–90	ruby laser	p	157
		0.7085*	20	Xe lamp	p	158–160
		0.7207	85–90	ruby laser	p	157
		0.7287	110–130	ruby laser	p	157
		0.7310	155	ruby laser	p	157
		0.745	210	ruby laser	p	157
SrF_2	${}^5D_0 \rightarrow {}^7F_1$	0.6969	4.2	Xe lamp	p	156

* Reference 159 includes whispering-galley-mode lasing.

Dysprosium (Dy^{2+}, $4f^{10}$)

Host crystal	Laser transition	Wavelength (μm)	Temp. (K)	Optical pump	Mode	Ref.
CaF_2	${}^5I_7 \rightarrow {}^5I_8$	2.35867	77	Hg lamp	p	906
		2.35867	4.2	Xe lamp	cw	907
		~2.358	4.2	W lamp	cw	908
		~2.358	77	sun	cw	909
		~2.358	4.2–77	W lamp	cw	910
		~2.358	4, 77	Xe lamp	p	911
SrF_2	${}^5I_7 \rightarrow {}^5I_8$	2.3659	20	Xe lamp	p	903

Thulium (Tm^{2+}, $4f^{13}$)

Host crystal	Laser transition	Wavelength (μm)	Temp. (K)	Optical pump	Mode	Ref.
CaF_2	${}^2F_{5/2} \rightarrow {}^2F_{7/2}$	1.116	4.2	Xe lamp	p	524
		1.116	2.7–4.2	Hg lamp	cw	525

1.1.6 Tables of Trivalent Lanthanide Ion Lasers

Table 1.1.11
Trivalent Lanthanide Ion Lasers

Optical pump

alex.	–	alexandrite (BeAl$_2$O$_4$:Cr) laser	NdL	–	neodymium laser
ArL	–	argon-ion laser	NdYAG	–	Nd:Y$_3$Al$_5$O$_{12}$ laser
CCL	–	color center laser	NdYLF	–	Nd:LiYF$_4$ laser
CeLiSAF	–	Ce:LiSrAlF$_6$ laser	NdYAP	–	Nd:YAlO$_3$ laser
CrLiSAF	–	Cr:LiSrAlF$_6$ laser	PT	–	pyrotechnical pumping
D	–	frequency doubled	Q	–	frequency quadrupled
DL	–	dye laser	RL	–	ruby laser
DyL	–	dysprosium laser	RS	–	Raman-shifted
ErGL	–	Er:glass laser	T	–	frequency tripled
Hg	–	mercury arc lamp	TiS	–	Ti:sapphire laser
Kr	–	krypton arc lamp	TmYAG	–	Tm:Y$_3$Al$_5$O$_{12}$ laser
KrFL	–	krypton fluoride excimer laser	W	–	tungsten arc lamp
LED	–	light emitting diode	qWI	–	quartz-tungsten-iodine lamp
LD	–	laser diode	Xe	–	Xe arc lamp
NdGL	–	Nd:glass laser	XeClL	–	XeCl laser

Mode of operation

AML	–	actively mode locked	qs	–	Q-switched
cas	–	cascade pumping/deactivation	pa	–	photon avalanche pumping
cw	–	continuous wave	SML	–	synchronously mode locked
p	–	pulsed	sml	–	soliton mode locked
PML	–	passively mode locked	sp	–	self pulsed
qcw	–	quasi-continuous wave	upc	–	upconversion pumping scheme

Cerium (Ce^{3+}, 4f^1)

Host crystal	Laser transition	Wavelength (μm)	Temp. (K)	Optical pump	Mode	Ref.
BaY$_2$F$_8$	5d → 4f	0.345	300	XeCl laser	p	14
LaF$_3$	5d → 4f	0.286	300	Kr laser	p	7
LiCaAlF$_6$	5d → 4f	0.278–0.335	300	QNdYAG	p	6
		0.2805–0.316	300	Cu laser	p	1128
		0.281–0.315	300	QNdYAG	p	238
		0.288	300	QNdYAG	p	8
LiLuF$_4$	5d → 4f	~0.308	300	Ce:LiSAF	p	1068
		0.323–0.335	300	Kr laser	p	12,
LiSrAlF$_6$	5d → 4f	0.275–0.339	300	QNdYAG	p	6
		0.283–0.313	300	QNdYAG	p	238
		0.285–0.297	300	QNdYAG	p	5
LiYF$_4$	5d → ^{2}F$_{5/2}$	0.306–0.315	300	Kr laser	p	9
	5d → ^{2}F$_{7/2}$	0.323–0.328	300	Kr laser	p	9
		0.3252	300	KrFL/ArFL	p	11

Praseodymium (Pr^{3+}, 4f^2)

Host crystal	Laser transition	Wavelength (μm)	Temp. (K)	Optical pump	Mode	Ref.
BaY$_2$F$_8$	$^3P_0 \to {}^3H_6$	0.6071	110, 300	Xe lamp	p	55, 56
	$^3P_0 \to {}^3F_2$	0.6388	300	Xe lamp	p	73
		0.6395	300	Xe lamp	p	1136
	$^3P_0 \to {}^3F_3$	0.6935–8	110, 190	Xe lamp	p	55, 56
	$^3P_0 \to {}^3F_4$	0.7191	300	Xe lamp	p	73
		0.7206	300	Xe lamp	p	1136
	$^3P_1 \to {}^1G_4$	0.9010	110	Xe lamp	p	245
	$^3P_0 \to {}^1G_4$	0.9148	110	Xe lamp	p, cas	241, 245
		0.9150	300	Xe lamp	p	245
	$^1G_4 \to {}^3H_5$	1.3345	110	Xe lamp	p, cas	241
BaYb$_2$F$_8$:Yb	$^1G_4 \to {}^3H_5$	1.3347	110	Xe lamp	p	241, 623
	$^1G_4 \to {}^3F_4$	3.6045	110	Xe lamp	p, cas	241
		3.6050	110	Xe lamp	p, cas	241, 623
Ca(NbO$_3$)$_2$	$^3P_0 \to {}^3H_6$	0.6105	110	Xe lamp	p	57
	$^1G_4 \to {}^3H_4$	1.04	77	Xe lamp	p	753
	$^1D_2 \to {}^3F_4$	1.0662	~110	Xe lamp	p	73, 1136
		1.0664	300	Xe lamp	p	73, 1136
CaWO$_4$	$^3P_0 \to {}^3F_2$	0.6497	110	Xe lamp	p	57
	$^1D_2 \to {}^3F_4$	1.0468	77	Xe lamp	p	310
		1.0495	300	Xe lamp	p	73, 1136
GdLiF$_4$	$^3P_1 \to {}^3H_5$	0.522	300	dye laser	p	29
	$^3P_0 \to {}^3H_5$	0.545	300	dye laser	p	29
	$^3P_0 \to {}^3H_6$	0.6045	300	dye laser	p	29
		0.607	300	dye laser	p	29
	$^3P_0 \to {}^3F_2$	0.639	300	dye laser	p	29
	$^3P_0 \to {}^3F_3$	0.697	300	dye laser	p	29
	$^3P_0 \to {}^3F_4$	0.720	300	dye laser	p	29
Gd$_2$SiO$_5$	$^3P_0 \to {}^3H_6$	0.660	300	dye laser	p	1033
KGd(WO$_4$)$_2$	$^1D_2 \to {}^3F_4$	1.0657	300	Xe lamp	qcw	1143
		1.0657	300	Xe lamp	p	1048
KY(WO$_4$)$_2$	$^1D_2 \to {}^3F_3$	1.0223	300	Xe lamp	p	1048, 1147
	$^1D_2 \to {}^3F_4$	1.0697	300	Xe lamp	p	1048, 1143
LaBGeO5	$^3P_0 \to {}^3H_6$	0.6080	~110	Xe lamp	p	330
LaBr$_3$	$^3P_1 \to {}^3H_5$	0.532	<300	dye laser	p	30
	$^3P_1 \to {}^3H_6$	0.621	<300	dye laser	p	30
	$^3P_2 \to {}^3F_3$	0.632	< 70	dye laser	p	30
	$^3P_0 \to {}^3F_2$	0.647	<300	dye laser	p	30

Praseodymium (Pr^{3+}, 4f^2)—*continued*

Host crystal	Laser transition	Wavelength (µm)	Temp. (K)	Optical pump	Mode	Ref.
LaCl$_3$	$^3P_0 \rightarrow {}^3H_4$	0.4892	5.5–14	dye laser	p	28
	$^3P_1 \rightarrow {}^3H_5$	0.5298	<300	dye laser	p	28
	$^3P_0 \rightarrow {}^3H_6$	0.6164	8, 65	dye laser	p	28
	$^3P_0 \rightarrow {}^3F_2$	0.644	80–210	dye laser	cw, pa	1038
		0.6452	300	dye laser	p	28
	$^3F_3 \rightarrow {}^3H_4$	1.611	130	TmYAG	p	761
		1.624	130	TmYAG	p	761
		1.644	130	TmYAG	p	761
	$^3F_3 \rightarrow {}^3H_6$	5.242	130–250	TmYAG	p	760
	$^3F_3 \rightarrow {}^3F_2$	7.141	~300	TmYAG	p	955
		7.152	~300	TmYAG	p	955
		7.24	148	TmYAG	p	950
		7.244	130	TmYAG	p	955
(La,Pr)Cl$_3$	$^3P_0 \rightarrow {}^3H_4$	0.4892	14	dye laser	p	26–28
	$^3P_1 \rightarrow {}^3H_5$	0.5298	12	dye laser	p	26–28
	$^3P_0 \rightarrow {}^3H_6$	0.6164	65, 300	dye laser	p	26–28
	$^3P_0 \rightarrow {}^3F_2$	0.6451	65, 300	dye laser	p	26–28
LaF$_3$	$^3P_0 \rightarrow {}^3H_6$	0.5985	77	Xe lamp	p	52
		0.5985	110	Xe lamp	p	50, 51
		0.6001	110	Xe lamp	p	50, 51
	$^3P_0 \rightarrow {}^3F_4$	0.7194	110	Xe lamp	p	50, 51
		0.7198	300	Xe lamp	p	50, 51, 161
	$^3P_0 \rightarrow {}^1G_4$	0.9185	110	Xe lamp	p	245
		0.9190	300	Xe lamp	p	245
(La,Pr)P$_5$O$_{14}$	$^3P_0 \rightarrow {}^3F_2$	0.637	300	dye laser	p	68
	$^3P_0 \rightarrow {}^3F_4$	0.717	300	dye laser	p	68
LiLuF$_4$	$^3P_0 \rightarrow {}^3H_5$	0.5380	110	Xe lamp	p	32, 33
	$^3P_0 \rightarrow {}^3H_6$	0.6042	110	Xe lamp	p	32, 33
		0.6071	110	Xe lamp	p	32, 33
	$^3P_0 \rightarrow {}^3F_2$	0.6399	110	Xe lamp	p	32, 33
		0.6401	300	Xe lamp	p	32, 33
	$^3P_0 \rightarrow {}^3F_3$	0.6958	110	Xe lamp	p	32, 33
		0.6977	110	Xe lamp	p	32
	$^3P_0 \rightarrow {}^3F_4$	0.7192	110	Xe lamp	p	32, 33
		0.7215	300	Xe lamp	p	32, 33
	$^3P_0 \rightarrow {}^1G_4$	0.9068	110–250	Xe lamp	p	246
LiPrP$_4$O$_{12}$	$^3P_0 \rightarrow {}^3H_6$	0.6048	300	dye laser	p	54
	$^3P_0 \rightarrow {}^3F_2$	0.6396	300	dye laser	p	74
	$^3P_0 \rightarrow {}^3F_4$	0.7204	300	dye laser	p	54
LiYF$_4$	$^3P_0 \rightarrow {}^3H_4$	0.479	300	Dy laser	p	23

Praseodymium (Pr^{3+}, 4f^2)—*continued*

Host crystal	Laser transition	Wavelength (μm)	Temp. (K)	Optical pump	Mode	Ref.
LiYF$_4$	$^3P_0 \rightarrow {}^3H_5$	0.5378	110	Xe lamp	p	31, 1136
	$^3P_0 \rightarrow {}^3H_6$	0.6044	300	Ar laser	cw	53
	$^3P_0 \rightarrow {}^3H_6$	0.6071	110–200	Xe lamp	p	31
		0.6072	300	Xe lamp	p	1052
		0.6073	300	Ar laser	cw	53
		0.6092	300	Ar laser	cw	53
		0.6130	300	Xe lamp	p, PML	53
	$^3P_1 \rightarrow {}^3F_2$	0.6158	300	Ar laser	cw	53
		0.6180	300	Ar laser	cw	53
		0.6201	300	Ar laser	cw	53
	$^3P_0 \rightarrow {}^3F_2$	0.6388	300	Ar laser	cw	53
		0.6395	110, 300	Xe lamp	p	31, 75, 1052
			300	Xe lamp	qcw	1047
			300	Xe lamp	q, AML	1049
		0.64	300	Xe lamp	p	73
		0.6444	300	Ar laser	cw	53
	$^3P_1 \rightarrow {}^3F_3$	0.6703	300	Ar laser	cw	53
	$^3P_0 \rightarrow {}^3F_3$	0.6954	110–180	Xe lamp	p	31
		0.6977	300	Ar laser	cw	53
	$^3P_1 \rightarrow {}^3F_4$	0.6994	300	Ar laser	cw	53
	$^3P_1 \rightarrow {}^3F_3$	0.7055	300	Ar laser	cw	53
	$^1I_6 \rightarrow {}^3F_4$	0.7082	300	Ar laser	cw	53
		0.7190	180–250	Xe lamp	p	31
	$^3P_0 \rightarrow {}^3F_4$	0.7206	300	Xe lamp	p	31, 168
		0.7209	300	Ar laser	cw	53
			300	Xe lamp	p	1052
		0.7222	300	Ar laser	cw	53
		0.7195	300	Ar laser	cw	53
	$^3P_0 \rightarrow {}^1G_4$	0.9066	110	Xe lamp	p, cas	241, 246
		0.9069	300	Xe lamp	p	246
		0.9145	110	Xe lamp	p	246
	$^1G_4 \rightarrow {}^3H_5$	1.3465	110	Xe lamp	p, cas	241
LiYbF$_4$	$^1G_4 \rightarrow {}^3H_5$	1.3465	110	Xe lamp	p	241
		1.3468	110	Xe lamp	p	241, 246
LuAlO$_3$	$^3P_0 \rightarrow {}^3H_6$	0.6155	110	Xe lamp	p	61, 62
	$^3P_0 \rightarrow {}^3F_3$	0.722	300	Xe lamp	p	169
	$^3P_0 \rightarrow {}^3F_4$	0.7496	300	Xe lamp	p	169
PrBr$_3$	$^3P_0 \rightarrow {}^3F_2$	0.6451	300	dye laser	p	30
PrCl$_3$	$^3P_0 \rightarrow {}^3H_6$	0.617	300	dye laser	p	27, 28
		0.620	300	dye laser	p	27, 28
		0.622	300	dye laser	p	27, 28

Praseodymium (Pr^{3+}, 4f^2)—*continued*

Host crystal	Laser transition	Wavelength (μm)	Temp. (K)	Optical pump	Mode	Ref.
PrCl$_3$	$^3P_0 \rightarrow {}^3H_6$	0.647	300	dye laser	p	27,28
PrF$_3$	$^3P_0 \rightarrow {}^3H_6$	0.5984	15	dye laser	p	49
PrP$_5$O$_{14}$	$^3P_0 \rightarrow {}^3F_2$	0.6374	300	Xe lamp	p	69–71
SrLaGa$_3$O$_7$	$^3P_0 \rightarrow {}^3F_2$	0.645	300	dye laser	p	24
	$^3P_0 \rightarrow {}^3H_4$	0.488	≤230	dye laser	p	24
SrMoO$_4$	$^1D_2 \rightarrow {}^3F_4$	1.04	—	—	p	290
YAlO$_3$	$^3P_0 \rightarrow {}^3H_6$	0.6139	110	Xe lamp	p	60–62
			300	Xe lamp	p	63
		0.6213	110	Xe lamp	p	61, 62
		0.6216	300	Xe lamp	p	63
		0.6624	300	Xe lamp	p	63
	$^3P_0 \rightarrow {}^3F_3$	0.7195	110	Xe lamp	p	60, 62
		0.7195	300	Xe lamp	p	63
		0.7197	300	Xe lamp	p	61, 62
		0.722	300	Xe lamp	p	169
		0.7437	300	Xe lamp	p	63
	$^3P_0 \rightarrow {}^3F_4$	0.7469	300	Xe lamp	p	60, 61, 63
		0.7537	300	Xe lamp	p	63
	$^3P_0 \rightarrow {}^1G_4$	0.9395	300	Xe lamp	p	63
		0.9660	300	Xe lamp	p	63
		0.9308	110	Xe lamp	p	62
		0.9312	300	Xe lamp	p	62
	$^1D_2 \rightarrow {}^3F_3$	0.9960	300	Xe lamp	p	62
Y$_3$Al$_5$O$_{12}$	$^3P_0 \rightarrow {}^3H_4$	0.4879	4–32	dye laser	p	657
	$^3P_0 \rightarrow {}^3H_6$	0.616	4–140	dye laser	p	657

Neodymium (Nd^{3+}, 4f^3)

Host crystal	Laser transition	Wavelength (μm)	Temp. (K)	Optical pump	Mode	Ref.
BaF$_2$	$^4F_{3/2} \rightarrow {}^4I_{11/2}$	1.0540	300	Xe lamp	p	445
		1.060	77	Xe lamp	p	463
	$^4F_{3/2} \rightarrow {}^4I_{13/2}$	1.3175	300	Xe lamp	p	541
		1.3270	300	Xe lamp	p	541
BaF$_2$-CeF$_3$	$^4F_{3/2} \rightarrow {}^4I_{11/2}$	1.0537	300	Xe lamp	p	440
		1.0543	300	Xe lamp	p	413
BaF$_2$-GdF$_3$	$^4F_{3/2} \rightarrow {}^4I_{11/2}$	1.0526	300	Xe lamp	p	413
BaF$_2$-LaF$_3$	$^4F_{3/2} \rightarrow {}^4I_{11/2}$	1.0534–1.0563	300–920	Xe lamp	p	435

Neodymium (Nd^{3+}, $4f^3$)—*continued*

Host crystal	Laser transition	Wavelength (μm)	Temp. (K)	Optical pump	Mode	Ref.
BaF_2-LaF_3	$^4I_{13/2} \rightarrow ^4I_{11/2}$	1.0538	77	Xe lamp	p	435
		1.0580	77	Xe lamp	p	435
	$^4F_{3/2} \rightarrow ^4I_{13/2}$	1.3185	300	Xe lamp	p	602
		1.3280	300	Xe lamp	p	602
		1.3290	77	Xe lamp	p	602
	$^4I_{13/2} \rightarrow ^4I_{11/2}$	5.15	300	Xe lamp	p, cas	959, 991
BaF_2-YF_3	$^4F_{3/2} \rightarrow ^4I_{11/2}$	1.0521	300	Xe lamp	p	413
	$^4F_{3/2} \rightarrow ^4I_{13/2}$	1.3200	300	Xe lamp	p	413
$BaGd_2(MoO_4)_4$	$^4F_{3/2} \rightarrow ^4I_{11/2}$	1.061	300	Xe lamp	p	584
$BaLaGa_3O_7$	$^4F_{3/2} \rightarrow ^4I_{11/2}$	1.059	300	—	p	513
		1.0595	300	Xe lamp	p	249
$BaLu_2F_8$	$^4F_{3/2} \rightarrow ^4I_{11/2}$	1.0483	300	Xe lamp	p	1142
		1.0483	300	laser diode	cw	1046
BaY_2F_8	$^4F_{3/2} \rightarrow ^4I_{11/2}$	1.0493	77	Xe lamp	p	366
		1.0495	300	Xe lamp	p	366
			300	laser diode	cw, qcw	1079
		1.0529	77	Xe lamp	p	422
		1.0530	300	Xe lamp	p	366
	$^4F_{3/2} \rightarrow ^4I_{13/2}$	1.3170	77	Xe lamp	p	366
		1.318	300	Xe lamp	p	366
$Ba_2MgGe_2O_7$	$^4F_{3/2} \rightarrow ^4I_{11/2}$	1.05436	300	Xe lamp	p	447
$Ba_{0.25}Mg_{2.75}$-$Y_2Ge_3O_{12}$	$^4F_{3/2} \rightarrow ^4I_{11/2}$	1.0615	300	Xe lamp	p	603
$Ba_2NaNb_5O_{15}$	$^4F_{3/2} \rightarrow ^4I_{11/2}$	1.0613	300	Xe lamp	p	595
		1.0630, 1.0710	300	laser diode	cw	1152
	$^4F_{3/2} \rightarrow ^4I_{13/2}$	~1.3336	300	laser diode	cw	1152
$Ba_2ZnGe_2O_7$	$^4F_{3/2} \rightarrow ^4I_{11/2}$	1.05437	300	Xe lamp	p	448
$Ba_3LaNb_3O_{12}$	$^4F_{3/2} \rightarrow ^4I_{9/2}$	0.9106	300	Xe lamp	p	249
	$^4F_{3/2} \rightarrow ^4I_{11/2}$	1.0591	77	Xe lamp	p	249
		1.0595	300	Xe lamp	p	249
$Ba_5(PO_4)_3F$	$^4F_{3/2} \rightarrow ^4I_{11/2}$	1.0555	300	Xe lamp	p	1027
		1.3209	300	Xe lamp	p	1027
$Bi_4Ge_3O_{12}$	$^4F_{3/2} \rightarrow ^4I_{11/2}$	1.0638	77	Xe lamp	p	291
		1.06425	77	Xe lamp	p	291, 661
		1.0644	300	Xe lamp	p	291, 661
	$^4F_{3/2} \rightarrow ^4I_{13/2}$	1.3418	300	Xe lamp	p	291, 661

Neodymium (Nd^{3+}, $4f^3$)—*continued*

Host crystal	Laser transition	Wavelength (μm)	Temp. (K)	Optical pump	Mode	Ref.
$Bi_4Si_3O_{12}$	$^4F_{3/2} \rightarrow {}^4I_{11/2}$	1.0629	77, 300	Xe lamp	p	625
	$^4F_{3/2} \rightarrow {}^4I_{13/2}$	1.3407	300	Xe lamp	p	625
$Bi_4(Si,Ge)_3O_{12}$	$^4F_{3/2} \rightarrow {}^4I_{11/2}$	1.0635	300	Xe lamp	p	356
$Bi_{12}SiO_{20}$	$^4F_{3/2} \rightarrow {}^4I_{11/2}$	≈1.0716	300	TiS laser	p	1144
$Ca_{0.25}Ba_{0.75}(NbO_3)_2$	$^4F_{3/2} \rightarrow {}^4I_{11/2}$	1.062	295	Xe lamp	p	291
$Ca(NbO_3)_2$	$^4F_{3/2} \rightarrow {}^4I_{11/2}$	1.0615	77	Xe lamp	p, cw	588
			300	Xe lamp	p	598, 600
		1.0615–1.0625	300–650	Xe lamp	p	994
		1.0626	77	Xe lamp	p	588
	$^4F_{3/2} \rightarrow {}^4I_{13/2}$	1.3370	77	Xe lamp	p	607
		1.3380	300	Xe lamp	p	607
		1.3415	77	Xe lamp	p	607
		1.3425	300	Xe lamp	p	607
$CaAl_4O_7$	$^4F_{3/2} \rightarrow {}^4I_{11/2}$	1.05895	77	Xe lamp	p	510
		1.0596	300	Xe lamp	p	510
		1.0638	300	Xe lamp	p	510
		1.06585	77	Xe lamp	p	510
		1.07655	77	Xe lamp	p	510
		1.0772	77	Xe lamp	p	510
		1.0786	300	Xe lamp	p	510
	$^4F_{3/2} \rightarrow {}^4I_{13/2}$	1.3400	77	Xe lamp	p	478
		1.3420	300	Xe lamp	p	478
		1.3675	77	Xe lamp	p	478
		1.3710	300	Xe lamp	p	478
$CaAl_{12}O_{19}$	$^4F_{3/2} \rightarrow {}^4I_{11/2}$	1.0497	300	Xe lamp	p	?
CaF_2	$^4F_{3/2} \rightarrow {}^4I_{11/2}$	1.0370	300	Xe lamp	p	137
		1.0448	50	Xe lamp	p	300–302
		1.0456	50	Xe lamp	p	301, 302
		1.0457	77	Xe lamp	p	301, 302
		1.0461–1.0468	300–530	Xe lamp	p	137
		1.0466	50	Xe lamp	p	301, 302
		1.0467	77	Xe lamp	p	301, 302
		1.0480	50	Xe lamp	p	301, 302
		1.0507	50	Xe lamp	p	301, 302

Neodymium (Nd^{3+}, 4f^3)—*continued*

Host crystal	Laser transition	Wavelength (µm)	Temp. (K)	Optical pump	Mode	Ref.
CaF$_2$	$^4F_{3/2} \rightarrow ^4I_{11/2}$	1.0623–1.0628	560–300	Xe lamp	p	137
		1.0648	50	Xe lamp	p	301, 302
		1.0657	300	Xe lamp	p	541
		1.0661	300	Xe lamp	p	300
		1.0885–1.0889	300–420	Xe lamp	p	137, 344
	$^4F_{3/2} \rightarrow ^4I_{13/2}$	1.3225	300	Xe lamp	p	541
CaF$_2$-CeF$_3$	$^4F_{3/2} \rightarrow ^4I_{11/2}$	1.0657–1.0640	300–700	Xe lamp	p	137
	$^4F_{3/2} \rightarrow ^4I_{13/2}$	1.3190	300	Xe lamp	p	537
CaF$_2$-CeO$_2$	$^4F_{3/2} \rightarrow ^4I_{11/2}$	~1.0885	300	Xe lamp	p	523
CaF$_2$-GdF$_3$	$^4F_{3/2} \rightarrow ^4I_{11/2}$	1.0654	300	Xe lamp	p	427
	$^4F_{3/2} \rightarrow ^4I_{13/2}$	1.3185	300	Xe lamp	p	434
CaF$_2$-LaF$_3$	$^4F_{3/2} \rightarrow ^4I_{11/2}$	1.0645	300	Xe lamp	p	427
	$^4F_{3/2} \rightarrow ^4I_{13/2}$	1.3190	300	Xe lamp	p	434
CaF$_2$-LuF$_3$	$^4F_{3/2} \rightarrow ^4I_{11/2}$	1.0530	300	Xe lamp	p	425
		1.0623	300	Xe lamp	p	425
		1.0635	300	Xe lamp	p	425
	$^4F_{3/2} \rightarrow ^4I_{13/2}$	1.330	300	Xe lamp	p	425
CaF$_2$-NdF$_3$	$^4F_{3/2} \rightarrow ^4I_{11/2}$	1.0654	300	Xe lamp	p	434
CaF$_2$-ScF$_3$	$^4F_{3/2} \rightarrow ^4I_{11/2}$	1.0500	300	Xe lamp	p	374
		1.0607	300	Xe lamp	p	374
		1.0618	300	Xe lamp	p	374
	$^4F_{3/2} \rightarrow ^4I_{13/2}$	1.3505	300	Xe lamp	p	374
CaF$_2$-SrF$_2$	$^4F_{3/2} \rightarrow ^4I_{11/2}$	1.0369	300	Xe lamp	p	286
CaF$_2$-SrF$_2$-BaF$_2$-YF$_3$-LaF$_3$	$^4F_{3/2} \rightarrow ^4I_{11/2}$	1.0535–1.0547	300–550	Xe lamp	p	137, 438
		1.0585	300–700	Xe lamp	p	137
		1.0623–1.10585	300	Xe lamp	p	438
CaF$_2$-YF$_3$	$^4F_{3/2} \rightarrow ^4I_{11/2}$	1.0461	300	y	p	308, 309
		1.0536	110	y	p	439
		1.0540	300	Xe lamp	p	308, 309
		1.0540	300	Xe lamp	p	439
		1.0603–1.0632	95–300	Xe lamp	p	137

Neodymium (Nd^{3+}, 4f^3)—*continued*

Host crystal	Laser transition	Wavelength (µm)	Temp. (K)	Optical pump	Mode	Ref.
CaF$_2$-YF$_3$	$^4F_{3/2} \rightarrow \,^4I_{11/2}$	1.0632	300	Xe lamp	p	308, 309, 439
		1.0671	110	Xe lamp	p	439
	$^4F_{3/2} \rightarrow \,^4I_{13/2}$	1.3255	77	Xe lamp	p	538
		1.3270	300	Xe lamp	p	538
		1.3370	300	Xe lamp	p	538
		1.3380	77	Xe lamp	p	538
		1.3585	300	Xe lamp	p	538
		1.3600	77	Xe lamp	p	538
CaF$_2$-YF$_3$-NdF$_3$	$^4F_{3/2} \rightarrow \,^4I_{11/2}$	1.0632	300	Xe lamp	p	238
CaGd$_4$(SiO$_4$)$_3$O	$^4F_{3/2} \rightarrow \,^4I_{11/2}$	~1.06	300	Xe lamp	p	551
	$^4F_{3/2} \rightarrow \,^4I_{13/2}$	1.37	300	NdYAG	p	701
CaLa$_4$(SiO$_4$)$_3$O	$^4F_{3/2} \rightarrow \,^4I_{11/2}$	~1.061	300	Xe lamp	p	1007
		1.0610	300	Xe lamp	p	497, 586, 644
			300	W	cw	497, 586
		1.0612	300	Xe lamp	q	590, 644
	$^4F_{3/2} \rightarrow \,^4I_{13/2}$	1.3354	300	Xe lamp	p	644
CaLu$_2$F$_8$	$^4F_{3/2} \rightarrow \,^4I_{11/2}$	1.0615	300	TiS laser	cw	1145
CaMoO$_4$	$^4F_{3/2} \rightarrow \,^4I_{11/2}$	1.0573	295	Xe lamp	p	291, 474, 479
		1.061	300	Xe lamp	p, cw	291, 474, 479
			300	Xe lamp	p	458
		1.067	300	Xe lamp	p	458
		1.0673	77	Xe lamp	p, cw	291, 474, 479
CaSc$_2$O$_4$	$^4F_{3/2} \rightarrow \,^4I_{11/2}$	1.0720	300	Xe lamp	p	323
		1.0730	77	Xe lamp	p	323
		1.0755	77	Xe lamp	p	323
		1.0867	77	Xe lamp	p	323
		1.0868	300	Xe lamp	p	323
	$^4F_{3/2} \rightarrow \,^4I_{13/2}$	1.3565	300	Xe lamp	p	323, 504
		1.3870	77	Xe lamp	p	323, 504
CaWO$_4$	$^4F_{3/2} \rightarrow \,^4I_{9/2}$	0.9145	77	Xe lamp	p	253

Neodymium (Nd^{3+}, 4f^3)—*continued*

Host crystal	Laser transition	Wavelength (µm)	Temp. (K)	Optical pump	Mode	Ref.
CaWO$_4$	$^4F_{3/2} \rightarrow {}^4I_{11/2}$	1.0582–1.0597	300–700	Xe lamp	p	137
		1.0582	300	Xe lamp	p	484–487
			300	Hg lamp	cw	483
		1.0587	300	Xe lamp	p	137
		1.0601	77	Xe lamp	p	137
		1.0634	77	Xe lamp	p	137
		1.0649	77	Xe lamp	p	137
		1.0649	77	Hg lamp	cw	137
		1.0650	77	Xe lamp	p	735
		1.0652	300	Xe lamp	p	484–486
	$^4F_{3/2} \rightarrow {}^4I_{13/2}$	1.3310	77	Hg lamp	cw	617
		1.3340	300	Xe lamp	p	253, 980
		1.3345	77	Xe lamp	p	253, 549
		1.3370	300	Xe lamp	p, cw	253, 980
		1.3372	77	Xe lamp	p	253, 980
		1.3390	300	Xe lamp	p	253,9 80
		1.3459	77	Xe lamp	p	253, 980
		1.3475	300	Xe lamp	p	253, 980
		1.3880	77	Xe lamp	p	253, 980
		1.3885	300	Xe lamp	p	549
CaY$_2$Mg$_2$Ge$_3$O$_{12}$	$^4F_{3/2} \rightarrow {}^4I_{9/2}$	0.941	300	alex. laser	cw	265
	$^4F_{3/2} \rightarrow {}^4I_{11/2}$	1.0584	300	Xe lamp	p	494
		1.05896	300	alex. laser	cw	265
			300	Xe lamp	p	511
		1.059	300	alex. laser	cw	265
		1.0649	300	Xe lamp	p	494
CaY$_4$(SiO$_4$)$_3$O	$^4F_{3/2} \rightarrow {}^4I_{11/2}$	1.0672	300	Xe lamp	p	497
CaYAlO$_4$	$^4F_{3/2} \rightarrow {}^4I_{11/2}$	1.0775–1.0845	300	laser diode	cw	287
		1.0806	300	laser diode	cw	422
Ca$_2$Al$_2$SiO$_7$	$^4F_{3/2} \rightarrow {}^4I_{11/2}$	1.061	300	—	p	314
Ca$_2$Ga$_2$Ge$_4$O$_{14}$	$^4F_{3/2} \rightarrow {}^4I_{11/2}$	1.0688	300	Xe lamp	p	188, 667
			300	laser diode	cw, qcw	1000
Ca$_2$Ga$_2$SiO$_7$	$^4F_{3/2} \rightarrow {}^4I_{11/2}$	1.0609	77	Xe lamp	p	536, 1002
		1.0610	300	laser diode	cw, qcw	544–545
				Xe lamp	p	1002

Neodymium (Nd^{3+}, 4f^3)—*continued*

Host crystal	Laser transition	Wavelength (μm)	Temp. (K)	Optical pump	Mode	Ref.
Ca$_2$Ga$_2$SiO$_7$	$^4F_{3/2} \rightarrow ^4I_{11/2}$	1.0780	77	laser diode	cw, qcw	544, 545
			77	Xe lamp	p	1002
		1.0788	300	Xe lamp	p	544–546
		1.0788	300	Xe lamp	p	1002
	$^4F_{3/2} \rightarrow ^4I_{13/2}$	1.3365	300	Xe lamp	p	544–546
		1.3375	110	Xe lamp	p	544–546
Ca$_2$Y$_5$F$_{19}$	$^4F_{3/2} \rightarrow ^4I_{11/2}$	1.0498	300	Xe lamp	p	137
	$^4F_{3/2} \rightarrow ^4I_{13/2}$	1.3190	300	Xe lamp	p	606
		1.3200	77	Xe lamp	p	606
		1.3525	300	Xe lamp	p	606
Ca$_3$(Nb,Ga)$_2$-(Ga$_3$O$_{12}$	$^4F_{3/2} \rightarrow ^4I_{11/2}$	1.0583	110	Xe lamp	p	439, 506
		1.0588	300	Xe lamp	p	439, 506
		1.0605	110	Xe lamp	p	439, 506
		1.0612	300	Xe lamp	p	439, 506
		1.0648	110	Xe lamp	p	439
	$^4F_{3/2} \rightarrow ^4I_{13/2}$	1.3270	300	Xe lamp	p	506
Ca$_3$(Nb,Ga)$_2$-(Ga$_3$O$_{12}$:Cr	$^4F_{3/2} \rightarrow ^4I_{11/2}$	1.053–1.062	300	Xe lamp	p	428
Ca$_3$(VO$_4$)$_2$	$^4F_{3/2} \rightarrow ^4I_{11/2}$	1.067	300	Xe lamp	p	767
Ca$_3$Ga$_2$Ge$_3$O$_{12}$	$^4F_{3/2} \rightarrow ^4I_{11/2}$	1.0590	77	Xe lamp	p	519
		1.0596	300	Xe lamp	p	519
		1.0597	300	Xe lamp	p	675
		1.0633	77	Xe lamp	p	519
		1.0638	300	Xe lamp	p	510
		1.0639	300	Xe lamp	p	675
		1.0642	300	Xe lamp	p	519
		1.2805	110	Xe lamp	p	519, 571
	$^4F_{3/2} \rightarrow ^4I_{13/2}$	1.3150	300	Xe lamp	p	519
		1.3315	300	Xe lamp	p	519
		1.3317	300	Xe lamp	p	675
		1.3335	110	Xe lamp	p	519, 571
		1.3545	110	Xe lamp	p	519
Ca$_3$Ga$_2$Ge$_4$O$_{14}$	$^4F_{3/2} \rightarrow ^4I_{11/2}$	1.0688	300	Xe lamp	p	667
		1.0690	77	Xe lamp	p	667

Neodymium (Nd^{3+}, $4f^3$)—*continued*

Host crystal	Laser transition	Wavelength (μm)	Temp. (K)	Optical pump	Mode	Ref.
$Ca_3Ga_2Ge_4O_{14}$	$^4F_{3/2} \rightarrow ^4I_{13/2}$	1.3493	300	Xe lamp	p	667
$Ca_3Ga_2SiO_7$	$^4F_{3/2} \rightarrow ^4I_{11/2}$	1.0688	300	Xe lamp	p	1016
$Ca_3Ga_4O_9$	$^4F_{3/2} \rightarrow ^4I_{11/2}$	1.0582	300	Xe lamp	p	482
	$^4F_{3/2} \rightarrow ^4I_{13/2}$	1.3320	300	Xe lamp	p	482
$Ca_4GdO(BO_3)_3$	$^4F_{3/2} \rightarrow ^4I_{11/2}$	1.060	300	TiS laser	cw	1033
$Ca_4La(PO_4)_3O$	$^4F_{3/2} \rightarrow ^4I_{11/2}$	1.0613	300	Xe lamp	p	497
$Ca_4YO(BO_3)_3$	$^4F_{3/2} \rightarrow ^4I_{11/2}$	1.060	300	TiS/LD	cw	1129
$Ca_5(PO_4)_3F$	$^4F_{3/2} \rightarrow ^4I_{11/2}$	1.0629	300	Xe lamp	p	487
			300	laser diode	cw	499
		1.0630	300	Xe lamp	p	632, 633
				W lamp	cw	632, 633
	$^4F_{3/2} \rightarrow ^4I_{13/2}$	1.3345	77	Xe lamp	p	606
		1.3347	300	Xe lamp	p	606
CdF_2	$^4F_{3/2} \rightarrow ^4I_{11/2}$	1.0666	300	Xe lamp	p	763
CdF_2-CeF_3	$^4F_{3/2} \rightarrow ^4I_{11/2}$	1.0667	300	Xe lamp	p	763
	$^4F_{3/2} \rightarrow ^4I_{13/2}$	1.3360	300	Xe lamp	p	763
CdF_2-GaF_3	$^4F_{3/2} \rightarrow ^4I_{13/2}$	1.3365	300	Xe lamp	p	742
CdF_2-GdF_3	$^4F_{3/2} \rightarrow ^4I_{11/2}$	1.0672	300	Xe lamp	p	439, 742
		1.0676	110	Xe lamp	p	439
		1.0694	110	Xe lamp	p	439
	$^4F_{3/2} \rightarrow ^4I_{13/2}$	1.3365	300	Xe lamp	p	742
		1.3520	300	Xe lamp	p	742
CdF_2-LaF_3	$^4F_{3/2} \rightarrow ^4I_{11/2}$	~1.0665	300	Xe lamp	p	627
		1.0668	300	Xe lamp	p	381
		1.0670	≤200	Xe lamp	p	380
	$^4F_{3/2} \rightarrow ^4I_{13/2}$	1.3365	300	Xe lamp	p	380
CdF_2-LuF_3	$^4F_{3/2} \rightarrow ^4I_{11/2}$	1.0652	300	Xe lamp	p	742
	$^4F_{3/2} \rightarrow ^4I_{13/2}$	1.3500	300	Xe lamp	p	742
CdF_2-ScF_3	$^4F_{3/2} \rightarrow ^4I_{11/2}$	1.0507	300	Xe lamp	p	380
		1.0510	≤200	Xe lamp	p	380
	$^4F_{3/2} \rightarrow ^4I_{13/2}$	1.3300	300	Xe lamp	p	360
CdF_2-YF_3	$^4F_{3/2} \rightarrow ^4I_{11/2}$	1.0629–1.0656	600–300	Xe lamp	p	626, 627

Neodymium (Nd^{3+}, $4f^3$)—*continued*

Host crystal	Laser transition	Wavelength (μm)	Temp. (K)	Optical pump	Mode	Ref.
CdF_2-YF_3	$^4F_{3/2} \rightarrow ^4I_{13/2}$	1.3165	77	Xe lamp	p	538
		1.3245	300	Xe lamp	p	538
CdF_2-YF_3-LaF_3	$^4F_{3/2} \rightarrow ^4I_{11/2}$	1.065	300	Xe lamp	p	627
$CeCl_3$	$^4F_{3/2} \rightarrow ^4I_{11/2}$	1.0647	300	Xe lamp	p	720
				Ar laser	cw	721
CeF_3	$^4F_{3/2} \rightarrow ^4I_{11/2}$	1.0404	77	Xe lamp	p	294, 295
		1.0410	300	Xe lamp	p	294, 295
		1.0638	300	Xe lamp	p	662, 663
		1.0639	77	Xe lamp	p	662, 663
	$^4F_{3/2} \rightarrow ^4I_{13/2}$	1.3240	77	Xe lamp	p	538
		1.3310	77	Xe lamp	p	538
		1.3320	300	Xe lamp	p	538
		1.3675	77	Xe lamp	p	538
		1.3690	300	Xe lamp	p	602
CeP_5O_{14}	$^4F_{3/2} \rightarrow ^4I_{11/2}$	1.051	300	Ar laser	p	384
$CsGd_2F_7$	$^4F_{3/2} \rightarrow ^4I_{11/2}$	1.0555	300	laser diode	cw	462
$CsLa(WO_4)_2$	$^4F_{3/2} \rightarrow ^4I_{11/2}$	1.0575	300	Xe lamp	p	241, 475
		1.0581	110	Xe lamp	p	241, 475
	$^4F_{3/2} \rightarrow ^4I_{13/2}$	1.3298	300	Xe lamp	p	241, 475
$CsNd(MoO_4)_2$	$^4F_{3/2} \rightarrow ^4I_{11/2}$	1.0658	300	ruby laser	p	737
CsY_2F_7	$^4F_{3/2} \rightarrow ^4I_{11/2}$	1.0533	300	LD, Xe lamp	p, cw	1138
CsY_2F_7	$^4F_{3/2} \rightarrow ^4I_{11/2}$	1.05499	300	Ar laser	cw	452
$Gd_2(MoO_4)_3$	$^4F_{3/2} \rightarrow ^4I_{11/2}$	1.0606	300	Xe lamp	p	568, 569
			300	laser diode	qcw, cw	1141
		1.0701	300	Xe lamp	p	568, 569
			300	laser diode	qcw, cw	1141
Gd_2O_3	$^4F_{3/2} \rightarrow ^4I_{11/2}$	1.0741	300	Xe lamp	p	372
		1.0776	77	Xe lamp	p	372
		1.0789	77, 300	Xe lamp	p	372
$Gd_2(WO_4)_3$	$^4F_{3/2} \rightarrow ^4I_{11/2}$	1.0603	300	RS-RL	p	562
$Gd_3Ga_5O_{12}$	$^4F_{3/2} \rightarrow ^4I_{11/2}$	1.054	300	Xe lamp	p	443
		1.0584	77	Xe lamp	p	490, 492
		1.0591	300	Xe lamp	p	522

Neodymium (Nd^{3+}, 4f^3)—*continued*

Host crystal	Laser transition	Wavelength (μm)	Temp. (K)	Optical pump	Mode	Ref.
Gd$_3$Ga$_5$O$_{12}$	$^4F_{3/2} \rightarrow {}^4I_{11/2}$	1.0599	77	Xe lamp	p	490, 492
		1.06	~120	Xe lamp	p	490, 492
		1.06	300	Kr lamp	cw	554–555
		1.0600	~120	Xe lamp	p	490, 492
		1.0606	300	Xe lamp	p	505
		1.0615	~120	Xe lamp	p	490, 492
		1.0621	300	Xe lamp	p	490, 492
			300	PT	p	610
	$^4F_{3/2} \rightarrow {}^4I_{13/2}$	1.3077	77	Xe lamp	p	490, 492
		1.319	300	Xe lamp	p	613
		1.330	300	Xe lamp	p	618
		1.3315	300	Xe lamp	p	1067
		1.338	300	Xe lamp	p	613
Gd$_3$Ga$_5$O$_{12}$:Cr	$^4F_{3/2} \rightarrow {}^4I_{11/2}$	1.060	300	Xe lamp	p	555, 557
Gd$_3$Sc$_2$Al$_3$O$_{12}$	$^4F_{3/2} \rightarrow {}^4I_{9/2}$	0.936	300	Xe lamp	p	258, 259
	$^4F_{3/2} \rightarrow {}^4I_{11/2}$	1.05915	77	Xe lamp	p	279, 527
		1.05995	300	Xe lamp	p	279, 527
		1.06	300	W lamp	cw	279, 527
			300	sun	qcw	1075
			213–303	sun	qcw	1076
		1.0620	300	Xe lamp	p	279, 527
	$^4F_{3/2} \rightarrow {}^4I_{13/2}$	1.3360	300	Xe lamp	p	279, 527
Gd$_3$Sc$_2$Al$_3$O$_{12}$:Cr	$^4F_{3/2} \rightarrow {}^4I_{9/2}$	1.058	300	Xe lamp	p	358
Gd$_3$Sc$_2$Ga$_3$O$_{12}$:Cr	$^4F_{3/2} \rightarrow {}^4I_{9/2}$	0.936	300	Xe lamp	p	258, 259
	$^4F_{3/2} \rightarrow {}^4I_{11/2}$	1.05755	77	Xe lamp	p	477
		1.0580	77	Xe lamp	p	477
		1.06045	77	Xe lamp	p	477
		1.061	300	W lamp	cw	314
			300	Kr laser	cw	542
			300	Xe lamp	p	314, 581
			300	laser diode	p	531
			300	Xe lamp	qs	583, 645, 653
			300	Xe lamp	self qs	358, 659
			300	Xe lamp	PML	730, 732
		1.0612	300	Xe lamp	p	477
		1.0613	300	Xe lamp	PML	596

Neodymium (Nd^{3+}, $4f^3$)—*continued*

Host crystal	Laser transition	Wavelength (μm)	Temp. (K)	Optical pump	Mode	Ref.
$Gd_3Sc_2Ga_3O_{12}$:Cr	$^4F_{3/2} \rightarrow {}^4I_{13/2}$	1.32	300	Xe lamp	p	565
$(Gd, Ca)_3(Ga,Mg, Zr)_5O_{12}$:Cr	$^4F_{3/2} \rightarrow {}^4I_{11/2}$	~1.06	300	Xe lamp	p	502, 503
$GdAlO_3$	$^4F_{3/2} \rightarrow {}^4I_{11/2}$	1.0689	77	Xe lamp	p	439, 546
		1.0690	300	Xe lamp	p	439, 546
		1.0759	77	Xe lamp	p	213, 214
		1.0760	300	Xe lamp	p	213, 214
GdF_3-CaF_2	$^4F_{3/2} \rightarrow {}^4I_{11/2}$	1.0495	300	Xe lamp	p	369
$GdGaGe_2O_7$	$^4F_{3/2} \rightarrow {}^4I_{11/2}$	1.0567	77	Xe lamp	p	467
		1.0570	300	Xe lamp	p	467, 472
		1.0601	300	Xe lamp	p	467, 472
		1.0659	300	Xe lamp	p	467, 472
	$^4F_{3/2} \rightarrow {}^4I_{13/2}$	1.3298	300	Xe lamp	p	467, 472
GdP_5O_{14}	$^4F_{3/2} \rightarrow {}^4I_{11/2}$	1.051	300	Ar laser	p	384
$GdScO_3$	$^4F_{3/2} \rightarrow {}^4I_{11/2}$	1.0840	200	Xe lamp	p	459
		1.08515	300	Xe lamp	p	363
$GdVO_4$	$^4F_{3/2} \rightarrow {}^4I_{11/2}$	~1.06	300	LD, ArL	cw	550
		1.065	300	—	cw, PML	725
	$^4F_{3/2} \rightarrow {}^4I_{13/2}$	~1.34	300	LD, ArL	cw	550
HfO_2-Y_2O_3	$^4F_{3/2} \rightarrow {}^4I_{11/2}$	1.0604	300	Xe lamp	p	563, 564
		1.0615	110	Xe lamp	p	439, 546
	$^4F_{3/2} \rightarrow {}^4I_{13/2}$	1.3305	300	Xe lamp	p	563, 564
$KGd(WO_4)_2$	$^4F_{3/2} \rightarrow {}^4I_{11/2}$	1.0672	300	Xe lamp	p,cw	76
			300	LED	p	136
			300	Xe lamp	PML	117, 139
			300	Xe lamp	p, qs	536
			300	PT	p	610
	$^4F_{3/2} \rightarrow {}^4I_{13/2}$	1.3510	300	Xe lamp	p	76
$KLa(MoO_4)_2$	$^4F_{3/2} \rightarrow {}^4I_{11/2}$	1.0580	110	Xe lamp	p	439
		1.0585	300	Xe lamp	p	498
		1.0587	300	Xe lamp	p	498, 501
		1.0645	110	Xe lamp	p	439
		1.0646	300	Xe lamp	PML	361
	$^4F_{3/2} \rightarrow {}^4I_{13/2}$	1.3342	300	Xe lamp	p	501

Neodymium (Nd^{3+}, $4f^3$)—*continued*

Host crystal	Laser transition	Wavelength (μm)	Temp. (K)	Optical pump	Mode	Ref.
$KLa(MoO_4)_2$	$^4F_{3/2} \rightarrow {}^4I_{13/2}$	1.3350	77, 300	Xe lamp	p	498
		1.3533	300	Xe lamp	p	501
		1.3630	300	Xe lamp	p	501
		1.3657	300	Xe lamp	p	501
$KLu(WO_4)_2$	$^4F_{3/2} \rightarrow {}^4I_{11/2}$	1.0701	300	PT	p	610
		1.0701–1.0706	77–600	Xe lamp	p	257
		1.0714	300	Xe lamp	p	257
		1.0716–1.0721	550–77	Xe lamp	p	257
		1.0721	300	Xe lamp	p	211
	$^4F_{3/2} \rightarrow {}^4I_{13/2}$	1.3482	300	Xe lamp	p	211, 272
		1.3533	300	Xe lamp	p	257
		1.3550	300	Xe lamp	p	257
$KNdP_4O_{12}$	$^4F_{3/2} \rightarrow {}^4I_{11/2}$	1.052	300	dye laser	cw	404–405
				Ar laser	cw	406
(waveguide)		1.05	300	Ar laser	cw	317
(waveguide)	$^4F_{3/2} \rightarrow {}^4I_{13/2}$	1.3	300	Ar laser	cw	317
		1.32	300	Ar laser	cw	382
$KY(MoO_4)_2$	$^4F_{3/2} \rightarrow {}^4I_{11/2}$	1.0669	300	Xe lamp	p	765, 766
	$^4F_{3/2} \rightarrow {}^4I_{13/2}$	1.3485	300	Xe lamp	p	549, 765
$KY(WO_4)_2$	$^4F_{3/2} \rightarrow {}^4I_{9/2}$	0.9137	77	Xe lamp	p	250, 251
	$^4F_{3/2} \rightarrow {}^4I_{11/2}$		300	laser diode	p	984
		1.0687–1.0690	77–600	Xe lamp	p	250, 251
		1.0688	300	Xe lamp	p	250, 251
		1.0706	300	Xe lamp	p	205
	$^4F_{3/2} \rightarrow {}^4I_{13/2}$	1.3515	77	Xe lamp	p	250, 251
		1.3525	300	Xe lamp	p	250, 251
		1.3545	77, 300	Xe lamp	p	250, 251
KY_3F_{10}	$^4F_{3/2} \rightarrow {}^4I_{11/2}$	1.0554	300	Xe lamp	p	461
	$^4F_{3/2} \rightarrow {}^4I_{13/2}$	1.3185	300	Xe lamp	p	461
KYF_4	$^4F_{3/2} \rightarrow {}^4I_{11/2}$	1.04	300	TiS laser	qs, cw	559
		1.0412	300	Cr:LiSAF	p	573
			300	TiS laser	cw	573
		1.0417	300	laser diode	qcw, cw	1138
	$^4F_{3/2} \rightarrow {}^4I_{13/2}$	1.302	300	Cr:LiSAF	p	573
				TiS laser	cw	573

Neodymium (Nd^{3+}, $4f^3$)—*continued*

Host crystal	Laser transition	Wavelength (µm)	Temp. (K)	Optical pump	Mode	Ref.
KYF_4	$^4F_{3/2} \rightarrow ^4I_{13/2}$	1.307	300	Cr:LiSAF	p	573
				TiS laser	cw	573
K_2YF_5	$^4F_{3/2} \rightarrow ^4I_{11/2}$	1.0479	300	laser diode	qcw, cw	1138
$K_3Nd(PO_4)_2$	$^4F_{3/2} \rightarrow ^4I_{11/2}$	1.055	300	dye	qcw	373, 454
$K_5(Bi,Nd)$-(MoO_4)	$^4F_{3/2} \rightarrow ^4I_{11/2}$	1.066	300	Xe lamp	p	758
$K_5Nd(MoO_4)_4$	$^4F_{3/2} \rightarrow ^4I_{11/2}$	1.066	300	Xe lamp	p	758
			300	Ar laser	qcw	751
$K_5(Nd,Ce)Li_2F_{10}$	$^4F_{3/2} \rightarrow ^4I_{11/2}$	1.048	300	dye laser	cw	357
		1.052	300	dye laser	cw	357
$K_5NdLi_2F_{10}$	$^4F_{3/2} \rightarrow ^4I_{11/2}$	1.052	300	dye laser	cw	357
$LaAlO_3$	$^4F_{3/2} \rightarrow ^4I_{11/2}$	1.0804	300	Xe lamp	p	213
$(La,Sr)(Al,Ta)O_3$	$^4F_{3/2} \rightarrow ^4I_{11/2}$	1.059	300	TiS laser	cw	512
$LaAl_{11}MgO_{19}$	$^4F_{3/2} \rightarrow ^4I_{11/2}$	1.054	300	Kr lamp	cw	431
			300	TiS laser	cw	326
			300	laser diode	p	1123
		1.054–1.086	300	laser diode	cw	444
		1.0530–1.0565	300	Ar laser	cw	431, 432
		1.0780–1.0860	300	Ar laser	cw	431
		1.0547	300	Ar laser	cw	318
			300	Xe lamp	p	450
		~110	300	—	PML	451
		1.0550	77	Xe lamp	p	439, 455
		1.0552	300	Xe lamp	p	439, 455
		1.0812	77	Xe lamp	p	439, 455
		1.0817	300	Xe lamp	p	439, 455
	$^4F_{3/2} \rightarrow ^4I_{13/2}$	1.3760	300	Xe lamp	p	455
$LaBGeO_5$	$^4F_{3/2} \rightarrow ^4I_{11/2}$	1.0475	~110	Xe lamp	p	330
		1.048	300	TiS laser	cw	1056
			300	laser diode	cw	1057
		1.0482	300	TiS laser	cw	1056
		1.0700	~110	Xe lamp	p	330
		1.0711	~110	Xe lamp	p	330

Neodymium (Nd^{3+}, $4f^3$)—*continued*

Host crystal	Laser transition	Wavelength (μm)	Temp. (K)	Optical pump	Mode	Ref.
$LaBGeO_5$	$^4F_{3/2} \rightarrow {}^4I_{13/2}$	1.3141	300	Xe lamp	p	330
		1.3868	300	Xe lamp	p	330
LaF_3	$5d \rightarrow {}^4I_{11/2}$	0.172	300	Kr_2^*	p	1, 2
			300	F_2 laser	p	3, 4
	$^4D_{3/2} \rightarrow {}^4I_{11/2}$	0.38006	20–77	dye laser	cw, upc	15, 1087
		0.38052	20–77	dye laser	cw	15
	$^4F_{3/2} \rightarrow {}^4I_{11/2}$	1.0400	77	Xe lamp	p	292–293
		1.04065	300	Xe lamp	p, cw	292–293
			300	laser diode	cw	985
		1.04065–1.0410	300–430	Xe lamp	p	137
		1.0451	77	Xe lamp	p	303–305
		1.0523	77	Xe lamp	p	292–293
		1.0583	77	Xe lamp	p	292–293
		1.0595–1.0613	380–820	Xe lamp	p	137
		1.06305	77	Xe lamp	p	292–293
		1.0632–1.0642	400–700	Xe lamp	p	137
		1.06335	300	Xe lamp	p	292–293
		1.0633–1.0638	300–650	Xe lamp	p	137
			300	laser diode	cw	985
		1.064	300	TiS laser	cw	326
		1.0670	77	Xe lamp	p	291–293
	$^4F_{3/2} \rightarrow {}^4I_{13/2}$	1.3125	77	Xe lamp	p	537
		1.3235	77	Xe lamp	p	537
		1.3305	77	Xe lamp	p	537
		1.3310	300	Xe lamp	p	537
		1.3670	77	Xe lamp	p	537
		1.3675	300	Xe lamp	p	537
LaF_3-SrF_2	$^4F_{3/2} \rightarrow {}^4I_{11/2}$	1.0486	300	Xe lamp	p	362
		1.0635	300	Xe lamp	p	362
	$^4F_{3/2} \rightarrow {}^4I_{13/2}$	1.3170	77	Xe lamp	p	549
		1.3275	77	Xe lamp	p	549
		1.3315	300	Xe lamp	p	549
		1.3325	77	Xe lamp	p	549
$LaGaGe_2O_7$	$^4F_{3/2} \rightarrow {}^4I_{11/2}$	1.0591	300	Xe lamp	p	525, 526
$LaMgAl_{11}O_{19}$	$^4F_{3/2} \rightarrow {}^4I_{11/2}$	1.0530–1.0565	300	Ar laser	cw	431, 432
		1.054	300	Kr lamp	cw	431

Neodymium (Nd^{3+}, $4f^3$)—*continued*

Host crystal	Laser transition	Wavelength (µm)	Temp. (K)	Optical pump	Mode	Ref.
$LaMgAl_{11}O_{19}$	$^4F_{3/2} \rightarrow {}^4I_{11/2}$		300	Xe lamp	p	313
		1.054–1.086	300	laser diode	cw	444
		1.0547	300	Ar laser	cw	318
			300	Xe lamp	p	450
			300	—	PML	451
		1.0550	77	Xe lamp	p	439, 455
		1.0552	300	Xe lamp	p	439, 455
		1.0780–1.086	300	Ar laser	cw	431
		1.0812	77	Xe lamp	p	439, 482
		1.0817	300	Xe lamp	p	439, 482
		1.082	300	Xe lamp	p	313
		1.082–1.084	300	Kr laser	cw	444
	$^4F_{3/2} \rightarrow {}^4I_{13/2}$	1.3760	300	Xe lamp	p	455
$LaNbO_4$	$^4F_{3/2} \rightarrow {}^4I_{11/2}$	1.0618	300	Xe lamp	p	611
		1.0624	300	Xe lamp	p	619
LaP_5O_{14}	$^4F_{3/2} \rightarrow {}^4I_{11/2}$	1.05	300	Xe lamp	p	376
$(La,Nd)P_5O_{14}$	$^4F_{3/2} \rightarrow {}^4I_{11/2}$	1.0505	300	ion laser	qcw	378
		1.051	300	Ar laser	cw	347, 385
		1.0511	300	ion laser	qcw	378
		1.0512	300	ion laser	qcw	378
		1.052	300	dye laser	p	391
			300	Ar laser	cw	391, 392
			300	dye laser	cw, qcw	393, 395
(fiber)			300	Kr laser	cw, qcw	394
			300	dye laser	cw	396
		1.063	300	DRL	p	635
	$^4F_{3/2} \rightarrow {}^4I_{13/2}$	1.32	300	Ar laser	cw	386
		1.323	300	LED	cw, qcw	574, 575
$LaSc_3(BO_3)_4$	$^4F_{3/2} \rightarrow {}^4I_{11/2}$	1.062	300	Xe lamp	p	612
$LaSr_2Ga_{11}O_{20}$	$^4F_{3/2} \rightarrow {}^4I_{11/2}$	1.0572	300	Xe lamp	p	234, 439
			300	laser diode	qcw, cw	1137
		1.0690	110	Xe lamp	p	234, 439
		1.0706	300	Xe lamp	p	234, 439
			300	laser diode	qcw, cw	1137
		1.0725	110	Xe lamp	p	234, 439
		1.0746	110	Xe lamp	p	234, 439
		1.0778	110	Xe lamp	p	234, 439

Neodymium (Nd^{3+}, $4f^3$)—*continued*

Host crystal	Laser transition	Wavelength (μm)	Temp. (K)	Optical pump	Mode	Ref.
$LaSr_2Ga_{11}O_{20}$	$^4F_{3/2} \rightarrow {}^4I_{13/2}$	1.3628	300	Xe lamp	p	· 234, 439
$La_2Be_2O_5$	$^4F_{3/2} \rightarrow {}^4I_{11/2}$	1.0698	300	Xe lamp	p, qs	270, 274, 314
			300	W lamp	cw	254, 255
		1.07	300	laser diode	cw	260
		1.0785	77	Xe lamp	p	270
		1.079	300	Xe lamp	qs	416
		1.0790	300	Xe lamp	p	270
	$^4F_{3/2} \rightarrow {}^4I_{13/2}$	1.351	300	Xe lamp	p	649
		1.3510	300	Xe lamp	p	270, 314
			300	Xe lamp	qs	314, 649
		1.354	300	Xe lamp	PML	151
		1.365	300	Xe lamp	p	663
La_2O_3	$^4F_{3/2} \rightarrow {}^4I_{11/2}$	1.079	77	Xe lamp	p	296
$7La_2O_3$-$9SiO_2$	$^4F_{3/2} \rightarrow {}^4I_{11/2}$	1.0610	300	Xe lamp	p	466
La_2O_2S	$^4F_{3/2} \rightarrow {}^4I_{11/2}$	1.075	300	Xe lamp	p	371
$La_2Si_2O_7$	$^4F_{3/2} \rightarrow {}^4I_{11/2}$	1.0566	300	Xe lamp	p	466
		1.0576	300	Xe lamp	p	466
$La_3Ga_{5.5}Nb_{0.5}O_{14}$	$^4F_{3/2} \rightarrow {}^4I_{11/2}$	1.0638	77	Xe lamp	p	667, 668
		1.0645	300	Xe lamp	p	667, 668
		1.067	300	laser diode	cw	1050
	$^4F_{3/2} \rightarrow {}^4I_{13/2}$	1.3707	300	Xe lamp	p	667, 668
$La_3Ga_{5.5}Ta_{0.5}O_{14}$	$^4F_{3/2} \rightarrow {}^4I_{11/2}$	1.0641	300	Xe lamp	p	705
	$^4F_{3/2} \rightarrow {}^4I_{13/2}$	1.3730	300	Xe lamp	p	667, 705
$La_3Ga_5GeO_{14}$	$^4F_{3/2} \rightarrow {}^4I_{11/2}$	1.0650	300	Xe lamp	p	736
		1.0675	300	laser diode	cw	1000
		1.0680	77	Xe lamp	p	736
	$^4F_{3/2} \rightarrow {}^4I_{13/2}$	1.3730	300	Xe lamp	p	667, 736
$(La,Lu)_3(Lu,Ga)_2$-Ga_3O_{12}	$^4F_{3/2} \rightarrow {}^4I_{11/2}$	—	300	laser diode	P	1041
$La_3Ga_5SiO_{14}$	$^4F_{3/2} \rightarrow {}^4I_{11/2}$	1.0640	300	Xe lamp	p	695–698
			300	PT	p	610
		1.0645	300	Xe lamp	p	695, 697–8
			300	PT	p	610

Neodymium (Nd^{3+}, $4f^3$)—*continued*

Host crystal	Laser transition	Wavelength (μm)	Temp. (K)	Optical pump	Mode	Ref.
$La_3Ga_5SiO_{14}$	$^4F_{3/2} \rightarrow {}^4I_{11/2}$	1.0670	300	Xe lamp	p	695–698
			300	Xe lamp	p, AML	1059
		1.0672	300	Xe lamp	p	695, 697–8
		1.0673	300	Xe lamp	p	695, 697–8
		1.0675	77	Xe lamp	p	695, 697
	$^4F_{3/2} \rightarrow {}^4I_{13/2}$	1.3730	300	Xe lamp	p	695, 697–8
$Li(Bi,Nd)P_4O_{12}$ (waveguide)	$^4F_{3/2} \rightarrow {}^4I_{11/2}$	1.048	300	Ar laser	qcw	332
$Li(Nd,Gd)P_4O_{12}$	$^4F_{3/2} \rightarrow {}^4I_{11/2}$	1.048	300	Ar laser	cw	273
$Li(Nd,La)P_4O_{12}$	$^4F_{3/2} \rightarrow {}^4I_{11/2}$	1.048	300	Ar laser	qcw	336–338, 353, 354
			300	dye laser	p	1060, 1062
			300	Ar laser	cw	349, 353, 354, 1065
			300	laser diode	cw	350, 1061
(waveguide)	$^4F_{3/2} \rightarrow {}^4I_{13/2}$	1.317	300	Ar laser	qcw	353, 576
			300	Ar laser	cw	353, 354, 382
			300	laser diode	cw	350
$LiGd(MoO_4)_2$	$^4F_{3/2} \rightarrow {}^4I_{11/2}$	1.0595	110	Xe lamp	p	439, 458
		1.0599	300	Xe lamp	p, cw	252, 257
	$^4F_{3/2} \rightarrow {}^4I_{13/2}$	1.3400	77, 300	Xe lamp	p	655
		1.3455	77	Xe lamp	p	655
$LiGdF_4$	$^4F_{3/2} \rightarrow {}^4I_{11/2}$	1.047	300	Xe lamp	p	313
			300	CrLiSAF	p	1040
			300	TiS laser	cw	1040
$LiKYF_5$	$^4F_{3/2} \rightarrow {}^4I_{11/2}$	1.0481	300	Xe lamp	P	359
			300	laser diode	cw, qcw	1053, 1138
		1.0532	300	Xe lamp	P	359
	$^4F_{3/2} \rightarrow {}^4I_{13/2}$	1.3163	300	Xe lamp	p	1053
			300	laser diode	cw	1053
$LiLa(MoO_4)_2$	$^4F_{3/2} \rightarrow {}^4I_{11/2}$	1.0585	300	Xe lamp	P	496
		1.0658	77	Xe lamp	P	496
	$^4F_{3/2} \rightarrow {}^4I_{13/2}$	1.3370	300	Xe lamp	P	655

Neodymium (Nd^{3+}, 4f^3)—*continued*

Host crystal	Laser transition	Wavelength (µm)	Temp. (K)	Optical pump	Mode	Ref.
LiLa(MoO$_4$)$_2$	$^4F_{3/2} \to {}^4I_{13/2}$	1.3375	77	Xe lamp	P	655
		1.3440	77	Xe lamp	P	655
LiLuF$_4$	$^4F_{3/2} \to {}^4I_{11/2}$	1.0472	300	Xe lamp	P	32, 33
			300	laser diode	cw, qcw	983
		1.0474	300	laser diode	cw, qcw	1139
		1.05	300	Xe lamp	P	32, 33
		1.0529	110	Xe lamp	P	32, 33
		1.0531	300	Xe lamp	P	32, 33
	$^4F_{3/2} \to {}^4I_{13/2}$	1.3133	300	Xe lamp	P	32, 33
		1.3172	110	Xe lamp	P	32
		1.3208	300	Xe lamp	P	32, 33
		1.3257	110	Xe lamp	P	32
LiNbO$_3$	$^4F_{3/2} \to {}^4I_{11/2}$	1.0782–1.0787	590–450	Xe lamp	P	626
		1.0829–1.0859	300	Kr laser	cw	988
(waveguide)		1.0829–1.0859	300	TiS laser	qcw	996
		1.0840	77	Xe lamp	P	457
		1.0846	300	Xe lamp	P	383, 449
		1.0922–1.0933	620–300	Xe lamp	P	626
		1.0933	300	Xe lamp	P	383, 449
			300	Kr laser	p	348
	$^4F_{3/2} \to {}^4I_{13/2}$	1.3745	300	Xe lamp	P	383, 538
		1.3870	300	Xe lamp	P	383, 538
LiNbO$_3$:Mg	$^4F_{3/2} \to {}^4I_{11/2}$	1.084	300	dye laser	cw	400
		1.085	300	Kr laser	cw	36
			300	laser diode	cw	471
			300	TiS laser	p	996
			363	TiS laser	qcw	996
		1.0829–1.0859	300	Kr laser	cw	988
		1.093	300	dye laser	cw	36, 480
			300	laser diode	cw	471
			300	dye laser	qs	36, 480
		~1.094	300	TiS laser	cw, qcw	1030
LiNdP$_4$O$_{12}$	$^4F_{3/2} \to {}^4I_{11/2}$	1.046–1.064	300	Ar laser	cw	306, 307
		1.047	300	Ar laser	cw	317, 319
			300	laser diode	cw	223, 306, 350

Neodymium (Nd^{3+}, 4f^3)—*continued*

Host crystal	Laser transition	Wavelength (μm)	Temp. (K)	Optical pump	Mode	Ref.
LiNdP$_4$O$_{12}$	$^4F_{3/2} \rightarrow {}^4I_{11/2}$	1.048	300	dye laser	cw, qcw	1064
(waveguide)			300	Ar laser	qcw	1060–1063
			300	Ar laser	cw	349
	$^4F_{3/2} \rightarrow {}^4I_{13/2}$	1.316–1.340	300	Ar laser	cw	307
		1.317	300	laser diode	cw	306, 560, 561
		1.319	300	Ar laser	cw	317, 319, 780
		1.32	300	Ar laser	cw	382
LiYF$_4$	$^2P_{3/2} \rightarrow {}^4I_{11/2}$	0.413	12–40	dye laser	cw, pa	77, 1087
	$^2P_{3/2} \rightarrow {}^2H_{9/2}$	0.72952	12–40	dye laser	cw, pa	77, 1087
	$^4F_{3/2} \rightarrow {}^4I_{11/2}$	1.047	300	W lamp	cw	314, 315
			300	Kr lamp	cw	315
			300	laser diode	cw	316
		1.0471	300	Xe lamp	p	314, 321, 322
			300	laser diode	p	1041
			300	Xe lamp	(A,P)ML	327
		1.0528	77	Xe lamp	p	32, 33
		1.053	300	W lamp	cw	314
			300	Kr lamp	cw	315
			300	laser diode	cw	316
			300	Xe lamp	p	32, 33
			300	Xe lamp	qs	314
			300	Kr lamp	AML	429
			300	Xe lamp	(A,P)ML	430
		1.0530	300	Xe lamp	p	321, 322
	$^4F_{3/2} \rightarrow {}^4I_{13/2}$	1.313	300	W lamp	cw	314
			300	Xe lamp	p	32, 33, 314
		1.3212	300	Xe lamp	p	32, 33
		1.3256	110	Xe lamp	p	32, 33
LuAlO$_3$	$^4F_{3/2} \rightarrow {}^4I_{11/2}$	1.0671	77	Xe lamp	p	74
		1.0675	300	Xe lamp	p	74
		1.0759	300	Xe lamp	p	74
		1.0831	120	Xe lamp	p	74
		1.0832	300	Xe lamp	p	74
	$^4F_{3/2} \rightarrow {}^4I_{13/2}$	1.3437	300	Xe lamp	p	74

Neodymium (Nd^{3+}, 4f^3)—*continued*

Host crystal	Laser transition	Wavelength (μm)	Temp. (K)	Optical pump	Mode	Ref.
LuPO$_4$	$^4F_{3/2} \rightarrow {}^4I_{11/2}$	1.0648	287–303	laser diode	cw	1135
LuScO$_3$	$^4F_{3/2} \rightarrow {}^4I_{11/2}$	1.0785	300	Xe lamp	p	363
Lu$_2$SiO$_5$	$^4F_{3/2} \rightarrow {}^4I_{11/2}$	1.0790	300	Xe lamp	p	466
		1.07925	300	Xe lamp	p	420
	$^4F_{3/2} \rightarrow {}^4I_{13/2}$	1.3585	300	Xe lamp	p	466
Lu$_3$Al$_5$O$_{12}$	$^4F_{3/2} \rightarrow {}^4I_{9/2}$	0.9473	77	Xe lamp	p	267
	$^4F_{3/2} \rightarrow {}^4I_{11/2}$	1.0535	300	Xe lamp	p	436
		1.0605	77	Xe lamp	p	267
		1.0615	300	Xe lamp	p	505
		1.0637–1.0672	120–900	Xe lamp	p	267
		1.06425	300	Xe lamp	p	267
			300	PT	p	610
	$^4F_{3/2} \rightarrow {}^4I_{13/2}$	1.3209	300	Xe lamp	p	538
		1.3319	77	Xe lamp	p	538
		1.3326	300	Xe lamp	p	538
		1.3333	77	Xe lamp	p	538
		1.3342	300	Xe lamp	p	538
		1.3376	77	Xe lamp	p	602
		1.3387	300	Xe lamp	p, cw	267
		1.3410	300	Xe lamp	p	267
		1.3499	77	Xe lamp	p	602
		1.3525	300	Xe lamp	p	267
		1.3532	300	Xe lamp	p	496
Lu$_3$Ga$_5$O$_{12}$	$^4F_{3/2} \rightarrow {}^4I_{11/2}$	1.0587	77	Xe lamp	p	505
		1.0594	300	Xe lamp	p	529
		1.06025	77	Xe lamp	p	505
		1.0609	300	Xe lamp	p	505
		1.0615	300	Xe lamp	p	505
		1.0623	300	Xe lamp	p	505
	$^4F_{3/2} \rightarrow {}^4I_{13/2}$	1.3315	300	Xe lamp	p	505
Lu$_3$Sc$_2$Al$_3$O$_{12}$	$^4F_{3/2} \rightarrow {}^4I_{11/2}$	1.0591	300	Xe lamp	p	279
		1.0599	300	Xe lamp	p	279
		1.0620	300	Xe lamp	p	460
	$^4F_{3/2} \rightarrow {}^4I_{13/2}$	1.3360	300	Xe lamp	p	460
Lu$_3$Sc$_2$Ga$_3$O$_{12}$	$^4F_{3/2} \rightarrow {}^4I_{11/2}$	1.0622	300	Xe lamp,LD	p, qcw	1140

Neodymium (Nd^{3+}, $4f^3$)—*continued*

Host crystal	Laser transition	Wavelength (μm)	Temp. (K)	Optical pump	Mode	Ref.
Na(Nd,Gd)-(WO$_4$)$_2$	$^4F_{3/2} \rightarrow {}^4I_{11/2}$	1.06	77	Xe lamp	p	476
Na$_{0.4}$Y$_{0.6}$F$_{2.2}$	$^4F_{3/2} \rightarrow {}^4I_{11/2}$	1.042–1.075	300	Ar laser	cw	298
β"-Na$_{1+x}$Mg$_x$-Al$_{11-x}$O$_{17}$	$^4F_{3/2} \rightarrow {}^4I_{11/2}$	1.059	300	dye laser	p, cw	517
Na$_2$(Nd,La)$_2$Pb$_2$-(PO$_4$)$_6$Cl$_2$	$^4F_{3/2} \rightarrow {}^4I_{11/2}$	1.059	300	Ar laser	cw	520, 521
		1.068	300	Ar laser	cw	520, 521
Na$_3$(La,Nd)-(PO$_4$)$_2$		1.055	300	dye laser	qcw	373
Na$_3$Nd(PO$_4$)$_2$	$^4F_{3/2} \rightarrow {}^4I_{11/2}$	1.055	300	dye laser	qcw	373
Na$_5$(Nd,La)-(WO$_4$)$_4$	$^4F_{3/2} \rightarrow {}^4I_{11/2}$	1.0670	300	Xe lamp	p	1159, 1160
(powder)		1.067–1.068	77	dye laser	p	1159, 1160
Na$_5$Nd(WO$_4$)$_4$	$^4F_{3/2} \rightarrow {}^4I_{11/2}$	1.063	300	dye laser	qcw	373, 628
NaBi(WO$_4$)$_2$	$^4F_{3/2} \rightarrow {}^4I_{11/2}$	1.0638	300	Xe lamp	P	665
		1.0642	300	Xe lamp	P	665
	$^4F_{3/2} \rightarrow {}^4I_{13/2}$	1.3342	300	Xe lamp	P	665
α-NaCaCeF$_6$	$^4F_{3/2} \rightarrow {}^4I_{11/2}$	1.0633–1.0653	920–300	Xe lamp	P	137
		1.0653	~100	Xe lamp	P	745
			300	LD	cw, qcw	983
	$^4F_{3/2} \rightarrow {}^4I_{13/2}$	1.3165	77	Xe lamp	P	538
		1.3190	300	Xe lamp	P	538
α-NaCaYF$_6$	$^4F_{3/2} \rightarrow {}^4I_{11/2}$	1.0539	300	Xe lamp	P	504
		1.0539–1.0549	300–550	Xe lamp	P	137
		1.0629	300	Xe lamp	P	504
		1.0629–1.0597	333–1000	Xe lamp	P	137
	$^4F_{3/2} \rightarrow {}^4I_{13/2}$	1.3260	77	Xe lamp	P	538
		1.3285	300	Xe lamp	P	538
		1.3375	300	Xe lamp	P	538
		1.3390	77	Xe lamp	P	538
		1.36000	300	Xe lamp	P	538
5NaF-9YF$_3$	$^4F_{3/2} \rightarrow {}^4I_{11/2}$	1.0505	300	Xe lamp	P	379
		1.0506	300	Xe lamp	P	379
		1.0595	300	Xe lamp	p	530

Neodymium (Nd^{3+}, 4f^3)—*continued*

Host crystal	Laser transition	Wavelength (μm)	Temp. (K)	Optical pump	Mode	Ref.
5NaF-9YF$_3$	$^4F_{3/2} \to {}^4I_{13/2}$	1.3070	300	Xe lamp	P	537
NaGaGe$_2$O$_7$	$^4F_{3/2} \to {}^4I_{11/2}$	1.0592	110	Xe lamp	P	470
		1.0608	300	Xe lamp	P	470
NaGd(MoO$_4$)$_2$	$^4F_{3/2} \to {}^4I_{11/2}$	1.0663	110	Xe lamp	P	439
		1.0667	300	Xe lamp	P	439, 699
	$^4F_{3/2} \to {}^4I_{13/2}$	1.3385	300	Xe lamp	P	655
NaGd(WO$_4$)$_2$	$^4F_{3/2} \to {}^4I_{11/2}$	1.06	300	Xe lamp	P	476
NaGdGeO$_4$	$^4F_{3/2} \to {}^4I_{11/2}$	1.0615	300	Xe lamp	P	539, 604, 605
			300	LD	cw	985
	$^4F_{3/2} \to {}^4I_{13/2}$	1.3334	300	Xe lamp	P	539, 604, 605
NaLa(MoO$_4$)$_2$	$^4F_{3/2} \to {}^4I_{11/2}$	1.0482	300	Xe lamp	PML	361
		1.0595	300	Xe lamp	P	533, 534
		1.06	300	Xe lamp	qs	536
		1.0650	110	Xe lamp	P	439
		1.0653	300	Xe lamp	p	533, 534
			300	Xe lamp	cw	532
		1.0653–1.0665	300–750	Xe lamp	P	137
	$^4F_{3/2} \to {}^4I_{13/2}$	1.3380	77	Xe lamp	P	655
		1.3430	77	Xe lamp	P	655
		1.3440	300	Xe lamp	P	655
		1.3755	77	Xe lamp	P	655
		1.3840	77	Xe lamp	P	655
NaLa(WO$_4$)$_2$	$^4F_{3/2} \to {}^4I_{11/2}$	1.0635	300	Xe lamp	P	654
	$^4F_{3/2} \to {}^4I_{13/2}$	1.3355	300	Xe lamp	P	655
NaLuGeO$_4$	$^4F_{3/2} \to {}^4I_{11/2}$	1.0596	77	Xe lamp	P	539, 540
		1.0604	300	Xe lamp	P	539, 540
			300	LD	cw	985
	$^4F_{3/2} \to {}^4I_{13/2}$	1.3300	77	Xe lamp	P	539, 540
		1.3310	300	Xe lamp	P	539, 540
NaNdP$_4$O$_{12}$	$^4F_{3/2} \to {}^4I_{11/2}$	1.049–1.077	300	Ar laser	cw	307
		1.051	300	Ar laser	cw	335, 381
	$^4F_{3/2} \to {}^4I_{13/2}$	1.311–1.334	300	Ar laser	cw	307

Neodymium (Nd^{3+}, 4f^3)—*continued*

Host crystal	Laser transition	Wavelength (μm)	Temp. (K)	Optical pump	Mode	Ref.
NaNdP$_4$O$_{12}$	$^4F_{3/2} \rightarrow {}^4I_{13/2}$	1.320	300	Ar laser	cw	382
NaY(MoO$_4$)$_2$	$^4F_{3/2} \rightarrow {}^4I_{11/2}$	1.0663	110	Xe lamp	P	439
		1.0674	300	Xe lamp	P	439, 699
NaY(WO$_4$)$_2$	$^4F_{3/2} \rightarrow {}^4I_{11/2}$	1.059	300	TiS laser	cw	516
NaYGeO$_4$	$^4F_{3/2} \rightarrow {}^4I_{11/2}$	1.0609	300	Xe lamp	P	576, 577
	$^4F_{3/2} \rightarrow {}^4I_{13/2}$	1.3325	300	Xe lamp	P	539, 577
NdAl$_3$(BO$_3$)$_4$	$^4F_{3/2} \rightarrow {}^4I_{11/2}$	1.06	300	Ar laser	cw	317
		1.063	300	Xe lamp	P	641
			300	Xe lamp	p, qs	642
		1.0635	300	laser diode	cw	630, 652
		1.065	300	dye laser	cw, qcw	405
			300	ion laser	cw, qcw	656
			300	Ar laser	qcw	658
			300	laser diode	qcw	640
	$^4F_{3/2} \rightarrow {}^4I_{13/2}$	1.3	300	Ar laser	cw	317
		1.341	300	ion laser	cw, qcw	656
		1.345	300	dye laser	P	648
		1.345	300	ion laser	cw, qcw	656
NdAl$_3$(BO$_3$)$_4$:Cr	$^4F_{3/2} \rightarrow {}^4I_{11/2}$	1.063	300	DNdYLF	p	643
	$^4F_{3/2} \rightarrow {}^4I_{13/2}$	1.3	300	Ar laser	cw	317
Nd(Ga,Cr)$_3$(BO$_3$)$_4$	$^4F_{3/2} \rightarrow {}^4I_{11/2}$	1.066	300	dye laser	P	631
Nd$_3$Ga$_5$O$_{12}$	$^4F_{3/2} \rightarrow {}^4I_{11/2}$	1.0592	110	Xe lamp	P	470
		1.0608	300	Xe lamp	P	563
NdGaGe$_2$O$_7$	$^4F_{3/2} \rightarrow {}^4I_{11/2}$	1.0566	110	Xe lamp	P	467
		1.0569	300	Xe lamp	P	467, 470
		1.0654	300	Xe lamp	P	467, 470
		1.0689	300	Xe lamp	P	467, 470
	$^4F_{3/2} \rightarrow {}^4I_{13/2}$	1.3303	300	Xe lamp	P	467, 470
Nd$_3$Ga$_5$GeO$_{14}$	$^4F_{3/2} \rightarrow {}^4I_{11/2}$	1.0675	300	Xe lamp	P	667, 736
		1.0680	300	Xe lamp	P	667, 736
Nd$_3$Ga$_5$SiO$_{14}$	$^4F_{3/2} \rightarrow {}^4I_{11/2}$	1.0675	300	Xe lamp	P	695, 697, 698
Nd$_x$La$_{1-x}$P$_5$O$_{14}$	$^4F_{3/2} \rightarrow {}^4I_{11/2}$	1.047–1.078	300	Ar laser	cw	307

Neodymium (Nd^{3+}, 4f^3)—*continued*

Host crystal	Laser transition	Wavelength (µm)	Temp. (K)	Optical pump	Mode	Ref.
Nd$_x$La$_{1-x}$P$_5$O$_{14}$	$^4F_{3/2} \rightarrow ^4I_{11/2}$	1.05	300	Xe lamp	P	376
		1.051	300	dye laser	p	388
			300	Xe lamp	p	347, 389
		1.0511	300	ion laser	qcw	378
		1.0512	300	ion laser	qcw	378
		1.0513	300	Xe lamp	P	387
		1.0515	300	Xe lamp	p	355, 388, 389
		1.052	300	dye laser	p	390
			300	dye laser	cw,qcw	639
(fiber)			300	Kr laser	cw	394
		1.053	300	Ar laser	cw	387
		1.063	300	Xe lamp	p, AML	396
			300	laser diode	qcw	639, 640
			300	flashlamp	p	634
			300	XeF laser	p	636
			300	Kr lamp	cw	638
			300	Xe lamp	p	621, 634
	$^4F_{3/2} \rightarrow ^4I_{13/2}$	1.3	300	ion laser	cw	656
		1.32	300	Xe lamp	p	575
		1.323	300	ion laser	cw	656
		1.324	300	laser diode	cw	566
		[1.304–1.372]	300	Ar laser	cw	306, 566
PbCl$_2$	$^4F_{3/2} \rightarrow ^4I_{11/2}$	1.0615	300	TiS laser	p	1044
PbMoO$_4$	$^4F_{3/2} \rightarrow ^4I_{11/2}$	1.0586	300	Xe lamp	p	291, 500
	$^4F_{3/2} \rightarrow ^4I_{13/2}$	1.3320	77	Xe lamp	p	655
		1.3340	300	Xe lamp	p	655
		1.3375	77	Xe lamp	p	655
		1.3425	300	Xe lamp	p	655
		1.3450	77	Xe lamp	p	655
		1.3780	77	Xe lamp	p	655
PbWO$_4$	$^4F_{3/2} \rightarrow ^4I_{11/2}$	1.0580	300	Xe lamp	p	1146
		1.0630	300	Xe lamp	p	1146
Pb$_5$Ge$_3$O$_{11}$	$^4F_{3/2} \rightarrow ^4I_{11/2}$	1.0789	77	Xe lamp	p	417, 418
		1.0799	77	Xe lamp	p	417, 418

Neodymium (Nd^{3+}, $4f^3$)—*continued*

Host crystal	Laser transition	Wavelength (μm)	Temp. (K)	Optical pump	Mode	Ref.
$Pb_5(PO_4)_3F$	$^4F_{3/2} \rightarrow ^4I_{11/2}$	1.0551	300	Xe lamp	p	456
$RbNd(WO_4)_2$	$^4F_{3/2} \rightarrow ^4I_{11/2}$	1.0650	300	ruby laser	p	737
Sc_2O_3	$^4F_{3/2} \rightarrow ^4I_{9/2}$	0.966	300	TiS laser	cw	1161
	$^4F_{3/2} \rightarrow ^4I_{11/2}$	1.082, 1.087	300	TiS laser	cw	1161
	$^4F_{3/2} \rightarrow ^4I_{13/2}$	1.486	300	TiS laser	cw	1161
$SrAl_2O_4$	$^4F_{3/2} \rightarrow ^4I_{11/2}$	1.0497	300	Xe lamp	p	370
$SrAl_4O_7$	$^4F_{3/2} \rightarrow ^4I_{11/2}$	1.0566	77	Xe lamp	p	473
		1.0568	77	Xe lamp	p	473
$SrAl_4O_7$	$^4F_{3/2} \rightarrow ^4I_{11/2}$	1.0576	300	Xe lamp	p	473
		1.0627	77	Xe lamp	p	473
		1.0828	300	Xe lamp	p	473
	$^4F_{3/2} \rightarrow ^4I_{13/2}$	1.3320	77	Xe lamp	p	478
		1.3345	300	Xe lamp	p	478
		1.3530	77	Xe lamp	p	478
		1.3665	300	Xe lamp	p	478
		1.3680	300	Xe lamp	p	602
$SrAl_{12}O_{19}$	$^4F_{3/2} \rightarrow ^4I_{11/2}$	1.0491	300	Xe lamp	p	364, 365
		1.0498	300	Xe lamp	p	313
		1.0618	300	Xe lamp	p	313
		1.0621	300	Xe lamp	P	615
		1.074	300	Xe lamp	p	313
	$^4F_{3/2} \rightarrow ^4I_{13/2}$	1.3065	300	Xe lamp	p	478, 324
$Sr_xBa_{1-x}(NbO_3)_2$	$^4F_{3/2} \rightarrow ^4I_{13/2}$	1.0626	300	TiS laser	cw	1053
SrF_2	$^4F_{3/2} \rightarrow ^4I_{11/2}$	1.0370–1.0395	300–530	Xe lamp	p	155
		1.0437	77	Xe lamp	p	155
		1.0445	300	Xe lamp	p	155
		1.0446	500–550	Xe lamp	p	155
		1.0596	300	Xe lamp	p	541
	$^4F_{3/2} \rightarrow ^4I_{13/2}$	1.3250	300	Xe lamp	p	541
SrF_2-CeF_3	$^4F_{3/2} \rightarrow ^4I_{11/2}$	1.0590	300	Xe lamp	p	439
		1.0590	300	Xe lamp	p	518
		1.0594	110	Xe lamp	p	439
		1.0636	110	Xe lamp	p	439
	$^4F_{3/2} \rightarrow ^4I_{13/2}$	1.3255	300	Xe lamp	p	518

Neodymium (Nd^{3+}, $4f^3$)—*continued*

Host crystal	Laser transition	Wavelength (μm)	Temp. (K)	Optical pump	Mode	Ref.
SrF_2-CeF_3-GdF_3	$^4F_{3/2} \rightarrow {}^4I_{11/2}$	1.0589	300	Xe lamp	p	421
SrF_2-GdF_3	$^4F_{3/2} \rightarrow {}^4I_{11/2}$	1.0528	300	Xe lamp	p	421
		1.3250	77	Xe lamp	p	421
	$^4F_{3/2} \rightarrow {}^4I_{13/2}$	1.3260	300	Xe lamp	p	421
SrF_2-LaF_3	$^4F_{3/2} \rightarrow {}^4I_{11/2}$	1.0597–1.0583	300–800	Xe lamp	p	137
	$^4F_{3/2} \rightarrow {}^4I_{13/2}$	1.3160	77	Xe lamp	p	537
		1.3235	77	Xe lamp	p	537
		1.3250	300	Xe lamp	p	537
		1.3355	77	Xe lamp	p	537
SrF_2-LuF_3	$^4F_{3/2} \rightarrow {}^4I_{11/2}$	1.0556	300	Xe lamp	p	463
		1.0560	300	Xe lamp	p	464, 465
	$^4F_{3/2} \rightarrow {}^4I_{13/2}$	1.3200	300	Xe lamp	p	427
SrF_2-ScF_3	$^4F_{3/2} \rightarrow {}^4I_{11/2}$	1.0543	300	Xe lamp	p	446
		1.0605	300	Xe lamp	p	446
	$^4F_{3/2} \rightarrow {}^4I_{13/2}$	1.3285	300	Xe lamp	p	446
SrF_2-YF_3	$^4F_{3/2} \rightarrow {}^4I_{11/2}$	1.0567	300	Xe lamp	p	468
	$^4F_{3/2} \rightarrow {}^4I_{13/2}$	1.3225	77	Xe lamp	p	537
		1.3300	77	Xe lamp	p	537
	$^4F_{3/2} \rightarrow {}^4I_{13/2}$	1.3320	77	Xe lamp	p	537
$SrGdGa_3O_7$	$^4F_{3/2} \rightarrow {}^4I_{9/2}$	0.911	31	laser diode	p	247
	$^4F_{3/2} \rightarrow {}^4I_{11/2}$	1.064–1.065	31	laser diode	p	247
$SrGd_4(SiO_4)_3O$	$^4F_{3/2} \rightarrow {}^4I_{13/2}$	1.44	300	NdYAG	p	701
$SrMoO_4$	$^4F_{3/2} \rightarrow {}^4I_{11/2}$	1.0576	295	Xe lamp	p	291, 479
		1.059	77	Xe lamp	p	291, 479
		1.0611	77	Xe lamp	p	291, 479
		1.0627	77	Xe lamp	p	291, 479
		1.0640	77	Xe lamp	p	291, 479
		1.0643	295	Xe lamp	p	291, 479
		1.0652	77	Xe lamp	p	291, 479
	$^4F_{3/2} \rightarrow {}^4I_{13/2}$	1.3300	77	Xe lamp	p	655
		1.3325	300	Xe lamp	p	655
		1.3440	77	Xe lamp	p	655
		1.3790	77	Xe lamp	p	655
$SrWO_4$	$^4F_{3/2} \rightarrow {}^4I_{11/2}$	1.0574	77	Xe lamp	p	291

Neodymium (Nd^{3+}, $4f^3$)—*continued*

Host crystal	Laser transition	Wavelength (μm)	Temp. (K)	Optical pump	Mode	Ref.
$SrWO_4$	$^4F_{3/2} \rightarrow ^4I_{11/2}$	1.0607	77	Xe lamp	p	291
		1.0627	77	Xe lamp	p	291
		1.063	295	Xe lamp	p	291
$SrWO_4$:Na	$^4F_{3/2} \rightarrow ^4I_{11/2}$	1.06265	77	Xe lamp	p	57
		1.0628	300	Xe lamp	p	57
	$^4F_{3/2} \rightarrow ^4I_{13/2}$	1.3347	300	Xe lamp	p	57
$Sr_2Y_5F_{19}$	$^4F_{3/2} \rightarrow ^4I_{11/2}$	1.0493	300	Xe lamp	p	367
	$^4F_{3/2} \rightarrow ^4I_{13/2}$	1.3190	300	Xe lamp	p	367
$Sr_3Ca_2(PO_4)_3F$	$^4F_{3/2} \rightarrow ^4I_{11/2}$	1.0607	300	laser diode	cw	499
$Sr_3Ga_2Ge_4O_{14}$	$^4F_{3/2} \rightarrow ^4I_{11/2}$	1.0688	77	Xe lamp	p	188, 667
		1.0694	300	Xe lamp	p	188, 667
		1.0757	300	Xe lamp	p	188, 667
	$^4F_{3/2} \rightarrow ^4I_{13/2}$	1.3510	300	Xe lamp	p	188, 667
$Sr_4Ca(PO_4)_3F$	$^4F_{3/2} \rightarrow ^4I_{11/2}$	1.0593	300	laser diode	cw	499
$Sr_5(PO_4)_3F$	$^4F_{3/2} \rightarrow ^4I_{11/2}$	1.0585	300	Xe lamp	p	497
		1.0586	300	laser diode	cw	499
		1.059	300	Cr:LiSAF	p	515
			300	TiS laser	cw	515
	$^4F_{3/2} \rightarrow ^4I_{13/2}$	1.328	300	Cr:LiSAF	p	515
			300	TiS laser	cw	515
$Sr_5(VO_4)_3Cl$	$^4F_{3/2} \rightarrow ^4I_{11/2}$	1.065	300	TiS laser	qcw	728
$Sr_5(VO_4)_3F$	$^4F_{3/2} \rightarrow ^4I_{11/2}$	1.065	300	Xe lamp	q	727
			300	dye laser	cw	727
Y_2O_3	$^4F_{3/2} \rightarrow ^4I_{11/2}$	1.073	77	Xe lamp	p	325, 593
		~1.0746	300	Kr laser	cw	325, 593
		1.078	77	Xe lamp	p	325, 593
		1.08	300	Xe lamp	p	427
	$^4F_{3/2} \rightarrow ^4I_{13/2}$	~1.358	300	Kr laser	cw	325, 593
Y_2O_3-ThO_2	$^4F_{3/2} \rightarrow ^4I_{11/2}$	~1.074	300	Xe lamp	p	781
$Y_3Al_3O_{12}$	$^4F_{3/2} \rightarrow ^4I_{9/2}$	0.8910	300	Ar laser	p	244
		0.8999	300	Ar laser	p	244
		0.9385	300	Ar laser	p	244
			300	TiS laser	cw	1151

Neodymium (Nd^{3+}, 4f^3)—*continued*

Host crystal	Laser transition	Wavelength (μm)	Temp. (K)	Optical pump	Mode	Ref.
Y$_3$Al$_3$O$_{12}$	$^4F_{3/2} \rightarrow {}^4I_{9/2}$	0.946	260–300	dye laser	cw	261–263
			300	laser diode	cw	261, 262
			300	TiS laser	cw	1151
		0.9460	300	Xe lamp	p	266
	$^4F_{3/2} \rightarrow {}^4I_{11/2}$	1.052	300	Xe lamp	PML	407, 412
			300	Xe lamp	qcw	594
		1.0521	300	W lamp	cw	414
		1.0610–1.0627	77–600	Xe lamp	p	137
		1.0612	77	Xe lamp	p	591
Y$_3$Al$_3$O$_{12}$	$^4F_{3/2} \rightarrow {}^4I_{11/2}$	1.0615	300	Xe lamp	PML	407
			300	W lamp	cw	414
		1.0637–1.0670	170–900	Xe lamp	p	137
		1.064	300	Xe lamp	PML	407, 676, 693, 694
			300	Kr lamp	qs, AML	691, 692
			300	Xe lamp	p	570
			300	W lamp	cw	314
			300	sun	cw	1077, 1078
			300	sun	cw	1043, 1155
			300	laser diode	p	686, 984
			300	laser diode	cw	685, 687, 1041
			300	Xe lamp	p	314, 492
			300	Xe lamp	qcw	594
			300	laser diode	qs	686, 688
			300	Xe lamp	qs	314, 689
			300	lamp	AML	690
		1.06415	300	W lamp	cw	414
			300	Xe lamp	p	591
		1.0646	300	Kr lamp	cw	375
		1.0682	300	Xe lamp	p	116
		1.073	300	Xe lamp	qcw	594
		1.0737	300	Xe lamp	cw	414
			300	Xe lamp	PML	407
		1.0780	300	Kr lamp	cw	375
		1.1054	300	Kr lamp	cw	375
		1.1119	300	W lamp	cw	414
		1.1158	300	W lamp	cw	414

Neodymium (Nd^{3+}, $4f^3$)—*continued*

Host crystal	Laser transition	Wavelength (μm)	Temp. (K)	Optical pump	Mode	Ref.
		1.1225	300	W lamp	cw	414
$Y_3Al_3O_{12}$	$^4F_{3/2} \rightarrow {}^4I_{13/2}$	1.318	300	Xe lamp	p	570
			300	Xe lamp	p	779
			300	Xe lamp	PML	151
			300	Xe lamp	qcw	594
			300	laser diode	AML	346
		1.3187	300	Xe lamp	qcw	599
			300	Xe lamp	p, qs	723
		1.3188	300	Kr lamp	cw	375
		1.319	300	DL, LD	cw	618
		1.3200	300	Kr lamp	cw	375
		1.3338	300	Kr lamp	cw	375
		1.3350	300	K lamp	cw	375
		1.338	300	DL, LD	cw	618
			300	W lamp	cw	314
			300	Xe lamp	p, qs	314
		1.338	300	Xe lamp	qcw	594
			300	Xe lamp	p, qs	723
		1.3382	300	Kr lamp	cw	375
		1.339	300	Xe lamp	p	673
		1.3410	300	Kr lamp	cw	375
		1.3533	300	Xe lamp	p	660
		1.3564	77	Kr lamp	cw	375
		1.3572	300	Xe lamp	p	660, 723
		1.358	300	Xe lamp	qcw	594
		1.4140	300	Kr lamp	cw	375
		1.4150	300	Xe lamp	p, qs	723
		1.4444	300	Kr lamp	cw	375
			300	Xe lamp	p	724
	$^4F_{3/2} \rightarrow {}^4I_{15/2}$	1.833	293	Kr lamp	p	807
$Y_3Al_5O_{12}$:Cr	$^4F_{3/2} \rightarrow {}^4I_{11/2}$	1.0612	77	Hg lamp	cw	592
			77	W lamp	cw	592
		1.0641	300	Xe lamp	p	592
			300	Hg lamp	cw	592
			300	W lamp	cw	592
$Y_3Al_5O_{12}$:Cr,Ce	$^4F_{3/2} \rightarrow {}^4I_{11/2}$	1.064	300	Kr lamp	cw	671

Neodymium (Nd^{3+}, $4f^3$)—*continued*

Host crystal	Laser transition	Wavelength (μm)	Temp. (K)	Optical pump	Mode	Ref.
$Y_3Al_5O_{12}$:Er	$^4F_{3/2} \rightarrow ^4I_{11/2}$	1.064	300	Xe lamp	p	672
$Y_3Al_5O_{12}$:Fe	$^4F_{3/2} \rightarrow ^4I_{11/2}$	1.064	300	Xe lamp	p	345
$Y_3Al_5O_{12}$:Ho	$^4F_{3/2} \rightarrow ^4I_{11/2}$	1.064	300	Xe lamp	p	673
$Y_3Al_5O_{12}$:Ti	$^4F_{3/2} \rightarrow ^4I_{11/2}$	1.064	300	—	cw	670
$(Y,Ce)_3Al_5O_{12}$	$^4F_{3/2} \rightarrow ^4I_{11/2}$	1.0638–1.0644	300	laser diode	cw	669
$(Y,Lu)_3Al_5O_{12}$	$^4F_{3/2} \rightarrow ^4I_{11/2}$	1.0608	77	Xe lamp	p	597
		1.0636	77	Xe lamp	p	597
$(Y,Lu)_3Al_5O_{12}$	$^4F_{3/2} \rightarrow ^4I_{11/2}$	1.0642	295	Xe lamp	p	597
		1.0726	77	Xe lamp	p	597
$Y_3Ga_5O_{12}$	$^4F_{3/2} \rightarrow ^4I_{11/2}$	1.0583	77	Xe lamp	p	490
		1.0589	300	Xe lamp	p	490
		1.05975	77	Xe lamp	p	490
		1.0603	300	Xe lamp	p	491
		1.0614	77	Xe lamp	p	435
		1.0625	300	Xe lamp	p	490, 491
	$^4F_{3/2} \rightarrow ^4I_{13/2}$	1.3305	300	Xe lamp	p	314, 490
Y_2SiO_5	$^4F_{3/2} \rightarrow ^4I_{9/2}$	0.911–0.912	300	alex. laser	qs	248
		0.912	300	TiS laser	qs	248
	$^4F_{3/2} \rightarrow ^4I_{11/2}$	1.0644	77	Xe lamp	p	713
		1.0710	77	Xe lamp	p	320
		1.0711	300	Xe lamp	p	466
		1.0715	300	Xe lamp	p	320
		1.074	300	Xe lamp	qs	360
		1.0740	77	Xe lamp	p	713
		1.0741	300	Xe lamp	p	466
		1.0742	300	Xe lamp	p	320
		1.0781	77	Xe lamp	p	320
		1.0782	300	Xe lamp	p	320, 466
	$^4F_{3/2} \rightarrow ^4I_{13/2}$	1.3585	300	Xe lamp	p	466
$Y_3Sc_2Al_3O_{12}$	$^4F_{3/2} \rightarrow ^4I_{11/2}$	1.0587	77	Xe lamp	p	476
		1.0622	300	Xe lamp	p	609
	$^4F_{3/2} \rightarrow ^4I_{13/2}$	1.3360	300	Xe lamp	p	465, 504
$Y_3Sc_2Ga_3O_{12}$	$^4F_{3/2} \rightarrow ^4I_{11/2}$	1.0575	77	Xe lamp	p	427
		1.0583	300	Xe lamp	p	427

Neodymium (Nd^{3+}, $4f^3$)—*continued*

Host crystal	Laser transition	Wavelength (μm)	Temp. (K)	Optical pump	Mode	Ref.
$Y_3Sc_2Ga_3O_{12}$	$^4F_{3/2} \rightarrow {}^4I_{11/2}$	1.0584	300	lamp	cw	493
		1.0615	300	Xe lamp	p	427
	$^4F_{3/2} \rightarrow {}^4I_{13/2}$	1.3310	300	Xe lamp	p	427
$Y_3Sc_2Ga_3O_{12}$:Cr	$^4F_{3/2} \rightarrow {}^4I_{11/2}$	1.0584	300	Xe lamp	p, qs	493
$YAl_3(BO_3)_4$	$^4F_{3/2} \rightarrow {}^4I_{11/2}$	1.06	300	dye laser	p	553
$YAlO_3$	$^4F_{3/2} \rightarrow {}^4I_{9/2}$	0.930	300	Xe lamp	p	256
			300	TiS laser	p, sml	1134
	$^4F_{3/2} \rightarrow {}^4I_{11/2}$	1.0585	300	Xe lamp	p	495
		1.064–1.110	300	laser diode	cw	666
		1.0641–1.0654	77–500	Xe lamp	p	626, 702
		1.0644	300	Xe lamp	p, qs	712
		1.0645	300	Kr lamp	cw	718
			300	Xe lamp	p	716
			300	Xe lamp	cw	717
		1.0652–1.0659	310–500	Xe lamp	p	626, 702
		1.0726–1.0730	77–490	Xe lamp	p	626, 702
		1.0726	300	Xe lamp	p	495
		1.0729	300	Kr lamp	p	712, 717
			300	Kr lamp	cw	718
		1.0782–1.0815	300	Kr lamp	cw	718
		1.0795	300	Xe lamp	p	377, 716
			300	Xe lamp	cw	995
			300	Kr lamp	cw	424, 717, 718
			300	laser diode	p	424
			300	Kr lamp	qs, AML	441
			300	—	PML	442
			300	Xe lamp	PML	481
			300	Xe lamp	p, qs	536
			300	Kr lamp	cw	423
		1.0795–1.0802	77–600	Xe lamp	p	626, 702
		1.0796	300	Xe lamp	qs	712
		1.0796–1.0803	600–700	Kr lamp	cw	718
		1.083	300	laser diode	cw	666
		1.0832–1.0855	300	Kr lamp	cw	718
		1.0845	300	Kr lamp	cw	718

Neodymium (Nd^{3+}, 4f^3)—*continued*

Host crystal	Laser transition	Wavelength (µm)	Temp. (K)	Optical pump	Mode	Ref.
YAlO$_3$	$^4F_{3/2} \to {}^4I_{11/2}$	1.0847	530	Xe lamp	p	626, 702
		1.0909	300	Kr lamp	p, cw	718
		1.0913	530	Xe lamp	p	626, 702
		1.0921	300	Kr lamp	cw	718
		1.0989	300	Kr lamp	cw	718
		1.0991	500	Xe lamp	p	627, 702
	$^4F_{3/2} \to {}^4I_{13/2}$	1.3391	77	Xe lamp	p	549, 717
YAlO$_3$	$^4F_{3/2} \to {}^4I_{13/2}$	1.3393	77	Xe lamp	p	599, 717
		1.3400	300	Kr lamp	p, cw	717, 980
		1.341	300	laser diode	P	424
		1.3410	300	Kr lamp	p, cw	599, 717
		1.3413	300	Xe lamp	p, qs	712
		1.3414	300	Xe lamp	p, cw	995
		1.3416	300	Xe lamp	p, cw	549, 717
		1.3512	300	Xe lamp	p	549
		1.3514	300	Xe lamp	p	717, 980
		1.3644	77	Xe lamp	p	549, 717
		1.3849	77	Xe lamp	p	549
		1.4026	~110	Xe lamp	p	549
YAlO$_3$:Cr	$^4F_{3/2} \to {}^4I_{11/2}$	1.0645	295	Xe lamp	p	715
			300	Kr lamp	cw	729
		1.0795	300	Kr lamp	cw	423
YAlO$_3$:Ce,Cr,Fe	$^4F_{3/2} \to {}^4I_{11/2}$	1.0795	300	Xe lamp	p	437
YF$_3$	$^4F_{3/2} \to {}^4I_{11/2}$	1.0521	300	Xe lamp	p	415
	$^4F_{3/2} \to {}^4I_{13/2}$	1.339	300	Xe lamp	p	415
(Y,Nd)P$_5$O$_{14}$	$^4F_{3/2} \to {}^4I_{11/2}$	1.052	300	dye laser	p	399, 401
			300	Ar laser	p	384
			300	Ar laser	qcw	601
			300	Ar laser	cw	403
	$^4F_{3/2} \to {}^4I_{13/2}$	1.319–1.322	300	Ar laser	qcw	601
YScO$_3$	$^4F_{3/2} \to {}^4I_{11/2}$	1.0770	77	Xe lamp	p	363
		1.0774	130	Xe lamp	p	439
		1.0837	130	Xe lamp	p	439
		1.0843	300	Xe lamp	p	363, 439
YVO$_4$	$^4F_{3/2} \to {}^4I_{11/2}$	1.0625	300	Xe lamp	p	620

Neodymium (Nd^{3+}, 4f^3)—*continued*

Host crystal	Laser transition	Wavelength (μm)	Temp. (K)	Optical pump	Mode	Ref.
YVO$_4$	$^4F_{3/2} \rightarrow ^4I_{11/2}$	1.0634	300	Ar laser	cw	778
		1.064	300	laser diode	cw	684
		1.0641	300	Xe lamp	p	620
		1.0648	300	Xe lamp	p	620
		1.0664–1.0672	300–690	Xe lamp	p	626, 628
		1.069	~90	Xe lamp	p	987
	$^4F_{3/2} \rightarrow ^4I_{13/2}$	1.34	300	Ar laser	cw	778
YVO$_4$	$^4F_{3/2} \rightarrow ^4I_{13/2}$	1.3415	77	Xe lamp	p	537
		1.3425	300	Xe lamp	p	537
ZrO$_2$-Y$_2$O$_3$	$^4F_{3/2} \rightarrow ^4I_{11/2}$	1.0608	300	Xe lamp	p	563
	$^4F_{3/2} \rightarrow ^4I_{13/2}$	1.3320	300	Xe lamp	p	564

Samarium (Sm^{3+}, 4f^5)

Host crystal	Laser transition	Wavelength (μm)	Temp. (K)	Optical pump	Mode	Ref.
TbF$_3$	$^4G_{5/2} \rightarrow ^6H_{7/2}$	0.5932	116	Xe lamp	p	48

Europium (Eu^{3+}, 4f^6)

Host crystal	Laser transition	Wavelength (μm)	Temp. (K)	Optical pump	Mode	Ref.
Y$_2$O$_3$	$^5D_0 \rightarrow ^7F_2$	0.6113	220	Xe lamp	p	58
YVO$_4$	$^5D_0 \rightarrow ^7F_2$	0.6193	90	Xe lamp	p	66

Terbium (Tb^{3+}, 4f^8)

Host crystal	Laser transition	Wavelength (μm)	Temp. (K)	Optical pump	Mode	Ref.
LiYF$_4$:Gd	$^5D_4 \rightarrow ^7F_5$	0.5445	300	Xe lamp	p	35

Dysprosium (Dy^{3+}, 4f^9)

Host crystal	Laser transition	Wavelength (μm)	Temp. (K)	Optical pump	Mode	Ref.
Ba(Y,Er)$_2$F$_8$	$^6H_{11/2} \rightarrow {}^6H_{15/2}$	3.02	300	NdGL	p	965
		3.022	77	Xe lamp	p	960
BaYb$_2$F$_8$	$^6H_{13/2} \rightarrow {}^6H_{15/2}$	3.40	300	NdYAG	p	1029
LaF$_3$	$^6H_{11/2} \rightarrow {}^6H_{15/2}$	2.97	300	NdGL	p	759
LiYF$_4$	$^6H_{11/2} \rightarrow {}^6H_{13/2}$	4.34	300	ErYLF	p	990

Holmium (Ho^{3+}, 4f^{10})

Host crystal	Laser transition	Wavelength (μm)	Temp. (K)	Optical pump	Mode	Ref.
BaEr$_2$F$_8$:Tm	$^5I_7 \rightarrow {}^5I_8$	2.086	300	Xe lamp	p	870, 871
			300	Xe lamp	qs	872
BaTm$_2$F$_8$	$^5I_7 \rightarrow {}^5I_8$	2.0560	110–230	Xe lamp	p	813
BaY$_2$F$_8$	$^5I_7 \rightarrow {}^5I_8$	2.065	77	Xe lamp	p	152
		2.074	20	Xe lamp	p	838
	$^5S_2 \rightarrow {}^5I_5$	2.362	77	Xe lamp	p	838
BaEr$_2$F$_8$:Tm	$^5I_7 \rightarrow {}^5I_8$	2.086	300	Xe lamp	p	870, 871
			300	Xe lamp	qs	872
BaTm$_2$F$_8$	$^5I_7 \rightarrow {}^5I_8$	2.0560	110–230	Xe lamp	p	813
BaY$_2$F$_8$	$^5I_7 \rightarrow {}^5I_8$	2.065	77	Xe lamp	p	152
		2.074	20	Xe lamp	p	838
	$^5S_2 \rightarrow {}^5I_5$	2.362	77	Xe lamp	p	838
BaY$_2$F$_8$		2.363	77	Xe lamp	p	838
		2.375	77	Xe lamp	p	838
		2.377	20	Xe lamp	p	838
BaY$_2$F$_8$:Er,Tm	$^5I_7 \rightarrow {}^5I_8$	2.0555	20	Xe lamp	p	838
		2.0644	85	W lamp	cw	838
		2.0746–2.076	85	Xe lamp	p	838
		2.074	20	Xe lamp	p	838
		2.065	77	Xe lamp	cw	838
		2.0866	77	Xe lamp	p	838
		2.171	295	Xe lamp	p	838
BaYb$_2$F$_8$	$^5I_6 \rightarrow {}^5I_8$	1.190	300	Xe lamp	p	469
	$^5S_2 \rightarrow {}^5I_5$	1.3865	110	Xe lamp	p	703
	$^5I_7 \rightarrow {}^5I_8$	2.0563	295	NGL	p, upc	846
		2.0665	110	Xe lamp	p	703, 848
		2.0715	110	Xe lamp	p	703, 848

Holmium (Ho^{3+}, 4f^{10})—*continued*

Host crystal	Laser transition	Wavelength (μm)	Temp. (K)	Optical pump	Mode	Ref.
BaYb$_2$F$_8$	$^5I_7 \rightarrow {}^5I_8$	2.0895	110	Xe lamp	p	703, 848
	$^5I_6 \rightarrow {}^5I_7$	2.9073	293	Nd laser	p	989
BaYb$_2$F$_8$:Yb	$^5I_6 \rightarrow {}^5I_7$	2.8575	300	Xe lamp	p	703
		2.9	300	Xe lamp	p	51, 161, 703, 964
		2.9054	300	NdYLF	p	966
Ba(Y,Yb)$_2$F$_8$	$^5S_2 \rightarrow {}^5I_8$	0.5515	77	Xe lamp	p, upc	45
			300	NdYAG	p, upc	966
Bi$_4$Ge$_3$O$_{12}$	$^5I_7 \rightarrow {}^5I_8$	2.087	77	Xe lamp	p	229
Ca$_5$(PO$_4$)$_3$F:Cr	$^5I_7 \rightarrow {}^5I_8$	2.075	77	Xe lamp	p	497
CaF$_2$	$^5S_2 \rightarrow {}^5I_8$	0.5512	77	Xe lamp	p	44
	$^5I_7 \rightarrow {}^5I_8$	2.092	77	Xe lamp	p	463
CaF$_2$-ErF$_2$	$^5I_7 \rightarrow {}^5I_8$	2.030	77	Xe lamp	p	835
CaF$_2$-ErF$_3$-	$^5I_7 \rightarrow {}^5I_8$	2.05	100	Xe lamp	p	837
TmF$_3$-YbF$_3$		2.060	100	Xe lamp	p	837
		2.06	298	Xe lamp	p	836
		2.1	65	Xe lamp	cw	836
		2.1	77	Xe lamp	p	836
CaF$_2$-HoF$_3$	$^5I_7 \rightarrow {}^5I_8$	2.1110	110	Xe lamp	p	791
CaF$_2$-YF$_3$	$^5I_7 \rightarrow {}^5I_8$	2.0318	77	Xe lamp	p	834
CaMoO$_4$	$^5I_7 \rightarrow {}^5I_8$	2.0556	77	Xe lamp	p	821
		2.0707	77	Xe lamp	p	821
		2.074	77	Xe lamp	p	821
CaWO$_4$	$^5I_7 \rightarrow {}^5I_8$	2.046	77	Xe lamp	p	291, 820
		2.059	77	Xe lamp	p	291, 820
CaY$_4$(SiO$_4$)$_3$O: Er,Tm	$^5I_7 \rightarrow {}^5I_8$	2.060	77	Xe lamp	p	497
(Er,Ho)F$_3$	$^5I_7 \rightarrow {}^5I_8$	2.090	77	ErGL	upc	863
Er$_2$O$_3$	$^5I_7 \rightarrow {}^5I_8$	2.121	145	Xe lamp	p	893
Er$_2$SiO$_5$	$^5I_7 \rightarrow {}^5I_8$	2.085	77	Xe lamp	p	856
ErAlO$_3$	$^5I_7 \rightarrow {}^5I_8$	2.089–2.102	77	Xe lamp	p	878
		2.0985–2.0997	110	Xe lamp	p, cas	848, 882
		2.1205	77	Xe lamp	p	840
	$^5I_6 \rightarrow {}^5I_7$	2.9230	110	Xe lamp	p	792
(Er,Lu)AlO$_3$	$^5I_7 \rightarrow {}^5I_8$	2.0010	77	Xe lamp	p	840

Holmium (Ho^{3+}, 4f^{10})—*continued*

Host crystal	Laser transition	Wavelength (μm)	Temp. (K)	Optical pump	Mode	Ref.
(Er,Lu)AlO$_3$	$^5I_7 \to {}^5I_8$	2.1205	77	Xe lamp	p	840
Er$_3$Al$_5$O$_{12}$	$^5I_7 \to {}^5I_8$	2.0985	110	Xe lamp	p, cas	848
Er$_3$Sc$_2$Al$_3$O$_{12}$	$^5I_7 \to {}^5I_8$	2.0985	77	Xe lamp	p	279
ErVO$_4$:Tm	$^5I_7 \to {}^5I_8$	2.0416	77	Xe lamp	p	833
GdAlO3	$^5I_7 \to {}^5I_8$	1.9925	90	Xe lamp	p	825
Gd$_3$Ga$_5$O$_{12}$	$^5I_6 \to {}^5I_8$	1.2085	~110	Xe lamp	p	722
	$^5S_2 \to {}^5I_5$	1.4040	~110	Xe lamp	p, cas	722
	$^5I_7 \to {}^5I_8$	2.0885	110	Xe lamp	p, cas	722, 851
	$^5I_6 \to {}^5I_7$	2.9	300	Xe lamp	p	918
		2.9619	110	Xe lamp	p, cas	887
Gd$_3$Sc$_2$Al$_3$O$_{12}$	$^5I_7 \to {}^5I_8$	2.09	300	Xe lamp	p	861
Gd$_3$(Sc,Ga)$_5$O$_{12}$: Cr,Tm	$^5I_7 \to {}^5I_8$	2.088	300	Xe lamp	p, cas	818, 873
			300	Xe lamp	p	341
GdVO$_4$:Tm	$^5I_7 \to {}^5I_8$	2.049	300	TiS laser	qcw	578
	$^5I_6 \to {}^5I_7$	2.8484	300	TiS laser	qcw	578
Ho$_3$Al$_5$O$_{12}$	$^5I_7 \to {}^5I_8$	2.1224	~90	Xe lamp	p	892
		2.1227	77	Xe lamp	p	713
		2.1294	~90	Xe lamp	p	892
		2.1297	77	Xe lamp	p	713
HoF$_3$	$^5I_7 \to {}^5I_8$	2.090	77	Xe lamp	p	1069
Ho$_3$Ga$_5$O$_{12}$	$^5I_7 \to {}^5I_8$	2.086	77	Xe lamp	p	713
		2.1135	77	Xe lamp	p	713
Ho$_3$Sc$_2$Al$_3$O$_{12}$	$^5I_7 \to {}^5I_8$	2.1170	77	Xe lamp	p	713
		2.1285	77	Xe lamp	p	713
K(Y,Er)(WO$_4$)$_2$: Tm	$^5S_2 \to {}^5I_5$	1.3908	110	Xe lamp	p	993
	$^5I_7 \to {}^5I_8$	2.0565	110	Xe lamp	p	848
		2.0720	110–220	Xe lamp	p	76
		2.0765	110	Xe lamp	p	848, 887
KGd(WO$_4$)$_2$	$^5S_2 \to {}^5I_5$	1.3982	110	Xe lamp	p	993
	$^5I_7 \to {}^5I_8$	2.0740	110	Xe lamp	p	76
	$^5I_6 \to {}^5I_7$	2.9342	300	Xe lamp	p	796
KGd(WO$_4$)$_2$: Er,Tm	$^5I_7 \to {}^5I_8$	2.074	110	Xe lamp	p	848
KLa(MoO$_4$)$_{12}$	$^5I_6 \to {}^5I_7$	2.8415	110	Xe lamp	p	501
	$^5S_2 \to {}^5I_5$	1.4	110	Xe lamp	p	501

Holmium (Ho^{3+}, 4f^{10})—*continued*

Host crystal	Laser transition	Wavelength (μm)	Temp. (K)	Optical pump	Mode	Ref.
KLa(MoO$_4$)$_{12}$	$^5I_6 \rightarrow {}^5I_7$	2.9700	300	Xe lamp	p	501
KLu(WO$_4$)$_2$	$^5I_7 \rightarrow {}^5I_8$	2.0790	110	Xe lamp	p	887
	$^5I_6 \rightarrow {}^5I_7$	2.9445	300	—	—	965
KY(WO$_4$)$_2$	$^5S_2 \rightarrow {}^5I_5$	1.3908	~110	Xe lamp	p	993
	$^5I_7 \rightarrow {}^5I_8$	2.0565	110	Xe lamp	p	848
		2.0765	110	Xe lamp	p	848
	$^5I_6 \rightarrow {}^5I_7$	2.9395	300	Xe lamp	p	959
KY(WO$_4$)$_2$: Er,Tm	$^5I_7 \rightarrow {}^5I_8$	2.0720	110, 220	Xe lamp	p	959
LaNbO$_4$	$^5I_7 \rightarrow {}^5I_8$	2.0725	110	Xe lamp	p	851
LaNbO$_4$:Er	$^5I_7 \rightarrow {}^5I_8$	2.07	90	Xe lamp	p	839
LaNbO$_4$:Er	$^5I_6 \rightarrow {}^5I_7$	2.8510	90	Xe lamp	p	958
Li(Y,Er)F$_4$	$^5I_7 \rightarrow {}^5I_8$	2.0654	300	Xe lamp	p, qs	853
		2.0656	300	Xe lamp	p	847
Li(Y,Er)F$_4$:Tm	$^5I_7 \rightarrow {}^5I_8$	2.0	77	W lamp	cw	841
		2.06	300	—	p	849
		2.065	220–300	Xe lamp	p	787
		2.1	77–124	laser diode	cw	884
LiErF$_4$	$^5I_7 \rightarrow {}^5I_8$	2.0610–2.0650	300	Xe lamp	p	843
LiErF$_4$:Tm	$^5I_7 \rightarrow {}^5I_8$	2.0490–2.0559	300	Xe lamp	p	843
LiHoF$_4$	$^5F_5 \rightarrow {}^5I_7$	0.979	~90	Xe lamp	p	268, 269
	$^5F_5 \rightarrow {}^5I_5$	2.352	90	Xe lamp	p	51
LiLuF$_4$	$^5S_2 \rightarrow {}^5I_7$	0.7501	110	Xe lamp	p	32
		0.7505	110	Xe lamp	p	32
	$^5S_2 \rightarrow {}^5I_6$	1.0183	110	Xe lamp	p	32
	$^5S_2,{}^5F_4 \rightarrow {}^5I_5$	1.3918	110	Xe lamp	p	32
		1.3920	300	Xe lamp	p	32
LiLuF$_4$:Tm	$^5I_7 \rightarrow {}^5I_8$	2.06	300	—	p	849
		2.055	300	laser diode	qs	831
LiNbO$_3$	$^5I_7 \rightarrow {}^5I_8$	2.0786	77	Xe lamp	p	370
LiYF$_4$	$^5S_2 \rightarrow {}^5I_7$	0.7498	90, 300	Xe lamp	p	167, 181
		0.7505	300	Xe lamp	p	187
		0.7516	116	Xe lamp	p	181
		0.7555	116	Xe lamp	p	181
	$^5F_5 \rightarrow {}^5I_7$	0.9794	90, 300	Xe lamp	p	167
	$^5S_2 \rightarrow {}^5I_6$	1.0143	90, 300	Xe lamp	p	167, 181
	$^5S_2,{}^5F_4 \rightarrow {}^5I_5$	1.392	300	DNdGL	p	719

Holmium (Ho^{3+}, 4f^{10})—*continued*

Host crystal	Laser transition	Wavelength (μm)	Temp. (K)	Optical pump	Mode	Ref.
LiYF$_4$	$^5S_2, ^5F_4 \rightarrow ^5I_5$	1.3960	116, 300	DNdGL	p	167, 181
	$^5F_5 \rightarrow ^5I_6$	1.486	90	Xe lamp	p	268, 269
		1.4862	90, 116	Xe lamp	p	749, 750
		1.4912	190	Xe lamp	p	181
	$^5I_5 \rightarrow ^5I_7$	1.673	115	Xe lamp	p	750
		1.673	300	DNdGL	p	719
	$^5I_7 \rightarrow ^5I_8$	2.0505	300	Xe lamp	p	847
		2.0534	300	Xe lamp	p	847
		2.0672	90	Xe lamp	p	852
	$^5F_5 \rightarrow ^5I_5$	2.3520	116	Xe lamp	p	181
		2.3524	116	Xe lamp	p	750
	$^5I_6 \rightarrow ^5I_7$	2.850	300	—	—	181
		2.952	116–300	Xe lamp	p	189
LiYF$_4$	$^5I_6 \rightarrow ^5I_7$	2.952	300	Xe lamp	p	750
		2.955	300	dye laser	p, cas	967
	$^5S_2 \rightarrow ^5F_5$	3.369	300	dye laser	p, cas	967
	$^5I_5 \rightarrow ^5I_6$	3.893	300	dye laser	p, cas	967
		3.914	300	DNdGL	p	915
LiYF$_4$:Er	$^5I_7 \rightarrow ^5I_8$	2.066	77	Xe lamp	p	854
LiYF$_4$:Tm	$^5I_7 \rightarrow ^5I_8$	2.05	243	laser diode	cw	1001
		2.052	300	laser diode	QS	831
		2.065	220	Xe lamp	p	787
		2.067	300	laser diode	cw	340
		2.1	77	laser diode	cw	340
LiYF$_4$:Er,Tm	$^5I_7 \rightarrow ^5I_8$	2.048–2.071	77	Xe lamp	p	842
		2.1	77	W lamp	cw	841
LiYbF$_4$	$^5I_7 \rightarrow ^5I_8$	2.065	300	Xe lamp	p	850
	$^5I_6 \rightarrow ^5I_7$	2.83	300	Nd laser	p, cas	850
Lu$_3$Al$_5$O$_{12}$	$^5I_7 \rightarrow ^5I_8$	2.1005	110	Xe lamp	p	881
		2.1250	110	Xe lamp	p	887
		2.1300	110	Xe lamp	p	887
Lu$_3$Al$_5$O$_{12}$:Tm	$^5I_6 \rightarrow ^5I_7$	2.1004	300	TiS laser	p	862
Lu$_3$Al$_5$O$_{12}$:Yb	$^5I_6 \rightarrow ^5I_8$	1.2160	110	Xe lamp	p	993
	$^5S_2, ^5F_4 \rightarrow ^5I_5$	1.4085	~110	Xe lamp	p	993
	$^5I_7 \rightarrow ^5I_8$	2.1005	110	Xe lamp	p	881
		2.1252	110	Xe lamp	p	881
		2.1300	110	Xe lamp	p	881
Lu$_3$Al$_5$O$_{12}$: Cr,Tm	$^5I_7 \rightarrow ^5I_8$	2.1020	110	Xe lamp	p	889
		2.1303	300	Xe lamp	p	881
		2.1008	110	Xe lamp	p	881

Holmium (Ho^{3+}, 4f^{10})—*continued*

Host crystal	Laser transition	Wavelength (μm)	Temp. (K)	Optical pump	Mode	Ref.
Lu$_3$Al$_5$O$_{12}$: Cr,Tm	$^5I_7 \rightarrow {}^5I_8$	2.1241	289	Xe lamp	p	824
		2.1241	300	Xe lamp	p	881
Lu$_3$Al$_5$O$_{12}$: Cr,Yb	$^5I_6 \rightarrow {}^5I_7$	2.9460	300	Xe lamp	p	129, 941
Lu$_3$Al$_5$O$_{12}$: Er,Tm	$^5I_7 \rightarrow {}^5I_8$	2.1020	77	Xe lamp	p, cw	794
Lu$_3$Al$_5$O$_{12}$: Er,Tm,Yb	$^5I_6 \rightarrow {}^5I_7$	2.1005	110	Xe lamp	p	887
		2.1020	77	Xe lamp	p, cw	794
Lu$_3$Al$_5$O$_{12}$:Tm	$^5I_6 \rightarrow {}^5I_7$	2.1004	300	TiS laser	p	862
LuAlO$_3$	$^5I_7 \rightarrow {}^5I_8$	2.1348	90	Xe lamp	p	795
α-NaCaErF$_6$	$^5I_7 \rightarrow {}^5I_8$	2.0312	77	Xe lamp	p	805
		2.0345	150	Xe lamp	p	805
		2.0377	77	Xe lamp	p	805
NaLa(MoO$_4$)$_2$: Er	$^5I_7 \rightarrow {}^5I_8$	2.050	90	Xe lamp	p	839
SrF$_2$-(Y,Er)F$_3$	$^5I_7 \rightarrow {}^5I_8$	2.053	120	Xe lamp	p	844
		2.053	300	Xe lamp	p	845
SrF$_2$-(Y,Er)F$_3$: Tm	$^5I_6 \rightarrow {}^5I_7$	2.0496	120, 300	Xe lamp	p	844
Tm$_3$Al$_5$O$_{12}$	$^5I_7 \rightarrow {}^5I_8$	2.0995	110	Xe lamp	p	848
(Y,Er)$_3$Al$_5$O$_{12}$	$^5I_7 \rightarrow {}^5I_8$	2.0907	300	Xe lamp	p	278
		2.0978	77	W lamp	cw	865
		2.0979	300	Xe lamp	p	278
			300	Hg, W	cw	278
		2.123	300	Xe lamp	p	278
Y,Er)$_3$Al$_5$O$_{12}$: Tm	$^5I_7 \rightarrow {}^5I_8$	2.0982	300	Xe lamp	p	817
		2.0990	300	Xe lamp	p	895
				W lamp	cw	895
		2.1227	300	W lamp	p, cw	817
		2.1285	300	Xe lamp	p	895
		~2.13	300	Xe lamp	p	896, 897
			77	W lamp	cw	897
YAlO$_3$	$^5S_2 \rightarrow {}^5I_7$	0.7577	110–300	Xe lamp	p	190
		0.7610	110–300	Xe lamp	p	190
	$^5S_2 \rightarrow {}^5I_6$	1.0311	110–300	Xe lamp	p, cas	190, 241
	$^5I_6 \rightarrow {}^5I_8$	1.2198	110	Xe lamp	p, cas	241
	$^5S_2, {}^5F_4 \rightarrow {}^5I_5$	1.3806	300	Xe lamp	p, cas	190, 241

Holmium (Ho^{3+}, 4f^{10})—*continued*

Host crystal	Laser transition	Wavelength (μm)	Temp. (K)	Optical pump	Mode	Ref.
YAlO$_3$	$^5S_2,^5F_4 \rightarrow {}^5I_5$	1.3900	110	Xe lamp	p	190, 241
		1.3950	110	Xe lamp	p	190
		1.4003	110	Xe lamp	p	190
		1.4028	110	Xe lamp	p	993
		1.4058	110–300	Xe lamp	p	190
	$^5I_7 \rightarrow {}^5I_8$	2.1185	110	Xe lamp	p	887
		2.1189	110	Xe lamp	p, cas	190, 241
		2.1193	110	Xe lamp	p	190
		2.12	233	Xe lamp	p	787
		2.1300	110	Xe lamp	p	190, 887
	$^5I_6 \rightarrow {}^5I_7$	2.8578	300	Xe lamp	p, cas	952
		2.9155	77, 300	Xe lamp	p, cas	783
		2.9180	300	Xe lamp	p	129, 930
		2.9185	110–300	Xe lamp	p, cas	190
YAlO$_3$	$^5I_6 \rightarrow {}^5I_7$	2.9200	~110	Xe lamp	p	1122
		3.0177	~110	Xe lamp	p	1122
YAlO$_3$:Er	$^5I_6 \rightarrow {}^5I_7$	2.9200	110	Xe lamp	p, cas	792, 930
		3.0132	300	Xe lamp	p	129, 930
		3.0157	110	Xe lamp	p	792, 910
		3.0165	110–300	Xe lamp	p	190
		3.0177	110	Xe lamp	p	792
(Y,Er)AlO$_3$:Tm	$^5I_7 \rightarrow {}^5I_8$	2.119	300	Xe lamp	p	130
		2.12	233	Xe lamp	p	787
		2.123	300	Xe lamp	p	130
Y$_2$SiO$_5$	$^5I_7 \rightarrow {}^5I_8$	2.092	~110	Xe lamp	p	856
		2.105	~110, 220	Xe lamp	p	856
Y$_3$Al$_5$O$_{12}$	$^5I_6 \rightarrow {}^5I_8$	1.2155	110	Xe lamp	p	993
	$^5S_2, \rightarrow {}^5I_5$	1.4072	~110	Xe lamp	p	993
	$^5I_7 \rightarrow {}^5I_8$	2.065	300	alex. laser	p	858
		2.097	300	alex. laser	p	858
		2.0977	~77	W lamp	cw	579
		2.098	77	Xe lamp	p	497
		2.101	300	alex. laser	p	858
		2.0914	77	Xe lamp	p	868
		2.0975	77	Xe lamp	p	278, 868
		2.1295	300	Xe lamp	p	881
	$^5I_6 \rightarrow {}^5I_7$	2.9403	300	Xe lamp	p	129
Y$_3$Al$_5$O$_{12}$:Cr	$^5I_7 \rightarrow {}^5I_8$	2.095	77	Xe lamp	p	278
			85	W lamp	cw	278
		2.1223	77	Xe lamp	p	278

Holmium (Ho^{3+}, $4f^{10}$)—*continued*

Host crystal	Laser transition	Wavelength (μm)	Temp. (K)	Optical pump	Mode	Ref.
$Y_3Al_5O_{12}$:Nd	$^5I_6 \rightarrow {}^5I_7$	2.940	300	Xe lamp	p	673
		3.011	300	Xe lamp	p	673
$Y_3Al_5O_{12}$:Tm	$^5I_7 \rightarrow {}^5I_8$	2	300	Xe lamp	p	814, 815
		2.060	300	alex. laser	p	858
		2.06	300	TiS laser	cw, cas	743
$Y_3Al_5O_{12}$: Cr,Tm	$^5I_7 \rightarrow {}^5I_8$	2.0982	110	Xe lamp	p	881
		2.09	300	Xe lamp	p	1034
		2.091	300	laser diode	cw	776
		2.0974	300	dye, LD	cw	879, 880
		2.1	290	Xe lamp	p	883
		2.12	215–330	Xe lamp	p	814, 891
		2.1223	77	Xe lamp	p	881
$Y_3Al_5O_{12}$: Er,Tm	$^5I_7 \rightarrow {}^5I_8$	2.0982	110	Xe lamp	p	817
			77	sun	qs, cw	1042
		2.0983	110	Xe lamp	p	881
		2.1	77	W lamp	cw	841
$Y_3Al_5O_{12}$:Tm	$^5I_7 \rightarrow {}^5I_8$	2.1	110	laser diode	cw	888
$Y_3Al_5O_{12}$:Yb	$^5I_6 \rightarrow {}^5I_8$	1.2155	110	Xe lamp	p	993
	$^5I_7 \rightarrow {}^5I_8$	2.1	300	TiS, LD	cw	1126
$(Y,Er)_3Al_5O_{12}$	$^5I_7 \rightarrow {}^5I_8$	2.123	77	Xe lamp	p	278
		2.0917	77	Xe lamp	p	278
		2.0979	77	Xe lamp	p	278
$(Y,Er)_3Al_3O_{12}$: Tm	$^5I_7 \rightarrow {}^5I_8$	2.089–2.102	77	Xe lamp	p,qs, AML	878
		2.0982	77	Xe lamp	p	817
		2.0990	77	Xe lamp	p, cw	817
		2.1	77	laser diode	cw	885, 886
		2.1227	77	Xe lamp	p, cw	817
		2.1285	77	W lamp	cw	895
		~2.13	300	Xe lamp	p	896
		~2.13	77	W lamp	cw	897
$(Y,Ho)_3Al_5O_{12}$	$^5I_7 \rightarrow {}^5I_8$	2.097	77	Xe lamp	p	893
		2.123	77	Xe lamp	p	893
$(Er,Tm,Yb)_3$-Al_5O_{12}	$^5I_7 \rightarrow {}^5I_8$	2.1010	77	Xe lamp	p	726
$Y_3Fe_5O_{12}$	$^5I_7 \rightarrow {}^5I_8$	2.089	77	Xe lamp	p	857
		2.107	77	Xe lamp	p	857
$Y_3Fe_5O_{12}$:Er,Tm	$^5I_7 \rightarrow {}^5I_8$	2.086	77	Xe lamp	p	857
$Y_3Ga_5O_{12}$	$^5I_7 \rightarrow {}^5I_8$	2.086	77	Xe lamp	p	857

Holmium (Ho^{3+}, 4f^{10})—*continued*

Host crystal	Laser transition	Wavelength (μm)	Temp. (K)	Optical pump	Mode	Ref.
$Y_3Ga_5O_{12}$	$^5I_7 \rightarrow {}^5I_8$	2.114	77	Xe lamp	p	857
$Y_3Ga_5O_{12}$:Fe	$^5I_6 \rightarrow {}^5I_7$	2.086	77	TiS laser	cw	860
		2.114	77–185	TiS laser	cw	860
$Y_3Sc_2Al_3O_{12}$	$^5I_7 \rightarrow {}^5I_8$	2.080–2.089	300	Xe lamp	p	859
$Y_3Sc_2Al_3O_{12}$: Cr,Tm	$^5I_7 \rightarrow {}^5I_8$	1.944	300	Kr laser	cw	776
		2.086	300	Kr laser	cw	869
		2.095	300	Kr laser	cw	869
$Y_3Sc_2Ga_3O_{12}$	$^5I_7 \rightarrow {}^5I_8$	2.086	300	Kr laser	cw	869
		2.088	300	Xe lamp	p, qs	874–876
$Y_3Sc_2Ga_3O_{12}$: Cr,Tm	$^5I_7 \rightarrow {}^5I_8$	1.924	300	Kr laser	cw	776
		2.010	300	Kr laser	cw	776
$Y_3Sc_2Ga_3O_{12}$: Cr,Tm	$^5I_7 \rightarrow {}^5I_8$	2.086	300	Kr laser	cw	776
(Y,Er)$_3$Sc$_2$-Ga$_3$O$_{12}$:Tm	$^5I_7 \rightarrow {}^5I_8$	2.1	77	Xe lamp	p	848
			77		cw	841
$Yb_3Al_5O_{12}$	$^5I_7 \rightarrow {}^5I_8$	2.0960	77	Xe lamp	p	866
		2.1000	110	Xe lamp	p	848
		2.101	300	alex. laser	cw	858
$Yb_3Al_5O_{12}$: Er,Tm	$^5I_7 \rightarrow {}^5I_8$	2.0998	110	Xe lamp	p	881
		2.1010	77	Xe lamp	p	726
$Yb_3Al_5O_{12}$:Tm,	$^5I_7 \rightarrow {}^5I_8$	2.1000	110	Xe lamp	p	848
YScO$_3$:Gd	$^5I_6 \rightarrow {}^5I_7$	2.8637	77	Xe lamp	p	783
YVO$_4$:Er,Tm	$^5I_7 \rightarrow {}^5I_8$	2.0412	77	Xe lamp	p	833
ZrO_2-Er_2O_3	$^5I_7 \rightarrow {}^5I_8$	2.115	77	Xe lamp	p	764

Erbium (Er^{3+}, 4f^{11})

Host crystal	Laser transition	Wavelength (μm)	Temp. (K)	Optical pump	Mode	Ref.
BaEr$_2$F$_8$	$^4S_{3/2} \rightarrow {}^4I_{13/2}$	0.8425	110	Xe lamp	p	115
		0.8538	104–123	Xe lamp	p	232
		0.8543	110	Xe lamp	p	115
	$^4S_{3/2} \rightarrow {}^4I_{11/2}$	1.2312	100–112	Xe lamp	p	232
		1.2320	110	Xe lamp	p	215
	$^2H_{11/2} \rightarrow {}^4I_{9/2}$	1.6455	110	Xe lamp	p	115
	$^4S_{3/2} \rightarrow {}^4I_{9/2}$	1.7350	102–112	Xe lamp	p	232

Erbium (Er^{3+}, 4f^{11})—*continued*

Host crystal	Laser transition	Wavelength (μm)	Temp. (K)	Optical pump	Mode	Ref.
BaEr$_2$F$_8$	$^4S_{3/2} \rightarrow {}^4I_{9/2}$	1.7355	110	Xe lamp	p	115
	$^4F_{9/2} \rightarrow {}^4I_{11/2}$	1.9975	110	Xe lamp	p	115
	$^4I_{11/2} \rightarrow {}^4I_{13/2}$	2.7417	110	Xe lamp	p	115, 927
		2.7595	110	Xe lamp	p	115, 927
		2.7980	110	Xe lamp	p	115, 927
BaLu$_2$F$_8$	$^4I_{11/2} \rightarrow {}^4I_{13/2}$	~2.79	300	laser diode	cw	1055
BaY$_2$F$_8$	$^2P_{3/2} \rightarrow {}^4I_{11/2}$	0.4703	10	TiS laser	cw, upc	1086, 1090
	$^4S_{3/2} \rightarrow {}^4I_{15/2}$	0.5449	10	TiS laser	cw, upc	1090
	$^4S_{3/2} \rightarrow {}^4I_{15/2}$	0.5512	10	TiS laser	cw, upc	1090
		0.5541	10	TiS laser	cw, upc	1090
		0.5517	10	TiS laser	cw, upc	1086
BaY$_2$F$_8$	$^2P_{3/2} \rightarrow {}^4I_{9/2}$	0.6172	20	TiS laser	cw, upc	1090
	$^2P_{3/2} \rightarrow {}^4I_{9/2}$	0.6185	10	TiS laser	cw, upc	1086
	$^4F_{9/2} \rightarrow {}^4I_{15/2}$	0.6688	20	TiS laser	cw, upc	1090
	$^2H_{9/2} \rightarrow {}^4I_{11/2}$	0.7015	10	TiS laser	cw, upc	1090
		0.7032	10	TiS laser	cw, upc	1090
	$^4I_{11/2} \rightarrow {}^4I_{13/2}$	2.7	300	LD, DL	cw	1072
Ba(Y,Er)$_2$F$_8$	$^4S_{3/2} \rightarrow {}^4I_{15/2}$	0.5540	77	Xe lamp	p	46
	$^2H_{9/2} \rightarrow {}^4I_{13/2}$	0.5617	77	Xe lamp	p	45, 46
	$^4F_{9/2} \rightarrow {}^4I_{15/2}$	0.6709	77	Xe lamp	p	46
	$^2H_{9/2} \rightarrow {}^4I_{11/2}$	0.7037	77	Xe lamp	p	46
Ba(Y,Er)$_2$F$_8$	$^4S_{3/2} \rightarrow {}^4I_{13/2}$	0.8640	300	Xe lamp	p	1074
	$^4S_{3/2} \rightarrow {}^4I_{11/2}$	1.2316	300	Xe lamp	p	1074
	$^4S_{3/2} \rightarrow {}^4I_{9/2}$	1.7332	300	Xe lamp	p	1074
	$^4I_{11/2} \rightarrow {}^4I_{13/2}$	2.704	300	Ar laser	qcw	1070
		2.711	300	Ar laser	qcw	1070
		2.7415	~110	Xe lamp	p, qcw	1074
		2.7585	~110	Xe lamp	p, qcw	1074
		2.798	300	Ar laser	qcw	1070
			300	LD, ArL	cw	1071
		2.7980	300–450	Xe lamp	p, qcw	1074
Ba(Y,Yb)$_2$F$_8$	$^4F_{9/2} \rightarrow {}^4I_{15/2}$	0.6700	77	Xe lamp	p	45
BaYb$_2$F$_8$	$^4F_{9/2} \rightarrow {}^4I_{15/2}$	0.67	300	NdGL,ErGL	p, upc	78
		0.6700	110	Xe lamp	p	45, 115
	$^4S_{3/2} \rightarrow {}^4I_{11/2}$	1.26	300	NdGL	p, upc	587
	$^4S_{3/2} \rightarrow {}^4I_{9/2}$	1.7360	110	Xe lamp	p	115
	$^4F_{9/2} \rightarrow {}^4I_{11/2}$	1.96	300	Xe lamp	p, upc	953, 1004
		1.965	300	Xe lamp	p, upc	817, 1004
		1.9654-5	300	NdGL	p, upc	548, 549

Erbium (Er^{3+}, $4f^{11}$)—*continued*

Host crystal	Laser transition	Wavelength (μm)	Temp. (K)	Optical pump	Mode	Ref.
$BaYb_2F_8$	$^4F_{9/2} \rightarrow {}^4I_{11/2}$	1.9965	110,300	Xe lamp	p	115,1051
		1.9925	110	Xe lamp	p	813
	$^4I_{11/2} \rightarrow {}^4I_{13/2}$	2.7980	300	laser diode	cw	1148, 1149
$BaYb_2F_8$:Tm	$^4F_{9/2} \rightarrow {}^4I_{11/2}$	1.9650	~110	Xe lamp	p	1051
		1.9655	—	Xe lamp	p	1051
		1.9925	110	Xe lamp	p	813
	$^4I_{11/2} \rightarrow {}^4I_{13/2}$	2.7060	300	Xe lamp	p, cas	1051
		2.7980	—	Xe lamp	p, cas	1051
$Bi_4Ge_3O_{12}$	$^4S_{3/2} \rightarrow {}^4I_{13/2}$	0.853	77	Xe lamp	p	229
	$^4I_{13/2} \rightarrow {}^4I_{15/2}$	1.5578	77	Xe lamp	p	229
		1.6645	77	Xe lamp	p	229
$Ca(NbO_3)_2$	$^4I_{13/2} \rightarrow {}^4I_{15/2}$	1.61	77	Xe lamp	p	753
$Ca(NbO_3)_2$:Ti	$^4S_{3/2} \rightarrow {}^4I_{9/2}$	1.7410	110	Xe lamp	p	57
	$^4I_{11/2} \rightarrow {}^4I_{13/2}$	2.7175	110	Xe lamp	p	57
$Ca_2Al_2SiO_7$:Yb	$^4I_{13/2} \rightarrow {}^4I_{15/2}$	1.530	300	TiS laser	cw	744
		1.550	300	TiS laser	cw	733, 744
$Ca_2Al_2SiO_7$: Yb,Ce	$^4I_{13/2} \rightarrow {}^4I_{15/2}$	1.555	300	TiS laser	cw	744
$Ca_3(NbLiGa)_5$-O_{12}	$^4I_{11/2} \rightarrow {}^4I_{13/2}$	2.71	300	Xe lamp	p	929
$Ca_3Ga_2Ge_3O_{12}$	$^4S_{3/2} \rightarrow {}^4I_{13/2}$	0.8471	110	Xe lamp	p	221
		0.8615	110	Xe lamp	p	221
$CaAl_4O_7$	$^4I_{13/2} \rightarrow {}^4I_{15/2}$	1.5500	77	Xe lamp	p	510
		1.5815	77	Xe lamp	p	510
CaF_2	$^4S_{3/2} \rightarrow {}^4I_{13/2}$	0.8546	77	Xe lamp	p	235
		0.8548	77	Xe lamp	p	235
		0.855	77	CCL	cw, upc	1082
	$^4S_{3/2} \rightarrow {}^4I_{11/2}$	1.26	77	Xe lamp	p	614
	$^4I_{13/2} \rightarrow {}^4I_{15/2}$	1.5298	77	Xe lamp	p	674
		1.5308	77	Xe lamp	p	674
		1.617	77	Xe lamp	p	760
	$^4S_{3/2} \rightarrow {}^4I_{9/2}$	1.696	77	Xe lamp	p	592
		1.715	77	Xe lamp	p	592
		1.726	77	Xe lamp	p	592
	$^4I_{11/2} \rightarrow {}^4I_{13/2}$	2.7307	300	Xe lamp	p	998
CaF_2-ErF_3	$^4S_{3/2} \rightarrow {}^4I_{9/2}$	1.6615	110	Xe lamp	p	791
	$^4I_{11/2} \rightarrow {}^4I_{13/2}$	2.75	300	Xe lamp	p, upc	923–926
		2.7955	300	Xe lamp	p	791

Erbium (Er^{3+}, 4f^{11})—*continued*

Host crystal	Laser transition	Wavelength (μm)	Temp. (K)	Optical pump	Mode	Ref.
CaF$_2$-ErF$_3$	$^4I_{11/2} \rightarrow {}^4I_{13/2}$	2.7985	300	Xe lamp	p	791, 944
		2.80	300	Xe lamp	p, upc	923–926
		2.7295	300	Xe lamp	p	791
		2.7307	300	Xe lamp	p	997
CaF$_2$-ErF$_3$:Ho	$^4I_{11/2} \rightarrow {}^4I_{13/2}$	2.7290	300	Xe lamp	p	791
CaF$_2$-ErF$_3$:Tm	$^4I_{11/2} \rightarrow {}^4I_{13/2}$	2.69	298	Xe lamp	p	936
		2.7490	300	Xe lamp	p	791
CaF$_2$-ErF$_3$: Ho,Tm	$^4I_{11/2} \rightarrow {}^4I_{13/2}$	2.7460	300	Xe lamp	p	791
CaF$_2$-HoF$_3$-ErF$_3$	$^4S_{3/2} \rightarrow {}^4I_{13/2}$	0.8456	77	Xe lamp	p	215
CaF$_2$-ErF$_3$-TmF$_3$	$^4I_{11/2} \rightarrow {}^4I_{13/2}$	2.69	100	Xe lamp	p	837
CaF$_2$-HoF$_3$- ErF$_3$-TmF$_3$	$^4S_{3/2} \rightarrow {}^4I_{13/2}$	0.8456	77	Xe lamp	p	215
CaF$_2$-YF$_3$	$^4S_{3/2} \rightarrow {}^4I_{13/2}$	0.8430	77	Xe lamp	p	215, 216
		0.8456	77	Xe lamp	p	215, 216
	$^4I_{13/2} \rightarrow {}^4I_{15/2}$	1.547	77	Xe lamp	p	215, 216
CaWO$_4$	$^4I_{13/2} \rightarrow {}^4I_{15/2}$	1.612	77	Xe lamp	p	757
ErAlO$_3$	$^4S_{3/2} \rightarrow {}^4I_{9/2}$	1.6632	110	Xe lamp	p	925
ErYF4	$^4S_{3/2} \rightarrow {}^4I_{9/2}$	1.732	90	Xe lamp	p	268, 545
Er(Y,Gd)AlO$_3$	$^4I_{13/2} \rightarrow {}^4I_{15/2}$	1.5542	77	Xe lamp	p	783
(Er,Gd)Al$_5$O$_{12}$	$^4I_{11/2} \rightarrow {}^4I_{13/2}$	2.80	110	Xe lamp	p	368
		2.86	110	Xe lamp	p	368
		2.94	300	Xe lamp	p	368
Er$_3$Al$_5$O$_{12}$	$^4S_{3/2} \rightarrow {}^4I_{13/2}$	0.8628	110	Xe lamp	p	218
	$^4S_{3/2} \rightarrow {}^4I_{9/2}$	1.7762	110	Xe lamp	p	790
	$^4I_{11/2} \rightarrow {}^4I_{13/2}$	2.86	300	Xe lamp	p	882
		2.8750	110	Xe lamp	p	882
		2.8868	110	Xe lamp	p	918
		2.9366	300	Xe lamp	p	918
		2.94	300	Xe lamp	p	882
Er$_3$Al$_5$O$_{12}$:Ho	$^4I_{11/2} \rightarrow {}^4I_{13/2}$	2.6970	110	Xe lamp	p, cas	51, 882
Er$_3$Al$_5$O$_{12}$:Tm	$^4I_{11/2} \rightarrow {}^4I_{13/2}$	2.9367	300	Xe lamp	p	882
		2.9367	300	Xe lamp	p	882
Er$_3$Al$_5$O$_{12}$:Yb	$^4I_{11/2} \rightarrow {}^4I_{13/2}$	2.8595	110	Xe lamp	p	882
		2.86	110	Xe lamp	p	465
		2.9367	300	Xe lamp	p	882

Erbium (Er^{3+}, $4f^{11}$)—*continued*

Host crystal	Laser transition	Wavelength (µm)	Temp. (K)	Optical pump	Mode	Ref.
$Er_3Al_5O_{12}$:Yb	$^4I_{11/2} \rightarrow {}^4I_{13/2}$	2.9395	110	Xe lamp	p	882
		2.94	110	Xe lamp	p	882
$Er_3Al_5O_{12}$: Tm,Yb	$^4I_{11/2} \rightarrow {}^4I_{13/2}$	2.6970	110	Xe lamp	p, cas	51, 882
		2.9367	300	Xe lamp	p	882
$(Er,Lu)_3Al_5O_{12}$	$^4I_{11/2} \rightarrow {}^4I_{13/2}$	2.8298	300	Xe lamp	p	1073
		2.9395	300	Xe lamp	p	268, 1073
$GdAlO_3$	$^4I_{13/2} \rightarrow {}^4I_{15/2}$	1.5646	77	Xe lamp	p	784
	$^4S_{3/2} \rightarrow {}^4I_{9/2}$	1.6714	77	Xe lamp	p	784
$Gd_3Al_5O_{12}$	$^4I_{11/2} \rightarrow {}^4I_{13/2}$	2.8128	300	Xe lamp	p	217
$Gd_3Ga_5O_{12}$	$^4I_{11/2} \rightarrow {}^4I_{13/2}$	2.7034	110	Xe lamp	p	877
		2.7188	110	Xe lamp	p	877
		2.8128	300	Xe lamp	p	938
		2.8218	300	TiSL, LD	cw	942
		2.8549	110	Xe lamp	p	877
$KEr(WO_4)_2$	$^4S_{3/2} \rightarrow {}^4I_{13/2}$	0.8624	110	Xe lamp	p	218
	$^4S_{3/2} \rightarrow {}^4I_{9/2}$	1.7372	300	Xe lamp	p	809
	$^4I_{11/2} \rightarrow {}^4I_{13/2}$	2.8070	300	Xe lamp	p	339
$KGd(WO_4)_2$	$^4S_{3/2} \rightarrow {}^4I_{13/2}$	0.8467	110	Xe lamp	p	218
$KGd(WO_4)_2$		0.8468	300	Xe lamp	p	218–220
		0.8610	110	Xe lamp	p	218
	$^4S_{3/2} \rightarrow {}^4I_{9/2}$	1.7155	300	Xe lamp	p	219, 796
		1.7325	300	Xe lamp	p	219, 796
		1.7330	300	Xe lamp	p	809
	$^4I_{11/2} \rightarrow {}^4I_{13/2}$	2.7222	300	Xe lamp	p	796
		2.7990	300	Xe lamp	p	796
$KLa(MoO_4)_2$	$^4S_{3/2} \rightarrow {}^4I_{9/2}$	1.7280	110	Xe lamp	p	501
		1.73	300	Xe lamp	p	501
	$^4I_{11/2} \rightarrow {}^4I_{13/2}$	2.7220	110	Xe lamp	p	501
		2.7575	300	Xe lamp	p	501
KYF_4	$^4S_{3/2} \rightarrow {}^4I_{15/2}$	0.562	300	TiS laser	p, upc	1081
$K(WO_4)_2$	$^4S_{3/2} \rightarrow {}^4I_{13/2}$	0.8474	110	Xe lamp	p	218
		0.8479	110	Xe lamp	p	218
		0.85	300	Xe lamp	p	207
		0.8621	300	Xe lamp	p	218, 236
		0.8631	110	Xe lamp	p	218
		0.86325	110	Xe lamp	p	218
		0.8633	300	Xe lamp	p	218
	$^4S_{3/2} \rightarrow {}^4I_{9/2}$	1.7178	300	Xe lamp	p	236
		1.7370	300	Xe lamp	p	300

Erbium (Er^{3+}, 4f^{11})—*continued*

Host crystal	Laser transition	Wavelength (μm)	Temp. (K)	Optical pump	Mode	Ref.
K(WO$_4$)$_2$	$^4S_{3/2} \to {}^4I_{9/2}$	1.7383	300	Xe lamp	p	809
		1.7390	300	Xe lamp	p	809
	$^4I_{11/2} \to {}^4I_{13/2}$	2.8070	300	Xe lamp	p	328
		2.8092	300	Xe lamp	p	938
K(Y,Er)(WO$_4$)$_2$	$^4S_{3/2} \to {}^4I_{9/2}$	1.7372	300	Xe lamp	p	809
	$^4I_{11/2} \to {}^4I_{13/2}$	2.6887	300–150	Xe lamp	p	76
		2.8070	300	Xe lamp	p	236
LaF$_3$	$^4I_{13/2} \to {}^4I_{15/2}$	1.6113	77	Xe lamp	p	810
LiErF$_4$	$^4S_{3/2} \to {}^4I_{13/2}$	0.8540	110	Xe lamp	p	233
	$^4S_{3/2} \to {}^4I_{11/2}$	1.228	110	Xe lamp	p	233
		1.2292	90–102	Xe lamp	p	223
	$^4S_{3/2} \to {}^4I_{9/2}$	1.7042	110	Xe lamp	p	233
		1.732	90	Xe lamp	p	268, 269
		1.7322	90	Xe lamp	p	223
	$^4F_{9/2} \to {}^4I_{11/2}$	2.0005	110	Xe lamp	p	233
	$^4I_{11/2} \to {}^4I_{13/2}$	2.8500	110	Xe lamp	p	233
LiGdF$_4$	$^4S_{3/2} \to {}^4I_{15/2}$	0.550	300	Ti:sapphire	cw, upc	1091
LiLuF$_4$	$^4S_{3/2} \to {}^4I_{13/2}$	0.8506	110	Xe lamp	p	32
		0.8507	300	Xe lamp	p	32, 33
		0.8542	300	Xe lamp	p	32, 33
		0.8543	110	Xe lamp	p	32
	$^4S_{3/2} \to {}^4I_{11/2}$	1.2196	110	Xe lamp	p	32
		1.2292	110	Xe lamp	p	32
		1.2295	300	Xe lamp	p	32, 33
	$^4S_{3/2} \to {}^4I_{9/2}$	1.7343	110	Xe lamp	p	32
		1.7345	300	Xe lamp	p	32, 33
LiNbO$_3$	$^4I_{13/2} \to {}^4I_{15/2}$	1.532	300	CCL	cw	747
(waveguide)		1.563	300	CCL	p, cw	752
		1.576	300	CCL	p, cw	752
LiYF$_4$	$^2P_{3/2} \to {}^4I_{11/2}$	0.4697	10–35	dye laser	cw, upc	22
		0.486	≤80	CCL	sp, upc	1083
	$^4S_{3/2} \to {}^4I_{11/2}$	0.544	≤95	CCL	cw, upc	1084
		0.544	≤95	CCL	ML, upc	1085
		0.544075	20	TiS laser	cw	34
		0.550965	49	TiS laser	cw	34
		0.551	40	dye laser	cw, upc	42, 43
		0.551	60	TiS laser	cw, upc	863
		0.551	300	TiS laser	p, upc	1081
		0.551	≤95	CCL	cw, upc	1084
		0.551	≤95	CCL	ML, upc	1085

Erbium (Er^{3+}, 4f^{11})—*continued*

Host crystal	Laser transition	Wavelength (μm)	Temp. (K)	Optical pump	Mode	Ref.
LiYF$_4$	$^4S_{3/2} \to {}^4I_{11/2}$	0.551	≤ 77	laser diode	sp, upc	1088
		0.551	≤ 85	dye laser	p, upc	40
		0.551	300	TiS laser	cw, upc	218
		0.551	300	dye laser	p	38
		0.551	300	dye laser	cw, upc	41
	$^2H_{9/2} \to {}^4I_{13/2}$	0.5606	20	TiS laser	cw	34
			5–40	laser diode	cw, upc	1088
	$^4F_{9/2} \to {}^4I_{15/2}$	0.671	≤ 60	dye laser	p, upc	40
		0.7015	300	CCL	cw	39
	$^4S_{3/2} \to {}^4I_{13/2}$	0.85	300	Xe lamp	p	220
		0.851	300	ErGL	p, upc	1035
		0.8503	300	Xe lamp	p	222, 224
		0.8535	110	Xe lamp	p	226
		0.8537	116, 300	Xe lamp	p	181, 231
	$^4S_{3/2} \to {}^4I_{11/2}$	1.2195	116, 300	Xe lamp	p	222, 231
		1.2290	110	Xe lamp	p, as	502
		1.2294	120, 300	Xe lamp	p	224
		1.23	300	ErGL	p, upc	1035
		1.2308	110, 300	Xe lamp	p	226
	$^4S_{3/2} \to {}^4I_{9/2}$	1.620	300	Kr laser	cw, cas	755
		1.6470	110	Kr laser	cw, cas	226
		1.6640	138–300	Xe lamp	p	223, 231
	$^4S_{3/2} \to {}^4I_{9/2}$	1.7036	116, 250	Xe lamp	p	223, 231
		≈ 1.73	300	ErGL	p, upc	1035
		1.730	300	Xe lamp	p, qs	808
		1.7312	116, 300	Xe lamp	p	224
		1.7320	110, 300	Xe lamp	p	226
	$^4I_{11/2} \to {}^4I_{13/2}$	2.66	300	laser diode	cw	943
		2.7170	110	Xe lamp	p, cas	788
		2.72	300	laser diode	cw	943
		2.747	300	Ar, Kr, LD	cw	776
		2.77	300	Ar laser	cw	943
		2.8	300	AL, LD	p, cw	961
		2.8085	110	Xe lamp	p, cas	788
		2.81	300	ErGL	p, upc	926
		2.810	300	Kr laser	cw, cas	755
		2.81	300	AL, LD	cw	943
		2.84	300	Ar laser,	cw	943
		2.85	300	Ar laser,	cw	943
		2.870	110, 300	Xe lamp	p	939
	$^4F_{9/2} \to {}^4I_{9/2}$	3.41	≤ 120	laser diode	cw	951
LiYF$_4$:Pr	$^4S_{3/2} \to {}^4I_{13/2}$	0.8503	110, 300	Xe lamp	p	224, 329

Erbium (Er^{3+}, 4f^{11})—*continued*

Host crystal	Laser transition	Wavelength (μm)	Temp. (K)	Optical pump	Mode	Ref.
LiYF$_4$:Pr	$^4I_{11/2} \rightarrow {}^4I_{13/2}$	≈2.8	300	laser diode	cw	1070
LiYF$_4$:Tb	$^4S_{3/2} \rightarrow {}^4I_{13/2}$	0.8503	110, 300	Xe lamp	p	224
LiYF$_4$:Yb	$^4S_{3/2} \rightarrow {}^4I_{15/2}$	0.551	300	TiS laser	upc	1030
	$^4S_{3/2} \rightarrow {}^4I_{11/2}$	1.234	300	TiS laser	cw	734
	$^4F_{9/2} \rightarrow {}^4I_{11/2}$	2.0025	300	NdYLF	p, upc	813
LiY$_{0.5}$Er$_{0.5}$F$_4$	$^4S_{3/2} \rightarrow {}^4I_{13/2}$	0.8501	113, 161	Xe lamp	p	181
		0.8535	113, 163	Xe lamp	p	181
LuAlO$_3$	$^4S_{3/2} \rightarrow {}^4I_{9/2}$	1.6675	90	Xe lamp	p	795
LuAlO$_3$:Tm	$^4I_{11/2} \rightarrow {}^4I_{13/2}$	2.7126	~110	Xe lamp	p	792
Lu$_3$Al$_5$O$_{12}$	$^4S_{3/2} \rightarrow {}^4I_{13/2}$	0.8631	110	Xe lamp	p	218
		0.86325	77, 300	Xe lamp	p	239
	$^4I_{13/2} \rightarrow {}^4I_{15/2}$	1.6525	77	Xe lamp	p	794
		1.6630	77	Xe lamp	p	794
	$^4S_{3/2} \rightarrow {}^4I_{9/2}$	1.7762	300	Xe lamp	p	790
Lu$_3$Al$_5$O$_{12}$	$^4I_{11/2} \rightarrow {}^4I_{13/2}$	2.7973	300	Xe lamp	p	918
		2.7998	110	Xe lamp	p	918
		2.829	300	ErGL	p, upc	1094
		2.8297–2.8302	300	Xe lamp	p	918
		2.8552–2.8590	110	Xe lamp	p	918
		2.8700	110	Xe lamp	p	918
		2.8748–2.8752	110	Xe lamp	p	918
		2.8760	110	Xe lamp	p	918
		2.8967–2.8979	110	Xe lamp	p	918
		2.9395	300	Xe lamp	p	239
		2.9395–2.9397	110	Xe lamp	p	918
		2.940	300	ErGL	p, upc	1094
		2.9401	110	Xe lamp	p	918
		2.9408	300	Xe lamp	p	239
Lu$_3$Al$_5$O$_{12}$:Ho,Tm	$^4I_{11/2} \rightarrow {}^4I_{13/2}$	2.6990	300	Xe lamp	p	239
Lu$_3$Al$_5$O$_{12}$:Cr,Yb	$^4I_{11/2} \rightarrow {}^4I_{13/2}$	2.8298	300	Xe lamp	p	941
		2.9395	300	Xe lamp	p	239
		2.9405	300	Xe lamp	p	941
(Lu,Er)$_3$Al$_5$O$_{12}$	$^4S_{3/2} \rightarrow {}^4I_{13/2}$	0.8625	300	Xe lamp	p	218
	$^4S_{3/2} \rightarrow {}^4I_{13/2}$	0.8631	110	Xe lamp	p	218
	$^4S_{3/2} \rightarrow {}^4I_{9/2}$	1.7767	110	Xe lamp	p	790
	$^4I_{11/2} \rightarrow {}^4I_{13/2}$	2.6990	300	—	—	239

Erbium (Er^{3+}, 4f^{11})—*continued*

Host crystal	Laser transition	Wavelength (μm)	Temp. (K)	Optical pump	Mode	Ref.
(Lu,Er)$_3$Al$_5$O$_{12}$	$^4I_{11/2} \to {^4I_{13/2}}$	2.7140 → 2.7143	110	Xe lamp	p	918
		2.7953	300	Xe lamp	p	964
		2.7973 → 2.7968	300	Xe lamp	p	918
		2.799	300	Xe lamp	p	51
		2.7998 → 2.7980	110	Xe lamp	p	918
		2.8298	300	Xe lamp	p	918
(Lu,Er)$_3$Al$_5$O$_{12}$	$^4I_{11/2} \to {^4I_{13/2}}$	2.8298 → 2.8301	300	Xe lamp	p	918
		2.830	300	Xe lamp	p	947, 986
		2.8302	300	Xe lamp	p	964
		2.8700 → 2.8592	110	Xe lamp	p	918
		2.8760 → 2.8751	110	Xe lamp	p	918
(Lu,Er)$_3$Al$_5$O$_{12}$		2.8868 → 2.8846	110	Xe lamp	p	918
		2.9365	300	Xe lamp	p	964
		2.9401 → 2.9397	110	Xe lamp	p	918
		2.9403 → 2.9366	300	Xe lamp	p	918
(Lu,Er)$_3$Al$_5$O$_{12}$: Tm	$^4I_{11/2} \to {^4I_{13/2}}$	2.94	300	Xe lamp	p	918
(Lu,Er)$_3$Al$_5$O$_{12}$: Yb	$^4I_{11/2} \to {^4I_{13/2}}$	2.6987 → 2.9403	300	Xe lamp	p	51, 941
(Lu,Er)$_3$Al$_5$O$_{12}$: Ho,Er,Tm	$^4I_{11/2} \to {^4I_{13/2}}$	2.699	300	Xe lamp	p	51, 919, 920
Sc$_2$SiO$_5$	$^4I_{13/2} \to {^4I_{15/2}}$	1.545	300	TiS laser	cw	994
		1.556	300	TiS laser	cw	994
		1.558	300	TiS laser	cw	994
Sc$_2$SiO$_5$:Yb	$^4I_{13/2} \to {^4I_{15/2}}$	1.551	300	TiS, LD	cw	994
SrF$_2$	$^4I_{11/2} \to {^4I_{13/2}}$	≈2.8	300	ErGL	p, upc	926
SrF$_2$-ErF$_3$	$^4I_{11/2} \to {^4I_{13/2}}$	2.7285	300	Xe lamp	p	921
		2.7450	300	Xe lamp	p	921
		2.7930	300	Xe lamp	p	921
		2.80	300	Xe, ErGL	p, upc	926

Erbium (Er^{3+}, $4f^{11}$)—*continued*

Host crystal	Laser transition	Wavelength (μm)	Temp. (K)	Optical pump	Mode	Ref.
SrLaGa$_3$O$_7$:Pr	$^4I_{11/2} \to {}^4I_{13/2}$	2.7	300	TiS laser	cw	733
SrY$_4$(SiO$_4$)$_3$O: Yb	$^4I_{13/2} \to {}^4I_{15/2}$	1.554	300	laser diode	cw	426
YAlO$_3$	$^4S_{3/2} \to {}^4I_{15/2}$	0.5496	30–77	dye laser	cw, upc	37
		0.5498	34	TiS laser	cw, upc	1089
		0.550	20	dye laser	cw, upc	1087
	$^4S_{3/2} \to {}^4I_{13/2}$	0.84965	77	Xe lamp	p	129, 130
YAlO$_3$	$^4S_{3/2} \to {}^4I_{13/2}$	0.84975	300	Xe lamp	p	129, 130
		0.85165	77	Xe lamp	p	129, 130
		0.8594	300	Xe lamp	p	129, 130
	$^4F_{9/2} \to {}^4I_{13/2}$	1.2342	110	Xe lamp	p	1037
		1.2390	110	Xe lamp	p	1037
		1.2392	300	Xe lamp	p	1037
	$^4I_{13/2} \to {}^4I_{15/2}$	1.5554	77	Xe lamp	p	783
YAlO$_3$	$^4S_{3/2} \to {}^4I_{9/2}$	1.66	300	Xe lamp	p	788, 789
		1.662–3	300	Xe lamp	p	777, 782
		1.6628	110	Xe lamp	p	788, 789
		1.6628	110	Xe lamp	p, cas	789, 792
	$^4I_{11/2} \to {}^4I_{13/2}$	2.7310	300	ErGL	p, upc	1093
YAlO$_3$:Ho	$^4I_{11/2} \to {}^4I_{13/2}$	2.7310	~110	Xe lamp	p	792
		2.7398	~110	Xe lamp	p	792
		2.7608	~110	Xe lamp	p	792
(Y,Er)AlO$_3$	$^4S_{3/2} \to {}^4I_{9/2}$	1.6631	110	Xe lamp	p	788, 789, 925
		1.6632	300	Xe lamp	p, cas	788, 789
		1.6632	300	Ar laser	cw	797
		1.677	300	Xe lamp	p	777
		1.6776	300	Ar laser	cw	797
		1.706	300	Xe lamp	p	777
		1.7061	300	Ar laser	cw	797
		1.726	300	Xe lamp	p	777
		1.7296	300	Ar laser	cw	797
	$^4I_{11/2} \to {}^4I_{13/2}$	[2.71–2.86]	300	Xe lamp	p	916
		2.7118	300	Xe lamp	p	922
		2.73	290–330	Xe lamp	p	968
		[2.73–2.92]	300	Xe lamp	p	917
		2.7305	300	Xe lamp	p, cas	788, 789
		2.7305–2.7307	300	Xe lamp	p	788, 789, 922

Erbium (Er^{3+}, $4f^{11}$)—*continued*

Host crystal	Laser transition	Wavelength (µm)	Temp. (K)	Optical pump	Mode	Ref.
(Y,Er)AlO$_3$	$^4I_{11/2} \rightarrow {}^4I_{13/2}$	2.7309	300	Xe lamp	p	129
		2.7310	110	Xe lamp	p, cas	788, 789, 792, 930
		2.7398	110	Xe lamp	p	788, 789, 792, 930
		2.7608	110	Xe lamp	p	788, 789, 792, 930
		2.7645	300	Xe lamp	p	922
		2.7698	300	Xe lamp	p	788, 789
		2.79	290–330	Xe lamp	p	968
(Y,Er)AlO$_3$	$^4I_{11/2} \rightarrow {}^4I_{13/2}$	2.7969	300	Xe lamp	p	788, 789
		2.7955–2.7957	300	Xe lamp	p	788, 789, 792, 930
		2.8230	300	Xe lamp	p	922
		2.8400	300	Xe lamp	p	922
		2.8665	300	Xe lamp	p	922
(Y,Er)AlO$_3$		2.8756	300	Xe lamp	p	922
		2.9195–2.9200	77	Xe lamp	p	788, 789, 922
	$^4I_{9/2} \rightarrow {}^4I_{11/2}$	4.75	110	Xe lamp	p, cas	959, 965, 991
(Y,Gd)AlO$_3$	$^4I_{13/2} \rightarrow {}^4I_{15/2}$	1.6600	300	Xe lamp	p	783
Y$_3$Al$_5$O$_{12}$	$^4S_{3/2} \rightarrow {}^4I_{15/2}$	0.561	300	dye laser	p	38
	$^4S_{3/2} \rightarrow {}^4I_{13/2}$	0.8624	77	Xe lamp	p	237
		0.8627	77, 300	Xe lamp	p	129, 238
	$^4F_{9/2} \rightarrow {}^4I_{13/2}$	1.245	77	Xe lamp	p	237
	$^4I_{13/2} \rightarrow {}^4I_{15/2}$	1.632	300	Xe lamp	p	738
		1.64	300	Xe lamp	p, q	756
		1.640	300	Kr kaser	cw	776
		1.644	300	Xe lamp	p	954
		1.6449	295	Xe lamp	p	738
		1.6452	77	Xe lamp	p	651
		1.6596	77	Xe lamp	p	237
		1.6602	77	Xe lamp	p	651
	$^4S_{3/2} \rightarrow {}^4I_{9/2}$	1.7757	300	Xe lamp	p	790, 802
	$^4I_{11/2} \rightarrow {}^4I_{13/2}$	2.766	300	Xe lamp	p	928
		2.795	300	Xe lamp	p	928
		2.8302	300	Xe lamp	p	940
		2.936	300	Xe lamp	p	731, 928
		2.9365	300	Xe lamp	p	940
		2.937	300	TiS, LD	cw	942
		2.939	300	Xe lamp	p	928

Erbium (Er^{3+}, 4f^{11})—*continued*

Host crystal	Laser transition	Wavelength (µm)	Temp. (K)	Optical pump	Mode	Ref.
$Y_3Al_5O_{12}$	$^4I_{11/2} \rightarrow {}^4I_{13/2}$	2.9403	300	Xe lamp	p	129
$Y_3Al_5O_{12}$:Cr	$^4I_{11/2} \rightarrow {}^4I_{13/2}$	2.7	300	Kr laser	cw	776
$Y_3Al_5O_{12}$:Tm	$^4I_{11/2} \rightarrow {}^4I_{13/2}$	2.6975	300	Xe lamp	p	938
		2.8302	300	Xe lamp	p	938
$Y_3Al_5O_{12}$:Yb	$^4I_{13/2} \rightarrow {}^4I_{15/2}$	1.6459	295	Xe lamp	p	793
		1.646	300	laser diode	cw	1121
	$^4I_{11/2} \rightarrow {}^4I_{13/2}$	2.6975	300	Xe lamp	p	938
		2.8302	300	Xe lamp	p	938
$Y_3Al_5O_{12}$: Cr,Tm	$^4I_{11/2} \rightarrow {}^4I_{13/2}$	2.62–2.94	300	Xe lamp	p	934
$Y_3Al_5O_{12}$: Tm,Yb	$^4I_{11/2} \rightarrow {}^4I_{13/2}$	2.9365	300	Xe lamp	p	938, 942
$Y_3Al_5O_{12}$: Ho,Tm,Yb	$^4I_{11/2} \rightarrow {}^4I_{13/2}$	2.6975	300	Xe lamp	p	938
$(Y,Er)_3Al_5O_{12}$	$^4I_{13/2} \rightarrow {}^4I_{15/2}$	1.6596	77	Xe lamp	p	237
	$^4S_{3/2} \rightarrow {}^4I_{9/2}$	1.776	77–110	Xe lamp	p	237, 925
	$^4I_{11/2} \rightarrow {}^4I_{13/2}$	2.6930	77	Xe lamp	p	237
		2.830	300	ErGL	p, upc	1094
		2.8552 → 2.8590	110	Xe lamp	p	918
		2.6975 → 2.6979	110	Xe lamp	p	918
		2.7156 → 2.7150	110	Xe lamp	p	918
		2.733 → 2.7328	110	Xe lamp	p	918
		2.7953 → 2.7958	300	Xe lamp	p	918
		2.7953 → 2.7958	110	Xe lamp	p	918
		2.80	300	Xe lamp	AML	969
		2.830	300	ErGL	p, upc	926
		2.8302	300	Xe lamp	p	964
		2.8302 → 2.8297	300	Xe lamp	p	918
		2.8748 → 2.8752	110	Xe lamp	p	918
		2.8967 → 2.8979	110	Xe lamp	p	918

Erbium (Er^{3+}, 4f^{11})—*continued*

Host crystal	Laser transition	Wavelength (µm)	Temp. (K)	Optical pump	Mode	Ref.
(Y,Er)$_3$Al$_5$O$_{12}$	$^4I_{11/2} \rightarrow \,^4I_{13/2}$	2.8970	77	Xe lamp	p	237
		2.936	300	ErGL	p, upc	1094
		2.9362 → 2.9366	300	Xe lamp	p	918, 964
		2.9364	300	Xe lamp	p	225, 926, 973–975
		2.9365	300	Xe lamp	p	964
		2.937	300	Xe lamp	p	926
			300	ErGL	p, upc	926
(Y,Er)$_3$Al$_5$O$_{12}$	$^4I_{11/2} \rightarrow \,^4I_{13/2}$	2.9395 → 2.9397	110	Xe lamp	p	918
		2.943	110	Xe lamp	p	926
			100	ErGL	p, upc	926
(Y,Er)$_3$Al$_5$O$_{12}$: Cr	$^4I_{11/2} \rightarrow \,^4I_{13/2}$	2.94	300	Xe lamp	AML	957, 992
(Y,Er)$_3$Al$_5$O$_{12}$: Nd	$^4I_{11/2} \rightarrow \,^4I_{13/2}$	2.94	300	Xe lamp	p	672
Y$_3$Ga$_5$O$_{12}$	$^4I_{13/2} \rightarrow \,^4I_{15/2}$	1.64	300	Kr laser	cw	976
Y$_3$Sc$_2$Al$_3$O$_{12}$	$^4I_{11/2} \rightarrow \,^4I_{13/2}$	2.7	300	Kr laser	cw	776
Y$_3$Sc$_2$Ga$_3$O$_{12}$	$^4I_{13/2} \rightarrow \,^4I_{15/2}$	1.640	300	Kr kaser	cw	776
		1.643	300	ErGL	p	1026
	$^4I_{11/2} \rightarrow \,^4I_{13/2}$	2.791	300	Xe lamp	AML	992
	$^4I_{11/2} \rightarrow \,^4I_{13/2}$	2.794	300	ErGL	p. upc	1092–1093
		2.797	300	TiS, LD	cw	942
		2.802	300	ErGL	p. upc	1092–1093
Y$_3$Sc$_2$Ga$_3$O$_{12}$: Cr	$^4I_{11/2} \rightarrow \,^4I_{13/2}$	2.707	300	Kr laser	cw	776
		2.79–2.80	300	Xe lamp	p	970–972
			300	Xe lamp	Q	956
Y$_3$Sc$_2$Ga$_3$O$_{12}$: Cr,Tm	$^4I_{11/2} \rightarrow \,^4I_{13/2}$	2.640	300	Kr kaser	cw	776
		2.707	300	Kr kaser	cw	776
YScO$_3$:Gd	$^4I_{13/2} \rightarrow \,^4I_{15/2}$	1.6437	77	Xe lamp	p	783
		1.6682	77	Xe lamp	p	783
	$^4I_{11/2} \rightarrow \,^4I_{13/2}$	2.86372	77	Xe lamp	p	783
Y$_2$SiO$_5$:Yb	$^4I_{13/2} \rightarrow \,^4I_{15/2}$	1.57	300	TiS laser	cw	1125
		1.545–1.617	300	laser diode	cw	1121
Yb$_3$Al$_5$O$_{12}$	$^4I_{13/2} \rightarrow \,^4I_{15/2}$	1.6615	77	Xe lamp	p	726

Erbium (Er^{3+}, 4f^{11})—*continued*

Host crystal	Laser transition	Wavelength (μm)	Temp. (K)	Optical pump	Mode	Ref.
ZrO_2-Er_2O_3	$^4I_{13/2} \rightarrow {}^4I_{15/2}$	1.620	77	Xe lamp	p	764

Thulium (Tm^{3+}, 4f^{12})

Host crystal	Laser transition	Wavelength (μm)	Temp. (K)	Optical pump	Mode	Ref.
$BaEr_2F_8$	$^3F_4 \rightarrow {}^3H_6$	1.9975	110	Xe lamp	p	115
$Ba(Y,Yb)_2F_8$:	$^1I_6 \rightarrow {}^3F_4$	0.3479	300	TiS laser	cw, upc	1095
	$^1I_6 \rightarrow {}^3H_4$	0.4574	≈77	TiS laser	cw, upc	21
		0.4581	≈77	TiS laser	cw, upc	21
	$^1D_2 \rightarrow {}^3F_4$	0.455	300	TiS laser	cw, upc	1096
		0.456	300	LD/TiS	qcw, upc	21
	$^1G_4 \rightarrow {}^3H_6$	0.482	300	LD/TiS	qcw, upc	21
	$^1D_2 \rightarrow {}^3H_5$	0.512	300	LD/TiS	qcw, upc	21
	$^1G_4 \rightarrow {}^3H_4$	0.649	300	TiS laser	cw, upc	78
	$^1G_4 \rightarrow {}^3H_5$	0.799	300	TiS laser	cw, upc	207
	$^3H_4 \rightarrow {}^3F_4$	1.480	300	TiS laser	cw, upc	21
BaY_2F_8:Er	$^3F_4 \rightarrow {}^3H_6$	1.9190	300	Xe lamp	p, upc	813
		1.965	300	Xe lamp	p	1004
		1.9654-5	300	Xe lamp	p	548, 549
		1.9965	110	Xe lamp	p	115
	$^3H_4 \rightarrow {}^3H_5$	2.2845	300	NdGL	p, upc	711
$BaYb_2F_8$	$^1G_4 \rightarrow {}^3F_4$	0.649	300	NdGL	p, upc	1097
		0.6490	300	NdYAG	p	79
	$^3H_4 \rightarrow {}^3F_4$	~1.48	300	Xe lamp	p	1098
		1.482	300	NdGL	p, upc	711
			300	Xe lamp	p, upc	714
	$^1G_4 \rightarrow {}^3F_3$	~1.58	300	NdGL	p, upc	1097
		1.5808	300	NdYAG	p	79
	$^3F_4 \rightarrow {}^3H_6$	1.96	300	Xe lamp	p	953, 1004
		1.9925	110	Xe lamp	p, upc	813
$BaYb_2F_8$:Tm	$^1G_4 \rightarrow {}^3H_4$	0.6490	300	Nd laser	p, upc	1097
	$^1G_4 \rightarrow {}^3F_3$	1.5808	300	Nd laser	p, upc	1097
$Ca(NbGa)_2$-Ga_3O_{12}:Cr	$^3F_4 \rightarrow {}^3H_6$	2.02	300	Xe lamp	p	830
$Ca(NbO_3)_2$	$^3F_4 \rightarrow {}^3H_6$	1.91	77	Xe lamp	p	753

Thulium (Tm^{3+}, $4f^{12}$)—*continued*

Host crystal	Laser transition	Wavelength (μm)	Temp. (K)	Optical pump	Mode	Ref.
CaF_2	$^3F_4 \rightarrow {}^3H_6$	~1.9	77	Xe lamp	p	755
CaF_2:Er	$^3F_4 \rightarrow {}^3H_6$	1.894	~100	Xe lamp	p	790, 936
				Xe lamp	cw	836
CaF_2-ErF_3	$^3F_4 \rightarrow {}^3H_6$	1.860	77	Xe lamp	p	798
CaF_2-HoF_3-ErF_3	$^3F_4 \rightarrow {}^3H_6$	1.860	77	Xe lamp	p	834
$CaMoO_4$:Er	$^3F_4 \rightarrow {}^3H_6$	1.9060	77	Xe lamp	p	821
		1.9115	77	Xe lamp	p	821
$CaWO_4$	$^3F_4 \rightarrow {}^3H_6$	1.911	77	Xe lamp	p	291
		1.916	77	Xe lamp	p	291, 822
$CaY_4(SiO_4)_3O$	$^3F_4 \rightarrow {}^3H_6$	1.94	300	TiS laser	cw	826
Er_2O_3	$^3F_4 \rightarrow {}^3H_6$	1.934	77	Xe lamp	p, cw	816
$ErAlO_3$	$^3F_4 \rightarrow {}^3H_6$	1.872	77, 150	Xe lamp	p	819
$(Er,Lu)AlO_3$	$^3F_4 \rightarrow {}^3H_6$	1.8845	77	Xe lamp	p	840
$Er_3Al_5O_{12}$	$^3F_4 \rightarrow {}^3H_6$	1.9–2.0	77	—	p	823
$(Er,Yb)_3Al_5O_{12}$	$^3F_4 \rightarrow {}^3H_6$	1.8850	77	Xe lamp	p	726
		2.0195	77	Xe lamp	p	726
$GdAlO_3$	$^3F_4 \rightarrow {}^3H_6$	1.8529	77	Xe lamp	p	804
		1.9925	90	Xe lamp	p	825
$Gd_3Sc_2Ga_3O_{12}$:Cr	$^3H_4 \rightarrow {}^3F_4$	1.48	300	Xe lamp	p	710
$Gd_3Sc_2Ga_3O_{12}$: Cr,Ho	$^3H_4 \rightarrow {}^3H_5$	2.335	300	Xe lamp	p	710
			300	Xe lamp	p, cas	873
$GdVO_4$	$^3F_4 \rightarrow {}^3H_6$	~1.95	300	TiS laser	cw	608
$KGd(WO_4)_2$:Er	$^3F_4 \rightarrow {}^3H_6$	1.93	~110	Xe lamp	p	1048,1147
$KY(WO_4)_2$:Er,Yb	$^3F_4 \rightarrow {}^3H_6$	1.92	110–200	Xe lamp	p	1048,1147
$LiNbO_3$		1.8532	77	Xe lamp	p	370
$LiYF_4$	$^1D_2 \rightarrow {}^3F_4$	0.4502	≤70	TiS laser	cw, upc	18
		0.4502	75	dye laser	p, upc	19
		0.4502	77	NdYAG	p, upc	1100
		0.4526	~100	Xe lamp	p	20
		0.453	300	flashlamp	p, upc	1100
	$^1G_4 \rightarrow {}^3H_6$	0.4835	≤160	dye laser	cw, upc, pa	18
	$^3H_4 \rightarrow {}^3F_4$	1.50	300	TiS laser	cw, cas	743
	$^3F_4 \rightarrow {}^3H_6$	1.88	300	ruby laser	p, cas	811
		1.8890	110	Xe lamp	p, cas	812

Thulium (Tm^{3+}, $4f^{12}$)—*continued*

Host crystal	Laser transition	Wavelength (μm)	Temp. (K)	Optical pump	Mode	Ref.
LiYF$_4$	$^3F_4 \rightarrow {}^3H_6$	1.9090	110	Xe lamp	p	812
	$^3H_4 \rightarrow {}^3H_5$	2.20–2.46	300	TiS laser	cw	912
		2.295–2.424	300	TiS laser	cw	913
		2.30	300	ruby laser	p, cas	811
		2.303	110	Xe lamp	p, cas	812
LiYF$_4$:Tb	$^3H_4 \rightarrow {}^3F_4$	1.449–1.455	300	laser diode	cw	700
LiYF$_4$:Yb	$^1G_4 \rightarrow {}^3F_4$	0.650	300	TiS laser	upc	80
	$^1G_4 \rightarrow {}^3H_5$	0.792	300	TiS laser	upc	80
	$^3H_4 \rightarrow {}^3H_6$	0.810	300	TiS laser	upc	80
LiYF$_4$:Yb	$^3H_4 \rightarrow {}^3F_4$	1.464	300	laser diode	upc	80
		1.500	300	laser diode	cw, upc	80
	$^1G_4 \rightarrow {}^3F_3$	1.568	300	laser diode	cw, upc	80
Lu$_3$Al$_5$O$_{12}$	$^3F_4 \rightarrow {}^3H_6$	1.8855	77	Xe lamp	p	794
		2.0240	77	Xe lamp	p	794
Lu$_3$Al$_5$O$_{12}$:Cr,Ho	$^3H_4 \rightarrow {}^3H_5$	2.3425	110	Xe lamp	p, cas	889
α-NaCaErF$_6$	$^3F_4 \rightarrow {}^3H_6$	1.8580	150	Xe /W lamp	p/cw	805
		1.8885	77, 150	Xe /W lamp	p/cw	805
ScBeAlO$_4$	$^1G_4 \rightarrow {}^3H_5$	0.792	300			206
SrF$_2$	$^3F_4 \rightarrow {}^3H_6$	1.972	77	Xe lamp	p, cas	463
SrLaGa$_3$O$_7$	$^1G_4 \rightarrow {}^3H_6$	0.488	≤230	dye laser	p	24
	$^1G_4 \rightarrow {}^3F_4$	0.645	300	dye laser	p	24
Y$_2$O$_3$	$^3F_4 \rightarrow {}^3H_6$	~2.0	300		cw	867
YAlO$_3$	$^3F_4 \rightarrow {}^3H_6$	1.883	90	Xe lamp	p	803
	$^3H_4 \rightarrow {}^3H_5$	2.348	300	Xe lamp	p	905
		2.349	300	Xe lamp	p	905
YAlO$_3$:Cr	$^3F_4 \rightarrow {}^3H_6$	1.856	90	Xe lamp	p	803
		1.9335	90	Xe lamp	p	803
	$^3H_4 \rightarrow {}^3H_5$	2.274	300	Kr laser	p	904
		2.318	300	Xe lamp	p	904
		~2.34	90, 300	Xe lamp	p	803
		2.353	300	Xe lamp	p	904
		2.354	300	Xe lamp	p	904
		2.355	300	Xe lamp	p	904
YAlO$_3$:Er	$^3F_4 \rightarrow {}^3H_6$	1.861	77	Xe lamp	p	130
(Y,Er)AlO$_3$	$^3F_4 \rightarrow {}^3H_6$	1.861	77	Xe lamp	p	130
Y$_3$Al$_5$O$_{12}$	$^1G_4 \rightarrow {}^3H_6$	0.486	≤30	dye lasers	sp, upc	1080
Y$_3$Al$_5$O$_{12}$	$^3F_4 \rightarrow {}^3H_6$	1.87–2.16	300	TiS laser	cw	800

Thulium (Tm^{3+}, $4f^{12}$)—*continued*

Host crystal	Laser transition	Wavelength (µm)	Temp. (K)	Optical pump	Mode	Ref.
$Y_3Al_5O_{12}$	$^3F_4 \rightarrow {}^3H_6$	1.87–2.16	300	TiS laser	cw	800
		1.8834	77	Xe lamp	p	278
	$^3H_4 \rightarrow {}^3H_5$	2.324	300	W lamp	cw	904
$Y_3Al_5O_{12}$:Cr	$^3F_4 \rightarrow {}^3H_6$	1.945–2.014	300	Xe lamp	p	828
		2.0132	77	Xe /W lamp	p/cw	278
		2.014	300	TiS /KrL	cw	977
		2.015	300	TiS /KrL	cw	978
		2.019	300	W lamp	cw	278
$Y_3Al_5O_{12}$:Er	$^3F_4 \rightarrow {}^3H_6$	1.880	77	Xe lamp	p	228
$Y_3Al_5O_{12}$:Er	$^3F_4 \rightarrow {}^3H_6$	1.884	77	Xe lamp	p	228
		2.014–2.019	77	Xe lamp	p	228
$(Y,Er)_3Al_5O_{12}$	$^3F_4 \rightarrow {}^3H_6$	1.880	77		cw, upc	80
		1.884	77	Xe lamp	p	794
		2.014	85	Xe lamp	p	278
$Y_3Sc_2Ga_3O_{12}$	$^3F_4 \rightarrow {}^3H_6$	1.85–2.14	300	TiS laser	cw	800
		2.018	300	Xe lamp	p	829
$Y_3Sc_2Ga_3O_{12}$:Cr	$^3F_4 \rightarrow {}^3H_6$	1.862	300	Kr laser	cw	776
YVO_4	$^3F_4 \rightarrow {}^3H_6$	1.94	300	TiS laser	cw	827
		~2.0	77	—	p	832
ZrO_2-Er_2O_3	$^3F_4 \rightarrow {}^3H_6$	~1.896	77	Xe lamp	p	764

Ytterbium (Yb^{3+}, $4f^{13}$)

Host crystal	Laser transition	Wavelength (µm)	Temp. (K)	Optical pump	Mode	Ref.
$BaCaBO_3F$	$^2F_{5/2} \rightarrow {}^2F_{7/2}$	1.034	300	TiS laser	p	275
$Ca_3Sr_2(PO_4)_3F$	$^2F_{5/2} \rightarrow {}^2F_{7/2}$	1.046	300	TiS laser	cw	17
$Ca_4Sr(PO_4)_3F$	$^2F_{5/2} \rightarrow {}^2F_{7/2}$	0.985	300	TiS laser	cw	17
		1.046	300	TiS laser	cw	17
		1.110	300	TiS laser	cw	17
$Ca_5(PO_4)_3F$	$^2F_{5/2} \rightarrow {}^2F_{7/2}$	1.043	300	TiS laser	cw	284, 311
CaF_2:Nd	$^2F_{5/2} \rightarrow {}^2F_{7/2}$	1.0336	120	Xe lamp	p	285
$Gd_3Ga_5O_{12}$:Nd	$^2F_{5/2} \rightarrow {}^2F_{7/2}$	1.0232	77	Xe lamp	p	276
$Gd_3Sc_2Al_3O_{12}$:Nd	$^2F_{5/2} \rightarrow {}^2F_{7/2}$	1.0299	77	Xe lamp	p	279

Ytterbium (Yb^{3+}, 4f^{13})—*continued*

Host crystal	Laser transition	Wavelength (μm)	Temp. (K)	Optical pump	Mode	Ref.
KGd(WO$_4$)$_2$	$^2F_{5/2} \to \,^2F_{7/2}$	1.026–1.044	300	laser diode	cw	277
KY(WO$_4$)$_2$	$^2F_{5/2} \to \,^2F_{7/2}$	1.026–1.042	300	laser diode	cw	277
LiNbO$_3$	$^2F_{5/2} \to \,^2F_{7/2}$	1.008	300	TiS laser	cw	999
(waveguide)		1.030	300	TiS laser	cw	999
		1.060	300	TiS laser	cw	999
Lu$_3$Al$_5$O$_{12}$	$^2F_{5/2} \to \,^2F_{7/2}$	1.0297	77	Xe lamp	p	276
		1.03	175	laser diode	cw	283
Lu$_3$Al$_5$O$_{12}$:Nd,Cr	$^2F_{5/2} \to \,^2F_{7/2}$	1.0294	77	Xe lamp	p	276
Lu$_3$Ga$_5$O$_{12}$:Nd	$^2F_{5/2} \to \,^2F_{7/2}$	1.0230	77	Xe lamp	p	276
Lu$_3$Sc$_2$Al$_3$O$_{12}$:Nd	$^2F_{5/2} \to \,^2F_{7/2}$	1.0299	77	Xe lamp	p	279
Sr$_5$(PO$_4$)$_3$F	$^2F_{5/2} \to \,^2F_{7/2}$	0.985	300	Cr:LiSAF	qcw	67
		1.047	300	TiS laser	p	311
			300	laser diode	p	312
Sr$_5$(VO$_4$)$_3$F	$^2F_{5/2} \to \,^2F_{7/2}$	1.044	300	TiS laser	p	17
Y$_3$Al$_5$O$_{12}$	$^2F_{5/2} \to \,^2F_{7/2}$	1.0293	77	Xe lamp	p	276
		1.0296	77	Xe lamp	p	278
(waveguide)		1.03	300	TiS laser	cw	280
		1.03	300	laser diode	cw	281
		1.031	300	TiS laser	p	311
Y$_3$Al$_5$O$_{12}$:Nd	$^2F_{5/2} \to \,^2F_{7/2}$	1.0297	200	Xe lamp	p	276
(waveguide)		1.03	300	laser diode	cw	282
Y$_3$Al$_5$O$_{12}$:Nd,Cr	$^2F_{5/2} \to \,^2F_{7/2}$	1.0298	210	Xe lamp	p	276
(Y,Yb)$_3$Al$_5$O$_{12}$	$^2F_{5/2} \to \,^2F_{7/2}$	1.0293	77	Xe lamp	p	276
YCa$_4$O(BO$_3$)$_3$	$^2F_{5/2} \to \,^2F_{7/2}$	1.018–1.087	300	TiS laser	cw	1149
Y$_3$Ga$_5$O$_{12}$:Nd	$^2F_{5/2} \to \,^2F_{7/2}$	1.0233	77	Xe lamp	p	276

1.1.7 Actinide Ion Lasers

Table 1.1.12
Actinide Lasers

<u>Optical pump</u>

Hg — mercury arc lamp

TiS — Ti:sapphire laser

Xe — Xe arc lamp

<u>Mode of operation</u>

cw — continuous wave

p — pulsed

Uranium (U^{3+}, 5f^3)

Host crystal	Laser transition	Wavelength (μm)	Temp. (K)	Optical pump	Mode	Ref.
BaF$_2$	$^4I_{11/2} \rightarrow {}^4I_{9/2}$	2.556	20	Xe lamp	p	901
CaF$_2$	$^4I_{11/2} \rightarrow {}^4I_{9/2}$	2.234	77	Xe lamp	p	898
		2.439	77	Xe lamp	p	899
		2.511	77	Xe lamp	p	899, 900
		2.571	77	Xe lamp	p	900
		2.6	4.2	Xe lamp	p	932
		2.613	20, 77, 90	Xe lamp	p	933
		2.613	77	Hg lamp	cw	933
LiYF$_4$	$^4I_{11/2} \rightarrow {}^4I_{9/2}$	2.827	300	TiS laser	cw	945
SrF$_2$	$^4I_{11/2} \rightarrow {}^4I_{9/2}$	2.407	20–90	Xe lamp	p	902

1.1.8 Other Ions and Crystals Exhibiting Gain

Table 1.1.13
Other Ions Exhibiting Gain

Ion	Host crystal	Transition	Wavelength (μm)	Temp. (K)	Ref.
Ag^+	KI, RbBr	$^1T_{1u} \rightarrow {}^1A_{1g}$	~0.337	5	1101–2
Cr^{3+}	$LiNbO_3$	$^4T_2 \rightarrow {}^4A_2$	0.910	300	1103
Cr^{3+}	$LiNbO_3$	$^4T_2 \rightarrow {}^4A_2$	0.927	300	1104
Cr^{4+}	$CaGd_4(SiO_4)_3O$	$^3T_2 \rightarrow {}^3A_2$	1.15–1.5	300	1105
Cu^+	Ag–β''–alumina	$4s \rightarrow 3d$	0.6328	300	1106
Cu^+	Na–β''–alumina[a]	$4s \rightarrow 3d$	0.5145	300	1106
Dy^{3+}	$LaCl_3$	$^6H_{9/2} \rightarrow {}^6H_{15/2}$	1.319, 1.338	300	1107
In^+	KCl	$6p \rightarrow 6s$	0.442	—	1108
Mn^{5+}	Ca_2PO_4Cl	$^1E \rightarrow {}^3A_2$	1.15	300	1109
Mn^{5+}	$Sr_5(PO_4)_3Cl$	$^1E \rightarrow {}^3A_2$	~1.2	300	1109
Nd^{3+}	$Gd_3(Sc,Ga)_5O_{12}$:Cr	$^4F_{3/2} \rightarrow {}^4I_{11/2}$	1.061[b]	300	1110
Nd^{3+}	$LiYF_4$	$5d \rightarrow {}^4I_{11/2}$	0.186[c]	300	1111
Nd^{3+}	ZnS film	$^4F_{3/2} \rightarrow {}^4I_{11/2}$	~1.080[d]	77	1112
Rh^{2+}	$RbCaF_3$	$^2T_1 \rightarrow {}^2E$	0.700–0.720	300	1113
Ti^{4+}	Li_2GeO_3	charge transfer	0.388–0.524	300	1114
Tl^+	CsI	$6p \rightarrow 6s$	0.407	—	1115
Tl^+	KI	$6p \rightarrow 6s$	0.420	—	1116
UO_2^{2+}	$Ca(UO_2)(PO_4) \cdot H_2O$	charge transfer	0.500–0.550	—	1117
V^{2+}	$KMgF_3$	$^4T_2 \rightarrow {}^4A_2$	1.064	—	1118

(a) Typical composition: $Na_{1.67}Mg_{0.67}Al_{10.33}O_{19}$.
(b) X-ray induced optical gain.
(c) Because of the presence of excited state absorption, a negative net-induced gain coefficient was measured.
(d) Direct current electroluminescence (DCEL) and cathodoluminescence.

1.1.9 Self-Frequency-Doubled Lasers

Table 1.1.14
Self-Frequency-Doubled Transition Metal Ion Lasers

Host crystal	Lasing ion	Primary transition	Frequency-doubled wavelength (μm)[a]	Temp. (K)	Ref.
$Ca_2Ga_2SiO_7$	Cr^{3+}	$^4T_2 \rightarrow {}^4A_2$	0.5	300	1017
$La_3Ga_{5.5}Nb_{0.5}O_{14}$	Cr^{3+}	$^4T_2 \rightarrow {}^4A_2$	0.53	300	1017
$La_3Ga_{5.5}Ta_{0.5}O_{14}$	Cr^{3+}	$^4T_2 \rightarrow {}^4A_2$	0.54	300	1017
$La_3Ga_5GeO_{14}$	Cr^{3+}	$^4T_2 \rightarrow {}^4A_2$	0.52	300	1017
$La_3Ga_5SiO_{14}$	Cr^{3+}	$^4T_2 \rightarrow {}^4A_2$	0.48	300	1017
$Sr_3Ga_2GeO_{14}$	Cr^{3+}	$^4T_2 \rightarrow {}^4A_2$	0.5	300	1017

(a) Center wavelength of tuning range.

Table 1.1.15
Self-Frequency-Doubled Lanthanide Ion Lasers

Host crystal	Lasing ion	Primary transition	Frequency-doubled wavelength (μm)	Temp. (K)	Ref.
$YAl_3(BO_3)_4$	Nd^{3+}	$^4F_{3/2} \rightarrow {}^4I_{9/2}$	~0.455	300	1018
$Ba_2NaNb_5O_{15}$	Nd^{3+}	$^4F_{3/2} \rightarrow {}^4I_{11/2}$	0.5355, 0.53065	300	1152
$Ca_4YO(BO_3)_3$	Nd^{3+}	$^4F_{3/2} \rightarrow {}^4I_{11/2}$	0.53	300	1129, 1137
$\beta'-Gd_2(MoO_4)_3$	Nd^{3+}	$^4F_{3/2} \rightarrow {}^4I_{11/2}$	0.5303	300	1058
$Ca_2Ga_2SiO_7$	Nd^{3+}	$^4F_{3/2} \rightarrow {}^4I_{11/2}$	0.5344	300	1016
$Ca_4GdO(BO_3)_3$	Nd^{3+}	$^4F_{3/2} \rightarrow {}^4I_{11/2}$	0.530	300	489
$Gd_2(MoO_4)_3$	Nd^{3+}	$^4F_{3/2} \rightarrow {}^4I_{11/2}$	0.5150	300	1154
$LaBGeO_5$	Nd^{3+}	$^4F_{3/2} \rightarrow {}^4I_{11/2}$	0.524	300	1056–1057
$LaBGeO_5$	Nd^{3+}	$^4F_{3/2} \rightarrow {}^4I_{11/2}$	0.5241	300	1013–1015
$La_3Ga_{5.5}Nb_{0.5}O_{14}$	Nd^{3+}	$^4F_{3/2} \rightarrow {}^4I_{11/2}$	0.5323	300	1016
$La_3Ga_5SiO_{14}$	Nd^{3+}	$^4F_{3/2} \rightarrow {}^4I_{11/2}$	0.5335	300	1016, 1023
$La_3Ga_5GeO_{14}$	Nd^{3+}	$^4F_{3/2} \rightarrow {}^4I_{11/2}$	0.5337	300	1016
$LiNbO_3:MgO$	Nd^{3+}	$^4F_{3/2} \rightarrow {}^4I_{11/2}$	0.5425	408–17	1006
$LiNbO_3:MgO$	Nd^{3+}	$^4F_{3/2} \rightarrow {}^4I_{11/2}$	0.5441	300	1009–1011
$LiNbO_3:MgO$	Nd^{3+}	$^4F_{3/2} \rightarrow {}^4I_{11/2}$	0.5457	300	1009–1011
$LiNbO_3:MgO$	Nd^{3+}	$^4F_{3/2} \rightarrow {}^4I_{11/2}$	0.5464	300	1009–1011
$LiNbO_3:MgO$	Nd^{3+}	$^4F_{3/2} \rightarrow {}^4I_{11/2}$	0.5465	~425	36, 1008
$LiNbO_3:MgO$	Nd^{3+}	$^4F_{3/2} \rightarrow {}^4I_{11/2}$	0.5467	~360	1005
$LiNbO_3:MgO$	Nd^{3+}	$^4F_{3/2} \rightarrow {}^4I_{11/2}$	~0.547	300	855
$LiNbO_3:MgO$	Nd^{3+}	$^4F_{3/2} \rightarrow {}^4I_{11/2}$	0.547	300	81, 1156
$LiNbO_3:Sc_2O_3$	Nd^{3+}	$^4F_{3/2} \rightarrow {}^4I_{11/2}$	~0.546	300	1012
$LiNbO_3:ZnO$	Nd^{3+}	$^4F_{3/2} \rightarrow {}^4I_{11/2}$	0.547	300	1133
$LiYF_4$	Nd^{3+}	$^4F_{3/2} \rightarrow {}^4I_{11/2}$	0.5236	300	316
$Sr_xBa_{1-x}(NbO_3)_2$	Nd^{3+}	$^4F_{3/2} \rightarrow {}^4I_{11/2}$	0.5313	300	1153
$Sr_3Ga_2GeO_{14}$	Nd^{3+}	$^4F_{3/2} \rightarrow {}^4I_{11/2}$	0.5347	300	1017
$YAl_3(BO_3)_4$	Nd^{3+}	$^4F_{3/2} \rightarrow {}^4I_{11/2}$	~0.53	300	1019, 1020
$YAl_3(BO_3)_4$	Nd^{3+}	$^4F_{3/2} \rightarrow {}^4I_{11/2}$	0.530	300	47
$YAl_3(BO_3)_4$	Nd^{3+}	$^4F_{3/2} \rightarrow {}^4I_{11/2}$	0.531	300	1156
$LaBGeO_5$	Nd^{3+}	$^4F_{3/2} \rightarrow {}^4I_{11/2}$	0.524	300	1156
$LaBGeO_5$	Nd^{3+}	$^4F_{3/2} \rightarrow {}^4I_{13/2}$	0.6571	300	1013–1015
$YAl_3(BO_3)_4$	Nd^{3+}	$^4F_{3/2} \rightarrow {}^4I_{13/2}$	~0.66	300	1021
$YAl_3(BO_3)_4$	Nd^{3+}	$^4F_{3/2} \rightarrow {}^4I_{13/2}$	0.669	300	1127
$LiNbO_3$	Tm^{3+}	$^3F_4 \rightarrow {}^3H_6$	0.9266	77	65

1.1.10 Commercial Crystalline Transition Metal Ion Lasers

Table 1.1.16
Commercial Crystalline Transition Metal Ion Lasers

Laser Type	Operation	Principal wavelengths (μm)	Output
Ruby ($Cr:Al_2O_3$)	pulsed (SH)	0.347	0.1–0.3 J
	cw	0.6943	7 W
	pulsed	0.6943	0.3–100 J
Alexandrite ($Cr:BeAl_2O_4$)	pulsed	0.36–0.4 (SH)	10–500 mJ
	cw	0.7–0.8	0.1–2 W
	pulsed	0.72–0.82	10 mJ–3 J
Ti: sapphire ($Ti:Al_2O_3$)	pulsed	0.25–0.30 (TH)	50 mJ–1 mJ
	cw	0.36–0.46 (SH)	0.01–0.2 mW
	pulsed	0.36–0.45 (SH)	0.3–25 mJ
	cw	0.67–1.13	0.25–5 W
	pulsed	0.7–1.1	10 mJ–3 J
Cobalt perovskite ($Co:MgF_2$)	pulsed, cw	1.75–2.5	20–25 mJ
Chromium:LiSAF ($Cr:LiSrAlF_6$)	pulsed	0.78–1.01	2 mJ
Chromium fluoride ($Cr:KZnF_3$)	cw	0.78–0.85	1 W
	pulsed	0.78–0.85	10 mJ
Forsterite ($Cr:Mg_2SiO_4$)	pulsed	0.58–0.66 (SH)	2 mJ
	pulsed	1.13–1.36	0.1–20 J

Abbreviations: SH - second harmonic, TH - third harmonic,

1.1.11 Commercial Crystalline Lanthanide Ion Lasers

Table 1.1.17
Commercial Crystalline Lanthanide Ion Lasers

Laser Type	Operation	Wavelengths (μm)	Output
Neodymium			
Nd:YAB [YAl$_3$(BO$_3$)$_4$]	cw	0.53 (SH)	10 mW
	cw	1.06	10–200 W
	cw (DP)	1.06	0.1–1 W
Nd:YAG (Y$_3$Al$_5$O$_{12}$)	pulsed	0.213 (FFH)	4–15 mJ
	cw	0.266 (FH)	0.02–0.6 W
	pulsed	0.266 (FH)	1–300 mJ
	cw	0.355 (TH)	0.01–1.5 W
	pulsed	0.355 (TH)	1–800 mJ
	cw	0.532 (SH)	0.1–60 W
	cw (DP)	0.532 (SH)	0.1–0.5 W
	pulsed	0.532 (SH)	0.1–100 J
	pulsed (DP)	0.532 (SH)	0.001–0.1 J
	cw	0.946	10 mW
	cw (multimode)	1.064	1–3000 W
	cw (TEM$_{00}$)	1.064	0.1–60 W
	cw (DP)	1.064	1 mW–20 W
	pulsed (multi.mode)	1.064	0.1–2000 J
	pulsed (TEM$_{00}$)	1.064	1–2.5 J
	pulsed (DP)	1.064	0.1–250 mJ
	cw	1.319	0.2–100 W
	cw (DP)	1.319	0.2–2 W
	pulsed	1.319	1–5 J
Nd:YLF (LiYF$_4$)	pulsed	0.209 (FFH)	0.2 mJ
	pulsed	0.263 (FH)	0.2–2 mJ
	pulsed	0.351 (TH)	0.3–2 mJ
	pulsed	0.523 (SH)	0.02–15 mJ
	pulsed	0.527 (SH)	1–15 mJ
	cw	1.047	0.5–6 W
	cw (DP)	1.047	0.5–2 W
	pulsed	1.047	0.5 J
	pulsed (DP)	1.047	0.01–0.15 J
	cw	1.053	2–45 W
	cw (DP)	1.053	0.5–5 W
	pulsed	1.053	0.1–10 J
	pulsed (DP)	1.053	≤ 1 mJ
	cw	1.313	1.5–3 W
	cw (DP)	1.313	0.04–0.8 W
	pulsed (DP)	1.313	≤ 10^{-5} J
	cw	1.321	40-200 mW

Table 1.1.17—*continued*
Commercial Crystalline Lanthanide Ion Lasers

Laser Type	Operation	Wavelengths (μm)	Output
Nd:YVO$_4$	pulsed (DP)	0.355 (TH)	30 mW
	cw	0.473 (SH)	1–100 mW
	cw (DP)	0.473 (SH)	20 mW
	cw	0.532 (SH)	0.01–5 W
	cw (DP)	0.532 (SH)	10–50 mW
	pulsed	1.064	≤ 150 mJ
	cw	1.064	2.5–10 W
Nd:GGG (Gd$_3$Ga$_5$O$_{12}$)	pulsed	1.062	14 J
Nd:YAP or YALO (YAlO$_3$)	cw, pulsed	1.079	≤ 60 W
Nd,Cr:GSGG (Gd$_3$Sc$_2$Ga$_3$O$_{12}$)	pulsed	1.061	0.5–40 J
Holmium			
Ho:YLF	cw	2.048–2.069	0.05–1 W
Ho:YAG (Y$_3$Al$_5$O$_{12}$)	cw	2.088–2.091	0.05–1 W
	pulsed	2.1	1–5 J
Ho:YSGG (Y$_3$Sc$_2$Ga$_3$O$_{12}$)	pulsed	2.088	3 J
Ho,Tm,Cr:YAG (Y$_3$Al$_5$O$_{12}$)	cw	2.09	0.05–1 W
	pulsed	2.09	0.5–2 J
Erbium			
Er:YAG (Y$_3$Al$_5$O$_{12}$)	cw	2.90, 2.94	2–10 W
	pulsed	2.90, 2.94	1–4 J
Er:YSGG (Y$_3$Sc$_2$Ga$_3$O$_{12}$)	pulsed	2.79	2 J
Thulium			
Tm:LuAG (Lu$_3$Al$_5$O$_{12}$)	pulsed (DP)	2.019, 2.033	0.01 J
	cw	2.019, 2.033	0.05–1 W
Tm:YAG (Y$_3$Al$_5$O$_{12}$)	pulsed	2.01	2 J
	cw	2.006–2.025	0.05–1 W

Abbreviations: SH - second harmonic, TH - third harmonic, FH - fourth harmonic, FFH - fifth harmonic, DP – diode pumped.

1.1.12 References

1. Waynant, R. W., Vacuum ultraviolet laser emission from Nd^{+3}:LaF$_3$, *Appl. Phys. B* 28, 205 (1982).
2. Waynant, R. W. and Klein, P. H., Vacuum ultraviolet laser emission from Nd^{3+}:LaF$_3$, *Appl. Phys. Lett.* 46, 14 (1985).
3. Dubinskii, M. A., Cefalas, A. C., and Nicolaides, C. A., Solid state LaF$_3$:Nd^{3+} vuv laser pumped by a pulsed discharge F$_2$-molecular laser at 157 nm, *Opt. Commun.* 88, 122 (1992).
4. Dubinskii, M. A., Cefalas, A. C., Sarantopouou, E., Spyrou, S. M., Nicolaides, C. A., Abdulsabirov, R. Yu., Korableva, S. L., and Semashjko, V. V., Efficient LaF$_3$:Nd^{3+}-based vacuum-ultraviolet laser at 172 nm, *J. Opt. Soc. Am. B* 9, 1148 (1992).
5. Pinto, J. F., Rosenblat, G. H., Esterowitz, L., Castillo, V., and Quarles, G. J., Tunable solid-state laser action in Ce^{3+}:LiSrAlF$_6$, *Electron. Lett.* 30, 240 (1994).
6. Marshall, C. D., Speth, J. A., Payne, S. A., Krupke, W. F., Quarles, G. J., Castillo, V., and Chai, B. H. T., Ultraviolet laser emission properties of Ce^{3+}-doped LiSrAlF$_6$ and LiCaAlF$_6$, *J. Opt. Soc. Am. B* 11, 2054 (1994).
7. Ehrlich, D. J., Moulton, P. F., and Osgood, R. M., Jr., Optically pumped Ce:LaF$_3$, laser at 286 nm, *Opt. Lett.* 5, 339 (1980).
8. Dubinskii, M. A., Semashko, V. V., Naumov, A. K., Abdulsabirov, R. Yu., and Korableva, S. L., Ce^{3+}-doped colquiriite: a new concept of all-solid-state tunable ultraviolet laser, *J. Mod. Optics* 40, 1 (1993).
9. Ehrlich, D. J., Moulton, P. F., and Osgood, R. M., Jr., Ultraviolet solid-state Ce:YLF laser at 325 nm, *Opt. Lett.* 4, 184 (1978) and unpublished data.
10. Azamatov, Z. T., Arsen'yev, P. A., and Chukichev, M. V., Spectra of gadolinium in YAG single crystals, *Opt. Spectrosc.* 28, 156 (1970).
11. Okada, F., Togawa, S., and Ohta, K., Solid-state ultraviolet tunable laser: a Ce^{3+} doped LiYF$_4$ crystal, *J. Appl. Phys.* 75, 49 (1994).
12. Sarukura, N., Liu, Z., Segawa, Y., Edamatsu, K., Suzuke, Y., Itoh, T. et al., Ce^{3+}:LiLuF$_4$ as a broadband ultraviolet amplification medium, *Opt. Lett.* 20, 294 (1995).
13. Schmitt, K., Stimulated C'-emission of Ag$^+$-centers in KI, RbBr, and CsBr, *Appl. Phys. A* 38 (1985).
14. Kaminskii, A., A., Kochubei, S. A., Naumochkin, K. N., Pestryakov, E. V., Trunov, V. I., and Uvarova, T. V., Amplification of ultraviolet radiation due to the 5d-4f configurational transition of the Ce^{3+} ion in BaY$_2$F$_8$, *Sov. J. Quantum Electron.* 19, 340 (1989).
15. Macfarlane, R. M., Tong, F., Silversmith, A. J., and Lenth, W., Violet CW neodymium upconversion laser, *Appl. Phys. Lett.* 52, 1300 (1988).
16. Kaminskii, A. A., Demchuk, M. I., Zhavaronkov, N. V., and Mikhailov, V. P., Ultrashort pulsed of stimulated emission by Nd^{3+}-doped (^{4}F$_{3/2}$–^{4}I$_{11/2}$ channel) anisotropic fluorides, acentric dispersed silicates, and gallium garnets, *Phys. Stat. Sol.* (a) 113, K257 (1989).
17. Payne, S. A., Smith, L. K., DeLoach, L. D., Kway, W. L., Tassano, J. B., and Krupke, W. F., Ytterbium-doped apatite-structure crystals: A new class of laser materials, *J. Appl. Phys.* 75, 497 (1994).
18. Hebert, T., Wannemacher, R., Macfarlane, R.M., and Lenth, W., Blue continuously pumped upconversion lasing in Tm:YLiF$_4$, *Appl. Phys. Lett.* 60, 2592 (1992).
19. Nguyen, D.C., Faulkner, G.E., and Dulick, M., Blue-green (450-nm) upconversion Tm^{3+}:YLF laser, *Appl. Optics*, 28, 3553 (1989).
20. Baer, J. W., Knights, M. G., Chicklis, E. P., and Jenssen, H. P., XeF-pumped laser operation of Tm:YLF at 452 nm, in *Proc. Topical Meeting on Excimer Lasers*, IEEE-OSA, Charleston, SC (1979).
21. Thrash, R. J. and Johnson, L. F., Upconversion laser emission from Yb^{3+}-sensitized Tm^{3+} in BaY$_2$F$_8$, *J. Opt. Soc. Am. B* 11, 881 (1994).
22. Hebert, T., Wannemacher, R., Lenth, W., and Macfarlane, R.M., Blue and green cw upconversion lasing in Er:YLiF$_4$, *Appl. Phys. Lett.* 57, 1727 (1990).

23.   Esterowitz, L., Allen, R., Kruer, M., Bartoli, F., Goldberg, L. S., Jensen, H. P., Linz, A., and Nicolai, V. O., Blue light emission by Pr:LiYF$_4$ laser operated at room temperature, *J. Appl. Phys.* 48, 650 (1977).

24.   Malinowsdki, M., Pracka, I., Surma, B., Lukasiewicz, T., Wolinski, W., and Wolski, R., Spectroscopic and laser properties of SrLaGa$_3$O$_7$:Pr^{3+} crystals, *Opt. Mater.* 6, 305 (1996).

25.   Smith, R. G., New room temperature CW laser transition in YAlG:Nd, *IEEE J. Quantum Electron.* QE-4, 505 (1968).

26.   Varsanyi, F., Surface lasers, *Appl. Phys. Lett.*, 19, 169 (1971).

27.   German, K. R., Kiel, A., and Guggenheim, H., Stimulated emission from PrCl$_3$, *Appl. Phys. Lett.*, 22, 87 (1973).

28.   German, K. R. and Kiel, A., Radiative and nonradiative transitions in LaCl$_3$: Pr and PrCl$_3$, *Phys. Rev. B* 8, 1846 (1973).

29.   Danger, T., Sandrock, T., Heumann, E., Huber, G., and Chai, B., Pulsed laser action of Pr:GdLiF$_4$ at room temperature, *Appl. Phys. B* 57, 239 (1993).

30.   German, K. R., Kiel, A., and Guggenheim, H. J., Radiative and nonradiative transitions of Pr^{3+} in trichloride and tribromide hosts, *Phys. Rev. B*, 11, 2436 (1975).

31.   Kaminskii, A. A., Visible lasing on five intermultiplet transitions of the ion Pr^{3+} in LiYF$_4$, *Sov. Phys. Dokl.* 28, 668 (1983).

32.   Kaminskii A. A., Stimulated emission spectroscopy of Ln^{3+} ions in tetragonal LiLuF$_4$ fluoride, *Phys. Stat. Sol. A* 97, K53 (1986).

33.   Kaminskii, A. A., Markosyan, A. A., Pelevin, A. V., Polyakova, Yu. A., Sarkisov, S. E., and Uvarova, T. V., Luminescence properties and stimulated emission from Pr^{3+}, Nd^{3+} and Er^{3+} ions in tetragonal lithium-lutecium fluoride, *Inorg. Mater. (USSR)* 22, 773 (1986).

34.   McFarlane, R.A., High-power visible upconversion laser, *Opt. Lett.* 16, 1397 (1991).

35.   Jenssen, H. P., Castleberry, D., Gabbe, D., and Linz, A., Stimulated emission at 5445 Å in Tb^{3+}: YLF, in *Digest of Technical Papers CLEA* (1973), IEEE/OSA, Washington, DC (1971), p. 47.

36.   Fan, T. Y., Cordova-Plaza, A., Digonnet, M. J. F., Byer, R. L., and Shaw, H. J., Nd:MgO:LiNbO$_3$ spectroscopy and laser devices, *J. Opt. Soc. Am. B* 3, 140 (1986).

37.   Silversmith, A. I., Lenth, W., and Macfarlane, R. M., Green infrared-pumped erbium upconversion laser, *Appl. Phys. Lett.* 51, 1977 (1987).

38.   Brede, R., Danger, T., Heumann, E., and Huber, G., Room temperature green laser emission of Er:LiYF$_4$, *Appl. Phys. Lett.* 63, 729 (1993).

39.   Xie, P. and Rand, S. C., Continuous-wave, fourfold upconversion laser, *Appl. Phys. Lett.* 63, 3125 (1993).

40.   McFarlane, R. A., Dual wavelength visible upconversion laser, *Appl. Phys. Lett.* 54, 2301 (1989).

41.   Heine, F., Heumann, E., Danger, T., Schweizer, T., Koetke, J., Huber, G., and Chai, B. H. T., Room temperature continuous wave upconversion Er:YLF laser at 551 nm, *OSA Proc. Adv. Solid State Lasers*, Fan, T. Y. and Chai, B. H. T., Eds., 20, 344 (1995).

42.   Tong, F., Risk, W. P., Macfarlane, R. M., and Lenth, W., 551 nm diode-laser-pumped upconversion laser, *Electron. Lett.* 25, 1389 (1989).

43.   Stephens, R. R. and McFarlane, R. A., Diode-pumped upconversion laser with 100-mW output power, *Opt. Lett.* 18, 34 (1993).

44.   Voron'ko, Yu. K., Kaminskii, A. A., Osiko, V. V., and Prokhorov, A. M., Stimulated emission from Ho^{3+} in CaF$_2$ at 5512 Å, *JETP Lett.* 1, 3 (1965).

45.   Johnson, L. F. and Guggenheim, H. J., Infrared-pumped visible laser, *Appl. Phys. Lett.* 19, 44 (1971).

46.   Johnson, L. F. and Guggenheim, H. J., New laser lines in the visible from Er^{3+} ions in BaY$_2$F$_8$, *Appl. Phys. Lett.* 20, 474 (1972).

47.   Lu, B., Wang, J., Pan, I., and Jiang, M., Excited emission and self-frequency-doubling effect of Nd$_x$Y$_{1-x}$Al$_3$(BO$_3$)$_4$ crystal, *Chin. Phys. Lett. (China)* 3, 413 (1986).

48.   Kazakov, B. N., Orlov, M. S., Petrov, M. V., Stolov, A. L., and Takachuk, A. M., Induced emission of Sm^{3+}- ions in the visible region of the spectrum, *Opt. Spectrosc. (USSR)* 47, 676 (1979).

49. Hegarty, J. and Yen, W. M., Laser action in PrF_3, *J. Appl. Phys.* 51, 3545 (1980).

50. Kaminskii, A. A., Stimulated radiation at the transitions $^3P_0 \rightarrow {}^3F_4$ and $^3P_0 \rightarrow {}^3H_6$ of Pr^{3+} ions in LaF_3 crystals, *Izv. Akad. Nauk. SSSR* 17, 185 (1981), (in Russian).

51. Kaminskii, A A., Some current trends in physics and spectroscopy of laser crystals, in *Proc. Int. Conf. Lasers '80*, Collins, C. B., Ed., STS Press, Mclean, VA (1981), p. 328.

52. Solomon, R. and Mueller, L., Stimulated emission at 5985 Å from Pr^{3+} in LaF_3, *Appl. Phys. Lett.* 3, 135 (1963).

53. Sutherland, J. M., French, P. M. W., Taylor, J. R., and Chai, B. H. T., Visible continuous-wave laser transitions in Pr^{3+}:YLF and femtosecond pulse generation. *Opt. Lett.* 21, 797 (1996).

54. Szafranski, C., Strek, W., and Jezowqka-Trzebiatowska, B., Laser oscillation of a $LiPrP_4O_{12}$ single crystal, *Opt. Commun.* 47, 268 (1983).

55. Kaminskii, A. A., Sobolev, B. P., Uvarova, T. V., and Chertanov, M. I., Visible stimulated emission of Pr^{3+} ions in BaY_2F_8, *Inorg. Mater. (USSR)* 20, 622 (1984).

56. Kaminskii, A. A. and Sarkisov, S. E., Stimulated-emission spectroscopy of Pr^{3+} ions in monoclinic BaY_2F_8 fluoride, *Phys. Stat. Solidi A* 97, K163 (1986).

57. Kaminskii, A. A., Petrosyan, A. G., and Ovanesyan, K. L., Stimulated emission of Pr^{3+}, Nd^{3+}, and Er^{3+} ions in crystals with complex anions, *Phys. Status. Solidi A* 83, K159 (1984).

58. Chang, N. C., Fluorescence and stimulated emission from trivalent europium in yttrium oxide, *J. Appl. Phys.* 34, 3500 (1963).

59. Kvapil, J., Koselja, M., Kvapil, J., Perner, B., Skoda, V., Kubeika, J., Hamal, K., and Kubecek, V., Growth and stimulated emission of YAP:Ti, *Czech. J. Phys. B.* 38, 237 (1988).

60. Kaminskii, A. A., Petrosyan, A. G., Ovanesyan, K. L., and Chertanov, M. I., Stimulated emission of Pr^{3+} ions in $YAlO_3$ crystals, *Phys. Stat. Sol. A:*, 77, K173 (1983).

61. Kaminskii, A. A., Petrosyan, A. G., and Ovanesyan, K. L., Stimulated emission spectroscopy of Pr^{3+} ions in $YAlO_3$ and $LuAlO_3$, *Sov. Phys. Dokl.* 32, 591 (1987).

62. Kaminskii, A. A., Kurbanov, K., Ovanesyan, K. L., and Petrosyan, A. G., Stimulated emission spectroscopy of Pr^{3+} ions in orthorhombic $YAlO_3$ single crystals, *Phys. Stat. Solidi A* 105, K155 (1988).

63. Bleckman, A., Heine, F., Meyn, J. P., Danger, T., Heumann, E., and Huber, G., CW-lasing of Pr:$YAlO_3$ at room temperature, in *Advanced Solid-State Lasers*, Pinto, A. A. and Fan, T. Y., Eds., Proceedings Vol. 15, Optical Society of America, Washington, DC (1993), p. 199.

64. Luo, Z., Jiang, A., Huang, Y., and Qiu, M., *Chin. Phys. Lett.* 6, 440 (1989).

65. Johnson, L. F. and Ballman, A. A., Coherent emission from rare-earth ions in electro-optic crystals, *J. Appl. Phys.* 40, 297 (1969).

66. O'Connor, J. R., Optical and laser properties of Nd^{3+}- and Eu^{3+}-doped YVO_4, *Trans. Metallurg. Soc. AIME* 239, 362 (1967).

67. Bayramian, A. J., Bibeau, C., Beach, R. J., Marshall, C. D., Payne, S. A., and Krupke, W. F., Three-level Q-switched laser opertion of ytterbium-doped $Sr_5(PO_4)_3F$ at 985 nm, *Optics Lett.* 125, 622 (2000).

68. Szymanski, M., Simultaneous operation at two different wavelengths of an $(Pr,La)P_5O_{14}$ laser, *Appl. Phys.* 24, 13 (1981).

69. Borkowski, B., Crzesiak, E., Kaczmarek, F., Kaluski, Z., Karolczak, J., and Szymanski, M., Chemical synthesis and crystal growth of laser quality praseodymium pentaphosphate, *J. Crystal Growth* 44, 320 (1978).

70. Szymanski, M., Karolczak, I., and Kaczmarek, F., Laser properties of praseodymium pentaphosphate single crystals, *Appl. Phys.* 19, 345 (1979).

71. Dornauf, H. and Heber, J., Fluorescence of Pr^{3+}-ions in $La_{1-x}Pr_xP_5O_{14}$, *J. Lumin.* 20, 271 (1979).

72. Kaminskii, A. A., New room-temperature stimulated-emission channels of Pr^{3+} ions in anisotropic laser crystals, *Phys. Stat. Solidi A* 125, K109 (1991).

73. Knowies, D. S., Zhang, Z., Gabbe, D., and Jenssen, H.B., Laser action of Pr^{3+} in $LiYF_4$ and spectroscopy of Eu^{3+}- sensitized Pr in BaY_2F_8, *IEEE J. Quantum Electron.* 24, 1118 (1988).

74. Kaminskii, A. A., Ivanov, A. O., Sarkisov, S. E. et al., Comprehensive investigations of the spectral and lasing characteristics of the $LuAlO_3$ crystal doped with Nd^{3+}, *Sov. Phys.-JETP* 44, 516 (1976).

75. Kaminskii, A. A., Eichler, H. J., Liu, B., and Meindl, P., $LiYF_4:Pr^{3+}$ laser at 639.5 nm with 30 J flashlamp pumping and 87 mJ output energy, *Phys. Stat. Sol. A* 138, K45 (1993).

76. Kaminskii, A. A., Pavlyuk, A. A., Klevtsov, P. V. et al., Stimulated radiation of monoclinic crystals of $KY(WO_4)_2$, and $KGd(WO_4)_2$ with Ln^{3+} ions, *Inorg. Mater. (USSR)*, 13, 482 (1977).

77. Macfarlane, R. M., Silversmith, A. J., Tong, F., and Lenth, W., CW upconversion laser action in neodymium and erbium doped solids, in *Proceedings of the Topical Meeting on Laser Materials and Laser Spectroscopy*, World Scientific, Singapore, (1988), p. 24.

78. Antipenko, B. M., Voronin, S. P., and Privaiova, T. A., Addition of optical frequencies by cooperative processes, *Opt. Spectrosc. (USSR)* 63, 768 (1987).

79. Antipenko, B. M., Voronin, S. P., and Privalova, T. A., New laser channels of the Tm^{3+} ion, *Opt. Spectrosc.* 68, 164 (1990).

80. Heine, F., Ostroumov, V., Heumann, E., Jensen, T., Huber, G., and Chai, B. H. T., CW Yb,Tm:$LiYF_4$ upconversion laser at 650 nm, 800 nm, and 1500 nm, *OSA Proc. Adv. Solid State Lasers*, Chai, B. H. T. and Payne, S. A., Eds., 24, 77 (1995).

81. Li, R., Xie, C., Wang, J., Liang, X., Peng, K., and Xu, G., CW Nd:MgO:$LiNbO_3$ self-frequency-doubling laser at room temperature, *IEEE J. Quantum Electron.* 29, 2419 (1993).

82. Albers, P., Stark, E., and Huber, G., Continuous-wave laser operation and quantum efficiency of titanium-doped sapphire, *J. Opt. Soc. Am. B*, 3, 134, 1986

83. Moulton, P.F., Spectroscopic and laser characteristics of Ti:Al_2O_3, *J. Opt. Soc. Am. B* 3, 125 (1986).

84. Kruglik, G. S., Skripko, G. A., Shkadarevich, A. P., Kondratyuk, N. V., Urbanovich, V. S., and Nazarenko, R N., Output of Al_2O_3:Ti^{3+} crystals in the continuous and quasicontinuous regimes, *J. Appl. Spectrosc. (USSR). (Eng. Transl.)* 45, 1031 (1986).

85. Birnbaum, M. and Pertica, A. J., Laser material characteristics of Ti:Al_2O_3, *J. Opt. Soc. Am. B.* 4, 1434 (1987).

86. Sanchez, A., Strauss, A. J., Aggarwal, R. L., and Fahey, R. E., Crystal growth, spectroscopy, and laser characteristics of Ti:Al_2O_3, *IEEE J. Quantum Electron.* 24, 995 (1988).

87. Sanchez, A., Fahey, R. E., Strauss, A. J., and Aggurwal, R. L., Room-temperature continuous-wave operation of a Ti:Al_2O_3 laser, *Opt. Lett.* 11, 363 (1986).

88. Sanchez, A., Fahey, R. E., Strauss, A. J., and Aggarwal, R. L., Room-termperature cw operation of the Ti:Al_2O_3 laser, in *Tunable Solid-State Lasers* II, Vol. S2, Budgor, A. B., Esterowitz, L., and DeShazer, L. G., Eds., Springer-Verlag, New York (1986), p. 202.

89. Albers, P., Jenssen, H. P., Hube, G., and Kokta, M., Continuous wave tunable laser operation of Ti^{3+}- doped sapphire at 300 K, in *Tunable Solid-State Lasers II*, Vol. 52, Budgor, A. B., Esterowitz, L., and DeShazer. L. G., Eds., Springer-Verlag, New York (1986), p. 208.

90. Albrecht, G. F., Eggleston, J. M., and Ewing, J. J., Measurements of Ti^{3+}:Al_2O_3 as a lasing material, in *Tunable Solid State Lasers*, Proc. Int. Conf., Vol. 47, Hammerling, P., Budgor, A. B. and Pinto, A., Eds., Springer-Verlag, New York (1985), p. 68.

91. Sevast'ganov, B. K., Bagdasarov, Kh. S., Fedorov, E. A., Semenov, V. B., Tsigler, I. N., Chirkina, K. P., Starostina, L. S., Chirkin, A. P., Minaev, A. A., Orekhova, V. P., Seregin, V. F., Koierov, A. N., and Vratskii, A. N., Tunable laser based on Al_2O_3:Ti^{3+} crystal, *Sov. Phys. Crystallogr.* 29, 566 (1984).

92. Kruglik, C. S., Skripko, G. A., Shkadurevich, A. P., Kondratyuk, N. V., and Zhdanov, E. A., Output characteristics of a coherently pumped laser utilizing an Al_2O_3:Ti^{3+} crystal, *Sov. J. Quantum Electron.* 16, 792 (1986).

93. Sevast'yanov, B. K., Budasarov, Kh. S., Fedorov, E. A., Semenov, V. B., Tsigler, I. N., Chirkina, K. P., Starostin, L. S., Chirkin, A. P., Minaev, A. A., Orekhova, V. P., Seregin, V. F., and Kobro, A. N., Spectral and lasing characteristics of corundum crystals activated by Ti^{3+} ions ($Al_2O_3:Ti^{3+}$), *Sov. Phys. Dokl.* 30, 508 (1985).

94. Muller, C. H., III, Lowenthal, D. D., Kangus, K. W., Hamil, R. A., and Tisone, G. C., 2.0-J Ti sapphire laser oscillator, *Opt. Lett.* 13, 380 (1988).

95. Schmid, F. and Khattak, C. P., Growth of Co MgF_2 and $Ti:Al_2O_3$ crystals for solid state laser applications, in *Tunable Solid State Lasers*, Proc. Int. Conf., Vol. 47, Hammerling, P., Budgor, A. B., and Pinto, A., Eds., Springer-Verlag, New York (1985), p. 122.

96. Moulton, P. F., Tunable paramagnetic-ion lasers, in *Laser Handbook*, Vol. 5, Bass, M., and Stitch, M. L., Eds., North-Holland, Amsterdam, The Netherlands (1985), p. 203.

97. Moulton, P. F., Recent advances in transition metal-doped lasers, in *Tunable Solid State Lasers*, Proc. Int. Conf., Vol 47, Hammerling, P., Budgor, A. B., and Pinto, A., Eds., Springer-Verlag, New York (1985), p. 4.

98. Moulton, P. F., Spectroscopic and laser characteristics of $Ti:Al_2O_3$. *J. Opt. Soc. Am.* B 3, 125 (1986).

99. Barnes, N. P., Williams, J. A., Barnes, J. C., and Lockard, G. E., A self-injection locked, Q-switched, line-narrowed $Ti:Al_2O_3$ laser, *IEEE J. Quantum Electron.* 24, 1021 (1988).

100. Moncorgé, R., Boulon, G., Vivien, D., Lejus, A. M., Collongues, R., Djevahirdjian, V., Djevahirdjian, K., and Cagnard, R., Optical properties and tunable laser action of Verneuil-grown single crystals of $Al_2O_3:Ti^{3+}$, *J. IEEE Quantum Electron.* 24, 1049 (1988).

101. Bagdasarov, Kh. S., Krasilov, Yu. I., Kuznetsov, N. T., Kuratev, I. I., Potemkin, A. V., Shestakov, A. V., Zverev, G. M., Siyuchenko, O. G., and Zhitnnyuk, V. A., Laser properties of α-$Al_2O_3:Ti^{3+}$ crystals, *Sov. Phys. Dokl.* 30, 473 (1985).

102. Rapoport, W. R. and Khattak, C. P., Titanium sapphire laser characteristics, *Appl. Opt.* 27, 2677 (1988).

103. DeShazer, L. G., Albrecht, C. F., and Seamans, J. F., Tunable titanium sapphire lasers, in Proc. Int. Soc. Opt. Eng., *High Power and Solid State Lasers*, Vol. 622, Simmons, W. W., Ed., SPIE, Bellingham, WA (1986), p. 133.

104. Eggleston, J. M., DeShazer, L. G., and Kangas, K. W., Characteristics and kinetics of laser-pumped Ti:Sapphire oscillators, *IEEE J. Quantum Electron.* 24, 1009 (1988).

105. Rapoport, W. R. and Khattak, C. P., Efficient tunable Ti:sapphire laser, in *Tunable Solid-State Lasers II*, Vol. 52, Budgor. A. B., Esterowitz. L., and DeShazer, L. G., Eds., Springer-Verlag, New York (1986), p. 212.

106. DeShazer, L. G., Eggleston, J. M., and Kangas, K. W., Oscillator and amplifier performance of Ti:sapphire, in *Tunable Solid-State Lasers II*, Vol. 52, Budgor, A. B., Esterowitz, L., and DeShazer L. G., Eds., Springer-Verlag, New York (1986), p. 228.

107. Schepier, K. L., Laser performance and temperature-dependent spectroscopy of titanium-doped crystals, in *Tunable Solid-State Lasers II*, Vol. 52, Budgor, A. B., Esterowitz, L., and DeShazer, L. G., Eds., Springer-Verlag, New York (1986), p. 235.

108. DeShazer, L. G. and Kangas, K. W., Extended infrared operation of a titanium sapphire laser presented at *Conf. on Lasers and Electrooptics*, Baltimore, MD, Apr. 26–May 1 (1987), p. 296.

109. Bagdasarov, Kh. S., Danilov, V. P., Murina, T. M., Novikov, E. G., Prokhorov, A. M., Semenov, V. B., and Fedorov, E. A., Tunable flashlamp-pumped $Al_2O_3:Ti^{3+}$ laser, *Sov. Tech. Phys. Lett.* 13, 152 (1987).

110. Schulz, P. A., Single-frequency $Ti:Al_2O_3$ ring laser, *IEEE J. Quantum Electron.* 24, 1039 (1988).

111. Lacovara, P., Esterowitz, L., and Allen, R., Flash-lamp-pumped $Ti:Al_2O_3$ laser using fluorescent conversion, *Opt. Lett.* 10, 273 (1985).

112. Esterowitz, L. and Allen, R., Stimulated emission from flashpumped $Ti:Al_2O_3$, in *Tunable Solid State Lasers*, Proc. Int. Conf., Vol. 47, Hammerling, P., Budgor, A. B., and Pinto, A., Eds., Springer-Verlag, New York (1985), p. 73.

113. Alshuler, G. B., Karasev, V. B., Kondratyuk, N. V., Krugiik, G. S., Okishev, A. V., Skripko, G. A., Urbanovich, V. S., and Shkadarevich, A. P., Generation of ultrashort pulses in a synchronously pumped Ti^{3+} laser, *Sov. Tech. Phys. Lett.* 13, 324 (1987).

114. Fox, A. M., Maciel, A. C., and Ryan, J. F., Efficient CW performance of a $Co:MgF_2$ laser operating at 1.5—2.0 μm, *Opt. Commun.* 59, 142 (1986).

115. Kaminskii, A. A., Sobolev, B. P., Sarkisov, S. E., Denlsenko, G. A., Ryabchenkov, V. V., Fedorov, V. A., and Uvarova, T. V., Physicochemical aspects of the preparation spectroscopy, and stimulated emission of single crystals of $BaLn_2F_8\text{-}Ln^{3+}$, *Inorg. Mater. (USSR)* 18, 402 (1982).

116. Marin, V. I., Nikitin, V. I., Soskin, M. S., and Khizhnyak, A. I., Superluminescence emitted by $YAG:Nd^{3+}$ crystals and stimulated emission due to weak transitions, *Sov. J. Quantum Electron.* 5, 732 (1975).

117. Nebdaev, N. Ya., Petrenko, R. A., Piskarskas, A. S., Sirutajtis, V. A., Smil'gyavichyus, V. I., and Yuozapavlchus, A. S., Peculiarities of stimulated emission of picosecond pulses from the potassium gadolinium tungstate laser, *Ukr. Fiz. Zh.* 33, 1165 (1988) (in Russian).

118. Bukin, G. V., Volkov, S. Yu., Matrosov, V. N., Sevast'yanov, B. K., and Timoshechkin, M. I., Stimulated emission from alexandrite ($BeAl_2O_4:Cr^{3+}$), *Sov. J. Quantum Electron.* 8, 671 (1978).

119. Weber, M. J., Bass, M., and Demars, G. A., Laser action and spectroscopic properties of Er^{3+} in $YAlO_3$, *J. Appl. Phys.* 42, 301 (1971).

120. Sevastyanov, B. K., Remigailo, Yu. I., Orekhova, V. P., Matrosov, V. P., Tsvetkov, E. G., and Bukin, G. V., Spectroscopics and lasing properties of alexandrite ($BeAl_2O_4:Cr^{3+}$), *Sov. Phys. Dokl.* 26, 62 (1981).

121. Walling, J. C., Jenssen, H. P., Morris, R. C., O'Dell, E. W., and Peterson, O. G., Tunable-laser performance in $BeAl_2O_4:Cr^{3+}$, *Opt. Lett.* 4, 182 (1979).

122. Walling, J. C. and Peterson, O. G., High gain laser performance in alexandrite, *IEEE J. Quantum Electron.* QE-16, 119 (1980).

123. Buchert, J., Katz, A., and Alfano, R. R., Laser action in emerald, *IEEE J. Quantum Electron.* QE-19, 1477 (1983).

124. Sevast'yanov, B. K., Bagdasarov, Kh. S., Pasternak, L. B., Volkov, S. Yu., and Drekhova, V. P., Stimulated emission from Cr^{3+} ions in YAG crystals, *JETP Lett.* 17, 47 (1973).

125. McClung, F. J., Schwarz, S. E., and Meyers, F. J., R_2 line optical maser action in ruby, *J. Appl. Phys.* 33, 3139 (1962).

126. Collins, R. J., Nelson, D. F., Schawlow, A. L., Bond, W., Garrett, C. G. B., and Kaiser, W., Coherence, narrowing, directionality, and relaxation oscillations in the light emission from ruby, *Phys. Rev. Lett.* 5, 303 (1960).

127. Nelson, D. F. and Boyle, W. S., A continuously operating ruby optical maser, *Appl. Optics*, 1, 181 (1962).

128. Birnbaum, M., Tucker, A. W., and Fincher, C. L., CW ruby laser pumped by an argon ion laser, *IEEE J. Quantum Electron.* QE-13, 808 (1977).

129. Kaminskii, A. A., Butaeva, T. I., Ivanov, A. O. et al., New data on stimulated emission of crystals containing Er^{3+} and Ho^{3+} ions, *Sov. Tech. Phys. Lett.* 2, 308 (1976).

130. Weber, M., Bass, M., Varitimos, T., and Bua, D., Laser action from Ho^{3+}, Er^{3+} and Tm^{3+} in $YAlO_3$, *IEEE J. Quantum Electron.* QE-9, 1079 (1973).

131. Maiman, T. H., Stimulated optical radiation in ruby, *Nature* 187, 493 (1960).

132. Maiman, T. H., Optical maser action in ruby, *Br. Commun. Electron.* 7, 674 (1960).

133. Roess, D., Analysis of room temperature CW ruby lasers, *IEEE J. Quantum Electron.* QE-2, 208 (1966).

134. Evtuhov, V. and Kneeland, J. K., Power output and efficiency of continuous ruby laser, *J. Appl. Phys.* 38, 4051 (1967).

135. Burrus, C. A. and Stone, J., Room-temperature continuous operation of a ruby fiber laser, *J. Appl. Phys.* 49, 3118 (1978).

136. Galkin, S. L., Zakgeim, A. L., Markhonov, V. M., Nikolaev, V. M., Pavlyuk, A. A., Petrovich, I. P., Petrun'kin, V. Yu., Shkadarevich, A. P., and Yarzhemkovskil, V. D.,

Crystalline KGd(WO$_4$)$_2$ laser with a semiconductor pump system, *J. Appl. Spectrosc. (USSR).* (Engl. Transl.) 37, 886 (1982).

137. Kaminskii, A., High-temperature spectroscopic investigation of stimulated emission from lasers based on crystals and glass activated with Nd^{3+} ions, *Sov. Phys.-JETP* 27, 388 (1968).

138. Kirkin, A. N., Leontovich, A. M., and Mozharovskii, A. M., Generation of high power ultrashort pulses in a low temperature ruby laser with a small active volume, *Sov. J. Quantum Electron.* 8, 1489 (1978).

139. Ivanyuk, A. M., Shakhverdov, P. A., Belyaev, V. D., Ter-Pogosyan, M. A., and Ermolaev, V. L., A picosecond neodymium laser on potassium-gadolinium tungstate with passive mode locking operating in the repetitive regime, *Opt. Spectrosc. (USSR)* 58, 589 (1985).

140. Walling, J. C., Peterson, O. G., and Morris, R. C., Tunable CW alexandrite laser, *IEEE J. Quantum Electron.* QE-16, 120 (1980).

141. Samelson, H., Walling, J. C., Wernikowski, T., and Harter, D. J., CW arc-lamp-pumped alexandrite lasers, *IEEE J. Quantum Electron.* 24, 1141 (1988).

142. Walling, J. C., Heller, D. F., Samelson, H., Harter, D. J., Pete, J. A., and Morris, R. C., Tunable alexandrite lasers: development and performance, *IEEE J. Quantum Electron.* QE-21, 1568 (1985).

143. Lai, S. T. and Shand, M. L., High efficiency cw laser-pumped tunable alexandrite laser, *J. Appl. Phys.* 54, 5642 (1983).

144. Rapoport, W. R. and Samebon, H., Alexandrite slab laser, in *Proc. Int. Conf. Lasers 85*, Wang, C. P., Ed., STS Press, McLean, VA (1986), p. 744.

145. Zhang, S. and Zhang, K., Experiment on laser performance of alexandrite crystals, *Chin. Phys.*, 4, 667 (1984).

146. Guch, S., Jr. and Jones, C. E., Alexandrite-laser performance at high temperature, *Opt. Lett.* 7, 608 (1982).

147. Walling, J. C., Peterson, O. G., Jenssen, H. P., Morris, R. C., and O'Dell, E. W., Tunable alexandrite lasers, *IEEE J. Quantum Electron.* QE-16, 1302 (1980).

148. Shand, M. L., Progress in alexandrite lasers, in *Proc. Int. Conf. Lasers 85*, Wang, C. P., Ed., STS Press, McLean, VA (1986), p. 732.

149. Zhang, G. and Ma, X., Improvement of lasing performance of alexandrite crystals, *Chin. Phys. Lasers*, 13, 816 (1986).

150. Jones, J. E., Dobbins, J. D., Butier, B. D., and Hinsley, R. J., Performance of a 250-Hz, 100-W alexandrite laser system, in *Proc. Int. Conf. Lasers 85*, Wang, C. P., Ed., STS Press, McLean, VA (1986), p. 738.

151. Lisitsyn, V. N., Matrosov, V. N., Pestryakov, E. V., and Trunov, V. I., Generation of picosecond pulses in solid-state lasers using new active media, *J. Sov. Laser Res.* 7, 364 (1986).

152. Jones, J. E., Dobbins, J. D., Butier, B. D., and Hinsley, R. J., Performance of a 250-Hz, 100-W alexandrite laser system, in *Proc. Int. Conf. Lasers 85*, Wang, C. P., Ed., STS Press, McLean, VA (1986), p. 738.

153. Schawlow, A. L. and Devlin, G. E., Simultaneous optical maser action in two ruby satellite lines, *Phys. Rev. Lett.* 6, 96 (1961).

154. Walling, J. C. and Sam, C. L., unpublished data (1980).

155. Kaminskii, A. A. and Li, L., Spectroscopic studies of stimulated emission in an SrF$_2$:Nd^{3+} crystal laser, *J. Appl. Spectrosc. (USSR)* 12, 29 (1970).

156. Sorokin, P. P., Stevenson, M. J., Lankard, J. R., and Pettit, G. D., Spectroscopy and optical maser action in SrF$_2$:Sm^{2+}, *Phys. Rev.* 127, 503 (1962).

157. Vagin, Yu. S., Marchenko, V. M., and Prokhorov, A. M., Spectrum of a laser based on electron-vibrational transitions in a CaF$_2$:Sm^{2+} crystal, *Sov. Phys.-JETP* 28, 904 (1969).

158. Sorokin, P. P. and Stevenson, M. J., Solid-state optical maser using divalent samarium in calcium fluoride, *IBM J. Res. Dev.* 5, 56 (1961).

159. Kaiser, W., Garrett, C. G. B., and Wood, D. L., Fluorescence and optical maser effects in CaF$_2$,:Sm^{2+}, *Phys. Rev.* 123, 766 (1961).

160. Anan'yev, Yu. A., Grezin, A. K., Mak, A. A., Sedov, B. M., and Yudina, Ye. N., A fluorite:samarium laser, *Sov. J. Opt. Technol.* 35, 313 (1968).

161. Kaminskii, A. A., Achievements in the fields of physics and spectroscopy of insulating laser crystals, in *Lasers and Applications, Pt. I, Proc.*, Ursu, I., and Brokhorov, A. M., Eds., CIP Press, Bucharest, Romania (1983), p. 97.

162. Payne, S. A., Chase, L. L., Newkirk, H. W., Smith, L. K., and Krupke, W. F., LiCaAlF$_6$:Cr^{3+}: a promising new solid-state laser material, *IEEE J. Quantum Electron.* 24, 2243 (1988).

163. Chase, L. L., Payne, S. A., Smith, L. K., Kway, W. L., Newkirk, H. W., Chai, B. H. T., and Long, M., Laser performance and spectroscopy of Cr^{3+} in LiCaAlF$_6$ and LiSrAlF$_6$:Cr^{3+}, in *Tunable Solid StateLasers*, Shand M. L., and Jenssen, H. P., Eds., Optical Society of America, Washington, DC (1989), p. 71.

164. Lai, S. T., Highly efficient emerald laser, *J. Opt. Soc. Am. B.*, 4, 1286 (1987).

165. Shand, M. L. and Lai, S. T., CW laser pumped emerald laser, *IEEE J. Quantum Electron.* QE-20, 105 (1984).

166. Shand, M. L. and Walllng, J. C., A tunable emerald laser, *IEEE J. Quantum Electron.* QE-18, 1829 (1982).

167. Podkolzina, I. G., Tkachuk, A. M., Fedorov, V. A., and Feofilov, P. P., Multifrequency generation of stimulated emission of Ho^{3+} ion in LiYF$_4$ crystals, *Opt. Spectrosc.* 40, 111 (1976).

168. Kaminskii, A. A. and Pelevin, A. V., Low-threshold lasing of LiYF$_4$:Pr^{3+} crystals in the 0.72 µm range as a result of flashlamp pumping at 300 K, *Sov. J. Quantum Electron.* 21, 819 (1991).

169. Kaminskii, A. A. and Petrosyan, A. G., New laser crystal for the excitation of stimulated radiation in the dark-red part of the spectrum at 300 K, *Sov. J. Quantum Electron.* 21, 486 (1991).

170. Pestryakov, E. V., Trunov, V. I., and Alimpiev, A. I., Generation of tunable radiation in a BeAl$_2$O$_4$:Ti^{3+} laser subjected to pulsed coherent pumping at a high repetition frequency, *Sov. J. Quantum Electron.* 17, 585 (1987).

171. Alimpiev, A. I., Bukin, G. V., Matrosov, V. N., Pestryakov, E. V., Soinbev, V. P., Trunov, V. I., Tsvetkov, E. G., and Chebobev, V. P., Tunable BeAl$_2$O$_4$:Ti^{3+} laser, *Sov. J. Quantum Electron.* 16, 579 (1986).

172. Segawa, Y., Sugimoto, A., Kim, P. H., Namba, S., Yamagishi, K., Anzai, Y., and Yamaguchi, Y., Optical properties and lasing of Ti^{3+} doped BeAl$_2$O$_4$, *Jpn. J. Appl. Phys.* 26, L291 (1987).

173. Huber, G. and Petermann, K., Laser action in Cr-doped garnet and tungstates, in *Tunable Solid-State Lasers*, Vol. 47, Hammerling, P., Budgor, A. B., and Pinto, A., Eds., Springer-Verlag, New York (1985), p. 11.

174. Struve, B. and Huber, G., Laser performance of Cr^{3+}:Gd(Sc,Ga) garnet, *J. Appl. Phys.* 57, 45 (1985).

175. Huber, G., Drube, J., and Struve, B., Recent developments in tunable Cr-doped garnet lasers, in *Proc. Int. Conf. Lasers 83*, Powell, R. C., Ed., STS Press, McLean, VA (1983), p. 143.

176. Struve, B., Huber, C., Laptev, V. V., Shcherbakov, I. A., and Zharikov, E. V., Tunable room-temperature cw laser action in Cr^{3+}:GdScGa-garnet, *Appl. Phys. B* 30, 117 (1983).

177. Zharikov, E. V., Il'ichev, N. N., Kalitin, S. P., Laptev, V. V., Malyutin, A. A., Osiko, V. V., Ostroumov, V. G., Pashinin, P. P., Prokhorov, A. M., Smirnov, V. A., Umyskov, A. F., and Shcherbakov, I. A., Tunable laser utilizing an electronic-vibrational transition in chromium in a gadolinium scandium gallium garnet crystal, *Sov. J. Quantum Electron.* 13, 1274 (1983).

178. Payne, M. J. P. and Evans, H. W., Laser action in flashlamp-pumped chromium:GSG-garnet, in *Tunable Solid-State Lasers II*, Vol. 52, Budgor, A. B., Esterowitz, L., and DeShazer, L., Eds., Springer-Verlag New York (1986), p. 126.

179. Walling, J. C., Peterson, D. G., and Morris, R. C., Tunable CW alexandrite laser, *IEEE J. Quantum Electron.* QE-16, 120 (1980).

180. Caird, J. A., Payne, S. A., Staver, P. R., Ramponi, A. J., Chase, L. L., and Krupke, W. F., Quantum electronic properties of the $Na_3Ga_2Li_3F_{12}:Cr^{3+}$ laser, *IEEE J. Quantum Electron.* 24, 1077 (1988).

181. Petrov, M. V., Tkachuk, A. M., and Feofilov, P. P., Multifrequency and cascade production of induced emission from Ho^{3+} and Er^{3+} in $LiYF_4$ crystals and delayed induced afterglow from Ho^{3+}, *Bull. Acad. Sci. USSR, Phys. Ser.* 45, 167 (1981).

182. Drube, J., Struve, B., and Huber, G., Tunable room-temperature CW laser action in Cr^{3+}:GdScAl-garnet, *Opt. Commun.*, 50, 45 (1984).

183. Drube, J., Huber, G., and Mateika, D., Flashlamp-pumped Cr^{3+}:GSAG and Cr^{3+}:GSGG: slope efficiency, resonator design color centers, and tunability in *Tunable Solid-State Lasers II*, Vol. 52, Budgor, A. B., Esterowitz, L., and DeShazer, L. G., Eds., Springer-Verlag, New York (1986), p. 118.

184. Meier, J. V., Barnes, N. P., Remelius, D. K., and Kokta, M. R., Flashlamp-pumped Cr^{3+} :GSAG laser, *IEEE J. Quantum Electron.* QE-22, 2058 (1986).

185. Struve, B., Fuhrberg, P., Luhs, W., and Litfin, G., Thermal lensing and laser operation of flashlamp-pumped Cr:GSAG, *Opt. Commun.* 65, 291 (1988).

186. Chai, B. H. T., Lefaucheur, J., Stalder, M., and Bass, M., $Cr:LiSr_{0.8}Ca_{0.2}AlF_6$ tunable laser, *Opt. Lett.* 17, 1584 (1992).

187. Chicklis, E. P., Naiman, C. S., Esterowitz, L., and Allen, R., Deep red laser emission in Ho:YLF, *IEEE J. Quantum Electron.* QE-13, 893 (1977).

188. Kaminskii, A. A., Belokoneva, E. L., Mill, B. V., Pisarevskii, Yu. V., Sarkisov, S. E., Silvestrova, I. M., Butashin, A. V., and Khodzhabagyan, G. G., Pure and Nd^{3+}-doped $Ca_3Ga_2Ge_4O_{14}$ and $Sr_3Ga_2Ge_4O_{14}$ single crystals, their structure, optical, spectral luminescence, electromagnetic properties, and stimulated emission. *Phys. Stat. Sol. A* 86, 345 (1984).

189. Sugimoto, A., Segawa, Y., Anzai, Y., Yamagishi, K., Kim, P.H., and Namba, S., Flash-lamp-pumped tunable $Ti:BeAl_2O_4$ laser, *Japan. J. Appl. Phys.* 29, 1136 (1990).

190. Kaminskii, A. A., Luminescence and multiwave stimulated emission of Ho^{3+} and Er^{3+} ions in orthorhombic $YAlO_3$ crystals, *Sov. Phys. Dokl.*, 31, 823 (1986).

191. Brauch, U. and Dürr, U., $KZnF_3:Cr^{3+}$—a tunable solid state NIR-laser, *Opt. Commun.* 49, 61 (1984).

192. Dürr, U., Brauch, U., Knierim, W., and Weigand, W., Vibronic solid state lasers: transition metal ions in perovskites. in *Proc. Int. Conf. Lasers 83*, Powell, R. C., Ed., STS Press, McLean, VA (1983), p. 42.

193. Brauch, U. and Dürr, U., Room-temperature operation of the vibronic $KZnF_3:Cr^{3+}$ laser, *Opt. Lett.* 9, 441 (1984).

194. Dubinskii, M. A., Kolerov, A. N., Mityagin, M. V., Silkin, N. I., and Shkadarevich, A. P., Quasi-continuous operation of a $KZnF_3:Cr^{3+}$ laser, *Sov. J. Quantum Electron.* 16, 1684 (1986).

195. Abdulsabirov, R. Yu., Dubinskii, M. A., Korableva, S. L., Mityagin, M. V., Silkin, N. I., Skripko, C. A., Shkadarevich, A. P., and Yagudin, Sh. I., Tunable laser based on $KZnF_3:Cr^{3+}$ crystal with nonselective pumping, *Sov. Phys. Crystallogr.* 31, 353 (1986).

196. Payne, S. A., Chase, L. L., Smith, L. K., Kway, W. L., and Newkirk, H. W., Laser performance of $LiSrAlF_6:Cr^{3+}$, *J. Appl. Phys.* 66, 1051 (1989).

197. Woodbury, E. J. and Ng, W. K., Ruby laser operation in the near IR, *Proc. IRE* 50, 2367 (1962).

198. Bazylev, A. G., Voitovich, A. P., Demidovich, A. A., Kalinov, V. S., Timoshechkin, M. I., and Shkadarevich, A. P., Laser performance of $Cr^{3+}:(Gd,Ca)_3(Ga,Mg,Zr)_2Ga_3O_{12}$, *Opt. Commun.* 94, 82 (1992).

199. Beaud, P., Chen, Y.-F., Chai, B. H. T., and Richardson, M. C., Gain properties of $LiSrAlF6:Cr^{3+}$, *Opt. Lett.* 17, 1064 (1992).

200. Dymott, M. J. P., Botheroyd, I. M., Hall, G. J., Lincoln, J. R., and Ferguson, A. J., All-solid-state actively mode-locked Cr:LiSAF laser, *Opt. Lett.* 19, 634 (1994).

201. Stalder, M., Chai, B.H.T., and Bass, M., Flashlamp pumped $Cr:LiSrAlF_6$ laser, *Appl. Phys. Lett.* 58, 216 (1991).

202. Lai, S. T., Chai, B. H. T., Long, M., and Morris, R. C., $ScBO_3$:Cr—a room temperature near-infrared tunable laser, *IEEE J. Quantum Electron.*, QE-22, 1931 (1986).

203. Lai, S. T., Chai, B. H. T., Long, M., Shinn, M. D., Caird, J. A., Marion, J. E., and Staver, P. R., A $ScBO_3$:Cr laser, in *Tunable Solid-State Lasers 11*, Vol. 52, Budgor, A. B., Esterowitz, L., and DeShazer, L.G., Eds., Springer-Verlag, New York (1986), p. 145.

204. Alimpiev, A. I., Pestryakov, E. V., Petrov, V. V., Solntsev, V. P., Trunov, V. I., and Matrosov, V. N., Tunable lasing due to the T_2-$^4A_2$ electronic-vibrational transition in Cr^{3+} ions in $BeAl_2O_4$, *Sov. J. Quantum Electron.* 18, 323 (1988).

205. Kaminskii, A. A., Sarkisov, S. E., Pavlyuk, A. A., and Lyubchenko, V. V., Anisotropy of luminescence properties of the laser crystals $KGd(WO_4)_2$ and $KY(WO_4)_2$ with Nd^{3+} ions, *Inorg. Mater. (USSR)* 16, 501 (1980).

206. Chai, B. H. T., Shinn, M. D., Long, M. N., Lai, S. T., Miller, H. H., and Smith, L. K., Laser and spectroscopic properties of Cr-doped $ScAlBeO_4$, *Bull. Am. Phys. Soc.* 33, 1631 (1988).

207. Kaminskii, A. A., Pavlyuk, A. A., Agamalyan, N. P., Bobovich, L. I., Lukin, A. V., and Lyubchenko, V. V., Stimulated radiation of $KLu(WO_4)_2$-Er^{3+} crystals at room temperature, *Inorg. Mater. (USSR)* 15, 1182 (1979).

208. Kaminskii, A. A., Shkadarevich, A. P., Mill, B. V., Koptev, V. G., and Demidovich, A. A., Wide-band tunable stimulated emission from a $La_3Ga_5SiO_{14}$-Cr^{3+} crystal, *Inorg. Mater. (USSR)* 23, 618 (1987).

209. Lai, S. T., Chai, B. H. T., Long, M., and Shinn, M. D., Room temperature near-infrared tunable $Cr:La_3Ga_5SiO_{14}$ laser, *IEEE J. Quantum Electron.* 24, 1922 (1988).

210. Petermann, K. and Mitzscherlich, P., Spectroscopic and laser properties of Cr^{3+}-doped $Al_2(WO_4)_3$ and $Sc_2(WO_4)_3$, *IEEE J. Quantum Electron.* QE-23, 1122 (1987).

211. Chicklis, E. P. and Naiman, C. S., A review of near-infrared optically pumped solid-state lasers, in *Proceedings of the First European Electro-Optics Markets and Technology Conference*, IPC Science and Technology Press. (1973), p. 77.

212. Smith, L. K., Payne, S. A., Krupke, W. F., Kway, W. L., Chase, L. L., and Chai, B. H. T., Investigation of the laser properties of Cr^{3+}:$LiSrGaF_6$, *IEEE J. Quantum Electron.* 28, 2612 (1992).

213. Bagdasarov, Kh. S., Bogomolova, G. A., Gritsenlco, M. M., Kaminskii, A. A., and Kervorkov, A. M., Spectroscopic study of the $LaAlO_3$:Nd^{3+} laser crystal, *Sov. Phys.-Crystallogr.* 17, 357 (1972).

214. Arsen'yev, P. A. and Bienert, K. E., Synthesis and optical properties of neodymium-doped gadolinium aluminate ($GdAlO_3$) single crystals, *J. Appl. Spectrosc. (USSR)* 17, 1623 (1972).

215. Kaminskii, A. A., Cascading lasers based on activated crystals, *Inorg. Mater. (USSR)* 7, 802 (1971).

216. Kaminskii, A. A., Spectroscopic studies of stimulated emission from Er^{3+} ions in CaF_2-YF_3 crystals, *Opt. Spectrosc.* 31, 507 (1971).

217. Kaminskii, A. A., Inorganic materials with Ln^{3+} ions for producing stimulated emission radiation in the 3-μm bands, *Inorg. Mater. (USSR)* 15, 809 (1979).

218. Heine, F., Heumann, E., Danger, T., Schweizer, T., and Huber, G., Green upconversion continuous wave Er^{3+}:$LiYF_4$ laser at room temperature, *Appl. Phys. Lett.* 65, 383 (1994).

219. Kaminskii, A. A., Pavlyuk, A. A., Butaeva, T. I., Fedorov, V. A., Balashov, I. F., Berenberg, V. A., and Lyubchenko, V. V., Stimulated emission by subsidiary transitions of Ho^{3+} and Er^{3+} ions in $KGd(WO_4)_2$ crystals, *Inorg. Mater. (USSR)* 13, 1251 (1977).

220. Pollock, S. A., Chang, D. B., and Birnbaum, M., Threefold upconversion laser at 0.85, 1.23 and 1.73 μm in Er:YLF pumped with a 1.53 μm Er glass laser, *Appl. Phys. Lett.* 54, 869 (1989).

221. Kaminskii, A. A., Butashin, A. V., Markabaev, A. K., Mill', B. V., Knab, G. G., and Ursovskaya, A. A., Garnet $Ca_3Ga_2Ge_3O_{12}$:optical properties, microhardness and stimulated emission of Er^{3+} ions, *Sov. Phys. Crystallogr.* 32, 413 (1987).

222. Chicklis, E. P., Naiman, C. S., and Linz, A., Stimulated emission at 0.85 μm in Er^{3+}:YLF, in *Digest of Technical Papers VII International Quantum Electronics Conference*, Montreal (1972), p. 17.

223. Kubodera, K. and Otsuka, K., Spike-mode oscillations in laser-diode pumped $LiNdP_4O_{12}$ lasers, *IEEE J. Quantum Electron.* QE-17, 1139 (1981).
224. Tkachuk, A. M., Petrov, M. V., Linmov, L. D., and Korablova, S. L., Pulsed-periodic 0.8503-μm YLF:Er^{3+}, Pr^{3+} laser, *Opt. Spectrosc. (USSR)*, 54, 667 (1983).
225. Zhekov, V. I., Lobachev, V. A., Murina, T. M., and Prokhorov, A. M., Efficient cross-relaxation laser emitting at λ = 2.94 μm, *Sov. J. Quantum Electron.* 13, 1235 (1983).
226. Petrov, M. V. and Tkachuk. A. M., Optical spectra and multifrequency stimulated emission of $LiYF_4$-Er^{3+} crystals, *Opt. Spectrosc.* 45, 81 (1978).
227. Jenssen, H. P. and Lai, S. T., Tunable-laser characteristics and spectroscopic properties of $SrAlF_5$:Cr, *J. Opt. Soc. Am. B* 3, 115 (1986).
228. Caird, J. A., Staver, P. R., Shinn, M. D., Guggenheim, H. J., and Bahnak, D., Laser-pumped laser measurements of gain and loss in $SrAlF_5$:Cr crystals, in *Tunable Solid-State Lasers II*, Vol. 52, Budgor, A. B., Esterowitz, L., and DeShazer, L., Eds., Springer-Verlag, New York (1986), p. 159.
229. Kaminskii, A. A., Sarkisov, S. E., Butaeva, T. I., Denisenko, G. A., Hermoneit, B., Bohm, J., Grosskreutz, W., and Schultze, D., Growth spectroscopy, and stimulated emission of cubic $Bi_4Ge_3O_{12}$ crystals doped with Dy^{3+}, Ho^{3+}, Er^{3+}, Tm^{3+}, or Yb^{3+} ions, *Phys. Stat. Sol. A*: 56, 725 (1979).
230. Andryunas, K., Vishchakas, Yu., Kabelka, V., Mochalov, I.V., Pavlyuk, A. A., and Syrus, V., Picosecond lasing in $KY(WO_4)_2Nd^{3+}$ crystals at pulse repetition frequencies up to 10 Hz, *Sov. J. Quantum Electron.* 15, 1144 (1985).
231. Korableva, S. L., Livanova, L. D., Petrov, M. V., and Tkachuk, A. M., Stimulated emission of Er^{3+} ions in $LiYF_4$ crystals, *Sov. Phys. Tech. Phys.* 26, 1521 (1981).
232. Tkachuk, A. M., Petrov, M. V., Podkolzina, I. G., and Semenova, T. S., Generation of stimulated emission in a concentrated barium-erbium double-fluoride crystal, *Opt. Spectrosc. (USSR)* 53, 235 (1982).
233. Kaminskii, A. A., Sarkisov, S. E., Seiranyan, K. B., and Fedorov, V. A., Generation of stimulated emission for the waves of five channels of Er^{3+} ions in a self-activated $LiErF_4$ crystal, *Inorg. Mater. (USSR)* 18, 527 (1981).
234. Kaminskii, A. A., Mill', B. V., Belokoneva, E. L., Butashin, A. V., Sarkisov, S. E., Kurbanov, K., and Khodzhabagyan, G. G., Crystal structure intensity characteristics of luminescence and stimulated emission of the disordered gallate $LaSr_2Ga_{11}O_{20}$-Nd^{3+}, *Inorg. Mater. (USSR)* 22, 1635 (1986).
235. Voronko, Yu. K. and Sychugov, V. I., The stimulated emission of Er^{3+} ions in CaF_2 at λ = 8456 Å and λ = 8548 Å, *Phys. Stat. Sol.* 25, K 119 (1968).
236. Kaminskii, A. A., Pavlyuk, A. A., Balashov, I. F. et al., Stimulated emission by $KY(WO_4)_2$-Er^{3+} crystals at 0.85, 1.73 and 2.8 μm at 300 K, *Inorg. Mater. (USSR)* 14, 1765 (1978).
237. Zhekov, V. I., Zubov, B. V., Lobchev, V. A., Murina, T. M., Prokhorov, A. M., and Shevd', A. F., Mechanism of a population inversion between the $^4I_{11/2}$ and $^4I_{13/2}$ levels of the Er^{3+} ion in $Y_3Al_5O_{12}$ crystals, *Sov. J. Quantum Electron.* 10, 428 (1980).
238. Pinto, J. F., Esterowitz, L., and Quarles, G. J., High performance Ce^{3+}:$LiSrAlF_6$/$LiCaAlF_6$ UV lasers with extended tunability, *Electron. Lett.* 31, 2009 (1995).
239. Kaminskii, A. A., Butaeva, T. I., Fedorov, V. A., Bagdasarov, Kh. S., and Petrosyan, A. G., Absorption, luminescence and stimulated emission investigations in $Lu_3Al_5O_{12}$-Er^{3+} crystals, *Phys. Stat. Sol.* 39a, 541 (1977).
240. Kaminskii, A. A., Shkadarevich, A. P., Mill, B. V., Koptev, V. G., Butashin, A. V., and Demidovich, A. A., Wide-band tunable stimulated emission of Cr^{3+} ions in the trigonal crystal $La_3Ga_{5.5}Nb_{0.5}O_{14}$, *Inorg. Mater. (USSR)* 23, 1700 (1987).
241. Kaminskii, A. A., Stimulated-emission spectroscopy of activated laser crystals with ordered and disordered structure: seven selected experimental problems, private communication (1988).

242. Kaminskii, A. A., Shkadarevich, A. P., Mill, B. V., Koptev, V. G., Butashin, A. V., and Demidovich, A., Tunable stimulated emission of Cr^{3+} ions and generation frequency self-multipication effect in acentric crystals of Ca-gallogermanate structure, *Inorg. Mater. (USSR)* 24, 579 (1988).

243. Smith, L. K., Payne, S. A., Kway, W. L., and Krupke, W. F., Laser emission from the transition-metal compound $LiSrCrF_6$, *Opt. Lett.* 18, 200 (1993).

244. Birnbaum, M., Tucker, A. W., and Pomphrey, P. I., New Nd:YAG laser transition $^4F_{3/2} \rightarrow {}^4I_{9/2}$, *IEEE J. Quantum Electron.* QE-8, 502 (1972).

245. Kaminskii, A. A., Kurbanov, K., Peievin, A.V., Bobakova, Yu. A., and Uvarova, T. V., Intermultiplet transition $^3P_0 \rightarrow {}^1G_4$—new channel for stimulated Pr^{3+}-ion emission in anisotropic fluoride crystals, *Inorg. Mater. (USSR)* 24, 439 (1988).

246. Kaminskii, A. A., Kurbanov, K., Peievin, A. V., Bobakova, Yu. A., and Uvarova, T. V., New channels for stimulated emission of Pr^{3+} ions in tetragonal fluorides $LiRF_4$ with the structure of scheelite, *Inorg. Mater. (USSR)* 23, 1702 (1987).

247. Hanson, F., Dick, D., Versun, H. R., and Kokta, M., Optical properties and lasing of $Nd:SrGdGa_3O_7$, *J. Opt. Soc. Am. B* 8, 1668 (1991).

248. Beach, R., Albrecht, G., Solarz, R., Krupke, W., Mitchell, S., Comaskey, B., Brandle, C., and Berkstresser, G., Q-switched laser at 912 nm using ground state depleted neodymium in yttrium orthosilicate, *Opt. Lett.* 15, 1020 (1990).

249. Antonov, V. A., Arsenev, P. A., Evdokimov, A. A., Koptsik, E. K., Starikov, A. M., and Tadzhi-Aglaev, Kh. G., Spectral-luminescence properties of $Ba_3LaNb_3O_{12}:Nd^{3+}$ single crystals, *Opt. Spectrosc. (USSR)* 60, 57 (1986).

250. Kaminskii, A. A., Klevtsov, P. V., Li, L., and Pavlyuk, A. A., Spectroscopic and stimulated emission studies of the new $KY(WO_4)_2:Nd^{3+}$ laser crystal, *Inorg. Mater. (USSR)* 8, 1896 (1972).

251. Kaminskii, A. A., Klevtsov, P. V., Li, L., and Pavlyuk, A. A., Laser $^4F_{3/2} \rightarrow {}^4I_{11/2}$ and $^4F_{3/2} \rightarrow {}^4I_{13/2}$ transitions in $KY(WO_4)_2$, *IEEE J. Quantum Electron.* QE-8, 457 (1972).

252. Kaminskii, A. A., Klevtsov, P. V., Bagdasarov, Kh. S., Mayyer, A. A., Pavlyuk, A. A., Petrosyan, A. G., and Provotorov, M. V., New cw crystal lasers, *JETP Lett.* 16, 387 (1972).

253. Johnson, L. F. and Thomas, R. A., Maser oscillations at 0.9 and 1.35 microns in $CaWO_4:Nd^{3+}$, *Phys. Rev.* 131, 2038 (1963).

254. Morris, R. C., Cline, C. F., and Begley, R. F., Lanthanum beryllate: a new rare-earth ion host, *Appl. Phys. Lett.* 27, 444 (1975).

255. Jenssen. H. P., Begley, R. F., Webb, R., and Morris, R. C., Spectroscopic properties and laser performance of Nd^{3+} in lanthanum beryllate, *J. Appl. Phys.* 47, 1496 (1976).

256. Birnbaum, M. and Tucker, A. W., Nd-YALO Oscillation at 0.95 μm at 300 K, *IEEE J. Quantum Electron.* QE-9, 46 (1973).

257. Kaminskii, A. A., Agamalyan, N. R., Pavlyuk, A. A., Bobovich, L. I., and Lyubchenko, V. V., Preparation and luminescence-generation properties of $KLu(WO4)2-Nd^{3+}$, *Inorg. Mater. (USSR)* 19, 885 (1983).

258. Prokhorov, A. M., A new generation of solid-state lasers, *Sov. Phys. Usp.* 29, 3 (1986).

259. Zharikov, E. V., Zhltnyuk, V. A., Kuratev, I. I., Lapaev, V. V., Smirnov, V. A., Shestakov, A. V., and Shcherbakov, I A., Laser based on a $GSGG:Cr^{3+}$, Nd^{3+} crystal, which operates on the $^4F_{3/2} \rightarrow {}^4I_{9/2}$ transition at room temperature, in *Bull. Acad. Sci. USSR Phys. Ser.* 48, 98 (1984).

260. Scheps, R., Myers, J., Schimitschek, E. J., and Heller, D. F., End-pumped Nd:BEL laser performance, *Opt. Engineer.* 27, 830 (1988).

261. Moulton, P. F., Tunable paramagnetic-ion lasers, in *Laser Handbook*, Vol. 5, Bass, M., and Stitch, M. L., Eds., North-Holland, Amsterdam, The Netherlands (1985), p. 203

262. Fan, T. Y. and Byer, R. L., Continuous-wave operation of a room-temperature diode-laser-pumped 946-nm Nd:YAG laser, *Opt. Lett.* 12, 809 (1987).

263. Risk, W. P. and Lenth, W., Room-temperature, continuous-wave, 946-nm Nd:YAG laser pumped by laser-diode arrays and intracavity frequency doubling to 473 nm, *Opt. Lett.* 12, 993 (1987).

264. Fan, T. Y. and Byer, R. L., Modeling and CW operation of a quasi-three-level 946-nm Nd:YAG laser, *IEEE J. Quantum Electron.* QE-23, 605 (1987).

265. Birnbaum, M., Tucker, A. W., and Fincher, C. L., CW room-temperature laser operation of Nd:CAMGAR at 0.941 and 1.059 μ, *J. Appl. Phys.* 49, 2984 (1978).

266. Wallace, R. W. and Harris, S. E., Oscillation and doubling of the 0.946 μm line in Nd^{3+}:YAG, *Appl. Phys. Lett.* 15, 111 (1969).

267. Kaminskii, A. A., Bogomolova, G. A., Bagdasarov, Kh. S., and Petrosyan, A. C., Luminescence. absorption and stimulated emission of $Lu_3Al_5O_{12}$-Nd^{3+}- crystals, *Opt. Spectrosc.* 39, 643 (1975).

268. Morozov, A. M., Podkolzina, I. A., Tkachuk, A. M., Fedorov, V. A., and Feofilov, P. P., Luminescence and induced emission lithium-erbium and lithium-holmium binary fluorides, *Opt. Spectrosc.* 39, 338 (1975).

269. Christensen, H. P., Spectroscopic analysis of $LiHoF_4$ and $LiErF_4$, *Phys. Rev. B* 19, 6564 (1979).

270. Kaminskii, A. A., Ngoc, T., Sarklsov, S. E., Matrosov, V. N., and Timoshechkin, M. I., Growth, spectral and laser properties of $La_2Be_2O_5$:Nd^{3+} crystals in the $^4F_{3/2} \rightarrow {}^4I_{11/2}$ and $^4F_{3/2} \rightarrow {}^4I_{13/2}$ transitions, *Phys. Stat. Sol. A*: 59, 121 (1980).

271. Kolbe, W., Petermann, K., and Huber, G., Broadband emission and laser action of Cr^{3+} doped zinc tungstate at 1 μm wavelength. *IEEE J. Quantum Electron.* QE-21, 1596 (1985).

272. Kaminskii, A. A., Pavlyuk, A. A., Agamalyan, N. R., Sarkisov, S. E., Bobovich, L. I., Lukin, A. V., and Lyubchenko, V. V., Stimulated radiation of Nd^{3+} and Ho^{3+} ions in monoclinic $KLu(WO_4)_2$ crystals at room temperature, *Inorg. Mater. (USSR)* 15, 1649 (1979).

273. Otsuka, K., Nakano, J., and Yamada, T., Laser emission cross section of the system $LiNd_{0.5}M_{0.5}P_4O_{12}$ (M = Gd,La), *J. Appl. Phys.* 46, 5297 (1975).

274. Golubev, P. G., Kandaurov, A. S., Lazarev, V. V., and Safronov, E. K., Lasing properties of neodymium-activated lanthanum beryllate. *Sov. J. Quantum Electron.* 15, 1213 (1985).

275. Schaffers, K. I., DeLoach, L. D., and Payne, S. A., Crystal growth, frequency doubling, and infrared laser performance of Yb^{3+}:$BaCaBO_3F$. *IEEE J. Quantum Electron.* 32, 741 (1996).

276. Bogomolova, G. A., Vylegzhanin, D. N., and Karninskii, A. A., Spectral and lasing investigations of garnets with Yb^{3+} ions, *Sov. Phys. -JETP* 42, 440 (1976).

277. Lagatsky, A. A., Kuleshov, N. V., and Mikhailov, V. P., Diode-pumped CW lasing of Yb:KYW and Yb:KGW, *Opt. Commun.* 165, 71 (1999).

278. Johnson, L. F., Geusic, J. E., and Van Uitert, L. G., Coherent oscillations from Tm^{3+}, Ho^{3+}, Yb^{3+} and Er^{3+}- ions in yttrium aluminum garnet, *Appl. Phys. Lett.* 7, 127 (1965).

279. Bagdasarov, Kh. S., Kaminskii, A. A., Kevorkov, A. M., and Prokhorov, A. M. Rare earth scandium-aluminum garnets with impurity of TR^{3+} ions as active media for solid state lasers, *Sov. Phys. Dokl.* 19, 671 (1975).

280. Hanna, D. C., Jones, J. K., Large, A. C., Shepherd, D. P., Tropper, A, C., Chandler, P. J., Rodman, M. J., Townsend, P. D., and Zhang, L., Quasi-three level 1.03 μm laser operation of a planar ion-implanted Yb:YAG waveguide, *Opt. Commun.* 99, 211 (1993).

281. Lacovara, P., Choi, H. K., Wang, C. A., Aggarwal, R. L., and Fan, T. Y., Room-temperature diode-pumped Yb:YAG laser, *Optics. Lett.* 16, 1089 (1991).

282. Sugimoto, N., Ohishi, Y., Katoh, Y., Tate, A., Shimokozono, M., and Sudo, S., A ytterbium- and neodymium-codoped yttrium aluminum garnet-buried channel waveguide laser pumped at 0.81 μm, *Appl. Phys. Lett.* 67, 582 (1995).

283. Sumida, D. S., Fan, T. Y., and Hutcheson, R., Spectroscopy and diode-pumped lasing of Yb^{3+}-doped $Lu_3Al_5O_{12}$ (Yb:LuAG), *OSA Proc. Adv. Solid State Lasers*, Chai, B. H. T., and Payne, S. A., Eds., 24, 348 (1995).

284. Payne, S. A., Smith, L. K., DeLoach, L. D., Kway, W. L., Tassano, J. B., and Krupke, W. F., Laser, optical and thermomechanical properties of Yb-doped fluoroapatite. *IEEE J. Quantum Electron.* 30, 170 (1994).

285. Robinson, M. and Asawa, C. K., Stimulated emission from Nd^{3+} and Yb^{3+} in noncubic sites of neodymium- and ytterbium-doped CaF_2, *J. Appl. Phys.* 38, 4495 (1967).

286. Kaminskii, A. A., Mikaelyan, R. C., and Zigler, I. N., Room-temperature induced emission of CaF_2-SrF_2 crystals containing Nd^{3+}, *Phys. Stat. Sol.* 31, K85 (1969).

287. Stephens, E., Schearer, L. D., and Verdun, H. R., A tunable $Nd:CaYAlO_4$ laser, *Opt. Commun.* 90, 79 (1992).

288. Danielmeyer, H. G., Jeser, J. P., Schonherr, E., and Stetter, W., The growth of laser quality NdP_5O_{14} crystals, *J. Cryst. Growth* 22, 298 (1974).

289. Miyazawa, S. and Kubodera, K., Fabrication of $KNdP_4O_{12}$ laser epitaxial waveguide, *J. Appl. Phys.* 49, 6197 (1978).

290. Johnson, L. F., Optically pumped pulsed crystal lasers other than ruby, in *Lasers, A Series of Advances*, Vol. I, Levine, A. K., Ed., Marcel Dekker, New York (1966), p.137.

291. Johnson, L. F. and Ballman, A. A., Coherent emission from rare-earth ions in electro-optic crystals, *J. Appl. Phys.* 40, 297 (1969).

292. Vylegzhanin, D. N. and Kaminskii, A. A., Study of electron-phonon interaction in $LaF_3:Nd^{3+}$ crystals, *Sov. Phys.-JETP* 35, 361 (1972).

293. Voron'ko, Yu. K., Dmitruk, M. V., Kaminskii. A. A., Osiko, V. V., and Shpakov, V. N., CW stimulated emission in an $LaF_3:Nd^{3+}$ laser at room temperature, *Sov. Phys.-JETP* 27, 400 (1968).

294. O'Connor, J. R. and Hargreaves, W. A., Lattice energy transfer and stimulated emission from $CeF_3:Nd^{3+}$, *Appl. Phys. Lett.* 4, 208 (1964).

295. Dmitruk, M. V., Kaminskii, A. A., and Shcherbakov, I. A., Spectroscopic studies of stimulated emission from a $CeF_3:Nd^{3+}$ laser, *Sov. Phys.-JETP* 27 900 (1968).

296. Hoskins, R. H., and Soffer, B. H., Fluorescence and stimulated emission from $La_2O_3:Nd^{3+}$, *J. Appl. Phys.* 36, 323 (1965).

297. Evtuhov, V. and Neeland, J. K., A continuously pumped repetitively Q-switched ruby laser and applications to frequency-conversion experiments, *IEEE J. Quantum Electron.* QE-5, 207 (1969).

298. Chou, H., Alpers, P., Cassanho, A., and Jenssen, H.P., CW tunable laser emission of $Nd^{3+}:Na_{0.4}Y_{0.6}F_{2.2}$ in *Tunable Solid-State Lasers II*, Vol. 52, Budgor, A. B., Esterowitz, L., and DeShazer, L. G., Eds., Springer-Verlag, New York (1986), p. 322.

299. Oldberg, L. S. and Bradford, J. N., Passive mode locking and picosecond pulse generation in Nd:lanthanum beryllate, *Appl. Phys. Lett.* 28, 585 (1976).

300. Robinson, M. and Asawa, C. K., Stimulated emission from Nd^{3+} and Yb^{3+} in noncubic sites of neodymium- and ytterbium-doped CaF_2, *J. Appl. Phys.* 38, 4495 (1967).

301. Johnson, L. F., Optical maser characteristics of Nd^{3+} in CaF_2, *J. Appl. Phys.* 33, 756 (1962).

302. Kaminskii, A. A., Korniyenko, L. S., and Prokhorov, A. M., Spectral study of stimulated emission from Nd^{3+} in CaF_2, *Sov. Phys.-JETP* 21, 318 (1965).

303. Moulton, P. F., Fahey, R. E., and Krupke, W. F., Advanced solid-state lasers, in 1981 Laser Program Annual Report, George, E. V., Ed., Lawrence Livermore National Laboratory, Livermore, CA UCRL- 50021 (1982), p. 7, available from National Technical Information Service.

304. Moulton, P. F., Advances in tunable transition-metal lasers, *Appl. Phys. B* 28, 233 (1982).

305. Emmett, J. L., Krupke, W. F., and Trenholme, J. B., Future development of high-power solid-state laser systems, *Sov. J. Quantum Electron.* 13, 1 (1983).

306. Telle, H. R., Tunable CW laser oscillation of stoichiometric Nd-materials, in *Proc. Int. Conf. Lasers '85*, Wang. C. P., Ed., STS Press, McLean, VA (1986) 460.

307. Otsuka, K., Li, H., and Telle, H. R., CW Nd-lasers with broad tuning range, *Opt. Commun.* 63, 57 (1987).

308. Bagdasarov, Kh. S., Voron'ko, Yu. K., Kaminskii, A. A., Osiko, V. V., and Prokhorov, A. M., Stimulated emission in yttro-fluorite:Nd^{3+} crystals at room temperature, *Sov. Phys.-Crystallogr.* 10, 626 (1966).

309. Kaminskii, A. A., Osiko, V. V., Prokhorov, A. M., and Voron'ko, Yu. K., Spectral investigation of the stimulated radiation of Nd^{3+} in CaF_2-YF_3, *Phys. Lett.* 22, 419 (1966).

310. Yariv, A., Porto, S. P. S., and Nassau, K., Optical laser emission from trivalent praseodymium in calcium tungstate, *J. Appl. Phys.* 33, 2519 (1962).

311. DeLoach, L. D., Payne, S. A., Smith, L. K., Kway, W. L., and Krupke, W. F., Laser and spectroscopic properties of $Sr_5(PO_4)_3F$:Yb, *J. Opt. Soc. Am. B* 11, 269 (1994).

312. Marshall, C. D., Smith, L. K, Beach, R. J. et al., Diode-pumped ytterbium-doped $Sr_5(PO_4)_3F$ laser performance, *IEEE J. Quantum Electron.* 32, 650 (1996).

313. Collonjgues, R., Lejus, A. M., Thery, J., and Vivien, D., Crystal growth and characterization of new laser materials, *J. Cryst. Growth* 128, 986 (1993).

314. Barnes, N. P., Gettemy, D. J., Esterowitz, L., and Allen, R. E., Comparison of Nd 1.06 and 1.33 μm operation in various hosts, *IEEE J. Quantum Electron.* QE-23, 1434 (1987).

315. Pollak, T. M., Wing, W. F., Grasso, R. J., Chicklis, E. P., and Jenssen, H. P., CW laser operation of Nd:YLF, *IEEE J. Quantum Electron.* QE-18, 159 (1982).

316. Fan, T. Y., Dixon, G. J., and Byer, R. L., Efficient GaAlAs diode-laser pumped operation of Nd:YLF at 1.047 μm with intracavity doubling to 523.6 nm, *Opt. Lett.* 11, 204 (1986).

317. Zverev, G. M., Kuratev, I. I., and Shestakov, A. V., Solid-state microlasers based on crystals with a high concentration of neodymium ions, *Bull. Acad. Sci. USSR Phys. Ser.* 46, 108 (1982).

318. Bagdasarov, Kh. S., Dorozhkin, L. M., Kevorkov, A. M., Krasilov, Yu. I., Potemkin, A. V., Shestakov, A. V., and Kuratev, I. I., Continuous lasing in $La_{1-x}Nd_xMgAl_{11}O_{19}$ crystals, *Sov. J. Quantum Electron.* 13, 639 (1983).

319. Krühler, W. W., Plättner, R. D., and Stetter, W., CW oscillation at 1.05 and 1.32 μm of $LiNd(PO_3)_4$ lasers in external resonator and in resonator with directly applied mirrors, *Appl. Phys.* 20, 329 (1979).

320. Bagdasarov, Kh. S., Kaminskii, A. A., Kevorkov, A. M. et al., Laser properties of Y_2SiO_5-Nd^{3+} crystals irradiated at the $^4F_{3/2} \rightarrow {}^4I_{11/2}$ and $^4F_{3/2} \rightarrow {}^4I_{13/2}$ transitions, *Sov. Phys.-Dokl.* 18, 664 (1974).

321. Harmer, A. L., Linz, A., and Gabbe, D. R., Fluorescence of Nd^{3+} in lithium yttrium fluoride, *J. Phys. Chem. Sol.* 30, 1483 (1969).

322. Le Coff, D., Bettinger, A., and Labadens, A., Etude d'un oscillateur a blocage de modes utilisant un cristal de $LiYF_4$ dope au neodyme, *Opt. Commun.* 26, 108 (1978), in French.

323. Bagdasarov, Kh. S., Kaminskii, A. A., Kevorkov, A. M., and Prokhorov, A. M., Investigation of the stimulated radiation emitted by Nd^{3+} ions in $CaSc_2O_4$ crystals, *Sov. J. Quantum Electron.* 4, 927 (1975).

324. Bagdasarov, Kh. S., Kaminskii, A. A., Kevorkov, A. M. et al., Stimulated emission of Nd^{3+} ions in an $SrAl_2O_4$ crystal at the transitions $^4F_{3/2} \rightarrow {}^4I_{11/2}$ and $^4F_{3/2} \rightarrow {}^4I_{13/2}$, *Sov. Phys.-Dokl.* 19, 350 (1974).

325. Hoskins, R. H. and Soffer, B. H., Stimulated emission from Y_2O_3:Nd^{3+}, *Appl. Phys. Lett.* 4, 22 (1964).

326. Fan, T. Y. and Kokta, M. R., End-pumped Nd:LaF_3 and Nd:$LaMgAl_{11}O_{19}$ lasers, *IEEE J. Quantum Electron.* 25, 1845 (1989).

327. Weston, J., Chiu, P. H., and Aubert, R., Ultrashort pulse active passive mode-locked Nd:YLF laser, *Opt. Commun.* 61, 208 (1987).

328. Kaminskii, A. A., Stimulated-emission spectroscopy of Er^{3+} ions in cubic $(Y,Ln)_3Al_5O_{12}$ and monoclinic $K(Y,Ln)W_2O_8$ single crystals, *Phys. Stat. Sol. A* 96 K175 (1986).

329. Antipenko, B. M., Rab., O. B., Seiranyan, K. B., and Sukhnreva, L. K., Quasi-continuous lasing of an $LiYF_4$:Er:Pr crystal at 0.85 μm, *Sov. J. Quantum Electron.* 13, 1237 (1983).

330. Kaminskii, A. A., Mill', B. V., and Butashin, A. V., New low-threshold noncentrosymmetric $LaBGeO_5$:Nd^{3+} laser crystal, *Sov. J. Quantum Electron.* 20, 875 (1990).

331. Kaminskii, A.A., Mill', B.V., and Butashin, A.V., Stimulated emission from Nd^{3+} ions in acentric $LaBGeO_5$ crystals, *Phys. Stat. Solidi A* 115, K59 (1990).

332. Nakano, J., Kubodera, K., Miyazawa, S., Kondo, S., and Koizumi, H., $LiBi_xNd_{1-x}P_4O_{12}$ waveguide laser layer epitaxially grown on $LiNdP_4O_{12}$ substrate, *J. Appl. Phys.* 50, 6546 (1979).

333. Yamada, T., Otsuka, K., and Nakano, J., Fluorescence in lithium neodymium ultraphosphate single crystals, *J. Appl. Phys.* 45, 5096 (1974).

334. Hong, H. Y-P. and Chinn, S. R., Influence of local-site symmetry on fluorescence lifetime in high Nd-concentration laser materials, *Mater. Res. Bull.* 11, 461 (1976).

335. Nakano, J., Kubodera, K., Yarnada, T., and Miyazawa, S., Laser-emission cross sections of $MeNdP_4O_{12}$ (Me = Li,Na,K) crystals, *J. Appl. Phys.* 50, 6492 (1979).

336. Otsuka, K. and Yamada, T., Transversely pumped LNP laser performance, *Appl. Phys. Lett.* 26, 311 (1975).

337. Chinn, S. R. and Hong, H. Y-P., Low-threshold cw $LiNdP_4O_{12}$ laser, *Appl. Phys. Lett.* 26, 649 (1975).

338. Otsuka, K., Yamada, T., Saruwatari, M., and Kimura, T., Spectroscopy and laser oscillation properties of lithium neodymium tetraphosphate, *IEEE J. Quantum Electron.* QE-I I, 330 (1975).

339. Kaminskii, A. A., Pavlyuk, A. A., Butaeva, T. I., Bobovieh, L. I., and Lyuhchenko, V. V., Stimulated emission in the 2.8-μm band by a self-activated crystal of $KEr(WO_4)_2$, *Inorg. Mater. (USSR)* 15, 424 (1979).

340. Hemmati, H., 2.07-μm CW diode-laser-pumped Tm, Ho:$YLiF_4$ room-temperature laser, *Opt. Lett.* 14, 435 (1989).

341. Antipenko, B. M., Glebov, A. S., Krutova, L. I., Solntsev, V. M., and Sukhareva, L. K., Active medium of lasers operating in the 2-μm spectra range and utilizing gadolinium scandium gallium garnet crystals, *Sov. J. Quantum Electron.* 16, 995 (1986).

342. Shcherbakov, I. A., Optically dense active media for solid-state lasers, *IEEE J. Quantum Electron.* 24, 979 (1988).

343. D'yakanov, G. I., Egorov, G. N., Zharikov, E. V., Mlkhailov, V. A., Pak, S. K., Prokhorov, A. M., and Shcherbakov, I. A., Chromium- and neodymium-activated yttrium scandium gallium garnet laser with an efficiency of 3.6% emitting linearly polarized radiation of energy of 0.46 J per single pulse and a pulse repetition frequency 50 Hz, *Sov. J. Quantum Electron.* 18, 43 (1988).

344. Voron'ko, Yu. K., Kaminskii, A. A., Korniyenko, L. S., Osiko, V. V., Prokhorov, A. M., and Udoven'chik, V. T., Investigation of stimulated emission from CaF_2:Nd^{3+}- (Type II) crystals at room temperature, *JETP Lett.* 1, 39 (1965).

345. Korzhik, M. V., Livshits, M. G., Bagdasarov, Kh. S., Kevorkov, A. M., Melkonyan, T. A., and Mellman, M. L., Efficient pumping of Nd^{3+} ions via charge transfer bands of Fe^{2+} ions in YAG, *Sov. J. Quantum Electron.*, 19, 344 (1989).

346. Keen, S. J., Maker, G. T., and Ferguson, A. J., Mode-locking of diode laser-pumped Nd:YAG laser at 1.3 μm, *Electron. Lett.* 25, 490 (1989).

347. Khurgin, J. and Zwicker, W. K., High efficiency nanosecond miniature solid-state laser, *Appl. Opt.* 24, 3565 (1985).

348. Kaminow, I. P. and Stulz, L. W., Nd:$LiNbO_3$ laser, *IEEE J. Quantum Electron.* QE-11, 306 (1975).

349. Kubodera, K., Otsuka, K., and Miyazawa, S., Stable $LiNdP_4O_{12}$ miniature laser, *Appl. Opt.* 18, 844 (1979).

350. Kubodera, K. and Otsuka, K., Efficient $LiNdP_4O_{12}$ lasers pumped with a laser diode, *Appl. Opt.* 18, 3882 (1979).

351. Saruwatari, M., Kimura, T., Yamada, T., and Nakano, J., $LiNdP_4O_{12}$ laser pumped with an Al_xGa_{1-x} As electroluminescent diode, *Appl. Phys. Lett.* 27, 682 (1975).

352. Saruwatari, M. and Kimura, T., LED pumped lithium neodymium tetraphosphate lasers, *IEEE J. Quantum Electron.* QE-12, 584 (1976).

353. Kubodera, K. and Otsuka, K., Laser performance of a glass-clad $LiNdP_4O_{12}$ rectangular waveguide, *J. Appl. Phys.* 50, 6707 (1979).

354. Krühler, W. W., Plättner, R. D., and Stetter, W., CW oscillation at 1.05 μm and 1.32 μm of $LiNd(PO_3)_4$ lasers in external resonator and in resonator with directly applied mirrors, *Appl. Phys.* 20, 329 (1979).

355. Chinn, S. R., Zwicker, W. K., and Colak, S., Thermal behavior of NdP_5O_{14} lasers, *J. Appl. Phys.* 53, 5471 (1982).

356. Sarkisov, S. E., Lomonov, V. A., Kaminskii, A. A. et al., Spectroscopic investigation of the stable crystalline mixed system $Bi_4(Ge_{1-x}Si_x)_3O_{12}$, in *Abstracts of Papers of Fifth All-Union Symposium on Spectroscopy of Crystals*, Kazan (1976) p. 195 (in Russian).

357. Lempicki, A., McCollum, B. C., and Chinn, S. R., Spectroscopy and lasing in $K_5NdLi_2F_{10}$ (KNLF), *IEEE J. Quantum Electron.* QE-15, 896 (1979).

358. Danilov, A. A., Zharikov, E. V., Zagumennyl, A. I., Lutts, G. B., Nlkolskii, M. Yu., Tsvetkov, V. B., and Shcherhakov, I. A., Self-Q-switched high-power laser utilizing gadolinium scandium gallium garnet activated with chromium and neodymium, *Sov. J. Quantum Electron.* 19, 315 (1989).

359. Kaminskii, A. A. and Khaidukov, N. M., New low-threshold $LiKYF_5:Nd^{3+}$ laser crystal, *Sov. J. Quantum Electron.* 22, 193 (1992) and *Phys. Stat. Sol. A* 129, K65 (1992).

360. Comaskey, B., Albrecht, G. F., Beach, R. J., Moran, B. D., and Solarz, R. W., Flash-lamp-pumped laser operation of $Nd^{3+}:Y_2SiO_5$ at 1.074 µm, *Opt. Lett.* 18, 2029 (1993).

361. Viscakas, J. and Syrusas, V., Stimulated raman scattering self-conversion of laser radiation and the potential for creating multifrequency solid state lasers, *Sov. Phys. Collect.* 27, 31 (1987).

362. Dmitruk, M. V., Kaminskii, A. A., Osiko, V. V., and Tevosyan, T. A., Induced emission of hexagonal $LaF_3-SrF_2:Nd^{3+}$ crystals at room-temperature, *Phys. Stat. Sol.* 25, K75 (1968).

363. Bagdasarov, Kh. S., Kaminskii, A. A., Kevorkov, A. M. et al., Investigation of the stimulated emission of cubic crystals of $YScO_3$ with Nd^{3+} ions, *Sov. Phys.-Dokl.* 20, 681 (1975).

364. Eggleston, J. M., DeShazer, L. G., and Kangas, K. W., Characteristics and kinetics of laser-pumped Ti:Sapphire oscillators, *IEEE J. Quantum Electron.* 24, 1009 (1988).

365. Shand, M. L., Progress in alexandrite lasers, in *Proc. Int. Conf. Lasers 85*, Wang, C. P., Ed., STS Press, McLean, VA (1986), p. 732.

366. Kaminskii, A. A. and Sobolev, B. N., Monoclinic fluoride $BaY_2F_8-Nd^{3+}$—a new low-threshold inorganic laser materials, *Inorg. Mater. (USSR)* 19, 1718 (1983).

367. Kaminskii, A. A., Sarkisov, S. E., Seiranyan, K. B., and Sobolev, B. P., Study of stimulated emission in $Sr_2Y_5F_{19}$ crystals with Nd^{3+} ions, *Sov. J. Quantum Electron.* 4, 112 (1974).

368. Kaminskii, A. A., Petrosyran, A. G., Ovanesyan, K. L. et al., Concentrational 3 µm stimulated emission tuning in the $(Gd_{1-x}Er_x)_3Al_5O_{12}$ crystal system, *Phys. Stat. Solidi A* 82, K185 (1984).

369. Kaminskii, A. A., Agamalyan, N. R., Denisenko, G. A., Sarkisov, S. E., and Fedorov, P. P., Spectroscopy and laser emission of disordered $GdF_3-CaF_2:Nd^{3+}$ trigonal crystals, *Phys. Stat. Sol. A* 70, 397 (1982).

370. Kevorkov, A. M., Kaminskii, A. A., Bagdasarov, Kh. S., Tevosyan, T. A., and Sarkisov, S. E., Spectroscopic properties of $SrAl_2O_4:Nd^{3+}$ crystals, *Inorg. Mater. (USSR)* 9, 1637 (1973).

371. Alves, R. V., Buchanan, R. A., Wickersheim, K. A., and Yates, E. A., Neodymium-activated lanthanum oxysulfide: a new high-gain laser material, *J. Appl. Phys.* 42, 3043 (1971).

372. Soffer, B. H. and Hoskins, R. H., Fluorescence and stimulated emission from $Gd_2O_3:Nd^{3+}$ at room temperature and 77 K, *Appl. Phys. Lett.* 4, 113 (1964).

373. Chinn, S. R. and Hong, H. Y.-P., Fluorescence and lasing properties of $NdNa_5(WO_4)_4$, $K_3Nd(PO_4)_2$ and $Na_3Nd(PO_4)_2$, *Opt. Commun.* 18, 87 (1976).

374. Kaminskii, A. A., Zhmurova, Z. I., Lomonov, V. A., and Sarkisov, S.E., Two stimulated emission $^4F_{3/2} \rightarrow ^4I_{11/2,13/2}$ channels of Nd^{3+} ions in crystals of the CaF_2-ScF_3 system, *Phys. Stat. Sol. A* 84, K81 (1984).

375. Marling, J. B., 1.05-1.44 µm tunability and performance of the CW $Nd^{3+}:YAG$ laser, *IEEE J. Quantum Electron.* QE-14, 56 (1978).

376. Chinn, S. R. and Zwicker, W. K., A comparison of flash-lamp-excited $Nd_xLa_{1-x}P_5O_{14}(x = 1.0, 0.75, 0.20)$ lasers, *J. Appl. Phys.* 52, 66 (1981).

377. Bagdasarov, Kh. S. and Kaminskii, A. A., RE^{3+}-doped $YAlO_3$ as an active medium for lasers, *JETP Lett.* 9, 303 (1969).

378. Huber, G., Krühler, W. W., Bludau, W., and Danielmeyer, H. G., Anisotropy in the laser performance of NdP_5O_{14}, *J. Appl. Phys.* 46, 3580 (1975).

379. Bagdasarov, Kh. S., Kaminskii, A. A., and Sobolev, B. P., Laser based on $5NaF\cdot 9YF_3:Nd^{3+}$ cubic crystals, *Sov. Phys.-Crystallogr.* 13, 779 (1969).

380. Kaminskii, A. A., Markosyan, A. A., Pelevin, A. V., Polyakova, Yu. A., and Uvarova, T. V., Single-crystal $Cd_{1-x}Sc_xF_{2+x}$, with Nd^{3+} ions and its stimulated emission, *Inorg. Mater. (USSR)* 22, 777 (1986).

381. Nakano, J., Otsuka, K., and Yamada, T., Fluorescence and laser-emission cross sections in $NaNdP_4O_{12}$, *J. Appl. Phys.* 47, 2749 (1976).

382. Otsuka, K., Miyazawa, S., Yamada, T., Iwasaki, H., and Nakano, J., CW laser oscillations in $MeNdP_4O_{12}$ (Me = Li,Na,K) at 1.32 μm, *J. Appl. Phys.* 48, 2099 (1977).

383. Belabaev, K. C., Kaminskii, A. A., and Sarkisov, S. E., Stimulated emission from ferroelectric $LiNbO_3$ crystals containing Nd^{3+} and Mg^{2+}- ions, *Phys. Stat. Sol.* 28a, K17 (1975).

384. Gualtieri, J. G. and Aucoin, T. R., Laser performance of large Nd-pentaphosphate crystals, *Appl. Phys. Lett.* 28, 189 (1976).

385. Szymanski, M., Karolczak, J., and Kaczmarek, F., Temporal studies of intensity dependent laser and spontaneous emission in $NdLaP_5O_{14}$ monocrystals, *Acta Phys. Pol. A* A60 95 (1981).

386. He, N. J., Lu, G. X., Li, Y. C., and Zhao, L. X., A CW $La_{0.1}Nd_{0.9}P_5O_{14}$ laser at 1.051 and 1.32 μm, *Chin. Phys.* 2, 455 (1982).

387. Winzer, G., Mockd, P. G., Oberbacher, R., and Vite, L., Laser emission from polished NdP_5O_{14} crystals with directly applied mirrors, *Appl. Phys.* 11, 121 (1976).

388. Grigor'yanb, V. V., Makovetskil, A. A., and Tishchenko, R. P., Kinetics of emission from a neodymium pentaphosphate microlaser pumped by short pulses, *Sov. J. Quantum Electron.* 10, 1286 (1980).

389. Liu, J., Wang, M., Zhao, X., Liang, Y., Wang, B., and Lu, B., A miniature pulsed NdP_5O_{14} laser, *Chin. Phys. Lasers* 14, 45 1987.

390. Weber, H. P., Damen, T. C., Danielmeyer, H. G., and Tofield, B. C., Nd-ultraphosphate laser, *Appl. Phys. Lett.* 22, 534 (1973).

391. Damen, T. C., Weber, H. P., and Tofield, B. C., NdLa pentaphosphate laser performance, *Appl. Phys. Lett.* 23, 519 (1973).

392. Danielmeyer, H. G., Huber, G., Krühler, W. W., and Jeser, J. P., Continuous oscillation of a (Sc,Nd) pentaphosphate laser with 4 milliwatts pump threshold, *Appl. Phys.* 2, 335 (1973).

393. Chinn, S. R., Pierce, J. W., and Heckscher, H., Low-threshold, transversely excited NdP_5O_{14} laser, *IEEE J. Quantum Electron.* QE-11, 747 (1975).

394. Weber, H. P., Liao, P. F., Tofield, B. C., and Bridenbaugh, P. M., CW fiber laser of NdLa pentaphosphate, *Appl. Phys. Lett.* 26, 692 (1975).

395. Chinn, S. R., Intracavity second-harmonic generation in a Nd pentaphosphate laser, *Appl. Phys. Lett.* 29, 176 (1976).

396. Chinn, S. R. and Zwicker, W. K., FM mode-locked $N_{0.5},La_{0.5}P_5O_{14}$ laser, *Appl. Phys. Lett.* 34, 847 (1979).

397. Budin, J. -P., Neubauer, M., and Rondot, M., Miniature Nd-pentaphosphate laser with bonded mirrors side pumped with low-current-density LED's, *Appl. Phys. Lett.* 33, 309 (1978).

398. Budin, J. -P., Neubauer, M., and Rondot, H., On the design of neodymium miniature lasers, *IEEE J. Quantum Electron.* QE-14, 831 (1978).

399. Krühler, W. W., Huber, G., and Danielmeyer, H. G., Correlations between site geometries and level energies in the laser system $Nd_xY_{1-x}P_5O_{14}$, *Appl. Phys.* 8, 261 (1975).

400. Krühler, W. W., and Plättner, R. D., Laser emission of (Nd,Y)-pentaphosphate at 1.32 μm, *Opt. Commun.* 28, 217 (1979).

401. Krühler, W. W., Jeser, J. P., and Danielmeyer, H. G., Properties and laser oscillation of the (Nd,Y) pentaphosphate system, *Appl. Phys.* 2, 329 (1973).

402. Gualtieri, J. G. and Aucoin, T. R., Laser performance of large Nd-pentaphosphate crystals, *Appl. Phys. Lett.* 28, 189 (1976).

403. Krühler, W. W., Plättner, R. D., Fabian, W., Mockel, P., and Grabmaier, J. G., Laser oscillation of $N_{0.14}Y_{0.86}P_5O_{14}$ layers epilaxially grown on $Gd_{0.33}Y_{0.67}P_5O_{14}$ substrates, *Opt. Commun.* 20, 354 (1977).

404. Gueugnon, C. and Budin, J. P., Determination of fluorescence quantum efficiency and laser emission cross sections of neodymium crystals: application to $KNdP_4O_{12}$, *IEEE J. Quantum Electron.* QE 16, 94 (1980).

405. Chinn, S. R. and Hong, H. Y-P., CW laser action in acentric $NdAl_3(BO_3)_4$ and $KNdP_4O_{12}$, *Opt. Commun.* 15, 345 (1975).

406. Kubodera, K., Miyazawa, S., Nakano, J., and Otsuka, K., Laser performance of an epitaxially grown $KNdP_4O_{12}$ waveguide, *Opt. Commun.* 27, 345 (1978).

407. Badalyan, A. A., Sapondzhyan, S. O., Sarkisyan, D. G., and Torosyan, G. A., Mode-locked Nd:YAG laser with output at 1052, 1061, 1064, and 1074 nm, *Sov. Tech. Phys. Lett.* 11, 513 (1985).

408. DiDomenico, M., Jr., Geusic, J. E., Marcos, H. M., and Smith, R. G., Generation of ultrashort optical pulses by mode locking the YAlG:Nd laser, *Appl. Phys. Lett.* 8, 180 (1966).

409. Osterink, L. M. and Foster, J. D., A mode-locked Nd:YAG laser, *J. Appl. Phys.* 39, 4163 (1968).

410. Clobes, A. R. and Brienza, M. J., Passive mode locking of a pulsed Nd:YAG laser, *Appl. Phys. Lett.* 14, 287 (1969).

411. Dewhurst, R. J. and Jacoby, D., A mode-locked unstable Nd:YAG laser, *Opt. Commun.*, 28, 107 (1979).

412. Reali, G. C., Operation of a mode-locked Nd:YAG oscillator at 1.05 μm, *Appl. Optics* 18, 3975 (1979).

413. Kaminskii, A. A., Sobolev, B. P., Bagdasarov, Kh. S. et al., Investigation of stimulated emission from crystals with Nd^{3+} ions, *Phys. Stat. Solidi* 23A, K135 (1974).

414. Smith, R. G., New room temperature CW laser transition in YAlG:Nd, *IEEE J. Quantum Electron.*, QE-4, 505 (1968).

415. Dubinskii, M. A., Kazakov, B. N., and Yagudin, Sh. I., Spectroscopy and stimulated emission of Nd^{3+} ions in yttrium trifluoride crystals, *Opt. Spectrosc. (USSR)* 63, 412 (1987).

416. Tucker, A. W., Birnbaum, M., and Fincher, C. L., Repetitive Q-switched operation of x-axis $Nd:La_2Be_2O_5$, *J. Appl. Phys.* 52, 5434 (1981).

417. Kaminskii, A. A., Kursten, G. D., and Shultze, D., A new laser ferroelectric, $Pb_5Ge_3O_{17}$-Nd^{3+}, *Sov. Phys. Dokl.* 28, 492 (1983).

418. Kaminskii, A. A., Kirsten, H. D., and Schultze, D., Stimulated emission of ferroelectric $Pb_5Ge_3O_{17}:Nd^{3+}$, *Phys. Stat. Sol. A* 81, K19 (1984).

419. Gualtieri, J. G. and Aucoin, T. R., Laser performance of large Nd-pentaphosphate crystals, *Appl. Phys. Lett.* 28, 189 (1976).

420. Korovkin, A. M., Morozova, L. G., Petrov, M. V., Tkachuk, A. M., and Feofilov, P. P., Spontaneous and induced emission of neodymium in the crystals of yttrium silicates and rare-earth silicates, *Digest of Technical Papers VI All-Union Conf. on Spectroscopy of Crystals*, September 21-25, 1979, Krasnodar, p. 156.

421. Kaminskii, A. A., Sarkisov, S. E., Seiranyan, K. B., and Sobolev, B. P., Stimulated emission from Nd^{3+}- ions in SrF_2-GdF_3 crystals, *Inorg. Mater. (USSR)* 9, 310 (1973).

422. Verdun, H.R. and Thomas, L.M., $Nd:CaYAlO_4$–A new crystal for solid-state lasers emitting at 1.08 μm, *Appl. Phys. Lett.* 56, 608 (1990).

423. Shen, H., Zhou, Y., Yu, G., Huang, X., Wu, C., and Ni, Y., Influences of thermal effects on high power CW outputs of b-axis Nd:YAP lasers, *Chin. Phys.* 3, 45 (1983).

424. Hanson, F., Laser-diode side-pumped $Nd:YAlO_3$ laser at 1.08 and 1.34 μm, *Opt. Lett.* 14, 674 (1989).

425. Kaminskii, A. A., Sobolev, B. R., Zhmurova, Z. R., and Sarkisov, S. E., Generation of stimulated radiation by Nd^{3+} ions in the disordered crystal of a solid solution of the system CaF_2—LuF_3, *Inorg. Mater. (USSR)* 20, 759 (1984).

426. Souriau, J. C., Romero, R., Borel, C., and Wyon, C., Room-temperature diode-pumped continuous-wave $SrY_4(SiO_4)_3O:Yb^{3+}$, Er^{3+} crystal laser at 1554 nm, *Appl. Phys. Lett.* 64, 1189 (1994).

427. Tsuiki, H., Masumoto, T., Kitazawa, K., and Fueki, K., Effect of point defects on laser oscillation properties of Nd-doped Y_2O_3, *Jpn. J. Appl. Phys.* 21, 1017 (1982).

428. Voron'ko, Yu. K., Gessen, S. B., Es'kov, N. A., Osiko, V. V., Sobol', A. A., Tlmoshechkin, M. I., Ushakov, S. N., and Tsymbal, L. I., Spectroscopic and lasing properties of calcium niobium gallium garnet activated with Cr^{3+} and Nd^{3+}, *Sov. J. Quantum Electron.* 18, 198 (1988).

429. Vanherzeele, H., Optimization of a CW mode-locked frequency-doubled Nd:LiYF4 laser, *Appl. Opt.* 27, 3608 (1988).

430. Loth, C. and Bruneau, D., Single-frequency active-passive mode-locked Nd:YLF oscillator at 1.053 μm, *Appl. Opt.* 21, 2091 (1982).

431. Schearer, L. D., Leduc, M., Vlvien, D., Lejus, A. M., and Thery, J., LNA: A new CW Nd laser tunable around 1.05 and 1.08 μm, *IEEE J. Quantum Electron.* QE-22, 713 (1986).

432. Vivien, D., Lejus, A. M., Thery, J., Collongues, R., Aubert, J. J., Moncorgé, R., and Auzel, F., Observation of the continuous laser effect in $La_{0.9}Nd_{0.1}MgAl_{11}O_{19}$ aluminate single crystals (LNA) grown by the Czochralski method, *C. R. Seances Acad. Sci. Ser.* 11, 298, 195 (1984).

433. Kaminskii, A. A., Mill, B. V., Belokoneva, E. L., and Butashin, A. V., Structure refinement and laser properties of orthorhombic chromium-containing $LiNbGeO_5$ crystals, *Neorgan. Mater.* 27, 1899 (1991).

434. Kaminskii, A. A., Sobolev, B. P., Bagdasarov, Kh. S. et al., Investigation of stimulated emission of the $^4F_{3/2} \rightarrow {}^4I_{13/2}$ transition of Nd^{3+} ions in crystals, *Phys. Stat. Sol.* 26A, K63 (1974).

435. Kaminskii, A. A., On the possibility of investigation of the 'Stark' structure of TR^{3+} ion spectra in disordered fluoride crystal systems, *Sov. Phys.-JETP* 31, 216 (1970).

436. Kaminskii, A. A., Klevtsov, P. V., Bagdararov, Kh. S., Maier, A. A., Pavlyuk, A. A., Petrosyan, A. G. and Provotorov, M. V., New CW crystal lasers, *JETP Lett.* 16, 387 (1972).

437. Kvapil, J., Perner, B., Kvapil, J., Manek, B., Hamal, K., Koselja, M., and Kuhecek, V., Laser properties of coactivated YAP:Nd free of colour centers, *Czech. J. Phys. B* 38, 1281 (1988).

438. Kaminskii, A. A., Dsiko, V. V., and Voron'ko, Yu. K., Five-component fluoride: a new laser material, *Sov. Phys.-Crystallogr.* 13, 267 (1968).

439. Kaminskii, A. A., On the laws of crystal-field disorder of Ln^{3+} ions in insulating crystals, *Phys. Stat. Sol. A* 102, 389 (1987).

440. Kaminskii, A. A., Sobolev, B. P., Bagdararov, Kh. S., Tkachenko, N. L., Sarkisov, S. E., and Seiranyan, K. B., Investigation of stimulated emission from crystals with Nd^{3+} ions, *Phys. Stat. Sol. (a)* 23, K135 (1974).

441. Portella, M. T., Montelmacher, P., Bourdon, A., Evesque, P., Duran, J., and Boltz, J. C., Characteristics of a Nd-doped yttrium-aluminum-perovskite picosecond laser, *J. Appl. Phys.* 61, 4928 (1987).

442. Chen, L., Chen, S., and Xie, Z., A passively mode-locked Nd:YAP laser, *Laser J. (China)* 8, 4 (1981), (in Chinese).

443. Honda, T., Kuwano, T., Masumoto, T., and Shiroki, K., Laser action of pulse-pumped Nd^{3+}:$Gd_3Ga_5O_{12}$ at 1.054 μm, *J. Appl. Phys.* 51, 896 (1980).

444. Hamel, J., Cassimi, A., Abu-Safia, H., Leduc, M., and Schearer, L. D., Diode pumping of LNA lasers for helium optical pumping, *Opt. Commun.* 63, 114 (1987).

445. Kaminskii, A. A. and Lomonov, V. A., Stimulated emission of $M_{1-x}Nd_xF_{2+x}$ solid solutions with the structure of fluorite. *Inorg. Mater. (USSR)* 20, 1799 (1984).

446. Kaminskii, A. A. and Lomonov, V. A., Low-threshold stimulated emission of Nd^{3+} ions in disordered SrF_2-ScF_3 crystals, *Sov. Phys. Dokl.* 30, 388 (1985).

447. Alam, M., Gooen, K. H., DiBartolo, B., Linz, A., Sharp, E., Gillespie, L. F., and Janney, G., Optical spectra and laser action of neodymium in a crystal $Ba_2MgGe_3O_{12}$, *J. Appl. Phys.* 39, 4728 (1968).

448. Horowitz, D. J., Gillespie, L. F., Miller, J. E., and Sharp, E. J., Laser action of Nd^{3+} in a crystal $Ba_2ZnGe_3O_{12}$, *J. Appl. Phys.* 43, 3527 (1972).

449. Dmitriev, V. G., Raevskii, E. V., Rubina, N. M., Rashkovieh, L. N., Silichev, O. O., and Fomichev, A. A., Simultaneous emission at the fundamental frequency and the second harmonic in an active nonlinear medium: neodymium-doped lithium metaniobate, *Sov. Tech. Phys. Lett.* 5, 590 (1979).

450. Bagdasarov, Kh. S., Dorozhkin, L. M., Ermakov, L. A., Kevorkov, A. M., Krasilov, Yu. I., Kuznetsov, N. T., Kurstev, I. I., Potemkin, A. V., Ralskaya, L. N., Tseitlin, P. A., and Shestakov, A. V., Spectroscopic and lasing properties of lanthanum neodymium magnesium hexaaluminate, *Sov. J. Quantum Electron.* 13, 1082 (1983).

451. Demehouk, M. I., Mikhailov, V. P., Gilev, A. K., Zabaznov, A. M., and Shkadarevieh, A. P., Investigation of the passive mode locking in a La-Nd-Mg hexaaluminate-doped laser, *Opt. Commun.* 55, 33 (1985).

452. Dubinskii, M. A., Khaidukov, N. M., Garipov, I. G., Naumov, A. K., and Semashko, V. V., Spectroscopy and stimulated emission of Nd^{3+} in an acentric CsY_2F_7 host, *Appl. Optics* 31, 4158 (1992).

453. Karapetyan, V. E., Korovkin, A. M., Morozova, L. G., Petrov. M. V., and Feofilov, P. P., Luminescence and stimulated emission of neodymium ions in scandium silicate single crystals, *Opt. Spectrosc. (USSR)* 49, 109 (1980).

454. Hong, H. Y-P. and Chinn, S. R., Crystal structure and fluorescence lifetime of potassium neodymium orthophosphate, $K_3Nd(PO_4)_2$, a new laser material, *Mater. Res. Bull.* 11, 421 (1976).

455. Garmash, V. M., Kaminskii, A. A., Polyakov, M. I., Sarkisov, S. E., and Filimonov, A. A., Luminescence and stimulated emission of Nd^{3+} ions in $LaMgAl_{11}O_{19}$ crystals in the $^4F_{3/2} \rightarrow {}^4I_{11/2}$ and $^4F_{3/2} \rightarrow {}^4I_{13/2}$ transitions, *Phys. Stat. Sol. A* 75, K111 (1983).

456. Morozov, A, M., Morozova, L. G., Fedorov, V. A., and Feofilov, P. P., Spontaneous and stimulated emission of neodymium in lead fluorophosphate crystals, *Opt. Spectrosc.*, 39, 343 (1975).

457. Johnson, L. F. and Ballman, A. A., Coherent emission from rare-earth ions in electro-optic crystals, *J. Appl. Phys.* 40, 297 (1969).

458. Blistanov, A. A., Gabgan, B. I., Denker, B.I., Ivleva, L. I., Osiko, V. V., Polozkov, N. M., and Sverchkov, Yu. E., Spectral and lasing characteristics of $CaMoO_4$:Nd^{3+} single crystals, *Sov. J. Quantum Electron.* 19, 747 (1989).

459. Arsenev, P. A., Bienert, K. E., and Sviridova, R. K., Spectral properties of neodymium ions in the lattice of $GdScO_3$ crystals, *Phys. Stat. Sol.* (a) 9, K103 (1972).

460. Bagdararov, Kh. S., Kaminskii, A. A., Kevorkov, A. M., and Prokhorov, A. M., Rare earth scandium-aluminum garnets with impurity TR^{3+} ions as active media for solid state lasers, *Sov. Phys. Dokl.* 19, 671 (1975).

461. Abdulsabirov, R. Yu., Dubinskii, M. A., Kazakov, B. N., Silkin, N. I., and Yagudln, Sh. I., New fluoride laser matrix, *Sov. Phys. Crystallogr.* 32, 559 (1987).

462. Kaminskii, A. A., Khadokov, N. M., Joubert, M. F., Boulin, G., and Makou, R., $CsGd_2F_7$:Nd^{3+} - a new laser crystal, *Phys. Stat. Sol. A* 142, K51 (1994).

463. Johnson, L. F., Optical maser characteristics of rare-earth ions, *J. Appl. Phys.* 34, 897 (1963).

464. Kaminskii, A. A., Sobolev, B. P., Bagdararov, Kh. S., Tkachenko, N. L., Sarkisov, S. E., and Seiranyan, K. B., Investigation of stimulated emission from crystals with Nd^{3+} ions, *Phys. Stat. Sol.* (a) 23, K135 (1974).

465. Kaminskii, A. A., Sobolev, B. P., Bagdararov, Kh. S., Kevorkov, A. M.. Fedorov, P. P., and Sarkisov, S. E., Investigation of stimulated emission in the $^4F_{3/2} \rightarrow {}^4I_{13/2}$ transition of Nd^{3+} ions in crystals (VII), *Phys. Stat. Sol.* (a) 26, K63 (1974).

466. Tkachuk, A. M., Przhevusskii, A. K., Morozova, L. G., Poletimova, A. V., Petrov, M. V., and Korovkin, A. M., Nd^{3+} optical centers in lutetium, yttrium, and scandium silicate crystals and their spontaneous and stimulated emission, *Opt. Spectrosc. (USSR)* 60, 176 (1986).

467. Kaminskii, A. A., Mill', B. V., Butashin, A. V., Belokoneva, E. L., and Kurbanov, K., Germanates with $NdAlGe_2O_7$-type structure, *Phys. Stat. Sol. A* 103, 575 1987.

468. Garashina, L. S., Kaminskii, A. A., Li, L., and Sobolev, B. P., Laser based on SrF_2-YF_3:Nd^{3+} cubic crystals, *Sov. Phys.-Crystallogr.* 14, 799 (1970).

469. Antipenko, B. M., Mak, A. A., Sinitsin, B. E., and Uvarova, T. V., Laser converter on the base of $BaYb_2F_8$:Ho^{3+}, *Digest of Technical Papers VI All-Union Conf. on Spectroscopy of Crystals*, September 21-25 (1979), Krasnodar, p. 30.

470. Kaminskii, A. A., Mill', B. V., Kurbanov, I., and Butashin, A. V., Concentration quenching of luminescence and stimulated emission of Nd^{3+} in a monoclinic $NdGaGe_2O_7$ crystal, *Inorg. Mater. (USSR)* 23, 530 (1987).

471. Cordova-Plaza, A., Fan, T. Y., DiRonnet, M. J. F., Byer, R. L., and Shaw, H. J., Nd:MgO:$LiNbO_3$ continuous-wave laser pumped by a laser diode, *Opt. Lett.* 13, 209 (1988).

472. Kaminskii, A. A., Mill', B. V., Butashin, A. V., and Dosmagambetov, E. S., Stimulated emission of $GdGaGe_2O_7$-Nd^{3+} crystals, *Inorg. Mater. (USSR)* 23, 626 (1987).

473. Kevorkov, A. M., Kaminskii, A. A., Bagdararov, Kh. S., Tevosyan, T. A., and Sarkisov, S. E., Spectroscopic properties of crystals of $SrAl_4O_7$ - Nd^{3+}, *Inorg. Mater. (USSR)* 9, 1637 (1973).

474. Duncan, R. C., Continuous room-temperature Nd^{3+}:$CaMoO_4$ laser, *J. Appl. Phys.* 36, 874 (1965).

475. Kaminskii, A. A., Pavlyuk, A. A., Kurbanov, K., Ivannikova, N. V., and Polyakova, L. A., Growth of $CsLa(WO_4)_2$-Nd^{3+} crystals and study of their spectral-generation properties, *Inorg. Mater. (USSR)* 24, 1144 (1988).

476. Peterson, G. E. and Bridenbaugh, P. M., Laser oscillation at 1.06 μ in the series $Na_{0.5}Gd_{0.5-x}Nd_xWO_4$, *Appl. Phys. Lett.* 4, 173 (1964).

477. Kaminskii, A. A., Bagdasarov, Kh. S., Bogomolova, G. A. et al., Luminescence and stimulated emission of Nd^{3+} ions in $Gd_3Sc_2Ga_5O_{12}$ crystals, *Phys. Stat. Sol.* 34a, K109 (1976).

478. Kaminskii, A. A., Sarkisov, S. E., and Bagdasarov, Kh. S., Study of stimulated emission from Nd^{3+} ions in crystals at the $^4F_{3/2} \rightarrow {}^4I_{13/2}$ transition. II, *Inorg. Mater. (USSR)* 9, 457 (1973).

479. Flournoy, P. A. and Brixner, L. H., Laser characteristics of niobium compensated $CaMoO_4$ and $SrMoO_4$, *J. Electrochem. Soc.* 112, 779 (1965).

480. Cordova-Plaza, A., Digonnet, M. J. F., and Shaw, H. J., Miniature CW and active internally Q-switched Nd:MgO:$LiNbO_3$ lasers, *IEEE J. Quantum Electron.* QE-23, 262 (1987).

481. Demchuk, M. I., Mikhallov, V. P., Gilev, A. K., Ishchenko, A. A., Kudinova, M. A., Slominskii, Yu. L., and Tolmachev, A. I., Optimization of the passive mode-locking state in an yttrium aluminate laser, *J. Appl. Spectrosc. (USSR)* (Engl. Transl.) 42, 477 (1985).

482. Kaminskii, A. A., Mill', B. V., Tamazyan, S. A., Sarkisov, S. E., and Kurbanov, K., Luminescence and stimulated emission from an acentric crystal of $Ca_3Ga_4O_9$-Nd^{3+}, *Inorg. Mater. (USSR)* 21, 1733 (1986).

483. Johnson, L. F., Boyd, G. D., Nassau, K., and Soden, R. R., Continuous operation of a solid-state optical maser, *Phys. Rev.* 126, 1406 (1962).

484. Johnson, L. F., Characteristics of the $CaWO_4$:Nd^{+3} optical maser, in *Quantum Electronics Proceedings of the Third International Congress*, Grivet, P. and Bloembergen, N., Eds., Columbia University Press, New York (1964), p. 1021.

485. Kaminskii, A. A., Korniyenko, L. S., Maksimova, G. V., Osiko, V. V., Prokhorov, A. M., and Shipulo, G. P., CW $CaWO_4$:Nd^{3+} laser operating at room temperature, *Sov. Phys.-JETP* 22, 22 (1966).

486. Kaminskii, A. A., Spectral composition of stimulated emission from a $CaWO_4$:Nd^{3+} laser, *Inorg. Mater. (USSR)* 6, 347 (1970).

487. Hopkins, R. H., Steinbruegge, K. B., Melamed, N. T. et al., Technical RPT. AFAL-TR-69-239, Air Force Avionics Laboratory (1969).
488. Johnson, L. F. and Guggenheim, H. J., Phonon-terminated coherent emission from V^{2+}-ions in MgF_2, *J. Appl. Phys.* 38, 4837 (1967).
489. Mougel, F., Aka, G., Kahn-Hararı, A., Hubert, H., Benitez, J. M., and Vivien, D., Infrared laser performance and self-frequency doubling of Nd^{3+}:$Ca_4GdO(BO_3)_3$ (Nd:GdCOB), *Opt. Mater.* 8, 161 (1997).
490. Bagdasarov, Kh. S., Bogomolova, G. A., Gritsenko, M. M., Kaminskii, A. A., Kevorkov, A. M., Prokhorov, A. M., and Sarkisov, S. E., Spectroscopy of stimulated emission from $Gd_3Ga_5O_{12}$:Nd^{3+} crystals, *Sov. Phys.-Dokl.* 19, 353 (1974).
491. Karninskii, A. A., *Laser Crystals,* Springer-Verlag, New York (1980).
492. Geusic, J. E., Marcos, H. M., and Van Uitert, L. G., Laser oscillations in Nd-doped yttrium aluminum, yttrium gallium and gadolinium garnets, *Appl. Phys. Lett.* 4, 182 (1964).
493. Danilov, A. A., Zharikov, E. V., Zavartsev, Yu. D., Noginov, M. A., Nikol'skii, M. Yu., Ostroumov, V. G., Smirnov, V. A., Studenikin, P. A., and Shcherbakov, I. A., YSGG:Cr^{3+}:Nd^{3+} as a new effective medium for pulsed solid-state lasers, *Sov. J. Quantum Electron.* 17, 1048 (1987).
494. Tucker, A.W. and Birnbaum, M., Energy levels and laser action in Nd:$CaY_2Mg_2Ge_3O_{12}$ (CAMGAR), *Proc. Internat. Conf. Laser '78,* Orlando (STS Press, McLean, VA, 1979), p 168.
495. Bagdarasov, Kh. S. and Kaminskii, A. A., $YAlO_3$ with TR^{3+} ion impurity as an active laser medium, *JETP Lett.* 9, 303 (1969).
496. Kaminskii, A. A., Mayer, A. A., Provotorov, M. V., and Sarkisov, S. E., Investigation of stimulated emission from $LiLa(MoO_4)_2$:Nd^{3+} crystal laser, *Phys. Stat. Sol.* 17a, K115 (1973).
497. Steinbruegge, K. B., Hennigsen, R. H., Hopkins, R., Mazelsky, R., Melamed, N. T., Riedd, E. P., and Roland, G. W., Laser properties of Nd^{3+} and Ho^{3+} doped crystals with the apatite structure, *Appl. Optics* 11, 999 (1972).
498. Kaminskii, A. A., Klevtsov, P. V., Li, L. et al., Stimulated emission of radiation by crystals of $KLa(MoO_4)_2$ with Nd^{3+} ions, *Inorg. Mater. (USSR)* 9, 1824 (1973).
499. Faure, N., Borel, C., Templier, R., Couchaud, M., Calvat, C., and Wyon, C., Optical properties and laser performance of neodymium doped fluoroapatites $Sr_xCa_{5-x}(PO_4)_3F$ (x= 0, 1, 2, 3, 4, and 5), *Opt. Mater.* 6, 293 (1996).
500. Kariss, Ya. E., Tolstoy, M. N., and Feofilov, P. P., Stimulated emission from neodymium in lead molybdate single crystals, *Opt. Spectrosc.* 18, 99 (1965).
501. Kaminskii, A. A., Kozeeva, L. P., and Pavlyuk, A. A., Stimulated emission of Er^{3+} and Ho^{3+} ions in $KLa(Mo_4)_2$ crystals, *Phys. Stat. Sol. A:* 83, K65 (1984).
502. Zhang, L., Liu, L., Liu, H., and Lin, C., Growth and investigation of substituted gadolinium gallium garnet laser crystals, *J. Cryst. Growth* 80, 257 (1987).
503. Xun, D., Zhu, H., Jin, F., Liu, H., and Zhang, L., Growth and measurement of GGG(Ca Mg,Zr):(Nd Cr) laser crystals, *Chin. Phys. Lasers* 13, 820 (1986).
504. Bagdasarov, Kh. S., Kaminskii, A. A., Lapsker, Ya. Ye., and Sobolev, B. P., Neodymium-doped α-gagarinite laser, *JETP Lett.* 5, 175 (1967).
505. Bagdasarov, Kh. S., Bogomolova, C. A., Kaminskii, A. A. et al., Study of the stimulated emission of $Lu_3Al_5O_{12}$ crystals containing Nd^{3+} ions at the transitions $^4F_{3/2} \rightarrow {}^4I_{11/2}$ and $^4F_{3/2} \rightarrow {}^4I_{13/2}$, *Sov. Phys.-Dokl.* 19, 584 (1975).
506. Kaminskii, A. A., Mill', B. V., Bulashin, A. V., Sarkisov, S. E., and Nikol'skaya, O. K., Two channels of stimulated emission of Nd^{3+} ions in $Ca_3(Nb, Ga)_2Ga_3O_{12}$ crystal, *Inorg. Mater. (USSR)* 21, 1834 (1985).
507. Seas, A., Petricevic, V., and Alfano, R. R., Continuous-wave mode-locked operation of a chromium-doped forsterite laser, *Opt. Lett.* 16, 1668 (1991).
508. Deka, C., Chai, B. H. T., Shimony, Y., Zhang, X. X., Munin, E., and Bass, M., Laser performance of Cr^{4+}:Y_2SiO_5, *Appl. Phys. Lett.* 61, 2141 (1992).
509. Merkle, L. D., Pinto, A., Verdun, H. R., and McIntosh, B., Laser action from Mn^{5+} in $Ba_3(VO_4)_2$, *Appl. Phys. Lett.* 61, 2386 (1992).

510. Kevorkov, A. M., Kaminskii, A. A., Bsgdasarov, Kh. S., Tevosyan, T. A., and Sarkisov, S. E., Spectroscopic properties of $CaAl_4O_7$:Nd^{3+} crystals, *Inorg. Mater. (USSR)* 9, 146 (1973).

511. Sharp, E. J., Mitler, J. E., Horowitz, D. J. et al., Optical spectra and laser action in Nd^{3+}-doped $CaY_2M_2Ge_3O_{12}$, *J. Appl. Phys.* 45, 4974 (1974).

512. Springer, J., Clausen, R., Huber, G., Petermann, K., and Mateika, D., New Nd-doped perovskite for diode-pumped solid-state lasers, *OSA Proc. Adv. Solid State Lasers*, Dube, G. and Chase, L., Eds., 10, 346 (1991).

513. Ryba-Romanowski, W., Jezowska-Trzeblalowska, B., Piekarczyk, W., and Berkowski, M., Optical properties and lasing of $BaLaGa_3O_7$ single crystals doped with neodymium, *J. Phys. Chem. Solids* 49, 199 (1988).

514. Carrig, T. J. and Pollock, C. R., Tunable, cw operation of a multiwatt forsterite laser, *Opt. Lett.* 16, 1662 (1991).

515. Zhang, X. X., Hong, P., Loutts, G. B., Lefaucheur, J., Bass, M., and Chai, B. H. T., Efficient laser performance of Nd^{3+}:$Sr_5(PO_4)_3F$ at 1.059 and 1.328 µm, *Appl. Phys. Lett.* 64, 3205 (1994).

516. Zhou, W.-L., Zhang, X. X., and Chai, B. T. H., Laser oscillation at 1059 nm of new laser crystal: Nd^{3+}-doped $NaY(WO_4)_2$, *Proceedings Advanced Solid State Lasers* (1997).

517. Jansen, M., Alfrey, A., Stafsudd, O. M., Dunn, B., Yang, D. L., and Farrington, G. C., Nd^{3+} beta alumina platelet laser, *Opt. Lett.* 10, 119 (1984).

518. Kaminskii, A. A., Sarkisov, S. E., and Sobolev, B. P., unpublished.

519. Kaminskii, A. A., Mill', B. V., and Butashin, A. V., Growth and stimulated emission spectroscopy of $Ca_3Ga_2Ge_3O_{12}$-Nd^{3+} garnet crystals, *Phys. Stat. Sol. A*: 78, 723 (1983).

520. Michel, J.-C., Morin, D., and Auzel, F., Intensite' de fluorescence et duree de vie du niveau $^4F_{3/2}$ de Nd^{3+} dans une chlorapatite fortement dopee. Comparison avec d'autres matériaux, *C. R. Acad. Sci. Ser. B* 281,445 (1975).

521. Budin, J.-P., Michel, J.-C., and Auzel, F., Oscillator strengths and laser effect in $Na_2Nd_2Pb_6(PO_4)_6Cl_2$ (chloroapatite), a new high-Nd-concentration laser material, *J. Appl. Phys.* 50, 641 (1979).

522. Bagdasarov, Kh. S., Bogomolova, G. A., Grotsenko, M. M., Kaminskii, A. A., Kevorkov, A. M., Prokhorov, A. M., and Sarkisov, S. E., Spectroscopy of the stimulated emission of $Gd_3Al_5O_{12}$-Nd^{3+} crystals, *Sov. Phys. Dokl.* 19, 353 (1974).

523. Kaminskii, A. A., Osiko, V. V., and Voron'ko, Yu. K., Mixed systems on the basis of fluorides as new laser materials for quantum electronics. The optical and emission parameters, *Phys. Stat. Sol.* 21, 17 (1967).

524. Kiss, Z. J. and Duncan, R. C., Optical maser action in CaF_2:Tm^{2+}, *Proc. IRE* 50, 1532 (1962).

525. Duncan, R. C. and Kiss, Z. J., Continuously operating CaF_2:Tm^{2+} Optical Maser, *Appl. Phys. Lett.* 3, 23 (1963).

526. Kaminskii, A. A., Mill', B. V., Bebkoneva, E. L., Tomazyan, S. A., Bubshin, A. V., Kurbanov, K., and Dosmagambetov, E. S., Germanates of the $NdAlCe_2O_7$ structure: synthesis, structure of La-$GaGe_2O_7$, absorption-luminscence properties, and stimulated emission with Nd^{3+} activator, *Inorg. Mater.* 22, 1763 (1986).

527. Brandle, C. D. and Vanderleeden, J. C., Growth, optical properties and CW laser action of neodymium-doped gadolinium scandium aluminum garnet, *IEEE J. Quantum Electron.* QE-10, 67 (1974).

528. Kaminskii, A. A., Pavlyuk, A. A., Chan, Ng. et al., 3 µm stimulated emission by Ho^{3+} ions in $KY(WO_4)_2$ crystals at 300 K, *Sov. Phys.-Dokl.* 24, 201 (1979).

529. Bagdararov, Kh. S., Bogomolova, G. A., Kaminskii, A. A., Kevorkov, A. M., Li, L., Prokhorov, A. M., and Sarkisov, S. E., Study of the stimulated emission of $Lu_3Al_5O_{12}$ crystals containing Nd^{3+} ions at the transitions $^4F_{3/2} \to {}^4I_{11/2}$ and $^4F_{3/2} \to {}^4I_{13/2}$, *Sov. Phys. Dokl.* 19, 584 (1975).

530. Bagdararov, Kh. S., Kaminskii, A. A., and Sobolev, B. P., Laser action in cubic $5NaF \cdot 9YF_3$-Nd^{3+} crystals, *Sov. Phys.-Crystallogr.* 13, 900 (1969).

531. Caffey, D. P., Utano, R. A., and Allik, T. H., Diode array side-pumped neodymium-doped gadolinium scandium gallium garnet rod and slab lasers, *Appl. Phys. Lett.* 56, 808 (1990).

532. Kaminskii, A. A., Kolodnyy, G. Ya., and Sergeyeva, N. I., CW NaLa(MoO$_4$)$_2$:Nd^{3+} crystal laser operating at 300 K, *J. Appl. Spectrosc. (USSR)* 9, 1275 (1968).

533. Morozov, A. M., Tolstoy, M. N., Feofilov, P. P., and Shapovalov, V. N., Fluorescence and stimulated emission in neodymium in lanthanum molybdate-sodium crystals, *Opt. Spectrosc. (USSR)* 22, 224 (1967).

534. Zverev, C. M. and Kolodnyy, G. Ya., Stimulated emission and spectroscopic studies of neodymium-doped binary lanthanum molybdate-sodium single crystals, *Sov. Phys.-JETP* 25, 217 (1967).

535. Johnson, L. F., Guggenheim, H. J., and Thomas, R. A., Phonon-terminated optical masers, *Phys. Rev.* 149, 179 (1966).

536. Zharikov, E. V., Zhitnyuk, V. A., Zverev, G. M., Kalitin, S. P., Kuratev, I. I., Leptev, V. V., Onishchenko, A. M., Osiko, V. V., Pashkov, V. A., Pimenov, A. S., Prokhorov, A. M., Smirnov, V. A., Stel'makh, M. F., Shestakov, A. V., and Shcherbakov, I. A., Active media for high-efficiency neodymium lasers with nonselective pumping, *Sov. J. Quantum Electron.* 12, 1652 (1982).

537. Kaminskii, A. A., Sarkisov, S. E., and Bagdararov, Kh. S., Stimulated emission by Nd^{3+} ions in crystals, due to the $^4F_{3/2} \rightarrow {}^4I_{13/2}$ transition, *Inorg. Mater. (USSR)* 9, 457 (1973).

538. Kaminskii, A. A. and Sarkisov, S. E., Study of stimulated emission from Nd^{3+} ions in crystals emitting at the $^4F_{3/2} \rightarrow {}^4I_{13/2}$ transition. I., *Inorg. Mater. (USSR)* 9, 453 (1973).

539. Kaminskii, A. A., Orthorhombic NaREGeO$_4$ crystals with Nd^{3+} ions; structure and formation. Luminescence properties and stimulated emission, *Appl. Phys. A.* 46, 173 (1988).

540. Kaminskii, A. A., Timofeevs, V. A., Bykov, A. B., and Agsmajyan, N. R., Low-threshold stimulated emission by Nd^{3+} ions in NaLuGeO$_4$. *Sov. Phys. Dokl.* 29, 220 (1984).

541. Kaminskii, A. A. and Lomonov, V. A., Stimulated emission of M$_{1-x}$Nd$_x$F$_{2+x}$ solid solutions with the structure of fluorite. *Inorg. Mater. (USSR)* 20, 1799 (1984).

542. Pruss, D., Huber, G., Beimowskl, A., Laptev, V. V., Shcherbakov, I. A., and Zharikov, Y. V., Efficient Cr^{3+} sensitized Nd^{3+}:GdScGa-garnet laser at 1.06 µm, *Appl. Phys. B* 28, 355 1982.

543. Moulton, P. F. and Mooradian, A., Broadly tunable CW operation of Ni:MgF$_2$ and Co:MgF$_2$ lasers, *Appl. Phys. Lett.* 35, 838 (1979).

544. Kaminskii, A. A., Belokoneva, E. L., Mill', B. V., Tamazyan, S. A., and Kurbanov, K., Crystal structure, spectral-luminescence properties and stimulated radiation of gallium gehlenite, *Inorg. Mater. (USSR)* 22, 993 (1986).

545. Kaminskii, A. A., Belokoneva, E. L., Mill', B. V., Sarkisov, S. E., and Kurbanov, K., Crystal structure absorption luminescence properties, and stimulated emission of Ga gehlenite (Ca$_{2-x}$Nd$_x$Ga$_{2+x}$Si$_{1-x}$O$_7$), *Phys. Stat. Sol. A* 97, 279 (1986).

546. Kaminskii, A. A., Laws of crystal-field disorderness of Ln^{3+} ions in insulating laser crystals, *J. Phys. (Paris)*, Colloq. 12, C7–359 (1987).

547. Kaminskii, A. A., Mayer, A. A., Nikonova, N. S., Provotorov, M. V., and Sarkisov, S. E., Stimulated emission from the new LiGd(MoO$_4$)$_2$:Nd^{3+} crystal laser, *Phys. Stat. Sol.* 12a, K73 (1972).

548. Antipenko, B. M., Mak, A. A., Nikolaev, B. V., Raba, O. B., Seiranyan, K. B., and Uvarova, T. V., Analysis of laser situations in BaYb$_2$F$_8$:Er^{3+} with stepwise pumping schemes, *Opt. Spectrsoc. (USSR)* 56, 296 (1984).

549. Kaminskii, A. A., Sarkisov, S. E., and Li, L., Investigation of stimulated emission in the $^4F_{3/2} \rightarrow {}^4I_{13/2}$ transition of Nd^{3+} ions in crystals (III), *Phys. Stat. Sol. (a)* 15, K141 (1973).

550. Zagumennyi, A. I., Ostroumov, V. G., Shcherbakov, I. A., Jensen, T., Meyen, J. P., and Huber, G., The Nd:GdVO$_4$ crystal: a new material for diode-pumped lasers, *Sov. J. Quantum Electron.* 22, 1071 (1992).

551. Aivea, A. F., Westinghouse, unpublished.

552. Petricevic, V., Gayen, S. K., and Alfano, R. R., Continuous-wave laser operation of chromium-doped forsterite, *Opt. Lett.* 14, 612 (1989).

553. Lu, B., Wang, J., Pan, I., and Jiang, M., Excited emission and self-frequency-doubling effect of $Nd_xY_{1-x}Al_3(BO_3)_4$ crystal, *Chin. Phys. Lett. (China)* 3, 413 (1986).

554. Hayakawa, H., Maeda, K., Ishlkawa, T., Yokoyama, T., and Yoshimasa, F., High average power $Nd:Gd_3Ga_5O_{12}$ slab laser, *Jpn. J. Appl. Phys.* 26, L1623 (1987).

555. Zhang, L., Lin, C., Liu, H., Liu, L., Zhu, H., and Lin, X., Investigation of growth and laser properties of CGG:(Nd,Cr) single crystals, *Chin. J. Phys.* (Engl. Transl.) 5, 136 (1985).

556. Caird, J. A., Shinn, M. D., Kirchoff, T. A., Smith, L. K., and Wilder, R. E., Measurements of losses and lasing efficiency in GSGG:Cr, Nd and YAG:Nd laser rods, *Appl. Opt.* 25, 4294 (1986).

557. Zharikov, E. V., Il'lchev, N. N., Laptev, V. V., Malyutin, A. A., Ostroumov, V. G., Pashlnin, P. P., and Shcherbakov, I. A., Sensitization of neodymium ion luminescence by chromium ions in a $Gd_3Ga_5O_{12}$ crystal, *Sov. J. Quantum Electron.* 12, 338 (1982).

558. Behrens, E.G., Jani, M.G., Powell, R.C., Verdon, H.R., and Pinto, A., Lasing properties of chromium-aluminum-doped forsterite pumped with an alexandrite laser, *IEEE J. Quantum Electron.* 27, 2042 (1991).

559. Allik, T., Merkle, L. D., Utano, R. A., Chai, B.H.T., Lefaucheur, J.-L. V., Voss, H., and Dixon, G, J., Crystal growth, spectroscopy, and laser performance of $Nd^{3+}:KYF_4$, *J. Opt. Soc. Am.* B 10, 633 (1993).

560. Kubodera, K. and Otsuka, K., Efficient $LiNdP_4O_{12}$ lasers pumped with a laser diode, *Appl. Opt.* 18, 3882 (1979).

561. Kubodera, K. and Noda, J., Pure single-mode $LiNdP_4O_{12}$ solid-state laser transmitter for 1.3-μm fiber-optic communications, *Appl. Opt.* 21, 3466 (1982).

562. Berenberg, V. A., Ivanov, A. O., Krutova, L. I., et al., Spectral-luminescent characteristics and stimulated emission of the Nd^{3+} ion in $Gd_{2-x}Nd_x(WO_4)_3$ crystals, *Opt. Spectrosc. (USSR)* 57, 274 (1984).

563. Aleksandrov, V. I., Voron'ko, Yu. K., Mikhalevich, V. G. et al., Spectroscopic properties and emission of Nd^{3+} in ZnO_2 and HfO_2 crystals, *Sov. Phys.-Dokl.* 16, 657 (1972).

564. Aleksandrov, V. I., Kaminskii, A. A., Maksimova, C. V. et al., Stimulated radiation of Nd^{3+} ions in crystals for the $^4F_{3/2} \to {}^4I_{13/2}$ transition, *Sov. Phys.-Dokl.* 18, 495 (1974).

565. Zharikov, E. V., Zabaznov, A. M., Prokhorov, A. M., Shkadarevich, A. P., and Shcherbakov, 1. A., Use of GSGG:Cr:Nd crystals with photochromic centers as active elements in solid lasers, *Sov. J. Quantum Electron.* 16, 1552 (1986).

566. Telle, H. R., Tunable CW laser oscillation of NdP_5O_{14} at 1.3 μm, *Appl. Phys.* B 35 195 (1984).

567. Avanesov, A. G., Denker, B. I., Galagan, B. I., Osiko, V. V., Shestakov, A. V., and Sverchkov, S. E., Room-temperature stimulated emission from chromium (IV)-activated yttrium orthosilicate, *Quantum Electron.* 24, 198 (1994).

568. Borchardl, H. J. and Bierstedl, P. E., $Gd_2(MoO_4)_3$: A ferro-electric laser host. *Appl. Phys. Lett.* 8, 50 (1966).

569. Kaminskii, A. A., Laser and spectroscopic properties of activated ferroelectrics, *Sov. Phys.-Crystallogr.* 17, 194 (1972).

570. Bethea, C. G., Megawatt power at 1.318 μ in $Nd^{3+}:YAG$ and simultaneous oscillation at both 1.06 and 1.318 μ, *IEEE J. Quantum Electron.* QE-9, 254 (1973).

571. Kaminskii, A. A., Mill, B. V., and Butashin, A. V., New possibilities for exciting stimulated emission in inorganic crystalline materials with the garnet structure, *Inorg. Mater. (USSR)* 19, 1808 (1983).

572. Kück, S., Petermann, K., and Huber, G., Near infrared $Cr^{4+}:Y_3Sc_xAl_{5-x}O_{12}$ lasers, *OSA Proc. Adv. Solid State Lasers*, Fan, T. Y. and Chai, B. H. T., Eds., 20, 180 (1995).

573. Zhang, X. X., Hong, P., Bass, M., and Chai, B. H. T., Multisite nature and efficient lasing at 1041 and 1302 nm in Nd^{3+} doped potassium yttrium fluoride, *Appl. Phys. Lett.* 66, 926 (1995).

574. Blatte, M., Danielmeyer, H. G., and Ulrich R., Energy transfer and the complete level system of NdUP, *Appl. Phys.* 1, 275 (1973).

575. Choy, M. M., Zwicker, W. K., and Chinn, S. R., Emission cross section and flashlamp-excited NdP_5O_{14} laser at 1.32 μm, *Appl. Phys. Lett.* 34, 387 (1979).

576. Saruwatari, M., Otsuka, K., Miyazawa, S., Yamada, T., and Kimura, T., Fluorescence and oscillation characteristics of $LiNdP_4O_{12}$ lasers at 1.317 μm, *IEEE J. Quantum Electron.* QE-13, 836 (1977).

577. Kaminskii, A. G., Timofeevs, V. A., Bykov, A. B., and Sarkisov, S. E., Luminescence and stimulated emission in the $^4F_{3/2} \rightarrow {}^4I_{11/2}$ and $^4F_{3/2} \rightarrow {}^4I_{13/2}$ channels of Nd^{3+} ions in orthorhombic $NaYGeO_4$ crystals, *Phys. Stat. Sol. A*: 83, K165 (1984).

578. Morris, P. J., Lüthy, W., Weber, H. P., Zavartsev, Yu. D., Studenikin, P. A., Shcherbakov, I., and Zaguminyi, A. I., Laser operation and spectroscopy of Tm:Ho:$GdVO_4$, *Opt. Commun.* 111, 493 (1994).

579. Beck, R. and Gurs, K., Ho laser with 50-W output and 6.5% slope efficiency, *J. Appl. Phys.* 46, 5224 (1975).

580. Viana, B., Saber, D., Lejus, A. M., Vivien, D., Borel, C., Romero, R., and Wyon, C., Nd^{3+}:$Ca_2Al_2SiO_7$ a new solid-state laser material for diode pumping, in *Advanced Solid-State Lasers*, Pinto, A. A. and Fan, T. Y., Eds., Proceedings Vol. 15, Optical Society of America, Washington, DC (1993), p. 244.

581. Zharikov, E. V., Il'ichev, N. N., Laptev, V. V., Malyutin, A. A., Ostroumov, V. G., Pashinin, P. P., Pimenov, A. S., Smirnov, V. A., and Shcherhakov, I. A., Spectral, luminescence, and lasing properties of gadolinium scandium gallium garnet crystals activated with neodymium and chromium ions, *Sov. J. Quantum Electron.* 13, 82 (1983).

582. Brauch, U. and Dürr, U., Vibronic laser action of V^{2+}:$CsCaF_3$, *Opt. Commun.* 55, 35 (1985).

583. Reed, E., A flashlamp-pumped, Q-switched Cr:Nd:GSGG laser, *IEEE J. Quantum Electron.* QE-21, 1625 (1985).

584. Balakireva, T. P., Briskin, Ch. M., Vakulyuk, V. V., Vasil'ev, E. V., Zolin, V. F., Maier, A. A., Markushev, V. M., Murashov, V. A., and Provotorov, M. V., Luminescence and stimulated emission from $BaGd_{2-x}Nd_x(MoO_4)_4$ single crystals, *Sov. J. Quantum Electron.* 11, 398 (1981).

585. Eilers, H., Dennis, W. M., Yen, W. M., Kück, S., Peterman, K., Huber, G., and Jia, W., Performance of a Cr:YAG laser, *IEEE J. Quantum Electron.* 30, 2925 (1994).

586. Steinbruegge, K. B., High average power characteristics of CaLaSOAP:Nd laser materials, in Digest of Technical Papers CLEA 1973 IEEE/OSA, Washington, DC (1973), p. 49.

587. Antipenko, B. M., Mak, A. A., Sinitsyn, B. V., Raba, O. B., and Uvurova, T. V., New excitation schemes for laser transitions, *Sov. Phys. Tech. Phys.* 27, 333 (1982).

588. Kaminskii, A. A. and Li, L., Spectroscopic and laser studies on crystalline compounds in the system CaO-Nb_2O_5, $Ca(NbO_3)_2$-Nd^{3+} crystals, *Inorg. Mater. (USSR)* 6, 254 (1970).

589. French, P. M. W., Rizvi, N. H., Taylor, J. R., and Shestakov, A. V., Continuous-wave mode-locked Cr^{4+}:YAG laser, *Optics. Lett.* 18, 39 (1992).

590. Steinbruegge, K. B. and Baldwin, G. D., Evaluation of CaLa SOAP:Nd for high-power flash-pumped Q-switched lasers, *Appl. Phys. Lett.* 25, 220 (1974).

591. Geusic, J. E., Marcos, H. M., and Van Uitert, L. G., Laser oscillations in Nd-doped yttrium aluminum, yttrium gallium and gadolinium garnets, *Appl. Phys. Lett.* 4, 182 (1964).

592. Kiss, Z. I. and Duncan, R. C., Cross-pumped Cr^{3+}-Nd^{3+}:YAG laser system, *Appl. Phys. Lett.* 5, 200 (1964).

593. Stone, I. and Burrus, C. A., Nd:Y_2O_3-single-crystal fiber laser: room-temperature CW operation at 1.07- and 1.35-μm wavelength, *J. Appl. Phys.* 49, 2281 (1978).

594. DeSerno, U., Röss, D., and Zeidler, G., Quasicontinuous giant pulse emission of $^4F_{3/2} \rightarrow {}^4I_{13/2}$ transition at 1.32 μm in YAG-Nd^{3+}, *Phys. Lett. A* 28, 422 (1968).

595. Kaminskii, A. A., Koptsik, V. A., Maskaek, Yu. A. et al., Stimulated emission from Nd^{3+} ions in ferroelectric $Ba_2NaNb_5O_{15}$ crystals (bananas), *Phys. Stat. Solidi* 28a, K5 (1975).

596. Babushkin, A. V., Vorob'ev, N. S., Zharakov, E. V., Kalitiin, S. P., Osiko, V. V., Prokhorov, A. N., Serdyuchenko, Yu. N., Shchelev, M. Ya., and Shcherbakov, I. A., Picosecond laser made of gadolinium scandium gallium garnet crystal doped with Cr and Nd, *Sov. J. Quantum Electron.* 16, 428 (1986).

597. Voron'ko, Yu. K., Maksimova, G. V., Mikhalevieh, V. G., Osiko, V. V., Sobol', A. A., Timosheekin, M. I., and Shipulo, G. P., Spectroscopic properties of and stimulated emission from yttrium-lutecium-aluminum garnet crystals, *Opt. Spectrosc.* 33, 376 (1972).

598. Kaminskii, A. A. and Li, L., Spectroscopic and stimulated emission studies of crystal compounds in a $CaO-Nb_2O_5$ system, $Ca(NbO_3)_2:Nd^{3+}$ crystals, *Inorg. Mater. (USSR)* 6, 254 (1970).

599. Kaminskii, A. A., Karlov, N. V., Sarkisov, S. E., Stelmakh, O. M., and Tukish, V. E., Precision measurement of the stimulated emission wavelength and continuous tuning of $YAlO_3:Nd^{3+}$ laser radiation due to $^4F_{3/2} \rightarrow {}^4I_{13/2}$ transition, *Sov. J. Quantum Electron.* 6, 1371 (1976).

600. Bagdasarov, Kh. S., Gritsenko, M. M., Zubkova, F. M., Kaminskii, A. A., Kevorkov, A. M., and Li, L., CW $Ca(NbO_3)_2:Nd^{3+}$ crystal laser, *Sov. Phys.-Crystallogr.* 15, 323 (1970).

601. Krühker, W. W. and Plättner, R. D., Laser emission of (Nd,Y)-pentaphosphate at 1.32 μm, *Opt. Commun.* 28, 217 (1979).

602. Kaminskii, A. A. and Sarkisov, S. E., Stimulated emission by Nd^{3+} ions in crystals, due to the $^4F_{3/2} \rightarrow {}^4I_{13/2}$ transition, *Inorg. Mater. (USSR)* 9, 453 (1973).

603. Miller, J. E., Sharp, F. J., and Horowitz, D. J., Optical spectra and laser action of neodymium in a crystal $BaO_{0.25}Mg_{2.75}Y_4Ge_3O_{12}$, *J. Appl. Phys.* 43, 462 (1972).

604. Kaminskii, A. A., Timoreeva, V. A., Agamalyan, N. R., and Bykov, A. B., Infrared laser radiation from $NaCdGeO_4-Nd^{3+}$ crystals growth from solution in a melt, *Sov. Phys. Crystallogr.* 27, 316 (1982).

605. Kaminskii, A. A., Timoreeva, V. A., Agamalyan, N. R., and Bykov, A. B., Stimulated emission by Nd^{3+} ions in $NaGdGeO_4$ by the $^4F_{3/2} \rightarrow {}^4I_{11/2}$ and $^4F_{3/2} \rightarrow {}^4I_{13/2}$ transitions at 300 K, *Inorg. Mater. (USSR)* 17, 1703 (1981).

606. Aleksandrov, V. I., Kaminskii, A. A., Maksimova, G. V., Prokhorov, A. M., Sarkisov, S. E., Sobol', A. A., and Tatarintsev, V. M., Study of stimulated emission from Nd^{3+} ions in crystals emitting at the $^4F_{3/2} \rightarrow {}^4I_{13/2}$ transition, *Sov. Phys.-Dokl.*, 18, 495 (1974).

607. Kaminskii, A. A., Sarkisov, S. E., and Li. L., Investigation of stimulated emission in the $^4F_{3/2} \rightarrow {}^4I_{13/2}$ transition of Nd^{3+} ions in crystals (III), *Phys. Stat. Sol.* 15a, K141 (1973).

608. Wyss, Chr. P., Lüthy, W., Weber, H. P. et al., Performance of a $Tm^{3+}:GdVO_4$ microchip laser at 1.9 μm, *Opt. Commun.* 153, 63 (1998).

609. Allik, T. H., Morrison, C. A., Gruber, J. B., and Kokta, M. R., Crystallography, spectroscopic analysis, and lasing properties of $Nd^{3+}:Y_3Sc_2Al_3O_{12}$, *Phys. Rev. B* 41, 21 (1990).

610. Kaminskii, A. A., Bodretsova, A. I., Petrosyan, A. G., and Pavlyuk, A. A., New quasi-CW pyrotechnically pumped crystal lasers, *Sov. J. Quantum Electron.* 13, 975 (1983).

611. Godina, N. A., Tolstoi, M. N., and Feofilov, P. P., Luminescence of neodymium in yttrium and lanthanum niobates and tantalates, *Opt. Spectrosc. (USSR)* 23, 411 (1967).

612. Kutovoi, S. A., Laptev, V. V., Lebedev, V. A., Matsnev, S. Yu., Pisarenko, V. F., and Chuev, Yu. M., Spectral luminescent and lasing properties of the new laser crystals lanthanum scandium borate doped with neodymium and chromium, *Zh. Prikl. Spektr.* 53, 370 (1990).

613. Doroshenko, M.E., Osiko, V.V., Sigachev, V.B., and Timoshechkin, M.I., Stimulated emission from a neodymium-doped gadolinium gallium garnet crystal due to the $^4F_{3/2} - {}^4I_{13/2}$ (λ = 1.33 μm) transition, *Sov. J. Quantum Electron.* 21, 266 (1991).

614. Voron'ko, Yu. K., Zverev, G. M., and Prokhorov, A. M., Stimulated emission from Er^{3+}-ions in CaF_2, *Sov. Phys.-JETP* 21, 1023 (1964).

615. Bagdararov, Kh. S., Kaminskii, A. A., Kevorkov, A. M., Li, L., Prokhorov, A. M., Sarkisov, S. E., and Tevosyan, T. A., Stimulated emission of Nd^{3+} ions in an $SrAl_{12}O_{19}$ crystal at the transitions $^4F_{3/2} \rightarrow {}^4I_{11/2}$ and $^4F_{3/2} \rightarrow {}^4I_{13/2}$, *Sov. Phys. Dokl.* 19, 350 (1974).

616. Chinn, S. R., Pierce, J. W., and Heckscher, H., Low-threshold transversely excited NdP_5O_{14} laser, *Appl. Optics* 15, 1444 (1976).

617. Chinn, S. R., Intracavity second-harmonic generation in a Nd pentaphosphate laser, *Appl. Phys. Lett.* 29, 176 (1976).

618. Trutna, W. R., Jr., Donald, D. K., and Nazarathy, M., Unidirectional diode-laser-pumped Nd:YAG ring laser with a small magnetic field. *Opt. Lett.* 12, 248 (1987).

619. Bakhsheyeva, G. F., Karapetyan, V. Ye., Morozov, A. M., Morozova, L. G., Tolstoy, M. N., and Feofilov, P. P., Optical constants, fluorescence and stimulated emission of neodymium-doped lanthanum niobate single crystals, *Opt. Spectrosc.* 28, 38 (1970).

620. Kaminskii, A. A., Bogomolova, G. A., and Li, L., Absorption, fluorescence, stimulated emission and splitting of the Nd^{3+} levels in a YVO_4 crystal, *Inorg. Mater. (USSR)* 5, 573 (1969).

621. Chinn, S. R. and Zwicker, W. K., Flash-lamp-excited NdP_5O_{14} laser, *Appl. Phys. Lett.* 31, 178 (1977).

622. Singh, S., Miller, D. C., Potopowicz, J. R., and Shick, L. K., Emission cross section and fluorescence quenching of Nd^{3+} lanthanum pentaphosphate, *J. Appl. Phys.* 46, 1191 (1975).

623. Kaminskii, A. A., Kurbanov, K., and Uvarova, T. V., Stimulated radiation from single crystals of $BaYb_2F_8-Pr^{3+}$, *Inorg. Mater. (USSR)* 23, 940 (1987).

624. Sandrock, T., Heumann, E., Huber, G., and Chai, B. H. T., Continuous-wave Pr,Yb:LiYF$_4$ upconversion laser in the red spectral range at room temperature, *OSA Trends in Optics and Photonics on Advanced Solid State Lasers*, Vol. 1, Payne, S. A., and Pollack, C. R., Eds., Optical Society of America, Washington, DC (1996), p. 550.

625. Kaminskii, A. A., Sarkisov, S. E., Maier, A. A. et al., Eulytine with TR^{3+} ions as a laser medium, *Sov. Tech. Phys. Lett.* 2, 59 (1976).

626. Kaminskii, A. A., High-temperature spectroscopic investigation of stimulated emission from lasers based on crystals activated with Nd^{3+} ions, *Phys. Stat. Sol.* 1a, 573 (1970).

627. Bagdasarov, Kh. S., Iwtova, O. Ye., Kaminskii, A. A., Li, L., and Sobolev, B. P., Optical and laser properties of mixed $CdF_2-YF_3:Nd^{3+}$- crystals, *Sov. Phys.-Dokl.* 14, 939 (1970).

628. Hong, H. Y-P., and Dwight, K., Crystal structure and fluorescence lifetime of a laser material $NdNa_5(WO_4)_4$, *Mater. Res. Bull.* 9, 775 (1974).

629. Petricevic, V., Bykov, A. B., Evans, J. M., and Alfano, R. R., Room-temperature near-infrared tunable laser operation of $Cr^{4+}:Ca_2GeO_4$, *Opt. Lett.* 21, 1750 (1996).

630. Hattendorff, H.-D., Huber, G., and Lutz, F., CW laser action in $Nd(Al,Cr)_3(BO_3)_4$, *Appl. Phys. Lett.* 34, 437 (1979).

631. Lutz, F., Ruppel, D., and Leiss, M., Epitaxial layers of the laser material $Nd(Ga,Cr)_3(BO_3)_4$, *J. Cryst. Growth* 48, 41 (1980).

632. Ohlmann, R. C., Steinbruegge, K. B., and Mazelsky, R., Spectroscopic and laser characteristics of neodymium-doped calcium fluorophosphate, *Appl. Optics* 7, 905 (1968).

633. Bruk, Z. M., Voron'ko, Yu. K., Maksimova, G. V., Osiko, V. V., Prokhorov, A. M., Shipilov, K. F., and Shcherbakov, I. A., Optical properties of a stimulated emission from Nd^{3+} in fluorapatite crystals, *JETP Lett.* 8, 221 (1968).

634. Chinn, S. R., Research studies on neodymium pentaphosphate miniature lasers, Final Report ESD TR-78-392, (DDC Number AD-A073140, Lincoln Laboratory, M. I. T., Lexington, MA (1978).

635. Kaczmarek, F. and Szymanski, M., Performance of NdLa pentaphosphate laser pumped by nanosecond pulses, *Appl. Phys.* 13, 55 (1977).

636. Wilson, J., Brown, D. C., and Zwicker, W. K., XeF excimer pumping of NdP_5O_{14}, *Appl. Phys. Lett.* 33, 614 (1978).

637. Gaiduk, M. I., Grigor'yants, V. V., Zhabotinskii, M. E., Makovestskii, A. A., and Tishchenko, R. P., Neodymium pentaphosphate microlaser pumped by the second harmonic of a YAG:Nd^{3+} laser, *Sov. J. Quantum Electron.* 9, 250 (1979).

638. Weber, H. P. and Tofield, B. C., Heating in a cw Nd-pentaphosphate laser, *IEEE J. Quantum Electron.* QE-11, 368 (1975).

639. Chinn, S. R., Pierce, J. W., and Heckscher H., Low-threshold transversely excited NdP$_5$O$_{14}$ laser, *Appl. Opt.* 15, 1444 (1976).

640. Chinn, S. R., Hong, H. Y-P., and Pierce. J. W., Spiking oscillations in diode-pumped NdP$_5$O$_{14}$ and NdAl$_3$(BO$_3$)$_4$ lasers, *IEEE J. Quantum Electron.* QE-12, 189 (1976).

641. Luo, Z., Jiang, A., Huang, Y., and Qui, M., Laser performance of large neodymium aluminum borate (NdAl$_3$(BO$_3$)$_4$) crystals, *Chin. Phys. Lett.* 3, 541 (1986).

642. Huang, Y., Qiu, M., Chen, G., Chen, J., and Luo, Z., Pulsed laser characteristics of neodymium aluminum borate [NdAl$_3$(BO$_3$)$_4$] (NAB) crystals, *Chin. Phys. Lasers* 14, 623 (1987).

643. Dianov, E. M., Dmitruk, M. V., Karasik, A. Ya., Kirpickenkova, E. O., Osiko, V. V., Ostrounov, V. G., Timoshechin, M. I., and Shcherbakov, I. A., Synthesis and investigation of spectral, luminescence, and lasing properties of aluminoborate crystals activated with chromium and neodymium ions, *Sov. J. Quantum Electron.* 10, 1222 (1980).

644. Ivanov, A. O., Morozova, L. G., Mochalov, I.V., and Fedorov, V. A., Spectra of a neodymium ion in Ca,LaSOAP and Ca,YSOAP crystals and stimulated emission in Ca,LaSOAP-Nd crystals, *Opt. Spectrosc. (USSR)* 42, 556 (1977).

645. Arkhipov, R. N., Evstigneev, V. L., Zharikov, E. V., Pabenichnikov, S. M., Shcherbakov, I. A., and Yumashev, V. E., Optimization of the conditions of utilization of the stored energy by Q switching active elements in the form of gadolinium scandium gallium garnet crystals activated with Cr and Nd, *Sov. J. Quantum Electron.* 16, 688 (1986).

646. Zharikov, E. V., Zhitkova, M. B., Zverev, G. M., Isaev, M. P., Kalitin, S. P., Kurabv, I. I., Kushir, V. R., Laptev, V. V., Osiko, V. V., Pashkov, V. A., Pimenov, A. S., Prokhorov, A. M., Smirnov, V. A., Stel'makh, M. F., Shectakov, A. M., and Shcherbakov, I. A., Output characteristics of a gadolinium scandium gallium garnet laser operating in the pulse-periodic regime, *Sov. J. Quantum Electron.* 13, 1306 (1983).

647. Shestakov, A.V., Borodin, N.I., Zhitnyuk, V.A., Ohrimtchyuk, A.G., and Gapontsev, V.P., Tunable Cr^{4+}:YAG lasers, *CPDP12-1*, 594 (1993).

648. Lutz, F., Leiss, M., and Muller, J., Epitaxy of NdAl$_3$(BO$_3$)$_4$ for thin film miniature lasers, *J. Cryst. Growth*, 47, 130 (1979).

649. Lazarev, V. V. and Kandaurov, A. S., Lasing properties of La$_2$Be$_2$O$_5$:Nd^{3+} at 1.35-μm wavelength, *Opt. Spectrosc. (USSR)* 63, 519 (1987).

650. Grigor'ev, V. N., Egorov, G. N., Zharikov, E. V., Mlkhailov, V. A., Pak, S. K., Pinskii, Yu. A., Shklovskii, E. I., and Shcherbakov, I. A., Prism-resonator CSGG Cr:Nd laser with polarization coupling out of radiation, *Sov. J. Quantum Electron.* 16, 1554 (1986).

651. Matrosov, V. I., Timosheckin, M. I., Tsvetkov, E. I. et al., Investigation of the conditions of crystallization of lanthanum beryllate in *Abstracts of the Fifth All-Union Conference on Crystal Growth*, Proc. Acad. Sci. Georgian Sov. Soc. Rep., Tiflis (1977), p. 167 (in Russian).

652. Hattendorff, H.-D., Huber, G., and Danielmeyer, H. G., Efficient cross pumping of Nd^{3+} by Cr^{3+} in Nd(Al, Cr)$_3$(BO$_3$)$_4$ lasers, *J. Phys. C* 11, 2399 (1978).

653. Gondra, A. D., Gradov, V. M., Danilov, A. A., Dybko, V. V., Zharlkov, E. V., Kondantlnov, B. A., Nlkol'skii, M. Yu., Rogal'skii, Yu. I., Smotryaev, S. A., Terent'ev, Yu. I., Shcherbakov, A. A., and Shcherbakov, I. A., Chromium- and neodymium-activated gadolinium scandium gallium garnet laser with efficient pumping and Q switching, *Sov. J. Quantum Electron.* 17, 582 (1987).

654. Belokrinitskiy, N. S., Belousov, N. D., Bonchkovskiy, V. I., Kobzaraklenko, V. A., Skorobogatov. B. S., and Soskin, M. S., Study of stimulated emission from Nd^{3+}-doped NaLa(WO$_4$)$_2$ single crystals. *Ukrainskiyl Fizicheskiy Zhurnal* 14, 1400, 1969 (in Russian).

655.   Kaminskii, A. A. and Sarkisov, S. E., Study of stimulated emission from Nd^{3+} ions in crystals at the $^4F_{3/2} \rightarrow {}^4I_{13/2}$ transition. 4, *Sov. J. Quantum Electron.* 3, 248 (1973).

656.   Huber, G. and Danielmeyer, H. G., NdP_5O_{14} and $NdAl_3(BO_3)_4$ lasers at 1.3 μm, *Appl. Phys.* 18, 77 (1979).

657.   Malinowski, M., Joubert, M. F., and Jacquier, B., Simultaneous laser action at blue and orange wavelengths in $YAG:Pr^{3+}$, *Phys. Stat. Sol. A* 140, K49 (1993).

658.   Winzer G., Mockel, P. G., and Krühler, W., Laser emission from miniaturized $NdAl_3(BO_3)_4$ crystals with directly applied mirrors, *IEEE J. Quantum Electron.* QE-14, 840 (1978).

659.   Danilov, A. A., Evstigneev, V. L., Il'lchev, N. N., Mdyutin, A. A., Nikol'skii, M. Yu., Umyhkov, A. F., and Shcherbakov, I. A., Compact $GSGG:Cr^{3+}:Nd^{3+}$ laser with passive Q switching, *Sov. J. Quantum Electron.* 17, 573 (1987).

660.   DeSerno, U., Ross, D., and Zeidler, G., Quasicontinuous giant pulse emission of $^4F_{3/2} \rightarrow {}^4I_{13/2}$ transition at 1.32 μm in $YAG:Nd^{3+}$, *Phys. Lett.* 28A, 422 (1968).

661.   Kaminskii, A. A., Schultze, D., Hermoneit, B. et al., Spectroscopic properties and stimulated emission in the $^4F_{3/2} \rightarrow {}^4I_{11/2}$ and $^4F_{3/2} \rightarrow {}^4I_{13/2}$ transitions of Nd^{3+} ions from cubic $Bi_4Ge_3O_{12}$ crystals, *Phys. Stat. Solidi* 33a, 737 (1976).

662.   Moulton, P. F., Mooradian, A., Chen, Y., and Abraham, M. M., unpublished.

663.   Richards, J., Fueloep, K., Seymour, R. S., Cashmore, D., Picone, P. J., and Horsburgh, M. A., Nd:BeL laser at 1356 nm, in *Tunable Solid State Lasers*, Shand, M. L. and Jenssen, H. P., Eds., Proceedings Vol. 5, Optical Society of America, Washington, DC (1989), p. 119.

664.   French, P. M. W., Rizvi, N. H., Taylor, J. R., and Shestakov, A. V., Continuous-wave mode-locked $Cr^{4+}:YAG$ laser, *Opt. Lett.* 18, 39 (1993).

665.   Kaminskii, A. A., Kholov, A., Klevstov, P. V., and Khafizov, S. K., Growth and generation properties of $NaBi(WO_4)_2–Nd^{3+}$ single crystals, *Neorgan. Mater.* 25, 1054, 1989

666.   Schearer, L. D. and Tin, P., Laser performance and tuning characteristics of a diode pumped $Nd:YAlO_3$ laser at 1083 nm, *Opt. Commun.* 71, 170 (1989).

667.   Baturina, O. A., Grechushnikov, B. N., Kaminskii, A. A., Konstantinova, A. F., Markosyan, A., A., Mill', B. V., and Khodzhabagyan, G. G., Crystal-optical investigations of compounds with the structure of trigonal Ca-gallogermanate $(Ca_3Ga_2Ge_4O_{14})$, *Sov. Phys. Crystallogr.* 32, 236 (1987).

668.   Kaminskii, A. A., Mill', B. V., Belokonevs, E. L., Sarkisov, S. E., Pastukhova, T. Yu., and Khodazhabagyan, G. G., Crystal structure and stimulated emission of $La_3Ga_{5.5}Nb_{0.5}O_{12}-Nd^{3+}$, *Inorg. Mater. (USSR)* 20, 1793 (1984).

669.   Gavrilovic, P., O'Neill, M. S., Meehan, K., Zarrabi, J. H., Singh, S., and Grodliewicz, W. H., Temperature-tunable, single frequency microcavity lasers fabricated from flux-grown YCeAG:Nd, *Appl. Phys. Lett.* 60, 1652 (1992).

670.   Kvapil, J., Kvapil, Jos., Kubelka, J., and Perner, B., Laser properties of YAG:Nd Ti, *Czech. J. Phys. B* 32, 817 (1982).

671.   Kvapil, J., Kvapil, Jos., Perner, B., Kubelka, J., Mamek, B., and Kubecek, V., Laser properties of YAG:Nd Cr,Ce, *Czech. J. Phys. B* 34, 581 (1984).

672.   Shi, W. Q., Kurtz, R., Machan, J., Bass, M., Birnbaum, M., and Kokta, M., Simultaneous, multiple wavelength lasing of $(Er,Nd):Y_3Al_5O_{12}$, *Appl. Phys. Lett.* 51, 1218 (1987).

673.   Machan, J., Kurtz, R., Bass, M., Birnbaum, M., and Kokta, M., Simultaneous multiple wavelength lasing of $(Ho,Nd):Y_3Al_5O_{12}$, *Appl. Phys. Lett.* 51, 1313 (1987).

674.   Forrester, P. A. and Simpson, D. F., A new laser line due to energy transfer from colour centers to erbium ions in CaF_2, *Proc. Phys. Soc.* 88, 199 (1966).

675.   Es'kov, N. A., Osiko, V. V., Sobol, A. A. et al., A new laser garnet $Ca_3Ca_2Ge_3O_{12}-Nd^{3+}$, *Inorg. Mater. (USSR)* 14, 1764 (1978).

676.   Bezrodnyi, V. I., Tikhonov, E. A., and Nedbaev, N. Ya., Generation of controlled-duration ultrashort pulses in a passively mode-locked $YAG:Nd^{3+}$ laser, *Sov. J. Quantum Electron.* 16, 796 (1986).

677.   Smith, R. J., Rice, R. R., and Aden, L. B., Jr., 100 mW laser diode pumped Nd:YAG laser, in *Proc. Soc. Photo. Opt. Instrum. Eng., Advances in Laser Engineering and Applicatons*, Vol. 247 Stitch, M. L., Ed., SPIE, Bellingham, WA (1980), p. 144.

678. Berger, J., Welch, D. F., Streifer, W., Scifres, D. R., Hoffman, N. J., Smith, J. J., and Radecki, D., Fiber-bundle coupled, diode end-pumped Nd:YAG laser, *Opt. Lett.* 13, 306 (1988).

679. Sipes, D. L., Highly efficient neodymium:yttrium aluminum garnet laser end pumped by a semiconductor laser array *Appl. Phys. Lett.* 47, 74 (1985).

680. Berger, J., Welch, D. F., Scifres, D. R., Streifer, W., and Cross, P. S., 370 mW, 1.06 μm, CW TEM_{00} output from an Nd:YAG laser rod end-pumped by a monolithic diode array, *Electron. Lett.* 23, 669 (1987).

681. Berger, J., Welch, D. F., Scifres, D. R., Strelfer, W., and Cross, P. S., High power, high efficient neodymium:yttrium aluminum garnet laser end pumped by a laser diode array, *Appl. Phys. Lett.* 51, 1212 (1987).

682. Zhou, B., Kane, T. J., Dixon, G. J., and Byer, R. L., Efficient, frequency-stable laser-diode-pumped Nd:YAG laser, *Opt. Lett.* 10, 62 (1985).

683. Kane, T. J., Nilsson, A. C., and Byer, R. L., Frequency stability and offset locking of a laser-diode-pumped Nd:YAG monolithic nonplanar ring oscillator, *Opt. Lett.* 12, 175 1987.

684. Fields, R. A., Birnbaum, M., and Fincher, C. L., Highly efficient Nd:YVO4 diode-laser end-pumped laser, *Appl. Phys. Lett.*, 51, 1885 (1987).

685. Allik, T. H., Hovis, W. W., Caffey, D. P., and King, V., Efficient diode-array-pumped Nd:YAG and Nd:Lu:YAG lasers, *Opt. Lett.* 14, 116 (1989).

686. Reed, M. K., Kozlovsky, W. J., Byer, R. L., Harmgel, G. L., and Cross, P. S., Diode-laser-array-pumped neodymium slab oscillators, *Opt. Lett.* 13, 204 (1988).

687. Berger, J., Harnagel, G., Welch, D. F., Scifres, D. R., and Strelfer, W., Direct modulation of a Nd:YAG laser by combined side and end laser diode pumping, *Appl. Phys. Lett.* 53, 268 (1988).

688. Maker, G. T. and Ferguson, A. I., Single-frequency Q-switched operation of a diode-laser-pumped Nd:YAG laser, *Opt. Lett.* 13, 461 (1988).

689. Denisov, N. N., Manenkov, A. A., and Prokhorov, A. M., Kinetics of generation and amplification of $YAG:Nd^{3+}$ laser radiation in a periodic Q-switched regime with pulsed pumping, *Sov. J. Quantum Electron.* 14, 597 (1984).

690. De Silvestri, S., Laporta, P., and Magni, V., 14-W continuous-wave mode-locked Nd:YAG laser, *Opt. Lett.* 11, 785 (1986).

691. Kuizenga, D. J., Short-pulse oscillator development for the Nd:Glass laser-fusion systems, *IEEE J. Quantum Electron.* QE-17, 1694 (1981).

692. Dawes, J. B. and Sceats, M. G., A high repetition rate pico-synchronous Nd:YAG laser, *Opt. Commun.* 65, 275 (1988).

693. Prokhorenko, V. I., Tikhonov, E. A., Yatskiv, D. Ya., and Bushmakin, E. N., Generation of ultra-short pulses in a $YAG:Nd^{3+}$ laser in a colliding pulse scheme, *Sov. J. Quantum Electron.* 17, 505 (1987).

694. Varnavskii, O. P., Leontovich, A. M., Mozharovskii, A. M., and Solomatin, I. I., Mode-locked $YAG:Nd^{3+}$ laser with a high output energy and brightness, *Sov. J. Quantum Electron.* 13, 1251 (1983).

695. Kaminskii, A. A., Silvestrova, I. M., Sarkisov, S. E., and Denisenko, G. A., Investigation of trigonal $(La_{1-x}Nd_x)_3Ga_5SiO_{14}$ crystals, *Phys. Stat. Sol. A* 80, 607 (1983).

696. Kaminskii, A. A., Sarkisov, S. E., Mill', B. V., and Khodzhabagyan, G. G., Generation of stimulated emission of Nd^{3+} ions in a trigonal acentric $La_3Ga_5SiO_{14}$ crystal, *Sov. Phys. Dokl.* 27, 403 (1982).

697. Kaminskii, A. A., Mill', B. V., Silvestrova, I. M., and Khodzhabagyan, G. G., The nonlinear active material $(La_{1-x}Nd_x)_3Ga_5SiO_{14}$, *Bull. Acad. Sci. USSR Phys. Ser.* 47, 25 (1983).

698. Kaminskii, A. A., Sarkisov, S. E., Mill', B. V., and Khodzhabagyan, G. G., New inorganic material with a high concentration of Nd^{3+} ions for obtaining stimulated emission at the $^4F_{3/2} \rightarrow {}^4I_{11/2}$ and $^4F_{3/2} \rightarrow {}^4I_{13/2}$ transitions, *Inorg. Mater. (USSR)* 18, 1189 (1982).

699. Kaminskii, A. A., Agamalyan, N. R., Kozeeva, L. P., Nesterenko, V. F., and Pavlyuk, A. A., New data on stimulated emission of Nd^{3+} ions in disordered crystals with scheelite structure, *Phys. Stat. Sol. A* 75, K 1 (1983).

700. Rosenblatt, G. H., Stoneman, R. C., and Esterowitz, L., Diode-pumped room-temperature cw 1.45-μm Tm;Tb:YLF laser, in *Advanced Solid-State Lasers*, Jenssen, H. P. and Dubé, G., Eds., Proceedings Vol. 6, Optical Society of America, Washington, DC (1990), p. 26.

701. Moncorgé, R., Manaa, H., Deghoul, F., Borel, C., and Wyon, Ch., Spectroscopic study and laser operation of Cr^{4+}-doped $(Sr,Ca)Gd_4(SiO_4)_3O$ single crystal, *Opt. Commun.* 116, 393 (1995).

702. Kaminskii, A. A., Temperature pulsations and multi-frequency laser action in $YAlO_3:Nd^{3+}$, *JETP Lett.* 14, 222 (1971).

703. Kaminskii, A. A., Sobolev, B. P., Sarkisov, S. E., Denisenko, G. A., Ryabchenkov, V. V., Fedorov, V. A., and Uvarova, T. V., Physicochemical aspects of the preparation spectroscopy, and stimulated emission of single crystals of $BaLn_2F_8-Ln^{3+}$, *Inorg. Mater. (USSR)* 18, 402 (1982).

704. Moulton, P. F., Pulse-pumped operation of divalent transition-metal lasers, *IEEE J. Quantum Electron.* QE-18, 1185 (1982).

705. Kaminskii, A. A., Kurbanov, K., Markosyan, A. A., Mill', B. V., Sarkisov, S. E., and Khodzhabagyan, G. G., Luminescence-absorption properties and low-threshold stimulated emission of Nd^{3+} ions in $La_3Ga_{5.5}Ta_{0.5}O_{14}$, *Inorg. Mater. (USSR)* 21, 1722 (1985).

706. Petricevic, V., Gayen, S. K., and Alfano, R. R., Laser action in chromium-activated forsterite for near-infrared excitation: is Cr^{3+} the lasing ion?, *Appl. Phys. Lett.* 53, 2590 (1988).

707. Petricevic, V., Gayen, S. K., and Alfano, R. R., Continuous-wave operation of chromium-doped forsterite. *Opt. Lett.* 14, 612 (1989).

708. Verdun, H. R., Thomas, L. M., Andrauskas, D. M., McCollum, T., and Pinto, A., Chromium-doped forsterite laser pumped with 1.06 μm radiation, *Appl. Phys. Lett.* 53, 2593 (1988).

709. Petricevic, V., Gayen, S. K., Alfano, R. R., Yamagishi, K., Anzai, H., and Yamaguchi, Y., Laser action in chromium-doped forsterite, *Appl. Phys. Lett.* 52, 1040 (1988).

710. Antipenko, B. M., Krutova, L. I., and Sukhareva, L. K., Dual-frequency lasing of GSGG-Cr^{3+} Tm crystals, *Opt. Spectrsoc. (USSR)* 60, 252 (1986).

711. Antipenko, B. M., Mak, A. A., Raba, O. B., Seiranyan, K. B., and Uvarova, T. V., New lasing transition in the Tm^{3+} ion, *Sov. J. Quantum Electron.* 13, 558 (1983).

712. Akmanov, A. G., Val'shin, A. M., and Yamaletdinov, A. G., Frequency-tunable $YAlO_3:Nd^{3+}$ laser, *Sov. J. Quantum Electron.* 15, 1555 (1985).

713. Kaminskii, A. A., Butaeva, T. I., Kevorkov, A. M., Fedorov, V. A., Petrosyan, A. G., and Gritsenko, M. M., New data on stimulated emission by crystals with high concentrations of Ln^{3+} ions, *Inorg. Mater. (USSR)* 12, 1238 (1976).

714. Antipenko, B. M., Dumbravyanu, R. V., Perlin, Yu. E., Raba, O. B., and Sukhareva, L. K., Spectroscopic aspects of the $BaYb_2F_8$ laser medium, *Opt. Spectrosc. (USSR)* 59, 377 1985.

715. Bass, M. and Weber, M. J., Nd, $Cr:YAlO_3$ laser tailored for high-energy Q-switched operation, *Appl. Phys. Lett.* 17, 395 (1970).

716. Weber, M. J., Bass, M., Andringa, K., Monchamp, R. R., and Comperchio, L., Czochralski growth and properties of $YAlO_3$ laser crystals, *Appl. Phys. Lett.* 15, 342 (1969).

717. Massey, G. A. and Yarborough, J. M., High average power operation and nonlinear optical generation with the $Nd:YAlO_3$ laser, *Appl. Phys. Lett.* 18, 576 (1971).

718. Schearer, L. and Leduc, M., Tuning characteristics and new laser lines in an Nd:YAP CW laser, *IEEE J. Quantum Electron.* QE-22, 756 (1986).

719. Esterowitz, L., Eckardt, R. C., and Allen, R. E., Long-wavelength stimulated emission via cascade laser action in Ho:YLF, *Appl. Phys. Lett.* 35, 236 (1979).

720. Singh, S., Van Uitert, L. G., Potopowicz, I. R., and Grodkiewicz, W. H., Laser emission at 1.065 μm from neodymium-doped anhydrous cerium trichloride at room temperature, *Appl. Phys. Lett.* 24, 10 (1974).

721. Singh, S., Chesler, R. B., Grodkiewicz, W. H. et al., Room temperature CW $Nd^{3+}:CdCl_2$ laser, *J. Appl. Phys.* 46, 436 (1975).

722. Kaminskii, A. A., Fedorov, V. A., Sarkisov, S. E. et al., Stimulated emission of Ho^{3+} and Er^{3+} ions in $Gd_3Ga_5O_{12}$ crystals and cascade laser action of Ho^{3+} ions over the $^5S_2-$ $^5I_5-^5I_6-^5I_8$ scheme, *Phys. Stat. Sol.* 53a, K219 (1979).

723. Wong, S.K., Mathieu, P., and Pace, P., Eye-safe Nd:YAG laser, *Appl. Phys. Lett.*, 57, 650 (1990).

724. Hodgson, N., Nighan, Jr., W. L., Golding, D. J., and Eisel, D., Efficient 100-watt Nd:YAG laser operating at the wavelength of 1.444 μm, *Opt. Lett.* 19, 1328 (1994).

725: Sorokin, E., Sorokina, I., Wintner, E., Zagumennyi, A. I., and Shcherbakov, I. A., CW passive mode-locking of a new Nd^{3+}:$GdVO_4$ crystal laser, in *Advanced Solid-State Lasers*, Pinto, A. A. and Fan, T. Y., Eds., Proceedings Vol. 15, Optical Society of America, Washington, DC (1993), p. 238.

726. Bagdasarov, Kh. S., Kaminskii, A. A., Kevorkov, A. M., Prokhorov, A. M., Sarkisov, S. E., and Tevosyan, T. A., Stimulated emission from RE^{3+} ions in YAG crystals, *Sov. Phys.-Dokl.* 19, 592 (1975).

727. Wang, Q., Zhao, S., and Zhang, X., Laser characterization of low-threshold high-efficiency $Nd:Sr_5(VO4)_3F$ crystal, *Opt. Lett.* 20, 1262 (1995).

728. DeLoach, L., Payne, S. A., Chai, B. H. T., and Loutts, G., Laser demonstration of neodymium-doped strontium chlorovandate, *Appl. Phys. Lett.* 65, 1208 (1994).

729. Bass, M. and Weber, M. J., YALO:Robust at age 2, *Laser Focus*, 34 (1971).

730. Demchouk, M. I., Gilev, A. K., Zabaznov, A. M., Mikhailov, V. P., Stavrov, A. A., and Shkadarevich, A. P., Lasing of ultrashort pulses by a Nd,Cr-doped gadolinium-scandium-gallium garnet laser, *Opt. Commun.* 55, 207 (1985).

731. Bagdasarov, Kh. S., Danilov, V. P., Zhekov, V. I. et al., Pulse-periodic $Y_3Al_5O_{12}$:Er^{3+} laser with high activator concentration, *Sov. J. Quantum Electron.* 8, 83 (1978).

732. Danelyus, R., Kuratev, I., Piskarskas, A., Sirutkaitis, V., Shvom, E., Yuozapavichyus, A., and Yankauskas, A., Generation of picosecond pulses by a gadolinium scandium gallium garnet laser, *Sov. J. Quantum Electron.* 15, 1160 (1985).

733. Simondi-Teisseire, B., Viana, B., and Vivien, D., Near-infrared Er^{3+} laser properties in melilite type crystals $Ca_2Al_2SiO_7$ and $SrLaGa_3O_7$, *Proceedings Advanced Solid State Lasers* (1997), p. 467.

734. Heumann, E., Mobert, P., and Huber, G., Room-temperature upconversion-pumped cw Yb,Er:YLF laser at 1.234 μm, *OSA Trends in Optics and Photonics on Advanced Solid State Lasers*, Vol. 1, Payne, S. A. and Pollack, C. R., Eds., Optical Society of America, Washington, DC (1996), p. 288.

735. Kaminskii, A. A., Spectral composition of laser light from neodymium-doped calcium tungstate crystals, *Inorg. Mater. (USSR)* 347 (1970).

736. Kaminskii, A. A., Mill', B. V., Belokoneva, E. L., and Khodzhabagyan, G. G., Growth and crystal structure of a new inorganic lasing material $La_3Ga_5GeO_{14}$-Nd^{3+} *Inorg. Mater. (USSR)* 19, 1559 1983.

737. Paylyuk, A. A., Kozeeva, L. I., Folin, K. G., Gladyshev, V. G., Gulyaev, V. S., Pivbov, V. S., and Kaminskii, A. A., Stimulated emission on the transition $^4F_{3/2} \rightarrow {}^4I_{11/2}$ of Nd^{3+} ions in $RbNd(WO_4)_2$ and $CsNd(MoO_4)_2$, *Inorg. Mater. (USSR)* 19, 767 (1983).

738. White, K. O. and Schlenser, S. A., Coincidence of Er:YAG laser emission with methane absorption at 1645.1 nm, *Appl. Phys. Lett.* 21, 419 (1972).

739. Johnson, L. F., Dietz, R. E., and Guggenheim, H. J., Optical maser oscillation from Ni^{2+} in MgF_2 involving simultaneous emission of phonons, *Phys. Rev. Lett.* 11, 318 (1963).

740. Johnson, L. F., Guggenheim, H. J., and Thomas, R. A., Phonon-terminated optical masers, *Phys. Rev.* 149, 179 (1966).

741. White, K. O. Watkins, W. R., and Schleusener, S. A., Holmium 2.06-mm laser spectral characteristics and absorption by CO2 gas, *Appl. Optics* 14, 16 1975).

742. Kaminskii, A. A., Kurbanov, K., Sattarova, M. A., and Fedorov, P. P., Stimulated IR emission of Nd^{3+} ions in nonstoichiometric cubic fluorides, *Inorg. Mater. (USSR)* 21, 609 (1985).

743. Stoneman, R.C. and Esterowitz, L., Continuous-wave 1.50-μm thulium cascade laser, *Opt. Lett.* 16, 232 (1991).

744.   Simondi-Teisseire, B., Viana, B., Lejus, A.-M. et al., Room-temperature CW laser operation at ~1.55 µm (eye-safe region) of Yb:Er and Yb:Er:Ce:$Ca_2Al_2SiO_7$ crystals, *IEEE J. Quantum Electron.* 32, 2004 (1996).

745.   Kaminskii, A. A., Lapsker, Ya. Ye., and Sobolev, B. P., Induced emission of $NaCaCeF_6$:Nd^{3+} at room temperature, *Phys. Stat. Sol.* 23, K5 (1967).

746.   Quaries, G.J., Rosenbaum, A., Marquardt, C.L., and Esterowitz, L., High-efficiency 2.09 µm flashlamp-pumped laser, *Appl. Phys. Lett.* 55, 1062 (1989).

747.   Brinkman, R., Sohler, W., and Suche, H., Continuous-wave erbium-diffused $LiNbO_3$ waveguide laser, *Electron. Lett.* 415 (1991).

748.   Klein, P. B., Furneaux, J. E., and Henry, R. L., Laser oscillation at 3.53 µm from Fe^{2+} in n-InP:Fe, *Appl. Phys. Lett.* 42, 638 (1983).

749.   Morozov, A. M., Pogkolzina, I. G., Tkachuk, A. M., Fedorov, V. A., and Feofilov, P. P., Luminescence and induced emission of lithium-erbium and lithium-holmium binary fluorides, *Opt. Spectrosc. (USSR)* 39, 605 (1975).

750.   Gifeisman, Sh. N., Tkachuk, A. M., and Prizmak, V. V., Optical spectra of Ho^{3+} ion in $LiYF_4$ crystals, *Opt. Spectrosc. (USSR)* 44, 68 (1978).

751.   Lenth, W., Hattendorff, H.-D., Huber, G., and Lutz, F., Quasi-cw laser action in $K_5Nd(MoO_4)_4$, *Appl. Phys.* 17, 367 (1978).

752.   Becker, P., Brinkmann, R., Dinand, M., Sohler, W., and Suche, H., Er-diffused Ti:$LiNbO_3$ waveguide laser of 1563 and 1576 nm emission wavelengths, *Appl. Phys. Lett.* 61, 1257 (1992).

753.   Ballman, A. A., Porto, S. P. S., and Yariv, A., Calcium niobate $Ca(NbO_3)_2$–A new laser host, *J. Appl. Phys.* 34, 3155 (1963).

754.   Moulton, P. F. and Mooradian, A., Broadly tunable CW operation of Ni:MgF_2, and Co:MgF_2, lasers, *Appl. Phys. Lett.* 35, 838 (1979). (Unpublished results which improve upon those indicated in this reference have been included.)

755.   Schmaul, B., Huber, G., Clausen, R., Chai, B., LiKamWa, P., and Bass, M., Er^{3+}:$YLiF_4$ continuous wave cascade laser operation at 1620 and 2810 nm at room temperature, *Appl. Phys. Lett.* 62, 541 (1993).

756.   Camargo, M. B., Stultz, R. D., and Birnbaum, M., Passive Q switching of the Er^{3+}:$Y_3Al_5O_{12}$ laser at 1.64 µm, *Appl. Phys. Lett.* (1995).

757.   Kiss, Z. J. and Duncan, R. C., Optical maser action in $CaWO_4$:Er^{3+}, *Proc. IRE* 50, 1531 (1962).

758.   Kaminskii, A. A., Sarkisov, S. E., Bohm, J. et al., Growth, spectroscopic and laser properties of crystals in the $K_5Bi_{1-x}Nd_x(MoO_4)_4$ system, *Phys. Stat. Sol.* 43a, 71 (1977).

759.   Antipenko, B. M., Asbkalunin, A. L., Mak, A. A., Sinitsyn, B. V., Tomashevlch, Yu. V., and Shakhkalamyan, G. S., Three-micron laser action in Dy^{3+}, *Sov. J. Quantum Electron.* 10, 560 (1980).

760.   Pollack, S. A., Stimulated emission in CaF_2:Er^{3+}, *Proc. IEEE* 51, 1793 (1963).

761.   Bowman, S. R., Ganem, J., Feldman, B. J., and Kueny, A. W., Infrared laser characteristics of praseodymium-doped lanthanum trichloride, *IEEE J. Quantum Electron.* 30, 2925 (1994).

762.   Schmid, F. and Khattak, C. R., Growth of Co:MgF_2 and Ti:Al_2O_3 crystals for solid state laser applications, in *Tunable Solid State Lasers,* Proc. Int. Conf., Vol. 47, Hammerling, P., Budgor, A. B., and Pinto, A., Eds., Springer-Verlag, New York (1985), p. 22.

763.   Kaminskii, A. A., Kurbanov, K., Sarkisov, S. E., Sattarova, M. M., Uvarova, T. V., and Fedorov, P. P., Stimulated emission of Nd^{3+} ions in non-stoichiometric $Cd_{1-x}Ce_xF_{2+x}$ and $Cd_{1-x}Nd_xF_{2+x}$ fluorides with fluorite structure. *Phys. Stat. Sol. A*: 90, K55 (1985).

764.   Aleksandrov, V. I., Murina, T. M., Zhekov, V. K., and Tatarintsev, V. M., Stimulated emission from Tm^{3+} and Ho^{3+} in zirconium dioxide crystals, in Sbornik. Kratkiye Soobshcheniya po Fizike, *An SSSR Fizicheskiy Institut im P. N. Lebedeva* (1973), No. 2, p. 17 (in Russian).

765.   Kaminskii, A. A., Klevtsov, P. V., and Pavlyuk, A. A., Stimulated emission from $KY(MoO_4)_2$:Nd^{3+}- crystal laser, *Phys. Stat. Sol.* 1a, K91 (1970).

766. Moulton, P. F., Recent advances in transition metal-doped lasers, in *Tunable Solid State Lasers*, Proc. Int. Conf., Vol. 47, Hammerling, P., Budgor, A. B., and Pinto, A., Eds., Springer-Verlag, New York (1985), p. 4.

767. Brixner, L. H. and Flournoy, A. P., Calcium orthovanadate $Ca_3(VO_4)_2$—a new laser host crystal, *J. Electrochem. Soc.* 112, 303 (1965).

768. Moulton, P. F., An investigation of the $Co:MgF_2$ laser system. *IEEE J. Quantum Electron.* QE-21, 1582 (1985).

769. Welford, D. and Moulton, P. F., Room-temperature operation of a $Co:MgF_2$ laser, *Opt. Lett.* 13, 975 (1988).

770. Lovold, S., Moulton, P. F., Kilhinger, D. K., and Menyuk, N., Frequency tuning characteristics of a Q-switched $Co:MgF_2$ laser, *IEEE J. Quantum Electron.* QE-21, 202 (1985).

771. Johnson, B. C., Moulton, P. F., and Mooradian, A., Mode-locked operation of $Co:MgF_2$ and $Ni:MgF_2$ lasers, *Opt. Lett.* 10, 116 (1984).

772. Johnson, B. C., Rosenbluh, M., Moulton, P. F., and Mooradian, A., High average power mode-locked $Co:MgF_2$ laser, in *Ultrafast Phenomena IV*, Proc., Vol. 38, Auston, D. H. and Eisenthal, K. B., Eds., Springer-Verlag, New York (1984), p. 35.

773. Muciel, A. C., Maly, P., and Ryan, J. F., Simultaneous modelocking and Q-switching of a $Co:MgF_2$ laser by loss-modulation frequency detuning, *Opt. Commun.* 61, 125 (1987).

774. Johnson, L. F., Guggenheim, H. J., and Bahnck, D., Phonon-terminated laser emission from Ni^{2+} ions in $KMgF_3$, *Opt. Lett.* 8, 371 (1983).

775. Breteau, J. M., Mekhenin, D., and Auzel, F., Study of the $Ni^{2+}:MgF_2$ tunable laser, *Rev. Phys. Appl.* 22, 1419, 1987 (in French).

776. Huber, G., Duczynski, E. W., and Petermann, K., Laser pumping of Ho-, Tm-, Er-doped garnet lasers at room temperature, *IEEE J. Quantum Electron.* 24, 920 (1988).

777. Dätwyler, M., Lüthy, W., and Weber, H. P., New wavelengths of the $YAlO_3$:Er laser, *IEEE J. Quantum Electron.* QE-23, 158 (1987).

778. Tucker, A. W., Birnbaum, M., Fincher, C. L., and DeShazer, L. G., Continuous-wave operation of $Nd:YVO_4$, at 1.06 and 1.34 μm, *J. Appl. Phys.* 47, 232 (1976).

779. Lisitsyn, V. N., Matrosov, V. N., Pestryakov, E. V., and Trunov, V. I., Generation of picosecond pulses in solid-state lasers using new active media, *J. Sov. Laser Res.* 7, 364 (1986).

780. Telle, H. R., Injection locking of a 1.3 μm laser diode to an $LiNdP_4O_{12}$ laser yields narrow linewidth emission, *Electron. Lett.*, 22, 150 (1986).

781. Greskovich, C. and Chernoch, I. P., Improved polycrystalline ceramic lasers, *J. Appl. Phys.* 45, 4495 (1974).

782. Tocho, J. O., Jaque, F., Sole, J. G., Camarillo, E., Cusso, F., and Munoz Santiuste, J. E., Nd^{3+} active sites in $Nd:MgO:LiNbO_3$ lasers, *Appl. Phys. Lett.* 60, 3206 (1992).

783. Arsen'yev, P. A., Potemkin, A. V., Fenin, V. V., and Senff, I., Investigation of stimulated emission of Er^{3+} ions in mixed crystals with perovskite structure, *Phys. Stat. Sol.* 43a, K 15 (1977).

784. Arsen'yev, P. A. and Bienert, K. E., Absorption, luminescence, and stimulated emission spectra of Er^{3+} ions in $GdAlO_3$ crystals, *Phys. Stat. Sol.* 10a, K85 (1972).

785. Künzel, W., Knierim, W., and Dürr, U., CW infrared laser action of optically pumped $Co^{2+}:KZnF_3$, *Opt. Commun.* 36, 383 (1981).

786. German, K. R., Dürr, U., and Künzel, W., Tunable single-frequency continuous-wave laser action in $Co^{2+}:KZnF_3$, *Opt. Lett.* 11, 12 (1986).

787. Dischler, B. and Wettling, W., Investigation of the laser materials $YAlO_3$:Er and $LiYF_4$:Ho, *J. Phys. D* 17, 1115 (1984).

788. Kaminskii, A. A. and Fedorov, V. A., Cascade stimulated emission in crystals with several metastable states of Ln^{3+} ions, in *Proc. 2d Int. Conf. Trends in Quantum Electron.*, Prokhorov, A. M. and Ursu, I., Eds., Springer-Verlag, New York (1986), p. 69.

789. Kaminskii, A. A., Cascade laser generation by Er^{3+} ions in $YAlO_3$ crystals by the scheme $^4S_{3/2} \rightarrow {}^4I_{11/2}$ and $^4S_{3/2} \rightarrow {}^4I_{13/2}$, *Sov. Phys. Dokl.* 27, 1039 (1982).

790. Kaminskii, A. A. and Osiko, V. V., Sensitization in a CaF_2-ErF_3:Tm^{3+} laser, *Inorg. Mater. (USSR)* 3, 519 (1967).

791. Kaminskii, A. A., Sarkisov, S. E., Rysbcbenkov, V. V., Arakelysn, A. Z., Seiranyan, K. B., and Sharkilatunyan, R. O., Growth of CaF_2-HoF_3 and CaF_2-ErF_3 crystals, and their laser properties, *Sov. Phys. Crystallogr.* 27, 118 (1982).

792. Kaminskii, A. A., Fedorov, V. A., and Mochalov, I. V., New data on the three-micron lasing of Ho^{3+} and Er^{3+} ions in aluminates having the perovskite structure, *Sov. Phys. Dokl.* 25, 744 (1980).

793. Thornton, J. R., Rushworth, P. M., Kelly, E. A., McMillan, R. W., and Harper, L. L., in *Proceedings 4th Conference Laser Technology*, Vol. 11, University of Michigan, Ann Arbor (1970), p. 1249.

794. Kaminskii, A. A., Bagdasarov, Kh. S., Petrosyan, A. G., and Sarkisov, S. E., Investigation of stimulated emission from $Lu_3Al_5O_{12}$ crystal with Ho^{3+}, Er^{3+} and Tm^{3+} ions, *Phys. Stat. Solidi* 18a, K31 (1973).

795. Ivanov, A. O., Mochalov, I. V., Petrov, M. V. et al., Spectroscopic properties of single crystals of rare-earth aluminum garnet and rare-earth-orthoaluminate activated by the ions Ho^{3+}, Er^{3+} and Tm^{3+}, in *Abstracts of Papers of Fifth All-Union Symposium on Spectroscopy of Crystals*, Kazan (1976), 195 (in Russian).

796. Kaminskii, A. A., Pavlyuk, A. A., Butaeva, T. I. et al., Stimulated emission by subsidiary transitions of Ho^{3+} and Er^{3+}- ions in $KGd(WO_4)_2$ crystals, *Inorg. Mater. (USSR)* 13, 1251 (1977).

797. Andreae, T., Meschede, D., and Hänsch, T.W., New cw laser lines in the Er:$YAlO_3$ crystal, *Optics Comm.* 79, 1062 (1989).

798. Kaminskii, A. A. and Osiko, V. V., Sensitization in optical quantum generators based on CaF_2-ErF_3-Tm^{3+}, *Inorg. Mater. (USSR)* 3, 519 (1967).

799. Rines, D. M., Moulton, P. F., Welford, D., and Rines, G. A., High-energy operation of a Co:MgF_2 laser, *Opt. Lett.* 19, 628 (1994).

800. Stoneman, R.C. and Esterowitz, L., Efficient, broadly tunable, laser-pumped Tm:YAG and Tm:YSGG cw lasers, *Opt. Lett.* 15, 486 (1990).

801. Johnson, L. F., Dietz, R. E., and Guggenheim, H. J., Spontaneous and stimulated emission from Co^{2+}- ions in MgF_2 and ZnF_2, *Appl. Phys. Lett.* 5, 21 (1964).

802. Zverev, G. M., Garmash, V. M., Onischenko, A. M. et al., Induced emission by trivalent erbium ions in crystals of yttrium-aluminum garnet, *J. Appl. Spectrosc. (USSR)* 21, 1467 (1974).

803. Ivanov, A. O., Mochalov, I. V., Tkachuk, A. M., Fedorov, V. A., and Feofilov, P. P., Spectral characteristics of the thulium ion and cascade generation of stimulated radiation in a $YAlO_3$:Tm^{3+}:Cr^{3+} crystal, *Sov. J. Quantum Electron.* 5, 117 (1975).

804. Arsenev, P. A. and Bienert, K. E., Absorption, luminescence and stimulated emission spectra of Tm^{3+} in $GdAlO_3$ crystals, *Phys. Stat. Sol.* 13a, K125 (1972).

805. Bagdasarov, Kh. S., Kaminskii, A. A., and Sobolev, B. P., Stimulated emission in lasers based on α-$NaCaErF_6$:Ho^{3+} and α-$NaCaErF_6$:Tm^{3+} crystals, *Inorg. Mater. (USSR)* 5, 527 (1969).

806. Moulton, P. F., Mooradian, A., and Reed, T. B., Efficient CW optically pumped Ni:MgF_2 laser, *Opt. Lett.* 3, 164 (1978).

807. Wallace, R. W., Oscillation of the 1.833 μm line in Nd^{3+}:YAG, *IEEE J. Quantum Electron.* QE-7, 203 (1971).

808. Barnes, N. P., Allen, R. E., Esterowitz, L., Chickis, E. P., Knights, M. G., and Jenssen, H. R., Operation of an Er YLF laser at 1.73 μm, *IEEE J. Quantum Electron.* QE-22, 337 (1986).

809. Kaminskii, A. A., Pavlyuk, A. A., Polyakov, A. I., and Lyubchenko, V. V., A new lasing channel in a self-activated erbium crystal $KEr(WO_4)_2$, *Sov. Phys. Dokl.* 28, 154 (1983).

810. Krupke, W. F. and Gruber, J. B., Energy levels of Er^{3+} in LaF_3 and coherent emission at 1.61 μm, *J. Chem. Phys.* 41, 1225 (1964).

811. Esterowitz, L., Allen, R., and Eckardt, R., Cascade laser action in Tm^{3+}:YLF, in *The Rare Earths in Modern Science and Technology*, Vol. 3, McCarthy, G. I., Silber, H. B., and Rhyne, J. J., Eds., Plenum Press, New York (1982), p. 159.

812. Kaminskii, A. A., Two lasing channels of Tm^{3+} ions in lithium-yttrium fluoride, *Inorg. Mater. (USSR)* 19, 1247 (1983).

813. Kaminskii, A. A., Sobolev, B., and Uvarova, T. V., Stimulated emission of Ho^{3+} in $BaTm_2F_8$ and Tm^{3+} ions in $BaYb_2F_8$:Er^{3+} crystals, *Phys. Stat. Sol. A* 78, K13 (1983).

814. Antipenko, B. M., Glebov, A. S., Kiseleva, T. I., and Plamennyi, V. A., A new spectroscopic scheme of an active medium for the 2-μm band, *Opt. Spectrosc. (USSR)* 60, 95 (1986).

815. Antipenko, B. M., Glebov, A. S., Kiseleva, T. I., and Plamennyi, V. A., Conversion of absorbed energy in YAG:Cr^{3+},Tm^{3+},Ho^{3+} crystals, *Opt. Spectrosc. (USSR)* 64, 221 (1988).

816. Soffer, B. H. and Hoskins, R. H., Energy transfer and CW laser action in Tm^{3+}- Er_2O_3, *Appl. Phys. Lett.* 6, 200 (1968).

817. Johnson, L. F., Geusic, J. E., and Van Uitert, L. G., Efficient, high-power coherent emission from Ho^{3+} ions in yttrium aluminum garnet, assisted by energy transfer, *Appl. Phys. Lett.* 8, 200 (1966).

818. Smirnov, V. A. and Shcherbakov, I. A., Rare-earth scandium chromium garnets as active media for solid-state lasers, *IEEE J. Quantum Electron.* 24, 949 (1988).

819. Giorbachov, V. A., Zhekov, V, I., Murina, T. M. et al., Spectroscopic and growth properties of erbium aluminate with Tr^{3+} impurity ions, *Short Communications in Physics* 4, 16 (1973).

820. Johnson, L. F., Boyd, G. D., and Nassau, K., Optical maser characteristics of Ho^{3+} in $CaWO_4$, *Proc. IRE* 50, 87 (1962).

821. Johnson, L. F., Van Uitert, L. A., Rubin, J. J., and Thomas, R. A., Energy transfer from Er^{3+} to Tm^{3+} and Ho^{3+} ions in crystals, *Phys. Rev.*, 133A, 494 (1964).

822. Johnson, L. F., Boyd, G. D., and Nassau, K., Optical maser characteristics of Tm^{3+} in $CaWO_4$, *Proc. IRE.* 50, 86 (1962).

823. Van Uitert, L. G., Grodkiewicz, W. H., and Dearborn, E. F., Growth of large optical-quality yttrium and rare-earth aluminum garnets, *J. Am. Ceram. Soc.* 48, 105 (1965).

824. Barnes, N. P., Murray, K. E., Jani, M. G., and Kokta, M., Flashlamp pumped Ho:Tm:Cr:LuAG laser, *OSA Proc. Adv. Solid State Lasers*, Chai, B. H. T. and Payne, S. A., Eds., 24, 352 (1995).

825. Arsenyev, P. A. and Bienert, K. E., Spectral properties of Ho^{3+} in $GdAlO_3$ crystals, *Phys. Stat. Sol.* 13a, K129 (1972).

826. Rosenblatt, G. H., Quarles, G. J., Esterowitz, E., Randles, M., Creamer, J., and Belt, R., Continuous-wave 1.94-μm Tm:$CaY_4(SiO_4)_3O$ laser, *Opt. Lett.* 18, 1523 (1993).

827. Saito, H., Chaddha, S., Chang, R. S. F., and Djeu, N., Efficient 1.94-μm Tm^{3+} laser in YVO_4 host, *Opt. Lett.* 17, 189 (1992).

828. Pinto, J. F. and Esterowitz, L., Tunable, flashlamp-pumped operation of a Cr,Tm:YAG laser between 1.945 and 2.014 μm, in *Advanced Solid-State Lasers*, Jenssen, H. P. and Dubé, G., Eds., Proceedings Vol. 6, Optical Society of America, Washington, DC (1990), p. 134.

829. Alpat'ev, A. N., Denisov, A. L., Zharikov, E. V., Zubenko, D. A., Kalitin, S. P., Noginov, M. A., Saidov, Z. S., Smirnov, V.A., Umyskov, A. F., and Shcherbakov, I. A., Crystal YSGG:Cr^{3+}:Tm^{3+} laser emitting in the 2-μm range, *Sov. J. Quantum Electron.* 20, 780 (1990).

830. Voron'ko, Yu. K., Gessen, S. B., Es'kov, N. A., Zverev, A. A., Ryabochkina, P. A., Sobol' A. A., Ushakov, S, N., and Tsymbal, L. I., Spectroscopic and lasing properties of calcium niobium gallium garnet activated with Tm^{3+} and Cr^{3+} ions, *Sov. J. Quantum Electron.* 22, 581 (1992).

831. Jani, M. G., Barnes, N. P., Murray, K. E., Hart, D. W., Quarles, G. J., and Castillo, V. K., Diode-pumped Ho:Tm:$LuLiF_4$ laser at room temperature, *IEEE J. Quantum Electron.* 32, 113 (1997).

832. Rubin, J. J. and Van Uitert, L. G., Growth of large yttrium vanadate single crystals for optical maser studies, *J. Appl. Phys.* 37, 2920 (1966).

833. Wunderlich, J. A., Sliney, J. A., and DeShazer, L. G., Stimulated emission at 2.04 μm in Ho^{3+}- doped $ErVO_4$ and YVO_4, *IEEE J. Quantum Electron.* QE-13, 69 (1977).

834. Dmitruk, M. V. and Kaminskii, A. A., Stimulated emission in a laser based on CaF_2-YF_3 crystals with Ho^{3+} and Er^{3+} ions, *Sov. Phys.-Crystallogr.* 14, 620 (1970).

835. Dmitruk, M. V., Kaminskii, A. A., Osiko, V. V., and Fursikov, M. M., Sensitization in ·CaF_2- ErF_3:Ho^{3+} lasers, *Inorg. Mater. (USSR)* 3, 516 (1967).

836. Voron'ko, Yu. K., Dmitruk, M. V., Murina, T. M., and Osiko, V. V., CW lasers based on mixed yttrofluorine-type crystals, *Inorg. Mater. (USSR)* 5, 422 (1969).

837. Robinson, M. and Devor, D. P., Thermal switching of laser emission of Er^{3+} at 2.69 μ and Tm^{3+} at 1.861 μ in mixed crystals of CaF_2:ErF_3:TmF_3, *Appl. Phys. Lett.* 10, 167 (1967).

838. Johnson, L. F. and Guggenheim, H. J., Electronic- and phonon-terminated laser emission from Ho^{3+} in BaY_2F_8, *IEEE J. Quantum Electron.* QE-10, 442 (1974).

839. Korovkin, A. M., Morozov, A. M., Tkachuk, A. M., Fedorov, A. A., Fedorov, V. A., and Feofilov, P. P., Spontaneous and stimulated emission from holmium in $NaLa(MoO_4)_2$ and $LaNbO_4$ crystals, in *Sbornik. Spektroskopiya Krisaliov*, Nauka, Moskova (1975).

840. Bagdasarov, Kh. S., Kaminskii, A. A., Kevorkov, A. M., Sarkisov, S. E., and Tevosyan, T. A., Stimulated emission from (Er,Lu)AlO_3 crystals with Ho^{3+} and Tm^{3+} ions, *Sov. Phys.-Crystallogr.* 18, 681 (1974).

841. Kalisky, Y., Kagan, J., Lotem, H., and Sagie, D., Continuous wave operation of multiply doped Ho:YLF and Ho:YAG laser, *Opt. Commun.* 65, 359 (1988).

842. Barnes, N. P., Eye-safe solid-state lasers for LIDAR applications, in Proc. Soc. Photo. Opt. Instrum. Eng., *Laser Radar Technology and Applications,* Vol. 663, Cruickshank, I. M. and Harney R. C., Eds., SPIE, Bellingham, WA (1986), p. 2.

843. Erbil, A. and Jenssen, H. P., Tunable Ho^{3+}:YLF laser at 2.06 μm, *Appl. Opt.* 19, 1729 (1980).

844. Petrov, M. V. and Tkachuk, A. M., Delayed stimulated afterglow from holmium ions in crystals with coactivators, *Sov. J. Quantum Electron.* 10, 1478 (1980).

845. Anan'eva, G. V., Baranov, E. N., Zarzhitskaya, M. N. et al., Growth and physico-chemical investigation of single crystals of tysonite solid solutions $(Y,Ln)_{1-x}Sr_xF_{3-x}$, *Inorg. Mater. (USSR)* 16, 52 (1980).

846. Antipenko, B. M., Vorykhalov, I. V., Sirilitsyn, B. V., and Uvarova, T. V., Laser frequency converter based on a $BaYb_2F_8$:Ho^{3+} crystal stimulated emission at 2 μm, *Sov. J. Quantum Electron.* 10, 114 (1980).

847. Gillespie, P. S., Armstrong, R. L., and White, K. O., Spectral characteristics and atmospheric CO_2 absorption of the Ho^{+3}:YLF laser at 2.05 μm, *Appl. Optics* 15, 865 (1976).

848. Kaminskii, A. A., Petrosyan, A. G., Federov, V. A., Ryabchenkov, V. V., Pavlyuk, A. A., Lyubachenko, V. V., and Lukin, A. V., Two-micron stimulated emission of radiation by Ho^{3+} based on the $^5I_7 \rightarrow {}^5I_8$ transition in sensitized crystals, *Inorg. Mater. (USSR)* 17, 1430 (1981).

849. Cockayne, B., Plant, J. G., and Clay, R. A., The Czochralski growth and laser characteristics of Li(Y,Er,Tm,Ho)F, and Li(Lu,Er,Tm,Ho)F_4 scheelite single crystals, *J. Cryst. Growth* 54, 407 (1981).

850. Antipenko, B. M., Podkolzina, I. G., and Tomashevich, Yu. V., Use of $LiYbF_4$:Ho^{3+} as an active medium in a laser frequency converter, *Sov. J. Quantum Electron.* 10, 370 1980.

851. Kaminskii, A. A., Fedorov, V. A., Ryabchenkov, V., Sarkisov, S. E., Schultze, D., Bohm, J., and Reiche, P., Cascade generation of Ho^{3+} ions in $Gd_3Ga_5O_{12}$ crystal by the scheme $^5I_6$—$^5I_7$—$^5I_8$, *Inorg. Mater. (USSR)* 17, 828 (1981).

852. Podkolizina, I. G., Tkachuk, A. M., Fedorov, V. A., and Feofilov, P. P., Multifrequency generation of stimulated emission of Ho^{3+} ion in $LiYF_4$ crystals, *Opt. Spectrosc. (USSR)* 40, 111 (1976).

853. Chicklis, E. P., Naiman, C. S., Folweiller, R. C., Gabbe, D. R., Jenssen, H. P., and Linz, A., High efficiency room-temperature 2.06-μm laser using sensitized Ho^{3+}:YLF, *Appl. Phys. Lett.* 19, 119 (1971).

854. Remski, R. L., James, L. T., Gooen, K. H., DiBartolo, B., and Linz, A., Pulsed laser action in $LiYF_4$:Er^{3+}, Ho^{3+} at 77°K, *IEEE J. Quantum Electron.* QE-5, 212 (1969).

855. Gong, G. Z., Xu, G., Han, K., and Zhai, G., $Nd:MgO:LiNbO_3$ self-frequency-doubled laser pumped by a flashlamp at room temperature, *Electron. Lett.* 26, 2063 (1990).

856. Morozov, A. M., Petrov, M. V., Startsev, V. R. et al., Luminescence and stimulated emission of holmium in yttrium- and erbium-oxyortho-silicate single crystals, *Opt. Spectrosc.* 41, 641 (1976).

857. Johnson, L. F., Dillon, J. F., and Remeika, J. P., Optical properties of Ho^{3+} ions in yttrium gallium garnet and yttrium iron garnet, *Phys. Rev.*, Bl, 1935 (1970).

858. Jani, M. G., Reeves, R. J., and Powell, R. C., Alexandrite-laser excitation of a Tm:Ho:$Y_3Al_5O_{12}$ laser, *J. Opt. Soc. Am. B* 8, 741, (1991).

859. Cha, S., Sugimoto, N., Chan, K., and Killinger, D. K., Tunable 2.1 μm Ho laser for DIAL remote sensing of atmospheric water vapor, in *Advanced Solid-State Lasers*, Jenssen, H. P. and Dubé, G., Eds., Proceedings Vol. 6, Optical Society of America, Washington, DC (1990), p. 165.

860. Dixon, G. J. and Johnson. L. F., Low-threshold 2-μm holmium laser excited by nonradiative energy transfer from Fe^{3+} in YGG, *Opt. Lett.* 17, 1782 (1992).

861. Alpat'ev, A. N., Zharikov, E. V., Zagumennyi, A. I., Zubenko, D. A., Kalitin, S. P., Lutts, G. B., Noginov, M. A., Smirnov, V. A., Umyskov, A. F., and Shcherbakov, I. A., Holmium GSAG:Cr^{3+}:Tm^{3+}:Ho^{3+} crystal laser (λ=2.09 μm) operating at room temperature, *Sov. J. Quantum Electron.* 19, 1400 (1989).

862. Barnes, N. P., Filer, E. D., Naranjo, F. L., Rodriguez, W. J., and Kokta, M. R., Spectroscopic and lasing properties of Ho:Tm:LuAG, *Opt. Lett.* 18, 708 (1993).

863. McFarlane, R. A., Robinson, M., Pollack, S. A., Chang, D. B., and Jenssen, H. P., Visible and infrared laser operation by upconversion pumping of erbium-doped fluorides in *Tunable Solid State Lasers*, Shand, M. L. and Jenssen H. P., Eds., Optical Society of America, Washington, DC (1989), p. 179.

864. Ashurov, M. Kh., Voron'ko, Yu. K., Zharikov, E. V., Kaminskii, A. A., Osiko, V. V., Sobol', A. A., Timoshechkin, M. I., Fedorov, V. A., and Shabaltai, A. A., Structure, spectroscopy, and stimulated emission of crystals of yttrium holmium aluminum garnets, *Inorg. Mater. (USSR)* 15, 979 (1979).

865. Beck, R. and Gurs, K., Ho laser with 50-W output and 6.5% slope efficiency, *J. Appl. Phys.* 46, 5224 (1975).

866. Arsen'yev, P. A., Spectral parameters of trivalent holmium in a YAG lattice, *Ukrainskiy Fizicheskiy Zhurnal* 15, 689, 1970 (in Russian).

867. Diening, A., Dicks, B.-M., Heumann, E., Meyn, J. P., Petermann, K. and Huber, G., Continuous wave lasing near 2 μm in Tm^{3+} doped Y_2O_3, *Proceedings Advanced Solid State Lasers* (1997), p. 221.

868. Johnson, L. F., Geusic, J. E., and Van Uitert, L. G., Coherent oscillations from Tm^{3+}, Ho^{3+}, Yb^{3+} and Er^{3+}- ions in yttrium aluminum garnet, *Appl. Phys. Lett.* 7, 127 (1965).

869. Duczynski, E. W., Huber, G., Ostroumov, V. G., and Shcherbakov, I. A., CW double cross pumping of the $^5I_7 - {}^5I_8$ laser transition in Ho^{3+}-doped garnets, *Appl. Phys. Lett.* 48, 1562 (1986).

870. Antipenko, B. M., Mak, A. A., and Sukhareva, L. K., Cross-relaxation $BaEr_2F_8$:Tm + Ho laser, *Sov. Tech. Phys. Lett.* 10, 217 (1984).

871. Antipenko, B. M., Glebov, A. S., Danbravyanu, R. V., Sobolev, B. P., and Uvarova, T. V., Spectroscopy and lasing characteristics of $BaEr_2F_8$:Tm:Ho crystals, *Sov. J. Quantum Electron.* 17, 424 (1987).

872. Antipenko, B. M., Glebov, A. S., and Dumbravyanu, R. V., Physics of energy storage in a $BaEr_2F_8$:Tm:Ho active medium, *Sov. J. Quantum Electron.* 18, 806 (1988).

873. Antipenko, B. M., Krutov, L. I., and Sulbsrevs, L. K., Cascade lasing of GSGG-Cr + Tm + Ho crystals, *Opt. Spectrosc. (USSR)* 61, 414 (1986).
874. Alpat'ev, A. N., Zharikov, E. V., Klifin, S. P., Laptev, V. V., Osiko, V. V., Ostroumov, V. G., Prokhorov, A. M., Salaov, Z. S., Smirnov, V. A., Sorokov, I. T., Umyskov, A. F., and Shcherbakov, I. A., Lasing of holmium ions as a result of the $^5I_7 \to ^5I_8$ transition at room temperature in an yttrium scandium gallium garnet crystal activated with chromium, thulium, and holmium ions, *Sov. J. Quantum Electron.* 16, 1404 (1986).
875. Alpat'ev, A. N., Zharikov, E. V., Kalitin, S. P., Umyskov, A. F., and Shcherbakov, I. A., Efficient room-temperature lasing ($\lambda = 2.088$ μ) of yttrium scandium gallium garnet activated with chromium, thulium, and holmium ions, *Sov. J. Quantum Electron.* 17, 587 (1987).
876. Alpat'ev, A. N., Zharikov, E. V., Kalitin, S. P., Smirnov, V. A., Umyskov, A. F., and Shcherbakov, I. A., Q-switched laser utilizing an yttrium scandium gallium garnet crystal activated with holmium ions, *Sov. J. Quantum Electron.* 18, 617 (1988).
877. Kostin, V, V., Kulevsky, L. A., Murina, T. M. et al., CaF_2:Dy^{2+} giant pulse laser with high repetition rate. *IEEE J. Quantum Electron.* QE-2, 611 (1966).
878. Barnes, N. R. and Gettemy, D. J., Pulsed Ho:YAG oscillator and amplifier, *IEEE J. Quantum Electron.* QE-17, 1303 (1981).
879. Fan, T. Y., Huber, G., Byer, R. L., and Mitzscherlich, P., Continuous-wave operation at 2.1 μm of a diode-laser-pumped, Tm-sensitized Ho:$Y_3Al_5O_{12}$ laser at 300 K *Opt. Lett.* 12, 678 (1987).
880. Fan, T. Y., Huber, G., Byer, R. L., and Mitzscherlich, P., Spectroscopy and diode laser-pumped operation of Tm Ho:YAG, *IEEE J. Quantum Electron.* 24, 924 (1988).
881. Kaminskii, A. A., Kurbanov, K., and Petrosyan, A. G., Spectral composition and kinetics of 2 μm stimulated emission of Ho^{3+} ions in sensitized $Y_3Al_5O_{12}$ and $Lu_3Al_5O_{12}$ single crystals, *Phys. Stat. Sol. A* 98, K57 (1987).
882. Kaminskii, A. A., Petrosyan, A. G., and Fedorov, V. A., Cross-cascade stimulated emission of Er^{3+}, Ho^{3+}, and Tm^{3+} ions in $Er_3Al_5O_{12}$, *Sov. Phys. Dokl.* 26, 309 (1981).
883. Storm, M. E., Laser characteristics of a Q-switched Ho:Tm:Cr:YAG, *Appl. Opt.* 27, 4170 (1988).
884. Hemmati, H., Efficient holmium:yttrium lithium fluoride laser longitudinally pumped by a semiconductor laser array, *Appl. Phys. Lett.* 51, 564 (1987).
885. Allen, R., Esterowitz, L., Goldberg, L., and Weiler, J. F., Diode-pumped 2 μm holmium laser, *Electron. Lett.* 22, 947 (1986).
886. Esterowitz, L., Allen, R., Goldberg, L., Weller, J. F., Sterm, M., and Abella, I., Diode-pumped 2 μm holmium laser, in *Tunable Solid-State Lasers II*, Vol. 52. Budgor, A. B., Esterowitz, L. and DeShazer, L. G., Eds., Springer-Verlag, New York (1986), p. 291.
887. Kaminskii, A. A., Petrosyan, A. G., Fedorov, V. A., Sarkisov, S. E., Ryabchenkov, V. V., Pavlyuk, A. A., Lyubchenko, V. V., and Mechalov, I. V., Two-micron stimulated emission by crystals with Ho^{3+} ions based on the transition $^5I_7$–$^5I_8$, *Sov. Phys. Dokl.* 26, 846 (1981).
888. Kintz, G. J., Esterowitz, L., and Allen, R., CW diode-pumped Tm^{3+}, Ho^{3+}:YAG 2.1 μm room-temperature laser, *Electron. Lett.* 23, 616 (1987).
889. Kaminskii, A. A., Petrosyan, A. G., and Ovanesyan, K. L., Cross-cascade generation of the stimulated emission of Tm^{3+} and Ho^{3+} ions in $Lu_3Al_5O_{12}$:Cr^{3+}-Tm^{3+}, Ho^{3+}, *Inorg. Mater. (USSR)* 19 1098 (1983).
890. Antipenko, B. M., Glebov, A. S., Kiseleva, T. I., and Pismennyi, V. A., 2.12-μm Ho:YAG laser *Sov. Tech. Phys. Lett.* 11, 284 (1985).
891. Antipenko, B. M., Glebov, A. S., Kiseleva, T. I., and Pismennyi, V. A., Interpretation of the temperature dependence of the YAG:Cr^{3+} Tm^{3+} ho lasing threshold, *Opt. Spectrosc. (USSR)* 63, 230 (1987).
892. Ivanov, A. O., Mochalov, I. V., Tkachuk, A. M., Fedorov, V. A., and Feofilov, P. P., Emission of $\lambda = 2$ μ stimulated radiation by holmium in aluminum holmium garnet crystals, *Sov. J. Quantum Electron.* 5, 115 (1975).

893. Hoskins, R. H. and Soffer, B. H., Energy transfer and CW laser action in Ho^{3+}:Er_2O_3, *IEEE J. Quantum Electron.* QE-2, 253 (1966).

894. Robertson, A., Knappe, R., and Wallenstein, R., Diode-pumped broadly tunable (809–910 nm) femtosecond Cr:LiSAF laser, *Opt. Commun.* 147, 294 (1998).

895. Bakradze, R. V., Zverev, G. M., Kolodnyi, G. Ya. et al., Sensitized luminescence and stimulated radiation from yttrium-aluminum garnet crystals, *Sov. Phys.-JETP* 26, 323 (1968).

896. Remski, R. L. and Smith, D. J., Temperature dependence of pulsed laser threshold in YAG: Er^{3+}, Tm^{3+}, Ho^3, *IEEE J. Quantum Electron.* QE-6, 750 (1970).

897. Hopkins, R. H., Melamed, N. T., Henningsen, T., and Roland, G. W., Technical Rpt. AFAL-TR-70-103 (1970), Air Force Avionics Laboratory, Dayton, OH.

898. Porto, S. P. S. and Yariv, A., Trigonal sites and 2.24 micron coherent emission of U^{3+} in CaF_2, *J. Appl. Phys.* 33, 1620 (1962).

899. Porto, S. P. S. and Yariv, A., Low lying energy levels and comparison of laser action of U^{3+} in CaF_2 in *Proceedings 3rd International Conference Quantum Electronics*, Grivet, P. and Bloembergen, N., Eds., Columbia University Press, New York (1964), p. 717.

900. Wittke, J. P., Kiss, Z. J., Duncan, R. C., and McCormick, J. J., Uranium-doped calcium fluoride as a laser material, *Proc. IEEE* 51, 56 (1963).

901. Porto, S. P. S. and Yariv, A., Optical maser characteristics BaF_2:U^{3+}, *Proc. IRE* 50, 1542 (1962).

902. Porto, S. P. S. and Yariv, A., Excitation, relaxation and optical maser action at 2.407 microns in SrF_2:U^{3+}, *Proc. IRE* 50, 1543 (1962).

903. Zolotov, Ye. M., Osiko, V. V., Prokhorov, A. M., and Shipulo, O. P., Study of fluorescence and laser properties of SrF_2:Dy^{2+} crystals, *J. Appl. Spectrosc. (USSR)* 8, 627 (1968).

904. Caird, J. A., DeShazer, L. G., and Nella, J., Characteristics of room-temperature 2.3 μm laser emission from Tm^{3+} in YAG and $YAlO_3$, *IEEE J. Quantum Electron.* QE-11, 874 (1975).

905. Hobrock, L. M., DeShazer, L. G., Krupke, W. F., Keig, G. A., and Witter, D. E., Four-level operation of Tm:Cr:$YAlO_3$ laser at 2.35 μm, in *Digest of Technical Papers VII International Quantum Electronics Conference*, Montreal (1972), p. 15.

906. Kiss, Z. J. and Duncan, R. C., Pulsed and continuous optical maser action in CaF_2:Dy^{2+}, *Proc. IRE* 50, 1531 (1962).

907. Yariv, A., Continuous operation of a CaF_2:Dy^{2+} optical maser, *Proc. IRE* 50, 1699 (1962).

908. Kiss, Z. J., The CaF_2-Tm^{2+} and the CaF_2-Dy^{2+} optical maser systems, in *Quantum Electronics Proceedings of the Third International Congress*, Grivet, P. and Bloembergen, N., Eds., Columbia University Press, New York (1964), p. 805.

909. Kiss, Z. J., Lewis, H. R., and Duncan, R. C., Sun pumped continuous optical maser, *Appl. Phys. Lett.* 2, 93 (1963).

910. Pressley, R. J. and Wittke, J. P., CaF_2:Dy^{2+}-lasers, *IEEE J. Quantum Electron.* QE-3, 116 (1967).

911. Hatch, S. E., Parsons, W. F., and Weagley, J. R., Hot-pressed polycrystalline CaF_2:Dy^{3+} laser, *Appl. Phys. Lett.* 5, 153 (1964).

912. Pinto, J. F., Esterowitz, L., and Rosenblatt, G. H., Tm^{3+}:YLF laser continuously tunable between 2.20 and 2.46 μm, *Opt. Lett.* 19, 883 (1994).

913. Stoneman, R. C., Esterowitz, L., and Rosenblatt, G. H., Tunable 2.3 μm Tm^{3+}:$LiYF_4$ laser, in *Tunable Solid State Lasers*, Shand, M. L. and Jenssen, H. P., Eds., Proceedings Vol. 5, Optical Society of America, Washington, DC (1989), p. 154.

914. Page, R. H., Schaffers, K. I., DeLoach, L. D., Wilke, G. D., Patel, F. D., Tassano, J. B., Payne, S. A., and Krupke, W. F., Cr^{2+}-doped chalcogenides as efficient, widely-tunable mid-infrared lasers, *IEEE J. Quantum Electron.* 33, 609 (1997).

915. DeLoach, L. D., Page, R. H., Wilke, G. D., Payne, S. A., and Krupke, W. F., Transition metal-doped zinc chalcogenides: spectroscopy and laser demonstration of a new class of gain media, *IEEE J. Quantum Electron.* 32, 885 (1996).

916. Arutyunyan, S. M., Kostanyan, R. B., Petrosyan, A. G., and Sanamyan, T. V., $YAlO_3$:Er^{3+} crystal laser, *Sov. J. Quantum Electron.* 17, 1010 (1987).

917. Frauchiger, J., Lüthy, W., Albers, P., and Weber, H. P., Laser properties of selectively excited YAlO$_3$:Er, *Opt. Lett.* 13, 964 (1988).

918. Kaminskii, A. A., Petrosyan, A. G., Denisenko, G. A., Butaeva, T. I., Fedorov, V. A., and Sarkisov, S. E., Spectroscopic properties and 3 μm stimulated emission of Er^{3+} ions in the $(Y^{3+},Er^{3+})_3Al_5O_{12}$ and $(Lu^{3+},Er^{3+})_3Al_5O_{12}$ garnet crystal systems, *Phys. Stat. Sol. A* 71, 291 (1982).

919. Andriasyan, M. A., Vardanyan, N. V., and Kostanyan, R. B., Influence of the absorption of the excitation energy from the $^4I_{13/2}$ level of erbium ions on the operation of Lu$_3$Al$_5$O$_{12}$:Er crystal lasers, *Sov. J. Quantum Electron.* 12, 804 (1982).

920. Kaminskii, A. A. and Petrosyan, A. G., New functional scheme for 3-μ crystal lasers, *Sov. Phys. Dokl.* 24, 363 (1979).

921. Kaminskii, A. A., Seiranyan, K. B., and Arakelysn, A. Z., Peculiarity of the 3-μm stimulated emission of Er^{3+} ions in disordered fluoride crystals, *Inorg. Mater. (USSR)* 18, 446 (1982).

922. Stalder, M., Lüthy, W., and Weber, H. P., Five new 3-μm laser lines in YAlO$_3$:Er, *Opt. Lett.* 12, 602 (1987).

923. Pollack, S. A., Chang, D. B., and Moise, N. L., Continuous wave and Q-switched infrared erbium laser, *Appl. Phys. Lett.* 49 1578 (1986).

924. Pollack, S. A., Chang, D. B., and Moise, N. L., Upconversion-pumped infrared erbium laser, *J. Appl. Phys.* 60, 4077 (1986).

925. Kaminskii, A. A. and Petrosyan, A. G., Stimulated emissions from Er^{3+} ions at 1.7 μm in self-activated oxygen-containing Er crystals, *Inorg. Mater. (USSR)* 18, 1645 (1982).

926. Pollack, S. A. and Chang, D. B., Ion-pair upconversion pumped laser emission in Er^{3+} ions in YAG, YLF, SrF$_2$, and CaF$_2$ crystals, *J. Appl. Phys.* 64, 2885 (1988).

927. Kaminskii, A. A., Sobolev, B. P., Sarkisov, S. E., Fedorov, V. A., Ryabchenkov, V. V., and Uvarova, T. V., A new self-activated crystal for producing three-micron stimulated emission, *Inorg. Mater. (USSR)* 17, 829 (1981).

928. Kurtz, R., Fathe, L., and Birnbaum, M., New laser lines of erbium in yttrium aluminum garnet, in *Advanced Solid-State Lasers*, Jenssen, H. P. and Dubé, G., Eds., Proceedings Vol. 6, Optical Society of America, Washington, DC (1990), p. 247.

929. Es'kov, N. A., Kulevskii, L. A., Lukashev, A. V., Pashinin, P. P., Randoshkin, V. V., and Timoshechkin, M. I., Lasing of a calcium niobium gallium garnet crystal activated with chromium and erbium (λ=2.71 μm), *Sov. J. Quantum Electron.* 20, 785 (1990).

930. Kaminskii, A. A., Fedorov, V. A., Ivanov, A. O., Mochalov, I. V., and Krutov, L. I., Three-micron lasers based on YAlO$_3$ crystals with a high concentration of Ho^{3+} and Er^{3+} ions, *Sov. Phys. Dokl.* 27, 725 (1982).

931. Petrov, M. V. and Tkachuk, A. M., Optical spectra and multifrequency generation of induced emission of LiYF$_4$-Er^{3+} crystals, *Opt. Spectrosc. (USSR)* 45, 147 (1978).

932. Sorokin. P. P. and Stevenson, M. J., Stimulated infrared emission from trivalent uranium, *Phys. Rev. Lett.* 5, 557 (1960).

933. Boyd, G. D., Collins, R. J., Porto, S. P. S., Yariv, A., and Hargreaves, W. A., Excitation, relaxation and continuous maser action in 2.613 μm transition of CaF$_2$:U^{3+}, *Phys. Rev. Lett.* 8, 269 (1962).

934. Antipenko, B. M. and Dolgoborodov, L. E., Y$_3$Al$_5$O$_{12}$:Cr,Tm-Er four-level laser medium, in *Advanced Solid-State Lasers*, Jenssen, H. P. and Dubé, G., Eds., Proceedings Vol. 6, Optical Society of America, Washington, DC (1990), p. 244.

935. Rabinovich, W. S., Bowman, S. R., Feldman, B. J., and Winings, M. J., Tunable laser pumped 3 μm Ho:YAlO$_3$ laser, *IEEE J. Quantum Electron.* 27, 895 (1991).

936. Robinson, M. and Devor, D. P., Thermal switching of laser emission of Er^{3+} at 2.69 μ and Tm^{3+} at 1.86 μ in mixed crystals of CaF$_2$:ErF$_3$:TmF$_3$, *Appl. Phys. Lett.* 10, 167 (1967).

937. Bagdasarov, S. Kh., Kulevskii, L. A., Prokhorov, A. M. et al., Erbium-doped CaF$_2$, crystal laser operating at room temperature, *Sov. J. Quantum Electron.* 4, 1469 (1975).

938. Kaminskii, A. A., Inorganic materials with Ln^{3+} ions for producing stimulated radiation in the 3 μ band, *Inorg. Mater. (USSR)* 15, 809 (1979).

939. Chicklis, E. P., Esterowitz, L., Allen, R., and Kruer, M., Stimulated emission at 2.81 μm in Er:YLF, in *Proceedings of LASERS '78*, Orlando, FL (1978).

940. Prokhorov, A. M., Kaminskii, A. A., Osiko, V. V. et al., Investigations of the 3 μm stimulated emission from Er^{3+} ions in aluminum garnets at room temperature, *Phys. Stat. Sol.* 40a, K69 (1977).

941. Kaminskii, A. A. and Petrosyan, A. G., Sensitized stimulated emission from self-saturating $^3+1$~m transitions of Ho^{3+} and Er^{3+} ions in $Lu_3Al_5O_{12}$ crystals, *Inorg. Mater. (USSR)* 15, 425 (1979).

942. Dinerman, B. J. and Moulton, P. F., 3-μm cw laser operations in erbium-doped YSGG, GGG, and YAG, *Opt. Lett.* 19, 1143 (1994).

943. Hubert, S., Meichenin, D., Zhou, B.W., and Auzel, F., Emission properties, oscillator strengths and laser parameters of Er^{3+} in $LiYF_4$ at 2.7 μm, *J. Lumin.* 50, 7 (1991).

944. Xie, P. and Rand, S.C., Continuous-wave, pair-pumped laser, *Opt. Lett.* 15, 848 (1990).

945. Meichenin, D., Auzel, F., Hubert, S., Simoni, E., Louis, M., and Gesland, J. Y., New room temperature CW laser at 2.82 μm:U^{3+}/$LiYF_4$, *Electron. Lett.* 30, 1309 (1994).

946. Kaminskii, A. A., Pavlyuk, A. A., Butaeva, T. I., Bobovich, L. I., and Lyubchenko, V. V., Stimulated emission in the 2.8 μm band by a self-activated crystal of $KEr(WO_4)_2$, *Inorg. Mater. (USSR)* 15, 424 (1979).

947. Zharikov, E. V., Zhekov, V. I., Murina, T. M., Osiko, V. V., Timoshechkin, M. I., and Shcherbakov, I. A., Cross section of the $^4I_{11/2} \rightarrow {}^4I_{13/2}$ laser transition in Er^{3+} ions in yttrium-erbium-aluminum garnet crystals, *Sov. J. Quantum Electron.* 7, 117 (1977).

948. Kaminskii, A. A., *Moderne Problem der Laserkristallphysik*, Physikalische Gesellschaft der DDR, Dresden (1977).

949. Basiev, T. T., Zharikov, E. V., Zhekov, V. I., Murina, T. M., Osiko, V. V., Prokhorov, A. M., Starikov, B. P., Timoshechkin, M. I., and Shcherbakov, I. A., Radiative and nonradiative transitions exhibited by Er^{3+} ions in mixed yttrium-erbium aluminum garnets, *Sov. J. Quantum Electron.* 6, 796 (1976).

950. Bowman, S. R., Shaw, L. B., Feldman, B. J., and Ganem, J., A. seven micron solid-state laser, *OSA Topical Meeting on Advanced Solid State Lasers* (1995).

951. Pinto, J. F., Rosenblatt, G. H., and Esterowitz, L., Continuous-wave laser action in Er^{3+}:YLF at 3.41 μm, *Electron. Lett.* 30, 1596 (1994).

952. Bowman, S. R., Rabinovich, W. S., Feldman, B. J., and Winings, M. J., Tuning the 3-μm Ho:$YAlO_3$ laser, in *Advanced Solid-State Lasers*, Jenssen, H. P. and Dubé, G., Eds., Proceedings Vol. 6, Optical Society of America, Washington, DC (1990), p. 254.

953. Antipenko, B. M., Buehenkov, V. A., Nikitiehev, A. A., Sobolev, B. P., Stepanov, A. I., Sukhareva, L. K., and Uvarova, T. V., Optimization of a $BaYb_2F_8$:Er active medium, *Sov. J. Quantum Electron.* 16, 759 1986.

954. Spariosu, K. and Birnbaum, M., Room-temperature 1.644-micron Er:YAG Lasers, in *Advanced Solid-State Lasers*, Chase, L. L. and Pinto, A. A., Eds., Proceedings Vol. 13, Optical Society of America, Washington, DC (1992), p. 127.

955. Bowman, S. R., Shaw, L. B., Feldman, B. J., and Ganem, J., A 7-μm praseodymium-based solid-state laser, *IEEE J. Quantum Electron.* 32, 646 (1996).

956. Breguet, J., Umyuskov, A. F., Lüthy, W. A. R., Shcherbakov, I. A., and Weber, H. P., Electrooptically q-switched 2.79 μm YSGG:Cr:Er laser with an intracavity polarizer, *IEEE J. Quantum Electron.* 27, 274 (1991).

957. Vodop'y nov, K. L., Kulevskii, L. A., Malyutin, A. A., Pashinin, P. P., and Prokhorov, A. M., Active mode locking in an yttrium erbium aluminum garnet crystal laser (λ = 2.94 μm), *Sov. J. Quantum Electron.* 12, 541 (1982).

958. Kaminskii, A. A., Fedorov, V. A., and Chan, Ng., Three-micron stimulated emission by Ho^{3+} ions in an $LaNbO_4$ crystal. *Inorg. Mater. (USSR)* 14, 1061 (1978).

959. Kaminskii, A. A., Advances in inorganic laser crystals. *Inorg. Mater. (USSR)* 20, 782 (1984).

960. Johnson, L. F. and Guggenheim, H. J., Laser emission at 3 μ from Dy^{3+} in BaY_2F_8, *Appl. Phys. Lett.* 23, 96 (1973).

961. Kintz, G. J., Alban, R., and Esterowitz, L., CW and pulsed 2.8 μm laser emission from diode-pumped Er^{3+}:LiYF, at room temperature, *Appl. Phys. Lett.* 50, 1553 (1987).

962. Vodop'ganov, K. L., Kulevskii, L. A., Pashlnin, P. P., Umyskov, A. F., and Shcherbakov, I. A., Bandwidth-limited picosecond pulses from a $YSGG:Cr^{3+}:Er^{3+}$ laser ($\lambda = 2.79$ μm) with active mode locking, *Sov. J. Quantum Electron.* 17, 776 (1987).

963. Chickis, E. P., Esterowitz, L., Allen, R., and Kruer, M., Stimulated emission at 2.8 μm in Er^{3+}:YLF. in *Proc. Int. Conf. Lasers*, Corcoran, V. L., Ed., STS Press, McLean, VA (1979), p. 172.

964. Kaminskii, A., Stimulated emission in the presence of strong nonradiative decay in crystals in *Proc. Conf. Int. Sch. At. Mol. Spectrosc. Radiationless Processes*, Vol. 62, DiBartolo, B., Ed., Plenum Press, New York (1980), p. 499.

965. Kaminskii, A. A., Modern tendencies in the development of the physics and spectroscopy of laser crystals, *Bull. Acad. Sci. USSR Phys. Ser.* 45, 106 (1981).

966. Gilliland, G. D. and Powell, R. C., Spectral and up-conversion dynamics and their relationship to the laser properties of $BaYb_2F_8:Ho^{3+}$, *Phys. Rev. B* 38, 9958 (1988).

967. Eckardt, R. C. and Esterowitz, L., Multiwavelength mid-IR laser emission in Ho:YLF, *Digest for Conf. on Lasers and Electrooptics*, IEEE, Piscataway, NJ (1982), p. 160.

968. Stalder, M. and Lüthy, W., Polarization of 3 μm laser emission in $YAlO_3$:Er, *Opt. Commun.* 61, 274 (1987).

969. Andreeva, L. I., Vodop'yanov, K. L., Kaidalov, S. A., Kalinin, Yu. M., Karasev, M. E., Kulevskii, L. A., and Lukashev, A. V., Picosecond erbium-doped YAG laser ($\lambda = 2.94$ μm) with active mode locking, *Sov. J. Quantum Electron.* 16, 326 (1986).

970. Zharikov, E. V., Il'ichev, N. N., Kalitin, S. P., Laptev, V. V., Malyutin, A. A., Osiko, V. V., Pashinin, P. P., Prokhorov, A. M., Saidov, Z. S., Smirnov, V. A., Umyskov, A. P., and Shcherbakov, I. A., Spectral luminescence, and lasing properties of a yttrium scandium gallium garnet crystal activated with chromium and erbium, *Sov. J. Quantum Electron.* 16, 635 (1986).

971. Moulton, P. F., Manni, J. G., and Rines, G. A., Spectroscopic and laser characteristics of Er,Cr:YSGG, *IEEE J. Quantum Electron.* 24, 960 (1988).

972. Al'bers, P., Ostroumov, V. G., Umyskov, A. F., Shnell, S., and Sbcherbakov, I. A., Low threshold YSGG:Cr:Er laser for the 3-μm range with a high pulse repetition frequency. *Sov. J. Quantum Electron.* 18, 558 (1988).

973. Bagdasarov, Kh. S., Zhekov, V. I., Lobabev, V. A., Murina, T. M., and Prokhorov, A. M., Steady-state emission from a $Y_3Al_5O_{12}:Er^{3+}$ laser ($\lambda = 2.94$ μm, T = 300 K), *Sov. J. Quantum Electron.* 13, 262 (1983).

974. Bagdasarov, Kh. S., Zbekov, V. I., Kukvskii, L. A., Murina, T. M., and Prokhorov, A. M., Giant laser radiation pulses from erbium-doped yttrium aluminum garnet crystals, *Sov. J. Quantum Electron.* 10, 1127 (1980).

975. Frauchiger, J. and Lüthy, W., Power limits of a YAG:Er laser, *Opt. Laser Technol.* 19, 312 (1987).

976. Strange, H., Petermann, K., Huber, G., and Duczynski, E. W., Continuous-wave 1.6 μm laser action in Er-doped garnets at room temperature, *Appl. Phys. B* 49, 269 (1989).

977. Becker, T., Clausen, R., Huber, G., Duczynski E. W., and Mitzscherlich, P., Spectroscopic and laser properties of Tm-doped YAG at 2 μm, in *Tunable Solid State Lasers*, Shand, M. L. and Jenssen H. P., Eds., Optical Society of America, Washington, DC (1989), p. 150.

978. Quarles, G. J., Rosenbaum, A., Marquardt, C. L., and Esterowitz, L., Efficient room-temperature operation of a flash-lamp-pumped Cr,Tm:YAG laser at 2.01 μm, *Opt. Lett.* 15, 42 (1990).

979. Duczynski, E. W., Huber, G., and Mitzscberlich, P., Laser action of Cr,Nd Tm,Ho-doped garnets in *Tunable Solid-State Lasers II*, Vol. 52, Budgor, A. B., Esterowitz, L. and DeShazer, L. C., Eds., Springer-Verlag, New York (1986), p. 282.

980. Kaminskii, A. A., Sarkisov, S. E., Klevtsov, P. V., Bagdasarov, Kh. S., Pavlyuk, A. A., and Petrosyan, A. G., Investigation of stimulated emission in the $^4F_{13/2}$ transition of Nd^{3+} ions in crystals. V, *Phys. Stat. Sol.* 17a, K75 (1973).

981. Shkadarevich, A. P., Recent advances in tunable solid state lasers in *Tunable Solid State Lasers*, Shand, M. L. and Jenssen, H. P., Eds., Optical Society of America, Washington, DC (1989), p. 66.

982. Zverev, G. M. and Shesttkov, A. V., Tunable near-infrared oxide crystal lasers, *ibid*.

983. Kaminskii, A. A., Ueda, K., Uehara, N., and Verdun, H. R., Room-temperature diode-laser-pumped efficient cw and quasi-cw single-mode lasers based on Nd^{3+}-doped cubic disordered α-NaCaYF$_6$ and tetragonal ordered LiLuF$_4$ crystals, *Phys. Stat. Sol. A* 140, K45 (1993).

984. Davydov, S.V., Kulak, I.I., Mit'kovets, A.I., Stavrov, A.A., and Sckadarevich, A.P., Lasing in YAG:Nd^{3+} and KGdW:Nd^{3+} crystals pumped with semiconductor lasers, *Sov. J. Quantum Electron.* 21, 16 (1991).

985. Kaminskii, A. A. and Verdun, H. R., New room-temperature diode-laser-pumped cw lasers based on Nd^{3+}-ion doped crystals, *Phys. Stat. Sol. A* 129, K119 (1992).

986. Dmitruk, M. V, Zhekov, V. I., Prokhorov, A. M., and Timoshechkin, M. I., Spectroscopic properties of Er$_3$Al$_{5-x}$Ga$_x$O$_{12}$ films obtained by liquid-phase epitaxy, *Inorg. Mater. (USSR)* 15, 976 (1979).

987. O'Connor, J. R., Unusual crystal-field energy levels and efficient laser properties of YVO$_4$:Nd^{3+}, *Appl. Phys. Lett.* 9, 407 (1966).

988. Schearer, L. D., Loduc, M., and Zachorowski, J., CW laser oscillations and tuning characteristics of neodymium-doped lithium niobate crystals, *IEEE J. Quantum Electron.* QE-23, 1996 (1987).

989. Antipenko, B. M., Sinitsyn, B. V., and Uvarova, T. V., Laser converter utilizing BaYb$_2$F$_8$:Ho^{3+} with a three-micron output, *Sov. J. Quantum Electron.* 10, 1168 (1980).

990. Barnes, N.P. and Allen, R.E., Room temperature Dy:YLF laser operation at 4.34 µm, *IEEE J. Quantum Electron.* 27, 277 (1991).

991. Kaminskii, A. A., New lasing channels of crystals with rare earth ions, in *Laser and Applications, Pt. 1, Proc.,* Ursu, I., and Prokhorov, A. M., Eds., CIP Press, Bucharest, Romania (1982), p. 587.

992. Vodop'pnov, K. L., Vorob'ev, N. S., Kulevskii, L. A., Prokhorov, A. M., and Shchelev, M. Ya., Image-converter recording of picosecond pulses emitted by an erbium laser (Λ = 2.94 µ) with active mode locking, *Sov. J. Quantum Electron.* 13, 272 (1983).

993. Kaminskii, A. A., Federev, V. A., Petrosyan, A. G., Pavlyuk, A. A., Bohm, I., Reiche, P., and Schulz, D., Stimulated emission by Ho^{3+} ions in oxygen-containing crystals at low temperatures, *Inorg. Mater. (USSR)* 15, 1180 (1979).

994. Fornasiero, L., Petermann, K., Heumann, E., and Huber, G., Spectroscopic properties and laser emission of Er^{3+} in scandium silicates near 1.5 µm, *Opt. Mater.* 10, 9 (1998).

995. Shen, H. Y., Lin, W. X., Zeng, R. R. et al., 1079.5 and 1341.4 nm: larger energy from a dual-wavelength Nd:YAlO$_3$ pulsed laser, *Appl. Optics* 32, 5952 (1993).

996. Field, S. J., Hanna, D. C., Shepherd, D. P., Tropper, A. C., Chandler, M. J., Townsend, P. D., and Zhang, L. Ion-implanted Nd:MgO:LiNbO$_3$ planar waveguide laser, *Opt. Lett.* 16, 481 (1991).

997. Batygov, S. Kh., Kulevskii, L. A., Lavrukhin, S. A. et al., Laser based on CaF$_2$-ErF$_3$ crystals, *Kurzfassungen Internat. Tagung Laser und ihre Anwendungen,* Dresden (1973), Teil 2, K97.

998. Batygov, S. Kh., Kulevskii, L. A., Prokhorov, A. M. et al., Erbium-doped CaF$_2$, crystal laser operating at room temperature, *Sov. J. Quantum Electron.* 4, 1469 (1975).

999. Jones, J. K., de Sandro, J. P., Hempstead, M., Shepherd, D. P., Large, A. C., Tropper, A. C., and Wilkinson, J. S., Channel waveguide laser at 1 μm in Yb-diffused LiNbO$_3$, *Opt. Lett.* 20, 1477 (1995).

1000. Kaminskii, A. A., Verdun, G. R., Mill', B. V., and Butashin, A. V., New diode-laser-pumped continuous lasers based on compounds having the structure of calcium gallogermanate with Nd^{3+} ions, *Neorgan. Mater. (USSR)* 28, 141 (1992).

1001. Koch, G. J., Deyst, J. P., and Storm, M. E., Single-frequency lasing of monolithic Ho,Tm:YLF, *Opt. Lett.* 18, 1235 (1993).

1002. Kaminskii, A. A., Karasev, V. A., Dubrov, V. D., Yakunin, V. P., Mill', B. V., and Butashin, A. V., New disordered Ca$_2$Ga$_2$SiO$_7$:Nd^{3+} crystal for high-power solid-state lasers, *Sov. J. Quantum Electron.* 22, 97 (1992).

1003. Manaa, H., Guyot, Y., and Moncorgé, R., Spectroscopic and tunable laser properties of Co^{2+}-doped single crystals, *Phys. Rev. B* 48, 3633 (1993).

1004. Antipenko, B. M., Mak, A. A., Raba, O. B., Sukhareva, L. K., and Uvarova, T. V., 2-μm-range rare earth laser, *Sov. Tech. Phys. Lett.* 9, 227 (1983).

1005. Dmitriev, V. G., Raevskii, E. V., Rubina, N. M., Rashkovich, L. N., Silichev, O. O., and Formichev, A. A., Simultaneous lasing in bands of a sequence in an optically pumped CO$_2$ laser, *Sov. Tech. Phys. Lett.* 4, 590 (1979).

1006. Li, M. J., Wang, L., Xie, C., Peng, K., and Xu, G., High efficiency Nd:MgO:LiNbO$_3$ self-frequency-doubling laser, *Proc. SPIE* 1726, 519 (1992).

1007. Eckhardt, R. C., DeRosa, J. L., and Letellier, J. P., Characteristics of an Nd:CaLaSOAP mode-locked oscillator, *IEEE J. Quantum Electron.* QE-10, 620 (1974).

1008. Cordova-Plaza, A., Fan, T. Y., Digonnet, M. J. F., Byer, R. L., and Shaw, H. J., Nd:MgO:LiNbO$_3$ continuous-wave laser pumped by a laser diode, *Opt. Lett.* 13, 209 (1988).

1009. de Micheli, M. P., in *Guided Wave Nonlinear Optics,* Ostrowsky, D. B., and Renisch, R., Eds., Kluwer Academic Publishers, Dordrecht (1992), p. 147.

1010. Li, M. J., de Micheli, M. P., He, Q., and Ostrowsky, D. B., Cerenhov configuration second harmonic generation in proton exchanged lithium niobate guides, *IEEE J. Quantum Electron.* QE-26, 1384 (1990).

1011. He, Q., de Micheli, M. P., Ostrowsky, D. B. et al., Self-frequency-doubled high Δn proton exchange Nd:LiNbO$_3$ waveguide laser, *Opt. Commun.* 89, 54 (1992).

1012. Yamamoto, J. K., Sugimoto, A., and Yamagishi, K., Self-frequency doubling in Nd,Sc$_2$O$_3$:LiNbO$_3$ at room temperature, *Opt. Lett.* 19, 1311 (1994).

1013. Kaminskii, A. A., Mill, B. V., and Butashin, A. V., Stimulated emission from Nd^{3+} ions in acentric LaBGeO$_5$ crystals, *Phys. Stat. Sol.* (a) 118, K59 (1990).

1014. Kaminskii, A. A., Butashin, A. V., Maslyanitsin, I. A. et al., Pure and Nd^{3+}, Pr^{3+}-ion doped trigonal acentric LaBGeO$_5$, *Phys. Stat. Sol.* (a) 125, 671 (1991).

1015. Kaminskii, A. A., Bagaev, S. N., Mill', B. V., and Butashin, A. V., New inorganic materials LaBGeO$_5$.Nd^{3+} for crystalline lasers with self-multiplied generation frequency, *Neorg. Mater. (Russia)* 29, 545 (1993).

1016. Kaminskii, A. A., Shkadarevich, A. P., Mill', B. V., Koptev, V. G., Butashin, A. V., and Demidovich, A. A., Tunable stimulated emission of Cr^{2+} ions and generation frequency self-multilplication effect in acentric crystal of Ca-gallogermanate structure, *Neorg. Mater. (Russia)* 24, 690 (1986).

1017. Kaminskii, A. A., Butashin, A. V., Demidovich, A. A., Koptev, V. G., Mill', B. V., and Shkadarevich, A. P., Broad-band tunable stimulated emission from octahedral Cr^{3+} ion in a new acentric crystal with Ca-gallogermanate structure, *Phys. Stat. Sol.* (a) 112, 197 (1989).

1018. Knappe, R., Bartschke, J., Becher, C., Beier, B., Scheidt, M., Boller, K. J., and Wallenstein, R., *Conf. Proc. 1994 IEEE Nonlinear Optics* (1994), p. 39.

1019. Lu, B., Wang, J., Pan, H., Jiang, M., Liu, E., and Hou, X., Stimulated radiation of monoclinic crystals of KY(WO$_4$)$_2$ and KGd(WO$_4$)$_2$ with Ln^{3+} ions, *Chin. Phys. Lett.* 3, 423 (1986).

1020. Osiko, V. V., Sigachev, V. B., Strelov, V. I., and Timoshechkin, M. I., Erbium gadolinium gallium garnet crystal laser, *Sov. J. Quantum Electron.* 21, 159 (1991).

1021. Dorozhkin, L. M., Kuratev, I. I., Leonyuk, N. I., Timochenko, T. I., and Shestakov, A. V., Optical second-harmonic operation in a new nonlinear active medium: neodymium-yttrium-aluminum borate crystals, *Sov. Tech. Phys. Lett.* 7, 555 (1981).

1022. Kaminskii, A. A., Butashin, A. V., Demidovich, M. I., Zhavaronkov, N. I., Mikhailov, V. P., and Shkadarevich, A. P., Generation of picosecond pulsed by acentric disordered silicate $La_3Ga_5SiO_{14}$, *Neorg. Mater. (Russia)* 24, 2075 (1986).

1023. Kaminskii, A. A., Butashin, A. V., and Bagaev, S. N., New Nd^{3+}:$BaLu_2F_8$ laser crystal, *Quantum Electron.* 26, 753 (1996).

1024. Kaminskii, A. A., New high-temperature induced transition of an optical quantum generator based on SrF_2-Nd^{3+} crystals (type I), *Inorg. Mater. (USSR)* 5, 525 (1969).

1025. Tin, P., and Schearer, L. D., A high power, tunable, arc-lamp pumped Nd-doped lanthanum-hexaluminate laser, *J. Appl. Phys.* 68, 950 (1990).

1026. Spariosu, K., Birnbaum, M., and Kokta, M., Room-temperature 1.643-μm Er^{3+}:$Y_3Sc_2Ga_3O_{12}$ (Er:YSGG) laser, *Appl. Optics* 34, 8272 (1995).

1027. Loutts, G. B., Bonner, C., Meegoda, C. et al., Crystal growth, spectroscopic characterization, and laser performance of a new efficient laser material $Nd:Ba_5(PO_4)_3F$, *Appl. Phys. Lett.* 71, 303 (1997).

1028. Antipenko, B. M., Mak, A. A., Sinitsyn, B. V., Raba, O. B., and Uvurova, T. V., New excitation schemes for laser transitions, *Sov. Phys. Tech. Phys.* 27, 333 (1982).

1029. Djeu, N., Hartwell, V. E., Kaminskii, A. A., and Butashin, A. V., Room-temperature 3.4-μm $Dy:BaYb_2F_8$ laser, *Opt. Lett.* 22, 997 (1997).

1030. Möbert, P. E.-A., Heumann, E., Huber, G., and Chai, B. H. T., Green Er^{3+}:$YLiF_4$ upconversion laser at 551 nm with Yb^{3+} codoping: a novel pumping scheme, *Opt. Lett.* 22, 1412 (1997).

1031. Hömmerich, U., Wu, X., and Davis, V. R., Demonstration of room-temperature laser action at 2.5 μm from Cr^{2+}:$Cd_{0.85}Mn_{0.15}Te$, *Opt. Lett.* 22, 1180 (1997).

1032. Evans, J. M., Petricevic, V., Bykov, A. B., Delgado, A., and Alfano, R. R., Direct diode-pumped continuous-wave near-infrared tunable laser operation of Cr^{4+}:forsterite and Cr^{4+}:Ca_2GeO_4, *Opt. Lett.* 22, 1171 (1997).

1033. Kuleshov, N. V., Shcherbitsky, V. G., Lagatsky, A. A. et al., Spectroscopy, excited-state absorption and stimulated emission in Pr^{3+}-doped Gd_2SiO_5 and Y_2SiO_5 crystals, *J. Lumin.* 71, 27 (1997).

1034. Quarles, G. J., Rosenbaum, A., Marquardt, C. L., and Esterowitz, L., High efficient 2.09 μm flashlamp-pumped laser, *Appl. Phys. Lett.* 55, 1062 (1989).

1035. Pollack, S. A., Chang, D. B., and Birnbaum, M., Threefold upconversion laser at 0.85, 1.23, and 1.73 μm in Er:YLF pumped with a 1.53 μm Er glass laser, *Appl. Phys. Lett.* 54, 869 (1989).

1037. Kaminskii, A. A., Luminescence and multiwave stimulated emission of Ho^{3+} and Er^{3+} ions in orthorhombic $YAlO_3$ crystals, *Sov. Phys. Dokl.*, 31, 823 (1986).

1038. Koch, M. E., Kueny, A. W., and Case, W. E., Photon avalanche upconversion laser at 644 nm, *Appl. Phys. Lett.* 56, 1083 (1990) and *J. Lumin.* 45, 351 (1990).

1039. Knowles, M. R. H. and Webb, C. E., Efficient high-power copper-vapor-laser-pumped Ti:Al_2O_3 laser, *Opt. Lett.* 18, 607 (1993).

1040. Zhang, X. X., Bass, M., Villaverde, A. B., Lefaucheur, J., Pham, A., and Chai, B. H. T., Efficient laser performance of Nd:$GdLiF_4$: a new laser crystal, *Appl. Phys. Lett.* 62, 1197 (1993).

1041. Hanson, F. and Haddock, D., Laser diode side pumping of neodymium laser rods, *Appl. Optics* 27, 80 (1988).

1042. Bermair, R. M. J., Kagan, J., Kalisky, Y., Noter, Y., Oron, M., Shimony, Y., and Yogev, A., Solar-pumped Er,Tm,Ho:YAG laser, *Opt. Lett.* 15, 36 (1990).

1043. Weksler, M. and Shwartz, J., Solar-pumped solid-state lasers, *IEEE J. Quantum Electron.* 24, 1222 (1988).

1044. Kaminskii, A. A., Eichler, H. J., Findeisen, J., and Barta, Ch., Room-temperature high-order stimulated Raman scattering and stimulated emission in ultra-low-phonon energy orthorhombic $PbCl_2:Nd^{3+}$ crystal, *Phys. Stat. Sol. (b)* 206, R3 (1998).

1045. Kaminskii, A. A., Eichler, H. J., Grebe, D. et al., Orthorhombic ($LiNbGeO_5$): efficient stimulated Ramon scattering and tunable near-infrared laser emission from chromium doping, *Opt. Mater.* 10, 269 (1998).

1046. Kaminskii, A. A., Eichler, H. J., Grebe, D., Macdonald, R., and Butashin, A. V., New fluoride crystal $BaLu_2F_8:Nd^{3+}$ for diode pumped lasers, *Phys. Stat. Sol. (a)* 158, K31 (1996).

1047. Kaminskii, A. A., Lyashenko, A. I., Isaev, N. P., Karlov, V. N., Pavlovich, V. L. Bagaev, S. N., Butashin, A. V., and Li, L. E., Quasi-cw $Pr^{3+}:LiYF_4$ laser with $\lambda = 0.6395$ μm and an average output power of 2.3 W, *Quantum. Electron.* 28, 187 (1998).

1048. Kaminskii, A. A., Li, L., Butashin, A. V., Mironov, V. S., Pavlyuk, A. A., Bagaev, S. N., and Ueda, K., New crystalline lasers on the base of monoclinic $KR(WO_4)_2:Ln^{3+}$ tungstates (R = Y and Ln), *Opt. Review* 4, 309 (1997).

1049. Eichler, H. J., Liu, B., Meindl, P., and Kaminskii, A. A., Generation of short pulses by a $Pr^{3+}:LiYF_4$ crystal with an efficient laser transition at 639.5 nm, *Opt. Mater.* 3, 163 (1994).

1050. Demidovich, A. A., Shkadarevich, A. P., Batay, L. E. et al., $Nd:La_3Ga_{5.5}Nb_{0.5}O_{14}$ laser properties under LD pumping, *Proc. SPIE* 3176, 269 (1997).

1051. Kaminskii, A. A., Butashin, A. V., Mironov, V. S., Bagaev, S. N., and Eichler, H. J., Efficient 2 μm stimulated emission in the $^4F_{9/2} \rightarrow {}^4I_{11/2}$ channel from monoclinic $BaYb_2F_8:Er^{3+}$ crystals, *Phys. Stat. Sol. (b)* 194, 319 (1996).

1052. Eichler, H. J., Liu, B., Lu, Z., and Kaminskii, A. A., Orange, red, and deep-red flashlamp-pumped $Pr^{3+}:LiYF_4$ laser with improved output energy and efficiency, *Appl. Phys. B* 58, 421 (1994); Eichler, H. J., Liu, B., Meindl, P., and Kaminskii, A. A., Generation of short pulses by a $Pr^{3+}:LiYF_4$ with an efficient laser transition at 639.5 nm, *Opt. Mater.* 3, 163 (1994).

1053. Kaminskii, A. A., Mironov, V. S., Bagaev, S. N., Khaidukov, N. M., Joubert, M. F., Jacquier, B., and Boulon, G., Spectroscopy and laser action of anisotropic single-centered $LiKYF_5:Nd^{3+}$ crystals grown by the hydrothermal method, *Phys. Stat. Sol. (a)* 145, 177 (1994).

1054. Kaminskii, A. A., Khaidukov, N. M., Koechner, W., and Verdun, H. R., New laser-diode-pumped CW crystalline lasers based on Nd^{3+}-ion doped hydrothermal fluorides, *Phys. Stat. Sol. (a)* 132, K105 (1992).

1055. Kaminskii, A. A., Butashin, A. V., Bagaev, S. N., Eichler, H. J., Findeisen, J., Tauber, U., and Liu, B., Three-micron cw stimulated emission from a new $Er^{3+}:BaLu_2F_8$, laser crystal subjected to laser-diode pumping, *Quantum Electron.* 28, 93 (1998).

1056. Capmany, J., Bausá, L. E., Jaque, D., García Solé, J., and Kaminskii, A. A., CW end-pumped $Nd^{3+}:LaBGeO_5$ minilaser for self-frequency-doubling, *J. Lumin.* 72-74, 816 (1997).

1057. Capmany, J., Bausá, L. E., Jaque, D., García Solé, J., and Kaminskii, A. A., Continuous wave laser radiation at 524 nm from a self-frequency-doubled laser of $LaBGeO_5:Nd^{3+}$, *Appl. Phys. Lett.* 72, 531 (1998).

1058. Kaminskii, A. A., Bagaev, S. N., Ueda, K., Pavlyuk, A. A., and Musha, M., Frequency self-doubling of the cw 1-μm laser of a ferroelectroic $Nd3+: \beta'-Gd_2(MoO_4)_3$ crsytal with laser-diode pumping, *Quantum Electron.* 27, 657 (1997).

1059. Eichler, H. J., Ashkenaasi, D., Jian, H., and Kaminskii, A. A., Accentric disordered $Nd^{3+}:La_3Ga_5SiO_{14}$ crystal. A broadband luminescence material with high thermal conductivity to generate picosecond laser pulses, *Phys. Stat. Sol. (a)* 146, 833 (1994).

1060. Otsuka, K., Yamada, T., Nakano, J., Kimura, T., and Saruwatari, M., Lithium neodymium tetraphosphate laser, *J. Appl. Phys.* 46, 4600 (1975).

1061. Saruwatari, M., Kimura, T., and Otsuka, K., Miniaturized cw $LiNdP_4O_{12}$ laser pumped with a semiconductor laser, *Appl. Phys. Lett.* 29, 291 (1978).

1062. Kubodera, K, Nakano, J., Otsuka, K., and Miyazawa, S., A stab waveguide laser formed of glass-clad $LiNdP_4O_{12}$, *J. Appl. Phys.* 49, 65 (1978).

1063. Kubodera, K., and Otsuka, K., Single-transverse-mode $LiNdP_4O_{12}$ slab waveguide laser, *J. Appl. Phys.* 50, 653 (1979).

1064. Chinn, S. R. and Hong, H. Y-P., Low-threshold cw $LiNdP_4O_{12}$ laser, *Appl. Phys. Lett.* 29, 291 (1978).

1065. Otsuka, K., Kubodera, K., and Nakano, J., Stabilized dual-polarization oscillation in a $LiNd_{.5}La_{.5}P_4O_{12}$ laser, *IEEE J. Quantum Electron.* QE-13, 398 (1977).

1066. Kopf, D., Prasad, A., Zhang, G., Moser, A., and Keller, U., Broadly tunable femtosecond Cr:SAF laser, *Opt. Lett.* 22, 621 (997).

1067. Kaminskii, A. A., Osiko, V. V., Sarkisov, S. E., Timoshechkin, M. I., Zhekov, E. V., Bohm, J., Reiche, P., and Schulzte, D., Growth, spectroscopic investigations, and some new stimulated emission data on $Gd_3Ga_5O_{12}$-Nd^{3+} single crystals, *Phys. Stat. Sol.* (a) 49, 305 (1978).

1068. Rambaldi, P., Moncorgé, R., Wolf, J. P., Pedrini, C. et al., Efficient and stable pulsed laser operation of Ce:LiLuF$_4$ around 308 nm., *Opt. Commun.* 146, 163 (1998).

1069. Devor, D. P., Soffer, B. H., and Robinson, M., Stimulated emission from Ho^{3+} at 2 μm in HoF_3, *Appl. Phys. Lett.* 18, 122 (1971).

1070. Knowles, D. S. and Jenssen, H. P., Upconversion versus Pr-deactivation for efficient 3 μm laser operation in Er, *IEEE J. Quantum Electron.* 28, 1197 (1992).

1071. Eichler, H. J., Findeisen, J., Liu, B., Kaminskii, A. A., Butachin, A. V., and Peuser, P., Highly efficient diode-pumped 3-μm Er^{3+}:BaY_2F_8 laser, *EEE J. Sel. Topics Quantum Electron.* 13, 90 (1997).

1072. Auzel, F., Kaminskii, A. A., and Meichenin, D., Diode and dye pumped cw laser in BaY_2F_8:Er^{3+} at 2.7 μm *Phys. Stat. Sol.* (a) 131, K63 (1992).

1073. Prokhorov, A. M., Kaminskii, A. A., Osiko, V. V., Timoshechkin, M.I., Zharokov, E. V. Butaeva, T. I., Sarkisov, S. E., and Petrosyan, SA. G., Investigations of 3-μm stimulated emission from Er^{3+} ions in aluminum garnets at room temperature, *Phys. Stat. Sol.* (a) 40, K69 (1977).

1074. Kaminskii, A. A. and Uvarova, T. V., Quasi-continuous 3-micron stimulated emission in monoclinic BaY_2F_8-Er^{3+} fluoride, *Inorg. Mater.* 24, 2088 (1988).

1075. Thompson, G. A., Krupkin, V., Yogev, A., and Oron, M., Solar-pumped Nd:Cr:GSGG parallel array laser, *Opt. Engin.* 31, 2644 (1992).

1076. Thompson, G. A., Krupkin, V., Reich, A., and Oron, M., Effects of coolant temperature and pump power on the power output on solar-pumped solid state lasers, *Opt. Engin.* 31, 2488 (1992).

1077. Young, C. G., A sun-pumped cw one-watt laser, *Appl. Optics* 5, 993 (1966).

1078. Arashi, H., Oka, Y., Sasahara, N., Kaimai, A., and Ishigame, M., A solar-pumped cw 18 W Nd:YAG laser, *Jap. J. Appl. Phys.* 23, 1051 (1984).

1079. Kaminskii, A. A., New room-temperature diode-laser-pumped efficient quasi-cw and cw single-mode laser based on monoclinic BaY_2F_8:Nd^{3+} crystal, *Phys. Stat. Sol. (a)* 137, K61 (1993).

1080. Scott, B. P., Zhao, F., Chang, R. S. F., and Djeu, N., Upconversion-pumped blue laser in Tm:YAG, *Opt. Lett.* 18, 113 (1993).

1081. Brede, R., Heumann, E., Roetke, J., Danger, T., Huber, G., and Chai, B., Green upconversion laser emission in Er-doped crystals at room temperature, *Appl. Phys. Lett.* 63, 2030 (1993).

1082. Xie, P. and Rand, S. C., Continuous-wave trio upconversion laser, *Appl. Phys. Lett.* 57, 1182 (1990).

1083. Macfarlane, R. M., Whittaker, E. A., and Lenth, W., Blue, green and yellow upconversion lasing in Er:YLF$_4$ using 1.5 μm pumping, *Electron. Lett.* 28, 2136 (1992).

1084. Xie, P. and Rand, S. C.,Visible cooperative upconversion laser in Er:LiYF$_4$, *Opt. Lett.* 17, 1198 (1992); Xie, P. and Rand, S.C., Erratum, *Opt. Lett.* 17, 1822 (1992).

1085. Xie, P. and Rand, S. C., Continuous-wave mode-locked visible upconversion laser, *Opt. Lett.* 17, 1116 (1992), Erratum, *Opt. Lett.* 17, 1822 (1992).

1086. McFarlane, R. A., Upconversion laser in BaY_2F_8:Er 5% pumped by ground-state and excited-state absorption, *J. Opt. Soc. Am. B* 11, 871 (1994).

1087. Lenth, W. and Macfarlane, R. M., Excitation mechanisms for upconversion lasers, *J. Lumin.* 45, 346 (1990).

1088. Hebert, T., Risk, W. P., Macfarlane, R. M., and Lenth, W., Diode-laser-pumped 551-nm upconversion laser in $YLiF_4$:Er^{3+}, *Proc. Adv. Sol. State Lasers*, Vol. 6, Jenssen, H. J. and Dube, G., Eds., Optical Society of America, Washington, DC (1990), p. 379.

1089. Scheps, R., Er^{3+}:$YAlO_3$ upconversion laser, *IEEE J. Quantum Electron.* 30, 2914 (1994)., *ibid.* 31, 309 (1995).

1090. McFarlane, R. A., Spectroscopic studies and upconversion laser operation of BaY_2F_8:Er 5%, *Proc. Adv. Sol. State Lasers*, Vol. 13, Chase, L. L. and Pinto, A. A., Eds., Optical Society of America, Washington, DC (1992), p. 275.

1091. Brede, R., Heumann, E., Danger, T., Huber, G., and Chai, B. H. T., Room temperature green upconversion lasing in erbium-doped fluorides, *OSA Proc. Adv. Solid State Lasers*, Fan, T. Y. and Chai, B. H. T., Eds., 20, 348 (1995).

1092. Pollack, S. A. and Chang, D. B., Upconversion-pumped 2.8–2.9 µm lasing of Er^{3+} ions in garnets, *J. Appl. Phys.* 70, 7227 (1991).

1093. Pollack, S. A. and Chang, D. B., Upconversion-pumped population kinetics for $^4I_{13/2}$ and $^4I_{11/2}$ laser states of Er^{3+} ion in several host crystals, *Opt. Quantum Electron.* 22, S75 (1992).

1094. Kaminskii, A. A., Today and tomorrow of laser crystal physics, *Phys. Stat. Sol.* (a) 148, 10 (1995).

1095. Thrash, R. J. and Johnson, L. F., UV upconversion laser emission from Yb^{3+} sensitized Tm^{3+} in BaY_2F_2, *OSA Proc. Adv. Solid State Lasers*, Fan, T. Y. and Chai, B. H. T., Eds., 20, 400 (1995).

1096. Thrash, R. J. and Johnson, L. F., in *Compact Blue-Green Lasers, Techn. Digest. Ser.* Optical Society of America, Washington, DC (1992), p. 17.

1097. Antipenko, B. M., Voronin, S. P., and Privalova, T. A., New laser channels of the Tm^{3+} ion, *Opt. Spectrosc.* (USSR) 68, 164 (1990).

1098. Antipenko, B. M., Dumbravyanu, R. V., Perlin, Yu. E., Raba, O. B., and Sukhareva, L. K., Spectroscopic aspects of the BaY_2F_8 laser medium, *Opt. Spectrosc.* (USSR) 59, 337 (1985).

1099. Mak, A. A. and Antipenko, B. M., Rare earth converters of neodymium laser radiation, *Zh. Prikl. Spektrosk.* 37, 1029 (1982).

1100. Nguyen, D. C., Faulkner, G. E., Weber, M. E., and Dulick, M., Blue upconversion thulium laser, *Proc. SPIE* 1223, 54 (1990).

1101. Schmitt, K., Stimulated C^+-emission of Ag^--centers in KI, RbBr, and CsBr, *Appl. Phys. A* 38, 61 (1985).

1102. Boutinaud, P., Monnier, A., and Bill, H., Ag^+ center in alkaline-earth fluorides: new UV solid state lasers?, *Rad. Eff. Def. Solids* 136, 69 (1995).

1103. Zhou, F., De La Rue, R. M., Ironside, C. N., Han, T. P. J., Hendersen, B., and Ferguson, A. I., Optical gain in proton-exchanged Cr:$LiNbO_3$ waveguides, *Electron. Lett.* 28, 204 (1992).

1104. Almeida, J. M., Leite, A. P., De La Rue, R. M., et al., Spectroscopy and optical amplification in Cr doped $LiNbO_3$, *OSA Trends in Optics and Photonics on Advanced Solid State Lasers*, Vol. 1, Payne, S. A. and Pollack, C. R., Eds., Optical Society of America, Washington, DC (1996), p. 478.

1105. Moncorgé, R., Manaa, H, Deghoul, F., Borel, C., and Wyon, Ch., Spectroscopic study and laser operation of Cr^{4+}-doped $(Sr,Ca)Gd_4(SiO_4)_3O$ single crystal, *Opt. Commun.* 116, 393 (1995).

1106. Barrie, J. D., Dunn, B., Stafsudd, O. M., and Nelson, P., Luminescence of Cu^+-β-alumina, *J. Lumin.* 37, 303 (1987).

1107. Schaffers, K. I., Page, R. H., Beach, R. J., Payne, S. A., and Krupke, W. F., Gain measurements of Dy^{3+}-doped $LaCl_3$: a potential 1.3 μm optical amplifier for tele-communications, *OSA Trends in Optics and Photonics on Advanced Solid State Lasers*, Vol. 1, Payne, S. A. and Pollack, C. R., Eds., Optical Society of America, Washington, DC (1996), p. 469; Page, R. H. et al. Dy-doped chlorides as gain media for 1.3 μm tele-communications amplifiers, *J. Lightwave Technol.* 15, 786 (1997).

1108. Shkadarevich, A. P., Recent advances in tunable solid state lasers, in *Tunable Solid State Lasers*, Shand, M. L. and Jenssen, H. P., Eds., Optical Society of America, Washington, DC (1989), p. 66.

1109. Capobianco, J. A., Cormier, G., Moncourgé, R., Manaa, H., and Bettinelli, M., Gain measurements of Mn^{5+} $(3d^2)$ doped $Sr_5(PO_4)_3Cl$ and Ca_2PO_4Cl, *Appl. Phys. Lett.* 60, 163 (1992).

1110. Brannon, P. J., X-ray induced optical gain in Cr,Nd:GSGG, *OSA Proc. Adv. Solid State Lasers*, Chai, B. H. T. and Payne, S. A., Eds., 24, 232 (1995).

1111. Cashmore, J. S., Hooker, S. M., and Webb, C. E., Vacuum ultraviolet gain measurements in optically pumped $LiYF_4$:Nd^{3+}, *Appl. Phys. B* 64, 293 (1997).

1112. Zhong, G. Z. and Bryant, F. J., Laser phenomena in DCEL of ZnS:Cu:Nd:Cl thin films, *Solid State Commun.* 39, 907 (1981).

1113. Powell, R. C., Quarles, G. L., Martin, J. J., Hunt, C. A., and Sibley, W. A., Stimulated emission and tunable gain from Rh^{2+} ion lin $RbCaF_3$ crystals, *Opt. Lett.* 10, 212 (1985).

1114. Loiacono, G. M., Shone, M. F., Mizell, G., Powell, R. C., Quarles, G. J., and Elonadi, B., Tunable single pass gain in titanium-activated lithium germanium oxide, *Appl. Phys. Lett.* 48, 622 (1986).

1115. Pazzi, G. P, Baldecchi, M. G., Fabeni, P., Linari, R., Ranfagni, A., Agresti, A., Cetica, M., and Simpkin, D. J., Amplified spontaneous emission in doped alkali-halides, *SPIE* 369, 338 (1982).

1116. Nagli, L. E. and Plyovin, I. K., Induced recombination emission of activated alkali halide crystals, *Opt. Spectrosc.* (USSR) 44, 79 (1978).

1117. Haley, L. V. and Koningstein, J. A., Time resolved stimulated fluorescence of the uranyl ion in the mineral metaautunite, *J. Phys. Chem. Solids* 44, 431 (1983).

1118. Moulton, P. F., Recent advances in solid-state laser materials, in *Materials Research Society Symposium Proceedings* 24, 393 (1984).

1119. Kaminskii, A. A., Pavlyuk, A. A., Klevtsov, P. V. et al., Stimulated radiation of monoclinic $KY(WO_4)_2$ and $KGd(WO_4)_2$ with Ln^{3+} ions, *Inorg. Mater. (USSR)* 13, 582 (1977).

1120. Kaminskii, A. A., Fedorov, V. A., and Mochalov, I. V., New data on the three-micron lasing of Ho^{3+} and Er^{3+} ions in aluminates having the perovskite structure, *Dokl. Akad. Nauk SSSR* 254, 604 (1980).

1121. Schweizer, T., Jensen, T., Heumann, E., and Huber, G., Spectroscopic properties and diode pumped 1.6 μm laser performance in Yb-codoped Er:$Y_3Al_5O_{12}$ and Er:Y_2SiO_5, *Opt. Commun.* 118, 557 (1995).

1122. Kaminskii, A. A., Fedorov, V. A., and Mochalov, I. V., New data on the three-micron lasing of Ho^{3+} and Er^{3+} ions in aluminates having the perovskite structure, *Sov. Phys. Dokl.* 25, 744 (1980).

1123. Hughes, D. W., Majdabadi, A., Barr, J. R. M., and Hanna, D. C., FM mode-locked, laser-diode-pumped $La_{1-x}Nd_xMgAl_{11}O_{19}$ laser, *Appl. Optics* 32, 5958 (1993).

1124. Wagner, G., Carrig, T. J., Page, R. H. et al., Continuous-wave broadly tunable Cr^{2+}:ZnSe laser, *Opt. Lett.* 24, 19 (1999).

1125. Li, C., Moncorgé, R., Souriau, J. C., Borel, C., and Wyon, Ch., Room temperature cw laser action of Y_2SiO_5:Yb^{3+}, Er^{3+} at 1.57 μm, *Opt. Commun.* 107, 61 (1994).

1126. Rothacher, Th., Lüthy, W., and Weber, H. P., Diode pumping and laser properties of Yb:Ho:YAG, *Opt. Commun.* 155, 68 (1998).

1127. Jaque, D., Capmany, J., and Garciá Solé, J., Continuous wave laser radiation at 669 nm from a self-frequency-doubled laser of $YAl_3(BO_3)_4$:Nd^{3+}, *Appl. Phys. Lett.* 74, 1788 (1999).

1128. McGonigle, A. J. S., Coutts, D. W., and Webb, C. E., 530-mW 7-kHz cerium LiCAF laser pumped by the sum-frequency-mixed output of a copper-vapor laser, *Opt. Lett.* 24, 232 (1999).

1129. Zhang, H. J., Meng, X. L., Zhu, L. et al., Growth and laser properties of Nd:Ca$_4$YO(BO$_3$)$_3$ crystal, *Opt. Commun.* 160, 273 (1999).

1130. Kaminskii, A. A., Petrosyan, A. G., and Fedorov, V. A., Cross-cascade stimulated emission of Er^{3+}, Ho^{3+}, and Tu^{3+} ions in Er$_3$Al$_5$O$_{12}$, *Dokl. Akad. Nauk SSSR* 257, 79 (1981).

1131. Kaminskii, A. A., Butaeva, T. I., Ivanov, A. O. et al., *Zh. Tekl. Fiz. Pis'ma* 2, 787 (1976).

1132. Kaminskii, A. A., Inorganic materials with Ln^{3+} ions for producing stimulated radiation in the 3-μm band, *Izv. Akad. Nauk. SSSR Neorg. Mater.* 15, 1028 (1979).

1133. Capmany, J., Jaque, D., Sanz García, J. A., and García Solé, J., Continuous wave laser radiation and self-frequency-doubling in ZnO doped LiNbO$_3$:Nd^{3+}, *Opt. Commun.* 161, 253 (1999).

1134. Kellner, T., Heine, F., Huber, G. et al., Soliton mode locked Nd:YAlO$_3$ laser at 930 nm, *J. Opt. Soc. Am,. B* 15, 1663 (1998).

1135. Rapaport, A., Moteau, O., Bass, M., Boatner, L. A., and Deka, C., Optical spectroscopy and lasing properties of neodymium-doped lutetium orthophosphate, *J. Opt. Soc. Am,. B* 16, 911 (1999).

1136. Kaminskii, A. A., New room-temperature stimulated-emission channels of Pr^{3+} ions in anisotropic laser crystals, *Phys. Stat. Sol. (b)* 125, K109 (1991).

1137. Chai, B. H. T. et al., Self-frequency doubled Nd:YCOB laser, in *OSA Trends in Optics and Photonics: Advanced Solid State Lasers*, Vol. 19, Bosenberg, W. R. and Fejer, M. M., Eds. (1998).

1138. Kaminskii, A. A., Verdun, H. R., and Khaidukov, N. M., New continuous-wave crystal lasers with semiconductor laser pumping based on anisotropic fluorides with Nd^{3+} ions, grown by the hydrothermal method, *Phys. Dokl.* 38, 39 (1993).

1139. Kaminskii, A. A., Ueda, K., and Uehara, N., New laser-diode-pumped CW laser based on Nd^{3+}-ion-doped tetragonal LiLuF$_4$ crystal, *Jpn. J. Appl. Phys.* 32, L586 (1993).

1140. Kaminskii, A. A., Boulon, G., Buoncristiani, M. et al., Spectroscopy of a new laser garnet Lu$_3$Sc$_2$Ga$_3$O$_{12}$:Nd^{3+}, *Phys. Stat. Sol. (b)* 141, 471 (1994).

1141. Kaminskii, A. A., New room-temperature laser-diode-pumped efficeint quasi-CW and CW single-mode laser based on ferroelectric and ferroelastic Gd$_2$(MoO$_4$)$_3$:Nd^{3+} crystal, *Phys. Stat. Sol. (b)* 149, K39 (1995).

1142. Kaminskii, A. A., Butashin, A. V., and Bagaev, S. N., New Nd^{3+}:BaLu$_2$F$_8$ laser crystal, *Quantum Electron.* 26, 753 (1996).

1143. Kaminskii, A. A., Bagaev, S. N., Li, L., Kuznetsov, F A., and Pavlyuk, A. A., New crystalline lasers for the 1-μm wavelength range, *Quantum Electron.* 26, 1 (1996).

1144. Kaminskii, A. A., Eichler, H. J., Garciá Solé, J. et al., Piezoelectric sillenite Bi$_{12}$SiO$_{20}$:Nd. A new laser and SRS-active crystal, *Phys. Stat. Sol. (b)* 210, R9 (1998).

1145. Kaminskii, A. A., Butashin, A. V., Li. L., Jaque, D., and Garciá Solé, J, New Nd^{3+}:CaLu$_2$F$_8$ laser crystal containing lutetium, *Quantum Electron.* 29, 377 (1999).

1146. Kaminskii, A. A., Li. L. E., Butashin, A. V. et al., Tetragonal Nd^{3+} doped and undoped PbWO$_4$ crystals are new laser- and SRS-active materials, Proceedings Second International Symposium on Modern Problems of Laser Physics, Bagayev, S. N., and Denisov, V. I., Eds., Novosibirsk, Russia (1997).

1147. Kaminskii, A. A., Li. L., Butashin, A. V., et al., New stimulated emission channels of Pr^{3+} and Tm^{3+} ions in monoclinic KR(WO$_4$)$_2$ type crystals with ordered structure (R=Y and Gd), , *Jpn. J. Appl. Phys.* 36, L107 (1997).

1148. Eichler, H. J., Liu, B., Findeisen, J., Kaminskii, A. A., Butachin, A. V., and Peuser, P., Diode-pumped 3 μm Er^{3+}:BaY_2F_8 cw laser with optimized Er^{3+}-concentration, *OSA TOPS Vol. 10 Advanced Solid State Lasers*, Pollack, C. R. and Bosenberg, W. R., Eds., (1997), p. 222.

1149. Eichler, H. J., Findeisen, J. Liu, B., Kaminskii, A. A., Butachin, A. V., and Peuser, P., Highly efficient diode-pumped 3 μm Er^{3+}:BaY_2F_8 laser, *IEEE J. Select. Topics Quantum Electron.* 3, 90 (1997).

1150. Šhah, L., Ye, Q., Eichenholz, J. M. et al., Laser tunability in Yb^{3+}:$YCa_4O(BO_3)_3$ {Yb:YCOB}, *Opt. Commun.* 167, 149 (1999).

1151. Wang, C. Q., Cow, Y. T., and Yuan, D. R., CW dual-wavelength Nd:YAG laser at 946 and 938.5 nm and intracavity nonlinear frequency conversion with a CMTC crystal, *Optics Commun.* 165, 2312 (1999).

1152. Kaminskii, A. A., Jaque, D., Bagaev, S. N., Ueda, K., García Solé, J., and Capmany, J., New nonlinear-laser properties of ferroelectric Nd^{3+}:$Ba_2NaNb_5O_{15}$—cw stimulated emission ($^4F_{3/2} \rightarrow {}^4I_{11/2}$ and $^4F_{3/2} \rightarrow {}^4I_{13/2}$), collinear and diffuse self-frequency doubling and summation, *Quantum Electron.* 29, 95 (1999).

1153. Kaminskii, A. A., García Solé, J., Bagaev, S. N., Jaque, D., and Capmany, J., Ferroelectric Nd^{3+}:$Sr_xBa_{1-x}(NbO_3)_2$—a new nonlinear laser crystal: cw 1-μm stimulated emission ($^4F_{3/2} \rightarrow {}^4I_{11/2}$) and diffuse self-frequency-doubling, collinear and diffuse self-frequency doubling and summation, *Quantum Electron.* 28, 1031 (1998).

1154. Kaminskii, A. A., Butasshin, A. V., Eichler, H.-J. et al., Orthorhombic ferroelectric and ferroelastic $Gd_2(MoO_4)_3$:crystal – a new many-purposed nonliner and optical material: efficient multiple stimulated Raman scattering and CW and tunable second harmonic generation, *Opt. Mater.* 7, 59 (1997).

1555. Lando, M., Shimony, Y., Benmair, R., Abramovich, D., Krupkin, V., and Yogev, A., Visible solar-pumped lasers, *Opt. Mater.* 13, 111 (1999).

1156. Jaque, D., Capmany, J., Sanz Garcia, J. A., Brenier, A., Boulon, G., and García Solé J., Nd^{3+} ion based self frequency doubling solid state lasers, *Opt. Mater.* 13, 147 (1999).

1157. Seo, J. T., Hömmerich, U., Zong, H. et al., Mid-infrared lasing from a novel optical material: chromium-doped $Cd_{0.55}Mn_{0.45}Te$, *Phys. Stat. Solidi* (a) 175, R3 (1999).

1158. Adams, J. J., Bibeau, C, Page, R. H., Krol, D. M., Furu, L. H., and Payne, S. A., 4.0–4.5-μm lasing of Fe:ZnSe below 180 K, a new mid-infrared laser material, *Optics Lett.* 24, 1720 (1999).

1159. Markushev, V. M., Zolin, V. F., and Brishina, Ch. M., Luminescence and stimulated emission of neodymium in sodium lanthanum molybdate powders, *Sov. J. Quantum Electron.* 16, 261 (1986) and references cited therein.

1160. Ter-Gabrielyan, N. E., Markushev, V. M., Belan, V. R., Brishina, Ch. M., and Zolin, V. F., Stimulated emission spectra of powders of double sodium and lanthanum tetramolybdate, *Sov. J. Quantum Electron.* 21, 32 (1991).

1161. Fornasiero, L., Mix, E., Peters, V., Heumann, E., Petermann, K., and Huber, G., Efficient laser operation of Nd:Sc_2O_3 at 966 nm, 1082 nm, and 1486 nm, *OSA Trends in Optics and Photonics: Advanced Solid State Lasers*, Vol. 26, Fejer, M. M., Injeyan, H., and Keller, U., Eds. (1999), p. 249.

Section 1.2
GLASS LASERS

1.2.1 Introduction

The past two decades have witnessed increased activity in glass lasers, in the form of both bulk materials and fiber and planar waveguides. Fibers, with their long interaction region, and heavy metal fluoride glasses, with their low vibrational frequencies and hence reduced probabilities for decay by nonradiative processes, have made possible many new lasing transitions and operation at longer wavelengths. Upconversion pumping schemes, laser diode pumping, and double-clad fibers have further extended the ulitization of glass lasers.

Glass lasers and amplifiers have been based almost exclusively on the trivalent lanthanide ions. Ions and host glasses that have been used for lasers are summarized in Table 1.2.1. The energy levels, lasing transitions, and approximate wavelengths of lanthanide-ion glass lasers are shown in Figures 1.2.1(a) and 1.2.1(b).

Table 1.2.1
Host Glass Types and Laser Ions

Glass Type	Pr^{3+}	Nd^{3+}	Pm^{3+}	Sm^{3+}	Tb^{3+}	Ho^{3+}	Er^{3+}	Tm^{3+}	Yb^{3+}
Oxide Glasses									
borate		•			•				•
germanate		•				•		•	
phosphate		•	•				•		
silica	•	•		•		•	•	•	•
silicate		•				•	•	•	•
aluminosilicate							•		
borosilicate		•					•		
germanosilicate									•
tellurite		•							
Halide Glasses									
fluoroberyllate		•							
fluorozirconate	•	•				•	•	•	•
Oxyhalide Glasses									
fluoroaluminate		•					•		
fluorophosphate		•					•		
Chalcogenide Glasses									
sulfide		•							

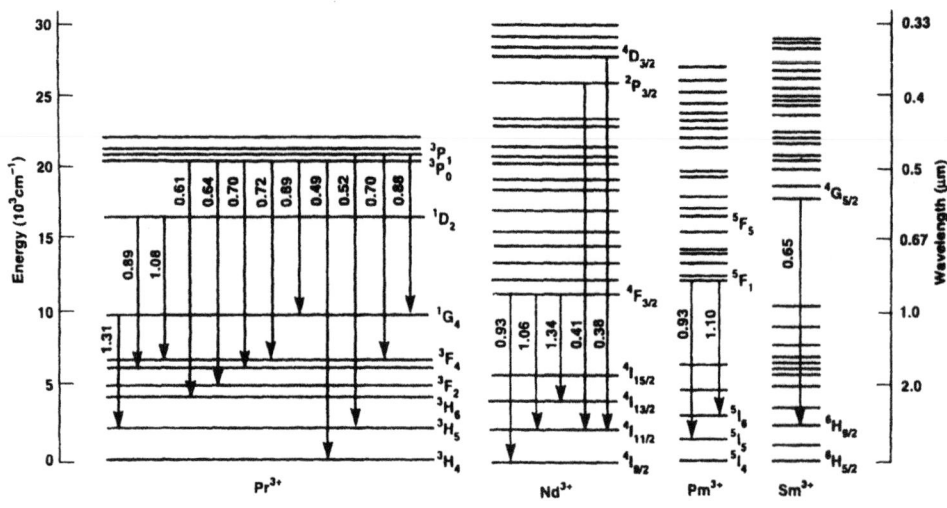

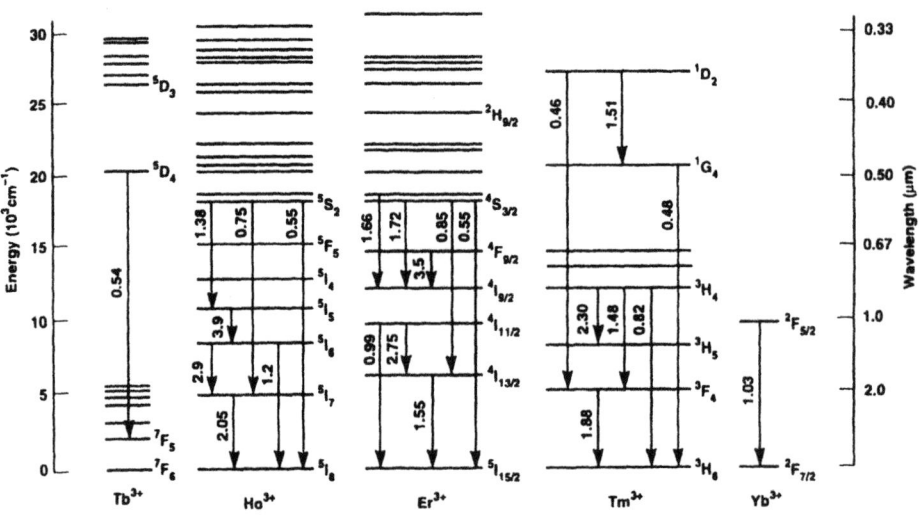

Figure 1.2.1 Energy levels, laser transitions, and approximate wavelengths (microns) of lanthanide-ion glass lasers.

As shown in Figure 1.2.2, the reported wavelengths of lanthanide-ion glass lasers range from 0.38 to 4 microns. The wavelength range is less than that of crystals at both the long and short wavelength extrema. The lasing wavelength could be extended to shorter wavelengths using glassy hosts with larger energy gaps such as beryllium fluoride and silica. Extension further into the infrared is limited by the vibrational frequencies associated with the glass network formers and nonradiative decay processes.

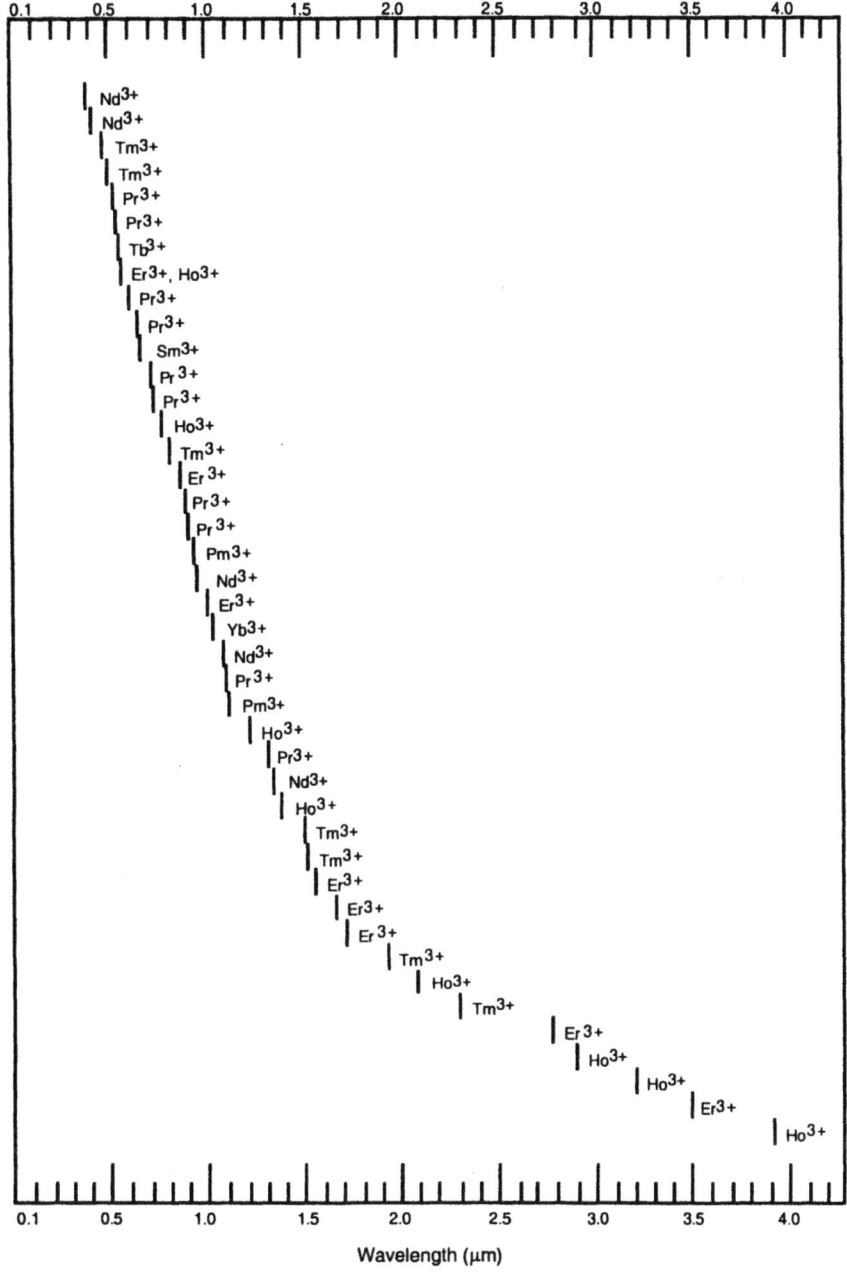

Figure 1.2.2 Approximate wavelengths of lanthanide-ion glass lasers; exact wavelengths are dependent on the glass composition and temperature and the specific Stark levels involved (from the *Handbook of Laser Wavelengths*, CRC Press, Boca Raton, FL, 1998, with permission).

Unlike crystals, which have a unique composition and structure, many glasses may be formed in broad compositional ranges with varying parentages of glass network formers (e.g., silicate, phosphate, borate) and network modifiers (e.g., alkali ions, alkaline earths ions). Compositional changes affect the stimulated emission cross sections, rates of radiative and nonradiative transitions, crystalline field splittings, and inhomogeneous broadening. Although trivially small compositional changes might technically constitute a new host material, the lasers listed in the tables in Section 1.2.2 are generally characterized by either significantly different host glass compositions or different operating properties, thus the tables are representative rather than exhaustive with respect to all glass lasers reported.

Because of site-to-site variations in the local fields in glass, there is a distribution of energy levels and transition frequencies that appear as inhomogeneous broadening and provide a small degree of tunability. Examples of reported tuning ranges of lanthanide-ion glass lasers are shown in Figure 1.2.3.

Upconversion excitation techniques involving multi-step absorption, energy transfer, excited state absorption, and photon avalanche have also been exploited for glass lasers. Examples of upconversion pumping schemes that have been used for fluorozirconate fiber lasers are given in Table 1.2.2.

In addition to glass laser operation involving a single transition, several cascade lasing schemes have been demonstrated, although not as many as for crystalline lasers (see Table 1.1.1). These schemes for glass lasers are summarized in Table 1.2.3. In all cases the lasers have utilized fluorozirconate fibers and have operated at room temperature.

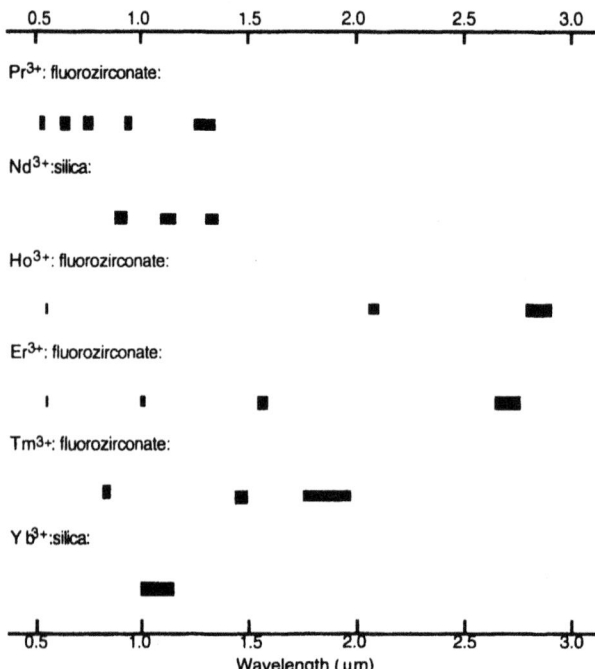

Figure 1.2.3 Reported tuning ranges of lanthanide-ion glass lasers (see Tables for specific wavelengths and host glasses).

Table 1.2.2
Upconversion Fluorozirconate Fiber Lasers

Laser wavelength(s) (μm)	Pumping scheme	Pump wavelength(s) (μm)	Temp. (K)	Mode	Ref.
Pr^{3+} lasers:					
0.491, 0.520, 0.605, 0.635	2-step absorption	0.835, 1.010	RT	cw	6
0.491, 0.520, 0.635	2-step absorption	1.020	RT	cw	228
Pr^{3+} (+ Yb^{3+}) lasers:					
0.491–0.493, 0.517–0.540, 0.605–0.622, 0.635–0.637	ET + ESA	0.860	RT	cw	8
0.635	ET + ESA	0.850	RT	cw	226
0.602, 0.635	ET + ESA	0.860	RT	cw	19
0.521, 0.635	ET + ESA	0.830, 1.016	RT	cw	10
0.635	ET + ESA	0.849	RT	cw	23
Nd^{3+} lasers:					
0.381, 0.412	2-step absorption	0.590	RT	p, cw	1, 2
Ho^{3+} lasers:					
0.540–0.553	2-step absorption	0.647	RT	cw	13
0.540–0.553	2-step absorption	0.643	RT	cw	227
Er^{3+} lasers:					
0.544	2-step absorption	0.801	RT	cw	229
0.544	2-step absorption	0.971	RT	cw	230
0.546, 0.850	2-step absorption	0.800	RT	cw	14
0.548	2-step absorption	0.800	RT	cw	16
Tm^{3+} lasers:					
0.455	2-step absorption	0.676, 0.647	77		3
0.455	2-step absorption	0.645, 1.064	RT	cw	4
0.480	2-step absorption	0.676, 0.647	77	sp	3
0.480	3-step absorption	1.12	RT	cw	5
0.810	2-step absorption	1.064	RT	cw	27
Tm^{3+} (+ Yb^{3+}) lasers:					
0.650	ET + ESA	1.120	RT		228

cw – continuous wave, ESA – excited state absorption, ET – energy transfer, sp – self-pulsed, RT – room temperature.

Table 1.2.3
Cascade Laser Schemes

Laser ion	Cascade transitions	Lasing wavelengths (μm)	Ref.
Ho^{3+}	$^5S_2 \rightarrow {}^5I_5 \Longrightarrow {}^5I_6 \rightarrow {}^5I_7$	1.4, 2.9	194
	$^5I_6 \rightarrow {}^5I_7 \rightarrow {}^5I_8$	1.2, 3.9	231
Er^{3+}	$^4S_{3/2} \rightarrow {}^4I_{9/2} \Longrightarrow {}^4I_{11/2} \rightarrow {}^4I_{13/2}$	1.7, 2.7	163
	$^4S_{3/2} \rightarrow {}^4I_{9/2} \Longrightarrow {}^4I_{11/2} \rightarrow {}^4I_{13/2} \rightarrow {}^4I_{15/2}$	1.7, 2.7, 1.55	232
Tm^{3+}	$^3H_4 \rightarrow {}^3H_5 \Longrightarrow {}^3F_4 \rightarrow {}^3H_6$	1.48, 1.88	115

$\rightarrow$ lasing transition $\Longrightarrow$ nonradiative transition

Further Reading

Becker, P. C., Olsson, N. A., and Simpson, J. R., *Erbium-Doped Fiber Amplifiers,* Academic Press, San Diego (1999).

Davey, S. T., Ainslie, B. J., and Wyatt, R., Waveguide Glasses, in *Handbook of Laser Science and Technology, Suppl. 2: Optical Materials,* CRC Press, Boca Raton, FL (1995), p. 635.

Desurvire, E., *Erbium-Doped Fiber Amplifiers,* John Wiley & Sons, New York, (1994).

Digonnet, M. J. F., Ed., *Selected Papers on Rare-Earth-Doped Fiber Laser Sources and Amplifiers,* SPIE Milestone Series, Vol. MS37, SPIE Optical Engineering Press, Bellingham, WA (1992).

Digonnet, M. J. F., Ed., *Rare Earth Doped Fiber Lasers and Amplifiers,* Marcel Dekker, New York (1993).

France, P. W., Ed., *Optical Fibre Lasers and Amplifiers,* Blackie and Sons, Ltd., Glasgow and London (1991).

Hall, D. W. and Weber, M. J., Glass Lasers, in *Handbook of Laser Science and Technology, Suppl. 1: Lasers,* CRC Press, Boca Raton, FL (1991), p. 137.

Miniscalco, W. J., Erbium-doped glasses for fiber amplifiers at 1500 nm, *J. Lightwave Techn.* 9, 234 (1991).

Rapp, C. F., Laser Glasses: Bulk Glasses, in *Handbook of Laser Science and Technology, Suppl. 2: Optical Materials,* CRC Press, Boca Raton, FL (1995), p. 619.

Stokowski, S. E., Glass Lasers, in *Handbook of Laser Science and Technology, Vol. I: Lasers and Masers,* CRC Press, Boca Raton, FL (1982), p. 215.

Weber, M. J., Science and technology of laser glass, *J. Non-Cryst. Solids,* 123, 208 (1991).

1.2.2 Tables of Glass Lasers

Data on specific lanthanide ion lasers are given in Table 1.2.4. The results are grouped by the form of the glass—bulk (rod, disk), fiber, planar waveguide, microcavity. Within each group, lasers are listed by the type of host glass that is specified by the glass network former (silicate, phosphate, fluorozirconate, etc.) and, if known, the principal glass network modifier cation(s). If codopants or sensitizing ions are added, they are listed in parentheses. Commercial glasses are identified by their commercial designation (in parentheses). For these lasers, the glass type is generally known but the detailed compositions are usually proprietary.

Other columns in Table 1.2.4 give the lasing transition and wavelength, pump wavelength or source (flashlamp, laser, sun), operating temperature, mode (pulsed, cw, quasi-cw; single- or multi-mode), together with primary references. Lasers involving upconversion schemes are noted in the mode column. Lasers that have been tuned over a range of wavelengths are listed by the lowest wavelength reported; the tuning range given is that for the configuration and experimental conditions used and may not represent the extremes possible.

Abbreviations used in Table 1.2.4:

Pump source			Mode of operation		
alex	—	alexandrite ($BeAl_2O_4:Cr$) laser	cw	—	continuous wave
Ar	—	argon ion laser	DFB	—	distributed feedback
CC	—	color center laser	m	—	multimode
D	—	frequency doubled	p	—	pulsed
LD	—	laser diode	qcw	—	quasi-continuous wave
NdYAG	—	$Nd:Y_3Al_5O_{12}$ laser	s	—	single mode
NdYLF	—	$Nd:LiYF_4$ laser	upc	—	upconversion pumping
scheme					
TiS	—	titanium doped sapphire (Al_2O_3) laser	wgm	—	whispering galley mode

Table 1.2.4
Glass lasers

Praseodymium (Pr^{3+}, $4f^2$) Lasers

Host glass	Lasing transition	Laser wavelength (μm)	Temp. (K)	Pump source [wavelength – μm]	Mode	Ref.
Fiber						
fluorozirconate	$^3P_0 \rightarrow {}^3H_4$	0.491	RT	Ar ion laser (0.476)	m	7
fluorozirconate	$^3P_0 \rightarrow {}^3H_4$	0.491	RT	TiS (0.835+1.01)	cw, upc	6
fluorozirconate	$^3P_0 \rightarrow {}^3H_4$	0.491	RT	Yb laser (1.020)	cw, upc	228
fluorozirconate	$^3P_0 \rightarrow {}^3H_4$	0.492	RT	TiS (0.835+1.017)	cw, upc	9
fluorozirconate	$^3P_0 \rightarrow {}^3H_4$	0.492	RT	LD (0.831+1.017)	cw, upc	225
fluorozirconate	$^3P_1 \rightarrow {}^3H_5$	0.520	RT	Ar ion laser (0.476)	cw, m	7
fluorozirconate	$^3P_1 \rightarrow {}^3H_5$	0.520	RT	LD (0.835+1.01)	cw	6
fluorozirconate	$^3P_1 \rightarrow {}^3H_5$	0.520	RT	Yb laser (1.020)	cw, upc	228
fluorozirconate	$^3P_0 \rightarrow {}^3H_6$	0.599–0.611	RT	Ar ion laser (0.476)	m	7
fluorozirconate	$^3P_0 \rightarrow {}^3H_6$	0.601–0.618	RT	Ar ion laser (0.476)	cw	18
fluorozirconate	$^3P_0 \rightarrow {}^3H_6$	0.605	RT	LD (0.835+1.01)	cw, upc	6
fluorozirconate	$^3P_0 \rightarrow {}^3F_2$	0.631–0.641	RT	Ar ion laser (0.476)	cw, m	7, 18
fluorozirconate	$^3P_0 \rightarrow {}^3F_2$	0.635	RT	Ar ion laser (0.476)	m	7
fluorozirconate	$^3P_0 \rightarrow {}^3F_2$	0.635	RT	LD (0.883+1.016)	cw, upc	10
fluorozirconate	$^3P_0 \rightarrow {}^3F_2$	0.635	RT	LD (0.835+1.01)	cw	6
fluorozirconate	$^3P_0 \rightarrow {}^3F_2$	0.635	RT	Yb laser (1.020)	cw, upc	228
fluorozirconate	$^3P_1 \rightarrow {}^3F_4$	0.690–0.703	RT	Ar ion laser (0.476)	cw	18
fluorozirconate	$^3P_1 \rightarrow {}^3F_4$	0.707–0.725	RT	Ar ion laser (0.476)	cw	18
fluororzirconate	$^3P_1 \rightarrow {}^3F_4$	0.715	RT	Ar ion laser (0.476)	m	7
fluororzirconate	$^3P_1 \rightarrow {}^3F_4$	0.715	RT	LD (0.835+1.01)	cw	6
fluorozirconate	$^3P_1 \rightarrow {}^1G_4$	0.880–0.886	RT	Ar ion laser (0.476)	cw	18
fluorozirconate	$^3P_0 \rightarrow {}^1G_4$	0.902–0.916	RT	Ar ion laser (0.476)	cw	18
fluorozirconate	$^1G_4 \rightarrow {}^3H_5$	1.24–1.34	RT	TiS laser (1.007)	s	94
fluorozirconate	$^1G_4 \rightarrow {}^3H_5$	1.294	RT	Nd:YAG laser (1.064)	cw	97

Host glass	Laser transition	Laser wavelength (μm)	Temp. (K)	Pump source [wavelength – μm]	Mode	Reference
fluorozirconate (Yb)	$^3P_0 \rightarrow {}^3H_4$	0.491–0.493	RT	TiS (0.780–0.880)	cw	8, 235
fluorozirconate (Yb)	$^3P_0 \rightarrow {}^3H_4$	0.491	RT	LD (0.840)	cw	235
fluorozirconate (Yb)	$^3P_1 \rightarrow {}^3H_5$	0.517–0.540	RT	TiS (0.780–0.880)	cw	8
fluorozirconate (Yb)	$^3P_1 \rightarrow {}^3H_5$	0.521	RT	LD (0.883+1.016)	cw, upc	10
fluorozirconate (Yb)	$^3P_0 \rightarrow {}^3H_6$	0.602	RT	laser diode (0.860)	cw	19
fluorozirconate (Yb)	$^3P_0 \rightarrow {}^3H_6$	0.605–0.622	RT	TiS (0.780–0.880)	cw	8
fluorozirconate (Yb)	$^3P_0 \rightarrow {}^3F_2$	0.635	RT	TiS (0.844, 0.849)	cw, upc	23
fluorozirconate (Yb)	$^3P_0 \rightarrow {}^3F_2$	0.635	RT	laser diode (0.860)	cw	19
fluorozirconate (Yb)	$^3P_0 \rightarrow {}^3F_2$	0.635	RT	(0.850)	cw, upc	226
fluorozirconate (Yb)	$^3P_0 \rightarrow {}^3F_2$	0.635–0.637	RT	TiS (0.780–0.880)	cw	8
silica	$^1D_2 \rightarrow {}^3F_4$	1.080	RT	dye laser (0.590)	s	201
silica	$^1D_2 \rightarrow {}^3F_3$	1.080	RT	Ar ion laser (0.488)	m	31
silica	$^1D_2 \rightarrow {}^3F_{3,4}$	1.000–1.085	RT	dye laser (0.592)	m	44
silica (Ge)	$^3P_0 \rightarrow {}^1G_4$	0.888	RT	dye laser (0.590)	s	31

Neodymium (Nd^{3+}, 4f^3) Lasers

Host glass	Laser transition	Laser wavelength (μm)	Temp. (K)	Pump source [wavelength – μm]	Mode	Reference
Bulk						
borate	$^4F_{3/2} \rightarrow {}^4I_{11/2}$	1.06	RT	Hg arc lamp	cw	68
La–Ba–Th borate	$^4F_{3/2} \rightarrow {}^4I_{13/2}$	1.37	RT	Xe flashlamp	p	110
Ba-In-Ga-Zn-fluoride	$^4F_{3/2} \rightarrow {}^4I_{11/2}$	1.0479	RT	flashlamp	p	217
fluoroberyllate	$^4F_{3/2} \rightarrow {}^4I_{11/2}$	1.047	RT	flashlamp	p	49
fluorophosphate	$^4F_{3/2} \rightarrow {}^4I_{11/2}$	1.051	RT	Xe flashlamp	p	54
glass ceramic	$^4F_{3/2} \rightarrow {}^4I_{11/2}$	1.06	RT	Xe flashlamp	p	70, 71

Neodymium (Nd^{3+}, 4f^3) Lasers—*continued*

Host glass	Laser transition	Laser wavelength (μm)	Temp. (K)	Pump source [wavelength – μm]	Mode	Reference
Li germanate	$^4F_{3/2} \rightarrow {}^4I_{11/2}$	1.06	RT	Ar ion laser (0.514)	p	73
phosphate (APG-1)	$^4F_{3/2} \rightarrow {}^4I_{11/2}$	1.0546	RT	Ar ion laser (0.514)	p	22
phosphate (LHG–5)	$^4F_{3/2} \rightarrow {}^4I_{11/2}$	1.054	RT	laser diode (0.800)	p, cw	95
phosphate (LHG–8)	$^4F_{3/2} \rightarrow {}^4I_{11/2}$	1.054	RT	laser diode (0.800)	cw	95
phosphate (LHG–8)	$^4F_{3/2} \rightarrow {}^4I_{11/2}$	1.054	RT	laser diode (0.800)	p	104
phosphate (LHG–8)	$^4F_{3/2} \rightarrow {}^4I_{11/2}$	1.054	RT	Ar ion laser (0.514)	cw	60
phosphate (LG-750)	$^4F_{3/2} \rightarrow {}^4I_{11/2}$	1.046–1.078	RT	Xe flashlamp	p	45
phosphate (LG-760)	$^4F_{3/2} \rightarrow {}^4I_{11/2}$	1.054	RT	laser diode (0.802)	cw	198
Li–La phosphate (Cr)	$^4F_{3/2} \rightarrow {}^4I_{11/2}$	1.06	RT	Kr laser (0.531, 0.647)	qcw	74
Li–Nd–La phosphate	$^4F_{3/2} \rightarrow {}^4I_{11/2}$	1.055	RT	Xe flashlamp	p	63
Li–Nd–La phosphate	$^4F_{3/2} \rightarrow {}^4I_{13/2}$	1.32	RT	Xe flashlamp	p	63
Na–K–Cd phosphate	$^4F_{3/2} \rightarrow {}^4I_{11/2}$	1.0535	RT	Ar ion laser (0.514)	p	58
Zn–Li phosphate	$^4F_{3/2} \rightarrow {}^4I_{11/2}$	1.054	RT	Xe flashlamp	p	61, 62
silica (Al)	$^4F_{3/2} \rightarrow {}^4I_{11/2}$	1.0635	RT	dye laser (0.588)	p	85
silica (+buffer)	$^4F_{3/2} \rightarrow {}^4I_{11/2}$	1.062	RT	Xe flashlamp	p	84
silicate (GLS-1)	$^4F_{3/2} \rightarrow {}^4I_{9/2}$	0.92	RT	dye laser (0.580)	p	35
silicate	$^4F_{3/2} \rightarrow {}^4I_{11/2}$	1.06	RT	Kr arc lamp	cw	75
silicate (AO-702)	$^4F_{3/2} \rightarrow {}^4I_{11/2}$	—	RT	sun	p	220
silicate (LG-680)	$^4F_{3/2} \rightarrow {}^4I_{11/2}$	1.052–1.076	RT	Xe flashlamp	p	45
K–Ba silicate	$^4F_{3/2} \rightarrow {}^4I_{11/2}$	1.061	RT	Xe flashlamp	p	82
Na–Ca silicate	$^4F_{3/2} \rightarrow {}^4I_{9/2}$	0.918	80	flashlamp	p	34
Ga–La sulfide	$^4F_{3/2} \rightarrow {}^4I_{11/2}$	1.08	RT	TiS laser (0.815)	cw	89
tellurite	$^4F_{3/2} \rightarrow {}^4I_{11/2}$	1.066	RT	TiS laser (0.804)	p	87
Li tellurite	$^4F_{3/2} \rightarrow {}^4I_{11/2}$	1.0605	RT	Ar ion laser (0.514)		58
Li tellurite	$^4F_{3/2} \rightarrow {}^4I_{11/2}$	1.0605	RT	Xe flashlamp	p	81

Fiber						
chalcogenide	$^4F_{3/2} \to {}^4I_{11/2}$	1.080	RT	TiS laser (0.815)	p, m	98
fluorophosphate	$^4F_{3/2} \to {}^4I_{13/2}$	1.323, 1.328, 1.355	RT	TiS laser (0.803)	s	102
fluorozirconate	$^4D_{3/2} \to {}^4I_{11/2}$	0.381	RT	dye laser (0.590)	m, upc	1
fluorozirconate	$^2P_{3/2} \to {}^4I_{11/2}$	0.412	RT	dye laser (0.581–0.593)	cw, upc	2
fluorozirconate	$^4F_{3/2} \to {}^4I_{11/2}$	1.049	RT	Ar ion laser (0.514)	p, m	105
fluorozirconate	$^4F_{3/2} \to {}^4I_{11/2}$	1.050	RT	Ar ion laser (0.514)	m	52
fluorozirconate	$^4F_{3/2} \to {}^4I_{13/2}$	1.31–1.36	RT	Ar ion laser (0.496)	cw, s	101
fluorozirconate	$^4F_{3/2} \to {}^4I_{13/2}$	1.340	RT	Ar ion laser (0.514)	p, m	105
fluorozirconate	$^4F_{3/2} \to {}^4I_{13/2}$	1.345	RT	laser diode (0.795)	cw, s	106
fluorozirconate	$^4F_{3/2} \to {}^4I_{13/2}$	1.350	RT	Ar ion laser (0.514)	cw, m	107
phosphate	$^4F_{3/2} \to {}^4I_{11/2}$	1.05	RT	laser diode (0.807)	s	50
phosphate	$^4F_{3/2} \to {}^4I_{13/2}$	1.363	RT	TiS, laser diode (0.815)	cw, s	108
phosphate (LHG–8)	$^4F_{3/2} \to {}^4I_{11/2}$	1.054	RT	Xe flashlamp	p	99
phosphate (LHG–8)	$^4F_{3/2} \to {}^4I_{11/2}$	1.054	RT	Ar arc lamp	cw	99
phosphate (LHG–8)	$^4F_{3/2} \to {}^4I_{11/2}$	1.054	RT	laser diode (0.807)	cw	109
phosphate (LHG–8)	$^4F_{3/2} \to {}^4I_{13/2}$	1.366	RT	laser diode (0.807)	cw	109
silica	$^4F_{3/2} \to {}^4I_{9/2}$	0.899–0.951	RT	Ar ion laser (0.514)	s	32
silica	$^4F_{3/2} \to {}^4I_{9/2}$	0.938	RT	dye laser (0.590)	s	?
silica	$^4F_{3/2} \to {}^4I_{9/2}$	0.938		laser diode (0.823)	s	38
silica	$^4F_{3/2} \to {}^4I_{9/2}$	0.900–0.945	RT	dye laser (0.590)	cw, s	33, 37
silica	$^4F_{3/2} \to {}^4I_{11/2}$	1.06	RT	Xe flashlamp	m	76
silica	$^4F_{3/2} \to {}^4I_{11/2}$	1.069–1.144	RT	Ar ion laser (0.514)	s	32
silica	$^4F_{3/2} \to {}^4I_{11/2}$	1.070–1.135	RT	dye laser (0.590)	cw, s	33, 37
silica	$^4F_{3/2} \to {}^4I_{11/2}$	1.0770–1.1386	RT	Ar ion laser (0.514)	m	88
silica (Er)	$^4F_{3/2} \to {}^4I_{9/2}$	0.908[a]	RT	Ar ion laser (0.514)	cw	140, 199
silica (Er)	$^4F_{3/2} \to {}^4I_{9/2}$	0.932[a]	RT	Ar ion laser (0.514)	cw	140

Neodymium (Nd^{3+}, 4f^3) Lasers—continued

Host glass	Laser transition	Laser wavelength (μm)	Temp. (K)	Pump source [wavelength – μm]	Mode	Reference
silica (Er)	$^4F_{3/2} \rightarrow {}^4I_{11/2}$	1.08[a]	RT	Ar ion laser (0.514)	cw	140, 199
silica (Al)	$^4F_{3/2} \rightarrow {}^4I_{11/2}$	1.060	RT	Ar (0.514), LD (0.590)	p	79
silica (Al)	$^4F_{3/2} \rightarrow {}^4I_{11/2}$	1.088	RT	laser diode (0.890)	cw	90
silica (Ge)	$^4F_{3/2} \rightarrow {}^4I_{11/2}$	1.088	RT	laser diode (0.820)	cw, s	91, 92
silica (P)	$^4F_{3/2} \rightarrow {}^4I_{11/2}$	1.050–1.075	RT	LD (0.790–0.850)	cw	53
silica (P)	$^4F_{3/2} \rightarrow {}^4I_{11/2}$	1.055	RT	Kr ion laser (0.752)	cw	64
silica (P)	$^4F_{3/2} \rightarrow {}^4I_{13/2}$	1.36	RT	Kr ion laser (0.752)	cw	64
silicate (GLS3)	$^4F_{3/2} \rightarrow {}^4I_{11/2}$	1.06	RT	Xe flashlamp	p	200
K–Ba silicate	$^4F_{3/2} \rightarrow {}^4I_{11/2}$	1.061	RT	Xe flashlamp	p	82
Pb silicate (F7)	$^4F_{3/2} \rightarrow {}^4I_{11/2}$	1.06	RT	LD (0.807)	s	72
tellurite	$^4F_{3/2} \rightarrow {}^4I_{11/2}$	1.061	RT	TiS laser (0.818)	s	83
Microsphere						
fluorozirconate	$^4F_{3/2} \rightarrow {}^4I_{11/2}$	1.051	RT	laser diode (0.800)	cw	197
fluorozirconate	$^4F_{3/2} \rightarrow {}^4I_{13/2}$	1.334	RT	laser diode (0.800)	cw	197
silica	$^4F_{3/2} \rightarrow {}^4I_{9/2}$	~0.930	RT	laser diode (0.807)	cw, wgm	217
silica	$^4F_{3/2} \rightarrow {}^4I_{11/2}$	~1.080	RT	laser diode (0.807)	cw, wgm	217
silicate	$^4F_{3/2} \rightarrow {}^4I_{9/2}$	0.860, 0.890	RT	Ar ion laser (0.514)	cw	219
Planar Waveguide						
borosilicate	$^4F_{3/2} \rightarrow {}^4I_{11/2}$	1.06	RT	dye laser (0.590)	p	69
borosilicate (BK–7)	$^4F_{3/2} \rightarrow {}^4I_{11/2}$	1.058	RT	TiS laser (0.807)	cw, s	66
phosphate	$^4F_{3/2} \rightarrow {}^4I_{9/2}$	0.905	RT	0.974	cw	21
phosphate	$^4F_{3/2} \rightarrow {}^4I_{11/2}$	1.057	RT	0.974	cw	21
phosphate	$^4F_{3/2} \rightarrow {}^4I_{13/2}$	1.356	RT	0.974	cw	21
phosphate (LG–760)	$^4F_{3/2} \rightarrow {}^4I_{11/2}$	1.05	RT	dye laser (0.590)	m	51

Host glass	Laser transition	Laser wavelength (μm)	Temp. (K)	Pump source [wavelength – μm]	Mode	Reference
phosphate (LHG-5)	$^4F_{3/2} \rightarrow {}^4I_{13/2}$	1.325, 1.355	RT	laser diode (0.802)	cw, m	103
silica	$^4F_{3/2} \rightarrow {}^4I_{11/2}$	1.053	RT	laser diode (0.805)	cw, m	57
silica (Al)	$^4F_{3/2} \rightarrow {}^4I_{11/2}$	1.062	RT	TiS laser (0.807)	cw	80
silica (P)	$^4F_{3/2} \rightarrow {}^4I_{11/2}$	1.0515	RT	TiS laser (0.80)	cw	55
silica (P)	$^4F_{3/2} \rightarrow {}^4I_{11/2}$	1.0525	RT	laser diode (0.805)	cw, m	56
silicate (GLS2, GLS3)	$^4F_{3/2} \rightarrow {}^4I_{11/2}$	~1.06	RT	flashlamp	p, m	77
silicate (LG–660)	$^4F_{3/2} \rightarrow {}^4I_{11/2}$	1.057	RT	Ar ion laser (0.528)	cw	65

(a) Simultaneous lasing of Er³⁺ at 1.53 μm.

Promethium (Pm³⁺, 4f⁴) Lasers

Host glass	Laser transition	Laser wavelength (μm)	Temp. (K)	Pump source [wavelength – μm]	Mode	Reference
Bulk						
Pb–In phosphate	$^5F_1 \rightarrow {}^5I_5$	0.933	RT	dye laser (0.572)	p	36
Pb–In phosphate	$^5F_1 \rightarrow {}^5I_6$	1.098	RT	dye laser (0.572)	p	36

Samarium (Sm³⁺, 4f⁵) Lasers

Host glass	Laser transition	Laser wavelength (μm)	Temp. (K)	Pump source [wavelength – μm]	Mode	Reference
Fiber						
silica	$^4G_{5/2} \rightarrow {}^6H_{9/2}$	0.651	RT	Ar ion laser (0.488)	s	24

Terbium (Tb³⁺, 4f⁹) Lasers

Host glass	Laser transition	Laser wavelength (μm)	Temp. (K)	Pump source [wavelength – μm]	Mode	Reference
Bulk						
borate	$^5D_4 \rightarrow {}^7F_5$	0.54[a]	RT	Xe flashlamp	p	11

(a) Evidence of lasing was based on the appearance of emission spikes.

Holmium (Ho^{3+}, 4f^{10}) Lasers

Host glass	Laser transition	Laser wavelength (μm)	Temp. (K)	Pump source [wavelength – μm]	Mode	Reference
Bulk						
germanate	$^5I_7 \to {}^5I_8$	2.09	RT	Xe flashlamp	p	196
Li–Mg–Al silicate	$^5I_7 \to {}^5I_8$	2.08	77	Xe flashlamp	p	179
Li–Mg–Al silicate (Er,Yb)	$^5I_7 \to {}^5I_8$	2.06–2.10	77	Xe flashlamp	p	178
Fiber						
fluorozirconate	$^5S_2 \to {}^5I_8$	~0.539–0.550	77, RT	dye laser (0.643)	m	17
fluorozirconate	$^5S_2 \to {}^5I_8$	0.540–0.553	RT	Kr ion laser (0.647)	cw, s, upc	13
fluorozirconate	$^5S_2 \to {}^5I_8$	0.540–0.553	RT	Kr ion laser (0.643)	cw, upc	227
fluorozirconate	$^5S_2 \to {}^5I_7$	0.7495–0.7545	RT	Kr ion laser (0.647)	cw, s, upc	13
fluorozirconate	$^5I_6 \to {}^5I_8$	1.19	77, RT	dye laser (0.639)	cw, m	127
fluorozirconate	$^5S_2, {}^5F_4 \to {}^5I_5$	1.38	RT	Ar ion laser (0.488)	m	111
fluorozirconate	$^5I_7 \to {}^5I_8$	2.05–2.06[a]	RT	TiS laser (0.890)	cw, m	231
fluorozirconate	$^5I_7 \to {}^5I_8$	2.08	RT	Ar ion laser (0.488)	m	111
fluorozirconate	$^5I_6 \to {}^5I_7$	2.83–2.95	RT	LD (0.640), TiS (0.750)	cw	191
fluorozirconate	$^5I_6 \to {}^5I_7$	2.83–2.95[a]	RT	TiS laser (0.890)	cw, m	231
fluorozirconate	$^5S_2 \to {}^5F_5$	~3.22	RT	DNd:YAG (0.532)	cw	221
fluorozirconate	$^5I_5 \to {}^5I_6$	3.95[b]	RT	dye laser (0.640)	qcw, m	194
fluorozirconate (Tm)	$^5I_7 \to {}^5I_8$	2.04	RT	laser diode (0.82)	cw, s	175
fluorozirconate (Tm)	$^5I_7 \to {}^5I_8$	2.024	RT	TiS laser (0.890)	cw	177
fluorozirconate (Tm)	$^5I_7 \to {}^5I_8$	2.076	RT	TiS laser (0.890)	cw	177
silica (Tm)	$^5I_7 \to {}^5I_8$	1.960–2.032	293–393	TiS laser (~0.812)	cw	173
silica (Ge)	$^5I_7 \to {}^5I_8$	2.04	RT	Ar ion laser (0.458)	s, cw	176

(a) Operates in cascade mode with simultaneous laser emission at 2.93 and 2.06 μm.
(b) Operates in cascade mode with simultaneous laser emission at ~1.2 μm.

Erbium (Er^{3+}, 4f^{11}) Lasers

Host glass	Laser transition	Laser wavelength (μm)	Temp. (K)	Pump source [wavelength – μm]	Mode	Reference
Bulk						
fluoroaluminate (Cr, Yb)	$^4I_{13/2} \rightarrow {}^4I_{15/2}$	1.6	RT	Kr ion laser (0.647)	cw	159
fluorophosphate (Yb)	$^4I_{13/2} \rightarrow {}^4I_{15/2}$	1.54	RT	Xe flashlamp	p	132
fluorophosphate (Cr, Yb)	$^4I_{13/2} \rightarrow {}^4I_{15/2}$	1.536–1.596	RT	Kr laser (0.647, 0.676)	cw	130
fluorozirconate	$^4I_{11/2} \rightarrow {}^4I_{13/2}$	2.70	RT	alex. laser (0.797)	p	167
fluorozirconate	$^4I_{11/2} \rightarrow {}^4I_{13/2}$	2.69–2.78	RT	Ar ion laser (0.514)	p, cw	184
fluorozirconate	$^4I_{11/2} \rightarrow {}^4I_{13/2}$	2.78	RT	Xe flashlamp	p	190
fluorozirconate	$^4I_{11/2} \rightarrow {}^4I_{13/2}$	2.79	RT	laser diode (~0.970)	cw, upc	233
phosphate	$^4I_{13/2} \rightarrow {}^4I_{15/2}$	1.54	RT	Nd:YAG laser (1.064)	p	134
phosphate (Yb)	$^4I_{13/2} \rightarrow {}^4I_{15/2}$	1.531–1.540	RT	laser diode (0.975)	cw	122
phosphate (Yb)	$^4I_{13/2} \rightarrow {}^4I_{15/2}$	1.5321–1.5348	RT	laser diode (0.980)	cw	123
phosphate (Yb)	$^4I_{13/2} \rightarrow {}^4I_{15/2}$	1.533	RT	laser diode (0.970)	p	124
phosphate (Yb)	$^4I_{13/2} \rightarrow {}^4I_{15/2}$	1.54	RT	LD (0.960–0.980)	cw	137
phosphate (Yb)	$^4I_{13/2} \rightarrow {}^4I_{15/2}$	1.54	RT	Nd:YAG laser (1.064)	p	138,139
phosphate (Yb)	$^4I_{13/2} \rightarrow {}^4I_{15/2}$	1.540	RT	laser diode (0.976)	s, cw	142
phosphate (Yb)	$^4I_{13/2} \rightarrow {}^4I_{15/2}$	1.545	RT	laser diode (0.970)	p	124
phosphate (Yb)	$^4I_{13/2} \rightarrow {}^4I_{15/2}$	1.549–1.563	RT	laser diode (0.975)	cw	122
Al–Zn phosphate (Yb)	$^4I_{13/2} \rightarrow {}^4I_{15/2}$	1.54	RT	Xe flashlamp	p	131
phosphate (Cr,Yb)	$^4I_{13/2} \rightarrow {}^4I_{15/2}$	1.54	RT	Xe flashlamp	p	136
Pb-Ba phosphate (Cr, Yb)	$^4I_{13/2} \rightarrow {}^4I_{15/2}$	1.54	RT	Xe flashlamp	p	133
Na–K Ba silicate (Yb)	$^4I_{13/2} \rightarrow {}^4I_{15/2}$	1.543	RT	Xe flashlamp	p	144
Na–K–Ba silicate	$^4I_{13/2} \rightarrow {}^4I_{15/2}$	1.55	RT	Xe flashlamp	p	146

Erbium (Er^{3+}, 4f^{11}) Lasers—continued

Host glass	Laser transition	Laser wavelength (µm)	Temp. (K)	Pump source [wavelength – µm]	Mode	Reference
Fiber						
fluorozirconate	$^4S_{3/2} \rightarrow {}^4I_{15/2}$	0.540–0.545	RT	TiS laser (0.97)	cw, upc	12
fluorozirconate	$^4S_{3/2} \rightarrow {}^4I_{15/2}$	0.544	RT	laser diode (0.971)	cw, upc	230
fluorozirconate	$^4S_{3/2} \rightarrow {}^4I_{15/2}$	0.544	RT	laser diode (0.801)	cw, upc	229
fluorozirconate	$^4S_{3/2} \rightarrow {}^4I_{15/2}$	0.546	RT	laser diode (0.801)	cw, upc	12,14
fluorozirconate	$^4S_{3/2} \rightarrow {}^4I_{15/2}$	0.548	RT	TiS laser (0.800)	cw, upc	16
fluorozirconate	$^4S_{3/2} \rightarrow {}^4I_{13/2}$	0.850	RT	laser diode (0.801)	cw, upc	29
fluorozirconate	$^4I_{11/2} \rightarrow {}^4I_{15/2}$	0.981–1.004	RT	Ar ion laser (0.514)	cw	41,42
fluorozirconate	$^4I_{11/2} \rightarrow {}^4I_{15/2}$	0.99	RT	Kr ion laser (0.647)	p	41
fluorozirconate	$^2H_{11/2} \rightarrow {}^4I_{9/2}$	1.660	RT	Ar ion laser (0.514)	p,m	162
fluorozirconate	$^2H_{11/2} \rightarrow {}^4I_{9/2}$	1.720	RT	Ar ion laser (0.514)	p,m	162
fluorozirconate	$^2H_{11/2} \rightarrow {}^4I_{9/2}$	1.724[a]	RT	TiS laser (0.791)	cw, s, upc	163
fluorozirconate	$^4I_{11/2} \rightarrow {}^4I_{13/2}$	2.702	RT	Ar laser (0.488, 0.514)	cw	186
fluorozirconate	$^4I_{11/2} \rightarrow {}^4I_{13/2}$	2.71	RT	laser diode (0.792)	s	187
fluorozirconate	$^4I_{11/2} \rightarrow {}^4I_{13/2}$	2.714	RT	Ar ion laser (0.476)	s	188
fluorozirconate	$^4I_{11/2} \rightarrow {}^4I_{13/2}$	2.715[a]	RT	TiS laser (0.791)	cw, s, upc	163
fluorozirconate	$^4I_{11/2} \rightarrow {}^4I_{13/2}$	2.75	RT	laser diode (0.792)	s	187
fluorozirconate	$^4I_{11/2} \rightarrow {}^4I_{13/2}$	2.701	RT	Ar ion laser (0.476)	m	185
fluorozirconate	$^4I_{11/2} \rightarrow {}^4I_{13/2}$	2.78	RT	flashlamp	s	187
fluorozirconate	$^4F_{9/2} \rightarrow {}^4I_{9/2}$	3.45	77	laser diode (0.655)	cw	192
fluorozirconate	$^4F_{9/2} \rightarrow {}^4I_{9/2}$	3.483	RT	laser diode (0.655)	cw	193
fluorozirconate	$^4F_{9/2} \rightarrow {}^4I_{9/2}$	3.535	RT	laser diode (0.655)	cw	193
fluorozirconate (Pr)	$^4I_{11/2} \rightarrow {}^4I_{13/2}$	2.65–2.77	RT	Kr ion laser (0.647)	cw	183
phosphate (LHG-8)	$^4I_{13/2} \rightarrow {}^4I_{15/2}$	1.535	RT	Ar ion laser (0.514)	cw	125
phosphate	$^4I_{13/2} \rightarrow {}^4I_{15/2}$	1.54	RT	Xe flashlamp	s, p	135
phosphosilicate (Al, Yb)	$^4I_{13/2} \rightarrow {}^4I_{15/2}$	1.535	RT	laser diode (0.980)	cw, DFB	218

Material	Transition	Temp	Wavelength (μm)	Pump	Mode	Ref
phosphosilicate (Al, Yb)	$^4I_{13/2} \rightarrow {}^4I_{15/2}$	RT	1.545	laser diode (0.980)	s, cw	222
phosphosilicate (Al, Yb)	$^4I_{13/2} \rightarrow {}^4I_{15/2}$	RT	1.534	laser diode (0.980)	cw	223
silica	$^4I_{13/2} \rightarrow {}^4I_{15/2}$	RT	1.529–1.554	Ar ion laser (0.514)	s	32
silica	$^4I_{13/2} \rightarrow {}^4I_{15/2}$	RT	1.55	Ar ion laser (0.514)	cw, p	147
silica	$^4I_{13/2} \rightarrow {}^4I_{15/2}$	RT	1.55	laser diode (0.811)	s	148
silica	$^4I_{13/2} \rightarrow {}^4I_{15/2}$	RT	1.553–1.603	Ar ion laser (0.514)	cw	151
silica (Al, Ge)	$^4I_{13/2} \rightarrow {}^4I_{15/2}$	RT	1.57–1.61	CC laser (1.55)	cw	157
silica (Al, P)	$^4I_{13/2} \rightarrow {}^4I_{15/2}$	RT	1.560	LD array (0.806)	cw	155
silica (Al, P, Yb)	$^4I_{13/2} \rightarrow {}^4I_{15/2}$	RT	1.56	dye laser (0.800–0.845)	cw	152,
silica (Al, P, Yb)	$^4I_{13/2} \rightarrow {}^4I_{15/2}$	RT	1.56	Nd:YAG laser (1.064) + dye laser (0.820)	s, cw	153
silica (Ge)	$^4I_{13/2} \rightarrow {}^4I_{15/2}$	RT	1.55	laser diode (0.807)	s, cw	149
silica (Ge)	$^4I_{13/2} \rightarrow {}^4I_{15/2}$	RT	1.566	Ar ion laser (0.514)	cw	156
silica (Ge, Al, P)	$^4I_{13/2} \rightarrow {}^4I_{15/2}$	RT	1.52–1.57	TiS laser (0.98)	cw	118
silica (Nd)	$^4I_{13/2} \rightarrow {}^4I_{15/2}$	RT	1.53[b]	Ar ion laser (0.514)	cw, s	199
silica (Nd)	$^4I_{13/2} \rightarrow {}^4I_{15/2}$	RT	1.552[c]	Ar ion laser (0.514)	cw, s	140
silica (Yb)	$^4I_{13/2} \rightarrow {}^4I_{15/2}$	RT	1.54	Nd:YAG (1.064)	cw, m	?
silica (Yb)	$^4I_{13/2} \rightarrow {}^4I_{15/2}$	RT	1.56	Nd:YAG (1.064), Nd:YLF (1.047)	cw	154
Microchip						
phosphate (Yb)	$^4I_{13/2} \rightarrow {}^4I_{15/2}$	RT	1.535	laser diode (0.980)	s	126
Planar Waveguide						
borosilicate (BK 7)	$^4I_{13/2} \rightarrow {}^4I_{15/2}$	RT	1.540	TiS laser (0.980)	cw	141
silica (P)	$^4I_{13/2} \rightarrow {}^4I_{15/2}$	RT	1.546	TiS laser (0.976)	cw	145
silica (P)	$^4I_{13/2} \rightarrow {}^4I_{15/2}$	RT	1.598, 1.604	TiS laser (0.98)	cw	158

(a) Co-lasing at 1.724 and 2.702 μm in a cascade mode.
(b) Simultaneous lasing of Nd^{3+} at 0.91 and 1.08 μm.
(c) Simultaneous lasing of Nd^{3+} at 0.908, 0.932, and 1.08 μm.

Thulium (Tm^{3+}, 4f^{12}) Lasers

Host glass	Lasing transition	Laser wavelength (μm)	Temp. (K)	Pump source [wavelength – μm]	Mode	Ref.
Bulk						
fluorozirconate (Tb)	$^3H_4 \rightarrow {}^3F_4$	1.47	RT	alex. laser (0.79)	p	113
fluorozirconate	$^3F_4 \rightarrow {}^3H_6$	1.88	RT	alex. laser (0.785)	p	167
fluorozirconate	$^3H_4 \rightarrow {}^3H_5$	2.25	RT	alex. laser (0.785)	p	167
Li–Mg–Al silicate	$^3F_4 \rightarrow {}^3H_6$	1.85, 2.015	RT	Xe flashlamp	p	166
Fiber						
fluorozirconate	$^1G_4 \rightarrow {}^3H_6$	0.482	RT	LD (1.135, 1.220)	m	15
fluorozirconate	$^1D_2 \rightarrow {}^3F_4$	0.455	RT	dye laser (0.645) + Nd:YAG laser (1.064)	cw, upc	4
fluorozirconate	$^1D_2 \rightarrow {}^3F_4$	0.455	77	Kr laser (0.676+0.647)	cw, upc	3
fluorozirconate	$^1G_4 \rightarrow {}^3H_6$	0.480	77	Kr laser (0.676+0.647)	cw, upc	3
fluorozirconate	$^1G_4 \rightarrow {}^3H_6$	0.480	RT	Nd:YAG laser (1.120)	cw, upc	5
fluorozirconate	$^3H_4 \rightarrow {}^3H_6$	0.803–0.816	RT	Nd:YAG laser (1.064)	m, upc	27
fluorozirconate	$^3H_4 \rightarrow {}^3H_6$	0.807–0.813	RT	TiS laser (0.778)	cw	224
fluorozirconate	$^3H_4 \rightarrow {}^3H_6$	~0.810	RT	TiS laser (0.780)	p, cw	25
fluorozirconate	$^3H_4 \rightarrow {}^3H_6$	0.815–0.825	RT	Kr ion laser (0.676)	cw	28
fluorozirconate	$^3H_4 \rightarrow {}^3F_4$	1.46–1.51	RT	Kr ion laser (0.676)	cw	28
fluorozirconate	$^3H_4 \rightarrow {}^3F_4$	~1.47	RT	TiS laser (0.790)	cw	112
fluorozirconate	$^3H_4 \rightarrow {}^3F_4$	1.475	RT	Nd:YAG laser (1.064)	cw, upc	114
fluorozirconate	$^3H_4 \rightarrow {}^3F_4$	1.475[b]	RT	laser diode (0.780)	cw, cascade	115
fluorozirconate	$^3H_4 \rightarrow {}^3H_6$	1.48	RT	Kr ion laser (0.676)	cw	28
fluorozirconate	$^1D_2 \rightarrow {}^1G_4$	1.51	77	Kr ion laser (0.647)	p, cw	3
fluorozirconate	$^3F_4 \rightarrow {}^3H_6$	1.82	RT	laser diode (0.790)	cw, m	165
fluorozirconate	$^3F_4 \rightarrow {}^3H_6$	1.84–1.94	RT	Kr ion laser (0.676)	cw	28
fluorozirconate	$^3H_4 \rightarrow {}^3F_4$	1.88[a]	RT	laser diode (0.780)	cw, cascade	115
fluorozirconate	$^3F_4 \rightarrow {}^3H_6$	~1.9	RT	TiS laser (0.790)	cw	112

Glass	Transition	Wavelength (μm)	Temp.	Pump	Mode	Ref.
fluorozirconate	$^3F_4 \to {}^3H_6$	1.925	RT	laser diode (0.790)	cw, m	165
fluorozirconate	$^3F_4 \to {}^3H_6$	1.972	RT	laser diode (0.795)	cw	174
fluorozirconate	$^3H_4 \to {}^3H_5$	2.25–2.50	RT	TiS laser (0.790)	s	180
fluorozirconate	$^3H_4 \to {}^3H_5$	2.27–2.40	RT	Kr ion laser (0.676)	cw	28
fluorozirconate	$^3H_4 \to {}^3H_5$	~2.3	RT	TiS laser (0.790)	cw	112
fluorozirconate	$^3H_4 \to {}^3H_5$	2.3	RT	laser diode (0.786)	cw, m	181
fluorozirconate	$^3H_4 \to {}^3H_5$	2.3	RT	alex. laser (0.790)	s, p	182
fluorozirconate	$^3H_4 \to {}^3H_5$	2.31(b)	RT	laser diode (0.790)	s, cw	165
fluorozirconate	$^3H_4 \to {}^3H_5$	2.35	RT	Kr ion laser (0.676)	cw	28
fluorozirconate (Pb)	$^3F_4 \to {}^3H_6$	1.818–1.858	RT	laser diode (0.790)	cw	165
fluorozirconate (Pb)	$^3F_4 \to {}^3H_6$	1.870–1.930	RT	laser diode (0.790)	cw	165
fluorozirconate (Tb)	$^3H_4 \to {}^3F_4$	1.481	RT	laser diode (0.786)	cw, m	116
fluorozirconate (Tb)	$^3H_4 \to {}^3F_4$	1.47	RT	laser diode (0.786)	cw, m	181
lead germanate	$^3F_4 \to {}^3H_6$	1.88	RT	TiS laser (0.794)	m	168
lead germanate	$^3F_4 \to {}^3H_6$	1.905	RT	laser diode (0.970)	m	168
silica	$^3F_4 \to {}^3H_6$	1.780–2.056		dye laser (0.800–0.840)	cw	164
silica	$^3F_4 \to {}^3H_6$	1.81–2.01	RT	dye laser (0.800)	s, cw	169
silica	$^3F_4 \to {}^3H_6$	1.84–1.90	RT	Er fiber laser (1.57)	cw	170
silica	$^3F_4 \to {}^3H_6$	1.937	RT	laser diode (825)	cw	171
silica	$^3F_4 \to {}^3H_6$	2.007	RT	laser diode (825)	cw	171
silica	$^3F_4 \to {}^3H_6$	2.049	RT	laser diode (825)	cw	171
silica	$^3F_4 \to {}^3H_6$	2.102	RT	laser diode (825)	cw	171
silica (Al)	$^3F_4 \to {}^3H_6$	1.70–2.00	RT	LD (0.786): Ti:S (830)	cw	160
silica (Ge)	$^3F_4 \to {}^3H_6$	1.65–1.86	RT	laser diode (0.786)	cw	160
silica (Ge)	$^3F_4 \to {}^3H_6$	1.94–1.96	RT	dye laser (0.797)	s, cw	172
silica (Ge)	$^3F_4 \to {}^3H_6$	1.88	RT	dye laser (~0.800)	s, cw	172

(a) Co-lasing at 1.475 and 1.88 μm in a cascade mode.
(b) Simultaneous operation at 1.82 and 2.31 μm.

Ytterbium (Yb^{3+}, 4f^{13}) Lasers

Host glass	Lasing transition	Laser wavelength (μm)	Temp. (K)	Pump source [wavelength – μm]	Mode	Ref.
Bulk						
Ca–Li borate (Nd)	$^2F_{5/2} \rightarrow {}^2F_{7/2}$	1.018	RT	Xe flashlamp	p	48
K–Ba silicate	$^2F_{5/2} \rightarrow {}^2F_{7/2}$	1.06	RT	Xe flashlamp	p	78
Li–Mg–Al silicate	$^2F_{5/2} \rightarrow {}^2F_{7/2}$	1.015	77	Xe flashlamp	p	45, 46
Li–Mg–Al silicate (Nd)	$^2F_{5/2} \rightarrow {}^2F_{7/2}$	1.015	RT	Xe flashlamp	p	IEE
Fiber						
fluorozirconate	$^2F_{5/2} \rightarrow {}^2F_{7/2}$	1.000–1.050	RT	TiS laser (0.911)	s, cw	43
germanosilicate	$^2F_{5/2} \rightarrow {}^2F_{7/2}$	1.115, 1.128	RT	Nd:YAG laser (1.064)	cw	195
silica	$^2F_{5/2} \rightarrow {}^2F_{7/2}$	0.974	RT	LD (0.800–0.850, 0.900)	s	39
silica	$^2F_{5/2} \rightarrow {}^2F_{7/2}$	1.010–1.162	RT	LD (0.800–0.850, 0.900)	s	39
silica	$^2F_{5/2} \rightarrow {}^2F_{7/2}$	1.015–1.140	RT	dye laser (0.840)	s, cw	47
silica	$^2F_{5/2} \rightarrow {}^2F_{7/2}$	1.028–1.064	RT	dye laser (0.822)	s, cw	47
silica (Al, P)	$^2F_{5/2} \rightarrow {}^2F_{7/2}$	0.980	RT	dye laser (0.890)	cw	40
silica (Ge)	$^2F_{5/2} \rightarrow {}^2F_{7/2}$	1.115	RT	Nd:YLF laser (1.047)	cw	93
silica (Ge)	$^2F_{5/2} \rightarrow {}^2F_{7/2}$	1.13	RT	Nd:YAG laser (1.064)	cw	93

1.2.3 Glass Amplifiers

Ion-glass systems used as amplifiers or in which only gain has been reported are listed in Table 1.2.5. The amplifying ion, host glass (if codopants or sensitizing ions are added, they are listed in parentheses), glass form, lasing wavelength, lasing transition, and pump wavelength are tabulated together with references.

Table 1.2.5
Ions in Glasses Exhibiting Optical Amplification

Ion	Host glass	Form	Wavelength (μm)	Transition	Pump (μm)	Ref.
CdSSe	silicate	bulk	0.625	—	0.532 (p)	30
Cu^+	aluminoborosilicate	bulk	0.560–0.585	$4s \rightarrow 3d$	0.266 (p)	128
Cu^+	fluorohafnate	bulk	0.633	$4s \rightarrow 3d$	0.265	202
Er^{3+}	aluminosilicate	fiber	1.527–1.560	$^4S_{3/2} \rightarrow {}^4I_{15/2}$	0.98	119
Er^{3+}	fluorozirconate	fiber	1.530–1.570	$^4S_{3/2} \rightarrow {}^4I_{15/2}$	1.481	121
Er^{3+}	fluorozirconate	fiber	1.543	$^4S_{3/2} \rightarrow {}^4I_{15/2}$	1.48	143
Er^{3+}	fluorozirconate	fiber	2.716	$^4I_{11/2} \rightarrow {}^4I_{132}$	0.642	189
Er^{3+}	fluorozirconate	fiber	0.546	$^4S_{3/2} \rightarrow {}^4I_{15/2}$	—	203
Er^{3+}	fluorozirconate	fiber	0.850	$^4S_{3/2} \rightarrow {}^4I_{15/2}$	0.476 (upc)	204
Er^{3+}	silica	fiber	1.50–1.70	$^4S_{3/2} \rightarrow {}^4I_{15/2}$	0.781	117
Er^{3+}	silica	fiber	1.531–1.556	$^4S_{3/2} \rightarrow {}^4I_{15/2}$	—	234
Er^{3+}	silica	fiber	1.5354	$^4I_{13/2} \rightarrow {}^4I_{15/2}$	0.820	205
Er^{3+}	silica	fiber	1.536,1.553	$^4I_{13/2} \rightarrow {}^4I_{15/2}$	1.48 (cw)	206
Er^{3+}	silica (Ge)	fiber	1.536, 1553	$^4S_{3/2} \rightarrow {}^4I_{15/2}$	0.827 (cw)	129
Er^{3+}	silica (Ge)	fiber	1.552	$^4S_{3/2} \rightarrow {}^4I_{15/2}$	0.810	150
Er^{3+}	silica (Ge, Ca, Al)	fiber	1.53–1.565	$^4S_{3/2} \rightarrow {}^4I_{15/2}$	0.813	120
Er^{3+}	silica (Ge)	fiber	1.536	$^4I_{13/2} \rightarrow {}^4I_{15/2}$	0.827	207
Er^{3+}	silica (Ge, Al)	fiber	~1.5–1.6	$^4I_{13/2} \rightarrow {}^4I_{15/2}$	1.48 (cw)	208
Er^{3+}	silica (Ge, Al)	fiber	~1.51–1.58	$^4I_{13/2} \rightarrow {}^4I_{15/2}$	0.98 (cw)	208
Er^{3+}	silica (Ge, Al)	fiber	1.57–1.62	$^4I_{13/2} \rightarrow {}^4I_{15/2}$	1.55 (cw)	157
Er^{3+}	silica (P)	fiber	1.54	$^4I_{13/2} \rightarrow {}^4I_{15/2}$	1.064 (cw)	209
Ho^{3+}	Al fluorozirconate (Tm)	bulk	2.05	$^5I_7 \rightarrow {}^5I_8$	0.79 (cw)	210
Ho^{3+}	fluorozircoaluminate (Yb)	fiber	0.543	$^5S_2 \rightarrow {}^5I_8$	0.974 (upc)	211
Nd^{3+}	Ba silicate	wg	1.06	$^4F_{3/2} \rightarrow {}^4I_{11/2}$	flashlamp	67
Nd^{3+}	Ba silicate	wg	1.064	$^4F_{3/2} \rightarrow {}^4I_{11/2}$	0.585 (cw)	86
Nd^{3+}	chalcogenide	fiber	1.083	$^4F_{3/2} \rightarrow {}^4I_{11/2}$	0.89 (cw)	212
Nd^{3+}	fluorozirconate	fiber	1.310–1.370	$^4F_{3/2} \rightarrow {}^4I_{13/2}$	0.795 (cw)	100
Nd^{3+}	fluorozirconate	fiber	1.338	$^4F_{3/2} \rightarrow {}^4I_{13/2}$	0.820	213
Pr^{3+}	fluorozirconate	fiber	0.6328	$^3P_0 \rightarrow {}^3F_2$	0.476	20
Pr^{3+}	fluorozirconate	fiber	1.290–1.320	$^1G_4 \rightarrow {}^3H_5$	1.047	97

Table 1.2.5—*continued*
Ions in Glasses Exhibiting Optical Amplification

Ion	Host glass	Form	Wavelength (μm)	Transition	Pump (μm)	Ref.
Pr^{3+}	fluorozirconate (Yb)	fiber	1.3	$^1G_4 \rightarrow ^3H_5$	0.98	214
Pr^{3+}	fluorozirconate (Yb)	fiber	1.31	$^1G_4 \rightarrow ^3H_5$	1.017	215
Tm^{3+}	fluorozirconate	fiber	0.800–0.830	$^3H_4 \rightarrow ^3H_6$	0.785	26
Tm^{3+}	silica	fiber	1.653–1.691	$^3F_4 \rightarrow ^3H_6$	0.781	161
Tm^{3+}	silica (Ta,Al,P)	fiber	1.91	$^3H_4 \rightarrow ^3H_6$	0.81–0.82	216

cw – continuous wave, p – pulsed, upc – upconversion excitation scheme, wg – waveguide.

1.2.4 Commercial Glass Lasers

Examples of commercial glass lasers and their mode of operation (cw or pulsed), principal wavelengths, and representative outputs are given in Table 1.2.6. These latter data are from recent (1997–1999) laser buyers' guides and manufacturers' literature and can be expected to change due to advances in technology.

Table 1.2.6
Commercial Glass Lasers

Laser Type	Operation	Principal wavelengths (μm)	Output
Nd:glass (phosphate)	pulsed	0.263 (FH)	0.04–4 J
	pulsed	0.351 (TH)	0.1–8 J
	pulsed	0.527 (SH)	0.2–22 J
	pulsed	1.054	1.0–80 J
Nd:glass (silicate)	pulsed	0.26 (FH)	0.1–0.8 J
	pulsed	0.35 (TH)	0.3–2 J
	pulsed	0.53 (SH)	0.1–5 J
	pulsed	1.06	0.2–20 J
	pulsed (DP)	1.060	
	cw (DP)	1.058	
Er:glass	pulsed (DP)	1.54	1.2 J
	cw	1.55	0.1 W
Er:glass (fiber)	cw	1.52–1.57	30 mW
	cw (DP)	1.54	50 mW
	pulsed (DP)	1.54	2 J
	pulsed (DP)	2.94	1 J
Er:doped fiber amplifier (EDFA)	cw	1.530–1.560	20 dB gain
Yb:glass	(DP)		

SH – second harmonic, TH – third harmonic, FH – fourth harmonic, DP – diode pumped

1.2.5 References

1. Funk, D. S., Carlson, J. W., and Eden, J. G., Ultraviolet (381 nm) room temperature laser in neodymium-doped fluorozirconate fibre, *Electron. Lett.* 30, 1859 (1994).
2. Funk, D. S., Carlson, J. W., and Eden, J. G., Room-temperature fluorozirconate glass fiber laser in the violet (412 nm), *Optics Lett.* 20, 1474 (1995).
3. Allain, J. Y., Monerie, M., and Poignant, H., Blue upconversion fluorozirconate fiber laser, *Electron. Lett.* 26, 166 (1990).
4. Le Flohic, M. P., Allain, J. Y., Stephan, G. M., and Maze, G., Room-temperature continuous-wave upconversion laser at 455 nm in a Tm^{3+} fluorozirconate fiber, *Optics Lett.* 19, 1982 (1994).
5. Grubb, S. G., Bennett, K. W., Cannon, R. S., and Humer, W. F., CW room-temperature blue upconversion fibre laser, *Electron. Lett.* 28, 1243 (1992).
6. Smart, R. G., Hanna, D. C., Tropper, A. C., Davey, S. T., Carter, S. F., and Szebesta, D., CW room temperature upconversion lasing at blue, green and red wavelengths in infrared-pumped Pr^{3+}-doped fluoride fiber, *Electron. Lett.* 27, 1307 (1991).
7. Smart, R. G., Carter, J. N., Tropper, A. C., Hanna, D. C., Davey, S. T., Carter S. F., and Szebesta, D., CW room temperature operation of praseodymium-doped fluorozirconate glass fiber lasers in the blue-green, green, and red spectral regions, *Optics Commun.* 86, 333 (1991).
8. Xie, P. and Gosnell, T. R., Room temperature upconversion fiber laser tunable in the red, orange, green and blue spectral regions, *Opt. Lett.* 20, 1014 (1995).
9. Zhao, Y. and Poole, S., Efficient blue Pr^{3+}-doped fluoride fibre upconversion laser, *Electron. Lett.* 30, 967 (1994).
10. Piehler, D., Carven, D., Kwong, N., and Zarem, H., Laser-diode-pumped red and green upconversion fibre lasers, *Electron. Lett.* 29, 1857 (1993).
11. Andreev, S. I., Bedilov, M. R., Karapetyan, G. O., and Likhachev, V. M., Stimulated radiation of glass activated by terbium, *Sov. J. Opt. Tech.* 34, 819 (1967; *Opt.-Mekh. Promst.* 34, 60 (1967).
12. Allain, J. Y., Monerie, M., and Poignant, H., Tunable green upconversion erbium fibre laser, *Electron. Lett.* 28, 111 (1992).
13. Allain, J. Y., Monerie, M., and Poignant, H., Room temperature cw tunable green upconversion holmium fiber laser, *Electron. Lett.* 26, 261 (1990).
14. Whitley, T. J., Millar, C. A., Wyatt, R., Brierley, M. C., and Szebesta, D., Upconversion pumped green lasing in erbium doped fluorozirconate fibre, *Electron. Lett.* 27, 1786 (1991).
15. Booth, I. J., Mackechnie, C. J., and Ventrudo, B. F., Operation of diode laser pumped Tm^{3+} ZBLAN upconversion fiber laser at 482 nm, *IEEE J. Quantum Electron.* 32, 118 (1996).
16. Hirao, K., Todoroki, S., and Soga, N., CW room temperature upconversion lasing in Er^{3+}-doped fluoride glass fiber, *J. Non-Cryst. Solids* 143, 40 (1992).
17. Funk, D. S., Stevens, S. B., Wu, S. S., and Eden, J. G., Tuning, temporal, spectral characteristics of the green (λ~549 nm), holmium-doped fluorozirconate glass fiber laser, *IEEE J. Quantum Electron.* 32, 638 (1996).
18. Allain, J. Y., Monerie, M., and Poignant, H., Tunable cw lasing around 610, 635, 695, 715, 885, and 910 nm in praseodymium-doped fluorozirconate fibre, *Electron. Lett.* 27, 189. (1991).
19. Baney, D. M., Yang, L., Ratcliff, J., and Chang, K. W., Red and orange Pr^{3+}/Yb^{3+} doped ZBLAN fibre upconversion lasers, *Electron. Lett.* 31, 1842 (1995).
20. Petreski, B. P., Murphy, M. M., Collins, S. F., and Booth, D. J., Amplification in Pr^{3+}-doped fluorozirconate optical fibre at 632.8 nm, *Electron. Lett.* 29, 1421 (1993).

21. Malone, K. J., Sanford, N. A., Hayden, J. S., and Sapak, D. L., Integrated optic laser emitting at 905, 1057, and 1356 nm, in *Advanced Solid-State Lasers*, Pinto, A. A. and Fan, T. Y., Eds., Proceedings Vol. 15, Optical Society of America, Washington, DC (1993), p. 286.

22. Payne, S. A., Marshall, C. D., Bayramian, A. J., Wilke, G. D., and Hayden, J. S., Properties of a new average power Nd-doped phosphate glass, *OSA Proc. Adv. Solid State Lasers*, Chai, B. H. T. and Payne, S. A., Eds., 24, 211(1995).

23. Allain, J. Y., Monerie, M., and Poignant, H., Red upconversion Yb-sensitised Pr doped fluoride fibre laser pumped in the 0.8 μm region, *Electron. Lett.* 27, 1156 (1991).

24. Farries, M. C., Morkel, P. R., and Townsend, J. E., Samarium^{3+}-doped glass laser operating at 651 nm, *Electron. Lett.* 24, 709 (1988).

25. Carter, J. N., Smart, R. G., Hanna, D. C., and Tropper, A. C., Lasing and amplification in the 0.8 μm region in thulium doped fluorozirconate fibers, *Electron. Lett.* 26, 1759 (1990).

26. Smart, R.G., Carter, J.N., Tropper, A.C., Hanna, D.C., Carter, S.F., and Szebesta, D., A 20 dB gain thulium-doped fluorozirconate fiber amplifier operating at around 0.8 μm, *Electron. Lett.* 27, 1123 (1991).

27. Dennis, M. L., Dixon, J. W., and Aggarwal, I., High power upconversion lasing at 810 nm in Tm:ZBLAN fibre, *Electron. Lett.* 30, 136 (1994).

28. Allain, J. Y., Monerie, M., and Poignant, H., Tunable cw lasing around 0.82, 1.48, 1.88, and 2.35 μm in a thulium-doped fluorozirconate fiber, *Electron. Lett.* 25, 1660 (1989).

29. Millar, C. A., Brierley, M. C., Hunt, M. H., and Carter, S. F., Efficient up-conversion pumping at 800 nm of an erbium-doped fluoride fiber laser operating at 850 nm, *Electron. Lett.* 26, 1876 (1990).

30. Zhou, F., Qin, W., Jin, C. et al., Optical gain of CdSSe-doped glass, *J. Lumin.* 60 & 61, 353 (1994).

31. Percival, R. M., Phillips, M. W., Hanna, D. C., and Tropper, A. C., Characterization of spontaneous and stimulated emission from praseodymium ions doped into a silica based monomode optical fiber, *IEEE J. Quantum Electron.* 25, 2119 (1989).

32. Reekie, L., Mears, R. J., Poole, S. B., and Payne, D. N., Tunable single-mode fiber lasers, *J. Lightwave Tech.* LT-4, 956 (1986).

33. Alcock, I. P., Ferguson, A. I., Hanna, D. C., and Tropper, A. C., Tunable, continuous-wave neodymium-doped monomode-fiber laser operating at 0.900–0.945 and 1.070–1.135 μm, *Optics Lett.* 11, 709 (1986).

34. Maurer, R. D., Operation of a Nd^{3+} glass optical maser at 9180 Å, *Appl. Opt.* 2, 87 (1963).

35. Artem'ev, E. P., Murzin, A. G., and Fromzel, V. A., Room-temperature laser action at 0.92 μm in neodymium glasses, *Sov. Phys. Tech. Phys.* 22, 274 (1977); *Zh. Tekh. Fiz.* 47, 456 (1977).

36. Krupke, W. F., Shinn, M. D., Kirchoff, T. A., Finch, D. B., and Boatner, L.A., Promethium-doped phosphate glass laser at 933 and 1098 nm, *Appl. Phys. Lett.* 51, 2186 (1987).

37. Alcock, I. P., Ferguson, A. I., Hanna, D. C., and Tropper, A. C., Continuous-wave oscillation of a monomode neodymium-doped fiber laser at 0.9 μm on the $^4F_{3/2} \rightarrow {}^4I_{9/2}$ transition, *Opt. Commun.* 58, 405 (1986).

38. Reekie, L., Jauncey, I. M., Poole, S. B., and Payne, D. N., Diode-laser-pumped Nd^{3+}-doped fiber laser operating at 938 nm, *Electron. Lett.* 23, 884 (1987).

39. Hanna, D. C., Percival, R. M., Perry, I. R., Smart, R. G., Suni, P. J., and Tropper, A. C., An ytterbium-doped monomode fiber laser: broadly tunable operation from 1.010 μm to 1.162 μm and three-level operation at 974 nm, *J. Mod. Opt.* 37, 517 (1990).

40. Armitage, J. R., Wyatt, R., Ainslie, B. J., and Craig-Ryan, S. P., Highly efficient 980 nm operation of an Yb^{3+}-doped silica fiber laser, *Electron. Lett.* 25, 298 (1989).

41. Allain, J. Y., Monerie, M., and Poignant, H., Q-switched 0.98 μm operation of erbium-doped fluorozirconate fiber laser, *Electron. Lett.* 25, 1082 (1989).

42. Allain, J. Y., Monerie, M., and Poignant, H., Lasing at 1.00 μm in erbium-doped fluorozirconate fibers, *Electron. Lett.* 25, 318 (1989).

43. Allain, J. Y., Monerie, M., and Poignant, H., Ytterbium-doped fluoride fibre laser operating at 1.02 μm, *Electron. Lett.* 28, 988 (1992).

44. Shi, Y., Poulsen, C. V., Sejka, M., Ibsen, M., and Poulsen, O., Tunable Pr^{3+}-doped silica-based fibre laser, *Electron. Lett.* 29, 1426 (1993).

45. Etzel, H. W., Gandy, H. W., and Ginther, R. J., Stimulated emission of infrared radiation from ytterbium-activated silicate glass, *Appl. Opt.* 1, 534 (1962).

46. Gandy, H. W. and Ginther, R. J., Simultaneous laser action of neodymium and ytterbium ions in silicate glass, *Proc. IRE* 50, 2114 (1962).

47. Hanna, D. C., Percival, R. M., Perry, I. R., Smart, R. G., Suni, P. J., Townsend, J. E., and Tropper, A. C., Continuous-wave oscillation of a monomode ytterbium-doped fiber laser, *Electron. Lett.* 24, 1111 (1988).

48. Pearson, A. D. and Porto, S.P.S., Non-radiative energy exchange and laser oscillation in Yb^{3+}- Nd^{3+}-doped borate glass, *Appl. Phys. Lett.* 4, 202 (1964).

49. Petrovksii, G. T., Tolstoi, M. N., Feofilov, P. P., Tsurikova, G. A., and Shapovalov, V. N., Luminescence and stimulated emission of neodymium in beryllium fluoride glass, *Opt. Spectrosc. (USSR)* 21, 72 (1966); *Opt. Spektrosk.* 21, 126 (1966).

50. Yamashita, T., Ammano, S., Masuda, I., Izumitani, T., and Ikushima, A. J., Nd and Er-doped phosphate glass fiber lasers, in *Conference on Lasers and Electro-Optics Technical Digest Series*, Opt. Soc. Amer., Washington, DC (1988), p. 320.

51. Robertson, G.R.F. and Jessop, P. E., Optical waveguide laser using an rf sputtered Nd:glass film, *Appl. Opt.* 30, 276 (1991).

52. Brierley, M. C. and France, P. W., Neodymium-doped fluorozirconate fiber laser, *Electron. Lett.* 23, 815 (1987).

53. Liu, K., Digonnet, M., Fesler, K., Kim, B. Y., and Shaw, H. J., Broadband diode-pumped fiber laser, *Electron. Lett.* 24, 838 (1988).

54. Stokowski, S. E., Martin, W. E., and Yarema, S. M., Optical and lasing properties of fluorophosphate glass, *J. Non-Cryst. Solids* 40, 48 (1980).

55. Hibino, Y., Kitagawa, T., Shimizu, M., Hanawa, F., and Sugita, A., Neodymium-doped silica optical waveguide laser on silicon substrate, *IEEE Phot. Tech. Lett.* 1, 349 (1990).

56. Kitagawa, T., Hattori, K., Hibino, Y., and Ohmori, Y., Neodymium-doped silica-based planar waveguide lasers, *J. Lightwave Techn.* 12, 436 (1994).

57. Hattori, K., Kitagawa, T., Ohmori, Y., and Kobayashi, M., Laser-diode pumping of waveguide laser based on Nd-doped silica planar lightwave circuit, *IEEE Phot. Techn. Lett.* 3, 882 (1991).

58. Michel, J. C., Morin, D., and Auzel, F., Properietes spectroscopiques et effet laser d'un verre tellurite et d'un verre phosphate fortement dopes en neodyme, *Rev. Phys. Appl.* 11, 859 (1978).

59. Aoki, H., Maruyama, O., and Asahara, Y., Glass waveguide laser, *IEEE Phot. Tech. Lett.* 2, 459 (1990).

60. Kishida, S., Washio, K., and Yoshikawa, S., CW oscillation in a Nd:phosphate glass laser, *Appl. Phys. Lett.* 34, 273 (1979).

61. Deutschbein, O., Pautrat, C., and Svirchevsky, I. M., Phosphate glasses, new laser materials, *Rev. Phys. Appl.* 1, 29 (1967).

62. Deutschbein, O. K. and Pautrat, C. C., CW laser at room temperature using vitreous substances, *IEEE J. Quantum Electron.* QE-4, 48 (1968).

63. Vodop'yanov, K. L., Denker, B. I., Maksimova, G. V., Malyutin, A. A., Osiko, V. V., Pashinin, P. P., and Prokhorov, A. M., Characteristics of simulated emission from LiNdLa phosphate glass, *Sov. J. Quantum Electron.* 8, 403 (1978).

64. Hakimi, F., Po, H., Tumminelli, R., McCollum, B. C., Zenteno, L., Cho, N. M., and Snitzer, E., Glass fiber laser at 1.36 μm from SiO_2:Nd, *Opt. Lett.* 14, 1060 (1989).

65. Sanford, M. A., Malone, K. J., and Larson, D. R., Integrated-optic laser fabricated by field-assisted ion exchange in neodymium-doped soda-lime-silicate glass, *Opt. Lett.* 15, 366 (1990).

66. Mwarania, E. K., Reekie, L., Wang, J., and Wilkinson, J. S., Low-threshold monomode ion-exchanged waveguide lasers in neodymium-doped BK-7 glass, *Electron. Lett.* 26, 1317 (1990).

67. Yajima, H., Kawase, S., and Sakimoto, Y., Amplification at 1.06 μm using a Nd-glass thin-film waveguide, *Appl. Phys. Lett.* 21, 407 (1972).

68. Young, C. G., Continuous glass laser, *Appl. Phys. Lett.* 2, 151 (1963).

69. Saruwatari, M. and Izawa, T., Nd-glass laser with three-dimensional optical waveguide, *Appl. Phys. Lett.* 24, 603 (1974).

70. Rapp, C. F. and Chrysochoos, J., Neodymium-doped glass-ceramic laser material, *J. Mater. Sci.* 7, 1090 (1972).

71. Müller, G. and Neuroth, N., Glass ceramic - a new host material, *J. Appl. Phys.* 44, 2315 (1973).

72. Wang, J., Reekie, L., Brocklesby, W. S., Chow, Y. T., and Payne, D. N., Fabrication, spectroscopy and laser performance of Nd^{3+}-doped lead-silicate glass fibers, *J. Non-Cryst. Solids* 180, 207 (1995).

73. Birnbaum, M., Fincher, C. L., Dugger, C. O., Goodrum, J., and Lipson, H., Laser characteristics of neodymium-doped lithium germanate glass, *J. Appl. Phys.* 41, 2470 (1970).

74. Härig, T., Huber, G., and Shcherbakov, I., Cr^{3+} sensitized Nd^{3+}:Li-La phosphate glass laser, *J. Appl. Phys.* 52, 4450 (1981).

75. Galaktionova, N. M., Garkavi, G. A., Zubkova, V. S., Mak, A. A., Soms, L. N., and Khaleev, M. M., Continuous Nd-glass laser, *Opt. Spectrosc. USSR* 37, 90 (1974); *Opt. Spektrosk.* 37, 162 (1974).

76. Koester, C. J. and Snitzer, E., Amplification in a fiber laser, *Appl. Opt.* 3, 1182 (1964).

77. Babukova, M. V., Berenberg, V. A., Glebov, L. B., Nikonorov, N. V., Petrovskii, G. T., and Terpugov, V. S., Investigation of neodymium silicate glass diffused waveguides, *Sov. J. Quantum Electron.* 15, 1304 (1985).

78. Snitzer, E., Laser emission at 1.06 μ from Nd^{3+}-Yb^{3+} glass, *IEEE J. Quantum Electron.* QE-2, 562 (1966).

79. Stone, J. and Burrus, C.A., Neodymium-doped silica lasers in end-pumped fiber geometry, *Appl. Phys. Lett.* 23, 388 (1973).

80. Tumminelli, R., Hakimi, F., and Haavisto, J., Integrated optic Nd glass laser fabricated by flame hydrolysis deposition using chelates, *Optics Lett.* 16, 1098 (1991).

81. Balashov, I. F., Berezin, B. G., Brachkovskaya, N. B., Volkova, V. V., Ivanov, V. N., Ovcharenko, N. V., Petrov, A. A., Przhevuskii, A. K., and Smirnova, T. V., Experimental study of tellurite laser glass doped with neodymium, *Zh. Prikl. Spekt.* 52, 781 (1989).

82. Snitzer, E., Optical maser action of Nd^{3+} in a barium crown glass, *Phys. Rev. Lett.* 7, 444 (1961).

83. Wang, J. S., Machewirth, D. P., Wu, F., Snitzer, E., and Vogel, E. M., Neodymium-doped tellurite single-mode fiber laser, *Optics Lett.* 19, 1448 (1994).

84. Galant, E. I., Kondrat'ev, Yu.N., Przhevuskii, A. K., Prokhorova, T. I., Tolstoi, M. N., and Shapovalov, V. N., Emission of neodymium ions in quartz glass, *Sov. JETP Lett.* 18, 372 (1973).

85. Thomas, I. M., Payne, S. A., and Wilke, G. D., Optical properties and laser demonstration of Nd-doped sol-gel silica glasses, *J. Non-Cryst. Solids* 151, 183 (1992).

86. Chen, B. -U. and Tang, C. L., Nd-glass thin-film waveguide in an active medium for Nd thin-film laser, *Appl. Phys. Lett.* 28, 435 (1976).

87. Lei, N., Xu, B., and Jiang, Z., Ti:sapphire laser pumped Nd:tellurite glass laser, *Optics Commun.* 127, 263 (1996).

88. Chaoyu, Y., Jiangde, P., and Bingkun, Z., Tunable Nd^{3+}-doped fiber ring laser, *Electron. Lett.* 25, 101 (1989).

89. Schweizer, T., Hewak, D. W., Payne, D. N., Jensen, T., and Huber, G., Rare-earth doped chalcogenide glass laser, *Electron. Lett.* 32, 666 (1996).

90. Stone, J. and Burrus, C. A., Neodymium-doped fiber lasers: room temperature cw operation with an injection laser pump, *Appl. Opt.* 13, 1256 (1974).

91. Poole, S. B., Payne, D. N., and Fermann, M. E., Fabrication of low-loss optical fibers containing rare-earth ions, *Electron. Lett.* 21, 737 (1985).

92. Mears, R. J., Reekie, L., Poole, S. B., and Payne, D. N., Neodymium-doped silica single-mode fiber lasers, *Electron. Lett.* 21, 738 (1985).

93. Mackechnie, C. J., Barnes, W. L., Hanna, D. C., and Townsend, J. E., High power ytterbium (Yb^{3+})-doped fibre laser operating in the 1.12 μm region, *Electron. Lett.* 29, 52 (1993).

94. Carter, S.F., Szebesta, D., Davey, S.T., Wyatt, R., Brierley, M.C., and France, P.W., Amplification at 1.3 μm in a Pr^{3+}-doped single-mode fluorozirconate fiber, *Electron. Lett.* 27, 628 (1991).

95. Kozlovsky, W. J., Fan, T. Y., and Byer, R. L., Diode-pumped continuous-wave Nd:glass laser, *Optics Lett.* 11, 788 (1986).

96. Whitley, T., Wyatt, R., Szebesta, D., Davey, S., and Williams, J. R., Quarter-watt output at 1.3 μm from a praseodymium-doped fluoride fibre amplifier pumped with a diode-pumped Nd:YLF laser, *IEEE Phot. Tech. Lett.* 5, 399 (1993).

97. Durteste, Y., Monerie, M., Allain, J.Y., and Poignant, H., Amplification and lasing at 1.3 μm in praseodymium-doped fluorozirconate fibers, *Electron. Lett.* 27, 626 (1991).

98. Schweizer, T., Samson, B. N., Moore, R. C., Hewak, D. W., and Payne, D. N., Rare-earth doped chalcogenide glass fibre laser, *Electron. Lett.* 33, 414 (1997).

99. Zapata, L. E., Continuous-wave 25-W Nd^{3+}:glass fiber bundle laser, *J. Appl. Phys.* 62, 3110 (1987).

100. Brierley, M., Carter, S., and France, P., Amplification in the 1300 nm telecommunications window in a Nd-doped fluoride fiber, *Electron. Lett.* 26, 329 (1990).

101. Miyajima, Y., Komuakai, T., and Sugawa, T., 1.31–1.36 μm optical amplification in Nd^{3+}-doped fluorozirconate fiber, *Electron. Lett.* 26, 194 (1990).

102. Ishikawa, E., Aoki, H., Yamashita, T., and Asahara, Y., Laser emission and amplication at 1.3 μm in neodymium-doped fluorophosphate fibres, *Electron. Lett.* 28, 1497 (1992).

103. Aoki, H., Maruyama, O., and Asahara, Y., Glass waveguide laser operated around 1.3 μm, *Electron. Lett.* 26, 1910 (1990).

104. Hanson, F. and Imthurn, G., Efficient laser diode side pumped neodymium glass slab laser, *IEEE J. Quantum Electron.* QE-24, 1811 (1988).

105. Miniscalco, W. J., Andrews, L. J., Thompson, B. A., Quimby, R. S., Vaca, L.J.B., and Drexhage, M. G., 1.3 μm fluoride fiber laser, *Electron. Lett.* 24, 28 (1988).

106. Millar, C. A., Fleming, S. C., Brierley, M. C., and Hunt, M. H., Single transverse mode operation at 1345 nm wavelength of a diode-laser pumped neodymium-ZBLAN multimode fiber laser, *IEEE Phot. Tech. Lett.* 2, 415 (1990).

107. Brierley, M. C. and Millar, C. A., Amplification and lasing at 1350 nm in a neodymium doped fluorozirconate fiber, *Electron. Lett.* 24, 438 (1988).

108. Grubb, S. G., Barnes, W. L., Taylor, E. R., and Payne, D. N., Diode-pumped 1.36 μm Nd-doped fiber laser, *Electron. Lett.* 26, 121 (1990).

109. Yamashita, T., Nd- and Er-doped phosphate glass for fiber laser, *SPIE Fiber Laser Sources and Amplifiers*, Vol. 1171, 291 (1989).

110. Mauer, P. B., Laser action in neodymium-doped glass at 1.37 microns, *Appl. Opt.* 3, 153 (1964).

111. Brierley, M. C., France, P. W., and Millar, C. A., Lasing at 2.08 μm and 1.38 μm in a holmium doped fluorozirconate fiber laser, *Electron. Lett.* 24, 539 (1988).

112. Smart, R.G., Carter, J.N., Tropper, A.C., and Hanna, D.C., Continuous-wave oscillation of Tm^{3+}-doped fluorozirconate fibre lasers at around 1.47 μm, 1.9 μm and 2.3 μm when pumped at 790 nm, *Optics Commun.* 82, 563 (1991).

113. Rosenblatt, G. H., Ginther, R. J., Stoneman, R. C., and Esterowitz, L., Laser emission at 1.47 μm from fluorozirconate glass doped with Tm^{3+} and Tb^{3+}, *Technical Digest - Tunable Solid State Laser Conference*, Opt. Soc. Amer., paper WE2-1 (1989).

114. Pericival, R. M., Szebesta, D., and Williams, J. R., Highly efficient 1.064 μm upconversion pumped 1.47 μm thulium doped fluoride fibre laser, *Electron. Lett.* 30, 1057 (1994).

115. Percival, R. M., Szebesta, D., and Davey, S. T., Highly efficient cw cascade operation of 1.47 and 1.82 μm transitions in Tm-doped fluoride fibre laser, *Electron. Lett.* 28, 1866 (1992).

116. Percival, R. M., Szebesta, D., and Davey, S. T., Thulium-doped terbium sensitised cw fluoride fibre laser operating on the 1.47 μm transition, *Electron. Lett.* 29, 1054 (1993).

117. Sankawa, I., Izumita, H., Furukawa, S., and Ishihara, M., An optical fiber amplifier for wide-band wavelength range around 1.65 μm, *IEEE Phot. Tech. Lett.* 2, 422 (1990).

118. Wyatt, R., High-power broadly tunable Er^{3+}-doped silica fiber laser, *Electron. Lett.* 25, 1498 (1989).

119. Yamada, M., Shimizu, M., Horiguchi, M., Okayasu, M., and Sugita, E., Gain characteristics of an Er^{3+}-doped multicomponent glass single-mode optical fiber, *IEEE Phot. Tech. Lett.* 2, 656 (1990).

120. Saifi, M. A., Andrejco, M. J., Way, W. I., Von Lehman, A., Yi-Yan, A., Lin, C., Bilodeau, F., and Hill, K. O., Er^{3+} doped GeO_2-CaO-Al_2O_3 silica core fiber amplifier pumped at 813 nm, *Optical Fiber Commun.* (OFC 91), Optical Soc. Am., San Diego, CA, paper FA6 (1991).

121. Spirit, D. M., Walker, G. R., France, P. W., Carter, S. F., and Szebesta, D., Characterization of diode-pumped erbium-doped fluorozirconate fibre optical amplifier, *Electron. Lett.* 26, 1218 (1990).

122. Taccheo, S., Laporta, P., and Svelto, O., Widely tunable single-frequency erbium-ytterbium phosphate glass laser, *Appl. Phys. Lett.* 68, 2621 (1996).

123. Laporta, P., Taccheo, S., and Svelto, O., High-power and high-efficiency diode-pumped Er:Yb:glass laser, *Electron. Lett.* 28, 490 (1992).

124. Hutchinson, J. A. and Allik, T. H., Diode array-pumped Er,Yb:phosphate glass laser, *Appl. Phys. Lett.* 60, 1424 (1992).

125. Yamashita, Y., Nd and Er doped phosphate glass for fiber laser, *Fiber Laser Sources and Amplifiers*, *SPIE* Vol. 1171, 291 (1989).

126. Laporta, P., Taccheo, S., Longhi, S., and Svelto, O., Diode-pumped microchip Er-Yb:laser, *Optics Lett.* 18, 1232 (1993).

127. Többen, H. and Wetenkamp, L., High-efficiency cw Ho-doped fluorozirconate fiber laser at 1.19 μm, in *Advanced Solid-State Lasers*, Chase, L. L. and Pinto, A. A., Eds., Proceedings Vol. 13, Optical Society of America, Washington, DC (1992), p. 119.

128. Kruglik, G. S., Skripko, G. A., Shkadarevich, A. P., Ermolenko, N. N., Gorodetskaya, O. G., Belokon, M. V., Shagov, A. A., and Zolotareva, L. E., Amplification of yellow-green light in copper-activated glass, *Opt. Spectrosc. (USSR)* 59, 439 (1985); Copper-doped aluminoborosilicate glass spectroscopic characteristics and stimulated emission, *J. Lumin.* 34, 343 (1986).

129. Horiguchi, H., Shimizu, M., Yamada, M., Yoshino, K., and Hanafusa, H., Highly efficient optical fiber amplifier pumped by a 0.8 nm band laser diode, *Electron. Lett.* 26, 1758 (1990).

130. Ledig, M., Heumann, E., Ehrt, D., and Seeber, W., Spectroscopic and laser properties of $Cr^{3+}:Yb^{3+}:Er^{3+}$ fluoride phosphate glass, *Opt. Quantum Electron.* 22, S107 (1990).

131. Snitzer, E., Woodcock, R. F., and Segre, J., Phosphate glass Er^{3+} laser, *IEEE J. Quantum Electron.* QE-4, 360 (1968).

132. Auzel, F., Stimulated emission of Er^{3+} in fluorophosphate glass, *C. R. Acad. Sci. B* 263, 765 (1966).

133. Lunter, S. G., Murzin, A. G., Tolstoi, M. N., Fedorov, Yu.K., and Fromzel, V. A., Possibility of improving the efficiency of lamp pumping of erbium-glass lasers, *Opt. Spectrosc. USSR* 55, 345 (1983; Energy parameters of lasers utilizing erbium glasses sensitized with ytterbium and chromium, *Sov. J. Quantum Electron.* 14, 66 (1984).

134. Maksimova, G. V., Sverchkov, S. E., and Sverchkov, Yu. E., Lasing tests on new ytterbium-erbium laser glass pumped by neodymium lasers, *Sov. J. Quantum Electron.* 21, 1324 (1991).

135. Astakhov, A. V., Butusov, M. M., Galkin, S. L., Ermakova, N. V., and Fedorov, Yu. K., Fiber laser with 1.54 μm radiation wavelength, *Opt. Spectrosc. USSR*, 62, 140 (1987).

136. Gapontsev, V. P., Gromov, A. K., Izyneev, A. A., Sadouskii, P. I., Stavrov, A. A., Tipenko, Yu. S., and Shkadarevich, A. P., Low-threshold erbium glass minilaser, *Sov. J. Quantum Electron.* 18, 447 (1989).

137. Laporta, P., De Silvestri, S., Magni, V., and Svelto, O., Diode-pumped bulk Er:Yb:glass laser, *Optics Lett.* 16, 1952 (1991).

138. Hanna, D. C., Kazer, A., and Shepherd, D. P., A 1.54 μm Er glass laser pumped by a 1.064 μm Nd:YAG laser, *Opt. Commun.* 63, 417 (1987).

139. Estie, D., Hanna, D. C., Kazer, A., and Shepherd, D. P., CW operation of Nd:YAG pumped Er:Yb phosphate glass laser at 1.54 μm, *Opt. Commun.* 69, 153 (1988).

140. Kimura, Y. and Nakazawa, M., Multiwavelength cw laser oscillation in a Nd^{3+} and Er^{3+} doubly doped fiber laser, *Appl. Phys. Lett.* 53, 1251 (1988).

141. Feuchter, T., Mwarania, E. K., Wang, J., Reekie, L., and Wilkinson, J. S., Erbium-doped ion-exchanged waveguide lasers in BK-7 glass, *IEEE Phot. Tech. Lett.* 4, 542 (1992).

142. Laporta, P., Longhi, S., Taccheo, S., Svelto, O., and Sacchi, G., Single-mode cw erbium-ytterbium glass laser at 1.5 μm, *Optics Lett.* 18, 31 (1993).

143. Ronarc'h, D., Guibert, M., Ibrahim, H., Monerie, M., Poignant, H., and Tromeur, A., 30 dB optical net gain at 1.543 μm in Er^{3+} doped fluoride fibre pumped around 1.48 μm, *Electron. Lett.* 27, 908 (1991).

144. Snitzer, E. and Woodcock, R., $Yb^{3+}-Er^{3+}$ glass laser, *Appl. Phys. Lett.* 6, 45 (1965).

145. Kitagawa, T., Bilodeau, F., Malo, B., Theriault, S., Albert, J., Jihnson, D. C., Hill, K. O., Hattori, K., and Hibino, Y., Single-frequency Er^{3+}-doped silica based planar waveguide laser with integrated photo-imprinted Bragg reflectors, *Electron. Lett.* 30, 1311 (1994).

146. Gandy, H. W., Ginther, R. J., and Weller, J. F., Laser oscillations in erbium activated silicate glass, *Phys. Lett.* 16, 266 (1965).

147. Mears, R. J., Reekie, L., Poole, S. B., and Payne, D. N., Low-threshold tunable CW and Q-switched fiber laser operating at 1.55 μm, *Electron. Lett.* 22, 159 (1986).

148. Reekie, L., Jauncey, I. M., Poole, S. B., and Payne, D. M., Diode-laser-pumped operations of an Er^{3+}-doped single mode fiber laser, *Electron. Lett.* 23, 1076 (1987).

149. Millar, C. A., Miller, I. D., Ainslie, B. J., Craig, S. P., and Armitage, J. R., Low-threshold cw operation of an erbium-doped fiber laser pumped at 807 nm wavelength, *Electron. Lett.* 23, 865 (1987).

150. Petersen, B., Zemon, S., and Miniscalco, W.J., Erbium doped fiber amplifiers pumped in the 800 nm band., *Electron. Lett.* 27, 1295 (1991).

151. Kimura, Y. and Nakazawa, M., Lasing characteristics of Er^{3+}-doped silica fibers from 1553 up to 1603 nm, *J. Appl. Phys.* 64, 516 (1988).

152. Hanna, D. C., Percival, R. M., Perry, I. R., Smart, R. G., and Tropper, A. C., Efficient operation of an Yb-sensitized Er fiber laser pumped in 0.8 μm region, *Electron. Lett.* 24, 1068 (1988).

153. Fermann, M. E., Hanna, D. C., Shepherd, D. P., Suni, P. J., and Townsend, J. E., Efficient operation of an Yb-sensitized Er fiber laser at 1.56 μm, *Electron. Lett.* 24, 1135 (1988).

154. Maker, G. T. and Ferguson, A. I., 1.56 μm Yb-sensitized Er fiber laser pumped by diode-pumped Nd:YAG and Nd:YLF lasers, *Electron. Lett.* 24, 1160 (1988).

155. Wyatt, R., Ainslie, B.J., and Craig, S.P., Efficient operation of an array pumped Er^{3+}-doped silica fiber laser at 1.5 μm, *Electron. Lett.* 24, 1362 (1988).

156. O'Sullivan, M.S., Chrostowski, J., Desurvire, E., and Simpson, J.R., High-power, narrow-linewidth Er^{3+}-doped fiber laser, *Optics Lett.* 14, 438 (1989).

157. Massicott, J.F., Armitage, J.R., Wyatt, R., Ainslie, B.J., and Craig-Ryan, S.P., High gain, broad bandwidth 1.6 μm Er^{3+} doped silica fiber amplifier, *Electron. Lett.* 26, 1645 (1990).

158. Kitagawa, T., Hattori, K., Shimizu, M., Ohmori, Y., and Kobayashi, M., Guided-wave laser based on erbium-doped silica planar lightwave circuit, *Electron. Lett.* 27, 334 (1991).

159. Heumann, E., Ledig, M., Ehrt, D., and Seeber, W., CW laser action of Er^{3+} in double sensitized fluoroaluminate glass at room temperature, *Appl. Phys. Lett.* 52, 255 (1988).

160. Barnes, W. L. and Townsend, J. E., Highly tunable and efficient diode pumped operation of Tm^{3+} doped fiber lasers, *Electron. Lett.* 26, 746 (1990).

161. Sankawa, I., Izumita, H., Furukawa, S., and Ishihara, K., An optical fiber amplifier for wide-band wavelength range around 1.65 μm, *IEEE Phot. Tech. Lett.* 2, 422 (1990).

162. Smart, R. G., Carter, J. N., Hanna, D. C., and Tropper, A. C., Erbium doped fluorozirconate fiber laser operating at 1.66 and 1.72 μm, *Electron. Lett.* 26, 649 (1990).

163. Ghisler, C., Pollnau, M., Bunea, G., Bunea, M., Lüthy, W., and Weber, H. P., Up-conversion cascade laser at 1.7 μm with simultaneous 2.7 μm lasing in erbium ZBLAN fibre, *Electron. Lett.* 31, 373 (1995); Pollnau, M., Ghisler, C., Bunea, G., Bunea, M., Lüthy, W., and Weber, H. P., 150 mW unsaturated output power at 3 μm from a single-mode-fiber erbium cascade laser, *Appl. Phys. Lett.* 66, 3564 (1995); Pollnau, M., Spring, R., Ghisler, C., Wittwer, S., Lüthy, W., and Weber, H. P., Efficiency of erbium 3-μm crystal and fiber lasers, *IEEE J. Quantum Electron.* 32, 657 (1996).

164. Hanna, C., Percival, R. M., Smart, R. G., and Tropper, A. C., Efficient and tunable operation of a Tm-doped fiber laser, *Opt. Commun.* 75, 283 (1990).

165. Percival, R. M., Szebesta, D., and Davey, S. T., Highly efficient and tunable operation of two colour Tm-doped fluoride fibre laser, *Electron. Lett.* 28, 671 (1992).

166. Gandy, H. W., Ginther, R. J., and Weller, J. F., Stimulated emission of Tm^{3+} radiation in silicate glass, *J. Appl. Phys.* 38, 3030 (1967).

167. Esterowitz, L., Allen, R., Kintz, G., Aggarwal, I., and Ginther, R. J., Laser emission in Tm^{3+} and Er^{3+}-doped fluorozirconate glass at 2.25, 1.88, and 2.70 μm, in *Conference on Lasers and Electro-Optics Technical Digest Series,* 7, Optical Society of America, Washington, DC (1988), p. 318.

168. Lincoln, J. R., Mackechnie, C. J., Wang, J., Brocklesby, W. S., Deol, R. S., Pearson, A., Hanna, D. C., and Payne, D. N., New class of fibre laser based on lead-germanate glass, *Electron. Lett.* 28, 1021 (1992); Wang, J., Lincoln, J. R., Brocklesby, W. S., Deol, R. S., Mackechnie, C. J., Pearson, A., Tropper, A. C., Hanna, D. C., and Payne, D. N., Fabrication and optical properties of lead-germanate glasses and a new class of optical fibers doped with Tm^{3+}, *J. Appl. Phys.* 73, 8066 (1993).

169. Hanna, D. C., Percival, R. M., Perry, I. R., Smart, R. G., Suni, P. J., and Tropper, A. C., Continuous-wave oscillation of a monomode thulium-doped silica fiber laser, *Technical Digest - Tunable solid State Laser Conference,* Optical Society of America (1989), p. 350.

170. Yamamoto, T., Miyajima, Y., and Komukai, T., 1.9 μm Tm-doped silica fibre laser pumped at 1.57 μm, *Electron. Lett.* 30, 220 (1994).

171. Boj, S., Delavaque, E., Allain, J. Y., Bayon, J. F., Niay, P., and Bernage, P., High efficiency diode pumped thulium-doped silica fibre lasers with intracore Bragg gratings in the 1.9-2.1 μm band, *Electron. Lett.* 30, 1019 (1994).

172. Hanna, D. C., Jauncey, I. M., Percival, R. M., Perry, I. R., Smart, R. G., Suni, P. J., Townsend, J. E., and Tropper, A. C., Continuous-wave oscillation of a monomode thulium-doped fluorozirconate fiber, *Electron. Lett.* 24, 935 (1988).

173. Ghisler, C., Luethy, W., and Weber, H. P., Tuning of a Tm^{3+}:Ho^{3+}:silica fiber laser at 2 μm, *IEEE J. Quantum Electron.* 31, 1877 (1995)

174. Carter, J.N., Smart, R.G., Hanna, D.C., and Tropper, A.C., CW diode-pumped operation of 1.97 μm thulium-doped fluorozirconate fiber laser, *Electron. Lett.* 26, 599 (1990).

175. Allain, J. Y., Monerie, M., and Poignant, H., High-efficiency cw thulium-sensitised holmium-doped fluoride fibre laser operating at 2.04 μm, *Electron. Lett.* 27, 1513 (1991).

176. Hanna, D. C., Percival, R. M., Smart, R. G., Townsend, J. E., and Tropper, A. C., Continuous wave oscillation of a holmium-doped silica fiber laser, *Electron. Lett.* 25, 593 (1989).

177. Percival, R. M., Szebesta, D., Davey, S. T., Swain, N. A., and King, T. A., High efficiency operation of 890-pumped holmium fluoride fibre laser, *Electron. Lett.* 28, 2064, (1992).

178. Veinberg, T. I., Zhmyreva, I. A., Kolobkov, V. P., and Kudryashov, P. I., Laser action of Ho^{3+} ions in silicate glasses coactivated by holmium, erbium, and ytterbium, *Opt. Spectrosc. USSR* 24, 441 (1968); *Opt. Specktrosk.* 24, 823 (1968).

179. Gandy, H. W. and Ginther, R. J., Stimulated emission from holmium activated silicate glass, *Proc. IRE* 50, 2113 (1962).

180. Percival, R. M., Carter, S. F., Szebesta, D., Davey, S. T., and Stallard, W. A., Thulium-doped monomode fluoride fibre laser broadly tunable from 2.25 to 2.5 μm, *Electron. Lett.* 27, 1912 (1991).

181. Allen, R. and Esterowitz, L., CW diode pumped 2.3 μm fiber laser, *Appl. Phys. Lett.* 55, 721 (1989).

182. Esterowitz, L., Allen, R., and Aggarwal, I., Pulsed laser emission at 2.3 μm in a thulium-doped fluorozirconate fiber, *Electron. Lett.* 24, 1104 (1988).

183. Allain, J. Y., Monerie, M., and Poignant, H., Energy transfer in Er^{3+}/Pr^{3+}-doped fluoride glass fibers and application to lasing at 2.7 μm, *Electron. Lett.* 27, 445 (1991).

184. Auzel, F., Meichenin, D., and Poignant, H., Tunable continuous-wave room-temperature Er^{3+}-doped ZrF_4-based glass laser between 2.69 and 2.78 μm, *Electron. Lett.* 24, 1463 (1988)

185. Brierley, M. C. and France, P.W., Continuous wave lasing at 2.7 μm in an erbium doped fluorozirconate fiber, *Electron. Lett.* 24, 935 (1988).

186. O'Sullivan, M. S., Chrostowski, J., Desurvire, E., and Simpson, J.R., High-power, narrow-linewidth Er^{3+}-doped fiber laser, *Optics Lett.* 14, 438 (1989).

187. Allen, R. and Esterowitz, L., Diode-pumped single-mode fluorozirconate fiber laser from the $^4I_{11/2} \rightarrow ^4I_{13/2}$ transition in erbium, *Appl. Phys. Lett.* 56, 1635 (1990).

188. Allain, J. Y., Monerie, M., and Poignant, H., Erbium-doped fluorozirconate single-mode fiber lasing at 2.71 μm, *Electron. Lett.* 25, 28 (1989).

189. Ronarch, D., Guibert, M., Auzel, F., Mechenin, D., Allain, J. Y., and Poignant, H., 35 dB optical gain at 2.716 μm in erbium doped ZBLAN fiber pumped at 0.642 μm, *Electron. Lett.* 27, 511 (1991).

190. Pollack, S. A. and Robinson, M., Laser emission of Er^{3+} in ZrF_4-based fluoride glass, *Electron. Lett.* 24, 320 (1988).

191. Wetenkamp, L., Efficient cw operation of a 2.9 μm Ho^{3+}-doped fluorozirconate fiber laser pumped at 640 nm, *Electron. Lett.* 26, 883 (1990).

192. Többen, H., CW lasing at 3.45 μm in erbium-doped fluorozirconate fibres, *Frequenz* 45, 250 (1991).

193. Többen, H., Room temperature cw fibre laser at 3.5 μm in Er^{3+}-doped ZBLAN glass, *Electron. Lett.* 28, 1361 (1992).

194. Schneider, J., Fluoride fibre laser operating at 3.9 μm, *Electron. Lett.* 31, 1250 (1995).

195. Mackechnie, C. J., Barnes, W. L., Carman, R. J., Townsend, J. E., and Hanna, D. C., High power ytterbium doped fiber laser operating at around 1.2 μm, in *Advanced Solid-State Lasers*, Pinto, A. A. and Fan, T. Y., Eds., Proceedings Vol. 15, Optical Society of America, Washington, DC (1993), p. 192.

196. Jiang, S., Myers, J., Belford, R., Rhonehouse, D., Myers, M., and Hamlin, S., Flashlamp pumped lasing of Ho:germanate glass at room temperature, *OSA Proc. Adv. Solid State Lasers*, Fan, T. Y. and Chai, B. H. T., Eds., Proceedings Vol. 20, 116 (1995).

197. Miura, K., Tanaka, K., and Hirao, K., CW laser oscillation on both the $^4F_{3/2}-^4I_{11/2}$ and $^4F_{3/2}-^4I_{13/2}$ transitions of Nd^{3+} ions using a fluoride glass microsphere, *J. Non-Cryst. Solids* 213&214, 276 (1997).

198. Basu, S. and Byer, R. L., Continuous-wave mode-locked Nd:glass laser pumped by a laser diode, *Optics Lett.* 13, 458 (1988).

199. Nakazawa, M. and Kimura, Y., Simultaneous oscillation at 0.91, 1.08, and 1.53 μm in a fusion-sliced fiber laser, *Appl. Phys. Lett.* 51, 1768 (1987).

200. Avakyants, L. I., Karpova, M. L., and Radchenko, V. V., Lasing properties of a multifiber Nd^{3+}-activated glass laser, *Sov. J. Quantum Electron.* 17, 553 (1987).

201. Reekie, L., Mears, R. J., Poole, S. B., and Payne, D. M., A Pr^{3+}-doped single-mode fibre laser, IEE Symp., May 1986.

202. DeShazer, L. G., Cuprous ion doped crystals for tunable lasers, in *Tunable Solid State Lasers*, Hammerling, P., Budgar, A. B., and Pinto, A., Eds., Springer-Verlag, Berlin (1985), p. 91.

203. Ugawa, T. S., Komukai, T., and Miyajuina, Y., Optical amplification in Er^{3+} doped single mode fluoride fiber, *IEEE Phot. Techn. Lett.* 2, 475 (1990).

204. Whitney, T. J., Millar, C. A., Brierley, M. C., and Carter, S. F., 23 dB gain upconversion pumped erbium doped fiber amplifier operating at 850 nm, *Electron. Lett.* 27, 189 (1991).

205. Nakazawa, M., Kimura, Y., and Suzuki, K., High gain erbium fiber amplifier pumped by 800 nm band, *Electron. Lett.* 26, 548 (1990).

206. Nakazawa, M., Kimura, Y., and Suzuki, K., Efficient Er^{3+}-doped optical fiber amplifier pumped by a 1.48 μm InGaAsP laser diode, *Appl. Phys. Lett.* 54, 295 (1989).

207. Horiguchi, H., Shimizu, M., Yamada, M., Yoshino, K., and Hanafusa, H., Highly efficient optical fiber amplifier pumped by a 0.8 nm band laser diode, *Electron. Lett.* 26, 1758 (1990).

208. Tachibana, M., Laming, R. I., Morkel, P. R., and Payne, D. N., Gain cross saturation and spectral hole burning in wideband erbium-doped fiber amplifiers, *Optics Lett.* 16, 1499 (1991).

209. Townsend, J. E., Barnes, W. L., Jedrzejewski, K. P., and Grubb, S. G., Yb^{3+} sensitised Er^{3+} doped silica optical fibre with ultrahigh transfer efficiency and gain, *Electron. Lett.* 27, 1958 (1991).

210. Doshida, M., Teraguchi. K., and Obara, M., Gain measurement and upconversion analysis in Tm^{3+}, Ho^{3+} co-doped alumino-zirco-fluoride glass, *IEEE J. Quantum Electron.* 31, 911 (1995).

211. Shikida, A., Yanagita, H., and Toratani, H., Ho-Yb fluoride glass fiber for green lasers, in *Advanced Solid-State Lasers*, Pinto, A. A. and Fan, T. Y., Eds., Proceedings Vol. 15, Optical Society of America, Washington, DC (1993), p. 261.

212. Mori, A., Ohishi, Y., Kanamori, T., and Sudo, S., Optical amplification with neodymium-doped chalcogenide glass fiber, *Appl. Phys. Lett.* 70, 1230 (1997).

213. Miyajima, Y., Sugawa, T., and Komukai, T., Efficient 1.3 μm-band amplification in a Nd^{3+}-doped single-mode fluoride fiber, *Electron. Lett.* 17, 1397 (1990).

214. Ohishi, Y., Kanamori, T., Temmyo, J. et al., Laser diode pumped Pr^{3+}-doped and Pr^{3+}-Yb^{3+}-codoped fluoride fibre amplifiers operating at 1.3 µm, *Electron. Lett.* 27, 1995 (1991); Ohishi, Y., Kanamori, T., Nishi, T., Takahashi, S., and Snitzer, E., Gain characteristics of Pr^{3+}-Yb^{3+} codoped fluoride fiber for 1.3 µm amplification, *IEEE Phot. Tech. Lett.* 3, 990 (1991).

215. Ohishi, Y., Kanamori, T., Kitagawa, T., Takahashi, S., Snitzer, E., and Sigel, G. H., Pr^{3+}-doped fluoride fiber superfluorescent laser, *Jap. J. Appl. Phys.* 30, L1282 (1991); Ohishi, Y. Kanamori, T., Kitagawa, T., Snitzer, E., and Sigel, Jr., G.H., Pr^{3+}-doped fluoride fiber amplifier operating at 1.31 µm, *Optics Lett.* 6, 1747 (1991).

216. Oh, K., Kilian, A., Reinhart, L., Zhang, Q., Morse, T. F., and Weber, P. M., Broadband superfluorescent emission of the $^3H_4 \rightarrow {}^3H_6$ transition in a Tm-doped multicomponent silicate fiber, *Optics Lett.* 19, 1131 (1994).

217. Sandoghdar, V., Treussart, F., Hare, J., Lefrvre-Seguin, V., Raimond, J.-M., and Haroche, S., Very low threshold whispering-galley-mode microsphere laser, *Phys. Rev. A* 54, R1777 (1996).

218. Kringlebotn, J. T., Archambault, J.-L., Reekie, L., and Payne, D. N., Er^{3+}:Yb^{3+}-codoped fiber distributed-feedback laser, *Optics Lett.* 19, 2101 (1994).

219. Wang, Y. Z., Lu, B. L., Li, Y. Q., and Liu, Y. S., Observation of cavity quantum-electrodynamic effects in a Nd:glass microsphere, *Optics Lett.* 20, 770 (1995).

220. Young, C. G., A sun-pumped cw one-watt laser, *Appl. Optics* 5, 993 (1966).

221. Carbonnier, C., Többen, H., and Unrau, U. B., Room temperature CW fibre laser at 3.22 µm, *Electron. Lett.* 34, 893 (1998).

222. Kringlebotn, J. T., Morkel, P. R., Reekie, L., Archambault, J.-L., and Payne, D. N., Efficient diode-pumped single-frequency erbium:ytterbium fiber laser, *IEEE Phot. Techn. Lett.* 5, 1162 (1993).

223. Kringlebotn, J. T., Archambault, J.-L., Reekie, L., Townsend, J. E., Vienne, G. G., and Payne, D. N., Highly-efficient, low-noise grating-feedback Er^{3+}:Yb^{3+}-codoped fibre laser, *Electron. Lett.* 30, 972 (1994).

224. Mejía B. E., Zenteno, L. A., Gavrilovic, P., and Goyal, A., High-efficiency lasing at 810 nm in single-mode Tm^{3+} doped fluorozirconate fiber pumped at 778 nm, *Opt. Eng.* 37, 2699 (1998).

225. Baney, D. M., Rankin, G., and Chang, K.-W., Blue Pr^{3+}-doped ZBLAN fiber upconversion laser, *Opt. Lett.* 21, 1372 (1996).

226. Scheife, H., Sandrock, T., Heumann, E., Huber, G., Pollock, C., and Bosenberg, W., Eds., *OSA TOPS Advanced Solid State Lasers*, Vol. 10 (Optical Society of America, Washington, DC, 1997).

227. Funk, D. S., Stevens, S. B., Wu, S. S., and Eden, J. G., *Proc. SPIE*, Vol. 2115, Bellingham, WA (1994), p. 108.

228. Pask, H. M., Tropper, A. C., and Hanna, A. C., A Pr^{3+}-doped ZBLAN fibre upconversion laser pumped by an Yb^{3+}-doped silica fibre laser, *Optics Commun.* 134 (1996).

229. Massicot, J. F., Brierley, M. C., Wyatt, R., Davey, S. T., and Szebesta, D., Low threshold, diode pumped operation of a green, Er^{3+} doped fluoride fibre laser, *Electron. Lett.* 29, 2119 (1993).

230. Piehler, D. and Graven, D., 11.7 mW green InGaAs-laser-pumped erbium fibre laser, *Electron. Lett.* 30 1759 (1994).

231. Sumiyoshi, T. and Sekita, H., Dual wavelength continuous-wave cascade oscillation at 3 and 2 µm with a holmium fluoride-glass fiber laser, *Optics Lett.* 23, 1837 (1998).

232. Schneider, J., Mid-infrared fluoride fiber lasers in multiple cascade operation, *IEEE Phot. Tech. Lett.* 7, 354 (1995).

233. Sandrock, T., Diening, A., and Huber, G., Laser emission of erbium-doped fluoride bulk glasses in the spectral range from 2.7 to 2.8 µm, *Optics Lett.* 24, 382 (1999).

234. Sun, Y., Judkins, J. B., Srivastava, A. K. et al., Transmission of 32-WDM 10-Gb/s channels over 640 km using broad-band, gain-flattened erbium-doped silica fiber amplifiers, *IEEE J. Quantum Electron.* 33, 1652 (1997).
235. Zellmer, H., Riedel, P., and Tünnermann, A., Visible upconversion lasers in praseodymium-ytterbium-doped fibers, *Appl. Phys. B* 69, 417 (1999).

Section 1.3
SOLID STATE DYE LASERS

1.3.1 Introduction

Solid state lasers based on organic molecules as the active element have utilized a variety of different host materials—plastics and polymers, organic and inorganic single crystals and glasses, gelatins, and biological materials—and have been operated in a wavelength range from 376 to 865 nm. One solid state dye laser has been offered commercially.

Data on the various types of solid state dye lasers are given in Tables 1.3.1 to 1.3.7 where the host material, dye, lasing wavelength, pump source and wavelength, and primary references to laser action are tabulated. The lasing wavelength and output of dye lasers depend on the characteristics of the optical cavity, the dye concentration, the optical pumping source and rate, and other operating conditions.[1] The original references should therefore be consulted for this information and its effect on the lasing wavelength. The references should also be consulted for details of the chemical composition and molecular structure of the dyes and host compounds.[2]

1. Schäfer, F. P., Ed., *Dye Lasers*, 3rd. edition, Springer-Verlag, Berlin (1990).
2. Maeda, M., *Laser Dyes*, Academic Press, New York (1984).

Further Reading

Bezrodnyi, V. I., Bondar, M. V., Kozak, G. Yu., Przhonskaya, O. V., and Tikhonov, E. A., Dye-activated polymer media for frequency-tunable lasers (review), *Zh. Prikl. Spektrosk. (USSR)* 50, 711 (1989).

Bezrodnyi, V. I., Przhonskaya, O. V., Tikhonov, E. A., Bondar, M. V., and Shpak, M. T., Polymer active and passive laser elements made of organic dyes, *Sov. J. Quantum Electron.* 12, 1602 (1982).

Dodabalapur, A., Chanddross, E. A., Berggren, M., and Slusher, R. L. Organic solid-state lasers: past and future, *Science* 277, 1787 (1997).

Dyumaev, K. M., Manenkov, A. A., Maslyukov, A. P., Matyushin, G. A., Nechitailo, V. S., and Prokhorov, A. M., Dyes in modified polymers: problems of photostability and conversion efficiency at high intensities, *J. Opt. Soc. Am. B* 9, 143 (1992).

Rahn, M. D. and King, T. A., Comparison of laser performance of dye molecules in sol-gel, polycom, ormosil, and poly(methyl methacryalte) host media, *Appl. Optics* 34, 8260 (1995).

Tagaya, A., Teramoto, S., Nihei, E., Sasake, K., and Koike, Y., High-power and high-gain organic dye-doped polymer optical fiber amplifiers: novel techniques for preparation and spectral investigation, *Appl. Optics* 36, 572 (1997).

Zink, J. I. and Dunn, B. S., Photonic materials by the sol-gel process, *J. Cer. Soc. Jpn.* 99, 878 (1991).

1.3.2 Dye Doped Organic Lasers

Table 1.3.1 summarizes the different polymer and molecular crystal hosts and organic dopants that have been used to demonstrate laser action. If the laser has been tuned over a range of wavelengths, the tuning range given is that for the experimental configuration and conditions used and may not represent the extremes possible. The lasing wavelength and output of dye lasers depend on the characteristics of the optical cavity, the dye concentration, the optical pumping source and rate, and other operating conditions. The original references should therefore be consulted for this information and its effect on the lasing wavelength.

Table 1.3.1
Dye Doped Organic Lasers

Host	Dye	Laser wavelength (nm)	Pump source (wavelength – nm)	Ref.
acrylic copolymer	pyrromethene 567	564	Nd:YAG laser (532)	94
acrylic copolymer	pyrromethene 580	570	Nd:YAG laser (532)	44
acrylic copolymer	pyrromethene 580	571	Nd:YAG laser (532)	95
acrylic copolymer	pyrromethene 597	587	Nd:YAG laser (532)	46, 94
acrylic monomers	sulforhodamine B	~600	Nd:YAG laser (532)	55
tris-(8-hydroxyquinoline) aluminum (Alq3)	DCM	589—635(a)	N2 laser (337) - ET	96
Alq3 (film)	DCM II	613	N2 laser (337) - ET	88
Alq3 (film)	DCM	645	N2 laser (337) - ET	89
Alq3	DCM	~655	N2 laser (337) - ET	97
4,4'-di(N-carbazole) biphenyl (CBP)	coumarin 47	460	N2 laser (337) - ET	99
4,4'-di(N-carbazole) biphenyl (CBP)	perylene	485	N2 laser (337) - ET	99
4,4'-di(N-carbazole) biphenyl (CBP)	coumarin 30	510	N2 laser (337) - ET	99
dibenzofurane (crystal)	anthracene	408—410	N2 laser (337)	9
dibenzyl (crystal)	β-naphthyl-p-biphenylethylene (β-BNE)	437, 461	XeCl laser (308)	7
dibenzyl (crystal 4.2 K)	tetracene	494, 521	Nd:YAG laser (355)	1
2,3-dimethylnaphthalene (crystal)	anthracene	408—410	N2 laser (337)	9

Host	Dye	Wavelength	Pump source	Ref.
diphenyl (crystal 4.2 K)	1,2-di-4-biphenylylethylene	418	Nd:YAG laser (355)	1,110
durol (crystal)	POPOP	469	XeCl laser (308)	7
fluorene (crystal)	anthracene	408—410	N$_2$ laser (337)	1,9,10,110
GPTMS/Ti(OBu)$_4$/MMA	rhodamine 6G	564—587	Nd:YAG laser (532)	102
HEMA/MMA (1:1)	coumarin 540A	515, 535	N$_2$ laser (337)	76
HEMA	ASPT	~600	Nd:YAG laser (1064) - TP	87
HEMA	DHASI	~606	Nd:YAG laser (1064) - TP	98
HEMA/MMA copolymer	rhodamine 640	654	N$_2$ laser (337)	70
HEMA/MMA copolymer	rhodamine 590	560	Nd:YAG laser (532)	107
HEMA/MMA (1:1)	rhodamine 6G		flashlamp	103
high temperature plastic	pyrromethene 567	522	Nd:YAG laser (532)	22
high temperature plastic	pyrromethene 597	528	Nd:YAG laser (532)	22
high temperature plastic	pyrromethene 580	550—570	Nd:YAG laser (532)	33
high temperature plastic	pyrromethene 597	565—590	Nd:YAG laser (532)	32
high temperature plastic	pyrromethene 570	566.3	Nd:YAG laser (532)	42
high temperature plastic	pyrromethene 597	587	Nd:YAG laser (532)	93
high temperature plastic	PM-IIMC	593	Nd:YAG laser (532)	22
high temperature plastic	PM-TEDC	598	Nd:YAG laser (532)	22
high temperature plastic	PM-HMC	615—635	Nd:YAG laser (532)	32
high temperature plastic	PM-TEDC	615—635	Nd:YAG laser (532)	32
high temperature plastic	PM 650	624	Nd:YAG laser (532)	46
methyl-cyclohexane/isopentane	coronene	443.9	flashlamp	14
methylmethacrylate	COP-2[b]	497	N$_2$ laser (337)	20
methylmethacrylate	COP-3[b]	511	N$_2$ laser (337)	20
methylmethacrylate[c]	1,3,5,7,8-pentamethyl-2,6-di-n-butylpyrromethene-BF$_2$	569.7, 571.0	Nd:YAG laser (532)	44
methylmethacrylate[c]	pyrromethene 567	571.4	Nd:YAG laser (532)	44

Table 1.3.1—*continued*
Dye Doped Organic Lasers

Host	Dye	Laser wavelength (nm)	Pump source (wavelength – nm)	Ref.
methylmethacrylate[c]	1,3,5,7-tetramethyl-8-cyanopyrro-methene-2,6-dicarboxylate-BF$_2$	617.9	Nd:YAG laser (532)	44
methylmethacrylate[c]	sulforhodamine B	618.2	Nd:YAG laser (532)	44
MMA/HPA	rhodamine 6G	595	coaxial flashlamp	
naphthalene (crystal 4.2 K)	4-phenylstilbene	376	Nd:YAG laser (355)	1
naphthalene (crystal 4.2 K)	1-(2-naphthyl)-2-phenylethylene	385	Nd:YAG laser (355)	1
naphthalene (crystal)	β-naphthyl-p-biphenylethylene (β-BNE)	395	XeCl laser (308)	7
naphthalene (crystal)	ββ-dinaphthylethylene (ββ-DNE)	395	XeCl laser (308), EB	6, 7
naphthalene (crystal 4.2 K)	1-(1-naphthyl)-2-(2-naphthyl)ethylene	397	Nd:YAG laser (355)	1
naphthalene (crystal 4.2 K)	1-(4-biphenyl)-2-(2-naphthyl)ethylene	397	Nd:YAG laser (355)	1
naphthalene (crystal 4.2 K)	1,2-di(2-naphthyl)ethylene	397	Nd:YAG laser (355)	1
naphthalene (crystal 4.2 K)	azulene	398	Nd:YAG laser (355)	1
naphthalene (crystal 4.2 K)	1-styryl-4-[1-(2-naphthyl)vinylbenzene	415, 442	Nd:YAG laser (355)	1
naphthalene (crystal 4.2 K)	1,4-bis(2-naphthyl)styrylbenzene	422, 449	Nd:YAG laser (355)	1
NAPOXA (film)	DCM II	~615	N$_2$ laser (337) - ET	90
para-terphenyl (crystal)	diphenylbutadiene (DPB)	383	electron beam	4
para-terphenyl (crystal)	diphenylbutadiene (DPB)	396	XeCl laser (308)	7
para-terphenyl (crystal)	naphthacene	528	XeCl laser (308)	7
para-terphenyl (crystal)	tetracene	530	N$_2$ laser (337)	9
para-terphenyl (crystal)	Cl-POPOP	589	XeCl laser (308)	7
para-terphenyl (crystal)	diphenylbutadiene (DPB)	589	XeCl laser (308)	7
para-terphenyl (crystal)	POPOP	589	XeCl laser (308)	7

Matrix	Dye	Pump source	Wavelength (nm)	Ref.
PBD (film)	coumarin 460	N_2 laser (337) - ET	~460	90
PBD (film)	PPV7	N_2 laser (337) - ET	575—590	91
PBD (photonic crystal)	coumarin 490, DCM	N_2 laser (337)	580, 596	109
PBD (film)	DCM II	N_2 laser (337) - ET	~600	90
PBD (film)	LDS 821	N_2 laser (337) - ET	~805	90
PHEMA (fiber)	ASPI	Nd:YAG laser (1064)- TP	~610	81
PHEMA+DEGMA	rhodamine 6G	N_2 laser (337)	580—620	51
P(HEMA:EDGMA 9:1)	rhodamine 6G	N_2 laser (337)	599	78
P(HEMA:MMA 1:1)	rhodamine 6G	Nd:YAG laser (532)	568	78
P(HEMA:MMA 3:7)	rhodamine 6G	N_2 laser (337)	584	78
P(HEMA:MMA 1:1)	rhodamine-Bz-MA	N_2 laser (337)	587	77
P(HEMA:MMA 7:3)	rhodamine 6G	N_2 laser (337)	587	78
P(HEMA:MMA 1:1)	rhodamine-Al	N_2 laser (337)	589	77
P(HEMA:MMA 1:1)	rhodamine 6G	N_2 laser (337)	593	78
P(HEMA:MMA 7:3)	rhodamine-Bz-MA	N_2 laser (337)	593	77
polyavylamide	rhodamine 6G	coaxial flashlamp	525—650	24
polyisobutylmethacrylate	α-NPO	N_2 laser (337)	~395	5
polymethylmethacrylate	α-NPO	Nd:YAG laser (355)	396	8
polymethylmethacrylate	2-(4-biphenylyl)-5-(p-styryl-phenyl)-1,3,4-oxadiazole	Nd:YAG laser (355)	400	8
polyisobutylmethacrylate	POPOP	N_2 laser (337)	~410	5
polymethylmethacrylate	2-(4-biphenylyl)-5-(1-naphthyl)-oxazole	Nd:YAG laser (355)	410	8
polymethylmethacrylate	5-phenyl-2-(p-styryl-phenyl)oxazole	Nd:YAG laser (355)	414	8
polymethylmethacrylate	POPOP	Nd:YAG laser (355)	415	8
polymethylmethacrylate	2-(1-naphthyl)-5-styryl-1,3,4-oxadiazole	Nd:YAG laser (355)	416	8
polymethylmethacrylate	2-phenyl-5-[p-(4-phenyl-1,3-bul-adrenyl)phenyl]-1,3,4-oxadiazole	Nd:YAG laser (355)	418	8
polyisobutylmethacrylate	dimethyl-POPOP	N_2 laser (337)	~420	5

Dye Doped Organic Lasers

Host	Dye	Laser wavelength (nm)	Pump source (wavelength – nm)	Ref.
polymethylmethacrylate	dimethyl-POPOP	423	Nd:YAG laser (355)	8
polymethylmethacrylate	5-phenyl-2-[p-(-phenylstyryl)-phenyl]oxazole	428	Nd:YAG laser (355)	8
polyisobutylmethacrylate	BBOT	~430	N_2 laser (337)	5
polymethylmethacrylate	5-(4-biphenylyl)-2-[p-(4-phenyl-1,3-baladienyl)phenyl]oxazole	443	Nd:YAG laser (355)	8
polymethylmethacrylate	1,2-bis(5-phenyloxazolyl)ethylene	444	N_2 laser (337)	5
polymethylmethacrylate	1,5-diphenyl-3-styryl-2-pyrazoline	450	Nd:YAG laser (355)	8
polymethylmethacrylate	1,4-bis[4-[5-(4-biphenylyl)-2-oxazolyl]styryl benzene	455	Nd:YAG laser (355)	8
polymethylmethacrylate	3-p-chlorostyryl-1,5-diphenyl-2-pyrazoline	457	Nd:YAG laser (355)	8
polymethylmethacrylate	(see reference)	478	Nd:YAG laser (355)	8
polymethylmethacrylate	2-[p-[2-(9-anthryl)vinyl]phenyl]-5-phenyloxazole	480	Nd:YAG laser (355)	8
polymethylmethacrylate (film)	CF_3-coumarin	484—529	N_2 laser (337)	18
polymethylmethacrylate	2-(2'-hydroxy-5'-fluorophenyl)benzimidazole	493	N_2 laser (337)	19
polymethylmethacrylate	5(6)-methoxycarbonyl-2(2'-hydroxphenyl)benzimidazole	494	N_2 laser (337)	20
polymethylmethacrylate	5(6)-methoxycarbonyl-12-(5'-fluoro-2'-hydroxyphenyl)benzimidazole	508	N_2 laser (337)	20
polymethylmethacrylate	fluoran	509—514	Nd:YAG laser (355)	8
polymethylmethacrylate	coumarin 540A	515	N_2 laser (337)	74
polymethylmethacrylate	rhodamine 6G	550	Nd:YAG laser (355)	8
polymethylmethacrylate	rhodamine 6G	552—575	Nd glass (530)	34

Host	Dye	Wavelength (nm)	Pump source	Ref.
polymethylmethacrylate	rhodamine 6G	552—595	Nd:YAG laser (532)	33
polymethylmethacrylate	rhodamine 6G	555—565	CdS laser (495)	36
polymethylmethacrylate	rhodamine 6G	562	Nd:YAG laser (532)	100
polymethylmethacrylate	rhodamine 6G	565.5, 570.9	Nd:YAG laser (532)	42
polymethylmethacrylate	rhodamine 6G	567—605	Ne laser (540)	43
polymethylmethacrylate	pyrromethene 570	569.6	Nd:YAG laser (532)	42
polymethylmethacrylate (WG)	DCM	570	dye laser (730)- TP	45
polymethylmethacrylate	rhodamine 590	573	Nd:YAG laser (532)	83
polymethylmethacrylate	rhodamine 590	573	Nd:YAG laser (355)	31
polymethylmethacrylate (film)	rhodamine 6G	577—590	N_2 laser (337)	49
polymethylmethacrylate	peryline orange (KF 241)	578	Nd:YAG laser (355)	31
polymethylmethacrylate	peryline orange (KF 241)	578	Nd:YAG laser (532)	83
polymethylmethacrylate	rhodamine 590	585	C-523 dye laser (525)	52
polymethylmethacrylate	rhodamine B	587.4, 594.0	N_2 laser (337) - ET	53
polymethylmethacrylate	rhodamine C	595	Nd:YAG laser (355)	8
polymethylmethacrylate	rhodamine 6G	601.0	Xe flashlamp	58
polymethylmethacrylate	peryline red (KF 856)	613	Nd:YAG laser (355)	31
polymethylmethacrylate	peryline red (KF 856)	613	Nd:YAG laser (532)	83
polymethylmethacrylate	rhodamine B	632.4	Xe flashlamp	58
PMMA(BBOT)(d)	perylene	455—475	N_2 laser (337)	16
PMMA(coumarin 1)(d)	acridine yellow	490—500	N_2 laser (337)	16
PMMA(coumarin 1)(d)	acriflavine	500—520	N_2 laser (337)	16
PMMA(coumarin 1)(d)	uranine	530—560	N_2 laser (337)	16
PMMA (rhodamine 6G)(d)	rhodamine B	600—620	N_2 laser (337)	16
PMMA (rhodamine 6G)(d)	cresyl violet	620—640	N_2 laser (337)	16
PMMA (rhodamine 6G)(d)	sulforhodamine 101	620—670	N_2 laser (337)	16
PMMA (rhodamine 6G)(d)	nile blue	695—720	N_2 laser (337)	16
PMMA (rh. 6G/nile blue)(d)	HITC	845—865	N_2 laser (337)	16

Table 1.3.1—*continued*
Dye Doped Organic Lasers

Host	Dye	Laser wavelength (nm)	Pump source (wavelength – nm)	Ref.
mPMMA	dye "11 B"	560—570	(not reported)	40
mPMMA	Rh-based 11 B	580	Nd:YAG laser (532)	111
mPMMA	rhodamine 6G	560—570	Nd glass (530)	39
mPMMA	rhodamine 6G chloride	560—570	(not reported)	40
mPMMA	Rh 6G-Cl⁻	569	Nd:YAG laser (532)	111
mPMMA	rhodamine 6G percholate	560—570	(not reported)	40
mPMMA	Rh 6G-ClO$_4$	569	Nd:YAG laser (532)	111
mPMMA	rhodamine 111	560—570	(not reported)	40
mPMMA	oxazine-17	620—640	(not reported)	40
mPMMA	oxazine-17	620—640	Nd glass (530)	39
polystyrene:TiO$_2$ (film)	MEH-PPV	~605	Nd:YAG laser (355)	92
proprietary polymer-gel	rhodamine 6G	565.1	Nd:YAG laser (532)	42
polystyrene	BBQ	381—394	N$_2$ laser (337)	3
polystyrene	α-NPO	411	Nd:YAG laser (355)	8
polystyrene	dimethyl-POPOP	414, 427	Nd:YAG laser (355)	8
polystyrene	POPOP	416—426	N$_2$ laser (337)	3
polystyrene	POPOP	420	Nd:YAG laser (355)	8
polystyrene	dimethyl-POPOP	424	Nd:YAG laser (355)	8
polystyrene	dimethyl-POPOP	426—445	N$_2$ laser (337)	3
polystyrene	dimethyl-POPOP	428—438	N$_2$ laser (337)	11
polystyrene	BBOT	440—472	N$_2$ laser (337)	3
polystyrene	BBOT + perylene	464—480	N$_2$ laser (337)	3
polystyrene	1,3-dimethylisobenzofuran	490—500	N$_2$ laser (337)	3
polystyrene	fluorescein (sodium salt)	553—564	N$_2$ laser (337)	11

Host material	Dye	Emission wavelength	Pump source (wavelength)	Ref.
polystyrene (microdisk laser)	pyrromethene dye	~565	dye laser (520)	108
polystyrene (microring laser)	pyrromethene dye	573, 577[e]	dye laser (520)	108
polystyrene (microring laser)	pyrromethene dye	~600—610[f]	dye laser (520)	108
polystyrene	rhodamine 640	~600	Nd:YAG laser (1064) - TP	56
polystyrene (WGM)	Nile Red	~610—620	dye laser (520)	102
polystyrene (WGM)	Nile Red	~605—620	Nd:YLF laser (523)	103
polystyrene (WG)	rhodamine 6G	614—624	N_2 laser (337)	11
polyurethane	rhodamine 6G	610—635	Ne laser (540)	62
polyurethane	rhodamine 6G	610—635	N_2 laser (337)	63, 64
polyurethane (film)	rhodamine B	632.8 (A)	N_2 laser (337)	68
polyurethane	rhodamine 110	643	Ar ion laser (514)	69
polyurethane	indodicarbocyanine (PK 643)	680 (A)	(532, 635, 670)	71
polyvinylxylene	dimethyl-POPOP	425	Nd:YAG laser (355)	8
Probimide 414[g] (WG)	cresyl violet 670	670	dye laser (590)	84
sym-octahydroanthracene (crystal)	anthracene	408—410	N_2 laser (337)	9

A — amplified wavelength, EB — electron beam, ET — energy transfer from host to dye, TP — two-photon pumped, WG — waveguide, WGM — whispering galley mode

(a) Laser emission varied by changing the thickness of the active organic layer.
(b) See Ref. 20 for molecular structure of copolymer.
(c) 16% hydroxypropyl acrylate/methyl methacrylate.
(d) Host materials were polymethylmethacrylate and polyvinylalcohol. Dye in parentheses serves as a donor dye in an energy transfer laser.
(e) Polymer droplets formed around a 125-µm silica fiber.
(f) Polymer droplets formed around a 17-µm silica fiber.
(g) A photosensitive benzophenone tetracarboxylic diahydride-alkylated diamine polyimide.

Abbreviations of host materials and dyes used in Tables 1.3.1:

Alq3 – tris-(8-hydroxyquinoline) aluminum, ASPI – trans-4-[p-(N-hydroxyethyl-N-methyl-amino)styryl]-N-methylpyridinium iodide, BBOT – 2,5-bis(5-tert-butyl-2-benzoxazoly)-thiophene BBQ – 4,4'''bis(2-butyloctyloxy)-*p*-quaterphenyl, CBP – 4,4'-di(N-carbazole) bi-phenyl, COP – cyclooctetraene, DCM – 4-dicyanomethylene-2-methyl-6-p-dimethylamino-styryl-4H-pyran, DEGMA – ethylene glycol dimethacrylate, DHASI – 4-[bis(2-hydroxy-ethyl)amino]-*N*-methylstilbazolium iodide, EGDMA—ethylene glycol dimethacrylate, GPTMS – glycidoxypropyltrimeth-oxysilane, HEMA – hydroxy ethyl meth-acrylate, HITC – 1,3,3,1',3',3'-hexamethyl-2,2'-indotricarbocyanine, HPA – hydroxypropyl acrylate, HTP – high temperature plastic (Korry Electronics), MEH – poly 2-methoxy-5-2'-ethylhexloxy, MMA – methylmethacrylate, mPMMA – modified polymethylmethacrylate, NAPOXA – 2-napthyl-4,5-bis(4-methoxyphenyl)-1,3-oxazole, NPO – 2-(1-naphthy)-5-phenyloxazole, PBD – 2-(4-biphenylyl)-5-(4-*t*-butylphenyl)-1,3,4-oxadiazole, PHEMA – poly (2-hydroxy-ethyl methacrylate), PMMA – polymethylmethacrylate, POPOP – 1,4-di[2-(5-phenyl-oxazolyl)] benzene, PPV – poly(1,4-phenylene vinylene), PS – polystyrene, PSI – proprietary polymer.

1.3.3 Silica and Silica-Gel Dye Lasers

The properties of dye doped silica and sol-gel silica lasers are summarized in Table 1.3.2. If the laser has been tuned over a range of wavelengths, the tuning range given is that for the experimental configuration and conditions used and may not represent the extremes possible. The lasing wavelength and output of dye lasers depend on the characteristics of the optical cavity, the dye concentration, the optical pumping source and rate, and other operating conditions. The original references should therefore be consulted for this information and its effect on the lasing wavelength.

Table 1.3.2
Silica and Silica-Gel Dye Lasers

Host	Dye	Laser wavelength (nm)	Pump source (wavelength – nm)	Reference
ORMOSIL	coumarin 153	498—574	C460 dye laser (460)	21
ORMOSIL	coumarin 540A	525	dye laser (460)	23
ORMOSIL	rhodamine 6G	557—598	C153 dye laser (539)	21
ORMOSIL	rhodamine 6G	559—587	C153 dye laser (539)	21
ORMOSIL	rhodamine 6G	562—590	Nd:YAG laser (532)	41
ORMOSIL	rhodamine 6G	568	dye laser (539)	23
ORMOSIL (film)	Lumogen LFO240	568—583	Nd:YAG laser (532)	54
ORMOSIL	pyrromethene 567	584	Nd:YAG laser (532)	46
ORMOSIL (film)	DCM	595—650	Nd:YAG laser (532)	54
ORMOSIL	rhodamine B	600—625	Nd:YAG laser (532)	41
ORMOSIL (film)	Lumogen LFR300	605—630	Nd:YAG laser (532)	54
ORMOSIL (film)	Lumogen LFR300	615—629	Nd:YAG laser (532)	65
ORMOSIL	nile blue	680—746	Nd:YAG laser (532)	72
ORMOSIL	LD 800	727—747	R640 dye laser (623)	73
ORMOSIL	HITC	819—844	R640 dye laser (630)	73
ORMOSIL (film)	rhodamine 610	585—635	Nd:YAG laser (532)	54
ORMOSIL (GPTA)	rhodamine 6G	605	N_2 laser (337) - ET	53
ORMOSIL (PMMA)	pyrromethene 567	572	Nd:YAG laser (532)	31
ORMOSIL (PMMA)	rhodamine 590	572	Nd:YAG laser (532)	31

Table 1.3.2—continued
Silica and Silica-Gel Dye Lasers

Host	Dye	Laser wavelength (nm)	Pump source (wavelength – nm)	Reference
ORMOSIL (PMMA)	BASF-241	575—590	Nd:YAG laser (532)	48
ORMOSIL (PMMA)	peryline orange	578	Nd:YAG laser (532)	31, 50
ORMOSIL (PMMA)	peryline orange (KF 241)	578	Nd:YAG laser (532)	31
ORMOSIL (PMMA)	peryline orange (KF 241)	578	Nd laser (532)	83
ORMOSIL (PMMA)	peryline red	604	Nd:YAG laser (532)	31
ORMOSIL (PMMA)	red perylimide dye	605—630	Nd:YAG laser (532)	48
ORMOSIL (PMMA)	peryline red	614	Nd:YAG laser (532)	50
ORMOSIL (PMMA)	peryline red (KF 856)	614	Nd laser (532)	83
ORMOSIL (TiO$_2$)	DCM	590.7—654.3	Nd laser (532)	82
ORMOSIL (VTEOS, MTEOS)	pyrromethene 567	542—606	Nd:YAG laser (532)	26
ORMOSIL (VTEOS, MTEOS)	pyrromethene 567	543—603	Nd laser (532)	27
ORMOSIL (VTEOS, MTEOS)	pyrromethene 580	554—584	Nd laser (532)	26
ORMOSIL (VTEOS, MTEOS)	peryline orange	567—594	Nd:YAG laser (532)	27
ORMOSIL (VTEOS, MTEOS)	pyrromethene 597	574—606	Nd:YAG laser (532)	26
ORMOSIL (VTEOS, MTEOS)	peryline orange	582—592	Nd:YAG laser (532)	26
ORMOSIL (VTEOS, MTEOS)	peryline red	595—640	Nd laser (532)	26
ORMOSIL (VTEOS, MTEOS)	peryline red	595—644	Nd:YAG laser (532)	27
ORMOSIL-γ-GLYMO (wg)	rhodamine B	~628	Nd:YAG laser (532)	66
silica	Exalite 377E	376	XeCl laser (308)	2
silica	LD 390	403	XeCl laser (308)	2
silica	coumarin 460	468—494	XeCl laser (308)	17
silica	rhodamine 6G	545—630	Nd:YAG laser (532)	28—30
silica	pyrromethene 567	549	Nd:YAG laser (532)	83
silica	pyrromethene 567	549	Nd:YAG laser (532)	31
silica	rhodamine 590	564	Nd:YAG laser (532)	83
silica	rhodamine 590	564	Nd:YAG laser (532)	31

Host	Dye	Pump laser	Wavelength	Ref.
silica	peryline orange	Nd:YAG laser (532)	585	83
silica	peryline orange	Nd:YAG laser (532)	585	31
silica-MMA	ASPI	Nd:YAG (532)	585—606	85
sol gel silica	Exalite 376	XeCl laser (308)	364	106
sol gel silica	coumarin 1	XeCl laser (308)	433—457	12
sol gel silica	stilbene	XeF laser (351)	440	80
sol gel silica	coumarin 460	XeF laser (351)	475	80
sol gel silica	coumarin 460	XeCl laser (308)	468—494	17
sol gel silica	coumarin 102	XeCl laser (308)	487—495	12
sol gel silica	coumarin 481	XeF laser (351)	535	79
sol gel silica	coumarin 153	XeCl laser (308)	545—572	12
sol gel silica	coumarin 521	XeF laser (351)	558	80
sol gel silica	rhodamine 6G	KrF laser (249)	560	37
sol gel silica	rhodamine 6G	Nd:YAG laser (532)	570—610	47, 86
sol gel silica	rhodamine 6G	dye laser (507)	577	37
sol gel silica	sulforhodamine 640	Nd:YAG laser (532)	600—650	57
sol gel silica	sulforhodamine 640	Nd:YAG laser (532)	605—648	59
sol gel silica	rhodamine B	XeCl laser (308)	610—620	12
sol gel silica	coumarin 560	XeF laser (351)	625	80
sol gel silica	coumarin 640	XeF laser (351)	655	80
sol-gel glass	pyrromethene 567	dye laser (506)	549	83
sol-gel glass	pyrromethene 567	Nd:YAG (532)	556	83
sol-gel glass	rhodamine 590	Nd:YAG (532)	568	83
sol-gel glass	rhodamine 590	dye laser (506)	580	83

wg — waveguide

Abbreviations of host materials and dyes used in Table 1.3.2:

ASPI – trans-4-[p-(N-hydroxyethyl-N-methylamino)styryl]-N-methylpyridinium iodide, DCM – 4-dicyanomethylene-2-methyl-6-p-dimethyl-aminostyryl-4H-pyran, GLYMO – glycidyloxyropyl trimethoxy silane, GPTA – glycerol propoxy triacrylate, HITC – 1,3,3,1',3',3'-hexamethyl-2,2'-indotricarbocyanine, MMA – methylmethacrylate, MTEOS – methyl-triethoxysilane, ORMOSIL – organically modified silicate, PMMA – poly-methylmethacrylate, VTEOS – vinyltriethoxysilane.

1.3.4 Dye Doped Inorganic Crystal Lasers

Table 1.3.3
Dye Doped Inorganic Crystal Lasers

Host	Dye	Wavelength (nm)	Pump Source (nm)	Ref.
Al_2O_3	rhodamine 6G	~560[a]	N_2 laser (337)	38
Al_2O_3	rhodamine B	~590[a]	N_2 laser (337)	38
Al_2O_3 [b]	rhodamine B	610–620	N_2 laser (337)	61
Al_2O_3	oxazine 4	~640[a]	N_2 laser (337)	38
$AlPO_4$-5 (zeolite)	Pyridine 2[c]	687[d]	Nd:YAG laser (532)	104, 105
K_2SO_4	pyrene	441, 541	Nd:YAG laser (355)	13
K_2SO_4	rhodamine	595–620	Nd:YAG laser (532)	13

(a) This is the wavelength of the fluorescence peak; lasing wavelengths were not cited.
(b) Alumina film doped with rhodamine 6G for an energy-transfer-type laser dye pair.
(c) 1-ethyl-4-[4-(*p*-dimethylaminophenyl)-1,3-butadienyl]-pyridinium perchlorate.
(d) Single mode, whispering gallery mode.

1.3.5 Dye Doped Glass Lasers

Table 1.3.4
Dye Doped Glass Lasers

Host	Dye	Wavelength (nm)	Pump Source (nm)	Ref.
aluminosilicate	coumarin 540A	557	dye laser (460)	23, 35
aluminosilicate	rhodamine 6G	570	dye laser (540)	23, 35
composite glass	perylimide (BASF 241)	530–630	Nd:YAG laser (532)	25
PFMPG[a]	pyrromethene 558	≈554	Nd:YAG laser (532)	75
PFMPG[a]	rhodamine 11B	≈563–570	Nd:YAG laser (532)	75
PFMPG[a]	pyrromethene 597	≈568, ≈571	Nd:YAG laser (532)	75
PFMPG[a]	pyrromethene 650	≈625	Nd:YAG laser (532)	75
polycom glass[b]	pyrromethene 567	572	Nd:YAG laser (532)	83
polycom glass[b]	rhodamine 590	572 (563)	Nd:YAG laser (532)	83
polycom glass[b]	rhodamine 590 chloride	573	dye laser (506)	83
polycom glass[b]	peryline orange (KF 241)	578	Nd:YAG laser (532)	83, 112
polycom glass[b]	perylene red (KF 856)	604	Nd:YAG laser (532)	83

(a) Polymer-filled microporous glass.
(b) Sol-gel glass–poly (methyl methacrylate) composite.

1.3.6 Dye Doped Gelatin Lasers

Table 1.3.5
Dye Doped Gelatin Lasers

Host	Dye	Wavelength (nm)	Pump Source (nm)	Ref.
Wratten[a] 22 filter	—	600–650[b]	N_2 laser (337)	15
Wratten[a] 29 filter	—	657	N_2 laser (337)	15
Knox[a] gelatin	methylumbelliferone	450–540[c]	N_2 laser (337)	15
Knox[a] gelatin	7-diethylamino-4-methylcoumarin	~453[c]	N_2 laser (337)	15
Knox[a] gelatin	Na fluorescein	550–570[c]	N_2 laser (337)	15
Knox[a] gelatin	rhodamine 6G	570–620[c]	XeCl laser (308)	16
Knox[a] gelatin	rhodamine B	~600–620[c]	N_2 laser (337)	15
gelatin	rhodamine 6G	~630	N_2 laser (337)	67

(a) Knox and Wratten are commercial product names.
(b) Output was stated to be in the orange-red region.
(c) Lasing wavelengths were not given but were stated to be close to those produced by the same dyes in liquid solutions.

1.3.7 Dye Doped Biological Material Lasers

Table 1.3.7
Dye Doped Biological Material Lasers

Host	Dye	Wavelength (nm)	Pump Source (nm)	Ref.
pig fat	rhodamine 640	609	Nd:YAG laser (532)	60
chicken tissue	rhodamine 640	613	Nd:YAG laser (532)	60

1.3.8 Commercial Solid State Dye Laser

Table 1.3.7
Commercial Solid State Dye Laser

Laser Type	Operation	Principal wavelengths (μm)	Output
polymeric host	pulsed	0.55–0.70[a]	≤ 150 mJ

(a) Tunable; several different polymer rods are needed to cover the wavelength range indicated.

1.3.9 References

1. Naboikin, Yu. V., Ogurtsova, L. A., Podgornyi, A. P., and Maikes, L. Ya., Stimulated emission of light from doped molecular crystals at 4.2 K, *Sov. J. Quantum Electron.* 8, 457 (1978).
2. Lam, K. S., Lo, D., and Wong, K. H., Observations of near-UV superradiance emission from dye-doped sol-gel silica, *Optics Comm.* 121, 121 (1995).
3. Muto, S., Ando, A., Yoda, O., Hanawa, T., and Ito, H., Dye laser by sheet of plastic fibers with wide tuning range, *Trans. IECE (Jpn)* E 70, 317 (1987).
4. Budakovskii, S. V., Gruzinskii, V. V., Davydov, S. V., Kulak, I. I., and Kolesnik, E. E., Stimulated emission from electron-beam-pumped diphenylbutadiene molecules imbedded in a crystalline matrix, *Sov. J. Quantum Electron.* 22, 16 (1992).
5. Muto, S., Ichikawa, A., Ando, A., Ito, C., and Inaba, H., Trial to plastic fiber dye laser, *Trans. IECE (Japan)* E 69, 374 (1986).
6. Bokhonov, A. F., Davydov, S. V., Kulam, I. I. et al., Spontaneous and stimulated-radiation of impurity molecular crystals with optical and electron pumping, *Zh. Prinkl. Spektrosk.* 50, 966 (1989).
7. Budakovskii, S. V., Gruzinskii, V. V., Davydov, S. V., and Kolesnik, E. E., Laser media on impurity organic crystals with pumping into an absorption band of the matrix, *Zh. Prinkl. Spektrosk.* 54, 79 (1991).
8. Naboikin, Yu. V., Ogurtsova, L. A., Podgornyi, A. P. et al., Spectral and energy characteristics of lasers based on organic molecules in polymers and toluol, *Opt. Spektrosk.* 28, 974 (1970); *Sov. Opt. Spectrosc.* 28, 528 (1970).
9. Karl, N., Laser emission from organic molecular crystals, mode selection and tunability, *J. Lumin.* 12-13, 851 (1976).
10. Karl, N., Laser emission from an organic molecular crystal, *Phys. Stat. Sol. A* 13, 651 (1972).
11. Tanuguchi, H., Fujiwara, T., Yamada, H., Tanosake, S., and Baba, M., Whispering-galley-mode dye lasers in blue, green, and orange regions using dye-doped, solid, small spheres, *Appl. Phys. Lett.* 62, 2155 (1993; see also, *J. Appl. Phys.* 73, 7957 (1993).
12. McKiernan, J. M., Yamanaka, S. A., Knoble, E. T., Pouxviel, J. C., Parvench, D. E., Dunn, S. B., and Zink, J. I., *J. Inorg. and Organomet. Polymers* 1, 87 (1991).
13. Rifani, M., Yin, Y.-Y., Elliott, D. S. et al., Solid state dye lasers from stereospecific host-guest interactions, *J. Am. Chem. Soc.* 117, 7572 (1995).
14. Kohlmannsperger, J., Ein organischer laser: coronen in MCH/IP bei 100 K, *Z. Naturf.* 24a, 1547 (1969).
15. Hänsch, T. W., Pernier, M., and Schawlow, A. L., Laser action of dyes in gelatin, *IEEE J. Quantum Electron.* QE-7, 45 (1971).
16. Muto, S., Shiba, T., Iijima, Y., Hattori, K., and Ito, C., Solid thin-film energy transfer dye lasers, *Trans. IECE (Japan)* J69-C, 25 (1986); [*Electron and Commun. Japan*, part 2, 70, 21 (1987)].
17. Lam, K.-S., Lo, D., and Wong, K.-H., Sol-gel silica laser tunable in the blue, *Appl. Optics* 34, 3380 (1995) and Wideband tuning of a XeCl laser-pumped dye-doped sol-gel silica laser, *IEEE Photon. Technol. Lett.* 7, 306 (1995)
18. Itoh, U., Takakusa, M., Moriya, T., and Saito, S., Optical gain of coumarin dye-doped thin film lasers, *Jpn. J. Appl. Phys.* 16, 1059 (1977).
19. Acuña, A. U., Amat-Guerri, F., Costela, A., Douhal, A., Figuera, J. M., Florido, F., and Sastre, R., Proton-transfer lasing from solid organic matrices, *Chem. Phys. Lett.* 187, 98 (1991).
20. Ferrer, M. L., Acuña, A. U., Amat-Guerri, F., Costela, A., Figuera, J. M., Florido, F. and Sastre, R., Proton-transfer lasers from solid polymeric chains with covalently bound 2-(2'-hydroxyphenyl)benzimidazole groups, *Appl. Optics* 33, 2266 (1994).
21. Knobbe, E. T., Dunn, B., Fuqua, P. D., and Nishida, F., Laser behavior and photostability characteristics of organic dye doped silicate gel materials, *Appl. Optics* 29, 2729 (1990).

22. Allik, T., Chandra, S., Robinson, T. R., Hutchinson, J. A., Sathyamoorthi, G., and Boyer, J. H., Laser performance and material properties of a high temperature plastic doped with pyrromethene-BF$_2$ dyes, Mat. Res. Soc. Proc. Vol. 329, *New Materials for Solid State Lasers* (1994), p. 291.

23. Dunn, B., Mackenzie, J. D., Zink, J. I., and Stafsudd, O. M., Solid-state tunable lasers based on dye-doped sol-gel materials, SPIE Vol. 1328, *Sol-Gel Optics* (1990), p. 174.

24. Kessler, W. J. and Davis, S. J., Novel solid state dye laser host, *Proceedings of the Solid State Dye Laser Workshop* (1994), p. 216.

25. Reisfeld, R., Brusilovsky, D., Eyal, M., Miron, E., Burstein, Z., and Ivri, J., A new solid-state tunable laser in the visible, *Chem. Phys. Lett.* 160, 43 (1989).

26. Canva, M., Dubois, A., Georges, P., Brum, A., Chaput, F., Ranger, A., and Boilot, J.-P., Perylene, pyrromethene and grafted rhodamine doped xerogels for tunable solid state laser, SPIE Vol. 2288, *Sol-Gel Optics III* (1994), p. 298.

27. Canva, M., Georges, P., Perelgritz, J.-F., Brum, A., Chaput, F., and Boilot, J.-P., Perylene- and pyrromethene-doped xerogel for a pulsed laser, *Appl. Optics* 34, 428 (1995).

28. Dul'nev, G. N., Zemskin, V. I., Krynetshi, B. B., Meshkovskii, I. K., Prokhorov, A. M., and Stelmakl, O. M., Tunable solid-state laser with a microcomposition matrix active medium, *Sov. Tech. Phys. Lett.* 4, 420 (1978).

29. Altshuler, G. B., Dulneva, E. G., Meshkovskii, I. K., and Krylov, K. I., Solid state active media based on dyes, (*Zh. Prikl. Spektrosk.*), *J. Appl. Spectrosc.* 36, 415 (1981).

30. Altshuler, G. B., Dulneva, E. G., Krylov, K. I., Meshkovskii, I. K., and Urbanovich, V. S., Output characteristics of a laser utilizing rhodamine 6G in microporous glass. *Sov. J. Quantum Electron.* 13, 784 (1983).

31. Rahn, M. D. and King, T. A., Lasers based on dye doped sol-gel composite glasses, SPIE Vol. 2288, *Sol-Gel Optics III* (1994), p. 382.

32. Chandra, S. and Allik, T., Compact high-brightness solid state dye laser, *Proceedings of the Solid State Dye Laser Workshop* (1994), p. 29.

33. Soffer, B. H. and McFarland, B. B., Continuously tunable, narrow band organic dye lasers, *Appl. Phys. Lett.* 10, 266 (1967).

34. Kaminow, I. P., Weber, H. P., and Chandross, E. A., Poly(methyl methacrylate) dye laser with internal diffraction grating resonator, *Appl. Phys. Lett.* 18, 497 (1971).

35. McKiernan, J. M., Yamanaka, S. A., Dunn, B., and Zink, J. I., *J. Phys. Chem.* 94, 5652, (1990).

36. Onstott, J. R., Short cavity dye laser excited by an electron beam, *Appl. Phys. Lett.* 31, 818 (1977).

37. Whitehurst, C., Shaw, D. J., and King, T. A., Sol-gel glass solid state lasers doped with organic molecules, SPIE Vol. 1328, *Sol-Gel Optics* (1990), p. 183.

38. Kobayashi, Y., Kurokawa, Y., and Imai, Y., A transparent alumina film doped with laser dye and its emission properties, *J. Non-Cryst. Solids* 105, 198 (1988).

39. Gromov, D. A., Dyumaev, K. M., Manenkov, A. A., Maslyukov, A. P., Matyushin, G. A., Nechitailo, V. S., and Prokhorov, A. M., Efficient plastic-host dye lasers, *J. Opt. Soc. Am. B* 2, 1028 (1985).

40. Manenkov, A. A., Maslyukov, A. P., Matyushin, G. A., and Nechitailo, V. S., Modified polymers - effective host materials for solid state dye lasers and laser beam control elements: a review, SPIE Vol. 2115, *Visible and UV Lasers* (1994), p. 136.

41. Altman, J. C., Stone, R. E., Dunn, B., and Nishida, F., Solid state laser using a rhodamine-doped silica gel compound, *IEEE Phot. Techn. Lett.* 3, 189 (1991).

42. Ewanizky, T. F. and Pearce, C. K., Solid state dye lasers in an unstable resonator configuration, *Proceedings of the Solid State Dye Laser Workshop* (1994), p. 154.

43. Fork, R. L., German, I. R., and Chandron, E. A., Photodimer distributed feedback laser, *Appl. Phys. Lett.* 20, 139 (1972).

44. Hermes, R. E, Allik, T. H., Chandra, S., and Hutchison, J. A., High-efficiency pyrromethene doped solid-state dye lasers, *Appl. Phys. Lett.* 63, 877 (1993).

45. Mukherjee, A., Two-photon pumped upconverted lasing in dye doped polymer waveguides, *Appl. Phys. Lett.* 62, 3423 (1993).

46. Boyer, J., Pyrromethene-BF$_2$ complexes (P-BF$_2$), *Proceedings of the Solid State Dye Laser Workshop* (1994), p. 66.
47. Altshuler, G. B., Bakhanov, V. A., Dulneva, E. G., Erofeev, A. V., Mazurin, O. V., Roskova, G. P., and Tsekhomskaya, T. S., Laser based on dye-activated silica gel, *Opt. Spectrosc. (USSR)* 62, 709 (1987).
48. Reisfeld, R., Film and bulk tunable lasers in the visible, SPIE Vol. 2288, *Sol-Gel Optics III* (1994), p. 563.
49. Schinke, D. P., Smith, R. G., Spencer, E. G., and Galvin, M. F., Thin-film distributed feedback laser fabricated by ion milling, *Appl. Phys. Lett.* 21, 494 (1972).
50. Rhan, M. D., King, T. A., Capozzi, C. A., and Seldon, A. B., Characteristics of dye doped ormosil lasers, SPIE Vol. 2288, *Sol-Gel Optics III* (1994), p. 364.
51. Amat-Guerri, F., Costela, A., Figuera, J. M., Florido, F., and Sastre, R., Laser action from rhodamine 6G-doped poly(2-hydroxyethyl methacrylate) matrices with different crosslinking degrees, *Chem. Phys. Lett.* 207, 352 (1993).
52. Mandl, A. and Klimek, D. E., 400 mJ long pulse (>1µs) solid state dye laser, *Proceedings of the Solid State Dye Laser Workshop* (1994), p. 192.
53. Wojcik, A. B. and Klein, L. C., Rhodamine 6G-doped inorganic/organic gels for laser and sensor applications, SPIE Vol. 2288, *Sol-Gel Optics III* (1994), p. 392.
54. Reisfeld, R., Film and bulk tunable lasers in the visible, *Proceedings of the Solid State Dye Laser Workshop* (1994), p. 56.
55. Hermes, R. E., McGrew, J. D., Wiswall, C. E., Monroe, S., and Kushina, M., A diode laser-pumped Nd:YAG-pumped polymeric host solid-state dye laser, *Appl. Phys. Commun.* 11, 1 (1992).
56. Misawa, H., Fujisawa, R., Sasaki, K., Kitamura, N., and Masuhara, H., Simultaneous manipulation and lasing of a polymer microparticle using a cw 1064 nm laser beam, *Jpn. J. Appl. Phys.* 32, L788 (1993).
57. Canva, M., Georges, P., Brun, A., Larrue, D., and Zarzycki, J., Impregnated SiO$_2$ gels used as dye laser matrix hosts, *J. Non-Cryst. Solids* 147&148, 636 (1992).
58. Peterson, O. G. and Snavely, B. B., Stimulated emission from flashlamp excited organic dyes in polymethyl methacrylate, *Appl. Phys. Lett.* 12, 238 (1968).
59. Salin, F., Le Saux, G., Georges, P., Brun, A., Bagnall, C., and Zarzycki, J., Efficient tunable solid-state laser near 630 nm using sulfo rhodamine 640-doped silica gel, *Opt. Lett.* 14, 785 (1989).
60. Siddique, M., Yang, L., Wang, Q. Z., and Alfano, R. R., Mirrorless laser action form optically pumped dye-treated animal tissues, *Optics Commun.* 117, 475 (1995).
61. Sasaki, H., Kobayashi, Y., Muto, S., and Kurokawa, Y., Preparation and photoproperties of a transparent alumina film doped with energy-transfer-type laser dye pairs, *J. Am. Ceram. Soc.* 73, 453 (1990).
62. Weber, H. P. and Ulrich, R., Unidirectional thin-film ring lasers, *Appl. Phys. Lett.* 20, 38 (1972).
63. Ulrich, R. and Weber, H. P., Solution-deposited thin films as passive and active light guides, *Appl. Optics* 11, 428 (1972).
64. Sasaki, K., Fukao, T., Saito, T., and Hamano, O., Thin film waveguide evanescent dye laser and its gain measurement, *J. Appl. Phys.* 51, 3090 (1980).
65. Shamrakov, D. and Reisfeld, R., Super radiant film laser operation of perylimide dyes doped silica-polymethylmethacrylate composite, *Opt. Mater.* 4, 103 (1994).
66. Reisfeld, R., Shamrakov, D., and Sorek, Y., Spectroscopic properties of thin glass films doped by laser dyes prepared by sol-gel method, *J. de Phys. IIII*, Vol. 4, Colloque C4, C4-487 (1994); Sorek, Y. and Reisfeld, R., Light amplification in a dye-doped glass planar waveguide, *Appl. Phys. Lett.* 66, 1169 (1995).
67. Kogelnik, H. and Shank, C. V., Stimulated emission in a periodic structure, *Appl. Phys. Lett.* 18, 152 (1971).
68. Chang, M. S., Burlamacchi, P., Hu, C., and Whinnery, J.R., Light amplification in a thin film, *Appl. Phys. Lett.* 20, 313 (1972).

69. Sriram, S., Jackson, H. E., and Boyd, J. T., Distributed-feedback dye laser integrated with a channel waveguide formed in silica, *Appl. Phys. Lett.* 36, 721 (1980).

70. Amat-Guerri, F., Costel, A., Figuera, J. M., Florido, F., Garcia-Moreno, I., and Sastre, R., Laser action from a rhodamine 640-doped copolymer of 2-hydroxyethyl methacrylate and methyl methacrylate, *Optics Comm.* 114, 442 (1995).

71. Bondar, M. V., Przhonskaya, O. V., and Tikhonov, E. A., Amplification of light by dyed polymers as the laser pumping frequency changes, *Opt. Spectrosc.* 74, 215 (1993).

72. Dunn, B., Nishida, F., Altman, J. C., and Stone, R. E., Spectroscopy and laser behavior of rhodamine-doped ORMOSILS, in *Chemical Processing of Advanced Materials*, Hench, L. L and West, J. K., Eds., Wiley, New York (1992), p. 941.

73. Dunn, B., Nishida, F., Toda, R., Zink, J. I., Allik, T., Chandra, S., and Hutchinson, J. A., Advances in dye-doped sol gel lasers, *Mat. Res. Soc. Proc.* Vol. 329, New Materials for Solid State Lasers (1994), p. 267.

74. Costela, A., Garcia-Moreno, I., Barroso, J., and Sastre, R., Laser performance of Coumarin 540A dye molecules in polymeric host media with different viscosities: from liquid solution to solid polymer matrix, *J. Appl. Phys.* 83, 650 (1998).

75. Aldag, H. R., Dolotov, S. M., Koldunov, M. F. et al., Efficient solid-state dye lasers based on polymer-filled microporous glass, SPIE Proceedings Vol. 3929 (in press).

76. Costela, A., Garcia-Moreno, I., Figuera, J. M., Amat-Guerri, F., Barroso, J., and Sastre, R., Solid-state dye laser based on coumarin 540A-doped polymeric matrices, *Opt. Commun.* 130, 44 (1996).

77. Costela, A., Garcia-Moreno, I., Figuera, J. M., Amat-Guerri, F., and Sastre, R., Solid-state dye lasers based on polymers incorporating covalently bonded modified rhodamine 6G, *Appl. Phys. Lett.* 68, 593 (1996).

78. Costela, A., Florido, F., Garcia-Moreno, I., Duchowicz, R., Amat-Guerri, F., Figuera, J. M., and Sastre, R., Solid-state dye lasers based on copolymers of 2-hydroxyethyl methacrylate and methyl methacrylate doped with rhodamine 6G, *Appl. Phys. B* 60, 383 (1995).

79. Lo. D., Parris, J. E., and Lawless, J. L., Multi-megawatt superradiant emissions from coumarin-doped sol-gel derived silica, *Appl. Phys. B* 55, 365 (1992).

80. Lo. D., Parris, J. E., and Lawless, J. L., Laser and fluorescence properties of dye-doped sol-gel silica from 400 nm to 800 nm, *Appl. Phys. B* 56, 385 (1993).

81. He, G. S., Bhawalkar, J. D., Zhao, C. F., and Park, C. K., Upconversion dye-doped polymer fiber laser, *Appl. Phys. Lett.* 68, 3549 (1996).

82. Hu, W., Chuangdong, H. Y., Jiang, Z., and Zhou, F., All-solid-state tunable dye laser pumped by a diode-pumped Nd:YAG laser, *Appl. Optics* 36, 579 (1997).

83. Rahn, M. D. and King, T. A., Comparison of laser performance of dye molecules in sol-gel, polycom, ormosil, and poly(methyl methancrylate) host media, *Appl. Optics* 34, 8260 (1995).

84. Weiss, M. N., Srivatava, R., Correia, R. R. B., Martins-Filho, J. F., and de Araujo, C. B., Measurement of optical gain at 670 nm in an oxazine-doped polyimide planar waveguide, *Appl. Phys. Lett.* 69, 3653 (1996).

85. Gvishi, R., Ruland, G., and Prasad, P. N., The influence of structure and environment on spectroscopic and lasing properties on dye-doped glasses, *Opt. Mater.* 8, 43 (1997).

86. Finkelstein, I., Ruschin, S., Sorek, Y., and Reisfeld, R., Waveguided visible lasing effects in a dye-doped sol-gel glass film, *Opt. Mater.* 7, 9 (1997).

87. He, G. S., Zhao, C. F., Bhawalkar, J. D., and Prasad, P. N., Two-photon pumped cavity lasing in novel dye doped bulk matrix rods, *Appl. Phys. Lett.* 67, 3703 (1995).

88. Berggren, M., Dodabalapur, A., and Slusher, R. E., Stimulated emission and lasing in dye-doped organic thin films with Forster transfer, *Appl. Phys. Lett.* 71, 2230 (1997).

89. Kozlov, V. G., Bulovic´, V., and Forrest, S. R., Temperature independent performance of organic semiconductor lasers, *Appl. Phys. Lett.* 71, 2575 (1997).

90. Berggren, M., Dodabalapur, A., Slusher, R. L., and Bao, Z., Light amplication in organic thin films using cascade energy transfer, *Nature* 389, 466 (1997).

91. Berggren, M., Dodabalapur, A., Bao, Z., and Slusher, R. E., Solid-state droplet laser made from an organic blend with a conjugated polymer emitter, *Adv. Mater.* 9, 968 (1997).

92. Hide, F., Schwartz, B. J., Diaz-Garcia, M. A., and Heeger, A. J., Laser emission from solutions and films containing semiconducting polymer and titanium dioxide nanocrystals, *Chem. Phys. Lett.* 256, 424 (1996).

93. Allik, T. H., Chandra, S., Robinson, T. R., Hutchinson, J. A., Sathyamoorthi, G., and Boyer, J. H., Laser performance and material properties of a high temperature plastic doped with pyrromethene-BF_2 dyes, *Mat. Res. Soc. Symp. Proc.* 329, 291 (1994).

94. Hermes, R.E., Lasing performance of pyrromethene-BF_2 laser dyes in a solid polymer host, SPIE Proceedings Vol. 2115: *Visible and UV Lasers* (1994), p. 240.

95. Allik, T. H., Chandra, S., Hermes, R. E., Hutchinson, J. A., Soong, M. L., and Boyer, J. H., Efficient and robust solid-state dye laser, OSA *Proc. on Adv. Solid-State Lasers* 15, 271 (1993).

96. Bulovic´, V., Kozlov, V. G., Khalfin, V. B., and Forrest, S. R., Transform-limited, narrow-linewidth lasing action in organic semiconductor microcavities, *Science* 279, 553 (1998).

97. Berggren, M., Dodabalapur, A., Slusher, R. E., Timko, A., and Nalamasu, O., Organic solid-state lasers with imprinted gratings on plastic substrates, *Appl. Phys. Lett.* 72, 410 (1998).

98. He, G. S., Kim, K.-S., Yuan, L., Cheng, N., and Prasad, P. N., Two-photon pumped partially cross-linked polymer laser, *Appl. Phys. Lett.* 71, 1619 (1997).

99. Kozlov, V. G., Parthasarathy, G., Burrows, P. E., Forrest, S. R., You, Y., and Thompson, M. E., Optically pumped blue organic semiconductor lasers, *Appl. Phys. Lett.* 72, 144 (1998).

100. Wang, H. H. L. and Gampel, L., A simple, efficient plastic dye laser, *Optics Commun.* 18, 444 (1976).

101. Finlayson, A. J., Peters, N., Kolinsky, P. V., and Venner, M. R. W., Flashlamp pumped polymer dye laser containing Rhodamine 6G, *Appl. Phys. Lett.* 72, 2153 (1997).

102. Kuwata-Gonokami, M., Takeda, K., Yasuda, H., and Ema, K., Laser emission from dye-doped polystrene microsphere, *Jpn. J. Appl. Phys.* 31, L99 (1992).

103. Kuwata-Gonokami, M. and Takeda, K., Polymer whispering gallery mode lasers, *Opt. Mater.* 9, 12 (1998).

104. Vietze, U., Krauss, O. Laeri, F. et al., Zeolite-dye microlasers, *Phys. Rev. Lett.* 81, 4628 (1998).

105. Ihlein, G., Schüth, F., Krauss, O., Vietze, U., and Laeri, F., Alignment of a laser dye in the channels of athe $AlPO_4$-5 molecular sieve, *Adv. Mater.* 10, 1117 (1998).

106. Wu, S. and Zhu, C., All-solid-state UV dye laser pumped by XeCl laser, *Opt. Mater.* 12, 99 (1999).

107. Giffin, S. M., McKinnie, I. T., Wadsworth, W. J. et al., Solid state dye laser based on 2-hydroxyethyl methacrylate and methyl methacrylate co-polymers, *Opt. Commun.* 161, 163 (1999).

108. Kuwata-Gonokami, M., Jordan, R. H., Dodabalapur, A., Katz, H. E., Schilling, M. L., and Slusher, R. E., Polymer microdisk and microring lasers, *Optics Lett.* 20, 3093 (1995).

109. Meier, M., Mekis, A., Dodabalapur, A., Timko, A. Slusher, R. E., Joannopoulos, J. D., and Nalamasu, O., Laser action from two-dimensional distributed feedback in photonic crystals, *Appl. Phys. Lett.* 74, 7 (1999).

110. Gorelik, V. S. and Zhabotubskii, A. M., Spontaneous and stimulated emission spectra of organic crystals upon two-photon excitation, *Optics Spectrosc.* 77, 218 (1994).

111. Maslyukov, A., Sokolov, S., Kaivola, M., Nyholm, K., and Popov, S., Solid-state dye laser with modified poly(methyl methacrylate)-doped active elements, *Appl. Optics* 34, 1516 (1995).

112. Rahn, M. D. and King, T. A., High-performance solid-state dye laser based on perylene-orange-doped polycom glass, *J. Mod. Optics* 45, 1259 (1998).

Section 1.4
COLOR CENTER LASERS

1.4.1 Introduction

The optically active centers in color center lasers are various types of point defects (i.e., color centers) in alkali halide and oxide crystals. The color centers are generally produced by ionizing radiation or are thermally induced. Additional ions may be present to stabilize the defect center and are included in the description of the active center. Other lasers in this category are based on vibrational transitions of molecular defects, such as CN^-.

Color center lasers are usually excited by optical pumping with broadband or laser radiation. Lasing involves allowed transitions between electronic energy levels, hence the gain can be high. Color center lasers have been reported that operate in the wavelength range from approximately 0.4 to 5 µm. Due to their large homogeneous emission bandwidths, color center lasers have varying degrees of tunability. The tuning ranges of some of the longer-lived alkali halide color center lasers are shown in Figure 1.4.1.

The output of color center lasers may be cw or pulsed. Representative average powers and average energies of cw and pulsed color center lasers in alkali halides are summarized in Ref. 73. As in the case of paramagnetic ion lasers, picosecond pulses can be obtained using various mode-locking techniques and femtoseconds pulses using saturable absorbers. The operative lifetimes of the color centers in these lasers depend on the temperature and can vary from hours to years. Many color center lasers require operation at low temperatures.

Further Reading

Baldacchini, G., Optical excitation and relaxation of solids with defects, in *Spectroscopy and Dynamics of Collective Excitations in Solids*, Di Bartolo, B. Ed., Plenum Press, New York (1997), p. 495. (This paper contains an interesting history of color center and color center laser research.)

Basiev, T. T. and Mirov, S. B., *Room Temperature Tunable Color Center Lasers*, Vol. 16 of Laser Science and Technology Series, Gordon & Breach, New York (1994), p. 1.

Basiev, T. T., Mirov, S. B., and Osiko, V. V., Room-temperature color center lasers, *IEEE J. Quantum Electron.* 24, 1052 (1988).

Gellermann, W., Color center lasers, *J. Phys. Chem. Solids* 52, 249 (1991).

German, K. R., Color Center Laser Technology, in *Handbook of Solid-State Lasers*, Cheo, P. K., Ed., Marcel Dekker, New York (1989), p. 457.

Mirov, S. B. and Basiev, T., Progress in color center lasers, in *Semiconductor Lasers, Selected Topics in Quantum Electronics* 1 (June 1995).

Mollenauer, L. F., Color Center Lasers, in *Handbook of Laser Science and Technology, Vol. I: Lasers and Masers*, CRC Press, Boca Raton, FL (1982), p. 171 and *Supplement 1: Lasers,* CRC Press, Boca Raton, FL (1991), p. 101.

Mollenauer, L. F., Color Center Lasers, in *Tunable Lasers*, 2nd. edition, Mollenauer, L. F., White, J. C., and Pollock, C. R., Eds., Springer-Verlag, Berlin (1992).

Pollock, C. R., Optical properties of laser-active color centers, *J. Lumin.* 35, 65 (1986).

Pollock, C. R., Color Center Lasers, in *Encyclopedia of Lasers and Optical Technology*, Meyers, R. A., Ed., Academic Press, San Diego (1991), p. 9.

Ter-Mikirtychev, V. V. and Tsuboi, T., Stable room-temperature tunable color center lasers and passive Q-switches, *Progr. Quantum Electron.* 20, 219 (1996).

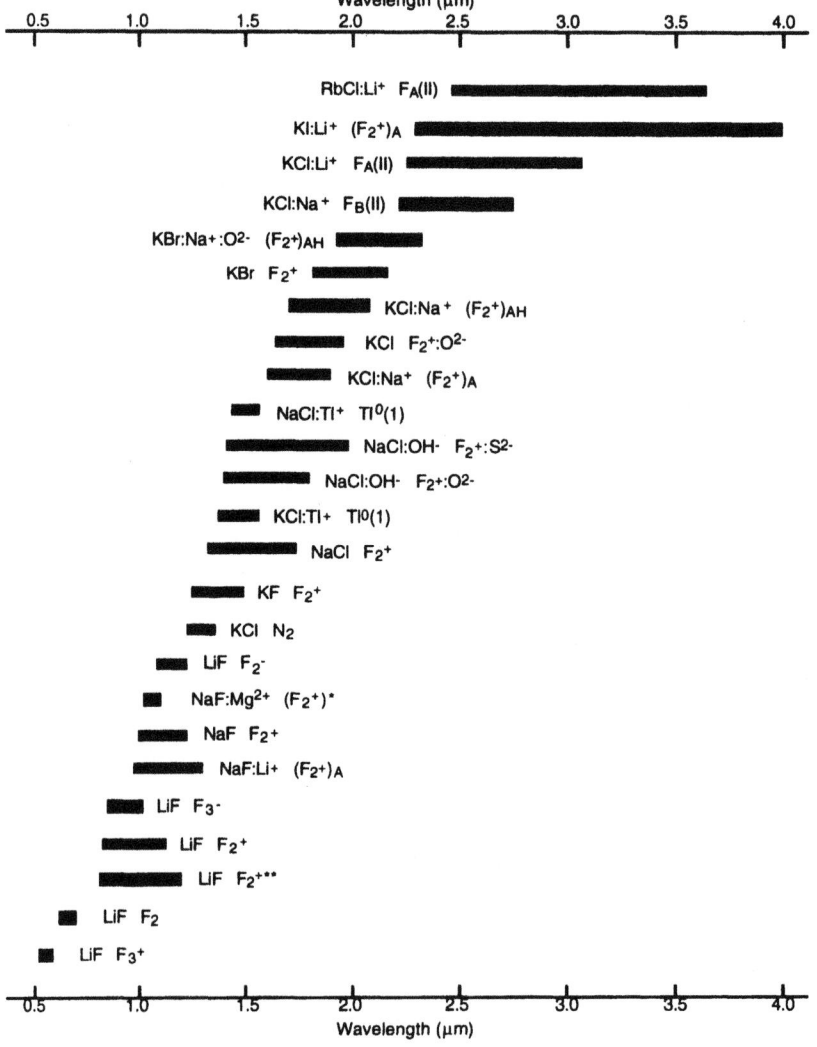

Figure 1.4.1 The reported tuning ranges for color center lasers in alkali halide hosts.

1.4.2 Centers and Crystals for Color Center Lasers

Color centers are formed by one or more vacancies in a crystal; the vacancy may also be accompanied by nearby foreign ions. The simple F center (an electron trapped at an anion vacancy) and several more complex aggregated centers are depicted by the structural models in Figure 1.4.2. Other centers, such as a F_3^+ center for example, consist of two electrons bound to three neighboring anion vacancies. Principal centers and crystalline hosts used for color center lasers are summarized in Table 1.4.1.

Table 1.4.1
Principal Centers and Crystals Used for Color Center Lasers

Center	Crystal[(a)]	Center	Crystal*
F_2	LiF	$(F_2^+)_{AH}$	$KCl:Na^+$
F_2^+	KBr	$(F_2)_A$	$CaF_2:Na^+$
	KCl		$MgF_2:Na^+$
	KF		$SrF_2:Na^+$
	LiF		
	$LiF:OH^-$	$F_2^+:O^{2-}$	$KCl:Na^+$
	$LiF:OH^-,Mg^{2+}$		$NaCl:OH^-$
	NaCl		
	$NaCl:OH^-$	$F_2^+:S^{2-}$	NaCl
	NaF		
		F_3^+	LiF
F_2^-	LiF		
		F_3^-	LiF
$(F_2^+)*$	$NaF:Mg^{2+}$		
		$F_A(II)$	$KCl:Li^+$
$(F_2^+)**$	LiF		$RbCl:Li^+$
$(F_2^+)_A$	$KCl:Li^+$	$F_A:Tl^0(1)$	$KCl:Tl^+$
	$KCl:Na^+$		
	$KI:Li^+$	$F_B(II)$	$KCl:Na^+$
	$NaF:Li^+$		
	$RbI:Li^+$	$Tl^0(1)$	$KBr:Tl^+$
			$KCl:Tl^+$
$(F_2^+)_H$	$NaCl:OH^-$		$KF:Tl^+$
			$NaCl:Tl^+$

(a) Dopants are given after the colon.

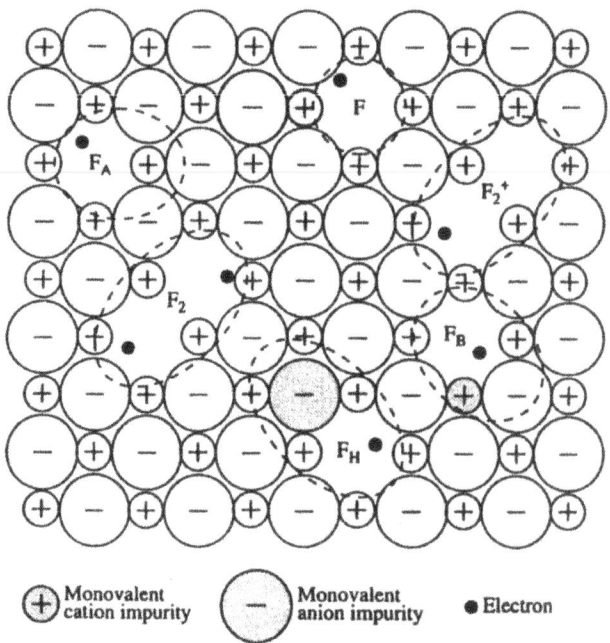

Figure 1.4.2 Structural models of the F center and other more complex aggregated color centers (from G. Baldacchini, Reference 73, with permission).

1.4.3 Table of Color Center Lasers

Color center lasers and their properties are listed by host crystal in Table 1.4.2. If the host contained dopants, they are included following the colon. The other columns in the table list the active center, laser wavelength, pump source and wavelength, operating temperature, and primary references. For lasers that have been tuned over a range of wavelengths, the tuning range given is that for the configuration and conditions used and may not represent the extremes possible.

The lasing wavelength and output power of color center lasers depend on the characteristics of the optical cavity, the temperature, the optical pump source, and other operating conditions. The original references should therefore be consulted for this information and its effect on the lasing wavelength. The original references should also be consulted for the method used to create the color centers and the operating lifetime of the laser.

Abbreviations used in Table 1.4.2:

Pump source
alex — alexandrite (BeAl$_2$O$_4$:Cr) laser
Ar — argon ion laser
CCL — color center laser
D — frequency doubled
DL — dye laser
Er:YLF — Er:LiYF$_4$ laser
Kr — krypton ion laser
NdYAG — Nd:Y$_3$Al$_5$O$_{12}$ laser
RL — ruby (Al$_2$O$_3$:Cr) laser
RS — Raman shifted
TiS — Ti sapphire (Al$_2$O$_3$) laser

Mode of operation
AML — actively mode-locked
cw — continuous wave
p — pulsed
qcw — quasi-continuous wave
PML — passively mode-locked
SML — synchronously mode-locked

Table 1.4.2
Color Center Lasers Arranged by Host Crystal

Host crystal	Active center	Laser wavelength (μm)	Pump source (wavelength-μm)	Mode	Temp. (K)	Ref.
Al$_2$O$_3$	unknown*	0.54–0.62	CCL (0.46)	p	RT	4
Al$_2$O$_3$	unknown*	0.75–0.95	ruby laser (0.694)	p	RT	5
Al$_2$O$_3$	unknown*	0.96–1.15	dye laser (0.83)	p	RT	5
Al$_2$O$_3$	unknown*	0.95–1.1	DL (0.825), RL(0.694)	p	RT	6
Al$_2$O$_3$:Mg	unknown*	0.51–0.59	dye laser (0.440)	p	RT	7

Table 1.4.2—*continued*
Color Center Lasers Arranged by Host Crystal

Host crystal	Active center	Laser wavelength (μm)	Pump source (wavelength-μm)	Mode	Temp. (K)	Ref.
C(diamond)	H_3	0.53(a)	Ar ion laser (0.488)	cw	RT	3
CaF_2:Na$^+$	$(F_2)_A$	0.76	dye laser (0.610)	p	RT	8–10
CaO	F^+	0.357–0.420(a)	Ar (0.351)/ N_2 (337)	p, cw	77	1
CsCl	$F_H(CN^-)$	4.88–5.00	Kr ion laser (0.647)	cw	77	49
KBr	CN^-	4.86	DCO_2 laser (4.81)	p	1.7	47
KBr	CN^-	4.86	CCL (2.42)	cw	1.7	48
KBr:O^{2-}	F_2^+	1.86–2.16	NdYAG laser (1.34)	cw	77	38, 41
KBr:O^{2-}	$(F_2^+)_H$	1.86–2.10	CCL (1.6)	cw, ML	77	42
KBr:Na$^+$:O^{2-}	$(F_2^+)_{AH}$	1.96–2.35	CCL (1.6)	cw, ML	77	42
KBr:Tl$^+$	$Tl^0(1)$	1.55–1.70	NdYAG laser (1.064)	cw	77	28, 76
KCl	N_2	1.23–1.35	NdYAG laser (1.064)	p	77	24
KCl	F_2^+	1.61–1.77	NdYAG (1.318/1.340)	cw	77	65
KCl	F_2^+:O^{2-}	1.66–1.97	NdYAG laser (1.34)	cw	77	38
KCl:Li$^+$	F_A	2.72	xenon flashlamp	p	77	74
KCl:Li$^+$	$(F_2^+)_A$	2.00–2.50	NdYAG laser (1.34)	cw	77	43
KCl:Li$^+$	$F_A(II)$	2.6–2.8	Kr laser (0.6471)	cw	77–200	75
KCl:Li$^+$	$F_A(II)$	2.30–3.10	dye laser (0.610)	cw	77	45
KCl:Li$^+$	$F_A(II)$	2.5–2.9	Ar (0.514), Kr (0.647)	cw	77	72
KCl:Na$^+$	$F_B(II)$	2.25–2.65	Ar (0.514), Kr (0.647)	cw	77	72
KCl:Li$^+$:Na$^+$	$F_A(II)$–$F_B(II)$	2.27–2.88	Ar (0.514), Kr (0.647)	cw	77	72
KCl:Na$^+$	$(F_2^+)_A$	1.62–1.91	NdYAG laser (1.34)	cw	77	37
KCl:Na$^+$	F_2^+:O^{2-}	~1.71–2.15	NdYAG (1.32/1.34)	cw	77	39
KCl:Na$^+$	F_2^+:O^{2-}	1.77–1.94	NdYAG laser (1.32)	p	77	66
KCl:Na$^+$	$(F_2^+)_{AH}$	1.73–2.10	NdYAG (1.32)	AML	77	40
KCl:Na$^+$	$F_B(II)$	2.22–2.75	dye laser (0.595)	cw	77	44, 45

Crystal	Center	Wavelength (μm)	Pump source (μm)	Mode	T (K)	Ref.
KCl:Tl$^+$	F$_A$:Tl^0(1)	1.40–1.60	NdYAG laser (1.064)	qcw, SML	77	27, 29
KCl:Tl$^+$	F$_A$:Tl^0(1)	1.40–1.60	NdYAG laser (1.064)	cw	77	28
KCl:Tl$^+$	Tl^0(1)	1.488–1.538	laser diode (0.96)	cw	84	36
KF	F$_2^+$	1.24–1.45	NdYAG laser (1.064)	SML	77	25
KF	F$_2^+$	1.26–1.48	NdYAG laser (1.064)	cw, SML	77	26
KF:Tl$^+$	Tl^0(1)	1.2–1.4	TiS laser (0.844)	cw	77	76
KI:Li$^+$	(F$_2^+$)$_A$	2.38–3.99	ErYLF laser (1.73)	p	77	46
KMgF$_3$:Cu$^+$	unknown	0.945–1.065	Kr ion laser (0.647)	cw	77	17
KMgF$_3$:Pb^{2+}	unknown	0.855–0.965	Kr ion laser (0.647)	cw	77	16, 17
LiF	F$_3^+$	0.51–0.58	dye laser (0.460)	p	RT	2, 60
LiF	F$_3^+$, F$_2$, unknown	0.519–0.722	dye laser (0.460)	p	RT	58
LiF	F$_3^+$	0.52–0.56	dye laser (0.432/0.44)	p	RT	50
LiF	F$_3^+$	0.543	dye laser (~0.460)	p	RT	56
LiF	F$_2$	0.64–0.71	dye laser (0.460)	p	RT	60
LiF	F$_2$	0.67	dye laser (0.436)	p	RT	11
LiF	F$_2$	0.67–0.74	—	p	RT	55
LiF	F$_2^+$	0.82–1.07	Kr ion laser (0.647)	cw	77	26
LiF	F$_2^+$	0.82–1.340	—	p	RT	70
LiF	F$_2^+$**	0.796–1.210	alex laser (0.740), RS D-NdYAG (0.683/0.633)	p	RT	68
LiF	F$_2^+$**	0.830–1.060	DNdYAG laser (0.532)	p	RT	57
LiF	F$_2$–F$_2^+$	0.84–1.10	DNdYAG laser (0.532)	p	RT	15
LiF	F$_3^-$	0.86–1.02	Ti sapphire laser	p	RT	18
LiF	F$_2^+$	0.88–0.995	laser diode (0.680)	p	RT	69
LiF	F$_2^+$	0.88–1.12	ruby laser (0.694)	p	RT	14
LiF(b)	F$_2^+$	0.936	DNdYAG laser (0.532)	p	RT	62
LiF	F$_2^-$	1.09–1.24	NdYAG laser (1.064)	p	RT	12, 53
LiF	F$_2^-$	1.1–1.2	NdYAG laser (1.064)	cw	313	67
LiF	F$_2^-$	1.122–1.201	laser diode (0.976)	p	RT	71
LiF	F$_2^-$	1.13–1.250	ruby laser (0.694)	p	RT	14

Table 1.4.2—*continued*
Color Center Lasers Arranged by Host Crystal

Host crystal	Active center	Laser wavelength (μm)	Pump source (wavelength-μm)	Mode	Temp. (K)	Ref.
LiF	F_2^-	1.15	Nd glass laser	SML	RT	13
LiF	F_2^-	1.150–1.172	NdYAG laser (1.064)	p	RT	59
LiF:OH⁻	F_2^+	0.84–1.13	ruby laser (0.694)	p	RT	52
LiF:OH⁻,Mg²⁺	F_2^+	0.85–1.040	DNdYAG laser (0.532)	p	RT	51
MgF₂:Na⁺	$(F_2)_A$	0.66	DNdYAG laser (0.532)	p	RT	9, 10
NaCl	F_2^+	1.40–1.56	NdYAG laser (1.064)	cw	77	65
NaCl	$F_2^+:S^{2-}$	1.43–2.00	NdYAG laser (1.064)	cw, ML	77	33
NaCl:K⁺(c)	$F_2^+:O^{2-}$	1.42–1.76	NdYAG laser (1.064)	cw	77	64
NaCl:OH⁻	F_2^+	1.37–1.77	NdYAG laser (1.064)	p	RT	63
NaCl:OH⁻	F_2^+	1.4–1.72	NdYAG laser (1.064)	p, qcw	RT	61
NaCl:OH⁻	$F_2^+:O^{2-}$ (d)	1.41–1.81	NdYAG laser (1.064)	cw	77	30, 31
NaCl:OH⁻	$F_2^+:O^{2-}$ (d)	1.41–1.81	NdYAG laser (1.064)	PML	77	32
NaCl:OH⁻	$(F_2^+)_H$	1.450–1.600(e)	NdYAG laser (1.064)	cw	30	34
NaCl:OH⁻	$(F_2^+)_H$	1.482–1.680	laser diode (0.99)	cw	≤140	35
NaCl:OH⁻	$(F_2^+)_H$	1.479–1.705	NdYAG laser (1.064)	cw	≤185	35
NaCl:OH⁻	$(F_2^+)_H$	~1.575	NdYAG laser (1.064)	PML	77	54
NaCl:Tl⁺	Tl⁰(1)	1.464–1.590	NdYAG laser (1.064)	cw	77	35
NaCl:Tl⁺	Tl⁰(1)	1.493–1.540	laser diode (0.993)	cw	77	35
NaF	F_2^+	0.99–1.22	CCL (0.870)	cw	77	20
NaF:Li⁺	$(F_2^+)_A$	0.95–1.3	ruby laser (0.694)	p	RT	19
NaF:Mg²⁺	$(F_2^+)^*$	1.03–1.12	laser diode (0.82)	cw	77	21
NaF:Mg²⁺	$(F_2^+)^*$	1.07	laser diode (0.82)	AML	77	22
NaF:OH⁻	F_2^+	1.10–1.30	CCL (0.90)	cw	77	23
RbCl:Li⁺	$F_A(II)$	2.48–3.64	Kr ion (0.647, 0.676)	cw	77	45
RbCl:Li⁺	$F_A(II)$	2.6–3.33	Kr ion (0.647, 0.752)	cw	77	72
RbCl:Na⁺	$F_B(II)$	2.5–2.9	Kr ion (0.647, 0.752)	cw	77	72

$RbCl:Li^+:Na^+$	$F_A(II)-F_B(II)$	2.5–3.15	Kr ion (0.647, 0.752)	cw	77	72
$RbI:Li^+$	$(F_2^+)_A$	2.84–3.68	Er:YLF (1.73)	p	77	77
$SrF_2:Na^+$	$(F_2)_A$	0.890	ruby laser (0.694)	p	RT	8

*   Color centers produced by neutron irradiation.
(a) Laser action requires further verification.
(b) LiF powder sample.
(c) Misidentified; the host crystal was actually $NaCl:OH^-$.
(d) The $NaCl:OH^-$ $F_2^+:O^{2-}$ laser and the $NaCl:OH^-$ $(F_2^+)_H$ laser are the same; there is no standard nomenclature for these centers.
(e) Emission of the (a) variety of $(F_2^+)_H$ center.

1.4.4 Commercial Color Center Lasers

Examples of commercial color center lasers are given in Table 1.4.3. General output properties are included. These data are taken from recent (1997–1999) buyers guides and manufacturers' literature and are representative rather than exhaustive. Performance figures may be expected to change due to changes and advances in technology.

Table 1.4.3
Commercial Color Center Lasers

Laser Type	Operation	Principal wavelengths (μm)	Output
LiF (F_2^+)	pulsed	1.09–1.27	£ 50 mJ
NaCl:OH (F_2^-)	cw	1.45–1.85	0.35 W
	pulsed	1.48–1.72	0.1 J
KCl:Na (F_B)	cw	2.30–2.55	1–100 mW
KCl:Li (F_A)	cw	2.45–2.80	100 mW
	pulsed	2.52–2.90	15 mJ
RbCl:Li(F_A)	cw	2.70–3.30	10 mW
	pulsed	2.73–3.18	15 mJ

1.4.5 References

1. Henderson, B., Tunable visible laser using F^+ centers in oxides, *Opt. Lett.* 6, 437 (1981).
2. Voytovich, A. P., Kalinov, V. S., Michonov, S. A., and Ovseichuk, S. I., Investigation of spectral and energy characteristics of green radiation generated in LiF with radiation color centers, *Kvant. Elektron.* 14, 1225 (1987); *Sov. J. Quantum Electron.* 17, 780 (1987).
3. Rand, S. C. and DeShazer, L. G., Visible color-center laser in diamond, *Opt. Lett.* 10, 481 (1985).
4. Martynovich, E. F., Baryshnikov, V. I., and Grigorov, V. A., Lasing in Al_2O_3 color centers at room temperature in the visible, *Opt. Comm.* 53, 257 (1985); *Sov. Tech. Phys. Lett.* 11, 81 (1985).
5. Martynovich, E. F., Tokarev, A. G., and Grigorov, V. A., Al_2O_3 color center lasing in near infrared at 300 K. *Opt. Commun.* 53, 254 (1985); *Sov. Phys. Tech. Phys.* 30, 243 (1985).
6. Voytovich, A. P., Grinkevich, V. A., Kalinov, V. S., Kononov, V. A., and Mikhnov, S. A., Spectroscopic and lasing characteristics of sapphire crystals containing color centers in the 1.0 μm range, *Sov. J. Quantum Electron.* 18, 202 (1988).
7. Boiko, B. B., Shakadarevich, A. P., Zdanov, E. A., Kalosha, I.I., Koptev, V. G., and Demidovich, A. A., Laser action of color centers in the Al_2O_3:Mg crystal, *Kvant. Electron.* 14, 914 (1987); *Sov. J. Quantum Electron.* 17, 581 (1987).
8. Arkhangel'skaya, V. A., Fedorov, A. A., and Feofilov, P. P., Spontaneous and stimulated emission of color centers in MeF_2-Na crystals, *Optica i Spectroskopiya* 44, 409 (1978); *Sov. Opt. Spectroscopy* 44, 240 (1978).
9. Arkhangel'skaya, V. A., Fedorov, A. A., and Feofilov, P. P., Luminescence and stimulated emission of M color centers in fluoride type crystals, *Izv. Akad. Nauk SSR, Ser Fiz.* 43, 1119 (1979); *Bull. Acad. Sci. USSR. Phys. Ser.* 43, 14 (1979).

10. Shkadarevich, A. P. and Yarmolkeich, A. P., New laser media in color centers of compound fluorides, *Inst. Phys. An BSSR, Minsk, USSR* (1985).

11. Gusev, Yu. L., Konoplin, S. N., and Marennikov, S. I., Laser radiation in F_2 color centers in an LiF single crystal, *Sov. J. Quantum Electron.* 7, 1157 (1977).

12. Basiev, T. T., Ershov, B. V., Kratstev, S. B., Mirov, Spiridonov, V. A., and Fedorov, V. B., Color center LiF laser with the oput energy of 100 J, *Sov. J. Quantum Electron.* 15, 745 (1985).

13. Babushkin, A. V., Basiev, T. T., Vorob'ev, N. S. et al., Generation and recording of high-power smoothly tunable ps radiation from $LiF:F_2^-$ laser, *Sov. J. Quantum Electron.* 16, 1492 (1986).

14. Gusev, Yu. L., Marennikov, S. I., and Chebotayev, V. P., Tunable laser via F_2^+ and F_2^- colour centers in the spectral region 0.88–1.25 μm, *Appl. Phys.* 14, 121 (1977).

15. Basiev, T. T., Karpushko, F. V., Kulaschik, S. M., Mirov, S. B., Morozov, V. P., Motkin, V. S., Saskevich, N. A., and Sinitsin, G. V., Automatic tunable MASLAN-201 laser, *Kvant. Elektron.* 14, 1726 (1987); *Sov. J. Quantum Electron.* 17, 1102 (1987).

16. Horsch, G. and Paus, H. J., A new color center laser on the basis of lead-doped $KMgF_3$, *Opt. Comm.* 60, 69 (1986).

17. Flassak, W., Goth, A., Horsch, G., and Paus, H. J., Tunable color center lasers with lead- and copper-doped $KMgF_3$, *IEEE J. Quantum Electron.* QE-24, 1070 (1988).

18. Shkadarevich, A. P., Demidovich, A. A., and Protassenya, A. L., Tunable lasers on F_3^--colour centers in LiF Crystals, *OSA Proc. Adv. Solid State Lasers*, Dubé, G. and Chase, L., Eds., 10, 153 (1991).

19. Gusev, Y. L., Kirpichnikov, S. N., Konoplin, S. N., and Marennikov, S. I., Laser operation in $(F_2^+)_A$ color centers in NaF crystal, *Sov. J. Quantum Electron.* 11, 833 (1981).

20. Mollenauer, L. F., Room-temperature-stable, F_2-like center yields cw laser tunable over the 0.9-1.22 μm range, *Opt. Lett.* 5, 188 (1980).

21. Doualan, J. L., Colour centre laser pumped by a laser diode, *Opt. Commun.* 70, 232 (1991).

22. Mazighi, K., Doualan, J. L., Hamel, J., Margerie, J., Mounier, D., and Ostrovsky, A., Active mode-locked operation of a diode pumped colour-centre laser, *Opt. Commun.* 85, 234 (1997).

23. Mollenauer, L. F., Laser-active defect stabilized F_2^+ center in NaF:OH and dynamics of defect-stabilized center formation, *Opt. Lett.* 6, 342 (1981).

24. Georgiou, E., Carrig, T. J., and Pollock, C. R., Stable, pulsed, color-center laser in pure KCl tunable from 1.23 to 1.35 μm, *Opt. Lett.* 13, 978 (1988).

25. Mollenauer, L. F. and Bloom, D. M., Color center laser generates picosecond pulses and several watts cw over the 1.24–1.45 μm range, *Opt. Lett.* 4, 247 (1979).

26. Mollenauer, L. F., Bloom, D. M., and DelGaudio, A. M., Broadly tunable cw lasers using F_2^+ centers for the 1.26–1.48 and 0.82–1.07 μm bands, *Opt. Lett.* 3, 48 (1978).

27. Pinto, J. F., Yakymyshyn, C. P., and Pollock, C. R., Acosto-optic mode-locked soliton laser, *Optics Lett.* 13, 383 (1988).

28. Gellerman, W., Luty, F., and Pollock, C. R., Optical properties and stable broadly tunable cw laser operation of new F_A-type centers in Tl^+-doped alkali-halides, *Opt. Commun.* 39, 391 (1981).

29. Mollenauer, L. F., Vieira, N. D., and Szeto, L., Mode locking by synchronous pumping using a gain medium with microsecond decay times, *Opt. Lett.* 7, 414 (1982).

30. Pinto, J. F., Georgiou, E., and Pollock, C. R., Stable color-center in OH^-doped NaCl operating in the 1.41—1.81-μm region, *Opt. Lett.* 11, 519 (1986).

31. Georgiou, E., Pinto, J. F., and Pollock, C. R., Optical properties and formation of oxygen-perturbed F_2^+ color center in NaCl, *Phys. Rev. B* 35, 7636 (1987).

32. Islam, M. N., Sunderman, E. R., Bar-Joseph, I., Sauer, N., and Chang, T. Y., Multiple quantum well passive mode locking of a NaCl color center laser, *Appl. Phys. Lett.* 54, 1203 (1989).

33. Möllmann, K. and Gellermann, W., Optical and laser properties of $(F_2^+)_H$ centers in sulfur-doped NaCl, *Opt. Lett.* 19, 804 (1994).

34. Konaté,, A., Doualan, J. L., Girard, S., and Margerie, J., Tunable cw laser emission of the (a) variety of $(F_2^+)_H$ centres in $NaCl:OH^-$, *Opt. Commun.* 133, 234 (1997).

35. Konaté, A., Donalan, J. I., Girard, S., Margerie, J., and Vicquelin, R., Diode-pumped colour-centre lasers tunable in the 1.5 μm range, *Appl. Phys. B* 62, 437 (1996).

36. Konaté, A., Doualan, J. I., and Margerie, J., Laser diode pumping of a colour centre laser with emission in the 1.5 μm wavelength domain, *Rad. Effects Def. Solids* 136, 61 (1995).
37. Schneider, I. and Marrone, M. J., Continuous-wave laser action of $(F_2^+)_A$ centers in sodium-doped KCl crystals, *Opt. Lett.* 4, 390 (1979).
38. Wandt, D., Gellerman, W., Luty, F., and Welling, H., Tunable cw laser action in the 1.45—2.16 μm range based on F_2^+-like center in O_2^-- doped NaCl, KCl, and KBr crystals, *J. Appl. Phys.* 61, 864 (1987).
39. Wandt, D. and Gellerman, W., Efficient cw color center laser operation in the 1.7 to 2.2 μm range based on F_2^+-like centers in $KCl:Na^+:O_2^-$ crystals, *Opt. Commun.* 61, 405 (1987).
40. Möllmann, K., Mitachke, F., and Gellermann, W., Optical properties and synchronously pumped mode locked 1.73-2.10 μm tunable laser operation of $(F_2^+)_{AH}$ centers in $KCl:Na^+:O_2^-$, *Opt. Commun.* 83, 177 (1991).
41. Doualan, J. L. and Gellerman, 4-W continuous-wave color-center laser pumps a $KBr:O_2^-$ $(F_A^+)_H$ center laser, in *Advanced Solid-State Lasers*, Jenssen, H. P. and Dubé, G., Eds., Proceedings Vol. 6 (Optical Society of America, Washington, DC (1990), p. 276.
42. Möllmann, K. Schrempel, M., Yu, B-K., and Gellermann, W., Subpicosecond and continuous-wave laser operation of $(F_A^+)_H$ and $(F_A^+)_{AH}$ color-center lasers in the 2-μm range, *Opt. Lett.* 19, 960 (1994).
43. Schneider, I. and Marquardt, C. L., Tunable, cw laser action using (F_2^+) centers in Li-doped KCl, *Opt. Lett.* 5, 214 (1980).
44. Litfin, G., Beigang, R., and Welling, H., Tunable cw laser operation in $F_B(II)$ type color center crystals, *Appl. Phys. Lett.* 31, 381 (1977).
45. German, K., Optimization of F_A (II) and F_B (II) color-center lasers, *J. Opt. Soc. Am.* B3, 149 (1986).
46. Schneider, I., Continuous tuning of a color-center laser between 2 and 4 μm, *Opt. Lett.* 7, 271 (1982).
47. Tkach, R. W., Gosnell, T. R., and Sievers, A. J., Solid-state vibrational laser—KBr :CN⁻, *Opt. Lett.* 9, 122, 1984).
48. Gosnell, T. R., Sievers, A. J., and Pollock, C. R., Continuous-wave operation of the KBr:CN⁻ solid-state vibration laser in the 5-μm region, *Opt. Lett.* 10, 125 (1985).
49. Gellerman, W., Yang, Y., and Luty, F., Laser operation near 5-μm of vibrationally excited F-center CN molecule defect pairs in CsCl crystals, pumped in the visible, *Opt. Commun.* 57, 196 (1986).
50. Tsuboi, T. and Ter-Mikirtychev, V. V., Characteristics of the $LiF:F_3^+$ color center laser, *Opt. Commun.* 116, 389 (1995).
51. Ter-Mikirtychev, V. V. and Tsuboi, T., Ultrabroadband $LiF:F_2^+$ color center laser using two-prism spatially-dispersive resonator, *Opt. Commun.* 137, 74 (1997).
52. Khulugurov, V. M. and Lobanov, B. D., Color-center lasing at 0.84–1.13 μm in a LiF–OH crystal at 300 K, *Sov. Tech. Phys. Lett.* 4, 595 (1978).
53. Basiev, T. T., Zverov, P. G., Fedorov, V. V., and Mirov, S. B., Multiline, superbroadband and sun-color oscillation of a $LiF:F_2^-$ color-center laser, *Appl. Opt.* 36, 2515 (1997).
54. Kennedy, G. T., Grant, R. S., and Sibbett, W., Self-mode-locked NaCl:OH⁻ color-center laser, *Opt. Lett.* 18, 1736 (1993).
55. Basiev, T. T., Mirov, S. B., and Osiko, V. V., Room-temperature color center lasers, *IEEE J. Quantum Electron.* 24, 1052 (1988).
56. Tsuboi, T. and Gu, H. E., Room-temperature-stable $LiF:F_3^+$ color-center laser with a two-mirror cavity, *Appl. Opt.* 33, 982 (1994).
57. Ter-Mikirtychev, V. V., Stable room-temperature $LiF:F_2^{+*}$ tunable color-center laser for the 830-1060-nm spectral range pumped by second-harmonic radiation from a neodymium laser, *Appl. Opt.* 34, 6114 (1995).
58. Gu, H.-E., Qi, L., and Wan, L.-F., Broadly tunable laser using some mixed centers in an LiF crystal for the 520-720 band, *Opt. Commun.* 67, 237 (1988).
59. Ter-Mikirtychau, V. V., Arestova, E. L., and Tsuboi, T., Tunable $LiF:F_2^-$ color center laser with an intracavity integrated-optic output coupler, *J. Lightwave Technol.* 14, 2353 (1996).

60. Gu, H.-E., Qi, L., Guo, S., and Wan, L.-F., A LiF crystal F_3^+–F_2 mixed color-center laser, *Chin. Phys.* 11, 148 (1991).
61. Matts, R. É, Stable laser based on color centers in the OH:NaCl crystal tunable in the range 1.4 to 1.7 μm and operating at room temperature, *Quantum Electron.* 23, 44 (1993).
62. Noginov, M. A., Noginova, N. E., Egarievwe, S. U., Caulfield, H. J., Venkateswarlu, P., Williams, A., and Mirov, S. B., Color-center powder laser: the effect of pulverization on color-center characteristics, *J. Opt. Soc. Am. B* 14, 2153 (1997).
63. Culpepper, C. F., Carrig, T. J., Pinto, J. F., Georgiou, E., and Pollock, C. R., Pulsed, room-temperature operation of a tunable NaCl color-center laser, *Opt. Lett.* 12, 882 (1987).
64. Pinto, J. F., Stratton, L. W., and Pollock, C. R, Stable color-center laser in K-doped NaCl tunable from 1.42 to 1.76 μm, *Opt. Lett.* 10, 384 (1985).
65. Gellermann, W., Lutz, F., Koch, K. P., and Littin, M., F_2^+ center stabilization and tunable laser operation in OH⁻ doped alkali halide, *Phys. Stat. Sol.* (a) 57, 411 (1980).
66. Steinmeyer, G., Morgner, U., Ostermeyer, M., Mitschke, F., and Welling, H., Subpicosecond pulses near 1.9 μm from a synchronously pumped color-center laser, *Optics Lett.* 18, 1544 (1993).
67. Basiev, T. T., Gusev, A. A., Kruzhalov, S. V., Mirov, S. B., and Petrun'kin, V. Yu., Continuous-wave ring LiF:F_2^- laser, *Sov. J. Quantum Electron.* 18, 315 (1988).
68. Dergachev, A. Yu. and Mirov, S. B., Efficient room temperature LiF:F_2^{+**} color center laser tunable in 820–1210 nm range, *Optics Commun.* 147, 107 (1998).
69. Ter-Mikirtychev, V. V., Diode-pumped LiF:F_2^{+*} color center laser tunable in 880–995-nm region at room temperature, *IEEE Phot. Techn. Lett.* 10, 1395 (1998).
70. Ter-Mikirtychev, V. V. and Tsuboi, T., Stable room-temperature tunable color center lasers and passive Q-switches, *Progr. Quantum Electron.* 20, 219 (1996).
71. Ter-Mikirtychev, V. V., Diode-pumped tunable room-temperature LiF:F_2^- color center laser, *Appl. Opt.* 37, 6442 (1998).
72. Koch, K.-P., Litfin, G., and Welling, H., Continuous-wave laser oscillation with extended tuning range in F_A(II)-F_B(II) color-center crystals, *Optics Lett.* 4, 387 (1979).
73. Baldacchini, G, New luminescent phenomena in color centers, *Physics and Chemistry of Luminescent Materials,* Struck, C. W., Mishra, K. C., and Di Bartolo, B., Eds., Electrochemical Society, Pennington, NJ (1999), p. 319.
74. Fritz, B. and Menke, E., Laser effect in KCl with F_A(Li) centers, *Solid State Commun.* 3, 61 (1965).
75. Mollenauer, L. F. and Olson, D. H., A broadly tunable cw laser using color centers, *Appl. Phys. Lett.* 24, 386 (1974).
76. Gellermann, W., Color center lasers, *J. Phys. Chem. Solids* 52, 249 (1991).
77. Foster, D. R. and Schneider, I., Recent progress in the development of $(F_2^+)_A$ color center lasers, in *Tunable Solid-State Lasers* II, Budgor, A. B., Esterowitz, L., and DeShazer, L. G., Eds., Springer-Verlag, New York (1986), p. 266.

Section 1.5
SEMICONDUCTOR LASERS

1.5.1 Introduction

Laser action in semiconductor diode lasers, in contrast to other solid state lasers, is associated with radiative recombination of electrons and holes at the junction of a n-type material (excess electrons) and a p-type material (excess holes). Excess charge is injected into the active region via an external electric field applied across a simple p-n junction (homojunction) or in a heterostructure consisting of several layers of semiconductor materials that have different band gap energies but are lattice matched. The ability to grow special structures one atomic layer at a time by liquid phase epitaxy (LPE), molecular bean epitaxy (MBE), and metal-organic chemical vapor deposition (MOCVD) has led to an explosive growth of activity and numerous new laser structures and configurations.

When the dimensions of the semiconductor material become <100 nm, quantum effects enter that modify the band gap. Quantum wells result from confinement in one dimension, quantum wires from confinement in two dimensions, and quantum dots or boxes from confinement in three dimensions. The wavelength of quantum well lasers can be changed by varying the quantum well thickness or the composition of the active material. By using materials of different lattice constants, thereby effectively straining the materials, one can further engineer the band gap.

The lasing material may be elemental, but more generally is a binary, ternary, or quaternary compound semiconductor. The latter includes II-VI, III-V, IV-VI, and other compounds. Figure 1.5.1 shows the elements that have been used as constituents to achieve laser action in elemental and compound semiconductor materials.

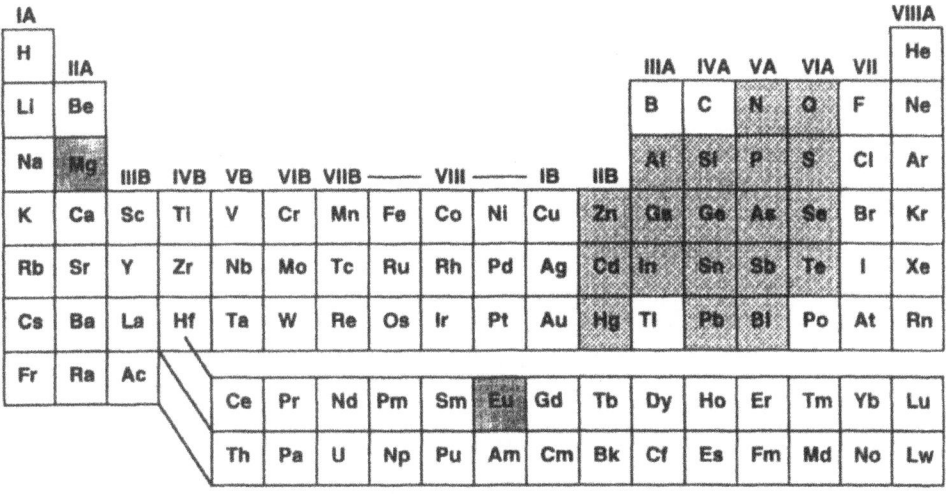

Figure 1.5.1   Periodic table of the elements showing the elements (shaded) that have been components of semiconductor laser materials.

Semiconductor lasers are divided by material type and listed in Tables 1.5.1–1.5.8. The lasers are furthered grouped by the method of excitation (injection, optically pumped, electron beam pumped). Quantum cascade and intersubband lasers are listed in a separate table (Table 1.5.9). Vertical cavity lasers are also listed in a separate table (Table 1.5.10). The tables include the lasing material, wavelength, structure, operating mode, temperature, and primary references. For lasers that have been tuned over a range of wavelengths, the tuning range given is that for the configuration and conditions used and may not represent the extremes possible. The lasing wavelength and output of semiconductor lasers depend on the chemical composition of the material, structural configuration, optical cavity, temperature, excitation rate, and other operating conditions. The original references should therefore be consulted for this information and its effect on the laser performance.

Because it is possible to vary the constituent elements and tailor the laser emission, the wavelength of semiconductor lasers is a less fundamental property than for other lasers involving transitions between specific atomic levels. Thus the tabulations generally include early pioneering papers and representative examples of different structures, preparation methods, and operating conditions rather than an exhaustive listing of all reported lasers.

The wavelength ranges of various types of semiconductor lasers are shown in Figure 1.5.2. The wavelength of quantum cascade lasers, unlike that of diode lasers, is determined by the active layer thickness rather than the band gap of the material. Multiple quantum well cascade lasers have been tailored to operate in the range ~3-17 μm, thereby extending the range of III-V compound lasers.

Only inorganic semiconducting materials are listed in this section. Dye-doped organic semiconductor lasers are included in Section 1.3; semiconducting polymer lasers are covered in Section 1.6. Commercial lasers are covered in Section 1.5.12.

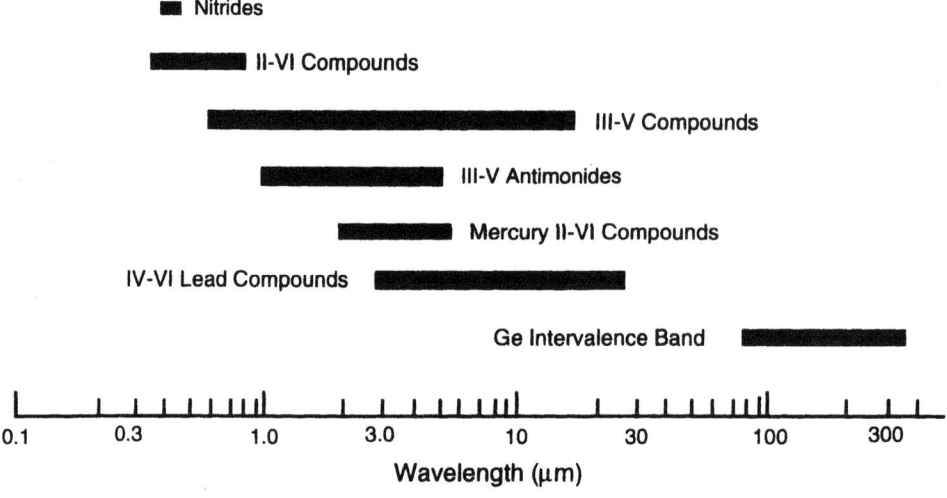

Figure 1.5.2 Reported ranges of output wavelengths of various types of semiconductor lasers. Quantum cascade lasers are included among the III-V compound lasers.

Further Reading

Amann, M.-C. and Buus, J., *Tunable Laser Diodes*, Artech House Publishers, Norwood, MA (1998).

Agrawal, G. P., Ed., *Semiconductor Lasers, Past, Present and Future*, AIP Press, Woodbury, NY (1995).

Botez, D. and Scifres, D. R., *Diode Laser Arrays*, Cambridge University Press, Cambridge (1994).

Carroll, J. E., Whiteaway, J. E. A., and Plumb, R. G. S., *Distributed Feedback Semiconductor Lasers*, IEE, Edison, NJ (1998).

Casey, H. C., Jr. and Parish, M. B., *Heterostucture Lasers, Part A: Fundamental Principles*, Academic Press, Orlando (1978).

Casey, H. C., Jr. and Parish, M. B., *Heterostucture Lasers, Part B: Materials and Operating Characteristics*, Academic Press, Orlando (1978).

Chang-Hasnain, C. J., Ed., Advances of VCSELs, *Optical Society of America Trends in Optics and Photonics Series*, Washington, DC (1997).

Chow, W. W., Koch, S. W., and Sargent, III, M., *Semiconductor Laser Physics*, 2nd. edition, Springer-Verlag, Berlin (1998).

Coleman, J. J., Ed., *Selected Papers on Semiconductor Diode Lasers*, SPIE Milestone Series, Vol. MS50, SPIE Optical Engineering Press, Bellingham, WA (1992).

Derry, P., Figueroa, L., and Hong, C. S., Semiconductor Lasers, in *Handbook of Optics*, Vol. 1, 2nd edition, McGraw-Hill, New York (1995), chapter 13.

Kapon, E., *Semiconductor Lasers I: Fundamentals* and *II: Materials and Structures*, Academic Press, New York (1998).

Manasreh, M.O., Ed., *Antimonide Related Heterostructures and Their Applications*, Gordon and Breach, New York (1997).

Nakamura, S. and Fasol, G., *The Blue Laser Diode: GaN Based Light Emitters and Lasers*, Springer-Verlag, Heidelburg (1997).

Nurmikko, A. V. and Gunshor, R. L., Physics and Device Science in II-VI Semiconductor Visible Light Emitters, in *Solid State Physics* 49, 205 (1995).

Partin, D. L., Lead salt quantum effect structures, *IEEE J. Quantum Electron.* 24, 1716 (1988).

Wilmsen, C., Temkin, H., and Coldren, L. A., Eds., *Vertical-Cavity Surface-Emitting Lasers: Design, Fabrication, and Applications*, Cambridge University Press, Cambridge, UK (1999).

Zory, Jr., P. S., Ed., *Quantum Well Lasers*, Academic Press, San Diego, CA (1993).

See, also, Far Infrared Semiconductor Lasers, special edition of the *Journal of Optical and Quantum Electronic* 23 (1991); Special Issue on Semiconductor Lasers, *IEEE Journal of Quantum Electronics*, (June 1993); Semiconductor Lasers, *Selected Topics in Quantum Electronics* 1 (June 1995); and Semiconductor Lasers, *Selected Topics in Quantum Electronics* 3 (April 1997).

MRS Internet Journal of Nitride Semiconductor Research (www.mrs.org)

Tables in this section are presented in the following order:

Type	Table
II-VI Compound Lasers	1.5.1
Mercury II-VI Compound Lasers	1.5.2
III-V Compound Lasers	1.5.3
III-V Compound Antimonide Lasers	1.5.4
Nitride Lasers	1.5.5
Lead IV-VI Compound Lasers	1.5.6
Germanium-Silicon Intervalence Band Lasers	1.5.7
Other Semiconductor Lasers	1.5.8
Quantum Cascade and Intersubband Lasers	1.5.9
Vertical Cavity Lasers	1.5.10

Abbreviations used to describe the laser structures and laser operation:

ADQW—asymmetric dual quantum well

BGSL—broken-gap superlattice

BH—buried heterostructure

BW—broad-waveguide

CMC—coupled microcavity

cw—continuous wave

DBR—distributed Bragg reflector

DFB—distributed feedback

DH—double heterostructure

DQW—double quantum well

GRIN—graded index

H—heterostructure

J—*p-n* junction

MC—microcavity or microcylinder

MD—microdisk

ML—monolayer or microlaser

MQW—multiple quantum well

p—pulsed

QB—quantum box

QC—quantum cascade

qcw—quasi-continuous wave

QD—quantum dot

QW—quantum well

RW—ridge waveguide

SB-BGSL — strain-balanced BGSL

SCBH—separate confinement buried H

SCH—separate confinement heterostructure

SL-MQW—strained-layer MQW

SLS—strained-layer superlattice

SL—superlattice

SQW—single quantum well

SSQW—strained SQW

T2QWL — type II quantum well laser

VCSEL—vertical cavity surface emitting laser

VC—vertical cavity

"W"—W (conduction band profile) active region

Section 1.5.2 II-VI Compound Lasers

Mercury-based II-VI compound lasers are tabulated in Table 1.5.2.

Table 1.5.1
II-VI Compound Lasers

Material	Wavelength (μm)	Structure	Mode	Temp. (K)	Ref.
Injection Lasers					
BeZnCdSe/BeZnSe/BeMgZnSe	0.507	SCH	p	77	268
BeMgZnSe/ZnCdSe	0.521	SCH QW	cw	300	353
CdZnS/ZnS	0.3755	SLS MQW	p	30	273
CdZnSe	0.490	SQW	p	77	38
CdZnSe/ZnSe	0.490–0.512	QW	cw	80	39
CdZnSe/ZnSe	0.508–0.535	QW	cw	300	39
(Zn,Cd)Se/Zn(S,Se)	~0.494	MQW	p	≤ 300	43
(Zn,Cd)Se/ZnSe	0.480–0.500	MQW	p	≤ 250	34
ZnCdSe/ZnSSe/ZnMgSSe	0.496	SCH	cw	85	269
ZnCdSe/ZnSSe/ZnMgSSe	0.5147	SQW SCH	cw	300	380
ZnCdSe/ZnSSe/ZnMgSSe	0.516	SCH	p	300–394	269
ZnCdSe/ZnSSe/ZnMgSSe	0.520	SCH	p	300	54
ZnCdSe/ZnSSe/ZnMgSSe	0.5235	SCH	p	300	270
ZnSe/ZnMgSSe	0.463	SCH	p	300	267
Optically Pumped Lasers					
CdS	0.4943	crystal	p	88	43
CdS	0.495–0.520	crystal	p	90–300	11
CdS	0.496, 0.502	crystal	p	77	46
CdSe	0.6917	crystal	p	77	90
CdSe	0.697	crystal	p	77	93
CdSe	0.698–0.752	crystal	p	90–300	11
CdSe/ZnSe	0.4647–0.4663	SL	p	80	29
CdSSe	0.4496–0.624	crystal	p	77	25
CdSSe	0.580–0.705	crystal	p	90–300	11
CdZnS/ZnS	0.333	QW	p	8	274
CdZnS/ZnS	0.357–0.390	SLS MQW	p	300	273
CdZnS/ZnS	0.3749	SLS	p	300	271
(Zn,Cd)Se	0.496	SCH VCSEL	p, qcw	300	45
ZnCdSe/ZnCdMgSe	0.512	GRINSCH	p	300	53
ZnCdSe/ZnSSe	0.504	MQW	p	300	51

Table 1.5.1—*continued*
II-VI Compound Lasers

Material	Wavelength (μm)	Structure	Mode	Temp. (K)	Ref.
ZnCdSe/ZnSe	0.498–0.517	MQW	p, cw	300–10	49
(Zn,Cd)Se/ZnSe	0.468	MQW	p	300	30
(Zn,Cd)Se/ZnSe	0.492	SQW	p	10	41
(Zn,Cd)Se/ZnSe	0.4963	SQW	p	150	41
(Zn,Cd)Se/ZnSe	0.50	SQW	p	250	41
ZnCdTe/ZnSe	0.49–0.56	SQW	p	300	36
ZnCdTe/ZnTe	0.575–0.602	SLS QW	cw	8–310	59
ZnMgSSe/ZnSSe/ZnCdSe	0.470–523	SCH QW	p	300	353
ZnO	0.375–0.400	crystal	p	80–300	11
ZnO	0.3775–0.3780	crystal layer	p	80–180	351
ZnO	0.3747	powder	p	80–180	351
ZnO	≈0.385	powder	p	300	352
ZnO	~0.38	film	p	300	16
ZnS	0.3497	crystal	p	300	3
ZnSe	0.430	crystal	p	≤ 200	23
ZnSe	0.469	crystal	p	300	31
ZnSe	0.469–0.475	crystal	p	300	32
ZnSe/ZnMgBeSe	0.444	SQW	p	300–473	388
ZnSe/ZnMnSe	~0.453	MQW	p	5.5–80	26
ZnSe/ZnSSe	0.445	H	p	< 400	272
ZnSe/ZnSSe	~0.445–0.455	SL	p	14–180	24
ZnSe/ZnSSe	0.462	H	p	≤ 260	28
ZnSSe	~0.438	crystal	p	≤ 200	23

Electron Beam Pumped Lasers

Material	Wavelength (μm)	Structure	Mode	Temp. (K)	Ref.
CdS	~0.49	crystal	p	90, 300	35
CdS	0.491	crystal	p	4.2, 77	40
CdS	0.4966	crystal	p	4	48
CdSe	~0.685	crystal	p	4.2	40
CdSSe	0.49–0.69	crystal	p	4.2, 77	37
CdTe	0.785	crystal	p	10–15	106,107
ZnCdSe	0.485	SQW	p	~20	290
ZnCdSe	0.585–0.620	crystal	p	10–310	64
ZnCdSe/ZnSe	0.4784	GRINSCH	p	83–225	33
ZnO	0.3757	crystal	p	77	12
ZnS	0.3245–0.3300	crystal	p	4.2, 77	1

Table 1.5.1—*continued*
II-VI Compound Lasers

Material	Wavelength (μm)	Structure	Mode	Temp. (K)	Ref.
ZnS	0.33	crystal	p	80	2
ZnSSe	~0.438	crystal	p	≤ 200	23
ZnSe	0.460	crystal	p	100	27
ZnSe/ZnSSE	0.474	SL	p	100	9
ZnSe/ZnSSe	0.454–0.474	SL	p	100	9
ZnTe	0.528	crystal	p	4	55
ZnTe	0.533	crystal	p	110	56

1.5.3 Mercury II-VI Compound Lasers

Table 1.5.2
Mercury II-VI Compound Lasers

Material	Wavelength (μm)	Structure	Mode	Temp. (K)	Ref.
Injection Lasers					
HgCdTe	2.86	DH	p	77	174
HgCdTe	2.9, 3.4, 3.9	DH	p	40–90	176,177
HgCdTe	3.4	DH	p	78	264
HgCdTe	3.6	layer	p	12–90	192
HgMnTe	5.3	H	p	77	220
Optically Pumped Lasers					
HgCdTe	1.89–1.97, 2.18–2.5	MQW SCH	p	300–10	277
HgCdTe	1.9–2.5	MQW GRINSCH	p	300	262
HgCdTe	2.0–2.2	SCH	p	150–300	276
HgCdTe	2.06–2.3, 3–3.3	MQW GRINSCH	p	300–10	277
HgCdTe	2.5	DH	p	110	263
HgCdTe	2.79	layer	p	~12	171
HgCdTe	~2.8	QW DH	p, cw	>60	173
HgCdTe	3.1	DH	p	120	266
HgCdTe	3.39–3.58	DH	p	78	278
HgCdTe	3.4	SCH	p	90	265
HgCdTe/CdTe	2.36	MQW SCH	p	12, 77	167
HgCdTe/CdZnTe	3.2	QW	p	80–154	275
HgZnTe	~5.4	layer	p	50–70	222

1.5.4 III-V Compound Lasers

Additional III-V compound lasers are included in the Section 1.5.10 on interband, intersubband, and cascade lasers (see Table 1.5.9). Antimonide III-V compound lasers are tabulated separately in Table 1.5.4.

Table 1.5.3
III-V Compound Lasers

Material	Wavelength (μm)	Structure	Mode	Temp. (K)	Ref.
Injection Lasers					
AlGaAs	0.680-0.700	DH	cw	300	84
AlGaAs	0.688–0.729	H	p, cw	100	88
AlGaAs	0.750	H	p	77, 273	99
AlGaAs	0.844–0.852	H	p	77, 300	123
AlGaAs	0.845	H	cw	100	88
AlGaAs/GaAs	0.674–0.681	SL QW RW	cw	300	83
AlGaAs/GaAs	0.750–0.855	GRINSCH SQW	cw	300	100
AlGaAs/InGaP	0.615	H	cw	300	71
AlGaAsP	~0.800–0.845	DH	p	77–300	109
AlGaInAs/AlInAs	1.008	DH	p	300	143
AlGaInP	0.5836	DH	cw	77	62
AlGaInP	0.621	mesa stripe	cw	273	72
AlGaInP/AlGaInP	0.6262	DH	p	300	74
AlGaInP	0.640	H	cw	293	79
AlGaInP/GaInP/AlGaInP	0.6897	DH	cw	300	89
GaAlAs/GaAs	0.843–0.8562	SH	cw	77	122
GaAs	0.69–0.79	SL QW	cw	300	83
GaAs	0.825	SQW	cw	300	83
GaAs	0.837–0.843	H	p	4.2, 77	114
GaAs	0.842	J	p	77	120
GaAs	0.843	H	p	77	121
GaAs	0.87	DH	cw	300	83
GaAs/AlGaAs	~0.765–0.855	ADQW	p, cw	300	385
GaAs/AlGaAs	0.7812	DH	p	77	105
GaAs/AlGaAs	0.8404	SCH	cw	300	94
GaAs/GaAlAs	0.680–0.785	SL SQW GRIN.	p, cw	300	85,86
GaAs/GaAlAs	0.848	SH DFB	p	77	298
GaAsP	0.710	J	p	77	98
GaAsP/InGaP	0.7010	H	cw	283	95

Table 1.5.3—*continued*
III-V Compound Lasers

Material	Wavelength (μm)	Structure	Mode	Temp. (K)	Ref
GaInP/AlInP	0.576–0.584	MQW	p	109–165	60
GaInP/AlInP	0.607	SSQW DFB	p	140	70
GaInP/GaAlInP	0.627–0.640	SQW GRINSCH	p	300	75
GaInP/AlGaInP	0.671	DH	cw	293	82
GaSb	1.55–1.60	J	p	78	141
GaSb	1.57	J	p	77	156
In(Al,GaP	~0.640	MQW H	cw	300	80
InAlAs	0.707	QD H	cw	77	97
InAs	0.982–0.992	GRINSCH QB	cw	79	139
InAs	1.05–1.24	QW QD[a]	p	295	375
InAs	3.112	J	p	4, 77	185
InAs/GaAs	1.00–1.05	SCH QD	cw	100–300	142
InAs/GaSb/GaInSb/GaSb	2.9	type-II "W" SL	p, cw	≤260	299
InAs/GaInSb	3.1	MQW	p	190	172
InAs/GaInSb/InAs/ AlGaInAsSb	2.7	SB-BGSL	p	180	279
InGa/GaAs	1.074	SL QW BH	p	300	144
InGaAlP	0.625	MQW	cw	300	73
InGaAlP	0.63–0.65	H	cw	295–323	76
InGaAlP	0.6378	SBR	cw	298	78
InGaAs	0.911	QD	p	80	130
InGaAs	0.979	SQW	cw	300	137
InGaAs	1.77, 2.07	J	cw	1.9	160
InGaAsPN	1.2–1.3	SQW	p	300	373
InGaAs/AlGaAs	0.956, 0.985	SCH[b]	p	300	382
InGaAs/AlGaAs	~0.955	SQW SCH	cw	300	374
InGaAs/GaAs	0.99	SL QW	cw	300	140
InGaAs/GaAs	1.028	SL SCH QD	cw	300	146
InGaAs/GaAs	1.14, 1.19	QD	cw	80–250	341
InGaAs/GaAs/InGaP	~0.978–0.984	RW QW	cw	300	136
InGaAs/GaAs/InGaP	0.99	SCH QW	qcw	278–288	343
InGaAs/GaInAsP	1.5	SL-MQW	cw	283–373	47
InGaAs/GaInAsP/GaInP	0.981–0.985	SCH QW	cw	300	348
InGaAs/InGaAs	2.051–2.056	MQW DFB	cw	298	358
InGaAs/InP	1.440–1.640	MQW GRINSCH	cw	300	151
InGaAs/InP	1.62	SQW GRINSCH	cw	300	159

Table 1.5.3—*continued*
III-V Compound Lasers

Material	Wavelength (μm)	Structure	Mode	Temp. (K)	Ref.
InGaAsP/GaAs	0.808	SQW	cw	300	110
InGaAsP/InaP[(c)]	1.281	SCH MQW	p	300	357
InGaAsP/InGaP	0.83	QW H	cw	300	112
InGaAsP/In(Ga,Al)P	0.735	SQW	cw	288	300
InGaAsP/In(Ga,Al)P	0.81	QW	cw	293	301
InGaAsP/InP	1.27	DH	p	300	148
InGaAsP/InP	1.31	BH	p	30	150
InGaP	0.761, 0.763	J	p	4.2, 77	101
InGaP/InP	1.43	BW SCH MQW	cw, qcw	~300	344
InGaP:N	0.5520	crystal	cw	77	58
InGaPAs/GaAs	~0.785	H	cw	300	108
InP	0.906–0.908	crystal	p	4, 77	128,129
InSb	5.085–5.28	J	p	10	215

Optically Pumped Lasers

Material	Wavelength (μm)	Structure	Mode	Temp. (K)	Ref.
AlGaAs/AlGaAs	~0.636	H	p	300	77
GaAs	0.83	crystal	p	300	111
GaAs	0.8365	crystal	p	77	113
GaAs/AlGaAs	~12.5	QW	p	77	244
GaAs/InAs	~0.84–0.86	SCH SSQW	p, cw	77	118
GaSb	1.541	crystal	p	4	154
In(Al,Ga)P	~0.640	MQW H	cw	300	80
InAs	2.94	crystal	p	20	154
InAs	~3.0	crystal	p	4	180
InAs/GaSb/InAs/AlSb	3.08–4.03	type "W"	p	100–360	349
InAs/GaSb/InAs/AlSb	2.72–3.55	type "W"	cw	78, 290	350
InAs/GaInSb/InAs/AlSb	5.4–7.4	type-II "W"	p	77–290	386
InAs/GaSb/InAs/AlSb	3.1–3.4	type-II "W"	p	300	377
InAs/GaInSb/InAs/AlAsSb	3.14–3.02	type "W"	cw	78, 275	350
InAs/GaInSb/InAs/AlSb	3.59–6.07	type-II "W"	cw	78–265	376
InAs/GaInSb/InAs/AlSb	5.4–7.1	type "W"	cw	78–265	350
InAs/GaInSb/InAs/ AlGaInAsSb	3.7, 5.2	SB-BGSL	p	300	279
InAs/GaInSb/InAs/AlSb	3.2, 3.4	T2QWL	p	350, 310	285
InAs/GaInSb/InAs/AlSb	3.4	T2QWL	p	310	285
InAs/GaInSb/InAs/AlSb	3.9–4.1	T2QWL	p	80–285	198
InAs/GaInSb/InAs/AlSb	4.2–4.5	T2QWL	p	100–310	203

Table 1.5.3—*continued*
III-V Compound Lasers

Material	Wavelength (μm)	Structure	Mode	Temp. (K)	Ref.
InAs/InAsSb.	3.3–3.4	type-II SL	cw	95	188
InAs/InAsSbP	3.1	DH	p	77–100	184
InAs/InAsSbP	3.1	DH	p	77–100	304
InGaAs	0.935	crystal	p	300	131
InGaAs/InAlAs	1.5–1.6	MQW	p	300	152
InGaAsP[c]	1.504	QW MC	p	143	356
InGaP	0.695	crystal	p	77	91
InP	0.850	crystal	p	77	126
InP	0.915	crystal	p	300	131
InPAs	1.602	crystal	p	77	158
InSb	5.16–5.32	crystal	cw[d]	20	201
InSb	5.258	crystal	p	4	312
InGaAs/GaAs	0.918–0.966, 0.940–0.978	QW CMC	qcw	300	302
InAs/GaAs	0.838–0.886	ML	cw	10	303
InAs/GaSb/InAs/AlSb	2.9	Type II	p	≤280	304
Electron Beam Pumped Lasers					
AlGaAs	0.696–0.760	layer	p	81	92
GaAs	~0.84	crystal	p	4	115–117
GaAsP	0.704	crystal	p	77	96
GaSb	1.51–1.53	crystal	p	4, 20	153
InAs	2.7–3.2	film	p	80–220	168
InGaP	0.549–0.562	crystal	p	10–150	57

(a) DWELL (dots in a well) design.
(b) Bipolar cascade laser.
(c) Photonic crystal.
(d) Two-photon (9.3 μm) pumped.

1.5.5 III-V Compound Antimonide Lasers

Table 1.5.4
III-V Compound Antimonide Lasers

Material	Wavelength (μm)	Structure	Mode	Temp. (K)	Ref.
Injection Lasers					
GaInAsSb/GaSb	~2.0	DH	cw	80	161
InAsSb/InAsSbP	3.6	DH	cw	77–100	296
InAsSb(P)/InAsSbP	2.7–3.9	DH	cw	80	196
InGaAsSb/AlGaAsSb	2.7	MQW	cw	170–234	283
InGaAsSb/AlGaAsSb	2.0–2.65	SCH QW	cw	300	
GaAsSb/AlGaAsSb	0.980	DH	p	300	138
GaInSb/InAs	2.8	MQW BGSL	p	225	172
GaInSb/InAs	3.1, 3.2	MQW	p	220, 255	172
GaInSb/InAs	3.28–3.90	MQW	p	170–84	187
GaInSb/InAs	3.4,–4.3	MQW	p	195–110	172
InAlSb/InSb	5.1	H	p	≤ 90	214
InAsSb	~3.17	crystal	p	77	186
InAsSb	3.5–3.6	MQW DH	p	77–135	287
InAsSb	~3.6	layer	p	77	184
InAsSb/InAs	3.8–3.9	MQW H	p	210	195
InGaAsSb/AlGaSb	2.023, 2.2	DH	p	140, 300	162
InGaAsSb/InPSb	3.06	DH	p	35	181
GaInAsSb/AlGaAsSb	2.1	DH	p, cw	300, 190	163
GaInAsSb/AlGaAsSb	2.02, 2.04	QW	p, cw	300	342
GaInAsSb/AlGaSb	~2.2	DH	p, cw	303	165
InAsSb/AlAsSb	3.97–3.985	DH	p, cw	155, 80	202
InAsSb/InAlAsSb	3.9	SQW	p, cw	165, 123	197
InAsSb/InAlAsSb	3.2–3.55	MQW	p, cw	225, 175	305
InAsSbP/InAsSb/InAs	~3.2	DH	p, cw	220, 77	288
InAsSbP/InGaAsSb	3.05–3.55	DH	p, cw	77	182
InGaAsSb	3.23	DH	p, cw	77	182
AlGaAsSb/InGaAsSb	2	SQW	qcw	283–288	289
Optically Pumped Lasers					
GaInAsSb/GaSb	3.08–3.30	DH	p	82–210	284
InAsSb/GaSb	3.86–3.97	DH	p	80–150	284
InAsSb/InAs	~3.6	J	p	77	186
AlGaSb/InAsSb/AlGaSb	3.9	DH	p	80–135	12

Table 1.5.4—*continued*
III-V Compound Antimonide Lasers

Material	Wavelength (μm)	Structure	Mode	Temp. (K)	Ref.
InAs/GaSb/InAs/AlSb	2.9	Type II "W"	p	≤280	306
InAsSbP/InAs	3.9	DH	p	77–125	184
InAsSb/InAsP	3.57–3.86	SLS	p	80–240	191
InAsSb	3.7	DH	qcw	77	307

Electron Beam Pumped Lasers

Material	Wavelength (μm)	Structure	Mode	Temp. (K)	Ref.
AlGaSb	1.10–1.60	crystal	p	83	147

1.5.6 Nitride Lasers

Table 1.5.5
Nitride Lasers

Material	Wavelength (μm)	Structure	Mode	Temp. (K)	Ref.
Injection Lasers					
InGaN	0.376	SQW SCH	p	300	199
InGaN	0.399–0.402	MQW	p, cw	300	291
InGaN	0.40583	MQW	cw	300	20
InGaN	0.407–0.411	MQW	p	300	297
InGaN	0.417	MQW	p	300	22
InGaN	0.419	MQW	p	300	44
Optically Pumped Lasers					
AlGaN/GaInN	0.389–0.399	DFB DH	p	300	295
AlGaN/GaInN	0.4025	DH	p	300	19
GaN	~0.356	SCH QD	p	20	4
GaN	0.359	crystal	p	2	5
GaN	0.362–0.381	crystal	p	10–375	7
GaN	0.3696	layer	p	300	10
GaN	0.376–0.378	layer VC	p	300	13,14
GaN	0.378	film	p	300	15
GaN	0.378	MD WGM	p	300	387
GaN	~0.385	ML	p	300	293
GaN	0.364–0.386	layer	p	77–450	334
GaN	—	powder	p	300	352
GaN/AlGaN	~0.3615	SCH	p	300	6

Table 1.5.5—*continued*
Nitride Lasers

Material	Wavelength (μm)	Structure	Mode	Temp. (K)	Ref.
GaN/AlGaN	0.3635	VCSEL	p	300	8
GaN/AlGaN	0.387	H	p	34	294
GaN/AlGaN	~0.359, ~0.365	DH	p	77, 295	243
GaN[a]	0.378–0.380[b]	film	p	1.8	335
GaN[a]	—[b]	film	p	300	336
InGaN	0.406	ML VC	p	300	296
InGaN/GaN	~0.341	MQW	p	10	337
InGaN/GaN	0.385	MQW	p	≤ 220	17
InGaN/GaN	0.415	VC	p	300	21
InGaN/GaN	0.427–0.437	MQW	p	175–575	50
InGaN/GaN/AlGaN`	0.401	VCSEL	p	300	384
InGaN/InGaN	0.410	MQW	p	300	292
InGaNAs/GaInP	1.17–1.26	DH	p	300	178,282

(a) Cubic gallium nitride.
(b) Gain was measured.

Section 1.5.7 Lead IV-VI Compound Lasers

Table 1.5.6
Lead IV-VI Compound Lasers

Material	Wavelength (μm)	Structure	Mode	Temp. (K)	Ref.
Injection Lasers					
PbCdS	3.5	J	cw	10–20	190
PbEuSe/PbSe/PbEuSe	5.55–7.81	BH	p, cw	30–160	223
PbEuSeTe	2.7–6.6	DH	cw	≤ 147	170
PbEuSeTe	2.6–6.6	DH	p	≤ 190	193
PbEuSeTe	3.8–6.6	DH	cw	≤ 147	193
PbEuSeTe/PbTe	4.19–6.49	SCBH	cw	215–20	280,281
PbEuSeTe/PbTe	4.2–6.4	BH	cw	90–203	204
PbEuSeTe/PbTe	3.97	SQW	p	260	170,200
PbEuSeTe/PbTe	4.41–6.45	SQW	cw	174–13	170,200
PbEuSeTe/PbTe	4.5362–5.7026	DH BH	cw	120–180	211
PbEuTe	3.4–6.5	DH	p	> 200	189
PbGeTe	4.40–6.50	J	cw	4	209
PbS	4.32	J	cw	4.2	207,208

Table 1.5.6—*continued*
Lead IV-VI Compound Lasers

Material	Wavelength (µm)	Structure	Mode	Temp. (K)	Ref.
PbSSe	6.1	SH, DH	cw	12	230
PbSe	8.5	J	p	4	240
PbSe/PbEuSe	2.88	DH	cw	100	175
PbSe/PbEuSe	5.7–7.8	DH	p, cw	220, 174	175
PbSe/PbSrSe	4.4–8.0	H	p, cw	290, 169	210
PbSnSe	10.2	J	p, cw	77	241
PbSnTe	6	MQW	p, cw	204, 130	229
PbSnTe	9.4, 12.7, 13.7	J	p, cw	12	241
PbSnTe/PbTeSe	5.90–8.55	MQW	p, cw	~10–204	228
PbSnTe	12.8	DH	p	77–188	189
PbSnTe/PbEuSeTe	4.77–7.18	BHG	cw	20–175	213
PbSnTe/PbTe	8.2–18.5	H	cw	12–80	225,236
PbSnTe/PbTeSe	~9	BH	cw	80	239
PbSrS/PbS	2.79–3.44	MQW	p	255–90	179
PbSrS/PbS	2.95–3.84	DH	p	250–90	179
PbSrS/PbS	2.97	DH	p, cw	245, 174	179
PbSrS/PbSrS	2.77–3.14	DH	p	180–90	179
PbSrSe/PbSePbSrSe	4.78	DH	p	300–333	345
PbTe	6.5	crystal	p	12	232

Optically Pumped Lasers

PbSSe	3.9–8.6	crystal	p[a]	2	18

Electron Beam Pumped Lasers

PbS	~4.3	crystal	p	4.2	206
PbSe	8.5	crystal	p	4.2	206
PbTe	6.41	crystal	p	4.2	206

(a) Optically pumped using a Nd:YAG laser (1.064 µm).

Section 1.5.8 Germanium-Silicon Intervalence Band Lasers

Table 1.5.7
Germanium-Silicon Intervalence Band Lasers

Material	Wavelength (µm)	Structure	Excitation	Temp. (K)[a]	Ref.
Ge(Be),Ge(Zn)	75–250	crystal	elect./mag. fields (p)	4.2	224, 226
Ge(Cu)	80–150	crystal	elect./mag. fields (p)	4.2	224, 226–7
Ge(Ga)	110–360	crystal	elect./mag. fields (p)	4.2	259–261
Ge(Ga)	≈86	crystal	electric field (p,cw)	4–8	389

Table 1.5.7—*continued*
Germanium-Silicon Intervalence Band Lasers

Material	Wavelength (μm)	Structure	Excitation	Temp. (K)[a]	Ref.
Ge(Ga)	90–125	crystal[b]	elect./mag. fields (p)	4.2	372
Ge(Ga),Ge(Al)	75–130	crystal	elect./mag. fields (p)	4.2–18	245–256
Ge(Ga),Ge(Al)	170–250	crystal	elect./mag. fields (p)	4.2–18	245–257
Ge(Tl)	85–110	crystal	elect./mag. fields (p)	4.2	309
Ge(Tl)	120–165	crystal	elect./mag. fields (p)	4.2	309
Si(B)	>50[c]	crystal	elect./mag. fields (p)	4.2	310

(a) For Ge lasers, during each laser pulse the crystal heats up and, while still lasing, reaches a temperature close to 20–25 K (E. Bründermann—private communication).
(b) Faraday configuration.
(c) Calculated optical gain spectrum; long wavelength limit is ~ 600 μm.

Section 1.5.9 Other Semiconductor Lasers

Table 1.5.8
Other Semiconductor Lasers

Material	Wavelength (μm)	Structure	Excitation (mode)	Temp. (K)	Ref.
Bi_xSb_{1-x}	~100	crystal	injection (cw)	4.2	258
Cd_3P_2	~2.12	crystal	optical-1064 nm (p)	4.2	164
$CdIn_2S_4$	0.765[a]	crystal	optical-495 nm[b] (p) optical-600 nm[c] (p)	100–300	102
$CdSiAs_2$	0.77	crystal	electron beam (p)	77	103
$CdSnP_2$	1.011	crystal	electron beam (p)	80	145
CuBr	~0.43	thin film	optical-337 nm (p)	10	308
CuCl	0.3914	quantum dot	optical-337,351 nm (p)	77–108	311
In_2Se	1.60	crystal	electron beam (p)	90	157
GaSe	0.59–0.60	crystal	electron beam (p)	77	65
GaSe	0.59–0.60	crystal	optical-1064 nm[d] (p)	77	66
GaSe	0.6010	crystal	optical-337 nm (p)	2	67,68
GaSe	0.602–0.604	crystal	optical-337 nm (p)	5	69
InSe	0.942, 0.945	crystal	optical-337 nm (p)	5	69
InSe	~0.945–0.985	crystal	optical-583 nm (p)	20, 90	133
InSe	0.97	crystal	electron beam (p)	90	134

(a) Optical gain was measured.
(b) Intrinsic.
(c) Extrinsic.
(d) Two-photon pumped.

1.5.10 Quantum Cascade and Intersubband Lasers

Interband-based cascade lasers (ICL), intersubband-based quantum cascade lasers (QCL), and non-cascaded intersubband lasers are included in this section. In the first, every injected electron generates multiple photons by making interband transitions at each step of the staircase-like quantum well structure; in the second an electron makes transitions between conduction subbands created by quantum confinement. A further distinction is that quantum cascade lasers are electrically pumped whereas quantum fountain lasers are optically pumped. In all cases the wavelength is determined by the layer thickness of the active region rather than the band gap, hence the lasers can be tailored to operate over a broad range of wavelengths in the mid-infrared.

Table 1.5.9
Quantum Cascade and Intersubband Lasers

Material	Wavelength (μm)	Structure	Mode	Temp. (K)	Ref.
AlInAs/GaInAs	4.26	MQW RW	p	10–90	205
AlInAs/GaInAs	5.37–5.44	MQW DFB	p	170–300	183
AlInAs/GaInAs	6.2–6.6	MQW RW	p, cw	10–280, 10–80	231
AlInAs/GaInAs	7.3	SL RW	p	10–300	364
AlInAs/GaInAs	~8	MQW RW	p, cw	10–300, 10–140	365
AlInAs/GaInAs	9.75–10.15	MQW RW	p	≤ 300	369
AlInAs/GaInAs	10.04–10.18	MQW DFB	p	80–300	367
AlInAs/GaInAs	8.3	SQW RW	p	≤220	361
AlInAs/GaInAs	8.47–8.54	MQW DFB	p, cw	300, 120	360
AlInAs/GaInAs	~10	DQW RW	p	≤220	361
AlInAs/GaInAs	~11.1	MQW RW	p, cw	10–200, 10–30	87,233
GaAs/AlGaAs	9.4	MQW RW	p	10–140	363
GaAs/AlGaAs	10	MQW MD	p	10–165	373
GaAs/AlGaAs	13	SL RW	p	10–50	366
GaAs/AlGaAs	14.1	QW	p[a]	77	339
GaAs/AlGaAs	14.5	MQW	p[b]	10–135	368
GaAs/AlGaAs	≈15.5	QW	p[b]	≤110	338
GaInAs/AlInAs	3.4–3.6	MQW RW	p, cw	10–280, 10–50	63
GaInAs/AlInAs	4.6	MQW RW	p, cw	10–200, 50–85	212
GaInAs/AlInAs	~5	MQW RW	p, cw	10–320, 10–140	87, 216-9
GaInAs/AlInAs	~5	MQW MD	p	10–150	52

Table 1.5.9—*continued*
Quantum Cascade and Intersubband Lasers

Material	Wavelength (μm)	Structure	Mode	Temp. (K)	Ref.
GaInAs/AlInAs	5.31–5.38	MQW DFB	p	110–315	61
GaInAs/AlInAs	6.33–6.28[c]	MQW RW	p	10–90	359
GaInAs/AlInAs	6.49–6.41[c]	MQW RW	p	10–77	359
GaInAs/AlInAs	≈6.76–6.60[d]	MQW RW	p	25–120	359
GaInAs/AlInAs	7.4–8.6	MQW RW	p cw	210 110	234
GaInAs/AlInAs	7.7	SL RW	p	10–240	235
GaInAs/AlInAs	7.78–7.93	MQW DFB	p	80–310	61
GaInAs/AlInAs	8.5	MQW RW	p	80–270	132
GaInAs/AlInAs	~8.5	MQW RW	p, cw	10–320, 10–110	87, 238
GaInAs/AlInAs	~9.3	MQW RW	p	10–220, 10–35	237
GaInAs/AlInAs	~9.5	MD	p	<140	242
GaInAs/AlInAs	9.75–10.15[e]	MQW RW	p	260–300	369
GaInAs/AlInAs	11.5	MQW MD	p	10	242
GaInAs/AlInAs	≈13	MQW RW	p	10–175	313
InAl/GaInSb	≈3.5	Type-II "W"	p	≤286	314
InAl/GaInSb	3.8	Type-II MQW	p	60–200	370
InAs/GaInSb	3.8	Type-II MQW	p	10–210	315
InAs/InGaSb/InAlSb	4	Type-II MQW	p	10–140	371
InAs/InGaSb/InAlSb	~3.9	Type-II MQW	p	40–170	124
InAs/InGaSb/InAlSb	3.79–3.8	Type-II MQW	p	170	194
InGaAs/AlInAs	≈7	SL RW	p	5–280	316
InGaAs/AlInAs	7.6	SL RW	p	300	354
InGaAs/AlInAs	7.22	SL RW	cw	<160	354
InGaAs/AlInAs	≈7.7	MQW RW	p	≤110	355
InGaAs/AlInAs	6.6, 7.3, 7.9	SL RW	p	≤300	340
InGaAs/InAlAs	~11.5	MQW RW	p	10–320	317
InGaAs/InAlAs	≈17	SL RW	p	10–150	362
InAsSb/InAsP	3.8–3.9	Type I MQW	p	80–170	318
InAs/GaInSb/InAs	2.80–2.99	Type II "W"	p	40–220	319

(a) No oscillation; gain measured using a pump-probe setup with a two-color free electron laser.
(b) Quantum fountain laser optically pumped using a CO_2 laser.
(c) Bidirectional laser.
(d) Symmetric bidirectional laser.
(e) Electrically tunable.

1.5.11 Vertical Cavity Lasers

The light output of vertical-cavity surface-emitting lasers (VCSELs), in contrast to edge-emitting diode lasers, is normal to the axis of the gain medium. The lasing wavelength is determined by the equivalent laser cavity thickness which can be varied by changing the thickness of either the wavelength spacer or the distributed Bragg reflector layers.

Table 1.5.10
Vertical Cavity Lasers

Material	Wavelength (µm)	Structure	Mode	Temp. (K)	Ref.
Injection Lasers					
AlGaAs	0.770	SL	cw	300	104
AlGaInAs/AlGaAs	~0.850	SL-MQW	cw	230–410	320
AlGaInAs	1.3	MQW	cw	300	149
AlGaInP/AlGaAs	0.670–0.690	QW	cw	300	81
GaAs	0.845	crystal	cw	300	125
GaAs	~0.900–0.962	layer	cw	300	127
InAlGaAs	0.9604	QD	cw	300	134
InGaAs	0.951–0.957	QW	cw	95–175	378
InGaAs	0.969, 0.986	MQW	cw	300	326
InGaAs	0.979	SQW	cw	300	327
InAsAs	0.958	MQW	p	300	322
InGaAs	0.9191–0.9507	MQW	p, cw	300	325
InGaAs/GaAs	~0.943–0.971	MQW	cw	300	328
InGaAs/GaAs	0.940–0.983[a]	SQW	p, cw	300	329
InGaAsP	1.542	QW	p, cw	300	381
InGaAsP	1.55	QW	p	300	155
InGaAsP	1.3	DH	p	300	331
InGaAsP	1.3	—	cw	273–343	379
InGaAsP/InP	1.18	DH	p	77	332
InGaN	0.381	3l cavity	p	77	346
InGaN	0.399	MQW	p	300	347
InSb	~5.2	J	p	~10	333
Optically Pumped Lasers					
InGaAs/GaAs	0.918–0.966 0.940–0.978	QW CMC	qcw	300	330
InAs/GaSb/InAs/AlSb	2.9	type II "W"	cw	78–160	324
CdHgTe/HgTe	3.06	DH	p	10–30	321
InAs/GaSb/InAs/AlSb	2.9	type II "W"	p	≤280	323

Table 1.5.10—*continued*
Vertical Cavity Lasers

Material	Wavelength (μm)	Structure	Mode	Temp. (K)	Ref.
CdHgTe/HgTe	3.06	DH	p	10–30	321
InAs/GaSb/InAs/AlSb	2.9	type II "W"	p	≤280	323
GaN	0.376–0.378	epilayer	p	300	13,14
GaN/AlGaN	0.363	layer	p	300	8
InGaN	0.406	ML	p	300	296
InGaN/GaN	0.415	H	p	300	21
InGaN/GaN/AlGaN`	0.401	ML DBR	p	300	384
InGaN/InGaN	0.410	MQW	p	300	292
(Zn,Cd)Se	0.496	QW SCH	p, qcw	300	45

(a) Rastered multiple wavelength (RMW) laser array.

1.5.12 Commercial Semiconductor Lasers

Diode lasers are available in a wide range of power levels and operating configurations according to the type of device and application; packing may include thermal control and fiber coupling. For the highest powers, diode laser arrays (bars) are stacked.

Table 1.5.11
Commercial Semiconductor Lasers

Laser Material	Operation	Principal wavelengths (µm)	Output
InGaN	cw	0.400	5 mW
GaN	pulsed	0.415	20 nJ
GaAlAs	cw	0.42, 0.43 (SH)	0.4–4.0 W
InGaAlP	cw	[0.63–0.68]	1–500 mW
	pulsed	0.68	≤ 10 J
GaAsP	cw	0.67	1–10 mW
	pulsed	0.67	3–10 J
GaAlAs	cw	[0.75–0.85]	1–200 mW
	pulsed	[0.75–0.85]	1–500 mJ
GaAlAs (arrays)	cw	[0.75–0.85]	10 W–60 W
	pulsed	[0.78–0.91]	0.1–30 J
GaAlAs (stacked arrays)	qcw, cw	[0.79–0.98]	≤ 3 kW
GaAs	pulsed	0.904	≤ 0.8 J
GaAs (arrays)	pulsed	0.904	≤ 5 J
InGaAs	cw	0.905–0.98	0.02–1.0 W
	pulsed	0.905–0.98	10^{-6}–1.0 J
InGaAs (arrays)	cw	0.91–0.98	30 W
InGaAsP	cw	1.27–1.33	0.1–3.0 W
	cw	1.52–1.58	0.5–100 mW
	cw	1.57–1.63	>2 mW
	pulsed	1.06–1.55	0.2 mJ
	pulsed	1.55	10^{-3}–0.6 J
InGaAsP (arrays)	pulsed	1.55	2.5 J
Pb salts	cw	3.3–27	0.1–25 mW
	pulsed	3.3–25	≤ 2 J
Pb salts (77 K)	cw	2.9–3.6	1–5 mW

SH – second harmonic

1.5.13 References

1. Hurwitz, C. E., Efficient ultraviolet laser emission in electron-beam-excited ZnS, *Appl. Phys. Lett.* 9, 116 (1966.

2. Bogdankevich, O. V., Zverev, M. M., Pechenov, A. N., and Sysoev, L. A., Recombination radiation of ZnS single crystals excited by a beam of fast electrons, *Sov. Phys. Sol. Stat.* 8, 2039 (1967).

3. Wang, S. and Chang, C. C., Coherent fluorescence from zinc sulphide excited by two-photon absorption, *Appl. Phys. Lett.* 12, 193 (1968).

4. Tanaka, S., Hirayama, H., Aoyagi, Y., Narukawa, Y., Kawakami, Y., Fujita, S., and Fujita, S., Stimulated emission form optically pumped GaN quantum dots, *Appl. Phys. Lett.* 71, 1299 (1997).

5. Dingle, R., Shaklee, K. L., Leheny, R. F., and Zenerstrom, R. B., Stimulated emission and laser action in gallium nitride, *Appl. Phys. Lett.* 19, 5 (1971).

6. Schmidt, T. J., Yang, X. H., Shan, W., Song, J. J., Salvador, A., Kim, W., Altas, Ö., Botchkarev, A., and Morkoc, H., Room-temperature stimulated emission in GaN/AlGaN separate confinement heterostructure grown by molecular beam epitaxy, *Appl. Phys. Lett.* 68, 1820 (1996).

7. Yang, X. H., Schmidt, T. J., Shan, W., and Song, J. J., Above room temperature near ultraviolet lasing from an optically pumped GaN film grown on sapphire, *Appl. Phys. Lett.* 66, 1 (1995).

8. Redwing, J. M., Loeber, D. A. S., Anderson, N. G., Tischler, M. A., and Flynn, J. S., An optically pumped GaN-AlGaN vertical cavity surface emitting laser, *Appl. Phys. Lett.* 69, 1 (1996).

9. Cammack, D. A., Dalby, R. J., Cornelissen, H. J., and Khurgin, J., *J. Appl. Phys.* 62, 3071 (1987).

10. Kurai, S., Naoi, Y., Abe, T., Ohmi, S., and Sakai, S., Photopumped stimulated emission from homoepitaxial GaN grown on bulk GaN prepared by sublimation method, *Jpn. J. Appl. Phys.* 35, L77 (1996).

11. Johnston, Jr., W. D., Characteristics of optically pumped platelet lasers of ZnO, CdS, CdSe and $CdS_{0.6}Se_{0.4}$ between 300 and 80 K, *J. Appl. Phys.* 42, 2731 (1971).

12. Nicoll, F. H., Ultraviolet ZnO laser pumped by an electron beam, *Appl. Phys. Lett.* 9, 13 (1966).

13. Khan. M. A., Olson, D. T., Van Hove, J. M., and Kuznia, J. N., Vertical-cavity, room-temperature stimulated emission from photopumped GaN films deposited over sapphire substrates using low-pressure metalorganic vapor deposition, *Appl. Phys. Lett.* 58, 1515 (1991).

14. Khan, M. A., Kuznia, J. N., Van Hove, J. M., Olson, D. T., Krishnankutty, S., and Kolbas, R. M., Growth of high optical and electrical quality GaN layers using low pressure metalorganic chemical vapor deposition, *Appl. Phys. Lett.* 58, 526 (1991).

15. Amano, H., Asahi, T., and Akasaki, I., Stimulated emission near ultraviolet at room temperature from a GaN film grown on sapphire by MOVPE using an AlN buffer layer, *Jpn. J. Appl. Phys.* 29, L205 (1990).

16. Segawa, Y., Ohtomo, A., Kawasaki, M., Koinuma, H., Tang, Z. K., Yu, P., and Wong, G. K. L., Growth of ZnO thin film by laser MBE: lasing of exciton at room temperature, *Phys. Stat. Sol. (b)* 202, 669 (1997).

17. Khan, M. A., Sun, C. J., Yang, J. W., Chen, Q., Lim, B. W., Anwar, M. Z., Osinsky, A., and Temkin, H., cleaved cavity optically pumped InGaN-GaN laser grown on spinel substrates, *Appl. Phys. Lett.* 69, 2418 (1996).

18. Mooradian, A., Strauss, A. J., and Rossi, J. A., Broad band laser emission from optically pumped $PbS_{1-x}Se_x$, *IEEE J. Quantum Electron.* QE-9, 347 (1973).

19. Amano, H., Tanaka, T., Kunii, Y., Kim, S. T., and Akasaki, I., Room-temperature violet stimulated emission from optically pumped AlGaN/GaInN double heterostructure, *Appl. Phys. Lett.* 64, 1377 (1994).

20. Nakamura, S., Senoh, M., Nagahame, S., Iwasa, N., Yamada, T., Matsushita, T., Sugimoto, Y., and Kiyoku. H., Room-temperature continuous-wave operation of InGaN multi-quantum-well structure diodes with a lifetime of 27 hours, *Appl. Phys. Lett.* 70, 1417 (1997).

21. Khan. M. A., Krishnankutty, S., Skogman, R. A., Kuznia, J. N., Olson, D. T., and George, T., Vertical-cavity stimulated emission from photopumped InGaN/GaN heterojunctions at room temperature, *Appl. Phys. Lett.* 65, 520 (1994).

22. Nakamura, S., Senoh, M., Nagahame, S., Iwasa, N., Yamada, T., Matsushita, T., Kiyoku. H., and Sugimoto, Y., InGaN-based multi-quantum-well-structure laser diodes, *Jpn. J. Appl. Phys.* 35, L74 (1996).

23. Yang, X. H., Hays, J. M., Shan, W., Song, J. J., and Cantwell, E., Two-photon pumped blue lasing in bulk ZnSe and ZnSSe, *Appl. Phys. Lett.* 62, 1071 (1993).

24. Suemune, I., Yamada, K., Masato, H., Kan, Y., and Yamanishi, M., Lasing in a ZnSSe/ZnSe multilayer structure with photopumping, *Appl. Phys. Lett.* 54, 981 (1989).

25. Brodin, M. S., VUrikhovskii, N. I., Zakrevskii, S. V., and Reznichenko, V. Ya., Generation in mixed CdS_x—$CdSe_{1-x}$ crystals excited with ruby laser radiation, *Sov. Phys. Sol. Stat.* 8, 2461 (1967).

26. Bylsma, R., Becker, W. M., Bonsett, T. C., Kolodziejski, L. A., Gunshor, R. L., Yamanishi, M., and Datta, S., Stimulated emission and laser oscillation in $ZnSe$-$Zn_{1-x}Mn_xSe$ multiple quantum wells at ~453 nm, *Appl. Phys. Lett.* 47, 1039 (1985).

27. Bogdankevich, O. V., Zverev, M. M., Krasiinikov, A. I., and Pechenov, A. N., Laser emission in electron-beam excited ZnSe, *Phys. Stat. Solid* 19, K56 (1967).

28. Nakanishi, K., Suemune, I., Masato, H., Kuroda, Y., and Yamanishi, M., Near-room-temperature photopumped blue lasers in ZnS_xSe_{1-x}/ZnSe multilayer structures, *Jpn. J. Appl. Phys.* 29, L2420 (1990).

29. Ledentsov, N. N., Krestnikov, I. L., Maximov, M. V. et al., Ground state exciton lasing in CdSe submonolayers inserted in a ZnSe matrix, *Appl. Phys. Lett.* 69, 1343 (1996).

30. Jeon, H., Ding, J., Nurmikko, A. V., Luo, H., Samarth, N., Furdyna, J. K., Bonner, W. A., and Nahory, R. E., Room-temperature blue lasing action in (Zn,Cd)Se/ZnSe optically pumped multiple quantum well structures on lattice-matched (Ga,In)As substrates, *Appl. Phys. Lett.* 57, 2413 (1990).

31. Zmudzinski, C. A., Guan, Y. and Zory, P. S., Room temperature photopumped ZnSe lasers, *IEEE Phot. Technol. Lett.* 2, 94 (1990).

32. Yang, X. H., Hays, J., Shan, W., Song, J. J., Cantwell, E., and Aldridge, J., Optically pumped lasing of ZnSe at room temperature, *Appl. Phys. Lett.* 59, 1681 (1991).

33. Hervé, D., Accomo, R., Molva, E., Vanzetti, L., Paggel, J. J., Sorba, L., and Franciose, A., Microgun-pumped blue lasers, *Appl. Phys. Lett.* 67, 2144 (1995).

34. Jeon, H., Ding, J., Patterson, W., Nurmikko, A. V., Xie, W., Grillo, D. C., Kobayashi, M., and Gunshor, R. L., Blue-green injection laser diodes in (Zn,Cd)Se/ZnSe quantum wells, *Appl. Phys. Lett.* 59, 3619 (1991).

35. Kurbatov, L. N., Mashchenko, V. E., and Mochalkin, N. N., Coherent radiation from cadmium sulfide single crystals excited by an electron beam, *Opt. Spectrosc.* 22, 232 (1967).

36. Ishihara, T., Ikemoto, Y., Goto, T., Tsujimura, A., Ohkawa, K., and Mitsuyu, T., Optical gain in an inhomogeneously broadened exciton system, *J. Lumin.* 58, 241 (1994).

37. Hurwitz, C. E., Efficient visible lasers of CdS_xSe_{1-x} by electron-beam excitation, *Appl. Phys. Lett.* 8, 243 (1966).

38. Haase, M. A., Qiu, J., DePuydt, J. M., and Cheng, H., Blue-green laser diodes, *Appl. Phys. Lett.* 59, 1272 (1991).

39. Walker, C. T., DePuydt, J. M., Haase, M. A., Qiu, J., and Cheng, H., Blue-green II-VI laser diodes, *Physica B* 185, 27 (1993).
40. Hurwitz, C. E., Electron beam pumped lasers of CdSe and CdS, *Appl. Phys. Lett.* 8, 121 (1966).
41. Ding, J., Jeon, H., Nurmikko, A. V., Luo, H., Samarth, N., and Furdyna, J. K., Laser action in the blue-green from optically pumped (Zn,Cd)Se/ZnSe single quantum well structures, *Appl. Phys. Lett.* 57, 2756 (1990).
42. Jeon, H., Ding, J., Nurmikko, A. V., Xie, W., Grillo, D. C., Kobayashi, M., Gunshor, R. L., Hua, G. C., and Otsuka, N., Blue and green diode lasers in ZnSe-based quantum wells, *Appl. Phys. Lett.* 60, 2045 (1992).
43. Konyukhov, V. K., Kulevskii, L. A., and Prokhorov, A. M., Optical oscillation in CdS under the action of two-photon excitation by a ruby laser, *Sov. Phys. Dokl.* 10, 943 (1966).
44. Nakamura, S., Senoh, M., Nagahame, S., Iwasa, N., Yamada, T., Matsushita, T., Kiyoku. H., and Sugimoto, Y., Characteristics of InGaN multi-quantum-well-structure laser diodes, *Appl. Phys. Lett.* 68, 3269 (1996).
45. Jeon, H., Kozlov, V., Kelkar, P. et al., Room-temperature optically pumped blue-green vertical cavity surface emitting laser, *Appl. Phys. Lett.* 67, 1668 (1995).
46. Basov, N. G., Grasyuk, A. Z., Zubarev, I. G., and Katuiin, V. A., Laser action in CdS induced by two-photon optical excitation from a ruby laser, *Sov. Phys. Sol. Stat.* 7, 2932 (1966).
47. Thijs, P. J. A., Binsma, J. J. M., Tiemeijer, L. F., Slootweg, R. W. M., van Roijen, R., and van Dongen, T., Sub-mA threshold operation of λ=1.5 μm strained InGaAs multiple quantum well lasers grown on (311) B InP substrates, *Appl. Phys. Lett.* 60, 3217 (1992).
48. Basov, N. G., Bogdankevid, O. V., and Devyatkov, A. G., Exciting a semiconductor quantum generator (laser) with a fast electron beam, *Sov. Phys. Dokl.* 9, 288 (1964).
49. Jeon, H., Ding, J., Nurmikko, A. V., Luo, H., Samarth, N., and Furdyna, J. K., Low threshold pulsed and continuous-wave laser action in optically pumped (Zn,Cd)Se/ZnSe multiple quantum well lasers in the blue-green, *Appl. Phys. Lett.* 59, 1293 (1991).
50. Bidnyk, S., Schmidt, T. J., Cho, Y. H., Gainer, G. H., and Song, J. J., High temperature stimulated emission in optically pumped InGaN/GaN multi-quantum wells, *Appl. Phys. Lett.* 72, 1623 (1998).
51. Kawakami, Y., Yamaguchi, S., Wu, Y., Ichino, K. Fujita, S., and Fujita, S., Optically pumped blue-green laser operation above room-temperature in $Zn_{0.80}Cd_{0.20}Se$-Zn $S_{0.08}Se_{0.92}$ multiple quantum well structures grown by metalorganic molecular beam epitaxy, *Jpn. J. Appl. Phys.* 30, L605 (1991).
52. Faist, J., Gmachl, C., Striccoli, M., Sirtori, C., Capasso, F., Sivco, D. L., and Cho, A. Y., Quantum cascade disk lasers, *Appl. Phys. Lett.* 69, 2456 (1996).
53. Guo, Y. Aizin, G., Chen, Y. C., Zeng, L., Cavus, A., and Tamargo, M., Photo-pumped ZnCdSe/ZnCdMgSe blue-green quantum well lasers grown on InP substrates, *Appl. Phys. Lett.* 70, 1351 (1997).
54. Yokogawa, T., Kamiyama, S., Yoshii, S., Ohkawa, K., Tsujimura, A., and Sasai, Y., Real-index guided blue-green laser diode with small beam astigmatism fabricated using ZnO buried structure, *Jpn. J. Appl. Phys.* 35, L314 (1996).
55. Hurwitz, C. E., Laser emission from electron beam excited ZnTe, *IEEE J. Quantum Electron.* QE-3, 333 (1967).
56. Vlasov, A. N., Kozina, G. S., and Fedorova, O. B., Stimulated emission from zinc telluride single crystals excited by fast electrons, *Sov. Phys. JETP* 25, 283 (1967).
57. Ermanov, O. N., Garba, L. S., Golvanov, Y. A., Sushov, V. P., and Chukichev, M. V., Yellow-green InGaP and InGaPAs LEDs and electron-beam-pumped lasers prepared by LPE and VPE, *IEEE Trans. Elect. Devices* ED-26, 1190 (1979).

58. Macksey, H. M., Holonyak, Jr., N., Dupuis, R. D., Campbell, J. C., and Zack, G. W., Crystal synthesis, electrical properties, and spontaneous and stimulated photoluminescence of $In_{1-x}Ga_xP$:N grown from solution, *J. Appl. Phys.* 44, 1333 (1973).

59. Glass, A. M., Tai, K., Bylsma, R B., Feldman, R. D., Olson, D. H., and Austin, R. F., Room-temperature optically pumped $Cd_{0.25}Zn_{0.75}Te$/ZnTe quantum well lasers grown on GaAs substrates, *Appl. Phys. Lett.* 53, 834 (1988).

60. Kaneko, Y., Kikuchi, A., Nomura, I., and Kishino, K., Yellow light (576 nm) lasing emission of GaInP/AlInP multiple quantum well lasers prepared by gas-source-molecular-beam-epitaxy, *Electron. Lett.* 26, 657(1990).

61. Faist, J., Gmachl, C., Capasso, F., Sirtori, C., Sivco, D. L., Baillargeon, J. N., and Cho, A. Y., Distributed feedback quantum cascade lasers, *Appl. Phys. Lett.* 70, 2670 (1997); Correction/addition, *ibid.*71, 986 (1997).

62. Hino, I., Kawat, S., Gomo, A., Kobayashi, K., and Suzuki, S., Continuous wave operation (77 K) of yellow (583.6 nm) emitting AlGaInP double heterostructure laser diodes, *Appl. Phys. Lett.* 48, 557 (1986).

63. Faist, J., Capasso, F., Sivco, D. L., Hutchinson. A. L., Chu, S. N. G., and Cho, A. Y., Short wavelength (λ~3.4 μm) quantum cascade laser based on strain compensated InGaAs/AlInAs, *Appl. Phys. Lett.* 72, 680 (1998) .

64. Khurgin, J., Fitzpatrick, B. J., and Seemungai, W., Cathodoluminiscence, gain, and stimulated emission in electron-beam-pumped ZnCdSe, *J. Appl. Phys.* 61, 1606 (1987).

65. Basov, N. G., Bogdankevich, O. V., and Abdullaev, A. N., Radiation in GaSe single crystals induced by excitation with fast electrons. *Sov. Phys. Dokl.* 10, 329 (1965).

66. Abdullaev, G. B., Aliev, M. Kh., and Mirzoev, B. R., Laser emission by GaSe under two-photon optical excitation conditions, *Sov. Phys. Semiconductor* 4, 1189 (1971).

67. Nahory, R. E., Shaklee, K. L., Leheny, R. F., and DeWinter, J. C., Stimulated emission and the type of bandgap in GaSe, *Solid State. Commun.* 9, 1107 (1971).

68. Catalano, I. M., Cingolani, A., Ferrara, M., and Minafra, A., Luminescence by exciton-exciton collision in GaSe, *Phys. Stat. Sol. (b)* 68, 341 (1975).

69. Abdullaev, G. B., Godhaev, I. O., Kakhramanov, N. B., and Suleimanov, R.A., Stimulated emission from indium selenide and gallium selenide layer semiconductors, *Sov. Phys. Solid State* 34, 39 (1992).

70. Jang, D.-H., Kaneko, Y., and Kishino, K., Shortest wavelength (607 nm) operations of GaInP/AlInP distributed Bragg reflector lasers, *Electron. Lett.* 28, 428 (1992).

71. Chang, L. B., and Shia, L.Z., Room-temperature continuous wave operation of a visible AlGaAs/InGaP transverse junction stripe laser grown by liquid phase epitaxy, *Appl. Phys. Lett.* 60, 1090 (1992).

72. Kawata, S., Kobayashi, K., Gomyo, A., Hino, I., and Suzuki, T., 621 nm cw operation (0°C) of AlGaInP visible semiconductor lasers, *Electron. Lett.* 22, 1265 (1986).

73. Rennie, J., Okajima, M., Watanabe, M., and Hatakoshi, G., Room temperature cw operation of orange light (625 nm) emitting InGaAlP laser, *Electron. Lett.* 28, 1950 (1992).

74. Kobayashi, K., Hino, I., and Suzuki, T., 626.2-nm pulsed operation (300 K) of an AlGaInP double heterostructure laser grown by metalorganic chemical vapor deposition, *Appl. Phys. Lett.* 46, 7 (1985).

75. Ou, S. S., Yang, J. J., Fu, R. J., and Hwang, C. J., High-power 630-640 nm GaInP/GaAlInP laser diodes, *Appl. Phys. Lett.* 61, 842 (1992).

76. Hatakoshi, G., Itaya, K., Ishikawa, M., Okajima, M., and Uematsu, Y., Short-wavelength InGaAlP visible laser diodes, *IEEE J. Quantum Electron.* 27, 1476 (1991).

77. Rinker, M., Kalt, H., and Köhler, K., Indirect stimulated emission at room temperature, *Appl. Phys. Lett.* 57, 584 (1990).

78. Ishikawa, M., Shiozawa, H., Tsuburai, Y., and Uematsu, Y., Short-wavelength (638 nm) room-temperature cw operation of InGaAlP laser diodes with quaternary active layer, *Electron. Lett.* 26, 211 (1990).

79. Kawata, S., Fujii, H., Kobayashi, K., Gomyo, A., Hino, I., and Suzuki, T., Room-temperature continuous-wave operation of a 640 nm AlGaInP visible-light semiconductor laser, *Electron. Lett.* 23, 1328 (1987).

80. Dallesasse, J. M., Nam, D. W., Deppe, D. G. et al., Short-wavelength (<6400Å) room-temperature continuous operation of *p-n* $In_{0.5}(Al_xGa_{1-x})_{0.5}P$ quantum well lasers, *Appl. Phys. Lett.* 53, 1826 (1988).

81. Schneider, Jr., R. P., Hagerott, M., Choquette, K. D., Lear, K. L., Kilcoyne, S. P., and Figiel, J. J., Improved AlGaInP-based red (670-690 nm) surface-emitting lasers with novel C-doped short-cavity epitaxial design, *Appl. Phys. Lett.* 67, 329 (1995).

82. Ikeda, M., Mori, Y., Sato, H., Kaneko, K., and Watanabe, N., Room-temperature continuous-wave operation of an AlGaInP double heterostructure laser grown by atmosphere pressure metalorganic chemical vapor deposition, *Appl. Phys. Lett.* 47, 1027 (1985).

83. Hayakawa, T., Suyama, T., Takahashi, K., Kondo, M., Yamamoto, S., and Hijikata, T., Low-threshold room-temperature cw operation of $(AlGaAs)_m(GaAs)_n$ superlattice quantum well lasers emitting at ~680 nm, *Appl. Phys. Lett.* 51, 70 (1987).

84. Yamamoto, S., Hayashi, H., Hayakawa, T., Miyauchi, N., Yano, S., and Hijikata, T., Room-temperature cw operation in the visible spectral range of 680-700 nm by AlGaAs double heterojunction lasers, *Appl. Phys. Lett.* 41, 796 (1982).

85. Hayakawa, T., Suyama, T., Takahashl, K., Kondo, M., Yamamoto, T., and Hijlkata, T., Low current and threshold AlGaAs visible laser diode with a $(AlGaAs)_m(GaAs)_n$ superlattice quantum well, *Appl. Phys. Lett.* 49, 637 (1986).

86. Derry, P. L., Yariv, A., Lau, K. Y., Bar-Chaim, N., Lee, K., and Rosenberg, J., Ultra-low-threshold graded-index separate-confinement single quantum well buried heterostructure (AlGa)As laser with high reflectivity coating, *Appl. Phys. Lett.* 50, 1773 (1987).

87. Capasso, F., Faist, J., Sirtori, C., and Cho, A. Y., Infrared (4-11 μm) quantum cascade laser, *Solid State Commun.* 102, 231 (1997).

88. Kressel, H. and Hawrylo, F. Z., Stimulated emission at 300 K and simultaneous lasing at two wavelengths in epitaxial $Al_xGa_{1-x}As$ injection lasers, *Proc. IEEE* 56, 1598 (1968).

89. Kobayashi, K., Kawata, S., Gomyo, A., Hino, I., and Suzuki, T., Room-temperature cw operation of AlGaInP double-heterostructure visible lasers, *Electron. Lett.* 23, 931 (1985).

90. Holonyak, Jr., N., Sirkis, H. D., Stillman, G. E., and Johnson, M. R., Laser operation of CdSe pumped with a Ga(AsP) laser diode. *Proc. IEEE* 54, 1068 (1966).

91. Burnham, R. D., Holonyak, Jr., N., Keune, D. L., Scrifres, D. R., and Dapkus, P. D., Stimulated emission in $In_{1-x}Ga_xP$, *Appl. Phys. Lett.* 17, 430 (1970).

92. Dolginov, L. M., Druzhinina, L. V., and Kryukova, I. V., Parameters of electron-beam-pumped $Al_xGa_{1-x}As$ lasers in the visible part of the spectrum, *Sov. J. Quantum Electron.* 4, 104 (1975).

93. Grasyuk, A. Z., Eiimkov, V. F., and Zubarev, I. G., Semiconductor CdSe laser with two-photon optical excitation, *Sov. Phys. Sol. Stat.* 8, 1548 (1966).

94. van der Ziel, J. P., Dupuis, R. D., Logan, R. A., Mikulyak, R. M., Pinzone, C. J., and Savage, A., Low threshold pulse and continuous laser oscillation from AlGaAs/GaAs double heterostructure grown by metalorganic chemical vapor deposition, *Appl. Phys. Lett.* 50, 454 (1987).

95. Kressel, H., Olssen, G. H., and Nuese, C. J., Visible $GaAs_{0.7}P_{0.3}$ cw heterostructure laser, *Appl. Phys. Lett.* 30, 249 (1977).

96. Basov, N. G., Bogdankevich, O. V., Eliseev, P. G., and Lavrushin, B. M., Electron beam excited lasers made from solid solutions of GaP_xAs_{1-x}, *Sov. Phys. Sol. Stat.* 8, 1073 (1966).

97. Fafard, S., Hinzer, K., Raymond, S., Dion, M., McCaffrey, J., Feng, Y., and Charbonneau, S., Red-emitting semiconductor quantum dot lasers, *Science* 274, 1350 (1996).

98. Holonyak, Jr., N. and Bevacqua, S. F., Coherent (visible) light emission from $Ga(As_{1-x}P_x)$ junctions, *Appl. Phys. Lett.* 1, 82 (1962).

99. Rupprecht, H., Woodall, J. M., and Pettit, G. D., Stimulated emission from Ga,Al,As diodes at 70 K, *IEEE J. Quantum Electron.* QE-4, 35 (1968).

100. Mehuys, D., Mittelstein, M., and Yariv, A., Optimised Fabry-Perot (AlGa)As quantum-well lasers tunable over 105 nm, *Electron. Lett.* 25, 143 (1989).

101. Macksey, H. M., Holonyak, Jr., N., Scifres, D. R., Dupuis, R. D., and Zack, G. W., $In_{1-x}Ga_xP$ p-n junction lasers, *Appl. Phys. Lett.* 19, 271 (1971).

102. Beauvais, J. and Fortin, E., Optical gain in $CdIn_2S_4$, *J. Appl. Phys.* 62, 1349 (1987).

103. Averkleva, G. K., Goryunova, N. A., and Prochukhan, V. D., Stimulated recombination radiation emitted by $CdSiAs_2$, *Sov. Phys. Semiconductor* 5, 151 (1971).

104. Lee, Y. H., Tell, B., Brown-Goebeler, K. F., Leibenguth, R. E., and Mattera, V. D., Deep-red continuous wave top-surface-emitting vertical-cavity AlGaAs superlattice lasers, *IEEE Phot. Technol. Lett.* 3, 108 (1991).

105. Windhorn, T. H., Metze, G. M., Tsaur, B.-Y., and Fan, J. C. C., AlGaAs double-heterostructure diode laser fabricated on a monolithic GaAs/Si substate, *Appl. Phys. Lett.* 45, 309 (1984).

106. Vavilov, V. S. and Nolle, E. L., Cadmium telluride laser with electron excitation, *Sov. Phys. Dokl.* 10, 827 (1966).

107. Vavilov, V. S., Nolle, E. L., and Egorov, V. D., New data on the electron-excited recombination radiation spectrum of cadmium telluride, *Sov. Phys. Sol. Stat.* 7, 749 (1965).

108. Mukai, S., Yajima, H., Mitsuhashi, Y., Shimada, J., and Kutsuwada, N., Continuous operating visible-light-emitting lasers using liquid-phase-epitaxial InGaPAs grown on GaAs substrates, *Appl. Phys. Lett.* 43, 24 (1983).

109. Burnham, R. D., Holonyak, N., Jr., Korb, H. W., Macksey, H. M., Scifres, D. R., and Woodhouse, J. B., Double heterostructure AlGaAsP quaternary laser, *Appl. Phys. Lett.* 19, 25 (1971).

110. Diaz, J., Eliashevich, I., He, X., Yi, H., Wang, L., Kolev, E., Garguzov, D., and Razeghi, M., High-power InGaAsP/GaAs 0.8-μm laser diodes and peculiarities of operational characteristics, *Appl. Phys. Lett.* 65, 1004 (1994).

111. Nakamura, M., Yariv, A., Yen, H. W., Somekh, S., and Garvin, H. L., Optically pumped GaAs surface laser with corrugation feedback, *Appl. Phys. Lett.* 22, (1973).

112. Wade, J. K., Mawst, L. J., Botez, D., Jansen, M., Fang, F., and Nabiev, R. F., High continuous wave power, 0.8 μm-band, Al-free active-region diode lasers, *Appl. Phys. Lett.* 70, 149 (1997).

113. Basov, N. G., Grasyuk, A. Z., and Katulin, V. A., Induced radiation in optically excited gallium arsenide, *Sov. Phys. Dokl.* 10, 343 (1965.

114. Quist, T. M., Rediker, R. H., and Keyes, R., Semiconductor maser of GaAs, *Appl. Phys. Lett.* 1, 91 (1962).

115. Hurwitz, C. E. and Keyes, R. J., Electron beam pumped GaAs laser, *Appl. Phys. Lett.* 5, 139 (1964).

116. Kurbatov, L, N., Kabanov, A. N., and Sigriyanskii, B. B., Generation of coherent radiation in gallium arsenide by electron excitation, *Sov. Phys. Dokl.* 10, 1059 (1966).

117. Cusano, D. A., Radiative recombination from GaAs directly excited by electron beams, *Solid State Commun.* 2, 353 (1964).

118. Lee, J. H., Hsieh, K. Y., and Kolbas, R. M., Photoluminescence and stimulated emission from monolayer-thick pseudomorphic InAs single-quantum-well heterostructures, *Phys. Rev. B* 41, 7678 (1990).

119. Chen, H. Z., Ghaflari, A., Wang, H., Morkoc, H., and Yariv, A., Continuous-wave operation of extremely low-threshold GaAs/AlGaAs broad-area injection lasers on (100) Si substrates at room temperature, *Optics Lett.* 12, 812 (1987).

120. Hall, R. N., Fenner, G. E., Kingley, J. D., Solbs, T. J., and Carlson, R. O., Coherent light emission from GaAs junctions, *Phys. Rev. Lett.* 9, 366 (1962).

121. Nathan, M. E., Dumke, W. P., Burns, C., Dill, Jr., F. H., and Lasher, G. J., Stimulated emission of radiation from GaAs p-n junction, *Appl. Phys. Lett.* 1, 62 (1962).

122. Scifres, D. R., Burnham, R. D., and Streifer, W., Distributed-feedback single heterojunction GaAs diode laser, *Appl. Phys. Lett.* 25, 203 (1974).

123. Susaki, W., Sogo, T., and Oku, T., Optical losses and efficiency in GaAs laser diodes, *IEEE J. Quantum Electron.* QE-4, 122 (1968).

124. Yang, R. Q., Yang, H. B., Lin, C.-H., Zhang, D., Murry, S. J., Wu, H., and Pei, S. S., High power mid-infrared interband cascade lasers based on type-II quantum wells, *Appl. Phys. Lett.* 71, 2409 (1997).

125. Tell, B., Lee, Y. H., Brown-Goebeler, K. F. et al., High-power cw vertical-cavity top surface-emitting GaAS quantum well lasers, *Appl. Phys. Lett.* 57, 1855 (1990).

126. Eliseev, P. G., Ismailo, I., and Mikhaillna, L. I., Coherent emission of InP optically excited by an injection laser, *JETP Lett.* 6, 15 (1967).

127. Yuen, W., Li, G. S., and Chang-Hasnain, C. J., Multiple-wavelength vertical-cavity surface-emitting laser arrays with a record wavelength span, *IEEE Phot. Technol. Lett.* 8, 4 (1996).

128. Weiser, K. and Levitt, R. S., Stimulated light emission from indium phosphide, *Appl. Phys. Lett.* 2, 178 (1963).

129. Basov, N. G., Eliseev, P. G., and Ismailov, I., Some properties of semiconductor lasers based on indium phosphide, *Sov. Phys. Solid State* 8, 2087 (1967).

130. Shoji, H., Ohtsuka, N., Sugawara, M., Uchida, T., and Ishikawa, H., Lasing at three-dimensionally quantum-confined sublevel of self-organized $In_{0.5}Ga_{0.5}As$ quantum dots by current injection, *IEEE Phot. Tech. Lett.* 7, 1385 (1995).

131. Rossi, J. A. and Chinn, S. R., Efficient optically pumped InP and $In_xGa_{1-x}As$ lasers, *J. Appl. Phys.* 43, 4806 (1972).

132. Slivken, S., Jelen, C., Rybaltowski, A., Diaz, J., and Razeghi, M., Gas-source molecular beam epitaxy growth of an 8.5 μm quantum cascade laser, *Appl. Phys. Lett.* 71, 2593 (1997).

133. Cingolani, A., Ferrara, M., Lugara, M., and Lévy, F., Stimulated photoluminescence in indium selenide, *Phys. Rev. B* 25, 1174 (1982).

134. Saito, H., Nishi, K., Ogura, I., Sugou, S., and Sugimoto, Y., Room-temperature lasing operation of a quantum-dot vertical-cavity surface-emitting laser, *Appl. Phys. Lett.* 69, 3140 (1996).

135. Kurbatov, L, N., Dirochka, A. I., and Britov, A. D., Stimulated emission of indium monoselenide subjected by electron bombardment, *Sov. Phys. Semiconductor* 5, 494 (1971).

136. Dutta, N. K., Hobson, W. S., Lopata, J., and Zydzik, G., Tunable InGaAs/GaAS/InGaP laser, *Appl. Phys. Lett.* 70, 1219 (1997).

137. Geels, R. S. and Coldren, L. A., Submilliamp threshold vertical-cavity laser diodes, *Appl. Phys. Lett.* 57, 1605 (1990).

138. Sugiyama K. and Saito, H., GaAsSb-AlGaAsSb double heterostructure laser, *Jpn. J. Appl. Phys.* 11, 1057 (1972).

139. Xie, Q., Kalburge, A., Chen, P., and Madhukar, A., Observation of lasing from vertically self-organized InAs three-dimensional island quantum boxes on GaAs (001), *IEEE Phot. Technol. Lett.* 8, 965 (1996).

140. Feketa, D., Chan, K. I., Ballantyne, J. M., and Eastman, L. F., Graded-index separate-confinement InGaAs/GaAs strained-layer quantum well laser grown by metalorganic chemical vapor deposition, *Appl. Phys. Lett.* 49, 1659 (1986).

141. Chipaux, C. and Eymard, R., Study of the laser effect in GaSb alloys, *Phys. Stat. Sol.* 10, 165 (1965).

142. Heinrichsdorff, F., Mao, M.-H., Kirstaedter, N., Krost, A., Bimberg, D., Kosogov, A. O., and Werner, P., Room-temperture continuous-wave lasing from stacked InAs/GaAs quantum dots grown by metalorganic chemical vapor deposition, *Appl. Phys. Lett.* 71, 22 (1997).

143. Chang-Hasnain, C. J., Bhat, R., Zah, C. E., Koza, M. A., Favire, F., and Lee, T. P., Novel AlGaInAs/AlInAs lasers emitting at 1 µm, *Appl. Phys. Lett.* 57, 2638 (1990).

144. York, P. K., Berenik, K., Fernandez, G. E., and Coleman, J. J., InGaAs-GaAs strained-layer quantum well buried heterostructure lasers ($\lambda > 1$ µm) by metalorganic chemical vapor deposition, *Appl. Phys. Lett.* 54, 499 (1989).

145. Berkovskil, F. M., Goryunova, N. A., and Ordov, V. M., CdSnP$_2$ laser excited with an electron beam, *Sov. Phys. Semiconductor* 2, 1027 (1969).

146. Kamath, K., Bhattacharya, P., Sosnowski, T., Norris, T., and Phillips, J., Room-temperature operation of In$_{0.4}$Ga$_{0.6}$As/GaAs self-organised quantum dot lasers, *Electron. Lett.* 32, 1374 (1996).

147. Akimov, Yu. A., Burov, A. A., and Zagarinskii, E. A., Electron-beam-pumped Al$_x$Ga$_{1-x}$Sb semiconductor laser, *Sov. J. Quantum Electron.* 5, 37 (1975).

148. Razeghi, M., Defour, M., Omnes, F., Maurel, Ph., Chazelas, J., and Brillouet, F., First GaInAsP-InP double-heterostructure laser emitting at 1.27 µm on a silicon substrate, *Appl. Phys. Lett.* 53, 725 (1988).

149. Qiun, Y., Zhu, Z. H., Lo, Y. H. et al., Long wavelength (1.3 µm) vertical-cavity surface-emitting lasers with a wafer-bonded mirror and an oxygen-implanted confinement region, *Appl. Phys. Lett.* 71, 25 (1997).

150. Wakao, K., Nakai, K., Sanada, T. et al., InGaAsP/InP planar buried heterostructure laser with semi-insulating InP current blocking layer grown by MOCVD, *IEEE J. Quantum Electron.* 23, 943 (1987).

151. Lidgard, A., Tnabun-Ek, T., Logan, R. A., Temkin, H., Wicht, K. W., and Olsson, N. A., External-cavity InGaAs/InP graded index multiquantum well laser with a 200 nm tuning range, *Appl. Phys. Lett.* 56, 816 (1990).

152. Temkin, H., Alavi, K., Wagner, W. R., Pearsall, T. P., and Cho, A. Y., 1.5–1.6-µm Ga$_{0.47}$In$_{0.53}$As/Al$_{0.48}$In$_{0.52}$As multiquantum will lasers grown by molecular beam epitaxy, *Appl. Phys. Lett.* 42, 845 (1983).

153. Benoit-a-la Guillaume, C., and Debever, J. M., Laser effect in gallium antimonide by electron bombardment, *Compt. Rend.* 259, 2200 (1964).

154. Benoit-a-la Guillaume, C., and Laurant, J. M., Laser effect in InAs and GaSb by optical excitation, *Compt. Rend.* 262, 275 (1966).

155. Babic´, D. I., Dudley, J. J., Strubel, K., Mirin, R. P., Bowers, J. E., and Hu, E. L., Double-fused 1.52-µm vertical-cavity lasers, *Appl. Phys. Lett.* 66, 1030 (1995).

156. Kryukova, I. V., Karnaukhov, V. G., and Paduchikh, L. I., Stimulated radiation from diffused p-n junction in gallium antimonide, *Sov. Phys. Sol. Stat.* 7, 2757 (1966).

157. Kurbatov, D. N., Dirochka, A. I., and Ogorodnik, A. D., Recombination radiation of In$_2$Se, *Sov. Phys. Semiconductor* 4, 1195 (1971).

158. Alexander, F. B., Bird, V. R., and Carpenter, D. B., Spontaneous and stimulated infrared emission from indium phosphide-arsenide diodes, *Appl. Phys. Lett.* 4, 13 (1964).

159. Zah, C.E., Bhat, R., Cheung, K. W. et al., Low-threshold (< 92 A/cm^2) 1.6 µm strained-layer single quantum well laser diodes optically pumped by a 0.8 µm laser diode, *Appl. Phys. Lett.* 57, 1608 (1990).

160. Melngailis, I., Strauss, A. J., and Rediker, R. H., Semiconductor $In_xGa_{1-x}As$ diode masers, *Proc. IEEE* 51, 1154 (1963).

161. Kano, H. and Sugiyama, K., 2.0 μm C.W. operation of GaInAsSb/GaSb D.H. lasers at 80 K, *Electron. Lett.* 16, 146 (1980).

162. Chiu, T. H., Tsang, W. T., Ditzenberger, J. A., and van der Ziel, J. P., Room-temperature operation of InGaAsSb/AlAsSb double-heterostructure lasers near 2.2 μm prepared by molecular beam epitaxy, *Appl. Phys. Lett.* 49, 1051 (1986).

163. Caneau, C., Srivastava, A. K., Zyskind, J. L., Sulhoff, J. W., Dentai, A. G., and Pollack, M. A., CW operation on GaInAsSb/AlGaAsSb laser up to 190 K, *Appl. Phys. Lett.* 49, 55 (1986).

164. Bishop, S. G., Moore, W. J., and Swiggard, E. M., Optically pumped Cd_3P_2 laser, *Appl. Phys. Lett.* 16, 459 (1970).

165. Choi, H. K. and Eglash, S. J., Room-temperature cw operation at 2.2 μm of GaInAsSb/AlGaAsSb diode lasers grown by molecular beam epitaxy, *Appl. Phys. Lett.* 59, 1165 (1991).

166. Gianordoli, S., Hvozara, L., Strasser, G., Schrenk, W., Unterrainer, K., and Gornik, E., GaAs/AlGaAs-based microcylinder lasers emitting at 10 μm, *Appl. Phys. Lett.* 75, 1045 (1999).

167. Mahavadi, K. K., Bleuse, J., Sivananthan, S., and Faurie, J. P., Stimulated emission from a CdTe/HgCdTe separate confinement heterostructure grown by molecular beam epitaxy, *J. Appl. Phys.* 56, 2077 (1990).

168. Kryukona, I. V., Leskovich, V. I., and Matveenko, E. V., Mechanisms of laser action in epitaxial InAs subjected to electron beam excitation, *Sov. J. Quantum Electron.* 9, 823 (1979).

169. Baranov, A. N., Imenkov, A. N., Sherstnev, V. V., and Yakovlev, Yu. P., 2.7-3.9 μm InAsSb(P)/InAsSbP low threshold diode lasers, *Appl. Phys. Lett.* 64, 2480 (1994).

170. Partin, D. L., Majkowski, R. F., and Swets, D. E., Quantum well diode lasers of lead-europium-selenide-telluride, *J. Vac. Sci. Technol. B* 3, 576 (1985).

171. Harman, T., Optically pumped LPE-grown $Hg_{1-x}Cd_xTe$ lasers, *J. Electron. Mater.* 8, 191 (1979).

172. Hasenberg, T. C., Miles, R. H., Kost, A. R., and West, L., Recent advances in Sb-based midwave-infrared lasers, *IEEE J. Quantum Electron.* 33, 1403 (1997).

173. Giles, N. C., Han, J. W., Cook, J. W., and Schetzina, J. F., Stimulated emission at 2.8 μm from Hg-based quantum well structures grown by photo-assisted molecular beam epitaxy, *J. Appl. Phys.* 55, 2026 (1989).

174. Zandian, M., Arias, J. M., Zucca, R., Gil, R. V., and Shin, S. H., HgCdTe double heterostructure injection lasers grown by molecular beam epitaxy, *J. Appl. Phys.* 59, 1022 (1991).

175. Tacke, M., Spanger, B., Lambrecht, A., Norton, P. R., and Böttner, H., Infrared double-heterostructure diode lasers made by molecular beam epitaxy of $Pb_{1-x}Eu_xSe$, *Appl. Phys. Lett.* 53, 2260 (1988).

176. Zucca, R., Zandian, M., Arias, J. M., and Gil, R. V., Mid-IR HgCdTe double heterostructure lasers, *SPIE* Vol. 1634, 161 (1992).

177. Zucca, R., Zandian, M., Arias, J. M., and Gil, R. V., HgCdTe double heterostructure diode lasers grown by molecular-beam epitaxy, *J. Vac. Sci. Technol. B* 10, 1587 (1992).

178. Sato, S., Osawa, Y., and Saitoh, Y., Room-temperature operation of GaInNAs/GaInP DH laser diodes grown by MOCVD, *Jpn. J. Appl. Phys.* 36(5A), 2671 (1997).

179. Ishida, A., Muramatsu, K., Takashiba, H., and Fuyiyasu, H., $Pb_{1-x}Sr_xS/PbS$ double-heterostructure lasers prepared by hot-wall expitaxy, *Appl. Phys. Lett.* 55, 430 (1989).

180. Melngailis, I., Optically pumped InAs laser, *IEEE J. Quantum Electron.* QE-1, 104 (1965).

181. Menna, R. J., Capewell, D. R., Martinelli, R. U., York, P. K., and Enstrom, R. E., 3 μm InGaAsSb/InPSb diode lasers grown by organometallic vapor-phase epitaxy, *SPIE* Vol. 1634, 174 (1992).

182. Aidaraliev, M., Zotova, N. V., Karndachev, S.A., Matseev, B. A., Stus', N. M., and Talalakin, G. N., Low-threshold lasers for the interval 3-3.5 μm based on InAsSbP/In$_{1-x}$Ga$_x$As$_{1-y}$Sb$_y$ double heterostructures, *Sov. Tech. Phys. Lett.* 15, 600 (1989).

183. Gmachl, C., Faist, J., Baillargeon, J. N., Capasso, F., Sirtori, C., Sivco, D. L., Chu, S. N. G., and Cho, A. Y., Complex-coupled quantum cascade distributed-feedback laser, *IEEE Phot. Tech. Lett.* 9, 1090 (1997).

184. van der Ziel, J. P., Logan, R. A., Mikulyak, R. M., and Ballman, A. A., Laser oscillation at 3-4 μm from optically pumped InAs$_{1-x-y}$Sb$_x$P$_y$, *IEEE J. Quantum Electron.* QE-21, 1827 (1985).

185. Melngailis, I., Maser action in InAs diodes, *Appl. Phys. Lett.* 2, 176 (1963).

186. Basov, N. G., Dudenkova, A. V., and Krasilnikov, A. I., Semiconductor p-n junction lasers in the InAs$_{1-x}$Sb$_x$ system, *Sov. Phys. Sol. Stat.* 8, 847 (1966).

187. Chow, D. H., Miles, R. H., Hasenberg, T. C., Kost, A. R., Zhang, Y.-H., Dunlap, H. L., and West, L., Mid-wave infrared diode lasers based on GaInSb/InAs and In As/AlSb superlattices, *Appl. Phys. Lett.* 67, 3700 (1995).

188. Zhang, Y.-H., Continuous wave operation of InAs/InAs$_x$Sb$_{1-x}$ midinfrared lasers, *Appl. Phys. Lett.* 66, 118 (1995).

189. Nishijima, Y., PbSnTe double-heterostructure lasers and PbEuTe double-heterostructure lasers by hot-wall epitaxy, *J. Appl. Phys.* 65, 935 (1989).

190. Nill, K. W., Strauss, A. J., and Blum, F. A., Tunable cw Pb$_{0.98}$Cd$_{0.02}$S diode laser emitting at 3.5 μm, *Appl. Phys. Lett.* 22, 677 (1973).

191. Kurtz, S. R., Allerman, A. A., and Biefeld, R. M., Midinfrared lasers and light-emitting diodes with InAsSb/InAsP strained-layer superlattice active regions, *Appl. Phys. Lett.* 70, 3188 (1997).

192. Ravid, A. and Zussman, A., Laser action and photoluminescence in an indium-doped *n*-type Hg$_{1-x}$Cd$_x$Te (x=0.375) layer grown by liquid phase epitaxy, *J. Appl. Phys.* 73, 3979 (1993).

193. Partin, D. L. and Thrush, C. M., Wavelength coverage of lead-europium-selenide-telluride diode lasers, *Appl. Phys. Lett.* 45, 193 (1984).

194. Lin, C.-H., Yang, R. Q., Zhang, D., Murry, S. J., Pei, S. S., Allerman, A. A., and Kurtz, S. R., Type-II interband quantum cascade laser at 3.8 μm, *Electron. Lett.* 33, 598 (1997).

195. Allerman, A. A., Biefeld, R. M., and Kurtz, S. R., InAsSb-based mid-infrared lasers (3.8-3.9 μm) and light-emitting diodes with AlAsSb claddings and semimetal electron injection, grown by metalorganic chemical vapor depostion, *Appl. Phys. Lett.* 69, 465 (1996).

196. van der Ziel, J. P., Chui, T. H., and Tsang, W. T., Optically pumped laser oscillation at 3.9 μm from Al$_{0.5}$Ga$_{0.5}$Sb/InAs$_{0.91}$Sb$_{0.09}$/Al$_{0.5}$Ga$_{0.5}$Sb double heterostructure grown by molecular beam epitaxy on GaSb, *Appl. Phys. Lett.* 48, 315 (1986).

197. Choi, H. K. and Turner, G. W., InAsSb/InAlAsSb strained quantum-well diode lasers emitting at 3.9 μm, *Appl. Phys. Lett.* 67, 332 (1995).

198. Malin, J. I., Meyer, J. R., Felix, C. L. et al., Type II mid-infrared quantum well laser, *Appl. Phys. Lett.* 68, 2976 (1996).

199. Akasaki, I., Sota, S., Sakai, H., Tanaka, T., Koike, M., and Amano, H., Shortest wavelength semiconductor laser diode, *Electron. Lett.* 32, 1105 (1996).

200. Partin, D. L., Single quantum well lead-europium-selenide-telluride diode lasers, *Appl. Phys. Lett.* 45, 487 (1984).

201. Yoshida, T., Miyazaki, K., and Fujisawa, K., Emission properties of two-photon pumped InSb laser under magnetic field, *Jpn. J. Appl. Phys.* 14, 1987 (1975).

202. Eglash, S. J. and Choi, H. K., InAsSb/AlAsSb double-heterostructure diode lasers emitting at 4 μm, *Appl. Phys. Lett.* 64, 833 (1994).

203. Felix, C. L., Meyer, J. R., Vurgaftman, I., Lin, C.-H., Murry, S. J., Zhang, D., and Pei, S.-S., High-temperature 4.5 μm type-II quantum-well laser with Auger suppression, *IEEE Phot. Techn. Lett.* 9, 734 (1997).

204. Feit, Z., Kostyk, D., Woods, R. J., and Mak, P., Single-mode molecular beam epitaxy grown PbEuSeTe/PbTe buried-heterostructure diode lasers for CO_2 high-resolution spectroscopy, *Appl. Phys. Lett.* 58, 343 (1991).

205. Faist, J., Capasso, F., Sivco, D. L., Sirtori, C., Hutchinson. A. L., and Cho, A. Y., Quantum cascade laser, *Science* 264, 553 (1994).

206. Hurwitz, C. E., Calawa, A. R., and Rediker, R. H., Electron beam pumped laser of PbS, PbSe and PbTe, *IEEE J. Quantum Electron.* QE-1, 102 (1965).

207. Butler, J. F. and Calawa, A. R., PbS diode laser, *J. Electrochem. Soc.* 112, 1056 (1965).

208. Ralston, R. W., Waipole, J. M., and Calawa, A. R., High CW output power in stripe-geometry PbS diode lasers, *J. Appl. Phys.* 45, 1323 (1974).

209. Anticliffe, G. A., Parker, S. G., and Bate, R. T., CW operation and nitric oxide spectroscopy using diode laser of $Pb_{1-x}Ge_xTe$, *Appl. Phys. Lett.* 21, 505 (1972).

210. Spanger, B., Schiesse, U., Lambrecht, A., Böttner, H., and Tacke, M., Naer-room-temperature operation of $Pb_{1-x}Sr_xSe$ infrared diode lasers using molecular beam epitaxy growth techniques, *Appl. Phys. Lett.* 53, 2583 (1988).

211. Feit, Z., Kostyk, D., Woods, R. J., and Mak, P., PbEuSeTe buried heterostructure lasers grown by molecular-beam epitaxy, *J. Vac. Sci. Tech. B* 8, 200 (1990).

212. Faist, J., Capasso, F., Sirtori, C., Sivco, D. L., Hutchinson. A. L., and Cho, A. Y., Continuous wave operation of a vertical transition quantum cascade laser above T=80 K, *Appl. Phys. Lett.* 67, 3057 (1995).

213. Feit, Z., Kostyk, D., Woods, R. J., and Mak, P., Molecular beam epitaxy-grown PbSnTe-PbEuSeTe buried heterostructure diode lasers, *IEEE Phot. Technol. Lett.* 2, 860 (1990).

214. Ashley, T., Elliott, C. T., Jefferies, R., Johnson, A. D., Pryce, G. J., and White, A. M., Mid-infrared $In_{1-x}Al_xSb$/InSb heterostructure diode lasers, *Appl. Phys. Lett.* 70, 931 (1997).

215. Melngailis, I., Phelan, R. J., and Rediker, R. H., Luminescence and coherent emission in large-volume injection plasma in InSb, *Appl. Phys. Lett.* 5, 99 (1964).

216. Faist, J., Capasso, F., Sirtori, C., Sivco, D. L., Hutchinson. A. L., and Cho, A. Y., Room temperature mid-infrared quantum cascade lasers, *Electron. Lett.* 32, 560 (1996).

217. Faist, J., Capasso, F., Sirtori, C., Sivco, D. L., Baillargeon, J. N., Hutchinson. A. L., Chu, S. N. G., and Cho, A. Y., High power mid-infrared (λ~5 μm) quantum cascade lasers operating above room temperature, *Appl. Phys. Lett.* 68, 3680 (1996).

218. Faist, J., Capasso, F., Sirtori, C., Sivco, D. L., Baillargeon, J. N., Hutchinson. A. L., Chu, S. N. G., and Cho, A. Y., High power mid-infrared quantum cascade lasers with a molecular beam epitaxy grown InP cladding operating above room temperature, *J. Cryst. Growth* 175/176, 22 (1997).

219. Faist, J., Tredicucci, A., Capasso, Sirtori, C. et al., High-power continuous-wave quantum cascade lasers, *IEEE J. Quantum Electron.* 34, 336 (1998).

220. Becla, P., HgMnTe light emitting diodes and laser heterostructures, *J. Vac. Sci. Technol. A* 6, 2725 (1988).

221. Kurbatov, L. N., Britov, A. D., and Dirochka, A. I., Stimulated radiation from solid solutions of chalcogenide of lead and tin in the range of 10 μm, *Quantum Electron.* Basov, N. G., Ed., *Soviet Radio* (1972), p. 97.

222. Ravid, A., Zussman, A., and Sher, A., Optically pumped $Hg_{1-x}Zn_xTe$ lasers grown by liquid phase epitaxy, *J. Appl. Phys.* 58, 337 (1991).

223. Schlereth, K.-H., Spanger, B., Böttner, H., Lambrecht, A., and Tacke, M., Buried waveguide double-heterostructure PbEuSe-lasers grown by MBE, *Infrared Phys.* 30, 449 (1990).

224. Bründermann, E., Linhart, A.M., Reichertz, L., Röser, H.P., Dubon, O.D., Hansen, W.L., Sirmain, G., and Haller, E.E., Double acceptor doped Ge: a new medium for inter-valence-band lasers, *Appl. Phys. Lett.* 68, 3075 (1996).

225. Groves, S. H., Nill, K. W., and Strauss, A. J., Double heterostructure $Pb_{1-x}Sn_xTe$-PbTe lasers with cw operation at 77 K, *Appl. Phys. Lett.* 25, 331 (1974).

226. Reichertz, L. A., Dubon, O. D., Sirmain, G., Bründermann, E., Hansen, W. L., Chamberlin, D. R., Linhart, A. M., Röser, H. P., and Haller, E. E., Stimulated far-infrared emission from combined cyclotron resonances in germanium, *Phys. Rev. B* 56, 12069 (1997).

227. Sirmain, G., Reichertz, L. A., Dubon, O. D., Haller, E. E., Hansen, W. L., Bründermann, E., Linhart, A. M., and Röser, H. P., Stimulated far-infrared emission from copper-doped germanium crystals, *Appl. Phys. Lett.* 70, 1659 (1997).

228. Ishida, A., Fujiyasu, H., Ebe, H., and Shinohara, K., Lasing mechanism of type-I' PbSnTe-PbTeSe multiquantum well laser with doping structure, *J. Appl. Phys.* 59, 3023 (1986).

229. Shinohara, K., Nishijima, Y., Ebe, H., Ishida, A., and Fujiyasu, H., PbSnTe multiple quantum well lasers for pulsed operation at 6 μm up to 204 K, *Appl. Phys. Lett.* 47, 1184 (1985).

230. McLane, G. F. and Sleger, K. J., Vacuum deposited epitaxial layers of $PbS_{1-x}Se_x$ for laser devices. *J. Electron. Mat.* 4, 465 (1975).

231. Faist, J., Capasso, F., Sirtori, C., Sivco, D. L., Hutchinson. A. L., and Cho, A. Y., Laser action by tuning the oscillator strength, *Nature* 387, 777 (1997).

232. Butler, J. F., Calawa, R. A., Phelan, R. J., Jarman, T. C., Strauss, A. J., and Rediker, R. H., PbTe diode laser, *Appl. Phys. Lett.* 5, 75 (1964).

233. Sirtori, C., Faist, J., Capasso, F., Sivco, D. L., Hutchinson, A. L., and Cho, A. Y., Long wavelength infrared ($\lambda \approx 11$ μm) quantum cascade lasers, *Appl. Phys. Lett.* 69, 2810 (1996).

234. Sirtori, C., Faist, J., Capasso, F., Sivco, D. L., Hutchinson, A. L., Chu, S. N. G., and Cho, A. Y., Continuous wave operation of midinfrared (7.4–8.6 μm) quantum cascade lasers up to 110 K temperature, *Appl. Phys. Lett.* 68, 1745 (1996).

235. Scamarcio, G., Capasso, F., Sirtori, C., Faist, J., Hutchinson, A. L., Sivco, D. L., and Cho, A. Y., High-power infrared (8-micrometer wavelength) superlattice lasers, *Science* 276, 773 (1997).

236. Zussman, A., Felt, Z., Eger, D., and Shahar, A., Long wavelength $Pb_{1-x}Sn_xTe$ homostructure diode lasers having a gallium doped cladding layer, *Appl. Phys. Lett.* 42, 344 (1983).

237. Sirtori, C., Faist, J., Capasso, F., Sivco, D. L., Hutchinson, A. L., and Cho, A. Y., Pulsed and continuous-wave operation of long wavelength infrared (λ=9.3 μm) quantum cascade lasers, *IEEE J. Quantum Electron.* 33, 89 (1997).

238. Sirtori, C., Faist, J., Capasso, F., Sivco, D. L., Hutchinson, A. L., and Cho, A. Y., Mid-infrared (8.5 μm) semiconductor lasers operating at room temperature, *IEEE Phot. Tech. Lett.* 9, 294 (1997).

239. Kasemset, D., Rotter, S., and Fonstad, C. G., $Pb_{1-x}Sn_xTe/PbTe_{1-y}Se_y$ lattice-matched buried heterostructure lasers with cw single mode output, *IEEE Electron. Device Lett.* EDL-1, 75 (1980).

240. Butler, J. F., Calawa, A. R., Phelan, R. J., Strauss, A. J., and Rediker, R. H., PbSe diode laser, *Solid State Commun.* 2, 303 (1964).

241. Butler, J. F., Calawa, A. R., and Harman, T. C., Diode lasers of $Pb_{1-x}Sn_xSe$ and $Pb_{1-x}Sn_xTe$, *Appl. Phys. Lett.* 9, 427 (1966).

242. Gmachl, C., Faist, J., Capasso, F., Sirtori, C., Sivco, D. L., and Cho, A. Y., Long-wavelength (9.5-11.5 μm) microdisk quantum-cascade lasers, *IEEE J. Quantum Electron.* 33, 1567 (1997).

243. Aggarwal, R. L., Maki, P. A., Molnar, R. J., Liau, Z.-L., and Melngailis, I., Optically pumped $GaN/Al_{0.1}Ga_{0.9}N$ double-heterostructure ultraviolet laser, *J. Appl. Phys.* 79, 2148 (1996).

244. Gauthier-Lafaye, O., Sauvage, S., Boucaud, P., Julien, F. H., Prazeres, R., Glotin, F., Ortega, J.-M., Thierry-Mieg, V., Planel, R., Leburton, J.-P., and Berrger, V., Intersubband stimulated emission in GaAs/AlGaAs quantum wells: pump-probe experiments using a two-color free-electron laser, *Appl. Phys. Lett.* 70, 3197 (1997).

245. Andronov, A. A., Zverev, I. V., Kozlov, V. A., Nozdrin, Yu. N., Pavlov, S. A., and Shastin, V. N., Stimulated emission in the long-wave IR region from hot holes in Ge in crossed electric and magnetic fields, *Sov. Phys.-JETP Lett.* 40, 804 (1984).

246. Komiyama, S., Iizuka, N., and Akasaka, Y., Evidence for induced far-infrared emission from p-Ge in crossed electric and magnetic fields, *Appl. Phys. Lett.* 47, 958 (1985).

247. Andronov, A. A., Nozdrin, Yu. N., and Shastin, V. N.,Tunable FIR lasers in semiconductors using hot holes, *Infrared Phys.* 27, 31 (1987).

248. Kuroda, S. and Komiyama, S., Far-infrared laser oscillation in p-type Ge: remarkably improved operation under uniaxial stress, *Infrared Phys.* 29, 361 (1989).

249. Kremser, C., Heiss, W., Unterrainer, K., Gornik, E., Haller, E. E., and Hansen, W.L., Stimulated emission from p-Ge due to transitions between light-hole Landau levels and excited states of shallow impurities, *Appl. Phys. Lett.* 60, 1785 (1992).

250. Komiyama, S., Morita, H., and Hosako, I., Continuous wavelength tuning of intervalence-band laser oscillation in p-type germanium over range of 80-120 µm, *Jpn. J. Appl. Phys.* 32, 4987 (1993).

251. Heiss, W., Kremser, C., Unterrainer, K., Strasser, G., Gornik, E., Meny, C., and Leotin, J., Influence of impurities on broadband p-type-Ge laser spectra under uniaxial stress, *Phys. Rev.* B47, 16586 (1993).

252. Bründermann, E., Röser, H. P., Muravjov, A. V., Pavlov, S. G., and Shastin, V. N., Mode fine structure of the FIR p-Ge intervalence band laser measured by heterodyne mixing spectroscopy with an optically pumped ring gas laser, *Infrared Phys. Technol.* 1, 59 (1995).

253. Bründermann, E., Röser, H. P., Heiss, W., Gornik, E., and Haller, E.E., Miniaturization of p-Ge lasers: progress toward continuous wave operation, *Appl. Phys. Lett.* 68, 1359 (1996).

254. Park, K., Peale, R. E., Weidner, H., and Kim, J. J., Submillimeter p-Ge laser using a Voigt-configured permanent magnet, *IEEE J. Quantum Electron.* 32, 1203 (1996).

255. Hovenier, J. N., Muravjov, A. V., Pavlov, S. G., Shastin, V.N., Strijbos, R. C., and Wenckebach, W. Th., Active mode locking of a p-Ge hot hole laser, *Appl. Phys. Lett.* 71, 443 (1997).

256. Bründermann, E. and Röser, H. P., First operation of a far-infrared p-Germanium laser in a standard closed-cycle machine at 15 Kelvin, *Infrared Phys. Technol.* 38, 201 (1997).

257. Hovenier, J. N., Klaassen, T. O., Wenckebach, W. Th., Muravjov, A. V., Pavlov, S. G., and Shastin, V.N., Gain of the mode locking p-Ge laser in the low field region, *Appl. Phys. Lett.* 72, 1140 (1998).

258. Aleksanyan, A. G., Kazaryan, R. K., and Khachatryan, A. M., Semiconductor laser made of $Bi_{1-x}Sb_x$, *Sov. J. Quantum Electron.* 14, 336 (1984).

259. Ivanov, Yu. L. and Vasil'ev, Yu. B., Submillimeter emission from hot holes in germanium in a transverse magnetic field, *Pis'ma v Zhurnal Tekhnicheskoi Fizika* 9, 613 (1983), Translation: *Sov. Tech. Phys. Lett.* 9, 264 (1983).

260. Mityagin, Yu. A., Murzin, V. N., Stepanov, O. N., and Stoklitsky, S. A., Cyclotron resonance submillimeter laser emission in hot hole Landau level system in uniaxially stressed p-germanium, *Physica Scripta* 49, 699 (1994).

261. Pfeffer, P., Zawadzki, W., Unterrainer, K., Kremser, C., Wurzer, C., Gornik, E., Murdin, B., and Pidgeon, C.R., p-type Ge cyclotron-resonance laser: Theory and experiment, *Phys. Rev.* B 47, 4522 (1993).

262. Bonnet-Gemard, J., Bleuse, J., Magnea, N., and Pautrat, J. L., Optical gain and laser emission in HgCdTe heterostructures, *J. Appl. Phys.* 78, 6908 (1995).

263. Ravid, A., Cinader, G., and Zussman, A., Optically pumped laser action in double-heterostructure HgCdTe grown by metalorganic chemical deposition on a CdTe substrate, *J. Appl. Phys.* 74, 15 (1993).

264. Million, A., Colin, T., Ferret, P., Zanattan, J. P., Bouchot, P., Destéfanis, G. L., and Bablet, J., HgCdTe double heterostructure for infrared injection laser, *J. Cryst. Growth* 127, 291 (1993).

265. Bleuse, J., Magnea, N., Pautrat, J. L., and Mriette, H., Cavity structure effects on CdHgTe photopumped heterostructure lasers, *Semicond. Sci. Technol.* 8, SC266 (1993).

266. Ravid, A., Sher, A., Cinader, G., and Zussman, A., Optically pumped laser action and photoluminescence in HgCdTe layer grown on (211) CdTe by metalorganic chemical vapor deposition, *J. Appl. Phys.* 73, 7102 (1993).

267. Grillo, D. C., Han, J., Ringle, M., Hua, G., Gunshor, R. L., Kelkar, P., Kozlov, V., Jeon, H., and Nurmikko, A. V., Blue ZnSe quantum-well diode laser, *Electron. Lett.* 30, 2131 (1994).

268. Waag, A., Fischer, F., Schüll, K. et al., Laser diodes based on beryllium-chalcogenides, *Appl. Phys. Lett.* 70, 280 (1997).

269. Gaines, J. M., Crenten, R. R., Haberern, K. W., Marshall, T., Mensz, P., and Petruzzello, J., Blue-green injection lasers containing pseudomorphic $Zn_{1-x}Mg_x-S_ySe_{1-y}$ cladding layers and operating up to 394 K, *Appl. Phys. Lett.* 62, 2462 (1993).

270. Nakayama, N., Itoh, S., Ohata, T., Nakano, K., Okuyama, H., Ozawa, M., Ishibashi, A., Ikeda, M., and Mori, Y., Room temperature continuous operation of blue-green laser diodes, *Electron. Lett.* 29, 1488 (1993).

271. Yamada, Y., Masumoto, Y., Mullins, J. T., and Taguchi, T., Ultraviolet stimulated emission and optical gain spectra in $Cd_xZn_{1-x}S$-ZnS strained-layer superlattices, *Appl. Phys. Lett.* 61, 2190 (1992).

272. Nakanichi, K., Suemune, I., Fujii, Y., Kuroda, Y., and Yamanishi, M., Extremely-low-threshold and high-temperature operation in a photopumped ZnSe/ZnSSe blue laser, *Appl. Phys. Lett.* 59, 1401 (1991).

273. Taguchi, T., Onodera, C., Yamada, Y., and Masumoto, Y., Band offsets in CdZnS/ZnS strained-layer quantum well and its application to UV laser diode, *Jpn. J. Appl. Phys.* 32, L1308 (1993).

274. Ozanyan, K. B., Nicholls, J. E., O'Neill, M., May, L., Hogg, J. H. C., Hagstom, W. E., Lunn, B., and Ashenford, D. E., Spectroscopic evidence for the excitonic lasing mechanism in ultraviolet ZnS/ZnCdS multiple quantum well lasers, *Appl. Phys. Lett.* 69, 4230 (1996).

275. Le, H. Q., Arias, J. M., Zandian, M., Zucca, R., and Liu, Y.-Z., High-power diode-laser-pumped midwave infrared HgCdTe/CdZnTe quantum-well lasers, *J. Appl. Phys.* 65, 810 (1994).

276. Bleuse, J., Magnea, N., Ulmer, L., Pautrat, J. L., and Mriette, H., Room-temperature laser emission near 2 µm from an optically pumped HgCdTe separate-confinement heterostructure, *J. Cryst. Growth* 117, 1046 (1992).

277. Bonnet-Gemard, J., Bleuse, J., Magnea, N., and Pautrat, J. L., Emission wavelength and cavity design dependence of laser behaviour in HgCdTe heterostructures, *J. Cryst. Growth* 159, 613 (1996).

278. Bouchut, P., Destefanis, G., Bablet, J., Million, A., Colin, T., and Ravetto, M., Mesa stripe transverse injection laser in HgCdTe, *J. Appl. Phys.* 61, 1561 (1992).

279. Flatte', M. E., Hasenberg, T. C., Olesberg, J. T., Anson, S. A., Boggess, T. F., Yan, C., and McDaniel, D. L. Jr., III-V interband 5.2 µm laser operating at 185 K, *Appl. Phys. Lett.* 71, 3764 (1997).

280. Feit, Z., Kostyk, D., Woods, R. J., and Mak, P., Molecular beam epitaxy-grown separate confinement buried heterostructure PbEuSeTe-PbTe diode lasers, *IEEE Phot. Technol. Lett.* 7, 1403 (1995).

281. Feit, Z., McDonald, M., Woods, R. J., Archambault, V., and Mak, P., Low-threshold PbEuSeTe/PbTe separate confinement buried heterostructure diode lasers, *Appl. Phys. Lett.* 68, 738 (1996).

282. Sato, S., Osawa, Y., and Saitoh, Y., Room-temperature operation of GaInNAs/GaInP DH laser diodes grown by MOCVD, *Jpn. J. Appl. Phys.* 36(5A), 2671 (1997).

283. Garbuzov, D. Z., Martinelli, R. U., Menna, R. J., York, P. K., Lee, H., Narayan, S. Y., and Connolly, J. C., 2.7-μm InGaAsSb/AlGaAsSb laser diodes with continuous-wave operation up to -39°C, *Appl. Phys. Lett.* 67, 1346 (1995).

284. Le, H. Q., Turner, G. W., Eglash, S. J., Choi, H. K., and Coppeta, D. A., High-power diode-laser-pumped InAsSb/GaSb and GaInAsSb/GaSb lasers emitting from 3 to 4 μm, *Appl. Phys. Lett.* 64, 152 (1994).

285. Malin, J. I., Felix, C. L., Meyer, J. R., Hoffman, C. A., Pinto, J. F., Lin, C.-H., Chang, P. C., Murry, S. J., and Pei, S. S., Type II mid-IR lasers operating above room temperature, *Electron. Lett.* 32, 1593 (1996).

286. Popov, A., Sherstnev, V., Yakovlev, Y., Mücke, R., and Werle, P., High power InAsSb/InAsSbP double heterostructure laser for continuous wave operation at 3.6 μm, *Appl. Phys. Lett.* 68, 2790 (1996).

287. Kurtz, S. R., Biefeld, R. M., Allerman, A. A., Howard, A. J., Crawford, M. H., and Pelczynski, M. W., Pseudomorphic InAsSb multiple quantum well injection laser emitting at 3.5 μm, *Appl. Phys. Lett.* 68, 1332 (1996).

288. Diaz, J., Yi, H., Rybaltowski, A., Lane, B., Lukas, G., Wu, D., Kim, S., Erdtmann, M., Kaas, E., and Razeghi, M., InAsSbP/InAsSb/InAs laser diodes (λ=3.2 μm) grown by low-pressure metal-organic chemical-vapor deposition, *Appl. Phys. Lett.* 70, 40 (1997).

289. Garbuzov, D. Z., Martinelli, R. U., Lee, H., Menna, R. J., York, P. K., DiMarco, L. A., Harvey, M. G., Matarese, R. J., Narayan, S. Y., and Connolly, J. C., 4 W quasi-continuous-wave output power from 2 μm AlGaAsSb/InGaAsSb single-quantum-well broadened waveguide laser diodes, *Appl. Phys. Lett.* 70, 2931 (1997).

290. Trager-Cowan, C., Bagnall, D. M., McGow, F. et al., Electron beam pumping of CdZnSe quantum well laser structures using a variable energy electron beam, *J. Cryst. Growth* 159, 618 (1996).

291. Nakamura, S., Senoh, M., Nagahama, S., Iwasa, N., Yamada, T., Matsusita, T., Sugimoto, Y., and Kiyoku, H., Subband emissions from InGaN multi-quantum-well laser diodes under room-temperature continuous wave operation, *Appl. Phys. Lett.* 70(20), 2753 (1997).

292. Hofstetter, D., Bour, D. P., Thorton, R. L., and Johnson, N. M., Excitation of a higher order transverse mode in an optically pumped $In_{0.15}Ga_{0.85}N/In_{0.05}Ga_{0.95}N$ multiquantum well laser structure, *Appl. Phys. Lett.* 70(13), 1650 (1997).

293. Gluschenkov, O., Myoung, J. M., Shim, K. H., Kim, K., Gigen, Z. G., Gao, J., and Eden, J. G., Stimulated emission at 300 K from photopumped GaN grown by plasma-assisted molecular beam epitaxy with an inductively coupled plasma source, *Appl. Phys. Lett.* 70, 811 (1997).

294. Nakadaira, A. and Tanaka, H., Stimulated emission at 34 K from an optically pumped cubic GaN/AlGaN heterostructure grown by metalorganic vapor-phase epitaxy, *Appl. Phys. Lett.* 71, 811 (1997).

295. Hofmann, R., Gauggel, H.-P., Griesinger, U. A., Gräbeldinger, H., Adler, F., Ernst, P., Bola, H., Härle, V., Scholz, F., Schweizer, H., and Pilkuhn, M. H., Realization of optically pumped second-order GaInN-distributed-feedback lasers, *Appl. Phys. Lett.* 69, 2068 (1996).

296. Kim, S. T., Amano, H., and Akasaki, I., Surface-mode stimulated emission from optically pumped GaInN at room temperature, *Appl. Phys. Lett.* 67, 267 (1996).

297. Nakamura, S., Senoh, M., Nagahame, S., Iwasa, N., Yamada, T., Matsushita, T., Sugimoto, Y., and Kiyoku. H., Ridge-geometry InGaN multi-quantum-well-structure laser diodes, *Appl. Phys. Lett.* 69, 1477 (1996).

298. Scifres, D. R., Burnham, R. D., and Streifer, W., Highly collimated laser beams from electrically pumped SH GaAs/GaAlAs distributed-feedback laser, *Appl. Phys. Lett.* 26, 48 (1975).

299. Bewley, W. W., Aifer, E. H., Felix, C. L., Vurgaftman, I., Meyer, J. R., Lin, C.-H., Murry, S. J., Zhang, D., and Pei, S.-S., High-temperature type-II superlattice diode laser at λ-2.9 μm, *Appl. Phys. Lett.* 71, 3607 (1997).

300. Al-Muhanna, A., Wade, J. K., Mawst, L. J., and Fu, L. J., 730-nm-emitting Al-free active-region diode lasers with compressively strained InGaAsP quantum wells, *Appl. Phys. Lett.* 72, 641 (1998).

301. Wade, J. K., Mawst, L. J., Botez, D., Nabiev, R. F., and Jansen, M., 5 W continuous wave power, 0.81 μm-emitting, Al-free active-region diode lasers, *Appl. Phys. Lett.* 71, 172 (1997).

302. Pellandini, P., Stanley, R. P., Houdré, J., Oesterle, U., Ilegems, M., and Weisbuch, C., Dual-wavelength laser emission form a coupled semiconductor microcavity, *Appl. Phys. Lett.* 71, 864 (1997).

303. Goni, A. R., Stroh, M., Thomsen, C., Heinrichsdorff, F., Türck, V., Krost, A., and Bimberg. D., High-gain excitonic lasing from a single InAs monolayer in bulk GaAs, *Appl. Phys. Lett.* 72, 1433 (1998).

304. Felix, C. L., Bewley, W. W., Vurgaftman, I. Meyer, J. R., Goldberg, L., Chow, D. H., and Selvig, E., Midinfrared vertical-cavity surface emitting laser, *Appl. Phys. Lett.* 71, 3483 (1997) .

305. Choi, H. K., Turner, G. W., Manfra, M. J., and Connors, M. K., 175 K continuous wave operation of InAsSb/InAlAsSb quantum-well diode lasers emitting at 3.5 μm, *Appl. Phys. Lett.* 68, 2936 (1996).

306. Felix, C. L., Bewley, W. W., Vurgaftman, I., Meyer, J. R., Goldberg, L., Chow, D. H., and Selvig, E., Midinfrared vertical-cavity surface emitting laser, *Appl. Phys. Lett.* 71, 3483 (1997) .

307. Le, H. Q., Turner, G. W., and Ochoa, J. R., High-power high-efficiency quasi-CW Sb-based mid-IR lasers sing 1.9-μm laser diode pmping, *IEEE Photon. Technol. Lett.* 10, 663 (1998).

308. Ichida, H., Nakayama, M., and Nishimura, H., Stimulated emission from exciton-exciton scattering in CuBr thin films, *J. Lumin.* 87-89 (2000).

309. Heiss, W., Unterrainer, K., Gornik, E., Hansen, W. L., and Haller, E. E., Influence of impurity absorption on germanium hot-hole laser spectra, *Semicond. Sci. Technol.* 9, B638 (1994).

310. Bründermann, E., Haller, E.E., and Muravjov, A. V., Terahertz emission of population-inverted hot-holes in single-crystalline silicon, *Appl. Phys. Lett.* 73, 723 (1998).

311. Masumoto, Y. and Kawamjra, T., Biexciton lasing in CuCl quantum dots, *Appl. Phys. Lett.* 62, 225 (1993); Masumoto, Y., Luminescence and lasing of CuCl nano-crystals, *J. Lumin.* 60/61, 256 (1994).

312. Phelan, R. J. and Redlker, R. H., Optically pumped semiconductor lasers, *Appl. Phys. Lett.* 6, 70 (1965).

313. Gmachl, C., Capasso, F., Tredicucci, A., Sivco, D. L. et al., Long wavelength (λ≈13 μm) quantum cascade lasers, *Electron. Lett.* 34, 1103 (1998).

314. Olafsen. L. J., Aifer, E. H., Vurgaftman, I., Bewley, W. W., Felix, C. L., Meyer, J. R., Zhang, D., Lin, C.-H., and Pei, S.-S., Near-room-temperature mid-infrared interband cascade laser, *Appl. Phys. Lett.* 72, 2370 (1998).

315. Bradshaw, J. L., Yang, R. Q., Bruno, J. D., Pham, J. T., and Wortman, D. E., High efficiency interband cascade lasers with peak powers exceeding 4W/facet, *Appl. Phys. Lett.* 75, 2362 (1999).

316. Tredicucci, A., Capasso, F., Gmachl, C., Sivco, D. L., Hutchinson, A. L., Cho, A. Y., Faist, J. F., and Scamarcio, G., High-power inter-miniband lasing in intrinsic superlattices, *Appl. Phys. Lett.* 72, 2388 (1998).

317. Faist, J., Sirtori, C., Capasso, F., Sivco, D. L. et al., High-power long-wavelength ($\lambda\sim11.5$ μm) quantum cascade lasers operating above room temperature, *IEEE Photon. Technol. Lett.* 10, 1100 (1998).

318. Allerman, A. A., Kurtz, S. R., Biefeld, R. M., and Baucom, K. C., 10-stage, 'cascaded' InAsSb quantum well laser at 3.9 μm, *Electron. Lett.* 34, 369 (1998).

319. Felix, C. L., Bewley, W. W., Vurgaftman, I., Meyer, J. R., Zhang, D., Lin, C.-H., Yang, R. Q., and Pei, S. S., Interband cascade laser emitting >1 photon per injected electron, *IEEE Photon. Technol. Lett.* 9, 1433 (1997).

320. Ko, J., Hegblom, E. R., Akulova, Y., Margalit, N. M., and Coldren, L. A., AlInGaAs/AsGaAs strained-layer 850 nm vertical-cavity lasers with very low thresholds, *Electron. Lett.* 33, 1551 (1997).

321. Hadji, E., Bleuse, J., Magnea, N., and Pautrat, J. L., Photopumped infrared vertical-cavity surface-emitting laser, *Appl. Phys. Lett.* 68, 2480 (1996).

322. Jewell, J. L., Scherer, A., McCall, S. L., Lee, Y. H., Walker, S., Harbison, J. P., and Florez, L. T., Low-threshold electrically pumped vertical-cavity surface-emitting mircrolasers, *Electron. Lett.* 25, 1123 (1989).

323. Felix, C. L., Bewley, W. W., Vurgaftman, I., Meyer, J. R., Goldberg, L., Chow, D. H., and Selvig, E., Midinfrared vertical-cavity surface emitting laser, *Appl. Phys. Lett.* 71, 3483 (1997) .

324. Bewley, W. W., Felix, C. L., Vurgaftman, I., Aifer, E. H., Meyer, J. R., Goldberg, L., Lindle, J. R., Chow, D. H., and Selvig, E., Continuous-wave mid-infrared VCSEL's, *IEEE Photon. Technol. Lett.* 10, 660 (1998).

325. Li, M. Y., Yuen, W., Li, G. S., and Chang-Hasnain, C. J., Top-emitting micromechanical VCSEL with a 31.6 nm tuning range, *Photon. Technol. Lett.* 10, 18 (1998) and *Electron. Lett.* 33, 1051 (1997).

326. Choquette, K. D., Schneider, Jr., R. P., Lear, K. L., and Geib, K. M., Low threshold voltage vertical cavity lasers fabricated by selective oxidation, *Electron. Lett.* 30, 2043 (1994).

327. Geels, R. S. and Coldren, L. A., Submilliamp threshold vertical-cavity laser diodes, *Appl. Phys. Lett.* 57, 1605 (1990).

328. Hu, S. Y., Hegblom, E. R., and Coldren, L. A., Multiple-wavelength top-emitting vertical-cavity photonic integrated emitter arrays for direct-coupled wavelength-division multiplexing applications, *Electron. Lett.* 34, 189 (1998).

329. Chang-Hasnain, C. J., Harbison, J. P., Zah, C.-E., Maeda, M. W., Florez, L. T., Stoffel, N. G., and Lee, T.-P., Multiple wavelength tunable surface emitting laser arrays, *IEEE J. Quantum Electron.* 27, 1368 (1991).; M.W. Maeda et al., Multi-gigabit/s operation of 16-wavelength vertical cavity surface emitting laser array, *IEEE Photon. Technol. Lett.* 3, 863 (1991).

330. Pellandini, P., Stanley, R. P., Houdré, J., Oesterle, U., Ilegems, M., and Weisbuch, C., Dual-wavelength laser emission form a coupled semiconductor microcavity, *Appl. Phys. Lett.* 71, 864 (1997).

331. Dudley, J. J., Babic´, D. I., Mirin, R. et al., Low threshold, wafer fused long wavlength vertical cavity lasers, *Appl. Phys. Lett.* 64, 1463 (1994) .

332. Soda, H., Iga, K., Kiahara, C., and Suematsu, Y., GaInAsP/InP surface emitting injection lasers, *Jpn. J. Appl. Phys.* 18, 2329 (1979).

333. Melngailis, I., Longitudinal injuection-plasma laser of InSb, *Appl. Phys. Lett.* 6, 59 (1995).

334. Zubrilov, A. S., Nikolaev, V. I., Tsvetkov, D. V. et al., Spontaneous and stimulated emission from photopumped GaN grown on SiC, *Appl. Phys. Lett.* 67, 533 (1995).

335. Holst, J., Eckey, L., Hoffmann, A. et al., Mechanisms of optical gain in cubic gallium nitrite, *Appl. Phys. Lett.* 72, 1439 (1998).

336. Klann, R., Brandt, O., Yang, H., Grahn, H. T., and Ploog, K. H., Optical gain in optically pumped cubic GaN at room temperature, *Appl. Phys. Lett.* 70, 1076 (1997).

337. Schmidt, T. J., Cho, Y.-H., Gainer, G. H., Song, J. J., Keller, S., Mishra, U. K., and DenBaars, S. P., Energy selective optically pumped stimulated emission from InGaN/GaN multiple quantum wells, *Appl. Phys. Lett.* 73, 560 (1998).

338. Gauthier-Lafaye, O., Boucaud, P., Julien, F. H., Sauvage, S., Cabaret, S., Lourtioz, J.-M., Thierry-Mieg, V., and Planel, R., Long-wavelength ($\approx$15.5 μm) uniploar semiconductor laser in GaAs quantum wells, *Appl. Phys. Lett.* 71, 3619 (1997).

339. Gauthier-Lafaye, O., Sauvage, S., Boucaud, P., Julien, F. H., Glotin, F., Prazeres, R., Ortega, J.-M., Thierry-Mieg, V., and Planel, R., Investigation of mid-infrared intersubband stimulated gain under optical pumping in GaAs/AlGaAs quantum wells, *J. Appl. Phys.* 83, 2920 (1998).

340. Tredicucci, A., Gmachl, C., Capasso, F., Sivco, D. L., Hutchinson, A. L., and Cho, A. Y., A multiwavelength semiconductor laser, *Nature* 396, 350 (1998).

341. Park, G., Shchekin, O. B., Huffaker, D. L., and Deppe, D. G., Lasing from InGaAs/GaAs quantum dots with extended wavelength and well-defined harmonic-oscillator energy levels, *Appl. Phys. Lett.* 73, 3351 (1998).

342. Choi, H. K., Walpol, J. N., Turner, G. W., Eglash, S. J., Missaggia, L. J., and Connors, M. K., GaInAsSb-AlGaAsSb tapered lasers emitting at 2 μm, *IEEE Photon. Technol. Lett.* 10, 1117 (1998).

343. Garbuzov, D. Z., Gokhale, M R., Dries, J. C., Studenkov, P., Martinelli, R. U., Connolly, J. C., and Forrest, S. R., 13.3W quasi-continuous operation of 0.99 μm wavelength SCH-QW InGaAs/GaAs/InGaP boradened waveguide lasers, *Electron. Lett.* 33, 1462 (1997).

344. Garbuzov, D. Z., Menna, R. J., Martinelli, R. U., Abeles, J. H., and Connolly, J. C., High power continuous and quasi-continuous wave InGaP/InP broad-waveguide separate confinement-heterostructure multiquantum well diode lasers, *Electron. Lett.* 33, 1635 (1997).

345. Schiebl, U. P. and Rohr, J., 60°C lead salt laser emission near 5-μm wavelength, *Infrared Phys. Technol.* 40, 325 (1999).

346. Someya, T., Tachibana, K., Lee, J., Kamiya, T., and Arakawa, Y., Lasing emission from an $In_{0.1}Ga_{0.9}N$ vertical cavity surface emitting laser. *Jpn. J. Appl. Phys.* 37, L1426 (1998).

347. Someya, T., Werner, R., Forchel, A., Catalano, M., Cingolani, R., and Arakawa, Y., Room temperature lasing at blue wavelengths in gallium nitrde micrcavities, *Science* 285, 1905 (1999).

348. Sidorin, Y. Blomberg, M., and Karioja, P., Demonstration of a tunable hybrid laser diode using an electrostatically tunable silicon micromachined Fabry-Perot interferometer device, *IEEE Photon. Technol. Lett.* 11, 18 (1999).

349. Bewley, W. W., Fleix, C. L., Aifer, E. H. et al., Above-room-temperature optically pumped midinfrared "W" lasers, *Appl. Phys. Lett.* 73, 3833 (1998).

350. Bewley, W. W., Fleix, C. L., Vurgaftman, I. et al., High-temperature continuous-wave 3-6.1 μm "W" lasers with diamond-pressure-bond heat sinking, *Appl. Phys. Lett.* 74, 1075 (1999).

351. Nikitenko, V. A., Tereshchnko, A. I., Kuzmina, I. P., and Lobachev, A. N., Stimulated emission in ZnO at a hgh level of one-photon excitation, *Opt. Spectrosc. (USSR)* 50, 331 (1981).

352. Cao, H., Zhao, Y. G., Ho, S. T., Seelig, E. W., Wang, Q. H., and Chang, R. P. H., Random laser action in semiconductor powder, *Phys. Rev. Lett.* 82, 2278 (1999).

353. Ivanov, S., Toropov, A., Sorokin, S. et al., ZnSe-based blue-green lasers with a short-period superlattice waveguide, *Appl. Phys. Lett.* 73, 2104 (1998).

354. Trediccuci, A., Capasso, F., Gmachi, C., Sivco, D. L., Hutchinson, A. L., and Cho, A. Y., High performance interminiband quantum cascade lasers with graded superlattices, *Appl. Phys. Lett.* 73, 2101 (1998).

355. Gmachl, C., Capasso, F., Trediccuci, A., Sivco, D. L., Hutchinson, A. L., Chu, S. N. G., and Cho, A. Y., Noncascaded intersubband injection lasers at $\lambda \approx 7.7$ μm, *Appl. Phys. Lett.* 73, 3830 (1998).

356. Painter, O., Lee, R. K., Scherer, A., Yariv, A., O'Brien, J. D., Dapkus, P. D., and Kim, I., Two-dimensional photonic band-gap defect mode laser, *Science* 284, 1819 (1999).

357. Imada, M., Noda, S., Chutinan, A., Tokuda, T., Murata, M., and Sasaki, G., Coherent two-dimensional lasing action in surface-emitting laser with triangular-lattice photonic crystal structure, *Appl. Phys. Lett.* 75, 316 (1999).

358. Mitsuhara, M., Ogasawara, M., Oishi, M., Sugiura, H., and Kasaya, K., 2.05-μm wavelength in InGaAs-InGaAs distributed-feedback multiquantum-well lasers with 10-mW output, *IEEE Photon. Technol. Lett.* 11, 33 (1999).

359. Gmachl, C., Trediccuci, A., Sivco, D. L., Hutchinson, Capasso, F., and Cho, A. Y., Bidirectional semiconductor laser, *Science* 286, 749 (1999).

360. Gmachl, C., Capasso, F., Faist, J., Hutchinson, A. L., Tredicucci, A., Sivco, D. L., Baillargeon, J. N., Chu, S.-N. G., and Cho, A. Y., Continuous-wave and high-power pulsed operation of index-coupled distributed feedback quantum cascade laser at ($\lambda \approx 8.5$ μm, *Appl. Phys. Lett.* 72, 1430 (1998).

361. Sirtori, C., Tredicucci, A., Capasso, Faist, J., F., Sivco, D. L., Hutchinson, A. L., and Cho, A. Y., Dual-wavelength emission from optically cascaded intersubband transitions, *Optics Lett.* 23, 463 (1998).

362. Tredicucci, A., Gmachl, C., Capasso, F., Sivco, D. L., Hutchinson, A. L., and Cho, A. Y., Long wavelength superlattice quantum cascade lasers at $\lambda \sim 17$ μm, *Appl. Phys. Lett.* 74, 638 (1999).

363. Sirtori, C., Kruck, P., Barbieri, Collot, P., Nagle, J., Beck, M., Faist, J., and Oesterle, U. Ga/Al$_x$Ga$_{1-x}$As mid-infrared ($\lambda=9.4$ μm) quantum cascade laser, *Appl. Phys. Lett.* 73, 3486 (1998).

364. Slivken, S., Matlis, A., Rybaltowski, A., Wu, Z., and Razeghi, M., Low-threshold 7.3 μm quantum cascade lasers grown by gas-source molecular beam epitaxy, *Appl. Phys. Lett.* 74, 2758 (1999).

365. Slivken, S., Matlis, A., Jelen, C., Rybaltowski, A., Diaz, J., and Razeghi, M., Low-threshold 7.3 μm quantum cascade lasers grown by gas-source molecular beam epitaxy, *Appl. Phys. Lett.* 74, 173 (1999).

366. Strasser, G., Gianordoli, S., Hvozara, L., Schrenk, W., Unterrainer, K. and Gornik, E., GaAs/AlGaAs superlattice quantum cascade lasers at ≈ 13 μm, *Appl. Phys. Lett.* 75, 1345 (1999).

367. Hofstetter, D., Faist, J., Beck, M., Müller, A., and Oesterle, U., Demonstration of high-performance 10.16 μm quantum cascade distributed feedback lasers fabricated without epitaxial regrowth, *Appl. Phys. Lett.* 75, 1345 (1999).

368. Gauthier-Lafaye, O., Julien, F. H., Cabaret, S., Lourtioz, M., Strasser, G., Gornik, E., Helm, M., and Bois, P., High-power GaAs/AlGaAs quantum fountain unipolar laser emitting at 14.5 μm with 2.5% tunability, *Appl. Phys. Lett.* 74, 1537 (1999).

369. Müller, A., Beck, M., Faist, J., Oesterle, U., and Ilegems, M., Electrically tunable room temperature quantum cascade lasers, *Appl. Phys. Lett.* 75, 1509 (1999).

370. Yang, R. Q., Bruno, J. D., Bradshaw, J. L., Pham, J. T., and Wortman, D. E., High power interband cascade lasers with quantum efficiency > 200%, *Electron. Lett.* 35, 1254 (1999).

371. Yang, B. H., Zhang, D., Yang, R. Q., Lin, C.-H., Murry, S. J., and Pei, S. S., Mid-infrared interband cascade lasers with quantum efficiencies > 200%, *Appl. Phys. Lett.* 72, 2220 (1998).

372. Muravjov, A. V., Withers, S. H., Strijbos, R. C. et al., Actively mode-locked *p*-Ge laser in Faraday configuration, *Appl. Phys. Lett.* 75, 2882 (1999).

373. Gokhale, M. R., Wei, J., Studenkov, P. V., Wang, H., and Forrest, S. R., High-performance long-wavelength ($\lambda \sim 1.3$ µm) InGaAsPN quantum-well lasers, *IEEE Photon. Techn. Lett.* 11, 952 (1999).

374. Höfling, E., Schafer, F., Reithmaier, J. P., and Forchel, A., Edge-emitting GaInAs-AlGaAs microlasers, *IEEE Photon. Techn. Lett.* 11, 943 (1999).

375. Lester, L. F., Stintz, A., Li, H., Newell, T. C., Pease, E. A., Fuchs, B. A., and Malloy, K. J., Optical characteristics of 1.24-µm InAs quantum-dot laser diodes, *IEEE Photon. Techn. Lett.* 11, 931 (1995).

376. Felix, C. L., Bewley, W. W., Olafsen, L. J. et al., Continuous-wave type-II "W" lasers emitting at $\lambda = 5.4$–7.1 µm, *IEEE Photon. Techn. Lett.* 11, 964 (1999).

377. Felix, C. L., Bewley, W. W., Vurgaftman, I. et al., High-efficiency midinfrared "W" laser with optical pumping injection cavity, *Appl. Phys. Lett.* 75, 2876 (1999).

378. Schmid, W., Wiedenmann, D., Grabherr, M., Jäger, R., Michalzik, R., and Ebeling, K. J., CW operation of a diode cascade InGaAs quantum well VCSEL, *Electron. Lett.* 34, 553 (1998).

379. Jayaraman, V., Geske, J. C., MacDougal, M. H., Peters, F. H., Lowes, T. D., and Char, T. T., Uniform threshold current, continuous-wave, singlemode 1300 nm vertical cavity lasers from 0 to 70°C, *Electron. Lett.* 34, 1405 (1998).

380. Taniguchi, S., Hino, T., Itoh, S., Nakano, K., Nakayama, N., Ishibashi, A., and Ikeda, M., 100h II-VI blue-green laser diode, *Electron. Lett.* 32, 552 (1996).

381. Babic', D. I., Streubel, K., Mirin, R. P. et al., Room-temperature continuous-wave operation of 1.54-µm vertical-cavity laser, *IEEE Photon. Techn. Lett.* 7, 1225 (1995).

382. Garcia, J. Ch., Rosencher, E., Collot, Ph., Laurent, N., Guyaux, J. L., Vinter, B., and Nagle, J., Epitaxially stacked lasers with Esaki junction: a bipolar cascade laser, *Appl. Phys. Lett.* 71, 3752 (1997).

383. Painter, O., Lee, R. K., Scherer, A., Yariv, A., O'Brien, J. D., Dapkus, P. D., and Kim, I. T., Room temperature photonic crystal defect lasers at near-infrared wavelengths in InGaAsP, *J. Lightwave Technol.* (1999).

384. Krestnikov, I. L., Lundin, W. V., Sakharov, A. V. et al., Room-temperature photopumped InGaN/GaN/AlGaN vertical-cavity surface-emitting laser, *Appl. Phys. Lett.* 75, 1192 (1999).

385. Lin, C.-F., Lee, B.-L., and Chen, M.-J., Broadly tunable semiconductor lasers using asymmetric dual quantum wells, *Opt. Commun.* 171, 271 (1999).

386. Stokes, D. W., Olafsen., L. J., Bewley, W. W. et al., Type-II quantum well "W" lasers emitting at $\lambda = 5.4$–7.3 µm, *J. Appl. Phys.* 86, 4729 (1999).

387. Chang, S., Rex, N. B., Chang, R. K., Chong, G., and Guido, L. J., Stimulated emission and lasing in whispering-gallery modes of GaN microdisk cavities, *Appl. Phys. Lett.* 75, 166 (1999).

388. Chang, J. H., Cho, M. W., Godo, K. et al., Low-threshold optically pumped lasing at 444 nm at room temperature with high characteristic temperature from Be-chalcogenide-based single-quantum-well laser structures, *Appl. Phys. Lett.* 75, 894 (1999).

389. Altukhov, I. V., Kagan, M. S., Gousev, Yu. P. et al., Continuous stimulated terahertz emission due to intra-center population inversion in uniaxially strained germanium, *Physica B*, 272, 458 (1999).

Section 1.6
POLYMER LASERS

1.6.1 Introduction

Polymer lasers encompass many different material compositions, active species (organic dyes), and forms (bulk, rods, fibers, films). In this section these are separated into pure polymer lasers, dye-doped polymer lasers (including host-to-dye energy tranfer), and rare-earth-doped polymer lasers. Liquid polymer lasers are covered in Section 2.3.

Further Reading—Pure Polymer Lasers

Dodabalapur, A., Chanddross, E. A., Berggren, M., and Slusher, R. L. Organic solid-state lasers: past and future, *Science* 277, 1787 (1997).

Friend, R. H., Denton, G. J., Halls, J. J. M. et al., Electronic excitations in luminescent conjugated polymers, *Solid State Commun.* 102, 249 (1997).

Frolov, S. V., Shkunov, M., Fujii, A., Yoshino, K., and Vardeny, Z. V., Lasing and stimulated emission in π-conjugated polymers, *IEEE J. Quantum Electron.* 36, 2 (2000).

Heeger, A. J., Light emission from semiconducting polymers: Light-emitting diodes, light-emitting electrochemical cells, lasers, and white light for the future, *Solid State Commun.* 107, 673 (1998).

Hide, F., Diaz-Garcia, M. A., Schwartz, B. J., Andersson, M. R., Pei, Q., and Heeger, A. J., Semiconducting polymers: a new class of solid-state laser materials, *Science* 273, 1833 (1996).

Jenekje, S. A. and Wynne, K. J., Eds., *Photonic and Optoelectronic Polymers*, American Chemical Society, Washington, DC (1997).

Lemmer, U., Stimulated emission and lasing in conjugated polymers, *Polym. Adv. Technol.* 9, 476 (1998).

Further Reading—Dye Doped Polymer Lasers

Bezrodnyi, V. I., Bondar, M. V., Kozak, G. Yu., Przhonskaya, O. V., and Tikhonov, E. A., Dye-activated polymer media for frequency-tunable lasers (review), *Zh. Prikl. Spektrosk. (USSR)* 50, 711 (1989).

Bezrodnyi, V. I., Przhonskaya, O. V., Tikhonov, E. A., Bondar, M. V., and Shpak, M. T., Polymer active and passive laser elements made of organic dyes, *Sov. J. Quantum Electron.* 12, 1602 (1982).

Dodabalapur, A., Chanddross, E. A., Berggren, M., and Slusher, R. L. Organic solid-state lasers: past and future, *Science* 277, 1787 (1997).

Dyumaev, K. M., Manenkov, A. A., Maslyukov, A. P., Matyushin, G. A., Nechitailo, V. S., and Prokhorov, A. M., Dyes in modified polymers: problems of photostability and conversion efficiency at high intensities, *J. Opt. Soc. Am. B* 9, 143 (1992).

Rahn, M. D. and King, T. A., Comparison of laser performance of dye molecules in sol-gel, polycom, ormosil, and poly(methyl methacrylte) host media, *Appl. Optics* 34, 8260 (1995).

Tagaya, A., Teramoto, S., Nihei, E., Sasake, K., and Koike, Y., High-power and high-gain organic dye-doped polymer optical fiber amplifiers: novel techniques for preparation and spectral investigation, *Appl. Optics* 36, 572 (1997).

1.6.2 Pure Polymer Lasers

Lasers based on neat and dilute blends of conjugated polymers are listed in Table 1.6.1. All experiments involved pulsed excitation and were performed at room temperature. The reported observations may be indicative of lasing or amplified spontaneous emission. The lasing material, solvent used, mode of photon confinement, and lasing and optical pumping wavelengths are listed together with the primary reference. For lasers that have been tuned over a range of wavelengths, the tuning range given is that for the experimental configuration and conditions used and may not represent the extremes possible. The references should be consulted for this information. The references should also be consulted for details of the chemical composition and molecular structure of the lasing compounds.

Abbreviations for the materials in Table 1.6.1:

BCHA	poly 2,5-bis(cholestanoxy)
BDOO-PF	poly[9,9-bis(3,6-dioxaoctyl)-fluorene-2,7-diyl]
BEH	poly 2,5-bis(2'-ethylhexyloxy)
BuEH	poly 2-butyl-5-2'-ethylhexyl
CB	chlorobenzene
CN-PPP	poly(2-(6'-mehylheptyloxy)-1,4-phenylene)
DCM/PS	4-(dicyanomethylene)-2-methyl-6-(4-dimethylaminostyry)-4*H*-pyran
DOO-PPV	2,5-dioctyloxy *p*-phenylene vinylene
m-EHOP	meta-(meta-2'-ethylhexoxyphenyl)
HEH-PF	poly(9-hexyl-9-2'-ethylhexyl)-fluorene-2,7-diyl)
LPPP	ladder-type poly(paraphenylene)
m-LPPP	methyl-substituted conjugated ladder-type poly(paraphenylene)
M3O	poly 2-methoxy-5-3'-octyloxy
MEH	poly 2-methoxy-5-2'-ethylhexloxy
NAPOXA	2-napthyl-4,5-bis(4-methoxyphenyl)-1,3-oxazole
Ooct-OPV5	5-ring *n*-octyloxy-substituted oligo[*p*-phenylene vinylene]
PBD	2-(4-biphenylyl)-5-(4-*t*-butylphenyl)-1,3,4-oxadiazole
PDAF8	di-octyl substituted polyfluorene
PMMA	polymethylmethacrylate
PPnVE	copolymers with phenylene, vinylene, and nonconjugated ethylidene units

PPPV	phenyl-substituted poly(p-phenylene vinylene)
PPV	p-phenylene vinylene
PS	polystyrene
Si-PPV	poly[dimethylsilylene-p-phene-vinylene -(2,5-di-n-octyl-p-phenylene)-PPV]
THF	tetrahydrofuran

Table 1.6.1
Pure Polymer Lasers

Material	Solvent	Photon confinement	Laser wavelength (nm)	Pump wavelength (nm)	Ref.
BCHA–PPV	p–xylene	wg	540, 630*	532	2
BCHA–PPV	xylene	wg	610	435	23
BDOO–PF	THF	wg	430, 450, 540	355	2
BEH–PPV	THF	wg	580, 625*	532	2
BEH–PPV	xylene	mr**	618–635	532	12
BEH–PPV	—	pc	626	555	15
BuEH-PPV	p–xylene	wg/mc	~550	435	8
BuEH–MEH(10:90)	p–xylene	wg	580, 625*	532	2
BuEH–MEH(70:30)	THF	wg	565, 600*	532	2
BuEH–MEH(90:10)	THF	wg	550, 580(sh)*	435	2
BuEH–MEH(95:5)	THF	wg	545, 580(sh)*	435	2
BuEH–MEH(97.5:2.5)	THF	wg	540, 570 (sh)*	435	2
BuEH–PPV	xylene	wg (DFB)	540–583	435	5, 13
BuEH–PPV	neat	wg	520–620	310	4
BuEH–PPV	THF	wg	~550	435	2
CN–PPP	THF	wg	420*	355	2
DCM/PS	THF	wg	640*	532	2
DOO-PPV	neat	wg/pc	625	532	9
m-EHOP–PPV	xylene	wg	550	435	23
4%BEH in m-EHOP	xylene	wg	675	435	23
4%BCHA in m-EHOP	xylene	wg	560	435	23
HEH–PF	THF	wg	425, 445*	355	2
LPPP	—	wg, s	487	400	18
m-LPPP	toluene	pc, s	490	390	17
m-LPPP	—	wg	~483–492	444	3
M3O–PPV	CB	wg	530, 620*	532	2
MEH–PPV	THF	wg	585, 625*	532	2
Ooct-OPV5		wg	~550	355	14
PBD	—	wg	392	337	1
PPnVE	b, t, cf	wg	473, 502, 637	337	20
PPPV–PMMA		res	480–545	450	11

Table 1.6.1—*continued*
Pure Polymer Lasers

Material	Solvent	Photon confinement	Laser wavelength (nm)	Pump wavelength (nm)	Ref.
PPV	neat	mc	~545	355	6
PPV	neat	mc	~545	355	7
PPV	neat	mc (DBR)	530, 550	325, 355	16
sexithiophene (6T)	—	sc	592.6	355	19
Si–PPV		wg	452	355	10

b—benzene, cf—chloroform, DBR—distributed Bragg reflectors, mc—microcavity, mr—microring, pc—planar cavity, res—resonator, s—single mode, sc—single crystal, sh—shoulder, t—toluene, wg—waveguide

* Peak(s) of photoluminescence; no lasing wavelength was reported.
** Whispering galley modes.

References—Table 1.6.1

1. Berggren, M., Dodabalapur, A., Slusher, R. L., and Bao, Z., Light amplification in organic thin films using cascade energy transfer, *Nature* 389, 466 (1997).
2. Hide, F., Diaz-Garcia, M. A., Schwartz, B. J., Andersson, M. R., Pei, Q., and Heeger, A. J., Semiconducting polymers: a new class of solid-state laser materials, *Science* 273, 1833 (1996).
3. Zenz, C., Graupner, W., Tasch, S., Leising, G., Müllen, K., and Scherf, U., Blue green stimulated emission from a high gain conjugated polymer, *Appl. Phys. Lett.* 71, 2566 (1997).
4. Schwartz, B. J., Hide, F., Andersson, M. R., and Heeger, A. J., Ultrafast studies of stimulated emission and gain in solid films of conjugated polymers, *Chem. Phys. Lett.* 265, 327 (1997).
5. McGehee, M. D., Diaz-Garcia, M. A., Hide, F., Gupta, R., Miller, E. K., Moses, D., and Heeger, A. J., Semiconducting polymer distributed feedback lasers, *Appl. Phys. Lett.* 72, 1536 (1998).
6. Tessler, N. Denton, G. J., and Friend, R. H., Lasing from conjugated-polymer microcavities, *Nature* 382, 695 (1996).
7. Friend, R. H., Denton, G. J., Halls, J. J. M. et al., Electronic excitations in luminescent conjugated polymers, *Solid State Commun.* 102, 249 (1997).
8. Diaz-Garcia, M. A., Hide, F., Schwartz, B. J., McGehee, M. D., Andersson, M. R., and Heeger, A. J., "Plastic" lasers: Comparison of gain narrowing with a soluble semiconducting polymer in waveguides and microcavities, *Appl. Phys. Lett.* 70, 3191 (1997).
9. Frolov, S. V., Gellermann, W., Ozaki, M., Yoshino, K., and Vardeny, Z. V., Cooperative emission in p-conjugated polymer thin films, *Phys. Rev. Lett.* 78, 729 (1997). See, also, Frolov, S. V., Shkunov, M., Vardeny, Z. V., and Yoshino, K., *Phys. Rev. B* 56, R4363 (1997) and Frolov, S. V., and Vardeny, Z. V., Cooperative and stimulated emission in poly(p-phenylene-vinylene) thin films and solutions, *Phys. Rev. B* 57, 9141 (1998).
10. Brouwer, H. J., Krasnikov, V., Hilberer, A., and Hadziioannou, G., Blue superradiance from neat semiconducting alternating copolymer films, *Adv. Mater.* 8, 935 (1996).
11. Wegmann, G., Giessen, H., Hertel, D., and Mahrt, R. F., Blue-green laser emission from a solid conjugated polymer, *Solid State Commun.* 104, 759 (1997); Giessen, H., Wegmann, G., Hertel, D., and Mahrt, R. F., A tunable blue-green laser from a solid conjugated polymer, *Phys. Stat. Sol. (b)* 206, 437 (1998); Wegmann, G., Giessen, H., Greiner, A., and Mahrt, R. F., Laser emission from a solid conjugated polymer: gain, tunability, and coherence, *Phys. Rev. B* 57, R4218 (1998).

12. Kawabe, Y., Spiegelberg, Ch., Schülzgen, A. et al., Whispering-galley-mode microring laser using a conjugated polymer, *Appl. Phys. Lett.* 72, 141 (1998).
13. McGehee, M. D., Gupta, R., Veenstra, S. Miller, E. K. Díaz-García, M. A., and Heeger, A. J., Amplified spontaneous emission from photopumped films of a conjugated polymer, *Phys. Rev. B.* 58, 7035 (1998).
14. Brouwer, H.-J., Krasnikov, V., V., Pham, T.-A., Gill, R. E., and Hadziioannou, G., Stimulated emission from vacuum-deposited thin films of a substituted oligo(*p*-phenylene vinylene), *Appl. Phys. Lett.* 73, 708 (1998).
15. Schülzgen, A., Spiegelberg, Ch., Morrell, M. M. et al., Near diffraction-limited laser emission from a polymer in a high finesse planar cavity, *Appl. Phys. Lett.* 72, 269 (1998).
16. Burns, S. E., Denton, G., Tessler, N., Stevens, M. A., Cacialli, F., and Friend, R. H., High finesse organic microcavities, *Opt. Mater.* 9, 18 (1998).
17. Stagira, S., Zavelani-Rossi, M., Nisoli, M. et al., Single-mode picosecond blue laser emission from a solid conjugated polymer, *Appl. Phys. Lett.* 73, 2860 (1998).
18. Kallinger, C., Hilmer, M., Haugeneder, A. et al., A flexible conjugated polymer laser, *Adv. Mater.* 10, 920 (1998) and Haugeneder, A., Hilmer, M., Kallinger, C. et al, Mechanism of gain narrowing in conjugated polymer thin films, *Appl. Phys. B* 66, 389 (1998).
19. Horowitz, G., Valat, P., Garnier, F., Kouki, F., and Wintgens, V., Photoinduced spontaneous and stimulated emission in sexithiophene single crystals, *Opt. Mater.* 9, 46 (1998).
20. Gelinck, G. H., Warman, J. M., Remmers, M., and Neher, D., Narrow-band emissions from conjugated-polymer films, *Chem. Phys. Lett.* 265, 320 (1997).
21. Shkunov, M., Österbacka, R., Fujii, A., Yoshino, K., and Vardeny, Z. V., Laser action in polydialkylfluorene films: influence of low-temperature thermal treatment, *Appl. Phys. Lett.* 74, 1648 (1999).
22. Fichou, D., Dumarcher, V., and Nunzi, J.-M., One- and two-photon stimulated emission in oligothiophenes single crystals, *Opt. Mater.* 12, 255 (1999).
23. Gupta, R., Stevenson, M., Dogariu, A. et al., Low-threshold amplified spontaneous emission in blends of conjugated polymers, *Appl. Phys. Lett.* 73, 3492 (1998).

1.6.3 Dye Doped Polymer Lasers

Table 1.6.2 summarizes the different polymer hosts and organic dye dopants that have been used to demonstrate laser action. If the laser has been tuned over a range of wavelengths, the tuning range given is that for the experimental configuration and conditions used and may not represent the extremes possible. The lasing wavelength and output of dye lasers depend on the characteristics of the optical cavity, the dye concentration, the optical pumping source and rate, and other operating conditions. The original references should therefore be consulted for this information and its effect on the lasing wavelength. The references should also be consulted for details of the chemical composition and molecular structure of the dyes and host compounds.

Table 1.6.2
Dye Doped Polymer Lasers

Host	Dye	Laser wavelength (nm)	Pump source (wavelength – nm)	Ref.
acrylic copolymer	pyrromethene 567	564	Nd:YAG laser (532)	1
acrylic copolymer	pyrromethene 580	570	Nd:YAG laser (532)	2
acrylic copolymer	pyrromethene 580	571	Nd:YAG laser (532)	3
acrylic copolymer	pyrromethene 597	587	Nd:YAG laser (532)	1, 4
acrylic monomers	sulforhodamine B	~600	Nd:YAG laser (532)	5
tris-(8-hydroxyquinoline) aluminum (Alq$_3$)	DCM	589–635(a)	N$_2$ laser (337) - ET	6
Alq$_3$ (film)	DCM II	613	N$_2$ laser (337) - ET	9
Alq$_3$ (film)	DCM	645	N$_2$ laser (337) - ET	7
Alq$_3$	DCM	~655	N$_2$ laser (337) - ET	8
4,4'-di(N-carbazole) biphenyl (film)	coumarin 47	460	N$_2$ laser (337) - ET	10
4,4'-di(N-carbazole) biphenyl (film)	perylene	485	N$_2$ laser (337) - ET	10
4,4'-di(N-carbazole) biphenyl (film)	coumarin 30	510	N$_2$ laser (337) - ET	10
GPTMS/Ti(OBu)$_4$/MMA	rhodamine 6G	564–587	Nd:YAG laser (532)	58
HEMA	ASPT	~600	Nd:YAG laser (1064) - TP	11
HEMA	DHASI	~606	Nd:YAG laser (1064) - TP	12

Host	Dye	Wavelength (nm)	Pump source	Ref.
HEMA/MMA (1:1)	coumarin 540A	515, 535	N_2 laser (337)	13
HEMA/MMA (1:1)	rhodamine 6G	—	flashlamp	59
HEMA/MMA copolymer	rhodamine 640	654	N_2 laser (337)	14
high temperature plastic	pyrromethene 567	522	Nd:YAG laser (532)	15
high temperature plastic	pyrromethene 597	528	Nd:YAG laser (532)	15
high temperature plastic	pyrromethene 580	550–570	Nd:YAG laser (532)	19
high temperature plastic	pyrromethene 597	565–590	Nd:YAG laser (532)	17
high temperature plastic	pyrromethene 570	566.3	Nd:YAG laser (532)	18
high temperature plastic	pyrromethene 597	587	Nd:YAG laser (532)	16
high temperature plastic	PM-HMC	593	Nd:YAG laser (532)	15
high temperature plastic	PM-TEDC	598	Nd:YAG laser (532)	15
high temperature plastic	PM-HMC	615–635	Nd:YAG laser (532)	17
high temperature plastic	PM-TEDC	615–635	Nd:YAG laser (532)	17
high temperature plastic	PM650	624	Nd:YAG laser (532)	4
PBD (film)	coumarin 460	~460	N_2 laser (337) - ET	20
PBD (film)	DCM II	~600	N_2 laser (337) - ET	20
PBD (film)	LDS821	~805	N_2 laser (337) - ET	20
PBD (film)	PPV7	575–590	N_2 laser (337) - ET	21
PBD (photonic crystal)	coumarin 490, DCM	580, 596	N_2 laser (337)	62
PHEMA (fiber)	ASPI	~610	Nd:YAG laser (1064) - TP	22
PHEMA+DEGMA	rhodamine 6G	580–620	N_2 laser (337)	23
P(HEMA:EDGMA 9:1)	rhodamine 6G	599	N_2 laser (337)	24
P(HEMA:MMA 1:1)	rhodamine 6G	568	Nd:YAG laser (532)	24
P(HEMA:MMA 1:1)	rhodamine-Bz-MA	587	N_2 laser (337)	25
P(HEMA:MMA 1:1)	rhodamine-Al	589	N_2 laser (337)	25
P(HEMA:MMA 1:1)	rhodamine 6G	593	N_2 laser (337)	24
P(HEMA:MMA 3:7)	rhodamine 6G	584	N_2 laser (337)	24
P(HEMA:MMA 7:3)	rhodamine 6G	587	N_2 laser (337)	24
P(HEMA:MMA 7:3)	rhodamine-Bz-MA	593	N_2 laser (337)	25

Table 1.6.2—*continued*
Dye Doped Polymer Lasers

Host	Dye	Laser wavelength (nm)	Pump source (wavelength – nm)	Ref.
methylmethacrylate	COP-2(b)	497	N$_2$ laser (337)	26
methylmethacrylate	COP-3(b)	511	N$_2$ laser (337)	26
methylmethacrylate(c)	1,3,5,7,8-pentamethyl-2,6-di-n-butylpyrromethene-BF$_2$	569.7, 571.0	Nd:YAG laser (532)	2
methylmethacrylate(c)	pyrromethene 567	571.4	Nd:YAG laser (532)	2
methylmethacrylate(c)	1,3,5,7-tetramethyl-8-cyanopyrromethene-2,6-dicarboxylate-BF$_2$	617.9	Nd:YAG laser (532)	2
methylmethacrylate(c)	sulforhodamine B	618.2	Nd:YAG laser (532)	2
MMA/HPA	rhodamine 6G	595	flashlamp	
NAPOXA (film)	DCM II	~615	N$_2$ laser (337) - ET	20
polyavylamide	rhodamine 6G	525–650	coaxial flashlamp	56
polyisobutylmethacrylate	α-NPO	~395	N$_2$ laser (337)	28
polyisobutylmethacrylate	BBOT	~430	N$_2$ laser (337)	28
polyisobutylmethacrylate	dimethyl-POPOP	~420	N$_2$ laser (337)	28
polyisobutylmethacrylate	POPOP	~410	N$_2$ laser (337)	28
polymethylmethacrylate	α-NPO	396	Nd:YAG laser (355)	29
polymethylmethacrylate	2-(4-biphenylyl)-5-(p-styryl-phenyl)-1,3,4-oxadiazole	400	Nd:YAG laser (355)	29
polymethylmethacrylate	2-(4-biphenylyl)-5-(1-naphthyl)-oxazole	410	Nd:YAG laser (355)	29
polymethylmethacrylate	5-phenyl-2-(p-styryl-phenyl)oxazole	414	Nd:YAG laser (355)	29
polymethylmethacrylate	POPOP	415	Nd:YAG laser (355)	29

Matrix	Dye	Wavelength	Pump source	Ref.
polymethylmethacrylate	2-(1-naphthyl)-5-styryl-1,3,4-oxadiazole	416	Nd:YAG laser (355)	29
polymethylmethacrylate	2-phenyl-5-[p-(4-phenyl-1,3-bul-adrenyl)phenyl]-1,3,4-oxadiazole	418	Nd:YAG laser (355)	29
polymethylmethacrylate	dimethyl-POPOP	423	Nd:YAG laser (355)	29
polymethylmethacrylate	2-[p-[2-(2-naphthyl)vinyl]phenyl]-5-phenyloxazole	425	Nd:YAG laser (355)	29
polymethylmethacrylate	5-phenyl-2-[p-(-phenylstyryl)-phenyl]oxazole	428	Nd:YAG laser (355)	29
polymethylmethacrylate	5-(4-biphenylyl)-2-[p-(4-phenyl-1,3-baladienyl)phenyl]oxazole	443	Nd:YAG laser (355)	29
polymethylmethacrylate	1,2-bis(5-phenyloxazolyl)ethylene	444	N_2 laser (337)	28
polymethylmethacrylate	1,5-diphenyl-3-styryl-2-pyrazoline	450	Nd:YAG laser (355)	29
polymethylmethacrylate	1,4-bis[4-[5-(4-biphenylyl)-2-oxazolyl]styryl benzene	455	Nd:YAG laser (355)	29
polymethylmethacrylate	3-p-chlorostyryl-1,5-diphenyl-2-pyrazoline	457	Nd:YAG laser (355)	29
polymethylmethacrylate	(see Reference 29)	478	Nd:YAG laser (355)	29
polymethylmethacrylate	2-[p-[2-(9-anthryl)vinyl]phenyl]-5-phenyloxazole	480	Nd:YAG laser (355)	29
polymethylmethacrylate (film)	CF$_3$-coumarin	484–529	N_2 laser (337)	31
polymethylmethacrylate	2-(2'-hydroxy-5'-fluorophenyl)benzimidazole	493	N_2 laser (337)	30
polymethylmethacrylate	5(6)-methoxycarbonyl-2(2'-hydroxphenyl)benzimidazole	494	N_2 laser (337)	26
polymethylmethacrylate	5(6)-methoxycarbonyl-12-(5'-fluoro-2'-hydroxyphenyl)benzimidazole	508	N_2 laser (337)	26
polymethylmethacrylate	fluoran	509–514	Nd:YAG laser (355)	29
polymethylmethacrylate	coumarin 540A	515	N_2 laser (337)	64
polymethylmethacrylate	rhodamine 6G	550	Nd:YAG laser (355)	29

Table 1.6.2—*continued*
Dye Doped Polymer Lasers

Host	Dye	Laser wavelength (nm)	Pump source (wavelength – nm)	Ref.
polymethylmethacrylate	rhodamine 6G	552–575	Nd glass (530)	36
polymethylmethacrylate	rhodamine 6G	552–595	Nd:YAG laser (532)	19
polymethylmethacrylate	rhodamine 6G	555–565	CdS (495)	37
polymethylmethacrylate	rhodamine 6G	562	Nd:YAG laser (532)	35
polymethylmethacrylate	rhodamine 6G	565.5, 570.9	Nd:YAG laser (532)	18
polymethylmethacrylate	rhodamine 6G	567–605	Ne laser (540)	38
polymethylmethacrylate	pyrromethene 570	569.6	Nd:YAG laser (532)	18
polymethylmethacrylate (wg)	DCM	570	dye laser (730) - TP	32
polymethylmethacrylate	rhodamine 590	573	Nd:YAG laser (532)	34
polymethylmethacrylate	rhodamine 590	573	Nd:YAG laser (355)	33
polymethylmethacrylate (film)	rhodamine 6G	577–590	N_2 laser (337)	39
polymethylmethacrylate	peryline orange (KF 241)	578	Nd:YAG laser (355)	33
polymethylmethacrylate	peryline orange (KF 241)	578	Nd:YAG laser (532)	34
polymethylmethacrylate	rhodamine 590	585	C-532 dye laser (525)	41
polymethylmethacrylate	rhodamine B	587.4, 594.0	N_2 laser (337) - ET	42
polymethylmethacrylate	rhodamine C	595	Nd:YAG laser (355)	29
polymethylmethacrylate	rhodamine 6G	601.0	Xe flashlamp	40
polymethylmethacrylate	peryline red (KF 856)	613	Nd:YAG laser (355)	33
polymethylmethacrylate	peryline red (KF 856)	613	Nd:YAG laser (532)	34
polymethylmethacrylate	rhodamine B	632.4	Xe flashlamp	40
PMMA (rhodamine 6G)(d)	rhodamine B	600–620	N_2 laser (337)	43
PMMA (rhodamine 6G)(d)	cresyl violet	620–640	N_2 laser (337)	43
PMMA (rhodamine 6G)(d)	sulforhodamine 101	620–670	N_2 laser (337)	43
PMMA (rhodamine 6G)(d)	nile blue	695–720	N_2 laser (337)	43
PMMA (rh. 6G/nile blue)(d)	HITC	845–865	N_2 laser (337)	43
PMMA(BBOT)(d)	pervlene	455–475	N_2 laser (337)	43

Host	Dye	Wavelength	Pump laser	Ref.
PMMA(coumarin 1)[d]	acridine yellow	490–500	N₂ laser (337)	43
PMMA(coumarin 1)[d]	acriflavine	500–520	N₂ laser (337)	43
PMMA(coumarin 1)[d]	uranine	530–560	N₂ laser (337)	43
mPMMA	dye "II B"	560–570	(not reported)	44
mPMMA	rhodamine 6G	560–570	Nd glass (530)	45
mPMMA	rhodamine 6G chloride	560–570	(not reported)	44
mPMMA	rhodamine 6G percholate	560–570	(not reported)	44
mPMMA	rhodamine III	560–570	(not reported)	44
mPMMA	oxazine-17	620–640	(not reported)	44
mPMMA	oxazine-17	620–640	Nd glass (530)	45
polystyrene:TiO₂ (film)	MEH-PPV	~605	Nd:YAG laser (532)	46
polystyrene	BBQ	381–394	N₂ laser (337)	47
polystyrene	α-NPO	411	Nd:YAG laser (355)	29
polystyrene	dimethyl-POPOP	414, 427	Nd:YAG laser (355)	29
polystyrene	POPOP	416–426	N₂ laser (337)	47
polystyrene	POPOP	420	Nd:YAG laser (355)	29
polystyrene	dimethyl-POPOP	424	Nd:YAG laser (355)	29
polystyrene	dimethyl-POPOP	426–445	N₂ laser (337)	47
polystyrene	dimethyl-POPOP	428–438	N₂ laser (337)	48
polystyrene	BBOT	440–472	N₂ laser (337)	47
polystyrene	BBOT + perylene	464–480	N₂ laser (337)	47
polystyrene	1,3-dimethylisobenzofuran	490–500	N₂ laser (337)	47
polystyrene	fluorescein (sodium salt)	553–564	N₂ laser (337)	48
polystyrene (microdisk laser)	pyrromethane dye	~565	dye laser (520)	63
polystyrene (microring laser)	pyrromethane dye	~573, 577[e]	dye laser (520)	63
polystyrene (microring laser)	pyrromethane dye	~600–610[f]	dye laser (520)	63
polystyrene	rhodamine 640	~600	Nd:YAG laser (1064) - TP	49
polystyrene (WGM)	Nile red	~605–620	Nd:YLF laser (523)	61
polystyrene (wg)	rhodamine 6G	614–624	N₂ laser (337)	48

Table 1.6.2—*continued*
Dye Doped Polymer Lasers

Host	Dye	Laser wavelength (nm)	Pump source (wavelength – nm)	Ref.
polyurethane	rhodamine 6G	610–635	Ne laser (540)	51
polyurethane	rhodamine 6G	610–635	N2 laser (337)	52, 53
polyurethane (film)	rhodamine B	632.8 (A)	He-Ne (632.8)	55
polyurethane	rhodamine 110	643	Ar laser (514)	54
polyurethane	indodicarbocyanine (PK 643)	680 (A)	(532, 635, 670)	50
polyvinylxylene	dimethyl-POPOP	425	Nd:YAG laser (355)	29
proprietary polymer-gel	rhodamine 6G	565.1	Nd:YAG laser (532)	18
Probimide 414(g) (wg)	cresyl violet 670	670	dye laser (590)	3

A – amplified wavelength, ET – energy tranfer from host to dye, TP – two-photon pumped, wg – waveguide, WGM — whispering-galley-mode

(a) Laser emission varied by changing the thickness of the active organic layer.
(b) See Ref. 26 for molecular structure of copolymer.
(c) 16% hydroxypropyl acrylate/methyl methacrylate.
(d) Host materials were polymethylmethacrylate and polyvinylalcohol. Dyes in parentheses serve as a donor dye in an energy transfer laser.
(e) Polymer droplets formed around a 125-μm silica fiber.
(f) Polymer droplets formed around a 17-μm silica fiber.
(g) A photosensitive benzophenone tetracarboxylic diahydride-alkylated diamine polyimide.

Abbreviations for hosts and dyes in Table 1.6.2:

ASPI – trans-4-[p-(N-hydroxyethyl-N-methylamino)styryl]-N-methylpyridinium iodide, BBOT – 2,5-bis(5-tert-butyl-2-benzoxazolyl)thiophene BBQ – 4,4'''bis(2-butyloctyloxy)-p-quaterphenyl, COP – cyclooctetraene, DCM – 4-dicyanomethylene-2-methyl-6-p-dimethylaminostyryl-4H-pyran, DHASI – 4-[bis(2-hydroxyethyl)amino]-N-methylstilbazolium iodide, GPTMS – glycidoxypropyl-trimethoxysilane, HITC – 1,3,3,1',3',3'-hexamethyl-2,2' indotri-carbocyanine, HPA – hydroxypropyl, MEH – poly 2-methoxy-5-2'-ethyl-hexloxy, MMA – methylmethacrylate, NAPOXA – 2-naphtyl-4,5-bis(4-methoxyphenyl)-1,3-oxazole, NPO – 2-(1-naphthy)-5-phenyloxazole, PBD – 2-(4-biphenylyl)-5-(4-t-butyl-phenyl)-1,3,4-oxadiazole, PMMA – polymethylmethacrylate, POPOP – 1,4-di[2-(5-phenyloxazolyl)] benzene, PPV – poly(1,4-phenylene vinylene).

References—Table 1.6.2

1. Hermes, R.E., Lasing performance of pyrromethene-BF$_2$ laser dyes in a solid polymer host, SPIE Vol. 2115: *Visible and UV Lasers* (1994), p. 240.
2. Hermes, R. E, Allik, T. H., Chandra, S., and Hutchison, J. A., High-efficiency pyrromethene doped solid-state dye lasers, *Appl. Phys. Lett.* 63, 877 (1993).
3. Allik, T. H., Chandra, S., Hermes, R. E., Hutchinson, J. A., Soong, M. L., and Boyer, J. H., Efficient and robust solid-state dye laser, OSA *Proc. on Adv. Solid-State Lasers* 15, 271 (1993).
4. Boyer, J., Pyrromethene-BF$_2$ complexes (P-BF$_2$), *Proceedings of the Solid State Dye Laser Workshop* (1994), p. 66.
5. Hermes, R. E., McGrew, J. D., Wiswall, C. E., Monroe, S., and Kushina, M., A diode laser-pumped Nd:YAG-pumped polymeric host solid-state dye laser, *Appl. Phys. Commun.* 11, 1 (1992).
6. Bulovic', V., Kozlov, V. G., Khalfin, V. B., and Forrest, S. R., Transform-limited, narrow-linewidth lasing action in organic semiconductor microcavities, *Science* 279, 553 (1998).
7. Kozlov, V. G., Bulovic', V., and Forrest, S. R., Temperature independent performance of organic semiconductor lasers, *Appl. Phys. Lett.* 71, 2575 (1997).
8. Berggren, M., Dodabalapur, A., Slusher, R. E., Timko, A., and Nalamasu, O., Organic solid-state lasers with imprinted gratings on plastic substrates, *Appl. Phys. Lett.* 72, 410 (1998).
9. Berggren, M., Dodabalapur, A., and Slusher, R. E., Stimulated emission and lasing in dye-doped organic thin films with Forster transfer, *Appl. Phys. Lett.* 71, 2230 (1997).
10. Kozlov, V. G., Parthasarathy, G., Burrows, P. E., Forrest, S. R., You, Y., and Thompson, M. E., Optically pumped blue organic semiconductor lasers, *Appl. Phys. Lett.* 72, 144 (1998).
11. He, G. S., Zhao, C. F., Bhawalkar, J. D., and Prasad, P. N., Two-photon pumped cavity lasing in novel dye doped bulk matrix rods, *Appl. Phys. Lett.* 67, 3703 (1995).
12. He, G. S., Kim, K.-S., Yuan, L., Cheng, N., and Prasad, P. N., Two-photon pumped partially cross-linked polymer laser, *Appl. Phys. Lett.* 71, 1619 (1997).
13. Costela, A., Garcia-Moreno, I., Figuera, J. M., Amat-Guerri, F., Barroso, J., and Sastre, R., Solid-state dye laser based on coumarin 540A-doped polymeric matrices, *Opt. Commun.* 130, 44 (1996).
14. Amat-Guerri, F., Costel, A., Figuera, J. M., Florido, F., Garcia-Moreno, I., and Sastre, R., Laser action from a rhodamine 640-doped copolymer of 2-hydroxyethyl methacrylate and methyl methacrylate, *Optics Comm.* 114, 442 (1995).
15. Allik, T., Chandra, S., Robinson, T. R., Hutchinson, J. A., Sathyamoorthi, G., and Boyer, J. H., Laser performance and material properties of a high temperature plastic doped with pyrromethene-BF$_2$ dyes, Mat. Res. Soc. Proc. Vol. 329, *New Materials for Solid State Lasers* (1994), p. 291.
16. Allik, T. H., Chandra, S., Robinson, T. R., Hutchinson, J. A., Sathyamoorthi, G., and Boyer, J. H., Laser performance and material properties of a high temperature plastic doped with pyrromethene-BF$_2$ dyes, *Mat. Res. Soc. Symp. Proc.* 329, 291 (1994).
17. Chandra, S. and Allik, T., Compact high-brightness solid state dye laser, *Proc. Solid State Dye Laser Workshop* (1994), p. 29.
18. Ewanizky, T. F. and Pearce, C. K., Solid state dye lasers in an unstable resonator configuration, *Proceedings of the Solid State Dye Laser Workshop* (1994), p. 154.
19. Soffer, B. H. and McFarland, B. B., Continuously tunable, narrow band organic dye lasers, *Appl. Phys. Lett.* 10, 266 (1967).
20. Berggren, M., Dodabalapur, A., Slusher, R. L., and Bao, Z., Light amplification in organic thin films using cascade energy transfer, *Nature* 389, 466 (1997).
21. Berggren, M., Dodabalapur, A., Bao, Z., and Slusher, R. E., Solid-state droplet laser made from an organic blend with a conjugated polymer emitter, *Adv. Mater.* 9, 968 (1997).
22. He, G. S., Bhawalkar, J. D., Zhao, C. F., and Park, C. K., Upconversion dye-doped polymer fiber laser, *Appl. Phys. Lett.* 68, 3549 (1996).

23. Amat-Guerri, F., Costela, A., Figuera, J. M., Florido, F., and Sastre, R., Laser action from rhodamine 6G-doped poly(2-hydroxyethyl methacrylate) matrices with different crosslinking degrees, *Chem. Phys. Lett.* 207, 352 (1993).
24. Costela, A., Florido, F., Garcia-Moreno, I., Duchowicz, R., Amat-Guerri, F., Figuera, J. M., and Sastre, R., Solid-state dye lasers based on copolymers of 2-hydroxyethyl methacrylate and methyl methacrylate doped with rhodamine 6G, *Appl. Phys. B* 60, 383 (1995).
25. Costela, A., Garcia-Moreno, I., Figuera, J. M., Amat-Guerri, F., and Sastre, R., Solid-state dye lasers based on polymers incorporating covalently bonded modified rhodamine 6G, *Appl. Phys. Lett.* 68, 593 (1996).
26. Ferrer, M. L., Acuña, A. U., Amat-Guerri, F., Costela, A., Figuera, J. M., Florido, F., and Sastre, R., Proton-transfer lasers from solid polymeric chains with covalently bound 2-(2'-hydroxyphenyl)benzimidazole groups, *Appl. Optics* 33, 2266 (1994).
27. Finlayson, A. J., Peters, N., Kolinsky, P. V., and Venner, M. R. W., Flashlamp pumped polymer dye laser containing Rhodamine 6G, *Appl. Phys. Lett.* 72, 2153 (1997).
28. Muto, S., Ichikawa, A., Ando, A., Ito, C., and Inaba, H., Trial to plastic fiber dye laser, *Trans. IECE (Japan)* E 69, 374 (1986).
29. Naboikin, Yu. V., Ogurtsova, L. A., Podgornyi, A. P. et al., Spectral and energy characteristics of lasers based on organic molecules in polymers and toluol, *Opt. Spektrosk.* 28, 974 (1970); *Sov. Opt. Spectrosc.* 28, 528 (1970).
30. Acuña, A. U., Amat-Guerri, F., Costela, A., Douhal, A., Figuera, J. M., Florido, F., and Sastre, R., Proton-transfer lasing from solid organic matrices, *Chem. Phys. Lett.* 187, 98 (1991).
31. Itoh, U., Takakusa, M., Moriya, T., and Saito, S., Optical gain of coumarin dye-doped thin film lasers, *Jpn. J. Appl. Phys.* 16, 1059 (1977).
32. Mukherjee, A., Two-photon pumped upconverted lasing in dye doped polymer waveguides, *Appl. Phys. Lett.* 62, 3423 (1993).
33. Rahn, M. D. and King, T. A., Lasers based on dye doped sol-gel composite glasses, SPIE Vol. 2288, *Sol-Gel Optics III* (1994), p. 382.
34. Rahn, M. D. and King, T. A., Comparison of laser performance of dye molecules in sol-gel, polycom, ormosil, and poly(methyl methancrylate) host media, *Appl. Optics* 34, 8260 (1995).
35. Wang, H. H. L. and Gampel, L., A simple, efficient plastic dye laser, *Optics Commun.* 18, 444 (1976).
36. Kaminow, I. P., Weber, H. P., and Chandross, E. A., Poly(methyl methacrylate) dye laser with internal diffraction grating resonator, *Appl. Phys. Lett.* 18, 497 (1971).
37. Onstott, J. R., Short cavity dye laser excited by an electron beam-pumped semiconductor laser, *Appl. Phys. Lett.* 31, 818 (1977).
38. Fork, R. L., German, I. R., and Chandron, E. A., Photodimer distributed feedback laser, *Appl. Phys. Lett.* 20, 139 (1972).
39. Schinke, D. P., Smith, R. G., Spencer, E. G., and Galvin, M. F., Thin-film distributed feedback laser fabricated by ion milling, *Appl. Phys. Lett.* 21, 494 (1972).
40. Peterson, O. G. and Snavely, B. B., Stimulated emission from flashlamp excited organic dyes in polymethyl methacrylate, *Appl. Phys. Lett.* 12, 238 (1968).
41. Mandl, A. and Klimek, D. E., 400 mJ long pulse (>1µs) solid state dye laser, *Proc. Solid State Dye Laser Workshop* (1994), p. 192.
42. Wojcik, A. B. and Klein, L. C., Rhodamine 6G-doped inorganic/organic gels for laser and sensor applications, SPIE Vol. 2288, *Sol-Gel Optics III* (1994), p. 392.
43. Muto, S., Shiba, T., Iijima, Y., Hattori, K., and Ito, C., Solid thin-film energy transfer dye lasers, *Trans. IECE (Japan)* J69-C, 25 (1986); [*Electron and Commun. Japan*, part 2, 70, 21 (1987)].
44. Manenkov, A. A., Maslyukov, A. P., Matyushin, G. A., and Nechitailo, V. S., Modified polymers - effective host materials for solid state dye lasers and laser beam control elements: a review, SPIE Vol. 2115, *Visible and UV Lasers* (1994), p. 136.
45. Gromov, D. A., Dyumaev, K. M., Manenkov, A. A., Maslyukov, A. P., Matyushin, G. A., Nechitailo, V. S., and Prokhorov, A. M., Efficient plastic-host dye lasers, *J. Opt. Soc. Am. B* 2, 1028 (1985).

46. Hide, F., Schwartz, B. J., Diaz-Garcia, M. A., and Heeger, A. J., Laser emission from solutions and films containing semiconducting polymer and titanium dioxide nanocrystals, *Chem. Phys. Lett.* 256, 424 (1996).
47. Muto, S., Ando, A., Yoda, O., Hanawa, T., and Ito, H., Dye laser by sheet of plastic fibers with wide tuning range, *Trans. IECE (Jpn)* E 70, 317 (1987).
48. Tanuguchi, H., Fujiwara, T., Yamada, H., Tanosake, S., and Baba, M., Whispering-galley-mode dye lasers in blue, green, and orange regions using dye-doped, solid, small spheres, *Appl. Phys. Lett.* 62, 2155 (1993; see also, *J. Appl. Phys.* 73, 7957 (1993).
49. Misawa, H., Fujisawa, R., Sasaki, K., Kitamura, N., and Masuhara, H., Simultaneous manipulation and lasing of a polymer microparticle using a cw 1064 nm laser beam, *Jpn. J. Appl. Phys.* 32, L788 (1993).
50. Bondar, M. V., Przhonskaya, O. V., and Tikhonov, E. A., Amplification of light by dyed polymers as the laser pumping frequency changes, *Opt. Spectrosc.* 74, 215 (1993).
51. Weber, H. P. and Ulrich, R., Unidirectional thin-film ring lasers, *Appl. Phys. Lett.* 20, 38 (1972).
52. Ulrich, R. and Weber, H. P., Solution-deposited thin films as passive and active light-guides, *Appl. Optics* 11, 428 (1972).
53. Sasaki, K., Fukao, T., Saito, T., and Hamano, O., Thin film waveguide evanescent dye laser and its gain measurement, *J. Appl. Phys.* 51, 3090 (1980).
54. Sriram, S., Jackson, H. E., and Boyd, J. T., Distributed-feedback dye laser integrated with a channel waveguide formed in silica, *Appl. Phys. Lett.* 36, 721 (1980).
55. Chang, M. S., Burlamacchi, P., Hu, C., and Whinnery, J.R., Light amplification in a thin film, *Appl. Phys. Lett.* 20, 313 (1972).
56. Kessler, W. J. and Davis, S. J., Novel solid state dye laser host, *Proc. Solid State Dye Laser Workshop* (1994), p. 216.
57. Weiss, M. N., Srivatava, R., Correia, R. R. B., Martins-Filho, J. F., and de Araujo, C. B., Measurement of optical gain at 670 nm in an oxazine-doped polyimide planar waveguide, *Appl. Phys. Lett.* 69, 3653 (1996).
58. Hu, L. amd Jiang, Z., Laser action in Rhodamine 6G doped titania-containing ormosils, *Optics Commun.* 148, 275 (1998).
59. Calderón, O. G., Guerra, J. M., Costela, A., García-Moreno, I., and Sastre, R., Laser emission of a flash-lamp pumped Rhodamine 6 G solid copolymer solution, *Appl. Phys. Lett.* 70, 25 (1997).
60. Kuwata-Gonokami, M., Takeda, K., Yasuda, H., and Ema, K., Laser emission from dye-doped polystrene microsphere, *Jpn. J. Appl. Phys.* 31, L99 (1992).
61. Kuwata-Gonokami, M. and Takeda, K., Polymer whispering gallery mode lasers, *Opt. Mater.* 9, 12 (1998).
62. Meier, M., Mekis, A., Dodabalapur, A., Timko,.A. Slusher, R. E., Joannopoulos, J. D., and Nalamasu, O., Laser action from two-dimensional distributed feedback in photonic crystals, *Appl. Phys. Lett.* 74, 7 (1999).
63. Kuwata-Gonokami, M., Jordan, R. H., Dodabalapur, A., Katz, H. E., Schilling, M. L., and Slusher, R. E., Polymer microdisk and microring lasers, *Optics Lett.* 20, 3093 (1995).
64. Costela, A., Garcia-Moreno, I., Barroso, J., and Sastre, R., Laser performance of Coumarin 540A dye molecules in polymeric host media with different viscosities: from liquid solution to solid polymer matrix, *J. Appl. Phys.* 83, 650 (1998).

1.6.4 Rare Earth Doped Polymer Lasers

Table 1.6.3 summarizes the laser properties of polymeric hosts and rare earth dopants that have been used to demonstrate stimulated emission.

Table 1.6.3
Rare Earth Doped Polymer Lasers

Host	Dopant	Transition	Laser wavelength (nm)	Pump source (wavelength – nm)	Ref.
polymethylmethacrylate (PMMA) - filaments	europium tris[4,4,4-trifluoro-1-(2-thienyl)-1,3 butanedione	Eu^{3+}: $^5D_0 \rightarrow ^7F_1$	~613	Xe flashlamp (~340)	1
polymethylmethacrylate (PMMA) - fiber	terbium thenoyltrifluoroacetonate	Tb^{3+}: $^5D_4 \rightarrow ^7F_5$	~545*	flashlamp (~320–370)	2
polymethylmethacrylate (PMMA) - fiber	neodymium octanoate	Nd^{3+}: $^4G_{5/2} \rightarrow ^4I_{9/2}$	575	Ar ion laser (514)	3

* No wavelength was given.

References—Table 1.6.3

1. Wolff, N. E. and Pressley, R. J., Optical laser action in an Eu^{+3}-containing organic matrix, *Appl. Phys. Lett.* 2, 152 (1963).
2. Huffman, E. H., Stimulated optical emission of a terbium ion chelate in a vinylic resin matrix, *Nature* 200, 158 (1963); Stimulated optical emission of a Tb^{3+} chelate in a vinylic resin matrix, *Phys. Lett.* 7, 237 (1963). For additional observations of probable stimulated emission of a terbium ion chelate in a vinylic resin matrix, see *Nature* 203, 1373 (1964). See, also, critique by Brecher, C., Lempicki, A. and Samelson, H, *Nature* 202, 580 (1964).
3. Zhang, Q. J., Wang, P., Sun, X. F., Zhai, Y., Dai, P., Yang, B., Hai, M. and Xie, J. P., Amplified spontaneous emission of an Nd^{3+}-doped poly(methyl methacrylate) optical fiber at ambient temperature, *Appl. Phys Lett.* 72, 407 (1998).

Section 1.7
SOLID STATE EXCIMER LASERS

Using matrix isolation techniques, it is possible to grow large, doped, rare-gas crystals. By introducing xenon and fluoride, XeF molecules can thus be formed by photodissociation of F_2 in Xe-F_2-Ar crystals. Optically pumped solid-state excimer laser action has been reported for XeF in Ar and Ne crystals. These results are summarized in Table 1.7.1.

Further Reading

Brau, C. A., Rare gas halide excimers, in *Excimer Lasers*, Rhodes, C. K., Ed., Springer Verlag, Berlin (1984).

Sliwinske, G. and Schwentner, N., Solid rare gas isolated ionic exciplexes for deep UV and VUV lasers, *J. Low Temp. Phys.* 111, 733 (1998).

Table 1.7.1
Solid State Excimer Lasers

Rare gas crystal	Composition	Transition	Lasing wavelength (nm)	Pump laser (nm)	Ref.
Ar (20 K)	Ar:Xe:F_2 (3000:1:1)	$B(^2\Sigma_{1/2}) \rightarrow X(^2\Sigma_{1/2})$	411	XeF (351)	1
		$C(^2\Pi_{3/2}) \rightarrow A(^2\Pi_{3/2,1/2})$	536[a]	XeF (351)	1
Ar (~10 K)	Ar:Xe:F_2 (1000:1:1)	$C(^2\Pi_{3/2}) \rightarrow A(^2\Pi_{3/2,1/2})$	~540	XeCl (308)	2
Ar (8 K)	Ar:Xe:F_2 (2000:1:1)	$C(^2\Pi_{3/2}) \rightarrow A(^2\Pi_{3/2,1/2})$	~540	XeCl (308)	3
Ar (15 K)	Ar:Xe:F_2 (2500:1:1)	$D(^2\Pi_{1/2}) \rightarrow X(^2\Sigma_{1/2})$	286	KrF (248)	4
Ar	Ar:Xe:F_2	$D(^2\Pi_{1/2}) \rightarrow X(^2\Sigma_{1/2})$	286	KrF (248)	5
Ne (6 K)	Ne:Xe:F_2 (6000:1:1)	$D(^2\Pi_{1/2}) \rightarrow X(^2\Sigma_{1/2})$	269.3	KrF (248)	6

(a) Gain measured from 520 to 590 nm.

References

1. Schwentner, N. and Apkarian, V. A., A solid state rare gas halide laser: XeF in crystalline argon, *Chem. Phys. Lett.* 154, 413 (1989).
2. Zerza, G., Sliwinski, G., and Schwentner, N., Relaxation-oscillations in the C→A laser emission of XeF in Ar crystals, *Appl. Phys. A* 54, 106 (1992).
3. Zerza, G., Sliwinski, G., and Schwentner, N., Threshold and saturation properties of a solid-state XeF (C-A) excimer laser, *Appl. Phys. B* 55, 331(1992).
4. Katz, A. I., Feld, J., and Apkarian, V. A., Solid-state XeF(D→X) laser at 286 nm, *Optics Lett.* 14, 441 (1989).
5. Zerza, G., Kometer, R., and Schwentner, N., A high gain tunable laser medium XeF doped Ar crystals, Orza, J.M., and Domingo, C., Eds., *Proc. SPIE* 1397, 107 (1991).
6. Zerza, G., Sliwinski, G., and Schwentner, N., Laser investigations at 269 nm for XeF (D-X) in Ne crystals, *Appl. Phys. A* 56, 156 (1993).

Section 1.8
RAMAN, BRILLOUIN, AND SOLITON LASERS

1.8.1 Introduction

Stimulated Raman scattering and stimulated Brillouin scattering are inelastic processes in which light at a laser or pump wavelength is converted into light at another wavelength accompanied by the excitation or deexcitation of an internal mode of the medium. The frequency of a laser v_L can be shifted to higher or lower energies through these light-matter interactions wherein vibrational energy of the material v_R is transferred to or is gained from the laser beam such that $v = v_L \pm v_R$, where the + and - signs denote anti-Stokes and Stokes transitions. Raman scattering is a third-order nonlinear optical process and it does not require phase matching. Many orders of stimulated Stokes radiation have been observed, thus providing multiwavelength emission.

Stimulated Raman scattering occurs in liquids, gases, and solids. Raman materials that have a high value of polarizability and vibrational modes that induce a large modulation of this polarizability have relatively high Raman scattering cross-sections, high Raman gains, and high optical frequency conversion efficiencies. Raman shifters are common accessories for lasers and have utilized solids (crystals and glasses), liquids, or gases as the active material.

Further Reading

Basiev, T. T. and Powell, R. C., Introduction, Special Issue on Solid State Raman Lasers, *Opt. Mat.* 11, 301 (1999).

Milanovich, F. P., Stimulated Raman Scattering, in *Handbook of Laser Science and Technology, Vol. III, Optical Materials*, CRC Press, Boca Raton, FL (1987), p. 283.

Mollenauer, L. F., The Soliton Laser, in *Handbook of Laser Science and Technology, Supplement 1: Lasers,* CRC Press, Boca Raton, FL (1991), p. 114.

Murray, J. T., Powell, R. C., and Peyghambarian, N., Properties of stimulated Raman scattering in crystals, *J. Lumin.* 66&67, 89 (1996).

Reinjtes, J. F., Stimulated Raman and Brillouin Scattering, in *Handbook of Laser Science and Technology, Supplement 1: Lasers,* CRC Press, Boca Raton, FL (1991), p. 334.

Stolen, R, H. and Lin, C., Fiber Raman Lasers, in *Handbook of Laser Science and Technology, Vol. I: Lasers and Masers*, CRC Press, Boca Raton, FL (1982), p. 171 and *Supplement 1: Lasers,* CRC Press, Boca Raton, FL (1991), p. 101.

Zverev, P. G., Basiev, T. T., and Prokhorov, A. M., Stimulated Raman scattering of laser radiation in Raman crystals, *Opt. Mat.* 11, 335 (1999).

1.8.2 Crystalline Raman Lasers

Crystals with high Raman gains have been used in high efficiency, compact, solid-state Raman lasers. With respect to the pump laser, Raman lasers can be designed with a shared resonator, coupled resonators, or an external resonator. The two-photon nature of Raman transitions results in a very small scattering cross section for the process and thus requires a very high pump density (about 1GW/cm^2), close to the laser damage threshold of many Raman crystals.

Table 1.8.1 lists Raman frequency shifts v_R of materials and the wavelengths that have been reported in the literature for crystalline Raman lasers. (This table was prepared by R. C. Powell, J. T. Murray, W. L. Austin and T. T.Basiev.) Most of the results have been obtained with a common pump laser such as ruby or Nd:YAG with the Raman laser operating in a first-Stokes configuration in which one quantum of vibrational energy is transferred from the pump photons to the material. It should be noted that many other laser wavelengths are implied by these results through the use of different pump lasers and higher- order Stokes or anti-Stokes processes. In addition, the spectroscopy of stimulated Raman scattering has been studied in many materials that have not yet been made into lasers (see Further Reading).

Table 1.8.1
Crystalline Raman Laser Wavelengths

Raman material	Pump wavelength (nm)	Raman wavelengths (nm)	Ref.
Ba(NO$_3$)$_2$	530	560, 598	5
v_R = 1048 cm^{-1}	1064	1197.4	14
	1100–1200	1243.3–1372.4, 1430–1600	14
	1064	1197, 1363, 598,5, 563	15
	1079	1216, 1394, 1632	16
	1318	1529	16
	1338	1556	16,17
	1318–1338	1535–1556, 1820–1860	18
	1047	1176, 588	19
	532	563, 599, 639, 504	20, 21
	1064	1197, 1369, 1598	16, 22
Ba(NO$_3$)$_2$ powder	532	563	101
Ba$_2$NaNb$_5$O$_{15}$	694.3	727.1	8
v_R = 650 cm^{-1}			
C (diamond)	694.3	765.1	10
v_R = 1332 cm^{-1}			
CaCO$_3$	694.3	694.26, 603.32, 566.23, 533.64	1
v_R = 1086 cm^{-1}	694.3	750.9	2, 3
	532	562, 599, 641	4
	530	560, 598	5
	532	562, 564.6	6, 7

Table 1.8.1—*continued*
Crystalline Raman Laser Wavelengths

Raman material	Pump wavelength (nm)	Raman wavelengths (nm)	Ref.
$CaWO_4$ $v_R = 911.3$ cm^{-1}	1064	1178, 1320, 589, 660	34
GaP $v_R = 403, 365$ cm^{-1}	1064	1112, 1164, 1221, 1284.3, 1107	49
	840	870	49
	895	929	49
$Gd_2(MoO_4)_3$ $v_R = 960, 943,$ $857, 100$ cm^{-1}	1064.15, 532.07	506.2–1185.2	37
β'-$Gd_2(MoO_4)_3$ $v_R = 768$ cm^{-1}	1064	1158.7	38
$InSb$ $v_R = 30–170$ cm^{-1}	10600	10900–13000	45
$KGd(WO_4)_2$ $v_R = 767, 901.5$ cm^{-1}	1067.2	973.5, 1180.8, 1321.5, 1500.2	23, 24
	1064	1176.95, 1158.6, 1293.7	25, 26
	1064	1176.9	14
	1064	1158.6	14
	1100–1200	1201.4–1321.7	14
	1069	1180	27, 28
	1068.8	1180.5, 1320.7	29, 30
	1064	1180, 1321.5, 1500.2, 1734.8	24, 26
	1064	973.5, 1162.3, 1276.1, 1414.5	24, 26
	532	554.8, 579.6, 606.6, 558.8	31
	496.5	516.2, 537.6; 519.7, 545.2	31
	1067.2	1180, 533.6, 561.9, 660.4, 590.3, 628.3	32
	1351	1538	33
KH_2PO_4 $v_R = 918$ cm^{-1}	532	559.3	39
$KY(WO_4)_2$ $v_R = 770, 901$ cm^{-1}	1068.8	1182.7, 1323.7, 974.9	28, 29
	532	554.8, 579.6, 606.6, 558.8	31
	496.5	516.2, 537.6; 519.7, 545.2	31
$Li_3Ba_2Gd_3(MoO_4)_8$ $v_R = 770$ cm^{-1}	532	554.7	35
$LiHCOO \cdot H_2O$ $v_R = 1372$ cm^{-1}	694.3	767	40
$LiIO_3$ $v_R = 760, 818$ cm^{-1}	1080	1180, 1310	41, 42
	1054	1143	43

Table 1.8.1—*continued*
Crystalline Raman Laser Wavelengths

Raman material	Pump wavelength (nm)	Raman wavelengths (nm)	Ref.
$LiNbO_3$	694.3	706.9	8
v_R = 152, 212–248,	694.3	726.4	8
491–628 cm^{-1}	694.3	701.7	9
	694.3	706.46–704.67	9
	694.3	718.8–725.9, 671.3	9
$LiTaO_3$	694.3	704.1	8
v_R = 200 cm^{-1}			
$NaClO_3$	1064	1181.8	44
v_R = 936 cm^{-1}			
$NaLa(MoO_4)_2$	496.5	519.7	31
v_R = 888, 320,	1064, 532	529.3–1185.2	36
100 cm^{-1}			
$NaNO_3$	532	564	6
v_R = 1068 cm^{-1}			
	530	560, 598	5
$Pb(NO_3)_2$	530	560, 598	5
v_R = 1045 cm^{-1}			
Si (silicon)	1064	1127	46, 47
v_R = 521 cm^{-1}		1008	48
SiO_2 (quartz)	694.3	717.6	8
v_R = 219–650 cm^{-1}	1064	1119.6	12
	532.0	545.6	12
	514.5	527.2	12
	496.5	508.3	12
	488.0	499.4	12
	476.5	487.3	12
	400–650	407.6–670.3	12
	1064,1090–1230	1116–1221, 995.4–1021.8	13
Stilbene	694.3	780.64	2
v_R = 997, 1593 cm^{-1}	694.3	745.93	2

1.8.3 Fiber Raman Lasers and Amplifiers

In its simplest form, a fiber Raman laser is an optical fiber with pump light focused into one end and stimulated Raman light appearing at the output. Pulsed single-pass or multipass oscillators pumped by either cw or pulsed lasers have been realized. Pump lasers have included Nd:YAG lasers, flashlamp- and nitrogen-laser-pumped dye lasers, ion lasers (Ar), and excimer lasers (XeCl). Cladding-pumped Raman fiber lasers have also been pumped by several multimode diode lasers. Output wavelengths have ranged from the near ultraviolet to the near infrared. The fiber Raman lasers in Table 1.8.2 have utilized fused silica (SiO_2), germanosilicate, or phosphosilicate optical fibers for which the Raman frequency shifts are ~440–490 cm^{-1}. Gas-in-glass fiber Raman lasers are listed in Table 1.8.3.

Table 1.8.2
Fiber Raman Lasers

Operating mode

CD	— cavity dumped	p	— pulsed
cw	— continuous wave	qcw	— quasi-continuous wave
ML	— mode locked	qs	— Q-switched

Pump (wavelength–μm)	Mode	Raman wavelengths (μm)	Ref.
XeCl laser (0.308)	p	0.312–0.331	64
XeCl laser (0.308)	p	0.3124, 0.3167, 0.3208, 0.3253, 0.3299, 0.3347, 0.3395, 0.34445, 0.3491	63
N$_2$ laser (0.337)	p	0.342	65
Ar laser (0.4880)	cw	0.4991	54
Ar laser (0.4880)	cw	0.4988, 0.5001	86
Ar laser (0.5145)	CD	1st–5th Stokes lines	58
Ar laser (0.5145)	cw	0.5153–0.5287	57
Ar laser (0.5145)	cw	0.5204–0.5285, 0.5302–0.5418, 0.5423–0.5540, 0.5620–0.5560	56
Ar laser (0.5145)	cw	0.521–0.5285	55
Ar laser (0.5145)	cw	0.5249–0.5279	53
Nd:YAG laser[a] (0.532)	p	0.545	11
dye laser (0.5992)	p	0.5975, 0.6036, 0.6143, 0.6334, 0.6506, 0.6573	66
Yb laser (1.06)	cw	1.24, 1.48[b]	103
Yb fiber laser (1.060)	cw	1.12, 1.18, 1.24[c]	99
Yb fiber laser (1.064)	cw	1.239, 1.484	107
Nd:YAG laser (1.064)	qs	1.12, 1.18, 1.23, 1.3	94
Nd:YAG laser (1.064)	qs	0.7–2.1	52
Nd:YAG laser (1.064)	cw	1.085–1.1175	62
Nd:YAG laser (1.064)	cw, ML	1.12–1.18, 1.24, 1.31, 1.07–1.32	60, 61
Nd:YAG laser (1.064)	ML, cw	1.101–1.125	58
Nd:YAG laser (1.064)	p	1.12, 1.18, 1.23, 1.3	50

(a) Frequency doubled
(b) SiO_2 plus 13 mol.% P_2O_5
(c) Germanium doped silica fiber

Table 1.8.3
Gas-in-Glass Fiber Raman Lasers

Gas diffused into SiO_2	Pump wavelength (μm)	Mode	Raman wavelength (μm)	Ref.
H_2	0.647	p	0.883	67
D_2	1.064	p, cw	1.56	68

Liquid-filled hollow fused silica fibers have also been used for Raman lasers; examples of liquids include carbon disulfide (Reference 51) and benzene (Reference 84).

Fiber Raman Amplifiers

Raman amplifiers can be constructed at virtually any wavelength by terminating the Raman laser one Stokes order less than where amplication is desired and then injecting a signal through the resonant structure. In the case of cascaded fiber Raman lasers, the desired upper wavelength limit is determined simply by the increased loss of the fiber at that wavelength and thus depends on what the acceptable efficiency is. The conversion efficiency is high if the cascaded Raman resonator is constructed with low-loss fiber Bragg gratings. The references below illustrate various methods that have been employed to obtain cascaded Raman generation and amplification.

Table 1.8.4
Fiber Raman Amplifiers

Pump (wavelength–μm)	Mode	Raman and amplifier wavelengths (μm)	Ref.
dye laser (0.540–0.590)	p	1st and higher order Stokes lines	85
Nd fiber laser (1.060)	cw	1.3085	97
Nd:YAG laser (1.064)	cw	1.117, 1.174, 1.24, 1.3	98
InGaAsP/InP (1.24)	cw	1.3[a]	100
Raman fiber laser (1.24)	cw	1.3	102
Color center laser (~1.47)	p	~1.58	76

(a) Fibers contained GeO_2 up to 25 mol%.

1.8.4 Fiber Soliton Lasers

In a solition laser, the output of a pulsed laser at some wavelength is coupled into a fiber. Reshaping of the laser pulse by the fiber forces the laser to produce much shorter pulses than it would by itself. The pulse width is controlled by the length of the fiber and can range from several picoseconds to a few tens of femtoseconds. Pulse compression effects in fiber Raman lasers result in the production of ultrafast optical pulses in both single-pass and oscillator configurations. A number of different cavity configurations and mode-locking techniques that have been used to generate femtosecond pulses are covered in the lasers listed in Table 1.8.5.

Table 1.8.5
Fiber Soliton Lasers

Pump (wavelength–µm)	Mode	Wavelengths (µm)	Pulse duration (fs)	Ref.
Nd:YAG laser (1.064)			800	79
Nd:YAG laser (1.064)	cw	1.076–1.12	400	77
Kr³⁺ ion laser	cw	~1.1[a]	6000	90
Nd:YAG laser (1.064)	qs, ML	1.6	18	78
Nd:YAG laser (1.064)	cw, ML	1.3841.400	100–200	74, 75
Pr:glass laser (1.3)	cw	~1.3	620	88
Nd:YAG laser (1.32)	ML	1.4	~100	80
Color center laser (~1.47)	p	~1.58	~240–300	76
Color center laser (~1.5)	ML	~1.6	210–2000	69
Color center laser (~1.5)			130, 50	70
Color center laser (~1.5)			60–600	71
Color center laser (~1.5)			50, 19	72
Er laser (1.550)	cw	1.585	1550	105
Yb-Er laser	cw	1.5390, 1.5436	1600, 12000	106
Yb-Er laser	cw	1.534–1.548	7000–8500	89
OPO[b] (1.54)	p	1.5–1.65	200	73

(a) Neodymium-doped fiber.
(b) Optical parmetric oscillator.

1.8.5 Fiber Brillouin Lasers

Stimulated Brillouin scattering involves the interaction of light with high-frequency sound waves in the medium. The scattered wave is almost always shifted to longer wavelengths than the pump. The results in Table 1.8.6 are for fused silica fibers.

Table 1.8.6
Fiber Brillouin Laser Wavelengths

Pump wavelength (µm)	Mode	Frequency shift (GHz)	Ref.
0.5145 (Ar ion laser)	cw	14 lines; total bandwidth – 476 GHz	83
0.5145 (Ar ion laser)	cw	34	82
0.5145 (Ar ion laser)	cw		91
0.5145 (Ar ion laser)	cw		92
0.5355	p	32.2	81
0.6328 (He-Ne laser)	cw		92
0.6328 (He-Ne laser)	cw	(spectral width ~2 kHz)	93
0.6328 (He-Ne laser)	cw	26,32, 26.85, 26.52, 27.05[a]	95
1.532 (Er fiber laser)	cw	10.35	96
1.561 (Er fiber laser)	cw	34 lines of Stokes and anti-Stokes waves[b]	104

(a) SiO_2 doped with 6 mol% GeO_2.
(b) Combined effect of stimulated Brillouin scattering and four-wave mixing.

1.8.6 References

1. Chiao, R. and Stoicheff, B. P., Angular dependence of maser-stimulated Raman radiation in calcite, *Phys. Rev. Lett* 12, 290 (1964).
2. Chirkov, V. A., Gorelik, V. S., Peregudov, G. V., and Sushinskii, M. M., Investigation of stimulated Raman scattering line width, *JETP Lett.* 10, 416 (1969).
3. Kudryavtseva, A. D., Morozova, E. A., and Moiseenko, M. M., Stimulated Raman scattering and distractions in calcite crystals, *Kratkie Soobshcheniya po Fizike* 10, 31 (1973).
4. Sukhareva, L. N. and Khazov, L. D., Ways to efficiency increase in solid-state SRS oscillators, *Sov. J. Quantum Electron.* 4, 13 (1977).
5. Eremenko, A. S., Karpukhin, S. N., and Stepanov, A. I., SRS of neodymium laser second harmonic in nitrate crystals, *Sov. J. Quantum Electron* 7, 196 (1980) .
6. Karpukhin, S. N. and Stepanov, A. I., Generation of radiation in a resonator under conditions of stimulated Raman scattering in $Ba(NO_3)_2$, $NaNO_3$, and $CaNO_3$ crystals, *Sov. J. Quantum Electron* 16, 1027 (1986); Karpukhin, S. N. and Yashin, V. E., Generation and amplification at SRS in crystals, *ibid* 14, 1337 (1984).
7. Ivanov, A. A. Mak, S. B. Papernyi, and Serebryakov, V. A., Picosecond pulses of back SRS, *Sov. J. Quantum Electron.* 13, 85 (1986) .
8. Johnston, Jr., W. D., Kaminow, I. P., and Bergmann, Jr., J. G., Stimulated Raman gain coefficients for Li^6NbO_3, $Ba_2NaNb_5O_{15}$, and other materials, *Appl. Phys. Lett.* 13, 190 (1968); Johnston, Jr., W. D. and Kaminow, I. P, Temperature dependence of Raman and Rayleigh scattering in $LiNbO_3$ and $LiTaO_3$, *Phys. Rev.* 168, 1045 (1968).
9. Gelbwachs, J., Pantell, R. H., Puthoff, H. E., and J. M. Yarborough, A tunable stimulated Raman oscillator, *Appl. Phys. Lett.* 14, 258 (1969).
10. McQuillan, A. K., Clements, W. R. C., and Stoicheff, B. P., Stimulated Raman emission in diamond: spectrum, gain, and angular distribution of intensity, *Phys. Rev. A* 1, 628 (1970).
11. Stolen, R. H., Ippen, E. P., and Tynes, A. R., Raman oscillation in glass optical waveguide, *Appl. Phys. Lett.* 20, 62 (1972).
12. Stolen, R. H., in *Fiber and Integrated Optics*, Ostrowsky, D. B., Ed., Plenum Press, N.Y., (1979).
13. Basiev, T. T., Dianov, E. M., Zharikov, E. A., Karasik, A.Ya., Mirov, S. B., and Prokhorov, A. M., Selective nonlinear spectroscopy of inhomogeneously broadened phonon resonances in a disordered medium, *JETP Lett.* 37, 2292 (1983).
14. Basiev, T. T., Voitsekhovskii, V. N., Zverev, P. G., Karpushko, F. V., Lyubimov, A. V., Mirov, S. B., Morozov, V. P., Mochalov, I. V., Pavlyuk, A. A., Sinitsyn, G. V., and Yakobson, V. E., Conversion of tunable radiation from a laser utilizing a LiF crystal containing F_2- color centers by stimulated Raman scattering in $Ba(NO_3)_2$ and $KGd(WO_4)_2$ crystals, *Sov. J. Quantum Electron.* 17, 1560 (1987).
15. Khulugurov, V. M., Ivanov, N. A., Inshakov, D. V., Oleinikov, E. A., Voitsekhovskii, V. N., and Yakobson, V. E., Amplification of $Ba(NO_3)_2$ SRS by $LiF{:}F_2^-$ crystal, *Sov. J. Quantum Electron* 19, 2, 162 (1992).
16. Zverev, P. G., and Basiev, T. T., Compact SRS laser on barium nitrate crystal,, *Proceedings of All Union Conference Laser Optics* 2, 363 (1993); Basiev, T. T., Sigachev, V. B., Doroshenko, M. E., Zverev, P. G., Osiko, V. V., and Prokhorov, A. M., *SPIE Proceeding* Vol. 2498, 171 (1994); Zverev, P. G., and Basiev, T. T., Barium nitrate Raman laser for near IR spectral region, *OSA Proceeding on Adv. Solid State Lasers*, Vol. 24 (1995), p. 288.
17. Murray, J. T., Powell, R. C., and Austin, E. L., High brightness, eye-safe lasers, *Optics and Photonic News* 6, 32 (1995).
18. Murray, J. T., Powell, R. C., Peyghambarian, N., Smith, D., Austin, W., and Stolzenberger, R. A., Generation of 1.5-μm radiation through intracavity solid-state Raman shifting in $Ba(NO_3)_2$ nonlinear crystals, *Optics Lett.* 20, 9, 1017 (1995).
19. Murray, J. T., Austin, W. L., Calmes, L. K., Powell, R. C., and G. J. Quarles, Nonlinear cavity-dumped intracavity solid-state Raman laser transmitters, *OSA TOPS on Advanced Solid State Lasers*, Vol. 20, Eds., Payne, S. A. and Pollock, C.K. (Optical Society of America, Washington, DC, 1997), p. 72.

20. Zverev, P. G., Murray, J. T., Powell, R. C., Reeves, R. J., and Basiev, T. T., Stimulated Raman scattering of picosecond pulses in barium nitrate crystals, *Opt. Commun.* 97, 59 (1993).

21. He, Ch. and Chyba, T. H., Solid-state barium nitrate raman laser in the visible region, *Opt. Commun.* 135, 273 (1997); McCray, C.L. and Chyba, T. H. *Opt. Mat.* 11, 335 (1999).

22. Kannari, F. and Takei, N., *Proc. Intern. Symp. on Laser and Nonlinear Optical Materials.* Singapore, 3-5 Nov. 1997, p. 289.

23. Andryunas, K., Vishakas, Yu., Kabelka, V., Mochalov, I.V., Pavlyuk, A.A., Petrovskii, G.T., and Syrus, V., Stimulated Raman self-conversion of Nd^{3+} laser light in double tungstate crystals, Pisíma v Zhurnal Eksper, i Teor. Fiz. 42, 333 (1985) (*JETP Lett.* 42, 410 (1986); Andryunas, K., Vishakas, Yu., Kabelka, V., Mochalov, I.V., Pavlyuk, A.A., Ionina, N.V., and Syrus, V., Method of picosecond pulse generation at a Raman frequency, USSR Patent # 1227074.

24. Berenberg, V. A., Karpukhin, S. N., and Mochalov, I. V., SRS of nanosecond pulses in $KGd(WO4)_2$ crystals, *Sov. J. Quantum Electron.* 14, 9, 1849-1850 (1987).

25. Mikhailov, A. V., Mochalov, I. V., and Lyubimov, A. V., Raman scattering in PGW crystals, *Lasers and Optical Nonlinearity*, Maldutis, E. K., Ed., Vilnyus (1987), p. 265 [in Russian].

26. Vishchakas, Yu. K., Mochalov, I. V., Mikhailov, A. V., Klevtsova, R. F., and Lyubimov, A. V., Crystal structure and Raman scattering in $KGd(WO_4)_2$ crystals, *Lietuvos fizikos rinkinys* 28, 2244 (1988)

27. Ivanyuk, A. M., Sandulenko, V. A., Ter-Pogosyan, M. A., Shakhverdov, P. A., Chervinskii, V. G., Lukin, A. V., and Ermolaev, V. L, Intracavity stimulated Raman scattering in a nanosecond neodymium laser based on potassium-gadolinium tungstate, *Optika i Spektroscopiya.* 62, 961 (1987) (*Soviet Opt. Spectr*).

28. Khulugurov, V. M., Ivanov, N. A., and Oleynikov, E. A., Nanosecond lasers based on Raman converters with resonant excitation, CLEO'94, abstract CTuK70.

29. Andryunas, K., Barila, A., Vishchakas, Yu., Mochalov, I. V., Petrovskii, G. T., and Syrus, V., Temporal characteristics of picosecond pulses of SRS self-conversion, *Optika i Spektroskopiya* 64, 2, 397 (1988) (*Sov. Opt. Spectr.*).

30. Andryunas, K., Barila, A.,Vishchakas, Yu., and Syrus, V., Investigation of pulse duration dynamics in a laser based on $KGd(WO4)_2$:Nd^{3+} and $KY(WO4)_2$:Nd^{3+}*Lasers and Optical Nonlinearity*, Maldutis, E. K., Ed., Vilnyus (1987), p. 43 [in Russian].

31. Kaminskii, A. A., H. Nishioka, Y. Kubota, K. Ueda, H. Takuma, Bagaev, S. N., and Pavlyuk, A. A., New optical phenomena in laser insulating crystal hosts with third-order nonlinear susceptibilities, *Phys. Stat. Solidi A.* 148 , 619 (1995).

32. Zverev, P. G. and Basiev, T. T., Intracavity stimulated Raman scattering in KGW:Nd laser passively Q-switched with $LiF:F_2^-$ color center crystal, in *Conf. Lasers and Electro-Optics*, Vol. 11, OSA Technical Digest Series (Optical Society of America, Washington, DC 1997), p. 379.

33. Ustimenko, N. S. and Gulin, A. V., 1.538 m wavelength radiation of $KGd(WO_4)_2$:Nd^{3+} lasers based on stimulated Raman scattering with self-conversion, *Pribory i Tekhnika Eksperimenta* 3, 99 (1998).

34. Murray, J. T., Austin, W. L., and Powell, R. C., End-pumped intracavity solid state Raman lasers, *OSA TOPS Vol. 19 Adv. Solid State. Lasers*, p. 129 (1998).

35. Belashenkov, N. R., Belayev, V. D., Gagarsky, S. V., and Titov, A. N., LBGM:Tr – The new laser Raman-active medium, *Advanced Solid State Laser Conference*, Memphis, TN (1995).

36. Kaminskii, A. A., Bagaev, S. N., Grebe, D., Eichler, G., Pavlyuk, A. A., and Macdonald, R., Efficient multi-wave Stokes and anti-Stokes generation of a Raman-parametric laser based on a $NaLa(MoO_4)_2$ tetragonal crystal, *Sov. J. Quantum Electron.* 23, 199 (1996).

37. Kaminskii, A. A., Bagaev, S. N., Grebe, D., Eichler, G., Pavlyuk, A. A., Kuznetsov, F. A., and Macdonald, R., Efficient multi-frequent Stokes and anti-Stokes Raman-parametric generation of $Gd_2(MoO_4)_3$ rhombic crystals, *Doklady Academii Nauk, Fizika* 348, 475 (1996).

38. Kaminskii, A. A., Butashin, A. V., Eichler, H. J., Grebe, D., Macdonald, R., Ueda, K., Nishioka, H., Odajima, W., Tateno, M., Song, J., Musha, M., Bageav, S. N., and Pavlyuk, A. A., Orthorhombic ferroelectric and ferroelastic $Gd_2(MoO_4)_3$ crystal a new many-purposed nonlinear and optical material: efficient multiple stimulated Raman scattering and CW and tunable second harmonic generation, *Opt. Mat.* 7, 59 (1997).

39. Milanovich, F. P., in *Handbook of Laser Science and Technology*, Vol. 3, Part 1, Ed. Weber, M. J. (CRC Press, Boca Raton, FL, 1987), p. 283.

40. Lai, K. K., Schüsslbauer, W., Silberbauer, H., Amler, H., Bogner, U., Maier, M., Jordan, M., and Jodl, H.-J., Stimulated Raman scattering in lithium formate monohydrate crystals at temperatures from 2 to 300 K, *Phys. Rev. B.* 42, 5834 (1990-II).

41. Ammann, E. O. and C. D. Decker, 0.9-W Raman oscillator, *J. Appl. Phys* 48, 1973 (1977).

42. Ammann, E. O., High-average-power Raman oscillator employing a shared-resonator configuration, *Appl. Phys. Lett.* 32, 52 (1978).

43. Grigoryan, G. G. and Sogomonyan, S.B., A synchronously pumped picosecond SRS-laser on a $LiIO_3$ crystal, *Sov. J. Quantum Electron.* 16, 2180 (1989).

44. Kaminskii, A. A., Hulliger, J., Eichler, H. J., Findeisen, J., Butashin, A. V., MacDonald, R., and Bagaev, S. N., Stimulated Raman scattering in cubic $NaClO_3$ crystal at room temperature under picosecond excitation, *Rap. Res. Not.* (b) 303, R9 (1997).

45. Patel, K. N. and Shaw, E. D., Tunable stimulated Raman scattering from mobile carriers in semiconductors, *Phys. Rev. B.* 3, 1279 (1971).

46. Ralston, J. M. and Chang, R. K., Spontaneous-Raman-scattering efficiency and stimulated scattering in silicon, *Phys. Rev. B.* 2, 1858 (1970).

47. Grassl, H.-P. and Maier, M., Efficient stimulated Raman scattering in silicon, *Opt. Commun.* 30, 253 (1979).

48. Kitazima, I. and Iwasawa, H., Stimulated Stokes and anti-Stokes emission from silicon by glass laser, *Opt. Commun.* 5, 18 (1972).

49. Suto, K. and Nishizawa, J.-I., *Semiconductor Raman Lasers*, Artech House Inc. (1994).

51. Ippen, E. P., Low power quasi-cw Raman oscillator, *Appl. Phys. Lett.* 16, 303 (1970).

52. Lin, C., Nguyen, V. T., and French, W. G., Wideband near-IR continuum (0.7-2.1 µm) generated in low-loss optical fibers, *Electron. Lett.* 14, 822 (1978).

53. Johnson, D. C., Hill, K. O., Kawasaki, B. S., and Kato, D., Tunable Raman fiber-optic laser, *Electron. Lett. 13,* 53 (1977).

54. Hill, K. O., Kawasaki, B. S., and Johnson, D. C., Low-threshold cw Raman laser, *Appl. Phys. Lett.* 29, 181 (1976).

55. Jain, R. K., Lin, C., Stolen, R. H., Pleibel, W., and Kaiser, P., A highly efficient tunable cw Raman oscillator, *Appl. Phys. Lett.* 30, 162 (1977).

56. Jain, R. K., Lin, C., Stolen, R. H., and Ashkin, A., A tunable multiple Stokes cw fiber Raman oscillator, *Appl. Phys. Lett.* 31, 89 (1977).

57. Stolen, R. H., Lin, C., and Jain, R. K., A time-dispersion-tuned fiber Raman oscillator, *Appl. Phys. Lett.* 30, 340 (1977).

58. Stolen, R. H., Lin, C., Shah, J., and Leheny, R. F., A fiber Raman ring laser, *IEEE J. Quantum Electron.* QE-14, 860 (1978).

59. Lin, C., Stolen, R. H., and Cohen, L. G., A tunable 1.1 µm fiber Raman oscillator, *Appl. Phys. Lett.* 31, 97 (1977).

60. Dianov, E. M., Isaev, S. K., Kornienko, L. S., Kravtsor, N. V., and Firsor, V. V., Raman Laser with optical fiber resonator, *Sov. J. Quantum Electron.* 8, 744 (1981).

61. Lin, C. and French, W. G., A near-infrared fiber Raman oscillator tunable from 1.07 to 1.32 µm, *Appl. Phys. Lett.* 34, 666 (1979).

62. Lin, C., Stolen, R. H., French, W. G., and Melone, T. G., A cw tunable near infrared (1.085–1.1175 µm) Raman oscillator, *Optics Lett.* 1, 96 (1977).

63. Pini, R., Salimbeni, R., Matera, M., and Lin, C., Wideband frequency conversion in the uv by nine orders of stimulated Raman scattering in a XeCl laser pumped multimode silica fiber, *Appl. Phys. Lett.* 43, 517 (1983).

64. Misunami, T., Miyazaki, T., and Takagi, K., Short-pulse ultraviolet fiber Raman laser pumped by a XeCl excimer laser, *J. Opt. Soc. Am.* B4, 498 (1987).

65. Rothschild, M. and Abad, H., Stimulated Raman scattering in fibers in the ultraviolet, *Optics Lett.* 8, 653 (1983).

66. Lin, C. H., Marshall, B. R., Nelson, M. A., and Theobald, J. K., Backward stimulated Raman scattering in multimode fiber, *Appl. Opt.* 17, 2486 (1978).

67. Stone, J., Chraplyvy, A. R., and Burrus, A. C., Gas-in-glass — a new Raman-gain medium: molecular hydrogen in solid-silica, *Optics Lett.* 7, 297 (1982.

68. Chraplyvy, A. R. and Stone, J., Single-pass modelocked or Q-switched operation of D_2 gas-in-glass fiber Raman lasers operating at 1.56 µm wavelength, *Optics Lett.* 10, 344 (1985).

69. Mollenauer, L. F. and Stolen, R. H., The soliton laser, *Optics Lett.* 29, 13 (1984).

70. Mollenauer, L. F., Solitons in optical fibres and the soliton laser, *Phil. Trans. Roy. Soc.* A315, 437 (1985).

71. Mitschke, F. M. and Mollenauer, L. F., Stabilizing the soliton laser, *IEEE J. Quantum Electron.* QE-22, 2242 (1986).

72. Mitschke, F. M. and Mollenauer, L. F., Ultrafast pulses fron the soliton laser, *Optics Lett.* 12, 407 (1987).

73. Dianov, E. M., Karasik, A. Va, Mamyshev, P. V., Prokhorov, A. M., Serkin, V. N., Stei'makh, M. F., and Fornichev, A. A., Stimulated-Raman conversion of multisoliton pulses in quartz optical fibers, *JETP Lett.* 41, 294 (1985).

74. Kafka, J. D. and Baer, T., Fiber Raman soliton laser pumped by a Nd:YAG laser, *Optics Lett.* 12, 181 (1987).

75. Gouveia-Neto, A. S., Gornes, A. S. L., and Taylor, J. R., Fermosecond soliton Raman ring laser, *Electron. Lett.* 23, 537 (1987).

76. Islam, M. N., Mollenauer, L. F., Stolen, R. H., Simpson, J. R., and Shang, H. T., Amplifier/compressor fiber Raman lasers, *Optics Lett.* 12, 814 (1987).

77. Dianov, E. M., Manryshev, P. V., Prokhorov, A. M., and Fursa, D. G., Tunable subpicosecond synchronously pumped fiber-optic stimulated-Raman laser, *JETP Lett.* 45, 599 (1987).

78. Grudinin, A. B., Dianov, E. M., Korohkin, D. B., Prokhorov, A. M., Serkin, V. N., and Khaidarov, D. V., Stimulated-Raman-scattering excitation of 18-fs pulses in the 1.6 µm region during pumping of a single-mode optical fiber by the beam from a Nd:YAG laser (λ = 1.064 µm), *JETP Lett.* 45, 260 (1987).

79. Kafka, J. D., Head, D. F., and Baer, T., Dispersion compensated fiber Raman oscillator, in *Ultrafast Phemomena, V,* Fleming, G. R. and Siegman, A. E., Eds., Springer-Verlag, Berlin (1986), p. 51.

80. Gouveia-Neto, A. S., Gomes, A. S. L., and Taylor, J. R., Higli-efficiency single-pass solitonlike compression of Raman radiation in an optical fiber around 1.4 µm, *Optics Lett.* 12, 1035 (1987.)

81. Ippen, E. P. and Stolen, R. H., Stimulated Brillouin scattering in optical fibers, *Appl. Phys. Lett.* 21, 539 (1972).

82. Hill, K. O., Kawasaki, B. S., and Johnson, D. C., cw Brillouin laser, *Appl. Phys. Lett.* 28, 608 (1976).

83. Hill, K. O., Johnson, D. C., and Kawasaki, B. S., CW generation of multiple Stokes and anti-Stokes Brillouin-shifted frequencies, *Appl. Phys. Lett.* 29, 185 (1977).

84. Stone, J., cw Raman fiber amplifier, *Appl. Phys. Lett.* 26, 163 (1975).

85. Lin, C., and Stolen, R. H., Backward Raman amplification and pulse steeping in silica fibers, *Appl. Phys. Lett.* 29, 428 (1976).

86. Johnson, D. C., Hill, K. O., and Kawasaki, B. S., Continuous-wave optical-fiber Raman oscillator employing a two-mirror resonator configuration, *Appl. Optics* 17, 3032 (1978).

87. Mitschke, F. M. and Mollenauer, L. F., Discovery of the soliton self-frequency shift, *Optics Lett.* 11, 659 (1986); Gordon, J. P., Theory of the soliton self-frequency shift, *Optics Lett.* 11, 662 (1986).

88. Guy, M. J., Noske, D. U., Boskovic, A., and Taylor, J. R., Femtosecond soliton generation in a praseodymium fluoride fiber laser, *Optics Lett.* 19, 828 (1994).

89. Romagnoli, M., Wabnitz, S., Franco, P., Midrio, M., Fontana, F., and Town, G. E., Tunable erbium-ytterbium fiber sliding-frequency soliton laser, *J. Opt. Soc. Am. B,* 12, 72 (1995).

90. Hofer, M., Ober, M. H., Hofer, R. et al., High-power neodymium soliton fiber laser that uses a chirped fiber grating, *Optics Lett.* 20, 1701 (1995).

91. Ponikvar, D. R. and Ezekiel, S., Stabilized single-frequency stimulated Brillouin fiber ring laser, *Optics Lett.* 6, 398 (1981).

92. Stokes, L. F., Chodorow, M., and Shaw, H. J., All-fiber stimulated Brillouin ring laser with submilliwatt pump threshold, *Optics Lett.* 7, 509 (1982).

93. Smith, S. P., Zarinetchi, F., and Ezekiel, S., Narrow-linewith stimulated Brillouin fiber laser and applications, *Optics Lett.* 16, 393 (1991).

94. Lin, C., Cohen, L. G., Stolen, R. H., Tasker, G. W., and French, W. G., Near-infrared sources in the 1–1.3 μm region by efficient stimulated Raman emission in glass fibers, *Opt. Commun.* 20, 426 (1977).

95. Kadiwar, R. K. and Giles, I. P., Optical fibre Brillouin laser gyroscope, *Electron. Lett.* 25, 1729 (1989).

96. Cowle, G. J. and Stepanov, D. Yu., Hybrid Brillouin/erbium fiber laser, *Optics Lett.* 21, 1250 (1996).

97. Hansen, P. B., Stentz, A. J., Eskilden, L., Grubb, S. G., Strasser, T. A., and Pedrazzani, J. R., High sensitivity 1.3 μm optically preamplified receiver using Raman amplification, *Electron. Lett.* 32, 2164 (1996).

98. Chernikov, S. V., Zhu, Y., Kashyap. R., and Taylor, J. R., High-gain, monolithic, cascaded fibre Raman amplifier operating at 1.3 μm, *Electron. Lett.* 34, 680 (1998).

99. Chernikov, S. V., Platonov, N. S., Gapontsev, D. V., Chang, Do Ill, Guy, M. J., and Taylor, J. R., Raman fibre laser operating at 1.24 μm, *Electron. Lett.* 31, 472 (1995).

100. Dianov, E. M., Fursa, D. G., Abramov, A. A. et al., Raman fibre-optic amplifier of signals at the wave length of 1.3 μm, *Quantum Electron.* 24, 749 (1994).

101. Perkins, A. E. and Lawandy, N. M., Light amplification in a disordered Raman medium, *Optics Commun.* 162, 191 (1999).

102. Gapontsev, D. V., Chernikov, S. V., and Taylor, J. R., Fibre Raman amplifiers for broadband operation at 1.3 μm, *Optics Commun.* 166, 85 (1999).

103. Karpov, V. I., Dianov, E. M., Paramonov, V. M. et al., Laser-diode-pumped phosphosilicate-fiber Raman laser with an output power of 1 W at 1.48 μm, *Optics Lett.* 24, 887 (1999).

104. Lim, D. S., Lee, H. K., Kim, K. H., Kang, S. B., Ahn, J. T., and Jeon. M.-Y., Generation of multiorder Stokes and anti-Stokes lines in a Brillouin erbium-fibre laser with a Sagnac loop mirror, *Optics Lett.* 323, 1671 (1998).

105. Matsas, V. J., Newson, T. P., Richardson, D. J., and Payne, D. N., Selfstarting passively mode-locked fibre ring soliton laser exploiting nonlinear polarisation rotation, *Electron. Lett.* 28, 1391 (1992).

106. Noske, D. U., Guy, M. J., Rottwitt, K., Kashyap, R., and Taylor, J. R., Dual-wavelength operation of a passively mode-locked "figure-of-eight" ytterbium-erbium fibre soliton laser, *Optics Commun.* 108, 297 (1994).

107. Kim, N. S., Prabhu, M., Li, C., Song, J., and Ueda, K., 1239/1484 nm cascaded phospho-silicate Raman fiber laser with CW output power of 1.36 W at 1484 nm pumped by CW Yb-doped double-clad fiber laser at 1064 nm and spectral continuum generation, *Optics Commun.* 176, 219 (2000).

Section 2: Liquid Lasers

Section 2.1
LIQUID ORGANIC DYE LASERS

2.1.1 Introduction

The most common and familiar liquid lasers are those based on strongly absorbing organic dye molecules in an organic solvent and involving allowed transitions of conjugate π electrons. Laser action has been reported for over 500 different dyes in the tables in Section 2.1.3. By the selection of the active dye and solvent, laser action spanning a wavelength range from the near-ultraviolet (336 nm) through the near-infrared (1.8 microns) has been achieved. General categories of dyes and spectral ranges of reported laser action are shown in Figure 2.1.1.

The very broad emission and gain spectra of organic dyes lead to tunable laser output—typically over several tens of nanometers. Because of this property, dye lasers are used extensively in wavelength-selective spectroscopy. Tuning curves for various commercial dyes and pumping sources are shown in Section 2.1.5.

In addition to standard dye laser configurations, organic dye laser action in strongly scattering media has been reported.[1,2] Lasing from dye-doped micrometer-size liquid droplets[3–5] and from evaporating layered microdroplets in the form of a glass core covered by liquid of dye in solution[6] has also been observed.

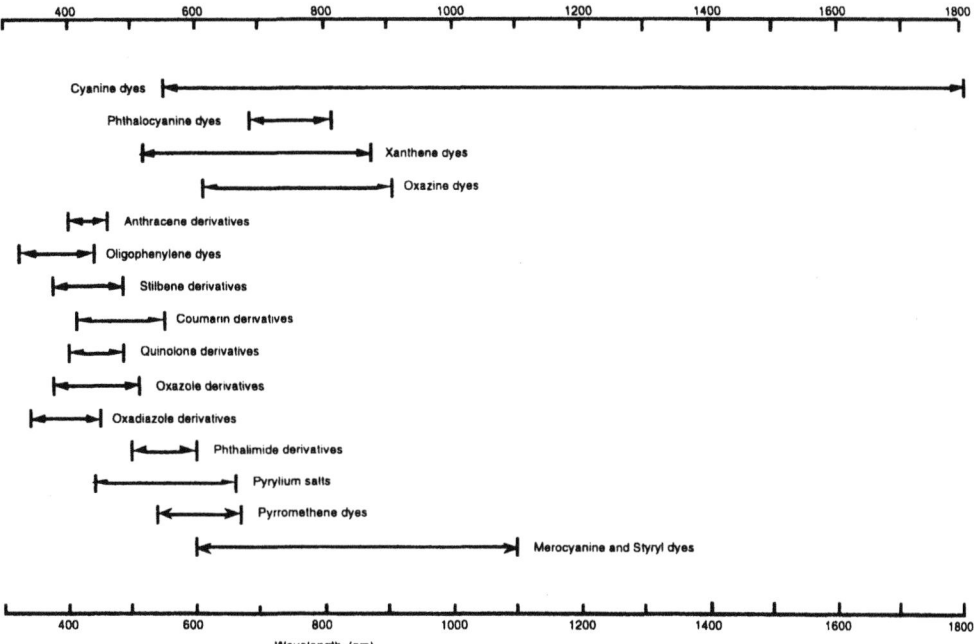

Figure 2.1.1   Reported ranges of output wavelengths of various types of organic dyes used in liquid lasers (from the *Handbook of Laser Wavelengths*, CRC Press, 1998).

Further Reading

Duarte, F. J., Ed., *High Power Dye Lasers*, Springer-Verlag, Berlin (1991).

Duarte, F. J., Ed., *Selected Papers on Dye Lasers*, SPIE Milestone Series, Vol. MS45, SPIE Optical Engineering Press, Bellingham, WA (1992).

Duarte, F. J. and Hillman, L. W., Eds., *Dye Laser Principles*, Academic Press, New York (1990).

Duarte, F. J., Paisner, J. A., and Penzkofer, A., Eds., Dye lasers, special issue of *Applied Optics* 31, 6977 (1992).

Maeda, M., *Laser Dyes*, Academic, New York (1984).

Schäfer, S. P., Ed., *Dye Lasers*, 3rd edition, Springer-Verlag, Berlin (1990).

Steppel, R. N., Organic Dye Lasers, in *Handbook of Laser Science and Technology, Vol. I: Lasers and Masers*, CRC Press, Boca Raton, FL (1982), p. 299.

Steppel, R. N., Organic Dye Lasers, in *Handbook of Laser Science and Technology, Suppl. 1: Lasers*, CRC Press, Boca Raton, FL (1991), p. 219.

Stuke, M., *Dye Lasers: 25 Years*, Springer-Verlag, Berlin (1992).

References

1. See, for example, Sha, W. L., Liu, C.-H., and Alfano, R. R., Spectral and temporal measurements of laser action of Rhodamine 640 dye in strongly scattering media, *Opt. Lett.* 19, 1922 (1994) and Siddique, R., Alfano, R. R., Berger, G. A., Kempe, M., and Genack, A. Z., Time-resolved studies of stimulated emission from colloidal dye solutions, *Opt. Lett.* 21, 450 (1996) and references cited therein.
2. Tanosaki, S. and Taniguchi, H., Microdroplet dye laser enhancing effects in dye-highly scattering intralipid mixture, *Appl. Phys. Lett.* 69, 719 (1996).
3. Biswas, A., Latifi, H., Armstrong, R. L., and Pinnick, R. G., Time-resolved spectroscopy of laser emission from dye-doped droplets, *Opt. Lett.* 14, 214 (1989).
4. Taniguchi, H. and Tanosaki, S., Three-color whispering-gallery-mode dye lasers using dye-doped liquid spheres, *Jpn. J. Appl. Phys.* 32, L1421 (1993).
5. Taniguchi, H. and Tomisawa, H., Suppression and enhancement of dye lasing and stimulated Raman scattering from various dye-doped liquid spheres, *Opt. Lett.* 19, 1403 (1994) and references cited therein.
6. Essien, M., Armstrong, R. L., and Gillespie, J. B., Lasing emission from an evaporating layered microdroplet, *Opt. Lett.* 18, 762 (1993).

2.1.2 Chemical Nomenclature

The names of various dyes found in the literature have in many instances been reduced to relatively simple pseudo acronyms by reducing the important parts of the name to a letter or by the assignment of a number to more complex molecules. The reader is referred to Steppel, R. N., Organic Dye Lasers, in *Handbook of Laser Science and Technology, Vol. 1: Lasers and Masers*, CRC Press, Boca Raton, FL (1982), p. 299; Steppel, R. N., Organic Dye Lasers, in *Handbook of Laser Science and Technology, Suppl. 1: Lasers*, CRC Press, Boca Raton, FL (1991), p. 219; and Maeda, M., *Laser Dyes*, Academic, New York (1984) for common names of the dyes and details of the dye structures and substitutent positions. The chemical names have been abbreviated according to the nomenclature system previously used by Fletcher. The shorthand notation for the parent compounds and substituents are as follows with numbers referring to the skeletal position (see numbered structure): AQ, azaquinolone; Q, quinolone, AC, azacoumarin; C, coumarin; A, amino; M, methyl; DM, dimethyl; DMA, dimethyl-amino; DEA, diethylamino; MO, methoxy; MOR, morpholino; TFM, trifluoromethyl; and OH, hydroxy. Thus 7A-M-AQ designates 7-amino-4-methylazaquinolone. Literature descriptions such as AQIF, QIF and C2H are not necessarily systematic. The first letter(s) refer to the family whereas the H and F refer to hydrogen and fluorine substitution.

Cyanine Dyes

The names of various dyes found in the literature have in many instances been reduced to relatively simple pseudo acronyms by reducing the important parts of the name to letters or by the assignment of a number to more complex molecules. The system used here for the cyanine dyes is an extension of that found in the literature with an attempt to further delineate the structure in a simple manner by designating the substituents independently of the basic chromophore. Thus, the parent molecule consists of the chromophore and standard substituents, and is represented by three or four capital letters. Substituents are indicated by letters and their position by numbers which generally precede the letters of the basic nucleus. For example, DTTC is the shorthand notation for 3,3'-diethylthiatricarbocyanine (considered to be the basic nucleus) and 4,5-Dbz DTTC is the shortened version of 3,3'-diethyl-2,2'-(4,5,4',5' dibenzo) thiatricarbocyanine. Both have the basic DTTC nucleus, but the second dye has the additional benzo substitution. The appropriate letters have been underlined to indicate the derivation of the acronym type abbreviation for the two dyes. Thus, the cyanine dyes have been listed in the table with an abbreviated form for the basic nucleus preceded by various substituents. For each of these cationic dyes a counter ion may be noted where the literature so designated. The counter-ion usually is iodide (I), bromide (Br), or perchorate (P) and may be noted following the dye name.

The basic cyanine dyes are the thiacarbocyanines(T), indocarbocyanines(I), oxacarbo-cyanines(O), quinocarbocyanines(Q), and those with mixed chromophores. In most cases the dyes have symmetrical structure about a vinylogous carbon chain. Numbers are used to indicate the position of substitution on the chromophore and primed numbers on the second chromophore. Where there are substituents on the carbon chain, their positions are indicated by the extension of the numbering sequence from the first chromophore.

With the exception of the quinocarbocyanines which have two possible sites of substitution to the carbon chain, the abbreviated names do not show the common sites of substitution, that is, -3,3'-diethyl and 2,2' carbon chain linkages to the chromophore. They are the same in each dye. For example, compare the names of various thiacarbocyanines.

However, the names and numbered positions of other substituents are indicated. In symmetrical cases the primed numbers of the substituents are dropped, that is, 5,6,5',6-dibenzo becomes 5,6-Dbz.

The letters used in the acronym type abbreviations are C (carbocyanine), T (thia), I (indo), O (oxa), Q (quino), D (diethyl), Dm (dimethyl), Dmo (dimethoxy), Dbz (dibenzo), T (thia or tri), Te (tetra), H (hexamethyl), and A (acetoxy). The counter-ions, I (iodide), Br (bromide), and P (perchorate) always appear at the end of the name or abbreviation. Thus I may be taken as indo or iodide; however, their relative position in the acronym allows one to discern the correct usage. Thus, the exact sequence of letters is of major importance in understanding the acronym. For example, HITC-I represents 1,3,3,1',3',3'-hexamethyl-2,2' indotricarbocyanine iodide. Similarly, T represents both thia and tri; however, only in the thiacarbocyanines does T represent thia while the next letter represents the number of carbo units making up the connecting chain. Thus in DTTC, the first T represents thia while the second represents tri and in DOTC the O represents oxa and again the T represents tri or the number or ethylene linkages between the two chromophores.

Counter-ions associated with each dye, where reported, are shown in the data listing but not below. Counter-ions such as perchlorate, bromide, and iodide are most frequently associated with the cyanine dye.

Commercial Laser Dyes

Commercial dyes may have different nomenclature depending upon the producer. Some synonyms for laser dyes are given below:

BBQ	BiBuQ
BPBD-365	Butyl PBD
Coumarin 440	Coumarin 120
Coumarin 450	Coumarin 2
Coumarin 456	Coumarin 4
Coumarin 460	Coumarin 1, Coumarin 47
Coumarin 461	Coumarin 311
Coumarin 478	Coumarin 106
Coumarin 480	Coumarin 102
Coumarin 481	Coumarin 152A, Coumarin 1F
Coumarin 485	Coumarin 152, Coumarin 2F
Coumarin 490	Coumarin 151, Coumarin 3F
Coumarin 500	Coumarin SA-28
Coumarin 503	Coumarin 307
Coumarin 504	Coumarin 314
Coumarin 515	Coumarin 30
Coumarin 519	Coumarin 343
Coumarin 521	Coumarin 334
Coumarin 522	Coumarin 8F
Coumarin 523	Coumarin 337
Coumarin 535	Coumarin 7
Coumarin 540	Coumarin 6

Coumarin 540A	Coumarin 153
Cresyl Violet 670 Perchlorate	Ozazine 9, Kresylviolett
Disodium Fluorescein	Uranin
DMOTC	Methyl DOTC (DmOTC-1)
DMT	BM-Terphenyl
Fluorescein 548	Fluorescein 27, Fluorescein
Fluorol 555	Fluorol 7GA
HIDC Iodide	Hexacyanin 2
HITC Iodide	Hexacyanin 3
IR-5	Q-switch 5
IR-26	Dye 26
Kiton Red 620	Sulforhodamine B
LD 390	Quinolon 390
LD 466	Coumarin 466
LD 490	Coumarin 6H
LD 690 Perchlorate	Oxazine 4 Perchlorate
LD 700 Perchlorate	Rhodamine 700
LD 800	Rhodamine 800
LDS 698	Pyridine 1
LDS 722	Pyridine 2
LDS 730 Perchlorate	Styryl 6
LDS 750	Styryl 7*
LDS 751	Styryl 8*
LDS 798	Styryl 11
LDS 821	Styryl 9/9M
LDS 925	Styryl 13
LDS 950	Styryl 14
Nile Blue 690 Perchlorate	Nile Blue A Perchlorate, Nileblau
Oxazine 725 Perchlorate	Oxazine 1 Perchlorate
Oxazine 720 Perchlorate	Oxazine 170 Perchlorate
p-Quaterphenyl	PQP
p-Terphenyl	PTP
Phenoxazone 660	Phenoxazone 9
Rhodamine 560 Chloride	Rhodamine 110
Rhodamine 590	Rhodamine 6G
Rhodamine 590 Perchloride	Rhodamine 6G Perchloride
Rhodamine 590 Tetrafluoroborate	Rhodamine 6G Tetrafluoroborate
Rhodamine 610 Chloride	Rhodamine B
Rhodamine 640 Perchlorate	Rhodamine 101
Stilbene 420	Stilbene 3
Sulforhodamine 640	Sulforhodamine 101

* These two dyes are the same as originally reported by Kato. Various sources supplying dyes should be consulted to verify that their products correspond to these dyes.

2.1.3 Tables of Liquid Organic Dye Lasers

Dye lasers are presented alphabetically in the following order:

Anthracenes	Perylenes
Benzimidazoles	Phosphorines
Benzoxazinones	Pteridines
Bimanes	Pyrazolines
Coumarins	Pyrromethenes
azacoumarins	
4,7-substituted	Pyrylium salts
3,7- and 3,4,7 substituted	azapyrylium salts
rigidized, bicyclic	thiopyrylium salts
rigidized, monocyclic	
	Quinolones
Cyanines	azaquinolones
indocarbocyanines	
oxacarbocyanines	Quinoxalinones
quinocarbocyanines	
thiocarbocyanines	Styryl type compounds
	stilbenes
Furans	styrylbenzenes
	styrylbiphenyls
Imitrines	styryldibenzenes
	triazinylstilbenes
Naphthalimides	triazolstilbenes
Oligophenylenes (polyphenyls)	Xanthenes
4-phenylpyridines	rhodamones
quaterphenyls	rosamones
terphenyls	
	Other dyes
Oxazines	
Oxazoles	
benzoxazoles	
bis-oxazoles	
oxadiazoles	

Specific dyes are listed by their chemical or commercial names or as designated in the original references or the tabulations in the *Handbook of Laser Science and Technology,* Volume 1 and Supplement 1. The latter should be consulted for additional information about the chemical composition, molecular structure, and substituent positions.

Tabulated data include the solvent used, wavelength of the absorption and fluorescence maxima, lasing wavelength and, if given, the tuning range, and the pump source and excitation wavelength. The wavelength, tuning range, and output of dye lasers depend on the characteristics of the dye, the dye concentration, solvent, optical pumping source and rate, optical cavity, and other operating conditions. The multiple listings of a given dye at several wavelengths reflect these differences. The original references should be consulted for the experimental conditions used and their effects on the lasing wavelength.

Table 2.1.1
Liquid Organic Dye Lasers

Excitation sources:

Ar (argon-ion laser), bb (broad band), coax (coaxial flashlamp), Cu (copper vapor laser), cw (continuous wave), FL (flashlamp), Kr (krypton-ion laser), KrF (krypton fluoride excimer laser), L (laser), N_2 (nitrogen laser), Nd:glass (neodymium:glass laser), Nd:YAG (neodymium:yttrium aluminum garnet laser), triax (triaxial flashlamp), XeCl (xenon chloride excimer laser).

Solvents:

Ar (argon), BuOAc (butylacetate), BzOH (benzyl alcohol), CCl_4 (carbon tetrachloride), CH_3CN (acetonitrile), CH_3COOH (acetic acid), COT (cyclooctetraene), DB (o-dichlorobenzene), DCE (1,2-dichloroethane), DEC (diethylcarbonate), DEE (diethyl ether), DMA (N,N-dimethylacetamide), DMA/EtOH (dimethylacetamide/ethanol), DMF (*N,N*-dimethylformamide), DMF/EtOH (dimethylformamide/ethanol), DMSO (dimethyl-sulfoxide), DPA (N,N-dipropylacetamide), DX (purified dioxane), EG (ethylene glycol), EtOH (ethanol), EtOH/H$_2$O (ethanol/water), G (glycerol), glyme (1,2-dimethoxyethane), H$_2$O (water), HCl (hydrochloric acid). HFIP (hexafluoroisopropanol), HFIP/H$_2$O (hexafluoroisopropanol-/water), LO (ammonyx LO), MC (methylene chloride), MCH (methylcyclohexane), MeOH (methanol), MeOH/H$_2$O (methanol/water), NB (nitrobenzene), NMP (N-methylpyrrolidone), PPH (1-phenoxy-2-propanol), 2-PrOH (2-propanol), TEA (triethylamine), TFE (trifluoroethanol), THF (tetrahydrofuran), a (acidic), ch (cyclohexane), d (DMF), e (ethanol), g (glycerol), m (DMA), p (p-dioxane). t (toluene), w (water).

2.1.3.1 Anthracene Dye Lasers

Solvent	Wavelength max. Abs. (nm)	Fluor. (nm)	Lasing wavelength (nm)	Pump source (nm)	Ref.
9-dichloroanthracene					
EtOH/MeOH:4/1		415.5/415.8		N_2 laser (337)	35
9-methylanthracene					
EtOH/MeOH:4/1		413.8/414.1		N_2 laser (337)	35
9-phenylanthracene					
EtOH/MeOH:4/1		417	N_2 laser (337)	35	
9,10-dimethylanthracene					
MCH/toluene:2/1			432.0/432.74	N_2 laser (337)	35
9,10-diphenylanthracene					
EtOH/MeOH:4/1	429.3/430.2			N_2 laser (337)	35
MCH/toluene:2/1			431.7/434.6	N_2 laser (337)	35
cyclohexane			432.6	ruby laser (347)	54
Diphenyl–anthracene					
ethanol, w, t			435-450	N_2 laser (337)	27

2.1.3.2 Benzimidazole Dye Lasers

Solvent	Wavelength max. Abs. (nm)	Fluor. (nm)	Lasing wavelength (nm)	Pump source (nm)	Ref.
2-(o-hydroxyphenyl) Benzimidazole					
dioxane			470 (460–495)	XeCl laser (308)	33
2-(2-hydroxy-5-fluoro) Benzimidazole (HOF-BIM)					
acetonitrile	326		487 (465–510)	N$_2$ laser (337)	128
2-(2-hydroxyphenyl) Benzimidazole (HOP-BIM)					
methanol	316		463 (450–480)	XeCl laser (308)	128
ethanol	317		465 (454–480)	XeCl laser (308)	128
p-dioxane	319		473 (460–490)	XeCl laser (308)	128
acetonitrile	316		474 (459–489)	XeCl laser (308)	128
DMF	318		477 (463–495)	XeCl laser (308)	128
2-(2-dihydroxyphenyl) Benzimidazole (DHOP-BIM)					
acetonitrile	300		524 (507–540)	N$_2$ laser (337)	128
ch/dioxane	301		524 (509–542)	N$_2$ laser (337)	128
p-dioxane	302		531 (515–549)	N$_2$ laser (337)	128
DMF	303		535 (515–553)	N$_2$ laser (337)	128
2-(2-hydroxy-5-methyl) Benzimidazole (HOMe-BIM)					
p-dioxane	326		495	N$_2$ laser (337)	128
2-(2-hydroxy-5-methoxy) Benzimidazole (OMeO-BIM)					
ch/dioxane	302/342		520 (509–537)	N$_2$ laser (337)	128
p-dioxane	301	341	530 (514–552)	N$_2$ laser (337)	128

2.1.3.3 Benzoxazinone Dye Lasers

Solvent	Wavelength max. Abs. (nm)	Fluor. (nm)	Lasing wavelength (nm)	Pump source (nm)	Ref.
Benzoxazinone 8-1-1					
ethanol	385	522	543	N$_2$ laser (337)	138
Benzoxazinone 8-1-2					
ethanol	423	530	563	N$_2$ laser (337)	138
Benzoxazinone 8-1-3					
ethanol	394	521	520	N$_2$ laser (337)	68
Benzoxazinone 8-1-4					
ethanol	403	551	570	N$_2$ laser (337)	68
Benzoxazinone 8-1-5					
ethanol	442	556	566	N$_2$ laser (337)	138
Benzoxazinone 8-1-6					
ethanol	415	540	554	N$_2$ laser (337)	138
Benzoxazinone 8-1-7					
ethanol	488	590	668	N$_2$ laser (337)	138

2.1.3.4 Bimane Dye Lasers

Solvent	Wavelength max. Abs. (nm)	Fluor. (nm)	Lasing wavelength (nm)	Pump source (nm)	Ref.
Bimane 5-1-1					
HFIP		480	501	flashlamp	134
H_2O	374	450	540	flashlamp	134
Bimane 5-1-2					
trifluoroethanol		474	516	flashlamp	137
Bimane 5-1-3					
HFIP/H_2O:4/1	344	473	507	flashlamp	50
Bimane 5-1-4					
H_2O		478	514	flashlamp	50
Bimane 5-1-5					
p–dioxane	380	437	435	flashlamp	50
H_2O	392	480	525	flashlamp	50

2.1.3.5 Coumarin Dye Lasers

Solvent	Wavelength max. Abs. (nm)	Fluor. (nm)	Lasing wavelength (nm)	Pump source (nm)	Ref.
4-methylcoumarin					
ethanol, w, t			445–490	N_2 laser (337)	27
Coumarin 6-1-1					
ethanol (air)			489	FL (triaxial)	38
ethanol (Ar)			493	FL (triaxial)	38
methanol (air)			493	FL (triaxial)	38
EG/EtOH (air)			497	FL (triaxial)	38
MeOH/H_2O (air)			497	FL (triaxial)	38
EtOH/H_2O (air)			501	FL (triaxial)	38
EtOH/H_2O (Ar)			501	FL (triaxial)	38
Coumarin 6-1-2					
hexane (Ar)			451–465	FL (triaxial)	38
benzene (Ar)			460–483	FL (triaxial)	38
toluene (Ar)			452–480	FL (triaxial)	38
toluene (air)			458–474	FL (triaxial)	38
benzene (air)			460–479	FL (triaxial)	38
DEC (air)			461–479	FL (triaxial)	38
butylacetate (air)			462–483	FL (triaxial)	38
DEC (Ar)			462–485	FL (triaxial)	38
butylacetate (Ar)			463–488	FL (triaxial)	38
glyme (air)			467–490	FL (triaxial)	38
THF (Ar)			467–491	FL (triaxial)	38
glyme (Ar)			467–497	FL (triaxial)	38
THF (air)			468–491	FL (triaxial)	38
acetone (air)			475–500	FL (triaxial)	38

Coumarin Dye Lasers—*continued*

Solvent	Wavelength max. Abs. (nm)	Fluor. (nm)	Lasing wavelength (nm)	Pump source (nm)	Ref.
Coumarin 6-1-2					
acetone (Ar)			475–504	FL (triaxial)	38
CH$_3$CN (Ar)			479–506	FL (triaxial)	38
DMA(Ar)			484–507	FL (triaxial)	38
2-PrOH (air)			486	FL (triaxial)	38
DMF (air)			486–508	FL (triaxial)	38
CH$_3$CN (air)			487–503	FL (triaxial)	38
DCE (air)			487–506	FL (triaxial)	38
DMA (air)			487–511	FL (triaxial)	38
DMF (Ar)			487–512	FL (triaxial)	38
DCE (Ar)			488–511	FL (triaxial)	38
NMP (air)			489–513	FL (triaxial)	38
ethanol (air)			491	FL (triaxial)	38
NMP (Ar)			491–513	FL (triaxial)	38
methanol (air)			494	FL (triaxial)	38
ethanol (Ar)			495	FL (triaxial)	38
EG/EtOH (air)			499	FL (triaxial)	38
MeOH/H$_2$O (air)			499	FL (triaxial)	38
EG/EtOH (Ar)			500	FL (triaxial)	38
EtOH/H$_2$O (air)			502	FL (triaxial)	38
MeOH (Ar)			502	FL (triaxial)	38
EtOH/H$_2$O (Ar)			503	FL (triaxial)	38
MeOH/H$_2$O (Ar)			505	FL (triaxial)	38
Coumarin 6-1-3					
ethanol (air)			489	FL (triaxial)	38
ethanol (Ar)			492	FL (triaxial)	38
methanol (air)			493	FL (triaxial)	38
MeOH (Ar)			497	FL (triaxial)	38
EG/EtOH (air)			498	FL (triaxial)	38
EG/EtOH (Ar)			498	FL (triaxial)	38
MeOH/H$_2$O (air)			498	FL (triaxial)	38
EtOH/H$_2$O (air)			501	FL (triaxial)	38
EtOH/H$_2$O (Ar)			502	FL (triaxial)	38
MeOH/H$_2$O (Ar)			504	FL (triaxial)	38
Coumarin 6-1-4					
MeOH/H$_2$O			502	flashlamp	135
MeOH (air)			503	FL (triaxial)	38
EtOH/H$_2$O (air)			512	FL (triaxial)	38
MeOH/H$_2$O (air)			515	FL (triaxial)	38
Coumarin 6-3-1					
ethanol (air)			483 (477–489)	FL (triaxial)	130
ethanol (Ar)	393	476	483 (478–489)	FL (triaxial)	130
EtOH/H$_2$O (Ar)	404	488	496 (484–508)	FL (triaxial)	130
EtOH/H$_2$O (air)			497 (485–508)	FL (triaxial)	130

Coumarin Dye Lasers—*continued*

Solvent	Wavelength max. Abs. (nm)	Fluor. (nm)	Lasing wavelength (nm)	Pump source (nm)	Ref.
Coumarin 6-3-10					
ethanol (air)	437	500	516 (511–522)	FL (triaxial)	130
ethanol (Ar)			517 (512–521)	FL (triaxial)	130
EtOH/H$_2$O (Ar)			526 (520–530)	FL (triaxial)	130
EtOH/H$_2$O (air)	445	508	541 (532–549)	FL (triaxial)	130
Coumarin 6-3-2					
ethanol (air)			534 (526–541)	FL (triaxial)	130
ethanol (Ar)	465	506	537 (531–544)	FL (triaxial)	130
EtOH/H$_2$O (Ar)	473	512	539 (530–548)	FL (triaxial)	130
EtOH/H$_2$O (air)			548 (542–554)	FL (triaxial)	130
Coumarin 6-3-3					
ethanol (Ar)	480	519	547 (540–555)	FL (triaxial)	130
ethanol (air)			547 (541–554)	FL (triaxial)	130
EtOH/H$_2$O (Ar)	490	526	552 (543–562)	FL (triaxial)	130
EtOH/H$_2$O (air)			555 (547–564)	FL (triaxial)	130
Coumarin 6-3-4					
ethanol (air)			533 (525–541)	FL (triaxial)	130
ethanol (Ar)	462	504	534 (525–543)	FL (triaxial)	130
EtOH/H$_2$O (air)			545 (537–553)	FL (triaxial)	130
EtOH/H$_2$O (Ar)	474	512	545 (537–553)	FL (triaxial)	130
Coumarin 6-3-5					
ethanol (Ar)	430	480	499 (492–507)	FL (triaxial)	130
ethanol (air)			500 (492–508)	FL (triaxial)	130
EtOH/H$_2$O (Ar)	442	488	513 (504–523)	FL (triaxial)	130
EtOH/H$_2$O (air)			514 (505–524)	FL (triaxial)	130
Coumarin 6-3-6					
ethanol (air)	415	492	504 (498–511)	FL (triaxial)	130
ethanol (Ar)			505 (498–512)	FL (triaxial)	130
EtOH/H$_2$O (air)	424	502	513 (505–520)	FL (triaxial)	130
EtOH/H$_2$O (Ar)			514 (505–523)	FL (triaxial)	130
Coumarin 6-3-7					
ethanol (air)	440	484	501 (493–508)	FL (triaxial)	130
ethanol (Ar)			502 (494–510)	FL (triaxial)	130
EtOH/H$_2$O (air)	450	492	514 (505–524)	FL (triaxial)	130
Coumarin 6-3-8					
ethanol (air)	435	494	512 (505–518)	FL (triaxial)	130
ethanol (Ar)			512 (507–517)	FL (triaxial)	130
EtOH/H$_2$O (Ar)			522 (518–525)	FL (triaxial)	130
EtOH/H$_2$O (air)	442	502	528 (520–536)	FL (triaxial)	130
Coumarin 6-3-9					
ethanol (air)	425	496	511 (504–518)	FL (triaxial)	130
ethanol (Ar)			512 (504–519)	FL (triaxial)	130
EtOH/H$_2$O (air)	433	506	520 (513–527)	FL (triaxial)	130
EtOH/H$_2$O (Ar)			520 (513–528)	FL (triaxial)	130

Coumarin Dye Lasers—*continued*

Solvent	Wavelength max. Abs. (nm)	Fluor. (nm)	Lasing wavelength (nm)	Pump source (nm)	Ref.
Coumarin 6-4-1					
ethanol			453–514	XeCl laser (308)	24
Coumarin 6-4-2					
ethanol			450–511	XeCl laser (308)	24
Coumarin 6-4-3					
ethanol			480–528	XeCl laser (308)	24
Coumarin 6-4-4					
ethanol			477–526	XeCl laser (308)	24
Coumarin 6-4-5					
ethanol			482–526	XeCl laser (308)	24
Coumarin 6-4-6					
ethanol			478–525	XeCl laser (308)	24
Coumarin 6-4-7					
ethanol			501–550	XeCl laser (308)	24
Coumarin 6-4-8					
ethanol			500–546	XeCl laser (308)	24
Coumarin 6-4-9					
ethanol			512–585	XeCl laser (308)	24
Coumarin 6-4-10					
ethanol			508–588	XeCl laser (308)	24
Coumarin 6-5-1					
1,2-DCE			518 (497–547)	XeCl laser (308)	132
Coumarin 6-5-2					
1,2-DCE			533 (510–558)	XeCl laser (308)	132
Coumarin 6-5-3					
1,2-DCE			529 (511–559)	XeCl laser (308)	132
Coumarin 6-5-4					
1,2-DCE			542 (520–570)	XeCl laser (308)	132
Coumarin 6-5-5					
1,2-DCE			603 (580–670)	XeCl laser (308)	132
Coumarin 6-5-6					
1,2-DCE			662 (600–706)	XeCl laser (308)	132
Coumarin 6-5-7					
1,2-DCE			610 (596–632)	XeCl laser (308)	132
1,2-DCE			674 (641–712)	XeCl laser (308)	132
Coumarin 6-5-8					
1,2-DCE			675 (628–712)	XeCl laser (308)	132
Coumarin 6-5-9					
ethanol			493 (475–518)	XeCl laser (308)	132
Coumarin 6-5-10					
ethanol		505	535 (512–562)	XeCl laser (308)	132

Coumarin Dye Lasers—*continued*

Solvent	Wavelength max.		Lasing	Pump source	Ref.
	Abs. (nm)	Fluor. (nm)	wavelength (nm)	(nm)	
Coumarin 6-7-5					
p-dioxane	440	485	528	N$_2$ laser (337)	139
p-dioxane+HOA	455		543–550	N$_2$ laser (337)	139
DMF+HOAc	440	535	532–550	N$_2$ laser (337)	139
Coumarin 6-7-6					
DMF+HOAc	480	552	560	N$_2$ laser (337)	139
Coumarin 6-8-1					
p-dioxane	460	525	510–535	Nd:YAG	136
EtOH+HOAc	433	510	530–570	N$_2$ laser (337)	136
Coumarin 6-8-2					
p-dioxane	455	545	550–580	Nd:YAG	136
EtOH+HOAc	446	525	565–610	N$_2$ laser (337)	136
Coumarin 6-9-1					
ethanol	549	572	590–616	L (520)	158
ethanol	571	594	595–617	L (520)	158
benzene	544	571	609	L (520)	158
CH$_3$CN	547	572	612	L (520)	158
ethanol	544	562	615–628	flashlamp	158
EtOH+HCl			620	L (520)	158
Coumarin 6-9-2					
methanol	582	603	605	L (520)	158
CH$_3$CN	567	586	607	L (520)	158
ethanol			608	L (520)	158
benzene	565	586	626	L (520)	158
ethanol	560	577	637–650	flashlamp	158
EtOH+HCl	561	587	654–669	flashlamp	158
EtOH+HCl			656	L (520)	158
Coumarin 6–7–5					
DMF+HOAc	440	535	532–550	N$_2$ laser (337)	139
DMF+HOAc	440	485	528	N$_2$ laser (337)	139
DMF+HOAc	455	525	543–550	Nd:YAG	139
Coumarin 6–7–6					
DMF+HOAc	480	552	560	Nd:YAG	139
Coumarin 138					
ethanol			~460	N$_2$ laser (337)	120
Coumarin 337					
ethanol	443	485	522	flashlamp	55
Coumarin 486					
water			486	flashlamp	55
Coumarin 488					
water			489	flashlamp	55

Azacoumarins

Solvent	Wavelength max. Abs. (nm)	Fluor. (nm)	Lasing wavelength (nm)	Pump source (nm)	Ref.
7MOR-4M-Azacoumarin					
ethanol			431	flashlamp	194
7OH-4M-Azacoumarin					
ethanol		415	431	flashlamp	194
7DMA-4M-Azacoumarin					
ethanol		420	434	flashlamp	194
AC2F-Azacoumarin					
ethanol			490	flashlamp	66,194

Coumarins: 4,7-substituted

Solvent	Wavelength max. Abs. (nm)	Fluor. (nm)	Lasing wavelength (nm)	Pump source (nm)	Ref.
7A-4MO-C; 7-amino-4-methoxycoumarin; 7-amino-4-methoxy-2H-1-benzopyran-2-one					
ethanol			417	flashlamp	43
7A-4M-C; 7-amino-4-methylcoumarin; 7-amino-4-methyl-2H-1-benzopyran-2-one					
ethanol		429	436	flashlamp	43
7DMA-4M-Coumarin					
ethanol		452	453	flashlamp	43
Coumarin 120, Coumarin 440, CSA-1; 7-amino-4-methyl-2H-1-benzopyran-2-one					
ethanol			436	flashlamp	43,44
ethanol	354	430	437 (420–457)	N_2 laser (337)	20
ethanol			438 (419–466)	N_2 laser (337)	56
methanol			440 (419–469)	flashlamp	59
ethanol			441	Nd:YAG(355)	60
ethanol			442	flashlamp	61
ethanol			442 (425–443)	flashlamp	25
methanol			442 (426–458)	Nd:YAG(355)	51
MeOH/H_2O:1/1			442 (435–455)	flashlamp	19
DPA,COT			450 (420–470)	Ar (351/364)	37
ethylene glycol			450 (427–477)	Ar laser (cw)	47
water			453	flashlamp	55
CSA-10; 7-pentylamino-4-methylcoumarin; 7-*n*-pentylamino-4-methoxy-2H-1-benzopyran-2-one					
ethanol			440	N_2 laser (337)	44
ethanol+HCl			485	N_2 laser (337)	44
CSA-4; 7(2-pyridylamino)-4-methylcoumarin; 7(2-pyridylamino)-2H-1-benzopyran-2-one					
ethanol			450	N_2 laser (337)	44
CSA-5; 7-pyrrolidino-4-methylcoumarin; 7-pyrrolidino-2H-1-benzopyran-2-one					
ethanol			450	N_2 laser (337)	44
Coumarin 175					
water	353		457	flashlamp	55

Coumarins: 4,7-substituted—*continued*

Solvent	Wavelength max.		Lasing	Pump source	Ref.
	Abs. (nm)	Fluor. (nm)	wavelength (nm)	(nm)	

Coumarin 445, CSA-6; 7-ethylamino-4-methylcoumarin; 7-ethylamino-4-methyl-2H-1-benzopyran-2-one

Solvent	Abs.	Fluor.	Lasing	Pump	Ref.
ethanol			445	N_2 laser (337)	44
MeOH/H_2O:1/1	363	430	445	flashlamp	26
ethylene glycol			460	Ar laser (cw)	63
ethanol+HCl			485 .	N_2 laser (337)	44

CSA-8; 7-propylamino-4-methylcoumarin; 7-n-propylamino-4-methoxy-2H-1-benzopyran-2-one

Solvent	Abs.	Fluor.	Lasing	Pump	Ref.
ethanol			440	N_2 laser (337)	44
ethanol+HCl			485	N_2 laser (337)	44

CSA-9; 7-butylamino-4-methylcoumarin; 7-n-butylamino-4-methoxy-2H-1-benzopyran-2-one

Solvent	Abs.	Fluor.	Lasing	Pump	Ref.
ethanol			440	N_2 laser (337)	44
ethanol+HCl			485	N_2 laser (337)	44

C3H; 7-aminocoumarin; 7-amino-2H-1-benzopyran-2-one

Solvent	Abs.	Fluor.	Lasing	Pump	Ref.
ethanol			450	flashlamp	191

CSA-11

Solvent	Abs.	Fluor.	Lasing	Pump	Ref.
ethanol			456	N_2 laser (337)	44

Coumarin 2, Coumarin 450; ; 7-ethylamino-4,6-dimethylcoumarin; 7-ethylamino-4,6-dimethyl-2H-1-benzopyran-2-one

Solvent	Abs.	Fluor.	Lasing	Pump	Ref.
ethanol	366	435	446	flashlamp	37
ethanol			446 (428–465)	N_2 laser (337)	20
ethanol/1.5%LO			449 (435–479)	flashlamp	25
methanol			450 (427–488)	flashlamp	59
ethylene glycol			450 (435–485)	Ar or Kr (uv)	164
ethylene glycol			452 (430–492)	Ar laser (cw)	47
methanol			454	flashlamp	61
methanol			454	Nd:YAG(355)	60
methanol			455 (433–474)	Nd:YAG(355)	51
methanol			458	flashlamp	191
20%aq. DPA			460 (430–480)	Ar (351/364)	192
MeOH/H_2O:4/6			460 (445–482)	flashlamp	19

Coumarin 311; 7-dimethylamino-4,-methylcoumarin; 7-dimethylamino-4-methyl-2H-1-benzopyran-2-one

Solvent	Abs.	Fluor.	Lasing	Pump	Ref.
ethanol			453	flashlamp	194
ethanol			457	flashlamp	58

Coumarin 1, Coumarin 460; 7-diethylamino-4,-methylcoumarin; 7-diethylamino-4,-methyl-2H-1-benzopyran-2-one

Solvent	Abs.	Fluor.	Lasing	Pump	Ref.
ethanol			457 (440–478)	N_2 laser (337)	20
ethanol	373	445	457 (450–484)	flashlamp	25
ethanol			460	Nd:YAG(355)	96
ethanol			460	flashlamp	43,61
ethanol			460	flashlamp	62
—			460 (442–490)	Nd:YAG(355)	53
methanol			461 (448–489)	flashlamp	25

Coumarins: 4,7-substituted—*continued*

Solvent	Wavelength max. Abs. (nm)	Fluor. (nm)	Lasing wavelength (nm)	Pump source (nm)	Ref.
Coumarin 1, Coumarin 460; 7-diethylamino-4,-methylcoumarin; 7-diethylamino-4,-methyl-2H-1-benzopyran-2-one					
ethylene glycol			470 (450–495)	Ar or Kr (uv)	164
—			471 (448–505)	Kr laser (uv)	52
ethylene glycol			472 (446–506)	Ar laser (cw)	47
Coumarin 360					
water	359		470	flashlamp	55
Coumarin 378					
water	361		468	flashlamp	55
Coumarin 379					
water	376		473	flashlamp	55
Coumarin 380					
water	362		470	flashlamp	55
Coumarin 381					
water	382		478	flashlamp	55
C2H; 7-dimethylaminocoumarin; 7-dimethylamino-2H-1-benzopyran-2-one					
ethanol			465	flashlamp	191
C1H, LD 466; 7-diethylaminocoumarin; 7-diethylamino-2H-1-benzopyran-2-one					
EtOH/p-dioxane			464 (446–492)	N$_2$ laser (337)	64
ethanol	368		465 (452–480)	N$_2$ laser (337)	1
ethanol			467 (459–477)	Nd:YAG(355)	65
methanol+LO			474 (462–490)	flashlamp	25
Coumarin 35, Coumarin 481, C1F, CSA-27; 7-diethylamino-4-trifluoromethylcoumarin; 7-diethylamino-4-trifluoromethyl-2H-1-benzopyran-2-one					
ethanol	390	465	480	N$_2$ laser (337)	44
ethanol			481	flashlamp	61,66
p-dioxane			481	flashlamp	66,67
p-dioxane			481	KrF laser (248)	6
p-dioxane			481 (460–518)	N$_2$ laser (337)	56
p-dioxane			481 (475–490)	flashlamp	25
p-dioxane			483 (463–516)	flashlamp	69
p-dioxane			483 (560–517)	N$_2$ laser (337)	20
p-dioxane			489	Nd:YAG(355)	60
dioxane/ethanol			507 (481–540)	N$_2$ laser (337)	69
ethanol			515 (492–545)	N$_2$ laser (337)	69
ethanol			516 (490–566)	N$_2$ laser (337)	56
Coumarin 151, Coumarin 490, C3F:CSA-29; 7-amino-4-trifluoromethylcoumarin; 7-amino-4-trifluoromethyl-2H-1-benzopyran-2-one					
ethanol	382	489	455	N$_2$ laser (337)	44
ethanol			484	flashlamp	66
ethanol			488	flashlamp	70
methanol			489 (467–510)	flashlamp	72
ethanol			490	flashlamp	61

Coumarins: 4,7-substituted—*continued*

Solvent	Wavelength max. Abs. (nm)	Fluor. (nm)	Lasing wavelength (nm)	Pump source (nm)	Ref.
Coumarin 151, Coumarin 490, C3F:CSA-29; 7-amino-4-trifluoromethylcoumarin; 7-amino-4-trifluoromethyl-2H-1-benzopyran-2-one					
methanol+LO			495 (477–515)	flashlamp	25
EtOH/H$_2$O			496	flashlamp	66
Coumarin 152, Coumarin 485, C2F; 7-dimethylamino-4-trifluoromethylcoumarin; 7-dimethylamino-4-trifluoromethyl-2H-1-benzopyran-2-one					
p-dioxane	397	510	479	flashlamp	66
ethanol			485	N$_2$ laser (337)	44
p-dioxane			500 (482–517)	N$_2$ laser (337)	73
ethanol			519	flashlamp	66,70
ethanol			520	flashlamp	61
ethanol			520 (490–562)	N$_2$ laser (337)	44
methanol			523	flashlamp	25
methanol			525 (502–573)	Nd:YAG(355)	53
Coumarin 307; Coumarin 503; 7-ethylamino-6-4-trifluoromethyl-2H-1-benzopyran-2-one					
ethanol	395	490	498 (477–531)	flashlamp	25
ethanol			502	flashlamp	61
MeOH+LO			504 (481–530)	flashlamp	25
Coumarin 316					
water	375		494	flashlamp	55
Coumarin 500, CSA-28; 7-ethylamino-4-trifluoromethylcoumarin; 7-ethylamino-4-trifluoromethyl-2H-1-benzopyran-2-one					
ethanol			500	KrF laser (248)	3
ethanol	392	495	500	N$_2$ laser (337)	44
ethanol			500 (473–547)	N$_2$ laser (337)	20
ethanol			500 (494–504)	Nd:YAG(355)	65
methanol			508 (481–573)	Nd:YAG(355)	53
methanol			514 (482–552)	Nd:YAG(355)	51
MeOH/H$_2$O:1/1			522 (500–548)	flashlamp	26

Coumarins: 3,7- and 3,4,7-substituted

Solvent	Wavelength max. Abs. (nm)	Fluor. (nm)	Lasing wavelength (nm)	Pump source (nm)	Ref.
Coumarins: 3,7-substituted:					
benzene			414	N$_2$ laser (337)	36
benzene			421	N$_2$ laser (337)	36
benzene			423	N$_2$ laser (337)	36
benzene			426	N$_2$ laser (337)	36
dichloromethane			430	N$_2$ laser (337)	36
benzene			434	N$_2$ laser (337)	36
benzene			437	N$_2$ laser (337)	36

Coumarins: 3,7- and 3,4,7-substituted—*continued*

Solvent	Wavelength max. Abs. (nm)	Fluor. (nm)	Lasing wavelength (nm)	Pump source (nm)	Ref.
Coumarins: 3,7-substituted:					
benzene			438	N_2 laser (337)	36
benzene			439	N_2 laser (337)	36
methanol			446	N_2 laser (337)	36
water			450	N_2 laser (337)	36
water			457	N_2 laser (337)	36
methanol			458	N_2 laser (337)	36
water			467	N_2 laser (337)	36
methanol			502	N_2 laser (337)	36
Coumarin 30, Coumarin 515; 7-(diethylamino)-3-(1-methyl-1H-benzimidazol-2-yl)-2H-1-benzopyran-2-one					
ethanol			482–507	N_2He(428)	68
aq.,DPA,COT	347	413	505 (495–515)	Ar (cw,458)	192
—			508 (477–548)	Kr (violet)	52
ethylene glycol			510 (492–550)	Ar laser (458)	48
ethylene glycol			515 (495–545)	Kr (400-420)	164
CSA-24; 3-[2-(6'-ethoxy)benzothiazolyl)-4-methyl)-7-(ethylamino)-2H-1-benzo-pyran-2-one					
ethanol	415	505	532	N_2 laser (337)	44
CSA-25; 3-(2-benzothiazolyl)-4-methyl)-7-(ethylamino)-2H-1-benzopyran-2-one					
ethanol	417	495	527	N_2 laser (337)	44
Coumarin 531, CSA-23; 3-[2-(6'-ethoxy)benzothiazolyl)-4-methyl)-7-(diethylamino)-2H-1-benzopyran-2-one					
ethanol	420	504	531	N_2 laser (337)	44
Coumarin 7, Coumarin 535; 3-(1H-benzothiazol-2-yl)-7-(diethylamino)-2H-1-benzopyran-2-one					
DPA+LO,COT	436	490	525 (500–575)	Ar (cw,477)	192
ethylene glycol			535 (500–565)	Ar laser (477)	164
ethylene glycol			535 (505–565)	Kr (400-420)	164
Coumarin 6, Coumarin 540; 3-(2-benzothiazoll)-7-(diethylamino)-2H-1-benzopyran-2-one					
ethanol	458	505	507–529	N_2He(428)	68
methanol			531 (510–556)	flashlamp	25
methanol			538 (521–551)	flashlamp	59
ethylene glycol			540 (515–566)	Ar laser (cw)	47
DPA+LO,COT			540 (515–585)	Ar laser (488)	76
methanol			544 (526–570)	flashlamp	26
EG/bz alcohol			560 (510–570)	Ar laser (488)	76
Coumarin; 3,7-substituted					
benzene			414, 421, 423,426	N_2 laser (337)	36
dichloromethane			430	N_2 laser (337)	36
benzene			434, 437, 438, 439	N_2 laser (337)	36
methanol			446, 458, 502	N_2 laser (337)	36
water			450, 457, 467	N_2 laser (337)	36

Coumarins: rigidized, bicyclic

Solvent	Wavelength max. Abs. (nm)	Fluor. (nm)	Lasing wavelength (nm)	Pump source (nm)	Ref.
Coumarin 478; Coumarin 106; 2,3,6,7,10,11-hexahydro-1H,5H-cyclopental[3,4][1]benzo-pyrano[6,7,8-ij]quinolizin-12(9H)-one					
ethanol	386	465	478	flashlamp	66
Coumarin 480; Coumarin 102; 2,3,6,7-tetrahydro-9-methyl-1H,5H,11H-[1[benzopyrano-[6,7,8-ij]-quinolizin-11-one					
ethanol	390	465	480	flashlamp	61
Coumarin C6H; LD 490; 2,3,6,7-tetrahydro-1H,5H,11H-[1[benzopyrano[6,7,8-ij]quinolizin-11-one					
—	396			flashlamp	191
Coumarin 504; Coumarin 314; 2,3,6,7-tetrahydro-11-oxo-l-1H,5H,11H-[1[benzopyrano[6,7,8-ij]quinolizine-10-carboxylic acid, ethyl ester					
methanol+LO			499 (484–537)	flashlamp	25
ethanol	437	480	504	flashlamp	61
MeOH/H$_2$O:1/1			505 (495–517)	flashlamp	26
MeOH/H$_2$O:1/1			520 (506–544)	flashlamp	19
Coumarin 217					
water	402		514	flashlamp	55
Coumarin 519; Coumarin 343; 2,3,6,7-tetrahydro-11-oxo-l-1H,5H,11H-[1[benzopyrano-[6,7,8-ij]quinolizine-10-carboxylic acid					
methanol	446	490	501 (490–513)	flashlamp	25
water	436		518	flashlamp	
ethanol	425		519	flashlamp	
Coumarin 521; Coumarin 334; 10-acetyl-2,3,6,7-tetrahydro-1H,5H,11H-[1[benzopyrano-[6,7,8-ij]quinolizin-11-one					
water	452	495	521	flashlamp	55
Coumarin 337, Coumarin 523; 2,3,6,7-tetrahydro-11-oxo-1H,5H,11H-[1[benzopyrano-[6,7,8-ij]quinolizine-10-carbonitrile					
ethanol	266	443	522	flashlamp	75
Coumarin 153, Coumarin 540A, C6F; 2,3,6,7-tetrahydro-9-(trifluoromethyl)-1H,5H,11H-[1[benzopyrano[6,7,8-ij]quinolizin-11-one					
ethanol			521	flashlamp	61
ethanol			522	flashlamp	61
p-dioxane			507	N$_2$ laser (337)	69
ethanol			536 (515–583)	flashlamp	25
ethanol			536 (517–576)	N$_2$ laser (337)	20
ethanol			538	flashlamp	66
ethanol			540	flashlamp	61
methanol			540 (516–590)	Nd:YAG(355)	51
methanol			541 (520–586)	flashlamp	25
ethanol			543	flashlamp	70
ethanol			562 (520–575)	Ar laser (476)	76

Coumarins: rigidized, monocyclic

Solvent	Wavelength max. Abs. (nm)	Fluor. (nm)	Lasing wavelength (nm)	Pump source (nm)	Ref.
Coumarin 522; C8H; 2,3,8,9-tetrahydro-9-methyl-2H-pyrano[3,2-g]quinolizin-2-one					
ethanol			475	flashlamp	37
C4H; 2,3,8,9-tetrahydro-2H-pyrano[3,2-g]quinolizin-2-one					
ethanol			477	flashlamp	191
Coumarin 386					
water	388		486	flashlamp	v1 -8
Coumarin 388					
water	390		489	flashlamp	v1 -8
Coumarin C4F; Coumarin 340; 6,7,8,9-tetrahydro-4-(trifluoromethyl)-2H-pyrano[3,2-g]-quinolizin-2-one					
ethanol	406	500	513	flashlamp	61
C340					
ethanol			522	flashlamp	66
Coumarin C8F; Coumarin 522; 6,7,8,9-tetrahydro-9-methyl-4-(trifluoromethyl)-2H-pyrano-[3,2-g]-quinolizin-2-one					
ethanol	412	515	520 (498–556)	flashlamp	25
ethanol			522 (500–572)	flashlamp	66
methanol+LO			526 (501–568)	flashlamp	25
DMF+MeOH			533 (515–570)	flashlamp	19
Coumarin 355; 6,7,8,9-tetrahydro-9-ethyl-4-(trifluoromethyl)-2H-pyrano-[3,2-g]-quinolizin-2-one					
ethanol	412		522	flashlamp	61

2.1.3.6 Cyanine Dye Lasers

Indocarbocyanines

Solvent	Wavelength max. Abs. (nm)	Fluor. (nm)	Lasing wavelength (nm)	Pump source (nm)	Ref.
6,7-Dbz-HITC-P (HDITC-P); 1,3,3,1',3',3'-hexamethyl-2,2'-(6,7,6',7'-dibenzo)-indotri-carbocyanine perchlorate					
DMSO+EG			920 (880–965)	Kr (752,799)	103
HIDC-1; 1,3,3,1',3',3'-hexarnethyl-2,2-indodicarbocyanine iodine					
methanol			720	Nd:YAG→585	95
methanol	641		675	Nd:YAG→585	95
DMSO			839 (826–850)	N_2 laser (337)	20
DMSO			770–830	FL→R620	88
DMSO			780–883	ruby (694.3)	28

Indocarbocyanines—*continued*

Solvent	Wavelength max. Abs. (nm)	Fluor. (nm)	Lasing wavelength (nm)	Pump source (nm)	Ref.
HIDC-1; 1,3,3,1',3',3'-hexarnethyl-2,2-indodicarbocyanine iodine					
DMSO			790–840	ruby (694.3)	89
ethylene glycol			800–882	FL→R640	25
DMSO	751		806 (788–832)	Nd:YAG(532)	51
acetone			819	ruby (694.3)	101
DMSO	743		822	Nd:YAG→700	110
EG/DMSO:3/1			832–911	Kr laser (647)	98
ethylene glycol			836	ruby (694.3)	112
DMSO			849	Nd:YAG(532)	111
DMSO			862	N$_2$ laser (337)	104
—			865 (825–912)	Kr laser (red)	52
ethylene glycol			869 (832–888)	Kr (752,799)	108
EG/DMSO:3/1			873 (819–937)	Kr laser (752)	113
HITC-P; 1,3,3,1',3',3'-hexamethyl-2,2'-indotricarbocyanine perchlorate					
DMSO+EG	751		870 (812–929)	Kr (752,799)	103
DMSO+EG			870 (828–909)	Kr (647,676)	103
EG/DMSO:84/16		875 (840–940)		Kr (647,676)	164
HITC-I; 1,3,3,1',3',3'-hexamethyl-2,2'-indotricarbocyanine iodine					
ethanol			807 (784–830)	Rh6G (587)	188
IR-125; 3,3,3',3'-tetramethyl-1,1'-di(4-sulfobutyl)4,5,4',5'-dibenzoindo-tricarbocyanine, monosodium salt					
DMSO			840–920	ruby (694.3)	89
DMSO	795		863 (846–907)	Nd:YAG	51
DMSO			903	Nd:YAG→700	101
DMSO			913	Nd:YAG(532)	111
DMSO			940	flashlamp	109
IR-144; anhydro-11-(4-ethoxycarbonyl-1-piperazinyl)-10,12-ethylene- 3,3,3',3'-tetramethyl-1,1'-di(3-sulfopropyl)-4,5,4',5'-dibenzoindotricarbocyanine hydroxide, triethylammonium salt					
ethylene glycol			800–870	FL→R640	109
EG/DMSO			834–892	Kr (752,799)	111
DMSO			835-890	ruby (694.3)	98
DMSO	745		863 (844–885)	Nd:YAG→700	25,110
DMSO			869	Nd:YAG(532)	105
DMSO			874	Nd:YAG(532)	114
DMSO			949–880	flashlamp	89

Oxacarbocyanines

Solvent	Wavelength max. Abs. (nm)	Fluor. (nm)	Lasing wavelength (nm)	Pump source (nm)	Ref.
DmOTC-I; 3,3'-dimethyl-2,2'-oxatricarbocyanine iodine					
acetone			711	ruby (694.3)	101
DMSO	682		725–780	ruby (694.3)	89
DMSO			750–810	ruby (694.3)	48
DMSO			788	N$_2$ laser (337)	104
DMSO+EG			~800 (750–864)	Kr (647,676)	103
DODC-I; 3,3'-diethyl-2,2'-oxadicarbocyanine iodine					
methanol	582		633	Nd:YAG(532)	89
DOTC-P (DEOTC); 3,3'-diethyl-2,2'-oxatricarbocyanine perchlorate					
DMSO+EG			795 (765–875)	Kr (647,676)	164
DOTC-I (DEOTC); 3,3'-diethyl-2,2'-oxatricarbocyanine iodine					
MeOH/N$_2$/COT	687		732	FL→R610	95
DMSO			725–765	ruby (694.3)	96
DMSO			740	Nd:YAG→585	95
acetone			742	ruby (694.3)	101
EG/DMSO:3/1			742–874	Kr laser (647)	98
DMSO			745	FL→R610	96
methanol			750–825	FL→R640	25
DMSO			770	Nd:YAG→585	95
DMSO			782	N$_2$ laser (337)	104
ethylene glycol			783 (750–833)	Kr (647,676)	108
DMSO+EG			800 (755–870)	Kr (647,676)	103
ethylene glycol			808 (756–871)	Kr laser (red)	52
DMSO	680		754	Nd:YAG(532)	105
DMSO			756 (736–793)	Nd:YAG(532)	51

Quinocarbocyanines

Solvent	Wavelength max. Abs. (nm)	Fluor. (nm)	Lasing wavelength (nm)	Pump source (nm)	Ref.
11-Br-D-2-QDC-I; 1,1'-diethyl-11-bromo-2,2'-quinodicarbocyanine iodine					
glycerin	694		815	ruby (694.3)	101
11-Br-Dm-2-QDC-I; ,1'-dimethyl-11-bromo-2,2'-quinodicarbocyanine iodine					
glycerin	691		745	ruby (694.3)	101
11-Br-Dm-4-QDC-I; 1,l'-diethyl-l l-bromo4,4'-quinodicarbocyanine iodine					
methanol	794		830	ruby (694.3)	101
D-2-QTC-I; 1,l'-diethyl-2,2'-quinotricarbocyanine iodine					
DMSO			865--920	ruby (694.3)	28
acetone	817		898	ruby (694.3)	101

Quinocarbocyanines—*continued*

Solvent	Wavelength max. Abs. (nm)	Fluor. (nm)	Lasing wavelength (nm)	Pump source (nm)	Ref.
D-2-QDC-I; 1,1'-diethyl-2,2'-quinodicarbocyanine iodine					
ethylene glycol	710		740–770	ruby (694.3)	89
D-4-QDC-I; 1,1'-diethyl-4,4'-quinodicarbocyanine iodine					
DMSO	820		845–920	ruby (694.3)	89
D-4-QTC-I; 1,1'-diethyl-4,4'-quinotricarbocyanine iodine					
DMSO			983–1081	ruby (694.3)	89
D-4-QTC-I; 1,1'-diethyl-4,4'-quinotricarbocyanine iodine					
acetone	923		1000	ruby (694.3)	101
Dm-4-QC-I; 1,1'-dimethyl-4,4'-quinocarbocyanine iodine					
glycerin	709		749	ruby (694.3)	101
Dm-4-QC-I; 1,1'-dimethyl-4,4'-quinocarbocyanine iodine					
glycerin	708		751	ruby (694.3)	101
Dm-4-QC-Br; 1,1'-dimethyl-4,4'-quinocarbocyanine bromine					
glycerin	710		754	ruby (694.3)	101

Thiacarbocyanines

Solvent	Wavelength max. Abs. (nm)	Fluor. (nm)	Lasing wavelength (nm)	Pump source (nm)	Ref.
Dmo-DTDC-I; 3,3'-diethyl-2,2'-(6,6'-dimethoxy)thiadicarbocyanine iodine					
DMSO	660		710–755	ruby (694.3)	89
DNDTPC-P; 3,3-diethyl-9,11,15,17-dineopentylene(6,7,6',7'-dibenzo) thiapentacarbo-cyanine perchlorate					
DMSO			1084–1125	Nd:YAG(1064)	116-7
DMSO			1231 (1192–1285)	Nd:YAG(1064)	119
5-Temo-DTTC-I; 3,3'-diethyl-2,2'-(5,6,5,6'-tetramethoxy)thiatricarbocyanine iodine					
DMSO			820–875	ruby (694.3)	89
9,11,15,17-Dnp-DTPC-P; 3,3'-diethyl-9,11,15,17-dineopentylene thiapentacarbocyanine perchlorate					
DMSO	~1060		1124 (1102–1148)	Nd:YAG(1064)	118
9,11,15,17-Dnp-5,6-Temo-DTPC-P; 3,3'-diethyl-9,11,15,17-dineopentylene (5,6,5',6'-tetramethoxy) thiapentacarbocyanine perchlorate					
—	~1060		1140 (1107–1187)	Nd:YAG(1064)	119
9,11,15,17-Dnp-6,7-Dbz-DTPC (DNDTPC)					
DMSO	~1060		1172 (1151–1198)	Nd:YAG(1064)	118
10-Cl-4,5-Dbz-DTDC-I; 3,3'-diethyl-10-chloro-2,2'(4,5,4',5'-dibenzo)-thiadicarbocyanine iodine					
DMSO	686		785	N_2 laser (337)	59

Thiacarbocyanines—*continued*

Solvent	Wavelength max. Abs. (nm)	Fluor. (nm)	Lasing wavelength (nm)	Pump source (nm)	Ref.
10-Cl-5,6-Dbz-DTDC-I; 3,3'-diethyl-10-chloro-2,2'-(5,6,5,'6'-dibenzo)-thiadicarbocyanine iodine					
acetone			714	ruby (694.3)	101
12A-DTTC-P; 3,3'-diethyl-12-acetoxy-2,2'-thiatetracarbocyanine perchlorate					
DMSO	872		920–950	ruby (694.3)	89
EG/DMSO:1/1			935–1019	Kr (752,799)	115
EG/DMSO:3/1			970 (915–1058)	Kr laser (752)	113
5-Dmo-DTTC-I; 3,3'-diethyl-2,2-(5,5'-dimethoxy)thiatricarbocyanine iodine					
DMSO			820–875	ruby (694.3)	89
5,6-Temo-DTTC-I; 3,3'-diethyl-2,2'-(5,6,5,6'-tetramethoxy)thiatricarbocyanine iodine					
acetone			853	ruby (694.3)	101
DMSO	793		855-885	ruby (694.3)	89
4,5-Dbz-DTTC-I; 3,3'-diethyl-2,2'-(4,5,4',5'-dibenzo)thiatricarbocyanine iodine					
DMSO	797		834–900	ruby (694.3)	28
acetone			860	ruby (694.3)	101
DMSO			928	flashlamp	104
DTDC-I; 3,3'-diethyl-2,2'-thiadicarbocyanine iodine					
DMSO	653		705–735	ruby (694.3)	89
acetone			711	ruby (694.3)	101
DMSO			720–775	Kr laser	103
DMSO			744	N$_2$ laser (337)	104
DTDC-P; 3,3'-diethyl-2,2'-thiadicarbocyanine perchlorate					
DMSO			746	N$_2$ laser (337)	104
DTTC-Br; 3,3'-diethyl-2,2'-thiatricarbocyanine bromine					
DMSO	760		810–830	ruby (694.3)	89
Dbz-DTTC-I; dibenzothiatricarbocyanine iodine					
acetone			774	ruby (694.3)	101
DTTC-I; 3,3'-diethylthiatricarbocyanine iodine					
methanol	757	798.5	796	ruby (694.3)	197
acetone	759	803	801.5	ruby (694.3)	197
1-propanol	764	807.5	807	ruby (694.3)	197
ethylene glycol	766	812	808	ruby (694.3)	197
DMF	766	812	808	ruby (694.3)	197
glycerin	775	815	809.5	ruby (694.3)	197
butyl alcohol	766	809	809.5	ruby (694.3)	197
DMSO	772.5	814	816	ruby (694.3)	197
ethanol	762.5	804.5	803	ruby (694.3)	197
DTTC-I; 3,3'-diethyl-2,2'-thiatricarbocyanine iodine					
DMSO			785	N$_2$ laser (337)	104
DMSO			790–871	ruby (694.3)	28
ethylene glycol			815–870	FL→R640	25
DMSO			820–900	FL→KR620	88

Thiacarbocyanines—*continued*

Solvent	Wavelength max. Abs. (nm)	Fluor. (nm)	Lasing wavelength (nm)	Pump source (nm)	Ref.
DTTC-I; 3,3'-diethyl-2,2'-thiatricarbocyanine iodine					
DMSO	763		828 (813–859)	Nd:YAG(532)	51
acetone			829	ruby (694.3)	101
ethanol			834	Nd:YAG→700	110
DMSO			840–870	Kr laser	59
DMSO			863 (816–855)	ruby (694.3)	89
DMSO			876	N_2 laser (337)	104
DMSO			889	flashlamp	151,152
IR-109-I; 3,3'-Diethyl-10,12-ethylene-11-morpholinothia-tricarbocyanine iodine					
DMSO			875	flashlamp	109
IR-116-I; 3,3'-diphenylthiatricarbocyanine iodine					
DMSO			885	flashlamp	109
IR-123-I; 3,3'-diethyl-9,11(oxy-o-phenylene)thiatricarbocyanine iodine					
DMSO			795–815	ruby (694.3)	89
DMSO	745		830	flashlamp	109
IR-132-P; 3,3'-di(3-acetoxypropyl)-11-diphenylamino-10,12-ethylene-5,6,5',6'-dibenzothiatricarbocyanine perchlorate					
EG/DMSO:3/1			863–1048	Kr laser (752)	113
DMSO	830		875–920	ruby (694.3)	89
DMSO			910	Nd:YAG(532)	111
EG/DMSO:3/1			916–984	Kr (752,799)	98
DMSO			972	flashlamp	109
IR-134-P; 11-(4-ethoxycarbonylpiperidino)-3,3'-diethyl-10,12-ethylene-4,5,4,5 dibenzothiatricarbocyanine perchlorate					
DMSO			888	flashlamp	109
IR-137-P); 3,3'-diethyl-10,12-ethylene-11-(N-methylanilino)-thiatricarbocyanine					
EG/DMSO:3/1			927 (855–1032)	Kr laser (752)	113
IR-137-P; 3,3'-diethyl-10,12-ethylene-11-(N-methylanilino)-thiatricarbocyanine perchlorate					
DMSO			950	flashlamp	109
IR-139-P; 11-dimethylamino-3,3-diethyl-10,12-ethylene-4,5,4',5'-di-benzothiatricarbocyanine perchlorate					
DMSO	868		883	flashlamp	109
IR-140-P; 5,5'-dichloro-11-diphenylamino-3,3'-diethyl-10,12-ethylenethiatricarbocyanine perchlorate					
ethanol			805–872	FL→Ox720	25
DMSO			850–930	ruby (694.3)	89
DMSO			875–916	FL→Ox720	25
EG/DMSO:3/1			887–986	Kr (752,799)	98
DMSO	823		893 (882–913)	Nd:YAG(532)	51
DMSO	776		898	Nd:YAG→700	101

Thiacarbocyanines—*continued*

Solvent	Wavelength max. Abs. (nm)	Fluor. (nm)	Lasing wavelength (nm)	Pump source (nm)	Ref.
IR-140-P; 5,5'-dichloro-11-diphenylamino-3,3'-diethyl-10,12-ethylenethiatricarbocyanine perchlorate					
DMSO	835		908	Nd:YAG(532)	105
DMSO			910	Nd:YAG(532)	111
EG/DMSO:3/1			927 (858–1030)	Kr laser(752)	113
DMSO			950	flashlamp	109
EG/DMSO:1/1			950 (862–1013)	Kr (752,799)	115
—			962	Kr laser (IR)	52
DMSO			946	flashlamp	109
R-141-I(DTTC-derivative); 5,5'-dichloro-3,3'-diethyl-10,12-ethylene-11-(N-methyl-anilino)thiatricarbocyanine iodine					
DMSO			946	flashlamp	109
IR-143-P(DTTC-derivative); 11-diphenylamino-3,3'-diethyl-10,12-ethylene-4,5,4',5'-dibenzothiatricarbocyanine perchlorate					
EG/DMSO:1/1			960 (913–1020)	Kr (752,799)	115
EG/DMSO:3/1			970 (894–1095)	Kr laser(752)	113
DMSO			972	flashlamp	109

The corresponding counter-ion has not been included in the name for any of the cyanine dyes (see reference cited in Section 2.1.2).

2.1.3.7 Furan Dye Lasers

Solvent	Wavelength maximum Abs. (nm)	Fluor. (nm)	Lasing wavelength (nm)	Pump source (nm)	Ref.
Furan 1-1-1					
toluene	349	423	415	N_2 laser (337)	32
Furan 1-1-2					
toluene	356	427	425	N_2 laser (337)	32
Furan 1-1-3					
toluene	342	412	407	N_2 laser (337)	32
Furan 1-1-4					
toluene	348	423	418	N_2 laser (337)	32
Furan 1-1-5					
toluene	348	425	420	N_2 laser (337)	32

2.1.3.8 Imitrine Dye Lasers

Solvent	Wavelength max. Abs. (nm)	Fluor. (nm)	Lasing wavelength (nm)	Pump source (nm)	Ref.
Imitrine 7-1-123					
ethanol	412	488	542 (530–554)	Nd:YAG(355)	133
Imitrine 7-1-124					
ethanol	395	490	507 (490–537)	Nd:YAG(355)	133
Imitrine 7-1-125					
ethanol	378	475	536 (522–548)	Nd:YAG(355)	133
Imitrine 7-1-127					
diethyl ether	406	532	565 (551–579)	Nd:YAG(355)	133
ethanol	402	535	570 (553–587)	Nd:YAG(355)	133
Imitrine 7-1-128					
ethanol	406	532	582 (561–593)	Nd:YAG(355)	133
Imitrine 7-1-129					
ethanol	406	532	576 (550–592)	Nd:YAG(355)	133
Imitrine 7-1-130					
ethanol	378	475	536 (522–546)	Nd:YAG(355)	133
Imitrine 7-1-131					
ethanol	390	475	491 (482–509)	Nd:YAG(355)	133
Imitrine 7-1-132					
ethanol	400	483	515 (505–535)	Nd:YAG(355)	133
Imitrine 7-1-133					
ethanol	395	472	508 (490–521)	Nd:YAG(355)	133
Imitrine 7-1-134					
ethanol	405	510	572 (560–584)	Nd:YAG(355)	133
Imitrine 7-1-135					
ethanol	405	512	540 (523–557)	Nd:YAG(355)	133
Imitrine 7-1-136					
ethanol	397	490	536 (526–546)	Nd:YAG(355)	133
Imitrine 7-1-137					
ethanol	396	505	546 (536–556)	Nd:YAG(355)	133
Imitrine 7-1-138					
ethanol	375	475	515 (502–531)	Nd:YAG(355)	133
Imitrine 7-1-139					
ethanol	455	510	544 (536–552)	Nd:YAG(355)	133
Imitrine 7-1-140					
ethanol	480	575	612 (606–618)	Nd:YAG(355)	133
Imitrine 7-1-141					
ethanol	435	520	540 (521–559)	Nd:YAG(355)	133
Imitrine 7-1-142					
ethanol	425	518	536 (515–557)	Nd:YAG(355)	133

2.1.3.9 Naphthalimide Dye Lasers

Solvent	Wavelength max. Abs. (nm)	Fluor. (nm)	Lasing wavelength (nm)	Pump source (nm)	Ref.
Brilliant Sulfaflavine; 6-amino-2,3-dihydro-2-(4-methylphenyl)-1,3-dioxo-1H-benz[de]-iso-quino-line-5-sulfonic acid, monosalt					
COT	443		562 (508–573	flashlamp	81,82
methanol			562 (522–618)	flashlamp	26
Fluorol 555; Fluorol 7GA: 6-n-butylamino-2,3-dihydro-2-(4-n-bytyl)-1,3-dioxo-1H-benz[de]-isoquinoline-5-sulfonic acid, monosal					
—	442		550–580	flashlamp	77
MeOH/LO			552	flashlamp	78
alcohol or water			555	flashlamp	69
ethanol			570 (540–590)	flashlamp	25
methanol			574 (535–590)	flashlamp	77
methanol			574 (542–592)	flashlamp	25

2.1.3.10 Oligophenylene Dye Lasers

Oligophenylenes

Solvent	Wavelength max. Abs. (nm)	Fluor. (nm)	Lasing wavelength (nm)	Pump source (nm)	Ref.
Oligophenylene 2-1-7					
ethanol			336 (331–342)	KrF laser (249)	2
Oligophenylene 2-1-9					
ethanol			336 (331–342)	KrF laser (249)	2
Oligophenylene 2-1-10					
ethanol	264	339	339 (332–346)	KrF laser (249)	2
Oligophenylene 2-3-15					
p-dioxane			386 (380–391)	XeCl laser (308)	21
Oligophenylene 2-4-7					
DMA/EtOH:4/1			386	flashlamp	17
ethanol			386	flashlamp	17
Oligophenylene 2-4-8					
DMF/EtOH:4/1			391	flashlamp	17
ethanol			391	flashlamp	17
Oligophenylene 2-5-4					
ethanol	310	366, 383	389	flashlamp	18
Oligophenylene 2-5-5					
DMA/EtOH:1/4			386	flashlamp	18
Oligophenylene 2-5-6					
DMA/EtOH:1/4			391	flashlamp	18

Oligophenylenes—*continued*

Solvent	Wavelength max. Abs. (nm)	Fluor. (nm)	Lasing wavelength (nm)	Pump source (nm)	Ref.
Oligophenylene 2-5-7					
DMA/EtOH:1/4	335	380, 399	395	flashlamp	18
Exalite 337E					
ethylene glycol			377 (365–400)	Ar laser	13
Exalite 392E					
ethylene glycol			393 (375–410)	Ar laser	30,31
Exalite 400E					
ethylene glycol			400 (385–425)	Ar laser	30,59

Phenylpyridines

Solvent	Wavelength max. Abs. (nm)	Fluor. (nm)	Lasing wavelength (nm)	Pump source (nm)	Ref.
4-phenylpyridine					
ethanol			418	N_2 laser (337)	44
ethanol			330–410	N_2 laser (337)	44
ethanol			330–408	N_2 laser (337)	44
ethanol			325–406	N_2 laser (337)	44

Quaterphenyls

Solvent	Wavelength max. Abs. (nm)	Fluor. (nm)	Lasing wavelength (nm)	Pump source (nm)	Ref.
3,3',2'',3''-tetramethyl-*p*-quaterphenyl					
cyclohexane			362 (354–388)	Nd:YAG (266)	10
***p*-quaterphenyl**					
DMF	294	365	362–390	flashlamp	8
toluene	297		374	N_2 laser (337)	3
4,4'''bis(2-butyloctyloxy)-*p*-quaterphenyl; BBQ					
cyclohexane	306	381	380	KrF laser (248)	5
butanol			381–389	flashlamp	15
cyclohexane			382 (373–391)	Nd:YAG (266)	10
—			386	KrF laser (248)	6
toluene/ethanol			386 (373–399)	N_2 laser (337)	20
DMF			389–395	flashlamp	15
DMF			390 (370–410)	flashlamp	25
EtOH/toluene:1/1			391 (380–410)	Nd:YAG (355)	28

Terphenyls

Solvent	Wavelength maximum Abs. (nm)	Fluor. (nm)	Lasing wavelength (nm)	Pump source (nm)	Ref.
2,2''-dimethly-p-terphenyl					
cyclohexane	251	336	332 (311–360)	KrF laser (248)	1
p-terphenyl					
cyclohexane	276	339	337	KrF laser (248)	3
ethanol		335	338	KrF laser (248)	3
—			338 (326–358)	KrF laser (248)	4
cyclohexane			339 (322–366)	KrF laser (248)	1
p-dioxane			340	KrF laser (248)	5,6
cyclohexane			340 (323–364)	KrF laser (248)	7
DMF			341	flashlamp	8
DMF			415	flashlamp	25
p,p'-N,N,N'',N''-tetraethydiamonoterphenyl					
ethanol (sat.)			417–427	N$_2$ laser (337)	36

2.1.3.11 Oxazine Dye Lasers

Solvent	Wavelength maximum Abs. (nm)	Fluor. (nm)	Lasing wavelength (nm)	Pump source (nm)	Ref.
Cresyl Violet 670: 5-imino-5H-benzo[a]phenoxazin-9-amine, monoperchlorate					
—	601	630	633 (615–655)	Nd:YAG (532)	83
methanol	594		637 (620–660)	Nd:YAG (532)	53
—			639 (620–670)	Nd:YAG (532)	84
methanol			639 (645–705)	flashlamp	81
MeOH/H$_2$O			640 (620–670)	Nd:YAG (532)	91
methanol			645–705	flashlamp	81
methanol			646 (6250660)	Nd:YAG (532)	91
—			647	Nd:YAG (532)	91
methanol			655 (646–697)	flashlamp	25
ethanol			659 (650–695)	flashlamp	25
ethanol			660 (641–687)	N$_2$ laser (337)	20
methanol			664 (631–705)	flashlamp	59
ethylene glycol			673 (650–696)	Ar laser (cw)	47
ethylene glycol			695 (675–708)	Ar (458-514)	164
LD 690 Perchlorate					
methanol	616		660	Nd:YAG (532)	94
ethanol			668 (655–705)	Nd:YAG (532)	93
DMSO/EtOH:2/1			670 (660–716)	N$_2$ laser (337)	93
ethylene glycol			696–780	Kr laser ((cw)	99

Oxazine Dye Lasers—*continued*

Solvent	Wavelength maximum		Lasing wavelength (nm)	Pump source (nm)	Ref.
	Abs. (nm)	Fluor. (nm)			

Nile Blue 690 Perchlorate, Nile Blue A Perchlorate,: 5-imino-9-(diethylamino)-benzo[a]-phenoxazin-7-amine, monoperchlorate

Solvent	Abs. (nm)	Fluor. (nm)	Lasing wavelength (nm)	Pump source (nm)	Ref.
—	624		681 (662–710)	Nd:YAG(532)	84
methanol	628		683	Nd:YAG(532)	97
ethanol			690	laser	80
ethanol			696 (683–710)	N_2 laser (337)	20
methanol			705	flashlamp	25
methanol			717 (689–750)	flashlamp	59
methanol			722	flashlamp	87
ethylene glycol			730 (692–782)	Kr laser (cw)	47
ethylene glycol			750 (710–790)	R590 (Ar)	164

Oxazine 720 Perchlorate; Oxazine 170 Perchlorate: 5-(ethylimino)-methyl-5H-benzo[a]-phenoxazin-9-amino, monoperchlorate

Solvent	Abs. (nm)	Fluor. (nm)	Lasing wavelength (nm)	Pump source (nm)	Ref.
methanol	627	650	668 (649–700)	Nd:YAG(532)	53
methanol	620		671 (613–708)	Nd:YAG(532)	51
ethanol			672	Nd:YAG(532)	81
methanol			692 (676–698)	flashlamp	25
methanol			698 (682–720)	flashlamp	26
ethanol			699 (675–711)	flashlamp	25
MeOH/H$_2$O			705 (675–730)	flashlamp	19
methanol			710 (690–740)	flashlamp	70

Oxazine 725 Perchlorate; Oxazine 1: 3,7-bis(diethylamino)phenoxazin-5-ium, perchlorate

Solvent	Abs. (nm)	Fluor. (nm)	Lasing wavelength (nm)	Pump source (nm)	Ref.
CH$_2$Cl$_2$	645	680	681	FL→R610	96
DMSO/EG/COT			687–826	Kr laser (647)	98
CH$_2$Cl$_2$			690	Nd:YAG(532)	97
DMSO			695	FL→R610	96
ethanol			715	flashlamp	80
—			723 (688–800	Kr laser (red)	52
methanol			724 (695–761)	Nd:YAG(532)	51
MeOH/R590			725 (705–745)	flashlamp	59
ethanol			725 (705–750)	N_2 laser (337)	20
CH$_2$Cl$_2$			740 (720–758)	flashlamp	25
DMSO/EGor G			745 (645–810)	Kr (647,676)	103
DMSO/EG			750 (695–801)	Kr (647,676)	164

Oxazine 750 Perchlorate and Oxazine 750 Chloride

Solvent	Abs. (nm)	Fluor. (nm)	Lasing wavelength (nm)	Pump source (nm)	Ref.
methanol	662	705	722 (704–786)	Nd:YAG(532)	51
ethanol			745 (700–785)	FL→R640	25
DMSO			760–775	Nd:YAG→585	95
EG/DMSO:4/1			770 (750–835)	Kr laser (647)	106
EG/DMSO:84/16			775 (747–885)	Kr (647,676)	164

Oxazine Dye Lasers—*continued*

Solvent	Wavelength maximum		Lasing	Pump source	Ref.
	Abs. (nm)	Fluor. (nm)	wavelength (nm)	(nm)	

Oxazine 750 Perchlorate and Oxazine 750 Chloride

Solvent	Abs. (nm)	Fluor. (nm)	wavelength (nm)	(nm)	Ref.
EG/DMSO:2/1			776 (747–801)	Kr (647,676)	107
ethylene glycol			780 (749–825)	Kr (647,676)	107

2.1.3.12 Oxazole Dye Lasers

Solvent	Wavelength max.		Lasing	Pump source	Ref.
	Abs. (nm)	Fluor. (nm)	wavelength (nm)	(nm)	

2-phenyl-5(4-difluoromethylsulphenyl) oxazole

Solvent	Abs. (nm)	Fluor. (nm)	wavelength (nm)	(nm)	Ref.
toluene			416 (403–437)	N_2 laser (337)	42
Oxazole 4-1-1					
H_2O (Ar)		476	494–512	flashlamp	129
Oxazole 4-1-2					
ethanol (air)		484	492–507	flashlamp	129
ethanol (Ar)		480	493–508	flashlamp	129
H_2O (air)		480	494–512	flashlamp	129
EtOH/H_2O (Ar)			495–511	flashlamp	129
EtOH/H_2O (air)			496–507	flashlamp	129
Oxazole 4-1-3					
2-PrOH (Ar)			559–582	flashlamp	129
ethanol (air)		567	560–583	flashlamp	129
ethanol (Ar)		578	567–587	flashlamp	129
methanol (Ar)			571–588	flashlamp	129
H_2O (Ar)			571–591	flashlamp	129
Oxazole 4-1-8					
ethanol (air)			480–497	flashlamp	129
MeOH (air)			482–509	flashlamp	129
MeOH (Ar)			484–512	flashlamp	129
EtOH/H_2O (air)			490–514	flashlamp	129
Oxazole 4-1-9					
ethanol (air)			492–504	flashlamp	129
MeOH (air)			496–508	flashlamp	129
Oxazole 4-2-1					
H_2O			490	flashlamp	45
H_2O	364	470	496	ruby (347)	45
Oxazole 4-2-2					
H_2O			504	flashlamp	45
H_2O	380	495	508	ruby (347)	45
Oxazole 4-2-3					
H_2O			501	flashlamp	45
H_2O	372	485	503	ruby (347)	45

Oxazole Dye Lasers—*continued*

Solvent	Wavelength max. Abs. (nm)	Wavelength max. Fluor. (nm)	Lasing wavelength (nm)	Pump source (nm)	Ref.
Oxazole 4-2-4					
H_2O	376	475	504	ruby (347)	45
Oxazole 4-2-5					
H_2O	380	490	503	ruby (347)	45
Oxazole 4-2-6					
H_2O	382	475	506	ruby (347)	45
Oxazole 4-2-7					
H_2O	380	447	452	ruby (347)	45
H_2O	376	500	503	ruby (347)	45
Oxazole 4-3-1					
MeOH (air)	379	482	492	flashlamp	131
ethanol (Ar)	379	482	493	flashlamp	131
H_2O (air)	371	480	494	flashlamp	131
H_2O (Ar)	371	480	494	flashlamp	131
EtOH/H_2O (Ar)	371	480	495	flashlamp	131
EtOH/H_2O (air)	371	480	496	flashlamp	131
Oxazole 4-3-3					
ethanol (air)	400	548	544	flashlamp	131
ethanol (Ar)	400	548	545	flashlamp	131
Oxazole 4-3-4					
methanol (air)			494	flashlamp	131
ethanol (Ar)	376	486	495	flashlamp	131
Oxazole 4-3-4					
EtOH/H_2O (Ar)	378	486	495	flashlamp	131
H_2O (Ar)			495	flashlamp	131
EG/H_2O (Ar)			496	flashlamp	131
MeOH/H_2O (air)			496	flashlamp	131
EG/H_2O (Ar)			498	flashlamp	131
MeOH/H_2O (Ar)			498	flashlamp	131
Oxazole 4-3-5					
EtOH (Ar)	376	486	508	flashlamp	131
H_2O (Ar)	374	490	510	flashlamp	131
Oxazole 4-3-7					
ethanol (Ar)	394	550	593	flashlamp	131
Oxazole 4-3-8					
ethanol (Ar)	388	492	499	flashlamp	131
Oxazole 4-3-9					
ethanol (Ar)	418	572	597	flashlamp	131
Oxazole 4-3-11					
ethanol (air)	422	560	559	flashlamp	131
ethanol (Ar)	422	560	566	flashlamp	131

Oxazole Dye Lasers—*continued*

Solvent	Wavelength max. Abs. (nm)	Fluor. (nm)	Lasing wavelength (nm)	Pump source (nm)	Ref.
Oxazole 4-3-12					
ethanol (Ar)	450	582	579	flashlamp	131
EtOH/H$_2$O (Ar)	444	588	582	flashlamp	131
H$_2$O (Ar)	435	580	582	flashlamp	131
Oxazole 4-3-13					
EtOH/H$_2$O (Ar)	420	517	519	flashlamp	131
ethanol (Ar)	420	517	520	flashlamp	131
H$_2$O (air)	409	515	520	flashlamp	131
EtOH/H$_2$O (air)	409	515	522	flashlamp	131
Oxazole 4-3-13					
H$_2$O (Ar)	420	517	523	flashlamp	131
ethanol (Ar)	420	517	524	flashlamp	131
Oxazole 4-3-14					
H$_2$O (air)	400	513	513	flashlamp	131
EtOH/H$_2$O (air)	400	513	514	flashlamp	131
H$_2$O (Ar)	395	511	514	flashlamp	131
ethanol (air)	395	511	515	flashlamp	131
EtOH/H$_2$O (Ar)	398	511	516	flashlamp	131
ethanol (Ar)	398	511	522	flashlamp	131
Oxazole 4-3-15					
ethanol (air)	397	525	494	flashlamp	131
Oxazole 4-3-16					
EtOH (Ar)	392	497	527	flashlamp	131
Oxazole 4-3-17					
ethanol (Ar)			537	flashlamp	131
EtOH/H$_2$O (Ar)			537	flashlamp	131
ethanol (Ar)			539	flashlamp	131
ethanol (Ar)			541	flashlamp	131
Oxazole 4-3-18					
EtOH (Ar)	428	604	635	flashlamp	131
2-propanol	399	555	559	flashlamp	131
ethanol (Ar)	410	567	567, 571	flashlamp	131
EtOH/H$_2$O (air)	39	562	570	flashlamp	131
Oxazole 4-3-20					
ethanol (Ar)	392	494	501	flashlamp	131
ethanol (air)	392	494	502	flashlamp	131
Oxazole 4-3-21					
ethanol (air)	390	497	502	flashlamp	131
ethanol (Ar)	390	497	502	flashlamp	131
EtOH/H$_2$O (air)	381	495	503	flashlamp	131
EtOH/H$_2$O (Ar)	381	495	503	flashlamp	131

Oxazole Dye Lasers—*continued*

Solvent	Wavelength max. Abs. (nm)	Fluor. (nm)	Lasing wavelength (nm)	Pump source (nm)	Ref.
Oxazole 4-3-21					
H$_2$O (air)	382	498	503	flashlamp	131
Oxazole 4-3-22					
ethanol (air)	418	580	573	flashlamp	131
ethanol (Ar)	418	580	574	flashlamp	131
Oxazole 4-4-1					
p-dioxane	343	390	388	XeCl laser (308)	22
Oxazole 4-4-2					
p-dioxane	304	390	390	XeCl laser (308)	22
Oxazole 4-5-1					
methanol			569	FL (coaxial)	150
methanol			581	FL (coaxial)	150
Oxazole 4-5-2					
methanol			569	FL (coaxial)	150
Oxazole 4-5-3					
methanol			569	FL (coaxial)	150
Oxazole 4-5-4					
methanol			569	FL (coaxial)	150
Oxazole 4-5-5					
methanol			569	FL (coaxial)	150
Oxazole 4-5-6					
methanol			573	FL (coaxial)	150
H$_2$O			582	FL (coaxial)	150
Oxazole 4-5-7					
MeOH			572	FL (coaxial)	150
H$_2$O			581	FL (coaxial)	150
Oxazole 4-5-8					
MeOH			568	FL (coaxial)	150
H$_2$O			580	FL (coaxial)	150
Oxazole 4-5-9					
methanol			565	FL (coaxial)	150
H$_2$O			578	FL (coaxial)	150
Oxazole 4-6-1					
ethanol	320	395	389–395	ruby (316)	28
H$_2$O/HOAc:95/5	368	475	462–489	ruby (316)	28
Oxazole 4-6-10					
H$_2$O/HOAc:1/1	401	556	590–620	ruby (316)	28
H$_2$O/HOAc:95/5	395	550	590–620	ruby (316)	28
H$_2$O	395	550	591–614	ruby (316)	28

Oxazole Dye Lasers—*continued*

Solvent	Wavelength max. Abs. (nm)	Fluor. (nm)	Lasing wavelength (nm)	Pump source (nm)	Ref.
Oxazole 4-6-11					
H_2O	394		568–598	ruby (316)	28
H_2O/HOAc:1/1			568–598	ruby (316)	28
H_2O/HOAc:95/5			568–598	ruby (316)	28
Oxazole 4-6-12					
ethanol	410	580	560–598	ruby (316)	28
H_2O	400	580	563–600	ruby (316)	28
Oxazole 4-6-13					
H_2O	403	585	570–600	ruby (316)	28
ethanol	420	580	581–598	ruby (316)	28
Oxazole 4-6-2					
ethanol	378	485	503–521	ruby (316)	28
Oxazole 4-6-3					
H_2O/HOAc:95/5	386	565	422–430	ruby (316)	28
ethanol	335	420	590–625	ruby (316)	28
Oxazole 4-6-4					
H_2O/HOAc:4/6	390	580	562–595	ruby (316)	28
Oxazole 4-6-5					
H_2O	375	485	492–511	ruby (316)	28
ethanol	375	480	494–504	ruby (316)	28
Oxazole 4-6-6					
H_2O	380	485	498–518	ruby (316)	28
ethanol	385	485	499–514	ruby (316)	28
Oxazole 4-6-7					
ethanol	380	485	510–522	ruby (316)	28
H_2O	374	496	510–527	ruby (316)	28
Oxazole 4-6-8					
H_2O/HOAc:95/5	378	490	509–531	ruby (316)	28
ethanol	385	485	529–536	ruby (316)	28
Oxazole 4-6-9					
H_2O/HOAc:1/1	395	560	590–614	ruby (316)	28
H_2O/HOAc:95/5	390	560	590–614	ruby (316)	28
H_2O	390	555	594–612	ruby (316)	28
4PyPO; 2-(4-pyridyl)-5-phenyloxazole					
ethanol	307,322	360	395–402	N_2 laser (337)	32
water+HCl (pH2)	244,364	470	493–512	N_2 laser (337)	32
water+HCl (pH2)			504	flashlamp	32
4PyPO-TS; 4[2-(5-phenyloxazole)]-1-methylpyridinium p-toluenesulfonate					
water	246,270	470	495–514	N_2 laser (337)	32
water			506	flashlamp	32
α-NPO; 2-(1-naphthy)-5-phenyloxazole					
cyclohexane			385–415	N_2 laser (337)	16

Oxazole Dye Lasers—*continued*

Solvent	Wavelength max. Abs. (nm)	Fluor. (nm)	Lasing wavelength (nm)	Pump source (nm)	Ref.
α-NPO; 2-(1-naphthy)-5-phenyloxazole					
ethanol			400	flashlamp	7
ethanol, w, t			390–398	N₂ laser (337)	27
ethanol	329	398		flashlamp	47
cyclohexane	332	391	385–415	N₂ laser (337)	51
POPOP: 1,4-bis[2-(5-phenyloxazolyl)] benzene					
vapor (Ar+N₂)			381	electron beam	14
vapor			393	N₂ laser (337)	29
tetrahydrofuran			415–430	N₂ laser (337)	16
toluene	358	410	419	flashlamp	46
ethanol			419–424	flashlamp	186
toluene			423 (414–442)	N₂ laser (337)	42
p-dioxane			427 (411–465)	N₂ laser (337)	40
ethanol, w, t			430–445	N₂ laser (337)	27
Dimethyl POPOP: 1,4-bis[2-(4-methyl-5-phenyloxazolyl)] benzene					
p-dioxane	266,286	428	~430	flashlamp	32
p-dioxane			430 (418–465)	N₂ laser (337)	11
7-diethylaminoPOPOP: 1,4-bis[2-(7-diethylamino-5-phenyloxazolyl)] benzene					
ethanol, w, t			390–445	N₂ laser (337)	27
PPO: 2,5-diphenyloxazole					
toluene	303	361	365 (359–391)	N₂ laser (337)	11
cyclohexane			372	KrF laser (248)	3
cyclohexane			381	flashlamp	8

Oxadiazoles

Solvent	Wavelength maximum Abs. (nm)	Fluor. (nm)	Lasing wavelength (nm)	Pump source (nm)	Ref.
2(2-fluorophenyl)-5-phenyl-1,3,4-oxadiazole					
cyclohexane	—	—	347	KrF laser (249)	9
2(3-fluorophenyl)-5-phenyl-1,3,4-oxadiazole					
ethanol	—	—	347	KrF laser (249)	9
2(4-fluorophenyl)-5-phenyl-1,3,4-oxadiazole					
ethanol	—	—	347	KrF laser (249)	9
di(2-fluorophenyl)-5-phenyl-1,3,4-oxadiazole					
ethanol	—	—	347	KrF laser (249)	9
di(3-fluorophenyl)-5-phenyl-1,3,4-oxadiazole					
ethanol	—	—	347	KrF laser (249)	9
2(4-bromophenyl)-5-phenyl-1,3,4-oxadiazole					
ethanol	—	—	353	KrF laser (249)	9

Oxadiazoles—*continued*

Solvent	Wavelength maximum Abs. (nm)	Fluor. (nm)	Lasing wavelength (nm)	Pump source (nm)	Ref.
2(4-chlorophenyl)-5-phenyl-1,3,4-oxadiazole					
ethanol	—	—	357	KrF laser (249)	9
2(3,4-dichlorophenyl)-5-phenyl-1,3,4-oxadiazole					
ethanol	—	—	357	KrF laser (249)	9
2(2,4-dichlorophenyl)-5-phenyl-1,3,4-oxadiazole					
ethanol	—	—	359	KrF laser (249)	9
3-methyl-alpha-naphthyloxadiazole					
toluene	—	—	372	N_2 laser (337)	9
2-fluoro-alpha-naphthyloxadiazole					
toluene	—	—	373	N_2 laser (337)	9
3-fluoro-alpha-naphthyloxadiazole					
toluene	—	—	375	N_2 laser (337)	9
4-bromo-alpha-naphthyloxadiazole					
toluene	—	—	377	N_2 laser (337)	9
4-methoxy-alpha-naphthyloxadiazole					
toluene	—	—	379	N_2 laser (337)	9
2,5-bis(4-biphenyl)-1,3,4-oxadiazole					
toluene	—	—	379	N_2 laser (337)	9
2,5-bis(4-biphenyl)-1,3,4-oxadiazole					
toluene	—	—	380 (372–406)	N_2 laser (337)	9
2,5-dinaphthyl-1,3,4-oxadiazole					
toluene	—	—	390 (385–417)	N_2 laser (337)	9
2,5-bis(4-diethylaminophenyl)l-1,3,4-oxadiazole					
methyl chloride	—	––	425	N_2 laser (337)	9

2.1.3.13 Perylene Dye Lasers

Solvent	Wavelength max. Abs. (nm)	Fluor. (nm)	Lasing wavelength (nm)	Pump source (nm)	Ref.
Perylene 11-1-1					
DMF	527	533	575 (569–585)	Nd:YAG (532)	155
Perylene BASF-241					
chloroform			577 (566–600)	Nd:YAG (532)	156

2.1.3.14 Phosphorine Dye Lasers

Solvent	Wavelength maximum		Lasing wavelength (nm)	Pump source (nm)	Ref.
	Abs. (nm)	Fluor. (nm)			
Phosphorine					
benzene	—	—	502	N_2 laser (337)	36
benzene	—	—	522	N_2 laser (337)	36
benzene	—	—	529	N_2 laser (337)	36
benzene	—	—	536	N_2 laser (337)	36

2.1.3.15 Pteridine Dye Lasers

Solvent	Wavelength max.		Lasing wavelength (nm)	Pump source (nm)	Ref.
	Abs. (nm)	Fluor. (nm)			
Pteridine					
methanol			375–380	N_2 laser (337)	10
MeOH, alkaline			460	N_2 laser (337)	10
water, alkaline			522	N_2 laser (337)	10

2.1.3.16 Pyrazoline Dye Lasers

Solvent	Wavelength maximum		Lasing wavelength (nm)	Pump source (nm)	Ref.
	Abs. (nm)	Fluor. (nm)			
2-pyrazoline					
methanol	—	—	516	N_2 laser (337)	36
methanol alkaline	—	—	544	N_2 laser (337)	36

2.1.3.17 Pyrromethene Dye Lasers

Solvent	Wavelength max.		Lasing wavelength (nm)	Pump source (nm)	Ref.
	Abs. (nm)	Fluor. (nm)			
Pyrromethene 546					
methanol			542 (523–580)	FL (triaxial)	143,144
DMA/MeOH			542 (532–565)	FL (coaxial)	142
ethanol			546 (bb)	flashlamp	189
Pyrromethene 556					
ethylene glycol			546 (527–583)	Ar laser (488)	145
ethylene glycol			547 (523–582)	Ar (699-1488)	23
methanol			548 (537–605)	FL (triaxial)	143
ethylene glycol			550 (527–584)	Ar laser (514.5)	145
ethylene glycol			553 (530–624)	Ar (458-514)	149
DMA/MeOH			555 (545–585)	FL (coaxial)	142
methanol			561 (540–580)	flashlamp	144-46

Pyrromethene Dye Lasers—*continued*

Solvent	Wavelength max. Abs. (nm)	Fluor. (nm)	Lasing wavelength (nm)	Pump source (nm)	Ref.
Pyrromethene 556; 1,3,5,7,8-pentamethyl pyrromethene-2,6 disulfonate-BF$_2$ complex					
ethylene glycol	500		560 (530–623)	Ar (458-524,cw)	124
Pyrromethene 567; 1,3,5,7-tetramethylpyrromethene-difluoroborate complex					
DMA/MeOH			540 (537–560)	FL (coaxial)	142
PPH			560 (543–584)	Ar laser (514.5)	145
PPH			566 (549–592)	Nd:YAG (532)	122
ethanol			567	flashlamp	123
PPH			568 (550–608)	Ar (458-524,cw)	124
methanol			570	flashlamp	144
NMP/PPH			571 (552–608)	Ar (all lines)	149
Pyrromethene 580; 1,3,5,7,8-pentamethyl 2,6-di-*n*-butylpyrromethene-difluoroborate complex					
methanol			552 (545–586)	Nd:YAG(532)	122
ethanol			552 (545–585)	Nd:YAG (532)	148
ethanol			569 (545–583)	Nd:YAG (532)	148
ethanol			570 (bb)	Nd:YAG (532)	153
PPH			575 (555–592)	Ar (all lines)	154
ethanol			580 (bb)	flashlamp	157
Pyrromethene 597; 1,3,5,7,8-pentamethyl 2,6-di-butylpyrromethene-difluoroborate complex					
ethanol			571	Nd:YAG (532)	148
ethanol			571 (560–600)	Nd:YAG(532)	122
ethanol			587 (bb)	Nd:YAG (532)	153
p-dioxane			593 (bb)	flashlamp	159
ethanol			597 (bb)	flashlamp	157
Pyrromethene 605					
ethanol			605 (bb)	flashlamp	157
Pyrromethene 650					
—			612 (604–630)	Nd:YAG (532)	160
xylene			631 (bb)	Nd:YAG (532)	153
Pyrromethene-BF$_2$					
ethanol			546	flashlamp	141
Pyrromethene-BF$_2$ 9-1-1; TMP-BF$_2$					
MeOH	505		533 (bb)	flashlamp	141-2
Pyrromethene-BF$_2$ 9-1-3					
methanol			551	flashlamp	141
Pyrromethene-BF$_2$(2)					
ethanol			571	flashlamp	125
Pyrromethene-BF$_2$(3)); 1,2,3,5,6,7,8-heptamethylpyrromethene-BF$_2$ complex					
ethanol			566	flashlamp	123
ethanol			593	flashlamp	125

Pyrromethene Dye Lasers—*continued*

Solvent	Wavelength max. Abs. (nm)	Fluor. (nm)	Lasing wavelength (nm)	Pump source (nm)	Ref.
Pyrromethene-BF$_2$(4); 1,3,5,7,8-pentamethyl-2,6-diethylpyrromethene-BF$_2$ complex					
methanol			567	flashlamp	123
ethanol			602	flashlamp	125
Pyrromethene-BF$_2$(5); 1,3,5,7,-tetramethyl-8-ethyl-2,6-dicarbethoxypyrromethene-BF$_2$					
complex					
methanol			556	flashlamp	123
ethanol			612	flashlamp	125
Pyrromethene-BF$_2$(6); 1,3,5,7,8-pentamethyl-2,6-dinitropyrromethene-BF$_2$ complex					
ethanol			559	flashlamp	123

2.1.3.18 Pyrylium Salt Dye Lasers

Solvent	Wavelength max. Abs. (nm)	Fluor. (nm)	Lasing wavelength (nm)	Pump source (nm)	Ref.
Pyrylium dye 7					
CH$_3$CN			516–550	N$_2$ laser (337)	190
Pyrylium dye 13					
CH$_3$CN			464–514	N$_2$ laser (337)	190
Pyrylium dye 14					
CH$_3$CN			458–503	N$_2$ laser (337)	190
Pyrylium dye 15					
CH$_3$CN			470–517	N$_2$ laser (337)	190
Pyrylium dye 16					
CH$_3$CN			532-609	N$_2$ laser (337)	190
Pyrylium dye 17					
CH$_3$CN			501–563	N$_2$ laser (337)	190
Pyrylium dye 18					
CH$_3$CN			500–562	N$_2$ laser (337)	190
Pyrylium dye 19					
CH$_3$CN			542–630	N$_2$ laser (337)	190
Pyrylium dye 20					
CH$_3$CN			532–622	N$_2$ laser (337)	190
Pyrylium dye 21					
CH$_3$CN			538–620	N$_2$ laser (337)	190
Pyrylium dye 22					
CH$_3$CN			526–596	N$_2$ laser (337)	190
Pyrylium dye 32					
CH$_3$CN			537–623	N$_2$ laser (337)	190
Pyrylium dye 34					
CH$_3$CN			503–557	N$_2$ laser (337)	190

Pyrylium Salt Dye Lasers—*continued*

Solvent	Wavelength max.		Lasing wavelength (nm)	Pump source (nm)	Ref.
	Abs. (nm)	Fluor. (nm)			
Pyrylium dye 35					
CH₃CN			558–632	N₂ laser (337)	190
Pyrylium dye 36					
CH₃CN			490–546	N₂ laser (337)	48
Pyrylium dye 37					
CH₃CN			516–550	N₂ laser (337)	190
Pyrylium dye 39					
CH₃CN			549–630	N₂ laser (337)	190
Pyrylium dye 40					
CH₃CN			556–629	N₂ laser (337)	190
Pyrylium dye 41					
CH₃CN			556–623	N₂ laser (337)	190
Pyrylium dye 43					
CH₃CN			558–603	N₂ laser (337)	190
Pyrylium dye 44					
CH₃CN			489–541	N₂ laser (337)	48
Pyrylium dye 45					
CH₃CN			499–547	N₂ laser (337)	190
Pyrylium dye 46					
CH₃CN			494–546	N₂ laser (337)	190
Pyrylium dye 47					
CH₃CN			553–655	N₂ laser (337)	190
Pyrylium dye 48					
CH₃CN			564–633	N₂ laser (337)	190
Pyrylium salt					
dichloromethane			442	N₂ laser (337)	36
Pyrylium salt					
dichloromethane			448	N₂ laser (337)	36
Pyrylium salt					
dichloromethane			468	N₂ laser (337)	36
Pyrylium salt					
dichloromethane			475	N₂ laser (337)	36
Pyrylium salt					
dichloromethane			478	N₂ laser (337)	36
Pyrylium salt					
dichloromethane			483	N₂ laser (337)	36
Pyrylium salt					
dichloromethane			488	N₂ laser (337)	36
Pyrylium salt					
dichloromethane			492	N₂ laser (337)	36

Pyrylium Salt Dye Lasers—*continued*

Solvent	Wavelength max. Abs. (nm)	Fluor. (nm)	Lasing wavelength (nm)	Pump source (nm)	Ref.
Pyrylium salt					
dichloromethane			497	N_2 laser (337)	36
Pyrylium salt					
dichloromethane			501	N_2 laser (337)	36
Pyrylium salt					
dichloromethane			505	N_2 laser (337)	36
Pyrylium salt					
dichloromethane			508	N_2 laser (337)	36
Pyrylium salt					
dichloromethane			512	N_2 laser (337)	36
Pyrylium salt					
dichloromethane			515	N_2 laser (337)	36
Pyrylium salt					
dichloromethane			522	N_2 laser (337)	36
Pyrylium salt					
dichloromethane			524	N_2 laser (337)	36
Pyrylium salt					
dichloromethane			526	N_2 laser (337)	36
Pyrylium salt					
dichloromethane			527	N_2 laser (337)	36
Pyrylium salt					
dichloromethane			532	N_2 laser (337)	36
Pyrylium salt					
dichloromethane			534	N_2 laser (337)	36
Pyrylium salt					
dichloromethane			545	N_2 laser (337)	36
Pyrylium salt					
dichloromethane			550	N_2 laser (337)	36
Pyrylium salt					
dichloromethane			560	N_2 laser (337)	36
Pyrylium salt					
dichloromethane			623	N_2 laser (337)	36
Pyrylium salt					
dichloromethane			627	N_2 laser (337)	36
Pyrylium salt					
dichloromethane			631	N_2 laser (337)	36
Pyrylium salt					
dichloromethane			670	N_2 laser (337)	36
2,6-di(4-CH$_3$-phenyl)-4-(2-chlorophenyl) Pyrylium Perchlorate					
acetonitride	433.0		515 (486–545)	N_2 laser (337)	196

Pyrylium Salt Dye Lasers—*continued*

Solvent	Wavelength max. Abs. (nm)	Fluor. (nm)	Lasing wavelength (nm)	Pump source (nm)	Ref.
2,6-di(4-F-phenyl)-4-(2-chlorophenyl) Pyrylium Perchlorate					
acetonitride	414.3		497 (480–518)	N$_2$ laser (337)	196
2,6-di(4-Cl-phenyl)-4-(2-chlorophenyl) Pyrylium Perchlorate					
acetonitride	421.3		505 (485–529)	N$_2$ laser (337)	196
2,6-di(4-Br-phenyl)-4-(3-chlorophenyl) Pyrylium Perchlorate					
acetonitride	425.6		512 (493–537)	N$_2$ laser (337)	196
2,6-di(4-CH$_3$-phenyl)-4-(3-chlorophenyl) Pyrylium Perchlorate					
acetonitride	433.0		512 (494–550)	N$_2$ laser (337)	196
2,6-di(4-OCH$_3$-phenyl)-4-(3-chlorophenyl) Pyrylium Perchlorate					
acetonitride	477.8		590 (545–640)	N$_2$ laser (337)	196
2,6-di(4-F-phenyl)-4-(3-chlorophenyl) Pyrylium Perchlorate					
acetonitride	415.7		500 (478–530)	N$_2$ laser (337)	196
2,6-di(4-Cl-phenyl)-4-(3-chlorophenyl) Pyrylium Perchlorate					
acetonitride	422.7		506 (485–540)	N$_2$ laser (337)	196
2,6-di(4-H-phenyl)-4-(3-chlorophenyl) Pyrylium Perchlorate					
acetonitride	410.9		495 (478–522)	N$_2$ laser (337)	196
2,6-di(4-H-phenyl)-4-(4-chlorophenyl) Pyrylium Perchlorate					
acetonitride	404.9		495 (477–518)	N$_2$ laser (337)	196
2,6-di(4-CH$_3$-phenyl)-4-(4-chlorophenyl) Pyrylium Perchlorate					
acetonitride	430.0		510 (488–512)	N$_2$ laser (337)	196
2,6-di(4-OCH$_3$-phenyl)-4-(4-chlorophenyl) Pyrylium Perchlorate					
acetonitride	473.1		569 (535–569)	N$_2$ laser (337)	196
2,6-di(4-F-phenyl)-4-(4-chlorophenyl) Pyrylium Perchlorate					
acetonitride	410.3		498 (480–525)	N$_2$ laser (337)	196
2,6-di(4-Cl-phenyl)-4-(4-chlorophenyl) Pyrylium Perchlorate					
acetonitride	417.1		510 (488–535)	N$_2$ laser (337)	196
7-(2-Br-phenyl)-5,6,8,9-tetrahydrodibenzo [c,h] Xanthylium Perchlorate					
acetonitride	450.1		503 (493–550)	N$_2$ laser (337)	196
7-(2-Cl-phenyl)-5,6,8,9-tetrahydrodibenzo [c,h] Xanthylium Perchlorate					
acetonitride	450.1		530 (493–555)	N$_2$ laser (337)	196
7-(2-F-phenyl)-5,6,8,9-tetrahydrodibenzo [c,h] Xanthylium Perchlorate					
acetonitride	452.6		520 (494–550)	N$_2$ laser (337)	196
7-(3-Br-phenyl)-5,6,8,9-tetrahydrodibenzo [c,h] Xanthylium Perchlorate					
acetonitride	448.5		525 (495–550)	N$_2$ laser (337)	196
7-(3-Cl-phenyl)-5,6,8,9-tetrahydrodibenzo [c,h] Xanthylium Perchlorate					
acetonitride	449.3		525 (495–555)	N$_2$ laser (337)	196
7-(3-F-phenyl)-5,6,8,9-tetrahydrodibenzo [c,h] Xanthylium Perchlorate					
acetonitride	448.5		525 (492–552)	N$_2$ laser (337)	196
7-(3-CH$_3$-phenyl)-5,6,8,9-tetrahydrodibenzo [c,h] Xanthylium Perchlorate					
acetonitride	444.53		520 (491–550)	N$_2$ laser (337)	196

Pyrylium Salt Dye Lasers—*continued*

Solvent	Wavelength max. Abs. (nm)	Fluor. (nm)	Lasing wavelength (nm)	Pump source (nm)	Ref.
7-(3-NO$_2$-phenyl)-5,6,8,9-tetrahydrodibenzo [c,h] Xanthylium Perchlorate					
acetonitride	451.8		525 (498–542)	N$_2$ laser (337)	196
7-(4-Br-phenyl)-5,6,8,9-tetrahydrodibenzo [c,h] Xanthylium Perchlorate					
acetonitride	448.4		525 (492–550)	N$_2$ laser (337)	196
7-(4-Cl-phenyl)-5,6,8,9-tetrahydrodibenzo [c,h] Xanthylium Perchlorate					
acetonitride	448.5		525 (493–550)	N$_2$ laser (337)	196
7-(4-F-phenyl)-5,6,8,9-tetrahydrodibenzo [c,h] Xanthylium Perchlorate					
acetonitride	446.9		520 (495–550)	N$_2$ laser (337)	196
7-(4-H-phenyl)-5,6,8,9-tetrahydrodibenzo [c,h] Xanthylium Perchlorate					
acetonitride	446.1		525 (492–553)	N$_2$ laser (337)	196
7-(4-CH$_3$-phenyl)-5,6,8,9-tetrahydrodibenzo [c,h] Xanthylium Perchlorate					
acetonitride	444.5		522 (490–548)	N$_2$ laser (337)	196
7-(4-NO$_2$-phenyl)-5,6,8,9-tetrahydrodibenzo [c,h] Xanthylium Perchlorate					
acetonitride	452.6		522 (500–542)	N$_2$ laser (337)	196
6-(3-chlorophenyl)-diindeno [1,2-b:2',1'-e] Pyrylium Perchlorate					
acetonitride	417.1		500 (480–527)	N$_2$ laser (337)	196
6-(4-chlorophenyl)-diindeno [1,2-b:2',1'-e] Pyrylium Perchlorate					
acetonitride	421.3		500 (481–527)	N$_2$ laser (337)	196
3,11-dimethoxy-7-(Cl-phenyl)-5,6,8,9-tetrahydrodibenzo [c,h] Xanthylium Perchlorate					
acetonitride	501.6		580 (562–640)	N$_2$ laser (337)	196
3,11-dimethoxy-7-(Cl-phenyl)-5,6,8,9-tetrahydrodibenzo [c,h] Xanthylium Perchlorate					
methanol			580 (562–640)	N$_2$ laser (337)	196
3,11-dimethoxy-7-(Cl-phenyl)-5,6,8,9-tetrahydrodibenzo [c,h] Xanthylium Perchlorate					
1,2-dichloroethane	501.6		590 (573–659)	N$_2$ laser (337)	196
3,11-dimethoxy-7-(Cl-phenyl)-5,6,8,9-tetrahydrodibenzo [c,h] Xanthylium Perchlorate					
1,1,2,2-tetrachloroethane			590 (576–655)	N$_2$ laser (337)	196
3,11-dimethoxy-7-(Cl-phenyl)-5,6,8,9-tetrahydrodibenzo [c,h] Xanthylium Perchlorate					
ethylenecyanohidrin			580 (570–646)	N$_2$ laser (337)	196
3,11-dimethoxy-5,6,8,9-tetrahydro dibenzo [c,h] Xanthylium Tetrafluoroborate					
trifluoroethanol	504		548–652	Cu vapor laser	121
5,6,8,9-tetrahydro bis-benzo-2,3-dihydrofuro [6,5-c:5',t'-h] Xanthylium Tetrafluoroborate					
acetonitrile	519		573–682	Cu vapor laser	121
3,11-dihydroxy-5,6,8,9-tetrahydro dibenzo [c,h] Xanthylium Tetrafluoroborate					
e/CH$_3$COOH	509		567–670	Cu vapor laser	121

Thiopyrylium salts

Solvent	Wavelength max. Abs. (nm)	Fluor. (nm)	Lasing wavelength (nm)	Pump source (nm)	Ref.
2,6-di(4-methylphenyl)-4-(3-chlorophenyl)-Thiopyrylium Perchlorate					
acetonitrile	434.0		520 (497–553)	N_2 laser (337)	196
3,11-dimethoxy-7-(chlorophenyl)-5,6,8,9-tetrahydrodibenzo [c,h] Thioxanthylium Perchlorate					
acetonitride	501.0		585 (562–647)	N_2 laser (337)	196
Thiopyrylium salt					
dichloromethane			513	N_2 laser (337)	36
Thiopyrylium salt					
dichloromethane			601	N_2 laser (337)	36
Thiopyrylium salt					
dichloromethane			612	N_2 laser (337)	36
Thiopyrylium salt					
dichloromethane			618	N_2 laser (337)	36
Dye 5, Q-switch					
—			1180–1530	Nd:YAG(1064)	183-4
DCE	1090	1170	1320 (1180–1400)	Nd:glass laser	176,181
Dye 26, IR-26					
DCE	1080	1140	1190 (1150–1240)	Nd:YAG laser	117,182
BzOH	1090	1170	1270 (1200–1320)	Nd:YAG laser	179
Dye 9860, Q-switch II					
DCE	1070	1110	1140	Nd:YAG(1.06)	176,180
Dye 301					
DCE	1270		1600	Nd glass (1054)	178
Dye 401					
DCE	1070	1110	1550	Nd glass (1054)	178
Dye 501					
DCE	1070	1110	1800	Nd glass (1054)	178

2.1.3.19 Quinolone Dye Lasers

Solvent	Wavelength maximum Abs. (nm)	Fluor. (nm)	Lasing wavelength (nm)	Pump source (nm)	Ref.
7DMA-1M-4MO-Quinolone					
water		396	409	flashlamp	191
7A-1,4M-DM-Quinolone					
water		413	409	flashlamp	191
7A-1M-Quinolone					
water		417	412	flashlamp	191
7A-4M-Quinolone					
water		417	412	flashlamp	191
7-amino-Quinolone					
water		425	418	flashlamp	191
7DMA-1,4DM-Quinolone					
ethanol		415	430	flashlamp	191

Quinolone Dye Lasers—*continued*

Solvent	Wavelength maximum Abs. (nm)	Fluor. (nm)	Lasing wavelength (nm)	Pump source (nm)	Ref.
7A-4,8DM-Quinolone					
water	429	420		flashlamp	191
7DEA-4M-Quinolone					
water		425		flashlamp	191
7DMA-4M-Quinolone					
water	444	426		flashlamp	191
7OH-4M-Quinolone					
Na_2CO_3/H_2O	429	441		flashlamp	191
7OH-3,4-DM-Quinolone					
Na_2CO_3/H_2O	437	447		flashlamp	191
7DMA-1M-Quinolone					
water	445	—		flashlamp	191
Q1F-Quinolone					
water		463		flashlamp	191
Q6F-Quinolone					
water		473		flashlamp	191
Q3F-Quinolone					
water		477		flashlamp	191

Azaquinolones

Solvent	Wavelength maximum Abs. (nm)	Fluor. (nm)	Lasing wavelength (nm)	Pump source (nm)	Ref.
7A-4M-Azaquinolone					
ethanol		386		flashlamp	43,194
7DMA-1M-4MO-Azaquinolone					
ethanol	383	390		flashlamp	43,194
7OH-4M-Azaquinolone					
ethanol		395		flashlamp	43,194
7OH-3,4DM-Azaquinolone					
ethanol		405		flashlamp	43,194
7DMA-1,4DM-Azaquinolone					
ethanol	400	407		flashlamp	43,194
7A-4TFM-Azaquinolone					
ethanol		437		flashlamp	43,194
AQ1F-Azaquinolone					
ethanol	430	452		flashlamp	43,194

2.1.3.20 Quinolinone Dye Lasers

Solvent	Wavelength maximum Abs. (nm)	Fluor. (nm)	Lasing wavelength (nm)	Pump source (nm)	Ref.
Quinoxalinone-TNH$_2$					
ethanol	—	—	489 (460–540)	N_2 laser (337)	71
Quinoxalinone-MeTNH$_2$					
ethanol	—	—	502 (474–556)	N_2 laser (337)	71

Quinolinone Dye Lasers—*continued*

Solvent	Wavelength maximum Abs. (nm)	Fluor. (nm)	Lasing wavelength (nm)	Pump source (nm)	Ref.
Quinoxalinone-TNMe₂					
ethanol	—	—	510 (478–570)	N₂ laser (337)	71
Quinoxalinone-MeTNH₂					
ethanol	—	—	511 (483–556)	N₂ laser (337)	71

2.1.3.21 Styryl Type Compound Dye Lasers

Solvent	Wavelength max. Abs. (nm)	Fluor. (nm)	Lasing wavelength (nm)	Pump source (nm)	Ref.
DCM; 4-(dicyanomethylene)-2-methyl-6-(*p*-dimethylaminostyryl) -4H-pyran					
methanol	482		640 (610–680)	Nd:YAG (532)	53
BzOH/EG			640 (605–680)	Ar laser (514)	90
DMF			649	flashlamp	86
DMSO			649 (615–688)	Cu (511,578)	85
DMSO			654 (601–716)	N₂ laser (337)	86
DMSO			655	flashlamp	86
DCM-OH; 4-dicyanomethylene-2-methyl-6-p-diethanolamino-styryl 4H-pyran					
methanol			600–700	flashlamp	127
methanol/H₂O			610–695	flashlamp	127
ASPI; trans-4-[p-(N-hydroxyethyl-N-methylamino)styryl]-N-methylpyridinium iodide					
ethanol			599–635	Nd:YAG(532)	33

Stilbenes

Solvent	Wavelength max. Abs. (nm)	Fluor. (nm)	Lasing wavelength (nm)	Pump source (nm)	Ref.
Stilbene 1					
ethylene glycol		347	415	Kr laser (UV)	39
Diphenystilbene, DPS					
p-dioxane	340	408	406 (396–416)	N₂ laser (337)	20
DMF			409	flashlamp	8
ethanol, w, t			400–420	N₂ laser (337)	27
Blankophor R					
methanol		340	416	N₂ laser (337)	40
Delft Weiss BSW					
methanol		351	434	N₂ laser (337)	40

Styrylbenzenes

Solvent	Wavelength max. Abs. (nm)	Fluor. (nm)	Lasing wavelength (nm)	Pump source (nm)	Ref.
Bis-dicyanostyryl-2-chlorobenzene					
p-dioxane, THF			414 (408–423)	N$_2$ laser (337)	36
1,4-distyrylbenzene					
toluene			419 (410–439)	N$_2$ laser (337)	34
Bis-sulfostyryl-2-chlorobenzene					
methanol			420 (413–431)	N$_2$ laser (337)	34
Bis-sulfostyrylbenzene					
methanol			423 (413–431)	N$_2$ laser (337)	34
Styryl 14					
PC/EG			980 (928–1084)	Ar laser	173
LDS 698/Pyridine 1					
methanol			684 (661–724)	Cu (511,578)	161
methanol			695 (658–738)	Nd:YAG(532)	26
DMSO			710 (670–760)	XeCl (308)	163
DMSO			718 (675–750)	N$_2$ (337)	165
PC/EG:15/85			726 (688–808)	Ar (458–514)	164
LDS 722/Pyridine 2					
methanol			722 (685–760)	Nd:YAG(532)	164
methanol			722 (687–755)	Cu (511,578)	161
DMSO			735 (700–780)	N$_2$ laser (337)	166
PC/EG			747 (682–810)	Ar (458-514)	167
LDS 730/Styryl 6					
methanol			716 (692–743)	Nd:YAG(532)	164
LDS 750/Styryl 7					
methanol			722 (698–743)	Nd:YAG(532)	164
LDS 751/Styryl 8					
methanol			750 (714–790)	Nd:YAG(532)	164
PC/EG			765 (715–840)	Ar (458–514)	164
LDS 765					
methanol			764 (738–800)	Nd:YAG(532)	164
methanol			767 (743–787)	Cu (511,578)	161
LDS 798/Styryl 11					
PC/EG:15/85			795 (768–850)	Ar laser	168
methanol			798 (765–845)	Nd:YAG(532)	164
LDS 821/Styryl 9 rigidized**					
methanol			815 (793–845)	Cu (511,578)	161
methanol			818 (785–850)	XeCl laser (308)	169
methanol			818 (785–851)	Nd:YAG(532)	164
PC			821 (802–852)	N$_2$ laser (337)	170
methanol			834 (817–842)	flashlamp	26
PC/EG			845 (780–960)	Ar (458–514)	164
PC/EG			880 (793–923)	Kr laser (647)	171-2

Styrylbenzenes—*continued*

Solvent	Wavelength max. Abs. (nm)	Fluor. (nm)	Lasing wavelength (nm)	Pump source (nm)	Ref.
LDS 867					
methanol			862 (851–890)	Cu(511,578)	161
methanol			866 (830–910)	Nd:YAG(532)	164
DMSO			946 (922–963)	flashlamp	26
LDS 925/Styryl 13					
DMSO			960 (902–1023)	XeCl laser (308)	174
PC/EG			925 (875–1050)	Ar laser	173
PC/EG			975 (930–1040)	Nd:YAG(532)	175

Styrylbiphenyls

Solvent	Wavelength max. Abs. (nm)	Fluor. (nm)	Lasing wavelength (nm)	Pump source (nm)	Ref.
Bis-dicyanostyrylbiphenyl					
p-dioxane, THF			413 (408–422)	N₂ laser (337)	34
Bis-dichlorestyrylbiphenyl					
p-dioxane, THF			425 (416–437)	N₂ laser (337)	34
Bis-sulfostyrylbiphenyl; Stilbene 420; Stilbene 3					
ethylene glycol		349	420–470	Ar laser (UV)	49
methanol			424 (411–436)	Nd:YAG(355)	51
—			425 (400–480)	Kr laser (UV)	167
methanol			425 (408–453)	N₂ laser (337)	34
methanol			429 (404–460)	Nd:YAG(355)	53
H₂O+NP-10			431 (415–458)	N₂ laser (337)	34
EG,methanol:9/1			432 (406–448)	Ar laser (UV)	162
H₂O			445 (421–468)	N₂ laser (337)	34

Triazinylstilbenes

Solvent	Wavelength max. Abs. (nm)	Fluor. (nm)	Lasing wavelength (nm)	Pump source (nm)	Ref.
4,4'-di(4-anilino-6-methoxytriazinyl)-2,2'-stilbene disulfonic acid					
methanol			425 (418–443)	N₂ laser (337)	52
NH₃-Uvitex CF					
methanol		348	429 (406–465)	N₂ laser (337)	40
methanol		351	430 (412–462)	N₂ laser (337)	40
4,4'-di(4-anilino-6-diethanolaminoxytriazinyl)-2,2'-stilbene disulfonic acid					
methanol			430 (420–438)	N₂ laser (337)	52
4,4'-di(4-sulfoanilino-6-*N*-methyl-*N*-ethanolaminoxytriazinyl)-2,2'-stilbene disulfonic acid					
methanol			430 (420–445)	N₂ laser (337)	52
4,4'-di(4-sulfoanilino-6-*N*-ethyltriazinyl)-2,2'-stilbene disulfonic acid					
methanol			431 (419–448)	N₂ laser (337)	52

Triazinylstilbenes—*continued*

Solvent	Wavelength max. Abs. (nm)	Fluor. (nm)	Lasing wavelength (nm)	Pump source (nm)	Ref.
4,4'-di(4-sulfoanilino-6-di-iso-propanolaminotriazinyl)-2,2'-stilbene disulfonic acid					
methanol			433 (420–447)	N$_2$ laser (337)	52
4,4'-di(4-sulfoanilino-6-diethanolaminotriazinyl)-2,2'-stilbene disulfonic acid-sodium salt					
methanol			433 (418–461)	N$_2$ laser (337)	52
H$_2$O, 1.5%NP10			447 (423–461)	N$_2$ laser (337)	52
4,4'-di(4-sulfoanilino-6-morpholinotriazinyl)-2,2'-stilbene disulfonic acid					
methanol			433 (418–448)	N$_2$ laser (337)	52
4,4'-di(4-sulfoanilino-6-*N*-methyl-*N*-ethanolaminoxytriazinyl)-2,2'-stilbene disulfonic acid					
methanol			433 (418–449)	N$_2$ laser (337)	52
Leukophor B					
methanol	277, 351		436 (410–462)	N$_2$ laser (337)	40

Triazolstilbenes

Solvent	Wavelength max. Abs. (nm)	Fluor. (nm)	Lasing wavelength (nm)	Pump source (nm)	Ref.
4,4'-di(4-phenyl-1,2,3-triazol-2-yl)-2,2'-stilbene disulfonic acid-potassium salt					
methanol	—	—	422 (412–432)	N$_2$ laser (337)	40
4,4'-di(4-phenyl-1,2,3-triazol-2-yl)-2,2'-stilbene disulfonic acid-potassium salt					
methanol	—	—	425 (414–438)	N$_2$ laser (337)	52
H$_2$O, 2.5%NP10	—	—	436 (417–455)	N$_2$ laser (337)	52
Heleofor BDC					
methanol	—	—	432(407–460)	N$_2$ laser (337)	40
Leukophor DC					
methanol		374	464 (447–510)	N$_2$ laser (337)	40
Tinopal GS					
methanol			440 (424–475)	N$_2$ laser (337)	40
Tinopal RBS					
MeOH/p–dioxane		360	434 (414–472)	N$_2$ laser (337)	40
Tinopal PCRP					
p-dioxane		371	427 (414–451)	N$_2$ laser (337)	40
benzene		371	430	N$_2$ laser (337)	40
Unitex NSI					
methanol	—	—	432 (412–464)	N$_2$ laser (337)	40
Unitex RS					
methanol	—	—	422 (412–432)	N$_2$ laser (337)	40
Unitex SFC					
methanol	—	—	427 (410–459)	N$_2$ laser (337)	40

2.1.3.22 Xanthene Dye Lasers

Solvent	Wavelength max. Abs. (nm)	Fluor. (nm)	Lasing wavelength (nm)	Pump source (nm)	Ref.
Fluorescein					
ethanol, w, t			520–600	N$_2$ laser (337)	27
Disodium Fluorescein; 3,6'-dihydroxy-spiro[isobenzofuran-1(3H),9'-[9H]xanthen]-3-one disodium salt					
EG, COT	501	531	522 (537–580)	Ar laser	47
ethylene glycol			552 (538–573)	Ar (458,514)	164
ethanol		527	yellow	ruby laser (347)	197
water		527	yellow	ruby laser (347)	197
ethanol			550	flashlamp	198
Kiton Red 620;Sulforhodamine B; N-[6-diethylamino-0-9-(2,4-disulfopheny)-3H-xanthen-3-ylidene]-N-ethyl-ethanaminium hydroxide, inner salt					
methanol	554	575	583 (570–604)	Nd:YAG(532)	53
trifluoroethanol			617 (595–639)	Cu (511,578)	85
methanol			620	flashlamp	88
ethanol			620 (580–630)	flashlamp	87
methanol+COT			621 (608–634)	flashlamp	26
Kiton Red 620;Sulforhodamine B; N-[6-diethylamino-0-9-(2,4-disulfopheny)-3H-xanthen-3-ylidene]-N-ethyl-ethanaminium hydroxide, inner salt					
ethanol+COT			623 (598–649)	flashlamp	25
methanol+COT			627 (595–629)	flashlamp	25
trifluoroethanol			628 (603–647)	N$_2$ laser (337)	86
methanol			631 (600–660)	flashlamp	59,87
ethylene glycol			636 (603–670)	flashlamp	16
DMSO			637	flashlamp	86
ethylene glycol			638 (610–670)	Ar laser (cw)	47
4%LO/H$_2$O			642 (622–665)	flashlamp	19
Rhodamine 640 Perchlorate: Rhodamine 101; 9-(2-carboxyphenyl)-2,3,6,7,12,13,16,17-octahydro-1H,5H,11H,15H-xanthen[2,3,4-i'j']diquinolizin-4-ium, perchlorate					
—		575	602 (589–623)	Nd:YAG(532)	83
methanol			602 (592–624)	Nd:YAG(532)	53
DMSO+HCl			671 (634–704)	N$_2$ laser (337)	86
methanol			611	Nd:YAG(532)	65
—			612 (598–640)	Nd:YAG(532)	84
methanol			613 (602–657)	Nd:YAG(532)	51
ethanol			630	flashlamp	80
ethanol			640	flashlamp	5
ethanol			640 (620–680)	N$_2$ laser (337)	92
methanol			642 (627–657)	flashlamp	26
ethanol			643 (623–657)	flashlamp	25
ethanol			644 (620–673)	N$_2$ laser (337)	93

Xanthene Dye Lasers—*continued*

Solvent	Wavelength max. Abs. (nm)	Fluor. (nm)	Lasing wavelength (nm)	Pump source (nm)	Ref.
Rhodamine 640 Perchlorate: Rhodamine 101; 9-(2-carboxyphenyl)-2,3,6,7,12,13,16,17-octahydro-1H,5H,11H,15H-xanthen[2,3,4-i'j']diquinolizin-4-ium, perchlorate					
ethylene glycol			645 (620–690)	Ar (458, 514)	164
—			648 (608–710)	Kr laser (568)	52
methanol			650	flashlamp	62
MeOH/H$_2$O:1/1			652 (620–687)	flashlamp	19
ethylene glycol			552 (536–602)	Ar (cw)	47
Rhodamine 6G; Rhodamine 590 as Chloride, Perchlorate, or Tetrafluoroborate					
methanol	530	560	550	Nd:YAG(532)	65
methanol			560 (548–580)	Nd:YAG(532)	53
methanol			562 (546–592)	Nd:YAG(532)	83
methanol			563 (550–590)	Nd:YAG(532)	84
methanol			564	Nd:YAG(532)	51
ethanol, w, t			565–620	N$_2$ laser (337)	27
ethanol			572 (564–600)	Cu (511,578)	85
ethanol			576 (555–618)	N$_2$ laser (337)	86
methanol			578 (565–612)	flashlamp	25
ethanol			579 (568–605)	N$_2$ laser (337)	20
ethanol			585	flashlamp	198
ethanol			580	KrF laser (248)	3
ethanol			584 (570–618)	flashlamp	25
methanol			586 (563–625)	flashlamp	59
ethanol			588–609 (MDR)	Ar laser (514)	195
methanol			590	flashlamp	62
p-dioxane			590	KrF laser (248)	36
ethylene glycol			590 (570–650)	Ar (458, 514)	164
MeOH/H$_2$O:1/3			596 (577–614)	flashlamp	26
MeOH/H$_2$O:1/1			598 (577–625)	flashlamp	19
Rhodamine 6G; Rhodamine 590 as Chloride, Perchlorate, or Tetrafluoroborate					
4%LO/H$_2$O			600	flashlamp	62
ethylene glycol			600 (567–657)	Ar laser (cw)	47
—			602 (560–654)	Kr (blue/green)	53
4%LO/H$_2$O			610 (585–633)	flashlamp	19
Rhodamine 560 Chloride, Rhodamine 110; 2-(6-amino-3-imino-3H-xanthen-9-yl)benzoic acid, monohydrochloride					
ethanol			560	flashlamp	80
methanol			563 (541–583)	flashlamp	26
methanol			565 (544–589)	flashlamp	25
ethanol			567 (546–587)	flashlamp	25
—			569 (533–600)	Kr (blue/green)	52

Xanthene Dye Lasers—*continued*

Solvent	Wavelength max. Abs. (nm)	Fluor. (nm)	Lasing wavelength (nm)	Pump source (nm)	Ref.
Rhodamine 560 Chloride, Rhodamine 110; 2-(6-amino-3-imino-3H-xanthen-9-yl)benzoic acid, monohydrochloride					
ethanol			570	flashlamp	80
ethylene glycol			570 (540–600)	Ar (458,514)	164
Rhodamine 575					
ethanol	518		575	flashlamp	80
ethanol	528		577 (563–602)	flashlamp	25
ethanol	518		585	flashlamp	80
MeOH/H$_2$O			590 (566–610)	flashlamp	19
Rhodamine B; Rhodamine 610 as Chloride or Perchlorate					
—	552	588	579 (570–596)	Nd:YAG(532)	83
—	554	580	586 (570–606)	Nd:YAG(532)	53
methanol			587 (579–601)	Nd:YAG(532)	65
—			590 (578–610)	Nd:YAG(532)	84
trifluoroethanol			591 (582–618)	Cu (511,578)	85
methanol			592 (578–629)	Nd:YAG(532)	51
—			609 (594–643)	N$_2$ laser (337)	20
methanol			613 (596–645)	flashlamp	25
ethanol			617 (598–647)	flashlamp	87
ethanol			620 (596–647)	flashlamp	25
ethylene glycol			630 (601–675)	Ar (458,514)	164
ethylene glycol			637 (608–682)	Ar (laser (cw)	47
Sulforhodamine 640; Sulforhodamine 101; 2',3',6',7',12',13',16',17'octahydrospirol[3H-2,1-benzoxathiole-3,9'-[1H,5H,9H,11H]xantheno[2,3,4-ij,5,6,7-i',j']diquinolizine]-6-sulfonic acid, 1,1-dioxide, sodium salt or free acid					
ethanol			590–640	Nd:YAG(532)	29
MeOH/H$_2$O:1/1			656	cw	71
MeOH/H$_2$O:1/1	576	602	662 (648–682)	flashlamp	26
ethylene glycol			668 (646–680)	Ar laser	90
LD 700 Perchlorate					
alcohol	647		690	Nd:YAG→585	96
LD 700 Perchlorate					
ethanol			706 (692–752)	N$_2$ laser (337)	93
ethanol			720 (698–758)	N$_2$ laser (337)	93
ethylene glycol			737 (700–810)	Kr laser	13
ethylene glycol			740 (700–820)	Kr (647,676)	13
Xanthene 10-1-1; SNH-8					
H$_2$O/MeOH:1/2	509.5		559 (548–588)	Nd:YAG(532)	147
Xanthene 10-1-2; SNH-8NH$_4$					
H$_2$O/MeOH:1/2	508		551 (545–559)	Nd:YAG(532)	147

Xanthene Dye Lasers—*continued*

Solvent	Wavelength max. Abs. (nm)	Fluor. (nm)	Lasing wavelength (nm)	Pump source (nm)	Ref.
Xanthene 10-1-3; SNH-19A					
H₂O/MeOH:1/2	523		560 (551–570)	Nd:YAG(532)	147
Xanthene 10-1-4; SNH-20A					
H₂O/MeOH:1/2	533		568 (559–580)	Nd:YAG(532)	70
Xanthene 10-1-5; SNH-25					
H₂O/MeOH:1/2	529		574	Nd:YAG(532)	147
Xanthene 10-1-6; CC-20					
H₂O/MeOH:1/2	538		563–580	Nd:YAG(532)	147

2.1.3.23 Other Dye Lasers

Solvent	Wavelength max. Abs. (nm)	Fluor. (nm)	Lasing wavelength (nm)	Pump source (nm)	Ref.
Acridine					
ethanol		437	438.5	ruby laser (347)	197
Acridine Red					
ethanol		580	orange	ruby laser (347)	197
ethanol		580	601.5	flashlamp	198
Carbazine 720; Carbazine 122; 7-hydroxy-2',3',5',6'-tetramethyl-spiro[acridine-9(2H),1'-[2,5]cyclohexadiene]-2,4'-dione, ion					
DMSO			740	flashlamp	73,100
EG (ethanol)			747 (687–811)	Kr (647,676)	102
Eosin					
ethanol		540	yellow	ruby laser (347)	197
Laser dye 13-1-1					
DMSO	360		416 (416–440)	N₂ laser (337)	41
Laser dye 13-1-2					
DMSO	380		447 (435–465)	N₂ laser (337)	41
MEH-PPV; poly 2-methoxy-5-2'-ethylhexloxy - poly(1,4-phenylene vinylene)					
xyl./chloroform			~598	Nd:YAG(532)	126
Salicylamide					
DMF	305	440	439	XeCl laser (308)	57
Sodium salicylate					
DMF, t, w	300	395	411	XeCl laser (308)	33

2.1.4 Commercial Dye Lasers

Table 2.1.2
Commercial Dye Lasers

Wavelength (μm)	Pump source	Output
CW Lasers:		
0.22–0.39 (SH)	Ar ion laser	0.01 W
0.38–1.0	Ar ion laser	0.1–2 W
Pulsed Lasers:		
0.2–0.4 (SH)	Nd:YAG, excimer lasers	1–60 mJ
0.25–0.4 (SH)	coaxial flashlamp	0.1–0.9 J
0.3–0.9	linear flashlamp	0.5–3 J
0.32–1.0	excimer laser	10–150 mJ
0.36–0.95	nitrogen laser	0.1–150 mJ
0.4–1.0	Nd:YAG laser	5–200 mJ
0.44–0.8	coaxial flashlamp	< 1–30 J
0.53–0.9	Cu vapor laser	0.1– 2 mJ
0.695–0.905	Ti:sapphire laser	≤0.15 J
0.9–4.5 (R)	Nd:YAG laser	1–10 mJ

R – Raman shifted, SH – second harmonic.

2.1.5 Dye Laser Tuning Curves

Lasing of organic dyes is dependent on the solvent, dye concentration, pumping source and rate, and other operating conditions. Relative energy outputs and tuning curves that may be obtained from commercially available pump sources and dyes are shown in Figures 2.1.2 – 2.1.13 (figures courtesy of Richard N. Steppel). The information is provided only as a guide and may not necessarily be extrapolated to systems other than those cited.

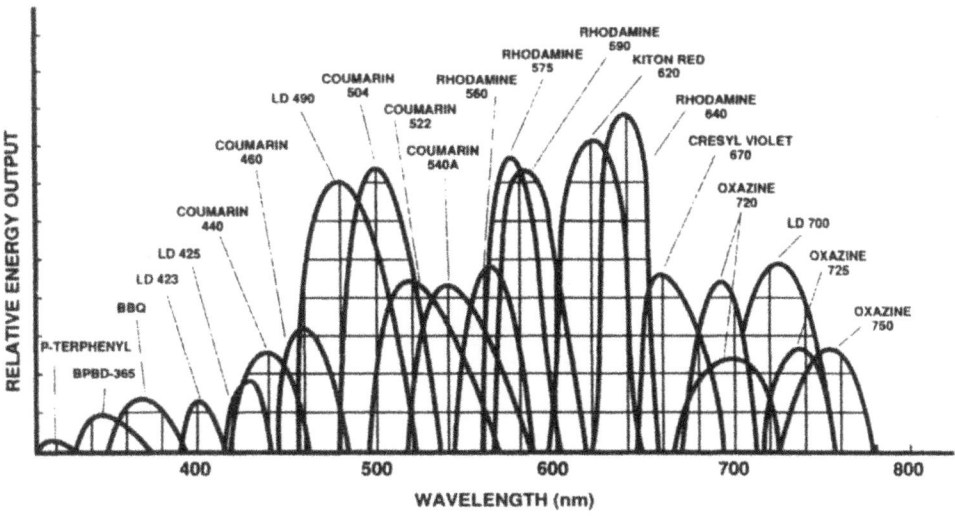

FIGURE 2.1.2   Tuning curves and relative energy outputs of various coaxial flashlamp pumped dyes. Data courtesy of Phase-R Corp., Box G-2, Old Bay Road, New Durham, NH.

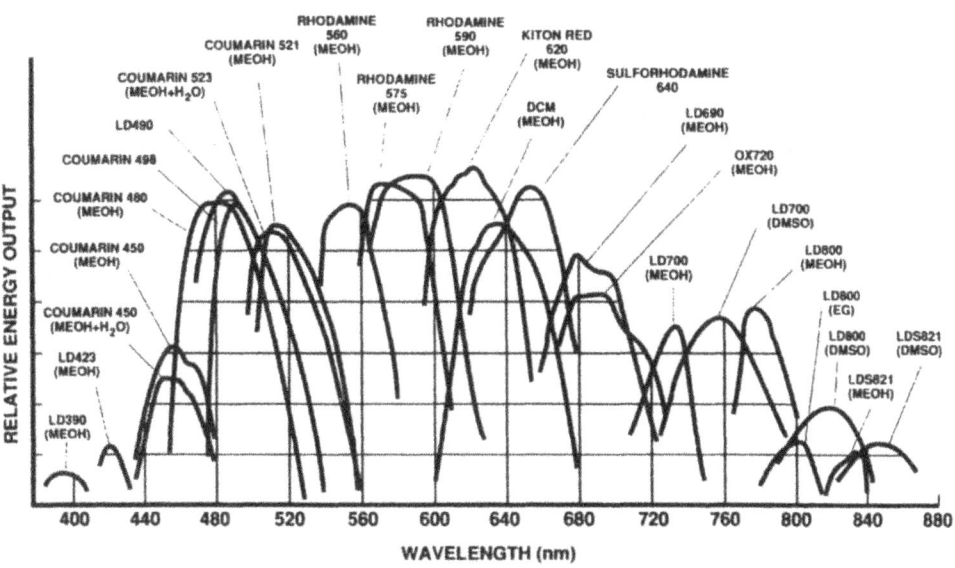

FIGURE 2.1.3   Tuning curves and relative energy outputs of various coaxial flashlamp pumped dyes. Data courtesy of Candela Laser Corp., 530 Boston Post Road, Wayland, MA.

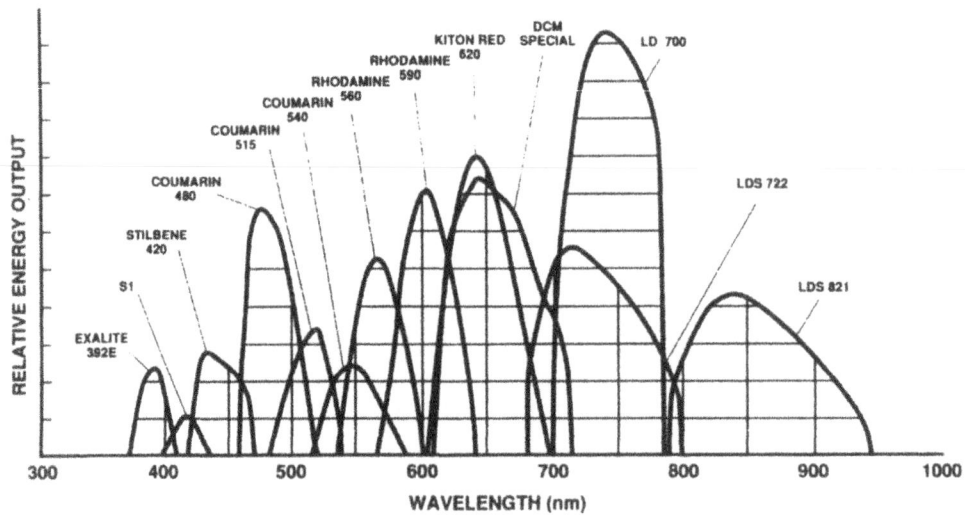

FIGURE 2.1.4 Tuning curves and relative energy outputs of various argon-ion and krypton-ion laser pumped dyes. Data courtesy of Coherent Inc., 3210 Porter Drive, Palo Alto, CA.

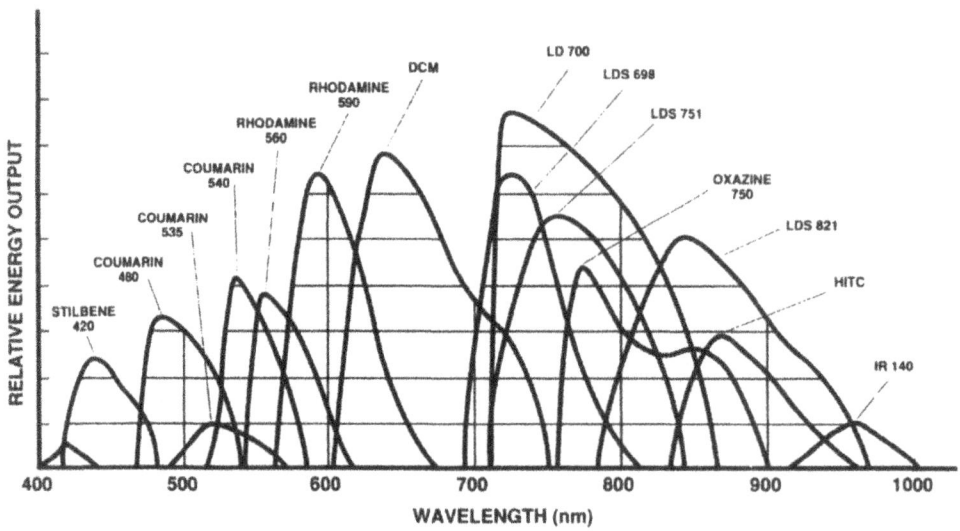

FIGURE 2.1.5 Tuning curves and relative energy outputs of various argon-ion and krypton-ion laser pumped dyes. Data Courtesy of Spectra-Physics Inc., 1250 Middlefield Road, Mountain View, CA.

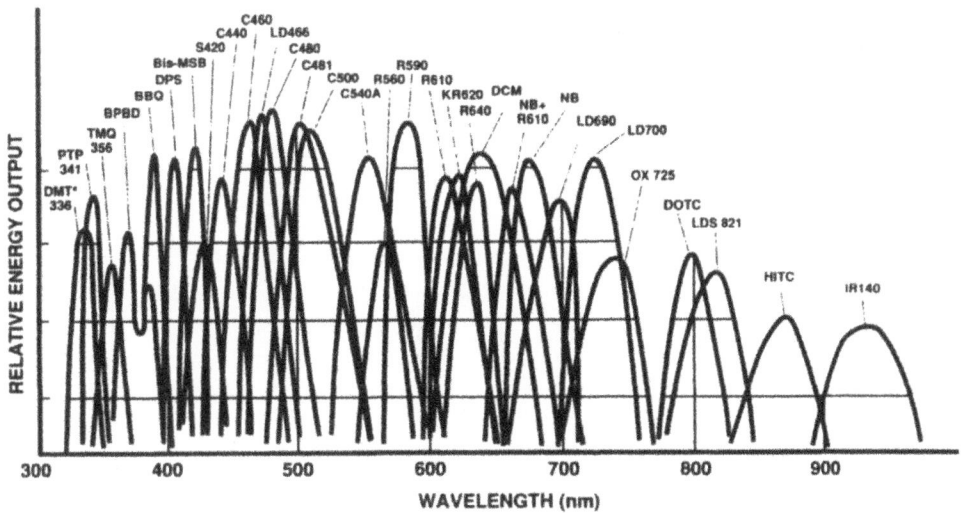

FIGURE 2.1.6   Tuning curves and relative energy outputs of various krypton fluoride and xenon chloride laser pumped dyes. Data courtesy of Lumonics, Inc., 105 Schneider Road, Kanata (Ottawa), Ontario, Canada.

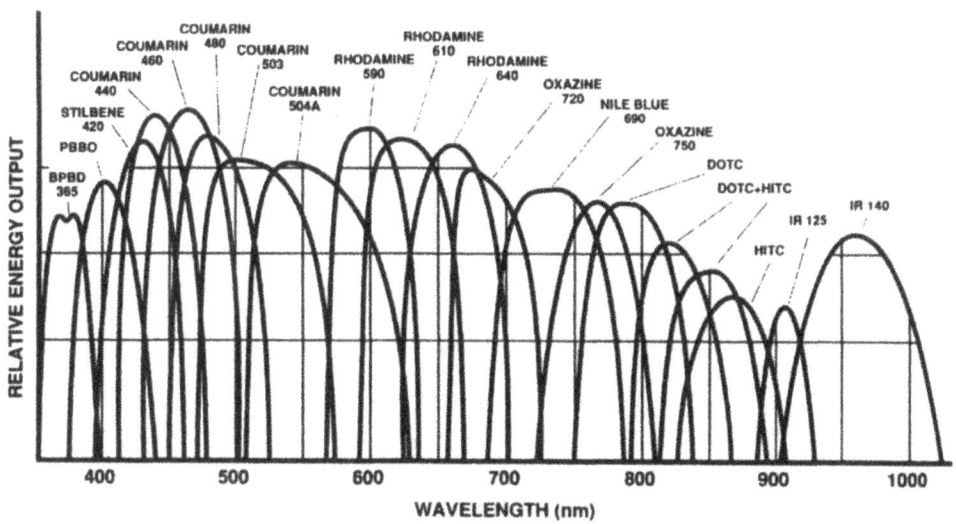

FIGURE 2.1.7   Tuning curves and relative energy outputs of various nitrogen laser pumped dyes. Data courtesy of Jobin Yvon, 16-18. rue du Canal B. P. 118, 91163 Longjumeau Cedex, France.

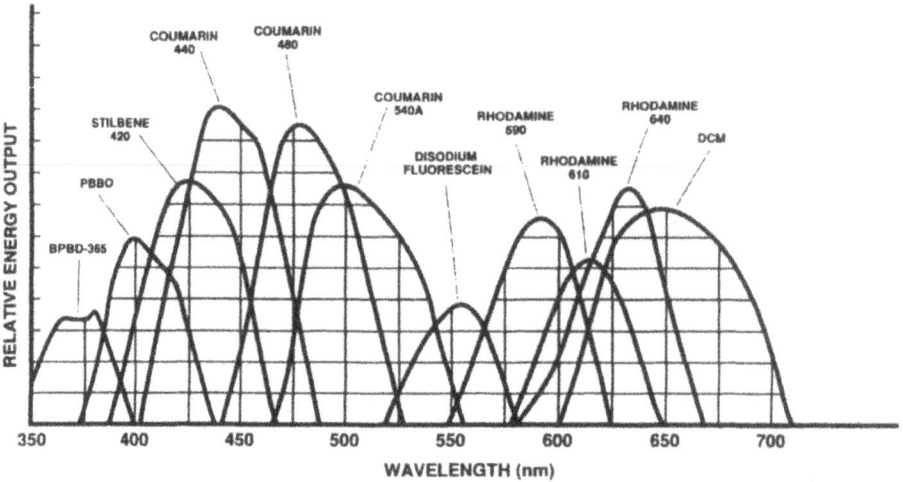

FIGURE 2.1.8 Tuning curves and relative energy outputs of various nitrogen laser pumped dyes. Data courtesy of Laser Science, Inc., 26 Landsowne Street, Cambridge, MA.

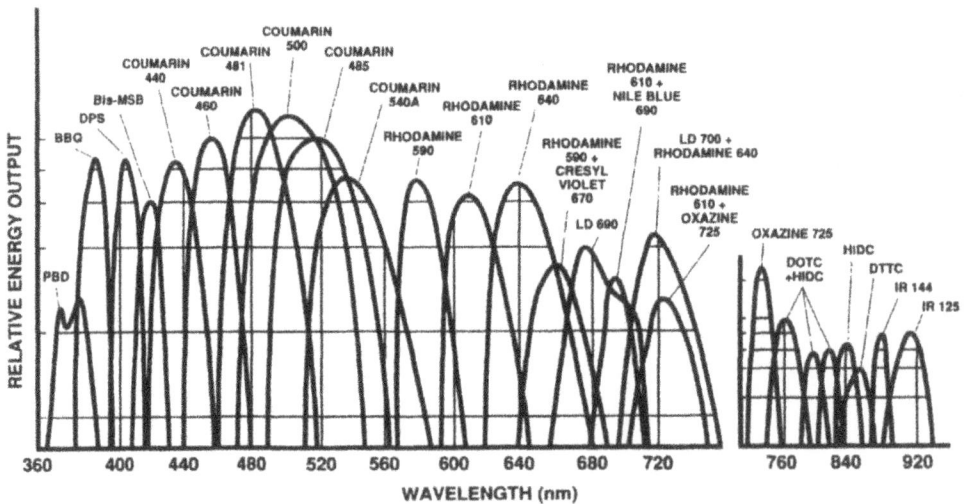

FIGURE 2.1.9 Tuning curves and relative energy outputs of various nitrogen laser pumped dyes. Data courtesy of Laser Photonics, Inc., 12351 Research Parkway, Orlando, FL.

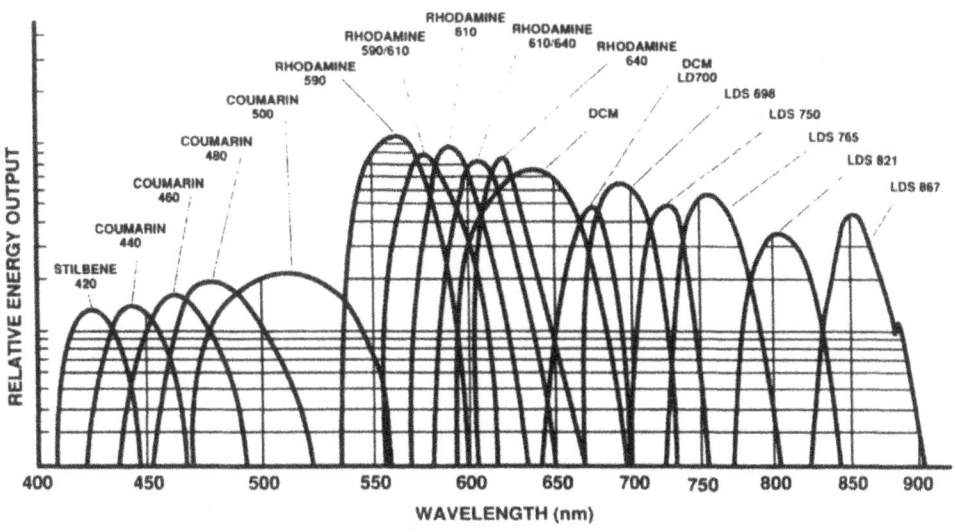

FIGURE 2.1.10   Tuning curves and relative energy outputs of various Nd:YAG laser pumped dyes. Data courtesy of Continuum, 3150 Central Expressway, Santa Clara, CA.

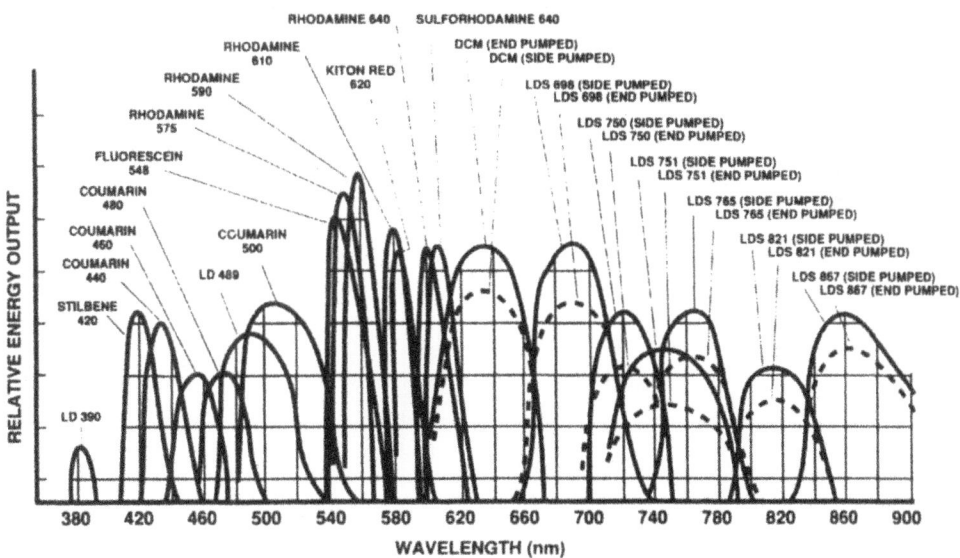

FIGURE 2.1.11   Tuning curves and relative energy outputs of various Nd:YAG laser pumped dyes. Data courtesy of Spectra-Physics/Quanta-Ray, 1250 Middlefield Road, Mountain View, CA.

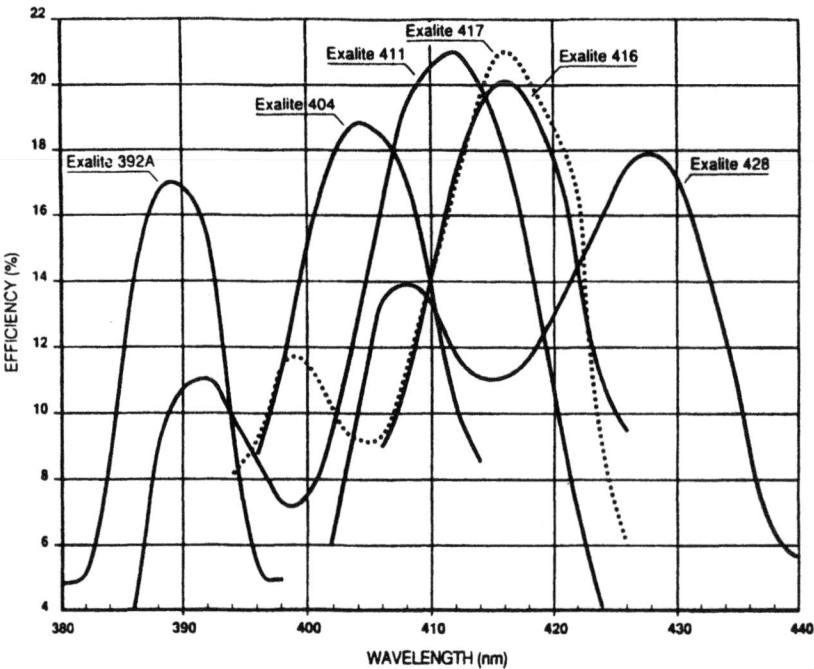

FIGURE 2.1.12   Tuning curves and efficiency of Exalite laser dyes (Exciton, Inc.) for Nd:YAG pumping at 355 nm. Data courtesy of Spectra-Physics/Quanta-Ray, 1250 Middlefield Road, Mountain View, CA.

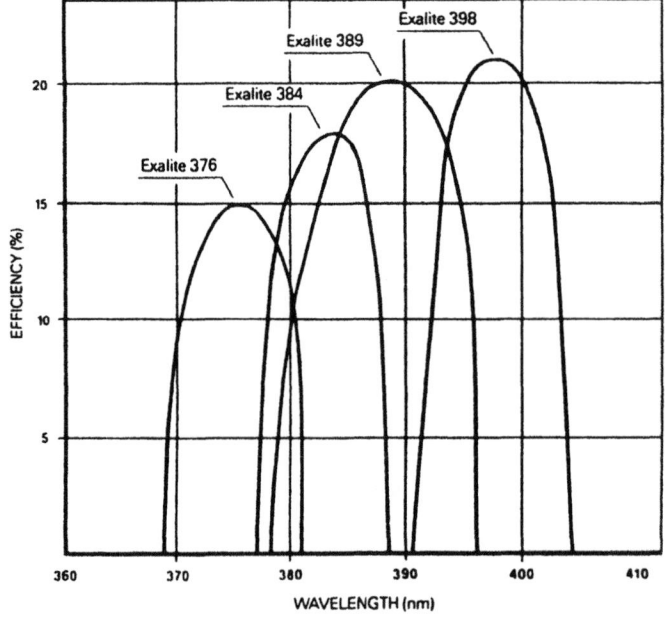

FIGURE 2.1.13   Tuning curves and efficiency of Exalite laser dyes (Exciton, Inc.) for Nd:YAG pumping at 355 nm. Data courtesy of Lumonics, Inc., 105 Schneider Road, Kanata (Ottawa), Ontario, Canada.

2.1.6 References

1. Zapka, W. and Brickmann, U., Shorter dye laser wavelengths from substituted p-terphenyl, *Appl. Phys.* 20, 283 (1979).
2. Gusten. H. and Rinke, M., Photophysical properties and laser performance of photostable uv laser dyes III. Sterically hindered p-quarterphenyls, *Appl. Phys.* B 45, 279 (1988).
3. Tomin, V. I., Alcock, A. J., Sarjeant, W. J., and Leopold, K. E., Some characteristics of efficient dye laser emission obtained by pumping at 248 nm with a high-power KrF* discharge laser, *Opt. Commun.* 26, 396 (1978).
4. Tomin, V. I., Alcock, A. J., Sarjeant, W. J., and Leopold, K. E., Tunable, narrow bandwidth, 2 MW dye laser pumped by a KrF* discharge laser. *Opt. Commun.* 28, 336 (1979).
5. McKee, T. J., Stoicheff, B. P., and Wallace, S. C., Tunable, coherent radiation in the Lyman-region (1210-1290 Å) using magnesium vapor, *Opt. Lett.* 3, 207 (1978).
6. McKee, T. J. and James, D. J., Characterization of dye laser pumping using a high-power KrF excimer laser at 248 nm, *Can. J. Phys.* (Sept. 1979).
7. Godard, B. and de Witte, O., Efficient laser emission in para-terphenyl tunable between 323 and 364 nm, *Opt. Commun.* 19, 325 (1976).
8. Furumoto, H. W. and Ceccon, H. L., Ultraviolet organic liquid lasers, *IEEE J. Quantum Electron.* QE-6, 262 (1970).
9. Rulliere, C., Morand, J. P., and de Witte, O., KrF laser pumps new dyes in the 3500 Å spectral range, *Opt. Commun.* 20, 339 (1977).
10. Ziegler, L. D. and Hudson, B. S., Tuning ranges of 266 nm pumped dyes in the near uv, *Opt. Commun.* 32, 119 (1980).
11. Dunning, F. B. and Stebbings, R. F., The efficient generation of tunable near UV radiation using an N_2 pumped dye laser, *Opt. Commun.* 11, 112 (1974).
12. Ducasse, L., Rayez, J. C., and Rulliere. C., Substitution effects enhancing the lasing ability of organic compounds, *Chem. Phys. Lett.* 57, 547 (1978).
13. Profitt, W., Coherent, Inc. (private communication, Steppel, R. N. 1980).
14. Marowsky, G., Cordray, R., Tittel, F. K., Wilson, W. L., and Collins, C. B., Intense laser emission from electron-beam-pumped ternary mixtures of Ar, N_2, and POPOP vapor, *Appl. Phys. Lett.* 33, 59 (1978).
15. Rulliere, C. and Joussat-Dubien, J., Dye laser action at 330 nm using benzoxazole: a new class of lasing dyes, *Opt. Commun.* 24, 38 (1978).
16. Myer, J. A., Itzkan, I., and Kierstead, E., Dye lasers in the ultraviolet, *Nature (London)* 225, 544, (1970).
17. Kelley, C. J., Ghiorghis, A., Neister, E., Armstrong, L., and Prause, P. R., Bridged quarterphenyls as flashlamp-pumpable laser dyes, *Laser Chem.* 7, 343 (1987).
18. Kauffman, J. M., Kelley, C. J., Ghiorghis, A., Neister, E., and Armstrong, L., Cyclic ether auxofluors on oligophenylene laser dyes, *Laser Chem.* 8, 335 (1988).
19. Chromatix, Inc., Sunnyvale, CA 94086.
20. Molectron Corporation, Sunnyvale, CA 94086.
21. Rinke, M., Gusten, H., and Ache. H. J., Photophysical properties and laser performance of photostable uv laser dyes II. Ring-bridged p-quarterphenyls, *J. Phys. Chem.* 90, 2666 (1986).
22. Gusten, H., Rinke, M., Kao, C., Zhou, Y., Wang, M., and Pan, J., New efficient laser dyes for operation in the near UV range, *Opt. Commun.* 59(5,6), 379 (1986).
23. Michelson, M. M. (private communication, Steppel, R. N., 1993).
24. Chen, C. H., Fox, J. L., Duarte, F. J., and Ehrlich, J. J., Lasing characteristics of new coumarin-analog dyes: broadband and narrow-linewidth performance, *Appl. Optics* 27, 443 (1988).
25. Phase-R Company, New Durham, NH.
26. Candela Corporation, Natick, MA.
27. Myer, J. A., Johnson, C. L., Kierstead, E., Sharma, R. D., and Itzkan, I., Dye laser stimulation with a pulsed N_2 laser line at 3371 Å. *Appl. Phys. Lett.* 16, 3 (1970).

28. Alekseeva, V. I., Afanasiadi, L. S., Volkov, V. M., Krasovitskii, B. M., Vernigor, E. M., Lebedev, S. A., Savvina, L. P., and Tur, I. N., Spectral luminescent and lasing properties of pyridylaryloxazoles, *J. Appl. Spectrosc. (USSR)* 44, 244 (1986): *Zh. Prikl. Spectrosc.* 44, 403 (1986).

29. Smith, P. W., Liao, P. F., Schank, C. V., Gustafson, T. K., Lin, C., and Maloney, P. J., Optically excited organic dye vapor laser, *Appl. Phys. Lett.* 25, 144 (1974).

30. Guggenheimer, S. C., Peterson, A. B., Knaak, L. E., and Steppel, R. N. (to be published).

31. Tully, F. P., and Durant, Jr., J. L., Exalite 392E: A new laser dye for efficient cw operation between 373 and 408 nm, *Appl. Optics* 27, 2096 (1988).

32. Gomez, M. S. and Guerra Perez, J. M., The ACDF. A new class of lasing dyes, *Optics Commun.* 40, 144 (1981).

33. Acuña, A. U., Amat, F., Catalan, J., Costel, A., Figuera, J. M., and Muñoz, J. M., Pulsed liquid lasers from proton transfer in the excited state, *Chem. Phys. Lett.* 132, 567 (1986).

34. Telle, H., Brinkmann, U., and Raue, R., Laser properties of bis-styryle compounds, *Opt. Commun.* 24, 248 (1978).

35. Ferguson, J. and Mau, A. W. H., Laser emission from meso-substituted anthracenes at low temperatures, *Chem. Phys. Lett.* 14, 245 (1972).

36. Basting, D., Schafer, F. P., and Steyer, B., New laser dyes, *Appl. Phys.* 3, 81 (1974).

37. Srinivasan, R., New materials for flash-pumped organic lasers, *IEEE J. Quantum Electron.* QE-5, 552 (1969).

38. Fletcher, A. N., Pietrak, M. E., and Bliss, D. E., Laser dye stability. Part II. The fluorinated azacoumarin dyes, *Appl. Phys.* B 42, 79 (1987).

39. Huffer, W., Schieder, R., Telle, H., Raue, R., and Brinkwerth, W., CW dye laser emission down to the near uv, *Opt. Commun.* 28, 353 (1979).

40. Majewski, W. and Krasinski, J., Laser properties of fluorescent brightening agents, *Opt. Commun.*, 18, 255 (1976).

41. Ebeid, E. M., Sabry, M. F., and El-Daly, S. A., 1,4-bis (β-pyridyl-2-vinyl)benzene (p2vb) and 2,5-distyryl-pyrazine (dsp) as blue laser dyes, *Laser Chem.* 5, 223 (1985).

42. Beterov, I. M., Ishchenko, V. N., Kogan, B. Ya., Krasovitskii, B. M., and Chernenko, A. A., Stimulated emission from 2-phenyl-5(4-difluoromethylsulfonyl-phenyl)oxazole pumped with nitrogen laser radiation, *Sov. J. Quantum Electron.* 7, 246 (1977).

43. Hammond, P. R., Fletcher, A. N., Henry, R. A., and Atkins, R. L., Search for efficient, near uv lasing dyes, *Appl. Phys.* 8, 311 (1975).

44. Srinivasan, R., von Gutfield, R. J., Angadiyavar, C. S., and Tynan, E. E., Photochemical studies on organic lasers, Air Force Materials Laboratory, Wright-Patterson Air Force Base, Dayton, Ohio, AFML-TR-74-110 (1974).

45. Dzyubenko, M. I., Krainov, I. P., and Maslov, V. V., Lasing properties of water-soluble dyes in the blue-green region, *Optics Spectrosc. (USSR)* 57, 58 (1984).

46. Furumoto, H. W. and Ceccon, H. L., Flashlamp pumped organic scintillator lasers, *J Appl. Phys.* 40, 4204 (1969).

47 Yarborough, J. M., CW laser emission spanning the visible spectrum, *Appl. Phys. Lett.* 24, 629 (1974).

48. Decker, C. D. and Tittel, P. K., Broadly tunable, narrow linewidth dye laser emission in the near infrared, *Opt. Commun.* 7, 155 (1974).

49. Eckstein, N. J., Ferguson, A. I., Hansch, T. W., Minard, C. A., and Chan, C. K., Production of deep blue tunable picosecond light pulses by synchronous pumping of a dye laser, *Opt. Commun.* 27, 466 (1978).

50. Pavlopoulos, T. G., Boyer, J. H., Politzer, I. R., and Lau, C. M., Syn-dioxabimanes as laser dyes, *Optics Commun.* 64, 367 (1987).

51. Quantel International, 928 Benecia Avenue, Sunnyvale, CA.

52. Telle, H., Brinkmann, U., and Raue, R., Laser properties of triazinyl stilbene compounds, *Opt. Commun.* 24, 33 (1978).

53. Quanta-Ray, 1250 Charleston Rd., Mountain View, CA.

54. Huth, B. G. and Farmer, G. I., Laser action in 9,10 diphenylanthracene, *IEEE J. Quantum Electron.* QE-4, 427 (1968).

55. Drexhage, K. H., Erickson, O. R.. Hawks, G. H., and Reynolds, G. A.. Water soluble coumarin dyes for flashlamp-pumped dye lasers, *Opt. Commun.* 15, 399 (1975).

56. Kittrell, C. (private communication, R. N. Steppel, 1977).

57. Acuña, A. U., Costel, A., and Muñoz, J. M., A proton-transfer laser, *J. Chem. Phys.* 90, 2807 (1986).

58. Fletcher, A. N. and Bliss, D. E., Laser dye stability. V. Efforts of chemical substituents of bicyclic dyes upon photodegradation parameters, *Appl. Phys.* 16, 289 (1978).

59. Marling, J. B., Hawley, J. H., Liston, E. M., and Grant, W. B., Lasing characteristics of seventeen visible-wavelength dyes using a coaxial-flashlamp-pumped laser, *Appl. Optics* 13, 2317 (1974).

60. Kato, K., 3547-Å pumped high power dye laser in the blue and violet, *IEEE J. Quantum Electron.* QE-II, 373 (1975).

61. Reynolds, G. A. and Drexhage, K. H., New coumarin dyes with rigidized structures for flashlamp pumped dye lasers, *Opt. Commun.* 13, 222 (1975).

62. Allain, J. Y., High energy pulsed dye lasers for atmospheric sounding, *Appl. Optics* 18, 287 (1979).

63. Roullard, F. P. (private communication, R. N. Steppel, 1976).

64. Williamson, A. (private communication, R. N. Steppel, 1977).

65. Green, W. R. (private communication, R. N. Steppel, 1977).

66. Schimitschek, E. J., Trias, J. A., Hammond, P. R., and Atkins, R. L., Laser performance and stability of fluorinated coumarin dyes, *Opt. Commun.* ll, 352 (1974).

67. Schimitschek, E. J., Trias, J. A., Taylor, M., and Celto, J. E., New improved laser dyes for the blue-green spectral region, *IEEE J. Quantum Electron.* QE-9, 781 (1974).

68. Collins, C. H., Taylor, K. N., and Lee, F. W., Dyes pumped by the nitrogen ion laser, *Opt. Commun.* 26, 101 (1978).

69. Halstead, J. A. and Reeves, R. R.. Mixed solvent systems for optimizing output from a pulsed dye laser, *Opt. Commun.* 27, 273 (1978).

70. Drexhage, K. H. and Reynolds, G. A., New highly efficient laser dyes, VII Int. Quantum Electronics Conf., Paper F. l, San Francisco, Calif. (1974); see also References 80 and 100.

71. Petty, B. W. and Morris, K., *Opt. Quantum Electron.* 8, 371 (1976).

72. Morton, R. O., Mack, M. E., and Itzkan, I., Efficient cavity dumped dye laser, *Appl. Opt.* 17, 3268 (1978).

73. Ledbetter, J. W. (private communication, R. N. Steppel, 1977).

74. Gacoin, P., Bokobza, A., Bos, F., LeBris, M. T., and Hayat, G., New class of high-efficiency laser dyes: the quinoxalinones, Conference on Laser and Electrooptical Systems, San Diego, (1978), S6, WEE6.

75. Fletcher, A. N., Laser dye stability. III. Bicyclic dyes in ethanol, *Appl. Phys.* 14, 295 (1977).

76. Blazy, J. (private communication, R. N. Steppel, 1978).

77. Lill, E., Schneider, S., and Dorr, F., Passive mode-locking of a flashlamp-pumped fluorol 7GA dye laser in the green spectral region, *Opt. Commun.* 20, 223 (1977).

78. Blazej, D. (private communication, R. N. Steppel, 1977).

79. Lambropoulos, M., Fluorol 7CA: An efficient yellow-green dye for flashlamp-pumped lasers, *Opt. Commun.* 15, 35 (1975).

80. Drexhage, K. H., What's ahead in laser dyes, *Laser Focus* 9, 35 (1974).

81. Marling, J. B., Wood, L. L., and Gregg, D. W., Long pulse dye laser across the visible spectrum, *IEEE J. Quantum Electron.* QE-7, 498 (1971).

82. Marling, J. B., Gregg, D. W., and Thomas, S. J., Effect of oxygen on flashlamp-pumped organic-dye lasers, *IEEE J. Quantum Electron.* QE-6, 570 (1970).

83. Hartig, W., A high power dye-laser pumped by the second harmonic of a Nd-YAG laser, *Opt. Commun.* 27, 447 (1978).

84. J. K. Lasers Ltd., Somers Road, Rugby, Warwickshire, U.K.

85. Hargrove, R. S. and Kan, T., Efficient, high average power dye amplifiers pumped by copper vapor lasers, *IEEE J. Quantum Electron.* QE-13, 28D (1977).

86. Hammond, P. R., Laser dye DCM, spectral properties, synthesis and comparison with other dyes in the red, *Opt. Commun.*, Preprint.

87. Drake, J. M., Steppel, R. N., and Young, D., Kiton red s and rhodamine b. The spectroscopy and laser performance of red laser dyes, *Chem. Phys. Lett.* 35, 181 (1975).

88. Passner, A. and Venkatesan, T., Inexpensive, pulsed, tunable ir dye laser pumped by a driven dye laser, *Rev. Sci. Instrum.* 49, 1413 (1978).

89. Oettinger, P. E. and Dewey, C. F., Lasing efficiency and photochemical stability of infrared laser dyes in the 710-1080 nm region, *IEEE J. Quantum Electron.* QE-12, 95 (1976).

90. Wayashita, M., Kasarnatsu, M., Kashiwagi, H., and Machida, K., The selective excitation of lithium isotopes by intracavity nonlinear absorption in a cw dye laser, *Opt. Commun.* 26, 343 (1978).

91. McDonald, J. (private communication, R. N. Steppel, 1974).

92. Woodruff, S. and Ahlgren, D. (private communication, R. N. Steppel, 1977).

93. Holton, G. (private communication, R. N. Steppel, 1978).

94. (a) Shirley, J., private communication (1977); (b) Hall, R. J., Shirley, J. A., and Eckbreth, A. C., Coherent anti-Stokes Raman spectroscopy: Spectra of water vapor in flames, *Opt. Lett.* 4, 87 (1979).

95. Drell, P. (private communication, R. N. Steppel, 1978).

96. Mahon, R., McIlrath, T. J., and Koopman, D. W., High-power TEM_{00} tunable laser system, *Appl. Optics* 18, 891 (1979).

97. Kato, K., A high-power dye laser at 6700-7700 Å, *Opt. Commun.* 19, 18 (1976).

98. Kuhl, J., Lambrich, R., and von der Linde, D., Generation of near-infrared picosecond pulses by mode locked synchronous pumping of a jet-stream dye laser, *Appl. Phys. Lett.* 31, 657 (1977).

99. Jarett, S., Spectra Physics (private communication, Steppel, R. N., 1980).

100. Drexhage, K. H., Structure and properties of laser dyes, in *Dye Lasers*, Vol. 1, Schafer, F. P., Ed., Springer-Verlag, Berlin (1973), p. 44 and references therein.

101. Miyazoe, Y. and Mitsuo, M., Stimulated emission from 19 polymethine dyes-laser action over the continuous range 710-1060 μm, *Appl. Phys. Lett.* 12, 206 (1968).

102. Szabo, A., National Research Council of Canada, private communication (1980), Jessop, P. E. and Scabo, A., Single frequency cw dye laser operation in the 690-700 nm gap, *IEEE J. Quantum Electron.* 16, 812 (1980).

103. Romanek, K. M., Hildebrand, O., and Gobel, E., High power CW dye laser emission in the near IR from 685 nm to 965 nm, *Opt. Commun.* 21, 16 (1977); *Spectra-Physics Laser Review* 4, April (1977).

104. Hildebrand, O., Nitrogen laser excitation of polymethine dyes for emission wave-length up to 9500 Å, *Opt. Commun.* 10, 310 (1974).

105. Moore, C. A. and Decker, C. D., Power-scaling effects in dye lasers under high power laser excitation, *J. Appl. Phys.* 49, 47, (1978).

106. Fehrenback, G. W., Oruntz, K. J., and Ulbrich, R. G., Subpicosecond light pulses from a synchronously mode-locked dye laser with composite gain and absorber medium, *Appl. Phys. Lett.* 33, 159 (1978).

107. Bryon, D, A., McDonnell Douglas Astronautics Company, (private communication, R. N. Steppel, 1979).

108. Donzel, A. and Weisbach, C., CW dye laser emission in the range 7540-8880 Å, *Opt. Commun.*, 17, 153 (1976).

109. Webb, J. P., Webster, F. G., and Plourde, B. E., Sixteen new infrared laser dyes excited by a simple, linear flashlamp, *IEEE J. Quantum Electron.* QE-11, 114 (1975).

110. Kato, K., Near infrared dye laser pumped by a carbazine 122 dye laser, *IEEE J. Quantum Electron.* QE-12, 442 (1976).

111. Decker, C. D., Excited state absorption and laser emission from infrared dyes optically pumped at 532 nm, *Appl. Phys. Lett.* 27, 607 (1975).

112. Miyazoe, Y. and Maeda, M., Polymethine dye lasers, *Opto Electronics* 2, 227 (1970).

113. Leduc, M., Synchronous pumping of dye lasers up to 1095 nm, *Opt. Commun.* 31, 66 (1979).

114. Ammann, E. O., Decker, C. D., and Falk, J., High-peak-power 532 nm pumped dye laser, *IEEE J. Quantum Electron.* QE-10, 463 (1974).

115. Leduc, M. and Weisbach, C., CW dye laser emission beyond 1000 nm, *Opt. Commun.* 26, 78 (1978).

116. Ferrario, A., 13 MW peak power dye laser tunable in the 1.1 μm range, *Opt. Commun.* 30, 83 (1979).

117. Ferrario, A., A picosecond dye laser tunable in the 1.1 μm region, *Opt. Commun.* 30, 85 (1979).

118. Kato, K., Nd:YAG laser pumped infrared dye laser, *IEEE J. Quantum Electron.* QE-14, 7 (1978).

119. Kato, K., Broadly tunable dye laser emission to 12850 Å, *Appl. Phys. Lett.* 33(16), 509 (1978).

120. Yenagi, J. V., Gorbal, M. R., Savadatti, M. I., and Naik, D. B., A new laser dye, *Opt. Commun.* 85, 223 (1991).

121. Doizi, D., Jaraudias, J., and Salvetat, G., Laser performance of dibenzoxanthylium salts, *Opt. Commun.* 99, 207 (1993).

122. O'Neil, M. P., Synchronously pumped visible laser dye with twice the efficiency of Rhodamine 6G, *Opt. Lett.* 18, 37 (1993).

123. Pavlopoulos, T. G., Boyer, J. H., Shah, M., Thangaraj, K., and Soong, M.-L., Laser action from 2,6,8-position trisubstituted 1,3,5,7-tetramethylpyrromethene-BF$_2$, *Appl. Optics* 29, 3885 (1990).

124. Guggenheimer, S. C., Boyer, J. H., Thangaraj, K., Shah, M., Soong, M.-L., and Pavlopoulos, T. G., Efficient laser action from two cw laser-pumped pyrromethene-BF$_2$ complexes, *Appl. Optics* 32, 3942 (1993).

125. Boyer, J. H., Haag, A., Soong, M.-L., Thangaraj, K., and Pavlopoulos, T. G., Laser action from 2,6,8-position trisubstituted 1,3,5,7-tetramethyl-pyrromethene-BF$_2$ complexes: part 2, *Appl. Optics* 30, 3788 (1991).

126. Moses, D., High quantum efficiency luminescence from a conducting polymer in solution: a novel polymer laser dye, *Appl. Phys. Lett.* 60, 3215 (1992).

127. Said, J. and Boquillon, J., Lasing characteristics of a new DCM derivative under flash-lamp pumping, *Opt. Commun.* 82, 51 (1991).

128. Costela, A., Amat, F., Catalan, J., Douhal, A., Figuera, J. M., Munoz, J. M., and Acuna, A. U., Phenylbenzimidazole proton-transfer laser dyes: spectral and operational properties, *Optics Commun.* 64, 457 (1987).

129. Fletcher, A. N., Henry, R. A., Kubin, R. F., and Hollins, R. A., Fluorescence and lasing characteristics of some long-lived flashlamp-pumpable, oxazole dyes, *Optics Commun.* 48, 352 (1984).

130. Fletcher, A. N., Bliss D. E., and Kauffman, J. M., Lasing and fluorescent characteristics of nine, new, flashlamp-pumpable, coumarin dyes in ethanol and ethanol:water, *Optics Commun.* 47, 57 (1983).

131. Fletcher, A. N., Henry, R. A., Pietrak, M. E., and Bliss D. E., Laser dye stability, part 12. The pyridinium salts, *Appl. Phys. B* 43, 155 (1987).

132. Raue, R., Harnisch, H., and Drexhage, K. H., Dyestuff lasers and light collectors—two new fields of application for fluorescent heterocyclic compounds, *Heterocycles* 21, 167 (1984).

133. Komel'kova, L. A., Kruglenko, V. P., Logunov, O. A., Povstyanoi, M. V., Startsev, A. V., Stoilov, Yu., and Timoshin, A. A., Imitrines. IV. New laser compounds in the imitrine class operating in the 482–618 nm range, *Sov. J. Quantum Electron.* 13, 549 (1983).

134. Pavlopoulos, T. G., Boyer, J. H., Politzer, I. R., and Lau, C. M., Laser action from syn-(methyl, methyl)bimane, *J. Appl. Phys.* 60, 4028 (1986).

135. Everett, P. N., Aldag, H. R., Ehrlich, J. J., Janes, G. S., Klimek, D. E., Landers, F. M., and Pacheco, D. P., Efficient 7-1 flashlamp-pumped dye laser at 500-nm wavelength, *Appl. Optics*, 25, 2142 (1986).

136. Asimov, M. M., Katarkevich, V. M., Kovalenko, A. N., Nikitchenko, V. M., Novikov, A. I., Rubinov, A. N., and Efendiev, T. Sh., Spectroluminescence and lasing characteristics of a new series of bifluorophoric laser dyes, *Opt. Spectrosc. (USSR)* 63, 356 (1987).

137. Pavlopoulos, T. G., McBee, C. J., Boyer, J. H., Politzer, I. R., and Lau, C. M., Laser action from syn-(methyl,chloro)bimane, *J. Appl. Phys.* 62, 36 (1987).

138. Dupuy, F., Rulliere, C., Le Bris, M. T., and Valeur, B., A new class of laser dyes: benzoxazinone derivatives, *Optics Commun.* 51, 36 (1984).

139. Asimov, M. M., Nikitchenko, V. M., Novikov, A. I., Rubinov, A. N., Bor, Zs., and Gaty, L., New high-efficiency biscoumarin laser dyes, *Chemical Phys. Lett.* 149, 140 (1988).

140. Pavlopoulos, T. G., Shah, M., and Boyer, J. H., Efficient laser action from 1,3,5,7,8-dentamethylpyrromethene-BF_2 complex and its disodium 2,6-di-sulfonate derivative, *Optics Commun.* 70, 425 (1989).

141. Pavlopoulos, T. G., Shah, M., and Boyer, J. H., Laser action from a tetramethyl-pyrromethene-BF_2 complex, *Appl. Optics* 27(24), 4998 (1988).

142. Neister, S.E. (private communication, Steppel, R. N.).

143. Davenport, W. E., Ehrlich, J. J., and Neister, S. E., Characterization of pyrromethene-BF_2 complexes as laser dyes, *Proceedings of the International Conferences on Lasers '89*, New Orleans, LA, (1989), p. 408.

144. Shah, M., Thangaraj, K., Soong, M. L., Wolford, L. T., Boyer, J. H., Politzer, I. R., and Pavlopoulos, T. G., Pyrromethene-BF_2 complexes as laser dyes: 1, *Heteroatom. Chem.* 1, 389 (1990).

145. Benson, M., Coherent Laser Group (private communication, R. N. Steppel, 1994).

146. Hsia, J., Candela Laser Corporation (private communication, R. N. Steppel, 1989).

147. Piechowski, A. P. and Bird, G. R., A new family of lasing dyes from an old family of fluors, *Optics Commun.* 50, 386 (1984).

148. Partridge, Jr., W. P., Laurendeau, N. M., Johnson, C.C., and Steppel, R. N., Performance of pyrromethene 580 and 597 in a commercial Nd:YAG-pumped dye laser system, *Optics Lett.* 19, 1630 (1994).

149. Guggenheimer, S. G., Boyer, J. H., Thangaraj, K., Shah, M., Soong, M. L., and Pavlopoulos, T. G., Efficient laser action from two cw laser pumped pyrromethene-BF_2 complexes, *Appl. Optics* 32(21), 3942 (1993).

150. Kauffman, J. M. and Bently, J. H., Effect of various anions and zwittenons on the lasing properties of a photostable cationic laser dye, *Laser Chem.* 8 (1988).

151. Maeda, M. and Miyazoe, Y., Flashlamp-excited organic liquid laser in the range from 342 to 889 nm, *Jpn. J. Appl. Phys.* 11, 692 (1972).

152. Loth, C. and Gacoin, P., Improvement of infrared flashlamp-pumped dye laser solution with a double effect additive, *Opt. Commun.* 15, 179 (1975).

153. Allik, T. H., Hermes, R. E., Sathyamoorthi, G., and Boyer, J. H., Spectroscopy and laser performance of new BF_2-complex dyes in solution, *SPIE Proceedings: Visible and UV Lasers* 2115, 240 (1994).

154. Shinn, M. D., Bryn Mawr College (private communication, R. N. Steppel, 1994).

155. Sadrai, M. and Bird, G. R., A new laser dye with potential for high stability and a broad band of lasing action: perylene-3,4,9,10 tetracarboxylic acid-bis-n,n' (2',6'xylidyl) diimide, *Optics Commun.* 51, 62 (1984).

156. Ivri, J. Burshtein, Z., Miron, E., Reisfeld, R., and Eyal, M., The perylene derivative BASF-241 solution as a new tunable dye laser in the visible, *IEEE J. Quantum Electron.* 26, 1516 (1990).

157. Boyer, J. H., Haag, A. M., Sathyamoorthi, G., Soong, M. L., and Thangara, K., Pyrromethene-BF_2 complexes as laser dyes: 2, *Heteroatom. Chem.* 4, 39 (1993).

158. Maslov, V. V., Dzyubenko, M. I., Kovalenko, S. N., Nikitchenko, V. M., and Nivikov, A. I., New efficient dyes for the red part of the lasing spectrum, *Sov. J. Quantum Electron.* 17, 998 (1987).

159. Boyer, J. H., Haag, A., Shah, M., Soong, M. L., Thangaraj, K., and Pavlopoulos, T. G., Laser action from 2,6,8-trisubstituted-1,3,5,7-tetramethyl-pyrromethene-BF_2 complexes: Part 2, *Appl. Optics* 30(27), 3788 (1991).

160. Richter, D. (private communication, Steppel, R. N., 1994).

161. Broyer, M., Chevaleyre, J., Delacretaz, G., and Woste, L., CVL-pumped dye laser for spectroscopic application, *Appl. Phys. B* 35, 31 (1984).

162. Kuhl, J., Telle, H., Scheider, R., and Brinkmann, U., New efficient and stable laser dyes for cw operation in the blue and violet spectral range, *Opt. Commun.* 24, 251 (1978).

163. Antonov, V. S. and Hohla, K. L., Dye stability under excimer-laser pumping II. visible and UV dyes, *Appl. Phys. B*. 32, 9 (1983).

164. Spectra-Physics, 1250 W. Middlefield Road, Mountain View, CA 94039.

165. Friedrich, D. M., Nitrogen pumped LDS 698, (private communication, Steppel, R. N., 1985).

166. Jasny, J., Novel method for wavelength tuning of distributed feedback dye lasers, *Optics Commun.* 53, 238 (1985).

167. Coherent Inc., 3210 Porter Dr., Palo Alto, CA 94304.

168. Hoffnagle, J., Roesch, L. Ph., Schlumpf, N., and Weis, A., CW operation of laser dyes Styryl-9 and Styryl-11, *Optics Commun.* 42, 267 (1982); K. Kato, see Reference 5 therein.

169. Lumonics Inc., 105 Schneider Road, Kanata (Ottawa), Ontario, Canada K2K IY3.

170. Klein, P. (private communication, Steppel, R. N., 1983).

171. Giberson, K. W., Jeys, T. H., and Dunning, F. B., Generation of tunable cw radiation near 875 nm, *Appl. Optics* 22(18), 2768 (1983).

172. Schellenberg, F. (private communication, Steppel, R. N., 1982).

173. Kato, K., Ar-ion-laser-pumped infrared dye laser at 875-1084 nm, *Optics Lett.* 9, 544 (1984).

174. Bloomfield, L. A., Excimer-laser pumped infrared dye laser at 907-1023 nm, *Optics Commun.* 70, 223 (1989).

175. Stark, T. S., Dawson, M. D., and Smirl, A. L., Synchronous and hybrid mode-locking of a Styryl 13 dye laser, *Optics Commun.* 68, 361 (1988).

176. Seilmeier, A., Kopainsky, B., and Kaiser, W., Infrared fluorescence and laser action of fast mode-locking dyes, *Appl. Phys.* 22, 355 (1980).

177. Reynolds, G. A. and Drexhage, K. H., Stable heptamethine pyrylium dyes that absorb in the infrared, *J. Org. Chem.* 42, 885 (1977).

178. Polland, H. J., Elsaesser, T., Seilmeier, A., Kaiser, W., Kussler, M., Marx, N. J., Sens, B., and Drexhage, K. H., Picosecond dye laser emission in the infrared between 1.4 and 1.81 μm, *Appl. Phys.* B32, 53 (1983).

179. Seilmeier, A., Kaiser, W., Sens, B., and Drexhage, K. H., Tunable picosecond pulses around 1.3 μm generated by a synchronously pumped infrared dye laser, *Optics Lett.* 8, 205 (1983).

180. Kopainsky, B., Kaiser, W., and Drexhage, K. H., New ultrafast saturable absorbers for Nd:lasers, *Optics Commun.* 32, 451 (1980).

181. Elsaesser, T., Polland, H. J., Seilmeier, A., and Kaiser, W., Narrow-band infrared picosecond pulses tunable between 1.2 and 1.4 μm generated by a traveling-wave dye laser, *IEEE J. Quantum Electron.* QE-20. 191 (1984).

182. Kopainsky, B., Qiu, P., Kaiser, W., Sens, B., and Drexhage, K. H., Lifetime, photostability, and chemical structure of ir heptamethine cyanine dyes absorbing beyond 1 μm, *Appl. Phys.* B 29, 15 (1982).

183. Looentanzer, H. and Polland, H. J., Generation of tunable picosecond pulses between 1.18 μm and 1.53 μm in a ring laser configuration using dye no. 5., *Optics Commun.* 62, 35 (1987).

184. Alfano, R. R., Schiller, N. H., and Reynolds, G. A., Production of picosecond pulses by mode locking an nd:glass laser with dye #5, *IEEE J. Quantum Electron.* QE-17, 290 (1981).

185. Rinke, M., Gusten, H., and Ache, H. J., Photophysical properties and laser performance of photostable uv laser dyes I. Substituted p-quarterphenyls, *J. Phys. Chem.* 90, 2661 (1986).

186. Schafer, F. P., Bor, Zs., Luttke, W., and Liphardt, B., Bifluorophoric laser dyes with intramolecular energy transfer, *Chem. Phys. Lett.* 56, 455 (1978).

187. Coherent Inc., 3210 Porter Dr., Palo Alto, CA.

188. Ivri, J., Burshtein, Z., and Miron, E., Characteristics of 1,1',3,3,3',3'-hexa-methyl-indotricarbocyanine iodide as a tunable dye laser in the near infrared, *Appl. Optics* 30, 2484 (1991).

189. Pavlopoulos, T. G., Shah, M., and Boyer, J. H., Efficient laser action from 1,3,5,7,8-pentamethylpyrromethene-BF_2 complex and its disodium 2,6- disulfonate derivative, *Opt. Commun.*, 70, 425 (1989).

190. Valat, P., Tascano, V., Kossanyi, J., and Bos, F., Laser effect of a series of variously substituted pyrylium and thiopyrylium salts, *J. Lumin.* 37, 149 (1987).

191. Schimitschek, E. J., Trias, J. A., Hammond, P. R., Henry, R. A., and Atkins, R. L., New laser dyes with blue-green emissions, *Opt. Commun.* 16, 313 (1976).

192. Tuccio, S. A., Drexhage, K. H., and Reynolds, G. A., CW laser emission from coumarin dyes in the blue and green, *Opt. Commun.* 7, 248 (1974).

193. Hoffnagle, J. (private communication, Steppel, R. N., 1987).

194. Hammond, P. R., Fletcher, A. N., Henry, R. A., and Atkins, R. L., Search for efficient, near uv lasing dyes. II. Aza substitution in bicyclic dyes, *Appl. Phys.* 8, 315 (1975).

195. Tzeng, H.-M., Wall, K. F., Long, M. B., and Chang, R. K., Laser emission from individual droplets at wavelengths corresponding to morphology-dependent resonances, *Optics Lett.* 9, 499 (1984).

196. Kotowski, T., Skubiszak, W., Soroka, J. A., Soroka, K. B., and Stacewicz, T., Pyrylium and thiopyrylium high efficiency laser dyes, *J. Lumin.* 50, 39 (19910.

197. Sorokin, P. P., Lankard, J. R., Hammond, E. C., and Moruzzi, V. L., Laser-pumped stimulated emission from organic dyes: experimental studies and analytical comparisons, *IBM Journal* 11, 130 (1967).

198. Sorokin, P. P. and Lankard, J. R., Flashlamp excitation of organic dye lasers: a short commun-ication, *IBM Journal* 11, 148(1967).

Section 2.2
RARE EARTH LIQUID LASERS

2.2.1 Introduction

Liquid lasers based on lanthanide ions have been of two types—rare earth chelate lasers and rare earth aprotic lasers. In rare earth chelate lasers, the rare earth is complexed with bidentate ligands such as β di-ketonate and carboxylate ions or organic phospate. Generally organic solvents are employed. These lasers are tabulated in Table 2.2.1

Rare earths have also been incorporated into inorganic aprotic solvents (no hydrogen anions). Thus far these have been oxyhalides or halides of the heavier elements such as phosphorous, sulfur, selenium, zirconium, tin, etc. Neodymium has been the active laser ion although other ions could undoubtedly be used. Neodymium aprotic liquid lasers and amplifiers have been operated in various pulsed modes; operating wavelengths are given in Tables 2.2.2 and 2.2.3.

Further Reading

Lempicki, A. and Samelson, H., Organic laser systems, in *Lasers*, Levine, A. D., Ed., Marcel Dekker, New York (1966), p. 181.

Lempicki, A., Samelson, H., and Brecher, C., Laser action in rare earth chelates, in *Applied Optics, Suppl. 2 of Chemical Lasers* (1965), p. 205.

Samelson, H., Inorganic Liquid Lasers, in *Handbook of Laser Science and Technology, Vol. I: Lasers and Masers*, CRC Press, Boca Raton, FL (1982), p. 397.

Samelson, H., Inorganic Liquid Lasers, in *Handbook of Laser Science and Technology, Suppl. 1: Lasers*, CRC Press, Boca Raton, FL (1991), p. 319.

2.2.2 Chelate Lasers

Rare earth chelate lasers are listed by lasing ion in Table 2.2.1. The ligand, cation, solvent, lasing wavelength, temperature, and references are included in the table. The lasers listed in this table were all excited by flashlamp discharge except for the following: Reference 12 (pulsed BBQ dye laser – 377 nm), Reference 14 (coumarin dye laser), and Reference 25 (pulsed N_2 laser – 337 nm). The neodymium chelate lasers are excited by direct pumping into levels of the rare earth ion. The excitation of the europium and terbium lasers is accomplished by energy transfer, the most effective absorption being in the singlet absorption band of the b-diketone ligand.

Table 2.2.1
Rare Earth Chelate Lasers[a]

Lasing ion	Ligand	Cation	Solvent	Wavelength (μm)	Temp. (K)	Ref.
Neodymium
Nd^{3+}	TBP_{d27}	(b)	HFB	1.054	300	27
Nd^{3+}	TBP_{d27}	(b)	CCL	1.054	300	27
Nd^{3+}	PFP1,10PHEN	—	$DSMO_{d6}$	1.057	300	28
Europium
Eu^{3+}	B	Na^+	EM (3:1)	0.6111	123–133	2,3
Eu^{3+}	B	Na^+	DMF	0.6111	133	2
Eu^{3+}	m-ClBTF	Me_2A^+	A	0.6117	300	3
Eu^{3+}	p-ClBTF	Me_2A^+	A	0.6117	300	3
Eu^{3+}	o-FBTF	Me_2A^+	A	0.6117	300	3
Eu^{3+}	p-FBTF	Me_2A^+	A	0.6117	300	3
Eu^{3+}	o-BrBTF	Me_2A^+	A	0.6117	300	3
Eu^{3+}	m-BrBTF	Me_2A^+	A	0.6117	300	3
Eu^{3+}	p-BrBTF	Me_2A^+	A	0.6117	300	3
Eu^{3+}	m-FBTF	Me_2A^+	A	0.6118	300	3
Eu^{3+}	NTF	P^+	A	0.6118	253	5
Eu^{3+}	o-ClBTF	Me_2A^+	A	0.6118	300	3,6
Eu^{3+}	BTF	I^+	A	0.6118	238	4
Eu^{3+}	BTF	Pyo^+	A	0.6118	300	8
Eu^{3+}	BTF	P^+	DMF (3:1:1)	0.6119	168	9
Eu^{3+}	BTF	P^+	A	0.6119	298	10
Eu^{3+}	TFA	Me_2A^+	A	0.6119	238	4
Eu^{3+}	D	P^+	DMF (9:3:2)	0.612	128	11
Eu^{3+}	BFA	—	—	0.612[c]	300	12
Eu^{3+}	TFA	NH_4^+	EM (3:1)	0.6122	123	13
Eu^{3+}	B	P^+	A	0.6123	300	14
Eu^{3+}	TTF	P^+	A	0.6123	300	14
Eu^{3+}	TTF	Me_2A^+	A	0.6125	238	4
Eu^{3+}	B	P^+	EM (3:1)	0.6129	110–150	15–23
Eu^{3+}	B	P^+	E	0.613	77–123	23,24
Eu^{3+}	DBM	phen	EG (~3:1)	0.613[d]	300	25
Eu^{3+}	B	NH_4^+	EM (3:1)	0.613	123	13
Eu^{3+}	B	P^+	DMF (3:1:1)	—	—	30
Eu^{3+}	B	P^+	N	—	—	30
Eu^{3+}	D	P^+	N	—	—	30
Eu^{3+}	BTF	P^+	ENAEO	—	168	31
Eu^{3+}	BTF	P^+_{d10}	ENAEO	—	168	31
Eu^{3+}	BTF_{d5}	P^+	ENAEO	—	168	31
Eu^{3+}	BTF_{d5}	P^+_{d10}	ENAEO	—	168	31
Eu^{3+}	p-FBTF	Me_2Pyr^+	ENAEO	—	168	5
Eu^{3+}	p-IBTF	Me_2Pyr^+	ENAEO	—	168	5

Table 2.2.1—*continued*
Rare Earth Chelate Lasers[a]

Lasing ion	Ligand	Cation	Solvent	Wavelength (µm)	Temp. (K)	Ref.
Eu^{3+}	p-CF_3BTF	Me_2Pyr^+	ENAEO	—	168	5
Eu^{3+}	p-CH_3OBTF	P^+	ENAEO	—	168	5
Eu^{3+}	m,p-Cl_2BTF	P^+	ENAEO	—	168	5
Terbium						
Tb^{3+}	TFA	[e]	A	0.547	300	1
Tb^{3+}	TFA	[e]	D	0.547	300	1

(a) Lasers listed in this table were all excited by flashlamp discharge except for the following: Reference 12 (pulsed dye laser – 377 nm), Reference 14 (coumarin laser), and Reference 25 (pulsed N_2 laser – 337 nm).
(b) In these lasers the active material seems to be a simple tris compound.
(c) Planar mircocavity.
(d) Morphology-dependent resonances.
(e) The active component is the dihydrate of a tris chelate and thus requires no cation.

Ligands in Table 2.2.1:

Symbol	Formula	Name
B	$C_6H_5COCHCOCH_3^-$	benzoylacetonate
BFA		4,4,4-trifluoro-1-phenyl-1,3-butanedione
D	$C_6H_5COCHCOC_6H_5^-$	dibenzoylmethide
DBM		dibenzoylmethane
BTF	$C_6H_5COCHCOCF_3^-$	benzoyltrifluoroacetonate
BTF_{d5}	$C_6D_5COCHCOCF_3^-$	deuterated benzoyltrifluoroacetonate
TTF	$C_4H_3SCOCHCOCF_3^-$	thenoyltrifluoroacetonate
TFA	$CF_3COCHCOCH_3^-$	trifluoroacetylacetonate
NTF	$C_{10}H_7COCHCOCF_3^-$	α-naphthoylltrifluoroacetonate
o-XBTF	$C_6H_4XCOCHCOCF_3^-$	o-halobenzoyltrifluoroacetonate[a]
m-XBTF	$C_6H_4XCOCHCOCF_3^-$	m-halobenzoyltrifluoroacetonate[a]
p-X8TF	$C_6H_4XCOCHCOCF_3^-$	p-halobenzoyltrifluoroacetonate[a]
3,4Cl2BTF	$C_6H_3Cl_2COCHCOCF_3$	m,p-dichlorobenzoyltrifluoroacetonate
p-RBTF	$C_6H_4RCOCHCOCF_3^-$	p-r benzoyltrifluoroacetonate[b]
TBP_{d27}	$(CD_3CD_2CD_2CD_2O)_3PO^-$	deuterotributyl phosphate
PFP1,10PHEN	$CF_3CF_2COO^-$ and$C_{12}H_8N_2$	pentafluoropropionate, 1,10 phenanthroline

(a) In the case of o- and m- substituents the halogens F, Cl, and Br have been used. In p-substitutions I has also been used.
(b) CH_3O-(methoxy) and CF_3 (trifluoromethyl) have been used as R.

Cations in Table 2.2.1:

Symbol	Formula	Name
P^+	$C_5H_{12}N^+$	piperidinium
$P^+{}_{d10}$	$C_5D_{10}NH_2{}^+$	deuterated piperidinium
Na^+	Na^+	sodium
M^+	$C_4H_8ONH_2$	morpholinium
I^+	$C_3H_5N_2{}^+$	imidazolium
Phen		phenanthroline
Pyo^+	$C_4H_8NO^+$	pyrrolidonium
Pyi^+	$C_4H_{10}N^+$	pyrrolidinium
Pyr^+	$C_5H_5NH^+$	pyridinium
Me_3Pyr^+	$(CH_3)_3C_5H_2NH^+$	2,4,6 trimethylpyridinium
Iq^+	$C_9H_8N^+$	isoquinolinium
$NH_4{}^+$	$NH_4{}^+$	ammonium
Me_2A^+	$(CH_3)_2NH_2{}^+$	dimethylammonium
Me_4A'	$(CH_3)_4N^+$	tetramethylammonium
Et_2A^+	$(C_2H_5)_2NH_2{}^+$	diethylammonium
Et_3A^+	$(C_2H_5)_3NH^+$	triethylammonium
Et_4A^+	$(C_2H_5)_4N^+$	tetraethylammonium
BA^+	$(C_2H_5)NH_3{}^+$	n-butylammonium
B_4A^+	$(C_4H_9)_4N^+$	tetra n-butylammonium
P_4A^+	$(C_3H_7)_4N^+$	tetra n-propylammonium
EOA^+	$HO(CH_2)_2NH_3{}^+$	2-hydroxyethylammonium
BeA^+	$C_7H_{10}N^+$	benzylammonium
Be_2A^+	$(C_7H_7)_2NH_2{}^+$	dibenzylammonium
Me_4G^+	$[N(C_2H_5)_2]_2CNH_2{}^+$	tetramethylguanidinium

Solvents in Table 2.2.1:

Symbol	Formula	Name
E	C_2H_5OH	ethanol
EG	C_2H_5OH	ethanol:glycerol
M	CH_3OH	methanol
A	CH_3CN	acetonitrile
DMSO	$(CH_3)_2SO$	dimethyl sulfoxide
DMSOd6	$(CD_3)_2SO$	deuterated dimethyl sulfoxide
DMF	$(CH_3)_2NCHO$	dimethyl formamide
EN	$CH_3CH(OC_2H_5)CN$	ethoxyproprionitrile
EO	$C_2H_5O(CH_2)_2OH$	ethoxyethanol
CCL	$CCl_4.$	carbon tetrachloride
HFB	C_6F_6	hexafluorobenzene

Solvents in Table 2.2.1—*continued*:

Symbol	Formula	Name
D	$C_4H_8O_2$	dioxane
PMMA		polymethylmethacrylate
EM		ethanol-methanol (3:1)
DMF		ethanol-methanol-dimethyl formamide
N		acetonitrile-proprionitrile-butyronitrile (1:1:1)
ENAEO		ethoxyproprionitrile-acetonitrile-ethoxy-ethanol (2:1:1)

References

1. Bjorklund, S., Kellermeyer, G., Hurt, C. R., McAvoy, N., and Filipescu, N., Laser action from terbium trifluoroacetylacetonate in p-dioxane and acetonitrile at room temperature, *Appl. Phys. Lett.* 10, 160 (1967).
2. Samelson, H., Brophy, V. A., Brecher, C., and Lempicki, A., Shift of laser emission of europium benzoylacetonate by inorganic ions, *J. Chem. Phys.* 41, 3998 (1964).
3. Meyer, Y., Astier, R., and Simon, J., Emission stimulee a 6111 Å dans le benzoylacetonate d'europium active au sodium, *Compt. Rend.* 259, 4604 (1964).
4. Schimitschek, E. J., Nehrich, R. B., and Trais, J.A., Fluorescence properties and stimulated emission in substituted europium chelates, *J. Chim. Phys.* 64, 173 (1967).
5. Riedel, E. P. and Charles, R. G., Spectroscopic and laser properties of europium naphthoyl-trifluoroacetonate in solution, *J. Chem. Phys.* 42 (1908 1966).
6. Schimitschek, E. J., Nehrich, R. B., and Trias, J. A., Recirculating liquid laser, *Appl. Phys. Lett.* 9, 103 (1966).
7. Schimitschek, E. J., Nehrich, R. B., and Trais, J. A., Laser action in fluorinated europium chelates in acetonitrile, *J. Chem. Phys.* 42, 788 (1965).
8. Schimitschek, E. J., Trais, J. A., and Nehrich, R. B., Stimulated emission in an europium chelate solution at room temperature, *J. Appl. Phys.* 36, 867 (1965).
9. Brecher, C., Samelson, H., and Lempicki, A., Laser phenomena in europium chelates, III: spectroscopic effects of chemical composition and molecular structure, *J. Chem. Phys.* 42, 1081 (1965).
10. Samelson, H., Lempicki, A., Brecher, C., and Brophy, V., Room temperature operation of a europium chelate liquid laser, *Appl. Phys. Lett.* 5, 173 1964).
11. Schimitschek, E. J. and Nehrich, R. B., Laser action in europium dibenzoylmethide, *J. Appl. Phys.* 35, 2786 (1964).
12. Ebina, K., Okadam Y., Yamasaki, A., and Ujihara, K., Spontaneous and stimulated emission by Eu-chelate in a planar microcavity, *Appl. Phys. Lett.* 66, 2783 (1995).
13. Nehrich, R. B., Schimitschek, E. J., and Tras, J. A., Laser action in europium chelates prepared with NH$_3$, *Phys. Lett.* 12, 198 (1964).
14. Malashkevich, G. E. and Kuznetsova, V. V., Laser excited lasing in solutions of some europium chelates, *J. Appl. Spectr.* 22, 170 (1975).
15. Bykov, V. P., Intramolecular energy transfer and quantum generators, *J. Exptl. Theor. Phys. (U.S.S.R.)* 43, 1634 (1962).
16. Charles, R. G. and Ohlmann, R. C., Europium thenoyl trifluoro acetonate, *J. Inor. Nucl. Chem.* 27, 255 (1965).
17. Metlay, M., Fluorescence lifetime of the europium dibenzoylmethides, *J. Phys. Chem.* 39, 491 (1963).

18. Bhaumik, M. L., Fletcher, P. C., Nugent, L. J., Lee, S. M., Higa, S., Telk, C. L., and Weinberg, M., Laser emission from a europium benzoylacetonate alcohol solution, *J. Phys. Chem.* 68, 1490 (1964).

19. Ohlmann, R. C. and Charles, R. G., Fluorescence properties of europium dibenzoylmethide and its complexes with Lewis bases, *J. Chem. Phys.* 41, 3131 (1964).

20. Aristov, A. V., Maslyukov, Yu. S., and Reznikova, I. I., Luminescence of a europium chelate solution under intense pulsed excitation, *Opt. Spectr.* 21, 286 (1966).

21. Lempicki, A., Samelson, H., and Brecher, C., Laser phenomena in europium chelates, IV. Characteristics of the europium benzoylacetonate laser, *J. Chem. Phys.* 41, 1214 (1964).

22. Lempicki, A., Samelson, H., and Brecher, C., Laser action in rare earth chelates, in *Applied Optics Supplement 2 of Chemical Lasers*, 205 (1965).

23. Aristov, A. V. and Maslyukov, Yu. S., Stimulated emission in europium benzoylacetonate solutions, *J. Appl. Spectr.* 8, 431 (1968).

24. Schimitschek, E. J., Stimulated emission in rare earth chelate (europium benzoylacetonate) in a capillary tube, *Appl. Phys. Lett.* 3, 117 (1963).

25. Taniguchi, H., Tomisawa, H., and Kido, J., Ultra-low-threshold europium chelate laser in morphology-dependent resonances, *Appl. Phys. Lett.* 66, 1578 (1995).

26. Lempicki, A. and Samelson, H., Optical maser action in europium-benzoylacetone, *Phys. Lett.* 4, 133 (1963).

27. Goryaeva, E. M., Shablya, A. V., and Serov, A. P., Luminescence and stimulated emission for solutions of complexes of neodymium nitrate with perdeutero-tributylphosphate, *J. Appl. Spectr.* 28, 55 (1976).

28. Heller, A., Fluorescence and room temperature laser action of trivalent neodymium in an organic liquid solution, *J. Am. Chem. Soc.* 89, 167 (1967).

29. Whittaker, B., Low threshold laser action of a rare earth chelate in liquid and solid host media, *Nature* 228, 157 (1970).

30. Samelson, H., Brecher, C., and Lempicki, A., Europium chelate lasers, *J. Chem. Phys.* 64, 165 (1967).

31. Ross, D. L., Blanc, J., and Pressley, R. J., Deuterium isotope effect on the performance of europium chelate lasers, *Appl. Phys. Lett.* 8, 101 (1966).

32. Ross, D. L. and Blanc, J., Europium chelates as laser materials, in *Advances in Chemistry Series*, No. 71, American Chemical Society, Washington, DC (1967), chapter 12.

2.2.3 Aprotic Liquid Lasers

Neodymium aprotic lasers are listed in order of increasing wavelength in Table 2.2.2 together with the solvent, mode of operation, and references. Neodymium aprotic laser amplifiers are listed separately in Table 2.2.3.

Table 2.2.2
Neodymium Aprotic Liquid Lasers

Wavelength (μm)	Solvent	Operation	Reference
1.050	$POCl_3$–$SnCl_4$–UO_2^{2+}	long-pulse	1
1.052	$POCl_3$–$SnCl_4$	long-pulse	2
1.053	$POCl_3$–$ZrCl_4$	long-pulse	3
1.053	$POCl_3$–$AlCl_3$	long-pulse	3
1.054	$POCl_3$–$AlCl_3$	long-pulse	3
1.054	$POCl_3$–BBr_3	long-pulse	3
1.054	$POCl_3$–$AlCl_3$	long-pulse	4
1.054	$POCl_3$–$SOCl_2$–$SnCl_4$	long-pulse	5
1.055	$SeOCl_2$–$SnCl_4$	long-pulse	6
1.055	$POCl_3$–$SnCl_4$	long-pulse	7,8
1.056	$SeOCl_2$–$SnCl_4$	long-pulse	4,9–12
1.056	$SeOCl_2$–$SnCl_4$	Q-switched	13–15
1.056	$SeOCl_2$–$SbCl_5$	long-pulse	4, 11
1.056	$SeOCl_2$–$SbCl_5$	Q-switched	16
1.056	$POCl_3$–$SnCl_4$	long-pulse	12,17
1.056	$POCl_3$–$ZrCl_4$	long-pulse	3,4,12,18,19
1.056	$POCl_3$–$ZrCl_4$	Q-switched	20–25
1.056	$POCl_3$–$AlCl_3$	long-pulse	4
1.058	$GaCl_3$–$SOCl_2$	long-pulse	26,27
1.058	$SeOCl_2$–$SnCl_4$	long-pulse	7,12
1.0585	$GaCl_3$–$SOCl_2$	long-pulse	39
1.060	$POCl_3$–$SnCl_4$	long-pulse	28
1.061	$GaCl_3$–CCl_3	long-pulse	29
1.066	PBr_3–$SbBr_3$–$AlBr_3$	long-pulse	30
1.330	$SeOCl_2$–$SnCl_4$	long-pulse	31

Table 2.2.3
Neodymium Aprotic Liquid Single-Pass Laser Amplifiers

Wavelength (μm)	Oscillator	Solvent	Reference
1.052	Nd:POCl$_3$–ZrCl$_4$	POCl$_3$–ZrCl$_4$	19
1.053	Nd:ethylene glyol	ethylene glyol	38
1.056	Nd:SeOCl$_2$–SnCl$_4$	SeOCl$_2$–SnCl$_4$	15
1.056	Nd:silicate glass	SeOCl$_2$–SnCl$_4$	12,15, 32
1.058	Nd:glass	POCl$_3$–ZrCl$_4$	35-37
1.062	Nd:POCl$_3$	POCl$_3$–ZrCl$_4$	18, 25, 33, 35

References

1. Dvachenko, P. P., Kalinin, V. V. Seregina, E. A. et al., Inorganic liquid laser doped with neodymium and uranyl, *Laser and Particle Beams* 11, 493 (1993).
2. Collier, F., Michon, M., and LeSergent, C., Parametres laser du systeme liquide Nd^{+3}-POCl$_3$-SnCl$_4$(H$_2$O) compares a ceux du YAG et du verre dope au neodyme, *Compt. Rend.* 272, 945 (1971).
3. Schimitschek, E. J., Laser emission of a neodymium salt dissolved in POCl$_3$, *J. Appl. Phys.* 39, 6120 (1968).
4. Weichselgartner, H. and Perchermeier, J., Anorganischer flüssigkeits laser, *Z. Naturforsch.* 25a, 1244 (1970).
5. Alekseev, N. E., Zhabotinski, M. E., Ivanova, E. B., Malashko, Ya. I., and Rudnitskii, Y. P., Effect of thionyl chloride on the laser characteristics of the liquid phosphor POCl$_3$ - SnCl$_4$ - Nd^{+3}, *Inorg. Mater.* 9, 215 (1973).
6. Samelson, H., Lempicki, A., and Brophy, V., Output properties of the Nd^{+3}:SeOCl$_2$ liquid lasers, *IEEE J. Quantum Electron.* QE-4, 849 (1968). See, also, Watson, W., Reich, S., Lempicki, A. and Lech, J., A circulating liquid laser system, *ibid.*, p. 842.
7. LeSergent, C., Michon, M., Rousseau, S., Collier, F., Dubost, H., and Raoult, G., Characteristics of the laser emission obtained with the solution POCl$_3$, SnCl$_4$, Nd$_2$O$_3$, *Compt. Rend.* 268, 1501 (1969).
8. Voronko, Yu. K., Krotova, L. V., Sychugov, V. A., and Shipulo, G. P., Lasers with liquid active materials based on POCl$_3$:Nd^{+3}, *J. Appl. Spect.* 10, 168 (1969).
9. Lempicki, A. and Heller, A., Characteristics of the Nd^{+3}:SeOCl$_y$ liquid laser, *Appl. Phys. Lett.* 9, 108 (1966).
10. Kato, D. and Shimoda, K., Liquid SeOCl$_2$:Nd^{+3} laser of high quality, *Jpn. J. Appl. Phys.* 7, 548 (1968).
11. Heller, A., A high gain, room-temperature liquid laser: trivalent neodymium in selenium oxychloride, *Appl. Phys. Lett.*, 9, 106 (1966), and Liquid lasers - design of neodymium based inorganic systems, *J. Mol. Spectrosc.* 28, 101 (1968).
12. Samelson, H., Kocher, R., Waszak, T., and Kellner, S., Oscillator and amplifier characteristics of lasers based on Nd^{+3} dissolved in aprotic solvents, *J. Appl. Phys.* 41, 2459 (1970).
13. Yamaguchi, G., Endo, F., Murakawa, S., Okamura, S., and Yamanaka, C., Room temperature, Q-switched liquid laser (SeOCl$_2$-Nd^{+3}), *Jpn. J. Appl. Phys.* 7, 179 (1968).
14. Samelson, H. and Lempicki, A., Q switching and mode locking of Nd^{+3}:SeOCl$_2$ liquid laser, *J. Appl. Phys.* 39, 6115 (1968).
15. Yamanaka, C., Yamanaka, T., Yamaguchi, G., Sasaki, T., and Nakai, S., Tandem amplifier systems of glass and SeOCl$_2$ liquid lasers doped with neodymium, *Nachrichten Tech. Fachberichte* 35, 791 (1968).

16. Lang, R. S., Die erzeugung von reisen impulsen durch einen aktiv und passiv geschalteten anorganischen neodym-flüssigkeits laser, *Z. Naturforsch.* 25a, 1354 (1970).

17. Zaretskii, A. I., Vladimirova, S. I., Kirillov, G. A., Kormes, S. B., Negiva, V. R., and Sukharov, S. A., Some characteristics of a $POCl_3$ + $SnCl_4$ + Nd^{+3} inorganic liquid laser, *Sov. J. Quantum Electron.* 4, 646 (1974).

18. Samelson, H. and Kocher, R., Final Technical Report, High Energy Liquid Lasers, Contract N0001468-C-0110 (1974).

19. Green, M. andReou, D., Little, V. I. and Selden, A. C., A multigigawatt liquid laser amplifier, *J. Phys. D* 9, 701 (1976).

20. Ueda, K., Hongyo, M., Sasaki, T. and Yamanaka, C., High power Nd^{+3} $POCl_3$ liquid laser system, *IEEE J. Quantum Electron.* QE-7, 291 (1971).

21. Brinkschulte, H., Fill, E. and Lang, R., Spectral output properties of an inorganic liquid laser, *J. Appl. Phys.* 43, 1807 (1972).

22. Brinkschulte, H., Perchermeier, J. and Schimitschek, E. J., A repetitively pulsed, Q-switched, inorganic liquid laser, *J. Phys.* D-7, 1361 (1974).

23. Andreou, D., Little, V., Selden, A. C. and Katzenstein, J., Output characteristics of a Q-switched laser system, Nd^{+3}:$POCl_3$:$ZrCl_4$, *J. Phys.* D-5, 59 (1972).

24. Fahlen, T. S., High average power Q-switched liquid laser, *IEEE J. Quantum Electron.* QE-9, 493 (1973).

25. Hongyo, M., Sasaki, T., Ngao, Y., Ueda, K. and Yamanaka, C., High power Nd^{+3}:$POCl_3$ liquid laser system, *IEEE J. Quantum Electron.* QE-8,192 (1972).

26. Mochalov, I. V., Bondareva, N. P., Bondareva, A. S. and Markosov, S. A., Spectral, luminescence and lasing properties of Nd^{3+} ions in systems utilizing $GaCl_3$-$SOCl_2$ and $AlCl_3$-$SOCl_2$ inorganic liquid media, *Sov. J. Quantum Electron.* 12, 647 (1982); Mokhova E. A. and Sviridov, V. V., Luminescence and lasing properties of $SOCl_2$-$GaCl_3$-Nd^{3+} inorganic laser liquids, *Zh. Prinkl. Spectrosk.* 50, 609 (1989).

27. Batyaev, I. M., Kabatskii, Yu. A. and Shilov, S. M., Luminescence spectrum and lasing parameters for Nd^{3+} in the $SOCl_2$-$GaCl_3$-$NdCl_3$ system, *Inorg. Mater.* 27, 1633 (1991).

28 Blumenthal, N., Ellis, C. B. and Grafstein, D., New room temperature liquid laser: Nd(III) in $POCl_3$-$SnCl_4$, *J. Chem. Phys.* 48, 5726 (1968).

29. Batyaev, I. M. and Kabatskii, Yu. A., Luminescence spectrum and lasing parameters for the CCl_3-$GaCl_3$-Nd^{3+} system, *Inorg. Mater.* 27, 1630 (1991).

30. Bondarev, A. S., Buchenkov, V. A., Volyukin, V. M., Mak, A. A., Pogodaev, A. K., Przhevaskii, A. K., Sidorenko, Yu. K., Soms, L. N. and Stepanov, A. I., New low toxicity inorganic Nd^{+3}-activated liquid medium for lasers, *Sov. J. Quantum Electron.* 6, 202 (1976).

31. Heller, A. and Brophy, V., Liquid lasers: stimulated emission of Nd^{+3} in selenium oxychloride solutions in the $^4F_{3/2}$ to $^4I_{13/2}$ transition, *J. Appl. Phys.* 39, 6120 (1968).

32. Sasaki, T., Yamanaka, T., Yamaguchi, G. and Yamanaka, C., A construction of the high power laser amplifier using glass and selenium oxychloride doped with Nd^{+3}, *Jpn. J. Appl. Phys.* 8, 1037-1045 (1969).

33. Andreou, D., A high power liquid laser amplifier, *J. Phys.* D-7, 1073 (1974).

34. Andreou, D., On the growth of stimulated Raman scattering in amplifying media, *Phys. Lett.* 57A, 250 (1976).

35. Fill, E. E., Ein Nd-$POCl_3$ laser verstärker, *Z. Angew. Phys.* 32. 356 (1972).

36. Andreou, D., Selden, A. C. and Little, V. I., Amplification of mode locked trains with a liquid laser amplifier, Nd^{+3}:$POCl_3$:$ZrCl_4$, *J. Phys.* D-5, 1405 (1972).

37. Andreou, D. and Little, V. I., The effect of frequency shifts on the power gain of a laser amplifier, *Opt. Commun.* 6, 180 (1972).

38. Han, K. G., Kong, H. J., Kim, H. S. and Um, G. Y., Nd^{3+}:ethylene glyol amplifier and its stimulated emission cross section, *Appl. Phys. Lett.* 67, 1501 (1995).

39. Batyaev, I. M., Kabatskii, Yu. A., Mokhova, E. A. and Sviridov, V. V., Luminescence and lasing properties of $SOCl_2$-$GaCl_3$-Nd^{3+} inorganic laser liquids, *Zh. Prik. Spektr.* 50, 609 (1988).

Section 2.3
LIQUID POLYMER LASERS

All of the liquid polymer lasers listed below were photopumped by pulsed lasers. The laser materials were contained in typical cuvettes or dye cell cavities at room temperature.

Table 2.3.1
Liquid Polymer Lasers

Material	Solvent	Laser wavelength (nm)	Pump laser (wavelength - nm)	Ref.
MEH-PPV	xylene	~598	Nd:YAG (532)	1
TOP-PPV	THF	438–456	Nd:YAG (355)	2
TOP-PPV	p-xylene	436–456	Nd:YAG (355)	2
TOP-PPV	hexane	414–452	Nd:YAG (355)	2

Abbreviations for the materials in Table 2.3.1:

MEH-PPV: poly[2-methoxy-5-92'-ethylhexloxy)-*p*-phenylene vinylene]

TOP-PPV: poly[2,5,2",5"-tetraoctyl)-*p*-terphenyl-4,4"-ylene vinylene-*p*-phenylene vinylene]

References

1. Moses, D., High quantum efficiency luminescence from a conducting polymer in solution: a novel polymer laser dye, *Appl. Phys. Lett.* 60, 3215 (1992).
2. Brouwer, H.-J., Krasnikov, V., Hilberer, A., Wildeman, J. and Hadziioannou, G., Novel high efficiency copolymer laser dye in the blue wavelength region, *Appl. Phys. Lett.* 66, 3404 (1996).

Section 2.4
LIQUID EXCIMER LASERS

Excimer lasers have been demonstrated in all phases of matter—gases, liquids, and solids. In liquid form, polar excimers are red shifted by large amounts, thus liquid excimer lasers are tunable both within the broad lasing line and by the choice of the solvent. Laser action has been excited by pulsed optical radiation and pulsed electron beams.

Further Reading

Excimer Lasers, Rhodes, C. K., Ed., Springer Verlag, Berlin (1984).

Table 2.4.1
Liquid Excimer Lasers

Excimer	Liquid host	Transition	Lasing wavelength (nm)	Excitation	Ref.
Kr_2	argon (104 K)	—	147	electron beam (1 MeV, 40 ns)	1
Xe_2	argon	—	175	electron beam (~ 1 MeV, 10 ns)	2
Xe_2	argon (104 K)	—	175	electron beam (1 MeV, 40 ns)	1
XeF	argon (~84 K)	$B \rightarrow X$	404	XeF laser (351 nm, 15 ns)	3
XeO	argon (104 K)	$^1S \rightarrow {}^1D$	547	electron beam (1 MeV, 40 ns)	4

References

1. Loree, T. R., Showalter, R. R., Johnson, T. M., Birmingham, B. S., and Hughes, W. M., Liquid excimers: lasing Xe_2 and Kr_2 in liquid argon, *Optics Lett.* 14, 1051 (1989).
2. Basov, N. G., Danilychev, V. A., Popov, Yu. M., and Khodevich, D. D., Laser operating in the vacuum region of the spectrum by excitation of liquid xenon with an electron beam, *JETP Lett.* 12, 493 (1970).
3. Shahidi, M., Jara, H., Pummer, H., Egger, H., and Rhodes, C. K., Optically excited XeF* excimer laser in liquid argon, *Optics Lett.* 10, 448 (1985).
4. Loree, T. R., Showalter, R. R., Johnson, T. M., Birmingham, B. S., and Hughes, W. M., Lasing XeO in liquid argon, *Optics Lett.* 11, 510 (1986).

Section 3: Gas Lasers

Section 3
GAS LASERS

3.0 Introduction

Gas lasers comprise the largest number of lasing transitions—over 12,000. Gas lasers may be categorized as neutral atom, ionic, or molecular. Molecular lasers can be further divided or characterized by the nature of the transitions involved in the stimulated emission process; that is, the transitions may be between electronic, vibrational, or rotational energy levels. The output of many lasers may consist of several lines of varying intensities.

Of the spectral ranges for lasers shown in the Preface, by far the largest range is that of gas lasers. The wavelength ranges of neutral atom, ionized, and molecular gas lasers extend from the extreme ultraviolet through the submillimeter. (Extreme ultraviolet and soft x-ray lasers are covered separately in Section 4.1.) Figure 3.0.1 shows the extremes of the wavelength ranges of different types of gas lasers. Neutral atom lasers emit throughout the ultraviolet, visible, and infrared. Ion lasers emit in the ultraviolet through the near infrared, the most important of which are based on the noble gas ions (Ar, Kr, Xe). These are operated in various states of ionization and in either pulsed or cw lasing modes. Metal vapor lasers, which may be either neutral atoms or ions, emit in the near-ultraviolet and visible and operate either pulsed or cw. Of these, cadmium, copper and gold are the most important examples.

Molecular lasers encompass a wide variety of molecules, operating conditions, and output wavelengths ranging from electronic transitions of the nitrogen (N_2) laser in the near ultraviolet, to the widely used vibrational-rotational transitions of the carbon dioxide (CO_2) laser in the mid-infrared, to the rotational transitions of various halide molecules lasing in the far infrared-submillimeter wavelength region. Electrically excited lasers such as H_2O, HCN, and DCN have transitions that extend well into the far infrared region.

Excimer lasers are based upon the formation in the gas phase of transient molecules such as XeCl, ArF, KrF, most of which emit in the ultraviolet or vacuum ultraviolet. These molecules, produced by collisions between rare gas ions or neutrals in excited states and

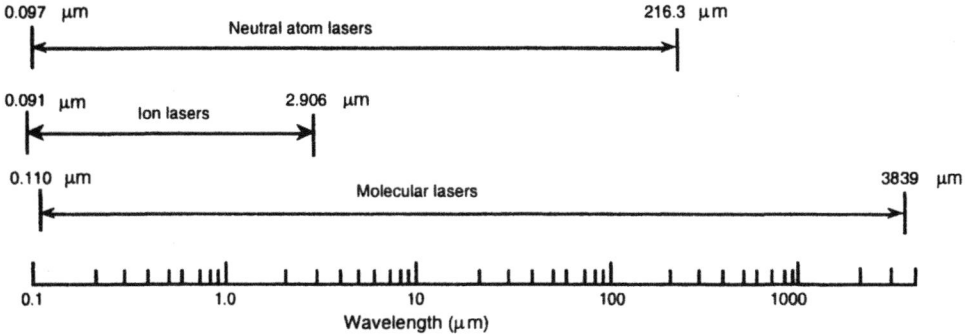

Figure 3.0.1   Reported extreme ranges of output wavelengths of various types of gas lasers (excluding extreme ultraviolet, soft x-ray, and optically pumped far infrared lasers).

halogen-containing molecular precursors, have strongly bound upper laser levels but ground states that are dissociative or weakly bound. They are generally produced in fast electrical discharges but can also be pumped optically or by intense electron or proton beams. Because the excited state lifetimes are short, ~10 ns, excimer lasers can emit powerful ultraviolet pulses of nanosecond duration.

In chemical lasers an inverted population is achieved—directly or indirectly—by an exothermic chemical reaction (for example, the exothermic reaction of H_2 and F_2 to yield vibrationally excited HF). Excitation processes include pumping in the course of photoinduced or electron-impact-induced chemical bond rupture, as well as by radiative association of atoms or molecules. In the oxygen-iodine laser, excited molecular oxygen transfers electronic energy to metastable levels of iodine. Chemical lasers operate in the near to middle infrared and have been operated pulsed and cw.

Gas lasers covered in this section are divided into four subsections: neutral atom gas lasers (Section 3.1), ionized gas lasers (Section 3.2), molecular gas lasers (Section 3.3), and far infrared and millimeter wave lasers (Section 3.4). For Section 3.4 one must decide on a definition of "far infrared". A perusal of numerous texts and handbooks reveals a variety of definitions beginning at 10 to 25 μm and extending to 300 to 1000 μm. Here we use 20 μm as the lower limit for the far infrared. By so doing we avoid the task of separating out the extremely numerous CO_2, N_2O, and other laser transitions in the 10 to 20 μm region which are covered in Section 3.3.2. The tabulation of lasing transitions in Section 3.4 extends to a few millimeters, thus overlapping millimeter wave masers.

As noted in the Preface, although there is a tremendous diversity of gas laser and transitions, only a few laser systems have achieved widespread use and commercial acceptance. Various types and properties of commercial gas lasers are listed in Section 3.5.

Comments about experimental conditions used, transition assignments, and other features associated with gas laser action are grouped together in Section 3.6.

The references with titles or descriptions of the contents are given in Section 3.7. The references generally include the original report of lasing plus other reports relevant to the identification of the lasing transition and operation.

Section 3.1
NEUTRAL ATOM GAS LASERS

3.1.1 Introduction

The tables in this section include laser lines originating from transitions between energy levels of elements in the neutral gas state. Figure 3.1.1 indicates by the shaded squares on the periodic table those elements that have produced neutral atom gas laser lines. Several different methods exist for exciting laser action in neutral species. These include weakly ionized DC and RF-excited discharges, pulsed discharges, pulsed electron beams, excitation in recombining plasmas, direct optical pumping, photodissociative optical pumping, excitation as a result of chemical reactions, and nuclear pumping.

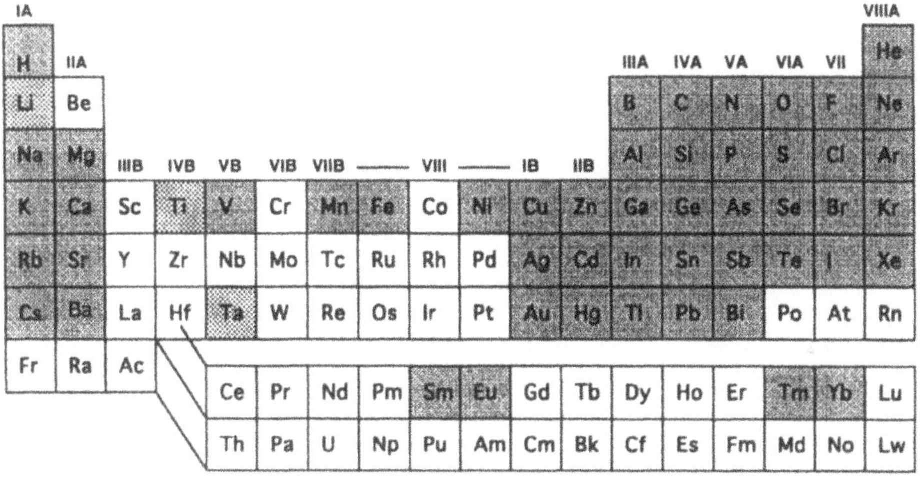

Figure 3.1.1 Periodic table of the elements showing those elements (shaded) which have exhibited laser oscillation in the neutral species.

The laser tables in this section are arranged by columns in the periodic table; within each column laser ions are further arranged in the order of increasing atomic number. Unless stated otherwise, and with the exception of the lanthanides, identified transitions are assigned and their wavelengths, *in vacuo* (in italics), given in accordance with energy level data given by C. E. Moore in *Atomic Energy Levels*.[1] Calculated wavelengths and spectral assignments for the lanthanides are taken from Martin, Zabubas, and Hagan, *Atomic Energy Levels—The Rare-Earth Elements*, NSRDS-NBS 60, 1978. Most modern energy level data appear to be accurate to $\leq 10^{-3}$ cm^{-1}; therefore several significance digits have been retained in calculated, *in vacuo*, wavelengths. One can easily convert these figures to air values by the use of tables of wavenumbers[2] or by direct computation from the Edlen's refractive index formula

$$n(v) = 1 + 6432.8 \times 10^{-8} + 2949810/(146 \times 10^8 - v^2) + 25540/(41 \times 10^8 - v^2),$$

where λ[microns, air] $= 10^4/n(v)v$[cm^{-1}, vacuum].

Figures of energy levels and laser transitions included in this section are from C. C. Davis, *Handbook of Laser Science and Technology, Vol. II: Gas Lasers.*

References include the first report of laser oscillation and representative selected papers on the laser system itself and the physical processes occurring within it. If several significantly different pumping conditions exist, these are included (the implication that laser action occurs only under the specific conditions indicated is not intended). Additional details about experimental conditions and results are given in the comments.

1. Moore, C. E., *Atomic Energy Levels*, Circular 467, Volumes 1, 2, 3, U.S. Government Printing Office, Washington, DC (1949, 1952, 1958).
2. Coleman, C. D., Bozman, W. R., and Meggers, W. F., *Table of Wave Numbers*, vols. 1 and 2, U.S. National Bureau of Standards Monograph 3 (1960).

Further Reading

Bennett, W. R., Jr., *Atomic Gas Laser Transition Data, A Critical Evaluation*, Plenum, New York (1979).

Davis, C. C., Neutral Gas Lasers, in *Handbook of Laser Science and Technology, Vol. II: Gas Lasers*, CRC Press, Boca Raton, FL (1982), p. 3.

Eden, J. G., Ed., *Selected Papers on Gas Laser Technology*, SPIE Milestone Series Vol. 159, SPIE Optical Engineering Press, Bellingham, WA (2000).

Goldhar, J., Neutral Gas Lasers, in *Handbook of Laser Science and Technology, Suppl. 1: Lasers*, CRC Press, Boca Raton, FL (1991), p. 325.

Waynant, R. W. and Ediger, M. N., Eds., *Selected Papers on UV, VUV, and X-Ray Lasers*, SPIE Milestone Series, Vol. MS71, SPIE Optical Engineering Press, Bellingham, WA (1993).

3.1.2 Tables of Neutral Atom Gas Lasers

The tables of neutral atom gas lasers are arranged in the following order:

Group IA – Table 3.1.1
- hydrogen
- lithium
- sodium
- potassium
- rubidium
- cesium

Group IB – Table 3.1.2
- copper
- silver
- gold

Group IIA – Table 3.1.3
- magnesium
- calcium
- strontium
- barium

Group IIB – Table 3.1.4
- zinc
- cadmium
- mercury

Group IIIA – Table 3.1.5
- boron
- aluminum
- gallium
- indium
- thallium

Group IVA – Table 3.1.6
- carbon
- silicon
- germanium
- tin
- lead

Group IVB – Table 3.1.7
- titanium

Group VA – Table 3.1.8
- nitrogen
- phosphorous
- arsenic
- antimony
- bismuth

Group VB – Table 3.1.9
- vanadium
- tantalum

Group VIA – Table 3.1.10
- oxygen
- sulfur
- selenium
- tellurium
- thulium

Group VII – Table 3.1.11
- fluorine
- chlorine
- bromine
- iodine
- ytterbium

Group VIIB – Table 3.1.12
- manganese

Group VIII – Table 3.1.13
- iron
- nickel
- samarium
- europium

Group VIIIA – Table 3.1.14
- helium
- neon
- argon
- krypton
- xenon

3.1.2.1 Group IA Lasers

Table 3.1.1
Group IA Lasers

Hydrogen (Figure 3.1.2)

Wavelength (μm)	Transition assignment	Comments	References
0.434046	$5 \rightarrow 2$(H-γ line, Balmer series)	130	1,3
0.486132	$4 \rightarrow 2$(H-β line, Balmer series)	130	1,3
0.656	$3D \rightarrow 2P$	6898,6899	1765,1766
1.875104	$4f\ ^2F^0_{7/2} \rightarrow 3d\ ^2D_{5/2}$ (P_a, first member of Paschen series, strongest fine-structure component)	130,339	1,4

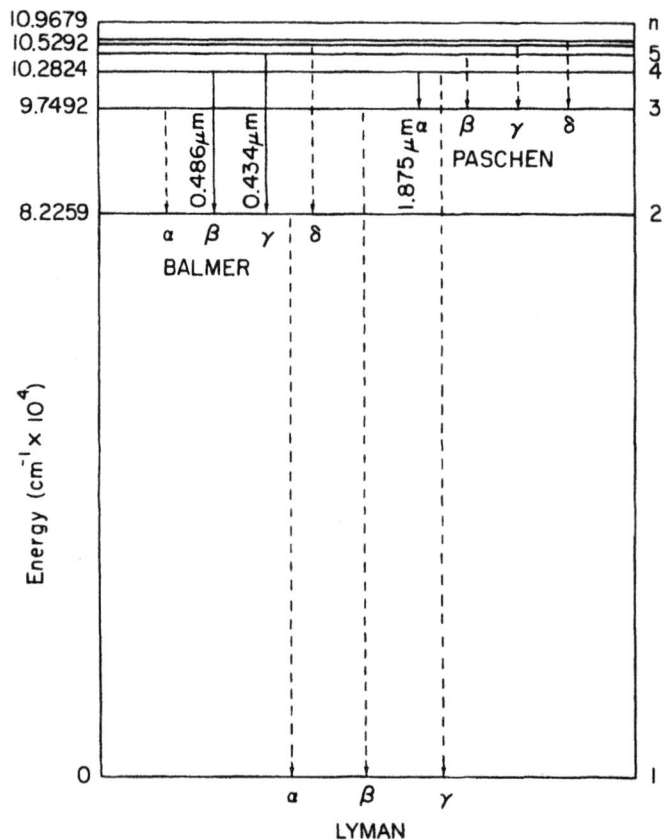

Figure 3.1.2 Partial energy level diagram of atomic hydrogen showing the reported laser transitions.

Lithium

Wavelength (μm)	Transition assignment	Comments	References
0.6707	$2p\ ^2P^0_{1/2,3/2} \rightarrow 2s\ ^2S_{1/2}$	6522,6524	1490
1.279	$5f\ ^2F^0 \rightarrow 3d\ ^2D$	6523	1499
1.870	$4f\ ^2F^0_{7/2} \rightarrow 3d\ ^2D_{5/2}$	6523	1499
2.689	$3p\ ^2P_{3/2} \rightarrow 3s\ ^2S_{1/2}$	6523	1499
4.029	$5f^2\ F^0 \rightarrow 4d\ ^2D$	6523	1499

Sodium (Figure 3.1.3)

Wavelength (μm)	Transition assignment	Comments	References
0.58899504	$3p\ ^2P^0_{3/2} \rightarrow 3s\ ^2S_{1/2}$	5,6,9,10,11	350
0.58959236	$3p\ ^2P^0_{1/2} \rightarrow 3s\ ^2S_{1/2}$	5,6,7,9,10	350,463
1.138145	$4s\ ^2S_{1/2} \rightarrow 3p\ ^2P^0_{1/2}$	5,8,12	573,679,772,785
1.140378	$4s\ ^2S_{1/2} \rightarrow 3p\ ^2P^0_{3/2}$	5,8,13	573,679,772,785
1.4553011	$4s\ ^4P_{5/2} \rightarrow 3p\ ^2D^0_{5/2}$	284,317	195,196
2.206	$4p\ ^2P^0_{3/2} \rightarrow 4s\ ^2S_{1/2}$	6525	1500
3.408	$5s\ ^2S_{1/2} \rightarrow 4p\ ^2P^0_{1/2}$	6526	1500
3.41	$5s\ ^2S_{1/2} \rightarrow 4p\ ^2P^0_{3/2}$	6527	1500

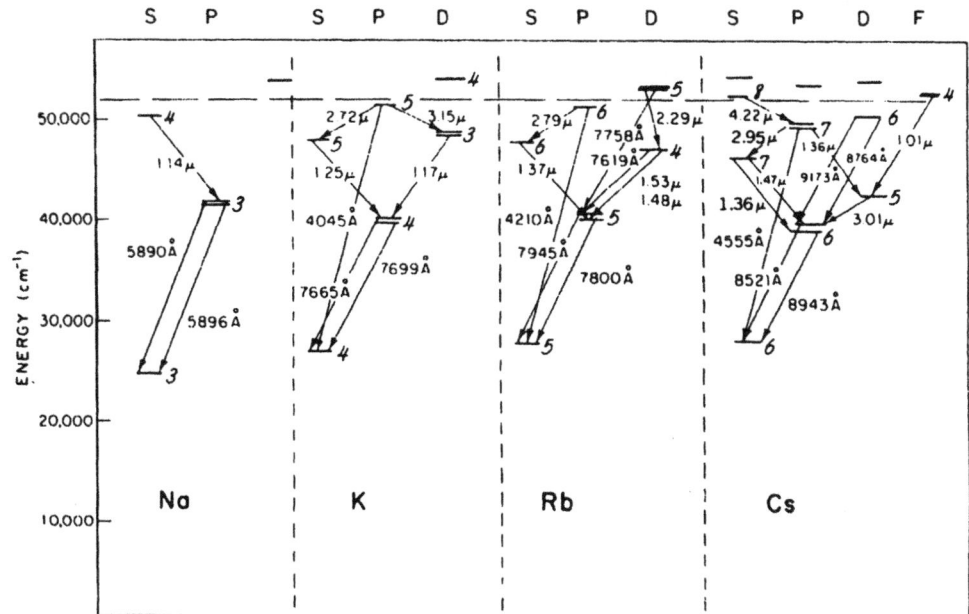

ALKALI DISSOCIATION LASERS

Figure 3.1.3 Laser transitions observed in neutral alkali metals following ArF laser dissociation of the alkali salts. The script numbers to the right of the levels are principal quantum numbers. (From Ehrlich, D. J. and Osgood, R. M., Jr., W. L., *Appl. Phys. Lett.* 34, 655, 1979. With permission.)

Potassium (Figure 3.1.3)

Wavelength (μm)	Transition assignment	Comments	References
0.4044136	5p $^2P^0_{3/2}$ → 4s $^2S_{1/2}$	14,15,20	350
0.40472602	5p $^2P^0_{1/2}$ → 4s $^2S_{1/2}$	14,15,21	350
0.6911	6s $^2S_{1/2}$ → 4p $^2P1/2$	1769	6903
0.7698959	4p $^2P^0_{1/2}$ → 4s $^2S_{1/2}$	14,23	350
0.7664899	4p $^2P^0_{3/2}$ → 4s $^2S_{1/2}$	14,22	350
1.177283	3d $^2D_{5/2}$ → 4p $^2P^0_{3/2}$	14,24	350
1.243224	5s $^2S_{1/2}$ → 4p $^2P^0_{1/2}$	14,19,25	772
1.252211	5s $^2S_{1/2}$ → 4p $^2P^0_{3/2}$	14,26	350,772
3.1415224	5p $^2P^0_{3/2}$ → 3d $^2D_{3/2}$	14,16,27	350,772
3.1601267	5p $^2P^0_{1/2}$ → 3d $^2D_{3/2}$	14,16,28	350,796
6.422525	6p $^2P^0_{3/2}$ → 6s $^2S_{1/2}$	14,17,29	131,142
6.4575288	6p $^2P^0_{1/2}$ → 6s $^2S_{1/2}$	14,17,30	131,142
7.8953393	7s $^2S_{1/2}$ → 6p $^2P^0_{3/2}$	14,31	131,142
9.1791962	6d $^2D_{3/2}$ → 5f $^2F^0$	14,18,32	131,142
12.568814	7p $^2P^0_{1/2}$ → 7s $^2S_{1/2}$	14,33	131,142
15.968063	6d $^2D_{3/2}$ → 7p $^2P^0_{1/2}$	14,34	131,142

Rubidium (Figure 3.1.3)

Wavelength (μm)	Transition assignment	Comments	References
0.420185	6p $^2P^0_{1/2}$ → 5s $^2S_{1/2}$	35,37,38	350
0.421556	6p $^2P^0_{3/2}$ → 5s $^2S_{1/2}$	35,37,39	350
0.7621029	5d $^2D_{3/2}$ → 5p $^2P^0_{1/2}$	37,42	350
0.7761570	5d $^2D_{3/2}$ → 5p $^2P^0_{3/2}$	37,43	350
0.7800268	5p $^2P^0_{3/2}$ → 5s $^2S_{1/2}$	37,40	350
0.7947603	5p $^2P^0_{1/2}$ → 5s $^2S_{1/2}$	37,41	350
1.32	6s $^2S_{1/2}$ → 5p $^2P^0_{1/2}$	6528	1501
1.366501	6s $^2S_{1/2}$ → 5p $^2P^0_{3/2}$	37,44	350
1.475241	4d $^2D_{3/2}$ → 5p $^2P^0_{1/2}$	37,45	350
1.528843	4d $^2D_{3/2}$ → 5p $^2P^0_{3/2}$	36,37,46	350
1.528948	4d $^2D_{5/2}$ → 5p $^2P^0_{3/2}$	36,37,47	350
2.252965	6p $^2P^0_{3/2}$ → 4d $^2D_{5/2}$	37,48	183,796
2.293247	6p $^2P^0_{1/2}$ → 4d $^2D_{3/2}$	37,50	183,350,796
2.73	6p $^2P^0_{3/2}$ → 6s $^2S_{1/2}$	6529	1501
2.790537	6p $^2P^0_{1/2}$ → 6s $^2S_{1/2}$	37,49	350

Cesium (Figure 3.1.3)

Wavelength (μm)	Transition assignment	Comments	References
0.0969	$5p^5 5d6s\ ^4D_{1/2} \to 5p^6 5d\ ^2D_{3/2}$	6924	903
0.4555276	$7p\ ^2P^0_{3/2} \to 6s\ ^2S_{1/2}$	51,52,53	350
0.8521133	$6p\ ^2P^0_{3/2} \to 6s\ ^2S_{1/2}$	52,54	350
0.87614150	$6d\ ^2D_{3/2} \to 6p\ ^2P^0_{1/2}$	52,55	350
0.8943468	$6p\ ^2P^0_{1/2} \to 6s\ ^2S_{1/2}$	51,52,56	350
0.91723217	$6d\ ^2D_{5/2} \to 6p\ ^2P^0_{3/2}$	52,57	350
1.01236025	$4f\ ^2F^0_{7/2} \to 5d\ ^2D_{5/2}$	52,58	350
1.358831	$7s\ ^2S_{1/2} \to 6p\ ^2P^0_{1/2}$	52,59	350
1.360257	$7p\ ^2P^0_{3/2} \to 5d\ ^2D_{5/2}$	52,60	796
1.375883	$7p\ ^2P^0_{1/2} \to 5d\ ^2D_{3/2}$	52,61	350,796
1.469493	$7s\ ^2S_{1/2} \to 6p\ ^2P^0_{3/2}$	52,62	350
1.47	$7\ ^2S_{1/2} \to 6\ ^2P_{3/2}$	6916	1764
2.9317981	$7p\ ^2P^0_{3/2} \to 7s\ ^2S_{1/2}$	52,63	350
3.01	$5\ ^2D_{3/2} \to 6\ ^2P_{1/2}$	6916	1764
3.01033	$5d\ ^2D_{3/2} \to 6p\ ^2P^0_{3/2}$	52,64	350
3.0111339	$5d\ ^2D_{3/2} \to 6p\ ^2P^0_{1/2}$	52,65	796
3.0961401	$7p\ ^2P^0_{1/2} \to 7s\ ^2S_{1/2}$	52,66	183,796
3.2050778	$8p\ ^2P^0_{1/2} \to 6d\ ^2D_{3/2}$	52,67	184,185,186
3.4909363	$5d\ ^2D_{5/2} \to 6p\ ^2P^0_{3/2}$	52,68	796
3.6140628	$5d\ ^2D_{5/2} \to 6p\ ^2P^0_{3/2}$	52,69	796
4.2181082	$8s\ ^2S_{1/2} \to 7p\ ^2P^0_{3/2}$	52,70	350
7.18537910	$8p\ ^2P^0_{1/2} \to 8s\ ^2S_{1/2}$	52,71	184,185,186,335

3.1.2.2 Group IB Lasers

Table 3.1.2
Group IB Lasers

Copper (Figure 3.1.4)

Wavelength (μm)	Transition assignment	Comments	References
0.510554	$4p\ ^2P^0_{3/2} \to 4s^2\ ^2D_{5/2}$	72,73,75	336–8,340–9, 351,356,367,378
0.570024	$4p\ ^2P^0_{3/2} \to 4s^2\ ^2D_{3/2}$	72,76	349
0.578213	$4p\ ^2P^0_{1/2} \to 4s^2\ ^2D_{3/2}$	72,73,77	336–8,340–9, 351,356,367,378
1.8199686	$4f\ ^2F^0_{5/2} \to 4d\ ^2D_{3/2}$	74,78	806
1.8234057	$4f\ ^2F^0_{7/2} \to 4d\ ^2D_{5/2}$	74,79	806
3.726	$4p\ ^2P^0_{1/2} \to 5s\ ^2S_{1/2}$	6530	1499
5.460	$7d\ ^2D_{3/2} \to 7p\ ^2p^0_{1/2}$	6531	1499

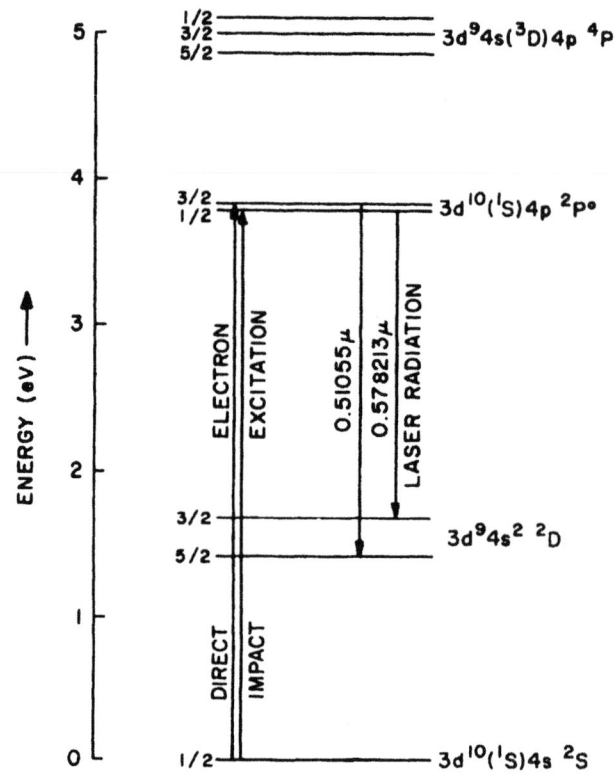

Figure 3.1.4 Partial energy level diagram of copper showing the strong self-terminating visible laser transitions.

Silver

Wavelength (μm)	Transition assignment	Comments	References
1.8380629	4f $^2F^0_{5/2,7/2}$ → 5d $^2D_{5/2}$	80,81	383,806
1.9371923	5s^2 $^2D_{3/2}$ → 5p $^2P^0_{1/2}$	82	806

Gold (Figure 3.1.5)

Wavelength (μm)	Transition assignment	Comments	References
0.3122784	6p $^2P^0_{3/2}$ → 6s^2 $^2D_{5/2}$	83,84	132,134,135
0.6278170	6p $^2P^0_{1/2}$ → 6s $^2D_{3/2}$	83,85	132–6,341,342

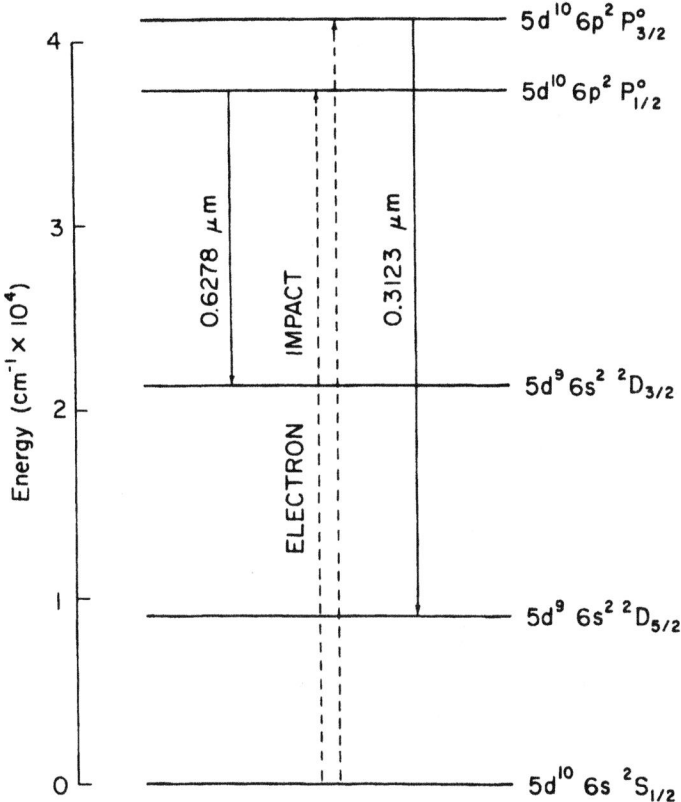

Figure 3.1.5 Partial energy level diagram of gold showing the two self-terminating laser transitions.

3.1.2.3 Group IIA Lasers

Table 3.1.3
Group IIA Lasers

Magnesium (Figure 3.1.6)

Wavelength (µm)	Transition assignment	Comments	References
1.502499	4p $^3P^0_2 \rightarrow$ 4s $^3S_1$	86,87,92,93	146
3.86573	7p $^3P^0\_?$ $\rightarrow$ 6s $^3S_1$	86,87,90,92,97	136
3.6789254	5d $^3D_1 \rightarrow$ 5p $^3P^0_1$	86,87,88,92,94	136
3.67895650	5d $^3D_2 \rightarrow$ 5p $^3P^0_1$	86,87,88,92,95	136
3.6825364	5d $^3D_{3,2,1} \rightarrow$ 5p $^3P^0_2$	86,87,89,96	136
4.2013276	5p $^3P^0_2 \rightarrow$ 5s $^3S_1$	86,87,92,98	136
4.3638859	5d $^3D_{3,2,1} \rightarrow$ 4f $^3F^0$	86,87,89,92,99	136

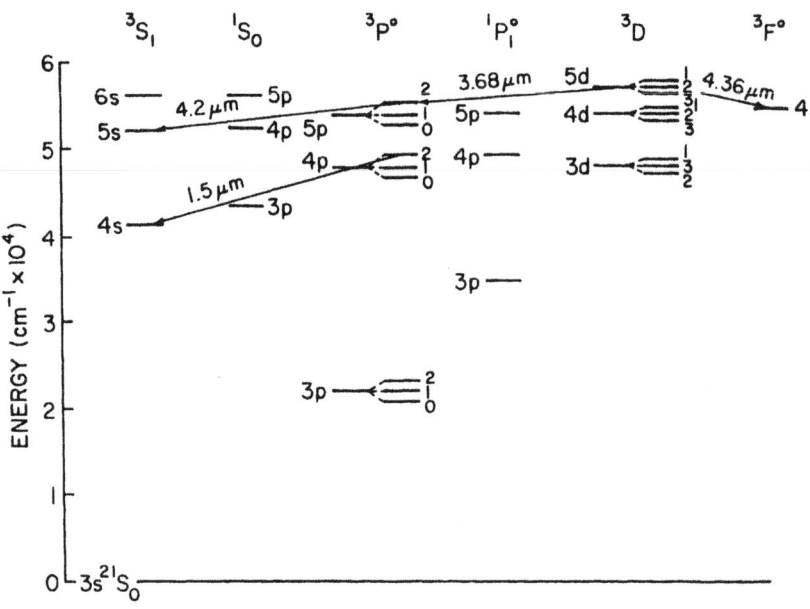

Figure 3.1.6 Partial energy level diagram of magnesium showing the majority of the reported laser transitions.

Calcium

Wavelength (μm)	Transition assignment	Comments	References
0.5349472	$4p'\ ^1F^0_3 \rightarrow 3d\ ^1D_2$	100,101,104	137
0.5857452	$4p^2\ ^1D_2 \rightarrow 4p\ ^1P^0_1$	100,101,105	137
0.644981	$4p'\ ^1D^0_2 \rightarrow 3d\ ^3D_1$	100,102,106	138
1.304	$3d\ ^3F_3 \rightarrow 4p'\ ^3F^0_3$	6532	1499
1.317	$3d\ ^3F_3 \rightarrow 4p'\ ^3F^0_2$	6533	1499
1.425	$5d\ ^3D_3 \rightarrow 4p'\ ^3F^0_2$	6534	1499
1.897	$3d^2\ ^3F_3 \rightarrow 4p'\ ^3D^0_2$	6535	1499
1.905	$3d^2\ ^3F_4 \rightarrow 4p'\ ^3D^0_3$	6536	1499
5.547327	$4p\ ^1P^0_1 \rightarrow 3d\ ^1D_2$	100,103,107	138–141

Strontium

Wavelength (μm)	Transition assignment	Comments	References
0.638075	$5p'\ ^1D^0_2 \rightarrow 4d\ ^3D_1$	108,110	138
3.0670208	$4d\ ^3D_1 \rightarrow 5p\ ^3P^0_2$	112	143,144
3.0118377	$4d\ ^3D_2 \rightarrow 5p\ ^3P^0_2$	111	143,144
6.4566866	$5p\ ^1P^0_1 \rightarrow 4d\ ^1D_2$	109,113	139,141

Barium (Figure 3.1.7)

Wavelength (μm)	Transition assignment	Comments	References
0.7120329	$6p'\ ^1D^0_2 \rightarrow 5d\ ^3D_1$	114,123,124	138,145
1.130304	$6p\ ^1P^0_1 \rightarrow 5d\ ^3D_2$	114–6,123,125	146,147,148
1.50004	$6p\ ^1P^0_1 \rightarrow 5d\ ^1D_2$	114–8,123,126	146–151,708
1.82041	$6p'\ ^1P^0_1 \rightarrow 5d2\ ^1D_2$	114,123,127	147
1.9022415	$6d\ ^1D_2 \rightarrow 6p'\ ^3D^0_3$	114,115,123,128	146
2.1573497	$6p'\ ^1P^0_1 \rightarrow 5d2\ ^3P_2$	114,115,123,129	146,147
2.32553	$6p\ ^3P^0_2 \rightarrow 5d\ ^3D_2$	114,123,130	146,147
2.4764593	$7s^1\ S_0 \rightarrow 6p'\ ^3D^0_1$	114,115,123,131	146
2.55157	$6p\ ^3P^0_2 \rightarrow 5d\ ^3D_3$	114,123,132	146,147
2.78	$6p\ ^3P^0_1 \rightarrow 5d\ ^3D_1$	6537	1502
2.9230381	$6p\ ^3P^0_1 \rightarrow 5d\ ^3D_2$	114,119,123,133	146,147
2.98	$6p\ ^3D^0_3 \rightarrow 5d2\ ^3F_4$	6538	1502
3.05	$6p'\ ^3D^0_3 \rightarrow 5d2\ ^3F_4$	6539	1502
3.9589222	$7s\ ^1S_0 \rightarrow 6p'\ ^3P^0_1$		146
4.0079678	$6p'\ ^3P^0_1 \rightarrow 5d2\ ^3P_0$	114,123,135	146
4.3285152	$7p\ ^1P^0_1 \rightarrow 6d\ ^1D_2$	114,123,136	147
4.6699795	$6d\ ^3D_1 \rightarrow 6p'\ ^1P^0_1$	114,121,123,137	146,147
4.7169143	$10d\ ^3D_2 \rightarrow 9p\ ^1P^0_1$	114,123,138	146
4.7184144	$6p\ ^3P^0_2 \rightarrow 5d\ ^1D_2$	114,123,139	146,147
5.0322846	$8p\ ^1P^0_1 \rightarrow 8s\ ^3S_1$	114,123,140	146
5.4798		114,122,123,141	146
5.5636		114,122,123,142	146
5.8899		114,122,123,143	146,147
6.4546		114,122,123,144	146

3.1.2.4 Group IIB Lasers

Table 3.1.4
Group IIB Lasers

Zinc

Wavelength (μm)	Transition assignment	Comments	References
0.4722	$5s\ ^3S_1 \rightarrow 4p\ ^3P^0_1$	6540	1503
0.4811	$5s\ ^3S_1 \rightarrow 4p\ ^3P^0_2$	6541	1503
1.305363	$5p\ ^3P^0_2 \rightarrow 5s\ ^3S_1$	145,146	383
1.315059	$5p\ ^3P^0_1 \rightarrow 5s\ ^3S_1$	145,147	383

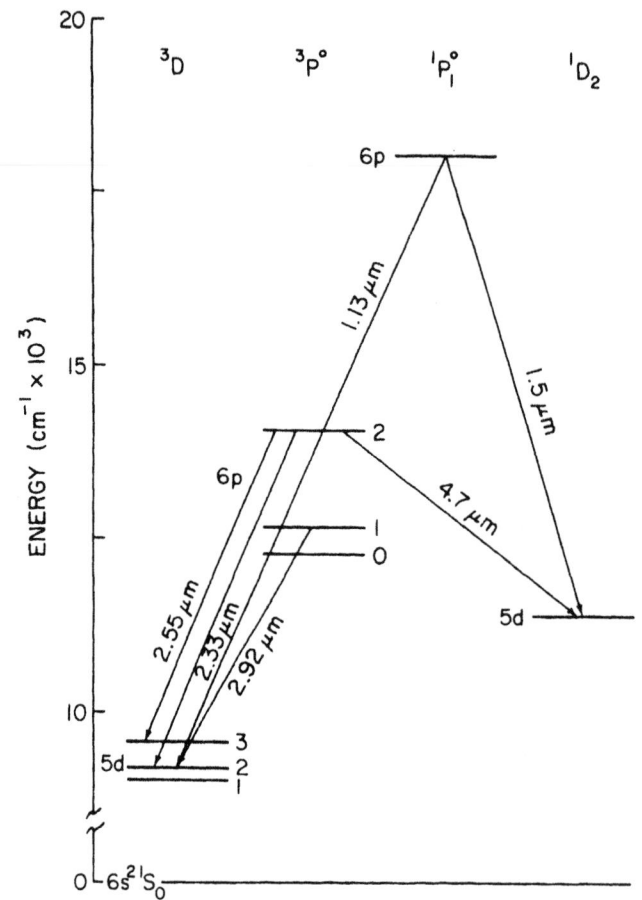

Figure 3.1.7 Partial energy level diagram of barium showing the laser transition between the lowest-lying levels.

Cadmium (Figure 3.1.8)

Wavelength (μm)	Transition assignment	Comments	References
0.4800	$6s\,^3S_1 \rightarrow 5p\,^3P^0_1$	6542	1503
0.5086	$6s\,^3S_1 \rightarrow 5p\,^3P^0_2$	6543	1503
0.7133974	$7p\,^1P^0_1 \rightarrow 6s\,^3S_1$	148,149,150	152
0.9838743	$9s\,^1S_0 \rightarrow 6p\,^3P^0_2$	155	152
1.0867911	$7d\,^1D_2 \rightarrow 6p\,^3P^0_2$	156	152
1.1485		157	152
1.1554		158	152
1.1663677	$8p\,^1P^0_1 \rightarrow 5d\,^3D_1$	148,159	152
1.1745636	$8s\,^1S_0 \rightarrow 6p\,^3P^0_0$	160	152
1.1874246	$6p\,^1P^0_1 \rightarrow 6s\,^3S_1$	161	152,153
1.3982714	$6p\,^3P^0_2 \rightarrow 6s\,^3S_1$	151	152,154,383
1.4331602	$6p\,^3P^0_1 \rightarrow 6s\,^3S_1$	153,154	152,153,154,383
1.4478302	$6p\,^3P^0_0 \rightarrow 6s\,^3S_1$	162	154,383,775

Cadmium (Figure 3.1.8)—*continued*

Wavelength (μm)	Transition assignment	Comments	References
1.6404449	4f $^3F^0_2$ → 5d $^3D_1$	163,164	152,154,383
1.6437081	4f $^3F^0_{2,3}$ → 5d $^3D_2$	165	152
1.6486189	4f $^3F^0_{2,3,4}$ → 5d $^3D_3$	166	152,153
1.9123124	6d $^1D_2$ → 6p $^1P^0_1$	167	152
13.188714	5d $^1D_2$ → 6p $^3P^0_1$	168	155
14.58202	6p $^1P^0_1$ → 5d $^1D_2$	169	155
1.1663677	8p $^1P^0_1$ → 5d $^3D_1$	148,159	152
1.1745636	8s $^1S_0$ → 6p $^3P^0_0$	160	152
1.1874246	6p $^1P^0_1$ → 6s $^3S_1$	161	152,153
1.3982714	6p $^3P^0_2$ → 6s $^3S_1$	151	152,154,383
1.4331602	6p $^3P^0_1$ → 6s $^3S_1$	153,154	152,153,154,383

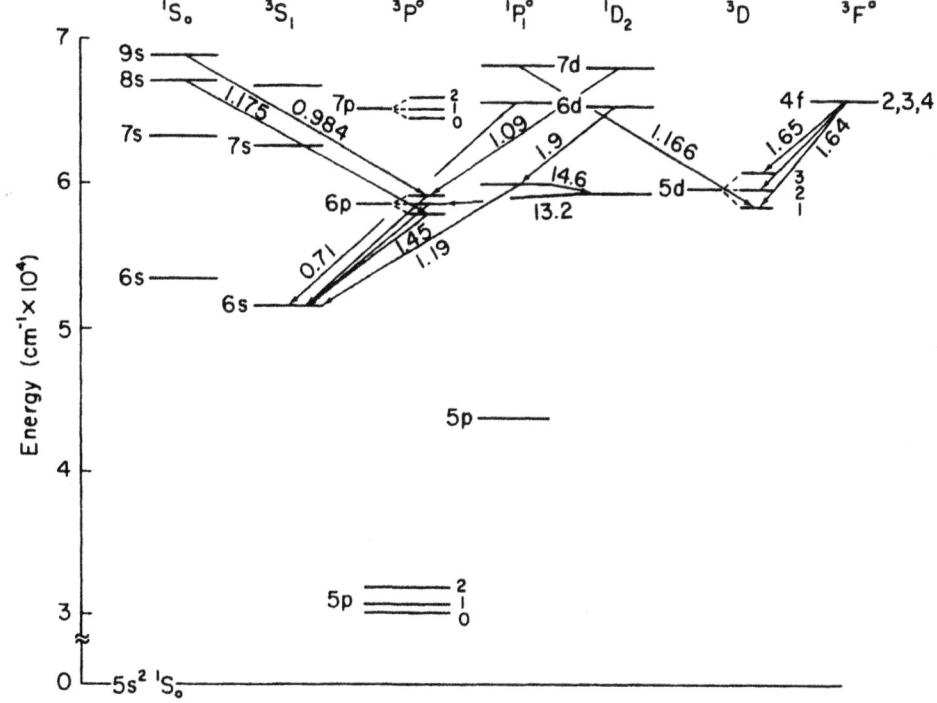

Figure 3.1.8 Partial energy level diagram of cadmium showing the majority of the laser transitions. Wavelengths are given in microns.

Mercury (Figure 3.1.9)

Wavelength (μm)	Transition assignment	Comments	References
0.365015	6d $^3D_3 \to$ 6p $^3P^0_2$	170,171,180,181	156
0.365483	6d $^3D_2 \to$ 6p $^3P^0_2$	170,171,180,182	156
0.366288	6d $^3D_1 \to$ 6p $^3P^0_2$	170,171,180,183	156
0.435835	7s $^3S_1 \to$ 6p $^3P^0_1$	170,173,180,185	156
0.546074	7s $^3S_1 \to$ 6p $^3P^0_2$	170,173,180,186	157–159
0.366328	6d $^1D_2 \to$ 6p $^3P^0_2$	170,171,180,184	156
0.404	7s $^3S_1 \to$ 6p $^3P^0_0$	6544	1504
0.579065	6d $^1D_2 \to$ 6p $^1P^0_1$	170,171,180,188	156
0.8677		175,180,189	160
0.576959	6d $^3D_2 \to$ 6p $^1P^0_1$	170,171,180,187	156
1.1179812	7p $^1P^0_1 \to$ 7s $^3S_1$	180,190	160,161
1.1290435	7p $^3P^0_2 \to$ 7s $^3S_1$	171,180,191	156
1.2222		180,192	162,163
1.2246		180,193	162,163
1.2545		180,194	160
1.2760		180,195	162,163
1.2981		180,196	160
1.3574217	7p $^1P^0_1 \to$ 7s $^1S_0$	180,197	156
1.3655	7p $^3P^0_1 \to$ 7s $^3S_1$ (probably)	171,180,198	160
1.3677207	7p $^3P^0_1 \to$ 7s $^3S_1$	171,180,199	156,161
1.3954389	7p $^3P^0_0 \to$ 7s $^3S_1$	171,180,200	156
1.529954	6p' $^3P^0_2 \to$ 7s $^3S_1$	180,201	160–174
1.6924775	5f $^1F^0_3 \to$ 6d $^1D_2$	180,202	160,161
1.6946636	5f $^3F^0_2 \to$ 6d $^3D_1$	180,203	160,161
1.7077438	5f $^3F^0_4 \to$ 6d $^3D_3$	180,204	160,161
1.7114554	5f $^3F^0_3 \to$ 6d $^3D_2$	180,205	160,161
1.7334185	7d $^1D_2 \to$ 7p $^1P^0_1$	180,206	161,167
1.8135329	6p' $^3F^0_4 \to$ 6d $^3D_3$	180,207	160,161,164–169
3.928361	6d $^3D_3 \to$ 6p' $^3P^0_2$ or 5g G $\to$ 5f F^0	176,177,180,208	161,167

3.1.2.5 Group IIIA Lasers

Table 3.1.5
Group IIIA Lasers

Boron

Wavelength (μm)	Transition assignment	Comments	References
3.601		213,214	176

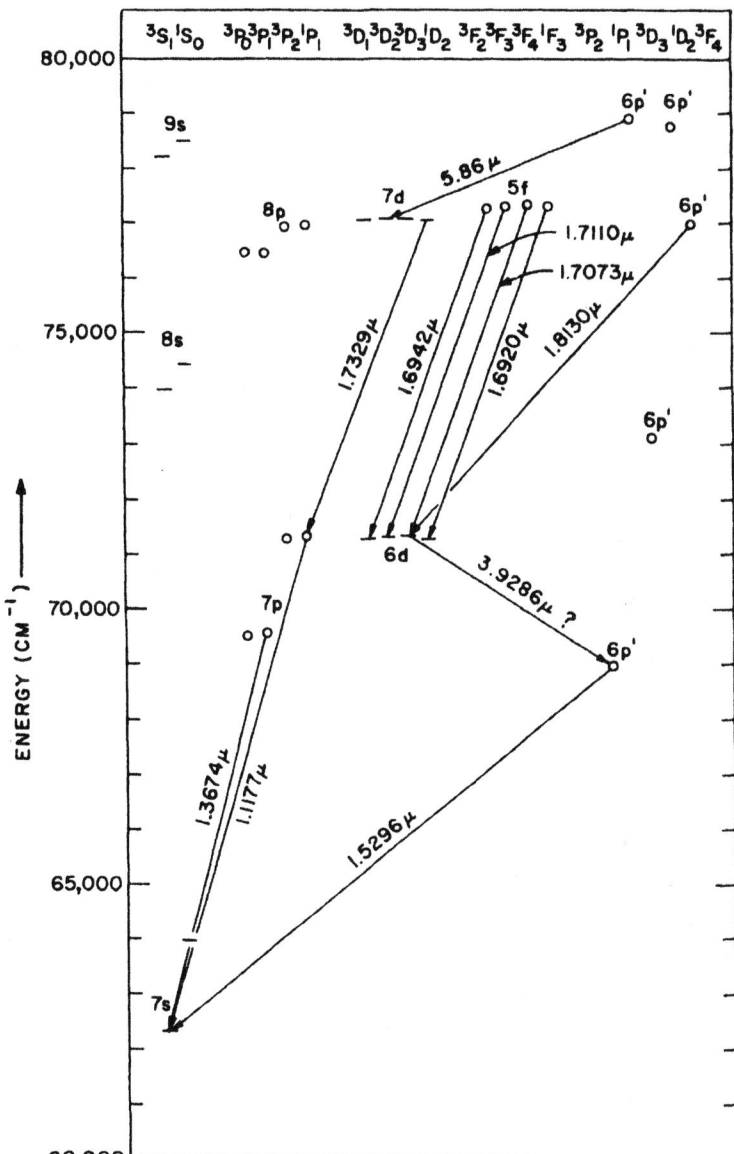

Figure 3.1.9 Partial term diagram of mercury showing the majority of the reported laser transitions. Circles represent odd levels; horizontal lines represent even levels. (From Bockasten, K., Garavaglia, M., Lengyel, B. A., and Lundholm, *J. Opt. Soc. Amer.* 55, 1051, 1965. With permission.)

Aluminum

Wavelength (μm)	Transition assignment	Comments	References
0.3962	$4s\ ^2S_{1/2} \rightarrow 3p\ ^2P_{3/2}$	6900	1771

Gallium (Figure 3.1.10)

Wavelength (μm)	Transition assignment	Comments	References
0.4032987	5s $^2S_{1/2}$ → 4p $^2P^0_{1/2}$	215,216,217	177
0.4172042	5s $^2S_{1/2}$ → 4p $^2P^0_{3/2}$	215,216,218	177
1.7367231	4p^2 $^4P_{5/2}$ → 5p $^2P^0_{3/2}$	219	175
5.754965	4d $^2D_{3/2}$ → 5p $^2P^0_{1/2}$	215,220	175
6.1477551	4d $^2D_{3/2}$ → 5p $^2P^0_{3/2}$	215,221	175

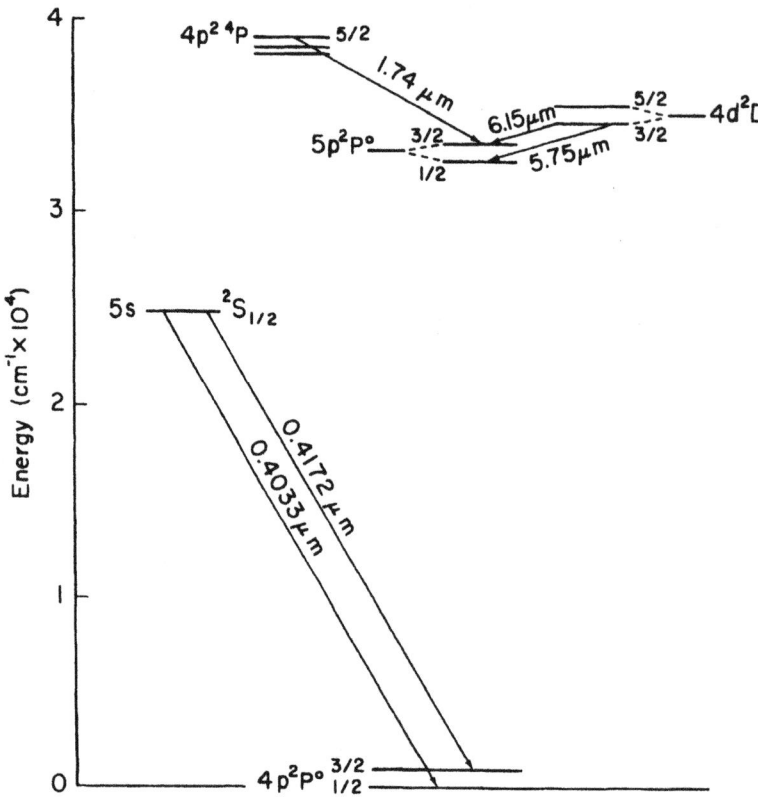

Figure 3.1.10 Partial energy level diagram of gallium showing the reported laser transitions. Wavelengths are given in microns.

Indium

Wavelength (μm)	Transition assignment	Comments	References
0.4511	$6\sigma\ ^2\Sigma_{1/2} \to 5\pi\ ^2P_{3/2}$	6911	1761
1.342996	$6p\ ^2P^0_{1/2} \to 6s\ ^2S_{1/2}$	222,225	383
1.431625	$6d\ ^2D_{5/2} \to 6p\ ^2P^0_{3/2}$	222,226	383
1.441920	$6d\ ^2D_{3/2} \to 6p\ ^2P^0_{3/2}$	222,227	383
1.8736732	$5p^2\ ^4P_{5/2} \to 6p\ ^2P^0_{3/2}$	222,228	175
2.3785794	$5p^2\ ^4P_{3/2} \to 6p\ ^2P^0_{1/2}$	222,229	175
0.4101745	$6s\ ^2S_{1/2} \to 5p\ ^2P^0_{1/2}$	222,223	178
0.4511299	$6s\ ^2S_{1/2} \to 5p\ ^2P^0_{3/2}$	222,224	178

Thallium

Wavelength (μm)	Transition assignment	Comments	References
0.377572	$7s\ ^2S_{1/2} \to 6p^2\ P^0_{1/2}$	230,233,234	180
0.535065	$7s\ ^2S_{1/2} \to 6p\ ^2P^0_{3/2}$	231–236	181,182,187–9
3.8135916	$8p\ ^2P^0_{1/2} \to 8s\ ^2S_{1/2}$	233,237	175
5.1072522	$6d\ ^2D_{3/2} \to 7p\ ^2P^0_{1/2}$	233,238	175
10.451505	$6d\ ^2D_{3/2} \to 7p\ ^2P^0_{3/2}$	233,239	175

3.1.2.6 Group IVA Lasers

Table 3.1.6
Group IVA Lasers

Carbon (Figure 3.1.11)

Wavelength (μm)	Transition assignment	Comments	References
0.8335149	$3p\ ^1S_0 \to 3s\ ^1P^0_1$	240,243	190
0.962	$3p\ ^3S_1 \to 3s\ ^3P^0_1$	6547	1505
0.9658	$3p\ ^3S_1 \to 3s\ ^3P^0_2$	6548	1505,1506
1.0683082	$3p\ ^3D_2 \to 3s\ ^3P^0_1$	240,241,245	192,191,193
1.0685345	$3p\ ^3D_1 \to 3s\ ^3P_0$	240,241,246	190,193
1.0691250	$3p\ ^3D_3 \to 3s\ ^3P^0_2$	240,241,247	190–1,193,194–6
1.0707333	$3p\ ^3D_1 \to 3s\ ^3P^0_1$	240,248	193
1.0730	$3p\ ^3D_2 \to 3s\ ^3P^0_2$	6549	1505
1.454250	$3p\ ^1P_1 \to 3s\ ^1P^0_1$	240,241,249	190–1,193,194–8, 383,403,405
2.0655993	$5d\ ^1D^0_2 \to 4p\ ^3P_2$	240,250	195
3.407422	$4d\ ^1D^0_2 \to 4p\ ^1P_1$	240,251	195
3.5117661	$6d\ ^3P^0_2 \to 5p\ ^3D_3$	240,252	195

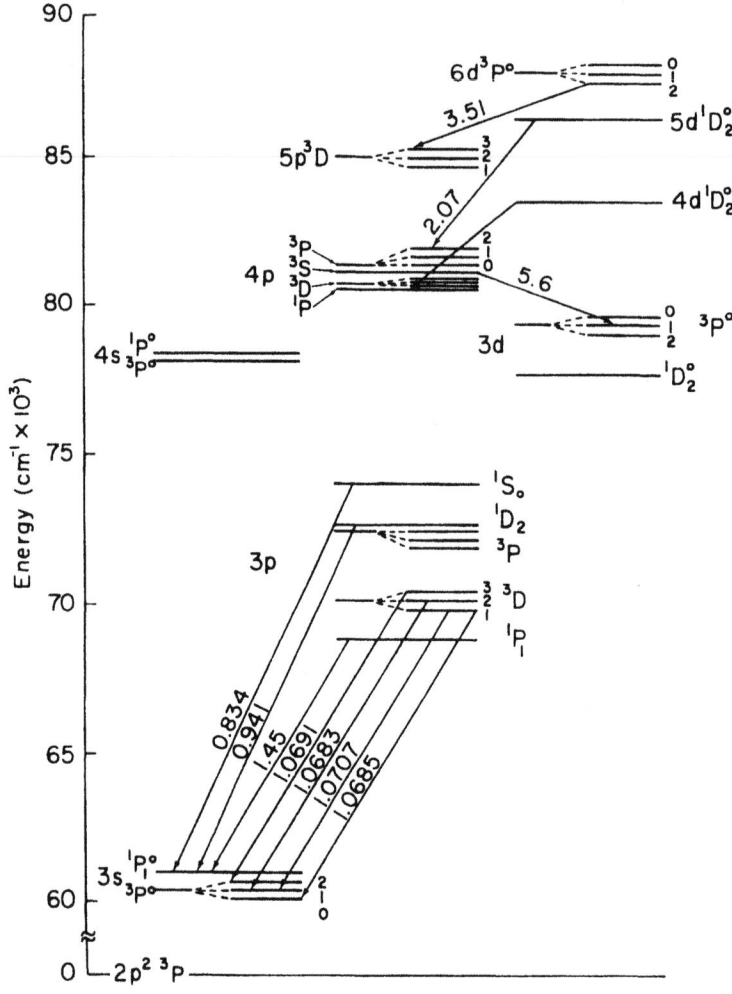

Figure 3.1.11 Partial energy level diagram of carbon showing the reported laser transitions. Wavelengths are given in microns.

Silicon

Wavelength (μm)	Transition assignment	Comments	References
1.1984187	$4p\ ^3D_2 \rightarrow 4s\ ^3P^0_1$	254,255	199
1.2031507	$4p\ ^3D_3 \rightarrow 4s\ ^3P^0_2$	254,256	199
1.5884410	$5s\ ^3P^0_1 \rightarrow 4p\ ^3D_1$	254,257	199,403

Germanium

Wavelength (μm)	Transition assignment	Comments	References
0.3269	$5s\ ^3P^0_1 \to 4p^2\ ^1D_2$	6550	1490
1.9814602	$6d\ ^3P^0_1 \to 6p\ ^3P_0$	258,260	175
2.020602	$6f\ ^3G_3 \to 5d\ ^3P^0_2$	258,259,261	175

Tin

Wavelength (μm)	Transition assignment	Comments	References
0.3801	$6s\ ^3P^0_1 \to 5p^2\ ^1D_2$	6551	1490
0.657903	$10d\ ^3D^0_2 \to 6p\ ^3P_1$?	262,263,264	200,201,202
1.0612556	$5p^3\ ^3D^0_1 \to 6p\ ^3P_2$	265	384
1.3612294	$6p\ ^1P_1 \to 6s\ ^1P^0_1$	266	383
4.6157396	$5d\ ^3D^0_1 \to 6p\ ^3P_1$	267	155

Lead (Figure 3.1.12)

Wavelength (μm)	Transition assignment	Comments	References
0.36395677	$7s(3/2,1/2)^0_1 \to 6p^2(3/2,3/2)_2$ $(7s\ ^3P^0_1 \to 6p^2\ ^3P_1)$	268,269,273	204,239
0.405780	$677s(1/2,1/2)^0_1 \to 6p^2(3/2,1/2)_2$ $(7s\ ^3P^0_1 \to 6p^2\ ^3P_2)$	268,269,274	204,207,239
0.40621360	$6d1/2[3/2]^0_1 \to 6p^2(3/2,3/2)_2$ $(6d\ ^3D^0_1 \to 6p^2\ ^1D_2)$	268,269,275	204,207,239
0.7228965	$37s(1/2,1/2)^0_1 \to 6p^2(3/2,3/2)_2$ $(7s^3P^0_1 \to 6p^2\ ^1D_2)$	268,269,270, 276	204–210
0.7229	$6p\ 7s\ ^3P_1 \to 6p^2\ ^1D_2$	6919	1776
1.2561370	$7p(1/2,1/2)_1 \to 7s(1/2,1/2)^0_0$ $(7p\ ^3P_1 \to 7s\ ^3P^0_0)$	268,277	211
1.3152769	$7d1/2[5/2]^0_3 \to 7p(1/2,3/2)_2$ $(7d\ ^3F^0_3 \to 7p\ ^3D_2)$	268,279	383
1.5335134	$7s(3/2,1/2)_1 \to 7p(1/2,1/2)_1$ $(7s\ ^1P^0_1 \to 7p\ ^3P_1)$	268,271,280	383
3.1748096	$6d1/2[3/2]^0_1 \to 7p(1/2,1/2)_1$ $(6d\ ^3D^0_1 \to 7p\ ^3P_1)$	268,281	175
7.1764192	$6d1/2[3/2]^0_1 \to 7p(1/2,3/2)_1$ $(6d\ ^3D^0_1 \to 7p\ ^3D_1)$	268,282	175
7.9423392	$6d1/2[3/2]^0_1 \to 7p(1/2,3/2)_2$ $(6d\ ^3D^0_1 \to 7p\ ^3D_2)$	268,283	175

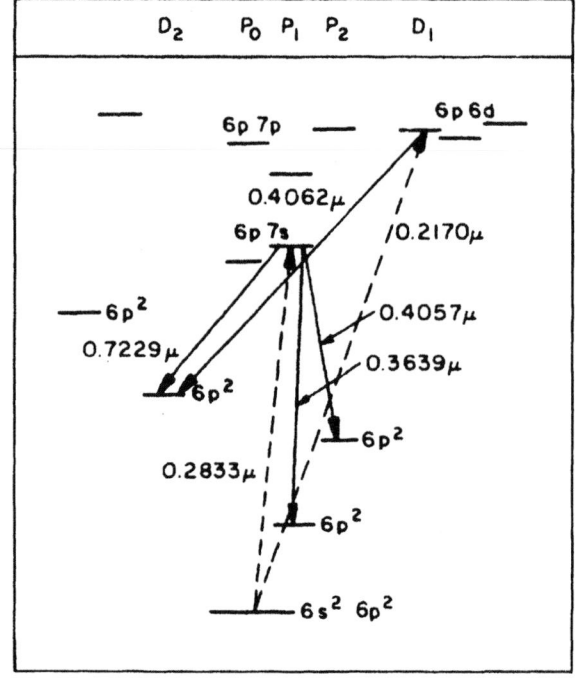

Figure 3.1.12 Partial energy level diagram of lead showing the self-terminating laser transitions. Wavelengths are given in microns.

3.1.2.7 Group IVB Lasers

Table 3.1.7
Group IVB Lasers

Titanium

Wavelength (μm)	Transition assignment	Comments	References
0.43148	$4p\ ^3D^0_3 \rightarrow 4s\ ^5F_4$	1779,1784	6906,6907
0.47102	$4p\ ^3D^0_1 \rightarrow 3d^2\ 4s^2\ ^3P_0$	1779,1784	6906,6907
0.47226	$4p\ ^3D^0_1 \rightarrow 3d^2\ 4s^2\ ^3P_1$	1779,1784	6906,6907
0.54742	$4p\ ^3G^0_5 \rightarrow 4s\ ^3F_4$	1779,1784	6906,6907
0.55125	$4p\ ^3D^0_3 \rightarrow 4s\ ^3F_4$	1779,1784	6906,6907
0.55144	$4p\ ^3D^0_1 \rightarrow 4s\ ^3F_2$	1779,1784	6906,6907
0.55145	$4p\ ^3D^0_2 \rightarrow 4s\ ^3F_3$	1779,1784	6906,6907
0.6258	$4p\ ^3G^0_4 \rightarrow 4s\ ^3F_3$	1779,1784	6906,6907

3.1.2.8 Group VA Lasers

Table 3.1.8
Group VA Lasers

Nitrogen (Figure 3.1.13)

Wavelength (μm)	Transition assignment	Comments	References
0.4120		284,285,295	212
0.4321334	$5p\ ^4P^0_{3/2} \to 2p^4\ ^4P_{3/2}$	284,286,296	213
0.4328395	$5p\ ^4D^0_{5/2} \to 2p^4\ ^4P_{5/2}$	284,286,297	213
0.4525		284,285,298	212
0.4750295	$4p\ ^2D^0_{3/2} \to 3s\ ^2P_{1/2}?$	284–286,299	212
0.5440		284,285,300	212
0.550042	$8s\ ^2P_{1/2} \to 3p\ ^2D^0_{3/2}?$	284–286,299	212
0.5540307	$5d\ ^4P_{3/2} \to 3p\ ^4D^0_{3/2}?$	284–286,299	212
0.8594005	$3p\ ^2P^0_{1/2} \to 3s\ ^2P_{1/2}$	284,303	190,214
0.8629238	$3p\ ^2P^0_{3/2} \to 3s\ ^2P_{3/2}$	284,287,304	190–2,214–6
0.8680	$3p\ ^4D^0_{7/2} \to 3s\ ^4P_{5/2}$	6552	1505
0.8683	$3p\ ^4D^0_{5/2} \to 3s\ ^4P_{3/2}$	6553	1505
0.8703	$3p\ ^4D^0_{1/2} \to 3s\ ^4P_{1/2}$	6555	1505
0.9045878	$3p'\ ^2F^0_{7/2} \to 3s'\ ^2D_{5/2}$	284,305	214
0.9187449	$3p'\ ^2D^0_{5/2} \to 3s'\ ^2D_{5/2}$	284,289,306	190
0.918784	$3p'\ ^2D^0_{5/2} \to 3s'\ ^2D_{3/2}$	284,307	190
0.9386805	$3p\ ^2D^0_{3/2} \to 3s\ ^2P_{1/2}$	284,287,308	190,191,215, 217,218
0.9392789	$3p\ ^2D^0_{5/2} \to 3s\ ^2P_{3/2}$	284,287,309	190–192,215, 216,218
1.0563328	$3d\ ^4D_{3/2} \to 3p\ ^4P^0_{5/2}$	284,290,310,311	219
1.0623177	$3d\ ^4P_{1/2} \to 3p\ ^4P^0_{1/2}$	284,290,311,312	219
1.0643981	$3d\ ^4P_{3/2} \to 3d\ ^4P^0_{1/2}$	284,290,311,313	219
1.342961	$3p\ ^2S^0_{1/2} \to 3s\ ^2P_{1/2}$	284,314	215
1.3581330	$3p\ ^2S^0_{1/2} \to 3s\ ^2P_{3/2}$	284,287,315	191,195,196,215
1.45423		284,292,316	215
1.4553011	$4s\ ^4P_{5/2} \to 3p\ ^2D^0_{5/2}$	284,317	195,196
3.7942		284,293,318	215
3.8154		284,294,319	215

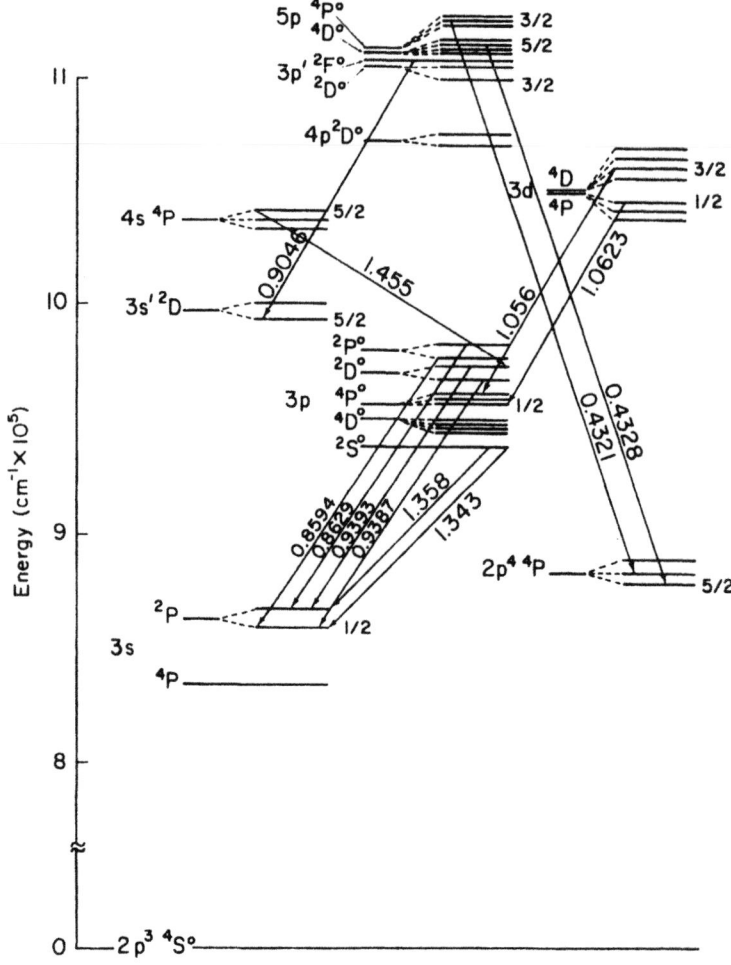

Figure 3.1.13 Partial energy level diagram of nitrogen showing the majority of the reported laser transitions. Wavelengths are given in microns.

Phosphorus (Figure 3.1.14)

Wavelength (μm)	Transition assignment	Comments	References
0.667193		320,321,323,324	220
1.008422	4p $^2P^0_{3/2}$ → 4s $^2P_{3/2}$	320,323,325	221
1.1163455	4p $^4S^0_{3/2}$ → 4 s$^2P_{1/2}$	320,323,326	221
1.11864700	4p $^2D^0_{5/2}$ → 4s $^2P_{3/2}$	320,323,327	221
1.1547277	4p $^4S^0_{3/2}$ → 4s $^2P_{3/2}$	320,323,328	221
01.1787698	4p $^4P^0_{3/2}$ → 4s $^2P_{1/2}$	320,323,329	221
1.5716351	4p $^2S^0_{1/2}$ → 4s $^2P_{1/2}$	320,323,330	221
1.648791	4p $^2S^0_{1/2}$ → 4s $^2P_{3/2}$	320,323,331	221
1.8943842	5s $^2P_{3/2}$ → 4p $^2P^0_{1/2}$	320,323,332	221
2.0962339	3d $^2D_{3/2}$ → 4p $^2P^0_{1/2}$	320,322,323,333	221

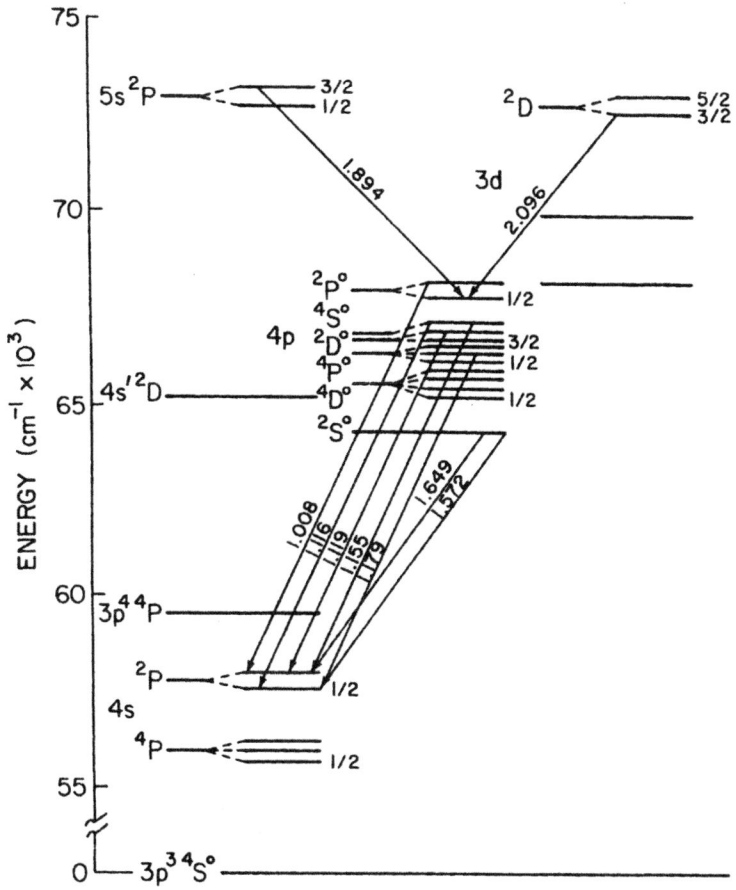

Figure 3.1.14 Partial energy level diagram of phosphorus showing all the classified laser transitions. Wavelengths are given in microns.

Arsenic (Figure 3.1.15)

Wavelength (μm)	Transition assignment	Comments	References
1.04559850	$5p\ ^2D^0_{5/2} \to 5s\ ^2P_{3/2}$	338	222
1.0617063	$5p\ ^2D^0_{3/2} \to 5s\ ^2P_{1/2}$	339	222
1.1247708	$5p\ ^4P^0_{1/2} \to 5s\ ^2P_{1/2}$	340	222
1.1522595	$6p\ ^2P^0_{1/2} \to 4d\ ^2P_{1/2}$	334,341	175,222
1.1524056	$5p\ ^4P^0_{5/2} \to 5\ s^2P_{3/2}$	342	175,222
1.2945989	$5p\ ^4D^0_{3/2} \to 5s\ ^2P_{1/2}$	343	222
1.4124892	$5p\ ^4D^0_{5/2} \to 5s\ ^2P_{3/2}$	335,344	222
1.4258622	$77121_{3/2,5/2} \to 5p'\ ^2D^0_{3/2}$	345	175
1.4259232	$6p\ ^4D^0_{3/2} \to 4d\ ^4P_{3/2}$	346	175
1.4629079	$5p\ ^2P^0_{1/2} \to 4p^4\ ^4P_{1/2}$	347	222
1.8053474	$6p\ ^4P^0_{1/2} \to 4d\ ^2D_{3/2}$	348	175
1.8068806	$5p\ ^4P^0_{3/2} \to 4p^4\ ^4P_{3/2}$	336,349	222
1.9754647	$75578.7_{3/2,5/2} \to 6p\ ^4D^0_{3/2}$	350	175

Arsenic (Figure 3.1.15)—*continued*

Wavelength (μm)	Transition assignment	Comments	References
2.0282741	6p $^2P^0_{3/2}$ → 6s $^2P_{1/2}$	337,351	175
2.4466627	4d $^2P_{3/2}$ → 5p $^4P^0_{3/2}$	352	175
2.9813368	5p $^2D^0_{5/2}$ → 5s' $^2D_{5/2}$	353	175
5.2879276	4d $^4P_{5/2}$ → 5p $^4D^0_{3/2}$	354	175

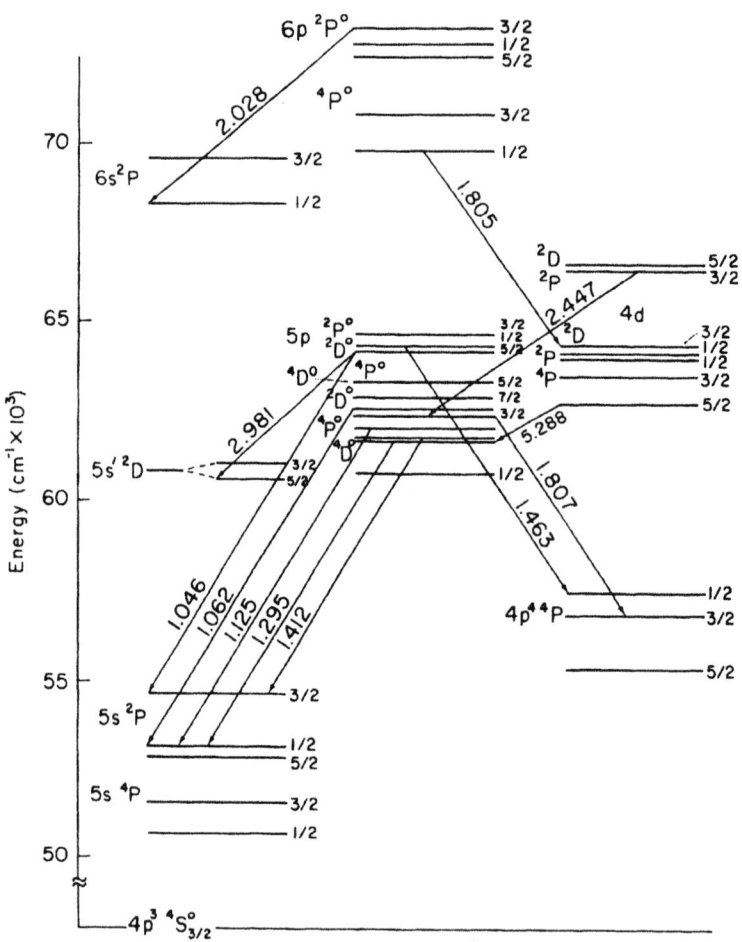

Figure 3.1.15 Partial energy level diagram of arsenic showing the majority of the reported laser transitions. Wavelengths are given in microns.

Antimony

Wavelength (μm)	Transition assignment	Comments	References
0.32676	$6s^2P_{1/2} \rightarrow 5p^3 {}^2P^0_{1/2}$	6556	1491
12.036591	$53443.3_{5/2} \rightarrow 52612.5^0_{3/2}$	355	175

Bismuth

Wavelength (μm)	Transition assignment	Comments	References
0.472252	$7s\ {}^4P_{1/2} \rightarrow 6p^3\ {}^2D^0_{3/2}$	356–359	223,386,769
1.1711	$6p\ {}^2({}^3P_0)7p_{1/2} \rightarrow 7s\ {}^4P_{1/2}$	6557	1492
5.3297517	$6d\ {}^2D_{5/2} \rightarrow 7p(?)^0_{3/2}$	356,361	175

3.1.2.9 VB Lasers

Table 3.1.9
Group VB Lasers

Vanadium

Wavelength (μm)	Transition assignment	Comments	References
0.4095	${}^4F^0_{7/2}$ - ${}^4D_{5/2}$	1787	6908
2.0200522	$f\ {}^6F_{9/2} \rightarrow v\ {}^4G^0_{7/2}$	362,363,364	155
2.4480996	$z\ {}^4P^0_{3/2} \rightarrow b\ {}^4D_{1/2}$	362,363,365	155

Tantalum

Wavelength (μm)	Transition assignment	Comments	References
0.2925	${}^6F^0_{1/2}$ - ${}^4P_{3/2}$	1778	6909
0.3227	${}^6F^0_{3/2}$ - ${}^4P_{5/2}$	1778	6909
0.3281	${}^6F^0_{3/2}$ - ${}^6D_{1/2}$	1778	6909
0.3304	${}^6F^0_{3/2}$ - ${}^6D_{3/2}$	1778	6909
0.351.4	${}^6F^0_{1/2}$ - ${}^2P_{1/2}$	1778	6909

3.1.2.10 Group VIA Lasers

Table 3.1.10
Group VIA Lasers

Oxygen (Figure 3.1.16)

Wavelength (μm)	Transition assignment	Comments	References
0.55788939	$2p^4\ ^1S_0 \rightarrow 2p^4\ ^1D_2$	366,373	224,225,387, 388,1486
0.844628	$3p\ ^3P_{0,2,1} \rightarrow 3\ s^3S^0_1$	368,374	226–239,1698,
0.844638	$3p\ ^3P_{0,2,1} \rightarrow 3\ s^3S^0_1$	368,376,377	226–240,1698,
0.844672	$3p\ ^3P_{0,2,1} \rightarrow 3s\ ^3S^0_1$	368,378	226–240,1698,
0.844680	$3p\ ^3P_{0,2,1} \rightarrow 3s\ ^3S^0_1$	368,380	226–240,1698,
0.88228702	$3p'\ ^1F_3 \rightarrow 3s'\ ^1D^0_2$	370,382	190
2.6513946	$4d\ ^5D^0 \rightarrow 4p\ ^5P$	371,372,383	236
2.8944397	$4p\ ^3P_{2,1,0} \rightarrow 4s\ ^3S^0_1$	384	1698,235
6.8175155	$3s'\ ^3D^0_2 \rightarrow 4p\ ^3P(2)$	387	215
6.8598868	$5p\ ^3P_{2,1,0} \rightarrow 5s\ ^3S^0_1$	388	235,1698
6.8745531	$3s'\ ^3D^0_3 \rightarrow 4p\ ^3P_2$	389	215
10.40312	$5p\ ^3P_{2,1,0} \rightarrow 4d\ ^3D^0_{3,2,1}$	385,390	1698,235

Sulfur (Figure 3.1.17)

Wavelength (μm)	Transition assignment	Comments	References
0.7726542	$3p^4\ ^1S_0 \rightarrow 3p^4\ ^1D_2$	391,395,396	238
1.0455451	$4p\ ^3P_2 \rightarrow 4s\ ^3S^0_1$	392,395,397	195,196,239
1.0635993	$4p'\ ^1F_3 \rightarrow 4s'\ ^1D^0_2$	392,395,398	195,196
1.4019620	$3s3p^5\ ^3P^0_2 \rightarrow 4p\ ^3P_2$	392,395,399	240
1.5422255	$4f\ ^3F_{4,(3)} \rightarrow 3d\ ^3D^0_3$	392,393,395,400	241
1.6542665	$5p\ ^5P_3 \rightarrow 3d\ ^5D^0_4$	392,395,401	241
2.436331	$5p\ ^3P_1 \rightarrow 3d\ ^3D^0_2$	392,395,403	241
2.2801247	$4s'\ ^3D^0_3 \rightarrow 4p\ ^5P_2$	395,402	242,403
3.389503	$4s'\ ^3D^0_3 \rightarrow 4p\ ^3P_2$	394,395,404	242

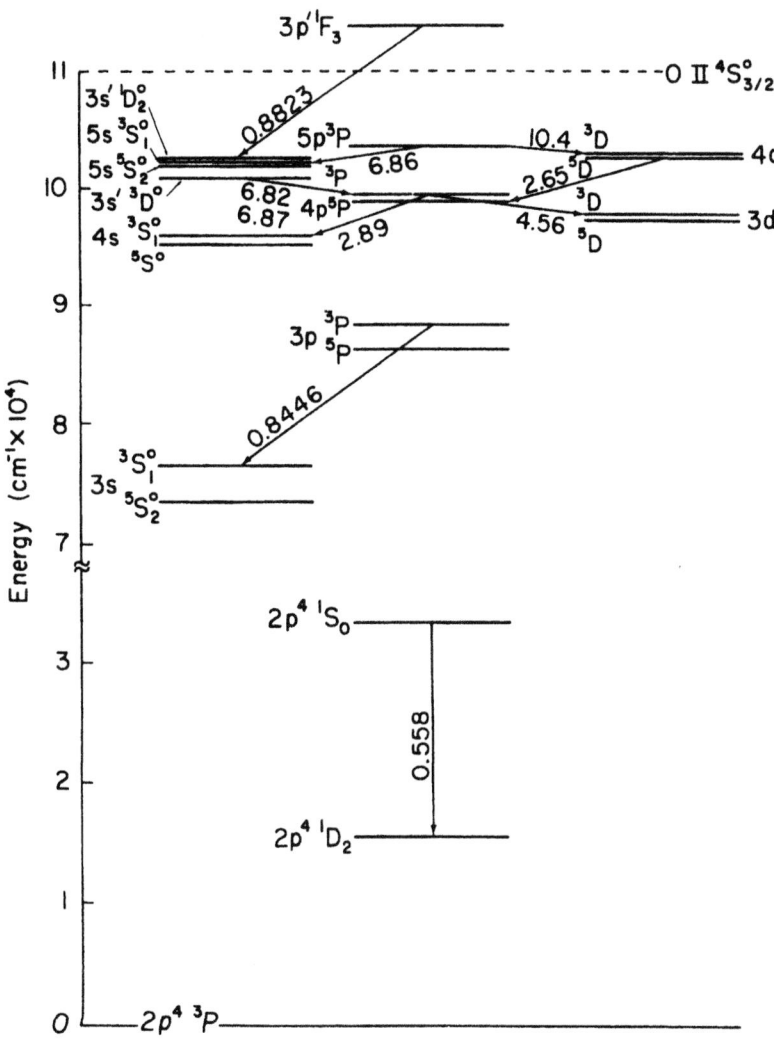

Figure 3.1.16 Partial energy level diagram of oxygen showing the majority of the reported laser transitions. Wavelengths are given in microns.

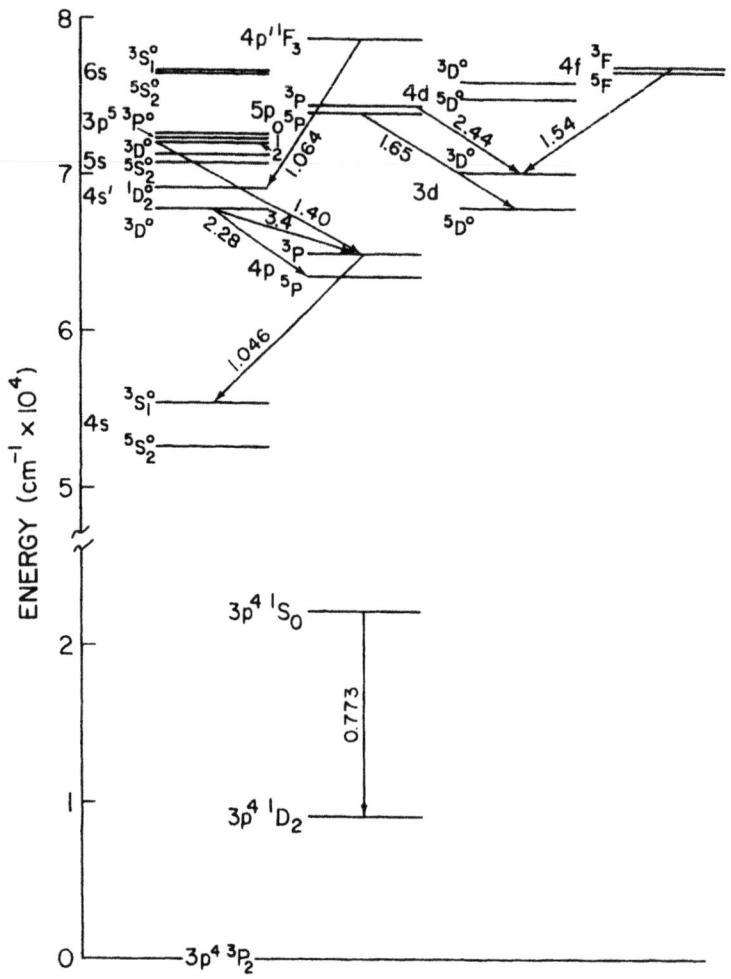

Figure 3.1.17 Partial energy level diagram of sulfur showing all of the reported laser transitions. Wavelengths are given in microns.

Selenium

Wavelength (μm)	Transition assignment	Comments	References
0.48884122	$4p^4\ ^1S_0 \rightarrow 4p^4\ ^3P_1$	405,406,407	245,390
0.77700379	$4p^4\ ^1S_0 \rightarrow 4p^4\ ^1D_2$	406,408	245,390
6.3687777	$5s'\ ^3D^0_3 \rightarrow 5p\ ^3P_2$	409	155

Tellurium

Wavelength (μm)	Transition assignment	Comments	References
7.8071691	5d $^5D^0_4 \rightarrow$ 6p $^5P_3$	410,412,419	155
6.7631543	5d $^5D^0_3 \rightarrow$ 6p $^5P_2$	410,412,418	155
3.1653653	5d $^3D^0_3 \rightarrow$ 6p $^3P_2$	410,412,417	155
0.63497		410,416	442
0.5640		410,415	443
0.5454		410,414	443

Thulium (Figure 3.1.18)

Wavelength (μm)	Transition assignment	Comments	References
0.106939	$6s^2$ 6p $\rightarrow 4f^{12}$ 5d $6s^2$ (J=7/2 $\rightarrow$ 5/2)	6913,6914	1785,1786
0.145304	$6s^2$ 6p $\rightarrow 4f^{12}$ 5d $6s^2$ (J=11/2 $\rightarrow$ 11/2)	6913,6914	1785,1786
0.149578	$4f^{13}$ 5d 6s $\rightarrow 4f^{12}$ 5d $6s^2$ (J=11/2 $\rightarrow$ 9/2)	6913,6914	1785,1786
0.3051		6559	1493
0.3079	$(3,5/2)_{7/2} \rightarrow {}^2F^0_{7/2}$	6559	1493
0.3497		6559	1493
0.3566	$(5/2,2)_{5/2} \rightarrow {}^2F^0_{7/2}$	6559	1493
0.3568	$(5/2,2)_{9/2} \rightarrow {}^2F^0_{7/2}$	6559	1493
0.3718	$(5,5/2)_{9/2} \rightarrow {}^2F^0_{7/2}$	6559	1493
0.3722	$(7/2,2)^0_{13/2} \rightarrow (6,3/2)_{13/2}$	6559	1493
0.4118	$(5/2,1)^0_{7/2} \rightarrow (7/2,0)_{7/2}$	6559	1493
0.4122		6559	1493
0.4151	$(7/2,0)^0_{7/2} \rightarrow (7/2,1)_{5/2}$	6559	1493
0.4360	$(5,3/2)_{5/2} \rightarrow {}^2F^0_{7/2}$	6559	1493
0.4415	$^1[11/2]^0_{5/2} \rightarrow (7/2,0)_{7/2}$	6559	1493
0.4662	$(3,3/2)^0_{7/2} \rightarrow (7/2,1)_{9/2}$	6559	1493
0.4881	$^3[9/2]^0_{11/2} \rightarrow (7/2,2)_{11/2}$	6559	
0.5061	$(7/2,2)_{7/2} \rightarrow {}^2F^0_{7/2}$	6559	1493
0.5115	$(7/2,2)_{5/2} \rightarrow {}^2F_{7/2}$	6559	1493
0.5307	$(6,5/2)_{9/2} \rightarrow {}^2F^0_{7/2}$	6559	1493
0.5634		6559	
0.5676	$(7/2,1)_{9/2} \rightarrow {}^2F^0_{7/2}$	6559	1493
0.5766	$(7/2,1)_{7/2} \rightarrow {}^2F^0_{7/2}$	6559	1493
0.5809	$(3,3/2)^0_{9/2} \rightarrow (4,5/2)_{11/2}$	6559	1493
0.58995	$(6,5/2)_{7/2} \rightarrow {}^2F^0_{7/2}$	6558,6559	1494
0.5971	$(7/2,0)_{7/2} \rightarrow {}^2F^0_{7/2}$	6559	1493
0.6119	$(6,5/2)^0_{11/2} \rightarrow (6,3/2)_{13/2}$	6559	1493

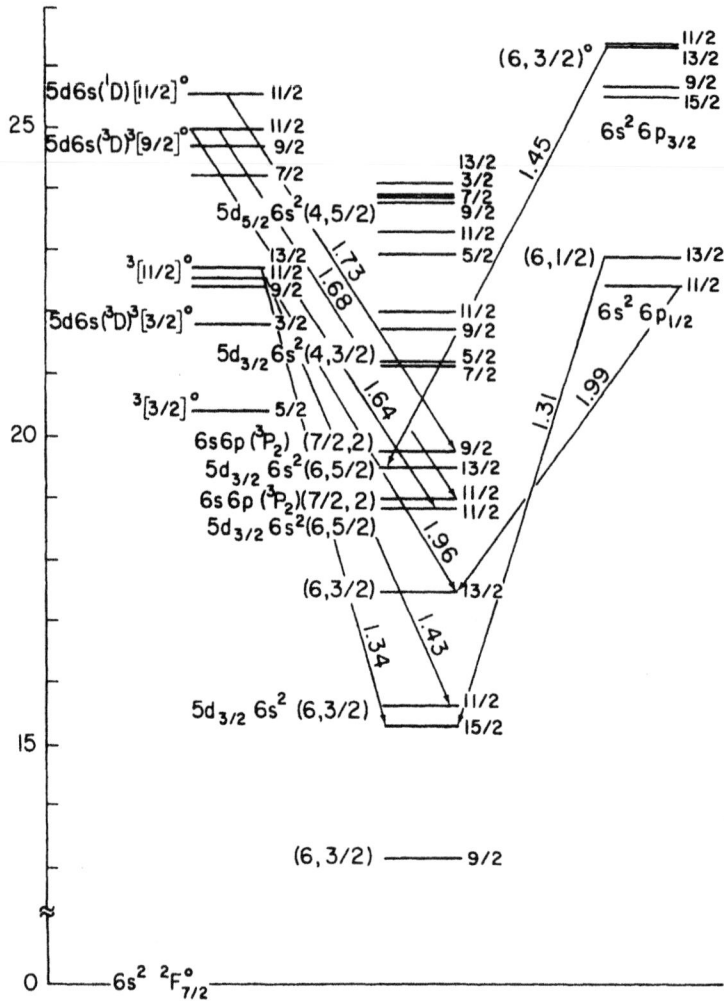

Figure 3.1.18 Partial energy level diagram of thulium showing the majority of the reported laser transitions. Wavelengths are given in microns.

Thulium—*continued*

Wavelength (μm)	Transition assignment	Comments	References
0.6170	$^1[3/2]^0{}_{9/2} \rightarrow (6,5/2)_{11/2}$	6559	1493
0.6404	$(4,1/2)^0{}_{9/2} \rightarrow (7/2,2)_{9/2}$	6559	1493
0.6461	$(7/2,1)^0{}_{9/2} \rightarrow (7/2,0)_{7/2}$	6559	1493
0.6519	$(2,5/2)_{3/2} \rightarrow {}^3[3/2]^0{}_{3/2}$	6559	1493
0.6589	$(7/2,2)_{11/2} \rightarrow {}^3[9/2]^0{}_{11/2}$	6559	1493
0.6620	$(4,5/2)_{7/2} \rightarrow {}^2F^0{}_{5/2}$	6559	1493
0.6658	$(7/2,1)^0{}_{7/2} \rightarrow (7/2,1)_{7/2}$	6559	1493
0.6723	$(6,9/2)^0{}_{5/2} \rightarrow (4,5/2)_{3/2}$	6559	1493
0.6782	$(7/2,1)^0{}_{7/2} \rightarrow (7/2,1)_{9/2}$	6559	1493
0.6845	$(7/2,1)^0{}_{7/2} \rightarrow (7/2,1)_{5/2}$	6559	1493
0.6848	$(7/2,1)^0{}_{9/2} \rightarrow (7/2,1)_{9/2}$	6559	1493

Thulium—*continued*

Wavelength (μm)	Transition assignment	Comments	References
0.6914	$(3,1/2)^0{}_{9/2} \rightarrow (4,5/2)_{11/2}$	6559	1493
0.6956	$(2,5/2)_{9/2} \rightarrow {}^3[5/2]^0{}_{7/2}$	6559	1493
0.6988		6559	1493
0.6996	$(7/2,1)_{9/2} \rightarrow {}^1[9/2]^0{}_{9/2}$	6559	1493
0.7038	$(7/2,0)^0{}_{7/2} \rightarrow (7/2,2)_{7/2}$	6559	1493
0.7339	$(4,1/2)^0{}_{9/2} \rightarrow (4,3/2)_{9/2}$	6559	1493
0.7342	$(2,5/2)_{7/2} \rightarrow {}^3[7/2]^0{}_{9/2}$	6559	1493
0.7392	$(7/2,1)^0{}_{7/2} \rightarrow (6,5/2)_{9/2}$	6559	1493
0.7406		6559	
0.7409	$(7/2,2)_{9/2} \rightarrow {}^3[5/2]^0{}_{7/2}$	6559	1493
0.7439		6559	
0.7448	$(4,3/2)_{7/2} \rightarrow {}^3[9/2]^0{}_{9/2}$	6559	1493
0.7481	$(4,1/2)^0{}_{9/2} \rightarrow (4,3/2)_{11/2}$	6559	1493
0.7522	$(7/2,0)^0{}_{3/2} \rightarrow (5/2,2)_{5/2}$	6559	1493
0.7526	$(7/2,1)^0{}_{11/2} \rightarrow (4,5/2)_{13/2}$	6559	1493
0.7532		6559	1493
0.7559		6559	1493
0.7577	$(7/2,0)^0{}_{7/2} \rightarrow (5/2,2)_{5/2}$	6559	1493
0.7669	${}^1[9/2]^0{}_{9/2} \rightarrow (5/2,2)_{7/2}$	6559	1493
0.7768	$(2,5/2)_{9/2} \rightarrow {}^3[9/2]^0{}_{9/2}$	6559	1493
0.7804		6559	1493
0.7845	${}^1[11/2]^0{}_{9/2} \rightarrow (5/2,1)_{7/2}$	6559	1493
0.7849		6559	1493
0.7929	$(7/2,1)^0{}_{7/2} \rightarrow (7/2,2)_{7/2}$	6559	1493
0.7931	$(7/2,1)^0{}_{7/2} \rightarrow (7/2,2)_{9/2}$	6559	1493
0.7983	$(2,5/2)_{3/2} \rightarrow {}^3[7/2]^0{}_{5/2}$	6559	1493
0.8020	$(7/2,1)^0{}_{9/2} \rightarrow (7/2,2)_{9/2}$	6559	1493
0.8170	${}^3[7/2]^0{}_{7/2} \rightarrow (4,3/2)_{5/2}$	6559	1493
0.8180	$(7/2,2)_{5/2} \rightarrow {}^1[7/2]^0{}_{7/2}$	6559	1493
0.8186	${}^1[7/2]^0{}_{11/2} \rightarrow (5,5/2)_{11/2}$	6559	1493
0.8215	$(3,1/2)^0{}_{11/2} \rightarrow (5,3/2)_{11/2}$	6559	1493
0.8233		6559	1493
0.8480	$(2,5/2)_{3/2} \rightarrow {}^1[3/2]^0{}_{3/2}$	6559	1493
0.8480		6559	1493
0.8482		6559	1493
0.8533	$(7/2,2)_{9/2} \rightarrow (5,1/2)^0{}_{9/2}$	6559	1493
0.8730	${}^3[9/2]_{11/2} \rightarrow (5/2,2)_{9/2}$	6559	1493
0.8803	$(7/2,1)^0{}_{5/2} \rightarrow (4,3/2)_{7/2}$	6559	1493
0.8836	$(7/2,1)^0{}_{5/2} \rightarrow (4,3/2)_{5/2}$	6559	1493
1.1101	$(6,3/2)^0{}_{15/2} \rightarrow (6,5/2)_{17/2}$	6558,6559	1494
1.3058983	$(4,5/2)'_{13/2} \rightarrow (6,1/2)^0{}_{11/2}$	420–24	246

Thulium—*continued*

Wavelength (μm)	Transition assignment	Comments	References
1.3104227	$(6,1/2)^0_{13/2} \to (6,3/2)_{15/2}$	420,423,425	246
1.3383700	$^3[11/2]^0_{13/2} \to (6,3/2)_{15/2}$	420,423,426	246
1.4343722	$^3[11/2]^0_{11/2} \to (6,3/2)_{11/2}$	420,423,427	246
1.4489080	$(6,3/2)^0_{11/2} \to (6,5/2)_{13/2}$	420,423,428	246
1.4998810	$^1[11/2]^0_{11/2} \to (6,5/2)_{11/2}$	420–23,429	246
1.5036834	$(2,3/2)_{5/2} \to {}^3[5/2]^0_{5/2}$	420–23,430	246
1.6383650	$^3[9/2]^0_{11/2} \to (6,5/2)_{11/2}$	420,421,423,431	246
1.6758663	$^3[9/2]^0_{11/2} \to (7/2,2)_{11/2}$	420,421,423,432	246
1.7323684	$^1[11/2]^0_{11/2} \to (7/2,2)_{9/2}$	420,423,433	247
1.9589851	$^3[11/2]^0_{1/2} \to (6,3/2)_{13/2}$	420,423,434	246
1.9722834	$(4,3/2)'_{9/2} \to {}^3[9/2]^0_{7/2}$	420,421,423,435	246
1.9947227	$(6,1/2)^0_{11/2} \to (6,3/2)_{13/2}$	420,423,436	
2.1059135	$(5,5/2)_{11/2} \to {}^3[11/2]^0_{13/2}$	420,421,423,437	246
2.1129476	$(4,1/2)^0_{9/2} \to (4,5/2)_{9/2}$	420,421,423,438	246
2.3851957	$^3[7/2]^0_{9/2} \to (7/2,2)_{9/2}$	420,421,423,439	246

3.1.2.11 Group VII Lasers

Table 3.1.11
Group VII Lasers

Fluorine (Figure 3.1.19)

Wavelength (μm)	Transition assignment	Comments	References
0.6239651	$3p\ ^4S^0_{3/2} \to 3\ s^4P_{5/2}$	440,442,444	248,249
0.6348508	$3p\ ^4S^0_{3/2} \to 3s\ ^4P_{3/2}$	440,442,445	248,249,250
0.6413651	$3p\ ^4S^0_{3/2} \to 3s\ ^4P_{1/2}$	440,442,446	248,249
0.6966349	$3p\ ^2P^0_{1/2} \to 3s\ ^2P_{3/2}$	440,442,447	250–252,440
0.7037469	$3p\ ^2P^0_{3/2} \to 3s\ ^2P_{3/2}$	440,442,448	197,249, 251–5,440
0.7127890	$3p\ ^2P^0_{1/2} \to 3s\ ^2P_{1/2}$	440,442,449	197,249,251, 253,255,440
0.7202360	$3p\ ^2P^0_{3/2} \to 3s\ ^2P_{1/2}$	440,442,450	250–253,440
0.7309033	$3p'\ ^2F_{7/2} \to 3s'\ ^2D_{5/2}$	440,442,451	249
0.7311019	$3p\ ^2S^0_{1/2} \to 3s\ ^2P_{3/2}$	440,442,452	250–252,255,440
0.7398688	$3p\ ^4P^0_{5/2} \to 3s\ ^4P_{5/2}$	440,442,453	248,250
0.7425645	$3p\ ^4P^0_{1/2} \to 3s\ ^4P_{3/2}$	440,442,454	256
0.7482723	$3p\ ^4P^0_{3/2} \to 3s\ ^4P_{3/2}$	440,442,455	256
0.7489155	$3p\ ^2S^0_{1/2} \to 3s\ ^2P_{1/2}$	440,442,456	250,252,440

Fluorine—*continued*

Wavelength (μm)	Transition assignment	Comments	References
0.7514919	3p $^4P^0_{1/2} \to$ 3s $^4P_{1/2}$	440,442,457	256
0.7552235	3p $^4P^0_{5/2} \to$ 3s $^4P_{3/2}$	440,442,458	250,248,249
0.7754696	3p $^2D^0_{5/2} \to$ 3s $^2P_{3/2}$	440,442,459	250,252,440
0.7800212	3p $^2D^0_{3/2} \to$ 3s $^2P_{1/2}$	440,442,460	250,252,254,440
7.435		440–42,461	155

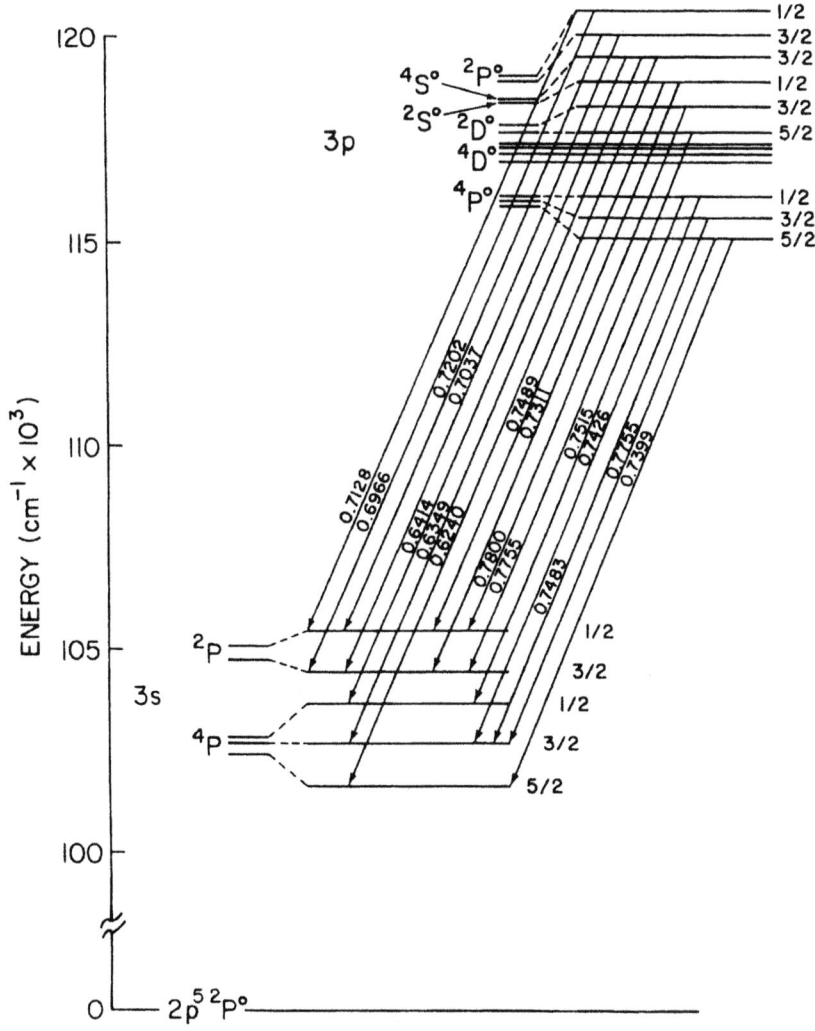

Figure 3.1.19 Partial energy level diagram of fluorine showing the majority of the reported laser transitions. Wavelengths are given in microns.

Chlorine (Figure 3.1.20)

Wavelength (μm)	Transition assignment	Comments	References
0.9452098	$4p\ ^2P^0_{3/2} \rightarrow 4s\ ^2P_{1/2}$	462,464,465	258,260
1.386330	$5s\ 2[2]_{5/2} \rightarrow 4p\ ^4D^0_{5/2}$	462,464,466	260
1.38931	$5s\ 2[2]_{3/2} \rightarrow 4p\ ^4D^0_{3/2}$	462,464,467	260
1.58697	$3d\ ^4F_{9/2} \rightarrow 4p\ ^4D^0_{7/2}$	462,464,468	197,261,391
1.5730	$3d\ ^4F_{7/2} \rightarrow 4p\ ^4D^0_{5/2}$	6560	1505
1.5970	$3d\ ^4F_{3/2} \rightarrow 4p\ ^4D^0_{1/2}$	6561	1505
1.97553	$3d\ ^4D_{7/2} \rightarrow 4p\ ^4P^0_{5/2}$	462,464,469	199,260–264
2.01994	$3d\ ^4D_{5/2} \rightarrow 4p\ ^4P_{3/2}$	462,464,470	199,260–263
2.44700	$3d\ ^4D_{7/2} \rightarrow 4p\ ^4D^0_{7/2}$	462,464,471	197,260,264
3.066080	$5p\ ^4D^0_{5/2} \rightarrow 5s\ '[1]_{3/2}$	462,463,464,472	264
3.543052	$5p\ ^4P^0_{5/2} \rightarrow 3d\ ^4P_{5/2}$	462,464,473	197
3.796602(	$^1D_2)4p^2P^0_{3/2} \rightarrow 5s[2]_{3/2}$	462,464,474	197

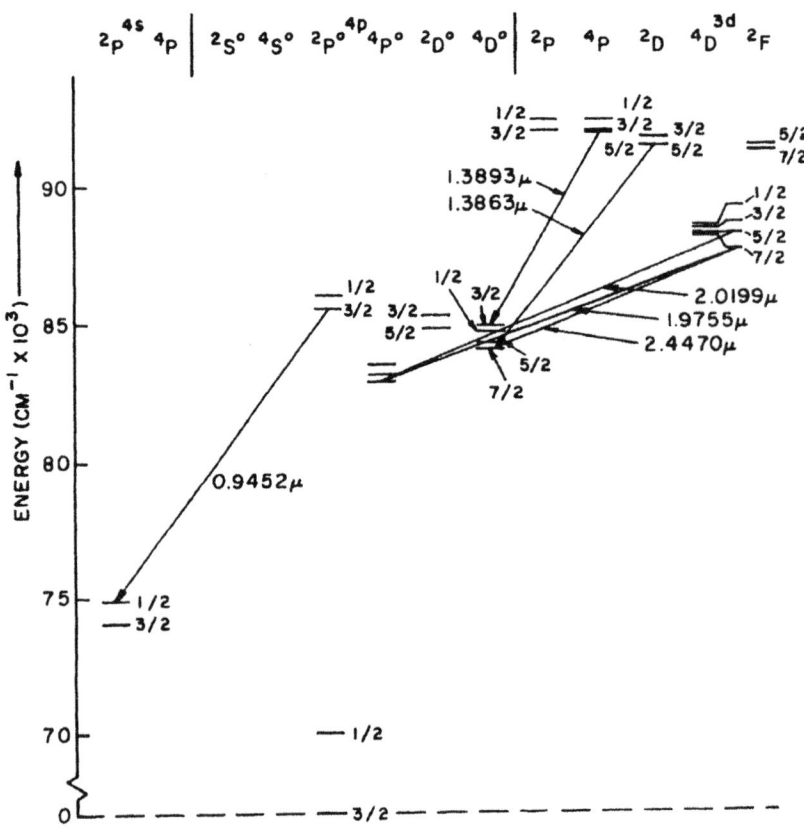

Figure 3.1.20 Partial energy level diagram of chlorine showing the majority of the reported laser transitions. Wavelengths are given in microns.

Bromine (Figure 3.1.21)

Wavelength (μm)	Transition assignment	Comments	References
0.8964	$5p\ ^2F^0_{5/2} \to 5s\ ^2D_{3/2}$	6562	1505
0.9173	$5p\ ^2P^0_{1/2} \to 5s\ ^2P_{1/2}$	6563	1505
0.9178	$5p\ ^4S^0_{1/2} \to 5s\ ^2P_{3/2}$	6564	1505
0.9793	$5p\ ^4S^0_{3/2} \to 5s\ ^2P_{1/2}$	6565	1505
1.1743	$5p\ ^4D^0_{5/2} \to 5s\ ^2P_{3/2}$	6566	1505
1.973362	$4d\ ^4F_{9/2} \to 5p\ ^4D^0_{7/2}$	475,478,479	261,403
2.286565	$4d\ ^4D_{7/2} \to 5p\ ^4P^0_{5/2}$	475,478,480	265
2.351215	$4d\ ^4D_{5/2} \to 5p\ ^4P^0_{3/2}$	475,478,481	265
2.7135274	$4p^5\ ^2P^0_{1/2} \to 4p^5\ ^2P^0_{3/2}$	475–8,482	437,438,439
2.837716	$4d\ ^4D_{7/2} \to 5p\ ^4D^0_{7/2}$	475,478,483	265

Iodine (Figure 3.1.22)

Wavelength (μm)	Transition assignment	Comments	References
0.98		484,485,490	266
1.01		484,485,491	266
1.03		484,485,492	266
1.06		484,485,493	266
1.3152443	$5p^5\ ^2P^0_{1/2} \to 5p^5\ ^2P_{3/2}$	484,486,494-95	267–283
1.4545941	$(^3P_0)6d[2]_{3/2} \to (^3P_2)7p[1]^0_{3/2}$	484,496	260
1.5533401	$(^3P_2)7s[2]_{3/2} \to (^3P_2)6p[1]^0_{1/2}$	484,487,497	400
2.598577	$5d'[2]_{3/2} \to 6p''[1]^0_{3/2}$	484,488,498	260
2.757298	$5d[0]_{1/2} \to 6p[2]^0_{3/2}$	484,488,499	265,401
3.036119	$5d[2]_{3/2} \to 6p[1]^0_{3/2}$	484,488,500	401,402
3.236285	$5d[2]_{5/2} \to 6p[1]^0_{3/2}$	484,488,501	164,263,265,401
3.429573	$5d[4]_{7/2} \to 6p[3]^0_{5/2}$	484,488-89,502	164,263,265, 401,402
4.3321362	$(^3P_2)5d[1]_{1/2} \to (^3P_2)6p[2]^0_{3/2}$	484,503	400
4.8584726	$(^3P_1)5d[1]_{3/2} \to (^3P_1)6p[2]^0_{5/2}$	484,504	401,402
4.8629144	$(^3P_2)5d[4]_{3/2} \to (^3P_2)6p[3]^0_{7/2}$	484,489,505	401,402
5.498705	$(^3P_0)5d[2]_{5/2} \to (^3P_0)6p[1]^0_{3/2}$	484,506	401,402
6.721966	$(^1D_2)6s[2]_{3/2} \to (^3P_2)6p[1]^0_{3/2}$	484,507	401,402
6.9035028	$(^3P_1)5d[3]_{7/2} \to (^3P_1)6p[2]^0_{5/2}$	484,508	401,402
9.0195724	$(^3P_2)5d[3]_{7/2} \to (^3P_2)6p[2]^0_{5/2}$	484,509	401,402

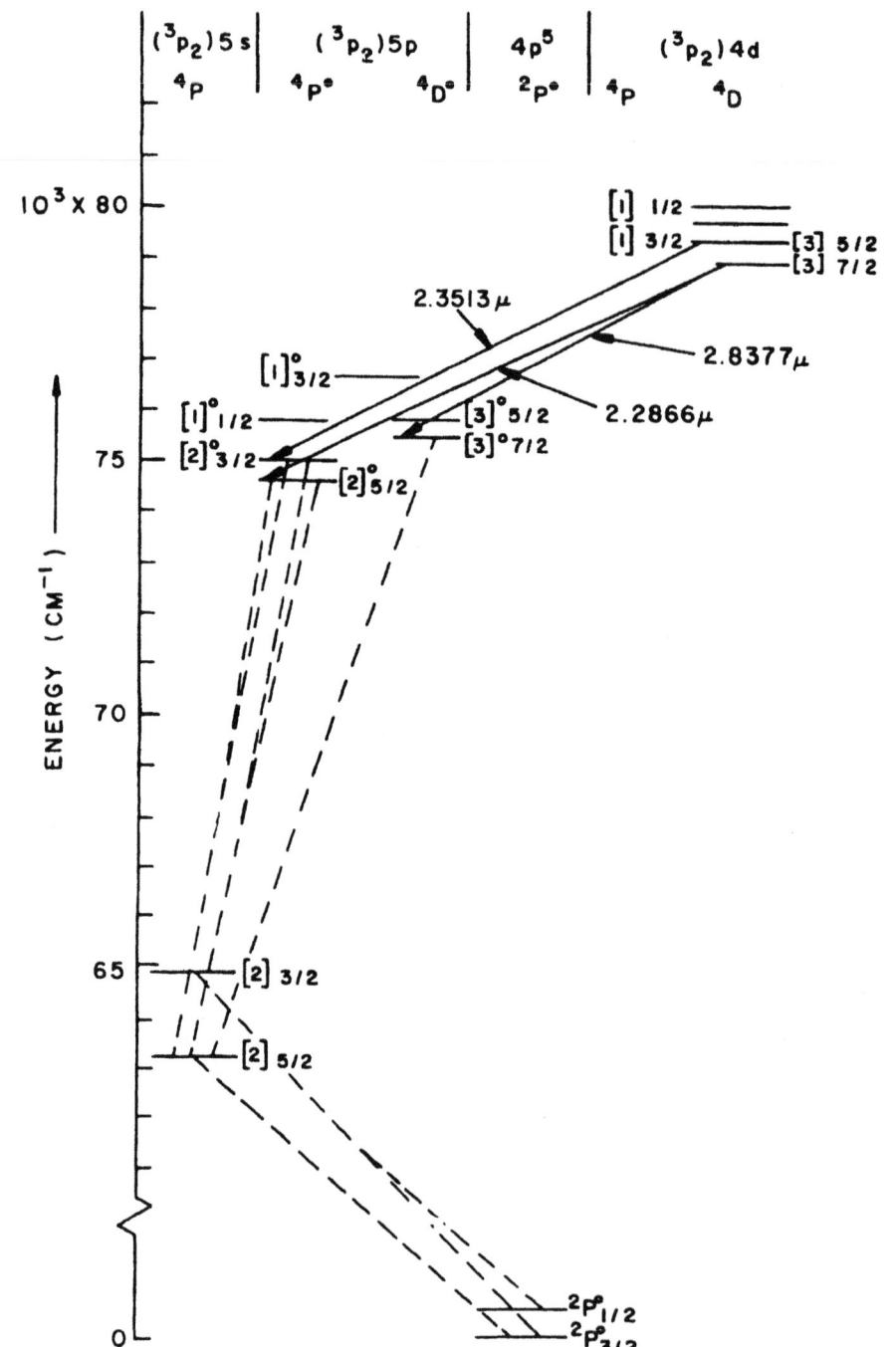

Figure 3.1.21 Partial energy level diagram of bromine showing the majority of the reported laser transitions. Wavelengths are given in microns.

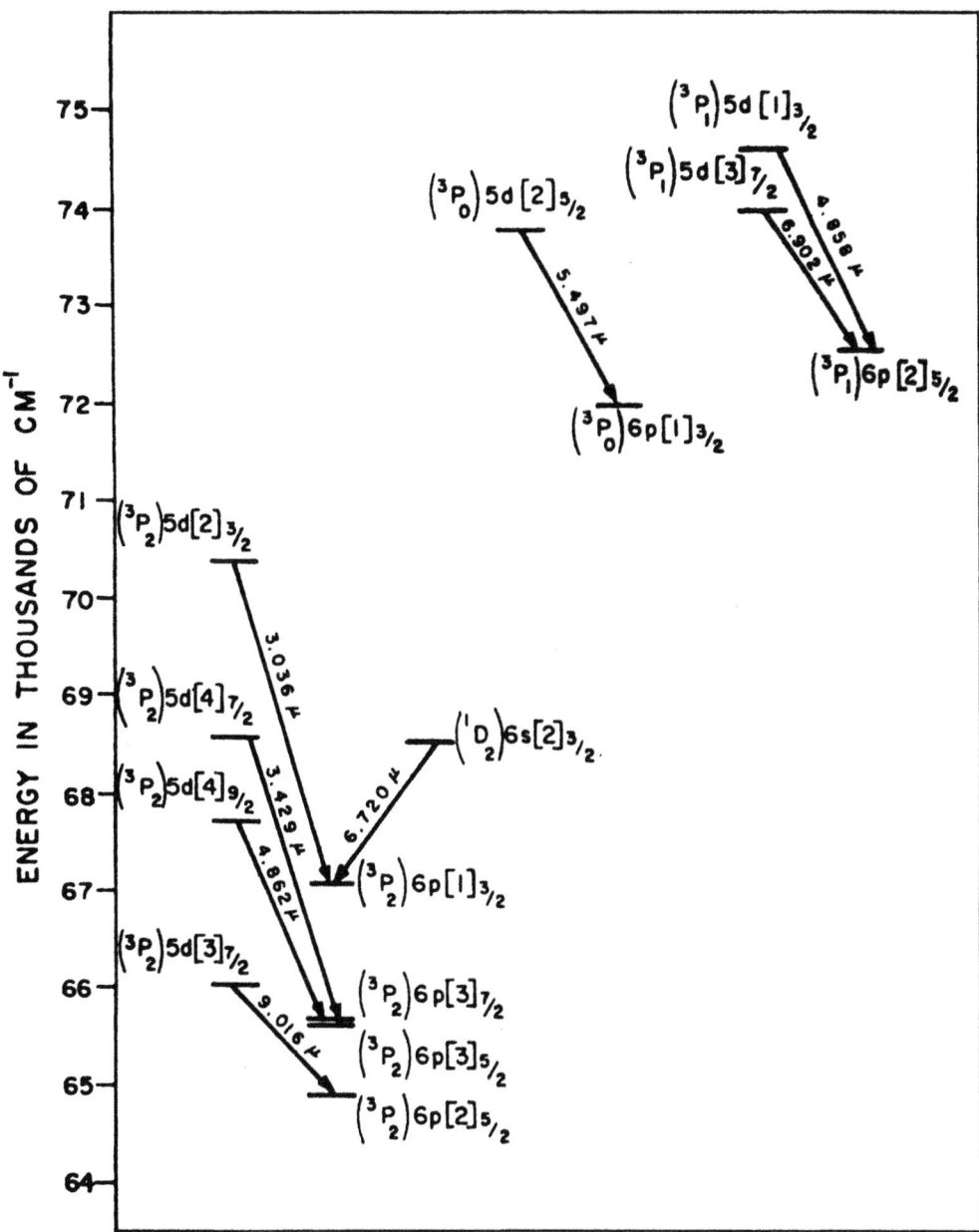

Figure 3.1.22 Partial energy level diagram of iodine showing the laser lines reported in CW gas discharge-excited operation. Wavelengths are given in microns. (From Kim, H., Paananen, R., and Hanst, P., *IEEE J. Quantum Electron.* QE-4, 385, 1968. With permission.)

Ytterbium (Figure 3.1.23)

Wavelength (μm)	Transition assignment	Comments	References
1.0324559	5d6s $^1D_2 \rightarrow$ 6s6p $^3P^0_1$	510,511,512	247
1.255136	5d6s $^1D_2 \rightarrow$ 6s6p $^3P^0_2$	510,511,513	247
1.4283698	6s^{2}6p $^1D_2 \rightarrow$ 5d6s^2 $^1D^0_2$	510,511,514	247
1.4793059	5d6s $^3D_2 \rightarrow$ 6s6p $^3P^0_1$	510,511,515	404
1.7459155	5d6s^2 $^3D^0_1 \rightarrow$ 6s7s $^3S_1$	510,511,516	247
1.798400	5d6s $^3D_3 \rightarrow$ 6s6p $^3P^0_2$	510,511,517	404
2.0041827	6s^{2}6p $^1D_2 \rightarrow$ 5d6s^2 $^1F^0_3$	510,511,519	246,247
2.1186997	5d6s^2 $^1P^0_1 \rightarrow$ 6s7s $^3S_1$	510,511,520	246,247
4.8021974	5d6s $^3D_3 \rightarrow$ 5d6s^2 $^3P^0_2$	510,511,521	247

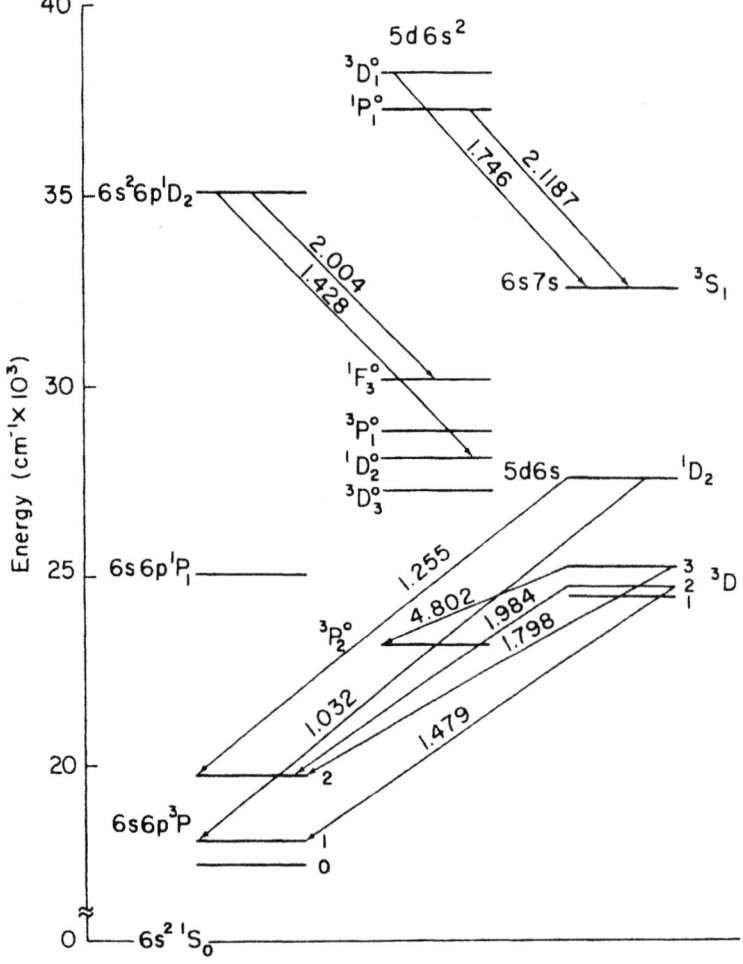

Figure 3.1.23 Partial energy level diagram of ytterbium showing all of the reported laser transitions. Wavelengths are given in microns.

3.1.2.12 Group VIIB Lasers

Table 3.1.12
Group VIIB Lasers

Manganese (Figure 3.1.24)

Wavelength (μm)	Transition assignment	Comments	References
0.5341065	y $^6P^0_{7/2} \to$ a $^6D_{9/2}$	522,523,524,525	336,406–414
0.5420368	y $^6P^0_{5/2} \to$ a $^6D_{7/2}$	522,526,527	336,406,407, 411–413
0.5470640	y $^6P^0_{5/2} \to$ a $^6D_{5/2}$	522,528	406,407,413
0.5481345	y $^6P^0_{3/2} \to$ a $^6D_{5/2}$	522,529	407,336
0.5516777	y $^6P^0_{3/2} \to$ a $^6D_{3/2}$	522,530	336,406,407.413
0.5537749	y $^6P^0_{3/2} \to$ a $^6D_{1/2}$	522,531	336,406,407.413
1.28997	z $^6P^0_{7/2} \to$ a $^6D_{9/2}$	522,532,533	336,406,407, 411,412,413
1.32941	z $^6P^0_{7/2} \to$ a $^6D_{7/2}$	522,534,535	336,406,407, 411,412,413
1.33179	z $^6P^0_{5/2} \to$ a $^6D_{7/2}$	522,536,537	336,406,407, 411,412,413
1.36257	z $^6P^0_{5/2} \to$ a$^6D_{1/2}$	522,538,539	336,406,407, 411,412,413
1.38638	z $^6P^0_{3/2} \to$ a $^6D_{3/2}$	522,540,541	336,406,407, 411,412,413
1.39970	z $^6P^0_{3/2} \to$ a $^6D_{1/2}$	522,542,543	336,406,407, 411,412,413

3.1.2.13 Group VIII Lasers

Table 3.1.13
Group VIII Lasers

Iron

Wavelength (μm)	Transition assignment	Comments	References
0.299951	x $^5F^0_5 \to$ a $^5F_5$	544,547	415
0.301618	x $^5F^0_1 \to$ a $^5F_2$	544,548	415
0.303164	x $^5F^0_1 \to$ a $^5F_1$	544,549	415
00.304043	x $^5F^0_5 \to$ a $^5F_5$	544,550	415
0.452862	x $^5D^0_4 \to$ a $^5P_3$	544,551	416
8.4927853	e $^5D_4 \to$ w $^5D^0_4$	546,553	155
6.8487052	w $^5P^0_3 \to$ e $^5D_4$	545,552	155

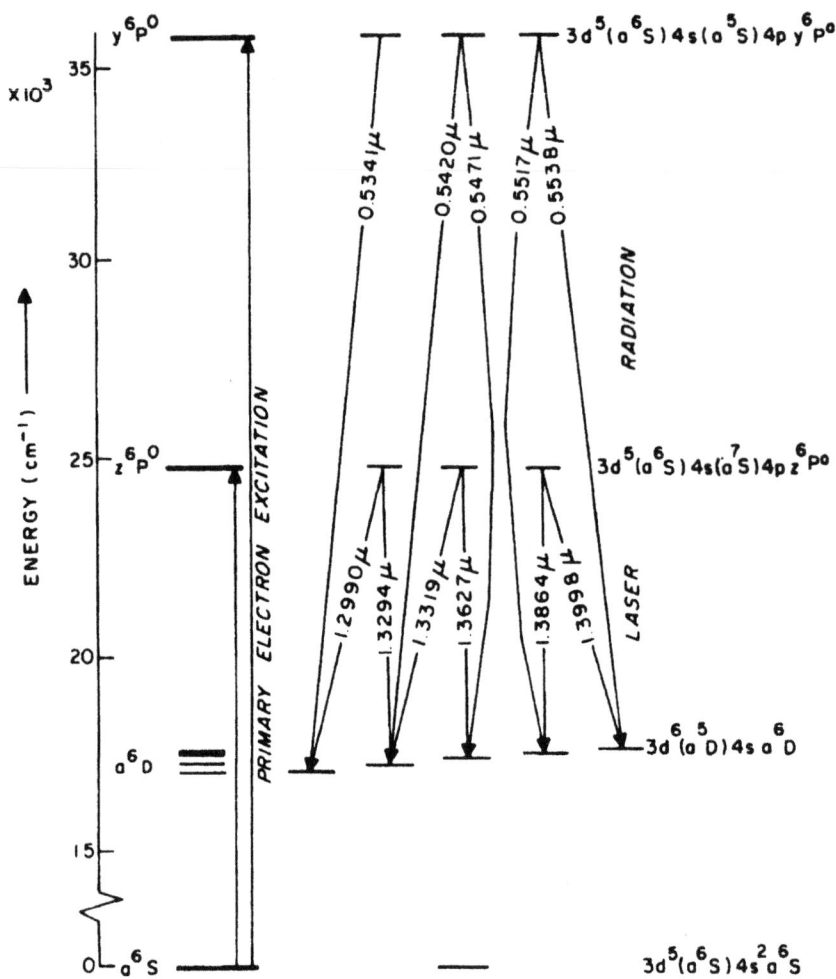

Figure 3.1.24 Partial energy level diagram of manganese showing the self-terminating laser transitions and the primary route for direct electron-impact excitation. Wavelengths are given in microns.

Nickel

Wavelength (μm)	Transition assignment	Comments	References
0.347	4p $^3D^0_3$ - 4s $^3D_2$	6905	1774
0.381	4p $^3D^0_3$ - 4s $^1D_2$	6905	1774
1.396710		554,556	417
1.4553719	h $^3F_2 \rightarrow$ w $^3P^0_1$	557	115

Samarium

Wavelength (μm)	Transition assignment	Comments	References
1.9124005	$5d6s^2\ ^7H^0_2 \rightarrow 5d(^8D)6s\ ^9D_3$	558,561	247
2.0482		559,562	247
2.7006079	$6s(^8F)7s\ ^9F_7 \rightarrow 5d6p(^3F^0)?\ ^9I_6?$	563	247
2.9663		564	247
3.4654		565	247
3.5361		560,566	247
4.1368		567	247
4.8656		568	247

Europium (Figure 3.1.25)

Wavelength (μm)	Transition assignment	Comments	References
0.545294	$z\ ^{10}D_{9/2} \rightarrow a\ ^{10}D^0_{7/2}$	569,571	421
0.545294	$z\ ^{10}D_{9/2} \rightarrow a\ ^{10}D^0_{7/2}$	569,571	421
0.557714	$z\ ^{10}D_{9/2} \rightarrow a\ ^{10}D^0_{11/2}$	569,572	421
0.557714	$z\ ^{10}D_{9/2} \rightarrow a\ ^{10}D^0_{11/2}$	569,572	421
0.605736	$z\ ^8F_{9/2} \rightarrow a\ ^8D^0_{9/2}$	569,573	420
1.7600985	$y\ ^8P_{9/2} \rightarrow a\ ^8D^0_{11/2}$	576	247,418
1.926	$5d^2\ ^8G^0_{11/2} \rightarrow z\ ^8F_{9/2}$	570,574	420
1.961	$5d^2\ ^8G^0_{9/2} \rightarrow z\ ^8F_{9/2}$	570,575	420
2.5818111	$b\ ^8D^0_{5/2} \rightarrow z\ ^8P_{5/2}$	577	247
2.7181668	$b\ ^8D^0_{9/2} \rightarrow z\ ^8P_{7/2}$	578	247
4.3213904	$y\ ^8P_{9/2} \rightarrow b\ ^8D^0_{11/2}$	579	247
4.6948356	$y\ ^8P_{9/2} \rightarrow b\ ^8P_{9/2}$	580	247
5.0660871	$y\ ^8P_{7/2} \rightarrow b\ ^8D^0_{9/2}$	581	247
5.28256430	$y\ ^8P_{7/2} \rightarrow b\ ^8D^0_{7/2}$	582	247
5.4306800	$y\ ^8P_{7/2} \rightarrow b\ ^8D^0_{5/2}$	583	247
5.7722389	$y\ ^8P_{5/2} \rightarrow b\ ^8D^0_{7/2}$	584	247
5.9495478	$y\ ^8P_{5/2} \rightarrow b\ ^8D^0_{5/2}$	585	247
6.0592473	$y\ ^8P_{5/2} \rightarrow b\ ^8D^0_{3/2}$	586	247

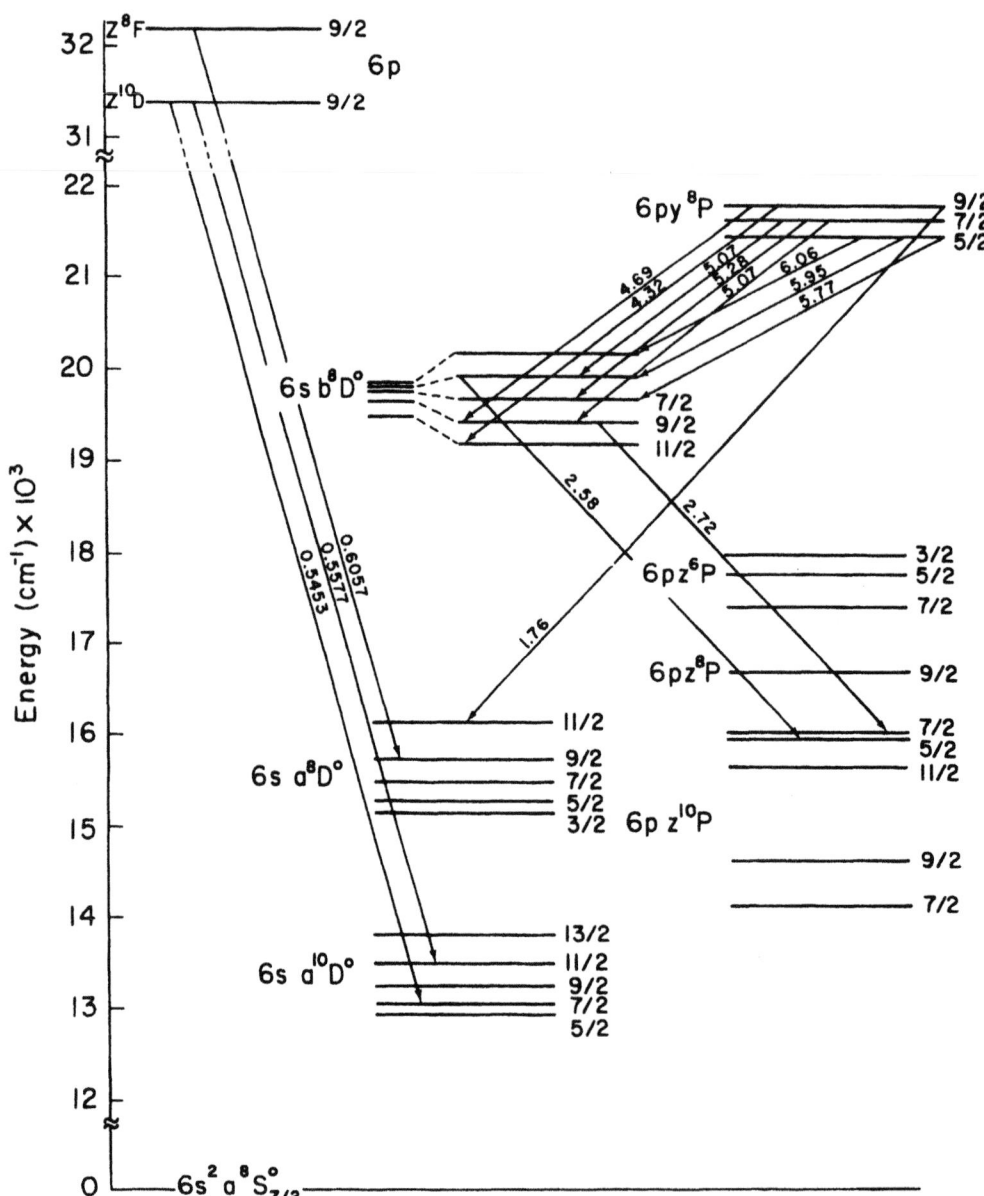

Figure 3.1.25 Partial energy level diagram of europium showing the majority of the reported laser transitions. Wavelengths are given in microns.

3.1.2.14 Group VIIIA Lasers

Table 3.1.14
Group VIIIA Lasers

Helium (Figure 3.1.26)

Wavelength (μm)	Transition assignment	Comments	References
0.6678	$3\ ^1D \to 2\ ^1P$	6917	1767
0.7067124	$3s^3\ S_1 \to 2p\ ^3P^0_2$	587,588,593	422,423
0.7067162	$3s\ ^3S_1 \to 2p\ ^3P^0_1$	587,588,594	422,423
0.7281	$3\ ^1S \to 2\ ^1P$	6917	1767
1.868596	$4f\ ^3F^0 \to 3d\ ^3D$	587,595	424,425
1.954313	$4p\ ^3P^0 \to 3d\ ^3D$	587,596	424–427
2.058130	$2p\ ^1P^0_1 \to 2s\ ^1S_0$	587,590,597	181,428
2.060755	$7d\ ^3D \to 4p\ ^3P^0$	587,591,598	185,429,431
4.60535	$5s\ ^1S_0 \to 4p\ ^1P^0_1$	587,592,599	432,433
4.60567(^{3}He)	$5s\ ^1S_0 \to 4p\ ^1P^0_1$	587,592,600	433
8.529294	$6s\ ^1S_0 \to 5p\ ^1P^0_1$	587,601	433
95.77994	$3p\ ^1P^0_1 \to 3d\ ^1D_2$	587,602	434,435,436
95.788		1998,1999	434,435
216.17882	$4p\ ^1P^1_1 \to 4d\ ^1D_2$	587,603	435,436
216.3		1996,1997	435

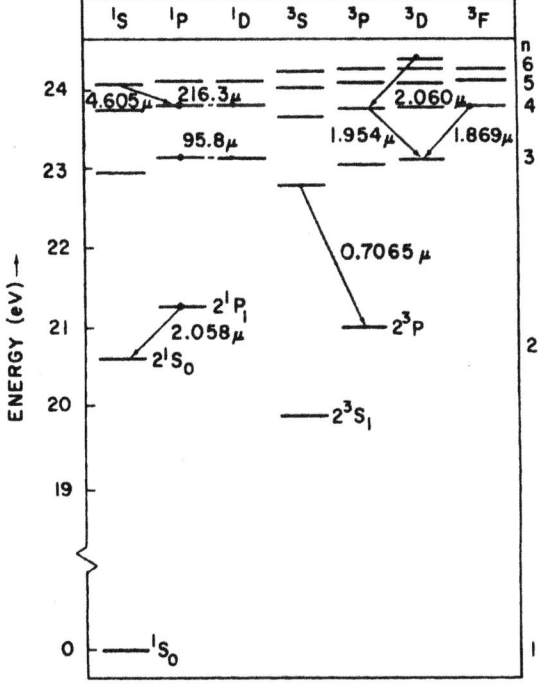

Figure 3.1.26 Partial energy level diagram of helium showing the reported laser transitions.

Neon (Figures 3.1.27–28)

Wavelength (µm)	Transition assignment Racah/Paschen	Comments	References
0.54005616	$3p'[1/2]_0 \rightarrow 3s[3/2]^0_1$ $2p_1 \rightarrow 1s_4$	604,605,621,627	188,336,444, 446,448,460
0.5435161	$5s'[1/2]^0_1 \rightarrow 3p[1/2]_1$ $3s_2 \rightarrow 2p_{10}$	604,621,628	449
0.58524878	$3p'[1/2]_0 \rightarrow 3s'[1/2]^0_1$ $2p_1 \rightarrow 1s_2$	604,605,621,629	185,450
0.58524878	$3p'[1/2]_0 \rightarrow 3s'[1/2]^0_1$ $2p_1 \rightarrow 1s_2$	604,605,621,629	450,185
0.58525	$3p'[1/2]_0 \rightarrow 3s'[1/2]^0_1$	6567,6568	1495–1497
0.59409633	$5s'[1/2]_1^0 \rightarrow 3p[5/2]_2$ $3s_2 \rightarrow 2p_8$	604,606,621, 630,631	185,212,451
0.59409633	$5s'[1/2]_1^0 \rightarrow 3p[5/2]_2$ $3s_2 \rightarrow 2p_8$	604,606,621, 630,631	185,212,451
0.59448342	$3p'[3/2]_2 \rightarrow 3s[3/2]^0_2$ $2p_4 \rightarrow 1s_5$	604,605,621,632	444,453–5
0.59448342	$3p'[3/2]_2 \rightarrow 3s[3/2]^0_2$ $2p_4 \rightarrow 1s_5$	604,605,621,632	444,453–455
0.60461348	$5s'[1/2]^0_1 \rightarrow 3p[3/2]_1$ $3s_2 \rightarrow 2p_7$	604,605,621,633	185,451
0.60461348	$5s'[1/2]^0_1 \rightarrow 3p[3/2]_1$ $3s_2 \rightarrow 2p_7$	604,605,621,633	451,185
0.61197087	$5s'[1/2]^0_1 \rightarrow 3p[3/2]_2$ $3s_2 \rightarrow 2p_6$	604,606,621,634	185,212,451,456
0.61197087	$5s'[1/2]^0_1 \rightarrow 3p[3/2]_2$ $3s_2 \rightarrow 2p_6$	604,606,621,634	185,212,451,456
0.61430623	$3p[3/2]_2 \rightarrow 3s[3/2]^0_2$ $2p_6 \rightarrow 1s_5$	604,605,621,635	444,447,448, 453–7,459–61
0.61430623	$3p[3/2]_2 \rightarrow 3s[3/2]^0_2$ $2p_6 \rightarrow 1s_5$	604,605,621,635	444,447,448, 453–7,459–61
0.62664950	$3p'[3/2]_1 \rightarrow 3s'[1/2]^0_0$ $2p_5 \rightarrow 1s_3$	604,605,621,636	462
0.62664950	$3p'[3/2]_1 \rightarrow 3s'[1/2]^0_0$ $2p_5 \rightarrow 1s_3$	604,605,621,636	462
0.62937447	$5s'[1/2]^0_1 \rightarrow 3p'[3/2]_1$ $3s_2 \rightarrow 2p_5$	604,605,621,637	451,456
0.62937447	$5s'[1/2]^0_1 \rightarrow 3p'[3/2]_1$ $3s_2 \rightarrow 2p_5$	604,605,621,637	451,456
0.63281646	$5s'[1/2]^0_1 \rightarrow 3p'[3/2]_2$ $3s_2 \rightarrow 2p_4$	604,605,621, 638,639	185,451,456,457, 459–462,465–74

Neon—*continued*

Wavelength (μm)	Transition assignment Racah/Paschen	Comments	References
0.63281646	$5s'[1/2]^0_1 \to 3p'[3/2]_2$ $3s_2 \to 2p_4$	604,605,621, 638,639	185,451,456,457, 459–462,465–74
0.63518618	$5s'[1/2]^0_1 \to 3p[1/2]_0$ $3s_2 \to 2p_3$	604–5,621, 640–41	451
0.63518618	$5s'[1/2]^0_1 \to 3p[1/2]_0$ $3s_2 \to 2p_3$	604–5,621, 640–41	451
0.64028455	$5s'[1/2]^0_1 \to 3p'[1/2]_1$ $3s_2 \to 2p_2$	604,621,642, 643,644	451,488,456, 489,523,524
0.64028455	$5s'[1/2]^0_1 \to 3p'[1/2]_1$ $3s_2 \to 2p_2$	604,621,642, 643,644	451,456,488, 489,523,524
0.6599	$3p'[1/2]_1 \to 3s'[1/2]^0_1$	6568	1496
0.6599	$3p'[1/2]_1 \to 3s'[1/2]^0_1$	6568	1496
0.70324	$3p[1/2]_1 \to 3s[3/2]^0_2$	6568	1495,1496
0.72452	$3p[1/2]_1 \to 3s[3/2]^0_1$	6568	1495,1496
0.73068569	$5s'[1/2]^0_1 \to 3p[1/2]_0$ $3s_2 \to 2p_1$	604,621, 645,646	451
0.73068569	$5s'[1/2]^0_1 \to 3p[1/2]_0$ $3s_2 \to 2p_1$	604,621, 645,646	451
0.84633569	$3d[3/2]^0_1 \to 3p[5/2]_2$ $3d_2 \to 2p_8$	604,605, 621,647	526
0.86376895	$3d'[3/2]^0_1 \to 3p'[3/2]_2$ $3s'_1 \to 2p_4$	604,621,648	526
0.86819216	$3d[3/2]^0_1 \to 3p[3/2]_1$ $3d_2 \to 2p_7$	604,605, 621,649	462
0.87740648	$3d'[3/2]^0_1 \to 3p'[1/2]_1$ $3s'_1 \to 2p_2$	604,621,650	526
0.88653057	$4s'[1/2]^0_1 \to 3p[1/2]_1$ $2s_2 \to 2p_{10}$	604,605, 621,651	490,526,527
0.89910237	$4s'[1/2]^0_0 \to 3p[1/2]_1$ $2s_3 \to 2p_{10}$	604,621,652	490,526
0.94892838	$4s[3/2]^0_1 \to 3p[1/2]_1$ $2s_4 \to 2p_{10}$	604,621,653	528
0.96680709	$4s[3/2]^0_2 \to 3p[1/2]_1$ $2s_5 \to 2p_{10}$	604,621,654	527,528
1.0298238	$4s'[1/2]^0_1 \to 3p[5/2]_2$ $2s_2 \to 2p_8$	604,621,655	490,526,527
1.0623574	$4s'[1/2]^0_1 \to 3p[3/2]_1$ $2s_2 \to 2p_7$	604,621,656	185,490,526, 531–533
1.0801000	$4s'[1/2]^0_0 \to 3p[3/2]_1$ $2s_3 \to 2p_7$	604,621,657	389,472,526, 529,534

Neon—*continued*

Wavelength (μm)	Transition assignment Racah/Paschen	Comments	References
1.0847447	$4s'[1/2]^0_1 \rightarrow 3p[3/2]_2$ $2s_2 \rightarrow 2p_6$	604,621,658	389,472,526, 529,534
1.1146071	$4s[3/2]^0_1 \rightarrow 3p[5/2]_2$ $2s_4 \rightarrow 2p_8$	604,621,659	389,472,526, 528,529,534-8
1.1180588	$4s[3/2]^0_2 \rightarrow 3p[5/2]_3$ $2s_5 \rightarrow 2p_9$	604,621,660	185,466,526-9, 532,535,537,539
1.1393552	$4s[3/2]^0_2 \rightarrow 3p[5/2]_2$ $2s_5 \rightarrow 2p_8$	604,621,661	389,466,526,528, 529,534
1.1412258	$4s'[1/2]^0_1 \rightarrow 3p'[3/2]_1$ $2s_2 \rightarrow 2p_5$	604,621,662	389,466,526, 528,529
1.1525900	$4s'[1/2]^0_1 \rightarrow 3p'[3/2]_2$ $2s_2 \rightarrow 2p_4$	604,621,672, 673	185,389,466, 470,472,527,529, 535-40,542-45
1.1528174	$4s[3/2]^0_1 \rightarrow 3p[3/2]_1$ $2s_4 \rightarrow 2p_7$	604,621,665	185,427,526, 528,529,535, 537,544,547
1.1604712	$4s'[1/2]^0_1 \rightarrow 3p[1/2]_0$ $2s_2 \rightarrow 2p_3$	604,621,666	389,466,526, 529,532,540
1.1617260	$4s[1/2]^0_0 \rightarrow 3p'[3/2]_1$ $2s_3 \rightarrow 2p_5$	604,621,667	540,337,389,466, 526,529,536,546
1.1770013	$4s'[1/2]^0_1 \rightarrow 3p'[1/2]_1$ $2s_2 \rightarrow 2p_2$	604,621,668	466,526,527-529, 535-538,550
1.1792270	$4s[3/2]^0_1 \rightarrow 3p[3/2]_2$ $2s_4 \rightarrow 2p_6$	604,621,669	490,526,527, 528,529
1.1988192	$4s'[1/2]^0_0 \rightarrow 3p'[1/2]_1$ $2s_3 \rightarrow 2p_2$	604,621,670	389,466,526, 529,536,539
1.2068179	$4s[3/2]^0_0 \rightarrow 3p[3/2]_2$ $2s_5 \rightarrow 2p_6$	604,621,671	389,526,528,529, 532,535-536, 546,550
1.2462797	$4s[3/2]^0_1 \rightarrow 3p'[3/2]_1$ $2s_4 \rightarrow 2p_5$	604,621,674	490,526,528,529
1.2588072	$5f[9/2]_5 \rightarrow 3d[7/2]^0_4$ $5V_5 \rightarrow 3d'_4$	604,621,675	553
1.2588088	$5f[9/2]_4 \rightarrow 3d[7/2]^0_4$ $5V_4 \rightarrow 3d'_4$	604,621,676	553
1.2598449	$4s[3/2]^0_1 \rightarrow 3p'[3/2]_2$ $2s_4 \rightarrow 2p_4$	604,621,677	528,529
1.2692672	$4s[3/2]^0_1 \rightarrow 3p[1/2]_0$ $2s_4 \rightarrow 2p_3$	604,621,678	426,434,526, 528,537,542,553

Neon—*continued*

Wavelength (μm)	Transition assignment Racah/Paschen	Comments	References
1.2773017	$4s[3/2]^0_2 \rightarrow 3p'[3/2]_1$ $2s_5 \rightarrow 2p_5$	604,621,679	526,528
1.2890684	$4s[3/2]^0_1 \rightarrow 3p'[1/2]^0_1$ $2s_4 \rightarrow 2p_2$	604,621,680	490
1.2915545	$4s[3/2]^0_2 \rightarrow 3p'[3/2]_2$ $2s_5 \rightarrow 2p_4$	604,621,681	434,490,526, 528,553
1.3222855	$?4s[3/2]^0_2 \rightarrow 3p'[1/2]_1$ $2s_5 \rightarrow 2p_2$	604,618, 621,682	528
1.4276		604,621,683	554
1.4304		604,621,684	554
1.4321		604,621,685	554
1.4330		604,621,686	554
1.4346		604,621,687	554
1.4368		604,621,688	554
1.4848636	$8p'[1/2]_1 \rightarrow 5s[3/2]^0_2$ $7p_2 \rightarrow 3s_5$	604,621,689	555
1.4873294	$7p[1/2]_1 \rightarrow 4p'[1/2]_0?$ $6p_{10} \rightarrow 3p_1?$	604,620,621,690	555
1.4876248	$5p'[1/2]_1 \rightarrow 3d[3/2]^0_1$ $4p_2 \rightarrow 3d_2$	604,621,691	555
1.4892012	$6s'[1/2]^0_1 \rightarrow 4p[3/2]_1$ $4s_2 \rightarrow 3p_7$	604,621,692	555
1.4903576	$6p[1/2]_1 \rightarrow 4p[3/2]_2?$ $5p_{10} \rightarrow 3p_6?$	604,620,621,693	555
1.4940304	$6s'[1/2]^0_1 \rightarrow 4p[3/2]_1$ $4s_3 \rightarrow 3p_7$	604,621,694	555
1.5234875	$4s'[1/2]^0_1 \rightarrow 3p'[1/2]_0$ $2s_2 \rightarrow 2p_1$	604,621,695	389,466,526, 529,534,550,556
1.6407031	$6s[3/2]^0_2 \rightarrow 4p[5/2]_3$ $4s_5 \rightarrow 3p_9$	604,621,696	551
1.7166616	$4s[3/2]^0_1 \rightarrow 3p'[1/2]_0$ $2s_4 \rightarrow 2p_1$	604,621,697	490,526
1.8215302	$4p'[1/2]_1 \rightarrow 4s[3/2]^0_1$ $3p_1 \rightarrow 2s_4$	604,621,698	526
1.8258313	$4f[5/2]_2 \rightarrow 3d[7/2]^0_3$ $4Y_2 \rightarrow 3d_4$	604,608, 621,699	185
1.8258357	$4f[5/2]_3 \rightarrow 3d[7/2]^0_3$ $4Y_3 \rightarrow 3d_4$	604,608, 621,700	185
1.827659	$4f[9/2]_{4,5} \rightarrow 3d[7/2]^0_4$ $4V \rightarrow 3d'_4$	604,609, 621,701	551,557–559

Neon—*continued*

Wavelength (µm)	Transition assignment Racah/Paschen	Comments	References
1.828258	$4f[9/2]_4 \rightarrow 3d[7/2]^0_3$ $4V \rightarrow 3d_4$	604,609, 621,702	551,557,558,559
1.830400	$4f[5/2]_{2,3} \rightarrow 3d[3/2]^0_2$ $4Y \rightarrow 3d_3$	604,609, 621,703	551,557,559
1.840316	$4f[5/2]_2 \rightarrow 3d[3/2]^0_1$ $4Y \rightarrow 3d_2$	604,609, 621,704	550,551,557,559
1.859112	$4f[7/2]_3 \rightarrow 3d[5/2]^0_2$ $4Z \rightarrow 3d''_1$	604,609, 621,705	557,551,559
1.859730	$4f[7/2]_{3,4} \rightarrow 3d[5/2]^0_3$ $4Z \rightarrow 3d'_1$	604,609, 621,706	556,557,559
1.95740	$4p'[3/2]_2 \rightarrow 4s'[3/2]^0_2$ $3p_4 \rightarrow 2s_5$	604,610, 621,707	185,526,542, 553,560
1.958248	$4p'[1/2]_1 \rightarrow 4s[3/2]^0_2$ $3p_2 \rightarrow 2s_5$	604,621,708	185,434,526,561
2.0355792	$4p'[3/2]_2 \rightarrow 4s[3/2]^0_1$ $3p_4 \rightarrow 2s_4$	604,621,709	185,426,434, 526,537,542,550, 560–562,564
2.0359432	$4p'[1/2]_1 \rightarrow 4s[3/2]^0_1$ $3p_2 \rightarrow 2s_4$	604,621,710	185,526,537, 542,564
2.1023345	$4d'[5/2]^0_2 \rightarrow 4p[3/2]_2$ $4s'''' \rightarrow 3p_6(?)$	604,621,711	185,435,526
2.10409	$4p'[1/2]^0_0 \rightarrow 4s'[1/2]^0_1$ $3p_1 \rightarrow 2s_2$	604,610, 621,712	185,434,526, 537,542,553,562
2.17074	$4p[1/2]_0 \rightarrow 4s[3/2]^0_1$ $3p_3 \rightarrow 2s_4$	604,610,621,713	185,434,526, 537,553
2.3266649	$4p[5/2]_2 \rightarrow 4s[3/2]^0_2$ $3p_8 \rightarrow 2s_5$	604,621,714	526
2.39579530	$4p'[3/2]_2 \rightarrow 4s[1/2]^0_1$ $3p_4 \rightarrow 2s_2$	604,611, 621,715	434,526,542, 556,560,565–568
2.3962995	$4p'[1/2]_1 \rightarrow 4s[1/2]^0_1$ $3p_2 \rightarrow 2s_2$	604,621,716	185,542,550, 567,568
2.4162547	$5s'[1/2]^0_1 \rightarrow 4p[1/2]_1$ $3s_2 \rightarrow 3p_{10}$	604,621,717	569
2.4225538	$4d[3/2]^0_1 \rightarrow 4p[5/2]_2$ $4d_2 \rightarrow 3p_8$	604,621,718	185,570
2.4256255	$4p'[3/2]_1 \rightarrow 4s'[1/2]^0_1$ $3p_5 \rightarrow 2s_2$	604,621,719	434,526,542
2.5400115	$4d[1/2]^0_1 \rightarrow 4p[3/2]_2$ $4d_5 \rightarrow 3p_6$	604,621,720	195,564

Neon—*continued*

Wavelength (μm)	Transition assignment Racah/Paschen	Comments	References
2.5531329	$4p[1/2]_1 \to 4s[3/2]^0_2$ $3p_\{10\} \to 2s_5$	604,621,721	526
2.7580982	$4d[3/2]^0_1 \to 4p[1/2]_0$ $4d_2 \to 3p_3$	604,621,722	261,195
2.7826380	$5s'[1/2]^0_0 \to 4p[3/2]_1$ $3s_3 \to 3p_7$	604,621,723	185,195,564
2.8633965	$8f'[7/2,5/2]_{2,3,4} \to 5d[7/2]^0_3?$ $8U \to 5d_4$	604,612,621,724	195
2.9455858	$5s[3/2]^0_1 \to 4p[1/2]_1$ $3s_4 \to 3p_{10}$	604,621,725	185,195,564
2.9676035	$4d[3/2]^0_1 \to 4p'[3/2]_1$ $4d_2 \to 3p_5$	604,621,726	195,564
2.9812503	$4d[3/2]^0_2 \to 4p'[3/2]_1$ $4d_3 \to 3p_5$	604,621,727	195,564
3.0267787	$4d[3/2]^0_2 \to 4p'[1/2]_1$ $4d_3 \to 3p_2$	604,621,728	185,195,564
3.0275836	$4d[3/2]^0_2 \to 4p'[3/2]_2$ $4d_3 \to 3p_4$	604,621,729	185,195,564
3.0720016	$5s'[1/2]^0_1 \to 4p[1/2]_0$ $3s_2 \to 3p_3$	604,621,730	571
3.3182141	$5s[3/2]^0_1 \to 4p[5/2]_2$ $3s_4 \to 3p_8$	604,621,731	195,542,559, 564,570
3.3341754	$5s'[1/2]^0_1 \to 4p'[3/2]_1$ $3s_2 \to 3p_5$	604,614, 621,732	195,559, 564,570
3.3361448	$5s[3/2]^0_2 \to 4p[5/2]_3$ $3s_5 \to 3p_9$	604,614, 621,733	195,559, 564,570
3.3510469	$6d[1/2]^0_1 \to 5p[1/2]_0$ $6d_5 \to 4p_3$	604,615, 621,734	432
3.3520466	$4p[1/2]_1 \to 4s'[1/2]^0_1$ $3p_{10} \to 2s_2$	604,615, 621,735	432
3.3813942	$7s'[1/2]^0_0 \to 5p'[3/2]_1$ $5s_3 \to 4p_5$	604,621,736	195,564
3.3849653	$7s'[1/2]^0_0 \to 5p'[1/2]_1$ $5s_3 \to 4p_2$	604,621,737	195,564
3.3912244	$5s'[1/2]^0_1 \to 4p'[1/2]_1$ $3s_2 \to 3p_2$	604,621,738	559,564,567, 570,578,579,580

Neon—*continued*

Wavelength (μm)	Transition assignment Racah/Paschen	Comments	References
3.3922348	$5s'[1/2]^0_1 \to 4p'[3/2]_2$ $3s_2 \to 3p_4$	604,621, 739,740	389,431,434,455, 456,466,468,470, 472,479,483,525, 556,564,567, 572,575
3.4480843	$5s[3/2]^0_1 \to 4p[3/2]_1$ $3s_4 \to 3p_7$	604,621,741	195,337,525,559, 564,570
3.4789495	$5s[3/2]^0_1 \to 4p[3/2]_2$ $3s_4 \to 3p_6$	604,621,742	432,525
3.5844556	$5s[3/2]^0_2 \to 4p[3/2]_2$ $3s_5 \to 3p_6$	604,621,743	185,559, 564,570
3.6164638	$9p[3/2]_{1,2} \to 5d[5/2]^0_2$ $8p_{7,6} \to 5d''_1$	604,621,744	432
3.7584321	$7s'[1/2]^0_1 \to 6p'[3/2]_2$ $5s_2 \to 5p_4$	604,617,621,800	564,559,195
3.7746325	$4p'[1/2]_0 \to 3d[3/2]^0_1$ $3p_1 \to 3d_2$	604,621,745	559,564
3.980630	$5s[3/2]^0_1 \to 4p[1/2]_0$ $3s_4 \to 3p_3$	604,610, 621,746	195,564
5.103		604,621,748	432
5.1711388	$5d'[5/2]^0_2 \to 5p'[3/2]_1$ $5s''''_1 \to 4p_5$	604,621,749	432
5.3258685	$5d[5/2]^0_2 \to 5p[3/2]_2$ $5d''_1 \to 4p_6$	604,615,621,750	432
5.3264331	$5d[3/2]^0_1 \to 5p[3/2]_1$ $5d_2 \to 4p_7$	604,615,621,751	432
5.4048094	$4p'[1/2]_0 \to 3d'[3/2]^0_1$ $3p_1 \to 3s'_1$	604,621,752	525,559,564, 570,582
5.515		604,621,753	432
5.6667372	$4p[1/2]_0 \to 3d[3/2]^0_1$ $3p_3 \to 3d_2$	604,621,754	195,564,570
5.7067951	$5p'[1/2]_1 \to 5s'[1/2]^0_0$ $4p_2 \to 3s_3$	604,621,755	432
5.7773913	$5p[5/2]_3 \to 5s[3/2]^0_2$ $4p_9 \to 3s_5$	604,621,756	432
5.8858082	$5p'[3/2]_1 \to 5s'[1/2]^0_1$ $4p_5 \to 3s_2$	604,621,757	432
5.9578742	$5p[5/2]_2 \to 5s[3/2]^0_1$ $4p_8 \to 3s_4$	604,621,758	432

Neon—*continued*

Wavelength (μm)	Transition assignment Racah/Paschen	Comments	References
6.7788016	$6s[3/2]^0_2 \rightarrow 5p[1/2]_1$ $4s_5 \rightarrow 4p_{10}$	604,621,759	432
6.8884755	$7d'[3/2]^0_1 \rightarrow 7p[1/2]_1$ $7s'_1 \rightarrow 6p_{10}$	604,621,760	432
6.9876797	$4p[3/2]_2 \rightarrow 3d[3/2]^0_2$ $3p_6 \rightarrow 3d_3$	604,621,761	432
7.098		604,621,762	432
7.3228367	$6s[3/2]^0_2 \rightarrow 5p[5/2]_3$ $4s_5 \rightarrow 4p_9$	604,621,763	195,570,564
7.405131	$6s'[1/2]^0_0 \rightarrow 5p'[3/2]_1?$ $4s_3 \rightarrow 4p_5$	604,616, 621,764	432
7.4222794	$6s'[1/2]^0_0 \rightarrow 5p'[1/2]_1$ $4s_3 \rightarrow 4p_2$	604,621,765	195,564
7.4235357	$5p'[1/2]_1 \rightarrow 4d[3/2]_2$ $4p_2 \rightarrow 4d_3$	604,621,766	195,564
7.4699904	$4p[3/2]_2 \rightarrow 3d[5/2]^0_2$ $3p_6 \rightarrow 3d''_1$	604,621,767	195
7.4799887	$4p[3/2]_2 \rightarrow 3d[5/2]^0_3$ $3p_6 \rightarrow 3d'_1$	604,621,768	559,564,570
7.4995237	$6s[3/2]^0_2 \rightarrow 5p[5/2]_2$ $4s_5 \rightarrow 4p_8$	604,621,769	185,195,559
7.5313000	$6s[3/2]^0_1 \rightarrow 5p[3/2]_1$ $4s_4 \rightarrow 4p_7$	604,621,770	185,570
7.5709798	$6d[1/2]^0_1 \rightarrow 5f[3/2]_2$ $6d_5 \rightarrow 5X_2$	604,621,771	432
7.5885028	$6d[1/2]^0_0 \rightarrow 5f[3/2]_1$ $6d_6 \rightarrow 5X_1$	604,621,772	432
7.6163805	$4p[3/2]_1 \rightarrow 3d[5/2]^0_2$ $3p_7 \rightarrow 3d''_1$	604,621,773	195,559, 564,570
7.6458386	$5p'[3/2]_1 \rightarrow 4d[5/2]^0_2$ $4p_5 \rightarrow 4d_1$	604,621,774	195
7.6511521	$4p[5/2]_2 \rightarrow 3d[7/2]^0_3$ $3p_8 \rightarrow 3d_4$	604,621,775	559,564,570
7.6926284	$4p'[3/2]_2 \rightarrow 3d'[5/2]^0_2$ $3p_4 \rightarrow 3s''''_1$	604,621,776	195
7.7016367	$4p'[3/2]_2 \rightarrow 3d'[5/2]^0_3$ $3p_4 \rightarrow 3s'''_1$	604,621,777	559,564,570
7.7408259	$4p[5/2]_2 \rightarrow 3d[3/2]^0_2$ $3p_8 \rightarrow 3d_3$	604,621,778	195

Neon—*continued*

Wavelength (μm)	Transition assignment Racah/Paschen	Comments	References
7.7655499	$4p'[1/2]_1 \rightarrow 3d'[3/2]^0_2$ $3p_2 \rightarrow 3s''_1$	604,621,779	559,570
7.7815561	$6s[3/2]^0_2 \rightarrow 5p[3/2]_1$ $4s_5 \rightarrow 4p_7$	604,621,780	195,564
7.809		604,621,781	432
7.8369230	$6s[3/2]^0_2 \rightarrow 5p[3/2]_2$ $4s_5 \rightarrow 4p_6$	604,621,782	195,559, 564,570
7.8716418	$8p[3/2]_2 \rightarrow 7s[3/2]^0_2$ $7p_6 \rightarrow 5s_5$	604,621,783	432
7.9429190	$8p[5/2]_2 \rightarrow 7s[3/2]^0_2$ $7p_8 \rightarrow 5s_5$	604,621,784	432
7.9846694	$8s'[1/2]^0_1 \rightarrow 7p[1/2]_1$ $6s_2 \rightarrow 6p_{10}$	604,621,785	432
8.0088892	$4p'[3/2]_1 \rightarrow 3d'[5/2]^0_2$ $3p_5 \rightarrow 3s''''_1$	604,621,786	195,559, 564,570
8.0621949	$4p[5/2]_3 \rightarrow 3d[7/2]^0_4$ $3p_9 \rightarrow 3d'_4$	604,621,787	195,559, 564,570
8.116		604,621,788	432
8.1736588	$4p[5/2]_3 \rightarrow 3d[3/2]_2$ $3p_9 \rightarrow 3d_3$	604,621,789	432
8.3371447	$4p[5/2]_2 \rightarrow 3d[5/2]^0_2$ $3p_8 \rightarrow 3d''_1$	604,615,621,790	195,564
8.3496011	$4p[5/2]_2 \rightarrow 3d[5/2]^0_3$ $3p_8 \rightarrow 3d'_1$	604,615,621,791	195,564
8.8414054	$4p[5/2]_3 \rightarrow 3d[5/2]^0_2$ $3p_9 \rightarrow 3d''_1$	604,621,792	195,564
8.8554154	$4p[5/2]_3 \rightarrow 3d[5/2]^0_3$ $3p_9 \rightarrow 3d'_1$	604,621,793	195,564
9.0894630	$6s[3/2]^0_1 \rightarrow 5p[1/2]_0$ $4s_4 \rightarrow 4p_3$	604,621,794	195,559, 564,570
10.063422	$4p[1/2]_1 \rightarrow 3d[1/2]^0_1$ $3p_{10} \rightarrow 3d_5$	604,621,795	195,564
10.981641	$4p[1/2]_1 \rightarrow 3d[3/2]^0_2$ $3p_{10} \rightarrow 3d_3$		195,559, 564,570
11.860553 0	$5p[1/2]_1 \rightarrow 5s'[1/2]^0_0$ $4p_{10} \rightarrow 3s_3$		564
11.902069	$5p[1/2]_0 \rightarrow 4d[3/2]^0_1$ $4p_3 \rightarrow 4d_2$		195
12.835	See Comments and References	2062	583,586,1452

Neon—*continued*

Wavelength (μm)	Transition assignment Racah/Paschen	Comments	References
12.835306	$5p'[1/2]_0 \rightarrow 4d'[3/2]^0_1$ $4p_1 \rightarrow 4s'_1$	604,621,799	195,564
13.758318	$4d'[5/2]^0_3 \rightarrow 4f[5/2]_3$ $4s'''_1 \rightarrow 4Y$	604,617,621, 801	195,559,564
13.759	See Comments and References	2061	583,586,1452
14.930		604,621,802	564
16.638		206	583,586,1452
16.638076	$5p[3/2]_2 \rightarrow 4d[5/2]^0_2$ $4p_6 \rightarrow 4d''_1$	604,621,803	195,564
16.667472	$5p[3/2]_2 \rightarrow 4d[5/2]^0_3$ $4p_6 \rightarrow 4d'_1$	604,621,804	564
16.668		2059	583,586,1452
16.893		2058	583,586,1452
16.893261	$5p[3/2]_1 \rightarrow 4d[5/2]^0_2$ $4p_7 \rightarrow 4d''_1$	604,621,805	559,564
16.946194	$5p[5/2]_2 \rightarrow 4d[7/2]^0_3$ $4p_3 \rightarrow 4d_4$	604,621,806	559,564,195
16.947		2057	583,586,1452
17.157279	$5p'[3/2]_2 \rightarrow 4d'[5/2]^0_3$ $4p_4 \rightarrow 4s'''_1$	604,621,807	559,564
17.158		2056	583,586,1452
17.188155	$5p'[3/2]_2 \rightarrow 4d'[3/2]^0_2$ $4p_4 \rightarrow 4s''_1$	604,621,808	564
17.189		2055	583,586,1452
17.803541	$5p'[1/2]_1 \rightarrow 4d'[3/2]^0_2$ $4p_2 \rightarrow 4s''_1$	604,621,809	559,564
17.804		2054	583,586,1452
17.839876	$5p'[3/2]_1 \rightarrow 4d'[5/2]^0_2$ $4p_5 \rightarrow 4s''''_1$	604,621,811	559,564
17.841		2053	583,586,1452
17.888		2052	583,586,1452
17.888255	$5p[5/2]_3 \rightarrow 4d[7/2]^0_4$ $4p_9 \rightarrow 4d'_4$	604,621,811	559,564,195
18.395067	$5p[5/2]_2 \rightarrow 4d[5/2]^0_2$ $4p_8 \rightarrow 4d''_1$	604,621,812	559,564,195
18.396		2051	583,586,1452
20.478165	$6p[1/2]_0 \rightarrow 5d[1/2]^0_1$ $5p_3 \rightarrow 5d_5$	604,621,813	559,564,195
20.48		2050	583,586,1452

Neon—*continued*

Wavelength (μm)	Transition assignment Racah/Paschen	Comments	References
21.750951	$6p[1/2]_0 \rightarrow 5d[3/2]^0_1$ $5p_3 \rightarrow 5d_2$	604,621,814	559,564
21.752		2049	583,586,1452
22.836		2048	583,586,1452
22.836211	$5p[1/2]_1 \rightarrow 4d[3/2]^0_2$ $4p_{10} \rightarrow 4d_3$	604,621,815	559,564
25.421617	$6p'[1/2]_0 \rightarrow 5d'[3/2]^0_1$ $5p_1 \rightarrow 5s'_1$	604,621,816	559,564
25.423		2047	583,586,1452
28.052222	$6p[3/2]_1 \rightarrow 5d[1/2]^0_1$ $5p_7 \rightarrow 6d_6$	604,621,817	559,564
28.053		2046	583,586,1452
31.552211	$6p[3/2]_2 \rightarrow 5d[5/2]^0_3$ $5p_6 \rightarrow 5d'_1$	604,621,818	564
31.553		2045	583,586,1452
31.928		2044	583,586,1452
31.930315	$6p[3/2]_1 \rightarrow 5d[5/2]^0_2$ $5p_7 \rightarrow 5d''_1$	604,621,819	583,585
32.015777	$6p[5/2]_2 \rightarrow 5d[7/2]^0_3$ $5p_8 \rightarrow 5d_4$	604,621,820	559
32.016		2043	583,586,1452
32.516		2042	583,586,1452
32.518956	$p'[3/2]_2 \rightarrow 5d'[5/2]^0_3$ $5p_4 \rightarrow 5s'''_1$	604,621,821	564
32.83		2041	583,586,1452
33.815657	$6p'[3/2]_1 \rightarrow 5d'[5/2]^0_2$ $5p_5 \rightarrow 5s''''_1$	604,621,822	564
33.834764	$6p[5/2]_3 \rightarrow 5d[7/2]^0_4$ $5p_9 \rightarrow 5d'_4$	604,621,823	564
34.550549	$6p'[1/2]_1 \rightarrow 5d'[3/2]^0_2$ $5p_2 \rightarrow 5s''_1$	604,621,824	564
34.552		2040	583,586,1452
34.678633	$6p[5/2]_2 \rightarrow 5d[5/2]^0_2$ $5p_8 \rightarrow 5d''_1$	604,621,825	583,585
34.679		2039	583,586,1452
35.602		2037,2038	583,586,1452
35.608858	$7p[1/2]_0 \rightarrow 6d[3/2]^0_1$ $6p_3 \rightarrow 6d_2$	604,621,826	337,583,585
37.230081	$7p'[1/2]_0 \rightarrow 6d'[3/2]^0_1$ $6p_1 \rightarrow 6s'_1$	604,621,827	583,585

Neon—*continued*

Wavelength (μm)	Transition assignment Racah/Paschen	Comments	References
37.231		2035,2036	583,586,1452
41.738317	$6p[1/2]_1 \to 5d[3/2]^0_2$ $5p_{10} \to 5d_3$	604,621,828	583,585
41.741		2034	583,586,1452
50.700169	$7p[3/2]_2 \to 6d[3/2]^0_2$ $6p_6 \to 6d_3$	604,621,829	583,586
50.705		2032,2033	583,586,1452
52.425		2030,2031	583,586,1452
52.429587	$7p'[1/2]_1 \to 6d'[3/2]^0_2$ $6p_2 \to 6s''_1$	604,621,830	586
53.478223	$7p[3/2]_2 \to 6d[5/2]^0_3$ $6p_6 \to 6d'_1$	604,621,831	583,585
53.486		2028,2029	583,586,1452
54.019		2026,2027	583,586,1452
54.041785	$7p[3/2]_1 \to 6d[5/2]^0_2$ $6p_7 \to 6d''_1$	604,621,832	583,585
54.106991	$7p[5/2]_2 \to 6d[7/2]^0_3$ $6p_8 \to 6d_4$	604,621,833	583,585
54.117		2024,2025	583,586,1452
55.537		2022,2023	583,586,1452
55.542287	$7p'[3/2]_1 \to 6d'[5/2]^0_2$ $6p_5 \to 6s''''_1$	604,621,834	586
57.355		2020,2021	583,586,1452
57.364118	$7p[5/2]_3 \to 6d[7/2]^0_4$ $6p_9 \to 6d'_4$	604,621,835	583,585
68.325612	$7p[1/2]_1 \to 6d[3/2]^0_2$ $6p_{10} \to 6d_3$	604,621,836	587,185
68.329		2019	583,586,1452
.72.108		2017,2018	583,586,1452
72.118851	$8p'[1/2]_0 \to 7d'[3/2]^0_1$ $7p_1 \to 7s'_1$	604,621,837	586
85.047		2016	586,1452
85.062223	$8p[3/2]_2 \to 7d[5/2]^0_3$ $7p_6 \to 7d'_1$	604,621,838	185,587
86.962		2014,2015	586,1452
86.977698	$8p'[3/2]_2 \to 7d'[5/2]^0_2$ $7p_4 \to 7s''''_1$	604,621,839	586
88.471		2012,2013	586,1452
88.487744	$8p[3/2]_1 \to 7d[5/2]^0_2$ $7p_7 \to 7d''_1$	604,621,840	185,586

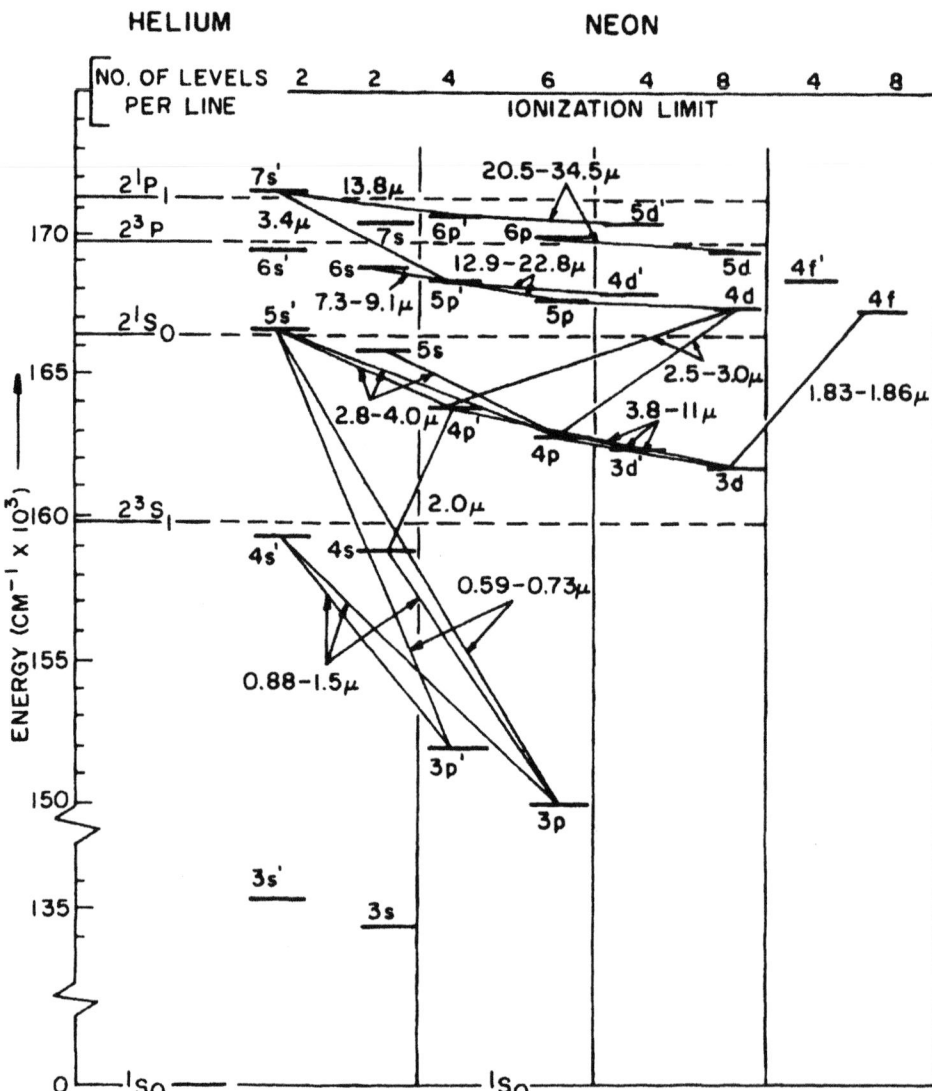

Figure 3.1.27 Partial energy level diagram of neon showing laser transition groups.

Neon—*continued*

Wavelength (μm)	Transition assignment Racah/Paschen	Comments	References
89.859		2010,2011	586,1452
89.872291	$8p[5/2]_3 \rightarrow 7d[7/2]^0_3$	604,621,841	586
	$7p_9 \rightarrow 7d_4$		
93.0		2008,2009	586,1452

Neon—*continued*

Wavelength (μm)	Transition assignment Racah/Paschen	Comments	References
93.02	?	604,621,842	586
106.07		2006,2007	586,1452
106.07828	$10p[1/2]_0 \rightarrow 9d[3/2]^0_1$ $9p_3 \rightarrow 9d_2$	604,621,843	185,586
124.4		2004,2005	586,1452
124.55626	$9p[3/2]_1 \rightarrow 8d[5/2]^0_2$ $8p_7 \rightarrow 8d''_1$	604,621,844	586
124.79564	$9p[3/2]_2 \rightarrow 8d[5/2]^0_3$ $8p_6 \rightarrow 8d'_1$	604,621,845	586
126.1	?	604,621,846, 2002,2003	586,1452
132.8	?	604,621,847, 2000,2001	586,1452

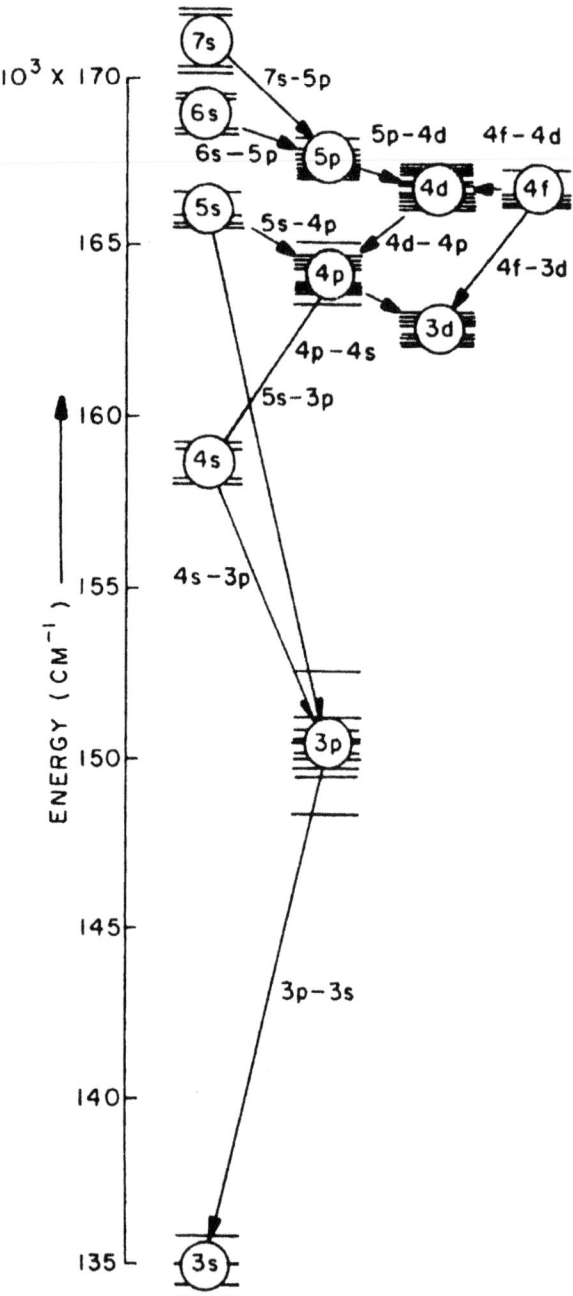

Figure 3.1.28 Energy level diagram of neon showing groups of laser transitions observed in a He-Ne mixture and energy coincidences with $He(2^3S_1, 2^1S_0, 2^3P_{2,1,0}$ and $2^1P_1)$ levels.

Argon (Figure 3.1.29)

Wavelength (μm)	Transition assignment	Comments	References
0.673000		1357	921
0.70691661	$4p'[3/2]_2 \rightarrow 4s[3/2]^0_2$ $2p_3 \rightarrow 1s_5$	848,857,860	454
0.75059341	$4p'[1/2]_0 \rightarrow 4s'[1/2]^0_1$ $2p_1 \rightarrow 1s_2$	848,857,861	450
0.7948	$4p'\ [3/2]_1 \rightarrow 4s'\ [1/2]s(o,o)$	6915	1762
0.912297	$4p[1/2]_1 \rightarrow 4s[3/2]^0_2$ $2p_{10} \rightarrow 1s_5$	848,857,862	248,588,589,722
0.965779	$4p[1/2]_1 \rightarrow 4s[3/2]^0_1$ $2p_{10} \rightarrow 1s_4$	848,857,863	588,589,593,722
1.04701	$4p[1/2]_1 \rightarrow 4s'[1/2]^0_0$ $2p_{10} \rightarrow 1s_3$	848,857,864	588
1.14881	$4p[1/2]_1 \rightarrow 4s'[1/2]^0_1$ $2p_{10} \rightarrow 1s_2$	848,857,865	588
1.2115637?	$3d[5/2]_3 \rightarrow 4p[5/2]_3$ $3d_1' \rightarrow 2p_9(?)$	848,849,857,866	532
1.21397378	$3d'[3/2]^0_1 \rightarrow 4p'[3/2]_1$ $3s'_1 \rightarrow 2p_4$	848,857,867	588
1.24028269	$3d[3/2]^0_1 \rightarrow 4p[3/2]_1$ $3d_2 \rightarrow 2p_7$	848,857,868	590,588
1.27022810	$3d'[3/2]_1 \rightarrow 4p'[1/2]_1$ $3s_1' \rightarrow 2p_2$	848,857,869	428,588, 590–593,722,806
1.28027391?	$3d[5/2]^0_2 \rightarrow 4p[5/2]_2$ $3d_1'' \rightarrow 2p_8$	848,850,857,870	532
1.3476544	$7d[3/2]^0_2 \rightarrow 5p[3/2]_2$ $7d_3 \rightarrow 3p_6$	848,857,871	594
1.40936399	$3d[3/2]^0_1 \rightarrow 4p[1/2]_0$ $3d_2 \rightarrow 2p_5$	848,857,872	538
1.504605	$3d'[3/2]^0_1 \rightarrow 4p'[1/2]_0$ $3s_1' \rightarrow 2p_1$	848,857,873	806
1.6180021	$5s[3/2]^0_2 \rightarrow 4p'[3/2]_2$ $2s_5 \rightarrow 2p_3$	848,857,874	389,429
1.65199	$3d[3/2]^0_2 \rightarrow 4p[3/2]_1$ $3d_3 \rightarrow 2p_7$	848,857,875	593
1.6940584	$3d[3/2]^0_2 \rightarrow 4p[3/2]_2$ $3d_3 \rightarrow 2p_6$	848,857,876	389,429,538, 588,593,806
1.7919615	$3d[1/2]^0_1 \rightarrow 4p[3/2]_2$ $3d_5 \rightarrow 2p_6$	848,851,857,877, 878	185,389,429,588, 591–593,596– 600,722,806
2.0616228	$3d[3/2]^0_2 \rightarrow 4p'[3/2]_2$ $3d_3 \rightarrow 2p_3$	848,857,879	389,429,596, 597,598,806

Argon—*continued*

Wavelength (μm)	Transition assignment	Comments	References
2.0986110	$3d[1/2]^0_1 \to 4p[1/2]_0$ $3d_5 \to 2p_5$	848,857,880	185,427,565,806
2.1332885	$3d[1/2]^0_1 \to 4p'[3/2]_1$ $3d_5 \to 2p_4$	848,857,881	195
2.1534205	$3d[3/2]^0_2 \to 4p'[1/2]_1$ $3d_3 \to 2p_2$	848,857,882	427,565,593
2.20771810	$3d[1/2]^0_1 \to 4p'[3/2]_2$ $3d_5 \to 2p_3$	848,852,857,883	195,264, 564,806
2.3133204	$3d[1/2]^0_1 \to 4p'[1/2]_1$ $3d_5 \to 2p_2$	848,857,884	195,538, 564,593
2.3966520	$3d[1/2]^0_0 \to 4p'[1/2]_1$ $3d_6 \to 2p_2$	848,857,885	195,264,428, 564,588,591
2.5014408	$6d'[3/2]^0_2 \to 6p[1/2]_1$ $6s''_1 \to 4p^{10}$	848,857,886	195,564
2.54946	$5p[5/2]_3 \to 3d[7/2]^0_3$ $3p_9 \to 3d_4$	848,857,887	195,564
2.5512187	$5p[1/2]_0 \to 5s[3/2]^0_1$ $3p_5 \to 2s_4$	848,857,888	195,564
2.5634025	$6d'[3/2]^0_2 \to 6p[5/2]_3$ $6s''_1 \to 4p_9$	848,857,889	195
2.56680230	$5p'[1/2]_0 \to 5s'[1/2]^0_1$ $3p_1 \to 3s_2$	848,857,890	564
2.6550282	$5p'[3/2]_1 \to 3d'[5/2]^0_2$ $3p_4 \to 3s''''_1$	848,857,891	601
2.6843026	$5p[3/2]_1 \to 3d[5/2]^0_2$ $3p_7 \to 3d_1''$	848,857,892	195,564
2.7152859	$5p[3/2]_1 \to 5s[3/2]^0_2$ $3p_7 \to 2s_5$	848,857,893	602
2.7363805	$5p'[1/2]_1 \to 3d'[3/2]^0_2$ $3p_2 \to 3s_1''$	848,857,894	195,564
2.8202417	$5p'[3/2]_1 \to 5s'[1/2]^0_0$ $3p_4 \to 2s_3$	848,853,857,895	195,564
2.8245953	$5p[3/2]_2 \to 5s[3/2]^0_1$ $3p_6 \to 2s_4$	848,853,857,896	195,564
2.8620231	$5p'[3/2]_2 \to 5s'[1/2]^0_1$ $3p_3 \to 2s_2$	848,857,897	598,603
2.8782932	$5p[5/2]_3 \to 5s[3/2]^0_2$ $3p_9 \to 2s_5$	848,857,898	195,564
2.8843088	$5p[3/2]_2 \to 3d[5/2]^0_3$ $3p_6 \to 3d'_1$	848,857,899	195,564

Argon—*continued*

Wavelength (µm)	Transition assignment	Comments	References
2.9134037	$5p'[3/2]_1 \rightarrow 5s'[1/2]^0_1$ $3p_4 \rightarrow 2s_2$	848,857,900	603
2.9280662	$5p[1/2]_0 \rightarrow 3d[3/2]^0_1$ $3p_5 \rightarrow 3d_2$	848,857,901	195,564
2.9796792	$5p[5/2]_2 \rightarrow 5s[3/2]^0_1$ $3p_8 \rightarrow 2s_4$	848,857,902	195,564
3.046207	$5p[5/2]_2 \rightarrow 3d[5/2]^0_3$ $3p_8 \rightarrow 3d_1'$	848,857,903	195,564
3.0996226	$5p[5/2]_3 \rightarrow 3d[5/2]^0_3$ $3p_9 \rightarrow 3d'_1$	848,857,904	195,564
3.13330280	$5p[1/2]_1 \rightarrow 5s[3/2]^0_2$ $3p10 \rightarrow 2s_5(?)$	848,857,905	564
3.1345761	$6p'[3/2]_2 \rightarrow 4d[5/2]^0_2$ $4p_3 \rightarrow 4d''_1$	848,857,906	195
3.631236	$6s'[1/2]^0_1 \rightarrow 5p'[3/2]_1$ $3s_2 \rightarrow 3p_4$	848,857,907	602
3.7013512	$6s'[1/2]^0_1 \rightarrow 5p'[1/2]_1$ $3s_2 \rightarrow 3p_2$	848,857,908	602
3.7086023	$4d[3/2]^0_1 \rightarrow 5p[3/2]_1$ $4d_2 \rightarrow 3p_7$	848,857,909	604
3.7143477	$6s'[1/2]^0_1 \rightarrow 5p'[3/2]_2$ $3s_2 \rightarrow 3p_3$	848,857,910	602
4.2044098	$5p[3/2]_2 \rightarrow 3d'[3/2]^0_2$ $3p_6 \rightarrow 3s''_1$	848,857,911	601
4.71516800	$5p[5/2]_3 \rightarrow 3d'[5/2]^0_3$ $3p_9 \rightarrow 3s'''_1$	848,857,912	601
4.9160256	$6p'[3/2]_2 \rightarrow 4d'[3/2]^0_2$ $4p_3 \rightarrow 4s''_1$	848,857,913	195,564
4.9207077	$5d[5/2]^0_2 \rightarrow 4f[7/2]^\wedge f_3$ $5d'_1 \rightarrow 4U$	848,857,914	195,564
5.0235584	$6p'[3/2]_1 \rightarrow 4d'[5/2]^0_2$ $4p_4 \rightarrow 4s''''_1$	848,857,915	602
12.140522	$4d'[3/2]^0_1 \rightarrow 4f[3/2]_1$ $4s' \rightarrow 4X_1$	848,857,932	195,559,564
12.146405	$4d'[3/2]^0_1 \rightarrow 4f[3/2]_2$ $4s'_1 \rightarrow 4X_2$	848,857,933	195,559,564
13.475		6572	1498
15.022		2067	1452
15.037		2066	1452
15.037067	$5d'[3/2]^0_2 \rightarrow 5f[5/2]_3$ $5s''_1 \rightarrow 5Y_3$	848,857,934	559,564

Argon—*continued*

Wavelength (μm)	Transition assignment	Comments	References
15.039		2065	1452
15.042133	$5d'[3/2]^0_2 \rightarrow 5f[5/2]_2$	848,857,935	559,564
	$5s''_1 \rightarrow 5Y_2$		
26.933		2064	1452
26.936		2063	1452
26.943974	$4d'[3/2]^0_2 \rightarrow 4f[5/2]_{,,}$	848,857,936	559,564
	$4s''_1 \rightarrow 4Y_3$		

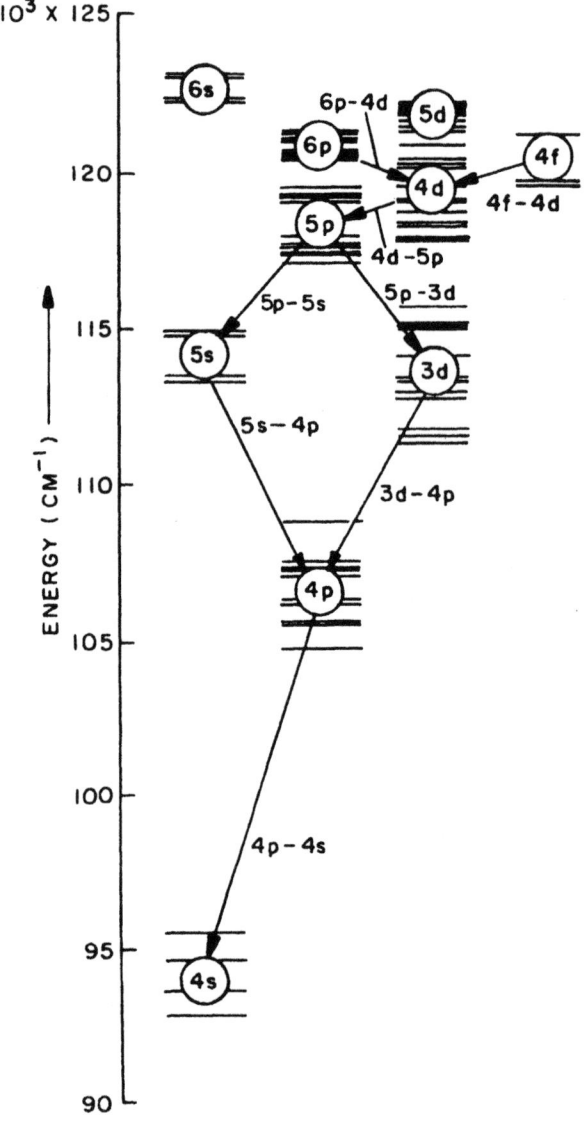

Figure 3.1.29 Partial energy level diagram of argon showing laser transition groups.

Krypton (Figure 3.1.30)

Wavelength (µm)	Transition assignment	Comments	References
0.760154393	$5p[3/2]_2 \rightarrow 5s[3/2]^0_2$ $2p_6 \rightarrow 1s_5$	937,945	604
0.810436392	$5p[5/2]_2 \rightarrow 5s[3/2]^0_2 2$ $p_8 \rightarrow 1s_5$	937,946	454,455,538
0.892869155	$5p[1/2]_1 \rightarrow 5s[3/2]^0_2$ $2p_{10} \rightarrow 1s_5$	937,947	588,589
1.14574813	$6s[3/2]^0_1 \rightarrow 5p[1/2]_1$ $2s_4 \rightarrow 2p_{10}$	937,948	538
1.31774118	$6s[3/2]^0_1 \rightarrow 5p[5/2]_2$ $2s_4 \rightarrow 2p_5$	937,949	538
1.36224153	$4d[3/2]^0_1 \rightarrow 5p[5/2]_2$ $3d_2 \rightarrow 2p_8$	937,950	538
1.44267933	$6s[3/2]^0_1 \rightarrow 5p[3/2]_1$ $2s_4 \rightarrow 2p_7$	937,951	538
1.47654720	$6s[3/2]^0_1 \rightarrow 5p[3/2]_2$ $2s_4 \rightarrow 2p_8$	937,952	538
1.49618939	$4d[3/2]^0_1 \rightarrow 5p[3/2]_1$ $3d_2 \rightarrow 2p_7$	937,953	605
1.53264796	$4d[3/2]^0_1 \rightarrow 5p[3/2]_2$ $3d_2 \rightarrow 2p_6$	937,954	605
1.68534881	$4d[7/2]^0_3 \rightarrow 5p[5/2]_3$ $3d_4 \rightarrow 2p_9$	937,955	538
1.68967525	$4d[1/2]^0_1 \rightarrow 5p[1/2]_1$ $3d_5 \rightarrow 2p_{10}$	937,956	389,429,538
1.69358061	$4d[5/2]^0_2 \rightarrow 5p[3/2]_1$ $3d''_1 \rightarrow 2p_7$	937,957	389,429
1.78427374	$4d[1/2]^0_0 \rightarrow 5p[1/2]_1$ $3d_6 \rightarrow 2p_{10}$	937,958	389,429
1.81673150	$4d[7/2]^0_4 \rightarrow 5p[5/2]_3$ $3d'_4 \rightarrow 2p_9$	937,959	427
1.81850539	$4d'[5/2]^0_2 \rightarrow 5p'[3/2]_2$ $3s''''_1 \rightarrow 2p_2$	937,960	389,429
1.9216572	$8s[3/2]^0_2 \rightarrow 6p[5/2]_2$ $4s_5 \rightarrow 3p_8$	937,961	389,429
2.11654709	$4d[3/2]^0_2 \rightarrow 5p[3/2]_1$ $3d_3 \rightarrow 2p_7$	937,962	185,389, 429,570
2.19025126	$4d[3/2]^0_2 \rightarrow 5p[3/2]_2$ $3d_3 \rightarrow 2p_6$	937,963	185,389,429, 570,607
2.24857754	$6p[3/2]_1 \rightarrow 4d[5/2]^0_2$ $3p_7 \rightarrow 3d''_1$	937,939,964	602

Krypton—*continued*

Wavelength (µm)	Transition assignment	Comments	References
2.42605059	$4d[1/2]^0_1 \to 5p[3/2]_1$ $3d_5 \to 2p_7$	937,965	427,565
2.52338198	$4d[1/2]^0_1 \to 5p[3/2]_2$ $3d_5 \to 2p_6$	937,966	389,428,538,559, 564,570,582,608
2.6266703	$4d[1/2]^0_0 \to 5p[3/2]_1$ $3d_6 \to 2p_7$	937,967	195,564
2.6288137	$7p[3/2]_2 \to 4d'[5/2]^0_2$ $4p_6 \to 3s''''_1$	937,968	195,564
2.8610550	$6p[5/2]_2 \to 6s[3/2]^0_2$ $3p_8 \to 2s_5$	937,969	195,538,559
2.8655717	$6p[5/2]_3 \to 6s[3/2]^0_2$ $3p_9 \to 2s_5$	937,970	195,538,559
2.9844656	$6p'[1/2]_1 \to 5d[5/2]^0_2$ $3p_3 \to 4d''_1$	937,941,971	195,564
2.9878091	$6p'[3/2]_1 \to 6s'[1/2]^0_0$ $3p_4 \to 2s_3$	937,941,972	195,564
3.0536574	$6p'[3/2]_1 \to 5d[5/2]^0_2$ $3p_4 \to 4d''_1$	937,973	195,564
3.0663542	$6p[1/2]_1 \to 6s[3/2]^0_2$ $3p_{10} \to 2s_5$	937,974	428,559,564,570
3.1514572	$6p'[1/2]_0 \to 5d[3/2]^0_1$ $3p_1 \to 4d_2$	937,975	195,564
3.34096354	$d[1/2]^0_1 \to 5p[1/2]_0$ $3d_5 \to 2p_5$	937,976	195,564,570
3.4679986	$7s[3/2]^0_1 \to 6p[1/2]_1$ $3s_4 \to 3p_{10}$	937,977	195,564
3.4882957	$6p'[1/2]_1 \to 7s[3/2]^0_2$ $3p_3 \to 3s_5$	937,978	195,564
3.4894892	$6p'[1/2]_1 \to 5d[3/2]^0_1$ $3p_3 \to 4d_2$	937,979	195,564
3.7742128	$7s[3/2]^0_1 \to 6p[5/2]_2$ $3s_4 \to 3p_8$	937,980	604
3.9557248	$5d[3/2]^0_1 \to 6p[5/2]^{\wedge}e_2$ $4d_2 \to 3p_8$	937,981	604
4.0685162	$7s[3/2]^0_1 \to 6p[3/2]_1$ $3s_4 \to 3p_7$	937,982	604
4.1526711	$7s[3/2]^0_1 \to 6p[3/2]_2$ $3s_4 \to 3p_6$	937,983	604
4.3747938	$5d[3/2]^0_1 \to 6p[3/2]_2$ $3s_4 \to 3p_6$	937,984	559,564,570,604

Krypton—*continued*

Wavelength (µm)	Transition assignment	Comments	References
4.3766712	$7s[3/2]^0_2 \to 6p[3/2]_2$	937,985	195
	$3s_5 \to 3p_6$		
4.8773393	$4d[3/2]^0_1 \to 5p'[3/2]_1$	937,986	195,564
	$3d_2 \to 2p_4$		
4.883133	$45d[5/2]^0_2 \to 6p[5/2]_3$	937,987	195,564
	$4d''_1 \to 3p_9$		
4.9996952	$4d'[3/2]^0_1 \to 6p[1/2]_1$	937,988	601,605
	$3s'_1 \to 3p_{10}$		
5.1311509	$6s[3/2]^0_1 \to 5p'[3/2]_2$	937,989	601
	$2s_4 \to 2p_2$		
5.2999768	$5d[3/2]^0_1 \to 6p[1/2]_0$	937,990	195,559,564
	$4d_2 \to 3p_5$		
5.3019553	$5d[3/2]^0_2 \to 6p[5/2]_2$	937,991	195,559,564
	$4d_3 \to 3p_8$		
5.5699805	$5d[7/2]^0_3 \to 6p[5/2]_2$	937,943,992	185,564,195
	$4d_4 \to 3p_8$		
5.5862769	$6d[7/2]^0_4 \to 4f[9/2]_5$	937,941,993	559,564,570
	$5d'_4 \to 4U$		
5.6305126	$6d[3/2]^0_2 \to 4f[5/2]_3$	937,994	559,564,570
	$5d_3 \to 4T$		
7.0580595	$4f[7/2]\_\{3,4\} \to 5d[7/2]^0_4$	937,995	195,564
	$4W \to 4d'_4$		
7.3625232	$4f[9/2]_5 \to 5d[7/2]^0_4$	937,996	601
	$4U_5 \to 4d'_4$		
8.1151	$8s[3/2]^0_1 \to 7p[5/2]_2$	6573	1498

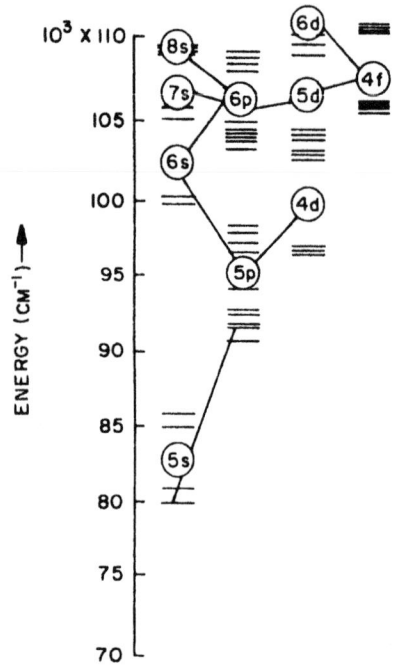

Figure 3.1.30 Partial energy level diagram of krypton showing laser transition groups.

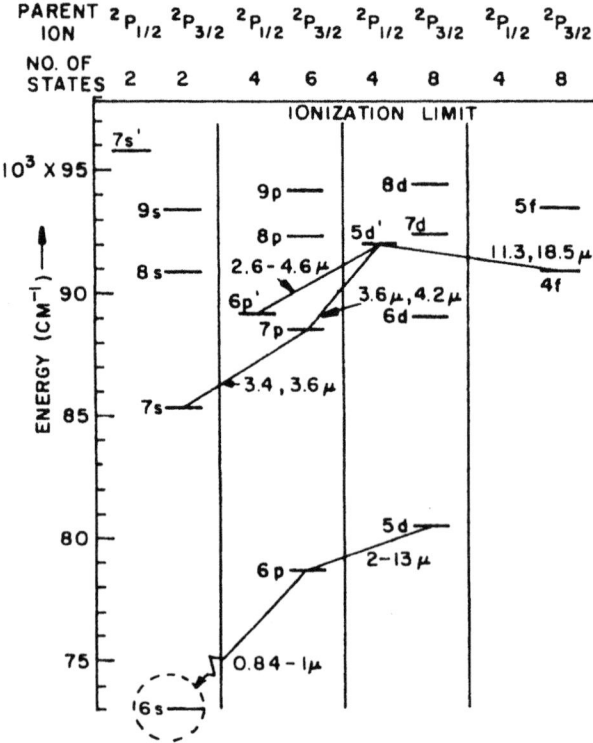

Figure 3.1.31 Partial energy level diagram of xenon showing groups of observed laser transitions. Wavelengths are given in microns.

Xenon (Figure 3.1.31)

Wavelength (μm)	Transition assignment	Comments	References
0.82316376	$6p[3/2]_2 \rightarrow 6s[3/2]^0_2$ $2p_6 \rightarrow 1s_5$	997,1006,1007	248,588,589
0.84091940	$6p[3/2]_1 \rightarrow 6s[3/2]^0_2$ $2p_7 \rightarrow 1s_5$	997,1006,1008	538,609
0.90454514	$6p[5/2]_2 \rightarrow 6s[3/2]^0_2$ $2p_9 \rightarrow 1s_5$	997,1006,1009	455,538
0.97997039	$6p[1/2]_1 \rightarrow 6s[3/2]^0_2$ $2p_{10} \rightarrow 1s_5$	997,1006,1010	455,588,589
1.0634		997,1000,1001, 1006,1011	609
1.0950		997,1000,1006, 1012	609
1.36570559	$7s[3/2]^0_1 \rightarrow 6p[5/2]_2$ $2s_4 \rightarrow 2p_9$	997,1006,1013	538
1.60576667	$s[3/2]^0_1 \rightarrow 6p[3/2]_2$ $2s_4 \rightarrow 2p_6$	997,1006,1014	538
1.7330499	$5d[3/2]^0_1 \rightarrow 6p[5/2]_2$ $3d_2 \rightarrow 2p_9$	997,1006,1015, 1016	538,588,589, 610–613,722
2.02622395	$5d[3/2]^0_1 \rightarrow 6p[3/2]_1$ $3d_2 \rightarrow 2p_7$	997,1006,1017, 1018,1019	185,429,389,538, 588,589,607,608, 610–616,618,806
2.31933328	$5d[5/2]^0_3 \rightarrow 6p[5/2]_2$ $3d'_1 \rightarrow 2p_9$	997,1006,1020	550,559,582,607
2.48247157	$5d[5/2]^0_3 \rightarrow 6p[5/2]_3$ $3d'_1 \rightarrow 2p_8$	997,1006,1021	610
2.5152702	$7d[7/2]^0_4 \rightarrow 7p[5/2]_3$ $5d'_4 \rightarrow 3p_9$	997,1006,1022	602
2.62690832	$5d[5/2]^0_2 \rightarrow 6p[5/2]_2$ $3d''_1 \rightarrow 2p_9$	997,1006,1023	559,582,607, 612,632
2.65108645	$5d[3/2]^0_1 \rightarrow 6p[1/2]_0$ $3d_2 \rightarrow 2p_5$	997,1006,1024	538,550,582,607, 610–613,623,625, 634,723
2.6608397	$5d'[3/2]^0_1 \rightarrow 6p'[1/2]_0$ $3s_1 \rightarrow 2p_1$	997,1002,1006, 1025	559,607
2.6672615	$7d[1/2]^0_1 \rightarrow 6p'[3/2]_1$ $3p_{10} \rightarrow 2p_4$	997,1006,1026	601
2.8590043	$7p[3/2]_2 \rightarrow 7s[3/2]^0_2$ $3p_6 \rightarrow 2s_5$	997,1006,1027	602
3.10692302	$5d[5/2]^0_3 \rightarrow 6p[3/2]_2$ $3d'_1 \rightarrow 2p_6$	997,1006,1028	550,559,582,607, 611,623,632

Xenon—*continued*

Wavelength (µm)	Transition assignment	Comments	References
3.27392788	$5d[3/2]^0_2 \to 6p[1/2]_1$ $3d_3 \to 2p_{10}$	997,1006,1029	427,550,565, 607,632
3.3094055	$8p[5/2]_3 \to 6d[5/2]^0_2$ $4p_8 \to 4d'_1$	997,1006,1030	601
3.36666991	$5d[5/2]^0_2 \to 6p[3/2]_1$ $3d''_1 \to 2p_7$	997,1006,1031	550,559,582, 607,611,613,623
3.4023945	$6p'[3/2]_1 \to 7s[3/2]^0_1$ $2p_4 \to 2s_4$	997,1006,1032	601
3.4344638	$7p[5/2]_2 \to 7s[3/2]^0_1$ $3p_9 \to 2s_4$	997,1003,1006, 1033,1034	185,559,564, 582,613,627,630
3.50702520	$5d[7/2]^0_3 \to 6p[5/2]_2$ $3d_4 \to 2p_9$	997,1004,1006, 1035,1036,1037	389,550,559,570, 582,607,611,612, 623–25,631–37
3.6219081	$5d'[3/2]^0_2 \to 7p[3/2]_2$ $3s''''_1 \to 3p_6$	997,1006,1038	559
3.6518315	$7p[1/2]_1 \to 7s[3/2]^0_2$ $3p_{10} \to 2s_5$	997,1006,1039	564,613,625, 627,630,631,723
3.6798859	$5d[1/2]^0_1 \to 6p[1/2]_1$ $3d_5 \to 2p_{10}$	997,1006,1040	582,559,632
3.6858866	$5d[5/2]^0_2 \to 6p[3/2]_2$ $3d'' \to 2p_6$	997,1006,1041	582,559,611,623
3.7265	$5d'[5/2]^0_2 \to 7p[5/2]_3$	6578	1498
3.8696535	$5d'[5/2]^0_3 \to 6p'[3/2]_2$ $3s''_1 \to 2p_3$	997,1006,1042	559,632
3.8950221	$5d[7/2]^0_3 \to 6p[5/2]_3$ $3d_4 \to 2p_8$	997,1006,1043	582,559,611
3.9966035	$5d[1/2]^0_0 \to 6p[1/2]_1$ $3d_6 \to 2p_{10}$	997,1006,1044	559,611,614, 623,624
5.3566973	$5d[1/2]^0_1 \to 6p[5/2]_2$ $3d_5 \to 2p_9$	997,1006,1053	571,601
5.4749269	$7d[5/2]^0_3 \to 4f[5/2]_3$ $5d'_1 \to 4U$	997,1006,1054	601
5.501		997,1005,1006, 1055	601
5.575472	$65d[7/2]^0_4 \to 6p[5/2]_3$ $3d'_4 \to 2p_8$	997,1006,1056	389,559,570,582, 611,623,632,647
5.6034328	$7d[7/2]^0_3 \to 4f[9/2]_4$ $5d_4 \to 4Z$	997,1006,1057	601
5.692		997,1005, 1006,1058	601

Xenon—*continued*

Wavelength (μm)	Transition assignment	Comments	References
5.913		997,1005, 1006,1059	601
6.132		997,1005, 1006,1060	601
6.3120378	$7d[7/2]^0_4 \rightarrow 4f[9/2]_5$ $5d'_4 \rightarrow 4T$	997,1006,1061	601
6.3154334	$7d[7/2]^0_4 \rightarrow 4f[9/2]_4$ $5d'_4 \rightarrow 4Z$	997,1006,1062	601
7.240		997,1005,1006, 1063	601
7.3168036	$5d[3/2]^0_2 \rightarrow 6p[3/2]_1$ $3d_3 \rightarrow 2p_7$	997,1006,1064	559,582, 623,632
7.4313142	$6d[3/2]^0_1 \rightarrow 7p[3/2]_2$ $4d_2 \rightarrow 3p_6$	997,1006,1065	601
7.7665	$6d[3/2]^0_1 \rightarrow 7p[3/2]_1$	6579	1498
7.7813	$5d'[3/2]^0_1 \rightarrow 8p[3/2]_1$	6580	1498
8.4042	$6d [3/2]^0_1 \rightarrow 7p[1/2]_0$	6581	1498
9.0067086	$5d[3/2]^0_2 \rightarrow 6p[3/2]_2$ $3d_3 \rightarrow 2p_6$	997,1006,1066	185,559,582, 611,623,632
9.7031910	$5d[1/2]^0_1 \rightarrow 6p[3/2]_1$ $3d_5 \rightarrow 2p_7$	997,1006,1067	582,559
11.298683	$5d'[5/2]^0_3 \rightarrow 4f[9/2]_4$ $3s'''_1 \rightarrow 4Z$		559,564
12.266	See Comments and References	2072	1452
12.917	See Comments and References	2071	1452
18.505324	$5d'[3/2]^0_2 \rightarrow 4f[5/2]_3$ $3s''''_1 \rightarrow 4U$	997,1006,1071	559,564
75.561687	$6p[1/2]_0 \rightarrow 5d[1/2]^0_1$ $2p_5 \rightarrow 3d_5$	997,1006,1072	652

Section 3.2
IONIZED GAS LASERS

3.2.1 Introduction

The tables in this section include laser lines originating from transitions between energy levels of the ionized states of atoms in gas discharges. Figure 3.2.1 indicates by the shaded squares on the periodic table those elements that have produced ionized gas laser lines. Most of the elements are those which occur as gases at room temperature or are easily vaporized. Some of the more refractory elements also exhibit ion laser oscillation when sputtered in a hollow-cathode discharge.

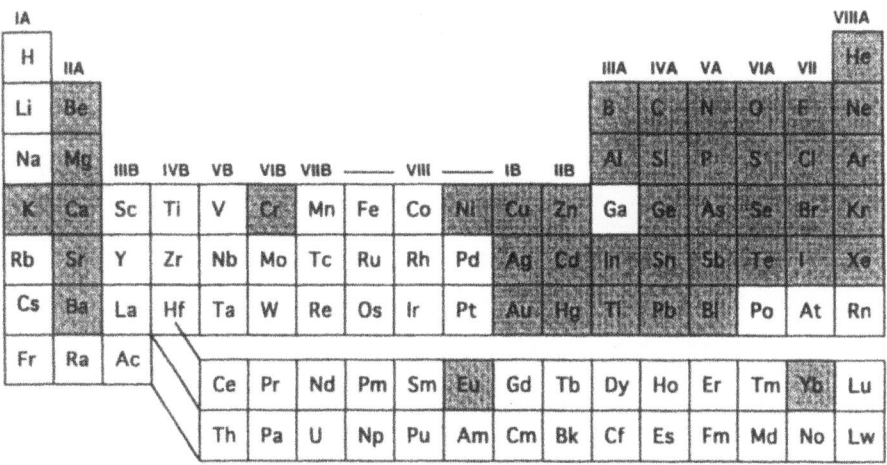

Figure 3.2.1 Periodic table of the elements showing those elements (shaded) whose ions have exhibited laser oscillation in gaseous or vapor form.

The laser tables in this section are arranged by columns in the periodic table; within each column laser ions are further arranged in the order of increasing atomic number. The tables include the observed wavelengths, element charge, transition assignments, comments, and references.

The Wavelength column lists the most accurate available wavelength for the given transition, in micrometers. In most cases, this value of wavelength has been derived from energy levels given in the spectroscopic literature, assuming the indicated classification for the line. The literature sources are given in the references; if no source is listed, the wavelength has been calculated from the energy levels listed by Moore in *Atomic Energy Levels*, National Bureau of Standards Circular 467.[1] The wavelength in air has been calculated from the vacuum energy levels using the Edlen formula:

$$n(v) = 1 + 6432.8 \times 10^{-8} + 2949810/(146 \times 10^8 - v^2) + 25540/(41 \times 10^8 - v^2)$$

where

$$\lambda[\text{microns, air}] = 10^4 /n(v)v[\text{cm}^{-1}, \text{vacuum}]$$

The calculations are accurate to $\pm 10^{-6}$ µm. For levels known only to the nearest cm^{-1}, the wavelength observed in spontaneous emission is more accurate and is listed as given by Harrison,[2] or some other spectroscopic source.

The Measured value column lists the most accurate measured wavelength of the actual laser output, together with the estimated error in the final digits as cited in the references.

The Transition column lists the levels of the atomic transition believed to be responsible for the laser emission. The level notation follows Reference 1, with the core configuration being given in parentheses. In cases where the configuration is unknown, the level energy rounded to the nearest cm^{-1} is given, with the J value as a subscript. In cases where the level configuration is partly known, either the partial designation in Reference 1 or the rounded energy designation is given. Identifications of the observed laser lines have generally been made only on the basis of the measured value of wavelength and theoretical plausibility. This can be done with some degree of confidence in the absence of other nearby spectral lines, but occasionally two equally plausible lines fall within the measurement error. In these cases, both identifications are given. Laser transitions can often be assigned to an ionization state on the basis of their behavior with discharge current relative to lines of known ionization state, even if they cannot be assigned to particular energy levels. A question mark (?) indicates that the classification or ionization state is uncertain or the existence of this laser line may be in doubt.

References include the first report of laser action on that line, although it may not represent the most accurate measurement of the laser wavelength. No attempt has been made to list all references in the case of lines which have received much attention; only references giving added data on gain or power or those in which new excitation techniques or other new features are mentioned.

1. Moore, C. E., *Atomic Energy Levels*, Circular 467, Volumes 1, 2, 3, U.S. Government Printing Office, Washington, DC (1949, 1952, 1958).
2. Harrison, G. R., *MIT Wavelength Tables*, John Wiley & Sons, New York (1952).

Further Reading

Bennett, W. R., Jr., *Atomic Gas Laser Transition Data,* Plenum, New York (1979).

Bridges, W. B., Ionized Gas Lasers, in *Handbook of Laser Science and Technology, Vol. II: Gas Lasers*, CRC Press, Boca Raton, FL (1982), p. 171.

King, D., Photoionization-Pumped Short Wavelength Lasers, in *Handbook of Laser Science and Technology, Suppl. 1: Lasers,* CRC Press, Boca Raton, FL (1991), p. 531.

Peterson, A. B., Ionized Gas Lasers in *Handbook of Laser Science and Technology, Suppl. 1: Lasers,* CRC Press, Boca Raton, FL (1991), p. 335.

3.2.2 Energy Level Diagrams

Figures 3.2.2 through 3.2.15 are energy level diagrams and laser transitions for selected laser ions and are from W. B. Bridges, *Handbook of Laser Science and Technology, Vol. II: Gas Lasers*, p. 171. The arrangement of the levels differs from diagram to diagram according to the complexity of the coupling scheme appropriate to the particular ion. Unlike conventional Grotrian diagrams where all levels and all observed spectral lines are usually indicated, not all levels are shown and only laser transitions are included.

Individual levels and laser transitions are not shown for Cu II, Ag II, and An II (Figures 3.2.2 to 3.2.4); however, the range of energies over which particular groups of levels occur is indicated by the vertical extent of the boxes. The number of laser transitions observed up to 1980 is given by the number beside each arrow. Both Penning and thermal charge-exchange collisional excitation processes are known to be responsible for various laser lines in Cu, Ag, An, Zn, Cd, and Hg; accordingly, the energies of the ions and metastables in He and Ne are indicated with respect to the neutral atom ground state in Figures 3.2.2 through 3.2.7.

Figures 3.2.8 through 3.2.15 for the noble gas ions are more complicated. Each known atomic energy state (up to the maximum energy indicated) appears as a horizontal line, and these states are classified in columns according to the orbital angular momentum, l, of the excited electron (l – 0, 1, 2 for s, p, d states). The configuration designations are given in the L-S (Russell-Saunders) coupling scheme, as listed in Reference 1, except as noted in the captions. Core configurations corresponding to unprimed and primed state designations follow Reference 1 and are explicitly given at the bottom of the figures. Higher orbital states (f, g, etc.) are not shown, since very few laser lines are known to involve any of these levels. In addition, levels not yet assigned to the s, p, or d systems are omitted, even though some laser lines involve these levels. The total angular quantum number J of each state appears to the right of the horizontal line denoting that state.

This arrangement of states in columns automatically separates states of even and odd parity, the latter being denoted by the symbol $^{\circ}$. Note that the only selection rules satisfied by all the laser lines are those of parity change and orbital momentum change $l = \pm 1$; L–S coupling is badly violated in many of these atoms, and changes in core configuration or total spin S are relatively common for laser transitions.

Vertical lines and dots are used to connect states having a common state designation, differing only in their total angular momentum, J.

The vertical scale represents energy of excitation above the ground state for the given ionic species. The left-hand axis is marked in thousands of cm^{-1}, and corresponding tic marks are indicated on the right-hand axis. In addition, the right-hand axis indicates energy in electron volts (1 eV = 8068 cm^{-1}). The energy placement of the atomic states is that given in Reference 1, with a few exceptions. The designation "I.P." in the upper corners of the figures gives the ionization potential of the particular ion considered, in centimeters^{-1} and in electron volts.

The diagonal lines drawn on the figures represent known ion laser lines. Their wavelengths (in air) are indicated in micrometers with four-place accuracy. A dotted line means that the indicated transition is one of two that could be responsible for the observed laser. For Ar II and Kr II, the laser lines ending on the s^{2}P energy states are listed in two separate columns, keyed to the corresponding upper energy levels; the transitions in these cases were so numerous that diagonal lines would have been difficult to distinguish.

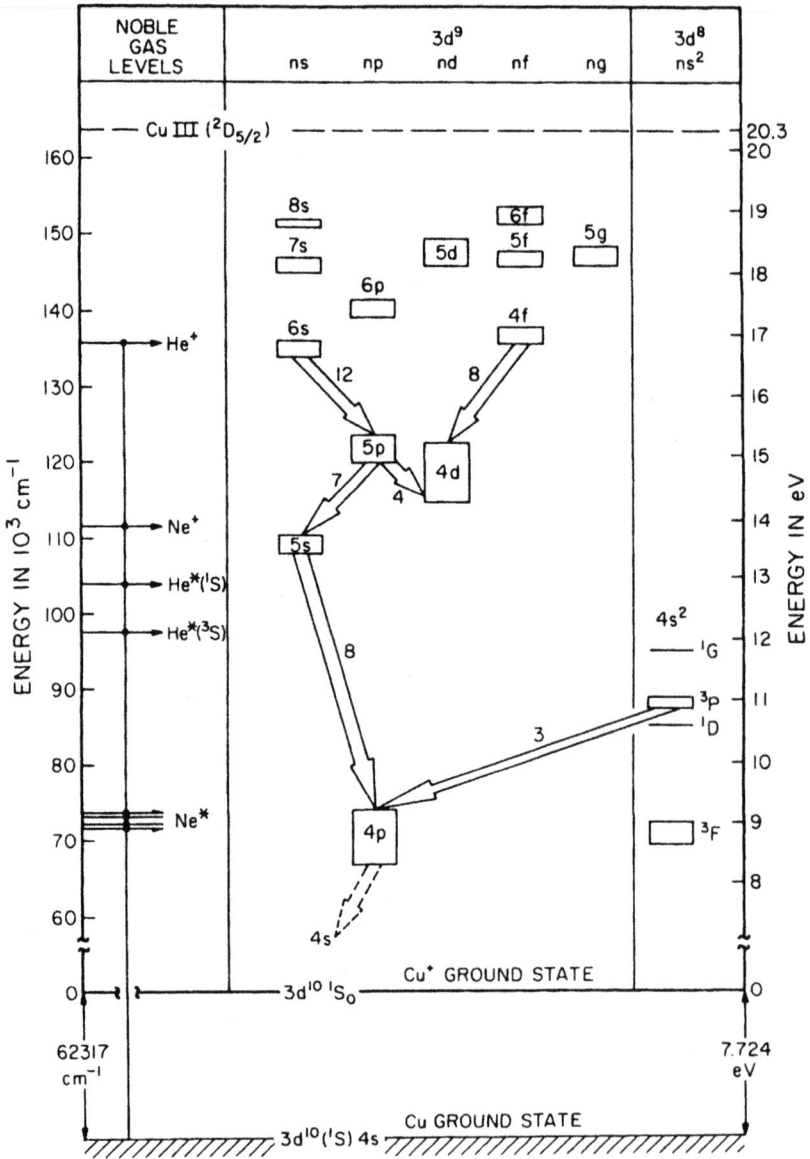

Figure 3.2.2 Simplified energy level diagram for singly ionized copper showing the number of laser lines in each multiplet. The energies of the He and Ne metastables and ground-state ions are also shown with respect to the ground state of the neutral copper atom. (From Bridges, W. B., *Methods of Experimental Physics*, Vol. 15A, Tang, C. L., Ed., Academic Press, New York, 1979. With permission.)

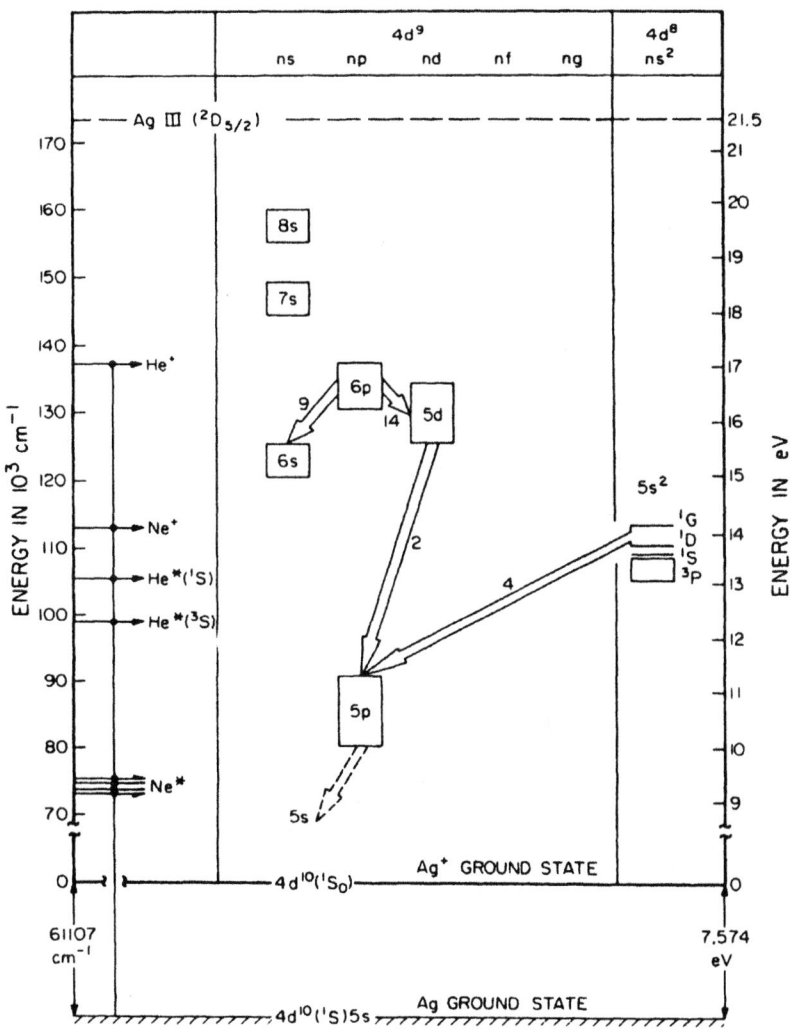

Figure 3.2.3 Simplified energy level diagram for singly ionized silver showing the number of laser lines in each supermultiplet. The energies of the He and Ne metastables and ground-state ions are also shown with respect to the ground state of the neutral silver atom.

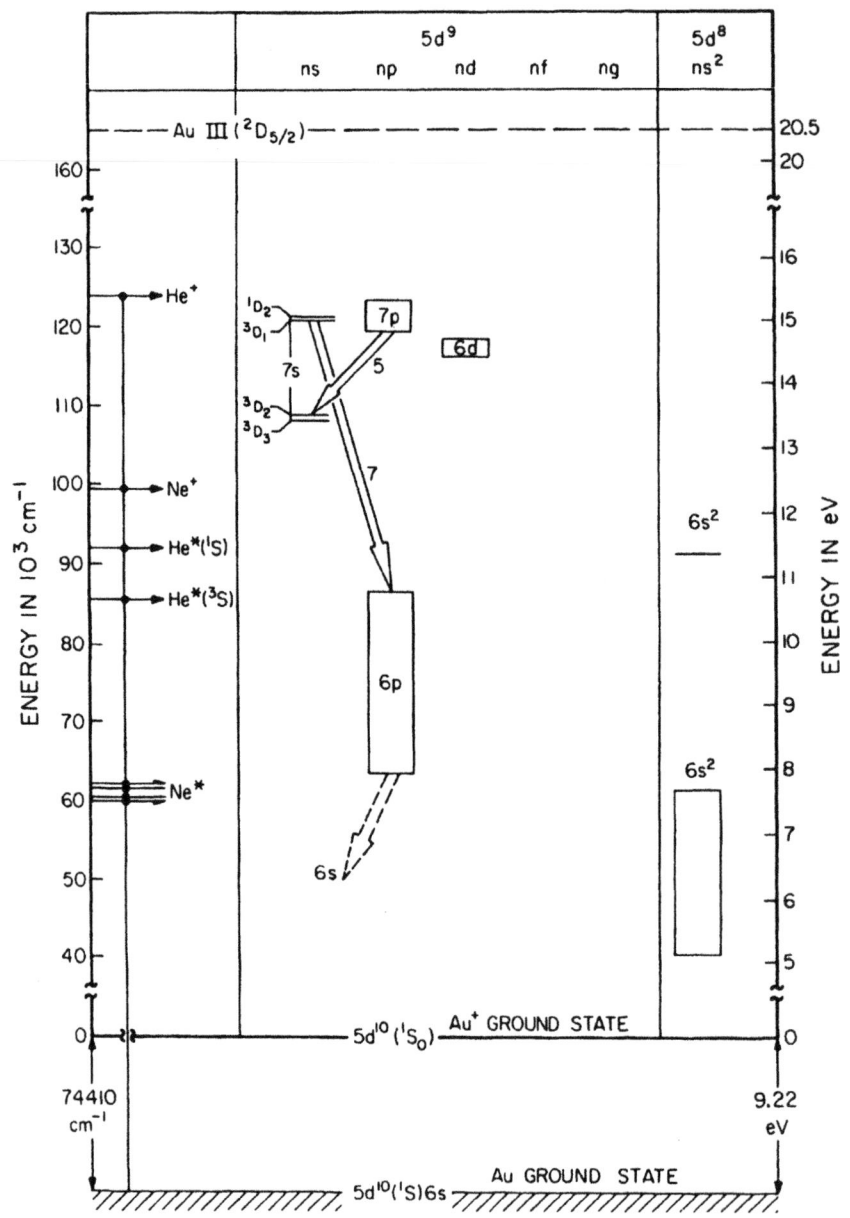

Figure 3.2.4 Simplified energy level diagram for singly ionized gold showing the number of laser lines in each supermultiplet. The energies of the He and Ne metastables and ground-state ions are also shown with respect to the ground state of the neutral gold atom.

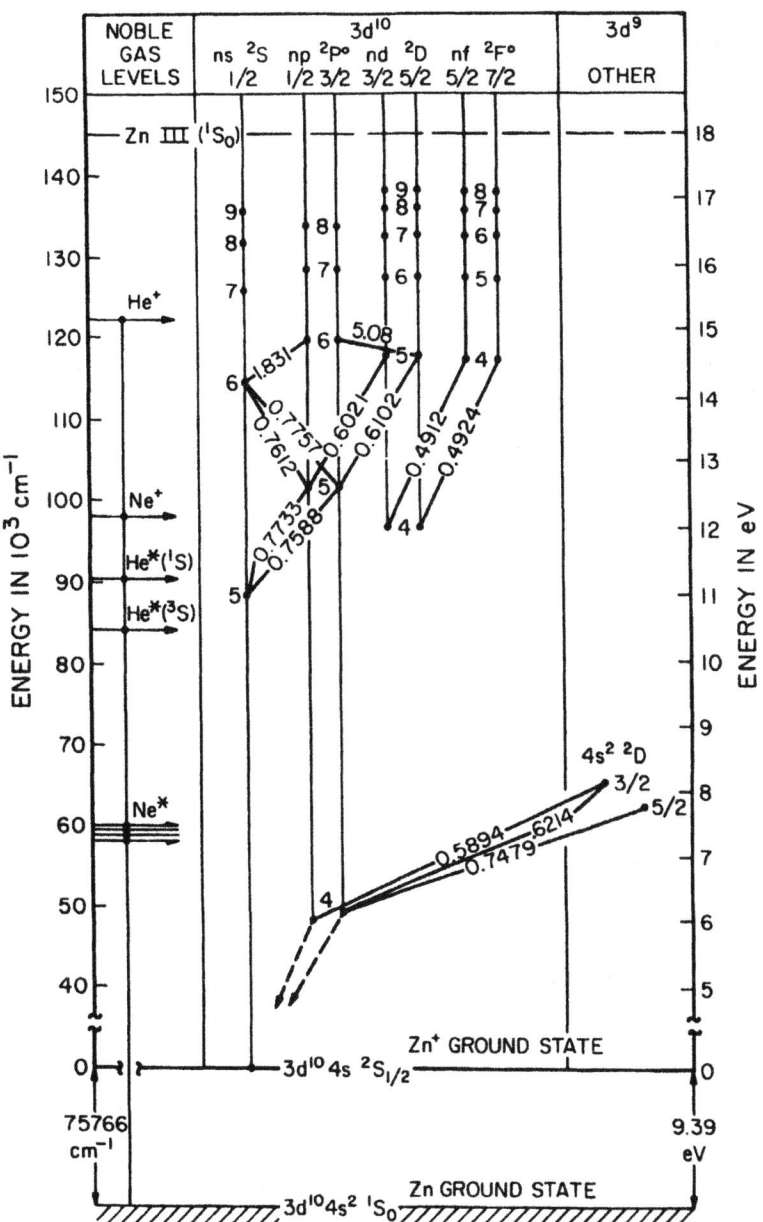

Figure 3.2.5 Simplified energy level diagram for singly ionized zinc showing the majority of observed Zn^+ laser lines. The energies of the He and Ne metastables and ground-state ions are also shown with respect to the ground state of the neutral zinc atom. (From Bridges, W. B., *Methods of Experimental Physics*, Vol. 15A, Tang, C. L., Ed., Academic Press, New York, 1979. With permission.)

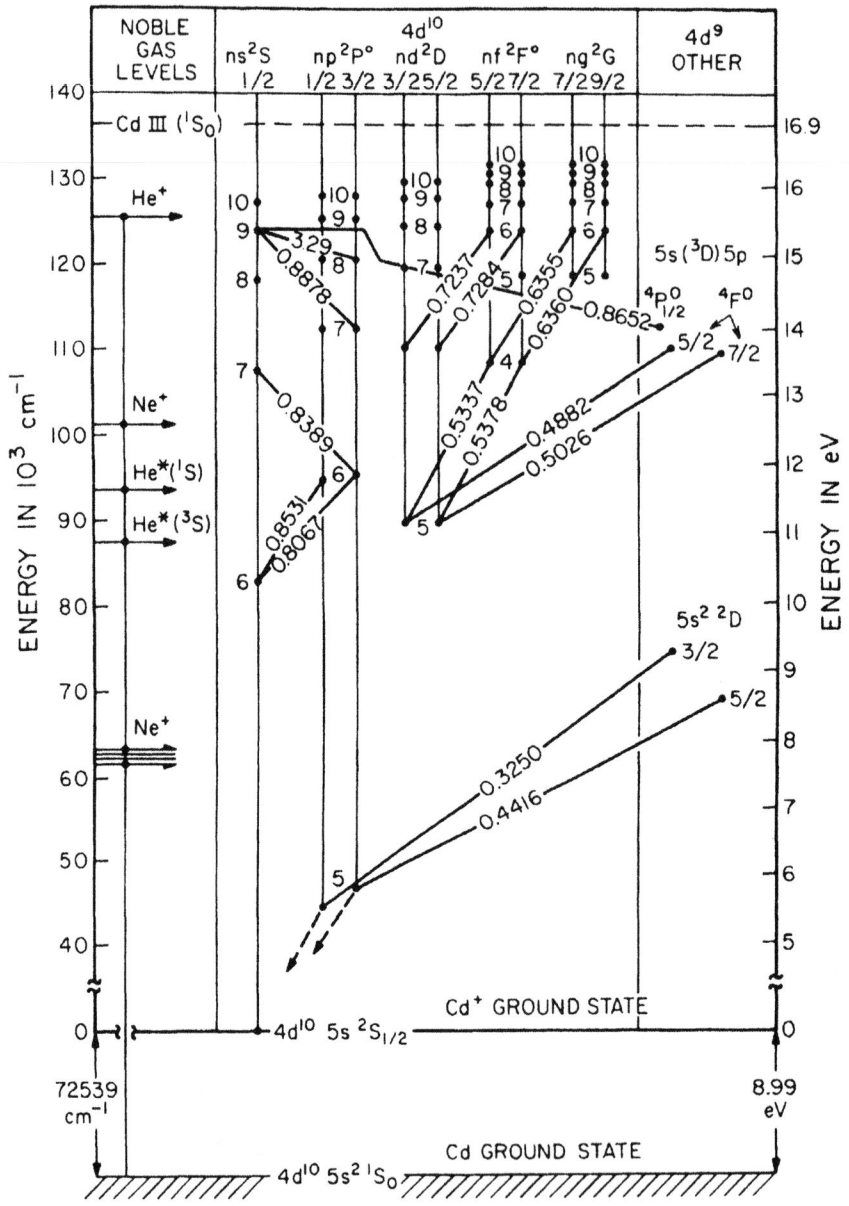

Figure 3.2.6 Simplified energy level diagram for singly ionized mercury showing the majority of observed Hg+ laser lines. The energies of the He and Ne metastables and ground-state ions are also shown with respect to the ground state of the neutral mercury atom. (From Bridges, W. B., *Methods of Experimental Physics*, Vol. 15A, Tang, C. L., Ed., Academic Press, New York, 1979. With permission.)

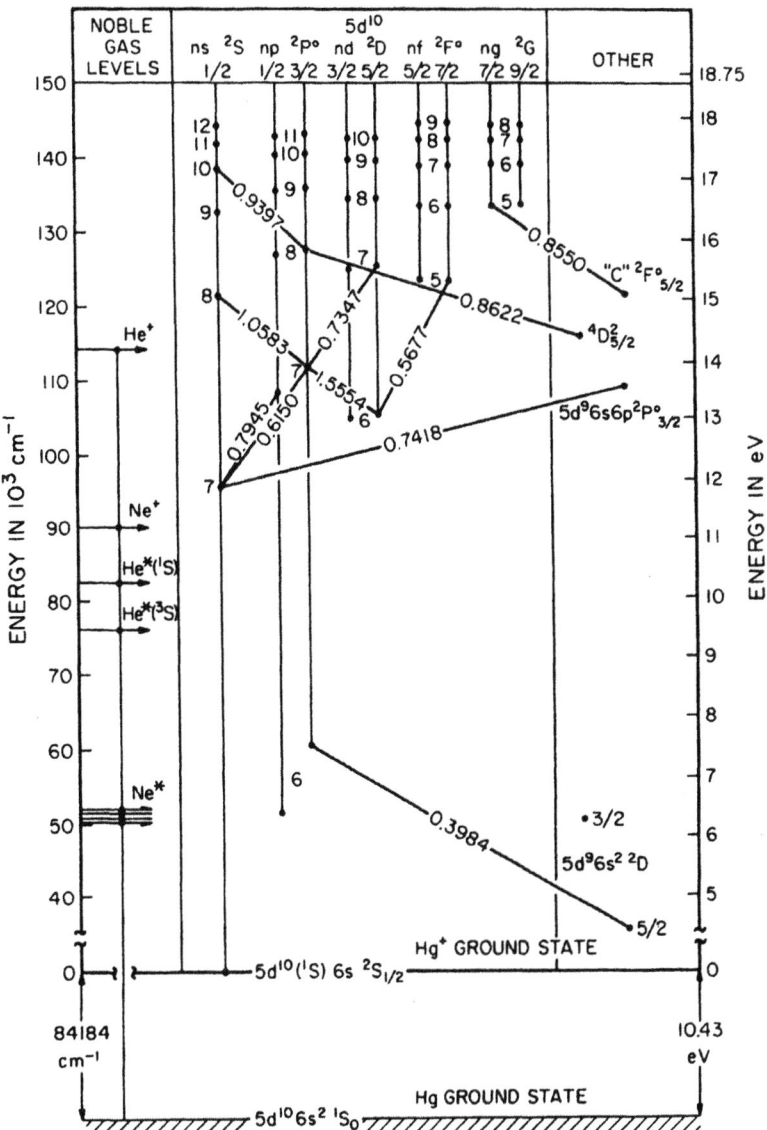

Figure 3.2.7 Simplified energy level diagram for singly ionized cadmium showing the majority of observed Cd$^+$ laser lines. The energies of the He and Ne metastables and ground-state ions are also shown with respect to the ground state of the neutral cadmium atom. (From Bridges, W. B., *Methods of Experimental Physics*, Vol. 15A, Tang, C. L., Ed., Academic Press, New York, 1979. With permission.)

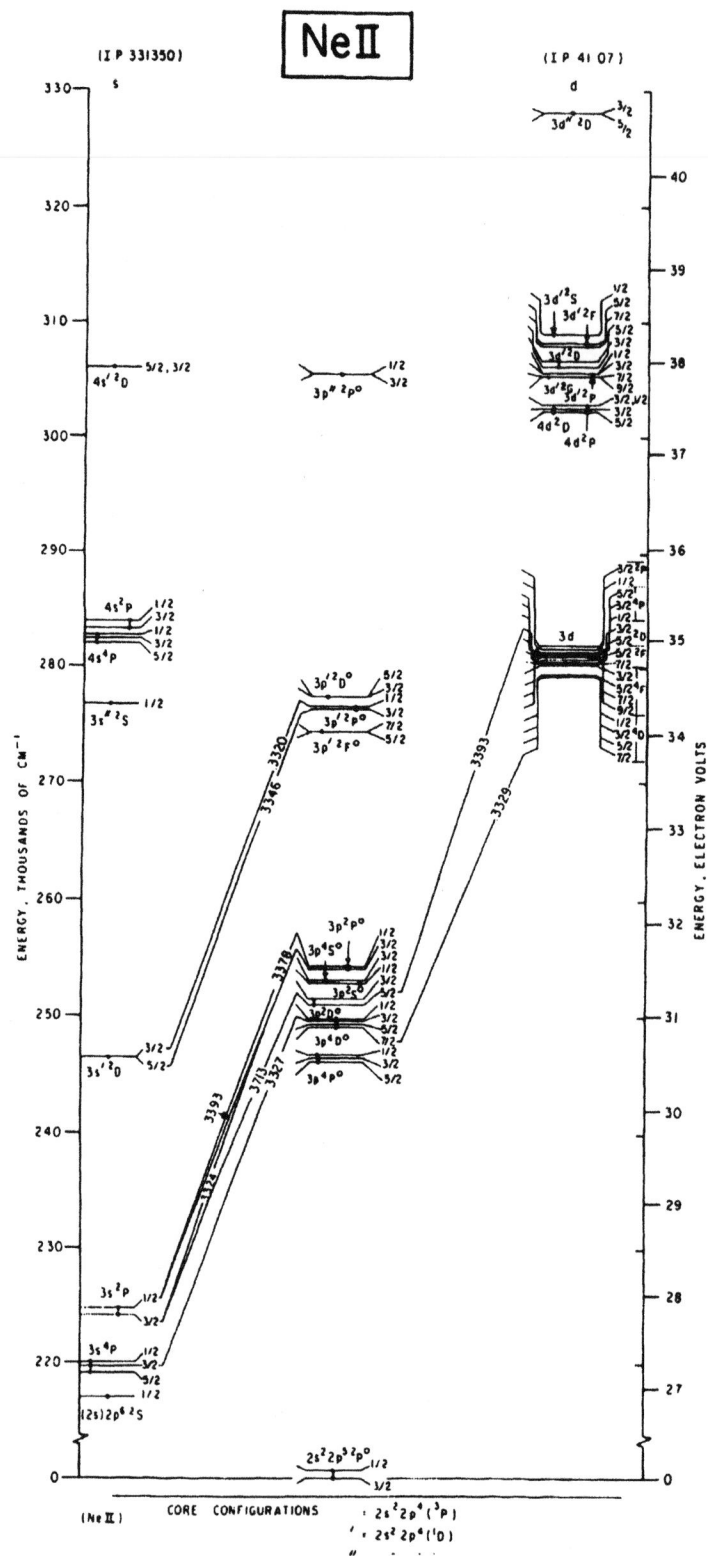

Figure 3.2.8 Energy level diagram for singly ionized neon showing the majority of Ne⁺ laser lines. Energy levels are from Moore, C. E., Reference 1757.

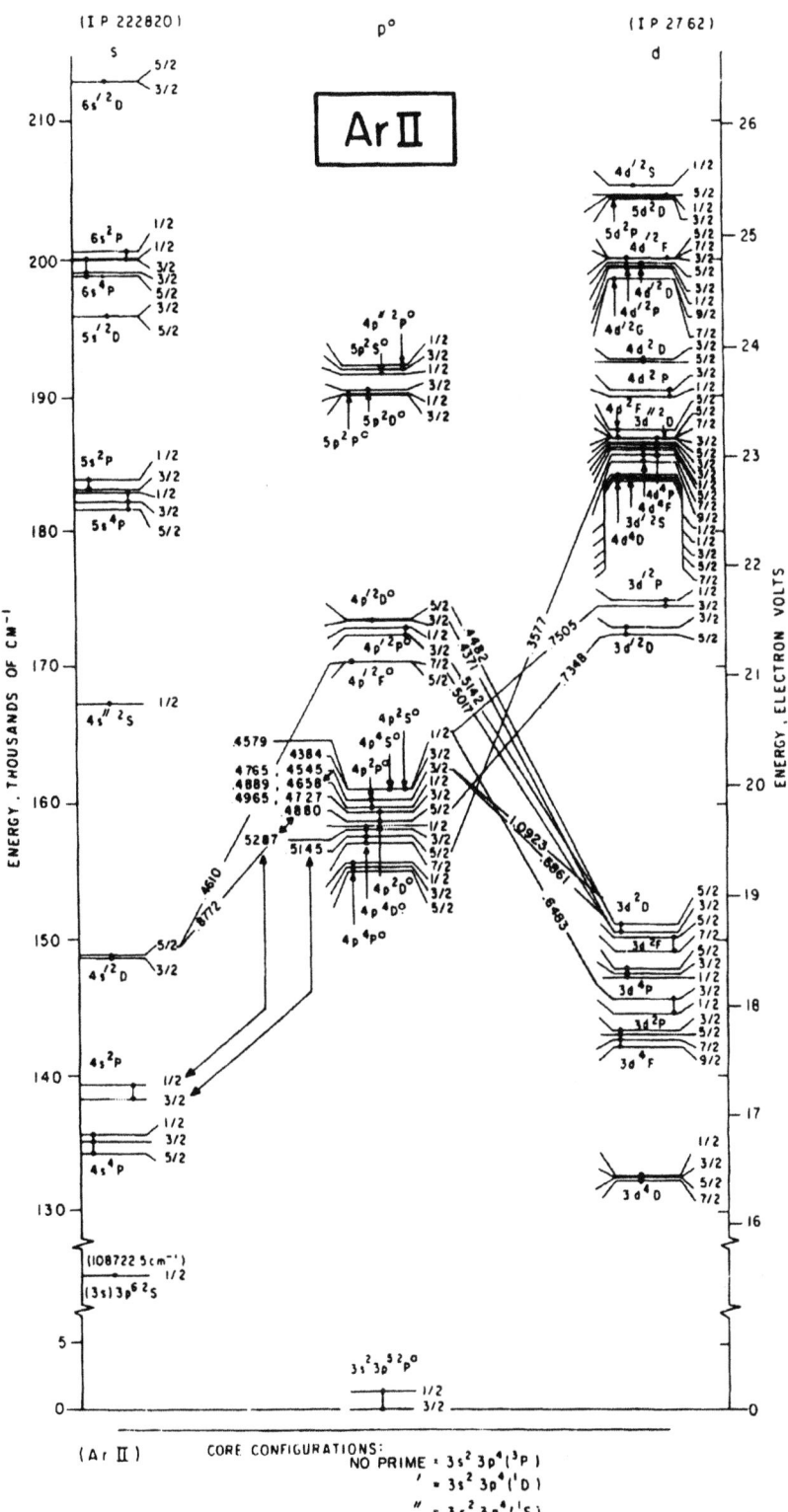

Figure 3.2.9 Energy level diagram for singly ionized argon showing the majority of Ar+ laser lines. Energy levels are from Moore, C. E., Reference 1757.

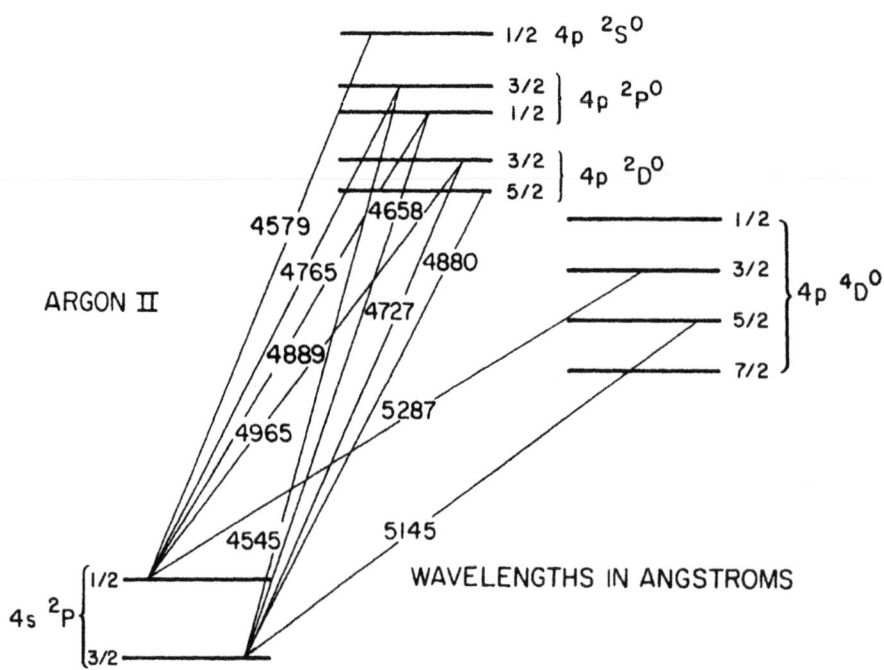

Figure 3.2.10 Simplified energy level diagram for Ar$^+$ showing the blue-green laser lines in the 4p → 4s supermultiplet. (From Bridges, W. B., *Appl. Phys. Lett.* 4, 128, 1964. With permission.)

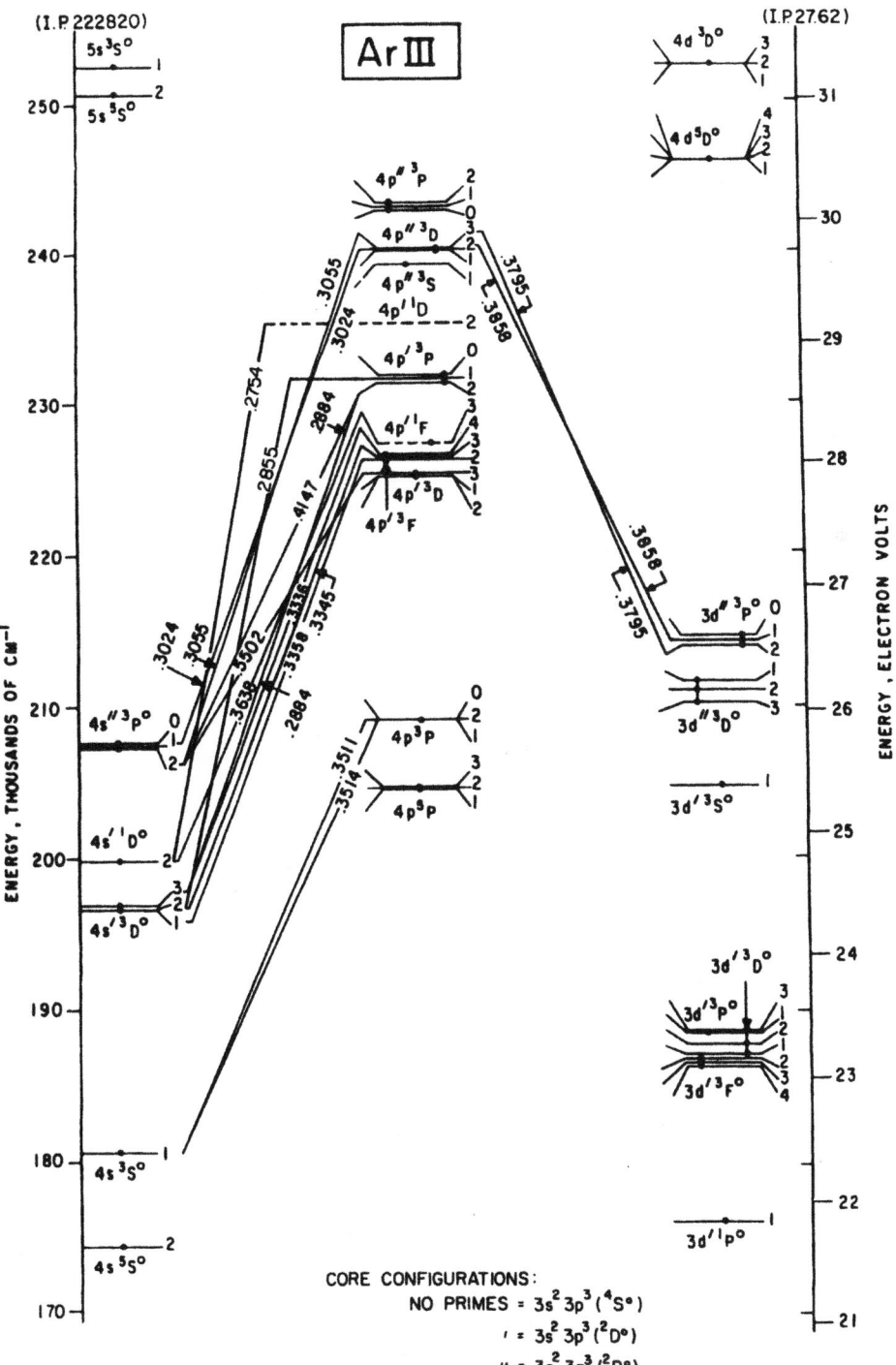

Figure 3.2.11 Energy level diagram for doubly ionized argon showing the majority of Ar^{2+} laser lines. Energy levels are from Moore, C. E., Reference 1757, except those shown dashed which are positioned through isoelectronic arguments by McFarlane[997] and Marling[989].

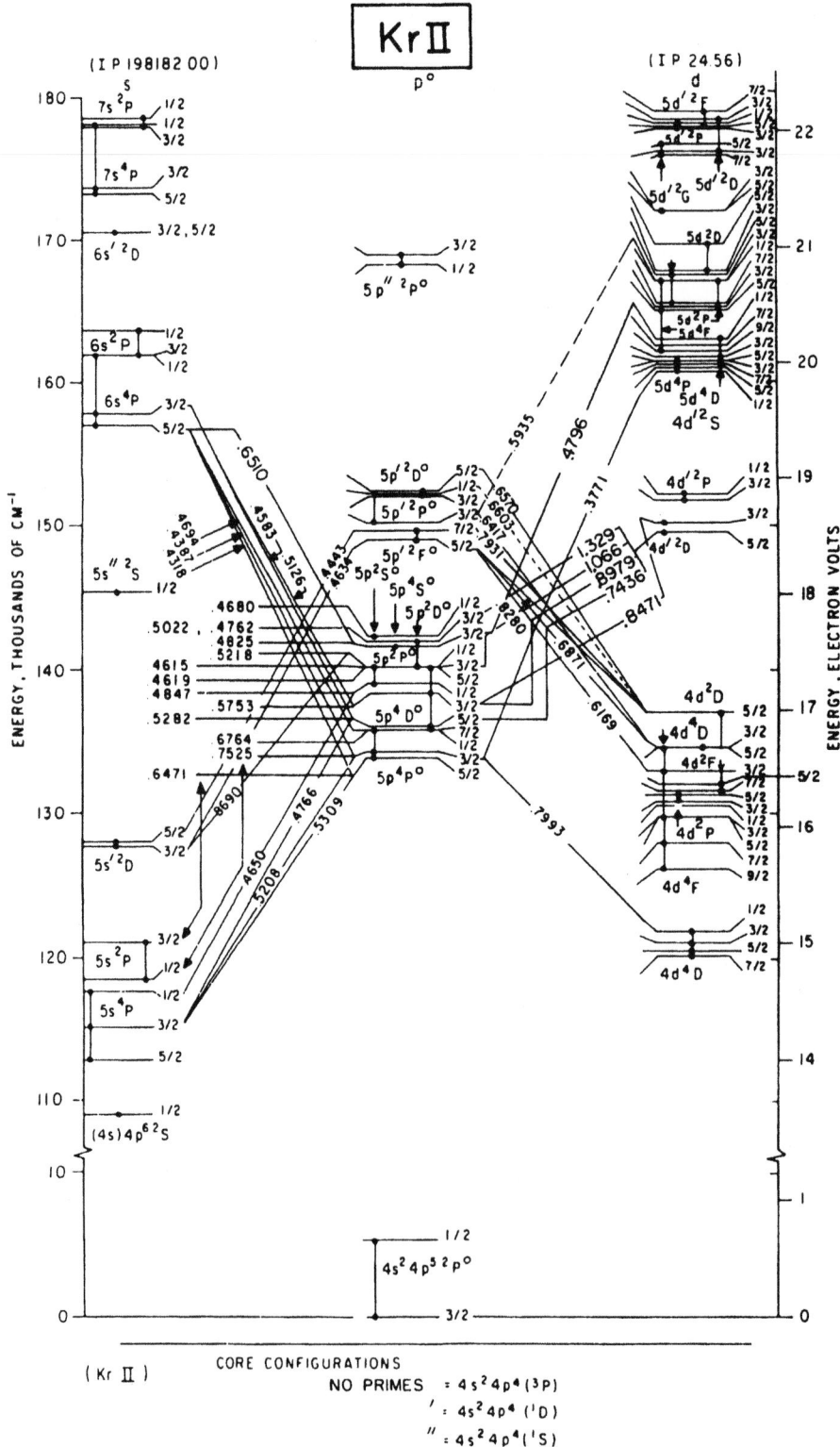

Figure 3.2.12 Energy level diagram for singly ionized krypton showing the majority of Kr⁺ lase lines. Energy levels are from Moore, C. E., Reference 1757; revisions from Minnhagen et al.[1007]

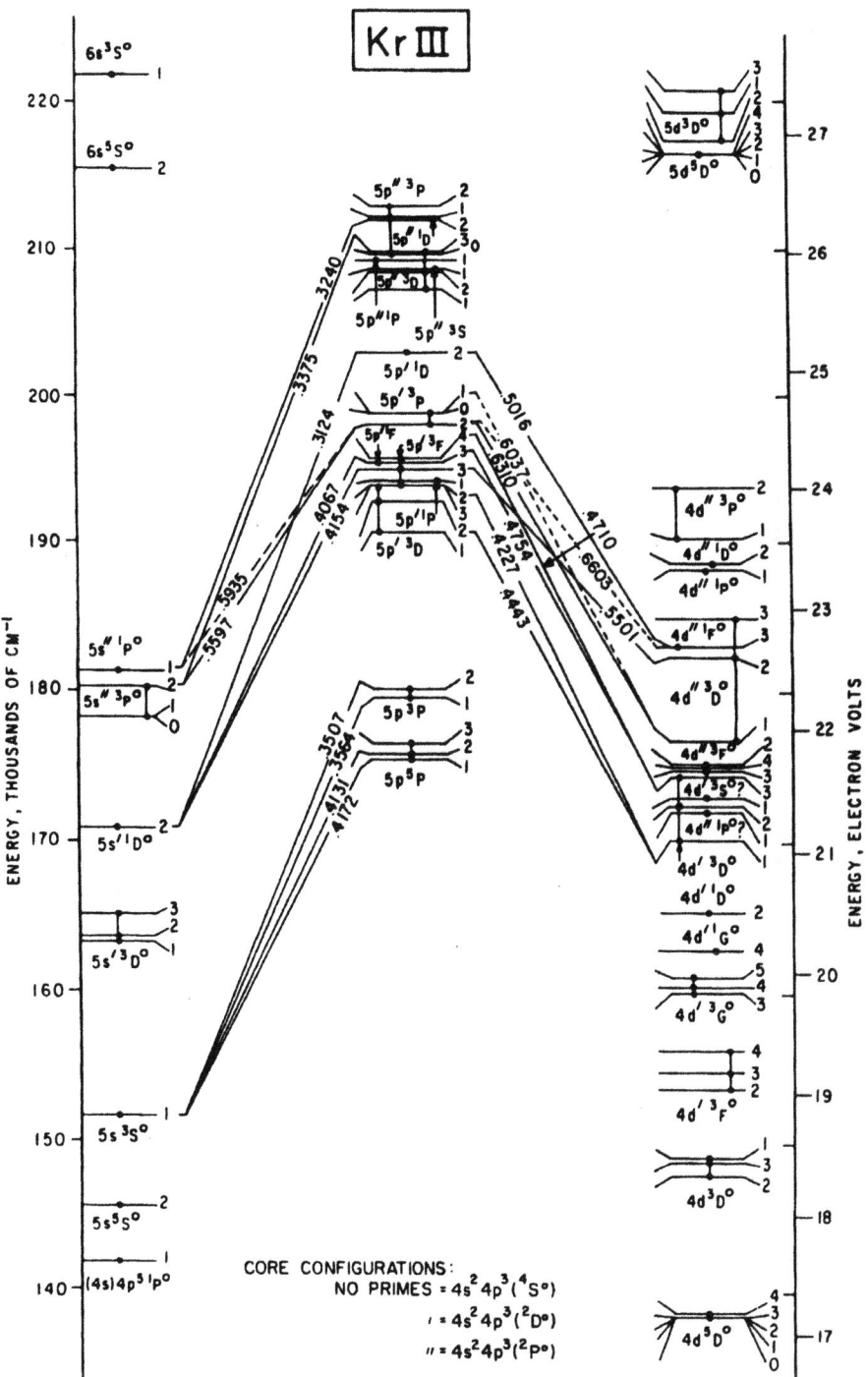

Figure 3.2.13 Energy level diagram for doubly ionized krypton showing the majority of Kr^{2+} laser lines. Energy levels are from Moore, C. E., Reference 1757.

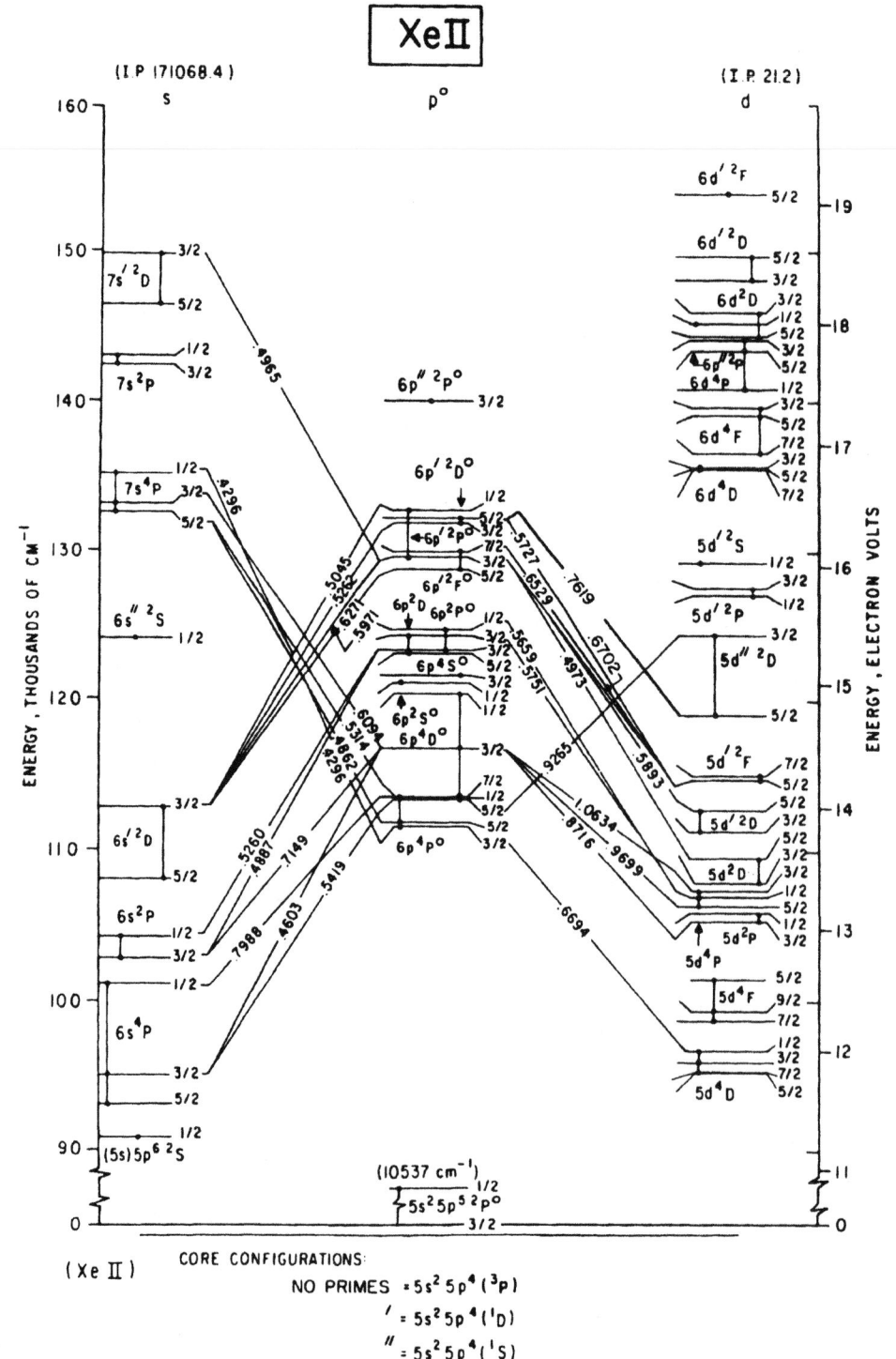

Figure 3.2.14 Energy level diagram for singly ionized xenon showing the majority of Xe⁺ laser lines. Energy levels are from Moore, C. E., Reference 1757, with revisions according to Minnhagen et al.[1007]

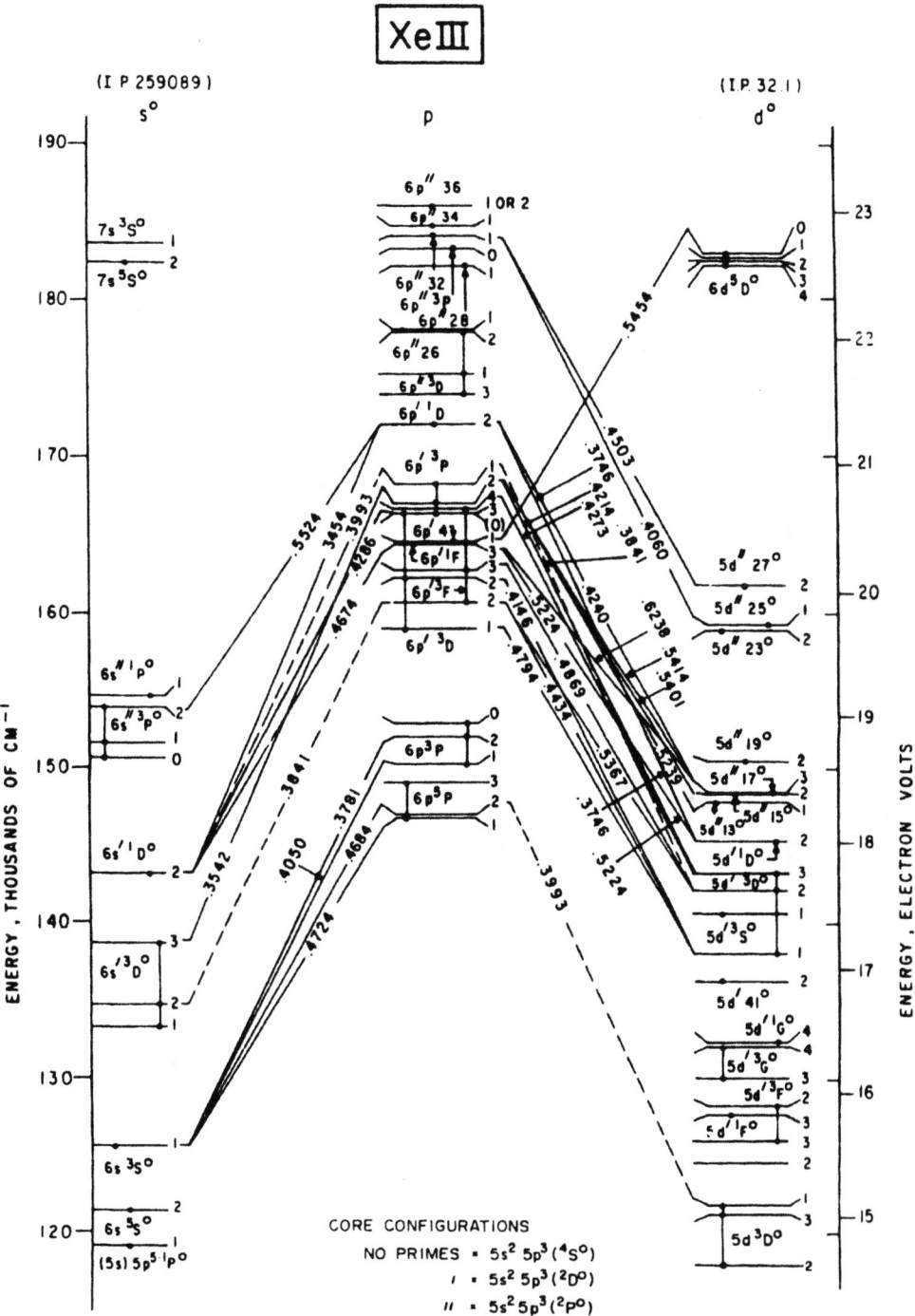

Figure 3.2.15 Energy level diagram for doubly ionized xenon showing the majority of Xe^{2+} laser lines. Energy levels are from Moore, C. E., Reference 1757, with revisions according to Gallardo et al.[875]

3.2.3 Tables of Ionized Gas Lasers

The tables of ionized gas lasers and elements are arranged in the following order:

<table>
<tr><td>

Group IA – Table 3.2.1
  potassium
</td><td>

Group IB – Table 3.2.2
  copper
  silver
  gold
</td><td>

Group IIA – Table 3.2.3
  beryllium
  magnesium
  calcium
  strontium
  barium
</td></tr>

<tr><td>

Group IIB – Table 3.2.4
  zinc
  cadmium
  mercury
</td><td>

Group IIIA – Table 3.2.5
  boron
  aluminum
  indium
  thallium
</td><td>

Group IVA – Table 3.2.6
  carbon
  silicon
  germanium
  tin
  lead
</td></tr>

<tr><td>

Group VA – Table 3.2.7
  nitrogen
  phosphorous
  arsenic
  antimony
  bismuth
</td><td>

Group VIA – Table 3.2.8
  oxygen
  sulfur
  selenium
  tellurium
  thulium
</td><td>

Group VIB – Table 3.2.9
  chromium
</td></tr>

<tr><td>

Group VII – Table 3.2.10
  fluorine
  chlorine
  bromine
  iodine
  ytterbium
</td><td>

Group VIII – Table 3.2.11
  nickel
  europium
</td><td>

Group VIIIA – Table 3.2.12
  helium
  neon
  argon
  krypton
  xenon
</td></tr>
</table>

3.2.3 Tables of Ionized Gas Lasers

3.2.3.1 Group IA Lasers

Table 3.2.1
Potassium

Wavelength (μm)	Measured value (μm)	Uncertainty	Charge	Transition	Comments	References
	0.612		1	$3p^54p\,^3D_3 \rightarrow 3p^53d\,^3F_4$	6921	847
	0.625		1	$3p^54p\,^3D_1 \rightarrow 3p^53d\,^3F_2$	6921	847
	0.631		1	$3p^54p\,^3D_2 \rightarrow 3p^53d\,^3F_3$	6921	847
	0.660		1	$3p^54p\,^3D_2 \rightarrow 3p^53d\,^3F_2$	6921	847

3.2.3.2 Group IB Lasers

Table 3.2.2
Copper

Wavelength (μm)	Measured value (μm)	Uncertainty	Charge	Transition	Comments	References
0.248579	0.248580	±0.000010	1	$(^2D_{3/2})5s\,^3D_1 \rightarrow (^2D_{3/2})4p\,^3F^0_2$	1080,1087,1090	937,999
0.250627	0.250650	±0.000010	1	$(^2D_{5/2})5s\,^3D_2 \rightarrow (^2D_{5/2})4p\,^3F^0_3$	1080,1087,1090	999
0.252930	0.252920	±0.000010	1	$(^2D_{3/2})5s\,^1D_2 \rightarrow (^2D_{5/2})4p\,^3D^0_3$	1080,1087,1090	937,1000
0.259052	0.259060	±0.000010	1	$(^2D_{3/2})5s\,^3D_1 \rightarrow (^2D_{5/2})4p\,^3D^0_2$	1080,1083,1087,1090	918,937,999
0.259881	0.259900	±0.000010	1	$(^2D_{5/2})5s\,^3D_2 \rightarrow (^2D_{3/2})4p\,^3F^0_2$	1080,1083,1087,1090	691,918,937,999
0.260027	0.260030	±0.000010	1	$(^2D_{3/2})5s\,^1D_2 \rightarrow (^2D_{3/2})4p\,^1F^0_3$	1080,1087,1090	691,937,1000
0.270097	0.270070	±0.000030	1	$(^2D_{3/2})5s\,^1D_2 \rightarrow (^2D_{3/2})4p\,^1D^0_2$	1080,1087,1090	1161
0.270318	0.270310	±0.000010	1	$(^2D_{3/2})5s\,^3D_1 \rightarrow (^2D_{3/2})4p\,^3D^0_1$	1080,1083,1087,1090	918,937,1000

Copper—continued

Wavelength (μm)	Measured value (μm)	Uncertainty	Charge	Transition	Comments	References
0.272168	0.2722		1	$(^2D_{3/2})5s\ ^3D_1 \rightarrow (^2D_{3/2})4p\ ^1D^0_2$	1080,1087,1090	937,938
0.450598	0.450660	±0.000030	1	$4s^2\ ^3P_1 \rightarrow (^2D_{5/2})4p\ ^3P^0_2$	1074,1087	1174
0.455591	0.455630	±0.000030	1	$4s^2\ ^3P_2 \rightarrow (^2D_{5/2})4p\ ^3P^0_2$	1074,1087	1174
0.467356	0.467320	±0.000030	1	$(^2D_{5/2})4f\ ^3P^0 \rightarrow (^2D_{5/2})4d\ ^3S_1$	1076,1087	1174
0.468198	0.468290	±0.000030	1	$(^2D_{5/2})4f\ ^3P^0_1 \rightarrow (^2D_{5/2})4d\ ^3S_1$	1076,1087	1174
0.485496	0.485580	±0.000030	1	$(^2D_{5/2})4\ f^3G^0_5 \rightarrow (^2D_{5/2})4d\ ^3G_5$	1076,1087	1174
0.490972	0.490970	±0.000030	1	$(^2D_{5/2})4f\ ^3H^0_6 \rightarrow (^2D_{5/2})4d\ ^3G_5$	1076,1083,1087	1174
0.493169	0.493170	±0.000030	1	$(^2D_{5/2})4f\ ^3H^0_5 \rightarrow (^2D_{5/2})4d\ ^3G_4$	1076,1083,1085, 1087,1106	975,1174
0.501262	0.501330	±0.000030	1	$(^2D_{5/2})4f\ ^3G^0_4 \rightarrow (^2D_{5/2})4d\ ^3F_3$	1076,1087	1174
0.502128	0.502190	±0.000030	1	$(^2D_{5/2})4f\ ^3D^0_3 \rightarrow (^2D_{5/2})4d\ ^3D_3$	1076,1087	1174
0.505177	0.505210	±0.000030	1	$(^2D_{5/2})4f\ ^3G^0_5 \rightarrow (^2D_{5/2})4d\ ^3F_4$	1076,1087	1174
0.506062	0.506050	±0.000030	1	$4s^2\ ^3P_1 \rightarrow (^2D_{3/2})4p\ ^3P^0_0$	1074,1087	1174
0.725576	0.725600	±0.000020	1	$(^2D_{3/2})6s\ ^1D_2 \rightarrow (^2D_{5/2})5p\ ^3D^0_2$	1076,1087	1077
0.739988	0.740020	±0.000020	1	$(^2D_{5/2})5p\ ^3D^0_3 \rightarrow (^2D_{5/2})5s\ ^3D_3$	1077,1087	1077
0.740434	0.740450	±0.000030	1	$(^2D_{5/2})6s^3D_3 \rightarrow (^2D_{5/2})5p^3\ P^0_2$	1076,1083,1084, 1087,1090,1092	934,1040,1051, 1174
0.743816	0.743900	±0.000020	1	$(^2D_{3/2})6s\ ^3D_1 \rightarrow (^2D_{3/2})5p^3\ P^0_0$	1078,1085,1087	1077
0.766466	0.766470	±0.000030	1	$(^2D_{5/2})6s\ ^3D_2 \rightarrow (^2D_{5/2})5p\ ^3F^0_3$	1076,1083,1084, 1087,1090,1092	934,1040,1051, 1174
0.773866	0.773870	±0.000030	1	$(^2D_{3/2})6s\ ^3D_1 \rightarrow (^2D_{3/2})5p\ ^3F^0_2$	1076,1083,1084, 1087,1090,1092	934,1040,1051, 1174
0.777879	0.777890	±0.000030	1	$(^2D_{5/2})6s\ ^3D_2 \rightarrow (^2F)4p^1D^0_2$	1076,1083,1087	1174
0.780523	0.780530	±0.000030	1	$(^2D_{5/2})6s\ ^3D_2 \rightarrow (^2D_{5/2})5p\ ^3P^0_1$	1076,1083,1087	934,1174
0.780767	0.780780	±0.000030	1	$(^2D_{5/2})6s\ ^3D_3 \rightarrow (^2D_{5/2})5p\ ^3F^0_4$	1076,1083,1084, 1087,1090,1092	691,934,1040, 1051,1149,1174
0.782565	0.782600	±0.000030	1	$(^2D_{5/2})5p\ ^3F^0_4 \rightarrow (^2D_{5/2})5s\ ^3D_3$	1076,1083,1087	934,1174

				Transition		
0.784508	0.784530	±0.000030	1	$(^2D_{3/2})\,6s\,^1D_2 \rightarrow (^2D_{3/2})\,5p\,^1F^0_3$	1076,1083,1087, 1090,1092	934,1051,
0.789584	0.789600	±0.000030	1	$(^2D_{3/2})5p\,^3F^0_2 \rightarrow (^2D_{3/2})\,5s\,^3D_1$	1076,1083,1087	934,1174
0.790254	0.790270	±0.000030	1	$(^2D_{3/2})5p\,^1F^0_3 \rightarrow (^2D_{3/2})\,5s\,^1D_2$	1076,1083,1087	935,1174,
0.794442	0.794480	±0.000030	1	$(^2D_{5/2})5p\,^3P^0_1 \rightarrow (^2D_{5/2})\,5s\,^3D_2$	1076,1083,1087	1935,1174,
0.798814	0.798820	±0.000030	1	$(^2D_{5/2})6s\,^3D_3 \rightarrow (^2F)4p\,^1F^0_3$	1076,1083,1087	934,1174
0.808862	0.808858		1	$(^2D_{3/2})6s\,^3D_1 \rightarrow (^2D_{3/2})5p\,^3D^0_1$	1076,1083,1087	935
0.809555	0.8096		1	$(^2D_{5/2})5p\,^3F^0_3 \rightarrow (^2D_{5/2})5s\,^3D_2$	1076,1083,1087	934
0.819223	0.819228		1	$(^2D_{3/2})6s\,^1D_2 \rightarrow (^2D_{3/2})5p\,^1D^0_2$	1076,1083, 1087,1108	934
0.819233	0.819228	±0.000200	1	$(^2D_{5/2})6s\,^3D_2 \rightarrow (^2D_{5/2})5p\,^3D^0_3$	1076,1083, 1087,1107	934
0.827758	0.827700	±0.000200	1	$(^2D_{5/2})5p\,^3P^0_2 \rightarrow (^2D_{5/2})5s\,^3D_3$	1076,1087	806
0.828317	0.828321		1	$(^2D_{5/2})6s\,^3D_3 \rightarrow (^2D_{5/2})5p\,^3D^0_3$	1076,1087	935
0.851108	0.851104		1	$(^2D_{5/2})6s\,^3D_2 \rightarrow (^2D_{5/2})5p\,^3D^0_2$	1076,1087	935
1.771	1.771000	±0.000200	1	$(^2D_{5/2})5p\,^3F^0_4 \rightarrow (^2D_{5/2})4d\,^3G_5$	1085,1087,1109	806
1.91495	1.915600	±0.000200	1	$(^2D_{3/2})5p\,^1F^0_3 \rightarrow (^2D_{3/2})4d\,^1G_4$	1075,1087	806
1.9473	1.948000	±0.000200	1	$(^2D_{3/2})5p\,^1D^0_2 \rightarrow (^2D_{3/2})4d\,^1F_3$	1077,1087	806
1.97071	1.971000	±0.000200	1	$(^2D_{3/2})5p\,^3F^0_2 \rightarrow (^2D_{3/2})4d\,^3G_3$	1073,1087	806
2.00002	2.000400	±0.000200	1		1077,1087	806

Silver

Wavelength (μm)	Measured value (μm)	Uncertainty	Charge	Transition	Comments	References
0.224357	0.224340	±0.000020	1	$(^2D_{3/2})5d\ ^1S_0 \rightarrow (^2D_{3/2})5p\ ^1P^0_1$	1076,1083,1087	937,1001
0.227743	0.227760	±0.000020	1	$(^2D_{5/2})5d\ ^3D_2 \rightarrow (^2D_{5/2})5p\ ^3P^0_1$	1076,1083,1087	1001
0.318070	0.318060	±0.000020	1	$5s^2\ ^1G_4 \rightarrow (^2D_{5/2})5p\ ^3F^0_3$	1080,1083,1087,1090,6927	691,1001
0.408590	0.408620	±0.000020	1	$5s^2\ ^1G_4 \rightarrow (^2D_{3/2})5p\ ^1F^0_3$	1080,1087,1090	691,948
0.478839	0.478840	±0.000020	1	$5s^2\ ^1D_2 \rightarrow (^2D_{3/2})5p\ ^1P^0_1$	1080,1087,1090,6928	691,948
0.502733	0.502720	±0.000020	1	$5s^2\ ^1D_2 \rightarrow (^2D_{3/2})5p\ ^1D^0_2$	1080,1087,1090	948
0.640529	0.640430	±0.000020	1	$(^2D_{3/2})6p\ ^3F^0_2 \rightarrow (^2D_{5/2})6s\ ^3D_2$	1076,1083,1087	1044
0.800517	0.800540	±0.000010	1	$(^2D_{5/2})6p\ ^3D^0_3 \rightarrow (^2D_{5/2})6s\ ^3D_3$	1076,1082,1083,1087,1090	691,954,1045
0.825482	0.825450	±0.000010	1	$(^2D_{5/2})6p\ ^3D^0_3 \rightarrow (^2D_{5/2})6s\ ^3D_2$	1076,1083,1087,1090	691,935,1044–45
0.826284	0.826300	±0.000200	1	$(^2D_{3/2})6p\ ^3D^0_1 \rightarrow (^2D_{3/2})6s\ ^3D_1$	1076,1087	806
0.832469	0.832480	±0.000010	1	$(^2D_{5/2})6p\ ^3D^0_2 \rightarrow (^2D_{5/2})6s\ ^3D_2$	1076,1083,1087	935,1044–45
0.837957	0.837950		1	$(^2D_{5/2})6p\ ^3D^0_1 \rightarrow (^2D_{5/2})6s\ ^3D_2$	1076,1083,1087	935,806
0.840384	0.840350	±0.000010	1	$(^2D_{5/2})6p\ ^3F^0_4 \rightarrow (^2D_{5/2})6s\ ^3D_3$	1076,1082,1083,1086,1087,1090	691,935,954,1044,1045
0.874833	0.874760	±0.000200	1	$(^2D_{3/2})6p\ ^1P^0_1 \rightarrow (^2D_{3/2})6s\ ^1D_2$	1076,1083,1087	806,935
0.877527	0.877300	±0.000200	1	$(^2D_{3/2})6p\ ^3F^0_2 \rightarrow (^2D_{3/2})6s\ ^3D_1$	1076,1083,1087	806
1.37538	1.375900	±0.000200	1	$(^2D_{5/2})6p\ ^3P^0_1 \rightarrow (^2D_{5/2})5d\ ^3S_1$	1076,1087	806
1.59759	1.598200	±0.000200	1	$(^2D_{5/2})6p\ ^3D^0_3 \rightarrow (^2D_{5/2})5d\ ^3P_2$	1076,1087	806
1.64593	1.646400	±0.000200	1	$(^2D_{5/2})6p\ ^3P^0_1 \rightarrow (^2D_{5/2})5d\ ^3P_1$	1076,1087	806
1.71983	1.720200	±0.000200	1	$(^2D_{5/2})6p\ ^3D^0_3 \rightarrow (^2D_{5/2})5d\ ^3D_3$	1076,1083,1087	806
1.73413	1.734600	±0.000200	1	$(^2D_{5/2})6p\ ^3F^0_4 \rightarrow (^2D_{5/2})5d\ ^3G_5$	1076,1083,1087	806
1.74749	1.748000	±0.000200	1	$(^2D_{3/2})6p\ ^3D^0_1 \rightarrow (^2D_{3/2})5d\ ^3D_1$	1076,1087	806
1.84051	1.841000	±0.000200	1	$(^2D_{5/2})6p\ ^3D^0_2 \rightarrow (^2D_{5/2})5d\ ^3F_3$	1076,1083,1087	806

Wavelength (μm)	Measured value (μm)	Uncertainty	Charge	Transition	Comments	References
1.84583	1.846400	±0.000200	1	$(^2D_{5/2})6p\ ^3D^0_3 \to (^2D_{5/2})5d\ ^3F_4$	1076,1083,1087	806
1.87200	1.872400	±0.000200	1	$(^2D_{5/2})6p\ ^3F^0_3 \to (^2D_{5/2})5d\ ^3G_4$	1076,1087	806
1.87896	1.879600	±0.000200	1	$(^2D_{5/2})6p\ ^3p^0_1 \to (^2D_{5/2})5d\ ^3D_2$	1076,1083,1087	806
1.89752	1.898000	±0.000200	1	$(^2D_{3/2})6p\ ^3F^0_2 \to (^2D_{3/2})5d\ ^3G_3$	1076,1083,1087	806
1.97086	1.971600	±0.000200	1	$(^2D_{5/2})6p\ ^3D^0_1 \to (^2D_{3/2})5d\ ^3F_2$	1076,1087	806
1.98177	1.982400	±0.000200	1	$(^2D_{3/2})6p\ ^1p^0_1 \to (^2D_{3/2})5d\ ^1D_2$	1076,1087	806
2.07907	2.079400	±0.000200	1	$(^2D_{5/2})6p\ ^3F^0_3 \to (^2D_{5/2})5d\ ^3D_3$	1076,1087	806

Gold

Wavelength (μm)	Measured value (μm)	Uncertainty	Charge	Transition	Comments	References
0.226362	0.226400		1	$(^2D_{3/2})7s\ ^3D_1 \to 76659^0_2$	1076,1087	938
0.253353	0.253350	±0.000020	1	$(^2D_{3/2})7s\ ^1D_2 \to 81659^0_1$	1076,1083,1087, 1090	673
0.261640	0.261640	±0.000020	1	$(^2D_{3/2})7s\ ^3D_1 \to 82613^0_0$	1076,1083,1087, 1090	673
0.282254	0.282220	±0.000020	1	$(^2D_{3/2})7s\ ^1D_2 \to 85699^0_3$	1076,1083,1087, 1090	673,691,937
0.284694	0.284720	±0.000020	1	$(^2D_{3/2})7s\ ^3D_1 \to 85707^0_1$	1076,1083,1087, 1090	673,691,937
0.289328	0.289350	±0.000020	1	$(^2D_{3/2})7s\ ^1D_2 \to 86565^0_2$	1076,1083,1087, 1090	673,691,937
0.291826	0.291810	±0.000020	1	$(^2D_{3/2})7s\ ^3D_1 \to 86565^0_2$	1076,1083,1087, 1090	673,691,937
0.755581	0.755600	±0.000020	1	$121862^0_1 \to (^2D_{5/2})7s\ ^3D_2$	1076,1083,1087	673
0.760050	0.760040	±0.000020	1	$121784^0_2 \to (^2D_{5/2})7s\ ^3D_2$	1076,1083,1087	673
0.827293	0.827290		1	$120256^0_3 \to (^2D_{5/2})7s\ ^3D_3$	1076,1087	935
0.859927	0.859900		1	$120256^0_3 \to (^2D_{5/2})7s\ ^3D_2$	1076,1087	938
0.886761	0.886760		1	$119446^0_2 \to (^2D_{5/2})7s\ ^3D_3$	1076,1087	935

3.2.3.3 Group IIA Lasers

Table 3.2.3
Beryllium

Wavelength (μm)	Measured value (μm)	Uncertainty	Charge	Transition	Comments	References
0.467346	0.4675		1	$4f^2\,{}^1F^0_{5/2,7/2} \rightarrow 3d^2\,{}^1D_{3/2,5/2}$	1076,1082,1086, 1087	1108
0.527032	0.5272		1	$4s\,{}^2S_{1/2} \rightarrow 3p\,{}^2P^0_{1/2}$	1076,1082	1108
0.527084	0.5272		1	$4s\,{}^2S_{1/2} \rightarrow 3p\,{}^2P^0_{3/2}$	1076,1082	1108
1.209298	1.2096		1	$3p\,{}^2P^0_{3/2} \rightarrow 3s\,{}^2S_{1/2}$	1080,1082,1087	1108
1.209561	1.2096		1	$3p\,{}^2P^0_{1/2} \rightarrow 3s\,{}^2S_{1/2}$	1080,1082,1087	1108

Table 3.2.4
Magnesium

Wavelength (μm)	Measured value (μm)	Uncertainty	Charge	Transition	Comments	References
0.921825	0.921800	±0.000150	1	$4p\,{}^2P^0_{3/2} \rightarrow 4s\,{}^2S_{1/2}$	1074,1076,1082, 1083,1086,1097	920,976,1046
0.924427	0.924400	±0.000150	1	$4p\,{}^2P^0_{1/2} \rightarrow 4s\,{}^2S_{1/2}$	1074,1080,1082, 1083,1086,1097	920,976,1046
1.091423	1.091500	±0.000150	1	$4p\,{}^2P^0_{3/2} \rightarrow 3d\,{}^2D_{5/2}$	1074,1083,1086, 1095	920,1046
1.091527	1.091500	±0.000150	1	$4p\,{}^2P^0_{3/2} \rightarrow 3d\,{}^2D_{3/2}$	1074,1083,1086, 1095	920,1046
1.095178	1.095200	±0.000150	1	$4p\,{}^2P^0_{1/2} \rightarrow 3d\,{}^2D_{3/2}$	1074,1083, 1086,1095	920,1046
2.40415	2.404160	±0.000300	1	$5p\,{}^2P^0_{3/2} \rightarrow 4d\,{}^2D_{5/2}$	1097	136,1046
2.41246	2.412520	±0.000300	1	$5p\,{}^2P^0_{1/2} \rightarrow 4d\,{}^2D_{3/2}$	1098	136,1046

Calcium

Wavelength (μm)	Measured value (μm)	Uncertainty	Charge	Transition	Comments	References
0.370602	0.3706		1	$(^1S)5s\ ^2S_{1/2} \rightarrow (^1S)4p\ ^2P^0_{1/2}$	1076,1082,1099	976,1109,1166
0.373690	0.3737		1	$(^1S)5s\ ^2S_{1/2} \rightarrow (^1S)4p\ ^2P^0_{3/2}$	1076,1082,1086, 1090,1100	976,1109,1166
0.854209	0.854180	±0.000060	1	$(^1S)4p\ ^2P^0_{3/2} \rightarrow (^1S)3d\ ^2D_{5/2}$	1086,1090,1091, 1101	336,1166
0.866214	0.866200	±0.000060	1	$(^1S)4p\ ^2P^0_{1/2} \rightarrow (^1S)3d\ ^2D_{3/2}$	1086,1090,1091, 1102	336,1166
0.992342	0.9940		1	$(^1S)6s\ ^2S_{1/2} \rightarrow (^1S)5p\ ^2P^0_{3/2}$		976

Strontium

Wavelength (μm)	Measured value (μm)	Uncertainty	Charge	Transition	Comments	References
0.407771	0.4078		1	$(^1S)5p\ ^2P^0_{3/2} \rightarrow (^1S)5s\ ^2S_{1/2}$	1091,1092	895
0.416179	0.4162		1	$(^1S)6s\ ^2S_{1/2} \rightarrow (^1S)5pP^0_{1/2}$	1076,1082	976,1109
0.430544	0.4305		1	$(^1S)6s\ ^2S_{1/2} \rightarrow (^1S)5p\ ^2P^0_{3/2}$	1076,1082,1086, 1090,1091	976,1109
0.868930	0.8700		1	$(^1S)6d\ ^2D_{5/2} \rightarrow (^1S)6p\ ^2P^0_{3/2}$	1076,1082,1086	976
1.03273	1.033050	±0.000050	1	$(^1S)5p\ ^2P^0_{3/2} \rightarrow (^1S)4d\ ^2D_{5/2}$	1082,1086,1091	139,881
1.09149	1.091450	±0.000050	1	$(^1S)5p\ ^2P^0_{1/2} \rightarrow (^1S)4d\ ^2D_{3/2}$	1082,1086,1091	139,881
1.122506	1.1230		1	$(^1S)7s\ ^2S_{1/2} \rightarrow (^1S)6p\ ^2P^0_{3/2}$	1082,1086	976

Barium

Wavelength (μm)	Measured value (μm)	Uncertainty	Charge	Transition	Comments	References
0.614171	0.614170	±0.000010	1	$(^1S)6p\ ^2P^0_{3/2} \rightarrow (^1S)5d\ ^2D_{5/2}$	1087,1090,1091, 1105	149,145
0.649690	0.649690	±0.000010	1	$(^1S)6p\ ^2P^0_{1/2} \rightarrow (^1S)5d\ ^2D_{3/2}$	1087,1090,1091	149,145
1.24748	1.2478		1	$(^1S)8s\ ^2S_{1/2} \rightarrow (^1S)7p\ ^2P^0_{3/2}$	1076,1082	976
2.59243	2.592300	±0.000150	1	$(^1S)7p\ ^2P^0_{3/2} \rightarrow (^1S)6d\ ^2D_{5/2}$		146,1133
2.90572	2.905900	±0.000200	1	$(^1S)7p\ ^2P^0_{1/2} \rightarrow (^1S)6d\ ^2D_{3/2}$		146,1133

3.2.3.4 Group IIB Lasers

Table 3.2.4
Zinc

Wavelength (μm)	Measured value (μm)	Uncertainty	Charge	Transition	Comments	References
	0.1270		2	$3d^94d\ ^1G_4 \rightarrow 3d^94p\ ^3F_3$	6923	850
	0.1306		2	$3d^84s\ ^1G_4 \rightarrow 3d^94p\ ^3D_3$	6923	850
	0.1319		2	$3d^94d\ ^3G_4 \rightarrow 3d^94p\ ^3F_3$	6923	850
0.491166	0.49116		1	$(^1S)4f\ ^2F^0_{5/2} \rightarrow (^1S)4d\ ^2D_{3/2}$	1076,1083,1084, 1087,1090,1110	905,944,945,1034 1036,1039,1059, 1083,1085,1145
0.492404	0.4925		1	$(^1S)4f\ ^2F^0_{7/2} \rightarrow (^1S)4d\ ^2D_{5/2}$	1076,1083,1084, 1087,1090,1111	868,905,953,1027 1034,1036,1039, 1083,1085,11145
0.589435	0.5894		1	$4s^2\ ^2D_{3/2} \rightarrow (^1S)4p\ ^2P^0_{1/2}$	1074,1083,1084, 1090,1112	905,945,1034, 1036,1039, 1085,1145

			Transition		
0.602126	0.6021	1	$(^1S)5d\ ^2D_{3/2} \rightarrow (^1S)5p\ ^2P^0_{1/2}$	1073,1083,1087, 1113	905,946,1145
0.610249	0.610280 ±0.000070	1	$(^1S)5d\ ^2D_{5/2} \rightarrow (^1S)5p\ ^2P^0_{3/2}$	1073,1083,1087, 1114	1064,1083,1085, 1145
0.621459	0.6214	1	$4s^2\ ^2D_{3/2} \rightarrow (^1S)4p\ ^2P^0_{3/2}$	1074,1083,1087, 1115	905,1036,1145
	0.7478	1	$3d^94s^2\ ^2D_{5/2} \rightarrow 3d^{10}4p\ ^2P^0_{3/2}$	6920	844
0.747879	0.747830 ±0.000160	1	$4s^2\ ^2D_{5/2} \rightarrow (^1S)4p\ ^2P^0_{3/2}$	1074,1083,1084, 1086,1088, 1090,1116	676,905,945,1023 1034,1036,1039, 1064,1085,1145
0.758848	0.758750 ±0.000160	1	$(^1S)5p\ ^2P^0_{3/2} \rightarrow (^1S)5s\ ^2S_{1/2}$	1076,1083,1084, 1087,1090,1117	676,905,945,953, 1034,1036,1039, 1064,1085,1145
0.76121	0.761118 ±0.000160	1	$9(^1S)6s\ ^2S_{1/2} \rightarrow (^1S)5p\ ^2P^0_{1/2}$	1118	905,1064,1145
0.773250	0.773250 ±0.000050	1	$(^1S)5p\ ^2P^0_{1/2} \rightarrow (^1S)5s\ ^2S_{1/2}$	1073,1083,1087, 1119	905,928,1145
0.775786	0.7757	1	$(^1S)6s\ ^2S_{1/2} \rightarrow (^1S)5p\ ^2P^0_{3/2}$	1076,1085,1120	868,905, 1139,1145
1.83083	1.831000 ±0.000100	1	$(^1S)6p\ ^2P^0_{3/2} \rightarrow (^1S)6s\ ^2S_{1/2}$	1084,1090,1121	155,1145
5.08483	5.086000 ±0.000400	1	$(^1S)6p\ ^2P^0_{3/2} \rightarrow (^1S)5d\ ^2D_{5/2}$	1084,1090,1122	155,1145

Cadmium

Wavelength (μm)	Measured value (μm)	Uncertainty	Charge	Transition	Comments	References
	0.2312		1	$4d^{10}5d\ ^2S_{1/2} \rightarrow 4d^{10}5p\ ^2P_{3/2}$	6923	852
	0.2573		1	$4d^{10}6s\ ^2S_{1/2} \rightarrow 4d^{10}5p\ ^2P_{1/2}$	6923	852
	0.2748		1	$4d^{10}6s\ ^2S_{1/2} \rightarrow 4d^{10}5p\ ^2P_{3/2}$	6923	852
	0.3250		1	$5s^2\ ^2D_{3/2} \rightarrow (^1S)5p\ ^2P^0_{1/2}$	6920	676,678
0.325029	0.3250		1	$5s^2\ ^2D_{3/2} \rightarrow (^1S)5p\ ^2P^0_{1/2}$	1074,1083,1086, 1087,1090	1034
0.441565	0.441560	±0.000070	1	$5s^2\ ^2D_{5/2} \rightarrow (^1S)5p\ ^2P^0_{3/2}$	1074,1081,1083, 1084,1086,1087, 1088,1090,1123	677,870,1023, 1034,1039,1064, 1065,1083,1085
	0.4416		1	$5s^2\ ^2D_{5/2} \rightarrow (^1S)5p\ ^2P^0_{3/2}$	6920	676–678
	0.4882		1	$(^3D)5p\ ^4F^0_{5/2} \rightarrow (^1S)5d\ ^2D_{3/2}$	1073,1083	1027
	0.50259		1	$(^3D)5p\ ^4F^0_{7/2} \rightarrow (^1S)5d\ ^2D_{5/2}$	1073,1083,1090	1027
	0.5337		1	$(^1S)4f\ ^2F^0_{5/2} \rightarrow (^1S)5d\ ^2D_{3/2}$	1073,1081,1083, 1084,1087,1088, 1090,1124	868,871,953,1023, 1034,1039,1057, 1058,1083,1085
0.537813	0.5338		1	$(^1S)4f\ ^2F^0_{7/2} \rightarrow (^1S)5d\ ^2D_{5/2}$	1073,1081,1083, 1084,1087,1088, 1090	868,871,953,1023, 1034,1039,1057, 1058,1083,1085
0.635480	0.63548		1	$(^1S)6g\ ^2{}^1G_{7/2} \rightarrow (^1S)4f\ ^2F^0_{5/2}$	1076,1083,1087, 1090	1034,1057,1058, 1084,1085
0.636004	0.63601		1	$(^1S)6g\ ^2G_{9/2} \rightarrow (^1S)4f\ ^2F^0_{7/2}$	1076,1083,1084, 1087,1090	871,1034,1039, 1057,1058,1082, 1084,1085
0.723689	0.72369		1	$(^1S)6f\ ^2F^0_{5/2} \rightarrow (^1S)6d\ ^2D_{3/2}$	1076,1083,1087, 1090	871,1034,1039, 1057,1058,1082, 1084,1085

Wavelength (μm)	Measured value (μm)	Uncertainty	Charge	Transition	Comments	References
0.728423	0.72843		1	$(^1S)6f\ ^2F^0_{7/2} \rightarrow (^1S)6d\ ^2D_{5/2}$	1076,1083,1084, 1087,1090	871,1034,1039, 1057,1082, 1084,1085
0.806687	0.80669		1	$(^1S)6p\ ^2P^0_{3/2} \rightarrow (^1S)6s\ ^2S_{1/2}$	1076,1083,1084, 1087,1090	871,953,1034, 1039,1059,1085
0.838889	0.8389		1	$(^1S)7s\ ^2S_{1/2} \rightarrow (^1S)6p\ ^2P^0_{3/2}$	1082,1086,1094	152,153
0.853026	0.85309		1	$(^1S)6p\ ^2P^0_{1/2} \rightarrow (^1S)6s\ ^2S_{1/2}$	1076,1083,1084, 1087,1090	871,1034,1039, 1059,1085
0.865190	0.8652		1	$(^1S)9s\ ^2S_{1/2} \rightarrow (^3D)5p\ ^4P^0_{1/2}$	1076,1087	938
0.887770	0.88778		1	$(^1S)9s\ ^2S_{1/2} \rightarrow (^1S)7p\ ^2P^0_{3/2}$	1076,1083,1084, 1087,1090	871,1039,1085
3.28825	3.289000	±0.000200	1	$(^1S)9s\ ^2S_{1/2} \rightarrow (^1S)8p\ ^2P^0_{3/2}$	1084,1090	155

Mercury

Wavelength (μm)	Measured value (μm)	Uncertainty	Charge	Transition	Comments	References
0.398398	0.398399	±0.000002	1	$(^1S)6p\ ^2P^0_{3/2} \rightarrow 6s^2\ ^2D_{5/2}$	1077,1091,1137	905,930
0.479701	0.479700	±0.000010	2	$5d^86s^2126468_4 \rightarrow 5d^96p105628_3$	1083,1086,1090, 1094	876,996,1131
0.521082	0.5210		2	$5d^86s^2122735_1 \rightarrow 5d^96p103549_2$	1083,1090	1131
0.567717	0.5678		1	$(^1S)5f\ ^2F^0_{7/2} \rightarrow (^1S)6d\ ^2D_{5/2}$	1081,1086,1090, 1092,1136	905,924,933
0.614947	0.6150		1	$(^1S)7p\ ^2P^0_{3/2} \rightarrow (^1S)7s\ ^2S_{1/2}$	1076,1081,1083, 1086,1087,1088, 1090,1094,1135	864,913,924,951, 1035,1060,1067, 1102,1106,1132
0.650138	0.6501		2	$5d^86s^2118926_2 \rightarrow 5d^96p103549_2$	1083,1090	1131
0.734637	0.7346		1	$(^1S)7d\ ^2D_{5/2} \rightarrow (^1S)7p\ ^2P^0_{3/2}$	1137	905,924
0.74181	0.74181		1	$5d^96s6p\ P^0_{3/2} \rightarrow (^1S)7s\ ^2S_{1/2}$	1086,1087,1138	886,905

Mercury—continued

Wavelength (μm)	Measured value (μm)	Uncertainty	Charge	Transition	Comments	References
0.794466	0.7945		1	$(^1S)7p\ ^2P^0_{1/2} \rightarrow (^1S)7s\ ^2S_{1/2}$	1076,1081,1083, 1086,1087,1139	905,951,1035, 1060,1067,1106
0.85482	0.8547		1	$(^1S)5g\ ^2G_{7/2} \rightarrow 121960\,^2F^0_{5/2}$	1130	160,905
0.86220?	0.8628		1	$(^1S)8p\ ^2P^0_{3/2} \rightarrow 116200_{5/2}$	1131	160
0.8677	0.8677		1		1128	160
0.93968	0.9396		1	$(^1S)10s\ ^2S_{1/2} \rightarrow (^1S)8p\ ^2P^0_{3/2}$		160
1.0584	1.0586		1	$(^1S)8s\ ^2S_{1/2} \rightarrow (^1S)7p\ ^2P^0_{3/2}$		160,924
1.2545	1.2545		1		1132	160
1.2981	1.2981		1		1131	160
1.5555	1.5550		1	$(^1S)7p\ ^2P^0_{3/2} \rightarrow (^1S)6d\ ^2D_{5/2}$		160,1102

3.2.3.5 Group IIIA Lasers

Table 3.2.5
Boron

Wavelength (μm)	Measured value (μm)	Uncertainty	Charge	Transition	Comments	References
0.345129	0.345132		1	$2p^2\ ^1D_2 \rightarrow (^2S)2p\ ^1P^0_1$		200

Aluminium

Wavelength (μm)	Measured value (μm)	Uncertainty	Charge	Transition	Comments	References
0.358655	0.358660	±0.000010	1	$(^2S)4f\,^3F^0_4 \rightarrow (^2S)3d\ ^3D_3$	1080,1087	423
0.358744	0.358740	±0.000020	1	$(^2S)4f\,^3F^0_2 \rightarrow (^2S)3d\ ^3D_1$	1080,1087,1090	878

Wavelength (μm)	Measured value (μm)	Uncertainty	Charge	Transition	Comments	References
0.691994	0.691998	±0.000010	1	$(^2S)5s\ ^1S_0 \rightarrow (^2S)4p\ ^1P^0_1$	1080,1083,1086, 1090,1092,1094	423,878,1050
0.704205	0.704209	±0.000010	1	$(^2S)4p\ ^3P^0_2 \rightarrow (^2S)4s\ ^3S_1$	1080,1083,1086, 1087,1090,1092, 1094	423,878,1050
0.705661	0.705640	±0.000020	1	$(^2S)4p\ ^3P^0_1 \rightarrow (^2S)4s\ ^3S_1$	1080,1087,1090	878
0.747145	0.747149	±0.000010	1	$(^2S)4f\ ^1F^0_3 \rightarrow (^2S)3d\ ^1D_2$	1080,1083,1086, 1087,1092,1094	423,878,1050

Indium

Wavelength (μm)	Measured value (μm)	Uncertainty	Charge	Transition	Comments	References
	0.1850		2	$4d^9 5s\ ^2D_{5/2} \rightarrow 4d^{10}5p\ ^2P_{3/2}$	6920	883
0.468111	0.468050	±0.000070	1	$(^2S)4f\ ^3F^0_4 \rightarrow (^2S)5d\ ^3D_3$		1064

Thallium

Wavelength (μm)	Measured value (μm)	Uncertainty	Charge	Transition	Comments	References
0.473705	0.4737		1	$(^2S)5f\ ^1F^0_3 \rightarrow (^2S)6d\ ^1D_2$	1073,1082,1086, 1140	905,931
0.498135	0.4981		1	$(^2S)5f\ ^3F^0_2 \rightarrow (^2S)6d\ ^3D_1$	1073,1082,1086, 1140	905,931
0.507854	0.5079		1	$(^2S)5f\ ^3F^0_3 \rightarrow (^2S)6d\ ^3D_2$	1073,1082,1086, 1141	905,931
0.515214	0.5152		1	$(^2S)5f\ ^3F^0_4 \rightarrow (^2S)6d\ ^3D_3$	1073,1080,1082, 1083,1086,1090, 1092,1142	905,931,977

Thallium—*continued*

Wavelength (μm)	Measured value (μm)	Uncertainty	Charge	Transition	Comments	References
0.594904	0.5949		1	$(^2S)7p\ ^3P^0_2 \rightarrow (^2S)7s\ ^3S_1$	1077,1080,1082, 1083,1084,1086, 1087,1090,1092, 1143	897,898,905, 931,977
0.69505	0.6951		1	$(^2S)7p\ ^1P^0_1 \rightarrow (^2S)7s\ ^1S_0$	1077,1080,1082, 1083–4,1086–7, 1090,1092,1144	898,905,931,977
0.9350	0.9350		1		1077,1082,1086	931
1.1350	1.1350		1		1073,1082,1086	931
1.1788	1.1750		1	$(^2S)6f\ ^3F^0_3 \rightarrow (^2S)7d\ ^3D_2$	1073,1082,1086	931

3.2.3.6 Group IVA Lasers

Table 3.2.6
Carbon

Wavelength (μm)	Measured value (μm)	Uncertainty	Charge	Transition	Comments	References
0.154819	0.154820	+\-0.000030	3	$2p\ ^2P^0_{3/2} \rightarrow 2s\ ^2S_{1/2}$	1084,1086,1090, 1092,1149	1048,1099
0.155077	0.155090	+\-0.000030	3	$2p\ ^2P^0_{1/2} \rightarrow 2s\ ^2S_{1/2}$	1084,1086,1090, 1092,1149	1048,1099
0.464742	0.464740	±0.000004	2	$(^2S)3p\ ^3P^0_2 \rightarrow (^2S)3s\ ^3S_1$	1084,1093,1094, 1147	450,664,874, 1038,1127
0.465025	0.465021	±0.000004	2	$(^2S)3p\ ^3P^0_1 \rightarrow (^2S)3s\ ^3S_1$	1084,1093,1094, 1146	450,664,874, 1038,1127

Wavelength (μm)	Measured value (μm)	Uncertainty	Charge	Transition	Comments	References
0.514516	0.514570	±0.000050	1	$(^3P^0)3p\ ^4P_{5/2} \rightarrow (^3P^0)3s\ ^4P_{5/2}$	1084,1090	1031
0.657803	0.657800	±0.000050	1	$(^1S)3p\ ^2P^0_{3/2} \rightarrow (^1S)3s\ S_{1/2}$	1084,1090	1031
0.678375	0.678360	±0.000050	1	$(^3P^0)3p\ ^4D_{7/2} \rightarrow (^3P^0)3s\ ^4P_{5/2}$	1084,1090	1031

Silicon

Wavelength (μm)	Measured value (μm)	Uncertainty	Charge	Transition	Comments	References
0.408885	0.408890	±0.000010	3	$4p\ ^2P^0_{3/2} \rightarrow 4s\ ^2S_{1/2}$	1093,1155	450,1090,1134
0.455262	0.455259	±0.000006	2	$(^2S)4p\ ^3P^0_2 \rightarrow (^2S)4s\ ^3S_1$	1085,1153	220,1020,1091
0.456782	0.456784	±0.000006	2	$(^2S)4p\ ^3P^0_1 \rightarrow (^2S)4s\ ^3S_1$	1085,1154	220,1020,1091
0.634710	0.634724	±0.000006	1	$(^1S)4p\ ^2P^0_{3/2} \rightarrow (^1S)4s\ ^2S_{1/2}$	1084,1085, 1150	220,1134, 1020,1062
0.637136	0.637148	±0.000006	1	$(^1S)4p\ ^2P^0_{1/2} \rightarrow (^1S)4s\ ^2S_{1/2}$	1085,1151	220,1020,1062
0.667188	0.667193	±0.000006	1	$(^3P^0)4p\ ^4D_{7/2} \rightarrow (^3P^0)4s\ ^4P^0_{5/2}$	1085,1152	220,1020,1062

Germanium

Wavelength (μm)	Measured value (μm)	Uncertainty	Charge	Transition	Comments	References
0.513175	0.513150	±0.000070	1	$(^1S)4f\ ^2F^0_{5/2} \rightarrow (^1S)4d\ ^2D_{3/2}$	1156	1063,1064
0.517865	0.517840	±0.000070	1	$(^1S)4f\ ^2F^0_{7/2} \rightarrow (^1S)4d\ ^2D_{5/2}$	1157	1063,1064

Tin

Wavelength (μm)	Measured value (μm)	Uncertainty	Charge	Transition	Comments	References
0.558943	0.5589		1	$(^1S)4f\ ^2F^0_{5/2} \rightarrow (^1S)5d\ ^2D_{3/2}$	1082,1162	202,905
0.579918	0.579870	±0.000070	1	$(^1S)4f\ ^2F^0_{7/2} \rightarrow (^1S)5d\ ^2D_{5/2}$	1082,1160	202,905,1064
0.645358	0.645300	±0.000070	1	$(^1S)6p\ ^2P^0_{3/2} \rightarrow (^1S)6s\ ^2S_{1/2}$	1074,1083, 1086,1161	676,905,1064
0.657926	0.657903	±0.000006	1	$(^1S)6p\ ^2P^0_{1/2} \rightarrow (^1S)6s\ ^2S_{1/2}$	1084,1093,1158	200–202,905
0.684420	0.684400	±0.000070	1	$(^1S)5f\ ^2F^0_{5/2} \rightarrow (^1S)6d\ ^2D_{3/2}$	1074,1083,1159	676,905,1064
1.06118	1.062		1	$(^1S)5f\ ^2F^0_{7/2} \rightarrow (^1S)6d\ ^2D_{5/2}$	1082,1085	202,384
1.07388	1.074		1		1082	202

Lead

Wavelength (μm)	Measured value (μm)	Uncertainty	Charge	Transition	Comments	References
0.53721	0.537210	±0.000070	1	$(^1S)5f\ ^2F^0_{7/2} \rightarrow 6s6p^2\ ^4P_{5/2}$	1163	905,1064
0.56088	0.560860	±0.000050	1	$(^1S)7p\ ^2P^0_{3/2} \rightarrow (^1S)7s\ ^2S_{1/2}$	1083,1164	905,1072
0.66600	0.666010	±0.000050	1	$(^1S)7p\ ^2P^0_{1/2} \rightarrow (^1S)7s\ ^2S_{1/2}$	1083,1165	905,1072
1.1592	1.159		1	$6p^2\ ^2D_{3/2} \rightarrow (^1S)7p\ ^2P_{1/2}$	1086	1086

3.2.3.7 Group VA Lasers

Table 3.2.7
Nitrogen

Wavelength (μm)	Measured value (μm)	Uncertainty	Charge	Transition	Comments	References
0.336734	0.336732	±0.000006	2	$(^3P^0)3p\ ^4P_{5/2} \rightarrow (^3P^0)3s\ ^4P^0_{5/2}$		653
0.347871	0.347876	±0.000005	3	$(^2S)3p\ ^3P^0_2 \rightarrow (^2S)3s\ ^3S_1$	1090,1092,1176	664,902,907,989

Wavelength (μm)	Measured value (μm)	Uncertainty	Charge	Transition	Comments	References
0.348299	0.348302	±0.000006	3	$(^2S)3p\ ^3P^0_1 \rightarrow (^2S)3s\ ^3S_1$	1177	653,902
0.399501	0.399499	±0.000001	1	$(^2P^0)3p\ ^1D_2 \rightarrow (^2P^0)3s\ ^1P^0_1$	1090,1175	1092,1170,1173
0.409732	0.409729	±0.000006	2	$(^1S)3p\ ^2P^0_{3/2} \rightarrow (^1S)3s\ ^2S_{1/2}$	1090,1092	664,907,989,1092
0.410338	0.410336	±0.000002	2	$(^1S)3p\ ^2P^0_{1/2} \rightarrow (^1S)3s\ ^2S_{1/2}$	1090,1094	664,989, 1092,1173
0.451088	0.451089	±0.000002	2	$(^3P^0)3p\ ^4D_{5/2} \rightarrow (^3P^0)3s\ ^4P^0_{3/2}$	1090,1094	664,989,1092
0.451487	0.451486	±0.000003	2	$(^3P^0)3p\ ^4D_{7/2} \rightarrow (^3P^0)3s\ ^4P^0_{5/2}$	1090,1094	664,989, 1092,1173
0.462140	0.462100	±0.000080	1	$(^2P^0)3p\ ^3P_0 \rightarrow (^2P^0)3s\ ^3P^0_1$	1090,1174	1016,1092,1170
0.463055	0.463051	±0.000002	1	$(^2P^0)3p\ ^3P_2 \rightarrow (^2P^0)3s\ ^3P^0_2$	1090,1094,1168	664,989,1092, 1170,1173
0.464310	0.464390	±0.000080	1	$(^2P^0)3p\ ^3P_1 \rightarrow (^2P^0)3s\ ^3P^0_2$	1090,1169	1016,1092,1170
0.501638	0.501639		1	$(^2P^0)3d\ ^3F_2 \rightarrow (^2P^0)3p\ ^3D_2$	1166	872,1170
0.566663	0.566662	±0.000003	1	$(^2P^0)3p\ ^3D_2 \rightarrow (^2P^0)3s\ ^3P^0_1$	1167	872,927,1170
0.567601	0.567603	±0.000003	1	$(^2P^0)3p\ ^3D_1 \rightarrow (^2P^0)3s\ ^3P^0_0$	1172	927,1170
0.567956	0.567953	±0.000003	1	$(^2P^0)3p\ ^3D_3 \rightarrow (^2P^0)3s\ ^3P^0_2$	1094,1173	212,450,872, 927,1170
0.568622	0.568690	±0.000080	1	$(^2P^0)3p\ ^3D_1 \rightarrow (^2P^0)3s\ ^3P^0_1$	1170	1016,1170
0.648209	0.648260	±0.000060	1	$(^2P^0)3p\ ^1P_1 \rightarrow (^2P^0)3s\ ^1P^0_1$	1171	986,1016,1170

Phosphorus

Wavelength (μm)	Measured value (μm)	Uncertainty	Charge	Transition	Comments	References
0.334769	0.334776	±0.000006	3	$(^2S)4p\ ^3P^0_2 \rightarrow (^2S)4s\ ^3S_1$		220
0.422208	0.422225	±0.000006	2	$(^1S)4p\ ^2P^0_{3/2} \rightarrow (^1S)4s\ ^2S_{1/2}$		220
0.602418	0.602427	±0.000006	1	$(^2P^0)4p\ ^3D_2 \rightarrow (^2P^0)4s\ ^3P^0_1$	1083,1092,1094, 1178	220,857,869,992
0.603404	0.603419	±0.000006	1	$(^2P^0)4p\ ^3D_1 \rightarrow (^2P^0)4s\ ^3P_0$	1179	220,992

Phosphorus—continued

Wavelength (μm)	Measured value (μm)	Uncertainty	Charge	Transition	Comments	References
0.604312	0.604322	±0.000006	1	$(^2P^0)4p\,^3D_3 \rightarrow (^2P^0)4s\,^3P^0_2$	1083,1092,1094, 1180	220,857,869,992
0.608782	0.608804	±0.000006	1	$(^2P^0)4p\,^3D_1 \rightarrow (^2P^0)4s\,^3P^0_1$	1181	220,992
0.616559	0.616574	±0.000006	1	$(^2P^0)4p\,^3D_2 \rightarrow (^2P^0)4s^3\,P^0_2$	1182	220,992
0.784563	0.7846		1	$(^2P^0)4p\,^1P_1 \rightarrow (^2P^0)4s\,^1P^0_1$	1183	869,992

Arsenic

Wavelength (μm)	Measured value (μm)	Uncertainty	Charge	Transition	Comments	References
0.538520	0.538510	±0.000040	1	$(^2P^0)6s\,^1P^0_1 \rightarrow (^2P^0)5p\,^3P_1$	1076,1083,1084, 1087,1184	981,1033,1041
0.549695	0.549680	±0.000040	1	$(^2P^0)6s\,^3P^0_2 \rightarrow (^2P^0)5p\,^3D_3$	1076,1083,1084, 1087,1185	981,1033,1041
0.549773	0.549760	±0.000040	1	$(^2P^0)5p\,^3D_1 \rightarrow (^2P^0)5s^3\,P^0_0$	1076,1083,1084, 1087,1090,1092, 1094,1186	941,981, 1033,1041
0.555809	0.555820	±0.000040	1	$(^2P^0)5p\,^3D_2 \rightarrow (^2P^0)5s\,^3P^0_1$	1076,1083,1084, 1087,1090,1092, 1094,1187	943,981,1033, 1041
0.565132	0.565200	±0.000100	1	$(^2P^0)5p\,^3D_3 \rightarrow (^2P^0)5s\,^3P^0_2$	1092,1188	943,981
0.583790	0.583800	±0.000040	1	$(^2P^0)6s\,^3P^0_2 \rightarrow (^2P^0)5p\,^3P_2$	1076,1083,1084, 1087,1189	981,1033,1041
0.617027	0.617020	±0.000040	1	$(^2P^0)5p\,^1P_1 \rightarrow (^2P^0)5s^3P^0_1$	1076,1083,1084, 1087,1092,1190	943,981, 1033,1041

Wavelength (μm)	Measured value (μm)	Uncertainty	Charge	Transition	Comments	References
0.651174	0.651180	±0.000040	1	$(^2P^0)6s\,^1P^0_1 \to (^2P^0)5p\,^1D_2$	1076,1083,1084, 1086,1087,1090, 1191	981,1033, 1041
0.710272	0.710250	±0.000040	1	$(^2P^0)5p\,^3P_1 \to 4p^3\,^3P^0_2$	1076,1083,1084, 1087,1192	981,1033, 1041

Antimony

Wavelength (μm)	Measured value (μm)	Uncertainty	Charge	Transition	Comments	References
0.613000	0.613000	±0.000100	1	$(^2P^0)6p\,^3D_3 \to (^2P^0)6s\,^3P^0_2$		442
0.699563	0.6996		1	$(^2P^0)6p\,^3D_1 \to (^2P^0)6s\,^3P^0_1$	1076,1086	932

Bismuth

Wavelength (μm)	Measured value (μm)	Uncertainty	Charge	Transition	Comments	References
0.456118	0.456070	±0.000010	2	$(^1S)7p\,^2P^0_{1/2} \to (^1S)7s\,^2S_{1/2}$	1086	957
0.571924	0.571920	±0.000010	1	$6p_{1/2}7p_{1/2}\,^3P_0 \to 6p_{1/2}7s\,^3P^0_1$	1086	957
0.75990	0.759870	±0.000050	2	$(^1S)6f\,^2F^0_{5/2} \to (^1S)7d\,^2D_{3/2}$	1086	957
0.806881	0.806920	±0.000050	2	$(^1S)6f\,^2F^0_{7/2} \to (^1S)7d\,^2D_{5/2}$	1086	957

3.2.3.8 Group VIA Lasers

Table 3.2.8
Oxygen

Wavelength (μm)	Measured value (μm)	Uncertainty	Charge	Transition	Comments	References
0.278104	0.278150	±0.000050	4	$(^2S)3p\ ^3P^o_2 \rightarrow (^2S)3s\ ^3S_1$	1092	908
0.298378	0.298389	±0.000006	2	$(^2P^o)3p\ ^1D_2 \rightarrow (^2P^o)3s\ ^1P^o_1$	1085,1090,1092, 1193	450,653, 908,989
0.304713	0.304715	±0.000006	2	$(^2P^o)3p\ ^3P_2 \rightarrow (^2P^o)3s\ ^3P^o_2$	1085,1090,1092	450,653,908,989
0.306342	0.306346	±0.000006	3	$(^1S)3p\ ^2P^o_{3/2} \rightarrow (^1S)3s\ ^2S_{1/2}$	1090,1092,1194	653,908,989,1130
0.338121	0.338134	±0.000006	3	$(^3P^o)3p\ ^4D_{5/2} \rightarrow (^3P^o)3s\ ^4P^o_{3/2}$	1196	653,1130
0.338130	0.338134	±0.000006	3	$(^3P^o)3p\ ^4D_{3/2} \rightarrow (^3P^o)3s\ ^4P^o_{1/2}$	1195	653,1130
0.338554	0.338554	±0.000006	3	$(^3P^o)3p\ ^4D_{7/2} \rightarrow (^3P^o)3s\ ^4P^o_{5/2}$	1092,1197	653,908,1130
0.372733	0.372711	±0.000050	1	$(^3P)3p\ ^4S^o_{3/2} \rightarrow (^3P)3s\ ^4P_{3/2}$	1092	908
0.374949	0.374947	±0.000004	1	$(^3P)3p\ ^4S^o_{3/2} \rightarrow (^3P)3s\ ^4P_{5/2}$	1090,1092,1198	450,664,908, 964,989
0.375467	0.375468	±0.000004	2	$(^2P^o)3p\ ^3D_2 \rightarrow (^2P^o)3s\ ^3P^o_1$	1090,1092,1199	450,664,908,964, 989,1026
0.375720	0.37572		2	$(^2P^o)3p\ ^3D_1 \rightarrow (^2P^o)3s\ ^3P^o_0$	1085,1090	989,1026,
0.375988	0.375989	±0.000005	2	$(^2P^o)3p\ ^3D_3 \rightarrow (^2P^o)3s\ ^3P^o_2$	1090,1092,1200	450,664,908, 964,989,1026
0.377399	0.37738		2	$(^2P^o)3p\ ^3D_1 \rightarrow (^2P^o)3s\ ^3P^o_1$		1026
0.434738	0.434738	±0.000004	1	$(^1D)3p\ ^2D^o_{3/2} \rightarrow (^1D)3s\ ^2D_{3/2}$	1090,1092,1201	450,664,908,1092
0.435128	0.435126	±0.000004	1	$(^1D)3p\ ^2D^o_{5/2} \rightarrow (^3P)3s\ ^2P_{3/2}$	1090,1202	450,664, 1026,1092
0.441488	0.441439	±0.000004	1	$(^3P)3p\ ^2D^o_{5/2} \rightarrow (^3P)3s\ ^2P_{3/2}$	1090,1203	450,664,1092
0.441697	0.441697	±0.000004	1	$(^3P)3p\ ^2D^o_{3/2} \rightarrow (^3P)3s\ ^2P_{1/2}$	1090,1204	450,664,1092
0.460552	0.460552	±0.000009	0		1205	664
0.464914	0.464908	±0.000010	1	$(^3P)3p\ ^4D^o_{7/2} \rightarrow (^3P)3s\ ^4P_{5/2}$		212,450

Wavelength (μm)	Measured value (μm)	Uncertainty	Charge	Transition	Comments	References
0.559237	0.559237	±0.000006	2	$(^2P^0)3p\ ^1P_1 \rightarrow (^2P^0)3s\ ^1P^0_1$	1086,1090,1206	450,664,872, 986,989,1026
0.664099	0.664020	±0.000100	1	$(^3P)3p\ ^2S^0_{1/2} \rightarrow (^3P)3s\ ^2P_{1/2}$	1090	986
0.666694	0.666694		1	$(^3P)4p\ ^2P^0_{1/2} \rightarrow (^3P)3d\ ^2P_{3/2}$	1207	872,1156
0.672136	0.672138	±0.000004	1	$(^3P)3p\ ^2S^0_{1/2} \rightarrow (^3P)3s\ ^2P_{3/2}$		664,872

Sulfur

Wavelength (μm)	Measured value (μm)	Uncertainty	Charge	Transition	Comments	References
0.263896	0.263896	±0.000001	2		1084,1090,1094, 1208	991,1489
0.263896	0.263896	±0.000001	3		1084,1090,1094, 1209	991,1489
0.332486	0.332486	±0.000001	2	$(^2P^0)4p\ ^3P_2 \rightarrow (^2P^0)3d\ ^3P^0_2$	1084,1090,1094	991,1489
0.349733	0.349733	±0.000001	2		1084,1090,1094	991,1489
0.370937	0.370935	±0.000001	2	$(^2P^0)4p\ ^3D_2 \rightarrow (^2P^0)3d\ ^3P^0_1$	1084,1090,1094	991,1489
0.492532	0.492560	±0.000006	1	$(^3P)4p\ ^4P^0_{3/2} \rightarrow (^3P)4s\ ^4P_{1/2}$	1084	1489
0.501401	0.501424	±0.000006	1	$(^3P)4p\ ^2P^0_{3/2} \rightarrow (^3P)4s\ ^2P_{3/2}$	1084	1489
0.503239	0.503262	±0.000006	1	$(^3P)4p\ ^4P^0_{5/2} \rightarrow (^3P)4s\ ^4P_{5/2}$	1084	1489
0.516032	0.516032	±0.000006	0		1084,1210	1489
0.521962	0.521962	±0.000006	0		1084,1211	1489
0.532070	0.532088	±0.000006	1	$(^1D)4p\ ^2F^0_{7/2} \rightarrow (^1D)4s\ ^2D_{5/2}$	1083,1084,1092, 1094,1212	857,869,1489
0.534567	0.534583	±0.000006	1	$(^1D)4p\ ^2F^0_{5/2} \rightarrow (^1D)4s\ ^2D_{3/2}$	1083,1084,1092, 1094,1213	857,869,1489
0.542864	0.542874	±0.000006	1	$(^3P)4p\ ^4D^0_{3/2} \rightarrow (^3P)4s\ ^4P_{1/2}$	1084,1094	869,1489
0.543274	0.543287	±0.000006	1	$(^3P)4p\ ^4D^0_{5/2} \rightarrow (^3P)4s\ ^4P_{3/2}$	1083,1084,1092, 1094,1214	857,869,1489

Sulfur—continued

Wavelength (μm)	Measured value (μm)	Uncertainty	Charge	Transition	Comments	References
0.545380	0.545388	±0.000006	1	$(^3P)4p\ ^4D^0_{7/2} \rightarrow (^3P)4s\ ^4P_{5/2}$	1083,1084,1092, 1094,1215	857,869,1489
0.547360	0.547374	±0.000006	1	$(^3P)4p\ ^4D^0_{1/2} \rightarrow (^3P)4s\ ^4P_{1/2}$	1084,1094	869,1489
0.550965	0.550990	±0.000006	1	$(^3P)4p\ ^4D^0_{3/2} \rightarrow (^3P)4s\ ^4P_{3/2}$	1084	1489
0.556491	0.556511	±0.000006	1	$(^3P)4p\ ^4D^0_{5/2} \rightarrow (^3P)4s\ ^4P_{5/2}$	1084	1489
0.563999	0.564012	±0.000006	1	$(^3P)4p\ ^2D^0_{5/2} \rightarrow (^3P)4s\ ^2P_{3/2}$	1083,1084,1092, 1094,1216	857,1869,489
0.564698	0.564716	±0.000006	1	$(^3P)4p\ ^2D^0_{3/2} \rightarrow (^3P)4s\ ^2P_{1/2}$	1084,1094	869,1489
0.581919	0.581935	±0.000006	1	$(^3P)4p\ ^2D^0_{3/2} \rightarrow (^3P)4s\ ^2P_{3/2}$	1084	1489

Selenium

Wavelength (μm)	Measured value (μm)	Uncertainty	Charge	Transition	Comments	References
0.446760	0.446800	±0.000050	1	$(^3P)5p\ ^2P^0_{1/2} \rightarrow (^3P)5s\ ^2P_{1/2}$	1076,1083,1086	963
0.460431	0.460460	±0.000050	1	$(^3P)5p\ ^2D^0_{5/2} \rightarrow (^3P)5s\ ^4P_{5/2}$	1076,1083,1087, 1090	915,1034,1069
0.461877	0.461910	±0.000050	1	$(^3P)5p\ ^4P^0_{5/2} \rightarrow (^3P)5s\ ^4P_{3/2}$	1076,1083,1086	963
0.464843	0.464860	±0.000050	1	$(^3P)5p\ ^4P^0_{3/2} \rightarrow (^3P)5s\ ^4P_{1/2}$	1076,1083,1090	915,1069
0.471782	0.471850	±0.000050	1	$?(^3P)5p\ ^4S^0_{3/2} \rightarrow 98118_{1/2}$	1076,1083,1086	963
0.474098	0.474060	±0.000050	1	$(^3P)5p\ ^2P^0_{3/2} \rightarrow 4s4p^4\ ^2P_{3/2}$	1076,1083,1086	963
0.476368	0.476410	±0.000050	1	$(^3P)5p\ ^2D^0_{3/2} \rightarrow (^3P)5s\ ^4P_{3/2}$	1076,1083	1069
0.476554	0.476510	±0.000050	1	$(^3P)5p\ ^2P^0_{1/2} \rightarrow 4s4p^4\ ^2P_{3/2}$	1076,1083,1086, 1090	915,963
0.484063	0.484060	±0.000050	1	$(^3P)5p\ ^2S^0_{1/2} \rightarrow (^3P)5s\ ^4P_{3/2}$	1076,1083	1069
0.484499	0.485500	±0.000050	1	$(^3P)5p\ ^4S^0_{3/2} \rightarrow (^3P)5s\ ^4P_{5/2}$	1076,1083,1086, 1087,1090	915,958, 1034,1069

Wavelength (μm)		±		Transition	References	References
0.497572	0.497610	±0.000050	1	$?$ $(^3P)5p\,^2D^0_{5/2} \to 4s4p^4\,^2P_{3/2}$	1076,1083,1086, 1087,1090,1217	915,958, 1034,1069
0.499279	0.499290	±0.000050	1	$(^3P)5p\,^4P^0_{3/2} \to (^3P)5s\,^4P_{3/2}$	1076,1083,1086, 1087,1090,1218	915,958, 1034,1069
0.506865	0.506870	±0.000050	1	$(^3P)5p\,^4P^0_{5/2} \to (^3P)5s\,^4P_{5/2}$	1076,1083,1086, 1087,1090,1219	915,958,963, 958,1034,1069
0.509650	0.509610	±0.000050	1	$?(^3P)5p\,^4D^0_{7/2} \to (^3P)4d\,^4F_{9/2}$	1076,1083,1086, 1092	915,943,958,1069
0.514215	0.514190	±0.000050	1	$(^3P)5p\,^4D^0_{3/2} \to (^3P)5s\,^4P_{1/2}$	1076,1083,1086	958,1069
0.517593	0.517600	±0.000050	1	$(^3P)5p\,^4D^0_{5/2} \to (^3P)5s\,^4P_{3/2}$	1076,1083,1086, 1087,1090,1220	915,943,958, 963,1034,1069
0.522749	0.522760	±0.000050	1	$(^3P)5p\,^4D^0_{7/2} \to (^3P)5s\,^4P_{5/2}$	1076,1083,1086, 1087,1088,1090, 1092,1221	915,943,958,963, 1034,1069
0.525309	0.525260	±0.000050	1	$(^3P)5p\,^2D^0_{5/2} \to (^3P)5s\,^2P_{3/2}$	1076,1083,1086, 1087	958,1034,1069
0.525367	0.525320	±0.000050	1	$(^3P)5p\,^4D^0_{1/2} \to (^3P)5s\,^4P^0_{1/2}$	1076,1083,1086, 1087	958,1069
0.527115	0.527130	±0.000050	1	$(^1D)5p\,^2D^0_{5/2} \to (^1D)5s\,^2D_{3/2}$	1076,1083,1086, 1222	958,1034,1069
0.527127	0.527130	±0.000050	1	$(^3P)5p\,^4D^0_{5/2} \to (^3P)4d\,^4F_{7/2}$	1076,1083,1086, 1223	958,1069
0.530539	0.530550	±0.000050	1	$(^3P)5p\,^2D^0_{3/2} \to (^3P)5s\,^2P_{1/2}$	1076,1083,1086, 1090	915,958,1069
0.552244	0.552280	±0.000050	1	$(^3P)5p\,^4P^0_{5/2} \to 4s4p^4\,^2P_{3/2}$	1076,1083,1086, 1087,1090,1225	915,943, 958,1069
0.552266	0.552280	±0.000050	1	$(^3P)5p\,^4P^0_{3/2} \to (^3P)5s\,^4P_{5/2}$	1076,1083,1086, 1087,1090,1224	915,943, 958,1069
0.556688	0.556710	±0.000050	1	$(^3P)5p\,^4D^0_{3/2} \to (^3P)5s\,^4P_{3/2}$	1076,1083,1086	963,958
0.559113	0.559160	±0.000050	1	$(^3P)5p\,^4P^0_{3/2} \to (^3P)5s\,^2P_{1/2}$	1076,1083,1090	915,1069

Selenium—*continued*

Wavelength (μm)	Measured value (μm)	Uncertainty	Charge	Transition	Comments	References
0.562315	0.562280	±0.000050	1	$(^3P)5p\ ^4P^0_{1/2} \rightarrow (^3P)5s\ ^4P_{1/2}$	1076,1083,1086	963
0.569782	0.569790	±0.000050	1	$(^3P)5p\ ^4D^0_{1/2} \rightarrow (^3P)5s\ ^4P_{3/2}$	1076,1083,1090	915,1069
0.574761	0.574790	±0.000050	1	$(^3P)5p\ ^4D^0_{5/2} \rightarrow (^3P)5s\ ^4P_{5/2}$	1076,1083	1069
0.584261	0.584280	±0.000050	1	$(^3P)5p\ ^2S_{1/2} \rightarrow 4s4p^4\ ^2P_{3/2}$	1076,1083,1086	963
0.586623	0.586670	±0.000050	1	$(^3P)5p\ ^4P^0_{5/2} \rightarrow (^3P)5s\ ^2P_{3/2}$	1076,1083,1086, 1090	915,1063
0.605592	0.605630	±0.000050	1	$(^3P)5p\ ^2P^0_{3/2} \rightarrow 104874_{3/2}$	1076,1083,1086, 1087,1090	915,958, 1034,1069
0.606573	0.606610	±0.000050	1	$(^3P)5p\ ^4P^0_{3/2} \rightarrow 4s4p^4\ ^2P_{3/2}$	1076,1083,1086	963
0.610198	0.610210	±0.000050	1	$(^3P)5p\ ^2D^0_{3/2} \rightarrow (^3P)5s\ ^2P_{3/2}$	1076,1083,1086	963
0.644426	0.644390	±0.000050	1	$(^3P)5p\ ^2D^0_{5/2} \rightarrow 104874_{3/2}$	1076,1083,1086, 1090	915,958, 1034,1069
0.649049	0.649010	±0.000050	1	$(^3P)5p\ ^4D^0_{1/2} \rightarrow (^3P)5s\ ^2P_{1/2}$	1076,1083,1086, 1090	915,958, 1034,1069
0.653440	0.653460	±0.000050	1	$(^3P)5p\ ^2P^0_{1/2} \rightarrow 105974_{1/2}$	1076,1083	1069
0.706389	0.706420	±0.000050	1	$(^3P)5p\ ^4P^0_{1/2} \rightarrow (^3P)5s\ ^2P_{1/2}$	1076,1083,1086	963
0.739206	0.739240	±0.000050	1	$(^3P)5p\ ^4P^0_{5/2} \rightarrow 104874_{3/2}$	1076,1083,1086	963
0.767472	0.767490	±0.000050	1	$(^3P)5p\ ^2P^0_{3/2} \rightarrow (^1D)5s\ ^2D_{5/2}$	1076,1083,1086	963
0.772406	0.772360	±0.000050	1	$(^3P)5p\ ^4D^0_{1/2} \rightarrow (^3P)5s\ ^2P_{3/2}$	1076,1083,1086	963
0.779610	0.779620	±0.000050	1	$(^3P)5p\ ^2P^0_{1/2} \rightarrow (^1D)5s\ ^2D_{3/2}$	1076,1083,1086	963
0.783877	0.783930	±0.000050	1	$(^3P)5p\ ^4P^0_{1/2} \rightarrow 4s4p^4\ ^2P_{3/2}$	1076,1083,1086	963
0.830930	0.830890	±0.000050	1	$(^3P)5p\ ^2D^0_{5/2} \rightarrow (^1D)5s\ ^2D_{5/2}$	1076,1083,1086	963
0.924930	0.924930	±0.000100	1		1076,1083,1086	963
0.995515	0.995470	±0.000100	1	$(^3P)5p\ ^4P^0_{5/2} \rightarrow (^1D)5s\ ^2D_{5/2}$	1076,1083,1086	963
1.04088	1.040940	±0.000100	1	$(^3P)5p\ ^4D^0_{1/2} \rightarrow 104694_{3/2}$	1076,1083,1086, 1226	963

| | 1.258790 | ±0.000100 | 1 | $(^3P)5p\ ^4P^0_{3/2} \to 108834_{5/2}$ | 1076,1083,1086, 1227 | 963 |

1.25867

Tellurium

Wavelength (μm)	Measured value (μm)	Uncertainty	Charge	Transition	Comments	References
0.48429	0.484330	±0.000040	1	$122888^0_{5/2} \to 102245_{3/2}$	1076,1083,1086, 1228	904,1073
0.502039	0.502000	±0.000040	1		1076,1083,1086, 1229	904,1073
0.525641	0.525640	±0.000040	1		1076,1083,1086, 1230	904,1073
0.544985	0.544980	±0.000040	1	$103936^0_{3/2} \to 85592_{5/2}$	1080,1083,1086, 1231	904,1073
0.54540	0.545400	±0.000050	1		1232	443,904
0.547905	0.547930	±0.000040	1	$105006^0_{3/2} \to 86760_{3/2}$	1080,1083,1086, 1233	904,1073
0.557634	0.557650	±0.000040	1	$5s^25p^2(^1D)6p^0_{7/2} \to 94860_{5/2}$	1076,1083,1086, 1094,1234	442,443, 904,1073
0.56405	0.564050	±0.000050	1		1235	443,904
0.566622	0.566610	±0.000040	1	$101221^0_{3/2} \to 83577_{1/2}$	1076,1083,1086, 1236	904,1073
0.570815	0.570850	±0.000050	1	$103106^0_{7/2} \to 85592_{5/2}$	1075,1083,1086, 1090,1094,1237	442,443,904, 1073,1098
0.574160	0.574150	±0.000040	1	$112272^0_{3/2} \to 94860_{5/2}$	1080,1083,1086, 1238	904,1073
0.575589	0.575570	±0.000040	1	$100112^0_{5/2} \to 82743_{3/2}$	1079,1083,1086, 1239	904,1073

Tellurium—*continued*

Wavelength (μm)	Measured value (μm)	Uncertainty	Charge	Transition	Comments	References
0.576528	0.576490	±0.000040	1	$112549_{5/2} \rightarrow 95208_{3/2}$	1076,1083,1086, 1240	904,1073
0.585105	0.585100	±0.000040	1	$111947^0_{5/2} \rightarrow 94860_{5/2}$	1079,1083,1086, 1241	904,1073
0.593618	0.593650	±0.000050	1	$99585^0_{3/2} \rightarrow 82743_{3/2}$	1080,1082,1083, 1086,1242	443,904,1073
0.597267	0.597230	±0.000040	1	$111947^0_{5/2} \rightarrow 95208_{3/2}$	1079,1083,1086, 1243	904,1073
0.597471	0.597430	±0.000040	1	$102324^0_{5/2} \rightarrow 85592_{5/2}$	1076,1083,1086, 1244	904,1073
0.601447	0.601470	±0.000040	1	$105583^0_{5/2} \rightarrow 88961_{3/2}$	1080,1083,1086, 1245	904,1073
0.60824	0.608240	±0.000040	1		1076,1083,1086, 1246	904,1073
0.623074	0.623040	±0.000040	1	$105006^0_{3/2} \rightarrow 88961_{3/2}$	1080,1083,1086, 1247	904,1073
0.624549	0.624550	±0.000050	1	$99585^0_{3/2} \rightarrow 83577_{1/2}$	1079,1082,1083, 1086,1248	443,904,1073
0.65850	0.658500	±0.000040	1		1076,1083,1086, 1249	904,1073
0.664859	0.664820	±0.000040	1	$97780^0_{1/2} \rightarrow 82743_{3/2}$	1080,1083,1086, 1250	904,1073
0.667602	0.667650	±0.000040	1	$103936^0_{3/2} \rightarrow 88961_{3/2}$	1080,1083,1086, 1251	904,1073
0.688508	0.688530	±0.000040	1	$100112^0_{5/2} \rightarrow 85592_{5/2}$	1080,1083,1086, 1252	904,1073

Wavelength (μm)	Uncertainty	Charge	Transition	Comments	References
0.703904	±0.000060	1	$97780^0_{1/2} \to 83577_{1/2}$	1080,1082,1083, 1086,1253	443,904,1073
0.780162	±0.000060	1	$105006^0_{3/2} \to 92192_{3/2}$	1080,1083,1086, 1254	904,1073
0.792155	±0.000060	1	$97780^0_{1/2} \to 85160_{3/2}$	1080,1083,1086, 1255	904,1073
0.86044	±0.000060	1		1076,1083,1086, 1256	904,1073
0.87343	±0.000060	1		1076,1083,1086, 1257	904,1073
0.897212	±0.000060	1	$103936^0_{3/2} \to 92793_{5/2}$	1080,1083,1086, 1258	904,1073
0.89982	±0.000060	1		1080,1083,1086, 1259	904,1073
0.937842	±0.000060	1	$99585^0_{3/2} \to 88925_{5/2}$	1080,1083,1086, 1260	904,1073

3.2.3.9 Group VIB Lasers

Table 3.2.9
Chromium

Wavelength (μm)	Measured value (μm)	Uncertainty	Charge	Transition	Comments	References
0.868347	0.868400	±0.000100	1	$(a^5D)4pz\ {}^4D^0_{3/2} \to (a^3D)4sb\ {}^2D_{3/2}$	1076,1087	939

3.2.3.10 Group VII Lasers

Table 3.2.10
Fluorine

Wavelength (μm)	Measured value (μm)	Uncertainty	Charge	Transition	Comments	References
0.275963	0.275959	±0.000006	2	$(^1D)3p\ ^2D^0_{5/2} \to (^1D)3s\ ^2D_{5/2}$	1085,1261	220,1022
0.282613	0.282608	±0.000006	3	$(^2P^0)3p\ ^3D_3 \to (^2P^0)3s\ ^3P^0_2$	1085	220
0.312154	0.312150	±0.000001	2	$(^3P)3p\ ^4D^0_{7/2} \to (^3P)3s\ ^4P_{5/2}$	1084,1090,1094,1262	220,991,1022
0.317417	0.317418	±0.000006	2	$(^3P)3p\ ^2D^0_{5/2} \to (^3P)3s\ ^2P_{3/2}$	1085,1263	220,1022
0.320275	0.320274	±0.000006	1	$(^2D^0)3p\ ^1D_2 \to (^2D^0)3s\ ^1D^0_2$	1264	220,1021
0.402472	0.402478	±0.000006	1	$(^4S^0)3p\ ^3P_2 \to (^4S^0)3s\ ^3S^0_1$	1085,1265	220,1021

Chlorine

Wavelength (μm)	Measured value (μm)	Uncertainty	Charge	Transition	Comments	References
0.263267	0.263269	±0.000001	2	$(^1D)4d\ ^2D_{5/2} \to (^1D)4p\ ^2F^0_{7/2}$	1084,1090,1094	220,991,1020
0.319146	0.319142	±0.000001	2	$(^3P)4p\ ^4S^0_{3/2} \to (^3P)4s\ ^4P_{5/2}$	1084,1085,1090,1094	220,991
0.339288	0.339286	±0.000001	2	$(^1D)4p\ ^2D^0_{3/2} \to (^1D)4s\ ^2D_{3/2}$	1084,1085,1090,1094	220,991
0.339345	0.339344	±0.000001	2	$(^1D)4p\ ^2D^0_{5/2} \to (^1D)4s\ ^2D_{5/2}$	1084,1085,1090,1094	220,991
0.353003	0.353002	±0.000001	2	$(^1D)4p\ ^2F^0_{7/2} \to (^1D)4s\ ^2D_{5/2}$	1084,1085,1090,1094	220,991
0.356068	0.356063	±0.000001	2	$(^1D)4p\ ^2F^0_{5/2} \to (^1D)4s\ ^2D_{3/2}$	1084,1085,1090,1094	220,991
0.360210	0.360210	±0.000006	2	$(^3P)4p\ ^4D^0_{7/2} \to (^3P)4s\ ^4P_{5/2}$	1085	220

Wavelength				Transition	References	
0.361282	±0.000006	0.361210	2	$(^3P)4p\ ^4D^o_{5/2} \rightarrow (^3P)4s\ ^4P_{3/2}$	1084,1085,1090,1094	220
0.362268	±0.000006	0.362269	2	$(^3P)4p\ ^4D^o_{3/2} \rightarrow (^3P)4s\ ^4P_{1/2}$	1085	220
0.372045	±0.000001	0.372044	2	$(^3P)4p\ ^2D^o_{5/2} \rightarrow (^3P)4s\ ^2P_{3/2}$	1085	220,991
0.374880	±0.000001	0.374877	2	$(^3P)4p\ ^2D^o_{3/2} \rightarrow (^3P)4s\ ^2P_{1/2}$	1084,1085,1090,1094	220,991
0.413250	±0.000010	0.413250	1	$(^2D^0)4p\ ^1D_2 \rightarrow (^2D^0)4s\ ^1D^0_2$	1083,1092	1107
0.474042	±0.000010	0.474040	1	$(^2P^0)4p\ ^1P_1 \rightarrow (^2P^0)3d\ ^1D^0_2$	1083,1092	1107
0.476870	±0.000006	0.476874	1	$(^2P^0)4p\ ^3D_2 \rightarrow (^2P^0)4s\ ^3P^0_1$	1083,1092	220,1107
0.478134	±0.000003	0.478134	1	$(^2P^0)4p\ ^3D_3 \rightarrow (^2P^0)4s\ ^3P^0_2$	1083,1092,1094	857,927,997
0.489685	±0.000003	0.489688	1	$(^2D^0)4p\ ^3F_4 \rightarrow (^2D^0)4s\ ^3D^0_3$	1083,1090,1092,1094	857,873,927,997
0.490482	±0.000003	0.490473	1	$(^2D^0)4p\ ^3F_3 \rightarrow (^2D^0)4s\ ^3D^0_2$	1083,1092,1094	857,927,997
0.491780	±0.000003	0.491766	1	$(^2D^0)4p\ ^3F_2 \rightarrow (^2D^0)4s\ ^3D^0_1$	1083,1092,1094	857,927,997
0.507828	±0.000003	0.507830	1	$(^2P^0)4p\ ^3D_3 \rightarrow (^2D^0)4s\ ^3D^0_3$	1083,1092,1094	857,927,997
0.510309	±0.000010	0.510310	1	$(^2D^0)4p\ ^3D_2 \rightarrow (^2D^0)4s\ ^3D^0_2$	1083,1092	1107
0.521792	±0.000003	0.521790	1	$(^4S^0)4p\ ^3P_2 \rightarrow (^4S^0)4s\ ^3S^0_1$	1083,1084,1090,1094	873,927,997,1094
0.522135	±0.000003	0.522130	1	$(^4S^0)4p\ ^3P_1 \rightarrow (^4S^0)4s\ ^3S^0_1$	1083,1084,1090,1092,1094	873,927,997,1094
0.539216	±0.000003	0.539215	1	$(^2D^0)4p\ ^1F_3 \rightarrow (^2D^0)4s\ ^1D^0_2$	1083,1090,1092,1094	857,873,927,997
0.6099472	±0.000003	0.609474	1	$^2D^0)4p\ ^1P_1 \rightarrow (^2D^0)4s\ ^1D^0_2$	1083,1092,1094	857,927,997

Bromine

Wavelength (μm)	Measured value (μm)	Uncertainty	Charge	Transition	Comments	References
	0.236246	±0.000001	3		1084,1090,1266	991
0.258125	0.258125	±0.000001	3		1084,1090,1267	991
0.278762	0.278762	±0.000001	2		1084,1090,1268	991
0.474270	0.474266	±0.000003	1	$(^2P^0)5p\,^3D_3 \rightarrow (^2P^0)5s\,^3P^0_2$	1269	955,1043
0.505465	0.505463	±0.000005	1	$(^2D^0)5p\,^3F_3 \rightarrow (^4S^0)4d\,^3D^0_2$	1270	955,1043
0.518227	0.518238	±0.000002	1	$(^4S^0)5p\,^3P_2 \rightarrow (^4S^0)5s\,^3S^0_1$	1083,1092,1271	857,943,955,1043
0.523823	0.523826	±0.000004	1	$(^4S^0)5p\,^3P_1 \rightarrow (^4S^0)5s\,^3S^0_1$	1083,1092,1272	857,955,1043
0.533205	0.533203	±0.000003	1	$(^2D^0)5p\,^1F_3 \rightarrow (^2D^0)5s\,^1D^0_2$	1083,1092,1273	857,943,955,1043
0.611761	0.611756	±0.000006	1	$(^4S^0)5p\,^5P_2 \rightarrow (^4S^0)5s\,^3S^0_1$	1084,1274	200,956,1043
0.616870	0.616878	±0.000006	1	$(^4S^0)5p\,^5P_1 \rightarrow (^4S^0)5s\,^3S^0_1$	1275	956,1043

Iodine

Wavelength (μm)	Measured value (μm)	Uncertainty	Charge	Transition	Comments	References
0.448855	0.448850	±0.000020	1	$(^2D^0)6p\,^3D_1 \rightarrow (^4S^0)5d\,^5D^0_1$	1076,1083,1087, 1276	993,1037
0.453379	0.453379	±0.000003	2, 3		1277	968
0.467440	0.467440	±0.000003	2, 3		1279	968
0.467553	0.467560	±0.000020	1	$(^2D^0)6p\,^1D_2 \rightarrow (^2D^0)6s\,^1D^0_2$	1076,1083,1087, 1281	993,1037
0.493467	0.493467	±0.000003	2, 3		1282	968
0.498692	0.498670	±0.000020	1	$(^2D^0)6p\,^3D_2 \rightarrow (^4S^0)5d\,^3D^0_1$	1076,1082,1083, 1086,1087, 1094,1284	266,993, 1037,1042
0.52140	0.521430	±0.000020	1	$(^2D^0)6p\,^3D_2 \rightarrow (^4S^0)5d\,^3D^0_3$	1076,1083,1087, 1285	993,1037

				Transition		
0.521626	0.521630	±0.000020	1	$(^2D^0)6p\ ^3F_2 \rightarrow (^4S^0)5d\ ^3D^0_1$	1076,1082,1083, 1086,1087,1088, 1094,1286	266,993,1037, 1042,1105
0.540737	0.540750	±0.000020	1	$(^2D^0)6p\ ^3D_2 \rightarrow (^2D^0)6s\ ^3D^0_2$	1076,1082,1083, 1086,1087,1088, 1090,1094,1287	266,866,867, 910,993,1032, 1037,1042,1105
0.55931	0.559310	±0.000020	1	$(^2D^0)6p\ ^3D_1 \rightarrow (^4S^0)5d\ ^3D^0_1$	1076,1083,1087, 1288	993,1037
0.562569	0.5625		1	$(^4S^0)6p\ ^3P_2 \rightarrow (^4S^0)6s\ ^3S^0_1$	1289	266,993
0.567807	0.567820	±0.000020	1	$(^2D^0)6p\ ^3F_2 \rightarrow (^2D^0)6s\ ^3D^0_2$	1076,1082,1083, 1086,1087,1088, 1090,1094,1290	266,862,866,911, 993,1032,1037, 1042,1105,1141
0.576071	0.576070	±0.000020	1	$(^2D^0)6p\ ^3D_2 \rightarrow (^2D^0)6s^3\ D^0_1$	1076,1082,1083, 1086,1087,1088, 1090,1094,1291	266,862,866,911, 993,1032,1037, 1042,1105,1141
0.606895	0.606900	±0.000020	1	$(^2D^0)6p\ ^3F_2 \rightarrow (^2D^0)6s\ ^3D_1$	1076,1083,1086, 1087,1088, 1094, 1292	993,1037, 1104,1105
0.612750	0.612740	±0.000020	1	$(^2D^0)6p\ ^3D_1 \rightarrow (^2D^0)6s^3\ D^0_2$	1076,1082–1084, 1086–1088,1090, 1094,1293	267,862,866,911, 993,1032,1037, 1042,1105,1141
0.620485	0.620490	±0.000020	1	$(^2D^0)6p\ ^1P_1 \rightarrow (^2D^0)6s\ ^1D^0_2$	1076,1083,1087, 1294	993,1037
0.634001	0.633990	±0.000020	1	$(^2D^0)6p\ ^3F_3 \rightarrow (^2D^0)6s\ ^3D^0_3$	1076,1083,1087, 1295	993,1037
0.635738	0.635737		1	$(^4S^0)8d\ ^5D^0_1 \rightarrow (^2P^0)6p\ ^3P_1$	1296	873,993
0.648899	0.648897		1	$(^4S^0)8d\ ^5D^0_4 \rightarrow (^4S^0)7p\ ^5P_3$	1076,1297	873,993
0.651615	0.651620	±0.000020	1	$(^2D^0)6p\ ^3F_2 \rightarrow 5s5p^5\ ^1P^0_1$	1076,1083,1087, 1088,1094,1298	993,1032,1037, 1103,1105

Iodine—continued

Wavelength (μm)	Measured value (μm)	Uncertainty	Charge	Transition	Comments	References
0.658520	0.658530	±0.000020	1	$(^2D^0)6p\ ^3D_1 \rightarrow (^2D^0)6s\ ^3D^0_1$	1076,1082,1083, 1086,1087,1088, 1090,1094,1299	266,866,867, 993,1032,1042, 1037,1104
0.662236	0.662250	±0.000020	1	$(^2D^0)6p\ ^3D_2 \rightarrow (^2D^0)5d\ ^3D^0_2$	1076,1083,1087, 1300	993,1037
0.667228	0.667227		1	$(^2P^0)6p\ ^3D_1 \rightarrow (^2D^0)5d\ ^1D^0_2$	1301	873,993
0.682520	0.682520	±0.000020	1	$(^2D^0)6p\ ^3F_2 \rightarrow (^2D^0)6s\ ^1D^0_3$	1076,1083,1087, 1302	993,1037,1104
0.690480	0.6904		1	$(^2D^0)6p\ ^3D_2 \rightarrow (^2D^0)6s\ ^1D^0_2$	1082,1085,1303	266,873,993
0.703299	0.703300	±0.000020	1	$(^2D^0)6p\ ^3F_2 \rightarrow (^2D^0)5d\ ^3D^0_2$	1073,1076,1082, 1083,1086,1087, 1088,1094,1304	266,866,867, 993,1032,1037, 1042,1104
0.713898	0.713900	±0.000020	1	$(^2D^0)6p\ ^3D_2 \rightarrow (^2D^0)5d\ ^3D^0_3$	1076,1083,1087, 1094,1305	993,1032, 1037,1104
0.761849	0.761850	±0.000020	1	$(^2D^0)6p\ ^3F_2 \rightarrow (^2D^0)5d\ ^3D^0_3$	1076,1083,1087, 1306	993,1037
0.773577	0.773580	±0.000020	1	$(^2D^0)6p\ ^3D_1 \rightarrow (^2D^0)5d\ ^3D^0_2$	1076,1083,1087, 1307	993,1037
0.817010	0.817020	±0.000020	1	$(^2D^0)6p\ ^3D_2 \rightarrow (^2D^0)5d\ ^3F^0_3$	1076,1083,1087, 1308	993,1037
0.825381	0.825390	±0.000020	1	$(^2D^0)6p\ ^3D_1 \rightarrow (^2D^0)5d\ ^3P^0_0$	1076,1082,1083, 1087,1094,1309	266,993, 1032,1037
0.880428	0.880428	±0.000020	1	$(^2D^0)6p\ ^3F_2 \rightarrow (^2D^0)5d\ ^3F^0_3$	1076,1082,1083, 1087,1094,1310	266,993, 1032,1037
0.887757	0.887740	±0.000020	1	$(^2D^0)6p\ ^3F_3 \rightarrow (^2D^0)5d\ ^3G^0_4$	1076,1083,1087, 1311	993,1037
1.04172	1.041720	±0.000060	2,3		1312	968

Ytterbium

Wavelength (μm)	Measured value (μm)	Uncertainty	Charge	Transition	Comments	References
1.26925	1.271400	±0.000100	1	$(^2F^0_{7/2})6s6p(^3P_0)_{7/2} \rightarrow$ $(^2F^0_{5/2})5d6s^3[9/2]$	1091,1466	404,1002
1.34527	1.345300	±0.000100	1	$(^1S)6p^2P^0_{3/2} \rightarrow (^1S)5d^2D_{3/2}$	1085,1091,1467	404,1002
1.64984	1.649800	±0.000200	1	$(^1S)6p^2P^0_{3/2} \rightarrow (^1S)5d^2D_{5/2}$	1086,1088,1091, 1468	247,1002, 1133
1.804031	1.8057		1	$(^2F^0_{7/2})6s6p(^3P_1)_{5/2} \rightarrow$ $(^2F^0_{7/2})5d6s^3[5/2]$	1469	404,1002
2.1480	2.1480		1 ?		1470	404,1002
2.43775	2.437700	±0.000200	1	$(^1S)6p\,^2\,^1P^0_{1/2} \rightarrow (^1S)5d^2\,^1D_{3/2}$	1088,1471	247,1002,1133

3.2.3.11 Group VIII Lasers

Table 3.2.11
Nickel

Wavelength (μm)	Measured value (μm)	Uncertainty	Charge	Transition	Comments	References
0.796213	0.796070	±0.000250	1	$(^1D)5p\,^2F^0_{7/2} \rightarrow (^1D)5s\,^2D_{5/2}$	1076,1087	936
0.797537	0.797480	±0.000250	1	$(^3P)5p\,^4D^0_{7/2} \rightarrow (^3P)5s\,^4P_{5/2}$	1076,1087	936

Europium

Wavelength (μm)	Measured value (μm)	Uncertainty	Charge	Transition	Comments	References
0.664506	0.6645		1	$z\,^9P_5 \rightarrow a\,^9D^0_6$	1073,1091,1460	1054,1087
0.989827	0.9898		1	$z\,^7P_3 \rightarrow a\,^7D^0_4$	1073,1090,1461	1054,1087
1.00195	1.0020		1	$z\,^7P_4 \rightarrow a\,^7D^0_5$	1073,1086,1090, 1462	1054,1078, 1087
1.01656	1.0166		1	$z\,^7P_4 \rightarrow a\,^7D^0_4$	1073,1086,1090, 1463	1054,1078, 1087
1.36070	1.3610		1	$z\,^9P_4 \rightarrow a\,^7D^0_5$	1073,1086,1090, 1464	1054,1078, 1087
1.47665	1.4770		1	$z\,^9P_3 \rightarrow a\,^7D^0_4$	1090,1465	1054,1087

3.2.3.12 Group VIIIA Lasers

Table 3.2.12
Helium

Wavelength (μm)	Measured value (μm)	Uncertainty	Charge	Transition	Comments	References
0.0164			1	$n=3 \rightarrow n=2$	6922	851

Neon

Wavelength (μm)	Measured value (μm)	Uncertainty	Charge	Transition	Comments	References
0.201844	0.201842	±0.000001	3	$(^1D)3p\,^2D^0_{3/2} \rightarrow (^1D)3s\,^2D_{3/2}$	1090,1314	989,1081
0.202219	0.202219	±0.000001	3	$(^1D)3p\,^2D^0_{5/2} \rightarrow (^1D)3s\,^2D_{5/2}$	1090,1315	989,990,1081
0.206530	0.206530	±0.000001	2, 3		1086,1090,1316	989,990

0.217771	0.217770	±0.000001	2	$(^2D^0)3p\ ^3P_1 \rightarrow (^2D^0)3s\ ^3D^0_2$	1090	989
0.218088	0.218086	±0.000001	2	$(^2D^0)3p\ ^3P_2 \rightarrow (^2D^0)3s\ ^3D^0_3$	1090	989
0.22657	0.226570		4	$0(^2P^0)3p\ ^3D_3 \rightarrow (^2P^0)3s\ ^3P^0_2$	1090	989
0.228579	0.228579	±0.000001	3	$(^1D)3p\ ^2F^0_{7/2} \rightarrow (^1D)3s\ ^2D_{5/2}$	1090	989
0.235252	0.235255	±0.000001	3	$(^3P)3p\ ^4D^0_{5/2} \rightarrow (^3P)3s\ ^4P_{3/2}$	1090	989
0.235796	0.235798	±0.000001	3	$(^3P)3p\ ^4D^0_{7/2} \rightarrow (^3P)3s\ ^4P_{5/2}$	1086,1090,1094	653,989
0.237320	0.237320	±0.000001	3 ?		1090,1318	989
0.247340	0.247340	±0.000001	2	$(^2D^0)3p\ ^1D_2 \rightarrow (^2D^0)3s\ ^1D^0_2$	1086,1090,1094, 1319	653,989
0.247340	0.2474	±0.0001	2	$(^2D^0)3p\ ^1D_2 \rightarrow (^2D^0)3s\ ^1D^0_2$	6583,6584	1757,1758
0.261003	0.260998	±0.000001	2	$(^2D^0)3p\ ^3F_4 \rightarrow (^2D^0)3s\ ^3D^0_3$	1090	989
0.261003	0.2610	±0.0001	2	$(^2D^0)3p\ ^3F_4 \rightarrow (^2D^0)3s\ ^3D^0_3$	6583,6584	1757,1758
0.261341	0.2613	±0.0001	2	$(^2D^0)3p\ ^3F_3 \rightarrow (^2D^0)3s\ ^3D^0_2$	6583,6584	1757,1758
0.261341	0.261340	±0.000010	2	$(^2D^0)3p\ ^3F_3 \rightarrow (^2D^0)3s\ ^3D^0_2$	1090	989
0.2677	0.2677		2	$(^4S^0)3p\ ^3P_{2,0} \rightarrow (^4S^0)3s\ 3s\ ^3S^0_1$	6583,6584	1757,1758
0.267790	0.267792	±0.000001	2	$(^4S^0)3p\ ^3P_{2,0} \rightarrow (^4S^0)3s\ ^3S^0_1$	1086,1090,1092, 1094	450,653, 907,989
0.267864	0.267869	±0.000001	2	$(^4S^0)3p\ ^3P_1 \rightarrow (^4S^0)3s\ 3s\ ^3S^0_1$	1086,1090,1092, 1094	450,653, 907,989
0.277765	0.277763	±0.000001	2	$(^2D^0)3p\ ^3D_3 \rightarrow (^2D^0)3s\ ^3D^0_3$	1090,1094	450,989
0.277765	0.2778	±0.0001	2	$(^2D^0)3p\ ^3D_3 \rightarrow (^2D^0)3s\ ^3D^0_3$	6583,6584	1757,1758
0.286673	0.286673	±0.000001	2	$(^2D^0)3p\ ^1F_3 \rightarrow (^2D^0)3s\ ^1D^0_2$	1090,1094,1320	450,653,989
0.331972	0.331975	±0.000001	1	$(^1D)3p\ ^2P^0_{1/2} \rightarrow (^1D)3s\ ^2D_{3/2}$	1090,1092, 1094,1321	653,907,989, 1030,1153
0.331972	0.3320	±0.0001	1	$(^1D)3p\ ^2P^0_{1/2} \rightarrow (^1D)3s\ ^2D_{3/2}$	6583,6584	1757,1758
0.332373	0.332375	±0.000001	1	$(^3P)3p\ ^2P^0_{3/2} \rightarrow (^3P)3s\ ^2P_{3/2}$	1083,1086,1090- 1092,1094,1322	450,454,907,987, 989,1018,1030, 1088,1137,1169
0.332715	0.332750	±0.000050	1	$(^3P)3p\ ^4D^0_{3/2} \rightarrow (^3P)3s\ ^4P_{3/2}$	1323	450,1030
0.332916	0.332902	±0.000010	1	$(^3P)3d\ ^4D_{7/2} \rightarrow (^3P)3p\ ^4D^0_{7/2}$	1324	1153,1030

Neon—continued

Wavelength (μm)	Measured value (μm)	Uncertainty	Charge	Transition	Comments	References
0.334545	0.333107	±0.000010	2	$(^2P^0)3p\ ^3D_2 \rightarrow (^2P^0)3s\ ^1P^0_1$		1153
0.334545	0.334545	±0.000002	1	$(^1D)3p\ ^2P^0_{3/2} \rightarrow (^1D)3s\ ^2D_{5/2}$	1083,1090,1091, 1092,1094,1325	454,653,907,989, 1030,1153,1169
0.337822	0.337826	±0.000001	1	$(^3P)3p\ ^2P^0_{1/2} \rightarrow (^3P)3s\ ^2P_{1/2}$	1083,1086,1090– 1092,1094,1326	450,454,907,989, 1018,1030,1169
0.339280	0.339280	±0.000001	1	$(^3P)3p\ ^2P^0_{3/2} \rightarrow (^3P)3s\ ^2P_{1/2}$	1083,1086,1090, 1094,1327	450,653,1169, 1018,1030
0.339318	0.339340	±0.000010	1	$(^3P)3d\ ^2D_{3/2} \rightarrow (^3P)3p\ ^2D^0_{5/2}$	1328	1030,1153
0.371308	0.371300	±0.000100	1	$(^3P)3p\ ^2D^0_{5/2} \rightarrow (^3P)3s\ ^2P_{3/2}$	1083,1086,1090, 1329	1030,1169,1172
0.372685	0.37268	±0.0001	1	$(^3P)3p\ ^2D^0_{3/2} \rightarrow (^3P)3s\ ^2P_{1/2}$		1757,1758
0.372710	0.37279	±0.0001	1	$(^3P)3p\ ^2D^0_{3/2} \rightarrow (^3P)3s\ ^2P_{3/2}$		1757,1758

Argon

Wavelength (μm)	Measured value (μm)	Uncertainty	Charge	Transition	Comments	References
0.501716	0.501717	±0.000002	1	$(^1D)4p\ ^2F^0_{5/2} \rightarrow (^3P)3d\ ^2D_{3/2}$	1081,1083,1086, 1351	888,965, 1005,1096,
0.184343	0.184343	±0.000003	4		1330	990
0.211398	0.211398	±0.000001	3		1090	989
0.224884	0.224884	±0.000001	3		1090	989
0.251328	0.251330	±0.000002	3	$(^3P)4p\ ^4S^0_{3/2} \rightarrow (^3P)4s\ ^4P_{5/2}$	1085,1090	989
0.262135	0.262138	±0.000001	3	$(^1D)4p\ ^2D^0_{3/2} \rightarrow (^1D)4s\ ^2D_{3/2}$	1090	989
0.262493	0.262488	±0.000001	3	$(^1D)4p\ ^2D^0_{5/2} \rightarrow (^1D)4s\ ^2D_{5/2}$	1090,1094	653,989
0.275392	0.275388	±0.000001	2	$[(^2D^0)4p\ ^1D_2 \rightarrow (^2D^0)4s\ ^1D_2]$	1081,1083,1089, 1090,1094,1331	450,653,905, 985,989,1128

λ (μm)	±	λ (μm)	Transition	Ref		
0.285521	±0.000001	0.285537	$(^2D^0)4p\,^3P_1 \rightarrow (^2D^0)4s\,^3D^0_2$	2	1090	989
0.288416	±0.000001	0.288422	$(^2D^0)4p\,^3P_2 \rightarrow (^2D^0)4s\,^3D^0_3$	2	1083,1090,1094	1153,653,989
0.291300	±0.000001	0.291292	$(^3P)4p\,^2D^0_{5/2} \rightarrow (^3P)4s\,^2P_{3/2}$	3	1086,1090, 1092,1094	450,653,907, 980,989
0.292627	±0.000001	0.292623	$(^3P)4p\,^2D^0_{3/2} \rightarrow (^3P)4s\,^2P_{1/2}$	3	1090,1094	450,653,989
0.300264	±0.000001	0.300264	$[(^2P^0)4p\,^1P_1 \rightarrow (^2P^0)3d\,^1D^0_2]$	2	1081,1083,1085, 1089,1090, 1092,1094	450,653,907, 989,985
0.302405	±0.000050	0.302400	$(^2P^0)4p\,^3D_3 \rightarrow (^2P^0)4s\,^3P^0_2$	2	1081,1083,1090	450,985
0.305484	±0.000050	0.305480	$(^2P^0)4p\,^3D_2 \rightarrow (^2P^0)4s\,^3P^0_1$	2	1081,1083,1090	450,985
0.333613	±0.000006	0.333621	$(^2D^0)4p\,^3F_4 \rightarrow (^2D^0)4s\,^3D^0_3$	2	1081,1083,1090	450,653985,1169
0.334472	±0.000006	0.334479	$(^2D^0)4p\,^3F_3 \rightarrow (^2D^0)4s\,^3D^0_2$	2	1081,1083,1090	450,653985,1169
0.335849	±0.000006	0.335852	$(^2D^0)4p\,^3F_2 \rightarrow (^2D^0)4s\,^3D^0_1$	2	1081,1083,1090	450,653985,1169
0.351112	±0.000006	0.351112	$(^4S^0)4p\,^3P_2 \rightarrow (^4S^0)4s\,^3S^0_1$	2	1081,1083,1086, 1090,1091,1092, 1332	450,454,929,985, 989,997,1018, 1088,1169
0.351418	±0.000006	0.351415	$(^4S^0)4p\,^3P_1 \rightarrow (^4S^0)4s\,^3S^0_1$	2	1083,1090,1092, 1094	450,907,989,997, 1092,1123
0.357661	±0.000050	0.357690	$(^3P)4d\,^4F_{7/2} \rightarrow (^3P)4p\,^4D^0_{5/2}$	1	1085,1333	450,1005
0.363789	±0.000004	0.363786	$[(^2D^0)4p\,^1F_3 \rightarrow (^2D^0)4s\,^1D^0_2]$	2	1081,1083,1086, 1089,1090,1092, 1334	450,905,907,985, 989,997,1088, 1092,1172
0.379532	±0.000006	0.379528	$(^2P^0)4p\,^3D_3 \rightarrow (^2P^0)3d\,^3P^0_2$	2	1083	450,1123
0.385829	±0.000006	0.385826	$(^2P^0)4p\,^3D_2 \rightarrow (^2P^0)3d\,^3P^0_1$	2	1083	450,1123
0.414671	±0.000004	0.414660	$(^2D^0)4p\,^3P_2 \rightarrow (^2P^0)4s\,^3P^0_2$	2	1083	450,1123
0.418298	±0.000006	0.418292	$[(^2D^0)4p\,^1P_1 \rightarrow (^2D^0)4s\,^1D^0_2]$	2	1083,1089, 1337	450,1081, 1128,1169
0.437075	±0.000006	0.437073	$(^1D)4p\,^2D^0_{3/2} \rightarrow (^3P)3d\,^2D_{3/2}$	1	1083,1338	450,1005,1118
0.438375	±0.000060	0.438360	$(^3P)4p\,^4S^0_{3/2} \rightarrow (^3P)4s\,^2P_{3/2}$	1	1339	1005,1016
0.448181	±0.000100	0.448200	$(^1D)4p\,^2D^0_{5/2} \rightarrow (^3P)3d\,^2D_{5/2}$	1	1083,1340	857,973,1005

Argon—continued

Wavelength (μm)	Measured value (μm)	Uncertainty	Charge	Transition	Comments	References
0.454505	0.454504	±0.000010	1	$(^3P)4p\ ^2P^0_{3/2} \rightarrow (^3P)4s\ ^2P_{3/2}$	1073,1081,1083, 1087,1094,1341	450,880,888, 1005,1096
0.457935	0.457936	±0.000016	1	$(^3P)4p\ ^2S^0_{1/2} \rightarrow (^3P)4s\ ^2P_{1/2}$	1073,1081,1083, 1086,1087,1342	450,880,888, 1005,1096
0.460956	0.460957	±0.000010	1	$(^1D)4p\ ^2F^0_{7/2} \rightarrow (^1D)4s\ ^2D_{5/2}$	1343	450,1005
0.465789	0.465795	±0.000002	1	$(^3P)4p\ ^2P^0_{1/2} \rightarrow (^3P)4s\ ^2P_{3/2}$	1081,1083,1086, 1344	888,965,1005, 1096
0.472686	0.472689	±0.000004	1	$(^3P)4p\ ^2D^0_{3/2} \rightarrow (^3P)4s\ ^2P_{3/2}$	1081,1083,1345	888,1005,1096
	0.4765		1	$(^3P)4p\ ^2P^0_{3/2} \rightarrow (^3P)4s\ ^2P_{1/2}$	6922	842
0.476486	0.476488	±0.000004	1	$(^3P)4p\ ^2P^0_{3/2} \rightarrow (^3P)4s\ ^2P_{1/2}$	1073,1081,1083, 1085,1086,1087, 1090-2,1346	880,888,907,929, 965,970,1005, 1096,1142
0.487986	0.487986	±0.000004	1	$(^3P)4p\ ^2D^0_{5/2} \rightarrow (^3P)4s\ ^2P_{3/2}$	1081,1083,1085, 1086,1090,1091, 1347	888,965, 1005,1096
0.488903	0.488906	±0.000006	1	$(^3P)4p\ ^2P^0_{1/2} \rightarrow (^3P)4s\ ^2P_{1/2}$	1083,1348	450,1005,1118
0.496507	0.496509	±0.000002	1	$(^3P)4p\ ^2D^0_{3/2} \rightarrow (^3P)4s\ ^2P_{1/2}$	1081,1083,1086, 1091,1349	888,965, 1005,1096
0.506204	0.506210	±0.000025	1	$(^3P)4p\ ^4P^0_{3/2} \rightarrow (^3P)4s\ ^4P_{1/2}$	1092,1352	975,1005,1111
0.514179	0.514180	±0.000005	1	$(^1D)4p\ ^2F^0_{7/2} \rightarrow (^3P)3d\ ^2D_{5/2}$	1083,1353	450,1118,1005
0.514532	0.514533	±0.000002	1	$(^3P)4p\ ^4D^0_{5/2} \rightarrow (^3P)4s\ ^2P_{3/2}$	1081,1083,1086, 1090,1354	888,965, 1005,1096
0.528690	0.528700	±0.000100	1	$(^3P)4p\ ^4D^0_{3/2} \rightarrow (^3P)4s\ ^2P_{1/2}$	1081,1083,1355	888,1005,1096
0.550220	0.550220	±0.000050	2	$(^2D^0)4p\,^3D_3 \rightarrow (^2P^0)4s\,^3P^0_2$		450
0.648308	0.648280	±0.000020	1	$(^3P)4p\ ^2S^0_{1/2} \rightarrow (^3P)3d\ ^2P_{3/2}$	1073,1083,1087, 1356	941,1005,1077
0.673000	0.673000	±0.000050	0		1357	921

Wavelength (μm)	Measured value (μm)	Uncertainty	Charge	Transition	Comments	References
0.686127		±0.000020	1	$(^3P)4p\ ^2P^0_{3/2} \rightarrow (^3P)3d\ ^2P_{3/2}$	1073,1083,1087, 1358	941,1005,1077
0.734805		±0.000005	1	$(^1D)3d\ ^2D_{5/2} \rightarrow (^3P)4p\ ^2D^0_{5/2}$	1091,1359	454,1005
0.750514		±0.000005	1	$(^1D)3d\ ^3P_{3/2} \rightarrow (^3P)4p\ ^2S^0_{1/2}$	1091,1360	454,1005
0.877186		±0.000300	1	$[(^3P)4p\ ^2P^0_{3/2} \rightarrow (^1D)4s\ ^2D_{5/2}]$	1073,1083,1087, 1089,1361	1005,1075, 1077,1117
1.092344		±0.000100	1	$(^3P)4p\ ^2P^0_{3/2} \rightarrow (^3P)3d\ ^2D_{5/2}$	1083,1090,1362	594,1005, 1117,1143

Krypton

Wavelength (μm)	Measured value (μm)	Uncertainty	Charge	Transition	Comments	References
	0.0907		2	$4s^04p^6\ ^1S_0 \rightarrow 4s^14p^5\ ^1P_1$	6925	892
0.175641	0.175641	±0.000003	3		1090	989,990
0.183243	0.183243	±0.000003	4		1090	989,990
0.195027	0.195027	±0.000003	3		1090	989,990
0.196808	0.196808	±0.000003	3		1090	989,990
0.205108	0.205108	±0.000001	3		1090	989,990
0.219192	0.219192	±0.000001	3		1090	989,990
0.225464	0.225464	±0.000001	3		1090	989
0.233848	0.233848	±0.000001	3		1090	989
0.241784	0.241784	±0.000001	2, 3		1090,1364	989
0.264936	0.264936	±0.000001	3		1085,1090,1094	653,989,1125
0.266440	0.266440	±0.000001	3		1085,1090,1094	653,989,1125
0.274138	0.274138	±0.000001	3		1090,1094	653,989
0.304970	0.304970	±0.000002	2		1090,1094	450,653,989
0.312436	0.312436	±0.000001	2		1081,1083, 1090,1094	653,985, 989

Krypton—continued

Wavelength (μm)	Measured value (μm)	Uncertainty	Charge	Transition	Comments	References
0.323951	0.323951	±0.000001	2	$(^2P^0)5p\ ^1D_2 \rightarrow (^2D^0)5s\ ^1P^0_1$	1081,1083,1090, 1094	450,653, 985,989
0.337496	0.337500	±0.000050	2	$(^2P^0)5p\ ^3D_3 \rightarrow (^2P^0)5s\ ^3P^0_2$	1081,1083,1090	450,985,1169
0.350742	0.350742	±0.000001	2	$(^4S^0)5p\ ^3P_2 \rightarrow (^4S^0)5s\ ^3S^0_1$	1081,1083,1086, 1090,1094,1366	450,985,989, 1018,1088,1092, 1169
0.356432	0.356420	±0.000006	2	$(^4S^0)5p\ ^3P_1 \rightarrow (^4S^0)5s\ ^3S^0_1$	1081,1083,1086, 1090,1094,1367	539,985,989, 1088,1166
0.377134	0.377134	±0.000005	1	$(^1D)4d\ ^2S_{1/2} \rightarrow (^3P)5p\ ^4P^0_{3/2}$	1090,1091,1368	454,1007
0.406737	0.406736	±0.000006	2	$(^2D^0)5p\ ^1F_3 \rightarrow (^2D^0)5s\ ^1D^0_2$	1081,1083,1090, 1369	450,974,1092, 1143,1169,1172
0.413133	0.413138	±0.000006	2	$(^4S^0)5p\ ^5P_2 \rightarrow (^4S^0)5s\ ^3S^0_1$	1081,1083,1090, 1370	450,974, 1092,1169
0.415444	0.415445	±0.000004	2	$(^2D^0)5p\ ^3F_3 \rightarrow (^2D^0)5s\ ^1D_2$	1081,1083	450,1123
0.417179	0.417181	±0.000010	2	$(^4S^0)5p\ ^5P_1 \rightarrow (^4S^0)5s\ ^3S^0_1$		450
0.422658	0.422651	±0.000006	2	$(^2D^0)5p\ ^3F_2 \rightarrow (^2D^0)4d\ ^3D^0_1$	1081	450
0.431780	0.431800	±0.000020	1	$(^3P)6s\ ^4P_{5/2} \rightarrow (^3P)5p\ ^4P^0_{5/2}$	1076,1082,1083, 1087,1371	880,978,1007, 1019,1077,1151
0.438653	0.438610	±0.000020	1	$(^3P)6s\ ^4P_{5/2} \rightarrow (^3P)5p\ ^4P^0_{3/2}$	1076,1082,1083, 1085,1087,1372	978,880,1007, 1077,1151
0.457720	0.457720	±0.000010	1	$(^1D)5p\ ^2F^0_{7/2} \rightarrow (^1D)5s\ ^2D_{5/2}$	1076,1083,1087, 1094,1373	450,861,1007, 1077,1116,
0.458285	0.458300	±0.000100	1	$(^3P)6s\ ^4P_{3/2} \rightarrow (^3P)5p\ ^4D^0_{5/2}$	1076,1082,1083, 1087,1374	941,978,1007, 1077,1128,1151
0.461528	0.461520	±0.000010	1	$(^3P)5p\ ^2P^0_{3/2} \rightarrow (^3P)5s\ ^2P_{3/2}$	1375	1007,1014
0.461915	0.461917	±0.000010	1	$(^3P)5p\ ^2D^0_{5/2} \rightarrow (^3P)5s\ ^2P_{3/2}$	1083,1094,1376	450,1007, 1116,1117

Wavelength (μm)	Uncertainty	Wavelength (μm)	No.	Transition	Ref.	Ref.
0.463388	±0.000006	0.463392	1	$(^1D)5p\ ^2F^0_{5/2} \rightarrow (^1D)5s\ ^2D_{3/2}$	1083,1094,1377	450,1007, 1116,1118
0.465016	±0.000010	0.465016	1	$(^3P)5p\ ^2P^0_{1/2} \rightarrow (^3P)5s\ ^4P_{1/2}$	1378	450,1007
0.468041	±0.000006	0.468045	1	$(^3P)5p\ ^2S^0_{1/2} \rightarrow (^3P)5s\ ^2P_{1/2}$	1081,1083,1094, 1379	450,1007, 1116,1117
0.469443	±0.000020	0.469410	1	$(^3P)6s\ ^4P_{5/2} \rightarrow (^3P)5p\ ^4D^0_{7/2}$	1076,1082,1083, 1087,1090,1092, 1380	940,941,978, 1007,1019,1049, 1077,1112,1151
0.471046	±0.000060	0.471030	2	$(^2D^0)5p\ ^3F_4 \rightarrow (^2D^0)4d\ ^3D^0_3$		1016
0.475447	±0.000030	0.475450	2	$(^2D^0)5p\ ^1F_3 \rightarrow (^2D^0)4d\ ^3D^0_3$		1016
0.476243	±0.000006	0.476244	1	$(^3P)5p\ ^2D^0_{3/2} \rightarrow (^3P)5s\ ^2P_{1/2}$	1081,1083,1092, 1094,1381	450,861,933, 1007,1116
0.476573	±0.000010	0.476571	1	$(^3P)5p\ ^4D^0_{5/2} \rightarrow (^3P)5s\ ^4P_{3/2}$	1083,1094,1382	450,1007, 1116,1117
0.479633	±0.000060	0.479630	1	$(^3P)5d\ ^4D_{1/2} \rightarrow (^3P)5p\ ^4S^0_{3/2}$	1383	1007,1016
0.482518	±0.000006	0.482518	1	$(^3P)5p\ ^4S^0_{3/2} \rightarrow (^3P)5s\ ^2P_{1/2}$	1081,1083,1094, 1384	450,861, 1007,1116
0.484660	±0.000006	0.484666	1	$(^3P)5p\ ^2P^0_{1/2} \rightarrow (^3P)5s\ ^2P_{3/2}$	1083,1385	450,1007,1117
0.501645	±0.000010	0.501640	2	$(^2D^0)5p\ ^1D_2 \rightarrow (^2P^0)4d\ ^1F^0_3$	1083,1386	857,1007,1118
0.502239	±0.000100	0.502200	1	$(^3P)5p\ ^4D^0_{3/2} \rightarrow (^3P)5s\ ^2P_{3/2}$	1387	1007,1016,1157
0.503747	±0.000060	0.503750	1			
0.512572	±0.000010	0.512600	1	$(^3P)6s\ ^4P_{3/2} \rightarrow (^3P)5p\ ^4D^0_{3/2}$	1076,1082,1083, 1087,1388	880,978, 1007,1151
0.520832	±0.000004	0.520832	1	$(^3P)5p\ ^4P^0_{3/2} \rightarrow (^3P)5s\ ^3P_{3/2}$	1081,1083,1092, 1094,1389	450,933,1007, 1116,1117
0.521793	±0.000040	0.521820	1	$(^3P)5p\ ^4D^0_{1/2} \rightarrow (^3P)5s\ ^2P_{1/2}$	1390	1007,1016
0.530865	±0.000004	0.530868	1	$(^3P)5p\ ^4P^0_{5/2} \rightarrow (^3P)5s\ ^4P_{3/2}$	1081,1083,1094, 1391	450,1007, 1116,1117
0.550143	±0.000050	0.550150	2	$(^2D^0)5p\ ^3F_3 \rightarrow (^2P^0)4d\ ^3D^0_2$	1085	1016
0.559732	±0.000100	0.559770	2	$(^2D^0)5p\ ^3P_2 \rightarrow (^2P^0)5s\ ^3P^0_2$	1085	1016

Krypton—continued

Wavelength (μm)	Measured value (μm)	Uncertainty	Charge	Transition	Comments	References
0.568188	0.568192	±0.000004	1	$(^3P)5p\ ^4D^0_{5/2} \rightarrow (^3P)5s\ ^2P_{3/2}$	1081,1083,1090,1092,1094,1392	450,861,933,1007,1116,1117
0.57529	0.575340	±0.000050	1	$(^3P)5d\ ^4D^0_{3/2} \rightarrow (^1D)5p\ ^2P^0_{3/2}$	1083,1094,1393	857,921,1007
0.593529	0.593530	±0.000060	1		1395	1007,1016
0.593506	0.593530	±0.000060	2	$(^2D^0)5p\ ^3P_2 \rightarrow (^2P^0)5s\ ^1P^0_1$	1394	1016
0.603811	0.603760	±0.000080	1		1397	1007,1015
0.603716	0.603760	±0.000080	2	$(^3P_2)4f[2]^0_{3/2} \rightarrow (^1D)4d\ ^2P_{3/2}$	1396	1015
0.6072	0.607200	±0.000100	2	$(^2D^0)5p\ ^3P_1 \rightarrow (^2P^0)4d\ ^3D^0_1$	1398	656
0.616880	0.616880	±0.000050	1	$(^1D)5p\ ^2F^0_{5/2} \rightarrow (^3P)4d\ ^4P_{3/2}$	1083,1399	1007,1118
0.631024	0.631030	±0.000080	2	$(^2D^0)5p\ ^3P_2 \rightarrow (^2P^0)4d\ ^3D^0_1$		1015
0.631276	0.631260	±0.000080	1	$(^1D)5p\ ^2P^0_{3/2} \rightarrow (^3P)4d\ ^2P_{3/2}$	1083,1400	921,926,1015
0.641660	0.641700	±0.000100	1	$(^3P)5p\ ^4P^0_{5/2} \rightarrow (^3P)5s\ ^2P_{3/2}$	1401	656,1007
0.647088	0.647100	±0.000050	1		1081,1083,1090,1402	1007,1116,1118
0.651016	0.651000	±0.000010	1	$(^3P)6s\ ^4P_{5/2} \rightarrow (^3P)5p\ ^4S^0_{3/2}$	1076,1083,1087,1403	880,1007,1077
0.657008	0.657000	±0.000050	1	$(^1D)5p\ ^2D^0_{5/2} \rightarrow (^3P)4d\ ^2D_{5/2}$	1083,1404	1007,1116,1118
0.660275	0.660280	±0.000080	1	$(^1D)5p\ ^2P^0_{1/2} \rightarrow (^3P)4d\ ^2D_{5/2}$	1083,1406	921,1007,1015
0.660293	0.660280	±0.000080	2	$[(^2D^0)5p\ ^3P_2 \rightarrow (^2P^0)4d\ ^1F^0_3]$	1083,1405	1015,921
0.676442	0.676457	±0.000010	1	$(^3P)5p\ ^4P^0_{1/2} \rightarrow (^3P)5s\ ^2P_{1/2}$	1081,1083,1094,1407	450,861,1007,1116
0.687085	0.687096	±0.000010	1	$(^1D)5p\ ^2F^0_{5/2} \rightarrow (^3P)4d\ ^2D_{3/2}$	1083,1094,1408	450,1007,1116,1118
0.743578	0.743576	±0.000001	1	$(^1D)4d\ ^2D_{5/2} \rightarrow (^3P)5p\ ^4D^0_{5/2}$	1090,1092,1094,1409	989,1007,1089
0.752546	0.752550	±0.000010	1	$(^3P)5p\ ^4P^0_{3/2} \rightarrow (^3P)5s\ ^2P_{1/2}$	1081,1083,1410	947,1007
0.79314	0.79314		1	$(^1D)5p\ ^2F^0_{7/2} \rightarrow (^3P)4d\ ^2F_{5/2}$	1081,1083,1411	973,1007
0.799322	0.799300	±0.000050	1	$(^3P)5p\ ^4P^0_{3/2} \rightarrow (^3P)4d\ ^4D_{1/2}$	1081,1083,1412	973,1007,1116

Wavelength (μm)	Measured value (μm)	Uncertainty	Charge	Transition	Comments	References
0.828034	0.828030	±0.000010	1	$(^1D)5p\ ^2F^0_{5/2} \to (^3P)4d\ ^2F_{5/2}$	1083,1094,1413	947,973,1007
0.8334	0.8334		1		1083,1414	1007,1152
0.847333	0.8473		1	$(^1D)4d\ ^2D_{3/2} \to (^3P)5p\ ^4D^0_{3/2}$	1415	1007,1089
0.8589	0.858900	±0.000300	2		1416	1075
0.869014	0.86901		1	$(^3P)5p\ ^2P^0_{1/2} \to (^1D)5s\ ^2D_{3/2}$	1083,1417	973,1007
0.897869	0.89784		1	$(^1D)4d\ ^2D_{5/2} \to (^3P)5p\ ^4D^0_{3/2}$	1085,1418	1007,1089
1.06596	1.06596		1	$(^1D)4d\ ^2D_{5/2} \to (^3P)5p\ ^2P^0_{3/2}$	1419	1007,1089
1.329404	1.3295		1	$(^1D)4d\ ^2D_{5/2} \to (^3P)5p\ ^2D^0_{3/2}$	1420	1007,1089

Xenon

Wavelength (μm)	Measured value (μm)	Uncertainty	Charge	Transition	Comments	References
0.1089			2	$5s^05p^6\ ^1S_0 \to 5s^15p^5\ ^1P_1$	6923	892,896,1756
0.223244	0.223244	±0.000001	2, 3		1081,1090,1421	989
0.231536	0.231536	±0.000001	2, 3		1081,1086,1090, 1423	989
0.24433	0.24433		3			1487
0.252666	0.252666	±0.000001	3		1090	989
0.25563	0.25563		3			1487
0.269194	0.269194	±0.000001	3		1089,1090,1094	653,874,983,989
0.276778	0.27680		2		6595	1487
0.281968	0.28199		2		6595	1487
0.282251	0.28227		2		6595	1487
0.287405	0.28741		2		6595	1487
0.295478	0.29547		2		6595	1487
0.297051	0.29707		2		6595	1487
0.301418	0.30142		2	$(^2P^0)6p\ ^3D_1 \to (^2D^0)5d\ ^3D^0_2$		
0.30438	0.30438		2		6595	1487

Xenon—continued

Wavelength (μm)	Measured value (μm)	Uncertainty	Charge	Transition	Comments	References
0.307974	0.307974	±0.000002	2		1081,1085,1089, 1090,1094	450,653,874, 983,989
0.310863	0.31089		3		6595	1487
0.312569	0.31260		2		6595	1487
0.324692	0.324692	±0.000001	3		1081,1085,1089, 1090,1094	653,983,929, 983,1024
0.330596	0.330596	±0.000001	3		1085,1089,1094	450,653,923, 983,989
0.330596	0.33060		3		6592,6593	1487
0.333087	0.333087	±0.000002	3		1081,1083,1089, 1090,1094,1426	450,653,874,983, 989,1024,1122
0.334974	0.334974	±0.000006	2		1089	653,874,923,983
0.345424	0.345425	±0.000001	2	$(^2D^0)6p\ ^1D_2 \rightarrow (^2D^0)6s\ ^1D_2$	1083,1090,1094, 1427	653,989, 1024,1169
0.348322	0.348331	±0.000003	3		1089,1094	450,653,874, 923,983
0.354233	0.354231	±0.000005	2	$(^2D^0)6p\ ^3P_2 \rightarrow (^2D^0)6s\ ^3D_3$	1091	454
0.359661	0.359600	±0.000100	2		1083,1089,1094	1169,923
0.364548	0.364548	±0.000001	3		1081,1085,1086, 1089,1090,1094, 1428	653,874,923,989, 983,1024,1025, 1092,1153
0.365461	0.36545		2	$(^2D^0)6p\ ^3P_1 \rightarrow (^2D^0)5d\ ^3S_1$		
0.365461	0.36545		2	$(^2D^0)6p\ ^3P_1 \rightarrow (^2P^0)5d\ ^3S_1$		
0.366920	0.366920	±0.000003	2		1083,1089	653,874,923, 983,1024,1123
0.367662	0.36766		2	$(^4S^0)6p\ ^3P_0 \rightarrow (^4S^0)6s\ ^3S_1$		

0.374571	0.374573	±0.000006	2	$(^2D^0)6p\,^1D_2 \rightarrow (^2D^0)5d\,^1D_2$	108i,1083,1090	653,1024, 1088,1169
0.375994	0.375994	±0.000003	3		1085,1089,1429	450,874,923, 983,1117,1127
0.376226	0.37622		2	$(^2P^0)6p\,^3D_2 \rightarrow (^2P^0)6s\,^3P_1$	6595	1487
0.377629	0.37765		2			1487
0.378097	0.378099	±0.000002	2	$(^4S^0)6p\,^3P_2 \rightarrow (^4S^0)6s\,^3S^0_1$	1083,1086,1090, 1094,1430	450,653,989, 1024,1088,1169
0.380329	0.380329	±0.000003	3		1089	653,874,923, 983,1024,1025
0.380329	0.38035		3			1487
0.384152	0.384100	±0.000100	2	$(^2D^0)6p\,^3F_2 \rightarrow (^2D^0)5d\,^3D^0_2$	1083,1431	1123
0.384186	0.384100	±0.000100	2	$(^2D^0)6p\,^3P_1 \rightarrow (^2D^0)5d\,^3D^0_2$	1083,1432	1123
0.397301	0.397301	±0.000003	3		1089,1433	653,874,923, 983,1024,1025
0.39732	0.39732		3		6592,6594	1487
0.399255	0.399300	±0.000100	2	$[(^4S^0)6p\,^5P_2 \rightarrow (^4S^0)5d\,^3D^0_1]$	1083,1089,1435	1016,1123
0.399285	0.399300	±0.000100	2	$(^2D^0)6p\,^3P_1 \rightarrow (^2D^0)6s\,^1D_2$	1083,1089,1434	1016,1123
0.405005	0.404990	±0.000020	2	$(^4S^0)6p\,^3P_1 \rightarrow (^4S^0)6s\,^3S^0_1$	1090	1015,1092
0.406041	0.406048	±0.000006	2	$(^2P^0)6p32_1 \rightarrow (^2P^0)5d25^0_1$	1083,1090,1436	450,989,1025, 1024,1092,1150, 1153,1169
0.414199	0.41420		2	$(^2P^0)26_1 \rightarrow (^2P^0)6s\,^3P^0_2$		1487
0.414572	0.414530	±0.000060	2	$(^2D^0)6p\,^3D_2 \rightarrow (^2D^0)5d\,^3D^0_1$	1083	1016,1123
0.421401	0.421405	±0.000006	2	$(^2D^0)6p\,^3P_2 \rightarrow (^2D^0)5d\,^3D^0_3$	1083,1086,1090, 1437	450,1169
0.424024	0.424026	±0.000010	2	$(^2D^0)6p\,^1D_2 \rightarrow (^2P^0)5d17^0_3$	1083,1090,1438	450,1169
0.427259	0.427260	±0.000006	2	$(^2D^0)6p\,^3F_4 \rightarrow (^2D^0)5d\,^3D^0_3$	1083,1090,1439	450,1169
0.428588	0.428592	±0.000006	2	$(^2D^0)6p\,^3D_3 \rightarrow (^2D^0)6s\,^1D^0_2$		450
0.429639	0.429633	±0.000005	1	$(^3P)7s\,^4P_{1/2} \rightarrow (^3P)6p\,^4P^0_{3/2}$	1090,1091	454,1092

Xenon—*continued*

Wavelength (μm)	Measured value (μm)	Uncertainty	Charge	Transition	Comments	References
0.430575	0.430575	±0.000003	3		1085,1086,1089, 1090	450,874,923,983, 989,1024,1092, 1117,1150
0.441314	0.441300	±0.000060	2		1089	923,1016
0.443415	0.443422	±0.000010	2	$(^2D^0)6p\ ^3F_2 \rightarrow (^2D^0)5d\ ^3D^0_1$	1083	450,921
0.450345	0.450350	±0.000060	2	$(^2P^0)6p\ 32_1 \rightarrow (^2P^0)5d\ 27^0_2$		1016
0.455874	0.455874	±0.000006	3		1089,1090	922,923,983,1092
0.460303	0.460302	±0.000004	1	$(^3P)6p\ ^4D^0_{3/2} \rightarrow (^3P)6s\ ^4P_{3/2}$	1083,1092,1094, 1440	450,884, 1116,1117
0.464740	0.464740	±0.000004	3	0.464740	1089,1093,1441	450,923,983,874, 1117,1127
0.465025	0.465025	±0.000001	2, 3	0.465025	1089,1093,1442	450,874,923, 983,1117,1127
0.467368	0.467373	±0.000006	2	$(^2D^0)6p\ ^1F_3 \rightarrow (^2D^0)6s\ ^1D^0_2$	1083,1086,1090	450,989,1092, 1101,1118
0.468354	0.468357	±0.000006	2	$(^4S^0)6p\ ^5P_2 \rightarrow (^4S^0)6s\ ^3S^0_1$	1083	450,1122
0.472360	0.472357	±0.000005	2	$(^4S^0)6p\ ^5P_1 \rightarrow (^4S^0)6s\ ^3S^0_1$	1089,1094	923,983,1014
0.474894	0.474894	±0.000001	2		1083,1089,1094	923,983,874,1118
0.479448	0.479450	±0.000060	2	$(^2D^0)6p\ ^3D_1 \rightarrow (^2D^0)5d\ ^3D^0_1$		1016
0.486249	0.486200	±0.000100	1	$(^3P)7s\ ^4P_{5/2} \rightarrow (^3P)6p\ ^4P^0_{5/2}$	1077,1080,1082, 1083,1087	979,1049,1151
0.486946	0.486948	±0.000006	2	$(^2D^0)6p\ ^3F_3 \rightarrow (^2D^0)5d\ ^3D^0_2$	1083	450,1118
0.488730	0.488700	±0.000100	1	$(^3P)6p\ ^2P_{3/2} \rightarrow (^3P)6s\ ^2P_{3/2}$	1083	857
0.495413	0.495418	±0.000003	3		1083,1086,1089, 1090,1444	450,874,923,983, 1024,1025,1074, 1092,1101,1122, 1150,1487

λ (µm)	λ (µm)	Uncertainty	Spectrum	Transition	References	References
0.496508	0.496508	±0.000006	1	$(^1D)7s\ ^2D_{3/2} \rightarrow (^1D)6p\ ^2P^0_{3/2}$	1083	450,857
0.497270	0.497271	±0.000005	1	$(^1D)6p\ ^2P^0_{3/2} \rightarrow (^3P)5d\ ^2D_{5/2}$	1091,1094	454,911,1150
0.500774	0.500780	±0.000003	3		1081,1083,1086, 1089,1090,1445	450,874,923,983, 1024,1025,1074, 1101,1122, 1150,1487
0.504492	0.504489	±0.000006	1	$(^1D)6p\ ^2P^0_{1/2} \rightarrow (^1D)6s^2\ D_{3/2}$	1083,1092,1094	450,884,1116–7
0.515704	0.515704	±0.000006	3		1089	922,923,983
0.515902	0.515908	±0.000003	3		1081,1083,1086, 1089,1090,1446	450,874,923,983, 1024,1025,1074, 1101,1122, 1150,1487
0.522364	0.522340	±0.000060	2	$(^2D^0)6p\ ^1F_3 \rightarrow (^2D^0)5d\ ^1D^0_2$		1016
0.523893	0.523889	±0.000006	2	$(^2D^0)6p\ ^3P_2 \rightarrow (^2P^0)5d\ 13^0_1$	1082,1083,1447	450,1074,1118
0.525630	0.525650	±0.000060	3		1089,1448	923,1016,1487
0.526017	0.526017	±0.000003	3		1081,1083,1085, 1086,1089,1090, 1449	450,874,922,923, 983,989,1010, 1024,1025,1074, 1101,1117,1124, 1127,1150
0.526043	0.526043	±0.000003	1	$(^3P)6p\ ^2P^0_{3/2} \rightarrow (^3P)6s\ ^2P_{1/2}$	1083,1089	874,923,983, 1117,1118,1127
0.526195	0.526150	±0.000100	1	$(^1D)6p\ ^2D^0_{3/2} \rightarrow (^1D)6s\ ^2D_{3/2}$	1077,1083,1087, 1092,1094,1450	450,884,1077, 1116,1117
0.531389	0.531400	±0.000100	1	$(^3P)7s\ ^4P_{5/2} \rightarrow (^3P)6p\ ^4D^0_{7/2}$	1077,1082,1083, 1086,1087	979,1049,1077, 1128,1151
0.534334	0.534334	±0.000005	3		1089,1094	922,923,983,1074

Xenon—continued

Wavelength (μm)	Measured value (μm)	Uncertainty	Charge	Transition	Comments	References
0.535290	0.535290	±0.000003	3		1081,1083,1086, 1089,1090,1451	450,874,922,929, 942,983,1024, 1025,1074,1101, 1122,1150,1487
0.536706	0.536700	±0.000060	2	$(^2D^0)6p\ ^3F_2 \rightarrow (^2D^0)5d\ ^3D^0_2$	1085	1016
0.539460	0.539460	±0.000003	3		1081,1083,1085, 1086,1089,1090, 1452	450,874,922,923, 929,942,983,1024 1025,1074,1101, 1122,1150,1487
0.540100	0.540090	±0.000030	2	$(^2D^0)6p\ ^3P_2 \rightarrow (^2P^0)5d\ 15^0_2$		1012
0.541353	0.541350	±0.000060	2	$(^2D^0)6p\ ^3P_2 \rightarrow (^2P^0)5d\ 17^0_3$		1016
0.541916	0.541915	±0.000006	1	$(^3P)6p\ ^4D^0_{5/2} \rightarrow (^3P)6s\ ^4P_{3/2}$	1083,1088,1092, 1453	450,884,1116, 1117,1167
0.545433	0.545460	±0.000060	2	$(^4S^0)6d\ ^5D^0_0 \rightarrow (^2D^0)6p\ 4_1$	1085,1086,1090	1016,1101
0.549942	0.549931	±0.000004	3		1089,1094	922,923,983,1074
0.552437	0.552450	±0.000050	1	$(^2D^0)6p\ ^1D_2 \rightarrow (^2P^0)6s\ ^3P_2$	1083	1118
0.559227	0.559235	±0.000005	3		1085,1089, 1094	904,923,967, 983,1074
0.565937	0.565900	±0.000100	1	$(^3P)6p\ ^2P^0_{1/2} \rightarrow (^3P)5d\ ^4P_{1/2}$	1083,1085,1089	857,923,983,1117
0.572690	0.572700	±0.000100	1	$(^1D)6p\ ^2D^0_{5/2} \rightarrow (^1D)5d\ ^2F_{5/2}$	1077,1083,1085, 1087,1089	857,1077, 1117,1151
0.575102	0.575100	±0.000100	1	$(^3P)6p\ ^2D^0_{3/2} \rightarrow (^3P)5d\ ^4P_{1/2}$	1083,1085	857,1117
0.589328	0.589330	±0.000003	1	$(^1D)6p\ ^2P^0_{3/2} \rightarrow (^1D)5d\ ^2D_{5/2}$	1089,1091,1094	454,874
0.595567	0.595567	±0.000003	3		1081,1086,1089, 1090,1454	450,874,922,923, 983,1010,1024, 1074,1101,1150
0.597111	0.597112	±0.000006	1	$(^1D)6p\ ^2P_{3/2} \rightarrow (^1D)6s\ ^2D_{3/2}$	1083,1092,1094	450,884,1116–17

λ	λ	Uncert.		Transition		
0.609361	0.609400	±0.000100	1	$(^3P)7s\ ^4P_{3/2} \rightarrow (^3P)6p\ ^4D^0_{3/2}$	1077,1082,1083, 1087	1077,1151
0.617615	0.617619	±0.000003	2		1083,1089,1094	874,921,923, 983,1015
0.623825(	0.623890	±0.000080	2	$(^2D^0)6p\ ^1F_3 \rightarrow (^2P^0)5d17^0_3$	1083	921,1015
0.627081	0.627090	±0.000010	1	$(^1D)6p\ ^2F^0_{5/2} \rightarrow (^1D)6s\ ^2D_{3/2}$	1083,1088,1092, 1094,1455	450,884,1116, 1117,1167
0.628641	0.628660	±0.000060	3		1089,1456	923,1016,1101
0.628641	0.628660	±0.000060	4		1089,1457	923,1016,1101
0.634343	0.634343	±0.000005	2		1089,1094	923,983,1101
0.652865	0.652850	±0.000050	1	$(^1D)6p\ ^2F^0_{7/2} \rightarrow (^1D)5d\ ^2F_{5/2}$	1083,1092	884,1118
0.669431	0.66943		1	$(^3P)6p\ ^4P^0_{3/2} \rightarrow (^3P)5d\ ^4D_{1/2}$	1083	973
0.669950	0.669950	±0.000030	3		1089	922,923
0.670225	0.670200	±0.000100	1	$(^1D)6p\ ^2P^0_{3/2} \rightarrow (^1D)5d\ ^2F_{5/2}$	1083,1092	884,1106
0.707234	0.70723		1	$37^0_{5/2} \rightarrow (^3P)6d\ ^4D_{5/2}$	1083	973
0.714903	0.714894	±0.000060	1	$(^3P)6p\ ^4D^0_{3/2} \rightarrow (^3P)6s\ ^2P_{3/2}$	1083,1085,1089, 1092	884,923,973, 983,1014
0.761859	0.761900	±0.000020	1	$(^1D)6p\ ^2D_{5/2} \rightarrow (^1S)5d\ ^2D_{5/2}$	1077,1083,1087	1077
0.7827633	0.782800	±0.000300	1	$5^0_{5/2} \rightarrow 16_{3/2}$		1075
0.798800	0.798900	±0.000300	1	$(^3P)6p\ ^4P^0_{1/2} \rightarrow (^3P)6s\ ^4P_{1/2}$		1075
0.833271	0.833000	±0.000300	1	$27^0_{5/2} \rightarrow (^3P)6d\ ^4D_{5/2}$		1075
0.844619	0.844300	±0.000300	1	$27^0_{5/2} \rightarrow (^3P)6d\ ^4D_{3/2}$		1075
0.8569	0.856900	±0.000300	2	$31^0_{3/2} \rightarrow 10_{5/2}$	1089,1458	874,923,983,1075
0.858251	0.858200	±0.000300	1	$(^3P)6p\ ^4D^0_{3/2} \rightarrow (^3P)5d\ ^2P_{3/2}$		1075
0.871617	0.871400	±0.000300	1	$27^0_{5/2} \rightarrow 16_{3/2}$	1083	973,1075
0.905930	0.906300	±0.000400	1	$(^1S)5d\ ^2D_{3/2} \rightarrow (^3P)6p\ ^4D^0_{5/2}$		1075
0.926539	0.926500	±0.000400	1	$13^0_{1/2} \rightarrow (^1D)5d\ ^2S_{1/2}$		1075
0.928854	0.928700	±0.000400	1	$(^3P)6p\ ^4D^0_{3/2} \rightarrow (^3P)5d\ ^4P_{5/2}$	1083	1075,1171
0.969859	0.969700	±0.000200	1	$(^3P)6p\ D^0_{3/2} \rightarrow (^3P)5d\ ^4P_{3/2}$		1075
1.063385	1.063400	±0.000600	1		1083	1075
1.0950	1.095000	±0.000600	1		1459	1075

Section 3.3
MOLECULAR GAS LASERS

3.3.1 Electronic Transition Gas Lasers

3.3.1.1 Introduction

Molecular gas lasers involving electronic transitions include a wide variety of systems such as the diatomic halogens species, metal halides, CO, H_2, N_2, alkali dimers, molecular ions, and rare earth complexes. Excitation by either electrical discharge or optical means is by far the most common. The former generally delivers the greater power, while the latter has much greater selectivity.

The tables of electronic transition molecular gas lasers are divided into subsections by the increasing number of atoms constituting the molecule: diatomic, triatomic, and poly-atomic. Within the tables, the ordering scheme is

1. alphabetical order of the chemical formulae,
2. increasing isotopic mass,
3. increasing band-center wavelength,
4. increasing lower vibrational state energy,
5. increasing transition wavelength within a given vibronic group.

The range of wavelengths of electronic transition molecular gas lasers extends from 109.82 nm (a para-H_2 transition) to 8210.2 nm (a N_2 transition). The laser wavelengths listed are in air (if in vacuum, wavelengths are in italics) and are followed by the transition assignment. The experimental conditions (pumping method, pump energy, and temperature and pressure of lasant and diluent species) and peak output are included in the comments in Section 3.6. References are grouped together in Section 3.7.

Further Reading

Davis, R. S. and Rhodes, C. K., Electronic Transition Lasers, in *Handbook of Laser Science and Technology, Vol. II: Gas Lasers*, CRC Press, Boca Raton, FL (1982), p. 273.

Eden, J. G., Electronic Transition Lasers, in *Handbook of Laser Science and Technology, Suppl. 1: Lasers*, CRC Press, Boca Raton, FL (1991), p. 341.

Gross, R. W. F. and Bott, J. F., *Handbook of Chemical Lasers*, John Wiley & Sons, New York (1975).

Hooker, S. M. and Webb, C. E., Progress in vacuum, ultraviolet lasers, *Progress in Quantum Electronics* 18, 227 (1994).

Rhodes, C. K. (Ed.), *Excimer Lasers*, 2nd edition, Springer-Verlag, Berlin (1984).

Electronic transition molecular gas lasers included in this section are presented in alphabetical order as follows:

Diatomic electronic transition lasers – Table 3.3.1.1:

Ar_2	para-H_2	NaRb
ArCl	HD	NO
ArF	HgBr	S_2
ArO	HgCl	Se_2
BH	HgI	SO
Bi_2	I_2	Te_2
BiF	ICl	$^{128}Te_2$
Br_2	IF	$^{130}Te_2$
BrF	K_2	$^{130}Te_2$
C_2	Kr_2	$^{126}Te^{128}Te$
CdBr	KrCl	$^{128}Te^{130}Te$
CdI	KrF	Xe_2
Cl_2	KrO	XeBr
ClF	Li_2	XeCl
CO	$^6Li_2$	XeF
CO^+	N_2	XeO
D_2	N_2^+	ZnI
F_2	Na_2	
H_2	NaK	

Triatomic electronic transition lasers – Table 3.3.1.2:

CS_2
Hg_3
Kr_2F
Xe_2Cl
Xe_2F

Polyatomic electronic transition lasers – Table 3.3.1.3:

coumarin 6
coumarin 7
coumarin 30
coumarin 153
7-dimethylamino-3(2'-benzoxazolyl) coumarin
$NdAl_3Cl_{12}$
Nd(thd)$_3$
p-phenylene-bis (5-phenyl-2-oxazole) (POPOP)
$TbAl_3Cl_{12}$

3.3.1.2 Diatomic Electronic Transition Lasers

Table 3.3.1.1
Diatomic Electronic Transition Lasers

Species	Wavelength (μm)	Transition assignment	Comments	References
Ar$_2$	0.124-0.128	AO$^+_u \rightarrow$ XO^+_g	6596,6597,6599	1523,1606,1622
	0.1261	$^1\Sigma^+_u \rightarrow X^1\Sigma^+_g$	1472–1475	1,40,51,62,73, 84,95,106,116
ArCl	0.1690	B$^2\Sigma^+_{1/2} \rightarrow$ X$^2\Sigma^+_{1/2}$	1476–1478	2,13,23, 33–36,116
	0.1750	B$^2\Sigma^+_{1/2} \rightarrow$ X$^2\Sigma^+_{1/2}$	1476–1478	2,13,23,33– 36,116
ArF	0.1933	B$^2\Sigma^+_{1/2} \rightarrow$ X$^2\Sigma^+_{1/2}$	1479,1480	13,33–35, 37,38,116
ArO	0.5580	O(^{1}S$_0$) $\rightarrow$ O(^{1}D$_2$)	1481–1483	2,39,41–47
BH	0.433	A-X (v' = 0 $\rightarrow$ v" = 0)	7252	1813
Bi$_2$	0.5929	A(O^-_u) $\rightarrow$ X(O^-_g)	1484,1486	48,49
	0.59293	A(O^+_u) $\rightarrow$ X(O^+_g) (v',v") = (17,17) R(254)	6600,6603	1524,1525
	0.61546	A(O^+_u) $\rightarrow$ X(O^+_g) (v',v") = (17,21) R(254)	6600	1524,1525
	0.6160	A(O^-_u) $\rightarrow$ X(O^-_g)	1484,1486	48,49
	0.61629	A(O^+_u) $\rightarrow$ X(O^+_g) (v',v") = (17,21) P(256)	6600,6603	1524,1525
	0.6239	A(O^-_u) $\rightarrow$ X(O^-_g)	1484,1486	48,49
	0.623987	A(O^+_u) $\rightarrow$ X(O^+_g) (v',v") = (16,22) P(199)	6600,6603	1524,1525
	0.629951	A(O^+_u) $\rightarrow$ X(O^+_g) (v',v") = (16,23) P(199)	6600,6603	1524,1525
	0.6300	A(O^-_u) $\rightarrow$ X(O^-_g)	1484,1486	48,49
	0.63308	A(O^+_u) $\rightarrow$ X(O^+_g) (v',v") = (17,24) P(254)	6600	1524,1525
	0.6339	A(O^-_u) $\rightarrow$ X(O^-_g)	1484,1486	48,49
	0.63396	A(O^+_u) $\rightarrow$ X(O^+_g) (v',v") = (17,24) P(256)	6600,6603	1524,1525
	0.6414	A(O^-_u) $\rightarrow$ X(O^-_g)	1484,1486	48,49
	0.641445	A(O^+_u) $\rightarrow$ X(O^+_g) (v',v") = (16,25) R(197)	6600,6603	1524,1525

Table 3.3.1.1—*continued*
Diatomic Electronic Transition Lasers

Species	Wavelength (μm)	Transition assignment	Comments	References
Bi_2	0.642134	$A(O^+_u) \rightarrow X(O^+_g)$ $(v',v'') = (16,25)$ P(199)	6600,6603	1524,1525
	0.6422	$A(O^-_u) \rightarrow X(O^-_g)$	1484,1486	48,49
	0.65183	$A(O^+_u) \rightarrow X(O^+_g)$ $(v',v'') = (9,21)$ R(198)	6600	1524,1525
	0.65264	$A(O^+_u) \rightarrow X(O^+_g)$ $(v',v'') = (9,21)$ P(200)	6600	1524,1525
	0.6576	$A(O^-_u) \rightarrow X(O^-_g)$	1484,1486	48,49
	0.65765	$A(O^+_u) \rightarrow X(O^+_g)$ $(v',v'') = (21,32)$ R(125)	6600	1524,1525
	0.65817	$A(O^+_u) \rightarrow X(O^+_g)$ $(v',v'') = (21,32)$ P(127)	6600	1524,1525
	0.6582	$A(O^-_u) \rightarrow X(O^-_g)$	1484,1486	48,49
	0.65847	$A(O^+_u) \rightarrow X(O^+_g)$ $(v',v'') = (9,22)$ R(198)	6600	1524,1525
	0.65923	$A(O^+_u) \rightarrow X(O^+_g)$ $(v',v'') = (9,22)$ P(200)	6600	1524,1525
	0.6603	$A(O^-_u) \rightarrow X(O^-_g)$	1484,1486	48,49
	0.660386	$A(O^+_u) \rightarrow X(O^+_g)$ $(v',v'') = (16,28)$ R(197)	6600,6603	1524,1525
	0.661110	$A(O^+_u) \rightarrow X(O^+_g)$ $(v',v'') = (16,28)$ P(199)	6600	1524,1525
	0.66405	$A(O^+_u) \rightarrow X(O^+_g)$ $(v',v'') = (21,33)$ R(125)	6600	1524,1525
	0.66406	$A(O^+_u) \rightarrow X(O^+_g)$ $(v',v'') = (17,29)$ R(254)	6600	1524,1525
	0.66457	$A(O^+_u) \rightarrow X(O^+_g)$ $(v',v'') = (21,33)$ P(127)	6600	1524,1525
	0.6650	$A(O^-_u) \rightarrow X(O^-_g)$	1484,1486	48,49
	0.66500	$A(O^+_u) \rightarrow X(O^+_g)$ $(v',v'') = (17,29)$ P(256)	6600,6603	1524,1525
	0.66510	$A(O^+_u) \rightarrow X(O^+_g)$ $(v',v'') = (9,23)$ R(198)	6600	1524,1525
	0.66594	$A(O^+_u) \rightarrow X(O^+_g)$ $(v',v'') = (9,23)$ P(200)	6600	1524,1525
	0.66661	$A(O^+_u) \rightarrow X(O^+_g)$ $(v',v'') = (19,32)$ R(10)	6600	1524,1525

Table 3.3.1.1—*continued*
Diatomic Electronic Transition Lasers

Species	Wavelength (µm)	Transition assignment	Comments	References
Bi_2	0.66665	$A(O^+_u) \rightarrow X(O^+_g)$ $(v',v'') = (19,32)$ P(12)	6600	1524,1525
	0.66710	$A(O^+_u) \rightarrow X(O^+_g)$ $(v',v'') = (19,32)$ P(58)	6600	1524,1525
	0.67191	$A(O^+_u) \rightarrow X(O^+_g)$ $(v',v'') = (9,24)$ R(198)	6600	1524,1525
	0.67272	$A(O^+_u) \rightarrow X(O^+_g)$ $(v',v'') = (9,24)$ P(200)	6600	1524,1525
	0.673448	$A(O^+_u) \rightarrow X(O^+_g)$ $(v',v'') = (16,30)$ R(197)	6600	1524,1525
	0.674241	$A(O^+_u) \rightarrow X(O^+_g)$ $(v',v'') = (16,30)$ P(199)	6600	1524,1525
	0.67704	$A(O^+_u) \rightarrow X(O^+_g)$ $(v',v'') = (17,31)$ R(254)	6600	1524,1525
	0.67804	$A(O^+_u) \rightarrow X(O^+_g)$ $(v',v'') = (17,31)$ P(256)	6600	1524,1525
	0.680156	$A(O^+_u) \rightarrow X(O^+_g)$ $(v',v'') = (16,31)$ R(197)	6600	1524,1525
	0.6809	$A(O^-_u) \rightarrow X(O^-_g)$	1484,1486	48,49
	0.680917	$A(O^+_u) \rightarrow X(O^+_g)$ $(v',v'') = (16,31)$ P(199)	6600,6603	1524,1525
	0.68366	$A(O^+_u) \rightarrow X(O^+_g)$ $(v',v'') = (17,32)$ R(254)	6600	1524,1525
	0.68465	$A(O^+_u) \rightarrow X(O^+_g)$ $(v',v'') = (17,32)$ P(256)	6600	1524,1525
	0.68651	$A(O^+_u) \rightarrow X(O^+_g)$ $(v',v'') = (19,35)$ R(10)	6600	1524,1525
	0.68656	$A(O^+_u) \rightarrow X(O^+_g)$ $(v',v'') = (19,35)$ P(12)	6600	1524,1525
	0.68687	$A(O^+_u) \rightarrow X(O^+_g)$ $(v',v'') = (19,35)$ R(56)	6600	1524,1525
	0.68710	$A(O^+_u) \rightarrow X(O^+_g)$ $(v',v'') = (19,35)$ P(58)	6600	1524,1525
	0.69010	$A(O^+_u) \rightarrow X(O^+_g)$ $(v',v'') = (20,36)$ R(170)	6600	1524,1525
	0.69084	$A(O^+_u) \rightarrow X(O^+_g)$ $(v',v'') = (20,36)$ P(172)	6600	1524,1525
	0.693677	$A(O^+_u) \rightarrow X(O^+_g)$ $(v',v'') = (16,33)$ R(197)	6600	1524,1525

Table 3.3.1.1—*continued*
Diatomic Electronic Transition Lasers

Species	Wavelength (μm)	Transition assignment	Comments	References
Bi_2	0.694482	$A(O^+_u) \rightarrow X(O^+_g)$	6600	1524,1525
		$(v',v'') = (16,33)$ P(199)		
	0.69667	$A(O^+_u) \rightarrow X(O^+_g)$	6600	1524,1525
		$(v',v'') = (20,37)$ R(170)		
	0.69742	$A(O^+_u) \rightarrow X(O^+_g)$	6600	1524,1525
		$(v',v'') = (20,37)$ P(172)		
	0.7006	$A(O^-_u) \rightarrow X(O^-_g)$	1484,1486	48,49
	0.700636	$A(O^+_u) \rightarrow X(O^+_g)$	6600,6603	1524,1525
		$(v',v'') = (16,34)$ R(197)		
	0.7013	$A(O^-_u) \rightarrow X(O^-_g)$	1484,1486	48,49
	0.701423	$A(O^+_u) \rightarrow X(O^+_g)$	6600,6603	1524,1525
		$(v',v'') = (16,34)$ P(199)		
	0.70403	$A(O^+_u) \rightarrow X(O^+_g)$	6600	1524,1525
		$(v',v'') = (17,35)$ R(254)		
	0.70507	$A(O^+_u) \rightarrow X(O^+_g)$	6600	1524,1525
		$(v',v'') = (17,35)$ P(256)		
	0.70725	$A(O^+_u) \rightarrow X(O^+_g)$	6600	1524,1525
		$(v',v'') = (19,38)$ R(10)		
	0.70730	$A(O^+_u) \rightarrow X(O^+_g)$	6600	1524,1525
		$(v',v'') = (19,38)$ P(12)		
	0.70759	$A(O^+_u) \rightarrow X(O^+_g)$	6600	1524,1525
		$(v',v'') = (19,38)$ R(56)		
	0.707677	$A(O^+_u) \rightarrow X(O^+_g)$	6600	1524,1525
		$(v',v'') = (16,35)$ R(197)		
	0.70784	$A(O^+_u) \rightarrow X(O^+_g)$	6600	1524,1525
		$(v',v'') = (19,38)$ P(58)		
	0.708506	$A(O^+_u) \rightarrow X(O^+_g)$	6600	1524,1525
		$(v',v'') = (16,35)$ P(199)		
	0.71065	$A(O^+_u) \rightarrow X(O^+_g)$	6600	1524,1525
		$(v',v'') = (20,39)$ R(170)		
	0.71096	$A(O^+_u) \rightarrow X(O^+_g)$	6600	1524,1525
		$(v',v'') = (17,36)$ R(254)		
	0.71141	$A(O^+_u) \rightarrow X(O^+_g)$	6600	1524,1525
		$(v',v'') = (20,39)$ P(172)		
	0.71199	$A(O^+_u) \rightarrow X(O^+_g)$	6600	1524,1525
		$(v',v'') = (17,36)$ P(256)		
	0.71714	$A(O^+_u) \rightarrow X(O^+_g)$	6600	1524,1525
		$(v',v'') = (21,41)$ R(102)		

Table 3.3.1.1—*continued*
Diatomic Electronic Transition Lasers

Species	Wavelength (μm)	Transition assignment	Comments	References
Bi_2	0.71756	$A(O^+_u) \to X(O^+_g)$ $(v',v'') = (21,41)\ P(104)$	6600	1524,1525
	0.71770	$A(O^+_u) \to X(O^+_g)$ $(v',v'') = (21,41)\ R(125)$	6600	1524,1525
	0.71775	$A(O^+_u) \to X(O^+_g)$ $(v',v'') = (20,40)\ R(170)$	6600	1524,1525
	0.71825	$A(O^+_u) \to X(O^+_g)$ $(v',v'') = (21,41)\ P(127)$	6600	1524,1525
	0.71848	$A(O^+_u) \to X(O^+_g)$ $(v',v'') = (20,40)\ P(172)$	6600	1524,1525
	0.721908	$A(O^+_u) \to X(O^+_g)$ $(v',v'') = (16,37)\ R(197)$	6600	1524,1525
	0.722799	$A(O^+_u) \to X(O^+_g)$ $(v',v'') = (16,37)\ P(199)$	6600	1524,1525
	0.72915	$A(O^+_u) \to X(O^+_g)$ $(v',v'') = (19,41)\ R(56)$	6600	1524,1525
	0.7292	$A(O^-_u) \to X(O^-_g)$	1484,1486	48,49
	0.729288	$A(O^+_u) \to X(O^+_g)$ $(v',v'') = (16,38)\ R(197)$	6600,6603	1524,1525
	0.72940	$A(O^+_u) \to X(O^+_g)$ $(v',v'') = (19,41)\ P(58)$	6600	1524,1525
	0.7301	$A(O^-_u) \to X(O^-_g)$	1484,1486	48,49
	0.730173	$A(O^+_u) \to X(O^+_g)$ $(v',v'') = (16,38)\ P(199)$	6600,6603	1524,1525
	0.73236	$A(O^+_u) \to X(O^+_g)$ $(v',v'') = (17,39)\ R(152)$	6600	1524,1525
	0.73238	$A(O^+_u) \to X(O^+_g)$ $(v',v'') = (17,39)\ R(254)$	6600	1524,1525
	0.73349	$A(O^+_u) \to X(O^+_g)$ $(v',v'') = (17,39)\ P(256)$	6600,6603	1524,1525
	0.7335	$A(O^-_u) \to X(O^-_g)$	1484,1486	48,49
	0.73355	$A(O^+_u) \to X(O^+_g)$ $(v',v'') = (17,39)\ P(154)$	6600	1524,1525
	0.73615	$A(O^+_u) \to X(O^+_g)$ $(v',v'') = (17,40)\ R(152)$	6600	1524,1525
	0.73623	$A(O^+_u) \to X(O^+_g)$ $(v',v'') = (19,42)\ R(10)$	6600,6603	1524,1525
	0.73629	$A(O^+_u) \to X(O^+_g)$ $(v',v'') = (19,42)\ P(12)$	6600,6603	1524,1525

Table 3.3.1.1—*continued*
Diatomic Electronic Transition Lasers

Species	Wavelength (μm)	Transition assignment	Comments	References
Bi₂	0.7364	$A(0^-_u) \rightarrow X(0^-_g)$	1484,1486	48,49
	0.73641	$A(0^+_u) \rightarrow X(0^+_g)$	6600	1524,1525
		(v',v") = (19,42) R(56)		
	0.7366	$A(0^-_u) \rightarrow X(0^-_g)$	1484,1486	48,49
	0.73667	$A(0^+_u) \rightarrow X(0^+_g)$	6600	1524,1525
		(v',v") = (19,42) P(58)		
	0.736693	$A(0^+_u) \rightarrow X(0^+_g)$	6600,6603	1524,1525
		(v',v") = (16,39) R(197)		
	0.73682	$A(0^+_u) \rightarrow X(0^+_g)$	6600	1524,1525
		(v',v") = (17,40) P(154)		
	0.737584	$A(0^+_u) \rightarrow X(0^+_g)$	6600,6603	1524,1525
		(v',v") = (16,39) P(199)		
	0.7376	$A(0^-_u) \rightarrow X(0^-_g)$	1484,1486	48,49
	0.73881	$A(0^+_u) \rightarrow X(0^+_g)$	6600	1524,1525
		(v',v") = (21,44) R(102)		
	0.73923	$A(0^+_u) \rightarrow X(0^+_g)$	6600	1524,1525
		(v',v") = (21,44) R(125)		
	0.73925	$A(0^+_u) \rightarrow X(0^+_g)$	6600	1524,1525
		(v',v") = (21,44) P(104)		
	0.73937	$A(0^+_u) \rightarrow X(0^+_g)$	6600	1524,1525
		(v',v") = (20,43) R(170)		
	0.73974	$A(0^+_u) \rightarrow X(0^+_g)$	6600,6603	1524,1525
		(v',v") = (17,40) R(254)		
	0.73978	$A(0^+_u) \rightarrow X(0^+_g)$	6600	1524,1525
		(v',v") = (21,44) P(127)		
	0.7398	$A(0^-_u) \rightarrow X(0^-_g)$	1484,1486	48,49
	0.74014	$A(0^+_u) \rightarrow X(0^+_g)$	6600	1524,1525
		(v',v") = (20,43) P(172)		
	0.7408	$A(0^-_u) \rightarrow X(0^-_g)$	1484,1486	48,49
	0.74084	$A(0^+_u) \rightarrow X(0^+_g)$	6600,6603	1524,1525
		(v',v") = (17,40) P(256)		
	0.74367	$A(0^+_u) \rightarrow X(0^+_g)$	6600	1524,1525
		(v',v") = (17,41) R(152)		
	0.74372	$A(0^+_u) \rightarrow X(0^+_g)$	6600	1524,1525
		(v',v") = (19,43) R(10)		
	0.74377	$A(0^+_u) \rightarrow X(0^+_g)$	6600	1524,1525
		(v',v") = (19,43) P(12)		
	0.74387	$A(0^+_u) \rightarrow X(0^+_g)$	6600	1524,1525
		(v',v") = (19,43) R(56)		

Table 3.3.1.1—*continued*
Diatomic Electronic Transition Lasers

Species	Wavelength (µm)	Transition assignment	Comments	References
Bi$_2$	0.7439	A(O^-_u) → X(O^-_g)	1484,1486	48,49
	0.74413	A(O^+_u) → X(O^+_g) (v',v") = (19,43) P(58)	6600	1524,1525
	0.744169	A(O^+_u) → X(O^+_g) (v',v") = (16,40) R(197)	6600	1524,1525
	0.74438	A(O^+_u) → X(O^+_g) (v',v") = (17,41) P(154)	6600	1524,1525
	0.745046	A(O^+_u) → X(O^+_g) (v',v") = (16,40) P(199)	6600	1524,1525
	0.74615	A(O^+_u) → X(O^+_g) (v',v") = (21,45) R(102)	6600	1524,1525
	0.74660	A(O^+_u) → X(O^+_g) (v',v") = (21,45) P(104)	6600	1524,1525
	0.74674	A(O^+_u) → X(O^+_g) (v',v") = (21,45) R(125)	6600	1524,1525
	0.74675	A(O^+_u) → X(O^+_g) (v',v") = (20,44) R(170)	6600,6603	1524,1525
	0.7468	A(O^-_u) → X(O^-_g)	1484,1486	48,49
	0.7471	A(O^-_u) → X(O^-_g)	1484,1486	48,49
	0.74713	A(O^+_u) → X(O^+_g) (v',v") = (17,41) R(254)	6600,6603	1524,1525
	0.74731	A(O^+_u) → X(O^+_g) (v',v") = (21,45) P(127)	6600	1524,1525
	0.7475	A(O^-_u) → X(O^-_g)	1484,1486	48,49
	0.74754	A(O^+_u) → X(O^+_g) (v',v") = (20,44) P(172)	6600,6603	1524,1525
	0.7482	A(O^-_u) → X(O^-_g)	1484,1486	48,49
	0.74829	A(O^+_u) → X(O^+_g) (v',v") = (17,41) P(256)	6600,6603	1524,1525
	0.74893	A(O^+_u) → X(O^+_g) (v',v") = (24,48) R(123)	6600	1524,1525
	0.74946	A(O^+_u) → X(O^+_g) (v',v") = (24,48) P(125)	6600	1524,1525
	0.75130	A(O^+_u) → X(O^+_g) (v',v") = (19,44) R(10)	6600	1524,1525
	0.75137	A(O^+_u) → X(O^+_g) (v',v") = (19,44) P(12)	6600	1524,1525
	0.75139	A(O^+_u) → X(O^+_g) (v',v") = (19,44) R(56)	6600	1524,1525

Table 3.3.1.1—*continued*
Diatomic Electronic Transition Lasers

Species	Wavelength (μm)	Transition assignment	Comments	References
Bi_2	0.75166	$A(O^+_u) \to X(O^+_g)$	6600	1524,1525
		$(v',v'') = (19,44)$ P(58)		
	0.751735	$A(O^+_u) \to X(O^+_g)$	6600	1524,1525
		$(v',v'') = (16,41)$ R(197)		
	0.752653	$A(O^+_u) \to X(O^+_g)$	6600	1524,1525
		$(v',v'') = (16,41)$ P(199)		
	0.75363	$A(O^+_u) \to X(O^+_g)$	6600	1524,1525
		$(v',v'') = (21,46)$ R(102)		
	0.75408	$A(O^+_u) \to X(O^+_g)$	6600	1524,1525
		$(v',v'') = (21,46)$ P(104)		
	0.75409	$A(O^+_u) \to X(O^+_g)$	6600	1524,1525
		$(v',v'') = (21,46)$ R(125)		
	0.75419	$A(O^+_u) \to X(O^+_g)$	6600,6603	1524,1525
		$(v',v'') = (20,45)$ R(170)		
	0.7543	$A(O^-_u) \to X(O^-_g)$	1484,1486	48,49
	0.75463	$A(O^+_u) \to X(O^+_g)$	6600	1524,1525
		$(v',v'') = (17,42)$ P(254)		
	0.75466	$A(O^+_u) \to X(O^+_g)$	6600	1524,1525
		$(v',v'') = (21,46)$ P(127)		
	0.75496	$A(O^+_u) \to X(O^+_g)$	6600,6603	1524,1525
		$(v',v'') = (20,45)$ P(172)		
	0.7551	$A(O^-_u) \to X(O^-_g)$	1484,1486	48,49
	0.75580	$A(O^+_u) \to X(O^+_g)$	6600	1524,1525
		$(v',v'') = (17,42)$ P(256)		
	0.75622	$A(O^+_u) \to X(O^+_g)$	6600	1524,1525
		$(v',v'') = (24,49)$ R(123)		
	0.75683	$A(O^+_u) \to X(O^+_g)$	6600	1524,1525
		$(v',v'') = (24,49)$ P(125)		
	0.76048	$A(O^+_u) \to X(O^+_g)$	6600	1524,1525
		$(v',v'') = (31,56)$ R(43)		
	0.76068	$A(O^+_u) \to X(O^+_g)$	6600	1524,1525
		$(v',v'') = (31,56)$ P(45)		
	0.76122	$A(O^+_u) \to X(O^+_g)$	6600	1524,1525
		$(v',v'') = (21,47)$ R(102)		
	0.76165	$A(O^+_u) \to X(O^+_g)$	6600	1524,1525
		$(v',v'') = (21,47)$ R(125)		
	0.76169	$A(O^+_u) \to X(O^+_g)$	6600	1524,1525
		$(v',v'') = (21,47)$ P(104)		

Table 3.3.1.1—*continued*
Diatomic Electronic Transition Lasers

Species	Wavelength (μm)	Transition assignment	Comments	References
Bi_2	0.76177	$A(O^+_u) \rightarrow X(O^+_g)$ $(v',v'') = (20,46)$ R(170)	6600	1524,1525
	0.76204	$A(O^+_u) \rightarrow X(O^+_g)$ $(v',v'') = (34,59)$ R(23)	6600	1524,1525
	0.76215	$A(O^+_u) \rightarrow X(O^+_g)$ $(v',v'') = (34,59)$ P(25)	6600	1524,1525
	0.76223	$A(O^+_u) \rightarrow X(O^+_g)$ $(v',v'') = (34,59)$ R(21)	6600	1524,1525
	0.76224	$A(O^+_u) \rightarrow X(O^+_g)$ $(v',v'') = (21,47)$ P(127)	6600	1524,1525
	0.76235	$A(O^+_u) \rightarrow X(O^+_g)$ $(v',v'') = (34,59)$ P(23)	6600	1524,1525
	0.76256	$A(O^+_u) \rightarrow X(O^+_g)$ $(v',v'') = (20,46)$ P(172)	6600	1524,1525
	0.76763	$A(O^+_u) \rightarrow X(O^+_g)$ $(v',v'') = (31,57)$ R(43)	6600	1524,1525
	0.76784	$A(O^+_u) \rightarrow X(O^+_g)$ $(v',v'') = (31,57)$ P(45)	6600	1524,1525
	0.76904	$A(O^+_u) \rightarrow X(O^+_g)$ $(v',v'') = (34,60)$ R(23)	6600	1524,1525
	0.76916	$A(O^+_u) \rightarrow X(O^+_g)$ $(v',v'') = (34,60)$ P(25)	6600	1524,1525
	0.76930	$A(O^+_u) \rightarrow X(O^+_g)$ $(v',v'') = (34,60)$ R(21)	6600	1524,1525
	0.76941	$A(O^+_u) \rightarrow X(O^+_g)$ $(v',v'') = (34,60)$ P(23)	6600	1524,1525
	0.77486	$A(O^+_u) \rightarrow X(O^+_g)$ $(v',v'') = (31,58)$ R(43)	6600	1524,1525
	0.77506	$A(O^+_u) \rightarrow X(O^+_g)$ $(v',v'') = (31,58)$ P(45)	6600	1524,1525
	0.77612	$A(O^+_u) \rightarrow X(O^+_g)$ $(v',v'') = (34,61)$ R(23)	6600	1524,1525
	0.77624	$A(O^+_u) \rightarrow X(O^+_g)$ $(v',v'') = (34,61)$ P(25)	6600	1524,1525
	0.77635	$A(O^+_u) \rightarrow X(O^+_g)$ $(v',v'') = (34,61)$ R(21)	6600	1524,1525
	0.77647	$A(O^+_u) \rightarrow X(O^+_g)$ $(v',v'') = (34,61)$ P(23)	6600	1524,1525

Table 3.3.1.1—*continued*
Diatomic Electronic Transition Lasers

Species	Wavelength (μm)	Transition assignment	Comments	References
Bi$_2$	0.78218	A(O^+_u) → X(O^+_g)	6600	1524,1525
		(v',v") = (31,59) R(43)		
	0.78237	A(O^+_u) → X(O^+_g)	6600	1524,1525
		(v',v") = (31,59) P(45)		
	0.78330	A(O^+_u) → X(O^+_g)	6600	1524,1525
		(v',v") = (34,62) R(23)		
	0.78342	A(O^+_u) → X(O^+_g)	6600	1524,1525
		(v',v") = (34,62) P(25)		
BiF	0.471	A-X, v' = 1 → v" = 4	7253	1812
Br$_2$	0.2915	E$^3\Pi_{2g}$ → B$^3\Pi_{2u}$	1487,1488	50,52–54
	0.55053	B$^3\Pi^+_{ou}$ → X$^1\Sigma^+_{og}$ (16,2)	6605,6607	1526
	0.58048	B$^3\Pi^+_{ou}$ → X$^1\Sigma^+_{og}$ (16,5)	6605,6607	1526
	0.58090	B$^3\Pi^+_{ou}$ → X$^1\Sigma^+_{og}$ (16,5)	6605,6607	1526
	0.61316	B$^3\Pi^+_{ou}$ → X$^1\Sigma^+_{og}$ (16,8)	6605,6607	1526
	0.61368	B$^3\Pi^+_{ou}$ → X$^1\Sigma^+_{og}$ (16,8)	6605,6607	1526
	0.63654	B$^3\Pi^+_{ou}$ → X$^1\Sigma^+_{og}$ (16,10)	6605,6607	1526
	0.63705	B$^3\Pi^+_{ou}$ → X$^1\Sigma^+_{og}$ (16,10)	6605,6607	1526
	0.67455	B$^3\Pi^+_{ou}$ → X$^1\Sigma^+_{og}$ (16,13)	6605,6607	1526
	0.67506	B$^3\Pi^+_{ou}$ → X$^1\Sigma^+_{og}$ (16,13)	6605,6607	1526
	0.74638	B$^3\Pi^+_{ou}$ → X$^1\Sigma^+_{og}$ (16,18)	6605,6607	1526
	0.74704	B$^3\Pi^+_{ou}$ → X$^1\Sigma^+_{og}$ (16,18)	6605,6607	1526
BrF	0.3542	D'(2$_g$) → A'(2$_u$)	6608	1527
	0.3545	D'(2$_g$) → A'(2$_u$)	6608	1527
C$_2$	0.5545	d$^3\Pi_g$ → a$^3\Pi_u$	6610	1528
		(v',v") = (1,2)		
	0.6075	d$^3\Pi_g$ → a$^3\Pi_u$	6610	1528
		(v',v") = (1,3)		
CdBr	0.811–0.816	B$^2\Sigma^+_{1/2}$ → X$^2\Sigma^+_{1/2}$	6612	1529
CdI	0.475	B$^2\Sigma^+_{1/2}$ → X$^2\Sigma^+_{1/2}$	6613,6616–19	1530
	0.6538	B$^2\Sigma^+_{1/2}$ → X$^2\Sigma^+_{1/2}$	6613,6616–19	1529,1530
	0.6553	B$^2\Sigma^+_{1/2}$ → X$^2\Sigma^+_{1/2}$	6613,6616–19	1529,1530,1531
	0.6568	B$^2\Sigma^+_{1/2}$ → X$^2\Sigma^+_{1/2}$ (1,62)	6613,6616–19	1531
	0.6571	B$^2\Sigma^+_{1/2}$ → X$^2\Sigma^+_{1/2}$	6613,6616–19	1529,1531
	0.6574	B$^2\Sigma^+_{1/2}$ → X$^2\Sigma^+_{1/2}$	6613,6616–19	1531

Table 3.3.1.1—*continued*
Diatomic Electronic Transition Lasers

Species	Wavelength (μm)	Transition assignment	Comments	References
CdI	0.6593	$B^2\Sigma^+_{1/2} \to X^2\Sigma^+_{1/2}$	6613,6616–19	1531
Cl_2	0.2580	$E^3\Pi_{2g} \to B^3\Pi_{2u}$	1489	55,56
ClF	0.2823-0.2832	$D'(2_g) \to A'(2_u)$	6620,6622	1534
	0.2840	$D'(2_g) \to A'(2_u)$	6620,6622	1527
	0.2844	$D'(2_g) \to A'(2_u)$	6620,6622	1527,1534
	0.2849	$D'(2_g) \to A'(2_u)$	6620,6622	1534
	0.2850	$3\Pi_{2g} \to {}^3\Pi_{2u}$	1490	56
	0.2860	$D'(2_g) \to A'(2_u)$	6620,6622	1534
CO	0.181085	$A'\Pi–X'\Sigma^+$ (2,6) Band, Q(5-13), R(2-9)	1498	55,56
	0.187831	$A'\Pi–X'\Sigma^+$ (2,6) Band, Q(5-13), R(2-9)		55,56
	0.189784	$A'\Pi–X'\Sigma^+$ (2,6) Band, Q(5-12), R(2-9)		55,56
	0.195006	$A'\Pi–X'\Sigma^+$ (2,6) Band, Q(5-11), R(2-9)		55,56
	0.197013	$A'\Pi–X'\Sigma^+$ (2,6) Band, Q(5-11)		55,56
	0.4210	$B^2\Sigma^+_u \to A^2\Pi$ (0-1)	1499	61
	0.55921	$B'\Sigma–A'\Pi$ (0-3) Band, Q(11)	1495,1496	59,60
	0.55949	$B'\Sigma–A'\Pi$ (0-3) Band, Q(10)	1496	59,60
	0.55975	$B'\Sigma–A'\Pi$ (0-3) Band, Q(9) or R(13)	1496	59,60
	0.55998	$B'\Sigma–A'\Pi$ (0-3) Band, Q(8)	1496	59,60
	0.56019	$B'\Sigma–A'\Pi$ (0-3) Band, Q(7)	1496	59,60
	0.56040	$B'\Sigma–A'\Pi$ (0-3) Band, Q(6)	1496	59,60
	0.56040	$B'\Sigma–A'\Pi$ (0-3) Band, Q(6)	1496	59,60
	0.56053	$B'\Sigma–A'\Pi$ (0-3) Band, Q(5)	1496	59,60

Table 3.3.1.1—*continued*
Diatomic Electronic Transition Lasers

Species	Wavelength (μm)	Transition assignment	Comments	References
CO	0.56053	B'Σ–A'Π (0-3) Band, Q(5)	1496	59,60
	0.60646	B'Σ–A'Π (0-4) Band, Q(9)	1496	59,60
	0.60674	B'Σ–A'Π (0-4) Band, Q(8)	1496	59,60
	0.60699	B'Σ–A'Π (0-4) Band, Q(7)	1496	59,60
	0.60722	B'Σ–A'Π (0-4) Band, Q(6)	1496	59,60
	0.60742	B'Σ–A'Π (0-4) Band, Q(5)	1496	59,60
	0.60759	B'Σ–A'Π (0-4) Band, Q(4)	1496	59,60
	0.65973	B'Σ–A'Π (0-5) Band, Q(10)	1496	59,60
	0.66013	B'Σ–A'Π (0-5) Band, Q(9)	1496	59,60
	0.66049	B'Σ–A'Π (0-5) Band, Q(8)orP(13)	1495,1496	59,60
	0.66082	B'Σ–A'Π (0-5) Band, Q(7)	1496	59,60
	0.66109	B'Σ–A'Π (0-5) Band, Q(6)	1496	59,60
CO^+	247.0	$B^2\Sigma^+_u$–$X^2\Sigma^+_u$ (0-2)	1499	61
	395.4	$B^2\Sigma^+_u$–$A^2\Pi^+_u$ (0-0)	1499	61
	421.0	$B^2\Sigma^+_u$–$A^2\Pi^+_u$ (0-1)	1499	61
D_2	0.11134	$C^1\Pi_u \rightarrow X^1\Sigma^+_g$ (1-4) R(0)	1515	66
	0.11377	$C^1\Pi_u \rightarrow X^1\Sigma^+_g$ (1-5) R(0)	1515	66
	0.11476	$C^1\Pi_u \rightarrow X^1\Sigma^+_g$ (1-5) P(2)	1515	66
	0.11565	$C^1\Pi_u \rightarrow X^1\Sigma^+_g$ (2-6) R(0)	1515	66
	0.11584	$C^1\Pi_u \rightarrow X^1\Sigma^+_g$ (2-6) P(2)	1515	66
	0.11881	$C^1\Pi_u \rightarrow X^1\Sigma^+_g$ (2-7) R(0)	1515	66
	0.11901	$C^1\Pi_u \rightarrow X^1\Sigma^+_g$ (2-7) P(2)	1515	66
	0.11975	$C^1\Pi_u \rightarrow X^1\Sigma^+_g$ (3-8) R(0)	1515	66
	0.11994	$C^1\Pi_u \rightarrow X^1\Sigma^+_g$ (3-8) P(2)	1515	66
	0.12064	$C^1\Pi_u \rightarrow X^1\Sigma^+_g$ (4-9) R(0)	1515	66

Table 3.3.1.1—*continued*
Diatomic Electronic Transition Lasers

Species	Wavelength (μm)	Transition assignment	Comments	References
D_2	0.12082	$C^1\Pi_u \to X^1\Sigma^+_g$ (4-9) P(2)	1515	66
	0.12280	$C^1\Pi_u \to X^1\Sigma^+_g$ (3-9) R(0)	1515	66
	0.12356	$C^1\Pi_u \to X^1\Sigma^+_g$ (4-10) R(0)	1515	66
	0.12424	$C^1\Pi_u \to X^1\Sigma^+_g$ (5-11) R(0)	1515	66
	0.12441	$C^1\Pi_u \to X^1\Sigma^+_g$ (5-11) P(2)	1515	66
	0.12483	$C^1\Pi_u \to X^1\Sigma^+_g$ (6-12) R(0)	1515	66
	0.12500	$C^1\Pi_u \to X^1\Sigma^+_g$ (6-12) P(2)	1515	66
	0.12533	$C^1\Pi_u \to X^1\Sigma^+_g$ (7-13) R(0)	1515	66
	0.13036	$B^1\Sigma^+_u \to X^1\Sigma^+_g$ (0-5) P(2)	1515	66,68
	0.13459	$B^1\Sigma^+_u \to X^1\Sigma^+_g$ (0-6) P(2)	1515	66,68
	0.13888	$B^1\Sigma^+_u \to X^1\Sigma^+_g$ (0-7) P(2)	1515	66,68
	0.14322	$B^1\Sigma^+_u \to X^1\Sigma^+_g$ (0-8) P(2)	1515	66,68
	0.15758	$B^1\Sigma^+_u \to X^1\Sigma^+_g$ (3-13) P(2)	1515	66,68
	0.15863	$B^1\Sigma^+_u \to X^1\Sigma^+_g$ (7-16) P(1)	1515	66,68
	0.15864	$B^1\Sigma^+_u \to X^1\Sigma^+_g$ (10-19) P(1)	1515	66,68
	0.15867	$B^1\Sigma^+_u \to X^1\Sigma^+_g$ (10-19) P(2)	1515	66,68
	0.15869	$B^1\Sigma^+_u \to X^1\Sigma^+_g$ (10-19) P(3)	1515	66,68
	0.15871	$B^1\Sigma^+_u \to X^1\Sigma^+_g$ (10-19) P(4)	1515	66,68
	0.15872	$B^1\Sigma^+_u \to X^1\Sigma^+_g$ (7-16) P(2)	1515	66,68
	0.15898	$B^1\Sigma^+_u \to X^1\Sigma^+_g$ (4-14) P(2)	1515	66,68
	0.15913	$B^1\Sigma^+_u \to X^1\Sigma^+_g$ (9-18) P(2)	1515	66,68
	0.15914	$B^1\Sigma^+_u \to X^1\Sigma^+_g$ (8-17) P(2)	1515	66,68
	0.15923	$B^1\Sigma^+_u \to X^1\Sigma^+_g$ (8-17) P(3)	1515	66,68
	0.15926	$B^1\Sigma^+_u \to X^1\Sigma^+_g$ (9-18) P(4)	1515	66,68
	0.16009	$B^1\Sigma^+_u \to X^1\Sigma^+_g$ (5-15) P(2)	1515	66,68
	0.16021	$B^1\Sigma^+_u \to X^1\Sigma^+_g$ (5-15) P(3)	1515	66,68
	0.16035	$B^1\Sigma^+_u \to X^1\Sigma^+_g$ (5-15) P(4)	1515,1516	66,68
	0.16058	$B^1\Sigma^+_u \to X^1\Sigma^+_g$ (9-19) R(0)	1515	66,68
	0.16065	$B^1\Sigma^+_u \to X^1\Sigma^+_g$ (9-19) P(1)	1515	66,68
	0.16068	$B^1\Sigma^+_u \to X^1\Sigma^+_g$ (9-19) P(2)	1515	66,68
	0.16077	$B^1\Sigma^+_u \to X^1\Sigma^+_g$ (6-16) P(1)	1515	66,68
	0.16085	$B^1\Sigma^+_u \to X^1\Sigma^+_g$ (6-16) P(2)	1515	66,68
	0.16096	$B^1\Sigma^+_u \to X^1\Sigma^+_g$ (6-16) P(3)	1515	66,68
	0.16107	$B^1\Sigma^+_u \to X^1\Sigma^+_g$ (7-17) R(0)	1515	66,68
	0.16108	$B^1\Sigma^+_u \to X^1\Sigma^+_g$ (6-16) P(4)	1515	66,68
	0.16115	$B^1\Sigma^+_u \to X^1\Sigma^+_g$ (8-18) P(1)	1515	66,68
	0.16117	$B^1\Sigma^+_u \to X^1\Sigma^+_g$ (7-17) P(1)	1515	66,68
	0.16120	$B^1\Sigma^+_u \to X^1\Sigma^+_g$ (8-18) P(2)	1515	66,68
	0.16124	$B^1\Sigma^+_u \to X^1\Sigma^+_g$ (7-17) P(2)	1515	66,68

Table 3.3.1.1—*continued*
Diatomic Electronic Transition Lasers

Species	Wavelength (μm)	Transition assignment	Comments	References
D_2	0.16126	$B^1\Sigma^+_u \rightarrow X^1\Sigma^+_g$ (8-18) P(3)	1515	66,68
	0.16132	$B^1\Sigma^+_u \rightarrow X^1\Sigma^+_g$ (7-17) P(3)	1515	66,68
	0.16141	$B^1\Sigma^+_u \rightarrow X^1\Sigma^+_g$ (7-17) P(4)	1515,1516	66,68
	0.16166	$B^1\Sigma^+_u \rightarrow X^1\Sigma^+_g$ (9-20) P(2)	1515	66,68
	0.82798	$E,F^1\Sigma^+_g \rightarrow B^1\Sigma^+_u$ (2-0) P(3)	1517	71,72,74
	0.95326	$E,F^1\Sigma^+_g \rightarrow B^1\Sigma^+_u$ (1-0) P(3)	1517	71,72,74
F_2	0.15671	$^3\Pi_g \rightarrow {}^3\Pi_u$	1500–1502	63–65
	0.15748	$^3\Pi_g \rightarrow {}^3\Pi_u$	1500–1502	63–65
	0.15759	$^3\Pi_g \rightarrow {}^3\Pi_u$	1500–1502	63–65
H_2	0.11020	$C'\Pi_u \rightarrow X'\Sigma^+_g$ (0-2) P(2)	1503	66–72,74
	0.11189	$C'\Pi_u \rightarrow X'\Sigma^+_g$ (1-3) P(2)	1503	66–72,74
	0.11486	$C'\Pi_u \rightarrow X'\Sigma^+_g$ (0-3) P(2)	1503	66–72,74
	0.11613	$C'\Pi_u \rightarrow X'\Sigma^+_g$ (1-4)Q(1)	1503	66–72,74
	0.11639	$C'\Pi_u \rightarrow X'\Sigma^+_g$ (1-4) P(2)	1503	66–72,74
	0.11662	$C'\Pi_u \rightarrow X'\Sigma^+_g$ (1-4) P(3)	1503	66–72,74
	0.11758	$C'\Pi_u \rightarrow X'\Sigma^+_g$ (2-5)Q(1)	1503	66–72,74
	0.11763	$C'\Pi_u \rightarrow X'\Sigma^+_g$ (2,5)Q(2)	6624,6626	1535
	0.11777	$C'\Pi_u \rightarrow X'\Sigma^+_g$ (2,5)Q(3)	6624,6625	1535
	0.11783	$C'\Pi_u \rightarrow X'\Sigma^+_g$ (2-5) P(2)	1503	66–72,74
	0.11805	$C'\Pi_u \rightarrow X'\Sigma^+_g$ (2-5) P(3)	1503	66–72,74
	0.11893	$C'\Pi_u \rightarrow X'\Sigma^+_g$ (3-6)Q(1)	1503	66–72,74
	0.12067	$C'\Pi_u \rightarrow X'\Sigma^+_g$ (1-5)Q(1)	1503	66–72,74
	0.12093	$C'\Pi_u \rightarrow X'\Sigma^+_g$ (1-5) P(2)	1503	66–72,74
	0.12173	$C'\Pi_u \rightarrow X'\Sigma^+_g$ (2-6) R(1)	1503	66–72,74
	0.12189	$C'\Pi_u \rightarrow X'\Sigma^+_g$ (2-6)Q(1)	1503	66–72,74
	0.12214	$C'\Pi_u \rightarrow X'\Sigma^+_g$ (2-6) P(2)	1503	66–72,74
	0.12236	$C'\Pi_u \rightarrow X'\Sigma^+_g$ (2-6) P(3)	1503	66–72,74
	0.12299	$C'\Pi_u \rightarrow X'\Sigma^+_g$ (3-7)Q(1)	1503	66–72,74
	0.12323	$C'\Pi_u \rightarrow X'\Sigma^+_g$ (3-7) P(2)	1503	66–72,74
	0.12394	$C'\Pi_u \rightarrow X'\Sigma^+_g$ (4-8)Q(1)	1503	66–72,74
	0.12417	$C'\Pi_u \rightarrow X'\Sigma^+_g$ (4-8) P(2)	1503	66–72,74
	0.12752	$B^1\Pi^+_u \rightarrow X^1\Sigma^+_g$ (0,3)R(0)	6624,6627	1535
	0.13336	$B^1\Pi^+_u \rightarrow X^1\Sigma^+_g$ (0,4)R(0)	6624,6627	1535
	0.1339	$B^1\Pi^+_u \rightarrow X^1\Sigma^+_g$ (0,4)P(2)	6624,6627	1535
	0.13423	$B^1\Pi^+_u \rightarrow X^1\Sigma^+_g$ (0-4) P(3)	1503	66–72,74
	0.13944	$B^1\Pi^+_u \rightarrow X^1\Sigma^+_g$ (0,5)R(0)	6624,6627	1535
	0.13998	$B^1\Pi^+_u \rightarrow X^1\Sigma^+_g$ (0,5)P(2)	6624,6627	1535

Table 3.3.1.1—*continued*
Diatomic Electronic Transition Lasers

Species	Wavelength (μm)	Transition assignment	Comments	References
H_2	0.14026	$B^1\Pi^+_u \rightarrow X^1\Sigma^+_g$ (0-5) P(3)	1503	66–72,74
	0.14187	$B^1\Pi^+_u \rightarrow X^1\Sigma^+_g$ (4-9) P(3)	1503	66–72,74
	0.14288	$B^1\Pi^+_u \rightarrow X^1\Sigma^+_g$ (1,6)R(2)	6624,6626	1535
	0.14362	$B^1\Pi^+_u \rightarrow X^1\Sigma^+_g$ (1-6) P(3)	1503	66–72,74
	0.14409	$B^1\Pi^+_u \rightarrow X^1\Sigma^+_g$ (1,6)P(4)	6624,6626	1535
	0.14409	$B^1\Pi^+_u \rightarrow X^1\Sigma^+_g$ (3-7) P(3)	1503	66–72,74
	0.14555	$B^1\Pi^+_u \rightarrow X^1\Sigma^+_g$ (0,6)R(0)	6624,6627	1535
	0.14609	$B^1\Pi^+_u \rightarrow X^1\Sigma^+_g$ (0,6)P(2)	6624,6627	1535
	0.14638	$B^1\Pi^+_u \rightarrow X^1\Sigma^+_g$ (0-6) P(3)	1503	66–72,74
	0.14670	$B^1\Pi^+_u \rightarrow X^1\Sigma^+_g$ (6-9) P(3)	1503	66–72,74
	0.14865	$B^1\Pi^+_u \rightarrow X^1\Sigma^+_g$ (1-7) R(0)	1503	66–72,74
	0.14876	$B^1\Pi^+_u \rightarrow X^1\Sigma^+_g$ (1,7)R(2)	6624,6626	1535
	0.14942	$B^1\Pi^+_u \rightarrow X^1\Sigma^+_g$ (11-14) P(3)	1504	66–72,74
	0.14952	$B^1\Pi^+_u \rightarrow X^1\Sigma^+_g$ (1-7) P(3)	1503	66–72,74
	0.14996	$B^1\Pi^+_u \rightarrow X^1\Sigma^+_g$ (1,7)P(4)	6624,6625	1535
	0.15233	$B^1\Pi^+_u \rightarrow X^1\Sigma^+_g$ (2-8) P(3)	1505	66–72,74
	0.15315	$B^1\Pi^+_u \rightarrow X^1\Sigma^+_g$ (2,8)P(5)	6624,6625	1535
	0.15449	$B^1\Pi^+_u \rightarrow X^1\Sigma^+_g$ (1-8) R(0)	1503	66–72,74
	0.15454	$B^1\Pi^+_u \rightarrow X^1\Sigma^+_g$ (1,8)R(2)	6624,6626	1535
	0.15534	$B^1\Pi^+_u \rightarrow X^1\Sigma^+_g$ (1-8) P(3)	1503	66,67,69–72,74
	0.15574	$B^1\Pi^+_u \rightarrow X^1\Sigma^+_g$ (1,8)P(4)	6624,6626	1535
	0.15655	$B^1\Pi^+_u \rightarrow X^1\Sigma^+_g$ (8-14) R(1)	1504	66,67,69–72,74
	0.15663	$B^1\Pi^+_u \rightarrow X^1\Sigma^+_g$ (8-14) R(0)	1503	66,67,69–72,74
	0.15673	$B^1\Pi^+_u \rightarrow X^1\Sigma^+_g$ (8-14) P(3)	1505,1506	66,67,69–72,74
	0.15708	$B^1\Sigma^+_u \rightarrow X^1\Sigma^+_g$ (2,9) R(3)	6624,6625	1535
	0.15720	$B^1\Pi^+_u \rightarrow X^1\Sigma^+_g$ (2-9) P(1)	1505,1506	66,67,69–72,74
	0.15774	$B^1\Pi^+_u \rightarrow X^1\Sigma^+_g$ (2-9) P(3)	1505,1506	66,67,69–72,74
	0.15777	$B^1\Pi^+_u \rightarrow X^1\Sigma^+_g$ (7-13) R(0)	1503	66,67,69–72,74
	0.15792	$B^1\Pi^+_u \rightarrow X^1\Sigma^+_g$ (7-13) P(1)	1505	66,67,69–72,74
	0.15800	$B^1\Pi^+_u \rightarrow X^1\Sigma^+_g$ (7-13) P(2)	1505	66,67,69–72,74
	0.15808	$B^1\Pi^+_u \rightarrow X^1\Sigma^+_g$ (7-13) P(3)	1506,1507	66,67,69–72,74
	0.15849	$B^1\Sigma^+_u \rightarrow X^1\Sigma^+_g$ (2,9) P(5)	6624,6625	1535
	0.15890	$B^1\Pi^+_u \rightarrow X^1\Sigma^+_g$ (3-10) R(0)	1503	66,67,69–72,74
	0.15913	$B^1\Pi^+_u \rightarrow X^1\Sigma^+_g$ (3-10) P(1)	1505,1506	66,67,69–72,74
	0.15934	$B^1\Pi^+_u \rightarrow X^1\Sigma^+_g$ (3-10) P(2)	1505	66,67,69–72,74
	0.15961	$B^1\Pi^+_u \rightarrow X^1\Sigma^+_g$ (3-10) P(3)	1506,1507	66,67,69–72,74
	0.16049	$B^1\Pi^+_u \rightarrow X^1\Sigma^+_g$ (4-11) P(1)	1506,1507	66,67,69–72,74
	0.16059	$B^1\Pi^+_u \rightarrow X^1\Sigma^+_g$ (6-13) R(0)	1503	66,67,69–72,74
	0.16062	$B^1\Pi^+_u \rightarrow X^1\Sigma^+_g$ (4-11) P(2)	1505	66,67,69–72,74

Table 3.3.1.1—*continued*
Diatomic Electronic Transition Lasers

Species	Wavelength (µm)	Transition assignment	Comments	References
H_2	0.16075	$B^1\Pi^+_u \to X^1\Sigma^+_g$ (6-13) P(1)	1506,1507	66,67,69–72,74
	0.16083	$B^1\Pi^+_u \to X^1\Sigma^+_g$ (6-13) P(2)	1503	66,67,69–72,74
	0.16084	$B^1\Pi^+_u \to X^1\Sigma^+_g$ (4-11) P(3)	1506,1507	66,67,69–72,74
	0.16090	$B^1\Pi^+_u \to X^1\Sigma^+_g$ (6-13) P(3)	1507	66,67,69–72,74
	0.16103	$B^1\Pi^+_u \to X^1\Sigma^+_g$ (5-12) P(1)	1507	66,67,69–72,74
	0.16117	$B^1\Pi^+_u \to X^1\Sigma^+_g$ (5-12) P(2)	1507	66,67,69–72,74
	0.16132	$B^1\Pi^+_u \to X^1\Sigma^+_g$ (5-12) P(3)	1506,1507	66,67,69–72,74
	0.16148	$B^1\Pi^+_u \to X^1\Sigma^+_g$ (5-12) P(4)	1504	66,67,69–72,74
	0.16165	$B^1\Pi^+_u \to X^1\Sigma^+_g$ (5-12) P(5)	1504	66,67,69–72,74
	0.16395	$B^1\Pi^+_u \to X^1\Sigma^+_g$ (4-11) R(0)	1503	66,67,69–72,74
	0.16415	$B^1\Pi^+_u \to X^1\Sigma^+_g$ (4-11) P(1)	1503	66,67,69–72,74
	0.16429	$B^1\Pi^+_u \to X^1\Sigma^+_g$ (4-11) P(2)	1503	66,67,69–72,74
	0.16444	$B^1\Pi^+_u \to X^1\Sigma^+_g$ (4-11) P(3)	1503	66,67,69–72,74
	0.75441	$E^1\Sigma^+_g \to B^1\Sigma^+_u$ (2,0) P(3)	6624,6626	1535
	0.83519	$E,F^1\Sigma^+_g \to B^1\Sigma^+_u$ (2-1) P(2)	1508,1509	71,72,74
	0.8370	$E^1\Sigma^+_g \to B^1\Sigma^+_u$ (2,1) P(3)	6624,6626	1535
	0.88787	$E,F^1\Sigma^+_g \to B^1\Sigma^+_u$ (1-0) P(4)	1508,1509	71,72,74
	0.89013	$E,F^1\Sigma^+_g \to B^1\Sigma^+_u$ (1-0) P(2)	1508,1509	71,72,74
	0.9222	$E^1\Sigma^+_g \to B^1\Sigma^+_u$ (2,2) P(4)	6624,6625	1535
	1.1165	$E,F^1\Sigma^+_g \to B^1\Sigma^+_u$ (0-0) P(4)	1508,1509	71,72,74
	1.1210	$E^1\Sigma^+_g \to B^1\Sigma^+_u$(0,0) P(1)	6624,6627	1535
	1.1225	$E,F^1\Sigma^+_g \to B^1\Sigma^+_u$ (0-0) P(2)	1508,1509	71,72,74
	1.3061	$E,F^1\Sigma^+_g \to B^1\Sigma^+_u$ (0-1) P(4)	1508,1509	71,72,74
	1.3166	$E,F^1\Sigma^+_g \to B^1\Sigma^+_u$ (0-1) P(2)	1508,1509	71,72,74
Para-H_2	0.10982	$C^2\Pi_u \to X^1\Sigma^+_g$ (0-2) R(0)	1510	66
	0.11152	$C^2\Pi_u \to X^1\Sigma^+_g$ (1-3) R(0)	1510	66
	0.11446	$C^2\Pi_u \to X^1\Sigma^+_g$ (0-3) R(0)	1510	66
	0.11600	$C^2\Pi_u \to X^1\Sigma^+_g$ (1-4) R(0)	1510	66
	0.11746	$C^2\Pi_u \to X^1\Sigma^+_g$ (2-5) R(0)	1510	66
	0.12054	$C^2\Pi_u \to X^1\Sigma^+_g$ (1-5) R(0)	1510	66
	0.12177	$C^2\Pi_u \to X^1\Sigma^+_g$ (2-6) R(0)	1510	66
	0.12287	$C^2\Pi_u \to X^1\Sigma^+_g$ (3-7) R(0)	1510	66
	0.12383	$C^2\Pi_u \to X^1\Sigma^+_g$ (4-8) R(0)	1510	66
	0.12462	$C^2\Pi_u \to X^1\Sigma^+_g$ (5-9) R(0)	1510	66
	0.12520	$C^2\Pi_u \to X^1\Sigma^+_g$ (6-10) R(0)	1510	66
	0.12795	$B^1\Sigma^+_u \to X^1\Sigma^+_g$ (0-3) P(2)	1510	66,68
	0.13386	$B^1\Sigma^+_u \to X^1\Sigma^+_g$ (0-4) P(2)	1510	66,68
	0.13598	$B^1\Sigma^+_u \to X^1\Sigma^+_g$ (4-6) P(2)	1510	66,68

Table 3.3.1.1—*continued*
Diatomic Electronic Transition Lasers

Species	Wavelength (μm)	Transition assignment	Comments	References
Para-H_2	0.13680	$B^1\Sigma^+_u \rightarrow X^1\Sigma^+_g$ (6-7) P(2)	1510	66,68
	0.13990	$B^1\Sigma^+_u \rightarrow X^1\Sigma^+_g$ (0-5) P(2)	1510	66,68
	0.14075	$B^1\Sigma^+_u \rightarrow X^1\Sigma^+_g$ (0-5) P(4)	1510	66,68
	0.14326	$B^1\Sigma^+_u \rightarrow X^1\Sigma^+_g$ (1-6) P(2)	1510	66,68
	0.14376	$B^1\Sigma^+_u \rightarrow X^1\Sigma^+_g$ (3-7) P(2)	1510	66,68
	0.14406	$B^1\Sigma^+_u \rightarrow X^1\Sigma^+_g$ (7-9) P(2)	1510	66,68
	0.14602	$B^1\Sigma^+_u \rightarrow X^1\Sigma^+_g$ (0-6) P(2)	1510	66,68
	0.14641	$B^1\Sigma^+_u \rightarrow X^1\Sigma^+_g$ (6-9) P(2)	1510	66,68
	0.14684	$B^1\Sigma^+_u \rightarrow X^1\Sigma^+_g$ (0-6) P(4)	1510	66,68
	0.14917	$B^1\Sigma^+_u \rightarrow X^1\Sigma^+_g$ (1-7) P(2)	1510	66,68
	0.15157	$B^1\Sigma^+_u \rightarrow X^1\Sigma^+_g$ (4-9) P(2)	1510	66,68
	0.15199	$B^1\Sigma^+_u \rightarrow X^1\Sigma^+_g$ (2-8) P(2)	1510	66,68
	0.15349	$B^1\Sigma^+_u \rightarrow X^1\Sigma^+_g$ (5-10) P(2)	1510	66,68
	0.15501	$B^1\Sigma^+_u \rightarrow X^1\Sigma^+_g$ (1-8) P(2)	1510	66,68
	0.15640	$B^1\Sigma^+_u \rightarrow X^1\Sigma^+_g$ (8-14) P(4)	1510	66,68
	0.15675	$B^1\Sigma^+_u \rightarrow X^1\Sigma^+_g$ (8-14) P(2)	1510	66,68
	0.15675	$B^1\Sigma^+_u \rightarrow X^1\Sigma^+_g$ (8-14) P(2)	1510	66,68
	0.15743	$B^1\Sigma^+_u \rightarrow X^1\Sigma^+_g$ (2-9) P(2)	1510	66,68
	0.15777	$B^1\Sigma^+_u \rightarrow X^1\Sigma^+_g$ (7-13) R(0)	1510,1511	66,68
	0.15800	$B^1\Sigma^+_u \rightarrow X^1\Sigma^+_g$ (7-13) P(2)	1510	66,68
	0.15811	$B^1\Sigma^+_u \rightarrow X^1\Sigma^+_g$ (2-9) P(4)	1510,1511	66,68
	0.15814	$B^1\Sigma^+_u \rightarrow X^1\Sigma^+_g$ (7-13) P(4)	1510,1511	66,68
	0.15890	$B^1\Sigma^+_u \rightarrow X^1\Sigma^+_g$ (3-10) R(0)	1510,1511	66,68
	0.15934	$B^1\Sigma^+_u \rightarrow X^1\Sigma^+_g$ (3-10) P(2)	1510	66,68
	0.15993	$B^1\Sigma^+_u \rightarrow X^1\Sigma^+_g$ (3-10) P(4)	1510	66,68
	0.16024	$B^1\Sigma^+_u \rightarrow X^1\Sigma^+_g$ (4-11) R(0)	1510,1511	66,68
	0.16059	$B^1\Sigma^+_u \rightarrow X^1\Sigma^+_g$ (6-13) R(0)	1510,1511	66,68
	0.16062	$B^1\Sigma^+_u \rightarrow X^1\Sigma^+_g$ (4-11) P(2)	1510,1511	66,68
	0.16083	$B^1\Sigma^+_u \rightarrow X^1\Sigma^+_g$ (6-13) P(2)	1510	66,68
	0.16096	$B^1\Sigma^+_u \rightarrow X^1\Sigma^+_g$ (6-13) P(4)	1510	66,68
	0.16103	$B^1\Sigma^+_u \rightarrow X^1\Sigma^+_g$ (5-12) P(1)	1510,1511	66,68
	0.16109	$B^1\Sigma^+_u \rightarrow X^1\Sigma^+_g$ (4-11) P(4)	1510,1511	66,68
	0.16117	$B^1\Sigma^+_u \rightarrow X^1\Sigma^+_g$ (5-12) P(2)	1510	66,68
	0.16132	$B^1\Sigma^+_u \rightarrow X^1\Sigma^+_g$ (5-12) P(3)	1510,1511	66,68
	0.16149	$B^1\Sigma^+_u \rightarrow X^1\Sigma^+_g$ (5-12) P(4)	1510	66,68
	0.16429	$B^1\Sigma^+_u \rightarrow X^1\Sigma^+_g$ (4-11) P(2)	1510	66,68
	0.16460	$B^1\Sigma^+_u \rightarrow X^1\Sigma^+_g$ (4-11) P(4)	1510	66,68
HD	0.11386	$C^1\Pi \rightarrow X^1\Sigma^+$ (1-4) R(0)	1512	66

Table 3.3.1.1—*continued*
Diatomic Electronic Transition Lasers

Species	Wavelength (μm)	Transition assignment	Comments	References
HD	0.11415	$C^1\Pi \rightarrow X^1\Sigma^+$ (1-4) P(2)	1512	66
	0.11520	$C^1\Pi \rightarrow X^1\Sigma^+$ (2-5) R(0)	1512	66
	0.11781	$C^1\Pi \rightarrow X^1\Sigma^+$ (1-5) R(0)	1512	66
	0.11900	$C^1\Pi \rightarrow X^1\Sigma^+$ (2-6) R(0)	1512	66
	0.11928	$C^1\Pi \rightarrow X^1\Sigma^+$ (2-6) P(2)	1512	66
	0.12010	$C^1\Pi \rightarrow X^1\Sigma^+$ (3-7) R(0)	1512	66
	0.12113	$C^1\Pi \rightarrow X^1\Sigma^+$ (4-8) R(0)	1512	66
	0.12284	$C^1\Pi \rightarrow X^1\Sigma^+$ (6-10) R(0)	1512	66
	0.12457	$B^1\Sigma^+_u \rightarrow X^1\Sigma^+_g$ (6-5) P(2)		66,68
	0.12528	$B^1\Sigma^+_u \rightarrow X^1\Sigma^+_g$ (0-3) P(2)		66,68
	0.13033	$B^1\Sigma^+_u \rightarrow X^1\Sigma^+_g$ (0-4) P(2)		66,68
	0.13551	$B^1\Sigma^+_u \rightarrow X^1\Sigma^+_g$ (0-5) P(2)		66,68
	0.14077	$B^1\Sigma^+_u \rightarrow X^1\Sigma^+_g$ (0-6) P(2)		66,68
	0.14884	$B^1\Sigma^+_u \rightarrow X^1\Sigma^+_g$ (1-8) P(2)		66,68
	0.15136	$B^1\Sigma^+_u \rightarrow X^1\Sigma^+_g$ (2-9) P(2)		66,68
	0.15299	$B^1\Sigma^+_u \rightarrow X^1\Sigma^+_g$ (5-11) P(2)		66,68
	0.15620	$B^1\Sigma^+_u \rightarrow X^1\Sigma^+_g$ (2-10) P(2)		66,68
	0.15713	$B^1\Sigma^+_u \rightarrow X^1\Sigma^+_g$ (9-16) R(0)		66,68
	0.15727	$B^1\Sigma^+_u \rightarrow X^1\Sigma^+_g$ (9-16) P(3)		66,68
	0.15743	$B^1\Sigma^+_u \rightarrow X^1\Sigma^+_g$ (9-16) P(2)		66,68
	0.15801	$B^1\Sigma^+_u \rightarrow X^1\Sigma^+_g$ (8-15) R(0)		66,68
	0.15809	$B^1\Sigma^+_u \rightarrow X^1\Sigma^+_g$ (3-11) P(2)		66,68
	0.15819	$B^1\Sigma^+_u \rightarrow X^1\Sigma^+_g$ (8-15) P(2)		66,68
	0.15825	$B^1\Sigma^+_u \rightarrow X^1\Sigma^+_g$ (8-15) P(3)		66,68
	0.15831	$B^1\Sigma^+_u \rightarrow X^1\Sigma^+_g$ (3-11) P(3)		66,68
	0.15938	$B^1\Sigma^+_u \rightarrow X^1\Sigma^+_g$ (4-12) P(1)		66,68
	0.15955	$B^1\Sigma^+_u \rightarrow X^1\Sigma^+_g$ (4-12) P(2)		66,68
	0.15974	$B^1\Sigma^+_u \rightarrow X^1\Sigma^+_g$ (4-12) P(3)		66,68
	0.16023	$B^1\Sigma^+_u \rightarrow X^1\Sigma^+_g$ (5-13) R(0)		66,68
	0.16037	$B^1\Sigma^+_u \rightarrow X^1\Sigma^+_g$ (5-13) P(1)		66,68
	0.16046	$B^1\Sigma^+_u \rightarrow X^1\Sigma^+_g$ (7-15) R(0)		66,68
	0.16052	$B^1\Sigma^+_u \rightarrow X^1\Sigma^+_g$ (5-13) P(2)		66,68
	0.16057	$B^1\Sigma^+_u \rightarrow X^1\Sigma^+_g$ (7-15) P(1)	1513	66,68
	0.16065	$B^1\Sigma^+_u \rightarrow X^1\Sigma^+_g$ (7-15) P(2)		66,68
	0.16067	$B^1\Sigma^+_u \rightarrow X^1\Sigma^+_g$ (6-14) R(0)		66,68
	0.16068	$B^1\Sigma^+_u \rightarrow X^1\Sigma^+_g$ (5-13) P(3)		66,68
	0.16069	$B^1\Sigma^+_u \rightarrow X^1\Sigma^+_g$ (7-15) P(3)		66,68
	0.16075	$B^1\Sigma^+_u \rightarrow X^1\Sigma^+_g$ (7-15) P(4)	1513	66,68
	0.16079	$B^1\Sigma^+_u \rightarrow X^1\Sigma^+_g$ (6-14) P(1)	1513	66,68

Table 3.3.1.1—*continued*
Diatomic Electronic Transition Lasers

Species	Wavelength (µm)	Transition assignment	Comments	References
HD	0.16083	$B^1\Sigma^+_u \to X^1\Sigma^+_g$ (5-13) P(4)	1513	66,68
	0.16091	$B^1\Sigma^+_u \to X^1\Sigma^+_g$ (6-14) P(2)		66,68
	0.16103	$B^1\Sigma^+_u \to X^1\Sigma^+_g$ (6-14) P(3)		66,68
	0.16113	$B^1\Sigma^+_u \to X^1\Sigma^+_g$ (6-14) P(4)		66,68
	0.9163	$E,F^1\Sigma^+_g \to B^1\Sigma^+_u$(1-0)	1514	71,72,74
HgBr	0.4990	$B^2\Sigma^+_{1/2} \to X^2\Sigma^+_{1/2}$ (0,21)	1518	75–78
	0.5018	$B^2\Sigma^+_{1/2} \to X^2\Sigma^+_{1/2}$ (0,22)	1519,1521	75–78
	0.5020-0.5026	$B^2\Sigma^+_{1/2} \to X^2\Sigma^+_{1/2}$	6629,6630,6634 6636,6638	1537–1545
	0.5023	$B^2\Sigma^+_{1/2} \to X^2\Sigma^+_{1/2}$ (0,22)	1519,1522	75–78
	0.5026	$B^2\Sigma^+_{1/2} \to X^2\Sigma^+_{1/2}$ (0,22)	1519,1522	75–78
	0.5039	$B^2\Sigma^+_{1/2} \to X^2\Sigma^+_{1/2}$ (0,23)	1519,1521	75–78
	0.5042	$B^2\Sigma^+_{1/2} \to X^2\Sigma^+_{1/2}$ (0,23)	1519,1521	75–78
	0.5046	$B^2\Sigma^+_{1/2} \to X^2\Sigma^+_{1/2}$ (0,23)	1519,1522	75–78
HgCl	0.5516	$B^2\Sigma_{1/2}^+ \to X^2\Sigma^+_{1/2}$ (0,21)	1525–528	75,78–83
	0.5523	$B^2\Sigma_{1/2}^+ \to X^2\Sigma^+_{1/2}$ (0,23)	1525–528	75,78–83
	0.5550	$B^2\Sigma_{1/2}^+ \to X^2\Sigma^+_{1/2}$ (0,22)	1525–528	75,78–83
	0.5576	$B^2\Sigma_{1/2}^+ \to X^2\Sigma^+_{1/2}$ (0,22)	1525–528	75,78–83
	0.558	$B^2\Sigma^+_{1/2} \to X^2\Sigma^+_{1/2}$	6639	1546
	0.5584	$B^2\Sigma_{1/2}^+ \to X^2\Sigma^+_{1/2}$ (1,23)	1525–528	75,78–83
	0.559	$B^2\Sigma^+_{1/2} \to X^2\Sigma^+_{1/2}$	6639	1546
	0.5590	$B^2\Sigma_{1/2}^+ \to X^2\Sigma^+_{1/2}$ (1,23)	1525–528	75,78–83
HgI	0.442-0.444	$B^2\Sigma^+_{1/2} \to X^2\Sigma^+_{1/2}$	6641	1547
	0.4430	$B^2\Sigma^+_{1/2} \to X^2\Sigma^+_{1/2}$ (0,15)	1529	78,79
	0.4450	$B^2\Sigma^+_{1/2} \to X^2\Sigma^+_{1/2}$ (0,17)	1529	78,79
I_2	0.3420	$^3\Pi_{2g} \to {}^3\Pi_{2u}$ (0-12), (2-15)	1531–1533	85–87
	0.3420-0.3428	$D'(2_g) \to A'(2_u)$	6643,6645,6648 6650,6652,6654	1548–51,1553
	0.3423	$^3\Pi_{2g} \to {}^3\Pi_{2u}$ (3-17)	1533	85–87
	0.3424	$^3\Pi_{2g} \to {}^3\Pi_{2u}$ (1-14)	1533	85–87
	0.3428	$^3\Pi_{2g} \to {}^3\Pi_{2u}$ (2-16), (0-13)	1533	85–87
	0.495-0.512		6655	1554,1555
	0.5443	$B^3\Pi_{2u} \to X^1\Sigma^+_g$	1534	88–92
	0.5550	$B^3\Pi_{2u} \to X^1\Sigma^+_g$	1535	88–92
	0.5567	$B^3\Pi_{2u} \to X^1\Sigma^+_g$	1534	88–92
	0.5680	$B^3\Pi_{2u} \to X^1\Sigma^+_g$	1535	88–92

Table 3.3.1.1—*continued*
Diatomic Electronic Transition Lasers

Species	Wavelength (µm)	Transition assignment	Comments	References
I_2	0.5697	$B^3\Pi_{2u} \rightarrow X^1\Sigma^+_g$ (43-9)	1536	88–92
	0.5745	$B^3\Pi_{2u} \rightarrow X^1\Sigma^+_g$	1535	88–92
	0.5764	$B^3\Pi_{2u} \rightarrow X^1\Sigma^+_g$	1534	88–92
	0.5815	$B^3\Pi_{2u} \rightarrow X^1\Sigma^+_g$	1535	88–92
	0.5830	$B^3\Pi_{2u} \rightarrow X^1\Sigma^+_g$ (43-11)	1536	88–92
	0.5880	$B^3\Pi_{2u} \rightarrow X^1\Sigma^+_g$	1535	88–92
	0.5905	$B^3\Pi_{2u} \rightarrow X^1\Sigma^+_g$	1534	88–92
	0.5968	$B^3\Pi_{2u} \rightarrow X^1\Sigma^+_g$ (43-13)	1536	88–92
	0.6025	$B^3\Pi_{2u} \rightarrow X^1\Sigma^+_g$	1535	88–92
	0.6048	$B^3\Pi_{2u} \rightarrow X^1\Sigma^+_g$	1534	88–92
	0.6111	$B^3\Pi_{2u} \rightarrow X^1\Sigma^+_g$ (43-15)	1536	88–92
	0.6175	$B^3\Pi_{2u} \rightarrow X^1\Sigma^+_g$	1535	88–92
	0.617520	$B^3\Pi_{2u} \rightarrow X^1\Sigma^+_g$ (34-13) R(84)	1535	88–92
	0.617730	$B^3\Pi_{2u} \rightarrow X^1\Sigma^+_g$ (35-13) R(107)	1535	88–92
	0.617900	$B^3\Pi_{2u} \rightarrow X^1\Sigma^+_g$ (33-13) R(58)	1535	88–92
	0.617970	$B^3\Pi_{2u} \rightarrow X^1\Sigma^+_g$ (34-13) R(89)	1535	88–92
	0.617990	$B^3\Pi_{2u} \rightarrow X^1\Sigma^+_g$ (34-13) P(86)	1535	88–92
	0.618245	$B^3\Pi_{2u} \rightarrow X^1\Sigma^+_g$ (33-13) P(60)	1535	88–92
	0.618325	$B^3\Pi_{2u} \rightarrow X^1\Sigma^+_g$ (35-13) P(109)	1535	88–92
	0.618490	$B^3\Pi_{2u} \rightarrow X^1\Sigma^+_g$ (34-13) P(91)	1535	88–92
	0.618580	$B^3\Pi_{2u} \rightarrow X^1\Sigma^+_g$ (33-13) P(65)	1535	88–92
	0.6198	$B^3\Pi_{2u} \rightarrow X^1\Sigma^+_g$	1534	88–92
	0.6258	$B^3\Pi_{2u} \rightarrow X^1\Sigma^+_g$ P(17), R(17)	1537	88–92
	0.6260	$B^3\Pi_{2u} \rightarrow X^1\Sigma^+_g$ (43-17)	1536	88–92
	0.6330	$B^3\Pi_{2u} \rightarrow X^1\Sigma^+_g$	1535	88–92
	0.6352	$B^3\Pi_{2u} \rightarrow X^1\Sigma^+_g$	1534	88–92
	0.6490	$B^3\Pi_{2u} \rightarrow X^1\Sigma^+_g$	1535	88–92
	0.6511	$B^3\Pi_{2u} \rightarrow X^1\Sigma^+_g$	1534	88–92
	0.6592	$B^3\Pi_{2u} \rightarrow X^1\Sigma^+_g$	1534	88–92
	0.6645	$B^3\Pi_{2u} \rightarrow X^1\Sigma^+_g$	1535	88–92

Table 3.3.1.1—*continued*
Diatomic Electronic Transition Lasers

Species	Wavelength (μm)	Transition assignment	Comments	References
I_2	0.6763	$B^3\Pi_{2u} \rightarrow X^1\Sigma^+_g$	1534	88–92
	0.6936	$B^3\Pi_{2u} \rightarrow X^1\Sigma^+_g$	1534	88–92
	0.7114	$B^3\Pi_{2u} \rightarrow X^1\Sigma^+_g$	1534	88–92
	0.8144	$B^3\Pi_{2u} \rightarrow X^1\Sigma^+_g$ P(38), R(38)	1537	88–92
	0.8358	$B^3\Pi_{2u} \rightarrow X^1\Sigma^+_g$ P(40), R(40)	1537	88–92
	0.8578	$B^3\Pi_{2u} \rightarrow X^1\Sigma^+_g$ P(42), R(42)	1537	88–92
	0.8579	$B^3\Pi_{2u} \rightarrow X^1\Sigma^+_g$ (43-42) P(17), R(15), R(11)	1536	88–92
	0.8804	$B^3\Pi_{2u} \rightarrow X^1\Sigma^+_g$ P(44), R(44)	1537	88–92
	0.8806	$B^3\Pi_{2u} \rightarrow X^1\Sigma^+_g$ (43-44) P(17), R(15)	1536	88–92
	0.8813	$B^3\Pi_{2u} \rightarrow X^1\Sigma^+_g$	1534	88–92
	0.9037	$B^3\Pi_{2u} \rightarrow X^1\Sigma^+_g$ P(46), R(46)	1537	88–92
	0.9038	$B^3\Pi_{2u} \rightarrow X^1\Sigma^+_g$ (43-46)	1536	88–92
	0.9047	$B^3\Pi_{2u} \rightarrow X^1\Sigma^+_g$	1534	88–92
	0.9060	$B^3\Pi_{2u} \rightarrow X^1\Sigma^+_g$	1534	88–92
	0.9274	$B^3\Pi_{2u} \rightarrow X^1\Sigma^+_g$ P(48), R(48)	1537	88–92
	0.9276	$B^3\Pi_{2u} \rightarrow X^1\Sigma^+_g$ (43-48) P(17), P(13), R(15), R(11)	1536	88–92
	0.9288	$B^3\Pi_{2u} \rightarrow X^1\Sigma^+_g$	1534	88–92
	0.9295	$B^3\Pi_{2u} \rightarrow X^1\Sigma^+_g$	1534	88–92
	0.9305	$B^3\Pi_{2u} \rightarrow X^1\Sigma^+_g$	1534	88–92
	0.9518	$B^3\Pi_{2u} \rightarrow X^1\Sigma^+_g$ P(50), R(50)	1537	88–92
	0.9520	$B^3\Pi_{2u} \rightarrow X^1\Sigma^+_g$ (43-50)	1536	88–92
	0.9545	$B^3\Pi_{2u} \rightarrow X^1\Sigma^+_g$	1534	88–92
	0.9555	$B^3\Pi_{2u} \rightarrow X^1\Sigma^+_g$	1534	88–92
	0.9766	$B^3\Pi_{2u} \rightarrow X^1\Sigma^+_g$ P(52), R(52)	1537	88–92
	0.9767	$B^3\Pi_{2u} \rightarrow X^1\Sigma^+_g$ (43-52) P(17), R(15)	1536	88–92
	0.9963	$B^3\Pi_{2u} \rightarrow X^1\Sigma^+_g$	1534	88–92
	0.9973	$B^3\Pi_{2u} \rightarrow X^1\Sigma^+_g$	1534	88–92
	1.0019	$B^3\Pi_{2u} \rightarrow X^1\Sigma^+_g$ (43-54) P(17), P(13), R(15), R(11)	1536	88–92
	1.0053	$B^3\Pi_{2u} \rightarrow X^1\Sigma^+_g$	1534	88–92
	1.016-1.340	$B^3\Pi_{u} \rightarrow X^1\Sigma^+_g$	6657	1556
	1.0225	$B^3\Pi_{2u} \rightarrow X^1\Sigma^+_g$	1534	88–92
	1.0245	$B^3\Pi_{2u} \rightarrow X^1\Sigma^+_g$	1534	88–92
	1.0255	$B^3\Pi_{2u} \rightarrow X^1\Sigma^+_g$	1534	88–92

Table 3.3.1.1—*continued*
Diatomic Electronic Transition Lasers

Species	Wavelength (μm)	Transition assignment	Comments	References
I_2	1.0274	$B^3\Pi_{2u} \rightarrow X^1\Sigma^+_g(43\text{-}56)$ P(17), P(13), R(15), R(11)	1536	88–92
	1.0534	$B^3\Pi_{2u} \rightarrow X^1\Sigma^+_g$	1534	88–92
	1.0775	$B^3\Pi_{2u} \rightarrow X^1\Sigma^+_g$	1534	88–92
	1.0788	$B^3\Pi_{2u} \rightarrow X^1\Sigma^+_g$	1534	88–92
	1.1066	$B^3\Pi_{2u} \rightarrow X^1\Sigma^+_g$	1534	88–92
	1.1073	$B^3\Pi_{2u} \rightarrow X^1\Sigma^+_g$	1534	88–92
	1.1206	$B^3\Pi_{2u} \rightarrow X^1\Sigma^+_g$ (11-44) R(40)	1538	88–92
	1.1214	$B^3\Pi_{2u} \rightarrow X^1\Sigma^+_g$ (11-44) P(42)	1538	88–92
	1.1216	$B^3\Pi_{2u} \rightarrow X^1\Sigma^+_g$ (11-44) R(58)	1538	88–92
	1.1226	$B^3\Pi_{2u} \rightarrow X^1\Sigma^+_g$ (11-44) P(60)	1538	88–92
	1.1255	$B^3\Pi_{2u} \rightarrow X^1\Sigma^+_g$	1534	88–92
	1.1328	$B^3\Pi_{2u} \rightarrow X^1\Sigma^+_g$ (12-45) R(127)	1538	88–92
	1.1334	$B^3\Pi_{2u} \rightarrow X^1\Sigma^+_g$ (13-46) R(84)	1538	88–92
	1.1347	$B^3\Pi_{2u} \rightarrow X^1\Sigma^+_g$ (12-45) P(129)	1538	88–92
	1.1348	$B^3\Pi_{2u} \rightarrow X^1\Sigma^+_g$ (13-46) P(86)	1538	88–92
	1.1350	$B^3\Pi_{2u} \rightarrow X^1\Sigma^+_g$	1534	88–92
	1.1453	$B^3\Pi_{2u} \rightarrow X^1\Sigma^+_g$ (12-46) R(61)	1538	88–92
	1.1464	$B^3\Pi_{2u} \rightarrow X^1\Sigma^+_g$ (12-46) P(63)	1538	88–92
	1.1502	$B^3\Pi_{2u} \rightarrow X^1\Sigma^+_g$ (13-47) P(57)	1539	88–92
	1.1510	$B^3\Pi_{2u} \rightarrow X^1\Sigma^+_g$ (13-47) R(57)	1539	88–92
	1.1515	$B^3\Pi_{2u} \rightarrow X^1\Sigma^+_g$ (13-47) P(82)	1539	88–92
	1.1522	$B^3\Pi_{2u} \rightarrow X^1\Sigma^+_g$ (13-47) R(75)	1539	88–92
	1.1529	$B^3\Pi_{2u} \rightarrow X^1\Sigma^+_g$ (13-47) R(82)	1539	88–92

Table 3.3.1.1—*continued*
Diatomic Electronic Transition Lasers

Species	Wavelength (µm)	Transition assignment	Comments	References
I_2	1.1698	$B^3\Pi_{2u} \to X^1\Sigma^+_g$ (13-48) P(75)	1539	88–92
	1.1703	$B^3\Pi_{2u} \to X^1\Sigma^+_g$ (13-48) P(82)	1539	88–92
	1.1711	$B^3\Pi_{2u} \to X^1\Sigma^+_g$ (13-48) R(75)	1539	88–92
	1.1718	$B^3\Pi_{2u} \to X^1\Sigma^+_g$ (13-48) R(82)	1539	88–92
	1.1740	$B^3\Pi_{2u} \to X^1\Sigma^+_g$ (14-49) P(62)	1539	88–92
	1.1750	$B^3\Pi_{2u} \to X^1\Sigma^+_g$ (14-49) R(63)	1539	88–92
	1.2170	$B^3\Pi_{2u} \to X^1\Sigma^+_g$ P(71), R(71)	1537	88–92
	1.2740	$B^3\Pi_{2u} \to X^1\Sigma^+_g$ P(76), R(76)	1537	88–92
	1.2870	$B^3\Pi_{2u} \to X^1\Sigma^+_g$	1534	88–92
	1.2925	$B^3\Pi_{2u} \to X^1\Sigma^+_g$	1534	88–92
	1.2940	$B^3\Pi_{2u} \to X^1\Sigma^+_g$ P(78), R(78)	1537	88–92
	1.3010	$B^3\Pi_{2u} \to X^1\Sigma^+_g$ (27-66) P(66)	1539	88–92
	1.3020	$B^3\Pi_{2u} \to X^1\Sigma^+_g$ (27-66) R(66)	1539	88–92
	1.3040	$B^3\Pi_{2u} \to X^1\Sigma^+_g$ P(79), R(79)	1537	88–92
	1.3069	$B^3\Pi_{2u} \to X^1\Sigma^+_g$ (29-68) P(64)	1539	88–92
	1.3080	$B^3\Pi_{2u} \to X^1\Sigma^+_g$ (26-68) R(64)	1539	88–92
	1.3153	$B^3\Pi_{2u} \to X^1\Sigma^+_g$	1534	88–92
	1.3192	$B^3\Pi_{2u} \to X^1\Sigma^+_g$	1534	88–92
	1.3200	$B^3\Pi_{2u} \to X^1\Sigma^+_g$ P(81), R(81)	1537	88–92
	1.3282	$B^3\Pi_{2u} \to X^1\Sigma^+_g$	1534	88–92
	1.3291	$B^3\Pi_{2u} \to X^1\Sigma^+_g$	1534	88–92
	1.3310	$B^3\Pi_{2u} \to X^1\Sigma^+_g$	1534	88–92
	1.3324	$B^3\Pi_{2u} \to X^1\Sigma^+_g$	1534	88–92
	1.3333	$B^3\Pi_{2u} \to X^1\Sigma^+_g$	1534	88–92

Table 3.3.1.1—*continued*
Diatomic Electronic Transition Lasers

Species	Wavelength (μm)	Transition assignment	Comments	References
I$_2$	1.3349	B$^3\Pi_{2u} \to$ X$^1\Sigma^+_g$	1534	88–92
	1.3380	B$^3\Pi_{2u} \to$ X$^1\Sigma^+_g$ P(83), R(83)	1537	88–92
	1.3406	B$^3\Pi_{2u} \to$ X$^1\Sigma^+_g$ (42-82) P(57)	1539	88–92
	1.3418	B$^3\Pi_{2u} \to$ X$^1\Sigma^+_g$ (42-82) R(57)	1539	88–92
	1.3421	B$^3\Pi_{2u} \to$ X$^1\Sigma^+_g$ (45-85) P(83) (44-84) P(75) (43-83) R(53)	1539	88–92
	1.3429	B$^3\Pi_{2u} \to$ X$^1\Sigma^+_g$ (45-85) R(74) (45-85) R(72) (44-84) R(62)	1539	88–92
ICl	0.430-0.437	D'(2$_g$) $\to$ A'(2$_u$)	6661,6663	1618,1557
IF	0.467	D'(2$_g$) $\to$ A'(2$_u$)	6665	1558
	0.4725	D'(2$_g$) $\to$ A'(2$_u$)	6665	1558
	0.4787	D'(2$_g$) $\to$ A'(2$_u$)	6665	1559
	0.4847	D'(2$_g$) $\to$ A'(2$_u$)	6665	1559
	0.485-0.491	D'(2$_g$) $\to$ A'(2$_u$)	6667	1565,1566
	0.4907	D'(2$_g$) $\to$ A'(2$_u$)	6665	1527,1559–61
	0.4965	D'(2$_g$) $\to$ A'(2$_u$)	6665	1559
	0.585	B$^3\Pi(0^+) \to$ X$^1\Sigma^+$ (v',v") = (0,3)	6669	1567
	0.60311	B$^3\Pi(0^+) \to$ X$^1\Sigma^+$ (v',v") = (0,4)	6669,6918	1567
	0.62293	B$^3\Pi(0^+) \to$ X$^1\Sigma^+$ (v',v") = (3,7) R(17)	6675	1568
	0.62297	B$^3\Pi(0^+) \to$ X$^1\Sigma^+$ (v',v") = (3,7) R(18)	6675	1568
	0.62369	B$^3\Pi(0^+) \to$ X$^1\Sigma^+$ (v',v") = (3,7) P(19)	6675	1568
	0.62378	B$^3\Pi(0^+) \to$ X$^1\Sigma^+$ (v',v") = (3,7) P(20)	6675	1568
	0.62378	B$^3\Pi(0^+) \to$ X$^1\Sigma^+$ (v',v") = (3,7) R(30)	6675	1568

Table 3.3.1.1—*continued*
Diatomic Electronic Transition Lasers

Species	Wavelength (µm)	Transition assignment	Comments	References
IF	0.62436	$B^3\Pi(0^+) \rightarrow X^1\Sigma^+$ $(v',v'') = (3,7)$ R(36)	6675	1568
	0.62493	$B^3\Pi(0^+) \rightarrow X^1\Sigma^+$ $(v',v'') = (0,5)$	6669	1567
	0.62508	$B^3\Pi(0^+) \rightarrow X^1\Sigma^+$ $(v',v'') = (3,7)$ P(32)	6675	1568
	0.62522	$B^3\Pi(0^+) \rightarrow X^1\Sigma^+$ $(v',v'') = (0,5)$ R(19)	6677	1569
	0.62522	$B^3\Pi(0^+) \rightarrow X^1\Sigma^+$ $(v',v'') = (0,5)$ R(20)	6677	1569
	0.62541	$B^3\Pi(0^+) \rightarrow X^1\Sigma^+$ $(v',v'') = (0,5)$ R(22)	6680,6683, 6686,6689	1569
	0.62541	$B^3\Pi(0^+) \rightarrow X^1\Sigma^+$ $(v',v'') = (0,5)$ R(23)	6680,6683, 6686,6689	1569
	0.62546	$B^3\Pi(0^+) \rightarrow X^1\Sigma^+$ $(v',v'') = (0,5)$ R(22)	6680,6683,6686 6689,6692	1569
	0.62546	$B^3\Pi(0^+) \rightarrow X^1\Sigma^+$ $(v',v'') = (0,5)$ R(23)	6680,6683,6686 6689,6692	1569
	0.62552	$B^3\Pi(0^+) \rightarrow X^1\Sigma^+$ $(v',v'') = (0,5)$ R(22)	6692	1569
	0.62592	$B^3\Pi(0^+) \rightarrow X^1\Sigma^+$ $(v',v'') = (3,7)$ P(38)	6675	1568
	0.62614	$B^3\Pi(0^+) \rightarrow X^1\Sigma^+$ $(v',v'') = (0,5)$ P(21)	6677	1569
	0.62626	$B^3\Pi(0^+) \rightarrow X^1\Sigma^+$ $(v',v'') = (0,5)$ P(22)	6677	1569
	0.62634	$B^3\Pi(0^+) \rightarrow X^1\Sigma^+$ $(v',v'') = (0,5)$ P(24)	6683	1569
	0.62639	$B^3\Pi(0^+) \rightarrow X^1\Sigma^+$ $(v',v'') = (0,5)$ P(24)	6683,6686, 6692	1569
	0.62639	$B^3\Pi(0^+) \rightarrow X^1\Sigma^+$ $(v',v'') = (0,5)$ P(25)	6683,6686, 6692	1569
	0.62645	$B^3\Pi(0^+) \rightarrow X^1\Sigma^+$ $(v',v'') = (0,5)$ P(24)	6680,6686, 6689	1569
	0.62645	$B^3\Pi(0^+) \rightarrow X^1\Sigma^+$ $(v',v'') = (0,5)$ P(25)	6680,6686, 6689	1569
	0.62651	$B^3\Pi(0^+) \rightarrow X^1\Sigma^+$ $(v',v'') = (0,5)$ P(25)	6689	1569

Table 3.3.1.1—*continued*
Diatomic Electronic Transition Lasers

Species	Wavelength (μm)	Transition assignment	Comments	References
IF	0.62657	$B^3\Pi(0^+) \to X^1\Sigma^+$ $(v',v'') = (0,5)\ P(25)$	6680,6689, 6692	1569
	0.62662	$B^3\Pi(0^+) \to X^1\Sigma^+$ $(v',v'') = (0,5)\ P(25)$	6686	1569
	0.64812	$B^3\Pi(0^+) \to X^1\Sigma^+$ $(v',v'') = (0,6)$	6669	1567
	0.65246	$B^3\Pi(0^+) \to X^1\Sigma^+$ $(v',v'') = (4,9)\ R(16)$	6695	1569
	0.65315	$B^3\Pi(0^+) \to X^1\Sigma^+$ $(v',v'') = (4,9)\ P(18)$	6695	1569
	0.6537	$B^3\Pi(0^+) \to X^1\Sigma^+$ $(v',v'') = (4,9)$	6696	1570
	0.65501	$B^3\Pi(0^+) \to X^1\Sigma^+$ $(v',v'') = (1,7)\ R(15)$	6698	1569
	0.65535	$B^3\Pi(0^+) \to X^1\Sigma^+$ $(v',v'') = (1,7)\ R(19)$	6698	1569
	0.65592	$B^3\Pi(0^+) \to X^1\Sigma^+$ $(v',v'') = (1,7)\ P(17)$	6698	1569
	0.65632	$B^3\Pi(0^+) \to X^1\Sigma^+$ $(v',v'') = (1,7)\ P(21)$	6698	1569
	0.675	$B^3\Pi(0^+) \to X^1\Sigma^+$ $(v',v'') = (0,7)$	6669	1567
	0.6762	$B^3\Pi(0^+) \to X^1\Sigma^+$ $(v',v'') = (4,10)$	6696	1570
	0.67686	$B^3\Pi(0^+) \to X^1\Sigma^+$ $(v',v'') = (4,10)\ R(22)$	6695	1569
	0.67793	$B^3\Pi(0^+) \to X^1\Sigma^+$ $(v',v'') = (4,10)\ P(24)$	6695	1569
	0.68036	$B^3\Pi(0^+) \to X^1\Sigma^+$ $(v',v'') = (1,8)\ R(20)$	6698	1569
	0.68081	$B^3\Pi(0^+) \to X^1\Sigma^+$ $(v',v'') = (1,8)\ R(27)$	6698	1569
	0.68137	$B^3\Pi(0^+) \to X^1\Sigma^+$ $(v',v'') = (1,8)\ P(22)$	6698	1569
	0.68210	$B^3\Pi(0^+) \to X^1\Sigma^+$ $(v',v'') = (1,8)\ P(29)$	6698	1569
	0.68420	$B^3\Pi(0^+) \to X^1\Sigma^+$ $(v',v'') = (5,11)\ R(19)$	6699	1569

Table 3.3.1.1—*continued*
Diatomic Electronic Transition Lasers

Species	Wavelength (μm)	Transition assignment	Comments	References
IF	0.68510	$B^3\Pi(0^+) \to X^1\Sigma^+$ $(v',v'') = (5,11)$ P(21)	6699	1569
	0.70730	$B^3\Pi(0^+) \to X^1\Sigma^+$ $(v',v'') = (1,9)$ R(26)	6698	1569
	0.70840	$B^3\Pi(0^+) \to X^1\Sigma^+$ $(v',v'') = (1,9)$ P(28)	6698	1569
	0.71406	$B^3\Pi(0^+) \to X^1\Sigma^+$ $(v',v'') = (2,10)$ R(11)	6700	1569
	0.7141	$B^3\Pi(0^+) \to X^1\Sigma^+$ $(v',v'') = (2,10)$	6696	1570
	0.71416	$B^3\Pi(0^+) \to X^1\Sigma^+$ $(v',v'') = (2,10)$ R(17)	6700	1569
	0.71466	$B^3\Pi(0^+) \to X^1\Sigma^+$ $(v',v'') = (2,10)$ P(13)	6700	1569
	0.71482	$B^3\Pi(0^+) \to X^1\Sigma^+$ $(v',v'') = (2,10)$ R(27)	6700	1569
	0.71515	$B^3\Pi(0^+) \to X^1\Sigma^+$ $(v',v'') = (2,10)$ P(19)	6700	1569
	0.71630	$B^3\Pi(0^+) \to X^1\Sigma^+$ $(v',v'') = (2,10)$ P(29)	6700	1569
	0.7215	$B^3\Pi(0^+) \to X^1\Sigma^+$ $(v',v'') = (3,11)$	6696	1570
	0.72156	$B^3\Pi(0^+) \to X^1\Sigma^+$ $(v',v'') = (3,11)$ R(12)	6701	1569
	0.72194	$B^3\Pi(0^+) \to X^1\Sigma^+$ $(v',v'') = (3,11)$ R(19)	6701	1569
	0.72205	$B^3\Pi(0^+) \to X^1\Sigma^+$ $(v',v'') = (3,11)$ R(22)	6701	1569
	0.72233	$B^3\Pi(0^+) \to X^1\Sigma^+$ $(v',v'') = (3,11)$ P(14)	6701	1569
	0.72294	$B^3\Pi(0^+) \to X^1\Sigma^+$ $(v',v'') = (3,11)$ P(21)	6701	1569
	0.72321	$B^3\Pi(0^+) \to X^1\Sigma^+$ $(v',v'') = (3,11)$ P(24)	6701	1569
	0.798523	$B^3\Pi(0^+) \to X^1\Sigma^+$ $(v',v'') = (5,15)$ R(18)	6669	1567
	0.799154	$B^3\Pi(0^+) \to X^1\Sigma^+$ $(v',v'') = (5,15)$ R(27)	6669	1567

Table 3.3.1.1—*continued*
Diatomic Electronic Transition Lasers

Species	Wavelength (μm)	Transition assignment	Comments	References
IF	0.799752	$B^3\Pi(0^+) \to X^1\Sigma^+$ $(v',v'') = (5,15)$ P(20)	6669	1567
	0.800991	$B^3\Pi(0^+) \to X^1\Sigma^+$ $(v',v'') = (5,15)$ P(29)	6669	1567
	0.832206	$B^3\Pi(0^+) \to X^1\Sigma^+$ $(v',v'') = (5,16)$ R(18)	6669	1567
	0.832819	$B^3\Pi(0^+) \to X^1\Sigma^+$ $(v',v'') = (5,16)$ R(27)	6669	1567
	0.833532	$B^3\Pi(0^+) \to X^1\Sigma^+$ $(v',v'') = (5,16)$ P(20)	6669	1567
	0.834767	$B^3\Pi(0^+) \to X^1\Sigma^+$ $(v',v'') = (5,16)$ P(29)	6669	1567
K_2	634.5	$B^1\Pi_u(v'=8) \to X^1\Sigma^+_g(v''=2)$	6901	1770
	638.1	$B^1\Pi_u(v'=8) \to X^1\Sigma^+_g(v''=3)$	6901	1770
	645.5	$B^1\Pi_u(v'=8) \to X^1\Sigma^+_g(v''=5)$	6901	1770
	671.7	$B^1\Pi_u(v'=8) \to X^1\Sigma^+_g(v''=12)$	6901	1770
	683.1	$B^1\Pi_u(v'=8) \to X^1\Sigma^+_g(v''=15)$	6901	1770
	686.9	$B^1\Pi_u(v'=8) \to X^1\Sigma^+_g(v''=16)$	6901	1770
	690.8	$B^1\Pi_u(v'=8) \to X^1\Sigma^+_g(v''=17)$	6901	1770
	0.6961	$B^1\Pi_u \to X^1\Sigma^+_g$	6702	1524
	0.6975	$B^1\Pi_u \to X^1\Sigma^+_g$	6702	1524
	0.6985	$B^1\Pi_u \to X^1\Sigma^+_g$	6702	1524
	0.6995	$B^1\Pi_u \to X^1\Sigma^+_g$	6702	1524
	0.7037	$B^1\Pi_u \to X^1\Sigma^+_g$	6702	1524
	0.7089	$B^1\Pi_u \to X^1\Sigma^+_g$	6702	1524
	0.145-0.146	$AO^+_u \to XO^+_g$	6704,6706	1606,1622,1571
	0.1457	$^1\Sigma^+_u \to X^1\Sigma^+_g$	1553–56	40,51,62,73,84,95, 106,116,103
Kr_2	145.7	$^1\Sigma^+_u \to X^1\Sigma^+_g$	1553–1556	40,51,62,73,84, 95,103,106,116
KrCl	0.222	$C^2\Sigma^+_{1/2} \to X^2\Sigma^+_{1/2}$	1540,1541,1542	23,33,36,93,94
	0.2235	$B \to X$	6708	1572
KrF	0.2484	$^2\Sigma^+_{1/2} \to X^2\Sigma^+_{1/2}$	1543,1545, 1547-9,1551-2	13,33–38, 96–102,116

Table 3.3.1.1—*continued*
Diatomic Electronic Transition Lasers

Species	Wavelength (μm)	Transition assignment	Comments	References
KrF	0.2491	$^2\Sigma^+_{1/2} \to X^2\Sigma^+_{1/2}$	1543,1545, 1547–9,1551–2	13,33–38, 96–102,116
KrO	0.55781	$0(^1S_0) \to 0(^1D_2)$	1557,1558	2,39,41–2,44,47
Li$_2$	0.5319	$B^1\Pi_u \to X^1\Sigma^+_g$	6726,6731	1577,1619
	0.5336	$B^1\Pi_u \to X^1\Sigma^+_g$	6726,6730	1577,1619
	0.5423	$B^1\Pi_u \to X^1\Sigma^+_g$	6726,6730	1577,1619
	0.5500	$B^1\Pi_u \to X^1\Sigma^+_g$	6726,6729	1577,1619
	0.5502	$B^1\Pi_u \to X^1\Sigma^+_g$	6726,6728	1577,1619
	0.5510	$B^1\Pi_u \to X^1\Sigma^+_g$	6726,6730	1577,1619
	0.5522	$B^1\Pi_u \to X^1\Sigma^+_g$	6726,6729	1577,1619
	0.5582	$B^1\Pi_u \to X^1\Sigma^+_g$	6726,6729	1577,1619
	0.5584	$B^1\Pi_u \to X^1\Sigma^+_g$	6726,6728	1577,1619
	0.5598	$B^1\Pi_u \to X^1\Sigma^+_g$	6726,6731	1577,1619
	0.5604	$B^1\Pi_u \to X^1\Sigma^+_g$	6726,6729	1577,1619
	0.7913	$A^1\Sigma^+_u \to X^1\Sigma^+_g$	6732	1577,1619
	0.8264	$A^1\Sigma^+_u \to X^1\Sigma^+_g$	6732	1577,1619
	0.8454	$A^1\Sigma^+_u \to X^1\Sigma^+_g$	6732	1577,1619
	0.8672	$A^1\Sigma^+_u \to X^1\Sigma^+_g$	6734	1577,1619
	0.8682	$A^1\Sigma^+_u \to X^1\Sigma^+_g$	6734	1577,1619
	0.8747	$A^1\Sigma^+_u \to X^1\Sigma^+_g$	6734	1577,1619
	0.8757	$A^1\Sigma^+_u \to X^1\Sigma^+_g$	6734	1577,1619
	0.8846	$A^1\Sigma^+_u \to X^1\Sigma^+_g$	6732	1577,1619
	0.886347	$A^1\Sigma^+_u \to X^1\Sigma^+_g$	6734	1577,1619
	0.8865	$A^1\Sigma^+_u \to X^1\Sigma^+_g$	6734	1577,1619
	0.887376	$A^1\Sigma^+_u \to X^1\Sigma^+_g$	6734	1577,1619
	0.8963	$A^1\Sigma^+_u \to X^1\Sigma^+_g$	6732	1577,1619
	0.8969	$A^1\Sigma^+_u \to X^1\Sigma^+_g$	6734	1577,1619
	0.8972	$A^1\Sigma^+_u \to X^1\Sigma^+_g$	6734	1577,1619
	0.8974	$A^1\Sigma^+_u \to X^1\Sigma^+_g$	6734	1577,1619
	0.8978	$A^1\Sigma^+_u \to X^1\Sigma^+_g$	6734	1577,1619
	0.8979	$A^1\Sigma^+_u \to X^1\Sigma^+_g$	6734	1577,1619
	0.8982	$A^1\Sigma^+_u \to X^1\Sigma^+_g$	6734	1577,1619
	0.8984	$A^1\Sigma^+_u \to X^1\Sigma^+_g$	6734	1577,1619
	0.8989	$A^1\Sigma^+_u \to X^1\Sigma^+_g$	6734	1577,1619
	0.8991	$A^1\Sigma^+_u \to X^1\Sigma^+_g$	6734	1577,1619
	0.9037	$A^1\Sigma^+_u \to X^1\Sigma^+_g$	6732	1577,1619
	0.9047	$A^1\Sigma^+_u \to X^1\Sigma^+_g$	6732	1577,1619

Table 3.3.1.1—*continued*
Diatomic Electronic Transition Lasers

Species	Wavelength (μm)	Transition assignment	Comments	References
Li_2	0.9064	$A^1\Sigma^+_u \to X^1\Sigma^+_g$	6734	1577,1619
	0.9068	$A^1\Sigma^+_u \to X^1\Sigma^+_g$	6732	1577,1619
	0.9074	$A^1\Sigma^+_u \to X^1\Sigma^+_g$	6734	1577,1619
	0.9122	$A^1\Sigma^+_u \to X^1\Sigma^+_g$	6732	1577,1619
$^6Li_2$	0.5237	$B^1\Pi_u \to S^1\Sigma^+_g$ $(v',J') \to (v'',J'')$: $(0,41) \to (4,40)$	6715,6723	1576
	0.5263	$B^1\Pi_u \to S^1\Sigma^+_g$ $(4,11) \to (7,11)$	6715,6718	1576
	0.5269	$B^1\Pi_u \to S^1\Sigma^+_g$ $(0,41) \to (4,42)$	6715,6723	1576
	0.5358	$B^1\Pi_u \to S^1\Sigma^+_g$ $(4,11) \to (8,11)$	6715,6718	1576
	0.5362	$B^1\Pi_u \to S^1\Sigma^+_g$ $(0,41) \to (5,42)$	6715,6723	1576
	0.5405	$B^1\Pi_u \to S^1\Sigma^+_g$	6715,6724	1576
	0.5417	$B^1\Pi_u \to S^1\Sigma^+_g$ $(4,31) \to (12,31)$	6715,6721	1576
	0.5417	$B^1\Pi_u \to S^1\Sigma^+_g$ $(7,8) \to (11,8)$	6715,6717	1576
	0.5424	$B^1\Pi_u \to S^1\Sigma^+_g$ $(0,41) \to (6,40)$	6715,6723	1576
	0.5446	$B^1\Pi_u \to S^1\Sigma^{\wedge+}_g$ $(5,41) \to (9,42)$	6715,6719	1576
	0.5451	$B^1\Pi_u \to S^1\Sigma^+_g$ $(4,11) \to (9,11)$	6715,6718	1576
	0.5459	$B^1\Pi_u \to S^1\Sigma^+_g$ $(0,41) \to (6,42)$	6715,6723	1576
	0.5496	$B^1\Pi_u \to S^1\Sigma^+_g$	6715,6724	1576
	0.5508	$B^1\Pi_u \to S^1\Sigma^+_g$ $(7,8) \to (12,8)$	6715,6717	1576
	0.5528	$B^1\Pi_u \to S^1\Sigma^+_g$ $(5,41) \to (10,42)$	6715,6719	1576
	0.5588	$B^1\Pi_u \to S^1\Sigma^+_g$	6715,6724	1576
	0.5593	$B^1\Pi_u \to S^1\Sigma^+_g$ $(5,41) \to (11,40)$	6715,6719	1576
	0.5599	$B^1\Pi_u \to S^1\Sigma^+_g$ $(7,8) \to (13,8)$	6715,6717	1576

Table 3.3.1.1—*continued*
Diatomic Electronic Transition Lasers

Species	Wavelength (μm)	Transition assignment	Comments	References
$^6Li_2$	0.5670	$B^1\Pi_u \rightarrow S^1\Sigma^+_g$ $(4,31) \rightarrow (15,31)$	6715,6721	1576
	0.5689	$B^1\Pi_u \rightarrow S^1\Sigma^+_g$ $(7,8) \rightarrow (14,8)$	6715,6717	1576
	0.5755	$B^1\Pi_u \rightarrow S^1\Sigma^+_g$} $(4,31) \rightarrow (16,31)$	6715,6717	1576
	0.5773	$B^1\Pi_u \rightarrow S^1\Sigma^+_g$ $(11,22) \rightarrow (18,21)$	6715,6717	1576
	0.5787	$B^1\Pi_u \rightarrow S^1\Sigma^+_g$ $(11,29) \rightarrow (18,28)$	6715,6722	1576
	0.5807	$B^1\Pi_u \rightarrow S^1\Sigma^+_g$	6715,6720	1576
	0.5809	$B^1\Pi_u \rightarrow S^1\Sigma^+_g$ $(11,29) \rightarrow (18,30)$	6715,6722	1576
	0.5867	$B^1\Pi_u \rightarrow S^1\Sigma^+_g$ $(11,29) \rightarrow (19,28)$	6715,6722	1576
	0.5889	$B^1\Pi_u \rightarrow S^1\Sigma^+_g$ $(11,29) \rightarrow (19,30)$	6715,6722	1576
N_2	0.3364903	$C^3\Pi_u \rightarrow B^3\Pi_g$ (0-0) Band $R_1(7)$	1559	104,105,109, 110,111
	0.3365474	$C^3\Pi_u \rightarrow B^3\Pi_g$ (0-0) Band $R_3(6)$	1559	104,105,109, 110,111
	0.3365537	$C^3\Pi_u \rightarrow B^3\Pi_g$ (0-0) Band $R_1(6)$	1559	104,105,109, 110,111
	0.3366156	$C^3\Pi_u \rightarrow B^3\Pi_g$ (0-0) Band $R_1(4)$	1559	104,105,109, 110,111
	0.3366211	$C^3\Pi_u \rightarrow B^3\Pi_g$ (0-0) Band $R_3(5)$	1559	104,105,109, 110,111
	0.3366682	$C^3\Pi_u \rightarrow B^3\Pi_g$ (0-0) Band $R_1(4)$	1559	104,105,109, 110,111
	0.3366911	$C^3\Pi_u \rightarrow B^3\Pi_g$ (0-0) Band $R_3(4)$	1559	104,105,109, 110,111
	0.3367218	$C^3\Pi_u \rightarrow B^3\Pi_g$ (0-0) Band $R_1(3)$	1559	104,105,109, 110,111
	0.3368432	$C^3\Pi_u \rightarrow B^3\Pi_g$ (0-0) Band $P'_3(20)$	1559	104,105,109, 110,111
	0.3368917	$C^3\Pi_u \rightarrow B^3\Pi_g$ (0-0) Band $P_3(19)$	1559	104,105,109, 110,111

Table 3.3.1.1—*continued*
Diatomic Electronic Transition Lasers

Species	Wavelength (μm)	Transition assignment	Comments	References
N_2	0.3369250	$C^3\Pi_u \to B^3\Pi_g$ (0-0) Band $P_1(1)$	1559	104,105,109, 110,111
	0.3369361	$C^3\Pi_u \to B^3\Pi_g$ (0-0) Band $P'_3(18)$	1559	104,105,109, 110,111
	0.3369502	$C^3\Pi_u \to B^3\Pi_g$ (0-0) Band $P'_2(18)$	1559	104,105,109, 110,111
	0.3369542	$C^3\Pi_u \to B^3\Pi_g$ (0-0) Band $Q_3(2)$	1559	104,105,109, 110,111
	0.3369555	$C^3\Pi_u \to B^3\Pi_g$ (0-0) Band $P'_2(2)$	1559	104,105,109, 110,111
	0.3369575	$C^3\Pi_u \to B^3\Pi_g$ (0-0) Band $P'_1(18)$	1559	104,105,109, 110,111
	0.3369760	$C^3\Pi_u \to B^3\Pi_g$ (0-0) Band $P_3(17)$	1559	104,105,109, 110,111
	0.3369838	$C^3\Pi_u \to B^3\Pi_g$ (0-0) Band $P_1(3)$	1559	104,105,109, 110,111
	0.3369852	$C^3\Pi_u \to B^3\Pi_g$ (0-0) Band $P_2(17)$	1559	104,105,109, 110,111
	0.3370081	$C^3\Pi_u \to B^3\Pi_g$ (0-0) Band $P'_1(4)$	1559	104,105,109, 110,111
	0.3370121	$C^3\Pi_u \to B^3\Pi_g$ (0-0) Band $P'_3(16)$	1559	104,105,109, 110,111
	0.3370138	$C^3\Pi_u \to B^3\Pi_g$ (0-0) Band $P'_1(16)$	1559	104,105,109, 110,111
	0.3370161	$C^3\Pi_u \to B^3\Pi_g$ (0-0) Band $P_1(16)$	1559	104,105,109, 110,111
	0.3370169	$C^3\Pi_u \to B^3\Pi_g$ (0-0) Band $P'_2(16)$	1559	104,105,109, 110,111
	0.3370297	$C^3\Pi_u \to B^3\Pi_g$ (0-0) Band $P_1(5)$	1559	104,105,109, 110,111
	0.3370316	$C^3\Pi_u \to B^3\Pi_g$ (0-0) Band $P_2(3)$	1559	104,105,109, 110,111
	0.3370360	$C^3\Pi_u \to B^3\Pi_g$ (0-0) Band $P'_1(15)$	1559	104,105,109, 110,111
	0.3370374	$C^3\Pi_u \to B^3\Pi_g$ (0-0) Band $P_1(15)$	1559	104,105,109, 110,111
	0.3370434	$C^3\Pi_u \to B^3\Pi_g$ (0-0) Band $P_2(15),P_3(15)$	1559	104,105,109, 110,111

Table 3.3.1.1—*continued*
Diatomic Electronic Transition Lasers

Species	Wavelength (μm)	Transition assignment	Comments	References
N_2	0.3370472	$C^3\Pi_u \rightarrow B^3\Pi_g$ (0-0) Band $P'_1(6)$	1559	104,105,109, 110,111
	0.3370529	$C^3\Pi_u \rightarrow B^3\Pi_g$ (0-0) Band $P'_1(14)$	1559	104,105,109, 110,111
	0.3370551	$C^3\Pi_u \rightarrow B^3\Pi_g$ (0-0) Band $P_1(14)$	1559	104,105,109, 110,111
	0.3370559	$C^3\Pi_u \rightarrow B^3\Pi_g$ (0-0) Band $P'_2(4)$	1559	104,105,109, 110,111
	0.3370614	$C^3\Pi_u \rightarrow B^3\Pi_g$ (0-0) Band $P'_1(7)$	1559	104,105,109, 110,111
	0.3370623	$C^3\Pi_u \rightarrow B^3\Pi_g$ (0-0) Band $P_1(7)$	1559	104,105,109, 110,111
	0.3370663	$C^3\Pi_u \rightarrow B^3\Pi_g$ (0-0) Band $P'_2(14)$	1559	104,105,109, 110,111
	0.3370682	$C^3\Pi_u \rightarrow B^3\Pi_g$ (0-0) Band $P_1(13)$	1559	104,105,109, 110,111
	0.3370716	$C^3\Pi_u \rightarrow B^3\Pi_g$ (0-0) Band $P'_1(8),P'_3(14)$	1559	104,105,109, 110,111
	0.3370731	$C^3\Pi_u \rightarrow B^3\Pi_g$ (0-0) Band $P_1(8)$	1559	104,105,109, 110,111
	0.3370757	$C^3\Pi_u \rightarrow B^3\Pi_g$ (0-0) Band $P'_1(12),P_3(3)$	1559	104,105,109, 110,111
	0.3370762	$C^3\Pi_u \rightarrow B^3\Pi_g$ (0-0) Band $P_2(5)$	1559	104,105,109, 110,111
	0.3370787	$C^3\Pi_u \rightarrow B^3\Pi_g$ (0-0) Band $P'_1(9)$	1559	104,105,109, 110,111
	0.3370803	$C^3\Pi_u \rightarrow B^3\Pi_g$ (0-0) Band $P_1(9)$	1559	104,105,109, 110,111
	0.3370821	$C^3\Pi_u \rightarrow B^3\Pi_g$ (0-0) Band $P'_1(10),P_1(11)$	1559	104,105,109, 110,111
	0.3370843	$C^3\Pi_u \rightarrow B^3\Pi_g$ (0-0) Band $P_2(13)$	1559	104,105,109, 110,111
	0.3370924	$C^3\Pi_u \rightarrow B^3\Pi_g$ (0-0) Band $P'_2(6)$	1559	104,105,109, 110,111
	0.3370941	$C^3\Pi_u \rightarrow B^3\Pi_g$ (0-0) Band $P_3(13)$	1559	104,105,109, 110,111
	0.3370990	$C^3\Pi_u \rightarrow B^3\Pi_g$ (0-0) Band $P'_2(12),P'_3(4)$	1559	104,105,109, 110,111

Table 3.3.1.1—*continued*
Diatomic Electronic Transition Lasers

Species	Wavelength (μm)	Transition assignment	Comments	References
N_2	0.3371042	$C^3\Pi_u \rightarrow B^3\Pi_g$ (0-0) Band $P_2(7)$	1559	104,105,109, 110,111
	0.3371082	$C^3\Pi_u \rightarrow B^3\Pi_g$ (0-0) Band $P_2(11)$	1559	104,105,109, 110,111
	0.3371120	$C^3\Pi_u \rightarrow B^3\Pi_g$ (0-0) Band $P'_2(8)$	1559	104,105,109, 110,111
	0.3371129	$C^3\Pi_u \rightarrow B^3\Pi_g$ (0-0) Band $P'_3(12)$	1559	104,105,109, 110,111
	0.3371141	$C^3\Pi_u \rightarrow B^3\Pi_g$ (0-0) Band $P'_2(10)$	1559	104,105,109, 110,111
	0.3371147	$C^3\Pi_u \rightarrow B^3\Pi_g$ (0-0) Band $P_2(9)$	1559	104,105,109, 110,111
	0.3371179	$C^3\Pi_u \rightarrow B^3\Pi_g$ (0-0) Band $P_3(5)$	1559	104,105,109, 110,111
	0.3371271	$C^3\Pi_u \rightarrow B^3\Pi_g$ (0-0) Band $P_3(11)$	1559	104,105,109, 110,111
	0.3371312	$C^3\Pi_u \rightarrow B^3\Pi_g$ (0-0) Band $P'_3(6)$	1559	104,105,109, 110,111
	0.3371371	$C^3\Pi_u \rightarrow B^3\Pi_g$ (0-0) Band $P'_3(10)$	1559	104,105,109, 110,111
	0.3371398	$C^3\Pi_u \rightarrow B^3\Pi_g$ (0-0) Band $P_3(7)$	1559	104,105,109, 110,111
	0.3371427	$C^3\Pi_u \rightarrow B^3\Pi_g$ (0-0) Band $P_3(9)$	1559	104,105,109, 110,111
	0.3371433	$C^3\Pi_u \rightarrow B^3\Pi_g$ (0-0) Band $P'_3(8)$	1559	104,105,109, 110,111
	0.3379898	$C^3\Pi_u \rightarrow B^3\Pi_g$ (0-0) Band $P_1(17)$	1559	104,105,109, 110,111
	0.3575980	$C^3\Pi_u \rightarrow B^3\Pi_g$ (0-1) Band $P_1(6)$	1559	104,105,109, 110,111
	0.3576194	$C^3\Pi_u \rightarrow B^3\Pi_g$ (0-1) Band $P_1(11)$	1559	104,105,109, 110,111
	0.3576250	$C^3\Pi_u \rightarrow B^3\Pi_g$ (0-1) Band $P_1(9)$	1559	104,105,109, 110,111
	0.3576320	$C^3\Pi_u \rightarrow B^3\Pi_g$ (0-1) Band $P_2(5)$	1559	104,105,109, 110,111
	0.3576571	$C^3\Pi_u \rightarrow B^3\Pi_g$ (0-1) Band $P_2(7)$	1559	104,105,109, 110,111

Table 3.3.1.1—*continued*
Diatomic Electronic Transition Lasers

Species	Wavelength (μm)	Transition assignment	Comments	References
N_2	0.3576613	$C^3\Pi_u \rightarrow B^3\Pi_g$ (0-1) Band $P_3(11)$	1559	104,105,109, 110,111
	0.3576778	$C^3\Pi_u \rightarrow B^3\Pi_g$ (0-1) Band $P_3(5)$	1559	104,105,109, 110,111
	0.3576899	$C^3\Pi_u \rightarrow B^3\Pi_g$ (0-1) Band $P_3(9)$	1559	104,105,109, 110,111
	0.3576955	$C^3\Pi_u \rightarrow B^3\Pi_g$ (0-1) Band $P_3(7)$	1559	104,105,109, 110,111
	0.4059	$C \rightarrow B$ $(v',v'') = (0,3)$	6737,6739,6740	1578–1580
	0.7482187	$B^3\Pi_g \rightarrow A^3\Sigma^+_u$ (4-2) Band $^QR_{23}(1)$	1560	105
	0.7485941	$B^3\Pi_g \rightarrow A^3\Sigma^+_u$ (4-2) Band	1560	105
	0.7486135	$B^3\Pi_g \rightarrow A^3\Sigma^+_u$ (4-2) Band $^PQ_{23}(3)$	1560	105
	0.7486253	$B^3\Pi_g \rightarrow A^3\Sigma^+_u$ (4-2) Band $^PQ_{23}(4)$	1560	105
	0.7486413	$B^3\Pi_g \rightarrow A^3\Sigma^+_u$ (4-2) Band $P_2(4)$	1560	105
	0.7487409	$B^3\Pi_g \rightarrow A^3\Sigma^+_u$ (4-2) Band $Q_1(9)$	1560	105
	0.7488046	$B^3\Pi_g \rightarrow A^3\Sigma^+_u$ (4-2) Band	1560	105
	0.7488246	$B^3\Pi_g \rightarrow A^3\Sigma^+_u$ (4-2) Band $P_2(6)$	1560	105
	0.7489107	$B^3\Pi_g \rightarrow A^3\Sigma^+_u$ (4-2) Band $^PQ_{23}(6)$	1560	105
	0.7489626	$B^3\Pi_g \rightarrow A^3\Sigma^+_u$ (4-2) Band $Q_1(8)$	1560	105
	0.7489809	$B^3\Pi_g \rightarrow A^3\Sigma^+_u$ (4-2) Band $P_2(13)$	1560	105
	0.7490096	$B^3\Pi_g \rightarrow A^3\Sigma^+_u$ (4-2) Band $P_2(12)$	1560	105
	0.7490317	$B^3\Pi_g \rightarrow A^3\Sigma^+_u$ (4-2) Band $^PQ_{23}(8)$	1560	105
	0.7491510	$B^3\Pi_g \rightarrow A^3\Sigma^+_u$ (4-2) Band	1560	105

Table 3.3.1.1—*continued*
Diatomic Electronic Transition Lasers

Species	Wavelength (μm)	Transition assignment	Comments	References
N_2	0.7491705	$B^3\Pi_g \to A^3\Sigma^+_u$ (4-2) Band $Q_1(7)$	1560	105
	0.7492379	$B^3\Pi_g \to A^3\Sigma^+_u$ (4-2) Band $^0P_{23}(4)$	1560	105
	0.7493082	$B^3\Pi_g \to A^3\Sigma^+_u$ (4-2) Band $^QR_{12}(6)$	1560	105
	0.7493716	$B^3\Pi_g \to A^3\Sigma^+_u$ (4-2) Band $Q_1(6)$	1560	105
	0.7493910	$B^3\Pi_g \to A^3\Sigma^+_u$ (4-2) Band	1560	105
	0.7495086	$B^3\Pi_g \to A^3\Sigma^+_u$ (4-2) Band $^0P_{23}(5)$	1560	105
	0.7495465	$B^3\Pi_g \to A^3\Sigma^+_u$ (4-2) Band $Q_1(5)$	1560	105
	0.7495660	$B^3\Pi_g \to A^3\Sigma^+_u$ (4-2) Band	1560	105
	0.7496024	$B^3\Pi_g \to A^3\Sigma^+_u$ (4-2) Band	1560	105
	0.7497256	$B^3\Pi_g \to A^3\Sigma^+_u$ (4-2) Band	1560	105
	0.7497524	$B^3\Pi_g \to A^3\Sigma^+_u$ (4-2) Band $Q_1(4)$	1560	105
	0.7497728	$B^3\Pi_g \to A^3\Sigma^+_u$ (4-2) Band	1560	105
	0.7498898	$B^3\Pi_g \to A^3\Sigma^+_u$ (4-2) Band	1560	105
	0.7499013	$B^3\Pi_g \to A^3\Sigma^+_u$ (4-2) Band	1560	105
	0.7499327	$B^3\Pi_g \to A^3\Sigma^+_u$ (4-2) Band $Q_1(3)$	1560	105
	0.7499593	$B^3\Pi_g \to A^3\Sigma^+_u$ (4-2) Band	1560	105
	0.7499825	$B^3\Pi_g \to A^3\Sigma^+_u$ (4-2) Band $P_1(13)$	1560	105
	0.7500071	$B^3\Pi_g \to A^3\Sigma^+_u$ (4-2) Band $^0P_{23}(7)$	1560	105
N_2	0.7500646	$B^3\Pi_g \to A^3\Sigma^+_u$ (4-2) Band $^PQ_{12}(12)$	1560	105

Table 3.3.1.1—*continued*
Diatomic Electronic Transition Lasers

Species	Wavelength (μm)	Transition assignment	Comments	References
N_2	0.7500734	$B^3\Pi_g \to A^3\Sigma^+_u$ (4-2) Band	1560	105
	0.7501056	$B^3\Pi_g \to A^3\Sigma^+_u$ (4-2) Band $Q_1(2)$	1560	105
	0.7501295	$B^3\Pi_g \to A^3\Sigma^+_u$ (4-2) Band	1560	105
	0.7501404	$B^3\Pi_g \to A^3\Sigma^+_u$ (4-2) Band	1560	105
	0.7501553	$B^3\Pi_g \to A^3\Sigma^+_u$ (4-2) Band $P_1(11)$	1560	105
	0.7502139	$B^3\Pi_g \to A^3\Sigma^+_u$ (4-2) Band $^PQ_{12}(9)$	1560	105
	0.7502729	$B^3\Pi_g \to A^3\Sigma^+_u$ (4-2) Band $Q_1(1)$	1560	105
	0.7502768	$B^3\Pi_g \to A^3\Sigma^+_u$ (4-2) Band $P_1(9)$	1560	105
	0.7503035	$B^3\Pi_g \to A^3\Sigma^+_u$ (4-2) Band $^PQ_{12}(7)$	1560	105
	0.7503371	$B^3\Pi_g \to A^3\Sigma^+_u$ (4-2) Band $^PQ_{12}(6)$	1560	105
	0.7503418	$B^3\Pi_g \to A^3\Sigma^+_u$ (4-2) Band $^PR_{13}(8)$	1560	105
	0.7503642	$B^3\Pi_g \to A^3\Sigma^+_u$ (4-2) Band $^PQ_{12}(5)$	1560	105
	0.7503669	$B^3\Pi_g \to A^3\Sigma^+_u$ (4-2) Band	1560	105
	0.7503697	$B^3\Pi_g \to A^3\Sigma^+_u$ (4-2) Band $P_1(7)$	1560	105
	0.7503838	$B^3\Pi_g \to A^3\Sigma^+_u$ (4-2) Band $^PQ_{12}(4)$	1560	105
	0.7503960	$B^3\Pi_g \to A^3\Sigma^+_u$ (4-2) Band $^PR_{13}(7)$	1560	105
	0.7503994	$B^3\Pi_g \to A^3\Sigma^+_u$ (4-2) Band $^PQ_{12}(3)$	1560	105
	0.7504106	$B^3\Pi_g \to A^3\Sigma^+_u$ (4-2) Band $^PQ_{12}(2)$	1560	105
	0.7504160	$B^3\Pi_g \to A^3\Sigma^+_u$ (4-2) Band $^PQ_{12}(1)$	1560	105

Table 3.3.1.1—*continued*
Diatomic Electronic Transition Lasers

Species	Wavelength (µm)	Transition assignment	Comments	References
N_2	0.7504184	$B^3\Pi_g \rightarrow A^3\Sigma^+_u$ (4-2) Band $^PR_{13}(6)$	1560	105
	0.7504274	$B^3\Pi_g \rightarrow A^3\Sigma^+_u$ (4-2) Band $Q_1(0)$	1560	105
	0.7504598	$B^3\Pi_g \rightarrow A^3\Sigma^+_u$ (4-2) Band $P_1(3)$	1560	105
	0.7504768	$B^3\Pi_g \rightarrow A^3\Sigma^+_u$ (4-2) Band $P_1(1)$	1560	105
	0.7505113	$B^3\Pi_g \rightarrow A^3\Sigma^+_u$ (4-2) Band $^PR_{13}(2)$	1560	105
	0.7505710	$B^3\Pi_g \rightarrow A^3\Sigma^+_u$ (4-2) Band	1560	105
	0.7505903	$B^3\Pi_g \rightarrow A^3\Sigma^+_u$ (4-2) Band	1560	105
	0.7506063	$B^3\Pi_g \rightarrow A^3\Sigma^+_u$ (4-2) Band	1560	105
	0.7506356	$B^3\Pi_g \rightarrow A^3\Sigma^+_u$ (4-2) Band $^0P_{23}(10)$	1560	105
	0.7508145	$B^3\Pi_g \rightarrow A^3\Sigma^+_u$ (4-2) Band $^0P_{23}(11)$	1560	105
	0.7509890	$B^3\Pi_g \rightarrow A^3\Sigma^+_u$ (4-2) Band	1560	105
	0.7510133	$B^3\Pi_g \rightarrow A^3\Sigma^+_u$ (4-2) Band $^0P_{12}(3)$	1560	105
	0.7510923	$B^3\Pi_g \rightarrow A^3\Sigma^+_u$ (4-2) Band	1560	105
	0.7511592	$B^3\Pi_g \rightarrow A^3\Sigma^+_u$ (4-2) Band $^0P_{12}(4)$	1560	105
	0.7512799	$B^3\Pi_g \rightarrow A^3\Sigma^+_u$ (4-2) Band	1560	105
	0.7513003	$B^3\Pi_g \rightarrow A^3\Sigma^+_u$ (4-2) Band $^0P_{12}(5)$	1560	105
	0.7513569	$B^3\Pi_g \rightarrow A^3\Sigma^+_u$ (4-2) Band	1560	105
	0.7514357	$B^3\Pi_g \rightarrow A^3\Sigma^+_u$ (4-2) Band $^0P_{12}(6)$	1560	105
	0.7515079	$B^3\Pi_g \rightarrow A^3\Sigma^+_u$ (4-2) Band	1560	105

Table 3.3.1.1—*continued*
Diatomic Electronic Transition Lasers

Species	Wavelength (μm)	Transition assignment	Comments	References
N_2	0.7515446	$B^3\Pi_g \to A^3\Sigma^+_u$ (4-2) Band	1560	105
	0.7515650	$B^3\Pi_g \to A^3\Sigma^+_u$ (4-2) Band $^0P_{12}(7)$	1560	105
	0.7517728	$B^3\Pi_g \to A^3\Sigma^+_u$ (4-2) Band $^0Q_{13}(8)$	1560	105
	0.7518013	$B^3\Pi_g \to A^3\Sigma^+_u$ (4-2) Band $^0P_{12}(9)$	1560	105
	0.7586439	$B^3\Pi_g \to A^3\Sigma^+_u$ (3-1) Band $Q_3(9)$	1560	105
	0.7587693	$B^3\Pi_g \to A^3\Sigma^+_u$ (3-1) Band $Q_3(7)$	1560	105
	0.7603477	$B^3\Pi_g \to A^3\Sigma^+_u$ (3-1) Band	1560	105
	0.7606374	$B^3\Pi_g \to A^3\Sigma^+_u$ (3-1) Band $^PQ_{23}(2)$	1560	105
	0.7607626	$B^3\Pi_g \to A^3\Sigma^+_u$ (3-1) Band $^PQ_{23}(3)$	1560	105
	0.7608801	$B^3\Pi_g \to A^3\Sigma^+_u$ (3-1) Band $^PQ_{23}(4)$	1560	105
	0.7609853	$B^3\Pi_g \to A^3\Sigma^+_u$ (3-1) Band $^PQ_{23}(5)$	1560	105
	0.7610759	$B^3\Pi_g \to A^3\Sigma^+_u$ (3-1) Band $^PQ_{23}(6)$	1560	105
	0.7611082	$B^3\Pi_g \to A^3\Sigma^+_u$ (3-1) Band $Q_1(8)$	1560	105
	0.7611514	$B^3\Pi_g \to A^3\Sigma^+_u$ (3-1) Band $^PQ_{23}(7)$	1560	105
	0.7612105	$B^3\Pi_g \to A^3\Sigma^+_u$ (3-1) Band $^PQ_{23}(8)$	1560	105
	0.7612528	$B^3\Pi_g \to A^3\Sigma^+_u$ (3-1) Band $^PQ_{23}(9)$	1560	105
	0.7613260	$B^3\Pi_g \to A^3\Sigma^+_u$ (3-1) Band $Q_1(7)$	1560	105
	0.7615347	$B^3\Pi_g \to A^3\Sigma^+_u$ (3-1) Band $Q_1(6)$	1560	105
N_2	0.7616994	$B^3\Pi_g \to A^3\Sigma^+_u$ (3-1) Band $^0P_{23}(5)$	1560	105

Table 3.3.1.1—*continued*
Diatomic Electronic Transition Lasers

Species	Wavelength (μm)	Transition assignment	Comments	References
	0.7617357	$B^3\Pi_g \rightarrow A^3\Sigma^+_u$ (3-1) Band $Q_1(5)$	1560	105
	0.7619288	$B^3\Pi_g \rightarrow A^3\Sigma^+_u$ (3-1) Band $Q_1(4)$	1560	105
	0.7620844	$B^3\Pi_g \rightarrow A^3\Sigma^+_u$ (3-1) Band	1560	105
	0.7620943	$B^3\Pi_g \rightarrow A^3\Sigma^+_u$ (3-1) Band	1560	105
	0.7621161	$B^3\Pi_g \rightarrow A^3\Sigma^+_u$ (3-1) Band $Q_1(3)$	1560	105
	0.7622235	$B^3\Pi_g \rightarrow A^3\Sigma^+_u$ (3-1) Band $^0P_{23}(7)$	1560	105
	0.7622565	$B^3\Pi_g \rightarrow A^3\Sigma^+_u$ (3-1) Band	1560	105
	0.7622959	$B^3\Pi_g \rightarrow A^3\Sigma^+_u$ (3-1) Band $Q_1(2)$	1560	105
	0.7623256	$B^3\Pi_g \rightarrow A^3\Sigma^+_u$ (3-1) Band	1560	105
	0.7623311	$B^3\Pi_g \rightarrow A^3\Sigma^+_u$ (3-1) Band	1560	105
	0.7623582	$B^3\Pi_g \rightarrow A^3\Sigma^+_u$ (3-1) Band $^pQ_{12}(10)$	1560	105
	0.7623686	$B^3\Pi_g \rightarrow A^3\Sigma^+_u$ (3-1) Band	1560	105
	0.7623918	$B^3\Pi_g \rightarrow A^3\Sigma^+_u$ (3-1) Band	1560	105
	0.7624220	$B^3\Pi_g \rightarrow A^3\Sigma^+_u$ (3-1) Band $^pQ_{12}(9)$	1560	105
	0.7624690	$B^3\Pi_g \rightarrow A^3\Sigma^+_u$ (3-1) Band $Q_1(1)$	1560	105
	0.7624924	$B^3\Pi_g \rightarrow A^3\Sigma^+_u$ (3-1) Band $P_1(9)$	1560	105
	0.7625115	$B^3\Pi_g \rightarrow A^3\Sigma^+_u$ (3-1) Band $^pQ_{12}(7)$	1560	105
	0.7625445	$B^3\Pi_g \rightarrow A^3\Sigma^+_u$ (3-1) Band $^pQ_{12}(6)$	1560	105
	0.7625709	$B^3\Pi_g \rightarrow A^3\Sigma^+_u$ (3-1) Band $^pQ_{12}(5)$	1560	105

Table 3.3.1.1—*continued*
Diatomic Electronic Transition Lasers

Species	Wavelength (μm)	Transition assignment	Comments	References
N_2	0.7625770	$B^3\Pi_g \rightarrow A^3\Sigma^+_u$ (3-1) Band	1560	105
	0.7625812	$B^3\Pi_g \rightarrow A^3\Sigma^+_u$ (3-1) Band $P_1(7)$	1560	105
	0.7625906	$B^3\Pi_g \rightarrow A^3\Sigma^+_u$ (3-1) Band $^PQ_{12}(4)$	1560	105
	0.7626007	$B^3\Pi_g \rightarrow A^3\Sigma^+_u$ (3-1) Band $^PR_{13}(7)$	1560	105
	0.7626044	$B^3\Pi_g \rightarrow A^3\Sigma^+_u$ (3-1) Band $^PQ_{12}(3)$	1560	105
	0.7626114	$B^3\Pi_g \rightarrow A^3\Sigma^+_u$ (3-1) Band $P_1(6)$	1560	105
	0.7626180	$B^3\Pi_g \rightarrow A^3\Sigma^+_u$ (3-1) Band $^PQ_{12}(2)$	1560	105
	0.7626207	$B^3\Pi_g \rightarrow A^3\Sigma^+_u$ (3-1) Band $^PQ_{12}(1)$	1560	105
	0.7626360	$B^3\Pi_g \rightarrow A^3\Sigma^+_u$ (3-1) Band $P_1(5)$	1560	105
	0.7626560	$B^3\Pi_g \rightarrow A^3\Sigma^+_u$ (3-1) Band $P_1(4)$	1560	105
	0.7626700	$B^3\Pi_g \rightarrow A^3\Sigma^+_u$ (3-1) Band $P_1(3)$	1560	105
	0.7626749	$B^3\Pi_g \rightarrow A^3\Sigma^+_u$ (3-1) Band $P_1(1)$	1560	105
	0.7626826	$B^3\Pi_g \rightarrow A^3\Sigma^+_u$ (3-1) Band $^0P_{23}(9)$	1560	105
	0.7628854	$B^3\Pi_g \rightarrow A^3\Sigma^+_u$ (3-1) Band $^0P_{23}(10)$	1560	105
	0.7629102	$B^3\Pi_g \rightarrow A^3\Sigma^+_u$ (3-1) Band	1560	105
	0.7630305	$B^3\Pi_g \rightarrow A^3\Sigma^+_u$ (3-1) Band	1560	105
	0.7631880	$B^3\Pi_g \rightarrow A^3\Sigma^+_u$ (3-1) Band	1560	105
	0.7632446	$B^3\Pi_g \rightarrow A^3\Sigma^+_u$ (3-1) Band $^0P_{12}(3)$	1560	105
	0.7633348	$B^3\Pi_g \rightarrow A^3\Sigma^+_u$ (3-1) Band	1560	105

Table 3.3.1.1—*continued*
Diatomic Electronic Transition Lasers

Species	Wavelength (µm)	Transition assignment	Comments	References
N_2	0.7633985	$B^3\Pi_g \rightarrow A^3\Sigma^+_u$ (3-1) Band $^0P_{12}(4)$	1560	105
	0.7634546	$B^3\Pi_g \rightarrow A^3\Sigma^+_u$ (3-1) Band	1560	105
	0.7634779	$B^3\Pi_g \rightarrow A^3\Sigma^+_u$ (3-1) Band	1560	105
	0.7635474	$B^3\Pi_g \rightarrow A^3\Sigma^+_u$ (3-1) Band $^0P_{12}(5)$	1560	105
	0.7636126	$B^3\Pi_g \rightarrow A^3\Sigma^+_u$ (3-1) Band	1560	105
	0.7636904	$B^3\Pi_g \rightarrow A^3\Sigma^+_u$ (3-1) Band $^0P_{12}(6)$	1560	105
	0.7637586	$B^3\Pi_g \rightarrow A^3\Sigma^+_u$ (3-1) Band	1560	105
	0.7638274	$B^3\Pi_g \rightarrow A^3\Sigma^+_u$ (3-1) Band $^0P_{12}(7)$	1560	105
	0.7639571	$B^3\Pi_g \rightarrow A^3\Sigma^+_u$ (3-1) Band $^0P_{12}(8)$	1560	105
	0.7639715	$B^3\Pi_g \rightarrow A^3\Sigma^+_u$ (3-1) Band	1560	105
	0.7640383	$B^3\Pi_g \rightarrow A^3\Sigma^+_u$ (3-1) Band	1560	105
	0.7640794	$B^3\Pi_g \rightarrow A^3\Sigma^+_u$ (3-1) Band $^0P_{12}(9)$	1560	105
	0.7641929	$B^3\Pi_g \rightarrow A^3\Sigma^+_u$ (3-1) Band $^0P_{12}(10)$	1560	105
	0.7642478	$B^3\Pi_g \rightarrow A^3\Sigma^+_u$ (3-1) Band	1560	105
	0.7644612	$B^3\Pi_g \rightarrow A^3\Sigma^+_u$ (3-1) Band	1560	105
	0.7743859	$B^3\Pi_g \rightarrow A^3\Sigma^+_u$ (2-0) Band $Q_1(5)$	1560	105
	0.7752354	$B^3\Pi_g \rightarrow A^3\Sigma^+_u$ (2-0) Band $P_1(8)$	1560	105
	0.7753652	$B^3\Pi_g \rightarrow A^3\Sigma^+_u$ (2-0) Band $P_1(2)$	1560	105
	0.8669223	$B^3\Pi_g \rightarrow A^3\Sigma^+_u$ (2-1) Band $Q_3(9)$	1560	105

Table 3.3.1.1—*continued*
Diatomic Electronic Transition Lasers

Species	Wavelength (µm)	Transition assignment	Comments	References
N_2	0.8671332	$B^3\Pi_g \rightarrow A^3\Sigma^+_u$ (2-1) Band $Q_3(7)$	1560	105
	0.8692580	$B^3\Pi_g \rightarrow A^3\Sigma^+_u$ (2-1) Band $^\wedge QP_{23}(1)$	1560	105
	0.8696366	$B^3\Pi_g \rightarrow A^3\Sigma^+_u$ (2-1) Band $^PQ_{23}(2)$	1560	105
	0.8697945	$B^3\Pi_g \rightarrow A^3\Sigma^+_u$ (2-1) Band $^PQ_{23}(3)$	1560	105
	0.8698263	$B^3\Pi_g \rightarrow A^3\Sigma^+_u$ (2-1) Band $Q_1(9)$	1560	105
	0.8699397	$B^3\Pi_g \rightarrow A^3\Sigma^+_u$ (2-1) Band $^PQ_{23}(4)$	1560	105
	0.8700670	$B^3\Pi_g \rightarrow A^3\Sigma^+_u$ (2-1) Band $^PQ_{23}(5)$	1560	105
	0.8700684	$B^3\Pi_g \rightarrow A^3\Sigma^+_u$ (2-1) Band $^PQ_{23}(5)$	1560	105
	0.8701481	$B^3\Pi_g \rightarrow A^3\Sigma^+_u$ (2-1) Band $Q_1(8)$	1560	105
	0.8701718	$B^3\Pi_g \rightarrow A^3\Sigma^+_u$ (2-1) Band $^PQ_{23}(6)$	1560	105
	0.8702451	$B^3\Pi_g \rightarrow A^3\Sigma^+_u$ (2-1) Band $^PQ_{23}(7)$	1560	105
	0.8702681	$B^3\Pi_g \rightarrow A^3\Sigma^+_u$ (2-1) Band $^0P_{23}(3)$	1560	105
	0.8703093	$B^3\Pi_g \rightarrow A^3\Sigma^+_u$ (2-1) Band $^PQ_{23}(8)$	1560	105
	0.8703457	$B^3\Pi_g \rightarrow A^3\Sigma^+_u$ (2-1) Band $^PQ_{23}(9)$	1560	105
	0.8704549	$B^3\Pi_g \rightarrow A^3\Sigma^+_u$ (2-1) Band $Q_1(7)$	1560	105
	0.8707478	$B^3\Pi_g \rightarrow A^3\Sigma^+_u$ (2-1) Band $Q_1(6)$	1560	105
	0.8710118	$B^3\Pi_g \rightarrow A^3\Sigma^+_u$ (2-1) Band $^0P_{23}(5)$	1560	105
	0.8710273	$B^3\Pi_g \rightarrow A^3\Sigma^+_u$ (2-1) Band $Q_1(5)$	1560	105
	0.8712956	$B^3\Pi_g \rightarrow A^3\Sigma^+_u$ (2-1) Band $Q_1(4)$	1560	105

Table 3.3.1.1—*continued*
Diatomic Electronic Transition Lasers

Species	Wavelength (μm)	Transition assignment	Comments	References
N_2	0.8713533	$B^3\Pi_g \to A^3\Sigma_u^+$ (2-1) Band $^0P_{23}(6)$	1560	105
	0.8715519	$B^3\Pi_g \to A^3\Sigma_u^+$ (2-1) Band $Q_1(3)$	1560	105
	0.8716718	$B^3\Pi_g \to A^3\Sigma_u^+$ (2-1) Band $^0P_{23}(7)$	1560	105
	0.8717377	$B^3\Pi_g \to A^3\Sigma_u^+$ (2-1) Band $P_1(11)$	1560	105
	0.8717970	$B^3\Pi_g \to A^3\Sigma_u^+$ (2-1) Band $Q_1(2)$	1560	105
	0.8718571	$B^3\Pi_g \to A^3\Sigma_u^+$ (2-1) Band $P_1(10)$	1560	105
	0.8718654	$B^3\Pi_g \to A^3\Sigma_u^+$ (2-1) Band $^PQ_{12}(9)$	1560	105
	0.8719537	$B^3\Pi_g \to A^3\Sigma_u^+$ (2-1) Band $^PQ_{12}(8)$	1560	105
	0.8719562	$B^3\Pi_g \to A^3\Sigma_u^+$ (2-1) Band $P_1(9)$	1560	105
	0.8719791	$B^3\Pi_g \to A^3\Sigma_u^+$ (2-1) Band	1560	105
	0.8720251	$B^3\Pi_g \to A^3\Sigma_u^+$ (2-1) Band $^PQ_{12}(7)$	1560	105
	0.8720284	$B^3\Pi_g \to A^3\Sigma_u^+$ (2-1) Band $Q_1(1)$	1560	105
	0.8720308	$B^3\Pi_g \to A^3\Sigma_u^+$ (2-1) Band	1560	105
	0.8720419	$B^3\Pi_g \to A^3\Sigma_u^+$ (2-1) Band $P_1(8)$	1560	105
	0.8720848	$B^3\Pi_g \to A^3\Sigma_u^+$ (2-1) Band $^PQ_{12}(6)$	1560	105
	0.8721155	$B^3\Pi_g \to A^3\Sigma_u^+$ (2-1) Band $P_1(7)$	1560	105
	0.8721327	$B^3\Pi_g \to A^3\Sigma_u^+$ (2-1) Band $^PQ_{12}(5)$	1560	105
	0.8721718	$B^3\Pi_g \to A^3\Sigma_u^+$ (2-1) Band $P_1(6)$	1560	105
	0.8721971	$B^3\Pi_g \to A^3\Sigma_u^+$ (2-1) Band	1560	105

Table 3.3.1.1—*continued*
Diatomic Electronic Transition Lasers

Species	Wavelength (µm)	Transition assignment	Comments	References
N_2	0.8722007	$B^3\Pi_g \to A^3\Sigma^+_u$ (2-1) Band $^PQ_{12}(3)$	1560	105
	0.8722220	$B^3\Pi_g \to A^3\Sigma^+_u$ (2-1) Band $P_1(5)$	1560	105
	0.8722341	$B^3\Pi_g \to A^3\Sigma^+_u$ (2-1) Band $^PQ_{12}(1)$	1560	105
	0.8722569	$B^3\Pi_g \to A^3\Sigma^+_u$ (2-1) Band $P_1(4)$	1560	105
	0.8722836	$B^3\Pi_g \to A^3\Sigma^+_u$ (2-1) Band $P_1(3)$	1560	105
	0.8723057	$B^3\Pi_g \to A^3\Sigma^+_u$ (2-1) Band $P_1(1)$	1560	105
	0.8726333	$B^3\Pi_g \to A^3\Sigma^+_u$ (2-1) Band $^0P_{12}(1)$	1560	105
	0.8728430	$B^3\Pi_g \to A^3\Sigma^+_u$ (2-1) Band $^0P_{12}(2)$	1560	105
	0.8730453	$B^3\Pi_g \to A^3\Sigma^+_u$ (2-1) Band $^0P_{12}(3)$	1560	105
	0.8732394	$B^3\Pi_g \to A^3\Sigma^+_u$ (2-1) Band $^0P_{12}(4)$	1560	105
	0.8734247	$B^3\Pi_g \to A^3\Sigma^+_u$ (2-1) Band $^0P_{12}(5)$	1560	105
	0.8735995	$B^3\Pi_g \to A^3\Sigma^+_u$ (2-1) Band $^0P_{12}(6)$	1560	105
	0.8737644	$B^3\Pi_g \to A^3\Sigma^+_u$ (2-1) Band $^0P_{12}(7)$	1560	105
	0.8739162	$B^3\Pi_g \to A^3\Sigma^+_u$ (2-1) Band $^0P_{12}(8)$	1560	105
	0.8740559	$B^3\Pi_g \to A^3\Sigma^+_u$ (2-1) Band $^0P_{12}(9)$	1560	105
	0.8742917	$B^3\Pi_g \to A^3\Sigma^+_u$ (2-1) Band $^0P_{12}(11)$	1560	105
	0.8845349	$B^3\Pi_g \to A^3\Sigma^+_u$ (1-0) Band $^SR_{32}(1)$	1560	105
	0.8856271	$B^3\Pi_g \to A^3\Sigma^+_u$ (1-0) Band $Q_3(9)$	1560	105
	0.8858470	$B^3\Pi_g \to A^3\Sigma^+_u$ (1-0) Band $Q_3(7)$	1560	105

Table 3.3.1.1—*continued*
Diatomic Electronic Transition Lasers

Species	Wavelength (µm)	Transition assignment	Comments	References
N_2	0.8880521	$B^3\Pi_g \rightarrow A^3\Sigma^+_u$ (1-0) Band $^QR_{23}(1)$	1560	105
	0.8884527	$B^3\Pi_g \rightarrow A^3\Sigma^+_u$ (1-0) Band $^PQ_{23}(2)$	1560	105
	0.8886204	$B^3\Pi_g \rightarrow A^3\Sigma^+_u$ (1-0) Band $^PQ_{23}(3)$	1560	105
	0.8886378	$B^3\Pi_g \rightarrow A^3\Sigma^+_u$ (1-0) Band $Q_1(9)$	1560	105
	0.8887756	$B^3\Pi_g \rightarrow A^3\Sigma^+_u$ (1-0) Band $^PQ_{23}(4)$	1560	105
	0.8889111	$B^3\Pi_g \rightarrow A^3\Sigma^+_u$ (1-0) Band $^PQ_{23}(5)$	1560	105
	0.8889738	$B^3\Pi_g \rightarrow A^3\Sigma^+_u$ (1-0) Band $Q_1(8)$	1560	105
	0.8890243	$B^3\Pi_g \rightarrow A^3\Sigma^+_u$ (1-0) Band $^PQ_{23}(6)$	1560	105
	0.8891133	$B^3\Pi_g \rightarrow A^3\Sigma^+_u$ (1-0) Band $^PQ_{23}(7)$	1560	105
	0.8891769	$B^3\Pi_g \rightarrow A^3\Sigma^+_u$ (1-0) Band $^PQ_{23}(8)$	1560	105
	0.8892149	$B^3\Pi_g \rightarrow A^3\Sigma^+_u$ (1-0) Band $^PQ_{23}(9)$	1560	105
	0.8892940	$B^3\Pi_g \rightarrow A^3\Sigma^+_u$ (1-0) Band $Q_1(7)$	1560	105
	0.8896001	$B^3\Pi_g \rightarrow A^3\Sigma^+_u$ (1-0) Band $Q_1(6)$	1560	105
	0.8898930	$B^3\Pi_g \rightarrow A^3\Sigma^+_u$ (1-0) Band $Q_1(5)$	1560	105
	0.8899078	$B^3\Pi_g \rightarrow A^3\Sigma^+_u$ (1-0) Band $^0P_{23}(5)$	1560	105
	0.8901733	$B^3\Pi_g \rightarrow A^3\Sigma^+_u$ (1-0) Band $Q_1(4)$	1560	105
	0.8902711	$B^3\Pi_g \rightarrow A^3\Sigma^+_u$ (1-0) Band $^0P_{23}(6)$	1560	105
	0.8904419	$B^3\Pi_g \rightarrow A^3\Sigma^+_u$ (1-0) Band $Q_1(3)$	1560	105
	0.8906097	$B^3\Pi_g \rightarrow A^3\Sigma^+_u$ (1-0) Band $^0P_{23}(7)$	1560	105

Table 3.3.1.1—*continued*
Diatomic Electronic Transition Lasers

Species	Wavelength (μm)	Transition assignment	Comments	References
N_2	0.8906649	$B^3\Pi_g \to A^3\Sigma^+_u$ (1-0) Band $P_1(11)$	1560	105
	0.8906994	$B^3\Pi_g \to A^3\Sigma^+_u$ (1-0) Band $Q_1(2)$	1560	105
	0.8907920	$B^3\Pi_g \to A^3\Sigma^+_u$ (1-0) Band $^PQ_{12}(9)$	1560	105
	0.8908808	$B^3\Pi_g \to A^3\Sigma^+_u$ (1-0) Band $^PQ_{12}(8)$	1560	105
	0.8908878	$B^3\Pi_g \to A^3\Sigma^+_u$ (1-0) Band $P_1(9)$	1560	105
	0.8909451	$B^3\Pi_g \to A^3\Sigma^+_u$ (1-0) Band $Q_1(1)$	1560	105
	0.8909527	$B^3\Pi_g \to A^3\Sigma^+_u$ (1-0) Band $^PQ_{12}(7)$	1560	105
	0.8909750	$B^3\Pi_g \to A^3\Sigma^+_u$ (1-0) Band $P_1(8)$	1560	105
	0.8910132	$B^3\Pi_g \to A^3\Sigma^+_u$ (1-0) Band $^PQ_{12}(6)$	1560	105
	0.8910480	$B^3\Pi_g \to A^3\Sigma^+_u$ (1-0) Band $P_1(7)$	1560	105
	0.8910612	$B^3\Pi_g \to A^3\Sigma^+_u$ (1-0) Band $^PQ_{12}(5)$	1560	105
	0.8911001	$B^3\Pi_g \to A^3\Sigma^+_u$ (1-0) Band $^PQ_{12}(4)$	1560	105
	0.8911063	$B^3\Pi_g \to A^3\Sigma^+_u$ (1-0) Band $P_1(6)$	1560	105
	0.8911280	$B^3\Pi_g \to A^3\Sigma^+_u$ (1-0) Band $^PQ_{12}(3)$	1560	105
	0.8911502	$B^3\Pi_g \to A^3\Sigma^+_u$ (1-0) Band $^PQ_{12}(2)$	1560	105
	0.8911538	$B^3\Pi_g \to A^3\Sigma^+_u$ (1-0) Band $P_1(5)$	1560	105
	0.8911608	$B^3\Pi_g \to A^3\Sigma^+_u$ (1-0) Band $^PQ_{12}(1)$	1560	105
	0.8911898	$B^3\Pi_g \to A^3\Sigma^+_u$ (1-0) Band $P_1(4)$	1560	105
	0.8912139	$B^3\Pi_g \to A^3\Sigma^+_u$ (1-0) Band $^0P_{23}(9)$	1560	105

Table 3.3.1.1—*continued*
Diatomic Electronic Transition Lasers

Species	Wavelength (µm)	Transition assignment	Comments	References
N_2	0.8918033	$B^3\Pi_g \to A^3\Sigma^+_u$ (1-0) Band $^pP_{12}(2)$	1560	105
	0.8920184	$B^3\Pi_g \to A^3\Sigma^+_u$ (1-0) Band $^0P_{12}(3)$	1560	105
	0.8922249	$B^3\Pi_g \to A^3\Sigma^+_u$ (1-0) Band $^0P_{12}(4)$	1560	105
	0.8924223	$B^3\Pi_g \to A^3\Sigma^+_u$ (1-0) Band $^0P_{12}(5)$	1560	105
	0.8926099	$B^3\Pi_g \to A^3\Sigma^+_u$ (1-0) Band $^0P_{12}(6)$	1560	105
	0.8927865	$B^3\Pi_g \to A^3\Sigma^+_u$ (1-0) Band $^0P_{12}(7)$	1560	105
	0.8929509	$B^3\Pi_g \to A^3\Sigma^+_u$ (1-0) Band $^0P_{12}(8)$	1560	105
	0.8931019	$B^3\Pi_g \to A^3\Sigma^+_u$ (1-0) Band $^0P_{12}(9)$	1560	105
	0.8933580	$B^3\Pi_g \to A^3\Sigma^+_u$ (1-0) Band $^0P_{12}(11)$	1560	105
	0.965389	$B^3\Pi_g \to A^3\Sigma^+_u$ (3-3) Band $P_2(11)$	1560	105
	0.965846	$B^3\Pi_g \to A^3\Sigma^+_u$ (3-3) Band $Q_1(7)$	1560	105
	0.966599	$B^3\Pi_g \to A^3\Sigma^+_u$ (3-3) Band $Q_1(5)$	1560	105
	0.967270	$B^3\Pi_g \to A^3\Sigma^+_u$ (3-3) Band $Q_1(3)$	1560	105
	0.967758	$B^3\Pi_g \to A^3\Sigma^+_u$ (3-3) Band $^pQ_{12}(7)$	1560	105
	0.967943	$B^3\Pi_g \to A^3\Sigma^+_u$ (3-3) Band $^pQ_{12}(5)$	1560	105
	0.968061	$B^3\Pi_g \to A^3\Sigma^+_u$ (3-3) Band $^pQ_{12}(3)$	1560	105
	0.969552	$B^3\Pi_g \to A^3\Sigma^+_u$ (3-3) Band $^0P_{12}(5)$	1560	105
	0.969879	$B^3\Pi_g \to A^3\Sigma^+_u$ (3-3) Band $^0P_{12}(7)$	1560	105
	3.29463	$a^1\Pi_g \to a'^1\Sigma^-_u$ (2-1) Band $Q(14)$	1561,1562	107,128

Table 3.3.1.1—*continued*
Diatomic Electronic Transition Lasers

Species	Wavelength (μm)	Transition assignment	Comments	References
N_2	3.30149	$a^1\Pi_g \rightarrow a'^1\Sigma^-_u$ (2-1) Band Q(12)	1561,1562	107,128
	3.30734	$a^1\Pi_g \rightarrow a'^1\Sigma^-_u$ (2-1) Band Q(10)	1561,1562	107,128
	3.30989	$a^1\Pi_g \rightarrow a'^1\Sigma^-_u$ (2-1) Band Q(9)	1561,1562	107,128
	3.31221	$a^1\Pi_g \rightarrow a'^1\Sigma^-_u$ (2-1) Band Q(8)	1561,1562	107,128
	3.31426	$a^1\Pi_g \rightarrow a'^1\Sigma^-_u$ (2-1) Band Q(7)	1562	107,128
	3.31607	$a^1\Pi_g \rightarrow a'^1\Sigma^-_u$ (2-1) Band Q(6)	1561,1562	107,128
	3.31760	$a^1\Pi_g \rightarrow a'^1\Sigma^-_u$ (2-1) Band Q(5)	1562	107,128
	3.31889	$a^1\Pi_g \rightarrow a'^1\Sigma^-_u$ (2-1) Band Q(4)	1561,1562	107,128
	3.32069	$a^1\Pi_g \rightarrow a'^1\Sigma^-_u$ (2-1) Band Q(2)	1562	107,128
	3.45184	$a^1\Pi_g \rightarrow a'^1\Sigma^-_u$ (1-0) Band Q(12)	1561,1562	107,128
	3.45832	$a^1\Pi_g \rightarrow a'^1\Sigma^-_u$ (1-0) Band Q(10)	1561,1562	107,128
	3.46114	$a^1\Pi_g \rightarrow a'^1\Sigma^-_u$ (1-0) Band Q(9)	1562	107,128
	3.46368	$a^1\Pi_g \rightarrow a'^1\Sigma^-_u$ (1-0) Band Q(8)	1561,1562	107,128
	3.46596	$a^1\Pi_g \rightarrow a'^1\Sigma^-_u$ (1-0) Band Q(7)	1562	107,128
	3.46795	$a^1\Pi_g \rightarrow a'^1\Sigma^-_u$ (1-0) Band Q(6)	1561,1562	107,128
	3.46967	$a^1\Pi_g \rightarrow a'^1\Sigma^-_u$ (1-0) Band Q(5)	1562	107,128
	3.47109	$a^1\Pi_g \rightarrow a'^1\Sigma^-_u$ (1-0) Band Q(4)	1561,1562	107,128
	3.62349	$w^1\Delta_u \rightarrow a^1\Pi_g$ (0-0) Band R(4)	1563	108
	3.62614	$w^1\Delta_u \rightarrow a^1\Pi_g$ (0-0) Band R(3)	1563	108

Table 3.3.1.1—*continued*
Diatomic Electronic Transition Lasers

Species	Wavelength (μm)	Transition assignment	Comments	References
N_2	3.62910	$w^1\Delta_u \to a^1\Pi_g$ (0-0) Band R(2)	1563	108
	3.64313	$w^1\Delta_u \to a^1\Pi_g$ (0-0) Band Q(4)	1563	108
	3.64472	$w^1\Delta_u \to a^1\Pi_g$ (0-0) Band Q(5)	1563	108
	3.64662	$w^1\Delta_u \to a^1\Pi_g$ (0-0) Band Q(6)	1563	108
	3.64883	$w^1\Delta_u \to a^1\Pi_g$ (0-0) Band Q(7)	1563	108
	3.65138	$w^1\Delta_u \to a^1\Pi_g$ (0-0) Band Q(8)	1563	108
	3.65424	$w^1\Delta_u \to a^1\Pi_g$ (0-0) Band Q(9)	1563	108
	3.65745	$w^1\Delta_u \to a^1\Pi_g$ (0-0) Band Q(10)	1563	108
	3.66095	$w^1\Delta_u \to a^1\Pi_g$ (0-0) Band Q(11)	1563	108
	3.66483	$w^1\Delta_u \to a^1\Pi_g$ (0-0) Band Q(12)	1563	108
	3.66899	$w^1\Delta_u \to a^1\Pi_g$ (0-0) Band Q(13)	1563	108
	3.67352	$w^1\Delta_u \to a^1\Pi_g$ (0-0) Band Q(14)	1563	108
	3.67834	$w^1\Delta_u \to a^1\Pi_g$ (0-0) Band Q(15)	1563	108
	8.1483	$a^1\Pi_g \to a'^1\Sigma^-_u$ (0-0) Band Q(10)	1562	107,128
	8.1827	$a^1\Pi_g \to a'^1\Sigma^-_u$ (0-0) Band Q(8)	1561,1562	107,128
	8.2102	$a^1\Pi_g \to a'^1\Sigma^-_u$ (0-0) Band Q(6)	1561,1562	107,128
N^+_2	0.391	$B \to X$ $(v',v'') = (0,0)$	6741	1581,1582
	0.3914	$B^2\Sigma^+_u \to X^2\Sigma^+_g$ (0-0)	1564–1567	112,114, 115,127
	0.42781	$B^2\Sigma^+_u \to X^2\Sigma^+_g$ (0-1)	1564–1567	112,114, 115,127

Table 3.3.1.1—*continued*
Diatomic Electronic Transition Lasers

Species	Wavelength (µm)	Transition assignment	Comments	References
N^+_2	0.428	$B \rightarrow X$	6741	1581,1582
		$(v',v'') = (0,1)$		
	0.4709	$B^2\Sigma^+_u \rightarrow X^2\Sigma^+_g(0\text{-}2)$	1564–1567	112,114, 115,127
	0.5228	$B^2\Sigma^+_u \rightarrow X^2\Sigma^+_g(0\text{-}3)$	1564–1567	112,114, 115,127
Na_2	0.3400-0.3600		6744–6746	1584
	0.4050-0.4600		6744–6746	
	0.430-0.452		6744–6746	1587
	0.4360	$^3\lambda_g \, (\lambda = \Sigma \text{ or } \Pi) \rightarrow a^3\Sigma^+_u$	6744–6746	1586
	0.5100	$B^1\Pi_u \rightarrow X^1\Sigma^+_g \, Q(9)$	6747,6749	1524,1588
	0.5137	$B^1\Pi_u \rightarrow X^1\Sigma^+_g \, Q(10)$	6747,6749	1524,1588
	0.5245	$B^1\Pi_u \rightarrow X^1\Sigma^+_g$	1568	117,118,119
	0.5251	$B^1\Pi_u \rightarrow X^1\Sigma^+_g \, Q(13)$	6747,6749	1524,1588
	0.52633	$B^1\Pi_u \rightarrow X^1\Sigma^+_g$	1569	117–119
	0.5274	$B^1\Pi_u \rightarrow X^1\Sigma^+_g$	1568	117–119
	0.5289	$B^1\Pi_u \rightarrow X^1\Sigma^+_g \, Q(14)$	6747,6749	1524,1588
	0.52982	$A^1\Sigma^+_u \rightarrow X^1\Sigma^+_g$	1569	117,118
	0.52995	$A^1\Sigma^+_u \rightarrow X^1\Sigma^+_g$	1569	117,118
	0.5310	$B^1\Pi_u \rightarrow X^1\Sigma^+_g \, R(17)$	6747,6751	1524,1588
	0.5319	$B^1\Pi_u \rightarrow X^1\Sigma^+_g \, P(17)$	6747,6749	1524,1588
	0.5321	$B^1\Pi_u \rightarrow X^1\Sigma^+_g$	1568	117–119
	0.5326	$B^1\Pi_u \rightarrow X^1\Sigma^+_g \, Q(15)$	6747,6749	1524,1588
	0.5333	$B^1\Pi_u \rightarrow X^1\Sigma^+_g$	1568	117–119
	0.5339	$B^1\Pi_u \rightarrow X^1\Sigma^+_g \, Q(15)$	6747,6752	1524,1588
	0.5340	$B^1\Pi_u \rightarrow X^1\Sigma^+_g$	1568	117–119
	0.53415	$A^1\Sigma^+_u \rightarrow X^1\Sigma^+_g$	1569	117,118
	0.53428	$A^1\Sigma^+_u \rightarrow X^1\Sigma^+_g$	1569	117,118
	0.5346	$B^1\Pi_u \rightarrow X^1\Sigma^+_g \, R(19)$	6747,6750	1524,1588
	0.5347	$B^1\Pi_u \rightarrow X^1\Sigma^+_g \, R(18)$	6747,6751	1524,1588
	0.53490	$B^1\Pi_u \rightarrow X^1\Sigma^+_g$	1569	117–119
	0.5352	$B^1\Pi_u \rightarrow X^1\Sigma^+_g \, P(19)$	6747,6750	1524,1588
	0.5355	$B^1\Pi_u \rightarrow X^1\Sigma^+_g \, P(18)$	6747,6751	1524,1588
	0.5363	$B^1\Pi_u \rightarrow X^1\Sigma^+_g \, Q(16)$	6747,6749	1524,1588
	0.53690	$B^1\Pi_u \rightarrow X^1\Sigma^+_g$	1569	117–119
	0.5370	$B^1\Pi_u \rightarrow X^1\Sigma^+_g$	1568	117–119
	0.5371	$B^1\Pi_u \rightarrow X^1\Sigma^+_g$	1568	117–119
	0.5374	$B^1\Pi_u \rightarrow X^1\Sigma^+_g \, Q(16)$	6747,6752	1524,1588

Table 3.3.1.1—*continued*
Diatomic Electronic Transition Lasers

Species	Wavelength (μm)	Transition assignment	Comments	References
Na$_2$	0.5376	B$^1\Pi_u \rightarrow$ X$^1\Sigma^+_g$	1568	117–119
	0.53781	B$^1\Pi_u \rightarrow$ X$^1\Sigma^+_g$	1569	117–119
	0.5382	B$^1\Pi_u \rightarrow$ X$^1\Sigma^+_g$ R(20)	6747,6750	1524,1588
	0.5383	B$^1\Pi_u \rightarrow$ X$^1\Sigma^+_g$ R(19)	6747,6751	1524,1588
	0.53850	A$^1\Sigma^+_u \rightarrow$ X$^1\Sigma^+_g$	1569	117,118
	0.53863	A$^1\Sigma^+_u \rightarrow$ X$^1\Sigma^+_g$	1569	117,118
	0.5388	B$^1\Pi_u \rightarrow$ X$^1\Sigma^+_g$ P(20)	6747,6750	1524,1588
	0.5389	B$^1\Pi_u \rightarrow$ X$^1\Sigma^+_g$	1568	117–119
	0.5391	B$^1\Pi_u \rightarrow$ X$^1\Sigma^+_g$ P(19)	6747,6751	1524,1588
	0.54024	B$^1\Pi_u \rightarrow$ X$^1\Sigma^+_g$	1569	117–119
	0.5409	B$^1\Pi_u \rightarrow$ X$^1\Sigma^+_g$ Q(17)	6747,6752	1524,1588
	0.54131	B$^1\Pi_u \rightarrow$ X$^1\Sigma^+_g$	1569	117–119
	0.5416	B$^1\Pi_u \rightarrow$ X$^1\Sigma^+_g$	1568	117–119
	0.5418	B$^1\Pi_u \rightarrow$ X$^1\Sigma^+_g$ R(21)	6747,6753,6554	1524,1588
	0.5424	B$^1\Pi_u \rightarrow$ X$^1\Sigma^+_g$ P(21)	6747,6753,6554	1524,1588
	0.54469	B$^1\Pi_u \rightarrow$ X$^1\Sigma^+_g$	1569	117–119
	0.5448	B$^1\Pi_u \rightarrow$ X$^1\Sigma^+_g$	1568	117–119
	0.5454	B$^1\Pi_u \rightarrow$ X$^1\Sigma^+_g$	1568	117–119
	0.5467	B$^1\Pi_u \rightarrow$ X$^1\Sigma^+_g$	1568	117–119
	0.5469	B$^1\Pi_u \rightarrow$ X$^1\Sigma^+_g$	1568	117–119
	0.5480	B$^1\Pi_u \rightarrow$ X$^1\Sigma^+_g$	1568	117–119
	0.5485	B$^1\Pi_u \rightarrow$ X$^1\Sigma^+_g$	1568	117–119
	0.54916	B$^1\Pi_u \rightarrow$ X$^1\Sigma^+_g$	1569	117–119
	0.5499	B$^1\Pi_u \rightarrow$ X$^1\Sigma^+_g$	1568	117–119
	0.5581	B$^1\Pi_u \rightarrow$ X$^1\Sigma^+_g$	1568	117–119
	0.5591	B$^1\Pi_u \rightarrow$ X$^1\Sigma^+_g$	1568	117–119
	0.5597	B$^1\Pi_u \rightarrow$ X$^1\Sigma^+_g$	1568	117–119
	0.7284	A$^1\Sigma^+_u \rightarrow$ X$^1\Sigma^+_g$ (v',v") = (34,34) P(51)	6755	1589,1590
	0.7563	A$^1\Sigma^+_u \rightarrow$ X$^1\Sigma^+_g$ (v',v") = (34,40) P(51)	6755	1589,1590
	0.7647	A$^1\Sigma^+_u \rightarrow$ X$^1\Sigma^+_g$ (v',v") = (34,42) P(51)	6755	1589,1590
	0.764852	A$^1\Sigma^+_u \rightarrow$ X$^1\Sigma^+_g$	6758	1589,1590
	0.765418	A$^1\Sigma^+_u \rightarrow$ X$^1\Sigma^+_g$	6758	1589,1590
	0.767324	A$^1\Sigma^+_u \rightarrow$ X$^1\Sigma^+_g$	6758	1589,1590
	0.7676	A$^1\Sigma^+_u \rightarrow$ X$^1\Sigma^+_g$ (v',v") = (34,43) R(49)	6755	1589,1590
	0.767606	A$^1\Sigma^+_u \rightarrow$ X$^1\Sigma^+_g$	6758	1589,1590

Table 3.3.1.1—*continued*
Diatomic Electronic Transition Lasers

Species	Wavelength (μm)	Transition assignment	Comments	References
Na$_2$	0.768487	$A^1\Sigma^+_u \to X^1\Sigma^+_g$	6758	1589,1590
	0.7687	$A^1\Sigma^+_u \to X^1\Sigma^+_g$	6755	1589,1590
		(v',v'') = (34,43) P(51)		
	0.773800	$A^1\Sigma^+_u \to X^1\Sigma^+_g$	6758	1589,1590
	0.775265	$A^1\Sigma^+_u \to X^1\Sigma^+_g$	6758	1589,1590
	0.775423	$A^1\Sigma^+_u \to X^1\Sigma^+_g$	6758	1589,1590
	0.784292	$A^1\Sigma^+_u \to X^1\Sigma^+_g$	6758	1589,1590
	0.78493	$A^1\Sigma^+_u \to X^1\Sigma^+_g$	1571	117,118
	0.7851	$A^1\Sigma^+_u \to X^1\Sigma^+_g$	6755	1589,1590
		(v',v'') = (34,48) R(49)		
	0.78569	$A^1\Sigma^+_u \to X^1\Sigma^+_g$	1571	117,118
	0.7879	$A^1\Sigma^+_u \to X^1\Sigma^+_g$	6755	1589,1590
		(v',v'') = (34,49) R(49)		
	0.7888	$A^1\Sigma^+_u \to X^1\Sigma^+_g$	6755	1589,1590
		(v',v'') = (34,49) P(51)		
	0.789204	$A^1\Sigma^+_u \to X^1\Sigma^+_g$	6758	1589,1590
	0.78974	$A^1\Sigma^+_u \to X^1\Sigma^+_g$	1571	117,118
	0.78979	$A^1\Sigma^+_u \to X^1\Sigma^+_g$	1571	117,118
	0.790912	$A^1\Sigma^+_u \to X^1\Sigma^+_g$	6758	1589,1590
	0.791003	$A^1\Sigma^+_u \to X^1\Sigma^+_g$	6758	1589,1590
	0.79178	$A^1\Sigma^+_u \to X^1\Sigma^+_g$	1571	117,118
	0.792346	$A^1\Sigma^+_u \to X^1\Sigma^+_g$	6758	1589,1590
	0.792346	$A^1\Sigma^+_u \to X^1\Sigma^+_g$	6758	1589,1590
	0.792517	$A^1\Sigma^+_u \to X^1\Sigma^+_g$	6758	1589,1590
	0.79295	$A^1\Sigma^+_u \to X^1\Sigma^+_g$	1571	117,118
	0.793253	$A^1\Sigma^+_u \to X^1\Sigma^+_g$	6758	1589,1590
	0.79370	$A^1\Sigma^+_u \to X^1\Sigma^+_g$	1571	117,118
	0.793903	$A^1\Sigma^+_u \to X^1\Sigma^+_g$	6758	1589,1590
	0.794072	$A^1\Sigma^+_u \to X^1\Sigma^+_g$	6758	1589,1590
	0.7957	$A^1\Sigma^+_u \to X^1\Sigma^+_g$	6755	1589,1590
		(v',v'') = (34,52) P(51)		
	0.795889	$A^1\Sigma^+_u \to X^1\Sigma^+_g$	6758	1589,1590
	0.796550	$A^1\Sigma^+_u \to X^1\Sigma^+_g$	6758	1589,1590
	0.7968	$A^1\Sigma^+_u \to X^1\Sigma^+_g$	6755	1589,1590
		(v',v'') = (34,53) R(49)		
	0.79747	$A^1\Sigma^+_u \to X^1\Sigma^+_g$	1571	117,118
	0.7975	$A^1\Sigma^+_u \to X^1\Sigma^+_g$	6755	1589,1590
		(v',v'') = (34,53) P(51)		
	0.797559	$A^1\Sigma^+_u \to X^1\Sigma^+_g$	6758	1589,1590

Table 3.3.1.1—*continued*
Diatomic Electronic Transition Lasers

Species	Wavelength (μm)	Transition assignment	Comments	References
Na$_2$	0.79766	A$^1\Sigma^+_u \to$ X$^1\Sigma^+_g$	1571	117,118
	0.798553	A$^1\Sigma^+_u \to$ X$^1\Sigma^+_g$	6758	1589,1590
	0.7989	A$^1\Sigma^+_u \to$ X$^1\Sigma^+_g$	6755	1589,1590
		(v',v") = (34,54) P(51)		
	0.79909	A$^1\Sigma^+_u \to$ X$^1\Sigma^+_g$	1571	117,118
	0.79966	A$^1\Sigma^+_u \to$ X$^1\Sigma^+_g$	1571	117,118
	0.80084	A$^1\Sigma^+_u \to$ X$^1\Sigma^+_g$	1571	117,118
	0.801329	A$^1\Sigma^+_u \to$ X$^1\Sigma^+_g$	6758	1589,1590
	0.802136	A$^1\Sigma^+_u \to$ X$^1\Sigma^+_g$	6758	1589,1590
	0.802345	A$^1\Sigma^+_u \to$ X$^1\Sigma^+_g$	6758	1589,1590
	0.803391	A$^1\Sigma^+_u \to$ X$^1\Sigma^+_g$	6758	1589,1590
	0.80365	A$^1\Sigma^+_u \to$ X$^1\Sigma^+_g$	1571	117,118
	0.80393	A$^1\Sigma^+_u \to$ X$^1\Sigma^+_g$	1571	117,118
	0.80445	A$^1\Sigma^+_u \to$ X$^1\Sigma^+_g$	1571	117,118
	0.804472	A$^1\Sigma^+_u \to$ X$^1\Sigma^+_g$	6758	1589,1590
	0.8050	A$^1\Sigma^+_u \to$ X$^1\Sigma^+_g$	6755,6760	1589,1590
	0.80537	A$^1\Sigma^+_u \to$ X$^1\Sigma^+_g$	1571	117,118
	0.80561	A$^1\Sigma^+_u \to$ X$^1\Sigma^+_g$	1571	117,118
	0.80694	A$^1\Sigma^+_u \to$ X$^1\Sigma^+_g$	1571	117,118
	0.80805	A$^1\Sigma^+_u \to$ X$^1\Sigma^+_g$	1571	117,118
	0.8175	A$^1\Sigma^+_u \to$ X$^1\Sigma^+_g$	6755,6760	1589,1590
	0.827-0.832	b$^3\Sigma^+_g \to$ x$^3\Sigma^+_u$	6761	1591
	0.8920	1$^3\Sigma^+_g \to$ 1$^3\Sigma^+_u$	6763	1592
	0.900	3$^1\Sigma^+_g \to$ A$^1\Sigma^+_u$	6765,6767	1593
	0.904	3$^1\Sigma^+_g \to$ A$^1\Sigma^+_u$	6765,6768	1593
		(v',v") = (6,5) P(44) P(46)		
	0.905	3$^3\Sigma^+_g \to$ A$^1\Sigma^+_u$	6770	1593
		(v',v") = (0,0) R(72)		
	0.907	3$^3\Sigma^+_g \to$ A$^1\Sigma^+_u$	6770	1593
		(v',v") = (0,0) P(74)		
	0.910	3$^1\Sigma^+_g \to$ A$^1\Sigma^+_u$	6765,6767	1593
		(v',v") = (4,4) R(33)		
	0.961	3$^1\Sigma^+_g \to$ A$^1\Sigma^+_u$	6765,6769	1593
	1.801	3$^1\Sigma^+_g \to$ B$^1\Pi_u$	6771	1593
		(v',v") = (4,2) R(11)		
	1.804	3$^1\Sigma^+_g \to$ B$^1\Pi_u$	6769,6771	1593
		(v',v") = (8,5) P(49)		
	1.844	3$^1\Sigma^+_g \to$ B$^1\Pi_u$	6767,6771	1593
		(v',v") = (4,3) Q(34) R(33)		

Table 3.3.1.1—*continued*
Diatomic Electronic Transition Lasers

Species	Wavelength (µm)	Transition assignment	Comments	References
Na$_2$	1.876	$3^1\Sigma^+_g \rightarrow B^1\Pi_u$ $(v',v'') = (0,0)$ R(72)	6770,6771	1593
	1.882	$3^1\Sigma^+_g \rightarrow B^1\Pi_u$ $(v',v'') = (0,0)$ Q(73)	6770,6771	1593
	1.888	$3^1\Sigma^+_g \rightarrow B^1\Pi_u$ $(v',v'') = (0,0)$ P(74)	6770,6771	1593
	1.925	$3^1\Sigma^+_g \rightarrow B^1\Pi_u$ $(v',v'') = (0,1)$ Q(73)	6770,6771	1593
	1.931	$3^1\Sigma^+_g \rightarrow B^1\Pi_u$ $(v',v'') = (4,5)$ P(35)	6767,6771	1593
	1.931	$3^1\Sigma^+_g \rightarrow B^1\Pi_u$ $(v',v'') = (5,6)$ R(33)	6767,6771	1593
	1.968	$3^1\Sigma^+_g \rightarrow B^1\Pi_u$ $(v',v'') = (4,6)$ Q(12)	6771	1593
	2.458	$C^1\Pi_u \rightarrow 3^1\Sigma_g$ $(v',v'') = (5,4)$ Q(34)	6767,6772	1593
	2.498	$C^1\Pi_u \rightarrow 3^1\Sigma_g$	6768,6772	1593
	2.503	$C^1\Pi_u \rightarrow 3^1\Sigma_g$	6769,6772	1593
	2.510	$C^1\Pi_u \rightarrow 3^1\Sigma_g$ $(v',v'') = (6,6)$ P(45) R(43)	6768,6772	1593
	2.517	$C^1\Pi_u \rightarrow 3^1\Sigma_g$ $(v',v'') = (5,5)$ Q(34)	6767,6772	1593
	2.520	$C^1\Pi_u \rightarrow 3^1\Sigma_g$ $(v',v'') = (4,4)$ Q(12)	6771,6772	1593
	2.524	$C^1\Pi_u \rightarrow 3^1\Sigma_g$ $(v',v'') = (1,1)$ R(67)	6771,6773	1593
	2.537	$C^1\Pi_u \rightarrow 3^1\Sigma_g$	6767,6772	1593
	2.538	$C^1\Pi_u \rightarrow 3^1\Sigma_g$ $(v',v'') = (0,0)$ Q(73)	6770,6772	1593
	2.557	$C^1\Pi_u \rightarrow 3^1\Sigma_g$	6772,6773	1593
	2.561	$C^1\Pi_u \rightarrow 3^1\Sigma_g$ $(v',v'') = (7,8)$ Q(48)	6769,6772	1593
NaK	0.520-0.565	$D^1\Pi(v'=0) \rightarrow X^1\Sigma^+$ $(v''=9\text{-}19)$	6902	1771
NaRb	0.666545		6774	1597
	0.668681		6774	1597
NO	218.1	$B'\,^2\Delta\,(v'=3) \rightarrow X^2\Pi\,(v''=10)$, $Q_{12}\,(7\frac{1}{2})$	6904	1772,1773

Table 3.3.1.1—*continued*
Diatomic Electronic Transition Lasers

Species	Wavelength (μm)	Transition assignment	Comments	References
NO	0.237	$A^2\Sigma^+ \to X^2\Pi$ (0,1)	6742	1583
	0.248	$A^2\Sigma^+ \to X^2\Pi$ (0,2)	6742	1583
S_2	0.362-0.570	$B^3\Sigma^-_u \to X^3\Sigma^-_g$	6777,6779	1598
	0.3620	$B^3\Sigma^-_u \to X^3\Sigma^-_g$	6778,6779	1599
		$(v',v'') = (2,7)$		
	0.365-0.570	$B^3\Sigma_u \to X^3\Sigma^-_g$	1572,1573	120,121
	0.3712	$B^3\Sigma^-_u \to X^3\Sigma^-_g$	6778,6779	1599
		$(v',v'') = (2,8)$		
	0.3807	$B^3\Sigma^-_u \to X^3\Sigma^-_g$	6778,6779	1599
		$(v',v'') = (2,9)$		
	0.3907	$B^3\Sigma^-_u \to X^3\Sigma^-_g$	6778,6779	1599
		$(v',v'') = (2,10)$		
	0.4011	$B^3\Sigma^-_u \to X^3\Sigma^-_g$	6778,6779	1599
		$(v',v'') = (2,11)$		
	0.4119	$B^3\Sigma^-_u \to X^3\Sigma^-_g$	6778,6779	1599
		$(v',v'') = (2,12)$		
	0.4233	$B^3\Sigma^-_u \to X^3\Sigma^-_g$	6778,6779	1599
		$(v',v'') = (2,13)$		
	0.4352	$B^3\Sigma^-_u \to X^3\Sigma^-_g$	6778,6779	1599
		$(v',v'') = (2,14)$		
	0.4477	$B^3\Sigma^-_u \to X^3\Sigma^-_g$	6778,6779	1599
		$(v',v'') = (2,15)$		
	0.4538	$B^3\Sigma^-_u \to X^3\Sigma^-_g$	6779,6780	1524
	0.4544	$B^3\Sigma^-_u \to X^3\Sigma^-_g$	6779,6780	1524
	0.4557	$B^3\Sigma^-_u \to X^3\Sigma^-_g$	6779,6780	1524
	0.4563	$B^3\Sigma^-_u \to X^3\Sigma^-_g$	6779,6780	1524
	0.4608	$B^3\Sigma^-_u \to X^3\Sigma^-_g$	6778,6779	1599
		$(v',v'') = (2,16)$		
	0.4671	$B^3\Sigma^-_u \to X^3\Sigma^-_g$	6779,6780	1524
	0.4677	$B^3\Sigma^-_u \to X^3\Sigma^-_g$	6779,6780	1524
	0.4690	$B^3\Sigma^-_u \to X^3\Sigma^-_g$	6779,6780	1524
	0.4697	$B^3\Sigma^-_u \to X^3\Sigma^-_g$	6779,6780	1524
	0.4745	$B^3\Sigma^-_u \to X^3\Sigma^-_g$	6778,6779	1599
		$(v',v'') = (2,17)$		
	0.4810	$B^3\Sigma^-_u \to X^3\Sigma^-_g$	6779,6780	1524
	0.4816	$B^3\Sigma^-_u \to X^3\Sigma^-_g$	6779,6780	1524
	0.4830	$B^3\Sigma^-_u \to X^3\Sigma^-_g$	6779,6780	1524
	0.4838	$B^3\Sigma^-_u \to X^3\Sigma^-_g$	6779,6780	1524

Table 3.3.1.1—*continued*
Diatomic Electronic Transition Lasers

Species	Wavelength (μm)	Transition assignment	Comments	References
S$_2$	0.4920	$B^3\Sigma_u^- \rightarrow X^3\Sigma_g^-$	6779,6780	1524
	0.4950	$B^3\Sigma_u^- \rightarrow X^3\Sigma_g^-$	6779,6780	1524
	1.0860	$^1\Sigma_g^+ \rightarrow X^3\Sigma_g^-$	1574	122
	1.0915	$^1\Sigma_g^+ \rightarrow X^3\Sigma_g^-$		122
	1.0917	$^1\Sigma_g^+ \rightarrow X^3\Sigma_g^-$		122
	1.0920	$^1\Sigma_g^+ \rightarrow X^3\Sigma_g^-$		122
	1.0923	$^1\Sigma_g^+ \rightarrow X^3\Sigma_g^-$		122
	1.0941	$^1\Sigma_g^+ \rightarrow X^3\Sigma_g^-$		122
	1.0946	$^1\Sigma_g^+ \rightarrow X^3\Sigma_g^-$		122
	1.0990	$^1\Sigma_g^+ \rightarrow X^3\Sigma_g^-$		122
	1.1000	$^1\Sigma_g^+ \rightarrow X^3\Sigma_g^-$		122
	1.1587	$^1\Sigma_g^+ \rightarrow X^3\Sigma_g^-$		122
Se$_2$	0.3869757	$BO_u^+ \rightarrow XO_g^+$ (v',v") = (11,7) R(10)	6781	1601
	0.3870360	$BO_u^+ \rightarrow XO_g^+$ (v',v") = (11,7) P(12)	6781	1601
	0.3925932	$BO_u^+ \rightarrow XO_g^+$ (v',v") = (11,8) R(10)	6781	1601
	0.3926552	$BO_u^+ \rightarrow XO_g^+$ (v',v") = (11,8) P(12)	6781	1601
	0.3983450	$BO_u^+ \rightarrow XO_g^+$ (v',v") = (11,9) R(10)	6781	1601
	0.3984086	$BO_u^+ \rightarrow XO_g^+$ (v',v") = (11,9) P(12)	6781	1601
	0.4042357	$BO_u^+ \rightarrow XO_g^+$ (v',v") = (11,10) R(10)	6781	1601
	0.4043010	$BO_u^+ \rightarrow XO_g^+$ (v',v") = (11,10) P(12)	6781	1601
	0.4164526	$BO_u^+ \rightarrow XO_g^+$ (v',v") = (11,12) R(10)	6781	1601
	0.4165214	$BO_u^+ \rightarrow XO_g^+$ (v',v") = (11,12) P(12)	6781	1601
	0.4227890	$BO_u^+ \rightarrow XO_g^+$ (v',v") = (11,13) R(10)	6781	1601
	0.4228597	$BO_u^+ \rightarrow XO_g^+$ (v',v") = (11,13) P(12)	6781	1601
	0.4359310	$BO_u^+ \rightarrow XO_g^+$ (v',v") = (11,15) R(10)	6781	1601

Table 3.3.1.1—*continued*
Diatomic Electronic Transition Lasers

Species	Wavelength (μm)	Transition assignment	Comments	References
Se$_2$	0.4360198	BO$^+_u \rightarrow$ XO^+_g	6781	1601
		(v',v") = (11,15) P(12)		
	0.4427765	BO$^+_u \rightarrow$ XO^+_g	6781	1601
		(v',v") = (11,16) R(10)		
	0.4428531	BO$^+_u \rightarrow$ XO^+_g	6781	1601
		(v',v") = (11,16) P(12)		
	0.4569772	BO$^+_u \rightarrow$ XO^+_g	6781	1601
		(v',v") = (11,18) R(10)		
	0.4570582	BO$^+_u \rightarrow$ XO^+_g	6781	1601
		(v',v") = (11,18) P(12)		
	0.4643593	BO$^+_u \rightarrow$ XO^+_g	6781	1601
		(v',v") = (11,19) R(10)		
	0.4644427	BO$^+_u \rightarrow$ XO^+_g	6781	1601
		(v',v") = (11,19) P(12)		
	0.4719389	BO$^+_u \rightarrow$ XO^+_g	6781	1601
		(v',v") = (11,20) R(10)		
	0.4720249	BO$^+_u \rightarrow$ XO^+_g	6781	1601
		(v',v") = (11,20) P(12)		
	0.4797236	BO$^+_u \rightarrow$ XO^+_g	6781	1601
		(v',v") = (11,21) R(10)		
	0.4798120	BO$^+_u \rightarrow$ XO^+_g	6781	1601
		(v',v") = (11,21) P(12)		
	0.4877210	BO$^+_u \rightarrow$ XO^+_g	6781	1601
		(v',v") = (11,22) R(10)		
	0.4878120	BO$^+_u \rightarrow$ XO^+_g	6781	1601
		(v',v") = (11,22) P(12)		
	0.4959397	BO$^+_u \rightarrow$ XO^+_g	6781	1601
		(v',v") = (11,23) R(10)		
	0.4960333	BO$^+_u \rightarrow$ XO^+_g	6781	1601
		(v',v") = (11,23) P(12)		
	0.5043880	BO$^+_u \rightarrow$ XO^+_g	6781	1601
		(v',v") = (11,24) R(10)		
	0.5044847	BO$^+_u \rightarrow$ XO^+_g	6781	1601
		(v',v") = (11,24) P(12)		
	0.5130754	BO$^+_u \rightarrow$ XO^+_g	6781	1601
		(v',v") = (11,25) R(10)		
	0.5131750	BO$^+_u \rightarrow$ XO^+_g	6781	1601
		(v',v") = (11,25) P(12)		

Table 3.3.1.1—*continued*
Diatomic Electronic Transition Lasers

Species	Wavelength (μm)	Transition assignment	Comments	References
Se$_2$	0.5220119	BO$^+_u \to$ XO^+_g	6781	1601
		$(v',v'') = (11,26)$ R(10)		
	0.5221138	BO$^+_u \to$ XO^+_g	6781	1601
		$(v',v'') = (11,26)$ P(12)		
	0.5312057	BO$^+_u \to$ XO^+_g	6781	1601
		$(v',v'') = (11,27)$ R(10)		
	0.5313116	BO$^+_u \to$ XO^+_g	6781	1601
		$(v',v'') = (11,27)$ P(12)		
	0.5387	BO$^+_u \to$ XO^+_g	6783	1601
	0.5406689	BO$^+_u \to$ XO^+_g	6781	1601
		$(v',v'') = (11,28)$ R(10)		
	0.5407782	BO$^+_u \to$ XO^+_g	6781	1601
		$(v',v'') = (11,28)$ P(12)		
	0.5504125	BO$^+_u \to$ XO^+_g	6781	1601
		$(v',v'') = (11,29)$ R(10)		
	0.5505254	BO$^+_u \to$ XO^+_g	6781	1601
		$(v',v'') = (11,29)$ P(12)		
	0.5604483	BO$^+_u \to$ XO^+_g	6781	1601
		$(v',v'') = (11,30)$ R(10)		
	0.5605648	BO$^+_u \to$ XO^+_g	6781	1601
		$(v',v'') = (11,30)$ P(12)		
	0.5685	BO$^+_u \to$ XO^+_g	6783	1601
	0.5707882	BO$^+_u \to$ XO^+_g	6781	1601
		$(v',v'') = (11,31)$ R(10)		
	0.5709091	BO$^+_u \to$ XO^+_g	6781	1601
		$(v',v'') = (11,31)$ P(12)		
	0.5788	BO$^+_u \to$ XO^+_g	6783	1601
	0.5814462	BO$^+_u \to$ XO^+_g	6781	1601
		$(v',v'') = (11,32)$ R(10)		
	0.5815369	BO$^+_u \to$ XO^+_g	6781	1601
		$(v',v'') = (11,32)$ P(12)		
	0.5924350	BO$^+_u \to$ XO^+_g	6781	1601
		$(v',v'') = (11,33)$ R(10)		
	0.5925637	BO$^+_u \to$ XO^+_g	6781	1601
		$(v',v'') = (11,33)$ P(12)		
	0.6037695	BO$^+_u \to$ XO^+_g	6781	1601
		$(v',v'') = (11,34)$ R(10)		
	0.6039026	BO$^+_u \to$ XO^+_g	6781	1601
		$(v',v'') = (11,34)$ P(12)		

Table 3.3.1.1—*continued*
Diatomic Electronic Transition Lasers

Species	Wavelength (μm)	Transition assignment	Comments	References
Se$_2$	0.6124	BO$^+_u \to$ XO^+_g	6783	1601
	0.6154650	BO$^+_u \to$ XO^+_g (v',v") = (11,35) R(10)	6781	1601
	0.6156027	BO$^+_u \to$ XO^+_g (v',v") = (11,35) P(12)	6781	1601
	0.6241	BO$^+_u \to$ XO^+_g	6783	1601
	0.6275378	BO$^+_u \to$ XO^+_g (v',v") = (11,36) R(10)	6781	1601
	0.6276804	BO$^+_u \to$ XO^+_g (v',v") = (11,36) P(12)	6781	1601
	0.6400050	BO$^+_u \to$ XO^+_g (v',v") = (11,37) R(10)	6781	1601
	0.6401526	BO$^+_u \to$ XO^+_g (v',v") = (11,37) P(12)	6781	1601
	0.6528847	BO$^+_u \to$ XO^+_g (v',v") = (11,38) R(10)	6781	1601
	0.6530380	BO$^+_u \to$ XO^+_g (v',v") = (11,38) P(12)	6781	1601
	0.6661968	BO$^+_u \to$ XO^+_g (v',v") = (11,39) R(10)	6781	1601
	0.6663315	BO$^+_u \to$ XO^+_g (v',v") = (11,39) P(12)	6781	1601
	0.6748596	BO$^+_u \to$ XO^+_g (v',v") = (13,41) R(52)	6784	1601
	0.6756099	BO$^+_u \to$ XO^+_g (v',v") = (13,41) P(54)	6784	1601
	0.6799589	BO$^+_u \to$ XO^+_g (v',v") = (11,50) R(10)	6781	1601
	0.6801233	BO$^+_u \to$ XO^+_g (v',v") = (11,40) P(12)	6781	1601
	0.6887340	BO$^+_u \to$ XO^+_g (v',v") = (13,42) R(52)	6781	1601
	0.6895119	BO$^+_u \to$ XO^+_g (v',v") = (13,42) P(54)	6781	1601
	0.6941961	BO$^+_u \to$ XO^+_g (v',v") = (11,41) R(10)	6781	1601
	0.6943668	BO$^+_u \to$ XO^+_g (v',v") = (11,41) P(12)	6781	1601

Table 3.3.1.1—*continued*
Diatomic Electronic Transition Lasers

Species	Wavelength (μm)	Transition assignment	Comments	References
Se$_2$	0.7089295	$BO^+_u \rightarrow XO^+_g$ $(v',v'') = (11,42)$ R(10)	6781	1601
	0.7091069	$BO^+_u \rightarrow XO^+_g$ $(v',v'') = (11,42)$ P(12)	6781	1601
SO	0.300-0.350	$B^3\Sigma^-$ $(v'=0) \rightarrow X^3\Sigma^-$ $(v''=8-11)$	6910	1777
Te$_2$	0.5571	$B(O^+_u) \rightarrow X(O^+_g)$	1575	48
	0.5575	$B(O^+_u) \rightarrow X(O^+_g)$	1575	48
	0.5578	$B(O^+_u) \rightarrow X(O^+_g)$	1575	48
	0.5579	$B(O^+_u) \rightarrow X(O^+_g)$	1575	48
	0.5626	$B(O^+_u) \rightarrow X(O^+_g)$	1575	48
	0.5638	$B(O^+_u) \rightarrow X(O^+_g)$	1575	48
	0.5642	$B(O^+_u) \rightarrow X(O^+_g)$	1575	48
	0.5643	$B(O^+_u) \rightarrow X(O^+_g)$	1575	48
	0.5646	$B(O^+_u) \rightarrow X(O^+_g)$	1575	48
	0.5647	$B(O^+_u) \rightarrow X(O^+_g)$	1575	48
	0.5649	$B(O^+_u) \rightarrow X(O^+_g)$	1575	48
	0.5650	$B(O^+_u) \rightarrow X(O^+_g)$	1575	48
	0.5696	$B(O^+_u) \rightarrow X(O^+_g)$	1575	48
	0.5701	$B(O^+_u) \rightarrow X(O^+_g)$	1575	48
	0.5711	$B(O^+_u) \rightarrow X(O^+_g)$	1575	48
	0.5714	$B(O^+_u) \rightarrow X(O^+_g)$	1575	48
	0.5715	$B(O^+_u) \rightarrow X(O^+_g)$	1575	48
	0.5719	$B(O^+_u) \rightarrow X(O^+_g)$	1575	48
	0.5720	$B(O^+_u) \rightarrow X(O^+_g)$	1575	48
	0.5721	$B(O^+_u) \rightarrow X(O^+_g)$	1575	48
	0.5721	$B(O^+_u) \rightarrow X(O^+_g)$	1575	48
	0.5724	$B(O^+_u) \rightarrow X(O^+_g)$	1575	48
	0.57284	$BO^+_u \rightarrow XO^+_g$ $(v',v'') = (16,31)$ R(36)	6795	1602
	0.5766	$B(O^+_u) \rightarrow X(O^+_g)$	1575	48
	0.5767	$B(O^+_u) \rightarrow X(O^+_g)$	1575	48
	0.5773	$B(O^+_u) \rightarrow X(O^+_g)$	1575	48
	0.5774	$B(O^+_u) \rightarrow X(O^+_g)$	1575	48
	0.5780	$B(O^+_u) \rightarrow X(O^+_g)$	1575	48
	0.5783	$B(O^+_u) \rightarrow X(O^+_g)$	1575	48
	0.5784	$B(O^+_u) \rightarrow X(O^+_g)$	1575	48

Table 3.3.1.1—*continued*
Diatomic Electronic Transition Lasers

Species	Wavelength (μm)	Transition assignment	Comments	References
Te$_2$	0.5785	$B(O^+_u) \rightarrow X(O^+_g)$	1575	48
	0.5787	$B(O^+_u) \rightarrow X(O^+_g)$	1575	48
	0.5789	$B(O^+_u) \rightarrow X(O^+_g)$	1575	48
	0.5790	$B(O^+_u) \rightarrow X(O^+_g)$	1575	48
	0.5793	$B(O^+_u) \rightarrow X(O^+_g)$	1575	48
	0.5794	$B(O^+_u) \rightarrow X(O^+_g)$	1575	48
	0.5797	$B(O^+_u) \rightarrow X(O^+_g)$	1575	48
	0.5798	$B(O^+_u) \rightarrow X(O^+_g)$	1575	48
	0.57989	$BO^+_u \rightarrow XO^+_g$ $(v',v'') = (16,32)$ R(36)	6795	1602
	0.58008	$BO^+_u \rightarrow XO^+_g$ $(v',v'') = (16,32)$ P(38)	6795	1602
	0.5841	$B(O^+_u) \rightarrow X(O^+_g)$	1575	48
	0.5849	$B(O^+_u) \rightarrow X(O^+_g)$	1575	48
	0.5857	$B(O^+_u) \rightarrow X(O^+_g)$	1575	48
	0.5859	$B(O^+_u) \rightarrow X(O^+_g)$	1575	48
	0.5865	$B(O^+_u) \rightarrow X(O^+_g)$	1575	48
	0.5869	$B(O^+_u) \rightarrow X(O^+_g)$	1575	48
	0.5870	$B(O^+_u) \rightarrow X(O^+_g)$	1575	48
	0.5874	$B(O^+_u) \rightarrow X(O^+_g)$	1575	48
	0.5924	$B(O^+_u) \rightarrow X(O^+_g)$	1575	48
	0.5927	$B(O^+_u) \rightarrow X(O^+_g)$	1575	48
	0.5934	$B(O^+_u) \rightarrow X(O^+_g)$	1575	48
	0.5936	$B(O^+_u) \rightarrow X(O^+_g)$	1575	48
	0.6002	$B(O^+_u) \rightarrow X(O^+_g)$	1575	48
	0.6004	$B(O^+_u) \rightarrow X(O^+_g)$	1575	48
	0.6005	$B(O^+_u) \rightarrow X(O^+_g)$	1575	48
	0.6008	$B(O^+_u) \rightarrow X(O^+_g)$	1575	48
	0.6009	$B(O^+_u) \rightarrow X(O^+_g)$	1575	48
	0.60190	$BO^+_u \rightarrow XO^+_g$ $(v',v'') = (16,35)$ R(36)	6795	1602
	0.60210	$BO^+_u \rightarrow XO^+_g$ $(v',v'') = (16,35)$ P(38)	6795	1602
	0.6082	$B(O^+_u) \rightarrow X(O^+_g)$	1575	48
	0.6083	$B(O^+_u) \rightarrow X(O^+_g)$	1575	48
	0.6085	$B(O^+_u) \rightarrow X(O^+_g)$	1575	48
	0.6087	$B(O^+_u) \rightarrow X(O^+_g)$	1575	48
	0.6089	$B(O^+_u) \rightarrow X(O^+_g)$	1575	48
	0.6162	$B(O^+_u) \rightarrow X(O^+_g)$	1575	48

Table 3.3.1.1—*continued*
Diatomic Electronic Transition Lasers

Species	Wavelength (μm)	Transition assignment	Comments	References
Te$_2$	0.6165	$B(O^+_u) \rightarrow X(O^+_g)$	1575	48
	0.6168	$B(O^+_u) \rightarrow X(O^+_g)$	1575	48
	0.6170	$B(O^+_u) \rightarrow X(O^+_g)$	1575	48
	0.6204	$B(O^+_u) \rightarrow X(O^+_g)$	1575	48
	0.62524	$BO^+_u \rightarrow XO^+_g$ $(v',v'') = (16,38)$ R(36)	6795	1602
	0.62545	$BO^+_u \rightarrow XO^+_g$ $(v',v'') = (16,38)$ P(38)	6795	1602
	0.6278	$B(O^+_u) \rightarrow X(O^+_g)$	1575	48
	0.6287	$B(O^+_u) \rightarrow X(O^+_g)$	1575	48
	0.6288	$B(O^+_u) \rightarrow X(O^+_g)$	1575	48
	0.6295	$B(O^+_u) \rightarrow X(O^+_g)$	1575	48
	0.63334	$BO^+_u \rightarrow XO^+_g$ $(v',v'') = (16,39)$ R(36)	6795	1602
	0.63355	$BO^+_u \rightarrow XO^+_g$ $(v',v'') = (16,39)$ P(38)	6795	1602
	0.6371	$B(O^+_u) \rightarrow X(O^+_g)$	1575	48
	0.6379	$B(O^+_u) \rightarrow X(O^+_g)$	1575	48
	0.6381	$B(O^+_u) \rightarrow X(O^+_g)$	1575	48
	0.6388	$B(O^+_u) \rightarrow X(O^+_g)$	1575	48
	0.6477	$B(O^+_u) \rightarrow X(O^+_g)$	1575	48
	0.6484	$B(O^+_u) \rightarrow X(O^+_g)$	1575	48
	0.6561	$B(O^+_u) \rightarrow X(O^+_g)$	1575	48
	0.6574	$B(O^+_u) \rightarrow X(O^+_g)$	1575	48
	0.6581	$B(O^+_u) \rightarrow X(O^+_g)$	1575	48
	0.66740	$BO^+_u \rightarrow XO^+_g$ $(v',v'') = (16,43)$ R(36)	6795	1602
	0.66764	$BO^+_u \rightarrow XO^+_g$ $(v',v'') = (16,43)$ P(38)	6795	1602
	0.71414	$BO^+_u \rightarrow XO^+_g$ $(v',v'') = (16,48)$ R(36)	6795	1602
	0.71441	$BO^+_u \rightarrow XO^+_g$ $(v',v'') = (16,48)$ P(38)	6795	1602
	0.72410	$BO^+_u \rightarrow XO^+_g$ $(v',v'') = (16,49)$ R(36)	6795	1602
	0.72437	$BO^+_u \rightarrow XO^+_g$ $(v',v'') = (16,49)$ P(38)	6795	1602
	0.73427	$BO^+_u \rightarrow XO^+_g$ $(v',v'') = (16,50)$ R(36)	6795	1602

Table 3.3.1.1—*continued*
Diatomic Electronic Transition Lasers

Species	Wavelength (μm)	Transition assignment	Comments	References
Te$_2$	0.73454	BO$^+_u \rightarrow$ XO^+_g (v',v") = (16,50) P(38)	6795	1602
	0.74466	BO$^+_u \rightarrow$ XO^+_g (v',v") = (16,51) R(36)	6795	1602
	0.74494	BO$^+_u \rightarrow$ XO^+_g (v',v") = (16,51) P(38)	6795	1602
^{128}Te$_2$	0.55374	BO$^+_u \rightarrow$ XO^+_g (v',v") = (16,28) R(56)	6798	1602
	0.55400	BO$^+_u \rightarrow$ XO^+_g (v',v") = (16,28) P(58)	6798	1602
	0.57434	BO$^+_u \rightarrow$ XO^+_g (v',v") = (16,31) R(56)	6798	1602
	0.57462	BO$^+_u \rightarrow$ XO^+_g (v',v") = (16,31) P(58)	6798	1602
	0.60372	BO$^+_u \rightarrow$ XO^+_g (v',v") = (16,35) R(56)	6798	1602
	0.60403	BO$^+_u \rightarrow$ XO^+_g (v',v") = (16,35) P(58)	6798	1602
	0.62733	BO$^+_u \rightarrow$ XO^+_g (v',v") = (16,38) R(56)	6798	1602
	0.62766	BO$^+_u \rightarrow$ XO^+_g (v',v") = (16,38) P(58)	6798	1602
	0.63552	BO$^+_u \rightarrow$ XO^+_g (v',v") = (16,39) R(56)	6798	1602
	0.63586	BO$^+_u \rightarrow$ XO^+_g (v',v") = (16,39) P(58)	6798	1602
	0.67001	BO$^+_u \rightarrow$ XO^+_g (v',v") = (16,43) R(56)	6798	1602
	0.67038	BO$^+_u \rightarrow$ XO^+_g (v',v") = (16,43) P(58)	6798	1602
	0.71737	BO$^+_u \rightarrow$ XO^+_g (v',v") = (16,48) R(56)	6798	1602
	0.71780	BO$^+_u \rightarrow$ XO^+_g (v',v") = (16,48) P(58)	6798	1602
	0.72746	BO$^+_u \rightarrow$ XO^+_g (v',v") = (16,49) R(56)	6798	1602
	0.72790	BO$^+_u \rightarrow$ XO^+_g (v',v") = (16,49) P(58)	6798	1602

Table 3.3.1.1—*continued*
Diatomic Electronic Transition Lasers

Species	Wavelength (μm)	Transition assignment	Comments	References
^{128}Te$_2$	0.73777	$BO^+_u \rightarrow XO^+_g$ $(v',v'') = (16,50)$ R(56)	6798	1602
	0.73822	$BO^+_u \rightarrow XO^+_g$ $(v',v'') = (16,50)$ P(58)	6798	1602
^{130}Te$_2$	0.46075	$BO^+_u \rightarrow XO^+_g$ $(v',v'') = (16,12)$ R(36)	6785	1602
	0.46087	$BO^+_u \rightarrow XO^+_g$ $(v',v'') = (16,12)$ P(38)	6785	1602
	0.46576	$BO^+_u \rightarrow XO^+_g$ $(v',v'') = (16,13)$ R(36)	6785	1602
	0.46588	$BO^+_u \rightarrow XO^+_g$ $(v',v'') = (16,13)$ P(38)	6785	1602
	0.47085	$BO^+_u \rightarrow XO^+_g$ $(v',v'') = (16,14)$ R(36)	6785	1602
	0.47098	$BO^+_u \rightarrow XO^+_g$ $(v',v'') = (16,14)$ P(38)	6785	1602
	0.47603	$BO^+_u \rightarrow XO^+_g$ $(v',v'') = (16,15)$ R(36)	6785	1602
	0.47616	$BO^+_u \rightarrow XO^+_g$ $(v',v'') = (16,15)$ P(38)	6785	1602
	0.48130	$BO^+_u \rightarrow XO^+_g$ $(v',v'') = (16,16)$ R(36)	6785	1602
	0.48143	$BO^+_u \rightarrow XO^+_g$ $(v',v'') = (16,16)$ P(38)	6785	1602
	0.48667	$BO^+_u \rightarrow XO^+_g$ $(v',v'') = (16,17)$ R(36)	6785	1602
	0.48680	$BO^+_u \rightarrow XO^+_g$ $(v',v'') = (16,17)$ P(38)	6785	1602
	0.49212	$BO^+_u \rightarrow XO^+_g$ $(v',v'') = (16,18)$ R(36)	6785	1602
	0.49226	$BO^+_u \rightarrow XO^+_g$ $(v',v'') = (16,18)$ P(38)	6785	1602
	0.49768	$BO^+_u \rightarrow XO^+_g$ $(v',v'') = (16,19)$ R(36)	6785	1602
	0.49782	$BO^+_u \rightarrow XO^+_g$ $(v',v'') = (16,19)$ P(38)	6785	1602
	0.50333	$BO^+_u \rightarrow XO^+_g$ $(v',v'') = (16,20)$ R(36)	6785	1602

Table 3.3.1.1—*continued*
Diatomic Electronic Transition Lasers

Species	Wavelength (μm)	Transition assignment	Comments	References
^{130}Te$_2$	0.50348	BO$^+_u \to$ XO^+_g $(v',v'') = (16,20)$ P(38)	6785	1602
	0.51495	BO$^+_u \to$ XO^+_g $(v',v'') = (16,22)$ R(36)	6785	1602
	0.51510	BO$^+_u \to$ XO^+_g $(v',v'') = (16,22)$ P(38)	6785	1602
	0.52092	BO$^+_u \to$ XO^+_g $(v',v'') = (16,23)$ R(36)	6785	1602
	0.52107	BO$^+_u \to$ XO^+_g $(v',v'') = (16,23)$ P(38)	6785	1602
	0.52504	BO$^+_u \to$ XO^+_g $(v',v'') = (18,24)$ R(172)	6785	1602
	0.52575	BO$^+_u \to$ XO^+_g $(v',v'') = (18,24)$ P(174)	6785	1602
	0.53318	BO$^+_u \to$ XO^+_g $(v',v'') = (16,25)$ R(36)	6785	1602
	0.53334	BO$^+_u \to$ XO^+_g $(v',v'') = (16,25)$ P(38)	6785	1602
	0.53949	BO$^+_u \to$ XO^+_g $(v',v'') = (16,26)$ R(36)	6785	1602
	0.53965	BO$^+_u \to$ XO^+_g $(v',v'') = (16,26)$ P(38)	6785	1602
	0.54355	BO$^+_u \to$ XO^+_g $(v',v'') = (18,27)$ R(172)	6785	1602
	0.54430	BO$^+_u \to$ XO^+_g $(v',v'') = (18,27)$ R(174)	6785	1602
	0.55245	BO$^+_u \to$ XO^+_g $(v',v'') = (16,28)$ R(36)	6785	1602
	0.55262	BO$^+_u \to$ XO^+_g $(v',v'') = (16,28)$ P(38)	6785	1602
	0.55645	BO$^+_u \to$ XO^+_g $(v',v'') = (18,29)$ R(172)	6785	1602
	0.55723	BO$^+_u \to$ XO^+_g $(v',v'') = (18,29)$ P(174)	6785	1602
	0.55912	BO$^+_u \to$ XO^+_g $(v',v'') = (16,29)$ R(36)	6785	1602
	0.55929	BO$^+_u \to$ XO^+_g $(v',v'') = (16,29)$ P(38)	6785	1602

Table 3.3.1.1—*continued*
Diatomic Electronic Transition Lasers

Species	Wavelength (μm)	Transition assignment	Comments	References
$^{130}Te_2$	0.56308	$BO^+_u \rightarrow XO^+_g$ $(v',v'') = (18,30)$ R(172)	6785	1602
	0.56388	$BO^+_u \rightarrow XO^+_g$ $(v',v'') = (18,30)$ P(174)	6785	1602
	0.57284	$BO^+_u \rightarrow XO^+_g$ $(v',v'') = (16,31)$ R(36)	6785	1602
	0.57302	$BO^+_u \rightarrow XO^+_g$ $(v',v'') = (16,31)$ P(38)	6785	1602
	0.57671	$BO^+_u \rightarrow XO^+_g$ $(v',v'') = (18,32)$ R(172)	6785	1602
	0.57754	$BO^+_u \rightarrow XO^+_g$ $(v',v'') = (18,32)$ P(174)	6785	1602
	0.57989	$BO^+_u \rightarrow XO^+_g$ $(v',v'') = (16,32)$ R(36)	6785	1602
	0.58008	$BO^+_u \rightarrow XO^+_g$ $(v',v'') = (16,32)$ P(38)	6785	1602
	0.58372	$BO^+_u \rightarrow XO^+_g$ $(v',v'') = (18,33)$ R(172)	6785	1602
	0.58457	$BO^+_u \rightarrow XO^+_g$ $(v',v'') = (18,33)$ P(174)	6785	1602
	0.59442	$BO^+_u \rightarrow XO^+_g$ $(v',v'') = (16,34)$ R(36)	6785	1602
	0.59461	$BO^+_u \rightarrow XO^+_g$ $(v',v'') = (16,34)$ P(38)	6785	1602
	0.59813	$BO^+_u \rightarrow XO^+_g$ $(v',v'') = (18,35)$ R(172)	6785	1602
	0.59902	$BO^+_u \rightarrow XO^+_g$ $(v',v'') = (18,35)$ P(174)	6785	1602
	0.60190	$BO^+_u \rightarrow XO^+_g$ $(v',v'') = (16,35)$ R(36)	6785	1602
	0.60210	$BO^+_u \rightarrow XO^+_g$ $(v',v'') = (16,35)$ P(38)	6785	1602
	0.60555	$BO^+_u \rightarrow XO^+_g$ $(v',v'') = (18,36)$ R(172)	6785	1602
	0.60645	$BO^+_u \rightarrow XO^+_g$ $(v',v'') = (18,36)$ P(174)	6785	1602
	0.62524	$BO^+_u \rightarrow XO^+_g$ $(v',v'') = (16,38)$ R(36)	6785	1602

Table 3.3.1.1—*continued*
Diatomic Electronic Transition Lasers

Species	Wavelength (μm)	Transition assignment	Comments	References
$^{130}Te_2$	0.62545	$BO^+_u \rightarrow XO^+_g$ $(v',v'') = (16,38)$ P(38)	6785	1602
	0.62866	$BO^+_u \rightarrow XO^+_g$ $(v',v'') = (18,39)$ R(172)	6785	1602
	0.62962	$BO^+_u \rightarrow XO^+_g$ $(v',v'') = (18,39)$ P(174)	6785	1602
	0.63334	$BO^+_u \rightarrow XO^+_g$ $(v',v'') = (16,39)$ R(36)	6785	1602
	0.63355	$BO^+_u \rightarrow XO^+_g$ $(v',v'') = (16,39)$ P(38)	6785	1602
	0.63666	$BO^+_u \rightarrow XO^+_g$ $(v',v'') = (18,40)$ R(172)	6785	1602
	0.63735	$BO^+_u \rightarrow XO^+_g$ $(v',v'') = (18,40)$ P(174)	6785	1602
	0.64418	$BO^+_u \rightarrow XO^+_g$ $(v',v'') = (18,42)$ P(174)	6785	1602
	0.65315	$BO^+_u \rightarrow XO^+_g$ $(v',v'') = (18,42)$ R(172)	6785	1602
	0.65862	$BO^+_u \rightarrow XO^+_g$ $(v',v'') = (16,42)$ R(36)	6785	1602
	0.65885	$BO^+_u \rightarrow XO^+_g$ $(v',v'') = (16,42)$ P(38)	6785	1602
	0.66163	$BO^+_u \rightarrow XO^+_g$ $(v',v'') = (18,43)$ P(172)	6785	1602
	0.66269	$BO^+_u \rightarrow XO^+_g$ $(v',v'') = (18,43)$ P(174)	6785	1602
	0.66740	$BO^+_u \rightarrow XO^+_g$ $(v',v'') = (16,43)$ R(36)	6785	1602
	0.66764	$BO^+_u \rightarrow XO^+_g$ $(v',v'') = (16,43)$ P(38)	6785	1602
	0.67029	$BO^+_u \rightarrow XO^+_g$ $(v',v'') = (18,44)$ R(172)	6785	1602
	0.67137	$BO^+_u \rightarrow XO^+_g$ $(v',v'') = (18,44)$ P(174)	6785	1602
	0.67637	$BO^+_u \rightarrow XO^+_g$ $(v',v'') = (16,44)$ R(36)	6785	1602
	0.67661	$BO^+_u \rightarrow XO^+_g$ $(v',v'') = (16,44)$ P(38)	6785	1602

Table 3.3.1.1—*continued*
Diatomic Electronic Transition Lasers

Species	Wavelength (μm)	Transition assignment	Comments	References
$^{130}Te_2$	0.68813	$BO^+_u \rightarrow XO^+_g$ $(v',v'') = (18,46)$ R(172)	6785	1602
	0.68926	$BO^+_u \rightarrow XO^+_g$ $(v',v'') = (18,46)$ P(174)	6785	1602
	0.69732	$BO^+_u \rightarrow XO^+_g$ $(v',v'') = (18,47)$ R(172)	6785	1602
	0.69848	$BO^+_u \rightarrow XO^+_g$ $(v',v'') = (18,47)$ P(174)	6785	1602
	0.70440	$BO^+_u \rightarrow XO^+_g$ $(v',v'') = (16,47)$ R(36)	6785	1602
	0.70466	$BO^+_u \rightarrow XO^+_g$ $(v',v'') = (16,47)$ P(38)	6785	1602
	0.70670	$BO^+_u \rightarrow XO^+_g$ $(v',v'') = (18,48)$ R(172)	6785	1602
	0.70789	$BO^+_u \rightarrow XO^+_g$ $(v',v'') = (18,48)$ P(174)	6785	1602
	0.71414	$BO^+_u \rightarrow XO^+_g$ $(v',v'') = (16,48)$ R(36)	6785	1602
	0.71748	$BO^+_u \rightarrow XO^+_g$ $(v',v'') = (16,48)$ P(38)	6785	1602
	0.72410	$BO^+_u \rightarrow XO^+_g$ $(v',v'') = (16,49)$ R(36)	6785	1602
	0.72437	$BO^+_u \rightarrow XO^+_g$ $(v',v'') = (16,49)$ P(38)	6785	1602
	0.73427	$BO^+_u \rightarrow XO^+_g$ $(v',v'') = (16,50)$ R(36)	6785	1602
	0.73454	$BO^+_u \rightarrow XO^+_g$ $(v',v'') = (16,50)$ P(38)	6785	1602
	0.73601	$BO^+_u \rightarrow XO^+_g$ $(v',v'') = (18,51)$ R(172)	6785	1602
	0.73728	$BO^+_u \rightarrow XO^+_g$ $(v',v'') = (18,51)$ P(174)	6785	1602
	0.74466	$BO^+_u \rightarrow XO^+_g$ $(v',v'') = (16,51)$ R(36)	6785	1602
	0.74494	$BO^+_u \rightarrow XO^+_g$ $(v',v'') = (16,51)$ P(38)	6785	1602
	0.74619	$BO^+_u \rightarrow XO^+_g$ $(v',v'') = (18,52)$ R(172)	6785	1602

Table 3.3.1.1—*continued*
Diatomic Electronic Transition Lasers

Species	Wavelength (μm)	Transition assignment	Comments	References
$^{130}Te_2$	0.74749	$BO^+_u \to XO^+_g$ $(v',v'') = (18,52)$ P(174)	6785	1602
	0.75527	$BO^+_u \to XO^+_g$ $(v',v'') = (16,52)$ R(36)	6785	1602
	0.75557	$BO^+_u \to XO^+_g$ $(v',v'') = (16,52)$ P(38)	6785	1602
	0.75657	$BO^+_u \to XO^+_g$ $(v',v'') = (18,53)$ R(172)	6785	1602
	0.75791	$BO^+_u \to XO^+_g$ $(v',v'') = (18,53)$ P(174)	6785	1602
	0.76613	$BO^+_u \to XO^+_g$ $(v',v'') = (16,53)$ R(36)	6785	1602
	0.76642	$BO^+_u \to XO^+_g$ $(v',v'') = (16,53)$ P(38)	6785	1602
	0.76718	$BO^+_u \to XO^+_g$ $(v',v'') = (18,54)$ R(172)	6785	1602
	0.76855	$BO^+_u \to XO^+_g$ $(v',v'') = (18,54)$ P(174)	6785	1602
	0.77722	$BO^+_u \to XO^+_g$ $(v',v'') = (16,54)$ R(36)	6785	1602
	0.77752	$BO^+_u \to XO^+_g$ $(v',v'') = (16,54)$ P(38)	6785	1602
	0.77801	$BO^+_u \to XO^+_g$ $(v',v'') = (18,55)$ R(172)	6785	1602
	0.77941	$BO^+_u \to XO^+_g$ $(v',v'') = (18,55)$ P(174)	6785	1602
	0.893	$AO^+_u \to b^2\Sigma^+_g$ $(v',v'') = (13,1)$ R(132)	6788	1603
	0.895	$AO^+_u \to b^2\Sigma^+_g$ $(v',v'') = (13,1)$ P(134)	6788	1603
	1.042	$BO^+_u \to b^1\Sigma^+_g$ $(v',v'') = (5,17)$ R(136), P(138)	6792	1603
	1.059	$BO^+_u \to b^1\Sigma^+_g$ $(v',v'') = (2,15)$ R(196), P(198)	6792	1603
	1.088	$BO^+_u \to b^1\Sigma^+_g$ $(v',v'') = (2,16)$ R(196), P(198)	6792	1603
	1.089	$BO^+_u \to b^1\Sigma^+_g$ $(v',v'') = (5,19)$ R(136), P(138)	6792	1603

Table 3.3.1.1—*continued*
Diatomic Electronic Transition Lasers

Species	Wavelength (μm)	Transition assignment	Comments	References
$^{130}Te_2$	1.113	$BO^+_u \rightarrow b^1\Sigma^+_g$ $(v',v'') = (2,17)$ R(196), P(198)	6792	1603
	1.134	$BO^+_u \rightarrow b^1\Sigma^+_g$ $(v',v'') = (2,18)$ R(196), P(198)	6792	1603
	1.144	$BO^+_u \rightarrow b^1\Sigma^+_g$ $(v',v'') = (5,21)$ R(136), P(138)	6792	1603
	1.161	$BO^+_u \rightarrow b^1\Sigma^+_g$ $(v',v'') = (2,19)$ R(196), P(198)	6792	1603
	1.171	$BO^+_u \rightarrow b^1\Sigma^+_g$ $(v',v'') = (5,22)$ R(136), P(138)	6792	1603
	1.191	$BO^+_u \rightarrow b^1\Sigma^+_g$ $(v',v'') = (5,23)$ R(136), P(138)	6792	1603
	1.224	$BO^+_u \rightarrow b^1\Sigma^+_g$ $(v',v'') = (5,24)$ R(136), P(138)	6792	1603
$^{126}Te^{128}Te$	0.56486	$BO^+_u \rightarrow XO^+_g$ $(v',v'') = (14,28)$ R(114)	6797	1602
	0.56541	$BO^+_u \rightarrow XO^+_g$ $(v',v'') = (14,28)$ P(116)	6797	1602
	0.58627	$BO^+_u \rightarrow XO^+_g$ $(v',v'') = (14,31)$ R(114)	6797	1602
	0.58686	$BO^+_u \rightarrow XO^+_g$ $(v',v'') = (14,31)$ P(116)	6797	1602
	0.59369	$BO^+_u \rightarrow XO^+_g$ $(v',v'') = (14,32)$ R(114)	6797	1602
	0.61685	$BO^+_u \rightarrow XO^+_g$ $(v',v'') = (14,35)$ R(114)	6797	1602
	0.61749	$BO^+_u \rightarrow XO^+_g$ $(v',v'') = (14,35)$ P(118)	6797	1602
	0.65000	$BO^+_u \rightarrow XO^+_g$ $(v',v'') = (14,39)$ R(114)	6797	1602
	0.65071	$BO^+_u \rightarrow XO^+_g$ $(v',v'') = (14,39)$ P(118)	6797	1602
	0.69549	$BO^+_u \rightarrow XO^+_g$ $(v',v'') = (14,44)$ R(114)	6797	1602
	0.69629	$BO^+_u \rightarrow XO^+_g$ $(v',v'') = (14,44)$ P(118)	6797	1602
	0.70518	$BO^+_u \rightarrow XO^+_g$ $(v',v'') = (14,45)$ R(114)	6797	1602

Table 3.3.1.1—*continued*
Diatomic Electronic Transition Lasers

Species	Wavelength (μm)	Transition assignment	Comments	References
$^{126}Te^{128}Te$	0.70600	$BO^+_u \rightarrow XO^+_g$ $(v',v'') = (14,45)$ P(118)	6797	1602
	0.71508	$BO^+_u \rightarrow XO^+_g$ $(v',v'') = (14,46)$ R(114)	6797	1602
	0.71592	$BO^+_u \rightarrow XO^+_g$ $(v',v'') = (14,46)$ P(118)	6797	1602
$^{128}Te^{130}Te$	0.54353	$BO^+_u \rightarrow XO^+_g$ $(v',v'') = (14,25)$ R(108)	6796	1602
	0.54402	$BO^+_u \rightarrow XO^+_g$ $(v',v'') = (14,25)$ P(110)	6796	1602
	0.56353	$BO^+_u \rightarrow XO^+_g$ $(v',v'') = (14,28)$ R(108)	6796	1602
	0.56405	$BO^+_u \rightarrow XO^+_g$ $(v',v'') = (14,28)$ P(110)	6796	1602
	0.58471	$BO^+_u \rightarrow XO^+_g$ $(v',v'') = (14,31)$ R(108)	6796	1602
	0.58526	$BO^+_u \rightarrow XO^+_g$ $(v',v'') = (14,31)$ P(110)	6796	1602
	0.59205	$BO^+_u \rightarrow XO^+_g$ $(v',v'') = (14,32)$ R(108)	6796	1602
	0.59261	$BO^+_u \rightarrow XO^+_g$ $(v',v'') = (14,32)$ P(110)	6796	1602
	0.61495	$BO^+_u \rightarrow XO^+_g$ $(v',v'') = (14,35)$ R(108)	6796	1602
	0.61555	$BO^+_u \rightarrow XO^+_g$ $(v',v'') = (14,35)$ P(110)	6796	1602
	0.62290	$BO^+_u \rightarrow XO^+_g$ $(v',v'') = (14,36)$ R(108)	6796	1602
	0.62351	$BO^+_u \rightarrow XO^+_g$ $(v',v'') = (14,36)$ P(110)	6796	1602
	0.64771	$BO^+_u \rightarrow XO^+_g$ $(v',v'') = (14,39)$ R(108)	6796	1602
	0.64837	$BO^+_u \rightarrow XO^+_g$ $(v',v'') = (14,39)$ P(110)	6796	1602
	0.65632	$BO^+_u \rightarrow XO^+_g$ $(v',v'') = (14,40)$ R(108)	6796	1602
	0.65700	$BO^+_u \rightarrow XO^+_g$ $(v',v'') = (14,40)$ P(110)	6796	1602

Table 3.3.1.1—*continued*
Diatomic Electronic Transition Lasers

Species	Wavelength (μm)	Transition assignment	Comments	References
	0.66512	$BO^+_u \rightarrow XO^+_g$ $(v',v'') = (14,41)$ R(108)	6796	1602
	0.66581	$BO^+_u \rightarrow XO^+_g$ $(v',v'') = (14,41)$ P(110)	6796	1602
	0.69263	$BO^+_u \rightarrow XO^+_g$ $(v',v'') = (14,44)$ R(108)	6796	1602
	0.69337	$BO^+_u \rightarrow XO^+_g$ $(v',v'') = (14,44)$ P(110)	6796	1602
	0.70220	$BO^+_u \rightarrow XO^+_g$ $(v',v'') = (14,45)$ R(108)	6796	1602
	0.70295	$BO^+_u \rightarrow XO^+_g$ $(v',v'') = (14,45)$ P(110)	6796	1602
	0.71197	$BO^+_u \rightarrow XO^+_g$ $(v',v'') = (14,46)$ R(108)	6796	1602
	0.71274	$BO^+_u \rightarrow XO^+_g$ $(v',v'') = (14,46)$ P(110)	6796	1602
	0.72195	$BO^+_u \rightarrow XO^+_g$ $(v',v'') = (14,47)$ R(108)	6796	1602
	0.72275	$BO^+_u \rightarrow XO^+_g$ $(v',v'') = (14,47)$ P(110)	6796	1602
Xe$_2$	0.170-0.176	$AO^+_u \rightarrow XO^+_g$	6799,6800	1606,1622
	0.1716	$^1\Sigma^+_u \rightarrow X^1\Sigma^+_g$	1590,1591	15–20,35,40, 51,62,73,84,93, 95,103,106,116
XeBr	0.2818	$B^1\Sigma^+_{1/2} \rightarrow X^1\Sigma^+_{1/2}$	1576,1577	33,34,35,116, 123–126
XeCl	0.3070	$B^2\Sigma^+_{1/2} \rightarrow X^2\Sigma^+_{1/2}(1\text{-}5)$	1578–1580	3–6,33–36, 116,126
	0.3073	$B^2\Sigma^+_{1/2} \rightarrow X^2\Sigma^+_{1/2}(1\text{-}6)$	1578–1580	3–6,33–36, 116,126
XeCl	0.30765	$B^2\Sigma^+_{1/2} \rightarrow X^2\Sigma^+_{1/2}$ (0-0)	1578–1580	3–6,33–36, 116,126
	0.30792	$B^2\Sigma^+_{1/2} \rightarrow X^2\Sigma^+_{1/2}$ (0-1)	1578–1580	3–6,33–36, 116,126
	0.30817	$B^2\Sigma^+_{1/2} \rightarrow X^2\Sigma^+_{1/2}(0\text{-}2)$	1578–1580	3–6,33–36, 116,126

Table 3.3.1.1—*continued*
Diatomic Electronic Transition Lasers

Species	Wavelength (μm)	Transition assignment	Comments	References
XeCl	0.30843	$B^2\Sigma^+_{1/2} \to X^2\Sigma^+_{1/2}(0\text{-}3)$	1578–1580	3–6,33–36, 116,126
XeF	0.3488	$B^2\Sigma^+_{1/2} \to X^2\Sigma^+_{1/2}$ (2-5)	1581–1583	7–9,12,37
	0.351	$B^2\Sigma \to X^2\Sigma$	6804	1607
	0.3511	$B^2\Sigma^+_{1/2} \to X^2\Sigma^+_{1/2}$ (0-2), (1-4)	1581–1583	7–9,12,37
	0.353	$B^2\Sigma \to X^2\Sigma$ (v'',v'') = (0,3)	6806,6808	1608,1609
	0.3531	$B^2\Sigma^+_{1/2} \to X^2\Sigma^+_{1/2}$ (0-3), (1-6)	1581–1583	7–9,12,37
	0.3540	$B^2\Sigma^+_{1/2} \to X^2\Sigma^+_{1/2}$ (0-4), (1-7)	1581,1585	7–9,12,37
	0.404	$B^2\Sigma \to X^2\Sigma$	6810	1610
	0.4830	$C^2\Pi_u \to A^2\Pi_g$	1586	10,11
	0.485	C(3/2) → A(3/2)	6812,6815,6817 6820,6822, 6824	1515–17,1609, 1611–1616
	0.4860	$C^2\Pi_u \to A^2\Pi_g$	1587	10,11
XeO	0.5376	$2^1\Sigma^+ \to 1^1\Sigma^+$ (0-5)	1588,1589	14,39,41
	0.5442	$2^1\Sigma^+ \to 1^1\Sigma^+$ (0-6)	1588,1589	14,39,41
	0.547	$2^1\Sigma^+ \to 1^1\Sigma^+$ (0,6)	6828	1519
ZnI	0.6011	$B^2S_{1/2}{}^{+1}$		
	0.6018	$B^2S_{1/2}{}^{+1}$		
	0.6025	$B^2S_{1/2}{}^{+1}$		
ZnI	0.6031	$B^2S_{1/2}{}^{+1}$		
	0.6039	$B^2S_{1/2}{}^{+1}$		

3.3.1.3 Triatomic Electronic Transition Lasers

Table 3.3.1.2
Triatomic Electronic Transition Lasers

Species	Wavelength (μm) air	Transition assignment	Comments	References
CS_2	0.4472	v=(0,10,0) $R^3B_2 \rightarrow$ v=(0,16,0) $X^1\Sigma^+_g$	6967	833
	0.4654	v=(0,10,0) $R^3B_2 \rightarrow$ v=(0,18,0) $X^1\Sigma^+_g$	6967	833
	0.4852	v=(0,10,0) $R^3B_2 \rightarrow$ v=(0,20,0) $X^1\Sigma^+_g$	6967	833
	0.5072	v=(0,10,0) $R^3B_2 \rightarrow$ v=(0,22,0) $X^1\Sigma^+_g$	6967	833
	0.5144	v=(0,10,0) $R^3B_2 \rightarrow$ v=(2,20,0) $X^1\Sigma^+_g$	6967	833
	0.5384	v=(0,10,0) $R^3B_2 \rightarrow$ v=(2,22,0) $X^1\Sigma^+_g$	6967	833
Hg_3	0.485-0.507		6628	1536
Kr_2F	0.430±0.25	$4^2\Gamma \rightarrow 1^2\Gamma, 2^2\Gamma$	6710,6712	1573
	0.4300		1592	129
	0.450±0.10	$4^2\Gamma \rightarrow 1^2\Gamma, 2^2\Gamma$	6710,6712	1574
Xe_2Cl	0.5180		1593,1594	21,22,34
	0.520–0.530	$4^2\Gamma \rightarrow 1^2\Gamma$	6801,6803	1604,1605
Xe_2F	0.580–0.635	$4^2\Gamma \rightarrow 1^2\Gamma$	6826	1518

3.3.1.4 Polyatomic Electronic Transition Lasers

Table 3.3.1.3
Polyatomic Electronic Transition Lasers

Species	Wavelength (μm) air	Transition assignment	Comments	References
coumarin 153	471–501		6895	1788,1789
7-diethylamino-3(2'-benzoxazolyl) coumarin	493–517		6895	1788,1789
coumarin 6	500–541		6895	1788,1789
coumarin 7	504–536		6895	1788,1789
coumarin 30	517–536		6895	1788,1789
$NdAl_5Cl_{12}$	1.060	$^4F_{3/2} \rightarrow {}^4I_{11/2}$	1595	24
$Nd(thd)_3$	1.0600	$^4F_{3/2} \rightarrow {}^4I_{11/2}$	1597	25
POPOP[a]	0.385	$S_1 \rightarrow S_0$	1600	32
$TbAl_3Cl_{12}$	0.435	$^5D_3 \rightarrow {}^7F_4$	1598	26,27
	0.545	$^5D_4 \rightarrow {}^7F_4$	1598	26,27

(a) POPOP – [*p*-phenylene-bis (5-phenyl-2-oxazole)]

3.3.2 Vibrational Transition Gas Lasers

3.3.2.1 Introduction

The tables in this section cover laser lines between 1.8 and 33 μm that are based on the "usual" vibrational-rotational transitions in molecular gases. Lasing may be excited by an electric discharge or by optically pumping. The spectral coverage overlaps partly that of the following section on far infrared lasers. Some of the far infrared transitions not covered in this section include transitions between resonantly mixed vibrational states (as in H_2O and HCN), inversion transitions (as in NH_3), transitions associated with torsional vibrations and internal rotations, and transitions due to ring puckering modes. These transitions are distinctly different from pure rotational transitions, but occur in the same spectral region as the latter. Consequently, the far infrared section is considered to be the proper section for their inclusion.

The tables and their contents are organized according to the following ordering scheme:

1. increasing number of atoms in the laser molecule,
2. alphabetical order of common chemical formulas,
3. increasing isotopic mass,
4. increasing band-center wavelength.

All wavelengths given are vacuum values. To obtain wavelengths in dry air, the reader may consult the monograph by Coleman et al.[1] Wavelength values are not given beyond 10^{-8} μm. In the rare event that more accuracy is needed for a truncated wavelength value, the full accuracy can be recovered by a simple inversion of the frequency value given.

Calculated values are used wherever highly accurate molecular constants from extensive polynomial fitting are available. Where accurate frequencies are known in units of Hertz, the values are converted to centimeter^{-1} unit by using the value of $c = 2.99792458 \times 10^8$ m/sec.

The estimated uncertainty of the listed wavelength is always less than 20 in the last two digits, and in most cases it is less than 10. When values of comparable accuracy were available from several different sources, the value used was chosen according to the following priority order:

1. direct laser measurement,
2. other spectroscopic measurements,
3. calculated value.

For several molecules more than one reference is given for the wavelength and frequency; the values listed are believed to be the most accurate ones (information provided by Tao-Yuan Chang).

All vibrational transitions in this section occur within the ground electronic state of the laser molecules. The spectroscopic notations used conform to the convention in Herzberg's books.[2,3]

The comments given in Section 3.6 include the experimental conditions used for the observation, the peak output power obtained, and selected references. The number of comments for a given line roughly reflects the relative importance of that line within a given band. For some laser lines that are optically pumped by another laser line, additional information is given to identify the transition of the pump laser.

1. Coleman, C. D., Bozman, W. R., Meggers, W. F., *Table of Wavenumbers II, National Bureau of Standards Monograph 3*, U.S. Government Printing Office, Washington, DC.
2. Herzberg, G., *Molecular Spectra and Molecular Structure, I.. Spectra of Diatomic Molecules*, 2nd edition, Van Nostrand Reinhold, New York (1950).
3. Herzberg, G., *Molecular Spectra and Molecular Structure, II. Infrared and Raman Spectra of Polyatomic Molecules*, Van Nostrand, Princeton, NJ (1945).

Further Reading

Chang, T.-Y., Vibrational Transition Lasers, in *Handbook of Laser Science and Technology, Vol. II: Gas Lasers*, CRC Press, Boca Raton, FL (1982), p. 313 and *Handbook of Laser Science and Technology, Suppl. 1: Lasers*, p. 387.

Cheo, P. K., Ed., *Handbook of Molecular Lasers*, Marcel Dekker Inc., New York (1987).

Evans, J. D., Ed., *Selected Papers on CO_2 Lasers*, SPIE Milestone Series, Vol. MS 24, SPIE Optical Engineering Press, Bellingham, WA (1990).

Witteman, W. J., *The CO_2 Laser*, Springer Verlag, Berlin (1987).

Vibrational transition molecular gas lasers included in this section:

Diatomic Lasers – Table 3.3.2.1

CN	$D^{37}Cl$	$H^{35}Cl$
$^{12}C^{16}O$	DF	$H^{37}Cl$
$^{13}C^{16}O$	HBr	HF
$D^{79}Br$	$H^{79}Br$	NO
$D^{81}Br$	$H^{81}Br$	OH
$D^{35}Cl$	HCl	

Triatomic Lasers – Table 3.3.2.2

$^{12}C^{16}O_2$	$^{14}C^{16}O_2$	H_2O
$^{12}C^{16}O^{18}O$	$^{14}C^{18}O_2$	$^{14}N_2O$
$^{12}C^{17}O_2$	^{12}CO	$^{14}N^{15}NO$
$^{12}C^{18}O_2$	^{13}CO	$^{15}N_2O$
$^{13}C^{16}O_2$	$^{12}CS_2$	$^{15}N^{14}NO$
$^{13}C^{16}O_2$	$^{12}CS_2$	NOCl
$^{13}C^{16}O^{18}$	$^{13}CS_2$	NSF
$^{13}C^{18}O_2$	HCN	

Four-Atom Lasers – Table 3.3.2.3

BCl_3	C_2H_2
C_2D_2	COF_2
$^{12}C_2D_2$	$^{14}NH_3$
$^{12}C_2DH$	$^{15}NH_3$

Five-Atom Lasers – Table 3.3.2.4

$^{12}CF_4$	$FClO_3$
$^{13}CF_4$	HCOOH
$^{14}CF_4$	SiF_4
CF_3I	SiH_4
CH_3F	

Six-Atom Lasers – Table 3.3.2.5

C_2H_4

Seven-Atom Lasers – Table 3.3.2.6

CH_3CCH	SF_6

3.3.2.2 Diatomic Vibrational Transition Lasers

Table 3.3.2.1
Diatomic Vibrational Transition Lasers

CN Laser

Wavelength (μm) vac	Frequency (cm^{-1})	Transition assignment	Comments
5.03723	1985.22	$v = 2 \rightarrow 1$ Band P(8)	1601,1602
5.04745	1981.20	$v = 2 \rightarrow 1$ Band P(9)	1601,1602
5.05779	1977.15	$v = 2 \rightarrow 1$ Band P(10)	1601,1602
5.06827	1973.06	$v = 2 \rightarrow 1$ Band P(11)	1601,1602
5.07887	1968.94	$v = 2 \rightarrow 1$ Band P(12)	1601,1602
5.08375	1967.05	$v = 3 \rightarrow 2$ Band P(6)	1601,1602
5.08960	1964.79	$v = 2 \rightarrow 1$ Band P(13)	1601,1602
5.09391	1963.13	$v = 3 \rightarrow 2$ Band P(7)	1601,1602
5.10418	1959.18	$v = 3 \rightarrow 2$ Band P(8)	1601,1602
5.11459	1955.19	$v = 3 \rightarrow 2$ Band P(9)	1601,1602
5.12513	1951.17	$v = 3 \rightarrow 2$ Band P(10)	1601,1602
5.13579	1947.12	$v = 3 \rightarrow 2$ Band P(11)	1601,1602
5.14221	1944.69	$v = 4 \rightarrow 3$ Band P(5)	1601,1602
5.14660	1943.03	$v = 3 \rightarrow 2$ Band P(12)	1601,1602
5.15241	1940.84	$v = 4 \rightarrow 3$ Band P(6)	1601,1602
5.15754	1938.91	$v = 3 \rightarrow 2$ Band P(13)	1601,1602
5.16273	1936.96	$v = 4 \rightarrow 3$ Band P(7)	1601,1602
5.16863	1934.75	$v = 3 \rightarrow 2$ Band P(14)	1601,1602
5.17320	1933.04	$v = 4 \rightarrow 3$ Band P(8)	1601,1602
5.18382	1929.08	$v = 4 \rightarrow 3$ Band P(9)	1601,1602
5.19454	1925.10	$v = 4 \rightarrow 3$ Band P(10)	1601,1602
5.20541	1921.08	$v = 4 \rightarrow 3$ Band P(11)	1601,1602
5.21643	1917.02	$v = 4 \rightarrow 3$ Band P(12)	1601,1602
5.22756	1912.94	$v = 4 \rightarrow 3$ Band P(13)	1601,1602
5.23366	1910.71	$v = 5 \rightarrow 4$ Band P(7)	1601,1602
5.23887	1908.81	$v = 4 \rightarrow 3$ Band P(14)	1601,1602
5.24431	1906.83	$v = 5 \rightarrow 4$ Band P(8)	1601,1602
5.25028	1904.66	$v = 4 \rightarrow 3$ Band P(15)	1601,1602
5.25511	1902.91	$v = 5 \rightarrow 4$ Band P(9)	1601,1602
5.26607	1925.10	$v = 5 \rightarrow 4$ Band P(10)	1601,1602
5.27713	1894.97	$v = 5 \rightarrow 4$ Band P(11)	1601,1602
5.28838	1890.94	$v = 5 \rightarrow 4$ Band P(12)	1601,1602
5.29973	1886.89	$v = 5 \rightarrow 4$ Band P(13)	1601,1602
5.31121	1882.81	$v = 5 \rightarrow 4$ Band P(14)	1601,1602
5.32286	1878.69	$v = 5 \rightarrow 4$ Band P(15)	1601,1602

$^{12}C^{16}O$ Laser

Wavelength (μm) vac	Frequency (cm^{-1})	Transition assignment	Comments
2.6886383	3719.35493	v = 12 → 10 Band P(5)	1603
2.6914188	3715.51240	v = 12 → 10 Band P(6)	1603
2.7261937	3668.11794	v = 13 → 11 Band P(5)	1603
2.7290264	3664.31044	v = 13 → 11 Band P(6)	1603
2.7319165	3660.43392	v = 13 → 11 Band P(7)	1603,1605
2.7348643	3656.48852	v = 13 → 11 Band P(8)	1605
2.7378700	3652.47438	v = 13 → 11 Band P(9)	1605
2.7409338	3648.39165	v = 13 → 11 Band P(10)	1605
2.7440560	3644.24048	v = 13 → 11 Band P(11)	1605
2.7472369	3640.02101	v = 13 → 11 Band P(12)	1605
2.7504767	3635.73339	v = 13 → 11 Band P(13)	1605
2.7537757	3631.37776	v = 13 → 11 Band P(14)	1605
2.7571343	3626.95427	v = 13 → 11 Band P(15)	1605
2.7605526	3622.46306	v = 13 → 11 Band P(16)	1605
2.7640311	3617.90428	v = 13 → 11 Band P(17)	1605
2.7646877	3617.04503	v = 14 → 12 Band P(5)	1603
2.7675700	3613.27807	v = 13 → 11 Band P(18)	1605
2.7675742	3613.27256	v = 14 → 12 Band P(6)	1603
2.7705197	3609.43108	v = 14 → 12 Band P(7)	1603
2.7711696	3608.58457	v = 13 → 11 Band P(19)	1605
2.7735245	3605.52073	v = 14 → 12 Band P(8)	1603
2.7765887	3601.54166	v = 14 → 12 Band P(9)	1605
2.7797128	3597.49402	v = 14 → 12 Band P(10)	1605
2.7828968	3593.37794	v = 14 → 12 Band P(11)	1605
2.7861412	3589.19359	v = 14 → 12 Band P(12)	1605
2.7894461	3584.94109	v = 14 → 12 Band P(13)	1605
2.7928120	3580.62060	v = 14 → 12 Band P(14)	1605
2.7962390	3576.23226	v = 14 → 12 Band P(15)	1605
2.7997275	3571.77621	v = 14 → 12 Band P(16)	1605
2.8032778	3567.25261	v = 14 → 12 Band P(17)	1605
2.8041538	3566.13815	v = 15 → 13 Band P(5)	1603
2.8068902	3562.66159	v = 14 → 12 Band P(18)	1605
2.8070958	3562.40070	v = 15 → 13 Band P(6)	1603
2.8100984	3558.59425	v = 15 → 13 Band P(7)	1603
2.8105651	3558.00329	v = 14 → 12 Band P(19)	1605
2.8143028	3553.27786	v = 14 → 12 Band P(20)	1605
2.8181037	3548.48545	v = 14 → 12 Band P(21)	1605
2.8219681	3543.62619	v = 14 → 12 Band P(22)	1605
2.8227207	3542.68141	v = 15 → 13 Band P(11)	1605
2.8258963	3538.70023	v = 14 → 12 Band P(23)	1605

$^{12}C^{16}O$ Laser—*continued*

Wavelength (μm) vac	Frequency (cm^{-1})	Transition assignment	Comments
2.8260306	3538.53215	v = 15 → 13 Band P(12)	1605
2.8294028	3534.31478	v = 15 → 13 Band P(13)	1605
2.8298888	3533.70771	v = 14 → 12 Band P(24)	1605
2.8328376	3530.02942	v = 15 → 13 Band P(14)	1605
2.8363353	3525.67623	v = 15 → 13 Band P(15)	1605
2.8398963	3521.25535	v = 15 → 13 Band P(16)	1605
2.8435208	3516.76691	v = 15 → 13 Band P(17)	1605
2.8446274	3515.39889	v = 16 → 14 Band P(5)	1603
2.8472093	3512.21108	v = 15 → 13 Band P(18)	1605
2.8476265	3511.69645	v = 16 → 14 Band P(6)	1603
2.8506881	3507.92503	v = 16 →·14 Band P(7)	1603
2.8509620	3507.58798	v = 15 → 13 Band P(19)	1605
2.8547793	3502.89776	v = 15 → 13 Band P(20)	1605
2.8569993	3500.17583	v = 16 → 14 Band P(9)	6951
2.8586616	3498.14057	v = 15 → 13 Band P(21)	1605
2.8626092	3493.31655	v = 15 → 13 Band P(22)	1605
2.8666225	3488.42584	v = 15 → 13 Band P(23)	1605
2.8703827	3483.85602	v = 16 → 14 Band P(13)	1605
2.8707019	3483.46858	v = 15 → 13 Band P(24)	1605
2.8738888	3479.60579	v = 16 → 14 Band P(14)	1605
2.8774596	3475.28774	v = 16 → 14 Band P(15)	1605
2.8810955	3470.90201	v = 16 → 14 Band P(16)	1605
2.8847967	3466.44874	v = 16 → 14 Band P(17)	1605
2.8885638	3461.92808	v = 16 → 14 Band P(18)	1605
2.8892040	3461.16098	v = 17 → 15 Band P(6)	1603
2.8923263	3457.42458	v = 17 → 15 Band P(7)	1603
2.8923969	3457.34017	v = 16 → 14 Band P(19)	1605
2.8955131	3453.61936	v = 17 → 15 Band P(8)	6951
2.8962965	3452.68516	v = 16 → 14 Band P(20)	1605
2.8987646	3449.74545	v = 17 → 15 Band P(9)	6951
2.9002630	3447.96318	v = 16 → 14 Band P(21)	1605
2.9020812	3445.80302	v = 17 → 15 Band P(10)	6951
2.9042967	3443.17439	v = 16 → 14 Band P(22)	1605
2.9083980	3438.31891	v = 16 → 14 Band P(23)	1605
2.9196501	3425.06792	v = 17 → 15 Band P(15)	1605
2.9233635	3420.71732	v = 17 → 15 Band P(16)	1605
2.9271441	3416.29920	v = 17 → 15 Band P(17)	1605
2.9287487	3414.42742	v = 18 → 16 Band P(5)	1603
2.9309924	3411.81371	v = 17 → 15 Band P(18)	1605
2.9318678	3410.79499	v = 18 → 16 Band P(6)	1603

$^{12}C^{16}O$ Laser—*continued*

Wavelength (μm) vac	Frequency (cm⁻¹)	Transition assignment	Comments
2.9349087	3407.26097	v = 17 → 15 Band P(19)	1605
2.9350529	3407.09359	v = 18 → 16 Band P(7)	1603,6951
2.9383044	3403.32339	v = 18 → 16 Band P(8)	6951
2.9388935	3402.64115	v = 17 → 15 Band P(20)	1605
2.9416225	3399.48452	v = 18 → 16 Band P(9)	6951
2.9429471	3397.95437	v = 17 → 15 Band P(21)	1605
2.9450075	3395.57713	v = 18 → 16 Band P(10)	6951
2.9470699	3393.20078	v = 17 → 15 Band P(22)	1605
2.9484597	3391.60136	v = 18 → 16 Band P(11)	6951
2.9512624	3388.38053	v = 17 → 15 Band P(23)	1605
2.9555249	3383.49374	v = 17 → 15 Band P(24)	1605
2.9598579	3378.54057	v = 17 → 15 Band P(25)	1605
2.9706038	3366.31896	v = 18 → 16 Band P(17)	1605
2.9724783	3364.19613	v = 19 → 17 Band P(5)	1603
2.9745362	3361.86861	v = 18 → 16 Band P(18)	1605
2.9756603	3360.59867	v = 19 → 17 Band P(6)	1603
2.9785387	3357.35102	v = 18 → 16 Band P(19)	1605
2.9789103	3356.93227	v = 19 → 17 Band P(7)	1603
2.9822285	3353.19706	v = 19 → 17 Band P(8)	6951
2.9826116	3352.76636	v = 18 → 16 Band P(20)	1605
2.9856154	3349.39321	v = 19 → 17 Band P(9)	6951
2.9867555	3348.11475	v = 18 → 16 Band P(21)	1605
2.9890712	3345.52084	v = 19 → 17 Band P(10	6951
2.9909705	3343.39635	v = 18 → 16 Band P(22)	1605
2.9925962	3341.58011	v = 19 → 17 Band P(11)	6951
2.9952574	3338.61128	v = 18 → 16 Band P(23)	1605
2.9996163	3333.75970	v = 18 → 16 Band P(24)	1605
3.0040479	3328.84174	v = 18 → 16 Band P(25)	1605
3.0085525	3323.85754	v = 18 → 16 Band P(26)	1605
3.0131307	3318.80724	v = 18 → 16 Band P(27)	1605
3.0173794	3314.13415	v = 20 → 18 Band P(5)	1603
3.0206264	3310.57166	v = 20 → 18 Band P(6)	1603
3.0239434	3306.94023	v = 20 → 18 Band P(7)	6951
3.0273307	3303.24001	v = 20 → 18 Band P(8)	6951
3.0274949	3303.06090	v = 19 → 17 Band P(20)	1605
3.0307887	3299.47114	v = 20 → 18 Band P(9)	6951
3.0317321	3298.44443	v = 19 → 17 Band P(21)	1605
3.0343177	3295.63377	v = 20 → 18 Band P(10)	6951
3.0360428	3293.76116	v = 19 → 17 Band P(22)	1605
3.0379180	3291.72805	v = 20 → 18 Band P(11)	6951

$^{12}C^{16}O$ Laser—*continued*

Wavelength (µm) vac	Frequency (cm⁻¹)	Transition assignment	Comments
3.0404274	3289.0i125	v = 19 → 17 Band P(23)	1605
3.0448864	3284.19483	v = 19 → 17 Band P(24)	1605
3.0494201	3279.31203	v = 19 → 17 Band P(25)	1605
3.0540291	3274.36301	v = 19 → 17 Band P(26)	1605
3.0587140	3269.34789	v = 19 → 17 Band P(27)	1605
3.0634751	3264.26682	v = 19 → 17 Band P(28)	1605
3.0668139	3260.71299	v = 21 → 19 Band P(6)	1603
3.0683130	3259.11994	v = 19 → 17 Band P(29)	1605
3.0702003	3257.11651	v = 21 → 19 Band P(7)	6950
3.0732282	3253.90737	v = 19 → 17 Band P(30)	1605
3.0736591	3253.45124	v = 21 → 19 Band P(8)	6950
3.0771907	3249.71734	v = 21 → 19 Band P(9)	6950
3.0779242	3248.94292	v = 20 → 18 Band P(21)	1605
3.0807955	3245.91494	v = 21 → 19 Band P(10)	6950
3.0823340	3244.29475	v = 20 → 18 Band P(22)	1605
3.0844737	3242.04420	v = 21 → 19 Band P(11)	6950
3.0868200	3239.57994	v = 20 → 18 Band P(23)	1605
3.0882257	3238.10525	v = 21 → 19 Band P(12)	6950
3.0913825	3234.79862	v = 20 → 18 Band P(24)	1605
3.0960223	3229.95094	v = 20 → 18 Band P(25)	1605
3.1007396	3225.03703	v = 20 → 18 Band P(26)	1605
3.1055351	3220.05703	v = 20 → 18 Band P(27)	1605
3.1104092	3215.01108	v = 20 → 18 Band P(28)	1605
3.1153625	3209.89932	v = 20 → 18 Band P(29)	1605
3.1177323	3207.45948	v = 22 → 20 Band P(7)	6951
3.1203956	3204.72188	v = 20 → 18 Band P(30)	1605
3.1212651	3203.82914	v = 22 → 20 Band P(8)	6951
3.1248729	3200.13017	v = 22 → 20 Band P(9)	6951
3.1255090	3199.47891	v = 20 → 18 Band P(31)	1605
3.1285561	3196.36271	v = 22 → 20 Band P(10)	6951
3.1323150	3192.52690	v = 22 → 20 Band P(11)	6951
3.1344855	3190.31623	v = 21 → 19 Band P(23)	1605
3.1361501	3188.62290	v = 22 → 20 Band P(12)	6951
3.1391557	3185.56995	v = 21 → 19 Band P(24)	1605
3.1439054	3180.75730	v = 21 → 19 Band P(25)	1605
3.1487351	3175.87843	v = 21 → 19 Band P(26)	1605
3.1536455	3170.93347	v = 21 → 19 Band P(27)	1605
3.1586370	3165.92256	v = 21 → 19 Band P(28)	1605
3.1637101	3160.84584	v = 21 → 19 Band P(29)	1605
3.1665944	3157.96680	v = 23 → 21 Band P(7)	6951

$^{12}C^{16}O$ Laser—*continued*

Wavelength (μm) vac	Frequency (cm^{-1})	Transition assignment	Comments
3.1688656	3155.70344	v = 21 → 19 Band P(30)	1605
3.1702038	3154.37137	v = 23 → 21 Band P(8)	6951
3.1738905	3150.70729	v = 23 → 21 Band P(9)	6951
3.1741039	3150.49550	v = 21 → 19 Band P(31)	1605
3.1776550	3146.97473	v = 23 → 21 Band P(10)	6951
3.1814976	3143.17382	v = 23 → 21 Band P(11)	6951
3.1834782	3141.21834	v = 22 → 20 Band P(23)	1605
3.1854187	3139.30470	v = 23 → 21 Band P(12)	6951
3.1882601	3136.50701	v = 22 → 20 Band P(24)	1605
3.1931240	3131.72931	v = 22 → 20 Band P(25)	1605
3.1980705	3126.88538	v = 22 → 20 Band P(26)	1605
3.2031002	3121.97537	v = 22 → 20 Band P(27)	1605
3.2082136	3116.99939	v = 22 → 20 Band P(28)	1605
3.2134114	3111.95760	v = 22 → 20 Band P(29)	1605
3.2168456	3108.63541	v = 24 → 22 Band P(7)	6951
3.2186940	3106.85013	v = 22 → 20 Band P(30)	1605
3.2205343	3105.07484	v = 24 → 22 Band P(8)	6951
3.2240622	3101.67711	v = 22 → 20 Band P(31)	1605
3.2243029	3101.44562	v = 24 → 22 Band P(9)	6951
3.2281517	3097.74790	v = 24 → 22 Band P(10)	6951
3.2320810	3093.98183	v = 24 → 22 Band P(11)	6951
3.2360914	3090.14755	v = 24 → 22 Band P(12)	6951
3.2685491	3059.46143	v = 25 → 23 Band P(7)	6951
3.2723202	3055.93567	v = 25 → 23 Band P(8)	6951
3.2761736	3052.34126	v = 25 → 23 Band P(9)	6951
3.2801099	3048.67834	v = 25 → 23 Band P(10)	6951
3.2841294	3044.94705	v = 25 → 23 Band P(11)	6951
3.2882324	3041.14753	v = 25 → 23 Band P(12)	6951
3.3217734	3010.44013	v = 26 → 24 Band P(7)	6951
3.3256299	3006.94914	v = 26 → 24 Band P(8)	6951
3.3295715	3003.38948	v = 26 → 24 Band P(9)	6951
3.3335986	2999.76129	v = 26 → 24 Band P(10)	6951
3.3377116	2996.06471	v = 26 → 24 Band P(11)	6951
3.3419110	2992.29989	v = 26 → 24 Band P(12)	6951
3.3461973	2988.46696	v = 26 → 24 Band P(13)	6951
3.3765921	2961.56587	v = 27 → 25 Band P(7)	6951
3.3845705	2954.58464	v = 27 → 25 Band P(9)	6951
3.3929023	2947.32916	v = 27 → 25 Band P(11	6951
3.3972019	2943.59895	v = 27 → 25 Band P(12)	6951
3.4015913	2939.80060	v = 27 → 25 Band P(13)	6951

$^{12}C^{16}O$ Laser—*continued*

Wavelength (μm) vac	Frequency (cm^{-1})	Transition assignment	Comments
3.4291380	2916.18480	v = 28 → 26 Band P(6)	6951
3.4330850	2912.83208	v = 28 → 26 Band P(7)	6951
3.4371224	2909.41048	v = 28 → 26 Band P(8)	6951
3.4412508	2905.92014	v = 28 → 26 Band P(9)	6951
3.4454705	2902.36120	v = 28 → 26 Band P(10)	6951
3.4497821	2898.73380	v = 28 → 26 Band P(11)	6951
3.4541860	2895.03809	v = 28 → 26 Band P(12)	6951
3.4586827	2891.27421	v = 28 → 26 Band P(13)	6951
3.4913383	2864.23117	v = 29 → 27 Band P(7)	6951
3.4954718	2860.84417	v = 29 → 27 Band P(8)	6951
3.4996993	2857.38837	v = 29 → 27 Band P(9)	6951
3.5040213	2853.86393	v = 29 → 27 Band P(10)	6951
3.5084383	2850.27099	v = 29 → 27 Band P(11)	6951
3.5129509	2846.60968	v = 29 → 27 Band P(12)	6951
3.5175595	2842.88015	v = 29 → 27 Band P(13)	6951
3.5222646	2839.08254	v = 29 → 27 Band P(14)	6951
3.5514460	2815.75446	v = 30 → 28 Band P(7)	6951
3.5556795	2812.40199	v = 30 → 28 Band P(8)	6951
3.5600103	2808.98067	v = 30 → 28 Band P(9)	6951
3.5644389	2805.49064	v = 30 → 28 Band P(10)	6951
3.5689659	2801.93203	v = 30 → 28 Band P(11)	6951
3.5735919	2798.30500	v = 30 → 28 Band P(12)	6951
3.5783172	2794.60969	v = 30 → 28 Band P(13)	6951
3.5831426	2790.84623	v = 30 → 28 Band P(14)	6951
3.6135103	2767.39217	v = 31 → 29 Band P(7)	6951
3.6178479	2764.07416	v = 31 → 29 Band P(8)	6951
3.6222865	2760.68721	v = 31 → 29 Band P(9)	6951
3.6268264	2757.23148	v = 31 → 29 Band P(10)	6951
3.6314683	2753.70710	v = 31 → 29 Band P(11)	6951
3.6362126	2750.11422	v = 31 → 29 Band P(12)	6951
3.6410600	2746.45297	v = 31 → 29 Band P(13)	6951
3.6460110	2742.72350	v = 31 → 29 Band P(14)	6951
3.6733000	2722.34774	v = 32 → 30 Band P(6)	6951
3.6776425	2719.13327	v = 32 → 30 Band P(7)	6951
3.6820890	2715.84964	v = 32 → 30 Band P(8)	6951
3.6866400	2712.49698	v = 32 → 30 Band P(9)	6951
3.6912962	2709.07544	v = 32 → 30 Band P(10)	6951
3.6960581	2705.58516	v = 32 → 30 Band P(11)	6951
3.7009263	2702.02628	v = 32 → 30 Band P(12)	6951
3.7059012	2698.39894	v = 32 → 30 Band P(13)	6951

$^{12}C^{16}O$ Laser—*continued*

Wavelength (μm) vac	Frequency (cm⁻¹)	Transition assignment	Comments
3.7109837	2694.70328	$v = 32 \to 30$ Band P(14)	6951
3.7161743	2690.93944	$v = 32 \to 30$ Band P(15)	6951
3.7395122	2674.14558	$v = 33 \to 31$ Band P(6)	6951
3.7439645	2670.96553	$v = 33 \to 31$ Band P(7)	6951
3.7485247	2667.71619	$v = 33 \to 31$ Band P(8)	6951
3.7531934	2664.39771	$v = 33 \to 31$ Band P(9)	6951
3.7579713	2661.01024	$v = 33 \to 31$ Band P(10)	6951
3.7628588	2657.55391	$v = 33 \to 31$ Band P(11)	6951
3.7678565	2654.02887	$v = 33 \to 31$ Band P(12)	6951
3.7729652	2650.43526	$v = 33 \to 31$ Band P(13)	6951
3.7781854	2646.77322	$v = 33 \to 31$ Band P(14)	6951
3.7835179	2643.04288	$v = 33 \to 31$ Band P(15)	6951
3.8080427	2626.02101	$v = 34 \to 32$ Band P(6)	6951
3.8126098	2622.87528	$v = 34 \to 32$ Band P(7)	6951
3.8172891	2619.66015	$v = 34 \to 32$ Band P(8)	6951
3.8220810	2616.37574	$v = 34 \to 32$ Band P(9)	6951
3.8269862	2613.02221	$v = 34 \to 32$ Band P(10)	6951
3.8320054	2609.59968	$v = 34 \to 32$ Band P(11)	6951
3.8371391	2606.10831	$v = 34 \to 32$ Band P(12)	6951
3.8423880	2602.54824	$v = 34 \to 32$ Band P(13)	6951
3.8477527	2598.91959	$v = 34 \to 32$ Band P(14)	6951
3.8532341	2595.22252	$v = 34 \to 32$ Band P(15)	6951
3.8790377	2577.95895	$v = 35 \to 33$ Band P(6)	6951
3.8837252	2574.84749	$v = 35 \to 33$ Band P(7)	6951
3.8885292	2571.66646	$v = 35 \to 33$ Band P(8)	6951
3.8934503	2568.41600	$v = 35 \to 33$ Band P(9)	6951
3.8984892	2565.09626	$v = 35 \to 33$ Band P(10)	6951
3.9036465	2561.70737	$v = 35 \to 33$ Band P(11)	6951
3.9089229	2558.24949	$v = 35 \to 33$ Band P(12)	6951
3.9143191	2554.72274	$v = 35 \to 33$ Band P(13)	6951
3.9198358	2551.12726	$v = 35 \to 33$ Band P(14)	6951
3.9254738	2547.46319	$v = 35 \to 33$ Band P(15)	6951
3.9526585	2529.94281	$v = 36 \to 34$ Band P(6)	6951
3.9574722	2526.86553	$v = 36 \to 34$ Band P(7)	6951
3.9624071	2523.71850	$v = 36 \to 34$ Band P(8)	6951
3.9674638	2520.50187	$v = 36 \to 34$ Band P(9)	6951
3.9726432	2517.21577	$v = 36 \to 34$ Band P(10)	6951
3.9779457	2513.86035	$v = 36 \to 34$ Band P(11)	6951
3.9833722	2510.43574	$v = 36 \to 34$ Band P(12)	6951
3.9889234	2506.94209	$v = 36 \to 34$ Band P(13)	6951

$^{12}C^{16}O$ Laser—*continued*

Wavelength (μm) vac	Frequency (cm^{-1})	Transition assignment	Comments
3.9946001	2503.37952	v = 36 → 34 Band P(14)	6951
4.0004029	2499.74818	v = 36 → 34 Band P(15)	6951
4.0391017	2475.79804	v = 37 → 35 Band P(8)	6951
4.0443011	2472.61510	v = 37 → 35 Band P(9)	6951
4.0496282	2469.36248	v = 37 → 35 Band P(10)	6951
4.0550837	2466.04033	v = 37 → 35 Band P(11)	6951
4.0606684	2462.64878	v = 37 → 35 Band P(12)	6951
4.0663829	2459.18798	v = 37 → 35 Band P(13)	6951
4.7451315	2107.42315	v = 1 → 0 Band P(9)	1606
4.7545021	2103.26967	v = 1 → 0 Band P(10)	1606
4.7639858	2099.08266	v = 1 → 0 Band P(11)	1606,1607
4.7735835	2094.86226	v = 1 → 0 Band P(12)	1606,1607
4.7768927	2093.41107	v = 2 → 1 Band P(6)	1606,1615
4.7832961	2090.60862	v = 1 → 0 Band P(13)	1606
4.7860765	2089.39410	v = 2 → 1 Band P(7)	1606,1607
4.7931243	2086.32187	v = 1 → 0 Band P(14)	1606
4.7953738	2085.34317	v = 2 → 1 Band P(8)	1606,1607
4.8030689	2082.00218	v = 1 → 0 Band P(15)	1606
4.8047854	2081.25840	v = 2 → 1 Band P(9)	1606,1607
4.8131309	2077.64969	v = 1 → 0 Band P(16)	1606
4.8143121	2077.13995	v = 2 → 1 Band P(10)	1606,1607
4.8233112	2073.26454	v = 1 → 0 Band P(17)	1606
4.8239547	2072.98796	v = 2 → 1 Band P(11)	1606,1607
4.8336105	2068.84688	v = 1 → 0 Band P(18)	1606
4.8337140	2068.80259	v = 2 → 1 Band P(12)	1606,1607
4.8435908	2064.58398	v = 2 → 1 Band P(13)	1606,1607
4.8440299	2064.39686	v = 1 → 0 Band P(19)	1606
4.8535861	2060.33227	v = 2 → 1 Band P(14)	1606
4.8545702	2059.91461	v = 1 → 0 Band P(20)	1606
4.8562333	2059.20912	v = 3 → 2 Band P(8)	1606,1615
4.8637006	2056.04761	v = 2 → 1 Band P(15)	1606
4.8652323	2055.40030	v = 1 → 0 Band P(21)	1606
4.8658027	2055.15938	v = 3 → 2 Band P(9)	1606,1607
4.8739353	2051.73016	v = 2 → 1 Band P(16)	1606
4.8754899	2051.07595	v = 3 → 2 Band P(10)	1606,1607
4.8842910	2047.38004	v = 2 → 1 Band P(17)	1606
4.8852957	2046.95900	v = 3 → 2 Band P(11)	1606,1607
4.8947688	2042.99742	v = 2 → 1 Band P(18)	1606
4.8952211	2042.80866	v = 3 → 2 Band P(12)	1606,1607
4.9052668	2038.62508	v = 3 → 2 Band P(13)	1606,1607

$^{12}C^{16}O$ Laser—*continued*

Wavelength (µm) vac	Frequency (cm⁻¹)	Transition assignment	Comments
4.9053694	2038.58244	v = 2 → 1 Band P(19)	1606
4.9088827	2037.12345	v = 4 → 3 Band P(7)	1607,1615
4.9154339	2034.40840	v = 3 → 2 Band P(14)	1606,1607
4.9160940	2034.13524	v = 2 → 1 Band P(20)	1606
4.9184943	2033.14256	v = 4 → 3 Band P(8)	1606,1607
4.9257231	2030.15878	v = 3 → 2 Band P(15)	1606
4.9282257	2029.12784	v = 4 → 3 Band P(9)	1606,1607
4.9361354	2025.87637	v = 3 → 2 Band P(16)	1606
4.9380779	2025.07944	v = 4 → 3 Band P(10)	1606,1607
4.9466717	2021.56129	v = 3 → 2 Band P(17)	1606
4.9480516	2020.99752	v = 4 → 3 Band P(11)	1606,1607
4.9573329	2017.21371	v = 3 → 2 Band P(18)	1606
4.9581478	2016.88221	v = 4 → 3 Band P(12)	1606,1607
4.9627718	2015.00300	v = 5 → 4 Band P(6)	1607,1615
4.9681201	2012.83377	v = 3 → 2 Band P(19)	1606
4.9683672	2012.73366	v = 4 → 3 Band P(13)	1606–8,1615
4.9724252	2011.09110	v = 5 → 4 Band P(7)	1607
4.9787110	2008.55203	v = 4 → 3 Band P(14)	1606–1608
4.9790342	2008.42162	v = 3 → 2 Band P(20)	1606
4.9822005	2007.14522	v = 5 → 4 Band P(8)	1606,1607
4.9891798	2004.33745	v = 4 → 3 Band P(15)	1606–1608
4.9900763	2003.97739	v = 3 → 2 Band P(21)	1606
4.9920987	2003.16553	v = 5 → 4 Band P(9)	1606,1607
4.9997748	2000.09007	v = 4 → 3 Band P(16)	1606
5.0012472	1999.50125	v = 3 → 2 Band P(22)	1606
5.0021205	1999.15216	v = 5 → 4 Band P(10)	1606–1608
5.0104969	1995.81004	v = 4 → 3 Band P(17)	1606
5.0122669	1995.10527	v = 5 → 4 Band P(11)	1606–8,1615
5.0125481	1994.99332	v = 3 → 2 Band P(23)	1606
5.0179407	1992.84936	v = 6 → 5 Band P(5)	1607,1615
5.0213470	1991.49751	v = 4 → 3 Band P(18)	1606
5.0225387	1991.02499	v = 5 → 4 Band P(12)	1606–1608
5.0239800	1990.45377	v = 3 → 2 Band P(24)	1606
5.0276354	1989.00658	v = 6 → 5 Band P(6)	1607
5.0323261	1987.15262	v = 4 → 3 Band P(19)	1606
5.0329368	1986.91149	v = 5 → 4 Band P(13)	1606–1608
5.0355441	1985.88273	v = 3 → 2 Band P(25)	1606,1615
5.0374542	1985.12969	v = 6 → 5 Band P(7)	1607,1609
5.0434353	1982.77552	v = 4 → 3 Band P(20)	1606
5.0434623	1982.76489	v = 5 → 4 Band P(14)	1606–1608

$^{12}C^{16}O$ Laser—*continued*

Wavelength (μm) vac	Frequency (cm^{-1})	Transition assignment	Comments
5.0472413	1981.28035	v = 3 → 2 Band P(26)	1606
5.0473980	1981.21884	v = 6 → 5 Band P(8)	1606,1607,1609
5.0541161	1978.58535	v = 5 → 4 Band P(15)	1606–1608
5.0546756	1978.36634	v = 4 → 3 Band P(21)	1606
5.0574676	1977.27417	v = 6 → 5 Band P(9)	1606,1607,1609
5.0590728	1976.64677	v = 3 → 2 Band P(27)	1606
5.0648990	1974.37302	v = 5 → 4 Band P(16)	1606,1608
5.0660480	1973.92525	v = 4 → 3 Band P(22)	1606
5.0676639	1973.29583	v = 6 → 5 Band P(10)	1606–1609
5.0710398	1971.98215	v = 3 → 2 Band P(28)	1606
5.0758122	1970.12805	v = 5 → 4 Band P(17)	1606,1608
5.0779878	1969.28397	v = 6 → 5 Band P(11)	1606–1609
5.0831434	1967.28662	v = 3 → 2 Band P(29)	1606
5.0841669	1966.89059	v = 7 → 6 Band P(5)	1607,1615
5.0868566	1965.85056	v = 5 → 4 Band P(18)	1606,1608,1610
5.0884403	1965.23874	v = 6 → 5 Band P(12)	1606–1609
5.0920025	1963.86391	v = 9 → 8 Band R(8)	1606,1615
5.0940286	1963.08282	v = 7 → 6 Band P(6)	1607
5.0953848	1962.56033	v = 3 → 2 Band P(30)	1606
5.0968272	1962.00493	v = 10 → 9 Band R(16)	1606,1615
5.0980333	1961.54073	v = 5 → 4 Band P(19)	1606,1610
5.0990223	1961.16027	v = 6 → 5 Band P(13)	1606–1609
5.1004303	1960.61891	v = 9 → 8 Band R(7)	1606
5.1040175	1959.24095	v = 7 → 6 Band P(7)	1606,1607,1609
5.1044124	1959.08935	v = 10 → 9 Band R(15)	1606
5.1077651	1957.80343	v = 3 → 2 Band P(31)	1606
5.1089803	1957.33776	v = 9 → 8 Band R(6)	1606,1615
5.1093433	1957.19868	v = 5 → 4 Band P(20)	1606,1610
5.1097348	1957.04872	v = 6 → 5 Band P(14)	1606–9,1612,1615
5.1121178	1956.13645	v = 10 → 9 Band R(14)	1606
5.1141344	1955.36511	v = 7 → 6 Band P(8)	1606,1607,1609
5.1199440	1953.14636	v = 10 → 9 Band R(13)	1606
5.1202856	1953.01606	v = 3 → 2 Band P(32)	1606
5.1205788	1952.90423	v = 6 → 5 Band P(15)	1606–1608
5.1207877	1952.82457	v = 5 → 4 Band P(21)	1606,1610
5.1227504	1952.07638	v = 11 → 10 Band R(22)	1606,1615
5.1243803	1951.45547	v = 7 → 6 Band P(9)	1606–1609
5.1278915	1950.11925	v = 10 → 9 Band R(12)	1606
5.1297237	1949.42274	v = 11 → 10 Band R(21)	1606
5.1315553	1948.72695	v = 6 → 5 Band P(16)	1606–1608,1610

$^{12}C^{16}O$ Laser—*continued*

Wavelength (μm) vac	Frequency (cm⁻¹)	Transition assignment	Comments
5.1323675	1948.41853	$v = 5 \rightarrow 4$ Band P(22)	1606,1610
5.1329476	1948.19836	$v = 3 \rightarrow 2$ Band P(33)	1606
5.1347561	1947.51216	$v = 7 \rightarrow 6$ Band P(10)	1606–1609
5.1359611	1947.05524	$v = 10 \rightarrow 9$ Band R(11)	1606
5.1368169	1946.73088	$v = 11 \rightarrow 10$ Band R(20)	1606
5.1426652	1944.51702	$v = 6 \rightarrow 5$ Band P(17)	1606–1608,1610
5.1440304	1944.00096	$v = 11 \rightarrow 10$ Band R(19)	1606
5.1440839	1943.98073	$v = 5 \rightarrow 4$ Band P(23)	1606,1610
5.1452628	1943.53534	$v = 7 \rightarrow 6$ Band P(11)	1606–1609
5.1513648	1941.23312	$v = 11 \rightarrow 10$ Band R(18)	1606
5.1519672	1941.00613	$v = 8 \rightarrow 7$ Band P(5)	1607,1615
5.1539097	1940.27460	$v = 6 \rightarrow 5$ Band P(18)	1606,1610
5.1559012	1939.52514	$v = 7 \rightarrow 6$ Band P(12)	1606–09,1612
5.1559380	1939.51129	$v = 5 \rightarrow 4$ Band P(24)	1606,1610
5.1588207	1938.42751	$v = 11 \rightarrow 10$ Band R(17)	1606
5.1620007	1937.23337	$v = 8 \rightarrow 7$ Band P(6)	1607
5.1652897	1935.99982	$v = 6 \rightarrow 5$ Band P(19)	1606,1610
5.1663986	1935.58429	$v = 11 \rightarrow 10$ Band R(16)	1606
5.1666724	1935.48171	$v = 7 \rightarrow 6$ Band P(13)	1606–1609,1612
5.1679309	1935.01038	$v = 5 \rightarrow 4$ Band P(25)	1606,1610
5.1721645	1933.42652	$v = 8 \rightarrow 7$ Band P(7)	1606,1607,1609
5.1740992	1932.70359	$v = 11 \rightarrow 10$ Band R(15)	1606
5.1768065	1931.69283	$v = 6 \rightarrow 5$ Band P(20)	1606,1610
5.1775774	1931.40521	$v = 7 \rightarrow 6$ Band P(14)	1606–9,1612,6950
5.1796042	1930.64944	$v = 12 \rightarrow 11$ Band R(24)	1606,1615
5.1800639	1930.47813	$v = 5 \rightarrow 4$ Band P(26)	1606,1610
5.1819229	1929.78557	$v = 11 \rightarrow 10$ Band R(14)	1606
5.1824596	1929.58571	$v = 8 \rightarrow 7$ Band P(8)	1606,1607,1609
5.1864328	1928.10750	$v = 12 \rightarrow 11$ Band R(23)	1606
5.1884610	1927.35379	$v = 6 \rightarrow 5$ Band P(21)	1606,1610
5.1886172	1927.29577	$v = 7 \rightarrow 6$ Band P(15)	1606–109,6950
5.1898705	1926.83037	$v = 11 \rightarrow 10$ Band R(13)	1606
5.1923380	1925.91468	$v = 5 \rightarrow 4$ Band P(27)	1606,1610
5.1928870	1925.71109	$v = 8 \rightarrow 7$ Band P(9)	1606–1609
5.1933833	1925.52706	$v = 12 \rightarrow 11$ Band R(22)	1606
5.1997928	1923.15354	$v = 7 \rightarrow 6$ Band P(16)	1606,1607,1609
5.2002545	1922.98282	$v = 6 \rightarrow 5$ Band P(22)	1606,1610
5.2004561	1922.90826	$v = 12 \rightarrow 11$ Band R(21)	1606
5.2034475	1921.80281	$v = 8 \rightarrow 7$ Band P(10)	1606–1609
5.2047546	1921.32019	$v = 5 \rightarrow 4$ Band P(28)	1606,1615

$^{12}C^{16}O$ Laser—*continued*

Wavelength (μm) vac	Frequency (cm^{-1})	Transition assignment	Comments
5.2076518	1920.25126	$v = 12 \to 11$ Band R(20)	1606
5.2111053	1918.97868	$v = 7 \to 6$ Band P(17)	1606–10,1615,6950
5.2113159	1918.90112	$v = 9 \to 8$ Band P(4)	1607
5.2121879	1918.58009	$v = 6 \to 5$ Band P(23)	1606,1610
5.2141422	1917.86102	$v = 8 \to 7$ Band P(11)	1606–1609
5.2149710	1917.55619	$v = 12 \to 11$ Band R(19)	1606
5.2173147	1916.69479	$v = 5 \to 4$ Band P(29)	1606
5.2213933	1915.19761	$v = 9 \to 8$ Band P(5)	1607
5.2224143	1914.82321	$v = 12 \to 11$ Band R(18)	1606
5.2225558	1914.77131	$v = 7 \to 6$ Band P(18)	1606–7,1610,6950
5.2242626	1914.14573	$v = 6 \to 5$ Band P(24)	1606,1610
5.2249720	1913.88586	$v = 8 \to 7$ Band P(12)	1606–1609,1612
5.2299820	1912.05246	$v = 12 \to 11$ Band R(17)	1606
5.2300198	1912.03864	$v = 5 \to 4$ Band P(30)	1606
5.2316034	1911.45986	$v = 9 \to 8$ Band P(6)	1606,1607
5.2341453	1910.53160	$v = 7 \to 6$ Band P(19)	1606,1610,6950
5.2359380	1909.87748	$v = 8 \to 7$ Band P(13)	1606–09,6950
5.2364797	1909.67989	$v = 6 \to 5$ Band P(25)	1606,1610
5.2376750	1909.24410	$v = 12 \to 11$ Band R(16)	1606
5.2419473	1907.68802	$v = 9 \to 8$ Band P(7)	1606,1607,1609
5.2428711	1907.35187	$v = 5 \to 4$ Band P(31)	1606
5.2454936	1906.39827	$v = 12 \to 11$ Band R(15)	1606
5.2458750	1906.25969	$v = 7 \to 6$ Band P(20)	1606,1610
5.2470411	1905.83602	$v = 8 \to 7$ Band P(14)	1606–110,6950
5.2488404	1905.18271	$v = 6 \to 5$ Band P(26)	1606,1610
5.2524257	1903.88223	$v = 9 \to 8$ Band P(8)	1606–1609
5.2558698	1902.63463	$v = 5 \to 4$ Band P(32)	1606
5.2577460	1901.95571	$v = 7 \to 6$ Band P(21)	1606,1610
5.2582825	1901.76164	$v = 8 \to 7$ Band P(15)	1606–110,6950
5.2613459	1900.65435	$v = 6 \to 5$ Band P(27)	1606,1610
5.2630398	1900.04263	$v = 9 \to 8$ Band P(9)	1606–1609
5.2690174	1897.88707	$v = 5 \to 4$ Band P(33)	1606
5.2696632	1897.65447	$v = 8 \to 7$ Band P(16)	1606–110,6950
5.2697594	1897.61982	$v = 7 \to 6$ Band P(22)	1606,1610
5.2737904	1896.16939	$v = 9 \to 8$ Band P(10)	1606–1609
5.2739975	1896.09494	$v = 6 \to 5$ Band P(28)	1606,1610
5.2811843	1893.51467	$v = 8 \to 7$ Band P(17)	1606–7,1610,6950
5.2819166	1893.25217	$v = 7 \to 6$ Band P(23)	1606,1610
5.2822426	1893.1353	$v = 10 \to 9$ Band P(4)	1607
5.2846787	1892.26263	$v = 9 \to 8$ Band P(11)	1606–1609,1612

$^{12}C^{16}O$ Laser—*continued*

Wavelength (μm) vac	Frequency (cm^{-1})	Transition assignment	Comments
5.2867965	1891.50463	$v = 6 \rightarrow 5$ Band P(29)	1606
5.2924984	1889.4668	$v = 10 \rightarrow 9$ Band P(5)	1607
5.2928469	1889.34237	$v = 8 \rightarrow 7$ Band P(18)	1606,1610
5.2942185	1888.85289	$v = 7 \rightarrow 6$ Band P(24)	1606,1610
5.2957055	1888.32251	$v = 9 \rightarrow 8$ Band P(12)	1606–1609,1612
5.2997441	1886.88356	$v = 6 \rightarrow 5$ Band P(30)	1606
5.3028905	1885.7640	$v = 10 \rightarrow 9$ Band P(6)	1606,1607
5.3046522	1885.13773	$v = 8 \rightarrow 7$ Band P(19)	1606,1610
5.3066666	1884.42214	$v = 7 \rightarrow 6$ Band P(25)	1606,1610
5.3068721	1884.34918	$v = 9 \rightarrow 8$ Band P(13)	1606–1609
5.3128417	1882.23188	$v = 6 \rightarrow 5$ Band P(31)	1606,1615
5.3134195	1882.0272	$v = 10 \rightarrow 9$ Band P(7)	1606,1607
5.3166012	1880.90089	$v = 8 \rightarrow 7$ Band P(20)	1606,1610
5.3181793	1880.34277	$v = 9 \rightarrow 8$ Band P(14)	1606–1609,1612
5.3192620	1879.96005	$v = 7 \rightarrow 6$ Band P(26)	1606,1610
5.3240867	1878.2564	$v = 10 \rightarrow 9$ Band P(8)	1606–1609
5.3260906	1877.54973	$v = 6 \rightarrow 5$ Band P(32)	1606
5.3286952	1876.63200	$v = 8 \rightarrow 7$ Band P(21)	1606,1610
5.3296283	1876.30344	$v = 9 \rightarrow 8$ Band P(15)	1606–1608,1610
5.3320059	1875.46677	$v = 7 \rightarrow 6$ Band P(27)	1606,1610
5.3348931	1874.4518	$v = 10 \rightarrow 9$ Band P(9)	1606–1609
5.3394923	1872.83725	$v = 6 \rightarrow 5$ Band P(33)	1606
5.3409354	1872.33119	$v = 8 \rightarrow 7$ Band P(22)	1606,1610
5.3412203	1872.23134	$v = 9 \rightarrow 8$ Band P(16)	1606–1610,6950
5.3432550	1871.51838	$v = 13 \rightarrow 12$ Band R(12)	1606,1615
5.3448998	1870.94246	$v = 7 \rightarrow 6$ Band P(28)	1606,1610
5.34583895	1870.61378	$v = 10 \rightarrow 9$ Band P(10)	1606–1609
5.3517174	1868.55904	$v = 13 \rightarrow 12$ Band R(11)	1606
5.3529563	1868.12660	$v = 9 \rightarrow 8$ Band P(17)	1606,1607,1610
5.3533230	1867.99863	$v = 8 \rightarrow 7$ Band P(23)	1606,1610
5.3546429	1867.53817	$v = 14 \rightarrow 13$ Band R(20)	1606,1615
5.3549016	1867.44796	$v = 11 \rightarrow 10$ Band P(4)	1607
5.3569272	1866.7418	$v = 10 \rightarrow 9$ Band P(11)	1606–1609,1615
5.3579449	1866.38724	$v = 7 \rightarrow 6$ Band P(29)	1606
5.3603122	1865.56297	$v = 13 \rightarrow 12$ Band R(10)	1606
5.3621811	1864.91278	$v = 14 \rightarrow 13$ Band R(19)	1606
5.3648374	1863.98938	$v = 9 \rightarrow 8$ Band P(18)	1606–7,1610,6950
5.3653410	1863.81445	$v = 11 \rightarrow 10$ Band P(5)	1607
5.3658592	1863.63444	$v = 8 \rightarrow 7$ Band P(24)	1606,1610
5.36815606	1962.83705	$v = 10 \rightarrow 9$ Band P(12)	1606–09,1612

$^{12}C^{16}O$ Laser—*continued*

Wavelength (μm) vac	Frequency (cm^{-1})	Transition assignment	Comments
5.3690401	1862.53033	$v = 13 \to 12$ Band R(9)	1606
5.3698498	1862.24948	$v = 14 \to 13$ Band R(18)	1606
5.3711425	1861.80127	$v = 7 \to 6$ Band P(30)	1606
5.3744699	1860.64862	$v = 15 \to 14$ Band R(28)	1606,1615
5.3759201	1860.14671	$v = 11 \to 10$ Band P(6)	1607
5.3768650	1859.81981	$v = 9 \to 8$ Band P(19)	1606,1610
5.3776497	1859.54843	$v = 14 \to 13$ Band R(17)	1606
5.3779018	1859.46126	$v = 13 \to 12$ Band R(8)	1606
5.3785453	1859.23878	$v = 8 \to 7$ Band P(25)	1606,1610
5.3795302	1858.8984	$v = 10 \to 9$ Band P(13)	1606–9,1612,6950
5.3810694	1858.36667	$v = 15 \to 14$ Band R(27)	1606
5.3844941	1857.18469	$v = 7 \to 6$ Band P(31)	1606
5.3855813	1856.80976	$v = 14 \to 13$ Band R(16)	1606
5.3866399	1856.44488	$v = 11 \to 10$ Band P(7)	1606,1607
5.3868980	1856.35591	$v = 13 \to 12$ Band R(7)	1606
5.3877986	1856.04562	$v = 15 \to 14$ Band R(26)	1606
5.3890400	1855.61805	$v = 9 \to 8$ Band P(20)	1606,1610
5.39104599	1854.92760	$v = 10 \to 9$ Band P(14)	1606–8,1610,6950
5.3913826	1854.81179	$v = 8 \to 7$ Band P(26)	1606,1610
5.3936454	1854.03364	$v = 14 \to 13$ Band R(15)	1606
5.3946580	1853.68562	$v = 15 \to 14$ Band R(25)	1606
5.3960296	1853.21444	$v = 13 \to 12$ Band R(6)	1606
5.3975014	1852.70912	$v = 11 \to 10$ Band P(8)	1606–1608
5.3980010	1852.53764	$v = 7 \to 6$ Band P(32)	1606
5.4013639	1851.38425	$v = 9 \to 8$ Band P(21)	1606,1610
5.4016481	1851.28682	$v = 15 \to 14$ Band R(24)	1606
5.4018425	1851.22021	$v = 14 \to 13$ Band R(14)	1606
5.4027105	1850.9228	$v = 10 \to 9$ Band P(15)	1606–8,1610,6950
5.4043724	1850.35362	$v = 8 \to 7$ Band P(27)	1606,1610
5.4052974	1850.03698	$v = 13 \to 12$ Band R(5)	1606
5.4085056	1848.93957	$v = 11 \to 10$ Band P(9)	1606–1608
5.4087695	1848.84937	$v = 15 \to 14$ Band R(23)	1606
5.4101733	1848.36961	$v = 14 \to 13$ Band R(13)	1606
5.4138377	1847.11853	$v = 9 \to 8$ Band P(22)	1606,1610
5.4145202	1846.8957	$v = 10 \to 9$ Band P(16)	1606–8,1610,6950
5.4147021	1846.82368	$v = 13 \to 12$ Band R(4)	1606,1615
5.4160225	1846.37342	$v = 15 \to 14$ Band R(22)	1606
5.4175160	1845.86440	$v = 8 \to 7$ Band P(28)	1606,1610
5.4186386	1845.48200	$v = 14 \to 13$ Band R(12)	1606
5.4196536	1845.13638	$v = 11 \to 10$ Band P(10)	1606–1608

$^{12}C^{16}O$ Laser—*continued*

Wavelength (µm) vac	Frequency (cm⁻¹)	Transition assignment	Comments
5.4234079	1843.85911	$v = 15 \rightarrow 14$ Band R(21)	1606
5.4242445	1843.57470	$v = 13 \rightarrow 12$ Band R(3)	1606
5.4264628	1842.82106	$v = 9 \rightarrow 8$ Band P(23)	1606,1610
5.4264777	1942.8160	$v = 10 \rightarrow 9$ Band P(17)	1606–7,1610,6950
5.4272390	1842.55753	$v = 14 \rightarrow 13$ Band R(11)	1606
5.4308149	1841.34429	$v = 8 \rightarrow 7$ Band P(29)	1606
5.4309260	1841.30661	$v = 15 \rightarrow 14$ Band R(20)	1606
5.4309464	1841.29970	$v = 11 \rightarrow 10$ Band P(11)	1606–1608,1612
5.4339257	1840.29018	$v = 13 \rightarrow 12$ Band R(2)	1606
5.4359752	1839.59633	$v = 14 \rightarrow 13$ Band R(10)	1606
5.4385776	1838.71605	$v = 15 \rightarrow 14$ Band R(19)	1606
5.4385843	1838.7138	$v = 10 \rightarrow 9$ Band P(18)	1606,1607,1610
5.4392405	1838.49197	$v = 9 \rightarrow 8$ Band P(24)	1606,1610
5.4399784	1838.24260	$v = 12 \rightarrow 11$ Band P(5)	1607,1615
5.4423852	1837.42966	$v = 11 \rightarrow 10$ Band P(12)	1606–1608
5.4442703	1836.79343	$v = 8 \rightarrow 7$ Band P(30)	1606
5.4448480	1836.59857	$v = 14 \rightarrow 13$ Band R(9)	1606
5.4463633	1836.08759	$v = 15 \rightarrow 14$ Band R(18)	1606
5.4507502	1834.60986	$v = 12 \rightarrow 11$ Band P(6)	1607
5.4508410	1834.5793	$v = 10 \rightarrow 9$ Band P(19)	1606,1610
5.4521720	1834.13142	$v = 9 \rightarrow 8$ Band P(25)	1606,1610
5.4539710	1833.52642	$v = 11 \rightarrow 10$ Band P(13)	1606–8,1610,6950
5.4542835	1833.42137	$v = 15 \rightarrow 14$ Band R(17)	1606
5.4578838	1832.21196	$v = 8 \rightarrow 7$ Band P(31)	1606
5.4616664	1830.94304	$v = 12 \rightarrow 11$ Band P(7)	1607
5.4623391	1830.71755	$v = 15 \rightarrow 14$ Band R(16)	1606
5.4632491	1830.4126	$v = 10 \rightarrow 9$ Band P(20)	1606,1610
5.4652587	1829.73953	$v = 9 \rightarrow 8$ Band P(26)	1606,1610
5.4657051	1829.59012	$v = 11 \rightarrow 10$ Band P(14)	1606–8,1610,6950
5.4705305	1827.97627	$v = 15 \rightarrow 14$ Band R(15)	1606
5.4716567	1827.60003	$v = 8 \rightarrow 7$ Band P(32)	1606
5.4727280	1827.24229	$v = 12 \rightarrow 11$ Band P(8)	1606–1608
5.4758098	1926.2139	$v = 10 \rightarrow 9$ Band P(21)	1606,1610
5.4775884	1825.62091	$v = 11 \rightarrow 10$ Band P(15)	1606–1608,1610
5.4785020	1825.31647	$v = 9 \rightarrow 8$ Band P(27)	1606,1610
5.4839361	1823.50776	$v = 12 \rightarrow 11$ Band P(9)	1606–1608
5.4855906	1822.95778	$v = 8 \rightarrow 7$ Band P(33)	1606
5.4885248	1821.9832	$v = 10 \rightarrow 9$ Band P(22)	1606,1610
5.4896223	1821.61893	$v = 11 \rightarrow 10$ Band P(16)	1606,1607,1610
5.4919033	1820.86237	$v = 9 \rightarrow 8$ Band P(28)	1606,1610

$^{12}C^{16}O$ Laser—*continued*

Wavelength (μm) vac	Frequency (cm^{-1})	Transition assignment	Comments
5.4952918	1819.73959	$v = 12 \rightarrow 11$ Band P(10)	1606–1608,1612
5.5013949	1817.7208	$v = 10 \rightarrow 9$ Band P(23)	1606,1610
5.5018080	1817.58434	$v = 11 \rightarrow 10$ Band P(17)	1606–7,1610,6950
5.5054638	1816.37738	$v = 9 \rightarrow 8$ Band P(29)	1606
5.5067961	1815.93794	$v = 12 \rightarrow 11$ Band P(11)	1606–1608,1612
5.5141466	1813.51727	$v = 11 \rightarrow 10$ Band P(18)	1606,1607,1610
5.5144216	1813.4268	$v = 10 \rightarrow 9$ Band P(24)	1606,1610
5.5164737	1812.75225	$v = 13 \rightarrow 12$ Band P(5)	1607
5.5184503	1812.10294	$v = 12 \rightarrow 11$ Band P(12)	1606–1608,6950
5.5191852	1811.86165	$v = 9 \rightarrow 8$ Band P(30)	1606
5.5266394	1809.41787	$v = 11 \rightarrow 10$ Band P(19)	1606,1610
5.5274439	1809.15449	$v = 13 \rightarrow 12$ Band P(6)	1607
5.5276064	1809.1013	$v = 10 \rightarrow 9$ Band P(25)	1606,1610
5.5302554	1808.23474	$v = 12 \rightarrow 11$ Band P(13)	1606–1608,6950
5.5330688	1807.31530	$v = 9 \rightarrow 8$ Band P(31)	1606
5.5385624	1805.52267	$v = 13 \rightarrow 12$ Band P(7)	1607,1608
5.5392876	1805.28628	$v = 11 \rightarrow 10$ Band P(20)	1606,1610
5.5409505	1804.7445	$v = 10 \rightarrow 9$ Band P(26)	1606,1610
5.5422127	1804.33350	$v = 12 \rightarrow 11$ Band P(14)	1606–1608,6950
5.5471162	1802.73850	$v = 9 \rightarrow 8$ Band P(32)	1606
5.5498302	1801.85693	$v = 13 \rightarrow 12$ Band P(8)	1607,1608
5.5520927	1801.12266	$v = 11 \rightarrow 10$ Band P(21)	1606,1610
5.5543233	1800.39935	$v = 12 \rightarrow 11$ Band P(15)	1606–1608,6950
5.5544555	1800.3565	$v = 10 \rightarrow 9$ Band P(27)	1610
5.5612484	1798.15742	$v = 13 \rightarrow 12$ Band P(9)	1606–1608
5.5613289	1798.13138	$v = 9 \rightarrow 8$ Band P(33)	1606
5.5650559	1796.92715	$v = 11 \rightarrow 10$ Band P(22)	1606,1610
5.5665884	1796.43244	$v = 12 \rightarrow 11$ Band P(16)	1606,1607
5.5728181	1794.42427	$v = 13 \rightarrow 12$ Band P(10)	1606–1608
5.5781785	1792.69988	$v = 11 \rightarrow 10$ Band P(23)	1606,1610
5.5790094	1792.43291	$v = 12 \rightarrow 11$ Band P(17)	1606,1610
5.5845404	1790.65764	$v = 13 \rightarrow 12$ Band P(11)	1606–1608
5.5914620	1788.44101	$v = 11 \rightarrow 10$ Band P(24)	1606,1610,1615
5.5915874	1788.40092	$v = 12 \rightarrow 11$ Band P(18)	1606–7,1610,6950
5.5948925	1787.34444	$v = 14 \rightarrow 13$ Band P(5)	1607
5.5948932	1787.3442	$v = 14 \rightarrow 13$ Band P(5)	1607
5.5964166	1786.85768	$v = 13 \rightarrow 12$ Band P(12)	1606–8,1612,6950
5.6043237	1784.33661	$v = 12 \rightarrow 11$ Band P(19)	1606,1610
5.6049077	1784.15069	$v = 11 \rightarrow 10$ Band P(25)	1606,1610
5.6060672	1783.78168	$v = 14 \rightarrow 13$ Band P(6)	1607

$^{12}C^{16}O$ Laser—*continued*

Wavelength (μm) vac	Frequency (cm⁻¹)	Transition assignment	Comments
5.6060681	1783.7814	v = 14 → 13 Band P(6)	1607
5.6084478	1783.02453	v = 13 → 12 Band P(13)	1606–8,1612,6950
5.6172198	1780.24012	v = 12 → 11 Band P(20)	1606,1610
5.6173942	1780.18485	v = 14 → 13 Band P(7)	1607,1608
5.6173950	1780.1846	v = 14 → 13 Band P(7)	1607,1608
5.6185171	1779.82904	v = 11 → 10 Band P(26)	1606
5.6206352	1779.15833	v = 13 → 12 Band P(14)	1606–1608,6950
5.6288744	1776.55411	v = 14 → 13 Band P(8)	1607,1608
5.6288754	1776.5518	v = 14 → 13 Band P(8)	1607,1608
5.6302769	1776.11159	v = 12 → 11 Band P(21)	1606,1610
5.6322917	1775.47623	v = 11 → 10 Band P(27)	1606
5.6329801	1775.25924	v = 13 → 12 Band P(15)	1606–8,1610,6950
5.6434963	1771.95118	v = 12 → 11 Band P(22)	1606,1610
5.6454837	1771.32739	v = 13 → 12 Band P(16)	1606,1610
5.6522995	1769.19147	v = 14 → 13 Band P(10)	1606–1608
5.6568796	1767.75903	v = 12 → 11 Band P(23)	1606,1610
5.6581474	1767.36294	v = 13 → 12 Band P(17)	1606,1607,1610
5.6642466	1765.45987	v = 14 → 13 Band P(11)	1606–1608,1612
5.6704281	1763.53528	v = 12 → 11 Band P(24)	1606,1610
5.6709724	1763.36603	v = 13 → 12 Band P(18)	1606,1607,1610
5.6763517	1761.69494	v = 14 → 13 Band P(12)	1606,1615
5.6839600	1759.33680	v = 13 → 12 Band P(19)	1606,1610
5.6841433	1759.28007	v = 12 → 11 Band P(25)	1606,1610
5.6866897	1758.49230	v = 15 → 14 Band P(6)	1607–8,1615
5.6886160	1757.89682	v = 14 → 13 Band P(13)	1606–1608
5.6971117	1755.27540	v = 13 → 12 Band P(20)	1606,1610
5.6980266	1754.99356	v = 12 → 11 Band P(26)	1606
5.6982315	1754.93046	v = 15 → 14 Band P(7)	1607,1608
5.7010408	1754.06567	v = 14 → 13 Band P(14)	1606–1608
5.7104288	1751.18197	v = 13 → 12 Band P(21)	1606,1610
5.7136274	1750.20163	v = 14 → 13 Band P(15)	1606–8,1610
5.7217888	1747.70521	v = 15 → 14 Band P(9)	1607,1608
5.7239128	1747.05667	v = 13 → 12 Band P(22)	1606,1610
5.7263770	1746.30485	v = 14 → 13 Band P(16)	1606–7,1610
5.7338066	1744.04209	v = 15 → 14 Band P(10)	1607–8,1612
5.7375651	1742.89962	v = 13 → 12 Band P(23)	1606,1610
5.7392911	1742.37546	v = 14 → 13 Band P(17)	1606–7,1610
5.7459855	1740.34550	v = 15 → 14 Band P(11)	1607,1608
5.7513871	1738.71099	v = 13 → 12 Band P(24)	1606,1610
5.7523709	1738.41362	v = 14 → 13 Band P(18)	1606–7,1610

$^{12}C^{16}O$ Laser—*continued*

Wavelength (µm) vac	Frequency (cm^{-1})	Transition assignment	Comments
5.7577821	1736.77986	v = 16 → 15 Band P(5)	1608,1615
5.7583267	1736.61559	v = 15 → 14 Band P(12)	1606–1608,6950
5.7653805	1734.49091	v = 13 → 12 Band P(25)	1606
5.7656179	1734.41948	v = 14 → 13 Band P(19)	1610
5.7708316	1732.85251	v = 15 → 14 Band P(13)	1606–1608,6950
5.7790335	1730.39317	v = 14 → 13 Band P(20)	1610
5.7795466	1730.23952	v = 13 → 12 Band P(26)	1606
5.7811482	1729.76019	v = 16 → 15 Band P(7)	1607
5.7835013	1729.05640	v = 15 → 14 Band P(14)	1606–8,1610,6950
5.7926190	1726.33484	v = 14 → 13 Band P(21)	1610
5.7930734	1726.19943	v = 16 → 15 Band P(8)	1607,1608
5.7938872	1725.95698	v = 13 → 12 Band P(27)	1606
5.7963373	1725.22741	v = 15 → 14 Band P(15)	1606–7,1610,6950
5.8051616	1722.60492	v = 16 → 15 Band P(9)	1606–1608
5.8063761	1722.24463	v = 14 → 13 Band P(22)	1610
5.8084037	1721.64342	v = 13 → 12 Band P(28)	1606
5.8093409	1721.36568	v = 15 → 14 Band P(16)	1606–7,1610,6950
5.8174142	1718.97680	v = 16 → 15 Band P(10)	1607,1608
5.8203061	1718.12270	v = 14 → 13 Band P(23)	1610
5.8225134	1717.47135	v = 15 → 14 Band P(17)	1610
5.8298322	1715.31522	v = 16 → 15 Band P(11)	1607,1608
5.8344106	1713.96918	v = 14 → 13 Band P(24)	1610
5.8358563	1713.54458	v = 15 → 14 Band P(18)	1610,6950
5.8424172	1711.62033	v = 16 → 15 Band P(12)	1607–08,1612
5.8486912	1709.78423	v = 14 → 13 Band P(25)	1610
5.8493710	1709.58551	v = 15 → 14 Band P(19)	1610
5.8542306	1708.16640	v = 17 → 16 Band P(6)	1608,1615
5.8551702	1707.89227	v = 16 → 15 Band P(13)	1607,1608,6950
5.8630590	1705.59429	v = 15 → 14 Band P(20)	1610
5.8662225	1704.67451	v = 17 → 16 Band P(7)	1608
5.8680928	1704.13119	v = 16 → 15 Band P(14)	1607,6950
5.8769218	1701.57105	v = 15 → 14 Band P(21)	1610
5.8783808	1701.14873	v = 17 → 16 Band P(8)	1607
5.8907066	1697.58921	v = 17 → 16 Band P(9)	1607,1608
5.8909609	1697.51594	v = 15 → 14 Band P(22)	1610
5.8944520	1696.51056	v = 16 → 15 Band P(16)	1610
5.9032014	1693.99608	v = 17 → 16 Band P(10)	1607–8,1612
5.9051778	1693.42911	v = 15 → 14 Band P(23)	1610
5.9078914	1692.65129	v = 16 → 15 Band P(17)	1610
5.9158663	1690.36950	v = 17 → 16 Band P(11)	1607,1608

$^{12}C^{16}O$ Laser—*continued*

Wavelength (μm) vac	Frequency (cm⁻¹)	Transition assignment	Comments
5.9195741	1689.31071	v = 15 → 14 Band P(24)	1610
5.9215060	1688.75958	v = 16 → 15 Band P(18)	1607,1610
5.9287028	1686.70962	v = 17 → 16 Band P(12)	1607,1608
5.9352973	1684.83557	v = 16 → 15 Band P(19)	1610
5.9492667	1680.87942	v = 16 → 15 Band P(20)	1610
5.9535374	1679.67367	v = 18 → 17 Band P(7)	1607–8,1615
5.9548958	1679.29051	v = 17 → 16 Band P(14)	1607,1608,6950
5.9634159	1676.89127	v = 16 → 15 Band P(21)	1610
5.9659362	1676.18285	v = 18 → 17 Band P(8)	1607,1608
5.9682551	1675.53159	v = 17 → 16 Band P(15)	1607
5.9777463	1672.87125	v = 16 → 15 Band P(22)	1610
5.9785074	1672.65830	v = 18 → 17 Band P(9)	1607,1608
5.9817917	1671.73994	v = 17 → 16 Band P(16)	1607,1610,6950
5.9955068	1667.91571	v = 17 → 16 Band P(17)	1610,6950
6.0041721	1665.50856	v = 18 → 17 Band P(11)	1607,1608
6.0094021	1664.05905	v = 17 → 16 Band P(18)	1610
6.0172683	1661.88367	v = 18 → 17 Band P(12)	1607,1608,6950
6.0234792	1660.17010	v = 17 → 16 Band P(19)	1610
6.0305424	1658.22563	v = 18 → 17 Band P(13)	1607,1608,6950
6.0377394	1656.24901	v = 17 → 16 Band P(20)	1610
6.0431812	1654.75761	v = 19 → 18 Band P(7)	1607,1615
6.0558284	1651.30175	v = 19 → 18 Band P(8)	1607,1608
6.0576299	1650.81066	v = 18 → 17 Band P(15)	1607,1610,6950
6.0686529	1647.81216	v = 19 → 18 Band P(9)	1607,1608
6.0714462	1647.05403	v = 18 → 17 Band P(16)	1607,1610,6950
6.0816560	1644.28898	v = 19 → 18 Band P(10)	1607,1608
6.0854464	1643.26483	v = 18 → i7 Band P(17)	1610,1615
6.0948392	1640.73236	v = 19 → 18 Band P(11)	1607,1608
6.0996319	1639.44321	v = 18 → 17 Band P(18)	1610
6.1082040	1637.14245	v = 19 → 18 Band P(12)	1607,1608,6950
6.1099818	1636.66608	v = 20 → 19 Band P(5)	1607,1615
6.1140043	1635.58930	v = 18 → 17 Band P(19)	1610
6.1217516	1633.51939	v = 19 → 18 Band P(13)	1607,1608,6950
6.1225250	1633.31306	v = 20 → 19 Band P(6)	1607,1608
6.1285653	1631.70325	v = 18 → 17 Band P(20)	1610,6950
6.1352478	1629.92602	v = 20 → 19 Band P(7)	1607,1608
6.1354838	1629.86333	v = 19 → 18 Band P(14)	1607,6950
6.1481516	1626.50510	v = 20 → 19 Band P(8)	1607,1608,6950
6.1494019	1626.17441	v = 19 → 18 Band P(15)	1607,6950
6.1612379	1623.05046	v = 20 → 19 Band P(9)	1607,1608

$^{12}C^{16}O$ Laser—*continued*

Wavelength (µm) vac	Frequency (cm^{-1})	Transition assignment	Comments
6.1745080	1619.56223	$v = 20 \rightarrow 19$ Band P(10)	1607,1608,6950
6.1879635	1616.04057	$v = 20 \rightarrow 19$ Band P(11)	1607,1608
6.1922880	1614.91198	$v = 19 \rightarrow 18$ Band P(18)	6950
6.2016057	1612.48562	$v = 20 \rightarrow 19$ Band P(12)	1607,1608,6950
6.2040571	1611.84847	$v = 21 \rightarrow 20$ Band P(5)	1607,1615
6.2069660	1611.09309	$v = 19 \rightarrow 18$ Band P(19)	6950
6.2154363	1608.89752	$v = 20 \rightarrow 19$ Band P(13)	1607,1608,6950
6.2168550	1608.53037	$v = 21 \rightarrow 20$ Band P(6)	1607,1608
6.2218381	1607.24207	$v = 19 \rightarrow 18$ Band P(20)	6950
6.2294567	1605.27643	$v = 20 \rightarrow 19$ Band P(14)	1607
6.2298377	1605.17825	$v = 21 \rightarrow 20$ Band P(7)	1607,1608
6.2430068	1601.79226	$v = 21 \rightarrow 20$ Band P(8)	1607,1608,6950
6.2436686	1601.62249	$v = 20 \rightarrow 19$ Band P(15)	1607,6950
6.2563637	1598.37255	$v = 21 \rightarrow 20$ Band P(9)	1607,1608,6950
6.2580735	1597.93584	$v = 20 \rightarrow 19$ Band P(16)	6950
6.2699099	1594.91926	$v = 21 \rightarrow 20$ Band P(10)	1607,1608,6950
6.2726732	1594.21664	$v = 20 \rightarrow 19$ Band P(17)	6950
6.2836468	1591.43253	$v = 21 \rightarrow 20$ Band P(11)	1607,1608
6.2874693	1590.46501	$v = 20 \rightarrow 19$ Band P(18)	6950
6.2975761	1587.91252	$v = 21 \rightarrow 20$ Band P(12)	1607,1608,6950
6.3007457	1587.11373	$v = 22 \rightarrow 21$ Band P(5)	1607-8,1615
6.3024636	1586.68112	$v = 20 \rightarrow 19$ Band P(19)	6950
6.3116993	1584.35937	$v = 21 \rightarrow 20$ Band P(13)	1607,6950
6.3138068	1583.83054	$v = 22 \rightarrow 21$ Band P(6)	1607,1608
6.3176578	1582.86510	$v = 20 \rightarrow 19$ Band P(20)	6950
6.3260181	1580.77323	$v = 21 \rightarrow 20$ Band P(14)	1607
6.3270583	1580.51333	$v = 22 \rightarrow 21$ Band P(7)	1607,1608
6.3405017	1577.16226	$v = 22 \rightarrow 21$ Band P(8)	1607-8,1615
6.3405339	1577.15424	$v = 21 \rightarrow 20$ Band P(15)	1607
6.3541385	1573.77746	$v = 22 \rightarrow 21$ Band P(9)	1607,1608,6950
6.3552487	1573.50255	$v = 21 \rightarrow 20$ Band P(16)	1607
6.3679703	1570.35908	$v = 22 \rightarrow 21$ Band P(10)	1607,1608
6.3701640	1569.81830	$v = 21 \rightarrow 20$ Band P(17)	1607
6.3819986	1566.90727	$v = 22 \rightarrow 21$ Band P(11)	1607,1608
6.3962250	1563.42218	$v = 22 \rightarrow 21$ Band P(12)	1607
6.4001617	1562.46053	$v = 23 \rightarrow 22$ Band P(5)	1607-8,1615
6.4134951	1559.21223	$v = 23 \rightarrow 22$ Band P(6)	1607
6.4252787	1556.35273	$v = 22 \rightarrow 21$ Band P(14)	1607
6.4270247	1555.92993	$v = 23 \rightarrow 22$ Band P(7)	1607,1608
6.4401094	1552.76866	$v = 22 \rightarrow 21$ Band P(15)	1607

$^{12}C^{16}O$ Laser—*continued*

Wavelength (μm) vac	Frequency (cm^{-1})	Transition assignment	Comments
6.4407519	1552.61374	v = 23 → 22 Band P(8)	1607,1608,6950
6.4546785	1549.26384	v = 23 → 22 Band P(9)	1607,1608,6950
6.4551449	1549.15189	v = 22 → 21 Band P(16)	1607
6.4688059	1545.88036	v = 23 → 22 Band P(10)	1607,1608,6950
6.4703872	1545.50257	v = 22 → 21 Band P(17)	1607
6.4831358	1542.46344	v = 23 → 22 Band P(11)	1607
6.5014990	1538.10682	v = 22 → 21 Band P(19)	6950
6.5024278	1537.88714	v = 24 → 23 Band P(5)	1607–8,1615
6.5124098	1535.52990	v = 23 → 22 Band P(13)	1607,6950
6.5160430	1534.67373	v = 24 → 23 Band P(6)	1607,1608
6.5273573	1532.01357	v = 23 → 22 Band P(14)	1607,6950
6.5298604	1531.42630	v = 24 → 23 Band P(7)	1607,1608
6.5425142	1528.46439	v = 23 → 22 Band P(15)	1607
6.5438816	1528.14500	v = 24 → 23 Band P(8)	1607,1608,6950
6.5578823	1524.88251	v = 23 → 22 Band P(16)	1607
6.5581082	1524.82997	v = 24 → 23 Band P(9)	1607,1608,6950
6.5871843	1518.09931	v = 24 → 23 Band P(11)	1607–8,1615
6.6020372	1514.68397	v = 24 → 23 Band P(12)	1607,1608,6950
6.6052724	1513.94211	v = 23 → 22 Band P(19)	6950
6.6076758	1513.39143	v = 25 → 24 Band P(5)	1607–8,1615
6.6171024	1511.23550	v = 24 → 23 Band P(13)	1607,1608
6.6215830	1510.21290	v = 25 → 24 Band P(6)	1607,1608
6.6356986	1507.00034	v = 25 → 24 Band P(7)	1607
6.6478767	1504.23970	v = 24 → 23 Band P(15)	1607,6950
6.6500243	1503.75389	v = 25 → 24 Band P(8)	1607,1608
6.6645619	1500.47371	v = 25 → 24 Band P(9)	1607,1608,6950
6.6793131	1497.15995	v = 25 → 24 Band P(10)	1607,1608,6950
6.6942795	1493.81274	v = 25 → 24 Band P(11)	1607,1608,6950
6.7020521	1492.08031	v = 26 → 25 Band P(4)	1607–8,1615
6.7094630	1490.43223	v = 25 → 24 Band P(12)	1607,1608,6950
6.7248655	1487.01858	v = 25 → 24 Band P(13)	1607
6.7302579	1485.82716	v = 26 → 25 Band P(6)	1607,1608
6.7404888	1483.57192	v = 25 → 24 Band P(14)	1607,6950
6.7446826	1482.64945	v = 26 → 25 Band P(7)	1607,1608
6.7563349	1480.09241	v = 25 → 24 Band P(15)	1607,1615,6950
6.7593242	1479.43785	v = 26 → 25 Band P(8)	1607,1608
6.7724057	1476.58018	v = 25 → 24 Band P(16)	1607
6.7892647	1472.91355	v = 26 → 25 Band P(10)	1607,1608,6950
6.8045673	1469.60115	v = 26 → 25 Band P(11)	1607,1608
6.8052295	1469.45817	v = 25 → 24 Band P(18)	6950

$^{12}C^{16}O$ Laser—*continued*

Wavelength (μm) vac	Frequency (cm^{-1})	Transition assignment	Comments
6.8200940	1466.25544	$v = 26 \rightarrow 25$ Band P(12)	1607,1608,6950
6.8358467	1462.87658	$v = 26 \rightarrow 25$ Band P(13)	1607,1608,6950
6.8389768	1462.20704	$v = 25 \rightarrow 24$ Band P(20)	6950
6.8422221	1461.51350	$v = 27 \rightarrow 26$ Band P(6)	1607–8,1615
6.8518273	1459.46470	$v = 26 \rightarrow 25$ Band P(14)	1607,6950
6.8680378	1456.01995	$v = 26 \rightarrow 25$ Band P(15)	1607
6.8719367	1455.19384	$v = 27 \rightarrow 26$ Band P(8)	1607
6.8844803	1452.54247	$v = 26 \rightarrow 25$ Band P(16)	1607
6.8871316	1451.98330	$v = 27 \rightarrow 26$ Band P(9)	1607,1608
6.9025539	1448.73914	$v = 27 \rightarrow 26$ Band P(10)	1607,1608
6.9182057	1445.46151	$v = 27 \rightarrow 26$ Band P(11)	1607,1608,6950
6.9340887	1442.15056	$v = 27 \rightarrow 26$ Band P(12)	1607,1608
6.9427938	1440.34236	$v = 28 \rightarrow 27$ Band P(5)	1607,1608
6.9428408	1440.3326	$v = 28 \rightarrow 27$ Band P(5)	1607–8,1615
6.9502052	1438.80644	$v = 27 \rightarrow 26$ Band P(13)	1607
6.9665570	1435.42929	$v = 27 \rightarrow 26$ Band P(14)	1607
6.9727211	1434.16033	$v = 28 \rightarrow 27$ Band P(7)	1607,1608
6.9831463	1432.01925	$v = 27 \rightarrow 26$ Band P(15)	1607,6950
6.9880306	1431.01835	$v = 28 \rightarrow 27$ Band P(8)	1607,1608,6950
7.0035732	1427.84258	$v = 28 \rightarrow 27$ Band P(9)	1607–8,1615
7.0193507	1424.63318	$v = 28 \rightarrow 27$ Band P(10)	1607,1608
7.0353653	1421.39030	$v = 28 \rightarrow 27$ Band P(11)	1607,1608
7.0516189	1418.11407	$v = 28 \rightarrow 27$ Band P(12)	1607,6950
7.0681136	1414.80464	$v = 28 \rightarrow 27$ Band P(13)	1607
7.0767012	1413.08777	$v = 29 \rightarrow 28$ Band P(6)	1607,1608,1615
7.0848516	1411.46217	$v = 28 \rightarrow 27$ Band P(14)	1607
7.0921255	1410.01453	$v = 29 \rightarrow 28$ Band P(7)	1607,1608
7.1018350	1408.08679	$v = 28 \rightarrow 27$ Band P(15)	1607
7.1077887	1406.90733	$v = 29 \rightarrow 28$ Band P(8)	1607,1608
7.1190660	1404.67865	$v = 28 \rightarrow 27$ Band P(16)	1607,6950
7.1236928	1403.76632	$v = 29 \rightarrow 28$ Band P(9)	1607,1608,6950
7.1398398	1400.59165	$v = 29 \rightarrow 28$ Band P(10)	1607,1608,6950
7.1562318	1397.38346	$v = 29 \rightarrow 28$ Band P(11)	1607,1608
7.1728710	1394.14191	$v = 29 \rightarrow 28$ Band P(12)	1607,1615,6950
7.1897594	1390.86713	$v = 29 \rightarrow 28$ Band P(13)	1607,6950
7.1995946	1388.96709	$v = 30 \rightarrow 29$ Band P(6)	1607–8,1615
7.2068993	1387.55928	$v = 29 \rightarrow 28$ Band P(14)	1607
7.2153787	1385.92864	$v = 30 \rightarrow 29$ Band P(7)	1607,1608
7.2242930	1384.21850	$v = 29 \rightarrow 28$ Band P(15)	1607,6950
7.2314099	1382.85620	$v = 30 \rightarrow 29$ Band P(8)	1607,1608

$^{12}C^{16}O$ Laser—*continued*

Wavelength (μm) vac	Frequency (cm⁻¹)	Transition assignment	Comments
7.2419428	1380.84493	v = 29 → 28 Band P(16)	1607
7.2476902	1379.74992	v = 30 → 29 Band P(9)	1607,1608,6950
7.2642218	1376.60994	v = 30 → 29 Band P(10)	1607,1608,6950
7.2810069	1373.43642	v = 30 → 29 Band P(11)	1607
7.2980475	1370.22950	v = 30 → 29 Band P(12)	1607
7.3106328	1367.87065	v = 31 → 30 Band P(5)	1607,1615
7.3153461	1366.98932	v = 30 → 29 Band P(13)	1607,6950
7.3265378	1364.90116	v = 31 → 30 Band P(6)	1607,1608
7.3329049	1363.71603	v = 30 → 29 Band P(14)	1607
7.3426965	1361.89750	v = 31 → 30 Band P(7)	1607,1608,6950
7.3507264	1360.40978	v = 30 → 29 Band P(15)	1607,6950
7.3591109	1358.85980	v = 31 → 30 Band P(8)	1607,1608
7.3688128	1357.07070	v = 30 → 29 Band P(16)	1607
7.3757832	1355.78822	v = 31 → 30 Band P(9)	1607,1608
7.3927156	1352.68291	v = 31 → 30 Band P(10)	1607,1608
7.4099103	1349.54401	v = 31 → 30 Band P(11)	1607
7.4273696	1346.37166	v = 31 → 30 Band P(12)	1607,6950
7.4414791	1343.81887	v = 32 → 31 Band P(5)	1607,1615
7.4450960	1343.16602	v = 31 → 30 Band P(13)	1607
7.4577655	1340.88421	v = 32 → 31 Band P(6)	1607,1608
7.4630918	1339.92724	v = 31 → 30 Band P(14)	6950
7.4743146	1337.91532	v = 32 → 31 Band P(7)	1607,1608
7.4911285	1334.91235	v = 32 → 31 Band P(8)	1607–8,1615,6950
7.5082096	1331.87545	v = 32 → 31 Band P(9)	1607,1608,6950
7.5255600	1328.80476	v = 32 → 31 Band P(10)	1607,1608,6950
7.5431822	1325.70044	v = 32 → 31 Band P(11)	1607,1608
7.5935345	1316.90981	v = 33 → 32 Band P(6)	1607,1615
7.5977040	1316.18711	v = 32 → 31 Band P(14)	6950
7.6104909	1313.97569	v = 33 → 32 Band P(7)	1607,1608
7.6277219	1311.00743	v = 33 → 32 Band P(8)	1607,1608
7.6354566	1309.67937	v = 32 → 31 Band P(16)	6950
7.6452297	1308.00518	v = 33 → 32 Band P(9)	1607,1608
7.6630168	1304.96908	v = 33 → 32 Band P(10)	1607,1608
7.6810857	1301.89930	v = 33 → 32 Band P(11)	1608,6950
7.7180788	1295.65922	v = 33 → 32 Band P(13)	6950
7.7341267	1292.97080	v = 34 → 33 Band P(6)	1608,1615
7.7515087	1290.07145	v = 34 → 33 Band P(7)	1608
7.7691754	1287.13789	v = 34 → 33 Band P(8)	1608,1615,6950
7.7871294	1284.17027	v = 34 → 33 Band P(9)	1608,6950
7.8053731	1281.16873	v = 34 → 33 Band P(10)	1608,6950

$^{12}C^{16}O$ Laser—*continued*

Wavelength (μm) vac	Frequency (cm⁻¹)	Transition assignment	Comments
7.8239092	1278.13344	v = 34 → 33 Band P(11)	1608
7.8427404	1275.06452	v = 34 → 33 Band P(12)	6950
7.8798521	1269.05935	v = 35 → 34 Band P(6)	1608,1615
7.8976792	1266.19476	v = 35 → 34 Band P(7)	1608
7.9158019	1263.29589	v = 35 → 34 Band P(8)	1608,6950
7.9529448	1257.39587	v = 35 → 34 Band P(10)	1608,6950
8.0310527	1245.16677	v = 36 → 35 Band P(6)	1608,1615
8.0493460	1242.33696	v = 36 → 35 Band P(7)	1608
8.0679465	1239.47277	v = 36 → 35 Band P(8)	1608
8.0868571	1236.57434	v = 36 → 35 Band P(9)	1608
8.1060805	1233.64183	v = 36 → 35 Band P(10)	1608
8.2068887	1218.48856	v = 37 → 36 Band P(7)	1608,1615
8.2259907	1215.65905	v = 37 → 36 Band P(8)	1608
8.2454152	1212.79520	v = 37 → 36 Band P(9)	1608
8.2651654	1209.89715	v = 37 → 36 Band P(10)	1608

$^{13}C^{16}O$ Laser

Wavelength (μm) vac	Frequency (cm⁻¹)	Transition assignment	Comments
5.3032817	1885.6249	v = 7 → 6 Band P(16)	1616
5.3145120	1881.6403	v = 7 → 6 Band P(17)	1616
5.3258763	1877.6253	v = 7 → 6 Band P(18)	1616
5.3284263	1876.7267	v = 8 → 7 Band P(12)	1616
5.3373755	1873.5900	v = 7 → 6 Band P(19)	1616
5.3393194	1872.8979	v = 8 → 7 Band P(13)	1616
5.3615063	1865.1475	v = 8 → 7 Band P(15)	1616
5.3728021	1861.2262	v = 8 → 7 Band P(16)	1616
5.3770194	1859.7664	v = 9 → 8 Band P(10)	1616
5.3842343	1857.2743	v = 8 → 7 Band P(17)	1616
5.3878364	1856.0326	v = 9 → 8 Band P(11)	1616
5.3958038	1853.2920	v = 8 → 7 Band P(18)	1616
5.3987880	1852.2676	v = 9 → 8 Band P(12)	1616
5.4075114	1849.2795	v = 8 → 7 Band P(19)	1616
5.4098751	1848.4715	v = 9 → 8 Band P(13)	1616
5.4324602	1840.7866	v = 9 → 8 Band P(15)	1616
5.4377867	1838.9835	v = 10 → 9 Band P(9)	1616
5.4439604	183&9980	v = 9 → 8 Band P(16)	1616
5.4486595	1835.3138	v = 10 → 9 Band P(10)	1616

$^{13}C^{16}O_2$ Laser—*continued*

Wavelength (µm) vac	Frequency (cm^{-1})	Transition assignment	Comments
5.4556237	1832.9710	v = 9 → 8 Band P(17)	1616
5.4596692	1831.6128	v = 10 → 9 Band P(11)	1616
5.4673804	1829.0295	v = 9 → 8 Band P(18)	1616
5.4708172	1827.8805	v = 10 → 9 Band P(12)	1616
5.4821042	1924.1171	v = 10 → 9 Band P(13)	1616
5.4913685	1821.0397	v = 9 → 8 Band P(20)	1616
5.4935308	1820.3229	v = 10 → 9 Band P(14)	1616
5.5000140	1818.1772	v = 11 → 10 Band P(8)	1616
5.5035779	1816.9998	v = 9 → 8 Band P(21)	1616
5.5050989	1816.4978	v = 10 → 9 Band P(15)	1616
5.5109423	1814.5717	v = 11 → 10 Band P(9)	1616
5.5168092	1812.6420	v = 10 → 9 Band P(16)	1616
5.5220103	1810.9347	v = 11 → 10 Band P(10)	1616
5.5286626	1808.7557	v = 10 → 9 Band P(17)	1616
5.5332186	1807.2664	v = 11 → 10 Band P(11)	1616
5.5406607	1804.8389	v = 10 → 9 Band P(18)	1616
5.5445687	1803.5668	v = 11 → 10 Band P(12)	1616
5.5528041	1800-8919	v = 10 → 9 Band P(19)	1616
5.5560612	1799.8362	v = 11 → 10 Band P(13)	1616
5.5676972	1796.0747	v = 11 → 10 Band P(14)	1616
5.5747348	1793.8073	v = 12 → 11 Band P(8)	1616
5.5794779	1792.2824	v = 11 → 10 Band P(15)	1616
5.5858604	1790.2345	v = 12 → 11 Band P(9)	1616
5.5914045	1788.4594	v = 11 → 10 Band P(16)	1616
5.5971292	1786.6302	v = 12 → 11 Band P(10)	1616
5.6034784	1784.6058	v = 11 → 10 Band P(17)	1616
5.6085416	1782.9947	v = 12 → 11 Band P(11)	1616
5.6200996	1779.3279	v = 12 → 11 Band P(12)	1616
5.6280714	1776.8076	v = 11 → 10 Band P(19)	1616
5.6436555	1771.9012	v = 12 → 11 Band P(14)	1616
5.6512718	1769.5132	v = 13 → 12 Band P(8)	1616
5.6626004	1765.9731	v = 13 → 12 Band P(9)	1616
5.6660928	1764.8846	v = 11 → 10 Band P(22)	1616
5.6678052	1764.3514	v = 12 → 11 Band P(16)	1616
5.6740756	1762.4016	v = 13 → 12 Band P(10)	1616
5.6801058	1760.5306	v = 12 → 11 Band P(17)	1616
5.6856987	1758.7988	v = 13 → 12 Band P(11)	1616
5.6974710	1755.1647	v = 13 → 12 Band P(12)	1616
5.7051640	1752.7980	v = 12 → 11 Band P(19)	1616
5.7214666	1747.8036	v = 13 → 12 Band P(14)	1616

$^{13}C^{16}O_2$ Laser—*continued*

Wavelength (μm) vac	Frequency (cm^{-1})	Transition assignment	Comments
5.7296871	1745.2960	v = 14 → 13 Band P(8)	1616
5.7308403	1744.9448	v = 12 → 11 Band P(21)	1616
5.7336925	1744.0768	v = 13 → 12 Band P(15)	1616
5.7412249	1741.7886	v = 14 → 13 Band P(9)	1616
5.7460720	1740.3193	v = 13 → 12 Band P(16)	1616
5.7529127	1738.2499	v = 14 → 13 Band P(10)	1616
5.7586063	1736.5313	v = 13 → 12 Band P(17)	1616
5.7712966	1732.7129	v = 13 → 12 Band P(18)	1616
5.7767459	1731.0784	v = 14 → 13 Band P(12)	1616
5.7841443	1728.8642	v = 13 → 12 Band P(19)	1616
5.7888926	1727.4461	v = 14 → 13 Band P(13)	1616
5.7971505	1724.9854	v = 13 → 12 Band P(20)	1616
5.7984477	1724.5995	v = 15 → 14 Band P(7)	1616
5.8011949	17217829	v = 14 → 13 Band P(14)	1616
5.8100469	1721.1565	v = 15 → 14 Band P(8)	1616
5.8103166	1721.0766	v = 13 → 12 Band P(21)	1616
5.8136537	1720.0887	v = 14 → 13 Band P(15)	1616
5.8262700	1716.3640	v = 14 → 13 Band P(16)	1616
5.8337074	1714.1758	v = 15 → 14 Band P(10)	1616
5.8390454	1712.6087	v = 14 → 13 Band P(17)	1616
5.8457708	1710.6394	v = 15 → 14 Band P(11)	1616
5.8519808	1708.8231	v = 14 → 13 Band P(18)	1616
5.8579913	1707.0698	v = 15 → 14 Band P(12)	1616
5.8650779	1705.0072	v = 14 → 13 Band P(19)	1616
5.8703698	1703.4702	v = 15 → 14 Band P(13)	1616
5.8924207	1697.0954	v = 16 → 15 Band P(8)	1616
5.9043954	1693.6535	v = 16 → 15 Band P(9)	1616
5.9084670	1692.4864	v = 15 → 14 Band P(16)	1616
5.9214909	1688.7639	v = 15 → 14 Band P(17)	1616
5.9288227	1686.6755	v = 16 → 15 Band P(11)	1616
5.9480336	1681.2279	v = 15 → 14 Band P(19)	1616
5.9538949	1679.5728	v = 16 → 15 Band P(13)	1616
5.9615550	1677.4147	v = 15 → 14 Band P(20)	1616
5.9648404	1676.4908	v = 17 → 16 Band P(7)	1616
5.9666761	1675.9750	v = 16 → 15 Band P(14)	1616
5.9752452	1673.5715	v = 15 → 14 Band P(21)	1616
5.9768819	1673.1132	v = 17 → 16 Band P(8)	1616
5.9796224	1672.3464	v = 16 → 15 Band P(15)	1616
5.9927349	1668.6872	v = 16 → 15 Band P(16)	1616
6.0014521	1666.2634	v = 17 → 16 Band P(10)	1616

$^{13}C^{16}O_2$ Laser—*continued*

Wavelength (μm) vac	Frequency (cm⁻¹)	Transition assignment	Comments
6.0060150	1664.9975	v = 16 → 15 Band P(17)	1616
6.0194643	1661.2774	v = 16 → 15 Band P(18)	1616
6.0266802	1659.2883	v = 17 → 16 Band P(12)	1616
6.0330839	1657.5271	v = 16 → 15 Band P(19)	1616
6.0395438	1655.7542	v = 17 → 16 Band P(13)	1616
6.0635085	1649.2102	v = 18 → 17 Band P(8)	1616
6.0657774	1648.5933	v = 17 → 16 Band P(15)	1616
6.0791497	1644.9669	v = 17 → 16 Band P(16)	1616
6.0885551	1642.4258	v = 18 → 17 Band P(10)	1616
6.0926946	1641.3099	v = 17 → 16 Band P(17)	1616
6.1013312	1638.9966	v = 18 → 17 Band P(11)	1616
6.1203065	1633.9051	v = 17 → 16 Band P(19)	1616
6.1273954	1632.0148	v = 18 → 17 Band P(13)	1616
6.1398711	1628.6987	v = 19 → 18 Band P(7)	1616
6.1406862	1628.4825	v = 18 → 17 Band P(14)	1616
6.1523828	1625.3865	v = 19 → 18 Band P(8)	1616
6.1541514	1624.9194	v = 18 → 17 Band P(15)	1616
6.1650658	1622.0427	v = 19 → 18 Band P(9)	1616
6.1677922	1621.3257	v = 18 → 17 Band P(16)	1616
6.1779214	1618.6674	v = 19 → 18 Band P(10)	1616
6.1816101	1617.7015	v = 18 → 17 Band P(17)	1616
6.1909503	1615.2609	v = 19 → 18 Band P(11)	1616
6.1956064	1614.0470	v = 18 → 17 Band P(18)	1616
6.2041540	1611.8233	v = 19 → 18 Band P(12)	1616
6.2097831	1610.3622	v = 18 → 17 Band P(19)	1616
6.2175344	1608.3546	v = 19 → 18 Band P(13)	1616
6.2308351	1604.9213	v = 20 → 19 Band P(7)	1616
6.2310925	1604.8550	v = 19 → 18 Band P(14)	1616
6.2435937	1601.6417	v = 20 → 19 Band P(8)	1616
6.2448297	1601.3247	v = 19 → 18 Band P(15)	1616
6.2565279	1598.3306	v = 20 → 19 Band P(9)	1616
6.2587474	1597.7639	v = 19 → 18 Band P(16)	1616
6.2696393	1594.9981	v = 20 → 19 Band P(10)	1616
6.2728473	1594.1724	v = 19 → 18 Band P(17)	1616
6.2829292	1591.6143	v = 20 → 19 Band P(11)	1616
6.2963990	1588.2094	v = 20 → 19 Band P(12)	1616
6.3015999	1586.8986	v = 19 → 18 Band P(19)	1616
6.3100504	1584.7734	v = 20 → 19 Band P(13)	1616
6.3238843	1581.3066	v = 20 → 19 Band P(14)	1616
6.3372336	1577.9756	v = 21 → 20 Band P(8)	1616

$^{13}C^{16}O_2$ Laser—*continued*

Wavelength (μm) vac	Frequency (cm^{-1})	Transition assignment	Comments
6.3379027	1577.8090	v = 20 → 19 Band P(15)	1616
6.3504272	1574.6972	v = 21 → 20 Band P(9)	1616
6.3521065	1574.2809	v = 20 → 19 Band P(16)	1616
6.3638031	1571.3874	v = 21 → 20 Band P(10)	1616
6.3664982	1570.7222	v = 20 → 19 Band P(17)	1616
6.3773627	1568.0463	v = 21 → 20 Band P(11)	1616
6.3810785	1567.1332	v = 20 → 19 Band P(18)	1616
6.4050388	1561.2708	v = 21 → 20 Band P(13)	1616
6.4468647	1551.1416	v = 22 → 21 Band P(9)	1616
6.4479667	1550.8765	v = 21 → 20 Band P(16)	1616
6.4605142	1547.8644	v = 22 → 21 Band P(10)	1616
6.4626595	1547.3506	v = 21 → 20 Band P(17)	1616
6.4743525	1544.5560	v = 22 → 21 Band P(11)	1616
6.4883811	1541.2165	v = 22 → 21 Band P(12)	1616
6.5026017	1537.8460	v = 22 → 21 Band P(13)	1616
6.5170156	1534.4447	v = 22 → 21 Band P(14)	1616
6.5316249	1531.0126	v = 22 → 21 Band P(15)	1616
6.5598797	1524.4182	v = 23 → 22 Band P(10)	1616
6.5614352	1524.0568	v = 22 → 21 Band P(17)	1616
6.6028484	1514.4979	v = 23 → 22 Band P(13)	1616
6.6175674	1511.1293	v = 23 → 22 Band P(14)	1616

$D^{79}Br$ Laser

Wavelength (μm) vac	Frequency (cm^{-1})	Transition assignment	Comments
5.56886	1795.70	v = 1 → 0 Band P(5)	1621
5.59785	1786.40	v = 1 → 0 Band P(6)	1621
5.62762	1776.95	v = 1 → 0 Band P(7)	1621
5.65822	1767.34	v = 1 → 0 Band P(8)	1621
5.68961	1757.59	v = 1 → 0 Band P(9)	1621
5.7110	1751.0	v = 2 → 1 Band P(5)	1621
5.7409	1741.9	v = 2 → 1 Band P(6)	1621
5.7717	1732.6	v = 2 → 1 Band P(7)	1621
5.8025	1723.4	v = 3 → 2 Band P(3)	1621
5.8035	1723.1	v = 2 → 1 Band P(8)	1621
5.8319	1714.7	v = 3 → 2 Band P(4)	1621
5.8360	1713.1	v = 2 → 1 Band P(9)	1621
5.86197	1705.91	v = 3 → 2 Band P(5)	1621,1622

D^{79}Br Laser—*continued*

Wavelength (μm) vac	Frequency (cm^{-1})	Transition assignment	Comments
5.89282	1696.98	v = 3 → 2 Band P(6)	1621,1622
5.92456	1687.89	v = 3 → 2 Band P(7)	1621,1622
5.95735	1678.60	v = 3 → 2 Band P(8)	1621,1622
5.9909	1669.2	v = 3 → 2 Band P(9)	1621
6.02090	1660.88	v = 4 → 3 Band P(5)	1621,1622
6.0255	1659.6	v = 3 → 2 Band P(10)	1621
6.05290	1652.10	v = 4 → 3 Band P(6)	1621,1622
6.08576	1643.18	v = 4 → 3 Band P(7)	1621,1622
6.11995	1634.00	v = 4 → 3 Band P(8)	1621,1622
6.15460	1624.80	v = 4 → 3 Band P(9)	1621,1622
6.19034	1615.42	v = 4 → 3 Band P(10)	1621,1622
6.22723	1605.85	v = 4 → 3 Band P(11)	1621,1622
6.25657	1598.32	v = 5 → 4 Band P(7)	1621,1622
6.2909	1589.41	v = 5 → 4 Band P(10)	1621
6.29164	1580.31	v = 5 → 4 Band P(8)	1621,1622
6.32787	1589.6	v = 5 → 4 Band P(9)	1621,1622

D^{81}Br Laser

Wavelength (μm) vac	Frequency (cm^{-1})	Transition assignment	Comments
5.57038	1795.21	v = 1 → 0 Band P(5)	1623
5.59939	1785.91	v = 1 → 0 Band P(6)	1623
5.62914	1776.47	v = 1 → 0 Band P(7)	1623
5.65973	1766.87	v = 1 → 0 Band P(8)	1623
5.69113	1757.12	v = 1 → 0 Band P(9)	1623
5.7127	1750.5	v = 2 → 1 Band P(5)	1623
5.7435	1741.1	v = 2 → 1 Band P(6)	1623
5.7733	1732.1	v = 2 → 1 Band P(7)	1623
5.8042	1722.9	v = 3 → 2 Band P(3)	1623
5.80528	1722.57	v = 2 → 1 Band P(8)	1623,1624
5.8336	1714.2	v = 3 → 2 Band P(4)	1623
5.8374	1713.1	v = 2 → 1 Band P(9)	1623
5.86362	1705.43	v = 3 → 2 Band P(5)	1623,1624
5.89442	1696.52	v = 3 → 2 Band P(6)	1623,1624
5.92610	1687.45	v = 3 → 2 Band P(7)	1623,1624
5.95901	1678.13	v = 3 → 2 Band P(8)	1623,1624
5.9927	1668.7	v = 3 → 2 Band P(9)	1623
6.02246	1660.45	v = 4 → 3 Band P(5)	1623,1624

D^{81}Br Laser—*continued*

Wavelength (μm) vac	Frequency (cm^{-1})	Transition assignment	Comments
6.0274	1659.1	v = 3 → 2 Band P(10)	1623
6.05440	1651.69	v = 4 → 3 Band P(6)	1623,1624
6.08732	1642.76	v = 4 → 3 Band P(7)	1623,1624
6.12156	1633.57	v = 4 → 3 Band P(8)	1623,1624
6.15616	1624.39	v = 4 → 3 Band P(9)	1623,1624
6.19184	1615.03	v = 4 → 3 Band P(10)	1623,1624
6.22374	1606.75	v = 5 → 4 Band P(6)	1623,1624
6.22886	1605.43	v = 4 → 3 Band P(11)	1623,1624
6.25810	1597.93	v = 5 → 4 Band P(7)	1623,1624
6.2925	1589.01	v = 5 → 4 Band P(10)	1623
6.29323	1579.92	v = 5 → 4 Band P(8)	1623,1624
6.32943	1589.2	v = 5 → 4 Band P(9)	1623,1624

D^{35}Cl Laser

Wavelength (μm) vac	Frequency (cm^{-1})	Transition assignment	Comments
5.00310	1998.76	v = 1 → 0 Band P(8)	1632
5.03438	1986.34	v = 1 → 0 Band P(9)	1632
5.04452	1982.35	v = 2 → 1 Band P(5)	1633
5.06673	1973.66	v = 1 → 0 Band P(10)	1632
5.07429	1970.72	v = 2 → 1 Band P(6)	1632,1633
5.10001	1960.78	v = 1 → 0 Band P(11)	1632
5.10491	1958.90	v = 2 → 1 Band P(7)	1632–1634
5.1222	1952.3	v = 3 → 2 Band P(3)	1632
5.13408	1947.77	v = 1 → 0 Band P(12)	1632
5.13627	1946.94	v = 2 → 1 Band P(8)	1632–1634
5.15105	1941.35	v = 3 → 2 Band P(4)	1632–1634
5.16884	1934.67	v = 2 → 1 Band P(9)	1632–1634
5.16908	1934.58	v = 1 → 0 Band P(13)	1632
5.18108	1930.10	v = 3 → 2 Band P(5)	1632–1634
5.2021	1922.3	v = 2 → 1 Band P(10)	1632
5.21178	1918.73	v = 3 → 2 Band P(6)	1632,1633
5.24348	1907.13	v = 3 → 2 Band P(7)	1632,1633
5.27601	1895.37	v = 3 → 2 Band P(8)	1632,1633
5.30969	1883.35	v = 3 → 2 Band P(9)	1632,1633
5.32445	1878.13	v = 4 → 3 Band P(5)	1632,1633
5.34431	1871.15	v = 3 → 2 Band P(10)	1632,1633
5.35616	1867.01	v = 4 → 3 Band P(6)	1632,1633

D^{35}Cl Laser—*continued*

Wavelength (μm) vac	Frequency (cm^{-1})	Transition assignment	Comments
5.37990	1858.77	v = 3 → 2 Band P(11)	1633
5.38892	1855.66	v = 4 → 3 Band P(7)	1632,1633
5.4227	1844.1	v = 4 → 3 Band P(8)	1632
5.45771	1832.27	v = 4 → 3 Band P(9)	1632,1633
5.49348	1820.34	v = 4 → 3 Band P(10)	1632,1633
5.50849	1815.38	v = 5 → 4 Band P(6)	1633
5.53036	1808.20	v = 4 → 3 Band P(11)	1632,1633
5.54228	1804.31	v = 5 → 4 Band P(7)	1633
5.57759	1792.89	v = 5 → 4 Band P(8)	1633
5.61369	1781.36	v = 5 → 4 Band P(9)	1633

D^{37}Cl Laser

Wavelength (μm) vac	Frequency (cm^{-1})	Transition assignment	Comments
5.00984	1996.07	v = 1 → 0 Band P(8)	1635
5.04124	1983.64	v = 1 → 0 Band P(9)	1635
5.05140	1979.65	v = 2 → 1 Band P(5)	1636
5.07357	1971.00	v = 1 → 0 Band P(10)	1635
5.08109	1968.08	v = 2 → 1 Band P(6)	1635,1636
5.10673	1958.20	v = 1 → 0 Band P(11)	1635
5.11182	1956.25	v = 2 → 1 Band P(7)	1635,1636
5.14070	1945.26	v = 1 → 0 Band P(12)	1635
5.14311	1944.35	v = 2 → 1 Band P(8)	1635,1636
5.1578	1938.8	v = 3 → 2 Band P(4)	1635
5.1757	1932.1	v = 2 → 1 Band P(9)	1635
5.18791	1927.56	v = 3 → 2 Band P(5)	1635,1636
5.2089	1919.8	v = 2 → 1 Band P(10)	1635
5.21855	1916.24	v = 3 → 2 Band P(6)	1635,1636
5.25025	1904.67	v = 3 → 2 Band P(7)	1635,1636
5.28287	1892.91	v = 3 → 2 Band P(8)	1635,1636
5.3163	1881.0	v = 3 → 2 Band P(9)	1635
5.36294	1864.65	v = 4 → 3 Band P(6)	1636
5.39561	1853.36	v = 4 → 3 Band P(7)	1636
5.42950	1841.79	v = 4 → 3 Band P(8)	1636

DF Laser

Wavelength (µm) vac	Frequency (cm⁻¹)	Transition assignment	Comments
1.836	5447	v = 3 → 1 Band P(4)	1649
1.844	5423	v = 3 → 1 Band P(5)	1649
1.854	5394	v = 3 → 1 Band P(6)	1649
3.49327	2862.65	v = 1 → 0 Band P(2)	1650
3.52140	2839.78	v = 1 → 0 Band P(3)	1650
3.55068	2816.36	v = 1 → 0 Band P(4)	1650
3.58110	2792.44	v = 1 → 0 Band P(5)	1650
3.61283	2767.91	v = 1 → 0 Band P(6)	1650
3.63630	2750.05	v = 2 → 1 Band P(3)	1652,1653
3.64560	2743.03	v = 1 → 0 Band P(7)	1650
3.66652	2727.38	v = 2 → 1 Band P(4)	1652,1653
3.67980	2717.54	v = 1 → 0 Band P(8)	1650,1651
3.69825	2703.98	v = 2 → 1 Band P(5)	1652,1653
3.71552	2691.41	v = 1 → 0 Band P(9)	1651,1652
3.73095	2680.28	v = 2 → 1 Band P(6)	1652,1653
3.75199	2665.25	v = 1 → 0 Band P(10)	1651,1652
3.75633	2662.17	v = 3 → 2 Band P(3)	1652,1653
3.76510	2655.97	v = 2 → 1 Band P(7)	1652,1653
3.78782	2640.04	v = 3 → 2 Band P(4)	1652,1653
3.79018	2611.10	v = 1 → 0 Band P(11)	1651,1652
3.80071	2631.09	v = 2 → 1 Band P(8)	1651–1653
3.82057	2617.41	v = 3 → 2 Band P(5)	1652,1653
3.82980	2583.7	v = 1 → 0 Band P(12)	1651–1653
3.83749	2605.87	v = 2 → 1 Band P(9)	1651–1653
3.8502	2597.3	v = 4 → 3 Band P(2)	1652
3.85471	2594.23	v = 3 → 2 Band P(6)	1652,1653
3.8704	2555.7	v = 1 → 0 Band P(13)	1652
3.87573	2580.19	v = 2 → 1 Band P(10)	1651–1653
3.8816	2576.2	v = 4 → 3 Band P(3)	1652
3.89028	2570.51	v = 3 → 2 Band P(7)	1652,1653
3.9128	2555.7	v = 1 → 0 Band P(14)	1652
3.9145	2554.6	v = 4 → 3 Band P(4)	1652
3.91547	2553.97	v = 2 → 1 Band P(11)	1651–1653
3.92716	2546.37	v = 3 → 2 Band P(8)	1651–1653
3.94867	2532.50	v = 4 → 3 Band P(5)	1652,1653
3.95653	2527.47	v = 2 → 1 Band P(12)	1651–1653
3.95717	2527.06	v = 1 → 0 Band P(15)	1652,1653

DF Laser—*continued*

Wavelength (μm) vac	Frequency (cm⁻¹)	Transition assignment	Comments
3.96541	2521.81	v = 3 → 2 Band P(9)	1651–1653
3.98429	2509.86	v = 4 → 3 Band P(6)	1652,1653
3.99949	2500.32	v = 2 → 1 Band P(13)	1652,1653
4.00317	2498.02	v = 1 → 0 Band P(16)	1652,1653
4.00543	2496.61	v = 3 → 2 Band P(10)	1651,1652,1653
4.02118	2486.83	v = 4 → 3 Band P(7)	1652,1653
4.0434	2473.2	v = 2 → 1 Band P(14)	1652
4.04639	2471.34	v = 3 → 2 Band P(11)	1651,1652,1653
4.0491	2469.7	v = 1 → 0 Band P(17)	1652
4.0594	2463.4	v = 4 → 3 Band P(8)	1652
4.0892	2445.5	v = 2 → 1 Band P(15)	1652
4.08949	2445.29	v = 3 → 2 Band P(12)	1652,1653
4.1336	2419.2	v = 3 → 2 Band P(13)	1652
4.13690	2417.27	v = 2 → 1 Band P(16)	1652,1653
4.17980	2392.46	v = 3 → 2 Band P(14)	1652,1653
4.18622	2388.79	v = 2 → 1 Band P(17)	1653

HBr Laser

Wavelength (μm) vac	Frequency (cm⁻¹)	Transition assignment	Comments
19.399	51.549		6960
19.988	50.030		6960
20.360	49.116		6960
20.896	47.773		6960
20.949	46.509		6960
21.501	46.412		6960
21.546	45.175		6960
22.136	44.992		6960
22.226	43.754		6960
22.855	42.669		6960
23.436	33.573		6960
29.786	33.846		6960
30.445	32.312		6960
30.948	31.880		6960
31.368	31.398		6960
31.849	30.799		6960
32.469	30.489		6960
32.799	29.932		6960
33.409	24.676		6960

H^{79}Br Laser

Wavelength (μm) vac	Frequency (cm^{-1})	Transition assignment	Comments
4.01703	2489.40	$v = 1 \rightarrow 0$ Band P(4)	1619,1620
4.04699	2470.97	$v = 1 \rightarrow 0$ Band P(5)	1619,1620
4.07825	2452.03	$v = 1 \rightarrow 0$ Band P(6)	1619,1620
4.11066	2432.70	$v = 1 \rightarrow 0$ Band P(7)	1619,1620
4.14424	2412.99	$v = 1 \rightarrow 0$ Band P(8)	1619
4.16531	2400.78	$v = 2 \rightarrow 1$ Band P(4)	1619,1620
4.19695	2382.68	$v = 2 \rightarrow 1$ Band P(5)	1619,1620
4.22947	2364.36	$v = 2 \rightarrow 1$ Band P(6)	1619,1620
4.26334	2345.58	$v = 2 \rightarrow 1$ Band P(7)	1619,1620
4.29880	2326.23	$v = 2 \rightarrow 1$ Band P(8)	1619,1620
4.32498	2312.15	$v = 3 \rightarrow 2$ Band P(4)	1619,1620
4.33539	2306.60	$v = 2 \rightarrow 1$ Band P(9)	1619,1620
4.35791	2294.68	$v = 3 \rightarrow 2$ Band P(5)	1619,1620
4.39250	2276.61	$v = 3 \rightarrow 2$ Band P(6)	1619,1620
4.42813	2258.29	$v = 3 \rightarrow 2$ Band P(7)	1619,1620
4.46524	2239.52	$v = 3 \rightarrow 2$ Band P(8)	1619,1620
4.50410	2220.20	$v = 3 \rightarrow 2$ Band P(9)	1619
4.53295	2206.07	$v = 4 \rightarrow 3$ Band P(5)	1619,1620
4.56911	2188.61	$v = 4 \rightarrow 3$ Band P(6)	1619,1620
4.60700	2170.61	$v = 4 \rightarrow 3$ Band P(7)	1619,1620
4.64626	2152.27	$v = 4 \rightarrow 3$ Band P(8)	1619

H^{81}Br Laser

Wavelength (μm) vac	Frequency (cm^{-1})	Transition assignment	Comments
4.01760	2489.05	$v = 1 \rightarrow 0$ Band P(4)	1617,1618
4.04755	2470.63	$v = 1 \rightarrow 0$ Band P(5)	1617,1618
4.07884	2451.68	$v = 1 \rightarrow 0$ Band P(6)	1617,1618
4.11123	2432.36	$v = 1 \rightarrow 0$ Band P(7)	1617,1618
4.14477	2412.68	$v = 1 \rightarrow 0$ Band P(8)	1617
4.16585	2400.47	$v = 2 \rightarrow 1$ Band P(4)	1617,1618
4.17962	2392.56	$v = 1 \rightarrow 0$ Band P(9)	1617
4.19754	2382.35	$v = 2 \rightarrow 1$ Band P(5)	1617,1618
4.26392	2345.26	$v = 2 \rightarrow 1$ Band P(7)	1617,1618
4.29937	2325.92	$v = 2 \rightarrow 1$ Band P(8)	1617,1618
4.32554	2311.85	$v = 3 \rightarrow 2$ Band P(4)	1617,1618
4.33595	2306.30	$v = 2 \rightarrow 1$ Band P(9)	1617,1618
4.35846	2294.39	$v = 3 \rightarrow 2$ Band P(5)	1617,1618
4.39306	2276.32	$v = 3 \rightarrow 2$ Band P(6)	1617,1618
4.42870	2258.00	$v = 3 \rightarrow 2$ Band P(7)	1617,1618

H^{81}Br Laser—*continued*

Wavelength (μm) vac	Frequency (cm^{-1})	Transition assignment	Comments
4.46576	2239.26	v = 3 → 2 Band P(8)	1617,1618
4.50467	2219.92	v = 3 → 2 Band P(9)	1617
4.53348	2205.81	v = 4 → 3 Band P(5)	1617,1618
4.56965	2188.35	v = 4 → 3 Band P(6)	1617,1618
4.60755	2170.35	v = 4 → 3 Band P(7)	1617,1618
4.64675	2152.04	v = 4 → 3 Band P(8)	1617

HCl Laser

Wavelength (μm) vac	Frequency (cm^{-1})	Transition assignment	Comments
13.872	720.88		6959
14.099	709.27		6959
14.343	697.20		6959
16.213	616.79		6959
16.644	600.82		6959
16.765	596.48		6960
17.034	587.06		6960
17.125	583.94		6959
17.492	571.69		6959
17.575	568.99		6960
17.98887	555.96		6959
17.997	555.56		6960
18.035	554.48		6960
18.522	539.90		6959
18.555	538.94		6960
18.593	577.84		6960
19.122	522.96		6960
19.145	522.33		6960
19.183	521.29		6960
19.700	507.61		6959
19.783	505.48		6960
19.821	504.52		6960
20.346	491.50		6959
20.411	489.93		6959
20.999	476.21		6959
21.047	475.13		6959
21.156	472.68		6959

HCl Laser—*continued*

Wavelength (μm) vac	Frequency (cm⁻¹)	Transition assignment	Comments
21.813	458.44		6959
21.971	455.14		6959
22.651	441.48		6959
22.864	437.37		6959
23.571	424.25		6959
23.849	419.30		6959
24.318	411.22		6959
24.583	406.78		6959
24.618	406.21		6959
24.937	401.01		6959
25.704	389.04		6959
26.146	382.47		6959
26.247	381.00		6960
27.508	363.53		6960

H^{35}Cl Laser

Wavelength (μm) vac	Frequency (cm⁻¹)	Transition assignment	Comments
3.572781	2798.940	v = 1 → 0 Band P(4)	1625
3.602617	2775.760	v = 1 → 0 Band P(5)	1625
3.633673	2752.036	v = 1 → 0 Band P(6)	1625
3.665989	2727.777	v = 1 → 0 Band P(7)	1625
3.699583	2703.007	v = 1 → 0 Band P(8)	1625,1626
3.70711	2697.52	v = 2 → 1 Band P(4)	1625–1628
3.734504	2677.732	v = 1 → 0 Band P(9)	1626
3.73830	2675.01	v = 2 → 1 Band P(5)	1625–1628
3.770787	2651.966	v = 1 → 0 Band P(10)	1626
3.77100	2651.82	v = 2 → 1 Band P(6)	1625–1628
3.80499	2628.13	v = 2 → 1 Band P(7)	1625–1628
3.808469	2625.727	v = 1 → 0 Band P(11)	1626
3.84011	2604.09	v = 2 → 1 Band P(8)	1625,1627–28
3.85091	2596.79	v = 3 → 2 Band P(4)	1625,1627–28
3.87684	2579.42	v = 2 → 1 Band P(9)	1625,1627–28
3.88395	2574.70	v = 3 → 2 Band P(5)	1625,1627–28
3.91491	2554.34	v = 2 → 1 Band P(10)	1625,1627–28
3.91810	2552.26	v = 3 → 2 Band P(6)	1625,1627–28
3.95365	2529.31	v = 3 → 2 Band P(7)	1625,1627–28
3.99093	2505.68	v = 3 → 2 Band P(8)	1625,1627–28

H^{35}Cl Laser—*continued*

Wavelength (μm) vac	Frequency (cm^{-1})	Transition assignment	Comments
4.00590	2496.32	v = 4 → 3 Band P(4)	1629
4.02951	2481.69	v = 3 → 2 Band P(9)	1625,1627–28
4.04042	2474.99	v = 4 → 3 Band P(5)	1629
4.07644	2453.12	v = 4 → 3 Band P(6)	1629
4.11399	2430.73	v = 4 → 3 Band P(7)	1629

H^{37}Cl Laser

Wavelength (μm) vac	Frequency (cm^{-1})	Transition assignment	Comments
3.636190	2750.131	v = 1 → 0 Band P(6)	1631
3.668485	2725.921	v = 1 → 0 Band P(7)	1631
3.702069	2701.192	v = 1 → 0 Band P(8)	1631
3.70971	2695.63	v = 2 → 1 Band P(4)	1631
3.736975	2675.961	v = 1 → 0 Band P(9)	1631
3.74079	2673.23	v = 2 → 1 Band P(5)	1630,1631
3.77347	2650.08	v = 2 → 1 Band P(6)	1630,1631
3.80742	2626.45	v = 2 → 1 Band P(7)	1630,1631
3.84249	2602.48	v = 2 → 1 Band P(8)	1630,1631
3.85361	2594.97	v = 3 → 2 Band P(4)	1631
3.88641	2573.07	v = 3 → 2 Band P(5)	1631
3.92049	2550.70	v = 3 → 2 Band P(6)	1630,1631
3.95602	2527.79	v = 3 → 2 Band P(7)	1630

HF Laser

Wavelength (μm) vac	Frequency (cm^{-1})	Transition assignment	Comments
1.31258	7618.6	v = 2 → 0 Band P(3)	6834
1.32125	7568.6	v = 2 → 0 Band P(4)	6833
1.33053	7515.8	v = 2 → 0 Band P(5)	6833
1.34043	7460.3	v = 2 → 0 Band P(6)	6833
1.35099	7402.0	v = 2 → 0 Band P(7)	6834
1.36219	7341.1	v = 2 → 0 Band P(8)	6834
1.37282	7284.3	v = 3 → 1 Band P(3)	6835
1.37406	7277.7	v = 2 → 0 Band P(9)	6834
1.38196	7236.1	v = 3 → 1 Band P(4)	6835,6836
1.39175	7185.2	v = 3 → 1 Band P(5)	6835

HF Laser—*continued*

Wavelength (μm) vac	Frequency (cm⁻¹)	Transition assignment	Comments
1.44693	6911.2	v = 4 → 2 Band P(4)	6835
1.45730	6862.0	v = 4 → 2 Band P(5)	6835
2.41381	4142.83	v = 1 → 0 Band R(4)	1637,1638
2.43312	4109.95	v = 1 → 0 Band R(3)	1637,1638
2.45381	4075.30	v = 1 → 0 Band R(2)	1637,1638
2.47588	4038.97	v = 1 → 0 Band R(1)	1637,1638
2.55083	3920.29	v = 1 → 0 Band P(1)	1637,1639
2.57885	3877.70	v = 1 → 0 Band P(2)	1637,1639
2.60848	3833.65	v = 1 → 0 Band P(3)	1637–1640
2.63976	3788.23	v = 1 → 0 Band P(4)	1637–1641
2.66679	3749.83	v = 2 → 1 Band P(1)	1637,1640
2.67274	3741.48	v = 1 → 0 Band P(5)	1637–1641
2.69625	3708.86	v = 2 → 1 Band P(2)	1637–40,1642,1646
2.70752377	3693.41171	v = 1 → 0 Band P(6)	1637–1642,1644
2.72749	3666.38	v = 2 → 1 Band P(3)	1637,1639,1640, 1642,1646
2.74412	3644.16	v = 1 → 0 Band P(7)	1637–39,1640–42
2.76036	3622.71	v = 2 → 1 Band P(4)	1637,1639,1640, 1641,1642,1646
2.78257	3593.90	v = 1 → 0 Band P(8)	1637–8,1640,1642
2.79023	3583.93	v = 3 → 2 Band P(1)	1637,1640
2.79527	3577.47	v = 2 → 1 Band P(5)	1637,1639,1640, 1641,1642,1646
2.82126	3544.51	v = 3 → 2 Band P(2)	1637,1640,1642
2.82310	3542.20	v = 1 → 0 Band P(9)	1637,1642
2.83181	3531.31	v = 2 → 1 Band P(6)	1637,1639,1640, 1641,1642,1646
2.85404	3503.80	v = 3 → 2 Band P(3)	1637,1640,1642
2.86567	3489.59	v = 1 → 0 Band P(10)	1637,1642
2.87057	3483.63	v = 2 → 1 Band P(7)	1637,1639,1640, 1641,1642,1646
2.88889	3461.54	v = 3 → 2 Band P(4)	1637,1640, 1642,1645
2.91026	3436.12	v = 1 → 0 Band P(11)	1637,1642
2.91106	3435.17	v = 2 → 1 Band P(8)	1637,1639,1642
2.92208	3422.22	v = 4 → 3 Band P(1)	1637,1640
2.92555	3418.16	v = 3 → 2 Band P(5)	1637,1640,1642
2.95391	3385.34	v = 2 → 1 Band P(9)	1637,1639,1642
2.95487	3384.24	v = 4 → 3 Band P(2)	1637,1640
2.95727	3381.50	v = 1 → 0 Band P(12)	1637,1642,1645

HF Laser—*continued*

Wavelength (µm) vac	Frequency (cm⁻¹)	Transition assignment	Comments
2.96432	3373.46	v = 3 → 2 Band P(6)	1637,1640,1642
2.98961	3344.92	v = 4 → 3 Band P(3)	1637,1640,1648
2.99891	3334.55	v = 2 → 1 Band P(10)	1637,1642,1645
3.00505	3327.73	v = 3 → 2 Band P(7)	1637,1640,1642
3.00642	3326.21	v = 1 → 0 Band P(13)	1637,1642,1645
3.02635	3304.31	v = 4 → 3 Band P(4)	1637,1648
3.04612	3282.86	v = 2 → 1 Band P(11)	1637,1642,1645
3.04819	3280.64	v = 3 → 2 Band P(8)	1637,1640,1642, 1645
3.05820	3269.90	v = 1 → 0 Band P(14)	1637,1642,1645
3.06517	3262.46	v = 4 → 3 Band P(5)	1637,1648
3.09350	3232.58	v = 3 → 2 Band P(9)	1637,1645
3.09580	3220.18	v = 2 → 1 Band P(12)	1637,1642,1645
3.09821	3227.67	v = 5 → 4 Band P(2)	1637,1646,1647
3.10616	3219.41	v = 4 → 3 Band P(6)	1637,1647
3.11255	3212.80	v = 1 → 0 Band P(15)	1637,1642,1645
3.13495	3189.84	v = 5 → 4 Band P(3)	1637,1646
3.14110	3183.60	v = 3 → 2 Band P(10)	1637,1645
3.14802	3176.60	v = 2 → 1 Band P(13)	1637,1642,1645
3.14939	3175.22	v = 4 → 3 Band P(7)	1637,1647
3.16962	3154.95	v = 1 → 0 Band P(16)	1637,1645
3.17387	3150.73	v = 5 → 4 Band P(4)	1637,1646
3.19117	3133.65	v = 3 → 2 Band P(11)	1637,1645
3.20293	3122.14	v = 2 → 1 Band P(14)	1637,1642,1645
3.21501	3110.41	v = 5 → 4 Band P(5)	1637,1646
3.22933	3096.62	v = 1 → 0 Band P(17)	1637,1645
3.24381	3082.79	v = 3 → 2 Band P(12)	1637,1645,1647
3.25849	3068.91	v = 5 → 4 Band P(6)	1637,1646,1648
3.26028	3067.22	v = 2 → 1 Band P(15)	1637,1642,1645
3.29196	3037.70	v = 1 → 0 Band P(18)	1637,1645
3.29913	3031.10	v = 3 → 2 Band P(13)	1637,1645
3.30438	3026.29	v = 5 → 4 Band P(7)	1637,1646
3.32059	3011.51	v = 2 → 1 Band P(16)	1637,1645
3.33348	299.87	v = 6 → 5 Band P(4)	1637,1646,1648
3.37720	2961.03	v = 6 → 5 Band P(5)	1637,1646,1648
10.198	980.58		6968
10.458	956.20		6968
10.582	945.00		6968
10.744	930.75		6968
10.812	924.90		6968

HF Laser—*continued*

Wavelength (μm) vac	Frequency (cm⁻¹)	Transition assignment	Comments
11.057	904.40		6968
11.403	876.96		6968
11.541	866.48		6968
11.785	848.54		6968
11.863	842.96		6960
12.208	819.13		6968
12.262	815.53		6968
12.678	788.77		6968
12.701	786.78		6968
13.188	758.26		6968
13.201	757.52		6968
13.221	756.64		6968
13.728	728.44		6968
13.784	725.48		6968
14.288	699.89		6968
14.441	692.47		6968
15.016	665.56		6968
16.022	624.14		6968
16.444	608.12		6960
16.655	600.42		6960
16.975	589.10		6960
17.095	584.97		6960
17.325	577.20		6960
17.645	566.73		6960
18.085	522.94		6960
18.801	531.89		6968
19.113	523.20		6968
20.134	496.67		6968
20.351	491.38		6968
20.939	477.76		6968
21.699	460.85		6968
21.789	458.95		6968

NO Laser

Wavelength (μm) vac	Frequency (cm⁻¹)	Transition assignment	Comments
5.504	1817	$^2\pi_{1/2}v = 2 \to 1$ Band	1654,1657
5.84618	1710.52	$^2\pi_{3/2}v = 6 \to 5$ Band P(7)	1655,1657
5.85487	1707.98	$^2\pi_{1/2}v = 6 \to 5$ Band P(8)	1655,1657
5.85840	1706.95	$^2\pi_{3/2}v = 6 \to 5$ Band P(8)	1655,1657
5.87058	1703.41	$^2\pi_{3/2}v = 6 \to 5$ Band P(9)	1655,1657
5.87893	1700.99	$^2\pi_{1/2}v = 6 \to 5$ Band P(10)	1655,1657
5.90361	1693.88	$^2\pi_{1/2}v = 6 \to 5$ Band P(12)	1655,1657
5.90831	1692.53	$^2\pi_{3/2}v = 6 \to 5$ Band P(12)	1655,1657
5.94234	1682084	$^2\pi_{3/2}v = 7 \to 6$ Band P(7)	1655,1657
5.95461	1697.37	$^2\pi_{3/2}v = 7 \to 6$ Band P(8)	1655,1657
5.9550	1679.3	$^2\pi_{1/2}v = 6 \to 5$ Band P(16)	1656,1657
5.96317	1676.96	$^2\pi_{1/2}v = 7 \to 6$ Band P(9)	1655–1657
5.96726	1675.81	$^2\pi_{3/2}v = 7 \to 6$ Band P(9)	1655,1657
5.97564	1673.46	$^2\pi_{1/2}v = 7 \to 6$ Band P(10)	1655–1657
5.97990	1672.27	$^2\pi_{3/2}v = 7 \to 6$ Band P(10)	1655,1657
5.98817	1669.96	$^2\pi_{1/2}v = 7 \to 6$ Band P(11)	1655,1657
5.99308	1668.59	$^2\pi_{3/2}v = 7 \to 6$ Band P(11)	1655,1657
6.00100	1666.39	$^2\pi_{1/2}v = 7 \to 6$ Band P(12)	1655–1657
6.00536	1665.18	$^2\pi_{3/2}v = 7 \to 6$ Band P(12)	1655,1657
6.01917	1661.36	$^2\pi_{3/2}v = 7 \to 6$ Band P(13)	1655,1657
6.02667	1659.29	$^2\pi_{1/2}v = 7 \to 6$ Band P(14)	1655–1657
6.03242	1657.71	$^2\pi_{3/2}v = 7 \to 6$ Band P(14)	1655–1657
6.0386	1656.0	$^2\pi_{1/2}v = 8 \to 7$ Band P(7)	1656,1657
6.04018	1655.58	$^2\pi_{1/2}v = 7 \to 6$ Band P(15)	1655–1657
6.04186	1655.12	$^2\pi_{3/2}v = 8 \to 7$ Band P(7)	1655,1657
6.05429	1651.72	$^2\pi_{3/2}v = 8 \to 7$ Band P(8)	1655,1657
6.06281	1649.40	$^2\pi_{1/2}v = 8 \to 7$ Band P(9)	1655–1657
6.06726	1648.19	$^2\pi_{3/2}v = 8 \to 7$ Band P(9)	1655,1657
6.08010	1644.71	$^2\pi_{3/2}v = 8 \to 7$ Band P(10)	1655,1657
6.08839	1642.47	$^2\pi_{1/2}v = 8 \to 7$ Band P(11)	1655–1657
6.09336	1641.13	$^2\pi_{3/2}v = 8 \to 7$ Band P(11)	1655–1657
6.10147	1638.95	$^2\pi_{1/2}v = 8 \to 7$ Band P(12)	1655,1657
6.12044	1633.87	$^2\pi_{3/2}v = 8 \to 7$ Band P(13)	1655,1657
6.14168	1628.22	$^2\pi_{1/2}v = 8 \to 7$ Band P(15)	1655–1657
6.15377	1625.02	$^2\pi_{1/2}v = 9 \to 8$ Band P(8)	1655,1657
6.1546	1624.8	$^2\pi_{1/2}v = 8 \to 7$ Band P(16)	1656,1657
6.15764	1624.00	$^2\pi_{3/2}v = 9 \to 8$ Band P(8)	1655,1657

NO Laser—*continued*

Wavelength (μm) vac	Frequency (cm^{-1})	Transition assignment	Comments
6.16629	1621.72	$^2\pi_{1/2}$v = 9 → 8 Band P(9)	1655,1657
6.17917	1618.34	$^2\pi_{1/2}$v = 9 → 8 Band P(10)	1655,1657
6.18383	1617.12	$^2\pi_{3/2}$v = 9 → 8 Band P(10)	1655,1657
6.19214	1614.95	$^2\pi_{1/2}$v = 9 → 8 Band P(11)	1655,1657
6.19721	1613.63	$^2\pi_{3/2}$v = 9 → 8 Band P(11)	1655,1657
6.1973	1613.6	$^2\pi_{1/2}$v = 8 → 7 Band P(19)	1656,1657
6.20548	1611.48	$^2\pi_{1/2}$v = 9 → 8 Band P(12)	1655,1657
6.21095	1610.06	$^2\pi_{3/2}$v = 9 → 8 Band P(12)	1655,1657
6.21914	1607.94	$^2\pi_{1/2}$v = 9 → 8 Band P(13)	1655,1657
6.22487	1606.46	$^2\pi_{3/2}$v = 9 → 8 Band P(13)	1655,1657
6.23807	1603.06	$^2\pi_{3/2}$v = 10 → 9 Band P(6)	1655,1657
6.25113	1599.71	$^2\pi_{3/2}$v = 10 → 9 Band P(7)	1655,1657
6.26021	1597.39	$^2\pi_{1/2}$v = 10 → 9 Band P(8)	1655,1657
6.26449	1596.30	$^2\pi_{3/2}$v = 10 → 9 Band P(8)	1655,1657
6.27782	1592.91	$^2\pi_{3/2}$v = 10 → 9 Band P(9)	1655,1657
6.28650	1590.71	$^2\pi_{1/2}$v = 10 → 9 Band P(10)	1655,1657
6.29129	1589.50	$^2\pi_{3/2}$v = 10 → 9 Band P(10)	1655,1657
6.29977	1587.36	$^2\pi_{1/2}$v = 10 → 9 Band P(11)	1655,1657
6.30509	1586.02	$^2\pi_{3/2}$v = 10 → 9 Band P(11)	1655,1657
6.31361	1583.88	$^2\pi_{1/2}$v = 10 → 9 Band P(12)	1655,1657
6.31912	1582.50	$^2\pi_{3/2}$v = 10 → 9 Band P(12)	1655,1657
6.32699	1580.53	$^2\pi_{1/2}$v = 10 → 9 Band P(13)	1655,1657
6.33356	1578.89	$^2\pi_{3/2}$v = 10 → 9 Band P(13)	1655,1657
6.37637	1568.29	$^2\pi_{3/2}$v = 11 → 10 Band P(8)	1655,1657
6.38941	1565.09	$^2\pi_{3/2}$v = 11 → 10 Band P(9)	1655,1657
6.39799	1562.99	$^2\pi_{1/2}$v = 11 → 10 Band P(10)	1655,1657
6.40311	1561.74	$^2\pi_{3/2}$v = 11 → 10 Band P(10)	1655,1657
6.42616	1556.14	$^2\pi_{1/2}$v = 11 → 10 Band P(12)	1655,1657
6.43207	1554.71	$^2\pi_{3/2}$v = 11 → 10 Band P(12)	1655,1657

OH Laser

Wavelength (μm) vac	Frequency (cm^{-1})	Transition assignment	Comments
2.935	3407	v = 1 → 0 Band P_1(4)	1658,1659
2.969	3368	v = 1 → 0 Band P_1(5)	1658,1659
3.078	3249	v = 2 → 1 Band P_1(4)	1658–1660
3.115	3210	v = 2 → 1 Band P_2(5)	1658–1660
3.157	3168	v = 2 → 1 Band P_1(6)	1658–1660
3.234	3092	v = 3 → 2 Band P_1(4)	1658–1660
3.274	3054	v = 3 → 2 Band P_1(5)	1658–1660

3.3.2.3 Triatomic Vibrational Transition Lasers

Table 3.3.2.2
Triatomic Vibrational Transition Lasers

$^{12}C^{16}O_2$ Laser

Wavelength (μm) vac	Frequency (cm^{-1})	Transition assignment	Comments
4.3131086	2318.5134	$10^01 \rightarrow 10^00$ Band P(10)	1661,1663
4.3162577	2316.8218	$10^01 \rightarrow 10^00$ Band P(12)	1661,1663
4.3178659	2315.9589	$02^01 \rightarrow 02^00$ Band P(14)	1661
4.32031	2314.65	$10^02 \rightarrow 10^01$ Band R(17)	1664
4.3211010	2314.2250	$02^01 \rightarrow 02^00$ Band P(16)	1661
4.3227096	2313.3638	$10^01 \rightarrow 10^00$ Band P(16)	1661,1663
4.3243851	2312.4675	$02^01 \rightarrow 02^00$ Band P(18)	1661
4.32492	2312.18	$10^02 \rightarrow 10^01$ Band R(13)	1664
4.3260126	2311.5975	$10^01 \rightarrow 10^00$ Band P(18)	1661,1663
4.32764	2310.73	$10^02 \rightarrow 10^01$ Band R(11)	1664
4.3277182	2310.6865	$02^01 \rightarrow 02^00$ Band P(20)	1661
4.3327731	2307.9907	$10^01 \rightarrow 10^00$ Band P(22)	1661,1663
4.3345326	2307.0538	$02^01 \rightarrow 02^00$ Band P(24)	1661
4.3362312	2306.1501	$10^01 \rightarrow 10^00$ Band P(24)	1661,1663
4.3380141	2305.2023	$02^01 \rightarrow 02^00$ Band P(26)	1661
4.3397413	2304.2848	$10^01 \rightarrow 10^00$ Band P(26)	1661,1663
4.3415456	2303.3272	$02^01 \rightarrow 02^00$ Band P(28)	1661
4.3433038	2302.3948	$10^01 \rightarrow 10^00$ Band P(28)	1661,1663
4.3451268	2301.4288	$02^01 \rightarrow 02^00$ Band P(30)	1661
4.3505861	2298.5409	$10^01 \rightarrow 10^00$ Band P(32)	1661,1663
4.35493	2296.25	$10^02 \rightarrow 10^01$ Band P(7)	1664
4.35802	2294.62	$10^02 \rightarrow 10^01$ Band P(9)	1664
4.36121	2292.94	$10^02 \rightarrow 10^01$ Band P(11)	1664
4.36443	2291.25	$10^02 \rightarrow 10^01$ Band P(13)	1664
4.36775	2289.51	$10^02 \rightarrow 10^01$ Band P(15)	1664
4.37109	2287.76	$10^02 \rightarrow 10^01$ Band P(17)	1664
4.37445	2286.00	$10^02 \rightarrow 10^01$ Band P(19)	1664
4.37790	2284.20	$10^02 \rightarrow 10^01$ Band P(21)	1664
4.38139	2282.38	$10^02 \rightarrow 10^01$ Band P(23)	1664
4.38495	2280.53	$10^02 \rightarrow 10^01$ Band P(25)	1664
9.0934949	1099.687205	$00^01 \rightarrow 02^00$ Band R(62)	1665,1666
9.0997598	1098.930108	$00^01 \rightarrow 02^00$ Band R(60)	1666,6954
9.1062290	1098.149406	$00^01 \rightarrow 02^00$ Band R(58)	1666,6954
9.1129052	1097.344890	$00^01 \rightarrow 02^00$ Band R(56)	1666,6954

$^{12}C^{16}O_2$ Laser—*continued*

Wavelength (μm) vac	Frequency (cm^{-1})	Transition assignment	Comments
9.1197910	1096.516358	$00^01 \rightarrow 02^00$ Band R(54)	1666,6954
9.1268888	1095.663613	$00^01 \rightarrow 02^00$ Band R(52)	1666,6954
9.1342014	1094.786462	$00^01 \rightarrow 02^00$ Band R(50)	1666,6954
9.1361235	1094.556136	$01^11 \rightarrow 03^10$ Band R(34)	6957
9.1377728	1094.358575	$01^11 \rightarrow 03^10$ Band R(33)	6957
9.1417311	1093.884721	$00^01 \rightarrow 02^00$ Band R(48)	1666,6954
9.1451984	1093.469990	$01^11 \rightarrow 03^10$ Band R(32)	6957
9.1494807	1092.958211	$00^01 \rightarrow 02^00$ Band R(46)	1666,6954
9.1545210	1092.356446	$01^11 \rightarrow 03^10$ Band R(30)	6957
9.1574525	1092.006758	$00^01 \rightarrow 02^00$ Band R(44)	1666,6954
9.1656492	1091.030197	$00^01 \rightarrow 02^00$ Band R(42)	1665,1666,6954
9.1740732	1090.028368	$00^01 \rightarrow 02^00$ Band R(40)	1666,6954
9.1773048	1089.644539	$01^11 \rightarrow 03^10$ Band R(25)	6957
9.1827270	1089.001119	$00^01 \rightarrow 02^00$ Band R(38)	1666,6954
9.1877699	1088.403401	$01^11 \rightarrow 03^10$ Band R(23)	6957
9.1916132	1087.948306	$00^01 \rightarrow 02^00$ Band R(36)	1666,6954
9.2007341	1086.869792	$00^01 \rightarrow 02^00$ Band R(34)	1666,6954
9.2049275	1086.374659	$01^11 \rightarrow 03^10$ Band R(20)	6957
9.2091716	1085.8740	$00^02 \rightarrow 02^01$ Band R(39)	1669
9.2100923	1085.765445	$00^01 \rightarrow 02^00$ Band R(32)	1666,6954
9.2177733	1084.8607	$00^02 \rightarrow 02^01$ Band R(37)	1669
9.2196902	1084.635145	$00^01 \rightarrow 02^00$ Band R(30)	1666,6954
9.2266150	1083.8211	$00^02 \rightarrow 02^01$ Band R(35)	1669
9.2295301	1083.478778	$00^01 \rightarrow 02^00$ Band R(28)	1666,6954
9.2356997	1082.7550	$00^02 \rightarrow 02^01$ Band R(33)	1669
9.2396145	1082.296238	$00^01 \rightarrow 02^00$ Band R(26)	1666,6954
9.2450296	1081.6623	$00^02 \rightarrow 02^01$ Band R(31)	1669
9.2499457	1081.087427	$00^01 \rightarrow 02^00$ Band R(24)	1666,6954
9.2546080	1080.5428	$00^02 \rightarrow 02^01$ Band R(29)	1669
9.2605261	1079.852256	$00^01 \rightarrow 02^00$ Band R(22)	1666,6954
9.2644362	1079.3965	$00^02 \rightarrow 02^01$ Band R(27)	1669
9.2713580	1078.590644	$00^01 \rightarrow 02^00$ Band R(20)	1666,6954
9.2745176	1078.2232	$00^02 \rightarrow 02^01$ Band R(25)	1669
9.2824437	1077.302520	$00^01 \rightarrow 02^00$ Band R(18)	1666,6954
9.2848536	1077.0229	$00^02 \rightarrow 02^01$ Band R(23)	1669
9.2937855	1075.987820	$00^01 \rightarrow 02^00$ Band R(16)	1666,6954
9.2954478	1075.7954	$00^02 \rightarrow 02^01$ Band R(21)	1669
9.3053856	1074.646490	$00^01 \rightarrow 02^00$ Band R(14)	1666,6954
9.3063026	1074.5406	$00^02 \rightarrow 02^01$ Band R(19)	1669
9.3172463	1073.278484	$00^01 \rightarrow 02^00$ Band R(12)	1666,6954

$^{12}C^{16}O_2$ **Laser**—*continued*

Wavelength (μm) vac	Frequency (cm^{-1})	Transition assignment	Comments
9.3174189	1073.2586	$00^02 \rightarrow 02^01$ Band R(17)	1669
9.3288003	1071.9492	$00^02 \rightarrow 02^01$ Band R(15)	1669
9.3293698	1071.883766	$00^01 \rightarrow 02^00$ Band R(10)	1666,6954
9.3404485	1070.6124	$00^02 \rightarrow 02^01$ Band R(13)	1669
9.341758159	1070.462308	$00^01 \rightarrow 02^00$ Band R(8)	1665,1666
9.3417582	1070.462308	$00^01 \rightarrow 02^00$ Band R(8)	1665,1666,6954
9.3523664	1069.2481	$00^02 \rightarrow 02^01$ Band R(11)	1669
9.3544136	1069.014093	$00^01 \rightarrow 02^00$ Band R(6)	1666,6954
9.3645559	1067.8563	$00^02 \rightarrow 02^01$ Band R(9)	1669
9.3673383	1067.539110	$00^01 \rightarrow 02^00$ Band R(4)	1666,6954
9.3743826	1066.736914	$01^11 \rightarrow 03^10$ Band P(6)	6957
9.3770190	1066.4370	$00^02 \rightarrow 02^01$ Band R(7)	1669
9.3805343	1066.037360	$00^01 \rightarrow 02^00$ Band R(2)	1666,6954
9.3897578	1064.9902	$00^02 \rightarrow 02^01$ Band R(5)	1669
9.3940036	1064.508853	$00^01 \rightarrow 02^00$ Band R(0)	1666,6954
9.3960044	1064.282176	$01^11 \rightarrow 03^10$ Band P(9)	6957
9.4027743	1063.5159	$00^02 \rightarrow 02^01$ Band R(3)	1669
9.4038519	1063.394035	$01^11 \rightarrow 03^10$ Band P(10)	6957
9.4108801	1062.599869	$01^11 \rightarrow 03^10$ Band P(11)	6957
9.4147246	1062.165965	$00^01 \rightarrow 02^00$ Band P(2)	1666,6954
9.4190258	1061.680923	$01^11 \rightarrow 03^10$ Band P(12)	6957
9.4260250	1060.892579	$01^11 \rightarrow 03^10$ Band P(13)	6957
9.4288861	1060.570666	$00^01 \rightarrow 02^00$ Band P(4)	1666,6954
9.4344952	1059.940124	$01^11 \rightarrow 03^10$ Band P(14)	6957
9.4433280	1058.948714	$00^01 \rightarrow 02^00$ Band P(6)	1666,6954
9.4502621	1058.171715	$01^11 \rightarrow 03^10$ Band P(16)	6957
9.4505542	1058.1390	$00^02 \rightarrow 02^01$ Band P(3)	1669
9.4580521	1057.300161	$00^01 \rightarrow 02^00$ Band P(8)	1666,6954
9.4648480	1056.5410	$00^02 \rightarrow 02^01$ Band P(5)	1669
9.4663283	1056.375783	$01^11 \rightarrow 03^10$ Band P(18)	6957
9.4730604	1055.625068	$00^01 \rightarrow 02^00$ Band P(10)	1666,6954
9.4794322	1054.9155	$00^02 \rightarrow 02^01$ Band P(7)	1669
9.4826960	1054.552423	$01^11 \rightarrow 03^10$ Band P(20)	6957
9.4883547	1053.923503	$00^01 \rightarrow 02^00$ Band P(12)	1666,6954
9.4943075	1053.2627	$00^02 \rightarrow 02^01$ Band P(9)	1669
9.4993669	1052.701738	$01^11 \rightarrow 03^10$ Band P(22)	6957
9.5039368	1052.195545	$00^01 \rightarrow 02^00$ Band P(14)	1666,6954
9.5094756	1051.5827	$00^02 \rightarrow 02^01$ Band P(11)	1669
9.5198086	1050.441282	$00^01 \rightarrow 02^00$ Band P(16)	1666,6954
9.5249398	1049.8754	$00^02 \rightarrow 02^01$ Band P(13)	1669

$^{12}C^{16}O_2$ **Laser**—*continued*

Wavelength (μm) vac	Frequency (cm^{-1})	Transition assignment	Comments
9.5359719	1048.660810	$00^01 \to 02^00$ Band P(18)	1666,6954
9.5407011	1048.1410	$00^02 \to 02^01$ Band P(15)	1669
9.5524283	1046.854234	$00^01 \to 02^00$ Band P(20)	1666,6954
9.5567603	1046.3797	$00^02 \to 02^01$ Band P(17)	1669
9.569179556	1045.021670	$00^01 \to 02^00$ Band P(22)	1666,1667
9.5691796	1045.021670	$00^01 \to 02^00$ Band P(22)	1666,1667,6954
9.5731220	1044.5913	$00^02 \to 02^01$ Band P(19)	1669
9.5747417	1044.414592	$01^11 \to 03^10$ Band P(31)	6957
9.586227377	1043.163239	$00^01 \to 02^00$ Band P(24)	1665–1667
9.5862274	1043.163239	$00^01 \to 02^00$ Band P(24)	1665–67,6954
9.5897854	1042.7762	$00^02 \to 02^01$ Band P(21)	1669
9.6035734	1041.279074	$00^01 \to 02^00$ Band P(26)	1666,1667,6954
9.6067533	1040.9344	$00^02 \to 02^01$ Band P(23)	1669
9.6109041	1040.484840	$01^11 \to 03^10$ Band P(35)	6957
9.6212192	1039.369315	$00^01 \to 02^00$ Band P(28)	1666,1667,6954
9.6240268	1039.0661	$00^02 \to 02^01$ Band P(25)	1669
9.6294208	1038.484060	$01^11 \to 03^10$ Band P(37)	6957
9.6391664	1037.434110	$00^01 \to 02^00$ Band P(30)	1666,1667,6954
9.6416089·	1037.1713	$00^02 \to 02^01$ Band P(27)	1669
9.6438773	1036.927335	$01^11 \to 03^10$ Band P(38)	6957
9.6482303	1036.459509	$01^11 \to 03^10$ Band P(39)	6957
9.6574165	1035.473616	$00^01 \to 02^00$ Band P(32)	1666,1667,6954
9.6594997	1035.2503	$00^02 \to 02^01$ Band P(29)	1669
9.6633616	1034.836576	$01^11 \to 03^10$ Band P(40)	6957
9.6673343	1034.411316	$01^11 \to 03^10$ Band P(41)	6957
9.6759711	1033.487999	$00^01 \to 02^00$ Band P(34)	1666,1667,6954
9.6777025	1033.3031	$00^02 \to 02^01$ Band P(31)	1669
9.6831675	1032.719929	$01^11 \to 03^10$ Band P(42)	6957
9.6867347	1032.339616	$01^11 \to 03^10$ Band P(43)	6957
9.6948316	1031.477431	$00^01 \to 02^00$ Band P(36)	1666,6954
9.6962175	1031.3300	$00^02 \to 02^01$ Band P(33)	1669
9.7032967	1030.577576	$01^11 \to 03^10$ Band P(44)	6957
9.7064333	1030.244545	$01^11 \to 03^10$ Band P(45)	6957
9.7139995	1029.442092	$00^01 \to 02^00$ Band P(38)	1666,6954
9.7150460	1029.3312	$00^02 \to 02^01$ Band P(35)	1669
9.7237511	1028.409709	$01^11 \to 03^10$ Band P(46)	6957
9.7264319	1028.126249	$01^11 \to 03^10$ Band P(47)	6957
9.7334763	1027.382172	$00^01 \to 02^00$ Band P(40)	1666,6954
9.7341914	1027.3067	$00^02 \to 02^01$ Band P(37)	1669
9.7445322	1026.216524	$01^11 \to 03^10$ Band P(48)	6957

$^{12}C^{16}O_2$ Laser—*continued*

Wavelength (μm) vac	Frequency (cm⁻¹)	Transition assignment	Comments
9.7467324	1025.984876	$01^11 \rightarrow 03^10$ Band P(49)	6957
9.7532633	1025.297865	$00^01 \rightarrow 02^00$ Band P(42)	1666,6954
9.7536530	1025.2569	$00^02 \rightarrow 02^01$ Band P(39)	1669
9.7733618	1023.189375	$00^01 \rightarrow 02^00$ Band P(44)	1666,6954
9.7935337	1021.0819	$00^02 \rightarrow 02^01$ Band P(43)	1669
9.7937734	1021.056912	$00^01 \rightarrow 02^00$ Band P(46)	1666,6954
9.8139549	1018.9572	$00^02 \rightarrow 02^01$ Band P(45)	1669
9.8144992	1018.900693	$00^01 \rightarrow 02^00$ Band P(48)	1666,6954
9.8355405	1016.720942	$00^01 \rightarrow 02^00$ Band P(50)	1666,6954
9.8568987	1014.517888	$00^01 \rightarrow 02^00$ Band P(52)	1666,6954
9.8785749	1012.291768	$00^01 \rightarrow 02^00$ Band P(54)	1666,6954
9.9005703	1010.042824	$00^01 \rightarrow 02^00$ Band P(56)	1666,6954
9.9228862	1007.771305	$00^01 \rightarrow 02^00$ Band P(58)	1665,1666,6954
9.922886254	1007.771302	$00^01 \rightarrow 02^00$ Band P(58)	1665,1666
9.9455238	1005.477463	$00^01 \rightarrow 02^00$ Band P(60)	1666,6954
9.9684841	1003.161558	$00^01 \rightarrow 02^00$ Band P(62)	1666,6954
9.9917683	1000.823853	$00^01 \rightarrow 02^00$ Band P(64)	1666,6954
10.0153774	998.464617	$00^01 \rightarrow 02^00$ Band P(66)	1666,6954
10.0259118	997.415520	$00^01 \rightarrow 10^00$ Band R(62)	1665,1666,6954
10.0334675	996.664412	$00^01 \rightarrow 10^00$ Band R(60)	1666,6954
10.0413232	995.884686	$00^01 \rightarrow 10^00$ Band R(58)	1666,6954
10.0494775	995.076610	$00^01 \rightarrow 10^00$ Band R(56)	1666,6954
10.0579292	994.240441	$00^01 \rightarrow 10^00$ Band R(54)	1666,6954
10.0666774	993.376426	$00^01 \rightarrow 10^00$ Band R(52)	1666,6954
10.0757210	992.484802	$00^01 \rightarrow 10^00$ Band R(50)	1666,6954
10.0850594	991.565797	$00^01 \rightarrow 10^00$ Band R(48)	1666,6954
10.0946920	990.619628	$00^01 \rightarrow 10^00$ Band R(46)	1666,6954
10.1046181	989.646505	$00^01 \rightarrow 10^00$ Band R(44)	1666,6954
10.1148375	988.646625	$00^01 \rightarrow 10^00$ Band R(42)	1666,6954
10.1253500	987.620180	$00^01 \rightarrow 10^00$ Band R(40)	1666,6954
10.12615	987.5419	$00^02 \rightarrow 10^01$ Band R(45)	1669
10.1361554	986.567351	$00^01 \rightarrow 10^00$ Band R(38)	1666,6954
10.13624	986.5588	$00^02 \rightarrow 10^01$ Band R(43)	1669
10.14662	985.5494	$00^02 \rightarrow 10^01$ Band R(41)	1669
10.1472538	985.488311	$00^01 \rightarrow 10^00$ Band R(36)	1666,6954
10.15730	984.5140	$00^02 \rightarrow 10^01$ Band R(39)	1669
10.1586453	984.383225	$00^01 \rightarrow 10^00$ Band R(34)	1665,1666,6954
10.16826	983.4527	$00^02 \rightarrow 10^01$ Band R(37)	1669
10.1703302	983.252249	$00^01 \rightarrow 10^00$ Band R(32)	1666,6954
10.17951	982.3657	$00^02 \rightarrow 10^01$ Band R(35)	1669

$^{12}C^{16}O_2$ Laser—*continued*

Wavelength (μm) vac	Frequency (cm^{-1})	Transition assignment	Comments
10.1823088	982.095531	$00^01 \to 10^00$ Band R(30)	1666,6954
10.19105	981.2531	$00^02 \to 10^01$ Band R(33)	1669
10.1945818	980.913211	$00^01 \to 10^00$ Band R(28)	1666,6954
10.20288	980.1150	$00^02 \to 10^01$ Band R(31)	1669
10.21501	978.9517	$00^02 \to 10^01$ Band R(29)	1669
10.2200135	978.472286	$00^01 \to 10^00$ Band R(24)	1666,6954
10.22742	977.7633	$00^02 \to 10^01$ Band R(27)	1669
10.2331739	977.213922	$00^01 \to 10^00$ Band R(22)	1666,6954
10.24013	976.5498	$00^02 \to 10^01$ Band R(25)	1669
10.2466319	975.930439	$00^01 \to 10^00$ Band R(20)	1666,6954
10.25314	975.3114	$00^02 \to 10^01$ Band R(23)	1669
10.2603888	974.621939	$00^01 \to 10^00$ Band R(18)	1666,6954
10.26643	974.0482	$00^02 \to 10^01$ Band R(21)	1669
10.2744457	973.288517	$00^01 \to 10^00$ Band R(16)	1666,6954
10.28002	972.7604	$00^02 \to 10^01$ Band R(19)	1669
10.2888041	971.930258	$00^01 \to 10^00$ Band R(14)	1666,6954
10.29391	971.4480	$00^02 \to 10^01$ Band R(17)	1669
10.30243	970.6452	$00^03 \to 10^02$ Band R(20)	1669
10.3034655	970.547244	$00^01 \to 10^00$ Band R(12)	1666,6954
10.30810	970.1111	$00^02 \to 10^01$ Band R(15)	1669
10.31616	969.3531	$00^03 \to 10^02$ Band R(18)	1669
10.3184315	969.139547	$00^01 \to 10^00$ Band R(10)	1666,6954
10.32258	968.7499	$00^02 \to 10^01$ Band R(13)	1669
10.33018	968.0370	$00^03 \to 10^02$ Band R(16)	1669
10.3337039	967.707233	$00^01 \to 10^00$ Band R(8)	1666,6954
10.33737	967.3643	$00^02 \to 10^01$ Band R(11)	1669
10.34451	966.6968	$00^03 \to 10^02$ Band R(14)	1669
10.3492846	966.250361	$00^01 \to 10^00$ Band R(6)	1666,6954
10.35246	965.9544	$00^02 \to 10^01$ Band R(9)	1669
10.35912	965.3326	$00^03 \to 10^02$ Band R(12)	1669
10.3651757	964.768981	$00^01 \to 10^00$ Band R(4)	1666,6954
10.36785	964.5204	$00^02 \to 10^01$ Band R(7)	1669
10.37404	963.9446	$00^03 \to 10^02$ Band R(10)	1669
10.3813793	963.263140	$00^01 \to 10^00$ Band R(2)	1666,6954
10.38355	963.0622	$00^02 \to 10^01$ Band R(5)	1669
10.38926	962.5328	$00^03 \to 10^02$ Band R(8)	1669
10.3978977	961.732874	$00^01 \to 10^00$ Band R(0)	1666,6954
10.39955	961.5800	$00^02 \to 10^01$ Band R(3)	1669
10.40477	961.0973	$00^03 \to 10^02$ Band R(6)	1669
10.41587	960.0737	$00^02 \to 10^01$ Band R(1)	1669

$^{12}C^{16}O_2$ Laser—*continued*

Wavelength (µm) vac	Frequency (cm^{-1})	Transition assignment	Comments
10.4232708	959.391745	$00^01 \to 10^00$ Band P(2)	1666,6954
10.4405872	957.800537	$00^01 \to 10^00$ Band P(4)	1666,6954
10.45803	956.2031	$00^02 \to 10^01$ Band P(3)	1669
10.4582274	956.184982	$00^01 \to 10^00$ Band P(6)	1666,6954
10.47545	954.6130	$00^02 \to 10^01$ Band P(5)	1669
10.4761945	954.545086	$00^01 \to 10^00$ Band P(8)	1666,6954
10.4853390	953.712608	$01^11 \to 11^10$ Band R(39)	6957
10.49319	952.9989	$00^02 \to 10^01$ Band P(7)	1669
10.4944915	952.880849	$00^01 \to 10^00$ Band P(10)	1666,6954
10.5081576	951.641608	$01^11 \to 11^10$ Band R(36)	6957
10.51126	951.3608	$00^02 \to 10^01$ Band P(9)	1669
10.5131217	951.192263	$00^01 \to 10^00$ Band P(12)	1666,1667,6954
10.5227783	950.319366	$01^11 \to 11^10$ Band R(33)	6957
10.52965	949.6988	$00^02 \to 10^01$ Band P(11)	1669
10.5320883	949.479313	$00^01 \to 10^00$ Band P(14)	1666,1667,6954
10.5455003	948.271744	$01^11 \to 11^10$ Band R(30)	6957
10.54838	948.0129	$00^02 \to 10^01$ Band P(13)	1669
10.5513950	947.741978	$00^01 \to 10^00$ Band P(16)	1666,1667,6954
10.5585931	947.095881	$01^11 \to 11^10$ Band R(28)	6957
10.5628423	946.714881	$01^11 \to 11^10$ Band R(27)	6957
10.56744	946.3030	$00^02 \to 10^01$ Band P(15)	1669
10.5710455	945.980229	$00^01 \to 10^00$ Band P(18)	1666,1667,6954
10.5767857	945.466827	$01^11 \to 11^10$ Band R(25)	6957
10.5857525	944.665959	$01^11 \to 11^10$ Band R(24)	6957
10.58684	944.5691	$00^02 \to 10^01$ Band P(17)	1669
10.5910255	944.195629	$01^11 \to 11^10$ Band R(23)	1670
10.5910435	944.194030	$00^01 \to 10^00$ Band P(20)	1666,1667
10.60658	942.8111	$00^02 \to 10^01$ Band P(19)	1669
10.6113931	942.383336	$00^01 \to 10^00$ Band P(22)	1666,1667,6954
10.62666	941.0291	$00^02 \to 10^01$ Band P(21)	1669
10.6320985	940.548098	$00^01 \to 10^00$ Band P(24)	1666,1667,6954
10.64710	939.2229	$00^02 \to 10^01$ Band P(23)	1669
10.6531641	938.688257	$00^01 \to 10^00$ Band P(26)	1666,1667,6954
10.66789	937.3926	$00^02 \to 10^01$ Band P(25)	1669
10.6745944	936.803747	$00^01 \to 10^00$ Band P(28)	1666,1667,6954
10.68565	935.8348	$00^03 \to 10^02$ Band P(24)	1669
10.68904	935.5380	$00^02 \to 10^01$ Band P(27)	1669
10.6963941	934.894495	$00^01 \to 10^00$ Band P(30)	1666,1667
10.70652	934.0103	$00^03 \to 10^02$ Band P(26)	1669
10.71055	933.6591	$00^02 \to 10^01$ Band P(29)	1669

$^{12}C^{16}O_2$ Laser—*continued*

Wavelength (μm) vac	Frequency (cm^{-1})	Transition assignment	Comments
10.7185683	932.960420	$00^01 \rightarrow 10^00$ Band P(32)	1665–1667,6954
10.72775	932.1620	$00^03 \rightarrow 10^02$ Band P(28)	1669
10.73243	931.7558	$00^02 \rightarrow 10^01$ Band P(31)	1669
10.7411220	931.001433	$00^01 \rightarrow 10^00$ Band P(34)	1666,1667,6954
10.74934	930.2897	$00^03 \rightarrow 10^02$ Band P(30)	1669
10.75468	929.8281	$00^02 \rightarrow 10^01$ Band P(33)	1669
10.7640606	929.017437	$00^01 \rightarrow 10^00$ Band P(36)	1666,1667
10.77130	928.3934	$00^03 \rightarrow 10^02$ Band P(32)	1669
10.77731	927.8757	$00^02 \rightarrow 10^01$ Band P(35)	1669
10.7873897	927.008324	$00^01 \rightarrow 10^00$ Band P(38)	1666,1667,6954
10.7890778	926.863273	$01^11 \rightarrow 11^10$ Band Q(11)	1670
10.79362	926.4731	$00^03 \rightarrow 10^02$ Band P(34)	1669
10.80032	925.8987	$00^02 \rightarrow 10^01$ Band P(37)	1669
10.8111149	924.973984	$00^01 \rightarrow 10^00$ Band P(40)	1666,6954
10.81632	924.5285	$00^03 \rightarrow 10^02$ Band P(36)	1669
10.82372	923.8969	$00^02 \rightarrow 10^01$ Band P(39)	1669
10.83941	922.5596	$00^03 \rightarrow 10^02$ Band P(38)	1669
10.84751	921.8702	$00^02 \rightarrow 10^01$ Band P(41)	1669
10.8510181	921.572511	$01^11 \rightarrow 11^10$ Band P(7)	6957
10.8597782	920.829122	$00^01 \rightarrow 10^00$ Band P(44)	1666,6954
10.8704334	919.926522	$01^11 \rightarrow 11^10$ Band P(9)	6957
10.87171	919.8185	$00^02 \rightarrow 10^01$ Band P(43)	1669
10.8808344	919.047160	$01^11 \rightarrow 11^10$ Band P(10)	6957
10.8847289	918.718330	$00^01 \rightarrow 10^00$ Band P(46)	1666,6954
10.8901847	918.258073	$01^11 \rightarrow 11^10$ Band P(11)	1670,6955
10.89631	917.7417	$00^02 \rightarrow 10^01$ Band P(45)	1669
10.9009647	917.350005	$01^11 \rightarrow 11^10$ Band P(12)	1670,6955
10.9101014	916.581769	$00^01 \rightarrow 10^00$ Band P(48)	1666,6954
10.92133	915.6396	$00^02 \rightarrow 10^01$ Band P(47)	1669
10.9214689	915.627747	$01^11 \rightarrow 11^10$ Band P(14)	1670,6955
10.9307080	914.853825	$01^11 \rightarrow 11^10$ Band P(15)	1670,6955
10.9359024	914.419282	$00^01 \rightarrow 10^00$ Band P(50)	1666,6954
10.9423512	913.880373	$01^11 \rightarrow 11^10$ Band P(16)	1670,6955
10.94676	913.5120	$00^02 \rightarrow 10^01$ Band P(49)	1669
10.9514867	913.118031	$01^11 \rightarrow 11^10$ Band P(17)	1670,6955
10.9621393	912.230701	$00^01 \rightarrow 10^00$ Band P(52)	1666,6954
10.9636156	912.107867	$01^11 \rightarrow 11^10$ Band P(18)	6957
10.9726149	931.359785	$01^11 \rightarrow 11^10$ Band P(19)	1670,6955
10.9852663	910.310206	$01^11 \rightarrow 11^10$ Band P(20)	1670,6955
10.9888196	910.015851	$00^01 \rightarrow 10^00$ Band P(54)	1666,6954

$^{12}C^{16}O_2$ Laser—*continued*

Wavelength (µm) vac	Frequency (cm⁻¹)	Transition assignment	Comments
10.9940962	909.579088	$01^11 \rightarrow 11^10$ Band P(21)	1670,6955
11.0073078	908.487360	$01^11 \rightarrow 11^10$ Band P(22)	1670
11.0159341	907.775947	$01^11 \rightarrow 11^10$ Band P(23)	1670
11.0159511	907.774544	$00^01 \rightarrow 10^00$ Band P(56)	1666,6954
11.0297447	906.639296	$01^11 \rightarrow 11^10$ Band P(24)	1670,6955
11.0381334	905.950279	$01^11 \rightarrow 11^10$ Band P(25)	1670,6955
11.0435420	905.506585	$00^01 \rightarrow 10^00$ Band P(58)	1666,6954
11.0525819	904.765972	$01^11 \rightarrow 11^10$ Band P(26)	1670,6955
11.0606972	904.102139	$01^11 \rightarrow 11^10$ Band P(27)	1670,6955
11.0716007	903.211767	$00^01 \rightarrow 10^00$ Band P(60)	1666,6954
11.0758242	902.867343	$01^11 \rightarrow 11^10$ Band P(28)	1670,6955
11.0836301	902.231481	$01^11 \rightarrow 11^10$ Band P(29)	1670,6955
11.0994769	900.943357	$01^11 \rightarrow 11^10$ Band P(30)	1670,6955
11.1001359	900.889873	$00^01 \rightarrow 10^00$ Band P(62)	1666,6954
11.1069364	900.338283	$01^11 \rightarrow 11^10$ Band P(31)	1670,6955
11.1235453	898.993956	$01^11 \rightarrow 11^10$ Band P(32)	1670,6955
11.1291567	898.540674	$00^01 \rightarrow 10^00$ Band P(64)	1666,6954
11.1306204	898.422514	$01^11 \rightarrow 11^10$ Band P(33)	1670,6955
11.1480349	897.019077	$01^11 \rightarrow 11^10$ Band P(34)	1670,6955
11.1546870	896.484143	$01^11 \rightarrow 11^10$ Band P(35)	6957
11.1586727	896.163933	$00^01 \rightarrow 10^00$ Band P(66)	1666,6954
11.1729515	895.018650	$01^11 \rightarrow 11^10$ Band P(36)	6957
11.1791407	894.523133	$01^11 \rightarrow 11^10$ Band P(37)	6957
11.1886935	893.759399	$00^01 \rightarrow 10^00$ Band P(68)	1666,6954
11.1983011	892.992599	$01^11 \rightarrow 11^10$ Band P(38)	6957
11.2039866	892.539445	$01^11 \rightarrow 11^10$ Band P(39)	6957
11.2240898	890.940842	$01^11 \rightarrow 11^10$ Band P(40)	6957
11.2292297	890.533034	$01^11 \rightarrow 11^10$ Band P(41)	6957
11.2503240	888.863292	$01^11 \rightarrow 11^10$ Band P(42)	6957
11.2548752	888.503851	$01^11 \rightarrow 11^10$ Band P(43)	6957
11.2770103	886.759853	$01^11 \rightarrow 11^10$ Band P(44)	6957
11.2809286	886.451844	$01^11 \rightarrow 11^10$ Band P(45)	6957
15.5936	641.289	$02^00 \rightarrow 01^10$ Band R(29)	1673
15.6311	639.752	$02^00 \rightarrow 01^10$ Band R(27)	1673
15.6688	638.212	$02^00 \rightarrow 01^10$ Band R(25)	1673
15.7067	636.671	$02^00 \rightarrow 01^10$ Band R(23)	1673
15.7449	635.127	$02^00 \rightarrow 01^10$ Band R(21)	1673
15.7833	4533.581	$02^00 \rightarrow 01^10$ Band R(19)	1673
15.822	632.033	$02^00 \rightarrow 01^10$ Band R(17)	1673
15.8609	630.483	$02^00 \rightarrow 01^10$ Band R(15)	1673

$^{12}C^{16}O_2$ Laser—*continued*

Wavelength (μm) vac	Frequency (cm^{-1})	Transition assignment	Comments
15.900	628.932	$02^00 \rightarrow 01^10$ Band R(13)	1673
15.9394	627.378	$02^00 \rightarrow 01^10$ Band R(11)	1673
15.979	625.823	$02^00 \rightarrow 01^10$ Band R(9)	1673
16.1819	617.974	$02^00 \rightarrow 01^10$ BandQ(8)	1673
16.1827	617.945	$02^00 \rightarrow 01^10$ BandQ(10)	1673
16.1836	617.908	$02^00 \rightarrow 01^10$ BandQ(12)	1673
16.1847	617.866	$02^00 \rightarrow 01^10$ BandQ(14)	1673
16.186	617.816	$02^00 \rightarrow 01^10$ BandQ(16)	1673
16.1874	617.7453	$02^00 \rightarrow 01^10$ BandQ(18)	1673
16.1891	617.701	$02^00 \rightarrow 01^10$ BandQ(20)	1673
16.1908	617.634	$02^00 \rightarrow 01^10$ BandQ(22)	1673
16.1928	617.559	$02^00 \rightarrow 01^10$ BandQ(24)	1673
16.1949	617.477	$02^00 \rightarrow 01^10$ BandQ(26)	1673
16.1972	617.389	$02^00 \rightarrow 01^10$ BandQ(28)	1673
16.367	610.986	$02^00 \rightarrow 01^10$ Band P(9)	1673
16.4091	609.419	$02^00 \rightarrow 01^10$ Band P(11)	1673
16.4515	607.848	$02^00 \rightarrow 01^10$ Band P(13)	1673
16.4941	606.278	$02^00 \rightarrow 01^10$ Band P(15)	1673
16.537	604.706	$02^00 \rightarrow 01^10$ Band P(17)	1673
16.5801	603.132	$02^00 \rightarrow 01^10$ Band P(19)	1673
16.586	602.96	$14^00 \rightarrow 13^10$, Q(?)	1671
16.597	602.51	$14^00 \rightarrow 13^10$, Q(?)	1671
16.6235	601.558	$02^00 \rightarrow 01^10$ Band P(21)	1673
16.6672	599.981	$02^00 \rightarrow 01^10$ Band P(23)	1673
16.7111	598.404	$02^00 \rightarrow 01^10$ Band P(25)	1673
16.7553	596.825	$02^00 \rightarrow 01^10$ Band P(27)	1673
16.7998	595.244	$02^00 \rightarrow 01^10$ Band P(29)	1673
17.023	587.43	$03^11 \rightarrow 02^21$, Q(?)	1671
17.029	587.25	$03^11 \rightarrow 02^21$, Q(?)	1671
17.036	587.00	$03^11 \rightarrow 02^21$, Q(?)	1671
17.036	587.00	$03^11 \rightarrow 02^21$, Q(?)	1671
17.048	586.59	$03^11 \rightarrow 02^21$, Q(?)	1671
17.37	575.71	$24^00 \rightarrow 23^10$, Q(?)	1671
17.376	575.49	$24^00 \rightarrow 23^10$, Q(?)	1671
17.39	575.05	$24^00 \rightarrow 23^10$, Q(?)	1671
18.42	543.0	$03^10 \rightarrow 10^00$ Band	6838

$^{12}C^{16}O^{18}O$ Laser

Wavelength (μm) vac	Frequency (cm^{-1})	Transition assignment	Comments
4.314	2318	$02^01 \rightarrow 02^00$ Band R(8)	1674
4.340	2304	$02^01 \rightarrow 02^00$ Band P(10)	1674
4.354	2297	$02^01 \rightarrow 02^00$ Band P(19)	1674
9.1378432636	1094.3501340	$00^01 \rightarrow 02^00$ Band R(33)	6839
9.137843272	1094.350133	$00^01 \rightarrow 02^00$ Band R(33)	1675
9.1423917999	1093.8056713	$00^01 \rightarrow 02^00$ Band R(32)	6839
9.142391802	1093.805671	$00^01 \rightarrow 02^00$ Band R(32)	1675
9.1469912065	1093.2556700	$00^01 \rightarrow 02^00$ Band R(31)	6839
9.146991207	1093.255670	$00^01 \rightarrow 02^00$ Band R(31)	1675
9.1516417199	1092.7001194	$00^01 \rightarrow 02^00$ Band R(30)	6839
9.151641723	1092.700119	$00^01 \rightarrow 02^00$ Band R(30)	1675
9.156343568	1092.139010	$00^01 \rightarrow 02^00$ Band R(29)	1675
9.1563435701	1092.1390098	$00^01 \rightarrow 02^00$ Band R(29)	6839
9.1610969896	1091.5723315	$00^01 \rightarrow 02^00$ Band R(28)	6839
9.161096994	1091.572331	$00^01 \rightarrow 02^00$ Band R(28)	1675
9.1659022105	1091.0000751	$00^01 \rightarrow 02^00$ Band R(27)	6839
9.165902211	1091.000075	$00^01 \rightarrow 02^00$ Band R(27)	1675
9.170759460	1090.422232	$00^01 \rightarrow 02^00$ Band R(26)	1675
9.1707594613	1090.4222319	$00^01 \rightarrow 02^00$ Band R(26)	6839
9.1756689718	1089.8387933	$00^01 \rightarrow 02^00$ Band R(25)	6839
9.175668974	1089.838793	$00^01 \rightarrow 02^00$ Band R(25)	1675
9.1806309692	1089.2497513	$00^01 \rightarrow 02^00$ Band R(24)	6839
9.180630972	1089.249751	$00^01 \rightarrow 02^00$ Band R(24)	1675
9.1856456819	1088.6550980	$00^01 \rightarrow 02^00$ Band R(23)	6839
9.185645690	1088.655097	$00^01 \rightarrow 02^00$ Band R(23)	1675
9.1907133355	1088.0548261	$00^01 \rightarrow 02^00$ Band R(22)	6839
9.190713345	1088.054825	$00^01 \rightarrow 02^00$ Band R(22)	1675
9.1958341547	1087.4489287	$00^01 \rightarrow 02^00$ Band R(21)	6839
9.195834169	1087.448927	$00^01 \rightarrow 02^00$ Band R(21)	1675
9.2010083628	1086.8373993	$00^01 \rightarrow 02^00$ Band R(20)	6839
9.201008374	1086.837398	$00^01 \rightarrow 02^00$ Band R(20)	1675
9.2062361840	1086.2202316	$00^01 \rightarrow 02^00$ Band R(19)	6839
9.206236198	1086.220230	$00^01 \rightarrow 02^00$ Band R(19)	1675
9.2115178379	1085.5974201	$00^01 \rightarrow 02^00$ Band R(18)	6839
9.211517856	1085.597418	$00^01 \rightarrow 02^00$ Band R(18)	1675
9.2168535462	1084.9689593	$00^01 \rightarrow 02^00$ Band R(17)	6839
9.216853566	1084.968957	$00^01 \rightarrow 02^00$ Band R(17)	1675
9.2222435271	1084.3348444	$00^01 \rightarrow 02^00$ Band R(16)	6839
9.222243539	1084.334843	$00^01 \rightarrow 02^00$ Band R(16)	1675

$^{12}C^{16}O^{18}O$ Laser—*continued*

Wavelength (μm) vac	Frequency (cm^{-1})	Transition assignment	Comments
9.2276879987	1083.6950709	$00^01 \rightarrow 02^00$ Band R(15)	6839
9.227688015	1083.695069	$00^01 \rightarrow 02^00$ Band R(15)	1675
9.2331871778	1083.0496347	$00^01 \rightarrow 02^00$ Band R(14)	6839
9.233187192	1083.049633	$00^01 \rightarrow 02^00$ Band R(14)	1675
9.2387412792	1082.3985322	$00^01 \rightarrow 02^00$ Band R(13)	6839
9.238741289	1082398531	$00^01 \rightarrow 02^00$ Band R(13)	1675
9.2443505168	1081.7417602	$00^01 \rightarrow 02^00$ Band R(12)	6839
9.244350527	1091.741759	$00^01 \rightarrow 02^00$ Band R(12)	1675
9.2500151042	1081.0793158	$00^01 \rightarrow 02^00$ Band R(11)	6839
9.250015111	1081.079315	$00^01 \rightarrow 02^00$ Band R(11)	1675
9.2557352512	1080.4111968	$00^01 \rightarrow 02^00$ Band R(10)	6839
9.255735258	1080.411196	$00^01 \rightarrow 02^00$ Band R(10)	1675
9.417384992	1061.865901	$00^01 \rightarrow 02^00$ Band P(14)	1675
9.4173850048	1061.8658996	$00^01 \rightarrow 02^00$ Band P(14)	6839
9.424619168	1061.050831	$00^01 \rightarrow 02^00$ Band P(15)	1675
9.4246191839	1061.0508292	$00^01 \rightarrow 02^00$ Band P(15)	6839
9.431914062	1060.230186	$00^01 \rightarrow 02^00$ Band P(16)	1675
9.4319140777	1060.2301842	$00^01 \rightarrow 02^00$ Band P(16)	6839
9.439269841	1059.403976	$00^01 \rightarrow 02^00$ Band P(17)	1675
9.4392698552	1059.4039744	$00^01 \rightarrow 02^00$ Band P(17)	6839
9.446686675	1058.572211	$00^01 \rightarrow 02^00$ Band P(18)	1675
9.4466866846	1058.5722099	$00^01 \rightarrow 02^00$ Band P(18)	6839
9.454164717	1057.734903	$00^01 \rightarrow 02^00$ Band P(19)	1675
9.4541647314	1057.7349014	$00^01 \rightarrow 02^00$ Band P(19)	6839
9.461704150	1056.892061	$00^01 \rightarrow 02^00$ Band P(20)	1675
9.4617041611	1056.8920598	$00^01 \rightarrow 02^00$ Band P(20)	6839
9.469305123	1056.043698	$00^01 \rightarrow 02^00$ Band P(21)	1675
9.4693051359	1056.0436966	$00^01 \rightarrow 02^00$ Band P(21)	6839
9.476967805	1055.189825	$00^01 \rightarrow 02^00$ Band P(22)	1675
9.4769678179	1055.1898236	$00^01 \rightarrow 02^00$ Band P(22)	6839
9.484692358	1054.330454	$00^01 \rightarrow 02^00$ Band P(23)	1675
9.4846923680	1054.3304529	$00^01 \rightarrow 02^00$ Band P(23)	6839
9.492478928	1053.465599	$00^01 \rightarrow 02^00$ Band P(24)	1675
9.4924789434	1053.4655973	$00^01 \rightarrow 02^00$ Band P(24)	6839
9.500327691	1052.595271	$00^01 \rightarrow 02^00$ Band P(25)	1675
9.5003277023	1052.5952697	$00^01 \rightarrow 02^00$ Band P(25)	6839
9.508238787	1051.719485	$00^01 \rightarrow 02^00$ Band P(26)	1675
9.5082387994	1051.7194836	$00^01 \rightarrow 02^00$ Band P(26)	6839
9.5162123445	1050.8392527	$00^01 \rightarrow 02^00$ Band P(27)	6839
9.516212378	1050.838254	$00^01 \rightarrow 02^00$ Band P(27)	1675

$^{12}C^{16}O^{18}O$ Laser—*continued*

Wavelength (μm) vac	Frequency (cm^{-1})	Transition assignment	Comments
9.524248619	1049.951592	$00^01 \rightarrow 02^00$ Band P(28)	1675
9.5242486252	1049.9515913	$00^01 \rightarrow 02^00$ Band P(28)	6839
9.532347648	1049.059515	$00^01 \rightarrow 02^00$ Band P(29)	1675
9.5323476576	1049.0595139	$00^01 \rightarrow 02^00$ Band P(29)	6839
9.540509632	1048.162036	$00^01 \rightarrow 02^00$ Band P(30)	1675
9.5405096371	1048.1620354	$00^01 \rightarrow 02^00$ Band P(30)	6839
9.5487347124	1047.2591711	$00^01 \rightarrow 02^00$ Band P(31)	6839
9.548734713	1047.259171	$00^01 \rightarrow 02^00$ Band P(31)	1675
9.557023028	1046.350937	$00^01 \rightarrow 02^00$ Band P(32)	1675
9.5570230286	1046.3509369	$00^01 \rightarrow 02^00$ Band P(32)	6839
9.5653747346	1045.4373485	$00^01 \rightarrow 02^00$ Band P(33)	6839
9.565374739	1045.437348	$00^01 \rightarrow 02^00$ Band P(33)	1675
9.5737899740	1044.5184224	$00^01 \rightarrow 02^00$ Band P(34)	6839
9.573789978	1044.518422	$00^01 \rightarrow 02^00$ Band P(34)	1675
9.5822688902	1043.5941753	$00^01 \rightarrow 02^00$ Band P(35)	6839
9.582268893	1043.594175	$00^01 \rightarrow 02^00$ Band P(35)	1675
9.5908116271	1042.6646241	$00^01 \rightarrow 02^00$ Band P(36)	6839
9.590811628	1042.664624	$00^01 \rightarrow 02^00$ Band P(36)	1675
9.599418318	1041.729787	$00^01 \rightarrow 02^00$ Band P(37)	1675
9.5994183247	1041.7297863	$00^01 \rightarrow 02^00$ Band P(37)	6839
9.608089110	1040.789681	$00^01 \rightarrow 02^00$ Band P(38)	1675
9.608089129	1040.799679	$00^01 \rightarrow 02^00$ Band P(38)	6839
10.087613	991.31477	$00^01 \rightarrow 10^00$ Band R(41)	1675
10.08761317	991.3147770	$00^01 \rightarrow 10^00$ Band R(41)	6839
10.092287	990.85569	$00^01 \rightarrow 10^00$ Band R(40)	1675
10.09228700	990.8556897	$00^01 \rightarrow 10^00$ Band R(40)	6839
10.097035	990.38971	$00^01 \rightarrow 10^00$ Band R(39)	1675
10.09703546	990.3897076	$00^01 \rightarrow 10^00$ Band R(39)	6839
10.10185856	989.9168499	$00^01 \rightarrow 10^00$ Band R(38)	6839
10.101859	989.91685	$00^01 \rightarrow 10^00$ Band R(38)	1675
10.1067563	989.437134	$00^01 \rightarrow 10^00$ Band R(37)	1675
10.10675630	989.4371353	$00^01 \rightarrow 10^00$ Band R(37)	6839
10.1117287	988.950581	$00^01 \rightarrow 10^00$ Band R(36)	1675
10.11172872	988.9505820	$00^01 \rightarrow 10^00$ Band R(36)	6839
10.1167758	988.457207	$00^01 \rightarrow 10^00$ Band R(35)	1675
10.11677584	998.4572079	$00^01 \rightarrow 10^00$ Band R(35)	6839
10.1218977	987.957029	$00^01 \rightarrow 10^00$ Band R(34)	1675
10.12189771	987.9570304	$00^01 \rightarrow 10^00$ Band R(34)	6839
10.12709436	987.4500664	$00^01 \rightarrow 10^00$ Band R(33)	6839
10.1270944	987.450065	$00^01 \rightarrow 10^00$ Band R(33)	1675

$^{12}C^{16}O^{18}O$ Laser—*continued*

Wavelength (μm) vac	Frequency (cm⁻¹)	Transition assignment	Comments
10.13236586	986.9363325	$00^01 \rightarrow 10^00$ Band R(32)	6839
10.1323659	986.936332	$00^01 \rightarrow 10^00$ Band R(32)	1675
10.13771226	986.4158449	$00^01 \rightarrow 10^00$ Band R(31)	6839
10.1377123	986.415844	$00^01 \rightarrow 10^00$ Band R(31)	1675
10.1431336	985.998618	$00^01 \rightarrow 10^00$ Band R(30)	1675
10.14313362	985.8886190	$00^01 \rightarrow 10^00$ Band R(30)	6839
10.1486300	985.354669	$00^01 \rightarrow 10^00$ Band R(29)	1675
10.14863003	985.3546703	$00^01 \rightarrow 10^00$ Band R(29)	6839
10.15420157	984.8140135	$00^01 \rightarrow 10^00$ Band R(28)	6839
10.1542016	984.814012	$00^01 \rightarrow 10^00$ Band R(28)	1675
10.1598483	984.266661	$00^01 \rightarrow 10^00$ Band R(27)	1675
10.15984832	984.2666631	$00^01 \rightarrow 10^00$ Band R(27)	6839
10.16557038	983.7126330	$00^01 \rightarrow 10^00$ Band R(26)	6839
10.1655704	983.712631	$00^01 \rightarrow 10^00$ Band R(26)	1675
10.17136785	983.1519369	$00^01 \rightarrow 10^00$ Band R(25)	6839
10.1713679	983.151934	$00^01 \rightarrow 10^00$ Band R(25)	1675
10.17724084	982.5845878	$00^01 \rightarrow 10^00$ Band R(24)	6839
10.1772409	982.584585	$00^01 \rightarrow 10^00$ Band R(24)	1675
10.18318948	982.0105996	$00^01 \rightarrow 10^00$ Band R(23)	6839
10.1831895	982.010596	$00^01 \rightarrow 10^00$ Band R(23)	1675
10.18921389	981.4299815	$00^01 \rightarrow 10^00$ Band R(22)	6839
10.1892139	981.429979	$00^01 \rightarrow 10^00$ Band R(22)	1675
10.1953142	980.942745	$00^01 \rightarrow 10^00$ Band R(21)	1675
10.19531420	980.8427486	$00^01 \rightarrow 10^00$ Band R(21)	6839
10.20149054	980.2489113	$00^01 \rightarrow 10^00$ Band R(20)	6839
10.2014906	980.248908	$00^01 \rightarrow 10^00$ Band R(20)	1675
10.20774308	979.6484808	$00^01 \rightarrow 10^00$ Band R(19)	6839
10.2077431	979.648477	$00^01 \rightarrow 10^00$ Band R(19)	1675
10.21407196	979.0414677	$00^01 \rightarrow 10^00$ Band R(18)	6839
10.2140720	979.041464	$00^01 \rightarrow 10^00$ Band R(18)	1675
10.22047734	978.4278825	$00^01 \rightarrow 10^00$ Band R(17)	6839
10.2204774	978.427879	$00^01 \rightarrow 10^00$ Band R(17)	1675
10.22695939	977.9077349	$00^01 \rightarrow 10^00$ Band R(16)	6839
10.2269594	977.807731	$00^01 \rightarrow 10^00$ Band R(16)	1675
10.43178618	958.6086056	$00^01 \rightarrow 10^00$ Band P(10)	6839
10.4317862	958.608605	$00^01 \rightarrow 10^00$ Band P(10)	1675
10.4405152	957.807139	$00^01 \rightarrow 10^00$ Band P(11)	1675
10.44051521	957.8071387	$00^01 \rightarrow 10^00$ Band P(11)	6839
10.4493292	956.999226	$00^01 \rightarrow 10^00$ Band P(12)	1675
10.44932925	956.9992260	$00^01 \rightarrow 10^00$ Band P(12)	6839

$^{12}C^{16}O^{18}O$ Laser—*continued*

Wavelength (μm) vac	Frequency (cm^{-1})	Transition assignment	Comments
10.45822869	956.1848657	$00^01 \to 10^00$ Band P(13)	6839
10.4582287	956.184866	$00^01 \to 10^00$ Band P(13)	1675
10.46721398	955.3640555	$00^01 \to 10^00$ Band P(14)	6839
10.4672140	955.364055	$00^01 \to 10^00$ Band P(14)	1675
10.4762855	954.536793	$00^01 \to 10^00$ Band P(15)	1675
10.47628554	954.5367929	$00^01 \to 10^00$ Band P(15)	6839
10.4854438	911.701074	$00^01 \to 10^00$ Band P(16)	1675
10.48544381	953.7030747	$00^01 \to 10^00$ Band P(16)	6839
10.4946892	952.862897	$00^01 \to 10^00$ Band P(17)	1675
10.49468924	952.9628975	$00^01 \to 10^00$ Band P(17)	6839
10.5040223	952.016257	$00^01 \to 10^00$ Band P(18)	1675
10.50402230	952.0162575	$00^01 \to 10^00$ Band P(18)	6839
10.51344346	951.1631505	$00^01 \to 10^00$ Band P(19)	6839
10.5134435	951.163150	$00^01 \to 10^00$ Band P(19)	1675
10.52295319	950.3035718	$00^01 \to 10^00$ Band P(20)	6839
10.5229532	950.303571	$00^01 \to 10^00$ Band P(20)	1675
10.53255199	949.4375165	$00^01 \to 10^00$ Band P(21)	6839
10.5325520	949.437516	$00^01 \to 10^00$ Band P(21)	1675
10.54224035	948.5649790	$00^01 \to 10^00$ Band P(22)	6839
10.5422404	948.564979	$00^01 \to 10^00$ Band P(22)	1675
10.5520188	947.685953	$00^01 \to 10^00$ Band P(23)	1675
10.55201880	947.6859535	$00^01 \to 10^00$ Band P(23)	6839
10.56188785	946.8004338	$00^01 \to 10^00$ Band P(24)	6839
10.5618879	946 800433	$00^01 \to 10^00$ Band P(24)	1675
10.5718480	945.908413	$00^01 \to 10^00$ Band P(25)	1675
10.57184804	945.9084132	$00^01 \to 10^00$ Band P(25)	6839
10.5818999	945.009884	$00^01 \to 10^00$ Band P(26)	1675
10.58189990	945.0098846	$00^01 \to 10^00$ Band P(26)	6839
10.59204399	944.1048407	$00^01 \to 10^00$ Band P(27)	6839
10.5920440	944.104840	$00^01 \to 10^00$ Band P(27)	1675
10.60228087	943.1932733	$00^01 \to 10^00$ Band P(28)	6839
10.6022809	943.191273	$00^01 \to 10^00$ Band P(28)	1675
10.6126111	942.275174	$00^01 \to 10^00$ Band P(29)	1675
10.61261113	942.2751743	$00^01 \to 10^00$ Band P(29)	6839
10.62303534	941,3505350	$00^01 \to 10^00$ Band P(30)	6839
10.6230354	941,350534	$00^01 \to 10^00$ Band P(30)	1675
10.6335541	940.419345	$00^01 \to 10^00$ Band P(31)	1675
10.63355411	940.4193461	$00^01 \to 10^00$ Band P(31)	6839
10.64416804	939.4815981	$00^01 \to 10^00$ Band P(32)	6839
10.6441681	939.481597	$00^01 \to 10^00$ Band P(32)	1675

$^{12}C^{16}O^{18}O$ Laser—*continued*

Wavelength (μm) vac	Frequency (cm^{-1})	Transition assignment	Comments
10.65487776	938.5372810	$00^01 \to 10^00$ Band P(33)	6839
10.6548778	938.537280	$00^01 \to 10^00$ Band P(33)	1675
10.6656839	937.586384	$00^01 \to 10^00$ Band P(34)	1675
10.66568390	937,5863843	$00^01 \to 10^00$ Band P(34)	6839
10.6765871	936.628897	$00^01 \to 10^00$ Band P(35)	1675
10.67658710	936.6288973	$00^01 \to 10^00$ Band P(35)	6839
10.6875880	935.664808	$00^01 \to 10^00$ Band P(36)	1675
10.68758802	935.6648086	$00^01 \to 10^00$ Band P(36)	6839
10.6986873	934.694106	$00^01 \to 10^00$ Band P(37)	1675
10.69868734	934.6941065	$00^01 \to 10^00$ Band P(37)	6839
10.7098857	933.716779	$00^01 \to 10^00$ Band P(38)	1675
10.70988572	933.7167788	$00^01 \to 10^00$ Band P(38)	6839
10.72118388	932.7328129	$00^01 \to 10^00$ Band P(39)	6839
10.721184	932.73281	$00^01 \to 10^00$ Band P(39)	1675
10.732582	931.74220	$00^01 \to 10^00$ Band P(40)	1675
10.73258251	931.7421959	$00^01 \to 10^00$ Band P(40)	6839
16.596	602.6	$02^00 \to 01^10$ Band R(6)	1674
16.76	596.7	$02^00 \to 01^10$ Band R(8)	1674
16.927	590.8	$02^00 \to 01^10$ Band P(9)	1674
16.97	589.3	$02^00 \to 01^10$ Band P(11)	1674

$^{12}C^{17}O_2$ Laser

Wavelength (μm) vac	Frequency (cm^{-1})	Transition assignment	Comments
9.0903830147	1100.0636589	$00^01 \to 02^00$ Band R(45)	6840
9.0943359524	1099.5855060	$00^01 \to 02^00$ Band R(44)	6840
9.0983367323	1099.1019891	$00^01 \to 02^00$ Band R(43)	6840
9.1023855930	1098,6130941	$00^01 \to 02^00$ Band R(42)	6840
9.1064827734	1098.1188071	$00^01 \to 02^00$ Band R(41)	6840
9.1106285129	1097.6191144	$00^01 \to 02^00$ Band R(40)	6840
9.1148230497	1097.1140027	$00^01 \to 02^00$ Band R(39)	6840
9.1190666207	1096.6034591	$00^01 \to 02^00$ Band R(38)	6840
9.1233594639	1096.0874708	$00^01 \to 02^00$ Band R(37)	6840
9.1277018147	1095.5660256	$00^01 \to 02^00$ Band R(36)	6840
9.1320939079	1095.0391116	$00^01 \to 02^00$ Band R(35)	6840
9.1365359806	1094.5067169	$00^01 \to 02^00$ Band R(34)	6840
9.1410282644	1D93.9688305	$00^01 \to 02^00$ Band R(33)	6840
9.1455709948	1093.4254412	$00^01 \to 02^00$ Band R(32)	6840

$^{12}C^{17}O_2$ Laser—*continued*

Wavelength (μm) vac	Frequency (cm^{-1})	Transition assignment	Comments
9.1501644027	10918765386	$00^01 \to 02^00$ Band R(31)	6840
9.1548087203	1092.3221124	$00^01 \to 02^00$ Band R(30)	6840
9.1595041789	1091.7621527	$00^01 \to 02^00$ Band R(29)	6840
9.1642510074	1091.1966501	$00^01 \to 02^00$ Band R(28)	6840
9.1690494356	1090.6255954	$00^01 \to 02^00$ Band R(27)	6840
9.1738996911	1090.0489799	$00^01 \to 02^00$ Band R(26)	6840
9.1788020012	1099.4667952	$00^01 \to 02^00$ Band R(25)	6840
9.1837565912	1089.8790334	$00^01 \to 02^00$ Band R(24)	6840
9.1887636870	1089.2856968	$00^01 \to 02^00$ Band R(23)	6840
9.1938235127	1087.6867482	$00^01 \to 02^00$ Band R(22)	6840
9.1989362908	1087.0822108	$00^01 \to 02^00$ Band R(21)	6840
9.2041022431	1086.4720682	$00^01 \to 02^00$ Band R(20)	6840
9.2093215901	1085.8563144	$00^01 \to 02^00$ Band R(19)	6840
9.2145945513	1085.2349438	$00^01 \to 02^00$ Band R(18)	6840
9.2199213456	1084.6079511	$00^01 \to 02^00$ Band R(17)	6840
9.2253021895	1083.9753316	$00^01 \to 02^00$ Band R(16)	6840
9.2307372990	10833370809	$00^01 \to 02^00$ Band R(15)	6840
9.2362268897	1082.6931949	$00^01 \to 02^00$ Band R(14)	6840
9.2417711746	1082.0436701	$00^01 \to 02^00$ Band R(13)	6840
9.2473703656	1081.3885034	$00^01 \to 02^00$ Band R(12)	6840
9.2530246740	1080.7276920	$00^01 \to 02^00$ Band R(11)	6840
9.2587343087	1080.0612337	$00^01 \to 02^00$ Band R(10)	6840
9.2644994798	1079.3891264	$00^01 \to 02^00$ Band R(9)	6840
9.2703203918	1078.7113689	$00^01 \to 02^00$ Band R(8)	6840
9.2761972527	1078.0279599	$00^01 \to 02^00$ Band R(7)	6840
9.2821302649	1077.3388990	$00^01 \to 02^00$ Band R(6)	6840
9.2881196322	1076.6441859	$00^01 \to 02^00$ Band R(5)	6840
9.3711993962	1067.0992663	$00^01 \to 02^00$ Band P(7)	6840
9.3779982255	1066.3256443	$00^01 \to 02^00$ Band P(8)	6840
9.3848562790	1065.5464189	$00^01 \to 02^00$ Band P(9)	6840
9.3917737361	1064.7615968	$00^01 \to 02^00$ Band P(10)	6840
9.3987507727	1063.9711853	$00^01 \to 02^00$ Band P(11)	6840
9.4057875665	1063.1751918	$00^01 \to 02^00$ Band P(12)	6840
9.4128842907	1062.3736244	$00^01 \to 02^00$ Band P(13)	6840
9.4200411203	1061.5664913	$00^01 \to 02^00$ Band P(14)	6840
9.4272582269	1060.7538013	$00^01 \to 02^00$ Band P(15)	6840
9.4345357789	1059.9355638	$00^01 \to 02^00$ Band P(16)	6840
9.4418739452	1059.1117884	$00^01 \to 02^00$ Band P(17)	6840
9.4492728943	1058.2824850	$00^01 \to 02^00$ Band P(18)	6840
9.4567327911	1057.4476641	$00^01 \to 02^00$ Band P(19)	6840

$^{12}C^{17}O_2$ Laser—*continued*

Wavelength (μm) vac	Frequency (cm^{-1})	Transition assignment	Comments
9.4642537986	1056.0073367	$00^01 \to 02^00$ Band P(20)	6840
9.4718360798	1055.7615140	$00^01 \to 02^00$ Band P(21)	6840
9.4794797973	1054.9102075	$00^01 \to 02^00$ Band P(22)	6840
9.4871851086	1054.0534295	$00^01 \to 02^00$ Band P(23)	6840
9.4949521722	1053.1911924	$00^01 \to 02^00$ Band P(24)	6840
9.5027811452	1052.3235090	$00^01 \to 02^00$ Band P(25)	6840
9.5106721823	1051,4503926	$00^01 \to 02^00$ Band P(26)	6840
9.5186254374	1050.5718568	$00^01 \to 02^00$ Band P(27)	6840
9.5266410620	1049.6879157	$00^01 \to 02^00$ Band P(28)	6840
9.5347192086	1048.7985835	$00^01 \to 02^00$ Band P(29)	6840
9.5428600243	1047.9038752	$00^01 \to 02^00$ Band P(30)	6840
9.5510636595	1047.0038057	$00^01 \to 02^00$ Band P(31)	6840
9.5593302580	1046.0983908	$00^01 \to 02^00$ Band P(32)	6840
9.5676599674	1045.1876461	$00^01 \to 02^00$ Band P(33)	6840
9.5760529309	1044 2715879	$00^01 \to 02^00$ Band P(34)	6840
9.5845092900	1043.3502329	$00^01 \to 02^00$ Band P(35)	6840
9.5930291871	1042.4235979	$00^01 \to 02^00$ Band P(36)	6840
9.6016127618	1041.4917002	$00^01 \to 02^00$ Band P(37)	6840
9.6102601521	1040.5545575	$00^01 \to 02^00$ Band P(38)	6840
9.6189714956	1039.6121877	$00^01 \to 02^00$ Band P(39)	6840
9.6277469294	1038.6646090	$00^01 \to 02^00$ Band P(40)	6840
9.6365865884	1037.7118400	$00^01 \to 02^00$ Band P(41)	6840
9.6454906048	1036.7538998	$00^01 \to 02^00$ Band P(42)	6840
9.6544591124	1035.7908075	$00^01 \to 02^00$ Band P(43)	6840
9.6634922422	1034.8225827	$00^01 \to 02^00$ Band P(44)	6840
9.6725901242	1033.8492453	$00^01 \to 02^00$ Band P(45)	6840
9.6817528890	1032.9708153	$00^01 \to 02^00$ Band P(46)	6840
9.6909806643	1031.8873132	$00^01 \to 02^00$ Band P(47)	6840
9.7002735768	1030.8987598	$00^01 \to 02^00$ Band P(48)	6840
9.7096317526	1029.9051761	$00^01 \to 02^00$ Band P(49)	6840
9.719055321	1028.906583	$00^01 \to 02^00$ Band P(50)	6840
9.728544397	1027.903003	$00^01 \to 02^00$ Band P(51)	6840
10.11386934	988.7412689	$00^01 \to 10^00$ Band R(37)	6840
10.11882628	988.2569109	$00^01 \to 10^00$ Band R(36)	6840
10.12385803	987.7657279	$00^01 \to 10^00$ Band R(35)	6840
10.12896464	987.2677371	$00^01 \to 10^00$ Band R(34)	6840
10.13414614	986.7629555	$00^01 \to 10^00$ Band R(33)	6840
10.13940260	986.2513992	$00^01 \to 10^00$ Band R(32)	6840
10.14473407	985.7330842	$00^01 \to 10^00$ Band R(31)	6840
10.15014061	985.2080262	$00^01 \to 10^00$ Band R(30)	6840

$^{12}C^{17}O_2$ Laser—*continued*

Wavelength (μm) vac	Frequency (cm⁻¹)	Transition assignment	Comments
10.15562232	984.6762402	$00^01 \rightarrow 10^00$ Band R(29)	6840
10.16117926	984.1377408	$00^01 \rightarrow 10^00$ Band R(28)	6840
10.16681153	983.5925424	$00^01 \rightarrow 10^00$ Band R(27)	6840
10.17251922	983.0406589	$00^01 \rightarrow 10^00$ Band R(26)	6840
10.17830245	982.4821037	$00^01 \rightarrow 10^00$ Band R(25)	6840
10.18416131	981.9168899	$00^01 \rightarrow 10^00$ Band R(24)	6840
10.19009593	981.3450300	$00^01 \rightarrow 10^00$ Band R(23)	6840
10.19610644	980.7665364	$00^01 \rightarrow 10^00$ Band R(22)	6840
10.20219297	980.1814209	$00^01 \rightarrow 10^00$ Band R(21)	6840
10.20835565	979.5896950	$00^01 \rightarrow 10^00$ Band R(20)	6840
10.21459464	978.9913695	$00^01 \rightarrow 10^00$ Band R(19)	6840
10.22091010	978.3864553	$00^01 \rightarrow 10^00$ Band R(18)	6840
10.22730217	977.7749625	$00^01 \rightarrow 10^00$ Band R(17)	6840
10.23377105	977.1569009	$00^01 \rightarrow 10^00$ Band R(16)	6840
10.24031689	976.5322800	$00^01 \rightarrow 10^00$ Band R(15)	6840
10.24693989	975.9011088	$00^01 \rightarrow 10^00$ Band R(14)	6840
10.25364024	9751633961	$00^01 \rightarrow 10^00$ Band R(13)	6840
10.26041813	974.6191499	$00^01 \rightarrow 10^00$ Band R(12)	6840
10.26727379	973.9683783	$00^01 \rightarrow 10^00$ Band R(11)	6840
10.27420741	973.3110886	$00^01 \rightarrow 10^00$ Band R(10)	6840
10.28121923	972.6472880	$00^01 \rightarrow 10^00$ Band R(9)	6840
10.44700792	957.2118716	$00^01 \rightarrow 10^00$ Band P(11)	6840
10.45581237	956.4058384	$00^01 \rightarrow 10^00$ Band P(12)	6840
10.46470237	955.5933503	$00^01 \rightarrow 10^00$ Band P(13)	6840
10.47367833	954.7744052	$00^01 \rightarrow 10^00$ Band P(14)	6840
10.48274069	953.9490005	$00^01 \rightarrow 10^00$ Band P(15)	6840
10.49188988	953.1171331	$00^01 \rightarrow 10^00$ Band P(16)	6840
10.50112635	952.2787997	$00^01 \rightarrow 10^00$ Band P(17)	6840
10.51045058	951.4339965	$00^01 \rightarrow 10^00$ Band P(18)	6840
10.51986303	950.5827192	$00^01 \rightarrow 10^00$ Band P(19)	6840
10.52936417	949.7249634	$00^01 \rightarrow 10^00$ Band P(20)	6840
10.53895450	948.8607241	$00^01 \rightarrow 10^00$ Band P(21)	6840
10.54863453	947.9899958	$00^01 \rightarrow 10^00$ Band P(22)	6840
10.55840475	947.1127728	$00^01 \rightarrow 10^00$ Band P(23)	6840
10.56826570	946.2297989	$00^01 \rightarrow 10^00$ Band P(24)	6840
10.57821790	945.3388175	$00^01 \rightarrow 10^00$ Band P(25)	6840
10.58826190	944.4420716	$00^01 \rightarrow 10^00$ Band P(26)	6840
10.59839824	943.5388038	$00^01 \rightarrow 10^00$ Band P(27)	6840
10.60862750	942.6290064	$00^01 \rightarrow 10^00$ Band P(28)	6840
10.61895025	941.7126710	$00^01 \rightarrow 10^00$ Band P(29)	6840

$^{12}C^{17}O_2$ Laser—*continued*

Wavelength (µm) vac	Frequency (cm^{-1})	Transition assignment	Comments
10.62936707	940.7897890	$00^01 \to 10^00$ Band P(30)	6840
10.63987856	939.8603514	$00^01 \to 10^00$ Band P(31)	6840
10.65048533	939.9243487	$00^01 \to 10^00$ Band P(32)	6840
10.66118800	937.9817711	$00^01 \to 10^00$ Band P(33)	6840
10.67198720	937,0326082	$00^01 \to 10^00$ Band P(34)	6840
10.68288358	936.0768492	$00^01 \to 10^00$ Band P(35)	6840
10.69387779	935.1144831	$00^01 \to 10^00$ Band P(36)	6840
10.70497050	934.1454983	$00^01 \to 10^00$ Band P(37)	6840

$^{12}C^{18}O_2$ Laser

Wavelength (µm) vac	Frequency (cm^{-1})	Transition assignment	Comments
4.346	2301	$02^01 \to 02^00$ Band R(8)	1676
4.371	2288	$02^01 \to 02^00$ Band P(10)	1676
4.376	2285	$02^01 \to 02^00$ Band P(14)	1676
4.382	2282	$02^01 \to 02^00$ Band P(18)	1676
4.384	2281	$02^01 \to 02^00$ Band P(20)	1676
4.392	2277	$02^01 \to 02^00$ Band P(24)	1676
4.398	2274	$02^01 \to 02^00$ Band P(28)	1676
8.9876734	1112.635002	$00^01 \to 02^00$ Band R(50)	1640,1677,1816
8.9949454	1111.735483	$00^01 \to 02^00$ Band R(48)	1640,1677,1816
9.0023806	1110.817287	$00^01 \to 02^00$ Band R(46)	1640,1677,1816
9.0099803	1109.880339	$00^01 \to 02^00$ Band R(44)	1640,1677,1816
9.0177460	1108.924564	$00^01 \to 02^00$ Band R(42)	1640,1677,1816
9.0256790	1107.949889	$00^01 \to 02^00$ Band R(40)	1640,1677,1816
9.0337807	1106.956249	$00^01 \to 02^00$ Band R(38)	1640,1677,1816
9.0420526	1105.943579	$00^01 \to 02^00$ Band R(36)	1640,1677,1816
9.0504960	1104.911817	$00^01 \to 02^00$ Band R(34)	1640,1677,1816
9.0591124	1103.860906	$00^01 \to 02^00$ Band R(32)	1640,1677,1816
9.0679030	1102.790794	$00^01 \to 02^00$ Band R(30)	1640,1677,1816
9.07505983	1101.92111	$00^02 \to 02^01$ Band R(39)	1678
9.0768694	1101.701430	$00^01 \to 02^00$ Band R(28)	1640,1677,1816
9.08316833	1100.93741	$00^02 \to 02^01$ Band R(37)	1678
9.0860128	1100.592768	$00^01 \to 02^00$ Band R(26)	1640,1677,1816
9.09145210	1099.93430	$00^02 \to 02^01$ Band R(35)	1678
9.0953347	1099.464768	$00^01 \to 02^00$ Band R(24)	1640,1677,1816
9.09991271	1098.91164	$00^02 \to 02^01$ Band R(33)	1678
9.1048363	1098.317391	$00^01 \to 02^00$ Band R(22)	1640,1677,1816

$^{12}C^{18}O_2$ Laser—*continued*

Wavelength (µm) vac	Frequency (cm^{-1})	Transition assignment	Comments
9.10855153	1097.86940	$00^02 \to 02^01$ Band R(31)	1678
9.1145190	1097.150604	$00^01 \to 02^00$ Band R(20)	1640,1677,1816
9.1173700	1096.80751	$00^02 \to 02^01$ Band R(29)	1678
9.1243841	1095.964378	$00^01 \to 02^00$ Band R(18)	1640,1677,1816
9.12636993	1095.72591	$00^02 \to 02^01$ Band R(27)	1678
9.1344331	1094.758689	$00^01 \to 02^00$ Band R(16)	1640,1677,1816
9.1446671	1093.533516	$00^01 \to 02^00$ Band R(14)	1640,1677,1816
9.14491873	1093.50343	$00^02 \to 02^01$ Band R(23)	1678
9.15447066	1092.36245	$00^02 \to 02^01$ Band R(21)	1678
9.1550876	1092.288843	$00^01 \to 02^00$ Band R(12)	1640,1677,1816
9.16420944	1091.20160	$00^02 \to 02^01$ Band R(19)	1678
9.1656957	1091.024659	$00^01 \to 02^00$ Band R(10)	1640,1677,1816
9.17413644	1090.02085	$00^02 \to 02^01$ Band R(17)	1678
9.1764928	1089.740958	$00^01 \to 02^00$ Band R(8)	1640,1677,1816
9.18425316	1088.82016	$00^02 \to 02^01$ Band R(15)	1678
9.1874801	1088.437737	$00^01 \to 02^00$ Band R(6)	1640,1677,1816
9.19456080	1087 59953	$00^02 \to 02^01$ Band R(13)	1678
9.1986588	1087.114999	$00^01 \to 02^00$ Band R(4)	1640,1677,1816
9.20506088	1086.35892	$00^02 \to 02^01$ Band R(11)	1678
9.2100304	1085.772751	$00^01 \to 02^00$ Band R(2)	1640,1677,1816
9.21575458	1085.09834	$00^02 \to 02^01$ Band R(9)	1678
9.22664333	1083.81777	$00^02 \to 02^01$ Band R(7)	1678
9.23772834	1082.51722	$00^02 \to 02^01$ Band R(5)	1678
9.2393103	1082.331865	$00^01 \to 02^00$ Band P(2)	1640,1677,1816
9.2513660	1080.921457	$00^01 \to 02^00$ Band P(4)	1640,1677,1816
9.2636198	1079.491631	$00^01 \to 02^00$ Band P(6)	1640,1677,1816
9.2760728	1078.042418	$00^01 \to 02^00$ Band P(8)	1640,1677,1816
9.2887264	1076.573858	$00^01 \to 02^00$ Band P(10)	1640,1677,1816
9.3015815	1075.085991	$00^01 \to 02^00$ Band P(12)	1640,1677,1816
9.3146394	1073.578866	$00^01 \to 02^00$ Band P(14)	1640,1677,1816
9.32725182	1072.12716	$00^02 \to 02^01$ Band P(9)	1678
9.3279011	1072.052534	$00^01 \to 02^00$ Band P(16)	1640,1677,1816
9.34005619	1070.65737	$00^02 \to 02^01$ Band P(11)	1678
9.3413677	1070.507051	$00^01 \to 02^00$ Band P(18)	1640,1677,1816
9.35306846	1069.16784	$00^02 \to 02^01$ Band P(13)	1678
9.3550403	1068.942477	$00^01 \to 02^00$ Band P(20)	1640,1677,1816
9.36628985	1067.65861	$00^02 \to 02^01$ Band P(15)	1678
9.3689201	1067.358878	$00^01 \to 02^00$ Band P(22)	1640,1677,1816
9.37972146	1066.12974	$00^02 \to 02^01$ Band P(17)	1678
9.3830079	1065.756323	$00^01 \to 02^00$ Band P(24)	1640,1677,1816

$^{12}C^{18}O_2$ Laser—*continued*

Wavelength (µm) vac	Frequency (cm^{-1})	Transition assignment	Comments
9.39336441	1064.58129	$00^02 \to 02^01$ Band P(19)	1678
9.3973049	1064.134886	$00^01 \to 02^00$ Band P(26)	1640,1677,1816
9.40721976	1063.01133	$00^02 \to 02^01$ Band P(21)	1678
9.4118121	1062.494644	$00^01 \to 02^00$ Band P(28)	1640,1677,1816
9.42128859	1061.42593	$00^02 \to 02^01$ Band P(23)	1678
9.4265306	1060.835680	$00^01 \to 02^00$ Band P(30)	1640,1677,1816
9.43557192	1059.81917	$00^02 \to 02^01$ Band P(25)	1678
9.4414613	1059.158080	$00^01 \to 02^00$ Band P(32)	1640,1677,1816
9.45007080	1058.19311	$00^02 \to 02^01$ Band P(27)	1678
9.4566052	1057.461932	$00^01 \to 02^00$ Band P(34)	1640,1677,1816
9.46478631	1056.54789	$00^02 \to 02^01$ Band P(29)	1678
9.4719633	1055.747333	$00^01 \to 02^00$ Band P(36)	1640,1677,1816
9.47971944	1054.88354	$00^02 \to 02^01$ Band P(31)	1678
9.4875366	1054.014377	$00^01 \to 02^00$ Band P(38)	1640,1677,1816
9.49487115	1051.20018	$00^02 \to 02^01$ Band P(33)	1678
9.5033261	1052.263168	$00^01 \to 02^00$ Band P(40)	1640,1677,1816
9.51024239	1051.49791	$00^02 \to 02^01$ Band P(35)	1678
9.5193326	1050.493809	$00^01 \to 02^00$ Band P(42)	1640,1677,1816
9.52583417	1049.77683	$00^02 \to 02^01$ Band P(37)	1678
9.5355573	1048.706409	$00^01 \to 02^00$ Band P(44)	1640,1677,1816
9.54164732	1048.03706	$00^02 \to 02^01$ Band P(39)	1678
9.5520009	1046.901078	$00^01 \to 02^00$ Band P(46)	1640,1677,1816
9.55768295	1046.27869	$00^02 \to 02^01$ Band P(41)	1678
9.5686644	1045.077932	$00^01 \to 02^00$ Band P(48)	1640,1677,1816
9.57394187	1044.50185	$00^02 \to 02^01$ Band P(43)	1678
9.5855488	1043.237087	$00^01 \to 02^00$ Band P(50)	1640,1677,1816
9.6026550	1041.378663	$00^01 \to 02^00$ Band P(52)	1640,1677,1816
9.6199838	1039.502785	$00^01 \to 02^00$ Band P(54)	1640,1677,1816
9.6375363	1037.609577	$00^01 \to 02^00$ Band P(56)	1640,1677,1816
9.6553134	1035.699167	$00^01 \to 02^00$ Band P(58)	1640,1677,1816
10.0880271	991.274097	$00^01 \to 10^00$ Band R(44)	1640,1677,1816
10.0960365	990.487702	$00^01 \to 10^00$ Band R(42)	1640,1677,1816
10.1043452	989.673231	$00^01 \to 10^00$ Band R(40)	1640,1677,1816
10.1129535	988.830810	$00^01 \to 10^00$ Band R(38)	1640,1677,1816
10.1218615	987.960561	$00^01 \to 10^00$ Band R(36)	1640,1677,1816
10.1310697	987.062599	$00^01 \to 10^00$ Band R(34)	1640,1677,1816
10.1405785	986.137035	$00^01 \to 10^00$ Band R(32)	1640,1677,1816
10.1503884	985.183973	$00^01 \to 10^00$ Band R(30)	1640,1677,1816
10.1605002	984.203513	$00^01 \to 10^00$ Band R(28)	1640,1677,1816
10.1709146	983.195749	$00^01 \to 10^00$ Band R(26)	1640,1677,1816

$^{12}C^{18}O_2$ Laser—*continued*

Wavelength (μm) vac	Frequency (cm^{-1})	Transition assignment	Comments
10.1816325	982.160770	$00^01 \to 10^00$ Band R(24)	1640,1677,1816
10.1926548	981.098661	$00^01 \to 10^00$ Band R(22)	1640,1677,1816
10.202035	980.19662	$00^02 \to 10^01$ Band R(23)	1678
10.2039827	980.009499	$00^01 \to 10^00$ Band R(20)	1640,1677,1816
10.213155	979.12939	$00^02 \to 10^01$ Band R(21)	1678
10.2156174	978.893358	$00^01 \to 10^00$ Band R(18)	1640,1677,1816
10.224579	978.03540	$00^02 \to 10^01$ Band R(19)	1678
10.2275601	977.750307	$00^01 \to 10^00$ Band R(16)	1640,1677,1816
10.236308	976.91470	$00^02 \to 10^01$ Band R(17)	1678
10.2398122	976.580410	$00^01 \to 10^00$ Band R(14)	1640,1677,1816
10.248345	975.76735	$00^02 \to 10^01$ Band R(15)	1678
10.2523753	975.383724	$00^01 \to 10^00$ Band R(12)	1640,1677,1816
10.260689	974.59339	$00^02 \to 10^01$ Band R(13)	1678
10.2652510	974.160302	$00^01 \to 10^00$ Band R(10)	1640,1677,1816
10.273344	973.39290	$00^02 \to 10^01$ Band R(11)	1678
10.2784410	972.910195	$00^01 \to 10^00$ Band R(8)	1640,1677,1816
10.2919471	971.633444	$00^01 \to 10^00$ Band R(6)	1640,1677,1816
10.3057713	970.330089	$00^01 \to 10^00$ Band R(4)	1640,1677,1816
10.3875886	962.687338	$00^01 \to 10^00$ Band P(6)	1640,1677,1816
10.4035359	961.211655	$00^01 \to 10^00$ Band P(8)	1640,1677,1816
10.4198197	959.709501	$00^01 \to 10^00$ Band P(10)	1640,1677,1816
10.4364429	958.180872	$00^01 \to 10^00$ Band P(12)	1640,1677,1816
10.441778	957.69129	$00^02 \to 10^01$ Band P(11)	1678
10.4534087	956.625760	$00^01 \to 10^00$ Band P(14)	1640,1677,1816
10.458482	956.16167	$00^02 \to 10^01$ Band P(13)	1678
10.4707201	955.044152	$00^01 \to 10^00$ Band P(16)	1640,1677,1816
10.475528	954.60579	$00^02 \to 10^01$ Band P(15)	1678
10.4883806	953.436031	$00^01 \to 10^00$ Band P(18)	1640,1677,1816
10.492919	953.02362	$00^02 \to 10^01$ Band P(17)	1678
10.5063938	951.801372	$00^01 \to 10^00$ Band P(20)	1640,1677,1816
10.510659	951.41515	$00^02 \to 10^01$ Band P(19)	1678
10.5247631	950.140148	$00^01 \to 10^00$ Band P(22)	1640,1677,1816
10.528750	949.78034	$00^02 \to 10^01$ Band P(21)	1678
10.5434925	948.452326	$00^01 \to 10^00$ Band P(24)	1640,1677,1816
10.547198	948.11915	$00^02 \to 10^01$ Band P(23)	1678
10.5625859	946.737867	$00^01 \to 10^00$ Band P(26)	1640,1677,1816
10.566004	946.43155	$00^02 \to 10^01$ Band P(25)	1678
10.5820472	944.996727	$00^01 \to 10^00$ Band P(28)	1640,1677,1816
10.585175	944.71749	$00^02 \to 10^01$ Band P(27)	1678
10.6018809	943.228859	$00^01 \to 10^00$ Band P(30)	1640,1677,1816

$^{12}C^{18}O_2$ Laser—*continued*

Wavelength (μm) vac	Frequency (cm^{-1})	Transition assignment	Comments
10.604713	942.97692	$00^00 2 \to 10^0 1$ Band P(29)	1678
10.6220912	941.434208	$00^0 1 \to 10^0 0$ Band P(32)	1640,1677,1816
10.624624	941.20978	$00^0 2 \to 10^0 1$ Band P(31)	1678
10.6426827	939.612715	$00^0 1 \to 10^0 0$ Band P(34)	1640,1677,1816
10.6636602	937.764315	$00^0 1 \to 10^0 0$ Band P(36)	1640,1677,1816
10.6850285	935.888938	$00^0 1 \to 10^0 0$ Band P(38)	1640,1677,1816
10.7067928	933.986510	$00^0 1 \to 10^0 0$ Band P(40)	1640,1677,1816
10.7289582	932.056948	$00^0 1 \to 10^0 0$ Band P(42)	1640,1677,1816
10.7515302	930.100166	$00^0 1 \to 10^0 0$ Band P(44)	1640,1677,1816
10.7745144	928.116073	$00^0 1 \to 10^0 0$ Band P(46)	1640,1677,1816
10.7979167	926.104569	$00^0 1 \to 10^0 0$ Band P(48)	1640,1677,1816
17.463	572.6	$02^0 1 \to 01^1 1$ Band R(8)	1676
17.730	564.0	$02^0 0 \to 01^1 0$ Band P(13)	1676
17.775	562.6	$02^0 0 \to 01^1 0$ Band P(15)	1676
17.821	561.1	$02^0 0 \to 01^1 0$ Band P(17)	1676
17.915	558.2	$02^0 0 \to 01^1 0$ Band P(21)	1676
17.962	556.7	$02^0 0 \to 01^1 0$ Band P(23)	1676
18.010	555.2	$02^0 0 \to 01^1 0$ Band P(25)	1676
18.053	553.9	$02^0 0 \to 01^1 0$ Band P(27)	1676

$^{13}C^{16}O_2$ Laser

Wavelength (μm) vac	Frequency (cm^{-1})	Transition assignment	Comments
4.39012	2277.84	$10^0 1 \to 10^0 0$ Band R(20)	6842
4.39267	2276.52	$10^0 1 \to 10^0 0$ Band R(18)	6842
4.39789	2273.82	$10^0 1 \to 10^0 0$ Band R(14)	6842
4.40056	2272.44	$10^0 1 \to 10^0 0$ Band R(12)	6842
4.44858	2247.91	$10^0 1 \to 10^0 0$ Band P(18)	6842
4.45210	2246.13	$10^0 1 \to 10^0 0$ Band P(20)	6842
4.45931	2242.50	$10^0 1 \to 10^0 0$ Band P(24)	6842
4.46299	2240.65	$10^0 1 \to 10^0 0$ Band P(26)	6842
4.46672	2238.78	$10^0 1 \to 10^0 0$ Band P(28)	6842
9.5500053	1,047.119835	$00^0 1 \to 02^0 0$ Band R(48)	6844,6957
9.5579535	1,046.249078	$00^0 1 \to 02^0 0$ Band R(46)	6844,6957
9.5661641	1,045.351083	$00^0 1 \to 02^0 0$ Band R(44)	6844,6957
9.5746401	1,044.425675	$00^0 1 \to 02^0 0$ Band R(42)	6844,6957
9.5833845	1,043.472687	$00^0 1 \to 02^0 0$ Band R(40)	6844,6957
9.5924001	1,042.491959	$00^0 1 \to 02^0 0$ Band R(38)	6844,6957

$^{13}C^{16}O_2$ Laser—*continued*

Wavelength (μm) vac	Frequency (cm^{-1})	Transition assignment	Comments
9.601689891	1041.4833369	$00^01 \to 02^00$ Band R(36)	1640,1679,6957
9.611256627	1040.4466766	$00^01 \to 02^00$ Band R(34)	1640,1679,6957
9.621103251	1039.3818408	$00^01 \to 02^00$ Band R(32)	1640,1679,6957
9.631232623	1038.2887004	$00^01 \to 02^00$ Band R(30)	1640,1679,6957
9.641647583	1037.1671347	$00^01 \to 02^00$ Band R(28)	1640,1679,6957
9.652350966	1036.0170314	$00^01 \to 02^00$ Band R(26)	1640,1679,6957
9.663345593	1034.8382866	$00^01 \to 02^00$ Band R(24)	1640,1679,6957
9.674634252	1033.6308053	$00^01 \to 02^00$ Band R(22)	1640,1679,6957
9.686219735	1032.3945014	$00^01 \to 02^00$ Band R(20)	1640,1679,6957
9.698104805	1031.1292979	$00^01 \to 02^00$ Band R(18)	1640,1679,6957
9.710292190	1029.8351269	$00^01 \to 02^00$ Band R(16)	1640,1679,6957
9.722784635	1028.5119297	$00^01 \to 02^00$ Band R(14)	1640,1679,6957
9.735584845	1027.1596570	$00^01 \to 02^00$ Band R(12)	1640,1679,6957
9.748695495	1025.7782690	$00^01 \to 02^00$ Band R(10)	1640,1679,6957
9.762119246	1024.3677355	$00^01 \to 02^00$ Band R(8)	1640,1679,6957
9.775858749	1022.9280357	$00^01 \to 02^00$ Band R(6)	1640,1679,6957
9.789916613	1021.4591587	$00^01 \to 02^00$ Band R(4)	1640,1679,6957
9.873040362	1012.8592241	$00^01 \to 02^00$ Band P(6)	1640,1679,6957
9.889229768	1011.2010991	$00^01 \to 02^00$ Band P(8)	1640,1679,6957
9.905756396	1009.5140248	$00^01 \to 02^00$ Band P(10)	1640,1679,6957
9.922622674	1007.7980729	$00^01 \to 02^00$ Band P(12)	1640,1679,6957
9.939830992	1006.0533246	$00^01 \to 02^00$ Band P(14)	1640,1679,6957
9.957383702	1064.2798709	$00^01 \to 02^00$ Band P(16)	1640,1679,6957
9.975283144	1002.4778119	$00^01 \to 02^00$ Band P(18)	1640,1679,6957
9.98393525	1001.60906	$00^02 \to 02^01$ Band P(17)	1680
9.993531627	1000.6472574	$00^01 \to 02^00$ Band P(20)	1640,1679,6957
10.002016	999.79840	$00^02 \to 02^01$ Band P(19)	1680
10.0121314	998.7883263	$00^01 \to 02^00$ Band P(22)	1640,1679,6957
10.0310849	996.9011466	$00^01 \to 02^00$ Band P(24)	1640,1679,6957
10.0503941	994.9858555	$00^01 \to 02^00$ Band P(26)	1640,1679,6957
10.0700615	993.0425989	$00^01 \to 02^00$ Band P(28)	1640,1679,6957
10.0900890	991.0715315	$00^01 \to 02^00$ Band P(30)	1640,1679,6957
10.1104791	989.0728168	$00^01 \to 02^00$ Band P(32)	1640,1679,6957
10.1312337	987.0466264	$00^01 \to 02^00$ Band P(34)	1640,1679,6957
10.1523550	984.9931404	$00^01 \to 02^00$ Band P(36)	1640,1679,6957
10.1738452	982.9125468	$00^01 \to 02^00$ Band P(38)	1640,1679,6957
10.19570615	980.8050417	$00^01 \to 02^00$ Band P(40)	6844,6957
10.21794020	978.6708287	$00^01 \to 02^00$ Band P(42)	6844,6957
10.24054929	976.5101189	$00^01 \to 02^00$ Band P(44)	6844,6957
10.26353546	974.3231306	$00^01 \to 02^00$ Band P(46)	6844,6957

$^{13}C^{16}O_2$ Laser—*continued*

Wavelength (μm) vac	Frequency (cm^{-1})	Transition assignment	Comments
10.28690074	972.1100894	$00^01 \to 02^00$ Band P(48)	6844,6957
10.6006284	943.3403026	$00^01 \to 10^00$ Band R(44)	1640,1679,6957
10.6131040	942.2314117	$00^01 \to 10^00$ Band R(42)	1640,1679,6957
10.6258494	941.1012389	$00^01 \to 10^00$ Band R(40)	1640,1679,6957
10.6388646	939.9499252	$00^01 \to 10^00$ Band R(38)	1640,1679,6957
10.6521502	938.7776043	$00^01 \to 10^00$ Band R(36)	1640,1679,6957
10.653167	938.68797	$00^02 \to 10^01$ Band R(41)	1680
10.6657064	937.5844035	$00^01 \to 10^00$ Band R(34)	1640,1679,6957
10.666054	937.55386	$00^02 \to 10^01$ Band R(39)	1680
10.679209	936.39893	$00^02 \to 10^01$ Band R(37)	1680
10.6795340	936.3704435	$00^01 \to 10^00$ Band R(32)	1640,1679,6957
10.692633	935.22332	$00^02 \to 10^01$ Band R(35)	1680
10.6936336	935.1358386	$00^01 \to 10^00$ Band R(30)	1640,1679,6957
10.706327	934.02715	$00^02 \to 10^01$ Band R(33)	1680
10.7080059	933.8806970	$00^01 \to 10^00$ Band R(28)	1640,1679,6957
10.720291	932.81054	$00^02 \to 10^01$ Band R(31)	1680
10.7226518	932.6051212	$00^01 \to 10^00$ Band R(26)	1640,1679,6957
10.734525	931.57361	$00^02 \to 10^01$ Band R(29)	1680
10.7375724	931.3092074	$00^01 \to 10^00$ Band R(24)	1640,1679,6957
10.749031	930.31645	$00^02 \to 10^01$ Band R(27)	1680
10.7527686	929.9930465	$00^01 \to 10^00$ Band R(22)	1640,1679,6957
10.763809	929.03916	$00^02 \to 10^01$ Band R(25)	1680
10.7682416	928.656723	$00^01 \to 10^00$ Band R(20)	1640,1679,6957
10.778861	927.74184	$00^02 \to 10^01$ Band R(23)	1680
10.7839929	927.300318	$00^01 \to 10^00$ Band R(18)	1640,1679,6957
10.794187	926.42457	$00^02 \to 10^01$ Band R(21)	1680
10.8000236	925.923906	$00^01 \to 10^00$ Band R(16)	1640,1679,6957
10.809789	925.08744	$00^02 \to 10^01$ Band R(19)	1680
10.8163353	924.527554	$00^01 \to 10^00$ Band R(14)	1640,1679,6957
10.825668	923.73053	$00^02 \to 10^01$ Band R(17)	1680
10.8329296	923.1113281	$00^01 \to 10^00$ Band R(12)	1640,1679,6957
10.841826	922.35389	$00^02 \to 10^01$ Band R(15)	1680
10.8498081	921.6752859	$00^01 \to 10^00$ Band R(10)	1640,1679,6957
10.858263	920.95760	$00^02 \to 10^01$ Band R(13)	1680
10.8669727	920.2194817	$00^01 \to 10^00$ Band R(8)	1640,1679,6957
10.874982	919.54173	$00^02 \to 10^01$ Band R(11)	1680
10.8844252	918.7439642	$00^01 \to 10^00$ Band R(6)	1640,1679,6957
10.891985	918.10632	$00^02 \to 10^01$ Band R(9)	1680
10.9021677	917.2487772	$00^01 \to 10^00$ Band R(4)	1640,1679,6957
10.909272	916.65142	$00^02 \to 10^01$ Band R(7)	1680

$^{13}C^{16}O_2$ **Laser**—*continued*

Wavelength (μm) vac	Frequency (cm⁻¹)	Transition assignment	Comments
10.9856555	910.2779562	$00^00 1 \rightarrow 10^00$ Band P(4)	1640,1679,6957
11.0050330	908.6751526	$00^00 1 \rightarrow 10^00$ Band P(6)	1640,1679,6957
11.009542	908.30102	$00^00 2 \rightarrow 10^01$ Band P(3)	1680
11.0247160	907.0528453	$00^00 1 \rightarrow 10^00$ Band P(8)	1640,1679,6957
11.028734	906.72238	$00^00 2 \rightarrow 10^01$ Band P(5)	1680
11.0447074	905.4110417	$00^00 1 \rightarrow 10^00$ Band P(10)	1640,1679,6957
11.048228	905.12249	$00^00 2 \rightarrow 10^01$ Band P(7)	1680
11.0650101	903.749742	$00^00 1 \rightarrow 10^00$ Band P(12)	1640,1679,6957
11.068028	903.50335	$00^00 2 \rightarrow 10^01$ Band P(9)	1680
11.0856271	902.068947	$00^00 1 \rightarrow 10^00$ Band P(14)	1640,1679,6957
11.088135	901.86496	$00^00 2 \rightarrow 10^01$ Band P(11)	1680
11.1065618	900.368647	$00^00 1 \rightarrow 10^00$ Band P(16)	1640,1679,6957
11.108552	900.20734	$00^00 2 \rightarrow 10^01$ Band P(13)	1680
11.1278173	898.648830	$00^00 1 \rightarrow 10^00$ Band P(18)	1640,1679,6957
11.129283	898.53047	$00^00 2 \rightarrow 10^01$ Band P(15)	1680
11.1493972	896.909477	$00^00 1 \rightarrow 10^00$ Band P(20)	1640,1679,6957
11.150331	896.83434	$00^00 2 \rightarrow 10^01$ Band P(17)	1680
11.1713050	895.150565	$00^00 1 \rightarrow 10^00$ Band P(22)	1640,1679,6957
11.171700	895.11894	$00^00 2 \rightarrow 10^01$ Band P(19)	1680
11.193392	893.38426	$00^00 2 \rightarrow 10^01$ Band P(21)	1680
11.1935445	893.372066	$00^00 1 \rightarrow 10^00$ Band P(24)	1640,1679,6957
11.215411	991.63025	$00^00 2 \rightarrow 10^01$ Band P(23)	1680
11.2161196	891.573944	$00^00 1 \rightarrow 10^00$ Band P(26)	1640,1679,6957
11.237762	889.85690	$00^00 2 \rightarrow 10^01$ Band P(25)	1680
11.2390343	889.756160	$00^00 1 \rightarrow 10^00$ Band P(28)	1640,1679,6957
11.260448	888.06417	$00^00 2 \rightarrow 10^01$ Band P(27)	1680
11.2622928	887.9186697	$00^00 1 \rightarrow 10^00$ Band P(30)	1640,1679,6957
11.283472	886.25201	$00^00 2 \rightarrow 10^01$ Band P(29)	1680
11.2858994	886.0614195	$00^00 1 \rightarrow 10^00$ Band P(32)	1640,1679,6957
11.306840	884.42038	$00^00 2 \rightarrow 10^01$ Band P(31)	1680
11.3098586	884.1943534	$00^00 1 \rightarrow 10^00$ Band P(34)	1640,1679,6957
11.330556	882.56922	$00^00 2 \rightarrow 10^01$ Band P(33)	1680
11.3341751	882.2874078	$00^00 1 \rightarrow 10^00$ Band P(36)	1640,1679,6957
11.354624	880.69848	$00^00 2 \rightarrow 10^01$ Band P(35)	1680
11.3588539	880.3705132	$00^00 1 \rightarrow 10^00$ Band P(38)	1640,1679,6957
11.379049	879.90808	$00^00 2 \rightarrow 10^01$ Band P(37)	1680
11.3838998	878.4335931	$00^00 1 \rightarrow 10^00$ Band P(40)	1640,1679,6957
11.403836	876.8979S	$00^00 2 \rightarrow 10^01$ Band P(39)	1680
11.4093182	876.4765648	$00^00 1 \rightarrow 10^00$ Band P(42)	1640,1679,6957
11.4161330	875.953358	$01^11 1 \rightarrow 11^10$ Band P(9)	6846

$^{13}C^{16}O_2$ Laser—*continued*

Wavelength (μm) vac	Frequency (cm^{-1})	Transition assignment	Comments
11.4273891	875.090530	$01^11 \to 11^10$ Band P(10)	6846
11.428989	874.96801	$00^02 \to 10^01$ Band P(41)	1680
11.435115	874.4993382	$00^01 \to 10^00$ Band P(44)	1640,1679,6957
11.4376785	874.303294	$01^11 \to 11^10$ Band P(11)	6846
11.4492864	873.416878	$01^11 \to 11^10$ Band P(12)	6846
11.454515	873.01819	$00^02 \to 10^01$ Band P(43)	1680
11.4595534	872.634356	$01^11 \to 11^10$ Band P(13)	6846
11.461294	872.5018106	$00^01 \to 10^00$ Band P(46)	1640,1679,6957
11.4715433	871.722294	$01^11 \to 11^10$ Band P(14)	6846
11.480419	871.04836	$00^02 \to 10^01$ Band P(45)	1680
11.4817608	870.946552	$01^11 \to 11^10$ Band P(15)	6846
11.487863	870.4838954	$00^01 \to 10^00$ Band P(48)	1640,1679,6957
11.4941634	870.006772	$001^11 \to 11^10$ Band P(16)	6846
11.5043041	869.239899	$01^11 \to 11^10$ Band P(17)	6846
11.506706	869.05845	$00^02 \to 10^01$ Band P(47)	1680
11.5171508	868.270301	$01^11 \to 11^10$ Band P(18)	6846
11.5271866	867.514370	$01^11 \to 11^10$ Band P(19)	6846
11.5405096	866.512865	$01^11 \to 11^10$ Band P(20)	6846
11.5504119	965.769994	$01^11 \to 11^10$ Band P(21)	6846
11.5642439	864.734444	$01^11 \to 11^10$ Band P(22)	6846
11.5739835	864.006760	$01^11 \to 11^10$ Band P(23)	6846
11.5883582	862.935011	$01^11 \to 11^10$ Band P(24)	6846
11.5979053	862.224662	$01^11 \to 11^10$ Band P(25)	6846
11.6128570	861.114537	$01^11 \to 11^10$ Band P(26)	6846
11.6221811	860.423692	$01^11 \to 11^10$ Band P(27)	6846
11.6377451	859.272985	$01^11 \to 11^10$ Band P(28)	6846
11.6468149	859.603840	$01^11 \to 11^10$ Band P(29)	6846
11.6630274	857.410317	$01^11 \to 11^10$ Band P(30)	6846
11.6718107	856.765092	$01^11 \to 11^10$ Band P(31)	6846
11.6887088	855.526487	$01^11 \to 11^10$ Band P(32)	6846
11.6971729	954.907431	$01^11 \to 11^10$ Band P(33)	6846
11.7147947	853.621445	$01^11 \to 11^10$ Band P(34)	6846
11.7229056	853.030839	$01^11 \to 11^10$ Band P(35)	6846
11.7412905	851.695137	$01^11 \to 11^10$ Band P(36)	6846
11.7490134	851.135294	$01^11 \to 11^10$ Band P(37)	6846
11.7682017	849.747504	$01^11 \to 11^10$ Band P(38)	6846
11.7755010	849.220770	$01^11 \to 11^10$ Band P(39)	6846

$^{13}C^{16}O^{18}O$ Laser

Wavelength (µm) vac	Frequency (cm⁻¹)	Transition assignment	Comments
4.473	2236	$02^01 \rightarrow 02^00$ Band	6847
4.528	2208	$02^01 \rightarrow 02^00$ Band	6847
9.6609222307	1035.0978676	$00^01 \rightarrow 02^00$ Band R(20)	6848
9.6665738784	1034.4926885	$00^01 \rightarrow 02^00$ Band R(19)	6848
9.6722905548	1033.8812656	$00^01 \rightarrow 02^00$ Band R(18)	6848
9.6780725408	1033.2635923	$00^01 \rightarrow 02^00$ Band R(17)	6848
9.6839201195	1032.6396621	$00^01 \rightarrow 02^00$ Band R(16)	6848
9.6898335717	1032.0094691	$00^01 \rightarrow 02^00$ Band R(15)	6848
9.6958131757	10313730080	$00^01 \rightarrow 02^00$ Band R(14)	6848
9.7018592092	1030.7302739	$00^01 \rightarrow 02^00$ Band R(13)	6848
9.7079719484	1030.0812624	$00^01 \rightarrow 02^00$ Band R(12)	6848
9.9236744988	1007.6912540	$00^01 \rightarrow 02^00$ Band P(17)	6848
9.9319829837	1006.8482816	$00^01 \rightarrow 02^00$ Band P(18)	6848
9.9403661154	1005.9991638	$00^01 \rightarrow 02^00$ Band P(19)	6848
9.9488241143	1005.1439130	$00^01 \rightarrow 02^00$ Band P(20)	6848
9.9573571956	1004 2825424	$00^01 \rightarrow 02^00$ Band P(21)	6848
9.9659655768	1003.4150653	$00^01 \rightarrow 02^00$ Band P(22)	6848
9.9746494693	1002.5414959	$00^01 \rightarrow 02^00$ Band P(23)	6848
9.9834090856	1001,6618486	$00^01 \rightarrow 02^00$ Band P(24)	6848
9.9922446352	1000.7761384	$00^01 \rightarrow 02^00$ Band P(25)	6848
10.00115632	999.8843809	$00^01 \rightarrow 02^00$ Band P(26)	6848
10.01014436	998.9865920	$00^01 \rightarrow 02^00$ Band P(27)	6848
10.01920894	998.0827884	$00^01 \rightarrow 02^00$ Band P(28)	6848
10.02835028	997.29970	$00^01 \rightarrow 02^00$ Band P(29)	6848
10.0375686	996.257205	$00^01 \rightarrow 02^00$ Band P(30)	6848
10.0468640	995.335462	$00^01 \rightarrow 02^00$ Band P(31)	6848
10.0562367	994.407775	$00^01 \rightarrow 02^00$ Band P(32)	6848
10.0656870	993.474164	$00^01 \rightarrow 02^00$ Band P(33)	6848
10.0752150	992.534648	$00^01 \rightarrow 02^00$ Band P(34)	6848
10.0848209	991.589248	$00^01 \rightarrow 02^00$ Band P(35)	6848
10.0945049	990.637986	$00^01 \rightarrow 02^00$ Band P(36)	6848
10.1042671	989.680882	$00^01 \rightarrow 02^00$ Band P(37)	6848
10.5758593	945.549642	$00^01 \rightarrow 10^00$ Band R(33)	6848
10.58205091	944.9963984	$00^01 \rightarrow 10^00$ Band R(32)	6848
10.58831203	944.4375994	$00^01 \rightarrow 10^00$ Band R(31)	6848
10.59464276	943.8732600	$00^01 \rightarrow 10^00$ Band R(30)	6848
10.60104316	943.3033950	$00^01 \rightarrow 10^00$ Band R(29)	6848
10.60751330	942.7280189	$00^01 \rightarrow 10^00$ Band R(28)	6848
10.61405328	942.1471454	$00^01 \rightarrow 10^00$ Band R(27)	6848
10.62066319	941.5607882	$00^01 \rightarrow 10^00$ Band R(26)	6848

$^{13}C^{16}O^{18}O$ Laser

Wavelength (μm) vac	Frequency (cm^{-1})	Transition assignment	Comments
10.62734311	940.9689602	$00^01 \rightarrow 10^00$ Band R(25)	6848
10.63409317	940.3716740	$00^01 \rightarrow 10^00$ Band R(24)	6848
10.64091348	939.7689418	$00^01 \rightarrow 10^00$ Band R(23)	6848
10.64780415	939.1607755	$00^01 \rightarrow 10^00$ Band R(22)	6848
10.65476531	938.5471863	$00^01 \rightarrow 10^00$ Band R(21)	6848
10.66179709	937.9281852	$00^01 \rightarrow 10^00$ Band R(20)	6848
10.66889965	937.3037828	$00^01 \rightarrow 10^00$ Band R(19)	6848
10.67607312	936.673494	$00^01 \rightarrow 10^00$ Band R(18)	6848
10.68331766	936.0388146	$00^01 \rightarrow 10^00$ Band R(17)	6848
10.69063344	935.3982680	$00^01 \rightarrow 10^00$ Band R(16)	6848
10.69802061	934.7523585	$00^01 \rightarrow 10^00$ Band R(15)	6848
10.70547937	934.1010950	$00^01 \rightarrow 10^00$ Band R(14)	6848
10.71300988	933.4444856	$00^01 \rightarrow 10^00$ Band R(13)	6848
10.92527598	915.3086856	$00^01 \rightarrow 10^00$ Band P(11)	6848
10.93476009	914.5148061	$00^01 \rightarrow 10^00$ Band P(12)	6848
10.94432366	913,7156674	$00^01 \rightarrow 10^00$ Band P(13)	6848
10.95396711	912.9112675	$00^01 \rightarrow 10^00$ Band P(14)	6848
10.96369084	912,1016042	$00^01 \rightarrow 10^00$ Band P(15)	6848
10.97349525	911.2866748	$00^01 \rightarrow 10^00$ Band P(16)	6848
10.98338078	910.4664763	$00^01 \rightarrow 10^00$ Band P(17)	6848
10.99334786	909.6410054	$00^01 \rightarrow 10^00$ Band P(18)	6848
11.00339692	908.8102582	$00^01 \rightarrow 10^00$ Band P(19)	6848
11.01352843	907.9742306	$00^01 \rightarrow 10^00$ Band P(20)	6848
11.02374283	907.1329182	$00^01 \rightarrow 10^00$ Band P(21)	6848
11.03404059	906.2863159	$00^01 \rightarrow 10^00$ Band P(22)	6848
11.04442221	905.4344185	$00^01 \rightarrow 10^00$ Band P(23)	6848
11.05488816	904.5772204	$00^01 \rightarrow 10^00$ Band P(24)	6848
11.06543894	903.7147154	$00^01 \rightarrow 10^00$ Band P(25)	6848
11.07607506	902 9468970	$00^01 \rightarrow 10^00$ Band P(26)	6848
11.08679705	901.9737584	$00^01 \rightarrow 10^00$ Band P(27)	6848
11.09760542	901.0952921	$00^01 \rightarrow 10^00$ Band P(28)	6848
11.10850073	900.2114905	$00^01 \rightarrow 10^00$ Band P(29)	6848
11.11948352	899.3223454	$00^01 \rightarrow 10^00$ Band P(30)	6848
11.13055436	898.4278481	$00^01 \rightarrow 10^00$ Band P(31)	6848
11.1417138	897.527990	$00^01 \rightarrow 10^00$ Band P(32)	6848
11.15296216	896.6227860	$00^01 \rightarrow 10^00$ Band P(33)	6848
11.1643009	895.712150	$00^01 \rightarrow 10^00$ Band P(34)	6848
11.1757298	894.796148	$00^01 \rightarrow 10^00$ Band P(35)	6848

$^{13}C^{18}O_2$ Laser

Wavelength (μm) vac	Frequency (cm^{-1})	Transition assignment	Comments
9.4686955	1056.111684	$00^01 \to 02^00$ Band R(52)	1682,6961
9.4762232	1055.272739	$00^01 \to 02^00$ Band R(50)	1682,6961
9.4839416	1054.413916	$00^01 \to 02^00$ Band R(48)	1682,6961
9.4918528	1053.535094	$00^01 \to 02^00$ Band R(46)	1682,6961
9.4999587	1052.636153	$00^01 \to 02^00$ Band R(44)	1682,6961
9.5082614	1051.716980	$00^01 \to 02^00$ Band R(42)	1682,6961
9.5167629	1050.777465	$00^01 \to 02^00$ Band R(40)	1682,6961
9.5254651	1049.817505	$00^01 \to 02^00$ Band R(38)	1682,6961
9.5343700	1048.837000	$00^01 \to 02^00$ Band R(36)	1682,6961
9.5434795	1047.835856	$00^01 \to 02^00$ Band R(34)	1682,6961
9.5527956	1046.813984	$00^01 \to 02^00$ Band R(32)	1682,6961
9.5623202	1045.771301	$00^01 \to 02^00$ Band R(30)	1682,6961
9.5720551	1044.707730	$00^01 \to 02^00$ Band R(28)	1682,6961
9.5820024	1043.623199	$00^01 \to 02^00$ Band R(26)	1682,6961
9.5921638	1042.517640	$00^01 \to 02^00$ Band R(24)	1682,6961
9.6025413	1041.390995	$00^01 \to 02^00$ Band R(22)	1682,6961
9.6131366	1040.243207	$00^01 \to 02^00$ Band R(20)	1682,6961
9.6239515	1039.074230	$00^01 \to 02^00$ Band R(18)	1682,6961
9.6349879	1037.884020	$00^01 \to 02^00$ Band R(16)	1682,6961
9.6462476	1036.672542	$00^01 \to 02^00$ Band R(14)	1682,6961
9.6577322	1035.439765	$00^01 \to 02^00$ Band R(12)	1682,6961
9.6694436	1034.185668	$00^01 \to 02^00$ Band R(10)	1682,6961
9.6813834	1032.910231	$00^01 \to 02^00$ Band R(8)	1682,6961
9.6935534	1031.613445	$00^01 \to 02^00$ Band R(6)	1682,6961
9.7059551	1030.295306	$00^01 \to 02^00$ Band R(4)	1682,6961
9.7646748	1024.099641	$00^01 \to 02^00$ Band P(4)	1682,6961
9.7783801	1022.664272	$00^01 \to 02^00$ Band P(6)	1682,6961
9.7923274	1021.207681	$00^01 \to 02^00$ Band P(8)	1682,6961
9.8065183	1019.729911	$00^01 \to 02^00$ Band P(10)	1682,6961
9.8209540	1018.231015	$00^01 \to 02^00$ Band P(12)	1682,6961
9.8356362	1016.711049	$00^01 \to 02^00$ Band P(14)	1682,6961
9.8505662	1015.170077	$00^01 \to 02^00$ Band P(16)	1682,6961
9.8657453	1013.608169	$00^01 \to 02^00$ Band P(18)	1682,6961
9.8811749	1012.025402	$00^01 \to 02^00$ Band P(20)	1682,6961
9.8968564	1010.421858	$00^01 \to 02^00$ Band P(22)	1682,6961
9.9127910	1008.797626	$00^01 \to 02^00$ Band P(24)	1682,6961
9.9289800	1007.152801	$00^01 \to 02^00$ Band P(26)	1682,6961
9.9454247	1005.487483	$00^01 \to 02^00$ Band P(28)	1682,6961
9.9621262	1003.801779	$00^01 \to 02^00$ Band P(30)	1682,6961
9.9790858	1002.095800	$00^01 \to 02^00$ Band P(32)	1682,6961

$^{13}C^{18}O_2$ Laser—*continued*

Wavelength (μm) vac	Frequency (cm^{-1})	Transition assignment	Comments
9.9963047	1000.369666	$00^01 \rightarrow 02^00$ Band P(34)	1682,6961
10.0137840	998.623500	$00^01 \rightarrow 02^00$ Band P(36)	1682,6961
10.0315248	996.857430	$00^01 \rightarrow 02^00$ Band P(38)	1682,6961
10.0495282	995.071591	$00^01 \rightarrow 02^00$ Band P(40)	1682,6961
10.0677953	993.266122	$00^01 \rightarrow 02^00$ Band P(42)	1682,6961
10.0863272	991.441167	$00^01 \rightarrow 02^00$ Band P(44)	1682,6961
10.1051249	989.596877	$00^01 \rightarrow 02^00$ Band P(46)	1682,6961
10.1241893	987.733404	$00^01 \rightarrow 02^00$ Band P(48)	1682,6961
10.1435216	985.850909	$00^01 \rightarrow 02^00$ Band P(50)	1682,6961
10.1631227	983.949553	$00^01 \rightarrow 02^00$ Band P(52)	1682,6961
10.1829934	982.029504	$00^01 \rightarrow 02^00$ Band P(54)	1682,6961
10.4739131	954.753005	$00^01 \rightarrow 10^00$ Band R(46)	1682,6961
10.4832265	953.904788	$00^01 \rightarrow 10^00$ Band R(44)	1682,6961
10.4928250	953.032193	$00^01 \rightarrow 10^00$ Band R(42)	1682,6961
10.5027083	952.135368	$00^01 \rightarrow 10^00$ Band R(40)	1682,6961
10.5128764	951.214452	$00^01 \rightarrow 10^00$ Band R(38)	1682,6961
10.5233296	950.269581	$00^01 \rightarrow 10^00$ Band R(36)	1682,6961
10.5340679	949.300882	$00^01 \rightarrow 10^00$ Band R(34)	1682,6961
10.5450918	948.308479	$00^01 \rightarrow 10^00$ Band R(32)	1682,6961
10.5564017	947.292486	$00^01 \rightarrow 10^00$ Band R(30)	1682,6961
10.5679980	946.253013	$00^01 \rightarrow 10^00$ Band R(28)	1682,6961
10.5798816	945.190165	$00^01 \rightarrow 10^00$ Band R(26)	1682,6961
10.5920530	944.104038	$00^01 \rightarrow 10^00$ Band R(24)	1682,6961
10.6045132	942.994725	$00^01 \rightarrow 10^00$ Band R(22)	1682,6961
10.6172631	941.862312	$00^01 \rightarrow 10^00$ Band R(20)	1682,6961
10.6303039	940.706880	$00^01 \rightarrow 10^00$ Band R(18)	1682,6961
10.6436367	939.528502	$00^01 \rightarrow 10^00$ Band R(16)	1682,6961
10.6572627	938.327247	$00^01 \rightarrow 10^00$ Band R(14)	1682,6961
10.6711835	937.103179	$00^01 \rightarrow 10^00$ Band R(12)	1682,6961
10.6854005	935.856356	$00^01 \rightarrow 10^00$ Band R(10)	1682,6961
10.6999154	934.586828	$00^01 \rightarrow 10^00$ Band R(8)	1682,6961
10.7147299	933.294644	$00^01 \rightarrow 10^00$ Band R(6)	1682,6961
10.8355947	922.884278	$00^01 \rightarrow 10^00$ Band P(8)	1682,6961
10.8355947	922.884278	$00^01 \rightarrow 10^00$ Band P(8)	1682,6961
10.8530426	921.400600	$00^01 \rightarrow 10^00$ Band P(10)	1682,6961
10.8708127	919.894429	$00^01 \rightarrow 10^00$ Band P(12)	1682,6961
10.8889078	918.3657S4	$00^01 \rightarrow 10^00$ Band P(14)	1682,6961
10.9073312	916.814558	$00^01 \rightarrow 10^00$ Band P(16)	1682,6961
10.9260861	915.240819	$00^01 \rightarrow 10^00$ Band P(18)	1682,6961
10.9451761	913.644507	$00^01 \rightarrow 10^00$ Band P(20)	1682,6961

$^{13}C^{18}O_2$ Laser—*continued*

Wavelength (μm) vac	Frequency (cm^{-1})	Transition assignment	Comments
10.9646046	912.025589	$00^01 \rightarrow 10^00$ Band P(22)	1682,6961
10.9843755	910.384024	$00^01 \rightarrow 10^00$ Band P(24)	1682,6961
11.0044927	908.719766	$00^01 \rightarrow 10^00$ Band P(26)	1682,6961
11.0249601	907.032164	$00^01 \rightarrow 10^00$ Band P(28)	1682,6961
11.0457819	905.322961	$00^01 \rightarrow 10^00$ Band P(30)	1682,6961
11.0669626	903.590293	$00^01 \rightarrow 10^00$ Band P(32)	1682,6961
11.0885067	901.934691	$00^01 \rightarrow 10^00$ Band P(34)	1682,6961
11.1104188	900.056078	$00^01 \rightarrow 10^00$ Band P(36)	1682,6961
11.1327039	898.254174	$00^01 \rightarrow 10^00$ Band P(38)	1682,6961
11.1553670	896.429491	$00^01 \rightarrow 10^00$ Band P(40)	1682,6961
11.1784134	894.581333	$00^01 \rightarrow 10^00$ Band P(42)	1682,6961
11.2018486	892.709801	$00^01 \rightarrow 10^00$ Band P(44)	1682,6961
11.2256780	890.814787	$00^01 \rightarrow 10^00$ Band P(46)	1682,6961
11.2499077	888.896178	$00^01 \rightarrow 10^00$ Band P(48)	1682,6961

$^{14}C^{16}O_2$ Laser

Wavelength (μm) vac	Frequency (cm^{-1})	Transition assignment	Comments
9.917877180	1008.280282	$00^01 \rightarrow 02^00$ Band R(40)	1683
9.9178771801	1008.2802820	$00^01 \rightarrow 02^00$ Band R(40)	6851
9.9273316267	1007.3200308	$00^01 \rightarrow 02^00$ Band R(38)	6851
9.927331635	1007.320030	$00^01 \rightarrow 02^00$ Band R(38)	1683
9.9370891511	1006.3309132	$00^01 \rightarrow 02^00$ Band R(36)	6851
9.937089163	1006.130912	$00^01 \rightarrow 02^00$ Band R(36)	1683
9.9471530250	1005.3127739	$00^01 \rightarrow 02^00$ Band R(34)	6851
9.947153044	1005.312772	$00^01 \rightarrow 02^00$ Band R(34)	1683
9.9575265191	1004.2654650	$00^01 \rightarrow 02^00$ Band R(32)	6851
9.957526539	1004.261461	$00^01 \rightarrow 02^00$ Band R(32)	1683
9.9682128982	1003.1888466	$00^01 \rightarrow 02^00$ Band R(30)	6851
9.968212914	1003.189845	$00^01 \rightarrow 02^00$ Band R(30)	1683
9.9792154208	1002.0827869	$00^01 \rightarrow 02^00$ Band R(28)	6851
9.979215440	1002.082785	$00^01 \rightarrow 02^00$ Band R(28)	1683
9.9905373357	1000.9471627	$00^01 \rightarrow 02^00$ Band R(26)	6851
9.990537353	1000.947161	$00^01 \rightarrow 02^00$ Band R(26)	1683
10.00218188	999.7818596	$00^01 \rightarrow 02^00$ Band R(24)	6851
10.0021819	999.781858	$00^01 \rightarrow 02^00$ Band R(24)	1683
10.01415228	998.5867724	$00^01 \rightarrow 02^00$ Band R(22)	6851
10.0141523	998.586771	$00^01 \rightarrow 02^00$ Band R(22)	1683

$^{14}C^{16}O_2$ Laser—*continued*

Wavelength (μm) vac	Frequency (cm^{-1})	Transition assignment	Comments
10.0264517	997.361804	$00^01 \rightarrow 02^00$ Band R(20)	1683
10.02645173	997.3618050	$00^01 \rightarrow 02^00$ Band R(20)	6851
10.03908344	996.1068712	$00^01 \rightarrow 02^00$ Band R(18)	6851
10.0390835	996.106870	$00^01 \rightarrow 02^00$ Band R(18)	1683
10.05205058	994.8218941	$00^01 \rightarrow 02^00$ Band R(16)	6851
10.0520506	994.821891	$00^01 \rightarrow 02^00$ Band R(16)	1683
10.0653563	993.506806	$00^01 \rightarrow 02^00$ Band R(14)	1683
10.06535630	993.5068073	$00^01 \rightarrow 02^00$ Band R(14)	6851
10.0790037	992.161553	$00^01 \rightarrow 02^00$ Band R(12)	1683
10.07900373	992.1615541	$00^01 \rightarrow 02^00$ Band R(12)	6851
10.09299597	990.7860883	$00^01 \rightarrow 02^00$ Band R(10)	6851
10.0929960	990.786087	$00^01 \rightarrow 02^00$ Band R(10)	1683
10.1073361	989.180371	$00^01 \rightarrow 02^00$ Band R(8)	1683
10.10733613	989.3803740	$00^01 \rightarrow 02^00$ Band R(8)	6851
10.12202725	987.9443859	$00^01 \rightarrow 02^00$ Band R(6)	6851
10.1220273	987 944385	$00^01 \rightarrow 02^00$ Bnd R(6)	1683
10.24370004	976.2097643	$00^01 \rightarrow 02^00$ Band P(8)	6851
10.2437001	976.209763	$00^01 \rightarrow 02^00$ Bnd P(8)	1683
10.2614943	974.516934	$00^01 \rightarrow 02^00$ Band P(10)	1683
10.26149432	974.5169354	$00^01 \rightarrow 02^00$ Band P(10)	6851
10.27966734	972.7941256	$00^01 \rightarrow 02^00$ Band P(12)	6851
10.2796674	972.794124	$00^01 \rightarrow 02^00$ Band P(12)	1683
10.29822186	971.0414219	$00^01 \rightarrow 02^00$ Band P(14)	6851
10.2982219	971.041421	$00^01 \rightarrow 02^00$ Band P(14)	1683
10.31716065	969.2589215	$00^01 \rightarrow 02^00$ Band P(16)	6851
10.3171607	969.258921	$00^01 \rightarrow 02^00$ Band P(16)	1683
10.3364864	967.446731	$00^01 \rightarrow 02^00$ Band P(18)	1683
10.33648641	967.4467319	$00^01 \rightarrow 02^00$ Band P(18)	6851
10.35620187	965.6049708	$00^01 \rightarrow 02^00$ Band P(20)	6851
10.3562019	965.604971	$00^01 \rightarrow 02^00$ Band P(20)	1683
10.37630968	963.7337658	$00^01 \rightarrow 02^00$ Band P(22)	6851
10.3763097	963.733766	$00^01 \rightarrow 02^00$ Band P(22)	1683
10.3968125	961.833254	$00^01 \rightarrow 02^00$ Band P(24)	1683
10.39681250	961.8332544	$00^01 \rightarrow 02^00$ Band P(24)	6851
10.41771296	959.9035834	$00^01 \rightarrow 02^00$ Band P(26)	6851
10.4177130	959.903583	$00^01 \rightarrow 02^00$ Band P(26)	1683
10.43901367	957.9449091	$00^01 \rightarrow 02^00$ Band P(28)	6851
10.4390137	957.944909	$00^01 \rightarrow 02^00$ Band P(28)	1683
10.4607172	955.957396	$00^01 \rightarrow 02^00$ Band P(30)	1683

$^{14}C^{16}O_2$ Laser—*continued*

Wavelength (μm) vac	Frequency (cm^{-1})	Transition assignment	Comments
10.46071722	955.9573967	$00^01 \rightarrow 02^00$ Band P(30)	6851
10.48282618	953 9412204	$00^01 \rightarrow 02^00$ Band P(32)	6851
10.4828262	953.941220	$00^01 \rightarrow 02^00$ Band P(32)	1683
10.5053431	951.996562	$00^01 \rightarrow 02^00$ Band P(34)	1683
10.50534311	951.8965628	$00^01 \rightarrow 02^00$ Band P(34)	6851
10.52827056	949.8236145	$00^01 \rightarrow 02^00$ Band P(36)	6851
10.5282706	949.823614	$00^01 \rightarrow 02^00$ Band P(36)	1683
11.106994	900.33358	$00^01 \rightarrow 10^00$ Band R(50)	1683
11.106994	900.33362	$00^01 \rightarrow 10^00$ Band R(50)	6851
11.119855	899.29229	$00^01 \rightarrow 10^00$ Band R(48)	6851
11.120968	899.20226	$00^01 \rightarrow 10^00$ Band R(48)	1683
11.1351978	898.053198	$00^01 \rightarrow 10^00$ Band R(46)	6851
11.135198	898.05318	$00^01 \rightarrow 10^00$ Band R(46)	1683
11.1496835	896.886443	$00^01 \rightarrow 10^00$ Band R(44)	6851
11.149684	896.88643	$00^01 \rightarrow 10^00$ Band R(44)	1683
11.164426	895.70211	$00^01 \rightarrow 10^00$ Band R(42)	1683
11.1644260	895.702120	$00^01 \rightarrow 10^00$ Band R(42)	6851
11.1794258	894.500320	$00^01 \rightarrow 10^00$ Band R(40)	6851
11.179426	894.50031	$00^01 \rightarrow 10^00$ Band R(40)	1683
11.194684	893.28113	$00^01 \rightarrow 10^00$ Band R(38)	1683
11.1946840	893.281132	$00^01 \rightarrow 10^00$ Band R(38)	6851
11.210201	892.04463	$00^01 \rightarrow 10^00$ Band R(36)	1683
11.2102013	892.044638	$00^01 \rightarrow 10^00$ Band R(36)	6851
11.2259788	890.790921	$00^01 \rightarrow 10^00$ Band R(34)	6851
11.225979	890.79092	$00^01 \rightarrow 10^00$ Band R(34)	1683
11.2420175	889.520056	$00^01 \rightarrow 10^00$ Band R(32)	6851
11.242018	889.52005	$00^01 \rightarrow 10^00$ Band R(32)	1683
11.2583176	888.232178	$00^01 \rightarrow 10^00$ Band R(30)	6851
11.258318	888.23212	$00^01 \rightarrow 10^00$ Band R(30)	1683
11.27488285	886.9271756	$00^01 \rightarrow 10^00$ Band R(28)	6851
11.274883	886.92718	$00^01 \rightarrow 10^00$ Band R(28)	1683
11.291712	885.60530	$00^01 \rightarrow 10^00$ Band R(26)	1683
11.29171205	885.6052965	$00^01 \rightarrow 10^00$ Band R(26)	6851
11.308807	884.26654	$00^01 \rightarrow 10^00$ Band R(24)	1683
11.30880736	884.2665442	$00^01 \rightarrow 10^00$ Band R(24)	6851
11.326170	882.91098	$00^01 \rightarrow 10^00$ Band R(22)	1683,1684
11.32617017	882.9109793	$00^01 \rightarrow 10^00$ Band R(22)	6851
11.34380199	881.5386594	$00^01 \rightarrow 10^00$ Band R(20)	6851
11.343802	891.53866	$00^01 \rightarrow 10^00$ Band R(20)	1683,1684
11.361704	880.14964	$00^01 \rightarrow 10^00$ Band R(18)	1683,1684

$^{14}C^{16}O_2$ Laser—*continued*

Wavelength (μm) vac	Frequency (cm^{-1})	Transition assignment	Comments
11.36170437	880.1496392	$00^01 \rightarrow 10^00$ Band R(18)	6851
11.37987894	878.7439704	$00^01 \rightarrow 10^00$ Band R(16)	6851
11.379879	878.74397	$00^01 \rightarrow 10^00$ Band R(16)	1683,1684
11.398327	877.32170	$00^01 \rightarrow 10^00$ Band R(14)	1683,1684
11.39832741	877.3217019	$00^01 \rightarrow 10^00$ Band R(14)	6851
11.41705156	875.8828799	$00^01 \rightarrow 10^00$ Band R(12)	6851
11.417052	875.88298	$00^01 \rightarrow 10^00$ Band R(12)	1683
11.436053	874.42754	$00^01 \rightarrow 10^00$ Band R(10)	1683
11.43605325	874.4275475	$00^01 \rightarrow 10^00$ Band R(10)	6851
11.455334	872.95574	$00^01 \rightarrow 10^00$ Band R(8)	1683
11.45533442	872.9557456	$00^01 \rightarrow 10^00$ Band R(8)	6851
11.474897	871.46751	$00^01 \rightarrow 10^00$ Band R(6)	1683
11.47489707	871.4675121	$00^01 \rightarrow 10^00$ Band R(6)	6851
11.494743	869.96288	$00^01 \rightarrow 10^00$ Band R(4)	1683
11.49474329	869.9628822	$00^01 \rightarrow 10^00$ Band R(4)	6851
11.609067	861.39566	$00^01 \rightarrow 10^00$ Band P(6)	1683
11.60906705	861.3956623	$00^01 \rightarrow 10^00$ Band P(6)	6851
11.63081284	859.7851359	$00^01 \rightarrow 10^00$ Band P(8)	6851
11.630813	859.78513	$00^01 \rightarrow 10^00$ Band P(8)	1683
11.652860	858.15839	$00^01 \rightarrow 10^00$ Band P(10)	1683
11.65286043	858.1583946	$00^01 \rightarrow 10^00$ Band P(10)	6851
11.67521261	856.5154514	$00^01 \rightarrow 10^00$ Band P(12)	6851
11.675213	856.51545	$00^01 \rightarrow 10^00$ Band P(12)	1683
11.697872	854.85631	$00^01 \rightarrow 10^00$ Band P(14)	1683,1684
11.69787227	854.8563165	$00^01 \rightarrow 10^00$ Band P(14)	6851
11.720842	853.18100	$00^01 \rightarrow 10^00$ Band P(16)	1683,1684
11.72084238	853.1809978	$00^01 \rightarrow 10^00$ Band P(16)	6851
11.744126	851.48950	$00^01 \rightarrow 10^00$ Band P(18)	1683,1684
11.74412603	851.4895000	$00^01 \rightarrow 10^00$ Band P(18)	6851
11.767726	849.78182	$00^01 \rightarrow 10^00$ Band P(20)	1683,1684
11.76772638	849.7818252	$00^01 \rightarrow 10^00$ Band P(20)	6851
11.79164670	848.0579728	$00^01 \rightarrow 10^00$ Band P(22)	6851
11.791647	848.05797	$00^01 \rightarrow 10^00$ Band P(22)	1683,1684
11.815890	846.11794	$00^01 \rightarrow 10^00$ Band P(24)	1683,1684
11.81589039	846.3179390	$00^01 \rightarrow 10^00$ Band P(24)	6851
11.84046091	844.5617172	$00^01 \rightarrow 10^00$ Band P(26)	6851
11.840461	844.56172	$00^01 \rightarrow 10^00$ Band P(26)	1683,1684
11.86536187	842.7892978	$00^01 \rightarrow 10^00$ Band P(28)	6851
11.865362	842.78930	$00^01 \rightarrow 10^00$ Band P(28)	1683

$^{14}C^{16}O_2$ Laser—*continued*

Wavelength (μm) vac	Frequency (cm^{-1})	Transition assignment	Comments
11.890597	841.00067	$00^00 1 \rightarrow 10^00 0$ Band P(30)	1683
11.8905970	841.000668	$00^00 1 \rightarrow 10^00 0$ Band P(30)	6851
11.916170	839.19581	$00^00 1 \rightarrow 10^00 0$ Band P(32)	1683
11.9161700	839.195812	$00^00 1 \rightarrow 10^00 0$ Band P(32)	6851
11.942085	837.37471	$00^00 1 \rightarrow 10^00 0$ Band P(34)	1683
11.9420850	837.374710	$00^00 1 \rightarrow 10^00 0$ Band P(34)	6851
11.968346	835.53734	$00^00 1 \rightarrow 10^00 0$ Band P(36)	1683
11.9683460	835.537340	$00^00 1 \rightarrow 10^00 0$ Band P(36)	6851
11.994957	833.68367	$00^00 1 \rightarrow 10^00 0$ Band P(38)	1683
11.9949572	833.683675	$00^00 1 \rightarrow 10^00 0$ Band P(38)	6851
12.0219229	831.813685	$00^00 1 \rightarrow 10^00 0$ Band P(40)	6851
12.021923	831.81368	$00^00 1 \rightarrow 10^00 0$ Band P(40)	1683
12.0492476	829.927338	$00^00 1 \rightarrow 10^00 0$ Band P(42)	6851
12.049248	829.92733	$00^00 1 \rightarrow 10^00 0$ Band P(42)	1683
12.076936	828.02458	$00^00 1 \rightarrow 10^00 0$ Band P(44)	1683
12.0769360	828.024594	$00^00 1 \rightarrow 10^00 0$ Band P(44)	6851
12.1049927	826.105414	$00^00 1 \rightarrow 10^00 0$ Band P(46)	6851
12.104993	826.10540	$00^00 1 \rightarrow 10^00 0$ Band P(46)	1683
12.1334228	824.169750	$00^00 1 \rightarrow 10^00 0$ Band P(48)	6851
12.133423	824.16974	$00^00 1 \rightarrow 10^00 0$ Band P(48)	1683

$^{14}C^{18}O_2$ Laser

Wavelength (μm) vac	Frequency (cm^{-1})	Transition assignment	Comments
9.95877	1004.1395781	$00^00 1 \rightarrow 02^00 0$ Band R(36)	6853
9.99854	1000.1461072	$00^00 1 \rightarrow 02^00 0$ Band R(28)	6853
9.89073556	1011.04715	$00^00 1 \rightarrow 02^00 0$ Band R(52)	1685
9.89073556	1011.047715	$00^00 1 \rightarrow 02^00 0$ Band R(52)	6852
9.898433863	1010.260827	$00^00 1 \rightarrow 02^00 0$ Band R(50)	6852
9.898433882	1010.260829	$00^00 1 \rightarrow 02^00 0$ Band R(50)	1685
9.906357727	1009.452744	$00^00 1 \rightarrow 02^00 0$ Band R(48)	6852
9.906357736	1009.452745	$00^00 1 \rightarrow 02^00 0$ Band R(48)	1685
9.914509593	1008.622757	$00^00 1 \rightarrow 02^00 0$ Band R(46)	1685
9.914509593	1008.622757	$00^00 1 \rightarrow 02^00 0$ Band R(46)	6852
9.922891926	1007.770725	$00^00 1 \rightarrow 02^00 0$ Band R(44)	6852
9.922891936	1007.770726	$00^00 1 \rightarrow 02^00 0$ Band R(44)	1685
9.9315071908	1006.896516	$00^00 1 \rightarrow 02^00 0$ Band R(42)	6852
9.931507202	1006.8965171	$00^00 1 \rightarrow 02^00 0$ Band R(42)	1685

$^{14}C^{18}O_2$ Laser—*continued*

Wavelength (μm) vac	Frequency (cm^{-1})	Transition assignment	Comments
9.94036	1006.0000032	$00^01 \rightarrow 02^00$ Band R(40)	6853
9.949446257	1006.0000032	$00^01 \rightarrow 02^00$ Band R(40)	1685
9.94945	1005.0810622	$00^01 \rightarrow 02^00$ Band R(38)	6853
9.958774894	1005.0810622	$00^01 \rightarrow 02^00$ Band R(38)	1685
9.968346125	1004.1395781	$00^01 \rightarrow 02^00$ Band R(36)	1685
9.96835	1003.1754410	$00^01 \rightarrow 02^00$ Band R(34)	6853
9.97816	1002.1885473	$00^01 \rightarrow 02^00$ Band R(32)	6853
9.978162343	1003.1754410	$00^01 \rightarrow 02^00$ Band R(34)	1685
9.988225899	1002.1885473	$00^01 \rightarrow 02^00$ Band R(32)	1685
9.98823	1001.1787996	$00^01 \rightarrow 02^00$ Band R(30)	6853
9.998539153	1001.1787996	$00^01 \rightarrow 02^00$ Band R(30)	1685
10.00910	999.0903858	$00^01 \rightarrow 02^00$ Band R(26)	6853
10.0091044	1000.1461072	$00^01 \rightarrow 02^00$ Band R(28)	1685
10.01992	998.0115579	$00^01 \rightarrow 02^00$ Band R(24)	6853
10.0199240	999.0903858	$00^01 \rightarrow 02^00$ Band R(26)	1685
10.03100	996.9095524	$00^01 \rightarrow 02^00$ Band R(22)	6853
10.0310003	998.0115579	$00^01 \rightarrow 02^00$ Band R(24)	1685
10.0423354	996.9095524	$00^01 \rightarrow 02^00$ Band R(22)	1685
10.04234	995.7843055	$00^01 \rightarrow 02^00$ Band R(20)	6853
10.05393	994.6357599	$00^01 \rightarrow 02^00$ Band R(18)	6853
10.0539317	995.7843055	$00^01 \rightarrow 02^00$ Band R(20)	1685
10.06579	993.4638655	$00^01 \rightarrow 02^00$ Band R(16)	6853
10.0657914	994.6357599	$00^01 \rightarrow 02^00$ Band R(18)	1685
10.0779166	993.4638655	$00^01 \rightarrow 02^00$ Band R(16)	1685
10.07792	992.2685790	$00^01 \rightarrow 02^00$ Band R(14)	6853
10.0903096	992.2685790	$00^01 \rightarrow 02^00$ Band R(14)	1685
10.09031	991.0498646	$00^01 \rightarrow 02^00$ Band R(12)	6853
10.1029726	989.8076932	$00^01 \rightarrow 02^00$ Band R(10)	1685
10.1029726	991.0498646	$00^01 \rightarrow 02^00$ Band R(12)	1685
10.10297260	989.807692	$00^01 \rightarrow 02^00$ Band R(10)	6852
10.1159076	989.5420431	$00^01 \rightarrow 02^00$ Band R(8)	1685
10.11590763	988.542042	$00^01 \rightarrow 02^00$ Band R(8)	6852
10.12911687	987.252899	$00^01 \rightarrow 02^00$ Band R(6)	6852
10.1291169	987.2528998	$00^01 \rightarrow 02^00$ Band R(6)	1685
10.14260239	985.940255	$00^01 \rightarrow 02^00$ Band R(4)	6852
10.1426024	985.9402562	$00^01 \rightarrow 02^00$ Band R(4)	1685
10.15636628	994.604111	$00^01 \rightarrow 02^00$ Band R(2)	6852
10.1563663	984.6041122	$00^01 \rightarrow 02^00$ Band R(2)	1685
10.20676128	979.742713	$00^01 \rightarrow 02^00$ Band P(4)	6852
10.2067613	979.7427147	$00^01 \rightarrow 02^00$ Band P(4)	1685

$^{14}C^{18}O_2$ Laser—*continued*

Wavelength (μm) vac	Frequency (cm^{-1})	Transition assignment	Comments
10.22180274	976.835945	$00^01 \to 02^00$ Band P(6)	6852
10.2218028	978.3010155	$00^01 \to 02^00$ Band P(6)	1685
10.2371335	976.8359469	$00^01 \to 02^00$ Band P(8)	1685
10.23713350	975.347555	$00^01 \to 02^00$ Band P(8)	6852
10.25275548	975.347555	$00^01 \to 02^00$ Band P(10)	6852
10.2527555	975.3475563	$00^01 \to 02^00$ Band P(10)	1685
10.26867054	973.835897	$00^01 \to 02^00$ Band P(12)	6852
10.2686706	973.8358986	$00^01 \to 02^00$ Band P(12)	1685
10.2848805	972.3010358	$00^01 \to 02^00$ Band P(14)	1685
10.28488054	972.301035	$00^01 \to 02^00$ Band P(14)	6852
10.3013873	970.7430371	$00^01 \to 02^00$ Band P(16)	1685
10.30138731	970.743036	$00^01 \to 02^00$ Band P(16)	6852
10.31819264	969.161978	$00^01 \to 02^00$ Band P(18)	6852
10.3181927	969.1619788	$00^01 \to 02^00$ Band P(18)	1685
10.3352983	967.5579443	$00^01 \to 02^00$ Band P(20)	1685
10.33529832	967.557944	$00^01 \to 02^00$ Band P(20)	6852
10.35270609	965.931024	$00^01 \to 02^00$ Band P(22)	6852
10.3527061	965.9310241	$00^01 \to 02^00$ Band P(22)	1685
10.37041767	964.281315	$00^01 \to 02^00$ Band P(24)	6852
10.3704177	964.2813157	$00^01 \to 02^00$ Band P(24)	1685
10.38843476	962.608923	$00^01 \to 02^00$ Band P(26)	6852
10.3884348	962.6089233	$00^01 \to 02^00$ Band P(26)	1685
10.4067590	960.9139580	$00^01 \to 02^00$ Band P(28)	1685
10.40675902	960.913958	$00^01 \to 02^00$ Band P(28)	6852
10.42539209	959.196538	$00^01 \to 02^00$ Band P(30)	6852
10.4253921	959.1965377	$00^01 \to 02^00$ Band P(30)	1685
10.4443356	957.4567868	$00^01 \to 02^00$ Band P(32)	1685
10.44433560	957.456787	$00^01 \to 02^00$ Band P(32)	6852
10.4635911	955.6948365	$00^01 \to 02^00$ Band P(34)	1685
10.46359112	955.694836	$00^01 \to 02^00$ Band P(34)	6852
10.4831602	953.9108240	$00^01 \to 02^00$ Band P(36)	1685
10.48316022	953.910824	$00^01 \to 02^00$ Band P(36)	6852
10.5030444	952.1048932	$00^01 \to 02^00$ Band P(38)	1685
10.50304443	952.104893	$00^01 \to 02^00$ Band P(38)	6852
10.52324528	950.277194	$00^01 \to 02^00$ Band P(40)	6852
10.5232453	950.2771940	$00^01 \to 02^00$ Band P(40)	1685
10.54376425	948.427882	$00^01 \to 02^00$ Band P(42)	6852
10.5437643	948.4278926	$00^01 \to 02^00$ Band P(42)	1685
10.56460279	946.557120	$00^01 \to 02^00$ Band P(44)	6852
10.5646028	946.5571209	$00^01 \to 02^00$ Band P(44)	1685

$^{14}C^{18}O_2$ Laser—*continued*

Wavelength (μm) vac	Frequency (cm^{-1})	Transition assignment	Comments
10.58576235	944.665076	$00^01 \to 02^00$ Band P(46)	6852
10.5857624	944.6650767	$00^01 \to 02^00$ Band P(46)	1685
10.6072443	942.7519236	$00^01 \to 02^00$ Band P(48)	1685
10.60724433	942.751923	$00^01 \to 02^00$ Band P(48)	6852
10.6290501	940.8178407	$00^01 \to 02^00$ Band P(50)	1685
10.62905014	940.817840	$00^01 \to 02^00$ Band P(50)	6852
10.6511811	938.8630124	$00^01 \to 02^00$ Band P(52)	1685
10.65118113	938.863011	$00^01 \to 02^00$ Band P(52)	6852
10.900267	917.40864	$00^01 \to 10^00$ Band R(48)	1685
10.900268	917.40868	$00^01 \to 10^00$ Band R(48)	6852
10.911233	916.486675	$00^01 \to 10^00$ Band R(46)	1685
10.9112334	916.48668	$00^01 \to 10^00$ Band R(46)	6852
10.922463	915.544400	$00^01 \to 10^00$ Band R(44)	1685
10.9224632	915.54439	$00^01 \to 10^00$ Band R(44)	6852
10.933957	914.581960	$00^01 \to 10^00$ Band R(42)	1685
10.9339572	914.58195	$00^01 \to 10^00$ Band R(42)	6852
10.945715	913.599495	$00^01 \to 10^00$ Band R(40)	1685
10.9457153	913.59949	$00^01 \to 10^00$ Band R(40)	6852
10.9577377	912.59713	$00^01 \to 10^00$ Band R(38)	6852
10.957738	912.597136	$00^01 \to 10^00$ Band R(38)	1685
10.970024	911,575009	$00^01 \to 10^00$ Band R(36)	1685
10.9700243	911.57501	$00^01 \to 10^00$ Band R(36)	6852
10.9825756	910.53323	$00^01 \to 10^00$ Band R(34)	6852
10.982576	910.533229	$00^01 \to 10^00$ Band R(34)	1685
10.9953918	909.47191	$00^01 \to 10^00$ Band R(32)	6852
10.995392	909.471908	$00^01 \to 10^00$ Band R(32)	1685
11.0084736	908.39115	$00^01 \to 10^00$ Band R(30)	6852
11.008474	908.391149	$00^01 \to 10^00$ Band R(30)	1685
11.021816	907.29104	$00^01 \to 10^00$ Band R(28)	6852
11.021822	907.291050	$00^01 \to 10^00$ Band R(28)	1685
11.035436	906.171705	$00^01 \to 10^00$ Band R(26)	1685
11.0354362	906.17169	$00^01 \to 10^00$ Band R(26)	6852
11.0493184	905.03319	$00^01 \to 10^00$ Band R(24)	6852
11.049319	905.033201	$00^01 \to 10^00$ Band R(24)	1685
11.063469	903.875620	$00^01 \to 10^00$ Band R(22)	1685
11.0634691	903.87560	$00^01 \to 10^00$ Band R(22)	6852
11.0778893	902.69902	$00^01 \to 10^00$ Band R(20)	6852
11.077890	902.699040	$00^01 \to 10^00$ Band R(20)	1685
11.0925799	901.50351	$00^01 \to 10^00$ Band R(18)	6852
11.092580	901.503533	$00^01 \to 10^00$ Band R(18)	1685

$^{14}C^{18}O_2$ Laser—*continued*

Wavelength (μm) vac	Frequency (cm^{-1})	Transition assignment	Comments
11.1075423	900.28914	$00^01 \rightarrow 10^00$ Band R(16)	6852
11.107543	900.289166	$00^01 \rightarrow 10^00$ Band R(16)	1685
11.1227776	899.05598	$00^01 \rightarrow 10^00$ Band R(14)	6852
11.122778	899.056002	$00^01 \rightarrow 10^00$ Band R(14)	1685
11.1382873	897.80408	$00^01 \rightarrow 10^00$ Band R(12)	6852
11.138288	897.80409	$00^01 \rightarrow 10^00$ Band R(12)	1685
11.1540728	896.53349	$00^01 \rightarrow 10^00$ Band R(10)	6852
11.154073	896.533510	$00^01 \rightarrow 10^00$ Band R(10)	1685
11.299632	884.98457	$00^01 \rightarrow 10^00$ Band P(6)	6852
11.299632	884.98458	$00^01 \rightarrow 10^00$ Band P(6)	1685
11.318130	883.53817	$00^01 \rightarrow 10^00$ Band P(8)	6852
11.318130	883.53819	$00^01 \rightarrow 10^00$ Band P(8)	1685
11.3369254	882.07336	$00^01 \rightarrow 10^00$ Band P(10)	6852
11.336926	882.073373	$00^01 \rightarrow 10^00$ Band P(10)	1685
11.356021	880.590133	$00^01 \rightarrow 10^00$ Band P(12)	1685
11.3560210	880.59011	$00^01 \rightarrow 10^00$ Band P(12)	6852
11.3754195	879.08844	$00^01 \rightarrow 10^00$ Band P(14)	6852
11.375420	879.088459	$00^01 \rightarrow 10^00$ Band P(14)	1685
11.395124	877.568337	$00^01 \rightarrow 10^00$ Band P(16)	1685
11.3951240	877.56831	$00^01 \rightarrow 10^00$ Band P(16)	6852
11.4151374	876.02973	$00^01 \rightarrow 10^00$ Band P(18)	6852
11.415138	876.029752	$00^01 \rightarrow 10^00$ Band P(18)	1685
11.435463	874.472679	$00^01 \rightarrow 10^00$ Band P(20)	1685
11.4354630	874.47266	$00^01 \rightarrow 10^00$ Band P(20)	6852
11.456104	872.897092	$00^01 \rightarrow 10^00$ Band P(22)	1685
11.4561042	872.89707	$00^01 \rightarrow 10^00$ Band P(22)	6852
11.477064	871.302959	$00^01 \rightarrow 10^00$ Band P(24)	1685
11.4770642	871.30294	$00^01 \rightarrow 10^00$ Band P(24)	6852
11.4983468	869.69023	$00^01 \rightarrow 10^00$ Band P(26)	6852
11.498347	869.690243	$00^01 \rightarrow 10^00$ Band P(26)	1685
11.5199556	868.05890	$00^01 \rightarrow 10^00$ Band P(28)	6852
11.519956	868.058901	$00^01 \rightarrow 10^00$ Band P(28)	1685
11.541894	866.408887	$00^01 \rightarrow 10^00$ Band P(30)	1685
11.5418945	866.40889	$00^01 \rightarrow 10^00$ Band P(30)	6852
11.5641676	864.74015	$00^01 \rightarrow 10^00$ Band P(32)	6852
11.564168	864.740147	$00^01 \rightarrow 10^00$ Band P(32)	1685
11.586779	863.052625	$00^01 \rightarrow 10^00$ Band P(34)	1685
11.5867790	863.05263	$00^01 \rightarrow 10^00$ Band P(34)	6852
11.609733	861.346255	$00^01 \rightarrow 10^00$ Band P(36)	1685
11.6097330	861.34626	$00^01 \rightarrow 10^00$ Band P(36)	6852
11.633034	859.620967	$00^01 \rightarrow 10^00$ Band P(38)	1685

$^{14}C^{18}O_2$ Laser—*continued*

Wavelength (μm) vac	Frequency (cm^{-1})	Transition assignment	Comments
11.6330341	859.62096	$00^01 \rightarrow 10^00$ Band P(38)	6852
11.656687	857.876685	$00^01 \rightarrow 10^00$ Band P(40)	1685
11.6566870	857.87668	$00^01 \rightarrow 10^00$ Band P(40)	6852
11.6806966	856.11331	$00^01 \rightarrow 10^00$ Band P(42)	6852
11.680697	856.113325	$00^01 \rightarrow 10^00$ Band P(42)	1685
11.7050679	854.33079	$00^01 \rightarrow 10^00$ Band P(44)	6852
11.705068	854.330795	$00^01 \rightarrow 10^00$ Band P(44)	1685
11.729806	852.528994	$00^01 \rightarrow 10^00$ Band P(46)	1685
11.7298063	852.52900	$00^01 \rightarrow 10^00$ Band P(46)	6852

^{12}COS Laser

Wavelength (μm) vac	Frequency (cm^{-1})	Transition assignment	Comments
8.2291	1215..2	$00^01 \to 10^00$ Band R(30)	1687
8.2318	1214.8	$00^01 \to 10^00$ Band R(29)	1687
8.2338	1214.5	$00^01 \to 10^00$ Band R(28)	1687
8.2366	1214.1	$00^01 \to 10^00$ Band R(27)	1687
8.23886	1211.76	$00^01 \to 10^00$ Band R(26)	1687,1688
8.24165	1213.35	$00^01 \to 10^00$ Band R(25)	1687,1688
8.24389	1211.02	$00^01 \to 10^00$ Band R(24)	1687,1688
8.2467	1212.6	$00^01 \to 10^00$ Band R(23)	1687
8.2495	1212.2	$00^01 \to 10^00$ Band R(22)	1687
8.25178	1211.86	$00^01 \to 10^00$ Band R(21)	1687,1688
8.25437	1211.48	$00^01 \to 10^00$ Band R(20)	1687,1688
8.25709	1211.09	$00^01 \to 10^00$ Band R(19)	1687,1688
8.25948	1210.73	$00^01 \to 10^00$ Band R(18)	1687,1688
8.26228	1210.32	$00^01 \to 10^00$ Band R(17)	1687,1688
8.26453	1209.99	$00^01 \to 10^00$ Band R(16)	1687,1688
8.26726	1209.59	$00^01 \to 10^00$ Band R(15)	1687,1688
8.2699	1209.2	$00^01 \to 10^00$ Band R(14)	1687
8.2727	1208.9	$00^01 \to 10^00$ Band R(13)	1687
8.2754	1208.4	$00^01 \to 10^00$ Band R(12)	1687
8.2781	1209.0	$00^01 \to 10^00$ Band R(11)	1687
8.2809	1207.6	$00^01 \to 10^00$ Band R(10)	1687
8.2836	1207.2	$00^01 \to 10^00$ Band R(9)	1687
8.2857	1206.9	$00^01 \to 10^00$ Band R(8)	1687
8.3278	1200.8	$00^01 \to 10^00$ Band P(6)	1687
8.3306	1200.4	$00^01 \to 10^00$ Band P(7)	1687
8.3333	1200.0	$00^01 \to 10^00$ Band P(8)	1687
8.3361	1199.6	$00^01 \to 10^00$ Band P(9)	1687
8.3389	1199.2	$00^01 \to 10^00$ Band P(10)	1687
8.3417	1199.8	$00^01 \to 10^00$ Band P(11)	1687
8.3452	1199.3	$00^01 \to 10^00$ Band P(12)	1687
8.3479	1197.9	$00^01 \to 10^00$ Band P(13)	1687
8.3507	1197.5	$00^01 \to 10^00$ Band P(14)	1687
8.3535	1197.1	$00^01 \to 10^00$ Band P(15)	1687
8.3563	1196.7	$00^01 \to 10^00$ Band P(16)	1687
8.3598	1196.2	$00^01 \to 10^00$ Band P(17)	1687
8.36246	1195.82	$00^01 \to 10^00$ Band P(18)	1687,1688
8.36540	1195.40	$00^01 \to 10^00$ Band P(19)	1687,1688
8.36855	1194.95	$00^01 \to 10^00$ Band P(20)	1687,1688
8.37156	1194.52	$00^01 \to 10^00$ Band P(21)	1687,1688

^{12}COS Laser—*continued*

Wavelength (μm) vac	Frequency (cm^{-1})	Transition assignment	Comments
8.37458	1194.09	$00^01 \rightarrow 10^00$ Band P(22)	1687,1688
8.37788	1193.62	$00^01 \rightarrow 10^00$ Band P(23)	1687,1688
8.38089	1193.19	$00^01 \rightarrow 10^00$ Band P(24)	1687,1688
8.38392	1192.76	$00^01 \rightarrow 10^00$ Band P(25)	1687,1688
8.38701	1192.32	$00^01 \rightarrow 10^00$ Band P(26)	1687,1688
8.39004	1191.89	$00^01 \rightarrow 10^00$ Band P(27)	1687,1688
8.39306	1191.46	$00^01 \rightarrow 10^00$ Band P(28)	1687,1688
8.39616	1191.02	$00^01 \rightarrow 10^00$ Band P(29)	1687,1688
8.39990	1190.49	$00^01 \rightarrow 10^00$ Band P(30)	1687,1688
8.40237	1190.14	$00^01 \rightarrow 10^00$ Band P(31)	1687,1688
8.40548	1199.70	$00^01 \rightarrow 10^00$ Band P(32)	1687,1688
8.40852	1189.27	$00^01 \rightarrow 10^00$ Band P(33)	1687,1688
8.41170	1188.82	$00^01 \rightarrow 10^00$ Band P(34)	1687,1688
8.41468	1188.40	$00^01 \rightarrow 10^00$ Band P(35)	1687,1688
8.41786	1187.95	$00^01 \rightarrow 10^00$ Band P(36)	1687,1688
8.42134	1187.46	$00^01 \rightarrow 10^00$ Band P(37)	1687,1688
8.42432	1187.04	$00^01 \rightarrow 10^00$ Band P(38)	1687,1688
8.4274	1186.6	$00^01 \rightarrow 10^00$ Band P(39)	1687
8.4310	1186.1	$00^01 \rightarrow 10^00$ Band P(40)	1687
8.4345	1185.6	$00^01 \rightarrow 10^00$ Band P(41)	1687
8.4374	1185.2	$00^01 \rightarrow 10^00$ Band P(42)	1687
8.4410	1184.7	$00^01 \rightarrow 10^00$ Band P(43)	1687
8.4438	1184.3	$00^01 \rightarrow 10^00$ Band P(44)	1687
8.4474	1183.8	$00^01 \rightarrow 10^00$ Band P(45)	1687
8.4509	1183.3	$00^01 \rightarrow 10^00$ Band P(46)	1687
8.4538	1182.9	$00^01 \rightarrow 10^00$ Band P(47)	1687
8.4574	1182.4	$00^01 \rightarrow 10^00$ Band P(48)	1687
18.983	526.8	$02^00 \rightarrow 01^10$ BandQ(4)	1689
19.057	524.7	$02^00 \rightarrow 01^10$ Band(5)	1689

^{13}COS Laser

Wavelength (μm) vac	Frequency (cm^{-1})	Transition assignment	Comments
8.6	1160	$00^01 \rightarrow 10^00$ Band	1690–1692

$^{12}CS_2$ Laser

Wavelength (μm) vac	Frequency (cm^{-1})	Transition assignment	Comments
6.6	1500	$10^01 \rightarrow 10^00$ Band P(60)	1693
11.476	871.11	$00^01 \rightarrow 10^00$ Band P(26)	1694
11.4823	870.90	$00^01 \rightarrow 10^00$ Band P(28)	1694,1695
11.4893	870 38	$00^01 \rightarrow 10^00$ Band P(30)	1694–1696
11.4962	869.85	$00^01 \rightarrow 10^00$ Band P(32)	1695
11.5031	869.33	$00^01 \rightarrow 10^00$ Band P(34)	1694–1696
11.5099	868.82	$00^01 \rightarrow 10^00$ Band P(36)	1695,1696
11.5166	868.31	$00^01 \rightarrow 10^00$ Band P(38)	1694,1695
11.5237	867.80	$00^01 \rightarrow 10^00$ Band P(40)	1695
11.5307	867.27	$00^01 \rightarrow 10^00$ Band P(42)	1694,1695
11.5376	866.73	$00^01 \rightarrow 10^00$ Band P(44)	1695,1696
11.5446	866.20	$00^01 \rightarrow 10^00$ Band P(46)	1694,1695
11.553	865.57	$00^01 \rightarrow 10^00$ Band P(48)	1694,1695
11.560	865.08	$00^01 \rightarrow 10^00$ Band P(50)	1694,1695
11.568	864.46	$00^01 \rightarrow 10^00$ Band P(52)	1694,1695
11.582	863.38	$00^01 \rightarrow 10^00$ Band P(56)	1694,1695

$^{13}CS_2$ Laser

Wavelength (μm) vac	Frequency (cm^{-1})	Transition assignment	Comments
6.9		$10^02 \rightarrow 10^01?$ or $02^02 \rightarrow 02^01?$	1697
11.959	836.2	$00^01 \rightarrow 10^00?$ Band R(40)	1698
11.963	835.9	$00^01 \rightarrow 10^00?$ Band R(38)	1698
11.983	834.5	$00^01 \rightarrow 10^00?$ Band R(30)	1698
12.214	818.7	$01^11 \rightarrow 11^10?$ Band P(23)	1698
12.237	817.2	$01^11 \rightarrow 11^10?$ Band P(30)	1698
12.247	816.5	$01^11 \rightarrow 11^10?$ Band P(32)	1698

HCN Laser

Wavelength (μm) vac	Frequency (cm^{-1})	Transition assignment	Comments
3.85	2600	$00^01 \rightarrow 01^10$ Band	1699
7.25	1380	$10^00 \rightarrow 01^10?$Band	1699
8.48	1180	$00^01 \rightarrow 10^00$ Band P(15)?	1699
12.85	778	$01^10 \rightarrow 00^00$ Band R(22)?	1700

H_2O Laser

Wavelength (µm) vac	Frequency (cm^{-1})	Transition assignment	Comments
2.2792	4387.6		1701
4.771	2096	$030(8_{-1}) \rightarrow 020(7_{-4})$?	1702
7.093	1410	$020 \rightarrow 010$ Band $(2_2) \rightarrow (3_2)$	1703
7.204	1388	$020 \rightarrow 010$ Band $(3_1) \rightarrow (4_1)$	1703
7.285	1373	$020 \rightarrow 010$ Band	1703
7.297	1371	$020 \rightarrow 010$ Band $(3_3) \rightarrow (4_3)$	1703
7.390	1353	$020 \rightarrow 010$ Band	1703
7.425	1347	$020 \rightarrow 010$ Band $(4_2) \rightarrow (5_2)$	1703
7.453	1342	$020 \rightarrow 010$ Band	1703
7.45879	1340.70	$020 \rightarrow 010$ Band $(4_3) \rightarrow (5_5)$	1704
7.543	1326	$020 \rightarrow 010$ Band	1703
7.590	1317	$020 \rightarrow 010$ Band	1703
7.59659	1316.38	$020 \rightarrow 010$ Band $(5_{-5}) \rightarrow (6_{-5})$	1704,1705
7.70897	1297.19	$020 \rightarrow 010$ Band $(6_{-5}) \rightarrow (7_{-7})$	1703–1705
7.740	1292	$020 \rightarrow 010$ Band $(6_4) \rightarrow (7_4)$	1703
9.39382	1064.53	$020 \rightarrow 010$ Band ?	1705
9.47472	1055.44	$020 \rightarrow 010$ Band ?	1705
9.56736	1045.22	$020 \rightarrow 010$ Band ?	1705
11.83	845.3	$020 \rightarrow 010$ Band ?	1702
11.96	836.1	$100 (7_1) \rightarrow 020 (6_{-3})$	1702
16.931	590.63		6969
23.359	428.10		6969
26.666	375.00		6969
27.971	356.46		6969
28.054	353.69		6969
28.273	303.68		6969
28.356	302.73		6969
32.929			6969
33.033			6969

$^{14}N_2O$ Laser

Wavelength (µm) vac	Frequency (cm^{-1})	Transition assignment	Comments
9.3073761	1,074.41667	$00^01 \rightarrow 02^00$ Band R(25)	6962
9.3127163	1,073.80056	$00^01 \rightarrow 02^00$ Band R(24)	6962
9.3181349	1,073.17613	$00^01 \rightarrow 02^00$ Band R(23)	6962
9.3236326	1,072.54334	$00^01 \rightarrow 02^00$ Band R(22)	6962
9.3292098	1,071.90214	$00^01 \rightarrow 02^00$ Band R(21)	6962

$^{14}N_2O$ Laser—*continued*

Wavelength (μm) vac	Frequency (cm^{-1})	Transition assignment	Comments
9.3348672	1,071.25252	$00^01 \rightarrow 02^00$ Band R(20)	6962
9.3406052	1,070.59444	$00^01 \rightarrow 02^00$ Band R(19)	6962
9.3464244	1,069.92788	$00^01 \rightarrow 02^00$ Band R(18)	6962
9.3523253	1,069.25280	$00^01 \rightarrow 02^00$ Band R(17)	6962
9.3583084	1,068.56918	$00^01 \rightarrow 02^00$ Band R(16)	6962
9.3643743	1,067.87700	$00^01 \rightarrow 02^00$ Band R(15)	6962
9.3705235	1,067.17624	$00^01 \rightarrow 02^00$ Band R(14)	6962
9.3767563	1,066.46687	$00^01 \rightarrow 02^00$ Band R(13)	6962
9.3830734	1,065.74888	$00^01 \rightarrow 02^00$ Band R(12)	6962
9.4025348	1,063.54299	$00^01 \rightarrow 02^00$ Band R(9)	6962
9.4091936	1,062.79034	$00^01 \rightarrow 02^00$ Band R(8)	6962
9.5442347	1,047.75294	$00^01 \rightarrow 02^00$ Band P(9)	6962
9.5526030	1,046.83509	$00^01 \rightarrow 02^00$ Band P(10)	6962
9.5610644	1,045.90865	$00^01 \rightarrow 02^00$ Band P(11)	6962
9.5696192	1,044.97366	$00^01 \rightarrow 02^00$ Band P(12)	6962
9.5782677	1,044.03013	$00^01 \rightarrow 02^00$ Band P(13)	6962
9.5870100	1,043.07808	$00^01 \rightarrow 02^00$ Band P(14)	6962
9.5958464	1,042.11756	$00^01 \rightarrow 02^00$ Band P(15)	6962
9.6047771	1,041.14858	$00^01 \rightarrow 02^00$ Band P(16)	6962
9.6138023	1,040.17117	$00^01 \rightarrow 02^00$ Band P(17)	6962
9.6229222	1,039.18538	$00^01 \rightarrow 02^00$ Band P(18)	6962
9.6321369	1,038.19123	$00^01 \rightarrow 02^00$ Band P(19)	6962
9.6414466	1,037.18876	$00^01 \rightarrow 02^00$ Band P(20)	6962
9.6508515	1,036.17800	$00^01 \rightarrow 02^00$ Band P(21)	6962
9.6603518	1,035.15899	$00^01 \rightarrow 02^00$ Band P(22)	6962
9.6699475	1,034.13178	$00^01 \rightarrow 02^00$ Band P(23)	6962
9.6796388	1,033.09640	$00^01 \rightarrow 02^00$ Band P(24)	6962
9.6894258	1,032.05290	$00^01 \rightarrow 02^00$ Band P(25)	6962
9.6993087	1,031.00132	$00^01 \rightarrow 02^00$ Band P(26)	6962
9.7092875	1,029.94170	$00^01 \rightarrow 02^00$ Band P(27)	6962
9.7193623	1,028.87408	$00^01 \rightarrow 02^00$ Band P(28)	6962
9.7295333	1,027.79852	$00^01 \rightarrow 02^00$ Band P(29)	6962
9.7398006	1,026.71507	$00^01 \rightarrow 02^00$ Band P(30)	6962
10.2582202	974.82797	$00^01 \rightarrow 10^00$ Band R(47)	6962
10.2652291	974.16238	$00^01 \rightarrow 10^00$ Band R(46)	6962
10.2722864	973.49311	$00^01 \rightarrow 10^00$ Band R(45)	6962
10.2793922	972.82016	$00^01 \rightarrow 10^00$ Band R(44)	6962
10.2865467	972.14355	$00^01 \rightarrow 10^00$ Band R(43)	6962
10.2937499	971.46328	$00^01 \rightarrow 10^00$ Band R(42)	6962
10.3010019	970.77936	$00^01 \rightarrow 10^00$ Band R(41)	6962

$^{14}N_2O$ Laser—*continued*

Wavelength (µm) vac	Frequency (cm^{-1})	Transition assignment	Comments
10.3083028	970.091800	$00^01 \rightarrow 10^00$ Band R(40)	1706,1707
10.3156527	969.400607	$00^01 \rightarrow 10^00$ Band R(39)	1707
10.3230518	968.705789	$00^01 \rightarrow 10^00$ Band R(38)	1707
10.3305000	968.007355	$00^01 \rightarrow 10^00$ Band R(37)	1707
10.3379976	967.305314	$00^01 \rightarrow 10^00$ Band R(36)	1707
10.3455446	966.599670	$00^01 \rightarrow 10^00$ Band R(35)	1707–08,6963–64
10.3531411	965.890435	$00^01 \rightarrow 10^00$ Band R(34)	1707–08,6963–64
10.3607873	965.177614	$00^01 \rightarrow 10^00$ Band R(33)	1707–08,6963–64
10.3684833	964.461213	$00^01 \rightarrow 10^00$ Band R(32)	1707–08,6963–64
10.3762292	963.741241	$00^01 \rightarrow 10^00$ Band R(31)	1707–08,6963–64
10.3840251	963.017705	$00^01 \rightarrow 10^00$ Band R(30)	1707–08,6963–64
10.3918711	962.290611	$00^01 \rightarrow 10^00$ Band R(29)	1707–08,6963–64
10.3997674	961.559966	$00^01 \rightarrow 10^00$ Band R(28)	1707–08,6963–64
10.4077141	960.825777	$00^01 \rightarrow 10^00$ Band R(27)	1707–08,6963–64
10.41105	960.518	$00^02 \rightarrow 10^01$ Band R(30)	1714
10.4157114	960.088050	$00^01 \rightarrow 10^00$ Band R(26)	1707–08,6963–64
10.41887	959.797	$00^02 \rightarrow 10^01$ Band R(29)	1714
10.4237593	959.346792	$00^01 \rightarrow 10^00$ Band R(25)	1707–08,6963–64
10.42674	959.072	$00^02 \rightarrow 10^01$ Band R(28)	1714
10.4318580	958.602009	$00^01 \rightarrow 10^00$ Band R(24)	1707–08,6963–64
10.43466	959.345	$00^02 \rightarrow 10^01$ Band R(27)	1714
10.4400076	957.853708	$00^01 \rightarrow 10^00$ Band R(23)	1707–08,6963–64
10.44263	957.613	$00^02 \rightarrow 10^01$ Band R(26)	1714
10.4482084	957.101893	$00^01 \rightarrow 10^00$ Band R(22)	1707–08,6963–64
10.45065	956.879	$00^02 \rightarrow 10^01$ Band R(25)	1714
10.4564603	956.346572	$00^01 \rightarrow 10^00$ Band R(21)	1707–08,6963–64
10.4587136	956.140531	$00^02 \rightarrow 10^01$ Band R(24)	1714
10.4647637	955.587750	$00^01 \rightarrow 10^00$ Band R(20)	1707–09,6963–64
10.46684	955.398	$00^02 \rightarrow 10^01$ Band R(23)	1714
10.4731186	954.825434	$00^01 \rightarrow 10^00$ Band R(19)	1707–09,6963–64
10.47501	954.653	$00^02 \rightarrow 10^01$ Band R(22)	1714
10.4815252	954.059628	$00^01 \rightarrow 10^00$ Band R(18)	1707–10,6963–64
10.48324	953.904	$00^02 \rightarrow 10^01$ Band R(21)	1714
10.4899836	953.290339	$00^01 \rightarrow 10^00$ Band R(17)	1707–10,6963–64
10.49151	953.152	$00^02 \rightarrow 10^01$ Band R(20)	1714
10.4984940	952.517571	$00^01 \rightarrow 10^00$ Band R(16)	1707–10,6963–64
10.49983	952.396	$00^02 \rightarrow 10^01$ Band R(19)	1714
10.5070566	951.741331	$00^01 \rightarrow 10^00$ Band R(15)	1707–10,6963–64
10.50821	951.637	$00^02 \rightarrow 10^01$ Band R(18)	1714
10.5156714	950.961625	$00^01 \rightarrow 10^00$ Band R(14)	1707–10,6963–64

$^{14}N_2O$ Laser—*continued*

Wavelength (μm) vac	Frequency (cm^{-1})	Transition assignment	Comments
10.51664	950.874	$00^02 \to 10^01$ Band R(17)	1714
10.5243388	950.178456	$00^01 \to 10^00$ Band R(13)	1707–10,6963–64
10.52512	950.108	$00^02 \to 10^01$ Band R(16)	1714
10.5330588	949.391831	$00^01 \to 10^00$ Band R(12)	1707–10,6963–64
10.53366	949.338	$00^02 \to 10^01$ Band R(15)	1714
10.5418316	948.601755	$00^01 \to 10^00$ Band R(11)	1707–10,6963–64
10.54224	948.565	$00^02 \to 10^01$ Band R(14)	1714
10.5506575	947.808232	$00^01 \to 10^00$ Band R(10)	1707–10,6ˆ63–64
10.5508756	947.788633	$00^02 \to 10^01$ Band R(13)	1714
10.5595364	947.011268	$00^01 \to 10^00$ Band R(9)	1707–08,6963–64
10.56830	946.226	$00^02 \to 10^01$ Band R(11)	1714
10.5684688	946.210868	$00^01 \to 10^00$ Band R(8)	1706–08,6963–64
10.57710	945.439	$00^02 \to 10^01$ Band R(10)	1714
10.5774546	945.407036	$00^01 \to 10^00$ Band R(7)	1707–08,6963–64
10.58595	944.649	$00^02 \to 10^01$ Band R(9)	1714
10.5864941	944.599778	$00^01 \to 10^00$ Band R(6)	1707–08,6963–64
10.59485	943.855	$00^02 \to 10^01$ Band R(8)	1714
10.5955875	943.789097	$00^01 \to 10^00$ Band R(5)	1707–08,6963–64
10.60380	943.058	$00^02 \to 10^01$ Band R(7)	1714
10.6047350	942.974999	$00^01 \to 10^00$ Band R(4)	1707–08,6963–64
10.61281	942.258	$00^02 \to 10^01$ Band R(6)	1714
10.6139368	942.157488	$00^01 \to 10^00$ Band R(3)	1707–08,6963–64
10.62187	941.454	$00^02 \to 10^01$ Band R(5)	1714
10.6231929	941.336569	$00^01 \to 10^00$ Band R(2)	1707–08,6963–64
10.6325038	940.512245	$00^01 \to 10^00$ Band R(1)	1707–08,6963–64
10.6418694	939.684522	$00^01 \to 10^00$ Band R(0)	1707–11,6963–64
10.6607661	938.018895	$00^01 \to 10^00$ Band P(1)	1711,6965
10.6702974	937.180999	$00^01 \to 10^00$ Band P(2)	1708,6966
10.6798844	936.339719	$00^01 \to 10^00$ Band P(3)	1707–08,6963–64
10.6895273	935.495062	$00^01 \to 10^00$ Band P(4)	1707–08,6963–64
10.6992262	934.647028	$00^01 \to 10^00$ Band P(5)	1707–08,6963–64
10.7089814	933.795623	$00^01 \to 10^00$ Band P(6)	1707–08,6963–64
10.7187931	932.940853	$00^01 \to 10^00$ Band P(7)	1707–08,6963–64
10.7286616	932.082716	$00^01 \to 10^00$ Band P(8)	1707–08,6963–64
10.73488	931.543	$00^02 \to 10^01$ Band P(6)	1714
10.7385869	931.221221	$00^01 \to 10^00$ Band P(9)	1707–08,6963–64
10.74466	930.695	$00^02 \to 10^01$ Band P(7)	1714
10.7485694	930.356369	$00^01 \to 10^00$ Band P(10)	1707–08,6963–64
10.75449	929.844	$00^02 \to 10^01$ Band P(8)	1714
10.7586093	929.488164	$00^01 \to 10^00$ Band P(11)	1707–08,6963–64

$^{14}N_2O$ Laser—*continued*

Wavelength (μm) vac	Frequency (cm^{-1})	Transition assignment	Comments
10.7643831	928.989601	$00^02 \rightarrow 10^01$ Band P(9)	1714
10.7687068	928.616611	$00^01 \rightarrow 10^00$ Band P(12)	1707–08,6963–64
10.77433	928.132	$00^02 \rightarrow 10^01$ Band P(10)	1714
10.7788621	927.741710	$00^01 \rightarrow 10^00$ Band P(13)	1707–08,6963–64
10.78434	927.271	$00^02 \rightarrow 10^01$ Band P(11)	1714
10.7890756	926.863466	$00^01 \rightarrow 10^00$ Band P(14)	1707–08,6963–64
10.79440	926.407	$00^02 \rightarrow 10^01$ Band P(12)	1714
10.7993473	925.981884	$00^01 \rightarrow 10^00$ Band P(15)	1707–08,6963–64
10.80452	925.539	$00^02 \rightarrow 10^01$ Band P(13)	1714
10.8096777	925.096964	$00^01 \rightarrow 10^00$ Band P(16)	1707–08,6963–64
10.81470	924.668	$00^02 \rightarrow 10^01$ Band P(14)	1714
10.820067	924.20871	$00^01 \rightarrow 10^00$ Band P(17)	1707–08,6963–64
10.82494	923.793	$00^02 \rightarrow 10^01$ Band P(15)	1714
10.830515	923.31713	$00^01 \rightarrow 10^00$ Band P(18)	1707–08,6963–64
10.8352265	922.915636	$00^02 \rightarrow 10^01$ Band P(16)	1714
10.841023	922.42221	$00^01 \rightarrow 10^00$ Band P(19)	1707–08,6963–64
10.84559	922.034	$00^02 \rightarrow 10^01$ Band P(17)	1714
10.851590	921.52398	$00^01 \rightarrow 10^00$ Band P(20)	1707–08,1710, 1712–13,6963–64
10.85600	921.150	$00^02 \rightarrow 10^01$ Band P(18)	1714
10.862216	920.62242	$00^01 \rightarrow 10^00$ Band P(21)	1707–08,1710, 1712–13,6963–64
10.86647	920.262	$00^02 \rightarrow 10^01$ Band P(19)	1714
10.872903	919.71754	$00^01 \rightarrow 10^00$ Band P(22)	1707–08,1710, 1712–13,6963–64
10.87700	919.371	$00^02 \rightarrow 10^01$ Band P(20)	1714
10.883651	918.80935	$00^01 \rightarrow 10^00$ Band P(23)	1707–08,1710, 1712–13,6963–64
10.88759	918.477	$00^02 \rightarrow 10^01$ Band P(21)	1714
10.89446	917.8978	$00^01 \rightarrow 10^00$ Band P(24)	1707–08,1710, 1712–13,6963–64
10.89824	917.579	$00^02 \rightarrow 10^01$ Band P(22)	1714
10.90533	916.9830	$00^01 \rightarrow 10^00$ Band P(25)	1707–08,1710, 1712–13,6963–64,
10.90895	916.678	$00^02 \rightarrow 10^01$ Band P(23)	1714
10.91626	916.0649	$00^01 \rightarrow 10^00$ Band P(26)	1707–08,1710, 1712–13,6963–64
10.91972	915.775	$00^02 \rightarrow 10^01$ Band P(24)	1714
10.92725	915.1434	$00^01 \rightarrow 10^00$ Band P(27)	1707–08,1710, 1712–13,6963–64

$^{14}N_2O$ Laser—*continued*

Wavelength (μm) vac	Frequency (cm^{-1})	Transition assignment	Comments
10.93055	914.867	$00^02 \rightarrow 10^01$ Band P(25)	1714
10.93830	914.2187	$00^01 \rightarrow 10^00$ Band P(28)	1707–08,1710, 1712–13,6963–64
10.94144	913.956	$00^02 \rightarrow 10^01$ Band P(26)	1714
10.94942	913.2906	$00^01 \rightarrow 10^00$ Band P(29)	1707–08,1710, 1712–13,6963–64
10.95240	913.042	$00^02 \rightarrow 10^01$ Band P(27)	1714
10.96059	912.3593	$00^01 \rightarrow 10^00$ Band P(30)	1707–08,1712, 6963–64
10.96341	912.125	$00^02 \rightarrow 10^01$ Band P(28)	1714
10.9718339	911.42466	$00^01 \rightarrow 10^00$ Band P(31)	1707–08,6963–64
10.97449	911.204	$00^02 \rightarrow 10^01$ Band P(29)	1714
10.9831367	910.48671	$00^01 \rightarrow 10^00$ Band P(32)	1707–08,6963–64
10.98314	910.487	$00^01 \rightarrow 10^00$ Band P(32)	1707–08,1712, 6963–64
10.98563	910.280	$00^02 \rightarrow 10^01$ Band P(30)	1714
10.9945027	909.54546	$00^01 \rightarrow 10^00$ Band P(33)	1707–08,6963–64
11.0059321	908.60091	$00^01 \rightarrow 10^00$ Band P(34)	1707–08,6963–64
11.0174254	907.65307	$00^01 \rightarrow 10^00$ Band P(35)	1707–08,6963–64
11.0289827	906.70194	$00^01 \rightarrow 10^00$ Band P(36)	6966
11.0406045	905.74751	$00^01 \rightarrow 10^00$ Band P(37)	6966
11.0522909	904.78979	$00^01 \rightarrow 10^00$ Band P(38)	6962
11.0640425	903.82878	$00^01 \rightarrow 10^00$ Band P(39)	6962
11.0758594	902.86448	$00^01 \rightarrow 10^00$ Band P(40)	6962
11.0877421	901.89688	$00^01 \rightarrow 10^00$ Band P(41)	6962
11.0996908	900.92600	$00^01 \rightarrow 10^00$ Band P(42)	6962
11.1117059	899.95182	$00^01 \rightarrow 10^00$ Band P(43)	6962
11.1237879	898.97435	$00^01 \rightarrow 10^00$ Band P(44)	6962
11.1359369	897.99360	P(45)	6962
11.1481534	897.00955	$00^01 \rightarrow 10^00$ Band P(46)	6962
11.1604377	896.02220	$00^01 \rightarrow 10^00$ Band P(47)	6962

$^{14}N^{15}NO$ Laser

Wavelength (μm) vac	Frequency (cm^{-1})	Transition assignment	Comments
4.6204	2164.3	$10^01 \rightarrow 10^00$ Band R(15)	1715
4.6812	2136.2	$10^01 \rightarrow 10^00$ Band P(17)	1715

$^{15}N_2O$ Laser

Wavelength (µm) vac	Frequency (cm^{-1})	Transition assignment	Comments
4.6		$10^01 \rightarrow 10^00$ Band	1716
10.99450	909.545	$00^01 \rightarrow 10^00$ Band P(33)	1707,1708,1712
11.00593	908.601	$00^01 \rightarrow 10^00$ Band P(34)	1707,1708,1712
11.01743	907.653	$00^01 \rightarrow 10^00$ Band P(35)	1707,1708,1712
11.02898	906.702	$00^01 \rightarrow 10^00$ Band P(36)	1708,1712
11.04061	905.747	$00^01 \rightarrow 10^00$ Band P(37)	1708,1712

$^{15}N^{14}NO$ Laser

Wavelength (µm) vac	Frequency (cm^{-1})	Transition assignment	Comments
4.5851	2181.0	$10^01 \rightarrow 10^00$ Band R(8)	1717
4.6189	2165.0	$10^01 \rightarrow 10^00$ Band P(10)	1717

NOCl Laser

Wavelength (µm) vac	Frequency (cm^{-1})	Transition assignment	Comments
16.4	608	$\nu_2+\nu_3 \rightarrow \nu_3$ Band	1718,1719
16.52	605.5	$\nu_2+\nu_3 \rightarrow \nu_3$ Band	1718,1720
16.57	603.4	$\nu_2+\nu_3 \rightarrow \nu_3$ Band	1718,1721
16.69	599.1	$\nu_2+\nu_3 \rightarrow \nu_3$ Band	1718,1722
16.7	598	$\nu_2+\nu_3 \rightarrow \nu_3$ Band	1718,1723
16.7	598	$\nu_2+\nu_3 \rightarrow \nu_3$ Band	1718,1724
16.75	597.0	$\nu_2+\nu_3 \rightarrow \nu_3$ Band	1718,1725
16.86	593.2	$\nu_2+\nu_3 \rightarrow \nu_3$ Band	1718,1726
16.9	590	$\nu_2+\nu_3 \rightarrow \nu_3$ Band	1718,1727
16.99	588.5	$\nu_2+\nu_3 \rightarrow \nu_3$ Band	1718,1728

NSF Laser

Wavelength (μm) vac	Frequency (cm⁻¹)	Transition assignment	Comments
15.19	658.3	$v_2+v_3 \rightarrow v_3$ Band	1729,1730
15.26	655.1	$v_2+v_3 \rightarrow v_3$ Band	1729,1731
15.34	651.7	$v_2+v_3 \rightarrow v_3$ Band	1729,1732
15.42	648.6	$v_2+v_3 \rightarrow v_3$ Band	1729,1733
15.47	646.5	$v_2+v_3 \rightarrow v_3$ Band	1729,1734
15.61	640.7	$v_2+v_3 \rightarrow v_3$ Band	1729,1735
15.61	640.6	$v_2+v_3 \rightarrow v_3$ Band	1729,1736
15.838	631.41	$v_2+v_3 \rightarrow v_3$ Band	1729,1737
15.89	629.2	$v_2+v_3 \rightarrow v_3$ Band	1729,1738
15.98	625.9	$v_2+v_3 \rightarrow v_3$ Band	1729,1739
16.03	623.9	$v_2+v_3 \rightarrow v_3$ Band	1729,1740
16.15	619.1	$v_2+v_3 \rightarrow v_3$ Band	1729,1741
16.19	617.6	$v_2+v_3 \rightarrow v_3$ Band	1729,1742

3.3.2.4 Four-Atom Vibrational Transition Lasers

Table 3.3.2.3
Four-Atom Vibrational Transition Lasers

BCl₃ Laser

Wavelength (μm) vac	Frequency (cm⁻¹)	Transition assignment	Comments
18.3	546.		1743
18.8	532.		1743
19.1	524.		1743
19.4	515.		1743
20.2	495.		1743
20.6	485.		1743
22.4	446.		1743
23.	435.		1743

C_2D_2 Laser

Wavelength (μm) vac	Frequency (cm^{-1})	Transition assignment	Comments
17.45	573.1	Hot Band ?R(23)?	1755
17.498	571050	$(\nu_4+\nu_5)\backslash\Sigma^+_u \to \nu_4$ Band R(22)	1747
17.56	569.6	$(\nu_4+\nu_5)\backslash\Sigma^+_u \to \nu_4$ Band R(21)	1747
17.61	567.8	$(\nu_4+\nu_5)\backslash\Sigma^+_u \to \nu_4$ Band R(16)	1753
17.610	567.87	$(\nu_4+\nu_5)\backslash\Sigma^+_u \to \nu_4$ Band R(20)	1747
17.665	566.08	$(\nu_4+\nu_5)\backslash\Sigma^+_u \to \nu_4$ Band R(19)	1747
17.722	564.28	$(\nu_4+\nu_5)\Sigma^+_u \to \nu_4$ Band R(18)	1747
17.778	562.48	$(\nu_4+\nu_5)\Sigma^+_u \to \nu_4$ Band R(17)	1747,1748
17.835	560.68	$(\nu_4+\nu_5)\Sigma^+_u \to \nu_4$ Band R(16)	1747
18.67	535.7	Hot Band ?Q(24)?	1756
18.79	532.2	$(2\nu_5+\nu_4)\pi_g \to$ $(\nu_5+\nu_4)\Sigma_u$ Band Q?	1758
18.79	532.3	$(\nu_4+\nu_5)\Sigma^+_u \to \nu_4$ Band ?	1747
18.84	530.7	$(\nu_4+\nu_5)\Sigma^+_u \to \nu_4$ Band ?	1747
18.85	530.6	$(\nu_4+\nu_5)\Sigma^+_u \to \nu_4$ Band Q(1)	1749
18.960	527.43	$(\nu_4+\nu_5)\Sigma^+_u \to \nu_4$ Band P(2)	1747,1750
18.97	527.1	$(2\nu_5+\nu_4)\pi_u \to 2\nu_4\Sigma_g$ Band Q?	1762
19.03	525.6	$(2\nu_5+\nu_4)\pi_g \to$ $(\nu_5+\nu_4)\Delta_u$ Band Q?	1760
19.03	525.6	$(2\nu_5+\nu_4)\pi_g \to$ $(\nu_5+\nu_4)\Sigma_u$ Band P(4)	1759
19.03	525.6	$(\nu_4+\nu_5)\Sigma^+_u \to \nu_4$ Band P(3)	1747
19.081	524.07	$(\nu_4+\nu_5)\Sigma^+_u \to \nu_4$ Band P(4)	1747
19.13	522.7	$(2\nu_5+\nu_4)\pi_u \to$ $2\nu_4\backslash\Sigma_g$ Band P(3)	1763
19.20	520.7	$(\nu_4+\nu_5)\Sigma^+_u \to \nu_4$ Band P(6)	1751
19.27	518.9	$(2\nu_5+\nu_4)\pi_g \to$ $(\nu_5+\nu_4)\Delta_u$ Band P(4)	1761
19.67	508.4	$(\nu_4+\nu_5)\Delta_u \to \nu_4$ Band P(18)	1754
19.947	501.33	$(\nu_4+\nu_5)\Sigma^+_u \to \nu_4$ Band P(18)	1747
20.010	499.75	$(\nu_4+\nu_5)\Sigma^+_u \to \nu_4$ Band P(19)	1747,1752
20.073	498.17	$(\nu_4+\nu_5)\Sigma^+_u \to \nu_4$ Band P(20)	1747
20.13	496.8	$(\nu_4+\nu_5)\Sigma^+_u \to \nu_4$ Band P(21)	1747
20.202	495.00	$(\nu_4+\nu_5)\Sigma^+_u \to \nu_4$ Band P(22)	1747
20.267	493.41	$(\nu_4+\nu_5)\Sigma^+_u \to \nu_4$ Band P(23)	1747
20.332	491.83	$(\nu_4+\nu_5)\Sigma^+_u \to \nu_4$ Band P(24)	1747
20.44	489.2	Hot Band ?P(25)?	1757

$^{12}C_2D_2$ Laser

Wavelength (μm) vac	Frequency (cm⁻¹)	Transition assignment	Comments
17.893	558.87	$(\nu_4+\nu_{-5})\Sigma^+_u \rightarrow \nu_4$ Band R(15)	6870
18.84	530.8	$(\nu_4+\nu_{-5})\Sigma^+_u \rightarrow \nu_4$ Band Q(?)	6870
19.511	512.84	$(\nu_4+\nu_{-5})\Sigma^+_u \rightarrow \nu_4$ Band P(11)	6870
19.634	509.32	$(\nu_4+\nu_{-5})\Sigma^+_u \rightarrow \nu_4$ Band P(13)	6870
19.758	506.12	$(\nu_4+\nu_{-5})\Sigma^+_u u \rightarrow \nu_4$ Band P(15)	6870
19.884	502.93	$(\nu_4+\nu_{-5})\Sigma^+_u u \rightarrow \nu_4$ Band P(17)	6870

$^{12}C_2DH$ Laser

Wavelength (μm) vac	Frequency (cm⁻¹)	Transition assignment	Comments
17.71	564.8	$(2\nu_4)\Sigma^+ \rightarrow \nu_4$ Band R(23)	6854,6855
18.09	552.8	HotbandsR(?)	6854,6863
18.45	542.1	$(2\nu_4)\Sigma^+ \rightarrow \nu_4$ Band R(12)	6854,6856
18.64	536.5	Hotbands R(10)	6854,6864
19.13	522.7	Hotbands Q(?)	6854,6865
19.18	521.4	$2\nu_4(\Delta_c) \rightarrow \nu_4$ Band Q(22)	6854,6862
19.33	517.4	Hotbands Q(?)	6854,6866
19.37	516.2	Hotbands?	6854,6867
19.4	515.5	$(2\nu_4)\Sigma^+ \rightarrow \nu_4$ Band Q(13)	6854,6858
19.4	515.4	$(2\nu_4)\Sigma^+ \rightarrow \nu_4$ Band Q(9)	6854,6857
19.48	513.4	Hotbands Q(11)	6854,6868
20.15	496.2	$(2\nu_4)\Sigma^+ \rightarrow \nu_4$ Band P(10)	6854,6859
20.38	490.7	Hotbands P(12)	6854,6869
20.47	488.5	$(2\nu_4)\Sigma^+ \rightarrow \nu_4$ Band P(14)	6854,6860
21.38	467.8	$(2\nu_4)\Sigma^+ \rightarrow \nu_4$ Band P(25)	6854,6861

C_2H_2 Laser

Wavelength (μm) vac	Frequency (cm⁻¹)	Transition assignment	Comments
8.03406	1244.70	$\nu_2 \rightarrow \nu^l_5$ Band Q(5)?	1744,1746
8.03523	1244.52	$\nu_2 \rightarrow \nu^l_5$ Band Q(7)?	1745,1746
8.03561	1244.46	$\nu_2 \rightarrow \nu^l_5$ Band Q?	1744
8.03781	1244.12	$\nu_2 \rightarrow \nu^l_5$ Band ?	1745
8.03800	1244.09	$\nu_2 \rightarrow \nu^l_5$ Band Q?	1744
8.04020	1243.75	$\nu_2 \rightarrow \nu^l_5$ Band Q(11)	1745,1746
8.04091	1243.64	$\nu_2 \rightarrow \nu^l_5$ Band Q?	1744

C_2H_2 Laser—*continued*

Wavelength (μm) vac	Frequency (cm⁻¹)	Transition assignment	Comments
8.04428	1243.12	$v_2 \to v^1_5$ BandQ(13)	1744,1746
8.04453	1243.08	$v_2 \to v^1_5$ Band Q?	1745
8.197	1220	$v_2 \to v^1_5$ Band P(10)	1746
8.299	1205	$v_2 \to v^1_5$ Band P(16)	1746

COF_2 Laser

Wavelength (μm) vac	Frequency (cm⁻¹)	Transition assignment	Comments
~10	~1000	0200 → 0100 Band ?	1764,1765

$^{14}NH_3$ Laser

Wavelength (μm) vac	Frequency (cm⁻¹)	Transition assignment	Comments
6.270	1595	$v_4 \to 0$ Band	6871
6.689	1495	$v_4 \to 0$ Band	6871
9.346	1070	$v_2 \to 0$ Band aR(6,K)	1766,1790
9.643	1037	$v_2 \to 0$ Band aR(4,K) or sR(3,K)	1766,1790
9.737	1027	$v_2 \to 0$ Band aR(3,1), aR(2,2), aR(2,1), aR(4,4), or aR(4,3)	1766,1790
9.921	1008	$v_2 \to 0$ Band aR(2,0), sR(1,1), sR(1,0), aR(3,3), or aR(3,2)	1766,1790
10.29016	971.882	$v_2 \to 0$ Band aR(1,1)	1766,1790
10.3376	967.346	$v_2 \to 0$ Band sQ(3,3)	6873
10.3423	966.905	$v_2 \to 0$ Band sQ(4,3)	6873
10.3589	965.354	$v_2 \to 0$ Band sQ(6,6)	6873
10.3670	964.596	$v_2 \to 0$ Band sQ(7,6)	6873
10.5067	951.776	$v_2 \to 0$ Band aR(0,0)	6873
10.5459	948.232	$v_2 \to 0$ Band sP(1,0)	6873
10.54594	948.232	$v_2 \to 0$ Band sP(1,0)	1766,1790
10.6	941	$v_2 \to 0$ Band aQ(9,3), aQ(10,4), aQ(9,2), aQ(10,3), or aQ(9,1)	1766,1790

$^{14}NH_3$ Laser—*continued*

Wavelength (μm) vac	Frequency (cm^{-1})	Transition assignment	Comments
10.7	931	$\nu_2 \to 0$ Band	1766,1790
		aQ(8,6), aQ(9,7), aQ(6,5),	
		aQ(3,3), aQ(5,4), aQ(5,5),	
		Q(7,6), aQ(4,4) or aQ(2,2)	
10.7182	932.992	$\nu_2 \to 0$ Band aQ(5,3)	6873
10.7322	931.774	$\nu_2 \to 0$ Band aQ(4,3)	6873
10.73729	931.334	$\nu_2 \to 0$ Band aQ(2,2)	1767,1790
10.74394	930.757	$\nu_2 \to 0$ Band aQ(3,3)	1767,1790
10.75387	929.898	$\nu_2 \to 0$ Band aQ(4,4)	1767,1790
10.7624	929.162	$\nu_2 \to 0$ Band aQ(7,6)	6873
10.76710	928.755	$\nu_2 \to 0$ Band aQ(5,5)	1767,1790
10.7837	927.323	$\nu_2 \to 0$ Band aQ(6,6)	6873,6874
10.8548	921.255	$\nu_2 \to 0$ Band aQ(9,9)	6873
11.01	908	$\nu_2 \to 0$ Band sP(3,1 or 2)	1769
11.01080	892.157	$\nu_2 \to 0$ Band sP(3,0)	1767,1790
11.0111	908.177	$\nu_2 \to 0$ Band sP(3,1)	6873
11.20879	892.157	$\nu_2 \to 0$ Band aP(2,0)	1767,1770,1790
11.2123	891.882	$\nu_2 \to 0$ Band aP(2,1)	6873
11.2603	888.079	$\nu_2 \to 0$ Band sP(4,1)	6873
11.261	887.99	$\nu_2 \to 0$ Band sP(4,1-2)	1770,1771,1790
11.2628	887.877	$\nu_2 \to 0$ Band sP(4,3)	6873
11.46044	872.567	$\nu_2 \to 0$ Band aP(3,1)	1766,1767,1770, 1771,1773,1790
11.47135	871.737	$\nu_2 \to 0$ Band aP(3,2)	1770
11.52074	868.000	$\nu_2 \to 0$ Band sP(5,0)	1766,1767,1770, 1771,1774,1790
11.5212	868.035	$\nu_2 \to 0$ Band sP(5,1)	6873
11.52446	867.719	$\nu_2 \to 0$ Band sP(5,3)	1790-1792
11.55466	865.452	$2\nu_2 \to \nu_2$ Band sP(4,1)	1783
11.71208	853.819	$\nu_2 \to 0$ Band aP(4,0)	1767,1770-1,1790
11.71582	853.547	$\nu_2 \to 0$ Band aP(4,1)	1766-7,1771,1790
11.72712	852.724	$\nu_2 \to 0$ Band aP(4,2)	1767,1770,1790
11.74637	851.327	$\nu_2 \to 0$ Band aP(4,3)	1767,1770,1790
11.7942	847.876	$\nu_2 \to 0$ Band sP(6,1)	6873
11.796	847.78	$\nu_2 \to 0$ Band sP(6,1-2)	1770
11.7983	847.578	$\nu_2 \to 0$ Band sP(6,3)	6873,6874
11.80167	847.338	$\nu_2 \to 0$ Band sP(6,4)	1770,1774
11.97859	834.823	$\nu_2 \to 0$ Band aP(5,1)	1766,1767,1770, 1771,1790

$^{14}NH_3$ Laser—*continued*

Wavelength (µm) vac	Frequency (cm^{-1})	Transition assignment	Comments
11.99025	834.011	$\nu_2 \to 0$ Band aP(5,2)	1766,1767,1770, 1771,1790
12.01008	832.634	$\nu_2 \to 0$ Band aP(5,3)	1766,1767,1770, 1771,1777,1790
12.03872	830.653	$\nu_2 \to 0$ Band aP(5,4)	1767,1770–1,1790
12.07912	827.875	$\nu_2 \to 0$ Band sP(7,0)	1766,1767,1770, 1771,1774,1778, 1779,1790,1792
12.0797	827.833	$\nu_2 \to 0$ Band s P(7,1)	6873
12.0997	826.470	$\nu_2 \to 0$ Band s P(7,6)	6873
12.11418	825.479	$2\nu_2 \to \nu_2$ Band sP(6,4)	1784
12.15575	822.656	$2\nu_2 \to \nu_2$ Band sP(6,3)	1785
12.18444	820.719	$2\nu_2 \to \nu_2$ Band sP(6,2)	1786
12.24521	816.646	$\nu_2 \to 0$ Band aP(6,0)	1766–7,1770,1771, 1774,1779,1790
12.24911	816.386	$\nu_2 \to 0$ Band aP(6,1)	1766,1767,1770, 1771,1790
12.26105	815.591	$\nu_2 \to 0$ Band aP(6,2)	1766,1767,1770, 1771,1790
12.28136	814.242	$\nu_2 \to 0$ Band aP(6,3)	1766,1767,1770, 1771,1783,1790
12.31072	812.300	$\nu_2 \to 0$ Band aP(6,4)	1766,1767,1770, 1771,1790
12.35002	809.715	$\nu_2 \to 0$ Band aP(6,5)	1766,1767,1770, 1771,1790
12.37821	807.871	$\nu_2 \to 0$ Band sP(8,1)	1770
12.38425	807.477	$\nu_2 \to 0$ Band sP(8,3)	1770
12.39560	806.738	$\nu_2 \to 0$ Band sP(8,5)	1770
12.4027	806.274	$\nu_2 \to 0$ Band sP(8,6)	6873
12.52781	798.224	$\nu_2 \to 0$ Band aP(7,1)	1766,1767,1770, 1771,1790–92
12.53999	797.449	$\nu_2 \to 0$ Band aP(7,2)	1766,1767,1770, 1771,1790,1792
12.56068	796.135	$\nu_2 \to 0$ Band aP(7,3)	1766,1767,1770, 1771,1790,1792
12.59	794.3	$2\nu_2 \to \nu_2$ Band aR(7,5)	6875
12.59059	794.244	$\nu_2 \to 0$ Band aP(7,4)	1766,1767,1770, 1771,1790,1792
12.63063	791.720	$\nu_2 \to 0$ Band aP(7,5)	1766,1767,1770, 1771,1790,1792

$^{14}NH_3$ Laser—*continued*

Wavelength (μm) vac	Frequency (cm^{-1})	Transition assignment	Comments
12.68213	788.511	$\nu_2 \to 0$ Band aP(7,6)	1766,1767,1770, 1771,1790
12.69716	787.578	$\nu_2 \to 0$ Band sP(9,3)	1770
12.7196	786.191	$\nu_2 \to 0$ Band sP(9,6)	6873
12.80959	780,665	$2\nu_2 \to \nu_2$ Band sP(8,5)	1787
12.81145	780.552	$\nu_2 \to 0$ Band aP(8,0)	1766,1767,1770, 1771,1781,1784, 1790,1792
12.81532	780.316	$\nu_2 \to 0$ Band aP(8,1)	1766,1767,1770, 1771,1790
12.82765	779.566	$\nu_2 \to 0$ Band aP(8,2)	1767,1770,1790, 1792
12.84863	778.293	$\nu_2 \to 0$ Band aP(8,3)	1766,1767,1770, 1771,1790,1792
12.87890	776.464	$\nu_2 \to 0$ Band aP(8,4)	1766,1767,1770, 1771,1790,1792
12.91946	774.026	$\nu_2 \to 0$ Band aP(8,5)	1766,1767,1770, 1771,1790,1792
12.97163	770.913	$\nu_2 \to 0$ Band aP(8,6)	1766,1767,1770, 1771,1790,1792
12.98352	770.207	$2\nu_2 \to \nu_2$ Band aR(6,5)	1787
13.0241	767.809	$\nu_2 \to 0$ Band sP(10,3)	6873
13.03715	767.039	$\nu_2 \to 0$ Band aP(8,7)	1766–67,1790
13.0505	766.252	$\nu_2 \to 0$ Band sP(10,6)	6873
13.11233	762.641	$\nu_2 \to 0$ Band aP(9,1)	1766,1790
13.12477	761.918	$\nu_2 \to 0$ Band aP(9,2)	1766,1790
13.14593	760.692	$\nu_2 \to 0$ Band aP(9,3)	1766–67,1790
13.17643	758.931	$\nu_2 \to 0$ Band aP(9,4)	1766,1790
13.21725	756.587	$\nu_2 \to 0$ Band aP(9,5)	1766,1790
13.23390	755.635	$2\nu_2 \to \nu_2$ Band aR(5,3)	1777
13.26978	753.592	$\nu_2 \to 0$ Band aP(9,6)	1766–67,1790
13.33580	749.861	$\nu_2 \to 0$ Band aP(9,7)	1766,1790
13.411	745.7	$\nu_2 \to 0$ Band aP(9,8), aP(10,1), or aP(10,0)	1766,1790
13.4153	745.420	$\nu_2 \to 0$ Band aP(10,0)	6873
13.4534	780.665	$\nu_2 \to 0$ Band aP(10,3)	6873
13.57749	736.513	$\nu_2 \to 0$ Band aP(10,6)	1766,1790
13.65533	732.315	$2\nu_2 \to \nu_2$ Band aR(4,3)	1777
13.72555	728.568	$2\nu_2 \to \nu_2$ Band aR(4,4)	1784
13.82608	723.271	$\nu_2 \to 0$ Band aP(10,9)	1766,1790

$^{14}NH_3$ Laser—*continued*

Wavelength (μm) vac	Frequency (cm^{-1})	Transition assignment	Comments
14.78	677	$2\nu_2 \to \nu_2$ Band	1788
15.04	665	$2\nu_2 \to \nu_2$ Band	1788
15.08	663	$2\nu_2 \to \nu_2$ Band	1788
15.41	649	$2\nu_2 \to \nu_2$ Band	1788
15.47	646	$2\nu_2 \to \nu_2$ Band	1788
15.68	637.8	$2\nu_2 \to \nu_2$ Band aQ(6,3)	6876
15.70	636.9	$2\nu_2 \to \nu_2$ Band aQ(8,5)	6875
15.78148	633.654	$2\nu_2 \to \nu_2$ Band aQ(5,3)	1777
15.81600	632.271	$2\nu_2 \to \nu_2$ Band aQ(7,5)	1787
15.85726	630.626	$2\nu_2 \to \nu_2$ Band aQ(4,3)	1777
15.87758	629.819	$2\nu_2 \to \nu_2$ Band aQ(5,4)	1777
15.91292	628.420	$2\nu_2 \to \nu_2$ Band aQ(6,5)	1787
15.913	628.40	$2\nu_2 \to \nu_2$ Band aQ(3,3)	6877
15.94637	627.102	$2\nu_2 \to \nu_2$ Band aQ(4,4)	1784
16.04	623.4	$2\nu_2 \to \nu_2$ Band ?	6878
16.936	590.45	$2\nu_2 \to \nu_2$ Band aP(2,0)	6877
16.95	590.0	$2\nu_2 \to \nu_2$ Band aP(2,1)	6879
18.046	554.14	$2\nu_2 \to \nu_2$ Band aP(4,0)	6877
18.203	549.37	$2\nu_2 \to \nu_2$ Band aP(4,3)	6877
18.21	549	$2\nu_2 \to \nu_2$ Band	1788
18.798	531.97	$2\nu_2 \to \nu_2$ Band aP(5,3)	6876,6877
18.92674	528.353	$2\nu_2 \to \nu_2$ Band aP(5,4)	1784
19.198	520.89	$2\nu_2 \to \nu_2$ Band aP(6,0)	6952
19.29	518.4	$2\nu_2 \to \nu_2$ Band aP(6,2)	6880
19.401	515.43	$2\nu_2 \to \nu_2$ Band aP(6,3)	6876,6877
19.55019	511.504	$2\nu_2 \to \nu_2$ Band aP(6,4)	1784
20.008	499.80	$2\nu_2 \to \nu_2$ Band aP(7,3)	6876,6877
20.358	491.20	$2\nu_2 \to \nu_2$ Band aP(8,0)	6877
20.38798	490.485	$2\nu_2 \to \nu_2$ Band aP(7,5)	1787
20.48	488.3	$2\nu_2 \to \nu_2$ Band aP(8,2)	6881
20.604	485.34	$2\nu_2 \to \nu_2$ Band aP(7,6)	6877
20.622	484.92	$2\nu_2 \to \nu_2$ Band aP(8,3)	6877
21.05409	474.967	$2\nu_2 \to \nu_2$ Band aP(8,5)	1787
21.228	471.08	$2\nu_2 \to \nu_2$ Band aP(9,3)	6952
21.333	468.75	$2\nu_2 \to \nu_2$ Band aP(8,6)	6877
21.471	465.74	$3\nu_2 \to 2\nu_2$ Band aP(2,0)?	1789
22.542	443.62	$3\nu_2 \to 2\nu_2$ Band aP(3,2)	1789
22.563	443.20	$3\nu_2 \to 2\nu_2$ Band aP(3,1)	1789
22.71	440	$3\nu_2 \to 2\nu_2$ Band	1788
23.675	422.39	$3\nu_2 \to 2\nu_2$ Band aP(4, or 1)	1789

$^{14}NH_3$ **Laser**—*continued*

Wavelength (μm) vac	Frequency (cm^{-1})	Transition assignment	Comments
23.86	419	$3\nu_2 \to 2\nu_2$ Band	1788
24.918	401.32	$3\nu_2 \to 2\nu_2$ Band aP(5,1)	1789
25.12	398	$3\nu_2 \to 2\nu_2$ Band	1788
26.282	380.49	$3\nu_2 \to 2\nu_2$ Band aP(6,0)	1789
30.69	326	$3\nu_2 \to 2\nu_2$ Band	1788
31.47	318	$3\nu_2 \to 2\nu_2$ Band	1788
31.951	312.98	$3\nu_2 \to 2\nu_2$ Band	1789
32.13	311	$3\nu_2 \to 2\nu_2$ Band	1788

$^{15}NH_3$ **Laser**

Wavelength (μm) vac	Frequency (cm^{-1})	Transition assignment	Comments
10.789	926.8	$\nu_2 \to 0$ Band aQ(3,3)	6882
10.829	923.45	$\nu_2 \to 0$ Band aQ(6,6)	6953
11.072	903.15	$\nu_2 \to 0$ Band sP(3,0)	6953
11.257	888.32	$\nu_2 \to 0$ Band aP(2,0)	6882
11.586	863.12	$\nu_2 \to 0$ Band sP(5,0)	6882,6883
11.763	850.11	$\nu_2 \to 0$ Band aP(4,0)	6882,6883
11.798	847.60	$\nu_2 \to 0$ Band aP(4,3)	6882
11.859	843.25	$\nu_2 \to 0$ Band sP(6,1)	6883
11.866	842.75	$\nu_2 \to 0$ Band aP(6,3)	6882
12.063	828.98	$\nu_2 \to 0$ Band aP(5,3)	6882,6883
12.148	823.21	$\nu_2 \to 0$ Band sP(7,0)	6882,6883
12.299	813.08	$\nu_2 \to 0$ Band aP(6,0)	6882,6883
12.336	810.66	$\nu_2 \to 0$ Band aP(6,3)	6882,6883
12.447	803.40	$\nu_2 \to 0$ Band sP(8,1)	6883
12.616	792.64	$\nu_2 \to 0$ Band aP(7,3)	6882,6883
12.739	785.00	$\nu_2 \to 0$ Band aP(7,6)	6882,6883
12.867	777.18	$\nu_2 \to 0$ Band aP(8,0)	6882,6883
12.905	774.88	$\nu_2 \to 0$ Band aP(8,3)	6882,6883
12.967	771.16	$\nu_2 \to 0$ aP(8,0)	6953
12.977	770.60	$\nu_2 \to 0$ Band aP(8,5)	6884
13.030	767.47	$\nu_2 \to 0$ Band aP(8,6)	6882,6883
13.204	757.37	$\nu_2 \to 0$ Band aP(9,3)	6882,6883
13.323	750.60	$\nu_2 \to 0$ Band aP(9,6)	6883
13.473	742.20	$\nu_2 \to 0$ Band aP(10,0)	6883
13.910	718.92	$2\nu_2 \to \nu_2$ Band sP(10,1)	6885
14.3	699		1793

$^{15}NH_3$ Laser—*continued*

Wavelength (μm) vac	Frequency (cm^{-1})	Transition assignment	Comments
14.8	676		1793
15.2	658		1793
15.7	637		1793
16.0	625		1793
17.8	562		1793

3.3.2.5 Five-Atom Vibrational Transition Lasers

Table 3.3.2.4
Five-Atom Vibrational Transition Lasers

$^{12}CF_4$ Laser

Wavelength (μm) vac	Frequency (cm^{-1})	Transition assignment	Comments
15.306	653.32	$\nu_2+\nu_4 \rightarrow \nu_2$ Band R(41)	1794–95,1804
15.40	649.3	$\nu_2+\nu_4 \rightarrow \nu_2$ Band R(33)	1794,1795
15.46	647.0	$\nu_2+\nu_4 \rightarrow \nu_2$ Band –	1797,1815
15.48	646.0	$\nu_2+\nu_4 \rightarrow \nu_2$ Band R(27)	1794–1799,1807
15.49	645.5	$\nu_2+\nu_4 \rightarrow \nu_2$ Band R(26)	1800,1808
15.547	643.23	$\nu_2+\nu_4 \rightarrow \nu_2$ Band R(22)	1797–1801,1809
15.57	642.4	$\nu_2+\nu_4 \rightarrow \nu_2$ Band R(21)	1794–96,1810
15.58	641.9	$\nu_2+\nu_4 \rightarrow \nu_2$ Band R(20)	1799,1811
15.60	640.9	$\nu_2+\nu_4 \rightarrow \nu_2$ Band R(18)	1799,1812
15.607	640.73	$\nu_2+\nu_4 \rightarrow \nu_2$ Band R(17)	1797–9,1801,1813
15.71	636.7	$\nu_2+\nu_4 \rightarrow \nu_2$ Band	1797,1816
15.71	636.6	$\nu_2+\nu_4 \rightarrow \nu_2$ Band R(9)	1797–99,1814
15.74	635.2	$\nu_2+\nu_4 \rightarrow \nu_2$ Band R(7)	1800,1817
15.83	631.8	$\nu_2+\nu_4 \rightarrow \nu_2$ Band Q(7)?	1799,1818
15.84	631.5	$\nu_2+\nu_4 \rightarrow \nu_2$ Band Q(14)	1795,182
15.84	631.3	$\nu_2+\nu_4 \rightarrow \nu_2$ Band Q(28)	1799,1819
15.844	631.15	$\nu_2+\nu_4 \rightarrow \nu_2$ Band Q(36)	1799,1801,1821
15.845	631.12	$\nu_2+\nu_4 \rightarrow \nu_2$ Band Q(31)	1794,1822
15.847	631.05	$\nu_2+\nu_4 \rightarrow \nu_2$ Band Q(29)	1799,1801,1823
15.85	631.0	$\nu_2+\nu_4 \rightarrow \nu_2$ Band	1795,1826
15.85	630.8	$\nu_2+\nu_4 \rightarrow \nu_2$ Band Q(11)	1800,1824
15.85	630.8	$\nu_2+\nu_4 \rightarrow \nu_2$ Band Q(35)	1795,1825

$^{12}CF_4$ Laser—*continued*

Wavelength (μm) vac	Frequency (cm⁻¹)	Transition assignment	Comments
15.85	630.8	$\nu_2+\nu_4 \rightarrow \nu_2$ Band Q(37)	1799,1827
15.91	628.5	$\nu_2+\nu_4 \rightarrow \nu_2$ Band P(5)	1800,1828
15.92	628.2	$\nu_2+\nu_4 \rightarrow \nu_2$ Band P(6)	1800,1829
15.94	627.3	$\nu_2+\nu_4 \rightarrow \nu_2$ Band P(8)	1800,1830
16.00	624.9	$\nu_2+\nu_4 \rightarrow \nu_2$ Band P(12)	1800,1831
16.02	624.4	$\nu_2+\nu_4 \rightarrow \nu_2$ Band P(14)	1796,1832
16.06	622.8	$\nu_2+\nu_4 \rightarrow \nu_2$ Band P(16)	1800,1833
16.07	622.4	$\nu_2+\nu_4 \rightarrow \nu_2$ Band P(17)	1800,1834
16.07	622.4	$\nu_2+\nu_4 \rightarrow \nu_2$ Band P(18)	1796,1835
16.09	621.4	$\nu_2+\nu_4 \rightarrow \nu_2$ Band P(19)	1800,1836
16.11	620.6	$\nu_2+\nu_4 \rightarrow \nu_2$ Band P(21)	1795,1796,1837
16.178	618.11	$\nu_2+\nu_4 \rightarrow \nu_2$ Band P(25)	1797–9,1801,1838
16.21	617.0	$\nu_2+\nu_4 \rightarrow \nu_2$ Band P(27)	1795,1839
16.24	615.7	$\nu_2+\nu_4 \rightarrow \nu_2$ Band P(30)	1799,1840
16.25	615.4	$\nu_2+\nu_4 \rightarrow \nu_2$ Band P(30)?	1800,1841
16.259	615.06	$\nu_2+\nu_4 \rightarrow \nu_2$ Band P(31)	1797–99,1801,1802
16.27	614.7	$\nu_2+\nu_4 \rightarrow \nu_2$ Band P(32)?	1799,1842
16.29	613.7	$\nu_2+\nu_4 \rightarrow \nu_2$ Band P(35)?	1799,1843
16.340	611.99	$\nu_2+\nu_4 \rightarrow \nu_2$ Band P(37)	1797–9,1801,1844
16.40	609.6	$\nu_2+\nu_4 \rightarrow \nu_2$ Band P(42)	1795,1845

$^{13}CF_4$ Laser

Wavelength (μm) vac	Frequency (cm⁻¹)	Transition assignment	Comments
15.29	654.2	$\nu_2+\nu_4 \rightarrow \nu_2$ Band R	1846,1852
15.32	652.9	$\nu_2+\nu_4 \rightarrow \nu_2$ Band R	1846,1853
15.42	648.4	$\nu_2+\nu_4 \rightarrow \nu_2$ Band R	1846,1854
15.44	647.8	$\nu_2+\nu_4 \rightarrow \nu_2$ Band R	1846,1855
15.45	647.1	$\nu_2+\nu_4 \rightarrow \nu_2$ Band R	1847,1856
15.47	646.6	$\nu_2+\nu_4 \rightarrow \nu_2$ Band R	1846,1857
15.53	644.0	$\nu_2+\nu_4 \rightarrow \nu_2$ Band R	1848,1858
15.54	643.3	$\nu_2+\nu_4 \rightarrow \nu_2$ Band R	1847,1848,1859
15.59	641.3	$\nu_2+\nu_4 \rightarrow \nu_2$ Band R	1846,1860
15.62	640.4	$\nu_2+\nu_4 \rightarrow \nu_2$ Band R	1846,1861
15.62	640.3	$\nu_2+\nu_4 \rightarrow \nu_2$ Band R	1848,1862
15.74	635.5	$\nu_2+\nu_4 \rightarrow \nu_2$ Band R	1849,1863
15.75	635.0	$\nu_2+\nu_4 \rightarrow \nu_2$ Band R	1850,1864

$^{13}CF_4$ Laser—*continued*

Wavelength (μm) vac	Frequency (cm^{-1})	Transition assignment	Comments
15.78	633.9	$v_2+v_4 \rightarrow v_2$ Band R	1846,1865
15.80	632.8	$v_2+v_4 \rightarrow v_2$ Band R	1851,1866
15.82	632.2	$v_2+v_4 \rightarrow v_2$ Band R	1850,1867
15.89	629.5	$v_2+v_4 \rightarrow v_2$ Band Q	1846,1868
15.89	629.5	$v_2+v_4 \rightarrow v_2$ Band Q	1846,1869
15.89	629-5	$v_2+v_4 \rightarrow v_2$ Band Q	1848,1870
16.00	625.0	$v_2+v_4 \rightarrow v_2$ Band P	1848,1871
16.09	621.4	$v_2+v_4 \rightarrow v_2$ Band P	1851,1872
16.10	621.0	$v_2+v_4 \rightarrow v_2$ Band	1851,1873
16.11	620.7	$v_2+v_4 \rightarrow v_2$ Band P	1851,1874
16.15	619.3	$v_2+v_4 \rightarrow v_2$ Band P	1847,1875
16.15	619.2	$v_2+v_4 \rightarrow v_2$ Band P	1849,1876
16.16	618.7	$v_2+v_4 \rightarrow v_2$ Band P	1848,1877
16.20	617.1	$v_2+v_4 \rightarrow v_2$ Band P	1848,1878
16.23	616.0	$v_2+v_4 \rightarrow v_2$ Band P	1846,1849,1879
16.25	615.4	$v_2+v_4 \rightarrow v_2$ Band P	1851,188
16.26	615.0	$v_2+v_4 \rightarrow v_2$ Band P	1846,1881
16.27	614.6	$v_2+v_4 \rightarrow v_2$ Band P	1851,1882
16.29	613.7	$v_2+v_4 \rightarrow v_2$ Band P	1848,1883
16.30	613.4	$v_2+v_4 \rightarrow v_2$ Band P	1848,1884
16.32	612.9	$v_2+v_4 \rightarrow v_2$ Band P	1846,1885
16.36	611.1	$v_2+v_4 \rightarrow v_2$ Band P	1846,1886
16.48	606.8	$v_2+v_4 \rightarrow v_2$ Band P	1846,1887

$^{14}CF_4$ Laser

Wavelength (μm) vac	Frequency (cm^{-1})	Transition assignment	Comments
15.39	649.8	$v_2 + v_4 \rightarrow v_2$ Band R	1888,1891
15.46	647.0	$v_2 + v_4 \rightarrow v_2$ Band R	1889,1892
15.51	644.9	$v_2 + v_4 \rightarrow v_2$ Band R	1889,1893
15.54	643.7	$v_2 + v_4 \rightarrow v_2$ Band R	1888,1894,1896
15.55	643.5	$v_2 + v_4 \rightarrow v_2$ Band R	1889,1895
15.58	642.0	$v_2 + v_4 \rightarrow v_2$ Band R	1889,1897
15.60	641.2	$v_2 + v_4 \rightarrow v_2$ Band R	1888,1898
15.61	640.5	$v_2 + v_4 \rightarrow v_2$ Band R	1888,1899
15.62	640.1	$v_2 + v_4 \rightarrow v_2$ Band R	1890,1900
15.64	639.2	$v_2 + v_4 \rightarrow v_2$ Band R	1889,1901

$^{14}CF_4$ Laser—*continued*

Wavelength (μm) vac	Frequency (cm⁻¹)	Transition assignment	Comments
15.65	638.8	$\nu_2 + \nu_4 \to \nu_2$ Band R	1890,1902
15.71	636.5	$\nu_2 + \nu_4 \to \nu_2$ Band R	1889,1903
15.79	633.5	$\nu_2 + \nu_4 \to \nu_2$ Band R	1889,1904,1905
15.82	632.1	$\nu_2 + \nu_4 \to \nu_2$ Band R	1890,1906
15.93	627.9	$\nu_2 + \nu_4 \to \nu_2$ Band Q	1889,1907,1908
15.94	627.4	$\nu_2 + \nu_4 \to \nu_2$ Band Q	1888,1909–1416
15.95	627.1	$\nu_2 + \nu_4 \to \nu_2$ Band Q	1888,1917
15.99	625.3	$\nu_2 + \nu_4 \to \nu_2$ Band P	1890,1918
16.04	623.6	$\nu_2 + \nu_4 \to \nu_2$ Band P	1890,1919
16.08	621.7	$\nu_2 + \nu_4 \to \nu_2$ Band P	1888,1920
16.14	619.7	$\nu_2 + \nu_4 \to \nu_2$ Band P	1890,1921
16.15	619.3	$\nu_2 + \nu_4 \to \nu_2$ Band P	1890,1922
16.21	616.8	$\nu_2 + \nu_4 \to \nu_2$ Band P	1888,1923
16.22	616.6	$\nu_2 + \nu_4 \to \nu_2$ Band P	1889,1924
16.25	615.2	$\nu_2 + \nu_4 \to \nu_2$ Band P	1888,1925
16.28	614.3	$\nu_2 + \nu_4 \to \nu_2$ Band P	1890,1926
16.30	613.6	$\nu_2 + \nu_4 \to \nu_2$ Band P	1889,1927
16.33	612.2	$\nu_2 + \nu_4 \to \nu_2$ Band P	1888,1928
16.35	611.8	$\nu_2 + \nu_4 \to \nu_2$ Band P	1888,1929
16.44	608.2	$\nu_2 + \nu_4 \to \nu_2$ Band P	1889,1930

CF_3I Laser

Wavelength (μm) vac	Frequency (cm⁻¹)	Transition assignment	Comments
13.54	738.4	$\nu_2 + \nu_3 \to \nu_3$ Band	1931,1932
13.57	736.8	$\nu_2 + \nu_3 \to \nu_3$ Band	1931,1933
13.63	733.6	$\nu_2 + \nu_3 \to \nu_3$ Band	1931,1934

CH_3F Laser

Wavelength (μm) vac	Frequency (cm⁻¹)	Transition assignment	Comments
9.747	1026	$2\nu_3 + \nu_3 \to \nu_3$ Band P(4,1)	1935

FClO$_3$ Laser

Wavelength (μm) vac	Frequency (cm^{-1})	Transition assignment	Comments
16.32	612.6	$\nu_5 + \nu_6 \rightarrow \nu_6$ Band	1936,1937
16.34	612.2	$\nu_5 + \nu_6 \rightarrow \nu_6$ Band	1936,1938
16.35	611.7	$\nu_5 + \nu_6 \rightarrow \nu_6$ Band	1936,1939
16.45	607.9	$\nu_5 + \nu_6 \rightarrow \nu_6$ Band	1936,1940
16.49	606.3	$\nu_5 + \nu_6 \rightarrow \nu_6$ Band	1936,1941,1942
16.5	606.1	$\nu_5 + \nu_6 \rightarrow \nu_6$ Band	1936,1943
16.52	605.4	$\nu_5 + \nu_6 \rightarrow \nu_6$ Band	1936,1944,1945
16.56	603.8	$\nu_5 + \nu_6 \rightarrow \nu_6$ Band	1936,1946,1947
16.61	601.9	$\nu_5 + \nu_6 \rightarrow \nu_6$ Band	1936,1948
16.66	600.3	$\nu_5 + \nu_6 \rightarrow \nu_6$ Band	1936,1949
16.73	597.8	$\nu_5 + \nu_6 \rightarrow \nu_6$ Band	1936,1950
16.75	596.9	$\nu_5 + \nu_6 \rightarrow \nu_6$ Band	1936,1951
16.76	596.8	$\nu_5 + \nu_6 \rightarrow \nu_6$ Band	1936,1952,1953
16.77	596.3	$\nu_5 + \nu_6 \rightarrow \nu_6$ Band	1936,1954
16.79	595.6	$\nu_5 + \nu_6 \rightarrow \nu_6$ Band	1936,1955
16.82	594.6	$\nu_5 + \nu_6 \rightarrow \nu_6$ Band	1936,1956
16.93	590.7	$\nu_5 + \nu_6 \rightarrow \nu_6$ Band	1936,1957
17.15	588.3	$\nu_5 + \nu_6 \rightarrow \nu_6$ Band	1936,1959
17.19	583.1	$\nu_5 + \nu_6 \rightarrow \nu_6$ Band	1936,1960
17.22	581.6	$\nu_5 + \nu_6 \rightarrow \nu_6$ Band	1936,1961
17.26	580.7	$\nu_5 + \nu_6 \rightarrow \nu_6$ Band	1936,1962
17.28	579.5	$\nu_5 + \nu_6 \rightarrow \nu_6$ Band	1936,1963
17.32	578.8	$\nu_5 + \nu_6 \rightarrow \nu_6$ Band	1936,1964
17.36	577.3	$\nu_5 + \nu_6 \rightarrow \nu_6$ Band	1936,1965
17.44	576.1	$\nu_5 + \nu_6 \rightarrow \nu_6$ Band	1936,1966
17.46	573.4	$\nu_5 + \nu_6 \rightarrow \nu_6$ Band	1936,1967
17.58	568.8	$\nu_5 + \nu_6 \rightarrow \nu_6$ Band	1936,1968
17.71	564.8	$\nu_5 + \nu_6 \rightarrow \nu_6$ Band	1936,1969

HCOOH Laser

Wavelength (μm) vac	Frequency (cm^{-1})	Transition assignment	Comments
5.7			1970,1971

SiF$_4$ Laser

Wavelength (μm) vac	Frequency (cm^{-1})	Transition assignment	Comments
24.78	403.5		1972,1973
25.31	395.1		1972,1974
25.36	394.3		1972,1975
25.40	393.7		1972,1976
25.67	389.5		1972,1977
25.67	389.5		1972,1978
25.68	389.4		1972,1979
25.77	388.0		1972,1980
25.79	387.7.		1972,1981
26.01	384.4		1972,1982
26.14	382.5		1972,1983

SiH$_4$ Laser

Wavelength (μm) vac	Frequency (cm^{-1})	Transition assignment	Comments
7.90202	1265.50		1984
7.92198	1262.31		1984
7.94900	1258.02		1984
7.95482	1257.10		1984
7.96902	1254.86		1984
7.99201	1251.25		1984

3.3.2.6 Six-Atom Vibrational Transition Lasers

Table 3.3.2.5
Six-Atom Vibrational Transition Lasers

C$_2$H$_4$ Laser

Wavelength (μm) vac	Frequency (cm^{-1})	Transition assignment	Comments
10.53	950		1985,1986
10.98	911		1985,1987

3.3.2.7 Seven-Atom Vibrational Transition Lasers

Table 3.3.2.6
Seven-Atom Vibrational Transition Lasers

CH$_3$CCH Laser

Wavelength (μm) vac	Frequency (cm^{-1})	Transition assignment	Comments
15.53	644.0	$\nu_9 + \nu_{10} \rightarrow \nu_{10}$ Band?	6886,6887
15.54	643.3	$\nu_9 + \nu_{10} \rightarrow \nu_{10}$ Band?	6886,6888
15.711	636.5	$\nu_9 + \nu_{10} \rightarrow \nu_{10}$ Band?	1988,1989
15.716	636.3	$\nu_9 + \nu_{10} \rightarrow \nu_{10}$ Band?	1988,1990
15.72	636.2	$\nu_9 + \nu_{10} \rightarrow \nu_{10}$ Band?	6886,6889
15.721	636.1	$\nu_9 + \nu_{10} \rightarrow \nu_{10}$ Band?	1988,1991
15.80	632.9	$\nu_9 + \nu_{10} \rightarrow \nu_{10}$ Band?	6886,6891
15.83	631.9	$\nu_9 + \nu_{10} \rightarrow \nu_{10}$ Band?	6886,6892
16.06	622.7	$\nu_9 + \nu_{10} \rightarrow \nu_{10}$ Band?	6886,6893
16.10	621.2	$\nu_9 + \nu_{10} \rightarrow \nu_{10}$ Band?	6886,6894
16.121	620.3	$\nu_9 + \nu_{10} \rightarrow \nu_{10}$ Band?	1988,1992
16.39	610	$\nu_9 + \nu_{10} \rightarrow \nu_{10}$ Band?	1988,1993
16.42	609	$\nu_9 + \nu_{10} \rightarrow \nu_{10}$ Band?	1988,1994

SF$_6$ Laser

Wavelength (μm) vac	Frequency (cm^{-1})	Transition assignment	Comments
15.905	628.74	$2\nu_3 \rightarrow \nu_2 + \nu_4$ Band?	1995

Section 3.4
FAR INFRARED AND MILLIMETER WAVE GAS LASERS

3.4.1 Introduction

Organic molecules used for far infrared and millimeter wave lasers number more than one hundred. Combined with the use of several different isotopes and the possibility of transitions between many different vibrational and rotational levels, reported lasing transitions are now numbered in the thousands. The molecular gas lasers in this section are either excited by an electrical discharge or are optically pumped by narrowband pump sources to excite molecules into a specific rotational state of an excited vibrational state; the latter are usually operated cw or quasi-cw.

As noted in the introduction, in this section 20 μm is used as the lower limit for the far infrared. Four noble gas lasers have transitions in this far infrared region; tables of these laser lines are given in Section 3.4.2.

Section 3.4.3 presents tables of molecular far infrared and millimeter wave laser lines; these are listed by molecule and for each molecule lines are listed in order of increasing wavelength. The uncertainty in the wavelength determination is noted in the second column. Accurate measurements, typically 1 part in 10^5 or better, refer to vacuum since they are calculated from frequency measurements. Interferometric wavelength measurements may refer to vacuum, the laser medium, or air but are of low accuracy, ranging from a few percent to (rarely) 1 part in 10^4. Thus within the measurement uncertainties almost all measurements may be considered to refer to vacuum. The third column lists the CO_2 laser pump transition and or may include various comments (given in Section 3.6) about the laser output power, relative polarization of the output radiation with respect to the pump radiation, and the pump transition. Pump transitions are usually those for a CO_2 laser; if an isotopic CO_2 or other pump laser was used or if the laser was pumped by an electric discharge, this is noted among the comments.

Laser output powers depend not only on molecular properties but also on the geometry and pump power, factors that vary with the design of the experiment. The reader is advised to consult the original references for this information and its effect on the laser performance.

For those interested in the most intense far-infrared and millimeter wave laser lines, a table of calibrated power measurements of over 150 lines between 40 μm and 2 mm having output powers of 1 mW or more is given in Douglas.[1]

References with titles or descriptions of the contents are given in Section 3.7. The references may include the original report of lasing and other reports relevant to the identification of the transition and laser operation; however, because of the huge literature, not all relevant measurements are noted.

1. Douglas, N. G., *Millimetre and Submillimetre Wavelength Lasers: A Handbook of cw Measurements*, Springer-Verlag, Berlin Heidelberg (1989).

Further Reading

Cheo, P. K., Ed., *Handbook of Molecular Lasers*, Marcel Dekker Inc., New York (1987).

Dodel, G., On the history of far-infrared (FIR) gas lasers: Thirty-five years of research and application, *Infrared Phys. & Technol.* 40, 127 (1999).

Inguscio, M., Moruzzi, G., Evenson, K. M., and Jennings, D. A., A review of frequency measurements of optically pumped lasers from 0.1 to 8 THz, *J. Appl. Phys.* 60, R161 (1986).

Jacobsson, S., Optically pumped far infrared lasers, *Infrared Phys.* 29, 853 (1989).

Knight, D. J. E., Far-Infrared CW Gas Lasers in *Handbook of Laser Science and Technology, Vol. II: Gas Lasers*, CRC Press, Boca Raton, FL (1982), p. 411 and *Handbook of Laser Science and Technology, Suppl. 1: Lasers*, CRC Press, Boca Raton, FL (1991), p. 415.

Moruzzi, G., Winnewisser, B. P., Winnewisser, M., Mukkopadhyay, I., and Strumia, F., *Microwave, Infrared and Laser Transitions of Methanol: Atlas of Assigned Lines from 0 to 1258 cm^{-1}*, CRC Press, Boca Raton, FL (1995).

Tobin, M. S., A review of optically pumped NMMW lasers, *Proc. IEEE* 73, 61 (1985).

For reports of new infrared lasers, see the *International Journal of Infrared and Millimeter Waves* (the proceedings of the Infrared and Millimeter Wave conference series are published in this journal), *Infrared Physics and Technology, Journal of Molecular Spectroscopy, Journal of Applied Physics, IEEE Journal of Quantum Electronics, and Quantum Electronics* (Russian).

3.4.2 Tables of Atomic Far Infrared Gas Lasers
(for other noble gas transitions, see the tables and references in Section 3.1.2.14.)

Table 3.4.1
Atomic Far Infrared Gas Lasers

Helium

Wavelength (μm)	Uncertainty (μm)	Comments	References
95.788	±0.0018	1998,1999,6929	434,435
216.3	±0.43	1996,1997,6929	435,436

Neon

Wavelength (μm)	Uncertainty (μm)	Comments	References
20.48	±0.0051	2050	583,586,1452
21.752	±0.0054	2049	583,586,1452
22.836	±0.0057	2048	583,586,1452
25.423	±0.0051	2047	583,586,1452
28.053	±0.0056	2046	583,586,1452
31.553	±0.0047	2045	583,586,1452
31.928	±0.0048	2044	583,586,1452
32.016	±0.0048	2043	583,586,1452
32.516	±0.0049	2042	583,586,1452
32.83	±0.0066	2041	583,586,1452
34.552	±0.0052	2040	583,586,1452
34.679	±0.0052	2039	583,586,1452
35.602	±0.0053	2037,2038	583,586,1452
37.231	±0.0056	2035,2036	583,586,1452
41.741	±0.0042	2034	583,586,1452
50.705	±0.0051	2032,2033	583,586,1452
52.425	±0.0052	2030,2031	583,586,1452
53.486	±0.0053	2028,2029	583,586,1452
54.019	±0.0054	2026,2027	583,586,1452
54.117	±0.0054	2024,2025	583,586,1452
55.537	±0.0056	2022,2023	583,586,1452
57.355	±0.0057	2020,2021	583,586,1452
68.329	±0.0068	2019	583,586,1452
72.108	±0.0072	2017,2018	583,586,1452
85.047	±0.0085	2016	583,586,1452
86.962	±0.0087	2014,2015	583,586,1452
88.471	±0.0088	2012,2013	583,586,1452

Neon—*continued*

Wavelength (μm)	Uncertainty (μm)	Comments	References
89.859	±0.0090	2010,2011	583,586,1452
93.0	±0.22	2008,2009	583,586,1452
106.07	±0.053	2006,2007	583,586,1452
124.4	±0.30	2004,2005	583,586,1452
126.1	±0.30	2002,2003	583,586,1452
132.8	±0.32	2000,2001	583,586,1452

Argon

Wavelength (μm)	Uncertainty (μm)	Comments	References
26.933	±0.0027	2064	1452
26.936	±0.0027	2063	1452

Xenon

Wavelength (μm)	Uncertainty (μm)	Comments	References
18.506	±0.0056	2070	1452
75.561687		997,1006,1072	652

3.4.3 Tables of Far Infrared and Millimeter Wave Gas Lasers

Molecular gas lasers included in this section are ordered as follows:

Formula	Name	Table
BCl_3	boron chloride	3.4.2
$^{10}BCl_3$	boron chloride – ^{10}B	3.4.3
$^{11}BCl_3$	boron chloride – ^{11}B	3.4.4
$C_2H_4O_2H_2$	$[C_2H_4(OH)_2]$ dihydroxyethane (ethylene glycol)	3.4.5
CDF_3	deuterotrifluoromethane	3.4.6
CD_2Cl_2	dideuterodichloromethane	3.4.7
CD_2F_2	dideuterodifluoromethane	3.4.8
CD_3Br	trideuterobromomethane	3.4.9
CD_3Cl	trideuterochloromethane	3.4.10
CD_3CN	trideuterocyanomethane	3.4.11
CD_3F	trideuterofluoromethane	3.4.12
$^{13}CD_3F$	trideuterofluoromethane – ^{13}C	3.4.13
CD_3I	trideuteroiodomethane	3.4.14
$^{13}CD_3I$	trideuteroiodomethane – ^{13}C	3.4.15
CD_3OD	deuterooxytrideuteromethanol	3.4.16
$^{13}CD_3OD$	deuteroxytrideuteromethane – ^{13}C	3.4.17
CD_3OH	trideuteromethanol	3.4.18
$^{13}CD_3OH$	trideuteromethanol – ^{13}C	3.4.19
CF_2Cl_2	dichlorodifluoromethane (fluorcarbon)	3.4.20
CF_3Br	bromotrifluoromethane	3.4.21
$CHCl_2F$	dichlorofluoromethane	3.4.22
$CHClF_2$	chlorodifluoromethane	3.4.23
CHD_2F	dideuterofluoromethane	3.4.24
CHD_2OH	dideuteromethanol	3.4.25
$CHFCHF$	cis 1,2-difluoroethene (cis 1,2-difluoroethylene)	3.4.26
$CHFO$	(HFCO, HCOF) formyl fluoride (fluoroformaldehyde)	3.4.27
CHF_2	difluoromethane	3.4.28
CH_2CF_2	1,1-difluoroethene (1,1-difluoroethylene)	3.4.29
CH_2CHBr	bromoethene (vinyl bromide)	3.4.30
CH_2CHCl	chloroethene (vinyl chloride)	3.4.31
CH_2CHCN	acrylonitrile, vinyl cyanide)	3.4.32
CH_2CHF	fluoroethene (vinyl fluoride)	3.4.33
CH_2Cl_2	dichloromethane	3.4.34
CH_2ClF	chlorofluoromethane	3.4.35
CH_2DOH	deuteromethanol	3.4.36
CH_2F_2	difluoromethane	3.4.37
$^{13}CH_2F_2$	difluoromethane – ^{13}C	3.4.38
CH_2NOH	formaldoxime	3.4.39
CH_3Br	bromomethane (methyl bromide)	3.4.40
$CH_3{}^{79}Br$	bromomethane (methyl bromide) – ^{79}Br	3.4.41
$CH_3{}^{81}Br$	bromomethane (methyl bromide) – ^{81}Br	3.4.42

Molecular gas lasers included in this section—*continued*

Formula	Name	Table
$^{13}CH_3Br$	bromomethane – ^{13}C	3.4.43
CH_3CCH	methyl acetylene	3.4.44
CH_3CD_2OH	1,1-dideuteroethanol	3.4.45
CH_3CF_3	1,1,1-trifluoroethane (methyl fluorofrom)	3.4.46
CH_3CHDOH	1-deuteroethanol	3.4.47
CH_3CHF_2	1,1-difluoroethane	3.4.48
CH_3CH_2Br	(C_2H_5Br) bromoethane (ethyl bromide)	3.4.49
CH_3CH_2Cl	(C_2H_5Cl) chloroethane (ethyl chloride)	3.4.50
CH_3CH_2F	(C_2H_5F) fluoroethane (ethyl fluoride)	3.4.51
CH_3CH_2I	(C_2H_5I) iodoethane (ethyl iodide)	3.4.52
CH_3CH_2OH	(C_2H_5OH) ethanol (ethyl alcohol)	3.4.53
CH_3CHO	ethanal (acetaldehyde)	3.4.54
CH_3Cl	chloromethane (methyl chloride)	3.4.55
CH_3CN	ethanenitrile (acetonitrile, cyanomethane, methyl cyanide)	3.4.56
CH_3COOD	deuterooxyethanoic acid	3.4.57
CH_3F	fluoromethane (methyl fluoride	3.4.58
$^{13}CH_3F$	fluoromethane – ^{13}C	3.4.50
CH_3I	iodomethane (methyl iodide)	3.4.60
$^{13}CH_3I$	iodomethane – ^{13}C	3.4.61
CH_3NC	methyl isocyanide	3.4.62
CH_3NH_2	aminomethane (methylamine)	3.4.63
CH_3NO_2	nitromethane	3.4.64
CH_3OCH_3	methoxymethane (dimethyl ether)	3.4.65
CH_3OD	deuterooxymethanol	3.4.66
CH_3OH	methanol (methyl alcohol)	3.4.67
$^{13}CH_3OH$	methanol – ^{13}C	3.4.68
$CH_3{}^{18}OH$	methanol – ^{18}O	3.4.69
CH_3SH	methyl mercaptan	3.4.70
ClO_2	chlorine dioxide	3.4.71
COF_2	carbonyl fluoride	3.4.72
COH_2	(H_2CO) methanal (formaldehyde)	3.4.73
DCN	deuterohydrocyanic acid (deuterium cyanide)	3.4.74
$DCOOD$	dideuteromethanic acid (dideutero formic acid)	3.4.75
$DCOOH$	deuteromethanic acid (deutero formic acid)	3.4.76
D_2CO	dideuteromethanal (dideuteroformaldehyde)	3.4.77
D_2O	deuterium oxide (heavy water)	3.4.78
$DFCO$	(CDFO, DCOF) deuterated formyl fluoride	3.4.79
FCN	cyanogen fluoride	3.4.80
HBr	hydrogen bromide	3.4.81
$HCCCH_2F$	$(FCH_2C:CH)$ 3-fluoropropyne (propargyl fluoride)	3.4.82
$HCCCHO$	$(HC:CCHO, C_3H_2O)$ propynal	3.4.83
$HCCF$	fluoroacetylene	3.4.84
HCl	hydrogen chloride (hydrochloric acid)	3.4.85

Molecular gas lasers included in this section—*continued*

Formula	Name	Table
HCN	hydrocyanic acid (hydrogen cyanide)	3.4.86
HCOOD	deuterooxymethanic acid (deuteroxy formic acid)	3.4.87
HCOOH	methanoic acid (formic acid)	3.4.88
$H^{13}COOH$	formic acid – ^{13}C	3.4.89
HDCO	deuteromethanal (deuteroformaldehyde)	3.4.90
HF	hydrogen fluoride (hydrofluoric acid)	3.4.91
H_2CO	methanal (formaldehyde)	3.4.92
$(H_2CO)_3$	trioxane (cyclic trimer of formaldehyde)	3.4.93
H_2O	water	3.4.94
H_2S	hydrogen sulfide	3.4.95
ND_3	trideuteroammonia (fully deuterated ammonia)	3.4.96
ND_2ND_2	(N_2D_4) dideuterohydrazine	3.4.97
NH2D	deuteroammonia	3.4.98
NH_3	ammonia	3.4.99
$^{15}NH_3$	ammonia – ^{15}N	3.4.100
NH_2NH_2	(N_2H_4) hydrazine	3.4.101
NH_2OH	hydroxylamine	3.4.102
O_3	ozone	3.4.103
OCS	(COS) carbonyl sulfide	3.4.104
PH_3	phosphine	3.4.105
SiF_4	silicon tetrafluoride	3.4.106
$SiHF_3$	$(SiHF_3)$ trifluorosilane (silyl fluoride)	3.4.107
SiH_2F_2	(SiH_2F_2) difluorosilane	3.4.108
SiH_3F	(SiH_3F) fluorosilane	3.4.109
SO_2	sulfur dioxide	3.4.110
SO_2	sulfur dioxide (isotopically substituted)	3.4.111

Table 3.4.2
Boron Chloride – BCl_3

Wavelength (µm)	Uncertainty (µm)	Comments	References
18.3		electric discharge	1182,1431
18.8		electric discharge	1182,1431
19.1		electric discharge	1182,1431
19.4		electric discharge	1182,1431
20.2		electric discharge	1182,1431
20.6		electric discharge	1182,1431
22.4		electric discharge	1182,1431
23.		electric discharge	1182,1431

Table 3.4.3
Boron Chloride – $^{10}BCl_3$

Wavelength (μm)	Uncertainty (μm)	Comments	References
18.3	±0.09	2586–2589	1431
19.1	±0.094	2582–2585	1431
19.4	±0.095	2578–2581	1431
22.4	±0.11	2574–2577	1431

Table 3.4.4
Boron Chloride – $^{11}BCl_3$

Wavelength (μm)	Uncertainty (μm)	Comments	References
18.8	±0.092	2602–2605	1431
20.2	±0.099	2598–2601	1431
20.6	±0.1	2594–2597	1431
23.	±.011	2590–2593	1431

Table 3.4.5
Dihydroxyethane (Ethylene Glycol) – $C_2H_4O_2H_2$

Wavelength (μm)	Uncertainty (μm)	CO_2 laser pump line	References
62.5		10R16	1799
69.1		10R16	1799
70.1		9P34	1799
75.2		9P32	1799
77.4		10R16	1799
90.8		9P32	1799
95.8		9R10	1799
109.1		9R16	1799
117.1		9P14	1799
118.		9R18	1799
118.		9P36	1799
118.9		9P34	1799
125.8		9P34	1799
132.		9P36	1799
132.		9P24	1799
135.		9P36	1799
164.		9R10	1799
164.		9P14	1799

Dihydroxyethane (Ethylene Glycol) – $C_2H_4O_2H_2$—*continued*

Wavelength (μm)	Uncertainty (μm)	CO_2 laser pump line	References
169.		9P36	1799
171.		9R08	1799
185.		9P34	1799
185.		9R18	1799
189.		9P34	1799
189.		9P36	1799
192.		9P36	1799
197.		9P36	1799
200.		9P36	1799
231.		9R10	1799
240.		9R10	1799
250.		9R18	1799
252.		9P34	1799
262.		9P34	1799
277.		9P36	1799
288.		9P12	1799
290.		9P36	1799
299.		9P34	1799
344.		9P22	1799
358.		9P34	1799
388.		9P36	1799
415.		9P14	1799
696.		9P34	1799

Table 3.4.6
Deuterotrifluoromethane – CDF_3

Wavelength (μm)	Uncertainty (μm)	CO_2 laser pump line	References
266.		10P14	1670
266.9		9P44	1670
286.3		10R48	1670
286.8		10R04	1670
286.8		10R10	1670
316.6		10R20	1670
330.019		10R08	1671
345.8		9P22	1670
361.231		10P24	1671
362.423		9R24	1671
388.273		10R32	1671
388.652		10R42	1671
420.311		10R26	1671

Deuterotrifluoromethane – CDF₃—*continued*

Wavelength (μm)	Uncertainty (μm)	CO₂ laser pump line	References
420.980		10R46	1671
432.987		10P20	1671
445.663		10P20	1671
459.4		10R18	1668
459.6		10R26	1668
488.528		10R38	1671
504.752		10R38	1671
521.237		10R24	1671
540.736		10R36	1671
560.703		10R36	1671
560.803		10R40	1671
581.984		10R26	1671
582.1		10R12	1671
605.6		10R20	1668
657.938		10P12	1671
657.989		10R10	1671
658.152		10P06	1671
687.837		10R10	1671
1008.558		10R12	1671
1080.537		10R12	1671
1260.561		10R16	1671
1377.		10R22	1668

Table 3.4.7
Dideuterodichloromethane – CD₂Cl₂

Wavelength (μm)	Uncertainty (μm)	Comments	References
171.2		6971,7099	1840
246.1		6971,7096	1840
249.	±1.2	3815–3817	1478
254.	±1.2	3812–3814	1478
287.6		6971,7084	1840
287.9		6971,7097	1840
292.5		6971,7088	1840
310.3		6971,7099	1840
328.2		6971,7089	1840
331.4		6971,7095	1840
340.9		6971,7098	1840
342.	±1.7	3809–3811	1478
348.1		6971,7086	1840
367.5		6971,7093	1840

Table 3.4.7
Dideuterodichloromethane – CD$_2$Cl$_2$—*continued*

Wavelength (μm)	Uncertainty (μm)	Comments	References
380.8		6971,7093	1840
396.7		6971,7087	1840
398.4		6971,7087	1840
410.2		6971,7092	1840
417.6		6971,7089	1840
437.4		6971,7097	1840
451.3		6971,7091	1840
469.	±2.3	3806–3808	1478
504.8		6971,7083	1840
520.	±2.5	3803–3805	1478
559.3		6971,7094	1840
618.3		6971,7099	1840
628.4		6971,7090	1840
631.	±3.1	3800–3802	1478
681.8		6971,7098	1840
726.8		6971,7085	1840
772.6		6971,7091	1840
829.	±4.1	3797–3799	1478

Table 3.4.8
Dideuterodifluoromethane – CD$_2$F$_2$

Wavelength (μm)	Uncertainty (μm)	CO$_2$ laser pump line	References
192.0		10R42	1710
207.835		10R38	1710
218.267		10R38	1710
303.8		10R48	1710
320.597		10R44	1710
456.2		10R42	1710

Table 3.4.9
Trideuterobromomethane – CD₃Br

Wavelength (μm)	Uncertainty (μm)	Comments	References
290.	±5.8	3976–3979	1438,1439
297.	±5.9	3972–3975	1438,1439
341.	±6.8	3968–3971	1438,1439
367.	±7.3	3964–3967	1438,1439
428.	±8.6	3960–3963	1438,1439
430.	±8.6	3956–3959	1438,1439
440.	±8.8	3952–3955	1438,1439
530.	±11.	3948–3951	1438,1439
550.	±11.	3944–3947	1438,1439
560.	±11.	3940–3943	1438,1439

Table 3.4.10
Trideuterochloromethane – CD₃Cl

Wavelength (μm)	Uncertainty (μm)	Comments	References
224.	±1.1	3598–3601	1388,1401
245.	±1.2	3594–3597	1388,1401
246.	±1.2	3590–3593	1388,1401
249.	±1.2	3586–3589	1388,1401
288.	±1.4	3578–3581	1388,1401
288.	±1.4	3582–3585	1388,1401
293.6480	±0.00059	3574–3577	1388,1401
318.	±1.6	3570–3573	1388,1401
383.2845	±0.00077	3566–3569	1388,1401
443.2645	±0.00089	3562–,3565	1388,1401
449.7997	±0.00090	3558–3561	1388,1401
464.7567	±0.00093	3554–3557	1388,1401
480.3101	±0.00096	3550–3553	1388,1401
519.303	±0.0010	3546–3549	1388,1401
698.555	±0.0014	3542–3545	1388,1401
735.130	±0.0015	3538–3541	1388,1401
792.	±3.9	3534–3537	1388,1401
883.598	±0.0018	3530–3533	1388,1401
1239.480	±0.0025	3526–3529	1388,1401
1990.757	±0.0040	3522–3525	1388,1401

Table 3.4.11
Trideuterocyanomethane – CD$_3$CN

Wavelength (µm)	Uncertainty (µm)	CO$_2$ laser pump line	References
455.073		9P08	1711
516.253		9P30	1711
529.880		9R04	1711

Table 3.4.12
Trideuterofluoromethane – CD$_3$F

Wavelength (µm)	Uncertainty (µm)	Comments	References
155.6	±0.45	2689–2692	1472,1485
172.8	±0.50	2685–2688	1472,1485
181.37		7250	1821
186.13		7239	1821
189.84		7222	1821
190.60		7248	1821
193.92		7247	1821
194.56		7221	1821
195.63		7247	1821
196.01		7247	1821
199.05		7218	1821
199.73		7220	1821
200.0	±0.48	2681–2684	1472,1485
201.30		7246	1821
201.5	±0.48	2677–2680	1472,1485
205.10		7219	1821
206.0	±0.49	2673–2676	1472,1485
206.20		7245	1821
206.36		7245	1821
210.85		7218	1821
212.01		7244	1821
212.32		7216	1821
216.46		7218	1821
217.21		7217	1821
218.30		7243	1821
218.40		7243	1821
218.49		7243	1821
218.59		7243	1821
218.64		7243	1821
219.11		7243	1821
219.21		7243	1821
223.82		7216	1821

Trideuterofluoromethane – CD$_3$F—*continued*

Wavelength (µm)	Uncertainty (µm)	Comments	References
224.02		7242	1821
224.63		7242	1821
224.93		7242	1821
225.23		7242	1821
225.59		7242	1821
229.21		7241	1821
229.84		7213	1821
230.69		7215	1821
231.86		7241	1821
231.92		7241	1821
232.03		7241	1821
232.08		7241	1821
232.13		7241	1821
232.19		7241	1821
232.35		7241	1821
237.71		7215	1821
239.18		7240	1821
239.30		7240	1821
239.47		7240	1821
239.55		7240	1821
239.99		7240	1821
240.33		7240	1821
245.95		7213	1821
246.01		7214	1821
246.80		7239	1821
246.89		7239	1821
247.22		7239	1821
247.29		7239	1821
247.3	±0.49	2669–2672	1472,1485
247.5	±0.50	2665–2668	1472,1485
247.53		7239	1821
247.59		7239	1821
247.71		7239	1821
247.90		7239	1821
254.20		7212	1821
255.69		7238	1821
255.76		7238	1821
256.09		7238	1821
256.41		7238	1821
256.68		7238	1821
263.51		7211	1821
264.69		7237	1821
264.76		7237	1821

Trideuterofluoromethane – CD₃F—*continued*

Wavelength (µm)	Uncertainty (µm)	Comments	References
264.83		7237	1821
264.97		7237	1821
265.	±5.0	2661–2664	1472,1485
265.33		7237	1821
265.54		7237	1821
265.68		7237	1821
272.93		7210	1821
274.58		7236	1821
274.65		7236	1821
275.03		7236	1821
275.33		7236	1821
283.37		7209	1821
284.74		7235	1821
285.06		7235	1821
285.14		7235	1821
285.23		7235	1821
285.31		7235	1821
285.47		7235	1821
285.71		7235	1821
285.80		7235	1821
286.29		7235	1821
296.21		7234	1821
296.38		7234	1821
296.47		7234	1821
296.73		7234	1821
308.26		7233	1821
308.54		7232	1821
308.73		7233	1821
308.83		7233	1821
308.92		7233	1821
309.21		7233	1821
309.31		7233	1821
309.59		7233	1821
322.16		7232	1821
322.26		7232	1821
322.37		7232	1821
323.3	±0.45	2657–2660	1472,1485
336.02		7231	1821
336.6	±0.44	2653–2656	1472,1485
336.69		7231	1821
336.81		7231	1821
337.26		7231	1821
337.38		7231	1821

Trideuterofluoromethane – CD₃F—*continued*

Wavelength (µm)	Uncertainty (µm)	Comments	References
349.0	±0.45	2649–2652	1472,1485
352.73		7230	1821
368.4	±0.48	2645–2648	1472,1485
370.26		7229	1821
370.46		7229	1821
371.19		7229	1821
371.47		7229	1821
371.60		7229	1821
384.7	±0.46	2641–2644	1472,1485
389.25		7228	1821
408.82		7208	1821
411.18		7227	1821
411.35		7227	1821
411.52		7227	1821
433.08		7207	1821
435.53		7226	1821
435.72		7226	1821
461.04		7206	1821
462.74		7225	1821
463.17		7225	1821
463.39		7225	1821
490.91		7205	1821
494.06		7224	1821
525.76		7204	1821
529.65		7223	1821
564.65		7201	1821
566.57		7203	1821
606.42		7200	1821
612.74		7202	1821
662.68		7202	1821
667.10		7199	1821
670.23		7201	1821
735.83		7200	1821
816.99		7198	1821
817.66		7199	1821
922.50		7251	1821
1042.75		7198	1821
1225.49		7197	1821
1450.	±19	2637–2640	1472,1485
1490.	±10	2633–2636	1472,1485

Table 3.4.13
Trideuterofluoromethane – $^{13}CD_3F$

Wavelength (μm)	Uncertainty (μm)	CO_2 laser pump line	References
183.4		9P36	1663
209.1		9P24	1663
249.9		10R34	1663
280.8		10P20	1663
299.5		9P48	1663
325.9		9R22	1663
336.5		hot band	1663
376.8		10P38	1663
470.065		10P34	1700
537.41		hot band	1700

Table 3.4.14
Trideuteroiodomethane – CD_3I

Wavelength (μm)	Uncertainty (μm)	Comments	References
272.	±1.3	4216–4219	1401
301.	±1.5	4212,–4215	1401
390.	±1.9	4208–4211	1401
433.1036	±0.00087	4204–4207	1401
444.3862	±0.00089	4200–4203	1401
460.5619	±0.00092	4196–4199	1401
487.2260	±0.00097	4192–4195	1401
490.3909	±0.00098	4188–4191	1401
523.406	±0.0010	4185–4187	1401
540.	±2.6	4181–4184	1401
556.876	±0.0011	4178–4180	1401
569.477	±0.0011	4175–4177	1401
599.550	±0.0012	4171–4174	1401
614.110	±0.0012	4167–4170	1401
640.	±3.1	4164–4166	1401
644.	±3.2	4161–4163	1401
660.582	±0.0013	4158–4160	1401
667.232	±0.0013	4155–4157	1401
668.5		6971,7053	1842
670.094	±0.0013	4152–4154	1401
670.114	±0.0013	4149–4151	1401
691.119	±0.0014	4145–4148	1401
730.323	±0.0015	4141–4144	1401
734.262	±0.0015	4137–4140	1401
740.2		7054	1842

Trideuteroiodomethane – CD₃I—*continued*

Wavelength (μm)	Uncertainty (μm)	Comments	References
745.	±3.7	4134–4136	1401
788.482	±0.0016	4131–4133	1401
895.	±4.4	4128–4130	1401
918.610	±0.0018	4124–4127	1401
953.880	±0.0019	4120–4123	1401
981.709	±0.0020	4117–4119	1401
1005.348	±0.0020	4114–4116	1401
1099.544	±0.0022	4111–4113	1401
1549.505	±0.0031	4107–4110	1401

Table 3.4.15
Trideuteroiodomethane – ^{13}CD₃I

Wavelength (μm)	Uncertainty (μm)	CO_2 laser pump line	References
554.7		$^{13}CO_2$ laser	1800
574.6		$C^{18}O_2$	1800
690.		10P10	1388
745.5		$^{13}CO_2$ laser	1800
806.		10P12	1388
901.3		$^{13}CO_2$ laser	1800
929.8		$^{13}CO_2$ laser	1800
1182.2		$^{13}CO_2$ laser	1800

Table 3.4.16
Deuterooxytrideuteromethanol – CD₃OD[a]

Wavelength (μm)	Uncertainty (μm)	Comments	References
38.1		6971,7007	1822
43.697		6970,7183	1838
45.362		6971,7182	1838
62.2		6970,7175	1836
76.6		6971,7013	1822
77.2		6970,7173	1836
79.257		6971,7182	1838
80.1		6971,7173	1836
81.557101	±4.1e-05	5561,5562,5563	1391,1453,1481
86.4	±0.42	5558,5559,5560	1391
100.0		6971,7173	1836

Deuterooxytrideuteromethanol – CD₃OD—*continued*

Wavelength (μm)	Uncertainty (μm)	Comments	References
102.6	±0.50	5555,5556,5557	1391
108.5		6971,7010	1822
108.66842	±5.4e-05	5554	1453,1481
112.3	±0.55	5551,5552,5553	1391
125.9		6970,7173	1836
128.7	±0.63	5548,5549,5550	1391
131.56276	±6.6e-05	5547	1453,1481
136.62721	±6.8e-05	5546	1453,1481
144.11787	±7.2e-05	5543,5544,5545	1391,1453,1481
146.1		6970,7177	1836
158.0	±0.77	5540,5541,5542	1410
174.0		6971,7018	1822
179.0	±0.88	5537,5538,5539	1410
180.74051	±9.0e-05	5534,5535,5536	1391,1453,1481
182.1		6970,7012	1822
182.56629	±9.1e-05	5533	1453,1481
184.0	±0.90	5530,5531,5532	1402
188.1		6971,7014	1822
188.42390	±9.4e-05	5529	1453,1481
189.1		6971,7010	1822
191.9	±0.94	5526,5527,5528	1391
195.85		6970,7174	1836
199.2		6971,7179	1836
200.9		6970,7173	1836
201.	±0.98	5523,5524,5525	1410
210.6		6971,7016	1822
214.6		6970,7008	1822
215.0812	±0.00011	5522	1453,1481
220.	±1.1	5519,5520,5521	1391
222.	±1.1	5516,5517,5518	1402
223.	±1.1	5513,5514,5515	1410
228.8		6971,7017	1822
232.	±1.1	5510,5511,5512	1402
236.	±1.2	5507,5508,5509	1410
238.	±1.2	5504,5505,5506	1391
246.2		6971,7180	1838
253.7196	±0.00013	5501,5502,5503	1391,1453,1481
258.4356	±0.00013	5498,5499,5500	1402,1453,1481
265.	±1.3	5495,5496,5497	1402
265.019		6971,7181	1838
266.	±1.3	5492,5493,5494	1402
267.	±1.3	5489,5490,5491	1402
268.	±1.3	5486,5487,5488	1402

Deuterooxytrideuteromethanol – CD₃OD—*continued*

Wavelength (μm)	Uncertainty (μm)	Comments	References
274.1		6971,7174	1836
276.7157	±0.00014	5483,5484,5485	1402,1453,1481
277.	±1.4	5480,5481,5482	1402
278.	±1.4	5477,5478,5479	1402
282.4		6970,7179	1836
286.1974	±0.00014	5474,5475,5476	1402,1453,1481
286.7242	±0.00014	5471,5472,5473	1402,1453,1481
287.3076	±0.00014	5468,5469,5470	1402,1453,1481
290.	±1.4	5465,5466,5467	1391
290.9		6970,7012	1822
291.2		6970,7011	1822
295.3		6971,7177	1836
297.	±1.5	5459,5460,5461	1402
297.	±1.5	5462,5463,5464	1402
299.	±1.5	5456,5457,5458	1402
309.	±1.5	5453,5454,5455	1402
310.	±1.5	5450,5451,5452	1402
312.9		6971,7177	1836
313.3		6970,7177	1836
316.0		6971,7017	1822
321.	±1.6	5447,5448,5449	1402
336.	±1.6	5441,5442,5443	1402
336.	±1.6	5444,5445,5446	1402
343.001		6971,7178	1836
343.2		6970,7009	1822
346.	±1.7	5438,5439,5440	1410
346.3		6970,7014	1822
350.	±1.7	5435,5436,5437	1402
351.	±1.7	5432,5433,5434	1402
352.	±1.7	5426,5427,5428	1402
352.	±1.7	5429,5430,5431	1402
353.	±1.7	5423,5424,5425	1402
357.1		6971,7015	1822
361.2		6970,7009	1822
370.	±1.8	5417,5418,5419	1402
370.	±1.8	5420,5421,5422	1402
385.	±1.9	5414,5415,5416	1402
386.	±1.9	5411,5412,5413	1402
391.1		6971,7177	1836
398.	±2.0	5408,5409,5410	1402
406.6		6971,7179	1836
407.	±2.0	5405,5406,5407	1402
407.3		6971,7177	1836

Deuterooxytrideuteromethanol – CD₃OD—*continued*

Wavelength (μm)	Uncertainty (μm)	Comments	References
408.0		66970,7177	1836
409.	±2.0	5402,5403,5404	1402
410.	±2.0	5399,5400,5401	1402
411.6		6971,7176	1836
412.	±2.0	5396,5397,5398	1402
418.7118	±0.00021	5393,5394,5395	1402,1453,1481
421.	±2.1	5390,5391,5392	1402
422.	±2.1	5387,5388,5389	1402
426.0		6971,7008	1822
435.	±2.1	5384,5385,5386	1410
455.	±2.2	5381,5382,5383	1402
455.2		6971,7013	1822
472.	±2.3	5378,5379,5380	1402
480.	±2.4	5375,5376,5377	1402
483.	±2.4	5369,5370,5371	1402
483.	±2.4	5372,5373,5374	1402
495.	±2.4	5366,5367,5368	1402
498.	±2.4	5363,5364,5365	1391
508.	±2.5	5360,5361,5362	1402
517.	±2.5	5357,5358,5359	1402
551.	±2.7	5354,5355,5356	1402
553.	±2.7	5348,5349,5350	1402
553.	±2.7	5351,5352,5353	1402
554.	±2.7	5345,5346,5347	1402
583.	±2.9	5342,5343,5344	1402
599.	±2.9	5339,5340,5341	1402
646.	±3.2	5336,5337,5338	1402
648.	±3.2	5333,5334,5335	1402
680.	±3.3	5330,5331,5332	1402
685.	±3.4	5327,5328,5329	1402
695.	±3.4	5324,5325,5326	1402
702.	±3.4	5321,5322,5323	1402
703.	±3.4	5318,5319,5320	1402
711.	±3.5	5315,5316,5317	1402
722.	±3.5	5312,5313,5314	1402
745.	±3.7	5309,5310,5311	1402
760.	±3.7	5306,5307,5308	1402
774.	±3.8	5303,5304,5305	1402
871.5850	±0.00044	5300,5301,5302	1402,1453,1481
968.	±4.7	5297,5298,5299	1402
1100.	±5.4	5294,5295,5296	1402
1146.	±5.6	5291,5292,5293	1402
1290.	±6.3	5288,5289,5290	1402

(a) See, also, reference 1849.

Table 3.4.17

Deuterooxytrideuteromethanol – $^{13}CD_3OD$

Wavelength (μm)	Uncertainty (μm)	CO_2 laser pump line	References
46.4		10R32*	1838
52.2		10R46	1838
54.159		10SR23	1838
57.3		10R24	453
62.6		10R08	1838
65.449		10R44	1838
70.947		10R40*	1838
72.194		10R32*	1838
74.817		10SR13	1838
75.275		9R24	1678
75.5		9R10	1678
81.8		10R06	453
82.1		9R34	1678
82.4		9R14	1678
84.4		10P12	1678
84.7		10P06	453
90.155		10R24	1838
92.9		10P12	453
93.6		10R16	1678
95.93		10SR17*	1838
100.506		10SR17*	1838
101.9		10P26	453
105.696		10R52	1838
109.3		1QR16	1832
109.926		10R16	1678
109.938		10R16	1678
110.6		10SR17*	1838
115.2		10P18	453
118.1		1QR20	1832
118.553		9R14	1678
120.9		1QR14	1832
121.1		1QR18	1832
123.7		10R38	453
124.253		10P24	1678
126.2		9P12	1678
128.1		10R16	1678
129.2		9P38	1678
132.7		10P24	453
132.862		10R53	1838
134.920		10R44	1838
138.040		10R08	1838
139.0		10P28	453

Deuterooxytrideuteromethanol – $^{13}CD_3OD$—*continued*

Wavelength (µm)	Uncertainty (µm)	Comments	References
140.354		10R44	1838
146.129		10P08	1838
146.326		10R26	1838
148.617		10R30	1678
149.2		10P08	453
150.3		9R32	1678
151.0		9P24	1678
151.8		9P28	1678
156.5		10R40	453
159.4		1QR20	1832
162.503		10P14	1824
164.142		10R40*	1838
170.9		10P26	453
173.637		10R20	1678
175.1		9R22	1678
192.846		10P14	1824
193.3		10P14	453
202.6		10P32	453
209.233		10R12	1678
211.0		1QR22	1832
211.6		1QR20	1832
215.005		10R54	1838
215.466		10R12	1838
216.3		1QR16	1832
216.356		10P24	1678
222.1		10P16	453
227.0		9R28	1678
235.3		10R36	453
241.4		1QR20	1832
241.6		9R20	1678
243.0		9P32	1678
247.0		9P12	1678
252.738		10R28	1838
261.2		10P30	453
272.958		10R26	1678
298.6		10P28	453
298.811		10R08	1838
300.5		10R36	453
314.1		10R06	453
321.410		10R12	1678
322.4		10P26	453
322.5		10R38	453
324.140		10R14	1678

Deuterooxytrideuteromethanol – $^{13}CD_3OD$—*continued*

Wavelength (μm)	Uncertainty (μm)	Comments	References
335.1		10P24	453
335.6		10P20	453
343.229		10P14	1824
343.3		10P14	453
347.0		10P18	453
351.8		10P26	453
353.1		9R26	1678
353.3		9R14	1678
358.4		9P32	1678
407.1		9P28	1678
417.3		9R20	1678
427.2		10R06	453
427.3		10R36	453
464.453		10R08	1824
464.7		10R08	1678
472.3		1QR16	1832
475.1		10P16	453
521.6		10R40	453
522.7		10P06	453
591.1		10P06	453
670.7		10R08	453
719.426		10R38	1838
783.3		10R10	453
1194.0		10P06	453

* Indicates different CO_2 laser frequency offsets.

Table 3.4.18
Trideuteromethanol – CD₃OH

Wavelength (μm)	Uncertainty (μm)	CO₂ laser pump line	References
27.7		9P34	1720
30.7		9P30	1720
31.1		9R20	1720
34.1		10R34	1719
34.2		10R38	1720
34.8		10P22	1391
35.500		9R06	826
35.7		10P26	1719
37.1		9P24	1720
37.6		10R34	1391
40.000		9R28	826
40.1		10P22	1391
41.25		10R08	1692
41.355		10R18	826
41.46		10R34	1806
41.50		10R08	1391
41.8		10R18	1391
42.5		9P38	1720
42.5		10R34	1719
42.6		9R28	1720
42.92		9P38	1806
43.697		10R18	826
43.7		10R48	1806
44.3		9R28	1720
44.55		10R32	1806
44.7		9P08	1806
44.8		10R08	1719
45.0		10R08	1806
45.66		10P08	1806
46.8		10HR30	1837
47.1		9P28	1720
47.2		10P24	1837
48.6		9R34	1720
48.7		9R06	1720
49.07		10R32	1806
49.78		10R20	1806
49.8		9R28	1410
50.000		9R26	826

Trideuteromethanol – CD₃OH—*continued*

Wavelength (μm)	Uncertainty (μm)	CO₂ laser pump line	References
50.000		10R38	826
50.1		10R34	1719
50.3		10R24	1806
51.1		10R38	1837
52.0		10P18	1837
52.8		10R20	1719
52.9		9R34	1410
53.10		9P44	1806
53.2		10R38	1837
53.3		9R34	1801
53.82		9R34	1692
54.1		9P28	1720
54.4		10R56	1837
54.7		10R18	1719
55.4		10R20	1801
55.56		9R28	1692
56.5		9R06	1720
56.7		10R30	1719
56.87		9P26	1806
59.6		10R40	1806
60.10		9R34	1692
60.7		10P18	1837
60.8		9R34	1410
61.4		10P08	1806
61.7		10R24	1719
65.87		10R32	1806
66.4		10R34	1806
66.8		9R18	1806
67.479		10R30	826
68.1		10R14	1801
68.45		9R06	1806
68.7		10R14	1806
68.8		10R46	826
69.18		10R36	1806
70.6		10R24	1801
70.989		10R08	826
71.4		10R08	1801
71.5		10R40	1806
71.7		10R10	1806
76.0		10R24	1806
76.1		10P32	1391
76.3		10P42	826
76.90		10R34	1806

Trideuteromethanol – CD₃OH—*continued*

Wavelength (μm)	Uncertainty (μm)	CO₂ laser pump line	References
76.93		10P48	1806
78.6		9R24	1806
78.78		9R42	1806
80.44		10R28	1806
80.9		10R16	1801
81.557		10R16	826
81.6		10HR30	1837
82.7		9P18	1719
83.6		10R32	1801
83.7		10R14	1806
83.9		10R32	1719
84.5		10P06	1719
86.3		10P56	1806
86.4		10R16	1391
86.7		10R16	1837
86.741		10R34	826
87.8		10P20	1719
87.9		9P28	1719
90.16		9P22	1806
92.7		9P48	1837
93.88		10R32	1806
94.9		10R32	1806
102.6		10R34	1391
103.0		10P14	1719
105.2		10P32	1837
107.2		10R14	1806
108.668		10P10	826
109.1		9P22	1719
111.4		10R28	1806
112.1		9R32	1806
112.3		10R34	1391
114.4		10R32	1806
116.5		9P48	1719
117.62		10R08	1806
118.8		9P38	1719
119.0		9R14	1801
119.9		10R34	1806
120.3		9R14	1719
120.45		9R40	1806
120.661		9R14	826
122.154		10R38	826
123.55		10R12	1806
124.93		10P10	1806

Trideuteromethanol – CD₃OH—*continued*

Wavelength (μm)	Uncertainty (μm)	CO_2 laser pump line	References
127.3		10P04	1719
128.034		10R34	826
128.7		10R34	1391
131.563		10R32	826
132.1		10P22	1806
133.7		9P14	1719
135.4		10R24	1719
136.5		9R06	1719
136.627		10R14	826
138.4		10R34	1806
140.0		9R10	1719
140.95		10R08	1806
142.2		10HR30	1837
143.4		$^{13}CO_2$ laser	1810
143.8		9P12	1719
143.80		10R12	1806
144.118		10P18	826
144.4		10R26	1806
145.7		9P08	1719
147.28		9R26	1806
147.349		10P32	826
147.65		10P12	1806
148.0		10P32	1719
148.940		9R32	1806
150.8		10R16	1719
151.3		9R32	1719
151.8		10R20	1806
153.7		10P46	1806
158.0		9R28	1410
158.9		9R18	1719
159.400		9R28	826
161.1		10R38	1806
162.85		10P24	1806
165.000		10R32	826
166.76		10R32	1806
168.083		10R34	826
172.62		10P12	1806
174.00		10R32	1806
176.8		9R32	1806
177.00		10R12	1806
177.4		9P30	1719
179.0		9R14	1410
180.741		10R34	826

Trideuteromethanol – CD₃OH—*continued*

Wavelength (μm)	Uncertainty (μm)	CO₂ laser pump line	References
180.75		10R08	1806
181.0		9R28	1801
181.711		9R28	826
182.566		9R14	826
184.0		9R08	1402
185.0		10R14	1719
187.05		10P24	1806
188.424		10P42	826
188.9		10P28	1719
189.73		9P36	826
190.0		10P28	826
191.356		10R34	826
196.6		10P08	1719
196.950		9R36	1806
198.6		9P22	1806
198.682		9P40	826
199.5		9P36	1806
199.81		10R44	1806
200.870		9R36	1806
201.0		9P40	1410
203.300		10R08	826
203.5		10P16	1719
205.8		10P36	1719
215.081		10P32	826
215.50		10R20	1806
215.6		10P36	1719
217.2		10R28	1806
219.0		10P12	1719
219.70		9R40	1806
219.9		10R18	1391
221.0		¹³CO₂ laser	1810
221.88		9P40	1806
221.9		9P08	1719
222.0		9P06	1402
222.217		10R34	826
222.7		10P48	1806
223.0		9P08	1410
225.0		9P20	1719
225.0		9R46	1719
225.8		9P06	1719
226.9		9R24	1719
228.30		10R34	1806
229.10		9P44	1806

Trideuteromethanol – CD₃OH—*continued*

Wavelength (μm)	Uncertainty (μm)	CO₂ laser pump line	References
231.1		10P40	1719
232.1		9R44	1719
234.7		¹³CO₂ laser	1810
234.8		10P40	1719
235.4		9P04	1806
235.8		10R30	1806
236.0		9R14	1410
237.1		10P24	1719
238.3		10P24	1391
239.65		10R34	1806
250.1		9P04	1719
251.4		10P24	1806
252.0		10R36	1801
252.3		9R24	1719
253.1		10P12	1719
253.720		10R36	826
253.8		10R38	1719
255.2		10R04	1719
257.0		10P22	1801
258.0		9P20	1801
258.30		9P46	1806
258.436		10P22	826
260.0		10P22	1402
264.7		10R04	1806
264.759		10R34	826
265.0		10R34	1719
265.3		9P14	1719
266.0		9P20	1719
267.0		10R14	1402
267.2		10P22	1806
268.0		9P14	1402
268.6		10R14	1719
272.3		10P12	1806
276.6		9R26	1719
276.716		10P28	826
276.9		10R24	1719
278.0		10R24	1402
282.80		10R32	1806
283.750		9P42	1806
284.3		9P40	1719
284.4		9R32	1806
285.0		10P24	1801
286.0		10P18	1801

Trideuteromethanol – CD₃OH—*continued*

Wavelength (μm)	Uncertainty (μm)	CO$_2$ laser pump line	References
286.197		9P40	826
286.2		10R24	1719
286.724		10P24	826
287.308		10P18	826
287.95		9P50	1806
290.0		10P18	1391
297.0		10R34	1402
297.1		9R20	1719
299.0		9R06	1402
308.5		10R28	1719
309.0		10P20	1402
310.0		10R28	1402
310.1		9P36	1806
310.35		10R26	1806
310.7		10P20	1719
310.8		10P48	1806
312.50		10R08	1806
312.9		9P48	1719
321.0		9R16	1402
322.1		10P12	1806
322.35		10P12	1806
323.5		9P24	1806
329.5		10P32	1719
333.9		9P36	1806
336.0		10R30	1402
336.5		9R32	1719
336.8		9P34	1719
337.3		10R30	1719
337.5		9P24	1719
346.0		9R14	1410
350.5		9P32	1719
351.2		10R30	1719
351.4		9R22	1719
351.4		9R32	1719
352.3		9R06	1719
352.503		9R14	826
362.8		9P28	1719
369.55		10P08	1806
369.7		10P38	1806
370.0		9P28	1402
370.483		9R28	826
372.36		10P12	1806
380.8		10R32	1719

Trideuteromethanol – CD₃OH—*continued*

Wavelength (μm)	Uncertainty (μm)	CO₂ laser pump line	References
385.7		9P30	1719
386.037		10R34	826
386.6		9P16	1719
386.9		9R22	1719
388.0		10R14	1719
396.4		10R28	1719
398.0		10R28	1402
407.0		9R44	1402
407.9		9R34	1719
409.1		10R40	1806
410.0		9P32	1402
412.0		10R12	1402
417.0		10R32	1806
418.1		10R38	1719
418.712		10R36	826
420.0		10R36	1402
420.3		10R32	1719
422.0		9P20	1402
430.927		9R34	826
431.4		10R34	1719
433.6		10P20	1719
435.0		9P28	1410
435.1		10R36	1806
435.30		10R34	1806
438.87		10R44	1806
452.90		10R20	1719
455.6		9P18	1719
472.4		9R18	1719
476.25		10R34	1806
477.3		10R34	1719
480.0		9P16	1402
482.7		9R26	1719
482.7		10P26	1719
495.0		10R18	1402
498.0		10R34	1391
498.700		9R26	826
509.5		9P06	1719
516.5		10P42	1719
520.3		10R18	1719
524.6		10P24	1719
530.4		13CO2	1810
550.2		9P22	1719
551.9		9R16	1719

Trideuteromethanol – CD₃OH—*continued*

Wavelength (μm)	Uncertainty (μm)	CO₂ laser pump line	References
553.0		9R34	1719
554.0		10R08	1719
562.4		10R36	1806
583.3		9R22	1719
593.1		10R06	1719
598.6		10P28	1719
599.0		10R16	1402
610.3		10R08	1719
645.2		10R36	1811
646.477		10R08	826
680.0		9P06	1402
684.5		10R34	1719
690.0		9P12	1719
695.0		9P12	1402
699.0		10R18	1719
701.5		9P24	1719
704.2		10R36	1811
706.6		9P10	1719
711.0		9P08	1402
722.0		10P20	1402
745.0		9R26	1402
760.0		10P18	1402
774.0		9P30	1402
812.6		9P06	1719
854.7		10R18	1811
858.254		10R18	826
861.1		10R12	1806
862.0		10R18	1719
871.585		9R14	1665
968.0		9R20	1402
988.1		10R18	1719
1092.8		9P12	1719
1100.0		9P12	1402
1146.0		9P24	1402
1155.5		10R38	1806
1290.0		10R20	1402
1676.		10R36	1719
1930.		10R36	1719
3030.		10R36	1719

Table 3.4.19
Trideuteromethanol – $^{13}CD_3OH$

Wavelength (μm)	Uncertainty (μm)	CO_2 laser pump line	References
~50		10R06	1843
52.1		9R32	1676
52.2		9R18	1676
53.4		10P10	1843
54.5		9P26	1843
55.8		9R32	1676
56.0		10P14	1843
58.1		10R30	1843
59.6		10R30	1843
61.5		9P26	1843
63.0		10R08	1823
64.4		9P22	1843
64.5		10R24	1823
65.4		9R10	1676
67.6		10R10	1843
67.8		10R12	1676
70.0		10P12	1843
71.7		10R10	1843
72.1		10P10	1843
72.9		9P10	1676
73.467		10R20	1676
75.7		10P42	1843
80.8		10P08	1843
82.5		10R24	1823
84.406		10R22	1676
84.7		10P12	1843
87.0		10P10	1843
87.0		10P10	1843
91.0		10R10	1843
91.8		9R26	1823
98.5		9R34	1676
99.8		9R24	1823
110.0		10R22	1676
112.3		10P14	1843
118.6		10P24	1827
119.1		10P42	1676
119.4		9R14	1676
124.3		10P22	1676
125.3		10P34	1843
126.1		9P10	1676
127.021		10P08	1676
127.2		10P10	1843

Trideuteromethanol – $^{13}CD_3OH$—*continued*

Wavelength (μm)	Uncertainty (μm)	CO₂ laser pump line	References
127.656		10R22	1676
128.3		10P08	1843
133.9		10P12	1843
142.0		10P10	1843
145.4		10P34	1843
145.563		10R24	1676
145.7		10R10	1843
146.326		10R26	1676
146.4		10P12	1843
148.3		10P42	1676
149.6		10P34	1843
150.2		9R30	1676
151.0		9R38	1676
151.6		10P10	1843
153.694		9R28	1676
156.0		9P38	1707
157.20		10R40	1843
162.2		10P14	1843
175.260		10P08	1809
177.6		9R40	1676
187.2		10P42	1843
188.0		10R40	1843
192.6		10P44	1843
196.2		9R08	1676
196.2		10P14	1823
197.046		10R26	1676
200.4		10P40	1843
205.4		10P40	1843
209.0		10R26	1676
214.0		10P24	1827
216.36		10P24	1827
221.0		9R08	1676
225.1		10P36	1843
227.88		10P24	1827
231.8		9P22	1843
241.6		10P34	1843
242.4		9R20	1823
255.7		10P30	1843
280.0		10P40	1843
280.28		10P24	1827
280.3		10P42	1823
280.7		10P24	1827

Trideuteromethanol – $^{13}CD_3OH$—*continued*

Wavelength (μm)	Uncertainty (μm)	CO_2 laser pump line	References
281.39		10R08	1843
281.7		10P26	1843
291.0		10P22	1676
309.7		9R30	1823
313.		10P22	1823
327.28		10R10	1843
333.261		10P16	1676
336.5		9R28	1676
336.9		10P22	1827
340.627		10P16	1676
387.2		9R38	1676
389.6		9R08	1676
393.3		10P40	1843
399.8		9P34	1676
414.6		10P14	1843
462.8		10P08	1676
468.965		10R26	1676
510.0		10P24	1827
561.8		9R14	1823
604.9		10P42	1843
655.0		10R26	1823
2615		10R26	1823

Table 3.4.20
Dichlorodifluoromethane (Fluorcarbon) – CF_2Cl_2

Wavelength (μm)	Uncertainty (μm)	Comments	References
614.3	±0.74	4020–4023	1442
638.4	±0.57	4016–4019	1442
684.7	±0.48	4012–4015	1442
684.7	±0.75	4008–4011	1442
751.4	±0.60	4004–4007	1442
765.2	±0.54	4000–4003	1442
858.7	±0.52	3996–3999	1442
980.	±9.8	3992–3995	1442
1025.	±1.0	3988–3991	1442
1164.	±2.3	3984–3987	1442
1205.	±2.4	3980–3983	1442

Table 3.4.21
Bromotrifluoromethane – CF₃Br

Wavelength (μm)	Uncertainty (μm)	Comments	References
824.0	±1.4	4256–4259	1443,1456,1444
883.0	±2.4	4252–4255	1443,1456,1444
1040.	±11.	4249,4250,4251	1443,1444,1456
1083.0	±4.3	4246,4247,4248	1443,1444,1456
1151.0	±4.6	4243,4244,4245	1443,1444,1456
1530.0	±11.	4240,4241,4242	1443,1444,1456
1550.0	±4.5	4237,4238,4239	1443,1444,1456
1692.0	±4.9	4234,4235,4236	1443,1444,1456
1890.0	±9.3	4231,4232,4233	1443,1444,1456
2140.	±19.	4228,4229,4230	1443,1444,1456

Table 3.4.22
Dichlorofluoromethane – CHCl₂F

Wavelength (μm)	Uncertainty (μm)	CO₂ laser pump line	References
340.300		9P20	1674
365.725		9P18	1674
375.980		9P16	1674
467.515		9R06	1674
470.386		9R34	1674
492.040		9R36	1674
495.963		9P08	1674
530.854		9R06	1674
547.529		9R10	1674
549.258		9R08	1674
561.028		9R40	1674
580.869		9R12	1674
661.153		9R30	1674
832.757		9R04	1674
905.428		9R04	1674

Table 3.4.23

Chlorodifluoromethane – CHClF$_2$

Wavelength (μm)	Uncertainty (μm)	CO$_2$ laser pump line	References
298.049		9R24	1696
301.		9P28	1696
306.053		9R24	1696
324.		9R08	1696
326.		9R12	1696
328.960		9R28	1696
335.467		9R16	1696
337.094		9R32	1696
345.		10P34	1696
360.606		9R14	1696
366.273		9R30	1696
370.		9R40	1696
372.870		9R22	1696
380.		9R30	1696
382.766		10R40	1696
385.687		9R18	1696
386.966		9R18	1696
387.		9R16	1696
388.		9R36	1696
396.		9R10	1696
414.351		9R36	1696
415.075		9R30	1696
427.807		9R26	1696
432.		10P06	1696
432.244		9R40	1696
433.438		10R34	1696
444.		9R40	1696
467.7		9R44	1672
476.		10R30	1696
481.452		9R30	1696
487.144		9R32	1696
533.137		10R16	1696
534.430		9R32	1696
556.097		9R32	1696
562.450		9R32	1696
590.		10P12	1696
591.130		10R24	1696
592.441		9R40	1696
615.329		10R14	1696
617.656		9R40	1696
665.885		10P18	1696

Chlorodifluoromethane – CHClF$_2$—*continued*

Wavelength (µm)	Uncertainty (µm)	CO$_2$ laser pump line	References
682.175		10P14	1696
747.050		9R10	1696
842.125		9R16	1696
899.384		9R38	1696

Table 3.4.24
Dideuterofluoromethane – CHD$_2$F

Wavelength (µm)	Uncertainty (µm)	CO$_2$ laser pump line	References
163.3		9P34	662
193.4		9P42	662
204.0		10P48	662
231.0		9P10	662
232.0		9P16	662
260.0		9R16	662
285.1		9R20	662
292.7		9R30	662
301.0		9P44	662
375.407		9R06	662
384.319		10P28	662
406.878		10P28	662
435.427		10R38	662
691.250		10R26	662
768.012		10R26	662
862.544		10R20	662
984.795		10P46	662
986.070		10R20	662

Table 3.4.25
Dideuteromethanol – CHD$_2$OH

Wavelength (µm)	Uncertainty (µm)	Comments	References
165.0	±0.81	5285–5287	1478
168.0	±0.82	5282–5284	1478
179.0	±0.88	5279–5281	1478
238.0	±1.2	5276–5278	1478
260.0	±1.3	5273–5275	1478
346.0	±1.7	5270–5272	1478

Dideuteromethanol – CHD$_2$OH—*continued*

Wavelength (μm)	Uncertainty (μm)	Comments	References
355.0	±1.7	5267–5269	1478
363.0	±1.8	5264–5266	1478
426.	±2.1	5261–5263	1478
483.	±2.4	5258–5260	1478
518.	±2.5	5255–5257	1478

Table 3.4.26
Cis 1,2-Difluoroethene (Cis 1,2-Difluoroethylene) – CHFCHF

Wavelength (μm)	Uncertainty (μm)	CO$_2$ laser pump line	References
161.8		10P10	665
185.0		9R08	1362
190.0		9R04	1362
196.3		10R04	665
198.0		10R04	1362
213.3		9P36	665
219.3		10P08	665
219.5		10R18	665
220.7		10P26	665
228.1		9P20	665
231.1		10R38	665
232.8		10R26	665
241.5		10P20	665
242.6		10R14	1801
260.1		10R20	1801
262.0		10R24	1362
272.1		10P14	665
284.6		10R18	665
286.5		9P28	1801
289.8		10R24	665
307.5		10R30	1801
310.0		10R34	1362
310.8		9P34	1801
326.6		10P06	665
339.0		10P26	665
360.0		10R38	665
360.5		9P30	665
376.7		10R16	665
386.5		10R40	665
389.3		10R20	665

Cis 1,2-Difluoroethene – CHFCHF—*continued*

Wavelength (μm)	Uncertainty (μm)	CO_2 laser pump line	References
411.1		9P16	665
422.5		9P28	665
424.0		10R14	665
433.8		9P16	665
437.6		10R06	665
442.1		10R30	665
543.2		9P34	665
546.8		9P20	665
549.5		10P06	665
557.0		9P40	665
583.7		9P42	665
687.2		9P38	665
705.0		10R16	665

Table 3.4.27
Formyl Fluoride (Fluoroformaldehyde) – CHFO

Wavelength (μm)	Uncertainty (μm)	CO_2 laser pump line	References
120.		9P24	1772
128.		9R22	1771
196.		9R06	1771,1772
220.		9R36	1771
258.		$C^{18}O_2$ laser	1771
260.		9P16	1771
280.		$C^{18}O_2$ laser	1771
282.		$C^{18}O_2$ laser	1771
306.		$C^{18}O_2$ laser	1771
432.		9R32	1771
654.		9P18	1771

Table 3.4.28
Difluoromethane – CHF_2

Wavelength (μm)	Uncertainty (μm)	Comments	References
185.	±1.9	5956–5959	1362,1389
190.	±1.9	5952–5955	1362,1389
198.	±2.0	5948–5951	1362,1389
228.	±2.3	5944–5947	1362,1389
231.0	±0.51	5940–5943	1389

Difluoromethane – CHF$_2$—*continued*

Wavelength (µm)	Uncertainty (µm)	Comments	References
242.6	±0.49	5936–5939	1389
260.1	±0.47	5932–5935	1362,1389
262.	±2.6	5928–5931	1362,1389
286.5	±0.46	5924–5927	1389
307.5	±0.49	5920–5923	1362,1389
310.	±3.1	5916–5919	1362,1389
310.8	±0.50	5912–5915	1389

Table 3.4.29
1,1-Difluoroethene (1,1-Difluoroethylene) – CH$_2$CF$_2$

Wavelength (µm)	Uncertainty (µm)	Comments	References
288.5	±0.58	6011–6013	1387,1388,1396,1465
325.0		7100	1839
354.0		7101	1839
375.5449	±0.00045	6007–6010	1387,1388,1396, 1465,1480,1483
407.2937	±0.00053	6003–6006	1387,1388,1420, 1465,1480,1483
458.0	±0.92	6000–6002	1387,1388,1396,1465
464.3	±0.93	5997–5999	1387,1388,1396,1465
520.	±10.	5993–5996	1387,1388,1459,1465
554.365	±0.0021	5989–5992	1387,1388,1396, 1465,1480,1483
570.	±17.	5985–5988	1387,1388,1420,1465
662.816	±0.0015	5981–5984	1387,1388,1396, 1465,1480,1483
764.	±3.7	5977–5980	1367,1387,1388,1465
770.	±15.	5973–5976	1387,1388,1465, 1480,1483
880.	±26.	5969–5972	1387,1388,1396,1465
890.	±1.8	5963–5965	1387,1388,1396,1465
890.	±1.8	596665968	1396,1387,1388,1465
990.	±2.0	5960–5962	1396,1387,1388,1465
1020.	±31.	5908–5911	1387,1388,1420,1465

Table 3.4.30
Bromoethene (Vinyl Bromide) – CH$_2$CHBr

Wavelength (μm)	Uncertainty (μm)	Comments	References
283.0	±0.57	6190–6193	1404
356.0	±0.71	6186,–6189	1404
370.0	±0.74	6182–6185	1404
396.0	±0.79	6178–6181	1404
411.0	±0.82	6174,–6177	1404
416.0	±0.83	6170,–6173	1404
419.0	±0.84	6166–6169	1404
424.0	±0.85	6162–6165	1404
427.0	±0.85	6158–6161	1404
438.5069	±0.00044	6154–6157	1404
443.5	±0.89	6150–6153	1404
445.0	±0.89	6146–6149	1404
448.0		7102	1839
482.9615	±0.00048	6142–6145	1404
490.0829	±0.00049	6138–6141	1404
506.	±1.0	6134–6137	1404
528.4965	±0.00053	6130–6133	1404
553.6962	±0.00055	6126–6129	1404
594.7286	±0.00059	6122–6125	1404
617.0		7103	1839
618.4462	±0.00062	6118–6121	1404
624.0958	±0.00062	6114–6117	1404
635.3548	±0.00064	6110–6113	1404
646.	±1.3	6106–6109	1404
649.4255	±0.00065	6102–6105	1404
680.5414	±0.00068	6098–6101	1404
693.1396	±0.00069	6094–6097	1404
707.2210	±0.00071	6090–6093	1404
712.	±1.4	6086–6089	1404
724.1399	±0.00072	6082–6085	1404
724.1399	±0.00072	6082–6085	1404
741.1149	±0.00074	6078–6081	1404
780.1330	±0.00078	6074–6077	1404
784.2681	±0.00078	6070–6073	1404
826.9443	±0.00083	6066–6069	1404
853.4380	±0.00085	6062–6065	1404
900.1338	±0.00090	6058–6061	1404
934.2230	±0.00093	6054–6057	1404
936.1590	±0.00094	6050–6053	1404
963.4873	±0.00096	6046–6049	1404
985.8588	±0.00099	6042–6045	1404
989.1904	±0.00099	6038–6041	1404

Bromoethene (Vinyl Bromide) – CH$_2$CHBr—*continued*

Wavelength (μm)	Uncertainty (μm)	Comments	References
990.6303	±0.00099	6034–6037	1404
1247.594	±0.0012	6030–6033	1404
1383.882	±0.0014	6026–6029	1404
1394.063	±0.0014	6022–6025	1404
1614.888	±0.0016	6018–6021	1404
1899.889	±0.0019	6014–6017	1404

Table 3.4.31
Chloroethene (Vinyl Chloride) – CH$_2$CHCl

Wavelength (μm)	Uncertainty (μm)	Comments	References
218.5		7104	1839
285.0		7104	1839
352.2		7105	1839
385.9092	±0.00046	5905–5907	1375,1480,1483
403.2		7106	1839
417.5		7107	1839
424.	±8.5	5902–5904	1459
442.1678	±0.00066	5899–5901	1480,1483
445.	±8.9	5896–5898	1459
477.3		7108	1839
487.	±9.7	5893–5895	1459
507.5840	±0.00081	5890–5892	1375,1480,1483
520.	±10.	5887–5889	1459
530.	±11.	5884–5886	1459
540.	±11.	5881–5883	1459
567.946	±0.0010	5878–5880	1459,1480,1483
601.897	±0.0012	5875–5877	1459,1480,1483
634.471	±0.0013	5872–5874	1375,1480,1483
640.	±13.	5869–5871	1459
669.9		7109	1839
700.	±14.	5866–5868	1459
710.	±14.	5863–5865	1459
830.	±17.	5860–5862	1459
940.	±19.	5857–5859	1459
987.8		7110	1839
988.5		7110	1839
1000.	±20.	5854–5856	1459
1040.	±21.	5851–5853	1459

Table 3.4.32
Acrylonitrile, Vinyl Cyanide) – CH$_2$CHCN

Wavelength (μm)	Uncertainty (μm)	Comments	References
270.6	±0.54	6413,6414	1396
361.0		7111	1839
415.3		7112	1839
489.	±9.8	6410–6412	1459
500.	±10.	6407–6409	1459
550.	±1.1	6405,6406	1396
574.	±1.1	6403,6404	1396
580.	±12.	6400–6402	1459
584.	±1.2	6398,6399	1396
587.	±1.2	6396,6397	1396
620.	±12.	6393–6395	1459
630.	±13.	6390–6392	1459
720.	±14.	6387,–6389	1459
740.	±15.	6384–6386	1459
780.	±16.	6381–6383	1459
790.	±16.	6378–6380	1459
830.	±17.	6375–6377	1459
910.	±18.	6372–6374	1459
940.	±19.	6369–6371	1459
1160.	±23.	6366–6368	1459
1180.	±24.	6363–6365	1459

Table 3.4.33
Fluoroethene (Vinyl Fluoride) – CH$_2$CHF

Wavelength (μm)	Uncertainty (μm)	CO$_2$ laser pump line	References
121.0		N$_2$O laser	1808
137.0		9R22	1839
142.6		10P54	1732
148.137		10R50	1846
148.2		10R50	1732
171.8		10P22	1653
194.2		10P54	1732
194.255		10P54	1846
201.9		sequence band	1732
203.		10R20	1658
222.3		10R50	1732
229.3		N$_2$O laser	1808
244.1		10P50	1732
263.5		10R12	1732

Fluoroethene (Vinyl Fluoride) – CH$_2$CHF—*continued*

Wavelength (μm)	Uncertainty (μm)	CO$_2$ laser pump line	References
269.0		10R02	1839
275.5		10P48	1732
275.839		10P48	1846
281.6		10R44	1732
285.0		10P32	1839
290.		10P36	1658
293.4		N$_2$O laser	1808
298.		10P08	1658
309.5		N$_2$O laser	1808
321.0		hot band	1732
322.8		10R26	1732
329.8		N$_2$O laser	1808
330.1		10R22	1732
332.363		10P06	1846
333.6		N$_2$O laser	1808
335.		10P06	1658
336.		10P38	1658
337.095		10P38	1846
344.8		N$_2$O laser	1808
345.5		10P38	1732
351.0		N$_2$O laser	1808
352.8		N$_2$O laser	1808
353.0		N$_2$O laser	1808
354.5		N$_2$O laser	1808
355.		10P20	1658
356.0		N$_2$O laser	1808
356.6		N$_2$O laser	1808
360.9		N$_2$O laser	1808
361.8		N$_2$O laser	1808
362.2		10P24	1732
362.8		10R28	1732
372.		10R20	1658
376.824		10P14	1846
377.4		10P14	1653
407.6		N$_2$O laser	1808
412.2		sequence band	1732
420.		10P40	1658
421.8		N$_2$O laser	1808
423.0		10R46	1732
429.9		N$_2$O laser	1808
430.		10P18	1658
433.		C^{18}O$_2$ laser	1729
433.5		N$_2$O laser	1808

Fluoroethene (Vinyl Fluoride) – CH$_2$CHF—*continued*

Wavelength (μm)	Uncertainty (μm)	CO$_2$ laser pump line	References
441.7		sequence band	1732
444.4		10R20	1653
444.5		10P10	1839
444.797		10R20	1846
445.2		N$_2$O laser	1808
445.831		10R50	1846
446.7		10R50	1732
446.7		N$_2$O laser	1808
447.3		10P14	1839
454.3		hot band	1732
456.		^{13}CO$_2$ laser	1729
458.		N$_2$O laser	1729
458.0		hot band	1732
459.8		10R20	1732
461.0		^{13}C^{18}O$_2$ laser	1732
461.0		hot band	1732
467.2		N$_2$O laser	1808
472.4		10P20	1653
476.0		N$_2$O laser	1808
477.		10P36	1658
483.8		10P10	1732
487.5		N$_2$O laser	1808
487.7		hot band	1732
487.8		10P20	1732
489.3		N$_2$O laser	1808
490.		10P22	1658
490.0		N$_2$O laser	1732
493.5		N$_2$O laser	1808
505.0		10P22	1732
506.3		hot band	1732
508.		10P38	1658
518.4		sequence band	1732
518.6		10P52	1732
519.6		N$_2$O laser	1808
538.6		10P56	1732
540.		10P32	1658
551.5		10P56	1732
557.		10P36	1658
563.		10R36	1658
565.		^{13}CO$_2$ laser	1729
573.		^{13}CO$_2$ laser	1729
579.		^{13}CO$_2$ laser	1729
582.5		10R32	1732

Fluoroethene (Vinyl Fluoride) – CH$_2$CHF—*continued*

Wavelength (μm)	Uncertainty (μm)	CO$_2$ laser pump line	References
583.1		N$_2$O laser	1808
585.8		sequence band	1732
586.		^{13}CO$_2$ laser	1729
605.0		10R02	1732
606.6		N$_2$O laser	1808
606.8		10P42	1732
617.0		hot band	1732
618.0		10P32	1732
655.0		N$_2$O laser	1808
660.0		10P24	1653
660.2		N$_2$O laser	1808
671.0		hot band	1732
671.128		10P36	1846
672.1		10P36	1653
700.3		10P36	1732
720.8		10P04	1732
758.7		10P50	1732
764.2		hot band	1732
774.9		10P22	1653
777.3		10P22	1839
777.5		10P22	1839
783.		10P32	1658
796.3		N$_2$O laser	1808
796.5		hot band	1732
838.2		10P10	1732
847.7		10P18	1732
875.		^{13}CO$_2$ laser	1729
894.0		hot band	1732
939.0		N$_2$O laser	1808
949.0		N$_2$O laser	1808
952.2		10P54	1732
961.		N$_2$O laser	1729
971.		^{13}CO$_2$ laser	1729
981.1		N$_2$O laser	1808
1011.0		10P38	1732
1063.0		10P32	1732
1082.6		10P48	1732
1083.1		10P32	1732
1094.0		10P24	1732
1165.7		10P46	1732
1167.6		10P28	1732
1170.7		10P32	1732
1170.7		N$_2$O laser	1808

Fluoroethene (Vinyl Fluoride) – CH₂CHF—*continued*

Wavelength (μm)	Uncertainty (μm)	CO_2 laser pump line	References
1234.3		N₂O laser	1808
1255.7		10P34	1732
1264.3		N₂O laser	1808
1523.0		hot band	1732
1612.		¹³CO₂ laser	1729
1733.		¹³CO₂ laser	1729
2453.0		sequence band	1732
2525.0		10P42	1732

Table 3.4.34
Dichloromethane – CH₂Cl₂

Wavelength (μm)	Uncertainty (μm)	CO_2 laser pump line	References
195.0		10P12	1736
208.3		10P18	1736
231.0		19P22	1736
235.5		10P24	1736
254.7		10P26	1736
294.6		10P26	1736
298.5		19P22	1736

Table 3.4.35
Chlorofluoromethane – CH₂ClF

Wavelength (μm)	Uncertainty (μm)	CO_2 laser pump line	References
176.		9P12	1669
218.		9P22	1669
221.		9R22	1669
244.		9P26	1669
246.		9R26	1669
284.		9R16	1669
292.		9R20	1669
296.		9R26	1669
308.		9R22	1669
324.		9R04	1669
344.		9P32	1669
349.		9R10	1669
1014.		9P08	1669

Table 3.4.36
Deuteromethanol – CH$_2$DOH

Wavelength (μm)	Uncertainty (μm)	Comments	References
87.10	±0.096	5111–5113	1467
87.90	±0.097	5108–5110	1467
90.40	±0.099	5105–5107	1467
100.00	±0.10	5102–5104	1467
102.02349	±5.1e-05	5099–5101	1467
108.81775	±5.4e-05	5096–5098	1467,1478
108.94124	±5.4e-05	5093–5095	1467
112.53224	±5.6e-05	5090–5092	1467
117.08507	±5.9e-05	5087–5089	1467
124.43170	±6.2e-05	5084–5086	1467,1478
135.17175	±6.8e-05	5081–5083	1467
135.17256	±6.8e-05	5078–5080	1467
135.83350	±6.8e-05	5075–5077	1467
140.30	±0.098	5072–5074	1467
147.6		6971,7047	1842
149.38792	±7.5e-05	5069–5071	1467
149.61284	±7.5e-05	5066–5068	1467
150.57167	±7.5e-05	5063–5065	1467
150.81629	±7.5e-05	5060–5062	1467,1478
152.7	±0.11	5057–5059	1467
159.21794	±8.0e-05	5054–5056	1467
162.70	±0.098	5051–5053	1467
164.74645	±8.2e-05	5048–5050	1467
167.35235	±8.4e-05	5045–5047	1467
167.54117	±8.4e-05	5042–5044	1467,1478
171.8	±0.10	5039–5041	1467
172.84620	±8.6e-05	5036–5038	1467,1478
182.10	±0.091	5033–5035	1467
183.62132	±9.2e-05	5030–5032	1467
188.41111	±9.4e-05	5027–5029	1467
189.30	±0.095	5024–5026	1467
195.49558	±9.8e-05	5021–5023	1467
196.10	±0.098	5018–5020	1467
200.0	±0.48	5015–5017	1467
206.6874	±0.00010	5012–5014	1467,1478
207.5		6970,7050	1842
212.5	±0.11	5010,5011	1467
216.8	±0.11	5007–5009	1467
218.0	±0.11	5004–5006	1467
219.0960	±0.00011	5001–5003	1467
224.2256	±0.00011	4998–5000	1467
226.2974	±0.00011	4995–4997	1467

Deuteromethanol – CH₂DOH—*continued*

$$\text{Deuteromethanol} - CH_2DOH$$

Wavelength (μm)	Uncertainty (μm)	Comments	References
245.1		6971,7052	1842
248.1220	±0.00012	4992–4994	1467
249.7204	±0.00012	4989,–4991	1467,1478
266.7352	±0.00013	4986–4988	1467
272.2516	±0.00014	4983–4985	1467,1478
273.0037	±0.00014	4980–4982	1467
275.6		6971,7051	1842
295.3967	±0.00015	4977–4979	1467,1478
295.6394	±0.00015	4974–4976	1467,1478
300.0	±0.48	4971–4973	1467
308.0405	±0.00015	4968–4970	1467,1478
308.2957	±0.00015	4965–4967	1467
322.4522	±0.00016	4962–4964	1467,1478
336.2461	±0.00017	4959–4961	1467
340.3566	±0.00017	4956–4958	1467
374.0861	±0.00019	4953–4955	1467,1478
379.1		6971,7046	1842
387.5591	±0.00019	4950–4952	1467
396.00	±0.099	4947–4949	1467,1478
413.5		6971,7049	1842
422.1512	±0.00021	4944–4946	1467
427.2	±0.13	4941–4943	1467
451.4754	±0.00023	4938–4940	1467
452.40	±0.10	4935–4937	1467
468.2359	±0.00023	4932–4934	1467,1478
509.3717	±0.00025	4929,–4931	1467
523.0914	±0.00026	4926–4928	1467
563.4		6971,7048	1842
616.3351	±0.00031	4923–4925	1467,1478
682.6	±0.10	4920–4922	1467
762.50	±0.092	4917–4919	1467

Table 3.4.37
Difluoromethane – CH₂F₂

$$\text{Difluoromethane} - CH_2F_2$$

Wavelength (μm)	Uncertainty (μm)	Comments	References
95.551057	±4.8e-05	3790–3793	1392,1394,1454,1466
97.6		6971,7040	1845
105.1		6971,7038	1845
105.51827	±5.3e-05	3786–3789	1392,1394,1454,1466

Difluoromethane – CH_2F_2—*continued*

Wavelength (μm)	Uncertainty (μm)	Comments	References
109.29579	±5.5e-05	3782–3785	1386,1392,1394, 1454,1466
117.72748	±5.9e-05	3778–3781	1386,1392,1394, 1454,1466
122.46551	±6.1e-05	3774–3777	1392,1394,1454,1466
122.46581	±6.1e-05	3770–3773	1392,1394,1454,1466
133.99765	±6.7e-05	3766–3769	1392,1394,1454,1466
135.26932	±6.8e-05	3762–3765	1386,1392,1394, 1454,1466
142.9		6971,7037	1845
158.51348	±7.9e-05	3758–,3761	1386,1392,1394, 1454,1466
158.96020	±8.0e-05	3754–3757	1392,1394,1454,1466
166.63105	±8.3e-05	3750–3753	1392,1394,1454,1466
166.67665	±8.3e-05	3746–3749	1392,1394,1454,1466
182.112		6970,7059	1835
184.30590	±9.2e-05	3742–3745	1386,1392,1394, 1454,1466
185.0		6970,7043	1845
191.84803	±9.6e-05	3738–3741	1392,1394,1454,1466
193.90445	±0.00010	3734–3737	1392,1394,1454,1466
194.418		6971,7040	1845
194.44761	±0.00010	3730–3733	1392,1394,1454,1466
194.484		6971,7044	1845
201.936		6971,7042	1845
202.4649	±0.00010	3726–3729	1392,1394,1454,1466
210.130		6971,7040	1845
214.5791	±0.00011	3722–3725	1386,1392,1394, 1454,1466
225.9		6971,7045	1845
227.6570	±0.00011	3718–3721	1392,1394,1454,1466
230.1059	±0.00012	3714–3717	1392,1394,1454,1466
235.6541	±0.00012	3710–3713	1386,1392,1394, 1454,1466
236.5915	±0.00012	3706–3709	1392,1394,1454,1466
236.6008	±0.00012	3702–3705	1392,1394,1454,1466
256.0270	±0.00013	3698–3701	1386,1392,1394, 1454,1466
261.7292	±0.00013	3694–3697	1386,1392,1394, 1454,1466
270.0055	±0.00014	3690–3693	1392,1394,1454,1466
272.3389	±0.00014	3686–3689	1386,1392,1394, 1454,1466

Difluoromethane – CH₂F₂—*continued*

Wavelength (μm)	Uncertainty (μm)	Comments	References
279.9		6971,7041	1845
287.6672	±0.00014	3682–3685	1386,1392,1394, 1454,1466
289.4999	±0.00014	3678–3681	1392,1394,1454,1466
293.9015	±0.00015	3674–3677	1392,1394,1454,1466
298.2910	±0.00015	3670–3673	1392,1394,1454,1466
326.4230	±0.00016	3666–3669	1392,1394,1454,1466
346.8		6970,7044	1845
355.1261	±0.00018	3662–3665	1392,1394,1454,1466
381.9956	±0.00019	3658–3661	1392,1394,1454,1466
382.6392	±0.00019	3654–3657	1392,1394,1454,1466
394.7009	±0.00020	3650–3653	1392,1394,1454,1466
418.2703	±0.00021	3646–3649	1392,1394,1454,1466
434.9514	±0.00022	3642–3645	1392,1394,1454,1466
464.4123	±0.00023	3638–3641	1392,1394,1454,1466
503.0567	±0.00025	3634–3637	1392,1394,1454,1466
511.4451	±0.00026	3630–3633	1392,1394,1454,1466
540.9864	±0.00027	3626–3629	1392,1394,1454,1466
567.4		6971,7036	1845
567.5316	±0.00028	3622–3625	1392,1394,1454,1466
588.0276	±0.00029	3618–3621	1392,1394,1454,1466
616.179		6971,7039	1845
642.5999	±0.00032	3614–3617	1392,1394,1454,1466
657.2391	±0.00033	3610–3613	1392,1394,1454,1466
724.9203	±0.00036	3606–3609	1392,1394,1454,1466
724.9203	±0.00036	3606–3609	1392,1394,1454,1466
1448.0958	±0.00072	3602–3605	1392,1394,1454,1466

Table 3.4.38
Difluoromethane – ¹³CH₂F₂

Wavelength (μm)	Uncertainty (μm)	CO_2 laser pump line	References
106.400		9R14	1666
112.000		9P44	1666
135.523		9R44	1666
138.281		9R22	1666
140.405		9R38	1666
164.656		9P12	1666
164.815		9R04	1666
180.600		9P34	1666
182.381		10R20	1666

Difluoromethane – $^{13}CH_2F_2$—*continued*

Wavelength (μm)	Uncertainty (μm)	CO_2 laser pump line	References
183.289		9P28	1666
186.043		9P38	1666
193.497		9P22	1666
195.158		9R20	1666
195.931		9HP12	1845
197.388		9P20	1666
198.548		9HP14	1845
200.295		9P16	1666
201.739		9R02	1845
206.043		10R18	1666
211.116		9R10	1845
213.351		9R44	1666
214.597		9R34	1665
221.284		9HP11	1845
223.973		9R34	1845
234.800		9R18	1666
245.652		9P28	1666
248.606		9R26	1666
254.802		9P04	1666
260.042		10R20	1666
260.248		9R50	1845
266.866		9R24	1666
271.212		9HP12	1845
273.764		9P26	1666
279.014		9P22	1666
299.5		9HP11	1845
300.233		9R36	1666
300.246		10R38	1666
301.654		10R04	1666
306.993		9R40	1666
311.213		9R08	1666
312.276		9P24	1666
316.329		9R24	1666
318.080		9R12	1666
333.926		9P36	1666
344.521		9R16	1666
355.126		9P08	1845
357.867		9R20	1666
359.362		9R22	1666
360.504		9P16	1666
377.718		9R10	1666
391.461		9R30	1666
399.288		9R34	1666

Difluoromethane – $^{13}CH_2F_2$—*continued*

Wavelength (μm)	Uncertainty (μm)	CO$_2$ laser pump line	References
403.777		9P32	1666
415.363		9P04	1666
421.053		9P14	1666
438.022		10R18	1666
452.425		9P16	1666
479.123		9R40	1666
496.660		9P14	1666
531.363		9P08	1666
551.100		9R06	1666
570.332		9R20	1666
618.896		9R08	1666
638.394		9P26	1666
674.061		9P04	1666
689.178		9R22	1666
734.959		9P36	1666
738.414		9R12	1666
740.000		9P28	1666
894.021		9R16	1666
897.521		9P14	1666
902.789		9R30	1666
935.604		9R10	1666
936.000		9R04	1666
1082.000		9R10	1666

Table 3.4.39
Formaldoxime – CH_2NOH

Wavelength (μm)	Uncertainty (μm)	CO$_2$ laser pump line	References
264.9		$^{13}CO_2$ laser	1740
278.3		$^{13}CO_2$ laser	1740
291.3		$^{13}CO_2$ laser	1740
301.2		10P10	1740
653.7		$^{13}CO_2$ laser	1740

Table 3.4.40
Bromomethane (Methyl Bromide) – CH₃Br

Wavelength (μm)	Uncertainty (μm)	Comments	References
245.0	±0.098	3937–3939	1378
279.8	±0.098	3934–3936	1378
294.3	±0.10	3931–3933	1378
311.07	±0.10	3928–3930	1378
311.10	±0.10	3925–3927	1378
311.21	±0.10	3922–3924	1378
332.86	±0.10	3919–3921	1378
333.15	±0.10	3916–3918	1378
352.75	±0.099	3913–3915	1378
368.9		6971,7193	1826
369.1		6970,7194	1826
380.02	±0.095	3910–3912	1378
407.7	±0.10	3907–3909	1378
417.9		6971,7192	1826
418.3	±0.10	3904–3906	1378
422.8	±0.10	3901–3903	1378
464.9		6971,7196	1826
466.2		6971,7195	1826
508.5	±0.10	3898–3900	1378
531.1	±0.10	3895–3897	1378
545.39	±0.093	3892–3894	1378
631.9	±0.10	3889–3891	1378
632.0	±0.10	3886–3888	1378
660.70	±0.099	3883–3885	1378
990.51	±0.099	3880–3882	1378

Table 3.4.41
Bromomethane (Methyl Bromide) – CH₃⁷⁹Br

Wavelength (μm)	Uncertainty (μm)	Comments	References
264.1	±0.10	3850–3853	1378,1450
414.98	±0.10	3846–3849	1378,1450
715.4	±0.10	3842–3845	1378,1450
749.29	±0.090	3838–3841	1378,1450
749.36	±0.090	3834–3837	1378,1450
925.5	±0.10	3830–3833	1378,1450
1310.4	±0.10	3826–3829	1378,1450
1572.64	±0.093	3822–3825	1378,1450
1965.34	±0.096	3818–3821	1378,1450

Table 3.4.42
Bromomethane (Methyl Bromide) – CH$_3$^{81}Br

Wavelength (μm)	Uncertainty (μm)	Comments	References
311.20	±0.10	3876–3879	1378,1450
545.21	±0.093	3872–3875	1378,1450
564.7	±0.096	3868–3871	1378,1450
585.72	±0.10	3864–3867	1378,1450
658.53	±0.099	3860–3863	1378,1450
831.13	±0.10	3856–3859	1378,1450
2650.	±13.	3854,3855	1450

Table 3.4.43
Bromomethane (Methyl Bromide) – ^{13}CH$_3$Br

Wavelength (μm)	Uncertainty (μm)	CO$_2$ laser pump line	References
660.882		10R20	1665

Table 3.4.44
Methyl Acetylene – CH$_3$CCH

Wavelength (μm)	Uncertainty (μm)	Comments	References
428.87	±0.099	6359–6362	1377,1378
516.77	±0.098	6355–6358	1377,1378
531.1	±0.10	6351–6354	1377,1378
566.4	±0.096	6347–6350	1377,1378
583.77	±0.099	6343–6346	1377,1378
647.89	±0.049	6340–6342	1377
675.3	±0.10	6336–6339	1377,1378
1097.11	±0.098	6332–6335	1377,1378
1174.87	±0.049	6328–6331	1377

Table 3.4.45
1,1-Dideuteroethanol – CH_3CD_2OH

Wavelength (μm)	Uncertainty (μm)	CO_2 laser pump line	References
491.8		9P40	1742
581.6		10R24	1742
1010.		10R20	1742

Table 3.4.46
1,1,1-Trifluoroethane (Methyl Fluorofrom) – CH_3CF_3

Wavelength (μm)	Uncertainty (μm)	CO_2 laser pump line	References
369.1		10R14	1684
383.2		10P14	1684
388.9		N_2O laser	1684
393.3		10P20	1684
393.3		10R14	1684
410.1		N_2O laser	1684
411.2		N_2O laser	1684
422.0		10R14	1684
454.6		N_2O laser	1684
454.8		10R52	1684
463.0		10R12	1684
471.2		N_2O laser	1684
477.1		10R50	1684
485.4		10P08	1684
485.6		10R38	1684
485.8		10P12	1684
486.1		10R32	1684
501.6		10R48	1684
510.4		N_2O laser	1684
510.7		10P18	1684
518.8		10R04	1684
519.2		N_2O laser	1684
558.8		N_2O laser	1684
580.6		10R46	1684
580.8		10R30	1684
606.8		10R26	1684
629.3		10R28	1684
633.4		N_2O laser	1684
634.0		10R32	1684
634.7		10P10	1684

1,1,1-Trifluoroethane – CH₃CF₃—*continued*

Wavelength (μm)	Uncertainty (μm)	CO₂ laser pump line	References
676.7		10P02	1684
709.2		N₂O laser	1684
709.5		10R36	1684
709.8		10R40	1684
766.6		10P06	1684
767.8		10R20	1684
782.7		sequence band	1684
783.2		N₂O laser	1684
784.5		10R30	1684
805.8		10P06	1684
833.3		10R18	1684
851.0		10R16	1684
852.5		10P10	1684
853.3		10R20	1684
878.1		10R08	1684
878.5		N₂O laser	1684
967.9		10R32	1684
968.9		sequence band	1684
969.0		N₂O laser	1684
973.0		N₂O laser	1684
1035.6		N₂O laser	1684
1264.3		N₂O laser	1684
1324.3		N₂O laser	1684
1613.0		10R14	1684

Table 3.4.47
1-Deuteroethanol – CH₃CHDOH

Wavelength (μm)	Uncertainty (μm)	CO₂ laser pump line	References
351.5		9P46	1742
379.5		9P32	1742
432.3		9P24	1742
889.0		9R12	1742

Table 3.4.48
1,1-Difluoroethane – CH₃CHF₂

Wavelength (μm)	Uncertainty (μm)	Comments	References
460.	±14.	6477,6478,6479	1420
530.	±16.	6474,6475,6476	1420

Table 3.4.49
Bromoethane (Ethyl Bromide) – CH₃CH₂Br

Wavelength (μm)	Uncertainty (μm)	CO₂ laser pump line	References
327.6		10R10	1685
453.6		10R34	1685
456.2		10R34	1834
520.7		10R30	1834
526.3		10R30	1834
527.9		10R30	1685
534.1		10R30	1834
555.7		10P20	1834
582.2		10P18	1834
641.1		10P16	1834
705.2		10P14	1834
707.8		10P14	1685
768.9		10R22	1834
769.8		10R22	1685
836.5		10R22	1834
837.4		10R20	1834
838.3		10R20	1685
896.5		10P10	1685
1035.3		10P08	1834
1056.7		10P08	1834
1059.0		10P08	1685

Table 3.4.50
Chloroethane (Ethyl Chloride) – CH_3CH_2Cl

Wavelength (μm)	Uncertainty (μm)	CO_2 laser pump line	References
299.2061		10R34	1829
299.5968		10R28	1829
423.6312		10P16	1829
447		10R28	1745
480.0346		10P14	1829
584.6650		10P08	1829
615.0399		10P06	1829
620.4012		10P06	1829
698		10R38	1745
720.3745		10R38	1829
845.3513		10R34	1829
963.7773		10R32	1829
1306		10R28	1745
1669		10R26	1745

Table 3.4.51
Fluoroethane (Ethyl Fluoride) – CH_3CH_2F

Wavelength (μm)	Uncertainty (μm)	Comments	References
264.7	±0.10	6460,6461,6462	1459,1474
336.7	±0.10	6457,6458,6459	1459,1474
404.	±0.97	6454,6455,6456	1459,1474
405.5044	±0.00053	6451,6452,6453	1459,1480,1483
438.5		7113	1839
452.	±9.0	6448,6449,6450	1480,1483
486.	±9.7	6445,6446,6447	1459
502.2623	±0.00080	6442,6443,6444	1459,1480,1483
503.2		7114	1839
519.075	±0.0013	6439,6440,6441	1459,1480,1483
551.0		7114	1839
593.506	±0.0012	6436,6437,6438	1459,1480,1483
620.40	±0.099	6433,6434,6435	1459,1474
660.	±13.	6430,6431,6432	1480,1483
851.9	±0.10	6427,6428,6429	1459,1474
1005.230	±0.0016	6424,6425,6426	1459,1480,1483
1070.	±21.	6421,6422,6423	1459
1440.	±29.	6418,6419,6420	1480,1483
1521.376	±0.0037	6415,6416,6417	1459,1480,1483

Table 3.4.52
Iodoethane (Ethyl Iodide) – CH_3CH_2I

Wavelength (μm)	Uncertainty (μm)	CO_2 laser pump line	References
493.0		10P34	1685
504.0		10P32	1685
542.0		10P30	1685
626.8		10R06	1685
660.328		10P26	1660
1044.4		10P20	1685
1049.810		10R04	1660

Table 3.4.53
Ethanol (Ethyl Alcohol) – CH_3CH_2OH

Wavelength (μm)	Uncertainty (μm)	Comments	References
285.3		6942	1742
311.9		6943	1742
388.060		6944	1802
396.	±4.0	6480–6482	1430
449.0		6945	1742
529.3		6946	1742
552.0		6943	1742
566.1		6930	1742
575.3		6947	1742
620.3		6948	1742

Table 3.4.54
Ethanal (Acetaldehyde) – CH_3CHO

Wavelength (μm)	Uncertainty (μm)	CO_2 laser pump line	References
176.		9R40	1679
319.154		9R38	1846
328.		9R22	1679
328.784		9R22	1846
335.282		9R30	1846
343.		9R30	1679
385.		9R30	1679
385.102		9R30	1846
415.		9R40	1679
415.356		9R40	1846

Ethanal (Acetaldehyde) – CH₃CHO—*continued*

Wavelength (μm)	Uncertainty (μm)	CO₂ laser pump line	References
470.967		9R50	1846
509.		9R36	1679
509.706		9R36	1846
528.277		9R50	1846

Table 3.4.55
Chloromethane (Methyl Chloride) – CH₃Cl

Wavelength (μm)	Uncertainty (μm)	Comments	References
227.15	±0.10	3518–3521	1378,1385
236.25	±0.097	3514–3517	1378,1385
240.98	±0.096	3510–3513	1378,1385
247.7		6971,7188	1826
253.2		6971,7186	1826
261.03	±0.099	3506–3509	1378,1385
264.6		6971,7184	1826
270.9		6971,7182	1826
271.3	±0.10	3502–3505	1378,1385
275.00	±0.096	3498–3501	1378,1385
275.09	±0.096	3494–3497	1378,1385
277.6		6971,7181	1826
281.7	±0.099	3490–3493	1378,1385
284.2		6971,7179	1826
284.3		6971,7136	1826
284.3		6971,7145	1826
284.3		6971,7190	1826
284.5		6971,7172	1826
286.8	±0.10	3486–3489	1378,1385
291.5		6971,7177	1826
299.3		6971,7159	1826
299.3		6971,7170	1826
299.5		6971,7176	1826
307.0		6971,7174	1826
307.1		6971,7189	1826
307.3		6971,7137	1826
307.3		6971,7158	1826
307.3		6971,7168	1826
307.65	±0.098	3482–3485	1378,1385
315.4		6971,7175	1826
315.5		6971,7156	1826
315.6		6971,7138	1826

Chloromethane (Methyl Chloride) – CH₃Cl—*continued*

Wavelength (µm)	Uncertainty (µm)	Comments	References
315.6		6971,7187	1826
315.8		6971,7166	1826
315.9		6971,7173	1826
324.9		6971,7165	1826
333.9		6971,7185	1826
334.0	±0.10	3478–3481	1378,1385
334.1		6971,7139	1826
334.1		6971,7164	1826
343.4		6971,7183	1826
344.4		6971,7140	1826
344.4		6971,7162	1826
349.3	±0.10	3474–3477	1378,1385,1430
354.7		6971,7146	1826
366.0		6971,7153	1826
366.2		6971,7147	1826
366.3		6971,7161	1826
366.3		6971,7180	1826
366.4		6971,7141	1826
378.2		6971,7155	1826
378.2		6971,7178	1826
378.4		6971,7171	1826
378.5		6971,7148	1826
378.5		6971,7160	1826
378.57	±0.095	3470–3473	1378,1385
378.6		6971,7142	1826
405.2		6971,7149	1826
405.2		6971,7169	1826
405.4		6971,7157	1826
405.5		6971,7143	1826
420.0		6971,7150	1826
420.2		6971,7167	1826
420.3		6971,7144	1826
453.5		6971,7151	1826
461.2	±0.10	3466–3469	1378,1385
472.4		6971,7152	1826
472.6		6971,7163	1826
476.0		6971,7191	1826
499.3		6971,7154	1826
511.9	±0.10	3462–3465	1378,1385
568.8	±0.097	3458–3461	1378,1385
870.80	±0.096	3454–3457	1378,1385
944.0	±0.10	3450–3453	1378,1385,1430
958.25	±0.096	3446–3449	1378,1385
1886.87	±0.092	3442–3445	1378,1385

Table 3.4.56
Ethanenitrile (Methyl Cyanide) – CH$_3$CN[a]

Wavelength (µm)	Uncertainty (µm)	Comments	References
281.18	±0.098	5778–5781	1364,1365,1377,1378
281.98	±0.099	5774–5777	1364,1365,1377,1378
286.9	±0.10	5770–5773	1364,1365,1377,1378
346.3	±0.10	5766–5769	1364,1365,1377,1378
372.87	±0.045	5762–5765	1364,1365,1377, 1459,1469
386.4	±0.097	5758–5761	1364,1365,1377,1378
387.3	±0.097	5754–5757	1364,1365,1377, 1378,1459
388.4	±0.097	5750–5753	1364,1365,1377,1378
427.04	±0.098	5746–5749	1364,1365,1377,1378
430.55	±0.052	5742–5745	1364,1365,1377
441.1	±0.10	5738–5741	1364,1365,1377,1378
453.3974	±0.00068	5734–5737	1364,1365,1377,1378 1459,1480,1483
466.25	±0.093	5730–5733	1364,1365,1377,1378
480.0	±0.096	5726–5729	1364,1365,1377,1378
494.6461	±0.00074	5722–5725	1364,1365,1377,1378 1459,1480,1483
510.2	±0.10	5718–5721	1364,1365,1377,1378
561.4	±0.095	5714–5717	1364,1365,1377,1378
652.68	±0.098	5710–5713	13641365,1377, 1378,1459
704.53	±0.099	5706–5709	1364,1365,1377,1378
713.72	±0.050	5702–5705	1364,1365,377
741.62	±0.089	5698–5701	1364,1365,1377,1378
750.	±15.	5694–5697	1364,1365,1377, 1480,1483
854.4	±0.10	5690–5693	1364,1365,1377,1378
1014.9	±0.10	5686–5689	1364,1365,1377,1378
1016.3	±0.10	5682–5685	1364,1365,1377,1378
1086.89	±0.097	5678–5681	1364,1365,1377,1378
1164.8	±0.10	5674–5677	1364,1365,1377,1378
1351.8	±0.11	5670–5673	1364,1365,1377,1378
1814.37	±0.049	5666–5669	1364,1365,1377

(a) See, also, reference 1848.

Table 3.4.57
Deuterooxyethanoic Acid – CH$_3$COOD

Wavelength (μm)	Uncertainty (μm)	CO$_2$ laser pump line	References
363.		10R28	1682
433.		10P20	1682
451.		9P32	1682
465.		10R20	1682
525.		10R12	1682
628.		9R06	1682
675.		10R22	1682
676.		10P18	1682
701.		10R18	1682
756.		10P18	1682

Table 3.4.58
Fluoromethane (Methyl Fluoride) – CH$_3$F

Wavelength (μm)	Uncertainty (μm)	Comments	References
192.78	±0.048	2625–2628	1363,1374,1377,1393
251.91	±0.050	2621–2624	1363,1374,1377,1393
372.68	±0.045	2617–2620	1363,1374,1377,1393
496.072	±0.0024	2614–2616	1363,1374,1375, 1377,1393
496.1009	±0.00040	2610–2613	1363,1374,1377, 1393,1434
992.	±4.9	2606–2609	1363,1374,1377,1393

Table 3.4.59
Fluoromethane (Methyl Fluoride) – $^{13}CH_3F$

Wavelength (μm)	Uncertainty (μm)	Comments	References
85.317287	±4.3e-05	4901–4904	1415,1417,1453,1481
85.79	±0.026	4898–4900	1415,1417
86.111788	±4.3e-05	4895–4897	1415,1417,1453,1481
87.90	±0.026	4892–4894	1415,1417
103.48079	±5.2e-05	4888–4891	1415,1417,1453,1481
103.58629	±5.2e-05	4885–4887	1415,1417,1453,1481
105.14719	±5.3e-05	4881–4884	1415,1417,1453,1481
110.43238	±5.5e-05	4877–4880	1415,1417,1453,1481
115.82318	±5.8e-05	4873–4876	1415,1417,1453,1481
118.01308	±5.9e-05	4869–4872	1415,1417,1453,1481
121.20	±0.036	4866–4868	1415,1417
123.26	±0.037	4863–4865	1415,1417
146.09738	±7.3e-05	4860–4862	1415,1417,1453,1481
147.97	±0.044	4857–4859	1415,1417
148.59041	±7.4e-05	4855,4856	1415,1417,1453,1481
149.27228	±7.5e-05	4851–4854	1415,1417,1453,1481
152.07569	±7.6e-05	4849,4850	1415,1417,1453,1481
157.92848	±7.9e-05	4845–4848	1415,1417,1453,1481
168.84	±0.051	4842–4844	1415,1417
171.75758	±8.6e-05	4838–4841	1415,1417,1453,1481
203.6358	±0.00010	4834–4837	1415,1417,1453,1481
208.4121	±0.00010	4831–4833	1415,1417,1453,1481
236.5303	±0.00012	4829,4830	1415,1417,1453,1481
237.5230	±0.00012	4826–4828	1415,1417,1453,1481
238.5227	±0.00012	4822–4825	1415,1417,1453,1481
268.5722	±0.00013	4818–4821	1415,1417,1453,1481
280.2183	±0.00014	4816,4817	1415,1417,1453,1481
280.2397	±0.00014	4814,4815	1415,1417,1453,1481
291.61	±0.087	4811–4813	1415,1417
307.78	±0.092	4807–4810	1415,1417,1453,1481
325.17	±0.098	4804–4806	1415,1417
332.6034	±0.00017	4802,4803	1415,1417,1453,1481
338.9638	±0.00017	4799–4801	1415,1417,1453,1481
358.9	±0.11	4796–4798	1415,1417
461.3847	±0.00023	4792–4795	1415,1417,1453,1481
629.8442	±0.00031	4789–4791	1415,1417,1453,1481
690.	±3.4	4224–4227	1388
806.	±3.9	4220–4223	1388
1221.79	±0.048	2629–2632	1377,1461

Table 3.4.60
Iodomethane (Methyl Iodide) – CH₃I

Wavelength (μm)	Uncertainty (μm)	Comments	References
377.45	±0.094	4103,4104,4105,4106	1378,1399,1409
390.5	±0.098	4099,4100,4101,4102	1378,1409
392.48	±0.098	4095,4096,4097,4098	1378,1409
447.1421	±0.00089	4091,4092,4093,4094	1378,1401,1409
451.5		6971,7057	1842
457.2	±0.10	4087,4088,4089,4090	1378,1409
459.2	±0.10	4083,4084,4085,4086	1378,1409
477.9	±0.096	4079,4080,4081,4082	1378,1409
508.4	±0.10	4075,4076,4077,4078	1378,1409
517.33	±0.098	4071,4072,4073,4074	1378,1409
525.32	±0.10	4067,4068,4069,4070	1378,1409
529.3	±0.10	4063,4064,4065,4066	1378,1409
542.99	±0.092	4059,4060,4061,4062	1378,1409
576.17	±0.098	4056,4057,4058	1378,1409
578.90	±0.098	4052,4053,4054,4055	1378,1409
580.1		6971,7056	1842
583.87	±0.099	4048,4049,4050,4051	1378,1409
639.7	±0.10	4044,4045,4046,4047	1378,1409
671.0	±0.10	4040,4041,4042,4043	1378,1409
719.3	±0.10	4036,4037,4038,4039	1378,1409
751.9		6971,7055	1842
964.	±4.7	4032,4033,4034,4035	1401,1409
1063.29	±0.095	4028,4029,4030,4031	1378,1409
1253.738	±0.0025	4024,4025,4026,4027	1378,1401,1409

Table 3.4.61
Iodomethane (Methyl Iodide) – ¹³CH₃I

Wavelength (μm)	Uncertainty (μm)	CO₂ laser pump line	References
293.13		sequence band	1675
346.67		hot band	1675
355.55		10P30	1675
366.92		hot band	1675
372.80		hot band	1675
395.00		sequence band	1675
521.11		hot band	1675
558.82		10P26	1675
573.75		10P26	1675
663.08		sequence band	1675

Iodomethane (Methyl Iodide) – $^{13}CH_3I$—*continued*

Wavelength (μm)	Uncertainty (μm)	CO_2 laser pump line	References
820.00		10P48	1675
848.00		hot band	1675
1245.71		hot band	1675
1624.00		N_2O laser	1675

Table 3.4.62
Methyl Isocyanide – CH_3NC

Wavelength (μm)	Uncertainty (μm)	CO_2 laser pump line	References
250.		10R22	1679
277.		10R24	1679
280.		10R24	681
284.		10R18	1679
288.		10R18	681
402.		10R12	1679
404.		10R04	681
454.		10R04	1679
481.		10P42	1679
823.		10P30	1680
938.		10P32	1680
2140.		10P14	1680

Table 3.4.63
Aminomethane (Methylamine) – CH_3NH_2

Wavelength (μm)	Uncertainty (μm)	CO_2 laser pump line	References
68.		9P14	1752
87.		9P34	1752
92.0		9R20	1752
99.5		9R14	1799
100.		9R14	1437
102.		9P18	1752
104.0		9P28	1799
105.0		9P28	1752
109.		9P40	1752
115.5		9P44	1799
116.		9P44	1437
118.		9P06	1752

Aminomethane (Methylamine) – CH₃NH₂—*continued*

Wavelength (μm)	Uncertainty (μm)	CO₂ laser pump line	References
118.0		9P08	1799
119.		9P08	1437
120.		10P24	1437
126.		10R06	1799
128.		10R12	1437
130.		9P20	1752
134.		9R14	1799
134.		9R18	1799
137.		9P18	1752
139.		9R14	1799
141.		10R22	1799
142.		10R20	1437
142.		10R22	1752
143.		9R14	1799
145.		9R14	1437
146.		9R08	1437
147.		10R36	1437
147.0		9P24	1799
147.845		9P24	1483
148.5		9P24	1805
150.		9P10	1752,6949
153.0		9P08	1799
159.0		9P24	1799
164.		9R18	1799
165.		9R18	1437
165.		10P24	1437
166.		10P24	1752,6949
166.0		9P32	1799
168.		9R22	1799
169.		9R22	1437
175.		10R06	1799
176.		10R32	1799
177.0		9R12	1799
178.		9P12	1752,6949
178.		9R12	1437
178.		10R32	1437
179.		9P22	1437
180.		9P46	1459
183.		9R14	1799
185.		9R14	1437
194.0		9R08	1799
197.940		9P24	1665
198.		9R20	1799

Aminomethane (Methylamine) – CH₃NH₂—*continued*

Wavelength (μm)	Uncertainty (μm)	CO₂ laser pump line	References
199.0		9R20	1752
201.		9R12	1799
203.		9P34	1752,6949
208.		9R12	1799
218.		9P24	1805
218.749		9P24	1752
219.0		9P32	1799
220.		9P32	1437
221.		10R20	1437
226.		9P14	1752
243.0		9P24	1799
244.890		9P24	1752
245.		9P34	1752,6949
246.		10P12	1437
250.000		9P22	1752
250.138		9P24	1665
251.		9R22	1437
251.180		9P24	1752
254.		10R40	1437
267.0		9P40	1799
268.		9R12	1799
270.		9R18	1437
271.		9R12	1437
281.		9R14	1752
283.		9P46	1437
288.		9R04	1459
314.847		9R04	1483
347.		10R20	1799
349.		10R18	1437
349.937		10R20	1665
351.		9P46	1437
377.		9P18	1752
387.		9P34	1752,6949
600.		9P40	1752,6949

Table 3.4.64
Nitromethane – CH_3NO_2

Wavelength (μm)	Uncertainty (μm)	CO_2 laser pump line	References
311.		10P16	1682
318.		10P18	1682
340.		9P08	1682
344.		10P20	1682
344.		10P22	1682
351.		10P20	1682
376.		9P06	1682
378.		10P22	1682
398.		9R36	1682
414.		9R04	1682
424.		10P24	1682
426.		9R22	1682
450.		9R34	1682
454.		10P26	1682
470.		9P14	1682
472.		9R08	1682
487.		9R06	1682
487.		9R12	1682
489.		10P28	1682
514.		9P28	1682
524.		9R28	1682
530.		9R16	1682
550.		10P30	1682
552.		9P06	1682
564.		10P30	1682
594.		9R28	1682
598.		9R16	1682
620.		9R18	1682
631.		10P32	1682
634.		9R30	1682
646.		9R18	1682
656.		9R20	1682
673.		9R30	1682
675.		9R14	1682
697.		9R30	1682
717.		9R14	1682
735.		10P34	1682
778.		9R40	1682
780.		9R32	1682
809.		10P40	1682
841.		9R10	1682
845.		9R42	1682
869.		10P36	1682
973.		9P08	1682
1001.		9R26	1682
1070.		9R34	1682

Table 3.4.65
Methoxymethane (Dimethyl Ether) – CH_3OCH_3

Wavelength (μm)	Uncertainty (μm)	CO_2 laser pump line	References
209.3		9R22	1686
220.3		9R32	1686
304.3		9R40	1686
338.9		9R22	1686
375.		10P20	1799
378.2		10P34	1686
441.3		10P08	1686
461.		10P34	1799
480.		10P34	1799
492.		10P34	1799
495.		10P12	1799
496.5		10P20	1686
497.4		9R26	1686
511.9		10P52	1686
520.		10P12	1799
526.3		10P12	1686
530.7		9R28	1686
564.7		10P16	1686
934.2		10P20	1686
64505		9R34	1686

Table 3.4.66
Deuterooxymethanol – CH_3OD

Wavelength (μm)	Uncertainty (μm)	Comments	References
42.6		6971,7145	1844
46.7		7140	1844
48.775		6971,7155	1844
50.2		6971,7153	1844
51.4		6970,7151	1844
51.9		6970,7161	1844
53.219		6971,7154	1844
53.3		7143	1844
54.8		6970,7171	1844
55.3		6971,7169	1844
58.3		6970,7163	1844
58.9		6971,7141	1844
60.622		6970,7155	1844
61.674		6971,7164	1844
62.133		6971,7160	1844

Deuterooxymethanol – CH₃OD—*continued*

Wavelength (μm)	Uncertainty (μm)	Comments	References
62.544		69717161	1844
65.3		6971,7162	1844
68.865		6971,7156	1844
69.6		6971,7158	1844
70.142		6971,7167	1844
73.691		6971,7160	1844
76.381		6970,7142	1844
77.442		6970,7167	1844
80.0	±0.16	5240,5241,5242	1397,1436
81.9	±0.40	5237,5238,5239	1451
85.2		6970,7168	1844
85.760		6971,7165	1844
87.222		6971,7144	1844
88.8		6971,7152	1844
89.6	±0.44	5234,5235,5236	1451
92.2		6971,7138	1844
92.738		6971,7170	1844
95.341		6970,7156	1844
101.378		6971,7158	1844
101.6	±0.50	5231,5232,5233	1436,1440,1451
103.12463	±1.8e-05	5228,5229,5230	1397,1480,1483
106.	±2.1	5219,5220,5221	1436
106.	±2.1	5222,5223,5224	1436
106.857		6971,7148	1844
108.4		6971,7172	1844
110.	±1.1	5225,5226,5227	1433,1436
110.7	±0.54	5216,5217,5218	1436,1451
112.912		6971,7166	1844
113.8	±0.56	5213,5214,5215	1451
117.22707	±2.2e-05	5210,5211,5212	1397,1480,1483
117.4		6970,7165	1844
118.1		6971,7157	1844
125.842		6971,7150	1844
126.208		6971,7146	1844
128.0	±0.26	5208,5209	1397
134.0	±0.54	5205,5206,5207	1425,1436,1440
135.	±1.4	5202,5203,5204	1367
136.	±1.4	5199,5200,5201	1425,1436
137.	±2.7	5196,5197,5198	1436
141.	±2.8	5193,5194,5195	1436
141.3		6971,7145	1844
145.6	±0.71	5187,5188,5189	1451
145.66171	±2.2e-05	5190,5191,5192	1370,1433

Deuterooxymethanol – CH₃OD—*continued*

Wavelength (μm)	Uncertainty (μm)	Comments	References
148.5		6971,7165	1844
152.366		6971,7149	1844
168.1	±0.34	5185,5186	1397
169.	±3.4	5182,5183,5184	1436
173.754		6971,7136	1844
173.785		6971,7139	1844
179.0	±0.36	5179,5180,5181	1397,1437
182.	±3.6	5176,5177,5178	1436
182.1	±0.89	5174,5175	1451
186.	±3.7	5171,5172,5173	1436
204.6		6971,7137	1844
207.179		6971,7147	1844
212.	±2.1	5168,5169,5170	1436,1440
215.37246	±3.2e-05	5165,5166,5167	1370,1433
224.	±4.5	5162,5163,5164	1436
225.	±2.3	5159,5160,5161	1436,1440
229.	±1.1	5156,5157,5158	1367
234.	±1.1	5153,5154,5155	1451
238.	±1.2	5150,5151,5152	1451
238.	±2.4	5147,5148,5149	1436,1440
241.	±1.2	5144,5145,5146	1436,1451
279.4	±0.56	5141,5142,5143	1397,1436
280.	±1.4	5138,5139,5140	1451
294.81098	±2.9e-05	5135,5136,5137	1370,1433
305.72611	±3.1e-05	5132,5133,5134	1370,1433
320.O	±0.64	5130,5131	1397
330.	±1.6	5127,5128,5129	1367
352.5	±0.71	5125,5126	1397
417.	±2.0	5122,5123,5124	1367
498.0	±1.0	5120,5121	1397

Table 3.4.67
Methanol (Methyl Alcohol) – CH₃OH[(a)]

Wavelength (μm)	Uncertainty (μm)	Comments	References
48.283		6970,7255	1855
48.4		6970,7187	1833
49.2		6970,7186	1833
52.0		6970,7190	1833
57.4		7254	1855
71.280		6971,7184	1833

Methanol (Methyl Alcohol) – CH₃OH—*continued*

Wavelength (μm)	Uncertainty (μm)	Comments	References
71.290		6971,7188	1833
75.616		6970,7184	1833
77.394		6970,7188	1833
80.	±2.4	4711,4712,4713,4714	1414,1416,1418, 1419,1420,1427, 1428,1429,1473
80.6	±0.39	4707,4708,4709,4710	1414,1416,1418, 1419,1427,1428, 1429,1473,1475
85.600931	±4.3e-05	4703,4704,4705,4706	1414,1416,1418, 1419,1427,1428, 1429,1453,1473, 1481
86.239385	±4.3e-05	4699,4700,4701,4702	1397,1414,1416, 1418,1419,1427, 1428,1429,1453, 1473,1481
92.543913	±4.6e-05	4695,4696,4697,4698	1414,1416,1418, 1419,1427,1428, 1429,1453,1473,1481
92.664287	±4.6e-05	4691,4692,4693,4694	1414,1416,1418, 1419,1427,1428, 1429,1453,1473,1481
96.522395	±1.5e-05	4688,4689,4690	1397,1414,14161418, 1419,1427,1428, 1429,1455,1473
97.518534	±4.9e-05	4684,4685,4686,4687	1414,1416,1418, 1419,1427,1428, 1429,1453,1473,1481
98.	±2.9	4680,4681,4682,4683	1414,1416,1418, 1419,1427,1428, 1429,1471,1473
100.80647	±5.0e-05	4676,4677,4678,4679	1414,1416,1418, 1419,1427,1428, 1429,1453,1473,1481
102.061		6971,7256	1855
104.3		7257	1855
111.704		6971,7258	1855
113.73188	±5.7e-05	4672,4673,4674,4675	1414,1416,1418, 1419,1427,1428, 1429,1453,1473,1481

Methanol (Methyl Alcohol) – CH₃OH—*continued*

Wavelength (μm)	Uncertainty (μm)	Comments	References
117.95948	±5.9e-05	4668,4669,4670,4671	1414,1416,1418, 1419,1427,1428, 1429,1453,1473,1481
118.117		6971,7185	1833
118.83409	±2.4e-05	4664,4665,4666,4667	1397,1414,1416, 1418,1419,1427, 1428,1455,1469,1473
121.	±2.8	4660,4661,4662,4663	1414,1416,1418, 1419,1427,1428, 1429,1471,1473
129.5497	±0.00013	4656,4657,4658,4659	1414,1433,1416,1418 1419,1427,1428, 1429,1473
130.6		7256	1855
133.1196	±0.00013	4653,4654,4655	1397,1414,1416, 1418,1419,1427, 1428,1429,1473
145.5	±0.71	4649,4650,4651,4652	1414,1416,1418, 1419,1427,1428, 1429,1473
151.25369	±7.6e-05	4645,4646,4647,4648	1414,1453,1416, 1418,1419,1427, 1428,1429,1473,1481
152.	±3.0	4641,4642,4643,4644	1414,1416,14181419, 1427,1428,1429, 1436,1473
159.2	±0.78	4637,4638,4639,4640	1414,1416,1418, 1419,1427,1428, 1429,1473,1475,
159.67569	±8.0e-05	4633,4634,4635,4636	1414,1416,1418, 1419,1427,1428, 1429,1453,1473,1481
163.03353	±4.6e-05	4629,4630,4631,4632	1414,1416,1418, 1419,1427,1428, 1429,1455,1459, 1473
164.0	±0.80	4625,4626,4627,4628	1414,1416,1418,1419 1427,1428,1429,1473
164.5076	±0.00016	4621,4622,4623,4624	1414,1416,14181419, 1427,1428,1429, 1471,1473

Methanol (Methyl Alcohol) – CH₃OH—*continued*

Wavelength (μm)	Uncertainty (μm)	Comments	References
164.56421	±8.2e-05	4617,4618,4619,4620	1414,1416,14181419, 1427,1428,1453, 1473,1429,1481
164.60038	±8.2e-05	4613,4614,4615,4616	1375,1414, 1416,1418,1419, 1427,1428,1429, 1453,1473,1481
164.69747	±8.2e-05	4609,4610,4611,4612	1414,1416,1418, 1419,1427,1428, 1429,1453,1473,1481
164.7832	±0.00016	4605,4606,4607,4608	1375,1414,1416, 1418,1419,1427, 1428,1429,1473,1481
167.58700	±8.4e-05	4601,4602,4603,4604	1414,1416,1418, 1419,1427,1428, 1429,1453,1473,1481
170.57637	±4.8e-05	4597,4598,4599,4600	1397,1414,1416, 1418,1419,1427, 1428,1429,1455,1473
171.3	±0.84	4593,4594,4595,4596	1414,1416,1418, 1419,1427,1428, 1429,1469,1473,1475
176.	±3.5	4589,4590,4591,4592	1414,1416,14181419, 1427,1428,1429, 1436,1473
178.	±3.6	4585,4586,4587,4588	1436,1414,1416, 1418,1419,1427, 1428,1429,1473
179.72791	±9.0e-05	4581,4582,4583,4584	1414,1416,1418, 1419,1427,1428, 1429,1453,1473,1481
180.4	±0.36	4578,4579,4580	1397,1414,1416, 1418,1419,1427, 1428,1429,1473
185.50040	±9.3e-05	4574,4575,4576,4577	1375,1414,14161418, 1419,1427,1428, 1429,1453,1473,1481
185.9	±0.37	4571,4572,4573	1397,1414,1416, 1418,1419,1427, 1428,1429,1473

Methanol (Methyl Alcohol) – CH₃OH—*continued*

Wavelength (μm)	Uncertainty (μm)	Comments	References
186.04219	±9.3e-05	4567,4568,4569,4570	1397,1414,1416, 1418,1419,1427, 1428,1429,1453, 1473,1481
190.72590	±9.5e-05	4563,4564,4565,4566	1397,1414,1416, 1418,1419,1427, 1428,1429,1453, 1473,1481
191.5	±0.38	4560,4561,4562	1397,1414,1416, 1418,1419,1427, 1428,1429,1473
191.61960	±9.6e-05	4557,4558,4559	1414,1416,1418, 1419,1427,1428, 1429,1453,1473,1481
193.14158	±9.7e-05	4553,4554,4555,4556	1397,1414,1416, 1418,1419,1427, 1428,1429,1453, 1473,1481
194.06320	±9.7e-05	4549,4550,4551,4552	1414,1416,1418, 1419,1427,1428,1429 1453,1473,1481
198.66433	±9.9e-05	4545,4546,4547,4548	1397,1414,1416, 1418,1419,1427, 1428,1429,1453, 1473,1481
202.40	±0.051	4542,4543,4544	1375,1414,1416, 1418,1419,1427, 1428,1429,1473
205.	±1.0	4540,4541	1414,1416,1418, 1419,1427,1428, 1429,1473
206.90	±0.062	4536,4537,4538,4539	1414,1416,1418, 1419,1427,1428, 1429,1453,1473,1481
208.	±2.1	4534,4535	1414,1416,1418, 1419,1427,1428, 1429,1473
209.9302	±0.00010	4530,4531,4532,4533	1414,1416, 1418,1419,1427, 1428,1429,1473,1481
211.2629	±0.00011	4526,4527,4528,4529	1414,1416,1418, 1419,1427,1428, 1429,1453,1473,1481

Methanol (Methyl Alcohol) – CH₃OH—*continued*

Wavelength (μm)	Uncertainty (μm)	Comments	References
211.3148	±0.00011	4523,4524,4525	1397,1414,1416, 1418,1419,1427, 1428,1429,1453, 1473,1481
213.4625	±0.00011	4519,4520,4521,4522	1397,1414,1416, 1418,1419,1427, 1428,1429,1453, 1473,1481
214.35	±0.064	4515,4516,4517,4518	1414,1416,1418, 1419,1427,1428, 1429,1453,1473,1481
218.22	±0.065	4511,4512,4513,4514	1414,1416,1418,1419 1427,1428,1473,1429
223.50	±0.056	4508,4509,4510	1375,1414,1416, 1418,1419,1427, 1428,14291473
225.5159	±0.00011	4504,4505,4506,4507	1397,1414,1416, 1418,1419,1427, 1428,1429,1453, 1473,1481
232.7	±0.47	4501,4502,4503	1397,1414,1416,1418 1419,1427,1428,1429 1473
232.7884	±0.00012	4497,4498,4499,4500	1414,1416,1418, 1419,1427,1428, 1429,1453,1473,1481
232.93906	±9.1e-05	4494,4495,4496	1397,1414,14161418, 1419,1427,1428, 1429,1455,1473
237.60	±0.048	4491,4492,4493	1375,1414,1416, 1418,1419,1427, 1428,1429,1473
242.4727	±0.00012	4487,4488,4489,4490	1414,1416,1418, 1419,1427,1428, 1429,1453,1473,1481
242.5	±0.49	4484,4485,4486	1397,1414,1416, 1418,1419,1427, 1428,1429,1473
242.79	±0.024	4480,4481,4482,4483	1414,1416,1418, 1419,1427,1428, 1429,1473,1474

Methanol (Methyl Alcohol) – CH₃OH—*continued*

Wavelength (μm)	Uncertainty (μm)	Comments	References
250.7813	±0.00010	4476,4477,4478,4479	1414,1416,1418, 1419,1427,1428, 1429,1455,1459,1473
251.	±2.8	4472,4473,4474,4475	1414,1416,1418, 1419,1427,1428, 1429,1471,1472,1473
251.1398	±0.00010	4468,4469,4470,4471	1414,14161418,1419, 1427,1428,1429, 1455,1459,1473
251.4324	±0.00013	4464,4465,4466,4467	1414,1416,1418,1419 1427,1428,1429, 1453,1473,1481
253.5530	±0.00013	4457,4458,4459,4460	1375,1414,14161418, 1419,1427,1428, 1429,1453,1473,1481
253.60	±0.051	4461,4462,4463	1375,1414,1416, 1418,1419,1427, 1428,1429,1473
261.	±5.2	4453,4454,4455,4456	1414,1416,1418, 1419,1427,1428, 1429,1436,1473
263.70	±0.053	4450,4451,4452	1375,1414,1416, 1418,1419,1427, 1428,1429,1473
264.5359	±0.00013	4446,4447,4448,4449	1375,1414,1416, 1418,1419,1427, 1428,1429,1453, 1473,1481
267.4432	±0.00013	4443,4444,4445	1414,1416,1418, 1419,1427,1428, 1429,1453,1473,1481
270.	±1.3	4439,4440,4441,4442	1414,1416,1418, 1419,1427,1428, 1429,1473
274.	±2.7	4435,4436,4437,4438	1414,1416,1418, 1419,1427,1428, 1429,1471,1473
278.8048	±0.00014	4431,4432,4433,4434	1375,1414,1416, 1418,1419,1427, 1428,1429,1453, 1473,1481

Methanol (Methyl Alcohol) – CH₃OH—*continued*

Wavelength (μm)	Uncertainty (μm)	Comments	References
280.9341	±0.00014	4427,4428,4429,4430	1375,1414,1416, 1418,1419,1427, 1428,1429,1453, 1473,1481
286.	±2.9	4423,4424,4425,4426	1414,1416,1418, 1419,1427,1428, 1429,1471,1473
290.62	±0.087	4420,4421,4422	1414,1416,1418, 1419,1427,1428, 1429,1473
292.1415	±0.00015	4416,4417,4418,4419	1375,1414,1416, 1418,1419,1427, 1428,1429,1453, 1473,1481
292.50	±0.050	4413,4414,4415	1375,1414,1416, 1418,1419,1427, 1428,1429,1473
293.8217	±0.00015	4409,4410,4411,4412	1414,1416,1418, 1419,1427,1428, 1429,1453,1473,1481
301.9943	±0.00015	4405,4406,4407,4408	1414,1416,1418, 14191427,1428, 1429,1453,1473,1481
311.2	±0.62	4402,4403,4404	1397,1414,1416, 1418,1419,1427, 1428,1429,1473
346.4875	±0.00017	4398,4399,4400,4401	1397,1414,1416, 1418,1419,1427, 1428,1429,1453, 1473,1481
369.1137	±0.00023	4395,4396,4397	1375,1414,1416, 1418,1419,1427, 1428,1429,1455,1473
386.3392	±0.00019	4391,4392,4393,4394	1414,1416,1418, 14191427,1428,1429, 1453,1473,1481
390.1	±0.78	4387,4388,4389,4390	1414,1416,1418, 1419,1427,1428, 1429,1473,1475
392.0687	±0.00026	4383,4384,4385,4386	1375,1469,1414, 1416,1418,1419, 1427,1428,1429, 1455,1473

Methanol (Methyl Alcohol) – CH₃OH—*continued*

Wavelength (μm)	Uncertainty (μm)	Comments	References
416.5223	±0.00042	4379,4380,4381,4382	1414,1471,1416, 1418,1419,1427, 1428,1429,1473
418.0827	±0.00021	4375,4376,4377,4378	1375,1414,1416, 1418,1419,1427, 1428,14291453, 1473,1481
453.697		6971,7189	1833
469.0233	±0.00037	4371,4372,4373,4374	1414,1416,1418, 1419,1427,1428, 1429,1459,1473
486.1	±0.97	4367,4368,4369,4370	1414,1416,1418, 1419,1427,1428, 1429,1455,1475,1473
495.	±2.9	4363,4364,4365,4366	1414,1416,1418, 1419,1427,1428, 1429,1453,1471, 1473,1481
570.5687	±0.00055	4360,4361,4362	1375,1414,1416,1418 1419,1427,1428, 1429,1455,1473
602.4870	±0.00030	4356,4357,4358,4359	1397,1414,1416, 1418,1419,1427, 1428,14291453, 1473,1481
614.2851	±0.00031	4352,4353,4354,4355	1397,1414,1416, 1418,1419,1427, 1428,1429,1453, 1473,1481
624.4301	±0.00031	4348,4349,4350,4351	1397,1414,1416, 1418,1419,1427, 1428,14291453, 1473,1481
694.	±2.8	4344,4345,4346,4347	1414,1416,1418, 1419,1427,1428, 1429,1471,1473
694.1893	±0.00035	4340,4341,4342,4343	1414,1416,1418, 1419,1427,1428, 1429,14531473,1481
695.3499	±0.00035	4336,4337,4338,4339	1414,1416,1418, 1419,1427,1428, 1429,1453,1473,1481

Methanol (Methyl Alcohol) – CH₃OH—*continued*

Wavelength (μm)	Uncertainty (μm)	Comments	References
699.4226	±0.00084	4333,4334,4335	1375,1414,1416, 1418,1419,1427, 1428,1429,1455,1473
1223.858	±0.0024	4329–4332	1414,1416,1418, 1419,1427,1428, 1429,1470,1473, 1480,1483

(a) See, also, references 1851–1853.

Table 3.4.68
Methanol (Methyl Alcohol) – ¹³CH₃OH[a]

Wavelength (μm)	Uncertainty (μm)	CO₂ laser pump line	References
34.790		10R22	1415
41.9		10P16	1415
44.00869		10R54	1856
44.92678		10R52	1856
46.92769		10SR17	1856
49.1		10R34	1856
50.48388		10SR11	1856
54.2		9R16	1828
58.8		9R40	1828
60.0		10R14	1707
63.096		9P12	1415
70.0		10SR19	1856
70.00		9P43	1707
71.70		9P32	1707
71.75232		10R06	1856
72.0		10R40	1707
77.489		10R26	1415
80.1809		10SR15	1856
80.3		10P34	1706
80.7375		10SR21	1856

Methanol (Methyl Alcohol) – $^{13}CH_3OH^{(a)}$

Wavelength (µm)	Uncertainty (µm)	CO_2 laser pump line	References
81.010		9P20	1707
81.72535		10R06	1856
83.82088		10R50	1856
85.317		9P22	1415
85.79		10R28	1415
86.112		9P10	1415
87.900		9P08	1415
88.73787		9P44	1856
89.00		9P44	1707
95.24		9P32	1707
98.0		9P28	1706
101.3		9R40	1706
103.00		9P24	1707
103.481		9P22	1415
103.586		10R26	1415
105.07		9P10	1707
105.147		10R18	1415
106.7		9R40	1828
107.8		9R20	1706
110.432		10R18	1415
110.50965		10R46	1856
110.9		10R46	1706
112.5		10SR17	1856
113.4		9P06	1706
113.60		9P30	1707
115.00		9P22	1415
115.823		10R16	1415
117.92		9P26	1707
118.013		9P22	1415
118.3846		9R32	1828
121.20		10R28	1415
122.0		10R02	1707
122.402		9P20	1803
122.885		10R30	1803
123.26		10P16	1415
125.9		10R42	1856
126.2243		10R28	1828
127.1		10R06	1856
128.9		10SR17	1856
133.7		10P12	1706
140.9		10P30	1706
142.1		9R32	1828
145.8727		10R44	1856

Methanol (Methyl Alcohol) – $^{13}CH_3OH$[a]

Wavelength (μm)	Uncertainty (μm)	CO$_2$ laser pump line	References
146.097		9P10	1415
146.61175		10R38	1856
147.97		9P30	1415
148.590		10R16	1417
149.272		9P22	1415
152.076		10R16	1417
155.0		10P08	1706
155.5		9R40	1828
156.6020		10R02	1828
157.929		9P12	1415
160.93747		10R46	1856
160.937765		10R46	1856
166.28		9P12	1707
168.84		9P40	1415
171.758		10R18	1415
176.3		9R24	1828
181.20		10P16	1706
184.4		9R22	1828
188.7859		9R30	1828
188.96		9P26	1707
190.3		9P04	1706
194.8		9R24	1828
198.79		10R30	1707
203.25239		10R44	1856
203.636		10R16	1415
203.96		9P20	1707
205.00		9P06	1707
208.412		9P10	1415
214.3		10P28	1706
216.5		9R32	1706
222.8		10R02	1706
230.40086		9HP20	1856
230.75554		9HP20	1856
236.530		9P10	1692
237.523		9P12	1415
238.523		9P12	1415
240.1		10R06	1706
247.4		10P10	1706
249.1		9P26	1706
253.50		10R20	1707
255.2		9R08	1828
268.572		10R16	1415
268.6		10R18	1706

Methanol (Methyl Alcohol) – $^{13}CH_3OH$[a]

Wavelength (µm)	Uncertainty (µm)	CO_2 laser pump line	References
269.9		10R14	1706
275.61		10R26	1707
280.218		10R16	1417
280.240		10R16	1417
281.0		9P32	1706
282.96		10R20	1707
291.62		9P36	1415
294.04		10R20	1707
306.500		9P20	1706
307.07		10R26	1707
307.780		9P22	1415
311.1		10P28	1706
313.65018		10R14	1856
319.7		10R14	1706
320.4		10R20	1706
321.3633		10R30	1828
325.17		9P36	1415
328.9		9P38	1706
332.603		10R16	1417
334.6		10R40	1706
338.964		9P22	1415
339.90		10R36	1706
340.00		10R32	1707
356.4		9R22	1828
358.92		9P40	1415
392.4634		10P16	1828
398.3		9R08	1828
400.1		9R20	1706
420.0230		10R02	1828
425.8		9P26	1706
452.4		9R36	1706
461.385		9P12	1415
496.3		10R16	1706
496.40		9P30	1706
629.844		9P12	1415
784.4		9P36	1706

(a) See, also, reference 1853.

Table 3.4.69
Methanol (Methyl Alcohol) – $CH_3{}^{18}OH$[a]

Wavelength (µm)	Uncertainty (µm)	CO_2 laser pump line	References
34.60		9P30	1741
35.00		10R36	1741
40.00		10P26	1741
43.70		9P30	1741
48.40		10R38	1741
49.50		9P16	1741
52.70		10R36	1741
53.60		10R30	1741
65.55		9P22	1741
69.90		10R12	1741
77.65		9P18	1741
78.20		10R24	1741
87.65		9R06	1741
90.97		10P42	1741
92.60		9P34	1741
93.40		9P22	1741
98.65		10R10	1741
99.14		9P30	1741
104.60		9P22	1741
109.3		10R04	1741
111.60		10P06	1741
114.20		9P32	1741
115.70		9P36	1741
115.80		9P40	1741
119.84		9P26	1741
123.85		9P34	1741
123.90		9P30	1741
127.77		10R30	1741
131.69		9R30	1741
134.60		9P30	1741
142.43		9P32	1741
142.80		10R26	1741
143.64		9P40	1741
144.18		10R16	1741
149.00		9P30	1741
151.65		9P22	1741
153.54		9P36	1741
165.10		9P10	1741
170.10		10P24	1741
170.18		9R26	1741
176.45		10R04	1741

Methanol (Methyl Alcohol) – $CH_3^{18}OH^{(a)}$—*continued*

Wavelength (μm)	Uncertainty (μm)	CO_2 laser pump line	References
179.80		9P18	1741
181.10		10R30	1741
181.20		9P32	1741
181.60		10R06	1741
182.19		9P14	1741
183.36		10P26	1741
184.80		10P24	1741
191.04		9P42	1741
193.25		9P16	1741
193.55		10R18	1741
199.90		10R26	1741
203.80		10P06	1741
206.60		9P20	1741
214.20		9P14	1741
215.80		9P06	1741
218.70		9P30	1741
219.80		10R20	1741
219.90		10R20	1741
220.27		10R18	1741
221.86		9P30	1741
222.50		9P26	1741
227.00		10P10	1741
229.40		9P44	1741
230.70		9P10	1741
232.65		10P06	1741
241.50		10P10	1741
241.75		10P06	1741
242.47		9P22	1741
251.90		9P32	1741
262.40		9P28	1741
268.30		9R38	1741
277.00		9P44	1741
284.15		9R10	1741
284.50		10R06	1741
284.90		9P30	1741
285.25		10R04	1741
294.30		9P06	1741
300.60		10R20	1741
307.20		9P42	1741
327.50		9P32	1741
342.80		10P42	1741

Methanol (Methyl Alcohol) – CH$_3$^{18}OH$^{(a)}$—*continued*

Wavelength (μm)	Uncertainty (μm)	CO$_2$ laser pump line	References
359.20		9P10	1741
362.65		10R20	1741
363.86		10R20	1741
364.30		10R18	1741
364.50		9P34	1741
382.88		10R18	1741
407.50		10P26	1741
434.95		10R16	1741
438.10		9P38	1741
465.50		9P18	1741
465.70		9R34	1741
482.12		9P14	1741
505.80		9P42	1741
506.25		9P36	1741
546.80		10R36	1741
555.75		10R20	1741
621.70		9R08	1741
653.22		9P14	1741

(a) See, also, reference 1850.

Table 3.4.70
Methyl Mercaptan – CH₃SH

Wavelength (μm)	Uncertainty (μm)	Comments	References
116.	±2.3	5848,5849,5850	1435
117.	±2.3	5845,5846,5847	1435
124.	±2.5	5842,5843,5844	1435
127.	±2.5	5839,5840,5841	1435
128.	±2.6	5836,5837,5838	1435
147.	±2.9	5833,5834,5835	1435
161.	±3.2	5830,5831,5832	1435
185.	±3.7	5827,5828,5829	1435
205.	±4.1	5824,5825,5826	1435
224.	±4.5	5821,5822,5823	1435
234.	±4.7	5818,5819,5820	1435
262.	±5.2	5815,5816,5817	1435
298.	±6.0	5812,5813,5814	1435
316.	±6.3	5809,5810,5811	1435
319.	±6.4	5806,5807,5808	1435
324.	±6.5	5803,5804,5805	1435
341.	±6.8	5800,5801,5802	1435
351.	±7.0	5797,5798,5799	1435
370.	±7.4	5794,5795,5796	1435
379.	±7.6	5791,5792,5793	1435
384.	±7.7	5788,5789,5790	1435
403.	±8.1	5785,5786,5787	1435
456.	±9.1	5782,5783,5784	1435

Table 3.4.71
Chlorine Dioxide – ClO$_2$

Wavelength (μm)	Uncertainty (μm)	CO$_2$ laser pump line	References
176.		C^{18}O$_2$ laser	681
196.		C^{18}O$_2$ laser	681
204.		C^{18}O$_2$ laser	681
207.		9R20	1008
215.		10R32	1008
216		C^{18}O$_2$ laser	681
233.		10R08	1008
247.		10R30	1008
255.		9R18	1008
264.		10R24	1008
264.		9R24	1008
285.		10R20	1008
300.		9R12	1008
337.		9R26	1008
340.		9R40	1008
380.		9R22	1008
409.		10P20	1008
418.		10R08	1008
459.886		10R24	1008
509.859		9R36	1008
525.		9R36	1008
775.		9R14	1008
914.721		9R14	1008
914.735		9R14	1008
914.755		9R14	1008
914.780		9R14	1008
949.685		10P16	1008
1134.113		10P16	1008
1310.748		10P14	1008
1827.424		10P20	1008

Table 3.4.72
Carbonyl Fluoride – COF$_2$

Wavelength (μm)	Uncertainty (μm)	CO$_2$ laser pump line	References
297.09		10P40	1798
301.37		10R54	1798
304.35		10R52	1798
305.24		10R50	1798
312.91		10P38	1798
335.85		N$_2$O laser	1798
339.		10R08	1658
345.50		N$_2$O laser	1798
354.63		N$_2$O laser	1798
357.		10P32	1658
358.111		10P32	1824
369.62		sequence band	1798
379.242		10R40	1765
379.59		N$_2$O laser	1798
384.916		10R16	1824
390.780		10R38	1765
393.33		10R32	1798
402.915		10P06	1765
424.13		N$_2$O laser	1798
430.91		N$_2$O laser	1798
437.		10P22	1658
440.		10P36	1658
444.745		10P36	1765
478.072		10P24	1765
485.27		10P22	1798
488.11		10P24	1798
505.829		10P22	1765
509.44		10R10	1798
516.382		10R08	1765
527.		10R22	1658
538.415		10P16	1765
539.10		N$_2$O laser	1798
552.94		N$_2$O laser	1798
572.51		10R12	1798
601.67		N$_2$O laser	1798
640.35		10R18	1798

Carbonyl Fluoride – COF$_2$—*continued*

Wavelength (μm)	Uncertainty (μm)	CO$_2$ laser pump line	References
650.70		N$_2$O laser	1798
665.70		N$_2$O laser	1798
765.42		N$_2$O laser	1798
799.17		N$_2$O laser	1798
817.50		N$_2$O laser	1798
837.27		10R14	1798
839.40		10R18	1798
867.27		N$_2$O laser	1798
1079.38		10P10	1798
1135.070		10R14	1765
1184.38		10P10	1798
1191.563		10P32	1700
1650.312		10P06	1765
1891.062		10R08	1765
1900.		10R08	1659

Table 3.4.73
Methanal (Formaldehyde) – COH$_2$

Wavelength (μm)	Uncertainty (μm)	Comments	References
101.9		6929	1734
119.6		6929	1734
122.8		6929	1734
125.9		6929	1734
155.1		6929	1734
157.6		6929	1734
159.5		6929	1734
163.8		6929	1734
170.2		6929	1734
184.4		6929	1734

Table 3.4.74
Deuterohydrocyanic Acid (Deuterium Cyanide) – DCN

Wavelength (μm)	Uncertainty (μm)	Comments	References
181.788			1477
189.9490	±0.00038	2129,2130	1373,1423,1447,1791
190.0090	±0.00038	2127,2128	1373,1423,1447,1791
194.7027	±0.00039	2125,2126	1373,1423,14471792
194.7644	±0.00039	2123,2124	1373,1423,1447,1792
204.3872	±0.00041	2122	1373,1423,1447,1792

Table 3.4.75
Dideuteromethanic Acid (Dideutero Formic Acid) – DCOOD

Wavelength (μm)	Uncertainty (μm)	Comments	References
218.0	±0.44	3438–3441	1384,1403
241.2	±0.48	3434–3437	1384,1403
266.1	±0.53	3430–3433	1384,1403
276.1	±0.55	3426–3429	1384,1403
283.1	±0.57	3422–3425	1384,1403
298.0	±0.60	3418–3421	1384,1403
304.0832	±0.00030	3414–3417	1384,1403
310.0	±0.62	3410–3413	1384,1403
323.1	±0.65	3406–3409	1384,1403
325.2	±0.65	3402–3405	1384,1403
335.7087	±0.00034	3398–3401	1384,1403
350.2	±0.70	3394–3397	1384,1403
351.9	±0.70	3390–3393	1384,1403
366.9	±0.73	3386–3389	1384,1403
380.5654	±0.00038	3382–3385	1384,1403
389.9070	±0.00039	3378–3381	1384,1403
395.1488	±0.00040	3374–3377	1384,1403
396.0	±0.79	3370–3373	1384,1403
397.1	±0.79	3366–3369	1384,1403
414.1	±0.83	3362–3365	1384,1403
415.2	±0.83	3358–3361	1384,1403
425.2	±0.85	3354–3357	1384,1403
442.8	±0.89	3350–3353	1384,1403
452.2	±0.90	3346–3349	1384,1403
457.3410	±0.00046	3342–3345	1384,1403
469.2	±0.94	3338–3341	1384,1403
478.9	±0.96	3334–3337	1384,1403
491.8906	±0.00049	3330–3333	1384,1403
508.	±1.0	3326–3329	1384,1403

Dideuteromethanic Acid – DCOOD—*continued*

Wavelength (μm)	Uncertainty (μm)	Comments	References
508.7911	±0.00051	3322–3325	1384,1403
514.9507	±0.00051	3318–3321	1384,1403
526.4856	±0.00053	3314–3317	1384,1403
527.2146	±0.00053	3310–3313	1384,1403
561.2939	±0.00056	3306–3309	1384,1403
567.8683	±0.00057	3302–3305	1384,1403
591.6157	±0.00059	3298–3301	1384,1403
593.	±1.2	3294–3297	1384,1403
645.	±1.3	3290–3293	1384,1403
666.	±1.3	3286–3289	1384,1403
726.9203	±0.00073	3282–3285	1384,1403
737.	±1.5	3278–3281	1384,1403
761.7617	±0.00076	3274–3277	1384,1403
779.8744	±0.00078	3270–3273	1384,1403
789.4203	±0.00079	3266–3269	1384,1403
795.	±1.6	3262–3265	1384,1403
812.	±1.6	3258–3261	1384,1403
835.	±1.7	3254–3257	1384,1403
843.2369	±0.00084	3250–3253	1384,1403
877.5481	±0.00088	3246–3249	1384,1403
927.9814	±0.00093	3242–3245	1384,1403
935.0095	±0.00094	3238–3241	1384,1403
936.6023	±0.00094	3234–3237	1384,1403
998.5140	±0.0010	3230–3233	1384,1403
1009.409	±0.0010	3226–3229	1384,1403
1070.231	±0.0011	3222–3225	1384,1403
1158.	±2.3	3218–3221	1384,1403
1281.649	±0.0013	3214–3217	1384,1403

Table 3.4.76
Deuteromethanic Acid (Deutero Formic Acid) – DCOOH

Wavelength (μm)	Uncertainty (μm)	Comments	References
265.1	±0.53	3211,3212,3213	1403
272.0	±0.54	3208,3209,3210	1403
312.0	±0.62	3205,3206,3207	1403
328.4570	±0.00033	3202,3203,3204	1403
341.8	±0.68	3199,3200,3201	1403
362.1	±0.72	3196,3197,3198	1403
365.2	±0.73	3193,3194,3195	1403
433.2	±0.87	3190,3191,3192	1403
433.2353	±0.00043	3187,3188,3189	1403
466.5461	±0.00047	3184,3185,3186	1403
479.9040	±0.00048	3181,3182,3183	1403
639.1282	±0.00064	3178,3179,3180	1403
647.3485	±0.00065	3175,3176,3177	1403
697.4552	±0.00070	3172,3173,3174	1403
710.	±1.4	3169,3170,3171	1403
713.1056	±0.00071	3166,3167,3168	1403
752.7485	±0.00075	3163,3164,3165	1403
971.8064	±0.00097	3160,3161,3162	1403
1047.579	±0.0010	3157,3158,3159	1403
1237.966	±0.0012	3154,3155,3156	1403

Table 3.4.77
Dideuteromethanal (Dideuteroformaldehyde) – D₂CO

Wavelength (μm)	Uncertainty (μm)	Comments	References
233.	±1.1	2570–2573	1380,1382
244.	±4.9	2566–2569	1382,1438,1439
245.	±1.2	2562–2565	1380,1382
245.	±4.9	2559,2560,2561	1382,1438,1439
256.	±5.1	2556,2557,2558	1382,1438,1439
279.	±1.4	2552–2555	1380,1382
294.	±5.9	2549,2550,2551	1382,1438,1439
320.	±6.4	2545–2548	1382,1438,1439
324.	±6.5	2541–2544	1382,1438,1439
341.	±6.8	2538,2539,2540	1382,1438,1439
346.	±6.9	2535,2536,2537	1382,1438,1439
733.5740	±0.00073	2531–2534	1380,1382
752.6808	±0.00075	2527–2530	1380,1382

Table 3.4.78
Deuterium Oxide (Heavy Water) – D$_2$O

Wavelength (μm)	Uncertainty (μm)	Comments	References
33.896		6929	1713
35.090		6929	1713
36.319		6929	1713
36.524		6929	1713
37.791		6929	1713
40.994		6929	1713
56.845		6929	1713
71.965		6929	1713
72.429		6929	1713
72.748		6929	1455
73.337		6929	1713
74.545		6929	1713
76.305		6929	1713
84.111		6929	1713
84.279		6929	1793
94.52	±0.46	2103–2106,6941	1368,1393
107.720		6929	1455
112.6	±0.55	2097–2100	1368,1393
171.670		6929	1794

Table 3.4.79
Deuterated Formyl Fluoride – DFCO

Wavelength (μm)	Uncertainty (μm)	CO$_2$ laser pump line	References
124.		9R12	684
144.		9R16	684
164.		9R12	684
198.		C^{18}O$_2$ laser	684
354.		C^{18}O$_2$ laser	684
358.		C^{18}O$_2$ laser	684
384.		9R12	684
384.		10P20	684
450.		9P12	684
514.		10P16	684
569.		C^{18}O$_2$ laser	684
608.		C^{18}O$_2$ laser	684

Deuterated Formyl Fluoride – DFCO—*continued*

Wavelength (μm)	Uncertainty (μm)	CO_2 laser pump line	References
664.		$C^{18}O_2$ laser	684
750.		9P20	684
788.		$C^{18}O_2$ laser	684
906.		$^{13}C^{18}O_2$ laser	684
1005.		$C^{18}O_2$ laser	684
2216.		$C^{18}O_2$ laser	684

Table 3.4.80
Cyanogen Fluoride – FCN

Wavelength (μm)	Uncertainty (μm)	CO_2 laser pump line	References
308.		$C^{18}O_2$ laser	681
988.		9R28	1729

Table 3.4.81
Hydrogen Bromide – HBr Laser

Wavelength (μm)	Uncertainty (μm)	Comments	References
20.360		6929	1183
20.896		6929	1183
20.949		6929	1183
21.501		6929	1183
21.546		6929	1183
22.136		6929	1183
22.226		6929	1183
22.855		6929	1183
23.436		6929	1183
29.786		6929	1183
30.445		6929	1183
30.948		6929	1183
31.368		6929	1183
31.849		6929	1183
32.469		6929	1183
32.799		6929	1183
33.409		6929	1183
40.526		6929	1183

Table 3.4.82
Fluoropropyne (Propargyl Fluoride) – $HCCCH_2F$

Wavelength (µm)	Uncertainty (µm)	CO_2 laser pump line	References
623.		9P24	1729
1006.		9P32	1729
1547.		9P18	1729

Table 3.4.83
Propynal – HCCCHO

Wavelength (µm)	Uncertainty (µm)	CO_2 laser pump line	References
148.		10P18	681
156.		10P22	681
366.		10P26	681
516.		10P14	681

Table 3.4.84
Fluoroacetylene – HCCF

Wavelength (µm)	Uncertainty (µm)	CO_2 laser pump line	References
590.		$C^{18}O_2$ laser	681
1028.		9R18	681

Table 3.4.85
Hydrogen Chloride (Hydrochloric Acid) – HCl

Wavelength (µm)	Uncertainty (µm)	Comments	References
20.346		6929	1255
20.411		6929	1255
20.999		6929	1255
21.047		6929	1255
21.156		6929	1255
21.813		6929	1255
21.971		6929	1255
22.651		6929	1255
22.864		6929	1255
23.571		6929	1255

Hydrogen Chloride (Hydrochloric Acid) – HCl—*continued*

Wavelength (μm)	Uncertainty (μm)	Comments	References
23.849		6929	1255
24.318		6929	1255
24.583		6929	1255
24.618		6929	1255
24.937		6929	1255
25.704		6929	1255
26.146		6929	1255
26.247		6929	1183
27.508		6929	1183

Table 3.4.86
Hydrocyanic Acid (Hydrogen Cyanide) – HCN

Wavelength (μm)	Uncertainty (μm)	Comments	References
71.899		6929	1445
73.101		6929	1445
76.093		6929	1445
77.001		6929	1445
81.554		6929	1445
96.401		6929	1445
98.693		6929	1445
101.257		6929	1445
110.240		6929	1445
112.066		6929	1445
113.311		6929	1445
116.132		6929	1445
126.164		6929	1447
128.629	±0.0063	2120,2121,6929	1422,1441, 1445–1449
130.839		6929	1447
134.933		6929	1445
138.768		6929	1445
165.150		6929	1445
201.059		6929	1445
211.00	±0.017	2118,2119,6929	1422,1441, 1445–1449
222.949		6929	1445
284.000		6929	1796

Hydrocyanic Acid (Hydrogen Cyanide) – HCN—*continued*

Wavelength (µm)	Uncertainty (µm)	Comments	References
309.714		6929	1796.
309.7140	±0.00031	2117,6929	1422,1441,1445, 1447–49,1796
310.8870	±0.00031	2115,2116,6929	1422,1441,1445, 1447–49,1797
335.1831	±0.00034	2114,6929	1422,1441,1445, 1447–49,1796
336.5578	±0.00034	2112,2113,6929	1422,1441,1445, 1447–49,1797
372.5282	±0.00037	2111,6929	1422,1441,1445, 1447–49,1796

Table 3.4.87
Deuterooxymethanic Acid (Deuteroxy Formic Acid) – HCOOD

Wavelength (µm)	Uncertainty (µm)	Comments	References
240.0	±0.48	3150–153	1384,1403
291.9	±0.58	3146–3149	1384,1403
304.1	±0.61	3142–3145	1384,1403
324.1	±0.65	3138–3141	1384,1403
325.9	±0.65	3134–3137	1384,1403
339.9	±0.68	3130–3133	1384,1403
347.0	±0.69	3126–3129	1384,1403
351.0	±0.70	3122–3125	1384,1403
351.9	±0.70	3118–3121	1384,1403
353.1	±0.71	3110–3113	1384,1403
353.1	±0.71	3114–3117	1384,1403
355.2	±0.71	3106–3109	1384,1403
356.0	±0.71	3102–3105	1384,1403
358.2	±0.72	3098–3101	1384,1403
361.2	±0.72	3094–3097	1384,1403
369.9678	±0.00037	3090–3093	1384,1403
372.0	±0.74	3082–3085	1384,1403
372.0	±0.74	3086–3089	1384,1403
373.8	±0.75	3078–3081	1384,1403
387.8	±0.78	3074–3077	1384,1403
391.6886	±0.00039	3070–3073	1384,1403
392.9	±0.79	3066–3069	1384,1403
395.0	±0.79	3058–3061	1384,1403
395.0	±0.79	3062–3065	1384,1403
395.7124	±0.00040	3054–3057	1384,1403

Deuterooxymethanic Acid – HCOOD—*continued*

Wavelength (μm)	Uncertainty (μm)	Comments	References
398.1	±0.80	3050–3053	1384,1403
411.2	±0.82	3046–3049	1384,1403
417.0	±0.83	3042–3045	1384,1403
429.6898	±0.00043	3038–3041	1384,1403
430.4380	±0.00043	3034–3037	1384,1403
433.2	±0.87	3030–3033	1384,1403
446.8	±0.89	3026–3029	1384,1403
450.1	±0.90	3022–3025	1384,1403
450.9799	±0.00045	3018–3021	1384,1403
461.2610	±0.00046	3014–3017	1384,1403
472.1	±0.94	3010–3013	1384,1403
472.9	±0.95	3006–3009	1384,1403
477.4	±0.95	3002–3005	1384,1403
493.1562	±0.00049	2998–3001	1384,1403
498.0	±1.0	2994–2997	1384,1403
513.7572	±0.00051	2934–2937	1384,1403
531.	±1.1	2990–2993	1384,1403
567.1065	±0.00057	2986–2989	1384,1403
582.5536	±0.00058	2982–2985	1384,1403
590.	±1.2	2978–2981	1384,1403
594.	±1.2	2974–2977	1384,1403
630.1661	±0.00063	2970–2973	1384,1403
657.	±1.3	2966–2969	1384,1403
660.	±1.3	2962–2965	1384,1403
668.	±1.3	2958–2961	1384,1403
689.9981	±0.00069	2954–2957	1384,1403
692.	±1.4	2950–2953	1384,1403
695.6720	±0.00070	2946–2949	1384,1403
727.9491	±0.00073	2942–2945	1384,1403
733.	±1.5	2938–2941	1384,1403
819.	±1.6	2930–2933	1384,1403
826.	±1.7	2926–2929	1384,1403
919.9355	±0.00092	2922–2925	1384,1403
926.2087	±0.00093	2918–2921	1384,1403
986.3125	±0.00099	2914–2917	1384,1403
1157.318	±0.0012	2910–2913	1384,1403
1161.676	±0.0012	2906–2909	1384,1403
1541.750	±0.0015	2902–2905	1384,1403
1730.833	±0.0017	2898–2901	1384,1403

Table 3.4.88
Methanoic Acid (Formic Acid) – HCOOH

Wavelength (µm)	Uncertainty (µm)	Comments	References
133.9	±0.27	2876–2879	1366,1381,1384, 1438,1439
196.5	±0.39	2872–2875	1366,1381,1384, 1438,1439
254.5	±0.51	2868–2871	1366,1381,1384, 1400,1403,1438,1439
278.5	±0.56	2864–2867	1366,1381,1384, 1400,1403,1438,1439
302.2781	±0.00030	2860–2863	1366,1379,1381, 1384,1403,1438,1439
309.5	±0.62	2857–2859	1366,1381,1384, 1438,1439
311.554	±0.0015	2853–2856	1366,1381,1384, 1400,1403,1406, 1438,1439
319.9	±0.64	2849–2852	1366,1381,1384, 1403,1438,1439
336.3	±0.67	2845–2848	1366,1381,1384, 1400,1403,1438,1439
359.9	±0.72	2841–2844	1366,1381,1384, 1403,1438,1439
393.6311	±0.00016	2837–2840	1366,1381,1379, 1384,1434,1438,1439
394.2	±0.79	2833–2836	1366,1381,1384, 1400,1438,1439
404.0	±0.81	2829–2832	1366,1381,1384, 1403,1438,1439
405.0	±0.81	2825–2828	1366,1381,1384, 1403,1400,1438,1439
405.5848	±0.00039	2821–2824	1366,1381,1384, 1403,1438,1439, 1480,1483
418.1	±0.84	2817–2820	1366,1381,1384, 1403,1438,1439
418.6129	±0.00042	2813–2816	1366,1379,1381, 1384,1403,1438,1439
420.3911	±0.00042	2809–2812	1366,1381,1384, 1403,1438,1439
432.1094	±0.00043	2805–2808	1366,1381,1384, 1403,1438,1439
432.6313	±0.00013	2801–2804	1366,1381,1384, 1403,1434,1438,1439

Methanoic Acid (Formic Acid) – HCOOH—*continued*

Wavelength (μm)	Uncertainty (μm)	Comments	References
432.6665	±0.00043	2797–2800	1366,1381,1384, 1403,1438,1439
437.4510	±0.00044	2793–2796	1366,1381,1384, 1403,1438,1439
444.8	±0.89	2789–2792	1366,1381,1384, 1403,1438,1439
445.8996	±0.00045	2785–2788	1366,1381,1384, 1403,1438,1439
446.5054	±0.00031	2781–2784	1366,1381,1384, 1403,1400,1438,1439
446.8730	±0.00045	2777–2780	1366,1381,1384, 1403,1438,1439
458.5229	±0.00069	2773–2776	1366,1379,1381, 1384,1403,1438,1439
513.0022	±0.00077	2769–2772	1366,1379,1381, 1384,1403,1438,1439
513.0157	±0.00051	2765–2768	1366,1381,1384, 1403,1438,1439
515.1695	±0.00052	2761–2764	1366,1381,1384, 1403,1438,1439
533.6783	±0.00053	2757–2760	1366,1381,1384, 1403,1438,1439
533.7006	±0.00053	2753–2756	1366,1381,1384, 1403,1438,1439
534.6		6971,6994	1830
535.	±1.1	2745–2748	1366,1381,1384, 1400,1438,1439
535.	±1.1	2749–2752	1366,1381,1384, 1400,1438,1439
580.3872	±0.00058	2741–2744	1366,1381,1384, 1438,1439
580.8010	±0.00058	2737–2740	1366,1381,1384, 1403,1438,1439
666.4		6971,6995	1830
669.531	±0.0010	2733–2736	1366,1379,1381, 1384,1403,1438,1439
705.	±1.4	2729–2732	1366,1381,1384, 1403,1438,1439
742.572	±0.0015	2725–2728	1366,1379,1381, 1384,1403,1438,1439
744.050	±0.0015	2721–2724	1366,1379,1381, 1384,1403,1438,1439

Methanoic Acid (Formic Acid) – HCOOH—*continued*

Wavelength (μm)	Uncertainty (μm)	Comments	References
760.	±15	2717–2720	1366,1381,1384,1438 1439,1459
786.1617	±0.00079	2713–2716	1366,1381,1384, 1438,1439,1459
786.9419	±0.00079	2709–2712	1366,1381,1384, 1403,1438,1439
789.8396	±0.00079	2705–2708	1366,1381,1384, 1403,1438,1439
930.	±19	2697–2700	1366,1381,1384, 1438,1483
930.	±19	2701–2704	1366,1381,1384, 1438,1439,1480,1483
1213.362	±0.0012	2693–2696	1366,1381,1384, 1403,1438,1439

Table 3.4.89
Methanoic Acid (Formic Acid) – H¹³COOH

Wavelength (μm)	Uncertainty (μm)	Comments	References
185.3		6971,7058	1841
231.7		6971,7072	1841
232.2		6971,6996	1830
255.7		6971,7070	1841
258.4		6971,7071	1841
260.	±1.3	2895–2897	1384
292.4		6971,7080	1841
311.1		6971,7080	1841
313.	±1.5	2892–2894	1384
351.0		6971,7078	1841
366.5		69701,7076	1841
393.5		6971,7074	1841
418.1		6971,7080	1841
448.5		6970,7062	1841
448.5335	±0.00045	2889–2891	1384
458.0		6971,7075	1841
464.8		6970,7062	1841
480.	±2.4	2886–2888	1384
513.1		6971,7061	1841
536.1		6971,7073	1841
536.7		6971,7065	1841
537.2		6971,6997	1830

Methanoic Acid (Formic Acid) – H^{13}COOH—*continued*

Wavelength (μm)	Uncertainty (μm)	Comments	References
572.5		6970,7059	1841
613.4		6971,7065	1841
613.8		6971,6997	1830
638.1		6971,7064	1841
639.4		6971,6998	1830
669.4		6971,7068	1841
740.1		6971,7079	1841
743.3		6970,7077	1841
745.8		6971,7063	1841
746.6		6971,6999	1830
788.9192	±0.00079	2883–2885	1384
789.0		6971,7069	1841
830.2		6971,7067	1841
1030.378	±0.0010	2880–2882	1384
1030.5		6971,7060	1841
1219.9		6970,7066	1841

Table 3.4.90
Deuteromethanal (Deuteroformaldehyde) – HDCO

Wavelength (μm)	Uncertainty (μm)	Comments	References
152.000		6940	1382
155.000		6939	1382
194.352		9P08	1382
195.0	±0.96	2523–2526,6933	1380
196.0	±0.96	2519–2522	1380
331.088		6937	1382
405.486		6938	1382

Table 3.4.91
Hydrogen Fluoride (Hydrofluoric Acid) – HF

Wavelength (μm)	Uncertainty (μm)	Comments	References
20.134		6929	1618
20.351		6929	1618
20.939		6929	1618
21.699		6929	1618
21.789		6929	1618

Table 3.4.92
Methanal (Formaldehyde) – H$_2$CO

Wavelength (μm)	Uncertainty (μm)	Comments	References
101.9		6929	1734
119.6		6929	1734
122.8		6929	1734
125.9		6929	1734
155.1		6929	1734
157.6		6929	1734
159.5		6929	1734
163.8		6929	1734
170.2		6929	1734
184.4		6929	1734

Table 3.4.93
Trioxane (Cyclic Trimer Of Formaldehyde) – (H$_2$CO)$_3$

Wavelength (μm)	Uncertainty (μm)	Comments	References
376.3		6971,7124	1834
384.	±1.9	6519–6521	1380
384.8		6971,7117	1834
391.6		6971,7134	1834
414.4		6971,7125	1834
419.9		6971,7127	1834
433.	±2.1	6516–6518	1380
439.7		6971,7133	1834
446.7		6971,7126	1834
453.3		6971,7128	1834
460.	±2.3	6513–6515	1380
482.8		6971,7129	1834
492.6		6971,7128	1834
512.	±2.5	6510–6512	1380
559.2		6971,7120	1834
619.	±3.0	6507–6509	1380
648.2		6971,7132	1834
662.9		6971,7122	1834
680.	±3.3	6504,6505,6506	1380
695.3		6971,7123	1834
696.	±3.4	6501,6502,6503	1380
712.	±3.5	6498,6499,6500	1380
729.6		6971,7123	1834
730.4		6971,7115	1834

Trioxane (Cyclic Trimer Of Formaldehyde) – $(H_2CO)_3$—*continued*

Wavelength (μm)	Uncertainty (μm)	Comments	References
731.1		6971,7135	1834
750.	±3.7	6495–6497	1380
789.5		6971,7118	1834
815.	±4.0	6492–6494	1380
815.3		6971,7131	1834
815.3		6971,7135	1834
837.3		6971,7119	1834
889.9		6971,7116	1834
890.	±4.4	6489–6491	1380
891.	±4.4	6486–6488	1380
948.4		6971,7123	1834
948.9247	±0.00095	6483–6485	1380
949.2		6971,7121	1834
1058.		6971,7130	1834
1139.2		6971,7117	1834
1139.2		6971,7121	1834

Table 3.4.94
Water – H_2O

Wavelength (μm)	Uncertainty (μm)	Comments	References
23.359		6929	1713
26.666		6929	1713
27.9707534	±2.5e-07	2093,2094,6929	1368,1369,1407
27.971		6929	1369
28.054		6929	1713
28.273		6929	1713
28.356		6929	1713
32.929		6929	1713
33.029	±0.0033	2092	1368
33.033		6929	1713
35.000		6929	1713
35.841		6929	1713
36.619		6929	1713
37.859		6929	1713
38.094		6929	1713
39.698		6929	1713
40.629		6929	1713
45.523		6929	1713

Water _ H₂O—*continued*

Wavelength (μm)	Uncertainty (μm)	Comments	References
47.244	±0.0047	2090,2091	1368
47.251		6929	1713
47.46315	±9.5e-05	2089,6929	1368,1383
47.687	±0.0048	2087,2088,6929	1368
47.693		6929	1383
48.677		6929	1383
53.609		6929	1383
55.077		6929	1383
55.088	±0.0055	2085,2086,6929	1368
57.660		6929	1383
67.177		6929	1383
73.401	±0.0073	2083,2084,6929	1368
73.402		6929	1383
78.443		6929	1407
78.443327	±2.4e-05	2081,2082,6929	1368,1455
79.091		6929	1455
79.09101	±2.4e-05	2079,2080,6929	1368,1455
89.775		6929	1455
115.32	±0.012	2077,2078,6929	1368
115.420		6929	1455
118.5910	±0.00012	2075,2076	1350,1368,1457
120.08		6929	1350
220.2279	±0.00022	2073,2074,6929	1368,1457
791.06		6929	1746

Table 3.4.95
Hydrogen Sulfide _ H₂S

Wavelength (μm)	Uncertainty (μm)	Comments	References
33.47		6929	1753
33.64		6929	1753
49.62		6929	1753
52.40		6929	1753
56.84		6929	1753
60.29		6929	1753
61.50		6929	1753
73.52		6929	1753
80.50		6929	1753

Hydrogen Sulfide – H₂S—*continued*

Wavelength (μm)	Uncertainty (μm)	Comments	References
83.43		6929	1753
87.47		6929	1753
92.00		6929	1753
96.38		6929	1753
103.3		6929	1753
108.8		6929	1753
116.8		6929	1753
126.2		6929	1753
129.1		6929	1753
130.8		6929	1753
135.5		6929	1753
140.6		6929	1753
162.4		6929	1753
192.9		6929	1753
225.3		6929	1753

Table 3.4.96
Trideuteroammonia (Fully Deuterated Ammonia) – ND₃

Wavelength (μm)	Uncertainty (μm)	Comments	References
86.90	±1.7	2515–2518	1437

Table 3.4.97
Dideuterohydrazine – ND₂ND₂

Wavelength (μm)	Uncertainty (μm)	CO₂ laser pump line	References
115.0		10P16	1770
134.0		10P18	1770
159.5		9P36	1770
217.0		10R14	1770
244.0		10P22	1770
249.0		9P22	1770
252.0		10R36	1770
275.0		10R12	1770
278.0		10P38	1770

Dideuterohydrazine – ND$_2$ND$_2$—*continued*

Wavelength (μm)	Uncertainty (μm)	CO$_2$ laser pump line	References
285.0		10R38	1770
285.5		10P38	1770
286.0		10P32	1770
290.0		10R08	1770
293.0		9P14	1770
296.0		10P32	1770
301.0		10R24	1770
311.0		10P30	1770
354.5		10P32	1770
386.5		10R18	1770
389.0		10P22	1770
434.0		9P14	1770
454.0		10R40	1770
533.0		10P38	1770
552.0		10R12	1770
587.5		10P34	1770
641.0		10P38	1770
658.5		9P22	1770
699.0		10R12	1770
724.0		10R30	1770

Table 3.4.98
Deuteroammonia – NH$_2$D

Wavelength (μm)	Uncertainty (μm)	Comments	References
85.90	±1.7	2503–2506	1437
107.8	±2.2	2499–2502	1437
113.1	±2.3	2495–2498	1437
123.9	±2.5	2491–2494	1437

Table 3.4.99
Ammonia – NH$_3$

Wavelength (μm)	Uncertainty (μm)	Comments	References
21.46		6929	1206
22.54		6929	1206
22.71		6929	1206
23.68		6929	1206
23.86		6929	1206
24.92		6929	1206
25.12		6929	1206
26.27		6929	1206
30.69		6929	1206
31.47		6929	1206
31.92		6929	1206
32.13		6929	1206
81.480	±0.020	2385–2390,2395–2397	1393,1461,1462,1463 1464,1484
81.500	±0.049	2391–2394	1376,1393,1461,1462 1463,1464,1469,1484
87.093	±0.022	2376–2384	1393,1461,1462,1463 1464,1484
87.41	±0.43	2372–2375	1393,1461,1462,1463 1464,1484
88.059	±0.018	2368–2371	1393,1461,1462,1463 1464,1484
90.934	±0.025	2365–2367	1393,1461,1462,1463 1464,1484
92.876	±0.026	2361–2364	1393,1461,1462,1463 1464,1484
94.447	±0.026	2352–2360	1393,1461,1462,1463 1464,1484
96.674	±0.028	2337–2351	1393,1461,1462,1463 1464,1484
105.35	±0.034	2325–2336	1393,1461,1462,1463 1464,1484
112.22	±0.038	2322–2324	1393,1461,1462,1463 1464,1479,1484
114.29	±0.040	2318–2321	1393,1461,1462,1463 1464,1479,1484
116.27	±0.041	2312–2317	1393,1461,1462,1463 1464,1484
119.02	±0.042	2303–2311	1393,1461,1462,1463 1464,1484

Ammonia – NH$_3$—*continued*

Wavelength (μm)	Uncertainty (μm)	Comments	References
147.15	±0.065	2299–2302	1393,1437,1461,1462 1463,1464,1484
151.49	±0.068	2285–2298	1393,1461,1462,1463 1464,1484
155.28	±0.071	2267–2284	1393,1461,1462,1463 1464,1484
215.01	±0.14	2264–2266	1393,1461,1462,1463 1464,1484
218.28	±0.14	2258–2263	1393,1461,1462,1463 1464,1484
223.91	±0.15	2246–2257	1393,1461,1462,1463 1464,1484
225.07	±0.15	2242–2245	1393,1461,1462,1463 1464,1484
250.06	±0.19	2236–2241	1393,1461,1462,1463 1464,1484
257.13	±0.20	2233–2235	1393,1461,1462,1463 1464,1484
263.40	±0.053	2229–2232	1376,1393,1461, 1462,1463,1464, 1469,1484
263.44	±0.21	2222–2228	1393,1461,1462,1463 1464,1484
268.82	±0.22	2213–2221	1393,1461,1462,1463 1464,1484
273.36	±0.22	220–2209	1393,1461,1462,1463 1464,1484
276.79	±0.23	2204–2206,2210–2212	1393,1461,1462,1463 1464,1484
279.32	±0.23	2195–2203	1393,1461,1462,1463 1464,1484
288.51	±0.25	2189–2194	1393,1461,1462,1463 1464,1484
289.35	±0.25	2180–2182	1393,1461,1462,1463 1464,1484
289.35	±0.25	2183–2188	1393,1461,1462,1463 1464,1484
290.2	±1.4	2176–2179	1393,1461,1462,1463 1464,1484
290.44	±0.25	2172–2175	1393,1461,1462,1463 1464,1484
290.95	±0.25	2159–2171	1393,1461,1462,1463 1464,1484

Ammonia – NH₃—*continued*

Wavelength (μm)	Uncertainty (μm)	Comments	References
301.3	±0.27	2155–2158	1393,1461,1462,1463 1464,1484
306.3	±0.28	2152–2154	1393,1461,1462,1463 1464,1484
309.5	±0.29	2149–2151	1393,1461,1462,1463 1464,1484
404.7	±0.45	2139–2148	1393,1461,1462,1463 1464,1484

Table 3.4.100
Ammonia – ^{15}NH₃

Wavelength (μm)	Uncertainty (μm)	Comments	References
111.9	±0.55	2487–2490	1390
152.9	±0.31	2483–2486	1390,1476,1477
218.0	±1.1	2479–2482	1390
375.0	±1.8	2475–2478	1390,1476

Table 3.4.101
Hydrazine – NH₂NH₂

Wavelength (μm)	Uncertainty (μm)	Comments	References
73.07		7002	1831
89.5		7000	1831
93.04		7034	1847
93.5		7001	1831
98.0		7020	1847
102.0		7003	1831
106.18		7035	1847
106.19		7021	1847
113.93		7025	1847
114.2		7029	1847
120.62		7028	1847
136.79		7027	1847
155.2		6971,6979	1825
157.58		7031	1847
157.8		6970,6987	1825
160.6		6970,6979	1825

Hydrazine – NH$_2$NH$_2$—*continued*

Wavelength (μm)	Uncertainty (μm)	Comments	References
161.28		7024	1847
163.4		6971,6981	1825
165.0		7004	1831
165.3		6971,6973	1825
178.6		6970,6976	1825
181.92643	±5.5e-05	4326–4328	1480,1483
192.9072	±0.00035	4323–4325	1398,1480,1483
206.8		6970,6980	1825
209.5		6971,6977	1825
209.8		6971,6977	1825
210.21		7032	1847
213.3		6970,6981	1825
218.84		7023	1847
220.6		6971,6977	1825
227.1		6971,6989	1825
232.74		7026	1847
233.8		6971,6985	1825
233.9157	±0.00019	4320–4322	1398,1480,1483
234.0	±0.47	4318,4319	1398
234.12		7005	1831
234.36		7006	1831
235.1		6971,6993	1825
235.2		6971,6992	1825
241.6		7030	1847
246.5	±0.49	4316,4317	1398
250.5	±0.50	4314,4315	1398
251.9		6970,6975	1825
257.5		7022	1847
262.0	±0.52	4312,4313	1398
262.1		6970,6972	1825
264.8		6971,6974	1825
264.8014	±0.00024	4309,4310,4311	1398,1480,1483
265.0	±0.53	4307,4308	1398
271.5	±0.54	4305,4306	1398
278.2		6970,6991	1825
283.8		6971,6976	1825
286.5		6970,6988	1825
287.4		6971,6977	1825
287.96		7033	1847
288.4		6971,6978	1825
301.1		6971,6985	1825
301.2754	±0.00045	4302,4303,4304	1398,1480,1483
301.3		6971,6982	1825

Hydrazine – NH₂NH₂—*continued*

Wavelength (μm)	Uncertainty (μm)	Comments	References
311.0747	±0.00031	4299–4301	1398,1480,1483
319.4		6971,6986	1825
327.0	±0.65	4297,4298	1398
331.5	±0.66	4295,4296	1398
331.6694	±0.00036	4292–4294	1398,1480,1483
335.1		6971,6984	1825
336.0	±0.67	4290,4291	1398
339.4		7019	1847
368.862	±0.0023	4288,4289	1398,1480,1483
372.5	±0.75	4286,4287	1398
373.0	±0.75	4284,4285	1398
394.0		6971,6974	1825
402.8		6971,6976	1825
419.7		6971,6983	1825
428.9		6970,6986	1825
435.7718	±0.00031	4281–4283	1398,1480,1483
461.0718	±0.00069	4278–4280	1398,1480,1483
461.2		6970,6988	1825
483.5	±0.97	4276,4277	1398
527.8730	±0.00090	4273–4275	1480,1483
533.5		6971,6990	1825
533.655	±0.0048	4271,4272	1480,1398,1483
669.3		6970,6978	1825
721.	±1.4	4269,4270	1398
734.1616	±0.00088	4266–4268	1480,1483
795.	±1.6	4264,4265	1398
802.	±1.6	4262,4263	1398
1007.	±2.0	4260,4261	1398

Table 3.4.102
Hydroxylamine – NH₂OH

Wavelength (μm)	Uncertainty (μm)	CO₂ laser pump line	References
277.		9R24	1768
290.		9P12	1768
292.		$^{13}CO_2$ laser	1768
545.		$C^{18}O_2$	1768
659.		$C^{18}O_2$	1768

Table 3.4.103
Ozone – O$_3$

Wavelength (µm)	Uncertainty (µm)	CO$_2$ laser pump line	References
121.00		9P14	1474
149.20		9P06	1798
163.61		9P40	1474
171.50		9P30	1474
217.83		9P30	1798
313.60		9P30	1798
489.038		9R32	1798

Table 3.4.104
Carbonyl Sulfide – OCS

Wavelength (µm)	Uncertainty (µm)	Comments	References
123.0		6929	1753
132.0		6929	1753
378.4	±0.76	2131–2134,6935	1436

Table 3.4.105
Phosphine – PH$_3$

Wavelength (µm)	Uncertainty (µm)	CO$_2$ laser pump line	References
83.77		10R34	1777
104.00		9R14	1777
135.94		9R12	1777
194.00		10P42	1777

Table 3.4.106
Silicon Tetrafluoride – SiF$_4$

Wavelength (μm)	Uncertainty (μm)	Comments	References
24.78		1972,1973	1219
25.31		1972,1974	1219
25.36		1972,1975	1219
25.40		1972,1976	1219
25.67		1972,1977,1978	1219
25.68		1972,1979	1219
25.77		1972,198	1219
25.79		1972,1981	1219
26.01		1972,198	1219
26.14		1972,1983	1219

Table 3.4.107
Trifluorosilane (Silyl Fluoride) – SiHF$_3$

Wavelength (μm)	Uncertainty (μm)	CO$_2$ laser pump line	References
149.0		C^{18}O$_2$ laser	1787
301.0		10R14	1787
322.5		10R32	1787
330.0		^{13}CO$_2$ laser	1787
334.0		10R12	1787
345.0		9P32	1787
355.5		^{13}CO$_2$ laser	1787
361.5		9P34	1787
412.0		^{13}CO$_2$ laser	1787
436.5		10R30	1787
439.0		^{13}CO$_2$ laser	1787
455.5		10R28	1787
465.0		10R16	1787
487.0		C^{18}O$_2$ laser	1787
488.0		^{13}CO$_2$ laser	1787
498.5		10R20	1787
523.5		C^{18}O$_2$ laser	1787

Table 3.4.108
Difluorosilane – SiH$_2$F$_2$

Wavelength (μm)	Uncertainty (μm)	CO$_2$ laser pump line	References
169.0		C^{18}O$_2$ laser	1787
175.5		9P20	1787
184.5		^{13}CO$_2$ laser	1787
190.5		C^{18}O$_2$ laser	1787
192.0		^{13}CO$_2$ laser	1787
193.0		C^{18}O$_2$ laser	1787
195.5		10R28	1787
261.5		C^{18}O$_2$ laser	1787
263.0		^{13}CO$_2$ laser	1787
317.5		C^{18}O$_2$ laser	1787
330.0		10P22	1787
343.0		10R14	1787
352.5		^{13}CO$_2$ laser	1787
355.0		C^{18}O$_2$ laser	1787
375.5		C^{18}O$_2$ laser	1787
443.0		^{13}CO$_2$ laser	1787
471.0		10R20	1787
494.0		10R18	1787
613.0		10R22	1787
1053.0		^{13}CO$_2$ laser	1787

Table 3.4.109
Fluorosilane – SiH$_3$F

Wavelength (μm)	Uncertainty (μm)	CO$_2$ laser pump line	References
187.0		10R10	1788
221.0		10R36	1788
236.0		10R06	1788
264.5		10R26	1788
280.5		10R38	1788
330.0		10R06	1788
340.0		10R22	1788
343.5		10R14	1788
369.0		10R16	1788
516.0		10R20	1788

Fluorosilane – SiH₃F—*continued*

Wavelength (μm)	Uncertainty (μm)	CO₂ laser pump line	References
622.0		$^{13}CO_2$ laser	1788
689.0		$^{13}CO_2$ laser	1788
1014.		10R30	1788
1056.		$^{13}CO_2$ laser	1788
1058.		$^{13}CO_2$ laser	1788
1286.		10R32	1788

Table 3.4.110
Sulfur Dioxide – SO₂

Wavelength (μm)	Uncertainty (μm)	Comments	References
128.1		6934	812
139.60		6932	1426
140.78		6929	1795
140.88		6929	1795
140.89	±0.042	2137,2138	1395,1411,1426,1468
142.00		6929	1426
142.1		6934	812
146.2		6932	812
149.7		6934	812
150.00		6929	1426
151.19		6929	1795
151.31		6929	1795
159.5		6930	811
165.2		6933	812
169.6		6930	812
171.4		6930	812
180.0		6931	812
182.0		6934	811
192.71		6929	1795
192.72	±0.058	2135,2136	1395,1411,1426,1468
193.1		6931	812
205.3		6932	812
206.40		6929	1426
208.0		6932	811
215.33		6929	1795
258.0		6930	811
282.1		6930	812
312.1		6931	811
349.1		6931	812

Table 3.4.111
Sulfur Dioxide – SO$_2$ (isotopically substituted)

Wavelength (μm)	Uncertainty (μm)	CO$_2$ laser pump line	References
134.2		9P10	1791
148.2		9P16	1653
166.8		9R18	1652
174.7		9R24	1652
184.1		9R22	1652
185.1		9R20	1652
192.0		9P10	1652
194.5		9R20	1652
208.8		9R18	1652
215.3		9P16	1652
218.2		9P32	1652
221.2		9R30	1652
232.9		9R30	1652
279.9		9P10	1652
298.9		9R20	1652
471.8		9R18	1653
505.0		9R24	1652
525.3		9R30	1652
570.3		9R16	1652
1570.2		9R16	1652

Section 3.5
COMMERCIAL GAS LASERS

Commercial gas laser types, mode of operation (cw or pulsed), wavelengths, and representative outputs are given in Table 3.5.1. The data were compiled from recent (1997–1999) laser buyers' guides and manufacturers' literature and may not be the only lasers available commercially nor may the lasers still be manufactured. Wavelengths enclosed in brackets denote the extremes of a group of discrete laser lines.

Further Reading

Eden, J. G., Ed., *Selected Papers on Gas Laser Technology*, SPIE Milestone Series Vol. 159, SPIE Optical Engineering Press, Bellingham, WA (2000).

Hecht, J., *The Laser Guidebook* (second edition), McGraw-Hill, New York (1992).

Laser Focus World Buyers Guide, Pennwalt Publishing Company, Tulsa, OK.

Table 3.5.1
Commercial Gas Lasers

Laser Type	Operation	Wavelength(s) (μm)	Output
Helium-neon (He-Ne)	cw	0.5435	0.1–3 mW
	cw	0.5941	0.5–7 mW
	cw	0.6119	0.5–7 mW
	cw	0.6328	0.5–50 mW
	cw	1.152	1–13 mW
	cw	1.523	0.5–1 mW
	cw	3.391	1–40 mW
Helium-cadmium (He-Cd)	cw	0.325	1–100 mW
	cw	0.4416	10–200 mW
Helium-silver (He-Ag$^+$)	cw	0.2243	1 mW
	pulsed	0.2243	0.1 J
Helium-gold (He-Au$^+$)	cw	[0.282–0.292]	3 mW
	pulsed	[0.282–0.292]	0.3 J
Iodine (I)	pulsed	1.315	≤ 1–3 J
Neon-copper (Ne-Cu$^+$)	cw	[0.248–0.270]	3 mW
	pulsed	[0.248–0.270]	0.3 J

Table 3.5.1—*continued*
Commercial Gas Lasers

Laser Type	Operation	Wavelength(s) (μm)	Output
Xenon-helium (Xe-He)	cw	2–4	1–600 mW
	pulsed	2–4	0.5 J
Molecular Lasers:			
Carbon dioxide (CO_2)[a]	cw	10.6	1 W–10 kW
	pulsed	10.6 (other lines from 9.2 to 11.4)	100 mJ–3 kJ
Carbon monoxide (CO)	cw, pulsed	several lines between 5 and 7	1–35 W
Nitrogen (N_2)	pulsed	0.3371	0.1–10 mJ
Nitrous oxide (N_2O)	cw	10.65 (other lines 10.3 to 11.1)	15 W
	pulsed	10.65 (other lines 10.3 to 11.1)	1 mJ
Metal Vapor Lasers:			
Copper (Cu)	cw	0.5105, 0.5782	100 W
	pulsed	0.5105, 0.5782	1–20 mJ
Gold (Au)	cw	0.628	2 W
	pulsed	0.628	0.2–0.6 mJ
Ion Lasers:			
Neon (Ne^+)	cw	0.3324 (other lines–0.3345, 0.3378, 0.3392, 0.3713, 0.373)	1 W
Argon (Ar^+)	cw, pulsed	0.4880, 0.5145 (other lines–0.351, 0.4545, 0.4579, 0.4765, 0.4965, 0.5017, 0.5287)	5 mW–50 W
Krypton (Kr^+)	cw, pulsed	0.6471 (other lines–0.3375, 0.3564,0.4762, 0.5208, 0.5309, 0.5682, 0.6764, 0.7525, 0.7993)	0.1–6 W
Argon-Krypton (Ar^+-Kr^+)	cw	many lines between 0.34–0.80	1–3 W
		several lines between 0.458–0.676	0.2–10 W
Xenon (Xe^{3+})	pulsed	0.5395	0.6 J

Table 3.5.1—*continued*
Commercial Gas Lasers

Laser Type	Operation	Wavelength(s) (μm)	Output
Excimer Lasers:			
Fluorine (F_2)	pulsed	0.157	1–60 mJ
Argon fluoride (ArF)	pulsed	0.193	3–700 mJ
Krypton chloride (KrCl)	pulsed	0.222	0.3–1.2 J
Krypton fluoride (KrF)	pulsed	0.248	5 mJ–2 J
Xenon chloride (XeCl)	pulsed	0.308	0.1–0.3 J
Xenon fluoride (XeF)	pulsed	0.351	2 mJ–0.5 J
Chemical Lasers:			
Hydrogen fluoride (HF)	cw	2.6-3.0	2–1000 W
	pulsed		50 mJ–3 J
Deuterium fluoride (DF)	cw	3.6-4.0	1–100 W
	pulsed	3.6-4.0	30 mJ–3 J
Far Infrared Lasers:			
Methanol (CH_3OH)	pulsed, cw	37.9, 70.5, 96.5, 118, 571, 699 other lines from 37 to 1224	< 1 W
Methyl fluoride (CH_3F)	pulsed, cw	496, 1222	< 1 W
Other molecules[b]	cw	lines from ~40 to 1000	0.1–1 W
	pulsed	lines from ~40 to 1200	≤ 750 mJ

(a) Operating configurations include axial gas flow (20 W–5 kW), transverse gas flow (500 W–15 kW), sealed tube (3 W–100 W), TEA (tranverse excited, atmospheric pressure), and waveguide (0.1–50 W).

(b) Methanol (fully deuterated) (CD_3OD): 41.0, 184, 229, 255 μm.
Methylamine (CH_3NH_2): 147.8 μm, other lines from 100 to 351 μm.
Methyl iodide (fully deuterated) (CD_3I): 461, 520 μm; other lines from 272 to 1550 μm.
Formic acid (HCOOH): 432.6 μm, other lines from 134 to 1213 μm.
Difluoromethane (CH_2F_2): 375, 889, 1018 μm.

Section 3.6
COMMENTS

1. Measured wavelengths were taken from Wiese, W. L., Smith, M. W., and Glennon, B. M., *Natl. Stand. Ref. Data Ser. Natl. Bur. Stand.*, NSRDS-NBS4 (1966).
2. Harrison, G. R., *MIT Wavelength Tables*, John Wiley & Sons, New York (1952).
3. As an impurity in a pulsed discharge in Ne at 1.5 torr; D = 25 mm; E/p = 140 V/cm torr.
4. Pulsed; as an impurity in 3.5 torr of He; optimum H pressure 0.01 torr; D = 7 mm.
5. Wavelength and spectral assignments were taken from Risberg, P., *Ark. Fys.*, 10, 583-606 (1956).
6. Resonance line.
7. Excitation results from the reaction NaI (hv)$\rightarrow$ Na(3p $^2P^0$) + I(5p^5 $^2P_{3/2}$).
8. Selective excitation occurs via the two-body recombination reaction; Na$^+$ + H$^-$ $\rightarrow$ Na(4s $^2S_{1/2}$) + H.
9– Pulsed; operates in an ASE mode following photodissociation of NaI with the fifth
10. harmonic of a Q-switched Nd/YAG laser at 0.2128 μm; NaI in a cell at 600 °C with 10 torr of Ar; NaI density 1.1 ± 0.3 x 10^{15} cm^{-3}. Also with ArF laser pumping (193 nm) of NaI or NaBr heated in an oven at temperatures up to 1000 °C, generally operated at 500-700 °C which corresponds to 10^{-4} - 5 x 10^{-1} torr vapor pressure; no buffer gas used.
11. Pulsed; NaI in a cell at 600 °C with 10 torr of Ar.
12. Pulsed; 0.001-0.003 torr of Na with 1-10 torr of H; D = 12 mm.
13. Pulsed; 0.001-0.003 torr of Na with 1-10 torr of H; D = 12 mm.
14. Wavelengths and spectral assignments taken from Risberg, P., *Ark. Fys.*, 10, 583-606 (1956).
15. Both components of this doublet probably oscillate, although this is not made clear in Reference 350.
16. Unclear whether both components of doublet were observed in Reference 350.
17. Unclear whether both components were observed.
18. Weak line in competition with the 15.97-μm transition.
19. Tube bore apparently 12 mm.
20. Pulsed; pumped with an ArF laser (193 nm) using KI or KBr heated in a cell to 500-700 °C without buffer gas.
21. Pulsed; pumped with an ArF laser (193 nm) using KI or KBr heated in a cell to 500-700 °C without buffer gas.
22. Pulsed; pumped with an ArF laser (193 nm) using KI or KBr heated in a cell to 500-700 °C without buffer gas.
23. Pulsed; pumped with an ArF laser (193 nm) using KI or KBr heated in a cell to 500-700 °C without buffer gas.
24. Pulsed; pumped with an ArF laser (193 nm) using KI or KBr heated in a cell to 500-700°C without buffer gas.
25. Pulsed; 0.1 torr of K with 3-5 torr of H.
26. Pulsed; pumped with an ArF laser (193 nm) using KI or KBr heated in a cell to 500-700°C without buffer gas; 0.1 torr of K with 3-5 torr of H.
27. Pulsed; K vapor excited with a Q-switched ruby laser (694.3 nm); also as for 1.177-μm line.

28. Pulsed; K vapor excited with a Q-switched ruby laser (694.3 nm); also as for 1.177-μm line.
29. Pulsed; K vapor discharge in a heat pipe at 370°C (1 torr vapor pressure) pumped with a flashlamp pumped coumarin dye laser (534.31 nm).
30. Pulsed; K vapor discharge in a heat pipe at 370 °C (1 torr vapor pressure) pumped with a flashlamp pumped coumarin dye laser (534.31 nm).
31. Pulsed; K vapor discharge in a heat pipe at 370°C (1 torr vapor pressure) pumped with a flashlamp pumped coumarin dye laser (534.31 nm).
32. Pulsed; K vapor discharge in a heat pipe at 370 °C (1 torr vapor pressure) pumped with a flashlamp pumped coumarin dye laser (534.3) nm).
33. Pulsed; K vapor discharge in a heat pipe at 370°C (1 torr vapor pressure) pumped with a flashlamp pumped coumarin dye laser (534.31 nm).
34. Pulsed; K vapor discharge in a heat pipe at 370°C (1 torr vapor pressure) pumped with a flashlamp pumped coumarin dye laser (534.31 nm).
35. Wavelengths taken from Meggers, W. F., Corliss, C. H., and Scribner, B. F., *Natl. Bur. Stand. U.S. Monogr.*, 145 (1) (1975). It is not clear whether one or both of these fine-structure components were observed in Reference 350.
36. It is not clear whether one or both of these fine-structure components were observed in Reference 350.
37. Unless otherwise indicated, wavelengths and spectral assignments are taken from data in Johansson, L., *Ark. Fys.*, 20, 135-146 (1961).
38. Pulsed; RbI or RbBr heated to 500-700 °C in a cell without buffer gas, pumped with an ArF laser (193 nm).
39. Pulsed; RbI or RbBr heated to 500-700 °C in a cell without buffer gas, pumped with an ArF laser (193 nm).
40. Pulsed; RbI or RbBr heated to 500-700 °C in a cell without buffer gas, pumped with an ArF laser (193 nm).
41. Pulsed; RbI or RbBr heated to 500-700 °C in a cell without buffer gas, pumped with an ArF laser (193 nm).
42. Pulsed; RbI or RbBr heated to 500-700 °C in a cell without buffer gas, pumped with an ArF laser (193 nm).
43. Pulsed; RbI or RbBr heated to 500-700 °C in a cell without buffer gas, pumped with an ArF laser (193 nm).
44. Pulsed; RbI or RbBr heated to 500-700 °C in a cell without buffer gas, pumped with an ArF laser (193 nm).
45. Pulsed; RbI or RbBr heated to 500-700 °C in a cell without buffer gas, pumped with an ArF laser (193 nm).
46. Pulsed; RbI or RbBr heated to 500-700 °C in a cell without buffer gas, pumped with an ArF laser (193 nm).
47. Pulsed; RbI or RbBr heated to 500-700 °C in a cell without buffer gas, pumped with an ArF laser (193 nm).
48. Pulsed; Rb vapor in a cell at about 400 °C with a He buffer, pumped with a Q-switched ruby laser.
49. Pulsed; RbI or RbBr heated to 500-700 °C in a cell without buffer gas, pumped with an ArF laser (193 nm).
50. Pulsed; RbI or RbBr heated to 500-700 °C in a cell without buffer gas, pumped with an ArF laser (193 nm).
51. Mean value for hyperfine structure components.
52. Wavelength and spectral assignments are taken from data in Johansson, L., *J. Opt. Soc. Am.*, 52, 441-447 (1962).
53. Pulsed; CsI or CsBr in a cell at 500-700 °C excited with an ArF laser (193 nm).

54. Pulsed; CsI or CsBr in a cell at 500-700 °C excited with an ArF laser (193 nm).
55. Pulsed; CsI or CsBr in a cell at 500-700 °C excited with an ArF laser (193 nm).
56. Pulsed; CsI or CsBr in a cell at 500-700 °C excited with an ArF laser (193 nm).
57. Pulsed; CsI or CsBr in a cell at 500-700 °C excited with an ArF laser (193 nm).
58. Pulsed; CsI or CsBr in a cell at 500-700 °C excited with an ArF laser (193 nm).
59. Pulsed; CsI or CsBr in a cell at 500-700 °C excited with an ArF laser (193 nm).
60. Pulsed; Cs vapor excited by a 765.8-nm nitrobenzene Raman laser; absence of He required
61. Pulsed; Cs vapor excited by a 765.8-nm nitrobenzene Raman laser; absence of He required; CsI or CsBr in a cell at 500-700 °C excited with an ArF laser (193 nm).
62. Pulsed; CsI or CsBr in a cell at 500-700 °C excited with an ArF laser (193 nm).
63. Pulsed; CsI or CsBr in a cell at 500-700 °C excited with an ArF laser (193 nm).
64. Pulsed; CsI or CsBr in a cell at 500-700 °C excited with an ArF laser (193 nm).
65. Pulsed; Cs vapor excited with a Q-switched 1.06-μm laser.
66. Pulsed; Cs vapor excited with 694.3, 765.8, 740–900 nm or 1.06-μm laser pulses; He buffer required
67. CW; vapor pressure of Cs at 175 °C optically pumped with a He lamp at 388.9 nm; D = 10 mm.
68. Pulsed; Cs vapor excited with a Q-switched 1.06-μm laser.
69. Pulsed; Cs vapor excited with a Q-switched 1.06-μm laser.
70. Pulsed; CsI or CsBr in a cell at 500-700 °C excited with an ArF laser (193 nm).
71. CW; vapor pressure of Cs at 175 °C optically pumped with a He lamp at 388.9 nm; D = 10 mm.
72. Measured wavelengths from Meggers, W. F., Corliss, C. H., and Scribner, B. F., *Natl. Bur. Stand. (U.S.) Monogr.,* 145(1) (1975).
73. Very strong self-terminating lines; gain can be greater than 42 dBm $^{-1}$.
74. Excitation believed to involve recombination of electrons with metal ions which are formed by election impact and charge-transfer during the current pulse.
75. Pulsed; short rise-time high-voltage single, double-, or multiple-pulse excitation of various Cu compounds or of Cu vapor at high temperature.
76. Pulsed; short rise-time high-voltage pulsed excitation of Cu iodide at moderately high temperature (600 °C), vapor pressure 1-10 torr; D = 9 mm.
77. Pulsed; short rise-time high-voltage single-, double-, or multiple-pulse excitation of various Cu compounds or of Cu vapor at high temperature; see text for further details.
78. Pulsed; lases in the afterglow of a slotted Cu hollow-cathode discharge, He or Ne buffer at 8-20 torr used.
79. Pulsed; lases in the afterglow of a slotted Cu hollow-cathode discharge, He or Ne buffer at 8-20 torr used
80– Pulsed; lases in the afterglow of a discharge in a slotted Ag hollow-cathode; He or
81. Ne buffer at 8-20 torr used; excitation by ion-electron recombination appears. likely; also in a segmented transversely excited device incorporating recombining silver plasmas in a few torr of He.
82. Pulsed; lases both during current pulse and in the afterglow of a discharge in a slotted Ag hollow-cathode; He or Ne buffer at 8-20 torr used.
83. Wavelengths and spectral assignments taken from Ehrhardt, J. C. and Davis, S. P., *J. Opt. Soc. Am.,* 61, 1342-1349 (1971).
84– Pulsed; Au-coated tube filled with 10 torr of Ne self-heated by repetitive high volt-
85. age short-pulse excitation; D = 16mm, however, optimum buffer pressure 25 torr of Ne or 30 torr of He; also operates with Ar or Xe buffer; D = 16.

86. No neutral Be laser transitions have been observed to date, three Be laser transitions listed by Beck et al. as neutral transitions are in fact singly ionized Be lines. Reference 1758.

87. Wavelengths and spectral assignments taken from Risberg, G., *Ark. Fys.,* 28, 381-395 (1965).

88. The wavelength resolution in Reference 136 was not great enough to determine whether one or both these fine-structure components at 3.6789254 or 3.6789565 μm were oscillating.

89. Mean calculated wavelength of fine-structure components.

90– Three possible assignments, slightly outside the stated error limits of the measured
91. wavelength reported in Reference 136, are 7p $^3P_2 \rightarrow$ 6s $^3S_1$ at 3.8670358 μm, 7p $^3P_1 \rightarrow$ 6s $^3S_1$ at 3.8681427 μm, 7p $^3P_0 \rightarrow$ 6s $^3S_1$ at 3.868638 μm.

92. Lines at 0.9218, 0.9244, 1.0952, and 1.0915 listed as neutral Mg lines in Reference 1758 are singly ionized Mg laser lines.

93. Pulsed; in a segmented, transversely excited device incorporating recombining Mg plasmas in a few torr of He.

94. CW; in Mg vapor above 450 °C in He, Ne, or Ar; D = 10 mm.

95. CW; in Mg vapor above 450 °C in He, Ne, or Ar; D = 10 mm.

96. CW; in Mg vapor above 450 °C in He, Ne, or Ar; D = 10 mm.

97. CW; in Mg vapor above 450 °C in He, Ne, or Ar; D = 10 mm.

98. CW; in Mg vapor above 450 °C in He, Ne, or Ar; D = 10 mm.

99. CW; in Mg vapor above 450 °C in He, Ne, or Ar; D = 10 mm.

100. Unless otherwise indicated, wavelengths and spectral assignments are taken from Risberg, G., *Ark. Fys.,* 37, 231-249 (1968).

101. Excitation of these transitions is believed to involve the absorption of a 249-nm photon by a Ca quasimolecule which gives some allowed character to the otherwise nonallowed $4s^2\ ^1S_0 \rightarrow$ 4s 6p $^1P^0_1$ resonance absorption at 239.856 nm.

102. Measured wavelength from Meggers, W. F., Corliss, C. H., and Scribner, B. F., *Natl. Bur. Stand. (U.S.) Monogr.,* 145(1) (1975).

103. When excited with short rise-time high-voltage pulses, this line can have a gain in excess of 300 dBm $^{-1}$.

104. Pulsed; Ca vapor with 0-1000 torr of noble gas in a heat-pipe oven at about 1200 K (Ca density about 5 x 10^{18} atoms cm^{-3}), excited with a KrF laser (249 nm).

105. Pulsed; Ca vapor with 0-1000 torr of noble gas in a heat-pipe oven at about 1200 K (Ca density about 5 x 10^{18} atoms cm^{-3}, excited with a KrF laser (249 nm).

106. Pulsed; in a hollow-cathode discharge with a DC trickle ionizing discharge; optimum buffer gas pressure ~25 m torr of Xe.

107. Short rise-time high-voltage pulsed excitation of Ca vapor at 460 °C with 3 torr of He in small-bore tubes; also CW in a 7:1:1 He-Ne-H_2 mixture at 1 torr with Ca vapor at temperatures from 590-650 °C; D = 4 mm.

108. Measured wavelength from Meggers, W. F., Corliss, C. H., and Scribner, B. F., *Natl. Bur. Stand. (U.S.) Monogr.,* 145(1) (1975).

109. The gain of this line when excited by short rise-time high-voltage pulses can be as great as 300 dBm^{-1}.

110. Pulsed; in a hollow-cathode discharge operated with a preionizing DC current, 18-85 mtorr of Xe buffer.

111. Pulsed; in a self-heated discharge tube containing pieces of Sr, optimum with 80 torr of He; D = 7 or 10 mm.

112. Pulsed; in a self-heated discharge tube containing pieces of Sr, optimum with 80 torr of He; D = 7 or 10 mm.

113. Pulsed; in a self-heated discharge tube containing pieces of Sr, optimum with 80 torr of He; D = 7 or 10 mm; also CW in ~ 10^{-2} torr of Ca vapor in a tube at ~ 600 °C with 0.1-5 torr of H.

114. Wavelengths and spectral assignments are taken from *Natl. Stand. Ref. Data Ser. Natl. Bur. Stand.*, NSRDS-NBS 34, Vol. 3 (1955), Russel, H. N. and Moore, C. E., *J. Res. Natl. Bur. Stand.*, 55, 299-306 (1955).

115. Lines which operate in an ASE mode.

116. A gain of 65 dBm^{-1} has been reported for this line. Reference 146.

117. A gain of 40 dBm^{-1} has been reported for this line. Reference 146.

118. Optically pumped in a Ba-Tl-Ar mixture by the following pair-absorption process: Ba $(6s^2\ {}^1S_0)$ + Tl $(6p\ {}^2P^0_{1/2})$ + hv (386.7 nm) $\rightarrow$ Ba $(6p\ {}^1P^0_1)$ + Tl$(6p\ {}^2P^0_{3/2})$. Reference 708.

119–
120. Two additional assignments are within the experimental error of the reported laser wavelength: 6d $^1D_2 \rightarrow$ 6p' $^1F^0_3$ at 2.9235303 μm and 6d' $^3P_1 \rightarrow$ 4f$^3F^0_2$ at 2.9236192 μm. The assignment given in the table has been suggested as the one most likely to be correct as it terminates on a metastable level in common with many other Ba-I laser lines.

121. Assignment suggested by C. C. Davis.

122. No energy level combination corresponding to these wavelengths could be found by searching *Atomic Energy Levels*, Vol. 3.

123. Note: Lines at 2.5924 and 2.9057 μm listed as neutral Ba lines in Reference 1758 are singly ionized Ba transitions.

124. Pulsed; in a hollow-cathode discharge with a sustainer DC discharge, optimum with 60 mtorr of Xe or 800-1000 torr of He; also in a discharge tube at 710-900 °C with 0.4 torr He, 0.1 to Ne, or 0.04-0.1 torr Xe; D = 2.8 cm

125. Pulsed; in Ba vapor in a tube at 500-850 °C with He, Ne, Ar, or H at 1-3 torr D = 5-10 mm.

126. Pulsed; in Ba vapor in a tube at 500-850 °C with He, Ne, Ar, or H at 1-3 torr D = 5-10 mm; also optically pumped

127. Pulsed; in a self-heated high repetition frequency (5-8 kHz) discharge in Ba vapor with 10-15 torr of Ne; D = 4 mm.

128. Pulsed; in Ba vapor in a tube at 500-850 °C with He, Ne, Ar, or H at 1-3 torr D = 5-10 mm.

129. Pulsed; in Ba vapor in a tube at 500-850 °C with He, Ne, Ar, or H at 1-3 torr D = 5-10 mm.

130. Pulsed; in Ba vapor in a tube at 500-850 °C with He, Ne, Ar, or H at 1-3 torr D = 5-10 mm.

131. Pulsed; in Ba vapor in a tube at 500-850 °C with He, Ne, Ar, or H at 1-3 torr D = 5-10 mm.

132. Pulsed; in Ba vapor in a tube at 500-850 °C with He, Ne, Ar, or H at 1-3 torr D = 5-10 mm.

133. Pulsed; in Ba vapor in a tube at 500-850 °C with He, Ne, Ar, or H at 1-3 torr D = 5-10 mm.

134. Pulsed; in Ba vapor in a tube at 500-850 °C with He, Ne, Ar, or H at 1-3 torr D = 5-10 mm.

135. Pulsed; in Ba vapor in a tube at 500-850 °C with He, Ne, Ar, or H at 1-3 torr D = 5-10 mm.

136. Pulsed; in a self-heated high repetition frequency (5-8kHz) discharge in Ba vapor with 10-15 torr of Ne; D = 4 mm.

137. Pulsed; in Ba vapor in a tube at 500-850 °C with He, Ne, Ar, or H at 1-3 torr D = 5-10 mm.

138. Pulsed; in Ba vapor in a tube at 500-850 °C with He, Ne, Ar, or H at 1-3 torr D = 5-10 mm.

139. Pulsed; in Ba vapor in a tube at 500-850 °C with He, Ne, Ar, or H at 1-3 torr D = 5-10 mm.

140. Pulsed; in Ba vapor in a tube at 500-850 °C with He, Ne, Ar, or H at 1-3 torr D = 5-10 mm.

141. Pulsed; in Ba vapor in a tube at 500-850 °C with He, Ne, Ar, or H at 1-3 torr D = 5-10 mm.

142. Pulsed; in Ba vapor in a tube at 500-850 °C with He, Ne, Ar, or H at 1-3 torr D = 5-10 mm.

143. Pulsed; in Ba vapor in a tube at 500-850 °C with He, Ne, Ar, or H at 1-3 torr D = 5-10 mm.

144. Pulsed; in Ba vapor in a tube at 500-850 °C with He, Ne, Ar, or H at 1-3 torr D = 5-10 mm.

145. Wavelength and spectral assignments taken from Johansson, I. and Contreras, R., *Ark. Fys.*, 37, 513-520 (1967).

146. Pulsed; in a segmented transversely excited device incorporating recombining Zn plasmas in a few torr of He.

147. Pulsed; in a segmented transversely excited device incorporating recombining Zn plasmas in a few torr of He.

148. Wavelength and spectral assignment taken from data in Burns, K. and Adams, K. B., *J. Opt. Soc. Am.*, 46, 94-99 (1956).

149. For those lines observed by Dubrovin et al., the optimum buffer gas pressures were low in the case of lines lasing in the rising edge of the current pulse and were 6 to 8 torr for the afterglow lines; the optimum Cd pressure was ~ 0.1 torr. Reference 152.

150. Pulsed; in 0.001-0.3 torr of Cd with 0.1-20 torr of He or Ne; discharge tube heated in a furnace; D = 15 mm; lases in rising edge of current pulse.

151– Pulsed; in 0.001-0.3 torr of Cd with 0.1-20 torr of He or Ne; discharge tube heated
152. in a furnace; D = 15 mm; lases only in the afterglow; also lases in a recombining plasma produced by vaporization of a Cd target with 1.06- or 10.6-μm laser radiation; 5 torr of He buffer used.

153– Pulsed; in 0.001-0.3 torr of Cd with 0.1-20 torr of He or Ne; discharge tube heated
154. in a furnace; D = 15 mm; lases only in the afterglow; also lases in a recombining plasma produced by vaporization of a Cd target with 1.06- or 10.6-μm laser radiation; 5 torr of He buffer used.

155. Pulsed; in 0.001-0.3 torr of Cd with 0.1-20 torr of He or Ne; discharge tube heated in a furnace; D = 15 mm; lases in rising edge of current pulse.

156. Pulsed; in 0.001-0.3 torr of Cd with 0.1-20 torr of He or Ne; discharge tube heated in a furnace; D = 15 mm; lases in rising edge of current pulse.

157. Pulsed; in 0.001-0.3 torr of Cd with 0.1-20 torr of He or Ne; discharge tube heated in a furnace; D = 15 mm; lases in rising edge of current pulse.

158. Pulsed; in 0.001-0.3 torr of Cd with 0.1-20 torr of He or Ne; discharge tube heated in a furnace; D = 15 mm; lases in rising edge of current pulse.

159. Pulsed; in 0.001-0.3 torr of Cd with 0.1-20 torr of He or Ne; discharge tube heated in a furnace; D = 15 mm; lases in rising edge of current pulse.

160. Pulsed; 0.001-0.3 torr of Cd with 0.1-20 torr of He or Ne; discharge tube heated in a furnace; D = 15 mm; lases in rising edge of current pulse.

161. Pulsed; 0.001-0.3 torr of Cd with 0.1-20 torr of He or Ne; discharge tube heated in a furnace; D = 15 mm; lases in rising edge of current pulse.

162.   Pulsed; lases in a recombining plasma produced by vaporization of a Cd target with 1.06- or 10.6-μ m laser radiation; 5 torr of He buffer used.

163–
164.   Pulsed; in 0.001-0.3 torr of Cd with 0.1-20 torr of He or Ne; discharge tube heated in a furnace; D = 15 mm; lases only in the afterglow; also lases in a recombining plasma produced by vaporization of a Cd target with 1.06- or 10.6-μm laser tradiation; 5 torr of He buffer used.

165.   Pulsed; in 0.001-0.3 torr of Cd with 0.1-20 torr of He or Ne; discharge tube heated in a furnace; D = 15 mm; lases only in the afterglow.

166.   Pulsed; in 0.001-0.3 torr of Cd with 0.1-20 torr of He or Ne; discharge tube heated in a furnace; D = 15 mm; lases in rising edge of current pulse.

167.   Pulsed; in 0.001-0.3 torr of Cd with 0.1-20 torr of He or Ne; discharge tube heated in a furnace; D = 15 mm; lases in rising edge of current pulse.

168.   Pulsed; by dissociation of 0.04 torr of $Cd(CH_3)_2$ in 1.3 torr of He in a transversely excited double-discharge laser.

169.   Pulsed; by dissociation of 0.04 torr of $Cd(CH_3)_2$ in 1.3 torr of He in a transversely excited double-discharge laser.

170.   Measured wavelength from Meggers, W. F., Corliss, C. H., and Scribner, B. F., *Natl. Bur. Stand. (U.S.) Monogr.*, 145 (1) (1975).

171–
172.   The excitation of the 6d ^3D, 7p $^3P^0$, 8s ^3S, 6d $^1D_2$, and 7p $^1P^0_1$ levels following optical pumping with 266-nm radiation involves dissociation of electronically excited Hg_2 molecules produced by the absorption of two pump photons. Reference 156.

173–
174.   Excitation of the 7s $^3S_1$ level when pulsed optical pumping with 266-nm radiation is used probably involves a collisional reaction between two excited Hg dimers. $Hg^*_2 + Hg^*_2 \rightarrow Hg\,(7s\ ^3S_1) + 3Hg$. Reference 156.

175.   May be a Hg-II line.

176.   Measured wavelength from Plyler, E. K., Blaine, L. R., and Tidwell, E. D., *J. Res. Natl. Bur. Stand.*, 55, 279-284 (1955).

177.   Assignment as given in Reference 161.

178.   The position of the 6p' $^1P^0_1$ is not very accurately known.

179.   Several additional assignments are possible for this transition; these other possibilities involve transitions between higher lying states than the ones listed here.

180.   A line at 3.34 μm reported in Reference 167 and assigned to Hg-I is a neutral Kr line.

181.   Pulsed; in an ASE mode following excitation of Hg vapor in a cell at 570 °C with the 266-nm fourth harmonic of a Nd/YAG laser.

182.   Pulsed; in a ASE mode following excitation of Hg vapor in a cell at 570 °C with the 266-nm fourth harmonic of a Nd/YAG laser.

183.   Pulsed; in a ASE mode following excitation of Hg vapor in a cell at 570 °C with the 266-nm fourth harmonic of a Nd/YAG laser.

184.   Pulsed; in a ASE mode following excitation of Hg vapor in a cell at 570 °C with the 266-nm fourth harmonic of a Nd/YAG laser.

185.   Pulsed; in a ASE mode following excitation of Hg vapor in a cell at 570 °C with the 266-nm fourth harmonic of a Nd/YAG laser.

186.   CW; optically pumped with a Hg lamp, in 10 to 120 torr of N (optimum 25 torr); D = 3 mm; also as for the 0.365-μm line above.

187.   Pulsed; in an ASE mode following excitation of Hg vapor in a cell at 570 °C with the 266-nm fourth harmonic of a Nd/YAG laser.

188.   Pulsed; in an ASE mode following excitation of Hg vapor in a cell at 570 °C with the 266-nm fourth harmonic of a Nd/YAG laser.

189. Pulsed; in 0.001 torr of Hg with 0.8-1.2 torr of He; D = 15 mm.
190. Pulsed; in 0.09-0.12 torr of Hg with 0.005-0.05 torr of He; D = 6 mm.
191. Pulsed; in an ASE mode following excitation of Hg vapor in a cell at 570 °C with the 266-nm fourth harmonic of a Nd/YAG laser.
192. Pulsed; 0.001 torr of Hg with 0.2 torr of Ar; D = 5 mm.
193. Pulsed; in a mixture of Hg and Ar; D = 5 mm.
194. Pulsed; 0.001 torr of Hg with 0.8-1.2 torr of He; D = 15 mm.
195. Pulsed; 0.001 torr of Hg with 0.2 torr of Ar; 0 = 5 mm.
196. Pulsed; 0.001 torr of Hg with 0.8 torr of Ar or 1.2 torr of He; D = 15 mm.
197. Pulsed; in an ASE mode following excitation of Hg vapor in a cell at 570 °C with the 266-nm fourth harmonic of a Nd/YAG laser.
198. Pulsed; 0.001 torr of Hg with 0.8-1.2 torr of He; D = 15 mm.
199. Pulsed; in 0.09-0.12 torr of Hg with 0.005-0.05 torr of He; D = 6 mm; in an ASE mode following excitation of Hg vapor in a cell at 570 °C with the 266-nm fourth harmonic of a Nd/YAG laser.
200. Pulsed; in an ASE mode following excitation of Hg vapor in a cell at 570 °C with the 266-nm fourth harmonic of a Nd/YAG laser.
201. CW; 0.09-0.12 torr of Hg with 0.1-1.0 torr of He, Ne, Kr, or Ar; D = 6-8 mm.
202. Pulsed; 0.09-0.12 torr of Hg with 0.005-0.05 torr of He; D = 6 mm.
203. Pulsed; 0.09-0.12 torr of Hg with 0.005-0.05 torr of He; D = 6 mm.
204. Pulsed; 0.09-0.12 torr of Hg with 0.005-0.05 torr of He; D = 6 mm.
205. Pulsed; in 0.09-0.12 torr of Hg with 0.005-0.05 torr of He; D = 6 mm.
206. Pulsed; in 0.09-0.3 torr of Hg with 0.005-0.1 torr of He, Ne, Kr, or air; D = 6 mm.
207. CW; in 0.09-0.3 torr of Hg with 0.005-0.1 torr of He, Ne, Kr, or air; D = 6 mm.
208. Pulsed; in 0.3 torr of Hg with 0.25 torr of Kr; D = 8 mm.
209. Pulsed; in 0.3 torr of Hg with 0.25 torr of Kr; D = 8 mm.
210. Pulsed; by dissociation of 60 torr of $Hg(CH_3)_2$ with 2.4 torr of He in a transversely excited double-discharge laser.
211. Pulsed; in 0.3 torr of Hg with 0.25 torr of Kr; D = 8 mm; may have been at 6.4887747 μm.
212. Pulsed; in 0.3 torr of Hg with 0.25 torr of Kr; D = 8 mm; may have been at 6.477439 μm.
213. A search of spectroscopic data on neutral revealed no likely assignment for this transition. Its positive identification as a neutral atomic B transition therefore remains uncertain.
214. Pulsed; in discharges in mixtures containing about 10 torr of He with about 0.025 torr of B-containing compounds such as B_2H_6, H_3B CO, B_5H_9, or BBr_3; lases in afterglow.
215. Wavelengths and spectral assignments taken from data in Johansson, I. and Litzen, U., *Ark. Fys.*, 34, 573-587 (1966).
216. Resonance transition. At the operating temperature some dissociation of GaI_3 into GaI occurs and the latter species may be the predominant one directly photodissociated to yield Ga 5s $^2S_{1/2}$ atoms.
217. Pulsed; by dissociation of GaI_3 with an ArF laser (193 nm); operates with 5-20 torr of GaI_3 with 300 torr of Ar at 160-210 °C
218. Pulsed; by dissociation of GaI_3 with an ArF laser (193 nm); operates with 5-20 torr of GaI_3 with 300 torr of Ar at 160-210 °C
219. Pulsed; by dissociation of 70 torr of $Ga(CH_3)_3$ with 2.8 torr of He in a transversely excited double-discharge laser.

220. Pulsed; by dissociation of 70 torr of Ga(CH$_3$)$_3$ with 2.8 torr of He in a transversely excited double-discharge laser.
221. Pulsed; by dissociation of 70 torr of Ga(CH$_3$)$_3$ with 22.8 torr of He in a transversely excited double-discharge laser.
222. Wavelengths and spectral assignments taken from data in Johansson, I. and Litzen, U., *Ark. Fys.*, 34, 573-587 (1966).
223. Pulsed; in an ASE mode following dissociation of InI in a cell at temperatures from 200-600 °C with an ArF laser (193 nm); optimum operating temperature 330 °C.
224. Pulsed; in an ASE mode following dissociation of InI in a cell at temperatures from 200-600 °C with an ArF laser (193 nm); optimum operating temperature 330 °C; far stronger than 0.4101745 μm line
225. Pulsed; in a segmented transversely excited device incorporating recombining In plasmas in a few torr of He.
226. Pulsed; in a segmented transversely excited device incorporating recombining In plasmas in a few torr of He.
227. Pulsed; in a segmented transversely excited device incorporating recombining In plasmas in a few torr of He.
228. Pulsed; by dissociation of 70 torr of In(CH$_3$)$_3$ with 2.8 torr of He in a transversely excited double-discharge laser.
229. Pulsed; by dissociation of 50 torr of In(CH$_3$)$_3$ with 2.8 torr of He in a transversely excited double-discharge laser.
230. Resonance line, measured wavelength from Meggers, W. F., Corliss, C. H., and Scribner, B. F., *Natl. Bur. Stand. (U.S.) Monogr.*, 145(1) (1975).
231. Measured wavelength and assignment from Seguie, J., *C. R. Acad. Sci. Ser. B*, 263B, 147-150 (1966).
232. Excitation follows dissociation of an exciplex such as (Tl-Hg); optimum operating conditions 0.01 torr Tl(~ 600 °C) with 400 torr Hg(~325 °C).
233. Lines at 0.5152, 0.5949, and 0.6950 μm listed as neutral Tl lines in Reference 1758 are ionized Tl laser lines.
234. Pulsed; by dissociation of 0.001-0.5 torr of Tl I vapor with an ArF laser (193 nm).
235– Pulsed; short rise-time high-voltage excitation of more than 0.01 torr of Tl with
236. several torr of Ne or He; D = 1.3, 2.0, or 3 mm, or of Tl I at 370-440 °C with added He, Ne, Ar, or Xe; D = 1.3 mm; also, in a Tl-Hg or Tl-Cd-Ar mixture excited with a N laser; D = 12 mm.
237. Pulsed; by dissociation of 120 torr of Tl(CH$_3$)$_3$ with 6 torr of He in a transversely excited, double-discharge laser.
238. Pulsed; by dissociation of 120 torr of Tl(CH$_3$)$_3$ with 6 torr of He in a transversely excited, double-discharge laser, but with 60 torr of Tl(CH$_3$)$_3$ and 2.4 torr of He.
239. Pulsed; by dissociation of 120 torr of Tl(CH$_3$)$_3$ with 6 torr of He in a transversely excited, double-discharge laser, but with 60 torr of Tl (CH$_3$)$_3$ and 2.4 torr of He.
240. Wavelengths and spectral assignment taken from data in Moore, C. E., *Natl. Stand. Ref. Dat. Ser. U.S. Natl. Bur. Stand.*, NSRDS-NBS3 (1975), St. 3.
241– The excitation mechanism for these transitions observed in He-CO and Ne-CO dis-
242. charges was originally thought to be due to dissociative excitation transfer involving He or Ne metastables, e.g., CO + Ne* (1s$_5$ or 1s$_3$) → C' + O + Ne. However, other work has indicated that, in fact, their excitation mechanism involves collisional-radiative ion-electron recombination, namely, C$^+$ + 2e → C' + e. Reference 191, 192.
243. Pulsed; in a mixture of CO$_2$ and Ne at 4 torr; D = 15 mm.
244. Pulsed; in a mixture of CO$_2$ and Ne at 4 torr; D = 15 mm and also in a mixture of 0.05 torr of CO with 2-16 torr of He, optimum 5 torr of He; D = 7 mm.

245. Pulsed; in a mixture of CO_2 and Ne at 4 torr; $D = 15$ mm and also in a mixture of 0.05 torr of CO with 2-16 torr of He, optimum 5 torr of He; $D = 7$ mm and also in a mixture of CO_2 and He; $D = 16$ mm.

246. Pulsed; in a mixture of 0.05 torr of CO with 2-16 torr of He, optimum 5 torr; $D = 7$ mm; also in CO_2 with He; $D = 16$ mm.

247. CW; in 0.01 torr of CO or CO_2 with 2 torr of He; $D = 5$ mm.

248. Pulsed; in a mixture of CO_2 and He; $D = 16$ mm.

249. CW; in 0.01 torr of CO or CO_2 with 2 torr of He; $D = 5$ mm; also pulsed in OCS and several other organic gases; also nuclear pumped by the reaction $^{10}B(n,$ alpha$)^7$Li in He-CO and He-CO_2 mixtures

250. CW; in 0.02 torr of CO with 1 torr of He; $D = 10$ mm.

251. CW; in 0.02 torr of CO with 1 torr of He; $D = 10$ mm.

252. CW; in 0.02 torr of CO with 1 torr of He; $D = 10$ mm.

253. CW; in 0.02 torr of CO with 1 torr of He; $D = 10$ mm.

254. Measured wavelengths and spectral assignments taken from Moore, C. E., *Natl. Stand. Ref. Data Ser. Natl. Bur. Stand.*, NSRDS-NBS 3 (1975), St. 2.

255. CW; 0.03 torr of $SiCl_4$ with 0.5 torr of Ne; $D = 6$ mm.

256. CW; in 0.04 torr of $SiCl_4$ with 0.5 torr of Ne; $D = 6$ mm.

257. CW; in 0.03-0.05 torr of $SiCl_4$ with 1-5 torr of Ne; $D = 6$ mm.

258. Calculated wavelengths and spectral assignments from Andrew, K. I. and Meissner, K. W., *J. Opt. Soc. Am.*, 49, 146-161 (1959).

259. The lower level of this transition is incorrectly designated $3P0_1$ in *Atomic Energy Levels*, Vol. 2.

260. Pulsed; by dissociation of 40 torr of $GaCl_4$ with 0.4 torr of He in a transversely excited double-discharge laser.

261. Pulsed; by dissociation of 40 torr of $GaCl_4$ with 0.4 torr of He in a transversely excited double-discharge laser.

262. Measured wavelength.

263. This assignment is by no means certain; this could be an ionized tin transition; it may be the same line as one at 0.657926 μm reported in *M.I.T. Wavelength Tables*. Reference 658.

264. Pulsed; in $SnCl_4$ vapor at room temperature; $D = 5.6$ mm; also in Sn vapor at 0.001-0.1 torr; $D = 7$ mm.

265. Pulsed; in a transversely excited pin laser with 0.05-3 torr of $SnCl_4$ with 85-250 torr of He; lases in afterglow.

266. Pulsed; in a segmented, transversely excited device incorporating recombining tin plasmas in a few torr of He.

267. Pulsed; by dissociation of 50 torr of $SnCl_4$ with 2 torr of He in a transversely excited double-discharge laser

268. Wavelengths and spectral assignments are taken from data in Wood, D. R. and Andrew, K. C., *J. Opt. Soc. Am.*, 58, 818-829 (1968). Level designations according to *Atomic Energy Levels* are in parentheses.

269. The vapor pressure of lead can be deduced from Honig, R. E., *R.C.A. Rev.*, 18, 195-204 (1957).

270. A single-pass gain of 600 dB m^{-1} has been reported for this line. Reference 206.

271– Appears the most likely assignment for this line; however, there are several other

  272. possible ones, namely, $5f1/2[7/2]_4 \rightarrow 6d1/2[5/2]0_3(5f\ ^3F_4 \rightarrow 6d\ ^3F0_3)$ at 1.53148 μm $5f1/2[7/2]_3 \rightarrow 6d1/2[5/2]0_3(5f\ ^3F_3 \rightarrow 6d\ ^3F0_3)$ at 1.53276 μm and $5f1/2[5/2]_2 \rightarrow 6d1/2[5/2]0_3(5f\ ^3F_2 \rightarrow 6d\ ^3F0_3)$ at 1.53310 μm. The laser wavelength reported in Reference 383 is 1.532 μm.

273.   Transient laser line requiring fast rise-time high-voltage pulsed excitation; in vapor pressure of Pb at a temperature of 800-900 °C with a He, Ne, or Ar buffer; D = 2 mm.

274.   Pulsed; transient laser line requiring fast rise-time high-voltage pulsed excitation; in vapor pressure of Pb at a temperature of 800-900 °C with a He, Ne, or Ar buffer; D = 2 mm.

275.   Pulsed; transient laser line requiring fast rise-time high-voltage pulsed excitation; in vapor pressure of Pb at a temperature of 800-900 °C with a He, Ne, or Ar buffer; D = 2 mm.

276.   Transient laser line requiring fast rise-time high-voltage pulsed excitation; in 0.2-2.0 torr vapor pressure of Pb with 3 torr of He; D = 10 mm.

277.   Pulsed; in Pb vapor in a heated tube at 1400 °C; D = 5-10 mm.

278.   Pulsed; in a segmented, transversely excited device incorporating recombining Pb plasmas in a few torr of He. May have been at 1.3152769 μm.

279.   Pulsed; in a segmented, transversely excited device incorporating recombining Pb plasmas in a few torr of He. May have been at 1.3103722 μm.

280.   Pulsed; in a segmeted, transversely excited device incorporating recombining Pb plasmas in a few torr of He.

281.   Pulsed; by dissociation of 0.5 torr of $Pb(CH_3)_4$ with 15 torr of He in a transversely excited double-discharge laser.

282.   Pulsed; by dissociation of 0.5 torr of $Pb(CH_3)_4$ with 15 torr of He in a transversely excited double-discharge laser.

283.   Pulsed; by dissociation of 0.06 torr of $Pb(CH_3)_4$ with 0.6 torr of He in a transversely excited double-discharge laser.

284.   Wavelengths and spectral assignments taken from data in Moore, C. E., *Natl. Stand. Ref. Data Ser. Natl. Bur. Stand.*, NSRDS-NBS 3(1975), St. 5.

285.   There is doubt about the accuracy of the determination of the wavelength of this line and the assignment here to neutral N is probably dubious.

286.   Tentative assignment made by C. C. Davis under the assumption that the reported wavelength is accurate.

287–   Excitation mechanism for this transition in a Ne-N or He-N discharge was original-
288.   ly thought to be due to dissociative excitation transfer involving He or Ne metastables, e.g., N_2 + Ne* ($1s_5$ or $1s_3$) →N' + N + Ne. However, recent work has indicated that in fact the excitation involves collisional-radiative ion-electron recombination, namely, N^+ + 2e → N' + e. References 191, 192.

289.   Alternative assignment suggested by C. C. Davis.

290–   This line, first observed and assigned by Sutton, is reported by him at a measured
291.   wavelength which agrees very well with the calculated wavelength of his assignment based on energy level values for NI reported in *Atomic Energy Levels*, Vol. 1. However, in view of the revision of the NI energy level scheme reported by Moore. Sutton's assignment is incorrect. The assignment given here is suggested by C. C. Davis on the basis of Reference 219.

292.   A possible assignment for this line suggested by C. C. Davis is 4s $^4P_{1/2}$ → 3p at 1.454855 μm.

293.   A possible assignment for this line, suggested by C. C. Davis is 4p $^4D^0_{3/2}$ → 4s $^2P_{1/2}$ at 3.7972035 μm.

294.   A possible assignment for this line, suggested by C. C. Davis is 4p $^4S^0_{3/2}$ → 3d $^4P_{5/2}$ at 3.816051 μm.

295.   Pulsed; in a Hg-N mixture at 0.001-0.02 torr; D = 3 mm.

296.   Pulsed; in a theta pinch discharge in 1 torr of N; D = 25 mm.

297.   Pulsed; in a theta pinch discharge in 1 torr of N; D = 25 mm.

298.   Pulsed; in a Hg-N mixture at 0.001-0.02 torr; D = 3 mm.

299.   Pulsed; in a Hg-N mixture at 0.001-0.02 torr; D = 3 mm.

300.   Pulsed; in a Hg-N mixture at 0.001-0.02 torr; D = 3 mm.

301.   Pulsed; in a Hg-N mixture at 0.001-0.02 torr; D = 3 mm.

302.   Pulsed; in a Hg-N mixture at 0.001-0.02 torr; D = 3 mm.

303.   Pulsed; in a mixture of N and He at about 4 torr; D = 15 mm.

304.   Pulsed; 0.2-0.7 torr of N with 3 torr of He or 0.15 torr of N only; D = 15 mm; also quasi-CW when nuclear pumped using 75-375 torr of a Ne-N_2 mixture with <0.001 torr of N_2; D = 2.5 cm.

305.   Pulsed; in 0.3 torr of N with 12 torr of He; D = 11 mm.

306.   Pulsed; in a mixture of N and He at 4 torr; D = 15 mm.

307.   Pulsed; in a mixture of N and He at 4 torr; D = 15 mm.

308.   CW; in 0.02-0.2 torr of N or nitrous oxide with 0.01-0.1 torr of O, H, He, or Ne; D = 3 mm; or in a mixture of N and He at 5 torr; D = 15 mm.

309.   CW; in 0.02-0.2 torr of N or nitrous oxide with 0.01-0.1 torr of O, H, He, or Ne; D = 3 mm; or in a mixture of N and He at 5 torr; D = 15 mm, also quasi-CW in a nuclear pumped 75-175 torr Ne-N_2 mixture with <0.01 torr of N_2; D = 2.5 cm.

310.   Pulsed; in 0.2 torr of N with 100 torr of He in a transversely excited pin laser.

311.   Line, first observed and assigned by Sutton (Reference 219), is reported by him at a measured wavelength that agrees very well with the calculated wavelength of his assignment based on energy level values for NI reported in *Atomic energy Levels*, Vol. 1. However, in view of the revision of the NI energy level values reported by Moore above, Sutton's assignment is incorrect. The assignment given here was suggested by C. C. Davis.

312.   Pulsed; in 0.2 torr of N with 100 torr of He in a transversely excited pin laser; Reference contains an incorrect wavelength and/or transition assignment; may be at 1.0643981 μm.

313.   Pulsed; in 0.2 torr of N with 100 torr of He in a transversely excited pin laser; Reference contains an incorrect wavelength and/or transition assignment; may be at 1.0623177 μm.

314.   Pulsed; in 0.7 torr of N or in 0.15 of N with 3 torr of He; D = 15 mm (or 75 mm?).

315.   CW; in 0.03 torr of nitric or nitrous oxide with 2 torr of He or 1 torr of Ne; D = 5 mm.

316.   Pulsed; in 0.2-0.7 torr of N or 0.15 torr of N with 3 torr of He; D = 15 mm (or 75 mm?).

317.   CW; in 0.03 torr of nitric or nitrous oxide with 2 torr of He or 1 torr of Ne.

318.   Pulsed; in 0.2-0.7 torr of N with 3 torr of He; D = 15 mm (or 75 mm?).

319.   Pulsed; in 0.15 torr of N with 3 torr of He or in 0.2-0.7 torr of N only; D = 15 mm (or 75 mm?).

320.   Wavelengths and spectral assignments taken from data in Martin, W. C., *J. Opt. Soc. Am.*, 49, 1071-1085 (1959).

321.   Probably an ionized P line.

322.   Assigned by C. C. Davis the assignment given in Reference 184, namely, 5s $^2P_{1/2}$ → 4p at 2.0596346 μm, is substantially outside the error of the measured wavelength. Reference 221.

323.   A line at 0.784563 μm listed in Reference 244 as a neutral transition is a singly ionized P transition.

324.   Pulsed; in 0.04 torr of P; D = 3 mm.

325.   Pulsed; in P vapor at 0.02-0.1 torr with 0.2-3.5 torr of He; D = 9 mm.

326. Pulsed; in P vapor at 0.02-0.1 torr with 0.2-3.5 torr of He or Ne, lases in afterglow; D = 9 mm.

327. Pulsed; in P vapor at 0.02-0.1 torr with 0.2-3.5 torr of He or Ne, lases in afterglow; D = 9 mm.

328. Pulsed; in P vapor at 0.02-0.1 torr with 0.2-3.5 torr of He or Ne, lases in afterglow; D = 9 mm.

329. Pulsed; in P vapor at 0.02-0.1 torr with 0.2-3.5 torr of He or Ne, lases in afterglow; D = 9 mm.

330. Pulsed; in P vapor at 0.02-0.1 torr with 0.2-3.5 torr of He or Ne, lases in afterglow; D = 9 mm.

331. Pulsed; in P vapor at 0.02-0.1 torr with 0.2-3.5 torr of He or Ne, lases in afterglow; D = 9 mm.

332. Pulsed; P vapor at 0.02-0.1 torr with 0.2-3.5 torr of He or Ne, lases in afterglow; D = 9 mm.

333. Pulsed; in P vapor at 0.02-0.1 torr with 0.2-3.5 torr of He or Ne, lases in afterglow; D = 9 mm.

334. Reference 222 lists only the second of these two assignments, however, the accuracy of the wavelength reported there is not sufficiently great to rule out the other possible assignment.

335. Wavelength reported in Reference 222 is 1.42 ± 0.01 μm, so this may be the transition immediately below whose laser wavelength was accurately measured. Reference 175.

336. Wavelength reported in Reference 222 is 1.80 ± 0.01, so this may be the transition immediately above whose laser wavelength was accurately measured. Reference 175.

337. Several additional assignments are possible for this line; these other possibilities involve transitions between higher-lying energy levels than the ones listed.

338. Pulsed; double-pulse excitation of Ar vapor at a pressure about 1 mtorr with 8 torr of He or Ne; D = 8 mm.

339. Pulsed; double-pulse excitation of Ar vapor at a pressure about 1 mtorr with 4 torr of He or Ne; D = 8 mm.

340. Pulsed; double-pulse excitation of Ar vapor at a pressure about 1 mtorr with 4 torr of He or Ne; D = 8 mm.

341. Pulsed; double-pulse excitation of Ar vapor at a pressure about 1 mtorr with 6 torr of He or 4 torr of Ne; D = 8 mm, also following dissociation of 0.04 torr of $AsCl_3$ with 1.6 torr of He in a double-discharge transversely excited laser.

342. Pulsed; double-pulse excitation of Ar vapor at a pressure about 1 mtorr with 6 torr of He or 4 torr of Ne; D = 8 mm, also following dissociation of 0.04 torr of $AsCl_3$ with 1.6 torr of He in a double-discharge transversely excited laser.

343. Pulsed; double-pulse excitation of Ar vapor at a pressure about 1 mtorr with 4 torr of He or Ne; D = 8 mm.

344. Pulsed; double-pulse excitation of Ar vapor at a pressure about 1 mtorr with 8 torr of He or Ne, and also with 1.4 torr of added Ar; D = 8 mm.

345. Pulsed; by dissociation of 0.04 torr of $AsCl_3$ with 1.6 torr of He in a double-discharge, transversely excited laser.

346. Pulsed; by dissociation of 0.04 torr of $AsCl_3$ with 1.6 torr of He in a double-discharge, transversely excited laser.

347. Pulsed; double-pulse excitation of Ar vapor at a pressure about 1 mtorr with 8 torr of added He; D = 8 mm.

348. Pulsed; by dissociation of 0.04 torr of $AsCl_3$ with 1.6 torr of He in a double-discharge, transversely excited laser.

349.   Pulsed; double-pulse excitation of Ar vapor at a pressure about 1 mtorr with 8 torr of added He or 4 torr of Ne; D = 8 mm.

350.   Pulsed; by dissociation of 0.04 torr of $AsCl_3$ with 1.6 torr of He in a double-discharge, transversely excited laser.

351.   Pulsed; by dissociation of 0.04 torr of $AsCl_3$ with 1.6 torr of He in a double-discharge, transversely excited laser.

352.   Pulsed; by dissociation of 0.04 torr of $AsCl_3$ with 1.6 torr of He in a double-discharge, transversely excited laser.

353.   Pulsed; by dissociation of 0.04 torr of $AsCl_3$ with 1.6 torr of He in a double-discharge, transversely excited laser.

354.   Pulsed; by dissociation of 0.04 torr of $AsCl_3$ with 1.6 torr of He in a double-discharge, transversely excited laser.

355.   Pulsed; by dissociation of 50 torr of Sb $(CH_3)_3$ with 1.5 torr of He in a transversely excited double-discharge laser. (First pulse originally intended for preionization, but in fact serves to dissociate metal complex. Second pulse excites metal atoms.

356.   Mean value of hyperfine component wavelengths listed by Meggers, W. F., Corliss, C. H., and Scribner, B. F., *Natl. Bur. Stand. (U.S.) Monogr.*, 145(1) (1975).

357.   This is a transient laser line; optical pumping results following absorption of XeCl laser photons on the bismuth resonance transition at 306.8 nm.

358.   A line at 0.475 μm listed in Reference 386 is not included as a neutral Bi laser line. This line was generated by stimulated Raman scattering of the 308-nm pump radiation.

359–   Pulsed; in an ASE mode in Bi vapor in a self-heated discharge tube with added He,
360.   Ne, or Ar; optimum with 32 torr of Ne; D = 8 or 40 mm; also by optically pumping Bi atoms at densities between 10^{16} and 10^{17} cm^{-3} with a XeCl laser (λ ~ 308 nm).

361.   Pulsed; by dissociation of 60 torr of $Bi(CH_3)_3$ with 2.4 torr of He in a transversely excited, double-discharge laser.

362.   Calculated wavelengths and spectral assignments from data in Davis, D. S. and Andrew, K. L., *J. Opt. Soc. Am.*, 68, 206-235 (1978); Davis, D. S., Andrew, K. L., and Verges, L., *J. Opt. Soc. Am.*, 28, 235-242 (1978).

363.   Several different assignments are possible for this line; these other possibilities involve transitions between higher-lying levels than the ones listed here.

364–   Pulsed; by dissociation of 90 torr of VCl_4 with 2.3 torr of He in a transversely
365.   excited double-discharge laser.

366–   This is the electric-dipole-forbidden and electric-quadrupole-allowed auroral line of
367.   atomic O which becomes weakly electric-dipole-allowed by virtue of a collision complex such as Ar-O($^1S_0$). This complex may be weakly bound, as in the case of Xe-O($^1S_0$), in which case this laser transition might equally well be referred to as a molecular laser transition.

368–   This is the O-I triplet, as shown in Reference 228. The quartet oscillation is due to
369.   the large Doppler width of the line (caused by excitation and dissociation of O molecules) and radiation trapping at the line center causing gain to occur only in the wings of the line.

370.   Tunitskii and Cherkasov have some reservations about the definite assignment of this laser line they observed to O I. Reference 190.

371.   Mean value of wavelength of fine structure components.

372.   Laser oscillation could have been occurring on several components of this transition, but were not capable of being separately resolved. Reference 236.

373.   Pulsed; 5-15 torr of O with several atmospheres of Ar, Kr, or Xe or 0.1-10 torr of N$_2$O with several atmospheres of Ar; excited in each case with a high-energy electron beam.

374–   CW; in approximately 0.01-0.04 torr of O with 0.35 torr of Ne, 1.4 torr of Ar, CO
375.   or CO$_2$, or He; D = 7 mm; as an impurity in Br with He or Ne, also in NO and in pure O at 0.1-2 torr; D = 10 mm; also in a transversely excited pin laser in 2 torr of O with 80 torr of He.

376–   CW; in approximately 0.01-0.04 torr of O with 0.35 torr of Ne, 1.4 torr of Ar, CO
377.   or CO$_2$, or He; D = 7 mm; as an impurity in Br with He or Ne, also in NO and in pure O at 0.1-2 torr; D = 10 mm; also in a transversely excited pin laser in 2 torr of O with 80 torr of He.

378–   CW; in approximately 0.01-0.04 torr of O with 0.35 torr of Ne, 1.4 torr of Ar, CO
379.   or CO$_2$, or He; D = 7 mm; as an impurity in Br with He or Ne, also in NO and in pure O at 0.1-2 torr; D = 10 mm; also in a transversely excited pin laser in 2 torr of O with 80 torr of He.

380–   CW; in approximately 0.01-0.04 torr of O with 0.35 torr of Ne, 1.4 torr of Ar, CO
381.   or CO$_2$, or He; D = 7 mm; as an impurity in Br with He or Ne, also in NO and in pure O at 0.1-2 torr; D = 10 mm; also in a transversely excited pin laser in 2 torr of O with 80 torr of He.

382.   Pulsed; in CO$_2$ and Ne at 4 torr; D = 15 mm.

383.   Pulsed; in a transversely excited pin laser in 2 torr of O with 80 torr of He.

384.   CW; 0.08 torr of O with 0.5-1.0 torr of He or Ne (probably D = 5 or 7 mm).

385.   CW; 0.08 torr of O with 0.5-1.0 torr of He or Ne (probably D = 5 or 7 mm), and also in 0.03-0.3 torr of O with several millitorr of water vapor; D=15 mm.

386.   CW; 0.08 torr of O with 0.5-1.0 torr of He or Ne (probably D = 5 or 7 mm).

387.   Pulsed; in pure O; D=15mm(or 75 mm?).

388.   CW; 0.08 torr of O with 0.5-1.0 torr of He or Ne (probably D = 5 or 7 mm).

389.   Pulsed; in pure O; D = 15 mm (or 75 mm?).

390.   CW; 0.08 torr of O with 0.5-1.0 torr of He or Ne (probably D = 5 or 7 mm).

391.   This transition is electric-dipole-forbidden, electric-quadrupole-allowed.

392.   Measured wavelengths and spectral assignments are from Jacobsson, L. R., *Ark. Fys.,* 34, 19-31 (1966).

393.   The 4f ^{3}F$_4$ and 4f ^{3}F$_3$ levels are separated by only 0.006 cm^{-1}, so these two possible assignments lie within a Doppler width of each other.

394.   Incorrectly assigned in Reference 242, assignment here made by present author.

395.   Lines listed in previous compilations at 0.516032 and 0.521962 μm are ionized sulfur lines. References 244, 659, 660.

396.   Pulsed; by photodissociation of OCS with a Kr*$_2$ laser (146 nm); optimum mix was 1.5 torr OCS, 25 torr SF$_6$and 25 torr N$_2$.

397.   CW; 0.03 torr of SF$_6$ or 0.03 torr of SF$_6$ with 2 torr of He; also in H$_2$S with He, Ne, or air; D = 5 mm.

398.   CW; 0.03 torr of SF$_6$ or 0.03 torr of SF$_6$ with 2 torr of He; also in H$_2$S with He, Ne, or air; D = 5 mm.

399.   Pulsed; in 0.2-0.8 torr of SF$_6$, CS$_2$, SO$_2$, H$_2$S, or OCS with 1-10 torr of He; optimum with 0.4 torr of SF$_6$ and 5 torr of He; D = 2.5 cm.

400.   Pulsed; in 5 torr of He containing <0.1 torr of SF$_6$.

401.   Pulsed; in 5 torr of He containing <0.1 torr of SF$_6$.

402.   Pulsed; in a low-pressure discharge in SO$_2$ at about 0.1 torr with or without added He.

403.   Pulsed; in 5 torr of He containing <0.1 torr of SF$_6$.

404. Pulsed; in a low-pressure discharge in SO_2 at about 0.1 torr with or without added He.

405. This transition is electric-dipole-forbidden; because $\Delta s \neq 0$, it becomes electric-quadrupole-allowed through deviations from true LS coupling.

406. CO buffer used to quench any population in the $Se(^3P_1)$ or $(^1D_2)$ lower laser levels produced either by direct photolysis or by quenching of higher-lying levels.

407. Pulsed; by photodissociation of carbonyl selenide with a Xe^*_2 laser (172 nm); operating mixture 1 torr of OCSe with 50 torr of CO.

408. Pulsed; by photodissociation of carbonyl selenide with a Xe^*_2 laser (172 nm); operating mixture 1 torr of OCSe with 50 torr of CO.

409. Pulsed; by dissociation of 80 torr of $Se(CH_3)_2$ with 2.0 torr of He in a transversely excited double-discharge laser.

410–
411. Spectral assignments and calculated wavelengths are from data in Morillon, C. and Verges, J., *Phys. Scripta*, 12, 129-144 (1975).

412–
413. Although Chou and Cool assigned this transition to Te I, they inadvertently listed a spectral assignment of Te II fairly close to the measured wavelength. The transition listed here is given by C. C. Davis as the likely correct assignment; its wavelength agrees very well with the measured wavelength given in Reference 155 and involves low-lying levels.

414. Pulsed; Te I at a temperature of 125-250 °C with 0.1-0.25 torr of Ne; D = 6 mm.

415. Pulsed; 0.001-0.002 torr of TeI with 0.2 torr of Ne; D = 3 mm.

416. CW; in a Te I-Ne mixture.

417. Pulsed; by dissociation of 60 torr of $Te(CH_3)_2$ with 1.6 torr of He in a transversely excited double-discharge laser; Reference contains an incorrect wavelength and/or transition assignment.

418. Pulsed; by dissociation of 40 torr of $Te(CH_3)_2$ with 1.0 torr of He in a transversely excited double-discharge laser; Reference contains an incorrect wavelength and/or transition assignment.

419. Pulsed; by dissociation of 40 torr of $Te(CH_3)_2$ with 1.0 torr of He in a transversely excited double-discharge laser; Reference contains an incorrect wavelength and/or transition assignment.

420. Calculated wavelengths and spectral assignments taken from data in Sugar, J., Meggers, W. F., and Cams, P., *J. Res. Natl. Bur. Stand.*, 77A, 1-43 (1973).

421. Difficult laser line to excite; wavelength not measured very accurately in Reference 246; spectral assignments seem likely to be correct, however, as it involves low-lying levels, and calculated wavelength is in good agreement with experimental values.

422. Assignment suggested by C. C. Davis.

423. Where a level designation is given with a prime, it is to indicate that this is the second-lowest energy level with this designation and has a different core configuration.

424. Pulsed; in Tm vapor above 800 °C (Tm vapor pressure ~ 0.1 torr) with 2.5 torr of He, 1.5 torr of Ne, or 0.8 torr of Ar.

425. Pulsed; in Tm vapor above 800 °C (Tm vapor pressure ~ 0.1 torr) with 2.5 torr of He, 1.5 torr of Ne, or 0.8 torr or Ar.

426. Pulsed; in Tm vapor above 800 °C (Tm vapor pressure ~ 0.1 torr) with 2.5 torr of He, 1.5 torr of Ne, or 0.8 torr of Ar.

427. Pulsed; in Tm vapor above 800 °C (Tm vapor pressure ~ 0.1 torr) with 2.5 torr of He, 1.5 torr of Ne, or 0.8 torr of Ar.

428. Pulsed; in Tm vapor above 800 °C (Tm vapor pressure ~ 0.1 torr) with 2.5 torr of He, 1.5 torr of Ne, or 0.8 torr of Ar.

429.   Pulsed; in Tm vapor above 800 °C (Tm vapor pressure ~ 0.01 torr) with 2.5 torr
of He, 1.5 torr of Ne, or 0.8 torr of Ar.

430.   Pulsed; in Tm vapor above 800 °C (Tm vapor pressure ~ 0.01 torr) with 2.5 torr
of He, 1.5 torr of Ne, or 0.8 torr of Ar.

431.   Pulsed; in Tm vapor above 800 °C (Tm vapor pressure ~ 0.1 torr) with 2.5 torr of
He, 1.5 torr of Ne, or 0.08 torr of Ar.

432.   Pulsed; in Tm vapor above 800 °C (Tm vapor pressure ~ 0.1 torr) with 2.5 torr of
He, 1.5 torr of Ne, or 0.8 torr of Ar.

433.   Pulsed; in Tm vapor above 800 °C (Tm vapor pressure ~ 0.1 torr) with 2.5 torr of
He, 1.5 torr of Ne, or 0.8 torr of Ar.

434.   Pulsed; in Tm vapor above 800 °C (Tm vapor pressure ~ 0.1 torr) with 2.5 torr of
He, 1.5 torr of Ne, or 0.8 torr of Ar.

435.   Pulsed; in Tm vapor above 800 °C (Tm vapor pressure ~ 0.1 torr with 2.5 torr of
He, 1.5 torr of Ne, or 0.8 torr of Ar.

436.   Pulsed; in Tm vapor above 800 °C (Tm vapor pressure ~ 0.1 torr) with 2.5 torr of
He, 1.5 torr of Ne, or 0.8 torr of Ar.

437.   Pulsed; in Tm vapor above 800 °C (Tm vapor pressure ~ 0.1 torr) with 2.5 torr of
He, 1.5 torr of Ne, or 0.8 torr of Ar.

438.   Pulsed; in Tm vapor above 800 °C (Tm vapor pressure ~ 0.1 torr) with 2.5 torr of
He, 1.5 torr of Ne, or 0.8 torr of Ar.

439.   Pulsed; in Tm vapor above 800 °C (Tm vapor pressure ~ 0.1 torr) with 2.5 torr of
He, 1.5 torr of Ne, or 0.8 torr of Ar.

440.   Measured wavelengths and spectral assignments are taken from Liden, K., The arc
spectrum of fluorine, *Ark., Fys.*, 1, 229-267 (1949).

441.   Measured wavelength from Reference 155, ± 0.004 μm.

442–   Reference 254 reports 20 unidentified lines, believed to be from atomic F, at wave-
443.   lengths between 1.5900 and 9.3462 μm, but lists no actual wavelengths. For
further details of the F laser, see the text of *CRC Handbook of Laser Science and
Technology, Vol. II,*, section 1.

444.   Pulsed; in a 100:1 He-NF$_3$ or He-F$_2$ mixture at pressures from 0.25-5 atm excited
in a TEA laser.

445.   Pulsed; in a vacuum UV photopreionized TEA laser with a 50:1 He-F$_2$ mixture at
160-1600 torr.

446.   Pulsed; in a 100:1 He-NF$_3$ or He-F$_2$ mixture at pressures from 0.25-5 atm excited
in a TEA laser.

447.   Pulsed; in a 100:1 He-NF$_3$ mixture at optimum pressures from 100-150 torr in a
transversely excited pin laser.

448.   Pulsed; in flowing CF$_4$, SF$_6$, or C$_2$F$_6$ at 0.03-0.1 torr with 2-10 torr He; D = 25
mm; also in 0.05 torr of HF with 0.3 torr of He; He essential with HF; also in a
vacuum-UV photopreionized TEA laser with a 50:1 He-F$_2$ mixture at 160-1600
torr.

449.   Pulsed; in flowing CF$_4$, SF$_6$, or C$_2$F$_6$ at 0.03-0.1 torr with 2-10 torr He; D = 25
mm; also in 0.05 torr of HF with 0.3 torr of He; He essential with HF; also in a
vacuum-UV photopreionized TEA laser with a 50:1 He-F$_2$ mixture at 160-1600
torr.

450.   Pulsed; in flowing CF$_4$, SF$_6$, or C$_2$F$_6$ at 0.03-0.1 torr with 2-10 torr He; D = 25
mm; also in 0.05 torr of HF with 0.3 torr of He; He essential with HF; also in a
vacuum-UV photopreionized TEA laser with a 50:1 He-F$_2$ mixture at 160-1600
torr.

451.   Pulsed; in a 100:1 He-NF$_3$ or He-F$_2$ mixture at high pressure, with optimum ~1.1
atom, excited in a TEA laser.

452. Pulsed; in a flowing He or He/Ar mixture at 5 torr containing a small amount of F.
453. Pulsed; in a vacuum-UV photopreionized TEA laser with 50:1 He-F_2-mixture at 160-1600 torr.
454. Pulsed; in a double-discharge TEA laser with a 0.1% F in He mixture at 0.5-2 atm.
455. Pulsed; in a double-discharge TEA laser with a 0.1% F in He mixture at 2.5 atm.
456. Pulsed; in a 100:1 He-F_2 mixture at 10-50 torr; D = 6 mm.
457. Pulsed; in a double-discharge TEA laser with a 0.1% F in He mixture at 1-3 atm.
458. Pulsed; in a vacuum-UV photopreionized TEA laser with 50:1 He-F_2-mixture at 160-1600 torr.
459. Pulsed; in a 100:1 He-F_2 mixture at 1-5 torr; D = 6 mm; also in a vacuum UV photopreionized TEA laser with a 50:1 He-F_2 mixture at 160-1600 torr.
460. Pulsed; in 0.05 of HF with 0.3 torr of He; D = 10 cm; also in a 100:1 He-F_2 mixture at 1-5 torr; D = 6 cm, or excited in a TEA laser.
461. Pulsed; in a mixture of UF_6 and He or WF_6 and He in a transversely excited double-discharge pin laser.
462. Measured wavelengths and spectral assignments are taken from Radziemski, L. J., Jr. and Kaufman, V., *J. Opt. Soc. Am.*, 59, 429-443 (1969); and Humphreys, C. S. and Paul, E., Jr., *J. Opt. Soc. Am.*, 62, 432-439 (1972).
463. The upper level of this laser transition appears to be excited selectively by excitation transfer from the 4s state of Ar I. Reference 264.
464. Lines at 1.589, 2.499, 2.535, 2.602, 2.784, and 3.801 μm observed to oscillate CW in discharge through He-Freon (CCl_2F_2) mixtures are also possibly Cl-I lines. Reference 261.
465. CW; in Cl at 0.01-0.08 torr with 0.3-3 torr of He or Ne; D = 6 mm; also in Freon at 0.001 torr with 0.8 torr of Ne; D = 7 mm.
466. Pulsed; in 0.3 torr of HCl with 0.1 torr of He or Ne; D = 14 mm.
467. Pulsed; in Cl at 0.01-0.08 torr with 0.3-3 torr of He or Ne; D = 6 mm; also in Freon at 0.001 torr of Ne; D = 7 mm; pulsed; in 0.3 torr of HCl with 0.1 torr of He or Ne; D = 14 mm.
468. CW; in a mixture of Freon (CCl_2F_2) and He at 3.3 torr.
469. CW; in 0.1 torr of Cl or in HCl or 0.3 torr of silicon tetrachloride with 0.1 torr of He or Ne
470. CW; in 0.1 torr of Cl or in HCl or 0.3 torr of silicon tetrachloride with 0.1 torr of He or Ne
471. Pulsed; in 0.3 torr of HCl with 0.1 torr of He or Ne; D = 14 mm, or CW in 0.09 torr of Cl with 1.5-7.2 torr of He; D = 25 mm.
472. CW; in 0.09 torr of Cl with 2.1 torr of Ar; D = 25 mm.
473. Pulsed; in 0.6 torr of Cl with 17-30 torr of He; D = 25 mm.
474. Pulsed; in 0.6 torr of Cl with 17-30 torr of He; D = 25 mm.
475. Unless otherwise indicated, measured wavelength and spectral assignments are taken from Humphreys, C. J. and Paul, E., Jr., *J. Opt. Soc. Am.*, 62 432-439 (1972).
476. Calculated vacuum wavelength from Tech, J. L., *J. Res. Natl. Bur. Stand.* 67A, 505-554 (1963).
477. Magnetic dipole transition.
478. Lines near 0.8446 μm originally thought to be Br lines are in fact O lines. References 196, 228.
479. CW; in a 14:1 $CBrF_3$-He mixture at 2.8 torr; also pulsed in a 1:100 Br-He mixture at 51 torr; D = 10 mm.
480. CW; in 0.3 torr of hydrogen bromide; D = 12 mm.

481. CW; in 0.3 torr of hydrogen bromide; D = 12 mm.
482. Pulsed; by flash photolysis of IBr at 0.5-5 torr; D = 8 mm; also by flash photolysis of CF_3Br, optimum pressure about 40 torr; D = 7 mm, and as a result of the chemical reaction: $I(5p^5\ ^2P^0_{1/2}) + Br_2 -> IBr + Br(4p^5\ ^2P^0_{1/2})$.
483. CW; in 0.3 torr of hydrogen bromide; D = 12 mm.
484. Unless otherwise indicated, calculated wavelengths and spectral assignments are taken from data in Minnhagen, L., *Ark. Fys.*, 21, 415-478 (1962).
485. This line may be an ionized I transition.
486. Magnetic dipole transition. For further discussion of laser systems based on this transition, see the text of *CRC Handbook of Laser Science and Technology, Vol. II*, section 1.
487. This assignment is quite likely to be correct, although Reference 400 lists others. For example, a possible alternative assignment is $8p[2]^0_{5/2} \rightarrow 7s[2]_{3/2}$ at 1.5533932 μm.
488. Measured wavelength and spectral assignment from Humphreys, C. J. and Paul, E., Jr., *J. Opt. Soc. Am.*, 62, 432-439 (1972).
489. Strongest transitions in CW gas discharge excitation.
490. Pulsed; in 0.1 torr of I with a few torr of He; D = 5 mm.
491. Pulsed; in 0.1 torr of I with a few torr of He; D = 5 mm.
492. Pulsed; in 0.1 torr of I with a few torr of He; D = 5 mm.
493. Pulsed; in 0.1 torr of I with a few torr of He; D = 5 mm.
494– Pulsed; by flash photolysis of tens of torr of various I-containing organic
  495. compounds, such as CF_3I, CH_3I, C_3F_7I, with or without a noble gas buffer; D is not critical; also in 0.18 torr of CF_3I with 70 torr of N in a vacuum-UV photopreionized TEA laser also operates CW as a result of a chemical reaction and with optical pumping.
496. Pulsed; in 0.3 torr of HI; D = 14 mm.
497. CW; in 0.05 torr of CH_2I_2 with or without added Ar; D ~ 12 mm.
498. CW; in 0.3 torr of HI with 0.3 torr of Ne; D = 14 mm.
499. CW; in HI; D = 12 mm or in 0.5 torr of I with 10 torr of He; D = 12.7 mm.
500. CW; in I vapor, CH_3I, CF_3I, or HI, with added He, Ar or Xe; optimum 0.5 torr of I with 5 torr of He; D = 2-8 cm.
501. CW; in HI or I vapor with or without added He; D = 13 mm.
502. CW; in I, CH_3I, CF_3I, or HI, with added Helium, Ar, or Xe; optimum 0.4 torr of I with 5 torr of He; D = 2-8 cm.
503. CW; in 0.05 torr of CH_2I_2 with added Ar; Ar was essential to obtain laser action; D ~ 12 mm.
504. CW; in I vapor, CH_3I, CF_3I, or HI, with added He, Ar, or Xe; optimum, 0.4 torr of I with 5 torr of He; D = 2-8 cm.
505. CW; in I vapor, CH_3I, CF_3, I, or HI, with added He, Ar, or Xe; optimum 0.4 torr of I with 5 torr of He; D = 2-8 cm.
506. CW; in I vapor; CH_3I, CF_3, I, or HI, with added He, Ar, or Xe; optimum 0.4 torr of I with 5 torr of He; D = 2-8 cm.
507. CW; in I vapor, CH_3I, CF_3, I, or HI, with added He, Ar, or Xe; optimum 0.4 torr of I with 5 torr of He; D = 2-8 cm.
508. CW; in I vapor, CH_3I, CF_3, I, or HI, with added He, Ar, or Xe; optimum 0.4 torr of I with 5 torr of He; D = 2-8 mm.
509. CW; in I vapor, CH_3I, CF_3, I, or HI, with added He, Ar, or Xe; optimum 0.4 torr of I with 5 torr of He; D = 2-8 mm.
510. Measured and calculated vacuum wavelengths and spectral assignments are from Wyart, J. F. and Cams, P., *Phys. Scripta*, 20, 43-59 (1979).

511.   A line at 2.7087 μm reported in References 196, 197 is in fact an YbII line at 2.4377 μm.

512.   Pulsed; in Yb vapor above 500 °C with 2.5 torr of He, 1.5 torr Ne, or 0.8 torr of Ar.

513.   Pulsed; in Yb vapor above 500 °C with 2.5 torr of He, 1.5 torr Ne, or 0.8 torr of Ar.

514.   Pulsed; in Yb vapor above 500 °C with 2.5 torr of He, 1.5 torr Ne, or 0.8 torr of Ar.

515.   Pulsed; in 0.001-1 torr of Yb vapor with from 0.01-760 torr of noble gas buffer - usually He; D = 7 mm.

516.   Pulsed; in Yb vapor above 500 °C with 2.5 torr of He, 1.5 torr Ne, or 0.8 torr of Ar.

517.   Pulsed; in 0.001-1 torr of Yb vapor with from 0.01-760 torr of noble gas buffer - usually He; D = 7 mm.

518.   Pulsed; in Yb vapor above 500 °C with 2.5 torr of He, 1.5 torr Ne, or 0.8 torr of Ar.

519.   Pulsed; in Yb vapor above 500 °C with 2.5 torr of He, 1.5 torr Ne, or 0.8 torr of Ar.

520.   Pulsed; in Yb vapor above 500 °C with 2.5 torr of He, 1.5 torr Ne, or 0.8 torr of Ar.

521.   Pulsed; in Yb vapor above 500 °C with 2.5 torr of He, 1.5 torr Ne, or 0.8 torr of Ar.

522.   Measured wavelengths and spectral assignments are from Catalan, M. A., Meggers, W. F., and Garcia-Riquelme, O., *J. Res. Natl. Bur. Stand.,* 68A, 9-59 (1964).

523.   Oscillation has been reported on six hyperfine components of this line. Reference 407.

524–   Short rise-time high-voltage pulsed excitation of 0.1-2.0 torr of Mn (at a tempera-
525.   ture of 1100-1300 °C) with 1-2 torr of He or Ne; D = 10 mm; also by single- or double-pulse excitation of MnCl$_2$ at 700-800 °C with up to 120 torr of Ne, 80 torr of He, or 20 torr of Xe; optimum is about 15 torr of Ne; D = 16 mm.

526–   Short rise-time high-voltage pulsed excitation of 0.1-2.0 torr of Mn (at a tempera-
527.   ture of 1100-1300 °C) with 1-2 torr of He or Ne; D = 10 mm; also by single- or double-pulse excitation of MnCl$_2$ at 700-800 °C with up to 120 torr of Ne, 80 torr of He, or 20 torr of Xe; optimum is about 15 torr of Ne; D = 16 mm.

528.   Pulsed; short rise-time high-voltage pulsed excitation of 0.1-2.0 torr of Mn (at a temperature of 1100-1300 °C) with 1-2 torr of He or Ne; D = 10 mm.

529.   Pulsed; short rise-time high-voltage pulsed excitation of 0.1-2.0 torr of Mn (at a temperature of 1100-1300 °C) with 1-2 torr of He or Ne; D = 10 mm.

530.   Pulsed; short rise-time high-voltage pulsed excitation of 0.1-2.0 torr of Mn (at a temperature of 1100-1300 °C) with 1-2 torr of He or Ne; D = 10 mm.

531.   Pulsed; short rise-time high-voltage pulsed excitation of 0.1-2.0 torr of Mn (at a temperature of 1100-1300 °C with 1-2 torr of He or Ne; D = 10 mm.

532–   Short rise-time high-voltage pulsed excitation of 0.1-2.0 torr of Mn (at a tempera-
533.   ture of 1100-1300 °C) with 1-2 torr of He or Ne; D = 10 mm; also by single- or double-pulse excitation of MnCl$_2$ at 700-800 °C with up to 120 torr of Ne, 80 torr of He, or 20 torr of Xe; optimum is about 15 torr of Ne; D = 16 mm.

534–   Short rise-time high-voltage pulsed excitation of 0.1-2.0 torr of Mn (at a tempera-
535.   ture of 1100-1300 °C) with 1-2 torr of He or Ne; D = 10 mm; also by single- or double-pulse excitation of MnCl$_2$ at 700-800 °C with up to 120 torr of Ne, 80 torr of He, or 20 torr of Xe; optimum is about 15 torr of Ne; D = 16 mm.

536– Short rise-time high-voltage pulsed excitation of 0.1-2.0 torr of Mn (at a tempera-
537. ture of 1100-1300 °C) with 1-2 torr of He or Ne; D = 10 mm; also by single- or double-pulse excitation of $MnCl_2$ at 700-800 °C with up to 120 torr of Ne, 80 torr of He, or 20 torr of Xe; optimum is about 15 torr of Ne; D = 16 mm.

538– Short rise-time high-voltage pulsed excitation of 0.1-2.0 torr of Mn (at a tempera-
539. ture of 1100-1300 °C) with 1-2 torr of He or Ne; D = 10 mm; also by single- or double-pulse excitation of $MnCl_2$ at 700-800 °C with up to 120 torr of Ne, 80 torr of He, or 20 torr of Xe; optimum is about 15 torr of Ne; D = 16 mm.

540 Short rise-time high-voltage pulsed excitation of 0.1-2.0 torr of Mn (at a tempera-
541. ture of 1100-1300 °C) with 1-2 torr of He or Ne; D = 10 mm; also by single- or double-pulse excitation of $MnCl_2$ at 700-800 °C with up to 120 torr of Ne, 80 torr of He, or 20 torr of Xe; optimum is about 15 torr of Ne; D = 16 mm.

542– Short rise-time high-voltage pulsed excitation of 0.1-2.0 torr of Mn (at a tempera-
543. ture of 1100-1300 °C) with 1-2 torr of He or Ne; D = 10 mm; also by single- or double-pulse excitation of $MnCl_2$ at 700-800 °C with up to 120 torr of Ne, 80 torr of He, or 20 torr of Xe; optimum is about 15 torr of Ne; D = 16 mm.

544. Measured wavelength from Meggers, W. F., Corliss, C. H., and Scribner, B. F., *Natl. Bur. Stand. (U.S.) Monogr.*, 145(1) (1975).

545. Alternative two-electron transition assignment is z $^3I^0_5 \rightarrow$ e $^5D_4$ at 6.8544324 μm. Measured wavelength of laser transition in Reference 155 was 6.847 μm.

546. Several additional assignments are possible for this line; these other possibilities involve transitions between higher-lying levels than the ones listed here. Measured wavelength in Reference 155 was 8.490 μm.

547. Pulsed; Fe atoms at a density of $\sim 10^{14}$ atoms cm^{-3}; produced by flash photolysis of $Fe(CO)_5$ in an Ar buffer or by a discharge through 0.1 torr of $Fe(CO)_5$ with 50 torr of Ne, optically pumped with a KrF laser (248 nm).

548. Pulsed; Fe atoms at a density of $\sim 10^{14}$ atoms cm^{-3}; produced by flash photolysis of $Fe(CO)_5$ in an Ar buffer or by a discharge through 0.1 torr of $Fe(CO)_5$ with 50 torr of Ne, optically pumped with a KrF laser (248 nm).

549. Pulsed; Fe atoms at a density of $\sim 10^{14}$ atoms cm^{-3}; produced by flash photolysis of $Fe(CO)_5$ in an Ar buffer or by a discharge through 0.1 torr of $Fe(CO)_5$ with 50 torr of Ne, optically pumped with a KrF laser (248 nm).

550. Pulsed; Fe atoms at a density of $\sim 10^{14}$ atoms cm^{-3}; produced by flash photolysis of $Fe(CO)_5$ in an Ar buffer or by a discharge through 0.1 torr of $Fe(CO)_5$ with 50 torr of Ne, optically pumped with a KrF laser (248 nm).

551. Pulsed; short rise-time high-voltage pulsed excitation of Fe vapor in a tube at 1680 °C with 1.5-3.5 torr of Ne; D = 16 mm.

552. Pulsed; by dissociation of 0.08 torr of $Fe(CO)_5$ with 2 torr of He in a transversely excited double-discharge laser.

553. Pulsed; by dissociation of 0.08 torr of $Fe(CO)_5$ with 2 torr of He in a transversely excited double-discharge laser.

554– Measured laser wavelength in Reference 417 was 1.3968 μm. This is probably the
555. unidentified transition at the wavelength listed above reported by Fisher, R. A., Knopf, W. C., Jr., and Kinney, F. E., *Astrophys. J.*, 130, 683-687 (1959). However, two nearby assigned transitions are e $^3D_2 \rightarrow$ y $^1F^0_3$ at 1.3984294 μm and g $^3D_2 \rightarrow$ w $^3D^0_1$ at 1.3987469 μm.

556. Pulsed; in Ni vapor produced by sputtering in a slotted hollow-cathode discharge.

557. Pulsed; by dissociation of 80 torr of $Ni(CO)_4$ with 2 torr of He in a transversely excited double-discharge laser.

558. Assignment suggested by C. C. Davis; involves low-lying levels and calculated wavelength is within reported error of measured wavelength in Reference 196.

559. Closest calculated line is 5d $6s^2$ $^5F^0_4 \rightarrow$ 5d(^6P)6s $^7F_3$ at 2.049928 μm.
560. Closest calculated line is $31893.780_3 \rightarrow 29006.022_2$ at 3.536368 μm.
561. Pulsed; in Sm vapor at about 0.1 torr with He, Ne, or Ar.
562. Pulsed; in Sm vapor at about 0.1 torr with He, Ne, or Ar.
563. Pulsed; in Sm vapor at about 0.1 torr with He, Ne, or Ar.
564. Pulsed; in Sm vapor about 0.1 torr with He, Ne, or Ar.
565. Pulsed; in Sm vapor about 0.1 torr with He, Ne, or Ar.
566. Pulsed; in Sm vapor about 0.1 torr with He, Ne, or Ar.
567. Pulsed; in Sm vapor about 0.1 torr with He, Ne, or Ar.
568. Pulsed; in Sm vapor about 0.1 torr with He, Ne, or Ar.
569. Measured wavelengths from Meggers, W. F., Corliss, C. H., and Scribner, B. F., *Natl. Bur. Stand. (U.S.) Monogr.*, 145(1) (1975).
570. Measured wavelength from Reference 420.
571. Pulsed; operates in a true superfluorescent mode.
572. Pulsed; operates in a true superfluorescent mode.
573. Pulsed; operates in a true superfluorescent mode following two-photon excitation of 7.5 torr of Eu vapor.
574. Pulsed; operates in a true superfluorescent mode following two-photon excitation of 7.5 torr of Eu vapor.
575. Pulsed; operates in a true superfluorescent mode following two-photon excitation of 7.5 torr of Eu vapor.
576. Pulsed; in about 0.1 torr of Eu vapor with He, Ne, or Ar; also in a self-heated repetitively pulsed discharge with Eu and 15-25 torr of He; D = 1.1 or 2 cm.
577. Pulsed; in about 0.1 torr of Eu vapor with He, Ne, or Ar.
578. Pulsed; in about 0.1 torr of Eu vapor with He, Ne, or Ar.
579. Pulsed; in about 0.1 torr of Eu vapor with He, Ne or Ar.
580. Pulsed; in about 0.1 torr of Eu vapor with He, Ne, or Ar.
581. Pulsed; in about 0.1 torr of Eu vapor with He, Ne, or Ar.
582. Pulsed; in about 0.1 torr of Eu vapor with He, Ne, or Ar.
583. Pulsed; in about 0.1 torr of Eu vapor with He, Ne, or Ar.
584. Pulsed; in about 0.1 torr of Eu vapor with He, Ne, or Ar.
585. Pulsed; in about 0.1 torr of Eu vapor with He, Ne, or Ar.
586. Pulsed; in about 0.1 torr of Eu vapor with He, Ne, or Ar.
587. Unless otherwise stated, indicated wavelengths and spectral assignments are taken from data in Martin, W. C., *J. Res. Natl. Bur. Stand.*, 64, 19-28 (1960).
588–589. It is not clear whether one or both lines near 0.7067 μm were observed in Reference 422. The excitation mechanism proposed for these transition involves a two-body recombination process H^- (1s $^1S_0$) + He^+ (1s $^2S_{1/2}$) $\rightarrow$ H (1s $^2S_{1/2}$) + He(1s 3s $^3S_1$).
590. This is a self-terminating laser transition.
591. The trace of argon or nitrogen is to depopulate the metastable He 2s $^3S_1$ level from which the 4p^{3}P lower laser level is otherwise excited by electron impact.
592. Measured wavelength from Reference 433.
593. Pulsed; in a 8:7 He/H_2 mix at 12.5 torr in a hollow-cathode discharge.
594. Pulsed; in a 8:7 He/H_2 mix at 12.5 torr in a hollow-cathode discharge.
595. CW; in 0.4 torr of He; D = 6 mm.
596. CW; in 0.4 torr of He; D = 6 mm.
597. Short rise-time, high-voltage pulsed excitation of 2.7 torr of He; D = 1.3 mm; operates in an ASE mode; also following pulsed excitation of 65 torr of He/NH_3 mixture containing 18% NH_3 in a transversely excited pin laser.
598. CW; in 8.0 torr of He with a trace of Ar or N; D = 7 mm.

599. CW; in 0.2-0.4 torr of He; D = 15 mm.
600. CW; in 0.2-0.4 torr of He; D = 15 mm.
601. CW; in 0.2-0.4 torr of He; D = 15 mm.
602. Pulsed; in 0.5 torr of He; D = 75 mm or CW in 0.1 torr of He; D = 60 mm.
603. CW; in 0.1 torr of He; D = 60 mm.
604. Except where indicated, wavelengths are calculated from accurate energy level values given by Kaufmann, V. and Minnhagen, L., *J. Opt. Soc. Am.*, 62, 92-95 (1972).
605. Measured wavelengths from Burns, K., Adams, K. B., and Longwell, J., *J. Opt. Soc. Am.*, 40, 339-344 (1950).
606– Although oscillation between the Ne-$3s_2$ and 2p levels is normally observed only in
607. a He-Ne mixture, it has been reported to have been observed on the 0.5941- and 0.6120-μm lines under pulsed conditions in pure Ne. In view of the excitation used and nonobservation of the 0.6328-μm line, it is likely that the two lines mentioned above should have been identified as the 0.5945- and 0.6143-μm transient laser lines. Reference 212.
608. This line appears in Reference 185, but no mention of it appears to have been made.
609. Measured wavelengths and classifications from Humphreys, C. J. and Kostkowski, H. J., *J. Res. Natl. Bur. Stand.*, 49, 73-84 (1952).
610. Measured wavelengths and assignments from Hepner, G., *Ann. de Phys.*, 13. (6), 744-750 (1961).
611. Originally measured with low precision and not correctly identified in Reference 556.
612– Originally assigned to $4d'[3/2]^0_1 \rightarrow 4p'[3/2]_2$ ($4s'_1 \rightarrow 3p_4$) in Reference 195; how-
613. ever, this assignment has a wavelength of 2.4395241 μm. Either the reported wavelength or assignment given in Reference 195 must be incorrect. The assignment given here was suggested by C. C. Davis on the assumption that the reported wavelength, 2.864 μm, is correct.
614. Both of these lines were observed. However, it is uncertain whether one or both were observed in oscillation in References 195, 559, 564.
615. Preferred assignment of the two alternates.
616. This transition was not observed and was incorrectly reported previously. However, it is possible that this is the transition observed at 7.400 μm in References 432 and previously unidentified. References 195, 432, 564, 659.
617. This is the most likely transition to give the 13.76-μm line; selective excitation of the Ne-$7s'[1/2]^0_1$ state occurs via excitation transfer from He-2p $^1P^0_1$ atoms. References 471, 564.
618– This transition is reported in Reference 458 where its wavelength is given as
619. 1.3912 μm. This wavelength does not correspond to the transition assigned, nor to any Ne 2p → 2s transition. Either the wavelength reported is incorrect, or the line was not observed at all, or some other unidentified transition of Ne or an impurity was observed. Reference 528.
620. Assignment given in Reference 555 seems unlikely to be correct, this is a no-parity change electric-dipole forbidden transition.
621– Several previous compilations of the neutral gas laser lines contain incorrect entries
626. corresponding to data taken from the work of Faust et al. In some early work of these latter authors, several lines were incorrectly or ambiguously assigned and in many cases more than one measured wavelength for the same transition was reported. These ambiguities were clearly indicated in Reference 559 and partially

eliminated in Reference 570. Excitation transfer is responsible for the selective excitation of the Ne $5s_2$, $3s_2$, and $2s_{2-4}$ levels via excited He atoms in the He $^1P_1$,He* $^1S_0$, and He* 2^3S_1 states, respectively. The 1/D gain relationship (under optimum discharge conditions exhibited by laser transitions from the Ne-$3s_2$ and $2s_2$ levels in the He-Ne l aser) is due to the populations of these levels following the concentration of He 2^1S_0 and 2^3S_1 metastables which follows a 1/D relationship. The 1/D gain relationship is not due to Ne-1s metastables, as stated throughout the laser review and book literature. This can be deduced from an analysis of the results of White and Gordon (see Labuda). The same 1/D gain relationship is also shown by the 0.4416- and 0.3250-μm laser lines in the He-Cd ion laser. The upper levels of these lines are also selectively excited (in a Penning reaction) by helium metastable (He* 2^3S_1) atoms. References 195, 468, 559, 564, 570, 659

627.   Transient line requiring short rise-time high voltage, high current pulses. In 3 torr of neon; D=5 mm in a longitudinal discharge system. In a transverse discharge system, 30-35 torr of neon with interelectrode separations of 2.5-10 cm.

628.   CW. In a 7:1 ^{3}He-^{20}Ne mixture at 1 torr; D=4mm.

629.   Transient line observed in a pulsed He-Ne mixture requiring trace of argon to destroy Ne-1s metastables. Total pressures 1-200 mtorr; D=3 mm.

630.   CW. In a 5:1 He-Ne mixture at 3.6 torr-mm. Need to suppress high gain. 3.39 μm ($3s_2 \rightarrow 3p_4$) line and strong 0.6328 μm ($3s_2 \rightarrow 2p_4$) line.

631.   The Ne $3s_2$ level is selectively excited mainly by He 2^1S_0 metastables in an endothermic excitation transfer reaction.

632.   Transient line requiring short rise-time, high voltage, high current pulses. In about 0.3 torr of neon, D=5 mm and pulse current of 120 A.

633.   CW. In a 5:1 He-Ne mixture at 3.6 torr-mm. Need to suppress high gain 3.39 μm line and the strong 0.6328 μm line.

634.   CW. In a 5:1 He-Ne mixture at 3.6 torr-mm; need to suppress high gain 3.39 μm line and the strong 0.6328 μm line

635.   Transient line requiring short rise time, high-voltage, high-current pulses; in about 0.3 torr of Ne; D = 5 mm and pulse current of 120 A.

636.   Pulsed; in an ASE mode; D = 1.3 mm; optimum Ne pressure ~ 2 torr.

637.   CW; in a 5:1 He-Ne mixture at 3.6 torr-mm, need to suppress 3.39 and 0.6328 μm lines.

638–
639.   CW; strongest of the $3s_2 \rightarrow 2p$ lines; observed in a 5:1 He-Ne mixture at 3.6 torr-mm; the He-Ne mixture ratio and pressure depend on the bore of the discharge tube; the Ne-$3s_3$ level is selectively excited mainly by He 2^1S_00 metastables in an endothermic excitation transfer reaction, as well by direct electron impact; also CW nuclear-pumped by the reaction ^{3}He(n,p) ^{3}H; threshold flux 2 x 10^{11} n cm^{-2} s^{-1}.

640–
641.   CW; strongest of the $3s_2 \rightarrow 2p$ lines; observed in a 5:1 He-Ne mixture at 3.6 torr-mm; the He-Ne mixture ratio and pressure depend on the bore of the discharge tube; the Ne-$3s_3$ level is selectively excited mainly by He 2^1S_0 metastables in an endothermic excitation transfer reaction, as well by direct electron impact; also CW nuclear-pumped by the reaction ^{3}He(n,p) ^{3}H; threshold flux 2 x $10^{10^{11}}$ n cm^{-2} s^{-1}; need to suppress the 3.39 and the 0.6328-μm lines.

642–
644.   CW; strongest of the $3s_2 \rightarrow 2p$ lines; observed in a 5:1 He-Ne mixture at 3.6 torr-mm; the He-Ne mixture ratio and pressure depend on the bore of the discharge tube; the Ne-$3s_3$ level is selectively excited mainly by He 2^1S_0 metastables in an endothermic excitation transfer reaction, as well by direct electron impact; also CW nuclear-pumped by the reaction ^{3}He(n,p) ^{3}H; threshold flux 2 x 10^{11} n cm^{-2} s^{-1}.

needed to suppress the 0.6328-μm line by the use of a prism or an unstable optical cavity and a discharge current more than optimal for the 0.6328-μm line.

645– CW; strongest of the $3s_2 \rightarrow 2p$ lines; observed in a 5:1 He-Ne mixture at 3.6 torr-
646. mm; the He-Ne mixture ratio and pressure depend on the bore of the discharge tube; the Ne-$3s_3$ level is selectively excited mainly by He 2^1S_0 metastables in an endothermic excitation transfer reaction, as well by direct electron impact; also CW nuclear-pumped by the reaction $^3He(n,p)$ ^3H; threshold flux 2×10^{11} n cm^{-2} s^{-1}; needed to suppress the 3.39 and the 0.6328-μm lines.

647. In an ASE mode following fast rise-time high-voltage pulsed excitation of 0.2 torr of Ne; D = 6 mm.

648. In an ASE mode following fast rise-time high-voltage pulsed excitation of 3 torr of Ne; D = 6 mm.

649. Pulsed; in an ASE mode; D = 1.3, 5, 6, or 6.5 mm; optimum pressure 2-4 torr of Ne, depending on tube diameter.

650. In an ASE mode following fast rise-time high-voltage pulsed excitation of 3 torr of Ne; D = 6 mm.

651. CW; observed in a very long discharge tube.

652. CW; observed in a very long discharge tube.

653. Pulsed; observed in a hollow-cathode discharge in a Ne-H_2 mixture; need to suppress oscillation on other $2s \rightarrow 2p$ transitions.

654. Pulsed; observed in a hollow-cathode discharge in a Ne-H_2 mixture; need to suppress oscillation on other $2s \rightarrow 2p$ transitions.

655. CW; observed in a very long discharge tube.

656. CW; in a He-Ne mixture, 0.15 torr partial pressure of Ne and 2.8 torr to tal pressure, D probably 5-10 mm; the gain on this transition is only a few tenths of a percent per meter even with oscillation suppressed on competing transitions.

657. CW; in a 10:1 He-Ne mixture at a pD of 7-14 torr-mm.

658. CW; in a 10:1 He-Ne mixture at a pD of 7-14 torr-mm.

659. CW; in a 10:1 He-Ne mixture at a pD of 7-14 torr-mm.; also observed in hollow-cathode discharge in pure Ne and in a mixture of Ne and H; H destroys the Ne-1s metastables.

660. CW; in a 10:1 He-Ne mixture at a pD of 7-14 torr-mm.; also observed in hollow-cathode discharge in pure Ne and in a mixture of Ne and H; H destroys the Ne-1s metastables.

661. CW; in a 10:1 He-Ne mixture at a pD of 7-14 torr-mm.

662– CW; in a 10:1 He-Ne mixture at a pD of 7-14 torr-mm. Lasing was also observed
664. in hollow-cathode discharge in pure Ne and in a mixture of Ne and H; H destroys the Ne-1s metastables.

665. CW; in a glow discharge in pure Ne, optimum pD is less than 0.5 torr-mm, He suppresses oscillation; observed also in hollow-cathode discharges in pure Ne and a Ne-H mixture.

666. CW; in a He-Ne mixture.

667. CW; in a He-Ne mixture.

668. CW; in a 10:1 He-Ne mixture at a pD of 7-14 torr-mm; observed also in hollow-cathode discharges in pure Ne and in mixtures of Ne and H and Ne and O.

669. CW; observed in a very long discharge tube, also in a hollow-cathode discharge in a mixture of Ne and H.

670. CW; in a He-Ne mixture.

671. CW; in a He-Ne mixture in glow and hollow-cathode discharges.

672–   CW; strongest of the 2s → 2p lines; in a 10:1 He-Ne mixture at about 11 torr-mm;
673.   the Ne-2s levels are selectively excited mainly by He 2^3S_1 metastables in an exothermic excitation transfer reaction, as well as by direct electron impact; observed also in hollow-cathode discharges in pure Ne and mixtures of Ne and H and Ne and O.

674.   CW; observed in a very long laser, also in a hollow-cathode discharge in a mixture of Ne and H.

675.   Pulsed, via optical pumping by a He lamp; 0.2 torr of Ne, 3-4 torr of He, D = 5.2 mm.

676.   Pulsed, via optical pumping by a He lamp; 0.2 torr of Ne, 3-4 torr of He, D = 5.2 mm.

677.   Pulsed, in a hollow-cathode discharge in a mixture of Ne and H.

678.   CW; observed in a very long laser, also in a hollow-cathode discharge in a mixture of Ne and H.

679.   CW; observed in a very long laser, also in a hollow-cathode discharge in a mixture of Ne and H.

680.   CW; observed in a very long discharge in a He-Ne mixture.

681.   CW; in 0.7-torr He with 0.07-torr Ne; D = 9 mm; also in a hollow-cathode discharge in a mixture of Ne and H.

682.   Pulsed, in a hollow-cathode discharge in a mixture of Ne and H.

683.   CW.

684.   CW.

685.   CW.

686.   CW.

687.   CW.

688.   CW.

689.   CW; in a 10:1 He to Ne mixture at 10 torr; D = 5 mm.

690.   CW; in a 10:1 He to Ne mixture at 10 torr; D = 5 mm.

691.   CW; in a 10:1 He to Ne mixture at 10 torr; D = 5 mm.

692.   CW; in a 10:1 He to Ne mixture at 10 torr; D = 5 mm.

693.   CW; in a 10:1 He to Ne mixture at 10 torr; D = 5 mm.

694.   CW; in a 10:1 He to Ne mixture a 10 torr; D = 5 mm.

695.   CW; in a 10:1 He-Ne mixture at a pD of 7-14 torr-mm.

696.   Pulsed; via optical pumping by a He lamp; 0.2 torr of Ne, 3-4 torr of He, D = 5.2 mm.

697.   CW; observed in a very long discharge in a He-Ne mixture.

698.   CW; observed in a very long discharge in a He-Ne mixture.

699.   CW.

700.   CW.

701.   CW; in a 10:1 or 100:1 He-Ne mixture at a pD of about 8 torr-mm.

702.   CW; in a 10:1 He-Ne mixture at a pD of about 8 torr-mm.

703.   CW; in a 10:1 He-Ne mixture at a pD of about 8 torr-mm.

704.   CW; in a 100:1 He-Ne mixture at a pD of about 8 torr-mm.

705.   CW; in a 10:1 or 100:1 He-Ne mixture at a pD of about 8 torr-mm.

706.   CW; in a 10:1 or 100:1 He-Ne mixture at a pD of about 8 torr-mm.

707.   CW; in a He-Ne mixture; oscillation is due to cascading from the high-gain 3.39-μm ($3s_2 \to 3p_4$) line; also in pure Ne in a 10-m long discharge tube at 0.01-0.05 torr; D = 10 mm.

708.   CW; in a He-Ne mixture; oscillation is due to cascade transitions from the well populated Ne-3s levels.

709. CW; in a He-Ne mixture; oscillation is due to cascading from the high-gain 3.39-µm ($3s_2 \rightarrow 3p_4$) line; also in pure Ne in a 10-m long discharge tube at 0.01-0.05 torr; D = 10 mm; also in pure Ne at 0.01-0.05 torr; D = 10 mm.

710. CW; in a He-Ne mixture; oscillation is due to cascade transitions from the well populated Ne-3s levels.

711. CW; observed in a very long discharge tube.

712. CW; in a He-Ne mixture; oscillation is due to cascade transitions from the well-populated Ne-3s levels; but also oscillates in 250 mtorr of pure Ne in an 8-m long, 10-mm bore discharge tube.

713. CW; in pure Ne at 0.05 torr; He suppresses oscillation by selectively populating the $2s_4$ lower laser level.

714. CW; observed in a very long discharge in pure Ne or a He-Ne mixture.

715. CW; in a He-Ne mixture, also in pure Ne; oscillation due to cascading through the high-gain $3s_2 \rightarrow 3p_4$ transition at 3.39 µm.

716. CW; in a 5:1 He-Ne mixture, total pressure 0.6 torr; D = 8 mm.

717. CW; in a 13:1 He-Ne mixture at a total pressure of 0.86 torr in a long discharge tube; D = 22 mm.

718. CW; in pure Ne at 0.15 torr; D = 15 mm.

719. CW; in a He-Ne mixture in cascade from the well-populated Ne-3p levels.

720. CW; in a He-Ne mixture, over ranges 0.01-0.2 torr of Ne 0.00-1.0 torr of Helium; D = 10 mm.

721. CW; in a long discharge tube, in He-Ne mixture in pure Ne; D = 10 mm.

722. CW; in a He-Ne mixture over ranges 0.01-0.2 torr of Ne, 0.00-1.0 torr of He; D = 10 mm.

723. CW; in a He-Ne mixture over ranges 0.01-0.2 torr of Ne, 0.00-1.0 torr of He; D = 10 mm.

724. CW; in a He-Ne mixture over ranges 0.01-0.2 torr of Ne, 0.00-1.0 torr of He; 0 = 10 mm.

725. CW; in a He-Ne mixture over ranges 0.01-0.2 torr of Ne, 0.00-1.0 torr of He; 0 = 10 mm.

726. CW; in a He-Ne mixture over ranges 0.01-0.2 torr of Ne, 0.00-1.0 torr of He; 0 = 10 mm.

727. CW; in a He-Ne mixture over ranges 0.01-0.2 torr of Ne, 0.00-1.0 torr of He; D = 10 mm.

728. CW; in a He-Ne mixture over ranges 0.01-0.2 torr of Ne, 0.00-1.0 torr of He; D = 10 mm.

729. CW; in a He-Ne mixture over ranges 0.01-0.2 torr of Ne, 0.00-1.0 torr of He; D = 10 mm.

730. CW; in a 12:1 He-Ne mixture at a total pressure of 0.65 torr; D = 10 mm; requires wavelength selection to suppress ASE mode operation of the 3.39-µm line.

731. CW; in a He-Ne mixture over ranges 0.01-0.2 torr of Ne, 0.00-1.0 torr of He; D = 10 mm.

732. CW; in a He-Ne mixture over ranges 0.01-0.2 torr of Ne, 0.00-1.0 torr of He; D = 10 mm.

733. CW; in a He-Ne mixture over ranges 0.01-0.2 torr of Ne, 0.00-1.0 torr of He; D = 10 mm.

734. CW; in He-Ne mixtures in ratios from 10:1 to 5:1 at total pressures between 0.3-0.5 torr; D = 15 mm.

735. CW; in He-Ne mixtures in ratios from 10:1 to 5:1 at total pressures between 0.3-0.5 torr; D = 15 mm.

736. CW; in a He-Ne mixture over ranges 0.01-0.2 torr of Ne; 0.00-1.0 torr of He; D = 10 mm.

737. CW; in a He-Ne mixture over ranges 0.01-0.2 torr of Ne; 0.00-1.0 torr of He; D = 10 mm.

738. CW; in a 5:1 He-Ne mixture at pD of 3.6 torr-mm; the Ne-$3s_2$ level is selectively excited mainly by excitation transfer from He 2^1S_0 metastables in an endothermic reaction, as well as by direct electron impact from the ground state.

739– CW; in a 5:1 He-Ne mixture at pD of 3.6 torr-mm; the Ne-$3s_2$ level is selectively
740. excited from He 2^1S_0 metastables in an endothermic react ion, as well as by direct electron impact from the ground state; also in 3 torr of pure Ne with high-voltage fast-pulse excitation; D = 6 mm; this line exhibits very gain.

741. CW; in a He-Ne mixture over ranges 0.01-0.2 torr of Ne, 0.00-1.0 torr of He; D = 10 mm; also in an ASE mode following high-voltage fast-pulse excitation of 3 torr of pure Ne; D = 6 mm.

742. CW; in He-Ne mixtures in ratios from 10:1 to 5:1 at pressures from 0.3-0.5 torr; D = 15 mm; also in an ASE mode following high-voltage fast-pulse excitation of 3 torr of pure Ne; D = 6 mm.

743. CW; in a He-Ne mixture over ranges 0.01-0.2 torr of Ne, 0.00-1.0 torr of He; D = 10 mm.

744. CW; in He-Ne mixtures in ratios from 10:1 to 5:1 at pressures from 0.3-0.5 torr; D = 15 mm.

745. CW; in a He-Ne mixture over ranges 0.01-0.2 torr of Ne, 0.00-1.0 torr of He; D = 10 mm.

746. CW; in a He-Ne mixture over ranges 0.01-0.2 torr of Ne, 0.00-1.0 torr of He; D = 10 mm.

747. CW; in a 12:1 He-Ne mixture at a total pressure of 0.65 torr; D = 10 mm; requires wavelength selection to suppress ASE mode operation of the 3.39-μm line.

748. CW; in 0.3 torr of pure Ne; D = 15 mm.

749. CW; in pure Ne at 0.3 torr or from 0.5-0.6 torr; D = 15 mm.

750. CW; in pure Ne at 0.3 torr or from 0.5-0.6, torr or in a He-Ne mixture in ratios from 10:1 to 5:1 at total pressures from 0.3-0.5 torr; D = 15 mm.

751. CW; in pure Ne at 0.3 torr or from 0.5-0.6, torr or in a He-Ne mixture in ratios from 10:1 to 5:1 at total pressures from 0.3-0.5 torr; D = 15 mm.

752. CW; in a He-Ne mixture over ranges 0.01-0.2 torr of Ne, 0-1.0 torr of He; D = 10 mm. also in an ASE mode following high-voltage fast-pulse excitation of 3 torr of pure Ne; D = 6 mm; Reference 525 contains an incorrect wavelength and/or transition assignment.

753. CW; in pure Ne at 0.3 torr or from 0.5-0.6 torr or in a He-Ne mixture in ratios from 10:1 to 5:1 at total pressures from 0.3-0.5 torr; D = 15 mm.

754. CW; in a He-Ne mixture over ranges 0.01-0.2 torr of Ne, 0-1.0 torr of He; D = 10 mm.

755. CW; in a He-Ne mixture having He to Ne ratios from 10:1 to 5:1 at 0.3 torr; D = 15 mm.

756. CW; in pure Ne at pressures of 0.3 or from 0.5-0.6 torr or in a He-Ne mixture in ratios from 10:1 to 5:1 at total pressures from 0.3-0.5 torr; D = 15 mm.

757. CW; in a He-Ne mixture in ratios from 10:1 to 5:1 at pressures from 0.3- 0.5 torr; D = 15 mm.

758. CW; in a He-Ne mixture in ratios from 10:1 to 5:1 at pressures from 0.3- 0.5 torr; D = 15 mm.

759. CW; in a He-Ne mixture in ratios from 10:1 to 5:1 at pressures from 0.3- 0.5 torr; D = 15 mm.

760. CW; in a He-Ne mixture in ratios from 10:1 to 5:1 at pressures from 0.3- 0.5 torr; D = 15 mm.

761. CW; in pure Ne at 0.3 or from 0.5-0.6 torr or in a He-Ne mixture in ratios from 10:1 to 5:1 at pressures from 0.3-0.5 torr; D = 15 mm.

762. CW; in pure Ne at 0.3 or from 0.5-0.6 torr or in a He-Ne mixture in ratios from 10:1 to 5:1 at 0.5 torr; D = 15 mm.

763. CW; in a He-Ne mixture over ranges 0.01-0.2 torr of Ne, 0-1.0 torr of He; D = 10 mm.

764. CW; in a He-Ne mixture in ratios from 10:1 to 5:1 at 0.5 torr; D = 15 mm.

765. CW; in a He-Ne mixture over ranges 0.01-0.2 torr of Ne, 0-1.0 torr of He; D = 10 mm.

766. CW; in a He-Ne mixture over ranges 0.01-0.2 torr of Ne, 0-1.0 torr of He; D = 10 mm.

767. CW; in a He-Ne mixture over ranges 0.01-0.2 torr of Ne, 0-1.0 torr of He; D = 10 mm.

768. CW; in a He-Ne mixture with 0.15 torr of Ne and 0.3 torr of He; D = 15 mm.

769. CW; in a He-Ne mixture with 0.15 torr of Ne and 0.3 torr of He.

770. CW; in a He-Ne mixture at 0.15 torr of Ne, 0.3 torr of He; D = 15 mm.

771. CW; in a He-Ne mixture in ratios from 10:1 to 5:1 at pressures from 0.3- 0.5 torr; D = 15 mm.

772. CW; in pure Ne at 0.3 or from 0.5-0.6 torr or in a He-Ne mixture in ratios from 10:1 to 5:1 at pressures from 0.3-0.5 torr; D = 15 mm.

773. CW; in a He-Ne mixture over ranges 0.01-0.2 torr of Ne, 0-1.0 torr of He; D = 10 mm.

774. CW; in a He-Ne mixture over ranges 0.01-0.2 torr of Ne, 0-1.0 torr of He; D = 10 mm.

775. CW; in a He-Ne mixture over ranges 0.01-0.2 torr of Ne, 0-1.0 torr of He; D = 10 mm.

776. CW; in a He-Ne mixture over ranges 0.01-0.2 torr of Ne, 0-1.0 torr of He; D = 10 mm.

777. CW; in a He-Ne mixture over ranges 0.01-0.2 torr of Ne, 0-1.0 torr of He; D = 10 mm.

778. CW; in a He-Ne mixture over ranges 0.01-0.2 torr of Ne, 0-1.0 torr of He; D = 10 mm.

779. CW; in a He-Ne mixture over ranges 0.01-0.2 torr of Ne, 0-1.0 torr of He; D = 10 mm.

780. CW; in a He-Ne mixture over ranges 0.01-0.2 torr of Ne, 0-1.0 torr of He; D = 10 mm.

781. CW; in pure Ne at 0.3 or from 0.5-0.6 torr or in a He-Ne mixture in ratios from 10:1 to 5:1 at pressures from 0.3-0.5 torr; D = 15 mm.

782. CW; in a He-Ne mixture over ranges 0.01-0.2 torr of Ne, 0-1.0 torr of He; D = 10 mm.

783. CW; in pure Ne at 0.3 or from 0.5-0.6 torr or in a He-Ne mixture in ratios from 10:1 to 5:1 at pressures from 0.3-0.5 torr; D = 15 mm.

784. CW; in He-Ne mixture in ratios from 10:1 to 5:1 at pressures from 0.3-0. 5 torr; D = 15 mm.

785. CW; in He-Ne mixtures in ratios from 10:1 to 5:1 at pressures from 0.3-0.5 torr D = 15 mm.

786. CW; in a He-Ne mixture over ranges 0.01-0.2 torr of Ne, 0-1.0 torr of He; D = 10 mm.

787. CW; in a He-Ne mixture over ranges 0.01-0.2 torr of Ne, 0-1.0 torr of He; D = 10 mm.

788. CW; in pure Ne at 0.3 or from 0.5-0.6 torr or in a He-Ne mixture in ratios from 10:1 to 5:1 at pressures from 0.3-0.5 torr; D = 15 mm.

789. CW; in pure Ne at 0.3 or from 0.5-0.6 torr or in a He-Ne mixture in ratios from 10:1 to 5:1 at pressures from 0.3-0.5 torr; D = 15 mm.

790. CW; in a He-Ne mixture over ranges 0.01-0.2 torr of Ne, 0-1.0 torr of He; D = 10 mm.

791. CW; in a He-Ne mixture over ranges 0.01-0.2 torr of Ne, 0-1.0 torr of He; D = 10 mm.

792. CW; in a He-Ne mixture over ranges 0.01-0.2 torr of Ne, 0-1.0 torr of He; D = 10 mm.

793. CW; in a He-Ne mixture over ranges 0.01-0.2 torr of Ne, 0-1.0 torr of He; D = 10 mm.

794. CW; in a He-Ne mixture over ranges 0.01-0.2 torr of Ne, 0-1.0 torr of He; D = 10 mm.

795. CW; in a He-Ne mixture over ranges 0.01-0.2 torr of Ne, 0-1.0 torr of He; D = 10 mm.

796. CW; in a He-Ne mixture over ranges 0.01-0.2 torr of Ne, 0-torr of He; D = 10 mm.

797. CW; in a He-Ne mixture over ranges 0.01-0.2 torr of Ne, 0-1.0 torr of He, D = 10 mm.

798 CW; in a He-Ne mixture over ranges 0.01-0.2 torr of Ne, 0-1.0 torr of He; D = 10 mm.

799. CW; in a He-Ne mixture over ranges 0.01-0.2 torr of Ne, 0-1.0 torr of He; D = 10 mm.

800. CW; in a He-Ne mixture over ranges 0.01-0.2 torr of Ne, 0-1.0 torr of He; D = 10 mm CW; in a He-Ne mixture over ranges 0.01-0.2 torr of Ne, 0- 1.0 torr of He; D = 10 mm.

801. CW; in a He-Ne mixture over ranges 0.01-0.2 torr of Ne, 0- 1.0 torr of He; D = 10 mm CW; in a He-Ne mixture over ranges 0.01-0.2 torr of Ne, 0- 1.0 torr of He; D = 10 mm.

802. CW; in a He-Ne mixture over ranges 0.01-0.2 torr of Ne, 0-1.0 torr of He; D = 10 mm.

803. CW; in a He-Ne mixture over ranges 0.01-0.2 torr of Ne, 0-1.0 torr of He; D = 10 mm.

804. CW; in a He-Ne mixture over ranges 0.01-0.2 torr of Ne, 0-1.0 torr of He; D = 10 mm.

805. CW; in a He-Ne mixture over ranges 0.01-0.2 torr of Ne, 0-1.0 torr of He; D = 10 mm.

806. CW; in a He-Ne mixture over ranges 0.01-0.2 torr of Ne, 0-1.0 torr of He; D = 10 mm.

807. CW; in a He-Ne mixture over ranges 0.01-0.2 torr of Ne, 0-1.0 torr of He; D = 10 mm.

808. CW; in a He-Ne mixture over ranges 0.01-0.2 torr of Ne, 0-1.0 torr of He; D = 10 mm.

809. CW: in A He-Ne mixture over ranges 0.01-0.2 torr of Ne, 0-1.0 torr of He; D = 10 mm.

810. CW; in a He-Ne mixture over ranges 0.01-0.2 torr of Ne, 0-1.0 torr of He; D = 10 mm.

811. CW; in a He-Ne mixture over ranges 0.01-0.2 torr of Ne, 0-1.0 torr of He; D = 10 mm.

812.   CW; in a He-Ne mixture over ranges 0.01-0.2 torr of Ne, 0-1.0 torr of He; D = 10 mm.

813.   CW; in a He-Ne mixture over ranges 0.01-0.2 torr of Ne, 0-1.0 torr of He; D = 10 mm.

814.   CW; in a He-Ne mixture over ranges 0.01-0.2 torr of Ne, 0-1.0 torr of He; D = 10 mm.

815.   CW; in a He-Ne mixture over ranges 0.01-0.2 torr of Ne, 0-1.0 torr of He; D = 10 mm.

816.   CW; in a He-Ne mixture over ranges 0.01-0.2 torr of Ne, 0-1.0 torr of He; D = 10 mm.

817.   CW; in a He-Ne mixture over ranges 0.01-0.2 torr of Ne, 0-1.0 torr of He; D = 10 mm.

818.   CW; in a He-Ne mixture over ranges 0.01-0.2 torr of Ne, 0-1.0 torr of He; D = 10 mm.

819.   CW; in pure Ne at 0.05 torr, D = 21 mm.

820.   CW; in a He-Ne mixture over ranges 0.01-0.2 torr of Ne, 0-1.0 torr of He; D = 10 mm.

821.   CW; in a He-Ne mixture over ranges 0.01-0.2 torr of Ne, 0-1.0 torr of He; D = 10 mm.

822.   CW; in a He-Ne mixture over ranges 0.01-0.2 torr of Ne, 0-1.0 torr of He; D = 10 mm.

823.   CW; in a He-Ne mixture over ranges 0.01-0.2 torr of Ne, 0-1.0 torr of He; D = 10 mm.

824.   CW; in a He-Ne mixture over ranges 0.01-0.2 torr of Ne, 0-1.0 torr of He; D = 10 mm.

825.   CW; in pure Ne at 0.05 torr, D = 21 mm.

826.   CW; in pure Ne at 0.05 torr, D = 21 mm.

827.   CW; in pure Ne at 0.05 torr, D = 21 mm.

828.   CW; in pure Ne at 0.05 torr, D = 21 mm.

829.   CW; in pure Ne at 0.05 torr, D = 21 mm.

830.   CW; in pure Ne at 0.02 torr, D = 47 mm.

831.   CW; in pure Ne at 0.05 torr, D = 21 mm.

832.   CW; in He-Ne mixture, 0.05 torr of Ne with 0.1 torr of He; D = 21 mm.

833.   CW; in He-Ne mixture, 0.05 torr of Ne with 0.1 torr of He; D = 21 mm.

834.   CW; in pure Ne at 0.02 torr, D = 47 mm.

835.   CW; in He-Ne mixture, 0.03 torr of Ne with 0.07 torr of He; D = 21 mm.

836.   CW; in pure Ne at 0.035 torr, D = 34 mm.

837.   CW; in pure Ne at 0.02 torr, D = 47 mm.

838.   CW; in pure Ne at 0.035 torr, D = 34 mm.

839.   CW; in pure Ne at 0.035 torr, D = 34 mm.

840.   CW; in pure Ne at 0.035 torr, D = 34 mm.

841.   CW; in pure Ne at 0.035 torr, D = 34 mm.

842.   CW; in pure Ne at 0.01 torr; D = 47 mm.

843.   CW; in pure Ne at 0.01 torr; D = 47 mm.

844.   CW; in pure Ne at 0.01 torr; D = 47 mm.

845.   CW; in pure Ne at 0.01 torr; D = 47 mm.

846.   CW; in pure Ne at 0.01 torr; D = 47 mm.

847.   CW; in pure Ne at 0.01 torr; D = 47 mm.

848.   Wavelengths and spectral assigments are from data in Minnhagen, L., *J. Opt. Soc. Am.*, 63, 1185-1198 (1973) and Norlen, G., *Phys. Scripta*, 8, 249- 268 (1973).

849.   Possibly the same as the Ar-I line at 1.21397 μm.

850. Possibly the same as the Ar-I line at 1.27023 μm.
851. This is a very strong line in a high-pressure He-Ar TEA laser. The assignment listed, rather than an alternate listed previously in the literature, was confirmed by Dauger and Stafsudd in line competition experiments. Reference 598.
852. The assignment listed, rather than an alternate listed previously in the literature was confirmed by Dauger and Stafsudd in line competition experiments. Reference 598.
853. Preferred assignment.
854– An alternate assignment of 5d $[7/2]^0_3 \rightarrow$ 4f$[9/2]_4$ (5d$'_4 \rightarrow$ 4V$_4$) at 5.1203927 μm
856. listed in previous compilations of neutral argon transitions has been eliminated; it is considerably outside the reported error of the measured wavelength 5.1218±10⁻⁴ μm. Reference 559.
857– Lines listed in previous compilations at 0.8780, 1.0935, and 1.8167 μm have been
859. omitted. The first of these is an Ar-II line, the second is almost certainly an Ar-II line, but is listed here under miscellaneous and unidentified possible neutral laser transitions, and the third line is a neutral Kr line. Unidentified laser lines observed in a pulsed Ar-Hg discharge at 1.222, 1.246, and 1.276 μm have been listed as unidentified neutral Hg transitions. References 162, 244, 659, 660, 662.
860. Transient line requiring short rise-time, high-voltage, high-current pulses.
861. Transient line as above. Observed in an Ar-Ne (He) discharge; favors very low Ar pressure.
862. Pulsed; in a high-pressure (>1 atm) He-Ar mixture excited in a vacuum UV photo-preionized TEA laser.
863. Pulsed; in a high-pressure (>1 atm) He-Ar mixture excited in a vacuum UV photo-preionized TEA laser.
864. Pulsed; in a high-pressure (>1 atm) He-Ar mixture excited in a vacuum UV photo-preionized TEA laser.
865. Pulsed; in a high-pressure (>1 atm) He-Ar mixture excited in a vacuum UV photo-preionized TEA laser.
866. Pulsed; in a mixture of Ar and He above 200 torr; D = 11 mm.
867. Pulsed; observed in an ASE mode in argon at 0.03 torr; D = 3 mm; also in a pulsed hollow-cathode laser with a 1:75 Ar-He mixture, optimum pressure 30 torr; also as for 0.9123-μm line.
868. Pulsed; observed in an ASE mode in argon at 0.03 torr; D = 3 mm; also as for 0.9123-μm line.
869. Pulsed; observed in an ASE mode in Ar at 0.03 torr; D = 3 mm; also in a pulsed hollow-cathode laser with a 1:75 Ar-He mixture, optimum pressure 30 torr; also as for 0.9123-μm line; also in pure Ar or Ar-He mixtures in a transversely excited pin laser.
870. Pulsed; in a mixture of Ar and He above 200 torr; D = 11 mm.
871. CW; in Ar at 0.25 torr; D = 2.2 mm.
872. Pulsed; in an ASE mode in Ar at 0.04 torr; D = 7 mm.
873. Pulsed; in a Cu hollow-cathode laser with a 1:75 Ar-He mixture; optimum pressure 30 torr.
874. CW; in Ar at 0.05 torr; 0 = 7 mm.
875. Pulsed; in 5 torr Ar with 3 torr SF$_6$ in a transversely excited pin laser.
876. CW; in 2.5 torr of Ar; D = 7 mm; also pulsed in pure Ar or Ar-SF$_6$ mix ture at low pressures, up to 5 torr; in transversely excited lasers and in a hollow-cathode laser with a 1:75 Ar-He mixture, optimum pressure 30 torr.
877. CW; in 0.035 torr of Ar; D = 7 mm; also in low pressure Ar or Ar-SF$_6$ mixtures in transversely excited lasers or in high-pressure (>1 atm) He-Ar.

878.   Mixtures excited in a vacuum-UV photopreionized TEA laser; also nuclear pumped by the reaction ^{3}He $(n,p)^3$H in a 9:1 ^{3}He-Ar mixture at pressures from 200-700 torr; D = 2 cm; ref 520 contains an incorrect wavelength and/or transition assignment.

879.   CW; in 0.035 torr of Ar; D = 7 mm; also pulsed, in a Cu hollow-cathode laser with a 1:75 Ar-He mixture; optimum pressure 30 torr.

880.   CW; in 0.012 torr of Ar; also pulsed in a Cu hollow-cathode laser with a 1:75 Ar-He mixture; optimum pressure 30 torr.

881.   CW; in 0.01-0.05 torr of Ar; D = 10 mm.

882.   CW; in 0.018 torr of Ar; also pulsed in 5.0 torr of Ar plus 3 torr of SF_6 in a transversely excited pin laser.

883.   CW; in 0.01-0.05 torr of Ar; D = 10 mm; also enhanced by the addition of Cl to an Ar-He mixture; also pulsed in a Cu hollow-cathode laser with a 1:75 Ar-He mixture; optimum pressure 30 torr.

884.   CW; in 0.01-0.05 torr of Ar; D = 10 mm; also pulsed in 5.0 torr of Ar plus 3 torr of SF_6 in a transversely excited pin laser.

885.   CW; in 0.01-0.05 torr of Ar; D = 10 mm; enhanced by the addition of Cl to an Ar-He mixture; also pulsed in pure Ar at low-pressure (<1 torr) or a high-pressure He-Ar mixture (>1 atm) excited in a vacuum UV photopreionized laser.

886.   CW; in 0.01-0.5 torr of Ar; D = 10 mm.

887.   CW; in 0.01-0.5 torr of Ar; D = 10 mm.

888.   CW; in 0.01-0.5 torr of Ar; D = 10 mm.

889.   CW; in 0.01-0.5 torr of Ar; D = 10 mm.

890.   CW; in 0.01-0.5 torr of Ar; D = 10 mm.

891.   CW; in 0.02 torr of Ar with 0.2 torr of He; D = 15 mm.

892.   CW; in 0.01-0.05 torr of Ar; D = 10 mm.

893.   Pulsed; in 0.008-0.014 torr of Ar; D = 4 mm.

894.   CW; in 0.01-0.05 torr of Ar; D = 10 mm.

895.   CW; in 0.01-0.05 torr of Ar; D = 10 mm.

896.   CW; in 0.01-0.05 torr of Ar; D = 10 mm.

897.   CW; in 0.09 torr of Ar with 3 torr of He.

898.   CW; in 0.01-0.05 torr of Ar; D = 10 mm.

899.   CW; in 0.01-0.05 torr of Ar; D = 10 mm.

900.   CW.

901.   CW; in 0.01-0.05 torr of Ar; D = 10 mm.

902.   CW; in 0.01-0.05 torr of Ar; D = 10 mm.

903.   CW; in 0.01-0.05 torr of Ar; D = 10 mm.

904.   CW; in 0.01-0.05 torr of Ar; D = 10 mm.

905.   CW; in 0.01-0.05 torr of Ar; D = 10 mm.

906.   CW; in 0.01-0.05 torr of Ar; D = 10 mm.

907.   Pulsed; in 0.008-0.014 torr of Ar; D = 4 mm.

908.   Pulsed; in 0.008-0.014 torr of Ar; D = 4 mm.

909.   Pulsed; in 0.01-0.015 torr of Ar; D = 8 mm.

910.   Pulsed; in 0.008-0.014 torr or Ar; D = 4 mm.

911.   CW; in 0.06 torr of Ar; D = 15 mm.

912.   CW; in 0.02 torr of Ar with 0.2 torr of He; D = 15 mm.

913.   CW; in 0.01-0.05 torr or Ar; D = 10 mm.

914.   CW; in 0.01-0.05 torr or Ar; D = 10 mm.

915.   Pulsed; in 0.008-0.014 torr of Ar; D = 4 mm.

916.   CW; in 0.01-0.05 torr of Ar; D = 10 mm.

917.   CW; in 0.06 torr of Ar; D = 15 mm.

918.   CW; in 0.05 torr of Ar; D = 15 mm.

919.   CW; in 0.05 torr of Ar; D = 15 mm.

920.   CW; in 0.02 torr of Ar with 0.2 torr of He; D = 15 mm; also pulsed in a 9:3 He-Ar mixture at 100 torr in a transversely excited pin laser.

921.   CW; in 0.05 torr of Ar; D = 15 mm.

922.   CW; in 0.01-0.05 torr of Ar; D = 10 mm.

923.   CW; in 0.01-0.05 torr of Ar; D = 10 mm.

924.   CW; in 0.02 torr of Ar with 0.2 torr of He; D = 15 mm.

925.   CW; in 0.01-0.05 torr of Ar; D = 10 mm.

926.   CW; in 0.01-0.05 torr of Ar; D = 10 mm.

927.   CW; in 0.05 torr of Ar; D = 10 mm.

928.   Pulsed; in a 7:3 He-Ar mixture at 100 torr in a transversely excited pin laser.

929.   CW; in 0.05 torr of Ar; D = 10 mm.

930.   CW; in 0.05 torr of Ar; D = 10 mm.

931.   CW; in 0.01-0.05 torr of Ar; D = 10 mm.

932.   CW; in 0.01-0.05 torr of Ar; D = 10 mm.

933.   CW; in 0.01-0.05 torr of Ar; D = 10 mm.

934.   CW; in 0.05 torr of Ar; D = 10 mm.

935.   CW; in 0.05 torr of Ar; D = 10 mm.

936.   CW; in 0.05 torr of Ar; D = 10 mm.

937–   Unless otherwise indicated, wavelengths and spectral assignment are taken from data

938.   on ^{86}Kr in Kaufman, V. and Humphreys, C. J., *J. Opt. Soc. Am.*, 59, 1614-1628 (1969). Although ^{86}Kr represents only 17.4% of natural isotopic abundance Kr, it is the spectral lines of this isotope that have been selected as wavelength standards by the International Astronomical Union and the International Committee of Weights and Measures.

939–   This assignment suggested by C. Davis as being much more likely to be correct

940.   than the two alternative assignments involving higher-lying levels listed by Linford. It is also closer to the measured wavelength. Reference 602.

941–   Calculated wavelength taken from data in Moore, C. E., *Natl. Stand. Ref. Data.*

942.   *Ser. Natl Bur. Stand.*, NSRDS-NBS 35, 2 (1971).

943–   The existence of this laser transition is doubtful; it is probably the Xe transition at

945.   5.5755 μm; Xe had been used in the same laser tube. Pulsed; in 0.01-0.015 torr of Kr; D = 8 mm.

946.   Transient line requiring short rise-time, high-voltage, high-current pulses; in about 0.1 torr of Kr; D = 3 mm; peak current 1000 A.

947.   Pulsed; in a 7 atm He-Kr mixture excited in a vacuum-UV photopreionized TEA laser.

948.   Pulsed; in an ASE mode in 0.04 torr of Kr; D = 7 mm.

949.   Pulsed; in an ASE mode in 0.06 torr of Kr; D = 7 mm.

950.   Pulsed; in an ASE mode in 0.03 torr of Kr; D = 7 mm.

951.   Pulsed; in an ASE mode in 0.08 torr of Kr; D = 7 mm.

952.   Pulsed; in an ASE mode in 0.2 torr of Kr; D = 7 mm.

953.   Pulsed; in an ASE mode in 0.01-0.015 torr of Kr; D = 2 or 4 mm.

954.   Pulsed; in an ASE mode in 0.01-0.015 torr of Kr; D = 2 or 4 mm.

955.   Pulsed; in an ASE mode in 0.07 torr of Kr; D = 7 mm.

956.   Pulsed; in an ASE mode in 0.08 torr of Kr; D = 7 mm.

957.   CW; in 0.05 torr of Kr; D = 7 mm.

958.   CW; in 0.07 torr of Kr; D = 7 mm.

959.   CW; in 0.015 torr of Kr; D = 9 mm.

960.   CW; in 0.07 torr of Kr; D = 7 mm.

961.   CW; in 0.035 torr of Kr; D = 7 mm; Reference 429 contains an incorrect wavelength and/or transition assignment.

962.   CW; in 0.035 torr of Kr; 0 = 7 mm.

963.   CW; in 0.035 torr of Kr; D = 7 mm.

964.   Pulsed; in 0.008-0.014 torr of Kr; D = 4 mm.

965.   CW.

966.   CW; in an ASE mode in 1.0 torr of Kr; D = 7 mm; also pulsed in a 93:7 He-Kr mixture at 760 torr in a transversely excited pin laser; also nuclear pumped by the reaction $^3He(n,p)^3H$.

967.   CW; in 0.02 torr of Kr; D = 10 mm.

968.   CW; in 0.02 torr of Kr; D = 10 mm.

969.   CW; in 0.02 torr of Kr; D = 10 mm.

970.   CW; in 0.02 torr of Kr; D = 10 mm.

971.   CW; in 0.02 torr of Kr; D = 10 mm.

972.   CW; in 0.02 torr of Kr; D = 10 mm.

973.   CW; in 0.02 torr of Kr; D = 10 mm.

974.   CW; in 0.03 torr of Kr; D = 15 mm; also pulsed in a 93:7 He-Kr mixture at 760 torr in a transversely excited pin laser.

975.   CW; in 0.02 torr of Kr; D = 10 mm.

976.   CW; in 0.02 torr of Kr; D = 10 mm; Reference 559 contains an incorrect wavelength and/or transition assignment

977.   CW; in 0.02 torr of Kr; D = 10 mm.

978.   CW; in 0.02 torr of Kr; D = 10 mm.

979.   CW; in 0.02 torr of Kr; D = 10 mm.

980.   Pulsed; in 0.01-0.015 torr of Kr; D = 8 mm.

981.   Pulsed; in 0.01-0.015 torr of Kr; D = 8 mm.

982.   Pulsed; in 0.01-0.015 torr of Kr; D = 8 mm.

983.   Pulsed; in 0.01-0.015 torr of Kr; D = 8 mm.

984.   CW; in 0.02 torr of Kr; D = 10 mm.

985.   CW; in 0.02 torr of Kr; D = 10 mm.

986.   CW; in 0.02 torr of Kr; D = 10 mm.

987.   CW; in 0.02 torr of Kr; D = 10 mm.

988.   CW; in pure Kr at 0.02 torr or a Kr-He mixture with 0.02 torr of Kr and 0.2 torr of He; D = 15 mm.

989.   CW; in pure Kr at 0.02 torr of a Kr-He mixture with 0.02 torr of Kr and 0.2 torr of He; D = 15 mm.

990.   CW; in 0.02 torr of Kr; D = 10 mm.

991.   CW; in 0.02 torr of Kr; D = 10 mm.

992.   CW; in 0.02 torr of Kr; D = 10 mm.

993.   CW; in 0.03 torr of Kr; D = 15 mm.

994.   CW; in 0.03 torr of Kr; D = 15 mm.

995.   CW; in 0.02 torr of Kr; D = 10 mm.

996.   CW; in pure Kr at 0.02 torr or in a Kr-He mixture with 0.02 torr of Kr and 0.2 torr of He; D = 15 mm.

997– 999.   Unless otherwise indicated, measured and calculated wavelengths are taken from data on ^{136}Xe in Humphreys, C. J. and Paul, E., Jr., *J. Opt. Soc. Am.*, 60, 1302-1310 (1970). ^{136}Xe is the heaviest isotope of Xe and the measured wavelengths for this isotope are the most accurately measured. In any case, the isotope shift in Xe is extremely small, e.g., 1500 kHz (2.10^{-8} μm) between the ^{134}Xe-^{136}Xe

transitions at 2.03 µm. Thus, the wavelength values given in the table probably only differ in the eighth decimal place between the various isotopes. Reference 626.

1000. Wavelength measured in Reference 609.

1001. May be an ionized Xe transition. Reference 244.

1002. Calculated wavelength from data in Moore, C. E., *Natl. Stand. Ref. Data. Ser. Natl. Bur. Stand.*, NSRDS-NBS 35, 3 (1971).

1003. This line was observed in Reference 627, but was incorrectly reported there as the transition at 3.507 µm. This latter transition is rarely, or only weakly observed in high-pressure transversely excited He-Xe lasers. References 627, 632.

1004. This line was observed in Reference 627, but was incorrectly reported there as the transition at 3.507 µm. This latter transition is rarely, or only weakly observed in high-pressure transversely excited He-Xe lasers. References 627, 632.

1005. Vacuum wavelength measured in Reference 601.

1006. Lines at 6.384 and 8.191 µm included by Willett in one compilation of neutral Xe laser transitions, but omitted in a second, are not included here. These lines were apparently only mentioned in a U.S. Goverment contract report and never reported in the literature.

1007. Pulsed; in a 7 atm He-Xe mixture in a vacuum-UV photopreionized TEA laser.

1008. Transient line which operates in an ASE mode in short rise-time, high-voltage, high-current pulsed discharges; in 0.04 torr of Xe; D = 7 mm.

1009. Transient line which operates in an ASE mode as above; in 0.12 torr of Xe; D = 7 mm.

1010. Pulsed; in 0.2-0.4 torr of Xe; also in a 7 atm He-Xe or a 1 atm Ar-Xe mixture in a vacuum-UV photopreionized TEA laser.

1011. Pulsed; in 0.001-0.02 torr of Xe; D = 2.7-4.0 mm.

1012. Pulsed in 0.001-0.02 torr of Xe; D = 2.7-4.0 mm.

1013. Pulsed; in an ASE mode in 0.04 torr of Xe; D = 7 mm.

1014. Pulsed; in an ASE mode in 0.1 torr of Xe; D = 7 mm.

1015–
1016. Pulsed; in ASE mode in 0.15 torr of Xe; D = 7 mm; also CW in 0.03-0.1 torr of Xe; D = 4 or 7 mm; also pulsed in a 1 atm He-Xe or Ar-Xe mixture in a vacuum-UV photopreionized TEA laser or in a high-pressure Ar-Xe mixture excited in a E-beam ionizer-sustainer mode laser.

1017–
1019. CW; operates in an ASE mode even in short discharge tubes; the gain in a 5-mm bore tube is more than 45 dBm^{-1}; in a few mtorr of Xe or in a 100:1 He-Xe mixture at a total pressure of about 10 torr; D = 5 mm. To avoid cataphoretic effects in a He-Xe mixture rf-excitation is necessary; clean-up of xenon is a problem; also pulsed in high-pressure (~1 atm) He-Xe, Ar-Xe, or Ne-Xe mixtures in a vacuum-UV photopreionized TEA laser; pulsed in a Cu hollow-cathode discharge; also pulsed, nuclear-pumped by the reaction ^{3}He(n,p)^{3}H; D = 10 mm.

1020. CW; in 0.01-0.04 torr of Xe, He also added to about 1 torr; D = 7 mm.

1021. CW; in 0.03-0.1 torr of Xe; D = 4 or 7 mm.

1022. Pulsed; in 0.005-0.011 torr of Xe; D = 4 mm.

1023. CW; in 0.01-0.04 torr of Xe, He also added to about 1 torr; D = 7 mm; also pulsed in a high-pressure Ar-Xe mixture containing <5% Xe excited in a E-beam ionizer-sustainer mode

1024. CW; in 0.01-0.04 torr of Xe, He also added to about 1 torr; D = 7 mm; also in high-pressure Ar-Xe or He-Xe mixtures containing <5% Xe excited in E-beam ionizer-sustainer and other types of pulsed transversely excited lasers

1025. CW; in 0.01-0.04 torr of Xe, He also added to about 1 torr; D = 7 mm.

1026. CW; in 0.01-0.06 torr of Xe; D = 15 mm.

1027. Pulsed; in 0.005-0.011 torr of Xe; D = 4 mm.

1028. CW; in 0.01-0.04 torr of Xe, He also added to about 1 torr; D = 7 mm.
1029. CW; in a 250:1 He-Xe mixture in an rf-discharge at about 0.4 torr; D = 11 mm.
1030. CW; in 0.01-0.06 torr of Xe; D = 15 mm.
1031. CW; in 0.01-0.04 torr of Xe, He added to 1 torr; D = 7 mm; also weakly in a pulsed high pressure Ar-Xe Mixture with <1% Xe excited in an E-beam ionizer-sustainer laser.
1032. CW; in pure Xe at 0.01-0.06 torr or in 0.015 torr of Xe with 0.3 torr of He; D = 15 mm.
1033. CW; in 0.01-0.04 torr of Xe, He added to 1 torr; D = 7 mm; also pulsed in a high repetition rate transversely excited He-Xe mixture containing <1% Xe or in a high-pressure He-Xe mixture excited in an E-beam ionizer-sustainer laser.
1034. Reference 627 contains an incorrect wavelength and/or transition assignment.
1035– CW; operates in ASE mode in short discharge tubes, gain as high as 400 dBm^{-1} in
1037. a 0.75-mm bore tube operates in a few tens of mtorr of Xe and with He to about 10 torr with D = 5 mm; clean-up of Xe is a problem; also pulsed in a high-pressure He-Xe mixture containing <1% Xe excited in an E-beam ionizer-sustainer laser or nuclear pumped by the reaction $^3He(n,p)^3H$ in a 10-mm bore tube or in a 200 torr 20:1 He-Xe mixture in a neutron activated ^{235}U lined tube; D = 19 mm.
1038. CW; in 0.01-0.04 torr of Xe, He added to 1.0 torr; D = 7 mm.
1039. CW; in 0.02 torr of Xe; D=7 mm. also pulsed in high-pressure He-Xe mixture with ~5% Xe excited in E-beam ionizer-sustainer and various other transversely excited lasers; also nuclear pumped by the reaction ^3He(n,p)^{3}H in a 10-mm bore tube.
1040. CW; in 0.01-0.04 torr of Xe, He added up to 1.0 torr; D = 7 mm.
1041. CW; in 0.01-0.04 torr of Xe, He added up to 1.0 torr; D = 7 mm.
1042. CW; in 0.01-0.4 torr of Xe, He added up to 1.0 torr; D = 7 mm.
1043. CW; in 0.01-0.4 torr of Xe, He added up to 1.0 torr; D = 7mm.
1044. CW; in 0.01-0.4 torr of Xe, He added up to 1.0 torr; D = 7 mm.
1045. CW; in 0.01-0.06 torr of Xe or in 0.015 torr of Xe with 0.3 torr of He; D = 15 mm; also pulsed in a high-pressure, high repetition rate TEA laser with a 200:1 He-Xe mixture at 300 torr.
1046. CW; in 0.01-0.04 torr of Xe, He added up to 1.0 torr; D = 7 mm.
1047. CW; in 0.01-0.06 torr of Xe or 0.015 torr Xe with 0.3 torr of He; D = 15 mm.
1048. CW; in 0.01-0.04 torr of Xe, He added to give total pressure of 1 torr; D = 7 mm, or in pure Xenon at 0.01-0.06 torr; D = 15 mm.
1049. CW; in 0.01-0.06 torr of Xe or 0.015 torr Xe with 0.3 torr of He; D = 15 mm.
1050. CW; in 0.01-0.06 torr of Xe or 0.015 torr Xe with 0.3 torr of He; D = 15. mm.
1051. CW; in 0.01-0.04 torr of Xe, He added to give total pressure of 1 torr; D = 7 mm, or in pure Xenon at 0.01-0.06 torr; D = 15 mm.
1052. CW; in 0.01-0.06 torr of Xe; D = 15 mm.
1053. CW; in pure Xe at 0.01-0.06 torr or in 0.015 torr of Xe with 0.3 torr of He; D = 15 mm; lasing is more easily achieved if the 3.507-μm line is suppressed
1054. CW; in 0.01-0.06 torr of Xe or 0.015 torr Xe with 0.3 torr of He; D = 15 mm.
1055. CW; in 0.01-0.06 torr of Xe or 0.015 torr Xe with 0.3 torr of He; D = 15 mm.
1056. CW; in 0.01-0.04 torr of Xe, He added up to 1 torr; D = 7 mm.
1057. CW; in 0.01-0.06 torr of Xe; D = 15 mm.
1058. CW; in 0.01-0.06 torr of Xe or 0.015 torr Xe with 0.3 torr of He; D = 15 mm.
1059. CW; in 0.015 torr of Xe with 0.3 torr of He; D = 15 mm.
1060. CW; in 0.01-0.06 torr of Xe; D = 15 mm.
1061. CW; in 0.01-0.06 torr of Xe; D = 15 mm.
1062. CW; in 0.01-0.06 torr of Xe; D = 15 mm.

1063.   CW; in 0.015 torr of Xe with 0.3 torr of He; D = 15 mm.

1064.   CW; in 0.01-0.04 torr of Xe, He added up to 1.0 torr; D = 7 mm.

1065.   CW; in 0.01-0.06 torr of Xe or 0.015 torr of Xe with 0.3 torr of He; D = 15 mm.

1066.   CW; in 0.01–0.04 torr of Xe, He added up to 1.0 torr; D = 7 mm.

1067.   CW; in 0.01–0.04 torr of Xe, He added up to 1.0 torr; D = 7 mm.

1068.   CW; in 0.01–0.04 torr of Xe, He added up to 1.0 torr; D = 7 mm.

1069.   CW; in 0.01–0.04 torr of Xe, He added up to 1.0 torr; D = 7 mm.

1070.   CW; in 0.01–0.04 torr of Xe, He added up to 1.0 torr; D = 7 mm.

1071.   CW; in 0.01–0.04 torr of Xe, He added up to 1.0 torr; D = 7 mm.

1072.   CW; in a 100:1 He-Xe mixture at 35 mtorr of Xe or in a 3:1 Kr-Xe mixture at 15-20 mtorr of Xe; D = 6 mm.

1073.   Collisions with He required for laser oscillation.

1074.   Excited by Penning collisions with metastable He.

1075.   Collisions with He,Ne required for laser oscillation.

1076.   Excited by charge-exchange collisions with ground-state ion of He.

1077.   Collisions with Ne required for laser oscillation.

1078.   Excited by charge-exchange collisions with ground-state ion of He.

1079.   Collisions with Ne,He required for laser oscillation.

1080.   Excited by charge-exchange collisions with ground-state ion of Ne.

1081.   See more complete discussion following the tables in the *CRC Handbook of Laser Science and Technology,* Vol. II, Gas Lasers, section 2.

1082.   Laser oscillation observed only in the afterglow of the discharge pulse.

1083.   Continuous oscillation reported.

1084.   Ion obtained by dissociation of molecular compound.

1085.   Error in classification or wavelength.

1086.   Gain measured.

1087.   Hollow-cathode excitation.

1088.   Hyperfine structure investigated.

1089.   Identification of line given or discussed.

1090.   Power output reported.

1091.   Superradiant operation reported.

1092.   Unique or unusual excitation method.

1093.   Existence of line discussed.

1094.   Accurate wavelength measurement

1095.   Or 1.091527 μm line and transition; Spectroscopic Reference: Mg II (1046).

1096.   Or 1.091423 μm line and transition; Spectroscopic Reference: Mg II (1046).

1097.   Spectroscopic Reference: Mg II (1046).

1098.   Spectroscopic Reference: Mg II (1046).

1099.   Spectroscopic Reference: Ca II (1166).

1100.   Spectroscopic Reference: Ca II (1166).

1101.   Spectroscopic Reference: Ca II (1166).

1102.   Spectroscopic Reference: Ca II (1166).

1103.   Spectroscopic Reference: Ca II (1166).

1104.   Spectroscopic Reference: Ca II (1166).

1105.   Strong or characteristic laser line in pure gas.

1106.   Reference 1849 changes previous data.

1107.   Or 0.819223 μm line and transition.

1108.   Or 0.819233 μm line and transition.

1109.   Classification or ionization state uncertain.

1110.   Spectroscopic Reference: Zn II (1145); wavelength from Harrison (1099.

1111.   Spectroscopic Reference: Zn II (1145); wavelength from Harrison (1099.

1112. Spectroscopic Reference: Zn II (1145); wavelength from Harrison (1099); strong or characteristic laser line in pure gas.
1113. Spectroscopic Reference: Zn II (1145); wavelength from Harrison (1099.
1114. Spectroscopic Reference: Zn II (1145).
1115. Spectroscopic Reference: Zn II (1145); wavelength from Harrison (1099.
1116. Spectroscopic Reference: Zn II (1145); wavelength from Harrison (1099); strong or characteristic laser line in pure gas.
1117. Spectroscopic Reference: Zn II (1145); wavelength from Harrison (1099.
1118. Spectroscopic Reference: Zn II (1145); wavelength from Harrison (1099.
1119. Spectroscopic Reference: Zn II (1145); wavelength from Harrison (1099.
1120. Spectroscopic Reference: Zn II (1145); wavelength from Harrison (1099.
1121. Spectroscopic Reference: Zn II (1145).
1122. Spectroscopic Reference: Zn II (1145).
1123. Strong or characteristic laser line in pure gas.
1124. Strong or characteristic laser line in pure gas.
1125. Strong or characteristic laser line in pure gas.
1126. Strong or characteristic laser line in pure gas.
1127. Strong or characteristic laser line in pure gas.
1128. Classification or ionization state uncertain
1129. Classification or ionization state uncertain
1130. Wavelength from Harrison, Reference 1099.
1131. Classification or ionization state uncertain.
1132. Classification or ionization state uncertain.
1133. Wavelength from Harrison, Reference 1099; strong or characteristic laser line in pure gas.
1134. Wavelength from Harrison, Reference 1099; strong or characteristic laser line in pure gas.
1135. Strong or characteristic laser line in pure gas; wavelength from Reference 1165.
1136. Wavelength from Harrison, Reference 1099; strong or characteristic laser line in pure gas.
1137. Wavelength from Harrison, Reference 1099.
1138. Wavelength from Harrison, Reference 1099.
1139. Wavelength from Harrison, Reference 1099; strong or characteristic laser line in pure gas.
1140. Wavelength from Harrison, Reference 1099.
1141. Wavelength from Harrison, Reference 1099.
1142. Wavelength from Harrison, Reference 1099.
1143. Wavelength from Harrison, Reference 1099.
1144. Wavelength from Harrison, Reference 1099.
1145. Wavelength from Harrison, Reference 1099.
1146. Spectroscopic Reference: C III (1038).
1147. Spectroscopic Reference: C III (1038).
1148. Spectroscopic Reference: C IV (1048); wavelength in vacuum.
1149. Spectroscopic Reference: C IV (1048); wavelength in vacuum.
1150. Spectroscopic Reference: Si II (294).
1151. Spectroscopic Reference: Si II (294).
1152. Spectroscopic Reference: Si II (294).
1153. Spectroscopic Reference: Si III (324).
1154. Spectroscopic Reference: Si III (324).
1155. Spectroscopic Reference: Si IV (323); classification or ionization state uncertain; [cf. Ar].

1156. Spectroscopic Reference: Ge II (295).
1157. Spectroscopic Reference: Ge II (295).
1158. Wavelength from Harrison Reference 1099; classification or ionization state uncertain.
1159. Wavelength from Harrison, Reference 1099.
1160. Wavelength from Harrison, Reference 1099.
1161. Wavelength from Harrison, Reference 1099.
1162. Wavelength from Harrison, Reference 1099.
1163. Wavelength from Harrison, Reference 1099.
1164. Wavelength from Harrison, Reference 1099.
1165. Wavelength from Harrison, Reference 1099.
1166. Spectroscopic Reference: N II (1170).
1167. Spectroscopic Reference: N II (1170).
1168. Spectroscopic Reference: N II (1170).
1169. Spectroscopic Reference: N II (1170).
1170. Spectroscopic Reference: N II (1170).
1171. Spectroscopic Reference: N II (1170).
1172. Spectroscopic Reference: N II (1170).
1173. Spectroscopic Reference: N II (1170); strong or characteristic laser line in pure gas.
1174. Spectroscopic Reference: N II (1170).
1175. Spectroscopic Reference: N II (1170).
1176. Spectroscopic Reference: N IV (902).
1177. Spectroscopic Reference: N IV (902).
1178. Spectroscopic Reference: P II (992); strong or characteristic laser line in pure gas.
1179. Spectroscopic Reference: P II (992).
1180. Spectroscopic Reference: P II (992); strong or characteristic laser line in pure gas.
1181. Spectroscopic Reference: P II (992).
1182. Spectroscopic Reference: P II (992).
1183. Spectroscopic Reference: P II (992).
1184. Spectroscopic Reference: (981).
1185. Spectroscopic Reference: (981).
1186. Spectroscopic Reference: (981).
1187. Spectroscopic Reference: (981).
1188. Spectroscopic Reference: (981).
1189. Spectroscopic Reference: (981).
1190. Spectroscopic Reference: (981).
1191. Spectroscopic Reference: (981).
1192. Spectroscopic Reference: (981).
1193. Strong or characteristic laser line in pure gas.
1194. Spectroscopic Reference: O IV (1130).
1195. Spectroscopic Reference: O IV (1130); or 0.338121 μm line and transition.
1196. Spectroscopic Reference: O IV (1130); or 0.338130 μm line and transition
1197. Spectroscopic Reference: O IV (1130).
1198. Strong or characteristic laser line in pure gas.
1199. Strong or characteristic laser line in pure gas.
1200. Strong or characteristic laser line in pure gas.
1201. Strong or characteristic laser line in pure gas.
1202. Strong or characteristic laser line in pure gas.
1203. Strong or characteristic laser line in pure gas.
1204. Strong or characteristic laser line in pure gas.
1205. Classification or ionization state uncertain.

1206.	Strong or characteristic laser line in pure gas.
1207.	Reference 1156 changes previous data.
1208.	Classification or ionization state uncertain.
1209.	Classification or ionization state uncertain.
1210.	Classification or ionization state uncertain.
1211.	Classification or ionization state uncertain.
1212.	Strong or characteristic laser line in pure gas.
1213.	Strong or characteristic laser line in pure gas.
1214.	Strong or characteristic laser line in pure gas.
1215.	Strong or characteristic laser line in pure gas.
1216.	Strong or characteristic laser line in pure gas.
1217.	Strong or characteristic laser line in pure gas.
1218.	Strong or characteristic laser line in pure gas.
1219.	Strong or characteristic laser line in pure gas.
1220.	Strong or characteristic laser line in pure gas.
1221.	Strong or characteristic laser line in pure gas.
1222.	Or 0.527127 μm line and transition.
1223.	Or 0.527115 μm line and transition.
1224.	Or 0.552244 μm line and transition.
1225.	Or 0.552266 μm line and transition.
1226.	Strong or characteristic laser line in pure gas.
1227.	Strong or characteristic laser line in pure gas.
1228.	Spectroscopic Reference: Te II (904).
1229.	Spectroscopic Reference: Te II (904); wavelength from Reference 904.
1230.	Spectroscopic Reference: Te II (904); wavelength from Reference 904.
1231.	Spectroscopic Reference: Te II (904).
1232.	Spectroscopic Reference: Te II (904); Classification or ionization state uncertain.
1233.	Spectroscopic Reference: Te II (904).
1234.	Spectroscopic Reference: Te II (904).
1235.	Spectroscopic Reference: Te II (904); Classification or ionization state uncertain.
1236.	Spectroscopic Reference: Te II (904).
1237.	Spectroscopic Reference: Te II (904).
1238.	Spectroscopic Reference: Te II (904).
1239.	Spectroscopic Reference: Te II (904).
1240.	Spectroscopic Reference: Te II (904).
1241.	Spectroscopic Reference: Te II (904).
1242.	Spectroscopic Reference: Te II (904).
1243.	Spectroscopic Reference: Te II (904).
1244.	Spectroscopic Reference: Te II (904).
1245.	Spectroscopic Reference: Te II (904).
1246.	Spectroscopic Reference: Te II (904).
1247.	Spectroscopic Reference: Te II (904).
1248.	Spectroscopic Reference: Te II (904).
1249.	Spectroscopic Reference: Te II (904).
1250.	Spectroscopic Reference: Te II (904).
1251.	Spectroscopic Reference: Te II (904).
1252.	Spectroscopic Reference: Te II (904).
1253.	Spectroscopic Reference: Te II (904).
1254.	Spectroscopic Reference: Te II (904).
1255.	Spectroscopic Reference: Te II (904).
1256.	Spectroscopic Reference: Te II (904).

1257.   Spectroscopic Reference: Te II (904).
1258.   Spectroscopic Reference: Te II (904).
1259.   Spectroscopic Reference: Te II (904).
1260.   Spectroscopic Reference: Te II (904).
1261.   Spectroscopic Reference: F III (1022).
1262.   Spectroscopic Reference: F III (1022).
1263.   Spectroscopic Reference: F III (1022).
1264.   Spectroscopic Reference: F II (1024).
1265.   Spectroscopic Reference: F II (1024).
1266.   Classification or ionization state uncertain.
1267.   Classification or ionization state uncertain.
1268.   Classification or ionization state uncertain.
1269.   Spectroscopic Reference: Br II (1043).
1270.   Spectroscopic Reference: Br II (1043).
1271.   Spectroscopic Reference: Br II (1043).
1272.   Spectroscopic Reference: Br II (1043).
1273.   Spectroscopic Reference: Br II (1043).
1274.   Spectroscopic Reference: Br II (1043).
1275.   Spectroscopic Reference: Br II (1043).
1276.   Spectroscopic Reference: I II (993).
1277.   Classification or ionization state uncertain.
1278.   Classification or ionization state uncertain.
1279.   Classification or ionization state uncertain.
1280.   Classification or ionization state uncertain.
1281.   Spectroscopic Reference: I II (993).
1282.   Classification or ionization state uncertain.
1283.   Classification or ionization state uncertain.
1284.   Spectroscopic Reference: I II (993).
1285.   Spectroscopic Reference: I II (993).
1286.   Spectroscopic Reference: I II (993).
1287.   Spectroscopic Reference: I II (993); strong or characteristic laser line in pure gas.
1288.   Spectroscopic Reference: I II (993).
1289.   Spectroscopic Reference: I II (993).
1290.   Spectroscopic Reference: I II (993); strong or characteristic laser line in pure gas.
1291.   Spectroscopic Reference: I II (993); strong or characteristic laser line in pure gas.
1292.   Spectroscopic Reference: I II (993).
1293.   Spectroscopic Reference: I II (993); strong or characteristic laser line in pure gas.
1294.   Spectroscopic Reference: I II (993).
1295.   Spectroscopic Reference: I II (993).
1296.   Spectroscopic Reference: I II (993).
1297.   Spectroscopic Reference: I II (993).
1298.   Spectroscopic Reference: I II (993).
1299.   Spectroscopic Reference: I II (993); strong or characteristic laser line in pure gas.
1300.   Spectroscopic Reference: I II (993).
1301.   Spectroscopic Reference: I II (993).
1302.   Spectroscopic Reference: I II (993).
1303.   Spectroscopic Reference: I II (993).
1304.   Spectroscopic Reference: I II (993).
1305.   Spectroscopic Reference: I II (993).
1306.   Spectroscopic Reference: I II (993).
1307.   Spectroscopic Reference: I II (993).

1308. Spectroscopic Reference: I II (993).
1309. Spectroscopic Reference: I II (993).
1310. Spectroscopic Reference: I II (993).
1311. Spectroscopic Reference: I II (993).
1312. Classification or ionization state of iodine IV uncertain.
1313. Classification or ionization state of iodine IV uncertain.
1314. Wavelength from Reference 1759.
1315. Wavelength from Reference 1759.
1316. Classification or ionization state uncertain.
1317. Classification or ionization state uncertain.
1318. Classification or ionization state uncertain.
1319. Reference 989 changes previous data.
1320. Reference 222 changes previous data.
1321. Spectroscopic Reference: Ne II (993).
1322. Spectroscopic Reference: Ne II (993); strong or characteristic laser line in pure gas.
1323. Spectroscopic Reference: Ne II (993).
1324. Spectroscopic Reference: Ne II (993).
1325. Spectroscopic Reference: Ne II (993); strong or characteristic laser line in pure gas.
1326. Spectroscopic Reference: Ne II (993); strong or characteristic laser line in pure gas.
1327. Spectroscopic Reference: Ne II (993); strong or characteristic laser line in pure gas.
1328. Spectroscopic Reference: Ne II (993).
1329. Spectroscopic Reference: Ne II (993); strong or characteristic laser line in pure gas.
1330. Classification or ionization state uncertain; vacuum wavelength.
1331. Wavelength from Harrison (Reference 1099).
1332. Strong or characteristic laser line in pure gas.
1333. Spectroscopic Reference: Ar II (1005).
1334. Strong or characteristic laser line in pure gas. Wavelength from Harrison (Reference 141).
1335. Existence of this laser line may be in doubt; classification or ionization state uncertain.
1336. cf. Si IV; Classification or ionization state uncertain.
1337. Strong or characteristic laser line in pure gas; wavelength from Reference 1759.
1338. Spectroscopic Reference: Ar II (1005).
1339. Spectroscopic Reference: Ar II (1005).
1340. Spectroscopic Reference: Ar II (1005).
1341. Spectroscopic Reference: Ar II (1005).
1342. Spectroscopic Reference: Ar II (1005); strong or characteristic laser line in pure gas.
1343. Spectroscopic Reference: Ar II (1005).
1344. Spectroscopic Reference: Ar II (1005).
1345. Spectroscopic Reference: Ar II (1005).
1346. Spectroscopic Reference: Ar II (1005); strong or characteristic laser line in pure gas.
1347. Spectroscopic Reference: Ar II (1005); strong or characteristic laser line in pure gas.
1348. Spectroscopic Reference: Ar II (1005).
1349. Spectroscopic Reference: Ar II (1005); strong or characteristic laser line in pure gas.
1350. Existence of this laser line may be in doubt; classification or ionization state uncertain.
1351. Spectroscopic Reference: Ar II (1005); strong or characteristic laser line in pure gas.
1352. Spectroscopic Reference: Ar II (1005).
1353. Spectroscopic Reference: Ar II (1005).
1354. Spectroscopic Reference: Ar II (1005); strong or characteristic laser line in pure gas.
1355. Spectroscopic Reference: Ar II (1005).

1356.   Spectroscopic Reference: Ar II (1005).
1357.   Existence of this laser line may be in doubt; classification or ionization state uncertain.
1358.   Spectroscopic Reference: Ar II (1005).
1359.   Spectroscopic Reference: Ar II (1005).
1360.   Spectroscopic Reference: Ar II (1005).
1361.   Spectroscopic Reference: Ar II (1005).
1362.   Spectroscopic Reference: Ar II (1005); strong or characteristic laser line in pure gas.
1363.   Classification or ionization state uncertain.
1364.   Classification or ionization state uncertain.
1365.   Classification or ionization state uncertain.
1366.   Strong or characteristic laser line in pure gas.
1367.   Strong or characteristic laser line in pure gas.
1368.   Spectroscopic Reference: Kr II (1007).
1369.   Strong or characteristic laser line in pure gas.
1370.   Strong or characteristic laser line in pure gas.
1371.   Spectroscopic Reference: Kr II (1007).
1372.   Spectroscopic Reference: Kr II (1007).
1373.   Spectroscopic Reference: Kr II (1007).
1374.   Spectroscopic Reference: Kr II (1007).
1375.   Spectroscopic Reference: Kr II (1007).
1376.   Spectroscopic Reference: Kr II (1007).
1377.   Spectroscopic Reference: Kr II (1007).
1378.   Spectroscopic Reference: Kr II (1007); cf. C III, Xe IV.
1379.   Spectroscopic Reference: Kr II (1007); strong or characteristic laser line in pure gas.
1380.   Spectroscopic Reference: Kr II (1007).
1381.   Spectroscopic Reference: Kr II (1007); strong or characteristic laser line in pure gas.
1382.   Spectroscopic Reference: Kr II (1007).
1383.   Spectroscopic Reference: Kr II (1007).
1384.   Spectroscopic Reference: Kr II (1007); strong or characteristic laser line in pure gas.
1385.   Spectroscopic Reference: Kr II (1007).
1386.   Spectroscopic Reference: Kr II (1007).
1387.   Spectroscopic Reference: Kr II (1007); wavelength from Reference 1157.
1388.   Spectroscopic Reference: Kr II (1007).
1389.   Spectroscopic Reference: Kr II (1007); strong or characteristic laser line in pure gas.
1390.   Spectroscopic Reference: Kr II (1007).
1391.   Spectroscopic Reference: Kr II (1007); strong or characteristic laser line in pure gas.
1392.   Spectroscopic Reference: Kr II (1007); strong or characteristic laser line in pure gas.
1393.   Spectroscopic Reference: Kr II (1007).
1394.   Or Kr II 0.593529 μm line and transition.
1395.   Or Kr III 0.593506 μm line and transition; spectroscopic Reference: Kr II (1007).
1396.   Or Kr II 0.603811 μm line and transition
1397.   Or Kr III 0.603716 μm line and transition; spectroscopic Reference: Kr II (1007).
1398.   Existence of this laser line may be in doubt.
1399.   Spectroscopic Reference: Kr II (1007).
1400.   Wavelength from Reference (162).
1401.   Spectroscopic Reference: Kr II (1007).
1402.   Spectroscopic Reference: Kr II (1007); strong or characteristic laser line in pure gas.
1403.   Spectroscopic Reference: Kr II (1007).
1404.   Spectroscopic Reference: Kr II (1007).
1405.   Or Kr II 0.660275 line and transition.

1406. Or Kr III 0.660293 line and transition; spectroscopic Reference: Kr II (1007).
1407. Spectroscopic Reference: Kr II (1007); strong or characteristic laser line in pure gas.
1408. Spectroscopic Reference: Kr II (1007).
1409. Spectroscopic Reference: Kr II (1007).
1410. Spectroscopic Reference: Kr II (1007); strong or characteristic laser line in pure gas.
1411. Spectroscopic Reference: Kr II (1007).
1412. Spectroscopic Reference: Kr II (1007).
1413. Spectroscopic Reference: Kr II (1007).
1414. Spectroscopic Reference: Kr II (1007); classification or ionization state uncertain.
1415. Spectroscopic Reference: Kr II (1007).
1416. Classification or ionization state uncertain.
1417. Spectroscopic Reference: Kr II (1007).
1418. Spectroscopic Reference: Kr II (1007).
1419. Spectroscopic Reference: Kr II (1007).
1420. Spectroscopic Reference: Kr II (1007).
1421. Classification or ionization state uncertain.
1422. Classification or ionization state uncertain.
1423. Strong or characteristic laser line in pure gas; classification or ionization state uncertain.
1424. Strong or characteristic laser line in pure gas; classification or ionization state uncertain.
1425. Classification or ionization state uncertain.
1426. Strong or characteristic laser line in pure gas.
1427. Strong or characteristic laser line in pure gas.
1428. Strong or characteristic laser line in pure gas.
1429. Classification or ionization state uncertain.
1430. Strong or characteristic laser line in pure gas.
1431. Or 0.384186 μm line and transition.
1432. Or 0.384152 μm line and transition.
1433. Classification or ionization state uncertain.
1434. Or 0.399255 μm line and transition.
1435. Or 0.399285 μm line and transition.
1436. Strong or characteristic laser line in pure gas.
1437. Strong or characteristic laser line in pure gas.
1438. Strong or characteristic laser line in pure gas.
1439. Strong or characteristic laser line in pure gas.
1440. Strong or characteristic laser line in pure gas.
1441. [cf. C III]
1442. Classification or ionization state uncertain; [cf. C III, Kr II].
1443. Classification or ionization state uncertain; [cf. C III, Kr II].
1444. Strong or characteristic laser line in pure gas; wavelength from Reference 163.
1445. Strong or characteristic laser line in pure gas; wavelength from Reference 163.
1446. Strong or characteristic laser line in pure gas; wavelength from Reference 163.
1447. Strong or characteristic laser line in pure gas.
1448. Wavelength from Reference (163).
1449. Strong or characteristic laser line in pure gas; wavelength from Reference (163).
1450. Strong or characteristic laser line in pure gas.
1451. Strong or characteristic laser line in pure gas; wavelength from Reference (163).
1452. Strong or characteristic laser line in pure gas; wavelength from Reference (163).
1453. Strong or characteristic laser line in pure gas.
1454. Strong or characteristic laser line in pure gas.

1455. Strong or characteristic laser line in pure gas.
1456. Classification or ionization state uncertain.
1457. Classification or ionization state uncertain.
1458. Classification or ionization state uncertain.
1459. Classification or ionization state uncertain.
1460. Spectroscopic Reference: Eu II (285).
1461. Spectroscopic Reference: Eu II (285).
1462. Spectroscopic Reference: Eu II (285).
1463. Spectroscopic Reference: Eu II (285).
1464. Spectroscopic Reference: Eu II (285).
1465. Spectroscopic Reference: Eu II (285).
1466. Spectroscopic Reference: Yb II (236).
1467. Spectroscopic Reference: Yb II (236).
1468. Spectroscopic Reference: Yb II (236).
1469. Spectroscopic Reference: Yb II (236).
1470. Spectroscopic Reference: Yb II (236); classification or ionization state uncertain.
1471. Spectroscopic Reference: Yb II (236).
1472. Electron-beam pumping (800 keV, 60 ns) of reagent grade Ar at pressures of 20 to 68 atm. Reference 1.
1473. Collisional processes in dense, electron beam-excited rare gases are reviewed. Reference 40.
1474. Spectral and temporal characteristics of high-current density (1.7 kA/cm^2) electron beam-excited argon at 0.2 to 65 atm. Reference 51.
1475. Microwave- and condensed discharge-excited emission continua of the rare gases were investigated with Reference to photodetection methods, order separation, and wavelength standards in the VUV. Reference 62.
1476. First rare gas-halide exciplex system found to lase after the advent of the rare gas lasers. Reference 2.
1477. ArCl exhibits predissociation as do KrCl and XeBr and is therefore an inefficient lasing material. Reference 13.
1478. Fast discharge excitation (80 to 110 kV, 2.5 ns rise time) of Cl$_2$ to Ar to He (1:15:84%) at atmospheric pressure produces a 10-ns pulse of energy ~ 0.2 mJ. Reference 23.
1479. Axial e-beam excitation (2 MeV, 55 kA, 55 ns) of 5-torr F$_2$ in Ar at ~ 2 atm produces a 55-ns pulse of energy 92 J with the peak power of 1.6 GW. Reference 35.
1480. UV-preionized transverse electric discharge (25 kV, 10 J, 40 ns) excitation of F$_2$ to Ar to He (0.3:30:69.7%) at a pressure; ~ 2 atm produces a 20- to 25-ns pulse with peak energy of 60 mJ and intrinsic efficiency of $\sim 1\%$. Reference 37.
1481. This system once appeared to be an attractive candidate for high-energy storage short-pulse laser applications, e.g., thermonuclear fusion. Excitation mechanisms, kinetic processes, and collisionally stimulated emission in Group VI elements are examined. References 39, 41, 42, 43.
1482. A theoretical expression for the emission-rate coefficient of ArO is developed, interatomic potentials and transition moments are discussed, and a theory is given which explains the main features of O(^{1}S) collision-induced emission in rare gases. References 44, 45.
1483. Vacuum UV photolysis of 2 torr N$_2$O in 12 to 41 atm Ar utilizing 192-nm Ar excimer radiation produced O(^{1}S$_0$) at high densities. The gain profile and stimulated emission cross section near 558 nm were determined. Reference 46.

1484– Optical pumping by a dye laser ($540 < \lambda < 580$ nm) of ~ 2 torr Bi_2 in 10 torr Ar at
1485.   1400 K produces a dense output spectra covering the entire region between 660 and 710 nm. Multiline output power up to 350 mW observed with several watts pumping power. Reference 49.

1486.   Optical pumping (Ar_2^*, 514.5 nm) of Bi_2 to Bi to Ar (6:22:10-55 torr) exhibits an optimum operating temperature of 1300 K. Reference 48.

1487.   Electron-beam excitation (235 keV, $5A/cm^2$, 50 mJ/cm^3) of 1 to 3 torr Br_2 in ~2 atm Ar produced maximum energy of 17 mJ with an efficiency ~0.1%. Reference 50.

1488.   Discharge pumping (11 kV/cm, 25 A/cm^2) in a high-energy electron beam of 0.4% Br_2 in Ar at 4 atm produced >200-ns pulses with energy 10 μJ. Reference 52.

1489.   Longitudinal electron-beam excitation of 2 torr Cl_2 in ~12 atm He yielded 96 mJ with an efficiency ~0.4%. Reference 55.

1490– Fluorescence and laser emission were observed utilizing coaxial electron beam (600
1491.   keV, 5 kA, 2 ns) excitation of Cl_2 to F_2 to Ne (0.025:0.025:99.5%) at 20 torr. Simultaneous fluorescing of F_2 (158 nm) and Cl_2 (258 nm) with similar band intensities occurred. Reference 56.

1492– Photodissociation and predissociation by flash photolysis (15 to 18 kV, 117 to 2.5
1493.   kV) of the parent molecules HCN, ClCN, BrCN, ICN,$(CN)_2$, CH_3CN, CF_3CN, and C_2F_5CN in Ar diluent. Methyl isocyanide to Ar (1:99%) at 50 torr produces the most intense emission. Not all lines are observed with each species. Reference 57.

1494.   Open discharge, electrical explosion of W wire pumping (50 kV, 60 kJ) of $(CN)_2C_4F_8$ to Ar (5:95%) at 200 torr produced 70 mJ. Reference 58.

1495.   Observed in a pulsed (25 pps) discharge (80 A peak) in a 1.17-m long, 10-mm I.D. tube. Reference 59.

1496. $\pm 10^{-5}$ μm.

1497– Superradiant 1.5-ns 6-W pulses produced in 0.05 x 1.2 x 120-cm tube containing
1498.   research grade CO (60 torr) with a high-current, Blumlein-circuit, parallel-plate discharge. Reference 60.

1499.   Electron-beam excitation (0.5 MeV, 10 kA, 3 ns) of 0.1 to 10.0 torr CO in 0.5 to 3 atm He. This laser may exhibit an efficiency approaching 12.7%.

1500.   Electron-beam excitation (1.9 MeV, 400 J, 50 ns) with magnetic field guidance (4.5 kG) of 10 torr F_2 in 10.2 atm He produced a 33-ns pulse of energy 0.25 J/cm^2 with a peak power of 7.6 MW/cm^2 and 2.6% efficiency. Reference 63.

1501.   Electron-beam pumping (850 keV, 20 kA, 300 ns) in a 3-KG field of 1 to 4 torr F_2 in 2 atm He produced a maximum energy of 22 mJ. Reference 64.

1502.   UV-preionized (20 kV, 10 J, 300 ns) fast-discharge pumping (35, 40 ns) produced power output with marked sensitivity toward oxygen impurity. Reference 65.

1503.   Electron-beam excitation (600 kV, 10 kA) of hydrogen isotopes at a pressure of ~ 8 torr in a liquid nitrogen-cooled stainless tube. Reference 66.

1504.   Optimum conditions in a TEA H and D laser, 0.1 torr and 5-kV discharge voltage produced 0.5 mJ. Reference 67.

1505.   Additional observations made with the same apparatus. Reference 68.

1506.   Observed in a pulsed discharge in a 1-m long, 12- x 3-mm tube. Maximum power output is several hundred kilowatts in a 1-ns pulse. Reference 69.

1507.   Observed in a pulsed discharge in a 1.2-m long, 12- x 0.4-mm tube; pressure, 20 to 150 torr H_2; maximum power output, 1.5 kW in a 2-ns pulse. Reference 70.

1508.   Observed in a pulsed discharge in a 1.02-m long, 7-mm I.D. tube; pressure, 3 torr H_2. Reference 71.

1509.   Observed in a pulsed (20 pps) discharge in a 1.45-m long, 15-mm I.D. tube; in H_2 or H_2 - D_2 mixtures, pressure 3 torr. References 72, 74.

1510.   Electron-beam excitation (600 kV, 10 kA) of hydrogen isotopes at a pressure of ~ 8 torr in a liquid nitrogen-cooled stainless tube. Reference 66.

1511.   Additional observations made with the same apparatus. Reference 68.

1512.   Electron-beam excitation (600 kV, 10 kA) of hydrogen isotopes at a pressure of ~ 8 torr in a liquid nitrogen-cooled stainless tube. Reference 66.

1513.   Additional observations made with the same apparatus. Reference 68.

1514.   Observed in a pulsed (20 pps) discharge in a 1.45-m long, 15-mm I.D. tube; in H_2 or H_2 - D_2 mixtures, pressure 3 torr. References 72, 74.

1515.   Electron-beam excitation (600 kV, 10 kA) of hydrogen isotopes at a pressure of ~ 8 torr in a liquid nitrogen-cooled stainless tube. Reference 66.

1516.   Additional observations made with the same apparatus. Reference 68.

1517.   Observed in a pulsed (20 pps) discharge in a 1.45-m long, 15-mm I.D. tube; in H_2 or H_2 - D_2 mixtures, pressure 3 torr. References 72,74.

1518.   Electron-beam sustained discharge pumping (25 kV, 2.5 A/cm^2, 800 ns) of HgBr to CCl_4 to Ar (0.2:05:99.3%) at ~ 2 atm produces ~ 1 mJ in an unoptimized system. Reference 75.

1519–   Electron-beam excitation (240 kV, and 100 A/cm^2, 150 ns) of a mixture of HBr to
1520.   Hg to Xe to Ar (0.8:2.0:10.8:86.4%) at an Ar density of 3 amagats produced 501.8-nm radiation of 3.2-mJ energy in a 60-ns pulse resulting in a peak power of 50 kW and an efficiency of 0.25%. Reference 76.

1521. $HgBr^{81}$ transition.

1522. $HgBr^{79}$ transition.

1523.   Optical pumping (ArF*, 193 nm) on the photodissociation of $HgBr_2$ at a vapor pressure of 50 torr produced 0.25 mJ.

1524.   UV-preionized transverse-electric-discharge photodissociation of $HgBr_2$ in N_2 to He (100:900 torr) produced 7.5 mJ with an efficiency of 0.1%.

1525.   First report of electron-beam excitation of CCl_4 to Hg to Xe to Ar (1.1:2.1:11.1:85.7%) at an Ar density of 3 amagats on the 557.6-nm line produced 138 kW with an efficiency of 0.5%. Reference 80.

1526.   Electron-beam (300 kV, 2 A/cm^2, 400 ns) controlled discharge pumping (2.2 kV, 2.5 kA) of Ar to Hg to Cl_2 (0.7:0.5:98%) at ~ 2 atm lased with 0.1 mJ energy and an efficiency of 0.01%. Reference 81.

1527.   UV-preionized transverse-discharge-initiated predissociation of $HgCl_2$ to N_2 to He (1:10:89%) resulted in an order of magnitude improvement in efficiency and output producing 6 J in a 50-ns pulse with an efficiency of 0.05%. Reference 78.

1528.   Photolyic predissociation (Xe_2, 172 nm) of $HgCl_2$ in He at 200 torr produces emission on the (0,22) and (1,23) lines of the B → X transition. Reference 82.

1529–   UV-preionized discharge (6 J in 50 cm^3, 50 ns) dissociation of HgI_2 (1 to 10 torr)
1530.   in He (350 to 1000 torr) produces maximal energy of 0.3 mJ in a mixture of HgI_2 to N_2 to He (1:10:89%). Energy output was an order of magnitude less without the nitrogen. The spectrum consists of six lines which are correlated to the (0,15) and (0,17) vibrational bands. References 78,79.

1531.   Electron-beam pumping (1 MeV, 20 kA, 20 ns) of 30 torr CF_3I in Ar at 10 atm produced 36 mJ in a 10-ns pulse corresponding to a peak power of 3.6 MW. Reference 85.

1532.   Electron-beam pumping of HI, CF_3I, or CH_3I in Ar at ~ 4 atm produced lasing on eight vibrational bands; maximum energy obtained for 8 torr HI in Ar at 5.3 atm was 1 J in a 40-ns pulse corresponding to a peak power of 25 MW. Reference 86.

1533. Exploding W wire (45 kV, 6 kJ, 5 μs) photoexcitation of 5 torr I_2 in 1.5 atm SF_6 produces a 5-μs pulse with an energy of 0.4 J and an overall efficiency of 0.1%. Reference 87.

1534. Optical pumping at 531.9 nm (Nd to YAG laser, 190-ns pulse) of I_2 in a Brewster cell exhibits pumping threshold of a few microjoules. Reference 88.

1535. Optical pumping at 530.6 nm (Nd to YAG laser, 190-ns pulse) of I_2 in a Brewster cell exhibits pumping threshold of a few microjoules Reference 88.

1536. CW-optical pumping at 514.5 nm (Ar^+ laser, 1 to 4 W) of I_2 vapor in a room-temperature 50-cm cell produces a maximum of 3 mW average power in ~ 50 μs pulses with a maximum conversion efficiency of 0.14%. Reference 89.

1537. CW-oscillation produced by Ar^+ laser pumping at 514.5 nm yields up to 250 mW with a pump power of 3 W. CW-oscillation in the spectral ranges 625.8 to 814.4 and 976.6 to 1217 nm should be possible with suitable mirrors. Reference 90.

1538. Optical pumping in the wavelength range 593.1 to 597.2 nm by a dye laser (1 to 2 kW, 1 μs) of I_2 in a 70-cm glass tube produced 20 to 40 W in a 0.2- to 0.6-μs pulse. At 125 °C an output power of 150 W is produced. Reference 91.

1539. Superfluorescent emission is produced via pumping with a broad-band dye laser over the range 515 to 583 nm. Threshold pump intensity is ~10.μJ at 532 nm. Rotational assignment is tentative. Reference 92.

1540. First demonstration of electron-beam pumping (600 keV, 760 J, 50 ns) of a mixture of Cl_2 to Kr to Ar (0.15:2.96.95%) at 4.5 atm produces 50 mJ in a ~ 30-ns pulse. Reference 93.

1541. Discharge pumping (80 to 110 kV) of mixtures of Cl_2 to Kr to He(1.0:10.89%) at ~ 1 atm produced superfluorescent lasing with 1.3 mJ in a 10-ns pulse corresponding to a peak power of 60 kW. Reference 23.

1542. Discharge excitation of mixtures of HCl to Kr to He (0.15:10:89.85%) at 3.3 atm produced 100 mJ per pulse. Reference 36.

1543– Ab initio configuration interaction calculations were performed; the main emission
1544. band corresponds to the $III_{1/2} \rightarrow I_{1/2}$ ionic to covalent transition, while emission features at 220 and 275 nm are assigned to the $IV_{1/2} \rightarrow I_{1/2}$ and $II_{3/2} \rightarrow I_{3/2}$ transitions, respectively. Reference 96.

1545– Axial electron-beam excitation (2 MeV, 55 kA, 55 ns) of a mixture of F_2 to Kr to
1546. Ar (0.3:7.1: 92.6%) produced 108 J of energy in a ~55-ns pulse corresponding to a peak power of ~ 1.96 W and an intrinsic efficiency of ~ 3%.

1547. Fast-discharge pumping (5 to 20 kV, 10 ns) of NF_3 to Kr to He (0.1:1.0:98.9%) at 1 to 4 atm produced 6 to 8 mJ in a ~100-ns pulse with an intrinsic efficiency of 0.4%. Reference 98.

1548. Electron-beam controlled-discharge pumping (350 kV, 2.2 kA, 100 ns) of a mixture of F_2 to Kr to Ar (0.1:2.1:97.9%) at 1 atm produced a maximum of 6.0 mJ energy in a 90-ns pulse corresponding to >100 kW and an efficiency of 0.2%. Reference 99.

1549– Electron-beam excitation (250 keV, 12 A/cm^2, 0.5–1.0 μs) of a mixture F_2 to Kr
1550. to Ar (0.2:4.0:95.8%) at 1.7 atm and a volume of 8.5 l produced 102 J with an efficiency of 9%. An accurate computer code is developed which explains the main features of the electron beam-produced KrF radiation. Reference 100.

1551. Discharge-pumping of F_2 to Kr to Ar (0.7:4.2:95.1%) produces 10 W of average power at a pulse rate of 1 kHz with an efficiency of 0.13%. Reference 101.

1552. An ultra-high spectral brightness KrF laser with a pulse energy of ~ 60 mJ and spectral width of 150 ± 30 MHz is tunable over 2 cm^{-1} and has beam divergence of 50 μrad. Reference 102.

1553.   First report of coherent oscillation in Kr_2 with a 10-ns pulse. The 147.0-nm resonance line of Xe impurity is a prominent absorptive feature in the spectrum. Threshold pressure for oscillation is ~ 250 psi. Reference 73.

1554.   The physics of electron beam-excited rare gases at high densities is reviewed in Reference 40.

1555.   Measurements were made of conversion reaction coefficients for the three-body conversion reactions of Kr and Xe to their molecular ions, Kr^+_2 and Xe^+_2, in electron afterglow plasmas. Reference 103.

1556.   An early dynamic model of high-pressure rare gas excimer lasers is given. Reference 84.

1557.   Electron-beam excitation (1 MeV, 50 ns, 2 x 10 cm) of 5 torr O_2 in 25 atm Kr produces an optimum output of 5 to 10 mJ in a 100-ns pulse corresponding to a peak power output of ~ 100 kW. Reference 39.

1558.   The metastable $^3P_2$ state of Kr strongly resembles Rb in its chemical properties. Reference 2.

1559.   Electrical-discharge (23 kV) in a 240- x 10- x 5-mm channel containing 70 torr of N_2 produces a 1-ns pulse of energy 0.4 mJ and a peak power of 400 kW with no mirror. High-resolution Czerny-Turner spectrograph employed. Reference 104.

1560.   Discharge pumping (50 kV) of 0.5 to 1 torr N_2 in 2.5 to 11 mm I.D., 30- to 120-mm pyrex tubes at liquid air temperature. Reference 105.

1561.   Observed in a pulsed discharge (several hundred amperes) in a 2.25-m, 15-mm I.D. tube containing 0.15 torr N_2 and 0.5 torr Ne. Reference 107.

1562.   Observed in a pulsed discharge (several hundred amperes) in a 2.25-m, 15-mm I.D. tube containing 0.15 torr N_2 and 0.5 torr Ne. Reference 128.

1563.   Observed in a pulsed (15 pps) discharge at pressure of 1 torr. Reference 108.

1564.   Preionized discharge excitation (60 kV) of N_2 to He (1:99%) at 3 to 11 atm produced superradiant power of 400 kW. Reference 112.

1565.   Electron-beam excitation (200 keV, 3kA, 50 ns) of 0.02 to 0.2% N_2 in He at 0.5 to 7 atm. Reference 127.

1566.   Spark-preionized discharge excitation (30 kV, 27 J, 15 ns) of 1 torr N_2 in 2.6 atm. He produces a 6-ns pulse with energy 3 mJ and peak power of 0.5 MW. Reference 114.

1567.   Electron-beam pumping (900 kV, 14 kA, 20 ns) of 10 torr N_2 in 7 atm He produces a 15-ns pulse of energy 0.27 J/l, power 0.5 MW, and an efficiency of 1.9%. Reference 115.

1568.   Optical pumping with the continuous Ar^+ - laser in the range 458 to 488 nm of Na_2 in a heat pipe produces continuous oscillation with an output power of up to 3 mW. Reference 117.

1569–
  1570.   Nd to YAG 473-nm radiation focused through antireflection windows into a stain-less steel pipe at 605 °C containing Na_2 to Na in a He buffer at 30 torr produces an average superfluorescent pulse energy of ~0.13 µJ with an intrinsic efficiency of ~ 0.07%. Reference 118.

1571.   A-band laser lines excited by 659-nm lines of apparatus in Comment 1569 produces only 2.4 nJ due to nonoptimum output coupling and losses at the Na vapor-fogged windows. Reference 118.

1572.   Line-tunable laser action is observed via frequency-doubled dye laser pumping of 1 to 10 torr S_2. Transitions corresponding to (3-0) to (7-0) were excited; intrinsic efficiency is ~ 2%. Reference 120.

1573.   Vibronic transitions of S_2 are given. Reference 121.

1574.   Photolysis of ~ 30 torr COS in 100 torr Xe diluent produced specific energy output of 0.5 mJ/cm^3. Reference 122.

1575.   CW-optical pumping with the 476.5-nm Ar laser line of Te_2 vapor in a heat pipe at 1000 K produces CW-output having maximum multiline power output of 20 mW with a pump power of 1 W. Reference 48.

1576.   Electron-beam pumping (433 keV, 36 J, 50 ns) of 0.10 to 4% analytical grade Br_2 in Xe at 500 to 1500 torr produced peak power of 200 W. Reference 123.

1577.   Electrical-discharge pumping (26 kV, ~ 100 kW, ~ 35 ns) of HBr to Xe to He (0.1:5:94.9%) at 3.3 atm overcomes the problem of Xe_2^+ absorption and produces 60 mJ per pulse. Simultaneous lasing of Br_2 at 291.3 nm is observed. Reference 124.

1578.   Electron-beam pumping (300 keV, 14 A/cm^2, 1.2 μs) of HCl to Xe to Ne (0.07:1.0:98.93%) at 4 atm produces 3.0 J/ l with an efficiency of 5%. Reference 3.

1579.   UV-preionized-discharge pumped HCl to Xe to He (0.2:3.0:96.8%) at 2.7 atm produced maximum output energy of 1110 mJ in a 30 ns pulse with an efficiency of >0.8%. Reference 4.

1580.   Electric-discharge excitation (48 kV, 150 kA) of HCl to Xe to He (0.2:5.0:94.8%) at ~ 3.3 atm produces peak energy of 180 mJ in a ~ 30 ns pulse and is amenable to continuous operation. Reference 36.

1581.   Fast-discharge Blumlein-type circuit electrical discharge with preionization (20 to 33 kV, ~ 1 kA, 8 ns) of NF_3 to Xe to He (1.2:97.6%) at 750 torr produced a 4-ns pulse of energy 100 mJ corresponding to peak power of 25 MW. Reference 7.

1582.   UV-preionized discharge-pumped (25 kV, 10 J, 40 ns) NF_3 to Xe to He (0.3:1.0:98.7%) at ~ 2 atm produced 65 mJ. Reference 37.

1583–   Electron-beam pumping (at 1 MeV, 20 kA, and 20 ns) of a ratio NF_3 to Xe to Ar
1584.   (0.36:10.0:89.4%) at 1.7 atm produces a 10-ns pulse of 5 mJ corresponding to 500 kW with an efficiency >1%. Coaxial excitation produced a 100-ns pulse of 80 mJ with an efficiency of 3%. Reference 8.

1585.   Electron-beam excitation of F_2 to Xe to Ar (0.1:0.3:99.6%) at <4 atm produces 6 kW. Reference 9.

1586.   Photodissociation (Xe_2^*, 172 nm, 10^5 W/cm^2) of ~ 2 torr XeF_2 in Xe at 6 atm produces output on the 353- and 483-nm bands of slightly more than 1 mJ. Reference 10.

1587.   Electron-beam pumping (1 MeV, 20 kA, 8 ns) of NF_3 to Xe to Ar (0.17:0.33:99.5%) at 5.9 atm produced a 20-ns pulse with peak power of 5 kW. Reference 11.

1588.   Pulsed electron-beam pumping (1 MeV, 50 kA, 50 ns) of 10 torr O_2 in 12 atm Xe produces a 60-ns pulse of energy ~ 10 mJ corresponding to a peak power of ~ 80 kW. Reference 39.

1589.   Open high-current discharge (45 kV, 20 μs) optical pumping (130 < λ < 145 nm) of 3.5 torr N_2O in Xe at 0.7 atm at a temperature of 170 °K produced 10 $μJ/cm^3$ in a 4- to 5-μs pulse. Reference 14.

1590.   First report of laser action on liquified Xe by electron-beam pumping (300 A/cm^2) produces peak power of ~ 1 kW on a line near 176 nm. Reference 15.

1591.   First report of lasing on high pressure (1 to 30 atm) gaseous Xe using electron-beam excitation (0.6 MeV, 7 kA, 2 ns) showed a threshold pressure of 13.6 atm and produced a 3-ns pulse with an intrinsic efficiency of ~ 20%. Reference 20.

1592.   First observation of lasing produced via e-beam-excited Ar/Kr/NF_3 (4 torr, 400 torr, 6 to 8 atm) yields ~ 10 kW in a 25-ns pulse. Reference 129.

1593.   First observation of laser action produced from e-beam-excited Ar/Xe/CCl_4 (4 to 9 atm, 100 to 750 torr, 0.5 to 10 torr). Peak laser power measured to be ~ 2 kW. Reference 21.

1594. Gain was measured on the blue-green band of Xe$_2$Cl and its characteristics were shown to be similar to those of the XeF(C → A) transition. Reference 22.

1595– First observation of optical gain for trivalent rare earth molecular vapors. A purified
1596. sample of NdCl$_3$/AlCl$_3$ in a Brewster-angle quartz cell at ~ 350 °C is excited by a pulsed dye laser (587.2 nm, ~ 1.5 torr). The existence of optical gain, etching, impurities, and Schlieren effects prevent demonstration of laser oscillation. Reference 24.

1597. Fluorescence decays were studied using the vapor-phase chelate 2,2,2,6-tetramethyl-3,5-heptanedione.

1598– A double Brewster-angle quartz cell containing TbCl$_3$/AlCl$_3$ (55.8 mg/354.8 mg)
1599. starting materials was situated in a cylindrical oven. The driver, a 15- to 20-mJ KrF* laser directed axially along the cell, produced excited-state densities of ~ 10J l sustained for ~ 10 μs. Reference 26.

1600. Stimulated emission from POPOP [p-phenylene-bis (5-phenyl-2-oxazole)] has been observed under electron beam excitation in a mixture involving a 4 to 5 atm argon buffer.

1601. The values of P(N) given here represent the calculated mean values of closely spaced spin doublets P$_1$ (J - 1/2) and P$_2$ (J + 1/2) which were not resolved in the experiment. Reference 58.

1602. Observed during flash photolysis of (CN)$_2$, CF$_3$CN, BrCN, or C$_2$F$_5$CN in up to 50 torr of Ar. All transitions listed here are in the ground electronic state. Electronic transitions give rise to laser lines in the 1- to 2-μm region. Reference 58.

1603– Observed in a supersonic flow laser using an electric-discharge-excited CO-He-O$_2$
1604. mixture. The transition frequencies are calculated from the best spectroscopic data available. A similar output extending to 2.35 μm has also been reported for a CS$_2$-O$_2$ chemical laser. However, the reported spectrum is inconsistent with the well established spectroscopic data on CO. References 1293, 1304, 1315.

1605. Observed in an electron-beam-controlled discharge in CO-N$_2$-He mixtures cooled to 100 K. The transition frequencies are calculated from the best spectroscopic data available. References 1304, 1324.

1606. Observed during flash photolysis of CS$_2$-O$_2$ mixtures. Reference 1176.

1607. Observed in a liquid-nitrogen cooled longitudinal discharge in CO-N$_2$-He mixtures. Reference 1187.

1608. Observed in a liquid-nitrogen cooled longitudinal discharge in CO-He-O$_2$ mixtures with Q-switching. Reference 1198.

1609. Observed in a pulsed longitudinal discharge in CO. Reference 1209.

1610– Observed in a CW discharge in CO-N$_2$ mixtures with a tube jacket cooled to either
1611. -78 °C or 15 °C. A multiline CW output of 25 W has been reported for a water-cooled tube using a He-CO-N$_2$-Hg-Xe mixture. A multiline pulsed output of 5 J has been reported for a transverse discharge tube using a CO-N$_2$-He mixture at -20 °C. References 1230, 1241, 1247.

1612– Observed in a liquid-nitrogen-cooled longitudinal discharge in CO-N$_2$-O$_2$-He mix-
1614. tures. A multiline peak output power of 7.7 kW and a greater number of laser lines were obtained in Q-switched operation. In CW mode, a multiline output of 70 W has been obtained with the addition of Xe. A discharge excited, supersonic expansion CO-He-O$_2$ laser has produced a 940 W CW output with a similar output spectrum. An output energy of 1600 J per pulse has been reported for an electron-beam sustained discharge CO-N$_2$ laser. Reference 1248, 1249, 1251, 1252:

1615. The transition frequencies given are either calculated values (eight digits) or observed values (nine digits) based on the best data available. Reference 1304.

1616.   Observed in a CW longitudinal discharge in $C^{13}O$-N-Xe-He mixtures cooled to -78 °C. The frequencies and wavelengths are calculated from best available values of Dunham coefficients. References 1253, 1254.

1617.   Observed in a pulsed longitudinal discharge in H_2-Br_2 mixtures. Reference 1255.

1618.   Observed in a pulsed transverse discharge in H_2-Br_2-He mixtures. A maximum energy of 550 mJ per pulse has been obtained by using a mixture of H_2, Br_2, Ar, and benzene. References 1256, 1257.

1619.   Observed in a pulsed longitudinal discharge in H_2-Br_2 mixtures. Reference 1255.

1620.   Observed in a pulsed transverse discharge in H_2-Br_2-He mixtures. A maximum energy of 550 mJ per pulse has been obtained by using a mixture of H_2, Br_2, Ar, and benzene. References 1256, 1257.

1621.   Observed in a pulsed transverse discharge in D_2-Br_2-He mixtures; peak output power ~6 kW. Frequencies and wavelengths are calculated from spectroscopic data. References 1256, 1258.

1622.   Observed in a pulsed longitudinal discharge in D_2-Br_2 mixtures. Reference 1255.

1623.   Observed in a pulsed transverse discharge in D_2-Br_2-He mixtures; peak output power ~6 kW. Frequencies and wavelengths are calculated from spectroscopic data. References 1256, 1258.

1624.   Observed in a pulsed longitudinal discharge in D_2-Br_2 mixtures. Reference 1255.

1625.   Observed in a pulsed transverse discharge in H_2-Cl_2-He mixtures; peak output power ~2.4 kW. Frequencies and wavelengths for the $v = 1 \rightarrow 0$ band are taken from absorption spectroscopy. Reference 1256, 1259, 1260.

1626.   Observed during a flash initiated reaction of H_2 and Cl_2. References 1262, 1263.

1627.   Observed in a pulsed longitudinal discharge in H_2-Cl_2 mixtures. Reference 1255.

1628.   Observed in a pulsed longitudinal discharge in H_2-Cl_2 or H_2-NOCl mixtures. Reference 1264.

1629.   Observed in a pulsed transverse discharge in H_2-ICl-He mixtures. Frequencies and wavelengths are calculated from spectroscopic data. References 1256, 1259.

1630.   Observed in a pulsed longitudinal discharge in H_2-Cl_2 mixtures. Reference 1255.

1631.   Observed in a pulsed transverse discharge in H_2-Cl_2-He mixtures; peak output power ~2.4 kW. Frequencies and wavelengths for the $v = 1 \rightarrow 0$ band are taken from absorption spectroscopy. References 1256, 1259, 1260.

1632.   Observed in a pulsed transverse discharge in D_2-Cl_2-He mixtures; peak output power ~5 kW. Frequencies and wavelengths for the $v = 1 \rightarrow 0$ band are taken from absorption spectroscopy. References 1256, 1265.

1633.   Observed in a pulsed longitudinal discharge in D_2-Cl_2 mixtures. Reference 1255.

1634.   Observed during a flash initiated reaction of D_2 and Cl_2 mixtures. Reference 1263.

1635.   Observed in a pulsed transverse discharge in D_2-Cl_2-He mixtures; peak output power ~5 kW. Frequencies and wavelengths for the $v = 1 \rightarrow 0$ band are taken from absorption spectroscopy. References 1256, 1265.

1636.   Observed in a pulsed longitudinal discharge in D_2-Cl_2 mixtures. Reference 1255.

1637.   The laser wavelengths and frequencies are either taken or calculated from absorption spectroscopy. Reference 1266.

1638.   Observed in a pulsed transverse discharge in H_2-SF_6 mixtures. References 1267, 1268.

1639.   Observed in a pulsed longitudinal discharge in H_2-SF_6 mixtures. Reference 1269.

1640.   Observed during flash photolysis of H_2-F_2-He mixtures. Reference 1270.

1641.   Observed in a continuous mixing flow of arc-heated N_2, SF_6, and H_2, with supersonic expansion. A maximum output power of 12.4 kW was obtained in a system using He, F_2, and H_2. References 1271, 1273.

1642– Observed in pulsed longitudinal discharge in H_2 - Freon mixtures. A similar spect-
1643. rum was observed in a transverse discharge laser using $He-SF_6-H_2$ mixtures. 25 A maximum output energy of 4.2 kJ per pulse was obtained by electron-beam excitation of $F_2-O_2-H_2$ mixtures. References 1256,1274, 1275.

1644. Frequency accurately measured against a CO laser line. Reference 1276.

1645. Observed in HF optically pumped by a pulsed HF laser. Reference 1277.

1646. Observed during flash photolysis of IF_5-H_2 mixtures. Reference 1278.

1647. Observed during flash photolysis of H_2-F_2 mixtures. Reference 1279.

1648. Observed in a pulsed transverse discharge in $HI-He-SF_6$ (or SO_2F_2) mixtures. Reference 1280.

1649. Observed during flash photolysis of $N_2F_4-CD_4$ mixtures. Reference 1281.

1650. Observed in a pulsed longitudinal discharge in D_2-SF_6-He mixtures. Reference 1282.

1651. Observed in a continuous mixing flow of arc-heated N_2, SF_6, and D_2, with supersonic expansion. A maximum output power of 340 W was obtained. References 1271, 1282.

1652. Observed in a pulsed transverse discharge in $He-SF_6-D_2$ mixtures. Output power ~24 kW. Reference 1256.

1653. Observed in a pulsed longitudinal discharge in D_2-Freon mixtures. Reference 1274.

1654. Observed in Br_2-NO-He mixtures during flash photolysis of Br_2. NO molecules are vibrationally excited by energy transfer from electronically excited Br atoms. Reference 1284.

1655. Observed in a pulsed longitudinal discharge in NOCl-He mixtures. Reference 1285.

1656. Observed during flash photolysis of NOCl-He mixtures. Output power 10 W. References 1286, 1287.

1657. Transition, $P(J-1/2)$.

1658. The subscript 1 refers to the F_1 component of the spin doublet which is expected to dominate over the F_2 component. For F_1 components, $J = K + 1/2$. Reference 58.

1659. Observed during flash photolysis of O_3-H_2 mixtures. Reference 1288.

1660. Observed in a pulsed transverse discharge in O_3-H_2 mixtures. Reference 1289.

1661– Observed in pulsed transverse discharge in CO_2-N_2-He mixtures which is simul-
1662. taneously stimulated by the output from a separate CO_2 laser operating on $00^0 2 \rightarrow 02^0 1$ or $00^0 2 \rightarrow 10^0 1$ sequence band. Frequencies and wavelengths are calculated from spectroscopic data. References 1294, 1295.

1663. Observed in Br_2-CO_2-He mixtures during flash photolysis of Br_2. CO_2 molecules are excited by energy transfer from electronically excited Br atoms. Frequencies and wavelengths are calculated from spectroscopic data. References 1295, 1296.

1664. Observed in a longitudinal discharge in $He-N_2-CO_2$ mixtures following Q-switched oscillation near 10.6 μm. Reference 1297.

1665. All wavelengths and frequencies given are calculated from accurately determined molecular constants. Reference 1298.

1666. Observed in a continuous longitudinal discharge in a $He-N_2-CO_2$ mixture. Reference 1299.

1667– Observed in a continuous longitudinal discharge in CO_2. A continuous output
1668. power of 27.2 kW has been reported for a transverse-flow, transverse-discharge $He-N_2-CO_2$ laser, and a quasicontinuous output of 60 kW has been reported for an N_2-CO_2 gas dynamic laser. In the pulse mode, an output power of 3 GW has been obtained from a UV preionized transverse discharge laser, while a 2-TW output beam has been produced by an electron-beam controlled transverse discharge laser. References 1300–1303.

1669.   Observed in a longitudinal-discharge He-N_2-CO_2 laser with an intracavity hot CO_2 cell. Reference 1306.

1670.   Observed in a longitudinal-discharge He-N_2-CO_2 laser with an intracavity CO_2 absorption cell. Reference 1307.

1671.   Observed in a pulsed longitudinal discharge of CO_2. Reference 1308.

1672.   Observed in a pulsed longitudinal discharge in cryogenically cooled CO_2-N_2-He mixtures which are also irradiated by a 10.6 μm beam from a TEA laser. Reference 1309. Frequencies and wavelengths are based on absorption spectroscopy. Reference 1310.

1673.   Observed in a pulsed longitudinal discharge in cryogenically cooled CO_2-HBr-Ar mixtures cooled to -80 °C which are optically pumped by the outputs from a pulsed HBr and a pulsed (10- or 9-μm) laser. Reference 1311. Frequencies and wavelengths are based on absorption spectroscopy. Reference 1310.

1674.   Observed in $C^{12}O^{16}O^{18}$ optically pumped by the output from an HF laser. Reference 1312.

1675.   Observed in a longitudinal discharge in a mixture of $C^{12}O^{16}O^{18}$-N_2-Xe-He-H_2. Reference 1298.

1676.   Observed in $C^{12}O_2^{18}$ optically pumped by the output from an HF laser. Reference 1312.

1677.   Observed in a longitudinal discharge in a mixture of $C^{12}O_2^{18}$-N_2-Xe-He-H_2 cooled to -60 °C. References 1298, 1313.

1678.   Observed in a longitudinal-discharge $C^{12}O_2^{18}$-N_2-Xe-He laser with an intracavity hot $C^{12}O_2^{18}$ cell. Reference 1237.

1679.   Observed in a longitudinal discharge in a mixture of $C^{13}O_2^{16}$-N_2-Xe-He-H_2. Reference 1298,1313.

1680–   Observed in a longitudinal-discharge $C^{13}O_2^{16}$-N_2-Xe-He laser with an intracavity
1681.   hot $C^{13}O_2^{16}$-cell. Reference 1238.

1682.   Observed in a longitudinal discharge in a mixture of $C^{13}O_2^{18}$-N_2-Xe-He-H_2 cooled to -60 °C. References 1298, 1313.

1683.   Observed in a longitudinal discharge in a mixture of $C^{14}O_2^{16}$-N_2-Xe-He-H_2. Reference 1298.

1684.   Observed in a longitudinal discharge in a mixture of $C^{14}O_2^{18}$-N_2-Xe-He-H_2. Reference 1298.

1685.   Observed in a longitudinal discharge in a mixture of $C^{14}O_2^{18}$-N_2-Xe-He-H_2. Reference 1298.

1686.   Observed in a pulsed longitudinal discharge in COS-He mixtures. Reference 1316.

1687.   Observed in COS and COS-CO mixtures optically pumped by a frequency doubled CO_2 laser. The frequencies and wavelengths are calculated from existing spectroscopic data. References 1317, 1318, 1338.

1688.   Observed in a pulsed longitudinal discharge in COS-He mixtures. Reference 1319.

1689.   Observed in COS optically pumped by a 9.57 μm CO_2 laser. Reference 1320.

1690.   Observed in C^{13}OS-CO mixtures optically pumped by a frequency doubled CO_2 laser. Reference 1317.

1691.   Observed in C^{13}OS optically pumped by a CO laser. References 1336, 1337.

1692.   Wavelength given is from a group of unresolved laser lines.

1693.   Observed in CS_2 optically pumped by the second harmonic of the 9.2-μm CO_2 R(30) laser line. Reference 1317.

1694.   Observed in optically pumped CO-CS_2 mixtures. A frequency-doubled TEA CO_2 laser is used to excite CO molecules which in turn transfer energy to CS_2. Reference 1322.

1695. Observed in continuously flowing mixtures of preexcited N_2 and CS_2. Reference 1323.

1696. Observed in an electron-beam-stabilized electric discharge in He-CO-CS_2 mixtures. Reference 1324.

1697. Observed in $C^{13}S_2$ vapor optically pumped by a pulsed HF laser. References 1325,1326.

1698. Observed in an electron-beam-stabilized electric discharge in mixtures of He-CO-CS_2. Reference 1323.

1699– Observed in Br_2-HCN-He mixtures during flash photolysis of Br_2. HCN molecules
1700. are vibrationally excited by energy transfer from electronically excited Br atoms. Reference 1327.

1701. Actual line at this location may be in question. Reference 1329.

1702. Observed in a pulsed longitudinal discharge in H_2O vapor. Reference 1328.

1703. Observed in Br_2-H_2O-He mixtures during flash photolysis of Br_2. H_2O molecules are vibrationally excited by energy transfer from electronically excited Br atoms. Reference 1284.

1704. Observed in a pulsed longitudinal discharge in H_2O-He mixtures. References 1329,1330.

1705. Observed in a pulsed transverse discharge in H_2O-He mixtures. Reference 1331.

1706. All values of frequencies and wavelengths given are calculated from accurately determined molecular constants. Reference 1335.

1707. Observed in a continuous longitudinal discharge in N_2O-N_2-He mixtures. Reference 1335.

1708. Observed in a continuous longitudinal discharge in N_2O-N_2 mixtures. Reference 1334.

1709. Observed in a pulsed radio-frequency discharge in N_2O-N_2 mixtures. Reference 1333.

1710. Observed in a pulsed transverse discharge in N_2O-He mixtures; output power ~6 kW. Reference 1331.

1711. Observed in a continuous longitudinal discharge in N_2O-N_2-He mixtures. Reference 1332.

1712. Observed in continuous flowing mixtures of N_2O and discharge-excited N_2. Reference 1177.

1713. Observed in a pulsed discharge in N_2O. Reference 1178.

1714. Observed in a CW, longitudinal-discharge, N_2O-N_2-He laser with an intracavity hot N_2O cell. Reference 1179.

1715. Observed in isotopic N_2O gas optically pumped by a pulsed HF laser.

1716. Observed in isotopic N_2O gas optically pumped by a pulsed HF laser.

1717. Observed in isotopic N_2O gas optically pumped by a pulsed HF laser.

1718. Observed in NOCl at 220 K when optically pumped by various 10.7-μm lines of a TEA CO_2 laser. References 1180,1181.

1719. The CO_2 laser transition = 10P(26).

1720. The CO_2 laser transition = 10P(28).

1721. The CO_2 laser transition = 10P(30).

1722. The CO_2 laser transition = 10P(34).

1723. The CO_2 laser transition = 10P(36).

1724. The CO_2 laser transition = 10P(38).

1725. The CO_2 laser transition = 10P(40).

1726. The CO_2 laser transition = 10P(42).

1727. The CO_2 laser transition = 10P(44).

1728. The CO_2 laser transition = 10P(34).

1729.   Observed in NSF between 220 and 250 K when optically pumped by various 9.7-μm lines of a TEA CO_2 laser. Reference 1239.

1730.   The CO_2 laser transition = 9P40.

1731.   The CO_2 laser transition = 9P36.

1732.   The CO_2 laser transition = 9P46.

1733.   The CO_2 laser transition = 9P48.

1734.   The CO_2 laser transition = 9P36.

1735.   The CO_2 laser transition = 9P44.

1736.   The CO_2 laser transition = 9P36.

1737.   The CO_2 laser transition = 9P44.

1738.   The CO_2 laser transition = 9P46.

1739.   The CO_2 laser transition = 9P42.

1740.   The CO_2 laser transition = 9P42.

1741.   The CO_2 laser transition = 9P40.

1742.   The CO_2 laser transition = 9P36.

1743.   Observed in a longitudinal discharge in CO_2-N_2-He-BCl_3 mixtures. However, this result is yet to be corroborated. The results of a subsequent study cast strong doubts to the claim that these emission lines are associated with the BCl_3 molecule. References 1182, 1183.

1744.   Observed in a pulsed longitudinal discharge in H_2-C_2H_2-He mixtures. Reference 1184.

1745.   Observed in CO-C_2H_2 mixtures in which CO is optically pumped by a frequency-doubled CO_2 TEA laser. References 1317, 1185.

1746.   Observed in electron beam-controlled discharges in CO-C_2H_2 mixtures. Reference 1186.

1747.   Observed in C_2D_2 optically pumped in the 9.4-μm R-branch of the $v_4 + v_5$ band by a high-pressure continuously tunable CO_2 laser. The transition frequencies given are either observed values (4 digits) or calculated values (5 digits). References 1240, 1242.

1748.   Observed in C_2D_2 optically pumped by a TEA CO_2 laser in the 9-μm band. The CO_2 laser transition = R(12).

1749.   Observed in C_2D_2 optically pumped by a TEA CO_2 laser in the 9-μm band. The CO_2 laser transition = P(24).

1750.   Observed in C_2D_2 optically pumped by a TEA CO_2 laser in the 9-μm band. The CO_2 laser transition = P(24).

1751.   Observed in C_2D_2 optically pumped by a TEA CO_2 laser in the 9-μm band. The CO_2 laser transition = P(36).

1752.   Observed in C_2D_2 optically pumped by a TEA CO_2 laser in the 9-μm band. The CO_2 laser transition = R(12).

1753.   Observed in C_2D_2 optically pumped by a TEA CO_2 laser in the 9-μm band. The CO_2 laser transition = R(20).

1754.   Observed in C_2D_2 optically pumped by a TEA CO_2 laser in the 9-μm band. The CO_2 laser transition = R(20).

1755.   Observed in C_2D_2 optically pumped by a TEA CO_2 laser in the 9-μm band. The CO_2 laser transition = R(14).

1756.   Observed in C_2D_2 optically pumped by a TEA CO_2 laser in the 9-μm band. The CO_2 laser transition = R(14).

1757.   Observed in C_2D_2 optically pumped by a TEA CO_2 laser in the 9-μm band. The CO_2 laser transition = R(14).

1758.   Observed in C_2D_2 optically pumped by a TEA CO_2 laser in the 9-μm band. The CO_2 laser transition = P(38).

1759.   Observed in C_2D_2 optically pumped by a TEA CO_2 laser in the 9-μm band. The CO_2 laser transition = P(38).

1760.   Observed in C_2D_2 optically pumped by a TEA CO_2 laser in the 9-μm band. The CO_2 laser transition = P(38).

1761.   Observed in C_2D_2 optically pumped by a TEA CO_2 laser in the 9-μm band. The CO_2 laser transition = P(38).

1762.   Observed in C_2D_2 optically pumped by a TEA CO_2 laser in the 9-μm band. The CO_2 laser transition = P(26).

1763.   Observed in C_2D_2 optically pumped by a TEA CO_2 laser in the 9-μm band. The CO_2 laser transition = P(26).

1764.   Observed in COF_2 optically pumped by a CO laser. References 1336, 1337.

1765.   Wavelength is approximate.

1766.   Observed in optically pumped $N^{14}H_3$-N_2 mixtures. The pump source is the 9.22 μm, R(30) line of a transverse-discharge, pulsed CO_2 laser which excites the sR(5,0) transition of the $0 \rightarrow v_2$ band of $N^{14}H_3$. Reference 1189.

1767.   Observed in optically pumped $N^{14}H_3$-N_2 mixtures. The pump source is the 9.22 μm, R(30) line of a transverse-discharge, pulsed CO_2 laser which excites the sR(5,0) transition of the $0 \rightarrow v_2$ and of $N^{14}H_3$. Reference 1190.

1768.   Wavelengths and frequencies of all positively identified transitions are based on the absorption spectrum.

1769.   Observed in NH_3 optically pumped by the 10.33 μm, R(8) line of a pulsed, transverse-discharge CO_2 laser. Reference 1181.

1770.   Observed in NH_3-He mixtures optically pumped by the 9.22 μm, R(30) line of a pulsed, transverse-discharge CO_2 laser. Wavelengths and frequencies of all positively identified transitions are based on the absorption spectrum. Reference 1191.

1771.   Observed in NH_3-N_2 mixtures optically pumped by the 9.22 μm, R(30) line of a pulsed, transverse-discharge CO_2 laser; maximum conversion efficiency = 20%. Reference 1192.

1772.   Wavelengths and frequencies of all positively identified transitions are based on the absorption spectrum.

1773.   Observed in NH_3 optically pumped by the 10.29 μm, R(14) line of a pulsed, transverse-discharge CO_2 laser. The absorbing transition in NH_3 is the aR(1,1) line of the $0 \rightarrow v_2$ band. References 1193, 1194, 1195.

1774.   Observed in NH_3 optically pumped by the 9.22 μm, R(30) line of a transverse-discharge, pulsed CO_2 laser. Reference 1195.

1777.   Observed in NH_3 optically pumped by two pulsed transverse-discharge CO_2 lasers, one of which is tuned to the 10.72 μm P(32) line and the other to the 9.59 μm P(32) line. Reference 1196.

1778.   Observed in NH_3 optically pumped by the 9.22 μm, R(30) line of a transverse-discharge, pulsed CO_2 laser. Peak output powers of up to 1 MW and energy conversion efficiencies of up to 10 % have been reported. References 1193, 1194, 1195, 1197.

1779.   Observed in NH_3 optically pumped by the 9.29 μm, R(16) line of a transverse-discharge, pulsed CO_2 laser. The absorbing transition in NH_3 is the aR(6,0) line of the $0 \rightarrow v_2$ band. Reference 1199.

1781.   Observed in NH_3 optically pumped by the 9.29 μm, R(16) line of a transverse-discharge, pulsed CO_2 laser. The absorbing transition in NH_3 is the aR(6,0) line of the $0 \rightarrow v_2$ band. References 1193, 1194, 1199, 1200.

1782.   Peak output powers of up to 1 MW and energy conversion efficiencies of up to 28% have been reported.

1783. Observed in NH_3 optically pumped simultaneously by the 10.29 μm, R(14) line of a TEA $C^{12}O_2$ laser and the 9.94 μm, P(14) line of a TEA $C^{13}O_2$ laser. Reference 1203.
1784. Observed in NH_3 optically pumped simultaneously by the 10.74 μm, P(34) line and the 10.57 μm, P(18) line of TEA CO_2 lasers. References 1203, 1204.
1785. Observed in NH_3 optically pumped simultaneously by the 9.89 μm, P(8) line of a TEA $C^{13}O_2$ laser and the 9.59 μm, P(24) line of a TEA $C^{12}O_2$ laser. Reference 1203.
1786. Observed in NH_3 optically pumped simultaneously by the 9.68 μm, P(34) line and the 10.63 μm, P(24) line of TEA CO_2 lasers. Reference 1203.
1787. Observed in NH_3 optically pumped simultaneously by the 9.34 μm, R(8) line and the 10.63 μm, P(24) line of TEA CO_2 lasers. Reference 1205.
1788. Observed in a pulsed longitudinal discharge in NH_3. Reference 1206.
1789. Observed in a pulsed longitudinal discharge in NH_3. References 1207, 1208.
1790. Wavelengths and frequencies of all positively identified transitions are based on the absorption spectrum. Reference 1210.
1791. Observed in NH_3 optically pumped by a TEA CO_2 laser operating in the 9.5-μm sequence band. Reference 1243.
1792. Observed in NH_3 optically pumped by a high-pressure, continuously tunable CO_2 laser. Reference 1244.
1793. Observed in $^{15}NH_3$ optically pumped by the P(6) line of the 3 – 2 band of a pulsed HF laser. Reference 1211.
1794. Observed in $^{12}CF_4$ at ~155 K when optically pumped by one of the 9.4-μm P(J) lines of a TEA $^{12}C^{16}O_2$ laser, where J = 4 to 12. Reference 1181.
1795. Observed in $^{12}CF_4$ at ~155 K when optically pumped by one of the 9.4-μm P(J) lines of a TEA $^{12}C^{16}O_2$ laser, where J = 4 to 14. Reference 1212.
1796. Observed in $^{12}CF_4$ at ~155 K when optically pumped by one of the 9.4-μm P(J) lines of a TEA $^{12}C^{16}O_2$ laser, where J = 4 to 10. Reference 1213.
1797. Observed in $^{12}CF_4$ at 100 to 200 K when optically pumped by one of the 9.3-μm P(J) lines of a TEA $^{12}C^{16}O_2$ laser, where J = 4 to 12. Reference 1181.
1798. Observed in $^{12}CF_4$ at 100 to 200 K when optically pumped by one of the 9.3-μm P(J) lines of a TEA $^{12}C^{16}O_2$ laser, where J = 10 to 24. Reference 1213.
1799. Observed in $^{12}CF_4$ at 100 to 200 K when optically pumped by one of the 9.3-μm P(J) lines of a TEA $^{12}C^{16}O_2$ laser, where J = 4 to 12. Reference 1212.
1800. Observed in $^{12}CF_4$ at ~ 155 K when optically pumped by one of the 9.4-μm P(J) lines of a TEA $^{12}C^{18}O_2$ laser, where J = 18 to 32. References 1213, 1488.
1801. Observed in $^{12}CF_4$ at ~155 K when optically pumped by one of the 9.3-μm P(J) lines of a TEA $^{12}C^{16}O_2$ laser, where J = 10 to 22. Reference 1181.
1802–
1803. Observed in $^{12}CF_4$ at 100 to 200 K when optically pumped by one of the 9.3-μm P(J) lines of a TEA $^{12}C^{16}O_2$ laser, where J = 12, which excites the $R^+(29)$ line of the $v_2 + v_4$ band of $^{12}CF_4$. An output energy of 65 mJ per pulse at an energy conversion efficiency of ~3% has been reported. References 1180, 1202, 1212.
1804. (J) of CO_2 = 10.
1805. (J) of CO_2 = 8.
1807. (J) of CO_2 = 6,24.
1808. (J) of CO_2 = 32.
1809. (J) of CO_2 = 20.
1810. (J) of CO_2 = 4.
1811. (J) of CO_2 = 18.
1812. (J) of CO_2 = 16.
1813. (J) of CO_2 = 16.

1814. (J) of CO_2 = 10.
1815. (J) of CO_2 = 8.
1816. (J) of CO_2 = 6.
1817. (J) of CO_2 = 26.
1818. (J) of CO_2 = 6.
1819. (J) of CO_2 = 18.
1820. (J) of CO_2 = 4.
1821. (J) of CO_2 = 22.
1822. (J) of CO_2 = 12.
1823. (J) of CO_2 = 18.
1824. (J) of CO_2 = 18.
1825. (J) of CO_2 = 14.
1826. (J) of CO_2 = 4.
1827. (J) of CO_2 = 22.
1828. (J) of CO_2 = 24.
1829. (J) of CO_2 = 24.
1830. (J) of CO_2 = 28.
1831. (J) of CO_2 = 20.
1832. (J) of CO_2 = 6.
1833. (J) of CO_2 = 26.
1834. (J) of CO_2 = 26.
1835. (J) of CO_2 = 8.
1836. (J) of CO_2 = 18.
1837. (J) of CO_2 = 10.
1838. (J) of CO_2 = 10.
1839. (J) of CO_2 = 14.
1840. (J) of CO_2 = 12.
1841. (J) of CO_2 = 28.
1842. (J) of CO_2 = 8.
1843. (J) of CO_2 = 14.
1844. (J) of CO_2 = 14.
1845. (J) of CO_2 = 4.
1846. Observed in $^{13}CF_4$ optically pumped by one of the 9.3-μm R(J) lines of a TEA $^{12}C^{16}O_2$ laser, where J = 2 to 18 References 1216, 1217.
1847. Observed in $^{13}CF_4$ optically pumped by one of the 9.4-μm P(J) $^{12}C^{16}O_2$ laser lines, where J = 8 to 12. Reference 1213.
1848. Observed in $^{13}CF_4$ optically pumped by one of the 9.4-μm P(J) $^{12}C^{16}O_2$ laser lines, where J = 4 to 18. References 1216, 1217.
1849. Observed in $^{13}CF_4$ optically pumped by one of the 9.3-μm R(J) lines of a TEA $^{12}C^{16}O_2$ laser, where J = 6 to 8. Reference 1213.
1850. Observed in $^{13}CF_4$ optically pumped by one of the 9.4-μm P(J) $^{12}C^{18}O_2$ laser lines, where J = 28 to 30. Reference 1213.
1851. Observed in $^{13}CF_4$ optically pumped by one of the 9.4-μm P(J) $^{12}C^{18}O_2$ laser lines, where J = 28 to 30. Reference 1488.
1852. (J) of CO_2 = (8).
1853. (J) of CO_2 = (8).
1854. (J) of CO_2 = (6).
1855. (J) of CO_2 = (6).
1856. (J) of CO_2 = (10).
1857. (J) of CO_2 = (6).
1858. (J) of CO_2 = (8).

1859.   (J) of CO_2 = (8).
1860.   (J) of CO_2 = (4).
1861.   (J) of CO_2 = (4).
1862.   (J) of CO_2 = (6).
1863.   (J) of CO_2 = (8).
1864.   (J) of CO_2 = (30).
1865.   (J) of CO_2 = (2)
1866.   (J) of CO_2 = (28).
1867.   (J) of CO_2 = (28).
1868.   (J) of CO_2 = (18).
1869.   (J) of CO_2 = (14).
1870.   (J) of CO_2 = (12).
1871.   (J) of CO_2 = (4).
1872.   (J) of CO_2 = (28).
1873.   (J) of CO_2 = (28).
1874.   (J) of CO_2 = (28).
1875.   (J) of CO_2 = (12).
1876.   (J) of CO_2 = (6).
1877.   (J) of CO_2 = (12).
1878.   (J) of CO_2 = (14).
1879.   (J) of CO_2 = (8).
1880.   (J) of CO_2 = (30).
1881.   (J) of CO_2 = (12).
1882.   (J) of CO_2 = (30).
1883.   (J) of CO_2 = (18).
1884.   (J) of CO_2 = (4).
1885.   (J) of CO_2 = (10).
1886.   (J) of CO_2 = (12).
1887.   (J) of CO_2 = (14).
1888.   Observed in $^{14}CF_4$ optically pumped by one of the 9.4-μm P(J) lines of a TEA $^{12}C^{16}O_2$ laser, where J = 4 to 20. References 1216, 1217.
1889.   Observed in $^{14}CF_4$ optically pumped by one of the 9.3-μm R(J) lines of a $^{12}C^{16}O_2$ laser, where J = 2 to 24. References 1216, 1217.
1890.   Observed in $^{14}CF_4$ optically pumped by one of the 9.4-μm P(J) lines of a $^{12}C^{18}O_2$ laser, where J = 24 to 30. Reference 1488.
1891.   (J) of CO_2 = (16).
1892.   (J) of CO_2 = (4).
1893.   (J) of CO_2 = (22).
1894.   (J) of CO_2 = (12).
1895.   (J) of CO_2 = (20).
1896.   (J) of CO_2 = (2).
1897.   (J) of CO_2 = (18).
1898.   (J) of CO_2 = (10).
1899.   (J) of CO_2 = (10).
1900.   (J) of CO_2 = (24).
1901.   (J) of CO_2 = (14).
1902.   (J) of CO_2 = (24).
1903.   (J) of CO_2 = (10).
1904.   (J) of CO_2 = (6).
1905.   (J) of CO_2 = (6).
1906.   (J) of CO_2 = (26).

1907.   (J) of CO_2 = (18).
1908.   (J) of CO_2 = (14).
1909.   (J) of CO_2 = (6).
1910.   (J) of CO_2 = (10).
1911.   (J) of CO_2 = (22).
1912.   (J) of CO_2 = (6).
1913.   (J) of CO_2 = (14).
1914.   (J) of CO_2 = (18).
1915.   (J) of CO_2 = (14).
1916.   (J) of CO_2 = (28).
1917.   (J) of CO_2 = (8).
1918.   (J) of CO_2 = (30).
1919.   (J) of CO_2 = (26).
1920.   (J) of CO_2 = (8).
1921.   (J) of CO_2 = (30).
1922.   (J) of CO_2 = (30).
1923.   (J) of CO_2 = (14).
1924.   (J) of CO_2 = (4).
1925.   (J) of CO_2 = (16).
1926.   (J) of CO_2 = (20).
1927.   (J) of CO_2 = (6).
1928.   (J) of CO_2 = (6).
1929.   (J) of CO_2 = (20).
1930.   (J) of CO_2 = (10).
1931.   Observed in CF_3I at 170 K when optically pumped by various 9.6-μm lines of a TEA CO_2 laser. Reference 1181.
1932.   (J) of CO_2 = 9P(36).
1933.   (J) of CO_2 = 9P(34).
1934.   (J) of CO_2 = 9P(30).
1935.   Observed in CH_3F optically pumped simultaneously by the 9.5-μm P(14) and the 9.6-μm P(30) lines of CO_2 TEA lasers. Reference 1218.
1936.   Observed in perchloryl fluoride ($FClO_3$) cooled to 160 K and optically pumped by a TEA CO_2 laser operating in the 10.2-μm R branch. Reference 1245.
1937.   (J) of CO_2 = 10R(22).
1938.   (J) of CO_2 = 10R(34).
1939.   (J) of CO_2 = 10R(40).
1940.   (J) of CO_2 = 10R(26).
1941.   (J) of CO_2 = 10R(30).
1942.   (J) of CO_2 = 10R(38).
1943.   (J) of CO_2 = 10R(42).
1944.   (J) of CO_2 = 10R(36).
1945.   (J) of CO_2 = 10R(22).
1946.   (J) of CO_2 = 10R(30).
1947.   (J) of CO_2 = 10R(34).
1948.   (J) of CO_2 = 10R(36).
1949.   (J) of CO_2 = 10R(38).
1950.   (J) of CO_2 = 10R(32).
1951.   (J) of CO_2 = 10R(32).
1952.   (J) of CO_2 = 10R(28).
1953.   (J) of CO_2 = 10R(34).
1954.   (J) of CO_2 = 10R(42).

1955. (J) of CO_2 = 10R(34).
1956. (J) of CO_2 = 10R(36).
1957. (J) of CO_2 = 10R(34).
1958. (J) of CO_2 = 10R(26).
1959. (J) of CO_2 = 10R(40).
1960. (J) of CO_2 = 10R(38).
1961. (J) of CO_2 = 10R(34).
1962. (J) of CO_2 = 10R(32).
1963. (J) of CO_2 = 10R(42).
1964. (J) of CO_2 = 10R(28).
1965. (J) of CO_2 = 10R(26).
1966. (J) of CO_2 = 10R(20).
1967. (J) of CO_2 = 10R(20).
1968. (J) of CO_2 = 10R(36).
1969. (J) of CO_2 = 10R(32).
1970. A group of lines. Observed in HCOOH vapor when optically pumped by the P(5) line of the $v = 3 \rightarrow 2$ band of a pulsed HF laser. References 1325, 1326.
1971. Wavelength is from a group of unresolved laser lines.
1972. Observed in SiF_4 when cooled to 125 K and optically pumped by various 9.6-μm lines of a TEA CO_2 laser. Reference 1219.
1973. (J) of CO_2 = P(8).
1974. (J) of CO_2 = P(18).
1975. (J) of CO_2 = P(24).
1976. (J) of CO_2 = P(20).
1977. (J) of CO_2 = P(24).
1978. (J) of CO_2 = P(34).
1979. (J) of CO_2 = P(18).
1980. (J) of CO_2 = P(26).
1981. (J) of CO_2 = P(18).
1982. (J) of CO_2 = P(30).
1983. (J) of CO_2 = P(18).
1984. Observed in CO-SiH_4 mixtures in which CO is optically pumped by a frequency-doubled CO_2 TEA laser. Reference 1186.
1985. Observed in C_2H_4 optically pumped by one of the 10.3-μm R(J) lines (identified in the remarks column) of a TEA CO_2 laser. Reference 1194.
1986. (J) of CO_2 = R(10).
1987. (J) of CO_2 = R(16).
1988. Observed in CH_3CCH (propyne, methylacetylene) when cooled to 180 K and optically pumped by various 10.4-μm lines of a TEA CO_2 laser. Reference 1246.
1989. The CO_2 laser line = 10P14.
1990. The CO_2 laser line = 10R4.
1991. The CO_2 laser line = 10R16.
1992. The CO_2 laser line = 10P14.
1993. The CO_2 laser line = 10P14.
1994. The CO_2 laser line = 10R16.
1995. Observed in SF_6 optically pumped simultaneously by 10.5-μm P(12) and P(14) lines of a TEA CO_2 laser. Reference 1221.
1996. For line assignments see Reference 1441.
1997. Output power: 0.1 mW.
1998. For line assignments see Reference 1441.
1999. Output power: 1 mW.

2000.   For line assignments see Reference 1452.
2001.   Output power: 1e-07 mW.
2002.   For line assignments see Reference 1452.
2003.   Output power: 1e-06 mW.
2004.   For line assignments see Reference 1452.
2005.   Output power: 1e-06 mW.
2006.   For line assignments see Reference 1452.
2007.   Output power: 1e-06 mW.
2008.   For line assignments see Reference 1452.
2009.   Output power: 1e-08 mW.
2010.   For line assignments see Reference 1452.
2011.   Output power: 1e-06 mW.
2012.   For line assignments see Reference 1452.
2013.   Output power: 1e-05 mW.
2014.   For line assignments see Reference 1452.
2015.   Output power: 1e-07 mW.
2016.   For line assignments see Reference 1452.
2017.   For line assignments see Reference 1452.
2018.   Output power: 1e-07 mW.
2019.   For line assignments see Reference 1452.
2020.   For line assignments see Reference 1452.
2021.   Output power: 0.0001 mW.
2022.   For line assignments see Reference 1452.
2023.   Output power: 1e-06 mW.
2024.   For line assignments see Reference 1452.
2025.   Output power: 0.0001 mW.
2026.   For line assignments see Reference 1452.
2027.   Output power: 0.001 mW.
2028.   For line assignments see Reference 1452.
2029.   Output power: 0.001 mW.
2030.   For line assignments see Reference 1452.
2031.   Output power: 1e-07 mW.
2032.   For line assignments see Reference 1452.
2033.   Output power: 1e-07 mW.
2034.   For line assignments see Reference 1452.
2035.   For line assignments see Reference 1452.
2036.   Output power: 0.0001 mW.
2037.   For line assignments see Reference 1452.
2038.   Output power: 0.0001 mW.
2039.   For line assignments see Reference 1452.
2040.   For line assignments see Reference 1452.
2041.   For line assignments see Reference 1452.
2042.   For line assignments see Reference 1452.
2043.   For line assignments see Reference 1452.
2044.   For line assignments see Reference 1452.
2045.   For line assignments see Reference 1452.
2046.   For line assignments see References 586, 1452, 1854.
2047.   For line assignments see References 586, 1452, 1854.
2048.   For line assignments see References 586, 1452, 1854.
2049.   For line assignments see References 586, 1452, 1854.
2050.   For line assignments see References 586, 1452, 1854.

2051.   For line assignments see References 586, 1452, 1854.
2052.   For line assignments see References 586, 1452, 1854.
2053.   For line assignments see References 586, 1452, 1854.
2054.   For line assignments see References 586, 1452, 1854.
2055.   For line assignments see References 586, 1452, 1854.
2056.   For line assignments see References 586, 1452, 1854.
2057.   For line assignments see References 586, 1452, 1854.
2058.   For line assignments see References 586, 1452, 1854.
2059.   For line assignments see References 586, 1452, 1854.
2060.   For line assignments see References 586, 1452, 1854.
2061.   For line assignments see References 586, 1452, 1854.
2062.   For line assignments see References 586, 1452, 1854.
2063.   For line assignments see Reference 1452.
2064.   For line assignments see Reference 1452.
2065.   For line assignments see Reference 1452.
2066.   For line assignments see Reference 1452.
2067.   For line assignments see Reference 1452.
2068.   For line assignments see Reference 1452.
2069.   For line assignments see Reference 1452.
2070.   For line assignments see Reference 1452.
2071.   For line assignments see Reference 1452.
2072.   For line assignments see Reference 1452.
2073.   Output power: 0.01 mW; laser has a Lamb dip.
2074.   For line assignments see Reference 1368.
2075.   For line assignments see Reference 1368.
2076.   Output power: 0.1 mW; laser has a Lamb dip.
2077.   For line assignments see Reference 1368.
2078.   Output power: 0.001 mW.
2079.   For line assignments see Reference 1368.
2080.   Output power: 10 mW; laser has a Lamb dip.
2081.   For line assignments see Reference 1368.
2082.   Output power: 10 mW; laser has a Lamb dip.
2083.   For line assignments see Reference 1368.
2084.   Output power: 0.001 mW.
2085.   For line assignments see Reference 1368.
2086.   Output power: 0.001 mW.
2087.   For line assignments see Reference 1368.
2088.   Output power: 0.1 mW.
2089.   For line assignments see Reference 1368.
2090.   For line assignments see Reference 1368.
2091.   Output power: 0.01 mW; laser has a Lamb dip.
2092.   For line assignments see Reference 1368.
2093.   For line assignments see Reference 1363.
2094.   Output power: 100 mW; laser has a Lamb dip.
2095.   For line assignments see References 1368, 1393.
2096.   Laser has a Lamb dip.
2097.   For line assignments see References 1368, 1393.
2098.   Output power: 10 mW.
2099.   -39 MHz offset from center of 9.32 μm CO_2 sequence band pump laser; R17 transition; perpendicular polarization.
2100.   Threshold pump power: 1000 mW.

2101. For line assignments see References 1368, 1393.
2102. Output power: 10 mW; Laser has a Lamb dip.
2103. For line assignments see References 1368, 1393.
2104. Output power: 0.1 mW.
2105. -39 MHz offset from center of 9.32 μm CO_2 Sequence band pump laser; R17 transition; parallel polarization.
2106. Threshold pump power: 1000 mW.
2107. For line assignments see References 1368, 1393.
2108. Output power: 10 mW; laser has a Lamb dip.
2109. For line assignments see References 1368, 1393.
2110. Output power: 10 mW; laser has a Lamb dip.
2111. For line assignments see References 1422, 1441, 1445, 1447, 1448, 1449.
2112. For line assignments see References 1422, 1441, 1445, 1447, 1448, 1449.
2113. Output power: 200 mW.
2114. For line assignments see References 1422, 1441, 1445, 1447, 1448, 1449.
2115. For line assignments see References 1422, 1441, 1445, 1447, 1448, 1449.
2116. Output power: 1 mW.
2117. For line assignments see References 1422, 1441, 1445, 1447, 1448, 1449.
2118. For line assignments see References 1422, 1441, 1445, 1447, 1448, 1449.
2119. Output power: 0.1 mW.
2120. For line assignments see References 1422, 1441, 1445, 1447, 1448, 1449.
2121. Output power: 0.1 mW.
2122. For line assignments see References 1423, 1447.
2123. For line assignments see References 1423, 1447.
2124. Output power: 100 mW.
2125. For line assignments see References 1423, 1447.
2126. Output power: 200 mW.
2127. For line assignments see References 1423, 1447.
2128. Output power: 100 mW.
2129. For line assignments see References 1423, 1447.
2130. Output power: 200 mW.
2131. For line assignments see Reference 1436.
2132. Output power: 0.02 mW.
2133. 9.34 μm CO_2 pump laser; R08 transition; parallel polarization.
2134. Threshold pump power: 5000 mW.
2135. For line assignments see References 1426, 1468.
2136. Output power: 0.1 mW.
2137. For line assignments see References 1426, 1468.
2138. Output power: 1 mW.
2139. For line assignments see References 1393, 1461–1464.
2140. 20.6 kV/cm Stark field 10.09 μm CO_2 pump laser; R48 transition; either parallel or perpendicular polarization.
2141. Threshold pump power: 10000 mW.
2142. For line assignments see References 1393, 1461–1464.
2143. 24.5 kV/cm Stark field 10.72 μm CO_2 pump laser; P32 transition; either parallel or perpendicular polarization.
2144. Threshold pump power: 10000 mW.
2145. For line assignments see References 1393, 1461–1464.
2146. Output power: 2 mW.
2147. 7.7 kV/cm Stark field 10.73 μm N_2O pump laser; P08 transition; either parallel or perpendicular polarization.

2148.   Threshold pump power: 1000 mW.

2149.   For line assignments see References 1393, 1461–1464.

2150.   12.3 kV/cm Stark field 9.22 μm CO_2 pump laser; R30 transition; either parallel or perpendicular polarization.

2151.   Threshold pump power: 10000 mW.

2152.   For line assignments see References 1393, 1461–1464.

2153.   21.0 kV/cm Stark field 9.49 μm CO_2 pump laser; P12 transition; parallel polarization.

2154.   Threshold pump power: 10000 mW.

2155.   For line assignments see References 1393, 1461–1464.

2156.   Output power: 0.1 mW.

2157.   14.2 kV/cm Stark field 10.35 μm CO_2 pump laser; R06 transition; either parallel or perpendicular polarization.

2158.   Threshold pump power: 10000 mW.

2159.   For line assignments see References 1393, 1461–1464.

2160.   42.2 kV/cm Stark field 9.55 μm CO_2 pump laser; P20 transition; either parallel or perpendicular polarization.

2161.   Threshold pump power: 10000 mW.

2162.   For line assignments see References 1393, 1461–1464.

2163.   47.5 kV/cm Stark field 9.57 μm CO_2 pump laser; P22 transition; either parallel or perpendicular polarization.

2164.   Threshold pump power: 10000 mW.

2165.   For line assignments see References 1393, 1461–1464.

2166.   Output power: 0.005 mW

2167.   12.4 kV/cm Stark field 10.72 μm CO_2 pump laser; P32.

2168.   Threshold pump power: 10000 mW.

2169.   For line assignments see References 1393, 1461–1464.

2170.   46.5 kV/cm Stark field 10.35 mm N_2O pump laser; R34 transition; either parallel or perpendicular polarization.

2171.   Threshold pump power: 1000 mW.

2172.   For line assignments see References 1393, 1461–1464.

2173.   Output power: 2 mW.

2174.   8.5 kV/cm Stark field 9.22 μm CO_2 pump laser; R30 transition; either parallel or perpendicular polarization.

2175.   Threshold pump power: 10000 mW 2176. For line assignments see References 1393, 1461–1464.

2177.   Output power: 0.05 mW.

2178.   -130 MHz offset from center of 9.56 μm CO_2 sequence band pump laser; P17 transition; perpendicular polarization.

2179.   Threshold pump power: 1000 mW.

2180.   For line assignments see References 1393, 1461–1464.

2181.   38.2 kV/cm Stark field 9.73 μm CO_2 pump laser; P40 transition; either parallel or perpendicular polarization.

2182.   Threshold pump power: 10000 mW.

2183.   For line assignments see References 1393, 1461–1464.

2184.   45.4 kV/cm Stark field 10.33 μm CO_2 pump laser; R08 transition; either parallel or perpendicular polarization.

2185.   Threshold pump power: 10000 mW.

2186.   For line assignments see References 1393, 1461–1464.

2187.   55.3 kV/cm Stark field 10.35 μm CO_2 pump laser; R06 transition; either parallel or perpendicular polarization.

2188.   Threshold pump power: 10000 mW.

2189.   For line assignments see References 1393, 1461–1464.
2190.   21.6 kV/cm Stark field 10.37 μm CO_2 pump laser; R04 transition; either parallel or perpendicular polarization.
2191.   Threshold pump power: 10000 mW.
2192.   For line assignments see References 1393, 1461–1464.
2193.   15.2 kV/cm Stark field 10.38 μm N_2O pump laser; R31 transition; perpendicular polarization.
2194.   Threshold pump power: 1000 mW.
2195.   For line assignments see References 1393, 1461–1464.
2196.   34.7 kV/cm Stark field 10.33 μm CO_2 pump laser; R08 transition; either parallel or perpendicular polarization.
2197.   Threshold pump power: 10000 mW.
2198.   For line assignments see References 1393, 1461–1464.
2199.   31.3 kV/cm Stark field 10.35 μm CO_2 pump laser; R06 transition; either parallel or perpendicular polarization.
2200.   Threshold pump power: 10000 mW.
2201.   For line assignments see References 1393, 1461–1464.
2202.   53.6 kV/cm Stark field 10.35 μm N_2O pump laser; R34 transition; either parallel or perpendicular polarization.
2203.   Threshold pump power: 1000 mW.
2204.   For line assignments see References 1393, 1461–1464.
2205.   17.0 kV/cm Stark field 10.35 μm N_2O pump laser; R34 transition; either parallel or perpendicular polarization.
2206.   Threshold pump power: 1000 mW.
2207.   For line assignments see References 1393, 1461–1464.
2208.   14.6 kV/cm Stark field 10.35 μm CO_2 pump laser; R06 transition; either parallel or perpendicular polarization.
2209.   Threshold pump power: 10000 mW.
2210.   For line assignments see References 1393, 1461–1464.
2211.   47.0 kV/cm Stark field 10.36 μm N_2O pump laser; R33 transition; parallel polarization.
2212.   Threshold pump power: 1000 mW.
2213.   For line assignments see References 1393, 1461–1464.
2214.   40.1 kV/cm Stark field 10.35 μm N_2O pump laser; R34 transition; perpendicular polarization.
2215.   Threshold pump power: 1000 mW.
2216.   For line assignments see References 1393, 1461–1464.
2217.   10.8 kV/cm Stark field 10.37 μm N_2O pump laser; R32 transition; either parallel or perpendicular polarization.
2218.   Threshold pump power: 1000 mW.
2219.   For line assignments see References 1393, 1461–1464.
2220.   54.3 kV/cm Stark field 10.38 μm N_2O pump laser; R31 transition; perpendicular polarization.
2221.   Threshold pump power: 1000 mW.
2222.   For line assignments see References 1393, 1461–1464.
2223.   28.0 kV/cm Stark field 10.38 μm CO_2 pump laser; R02 transition; perpendicular polarization.
2224.   Threshold pump power: 10000 mW.
2225.   For line assignments see References 1393, 1461–1464.
2226.   Output power: 0.005 mW.
2227.   8.4 kV/cm Stark field 10.37 μm N_2O pump laser; R32 transition; either parallel or perpendicular polarization.
2228.   Threshold pump power: 200 mW.
2229.   For line assignments see References 1393, 1461–1464.

2230. Output power: 0.5 mW.
2231. 10.78 μm N_2O pump laser; P13 transition; parallel polarization.
2232. Threshold pump power: 50 mW.
2233. For line assignments see References 1393, 1461–1464.
2234. 20.3 kV/cm Stark field 10.39 μm N_2O pump laser; R29 transition; perpendicular polarization.
2235. Threshold pump power: 1000 mW.
2236. For line assignments see References 1393, 1461–1464.
2237. 15.8 kV/cm Stark field 10.39 μm N_2O pump laser; R29 transition; perpendicular polarization.
2238. Threshold pump power: 1000 mW.
2239. For line assignments see References 1393, 1461–1464.
2240. 52.4 kV/cm Stark field 10.41 μm N_2O pump laser; R27 transition; parallel polarization.
2241. Threshold pump power: 1000 mW.
2242. For line assignments see References 1393, 1461–1464.
2243. Output power: 0.1 mW.
2244. 21.3 kV/cm Stark field 10.44 μm N_2O pump laser; R23 transition; parallel polarization.
2245. Threshold pump power: 1000 mW.
2246. For line assignments see References 1393, 1461–1464.
2247. 26.3 kV/cm Stark field 9.49 μm CO_2 pump laser; P12 transition; parallel polarization.
2248. Threshold pump power: 10000 mW.
2249. For line assignments see References 1393, 1461–1464 .
2250. 48.8 kV/cm Stark field 10.72 μm CO_2 pump laser; P32 transition; either parallel or perpendicular polarization.
2251. Threshold pump power: 10000 mW.
2252. For line assignments see References 1393, 1461–1464.
2253. 47.3 kV/cm Stark field 10.72 μm N_2O pump laser; P07 transition; either parallel or perpendicular polarization.
2254. Threshold pump power: 1000 mW.
2255. For line assignments see References 1393, 1461–1464.
2256. 56.7 kV/cm Stark field 10.74 μm N_2O pump laser; P09 transition; either parallel or perpendicular polarization.
2257. Threshold pump power: 1000 mW.
2258. For line assignments see References 1393, 1461–1464.
2259. 42.1 kV/cm Stark field 10.72 μm CO_2 pump laser; P32 transition; either parallel or perpendicular polarization.
2260. Threshold pump power: i0000 mW.
2261. For line assignments see References 1393, 1461–1464.
2262. 34.3 kV/cm Stark field 10.72 μm N_2O pump laser; P07 transition; either parallel or perpendicular polarization.
2263. Threshold pump power: 1000 mW.
2264. For line assignments see References 1393, 1461–1464.
2265. 37.7 kV/cm Stark field 10.71 μm N_2O pump laser; P06 transition; either parallel or perpendicular polarization.
2266. Threshold pump power: 1000 mW.
2267. For line assignments see References 1393, 1461–1464.
2268. 15.4 kV/cm Stark field; 9.22 μm CO_2 pump laser; R30 transition; parallel polarization.
2269. Threshold pump power: 10000 mW.
2270. For line assignments see References 1393, 1461–1464.

2271. 26.3 kV/cm Stark field; 9.69 μm CO_2 pump laser; P36 transition; either parallel or perpendicular polarization.
2272. Threshold pump power: 10000 mW.
2273. For line assignments see References 1393, 1461–1464.
2274. 66.7 kV/cm Stark field; 9.71 μm CO_2 pump laser; P38 transition; either parallel or perpendicular polarization.
2275. Threshold pump power: 10000 mW.
2276. For line assignments see References 1393, 1461–1464.
2277. 24.4 kV/cm Stark field; 10.74 μm CO_2 pump laser; P34 transition; either parallel or perpendicular polarization.
2278. Threshold pump power: 10000 mW.
2279. For line assignments see References 1393, 1461–1464.
2280. 22.6 kV/cm Stark field; 10.73 μm N_2O pump laser; P08 transition; perpendicular polarization.
2281. Threshold pump power: 1000 mW.
2282. For line assignments see References 1393, 1461–1464.
2283. 68.6 kV/cm Stark field; 10.75 μm N_2O pump laser; P10 transition; either parallel or perpendicular polarization.
2284. Threshold pump power: 1000 mW.
2285. For line assignments see References 1393, 1461–1464.
2286. 62.0 kV/cm Stark field; 9.68 μm CO_2 pump laser; P34 transition; either parallel or perpendicular polarization.
2287. Threshold pump power: 10000 mW.
2288. For line assignments see References 1393, 1461–1464.
2289. Output power: 2 mW.
2290. 12.4 kV/cm Stark field; 10.72 μm CO_2 pump laser; P32 transition; either parallel or perpendicular polarization.
2291. Threshold pump power: 10000 mW.
2292. For line assignments see References 1393, 1461–1464.
2293. 22.3 kV/cm Stark field; 10.71 μm N_2O pump laser; P06 transition; either parallel or perpendicular polarization.
2294. Threshold pump power: 1000 mW.
2295. For line assignments see References 1393, 1461–1464.
2296. Output power: 0.1 mW.
2297. 16.0 kV/cm Stark field; 10.72 μm N_2O pump laser; P07 transition; either parallel or perpendicular polarization.
2298. Threshold pump power: 1000 mW.
2299. For line assignments see References 1393, 1461–1464.
2300. Output power: 0.1 mW; cascade transition
2301. 0.0 kV/cm Stark field; 9.22 μm CO_2 pump laser; R30 transition; parallel polarization.
2302. Threshold pump power: 10000 mW.
2303. For line assignments see References 1393, 1461–1464.
2304. 27.1 kV/cm Stark field; 9.52 μm CO_2 pump laser; P16 transition; parallel polarization.
2305. Threshold pump power: 10000 mW.
2306. For line assignments see References 1393, 1461–1464.
2307. 48.7 kV/cm Stark field; 9.54 μm CO_2 pump laser; P18 transition; parallel polarization.
2308. Threshold pump power: 10000 mW.
2309. For line assignments see References 1393, 1461–1464.

2310. 21.8 kV/cm Stark field; 10.74 μm N_2O pump laser; P09 transition; perpendicular polarization.
2311. Threshold pump power: 1000 mW.
2312. For line assignments see References 1393, 1461–1464.
2313. 52.7 kV/cm Stark field; 10.71 μm N_2O pump laser; P06 transition; perpendicular polarization.
2314. Threshold pump power: 1000 mW.
2315. For line assignments see References 1393, 1461–1464.
2316. 58.0 kV/cm Stark field; 10.73 μm N_2O pump laser; P08 transition; either parallel or perpendicular polarization.
2317. Threshold pump power: 1000 mW.
2318. For line assignments see References 1393, 1461–1464.
2319. Output power: 2 mW.
2320. 21.0 kV/cm Stark field; 9.49 μm CO_2 pump laser; P12 transition; Unspecified polarization.
2321. Threshold pump power: 10000 mW.
2322. For line assignments see References 1393, 1461–1464.
2323. 46.6 kV/cm Stark field; 9.47 μm CO_2 pump laser; P10 transition; parallel polarization.
2324. Threshold pump power: 10000 mW.
2325. For line assignments see References 1393, 1461–1464.
2326. 38.2 kV/cm Stark field; 9.73 μm CO_2 pump laser; P40 transition; either parallel or perpendicular polarization.
2327. Threshold pump power: 10000 mW.
2328. For line assignments see References 1393, 1461–1464.
2329. 45.4 kV/cm Stark field; 10.33 μm CO_2 pump laser; R08 transition; either parallel or perpendicular polarization.
2330. Threshold pump power: 10000 mW.
2331. For line assignments see References 1393, 1461–1464.
2332. 55.3 kV/cm Stark field; 10.35 μm CO_2 pump laser; R06 transition; either parallel or perpendicular polarization.
2333. Threshold pump power: 10000 mW.
2334. For line assignments see References 1393, 1461–1464.
2335. 42.1 kV/cm Stark field; 10.72 μm CO_2 pump laser; P32 transition; either parallel or perpendicular polarization.
2336. Threshold pump power: 10000 mW.
2337. For line assignments see References 1393, 1461–1464.
2338. 22.0 kV/cm Stark field; 9.35 μm CO_2 pump laser; R06 transition; either parallel or perpendicular polarization.
2339. Threshold pump power: 10000 mW.
2340. For line assignments see References 1393, 1461–1464.
2341. 34.9 kV/cm Stark field; 9.37 μm CO_2 pump laser; R04 transition; either parallel or perpendicular polarization.
2342. Threshold pump power: 10000 mW.
2343. For line assignments see References 1393, 1461–1464.
2344. 19.5 kV/cm Stark field; 10.76 μm CO_2 pump laser; P36 transition; either parallel or perpendicular polarization.
2345. Threshold pump power: 10000 mW.
2346. For line assignments see References 1393, 1461–1464.
2347. 39.4 kV/cm Stark field; 10.75 μm N_2O pump laser; P10 transition; either parallel or perpendicular polarization.

2348.   Threshold pump power: 1000 mW.
2349.   For line assignments see References 1393, 1461–1464.
2350.   45.1 kV/cm Stark field; 10.77 μm N_2O pump laser; P12 transition; either parallel or perpendicular polarization.
2351.   Threshold pump power: 1000 mW.
2352.   For line assignments see References 1393, 1461–1464.
2353.   47.9 kV/cm Stark field; 9.33 μm CO_2 pump laser; R10 transition; either parallel or perpendicular polarization.
2354.   Threshold pump power: 10000 mW.
2355.   For line assignments see References 1393, 1461–1464.
2356.   19.1 kV/cm Stark field; 9.34 μm CO_2 pump laser; R08 transition; either parallel or perpendicular polarization.
2357.   Threshold pump power: 10000 mW.
2358.   For line assignments see References 1393, 1461–1464.
2359.   37.9 kV/cm Stark field; 10.72 μm CO_2 pump laser; P32 transition; either parallel or perpendicular polarization.
2360.   Threshold pump power: 10000 mW.
2361.   For line assignments see References 1393, 1461–1464.
2362.   Output power: 0.1 mW.
2363.   8.8 kV/cm Stark field; 10.70 μm CO_2 pump laser; P30 transition; either parallel or perpendicular polarization.
2364.   Threshold pump power: 10000 mW.
2365.   For line assignments see References 1393, 1461–1464.
2366.   49.5 kV/cm Stark field; 9.29 μm CO_2 pump laser; R16 transition; either parallel or perpendicular polarization.
2367.   Threshold pump power: 10000 mW.
2368.   For line assignments see References 1393, 1461–1464.
2369.   Output power: 2 mW.
2370.   7.1 kV/cm Stark field; 10.35 μm CO_2 pump laser; R06 transition; either parallel or perpendicular polarization.
2371.   Threshold pump power: 10000 mW.
2372.   For line assignments see References 1393, 1461–1464.
2373.   Output power: 1 mW.
2374.   -130 MHz offset from center of 9.56 μm CO_2. Sequence band pump laser; P17 transition; parallel polarization.
2375.   Threshold pump power: 1000 mW.
2376.   For line assignments see References 1393, 1461–1464.
2377.   42.2 kV/cm Stark field; 9.55 μm CO_2 pump laser; P20 transition; either parallel or perpendicular polarization.
2378.   Threshold pump power: 10000 mW.
2379.   For line assignments see References 1393, 1461–1464.
2380.   47.5 kV/cm Stark field; 9.57 μm CO_2 pump laser; P22 transition; either parallel or perpendicular polarization.
2381.   Threshold pump power: 10000 mW.
2382.   For line assignments see References 1393, 1461–1464.
2383.   12.3 kV/cm Stark field; 10.72 μm CO_2 pump laser; P32 transition; perpendicular polarization.
2384.   Threshold pump power: 10000 mW.
2385.   For line assignments see References 1393, 1461–1464.
2386.   53.7 kV/cm Stark field; 9.19 μm CO_2 pump laser; R36 transition; parallel polarization.

2387.   Threshold pump power: 10000 mW.

2388.   For line assignments see References 1393, 1461–1464.

2389.   39.8 kV/cm Stark field; 9.21 μm CO_2 pump laser; R32 transition; parallel polarization.

2390.   Threshold pump power: 10000 mW.

2391.   For line assignments see References 1393, 1461–1464.

2392.   Output power: 50 mW.

2393.   10.78 μm N_2O pump laser; P13 transition; perpendicular polarization.

2394.   Threshold pump power: 20 mW.

2395.   For line assignments see References 1393, 1461–1464.

2396.   15.6 kV/cm Stark field; 10.77 μm N_2O pump laser; P12 transition; perpendicular polarization.

2397.   Threshold pump power: 1000 mW.

2398.   For line assignments see References 1393, 1461–1464.

2399.   67.2 kV/cm Stark field; 9.17 μm CO_2 pump laser; R42 transition; either parallel or perpendicular polarization.

2400.   Threshold pump power: 10000 mW.

2401.   For line assignments see References 1393, 1461–1464.

2402.   34.6 kV/cm Stark field; 9.18 μm CO_2 pump laser; R38 transition; either parallel or perpendicular polarization.

2403.   Threshold pump power: 10000 mW.

2404.   For line assignments see References 1393, 1461–1464.

2405.   Output power: 0.005 mW.

2406.   18.9 kV/cm Stark field; 10.74 μm CO_2 pump laser; P34 transition; either parallel or perpendicular polarization.

2407.   Threshold pump power: 10000 mW.

2408.   For line assignments see References 1393, 1461–1464.

2409.   31.3 kV/cm Stark field; 10.73 μm N_2O pump laser; P08 transition; perpendicular polarization.

2410.   Threshold pump power: 1000 mW.

2411.   For line assignments see References 1393, 1461–1464.

2412.   62.8 kV/cm Stark field; 10.75 μm N_2O pump laser; P10 transition; either parallel or perpendicular polarization.

2413.   Threshold pump power: 1000 mW.

2414.   For line assignments see References 1393, 1461–1464.

2415.   Output power: 0.005 mW.

2416.   10.2 kV/cm Stark field; 10.71 μm N_2O pump laser; P06 transition; either parallel or perpendicular polarization.

2417.   Threshold pump power: 1000 mW.

2418.   For line assignments see References 1393, 1461–1464.

2419.   26.7 kV/cm Stark field; 10.72 μm CO_2 pump laser; P32 transition; perpendicular polarization.

2420.   Threshold pump power: 10000 mW.

2421.   For line assignments see References 1393, 1461–1464.

2422.   Output power: 0.1 mW.

2423.   0.0 kV/cm Stark field; 9.22 μm CO_2 pump laser; R30 transition; parallel polarization.

2424.   Threshold pump power: 2000 mW.

2425.   For line assignments see References 1393, 1461–1464.

2426.   12.3 kV/cm Stark field; 9.22 μm CO_2 pump laser; R30 transition; either parallel or perpendicular polarization.

2427.   Threshold pump power: 10000 mW.

2428.   For line assignments see References 1393, 1461–1464.

2429.   28.8 kV/cm Stark field; 10.37 μm CO_2 pump laser; R04 transition; either parallel or perpendicular polarization.

2430.   Threshold pump power: 10000 mW.

2431.   For line assignments see References 1393, 1461–1464.

2432.   33.1 kV/cm Stark field; 10.35 μm N_2O pump laser; R34 transition; perpendicular polarization.

2433.   Threshold pump power: 1000 mW.

2434.   For line assignments see References 1393, 1461–1464.

2435.   Output power: 2 mW.

2436.   8.5 kV/cm Stark field; 9.22 μm CO_2 pump laser; R30 transition; either parallel or perpendicular polarization.

2437.   Threshold pump power: 10000 mW.

2438.   For line assignments see References 1393, 1461–1464.

2439.   45.0 kV/cm Stark field; 10.35 μm N_2O pump laser; R34 transition; perpendicular polarization.

2440.   Threshold pump power: 1000 mW.

2441.   For line assignments see References 1393, 1461–1464.

2442.   31.1 kV/cm Stark field; 10.37 μm N_2O pump laser; R32 transition; perpendicular polarization.

2443.   Threshold pump power: 1000 mW.

2444.   For line assignments see References 1393, 1461–1464.

2445.   17.2 kV/cm Stark field; 10.36 μm N_2O pump laser; R33 transition; either parallel or perpendicular polarization.

2446.   Threshold pump power: 1000 mW.

2447.   For line assignments see References 1393, 1461–1464.

2448.   41.3 kV/cm Stark field; 10.36 μm N_2O pump laser; R33 transition; perpendicular polarization.

2449.   Threshold pump power: 1000 mW.

2450.   For line assignments see References 1393, 1461–1464.

2451.   40.2 kV/cm Stark field; 10.38 μm N_2O pump laser; R31 transition; perpendicular polarization.

2452.   Threshold pump power: 1000 mW.

2453.   For line assignments see References 1393, 1461–1464.

2454.   21.6 kV/cm Stark field; 10.37 μm CO_2 pump laser; R04 transition; either parallel or perpendicular polarization.

2455.   Threshold pump power: 10000 mW.

2456.   For line assignments see References 1393, 1461–1464.

2457.   48.1 kV/cm Stark field; 10.38 μm CO_2 pump laser; R02 transition; perpendicular polarization.

2458.   Threshold pump power: 10000 mW.

2459.   For line assignments see References 1393, 1461–1464.

2460.   15.2 kV/cm Stark field; 10.38 μm N_2O pump laser; R31 transition; perpendicular polarization.

2461.   Threshold pump power: 1000 mW.

2462.   For line assignments see References 1393, 1461–1464.

2463.   Output power: 2 mW.

2464.   7.0 kV/cm Stark field; 10.79 μm N_2O pump laser; P14 transition; either parallel or perpendicular polarization.

2465.   Threshold pump power: 1000 mW.

2466.   For line assignments see References 1393, 1461–1464.

2467.   25.2 kV/cm Stark field; 10.38 μm CO_2 pump laser; R02 transition; perpendicular polarization.

2468.   Threshold pump power: 10000 mW.

2469.   For line assignments see References 1393, 1461–1464.

2470.   16.0 kV/cm Stark field; 10.38 μm N_2O pump laser; R30 transition; perpendicular polarization.

2471.   Threshold pump power: 1000 mW.

2472.   For line assignments see References 1393, 1461–1464.

2473.   11.4 kV/cm Stark field; 10.38 μm N_2O pump laser; R30 transition; perpendicular polarization.

2474.   Threshold pump power: 1000 mW.

2475.   For line assignments see Reference 1390.

2476.   Output power: 5 mW.

2477.   10.11 μm CO_2 pump laser; R42 transition; parallel polarization.

2478.   Threshold pump power: 1000 mW.

2479.   For line assignments see Reference 1390.

2480.   Output power: 0.002 mW.

2481.   10.78 μm CO_2 sequence band pump laser; P35 transition; perpendicular polarization.

2482.   Threshold pump power: 1000 mW.

2483.   For line assignments see Reference 1390.

2484.   Output power: 200 mW.

2485.   10.78 μm CO_2 pump laser; R18 transition; unspecified polarization.

2486.   Threshold pump power: 10 mW.

2487.   For line assignments see Reference 1390.

2488.   Output power: 0.002 mW.

2489.   10.73 μm CO_2 sequence band pump laser; P31 transition; perpendicular polarization.

2490.   Threshold pump power: 1000 mW.

2491.   For line assignments see Reference 1437.

2492.   Output power: 0.05 mW.

2493.   10.29 μm CO_2 pump laser; R14 transition; parallel polarization.

2494.   Threshold pump power: 2000 mW.

2495.   For line assignments see Reference 1437.

2496.   Output power: 0.05 mW.

2497.   10.81 μm CO_2 pump laser; P40 transition; parallel polarization.

2498.   Threshold pump power: 2000 mW.

2499.   For line assignments see Reference 1437.

2500.   Output power: 0.02 mW.

2501.   10.21 μm CO_2 pump laser; R26 transition; perpendicular polarization.

2502.   Threshold pump power: 10000 mW.

2503.   For line assignments see Reference 1437.

2504.   Output power: 0.1 mW.

2505.   10.81 μm CO_2 pump laser; P40 transition; parallel polarization.

2506.   Threshold pump power: 500 mW.

2507.   For line assignments see Reference 1437.

2508.   Output power: 0.05 mW.

2509.   10.29 μm CO_2 pump laser; R14 transition; parallel polarization.

2510.   Threshold pump power: 2000 mW.

2511.   For line assignments see Reference 1437.

2512.   Output power: 0.05 mW.
2513.   10.18 μm CO_2 pump laser; R30 transition; parallel polarization.
2514.   Threshold pump power: 5000 mW.
2515.   For line assignment see Reference 1437.
2516.   Output power: 0.05 mW.
2517.   9.17 μm CO_2 pump laser; R40 transition; parallel polarization.
2518.   Threshold pump power: 2000 mW.
2519.   For line assignments see Reference 1382.
2520.   Output power: 0.05 mW.
2521.   9.46 μm CO_2 pump laser; P08 transition; unspecified polarization.
2522.   Threshold pump power: 1000 mW.
2523.   For line assignments see Reference 1382.
2524.   Output power: 0.05 mW.
2525.   9.24 μm CO_2 pump laser; R26 transition; unspecified polarization.
2526.   Threshold pump power: 1000 mW.
2527.   For line assignments see Reference 1382.
2528.   Output power: 0.05 mW.
2529.   9.21 μm CO_2 pump laser; R32 transition; unspecified polarization.
2530.   Threshold pump power: 1000 mW.
2531.   For line assignments see Reference 1382.
2532.   Output power: 0.05 mW.
2533.   9.66 μm CO_2 pump laser; P32 transition; unspecified polarization.
2534.   Threshold pump power: 1000 mW.
2535.   For line assignments see Reference 1382.
2536.   Output power: 0.01 mW; cascade transition.
2537.   10.61 μm CO_2 pump laser; P22 transition; perpendicular polarization.
2538.   For line assignments see Reference 1382.
2539.   Output power: 0.01 mW; cascade transition.
2540.   9.68 μm CO_2 pump laser; P34 transition; parallel polarization.
2541.   For line assignments see Reference 1382.
2542.   Output power: 0.1 mW.
2543.   10.61 μm CO_2 pump laser; P22 transition; perpendicular polarization.
2544.   Threshold pump power: 1000 mW.
2545.   For line assignments see Reference 1382.
2546.   Output power: 0.1 mW.
2547.   9.68 μm CO_2 pump laser; P34 transition; parallel polarization.
2548.   Threshold pump power: 2000 mW.
2549.   For line assignments see Reference 1382.
2550.   Output power: 0.01 mW; cascade transition.
2551.   10.48 μm CO_2 pump laser; P08 transition; parallel polarization.
2552.   For line assignments see Reference 1382.
2553.   Output power: 0.05 mW.
2554.   10.48 μm CO_2 pump laser; P08 transition; unspecified polarization.
2555.   Threshold pump power: 1000 mW.
2556.   For line assignments see Reference 1382.
2557.   Output power: 0.01 mW; cascade transition.
2558.   9.25 μm CO_2 pump laser; R24 transition; parallel polarization.
2559.   For line assignments see Reference 1382.
2560.   Output power: 0.01 mW; cascade transition.
2561.   9.31 μm CO_2 pump laser; R14 transition; parallel polarization.
2562.   For line assignments see Reference 1382.

2563.   Output power: 0.2 mW.
2564.   9.25 μm CO_2 pump laser; R24 transition; unspecified polarization.
2565.   Threshold pump power: 1000 mW.
2566.   For line assignments see Reference 1382.
2567.   Output power: 0.05 mW.
2568.   10.55 μm CO_2 pump laser; P16 transition; perpendicular polarization.
2569.   Threshold pump power: 2000 mW.
2570.   For line assignments see Reference 1382.
2571.   Output power: 0.05 mW.
2572.   9.31 μm CO_2 pump laser; R14 transition; unspecified polarization.
2573.   Threshold pump power: 1000 mW.
2574.   For line assignments see Reference 1431.
2575.   Output power: 20 mW.
2576. CO_2 pump laser; unspecified polarization.
2577.   Threshold pump power: 10000 mW.
2578.   For line assignments see Reference 1431.
2578.   For line assignments see Reference 1431.
2579.   Output power: 50 mW.
2580. CO_2 pump laser; unspecified polarization.
2581.   Threshold pump power: 10000 mW.
2582.   For line assignments see Reference 1431.
2583.   Output power: 50 mW.
2584. CO_2 pump laser; unspecified polarization.
2585.   Threshold pump power: 10000 mW.
2586.   For line assignments see Reference 1431.
2587.   Output power: 50 mW.
2588. CO_2 pump laser; unspecified polarization.
2589.   Threshold pump power: 10000 mW.
2590.   For line assignments see Reference 1431.
2591.   Output power: 50 mW.
2592. CO_2 pump laser; unspecified polarization.
2593.   Threshold pump power: 10000 mW.
2594.   For line assignments see Reference 1431.
2595.   Output power: 100 mW.
2596. CO_2 pump laser; unspecified polarization.
2597.   Threshold pump power: 10000 mW.
2598.   For line assignments see Reference 1431.
2599.   Output power: 100 mW.
2600. CO_2 pump laser; unspecified polarization.
2601.   Threshold pump power: 10000 mW.
2602.   For line assignments see Reference 1431.
2603.   Output power: 20 mW.
2604. CO_2 pump laser; unspecified polarization.
2605.   Threshold pump power: 10000 mW.
2606.   For line assignments see References 1363, 1374, 1377, 1393.
2607.   Output power: 1 mW.
2608.   +30 MHz offset from center of 9.54 μm CO_2 Sequence band pump laser; P15 transition; perpendicular polarization.
2609.   Threshold pump power: 1000 mW.
2610.   For line assignments see References 1363, 1374, 1377, 1393.
2611.   Output power: 2 mW.

2612. +44 MHz offset from center of 9.55 μm CO_2 pump laser; P20 transition; perpendicular polarization.
2613. Threshold pump power: 500 mW.
2614. For line assignments see References 1363, 1374, 1377, 1393.
2615. Output power: 1 mW.
2616. -50 MHz offset from center of 9.55 μm CO_2 pump laser; P20 transition; perpendicular polarization.
2617. For line assignments see References 1363, 1374, 1377, 1393.
2618. Output power: 10 mW.
2619. -50 MHz offset from center of 9.84 μm CO_2 pump laser; P50 transition; parallel polarization.
2620. Threshold pump power: 1000 mW.
2621. For line assignments see References 1363, 1374, 1377, 1393.
2622. Output power: 0.1 mW.
2623. +25 MHz offset from center of 10.16 μm CO_2 pump laser; R34 transition; parallel polarization.
2624. Threshold pump power: 5000 mW.
2625. For line assignments see References 1363, 1374, 1377, 1393.
2626. Output power: 10 mW.
2627. 0 MHz offset from center of 10.17 μm CO_2 pump laser; R32 transition; parallel polarization.
2628. Threshold pump power: 5000 mW.
2629. For line assignments see References 1377, 1460.
2630. Output power: 1 mW.
2631. -26 MHz offset from center of 9.66 μm CO_2 pump laser; P32 transition; parallel polarization.
2632. Threshold pump power: 10000 mW.
2633. For line assignments see Reference 1472.
2634. Output power: 0.01 mW.
2635. 10.88 μm CO_2 pump laser; P46 transition; unspecified polarization.
2636. Threshold pump power: 10000 mW.
2637. For line assignments see Reference 1472.
2638. Output power: 0.001 mW.
2639. 9.19 μm CO_2 pump laser; R36 transition; unspecified polarization.
2640. Threshold pump power: 10000 mW.
2641. For line assignments see Reference 1472.
2642. Output power: 0.01 mW.
2643. 10.67 μm CO_2 pump laser; P28 transition; unspecified polarization.
2644. Threshold pump power: 10000 mW.
2645. For line assignments see Reference 1472.
2646. Output power: 0.01 mW.
2647. 10.09 μm CO_2 pump laser; R48 transition; perpendicular polarization.
2648. Threshold pump power: 10000 mW.
2649. For line assignments see Reference 1472.
2650. Output power: 0.01 mW.
2651. 9.68 μm CO_2 pump laser; P34 transition; unspecified polarization.
2652. Threshold pump power: 10000 mW.
2653. For line assignments see Reference 1472.
2654. Output power: 0.01 mW.
2655. 10.94 μm CO_2 pump laser; P50 transition; perpendicular polarization.
2656. Threshold pump power: 10000 mW.

2657.   For line assignments see Reference 1472.
2658.   Output power: 0.01 mW.
2659.   10.48 μm CO_2 pump laser; P08 transition; unspecified polarization.
2660.   Threshold pump power: 10000 mW.
2661.   For line assignments see Reference 1472.
2662.   Output power: 0.01 mW.
2663.   9.86 μm CO_2 pump laser; P52 transition; parallel polarization.
2664.   Threshold pump power: 10000 mW.
2665.   For line assignments see Reference 1472.
2666.   Output power: 1 mW.
2667.   9.33 μm CO_2 pump laser; R10 transition; parallel polarization.
2668.   Threshold pump power: 10000 mW.
2669.   For line assignments see Reference 1472.
2670.   Output power: 0.01 mW.
2671.   10.88 μm CO_2 pump laser; P46 transition; parallel polarization.
2672.   Threshold pump power: 10000 mW.
2673.   For line assignments see Reference 1472.
2674.   Output power: 2 mW.
2675.   9.52 μm CO_2 pump laser; P16 transition; parallel polarization.
2676.   Threshold pump power: 10000 mW.
2677.   For line assignments see Reference 1472.
2678.   Output power: 0.1 mW.
2679.   10.51 μm CO_2 pump laser; P12 transition; parallel polarization.
2680.   Threshold pump power: 10000 mW.
2681.   For line assignments see Reference 1472.
2682.   Output power: 0.01 mW.
2683.   9.62 μm CO_2 pump laser; P28 transition; unspecified polarization.
2684.   Threshold pump power: 10000 mW.
2685.   For line assignments see Reference 1472.
2686.   Output power: 0.01 mW.
2687.   10.35 μm CO_2 pump laser; R06 transition; unspecified polarization.
2688.   Threshold pump power: 10000 mW.
2689.   For line assignments see Reference 1472.
2690.   Output power: 0.01 mW.
2691.   10.49 μm CO_2 pump laser; P10 transition; parallel polarization.
2692.   Threshold pump power: 10000 mW.
2693.   For line assignments see References 1366, 1381, 1384, 1439.
2694.   Output power: 0.1 mW.
2695.   9.62 μm CO_2 pump laser; P28 transition; parallel polarization.
2696.   Threshold pump power: 2000 mW.
2697.   For line assignments see References 1366, 1381, 1384, 1439.
2698.   Output power: 0.001 mW.
2699.   10.29 μm CO_2 pump laser; R14 transition; unspecified polarization.
2700.   Threshold pump power: 1000 mW.
2701.   For line assignments see References 1366, 1381, 1384, 1439.
2702.   Output power: 0.001 mW.
2703.   10.17 μm CO_2 pump laser; R32 transition; unspecified polarization.
2704.   Threshold pump power: 1000 mW.
2705.   For line assignments see References 1366, 1381, 1384, 1439.
2706.   Output power: 0.1 mW.
2707.   9.19 μm CO_2 pump laser; R36 transition; parallel polarization.

2708.   Threshold pump power: 2000 mW.
2709.   For line assignments see References 1366, 1381, 1384, 1439.
2710.   Output power: 0.2 mW.
2711.   9.21 μm CO_2 pump laser; R32 transition; parallel polarization.
2712.   Threshold pump power: 2000 mW.
2713.   For line assignments see References 1366, 1381, 1384, 1439.
2714.   Output power: 0.05 mW.
2715.   9.17 μm CO_2 pump laser; R40 transition; unspecified polarization.
2716.   Threshold pump power: 500 mW.
2717.   For line assignments see References 1366, 1381, 1384, 1439.
2718.   Output power: 0.0005 mW.
2719.   9.25 μm CO_2 pump laser; R24 transition; unspecified polarization.
2720.   Threshold pump power: 2000 mW.
2721.   For line assignments see References 1366, 1381, 1384, 1439.
2722.   Output power: 0.2 mW.
2723.   9.25 μm CO_2 pump laser; R24 transition; parallel polarization.
2724.   Threshold pump power: 1000 mW.
2725.   For line assignments see References 1366, 1381, 1384, 1439.
2726.   Output power: 0.5 mW.
2727.   9.17 μm CO_2 pump laser; R40 transition; parallel polarization.
2728.   Threshold pump power: 200 mW.
2729.   For line assignments see References 1366, 1381, 1384, 1438.
2730.   Output power: 0.1 mW.
2731.   9.35 μm CO_2 pump laser; R06 transition; parallel polarization.
2732.   Threshold pump power: 2000 mW.
2733.   For line assignments see References 1366, 1381, 1384, 1438.
2734.   Output power: 0.2 mW.
2735.   9.22 mum CO_2 pump laser; R30 transition; parallel polarization.
2736.   Threshold pump power: 1000 mW.
2737.   For line assignments see References 1366, 1381, 1384, 1438.
2738.   Output power: 1 mW.
2739.   9.71 μm CO_2 pump laser; P38 transition; perpendicular polarization.
2740.   Threshold pump power: 2000 mW.
2741.   For line assignments see References 1366, 1381, 1384, 1438.
2742.   Output power: 0.05 mW.
2743.   9.26 μm CO_2 pump laser; R22 transition; unspecified polarization.
2744.   Threshold pump power: 2000 mW.
2745.   For line assignments see References 1366, 1381, 1384, 1438.
2746.   Output power: 0.2 mW.
2747.   9.25 μm CO_2 pump laser; R24 transition; unspecified polarization.
2748.   Threshold pump power: 2000 mW.
2749.   For line assignments see References 1366, 1381, 1384, 1438.
2750.   Output power: 0.02 mW.
2751.   9.54 μm CO_2 pump laser; P18 transition; unspecified polarization.
2752.   Threshold pump power: 2000 mW.
2753.   For line assignments see References 1366, 1381, 1384, 1438.
2754.   Output power: 1 mW.
2755.   9.23 μm CO_2 pump laser; R28 transition; parallel polarization.
2756.   Threshold pump power: 2000 mW.
2757.   For line assignments see References 1366, 1381, 1384, 1438.
2758.   Output power: 1 mW.

2759.   9.52 μm CO_2 pump laser; P16 transition; parallel polarization.
2760.   Threshold pump power: 2000 mW.
2761.   For line assignments see References 1366, 1381, 1384, 1438.
2762.   Output power: 1 mW.
2763.   9.52 μm CO_2 pump laser; P16 transition; perpendicular polarization.
2764.   Threshold pump power: 2000 mW.
2765.   For line assignments see References 1366, 1381, 1384, 1438.
2766.   Output power: 5 mW.
2767.   9.23 μm CO_2 pump laser; R28 transition; parallel polarization.
2768.   Threshold pump power: 2000 mW.
2769.   For line assignments see References 1366, 1381, 1384, 1438.
2770.   Output power: 1 mW.
2771.   9.23 μm CO_2 pump laser; R28 transition; parallel polarization.
2772.   Threshold pump power: 500 mW.
2773.   For line assignments see References 1366, 1381, 1384, 1438.
2774.   Output power: 0.5 mW.
2775.   9.18 μm CO_2 pump laser; R38 transition; perpendicular polarization.
2776.   Threshold pump power: 500 mW.
2777.   For line assignments see References 1366, 1381, 1384, 1438.
2778.   Output power: 0.2 mW.
2779.   9.29 μm CO_2 pump laser; R16 transition; parallel polarization.
2780.   Threshold pump power: 2000 mW.
2781.   For line assignments see References 1366, 1381, 1384, 1438.
2782.   Output power: 0.1 mW.
2783.   9.26 μm CO_2 pump laser; R22 transition; unspecified polarization.
2784.   Threshold pump power: 2000 mW.
2785.   For line assignments see References 1366, 1381, 1384, 1438.
2786.   Output power: 0.2 mW.
2787.   9.27 μm CO_2 pump laser; R20 transition; parallel polarization.
2788.   Threshold pump power: 2000 mW.
2789.   For line assignments see References 1366, 1381, 1384, 1438.
2790.   Output power: 0.2 mW.
2791.   10.53 μm CO_2 pump laser; P14 transition; perpendicular polarization.
2792.   Threshold pump power: 2000 mW.
2793.   For line assignments see References 1366, 1381, 1384, 1438.
2794.   Output power: 0.2 mW.
2795.   9.52 μm CO_2 pump laser; P16 transition; perpendicular polarization.
2796.   Threshold pump power: 2000 mW.
2797.   For line assignments see References 1366, 1381, 1384, 1438.
2798.   Output power: 2 mW.
2799.   9.27 μm CO_2 pump laser; R20 transition; parallel polarization.
2800.   Threshold pump power: 2000 mW.
2801.   For line assignments see References 1366, 1381, 1384, 1438.
2802.   Output power: 2 mW.
2803.   9.27 μm CO_2 pump laser; R20 transition; parallel polarization.
2804.   Threshold pump power: 1000 mW.
2805.   For line assignments see References 1366, 1381, 1384, 1438.
2806.   Output power: 1 mW.
2807.   9.26 μm CO_2 pump laser; R22 transition; parallel polarization.
2808.   Threshold pump power: 2000 mW.
2809.   For line assignments see References 1366, 1381, 1384, 1438.

2810.   Output power: 0.2 mW.
2811.   9.34 μm CO_2 pump laser; P08 transition; parallel polarization.
2812.   Threshold pump power: 2000 mW.
2813.   For line assignments see References 1366, 1381, 1384, 1438.
2814.   Output power: 2 mW.
2815.   9.26 μm CO_2 pump laser; R22 transition; parallel polarization.
2816.   Threshold pump power: 500 mW.
2817.   For line assignments see References 1366, 1381, 1384, 1438.
2818.   Output power: 0.05 mW.
2819.   9.25 μm CO_2 pump laser; R24 transition; parallel polarization.
2820.   Threshold pump power: 2000 mW.
2821.   For line assignments see References 1366, 1381, 1384, 1438.
2822.   Output power: 1 mW.
2823.   9.28 μm CO_2 pump laser; R18 transition; parallel polarization.
2824.   Threshold pump power: 500 mW.
2825.   For line assignments see References 1366, 1381, 1384, 1438.
2826.   Output power: 0.2 mW.
2827.   9.60 μm CO_2 pump laser; P26 transition; parallel polarization.
2828.   Threshold pump power: 2000 mW.
2829.   For line assignments see References 1366, 1381, 1384, 1438.
2830.   Output power: 0.2 mW.
2831.   10.11 μm CO_2 pump laser; R42 transition; perpendicular polarization.
2832.   Threshold pump power: 2000 mW.
2833.   For line assignments see References 1366, 1381, 1384, 1438.
2834.   Output power: 2 mW.
2835.   9.29 μm CO_2 pump laser; P16 transition; unspecified polarization.
2836.   Threshold pump power: 2000 mW.
2837.   For line assignments see References 1366, 1381, 1384, 1438.
2838.   Output power: 2 mW.
2839.   9.28 μm CO_2 pump laser; R18 transition; parallel polarization.
2840.   Threshold pump power: 1000 mW.
2841.   For line assignments see References 1366, 1381, 1384, 1438.
2842.   Output power: 0.1 mW.
2843.   9.20 μm CO_2 pump laser; R34 transition; parallel polarization.
2844.   Threshold pump power: 2000 mW.
2845.   For line assignments see References 1366, 1381, 1384, 1438.
2846.   Output power: 0.2 mW.
2847.   9.31 μm CO_2 pump laser; R14 transition; parallel polarization.
2848.   Threshold pump power: 2000 mW.
2849.   For line assignments see References 1366, 1381, 1384, 1438.
2850.   Output power: 0.02 mW.
2851.   10.22 μm CO_2 pump laser; R24 transition; parallel polarization.
2852.   Threshold pump power: 2000 mW.
2853.   For line assignments see References 1366, 1381, 1384, 1438.
2854.   Output power: 0.5 mW.
2855.   10.23 μm CO_2 pump laser; R22 transition; parallel polarization.
2856.   Threshold pump power: 2000 mW.
2857.   For line assignments see References 1366, 1381, 1384, 1438.
2858.   Output power: 0.01 mW; cascade transition.
2859.   9.37 μm CO_2 pump laser; R04 transition; parallel polarization.
2860.   For line assignments see References 1366, 1381, 1384, 1438.

2861.   Output power: 5 mW.
2862.   9.37 µm CO_2 pump laser; R04 transition; parallel polarization.
2863.   Threshold pump power: 500 mW.
2864.   For line assignments see References 1366, 1381, 1384, 1439.
2865.   Output power: 0.2 mW.
2866.   9.64 µm CO_2 pump laser; P30 transition; parallel polarization.
2867.   Threshold pump power: 2000 mW.
2868.   For line assignments see References 1366, 1381, 1384, 1439.
2869.   Output power: 0.2 mW.
2870.   9.55 µm CO_2 pump laser; P20 transition; parallel polarization.
2871.   Threshold pump power: 2000 mW.
2872.   For line assignments see References 1366, 1381, 1384, 1439.
2873.   Output power: 0.02 mW.
2874.   9.26 µm CO_2 pump laser; R22 transition; perpendicular polarization.
2875.   Threshold pump power: 10000 mW.
2876.   For line assignments see References 1366, 1381, 1384, 1439.
2877.   Output power: 0.05 mW.
2878.   9.26 µm CO_2 pump laser; R22 transition; parallel polarization.
2879.   Threshold pump power: 2000 mW.
2880.   Output power: 0.05 mW.
2881.   9.22 µm CO_2 pump laser; R30 transition; unspecified polarization.
2882.   Threshold pump power: 2000 mW.
2883.   Output power: 0.2 mW.
2884.   9.49 µm CO_2 pump laser; P12 transition; unspecified polarization.
2885.   Threshold pump power: 2000 mW.
2886.   Output power: 0.05 mW.
2887.   10.09 µm CO_2 pump laser; R46 transition; unspecified polarization.
2888.   Threshold pump power: 2000 mW.
2889.   Output power: 0.2 mW.
2890.   9.26 µm CO_2 pump laser; R22 transition; unspecified polarization.
2891.   Threshold pump power: 2000 mW.
2892.   Output power: 0.05 mW.
2893.   9.44 µm CO_2 pump laser; P06 transition; unspecified polarization.
2894.   Threshold pump power: 2000 mW.
2895.   Output power: 0.05 mW.
2896.   9.52 µm CO_2 pump laser; P16 transition; unspecified polarization.
2897.   Threshold pump power: 2000 mW.
2898.   For line assignments see Reference 1384 (pump).
2899.   Output power: 0.2 mW.
2900.   10.22 µm CO_2 pump laser; 824 transition; parallel polarization.
2901.   Threshold pump power: 2000 mW.
2902.   For line assignments see Reference 1384 (pump).
2903.   Output power: 0.2 mW.
2904.   9.64 µm CO_2 pump laser; P30 transition; parallel polarization.
2905.   Threshold pump power: 2000 mW.
2906.   For line assignments see Reference 1384 (pump).
2907.   Output power: 0.1 mW.
2908.   10.25 µm CO_2 pump laser; R20 transition; parallel polarization.
2909.   Threshold pump power: 2000 mW.
2910.   For line assignments see Reference 1384 (pump).
2911.   Output power: 0.05 mW.

2912. 10.14 μm CO_2 pump laser; R38 transition; perpendicular polarization.
2913. Threshold pump power: 2000 mW.
2914. For line assignments see Reference 1384 (pump).
2915. Output power: 0.5 mW.
2916. 10.17 μm CO_2 pump laser; R32 transition; parallel polarization.
2917. Threshold pump power: 2000 mW.
2918. For line assignments see Reference 1384 (pump).
2919. Output power: 10 mW.
2920. 10.29 μm CO_2 pump laser; R14 transition; perpendicular polarization.
2921. Threshold pump power: 2000 mW.
2922. For line assignments see Reference 1384 (pump).
2923. Output power: 10 mW.
2924. 10.17 μm CO_2 pump laser; R32 transition; parallel polarization.
2925. Threshold pump power: 2000 mW.
2926. For line assignments see Reference 1384 (pump).
2927. Output power: 0.1 mW.
2928. 9.49 μm CO_2 pump laser; P12 transition; parallel polarization.
2929. Threshold pump power: 2000 mW.
2930. For line assignments see Reference 1384 (pump).
2931. Output power: 0.002 mW.
2932. 10.76 μm CO_2 pump laser; P36 transition; parallel polarization.
2933. Threshold pump power: 2000 mW.
2934. For line assignments see Reference 1384 (pump).
2935. Output power: 0.05 mW.
2936. 9.49 μm CO_2 pump laser; P12 transition; parallel polarization.
2937. Threshold pump power: 2000 mW.
2938. For line assignments see Reference 1384 (pump).
2939. Output power: 0.1 mW.
2940. 10.15 μm CO_2 pump laser; R36 transition; parallel polarization.
2941. Threshold pump power: 2000 mW.
2942. For line assignments see Reference 1384 (pump).
2943. Output power: 0.2 mW.
2944. 10.11 μm CO_2 pump laser; R42 transition; parallel polarization.
2945. Threshold pump power: 2000 mW.
2946. For line assignments see Reference 1384 (pump).
2947. Output power: 0.5 mW.
2948. 10.15 μm CO_2 pump laser; R36 transition; parallel polarization.
2949. Threshold pump power: 2000 mW.
2950. For line assignments see Reference 1384 (pump).
2951. Output power: 0.05 mW.
2952. 10.30 μm CO_2 pump laser; R12 transition; perpendicular polarization.
2953. Threshold pump power: 2000 mW.
2954. For line assignments see Reference 1384 (pump).
2955. Output power: 2 mW.
2956. 10.21 μm CO_2 pump laser; R26 transition; parallel polarization.
2957. Threshold pump power: 2000 mW.
2958. For line assignments see Reference 1384 (pump).
2959. Output power: 0.1 mW.
2960. 9.17 μm CO_2 pump laser; R40 transition; parallel polarization.
2961. Threshold pump power: 2000 mW.
2962. For line assignments see Reference 1384 (pump).

2963.   Output power: 0.05 mW.

2964.   10.30 μm CO_2 pump laser; R12 transition; perpendicular polarization.

2965.   Threshold pump power: 2000 mW.

2966.   For line assignments see Reference 1384 (pump).

2967.   Output power: 0.05 mW.

2968.   9.57 μm CO_2 pump laser; P22 transition; parallel polarization.

2969.   Threshold pump power: 2000 mW.

2970.   For line assignments see Reference 1384 (pump).

2971.   Output power: 0.5 mW.

2972.   10.32 μm CO_2 pump laser; R10 transition; perpendicular polarization.

2973.   Threshold pump power: 2000 mW.

2974.   For line assignments see Reference 1384 (pump).

2975.   Output power: 0.1 mW.

2976.   9.22 μm CO_2 pump laser; R30 transition; perpendicular polarization.

2977.   Threshold pump power: 2000 mW.

2978.   For line assignments see Reference 1384 (pump).

2979.   Output power: 0.1 mW.

2980.   10.53 μm CO_2 pump laser; P14 transition; parallel polarization.

2981.   Threshold pump power: 2000 mW.

2982.   For line assignments see Reference 1384 (pump).

2983.   Output power: 0.2 mW.

2984.   9.54 μm CO_2 pump laser; P18 transition; perpendicular polarization.

2985.   Threshold pump power: 2000 mW.

2986.   For line assignments see Reference 1384 (pump).

2987.   Output power: 1 mW.

2988.   10.53 μm CO_2 pump laser; P14 transition; parallel polarization.

2989.   Threshold pump power: 2000 mW.

2990.   For line assignments see Reference 1384 (pump).

2991.   Output power: 0.02 mW.

2992.   9.73 μm CO_2 pump laser; P40 transition; parallel polarization.

2993.   Threshold pump power: 2000 mW.

2994.   For line assignments see Reference 1384 (pump).

2995.   Output power: 0.05 mW.

2996.   9.77 μm CO_2 pump laser; P44 transition; parallel polarization.

2997.   Threshold pump power: 2000 mW.

2998.   For line assignments see Reference 1384 (pump).

2999.   Output power: 0.2 mW.

3000.   10.13 μm CO_2 pump laser; R40 transition; parallel polarization.

3001.   Threshold pump power: 2000 mW.

3002.   For line assignments see Reference 1384 (pump).

3003.   Output power: 0.02 mW.

3004.   9.50 μm CO_2 pump laser; P14 transition; parallel polarization.

3005.   Threshold pump power: 2000 mW.

3006.   For line assignments see Reference 1384 (pump).

3007.   Output power: 0.1 mW.

3008.   9.18 μm CO_2 pump laser; R38 transition; perpendicular polarization.

3009.   Threshold pump power: 2000 mW.

3010.   For line assignments see Reference 1384 (pump).

3011.   Output power: 1 mW.

3012.   10.55 μm CO_2 pump laser; P16 transition; parallel polarization.

3013.   Threshold pump power: 2000 mW.

3014. For line assignments see Reference 1384 (pump).
3015. Output power: 10 mW.
3016. 10.55 μm CO_2 pump laser; P16 transition; parallel polarization.
3017. Threshold pump power: 2000 mW.
3018. For line assignments see Reference 1384 (pump).
3019. Output power: 0.5 mW.
3020. 10.51 μm CO_2 pump laser; P12 transition; perpendicular polarization.
3021. Threshold pump power: 2000 mW.
3022. For line assignments see Reference 1384 (pump).
3023. Output power: 0.1 mW.
3024. 10.21 μm CO_2 pump laser; R26 transition; perpendicular polarization.
3025. Threshold pump power: 2000 mW.
3026. For line assignments see Reference 1384 (pump).
3027. Output power: 0.002 mW.
3028. 9.62 μm CO_2 pump laser; P28 transition; parallel polarization.
3029. Threshold pump power: 2000 mW.
3030. For line assignments see Reference 1384 (pump).
3031. Output power: 0.02 mW.
3032. 10.27 μm CO_2 pump laser; R16 transition; parallel polarization.
3033. Threshold pump power: 2000 mW.
3034. For line assignments see Reference 1384 (pump).
3035. Output power: 1 mW.
3036. 10.46 μm CO_2 pump laser; P06 transition; parallel polarization.
3037. Threshold pump power: 2000 mW.
3038. For line assignments see Reference 1384 (pump).
3039. Output power: 1 mW.
3040. 10.63 μm CO_2 pump laser; P24 transition; parallel polarization.
3041. Threshold pump power: 2000 mW.
3042. For line assignments see Reference 1384 (pump).
3043. Output power: 1 mW.
3044. 9.26 μm CO_2 pump laser; R22 transition; parallel polarization.
3045. Threshold pump power: 2000 mW.
3046. For line assignments see Reference 1384 (pump).
3047. Output power: 0.1 mW.
3048. 9.73 μm CO_2 pump laser; P40 transition; parallel polarization.
3049. Threshold pump power: 2000 mW.
3050. For line assignments see Reference 1384 (pump).
3051. Output power: 0.05 mW.
3052. 10.23 μm CO_2 pump laser; R22 transition; parallel polarization.
3053. Threshold pump power: 2000 mW.
3054. For line assignments see Reference 1384 (pump).
3055. Output power: 2 mW.
3056. 10.30 μm CO_2 pump laser; R12 transition; perpendicular polarization.
3057. Threshold pump power: 2000 mW.
3058. For line assignments see Reference 1384 (pump).
3059. Output power: 0.1 mW.
3060. 9.52 μm CO_2 pump laser; P16 transition; parallel polarization.
3061. Threshold pump power: 2000 mW.
3062. For line assignments see Reference 1384 (pump).
3063. Output power: 0.1 mW.
3064. 9.47 μm CO_2 pump laser; P10 transition; parallel polarization.

3065.   Threshold pump power: 2000 mW.
3066.   For line assignments see Reference 1384 (pump).
3067.   Output power: 0.2 mW.
3068.   9.29 μm CO_2 pump laser; R16 transition; parallel polarization.
3069.   Threshold pump power: 2000 mW.
3070.   For line assignments see Reference 1384 (pump).
3071.   Output power: 0.2 mW.
3072.   10.14 μm CO_2 pump laser; R38 transition; perpendicular polarization.
3073.   Threshold pump power: 2000 mW.
3074.   For line assignments see Reference 1384 (pump).
3075.   Output power: 0.2 mW.
3076.   10.37 μm CO_2 pump laser; R04 transition; perpendicular polarization.
3077.   Threshold pump power: 2000 mW.
3078.   For line assignments see Reference 1384 (pump).
3079.   Output power: 0.01 mW.
3080.   9.68 μm CO_2 pump laser; P34 transition; parallel polarization.
3081.   Threshold pump power: 2000 mW.
3082.   For line assignments see Reference 1384 (pump).
3083.   Output power: 0.05 mW.
3084.   10.15 μm CO_2 pump laser; R36 transition; perpendicular polarization.
3085.   Threshold pump power: 2000 mW.
3086.   For line assignments see Reference 1384 (pump).
3087.   Output power: 0.1 mW.
3088.   10.65 μm CO_2 pump laser; P26 transition; parallel polarization.
3089.   Threshold pump power: 2000 mW.
3090.   For line assignments see Reference 1384 (pump).
3091.   Output power: 1 mW.
3092.   10.19 μm CO_2 pump laser; R28 transition; parallel polarization.
3093.   Threshold pump power: 2000 mW.
3094.   For line assignments see Reference 1384 (pump).
3095.   Output power: 0.5 mW.
3096.   9.71 μm CO_2 pump laser; P38 transition; parallel polarization.
3097.   Threshold pump power: 2000 mW.
3098.   For line assignments see Reference 1384 (pump).
3099.   Output power: 0.005 mW.
3100.   10.27 μm CO_2 pump laser; R16 transition; parallel polarization.
3101.   Threshold pump power: 2000 mW.
3102.   For line assignments see Reference 1384 (pump).
3103.   Output power: 0.2 mW.
3104.   10.70 μm CO_2 pump laser; P30 transition; parallel polarization.
3105.   Threshold pump power: 2000 mW.
3106.   For line assignments see Reference 1384 (pump).
3107.   Output power: 0.2 mW.
3108.   9.71 μm CO_2 pump laser; P38 transition; parallel polarization.
3109.   Threshold pump power: 2000 mW.
3110.   For line assignments see Reference 1384 (pump).
3111.   Output power: 0.5 mW.
3112.   10.67 μm CO_2 pump laser; P28 transition; parallel polarization.
3113.   Threshold pump power: 2000 mW.
3114.   For line assignments see Reference 1384 (pump).
3115.   Output power: 2 mW.

3116.   10.35 μm CO_2 pump laser; R06 transition; parallel polarization.
3117.   Threshold pump power: 2000 mW.
3118.   For line assignments see Reference 1384 (pump).
3119.   Output power: 0.2 mW.
3120.   10.23 μm CO_2 pump laser; R22 transition; parallel polarization.
3121.   Threshold pump power: 2000 mW.
3122.   For line assignments see Reference 1384 (pump).
3123.   Output power: 0.1 mW.
3124.   9.69 μm CO_2 pump laser; P36 transition; parallel polarization.
3125.   Threshold pump power: 2000 mW.
3126.   For line assignments see Reference 1384 (pump).
3127.   Output power: 0.1 mW.
3128.   10.33 μm CO_2 pump laser; R08 transition; perpendicular polarization.
3129.   Threshold pump power: 2000 mW.
3130.   For line assignments see Reference 1384 (pump).
3131.   Output power: 0.2 mW.
3132.   9.62 μm CO_2 pump laser; P28 transition; parallel polarization.
3133.   Threshold pump power: 2000 mW.
3134.   For line assignments see Reference 1384 (pump).
3135.   Output power: 0.05 mW.
3136.   10.18 μm CO_2 pump laser; R30 transition; parallel polarization.
3137.   Threshold pump power: 2000 mW.
3138.   For line assignments see Reference 1384 (pump).
3139.   Output power: 0.05 mW.
3140.   10.32 μm CO_2 pump laser; R10 transition; parallel polarization.
3141.   Threshold pump power: 2000 mW.
3142.   For line assignments see Reference 1384 (pump).
3143.   Output power: 0.05 mW.
3144.   9.35 μm CO_2 pump laser; R06 transition; parallel polarization.
3145.   Threshold pump power: 2000 mW.
3146.   For line assignments see Reference 1384 (pump).
3147.   Output power: 0.5 mW.
3148.   10.72 μm CO_2 pump laser; P32 transition; parallel polarization.
3149.   Threshold pump power: 2000 mW.
3150.   For line assignments see Reference 1384 (pump).
3151.   Output power: 0.2 mW.
3152.   10.29 μm CO_2 pump laser; R14 transition; perpendicular polarization.
3153.   Threshold pump power: 2000 mW.
3154.   Output power: 0.05 mW.
3155.   10.22 μm CO_2 pump laser; R24 transition; parallel polarization.
3156.   Threshold pump power: 2000 mW.
3157.   Output power: 0.5 mW.
3158.   10.30 μm CO_2 pump laser; R12 transition; perpendicular polarization.
3159.   Threshold pump power: 2000 mW.
3160.   Output power: 0.05 mW.
3161.   10.19 μm CO_2 pump laser; R28 transition; parallel polarization.
3162.   Threshold pump power: 2000 mW.
3163.   Output power: 1 mW.
3164.   10.16 μm CO_2 pump laser; R34 transition; parallel polarization.
3165.   Threshold pump power: 2000 mW.
3166.   Output power: 2 mW.

3167.   10.16 μm CO_2 pump laser; R34 transition; parallel polarization.
3168.   Threshold pump power: 2000 mW.
3169.   Output power: 0.01 mW.
3170.   10.46 μm CO_2 pump laser; P06 transition; perpendicular polarization.
3171.   Threshold pump power: 2000 mW.
3172.   Output power: 0.2 mW.
3173.   10.15 μm CO_2 pump laser; R36 transition; parallel polarization.
3174.   Threshold pump power: 2000 mW.
3175.   Output power: 0.02 mW.
3176.   10.18 μm CO_2 pump laser; R30 transition; parallel polarization.
3177.   Threshold pump power: 2000 mW.
3178.   Output power: 1 mW.
3179.   10.48 μm CO_2 pump laser; P08 transition; perpendicular polarization.
3180.   Threshold pump power: 2000 mW.
3181.   Output power: 1 mW.
3182.   10.53 μm CO_2 pump laser; P14 transition; parallel polarization.
3183.   Threshold pump power: 2000 mW.
3184.   Output power: 0.5 mW.
3185.   10.53 μm CO_2 pump laser; P14 transition; perpendicular polarization.
3186.   Threshold pump power: 2000 mW.
3187.   Output power: 0.5 mW.
3188.   10.29 μm CO_2 pump laser; R14 transition; perpendicular polarization.
3189.   Threshold pump power: 2000 mW.
3190.   Output power: 0.02 mW.
3191.   10.55 μm CO_2 pump laser; P16 transition; parallel polarization.
3192.   Threshold pump power: 2000 mW.
3193.   Output power: 0.02 mW.
3194.   10.30 μm CO_2 pump laser; R12 transition; parallel polarization.
3195.   Threshold pump power: 2000 mW.
3196.   Output power: 0.1 mW.
3197.   10.59 μm CO_2 pump laser; P20 transition; parallel polarization.
3198.   Threshold pump power: 2000 mW.
3199.   Output power: 0.1 mW.
3200.   10.27 μm CO_2 pump laser; R16 transition; perpendicular polarization.
3201.   Threshold pump power: 2000 mW.
3202.   Output power: 1 mW.
3203.   10.61 μm CO_2 pump laser; P22 transition; parallel polarization.
3204.   Threshold pump power: 2000 mW.
3205.   Output power: 0.1 mW.
3206.   10.22 μm CO_2 pump laser; R24 transition; parallel polarization.
3207.   Threshold pump power: 2000 mW.
3208.   Output power: 0.2 mW.
3209.   10.70 μm CO_2 pump laser; P30 transition; parallel polarization.
3210.   Threshold pump power: 2000 mW.
3211.   Output power: 0.01 mW.
3212.   10.25 μm CO_2 pump laser; R20 transition; perpendicular polarization.
3213.   Threshold pump power: 2000 mW.
3214.   For line assignments see Reference 1384 (pump).
3215.   Output power: 0.05 mW.
3216.   9.71 μm CO_2 pump laser; P38 transition; parallel polarization.
3217.   Threshold pump power: 2000 mW.

3218. For line assignments see Reference 1384 (pump).
3219. Output power: 0.02 mW.
3220. 10.48 μm CO_2 pump laser; P08 transition; parallel polarization.
3221. Threshold pump power: 2000 mW.
3222. For line assignments see Reference 1384 (pump).
3223. Output power: 0.005 mW.
3224. 9.49 μm CO_2 pump laser; P12 transition; parallel polarization.
3225. Threshold pump power: 2000 mW.
3226. For line assignments see Reference 1384 (pump).
3227. Output power: 1 mW.
3228. 10.26 μm CO_2 pump laser; R18 transition; parallel polarization.
3229. Threshold pump power: 2000 mW.
3230. For line assignments see Reference 1384 (pump).
3231. Output power: 0.2 mW.
3232. 9.49 μm CO_2 pump laser; P12 transition; parallel polarization.
3233. Threshold pump power: 2000 mW.
3234. For line assignments see Reference 1384 (pump).
3235. Output power: 0.05 mW.
3236. 10.65 μm CO_2 pump laser; P26 transition; parallel polarization.
3237. Threshold pump power: 2000 mW.
3238. For line assignments see Reference 1384 (pump).
3239. Output power: 1 mW.
3240. 9.52 μm CO_2 pump laser; P16 transition; parallel polarization.
3241. Threshold pump power: 2000 mW.
3242. For line assignments see Reference 1384 (pump).
3243. Output power: 0.1 mW.
3244. 10.59 μm CO_2 pump laser; P20 transition; parallel polarization.
3245. Threshold pump power: 2000 mW.
3246. For line assignments see Reference 1384 (pump).
3247. Output power: 2 mW.
3248. 10.65 μm CO_2 pump laser; P26 transition; parallel polarization.
3249. Threshold pump power: 2000 mW.
3250. For line assignments see Reference 1384 (pump).
3251. Output power: 0.01 mW.
3252. 9.49 μm CO_2 pump laser; P12 transition; parallel polarization.
3253. Threshold pump power: 2000 mW.
3254. For line assignments see Reference 1384 (pump).
3255. Output power: 0.05 mW.
3256. 10.25 μm CO_2 pump laser; R20 transition; parallel polarization.
3257. Threshold pump power: 2000 mW.
3258. For line assignments see Reference 1384 (pump).
3259. Output power: 0.2 mW.
3260. 10.44 μm CO_2 pump laser; P04 transition; parallel polarization.
3261. Threshold pump power: 2000 mW.
3262. For line assignments see Reference 1384 (pump).
3263. Output power: 0.02 mW.
3264. 10.44 μm CO_2 pump laser; P04 transition; parallel polarization.
3265. Threshold pump power: 2000 mW.
3266. For line assignments see Reference 1384 (pump).
3267. Output power: 2 mW.
3268. 10.25 μm CO_2 pump laser; R20 transition; parallel polarization.

3269.   Threshold pump power: 2000 mW.
3270.   For line assignments see Reference 1384 (pump).
3271.   Output power: 0.05 mW.
3272.   10.65 μm CO_2 pump laser; P26 transition; parallel polarization.
3273.   Threshold pump power: 2000 mW.
3274.   For line assignments see Reference 1384 (pump).
3275.   Output power: 0.1 mW.
3276.   10.49 μm CO_2 pump laser; P10 transition; perpendicular polarization.
3277.   Threshold pump power: 2000 mW.
3278.   For line assignments see Reference 1384 (pump).
3279.   Output power: 0.02 mW.
3280.   10.44 μm CO_2 pump laser; P04 transition; parallel polarization.
3281.   Threshold pump power: 2000 mW.
3282.   For line assignments see Reference 1384 (pump).
3283.   Output power: 0.5 mW.
3284.   10.49 μm CO_2 pump laser; P10 transition; perpendicular polarization.
3285.   Threshold pump power: 2000 mW.
3286.   For line assignments see Reference 1384 (pump).
3287.   Output power: 0.1 mW.
3288.   10.70 μm CO_2 pump laser; P30 transition; parallel polarization.
3289.   Threshold pump power: 2000 mW.
3290.   For line assignments see Reference 1384 (pump).
3291.   Output power: 0.05 mW.
3292.   10.22 μm CO_2 pump laser; R24 transition; unspecified polarization.
3293.   Threshold pump power: 2000 mW.
3294.   For line assignments see Reference 1384 (pump).
3295.   Output power: 0.2 mW.
3296.   10.57 μm CO_2 pump laser; P18 transition; perpendicular polarization.
3297.   Threshold pump power: 2000 mW.
3298.   For line assignments see Reference 1384 (pump).
3299.   Output power: 1 mW.
3300.   10.21 μm CO_2 pump laser; R26 transition; parallel polarization.
3301.   Threshold pump power: 2000 mW.
3302.   For line assignments see Reference 1384 (pump).
3303.   Output power: 5 mW.
3304.   10.21 μm CO_2 pump laser; R26 transition; parallel polarization.
3305.   Threshold pump power: 2000 mW.
3306.   For line assignments see Reference 1384 (pump).
3307.   Output power: 2 mW.
3308.   10.59 μm CO_2 pump laser; P20 transition; perpendicular polarization.
3309.   Threshold pump power: 2000 mW.
3310.   For line assignments see Reference 1384 (pump).
3311.   Output power: 2 mW.
3312.   10.74 μm CO_2 pump laser; P34 transition; parallel polarization.
3313.   Threshold pump power: 2000 mW.
3314.   For line assignments see Reference 1384 (pump).
3315.   Output power: 2 mW.
3316.   10.74 μm CO_2 pump laser; P34 transition; parallel polarization.
3317.   Threshold pump power: 2000 mW.
3318.   For line assignments see Reference 1384 (pump).
3319.   Output power: 2 mW.

3320.   10.74 μm CO_2 pump laser; P34 transition; parallel polarization.
3321.   Threshold pump power: 2000 mW.
3322.   For line assignments see Reference 1384 (pump).
3323.   Output power: 1 mW.
3324.   10.48 μm CO_2 pump laser; P08 transition; perpendicular polarization.
3325.   Threshold pump power: 2000 mW.
3326.   For line assignments see Reference 1384 (pump).
3327.   Output power: 0.01 mW.
3328.   10.19 μm CO_2 pump laser; R28 transition; parallel polarization.
3329.   Threshold pump power: 2000 mW.
3330.   For line assignments see Reference 1384 (pump).
3331.   Output power: 1 mW.
3332.   10.48 μm CO_2 pump laser; P08 transition; perpendicular polarization.
3333.   Threshold pump power: 2000 mW.
3334.   For line assignments see Reference 1384 (pump).
3335.   Output power: 0.02 mW.
3336.   10.37 μm CO_2 pump laser; R04 transition; parallel polarization.
3337.   Threshold pump power: 2000 mW.
3338.   For line assignments see Reference 1384 (pump).
3339.   Output power: 0.1 mW.
3340.   10.51 μm CO_2 pump laser; P12 transition; perpendicular polarization.
3341.   Threshold pump power: 2000 mW.
3342.   For line assignments see Reference 1384 (pump).
3343.   Output power: 0.2 mW.
3344.   10.70 μm CO_2 pump laser; P30 transition; parallel polarization.
3345.   Threshold pump power: 2000 mW.
3346.   For line assignments see Reference 1384 (pump).
3347.   Output power: 0.1 mW.
3348.   10.32 μm CO_2 pump laser; R10 transition; parallel polarization.
3349.   Threshold pump power: 2000 mW.
3350.   For line assignments see Reference 1384 (pump).
3351.   Output power: 0.05 mW.
3352.   10.53 μm CO_2 pump laser; P14 transition; perpendicular polarization.
3353.   Threshold pump power: 2000 mW.
3354.   For line assignments see Reference 1384 (pump).
3355.   Output power: 0.5 mW.
3356.   10.57 μm CO_2 pump laser; P18 transition; perpendicular polarization.
3357.   Threshold pump power: 2000 mW.
3358.   For line assignments see Reference 1384 (pump).
3359.   Output power: 0.1 mW.
3360.   10.29 μm CO_2 pump laser; P14 transition; parallel polarization.
3361.   Threshold pump power: 2000 mW.
3362.   For line assignments see Reference 1384 (pump).
3363.   Output power: 0.2 mW.
3364.   10.86 μm CO_2 pump laser; P44 transition; parallel polarization.
3365.   Threshold pump power: 2000 mW.
3366.   For line assignments see Reference 1384 (pump).
3367.   Output power: 0.5 mW.
3368.   10.23 μm CO_2 pump laser; R22 transition; parallel polarization.
3369.   Threshold pump power: 2000 mW.
3370.   For line assignments see Reference 1384 (pump).

3371.   Output power: 1 mW.
3372.   10.25 μm CO_2 pump laser; R20 transition; parallel polarization.
3373.   Threshold pump power: 2000 mW.
3374.   For line assignments see Reference 1384 (pump).
3375.   Output power: 1 mW.
3376.   10.32 μm CO_2 pump laser; R10 transition; parallel polarization.
3377.   Threshold pump power: 2000 mW.
3378.   For line assignments see Reference 1384 (pump).
3379.   Output power: 1 mW.
3380.   10.30 μm CO_2 pump laser; R12 transition; parallel polarization.
3381.   Threshold pump power: 2000 mW.
3382.   For line assignments see Reference 1384 (pump).
3383.   Output power: 10 mW.
3384.   10.30 μm CO_2 pump laser; R12 transition; parallel polarization.
3385.   Threshold pump power: 2000 mW.
3386.   For line assignments see Reference 1384 (pump).
3387.   Output power: 0.1 mW.
3388.   9.37 μm CO_2 pump laser; R04 transition; parallel polarization.
3389.   Threshold pump power: 2000 mW.
3390.   For line assignments see Reference 1384 (pump).
3391.   Output power: 1 mW.
3392.   10.29 μm CO_2 pump laser; R14 transition; parallel polarization.
3393.   Threshold pump power: 2000 mW.
3394.   For line assignments see Reference 1384 (pump).
3395.   Output power: 1 mW.
3396.   10.13 μm CO_2 pump laser; R40 transition; parallel polarization.
3397.   Threshold pump power: 2000 mW.
3398.   For line assignments see Reference 1384 (pump).
3399.   Output power: 0.5 mW.
3400.   9.34 μm CO_2 pump laser; R08 transition; parallel polarization.
3401.   Threshold pump power: 2000 mW.
3402.   For line assignments see Reference 1384 (pump).
3403.   Output power: 0.02 mW.
3404.   10.19 μm CO_2 pump laser; R28 transition; parallel polarization.
3405.   Threshold pump power: 2000 mW.
3406.   For line assignments see Reference 1384 (pump).
3407.   Output power: 0.5 mW.
3408.   10.18 μm CO_2 pump laser; R30 transition; parallel polarization.
3409.   Threshold pump power: 2000 mW.
3410.   For line assignments see Reference 1384 (pump).
3411.   Output power: 2 mW.
3412.   10.22 μm CO_2 pump laser; R24 transition; parallel polarization.
3413.   Threshold pump power: 2000 mW.
3414.   For line assignments see Reference 1384 (pump).
3415.   Output power: 5 mW.
3416.   10.22 μm CO_2 pump laser; R24 transition; parallel polarization.
3417.   Threshold pump power: 2000 mW.
3418.   For line assignments see Reference 1384 (pump).
3419.   Output power: 0.2 mW.
3420.   10.35 μm CO_2 pump laser; R06 transition; parallel polarization.
3421.   Threshold pump power: 2000 mW.

3422.   For line assignments see Reference 1384 (pump).

3423.   Output power: 0.2 mW.

3424.   9.29 μm CO_2 pump laser; R16 transition; parallel polarization.

3425.   Threshold pump power: 2000 mW.

3426.   For line assignments see Reference 1384 (pump).

3427.   Output power: 0.5 mW.

3428.   9.52 μm CO_2 pump laser; P16 transition; parallel polarization.

3429.   Threshold pump power: 2000 mW.

3430.   For line assignments see Reference 1384 (pump).

3431.   Output power: 0.2 mW.

3432.   10.17 μm CO_2 pump laser; R32 transition; parallel polarization.

3433.   Threshold pump power: 2000 mW.

3434.   For line assignments see Reference 1384 (pump).

3435.   Output power: 0.1 mW.

3436.   10.15 μm CO_2 pump laser; R36 transition; parallel polarization.

3437.   Threshold pump power: 2000 mW.

3438.   For line assignments see Reference 1384 (pump).

3439.   Output power: 5 mW.

3440.   10.25 μm CO_2 pump laser; R20 transition; parallel polarization.

3441.   Threshold pump power: 2000 mW.

3442.   For line assignments see Reference 1385.

3443.   Output power: 2 mW.

3444.   +20 MHz offset from center of 9.60 μm CO_2 pump laser; P26 transition; perpendicular polarization.

3445.   Threshold pump power: 10000 mW.

3446.   For line assignments see Reference 1385.

3447.   Output power: 0.5 mW.

3448.   -50 MHz offset from center of 9.71 mm CO_2 pump laser; P38 transition; parallel polarization.

3449.   Threshold pump power: 20000 mW.

3450.   For line assignments see Reference 1385.

3451.   Output power: 10 mW.

3452.   -30 MHz offset from center of 9.32 μm CO_2 pump laser; R12 transition; parallel polarization.

3453.   Threshold pump power: 2000 mW.

3454.   For line assignments see Reference 1385.

3455.   Output power: 1 mW.

3456.   +30 MHz offset from center of 9.86 μm CO_2 pump laser; P52 transition; parallel polarization.

3457.   Threshold pump power: 5000 mW.

3458.   For line assignments see Reference 1385.

3459.   Output power: 0.02 mW.

3460.   +5 MHz offset from center of 10.21 μm CO_2 pump laser; R26 transition; parallel polarization.

3461.   Threshold pump power: 50000 mW.

3462.   For line assignments see Reference 1385.

3463.   Output power: 0.05 mW.

3464.   0 MHz offset from center of 10.07 μm CO_2 pump laser; R52 transition; perpendicular polarization.

3465.   Threshold pump power: 5000 mW.

3466.   For line assignments see Reference 1385.

3467. Output power: 0.02 mW.
3468. -25 MHz offset from center of 9.17 μm CO_2 pump laser; R42 transition; perpendicular polarization.
3469. Threshold pump power: 2000 mW.
3470. For line assignments see Reference 1385.
3471. Output power: 0.2 mW.
3472. +50 MHz offset from center of 9.29 μm CO_2 pump laser; R16 transition; perpendicular polarization.
3473. Threshold pump power: 20000 mW.
3474. For line assignments see Reference 1385.
3475. Output power: 10 mW.
3476. -5 MHz offset from center of 10.26 μm CO_2 pump laser; R18 transition; parallel polarization.
3477. Threshold pump power: 2000 mW.
3478. For line assignments see Reference 1385.
3479. Output power: 50 mW.
3480. +40 MHz offset from center of 9.75 μm CO_2 pump laser; P42 transition; perpendicular polarization.
3481. Threshold pump power: 1000 mW.
3482. For line assignments see Reference 1385.
3483. Output power: 0.5 mW.
3484. -50 MHz offset from center of 10.97 μm CO_2 pump laser; P19 transition; parallel polarization.
3485. Threshold pump power: 2000 mW.
3486. For line assignments see Reference 1385.
3487. Output power: 0.2 mW.
3488. -35 MHz offset from center of 10.16 μm CO_2 pump laser; R34 transition; perpendicular polarization.
3489. Threshold pump power: 20000 mW.
3490. For line assignments see Reference 1385.
3491. Cascade transition.
3492. +45 MHz offset from center of 9.31 μm CO_2 pump laser; R14 transition; parallel polarization.
3493. Threshold pump power: 5000 mW.
3494. For line assignments see Reference 1385.
3495. Output power: 0.05 mW.
3496. +10 MHz offset from center of 9.19 μm CO_2 pump laser; R36 transition; parallel polarization.
3497. Threshold pump power: 5000 mW.
3498. For line assignments see Reference 1385.
3499. Output power: 2 mW.
3500. +45 MHz offset from center of 9.31 μm CO_2 pump laser; R14 transition; parallel polarization.
3501. Threshold pump power: 5000 mW.
3502. For line assignments see Reference 1385.
3503. Output power: 0.5 mW.
3504. -35 MHz offset from center of 10.59 μm CO_2 pump laser; P20 transition; perpendicular polarization.
3505. Threshold pump power: 10000 mW.
3506. For line assignments see Reference 1385.
3507. Output power: 0.05 mW.

3508. -20 MHz offset from center of 10.74 μm CO_2 pump laser; P34 transition; parallel polarization.
3509. Threshold pump power: 20000 mW.
3510. For line assignments see Reference 1385.
3511. Output power: 1 mW.
3512. -30 MHz offset from center of 10.49 μm CO_2 pump laser; P10 transition; parallel polarization.
3513. Threshold pump power: 10000 mW.
3514. For line assignments see Reference 1385.
3515. Output power: 0.1 mW.
3516. -10 MHz offset from center of 9.38 μm CO_2 pump laser; R02 transition; perpendicular polarization.
3517. Threshold pump power: 2000 mW.
3518. For line assignments see Reference 1385.
3519. Output power: 0.1 mW.
3520. +50 MHz offset from center of 9.81 μm CO_2 pump laser; P48 transition; parallel polarization.
3521. Threshold pump power: 5000 mW.
3522. For line assignments see Reference 1388.
3523. Output power: 1 mW.
3524. 9.50 μm CO_2 pump laser; P14 transition; parallel polarization.
3525. Threshold pump power: 2000 mW.
3526. For line assignments see Reference 1388.
3527. Output power: 1 mW.
3528. 9.49 μm CO_2 pump laser; P12 transition; parallel polarization.
3529. Threshold pump power: 2000 mW.
3530. For line assignments see Reference 1388.
3531. Output power: 1 mW.
3532. 9.68 μm CO_2 pump laser; P34 transition; parallel polarization.
3533. Threshold pump power: 2000 mW.
3534. For line assignments see Reference 1388.
3535. Output power: 2 mW.
3536. -35 MHz offset from center of 9.62 μm CO_2 pump laser; P28 transition; unspecified polarization.
3537. Threshold pump power: 1000 mW.
3538. For line assignments see Reference 1388.
3539. Output power: 0.1 mW.
3540. 9.44 μm CO_2 pump laser; P06 transition; perpendicular polarization.
3541. Threshold pump power: 2000 mW.
3542. For line assignments see Reference 1388.
3543. Output power: 1 mW.
3544. 9.44 μm CO_2 pump laser; P06 transition; perpendicular polarization.
3545. Threshold pump power: 2000 mW.
3546. For line assignments see Reference 1388.
3547. Output power: 1 mW.
3548. 9.69 μm CO_2 pump laser; P36 transition; parallel polarization.
3549. Threshold pump power: 2000 mW.
3550. For line assignments see Reference 1388.
3551. Output power: 1 mW.
3552. 9.69 μm CO_2 pump laser; P36 transition; perpendicular polarization.
3553. Threshold pump power: 2000 mW.

3554.   For line assignments see Reference 1388.
3555.   Output power: 10 mW.
3556.   10.25 μm CO_2 pump laser; R20 transition; parallel polarization.
3557.   Threshold pump power: 2000 mW.
3558.   For line assignments see Reference 1388.
3559.   Output power: 10 mW.
3560.   10.25 μm CO_2 pump laser; R20 transition; parallel polarization.
3561.   Threshold pump power: 2000 mW.
3562.   For line assignments see Reference 1388.
3563.   Output power: 10 mW.
3564.   9.47 μm CO_2 pump laser; P10 transition; parallel polarization.
3565.   Threshold pump power: 2000 mW.
3566.   For line assignments see Reference 1388.
3567.   Output power: 10 mW.
3568.   9.20 μm CO_2 pump laser; R34 transition; parallel polarization.
3569.   Threshold pump power: 2000 mW.
3570.   For line assignments see Reference 1388.
3571.   Output power: 0.1 mW.
3572.   10.19 μm CO_2 pump laser; R28 transition; parallel polarization.
3573.   Threshold pump power: 2000 mW.
3574.   For line assignments see Reference 1388.
3575.   Output power: 10 mW.
3576.   9.59 μm CO_2 pump laser; P24 transition; perpendicular polarization.
3577.   Threshold pump power: 2000 mW.
3578.   For line assignments see Reference 1388.
3579.   Output power: 0.1 mW.
3580.   9.52 μm CO_2 pump laser; P16 transition; parallel polarization.
3581.   Threshold pump power: 2000 mW.
3582.   For line assignments see Reference 1388.
3583.   Output power: 0.1 mW.
3584.   10.26 μm CO_2 pump laser; R18 transition; parallel polarization.
3585.   Threshold pump power: 2000 mW.
3586.   For line assignments see Reference 1388.
3587.   Output power: 0.1 mW.
3588.   9.71 μm CO_2 pump laser; P38 transition; perpendicular polarization.
3589.   Threshold pump power: 2000 mW.
3590.   For line assignments see Reference 1388.
3591.   Output power: 1 mW.
3592.   10.29 μm CO_2 pump laser; R14 transition; parallel polarization.
3593.   Threshold pump power: 2000 mW.
3594.   For line assignments see Reference 1388.
3595.   Output power: 1 mW.
3596.   9.66 μm CO_2 pump laser; P32 transition; perpendicular polarization.
3597.   Threshold pump power: 2000 mW.
3598.   For line assignments see Reference 1388.
3599.   Output power: 0.1 mW.
3600.   9.23 μm CO_2 pump laser; R28 transition; parallel polarization.
3601.   Threshold pump power: 2000 mW.
3602.   For line assignments see References 1392 (pump), 1394.
3603.   Output power: 0.005 mW.

3604. 9.16 μm CO_2 pump laser; R44 transition; either parallel or perpendicular polarization.
3605. Threshold pump power: 5000 mW.
3606. For line assignments see References 1392 (pump), 1394.
3607. Output power: 0.5 mW.
3608. 9.43 μm CO_2 pump laser; P04 transition; perpendicular polarization.
3609. Threshold pump power: 10000 mW.
3610. For line assignments see References 1393 (pump), 1394.
3611. Output power: 0.5 mW.
3612. 9.47 μm CO_2 pump laser; P10 transition; perpendicular polarization.
3613. Threshold pump power: 10000 mW.
3614. For line assignments see References 1393 (pump), 1394.
3615. Output power: 0.2 mW.
3616. 9.16 μm CO_2 pump laser; R44 transition; parallel polarization.
3617. Threshold pump power: 5000 mW.
3618. For line assignments see References 1393 (pump), 1394.
3619. Output power: 0.1 mW.
3620. 9.15 μm CO_2 pump laser; R46 transition; either parallel or perpendicular polarization.
3621. Threshold pump power: 5000 mW.
3622. For line assignments see References 1393 (pump), 1394.
3623. Output power: 0.02 mW.
3624. 9.23 μm CO_2 pump laser; R28 transition; either parallel or perpendicular polarization.
3625. Threshold pump power: 10000 mW.
3626. For line assignments see References 1393 (pump), 1394.
3627. Output power: 0.1 mW.
3628. 9.17 μm CO_2 pump laser; R42 transition; parallel polarization.
3629. Threshold pump power: 10000 mW.
3630. For line assignments see References 1393 (pump), 1394.
3631. Output power: 0.1 mW.
3632. 9.23 μm CO_2 pump laser; R28 transition; perpendicular polarization.
3633. Threshold pump power: 2000 mW.
3634. For line assignments see References 1393 (pump), 1394.
3635. Output power: 0.5 mW.
3636. 9.35 μm CO_2 pump laser; R06 transition; perpendicular polarization.
3637. Threshold pump power: 10000 mW.
3638. For line assignments see References 1393 (pump), 1394.
3639. Output power: 0.2 mW.
3640. 9.44 μm CO_2 pump laser; P06 transition; parallel polarization.
3641. Threshold pump power: 10000 mW.
3642. For line assignments see References 1393 (pump), 1394.
3643. Output power: 0.5 mW.
3644. 9.35 μm CO_2 pump laser; R06 transition; perpendicular polarization.
3645. Threshold pump power: 2000 mW.
3646. For line assignments see References 1393 (pump), 1394.
3647. Output power: 0.01 mW.
3648. 9.32 μm CO_2 pump laser; R12 transition; parallel polarization.
3649. Threshold pump power: 2000 mW.
3650. For line assignments see References 1393 (pump), 1394.
3651. Output power: 1 mW.

3652.   9.44 μm CO_2 pump laser; P06 transition; parallel polarization.
3653.   Threshold pump power: 10000 mW.
3654.   For line assignments see References 1393 (pump), 1394.
3655.   Output power: 1 mW.
3656.   9.47 μm CO_2 pump laser; P10 transition; parallel polarization.
3657.   Threshold pump power: 10000 mW.
3658.   For line assignments see References 1393 (pump), 1394.
3659.   Output power: 0.1 mW.
3660.   9.19 μm CO_2 pump laser; R36 transition; parallel polarization.
3661.   Threshold pump power: 10000 mW.
3662.   For line assignments see References 1393 (pump), 1394.
3663.   Output power: 0.02 mW.
3664.   9.46 μm CO_2 pump laser; P08 transition; parallel polarization.
3665.   Threshold pump power: 10000 mW.
3666.   For line assignments see References 1393 (pump), 1394.
3667.   Output power: 0.02 mW.
3668.   9.31 μm CO_2 pump laser; R14 transition; perpendicular polarization.
3669.   Threshold pump power: 2000 mW.
3670.   For line assignments see References 1393 (pump), 1394.
3671.   Output power: 0.02 mW.
3672.   9.19 μm CO_2 pump laser; R36 transition; parallel polarization.
3673.   Threshold pump power: 10000 mW.
3674.   For line assignments see References 1393 (pump), 1394.
3675.   Output power: 0.01 mW.
3676.   9.55 μm CO_2 pump laser; P20 transition; parallel polarization.
3677.   Threshold pump power: 20000 mW.
3678.   For line assignments see References 1393 (pump), 1394.
3679.   Output power: 2 mW.
3680.   9.43 μm CO_2 pump laser; P04 transition; parallel polarization.
3681.   Threshold pump power: 10000 mW.
3682.   For line assignments see References 1393 (pump), 1394.
3683.   Output power: 10 mW.
3684.   9.20 μm CO_2 pump laser; R34 transition; parallel polarization.
3685.   Threshold pump power: 10000 mW.
3686.   For line assignments see References 1393 (pump), 1394.
3687.   Output power: 0.5 mW.
3688.   9.47 μm CO_2 pump laser; P10 transition; perpendicular polarization.
3689.   Threshold pump power: 10000 mW.
3690.   For line assignments see References 1393 (pump), 1394.
3691.   Output power: 0.1 mW.
3692.   9.26 μm CO_2 pump laser; R22 transition; perpendicular polarization.
3693.   Threshold pump power: 20000 mW.
3694.   For line assignments see References 1393 (pump), 1394.
3695.   Output power: 0.2 mW.
3696.   9.71 μm CO_2 pump laser; P38 transition; parallel polarization.
3697.   Threshold pump power: 10000 mW.
3698.   For line assignments see References 1393 (pump), 1394.
3699.   Output power: 2 mW.
3700.   9.59 μm CO_2 pump laser; P24 transition; perpendicular polarization.
3701.   Threshold pump power: 10000 mW.
3702.   For line assignments see References 1393 (pump), 1394.

3703.   Output power: 10 mW.
3704.   9.35 μm CO_2 pump laser; R06 transition; parallel polarization.
3705.   Threshold pump power: 2000 mW.
3706.   For line assignments see References 1393 (pump), 1394.
3707.   Output power: 10 mW.
3708.   9.35 μm CO_2 pump laser; 906 transition; parallel polarization.
3709.   Threshold pump power: 2000 mW.
3710.   For line assignments see References 1393 (pump), 1394.
3711.   Output power: 5 mW.
3712.   9.21 μm CO_2 pump laser; R32 transition; parallel polarization.
3713.   Threshold pump power: 10000 mW.
3714.   For line assignments see References 1393 (pump), 1394.
3715.   Output power: 0.5 mW.
3716.   9.17 μm CO_2 pump laser; R42 transition; perpendicular polarization.
3717.   Threshold pump power: 10000 mW.
3718.   For line assignments see References 1393 (pump), 1394.
3719.   Output power: 0.01 mW.
3720.   9.54 μm CO_2 pump laser; P18 transition; perpendicular polarization.
3721.   Threshold pump power: 20000 mW.
3722.   For line assignments see References 1393 (pump), 1394.
3723.   Output power: 200 mW.
3724.   9.20 μm CO_2 pump laser; R34 transition; perpendicular polarization.
3725.   Threshold pump power: 10000 mW.
3726.   For line assignments see References 1393 (pump), 1394.
3727.   Output power: 0.5 mW.
3728.   9.35 μm CO_2 pump laser; R06 transition; parallel polarization.
3729.   Threshold pump power: 2000 mW.
3730.   For line assignments see References 1392 (pump), 1394.
3731.   Output power: 0.02 mW.
3732.   9.32 μm CO_2 pump laser; R12 transition; perpendicular polarization.
3733.   Threshold pump power: 2000 mW.
3734.   For line assignments see References 1392 (pump), 1394.
3735.   Output power: 0.5 mW.
3736.   9.26 μm CO_2 pump laser; R22 transition; parallel polarization.
3737.   Threshold pump power: 20000 mW.
3738.   For line assignments see References 1392 (pump), 1394.
3739.   Output power: 1 mW.
3740.   9.57 μm CO_2 pump laser; P22 transition; parallel polarization.
3741.   Threshold pump power: 10000 mW.
3742.   For line assignments see References 1392 (pump), 1394.
3743.   Output power: 200 mW.
3744.   9.21 μm CO_2 pump laser; R32 transition; perpendicular polarization.
3745.   Threshold pump power: 10000 mW.
3746.   For line assignments see References 1392 (pump), 1394.
3747.   Output power: 1 mW.
3748.   9.26 μm CO_2 pump laser; R22 transition; parallel polarization.
3749.   Threshold pump power: 2000 mW.
3750.   For line assignments see References 1392 (pump), 1394.
3751.   Output power: 50 mW.
3752.   9.27 μm CO_2 pump laser; R20 transition; parallel polarization.
3753.   Threshold pump power: 2000 mW.

3754.   For line assignments see References 1392 (pump), 1394.

3755.   Output power: 0.01 mW.

3756.   9.55 μm CO_2 pump laser; P20 transition; parallel polarization.

3757.   Threshold pump power: 20000 mW.

3758.   For line assignments see References 1392 (pump), 1394.

3759.   Output power: 20 mW.

3760.   9.47 μm CO_2 pump laser; P10 transition; parallel polarization.

3761.   Threshold pump power: 10000 mW.

3762.   For line assignments see References 1392 (pump), 1394.

3763.   Output power: 5 mW.

3764.   9.59 μm CO_2 pump laser; P24 transition; parallel polarization.

3765.   Threshold pump power: 10000 mW.

3766.   For line assignments see References 1392 (pump), 1394.

3767.   Output power: 1 mW.

3768.   9.57 μm CO_2 pump laser; P22 transition; perpendicular polarization.

3769.   Threshold pump power: 10000 mW.

3770.   For line assignments see References 1392 (pump), 1394.

3771.   Output power: 10 mW.

3772.   9.26 μm CO_2 pump laser; R22 transition; perpendicular polarization.

3773.   Threshold pump power: 2000 mW.

3774.   For line assignments see References 1392 (pump), 1394.

3775.   Output power: 0.01 mW.

3776.   9.46 μm CO_2 pump laser; P08 transition; parallel polarization.

3777.   Threshold pump power: 10000 mW.

3778.   For line assignments see References 1392 (pump), 1394.

3779.   Output power: 100 mW.

3780.   9.27 μm CO_2 pump laser; R20 transition; perpendicular polarization.

3781.   Threshold pump power: 20000 mW.

3782.   For line assignments see References 1392 (pump), 1394.

3783.   Output power: 2 mW.

3784.   9.59 μm CO_2 pump laser; P24 transition; parallel polarization.

3785.   Threshold pump power: 10000 mW.

3786.   For line assignments see References 1392 (pump), 1394.

3787.   Output power: 0.2 mW.

3788.   9.52 μm CO_2 pump laser; P16 transition; parallel polarization.

3789.   Threshold pump power: 20000 mW.

3790.   For line assignments see References 1392 (pump), 1394.

3791.   Output power: 0.05 mW.

3792.   9.32 μm CO_2 pump laser; R12 transition; perpendicular polarization.

3793.   Threshold pump power: 2000 mW.

3794.   Output power: 0.1 mW.

3795.   10.65 μm CO_2 pump laser; P26 transition; unspecified polarization.

3796.   Threshold pump power: 10000 mW.

3797.   Output power: 0.5 mW.

3798.   10.46 μm CO_2 pump laser; P06 transition; unspecified polarization.

3799.   Threshold pump power: 2000 mW.

3800.   Output power: 0.5 mW.

3801.   10.26 μm CO_2 pump laser; R18 transition; unspecified polarization.

3802.   Threshold pump power: 2000 mW.

3803.   Output power: 0.5 mW.

3804.   10.30 μm CO_2 pump laser; R12 transition; unspecified polarization.

3805.   Threshold pump power: 2000 mW.

3806.   Output power: 0.2 mW.

3807.   10.37 μm CO_2 pump laser; R04 transition; unspecified polarization.

3808.   Threshold pump power: 2000 mW.

3809.   Output power: 0.5 mW.

3810.   10.55 μm CO_2 pump laser; P16 transition; unspecified polarization.

3811.   Threshold pump power: 2000 mW.

3812.   Output power: 0.2 mW.

3813.   10.15 μm CO_2 pump laser; R36 transition; unspecified polarization.

3814.   Threshold pump power: 2000 mW.

3815.   Output power: 2 mW.

3816.   10.27 μm CO_2 pump laser; R16 transition; unspecified polarization.

3817.   Threshold pump power: 2000 mW.

3818.   For line assignments see Reference 1450.

3819.   Output power: 1 mW.

3820.   0 MHz offset from center of 10.67 μm CO_2 pump laser; P28 transition; parallel polarization.

3821.   Threshold pump power: 20000 mW.

3822.   For line assignments see Reference 1450.

3823.   Output power: 2 mW.

3824.   0 MHz offset from center of 10.44 μm CO_2 pump laser; P04 transition; parallel polarization.

3825.   Threshold pump power: 5000 mW.

3826.   For line assignments see Reference 1450.

3827.   Output power: 5 mW.

3828.   -30 MHz offset from center of 10.37 μm CO_2 pump laser; R04 transition; perpendicular polarization.

3829.   Threshold pump power: 10000 mW.

3830.   For line assignments see Reference 1450.

3831.   Output power: 10 mW.

3832.   -35 MHz offset from center of 10.09 mm CO_2 pump laser; R46 transition; parallel polarization.

3833.   Threshold pump power: 5000 mW.

3834.   For line assignments see Reference 1450.

3835.   Cascade transition

3836.   -15 MHz offset from center of 10.29 mm CO_2 pump laser; R14 transition; perpendicular polarization.

3837.   Threshold pump power: 10000 mW.

3838.   For line assignments see Reference 1450.

3839.   Output power: 2 mW.

3840.   +5 MHz offset from center of 10.53 mm CO_2 pump laser; P14 transition; perpendicular polarization.

3841.   Threshold pump power: 10000 mW.

3842.   For line assignments see Reference 1450.

3843.   Output power: 2 mW.

3844.   -15 MHz offset from center of 10.29 μm CO_2 pump laser; R14 transition; perpendicular polarization.

3845.   Threshold pump power: 10000 mW.

3846.   For line assignments see Reference 1450.

3847.   Output power: 20 mW.

3848. +45 MHz offset from center of 10.38 μm CO_2 pump laser; R02 transition; parallel polarization.

3849. Threshold pump power: 10000 mW.

3850. For line assignments see Reference 1450.

3851. Output power: 5 mW.

3852. +10 MHz offset from center of 10.32 μm CO_2 pump laser; R10 transition; parallel polarization.

3853. Threshold pump power: 20000 mW.

3854. For line assignments see Reference 1450.

3855. 10.49 mm CO_2 pump laser; P10 transition; unspecified polarization.

3856. For line assignments see Reference 1450.

3857. Output power: 5 mW.

3858. +10 MHz offset from center of 10.67 mm CO_2 pump laser; P28 transition; parallel polarization.

3859. Threshold pump power: 20000 mW.

3860. For line assignments see Reference 1450.

3861. Output power: 5 mW.

3862. 0 MHz offset from center of 9.90 μm CO_2 pump laser; P56 transition; parallel polarization.

3863. Threshold pump power: 10000 mW.

3864. For line assignments see Reference 1450.

3865. Output power: 5 mW.

3866. +25 MHz offset from center of 9.73 μm CO_2 pump laser; P40 transition; parallel polarization.

3867. Threshold pump power: 5000 mW.

3868. For line assignments see Reference 1450.

3869. Cascade transition.

3870. -5 MHz offset from center of 10.79 μm CO_2 pump laser; P38 transition; unspecified polarization.

3871. Threshold pump power: 2000 mW.

3872. For line assignments see Reference 1450.

3873. Output power: 10 mW.

3874. -5 MHz offset from center of 10.79 μm CO_2 pump laser; P38 transition; parallel polarization.

3875. Threshold pump power: 2000 mW.

3876. For line assignments see Reference 1450.

3877. Output power: 1 mW.

3878. +45 MHz offset from center of 10.81 μm CO_2 pump laser; P40 transition; perpendicular polarization.

3879. Threshold pump power: 5000 mW.

3880. Output power: 1 mW.

3881. 0 MHz offset from center of 10.49 mm CO_2 pump laser; P10 transition; parallel polarization.

3882. Threshold pump power: 50000 mW.

3883. Output power: 20 mW.

3884. +25 MHz offset from center of 10.25 μm CO_2 pump laser; R20 transition; perpendicular polarization.

3885. Threshold pump power: 10000 mW.

3886. Output power: 0.02 mW.

3887. 0 MHz offset from center of 10.61 μm CO_2 pump laser; P22 transition; perpendicular polarization.

3888. Threshold pump power: 10000 mW.
3889. Output power: 0.002 mW.
3890. +20 MHz offset from center of 10.55 μm CO_2 pump laser; P16 transition; parallel polarization.
3891. Threshold pump power: 50000 mW.
3892. Output power: 10 mW.
3893. +20 MHz offset from center of 10.17 μm CO_2 pump laser; R32 transition; parallel polarization.
3894. Threshold pump power: 5000 mW.
3895. Output power: 5 mW.
3896. -5 MHz offset from center of 10.63 μm CO_2 pump laser; P24 transition; parallel polarization.
3897. Threshold pump power: 20000 mW.
3898. Output power: 2 mW.
3899. +30 MHz offset from center of 10.11 μm CO_2 pump laser; R42 transition; parallel polarization.
3900. Threshold pump power: 5000 mW.
3901. Output power: 0.01 mW.
3902. +50 MHz offset from center of 10.21 μm CO_2 pump laser; R26 transition; parallel polarization.
3903. Threshold pump power: 50000 mW.
3904. Output power: 0.5 mW.
3905. +10 MHz offset from center of 10.65 μm CO_2 pump laser; P26 transition; parallel polarization.
3906. Threshold pump power: 20000 mW.
3907. Output power: 0.01 mW.
3908. -40 MHz offset from center of 9.62 μm CO_2 pump laser; P28 transition; parallel polarization.
3909. Threshold pump power: 50000 mW.
3910. Output power: 5 mW.
3911. +50 MHz offset from center of 10.26 μm CO_2 pump laser; R18 transition; perpendicular polarization.
3912. Threshold pump power: 10000 mW.
3913. Output power: 2 mW.
3914. +25 MHz offset from center of 9.54 μm CO_2 pump laser; P18 transition; perpendicular polarization.
3915. Threshold pump power: 20000 mW.
3916. Output power: 0.01 mW.
3917. +35 MHz offset from center of 10.48 μm CO_2 pump laser; P08 transition; parallel polarization.
3918. Threshold pump power: 50000 mW.
3919. Output power: 0.5 mW.
3920. 0 MHz offset from center of 10.35 μm CO_2 pump laser; R06 transition; perpendicular polarization.
3921. Threshold pump power: 10000 mW.
3922. Output power: 0.5 mW.
3923. +50 MHz offset from center of 10.08 μm CO_2 pump laser; R50 transition; perpendicular polarization.
3924. Threshold pump power: 2000 mW.
3925. Output power: 1 mW.

3926.   +10 MHz offset from center of 10.59 μm CO_2 pump laser; P20 transition; parallel polarization.

3927.   Threshold pump power: 10000 mW.

3928.   Output power: 0.02 mW.

3929.   +30 MHz offset from center of 10.30 μm CO_2 pump laser; R12 transition; perpendicular polarization.

3930.   Threshold pump power: 10000 mW.

3931.   Output power: 1 mW.

3932.   -40 MHz offset from center of 10.19 μm CO_2 pump laser; R28 transition; parallel polarization.

3933.   Threshold pump power: 20000 mW.

3934.   Output power: 0.01 mW.

3935.   0 MHz offset from center of 10.07 μm CO_2 pump laser; R52 transition; parallel polarization.

3936.   Threshold pump power: 10000 mW.

3937.   Output power: 1 mW.

3938.   -40 MHz offset from center of 9.62 μm CO_2 pump laser; P28 transition; parallel polarization.

3939.   Threshold pump power: 10000 mW.

3940.   For line assignments see Reference 1439.

3941.   Output power: 0.05 mW.

3942.   9.22 μm CO_2 pump laser; R30 transition; parallel polarization.

3943.   Threshold pump power: 2000 mW.

3944.   For line assignments see Reference 1401.

3945.   Output power: 0.02 mW.

3946.   9.66 μm CO_2 pump laser; P32 transition; perpendicular polarization.

3947.   Threshold pump power: 10000 mW.

3948.   For line assignments see Reference 1401.

3949.   Output power: 0.05 mW.

3950.   10.32 μm CO_2 pump laser; R10 transition; parallel polarization.

3951.   Threshold pump power: 2000 mW.

3952.   For line assignments see Reference 1401.

3953.   Output power: 0.05 mW.

3954.   9.28 μm CO_2 pump laser; R18 transition; parallel polarization.

3955.   Threshold pump power: 2000 mW.

3956.   For line assignments see Reference 1401.

3957.   Output power: 0.05 mW.

3958.   9.54 μm CO_2 pump laser; P18 transition; perpendicular polarization.

3959.   Threshold pump power: 2000 mW.

3960.   For line assignments see Reference 1439.

3961.   Output power: 0.05 mW.

3962.   10.38 μm CO_2 pump laser; R02 transition; parallel polarization.

3963.   Threshold pump power: 200 mW.

3964.   For line assignments see Reference 1439.

3965.   Output power: 0.05 mW.

3966.   9.50 μm CO_2 pump laser; P14 transition; parallel polarization.

3967.   Threshold pump power: 1000 mW.

3968.   For line assignments see Reference 1439.

3969.   Output power: 0.005 mW.

3970.   9.69 μm CO_2 pump laser; P36 transition; perpendicular polarization.

3971.   Threshold pump power: 5000 mW.

3972. For line assignments see Reference 1439.
3973. Output power: 0.05 mW.
3974. 9.66 μm CO_2 pump laser; P32 transition; parallel polarization.
3975. Threshold pump power: 2000 mW.
3976. For line assignments see Reference 1439.
3977. Output power: 0.05 mW.
3978. 10.57 μm CO_2 pump laser; P18 transition; parallel polarization.
3979. Threshold pump power: 1000 mW.
3980. For line assignments see Reference 1442.
3981. Output power: 0.02 mW.
3982. -31 MHz offset from center of 10.76 μm CO_2 pump laser; P36 transition; unspecified polarization.
3983. Threshold pump power: 10000 mW.
3984. For line assignments see Reference 1442.
3985. Output power: 0.02 mW.
3986. -31 MHz offset from center of 10.76 mm CO_2 pump laser; P36 transition; unspecified polarization.
3987. Threshold pump power: 10000 mW.
3988. For line assignments see Reference 1442.
3989. Output power: 0.02 mW.
3990. +17 MHz offset from center of 10.74 mm CO_2 pump laser; P34 transition; unspecified polarization.
3991. Threshold pump power: 10000 mW.
3992. For line assignments see Reference 1442.
3993. Output power: 0.02 mW.
3994. 10.76 μm CO_2 pump laser; P36 transition; unspecified polarization.
3995. Threshold pump power: 10000 mW.
3996. For line assignments see Reference 1442.
3997. Output power: 0.05 mW.
3998. +12 MHz offset from center of 10.74 mm CO_2 pump laser; P34 transition; unspecified polarization.
3999. Threshold pump power: 10000 mW.
4000. For line assignments see Reference 1442.
4001. Output power: 0.1 mW.
4002. +28 MHz offset from center of 10.84 μm CO_2 pump laser; P42 transition; unspecified polarization.
4003. Threshold pump power: 5000 mW.
4004. For line assignments see Reference 1442.
4005. Output power: 0.1 mW.
4006. +10 MHz offset from center of 10.70 μm CO_2 pump laser; P30 transition; unspecified polarization.
4007. Threshold pump power: 20000 mW.
4008. For line assignments see Reference 1442.
4009. Output power: 0.02 mW.
4010. -22 MHz offset from center of 10.74 μm CO_2 pump laser; P34 transition; unspecified polarization.
4011. Threshold pump power: 10000 mW.
4012. For line assignments see Reference 1442.
4013. Output power: 0.2 mW.
4014. -13 MHz offset from center of 10.84 μm CO_2 pump laser; P42 transition; unspecified polarization.

4015.   Threshold pump power: 5000 mW.
4016.   For line assignments see Reference 1442.
4017.   Output power: 0.05 mW.
4018.   +35 MHz offset from center of 10.76 μm CO_2 pump laser; P36 transition; unspecified polarization.
4019.   Threshold pump power: 10000 mW.
4020.   For line assignments see Reference 1442.
4021.   Output power: 0.05 mW.
4022.   +33 MHz offset from center of 10.72 μm CO_2 pump laser; P32 transition; unspecified polarization.
4023.   Threshold pump power: 10000 mW.
4024.   For line assignments see Reference 1419.
4025.   Output power: 10 mW.
4026.   +25 MHz offset from center of 10.72 μm CO_2 pump laser; P32 transition; parallel polarization.
4027.   Threshold pump power: 5000 mW.
4028.   For line assignments see Reference 1419.
4029.   Output power: 5 mW.
4030.   +5 MHz offset from center of 10.78 μm CO_2 pump laser; P38 transition; parallel polarization.
4031.   Threshold pump power: 20000 mW.
4032.   For line assignments see Reference 1419.
4033.   Output power: 0.1 mW.
4034.   10.61 μm CO_2 pump laser; P22 transition; parallel polarization.
4035.   Threshold pump power: 2000 mW.
4036.   For line assignments see Reference 1419.
4037.   Output power: 0.5 mW.
4038.   +25 MHz offset from center of 10.61 μm CO_2 pump laser; P22 transition; parallel polarization.
4039.   Threshold pump power: 20000 mW.
4040.   For line assignments see Reference 1419.
4041.   Output power: 1 mW.
4042.   +20 MHz offset from center of 10.67 μm CO_2 pump laser; P28 transition; parallel polarization.
4043.   Threshold pump power: 20000 mW.
4044.   For line assignments see Reference 1419.
4045.   Output power: 0.2 mW.
4046.   +20 MHz offset from center of 9.44 μm CO_2 pump laser; P06 transition; parallel polarization.
4047.   Threshold pump power: 20000 mW.
4048.   For line assignments see Reference 1419.
4049.   Output power: 0.1 mW.
4050.   0 MHz offset from center of 9.42 μm CO_2 pump laser; P04 transition; parallel polarization.
4051.   Threshold pump power: 5000 mW.
4052.   For line assignments see Reference 1419.
4053.   Output power: 1 mW.
4054.   -30 MHz offset from center of 10.16 μm CO_2 pump laser; R34 transition; parallel polarization.
4055.   Threshold pump power: 50000 mW.
4056.   For line assignments see Reference 1419.

4057.   Output power: 2 mW.
4058.   -30 MHz offset from center of 10.55 μm CO_2 pump laser; P16 transition; parallel polarization.
4059.   For line assignments see Reference 1419.
4060.   Output power: 0.002 mW.
4061.   +10 MHz offset from center of 10.65 μm CO_2 pump laser; P26 transition; parallel polarization.
4062.   Threshold pump power: 10000 mW.
4063.   For line assignments see Reference 1419.
4064.   Output power: 5 mW.
4065.   0 MHz offset from center of 10.76 μm CO_2 pump laser; P36 transition; parallel polarization.
4066.   Threshold pump power: 20000 mW.
4067.   For line assignments see Reference 1419.
4068.   Output power: 0.1 mW.
4069.   -25 MHz offset from center of 9.42 μm CO_2 pump laser; P04 transition; parallel polarization.
4070.   Threshold pump power: 5000 mW.
4071.   For line assignments see Reference 1419.
4072.   Output power: 0.5 mW.
4073.   +30 MHz offset from center of 10.53 μm CO_2 pump laser; P14 transition; parallel polarization.
4074.   Threshold pump power: 50000 mW.
4075.   For line assignments see Reference 1419.
4076.   Output power: 10 mW.
4077.   0 MHz offset from center of 9.68 μm CO_2 pump laser; P34 transition; parallel polarization.
4078.   Threshold pump power: 10000 mW.
4079.   For line assignments see Reference 1409.
4080.   Output power: 0.02 mW.
4081.   +10 MHz offset from center of 9.60 μm CO_2 pump laser; P26 transition; parallel polarization.
4082.   Threshold pump power: 50000 mW.
4083.   For line assignments see Reference 1409.
4084.   Output power: 0.2 mW.
4085.   0 MHz offset from center of 10.47 μm CO_2 pump laser; P08 transition; perpendicular polarization.
4086.   Threshold pump power: 50000 mW.
4087.   For line assignments see Reference 1409.
4088.   Cascade transition.
4089.   0 MHz offset from center of 10.57 μm CO_2 pump laser; P18 transition; parallel polarization.
4090.   Threshold pump power: 1000 mW.
4091.   For line assignments see Reference 1409.
4092.   Output power: 20 mW.
4093.   0 MHz offset from center of 10.57 μm CO_2 pump laser; P18 transition; parallel polarization.
4094.   Threshold pump power: 1000 mW.
4095.   For line assignments see Reference 1409.
4096.   Output power: 0.02 mW.

4097.   +5 MHz offset from center of 9.30 μm CO_2 pump laser; R14 transition; parallel polarization.
4098.   Threshold pump power: 20000 mW.
4099.   For line assignments see Reference 1409.
4100.   Output power: 2 mW.
4101.   +25 MHz offset from center of 10.84 μm CO_2 pump laser; P42 transition; parallel polarization.
4102.   Threshold pump power: 10000 mW.
4103.   For line assignments see Reference 1409.
4104.   Output power: 5 mW.
4105.   +5 MHz offset from center of 9.29 μm CO_2 pump laser; R16 transition; parallel polarization.
4106.   Threshold pump power: 10000 mW.
4107.   For line assignments see Reference 1401.
4108.   Output power: 0.1 mW.
4109.   9.33 μm CO_2 pump laser; R10 transition; parallel polarization.
4110.   Threshold pump power: 2000 mW.
4111.   For line assignments see Reference 1401.
4112.   10.61 μm CO_2 pump laser; P22 transition; parallel polarization.
4113.   Threshold pump power: 2000 mW.
4114.   For line assignments see Reference 1401.
4115.   10.74 μm CO_2 pump laser; P34 transition; parallel polarization.
4116.   Threshold pump power: 2000 mW.
4117.   For line assignments see Reference 1401.
4118.   10.61 μm CO_2 pump laser; P22 transition; parallel polarization.
4119.   Threshold pump power: 2000 mW.
4120.   For line assignments see Reference 1401.
4121.   Output power: 1 mW.
4122.   9.23 μm CO_2 pump laser; R28 transition; parallel polarization.
4123.   Threshold pump power: 2000 mW.
4124.   For line assignments see Reference 1401.
4125.   Output power: 10 mW.
4126.   9.23 μm CO_2 pump laser; R28 transition; parallel polarization.
4127.   Threshold pump power: 2000 mW.
4128.   For line assignments see Reference 1401.
4129.   10.70 μm CO_2 pump laser; P30 transition; parallel polarization.
4130.   Threshold pump power: 2000 mW.
4131.   For line assignments see Reference 1401.
4132.   10.51 μm CO_2 pump laser; P12 transition; parallel polarization.
4133.   Threshold pump power: 2000 mW.
4134.   For line assignments see Reference 1401.
4135.   10.48 μm CO_2 pump laser; P08 transition; parallel polarization.
4136.   Threshold pump power: 2000 mW.
4137.   For line assignments see Reference 1401.
4138.   Output power: 1 mW.
4139.   9.57 μm CO_2 pump laser; P22 transition; parallel polarization.
4140.   Threshold pump power: 2000 mW.
4141.   For line assignments see Reference 1401.
4142.   Output power: 1 mW.
4143.   9.23 μm CO_2 pump laser; R28 transition; parallel polarization.
4144.   Threshold pump power: 2000 mW.

4145. For line assignments see Reference 1401.
4146. Output power: 0.1 mW.
4147. 9.27 μm CO_2 pump laser; R20 transition; parallel polarization.
4148. Threshold pump power: 2000 mW.
4149. For line assignments see Reference 1401.
4150. 10.33 μm CO_2 pump laser; R08 transition; parallel polarization.
4151. Threshold pump power: 2000 mW.
4152. For line assignments see Reference 1401.
4153. 10.33 μm CO_2 pump laser; R08 transition; parallel polarization.
4154. Threshold pump power: 2000 mW.
4155. For line assignments see Reference 1401.
4156. 10.49 μm CO_2 pump laser; P10 transition; parallel polarization.
4157. Threshold pump power: 2000 mW.
4158. For line assignments see Reference 1401.
4159. 10.88 μm CO_2 pump laser; P46 transition; parallel polarization.
4160. Threshold pump power: 2000 mW.
4161. For line assignments see Reference 1401.
4162. 10.55 μm CO_2 pump laser; P16 transition; perpendicular polarization.
4163. Threshold pump power: 2000 mW.
4164. For line assignments see Reference 1401.
4165. 10.26 μm CO_2 pump laser; R18 transition; perpendicular polarization.
4166. Threshold pump power: 2000 mW.
4167. For line assignments see Reference 1401.
4168. Output power: 10 mW.
4169. 10.23 μm CO_2 pump laser; R22 transition; parallel polarization.
4170. Threshold pump power: 2000 mW.
4171. For line assignments see Reference 1401.
4172. Output power: 10 mW.
4173. 10.23 μm CO_2 pump laser; R22 transition; parallel polarization.
4174. Threshold pump power: 2000 mW.
4175. For line assignments see Reference 1401.
4176. 10.76 μm CO_2 pump laser; P36 transition; parallel polarization.
4177. Threshold pump power: 2000 mW.
4178. For line assignments see Reference 1401.
4179. 10.76 μm CO_2 pump laser; P36 transition; parallel polarization.
4180. Threshold pump power: 2000 mW.
4181. For line assignments see Reference 1401.
4182. Output power: 0.1 mW.
4183. 9.35 μm CO_2 pump laser; R06 transition; perpendicular polarization.
4184. Threshold pump power: 2000 mW.
4185. For line assignments see Reference 1401.
4186. 10.79 μm CO_2 pump laser; P38 transition; parallel polarization.
4187. Threshold pump power: 2000 mW.
4188. For line assignments see Reference 1401.
4189. Output power: 10 mW.
4190. 9.26 μm CO_2 pump laser; R22 transition; parallel polarization.
4191. Threshold pump power: 2000 mW.
4192. For line assignments see Reference 1401.
4193. Output power: 1 mW.
4194. 9.47 μm CO_2 pump laser; P10 transition; parallel polarization.
4195. Threshold pump power: 2000 mW.

4196.   For line assignments see Reference 1401.
4197.   Output power: 10 mW.
4198.   9.32 μm CO_2 pump laser; R12 transition; perpendicular polarization.
4199.   Threshold pump power: 2000 mW.
4200.   For line assignments see Reference 1401.
4201.   Output power: 10 mW.
4202.   9.21 μm CO_2 pump laser; R32 transition; perpendicular polarization.
4203.   Threshold pump power: 2000 mW.
4204.   For line assignments see Reference 1401.
4205.   Output power: 0.1 mW.
4206.   9.62 μm CO_2 pump laser; P28 transition; parallel polarization.
4207.   Threshold pump power: 2000 mW.
4208.   For line assignments see Reference 1401.
4209.   Output power: 0.01 mW.
4210.   9.60 μm CO_2 pump laser; P26 transition; parallel polarization.
4211.   Threshold pump power: 2000 mW.
4212.   For line assignments see Reference 1401.
4213.   Output power: 1 mW.
4214.   9.24 μm CO_2 pump laser; R26 transition; parallel polarization.
4215.   Threshold pump power: 2000 mW.
4216.   For line assignments see Reference 1401.
4217.   Output power: 0.1 mW.
4218.   9.49 μm CO_2 pump laser; P12 transition; perpendicular polarization.
4219.   Threshold pump power: 2000 mW.
4220.   For line assignments see Reference 1388 (pump).
4221.   Output power: 1 mW.
4222.   10.51 μm CO_2 pump laser; P12 transition; unspecified polarization.
4223.   Threshold pump power: 1000 mW.
4224.   For line assignments see Reference 1388 (pump).
4225.   Output power: 5 mW.
4226.   10.49 μm CO_2 pump laser; P10 transition; unspecified polarization.
4227.   Threshold pump power: 1000 mW.
4228.   For line assignments see References 1444, 1456.
4229.   9.20 μm CO_2 pump laser; R34 transition; unspecified polarization.
4230.   Threshold pump power: 10000 mW.
4231.   For line assignments see References 1444, 1456.
4232.   9.27 μm CO_2 pump laser; R20 transition; unspecified polarization.
4233.   Threshold pump power: 10000 mW.
4234.   For line assignments see References 1444, 1456.
4235.   9.17 μm CO_2 pump laser; R40 transition; unspecified polarization.
4236.   Threshold pump power: 10000 mW.
4237.   For line assignments see References 1444, 1456.
4238.   9.18 μm CO_2 pump laser; R38 transition; unspecified polarization.
4239.   Threshold pump power: 10000 mW.
4240.   For line assignments see References 1444, 1456.
4241.   9.29 μm CO_2 pump laser; R16 transition; unspecified polarization.
4242.   Threshold pump power: 10000 mW.
4243.   For line assignments see References 1444, 1456.
4244.   9.20 μm CO_2 pump laser; R34 transition; unspecified polarization.
4245.   Threshold pump power: 10000 mW.
4246.   For line assignments see References 1444, 1456.

4247.   9.23 μm CO_2 pump laser; R28 transition; unspecified polarization.
4248.   Threshold pump power: 10000 mW.
4249.   For line assignments see References 1444, 1456.
4250.   9.32 μm CO_2 pump laser; R12 transition; unspecified polarization.
4251.   Threshold pump power: 10000 mW.
4252.   For line assignments see References 1444, 1456.
4253.   Output power: 0.02 mW.
4254.   9.33 μm CO_2 pump laser; R10 transition; unspecified polarization.
4255.   Threshold pump power: 5000 mW.
4256.   For line assignments see References 1444, 1456.
4257.   Output power: 0.1 mW.
4258.   9.34 μm CO_2 pump laser; R08 transition; unspecified polarization.
4259.   Threshold pump power: 5000 mW.
4260.   Output power: 0.1 mW.
4261.   10.61 μm CO_2 pump laser; P22 transition; perpendicular polarization.
4262.   Output power: 0.1 mW.
4263.   10.22 μm CO_2 pump laser; R24 transition; parallel polarization.
4264.   Output power: 0.1 mW.
4265.   10.72 μm CO_2 pump laser; P32 transition; perpendicular polarization.
4266.   Output power: 0.01 mW.
4267.   10.14 μm CO_2 pump laser; R38 transition; unspecified polarization.
4268.   Threshold pump power: 1000 mW.
4269.   Output power: 0.1 mW.
4270.   10.51 μm CO_2 pump laser; P12 transition; parallel polarization.
4271.   Output power: 1 mW.
4272.   10.33 μm CO_2 pump laser; R08 transition; perpendicular polarization.
4273.   Output power: 0.01 mW.
4274.   9.49 μm CO_2 pump laser; P12 transition; unspecified polarization.
4275.   Threshold pump power: 1000 mW.
4276.   Output power: 1 mW.
4277.   9.55 μm CO_2 pump laser; P20 transition; parallel polarization.
4278.   Output power: 0.1 mW.
4279.   10.55 μm CO_2 pump laser; P16 transition; perpendicular polarization.
4280.   Threshold pump power: 1000 mW.
4281.   Output power: 1 mW.
4282.   10.63 μm CO_2 pump laser; P24 transition; perpendicular polarization.
4283.   Threshold pump power: 2000 mW.
4284.   Output power: 0.1 mW.
4285.   10.30 μm CO_2 pump laser; R12 transition; perpendicular polarization.
4286.   Output power: 0.01 mW.
4287.   10.57 μm CO_2 pump laser; P18 transition; perpendicular polarization.
4288.   Output power: 0.1 mW.
4289.   9.28 μm CO_2 pump laser; R18 transition; perpendicular polarization.
4290.   Output power: 0.1 mW.
4291.   10.63 μm CO_2 pump laser; P24 transition; perpendicular polarization.
4292.   Output power: 0.1 mW.
4293.   9.49 μm CO_2 pump laser; P12 transition; perpendicular polarization.
4294.   Threshold pump power: 2000 mW.
4295.   Output power: 0.01 mW.
4296.   9.64 μm CO_2 pump laser; P30 transition; parallel polarization.
4297.   Output power: 0.1 mW.

4298.   9.26 μm CO_2 pump laser; R22 transition; parallel polarization.
4299.   Output power: 1 mW.
4300.   9.55 μm CO_2 pump laser; P20 transition; parallel polarization.
4301.   Threshold pump power: 1000 mW.
4302.   Output power: 1 mW.
4303.   10.30 μm CO_2 pump laser; R12 transition; parallel polarization.
4304.   Threshold pump power: 1000 mW.
4305.   Output power: 0.01 mW.
4306.   10.57 μm CO_2 pump laser; P18 transition; parallel polarization.
4307.   Output power: 0.01 mW.
4308.   10.19 μm CO_2 pump laser; R28 transition; parallel polarization.
4309.   Output power: 1 mW.
4310.   10.25 μm CO_2 pump laser; R20 transition; parallel polarization.
4311.   Threshold pump power: 1000 mW.
4312.   Output power: 0.01 mW.
4313.   10.67 μm CO_2 pump laser; P28 transition; perpendicular polarization.
4314.   Output power: 0.1 mW.
4315.   9.34 μm CO_2 pump laser; R08 transition; parallel polarization.
4316.   Output power: 0.1 mW.
4317.   10.46 μm CO_2 pump laser; P06 transition; parallel polarization.
4318.   Output power: 1 mW.
4319.   10.16 μm CO_2 pump laser; R34 transition; parallel polarization.
4320.   Output power: 0.1 mW.
4321.   10.33 μm CO_2 pump laser; R08 transition; parallel polarization.
4322.   Threshold pump power: 200 mW.
4323.   Output power: 1 mW.
4324.   10.63 μm CO_2 pump laser; P24 transition; perpendicular polarization.
4325.   Threshold pump power: 500 mW.
4326.   Output power: 0.01 mW.
4327.   10.46 μm CO_2 pump laser; P06 transition; parallel polarization.
4328.   Threshold pump power: 1000 mW.
4329.   For line assignments see References 1414, 1416, 1418, 1419, 1427, 1428, 1473.
4330.   Output power: 0.1 mW.
4331.   9.52 μm CO_2 pump laser; P16 transition; parallel polarization.
4332.   Threshold pump power: 2000 mW.
4333.   For line assignments see References 1414, 1416, 1418, 1419, 1427, 1428, 1473.
4334.   Output power: 0.1 mW.
4335.   9.68 μm CO_2 pump laser; P34 transition; perpendicular polarization.
4336.   For line assignments see References 1414, 1416, 1418, 1419, 1427, 1428, 1473.
4337.   Output power: 0.02 mW.
4338.   0 MHz offset from center of 10.27 μm CO_2 pump laser; R16 transition; parallel polarization.
4339.   Threshold pump power: 10000 mW.
4340.   For line assignments see References 1414, 1416, 1418, 1419, 1427, 1428, 1473.
4341.   Output power: 0.05 mW.
4342.   0 MHz offset from center of 9.59 μm CO_2 pump laser; P24 transition; parallel polarization.
4343.   Threshold pump power: 10000 mW.
4344.   For line assignments see References 1414, 1416, 1418, 1419, 1427, 1428, 1473.
4345.   Output power: 0.01 mW.
4346.   10.27 μm CO_2 pump laser; R16 transition; unspecified polarization.

4347. Threshold pump power: 2000 mW.
4348. For line assignments see References 1414, 1416, 1418, 1419, 1427, 1428, 1473.
4349. Output power: 0.01 mW.
4350. 9.71 μm CO_2 pump laser; P38 transition; unspecified polarization.
4351. Threshold pump power: 10000 mW.
4352. For line assignments see References 1414, 1416, 1418, 1419, 1427, 1428, 1473.
4353. Output power: 0.02 mW.
4354. 0 MHz offset from center of 9.59 μm CO_2 pump laser; P24 transition; parallel polarization.
4355. Threshold pump power: 10000 mW.
4356. For line assignments see References 1414, 1416, 1418, 1419, 1427, 1428, 1473.
4357. Output power: 0.1 mW.
4358. 0 MHz offset from center of 9.59 μm CO_2 pump laser; P24 transition; parallel polarization.
4359. Threshold pump power: 10000 mW.
4360. For line assignments see References 1414, 1416, 1418, 1419, 1427, 1428, 1473.
4361. Output power: 2 mW.
4362. 9.52 μm CO_2 pump laser; P16 transition; parallel polarization.
4363. For line assignments see References 1414, 1416, 1418, 1419, 1427, 1428, 1473.
4364. Output power: 0.01 mW.
4365. 10.37 μm CO_2 pump laser; R04 transition; perpendicular polarization.
4366. Threshold pump power: 2000 mW.
4367. For line assignments see References 1414, 1416, 1418, 1419, 1427, 1428, 1473.
4368. Output power: 5 mW.
4369. 9.68 μm CO_2 Sequence band pump laser; P31 transition; unspecified polarization.
4370. Threshold pump power: 2000 mW.
4371. For line assignments see References 1414, 1416, 1418, 1419, 1427, 1428, 1473.
4372. Output power: 0.2 mW.
4373. 10.14 μm CO_2 pump laser; R38 transition; unspecified polarization.
4374. Threshold pump power: 1000 mW.
4375. For line assignments see References 1414, 1416, 1418, 1419, 1427, 1428, 1473.
4376. Output power: 1 mW; cascade transition.
4377. 9.69 μm CO_2 pump laser; P36 transition; perpendicular polarization.
4378. Threshold pump power: 10000 mW.
4379. For line assignments see References 1414, 1416, 1418, 1419, 1427, 1428, 1473.
4380. Output power: 0.01 mW.
4381. -15 MHz offset from center of 9.50 μm CO_2 pump laser; P14 transition; unspecified polarization.
4382. Threshold pump power: 1000 mW.
4383. For line assignments see References 1414, 1416, 1418, 1419, 1427, 1428, 1473.
4384. Output power: 5 mW.
4385. 9.69 μm CO_2 pump laser; P36 transition; perpendicular polarization.
4386. Threshold pump power: 200 mW.
4387. For line assignments see References 1414, 1416, 1418, 1419, 1427, 1428, 1473.
4388. Output power: 0.5 mW.
4389. 9.52 μm CO_2 sequence band pump laser; P13 transition; unspecified polarization.
4390. Threshold pump power: 2000 mW.
4391. For line assignments see References 1414, 1416, 1418, 1419, 1427, 1428, 1473.
4392. Output power: 0.05 mW.
4393. -15 MHz offset from center of 9.50 μm CO_2 pump laser; P14 transition; parallel polarization.

4394.   Threshold pump power: 10000 mW.
4395.   For line assignments see References 1414, 1416, 1418, 1419, 1427, 1428, 1473.
4396.   Output power: 2 mW.
4397.   9.52 μm CO_2 pump laser; P16 transition; parallel polarization.
4398.   For line assignments see References 1414, 1416, 1418, 1419, 1427, 1428, 1473.
4399.   Output power: 0.05 mW.
4400.   9.57 μm CO_2 pump laser; P22 transition; parallel polarization.
4401.   Threshold pump power: 10000 mW.
4402.   For line assignments see References 1414, 1416, 1418, 1419, 1427, 1428, 1473.
4403.   Output power: 0.2 mW.
4404.   9.59 μm CO_2 pump laser; P24 transition; perpendicular polarization.
4405.   For line assignments see References 1414, 1416, 1418, 1419, 1427, 1428, 1473.
4406.   Output power: 0.05 mW.
4407.   -15 MHz offset from center of 9.50 μm CO_2 pump laser; P14 transition; perpendicular polarization.
4408.   Threshold pump power: 10000 mW.
4409.   For line assignments see References 1414, 1416, 1418, 1419, 1427, 1428, 1473.
4410.   Output power: 1 mW.
4411.   10.32 μm CO_2 pump laser; R10 transition; parallel polarization.
4412.   Threshold pump power: 10000 mW.
4413.   For line assignments see References 1414, 1416, 1418, 1419, 1427, 1428, 1473.
4414.   Output power: 0.1 mW.
4415.   9.68 μm CO_2 pump laser; P34 transition; perpendicular polarization.
4416.   For line assignments see References 1414, 1416, 1418, 1419, 1427, 1428, 1473.
4417.   Output power: 0.1 mW.
4418.   9.71 μm CO_2 pump laser; P38 transition; parallel polarization.
4419.   Threshold pump power: 10000 mW.
4420.   For line assignments see References 1414, 1416, 1418, 1419, 1427, 1428, 1473.
4421.   -5 MHz offset from center of 9.49 μm CO_2 pump laser; P12 transition; perpendicular polarization.
4422.   Threshold pump power: 10000 mW.
4423.   For line assignments see References 1414, 1416, 1418, 1419, 1427, 1428, 1473.
4424.   Output power: 1 mW.
4425.   10.09 μm CO_2 pump laser; R48 transition; unspecified polarization.
4426.   Threshold pump power: 500 mW.
4427.   For line assignments see References 1414, 1416, 1418, 1419, 1427, 1428, 1473.
4428.   Output power: 0.05 mW.
4429.   +5 MHz offset from center of 9.28 μm CO_2 pump laser; R18 transition; parallel polarization.
4430.   Threshold pump power: 10000 mW.
4431.   For line assignments see References 1414, 1416, 1418, 1419, 1427, 1428, 1473.
4432.   Output power: 0.1 mW.
4433.   9.71 μm CO_2 pump laser; P38 transition; parallel polarization.
4434.   Threshold pump power: 10000 mW.
4435.   For line assignments see References 1414, 1416, 1418, 1419, 1427, 1428, 1473.
4436.   Output power: 0.05 mW.
4437.   10.10 μm CO_2 pump laser; R46 transition; unspecified polarization.
4438.   Threshold pump power: 10000 mW.
4439.   For line assignments see References 1414, 1416, 1418, 1419, 1427, 1428, 1473.
4440.   Output power: 0.05 mW.
4441.   9.66 μm CO_2 pump laser; P32 transition; unspecified polarization.

4442. Threshold pump power: 20000 mW.

4443. For line assignments see References 1414, 1416, 1418, 1419, 1427, 1428, 1473.

4444. -10 MHz offset from center of 10.16 μm CO_2 pump laser; R34 transition; perpendicular polarization.

4445. Threshold pump power: 10000 mW.

4446. For line assignments see References 1414, 1416, 1418, 1419, 1427, 1428, 1473.

4447. Output power: 1 mW.

4448. 9.68 μm CO_2 pump laser; P34 transition; parallel polarization.

4449. Threshold pump power: 10000 mW.

4450. For line assignments see References 1414, 1416, 1418, 1419, 1427, 1428, 1473.

4451. Output power: 1 mW.

4452. 9.68 μm CO_2 pump laser; P34 transition; parallel polarization.

4453. For line assignments see References 1414, 1416, 1418, 1419, 1427, 1428, 1473.

4454. Output power: 0.1 mW.

4455. 9.38 μm CO_2 pump laser; R02 transition; perpendicular polarization.

4456. Threshold pump power: 1000 mW.

4457. For line assignments see References 1414, 1416, 1418, 1419, 1427, 1428, 1473.

4458. Output power: 1 mW.

4459. 9.68 μm CO_2 pump laser; P34 transition; parallel polarization.

4460. Threshold pump power: 10000 mW.

4461. For line assignments see References 1414, 1416, 1418, 1419, 1427, 1428, 1473.

4462. Output power: 0.5 mW.

4463. 9.68 μm CO_2 pump laser; P34 transition; parallel polarization.

4464. For line assignments see References 1414, 1416, 1418, 1419, 1427, 1428, 1473.

4465. Output power: 0.05 mW.

4466. +5 MHz offset from center of 9.28 μm CO_2 pump laser; R18 transition; parallel polarization.

4467. Threshold pump power: 10000 mW.

4468. For line assignments see References 1414, 1416, 1418, 1419, 1427, 1428, 1473.

4469. Output power: 0.5 mW.

4470. 10.14 μm CO_2 pump laser; R38 transition; unspecified polarization.

4471. Threshold pump power: 500 mW.

4472. For line assignments see References 1414, 1416, 1418, 1419, 1427, 1428, 1473.

4473. Output power: 0.01 mW.

4474. 10.11 μm CO_2 pump laser; R44 transition; unspecified polarization.

4475. Threshold pump power: 5000 mW.

4476. For line assignments see References 1414, 1416, 1418, 1419, 1427, 1428, 1473.

4477. Output power: 0.2 mW.

4478. 10.16 μm CO_2 pump laser; R3 transition; unspecified polarization.

4479. Threshold pump power: 5000 mW.

4480. For line assignments see References 1414, 1416, 1418, 1419, 1427, 1428, 1473.

4481. Output power: 2 mW.

4482. -45 MHz offset from center of 10.17 μm CO_2 pump laser; R32 transition; parallel polarization.

4483. Threshold pump power: 20000 mW.

4484. For line assignments see References 1414, 1416, 1418, 1419, 1427, 1428, 1473.

4485. Output power: 1 mW.

4486. 9.20 μm CO_2 pump laser; R transition; parallel polarization.

4487. For line assignments see References 1414, 1416, 1418, 1419, 1427, 1428, 1473.

4488. Output power: 0.2 mW.

4489.   -10 MHz offset from center of 10.16 μm CO_2 pump laser; R34 transition; parallel polarization.

4490.   Threshold pump power: 10000 mW.

4491.   For line assignments see References 1414, 1416, 1418, 1419, 1427, 1428, 1473.

4492.   Output power: 0.5 mW.

4493.   9.68 μm CO_2 pump laser; P34 transition; parallel polarization.

4494.   For line assignments see References 1414, 1416, 1418, 1419, 1427, 1428, 1473.

4495.   Output power: 0.05 mW.

4496.   9.33 μm CO_2 pump laser; R10 transition; parallel polarization.

4497.   For line assignments see References 1414, 1416, 1418, 1419, 1427, 1428, 1473.

4498.   Output power: 0.02 mW.

4499.   +15 MHz offset from center of 9.26 μm CO_2 pump laser; R22 transition; parallel polarization.

4500.   Threshold pump power: 10000 mW.

4501.   For line assignments see References 1414, 1416, 1418, 1419, 1427, 1428, 1473.

4502.   Output power: 0.2 mW.

4503.   9.20 μm CO_2 pump laser; R transition; parallel polarization.

4504.   For line assignments see References 1414, 1416, 1418, 1419, 1427, 1428, 1473.

4505.   Output power: 0.02 mW.

4506.   9.34 μm CO_2 pump laser; R08 transition; parallel polarization.

4507.   Threshold pump power: 10000 mW.

4508.   For line assignments see References 1414, 1416, 1418, 1419, 1427, 1428, 1473.

4509.   Output power: 1 mW.

4510.   9.52 μm CO_2 pump laser; P16 transition; parallel polarization.

4511.   For line assignments see References 1414, 1416, 1418, 1419, 1427, 1428, 1473.

4512.   Output power: 0.002 mW.

4513.   -20 MHz offset from center of 9.47 μm CO_2 pump laser; P10 transition; parallel polarization.

4514.   Threshold pump power: 10000 mW.

4515.   For line assignments see References 1414, 1416, 1418, 1419, 1427, 1428, 1473.

4516.   Output power: 0.002 mW.

4517.   -20 MHz offset from center of 9.47 μm CO_2 pump laser; P10 transition; parallel polarization.

4518.   Threshold pump power: 10000 mW.

4519.   For line assignments see References 1414, 1416, 1418, 1419, 1427, 1428, 1473.

4520.   Output power: 0.1 mW.

4521.   9.57 μm CO_2 pump laser; P22 transition; parallel polarization.

4522.   Threshold pump power: 10000 mW.

4523.   For line assignments see References 1414, 1416, 1418, 1419, 1427, 1428, 1473.

4524.   -5 MHz offset from center of 9.49 μm CO_2 pump laser; P12 transition; parallel polarization.

4525.   Threshold pump power: 10000 mW.

4526.   For line assignments see References 1414, 1416, 1418, 1419, 1427, 1428, 1473.

4527.   Output power: 0.2 mW.

4528.   10.37 μm CO_2 pump laser; R04 transition; parallel polarization.

4529.   Threshold pump power: 2000 mW.

4530.   For line assignments see References 1414, 1416, 1418, 1419, 1427, 1428, 1473.

4531.   Output power: 0.05 mW.

4532.   9.31 μm CO_2 pump laser; R14 transition; parallel polarization.

4533.   Threshold pump power: 10000 mW.

4534.   For line assignments see References 1414, 1416, 1418, 1419, 1427, 1428, 1473.

4535. +140 MHz offset from center of 9.68 μm CO_2 pump laser; P34 transition; either parallel or perpendicular polarization.
4536. For line assignments see References 1414, 1416, 1418, 1419, 1427, 1428, 1473.
4537. Output power: 0.005 mW.
4538. -5 MHz offset from center of 9.49 μm CO_2 pump laser; P12 transition; parallel polarization.
4539. Threshold pump power: 10000 mW.
4540. For line assignments see References 1414, 1416, 1418, 1419, 1427, 1428, 1473.
4541. +120 MHz offset from center of 9.68 μm CO_2 pump laser; P34 transition; either parallel or perpendicular polarization.
4542. For line assignments see References 1414, 1416, 1418, 1419, 1427, 1428, 1473.
4543. Output power: 1 mW.
4544. 9.69 μm CO_2 pump laser; P36 transition; parallel polarization.
4545. For line assignments see References 1414, 1416, 1418, 1419, 1427, 1428, 1473.
4546. Output power: 0.5 mW.
4547. 9.71 μm CO_2 pump laser; P38 transition; perpendicular polarization.
4548. Threshold pump power: 10000 mW.
4549. For line assignments see References 1414, 1416, 1418, 1419, 1427, 1428, 1473.
4550. Output power: 0.1 mW.
4551. 9.31 μm CO_2 pump laser; R14 transition; perpendicular polarization.
4552. Threshold pump power: 2000 mW.
4553. For line assignments see References 1414, 1416, 1418, 1419, 1427, 1428, 1473.
4554. Output power: 0.1 mW.
4555. 9.71 μm CO_2 pump laser; P38 transition; perpendicular polarization.
4556. Threshold pump power: 10000 mW.
4557. For line assignments see References 1414, 1416, 1418, 1419, 1427, 1428, 1473.
4558. -15 MHz offset from center of 10.32 μm CO_2 pump laser; R10 transition; parallel polarization.
4559. Threshold pump power: 10000 mW.
4560. For line assignments see References 1414, 1416, 1418, 1419, 1427, 1428, 1473.
4561. Output power: 0.1 mW.
4562. 9.20 μm CO_2 pump laser; R transition; parallel polarization.
4563. For line assignments see References 1414, 1416, 1418, 1419, 1427, 1428, 1473.
4564. Output power: 0.1 mW.
4565. 9.68 μm CO_2 pump laser; P34 transition; perpendicular polarization.
4566. Threshold pump power: 10000 mW.
4567. For line assignments see References 1414, 1416, 1418, 1419, 1427, 1428, 1473.
4568. Output power: 0.2 mW.
4569. +5 MHz offset from center of 9.28 μm CO_2 pump laser; R18 transition; parallel polarization.
4570. Threshold pump power: 10000 mW.
4571. For line assignments see References 1414, 1416, 1418, 1419, 1427, 1428, 1473.
4572. Output power: 0.2 mW.
4573. 9.20 μm CO_2 pump laser; R transition; parallel polarization.
4574. For line assignments see References 1414, 1416, 1418, 1419, 1427, 1428, 1473.
4575. Output power: 0.2 mW.
4576. 9.68 μm CO_2 pump laser; P34 transition; perpendicular polarization.
4577. Threshold pump power: 10000 mW.
4578. For line assignments see References 1414, 1416, 1418, 1419, 1427, 1428, 1473.
4579. Output power: 1 mW.
4580. 9.68 μm CO_2 pump laser; P34 transition; perpendicular polarization.

4581.   For line assignments see References 1414, 1416, 1418, 1419, 1427, 1428, 1473.
4582.   Output power: 0.2 mW.
4583.   10.37 μm CO_2 pump laser; R04 transition; parallel polarization.
4584.   Threshold pump power: 10000 mW.
4585.   For line assignments see References 1414, 1416, 1418, 1419, 1427, 1428, 1473.
4586.   Output power: 0.005 mW.
4587.   10.38 μm CO_2 pump laser; R02 transition; parallel polarization.
4588.   Threshold pump power: 2000 mW.
4589.   For line assignments see References 1414, 1416, 1418, 1419, 1427, 1428, 1473.
4590.   Output power: 0.1 mW.
4591.   9.38 μm CO_2 pump laser; R02 transition; perpendicular polarization.
4592.   Threshold pump power: 1000 mW.
4593.   For line assignments see References 1414, 1416, 1418, 1419, 1427, 1428, 1473.
4594.   Output power: 0.5 mW.
4595.   9.59 μm CO_2 Sequence band pump laser; P21 transition; unspecified polarization.
4596.   Threshold pump power: 2000 mW.
4597.   For line assignments see References 1414, 1416, 1418, 1419, 1427, 1428, 1473.
4598.   Output power: 10 mW.
4599.   9.69 μm CO_2 pump laser; P36 transition; parallel polarization.
4600.   Threshold pump power: 200 mW.
4601.   For line assignments see References 1414, 1416, 1418, 1419, 1427, 1428, 1473.
4602.   Output power: 0.1 mW.
4603.   10.13 μm CO_2 pump laser; R40 transition; parallel polarization.
4604.   Threshold pump power: 10000 mW.
4605.   For line assignments see References 1414, 1416, 1418, 1419, 1427, 1428, 1473.
4606.   Output power: 10 mW.
4607.   9.33 μm CO_2 pump laser; R10 transition; perpendicular polarization.
4608.   Threshold pump power: 1000 mW.
4609.   For line assignments see References 1414, 1416, 1418, 1419, 1427, 1428, 1473.
4610.   Output power: 0.1 mW.
4611.   10 MHz offset from center of 9.59 μm CO_2 pump laser; P24 transition; perpendicular polarization.
4612.   Threshold pump power: 10000 mW.
4613.   For line assignments see References 1414, 1416, 1418, 1419, 1427, 1428, 1473.
4614.   Output power: 1 mW.
4615.   9.52 μm CO_2 pump laser; P16 transition; perpendicular polarization.
4616.   Threshold pump power: 10000 mW.
4617.   For line assignments see References 1414, 1416, 1418, 1419, 1427, 1428, 1473.
4618.   Output power: 0.2 mW.
4619.   9.50 μm CO_2 pump laser; P14 transition; perpendicular polarization.
4620.   Threshold pump power: 10000 mW.
4621.   For line assignments see References 1414, 1416, 1418, 1419, 1427, 1428, 1473.
4622.   Output power: 0.05 mW.
4623.   9.50 μm CO_2 pump laser; P14 transition; unspecified polarization.
4624.   Threshold pump power: 2000 mW.
4625.   For line assignments see References 1414, 1416, 1418, 1419, 1427, 1428, 1473.
4626.   Output power: 0.5 mW.
4627.   10.09 μm CO_2 pump laser; R48 transition; parallel polarization.
4628.   Threshold pump power: 20000 mW.
4629.   For line assignments see References 1414, 1416, 1418, 1419, 1427, 1428, 1473.
4630.   Output power: 1 mW.

4631. 10.14 μm CO_2 pump laser; R38 transition; unspecified polarization.
4632. Threshold pump power: 500 mW.
4633. For line assignments see References 1414, 1416, 1418, 1419, 1427, 1428, 1473.
4634. Output power: 1 mW.
4635. 9.24 μm CO_2 pump laser; R26 transition; parallel polarization.
4636. Threshold pump power: 10000 mW.
4637. For line assignments see References 1414, 1416, 1418, 1419, 1427, 1428, 1473.
4638. Output power: 1 mW.
4639. 9.68 μm CO_2 Sequence band pump laser; P31 transition; unspecified polarization.
4640. Threshold pump power: 2000 mW.
4641. For line assignments see References 1414, 1416, 1418, 1419, 1427, 1428, 1473.
4642. Output power: 0.1 mW.
4643. 9.38 μm CO_2 pump laser; R02 transition; perpendicular polarization.
4644. Threshold pump power: 500 mW.
4645. For line assignments see References 1414, 1416, 1418, 1419, 1427, 1428, 1473.
4646. Output power: 1 mW.
4647. 9.24 μm CO_2 pump laser; R26 transition; parallel polarization.
4648. Threshold pump power: 10000 mW.
4649. For line assignments see References 1414, 1416, 1418, 1419, 1427, 1428, 1473.
4650. Output power: 2 mW.
4651. -45 MHz offset from center of 10.17 μm CO_2 pump laser; R32 transition; perpendicular polarization.
4652. Threshold pump power: 20000 mW.
4653. For line assignments see References 1414, 1416, 1418, 1419, 1427, 1428, 1473.
4654. Output power: 1 mW.
4655. 0 MHz offset from center of 9.59 μm CO_2 pump laser; P24 transition; parallel polarization.
4656. For line assignments see References 1414, 1416, 1418, 1419, 1427, 1428, 1473.
4657. Output power: 0.1 mW.
4658. -10 MHz offset from center of 10.16 μm CO_2 pump laser; R34 transition; parallel polarization.
4659. Threshold pump power: 1000 mW.
4660. For line assignments see References 1414, 1416, 1418, 1419, 1427, 1428, 1473.
4661. Output power: 0.01 mW.
4662. 10.11 μm CO_2 pump laser; R44 transition; unspecified polarization.
4663. Threshold pump power: 10000 mW.
4664. For line assignments see References 1414, 1416, 1418, 1419, 1427, 1428, 1473.
4665. Output power: 20 mW.
4666. 9.69 μm CO_2 pump laser; P36 transition; unspecified polarization.
4667. Threshold pump power: 200 mW.
4668. For line assignments see References 1414, 1416, 1418, 1419, 1427, 1428, 1473.
4669. Output power: 0.2 mW.
4670. -15 MHz offset from center of 9.50 μm CO_2 pump laser; P14 transition; parallel polarization.
4671. Threshold pump power: 10000 mW.
4672. For line assignments see References 1414, 1416, 1418, 1419, 1427, 1428, 1473.
4673. Output power: 0.02 mW.
4674. 9.34 μm CO_2 pump laser; R08 transition; perpendicular polarization.
4675. Threshold pump power: 10000 mW.
4676. For line assignments see References 1414, 1416, 1418, 1419, 1427, 1428, 1473.
4677. Output power: 0.2 mW.

4678. 9.31 μm CO_2 pump laser; R14 transition; parallel polarization.
4679. Threshold pump power: 10000 mW.
4680. For line assignments see References 1414, 1416, 1418, 1419, 1427, 1428, 1473.
4681. Output power: 0.05 mW.
4682. 10.13 μm CO_2 pump laser; R40 transition; unspecified polarization.
4683. Threshold pump power: 10000 mW.
4684. For line assignments see References 1414, 1416, 1418, 1419, 1427, 1428, 1473.
4685. Output power: 0.2 mW.
4686. 0 MHz offset from center of 10.13 μm CO_2 pump laser; R40 transition; perpendicular polarization.
4687. Threshold pump power: 10000 mW.
4688. For line assignments see References 1414, 1416, 1418, 1419, 1427, 1428, 1473.
4689. Output power: 20 mW.
4690. 9.33 μm CO_2 pump laser; R10 transition; parallel polarization.
4691. For line assignments see References 1414, 1416, 1418, 1419, 1427, 1428, 1473.
4692. Output power: 0.02 mW.
4693. -10 MHz offset from center of 10.16 μm CO_2 pump laser; R34 transition; parallel polarization.
4694. Threshold pump power: 10000 mW.
4695. For line assignments see References 1414, 1416, 1418, 1419, 1427, 1428, 1473.
4696. Output power: 0.1 mW.
4697. 0 MHz offset from center of 9.59 μm CO_2 pump laser; P24 transition; parallel polarization.
4698. Threshold pump power: 10000 mW.
4699. For line assignments see References 1414, 1416, 1418, 1419, 1427, 1428, 1473.
4700. Output power: 0.02 mW.
4701. 9.34 μm CO_2 pump laser; R08 transition; parallel polarization.
4702. Threshold pump power: 10000 mW.
4703. For line assignments see References 1414, 1416, 1418, 1419, 1427, 1428, 1473.
4704. Output power: 0.002 mW.
4705. -10 MHz offset from center of 9.73 μm CO_2 pump laser; P40 transition; parallel polarization.
4706. Threshold pump power: 10000 mW.
4707. For line assignments see References 1414, 1416, 1418, 1419, 1427, 1428, 1473.
4708. Output power: 10 mW.
4709. 9.59 μm CO_2 sequence band pump laser; P21 transition; unspecified polarization.
4710. Threshold pump power: 2000 mW.
4711. For line assignments see References 1414, 1416, 1418, 1419, 1427, 1428, 1473.
4712. Output power: 0.1 mW.
4713. 9.68 μm CO_2 pump laser; P34 transition; unspecified polarization.
4714. Threshold pump power: 1000 mW.
4715. For line assignments see References 1414, 1416, 1418, 1419, 1427, 1428, 1473.
4716. Output power: 0.2 mW.
4717. 0 MHz offset from center of 10.27 μm CO_2 pump laser; R16 transition; parallel polarization.
4718. Threshold pump power: 10000 mW.
4719. For line assignments see References 1414, 1416, 1418, 1419, 1427, 1428, 1473.
4720. Output power: 0.1 mW.
4721. 9.28 μm CO_2 pump laser; R08 transition; parallel polarization.
4722. Threshold pump power: 10000 mW.
4723. For line assignments see References 1414, 1416, 1418, 1419, 1427, 1428, 1473.

4724. Output power: 0.002 mW.
4725. -10 MHz offset from center of 9.73 μm CO_2 pump laser; P40 transition; parallel polarization.
4726. Threshold pump power: 10000 mW.
4727. For line assignments see References 1414, 1416, 1418, 1419, 1427, 1428, 1473.
4728. Output power: 50 mW.
4729. +26 MHz offset from center of 9.68 μm CO_2 pump laser; P34 transition; perpendicular polarization.
4730. For line assignments see References 1414, 1416, 1418, 1419, 1427, 1428, 1473.
4731. Output power: 0.1 mW.
4732. 0 MHz offset from center of 10.27 μm CO_2 pump laser; R16 transition; perpendicular polarization.
4733. Threshold pump power: 10000 mW.
4734. For line assignments see References 1414, 1416, 1418, 1419, 1427, 1428, 1473.
4735. Output power: 0.01 mW.
4736. 9.28 μm CO_2 pump laser; R18 transition; perpendicular polarization.
4737. Threshold pump power: 10000 mW.
4738. For line assignments see References 1414, 1416, 1418, 1419, 1427, 1428, 1473.
4739. Output power: 0.5 mW.
4740. 9.68 μm CO_2 pump laser; P34 transition; unspecified polarization.
4741. Threshold pump power: 1000 mW.
4742. For line assignments see References 1414, 1416, 1418, 1419, 1427, 1428, 1473.
4743. Output power: 0.2 mW.
4744. 10.27 μm CO_2 pump laser; R16 transition; parallel polarization.
4745. Threshold pump power: 10000 mW.
4746. For line assignments see References 1414, 1416, 1418, 1419, 1427, 1428, 1473.
4747. Output power: 0.05 mW.
4748. 9.28 μm CO_2 pump laser; R18 transition; perpendicular polarization.
4749. Threshold pump power: 10000 mW.
4750. For line assignments see References 1414, 1416, 1418, 1419, 1427, 1428, 1473.
4751. Output power: 0.01 mW.
4752. -10 MHz offset from center of 9.73 μm CO_2 pump laser; P40 transition; parallel polarization.
4753. Threshold pump power: 10000 mW.
4754. For line assignments see References 1414, 1416, 1418, 1419, 1427, 1428, 1473.
4755. Output power: 0.02 mW.
4756. -10 MHz offset from center of 9.73 μm CO_2 pump laser; P40 transition; perpendicular polarization.
4757. Threshold pump power: 10000 mW.
4758. For line assignments see References 1414, 1416, 1418, 1419, 1427, 1428, 1473.
4759. Output power: 2 mW.
4760. +23 MHz offset from center of 10.15 μm CO_2 pump laser; R36 transition; perpendicular polarization.
4761. Threshold pump power: 20000 mW.
4762. For line assignments see References 1414, 1416, 1418, 1419, 1427, 1428, 1473.
4763. -10 MHz offset from center of 10.16 μm CO_2 pump laser; R34 transition; parallel polarization.
4764. Threshold pump power: 10000 mW.
4765. For line assignments see References 1414, 1416, 1418, 1419, 1427, 1428, 1473.
4766. Output power: 2 mW.

4767.   +23 MHz offset from center of 10.15 μm CO_2 pump laser; R36 transition; unspecified polarization.

4768.   Threshold pump power: 20000 mW.

4769.   For line assignments see References 1414, 1416, 1418, 1419, 1427, 1428, 1473.

4770.   Output power: 0.1 mW.

4771.   +38 MHz offset from center of 9.68 μm CO_2 pump laser; P34 transition; unspecified polarization.

4772.   Threshold pump power: 1000 mW.

4773.   For line assignments see References 1414, 1416, 1418, 1419, 1427, 1428, 1473.

4774.   Output power: 10 mW.

4775.   -16 MHz offset from center of 9.66 μm CO_2 pump laser; P32 transition; parallel polarization.

4776.   Threshold pump power: 1000 mW.

4777.   For line assignments see References 1414, 1416, 1418, 1419, 1427, 1428, 1473.

4778.   Output power: 1 mW.

4779.   10 MHz offset from center of 9.68 μm CO_2 pump laser; P34 transition; perpendicular polarization.

4780.   Threshold pump power: 1000 mW.

4781.   For line assignments see References 1414, 1416, 1418, 1419, 1427, 1428, 1473.

4782.   Output power: 1 mW.

4783.   -16 MHz offset from center of 9.66 μm CO_2 pump laser; P32 transition; unspecified polarization.

4784.   Threshold pump power: 1000 mW.

4785.   For line assignments see References 1414, 1416, 1418, 1419, 1427, 1428, 1473.

4786.   Output power: 0.002 mW.

4787.   9.50 μm CO_2 pump laser; P14 transition; unspecified polarization.

4788.   Threshold pump power: 10000 mW.

4789.   For line assignments see References 1415, 1417.

4790.   +25 MHz offset from center of 9.49 μm CO_2 pump laser; P12 transition; parallel polarization.

4791.   Threshold pump power: 10000 mW.

4792.   For line assignments see References 1415, 1417.

4793.   Output power: 2 mW.

4794.   -20 MHz offset from center of 9.49 μm CO_2 pump laser; P12 transition; perpendicular polarization.

4795.   Threshold pump power: 10000 mW.

4796.   For line assignments see References 1415, 1417.

4797.   +35 MHz offset from center of 9.73 μm CO_2 pump laser; P40 transition; perpendicular polarization.

4798.   Threshold pump power: 10000 mW.

4799.   For line assignments see References 1415, 1417.

4800.   +15 MHz offset from center of 9.57 μm CO_2 pump laser; P22 transition; parallel polarization.

4801.   Threshold pump power: 10000 mW.

4802.   For line assignments see References 1415, 1417.

4803.   10.27 μm CO_2 pump laser; R16 transition; unspecified polarization.

4804.   For line assignments see References 1415, 1417.

4805.   -20 MHz offset from center of 9.69 μm CO_2 pump laser; P36 transition; parallel polarization.

4806.   Threshold pump power: 10000 mW.

4807.   For line assignments see References 1415, 1417.

4808. Output power: 0.5 mW.
4809. -15 MHz offset from center of 9.57 μm CO_2 pump laser; P22 transition; parallel polarization.
4810. Threshold pump power: 10000 mW.
4811. For line assignments see References 1415, 1417.
4812. -20 MHz offset from center of 9.69 μm CO_2 pump laser; P36 transition; parallel polarization.
4813. Threshold pump power: 10000 mW.
4814. For line assignments see References 1415, 1417.
4815. 10.27 μm CO_2 pump laser; R16 transition; unspecified polarization.
4816. For line assignments see References 1415, 1417.
4817. 10.27 μm CO_2 pump laser; R16 transition; unspecified polarization.
4818. For line assignments see References 1415, 1417.
4819. Output power: 0.5 mW.
4820. +20 MHz offset from center of 10.27 μm CO_2 pump laser; R16 transition; parallel polarization.
4821. Threshold pump power: 10000 mW.
4822. For line assignments see References 1415, 1417.
4823. Output power: 2 mW.
4824. -20 MHz offset from center of 9.49 μm CO_2 pump laser; P12 transition; parallel polarization.
4825. Threshold pump power: 10000 mW.
4826. For line assignments see References 1415, 1417.
4827. +25 MHz offset from center of 9.49 μm CO_2 pump laser; P12 transition; parallel polarization.
4828. Threshold pump power: 10000 mW.
4829. For line assignments see References 1415, 1417.
4830. 9.49 μm CO_2 pump laser; P10 transition; unspecified polarization.
4831. For line assignments see References 1415, 1417.
4832. +25 MHz offset from center of 9.47 μm CO_2 pump laser; P10 transition; either parallel or perpendicular polarization.
4833. Threshold pump power: 10000 mW.
4834. For line assignments see References 1415, 1417.
4835. Output power: 2 mW.
4836. +20 MHz offset from center of 10.27 μm CO_2 pump laser; R16 transition; either parallel or perpendicular polarization.
4837. Threshold pump power: 10000 mW.
4838. For line assignments see References 1415, 1417.
4839. Output power: 1 mW.
4840. +25 MHz offset from center of 10.26 μm CO_2 pump laser; R18 transition; perpendicular polarization.
4841. Threshold pump power: 10000 mW.
4842. For line assignments see References 1415, 1417.
4843. +35 MHz offset from center of 9.73 μm CO_2 pump laser; P40 transition; parallel polarization.
4844. Threshold pump power: 10000 mW.
4845. For line assignments see References 1415, 1417.
4846. Output power: 1 mW.
4847. -20 MHz offset from center of 9.49 μm CO_2 pump laser; P12 transition; parallel polarization.
4848. Threshold pump power: 10000 mW.

4849.   For line assignments see References 1415, 1417.

4850.   10.27 μm CO_2 pump laser; R16 transition; unspecified polarization.

4851.   For line assignments see References 1415, 1417.

4852.   Output power: 0.1 mW.

4853.   +15 MHz offset from center of 9.57 μm CO_2 pump laser; P22 transition; perpendicular polarization.

4854.   Threshold pump power: 10000 mW.

4855.   For line assignments see References 1415, 1417.

4856.   10.27 μm CO_2 pump laser; R16 transition; unspecified polarization.

4857.   For line assignments see References 1415, 1417.

4858.   -25 MHz offset from center of 9.64 μm CO_2 pump laser; P30 transition; perpendicular polarization.

4859.   Threshold pump power: 10 W.

4860.   For line assignments see References 1415, 1417.

4861.   +25 MHz offset from center of 9.47 μm CO_2 pump laser; P10 transition; perpendicular polarization.

4862.   Threshold pump power: 10000 mW.

4863.   For line assignments see References 1415, 1417.

4864.   +40 MHz offset from center of 10.55 μm CO_2 pump laser; P16 transition; parallel polarization.

4865.   Threshold pump power: 10000 mW.

4866.   For line assignments see References 1415, 1417.

4867.   -5 MHz offset from center of 10.19 μm CO_2 pump laser; R28 transition; parallel polarization.

4868.   Threshold pump power: 10000 mW.

4869.   For line assignments see References 1415, 1417.

4870.   Output power: 2 mW.

4871.   -15 MHz offset from center of 9.57 μm CO_2 pump laser; P22 transition; perpendicular polarization.

4872.   Threshold pump power: 10000 mW.

4873.   For line assignments see References 1415, 1417.

4874.   Output power: 1 mW.

4875.   +20 MHz offset from center of 10.27 μm CO_2 pump laser; R16 transition; parallel polarization.

4876.   Threshold pump power: 10000 mW.

4877.   For line assignments see References 1415, 1417.

4878.   Output power: 2 mW.

4879.   +25 MHz offset from center of 10.26 μm CO_2 pump laser; R18 transition; either parallel or perpendicular polarization.

4880.   Threshold pump power: 10000 mW.

4881.   For line assignments see References 1415, 1417.

4882.   Output power: 1 mW.

4883.   +25 MHz offset from center of 10.26 μm CO_2 pump laser; R18 transition; either parallel or perpendicular polarization.

4884.   Threshold pump power: 10000 mW.

4885.   For line assignments see References 1415, 1417.

4886.   +40 MHz offset from center of 10.21 μm CO_2 pump laser; R26 transition; either parallel or perpendicular polarization.

4887.   Threshold pump power: 10000 mW.

4888.   For line assignments see References 1415, 1417.

4889.   Output power: 1 mW.

4890.   +15 MHz offset from center of 9.57 μm CO_2 pump laser; P22 transition; parallel polarization.
4891.   Threshold pump power: 10000 mW.
4892.   For line assignments see References 1415, 1417.
4893.   9.46 μm CO_2 pump laser; P08 transition; perpendicular polarization.
4894.   Threshold pump power: 10000 mW.
4895.   For line assignments see References 1415, 1417.
4896.   +25 MHz offset from center of 9.47 μm CO_2 pump laser; P10 transition; parallel polarization.
4897.   Threshold pump power: 10000 mW.
4898.   For line assignments see References 1415, 1417.
4899.   -5 MHz offset from center of 10.19 μm CO_2 pump laser; R28 transition; perpendicular polarization.
4900.   Threshold pump power: 10000 mW.
4901.   For line assignments see References 1415, 1417.
4902.   Output power: 1 mW.
4903.   -15 MHz offset from center of 9.57 μm CO_2 pump laser; P22 transition; parallel polarization.
4904.   Threshold pump power: 10000 mW.
4905.   For line assignments see References 1415, 1417.
4906.   +40 MHz offset from center of 10.21 μm CO_2 pump laser; R26 transition; parallel polarization.
4907.   Threshold pump power: 10000 mW.
4908.   For line assignments see References 1415, 1417.
4909.   +25 MHz offset from center of 9.49 μm CO_2 pump laser; P12 transition; parallel polarization.
4910.   Threshold pump power: 10000 mW.
4911.   For line assignments see References 1415, 1417.
4912.   -10 MHz offset from center of 10.55 μm CO_2 pump laser; P16 transition; perpendicular polarization.
4913.   Threshold pump power: 10000 mW.
4914.   For line assignments see References 1415, 1417.
4915.   10.23 μm CO_2 pump laser; R22 transition; parallel polarization.
4916.   Threshold pump power: 10000 mW.
4917.   Output power: 0.05 mW.
4918.   9.54 μm CO_2 pump laser; P18 transition; parallel polarization.
4919.   Threshold pump power: 20000 mW.
4920.   Output power: 0.02 mW.
4921.   9.25 μm CO_2 pump laser; R24 transition; perpendicular polarization.
4922.   Threshold pump power: 20000 mW.
4923.   Output power: 1 mW.
4924.   9.60 μm CO_2 pump laser; P26 transition; parallel polarization.
4925.   Threshold pump power: 20000 mW.
4926.   Output power: 0.2 mW.
4927.   9.73 μm CO_2 pump laser; P40 transition; parallel polarization.
4928.   Threshold pump power: 10000 mW.
4929.   Output power: 0.005 mW.
4930.   10.88 μm CO_2 pump laser; P46 transition; unspecified polarization.
4931.   Threshold pump power: 2000 mW.
4932.   Output power: 0.5 mW.
4933.   9.60 μm CO_2 pump laser; P26 transition; parallel polarization.

4934.   Threshold pump power: 20000 mW.

4935.   Output power: 0.5 mW.

4936.   9.79 μm CO_2 pump laser; P46 transition; perpendicular polarization.

4937.   Threshold pump power: 2000 mW.

4938.   Output power: 0.1 mW.

4939.   9.66 μm CO_2 pump laser; P32 transition; perpendicular polarization.

4940.   Threshold pump power: 20000 mW.

4941.   Output power: 0.005 mW.

4942.   10.76 μm CO_2 pump laser; P36 transition; perpendicular polarization.

4943.   Threshold pump power: 10000 mW.

4944.   Output power: 0.02 mW.

4945.   9.34 μm CO_2 pump laser; R08 transition; perpendicular polarization.

4946.   Threshold pump power: 10000 mW.

4947.   Output power: 0.05 mW.

4948.   9.54 μm CO_2 pump laser; P18 transition; parallel polarization.

4949.   Threshold pump power: 2000 mW.

4950.   Output power: 0.2 mW.

4951.   9.73 μm CO_2 pump laser; P40 transition; parallel polarization.

4952.   Threshold pump power: 10000 mW.

4953.   Output power: 0.1 mW.

4954.   10.88 μm CO_2 pump laser; P46 transition; parallel polarization.

4955.   Threshold pump power: 2000 mW.

4956.   Output power: 1 mW.

4957.   10.17 μm CO_2 pump laser; R32 transition; parallel polarization.

4958.   Threshold pump power: 20000 mW.

4959.   Output power: 0.5 mW.

4960.   9.69 μm CO_2 pump laser; P36 transition; parallel polarization.

4961.   Threshold pump power: 10000 mW.

4962.   Output power: 0.5 mW.

4963.   9.49 μm CO_2 pump laser; P12 transition; parallel polarization.

4964.   Threshold pump power: 10000 mW.

4965.   Output power: 0.5 mW.

4966.   10.16 μm CO_2 pump laser; R34 transition; unspecified polarization.

4967.   Threshold pump power: 10000 mW.

4968.   Output power: 0.5 mW.

4969.   9.50 μm CO_2 pump laser; P14 transition; parallel polarization.

4970.   Threshold pump power: 10000 mW.

4971.   Output power: 0.01 mW.

4972.   10.27 μm CO_2 pump laser; R16 transition; parallel polarization.

4973.   Threshold pump power: 20000 mW.

4974.   Output power: 1 mW.

4975.   10.16 μm CO_2 pump laser; R34 transition; parallel polarization.

4976.   Threshold pump power: 20000 mW.

4977.   Output power: 0.5 mW.

4978.   9.47 μm CO_2 pump laser; P10 transition; parallel polarization.

4979.   Threshold pump power: 10000 mW.

4980.   Output power: 0.02 mW.

4981.   9.44 μm CO_2 pump laser; P06 transition; parallel polarization.

4982.   Threshold pump power: 10000 mW.

4983.   Output power: 1 mW.

4984.   9.25 μm CO_2 pump laser; R24 transition; parallel polarization.

4985. Threshold pump power: 20000 mW.
4986. Output power: 0.1 mW.
4987. 9.66 μm CO_2 pump laser; P32 transition; parallel polarization.
4988. Threshold pump power: 20000 mW.
4989. Output power: 0.5 mW.
4990. 10.74 μm CO_2 pump laser; P34 transition; parallel polarization.
4991. Threshold pump power: 10000 mW.
4992. Output power: 1 mW.
4993. 10.74 μm CO_2 pump laser; P34 transition; perpendicular polarization.
4994. Threshold pump power: 10000 mW.
4995. Output power: 0.5 mW.
4996. 9.79 μm CO_2 pump laser; P46 transition; perpendicular polarization.
4997. Threshold pump power: 10000 mW.
4998. Output power: 0.05 mW.
4999. 10.76 μm CO_2 pump laser; P36 transition; parallel polarization.
5000. Threshold pump power: 10000 mW.
5001. Output power: 0.2 mW.
5002. 9.25 μm CO_2 pump laser; R24 transition; parallel polarization.
5003. Threshold pump power: 20000 mW.
5004. Output power: 0.02 mW.
5005. 9.26 μm CO_2 pump laser; R22 transition; parallel polarization.
5006. Threshold pump power: 20000 mW.
5007. Output power: 0.05 mW.
5008. 9.29 μm CO_2 pump laser; R16 transition; parallel polarization.
5009. Threshold pump power: 20000 mW.
5010. 10.27 μm CO_2 pump laser; R16 transition; parallel polarization.
5011. Threshold pump power: 20000 mW.
5012. Output power: 0.5 mW.
5013. 9.50 μm CO_2 pump laser; P14 transition; parallel polarization.
5014. Threshold pump power: 10000 mW.
5015. Output power: 0.02 mW.
5016. 9.71 μm CO_2 pump laser; P38 transition; perpendicular polarization.
5017. Threshold pump power: 10000 mW.
5018. Output power: 0.02 mW.
5019. 10.67 μm CO_2 pump laser; P28 transition; parallel polarization.
5020. Threshold pump power: 20000 mW.
5021. Output power: 0.5 mW.
5022. 9.69 μm CO_2 pump laser; P36 transition; parallel polarization.
5023. Threshold pump power: 10000 mW.
5024. Output power: 0.02 mW.
5025. 10.67 μm CO_2 pump laser; P28 transition; parallel polarization.
5026. Threshold pump power: 20000 mW.
5027. Output power: 0.5 mW.
5028. 10.65 μm CO_2 pump laser; P26 transition; parallel polarization.
5029. Threshold pump power: 20000 mW.
5030. Output power: 1 mW.
5031. 9.47 μm CO_2 pump laser; P10 transition; parallel polarization.
5032. Threshold pump power: 10000 mW.
5033. Output power: 0.05 mW.
5034. 9.26 μm CO_2 pump laser; R22 transition; parallel polarization.
5035. Threshold pump power: 20000 mW.

5036.   Output power: 0.2 mW.
5037.   9.49 μm CO_2 pump laser; P12 transition; perpendicular polarization.
5038.   Threshold pump power: 10000 mW.
5039.   Output power: 0.01 mW.
5040.   09.26 μm CO_2 pump laser; R22 transition; parallel polarization.
5041.   Threshold pump power: 20000 mW.
5042.   Output power: 0.5 mW.
5043.   9.54 μm CO_2 pump laser; P18 transition; parallel polarization.
5044.   Threshold pump power: 10000 mW.
5045.   Output power: 0.5 mW.
5046.   9.66 μm CO_2 pump laser; P32 transition; parallel polarization.
5047.   Threshold pump power: 20000 mW.
5048.   Output power: 0.05 mW.
5049.   9.34 μm CO_2 pump laser; R08 transition; parallel polarization.
5050.   Threshold pump power: 10000 mW.
5051.   Output power: 0.002 mW.
5052.   10.70 μm CO_2 pump laser; P30 transition; perpendicular polarization.
5053.   Threshold pump power: 20000 mW.
5054.   Output power: 1 mW.
5055.   10.16 μm CO_2 pump laser; R34 transition; parallel polarization.
5056.   Threshold pump power: 20000 mW.
5057.   Output power: 0.2 mW.
5058.   9.25 μm CO_2 pump laser; R24 transition; parallel polarization.
5059.   Threshold pump power: 20000 mW.
5060.   Output power: 2 mW.
5061.   10.16 μm CO_2 pump laser; R34 transition; parallel polarization.
5062.   Threshold pump power: 20000 mW.
5063.   Output power: 0.5 mW.
5064.   10.65 μm CO_2 pump laser; P26 transition; perpendicular polarization.
5065.   Threshold pump power: 20000 mW.
5066.   Output power: 0.05 mW.
5067.   10.17 μm CO_2 pump laser; R32 transition; parallel polarization.
5068.   Threshold pump power: 20000 mW.
5069.   Output power: 0.02 mW.
5070.   10.76 μm CO_2 pump laser; P36 transition; parallel polarization.
5071.   Threshold pump power: 10000 mW.
5072.   Output power: 0.02 mW.
5073.   9.55 μm CO_2 pump laser; P20 transition; perpendicular polarization.
5074.   Threshold pump power: 20000 mW.
5075.   Output power: 0.05 mW.
5076.   9.34 μm CO_2 pump laser; R08 transition; perpendicular polarization.
5077.   Threshold pump power: 10000 mW.
5078.   Output power: 0.2 mW.
5079.   10.17 μm CO_2 pump laser; R32 transition; parallel polarization.
5080.   Threshold pump power: 20000 mW.
5081.   Output power: 0.1 mW.
5082.   10.17 μm CO_2 pump laser; R32 transition; parallel polarization.
5083.   Threshold pump power: 20000 mW.
5084.   Output power: 2 mW.
5085.   10.74 μm CO_2 pump laser; P34 transition; parallel polarization.
5086.   Threshold pump power: 10000 mW.

5087. Output power: 0.05 mW.
5088. 9.66 μm CO_2 pump laser; P32 transition; perpendicular polarization.
5089. Threshold pump power: 20000 mW.
5090. Output power: 1 mW.
5091. 9.49 μm CO_2 pump laser; P12 transition; parallel polarization.
5092. Threshold pump power: 10000 mW.
5093. Output power: 0.05 mW.
5094. 9.66 μm CO_2 pump laser; P32 transition; perpendicular polarization.
5095. Threshold pump power: 20000 mW.
5096. Output power: 0.5 mW.
5097. 9.49 μm CO_2 pump laser; P12 transition; parallel polarization.
5098. Threshold pump power: 10000 mW.
5099. Output power: 0.1 mW.
5100. 9.52 μm CO_2 pump laser; P16 transition; perpendicular polarization.
5101. Threshold pump power: 20000 mW.
5102. Output power: 0.2 mW.
5103. 9.54 μm CO_2 pump laser; P18 transition; parallel polarization.
5104. Threshold pump power: 20000 mW.
5105. Output power: 0.2 mW.
5106. 10.70 μm CO_2 pump laser; P30 transition; parallel polarization.
5107. Threshold pump power: 20000 mW.
5108. Output power: 0.1 mW.
5109. 9.73 μm CO_2 pump laser; P40 transition; perpendicular polarization.
5110. Threshold pump power: 10000 mW.
5111. Output power: 0.2 mW.
5112. 9.54 μm CO_2 pump laser; P18 transition; perpendicular polarization.
5113. Threshold pump power: 20000 mW.
5114. Output power: 0.1 mW.
5115. 9.64 μm CO_2 pump laser; P30 transition; parallel polarization.
5116. Threshold pump power: 20000 mW.
5117. Output power: 0.02 mW.
5118. 9.71 μm CO_2 pump laser; P38 transition; perpendicular polarization.
5119. Threshold pump power: 10000 mW.
5120. Output power: 0.1 mW.
5121. 9.66 μm CO_2 pump laser; P32 transition; parallel polarization.
5122. Output power: 0.2 mW.
5123. 9.44 μm CO_2 pump laser; P06 transition; perpendicular polarization.
5124. Threshold pump power: 2000 mW.
5125. Output power: 0.1 mW.
5126. 9.64 μm CO_2 pump laser; P30 transition; perpendicular polarization.
5127. Output power: 0.2 mW.
5128. 9.37 μm CO_2 pump laser; R04 transition; perpendicular polarization.
5129. Threshold pump power: 2000 mW.
5130. Output power: 0.1 mW.
5131. 9.64 μm CO_2 pump laser; P30 transition; parallel polarization.
5132. Output power: 10 mW.
5133. 9.34 μm CO_2 pump laser; R08 transition; parallel polarization.
5134. Threshold pump power: 2000 mW.
5135. Output power: 10 mW.
5136. 9.34 μm CO_2 pump laser; R08 transition; parallel polarization.
5137. Threshold pump power: 2000 mW.

5138.   Output power: 0.005 mW.
5139.   10.57 μm CO_2 pump laser; P18 transition; parallel polarization.
5140.   Threshold pump power: 5000 mW.
5141.   Output power: 0.1 mW.
5142.   9.66 μm CO_2 pump laser; P32 transition; perpendicular polarization.
5143.   Threshold pump power: 2000 mW.
5144.   Output power: 0.05 mW.
5145.   10.10 μm CO_2 pump laser; R44 transition; parallel polarization.
5146.   Threshold pump power: 2000 mW.
5147.   Output power: 0.1 mW.
5148.   9.31 μm CO_2 pump laser; R14 transition; parallel polarization.
5149.   Threshold pump power: 5000 mW.
5150.   Output power: 0.005 mW.
5151.   10.16 μm CO_2 pump laser; R34 transition; parallel polarization.
5152.   Threshold pump power: 5000 mW.
5153.   Output power: 0.01 mW.
5154.   9.31 μm CO_2 pump laser; R14 transition; parallel polarization.
5155.   Threshold pump power: 5000 mW.
5156.   Output power: 0.5 mW.
5157.   9.44 μm CO_2 pump laser; P06 transition; parallel polarization.
5158.   Threshold pump power: 2000 mW.
5159.   Output power: 0.5 mW.
5160.   9.35 μm CO_2 pump laser; R06 transition; parallel polarization.
5161.   Threshold pump power: 5000 mW.
5162.   Output power: 0.02 mW.
5163.   10.23 μm CO_2 pump laser; R22 transition; parallel polarization.
5164.   Threshold pump power: 10000 mW.
5165.   Output power: 0.1 mW.
5166.   9.31 μm CO_2 pump laser; R14 transition; parallel polarization.
5167.   Threshold pump power: 2000 mW.
5168.   Output power: 1 mW.
5169.   9.37 μm CO_2 pump laser; R04 transition; parallel polarization.
5170.   Threshold pump power: 5000 mW.
5171.   Output power: 0.02 mW.
5172.   10.23 μm CO_2 pump laser; R28 transition; parallel polarization.
5173.   Threshold pump power: 5000 mW.
5174.   9.44 μm CO_2 pump laser; P06 transition; unspecified polarization.
5175.   Threshold pump power: 5000 mW.
5176.   Output power: 0.02 mW.
5177.   9.24 μm CO_2 pump laser; R26 transition; parallel polarization.
5178.   Threshold pump power: 5000 mW.
5179.   Output power: 0.5 mW.
5180.   9.66 μm CO_2 pump laser; P32 transition; perpendicular polarization.
5181.   Threshold pump power: 2000 mW.
5182.   Output power: 0.05 mW.
5183.   9.26 μm CO_2 pump laser; R22 transition; parallel polarization.
5184.   Threshold pump power: 5000 mW.
5185.   Output power: 0.1 mW.
5186.   9.64 μm CO_2 pump laser; P30 transition; perpendicular polarization.
5187.   Output power: 0.005 mW.
5188.   9.66 μm CO_2 pump laser; P32 transition; parallel polarization.

5189.   Threshold pump power: 5000 mW.

5190.   Output power: 0.1 mW.

5191.   9.64 μm CO_2 pump laser; P30 transition; parallel polarization.

5192.   Threshold pump power: 2000 mW.

5193.   Output power: 0.02 mW.

5194.   10.11 μm CO_2 pump laser; R42 transition; perpendicular polarization.

5195.   Threshold pump power: 2000 mW.

5196.   Output power: 0.05 mW.

5197.   10.23 μm CO_2 pump laser; R22 transition; parallel polarization.

5198.   Threshold pump power: 5000 mW.

5199.   Output power: 0.05 mW.

5200.   9.59 μm CO_2 pump laser; P24 transition; parallel polarization.

5201.   Threshold pump power: 2000 mW.

5202.   Output power: 0.2 mW.

5203.   9.44 μm CO_2 pump laser; P06 transition; perpendicular polarization.

5204.   Threshold pump power: 2000 mW.

5205.   Output power: 2 mW.

5206.   9.47 μm CO_2 pump laser; P10 transition; either parallel or perpendicular polarization.

5207.   Threshold pump power: 5000 mW.

5208.   Output power: 0.2 mW.

5209.   9.57 μm CO_2 pump laser; P22 transition; perpendicular polarization.

5210.   Output power: 20 mW.

5211.   9.60 μm CO_2 pump laser; P26 transition; parallel polarization.

5212.   Threshold pump power: 1000 mW.

5213.   Output power: 0.01 mW.

5214.   9.66 μm CO_2 pump laser; P32 transition; parallel polarization.

5215.   Threshold pump power: 5000 mW.

5216.   Output power: 0.05 mW.

5217.   9.66 μm CO_2 pump laser; P32 transition; either parallel cr perpendicular polarization.

5218.   Threshold pump power: 5000 mW.

5219.   Output power: 0.02 mW.

5220.   9.24 μm CO_2 pump laser; R26 transition; parallel polarization.

5221.   Threshold pump power: 5000 mW.

5222.   Output power: 0.05 mW.

5223.   9.38 μm CO_2 pump laser; R02 transition; parallel polarization.

5224.   Threshold pump power: 2000 mW.

5225.   Output power: 0.1 mW.

5226.   10.11 μm CO_2 pump laser; R44 transition; parallel polarization.

5227.   Threshold pump power: 1000 mW.

5228.   Output power: 10 mW.

5229.   9.64 μm CO_2 pump laser; P30 transition; perpendicular polarization.

5230.   Threshold pump power: 500 mW.

5231.   Output power: 1 mW.

5232.   9.60 μm CO_2 pump laser; P26 transition; either parallel or perpendicular polarization.

5233.   Threshold pump power: 2000 mW.

5234.   Output power: 1 mW.

5235.   9.64 μm CO_2 pump laser; transition; parallel polarization.

5236.   Threshold pump power: 5000 mW.

5237.   Output power: 1 mW.
5238.   9.64 μm CO_2 pump laser; P30 transition; perpendicular polarization.
5239.   Threshold pump power: 5000 mW.
5240.   Output power: 1 mW.
5241.   9.66 μm CO_2 pump laser; P32 transition; perpendicular polarization.
5242.   Threshold pump power: 2000 mW.
5243.   Output power: 0.2 mW.
5244.   9.29 μm CO_2 pump laser; R16 transition; parallel polarization.
5245.   Threshold pump power: 5000 mW.
5246.   Output power: 0.1 mW.
5247.   9.35 μm CO_2 pump laser; R06 transition; unspecified polarization.
5248.   Threshold pump power: 5000 mW.
5249.   Output power: 1 mW.
5250.   9.34 μm CO_2 pump laser; R08 transition; perpendicular polarization.
5251.   Threshold pump power: 1000 mW.
5252.   Output power: 0.5 mW.
5253.   9.34 μm CO_2 pump laser; R08 transition; unspecified polarization.
5254.   Threshold pump power: 2000 mW.
5255.   Output power: 0.5 mW.
5256.   9.64 μm CO_2 pump laser; P30 transition; unspecified polarization.
5257.   Threshold pump power: 2000 mW.
5258.   Output power: 0.2 mW.
5259.   9.44 μm CO_2 pump laser; P06 transition; unspecified polarization.
5260.   Threshold pump power: 2000 mW.
5261.   Output power: 0.5 mW.
5262.   10.14 μm CO_2 pump laser; R38 transition; unspecified polarization.
5263.   Threshold pump power: 2000 mW.
5264.   Output power: 2 mW.
5265.   10.27 μm CO_2 pump laser; R16 transition; unspecified polarization.
5266.   Threshold pump power: 2000 mW.
5267.   Output power: 0.5 mW.
5268.   10.57 μm CO_2 pump laser; P18 transition; unspecified polarization.
5269.   Threshold pump power: 2000 mW.
5270.   Output power: 0.2 mW.
5271.   19.55 μm CO_2 pump laser; P20 transition; unspecified polarization.
5272.   Threshold pump power: 2000 mW.
5273.   Output power: 2 mW.
5274.   10.25 μm CO_2 pump laser; R20 transition; unspecified polarization.
5275.   Threshold pump power: 2000 mW.
5276.   Output power: 2 mW.
5277.   10.57 μm CO_2 pump laser; P18 transition; unspecified polarization.
5278.   Threshold pump power: 2000 mW.
5279.   Output power: 0.5 mW.
5280.   10.27 μm CO_2 pump laser; R16 transition; unspecified polarization.
5281.   Threshold pump power: 2000 mW.
5282.   Output power: 0.5 mW.
5283.   10.14 μm CO_2 pump laser; R38 transition; unspecified polarization.
5284.   Threshold pump power: 2000 mW.
5285.   Output power: 0.5 mW.
5286.   9.28 μm CO_2 pump laser; R18 transition; unspecified polarization.
5287.   Threshold pump power: 2000 mW.

5288.   Output power: 0.1 mW.
5289.   10.25 µm CO_2 pump laser; R20 transition; parallel polarization.
5290.   Threshold pump power: 2000 mW.
5291.   Output power: 0.1 mW.
5292.   9.59 µm CO_2 pump laser; P24 transition; parallel polarization.
5293.   Threshold pump power: 2000 mW.
5294.   Output power: 0.2 mW.
5295.   9.49 µm CO_2 pump laser; P12 transition; parallel polarization.
5296.   Threshold pump power: 2000 mW.
5297.   Output power: 0.1 mW.
5298.   9.27 µm CO_2 pump laser; R20 transition; parallel polarization.
5299.   Threshold pump power: 2000 mW.
5300.   Output power: 5 mW.
5301.   10.26 µm CO_2 pump laser; R18 transition; parallel polarization.
5302.   Threshold pump power: 2000 mW.
5303.   Output power: 0.01 mW.
5304.   9.64 µm CO_2 pump laser; P30 transition; parallel polarization.
5305.   Threshold pump power: 2000 mW.
5306.   Output power: 0.1 mW.
5307.   10.57 µm CO_2 pump laser; P18 transition; parallel polarization.
5308.   Threshold pump power: 2000 mW.
5309.   Output power: 0.1 mW.
5310.   9.24 µm CO_2 pump laser; R26 transition; parallel polarization.
5311.   Threshold pump power: 2000 mW.
5312.   Output power: 0.05 mW.
5313.   10.59 µm CO_2 pump laser; P20 transition; perpendicular polarization.
5314.   Threshold pump power: 2000 mW.
5315.   Output power: 0.1 mW.
5316.   9.46 µm CO_2 pump laser; P08 transition; parallel polarization.
5317.   Threshold pump power: 2000 mW.
5318.   Output power: 1 mW.
5319.   10.15 µm CO_2 pump laser; R36 transition; perpendicular polarization.
5320.   Threshold pump power: 2000 mW.
5321.   Output power: 0.5 mW.
5322.   9.59 µm CO_2 pump laser; P24 transition; parallel polarization.
5323.   Threshold pump power: 2000 mW.
5324.   Output power: 1 mW.
5325.   9.47 µm CO_2 pump laser; P10 transition; parallel polarization.
5326.   Threshold pump power: 2000 mW.
5327.   Output power: 0.1 mW.
5328.   10.16 µm CO_2 pump laser; R34 transition; parallel polarization.
5329.   Threshold pump power: 2000 mW.
5330.   Output power: 0.02 mW.
5331.   9.44 µm CO_2 pump laser; P06 transition; parallel polarization.
5332.   Threshold pump power: 2000 mW.
5333.   Output power: 1 mW.
5334.   10.33 µm CO_2 pump laser; R08 transition; parallel polarization.
5335.   Threshold pump power: 2000 mW.
5336.   Output power: 1 mW.
5337.   10.15 µm CO_2 pump laser; R36 transition; perpendicular polarization.
5338.   Threshold pump power: 2000 mW.

5339.   Output power: 0.1 mW.
5340.   10.27 μm CO_2 pump laser; R16 transition; parallel polarization.
5341.   Threshold pump power: 2000 mW.
5342.   Output power: 1 mW.
5343.   9.26 μm CO_2 pump laser; R22 transition; parallel polarization.
5344.   Threshold pump power: 2000 mW.
5345.   Output power: 0.5 mW.
5346.   9.21 μm CO_2 pump laser; R32 transition; parallel polarization.
5347.   Threshold pump power: 2000 mW.
5348.   Output power: 2 mW.
5349.   10.33 μm CO_2 pump laser; R08 transition; parallel polarization.
5350.   Threshold pump power: 2000 mW.
5351.   Output power: 5 mW.
5352.   9.31 μm CO_2 pump laser; R14 transition; parallel polarization.
5353.   Threshold pump power: 2000 mW.
5354.   Output power: 0.2 mW.
5355.   9.57 μm CO_2 pump laser; P22 transition; perpendicular polarization.
5356.   Threshold pump power: 2000 mW.
5357.   Output power: 0.1 mW.
5358.   10.84 μm CO_2 pump laser; P42 transition; parallel polarization.
5359.   Threshold pump power: 2000 mW.
5360.   Output power: 0.1 mW.
5361.   9.46 μm CO_2 pump laser; P08 transition; parallel polarization.
5362.   Threshold pump power: 2000 mW.
5363.   Output power: 0.1 mW.
5364.   10.16 μm CO_2 pump laser; R34 transition; parallel polarization.
5365.   Threshold pump power: 2000 mW.
5366.   Output power: 1 mW.
5367.   10.26 μm CO_2 pump laser; R18 transition; parallel polarization.
5368.   Threshold pump power: 2000 mW.
5369.   Output power: 0.1 mW.
5370.   10.65 μm CO_2 pump laser; P26 transition; parallel polarization.
5371.   Threshold pump power: 2000 mW.
5372.   Output power: 0.5 mW.
5373.   9.26 μm CO_2 pump laser; R22 transition; parallel polarization.
5374.   Threshold pump power: 2000 mW.
5375.   Output power: 0.02 mW.
5376.   9.52 μm CO_2 pump laser; P16 transition; perpendicular polarization.
5377.   Threshold pump power: 2000 mW.
5378.   Output power: 0.02 mW.
5379.   9.29 μm CO_2 pump laser; R16 transition; parallel polarization.
5380.   Threshold pump power: 2000 mW.
5381.   Output power: 0.2 mW.
5382.   9.54 μm CO_2 pump laser; P18 transition; parallel polarization.
5383.   Threshold pump power: 2000 mW.
5384.   Output power: 0.001 mW.
5385.   9.62 μm CO_2 pump laser; P28 transition; perpendicular polarization.
5386.   Threshold pump power: 2000 mW.
5387.   Output power: 0.5 mW.
5388.   9.55 μm CO_2 pump laser; P20 transition; parallel polarization.
5389.   Threshold pump power: 2000 mW.

5390.   Output power: 0.5 mW.
5391.   10.17 μm CO_2 pump laser; R32 transition; perpendicular polarization.
5392.   Threshold pump power: 2000 mW.
5393.   Output power: 5 mW.
5394.   10.15 μm CO_2 pump laser; R36 transition; perpendicular polarization.
5395.   Threshold pump power: 2000 mW.
5396.   Output power: 0.1 mW.
5397.   10.30 μm CO_2 pump laser; R12 transition; parallel polarization.
5398.   Threshold pump power: 2000 mW.
5399.   Output power: 0.2 mW.
5400.   9.66 μm CO_2 pump laser; P32 transition; parallel polarization.
5401.   Threshold pump power: 2000 mW.
5402.   Output power: 0.05 mW.
5403.   9.20 μm CO_2 pump laser; R34 transition; perpendicular polarization.
5404.   Threshold pump power: 2000 mW.
5405.   Output power: 0.2 mW.
5406.   9.16 μm CO_2 pump laser; R44 transition; parallel polarization.
5407.   Threshold pump power: 2000 mW.
5408.   Output power: 0.5 mW.
5409.   10.19 μm CO_2 pump laser; R28 transition; parallel polarization.
5410.   Threshold pump power: 2000 mW.
5411.   Output power: 0.5 mW.
5412.   9.52 μm CO_2 pump laser; P16 transition; perpendicular polarization.
5413.   Threshold pump power: 2000 mW.
5414.   Output power: 0.01 mW.
5415.   9.64 μm CO_2 pump laser; P30 transition; perpendicular polarization.
5416.   Threshold pump power: 2000 mW.
5417.   Output power: 2 mW.
5418.   9.24 μm CO_2 pump laser; R26 transition; parallel polarization.
5419.   Threshold pump power: 2000 mW.
5420.   Output power: 0.2 mW.
5421.   9.62 μm CO_2 pump laser; P28 transition; parallel polarization.
5422.   Threshold pump power: 2000 mW.
5423.   Output power: 0.2 mW.
5424.   9.35 μm CO_2 pump laser; R06 transition; parallel polarization.
5425.   Threshold pump power: 2000 mW.
5426.   Output power: 0.1 mW.
5427.   9.23 μm CO_2 pump laser; R28 transition; parallel polarization.
5428.   Threshold pump power: 2000 mW.
5429.   Output power: 1 mW.
5430.   9.31 μm CO_2 pump laser; R14 transition; parallel polarization.
5431.   Threshold pump power: 2000 mW.
5432.   Output power: 0.1 mW.
5433.   10.18 μm CO_2 pump laser; R30 transition; perpendicular polarization.
5434.   Threshold pump power: 2000 mW.
5435.   Output power: 0.5 mW.
5436.   9.66 μm CO_2 pump laser; P32 transition; parallel polarization.
5437.   Threshold pump power: 2000 mW.
5438.   Output power: 0.1 mW.
5439.   9.31 μm CO_2 pump laser; R14 transition; parallel polarization.
5440.   Threshold pump power: 2000 mW.

5441.   Output power: 0.02 mW.
5442.   9.21 μm CO_2 pump laser; R32 transition; perpendicular polarization.
5443.   Threshold pump power: 2000 mW.
5444.   Output power: 0.1 mW.
5445.   10.18 μm CO_2 pump laser; R30 transition; parallel polarization.
5446.   Threshold pump power: 2000 mW.
5447.   Output power: 0.05 mW.
5448.   9.29 μm CO_2 pump laser; R16 transition; perpendicular polarization.
5449.   Threshold pump power: 2000 mW.
5450.   Output power: 0.5 mW.
5451.   10.19 μm CO_2 pump laser; R28 transition; perpendicular polarization.
5452.   Threshold pump power: 2000 mW.
5453.   Output power: 0.05 mW.
5454.   10.59 μm CO_2 pump laser; P20 transition; parallel polarization.
5455.   Threshold pump power: 2000 mW.
5456.   Output power: 0.02 mW.
5457.   9.35 μm CO_2 pump laser; R06 transition; parallel polarization.
5458.   Threshold pump power: 2000 mW.
5459.   Output power: 0.1 mW.
5460.   9.28 μm CO_2 pump laser; R18 transition; perpendicular polarization.
5461.   Threshold pump power: 2000 mW.
5462.   Output power: 0.5 mW.
5463.   10.16 μm CO_2 pump laser; R34 transition; parallel polarization.
5464.   Threshold pump power: 2000 mW.
5465.   Output power: 0.1 mW.
5466.   10.57 μm CO_2 pump laser; P18 transition; perpendicular polarization.
5467.   Threshold pump power: 2000 mW.
5468.   Output power: 5 mW.
5469.   10.57 μm CO_2 pump laser; P18 transition; parallel polarization.
5470.   Threshold pump power: 2000 mW.
5471.   Output power: 2 mW.
5472.   10.63 μm CO_2 pump laser; P24 transition; parallel polarization.
5473.   Threshold pump power: 2000 mW.
5474.   Output power: 0.2 mW.
5475.   9.73 μm CO_2 pump laser; P40 transition; parallel polarization.
5476.   Threshold pump power: 2000 mW.
5477.   Output power: 0.1 mW.
5478.   10.22 μm CO_2 pump laser; R24 transition; parallel polarization.
5479.   Threshold pump power: 2000 mW.
5480.   Output power: 0.1 mW.
5481.   9.24 μm CO_2 pump laser; R26 transition; parallel polarization.
5482.   Threshold pump power: 2000 mW.
5483.   Output power: 1 mW.
5484.   10.67 μm CO_2 pump laser; P28 transition; parallel polarization.
5485.   Threshold pump power: 2000 mW.
5486.   Output power: 0.1 mW.
5487.   9.50 μm CO_2 pump laser; P14 transition; parallel polarization.
5488.   Threshold pump power: 2000 mW.
5489.   Output power: 0.1 mW.
5490.   10.29 μm CO_2 pump laser; R14 transition; parallel polarization.
5491.   Threshold pump power: 2000 mW.

5492. Output power: 0.2 mW.
5493. 9.55 μm CO_2 pump laser; P20 transition; parallel polarization.
5494. Threshold pump power: 2000 mW.
5495. Output power: 0.5 mW.
5496. 10.16 μm CO_2 pump laser; R34 transition; parallel polarization.
5497. Threshold pump power: 2000 mW.
5498. Output power: 1 mW.
5499. 10.61 μm CO_2 pump laser; P22 transition; parallel polarization.
5500. Threshold pump power: 2000 mW.
5501. Output power: 0.5 mW.
5502. 10.15 μm CO_2 pump laser; R36 transition; perpendicular polarization.
5503. Threshold pump power: 2000 mW.
5504. Output power: 0.01 mW.
5505. 10.63 μm CO_2 pump laser; P24 transition; perpendicular polarization.
5506. Threshold pump power: 2000 mW.
5507. Output power: 0.1 mW.
5508. 9.31 μm CO_2 pump laser; R14 transition; parallel polarization.
5509. Threshold pump power: 2000 mW.
5510. Output power: 0.5 mW.
5511. 9.20 μm CO_2 pump laser; R34 transition; parallel polarization.
5512. Threshold pump power: 2000 mW.
5513. Output power: 0.01 mW.
5514. 9.46 μm CO_2 pump laser; P08 transition; perpendicular polarization.
5515. Threshold pump power: 2000 mW.
5516. Output power: 0.1 mW.
5517. 9.44 μm CO_2 pump laser; P06 transition; parallel polarization.
5518. Threshold pump power: 2000 mW.
5519. Output power: 0.01 mW.
5520. 10.26 μm CO_2 pump laser; R18 transition; parallel polarization.
5521. Threshold pump power: 2000 mW.
5522. 10.72 μm CO_2 pump laser; P52 transition; unspecified polarization.
5523. Output power: 0.1 mW.
5524. 9.73 μm CO_2 pump laser; P40 transition; parallel polarization.
5525. Threshold pump power: 2000 mW.
5526. Output power: 0.1 mW.
5527. 10.16 μm CO_2 pump laser; R34 transition; parallel polarization.
5528. Threshold pump power: 2000 mW.
5529. 10.84 μm CO_2 pump laser; P42 transition; unspecified polarization.
5530. Output power: 0.2 mW.
5531. 9.34 μm CO_2 pump laser; R08 transition; perpendicular polarization.
5532. Threshold pump power: 2000 mW.
5533. 9.31 μm CO_2 pump laser; R14 transition; unspecified polarization.
5534. Output power: 0.01 mW.
5535. 10.16 μm CO_2 pump laser; R34 transition; perpendicular polarization.
5536. Threshold pump power: 2000 mW.
5537. Output power: 0.5 mW.
5538. 9.31 μm CO_2 pump laser; R14 transition; perpendicular polarization.
5539. Threshold pump power: 2000 mW.
5540. Output power: 0.5 mW.
5541. 9.23 μm CO_2 pump laser; R28 transition; parallel polarization.
5542. Threshold pump power: 2000 mW.

5543.   Output power: 0.1 mW.
5544.   10.57 μm CO_2 pump laser; P18 transition; parallel polarization.
5545.   Threshold pump power: 2000 mW.
5546.   10.29 μm CO_2 pump laser; R14 transition; unspecified polarization.
5547.   10.17 μm CO_2 pump laser; R32 transition; unspecified polarization.
5548.   Output power: 0.001 mW.
5549.   10.16 μm CO_2 pump laser; R34 transition; parallel polarization.
5550.   Threshold pump power: 2000 mW.
5551.   Output power: 0.001 mW.
5552.   10.16 μm CO_2 pump laser; R34 transition; perpendicular polarization.
5553.   Threshold pump power: 2000 mW.
5554.   10.49 μm CO_2 pump laser; P10 transition; unspecified polarization.
5555.   Output power: 0.01 mW.
5556.   10.16 μm CO_2 pump laser; R34 transition; perpendicular polarization.
5557.   Threshold pump power: 2000 mW.
5558.   Output power: 0.01 mW.
5559.   10.27 μm CO_2 pump laser; R16 transition; perpendicular polarization.
5560.   Threshold pump power: 2000 mW.
5561.   Output power: 0.1 mW.
5562.   10.27 μm CO_2 pump laser; R16 transition; parallel polarization.
5563.   Threshold pump power: 2000 mW.
5564.   Output power: 0.1 mW.
5565.   10.72 μm CO_2 pump laser; P32 transition; parallel polarization.
5566.   Threshold pump power: 2000 mW.
5567.   Output power: 0.1 mW.
5568.   10.33 μm CO_2 pump laser; R08 transition; parallel polarization.
5569.   Threshold pump power: 2000 mW.
5570.   10.18 μm CO_2 pump laser; R30 transition; unspecified polarization.
5571.   Output power: 1 mW.
5572.   9.20 μm CO_2 pump laser; R34 transition; parallel polarization.
5573.   Threshold pump power: 2000 mW.
5574.   9.23 μm CO_2 pump laser; R28 transition; perpendicular polarization.
5575.   Threshold pump power: 2000 mW.
5576.   Output power: 1 mW.
5577.   9.20 μm CO_2 pump laser; R34 transition; perpendicular polarization.
5578.   Threshold pump power: 2000 mW.
5579.   Output power: 1 mW.
5580.   9.23 μm CO_2 pump laser; R28 transition; perpendicular polarization.
5581.   Threshold pump power: 2000 mW.
5582.   Output power: 1 mW.
5583.   10.26 μm CO_2 pump laser; R18 transition; parallel polarization.
5584.   Threshold pump power: 2000 mW.
5585.   Output power: 0.1 mW.
5586.   10.33 μm CO_2 pump laser; R08 transition; perpendicular polarization.
5587.   Threshold pump power: 2000 mW.
5588.   Output power: 1 mW.
5589.   10.26 μm CO_2 pump laser; R18 transition; perpendicular polarization.
5590.   Threshold pump power: 2000 mW.
5591.   10.33 μm CO_2 pump laser; R08 transition; parallel polarization.
5592.   Threshold pump power: 2000 mW.
5593.   Output power: 0.1 mW.

5594.   10.61 μm CO_2 pump laser; P22 transition; perpendicular polarization.

5595.   Threshold pump power: 2000 mW.

5596.   Output power: 0.1 mW.

5597.   10.16 μm CO_2 pump laser; R34 transition; parallel polarization.

5598.   Threshold pump power: 2000 mW.

5599.   Output power: 0.1 mW.

5600.   10.61 μm CO_2 pump laser; P22 transition; parallel polarization.

5601.   Threshold pump power: 2000 mW.

5602.   For line assignments see Reference 1425.

5603.   Output power: 0.1 mW.

5604.   -20 MHz offset from center of 10.26 μm CO_2 pump laser; R18 transition; unspecified polarization.

5605.   Threshold pump power: 2000 mW.

5606.   For line assignments see Reference 1425.

5607.   Output power: 1 mW.

5608.   10.22 μm CO_2 pump laser; R24 transition; parallel polarization.

5609.   Threshold pump power: 10000 mW.

5610.   For line assignments see Reference 1425.

5611.   Output power: 1 mW.

5612.   10.30 μm CO_2 pump laser; R12 transition; perpendicular polarization.

5613.   Threshold pump power: 10000 mW.

5614.   For line assignments see Reference 1425.

5615.   Output power: 1 mW; laser has a Lamb dip.

5616.   10.30 μm CO_2 pump laser; R12 transition; perpendicular polarization.

5617.   Threshold pump power: 1000 mW.

5618.   For line assignments see Reference 1425.

5619.   Output power: 0.5 mW.

5620.   10.27 μm CO_2 pump laser; R16 transition; unspecified polarization.

5621.   Threshold pump power: 10000 mW.

5622.   For line assignments see Reference 1425.

5623.   Output power: 1 mW; laser has a Lamb dip.

5624.   10.37 μm CO_2 pump laser; R04 transition; perpendicular polarization.

5625.   Threshold pump power: 1000 mW.

5626.   For line assignments see Reference 1425.

5627.   10.32 μm CO_2 pump laser; R10 transition; perpendicular polarization.

5628.   Threshold pump power: 1000 mW.

5629.   For line assignments see Reference 1425.

5630.   Output power: 1 mW; laser has a Lamb dip.

5631.   10.22 μm CO_2 pump laser; R24 transition; perpendicular polarization.

5632.   Threshold pump power: 1000 mW.

5633.   For line assignments see Reference 1425.

5634.   10.15 μm CO_2 pump laser; R36 transition; perpendicular polarization.

5635.   Threshold pump power: 1000 mW.

5636.   For line assignments see Reference 1425.

5637.   10.32 μm CO_2 pump laser; R10 transition; perpendicular polarization.

5638.   Threshold pump power: 1000 mW.

5639.   For line assignments see Reference 1425.

5640.   Output power: 1 mW; laser has a Lamb dip.

5641.   10.22 μm CO_2 pump laser; R24 transition; perpendicular polarization.

5642.   Threshold pump power: 1000 mW.

5643.   For line assignments see Reference 1425.

5644.   Output power: 0.05 mW.
5645.   10.23 μm CO_2 pump laser; R22 transition; unspecified polarization.
5646.   Threshold pump power: 10000 mW.
5647.   For line assignments see Reference 1425.
5648.   Output power: 0.1 mW.
5649.   10.18 μm CO_2 pump laser; R30 transition; unspecified polarization.
5650.   Threshold pump power: 10000 mW.
5651.   For line assignments see Reference 1425.
5652.   Output power: 0.2 mW.
5653.   10.21 μm CO_2 pump laser; R26 transition; unspecified polarization.
5654.   Threshold pump power: 10000 mW.
5655.   For line assignments see Reference 1425.
5656.   Output power: 2 mW.
5657.   10.49 μm CO_2 pump laser; P10 transition; unspecified polarization.
5658.   Threshold pump power: 10000 mW.
5659.   For line assignments see Reference 1425.
5660.   10.26 μm CO_2 pump laser; R18 transition; perpendicular polarization.
5661.   Threshold pump power: 1000 mW.
5662.   For line assignments see Reference 1425.
5663.   Output power: 5 mW.
5664.   10.19 μm CO_2 pump laser; R28 transition; unspecified polarization.
5665.   Threshold pump power: 10000 mW.
5666.   For line assignments see References 1364, 1377.
5667.   Output power: 1 mW.
5668.   0 MHz offset from center of 10.88 μm CO_2 pump laser; P46 transition; parallel polarization.
5669.   Threshold pump power: 10000 mW.
5670.   For line assignments see References 1364, 1377.
5671.   Output power: 0.02 mW.
5672.   -15 MHz offset from center of 9.27 μm CO_2 pump laser; R20 transition; perpendicular polarization.
5673.   Threshold pump power: 50000 mW.
5674.   For line assignments see References 1364, 1377.
5675.   Output power: 0.05 mW.
5676.   +45 MHz offset from center of 9.47 μm CO_2 pump laser; P10 transition; parallel polarization.
5677.   Threshold pump power: 20000 mW.
5678.   For line assignments see References 1364, 1377.
5679.   Output power: 0.2 mW.
5680.   +40 MHz offset from center of 9.73 μm CO_2 pump laser; P40 transition; perpendicular polarization.
5681.   Threshold pump power: 10000 mW.
5682.   For line assignments see References 1364, 1377.
5683.   Output power: 0.05 mW.
5684.   -40 MHz offset from center of 9.46 μm CO_2 pump laser; P08 transition; parallel polarization.
5685.   Threshold pump power: 5000 mW.
5686.   For line assignments see References 1364, 1377.
5687.   Output power: 0.02 mW.
5688.   -20 MHz offset from center of 9.31 μm CO_2 pump laser; R14 transition; perpendicular polarization.

5689.   Threshold pump power: 20000 mW.
5690.   For line assignments see References 1364, 1377.
5691.   Output power: 0.1 mW.
5692.   -10 MHz offset from center of 9.52 μm CO_2 pump laser; P16 transition; parallel polarization.
5693.   Threshold pump power: 20000 mW.
5694.   For line assignments see References 1364, 1377.
5695.   Output power: 0.001 mW.
5696.   9.52 μm CO_2 pump laser; P16 transition; unspecified polarization.
5697.   Threshold pump power: 2000 mW.
5698.   For line assignments see References 1364, 1377.
5699.   Output power: 0.2 mW.
5700.   +35 MHz offset from center of 9.34 μm CO_2 pump laser; R08 transition; perpendicular polarization.
5701.   Threshold pump power: 5000 mW.
5702.   For line assignments see References 1364, 1377 (pump).
5703.   Output power: 1 mW.
5704.   -40 MHz offset from center of 10.72 μm CO_2 pump laser; P32 transition; parallel polarization.
5705.   Threshold pump power: 10000 mW.
5706.   For line assignments see References 1364, 1377 (pump).
5707.   Output power: 0.1 mW.
5708.   -15 MHz offset from center of 9.20 μm CO_2 pump laser; R34 transition; parallel polarization.
5709.   Threshold pump power: 10000 mW.
5710.   For line assignments see References 1364, 1377 (pump).
5711.   Output power: 0.2 mW.
5712.   -15 MHz offset from center of 9.64 μm CO_2 pump laser; P30 transition; parallel polarization.
5713.   Threshold pump power: 20000 mW.
5714.   For line assignments see References 1364, 1377 (pump).
5715.   Output power: 0.2 mW.
5716.   +40 MHz offset from center of 9.34 μm CO_2 pump laser; R08 transition; parallel polarization.
5717.   Threshold pump power: 20000 mW.
5718.   For line assignments see References 1364, 1377 (pump).
5719.   Output power: 0.1 mW; cascade transition.
5720.   -10 MHz offset from center of 9.44 μm CO_2 pump laser; P06 transition; perpendicular polarization.
5721.   Threshold pump power: 5000 mW.
5722.   For line assignments see References 1364, 1377.
5723.   Output power: 5 mW.
5724.   -10 MHz offset from center of 9.44 μm CO_2 pump laser; P06 transition; perpendicular polarization.
5725.   Threshold pump power: 1000 mW.
5726.   For line assignments see References 1364, 1377.
5727.   Output power: 0.1 mW; cascade transition.
5728.   0 MHz offset from center of 9.29 μm CO_2 pump laser; R16 transition; perpendicular polarization.
5729.   Threshold pump power: 5000 mW.
5730.   For line assignments see References 1364, 1377.

5731. Output power: 0.5 mW; cascade transition.
5732. 0 MHz offset from center of 9.29 μm CO_2 pump laser; R16 transition; perpendicular polarization.
5733. Threshold pump power: 5000 mW.
5734. For line assignments see References 1364, 1377.
5735. Output power: 1 mW.
5736. 0 MHz offset from center of 9.29 μm CO_2 pump laser; R16 transition; perpendicular polarization.
5737. Threshold pump power: 2000 mW.
5738. For line assignments see References 1364, 1377.
5739. Output power: 0.1 mW; cascade transition.
5740. 0 MHz offset from center of 9.29 μm CO_2 pump laser; R16 transition; perpendicular polarization.
5741. Threshold pump power: 5000 mW.
5742. For line assignments see References 1364, 1377.
5743. Output power: 0.1 mW.
5744. -15 MHz offset from center of 10.57 μm CO_2 pump laser; P18 transition; parallel polarization.
5745. Threshold pump power: 10000 mW.
5746. For line assignments see References 1364, 1377.
5747. Output power: 0.02 mW.
5748. +15 MHz offset from center of 9.60 μm CO_2 pump laser; P26 transition; perpendicular polarization.
5749. Threshold pump power: 10000 mW.
5750. For line assignments see References 1364, 1377.
5751. Output power: 0.005 mW.
5752. +30 MHz offset from center of 9.57 μm CO_2 pump laser; P22 transition; parallel polarization.
5753. Threshold pump power: 50000 mW.
5754. For line assignments see References 1364, 1377.
5755. Output power: 0.5 mW.
5756. +15 MHz offset from center of 9.32 μm CO_2 pump laser; R12 transition; perpendicular polarization.
5757. Threshold pump power: 10000 mW.
5758. For line assignments see References 1364, 1377.
5759. Output power: 0.002 mW.
5760. -15 MHz offset from center of 9.79 μm CO_2 pump laser; P46 transition; parallel polarization.
5761. Threshold pump power: 20000 mW.
5762. For line assignments see References 1364, 1377.
5763. Output power: 0.2 mW.
5764. -30 MHz offset from center of 10.59 μm CO_2 pump laser; P20 transition; unspecified polarization.
5765. Threshold pump power: 1000 mW.
5766. For line assignments see References 1364, 1377.
5767. Output power: 0.5 mW.
5768. -10 MHz offset from center of 9.52 μm CO_2 pump laser; P16 transition; parallel polarization.
5769. Threshold pump power: 10000 mW.
5770. For line assignments see References 1364, 1377.
5771. Output power: 0.1 mW; cascade transition.

5772.   +50 MHz offset from center of 9.84 μm CO_2 pump laser; P50 transition; parallel polarization.
5773.   Threshold pump power: 2000 mW.
5774.   For line assignments see References 1364, 1377.
5775.   Output power: 0.5 mW.
5776.   +50 MHz offset from center of 9.84 μm CO_2 pump laser; P50 transition; parallel polarization.
5777.   Threshold pump power: 2000 mW.
5778.   For line assignments see References 1364, 1377.
5779.   Output power: 0.02 mW.
5780.   -20 MHz offset from center of 9.68 μm CO_2 pump laser; P34 transition; perpendicular polarization.
5781.   Threshold pump power: 10000 mW.
5782.   Output power: 0.05 mW.
5783.   9.77 μm CO_2 pump laser; P44 transition; parallel polarization.
5784.   Threshold pump power: 5000 mW.
5785.   Output power: 0.05 mW.
5786.   9.49 μm CO_2 pump laser; P12 transition; perpendicular polarization.
5787.   Threshold pump power: 5000 mW.
5788.   Output power: 0.1 mW.
5789.   9.52 μm CO_2 pump laser; P16 transition; parallel polarization.
5790.   Threshold pump power: 5000 mW.
5791.   Output power: 0.1 mW.
5792.   9.57 μm CO_2 pump laser; P22 transition; parallel polarization.
5793.   Threshold pump power: 2000 mW.
5794.   Output power: 0.05 mW.
5795.   10.16 μm CO_2 pump laser; R3 transition; parallel polarization.
5796.   Threshold pump power: 10000 mW.
5797.   Output power: 0.1 mW.
5798.   9.23 μm CO_2 pump laser; R28 transition; perpendicular polarization.
5799.   Threshold pump power: 5000 mW.
5800.   Output power: 0.005 mW.
5801.   9.52 μm CO_2 pump laser; P16 transition; perpendicular polarization.
5802.   Threshold pump power: 10000 mW.
5803.   Output power: 0.02 mW.
5804.   10.27 μm CO_2 pump laser; R16 transition; perpendicular polarization.
5805.   Threshold pump power: 10000 mW.
5806.   Output power: 0.1 mW.
5807.   9.71 μm CO_2 pump laser; P38 transition; parallel polarization.
5808.   Threshold pump power: 5000 mW.
5809.   Output power: 0.005 mW.
5810.   9.77 μm CO_2 pump laser; P44 transition; unspecified polarization.
5811.   Threshold pump power: 20000 mW.
5812.   Output power: 0.05 mW.
5813.   9.64 μm CO_2 pump laser; P30 transition; parallel polarization.
5814.   Threshold pump power: 5000 mW.
5815.   Output power: 0.02 mW.
5816.   9.71 μm CO_2 pump laser; P38 transition; parallel polarization.
5817.   Threshold pump power: 10000 mW.
5818.   Output power: 0.005 mW.
5819.   10.16 μm CO_2 pump laser; R34 transition; perpendicular polarization.

5820. Threshold pump power: 10000 mW.
5821. Output power: 0.005 mW.
5822. 9.77 μm CO_2 pump laser; P44 transition; parallel polarization.
5823. Threshold pump power: 10000 mW.
5824. Output power: 0.02 mW.
5825. 9.28 μm CO_2 pump laser; R18 transition; parallel polarization.
5826. Threshold pump power: 5000 mW.
5827. Output power: 0.1 mW.
5828. 9.28 μm CO_2 pump laser; R18 transition; perpendicular polarization.
5829. Threshold pump power: 2000 mW.
5830. Output power: 0.02 mW.
5831. 9.22 μm CO_2 pump laser; R30 transition; parallel polarization.
5832. Threshold pump power: 10000 mW.
5833. Output power: 0.005 mW.
5834. 10.16 μm CO_2 pump laser; R34 transition; unspecified polarization.
5835. Threshold pump power: 20000 mW.
5836. Output power: 0.005 mW.
5837. 9.59 μm CO_2 pump laser; P24 transition; parallel polarization.
5838. Threshold pump power: 5000 mW.
5839. Output power: 0.005 mW.
5840. 9.54 μm CO_2 pump laser; P18 transition; perpendicular polarization.
5841. Threshold pump power: 10000 mW.
5842. Output power: 0.02 mW.
5843. 10.22 μm CO_2 pump laser; R24 transition; either parallel or perpendicular polarization.
5844. Threshold pump power: 10000 mW.
5845. Output power: 0.005 mW.
5846. 10.16 μm CO_2 pump laser; R34 transition; parallel polarization.
5847. Threshold pump power: 5000 mW.
5848. Output power: 0.005 mW.
5849. 9.20 μm CO_2 pump laser; R34 transition; parallel polarization.
5850. Threshold pump power: 10000 mW.
5851. Output power: 0.002 mW.
5852. 10.15 μm CO_2 pump laser; R36 transition; unspecified polarization.
5853. Threshold pump power: 5000 mW.
5854. Output power: 0.002 mW.
5855. 10.21 μm CO_2 pump laser; R26 transition; unspecified polarization.
5856. Threshold pump power: 2000 mW.
5857. Output power: 0.0005 mW.
5858. 10.89 μm CO_2 pump laser; P46 transition; unspecified polarization.
5859. Threshold pump power: 1000 mW.
5860. Output power: 0.002 mW.
5861. 9.59 μm CO_2 pump laser; P24 transition; unspecified polarization.
5862. Threshold pump power: 2000 mW.
5863. Output power: 0.01 mW.
5864. 9.54 μm CO_2 pump laser; P18 transition; unspecified polarization.
5865. Threshold pump power: 5000 mW.
5866. Output power: 0.0005 mW.
5867. 9.57 μm CO_2 pump laser; P22 transition; unspecified polarization.
5868. Threshold pump power: 5000 mW.
5869. Output power: 0.002 mW.

5870.   10.46 µm CO_2 pump laser; P06 transition; unspecified polarization.
5871.   Threshold pump power: 2000 mW.
5872.   Output power: 1 mW.
5873.   9.55 µm CO_2 pump laser; P20 transition; parallel polarization.
5874.   Threshold pump power: 500 mW.
5875.   Output power: 0.02 mW.
5876.   10.79 µm CO_2 pump laser; P38 transition; unspecified polarization.
5877.   Threshold pump power: 1000 mW.
5878.   Output power: 0.01 mW.
5879.   10.55 µm CO_2 pump laser; P16 transition; unspecified polarization.
5880.   Threshold pump power: 1000 mW.
5881.   Output power: 0.0001 mW.
5882.   10.37 µm CO_2 pump laser; R04 transition; unspecified polarization.
5883.   Threshold pump power: 5000 mW.
5884.   Output power: 0.001 mW.
5885.   9.52 µm CO_2 pump laser; P16 transition; unspecified polarization.
5886.   Threshold pump power: 5000 mW.
5887.   Output power: 0.0005 mW.
5888.   10.74 µm CO_2 pump laser; P34 transition; unspecified polarization.
5889.   Threshold pump power: 5000 mW.
5890.   Output power: 1 mW.
5891.   10.61 µm CO_2 pump laser; P22 transition; perpendicular polarization.
5892.   Threshold pump power: 2000 mW.
5893.   Output power: 5e-05 mW.
5894.   9.47 µm CO_2 pump laser; P10 transition; unspecified polarization.
5895.   Threshold pump power: 5000 mW.
5896.   Output power: 0.001 mW.
5897.   10.26 µm CO_2 pump laser; R18 transition; unspecified polarization.
5898.   Threshold pump power: 5000 mW.
5899.   Output power: 0.01 mW.
5900.   10.55 µm CO_2 pump laser; P16 transition; unspecified polarization.
5901.   Threshold pump power: 1000 mW.
5902.   Output power: 0.002 mW.
5903.   10.20 µm CO_2 pump laser; R28 transition; unspecified polarization.
5904.   Threshold pump power: 5000 mW.
5905.   Output power: 0.5 mW.
5906.   10.61 µm CO_2 pump laser; P22 transition; perpendicular polarization.
5907.   Threshold pump power: 1000 mW.
5908.   For line assignments see References 1387, 1388, 1465.
5909.   Output power: 0.1 mW.
5910.   10.53 µm CO_2 pump laser; P14 transition; unspecified polarization.
5911.   Threshold pump power: 1000 mW.
5912.   For line assignments see Reference 1389 (pump).
5913.   Output power: 5 mW.
5914.   9.68 µm CO_2 pump laser; P34 transition; perpendicular polarization.
5915.   Threshold pump power: 10000 mW.
5916.   For line assignments see Reference 1389 (pump).
5917.   Output power: 1 mW.
5918.   10.16 µm CO_2 pump laser; R34 transition; unspecified polarization.
5919.   Threshold pump power: 2000 mW.
5920.   For line assignments see Reference 1389 (pump).
5921.   Output power: 10 mW.

5922.   10.18 μm CO_2 pump laser; R30 transition; parallel polarization.
5923.   Threshold pump power: 10000 mW.
5924.   For line assignments see Reference 1389 (pump).
5925.   Output power: 2 mW.
5926.   9.62 μm CO_2 pump laser; P28 transition; perpendicular polarization.
5927.   Threshold pump power: 10000 mW.
5928.   For line assignments see Reference 1389 (pump).
5929.   Output power: 0.1 mW.
5930.   10.22 μm CO_2 pump laser; R24 transition; unspecified polarization.
5931.   Threshold pump power: 2000 mW.
5932.   For line assignments see Reference 1389 (pump).
5933.   Output power: 10 mW.
5934.   10.25 μm CO_2 pump laser; R20 transition; parallel polarization.
5935.   Threshold pump power: 10000 mW.
5936.   For line assignments see Reference 1389 (pump).
5937.   Output power: 5 mW.
5938.   10.29 μm CO_2 pump laser; R14 transition; parallel polarization.
5939.   Threshold pump power: 10000 mW.
5940.   For line assignments see Reference 1389 (pump).
5941.   Output power: 5 mW.
5942.   10.14 μm CO_2 pump laser; R38 transition; perpendicular polarization.
5943.   Threshold pump power: 10000 mW.
5944.   For line assignments see Reference 1389 (pump).
5945.   Output power: 0.1 mW.
5946.   9.55 μm CO_2 pump laser; P20 transition; unspecified polarization.
5947.   Threshold pump power: 2000 mW.
5948.   For line assignments see Reference 1389 (pump).
5949.   Output power: 0.2 mW.
5950.   10.37 μm CO_2 pump laser; R04 transition; unspecified polarization.
5951.   Threshold pump power: 2000 mW.
5952.   For line assignments see Reference 1389 (pump).
5953.   Output power: 0.1 mW.
5954.   9.37 μm CO_2 pump laser; R04 transition; unspecified polarization.
5955.   Threshold pump power: 2000 mW.
5956.   For line assignments see Reference 1389 (pump).
5957.   Output power: 1 mW.
5958.   9.34 μm CO_2 pump laser; R08 transition; unspecified polarization.
5959.   Threshold pump power: 2000 mW.
5960.   For line assignments see References 1387, 1388, 1465.
5961.   Output power: 0.1 mW.
5962.   10.61 μm CO_2 pump laser; P22 transition; parallel polarization.
5963.   For line assignments see References 1387, 1388, 1465.
5964.   Output power: 0.1 mW.
5965.   10.61 μm CO_2 pump laser; P22 transition; parallel polarization.
5966.   For line assignments see References 1387, 1388, 1465.
5967.   Output power: 0.1 mW.
5968.   10.61 μm CO_2 pump laser; P22 transition; parallel polarization.
5969.   For line assignments see References 1387, 1388, 1465.
5970.   Output power: 1 mW.
5971.   10.51 μm CO_2 pump laser; P12 transition; unspecified polarization.
5972.   Threshold pump power: 1000 mW.

5973. For line assignments see References 1387, 1388, 1465.
5974. Output power: 0.001 mW.
5975. 10.49 μm CO_2 pump laser; P10 transition; unspecified polarization.
5976. Threshold pump power: 2000 mW.
5977. For line assignments see References 1387, 1388, 1465.
5978. Output power: 0.2 mW.
5979. 10.49 μm CO_2 pump laser; P10 transition; perpendicular polarization.
5980. Threshold pump power: 2000 mW.
5981. For line assignments see References 1387, 1388, 1465.
5982. Output power: 0.1 mW.
5983. 10.63 μm CO_2 pump laser; P24 transition; parallel polarization.
5984. Threshold pump power: 2000 mW.
5985. For line assignments see References 1387, 1388, 1465.
5986. Output power: 0.1 mW.
5987. 10.63 μm CO_2 pump laser; P24 transition; unspecified polarization.
5988. Threshold pump power: 1000 mW.
5989. For line assignments see References 1387, 1388, 1465.
5990. Output power: 1 mW.
5991. 10.53 μm CO_2 pump laser; P14 transition; perpendicular polarization.
5992. Threshold pump power: 2000 mW.
5993. For line assignments see References 1387, 1388, 1465.
5994. Output power: 0.01 mW.
5995. 10.55 μm CO_2 pump laser; P16 transition; unspecified polarization.
5996. Threshold pump power: 5000 mW.
5997. For line assignments see References 1387, 1388, 1465.
5998. Output power: 0.1 mW.
5999. 10.25 μm CO_2 pump laser; R20 transition; parallel polarization. Pump power: 2000 mW.
6000. For line assignments see References 1387, 1388, 1465.
6001. Output power: 0.1 mW.
6002. 10.70 μm CO_2 pump laser; P30 transition; parallel polarization.
6003. For line assignments see References 1387, 1388, 1465.
6004. Output power: 0.1 mW.
6005. 10.53 μm CO_2 pump laser; P14 transition; parallel polarization.
6006. Threshold pump power: 2000 mW.
6007. For line assignments see References 1387, 1388, 1465.
6008. Output power: 1 mW.
6009. 10.51 μm CO_2 pump laser; P12 transition; parallel polarization.
6010. Threshold pump power: 2000 mW.
6011. For line assignments see References 1387, 1388, 1465.
6012. Output power: 0.1 mW.
6013. 10.51 μm CO_2 pump laser; P12 transition; parallel polarization.
6014. $^{79}Br^{81}Br$ mixed 51%, 49%; for line assignments see Reference 1424.
6015. Output power: 0.5 mW.
6016. 10.59 μm CO_2 pump laser; P20 transition; parallel polarization.
6017. Threshold pump power: 2000 mW.
6018. $^{79}Br^{81}Br$ mixed 51%, 49%; for line assignments see Reference 1424.
6019. Output power: 0.2 mW.
6020. 10.65 μm CO_2 pump laser; P26 transition; parallel polarization.
6021. Threshold pump power: 2000 mW.
6022. $^{79}Br^{81}Br$ mixed 51%, 49%; for line assignments see Reference 1424.

6023.   Output power: 0.5 mW.
6024.   10.25 μm CO_2 pump laser; R20 transition; parallel polarization.
6025.   Threshold pump power: 2000 mW.
6026. $^{79}Br^{81}Br$ mixed 51%, 49%; for line assignments see Reference 1424.
6027.   Output power: 0.2 mW.
6028.   10.63 μm CO_2 pump laser; P24 transition; parallel polarization.
6029.   Threshold pump power: 2000 mW.
6030. $^{79}Br^{81}Br$ mixed 51%, 49%; for line assignments see Reference 1424.
6031.   Output power: 0.2 mW.
6032.   10.30 μm CO_2 pump laser; R12 transition; parallel polarization.
6033.   Threshold pump power: 2000 mW.
6034. $^{79}Br^{81}Br$ mixed 51%, 49%; for line assignments see Reference 1424.
6035.   Output power: 0.05 mW.
6036.   10.37 μm CO_2 pump laser; R04 transition; parallel polarization.
6037.   Threshold pump power: 1000 mW.
6038. $^{79}Br^{81}Br$ mixed 51%, 49%; for line assignments see Reference 1424.
6039.   Output power: 0.5 mW.
6040.   10.55 μm CO_2 pump laser; P16 transition; perpendicular polarization.
6041.   Threshold pump power: 2000 mW.
6042. $^{79}Br^{81}Br$ mixed 51%, 49%; for line assignments see Reference 1424.
6043.   Output power: 0.1 mW.
6044.   10.38 μm CO_2 pump laser; R02 transition; perpendicular polarization.
6045.   Threshold pump power: 1000 mW.
6046 $^{79}Br^{81}Br$ mixed 51%, 49%; for line assignments see Reference 1424.
6047.   Output power: 5 mW.
6048.   10.49 μm CO_2 pump laser; P10 transition; perpendicular polarization.
6049.   Threshold pump power: 2000 mW.
6050. $^{79}Br^{81}Br$ mixed 51%, 49%; for line assignments see Reference 1424.
6051.   Output power: 0.2 mW.
6052.   10.17 μm CO_2 pump laser; P32 transition; perpendicular polarization.
6053.   Threshold pump power: 2000 mW.
6054. $^{79}Br^{81}Br$ mixed 51%, 49%; for line assignments see Reference 1424.
6055.   Output power: 0.1 mW.
6056.   9.62 μm CO_2 pump laser; P28 transition; parallel polarization.
6057.   Threshold pump power: 2000 mW.
6058. $^{79}Br^{81}Br$ mixed 51%, 49%; for line assignments see Reference 1424.
6059.   Output power: 0.1 mW.
6060.   10.26 μm CO_2 pump laser; R18 transition; perpendicular polarization.
6061.   Threshold pump power: 2000 mW.
6062. $^{79}Br^{81}Br$ mixed 51%, 49%; for line assignments see Reference 1424.
6063.   Output power: 0.2 mW.
6064.   10.49 μm CO_2 pump laser; P10 transition; perpendicular polarization.
6065.   Threshold pump power: 2000 mW.
6066. $^{79}Br^{81}Br$ mixed 51%, 49%; for line assignments see Reference 1424.
6067.   Output power: 0.02 mW.
6068.   10.61 μm CO_2 pump laser; P22 transition; perpendicular polarization.
6069.   Threshold pump power: 2000 mW.
6070. $^{79}Br^{81}Br$ mixed 51%, 49%; for line assignments see Reference 1424.
6071.   Output power: 0.05 mW.
6072.   10.63 μm CO_2 pump laser; P24 transition; parallel polarization.
6073.   Threshold pump power: 2000 mW.

6074. $^{79}Br^{81}Br$ mixed 51%, 49%; for line assignments see Reference 1424.
6075. Output power: 0.2 mW.
6076. 10.29 μm CO_2 pump laser; P14 transition; parallel polarization.
6077. Threshold pump power: 2000 mW.
6078. $^{79}Br^{81}Br$ mixed 51%, 49%, for line assignments see Reference 1424.
6079. Output power: 0.02 mW.
6080. 10.59 μm CO_2 pump laser; P20 transition; parallel polarization.
6081. Threshold pump power: 2000 mW.
6082. $^{79}Br^{81}Br$ mixed 51%, 49%, for line assignments see Reference 1424.
6083. Output power: 0.2 mW.
6084. 10.53 μm CO_2 pump laser; P14 transition; parallel polarization.
6085. Threshold pump power: 2000 mW.
6086. $^{79}Br^{81}Br$ mixed 51%, 49%; for line assignments see Reference 1424.
6087. Output power: 0.005 mW.
6088. 10.32 μm CO_2 pump laser; R10 transition; parallel polarization.
6089. Threshold pump power: 2000 mW.
6090. $^{79}Br^{81}Br$ mixed 51%, 49%; for line assignments see Reference 1424.
6091. Output power: 0.05 mW.
6092. 10.22 μm CO_2 pump laser; R24 transition; parallel polarization.
6093. Threshold pump power: 2000 mW.
6094. $^{79}Br^{81}Br$ mixed 51%, 49%; for line assignments see Reference 1424.
6095. Output power: 1 mW.
6096. 10.27 μm CO_2 pump laser; R16 transition; parallel polarization.
6097. Threshold pump power: 2000 mW.
6098. $^{79}Br^{81}Br$ mixed 51%, 49%; for line assignments see Reference 1424.
6099. Output power: 2 mW.
6100. 10.27 μm CO_2 pump laser; R16 transition; parallel polarization.
6101. Threshold pump power: 2000 mW.
6102. $^{79}Br^{81}Br$ mixed 51%, 49%; for line assignments see Reference 1424.
6103. Output power: 2 mW.
6104. 10.57 μm CO_2 pump laser; P18 transition; parallel polarization.
6105. Threshold pump power: 2000 mW.
6106. $^{79}Br^{81}Br$ mixed 51%, 49%; for line assignments see Reference 1424.
6107. Output power: 0.02 mW.
6108. 10.21 μm CO_2 pump laser; R26 transition; perpendicular polarization.
6109. Threshold pump power: 2000 mW.
6110. $^{79}Br^{81}Br$ mixed 51%, 49%; for line assignments see Reference 1424.
6111. Output power: 1 mW.
6112. 10.21 μm CO_2 pump laser; R26 transition; perpendicular polarization.
6113. Threshold pump power: 2000 mW.
6114. $^{79}Br^{81}Br$ mixed 51%, 49%; for line assignments see Reference 1424.
6115. Output power: 0.5 mW.
6116. 10.26 μm CO_2 pump laser; R18 transition; parallel polarization.
6117. Threshold pump power: 2000 mW.
6118. $^{79}Br^{81}Br$ mixed 51%, 49%; for line assignments see Reference 1424.
6119. Output power: 0.1 mW.
6120. 10.18 μm CO_2 pump laser; R30 transition; parallel polarization.
6121. Threshold pump power: 2000 mW.
6122. $^{79}Br^{81}Br$ mixed 51%, 49%; for line assignments see Reference 1424.
6123. Output power: 0.05 mW.
6124. 10.72 μm CO_2 pump laser; P32 transition; perpendicular polarization.

6125.   Threshold pump power: 2000 mW.
6126. $^{79}Br^{81}Br$ mixed 51%, 49%; for line assignments see Reference 1424.
6127.   Output power: 0.2 mW.
6128.   10.81 μm CO_2 pump laser; P40 transition; perpendicular polarization.
6129.   Threshold pump power: 1000 mW.
6130. $^{79}Br^{81}Br$ mixed 51%, 49%; for line assignments see Reference 1424.
6131.   Output power: 0.2 mW.
6132.   10.13 μm CO_2 pump laser; R40 transition; perpendicular polarization.
6133.   Threshold pump power: 1000 mW.
6134. $^{79}Br^{81}Br$ mixed 51%, 49%; for line assignments see Reference 1424.
6135.   Output power: 0.1 mW.
6136.   10.14 μm CO_2 pump laser; R38 transition; parallel polarization.
6137.   Threshold pump power: 1000 mW.
6138. $^{79}Br^{81}Br$ mixed 51%, 49%; for line assignments see Reference 1424.
6139.   Output power: 0.5 mW.
6140.   10.55 μm CO_2 pump laser; P16 transition; parallel polarization.
6141.   Threshold pump power: 2000 mW.
6142. $^{79}Br^{81}Br$ mixed 51%, 49%; for line assignments see Reference 1424.
6143.   Output power: 0.2 mW.
6144.   10.65 μm CO_2 pump laser; P26 transition; perpendicular polarization.
6145.   Threshold pump power: 2000 mW.
6146. $^{79}Br^{81}Br$ mixed 51%, 49%; for line assignments see Reference 1424.
6147.   Output power: 0.2 mW.
6148.   10.61 μm CO_2 pump laser; P22 transition; perpendicular polarization.
6149.   Threshold pump power: 2000 mW.
6150. $^{779}Br^{81}Br$ mixed 51%, 49%; for line assignments see Reference 1424.
6151.   Output power: 0.2 mW.
6152.   10.63 μm CO_2 pump laser; P24 transition; perpendicular polarization.
6153.   Threshold pump power: 2000 mW.
6154. $^{79}Br^{81}Br$ mixed 51%, 49%; for line assignments see Reference 1424.
6155.   Output power: 0.2 mW.
6156.   10.67 μm CO_2 pump laser; P28 transition; parallel polarization.
6157.   Threshold pump power: 2000 mW.
6158. $^{79}Br^{81}Br$ mixed 51%, 49%; for line assignments see Reference 1424.
6159.   Output power: 0.005 mW.
6160.   10.22 μm CO_2 pump laser; R24 transition; parallel polarization.
6161.   Threshold pump power: 2000 mW.
6162. $^{79}Br^{81}Br$ mixed 51%, 49%; for line assignments see Reference 1424.
6163.   Output power: 0.2 mW.
6164.   10.59 μm CO_2 pump laser; P20 transition; parallel polarization.
6165.   Threshold pump power: 2000 mW.
6166. $^{79}Br^{81}Br$ mixed 51%, 49%; for line assignments see Reference 1424.
6167.   Output power: 0.02 mW.
6168.   10.17 μm CO_2 pump laser; R32 transition; parallel polarization.
6169.   Threshold pump power: 2000 mW.
6170. $^{79}Br^{81}Br$ mixed 51%, 49%; for line assignments see Reference 1424.
6171.   Output power: 0.1 mW.
6172.   10.23 μm CO_2 pump laser; R22 transition; perpendicular polarization.
6173.   Threshold pump power: 2000 mW.
6174. $^{79}Br^{81}Br$ mixed 51%, 49%; for line assignments see Reference 1424.
6175.   Output power: 0.02 mW.

6176.   10.21 μm CO_2 pump laser; R26 transition; parallel polarization.
6177.   Threshold pump power: 2000 mW.
6178. $^{79}Br^{81}Br$ mixed 51%, 49%; for line assignments see Reference 1424.
6179.   Output power: 0.02 mW.
6180.   9.68 μm CO_2 pump laser; P34 transition; parallel polarization.
6181.   Threshold pump power: 2000 mW.
6182. $^{79}Br^{81}Br$ mixed 51%, 49%; for line assignments see Reference 1424.
6183.   Output power: 0.01 mW.
6184.   10.67 μm CO_2 pump laser; P28 transition; parallel polarization.
6185.   Threshold pump power: 2000 mW.
6186. $^{779}Br^{81}Br$ mixed 51%, 49%; for line assignments see Reference 1424.
6187.   Output power: 0.05 mW.
6188.   10.25 μm CO_2 pump laser; R20 transition; perpendicular polarization.
6189.   Threshold pump power: 2000 mW.
6190. $^{79}Br^{81}Br$ mixed 51%, 49%;.For line assignments see Reference 1424.
6191.   Output power: 0.05 mW.
6192.   10.25 μm CO_2 pump laser; R20 transition; parallel polarization.
6193.   Threshold pump power: 2000 mW.
6194.   For line assignments see Reference 1437.
6195.   Output power: 0.05 mW.
6196.   9.79 μm CO_2 pump laser; P46 transition; parallel polarization.
6197.   Threshold pump power: 5000 mW.
6198.   For line assignments see Reference 1437.
6199.   Output power: 0.1 mW.
6200.   10.26 μm CO_2 pump laser; R18 transition; parallel polarization.
6201.   Threshold pump power: 5000 mW.
6202.   For line assignments see Reference 1437.
6203.   Output power: 0.02 mW.
6204.   9.37 μm CO_2 pump laser; R04 transition; unspecified polarization.
6205.   Threshold pump power: 1000 mW.
6206.   For line assignments see Reference 1437.
6207.   Output power: 0.01 mW.
6208.   9.37 μm CO_2 pump laser; R04 transition; unspecified polarization.
6209.   Threshold pump power: 2000 mW.
6210.   For line assignments see Reference 1437.
6211.   Output power: 0.1 mW.
6212   9.79 μm CO_2 pump laser; P46 transition; perpendicular polarization.
6213.   Threshold pump power: 2000 mW.
6214.   For line assignments see Reference 1437.
6215.   Output power: 0.02 mW.
6216.   9.32 μm CO_2 pump laser; R12 transition; perpendicular polarization.
6217.   Threshold pump power: 5000 mW.
6218.   For line assignments see Reference 1437.
6219.   Output power: 0.05 mW.
6220.   9.28 μm CO_2 pump laser; R18 transition; perpendicular polarization.
6221.   Threshold pump power: 5000 mW.
6222.   For line assignments see Reference 1437.
6223.   Output power: 0.1 mW.
6224.   10.13 μm CO_2 pump laser; R40 transition; parallel polarization.
6225.   Threshold pump power: 2000 mW.
6226.   For line assignments see Reference 1437.

6227.   9.59 μm CO_2 pump laser; P24 transition; parallel polarization.
6228.   For line assignments see Reference 1437.
6229.   Output power: 0.05 mW.
6230.   9.26 μm CO_2 pump laser; R22 transition; perpendicular polarization.
6231.   Threshold pump power: 10000 mW.
6232.   For line assignments see Reference 1437.
6233.   Output power: 0.05 mW.
6234.   10.51 μm CO_2 pump laser; P12 transition; parallel polarization.
6235.   Threshold pump power: 10000 mW.
6236.   For line assignments see Reference 1437.
6237.   Output power: 0.02 mW.
6238.   10.25 μm CO_2 pump laser; R20 transition; parallel polarization.
6239.   Threshold pump power: 10000 mW.
6240.   For line assignments see Reference 1437.
6241.   Output power: 0.02 mW.
6242   9.66 μm CO_2 pump laser; P32 transition; parallel polarization.
6243.   Threshold pump power: 10000 mW.
6244.   For line assignments see Reference 1437.
6245.   9.59 μm CO_2 pump laser; P24 transition; parallel polarization.
6246.   For line assignments see Reference 1437.
6247.   9.59 μm CO_2 pump laser; P24 transition; parallel polarization.
6248.   For line assignments see Reference 1437.
6249.   Output power: 0.05 mW.
6250.   9.31 μm CO_2 pump laser; R14 transition; perpendicular polarization.
6251.   Threshold pump power: 5000 mW.
6252.   For line assignments see Reference 1437.
6253.   Output power: 0.0001 mW.
6254.   9.79 μm CO_2 pump laser; P46 transition; unspecified polarization.
6255.   Threshold pump power: 2000 mW.
6256.   For line assignments see Reference 1437.
6257.   Output power: 0.02 mW.
6258   9.57 μm CO_2 pump laser; P22 transition; perpendicular polarization.
6259.   Threshold pump power: 5000 mW.
6260.   For line assignments see Reference 1437.
6261.   Output power: 0.1 mW.
6262.   10.17 μm CO_2 pump laser; R32 transition; perpendicular polarization.
6263.   Threshold pump power: 2000 mW.
6264.   For line assignments see Reference 1437.
6265.   Output power: 0.1 mW.
6266.   9.32 μm CO_2 pump laser; R12 transition; parallel polarization.
6267.   Threshold pump power: 2000 mW.
6268.   For line assignments see Reference 1437.
6269.   Output power: 0.1 mW.
6270.   9.26 μm CO_2 pump laser; R22 transition; parallel polarization.
6271.   Threshold pump power: 5000 mW.
6272.   For line assignments see Reference 1437.
6273.   Output power: 0.02 mW.
6274.   9.66 μm CO_2 pump laser; P32 transition; perpendicular polarization.
6275.   Threshold pump power: 5000 mW.
6276.   For line assignments see Reference 1437.
6277.   Output power: 0.02 mW.

6278.   10.63 μm CO_2 pump laser; P24 transition; unspecified polarization.
6279.   Threshold pump power: 20000 mW.
6280.   For line assignments see Reference 1437.
6281.   Output power: 0.1 mW.
6282   9.28 μm CO_2 pump laser; R18 transition; parallel polarization.
6283.   Threshold pump power: 2000 mW.
6284.   For line assignments see Reference 1437.
6285.   Output power: 10 mW.
6286.   9.59 μm CO_2 pump laser; P24 transition; perpendicular polarization.
6287.   Threshold pump power: 2000 mW.
6288.   For line assignments see Reference 1437.
6289.   Output power: 0.02 mW.
6290.   10.15 μm CO_2 pump laser; R36 transition; parallel polarization.
6291.   Threshold pump power: 5000 mW.
6292.   For line assignments see Reference 1437.
6293.   Output power: 0.02 mW.
6294.   9.34 μm CO_2 pump laser; R08 transition; parallel polarization.
6295.   Threshold pump power: 5000 mW.
6296.   For line assignments see Reference 1437.
6297.   Output power: 0.05 mW.
6298.   9.31 μm CO_2 pump laser; R14 transition; parallel polarization.
6299.   Threshold pump power: 5000 mW.
6300.   For line assignments see Reference 1437.
6301.   Output power: 0.02 mW.
6302.   10.25 μm CO_2 pump laser; R20 transition; perpendicular polarization.
6303.   Threshold pump power: 5000 mW.
6304.   For line assignments see Reference 1437.
6305.   Output power: 0.05 mW.
6306.   10.30 μm CO_2 pump laser; R12 transition; parallel polarization.
6307.   Threshold pump power: 5000 mW.
6308.   For line assignments see Reference 1437.
6309.   Output power: 0.02 mW.
6310.   10.35 μm CO_2 pump laser; R06 transition; perpendicular polarization.
6311.   Threshold pump power: 5000 mW.
6312.   For line assignments see Reference 1437.
6313.   Output power: 0.02 mW.
6314.   10.63 μm CO_2 pump laser; P24 transition; parallel polarization.
6315.   Threshold pump power: 10000 mW.
6316.   For line assignments see Reference 1437.
6317.   Output power: 0.02 mW.
6318.   9.46 μm CO_2 pump laser; P08 transition; perpendicular polarization.
6319.   Threshold pump power: 5000 mW.
6320.   For line assignments see Reference 1437.
6321.   Output power: 0.05 mW.
6322.   9.77 μm CO_2 pump laser; P44 transition; parallel polarization.
6323.   Threshold pump power: 2000 mW.
6324.   For line assignments see Reference 1437.
6325.   Output power: 0.02 mW.
6326.   9.31 μm CO_2 pump laser; R14 transition; parallel polarization.
6327.   Threshold pump power: 5000 mW.
6328.   For line assignments see Reference 1377.

6329.   Output power: 1 mW.

6330.   -15 MHz offset from center of 10.86 μm CO_2 pump laser; P44 transition; parallel polarization.

6331.   Threshold pump power: 2000 mW.

6332.   For line assignments see Reference 1377.

6333.   Output power: 0.5 mW.

6334.   -20 MHz offset from center of 9.46 μm CO_2 pump laser; P08 transition; parallel polarization.

6335.   Threshold pump power: 10000 mW.

6336.   For line assignments see Reference 1377.

6337.   Output power: 0.5 mW.

6338.   0 MHz offset from center of 9.73 μm CO_2 pump laser; P40 transition; parallel polarization.

6339.   Threshold pump power: 20000 mW.

6340.   For line assignments see Reference 1377.

6341.   Output power: 10 mW.

6342.   0 MHz offset from center of 10.53 μm CO_2 pump laser; P14 transition; parallel polarization.

6343.   For line assignments see Reference 1377.

6344.   Output power: 0.005 mW.

6345.   -15 MHz offset from center of 9.55 μm CO_2 pump laser; P20 transition; parallel polarization.

6346.   Threshold pump power: 10000 mW.

6347.   For line assignments see Reference 1377.

6348.   Output power: 0.5 mW.

6349.   +25 MHz offset from center of 9.54 μm CO_2 pump laser; P18 transition; parallel polarization.

6350.   Threshold pump power: 50000 mW.

6351.   For line assignments see Reference 1377.

6352.   Output power: 0.005 mW.

6353.   0 MHz offset from center of 9.44 μm CO_2 pump laser; P06 transition; parallel polarization.

6354.   Threshold pump power: 10000 mW.

6355.   For line assignments see Reference 1377.

6356.   Output power: 0.005 mW.

6357.   +35 MHz offset from center of 9.32 μm CO_2 pump laser; R12 transition; parallel polarization.

6358.   Threshold pump power: 20000 mW.

6359.   For line assignments see Reference 1377.

6360.   Output power: 0.01 mW.

6361.   +35 MHz offset from center of 9.18 μm CO_2 pump laser; R38 transition; parallel polarization.

6362.   Threshold pump power: 10000 mW.

6363.   Output power: 0.0005 mW.

6364.   10.14 μm CO_2 pump laser; R38 transition; unspecified polarization.

6365.   Threshold pump power: 2000 mW.

6366.   Output power: 5e-05 mW.

6367.   10.65 μm CO_2 pump laser; P26 transition; unspecified polarization.

6368.   Threshold pump power: 5000 mW.

6369.   Output power: 0.0001 mW.

6370.   10.68 μm CO_2 pump laser; P28 transition; unspecified polarization.

6371.   Threshold pump power: 5000 mW.
6372.   Output power: 0.0001 mW.
6373.   10.30 μm CO_2 pump laser; R12 transition; unspecified polarization.
6374.   Threshold pump power: 10000 mW.
6375.   Output power: 0.0001 mW.
6376.   10.26 μm CO_2 pump laser; R18 transition; unspecified polarization.
6377.   Threshold pump power: 5000 mW.
6378.   Output power: 0.0001 mW.
6379.   10.13 μm CO_2 pump laser; R40 transition; unspecified polarization.
6380.   Threshold pump power: 5000 mW.
6381.   Output power: 0.0001 mW.
6382.   10.12 μm CO_2 pump laser; R42 transition; unspecified polarization.
6383.   Threshold pump power: 2000 mW.
6384.   Output power: 5e-05 mW.
6385.   10.55 μm CO_2 pump laser; P16 transition; unspecified polarization.
6386.   Threshold pump power: 5000 mW.
6387.   Output power: 5e-05 mW.
6388.   10.84 μm CO_2 pump laser; P42 transition; unspecified polarization.
6389.   Threshold pump power: 5000 mW.
6390.   Output power: 0.0001 mW.
6391.   10.35 μm CO_2 pump laser; R06 transition; unspecified polarization.
6392.   Threshold pump power: 5000 mW.
6393.   Output power: 0.0001 mW.
6394.   10.30 μm CO_2 pump laser; R12 transition; unspecified polarization.
6395.   Threshold pump power: 10000 mW.
6396.   Output power: 0.1 mW.
6397.   10.59 μm CO_2 pump laser; P20 transition; parallel polarization.
6398.   Output power: 0.1 mW.
6399.   10.51 μm CO_2 pump laser; P12 transition; perpendicular polarization.
6400.   Output power: 0.001 mW.
6401.   10.29 μm CO_2 pump laser; R14 transition; unspecified polarization.
6402.   Threshold pump power: 2000 mW.
6403.   Output power: 0.1 mW.
6404.   10.27 μm CO_2 pump laser; R16 transition; perpendicular polarization.
6405.   Output power: 0.1 mW.
6406.   10.53 μm CO_2 pump laser; P14 transition; parallel polarization.
6407.   Output power: 5e-05 mW.
6408.   9.32 μm CO_2 pump laser; R12 transition; unspecified polarization.
6409.   Threshold pump power: 10000 mW.
6410.   Output power: 5e-05 mW.
6411.   10.48 μm CO_2 pump laser; P08 transition; unspecified polarization.
6412.   Threshold pump power: 5000 mW.
6413.   Output power: 0.1 mW.
6414.   10.63 μm CO_2 pump laser; P24 transition; perpendicular polarization.
6415.   Output power: 0.01 mW.
6416.   9.47 μm CO_2 pump laser; P10 transition; unspecified polarization.
6417.   Threshold pump power: 5000 mW.
6418.   Output power: 0.001 mW.
6419.   9.46 μm CO_2 pump laser; P08 transition; unspecified polarization.
6420.   Threshold pump power: 2000 mW.
6421.   Output power: 0.002 mW.

6422.   9.33 μm CO_2 pump laser; R10 transition; unspecified polarization.
6423.   Threshold pump power: 2000 mW.
6424.   Output power: 0.01 mW.
6425.   9.62 μm CO_2 pump laser; P28 transition; unspecified polarization.
6426.   Threshold pump power: 2000 mW.
6427.   Output power: 0.002 mW.
6428.   9.64 μm CO_2 pump laser; P30 transition; unspecified polarization.
6429.   Threshold pump power: 2000 mW.
6430.   Output power: 0.001 mW.
6431.   9.29 μm CO_2 pump laser; R16 transition; unspecified polarization.
6432.   Threshold pump power: 5000 mW.
6433.   Output power: 0.001 mW.
6434.   9.57 μm CO_2 pump laser; P22 transition; unspecified polarization.
6435.   Threshold pump power: 2000 mW.
6436.   Output power: 0.01 mW.
6437.   9.69 μm CO_2 pump laser; P36 transition; unspecified polarization.
6438.   Threshold pump power: 1000 mW.
6439.   Output power: 0.01 mW.
6440.   9.37 μm CO_2 pump laser; R04 transition; unspecified polarization.
6441.   Threshold pump power: 2000 mW.
6442.   Output power: 0.01 mW.
6443.   9.25 μm CO_2 pump laser; R24 transition; unspecified polarization.
6444.   Threshold pump power: 1000 mW.
6445.   Output power: 0.005 mW.
6446.   9.25 μm CO_2 pump laser; R24 transition; unspecified polarization.
6447.   Threshold pump power: 2000 mW.
6448.   Output power: 0.001 mW.
6449.   9.26 μm CO_2 pump laser; R22 transition; unspecified polarization.
6450.   Threshold pump power: 2000 mW.
6451.   Output power: 0.01 mW.
6452.   9.22 μm CO_2 pump laser; R30 transition; unspecified polarization.
6453.   Threshold pump power: 1000 mW.
6454.   Output power: 0.0005 mW.
6455.   9.68 μm CO_2 pump laser; P34 transition; unspecified polarization.
6456.   Threshold pump power: 5000 mW.
6457.   Output power: 0.002 mW.
6458.   9.29 μm CO_2 pump laser; R16 transition; unspecified polarization.
6459.   Threshold pump power: 5000 mW.
6460.   Output power: 0.01 mW.
6461.   9.54 μm CO_2 pump laser; P18 transition; unspecified polarization.
6462.   Threshold pump power: 2000 mW.
6463.   Output power: 0.1 mW.
6464.   10.20 μm CO_2 pump laser; R28 transition; unspecified polarization.
6465.   Threshold pump power: 10000 mW.
6466.   10.14 μm CO_2 pump laser; R38 transition; unspecified polarization.
6467.   Threshold pump power: 10000 mW.
6468.   Output power: 0.01 mW.
6469.   10.18 μm CO_2 pump laser; R30 transition; unspecified polarization.
6470.   Threshold pump power: 10000 mW.
6471.   Output power: 0.1 mW.
6472.   10.18 μm CO_2 pump laser; R30 transition; unspecified polarization.

6473.   Threshold pump power: 10000 mW.
6474.   Output power: 0.1 mW.
6475.   10.59 μm CO_2 pump laser; P20 transition; unspecified polarization.
6476.   Threshold pump power: 1000 mW.
6477.   Output power: 0.1 mW.
6478.   10.59 μm CO_2 pump laser; P20 transition; unspecified polarization.
6479.   Threshold pump power: 1000 mW.
6480.   Output power: 0.01 mW.
6481.   9.66 μm CO_2 pump laser; P32 transition; unspecified polarization.
6482.   Threshold pump power: 10000 mW.
6483.   Output power: 0.05 mW.
6484.   9.25 μm CO_2 pump laser; R24 transition; unspecified polarization.
6485.   Threshold pump power: 1000 mW.
6486.   Output power: 0.05 mW.
6487.   9.27 μm CO_2 pump laser; R20 transition; unspecified polarization.
6488.   Threshold pump power: 1000 mW.
6489.   Output power: 0.05 mW.
6490.   9.24 μm CO_2 pump laser; R26 transition; unspecified polarization.
6491.   Threshold pump power: 1000 mW.
6492.   Output power: 0.05 mW.
6493.   9.66 μm CO_2 pump laser; P32 transition; unspecified polarization.
6494.   Threshold pump power: 1000 mW.
6495.   Output power: 0.05 mW.
6496.   10.57 μm CO_2 pump laser; P18 transition; unspecified polarization.
6497.   Threshold pump power: 1000 mW.
6498.   Output power: 0.05 mW.
6499.   9.21 μm CO_2 pump laser; R32 transition; unspecified polarization.
6500.   Threshold pump power: 1000 mW.
6501.   Output power: 0.05 mW.
6502.   9.29 μm CO_2 pump laser; R16 transition; unspecified polarization.
6503.   Threshold pump power: 1000 mW.
6504.   Output power: 0.05 mW.
6505.   10.74 μm CO_2 pump laser; P34 transition; unspecified polarization.
6506.   Threshold pump power: 1000 mW.
6507.   Output power: 0.05 mW.
6508.   10.23 μm CO_2 pump laser; R22 transition; unspecified polarization.
6509.   Threshold pump power: 1000 mW.
6510.   Output power: 0.05 mW.
6511.   10.81 μm CO_2 pump laser; P40 transition; unspecified polarization.
6512.   Threshold pump power: 1000 mW.
6513.   Output power: 0.05 mW.
6514.   9.27 μm CO_2 pump laser; R22 transition; unspecified polarization.
6515.   Threshold pump power: 1000 mW.
6516.   Output power: 0.05 mW.
6517.   10.86 μm CO_2 pump laser; P44 transition; unspecified polarization.
6518.   Threshold pump power: 1000 mW.
6519.   Output power: 0.05 mW.
6520.   9.22 μm CO_2 pump laser; R30 transition; unspecified polarization.
6521.   Threshold pump power: 1000 mW.
6522.   It is not clear if one or both components of the doublet were observed.
6523.   Pulsed. Recombination in segmented plasma excitation recombination device.

6524. Pulsed. Photodissociation of LiI by ArF laser at 193 nm.

6525. Pulsed. Dye laser excitation of sodium vapor.

6526. Pulsed. Dye laser excitation of sodium vapor.

6527. Pulsed. Dye laser excitation of sodium vapor.

6528. CW. Optical pumping of rubidium vapor by dye laser.

6529. CW. Optical pumping of rubidium vapor by dye laser.

6530. Pulsed, recombination laser in segmented plasma excitation recombination device.

6531. Pulsed, recombination laser in segmented plasma excitation recombination device.

6532. Pulsed, recombination laser in segmented plasma excitation recombination device.

6533. Pulsed, recombination laser in segmented plasma excitation recombination device.

6534. Pulsed, recombination laser in segmented plasma excitation recombination device.

6535. Pulsed, recombination laser in segmented plasma excitation recombination device.

6536. Pulsed, recombination laser in segmented plasma excitation recombination device.

6537. Pulsed, longitudinal discharge in Ba vapor with He or Ne buffer.

6538. Pulsed, longitudinal discharge in Ba vapor with He or Ne buffer.

6539. Pulsed, longitudinal discharge in Ba vapor with He or Ne buffer.

6540. Pulsed. Photodissociation of ZnI_2 with KrF laser at 248 nm.

6541. Pulsed. Photodissociation of ZnI_2 with KrF laser at 248 nm.

6542. Pulsed. Photodissociation of CdI_2 with KrF laser at 248 nm.

6543. Pulsed. Photodissociation of CdI_2 with KrF laser at 248 nm.

6544. Pulsed. Photodissociation of $HgBr_2$ by ArF laser at 193 nm.

6545. Pulsed. Recombination laser in segmented plasma excitation recombination device.

6546. Pulsed. Recombination laser in segmented plasma excitation recombination device.

6547. Pulsed. Afterglow in a hollow cathode discharge in CO_2.

6548. Pulsed. Afterglow in a hollow cathode discharge in CO_2.

6549. Pulsed. Afterglow in a hollow cathode discharge in CO_2.

6550. Pulsed. Photodissociation of GeI_4 by ArF laser at 193 nm.

6551. Pulsed. Photodissociation of SnI_2 by ArF laser at 193 nm.

6552. Pulsed. Hollow cathode discharge in N_2.

6553. Pulsed. Hollow cathode discharge in N_2.

6554. Pulsed. Hollow cathode discharge in N_2.

6555. Pulsed. Hollow cathode discharge in N_2.

6556. Pulsed. Photodissociation of SbI_3 by ArF laser at 193 nm.

6557. Pulsed. Recombination in afterglow of hollow cathode discharge in He and Ar.

6558. These two lines were generated in a pulsed longitudinal discharge in Tm vapor with noble gas buffer gas.

6559. Pulsed. Optical excitation of Tm vapor by fourth harmonic of Nd:YAG laser at 266 nm.

6560. Pulsed. Afterglow in hollow cathode discharge in Cl_2.

6561. Pulsed. Afterglow in hollow cathode discharge in Cl_2.

6562. Pulsed. Afterglow in hollow cathode discharge in Br_2.

6563. Pulsed. Afterglow in hollow cathode discharge in Br_2.

6564. Pulsed. Afterglow in hollow cathode discharge in Br_2.

6565. Pulsed. Afterglow in hollow cathode discharge in Br_2.

6566. Pulsed. Afterglow in hollow cathode discharge in Br_2.

6567. 0.58525 μm line was previously observed only in low pressure discharge. Reference 450.

6568. Pulsed, Quasi-cw operation in high pressured He : Ne : Kr(Ar) mixture pumped by electron beam.

6569. Pulsed. Longitudinal discharge.

6570. Pulsed. Longitudinal discharge.

6571. Pulsed. Longitudinal discharge.
6572. Pulsed. Longitudinal discharge.
6573. Pulsed. Longitudinal discharge.
6574. Pulsed. Longitudinal discharge.
6575. Pulsed. Longitudinal discharge.
6576. Pulsed. Longitudinal discharge.
6577. Pulsed. Longitudinal discharge.
6578. Pulsed. Longitudinal discharge.
6579. Pulsed. Longitudinal discharge.
6580. Pulsed. Longitudinal discharge.
6581. Pulsed. Longitudinal discharge.
6582. Pulsed. Longitudinal discharge.
6583. Previously observed in pulsed operation only (see Reference 1507).
6584. Accuracy of measurement ± 0.1 nm.
6586. Previously observed in pulsed operation only (see Reference 1507).
6587. Accuracy of measurement ± 0.1 nm.
6588. Classification or ionization state uncertain.
6589. Previously observed in pulsed operation only (see Reference 1507).
6590. Wavelength from Reference 1585.
6591. Classification or ionization state uncertain.
6592. Previously observed in pulsed operation only (see Reference 1507).
6593. Wavelength from Reference 1585.
6594. Wavelength from Reference 1595.
6595. Classification or ionization state uncertain.
6596. The excited states of this molecule have been shown (Reference 1595) to obey Hund's case (c) coupling and, therefore, the lasing transition is properly denoted as shown.
6597. Similar results were obtained for Kr_2 (145 nm) and Xe_2 (172 nm).
6598– This laser has been tuned over the 124–128 nm region in a coaxial electron beam
6599. pumping configuration. A maximum output power of 16 MW was obtained at 126 nm.
6600– CW oscillation on the A → X band of Bi_2 was obtained by pumping several ind-
6602. ividual vibrational-rotational lines of the A → X band with various single mode lines from an Ar^+ laser. The pump lines, ranging in wavelength from 472.69 to 514.53 nm, correspond to the excitation of $A(O^+_u)$ vibrational levels from v' = 16 to 34. Power conversion efficiencies up to 15% were measured for broadband laser emission and up to 2.5% for a single line. Pump thresholds as low as 40 mW for a single line were observed.
6603– Laser transition first reported by B. Wellegehausen, D. Friede, G. Steger. Optically
6604. pumped continuous Bi_2 and Te_2 lasers, *Opt. Commun.*, 26, 391 (1978); also see W. P. West and H. P. Broida, Optically pumped vapor phase Bi_2 laser, *Chem. Phys.*, 56, 283-285 (1978).
6605– Lasing observed upon exciting 1.5 to 22.3 Torr of bromine vapor with frequency-
6606. doubled Nd:YAG (532 nm) radiation. The pump wavelength could be scanned over ~ 0.1 nm with an intracavity etalon. Although not published, it appears that output wavelengths as long as ~ 3.5 μm are possible with this system.
6607. λ_{pump} = 531.903 nm.
6608– Lasing was obtained in gas mixture composed of 2.7 bar He, 2 mb HBr and 2 mb
6609. NF_3 that was excited by a UV-preionized transverse discharge 54 cm in length. This same device was also used in observing stimulated emission in ClF and IF.

Strongest emission occurred at 354.5 nm but a much weaker line was also observed at 354.2 nm. The output coupling was 30%.

6610– Lasing obtained by two color excitation of C_2 in a (C_2H_2:Ar; T = 4.2 K) matrix.
6611. The first photon (v = 17658 cm^{-1}) pumped the A$^1\pi_u \rightarrow$ X$^1\Sigma^+_g$(6,0) transition and the second (v = 19632 cm^{-1}) excited the d$^3 \pi_g \rightarrow$ a$^3\pi_u$ (1,1) line.

6612. Lasing was observed upon photodissociating $CdBr_2$ (in the presence of 5 to 500 Torr of Ar buffer) at 193 nm (ArF laser). Strongest emission was observed at 811 nm with weaker features at 813 and 816 nm.

6613– By photodissociating CdI_2 (natural abundance) and $^{114}CdI_2$ with an ArF laser, las-
6615. ing was observed at several wavelengths in the vicinity of 655 nm. For natural abundance CdI_2, strongest lasing occurred on two closely spaced lines at 655.0 and 655.3 nm. With $^{114}CdI_2$, however, the spectrum changes dramatically and now consists of two transitions of nearly equal intensity at 655.3 and 657.1 nm. The laser output energy quadruples when using the isotopically enriched salt. Neither laser spectrum was rotationally resolved but the strongest features in the ^{114}CdI spectrum were attributed to B $\rightarrow$ X, v' = 0-2 $\rightarrow$ v" = 61, 62 transitions.

6616. Experiments in Reference 1530 also photodissociated CdI_2 with an ArF laser, demonstrating lasing in the red and blue (475 nm) from CdI. Both laser transitions were attributed to B$^2\Sigma^+_{1/2} \rightarrow$ X$^2\Sigma^+_{1/2}$.

6617– Lasing was observed in a transverse discharge of 50 cm excitation length and con-
6618. taining He, N_2 and CdI_2 or $^{114}CdI_2$ vapor. Natural abundance CdI_2 yielded over 30 distinguishable lines between 655 and 660 nm. References 1531, 1623.

6619. Injection locking experiments with a ^{114}CdI discharge laser showed that the slave oscillator could be completely locked over the 655 to 660 nm region with injected intensities of 5 W cm^{-2}. Reference 1533.

6620– Mixtures of Ne (< 20 bar), 10 mb F_2, and 10 mb Cl_2 were excited by an e-beam in
6621. a coaxial diode. Strongest lasing was observed from two lines of nearly equal intensity at 284.4 and 284.9 nm. Weaker transitions were also observed at 286.0 and in the 282 to 283.5 nm region. The e-beam excitation pulse width was 3 ns.

6622– Lasing at 284.4 nm (and very weak emission at 284.0 nm) was obtained in 3.3 bar
6623. He/2.7 mb Cl_2/4 mb F_2 mixtures excited in a UV-preionized transverse discharge having an active length of 54 cm. Other halogen donors (ClIF and NF_3) and the substitution of Ne for He also successfully yielded lasing but the results were less satisfactory than those obtained with the He/Cl_2/F_2 mixtures.

6624. Observed following two-photon excitation of H_2 at 193 nm with a tunable ($\Delta v \sim$ 5 cm^{-1}) ArF laser. The two-photon transition is E, F $^1\Sigma^+_g \rightarrow$ X$^1\Sigma^+_g$.

6625. Transition observed upon pumping the Q(3) transition of the E, F $^1\Sigma^+_g$ (0,2) band (v = 103282 cm^{-1}) in ~20 Torr of H_2.

6626. Transition observed upon pumping the Q(2) transition of the E, F $^1\Sigma^+_g$ (0,2) band (v = 103328 cm^{-1}) in 20 Torr of H_2.

6627. Transition observed at high (>600 Torr) H_2 pressure.

6628. Lasing from the Hg_3 trimer was observed in the blue-green by photodissociating $HgBr_2$ with an F_2 (λ = 158 nm) laser. Peak output occurred at 495 nm and the small signal gain coefficient was estimated to be 0.4% cm^{-1}.

6629. ~ 2 μs pulses of stimulated emission on the B $\rightarrow$ X band of HgBr were obtained by exploding a 0.1 mm diameter tungsten wire (50 cm in length) in $HgBr_2$ vapor. Specific output energies as large as 15 J l^{-1} were obtained. References 1537, 1538.

6630– Lasing on the B $\rightarrow$ X band of HgBr radicals was demonstrated by first dissociating
6633. $HgBr_2$ (in a discharge) to obtain HgBr(X) molecules and subsequently pumping the B $\rightarrow$ X absorption band (peaking at ~ 350 nm, FWHM ~ 60 nm) with a frequency-tripled Nd:YAG (355 nm) or XeF (351 nm) laser. The energy and

photon conversion efficiencies were measured to be 22 ± 1% and 33%, respectively. Recently, excited state (B$^2\Sigma$) absorption in the near-UV (350 nm) was shown to be the obstacle to long pulse operation of HgBr when pumped in this manner. Previously, this blue-green laser had been excited by e-beam and transverse discharge (see *CRC Handbook of Laser Science and Technology, Vol. II*, for References). Also, lasing has been obtained by photodissociating HgBr$_2$ with a laser (ArF) of λ <200 nm.

6634–
6635.
Isotopically enriched HgBr$_2$ salts (i.e., ^{200}Hg^{81}Br$_2$, ^{200}Hg^{79}Br$_2$ and Hg^{79}Br$_2$) were shown to significantly improve the efficiencies of both the HgBr (B $\rightarrow$ X) photodissociation and discharge-pumped lasers. Improvements in output energy ranging from 15 to 25% were reported. References 1541, 1542.

6636–
6637.
By adding ~ 7.5 Torr of SF$_6$ to the mixture of Ne, N$_2$, and HgBr$_2$ (or HgCl$_2$) vapor (1.8 atm Ne, 100 Torr N$_2$, 6.3 Torr HgBr$_2$ vapor), the blue-green output energy increased by 50% Similar improvement (40%) was observed for the HgCl laser. Reference 1543.

6638.
Simultaneous multi-wavelength operation of the mercury-halide lasers (HgI, HgBr, and HgCl) has been achieved by intense optical pumping or with a fast transverse discharge. References 1544, 1545.

6639–
6640.
Lasing on the B $\rightarrow$ X band of HgCl was obtained by photodissociating HgCl$_2$ vapor with the broadband optical radiation from a high current surface discharge (estimated brightness temperature of 2.5 x 10^4 K). The HgCl$_2$ vapor partial pressure was ~1 Torr and the buffer consisted of Ar/H$_2$ in the ratio 1:1.5 at a total pressure of 2.5 atm. Output energies as high as 2.1 J (or 7.6 J l^{-1}) were extracted in pulses ~ 3 to 7.6 µs in duration.

6,641–
6642.
Oscillation in the violet was obtained from the HgI radicals when HgI$_2$ vapor was photodissociated by the incoherent radiation from a surface discharge (brightness temperature ~2.5 X 10^4 K). Energies up to 0.5 J (for a specific output energy of ~ 0.6 J l^{-1}) were obtained in ~ 3.5 µs pulses.

6643–
6644.
Transverse pumping of I$_2$/SF$_6$(or Ar) mixtures (0.2 Torr I$_2$, ~1 atm buffer) by an ArF laser yielded strong lasing at 342 nm (cf. Reference 1549). An output energy of 230 mJ was reported which corresponds to an energy conversion efficiency of 30% or a photon conversion efficiency exceeding 50%.

6645–
6647.
Single pulse energies of 13J in 13 µs pulses (~ 10 J l^{-1} volumetric energy extraction) were obtained by pumping 2 to 4 Torr I$_2$ and 1.5 to 3 atm of buffer (CF$_4$ or SF$_6$) with radiation from an exploding wire. Strongest lasing was observed on the (v',v") = (1,14) and (2,15) transitions of the D' $\rightarrow$ A' band and, while the overall efficiency was ~ 1%, the quantum efficiency was estimated to be 11%. Also, the blackbody color temperature was 30 X 10^3 K. Reference 1550.

6648–
6649.
Improved output energies of 27 J/pulse and 13 µs in duration (18 J/l specific output energy) were obtained in CF$_4$/I$_2$ gas mixtures by broadband optical pumping. The efficiency, defined as the output energy normalized to the energy stored in the capacitors, was 0.27%. Reference 1551.

6650–
6651.
Operation of this UV laser at a pulse repetition frequency of 0.5 Hz was achieved by photoexciting mixtures composed of 3 to 5 Torr I$_2$ and 1 to 2 atm of perfluoromethane with quartz flashlamps. Output energies of 10 to 50 mJ in 3 to 5 µs FWHM pulses were obtained. Reference 1553.

6652–
6653.
Originally attributed to a $^3\Pi_{2g} \rightarrow ^3\Pi_{2u}$ transition, this laser has now been clearly shown by J. Tellinghuisen, *Chem. Phys. Lett.*, 49, 485 (1977), to be assigned D' $\rightarrow$ A' where D' is the lowest-lying excited state in the first tier of ion pair (I$^+$I$^-$) states of the molecule.

6654. Other References for this laser system may be found in Section 3.1 of the *Handbook of Laser Science and Technology, Vol. II.*

6655– Gain was observed over the indicated spectral region in the blue-green in e-beam
6656. pumped Ar (or Ne)/0.3% HI gas mixtures. The rare gas buffer pressure was 5930 Torr and the peak small signal gain coefficient was measured to be ~ 1.1% cm^{-1} at 506 nm. The electronic transition involved has not yet been positively identified. In Ar, the FWHM of the gain spectrum is 13 nm.

6657– Lasing was observed in ~ 40 ns pulses on several transitions of theB → X band of
6660. I_2 when iodine vapor was pumped by the focussed radiation from a copper vapor laser (λ = 510.6 and 578.2 nm). Characteristic doublets arising from P and R transitions and J' = 24, 30, 52, and 81 were observed. The strongest vibronic bands were (v',v") = (14,44), (14,48), (14,49), (16,47), (16,51), and (16,52). Excitation of the upper laser level occurred most efficiently via the v" = 0, J" = 51 and v" = 1, J" = 31 ground state levels which were pumped by the 578.2 nm Cu vapor laser line. The maximum conversion efficiency (into all lines) was ~ 1% for a yellow input power of 0.7 W.

6661– Transient gain was observed in Ar/5 Torr ICl mixtures pumped by a relativistic
6662. electron beam having an excitation length of 50 cm. Maximum gain of 0.3 to 0.7% cm^{-1} was measured at 431.3 nm and amplification was observed from 430.8 to 432.8 nm.

6663– Optical gain as high as 1.3% cm^{-1} was observed at 431.3 nm in a transverse dis-
6664. charge having an excitation length of 100 cm. Amplification (Γ_0 exceeding 0.4% cm^{-1}) of a dye laser probe was observed over the range 430 to 437 nm in $He/CF_3I/CCl_4$ discharges.

6665– Laser oscillation was realized in transverse discharges (UV-preionized) in $He/NF_3/$
6666. CF_3I gas mixtures; the 490.7 nm line was also observed by electron beam pumping in Reference 1527. The maximum instantaneous power measured to date is 140 kW or an energy per pulse of ~ 2 mJ.

6667– 100 mJ, 4 µs pulseswere obtained in the blue-green by pumping the $Ar/NF_3/CF_3I$
6668. mixtures ($Ar:NF_3:CF_3I$ = 1870:2:1) with the VUV radiation produced by an exploding tungsten wire. A total of nine lasing transitions were observed— two near 485 nm and the remainder in the vicinity of 491 nm.

6669– Gaseous mixtures of CF_3I, NF_2 and He were optically pumped by a pulsed dye
6674. laser to yield lasing on various ro-vibrational transitions of the IF(B → X) band. NF_2 radicals were produced by the pyrolysis of N_2F_4 at 170 °C. Subsequently, 10 ns pulses from a quadrupled Nd-glass laser (λ = 264 nm) photodissociated the NF_2 species to generate free F atoms. IF molecules in the ground (X) state were then formed by the reaction of F with CF_3I. The $B^3 Pi^+_0 → X^1\Sigma^+$ lines selected for dye laser pumping were the P(20) and R(27) transitions of the v' = 5 → v" = 0 vibrational band at 478.84 nm. Lasing occurred on a variety of B → X transitions, depending upon the He buffer pressure. For 3 < p_{He} < 30 Torr, lasing originated solely from v' = 5 states and the terminal vibrational levels were v" = 11, 15, and 16. The rotational lines observed were (as indicated above) R(18), R(27), P(20), and P(29). For He pressures above ~ 70 Torr, lasing was also observed from v' = 0 and 1 states. Transitions recorded were the P(14) - P(40) and R(14) - R(37) lines of the v' = 0 → v" = 4 to 6 vibrational bands. When p_{He}= 1000 Torr, only the (v', v") = (0,4), (0,5), and (0,6) transitions were observed to lase.

6675– CW oscillation on a series of ro-vibronic lines of the IF(B → X) band, originating
6676. from v' = 2 to 5, was observed when the molecule was pumped by a ring dye laser (~ 150 mW). For the indicated laser transitions, the pump was tuned to the R(17), R(18), P(32), or R(36) lines of the v' = 3 → v" = 0 band. Lasing was observed

only from the (v', J') level excited directly by the dye laser and oscillation could not be obtained for gas pressures in excess of 1.6 Torr.

6677– Pulsed lasing was reported from v' = 1 to 6 levels of the IF(B) state which were
6679. excited by a flashlamp-pumped dye laser. The transitions indicated were observed to lase upon pumping the v' = 1 → v" = 0 transition at 516.64 nm (P(6) or R(15)) or 517.17 nm (P(17) or R(26)). IF molecules in the X state were formed by reacting I$_2$ and F$_2$ in the presence of He buffer. For these transitions, the He partial pressure was maintained between 8.0 and 20.0 Torr.

6680– Pulsed lasing was reported from v' = 1 to 6 levels of the IF(B) state which were
6682. excited by a flashlamp-pumped dye laser. The transitions indicated were observed to lase upon pumping the v' = 2 → v" = 0 transition at 506.34 nm (P(10) or R(18)) or 506.71 nm (P(17) or R(25)). IF molecules in the X state were formed by reacting I$_2$ and F$_2$ in the presence of He buffer. For these transitions, the He partial pressure was maintained between 8.0 and 20.0 Torr.

6683– Pulsed lasing was reported from the v' = 1 to 6 levels of the IF(B) state which were
6685. excited by a flashlamp-pumped dye laser. The transitions indicated were observed to lase upon pumping the v' = 3 → v" = 0 transition at 496.40 nm (P(9) or R(17)) or 497.15 nm (P(22) or R(30)). IF molecules in the X state were formed by reacting I$_2$ and F$_2$ in the presence of He buffer. For these transitions, the He partial pressure was maintained between 8.0 and 20.0 Torr.

6686– Pulsed lasing was reported fromthe v' = 1 to 6 levels of the IF(B) state which were
6688. excited by a flashlamp-pumped dye laser. The transitions indicated were observed to lase upon pumping the v' = 5 → v" = 0 transition at 478.21 nm (P(8) or R(15)) or 479.45 nm (P(28) or R(35)) or 480.63 nm (P(39) or R(46)). IF molecules in the X state were formed by reacting I$_2$ and F$_2$ in the presence of He buffer. For these transitions, the He partial pressure was maintained between 8.0 and 20.0 Torr.

6689– Pulsed lasing was reported from the v' = 1 to 6 levels of the IF(B) state which were
6691. excited by a flashlamp-pumped dye laser. The transitions indicated were observed to lase upon pumping the v' = 6 → v" = 0 transition at 470.14 nm (P(13) or R(20)) or 470.44 nm (P(19) or R(25)) or 471.38 nm (P(30) or R(37)). IF molecules in the X state were formed by reacting I$_2$ and F$_2$ in the presence of He buffer. For these transitions, the He partial pressure was maintained between 8.0 and 20.0 Torr.

6692– Pulsed lasing was reported from the v' = 1 to 6 levels of the IF(B) state which were
6694. excited by a flashlamp-pumped dye laser. The transitions indicated were observed to lase upon pumping the v' = 4 → v" = 0 transition at 487.27 nm (P(14) or R(21)) or 488.51 nm (P(31) or R(3)) or 489.50 nm (P(39) or R(4)). IF molecules in the X state were formed by reacting I$_2$ and F$_2$ in the presence of He buffer. For these transitions, the He partial pressure was maintained between 8.0 and 20.0 Torr.

6695. The indicated transitions were observed to lase upon pumping the v' = 4 → v" = 0 transitions at 487.33 (P(18) or 487.21 nm (R(22). Furthermore, the He partial pressure was ~ 0.7 Torr.

6696– This line was reported by Davis and Hanko (Reference 1570) in 1980 who were the
6697. first to observe lasing on the B → X band of IF under optical pumping. Ground state IF molecules, produced chemically by the reaction of I$_2$ with F$_2$, were excited in a longitudinal configuration by a broadband dye laser. The B → X transitions pumped were (2,0), (3,0) and (4,0) and the small signal gain coefficient was estimated to be 1.8% cm^{-1}.

6698. The He partial pressure was ~ 0.7 Torr (low pressure regime of Reference 1569) but the pump laser was tuned to the v' = 1 → v" = 0 transition at 516.77 (R(15)), 516.92 (R(26)), 517.08 nm (R(19)), 517.23 nm (R(27)), or 517.53 nm (P(22)).

6699. v' = 4 → v" = 0 line pumped at 478.47 nm (R(19)); helium partial pressure ~ 0.7 Torr.

6700. Pump transition: v' = 2 → v" = 0 at 506.17 nm (R(11)), 506.29 nm (R(17)), or 507.02 nm (R(27)); p_{He} ~ 0.7 Torr.

6701. Transition pumped by the dye laser was v' = 3 → v" = 0 at 496.29 nm (R(12)), 497.14 nm (P(21)) or 497.39 nm (P(24)) and the He pressure was again 0.7 Torr.

6702– Lasing was demonstrated by pumping the (v', v") = (5, 1) band of the B → X
6703. transition of the dimer with the 647.1 nm line of the Kr^+ ion laser. The wavelengths given above have a stated accuracy (cf. Reference 1524) of ± 0.2 nm. Lasing occurred only at low temperatures (600 K, which corresponds to a K_2 vapor pressure of 4 X 10^{-3} mb).

6704– Rare gas polycrystals (~ 1 cm^3 in volume) excited by a relativistic, 3 ns pulse. A
6705. FWHM electron beam. Excited state densities estimated to be $5x10^{18}$ cm^{-3} and a gain coefficient of 50 cm^{-1} was obtained yielding ~ 10^7 W of peak power at 145 nm. See also the entries for Ar_2 and Xe_2.

6706– 6.6 MW of peak power (pressure of 15 atm) was obtained at 146 nm by exciting
6707. Kr with a coaxial electron beam (700 kV). Spectral widths as small as 0.1 nm were obtained at 145.7 nm (and peak output power = 3.5 MW) with an intracavity MgF_2 prism.

6708– Optical pumping of the KrCl excimer laser was achieved by exploding a tungsten
6709. wire in a mixture of 0.5 atm Kr and 6 Torr Cl_2. With 6% output coupling at 223 nm, >1 μ J of output energy was obtained in ~ 1 μs (FWHM) pulses for 380 Torr Kr/6 Torr Cl_2 mixtures. For earlier work on this laser, the reader is referred to the *CRC Handbook of Laser Science and Technology*, Vol. II.

6710– Mixtures of Ar, Kr, and NF_3 were excited by a relativistic e-beam, producing broad
6711. band lasing in the violet. Peak output powers of 5 kW in ~ 30 ns were obtained. For a thorough discussion of the properties of this laser, the reader is referred to the excellent discussion by D. L. Huestis, G. Marowsky, and F. K. Tittel in *Excimer Laser*, 2nd edition, C. K. Rhodes, Ed. (Springer-Verlag, Berlin, 1984), 181.

6712– Mixtures of KrF_2 and N_2 (1:1000; total pressure = 2 atm) were optically pumped
6714. by vacuum ultraviolet radiation generated by an open, high-current discharge (two tungsten wires exploding in the gas and located ~2.5 cm from the laser's optical axis). Output pulses ~ 1 μs in length (FWHM) were observed at 450 ± 10 nm which is ~20 nm to the red of the peak wavelength at which lasing occurs when Kr_2F is pumped by a pulsed e-beam. (For further details, the reader is referred to the Xe_2Cl and KrCl entries and to the References given in *CRC Handbook of Laser Science and Technology*, Vol. II.) Lasing was also obtained in $KrF_2/CF_4/Kr$ (1:200:800; total pressure = 2 atm) gas mixtures.

6715– Lasing was obtained by exciting about 4 Torr of ^6Li vapor in a ~ 120-cm long heat
6716. pipe oven with various lines from an argon ion laser. Oscillation was observed at 27 discrete wavelengths in the 523.7 to 588.9 nm region utilizing a ring cavity configuration. In all cases, the molecule was pumping by exciting different rovibronic transitions of the $B^1\Pi_u \to S^1\Sigma_g^+$ band.

6717. For these lasing transitions, the pump wavelength was 457.9 nm, corresponding to excitation of the (v',J') = (7,8) → (v",J") = (1,8) line of the B → X transition.

6718. The pump wavelength was 465.8 nm [(4,11) → (0,11) line of the B → X band].

6719. λ_{pump} = 465.8 nm but the pumping transition is (5,41) → (0,41).

6720. Pump wavelength is 472.7 nm, corresponding to the (11,22) → (5,21) transition.

6721. The Ar$^+$ laser wavelength was 476.5 nm and the (11,31) → (4,31) transition was excited.

6722. λ_{pump} again = 476.5 nm which also overlaps the (11,29) → (5,30) transition.

6723. Pump wavelength is 488.0 nm: (0,41) → (0,40) transition.

6724. The Ar$^+$ laser was tuned to 496.5 nm but the pumping and lasing transitions are unidentified.

6725– CW oscillation at the indicated wavelengths was obtained by pumping Li vapor
6727. (50 Torr in a heat pipe operated at 1000 °C) with individual lines from an Ar$^+$ ion laser. Pumping thresholds ranged from <50 to <200 mW and output powers (all lines) were measured to be in the 1 to 15 mW region.

6728. Oscillation obtained by pumping with the 457.9 nm line of the Ar$^+$ laser.

6729. λ_{pump} = 472.7 nm.

6730. The pump wavelength for this line was 476.5 nm.

6731. λ_{pump} = 488.0 nm.

6732– CW oscillation at the indicated wavelengths was obtained by pumping Li vapor
6733. (50 Torr in a heat pipe operated at 1000 °C) with the 647.1 nm line of a Kr ion laser operated at 1 W. Output power on this band (all lines) was ~ 30 mW and the threshold pumping power was <200 mW.

6734– Lithium vapor, produced in a stainless steel heatpipe, was excited by the 578.2 nm
6736. line from a copper vapor laser operated at a pulse repetition frequency of 5 kHz. The Li vapor pressure was 70 Torr, corresponding to a dimer partial pressure of 5.5 Torr. Laser threshold was observed at a pump power of 10 mW and the average laser output power was 8 mW for 190 mW of pump power. The maximum energy conversion efficiency was measured to be 7% and the strongest transitions occurred at 886.347 and 887.376 nm.

6737– By bottlenecking the (v',f") = (0,0), (0,1), and (0,2) transitions of the C → B band
6738. at 337.1, 357.7, and 380.5 nm, respectively, lasing on the (0,3) transition was observed. Mixtures of Ar (2000 Torr) and N$_2$ (100 Torr) were pumped by a coaxial e-beam of 50 cm length, yielding a violet laser pulse having a temporal width of ~ 200 ns (FWHM).

6739. Longitudinal pumping of 85% Ar/15% N$_2$ (2 atmospheres total pressure) mixtures yielded 0.3 J/l of extracted energy in 20 ns pulses.

6740. The addition of excess Ne and He into e-beam pumped Ar/N$_2$ mixtures resulted in obtaining peak intensities of 56, 44, and 66 kW cm^{-2} at 357.7, 380.5, and 405.9 nm, respectively. Pulse lengths up to 400 ns were obtained in the violet.

6741. The addition of H$_2$ to electron beam-pumped He/N$_2$ mixtures efficiently quenches the lower laser levels, leading to improved gain and output power at both wavelengths.

6742– The (0,0) transition of the $A^2\Sigma^+$ → $X^2\Pi$ band (Γ band) of NO was pumped with
6743. 227 nm radiation generated by summing a frequency-doubled dye laser beam (λ ~ 288.5 nm) with the fundamental of Nd:YAG (1.06 µm). The pump energy per pulse was ~ 1.3 mJ and the pump beam was directed longitudinally into the laser cavity. The NO pressure was 0.5 Torr.

6744. Stimulated emission on the so-called "diffuse" band of Na$_2$ was observed upon exciting the 3p → 3s or (two-photon) 4d → 3s transitions of Na with an Nd:YAG-pumped dye laser.

6745. ASE on the violet band of Na$_2$ was observed at 436 nm by two-photon excitation of the vapor with a dye laser operating in the 570 to 595 nm region.

6746. Na vapor was optically pumped at 350.7 nm with a Kr ion laser, yielding gain on the excimer (bound → free) band of Na$_2$ in the violet (430 < λ < 452 nm). The Na vapor pressure was typically 1 Torr.

6747–   A ring laser system pumped by a single mode Ar^+ laser at 488 nm yielded the new
6748.   new laser lines. The heatpipe temperatures ranged from 650 to 850 K and buffer gas
        pressures up to 150 mb were used. Also, the laser could be tuned slightly by
        means of an intracavity etalon.

6749.   These lines were obtained by pumping the $(v',J') = (6,43) \rightarrow (3,43)$ transition of
        the $B \rightarrow X$ band with 2 W of single mode power. The output powers obtained on
        individual lines ranged from 30 to 200 mW.

6750.   Pump transition was $(v',J') = (10,42) \rightarrow (6,41)$ of the $B^1\Pi_u \rightarrow X^1\Sigma^+_g$ band.

6751.   Line pumped was $(v',J') = (9,56) \rightarrow (5,55)$.

6752.   These lines were observed by pumping the $(7,98) \rightarrow (2,98)$ transition.

6753–   Lasing on the $B^1\Pi_u$ $(v' = 6) \rightarrow X^1\Sigma^+_g$ $(v'' = 13–15)$ transitions was also realized
6754.   by pumping the ground state dimers produced in a supersonic expansion.
        Threshold pumping powers below 1 mW and output powers exceeding 0.6 mW
        were reported in reference 1587.

6755– Na_2 ground state molecules in a heat pipe oven were photopumped to the $A^1\Sigma^+_u$
6757.   state with a Kr ion laser operating on a single longitudinal mode ($\Delta v \sim 6$ MHz) at
        568.188 nm with ~ 0.9 W of output power. The transition pumped was $A^1\Sigma^+_u$ $(v'$
        $= 34, J' = 50) \rightarrow X^1\Sigma^+_g$ $(v'' = 3, J'' = 51)$ and lasing in all cases originated from
        the pumped state $(v' = 34, J' = 50)$ and terminated on X state vibrational levels for
        which $34 < v'' < 54$. Optimum Na vapor pressure was found to be ~ 0.65 Torr.

6758– Na_2 laser emission over the 0.765 to 0.804 μm region was observed by pumping
6759.   sodium vapor with the 578.2 nm line from a copper vapor laser. The pump pulse
        repetition frequency was 3.3 kHz and the Na vapor pressure was 6 Torr ($T \sim 0.2$
        Torr). The maximum average output power obtained for this laser (all lines) was
        10 mW which corresponds to an energy conversion efficiency of $\sim 3\%$.

6760.   This laser is a bound $\rightarrow$ free transition and was assigned as $(v' = 34, J' = 50) \rightarrow$
        $(k'',J'' = 49,51)$.

6761–   Bound $\rightarrow$ free (excimer) laser emission, centered at about 830 nm, was obtained by
6762.   photopumping Na vapor with an N_2-pumped dye laser. The emission occurs in the
        827 to 832 nm region and partially resolved structure is observed. All transitions
        were attributed to $v' = 0$.

6763–   Lasing was observed in the near-infrared by pumping the $2^3\Pi_u$ state from ground
6764.   state in the 337.0 to 342.0 nm spectral region. The Na pressure was ~ 5 Torr,
        pumping threshold was ~ 70 μJ, and the divergence angle of the output radiation
        was ~ 6 mr.

6765–   Continuous lasing was obtained by pumping the various transitions of the $C^1\Pi_u \rightarrow$
6766. $X^1\Sigma^+_g$ band of the dimer with a single frequency Ar^+ or Kr^+ ion laser. The stated
        accuracy of the wavelengths of all lasing lines is given as ± 1 nm. The length of the
        heat pipe vapor zone was 15 cm. $T \sim 500$ °C and threshold pump powers as low as
        1 mW were measured.

6767.   The pump was an Ar^+ ion laser having a wavelength of 351.1 nm and the
        transition excited was $(v',v'') = (5,11), J' = 34 \rightarrow J'' = 34$ (Q(34)).

6768.   An Ar^+ laser pumped the (6,2)R(43) transition at 333.5 nm.

6769.   The Ar^+ laser pump wavelength was 333.6 nm which corresponds to excitation of
        the (7,2)Q(48) transition of the $C^1\Pi_u \rightarrow X^1\Sigma^+_g$ band.

6770.   These lasing transitions were observed by pumping the $C^1\Pi_u \rightarrow X^1\Sigma^+_g$
        (0,6)Q(1145) transition with the 350.7 nm line from a Kr^+ ion laser.

6771.   Lasing was observed upon pumping the (4,1)Q(12) transition of the $C^1\Pi_u \rightarrow$
        $X^1\Sigma^+_g$ band at 333.5 nm with a single frequency Ar^+ laser.

6772.   Pumping of the (1,7)P(69) transition of the $C^1\Pi_u \rightarrow X^1\Sigma^+_g$ band at 351.1 nm
        with a tuned Ar^+ laser resulted in lasing at this wavelength.

6773. Pulsed oscillation on this band has also been obtained by pumping the $C^1\Pi \to$ $X^1\Sigma^+_g$ transition with N_2, XeF and dye lasers (cf., References 104, 105, 107).

6774– 7 Torr of rubidium vapor (corresponding to ~ 0.1 Torr of Rb dimers) and
6776. containing sodium as an impurity (concentration unknown) was pumped by the 510.6 nm line of the copper vapor laser. Two laser lines were observed of which the 666.5 nm line was the stronger one. The observed resonance fluorescence spectrum identified the molecule as NaRb. Although a definitive assignment was not reported, the lasing electronic transition was believed to be $D^1\Pi$ (or possibly C) $\to X^1\Sigma^+$.

6777. $(v',v'') = (3,0)$ to $(7,0)$ transitions of the $B \to X$ band of S_2 were pumped by a frequency-doubled dye laser.

6778. $B \to X$ laser transitions ranging in wavelength from ~ 370 to 490 nm were observed upon exciting the $v' = 2 \to v'' = 4$ line with an N_2 (337.1 nm) laser. Superfluorescence was observed on $v' = 2 \to v'' = 7$-11, 13-17 transitions.

6779. The $B \to X$ system $(v',v'' = 4.1)$ was pumped by a free-running XeCl laser. The $v' = 4$ state was excited directly by the excimer laser but gain was measured on the (2, 17) and (3, 18) transitions (cf. Reference 1600).

6780. The UV lines from an Ar^+ laser resulted in CW oscillation on this line. The pump wavelengths were 351.1, 351.4, 363.8, and 379.5 nm, with 90% of the overall power contained in the 363.8 nm line.

6781– Unless indicated otherwise, all laser transitions given in this table were pumped by
6782. exciting $^{80}Se_2$ in a quartz cell with a single frequency Ar^+ ion laser operating at 351.1 nm. The line excited by the pump is he $(v',v'') = (11,0)$, P(12) transition of the Se_2 $BO^+_u \to XO^+_g$ band. The length of the vapor zone is ~ 7 cm and the optimum cell temperature was ~ 380 °C.

6783. Lasing was obtained by transversely pumping the quartz cell with XeF laser (351.1 nm) radiation.

6784. For these laser lines, the pumped transition was the (13,1) R(52) line.

6785– Laser transitions in $^{130}Te_2$ were pumped on BO^+_u $(v' = 16, J' = 37) \to XO^+_g$ $(v''$
6787. L= 0, J'' = 36), and B^+_u $(v' = 18, J' = 173) \to XO^+_g$ $(v'' = 0, J'' = 172)$ transitions by the 406.7 nm line of the Kr ion laser. As indicated in the table, the lasing transitions are BO^+_u $(v' = 16, J' = 37) \to XO^+_g$ $(12 < v'' < 54, J'' = 36,38)$ or BO^+_u $(v' = 18, J' = 173) \to XO^+_g$ $(24 < v'' < 55, J'' = 172,174)$. The pump laser power was 15 W and the quartz optical cell and reservoir temperatures were 700 and 570 °C, respectively. Small-signal gain coefficients were estimated to be as large as 4% cm^{-1}.

6788– Lasing of the $AO^+_u \to b^1\Sigma^+_g$ transition of the dimer was obtained by pumping the
6791. LAO^+_u $(v' = 13, J' = 133) \to XO^+_g(v'' = 0, J'' = 134)$ transition with 0.5 W on the 476.49 nm line of the Ar^+ laser (operating single mode). Between the $A \to b$ and $BO^+_u \to b^1\Sigma^+_g$ transitions (latter also shown above), 13 laser lines in the 0.8 to 1.3 μm spectral range were reported in Reference 127. All lines operated CW. The 893 and 895 nm lines yielded output powers of 1.0 to 1.3 mW and the threshold pumping power in each case was <140 mW. The cell and reservoir temperatures were 700 and 650° C, respectively, and the optical cell length was 7 cm.

6792– Eleven CW laser lines between 1.059 and 1.224 μm were observed by pumping
6794. the BO^+_u $(v' = 2, J' = 197) \to XO^+_g$ $(v'' = 5, J'' = 198)$ transition at 476.49 nm or the BO^+_u $(v' = 5, J' = 137) \to XO^+_g$ $(v'' = 4, J'' = 138)$ transition at 457.94 nm. The maximum pump powers available were 0.5 and 0.4 W, respectively, and the output power for each line was ~ 1.0 mW. Also, the threshold pumping powers

were <70 and <120 mW, respectively. Other experimental parameters are given under footnotes for the 893 and 895 nm lines.

6795.   Transitions similar to those for $^{130}Te_2$ but obtained with natural abundance Te_2. Once again, the pump is the 406.7 nm line of Kr^{2+} laser.

6796.   B → X laser transitions of $^{130}Te^{128}Te$ were obtained by pumping the molecule at 413.1 nm (Kr ion laser). The transition excited is BO^+_u (v' = 14, J' = 109) → XO^+_g (v" = 0, J" = 110).

6797. $^{128}Te^{126}Te$ laser transitions were excited by the BO^+_u (v' = 14, J' = 115) → XO^+_g (v" = 0, J" = 116) transition of the dimer. The pump is the 413.1 nm line of the Kr^{2+} laser.

6798.   Laser transitions on the BO^+_u → XO^+_g band of $^{128}Te_2$ were pumped at 406.7 nm (by a Kr^{2+} ion laser) which coincides with the BO^+_u (v' = 16, J' = 57) → XO^+_g (v" = 0, J" = 56) transition.

6799.   Electron beam excitation of rare gas crystals resulted in ~ 10^7 W of peak power output at 172 nm. Similar experiments were carried out for Ar_2 and Kr_2.

6800.   400 MW of peak power was obtained by exciting Xe gas with a coaxial e-beam; the laser was also tuned over the 170 to 176 nm interval while maintaining a peak output power of 1 MW and a spectral width (FWHM) of 0.1 nm.

6801–   Mixtures of $Cl_2/Xe/N_2$ (1:250:250; total pressure 2 atm) were photoexcited by the
6802.   radiation from an open current discharge (tungsten wire exploding in the gas mixture ~ 1 cm from the resonator axis). The length of the active medium was 80 cm and output pulses were ~ 3 μs in length. The lasing wavelengths (520 to 530 nm) lie slightly to the red of those observed when Xe_2Cl is pumped by an e-beam (see *CRC Handbook of Laser Science and Technology*, Vol. II, for further References).

6803.   The addition of N_2 to an electron beam-excited mixture composed of Ar, Xe, and CCl^4 yielded a threefold increase in the laser output power. The optimum N_2 partial pressure was found to be 200 Torr.

6804–   A high power, optically pumped XeF (B→X) laser was demonstrated by photodis-
6805.   sociating XeF_2 with the VUV radiation produced by sliding (surface) discharges. The gas mixture was composed of 450 Torr Ar, 50 Torr N_2, and 12 Torr XeF_2. Eight J, 5 μs pulses were obtained, yielding an average intensity of 130 kW cm^{-2}.

6806–   Lasing at 353 nm was observed upon photodissociating $KrF_2/Xe/N_2/Ar$ mixtures
6807.   (KrF_2:Xe:N_2:Ar = 4:6:160:1500 Torr) with the VUV radiation from an exploding tungsten wire of 78 cm length. Output energies of 1 μ J in 0.5 μs pulses were reported (cf. Reference 1608).

6808–   14.5 J laser pulses were measured on the ZeF(B → X) band when $XeF_2/N_2/Ar$ gas
6809.   mixtures (XeF_2:N_2:Ar = 3.4:311:1000 Torr) were excited optically by two 75 cm long exploding tungsten wires. The maximum measured instantaneous efficiency was 0.8%. For a XeF_2:N_2Ar = 9.0:311:1000 Torr mixture, single pulse energies as high as 28 J were obtained, which correspond to a specific output energy of 18 J l^{-1} and a maximum efficiency of 0.55%.

6810–   Mixtures of Xe and F_2 (1% of each) dissolved in liquid argon were optical pumped
6811.   (transversely) at 351 nm, resulting in stimulated emission at 404 nm. The output energy was measured to be ~ 70 μ J in a 5-ns FWHM pulse. The small-signal gain coefficient was ~ 26% cm^{-1} and the spectral bandwidth of the output radiation was ~ $60cm^{-1}$.

6812–   Since early work referenced in *CRC Handbook of Laser Science and Technology*,
6814.   Vol. II, lasing on the C → A band of XeF in the blue-green has also been achieved in transverse discharges. Peak gain was measured for gas mixtures consisting of 3 atm He, 3 Torr Xe and 2 Torr NF_3 and lasing occurred over the ~ 470-510 nm

region. Intracavity absorption due to several gaseous species is clearly evident in the laser spectrum. Output energies of 50 μJ in ~ 70 ns FWHM pulses were obtained and peak gain coefficients of 0.8 to 1.0% cm^{-1} were measured. References 1611, 1612.

6815– Tailoring the gas mixture so as to reduce excited state absorption has significantly
6816. improved the performance of the e-beam pumped XeF (C → A) laser. The use of F$_2$/NF$_3$ mixtures rather than F$_2$ as the fluorine donor alone, for example, raised the volumetric energy extraction for the device from 2 X 10^{-3} J l^{-1} to 100 mJ l^{-1} (cf. Reference 1613).

6817– The performance of the XeF (C → A) laser (e-beam pumped) was further improved
6819. by the addition of Kr to Ar/Xe/NF$_3$/F$_2$ gas mixtures. References 1515, 1614, 1615, 1616.

6820– Long pulse operation of the XeF (C → A) laser was demonstrated by electron beam
6821. pumping of Ar/Kr/Xe/NF$_3$/F$_2$ gas mixtures at moderate pump rates (~ 250 kW/cm^3). One joule, 400 ns pulses were obtained at an intrinsic efficiency of 0.7%. Reference 1516.

6822– 14.5 J blue-green pulses of 1.8 μs (FWHM) duration were obtained by photoexcit-
6823. ing XeF$_2$/N$_2$/Ar gas mixtures (XeF$_2$:N$_2$:Ar = 3.4:311:1000 Torr) with VUV radiation from two exploding tungsten wires. The output energy was 10 J l^{-1}, the maximum instantaneous efficiency was ~ 1%, and the cavity output coupling 6%. The resonator diameter was 5.4 cm and the peak output power recorded was 8 MW (cf. Reference 1609).

6824– A surface discharge having an active length >1 m was employed to optically pump
6825. aXeF$_2$/N$_2$/Kr gas mixture (composition - XeF$_2$:N$_2$:Kr = 1:100:300, P$_{total}$ = 1.1 atm). With an active length of 1.5 m, 45 J/pulse was obtained in the visible, corresponding to a specific output energy of 5 J l^{-1}. The optical aperture for this laser was 10 cm, the laser pulse width was ~ 5 μs FWHM and the energy conversion efficiency exceeded 5% (cf. Reference 1517).

6826– Lasing was obtained in a UV-preionized, transverse discharge in a mixture of Ne (2
6827. to 3.5 atm), 10 to 15 Torr Xe, and 3.5 Torr NF$_3$. The laser cavity was formed by two mirrors (99.5 and 97.5% reflectivities) separated by 60 cm. Peak lasing was observed at ~ 600 nm and the FWHM of the spectrum was ~ 20 nm. With increasing Ne pressure, the Xe$_2$F fluorescence spectrum shifts to the red (from λ_{peak} ~ 600 nm at 1 atm to 615 nm for p_{Ne} = 6 atm).

6828– Irradiating mixtures of Xe and N$_2$O (or CO$_2$) in liquid Ar with a relativistic e-beam
6829. yielded lasing a 547 nm from XeO. The gain coefficient was estimated to be ~23% cm^{-1} with an output coupling of 88%, 0.1 mJ of energy.

6830– Lasing on the B → X band of ZnI was demonstrated by photodissociating ZnI$_2$ or
6831. ^{64}ZnI$_2$ with an ArF laser. The natural abundance ZnI spectrum was dominated by lines at 602.5 and 603.1 nm whereas for ^{64}ZnI the most intense transition was further to the red (λ = 603.9 nm). The ZnI photodissociation laser efficiency doubled with the use of the isotopically enriched salt.

6832. Gain on the orange ZnI band was measured in a UV-preionized, transverse discharge having an active length of 13 or 25 cm. Peak gain was found to be ~ 0.7% cm^{-1} at 603 nm in He/N$_2$/ZnI$_2$ vapor discharges. References 1521, 1522.

6833. Observed in a chemical laser with H$_2$ injection into a flowing mixture of discharge dissociated He-0$_2$0-SF$_6$. The maximum CW output power obtained is 56 W. Reference 1624.

6834. Observed in a chemical laser with H$_2$ injection into a flowing mixture of discharge dissociated He-0$_2$0-SF$_6$. The maximum CW output power obtained is 56 W. Reference 1645.

6835. Observed in a pin discharge chemical laser containing SF_6 and HBr. Reference 1638.

6836. Observed in a mixture of SF_6, H_2S, and He excited by a longitudinal pulse discharge. Reference 1639.

6837. The laser lines in these bands are not new. However, compared to the old list in *CRC Handbook of Laser Science and Technology*, Vol. II, their frequencies are given here with an order of magnitude improvement in precision. Reference 1640.

6838. Observed in CO_2-Ne[6] and CO_2-Ar[7] gas dynamic lasers. The output power was 6 W-cw. Reference 1642.

6839. The laser lines in these bands are not new. However, compared to the old list in *CRC Handbook of Laser Science and Technology*, Vol. II, their frequencies are given here with an order of magnitude improvement in precision. Reference 1640.

6840. Observed in a grating tuned longitudinal discharge laser. The sealed gain tube was cooled to -60° C and contained a mixture of CO_2-N_2-Xe-He-H_2. Reference 1640.

6841. The laser lines in these bands are not new. However, compared to the old list in *CRC Handbook of Laser Science and Technology*, Vol. II, their frequencies are given here with an order of magnitude improvement in precision. Reference 1640.

6842– Observed in a compound cavity Q-switched laser with simultaneous output in the
6843. $00^02 \rightarrow 10^01$ sequence band. The laser cavity contains a sealed gain tube filled with a mixture of He-N_2 $C^{13}O^{16}_2$-Xe and an absorption cell filled with hot $C^{13}O^{16}_2$. References 1643, 1644.

6844. Observed in a grating tuned longitudinal discharge laser. The sealed gain tube was cooled to -60° C and contained a mixture of CO_2-N_2- Xe-He-H_2. Reference 1640.

6845. These laser lines are not new. However, compared to the old list in *CRC Handbook of Laser Science and Technology*, Vol. II, their frequencies are given here with an order of magnitude improvement in precision. Reference 1640.

6846. Observed in a sealed longitudinal discharge laser using a He-N_2-$C^{13}O_2^{16}$-H_2 mixture. Reference 1625.

6847. Observed in a CO_2 laser optically pumped by a pulsed HF laser. Reference 1626.

6848. Observed in a grating tuned longitudinal discharge laser. The sealed gain tube was cooled to -60 °C and contained a mixture of CO_2-N_2-Xe-He-H_2. Reference 1640.

6849. Observed in a longitudinal discharge laser using a mixture of $C^{13}O_2^{18}$-N_2-Xe-He-H_2 cooled to -60 °C; see References 1298 and 1313.

6850. These lines are not new. However, compared to the old list in *CRC Handbook of Laser Science and Technology*, Vol. II, their frequencies are given here with an order of magnitude improvement in precision. Reference 1640.

6851. The laser lines in these bands are not new. However, compared to the old list in *CRC Handbook of Laser Science and Technology*, Vol. II, their frequencies are given here with an order of magnitude improvement in precision.

6852. The laser lines in these bands are not new. However, compared to the old list in *CRC Handbook of Laser Science and Technology*, Vol. II, their frequencies are given here with an order of magnitude improvement in precision. Reference 1640.

6853. The data for these lines were in error in *CRC Handbook of Laser Science and Technology*, Vol. II, due to misalignments in the table.

6854. Optically pumped by 1-5J pulses from a CO_2 laser operating in the $00^01 \rightarrow 02^00$ band. The CO_2 lines are identified in the remarks column. Also given is the relative polarization between the pump and the output beams. Reference 1627.

6855. Parallel polarization, R(28).

6856. Parallel polarization, P(4).

6857. Perpendicular polarization, P(14).

6858 Perpendicular polarization, P(4).

6859. Parallel polarization, P(14).
6860. Parallel polarization, P(4).
6861. Parallel polarization, R(28).
6862 R(32).
6863. Parallel polarization, R(12).
6864. Parallel polarization, P(12).
6865. Perpendicular polarization, R(12).
6866. P(46).
6867. Parallel polarization, P(18).
6868. Parallel polarization, P(12).
6869. Parallel polarization, P(12).
6870– Optically pumped by 2 J pulses from a CO_2 laser operating on the R(12) line of
6871. the $00^01 \rightarrow 02^00$ band. The gas cell containing a C_2D_2-He mixture is cooled to 238 K and below. Reference 1628.
6872. Observed in NH_3 optically pumped simultaneously by the 10.74 μm, P(34) line and the 10.57 μm, P(18) line of TEA CO_2 lasers. References 1203, 1204.
6873. Observed in cooled NH_3-N_2 and NH_3-Ar mixtures optically pumped by a cw CO_2 laser operating on the 9.22 μm, R(30) line and frequency shifted by acoustooptic modulators. Reference 1629.
6874. Observed in a NH_3-N_2 mixture optically pumped by a cw CO_2 laser operating on the 9.22 μm, R(30) line and frequency shifted by acoustooptic modulator. References 1630, 1629.
6875. Observed in NH_3 optically pumped simultaneously by the 9.34 μm, R(8) line and the 10.63 μm, P(24) line of TEA CO_2 lasers. Reference 1631.
6876. Observed in NH_3 optically pumped simultaneously by the 10.72 μm, P(32) line and the 9.59 μm, P(24) line of TEA CO_2 lasers. Reference 1631.
6877. Observed in a NH_3-N_2 mixture optically pumped simultaneously by the 9.22 μm, R(30) line and the 9.59 μm, P(24) line of TEA CO_2 lasers. Reference 1632.
6878. Observed in NH_3 optically pumped by a TEA CO_2 laser operating on the 9.29 μm, R(16) line. Reference 1633.
6879. Observed in NH_3 optically pumped by a TEA CO_2 laser operating on the 10.63 μm, P(24) line. Reference 1634.
6880. Observed in NH_3 optically pumped by a TEA CO_2 laser operating on the 9.57 μm, P(22) line. Reference 1634.
6881. Observed in NH_3 optically pumped by a TEA CO_2 laser operating on the 9.25 μm, R(24) line. Reference 1634.
6882. Observed in dry-ice cooled $N^{15}H_3$-N_2 mixture optically pumped by a cw CO_2 laser operating on the 10.11 μm, R(42) line. Reference 1629.
6883. Observed in a mixture of $N^{15}H_3$-N_2 optically pumped by a pulsed CO_2 laser operating on the 9.33 μm, R(10) line. Reference 1635.
6884. Observed in dry-ice cooled $N^{15}H_3$ optically pumped by a cw CO_2 laser operating on the 9.38 μm, R(7) sequence line. Reference 1629.
6885. Observed in $N^{15}H_3$ optically pumped by a TEA CO_2 laser operating on the 9.15 μm, R(14) line. Reference 1636.
6886. Observed in CH_3CCH (propyne, methylacetylene) optically pumped by a TEA CO_2 laser operating in the $00^01 \rightarrow 10^00$ band. The specific CO_2 laser lines are identified in the comments. Reference 1637.
6887. 10P24 CO_2 laser line.
6888. 10R8 CO_2 laser line.
6889. 10P18 CO_2 laser line.
6890. 10P14 CO_2 laser line.

6891.   10R18 CO_2 laser line.

6892.   10R8 CO_2 laser line.

6893.   10R32 CO_2 laser line.

6894.   10P14 CO_2 laser line.

6895.   Excitation by third harmonic of a Nd laser (355 nm). Buffer gas: ether; temperature of vapor: ~200°C; pressure ~35 atm.

6896.   140 mW of CW output power was obtained by exciting silver metal vapor and He with dual electron guns. The threshold electron beam current for laser oscillation was 45 mA and a magnetic field ($\leq$ 3.5 kG) confined the electron beam. The silver vapor pressure was ~75 mtorr and the optimal magnetic field strength was ~2.8 kG.

6897.   The $2s4p^1P^o \leftarrow 2s^2\,^1S$ transition of doubly ionized carbon is resonantly pumped by 310 Å radiation from Mn VI ions formed in a laser-produced Mn plasma. CO_2 laser pulses (15 J, 50 ns) both generate the Mn plasma and trigger the vacuum arc discharge that produces the carbon ions. Gain on the 216.3 and 217.7 nm transitions was evident when an optical cavity was installed.

6898.   Stimulated emission on the n = 3 $\rightarrow$ n = 2 (Balmer-α) transition of hydrogen was generated in a low pressure flame by exciting the two photon 3S,D $\leftarrow$ 1S transition of the atom at 205 nm. Pumping threshold for the oscillator occurred at ~ 150 μJ/pulse and the red beam was visible to the eye.

6899.   The 205 nm pump beam can serve to both produce H atoms by photodissociation of H_2S and pump the two photon n = 3 $\leftarrow$ n = 1 transition. An injection-locked ArF laser, Raman shifted in D_2 (at 3 atm), produced the 5 mJ pump pulses. The red laser output energy was increased by more than a factor of two by adding ~ 100 Torr of helium to (typically) 700 mtorr of H_2S.

6900.   Amplified spontaneous emission was observed when flat aluminum targets were irradiated by a focussed ArF excimer (193 nm) laser beam. The pump energy ranged from 450 to 600 mJ/pulse and the presence of $5 \times 10^{18} - 10^{19}$ cm^{-3} of background H_2 was determined to be necessary for amplification. The characteristics of the system point to photoionization, followed by recombination in the expanding plasma as the dominant pumping processes.

6901.   Upon pumping potassium dimers produced in a supersonic jet with a CW dye laser tuned to the Q (11) transition of the K_2 $B^1\Pi_u$ (v'=8) $\leftarrow$ $X^1\Sigma^+_g$ (v''=0) band at 627.298 nm, lasing was observed on several B $\rightarrow$ X transitions in the red. Strongest emission was obtained on the v'=8 $\rightarrow$ v''=16 line at 686.9 nm with a folded, four-mirror optical cavity (output coupling of 2%). The highest power outputs and lowest pump threshold (< 1 mW) were observed for an oven temperature of 848 K.

6902.   Amplified spontaneous emission (ASE) on several vibrational transitions of the NaK (D $\rightarrow$ X) band was observed by pumping a mixture of Na and K vapor in a heat pipe with the 510.6 nm line of a 10 W copper vapor laser. The green line of the pump laser overlaps the D $\leftarrow$ X absorption band and strong ASE on several $D^1\Pi \rightarrow X^1\Sigma^+$ (v'=0 $\rightarrow$ v''=9-19) transitions was observed. Output intensity peaked at ~ 542 nm (v'=0 $\rightarrow$ v''=14) for an oven temperature of approximately 700 K.

6903.   Stimulated emission in the deep red was obtained by photoexciting mixtures of sodium and potassium metal vapors with a tunable dye laser. The authors attribute the pumping process to excitation of the $B^1\Pi \leftarrow X^1\Sigma^+$ transition of the heteronuclear NaK molecule, followed by collisional transfer to the $K(4^2P)$ states and culminated by photoexcitation of the K 5^2D and 7^2S states. Pumping resonances were observed at 578.2 nm ($7^2S \leftarrow 4^2P_{1/2}$), 580.2 nm ($7^2S \leftarrow 4^2P_{3/2}$), and 583.2 nm ($5^2D \leftarrow 4^2P_{3/2}$).

6904.   The B' $^2\Delta$ (v'=3) ← X$^2\prod$ (v'=0), Q$_{11}$ (7(1/2)) transition of NO was pumped by an F$_2$ laser (158 nm, 30 mJ/pulse). The mismatch between the F$_2$ laser wavelength and the absorption line (0.3-0.4 cm^{-1}) was offset by Zeeman shifting NO lines into resonance with a pulsed, 1.3 T magnetic field. For an NO pressure of 20 mb, the NO laser output energy was measured to be 87 μJ.

6905.   Amplification was observed on the indicated transitions by vaporizing a portion of a flat nickel plate with an XeCl laser and near-resonantly exciting Ni atoms with a time-delayed N$_2$ laser (337 nm) pulse. The pumping laser (N$_2$) energy fluence was 0.9 mJ/cm^2 and the maximum gain coefficients recorded were 0.65 dB - cm^{-1} (347 nm) and 1.03 dB - cm^{-1} (381 nm).

6906.   Stimulated emission was observed in double optical pulse experiments in which an initial laser pulse (either N$_2$, 337 nm, or XeCl, 308 nm) produced Ti vapor from a flat metal target. A second laser beam (either from an N$_2$ or pulsed dye laser), time-delayed with respect to the first pulse by 5-50 μs, pumped the upper laser level. The optical cavity consisted of a high reflector and an output coupler having a reflectivity between 6 and 97%. Maximum output pulse energies (on the green line) of 100 nJ were obtained for an output coupling of 74%, a pumping laser fluence of 0.9 mJ - cm^{-2}, and a time delay between the two laser pulses of 20 μs. Peak population inversions of ~ 1.5 x 10^{11} cm^{-3} were reported in Reference 25 for the 551.4 nm transition.

6907.   When an Nd:YAG laser pulse (14 J) is used to produce Ti vapor from the target, six new laser transitions are observed (431.5, 471.0, 472.3, 547.4, 551.3, and 551.45 nm) - Reference 113. With no He buffer, oscillation is observed on the 551.44, 472.3 and 471.0 nm lines but obtaining stimulated emission at the other wavelengths requires the addition of 5 to ~ 60 Torr He. The output laser pulse widths are ~ 3 ns (FWHM).

6908.   Vanadium vapor was produced by irradiating a flat V target with a 2.2 J Nd:YAG laser pulse. A time-delayed XeCl (308 nm) pulse photoexcites ground state V atoms to the upper laser level (4p ^{4}F$^0_{7/2}$). The temporal width of the violet laser output pulse is 4 ns and the peak power was measured to be ~ 1 W.

6909.   Lasing was obtained by first irradiating a tantalum plate with a 2J Nd:YAG pulse and subsequently exciting the metal vapor with KrF laser photons (248 nm, 10 mJ/pulse). Under these conditions, gain coefficients for the transitions listed above ranged from 9 to 35% - cm^{-1}. The maximum output power recorded (all lines) was 48 W for a He buffer gas pressure of 2.5 Torr.

6910.   Stimulated emission on four transitions of the B$^3\Sigma^-$ (v'=0) → X$^3\Sigma^-$ (v") band was observed at wavelengths in the 300-350 nm interval. Ground state SO was generated by photodissociating SO$_2$ with ArF laser pulses (193 nm, typically 40 mJ/pulse). Subsequent photoexcitation of the SO B$^3\Sigma^-$ (v'=0) ← X$^3\Sigma^-$ (v"=2) transition with a frequency-doubled dye laser results in gain coefficients of 5-6%/cm on the B → X (v'=0 → v"=8-11) transitions. Peak output power was measured to be 1.6 W (with 30% output coupling) or single pulse energies of 8-16 nJ. The optimal SO$_2$ pressure was ~ 2 Torr.

6911.   Flat indium targets were irradiated by a focussed excimer laser beam (ArF, 193 nm). Amplified spontaneous emission in the blue was observed with ~ 10^{19} cm^{-3} of background H$_2$.

6912.   Stimulated emission was observed in a pulsed longitudinal discharge (50 cm in length, 3 mm dia.) in O$_2$/He gas mixtures. Peak current was ~ 2.5 kA and the pulse width was ~ 1 μs. Typical gas fill pressures were 5 x 10^{-3} - 10^{-2} Torr and 0.1-1 Torr of O$_2$ and He, respectively. All of the transitions were pumped by recombination and had previously been observed to lase by W. B. Bridges and A.

N. Chester [*Appl. Opt.* 4, 573 (1965)] when the upper laser levels were pumped by electron impact ionization.

6913.   Eighteen transitions of atomic thulium were observed to lase in a pulsed discharge. Many of the lines had been reported previously but three new transitions were recorded when the BeO discharge tube temperature was in the 800-1250 °C range which correlates with a thulium vapor pressure of 4×10^{-3} to 6 Torr. For all measurements, 0.5 to 10 Torr of He was also present. Pulse widths were typically 10-15 ns FWHM (Reference 1785).

6914.   In addition to the transitions indicated above, lasing was observed on the 1448 nm line of neutral Tm when $TmBr_3$ vapor was dissociated in a pulsed discharge (cf. Reference 1786).

6915.   A mixture of Ar and NF_3 was pumped longitudinally by a 150 keV electron beam. The beam current density was varied between 0.1 and 0.5 kA - cm^{-2} and the active length of the laser was 50 cm. The optimal gas mixture was determined to be 10% NF_3 in Ar and an Ar pressure of ~ 90 Torr. Peak output power was ~ 2.5 W and the laser pulse width was 18 ns (FWHM).

6916. Cs_2 dimers produced in a heat pipe were irradiated with a pulsed 476 nm dye laser beam which photoexcited the E $^1\Sigma_u^+ \leftarrow$ X $^1\Sigma_g^+$ band of the molecule. Predissociation of the E $^1\Sigma_u^+$ state resulted in lasing on the $5\ ^2D_{3/2} \rightarrow 6\ ^2P_{1/2}$ line of the atom ($\lambda = 3.01$ μm). Also, collisionally induced dissociation of the E state (i.e., Cs_2^* (E) + Cs (6S) $\rightarrow Cs_2$ (X $^1\Sigma_g^+$) + Cs* (7S)) populates the $7\ ^2S_{1/2}$ state of Cs and results in stimulated emission on the $7\ ^2S_{1/2} \rightarrow 6\ ^2P_{3/2}$ transition at 1.47 μm. Peak 1.47 μm output was obtained for a pump laser wavelength of ~ 477 nm while, for 3.01 μm output, the optimal pump wavelength was ~ 475 nm. The Cs vapor pressure in the heat pipe was typically 6-10 Torr and the maximum pulse energy obtained at 3.01 μm was 0.5 mJ. The threshold pump energy was approximately 10 μJ/pulse.

6917.   In addition to the previously reported $3\ ^3S \rightarrow 2\ ^3P$ transition at 706.5 nm, lasing was obtained at 667.8 and 728.1 nm in a pulsed discharge between two cylindrical electrodes mounted in a coaxial configuration. The active length of the discharge was 40 cm and discharging a capacitor (V = 1-6 kV) across the electrodes produced 100-140 A pulses that were 0.5-1 μs in duration (FWHM). The optimal He/H_2 gas mixture was found to be a total pressure of 8 Torr and a He:H_2 ratio of 2:1. For a pulse repetition frequency of 1 kHz, the energy per pulse emitted by all lines (667.8, 706.5, and 728.1 nm) was 50 μJ and the gain was measured to be 1% - cm^{-1}. The laser pulse width was 3-4 μs FWHM.

6918.   Lasing was obtained on the B $\rightarrow$ X band of the heteronuclear upon optically pumping IF molecules in the ground state that were produced by a chemical reaction chain that was triggered by ultraviolet photolysis of NF_2 radicals of 264 nm. The gas mixture was composed of CF_3I, N_2F_4 and He and the optical cell was heated to 170°C so as to thermally dissociate N_2F_4 to yield NF_2. Iodine monofluoride molecules in the ground state (X $^1\Sigma^+$) were produced by the reactions following the photodissociation of NF_2 (to produce F atoms). A time-delayed dye laser pulse then pumped IF(X) to v'=5 of the B state. For He partial pressures between 3 and 30 Torr, lasing on the B $\rightarrow$ X band of IF occurred exclusively from the v'=5 state. For P_{He} = 70-100 Torr, stimulated emission was also observed from the v'=0 and 1 levels and, in particular, lasing was recorded on the P (14) - P (40) and R (14) - R (37) lines of the v'=0 $\rightarrow$ v"=4 - 6 bands. For He pressures above 450 Torr, lasing was also observed at 603.11, 624.93 and 648.12 nm. With 300 Torr of He in the gas mixture, the lasing spectrum was tuned with an intracavity prism over the following bands: v'=0 $\rightarrow$ v"=3 (585 nm) 0 $\rightarrow$ 4 (605 nm), $0 \to 5$ (625 nm), $0 \to 6$ (650 nm) and $0 \to 7$ (675 nm). Pulse energies of 0.2 mJ were obtained on the $v'=0 \to v''=4$ - 6 bands (all lines) for a pump energy of 0.2 J (1.6 µs pulse width [FWHM]).

6919. By photodissociating PbI_2 at 405.8 nm, stimulated emission was observed on the $^3P_1 \to {}^1D_2$ transition of atomic lead. The pump pulse energy and pulse width were 1 mJ and 20 ns, respectively. Maximum pulse energy at 722.9 nm was obtained for PbI_2 number densities of 1-5 x 10^{16} cm^{-3} which correspond to source temperatures of ~ 500-540 °C. Output pulse energies of 0.4 mJ were obtained for pumping energies of 15 mJ for a conversion efficiency of 4%.

6920. Direct photoionization.

6921. Resonance excitation followed by photoionization.

6922. Photoionization with shakeup.

6923. Photoionization then Auger decay.

6924. Collisions with photoelectrons.

6925. This laser transition reaches saturation before it achieves this gain.

6926. High-gain VUV laser.

6927. In Reference 1790 cw laser oscillation was obtained in a neon-silver vapor mixture excited by glow discharge electon beam. Output power was 14 mW using a nonoptimized optical cavity.

6928. In Reference 1790 cw laser oscillation was obtained in a neon-silver vapor mixture excited by glow discharge electon beam. Output power was 60 mW using a nonoptimized optical cavity.

6929. Electric discharge pumped.

6930. CO_2 laser pump line 9R28.

6931. CO_2 laser pump line 9R40.

6932. CO_2 laser pump line 9R14

6933. CO_2 laser pump line 9R26.

6934. CO_2 laser pump line 9R18.

6935. CO_2 laser pump line 9R08.

6936. CO_2 laser pump line 9P08.

6937. CO_2 laser pump line 10P30.

6938. CO_2 laser pump line 9P16.

6939. CO_2 laser pump line 10R34.

6940. CO_2 laser pump line 9R22.

6941. Sequence band CO_2 pump laser.

6942. CO_2 laser pump line 9P40.

6943. CO_2 laser pump line 9P26.

6944. CO_2 laser pump line 9P32.

6945. CO_2 laser pump line 9P22.

6946. CO_2 laser pump line 9R04.

6947. CO_2 laser pump line 9P34.

6948. CO_2 laser pump line 9R12.

6949. Molecule may have been partially deuterated, i.e., CH_3NHD or CH_3ND_2.

6950. See Reference 1819.

6951. See Reference 1818.

6952. See Reference 1815.

6953. See Reference 817.

6954. See References 1640 and 1816.

6955. See References 1307 and 1816.

6956. See Reference 1307.

6957. See Reference 1816.
6958. See Reference 18179.
6959. Observed in an electric discharge; see Reference 1255.
6960. Observed in an electric discharge; see Reference 1183.
6961. See References 1640 and 1816 for present values.
6962. See Reference 1820.
6963. See References 1335 and 1820.
6964. See Reference 1834.
6965. See Reference 1834.
6966. See References 1334 and 1820.
6967. Pulsed laser excitation at 343.6 nm; coherent emission is reported without need of a cavity. Gas pressure ranged from 350 to 450 mtorr.
6968. Observed in an electric discharge; see Reference 1618.
6969. Observed in an electric discharge; see Reference 1713.
6970. Relative polarization: perpendicular.
6971. Relative polarization: parallel.
6972. CO_2 laser pump line 9R24.
6973. CO_2 laser pump line 9R12.
6974. CO_2 laser pump line 10R34.
6975. CO_2 laser pump line 10R28.
6976. CO_2 laser pump line 10R26.
6977. CO_2 laser pump line 10R24.
6978. CO_2 laser pump line 10R22.
6979. CO_2 laser pump line 10R20.
6980. CO_2 laser pump line 10R18.
6981. CO_2 laser pump line 10R16.
6982. CO_2 laser pump line 10R14.
6983. CO_2 laser pump line 10R12.
6984. CO_2 laser pump line 10R10.
6985. CO_2 laser pump line 10R08.
6986. CO_2 laser pump line 10R06.
6987. CO_2 laser pump line 10P12.
6988. CO_2 laser pump line 10P16.
6989. CO_2 laser pump line 10P30.
6990. CO_2 laser pump line 10P32.
6991. CO_2 laser pump line 10P34.
6992. CO_2 laser pump line 10P42.
6993. CO_2 laser pump line 10P56.
6994. CO_2 laser pump line 9P18.
6995. CO_2 laser pump line 9P14.
6996. CO_2 laser pump line 9P22.
6997. CO_2 laser pump line 9R18.
6998. CO_2 laser pump line 9R20.
6999. CO_2 laser pump line 9R24.
7000. CO_2 laser pump line 10P44.
7001. CO_2 laser pump line 10HP19.
7002. CO_2 laser pump line 9P22.
7003. CO_2 laser pump line 9P34.
7004. CO_2 laser pump line 10R20.
7005. CO_2 laser pump line 10R12.
7006. CO_2 laser pump line 10SR11.

7007. CO_2 laser pump line 10R28.
7008. CO_2 laser pump line 10R24.
7009. CO_2 laser pump line 10R22.
7010. CO_2 laser pump line 10R20.
7011. CO_2 laser pump line 10R16.
7012. CO_2 laser pump line 10R14.
7013. CO_2 laser pump line 10R10.
7014. CO_2 laser pump line 10P10.
7015. CO_2 laser pump line 10P14.
7016. CO_2 laser pump line 10P18.
7017. CO_2 laser pump line 10P20.
7018. CO_2 laser pump line 10P22.
7019. N_2O laser pump line 10R38.
7020. N_2O laser pump line 10R36.
7021. N_2O laser pump line 10R25.
7022. N_2O laser pump line 10R24.
7023. N_2O laser pump line 10R04.
7024. N_2O laser pump line 10P11.
7025. N_2O laser pump line 10R15; different N_2O laser frequency offset from line center.
7026. N_2O laser pump line 10R15; different N_2O laser frequency offset from line center.
7027. N_2O laser pump line 10P16.
7028. N_2O laser pump line 10P24.
7029. N_2O laser pump line 10P26.
7030. N_2O laser pump line 10P29.
7031. N_2O laser pump line 10P30.
7032. N_2O laser pump line 10P34.
7033. N_2O laser pump line 10P34
7034. N_2O laser pump line 10P32.
7035. N_2O laser pump line 10P45.
7036. CO_2 laser pump line 9R60.
7037. CO_2 laser pump line 9R58.
7038. CO_2 laser pump line 9R56.
7039. CO_2 laser pump line 9R54.
7040. CO_2 laser pump line 9R52; different laser frequency offsets.
7041. CO_2 laser pump line 9R50.
7042. CO_2 laser pump line 9R48.
7043. CO_2 laser pump line 9SR05.
7044. CO_2 laser pump line 9R02.
7045. CO_2 laser pump line 9HP16.
7046. CO_2 laser pump line 9R38.
7047. CO_2 laser pump line 9P10.
7048. CO_2 laser pump line 9P12.
7049. CO_2 laser pump line 9P16.
7050. CO_2 laser pump line 9P38.
7051. CO_2 laser pump line 9P38.
7052. CO_2 laser pump line 10R18.
7053. CO_2 laser pump line 10R34.
7054. CO_2 laser pump line 10P08.
7055. CO_2 laser pump line 9P24.
7056. CO_2 laser pump line 9P30.
7057. CO_2 laser pump line 9P38.

7058. CO_2 laser pump line 9R34.
7059. CO_2 laser pump line 9R32.
7060. CO_2 laser pump line 9R30.
7061. CO_2 laser pump line 9R28.
7062. CO_2 laser pump line 9R26.
7063. CO_2 laser pump line 9R24.
7064. CO_2 laser pump line 9R20.
7065. CO_2 laser pump line 9R18.
7066. CO_2 laser pump line 9R18.
7067. CO_2 laser pump line 9R10.
7068. CO_2 laser pump line 9P10.
7069. CO_2 laser pump line 9P12.
7070. CO_2 laser pump line 9P14.
7071. CO_2 laser pump line 9P16.
7072. CO_2 laser pump line 9P22.
7073. CO_2 laser pump line 9P24.
7074. CO_2 laser pump line 9P32.
7075. CO_2 laser pump line 9P40.
7076. CO_2 laser pump line 9P48.
7077. CO_2 laser pump line 10R46.
7078. CO_2 laser pump line 10R40.
7079. CO_2 laser pump line 10R36.
7080. CO_2 laser pump line 10R28.
7081. CO_2 laser pump line 10P08.
7082. CO_2 laser pump line 10P42.
7083. CO_2 laser pump line 10R22.
7084. CO_2 laser pump line 10R18.
7085. CO_2 laser pump line 10R16.
7086. CO_2 laser pump line 10R14.
7087. CO_2 laser pump line 10R12.
7088. CO_2 laser pump line 10R10.
7089. CO_2 laser pump line 10R08.
7090. CO_2 laser pump line 10R06.
7091. CO_2 laser pump line 10R04.
7092. CO_2 laser pump line 10R02.
7093. CO_2 laser pump line 10P06.
7094. CO_2 laser pump line 10P10.
7095. CO_2 laser pump line 10P12.
7096. CO_2 laser pump line 10P14.
7097. CO_2 laser pump line 10P18.
7098. CO_2 laser pump line 10P20.
7099. CO_2 laser pump line 10P22.
7100. CO_2 laser pump line 10P10.
7101. CO_2 laser pump line 10P12.
7102. CO_2 laser pump line 10P18.
7103. CO_2 laser pump line 10P30.
7104. CO_2 laser pump line 10P36.
7105. CO_2 laser pump line 10P22.
7106. CO_2 laser pump line 10P24.
7107. CO_2 laser pump line 10P16.
7108. CO_2 laser pump line 10P36.

7109.　CO_2 laser pump line 10P20.
7110.　CO_2 laser pump line 10R26.
7111.　CO_2 laser pump line 10R12.
7112.　CO_2 laser pump line 10P12.
7113.　CO_2 laser pump line 9R24.
7114.　CO_2 laser pump line 9P18.
7115.　CO_2 laser pump line 9R38.
7116.　CO_2 laser pump line 9R30.
7117.　CO_2 laser pump line 9R28.
7118.　CO_2 laser pump line 9R08.
7119.　CO_2 laser pump line 9P08.
7120.　CO_2 laser pump line 9P20.
7121.　CO_2 laser pump line 9P28.
7122.　CO_2 laser pump line 9P34.
7123.　CO_2 laser pump line 10R22.
7124.　CO_2 laser pump line 10R12.
7125.　CO_2 laser pump line 10R10.
7126.　CO_2 laser pump line 10R08.
7127.　CO_2 laser pump line 10R06.
7128.　CO_2 laser pump line 10R04.
7129.　CO_2 laser pump line 10P06.
7130.　CO_2 laser pump line 10P08.
7131.　CO_2 laser pump line 10P32.
7132.　CO_2 laser pump line 10P36.
7133.　CO_2 laser pump line 10P42.
7134.　CO_2 laser pump line 10P46.
7135.　CO_2 laser pump line 10P04.
7136.　CO_2 laser 10P pump band; resonance pump transition $^PP_6(41)$.
7137.　CO_2 laser 10P pump band; resonance pump transition $^PP_6(38)$.
7138.　CO_2 laser 10P pump band; resonance pump transition $^PP_6(37)$.
7139.　CO_2 laser 10P pump band; resonance pump transition $^PP_6(35)$.
7140.　CO_2 laser 10P pump band; resonance pump transition $^PP_6(34)$.
7141.　CO_2 laser 10P pump band; resonance pump transition $^PP_6(32)$.
7142.　CO_2 laser 10P pump band; resonance pump transition $^PP_6(31)$.
7143.　CO_2 laser 10P pump band; resonance pump transition $^PP_6(29)$.
7144.　CO_2 laser 10P pump band; resonance pump transition $^PP_6(28)$.
7145.　CO_2 laser 10R pump band; resonance pump transition $^PP_2(41)$.
7146.　CO_2 laser 10R pump band; resonance pump transition $^PP_3(33)$.
7147.　CO_2 laser 10R pump band; resonance pump transition $^PP_3(32)$.
7148.　CO_2 laser 10R pump band; resonance pump transition $^PP_3(31)$.
7149.　CO_2 laser 10R pump band; resonance pump transition $^PP_3(29)$.
7150.　CO_2 laser 10R pump band; resonance pump transition $^PP_3(28)$.
7151.　CO_2 laser 10R pump band; resonance pump transition $^PP_3(26)$.
7152.　CO_2 laser 10R pump band; resonance pump transition $^PP_3(25)$.
7153.　CO_2 laser 10R pump band; resonance pump transition $^PP_2(32)$.
7154.　CO_2 laser 10R pump band; resonance pump transition $^PP_3(24)$.
7155.　CO_2 laser 10R pump band; resonance pump transition $^PP_2(31)$.
7156.　CO_2 laser 10R pump band; resonance pump transition $^PP_1(37)$.
7157.　CO_2 laser 9P pump band; resonance pump transition $^RR_0(27)$.
7158.　CO_2 laser 9P pump band; resonance pump transition $^RR_1(36)$.
7159.　CO_2 laser 9P pump band; resonance pump transition $^RR_1(37)$.

7160. CO_2 laser 9P pump band; resonance pump transition $^RR_0(29)$.
7161. CO_2 laser 9P pump band; resonance pump transition $^RR_0(30)$.
7162. CO_2 laser 9P pump band; resonance pump transition $^RR_0(32)$.
7163. CO_2 laser 9P pump band; resonance pump transition $^RR_1(23)$.
7164. CO_2 laser 9P pump band; resonance pump transition $^RR_0(33)$.
7165. CO_2 laser 9P pump band; resonance pump transition $^RR_0(34)$.
7166. CO_2 laser 9P pump band; resonance pump transition $^RR_1(35)$.
7167. CO_2 laser 9P pump band; resonance pump transition $^RR_1(26)$.
7168. CO_2 laser 9P pump band; resonance pump transition $^RR_0(36)$.
7169. CO_2 laser 9P pump band; resonance pump transition $^RR_0(27)$.
7170. CO_2 laser 9P pump band; resonance pump transition $^RR_0(37)$.
7171. CO_2 laser 9P pump band; resonance pump transition $^RR_1(29)$.
7172. CO_2 laser 9P pump band; resonance pump transition $^RR_0(39)$.
7173. CO_2 laser 9R pump band; resonance pump transition $^RR_3(35)$.
7174. CO_2 laser 9R pump band; resonance pump transition $^RR_3(36)$.
7175. CO_2 laser 9R pump band; resonance pump transition $^RR_3(36)$.
7176. CO_2 laser 9R pump band; resonance pump transition $^RR_3(37)$.
7177. CO_2 laser 9R pump band; resonance pump transition $^RR_3(38)$.
7178. CO_2 laser 9R pump band; resonance pump transition $^RR_2(29)$.
7179. CO_2 laser 9R pump band; resonance pump transition $^RR_3(39)$.
7180. CO_2 laser 9R pump band; resonance pump transition $^RR_4(30)$.
7181. CO_2 laser 9R pump band; resonance pump transition $^RR_3(40)$.
7182. CO_2 laser 9R pump band; resonance pump transition $^RR_3(41)$.
7183. CO_2 laser 9R pump band; resonance pump transition $^RR_4(32)$.
7184. CO_2 laser 9R pump band; resonance pump transition $^RR_3(42)$.
7185. CO_2 laser 9R pump band; resonance pump transition $^RR_4(33)$.
7186. CO_2 laser 9R pump band; resonance pump transition $^RR_3(44)$.
7187. CO_2 laser 9R pump band; resonance pump transition $^RR_4(35)$.
7188. CO_2 laser 9R pump band; resonance pump transition $^RR_3(45)$.
7189. CO_2 laser 9R pump band; resonance pump transition $^RR_4(36)$.
7190. CO_2 laser 9R pump band; resonance pump transition $^RR_4(39)$.
7191. CO_2 laser 9R pump band; resonance pump transition $^RR_6(23)$.
7192. CO_2 laser 10P pump band; resonance pump transition $^RP_3(39)$.
7193. CO_2 laser 10P pump band; resonance pump transition $^RP_3(44)$.
7194. CO_2 laser 10R pump band; resonance pump transition $^RQ_3(43)$.
7195. CO_2 laser 10R pump band; resonance pump transition $^RR_0(33)$.
7196. CO_2 laser 10R pump band; resonance pump transition $^PR_1(39)$.
7197. CO_2 laser pump line 10P07.
7198. CO_2 laser pump line 10P08.
7199. CO_2 laser pump line 10P10.
7200. CO_2 laser pump line 10P11.
7201. CO_2 laser pump line 10P12.
7202. CO_2 laser pump line 10P13.
7203. CO_2 laser pump line 10P14.
7204. CO_2 laser pump line 10P15.
7205. CO_2 laser pump line 10P16.
7206. CO_2 laser pump line 10P17.
7207. CO_2 laser pump line 10P18.
7208. CO_2 laser pump line 10P19.
7209. CO_2 laser pump line 10P27.
7210. CO_2 laser pump line 10P28.

7211. CO_2 laser pump line 10P29.
7212. CO_2 laser pump line 10P30.
7213. CO_2 laser pump line 10P31.
7214. CO_2 laser pump line 10P32.
7215. CO_2 laser pump line 10P33.
7216. CO_2 laser pump line 10P34.
7217. CO_2 laser pump line 10P35.
7218. CO_2 laser pump line 10P36.
7219. CO_2 laser pump line 10P37.
7220. CO_2 laser pump line 10P38.
7221. CO_2 laser pump line 10P39.
7222. CO_2 laser pump line 10P40.
7223. CO_2 laser pump line 10R13.
7224. CO_2 laser pump line 10R14.
7225. CO_2 laser pump line 10R15.
7226. CO_2 laser pump line 10R16.
7227. CO_2 laser pump line 10R17.
7228. CO_2 laser pump line 10R18.
7229. CO_2 laser pump line 10R19.
7230. CO_2 laser pump line 10R20.
7231. CO_2 laser pump line 10R21.
7232. CO_2 laser pump line 10R22.
7233. CO_2 laser pump line 10R23.
7234. CO_2 laser pump line 10R24.
7235. CO_2 laser pump line 10R25.
7236. CO_2 laser pump line 10R26.
7237. CO_2 laser pump line 10R27.
7238. CO_2 laser pump line 10R28.
7239. CO_2 laser pump line 10R29.
7240. CO_2 laser pump line 10R30.
7241. CO_2 laser pump line 10R31.
7242. CO_2 laser pump line 10R32.
7243. CO_2 laser pump line 10R33.
7244. CO_2 laser pump line 10R34.
7245. CO_2 laser pump line 10R35.
7246. CO_2 laser pump line 10R36.
7247. CO_2 laser pump line 10R37.
7248. CO_2 laser pump line 10R38.
7249. CO_2 laser pump line 10R39.
7250. CO_2 laser pump line 10R40.
7251. CO_2 laser pump line 10P09.
7252.   Excitation of a mixture of FN_3, B_2H_6, SF_6, and He by a pulsed CO_2 laser.
7253.   Stimulated emission on the P(43) and R(70) transitions was demonstrated using the reaction of premixed FN_3 and $Bi(CH_3)_3$. The transient chemistry was thermally initated by a pulsed CO_2 laser using SF_6 as a sensitizer. The peak gain was estimated at 3.6×10^{-4}/cm.
7254. CO_2 laser pump line 10R04.
7255. CO_2 laser pump line 10R32.
7256. CO_2 laser pump line 10R50.
7257. CO_2 laser pump line 10R52.
7258. CO_2 laser pump line 9P34.

Section 3.7
REFERENCES

1. Hughes, W. M., Shannon, J., Hunter, R., *Appl. Phys. Lett.*, 24, 488 (1974), 126.1 nm argon laser.
2. Golde, M. F., Thrush, B. A., *Chem. Phys. Lett.*, 29, 486 (1974), Vacuum UV emission from reactions of metastable inert-gas atoms: chemiluminescence of ArO and ArCl.
3. Champagne, L. F., *Appl. Phys. Lett.*, 33, 523 (1978), Efficient operation of the electron-beam-pumped XeCl laser.
4. Burnham, R., *Opt. Commun.*, 24, 161 (1978), Improved performance of the discharge-pumped XeCl laser.
5. Bichkov, Y. I., Gorbatenko, A. I., Mesyats, G. A., Tarasenko, V. F., *Opt. Commun.*, 30, 224 (1979), Effective XeCl-laser performance conditions with combined pumping.
6. Sur, A., Hui, A. K., Tellinghuisen, J., *J. Mol. Spectrosc.*, 74, 465 (1979), Noble gas halides.
7. Burnham, R., Powell, F. X., Djeu, N., *Appl. Phys. Lett.*, 29, 30 (1976), Efficient electric discharge lasers in XeF and KrF.
8. Ault, E. R., Bradford, R. S., Jr., Bhaumik, M. L., *Appl. Phys. Lett.*, 27, 413 (1975), High-power xenon fluoride laser.
9. Brau, C. A., Ewing, J. J., *Appl. Phys. Lett.*, 27, 435 (1975), 354-nm laser action on XeF.
10. Bischel, W. K., Nakano, H. H., Eckstrom, D. J., Hill, R. M., Huestis, D., *Appl. Phys. Lett.*, 34, 565 (1979), A new blue-green excimer laser in XeF.
11. Ernst, W. E., Tittel, F. K., *Appl. Phys. Lett.*, 35, 36 (1979), A new electron-beam pumped XeF laser at 486 nm.
12. Tellinghuisen, J., Tellinghuisen, P. C., Tisone, G. C., Hoffman, J. M., *J. Chem. Phys.*, 68, 5177 (1978), Spectroscopic studies of diatomic noble gas halides. III. Analysis of XeF 3500 Å, band system.
13. Velazco, J. E., Kolts, J. H., Setser, D. W., *J. Chem. Phys.*, 65, 3468 (1976), Quenching rate constants for metastable argon, krypton, and xenon atoms, by fluorine containing molecules and branching ratios for XeF* and KrF*.
14. Basov, N. G., Babeiko, Y. A., Zuev, V. S., Mesyats, G. A., Orlov, V. K., *Sov. J. Quantum Electron.*, 6, 505 (1976), Laser emission from the XeO molecule under optical pumping conditions.
15. Basov, N. G., Danilychev, V. A., Popov, Y. M., *Sov. J. Quantum Electron.*, 1, 18 (1971), Stimulated emission in the vacuum-ultraviolet region.
16. Turner, C. E., Jr., *Appl. Phys. Lett.*, 31, 659 (1977), Near-atmospheric-pressure xenon excimer laser.
17. Hughes, W. M., Shannon, J., Kolb, A., Ault, E., Bhaumik, M., *Appl. Phys. Lett.*, 23, 385 (1973), High-power ultraviolet laser radiation from molecular xenon.
18. Hoff, P. W., Swingle, J. C., Rhodes, C. K., *Opt. Commun.*, 8, 128 (1973), Demonstration of temporal coherence, spatial coherence, and threshold effects in the molecular xenon laser.
19. Bradley, D. J., Hull, D. R., Hutchinson, M. H. R., McGeoch, M. W., *Opt. Commun.*, 14, 1 (1975), Co-axially pumped, narrow band, continuously tunable, high power VUV xenon laser.
20. Koehler, H. A., Ferderber, L. J., Redhead, D. L., Ebert, P. J., *Appl. Phys. Lett.*, 21, 198 (1972), Stimulated VUV emission in high-pressure xenon excited by relativistic electron beams.

21. Tittel, F. K., Wilson, W. L., Stickel, R. E., Marowsky, G., Ernst, W. E., *Appl. Phys. Lett.*, 36, 405 (1980), A triatomic Xe_2Cl excimer laser in the visible.

22. Tang, K. Y., Lorents, D. C., Huestis, D. L., *Appl. Phys. Lett.*, 36, 347 (1980), Gain measurements on the triatomic excimer Xe_2Cl.

23. Waynant, R. W., *Appl. Phys. Lett.*, 30, 234 (1977), A discharge-pumped ArCl superfluorescent laser at 175.0 nm.

24. Jacobs, R. R., Krupke, W. F., *Appl. Phys. Lett.*, 32, 31 (1978), Optical gain at 1.06 μm in the neodymium chloride-aluminum chloride vapor complex.

25. Jacobs, R. R., Krupke, W. F., *Appl. Phys. Lett.*, 34, 497 (1979), Excited state kinetics for $Nd(thd)_3$ and $Tb(thd)_3$ chelate vapors and prospects as fusion laser media.

26. Jacobs, R. R., Krupke, W. F., *Appl. Phys. Lett.*, 35, 126 (1979), Kinetics and fusion laser potential for the terbium aluminum chloride vapor complex.

27. Jacobs, R. R., Weber, M. J., Pearson, R. K., *Chem. Phys. Lett.*, 34, 80 (1975), Nonradiative intramolecular deactivation of Tb^{+3} fluorescence in a vapor phase, terbium (III) complex.

28. Ginter, M. L., Battino, R., *J. Chem. Phys.*, 52, 4469 (1970), Potential-energy curves for the He_2 molecule.

29. Ninomiya, H., Hirata, K., *J. Appl. Phys.* 68, 5378 (1990), Visible laser action in N_2 laser pumped Ti vapor.

30. Hirata, K., Yoshino, S., Ninomiya, H., *J. Appl. Phys.* 68, 1460 (1990), Characteristics of an optically pumped titanium vapor laser.

31. Hirata, K., Yoshino, S., Ninomiya, H., *J. Appl. Phys.* 67, 45 (1990), Optically pumped titanium laser at 551.4 nm.

32. Johnson, R. O., Perram, G. P., Roh, W. B., *Appl. Phys. B* 65, 5 (1997), Dynamics of a $Br(4^2P_{1/2} \rightarrow 4^2P_{3/2})$ pulsed laser and a $Br (^2P_{1/2})$–NO(n=2 $\rightarrow$ n=1) transfer laser driven by photolysis of iodine monobromine.

33. Golde, M. F., *J. Mol. Spectrosc.*, 58, 261 (1975), Interpretation of the oscillatory spectra of the inert-gas halides.

34. Lorents, D. C., Huestis, D. L., McCusker, R. V., Nakano, H. H., Hill, R., *J. Chem. Phys.*, 68, 4657 (1978), Optical emissions of triatomic rare gas halides.

35. Hoffman, J. M., Hays, A. K., Tisone, G. C., *Appl. Phys. Lett.*, 28, 538 (1976), High-power noble-gas-halide lasers.

36. Sze, R. C., Scott, P. B., *Appl. Phys. Lett.*, 33, 419 (1978), Intense lasing in discharge-excited noble-gas monochlorides.

37. Burnham, R., Djeu, N., *Appl. Phys. Lett.*, 29, 707 (1976), Ultraviolet-preionized discharge-pumped lasers in XeF, KrF, and ArF.

38. Rokni, M., Jacob, J. H., Mangano, J. H., *Phys. Rev. A*, 16, 2216 (1977), Dominant formation and quenching processes in e-beam pumped ArF* and KrF* lasers.

39. Powell, H. T., Murray, J. R., Rhodes, C. K., *Appl. Phys. Lett.*, 25, 730 (1974), Laser oscillation on the green bands of XeO and KrO.

40. Lorents, D. C., *Physica*, 82C, 19 (1976), The physics of electron-beam excited rare-gases at high densities.

41. Murray, J. R., Rhodes, C. K., *J. Appl. Phys.*, 47, 5041 (1976), The possibility of high-energy-storage lasers using the auroral and transauroral transitions of column-VI elements.

42. Rockwood, S. D., LAUR 73-1031(1973), Mechanisms for Achieving Lasing on the 5577 Å Line of Atomic Oxygen, Los Alamos Scientific Laboratory.

43. Murray, J. R., Powell, H. T., Schlitt, L. G., Toska, J., UCRL-50021-76(1976), Laser Program Annual Report, Lawrence Livermore Laboratory.

44. Julienne, P. S., Krauss, M., Stevens, W., *Chem. Phys. Lett.*, 38, 374 (1976), Collision-induced O^1D_2 - $^1S_0$ emission near 5577 Å in argon.

45.   Krauss, M., Mies, F. H., *Excimer Lasers* (1979), Springer-Verlag, New York, Electronic structure and radiative transitions of excimer systems.

46.   Hughes, W. M., Olson, N. T., Hunter, R., *Appl. Phys. Lett.*, 28, 81 (1976), Experiments on the 558 nm argon oxide laser system.

47.   Cunningham, D. L., Clark, K. C., *J. Chem. Phys.* 61, 1118 (1974), Rates of collision-induced emission from metastable O^1S atoms.

48.   Wellegehausen, B., Friede, D., Steger, G., *Opt. Commun.*, 26, 391 (1978), Optically pumped continuous Bi_2 and Te_2 lasers.

49.   West, W. P., Broida, H. P., *Chem. Phys. Lett.*, 56, 283 (1978), Optically pumped vapor phase Bi_2 laser.

50.   Murray, J. R., Swingle, J. C., Turner, C. E., Jr., *Appl. Phys. Lett.*, 28, 530 (1976), Laser oscillation of the 292 nm band system of Br_2.

51.   Koehler, H. A., Ferderber, L. J., Redhead, D. L., Ebert, P. J., *Phys. Rev.* A, 9, 768 (1974), Vacuum-ultraviolet emission from high-pressure xenon and argon excited by high-current relativistic electron-beams.

52.   Ewing, J. J., Jacob, J. H., Mangano, J. A., Brown, H. A., *Appl. Phys. Lett.*, 28, 656 (1976), Discharge pumping of the Br_2^* laser.

53.   Hunter, R. O., unpublished (1975), ARPA Review Meeting, Stanford Research Institute.

54.   Veukateswarlu, P., Verma, R. D., *Proc. Indian Acad. Sci.*, 46, 251 (1957), Emission spectrum of bromine excited in the presence of argon - Part I.

55.   Hays, A. K., *Laser Focus*, 14, 28 (1978), Cl_2 laser emitting 96 mJ at 258 nm seen promising for iodine-laser pump.

56.   Diegelmann, M., Hohla, K., Kompa, K. L., *Opt. Commun.*, 29, 334 (1979), Interhalogen UV laser on the 285 nm line band of ClF^*.

57.   Baboshii, V. N., Dobychin, S. L., Zuev, V. S., Mikheev, L. D., Pavlov, A., *Sov. J. Quantum Electron.*, 7, 1183 (1977), Laser utilizing an electronic transition in CN radicals pumped by radiation from an open, high current discharge.

58.   West, G. A., Berry, M. J., *J. Chem. Phys.*, 61, 4700 (1974), CN photodissociation and predissociation chemical lasers: molecular electronic and vibrational laser emissions.

59.   Mathias, L. E. S., Crocker, A., *Phys. Lett.*, 7, 194 (1963), Visible laser oscillations from carbon monoxide.

60.   Hodgson, R. T., *J. Chem. Phys.*, 55, 5378 (1971), Vacuum-ultraviolet lasing observed in CO: 1800-2000 Å.

61.   Waller, R. A., Collins, C. B., Cunningham, A. J., *Appl. Phys. Lett.*, 27, 323 (1975), Stimulated emission from CO^+ pumped by charge transfer from He_2^+ in the afterglow of an e-beam discharge.

62.   Wilkinson, P. G., Byram, E. T., *Appl. Opt.*, 4, 581 (1965), Rare-gas light sources for the vacuum ultraviolet.

63.   Rice, J. K., Hays, A. K., Woodworth, J. R., *Appl. Phys. Lett.*, 31, 31 (1977), Vacuum-UV emissions from mixtures of F_2 and the noble gases - a molecular F_2 laser at 1575 Å.

64.   Pummer, H., Hohla, H., Diegelmann, M., Reilly, J. P., *Opt. Commun.*, 28, 104 (1979), Discharge pumped F_2 laser at 1580 Å.

65.   Woodworth, J. K., Rice, J. K., *J. Chem. Phys.*, 69, 2500 (1978), An efficient high-power F_2 laser near 157 nm.

66.   Dreyfus, R. W., *Phys. Rev.* A, 9, 2635 (1974), Molecular hydrogen laser: 1098-1613 Å.

67.   Knyazev, I. N., Letokhov, V. S., Movshev, V. G., *IEEE J. Quantum Electron.*, QE-11, 805 (1975), Efficient and practical hydrogen vacuum ultraviolet laser.

68.   Waynant, R. W., Ali, A. W., Julienne, P. S., *J. Appl. Phys.*, 42, 3406 (1971), Experimental observations and calculated bond strengths for the D_2 Lyman band laser.

69.   Waynant, R. W., Shipman, J. D., Jr., Elton, R. C., Ali, A. W., *Appl. Phys. Lett.*, 17, 383 (1970), VUV laser emission from molecular hydrogen.

70. Hodgson, R. T., *Phys. Rev. Lett.*, 25, 494 (1970), VUV laser action observed in the Lyman bands of molecular hydrogen.

71. Bockasten, K., Lundholm, T., Andrede, D., *J. Opt. Soc. Am.*, 56, 1260 (1966), Laser lines in atomic and molecular hydrogen.

72. Bazhulin, P. A., Knyazev, I. N., Petrash, G. G., *Sov. Phys. JETP*, 20, 1068 (1965), Pulsed laser action in molecular hydrogen.

73. Hoff, P. W., Swingle, J. C., Rhodes, C. K., *Appl. Phys. Lett.*, 23, 245 (1973), Observations of stimulated emission from high-pressure krypton and argon xenon mixtures.

74. Bazhulin, P. A., Knyazev, I. N., Petrash, G. G., *Sov. Phys. JETP*, 22, 11 (1966), Stimulated emission from hydrogen and deuterium in the infrared.

75. Whitney, W. T., *Appl. Phys. Lett.*, 32, 239 (1978), Sustained discharge excitation of HgCl and HgBr $1B^2S^+_{1/2} \rightarrow X^2S^+_{1/2}$ lasers.

76. Parks, J. H., *Appl. Phys. Lett.*, 31, 297 (1977), Laser action on the $B^2S^+_{1/2} \rightarrow X^2S^+_{1/2}$ band of HgBr at 5018 Å.

77. Schimitschek, E. J., Celto, J. E., Trias, J. A., *Appl. Phys. Lett.*, 31, 608 (1977), Mercuric bromide photodissociation laser.

78. Burnham, R., *Appl. Phys. Lett.*, 33, 156 (1978), Discharge pumped mercuric halide dissociation lasers.

79. Maya, J., *J. Chem. Phys.*, 67, 4976 (1977), Ultraviolet absorption cross-sections of HgI_2, $HgBr_2$, and tin (II) halide vapors.

80. Parks, J. H., *Appl. Phys. Lett.*, 31, 192 (1977), Laser action on the $B^2S^+_{1/2} \, X^2S^+_{1/2}$ band of HgCl at 5576 Å.

81. Tang, K. Y., Hunter, R. O., Oldenettel, J., Howton, C., Huestis, D., Eckstrom, D., Perry, B.., McCusker, M., *Appl. Phys. Lett.*, 32, 226 (1978), Electron-beam controlled HgCl* laser.

82. Eden, J. G., *Appl. Phys. Lett.*, 33, 495 (1978), VUV-pumped HgCl laser.

83. Eden, J. G., *Appl. Phys. Lett.*, 31, 448 (1977), Green HgCl ($B^2S^+ \rightarrow X^2S^+$) laser.

84. Werner, C. W., George, E. V., Hoff, P. W., Rhodes, C. K., *Appl. Phys. Lett.*, 25, 235 (1974), Dynamic model of high-pressure rare-gas excimer lasers.

85. Bradford, R. S., Jr., Ault, E. R., Bhaumik, M. L., *Appl. Phys. Lett.*, 27, 546 (1975), High-power I_2 laser in the 342 nm band system.

86. Hays, A. K., Hoffman, J. M., Tisone, G. C., *Chem. Phys. Lett.*, 39, 353 (1976), Molecular iodine laser.

87. Mikheev, L. D., Shirokikh, A. P., Startsev, A. V., Zuev, V. S., *Opt. Commun.*, 26, 237 (1978), Optically pumped molecular iodine laser on the 342 nm band.

88. Byer, R. L., Herbst, R. L., Kildal, H., *Appl. Phys. Lett.*, 20, 463 (1972), Optically pumped molecular iodine vapor-phase laser.

89. Koffend, J. B., Field, R. W., *J. Appl. Phys.*, 48, 4468 (1977), CW optically pumped molecular iodine laser.

90. Wellegehausen, B., Stephan, K. H., Friede, D., Welling, H., *Opt. Commun.*, 23, 157 (1977), Optically pumped continuous I_2 molecular laser.

91. Hartmann, B., Kleman, B., Steinvall, O., *Opt. Commun.*, 21, 33 (1977), Quasi-tunable I_2-laser for absorption measurements in the near infrared.

92. Hanko, L., Benard, D. J., Davis, S. J., *Opt. Commun.*, 30, 63 (1979), Observation of super-fluorescent emission of the B-X system in I_2.

93. Murray, J. R., Powell, H. T., *Appl. Phys. Lett.*, 29, 252 (1976), KrCl laser oscillation at 222 nm.

94. Eden, J. G., Searles, S. K., *Appl. Phys. Lett.*, 29, 350 (1976), Observation of stimulated emission in KrCl.

95.   Gedanken, A., Jortner, J., Raz, B., Szoke, A., *J. Chem. Phys.*, 57, 3456 (1972), Electronic energy transfer phenomena in rare gases.

96.   Hay, P. J., Dunning, T. H., Jr., *J. Chem. Phys.*, 66, 1306 (1977), The electronic states of KrF.

97.   Tisone, G. C., Hays, A. K., Hoffman, J. M., *Opt. Commun.*, 15, 188 (1975), 100 mW, 248.8 nm KrF laser excited by an electron beam.

98.   Ewing, J. J., Brau, C. A., *Appl. Phys. Lett.*, 27, 350 (1975), Laser action on the $2S^+{}_{1/2} \rightarrow 2S^+{}_{1/2}$ bands of KrF and XeCl.

99.   Mangano, J. A., Jacob, J. H., *Appl. Phys. Lett.*, 27, 495 (1975), Electron-beam-controlled discharge pumping of the KrF laser.

100.   Jacob, J. H., Hsia, J. C., Mangano, J. A., Rokni, M., *J. Appl. Phys.*, 50, 5130 (1979), Pulse shape and laser-energy extraction from e-beam-pumped KrF.

101.   Fahlen, T. S., *J. Appl. Phys.*, 49, 455 (1978), High-pulse-rate 10-W KrF laser.

102.   Hawkins, R. T., Egger, H., Bokor, J., Rhodes, C. K., *Appl. Phys. Lett.*, 36, 391 (1980), A tunable, ultrahigh spectral brightness KrF* excimer laser source.

103.   Smith, D., Dean, A. G., Plumb, I. C., *J. Phys. B*, 5, 2134 (1972), Three-body conversion reactions in pure rare gases.

104.   Petit, A., Launay, F., Rostas, J., *Appl. Opt.*, 17, 3081 (1978), Spectroscopic analysis of the transverse excited $C^3P_u \rightarrow B$, ^3P_g (0,0) UV laser band of N_2 at room temperature.

105.   Massone, C. A., Garavaglia, M., Gallardo, M., Calatroni, J. A. E., Tagliaferri, A. A., *Appl. Opt.*, 11, 1317 (1972), Investigation of a pulsed molecular nitrogen laser at low temperature.

106.   Verkovtseva, E. T., Ovechkin, A. E., Fogel, Y. M., *Chem. Phys. Lett.*, 30, 120 (1975), The vacuum-uv spectra of supersonic jets of Ar-Kr-Xe mixtures excited by an electron beam.

107.   McFarlane, R. A., *Phys. Rev.*, 140, 1070 (1965), Observation of a^1P - $^1S^-$ transition in the nitrogen molecule.

108.   McFarlane, R. A., *IEEE J. Quantum Electron.*, QE-2, 229 (1966), Precision spectroscopy of new infrared emission systems of molecular nitrogen.

109.   Ault, E. R., Bhaumik, M. L., Olson, N. T., *IEEE J. Quantum Electron.*, QE-10, 624 (1974), High-power Ar-N_2 transfer laser at 3577 Å.

110.   Black, G., Sharpless, R. L., Slanger, T. G., Lorents, D. C., *J. Chem. Phys.*, 62, 4266 (1975), Quantum yields for the production of $O(^1S)$, $N(^2D)$, and $N_2(A^2S^+{}_u)$ from VUV, photolysis of N_2O.

111.   Searles, S. K., Hart, G. A., *Appl. Phys. Lett.*, 25, 79 (1974), Laser emission at 3577 and 3805 Å in electron-beam-pumped Ar-N_2 mixtures.

112.   Ischenko, V. N., Lisitsyn, V. N., Razhev, A. M. et al., *Opt. Commun.*, 13, 231 (1975), The $N^+{}_2$ laser.

113.   Ninomiya, H., Takashima, N., Hirata, K., *J. Appl. Phys.* 69, 67 (1991), Measurement of a population inversion on the 551.4 nm transition in an optically pumped Ti vapor laser.

114.   Rothe, D. E., Tan, K. O., *Appl. Phys. Lett.*, 30, 152 (1977), High-power $N^+{}_2$ laser pumped by charge transfer in a high-pressure pulsed glow discharge.

115.   Collins, C. B., Cunningham, A. J., Stockton, M., *Appl. Phys. Lett.*, 25, 344 (1974), A nitrogen ion laser pumped by charge transfer.

116.   McDaniel, E. W., Flannery, M. R., Ellis, H. W., Eisele, F. L., Pope, W., Tech. Rep. H-78-1(1978), U.S. Army Missile Research and Development Command, Compilation of data relevant to rare gas-rare gas and rare gas-monohalide, excimer lasers.

117.   Wellegehauser, B., Shahdin, S., Friede, D., Welling, H., *IEEE J. Quantum Electron.*, QE-13, 65D (1977), Continuous laser oscillation in alkali dimers.

118.   Henesian, M. A., Herbst, R. L., Byer, R. L., *J. Appl. Phys.*, 47, 1515 (1976), Optically pumped superfluorescent Na_2 molecular laser.

119. Itoh, H., Uchiki, H., Matsuoka, M., *Opt. Commun.*, 18, 271 (1976), Stimulated emission from molecular sodium.

120. Leone, S. R., Kosnik, K. G., *Appl. Phys. Lett.*, 30, 346 (1977), A tunable visible and ultraviolet laser on S_2.

121. Gerasimov, V. A., Yunzhakov, B. P., *Sov. J. Quantum. Electron.*, 19, 1532 (1989), Investigation of a thulium vapor laser.

122. Zuev, V. S., Mikheev, L. D., Yalovi, V. I., *Sov. J. Quantum Electron.*, 5, 442 (1975), Photochemical laser utilizing the $^1S^+_g - {}^3S^-_g$ vibronic transition in S_2.

123. Searles, S. K., Hart, G. A., *Appl. Phys. Lett.*, 27, 243 (1975), Stimulated emission at 281.8 nm from XeBr.

124. Sze, R. C., Scott, P. B., *Appl. Phys. Lett.*, 32, 479 (1978), High-energy lasing of XeBr in an electric discharge.

125. Tellinghuisen, J., Hays, A. K., Hoffman, J. M., Tisone, G. C., *J. Chem. Phys.*, 65, 4473 (1976), Spectroscopic studies of diatomic noble gas halides. II. Analysis of bound-free emission, from XeBr, XeI, and KrF.

126. Velazco, J. E., Setser, D. W., *J. Chem. Phys.*, 62, 1990 (1975), Bound-free emission spectra of diatomic xenon halides.

127. Basov, N. G., Vasil'ev, L. A., Danilychev, V. A., Dolgov-Saval'ev, G. G., *Sov. J. Quantum Electron.*, 5, 869 (1975), High-pressure N_2^+ laser emitting violet radiation.

128. McFarlane, R. A., unpublished work.

129. Tittel, F. K., unpublished (private communication).

130. Dezenberg, G. J., Willett, C. S., *IEEE J. Quantum Electron.*, QE-7, 491-493 (1971), New unidentified high-gain oscillation at 486.1 and 434.0 nm in the presence of neon.

131. Grishkowsky, D. R., Sorokin, P. P., Lankard, J. R., *Opt. Commun.*, 18, 205-206 (1977), An atomic 16 micron laser.

132. Isaev, A. A., Kazaryan, M. A., Petrash, G. G., *Kratk. Soobshch. Fiz.*, 3, 3 (1972), unspecified.

133. Fahlen, T. S., *IEEE J. Quantum Electron.*, QE-12, 200-201 (1976), Self-heated, multiple-metal-vapor laser.

134. Markova, S. V., Cherezov, V. M., *Sov. J. Quantum Electron.*, 7, 339-342 (1977), Investigation of pulse stimulated emission from gold vapor.

135. Markova, S. V., Petrash, G. G., Cherezov, V. M., *Sov. J. Quantum Electron.*, 8, 904-906 (1978), Ultraviolet-emitting gold vapor laser.

136. Cahuzac, P., *IEEE J. Quantum Electron.*, QE-8, 500 (1972), New infrared laser lines in Mg vapor.

137. Trainor, D. W., Mani, S. A., *Appl. Phys. Lett.*, 33, 648-650 (1978), Atomic calcium laser: pumped via collision-induced absorption.

138. Baron, K. U., Stadler, B., unpublished (June, 1976), Hollow cathode-excited laser transitions in calcium, strontium and barium.

139. Deech, J. S., Sanders, J. H., *IEEE J. Quantum Electron.*, QE-4, 474 (1968), New self-terminating laser transitions in calcium and stronium.

140. Klimkin, V. M., Kolbycheva, P. D., *Sov. J. Quantum Electron.*, 7, 1037-1039 (1977), Tunable single-frequency calcium-hydrogen laser emitting at 5.54 μm.

141. Klimkin, V. M., Monastyrev, S. S., Prokop'ev, V. E., *JETP Lett.*, 20, 110-111 (1974), Selective relaxation of long-lived states of metal atoms in a gas discharge plasma. Stationary generation on $^1P^0_1-{}^1D_2$ transitions.

142. Grishkowsky, D. R., Lankard, J. R., Sorokin, P. P., *IEEE J. Quantum Electron.*, QE-13, 392-396 (1977), An atomic Rydberg state 16-m laser.

143. Platanov, A. V., Soldatov, A. N., Filonov, A. G., *Sov. J. Quantum Electron.*, 8, 120-121 (1978), Pulsed strontium vapor laser.

144. Bokhan, P. A., Burlakov, V. D., *Sov. J. Quantum Electron.*, 9, 374-376 (1979), Mechanism of laser action due to $4d^3D_{1/2} \rightarrow 5p^3P^0_2$ transitions in a strontium atom.

145. Baron, K. U., Stadler, B., *IEEE J. Quantum Electron.*, QE-11, 852-853 (1975), New visible laser transitions in Ba I and Ba II.

146. Cahuzac, P., *Phys. Lett.*, 32a, 150-151 (1970), New infrared laser lines in barium vapor.

147. Isaev, A. A., Kazaryan, M. A., Markova, S. V., Petrash, G. G., *Sov. J. Quantum Electron.*, 5, 285-287 (1975), Investigation of pulse infrared stimulated emission from barium vapor.

148. Bricks, B. G., Karras, T. W., Anderson, R. S., *J. Appl. Phys.*, 49, 38-40 (1978), An investigation of a discharge-heated barium laser.

149. Isaev, A. A., Kazaryan, M. A., Petrash, G. G., *Sov. J. Quantum Electron.*, 3, 358-359 (1974), Emission of laser pulses due to transitions from a resonance to a metastable level in barium vapor.

150. Cross, L. A., Gokay, M. C., *IEEE J. Quantum Electron.*, QE-14, 648 (1978), A pulse repetition frequency scaling law for the high repetition rate neutral barium laser.

151. Bokhan, P. A., Solomonov, V. I., *Sov. Tech. Phys. Lett.*, 4, 486-487 (1978), Barium vapor laser with a high average output power.

152. Dubrovin, A. N., Tibilov, A. S., Shevtsov, M. K., *Opt. Spectrosc.*, 32, 685 (1972), Lasing on Cd, Zn and Mg lines and possible applications.

153. Tibilov, A. S., *Opt. Spectrosc.*, 19, 463-464 (1965), Generation of radiation in He-Cd and Ne-Cd mixtures.

154. Silfvast, W. T., Szeto, L. H., Wood, O. R., II, *Opt. Lett.*, 4, 271-273 (1979), Recombination lasers in Nd and CO_2 laser-produced cadmium plasmas.

155. Chou, M. S., Cool, T. A., *J. Appl. Phys.*, 48, 1551-1555 (1977), Laser operation by dissociation of metal complexes. II. New transitions in Cd, Fe, Ni, Se, Sn, Te, V and Zn.

156. Komine, H., Byer, R. L., *J. Appl. Phys.*, 48, 2505-2508 (1977), Optically pumped atomic mercury photodissociation laser.

157. Djeu, N., Burnham, R., *Appl. Phys. Lett.*, 25, 350-351 (1974), Optically pumped CW Hg laser at 546.1 nm.

158. Artusy, M., Holmes, N., Siegman, A. E., *Appl. Phys. Lett.*, 28, 1331-1334 (1976), D.C.-excited and sealed-off operation of the optically pumped 546.1 nm Hg laser.

159. Holmes, N. C., Siegman, A. E., *J. Appl. Phys.*, 49, 3155-3170 (1978), The optically pumped mercury vapor laser.

160. Bloom, A. L., Bell, W. E., Lopez, F. O., *Phys. Rev.*, 135, A578-A579 (1964), Laser spectroscopy of a pulsed mercury-helium discharge.

161. Bockasten, K., Garavaglia, M., Lengyel, B. A., Lundholm, T., *J. Opt. Soc. Am.*, 55, 1051-1053 (1965), Laser lines in Hg I.

162. Heard, H. G., Peterson, J., *Proc. IEEE*, 52, 414 (1964), Laser action in mercury rare gas mixtures.

163. Heard, H. G., Peterson, J., *Proc. IEEE*, 52, 1049-1050 (1964), Mercury-rare gas visible-UV laser.

164. Rigden, J. D., White, A. D., *Nature (London)*, 198, 774 (1963), Optical laser action in iodine and mercury discharges.

165. Paananen, R. A., Tang, C. L., Horrigan, F. A., Statz, H., *J. Appl. Phys.*, 34, 3148-3149 (1963), Optical laser action in He-Hg rf discharges.

166. Armand, M., Martinot-Lagarde, P., *C. R. Acad. Sci. Ser.* B, 258, 867-868 (1964), Effect laser sur la vapeur de mercure dans un melange He-Hg.

167. Doyle, W. M., *J. Appl. Phys.*, 35, 1348-1349 (1964), Use of time resolution in identifying laser transitions in mercury rare gas discharge.

168. Chebotayev, V. P., *Opt. Spectrosc.*, 25, 267-268 (1968), Isotopic structure of the 1.5295 millimicron laser line of mercury.

169. Convert, G., Armand, M., Martinot-Lagarde, P., *C. R. Acad. Sci. Ser.* B, 257, 3259-3260 (1964), Effect laser dans des melanges mercure-gas rares.

170. Beterov, I. M., Klement'ev, V. M., Chebotaev, V. P., *Radio Eng. Electron. Phys.* (USSR), 14, 1790-1792 (1969), A mercury laser secondary frequency standard in the microwave region.

171. Bikmukhametov, K. A., Klement'ev, V. M., Chebotaev, V. P., *Sov. J. Quantum Electron.*, 2, 254-256 (1972), Investigation of the stability of the oscillation frequency of a mercury laser emitting at λ= 1.53 μ.

172. Bikmukhametov, K. A., Klement'ev, V. M., Chebotaev, V. P., *Opt. Spectrosc.*, 34, 616-617 (1973), Collision broadening of the 1.53 μm line of mercury in an Hg-He, Hg-Ne mixture.

173. Klement'ev, V. M., Solov'ev, M. V., *J. Appl. Spectrosc.*, 18, 29-32 (1973), Mercury-vapor laser.

174. Bikmukhametov, K. A., Klement'ev, V. M., Chebotaev, V. P., *Sov. J. Quantum Electron.*, 5, 278-281 (1975), Experimental investigation of the dependencies of the collision broadening and shift of the emission line of a mercury laser on He and Ne pressure.

175. Chou, M. S., Cool, T. A., *J. Appl. Phys.*, 47, 1055-1061 (1976), Laser operation by dissociation of metal complexes: new transitions in As, Bi, Ga, Hg, In, Pb, Sb, and Tl.

176. Stricker, J., Bauer, S. H., *IEEE J. Quantum Electron.*, QE-11, 701-702 (1975), An atomic boron laser-pumping by incomplete autoionization or ion-electron.

177. Hemmati, H., Collins, G. J., *Appl. Phys. Lett.*, 34, 844-845 (1979), Atomic gallium photodissociation laser.

178. Burnham, R., *Appl. Phys. Lett.*, 30, 132-133 (1977), Atomic indium photodissociation laser at 451 nm.

179. Gerasimov, V. A., Yunzhakov, B. P., *Sov. J. Quantum. Electron.*, 19, 1323 (1989), Stimulated emission from thulium bromide vapor.

180. Ehrlich, D. J., Maya, J., Osgood, R. M., Jr., *Appl. Phys. Lett.*, 33, 931-933 (1978), Efficient thallium photodissociation laser.

181. Isaev, A. A., Ischenko, P. I., Petrash, G. G., *JETP Lett.*, 6, 118-121 (1967), Super-radiance at transitions terminating at metastable levels of helium and thallium.

182. Isaev, A. A., Petrash, G. G., *JETP Lett.*, 7, 156-158 (1968), Pulsed superradiance at the green line of thallium in TlI-vapor.

183. Sorokin, P. P., Lankard, J. R., *J. Chem. Phys.*, 51, 2929-2931 (1969), Infrared lasers resulting from photodissociation of Cs_2 and Rb_2.

184. Jacobs, S., Rabinowitz, P., Gould, G., *Phys. Rev. Lett.*, 7, 415-417 (1961), Coherent light amplification in optically pumped Cs vapor.

185. Bennett, W. R., Jr., *Appl. Opt., Suppl. Chem. Lasers*, 3-33 (1965), Inversion mechanisms in gas lasers.

186. Rabinowitz, P., Jacobs, S., *Quantum Electronics III* (1964), 489-498, Columbia University Press, New York, The optically pumped cesium laser.

187. Isaev, A. A., Kazaryan, M. A., Petrash, G. G., *Opt. Spectrosc.*, 31, 180-183 (1971), Mechanism of pulsed lasing of the green thallium line in a thallium iodide vapor discharge.

188. Korolev, F. A., Odintsov, A. I., Turkin, N. G., Yakunin, V. P., *Sov. J. Quantum Electron.*, 5, 237-239 (1975), Spectral structure of pulse superluminescence lines of gases.

189. Chilukuri, S., *Appl. Phys. Lett.*, 34, 284-286 (1979), Selective optical excitation and inversions via the excimer channel: superradiance at the thallium green line.

190. Tunitskii, L. N., Cherkasov, E. M., *Sov. Phys. Tech. Phys.*, 13, 1696-1697 (1969), New oscillation lines in the spectra of NI and Cl.

191. Atkinson, J. B., Sanders, J. H., *J. Phys. B*, 1, 1171-1179 (1968), Laser action in C and N following dissociative excitation transfer.

192. Cooper, G. W., Verdeyen, J. T., *J. Appl. Phys.*, 48, 1170-1175 (1977), Recombination pumped atomic nitrogen and carbon afterglow lasers.

193. Voitovich, A. P., Dubovik, M. V., *J. Appl. Spectrosc.*, 27, 1399-1403 (1978), Time and power characteristics of pulsed atomic gas lasers in a magnetic field.

194. Boot, H. A. H., Clunie, D. M., *Nature (London)*, 197, 173-174 (1963), Pulsed gaseous maser.

195. Patel, C. K. N., McFarlane, R. A., Faust, W. L., *Quantum Electronics III* (1964), 561-572, Columbia University Press, New York, Further infrared spectroscopy using stimulated emission techniques.

196. Patel, C. K. N., McFarlane, R. A., Faust, W. L., *Phys. Rev.*, 133, A1244-A1248 (1964), Optical maser action in C, N, O, S and Br on dissociation of diatomic and polyatomic molecules.

197. English, J. R., II, Gardner, H. C., Merritt, J. A., *IEEE J. Quantum Electron.*, QE-8, 843-844 (1972), Pulsed stimulated emission from N, C, Cl and F atoms.

198. DePoorter, G. C., Balog, G., *IEEE J. Quantum Electron.*, QE-8, 917-918 (1972), New infrared laser line in OCS and new method for C atom lasing.

199. Shimazu, M., Suzaki, Y., *Jpn. J. Appl. Phys.*, 4, 819 (1965), Laser oscillations in silicon tetrachloride vapor.

200. Cooper, H. G., Cheo, P. K., *IEEE J. Quantum Electron.*, QE-2, 785 (1966), Laser transitions in BII, BrII, and Sn.

201. Carr, W. C., Grow, R. W., *Proc. IEEE*, 55, 1198 (1967), A new laser line in tin using stannic chloride vapor.

202. Zhukov, V. V., Latush, E. L., Mikhalevskii, V. S., Sem, M. F., *Sov. J. Quantum Electron.*, 5, 468-469 (1975), New laser transitions in the spectrum of tin and population-inversion mechanism.

203. Xing, D., Ueda, K., Takuma, H., *Appl. Phys. Lett.*, 60, 2961 (1992), K_2 yellow-band and Rb_2 orange-band excimer emissions by electron-beam excitation.

204. Isaev, A. A., Petrash, G. G., *JETP Lett.*, 10, 119-121 (1969), New generation and superradiance lines of lead vapor.

205. Fowles, G.R., Silfvast, W. T., *Appl. Phys. Lett.*, 6, 236-237 (1965), High gain laser transition in lead vapor.

206. Silfvast, W. T., Deech, J. S., *Appl. Phys. Lett.*, 11, 97-99 (1967), Six db/cm single pass gain at 7229 A in lead vapor.

207. Anderson, R. S., Bricks, B. G., Karras, T. W., Springer, L. W., *IEEE J. Quantum Electron.*, QE-12, 313-315 (1976), Discharge-heated lead vapor laser.

208. Kirilov, A. E., Kukharev, V. N., Soldatov, A. N., Tarasenko, V. F., *Sov. Phys. J.*, 20, 1381-1384 (1977), Lead vapor lasers.

209. Feldman, D. W., Liu, C. S., Pack, J. L., Weaver, L. A., *J. Appl. Phys.*, 49, 3679-3683 (1978), Long-lived lead-vapor lasers.

210. Kirilov, A. E., Kukharev, V. N., Soldatov, V. N., *Sov. J. Quantum Electron.*, 9, 285-287 (1979), Investigation of a pulsed $\lambda = 722.9$ nm Pb laser with a double-section gas-discharge tube.

211. Piltch, M., Gould, G., *Rev. Sci. Instrum.*, 37, 925-927 (1966), High temperature alumina discharge tube for pulsed metal vapor lasers.

212. Heard, H. G., Peterson, J., *Proc. IEEE*, 52, 1258 (1964), Visible laser transitions in ionized oxygen, nitrogen and carbon monoxide.

213. Hitt, J. S., Haswell, W. T., II, *IEEE J. Quantum Electron.*, QE-2, xlii (1966), Stimulated emission in the theta pinch discharge.

214. Chou, M. S., Zawadzkas, G. A., *Opt. Commun.*, 26, 92 (1978), Observation of new atomic nitrogen laser transition at 9046 A.

215. McFarlane, R. A., *Physics of Quantum Electronics* (1966), 655-663, McGraw-Hill, New York, Stimulated emission spectroscopy of some diatomic molecules.

216. DeYoung, R. J., Wells, W. E., Miley, G. H., Verdeyen, J. T., *Appl. Phys. Lett.*, 28, 519-521 (1976), Direct nuclear pumping of a Ne-N_2 laser.

217. Janney, G. M., *IEEE J. Quantum Electron.*, QE-3, 133 (1967), New infrared laser oscillations in atomic nitrogen.

218. Janney, G. M., *IEEE J. Quantum Electron.*, QE-3, 339 (1) (1967), Correction to near infrared laser oscillations in atomic nitrogen.

219. Sutton, D. G., *IEEE J. Quantum Electron.*, QE-12, 315-316 (1976), New laser oscillation in the N atom quartet manifold.

220. Cheo, P. K., Cooper, H. G., *Appl. Phys. Lett.*, 7, 202-204 (1965), UV and visible laser oscillations in fluorine, phosphorus and chlorine.

221. Fowles, G.R., Zuryk, J. A., Jensen, R. C., *IEEE J. Quantum Electron.*, QE-10, 394-395 (1974), Infrared laser lines in neutral atomic phosphorus.

222. Fowles, G.R., Zuryk, J. A., Jensen, R. C., *IEEE J. Quantum Electron.*, QE-10, 849 (1974), Infrared laser lines in arsenic vapor.

223. Markova, S. V., Petrash, G. G., Cherezov, V. M., *Sov. J. Quantum Electron.*, 7, 657 (1977), Pulse stimulated emission of the 472.2 nm line of the bismuth atom.

224. Powell, H. T., Murray, J. R., Rhodes, C. K., *Appl. Phys. Lett.*, 25, 730-732 (1974), Laser oscillation on the green bands of XeO and KrO.

225. Hughes, W. M., Olson, N. T., Hunter, R., *Appl. Phys. Lett.*, 28, 81-83 (1976), Experiments on 558-nm argon oxide laser systems.

226. Bennett, W. R., Jr., Faust, W. L., McFarlane, R. A., Patel, C. K. N., *Phys. Rev. Lett.*, 8, 470-473 (1962), Dissociative excitation transfer and optical maser oscillation in Ne-O_2 and Ar-O_2 rf discharges.

227. Tunitskii, L. N., Cherkasov, E. M., *Sov. Phys. Tech. Phys.*, 12, 1500-1501 (1968), Method for varying the frequency of a gas laser.

228. Tunitskii, L. N., Cherkasov, E. M., *J. Opt. Soc. Am.*, 56, 1783-1784 (1966), Interpretation of oscillation lines in Ar-Br_2 laser.

229. Rautian, S. G., Rubin, P. L., *Opt. Spectrosc.*, 18, 180-181 (1965), On some features of gas lasers containing mixtures of oxygen and rare gases.

230. Tunitskii, L. N., Cherkasov, E. M., *Opt. Spectrosc.*, 27, 344-346 (1969), Pulsed mode generation in an argon-oxygen laser.

231. Feld, M. S., Feldman, B. J., Javan, A., *Bull. Am. Phys. Soc.*, 12, 15 (1967), Frequency shifts of the fine structure oscillations of the 8446-Å atomic oxygen laser.

232. Tunitskii, L. N., Cherkasov, E. M., *Opt. Spectrosc.*, 23, 154-157 (1967), The mechanism of laser action in oxygen-inert gas mixtures.

233. Tunitskii, L. N., Cherkasov, E. M., *Sov. Phys. Tech. Phys.*, 13, 993-994 (1969), Pure oxygen laser.

234. Kolpakova, I. V., Redko, T. P., *Opt. Spectrosc.*, 23, 351-352 (1967), Some remarks on the operation of the neon-oxygen gas laser.

235. Feld, M. S., Feldman, B. J., Javan, A., Domash, L. H., *Phys. Rev.*, A7, 257-262 (1973), Selective reabsorption leading to multiple oscillations in the 8446Å atomic oxygen laser.

236. Sutton, D. G., Galvan, L., Suchard, S. N., *IEEE J. Quantum Electron.*, QE-11, 92 (1975), New laser oscillation in the oxygen atom.

237. Powell, F. X., Djeu, N. I., *IEEE J. Quantum Electron.*, QE-7, 176-177 (1971), CW atomic oxygen laser at 4.56 m.

238. Powell, H. T., Prosnitz, D., Schleicher, B. R., *Appl. Phys. Lett.*, 34, 571-573 (1979), Sulfur $^1S_0$-$^1D_2$ laser by OCS photodissociation.

239. Martinelli, R. U., Gerritsen, H. J., *J. Appl. Phys.*, 37, 444-445 (1966), Laser action in sulphur using hydrogen sulphide.

240. Ultee, C. J., *J. Appl. Phys.*, 44, 1406 (1973), Infrared laser emission from discharges through gaseous sulfur compounds.

241. Hocker, L. O., *J. Appl. Phys.*, 48, 3127-3128 (1977), New infrared laser transitions in neutral sulfur.

242. Hubner, G., Wittig, C., *J. Opt. Soc. Am.*, 61, 415-416 (1971), Some new infrared laser transitions in atomic oxygen and sulfur.

243. Cooper, H. G., Cheo, P. K., *Physics of Quantum Electronics* (1966), 690-697, McGraw-Hill, New York, Ion laser oscillations in sulphur.

244. Davis, C. C., King, T. A., *Advances in Quantum Electronics*, Vol. 3 (1977), 169-454, Academic Press, London, Gaseous ion lasers.

245. Powell, H. T., Ewing, J. J., *Appl. Phys. Lett.*, 33, 165-167 (1978), Photodissociation lasers using forbidden transitions of selenium atoms.

246. Cahuzac, P., *Phys. Lett.*, 27A, 473-474 (1968), Emission laser infrarouges dans les vapeurs de thulium et d'ytterbium.

247. Cahuzac, P., *Phys. Lett.*, 31A, 541-542 (1970), Infrared laser emission from rare-earth vapors.

248. Chapovsky, P. L., Kochubei, S. A., Lisitsyn, V. N., Razhev, A. M., *Appl. Phys.*, 14, 231-233 (1977), Excimer ArF/XeF lasers providing high-power stimulated radiation in Ar/Xe and F lines.

249. Lisitsyn, V. N., Razhev, A. M., *Sov. Tech. Phys. Lett.*, 3, 350-351 (1977), High-power, high-pressure laser based on red fluorine lines.

250. Loree, T. R., Sze, R. C., *Opt. Commun.*, 21, 255-257 (1977), The atomic fluorine laser: spectral pressure dependence.

251. Bigio, I. J., Begley, R. F., *Appl. Phys. Lett.*, 28, 263-264 (1976), High power visible laser action in neutral atomic fluorine.

252. Hocker, L. O., Phi, T. B., *Appl. Phys. Lett.*, 29, 493-494 (1976), Pressure dependence of the atomic fluorine transition intensities.

253. Kovacs, M. A., Ultee, C. J., *Appl. Phys. Lett.*, 17, 39-40 (1970), Visible laser action in fluorine I.

254. Jeffers, W. Q., Wiswall, C. E., *Appl. Phys. Lett.*, 17, 444-447 (1970), Laser action in atomic fluorine based on collisional dissociation of HF.

255. Florin, A. E., Jensen, R. J., *IEEE J. Quantum Electron.*, QE-7, 472 (1971), Pulsed laser oscillation at $0.7311\ \mu$ from F atoms.

256. Sumida, S., Obara, M., Fujioka, T., *J. Appl. Phys.*, 50, 3884-3887 (1979), Novel neutral atomic fluorine laser lines in a high-pressure mixture of F_2 and He.

257. Lawler, J. E., Parker, J. W., Anderson, L. W., Fitzsimmons, W. A., *IEEE J. Quantum Electron.*, QE-15, 609-613 (1979), Experimental investigation of the atomic fluorine laser.

258. Paananen, R. A., Horrigan, F. A., *Proc. IEEE*, 52, 1261-1262 (1964), Near infra-red lasering in $NeCl_2$ and $He-Cl_2$.

259. Shimazu, M., Suzaki, Y., *Jpn. J. Appl. Phys.*, 4, 381-382 (1965), Laser oscillation in the mixtures of freon and rare gases.

260. Jarrett, S. M., Nunez, J., Gould, G., *Appl. Phys. Lett.*, 8, 150-151 (1966), Laser oscillation in atomic Cl in HCl and HI gas discharges.

261. Trusty, G. L., Yin, P. K., Koozekanani, S. K., *IEEE J. Quantum Electron.*, QE-3, 368 (1967), Observed laser lines in freon-helium mixtures.

262. Paananen, R. A., Tang, C. L., Horrigan, F. A., *Appl. Phys. Lett.*, 3, 154-155 (1963), Laser action in Cl_2 and $He-Cl_2$.

263. Bockasten, K., *Appl. Phys. Lett.*, 4, 118-119 (1964), On the classification of laser lines in chlorine and iodine.

264. Dauger, A. B., Stafsudd, O. M., *IEEE J. Quantum Electron.*, QE-6, 572-573 (1970), Observation of CW laser action in chlorine, argon and helium gas mixtures.

265. Jarrett, S. M., Nunez, J., Gould, G., *Appl. Phys. Lett.*, 7, 294-296 (1965), Infrared laser oscillation in HBr and HI gas discharges.

266. Jensen, R. C., Fowles, G.R., *Proc. IEEE*, 52, 1350 (1964), New laser transitions in iodine-inert-gas mixtures.

267. Kasper, J. V. V., Pimentel, G. C., *Appl. Phys. Lett.*, 5, 231-233 (1964), Atomic iodine photodissociation laser.

268. Kasper, J. V. V., Parker, J. H., Pimentel, G. C., *J. Chem. Phys.*, 43, 1827-1828 (1965), Iodine-atom laser emission in alkyl iodide photolysis.

269. Pollack, M. A., *Appl. Phys. Lett.*, 8, 36-38 (1966), Pressure dependence of the iodine photodissociation laser peak output.

270. Andreeva, T. L., Dudkin, V. A., Malyshev, V. I., Mikhailov, G. V., Sorokin, V. N., *Sov. Phys. JETP*, 22, 969-970 (1966), Gas laser excited in the process of photodissociation.

271. DeMaria, A. J., Ultee, C. J., *Appl. Phys. Lett.*, 9, 67-69 (1966), High-energy atomic iodine photodissociation laser.

272. Gregg, D. W., Kidder, R. E., Dobler, C. V., *Appl. Phys. Lett.*, 13, 297-298 (1968), Zeeman splitting used to increase energy from a Q-switched laser.

273. Ferrar, C. M., *Appl. Phys. Lett.*, 12, 381-383 (1968), Q-switching and mode locking of a CF_3I photolysis laser.

274. Zalesskii, V. Yu, Venediktov, A. A., *Sov. Phys. JETP*, 28, 1104-1107 (1969), Mechanism of generation termination at the $5^2P_{1/2}$–$5^2P_{3/2}$ transition.

275. O'Brien, D. E., Bowen, J. R., *J. Appl. Phys.*, 40, 4767-4769 (1969), Kinetic model for the iodine photodissociation laser.

276. Andreeva, T. L., Malyshev, V. I., Maslov, A. I., Sobel'man, I. I., Sorokin, V. N., *JETP Lett.*, 10, 271-274 (1969), Possibility of obtaining excited iodine atoms as a result of chemical reactions.

277. Zalesskii, V. Yu., Moskalev, E. I., *Sov. Phys. JETP*, 30, 1019-1023 (1970), Optical probing of a photodissociation laser.

278. Belousova, I. M., Danilov, O. B., Sinitsina, I. A., Spiridonov, V. V., *Sov. Phys. JETP*, 31, 791-793 (1970), Investigation of the optical inhomogeneities of the active medium of a CF_3I photodissociation laser.

279. Velikanov, S. D., Kormer, S. B., Nikolaev, V. D., Sinitsyn, M. V., Solov'ev, Yu A., Urlin, V. D., *Sov. Phys. Dokl.*, 15, 478-480 (1970), Lower limit of the luminescence spectral linewidth of the $5^2P_{1/2}$ – $5^2P_{3/2}$ transition in atomic iodine in a photodissociation laser.

280. Gensel, P., Hohla, K., Kompa, K. L., *Appl. Phys. Lett.*, 18, 48-50 (1971), Energy storage of CF_3I photodissociation laser.

281. Belousova, I. M., Danilov, O. B., Kladovikova, N. S., Yachnev, I. L., *Sov. Phys. Tech. Phys.*, 15, 1212-1213 (1971), Quenching of excited atoms in a photodissociation laser.

282. O'Brien, D. E., Bowen, J. R., *J. Appl. Phys.*, 42, 1010-1015 (1971), Parametric studies of the iodine photodissociation laser.

283. DeWolf Lanzerotti, M. Y., *IEEE J. Quantum Electron.*, QE-7, 207-208 (1971), Iodine-atom laser emission in 2-2-2 trifluoroethyliodide.

284. Zalesskii, V. Yu., Krupenikova, T. I., *Opt. Spectrosc.*, 30, 439-443 (1971), Deactivation of metastable iodine atoms by collision with perfluoroalkyl iodide molecules.

285. Andreeva, T. L., Kuznetsova, S. V., Maslov, A. I., Sobel'man, I. I., Sorokin, V. N., *JETP Lett.*, 13, 449-452 (1971), Investigation of reactions of excited iodine atoms with the aid of a photodissociation laser.

286. Hohla, K., IPP Report IV/3(Dec, 1971), Max-Planck-Institut fur Plasma Physik, Photochemical iodine laser: kinetic foundations for giant pulse operation.

287. Belousova, I. M., Kiselev, V. M., Kurzenkov, V. N., *Opt. Spectrosc.*, 33, 112-114 (1972), Induced emission spectrum of atomic iodine due to the hyperfine structure of the transition $^2P_{1/2} - {}^2P_{3/2}$ (7603 cm^{-1}).

288. Belousova, I. M., Kiselev, V. M., Kurzenkov, V. N., *Opt. Spectrosc.*, 33, 115-116 (1972), Line width for induced emission due to the $^2P_{1/2} - {}^2P_{3/2}$ transition of atomic iodine.

289. Zalesskii, V. Yu., *Sov. Phys. JETP*, 34, 474-480 (1972), Kinetics of a CF_3I photodissociation laser.

290. Hwang, W. C., Kasper, J. V. V., *Chem. Phys. Lett.*, 13, 511-514 (1972), Zeeman effects in the hyperfine structure of atomic iodine photodissociation laser emission.

291. Hohla, K., Kompa, K. L., *Chem. Phys. Lett.*, 14, 445-448 (1972), Energy transfer in a photochemical iodine laser.

292. Hohla, K., Kompa, K. L., *Z. Naturforsch.*, 27a, 938-947 (1972), Kinetische prozesse in einem photochemischen jodlaser.

293. Filyukov, A. A., Karpov, Ya., *Sov. Phys. JETP*, 35, 63-65 (1972), A criterion for probable laser quenching.

294. Gavrilina, L. K., Karpov, V. Ya., Leonov, Yu. S., Sautkin, V. A., Filyukov, A. A., *Sov. Phys. JETP*, 35, 258-259 (1972), Selective pumping effect of a photodissociative laser.

295. Hohla, K., Kompa, K. L., *Appl. Phys. Lett.*, 22, 77-78 (1973), Gigawatt photochemical laser.

296. Aldridge, F. T., *Appl. Phys. Lett.*, 22, 180-182 (1973), High-pressure iodine photodissociation laser.

297. Alekseev, V. A., Andreeva, T. L., Volkov, V. N., Yukov, E. A., *Sov. Phys. JETP*, 36, 238-242 (1973), Kinetics of the generation spectrum of a photodissociation laser.

298. Yukov, E. A., *Sov. J. Quantum Electron.*, 3, 117-120 (1973), Elementary processes in the active medium of an iodine photodissociation laser.

299. Gusinow, M. A., Rice, J. K., Padrick, T. D., *Chem. Phys. Lett.*, 21, 197-199 (1973), The apparent late-time gain in a photodissociation iodine laser.

300. Hohla, K., *Laser Interaction and Related Plasma Phenomena*, Vol. 3A(1974), Plenum Press, New York, The iodine laser, a high power gas laser.

301. Birich, G. N., Drozd, G. I., Sorokin, V. N., Struk, I. I., *JETP Lett.*, 19, 27-29 (1974), Photodissociation iodine laser using compounds containing group-V atoms.

302. Belousova, I. M., Gorshkov, N. G., Danilov, O. B., Zalesskii, V. Yu., Yachnev, I. L., *Sov. Phys. JETP*, 38, 254-257 (1974), Accumulation of iodine molecules in flash photolysis of CF_3I and n-C_3F_7I vapor.

303. Belousova, I. M., Bobrov, B. D., Kiselev, V. M., Kurzenkov, V. N., Krepostnov, P. I., *Sov. Phys. JETP*, 38, 258-263 (1974), Photodissociative I^{127} laser in a magnetic field.

304. Basov, N. G., Golubev, L. E., Zuev, V. S., Katulin, V. A., Netemin, V. N., *Sov. J. Quantum Electron.*, 3, 524 (1974), Iodine laser emitting short pulses of 50 J energy and 5 nsec duration.

305. Golubev, L. E., Zuev, V. S., Katulin, V. A., Nosach, V. Yu., Nosach, O., *Sov. J. Quantum Electron.*, 3, 464-467 (1974), Investigation of optical inhomogeneities which appear in an active medium of a photodissociation laser during coherent emission.

306. Kuznetsova, S. V., Maslov, A. I., *Sov. J. Quantum Electron.*, 3, 468-471 (1974), Investigation of the reactions of atomic iodine in a photodissociation laser using n-C_3F_7I and i-C_3F_7I molecules.

307. Palmer, R. E., Gusinow, M. A., *J. Appl. Phys.*, 45, 2174-2178 (1974), Late-time gain of the CF_3I iodine photodissociation laser.

308. Belousova, I. M., Bobrov, B. D., Kiselev, V. M., Kurzenkov, V. N., Krepostnov, P. I., *Opt. Spectrosc.*, 37, 20-24 (1974), I^{127} atom in a magnetic field.

309. Palmer, R. E., Gusinow, M. A., *IEEE J. Quantum Electron.*, QE-10, 615-616 (1974), Gain versus time in the CF_3I iodine photodissociation laser.

310. Silfvast, W. T., Szeto, L. H., Wood, O. R., II, *Appl. Phys. Lett.*, 25, 593-595 (1974), C_3F_7I photodissociation laser initiated by a CO_2-laser-produced plasma.

311. Belousova, I. M., Bobrov, B. D., Kiselev, V. M., Kurzenkov, V. N., *Sov. J. Quantum Electron.*, 4, 767-769 (1974), Characteristics of the stimulated emission from iodine atoms in pulsed magnetic fields.

312. Hohla, K., Fuss, W., Volk, R., Witte, K. J., *Opt. Commun.*, 13, 114-116 (1975), Iodine laser oscillator in gain switch mode for ns pulses.

313. Hohla, K., Brederlow, G., Fuss, W., Kompa, K. L., Raeder, J., Volk, R., *J. Appl. Phys.*, 46, 808-809 (1975), 60J 1-nsec iodine laser.

314. Zalesskii, V. Yu., *Sov. J. Quantum Electron.*, 4, 1009-1014 (1975), Analytic estimate of the maximum duration of stimulated emission from a CF_3I photodissociation laser.

315. Butcher, R. J., Donovan, R. J., Fotakis, C., Fernie, D., Rae, A. G. A., *Chem. Phys. Lett.*, 30, 398-402 (1975), Photodissociation laser isotope effects.

316. Baker, H. J., King, T. A., *J. Phys. D*, 8, L31-L33 (1975), Mode-beating in gain-switch iodine photodissociation laser.

317. Baker, H. J., King, T. A., *J. Phys. D*, 8, 609-619 (1975), Iodine photodissociation laser oscillator characteristics.

318. Aldridge, F. T., *IEEE J. Quantum Electron.*, QE-11, 215-217 (1975), Stimulated emission cross section and inversion lifetime in a three-atmosphere iodine photodissociation laser.

319. Antonov, A. V., Basov, N. G., Zuev, V. S., Katulin, V. A., Korol'kov, K., *Sov. J. Quantum Electron.*, 5, 123 (1975), Amplifier with a stored energy over 700 J designed for a short-pulse iodine laser.

320. Ishii, S., Ahlborn, B., Curzon, F. L., *Appl. Phys. Lett.*, 27, 118-119 (1975), Gain switching and Q spoiling of iodine laser with a shock wave.

321. Borovich, B. L., Zuev, V. S., Katulin, V. A., Nosach, V. Yu., Nosach, O., *Sov. J. Quantum Electron.*, 5, 695-702 (1975), Characteristics of iodine laser short-pulse amplifier.

322. Zalesskii, V. Yu., Polikarpov, S. S., *Sov. J. Quantum Electron.*, 5, 826-831 (1975), Investigation of the conditions governing the stimulated emission threshold of a CF_3I (iodine) laser.

323. Pirkle, R. J., Davis, C. C., McFarlane, R. A., *J. Appl. Phys.*, 46, 4083-4085 (1975), Self-mode-locking of an iodine photodissociation laser.

324. Pleasance, L. D., Weaver, L. A., *Appl. Phys. Lett.*, 27, 407-409 (1975), Laser emission at 1.32 μm from atomic iodine produced by electrical dissociation of CF_3I.

325. Beverly, R. E., II, *Opt. Commun.*, 15, 204-208 (1975), Pressure-broadened iodine-laser-amplifier kinetics and a comparison of diluent effectiveness.

326. Gusinow, M. A., *Opt. Commun.*, 15, 190-192 (1975), The enhancement of the near UV flashlamp spectra with special emphasis on the iodine photodissociation laser.

327. Skribanowitz, N., Kopainsky, B., *Appl. Phys. Lett.*, 27, 490-492 (1975), Pulse shortening and pulse deformation in a high-power iodine laser amplifier.

328. Pirkle, R. J., Jr., Davis, C. C., McFarlane, R. A., *Chem. Phys. Lett.*, 36, 805-807 (1975), Comparative performance of CF_3I, CD_3I and CH_3I in an atomic iodine photodissociation laser.

329. Basov, N. G., Zuev, V. S., *Nuovo Cimento*, 31, 129-151 (1976), Short-pulsed iodine laser.

330.   Brederlow, G., Witte, K. J., Fill, E., Hohla, K., Volk, R., *IEEE J. Quantum Electron.*, QE-12, 152-155 (1976), The Asterix III pulsed high-power iodine laser.

331.   Jones, E. D., Palmer, M. A., Franklin, F. R., *Opt. Quantum Electron.*, 8, 231-235 (1976), Subnanosecond high-pressure iodine photodissociation laser oscillator.

332.   Katulin, V. A., Nosach, V. Yu., Petrov, A. L., *Sov. J. Quantum Electron.*, 6, 205-208 (1976), Iodine laser with active Q switching.

333.   Ishii, S., Ahlborn, B., *J. Appl. Phys.*, 47, 1076-1078 (1976), Elimination of compression waves induced by pump light in iodine lasers.

334.   Swingle, J. C., Turner, C. E., Jr., Murray, J. R., George, E. V., Krupke, W. F., *Appl. Phys. Lett.*, 28, 387-388 (1976), Photolytic pumping of the iodine laser by XeBr*.

335.   Rabinowitz, P., Jacobs, S., Gould, G., *Appl. Opt.*, 1, 511-516 (1962), Continuously optically pumped Cs laser.

336.   Walter, W. T., Solimene, N., Piltch, M., Gould, G., *IEEE J. Quantum Electron.*, QE-2, 474-479 (1966), Efficient pulsed gas discharge lasers.

337.   Walter, W. T., Solimene, N., Piltch, M., Gould, G., *Bull. Am. Phys. Soc.*, 11, 113 (1966), Pulsed-laser action in atomic copper vapor.

338.   Walter, W. T., *Bull. Am. Phys. Soc.*, 12, 90 (1967), 40-kW pulsed copper laser.

339.   Bockasten, K., Lundholm, T., Andrade, O., *J. Opt. Soc. Am.*, 56, 1260-1261 (1966), Laser lines in atomic and molecular hydrogen.

340.   Leonard, D. A., *IEEE J. Quantum Electron.*, QE-3, 380-381 (1967), A theoretical description of the 5106-A pulsed copper vapor laser.

341.   Walter, W. T., *IEEE J. Quantum Electron.*, QE-4, 355-356 (1968), Metal vapor lasers.

342.   Asmus, J. E., Moncur, N., *Appl. Phys. Lett.*, 13, 384-385 (1968), Pulse broadening in a MHD copper vapor laser.

343.   Isaev, A. A., Kazaryan, M. A., Petrash, G. G., *JETP Lett.*, 16, 27-29 (1972), Effective pulsed copper-vapor laser with high average generation power.

344.   Russell, G. R., Nerheim, M. M., Pivirotto, T. J., *Appl. Phys. Lett.*, 21, 656-657 (1972), Supersonic electrical-discharge copper vapor laser.

345.   Liu, C. S., Sucov, E. W., Weaver, L. A., *Appl. Phys. Lett.*, 23, 92 (1973), Copper superradiant emission from pulsed discharges in copper iodide vapor.

346.   Ferrar, C. M., *IEEE J. Quantum Electron.*, QE-9, 856-857 (1973), Copper-vapor laser with closed-cycle transverse vapor flow.

347.   Chen, C. J., Nerheim, N. M., Russell, G. R., *Appl. Phys. Lett.*, 23, 514-515 (1973), Double-discharge copper vapor laser with copper chloride as a lasant.

348.   Isaev, A. A., Kazaryan, M. A., Petrash, G. G., *Opt. Spectrosc.*, 35, 307-308 (1973), Copper vapor pulsed laser with a repetition frequency of 10 kHz.

349.   Weaver, L. A., Liu, C. S., Sucov, E. W., *IEEE J. Quantum Electron.*, QE-10, 140-147 (1974), Superradiant emission at 5106, 5700 and 5782 Å in pulsed copper iodide discharges.

350.   Ehrlich, D. J., Osgood, R. M., *Appl. Phys. Lett.*, 34, 655-658 (1979), Alkali-metal resonance-line lasers based on photodissociation.

351.   Fahlen, T. S., *J. Appl. Phys.*, 45, 4132-4133 (1974), Hollow-cathode copper-vapor laser.

352.   Gal'pern, M. G., Gorbachev, V. A., Katulin, V. A. et al., *Sov. J. Quantum Electron.*, 5, 1384-1385 (1975), Bleachable filter for the iodine laser emitting at $\lambda = 1.35$ μm.

353.   Brederlow, G., Fill, E., Fuss, W., Hohla, K., Volk, R., Witte, K. J., *Sov. J. Quantum Electron.*, 6, 491-495 (1976), High-power iodine laser development at the Institut fur Plasmaplysik, Garching.

354.   Davis, C. C., Pirkle, R. J., McFarlane, R. A., Wolga, G. J., *IEEE J. Quantum Electron.*, QE-12, 334-352 (1976), Output mode spectra, comparative parametric operation, quenching, photolytic reversibility, and short-pulse generation in atomic iodine photodissociation laser.

355. Arkhipova, E. V., Borovich, B. L., Zapol'skii, A. K., *Sov. J. Quantum Electron.*, 6, 686-696 (1976), Accumulation of excited iodine atoms in iodine photodissociation laser. Analysis of kinetic equations.

356. Liberman, I., Babcock, R. V., Liu, C. S., George, T. V., Weaver, L. A., *Appl. Phys. Lett.*, 25, 334-335 (1974), High-repetition-rate copper iodide laser.

357. Nosach, O. Yu., Orlov, E. P., *Sov. J. Quantum Electron.*, 6, 770-777 (1976), Some features of formation of the angular spectrum of the stimulated radiation emitted from iodine laser.

358. Andreeva, T. L., Birich, G. N., Sorokin, V. N., Struk, I. I., *Sov. J. Quantum Electron.*, 6, 781-789 (1976), Investigations of photodissociation iodine lasers utilizing molecules with bonds between iodine atoms and group V elements. I. Experimental investigation.

359. Zalesskii, V. Yu., Kokushkin, A. M., *Sov. J. Quantum Electron.*, 6, 813-817 (1976), Tracing chemical changes in the active media of iodine photodissociation laser.

360. Katulin, V. A., Nosach, V. Yu., Petrov, A. L., *Sov. J. Quantum Electron.*, 6, 998-999 (1976), Nanosecond iodine laser with an output energy of 200J.

361. Fuss, W., Hohla, K., *Z. Naturforsch.* Teil A., 31, 569-577 (1976), Pressure broadening of the 1.3 μm iodine laser line.

362. Fuss, W., Hohla, K., *Opt. Commun.*, 18, 427-430 (1976), A closed cycle iodine laser.

363. Fill, E., Hohla, K., *Opt. Commun.*, 18, 431-436 (1976), A saturable absorber for the iodine laser.

364. Kamrukov, A. S., Kashnikov, G. N., Kozlov, N. P., Malashchenko, V. A. et al., *Sov. J. Quantum Electron.*, 6, 1101-1104 (1976), Investigation of an iodine laser excited optically by high-current plasmadynamic discharges.

365. Mukhtar, E. S., Baker, H. J., King, T. A., *Opt. Commun.*, 19, 193-196 (1976), Selection of oscillation frequency in the 1.315 μm iodine laser.

366. Alekhin, B. V., Lazhintsev, B. V., Norarevyan, V. A., Petrov, N. N. et al., *Sov. J. Quantum Electron.*, 6, 1290-1292 (1976), Short-pulse photodissociation laser with magnetic-field modulation of gain.

367. Bokhan, P. A., Solomonov, V. I., *Sov. J. Quantum Electron.*, 3, 481-483 (1974), Mechanism of laser action in copper vapor.

368. Baker, H. J., King, T. A., *J. Phys. D*, 9, 2433-2445 (1976), Line broadening and saturation parameters for short-pulse high-pressure iodine photodissociation laser systems.

369. Padrick, T. D., Palmer, R. E., *J. Appl. Phys.*, 47, 5109-5110 (1976), Use of titanium doped quartz to eliminate carbon deposits in an atomic iodine photodissociation laser.

370. Fill, E., Hohla, K., Schappert, G. T., Volk, R., *Appl. Phys. Lett.*, 29, 805-807 (1976), 100-ps pulse generation and amplification in the iodine laser.

371. Olsen, J. N., *J. Appl. Phys.*, 47, 5360-5364 (1976), Pulse shaping in the iodine laser.

372. Belousova, I. M., Bobrov, B. D., Grenishin, A. S., Kiselev, V. M., *Sov. J. Quantum Electron.*, 7, 249-255 (1977), Magnetic-field control of the duration of pulses emitted from an iodine, photodissociation laser.

373. Andreeva, T. L., Birich, G. N., Sobel'man, I. I., Sorokin, V. N., et al., *Sov. J. Quantum Electron.*, 7, 1230-1234 (1977), Continuously pumped continuous-flow iodine laser.

374. Mukhtar, E. S., Baker, H. J., King, T. A., *Opt. Commun.*, 24, 167-169 (1978), Pressure-induced frequency shifts in the atomic iodine laser.

375. Kiselev, V. M., Bobrov, B. D., Grenishin, A. S., Kotlikova, T. N., *Sov. J. Quantum Electron.*, 8, 181-184 (1978), Faraday rotation in the active medium of an iodine photodissociation laser oscillator or amplifier.

376. Babkin, V. I., Kuznetsova, S. V., Maslov, A. I., *Sov. J. Quantum Electron.*, 8, 285-289 (1978), Simple method for determination of stimulated emission cross section of $^2P_{1/2}(F = 3) \rightarrow {}^2P_{3/2}$ (F' = 4) transition in atomic iodine.

377. Akitt, D. P., Wittig, C. F., *J. Appl. Phys.* 40, 902 (1969), Laser emission in ammonia.

378. Ferrar, C. M., *IEEE J. Quantum Electron.*, QE-10, 655-656 (1974), Buffer gas effects in a rapidly pulsed copper vapor laser.

379. Saito, H., Uchiyama, T., Fujioka, T., *IEEE J. Quantum Electron.*, QE-14, 302-309 (1978), Pulse propagation in the amplifier of a high-power iodine laser.

380. Gaidash, V. A., Mochalov, M. R., Shemyakin, V. I., Shurygin, V. K., *Sov. J. Quantum Electron.*, 8, 530-531 (1978), Modulation of the transmission of an atomic iodine switch.

381. McDermott, W. E., Pchelkin, N. R., Benard, D. G., Bousek, R. R., *Appl. Phys. Lett.*, 32, 469-470 (1978), An electronic transition chemical laser.

382. Antonov, A. S., Belousova, I. M., Gerasimov, V. A. et al., *Sov. Tech. Phys. Lett.*, 4, 459 (1978), Flashlamp-excited photodissociation 1000 J laser with 1.4 percent efficiency.

383. Silfvast, W. T., Szeto, L. H., Wood, O. R., II, *Appl. Phys. Lett.*, 36, 615 (1980), Simple metal vapor recombination lasers using segmented plasma excitation.

384. Sutton, D. G., Galvan, L., Suchard, S. N., *IEEE J. Quantum Electron.*, QE-11, 312 (1975), Two-electron laser transition in Sn(I)?.

385. Katulin, V. A., Nosach, V. Yu., Petrov, A. L., *Sov. J. Quantum Electron.*, 8, 380-382 (1978), Q-Switched iodine laser emitting at two frequencies.

386. Burnham, R., unpublished (May, 1978), Optically pumped bismuth lasers at 472 and 475 nm.

387. Murray, J. R., Powell, H. T., Rhodes, C. K., unpublished (June 1974), Inversion of the auroral green transition of atomic oxygen by argon excimer transfer to nitrous oxide.

388. Powell, H. T., Murray, J. R., Rhodes, C. K., unpublished (May, 1975), Collision-induced auroral line lasers.

389. Bennett, W. R., Jr., *Appl. Optics Supplement on Optical Masers* (1962), 24-61, Gaseous optical masers.

390. Powell, H. T., Ewing, J. J., unpublished (May, 1978), Forbidden transition selenium atom photodissociation lasers.

391. Pollack, M. A., unpublished (private communication to C. S. Willett).

392. Vinokurov, G. N., Zalesskii, V. Yu., *Sov. J. Quantum Electron.*, 8, 1191-1197 (1978), Chemical kinetics and gas dynamics of a Q-switch iodine laser with an optically thick active medium.

393. Benard, D. J., McDermott, W. E., Pchelkin, N. R., Bousek, R. R., *Appl. Phys. Lett.*, 34, 40-41 (1979), Efficient operation of a 100-W transverse-flow oxygen-iodine chemical laser.

394. Richardson, R. J., Wiswall, C. E., *Appl. Phys. Lett.*, 35, 138-139 (1979), Chemically pumped iodine laser.

395. Witte, K. J., Burkhard, P., Luthi, H. R., *Opt. Commun.*, 28, 202-206 (1979), Low pressure mercury lamp pumped atomic iodine laser of high efficiency.

396. Riley, M. E., Padrick, T. D., Palmer, R. E., *IEEE J. Quantum Electron.*, QE-15, 178-189 (1979), Multilevel paraxial Maxwell-Bloch equation description of short pulse amplification in the atomic iodine laser.

397. Zuev, V. S., Netemin, V. N., Nosach, O. Yu., *Sov. J. Quantum Electron.*, 9, 522-524 (1979), Wavefront instability of iodine laser radiation and dynamics of optical inhomogeneity evolution in the active medium.

398. Anderson, R. S., Springer, L., Bricks, B. G., Karras, T. W., *IEEE J. Quantum Electron.*, QE-11, 172-174 (1975), A discharge heated copper vapor laser.

399. Palmer, R. E., Padrick, T. D., Palmer, M. A., *Opt. Quantum Electron.*, 11, 61-70 (1979), Diffraction-limited atomic iodine photodissociation laser.

400. Djeu, N., Powell, F. X., *IEEE J. Quantum Electron.*, QE-7, 537-538 (1971), More infrared laser transitions in atomic iodine.

401. Kim, H. H., Marantz, H., *Appl. Opt.*, 9, 359-368 (1970), A study of the neutral atomic iodine laser.

402. Kim, H., Paananen, R., Hanst, P., *IEEE J. Quantum Electron.*, QE-4, 385-386 (1968), Iodine infrared laser.

403. Brandelik, J. E., Smith, G. A., *IEEE J. Quantum Electron.*, QE-16, 7-10 (1980), Br, C, Cl, S and Si laser action using a pulsed microwave discharge.

404. Klimkin, V. M., *Sov. J. Quantum Electron.*, 5, 326-329 (1975), Investigation of an ytterbium vapor laser.

405. Prelas, M. A., Akerman, M. A., Boody, F. P., Miley, G. H., *Appl. Phys. Lett.*, 31, 428-430 (1977), A direct nuclear pumped 1.45 μ atomic carbon laser in mixtures of He-CO and He-CO$_2$.

406. Piltch, M., Walter, W. T., Solimene, N., Gould, G., Bennett, W. R., Jr., *Appl. Phys. Lett.*, 7, 309-310 (1965), Pulsed laser transitions in manganese vapor.

407. Silfvast, W. T., Fowles, G. R., *J. Opt. Soc. Am.*, 56, 832-833 (1966), Laser action on several hyperfine transitions in MnI.

408. Chen, C. J., Russell, G. R., *Appl. Phys. Lett.*, 26, 504-505 (1975), High-efficiency multiply pulsed copper vapor laser utilizing copper chloride as a lasant.

409. Chen, C. J., *J. Appl. Phys.*, 44, 4246-4247 (1973), Manganese laser.

410. Chen, C. J., *Appl. Phys. Lett.*, 24, 499-500 (1974), Manganese laser using manganese chloride as lasant.

411. Isakov, V. K., Kapugin, M. M., Potapov, S. E., *Sov. Tech. Phys. Lett.*, 2, 292-293 (1976), MnCl$_2$-vapor laser (energy characteristics).

412. Isakov, V. K., Kalugin, M. M., Potapov, S. E., *Sov. Tech. Phys. Lett.*, 4, 333-334 (1978), Output spectrum of a manganese-chloride laser.

413. Bokhan, P. A., Burlakov, V. D., Gerasimov, V. A., Solomonov, V. I., *Sov. J. Quantum Electron.*, 6, 672-675 (1976), Stimulated emission mechanism and energy characteristics of manganese vapor laser.

414. Isaev, A. A., Kazaryan, M. A., Petrash, G. G., Cherezov, V. M., *Sov. J. Quantum Electron.*, 6, 978-980 (1976), An investigation of pulse manganese vapor laser.

415. Trainor, D. W., Mani, S. A., *J. Chem. Phys.*, 68, 5481-5485 (1978), Pumping iron: a KrF laser pumped atomic iron laser.

416. Linevsky, M. J., Karras, T. W., *Appl. Phys. Lett.*, 33, 720-721 (1978), An iron-vapor laser.

417. Solanki, R., Collins, G. J., Fairbank, W. M., Jr., *IEEE J. Quantum Electron.*, QE-15, 525 (1979), IR laser transitions in a nickel hollow cathode discharge.

418. Bokhan, P. A., Klimkin, V. M., Prokop'ev, V. E., Solomonov, V. I., *Sov. J. Quantum Electron.*, 7, 81-82 (1977), Investigation of a laser utilizing self-terminating transitions in europium atoms and ions.

419. Bokhan, P. A., Nikolaev, V. N., Solomonov, V. I., *Sov. J. Quantum Electron.*, 5, 96-98 (1975), Sealed copper vapor laser.

420. Cahuzac, P., Sontag, H., Toschek, P. E., *Opt. Commun.*, 31, 37-41 (1979), Visible superfluorescence from atomic europium.

421. Breichignac, C., Cahuzac, P., (unpublished).

422. Pixton, R. M., Fowles, G. R., *Phys. Lett.*, 29A, 654-655 (1969), Visible laser oscillations in helium at 7065 Å.

423. Schuebel, W. K., *Appl. Phys. Lett.*, 30, 516-519 (1977), Laser action in AlII and HeI in a slot cathode discharge.

424. Abrams, R. L., Wolga, G. J., *IEEE J. Quantum Electron.*, QE-5, 368 (1967), Near infrared laser transitions in pure helium.

425. Abrams, R. L., Wolga, G. J., *Phys. Rev. Lett.*, 19, 1411-1414 (1967), Direct demonstration of the validity of the Wigner spin rule for helium-helium collisions.

426. Cagnard, R., der Agobian, R., Otto, J. L., Echard, R., *C. R. Acad. Sci. Ser.* B, 257, 1044-1047 (1963), L'emission stimulee de quelque transitions infrarouges de l'helium et du neon.

427. der Agobian, R., Otto, J. L., Cagnard, R., Echard, R., *J. Phys.*, 25, 887-897 (1964), Emission stimulee de nouvelles transitions infrarouges dans les gaz rares.

428. Wood, O. R., Burkhardt, E. G., Pollack, M. A., Bridges, T. J., *Appl. Phys. Lett.*, 18, 261-264 (1971), High pressure laser action in 13 gases with transverse excitation.

429. Patel, C. K. N., Bennett, W. R., Jr., Faust, W. L., McFarlane, R. A., *Phys. Rev. Lett.*, 9, 102-104 (1962), Infrared spectroscopy using stimulated emission techniques.

430. Smilanski, I., Levin, L. A., Erez, G., *IEEE J. Quantum Electron.*, QE-11, 919-920 (1975), A copper laser using CuI vapor.

431. Bennett, W. R., Jr., Kindlmann, P. J., *Bull. Am. Phys. Soc.*, 8, 87 (1963), Collision cross sections and optical maser considerations for helium.

432. Brochard, J., Liberman, S., *C. R. Acad. Sci. Ser.* B, 260, 6827-6829 (1965), Emission stimulee de nouvelles transitions infrarouge de l'helium et du neon.

433. Brochard, J., Lespritt, J. F., Liberman, L., *C. R. Acad. Sci. Ser.* B, 600-602 (1970), Measurement of the isotopic separation of two infrared laser lines of HeI.

434. Mathias, L. E. S., Crocker, A., Wills, M. S., *IEEE J. Quantum Electron.*, QE-3, 170 (1967), Pulsed laser emission from helium at 95 mm.

435. Levine, J. S., Javan, A., *Appl. Phys. Lett.*, 14, No. 11, 348-350 (1969), Far infra-red continuous wave oscillation in pure helium.

436. Turner, R., Murphy, R. A., *Infrared Phys.*, 16, 197-200 (1976), The far infrared helium laser.

437. Giuliano, C. R., Hess, L. D., *J. Appl. Phys.*, 40, 2428-2430 (1969), Reversible photodissociative laser system.

438. Campbell, J. D., Kasper, J. V. V., *Chem. Phys. Lett.*, 10, 436-437 (1971), Hyperfine structure in the CF_3Br photodissociation laser.

439. Spencer, D. J., Wittig, C., *Opt. Lett.*, 4, 1-3 (1979), Atomic bromine electronic-transition chemical laser.

440. Hocker, L. O., *J. Opt. Soc. Am.*, 68, 262-265 (1978), High resolution study of the helium-fluorine laser.

441. Shukhtin, M., Fedotov, G. A., Mishakov, V. G., *Opt. Spectrosc.*, 39, 681 (1976), Lasing with CuI lines using copper bromide vapor.

442. Bell, W. E., Bloom, A. L., Goldsborough, J. P., *IEEE J. Quantum Electron.*, QE-2, 154 (1966), New laser transitions in antimony and tellurium.

443. Webb, C. E., *IEEE J. Quantum Electron.*, QE-4, 426-427 (1968), New pulsed laser transitions in TeII.

444. Clunie, D. M., Thorn, R. S. A., Trezise, K. E., *Phys. Lett.*, 14, 28-29 (1965), Asymmetric visible super-radiant emission from a pulsed neon discharge.

445. Leonard, D. A., Neal, R. A., Gerry, E. T., *Appl. Phys. Lett.*, 7, 175 (1965), Observation of a super-radiant self-terminating green laser transition in neon.

446. Leonard, D. A., *IEEE J. Quantum Electron.*, QE-3, 133-135 (1967), The 5401 Å pulsed neon laser.

447. Isaev, A. A., Kazaryan, M. A., Petrash, G. G., *Sov. J. Quantum Electron.*, 2, 49-51 (1972), Shape and duration of superradiance pulses corresponding to neon lines.

448. Magda, I. I., Tkach, Yu. V., Lemberg, E. A., Skachek, G. V., Gadetskii, N., *Sov. J. Quantum Electron.*, 3, 260-261 (1973), High power nitrogen and neon pulsed gas lasers.

449. Perry, D. L., *IEEE J. Quantum Electron.*, QE-7, 102 (1971), CW laser oscillation at 5433 Å in neon.

450. Bridges, W. B., Chester, A. N., *Appl. Opt.*, 4, 573-580 (1965), Visible and uv laser oscillation at 118 wavelengths in ionized neon, argon, krypton, xenon, oxygen and other gases.

451. White, A. D., Rigden, J. D., *Appl. Phys. Lett.*, 2, 211-212 (1963), The effect of super-radiance at 3.39 μ on the visible transitions in the He-Ne maser.

452. Sovero, E., Chen, C. J., Culick, F. E. C., *J. Appl. Phys.*, 47, 4538-4542 (1976), Electron temperature measurements in a copper chloride laser utilizing a microwave radiometer.

453. Viscovini, R. C., Scalabrin, A., Pereira, D., *IEEE J. Quantum Electron.*, 33, 916 (1997), $^{13}CD_3OD$ optically pumped by a waveguide CO_2 laser: new FIR laser lines.

454. Ericsson, K. G., Lidholt, L. R., *IEEE J. Quantum Electron.*, QE-3, 94 (1967), Super-radiant transitions in argon, krypton and xenon.

455. Rosenberger, D., *Phys. Lett.*, 14, 32 (1965), Superstrahlung in gepulsten argon-, krypton- und xenon-entladungen.

456. Bloom, A. L., *Appl. Phys. Lett.*, 2, 101-102 (1963), Observation of new visible gas laser transition by removal of dominance.

457. Ericsson, G., Lidholt, R., *Ark. Fys.*, 37, 557-568 (1967), Generation of short light pulses by superradiance in gases.

458. Rosenberger, D., *Phys. Lett.*, 13, 228-229 (1964), Laser-ubergange and superstrahlung bei 6143 A in einer gepulsten Neon-entladungen.

459. Abrosimov, G. V., *Opt. Spectrosc.*, 31, 54-56 (1971), Spatial and temporal coherence of the radiation of pulsed neon and thallium gas lasers.

460. Isaev, A. A., Kazaryan, M. A., Petrash, G. G., *Sov. J. Quantum Electron.*, 2, 49-52 (1972), Shape and duration of superradiance pulses corresponding to neon lines.

461. Odintsov, A. I., Turkin, N. G., Yakunin, V. P., *Opt. Spectrosc.*, 38, 244-245 (1975), Spatial coherence and angular divergence of pulsed superradiance of neon.

462. Isaev, A. A., Petrash, G. G., *Sov. Phys. JETP*, 29, 607-614 (1969), Mechanism of pulsed superradiance from 2p–1s transitions in neon.

463. White, J. C., *Appl. Phys. Lett.*, 33, 325-327 (1978), Inversion of the Na resonance line by selective photodissociation of NaI.

464. Shukhtin, A. M., Fedotov, G. A., Mishakov, V. G., *Opt. Spectrosc.*, 40, 237-238 (1976), Stimulated emission on copper lines during pulsed production of vapor without the use of a heating element.

465. White, A. D., Rigden, J. D., *Proc. IEEE*, 50, 1796 (1962), Continuous gas maser operation in the visible.

466. Rigden, J. D., White, A. D., *Proc. IEEE*, 51, 943 (1963), The interaction of visible and infrared maser transitions in the helium-neon system.

467. White, A. D., Gordon, E. I., *Appl. Phys. Lett.*, 3, 197-198 (1963), Excitation mechanism and current dependence of population inversion in He-Ne lasers.

468. Gordon, E. I., White, A. D., *Appl. Phys. Lett.*, 3, 199-201 (1963), Similarity laws for the effect of pressure and discharge diameter on gain of He-Ne lasers.

469. Labuda, E. F., Gordon, E. I., *J. Appl. Phys.*, 35, 1647-1648 (1964), Microwave determination of average electron energy and density in He-Ne discharges.

470. Young, R. T., Willett, C. S., Maupin, R. T., *J. Appl. Phys.*, 41, 2936-2941 (1970), The effect of helium on population inversion in the He-Ne laser.

471. Young, R. T., Jr., Willett, C. S., Maupin, R. T., *Bull. Am. Phys. Soc.*, 13, 206 (1968), An experimental determination of the relative contributions of resonance and electron impact collision to the excitation of Ne atom in He-Ne laser.

472. Jones, C. R., Robertson, W. W., *Bull. Am. Phys. Soc.*, 13, 198 (1968), Temperature dependence of reaction of the helium metastable atom.

473. Korolev, F. A., Odintsov, A. I., Mitsai, V. N., *Opt. Spectrosc.*, 19, 36-39 (1965), A study of certain characteristics of a He-Ne laser.

474. Suzuki, N., *Jpn. J. Appl. Phys.*, 4, 285-291 (1965), Vacuum uv measurements of helium-neon laser discharge.

475. Abrosimov, G. V., Vasil'tsov, V. V., Voloshin, V. N., Korneev, A. V. et al., *Sov. Tech. Phys. Lett.*, 2, 162-163 (1976), Pulsed laser action on self-limiting transitions of copper atoms in copper halide vapor.

476. Field, R. L., Jr., *Rev. Sci. Instrum.*, 38, 1720-1722 (1967), Operating characteristics of dc-excited He-Ne gas lasers.

477. Suzuki, N., *Jpn. J. Appl. Phys.*, 4, 642-647 (1965), Spectroscopy of He-Ne laser discharges.

478. Gonchukov, G. A., Ermakov, G A., Mikhenko, G. A., Protsenko, E. D., *Opt. Spectrosc.*, 20, 601-602 (1966), Temperature effects in the He-Ne laser.

479. Alekseeva, A. N., Gordeev, D. V., *Opt. Spectrosc.*, 23, 520-524 (1967), The effect of longitudinal magnetic field on the output of a helium-neon laser.

480. Smith, P. W., *IEEE J. Quantum Electron.*, QE-2, 62-68 (1966), The output power of a 6328-Å He-Ne gas laser.

481. Smith, P. W., *IEEE J. Quantum Electron.*, QE-2, 77-79 (1966), On the optimum geometry of a 6328 Å laser oscillator.

482. Herziger, G., Holzapfel, W., Seelig, W., *Z. Phys.*, 189, 385-400 (1966), Verstarkung einer He-Ne gasentladung fur die laser wellenlange, l=6328 Å.

483. Bell, W. E., Bloom, A. L., *Appl. Opt.*, 3, 413-415 (1964), Zeeman effect at 3.39 microns in a helium-neon laser.

484. Belusova, I. M., Danilov, O. B., Elkina, I. A., Kiselev, K. M., *Opt. Spectrosc.*, 16, 44-47 (1969), Investigation of the causes of gas temperature effects on the generation of power of a He-Ne laser at 6328 Å.

485. Landsberg, B. M., *IEEE Quantum Electron.*, QE-16, 684 (1980), Optically pumped CW submillimeter emission lines from methyl mercaptan CH_3SH.

486. Akirtava, O. S., Dzhikiya, V. L., Oleinik, Yu. M., *Sov. J. Quantum Electron.*, 5, 1001-1002 (1976), Laser utilizing CuI transitions in copper halide vapors.

487. Vasconcellos, E. C. C., Wyss, J. C., Petersen, F. R., Evenson, K. M., *Int. J. Infrared Millimeter Waves*, 4, 401 (1983), Frequency measurements of far infrared cw lasing lines in optically pumped $CHCl_2F$.

488. White, A. D., *Proc. IEEE*, 52, 721 (1964), Anomalous behaviour of the 6402.84 Å gas laser.

489. Bloom, A. L., Hardwick, D. L., *Phys. Lett.*, 20, 373-375 (1966), Operation of He-Ne lasers in the forbidden resonator region.

490. Zitter, R. N., *Bull. Am. Phys. Soc.*, 9, 500 (1964), 2s–2p and 3s–3p neon transitions in a very long laser.

491. Patel, C. K. N., *Lasers*, Vol. 1 (1968), 39-50, Marcel Dekker, New York, Gas lasers.

492. Henningsen, J. O., *Int. J. Infrared Millimeter Waves*, 7, 1605 (1986), Methanol laser lines from torsionally excited CO stretch states and from OH-bend, CH_3-rock, and CH_3-deformation states.

493. Uchida, T., *Appl. Opt.*, 4, 129-131 (1965), Frequency spectra of the He-Ne optical masers with external concave mirrors.

494. Massey, G. A., Oshman, M. K., Targ, R., *Appl. Phys. Lett.*, 6, 10-11 (1965), Generation of single-frequency light using the FM laser.

495. Steier, W. H., *Proc. IEEE*, 54, 1604-1606 (1966), Coupling of high peak power pulses from He-Ne lasers.

496. Kuznetsov, A. A., Mash, D. I., Skuratova, N. V., *Radio Eng. Electron. Phys.* (USSR), 12, 140-143 (1967), Effect of an axial magnetic field on the output power of a neon-helium laser.

497. Subotinov, N. V., Kalchev, S. D., Telbizov, P. K., *Sov. J. Quantum Electron.*, 5, 1003-1004 (1976), Copper vapor laser operating at a high pulse repetition frequency.

498. Lee, P. H., Skolnick, M. L., *Appl. Phys. Lett.*, 10, 303-305 (1967), Saturated neon absorption inside a 6328 Å laser.

499. Hochuli, U., Haldemann, P., Hardwick, D., *IEEE J. Quantum Electron.*, QE-3, 612-614 (1967), Cold cathodes for He-Ne gas lasers.

500. Carlson, F. P., *IEEE J. Quantum Electron.*, QE-4, 98-99 (1968), On the optimal use of dc and RF excitation in He-Ne lasers.

501. Suzuki, T., *Jpn. J. Appl. Phys.*, 9, 309-310 (1970), Discharge current noise in He-Ne laser and its suppression.

502. Ehlers, K. W., Brown, I. G., *Rev. Sci. Instrum.*, 41, 1505-1506 (1970), Regeneration of helium neon lasers.

503. Leontov, V. G., Ostapchenko, E. P., Sedov, G. S., *Opt. Spectrosc.*, 31, 418-419 (1971), Optimum lasing conditions of a He-Ne laser operating in the TEM_{00} axial mode.

504. Sakurai, T., *Jpn. J. Appl. Phys.*, 11, 1832-1836 (1972), Discharge current dependence of the saturation parameter of a He-Ne gas laser.

505. Wang, S. C., Siegman, A. E., *Appl. Phys.*, 2, 143-150 (1973), Hollow-cathode transverse discharge He-Ne and $He\text{-}Cd^+$ lasers.

506. Zborovskii, V. A., Molchanov, M. I., Turkin, A. A., Yaroshenko, N. G., *Opt. Spectrosc.*, 34, 704-705 (1973), Measurement of the natural line width of a travelling-wave He-Ne laser in the 0.63 μm region.

507. Dote, T., Yamaguchi, N., Nakamura, T., *Phys. Lett.*, 45A, 29-30 (1973), Effects of RF electric field on He-Ne laser output.

508. Aleksandrov, I. S., Babeiko, Yu. A., Babaev, A. A., Buzhinskii, O. I. et al., *Sov. J. Quantum Electron.*, 5, 1132-1133 (1976), Stimulated emission from a transverse discharge in copper vapor.

509. Belousova, I. M., Znamemskii, V. B., *J. Appl. Spectrosc.*, 25, 1109-1114 (1976), Properties of the lasing mechanism of a helium-neon mixture excited by a discharge in a hollow cathode.

510. Young, R. T., Jr., *J. Appl. Phys.*, 36, 2324-2325 (1965), Calculation of average electron energies in He-Ne discharges.

511. Crisp, M. D., *Opt. Commun.*, 19, 316-319 (1976), A magnetically polarized He-Ne laser.

512. Ihjima, T., Kuroda, K., Ogura, I., *J. Appl. Phys.*, 48, 437-439 (1977), Radial profiles of upper- and lower-laser-level emission in an oscillating He-Ne laser.

513. Muller, Ya. N., Geller, V. M., Lisitsyna, L. I., Grif, G. I., *Sov. J. Quantum Electron.*, 7, 1013-1015 (1977), Microwave-pumped helium-neon laser.

514. Leontov, V. G., Ostapchenko, E. P., *Opt. Spectrosc.*, 43, 321-325 (1977), Effect of excitation conditions on the radial distribution of population inversion in the active element of a He-Ne laser.

515. Honda, T., Endo, M., *IEEE J. Quantum Electron.*, QE-14, 213-214 (1978), International intercomparison of laser power at 633 nm.

516. Otieno, A. V., *Opt. Commun.*, 26, 207-210 (1978), Homogeneous saturation of the 6328 Å neon laser transition due to collisions in the weak collision model.

517. Ferguson, J. B., Morris, R. H., *Appl. Opt.*, 17, 2924-2929 (1978), Single-mode collapse in 6328 Å He-Ne lasers.

518. Chance, D. A., Chastang, J.-C. A., Crawford, V. S., Horstmann, R. E., *IBM J. Res. Develop.*, 23, 108-118 (1979), HeNe parallel plate laser development.

519. Alaev, M. A., Baranov, A. I., Vereshchagin, N. M., Gnedin, I. N. et al., *Sov. J. Quantum Electron.*, 6, 610-611 (1976), Copper vapor laser with a pulse repetition frequency of 100 kHz.

520. Chance, D. A., Brusic, V., Crawford, V. S., Macinnes, R. D., *IBM J. Res. Develop.*, 23, 119-127 (1979), Cathodes for HeNe lasers.

521. Ahearn, W. E., Horstmann, R. E., *IBM J. Res. Develop.*, 23, 128-131 (1979), Nondestructive analysis for HeNe lasers.

522. Schuocker, D., Reif, W., Lagger, H., *IEEE J. Quantum Electron.*, QE-15, 232-239 (1979), Theoretical description of discharge plasma and calculation of maximum output intensity of He-Ne waveguide lasers as a function of discharge tube.

523. Tobias, I., Strouse, W. M., *Appl. Phys. Lett.*, 10, 342-344 (1967), The anomalous appearance of laser oscillation at 6401 Å.

524. Schlie, L. A., Verdeyn, J. T., *IEEE J. Quantum Electron.*, QE-5, 21-29 (1969), Radial profile of Ne $1s_5$ atoms in a He-Ne discharge and their lens effect on lasing at 6401 Å.

525. Sanders, J. H., Thomson, J. E., *J. Phys. B*, 6, 2177-2183 (1973), New high-gain laser transitions in neon.

526. Zitter, R. N., *J. Appl. Phys.*, 35, 3070-3071 (1964), 2s–2p and 3p–2s transitions of neon in a laser ten meters long.

527. Schearer, L. D., *Phys. Lett.*, 27a, 544-545 (1968), Polarization transfer between oriented metastable helium atoms and neon atoms.

528. Chebotayev, V. P., Vasilenko, L. S., *Sov. Phys. JETP*, 21, 515-516 (1965), Investigation of a neon-hydrogen laser at large discharge current.

529. Chebotayev, V. P., *Radio Eng. Electron. Phys.* (USSR), 10, 316-318 (1965), Effect of hydrogen and oxygen on the operation of a neon maser.

530. Anderson, R. S., Bricks, B. G., Karras, T. W., *Appl. Phys. Lett.*, 29, 187-189 (1976), Copper oxide as the metal source in a discharge heated copper vapor laser.

531. McClure, R. M., Pizzo, R., Schiff, M., Zarowin, C. B., *Proc. IEEE*, 52, 851 (1964), Laser oscillation at 1.06 microns in He-Ne.

532. Shtyrkov, E. I., Subbes, E. V., *Opt. Spectrosc.*, 21, 143-144 (1966), Characteristics of pulsed laser action in helium-neon and helium-argon mixtures.

533. Itzkan, I., Pincus, G., *Appl. Opt.*, 5, 349 (1966), 1.0621-μ He-Ne laser.

534. McFarlane, R. A., Patel, C. K. N., Bennett, W. R., Jr., Faust, W. L., *Proc. IEEE*, 50, 2111-2112 (1962), New helium-neon optical maser transitions.

535. Chebotayev, V. P., Pokasov, V. V., *Radio Eng. Electron. Phys.* (USSR), 10, 817-819 (1965), Operation of a laser on a mixture of He-Ne with discharge in a hollow cathode.

536. Petrash, G. G., Knyazev, I. N., *Sov. Phys. JETP*, 18, 571-575 (1964), Study of pulsed laser generation in neon and mixtures of neon and helium.

537. der Agobian, R., Otto, J. L., Cagnard, R., Echard, R., *C. R. Acad. Sci. Ser.* B, 259, 323-326 (1964), Cascades de transitions stimulees dans le neon pur.

538. Andrade, O., Gallardo, M., Bockasten, K., *Appl. Phys. Lett.*, 11, 99-100 (1967), High-gain laser lines in noble gases.

539. Javan, A., Bennett, W. R., Jr., Herriott, D. R., *Phys. Rev. Lett.*, 6, 106-110 (1961), Population inversion and continuous optical maser oscillation in a gas discharge containing a He-Ne mixture.

540. Herriott, D. R., *J. Opt. Soc. Am.*, 52, 31-37 (1962), Optical properties of a continuous helium-neon optical maser.

541. Isaev, A. A., Kazaryan, M. A., Lemmerman, G. Yu., Petrash, G. G. et al., *Sov. J. Quantum Electron.*, 6, 976-977 (1976), Pulse stimulated emission due to transitions in copper atoms excited by discharges in cuprous bromide and chloride vapors.

542. der Agobian, R., Cagnard, R., Echard, R., Otto, J. L., *C. R. Acad. Sci. Ser.* B, 258, 3661-3663 (1964), Nouvelle cascade de transitions stimulee du neon.

543. Bennett, W. R., Jr., Knutson, J. W., Jr., *Proc. IEEE*, 52, 861-862 (1964), Simultaneous laser oscillation on the neon doublet at 1.1523 m.

544.   Patel, C. K. N., *J. Appl. Phys.*, 33, 3194-3195 (1962), Optical power output in He-Ne and pure Ne maser.

545.   Chebotayev, V. P., *Radio Eng. Electron. Phys.* (USSR), 10, 314-316 (1965), Operating condition of an optical maser containing a helium-neon mixture.

546.   Boot, H. A. H., Clunie, D. M., Thorn, R. S. A., *Nature (London)*, 198, 773-774 (1963), Pulsed laser operation in a high-pressure helium-neon mixture.

547.   Smith, J., *J. Appl. Phys.*, 35, 723-724 (1964), Optical maser action in the negative glow region of a cold cathode glow discharge.

548.   Cool, T. A., *Appl. Phys. Lett.*, 9, 418-420 (1966), A fluid mixing laser.

549.   Mustafin, K. S., Seleznev, V. A., Shtyrkov, E. I., *Opt. Spectrosc.*, 21, 429-430 (1966), Stimulated emission in the negative glow region of a glow discharge.

550.   Kuznetsov, A. A., Mash, D. I., Milinkis, B. M., Chirina, L. P., *Radio Eng. Electron. Phys.* (USSR), 9, 1576 (1964), Operating conditions of an optical quantum generator (laser) in helium-neon and xenon-helium gas mixtures.

551.   Lisitsyn, V. N., Fedchenko, A. I., Chebotayev, V. O., *Opt. Spectrosc.*, 27, 157-161 (1969), Generation due to upper neon transitions in a He-Ne discharge optically pumped by a helium lamp.

552.   Babeiko, Yu. A., Vasil'ev, L. A., Orlov, V. K., Sokolov, A. V. et al., *Sov. J. Quantum Electron.*, 6, 1258-1259 (1976), Stimulated emission from copper vapor in a radial transverse discharge.

553.   der Agobian, R., Otto, J. L., Echard, R., Cagnard, R., *C. R. Acad. Sci. Ser.* B, 257, 3844–3847 (1963), Emission stimulee de nouvelles transitions infrarouges du neon.

554.   Blau, E. J., Hochheimer, B. F., Massey, J. T., Schulz, A. G., *J. Appl. Phys.*, 34, 703 (1963), Identification of lasing energy levels by spectroscopic techniques.

555.   Smith, D. S., Riccius, H. D., *IEEE J. Quantum Electron.*, QE-13, 366 (1977), Observation of new helium-neon laser transitions near 1.49 µm.

556.   Gires, F., Mayer, H., Pailette, M., *C. R. Acad. Sci. Ser.* B, 256, 3438–3439 (1963), Sur quelques transitions presentant l'effet laser dans le melange helium-neon.

557.   McFarlane, R. A., Faust, W. L., Patel, C. K. N., *Proc. IEEE*, 51, 468 (1963), Oscillation of f–d transitions in neon in a gas optical maser.

558.   Lisitsyn, V. N., Chebotayev, V. P., *Opt. Spectrosc.*, 20, 603–604 (1966), The generation of laser action in the 4f–3d transitions of neon by optical pumping of a He-Ne discharge with helium lamp.

559.   McFarlane, R. A., Faust, W. L., Patel, C. K. N., Garrett, C. G. B., *Quantum Electronics III* (1964), 573–586, Columbia University Press, New York, Gas maser operation at wavelengths out to 28 micron.

560.   Rosenberger, D., *Phys. Lett.*, 9, 29–31 (1964), Oscillation of three 3p–2s transitions in a He-Ne mixtures.

561.   Bennett, W. R., Jr., Pawilkowski, A. T., *Bull. Am. Phys. Soc.*, 9, 500 (1964), Additional cascade laser transitions in He-Ne mixtures.

562.   Smiley, V. N., *Appl. Phys. Lett.*, 4, 123–124 (1964), New He-Ne and Ne laser lines in the infra-red.

563.   Fedorov, A. I., Sergeenko, V. P., Tarasenko, V. F., Sedoi, V. S., *Sov. Phys. J.*, 20, 251–253 (1976), Copper vapor laser with pulse production of the vapor.

564.   Faust, W. L., McFarlane, R. A., Patel, C. K. N., Garrett, C. G. B., *Phys. Rev.*, 133, A1476–A1478 (1964), Noble gas optical maser lines at wavelengths between 2 and 35 µ.

565.   Otto, J. L., Cagnard, R., Echard, R., der Agobian, R., *C. R. Acad. Sci. Ser.* B, 258, 2779–2780 (1964), Emission stimulee de nouvelles transition infrarouges dan les gas rares.

566.   Grudzinski, R., Pailette, M. R., Becrelle, J., *C. R. Acad. Sci. Ser.* B, 258, 1452–1454 (1964), Etude des transitions laser complees dans un melange helium-neon.

567. Bergman, K., Demtroder, W., *Phys. Lett.*, 29a, 94–95 (1969), A new cascade laser transition in He-Ne mixture.

568. Gerritsen, H. J., Goedertier, P. V., *Appl. Phys. Lett.*, 4, 20–21 (1964), A gaseous (He-Ne) cascade laser.

569. Smiley, V. N., *Quantum Electronics III*(1964), 587–591, Columbia University Press, New York, A long gas phase optical maser cell.

570. McMullin, P. G., *Appl. Opt.*, 3, 641–642 (1964), Precise wavelength measurement of infrared optical maser lines.

571. Brunet, H., Laures, P., *Phys. Lett.*, 12, 106–107 (1964), New infrared gas laser transitions by removal of dominance.

572. Bloom, A. L., Bell, W. E., Rempel, R. C., *Appl. Opt.*, 2, 317–318 (1963), Laser action at 3.39 μ in a helium-neon mixtures.

573. Tibilov, A. S., Shukhtin, A. M., *Opt. Spectrosc.*, 21, 69-70 (1966), Laser action with sodium lines.

574. Smilanski, I., Kerman, A., Levin, L. A., Erez, G., *IEEE J. Quantum Electron.*, QE-13, 24-36 (1977), A hollow-cathode copper halide laser.

575. Herceg, J. E., Miley, G. H., *J. Appl. Phys.*, 39, 2147-2149 (1968), A laser utilizing a low-voltage arc discharge in helium-neon.

576. Konovalov, I. P., Popov, A. I., Protsenko, E. D., *Opt. Spectrosc.*, 33, 6-10 (1972), Measurement of spectral characteristics of the 5s'[1/2]$^0_1$ → 4p[3/2]$_2$ Ne (3.39 μm) transition.

577. Mazanko, I. P., Ogurok, N. D. D., Sviridov, M. V., *Opt. Spectrosc.*, 35, 327-328 (1973), Measurement of the saturation parameter of a neon-helium mixture at the 3.39-μm wavelength.

578. Watanabe, S., Chihara, M., Ogura, I., *Jpn. J. Appl. Phys.*, 13, 164-169 (1974), Decay rate measurements of upper laser levels in He-Ne and He-Se lasers.

579. Balakin, V. A., Konovalov, I. P., Ocheretyanyi, A. I., Popov, A. I. et al., *Sov. J. Quantum Electron.*, 5, 230-231 (1975), Switching of the emission wavelength of a helium-neon laser in the 3.39 μ region.

580. Balakin, V. A., Konovalov, I. P., Protsenko, E. D., *Sov. J. Quantum Electron.*, 5, 581-583 (1975), Measurement of the spectral characteristics of the 3.3912 μ (3s$_2$-3p$_2$) Ne line.

581. Popov, A. I., Protsenko, E. D., *Sov. J. Quantum Electron.*, 5, 1153-1154 (1976), Laser gain due to 5s'[1/2]$^0_1$ – 4p'[3/2]$_2$ transition in neon at λ= 3.39 μ.

582. Faust, W. L., McFarlane, R. A., Patel, C. K. N., Garrett, C. G. B., *Appl. Phys. Lett.*, 4, 85-88 (1962), Gas maser spectroscopy in the infrared.

583. Patel, C. K. N., McFarlane, R. A., Garrett, C. G. B., *Appl. Phys. Lett.*, 4, 18-19 (1964), Laser action up to 57.355 μ in gaseous discharges (Ne, He-Ne).

584. Liu, G. S., Feldman, D. W., Pack, J. C., Weaver, L. A., *J. Appl. Phys.*, 48, 194-195 (1977), Axial cataphoresis effects in continuously pulsed copper halide lasers.

585. Patel, C. K. N., McFarlane, R. A., Garrett, C. G. B., *Bull. Am. Phys. Soc.*, 9, 65 (1964), Optical-maser action up to 57.355 μm in neon.

586. Patel, C. K. N., Faust, W. L., McFarlane, R. A., Garrett, C. G. B., *Proc. IEEE*, 52, No. 6, 713 (1964), CW optical-maser action up to 133 μm (0.133 mm) in neon discharges.

587. McFarlane, R. A., Faust, W. L., Patel, C. K. N., Garrett, C. G. B., *Proc. IEEE*, 52, 318 (1964), Neon gas maser lines at 68.329 μ and 85.047 μ.

588. Chapovsky, P. L., Lisitsyn, V. N., Sorokin, A. R., *Opt. Commun.*, 26, 33-36 (1976), High-pressure gas lasers on ArI, XeI and KrI transitions.

589. Kochubei, S. A., Lisitsyn, V. N., Sorokin, A. R., Chapovskii, P. L., *Sov. J. Quantum Electron.*, 7, 1142-1144 (1978), High-pressure tunable atomic gas lasers.

590. Bockasten, K., Lundholm, T., Andrade, O., *Phys. Lett.*, 22, 145-146 (1966), New near infrared laser lines in argon I.

591. Brisbane, A. D., *Nature (London)*, 214, 75 (1967), High gain pulsed laser.

592. Bockasten, K., Andrade, O., *Nature (London)*, 215, 382 (1967), Identification of high gain laser lines in argon.

593. Sutton, D. G., Galvan, L., Valenzuela, P. R., Suchard, S. N., *IEEE J. Quantum Electron.*, QE-11, 54-57 (1975), Atomic laser action in rare gas-SF_6 mixture.

594. Horrigan, F. A., Koozekanani, S. H., Paananen, R. A., *Appl. Phys. Lett.*, 6, 41-43 (1965), Infrared laser action and lifetimes in argon II.

595. Nerheim, N. M., *J. Appl. Phys.*, 48, 1186-1190 (1977), A parametric study of the copper chloride laser.

596. Dauger, A. B., Stafsudd, O. M., *Appl. Opt.*, 10, 2690-2697 (1971), Characteristics of the continuous wave neutral argon laser.

597. Willett, C. S., *Appl. Opt.*, 11, 1429-1431 (1972), Comments on characteristics of the continuous wave neutral argon laser.

598. Dauger, A. B., Stafsudd, O. M., *IEEE J. Quantum Electron.*, QE-8, 912-913 (1972), Line competition in the neutral argon laser.

599. Jalufka, N. W., DeYoung, R. J., Hohl, F., Williams, M. D., *Appl. Phys. Lett.*, 29, 188-190 (1976), Nuclear-pumped ^{3}He-Ar laser excited by the ^{3}He(n,p)^{3}H reaction.

600. Wilson, J. W., DeYoung, R. J., Harries, W. L., *J. Appl. Phys.*, 50, 1226-1234 (1979), Nuclear-pumped ^{3}He-Ar laser modelling.

601. Liberman, S., *C. R. Acad. Sci. Ser. B*, 261, 2601-2604 (1965), Emission stimulee de nouvelles transitions infrarouge de l'argon, du krypton et du xenon.

602. Linford, G. J., *IEEE J. Quantum Electron.*, QE-9, 611-612 (1973), New pulsed and CW laser lines in the heavy noble gases.

603. Brochard, J., Cahuzac, P., Vetter, R., *C. R. Acad. Sci. Ser. B*, 265, 467-470 (1967), Mesure des ecouts isotopiques de six raies laser infrarouges dans l'argon.

604. Linford, G. J., *IEEE J. Quantum Electron.*, QE-8, 477-482 (1972), High-gain neutral laser lines in pulsed noble-gas discharges.

605. Linford, G. J., *IEEE J. Quantum Electron.*, QE-9, 610-611 (1973), New pulsed laser lines in krypton.

606. Anderson, R. S., Bricks, B. G., Karras, T. W., *IEEE J. Quantum Electron.*, QE-13, 115-117 (1977), Steady multiply pulsed discharge-heated copper-vapor laser with copper halide lasant.

607. Walter, W. T., Jarrett, J. M., *Appl. Opt.*, 3, 789-790 (1964), Strong 3.27 μ oscillation in xenon.

608. DeYoung, R. J., Jalufka, N. W., Hohl, F., *Appl. Phys. Lett.*, 30, 19-21 (1977), Nuclear-pumped lasing of ^{3}He-Xe and ^{3}He-Kr.

609. Sinclair, D. C., *J. Opt. Soc. Am.*, 55, 571 (1965), Near-infrared oscillations in pulsed noble-gas-ion lasers.

610. Courville, G. E., Walsh, P. J., Wasko, J. H., *J. Appl. Phys.*, 35, 2547-2548 (1964), Laser action in Xe in two distinct current regions of ac and dc discharges.

611. Clark, P. O., *Phys. Lett.*, 17, 190-192 (1965), Pulsed operation of the neutral xenon laser.

612. Newman, L. A., DeTemple, T. A., *Appl. Phys. Lett.*, 27, 678-680 (1975), High-pressure infrared Ar-Xe laser system: ionizer-sustainer mode of excitation.

613. Lawton, S. A., Richards, J. B., Newman, L. A., Specht, L., DeTemple, T., *J. Appl. Phys.*, 50, 3888-3898 (1979), The high-pressure neutral infrared xenon laser.

614. Patel, C. K. N., Faust, W. L., McFarlane, R. A., *Appl. Phys. Lett.*, 1, 84-85 (1962), High gain gaseous (Xe-He) optical maser.

615. Tang, C. L., *Proc. IEEE,* 219-220 (1963), Relative probabilities for the xenon laser transitions.

616. Fork, R. L., Patel, C. K. N., *Appl. Phys. Lett.,* 2, 180-181 (1963), Broadband magnetic field tuning of optical masers.

617. Vetter, A. A., Nerheim, N. M., *Appl. Phys. Lett.,* 30, 405-407 (1977), Addition of HCl to the double-pulse copper chloride laser.

618. Aisenberg, S., *Appl. Phys. Lett.,* 2, 187-189 (1963), The effect of helium on electron temperature and electron density in rare gas lasers.

619. Patel, C. K. N., *Phys. Rev.,* 131, 1582-1584 (1963), Determination of atomic temperature and Doppler broadening in a gaseous discharge with population inversion.

620. Patel, C. K. N., McFarlane, R. A., Faust, W. L., *Quantum Electronics III* (1964), 507-514, Columbia University Press, New York, High gain medium for gaseous optical masers.

621. Faust, W. L., McFarlane, R. A., *J. Appl. Phys.,* 35, 2010-2015 (1964), Line strengths for noble-gas maser transitions; calculations of gain/inversion at various wavelengths.

622. Smiley, V. N., Lewis, A. L., Forbes, D. K., *J. Opt. Soc. Am.,* 55, 1552-1553 (1965), Gain and bandwidth narrowing in a regenerative He-Xe laser amplifier.

623. Kuznetsov, A. A., Mash, D. I., *Radio Eng. Electron. Phys. (USSR),* 10, 319-320 (1965), Operating conditions of an optical maser with a helium-xenon mixture in the middle infrared region of the spectrum.

624. Moskalenko, V. F., Ostapchenko, E. P., Pugnin, V. I., *Opt. Spectrosc.,* 23, 94-95 (1967), Mechanism of xenon-level population inversion in the positive column of a helium-xenon mixture.

625. Schwarz, S. E., DeTemple, T. A., Targ, R., *Appl. Phys. Lett.,* 17, 305-306 (1970), High-pressure pulsed xenon laser.

626. Shafer, J. H., *Phys. Rev.,* A3, 752-757 (1971), Optical heterodyne measurement of xenon isotope shifts.

627. Targ, R., Sasnett, M. W., *IEEE J. Quantum Electron.,* QE-8, 166-169 (1972), High-repetition-rate xenon laser with transverse excitation.

628. Abrosimov, G. V., Vasil'tsov, V. V., *Sov. J. Quantum Electron.,* 7, 512-513 (1977), Stimulated emission due to transitions in copper atoms formed in transverse discharge in copper halide vapors.

629. Dandawate, V. D., Thomas, G. C., Zembrod, A., *IEEE J. Quantum Electron.,* QE-8, 918-919 (1972), Time behavior of a TEA xenon laser.

630. Fahlen, T. S., Targ, R., *IEEE J. Quantum Electron.,* QE-9, 609 (1973), High-average-power xenon laser.

631. Mansfield, C. R., Bird, P. F., Davis, J. F., Wimett, T. F., Helmick, H., *Appl. Phys. Lett.,* 30, 640-641 (1977), Direct nuclear pumping of a ^{3}He-Xe laser.

632. Liberman, L., *C. R. Acad. Sci. Ser.* B, 266, 236-239 (1968), Sur la structure hyperfine de quelques raies laser infrarouges de xenon 129.

633. Armstrong, D. R., *IEEE J. Quantum Electron.,* QE-4, 968-969 (1968), A method for the control of gas pressure in the xenon laser.

634. Culshaw, W., Kannelaud, J., *Phys. Rev.,* 156, 308-319 (1967), Mode interaction in a Zeeman laser.

635. Paananen, R. A., Bobroff, D. L., *Appl. Phys. Lett.,* 2, 99-100 (1963), Very high gain gaseous (Xe-He) optical maser at 3.5 μ.

636. Bridges, W. B., *Appl. Phys. Lett.,* 3, 45-47 (1963), High optical gain at 3.5 μ in pure xenon.

637. Markin, E. P., Nikitin, V. V., *Opt. Spectrosc.,* 17, 519 (1964), The 3.5 μ Xe-Ne laser.

638. Clark, P. O., *IEEE J. Quantum Electron.,* QE-1, 109-113 (1965), Investigation of the operating characteristics of the 3.5 μ xenon laser.

639. Gabay, S., Smilanski, I., Levin, L. A., Erez, G., *IEEE J. Quantum Electron.*, QE-13, 364-366 (1977), Comparison of CuCl, CuBr, and CuI as lasants for copper-vapor lasers.

640. Kluver, J. W., *J. Appl. Phys.*, 37, 2987-2999 (1966), Laser amplifier noise at 3.5 microns in helium-xenon.

641. Freiberg, R. J., Weaver, L. A., *J. Appl. Phys.*, 38, 250-262 (1967), Effects of lasering upon the electron gas and excited-state populations in xenon discharges.

642. Aleksandrov, E. B., Kulyasov, V. N., *Sov. Phys. JETP*, 28, 396-400 (1969), Determination of the elementary-emitter spectrum latent in an inhomogeneously broadened spectral line.

643. Fork, R. L., Dienes, A., Kluver, J. W., *IEEE J. Quantum Electron.*, QE-5, 607-616 (1969), Effects of combined RF and optical fields on a laser medium.

644. Wang, S. C., Byer, R. L., Siegman, A. E., *Appl. Phys. Lett.*, 17, 120-122 (1970), Observation of an enhanced Lamb dip with a pure Xe gain cell inside a 3.51 μ He-Xe laser.

645. Kasuya, T., *Appl. Phys.*, 2, 339-343 (1973), Broad band frequency tuning of a He-Xe laser with a superconducting solenoid.

646. Linford, G. J., Peressini, E. R., Sooy, W. R., Spaeth, M. L., *Appl. Opt.*, 13, 379-390 (1974), Very long lasers.

647. Aleksandrov, E. B., Kulyasov, V. N., Kharnang, K., *Opt. Spectrosc.*, 38, 439-440 (1975), Tunable single-frequency xenon laser operating at the two infrared transitions, λ = 5.57 μm and 3.51 μm.

648. Wolff, P. A., Abraham, N. B., Smith, S. R., *IEEE J. Quantum Electron.*, QE-13, 400-403 (1977), Measurement of radial variation of 3.51 μm gain in xenon discharge tubes.

649. Vetter, R., *Phys. Lett.*, 42A, 231-232 (1972), Ecarts isotopiques dans la transition laser a l = 3.99 μm du xenon, Accroissement de l'effet de volume pour les couples.

650. Fahlen, T. S., *IEEE J. Quantum Electron.*, QE-13, 546-547 (1977), High pulse rate, mode-locked copper vapor laser.

651. Olson, R. A., Bletzinger, P., Garscadden, A., *IEEE J. Quantum Electron.*, QE-12, 316-317 (1976), New pulsed Xe-neutral laser line.

652. Petrov, Yu. N., Prokhorov, A. M., *Sov. Phys. Tech. Phys.*, 1, 24-25 (1965), 75-micron laser.

653. Cheo, P. K., Cooper, H. G., *J. Appl. Phys.*, 36, 1862-1865 (1965), Ultraviolet ion laser transitions between 2300 Å and 4000 Å.

654. Dreyfus, R. W., Hodgson, R. T., *IEEE J. Quantum Electron.*, QE-8, 537-538 (1972), Electron-beam gas laser excitation.

655. Dana, L., Laures, P., Rocherolles, R., *C. R.*, 260, 481-484 (1965), Raies laser ultraviolettes dans le neon, l'argon et le xenon.

656. Cottrell, T. H. E., Sinclair, D. C., Forsyth, J. M., *IEEE J. Quantum Electron.*, QE-2, 703 (1966), New laser wavelengths in krypton.

657. Akitt, D. P., Wittig, C. F., *J. Appl. Phys.*, 40, 902-903 (1969), Laser emission in ammonia.

658. Turner, R., Murphy, R. A., *Infrared Phys.* 16, 197 (1976), The far infrared helium laser.

659. Willett, C. S., *Handbook of Lasers*, R. Pressley, Ed., CRC Press, Boca Raton, FL (1971), Neutral gas lasers.

660. Willett, C. S., *Progress in Quantum Electronics*, Vol. 1, Pergamon Press, New York (1971), Laser lines in atomic species.

661. Isaev, A. A., Lemmerman, G. Yu., *Sov. J. Quantum Electron.*, 7, 799-801 (1977), Investigation of a copper vapor pulsed laser at elevated powers.

662. Tobin, M. S., *IEEE Quantum Electron.*, QE-20, 985 (1984), SMMW laser emission and frequency measurements in doubly deuterated methyl fluoride (CHD_2F)O.

663.   Facin, J. A., Pereira, D., Vasconcellos, E. C. C., Scalabrin, A., Ferrari, C. A., *Appl. Phys. B*, 48, 245 (1989), New FIR laser lines from CHD_2OH optically pumped by a cw CO_2 laser.

664.   McFarlane, R. A., *Appl. Phys. Lett.*, 5, 91-93 (1964), Laser oscillation on visible and ultraviolet transitions of singly and multiply ionized oxygen, carbon and nitrogen.

665.   Bennett, A. S., Herman, H., *IEEE Quantum Electron.*, QE-18, 323 (1982), Optically pumped far-infrared emission in the cis 1,2-difluoroethene laser.

666.   Beaulieu, A. J., *Appl. Phys. Lett.*, 16, 504-505 (1970), Transversely excited atmospheric pressure CO_2 laser.

667.   Dumanchin, R., Michon, M., Farcy, J. C., Boudinet, G., Rocca-Serra, J., *IEEE J. Quantum Electron.*, QE-8, 163-165 (1972), Extension of TEA CO_2 laser capabilities.

668.   Petrov, Yu. N., Prokhorov, A. M., *JETP Lett.*, 1, 24 (1965), 75- micron laser.

669.   Batenin, V. M., Burmakin, V. A., Vokhmin, P. A., Evtyunin, A. I., Klimov, *Sov. J. Quantum Electron.*, 7, 891-893 (1977), Time dependence of the electron density in a copper vapor laser.

670.   Lamberton, H. M., Pearson, P. R., *Electron. Lett.*, 7, 141-142 (1971), Improved excitation techniques for atmospheric pressure CO_2 lasers.

671.   Pearson, P. R., Lamberton, H. M., *IEEE J. Quantum Electron.*, QE-8, 145-149 (1972), Atmospheric pressure CO_2 lasers giving high output energy per unit volume.

672.   Richardson, M. C., Alcock, A. J., Leopold, K., Burtyn, P., *IEEE J. Quantum Electron.*, QE-9, 236-243 (1973), A 300-J multigigawatt CO_2 laser.

673.   Reid, R. D., McNeil, J. R., Collins, G. J., *Appl. Phys. Lett.*, 29, 666-668 (1976), New ion laser transitions in He-Au mixtures.

674.   Borodin, V. S., Kagan, Yu. M., *Sov. Phys. Tech. Phys.*, 11, 131-134 (1966), Investigations of hollow-cathode discharges. I. Comparison of the electrical characteristics of a hollow cathode and a positive column.

675.   Hodgson, R. T., Dreyfus, R. W., *Phys. Rev. Lett.*, 28, 536-539 (1972), Vacuum-uv laser action observed in H_2 Werner Bands: 1161-1240 Å.

676.   Silfvast, W. T., *Appl. Phys. Lett.*, 15, 23-25 (1969), Efficient CW laser oscillation at 4416 Å in Cd(II).

677.   Goldsborough, J. P., *Appl. Phys. Lett.*, 15, 159-161 (1969), Stable long life CW excitation of helium-cadmium lasers by dc cataphoresis.

678.   Goldsborough, J. P., *IEEE J. Quantum Electron.*, QE-5, 133 (1969), Continuous laser oscillation at 3250 Å in cadmium ion.

679.   Pogorelyi, P. A., Tibilov, A. S., *Opt. Spectrosc.*, 25, 301-305 (1968), On the mechanism of laser action in Na-H_2 mixtures.

680.   Bokhan, P. A., Solomonov, V. I., Shcheglov, V. B., *Sov. J. Quantum Electron.*, 7, 1032-1033 (1977), Investigation of the energy characteristics of a copper vapor laser with a longitudinal discharge.

681.   Davies, P. B., Jones, H., *Appl. Phys.*, 22, 53 (1980), New cw far infrared molecular lasrs from ClO_2, HCCF, FCN, CH_3NC, CH_3F and propynal.

682.   Telle, J., *IEEE Quantum Electron.*, QE-19, 1469 (1983), Continuous wave 16 μm CF_4 laser.

683.   Dangoisse, E. J., Glorieux, P., *J. Mol. Spec.* 92, 283 (1982), Optically pumped continuous wave submillimeter emission from $H^{13}COOH$: measurements and assignments.

684.   Jones, H., Davies, P. B., Lewis-Bevan, W., *Appl. Phys.* 30, 1 (1983), New FIR laser lines from optically pumped DCOF.

685.   Andriakin, V. M., Velikhov, E. P., Golubev, S. A., Krasil'nikov, S. S., *JETP Lett.*, 8, 214 (1968), Increase of CO_2 laser power under the influence of a beam of fast protons.

686.   Gancey, T., Verdeyen, J. T., Miley, G. H., *Appl. Phys. Lett.*, 18, 568-569 (1971), Enhancement of CO_2 laser power and efficiency by neutron irradiation.

687. Rhoads, H. S., Schneider, R. T., *Trans. Am. Nucl. Soc.*, 14, 429 (1971), Nuclear enhancement of CO_2 laser output.

688. DeYoung, R. J., Wells, W. E., Miley, G. H., *Trans. Am. Nucl. Soc.*, 19, 66 (1974), Enhancement of He-Ne lasers by nuclear radiation.

689. McArthur, D. A., Tollefsrud, P. B., *Appl. Phys. Lett.*, 26, 187-190 (1975), Observation of laser action in CO gas excited only by fission fragments.

690. Helmick, H. K., Fuller, J. L., Schneider, R. T., *Appl. Phys. Lett.*, 26, 327-328 (1975), Direct nuclear pumping of a helium-xenon laser.

691. Warner, B. E., Gerstenberger, D. C., Reid, R. D., McNeil, J. R. et al. *IEEE J. Quantum Electron.*, QE-14, 568-570 (1978), 1 W operation of singly ionized silver and copper lasers.

692. Dorgela, H. B., Alting, H., Boers, J., *Physica*, 2, 959-967 (1935), Electron temperature in the positive column in mixtures of neon and argon or mercury.

693. Chen, C. H., Haberland, H., Lee, Y. T., *J. Chem. Phys.*, 61, 3095-3103 (1974), Interaction potential and reaction dynamics of He(2^1S, 2^3S) + Ne, Ar by the crossed molecular beam method.

694. Leasure, E. L., Mueller, C. R., *J. Appl. Phys.*, 47, 1062-1064 (1976), Crossed-molecular beams investigation of the excitation of ground-state neon atoms by 4.6-eV helium metastables.

695. Brion, C. E., Olsen, L. A. R., *J. Phys. B*, 3, 1020-1033 (1970), Threshold electron impact excitation of the rare gases.

696. Holstein, T., *Phys. Rev.*, 72, 1212-1233 (1947), Imprisonment of resonance radiation in gases.

697. Holstein, T., *Phys. Rev.*, 83, 1159-1168 (1951), Imprisonment of resonance radiation in gases. II.

698. Donohue, T., Wiesenfeld, J. R., *Chem. Phys. Lett.*, 33, 176-180 (1975), Relative yields of electronically excited iodine atoms, I $5^2P_{1/2}$, in the photolysis of alkyl iodides.

699. Donohue, T., Wiesenfeld, J. R., *J. Chem. Phys.*, 63, 3130-3135 (1975), Photo-dissociation of alkyl iodides.

700. Ershov, L. S., Zaleskii, V. Yu., Sokolov, V. N., *Sov. J. Quantum Electron.*, 8, 494-501 (1978), Laser photolysis of perfluoroalkyl iodides.

701. Derwent, R. G., Thrush, B. A., *Faraday Disc. Chem. Soc.*, 53, 16-167 (1972), Excitation of iodine by singlet molecular oxygen.

702. Pirkle, R. J., Wiesenfeld, J. R., Davis, C. C., Wolga, G. J., et al., *IEEE J. Quantum Electron.*, QE-11, 834-838 (1975), Production of electronically excited iodine atoms, I ($^2P_{1/2}$) following injection of HI into a flow of discharged oxygen.

703. Beaty, E. C., Patterson, P. L., *Phys. Rev.*, A137, 346-357 (1965), Mobilities and reaction rates of ions in helium.

704. Deese, J. E., Hassan, H. A., *AIAA J.*, 14, 1589-1597 (1976), Analysis of nuclear induced plasmas.

705. DeYoung, R. J., Winters, P. A., *J. Appl. Phys.*, 48, 3600-3602 (1977), Power deposition in He from the volumetric ^{3}He(n,p)^{3}H reaction.

706. Wilson, J. W., DeYoung, R. J., *J. Appl. Phys.*, 49, 980-988 (1978), Power density in direct nuclear-pumped ^{3}He lasers.

707. Carter, B. D., Rowe, M. J., Schneider, R. T., *Appl. Phys. Lett.*, 36, 115-117 (1980), Nuclear-pumped CW lasing of the ^{3}He-Ne system.

708. Falcone, R. W., Zdasiuk, G. A., *Opt. Lett.*, 5, 155-157 (1980), Pair-absorption-pumped barium laser.

709. Ramsay, J. V., Tanaka, K., *Jpn. J. Appl. Phys.*, 5, 918-923 (1966), Construction of single-mode dc operated He/Ne lasers.

710. Sakurai, T., *Jpn. J. Appl. Phys.*, 11, 1826-1831 (1972), Dependence of He-Ne laser output power on discharge current, gas pressure and tube radius.

711. Brunet, H., *Appl. Opt.*, 4, 1354 (1965), Laser gain measurements in a xenon-krypton discharge.

712. Mash, D. I., Papulovskii, V. F., Chirina, L. P., *Opt. Spectrosc.*, 17, 431-432 (1964), On the operation of a xenon-krypton laser.

713. Allen, L., Jones, D. G. C., Schofield, D. G., *J. Opt. Soc. Am.*, 59, 842-847 (1969), Radiative lifetimes and collisional cross sections for XeI and II.

714. Allen, L., Peters, G. I., *Phys. Lett.*, 31A, 95-96 (1970), Superradiance, coherence brightening and amplified spontaneous emission.

715. Shukhtin, A. M., Mishakov, V. G., Fedotov, G. A., *Sov. Tech. Phys. Lett.*, 3, 304-305 (1977), Production of Cu vapor from Cu_2O dust in a pulsed discharge.

716. Peters, G. I., Allen, L., *J. Phys. A*, 4, 238-243 (1971), Amplified spontaneous emission. I. The threshold condition.

717. Allen, L., Peters, G. I., *J. Phys. A*, 4, 377-381 (1971), Amplified spontaneous emission. II. The connection with laser theory.

718. Allen, L., Peters, G. I., *J. Phys. A*, 4, 564-573 (1971), Amplified spontaneous emission. III. Intensity and saturation.

719. Peters, G. I., Allen, L., *J. Phys. A*, 5, 546-554 (1972), Amplified spontaneous emission. IV. Beam divergence and spatial coherence.

720. Shelton, R. A. J., *Trans. Faraday. Soc.*, 57, 2113-2118 (1961), Vapour pressures of the solid copper (I) halides.

721. Tobin, R. C., *Opt. Commun.*, 32, 325-330 (1980), Rapid differential decay of metastable populations in a copper halide laser.

722. Rothe, D. E., Tan, K. O., *Appl. Phys. Lett.*, 30, 152-154 (1977), High power N_2^+ laser pumped by change transfer in a high-pressure pulsed glow discharge.

723. Olson, R. A., Grosjean, D., Sarka, B., Jr., Garscadden, A. et al., *Rev. Sci. Instrum.*, 47, 677-683 (1976), High-repetition-rate closed-cycle rare gas electrical discharge laser.

724. Hasle, E. K., *Opt. Commun.*, 31, 206-210 (1979), Polarization properties of He-Ne lasers.

725. Nerheim, N. M., *J. Appl. Phys.*, 48, 3244-3250 (1977), Measurement of copper ground-state and metastable level population densities in copper-chloride laser.

726. Majer, J. R., Simons, J. P., *Adv. Photochem.*, 2, 137-182 (1964), Photochemical process in halogenated compounds.

727. Riley, S. J., Wilson, K. R., *Faraday Disc. Chem. Soc.*, 53, 132-146 (1972), Excited fragments from excited molecules: energy partitioning in the photodissociation of alkyl iodides.

728. Harris, G. M., Willard, J. E., *J. Am. Chem. Soc.*, 76, 4678-4687 (1954), Photochemical reactions in the system methyl iodide-iodine-methane: the reaction, $C^{14}H_3 + CH_4 - C^{14}H_4 + CH_3$.

729. Jaseja, T. S., Javan, A., Townes, C. H., *Phys. Rev. Lett.*, 10, 165-167 (1963), Frequency stability of He-Ne masers and measurements of length.

730. Jaseja, T. S., Javan, A., Murray, J., Townes, C. H., *Phys. Rev. A*, 133, 1221-1225 (1964), Test of special relativity or of the isotropy of space by use of infrared masers.

731. McFarlane, R. A., Bennett, W. R., Jr., Lamb, W. E., Jr., *Appl. Phys. Lett.*, 2, 189-190 (1963), Single mode tuning dip in the power output of an He-Ne optical maser.

732. Tobias, I., Skolnick, M., Wallace, R. A., Polanyi, T. G., *Appl. Phys. Lett.*, 6, 198-201 (1965), Deviation of a frequency-sensitive signal from a gas laser in an axial magnetic field.

733. Andrews, A. J., Webb, C. E., Tobin, R. C., Denning, R. G., *Opt. Commun.*, 22, 272-274 (1977), A copper vapor laser operating at room temperature.

734. Wallard, A. J., *J. Phys. E*, 6, 793-807 (1973), The frequency stabilization of gas lasers.

735. Shevirev, A. S., Dyubko, S. F., Efimenko, M. N., Fesenko, L. D., *Zh. Prikl. Specktosk.* 42, 480 (1985).

736. White, A. D., *IEEE J. Quantum Electron.*, QE-1, 349-357 (1965), Frequency stabilization of gas lasers.

737. Hochuli, U. E., Haldemann, P., Li, H. A., *Rev. Sci. Instrum.*, 45, 1378-1381 (1974), Factors influencing the relative frequency stability of He-Ne laser structures.

738. Hanes, G. R., Dahlstrom, C. E., *Appl. Phys. Lett.*, 14, 362-364 (1969), Iodine hyperfine structure observed in saturated absorption at 633 nm.

739. VanOorschot, B. D. J., VanderHoeven, C. J., *J. Phys. E*, 12, 51-55 (1979), A recently developed iodine-stabilized laser.

740. Schweitzer, W. G., Jr., Kessler, E. G., Jr., Deslattes, R. D., Layer, H. P., *Appl. Opt.*, 12, 2927-2938 (1973), Description, performance and wavelengths of iodine stabilized lasers.

741. Cerez, P., Brillet, A., Hartmann, F., *IEEE Trans. Inst. Meth.*, IM-23, 526-528 (1974), Metrological properties of the R(127) line of iodine studied by laser saturated absorption.

742. Fedorov, A. J., Sergeenko, V. P., Tarasenko, V. F., *Sov. J. Quantum Electron.*, 7, 1166-1167 (1977), Apparatus for investigating stimulated emission from explosively formed metal vapors.

743. Helmcke, J., Bayer-Helms, F., *IEEE Trans. Inst. Meth.*, IM-23, 529-531 (1974), He-Ne laser stabilized by saturated absorption in I_2.

744. Wallard, A. J., *IEEE Trans. Inst. Meth.*, IM-23, 52-535 (1974), The reproducibility of 633 nm lasers stabilized by $^{127}I_2$.

745. Cole, J. B., Bruce, C. F., *Appl. Opt.*, 14, 1303-1310 (1975), Iodine stabilized laser with three internal mirrors.

746. Cerez, P., Brillet, A., Hajdukovic, S., Man, N., *Opt. Commun.*, 21, 332-336 (1977), Iodine stabilized He-Ne laser with a hot wall iodine cell.

747. Melnikov, N. A., Privalov, V. E., Fofanov, Ya. A., *Opt. Spectrosc.*, 42, 425-428 (1977), Experimental investigation of He-Ne lasers stabilized by saturation absorption in iodine.

748. Layer, H. P., *Proc. Soc. Photo Opt. Instrum. Eng.*, 129, 9-11 (1977), The iodine stabilized laser as a realization of the length unit.

749. Chartier, J. M., Helmcke, J., Wallard, A. J., *IEEE Trans. Inst. Meth.*, IM-25, 450-453 (1976), International intercomparison of the wavelength of iodine-stabilized lasers.

750. Bagaev, S. N., Baklanov, E. V., Chebotaev, V. O., *JETP Lett.*, 16, 243-246 (1972), Anomalous decrease of the shift of the center of the Lamb dip in low-pressure, molecular gases.

751. Koshelyaevskii, N. B., Tatarenkov, V. M., Titov, A. N., *Sov. J. Quantum Electron.*, 6, 222-226 (1976), Power shift of the frequency of an He-Ne-CH_4 laser.

752. Liu, C. S., Feldman, D. W., Pack, J. L., Weaver, L. A., *IEEE J. Quantum Electron.*, QE-13, 744-751 (1977), Kinetic processes in continuously pulsed copper halide lasers.

753. Jolliffe, B. W., Kramer, G., Chartier, J. M., *IEEE Trans. Inst. Meth.*, IM-25, 447-450 (1976), Methane-stabilized He-Ne laser intercomparisons 1976.

754. Alekseev, V. A., Malyugin, A. V., *Sov. J. Quantum Electron.*, 7, 1075-1081 (1977), Influence of the hyperfine structure on the frequency reproducibility of an He-Ne laser with a methane absorption cell.

755. Nakazawa, M., Musha, T., Tako, T., *J. Appl. Phys.*, 50, 2544-2547 (1979), Frequency-stabilized 3.39 µm He-Ne laser with no frequency modulation.

756. Evenson, K. M., Day, G. W., Wells, J. S., Mullen, L. O., *Appl. Phys. Lett.*, 20, 133-134 (1972), Extension of absolute frequency measurements to the CW He-Ne laser at 88 THz (3.39 μ).

757. Evenson, K. M., Wells, J. S., Petersen, F. R., Danielson, B. L., Day, G., *Appl. Phys. Lett.*, 22, 192-195 (1973), Absolute frequencies of molecular transitions used in laser stabilization: the 3.39 μm transition in CH_4 and the 9.33- and 10.18-μm.

758. Barger, R. L., Hall, J. L., *Appl. Phys. Lett.*, 22, 196-199 (1973), Wavelength of the 3.39 μm laser-saturated absorption line of methane.

759. Jennings, D. A., Petersen, F. R., Evenson, K. M., *Opt. Lett.*, 4, 129-130 (1979), Frequency measurements of the 260-THz (1.15 μm) He-Ne laser.

760. Baird, K. M., Evenson, K. M., Hanes, G. R., Jennings, D. A., Peterson, F., *Opt. Lett.*, 4, 263-264 (1979), Extension of absolute-frequency measurements to the visible: frequencies of ten, hyperfine components of iodine.

761. Jennings, D. A., Petersen, F. R., Evenson, K. M., *Appl. Phys. Lett.*, 510-511 (1975), Extension of absolute frequency measurements to 148THz: frequencies of the 2.0- and, 3.5-μ Xe laser.

762. Gokay, M. C., Jenkins, R. S., Cross, L. A., *J. Appl. Phys.*, 48, 4395-4396 (1977), Output characteristics of the CuCl double-pulse laser at small pumping pulse delays.

763. Baird, K. M., Smith, D. S., Whitford, B. G., *Opt. Commun.*, 31, 367-368 (1979), Confirmation of the currently accepted value 299792458 metres per second for the speed of light.

764. Evenson, K. M., Wells, J. S., Petersen, F. R., Danielson, B. L., Day, G., *Phys. Rev. Lett.*, 29, 1346-1349 (1972), Speed of light from direct frequency and wavelength measurements of the methane-stabilized laser.

765. Knight, T. G., Rowley, W. R. C., Shotton, K. C., Woods, P. T., *Nature (London)*, 251, 46 (1974), Measurement of the speed of light.

766. Blaney, T. G., Bradley, C. C., Edwards, G. J., Knight, D. J. E., Woods, P. T., *Nature (London)*, 244, 504 (1973), Absolute frequency measurement of the R(12) transition of CO_2 at 9.3 mm.

767. Jolliffe, B. W., Rowley, W. R. C., Shotton, K. C., Wallard, A. J., Woods, P. T., *Nature (London)*, 251, 46-47 (1974), Accurate wavelength measurement on up-converted CO_2 laser radiation.

768. Woods, P. T., Shotton, K. C., Rowley, W. R. C., *Appl. Opt.*, 17, 1048-1054 (1978), Frequency determination of visible laser light by interferometric comparison.

769. Markova, S. V., Petrash, G. G., Cherezov, V. M., *Sov. J. Quantum Electron.*, 9, 707-711 (1979), Investigation of the stimulated emission mechanism in a pulsed bismuth vapor laser.

770. Fill, E. E., Thieme, W. H., Volk, R., *J. Phys. D.*, 12, L41-L45 (1979), A tunable iodine laser.

771. Silfvast, W. T., Szeto, L. H., Wood, O. R., II, *Appl. Phys. Lett.*, 31, 334-337 (1977), Recombination lasers in expanding CO_2 laser-produced plasmas of argon, krypton, and xenon.

772. Tibilov, A. S., Shukhtin, A. M., *Opt. Spectrosc.*, 25, 221-224 (1968), Investigation of generation of radiation in the Na-H_2 mixture.

773. Vetter, A. A., *IEEE J. Quantum Electron.*, QE-13, 889-891 (1977), Quantitative effect of initial current rise on pumping the double-pulsed copper chloride.

774. Silfvast, W. T., Szeto, L. H., Wood, O. R., II, *Appl. Phys. Lett.*, 34, 213-215 (1979), Ultra-high-gain laser-produced plasma laser in xenon using periodic pumping.

775. Silfvast, W. T., Szeto, L. H., Wood, O. R., II, *Appl. Phys. Lett.*, 36, 500-502 (1980), Power output enhancement of a laser-produced Cd plasma recombination laser by plasma confinement.

776.   Kneipp, H., Rentsch, M., *Sov. J. Quantum Electron.*, 7, 1454-1455 (1977), Discharge-heated copper vapor laser.

777.   Vetter, A. A., Nerheim, N. M., *IEEE J. Quantum Electron.*, QE-14, 73-74 (1978), Effect of dissociation pulse circuit inductance on the CuCl laser.

778.   Cross, L. A., Jenkins, R. S., Gokay, M. C., *J. Appl. Phys.*, 49, 453-454 (1978), The effects of a weak axial magnetic field on the total energy output of the CuCl, double-pulse laser.

779.   Nerheim, N. M., Vetter, A. A., Russell, G. R., *J. Appl. Phys.*, 49, 12-15 (1978), Scaling a double-pulsed copper chloride laser to 10 mJ.

780.   Gordon, E. B., Egorov, V. G., Pavcenko, V. S., *Sov. J. Quantum Electron.*, 8, 266-268 (1978), Excitation of metal vapor lasers by pulse trains.

781.   Bokhan, P. A., Shcheglov, V. B., *Sov. J. Quantum Electron.*, 8, 219-222 (1978), Investigation of a transversely excited pulsed copper vapor laser.

782.   Zemskov, K. I., Kazaryan, M. A., Mokerov, V. G., Petrash, G. G. et al., *Sov. J. Quantum Electron.*, 8, 245-247 (1978), Coherent properties of a copper vapor laser and dynamic holograms in vanadium dioxide films.

783.   Burmakin, V. A., Evtyunin, A.N., Lesnoi, M. A., Bylkin, V.I., *Sov. J. Quantum Electron.*, 8, 574-576 (1978), Long-life sealed copper vaper laser.

784.   Gridnev, A. G., Gorblinova, T. M., Elaev, V. F., Evtushenko, G. S. et al., *Sov. J. Quantum Electron.*, 8, 656-658 (1978), Spectroscopic investigation of a gas discharge plasma of a Cu + Ne laser.

785.   Mishakov, V. G., Tibilov, A. S., Shukhtin, A. M., *Opt. Spectrosc.*, 31, 176-177 (1971), Laser action in Na-H_2 and K-H_2 mixtures with pulsed injection of metal vapor into a gas-discharge plasma.

786.   Tenenbaum, J., Smilanski, I., Gabay, S., Erez, G., Levin, L. A., *J. Appl. Phys.*, 49, 2662-2665 (1978), Time dependence of copper-atom concentration in ground and metastable states in a pulsed CuCl laser.

787.   Piper, J. A., *IEEE J. Quantum Electron.*, QE-14, 405-407 (1978), A transversely excited copper halide laser with large active volume.

788.   Chen, C. J., Bhanji, A. M., Russell, G. R., *Appl. Phys. Lett.*, 33, 146-148 (1978), Long duration high-efficiency operation of a continuously pulsed copper laser utilizing copper bromide as a lasant.

789.   Gokay, M. C., Soltanoalkotabi, M., Cross, L. A., *J. Appl. Phys.*, 49, 4357-4358 (1978), Copper acetylacetonate as a source in the 5106 - A neutral copper laser.

790.   Nerheim, N. M., Bhanji, A. M., Russell, G. R., *IEEE J. Quantum Electron.*, QE-14, 686-693 (1978), A continuously pulsed copper halide laser with a cable-capacitor Blumlein discharge circuit.

791.   Babeiko, Yu. A., Vasil'ev, L. A., Sokolov, A. V., Sviridov, A. V., et al., *Sov. J. Quantum Electron.*, 8, 1153-1154 (1978), Coaxial copper-vapor laser with a buffer gas at above atmospheric pressure.

792.   Kushner, M. J., Culick, F. E. C., *Appl. Phys. Lett.*, 33, 728-731 (1978), Extrema of electron density and output pulse energy in a CuCl/Ne discharge and a Cu/CuCl double-pulsed laser.

793.   Bokhan, P. A., Gerasimov, V. A., Solomonov, V. I., Shcheglov, V. B., *Sov. J. Quantum Electron.*, 8, 1220-1227 (1978), Stimulated emission mechanism of a copper vapor laser.

794.   Kazaryan, M. A., Trofimov, A. N., *Sov. J. Quantum Electron.*, 8, 1390-1391 (1978), Gas-discharge tubes for metal halide vapor lasers.

795.   Bokhan, P. A., Gerasimov, V. A., *Sov. J. Quantum Electron.*, 9, 273-275 (1979), Optimization of the excitation conditions in a copper vapor laser.

796.   Sorokin, P. P., Lankard, J. R., *J. Chem. Phys.*, 54, 2184-2190 (1971), Infrared lasers resulting from giant-pulse laser excitation of alkali metal molecules.

797. Smilanski, I., Erez, G., Kerman, A., Levin, L. A., *Opt. Commun.*, 30, 70-74 (1979), High-power, high-pressure, discharge-heated copper vapor laser.

798. Kushner, M. J., Culick, F. E. C., *IEEE J. Quantum Electron.*, QE-15, 835-837 (1979), A continuous discharge improves the performance of the Cu/CuCl double pulse laser.

799. Tennenbaum, J., Smilanski, I., Gabay, S., Levin, L. A., Erez, G., *J. Appl. Phys.*, 50, 57-61 (1979), Laser power variation and time dependence of populations in a burst-mode CuBr laser.

800. Miller, J. L., Kan, T., *J. Appl. Phys.*, 50, 3849-3851 (1979), Metastable decay rates in a Cu-metal-vapor laser.

801. Kan, T., Ball, D., Schmitt, E., Hill, J., *Appl. Phys. Lett.*, 35, 676-677 (1979), Annular discharge copper vapor laser.

802. Gokay, M. C., Cross, L. A., *IEEE J. Quantum Electron.*, QE-15, 65-66 (1979), Comparison of copper acetylacetonate, copper (II) acetate, and copper chloride as lasants for copper vapor lasers.

803. Kazaryan, M. A., Trofimov, A. N., *Sov. J. Quantum Electron.*, 9, 148-152 (1979), Kinetics of metal salt vapor lasers.

804. Babeiko, Yu. A., Vasil'ev, L. A., Sviridov, A. V., Sokolov, A. V. et al., *Sov. J. Quantum Electron.*, 9, 651-653 (1979), Efficiency of a copper vapor laser.

805. Hargrove, R. S., Grove, R., Kan, T., *IEEE J. Quantum Electron.*, QE-15, 1228-1233 (1979), Copper vapor laser unstable resonator oscillator and oscillator-amplifier characteristics.

806. Solanki, R., Latush, E. L., Fairbank, W. M., Jr., Collins, G. J., *Appl. Phys. Lett.*, 34, 568-570 (1979), New infrared laser transitions in copper and silver hollow cathode discharges.

807. Landsberg, B. M., Shafki, M. S., Butcher, R. J., *Int. J. Infrared Millimeter Waves*, 2, 49 (1981).

808. Ioli, N., Moretti, A., Moruzzi, G., Strumia, F., D'Amato, F., New CH_2F_2 FIR laser lines pumped by a tunable WG CO_2 laser, *Int. J. Infrared Millimeter Waves*, 6, 1017 (1985).

809. Dangoisse, E. J., Wasscat, J., Colmont, J. M., Assignment of laser lines in an optically pumped submillimeter and near millimeter laser $(H_2CO)_3$, *Int. J. Infrared Millimeter Waves*, 2, 1177 (1981).

810. Bugaev, V. A., Shliteris, E. P., *Sov. J. Quantum Electron.*, 14, 1331 (1984), Submillimeter lasing transitions in isotopic modifications of SO_2.

811. Bugaev, V. A., Shliteris, E. P., *Sov. J. Quantum Electron.*, 11, 742 (1981), Sulfur dioxide laser pumped by CO_2 laser radiation and identification of the transitions.

812. Sattler, J. P., Lafferty, W. J., *Reviews of Infrared and Millimeter Waves*, 2, Button, K. J., Inguscio, M., Strumia, F., Eds., Plenum Press, New York.

813. Davies, P. B., Stern, D. P., *Int. J. Infrared Millimeter Waves*, 3, 909 (1982), New cw FIR laser lines from optically pumped silyl fluoride (SiH_3F).

814. Davies, P. B., Ferguson, A. H., Stern, D. P., *Infrared Phys.* 25, 87 (1985), New optically pumped lasers containing silicon.

815. Shafik, S., Crocker, D., Landsberg, B. M., Butcher, R. J., *IEEE Quantum Electron.*, QE-17, 115 (1981), Phosphine far-infrared CW laser transitions: optical pumping at more than 11 MHz from resonance.

816. Gerasimov, W. G., Dyubkko, S. F., Efimenko, L. D., Fesenko, L. D., Jarcev, W. I., *Ukr. Fiz. Zh.* 28, 1323 (1983).

817. Siemsen, K. J., Reid, J., Danagher, D. J., *Appl. Opt.*, 25, 86 (1986), Improved cw lasers in the 11-13 μm wavelength region produced by optically pumping NH_3.

818. Kroeker, D. F., Reid, J., *Appl. Opt.*, 25, 2929 (1986), Line-tunble cw *ortho-* and *para-*NH_3 lasers operating at wavelengths of 11 and 14 μm.

819. Jones, H., Taubmann, G., Takami, M., *IEEE Quantum Electron.*, QE-18, 1997 (1982), The optically pumped hydrazin FIR lasr: assignments and new laser lines.

820. Landsberg, B. M., *IEEE Quantum Electron.*, QE-16,704 (1980), New optically pumped CW submillimeter emission lines from OCS, CH_3OH, and CH_3OD.

821. Davies, P. B., Jones, H., *IEEE Quantum Electron.*, QE-17, 13 (1981), A powerful new optically pumped FIR laser—formylfluoride.

822. Dumanchin, R., Rocca-Serra, J., unpublished (Sept. 1970), High power density pulsed molecular laser.

823. Tang, F., Olafsson, A., Hennignsen, J. O., *Appl. Phys. B*, 47, 47 (1988), A study of the methanol laser with a 500 MHz tunable CO_2 pump laser.

824. Allario, F., Schneider, R. T., *Research on Uranium Plasmas*, NASA SP-236, No. 236(1971), Enhancement of laser output by nuclear reactions, National Aeronautics and Space Administration.

825. Horrigan, F. A., unpublished (1966), Raytheon Company, Estimated lifetimes in Neon I and Xenon I.

826. Saykally, R. J., Evenson, K., Jennings, D. A., Zink, L. R., Scalabrin, A., New FIR laser line and frequency measurement for optically pumped CD_3OH, *Int. J. Infrared Millimeter Waves*, 8, 653 (1987).

827. Baldacci, S. A., Ghersetti, S. H., Jurlock, S. C., Rao, K. N., *J. Mol. Spectrosc.* 42, 327 (1972), Spectrum of dideuteroacetylene near 18.6 microns.

828. Gastaud, C., Redon, M., Fourrier, M., *Appl. Phys. B*, 47, 303 (1988), New strong CW FIR laser actions and assignments in COF_2.

829. Petersen, J. C., Duxbury G., *Appl. Phys. B*, 27, 19 (1982), Observation and assignment of submillimetre laser lines from CH_3OH pumped by isotopic CO_2 lasers.

830. Petersen, J. C., Duxbury G., *Appl. Phys. B*, 34, 17 (1984), Submillimetre laser lines from CH_3OH pumped by a $^{13}C^{18}O_2$ pump laser: observations and assignments.

831. Coleman, C. D., Bozman, W. R., Meggers, W. F., Monograph, Vol. 1 & 2, No. 3 (1960), *Table of Wave Numbers*, U. S. National Bureau of Standards .

832. Jones, D. R., Little, C. E., *IEEE J. Quantum Electron.*, 28, 590 (1992), A lead bromide laser operating at 722.9 and 406.2 nm.

833. Liou, H.-T., Yang, H., Dan, P., *Appl. Phys. B*, 54, 221 (1992), Laser induced lasing in CS_2 vapor.

834. Vasconcellos, E. C. C., Wyss, J., Evenson, K. M., Frequency measurements of far infrared $^{13}CH_3OH$ laser lines, *Int. J. Infrared Millimeter Waves*, 8, 647 (1987).

835. Rutt, H. N., Green, J. M., *Opt. Commun.* 19, 320 (1978), Optically pumped laser action in dideuteroacetylene.

836. Deka, B. K., Dyer, P. E., Winfield, R. J., *Opt. Lett.*, 5, 194 (1980), New 17-21 μm laser lines in C_2D_2 using a continuously tunable CO_2 laser pump.

837. Hall, J. C., *Atomic Physics*, 615(1973), Plenum Press, New York, Saturated absorption spectroscopy with applications to the 3.39 μm methane transition.

838. Jones, D. R., Little, C. E., *Opt. Commun.* 91, 223 (1992), A 472.2 nm bismuth halide laser.

839. Thom, K., Schneider, R. T., *AIAA J.*, 10, 400-406 (1972), Nuclear pumped gas lasers.

840. Birnbaum, G., *Proc. IEEE*, 55, 1015-1026 (1967), Frequency stabilization of gas lasers.

841. Silfvast, W. T., Macklin, J. J., Wood II, O. R., *Opt. Lett.*, 8, 551 (1983), High-gain inner-shell photoionization laser in Cd vapor pumped by soft-X-ray radiation from a laser-produced plasma source.

842. Silfvast, W. T., Wood II, O. R., *J. Opt. Soc. Am. B*, 4, 609 (1987), Photoionization lasers pumped by broadband soft-X-ray flux from laser-produced plasmas.

843.   Hube, M., Kumkar, M., Dieckmann, M., Beigang, R., Willegehausen, B., *Optics Commun.* 66, 107 (1988), Potassium photoionization laser produced by innershell ionization of excited potassium.

844.   Lundberg, H., Macklin, J. J., Silfvast, W. T., Wood II, O. R., *Appl. Phys. Lett.*, 45, 335 (1984), High-gain soft-x-ray-pumped photoionization laser in zinc vapor.

845.   Hooker, S. M., Haxell, A. M., Webb, C. E., *Appl. Phys. B* 54, 119 (1992), Observation of new laser transitions and saturation effects in optically pumped NO.

846.   Berkeliev, B. M., Dolgikh, V. A., Rudoi, I. G., Soroka, A. M., *Sov. J. Quantum Electron.*, 21, 250 (1991), Efficient stimulated emission of ultraviolet and infrared radiation from a mixture of nitrogen and light rare gases; simultaneous lasing at 1790 and 358 nm.

847.   Willett, C. S., Gleason, T. J., Kruger, J. S., unpublished (1970).

848.   Hochuli, U. E., Haldemann, P., private communication.

849.   Laures, P., *Onde Electr.*, No. 469, (1966), Stabilisation de la frequence des lasers a gaz.

850.   Walker, D. J., Barty, C. P. J., Yin, G. Y., Young, J. F., Harris, S. E., *Opt. Lett.*, 12, 894 (1987), Observation of super Coster-Kronig-pumped gain in Zn III.

851.   Silfvast, W. T., Wood II, O. R., Al-Salameh, D. Y., In *Short Wavelength Coherent Radiation: Generation and Applications* AIP Conf. Proc. 147, Attwood, D. T., Bokor, J., Eds., AIP, New York (1986), p. 134., Direct photoionization pumping of VUV, UV and visible inversions in helium, cadmium and argon via two-electron shakeup and of sodium via the output from the LLNL soft-X-ray laser.

852.   Silfvast, W. T., Wood II, O. R., Lundberg, H., Macklin, J. J., *Opt. Lett.*, 10, 122 (1985), Stimulated emission in the ultraviolet by optical pumping from photoionization-produced inner-shell states in Cd^+.

853.   Hube, M., Brinkmann, R., Willing, H., Beigang, R., Willegehausen, B., *Appl. Phys. B* 45, 197(1988), Potassium photoionization laser produced by laser induced plasma radiation from a multi foci device.

854.   Klimovski, I. I., Vokhmin, P. A., *Proc. 13th Int. Conf. Phenomena in Ionized Gases* (Sept. 1977), Connection of the copper vapor laser emission pulse characteristics with plasma parameters, East German Physical Society .

855.   Rodin, A. V., Zemtsov, Yu. L., *Proc. 13th Int. Conf. Phenomenin Ionized Gases* (Sept. 1977), Electron energy distribution function for copper vapor, East German Physical Society.

856.   Elaev, V. F., Kirilov, A. E., Polunin, Yu. P., Soldatov, A. N., Fedorov, A. I., *Proc. 13th Int. Conf. Phenomenin Ionized Gases* (Sept. 1977), Experimental investigation of the pulse discharge in Cu + Ne mixture in high repetition rate regime, East German Physical Society.

857.   Bloom, A. L., unpublished, private communication.

858.   Russell, G. R., *Research on Uranium Plasmas*, NASA SP-236, No. 236 (1971), National Aeronautics and Space Administration, Feasibility of a nuclear laser excited by fission fragments produced in pulsed nuclear reactor.

859.   DeShong, J. A., ANL Report, No. 7310(1966), Argonne National Lab, Nuclear Pumped Carbon Dioxide Gas Lasers - Model I Experiment.

860.   Boguslovskii, A. A., Guryev, T. T., Didrikell, L. N., Novikova, V. A. K., *Electronnaya Tekhn.*, 8 (1967), unspecified.

861.   der Agobian, R., Otto, J. L., Cagnard, R., Barthelemy, J., Echard, R., *C.R. Acad. Sci. Paris*, 260, 6327-6329 (1965), Emission stimulee en regime permanent dans le spectre visible du krypton ionise.

862.   Banse, K., Herziger, G., Schafer, G., Seelig, W., *Phys. Lett.*, 27A, 682-683 (1968), Continuous UV-laser power in the watt range.

863.   Fendley, J. R. Jr., O'Grady, J. J., Tech. Rep. ECOM-0246-F (1970), RCA, Development, construction, and demonstration of a 100W cw argon-ion laser.

864. Ferrario, A., *Opt. Commun.*, 7, 375-378 (1973), Excitation mechanism in Hg⁺ ion laser.

865. Fowles, G.R., Jensen, R. C., *Proc. IEEE*, 52, 851-852 (1964), Visible laser transitions in the spectrum of singly ionized iodine.

866. Fowles, G.R., Jensen, R. C., *Appl. Opt.*, 3, 1191-1192 (1964), Visible laser transitions in ionized iodine.

867. Fowles, G.R., Jensen, R. C., *Phys. Rev. Lett.*, 14, 347-348 (1965), Laser oscillation on a single hyperfine transition in iodine.

868. Fowles, G.R., Silfvast, W. T., *IEEE J. Quantum Electron.*, QE-1, 131 (1965), Laser action in the ionic spectra of zinc and cadmium.

869. Fowles, G.R., Silfvast, W. T., Jensen, R. C., *IEEE J. Quantum Electron.*, QE-1, 183-184 (1965), Laser action in ionized sulfur and phosphorus.

870. Fowles, G.R., Hopkins, B. D., *IEEE J. Quantum Electron.*, QE-3, 419 (1967), CW laser oscillation at 4416 Å in cadmium.

871. Fukuda, S., Miya, M., *Jpn. J. Appl. Phys.*, 13, 667-674 (1974), A metal-ceramic He-Cd II laser with sectioned hollow cathodes and output power characteristics of simultaneous oscillations.

872. Gadetskii, N. P., Tkach, Yu V., Slezov, V. V., Bessarab, Ya. et al., *JETP Lett.*, 14, 101-105 (1971), New mechanism of coherent-radiation generation in the visible region of the spectrum in ionized oxygen and nitrogen.

873. Gadetskii, N. P., Tkach, Yu V., Bessarab, Ya. Ya., Sidel'nikova, A. V., *Sov. J. Quantum Electron.*, 3, 168-169 (1973), Stimulated emission of visible light due to transitions in singly ionized chlorine and iodine atoms.

874. Gallardo, M., Garavaglia, M., Tagliaferri, A. A., Gallego Lluesma, E., *IEEE J. Quantum Electron.*, QE-6, 745-747 (1970), About unidentified ionized Xe laser lines.

875. Gallardo, M., Massone, C. A., Tagliaferri, A. A., Garavaglia, M., *Phys. Scripta*, 19, 538-544 (1979), $5s^25p^3(^4S)nl$ Levels of Xe III.

876. Gerritsen, H. J., Goedertier, P. V., *J. Appl. Phys.*, 35, 3060-3061 (1964), Blue gas laser using Hg^{2+}.

877. Birnbaum, M., Stocker, T. L., unpublished (1965), private communication.

878. Gerstenberger, D. C., Reid, R. D., Collins, G. J., *Appl. Phys. Lett.*, 30, 466-468 (1977), Hollow-cathode aluminum ion laser.

879. Gill, P., Webb, C. E., *J. Phys. D.*, 11, 245-254 (1978), Radial profiles of excited ions and electron density in the hollow cathode He/Zn laser.

880. Gilles, M., *Ann. Phys.*, 15, 267-410 (1931), Recherches sur la Structure des Spectres du Soufre.

881. Cem Gokay, M., Soltanolkotabi, M., Cross, L. A., *IEEE J. Quantum Electron.*, QE-14, 1004-1007 (1978), Single- and double-pulse experiments on the Sr⁺ cyclic ion laser.

882. Goldsborough, J. P., Hodges, E. B., Bell, W. E., *Appl. Phys. Lett.*, 8, 137-139 (1966), RF induction excitation of CW visible laser transitions in ionized gases.

883. Lacy, R. A., Byer, R. L., Silfvast, W. T., Wood II, O. R., Svanberg, S., in *Short Wavelength Coherent Radiation: Generation and Applications*, AIP Conf. Proc. 147, Attwood, D. T., Bokor, J., Eds., AIP, New York (1986), p. 96., Optical gain at 185 nm in a laser ablated, inner-shell ionization-pumped indium plasma.

884. Goldsborough, J. P., Bloom, A. L., *IEEE J. Quantum Electron.*, QE-3, 96 (1) (1967), New CW ion laser oscillation in microwave-excited xenon.

885. Goldsborough, J. P., Hodges, E. B., *IEEE J. Quantum Electron.*, QE-5, 361-367 (1969), Stable long-life operation of helium-cadmium lasers at 4416 Å and 3250 Å.

886. Goldsborough, J. P., Bloom, A. L., *IEEE J. Quantum Electron.*, QE-5, 459-460 (1969), Near-infrared operating characteristics of the mercury ion laser.

887. Goldsmith, S., Kaufman, A. S., *Proc. Phys. Soc.*, 81, 544-552 (1963), The spectra of Ne IV, Ne V, and Ne VI: a further analysis.

888. Gordon, E. I., Labuda, E. F., Bridges, W. B., *Appl. Phys. Lett.*, 4, 178-180 (1964), Continuous visible laser action in singly ionized argon, krypton, and xenon.

889. Gordon, E. I., Labuda, E. F., Miller, R. C., Webb, C. E., *Proc. Phys. of QE Conf.*, 664-673 (1966), McGraw-Hill, New York, Excitation mechanisms of the argon-ion laser.

890. Gorog, I., Spong, F. W., *Appl. Phys. Lett.*, 9, 61-63 (1966), High pressure, high magnetic field effects in continuous argon ion lasers.

891. Goto, T., Kano, H., Yoshino, N., Mizeraczyk, J. K., Hattori, S., *J. Phys. B*, 10, 292-295 (1977), Construction of a practical sealed-off He-I$^+$ laser device.

892. Kapteyn, H. C., Falcone, R. W., *Phys. Rev. A* 37, 2033 (1988), Auger-pumped short-wavelength lasers in xenon and krypton.

893. Green, J. M., Collins, G. J., Webb, C. E., *J. Phys. B*, 6, 1545-1555 (1973), Collisional excitation and destruction of excited Zn II levels in a helium afterglow.

894. Green, J. M., Webb, C. E., *J. Phys. B*, 7, 1698-1711 (1974), The production of excited metal ions in thermal energy charge transfer and Penning reactions.

895. Green, W. R., Falcone, R. W., *Opt. Lett.*, 2, 115-116 (1978), Inversion of the resonance line of Sr$^+$ produced by optically pumped Sr atoms.

896. Sher, M. H., Macklin, J. J., Young, J. F., Harris, S. E., *Optics Lett.*, 12, 89 (1987), Saturation of the Xe III 109-nm laser using traveling-wave laser-produced-plasma, excitation.

897. Grozeva, M. G., Sabotinov, N. V., Vuchkov, N. K., *Opt. Commun.*, 29, 339-340 (1979), CW laser generation on Tl II in a hollow-cathode Ne-Tl discharge.

898. Grozeva, M. G., Sabotinov, N. V., Telbizov, P. K., Vuchkov, N. K., *Opt. Commun.*, 31, 211-213 (1979), CW laser oscillation on transitions of Tl in a hollow-cathode Ne-Tl halide discharge.

899. Gunderson, M., Harper, C. D., *IEEE J. Quantum Electron.*, QE-9, 1160 (1973), A high-power pulsed xenon ion laser.

900. Hagen, L., Martin, W. C., Special Publication 363(1972), *Bibliography on Atomic Energy Levels and Spectra*, July 1968 through June 1971, U. S. National Bureau of Standards.

901. Hagen, L., Special Publication 363, Supplement 1(1977), *Bibliography on Atomic Energy Levels and Spectra*, July 1971 through June 1975, U. S. National Bureau of Standards.

902. Hallin, R., *Ark. Fys.*, 32, 201-210 (1966), The spectrum of N IV.

903. Barty, C. P. J., King, D. A., Yin, G. Y., Hahn, K. H., Field, J. E., Young, J. F., Harris, S. E., *Phys. Rev. Lett.*, 61, 2201 (1988), 12.8-eV laser in neutral cesium.

904. Handrup, M. B., Mack, J. E., *Physica*, 30, 1245-1275 (1964), On the spectrum of ionized tellurium, Te II.

905. Tachikawa, M., Evenson, K. M., *Opt. Lett.*, 21, 1247 (1996), Sequential optical pumping of a far-infrared ammonia laser.

906. Harper, C. D., Gunderson, M., *Rev. Sci. Instrum.*, 45, 400-402 (1974), Construction of a high power xenon ion laser.

907. Hashino, Y., Katsuyama, Y., Fukuda, K., *Jpn. J. Appl. Phys.*, 11, 907 (1972), Laser oscillation of multiply ionized Ne, Ar, and N ions in a Z-pinch discharge.

908. Hashino, Y., Katsuyama, Y., Fukuda, K., *Jpn. J. Appl. Phys.*, 12, 470 (1973), Laser oscillation of O V in Z-pinch discharge.

909. Hattori, S., Kano, H., Tokutome, K., Collins, G. J., Goto, T., *IEEE J. Quantum Electron.*, QE-10, 530-531 (1974), CW iodine-ion laser in a positive-column discharge.

910. Hattori, S., Kano, H., Goto, T., *IEEE J. Quantum Electron.*, QE-10, 739-740 (1974), A continuous positive-column He-I$^+$ laser using a sealed-off tube.

911. Heard, H. G., Peterson, J., *Proc. IEEE*, 52, 1050 (1964), Orange through blue-green transitions in a pulsed-CW xenon gas laser.

912. Hernqvist, K. G., Pultorak, D. C., *Rev. Sci. Instrum.*, 41, 696-697 (1970), Simplified construction and processing of a helium-cadmium laser.
913. Bell, W. E., unpublished (1964), private communication.
914. Hernqvist, K. G., *Appl. Phys. Lett.*, 16, 464-466 (1970), Stabilization of He-Cd laser.
915. Hernqvist, K. G., Pultorak, D. C., *Rev. Sci. Instrum.*, 43, 290-292 (1972), Study of He-Se laser performance.
916. Hernqvist, K. G., *IEEE J. Quantum Electron.*, QE-8, 740-743 (1972), He-Cd lasers using recirculation geometry.
917. Hernqvist, K. G., *RCA Rev.*, 34, 401-407 (1973), Vented-bore He-Cd lasers.
918. Hernqvist, K. G., *IEEE J. Quantum Electron.*, QE-13, 929 (1977), Continuous laser oscillation at 2703 A in copper ion.
919. Herziger, G., Seelig, W., *Z. Phys.*, 219, 5-31 (1969), Ionenlaser hober leistung.
920. Hodges, D. T., *Appl. Phys. Lett.*, 18, 454-456 (1971), CW laser oscillation in singly ionized magnesium.
921. Hodges, D. T., Tang, C. L., *IEEE J. Quantum Electron.*, QE-6, 757-758 (1970), New CW ion laser transitions in argon, krypton, and xenon.
922. Hoffmann, V., Toschek, P., *IEEE J. Quantum Electron.*, QE-6, 757 (1970), New laser emission from ionized xenon.
923. Hoffmann, V., Toschek, P. E., *J. Opt. Soc. Am.*, 66, 152-154 (1976), On the ionic assignment of xenon laser lines.
924. Bell, W. E., *Appl. Phys. Lett.*, 4, 34-35 (1964), Visible laser transitions in Hg^+.
925. Humphreys, C. J., Meggers, W. F., de Bruin, T. L., *J. Res. Natl. Bur. Stand.*, 23, 683-699 (1939), Zeeman effect in the second and third spectra of xenon.
926. Humphreys, C. J., unpublished, Line list for ionized krypton.
927. McFarlane, R. A., unpublished (private communication).
928. Iijima, T., Sugawara, Y., *J. Appl. Phys.*, 45, 5091-5092 (1974), New CW laser oscillation in He-Zn hollow cathode laser.
929. Illingworth, R., *Appl. Phys.*, 3, 924-930 (1970), Laser action and plasma properties of an argon Z-pinch discharge.
930. Isaev, A. A., Petrash, G. G., *J. Appl. Spectrosc. (USSR)*, 12, 835-837 (1970), New superradiance on the violet line of the mercury ion.
931. Ivanov, I. G., Sem, M. F., *J. Appl. Spectrosc. (USSR)*, 19, 1092-1093 (1973), New lasing lines in thallium.
932. Ivanov, I. G., Il'yushko, V. G., Sem, M. F., *Sov. J. Quantum Electron.*, 4, 589-593 (1974), Dependences of the gain of cataphoretic lasers on helium pressure and discharge-tube diameter.
933. Bell, W. E., *Appl. Phys. Lett.*, 7, 190-191 (1965), Ring discharge excitation of gas ion lasers.
934. Jain, K., *Opt. Commun.*, 28, 207-208 (1979), Cw laser oscillation at 8096 Å in Cu II in a hollow cathode discharge.
935. Jain, K., *Appl. Phys. Lett.*, 34, 398-399 (1979), New ion laser transitions in copper, silver, and gold.
936. Jain, K., *Appl. Phys. Lett.*, 34, 845-846 (1979), A nickel-ion laser.
937. Jain, K., *Appl. Phys. Lett.*, 36, 10-11 (1980), A milliwatt-level cw laser source at 224 nm.
938. Jain, K., *IEEE J. Quantum Electron.*, QE-16, 387-388 (1980), New UV and IR transitions in gold, copper, and cadmium hollow cathode lasers.
939. Jain, K., *Appl. Phys. Lett.*, 37, 362-364 (1980), Laser action in chromium vapor.
940. Janossy, M., Csillag, L., Rozsa, K., Salamon, T., *Phys. Lett.*, 46A, 379-380 (1974), CW laser oscillation in a hollow cathode He-Kr discharge.
941. Janossy, M., Rozsa, K., Csillag, L., Bergou, J., *Phys. Lett.*, 68A, 317-318 (1978), New cw laser lines in a noble gas mixture high voltage hollow cathode discharge.

942.   Jarrett, S. M., Barker, G. C., *IEEE J. Quantum Electron.*, QE-5, 166 (1969), High-power output at 5353 Å from a pulsed xenon ion laser.

943.   Bell, W. E., Bloom, A. L., Goldsborough, J. P., *IEEE J. Quantum Electron.*, QE-1, 400 (1965), Visible laser transitions in ionized selenium, arsenic, and bromine.

944.   Jensen, R. C., Bennett, W. R., Jr., *IEEE J. Quantum Electron.*, QE-4, 356 (1968), Role of charge exchange in the zinc ion laser.

945.   Jensen, R. C., Collins, G. J., Bennett, W. R., Jr., *Phys. Rev. Lett.*, 23, 363-367 (1969), Charge-exchange excitation and cw oscillation in the zinc-ion laser.

946.   Jensen, R. C., Collins, G. J., Bennett, W. R., Jr., *Appl. Phys. Lett.*, 18, 50-51 (1971), Low-noise CW hollow-cathode zinc-ion laser.

947.   Johnson, A. M., Webb, C. E., *IEEE J. Quantum Electron.*, QE-3, 369 (1967), New CW laser wavelength in Kr II.

948.   Johnson, W. L., McNeil, J. R., Collins, G. J., Persson, K. B., *Appl. Phys. Lett.*, 29, 101-102 (1976), CW laser action in the blue-green spectral region from Ag II.

949.   Johnston, T. F., Jr., Kolb, W. P., *IEEE J. Quantum Electron.*, QE-12, 482-493 (1976), The self-heated 442-nm He-Cd laser: optimizing the power output, and the origin of beam noise.

950.   Kano, H., Goto, T., Hattori, S., *IEEE J. Quantum Electron.*, QE-9, 776-778 (1973), Electron temperature and density in the He-CdI$_2$ positive column used for an I$^+$ laser.

951.   Kano, H., Goto, T., Hattori, S., *J. Phys. Soc. Jpn.*, 38, 596 (1975), CW laser oscillation of visible and near-infrared Hg(II) lines in a He-Hg positive, column discharge.

952.   Kano, H., Shay, T., Collins, G. J., *Appl. Phys. Lett.*, 27, 610-612 (1975), A second look at the excitation mechanism of the He-Hg$^+$ laser.

953.   Karabut, E. K., Mikhalevskii, V. S., Papakin, V. F., Sem, M. F., *Sov. Phys. Tech. Phys.*, 14, 1447-1448 (1970), Continuous generation of coherent radiation in a discharge in Zn and Cd vapors obtained by cathode sputtering.

954.   Karabut, E. K., Kravchenko, V. F., Papakin, V. F., *J. Appl. Spectrosc. (USSR)*, 19, 938-939 (1973), Excitation of the Ag II Lines by a pulsed discharge in a mixture of silver vapor and helium.

955.   Keeffe, W. M., Graham, W. J., *Appl. Phys. Lett.*, 7, 263-264 (1965), Laser oscillation in the visible spectrum of singly ionized pure bromine vapor.

956.   Keeffe, W. M., Graham, W. J., *Phys. Lett.*, 20, 643 (1966), Observation of new Br II laser transitions.

957.   Keiden, V. F., Mikhalevskii, V. S., *J. Appl. Spectrosc. (USSR)*, 9, 1154 (1968), Pulsed generation in bismuth vapor.

958.   Keiden, V. F., Mikhalevskii, V. S., Sem, M. F., *J. Appl. Spectrosc. (USSR)*, 15, 1089-1090 (1971), Generation from ionic transitions of selenium.

959.   Kiess, C. C., de Bruin, T. L., *J. Res. Natl. Bur. Stand.*, 23, 443-470 (1939), Second spectrum of chlorine and its structure.

960.   Kitaeva, V. F., Odintsov, A. N., Sobolev, N. N., *Sov. Phys. Usp.*, 12, 699-730 (1970), Continuously operating argon ion lasers.

961.   Klein, M. B., Ph.D. dissertation (1969), University of California, Radiation Trapping Processes in the Pulsed Ion Laser.

962.   Klein, M. B., *Appl. Phys. Lett.*, 17, 29-32 (1970), Time-resolved temperature measurements in the pulsed argon ion laser.

963.   Klein, M. B., Silfvast, W. T., *Appl. Phys. Lett.*, 18, 482-485 (1971), New CW laser transitions in Se II.

964.   Akirtava, O. S., Bogus, A. M., Dzhilkiya, V. L., Oleinik, Yu. M., *Sov. J. Quantum Electron.*, 3, 519-520 (1974), Quasicontinuous emission from ion lasers in electrodeless high-frequency discharges.

965. Bennett, W. R., Jr., Knutson, J. W., Jr., Mercer, G. N., Detch, J. L., *Appl. Phys. Lett.*, 4, 180-182 (1964), Super-radiance, excitation mechanisms, and quasi-cw oscillation in the visible Ar+ laser.

966. Kobayashi, S., Izawa, T., Kawamura, K., Kamiyama, M., *IEEE J. Quantum Electron.*, QE-2, 699-700 (1966), Characteristics of a pulsed Ar II ion laser using the external spark gap.

967. Kobayashi, S., Kurihara, K., Kamiyama, M., *Oyo Butsuri*, 38, 766-768 (1969), New laser oscillation in ionized xenon at 5592 Å.

968. Koval'chuk, V. M., Petrash, G. G., *JETP Lett.*, 4, 144-146 (1966), New generation lines of a pulsed iodine-vapor laser.

969. Kruithof, A. A., Penning, F. M., *Physica*, 4, 430-449 (1937), Determination of the Townsend ionization coefficient alpha for mixtures of neon and argon.

970. Kulagin, S. G., Likhachev, V. M., Markuzon, E. V., Rabinovich, M. S. et al., *JETP Lett.*, 3, 6-8 (1966), States with population inversion in a self-compressed discharge.

971. Labuda, E. F., Gordon, E. I., Miller, R. C., *IEEE J. Quantum Electron.*, QE-1, 273-279 (1965), Continuous-duty argon ion lasers.

972. Labuda, E. F., Webb, C. E., Miller, R. C., Gordon, E. I., *Bull. Am. Phys. Soc.*, 11, 497 (1966), A study of capillary discharges in noble gases at high current densities.

973. Labuda, E. F., Johnson, A. M., *IEEE J. Quantum Electron.*, QE-2, 700-701 (1966), Threshhold properties of continuous duty rare gas ion laser transitions.

974. Latimer, I. D., *Appl. Phys. Lett.*, 13, 333-335 (1968), High power quasi-CW ultra-violet ion laser.

975. Schneider, M., Evenson, K. M., Johns, J. W. C., *Opt. Lett.*, 21, 1038 (1996), Far-infrared continuous-wave laser emission from H_2O and from NH_3 optically pumped by a CO laser.

976. Latush, E. L., Sem, M. F., *Sov. J. Quantum Electron.*, 3, 216-219 (1973), Stimulated emission due to transitions in alkaline-earth metal ions.

977. Latush, E. L., Mikhalevskii, V. S., Sem, M. F., Tolmachev, G. N. et al., *JETP Lett.*, 24, 69-71 (1976), Metal-ion transition lasers with transverse HF excitation.

978. Laures, P., Dana, L., Frapard, C., *C.R. Acad. Sci. Paris*, 258, 6363-6365 (1964), Nouvelles transitions laser dans le domaine 0.43-0.52 μ obtenues a partir du spectre du krypton ionise.

979. Laures, P., Dana, L., Frapard, C., *C.R. Acad. Sci. Paris*, 259, 745-747 (1964), Nouvelles raies laser visibles dans le xenon ionise.

980. Levinson, G. R., Papulovskiy, V. F., Tychinskiy, V. P., *Radio Eng. Electron. Phys. (USSR)*, 13, 578-582 (1968), The mechanism of inversion of the populations of the various levels in multivalent argon ions.

981. Li, H., Andrew, K. L., *J. Opt. Soc. Am.*, 61, 96-109 (1971), First spark spectrum of arsenic.

982. Littlewood, I. M., Piper, J. A., Webb, C. E., *Opt. Commun.*, 16, 45-49 (1976), Excitation mechanisms in CW He-Hg lasers.

983. Lluesma, E. G., Tagliaferri, A. A., Massone, C. A., Garavaglia, A. et al., *J. Opt. Soc. Am.*, 63, 362-364 (1973), Ionic assignment of unidentified xenon laser lines.

984. Luthi, H. R., Seelig, W., Steinger, J., Lobsiger, W., *IEEE J. Quantum Electron.*, QE-13, 404-405 (1977), Continuous 40-W UV laser.

985. Luthi, H. R., Seelig, W., Steinger, J., *Appl. Phys. Lett.*, 31, 670-672 (1977), Power enhancement of continuous ultraviolet lasers.

986. Birnbaum, M., Tucker, A. W., Gelbwachs, J. A., Fincher, C. L., *IEEE J. Quantum Electron.*, QE-7, 208 (1971), New O II 6640 Å laser line.

987. Luthi, H. R., Steinger, J., *Opt. Commun.*, 27, 435-438 (1978), Continuous operation of a high power neon ion laser.

988. Manley, J. H., Duffendack, O. S., *Phys. Rev.*, 47, 56-61 (1935), Collisions of the second kind between magnesium and neon.

989. Marling, J. B., *IEEE J. Quantum Electron.*, QE-11, 822-834 (1975), Ultraviolet ion laser performance and spectroscopy. I. New strong noble-gas transitions below 2500 Å.

990. Marling, J. B., Lang, D. B., *Appl. Phys. Lett.*, 31, 181-184 (1977), Vacuum ultraviolet lasing from highly ionized noble gases.

991. Marling, J. B., *IEEE J. Quantum Electron.*, QE-14, 4-6 (1978), Ultraviolet ion laser performance and spectroscopy for sulfur, fluorine, chlorine, and bromine.

992. Martin, W. C., *J. Opt. Soc. Am.*, 49, 1071-1085 (1959), Atomic energy levels and spectra of neutral and singly ionized phosphorus (P I and P II).

993. Martin, W. C., Corliss, C. H., *J. Res. Natl. Bur. Stand.*, 64A, 443-477 (1960), The spectrum of singly ionized atomic iodine (I.II).

994. Martin, W. C., Kaufman, V., *J. Res. Natl. Bur. Stand.*, 74A, 11-22 (1970), New vacuum ultraviolet wavelengths and revised energy levels in the second spectrum, of zinc (Zn II).

995. Davies, P. B., Jones, H., *Appl. Phys.* 22, 53 (1980), New cw far infrared molecular lasers from ClO_2, HCCF, FCN, CH_3NC, CH_3F, and propynal.

996. Gerritsen, H. J., unpublished (1965), private communication.

997. McFarlane, R. A., *Appl. Opt.*, 3, 1196 (1964), Optical maser oscillation on iso-electronic transitions in Ar III and Cl II.

998. Davis, I. H., Pharaoh, K. I., Knight, D. J. E., *Int. J. Infrared Millimeter Waves*, 8, 765 (1987), Frequency measurements on far-infrared laser emissions of carbonyl fluoride (COF_2).

999. McNeil, J. R., Collins, G. J., Persson, K. B., Franzen, D. L., *Appl. Phys. Lett.*, 28, 207-209 (1976), Ultraviolet laser action from Cu II in the 2500 Å region.

1000. McNeil, J. R., Collins, G. J., *IEEE J. Quantum Electron.*, QE-12, 371-372 (1976), Additional ultraviolet laser transitions in Cu II.

1001. McNeil, J. R., Johnson, W. L., Collins, G. J., Persson, K. B., *Appl. Phys. Lett.*, 29, 172-174 (1976), Ultraviolet laser action in He-Ag and Ne-Ag mixtures.

1002. Meggers, W. F., *J. Res. Natl. Bur. Stand.*, 71A, 396-544 (1967), The second spectrum of ytterbium (Yb II).

1003. Petersen, A. B., *Handbook of Laser Science and Technology*, Vol. II, Gas Lasers, CRC Press, Boca Raton, FL (1991), p. 335, Ionized gas lasers.

1004. Minnhagen, L., Stigmark, L., *Ark. Fys.*, 13, 27-36 (1957), The excitation of ionic spectra by 100 kw high frequency pulses.

1005. Minnhagen, L., *Ark. Fys.*, 25, 203-284 (1963), The spectrum of singly ionized argon, Ar II.

1006. Ninomiya, H., Abe, M., Takashima, N., *Appl. Phys. Lett.*, 58, 18191 (1991), Laser action of optically pumped atomic vanadium vapor.

1007. Minnhagen, L., Strihed, H., Petersson, B., *Ark. Fys.*, 39, 471-493 (1969), Revised and extended analysis of singly ionized krypton, Kr II.

1008. Dyubko, S. F., Svich, V. A., Fesenko, L. D., *Sov. Tech. Phys. Lett*, 1, 192 (1975), Magnetically tunable submillimeter laser based on paramagnetic ClO_2.

1009. Moore, C. E., Merrill, P. W., *NSRDS-NBS*23(1968), Partial Grotrian Diagrams of AstroPhysical Interest, U. S. National Bureau of Standards.

1010. Moskalenko, V. F., Ostapchenko, E. P., Perchurina, S. V., Stepanov, V. A, *Opt. Spectrosc.*, 30, 201-202 (1971), Radiation of a pulsed ion laser.

1011. Myers, R. A., Wieder, H., Pole, R. V., *IEEE J. Quantum Electron.*, QE-2, 270-275 (1966), 9A5 - Wide field active imaging.

1012. Neusel, R. H., *IEEE J. Quantum Electron.*, QE-2, 70 (1966), A new xenon laser oscillation at 5401 Å.

1013. Neusel, R. H., *IEEE J. Quantum Electron.*, QE-2, 106 (1966), A new krypton laser oscillation at 5016.4 Å.

1014. Neusel, R. H., *IEEE J. Quantum Electron.*, QE-2, 334 (1) (1966), New laser oscillations in krypton and xenon.

1015. Neusel, R. H., *IEEE J. Quantum Electron.*, QE-2, 758 (1966), New laser oscillations in xenon and krypton.

1016. Neusel, R. H., *IEEE J. Quantum Electron.*, QE-3, 207-208 (1967), New laser oscillations in Ar, Kr, Xe, and N.

1017. Olme, A., *Phys. Scripta*, 1, 256-260 (1970), The spectrum of singly ionized boron, B II.

1018. Paananen, R., *Appl. Phys. Lett.*, 9, 34-35 (1966), Continuously operated ultraviolet laser.

1019. Pacheva, Y., Stefanova, M., Pramatarov, P., *Opt. Commun.*, 27, 121-122 (1978), Cw laser oscillations on the Kr II 4694 Å and Kr II 4318 Å lines in a hollow-cathode, He-Kr discharge.

1020. Palenius, H. P., *Appl. Phys. Lett.*, 8, 82 (1966), The identification of some Si and Cl laser lines observed by Cheo and Cooper.

1021. Palenius, H. P., *Ark. Fys.*, 39, 15-64 (1969), Spectrum and term system of singly ionized fluorine, F II.

1022. Palenius, H. P., *Phys. Scripta*, 1, 113-135 (1970), Spectrum and term system of doubly ionized fluorine, F III.

1023. Papakin, V. F., Sem, M. F., *Sov. Phys. J.*, 13, 230-231 (1970), Use of isotopes in cadmium and zinc vapor lasers.

1024. Papayoanou, A., Buser, R. G., Gumeiner, I. M., *IEEE J. Quantum Electron.*, QE-9, 580-585 (1973), Parameters in a dynamically compressed xenon plasma laser.

1025. Papayoanou, A., Gumeiner, I., *Appl. Phys. Lett.*, 16, 5-8 (1970), High power xenon laser action in high current pinched discharges.

1026. Pappalardo, R., *J. Appl. Phys.*, 45, 3547-3553 (1974), Observation of afterglow character and high gain in the laser lines of O(III).

1027. Bloom, A. L., Goldsborough, J. P., *IEEE J. Quantum Electron.*, QE-6, 164 (1970), New CW laser transitions in cadmium and zinc ion.

1028. Pappalardo, R., *IEEE J. Quantum Electron.*, QE-10, 897-898 (1974), Some observations on multiple ionized xenon laser lines.

1029. Penning, F. M., *Physica*, 1, 1028-1044 (1934), The starting potential of the glow discharge in neon-argon mixtures between large parallel plates.

1030. Persson, W., *Phys. Scripta*, 3, 133-155 (1971), The spectrum of singly ionized neon, Ne II.

1031. Petersen, A. B., Birnbaum, M., *IEEE J. Quantum Electron.*, QE-10, 468 (1974), The singly ionized carbon laser at 6783, 6578, and 5145 Å.

1032. Piper, J. A., Collins, G. J., Webb, C. E., *Appl. Phys. Lett.*, 21, 203-205 (1972), CW laser oscillation in singly ionized iodine.

1033. Piper, J. A., Webb, C. E., *J. Phys. B*, 6, L116-L120 (1973), Continuous-wave laser oscillation in singly ionized arsenic.

1034. Piper, J. A., Webb, C. E., *J. Phys. D*, 6, 400-407 (1973), A hollow cathode device for CW helium-metal vapor laser systems.

1035. Piper, J. A., Webb, C. E., *Opt. Commun.*, 13, 122-125 (1975), Power limitations of the CW He-Hg laser.

1036. Piper, J. A., Gill, P., *J. Phys. D*, 8, 127-134 (1975), Output characteristics of the He-Zn laser.

1037. Piper, J. A., Webb, C. E., *IEEE J. Quantum Electron.*, QE-12, 21-25 (1976), High-current characteristics of the continuous-wave hollow-cathode He-I_2 laser.

1038.   Bockasten, K., *Ark. Fys.*, 9, 457-481 (1955), A study of C III by means of a sliding vacuum spark.

1039.   Piper, J. A., Brandt, M., *J. Appl. Phys.*, 48, 4486-4494 (1977), Cw laser oscillation on transitions of Cd^+ and Zn^+ in He-Cd-halide and He-Zn-halide discharges.

1040.   Piper, J. A., Neely, D. F., *Appl. Phys. Lett.*, 33, 621-623 (1978), Cw laser oscillation on transitions of Cu^+ in He-Cu-halide gas discharges.

1041.   Piper, J. A., *Opt. Commun.*, 31, 374-376 (1979), CW laser oscillation on transitions of As^+ in $He-AsH_3$ gas discharges.

1042.   Pugnin, V. I., Rudelev, S. A., Stepanov, A. F., *J. Appl. Spectrosc. (USSR)*, 18, 667-668 (1973), Laser generation with ionic transitions of iodine.

1043.   Rao, Y. B., *Indian J. Phys.*, 32, 497-515 (1958), Structure of the spectrum of singly ionized bromine.

1044.   Reid, R. D., Johnson, W. L., McNeil, J. R., Collins, G. J., *IEEE J. Quantum Electron.*, QE-12, 778-779 (1976), New infrared laser transitions in Ag II.

1045.   Reid, R. D., Gerstenberger, D. C., McNeil, J. R., Collins, G. J., *J. Appl. Phys.*, 48, 3994 (1977), Investigations of unidentified laser transitions in Ag II.

1046.   Risberg, P., *Ark. Fys.*, 9, 483-494 (1955), The spectrum of singly ionized magnesium, Mg II.

1047.   Riseberg, L. A., Schearer, L. B., *IEEE J. Quantum Electron.*, QE-7, 40-41 (1971), On the excitation mechanism of the He-Zn laser.

1048.   Bockasten, K., *Ark. Fys.*, 10, 567-582 (1956), A study of C IV: term values, series formulae, and stark effect.

1049.   Rozsa, K., Janossy, M., Bergou, J., Csillag, L., *Opt. Commun.*, 23, 15-18 (1977), Noble gas mixture CW hollow cathode laser with internal anode system.

1050.   Rozsa, K., Janossy, M., Csillag, L., Bergou, J., *Phys. Lett.*, 63A, 231-232 (1977), CW aluminum ion laser in a high voltage hollow cathode discharge.

1051.   Rozsa, K., Janossy, M., Csillag, L., Bergou, J., *Opt. Commun.*, 23, 162-164 (1977), CW Cu II laser in a hollow anode-cathode discharge.

1052.   Rudko, R. I., Tang, C. L., *Appl. Phys. Lett.*, 9, 41-44 (1966), Effects of cascade in the excitation of the Ar II laser.

1053.   Rudko, R. I., Tang, C. L., *J. Appl. Phys.*, 38, 4731-4739 (1967), Spectroscopic studies of the Ar^+ laser.

1054.   Russell, H. N., Albertson, W., Davis, D. N., *Phys. Rev.*, 60, 641-656 (1941), The spark spectrum of europium, Eu II.

1055.   Schearer, L. D., Padovani, F. A., *J. Chem. Phys.*, 52, 1618-1619 (1970), De-excitation cross section of metastable helium by Penning collisions with cadmium atoms.

1056.   Schearer, L. D., *IEEE J. Quantum Electron.*, QE-11, 935-937 (1975), A high-power pulsed xenon ion laser as a pump source for a tunable dye laser.

1057.   Schuebel, W. K., *Appl. Phys. Lett.*, 16, 470-472 (1970), New cw Cd-vapor laser transitions in a hollow-cathode structure.

1058.   Schuebel, W. K., *IEEE J. Quantum Electron.*, QE-6, 574-575 (1970), Transverse-discharge slotted hollow-cathode laser.

1059.   Schuebel, W. K., *IEEE J. Quantum Electron.*, QE-6, 654-655 (1970), New CW laser transitions in singly-ionized cadmium and zinc.

1060.   Schuebel, W. K., *IEEE J. Quantum Electron.*, QE-7, 39-40 (1971), Continuous visible and near-infrared laser action in Hg II.

1061.   Shay, T., Kano, H., Hattori, S., Collins, G. J., *J. Appl. Phys.*, 48, 4449-4453 (1977), Time-resolved double-probe study in a He-Hg afterglow plasma.

1062.   Shenstone, A. G., *Proc. Roy. Soc.*, A261, 153-174 (1961), The second spectrum of silicon.

1063. Shenstone, A. G., *Proc. Roy. Soc.*, A276, 293-307 (1963), The second spectrum of germanium.

1064. Silfvast, W. T., Fowles, G.R., Hopkins, B. D., *Appl. Phys. Lett.*, 8, 318-319 (1966), Laser action in singly ionized Ge, Sn, Pb, In, Cd, and Zn.

1065. Silfvast, W. T., *Appl. Phys. Lett.*, 13, 169-171 (1968), Efficient cw laser oscillation at 4416 Å in Cd(II).

1066. Silfvast, W. T., Szeto, L. H., *Appl. Opt.*, 9, 1484-1485 (1970), A simple high temperature system for CW metal vapor lasers.

1067. Aleinikov, V. S., *Opt. Spectra*, 28, 15-17 (1970), Use of an electron gun to determine the nature of collisions of the second kind in a mercury-helium mixture.

1068. Boersch, H., Boscher, J., Hoder, D., Schafer, G., *Phys. Lett.*, 31A, 188-189 (1970), Saturation of laser power of CW ion laser with large bored tubes and high power CW UV.

1069. Silfvast, W. T., Klein, M. B., *Appl. Phys. Lett.*, 17, 400-403 (1970), CW laser action on 24 visible wavelengths in Se II.

1070. Silfvast, W. T., Szeto, L. H., *Appl. Phys. Lett.*, 19, 445-447 (1971), Simplified low-noise He-Cd laser with segmented bore.

1071. Silfvast, W. T., *Phys. Rev. Lett.*, 27, 1489-1492 (1971), Penning ionization in a He-Cd DC discharge.

1072. Silfvast, W. T., unpublished (1971), private communication.

1073. Silfvast, W. T., Klein, M. B., *Appl. Phys. Lett.*, 20, 501-504 (1972), CW laser action on 31 transitions in tellurium vapor.

1074. Simmons, W. W., Witte, R. S., *IEEE J. Quantum Electron.*, QE-6, 466-469 (1970), High power pulsed xenon ion lasers.

1075. Sinclair, D. C., *J. Opt. Soc. Am.*, 55, 571-572 (1965), Near-infrared oscillation in pulsed noble-gas-ion lasers.

1076. Sinclair, D. C., *J. Opt. Soc. Am.*, 56, 1727-1731 (1966), Polarization characteristics of an ionized-gas laser in a magnetic field.

1077. Solanki, R., Latush, E. L., Gerstenberger, D. C., Fairbank, W. M., Jr., *Appl. Phys. Lett.*, 35, 317-319 (1979), Hollow-cathode excitation of ion laser transitions in noble-gas mixtures.

1078. Bokhan, P. A., Klimkin, V. M., Prokop'ev, V. E., *JETP Lett.*, 18, 44-45 (1973), Gas laser using ionized europium.

1079. Sosnowski, T. P., *J. Appl. Phys.*, 40, 5138-5144 (1969), Cataphoresis in the helium-cadmium laser discharge tube.

1080. Sosnowski, T. P., Klein, M. B., *IEEE J. Quantum Electron.*, QE-7, 425-426 (1971), Helium cleanup in the helium-cadmium laser discharge.

1081. Sattler, J. P., Worchesky, T. L., Tobin, M. S., Ritter, K. J., Daley, T. W., *Int. J. Infrared Millimeter Waves*, 1, 127 (1980), Submillimeter emission assignments for 1,1-difluoroethylene.

1082. Sugawara, Y., Tokiwa, Y., *Technology Reports of Seikei University*, 9 (1970), Seikei University, CW hollow cathode laser oscillation in Zn^+ and Cd^+.

1083. Sugawara, Y., Tokiwa, Y., *Jpn. J. Appl. Phys.*, 9, 588-589 (1970), CW laser oscillations in Zn II and Cd II in hollow cathode discharges.

1084. Sugawara, Y., Tokiwa, Y., Iijima, T., *Int. Quantum Electronics Conf.*, 320-321(1970), Excitation mechanisms of CW laser oscillations in Zn II and Cd II in hollow cathode.

1085. Sugawara, Y., Tokiwa, Y., Iijima, T., *Jpn. J. Appl. Phys.*, 9, 1537 (1970), New CW laser oscillations in Cd-He and Zn-He hollow cathode lasers.

1086. Szeto, L. H., Silfvast, W. T., Wood, O. R., II, *IEEE J. Quantum Electron.*, QE-15, 1332-1334 (1979), High-gain laser in Pb^+ populated by direct electron excitation from the neutral ground states.

1087.   Bokhan, P. A., Klimkin, V. M., Prokop'ev, V. E., *Sov. J. Quantum Electron.*, 4, 752-754 (1974), Collision gas-discharge laser utilizing europium vapor. I. Observation of self-terminating oscillations and transition from cyclic to quasicontinuous.

1088.   Tio, T. K., Luo, H. H., Lin, S.-C., *Appl. Phys. Lett.*, 29, 795-797 (1976), High cw power ultraviolet generation from wall-confined noble gas ion lasers.

1089.   Tolkachev, V. A., *J. Appl. Spectrosc. (USSR)*, 8, 449-451 (1968), Super-radiant transitions in Ar and Kr.

1090.   Toresson, Y. G., *Ark. Fys.*, 17, 179-192 (1959), Spectrum and term system of trebly ionized silicon, Si IV.

1091.   Toresson, Y. G., *Ark. Fys.*, 18, 389-416 (1960), Spectrum and term system of doubly ionized silicon, Si III.

1092.   Tucker, A. W., Birnbaum, M., *IEEE J. Quantum Electron.*, QE-10, 99-100 (1974), Pulsed-ion laser performance in nitrogen, oxygen, krypton, xenon, and argon.

1093.   Turner-Smith, A. R., Green, J. M., Webb, C. E., *J. Phys. B*, 6, 114-130 (1973), Charge transfer into excited states in thermal energy collisions.

1094.   Vuchkov, N. K., Sabotinov, N. V., *Opt. Commun.*, 25, 199-200 (1978), Pulse generation of Cl II in vapours of CuCl and $FeCl_2$.

1095.   Wang, C. P., Lin, S.-C., *J. Appl. Phys.*, 43, 5068-5073 (1972), Experimental study of argon ion laser discharge at high current.

1096.   Bridges, W. B., *Appl. Phys. Lett.*, 4, 128-130 (1964), Laser oscillation in singly ionized argon in the visible spectrum.

1097.   Wang, C. P., Lin, S.-C., *J. Appl. Phys.*, 44, 4681-4682 (1973), Performance of a large-bore high-power argon ion laser.

1098.   Watanabe, S., Chihara, M., Ogura, I., *Jpn. J. Appl. Phys.*, 11, 600 (1972), New continuous oscillation at 5700 Å in He-Te laser.

1099.   Waynant, R. W., *Appl. Phys. Lett.*, 22, 419-420 (1973), Vacuum ultraviolet laser emission from C IV.

1100.   Webb, C. E., Turner-Smith, A. R., Green, J. M., *J. Phys. B*, 3, L134-L138 (1970), Optical excitation in charge transfer and Penning ionization.

1101.   Wheeler, J. P., *IEEE J. Quantum Electron.*, QE-7, 429 (1971), New xenon laser line observed.

1102.   Wieder, H., Myers, R. A., Fisher, C. L., Powell, C. G., Colombo, J., *Rev. Sci. Instrum.*, 38, 1538-1546 (1967), Fabrication of wide bore hollow cathode Hg^+ lasers.

1103.   Willett, C. S., Heavens, O. S., *Opt. Acta*, 13, 271-274 (1966), Laser transition at 651.6 nm in ionized iodine.

1104.   Willett, C. S., *IEEE J. Quantum Electron.*, QE-3, 33 (1967), New laser oscillations in singly-ionized iodine.

1105.   Willett, C. S., Heavens, O. S., *Opt. Acta*, 14, 195-197 (1967), Laser oscillation on hyperfine transitions in ionized iodine.

1106.   Willett, C. S., *IEEE J. Quantum Electron.*, QE-6, 469-471 (1970), Note on near-infrared operating characteristics of the mercury ion laser.

1107.   Zarowin, C. B., *Appl. Phys. Lett.*, 9, 241-242 (1966), New visible CW laser lines in singly ionized chlorine.

1108.   Zhukov, V. V., Il'yushko, V. G., Latush, E. L., Sem, M. F., *Sov. J. Quantum Electron.*, 5, 757-760 (1975), Pulse stimulated emission from beryllium vapor.

1109.   Zhukov, V. V., Kucherov, V. S., Latush, E. L., Sem, M. F., *Sov. J. Quantum Electron.*, 7, 708-714 (1977), Recombination lasers utilizing vapors of chemical elements II. Laser action due to transitions in metal ions.

1110.   Toyoda, K., Kobiyama, M., Namba, S., *Jpn. J. Appl. Phys.*, 15, 2033-2034 (1976), Laser oscillation at 5378 Å by laser-produced cadmium plasma.

1111. Bennett, J., Bennett, W. R., Jr., *IEEE J. Quantum Electron.*, QE-15, 842-843 (1979), CW oscillation on a new argon ion laser line at 5062 Å and relation to laser Raman spectroscopy.

1112. Kato, I., Nakaya, M., Satake, T., Shimizu, T., *Jpn. J. Appl. Phys.*, 14, 2001-2004 (1975), Output power characteristics of microwave-pulse-excited He-Kr$^+$ ion laser.

1113. Radford, H. E., *IEEE J. Quantum Electron.*, 11, 213 (1992), New cw lines from a submillimeter waveguide laser.

1114. Anon., unpublished, LGI-37 High Power Pulsed Optical Quantum Oscillator.

1115. Bridges, W. B., unpublished (1964), private communication.

1116. Bridges, W. B., *Proc. IEEE*, 52, 843-844 (1964), Laser action in singly ionized krypton and xenon.

1117. Bridges, W. B., Chester, A. N., *IEEE J. Quantum Electron.*, QE-1, 66-84 (1965), Spectroscopy of ion lasers.

1118. Bridges, W. B., Halsted, A. S., *IEEE J. Quantum Electron.*, QE-2, 84 (1966), New CW laser transitions in argon, krypton, and xenon.

1119. Bridges, W. B., Clark, P. O., Halsted, A. S., Tech. Rep. AFAL-TR-66-369, DDC No. AD-807363(1967), Hughes Research Laboratories, High power gas laser research.

1120. Alferov, G. N., Donin, V. I., Yurshin, B. Ya., *JETP Lett.*, 18, 369-370 (1973), CW argon laser with 0.5 kW output power.

1121. Bridges, W. B., Halsted, A. S., Tech. Rep. AFAL-TR-67-89; DDC No. AD-814897(1967), Hughes Research Laboratories, Gaseous Ion Laser Research.

1122. Bridges, W. B., Mercer, G. N., *IEEE J. Quantum Electron.*, QE-5, 476-477 (1969), CW operation of high ionization states in a xenon laser.

1123. Bridges, W. B., Mercer, G. N., Tech. Rep. ECOM-0229-F, DDC No.AD-861927(1969), Hughes Research Laboratories, Ultraviolet Ion Laser.

1124. Bridges, W. B., Chester, A. N., unpublished data (1969 and 1970), private communication.

1125. Bridges, W. B., Chester, A. N., *Handbook of Lasers*, R. Pressley, Ed., CRC Press, Boca Raton, FL (1971), 242-297. Ionized gas lasers.

1126. Bridges, W. B., Chester, A. N., Halsted, A. S., Parker, J. V., *Proc. IEEE*, 59, 724-737 (1971), Ion laser plasmas.

1127. Bridges, W. B., Chester, A. N., *IEEE J. Quantum Electron.*, QE-7, 471-472 (1971), Comments on the identification of some xenon ion laser lines.

1128. Bridges, W. B., unpublished (1975), private communication.

1129. Bridges, W. B., *Methods of Experimental Physics*, Vol. 15A (1979), Academic Press, London, Atomic and ionic gas lasers.

1130. Bromander, J., *Ark. Fys.*, 40, 257-274 (1969), The spectrum of triply-ionized oxygen, O IV.

1131. Burkhard, P., Luthi, H. R., Seelig, W., *Opt. Commun.*, 18, 485-487 (1976), Quasi-CW laser action from Hg-III lines.

1132. Byer, R. L., Bell, W. E., Hodges, E., Bloom, A. L., *J. Opt. Soc. Am.*, 55, 1598-1602 (1965), Laser emission in ionized mercury: isotope shift, linewidth, and precise wavelength.

1133. Cahuzac, Ph., *J. Phys.*, 32, 499-505 (1971), Raies laser infrarouges dans les vapeurs de terres rares et D'Alcalino-Terreux.

1134. Carr, W. C., Grow, R. W., *Proc. IEEE*, 55, 726 (1967), Silicon and chlorine laser oscillations in SiCl$_4$.

1135. Chester, A. N., *Phys. Rev.*, 169, 172-184 (1968), Gas pumping in discharge tubes.

1136. Chester, A. N., *Phys. Rev.*, 169, 184-193 (1968), Experimental measurements of gas pumping in an argon discharge.

1137. Clunie, D. M., unpublished (1966), private communication.

1138. Collins, G. J., Ph.D. thesis (1970), Cw Oscillation and Charge Exchange Excitation in the Zinc-Ion Laser, Yale University.

1139. Collins, G. J., Jensen, R. C., Bennett, W. R., Jr., *Appl. Phys. Lett.*, 18, 282-284 (1971), Excitation mechanisms in the zinc ion laser.

1140. Collins, G. J., Jensen, R. C., Bennett, W. R., Jr., *Appl. Phys. Lett.*, 19, 125-130 (1971), Charge-exchange excitation in the He-Cd laser.

1141. Collins, G. J., Kuno, H., Hattori, S., Tokutome, K., Ishikawa, M., Kamid, *IEEE J. Quantum Electron.*, QE-8, 679-680 (1972), Cw laser oscillation at 6127 Å in singly ionized iodine.

1142. Convert, G., Armand, M., Martinot-Lagrade, P., *C.R. Acad. Sci. Paris*, 258, 3259-3260 (1964), Effet laser dans des melanges mercure-gaz rares.

1143. Wernsman, B., Prabhuram, T., Lewis, K., Gonzalez, F., Villagran, M., Rocca, J. J., *IEEE J. Quantum Electron.*, 24, 1554 (1988), CW silver ion laser with electron beam excitation.

1144. Landberg, B. M., Shafik, M. S., Butcher, J. R., CW optically pumped far-infrared mission from acetaldehyde, vinyl chloride, and methyl isocyanide, *IEEE J. Quantum Electron.*, 17, 828 (1992).

1145. Crooker, A. M., Dick, K. A., *Can. J. Phys.*, 46, 1241-1251 (1968), Extensions to the spark spectra of Zinc I, Zinc II and Zinc IV.

1146. Csillag, L., Janossy, M., Kantor, K., Rozsa, K., Salamon, T., *Appl. Phys.*, 3, 64-68 (1970), Investigation on a continuous wave 4416 Å cadmium ion laser.

1147. Csillag, L., Janossy, M., Salamon, T., *Phys. Lett.*, 31A, 532-533 (1970), Time delay of laser oscillation of the green transition of a pulsed He-Cd laser.

1148. Csillag, L., Itagi, V. V., Janossy, M., Rozsa, K., *Phys. Lett.*, 34A, 110-111 (1971), Laser oscillation at 4416 Å in a Ne-Cd discharge.

1149. Csillag, L., Janossy, M., Rozsa, K., Salamon, T., *Phys. Lett.*, 50A, 13-14 (1974), Near infrared Cw laser oscillation in Cu II.

1150. Dahlquist, J. A., *Appl. Phys. Lett.*, 6, 193-194 (1965), New line in a pulsed xenon laser.

1151. Dana, L., Laures, P., *Proc. IEEE*, 53, 78-79 (1965), Stimulated emission in krypton and xenon ions by collisions with metastable atoms.

1152. Kielkopf, J., *J. Opt. Soc. Am. B* 8, 212 (1991), Lasing in the aluminum and indium resonance lines following photoionization, and recombination in the presence of H_2.

1153. Dana, L., Laures, P., Rocherolles, R., *C. R. Acad. Sci. Paris*, 260, 481-484 (1965), Raies laser ultraviolettes dans le neon, l'argon, et le xenon.

1154. Davis, C. C., King, T. A., *Phys. Lett.*, 36A, 169-170 (1971), Time-resolved gain measurements and excitation mechanisms of the pulsed argon ion laser.

1155. Davis, C. C., King, T. A., *IEEE J. Quantum Electron.*, QE-8, 755-757 (1972), Laser action on unclassified xenon transitions in a highly ionized plasma.

1156. Davis, C. C., King, T. A., *Advances in Quantum Electronics*, Vol. 3(1975), 169, Academic Press, London, Gaseous ion lasers.

1157. deBruin, T. L., Humphreys, C. J., Meggers, W. F., *J. Res. Natl. Bur. Stand.*, 11, 409-440 (1933), The second spectrum of krypton.

1158. Demtroder, W., *Phys. Lett.*, 22, 436-438 (1966), Excitation mechanisms of pulsed argon ion lasers at 4880 Å.

1159. Donin, V. I., *Sov. Phys. JETP*, 35, 858-864 (1972), Output power saturation with a discharge current in powerful continuous argon lasers.

1160. Donin, V. I., Shipilov, A. F., Grigor'ev, V. A., *Sov. J. Quantum Electron.*, 9, 210-212 (1979), High-power cw ion lasers with an improved service life.

1161. Auschwitz, B., Eichler, H. J., Wittwer, W., *Appl. Phys. Lett.*, 36, 804-805 (1980), Extension of the operating period of an UV Cu II-laser by a mixture of argon.

1162. Duffendack, O. S., Thomson, K., *Phys. Rev.*, 43, 106-111 (1933), Some factors affecting action cross section for collisions of the second kind between atoms and ions.

1163. Duffendack, O. S., Gran, W. H., *Phys. Rev.*, 51, 804-809 (1937), Regularity along a series in the variation of the action cross section with energy discrepancy in impacts of the second kind.

1164. Dunn, M. H., Ross, J. N., *Progr. Quantum Electron.*, 4, 233-269 (1976), The argon ion laser.

1165. Dyson, D. J., *Nature (London)*, 207, 361-363 (1965), Mechanism of population inversion at 6149 Å in the mercury ion laser.

1166. Edlen, B., Risberg, P., *Ark. Fys.*, 10, 553-566 (1956), The spectrum of singly-ionized calcium, Ca II.

1167. Engelhard, E., Spieweck, F., *Z. Naturforsch.*, 25a, 156 (1969), Ein ionen-laser fur metrologische anwendungen mitteilung aus der physika, lisch-technischen bundesanstalt braunschweig.

1168. Fendley, J. R. Jr., Gorog, I., Hernqvist, K. G., Sun, C., *IEEE J. Quantum Electron.*, QE-6, 8 (1970), Characteristics of a sealed-off He^3-Cd^{114} laser.

1169. Fendley, J. R. Jr., *IEEE J. Quantum Electron.*, QE-4, 627-631 (1968), Continuous UV lasers.

1170. Eriksson, K. B. S., *Ark. Fys.*, 13, 303-329 (1958), The spectrum of the singly-ionized nitrogen atom.

1171. Tell, B., Martin, R. J., McNair, D., *IEEE J. Quantum Electron.*, QE-3, 96 (2) (1967), CW laser oscillation in ionized xenon at 9697 Å.

1172. Bridges, W. B., Freiberg, R. J., Halsted, A. S., *IEEE J. Quantum Electron.*, QE-3, 339 (2) (1967), New continuous UV ion laser transitions in neon, argon, and krypton.

1173. Allen, R. B., Starnes, R. B., Dougal, A. A., *IEEE J. Quantum Electron.*, QE-2, 334 (2) (1966), A new pulsed ion laser transition in nitrogen at 3995 Å.

1174. McNeil, J. R., Collins, G. J., Persson, K. B., Franzen, D. L., *Appl. Phys. Lett.*, 27, 595-598 (1975), CW laser oscillation in Cu II.

1175. Coleman, C. D., Bozman, W. R., Meggers, W. F., Monograph, No. 3(1960), *Table of Wavenumbers II*, U. S. National Bureau of Standards.

1176. Gregg, D. W., Thomas, S. J., *J. Appl. Phys.*, 39, 4399 (1968), Analysis of the CS_2-O_2 chemical laser showing new lines and selective excitation.

1177. Patel, C. K. N., *Appl. Phys. Lett.*, 6, 12 (1965), CW laser action in N_2O (N_2-N_2O system).

1178. Mathias, L. E. S., Crocker, A., Wills, M. S., *Phys. Lett.*, 13, 303 (1964), Laser oscillations from nitrous oxide at wavelengths around 10.9 micrometers.

1179. Siemsen, K., Reid, J., *Opt. Commun.*, 20, 284 (1977), New N_2O laser band in the 10-μm wavelength region.

1180. Tiee, J. J., Wittig, C., *Appl. Phys. Lett.*, 30, 420 (1977), CF_4 and NOCl molecular lasers operating in the 16-μm region.

1181. Tiee, J. J., Wittig, C., *J. Appl. Phys.*, 49, 61 (1978), Optically pumped molecular lasers in the 11- to 17-μm region.

1182. Karlov, N. V., Konev, Yu. B., Petrov, Yu. N., Prokhorov, A. M. et al., *JETP Lett.*, 8, 12 (1968), Laser based on boron trichloride.

1183. Akitt, D. P., Yardley, J. T., *IEEE J. Quantum Electron.*, QE-6, 113 (1970), Far-infrared laser emission in gas discharges containing boron trihalides.

1184. Shelton, C. F., Byrne, F. T., *Appl. Phys. Lett.*, 17, 436 (1970), Laser emission near 8 μ from a H_2-C_2H_2-He mixture.

1185. Kildal, H., Deutsch, T. F., *Appl. Phys. Lett.*, 27, 500 (1975), Optically pumped infrared V-V transfer lasers.

1186. Nelson, L. Y., Fisher, C. H., Hoverson, S. J., Byron, S. R., O'Neill, F., *Appl. Phys. Lett.*, 30, 192 (1977), Electron-beam-controlled discharge excitation of a CO-C_2H_2 energy transfer laser.

1187. Puerta, J., Herrmann, W., Bourauel, G., Urban, W., *Appl. Phys.*, 19, 439 (1979), Extended spectral distribution of lasing transitions in a liquid-nitrogen cooled CO-laser.

1188. Rutt, H. N., Green, J. M., *Opt. Commun.*, 26, 422 (1978), Optically pumped laser action in dideuteroacetylene.

1189. Fry, S. M., *Opt. Commun.*, 19, 320 (1976), Optically pumped multiline NH_3 laser.

1190. Yamabayashi, N., Yoshida, T., Myazaki, K., Fujisawa, K., *Opt. Commun.*, 30, 245 (1979), Infrared multiline NH_3 laser and its application for pumping an InSb laser.

1191. Tashiro, H., Suzuki, K., Toyoda, K., Namba, S., *Appl. Phys.*, 21, 237 (1980), Wide-range line-tunable oscillation of an optically pumped NH_3 laser.

1192. Grasiuk, A. Z., *Appl. Phys.*, 21, 173 (1980), High-power tunable ir Raman and optically pumped molecular lasers for spectroscopy.

1193. Danielewicz, E. J., Malk, E. G., Coleman, P. D., *Appl. Phys. Lett.*, 29, 557 (1976), High-power vibration-rotation emission from $^{14}NH_3$ optically pumped off resonance.

1194. Chang, T. Y., McGee, J. D., *Appl. Phys. Lett.*, 29, 725 (1976), Off-resonant infrared laser action in NH_3 and C_2H_4 without population inversion.

1195. Mochizuki, T., Yamanaka, M., Morikawa, M., Yamanaka, C., *Jpn. J. Appl. Phys.*, 17, 1295 (1978), unspecified.

1196. Bobrovskii, A. N., Vedenov, A. A., Kozhevnikov, A. V., Sobolenko, D. N., *JETP Lett.*, 29, 537 (1979), NH_3 laser pumped by two CO_2 lasers.

1197. Baranov, V. Yu., Kazakov, S. A., Pis'mennyi, V. D., Starodubtsev, A. I., *Appl. Phys.*, 17, 317 (1978), Multiwatt optically pumped ammonia laser operation in the 12- to 13-µm region.

1198. Roh, W. B., Rao, K. N., *J. Mol. Spectrosc.*, 49, 317 (1974), CO laser spectra.

1199. Yoshida, T., Yamabayashi, N., Miyazaki, K., Fujisawa, K., *Opt. Commun.*, 26, 410 (1978), Infrared and far-infrared laser emissions from a TE CO_2 laser pumped NH_3 gas.

1200. Chang, T. Y., McGee, J. D., *Appl. Phys. Lett.*, 28, 526 (1976), Laser action at 12.812 mm in optically pumped NH_3.

1201. Shaw, E. D., Patel, C. K. N., *Opt. Commun.*, 27, 419 (1978), Improved pumping geometry for high-power NH_3 lasers.

1202. Gupta, P. K., Kar, A. K., Taghizadeh, M. R., Harrison, R. G., *Appl. Phys. Lett.*, 39, 32 (1981), 12.8-m NH_3 laser emission with 40-60% power conversion and up to 28% energy conversion efficiency.

1203. Lee, W., Kim, D., Malk, E., Leap, J., *IEEE J. Quantum Electron.*, QE-15, 838 (1979), Hot-band lasing in NH_3.

1204. Jacobs, R. R., Prosnitz, D., Bischel, W. K., Rhodes, C. K., *Appl. Phys. Lett.*, 29, 710 (1976), Laser generation from 6 to 35 µm following two-photon excitation of ammonia.

1205. Eggleston, J., Dallarosa, J., Bischel, W. K., Bokor, J., Rhodes, C. K., *J. Appl. Phys.*, 50, 3867 (1979), Generation of 16-µm radiation in $^{14}NH_3$ by two-quantum excitation of the $2n^{-2}(7.5)$ state.

1206. Akitt, D. P., Wittig, C. F., *J. Appl. Phys.*, 40, 902 (1969), Laser emission in ammonia.

1207. Mathias, L. E. S., Crocker, A., Wills, M. S., *Phys. Lett.*, 14, 33 (1965), Laser oscillations at wavelengths between 21 and 32 µ from a pulsed discharge through ammonia.

1208. Lide, D. R., Jr., *Phys. Lett.*, A, 24, 599 (1967), Interpretation of the far-infrared laser oscillation in ammonia.

1209. Patel, C. K. N., Kerl, R. J., *Appl. Phys. Lett.*, 5, 81 (1964), Laser oscillation on X'S$^+$ vibrational rotational transitions of CO.

1210. Urban, S., Spirko, V., Papousek, D., McDowell, R S., Nereson, N. G. et al., *J. Mol. Spectrosc.* 79, 455 (1980), Coriolis and l-type interactions in the n_2, $2n_2$, and n_4 states of $^{14}NH_3$.

1211. Jones, C. R., Buchwald, M. I., Gundersen, M., Bushnell, A. H., *Opt. Commun.*, 24, 27 (1978), Ammonia laser optically pumped with an HF laser.

1212. McDowell, R S., Patterson, C. W., Jones, C. R., Buchwald, M. I., Telle, J. M., *Opt. Lett.*, 4, 274 (1979), Spectroscopy of the CF_4 laser.

1213. Averim, V. G., Alimpiev, S. S., Baronov, G. S., Karlov, N. V. et al., *Sov. Tech. Phys. Lett.*, 4, 527 (1978), Spectroscopic characteristics of an optically pumped carbon tetrafluoride laser.

1214. Alimpiev, S. S., Baronov, G. S., Karlov, N. V., Karchevskii, A. I. et al., *Sov. Tech. Phys. Lett.*, 4, 69 (1978), Tuning and stabilization of an optically pumped carbon tetrafluoride laser.

1215. Lomaev, M. I., Nagornyi, D. Yu., Tarasenko, V. F., Fedenev, A. V., Kirillin, G. V., *Sov. J. Quantum Electron.*, 19, 1321 (1989), Lasing due to atomic transitions of rare gases in mixtures with NF_3.

1216. Jones, C. R., Telle, J. M., Buchwald, M. I., *J. Opt. Soc. Am.*, 68, 671 (1978), Optically pumped isotopic CF_4 lasers.

1217. Knyazev, I. N., Letokhov, V. S., Lobko, V. V., *Opt. Commun.*, 29, 73 (1979), Weakly forbidden vibration-rotation transitions DR ≠ 0 in CF_4 laser.

1218. Prosnitz, D., Jacobs, R. R., Bischel, W. K., Rhodes, C. K., *Appl. Phys. Lett.*, 32, 221 (1978), Stimulated emission at 9.75 μm following two-photon excitation of methyl fluoride.

1219. Green, J. M., Rutt, H. N., *Proc. 2nd Int. Conf. on Infrared Physics*, 205 (Mar, 1979), Optically pumped laser action in silicon tetrafluoride.

1220. Patel, C. K. N., *Phys. Rev.*, 141, 71 (1966), Vibrational-rotational laser action in carbon monoxide.

1221. Barch, W. E., Fetterman, H. R., Schlossberg, H. R., *Opt. Commun.*, 15, 358 (1975), Optically pumped 15.90 μm SF_6 laser.

1222. Dunham, J. L., *Phys. Rev.*, 41, 721 (1932), The energy levels of a rotating vibrator.

1223. Osgood, R. M., Jr., Eppers, W. C., Jr., *Appl. Phys. Lett.*, 13, 409 (1968), High-power CO-N_2-He laser.

1224. Treanor, C. E., Rich, J. W., Rehm, R. G., *J. Chem. Phys.*, 48, 1798 (1968), unspecified.

1225. Jeffers, W. Q., Wiswall, C. E., *J. Appl. Phys.*, 42, 5059 (1971), Analysis of pulsed CO lasers.

1226. Rich, J. W., Thompson, H. M., *Appl. Phys. Lett.*, 19, 3 (1971), Infrared sidelight studies in the high-power carbon monoxide laser.

1227. Amat, G., *J. Chim. Phys.*, 64, 91 (1967), Discussion following a paper by C. K. N. Patel.

1228. Tyte, D. C., *Advances in Quantum Electronics*, Vol. 1 (1970), Academic Press, London, Carbon dioxide lasers.

1229. Cheo, P. K., *Lasers*, Vol. 3 (1971), Marcel Dekker, New York, CO_2 lasers.

1230. Patel, C. K. N., *Appl. Phys. Lett.*, 7, 246 (1965), CW laser on vibrational-rotational transitions of CO.

1231. Nigham, W. L., *Phys. Rev.* A, 2, 1989 (1970), Electron energy distribution and collision rates in electrically excited N_2, CO, and CO_2.

1232. Burkhardt, E. G., Bridges, T. J., Smith, P. W., *Opt. Commun.*, 6, 193 (1972), BeO capillary CO_2 waveguide laser.

1233. Wood, O. R., II, *Proc. IEEE*, 62, 355 (1974), High-pressure pulsed molecular lasers.

1234. Taylor, R. S., Alcock, A. J., Sarjeant, W. J., Leopold, K. E., *IEEE J. Quantum Electron.*, QE-15, 1131 (1979), Electrical and gain characteristics of a multiatmosphere uv-preionized CO_2 laser.

1235. Chang, T. Y., Wood, O. R., II, *IEEE J. Quantum Electron.*, QE-13, 907 (1977), Optically pumped continuously tunable high-pressure molecular lasers.

1236. Javan, A., *Phys. Rev.*, 107, 1579 (1957), Theory of a three-level maser.

1237. Siemsen, K. J., *Opt. Lett.*, 6, 114 (1981), Sequence bands of the isotope $^{12}C^{18}O_2$ laser.

1238. Siemsen, K. J., *Opt. Commun.*, 34, 447 (1980), The sequence bands of the carbon-13 isotope CO_2 laser.

1239. Fischer, T. A., Tiee, J. J., Wittig, C., *Appl. Phys. Lett.*, 37, 592 (1980), Optically pumped NSF molecular laser.

1240. Deka, B. K., Dyer, P. E., Winfield, R. J., *Opt. Lett.*, 5, 194 (1980), New 17-21-μm laser lines in C_2D_2 using a continuously tunable CO_2 laser pump.

1241. Bhaumik, M. L., *Appl. Phys. Lett.*, 17, 188 (1970), High efficiency CO laser at room temperature.

1242. Baldacci, A., Ghersetti, S., Hurlock, S. C., Rao, K. N., *J. Mol. Spectrosc.*, 42, 327 (1972), Spectrum of dideuteroacetylene near 18.6 microns.

1243. Znotins, T. A., Reid, J., Garside, B. K., Ballik, E. A., *Opt. Lett.*, 5, 528 (1980), 12-μm NH_3 laser pumped by a sequence CO_2 laser.

1244. Deka, B. K., Dyer, P. E., Winfield, R. J., *Opt. Commun.*, 33, 206 (1980), Optically pumped NH_3 laser using a continuously tunable CO_2 laser.

1245. Rutt, H. N., *Opt. Commun.*, 34, 434 (1980), Optically pumped laser action in perchloryl fluoride.

1246. Fischer, T. A., Wittig, C., *Appl. Phys. Lett.*, 39, 6 (1981), 16-mm laser oscillation in propyne.

1247. Schmid, W. E., *High-Power Lasers and Applications* (1979), 148, Springer-Verlag, New York, A simple high energy TEA CO laser.

1248. Osgood, R. M., Jr., Nichols, E. R., Eppers, W. C., Jr., Petty, R. D., *Appl. Phys. Lett.*, 15, 69 (1969), Q switching of the carbon monoxide laser.

1249. Bhaumik, M. L., Lacina, W. B., Mann, M. M., *IEEE J. Quantum Electron.*, QE-6, 576 (1970), Enhancement of CO laser efficiency by addition of Xenon.

1250. Qi, N., Krishnan, M., *Phys. Rev. Lett.*, 59, 2051 (1987), Photopumping of a C III ultraviolet laser by Mn VI line radiation.

1251. Daiber, J. W., Thompson, H. M., *IEEE J. Quantum Electron.*, QE-13, 10 (1977), Performance of a large, cw, preexcited CO supersonic laser.

1252. Boness, M. J. W., Center, R. E., *J. Appl. Phys.*, 48, 2705 (1977), High-pressure pulsed electrical CO laser.

1253. Johns, J. W. C., McKellar, A. R. W., Weitz, D., *J. Mol. Spectrosc.*, 51, 539 (1974), Wavelength measurements of $^{13}C^{16}O$ laser transitions.

1254. Ross, A. H. M., Eng, R. S., Kildal, H., *Opt. Commun.*, 12, 433 (1974), Heterodyne measurements of $^{12}C^{18}O$, $^{13}C^{16}O$, and $^{13}C^{18}O$ laser frequencies; mass dependence of Dunham coefficients.

1255. Deutsch, T. F., *IEEE J. Quantum Electron.*, QE-3, 419 (2) (1967), New infrared laser transitions in HCl, HBr, DCl, and DBr.

1256. Wood, O. R., Chang, T. Y., *Appl. Phys. Lett.*, 20, 77 (1972), Transverse-discharge hydrogen halide lasers.

1257. Rutt, H. N., *J. Phys. D.*, 12, 345 (1979), A high-energy hydrogen bromide laser.

1258. Keller, F. L., Nielsen, A. H., *J. Chem. Phys.*, 22, 294 (1954), The infrared spectrum and molecular constants of DBr.

1259. Rank, D. H., Rao, B. S., Wiggins, T. A., *J. Mol. Spectrosc.*, 17, 122 (1965), Molecular constants of HCl^{35}.

1260. Webb, D. U., Rao, K. N., *Appl. Opt.*, 5, 1461 (1966), A heated absorption cell for studying infrared absorption bands.

1261. Diemer, U., Demtröder, W., *Chem. Phys. Lett.*, 175, 135 (1991), Infrared atomic Cs laser based on optical pumping of Cs_2 molecules.

1262. Kasper, J. V. V., Pimentel, G. C., *Phys. Rev. Lett.*, 14, 352 (1965), HCl chemical laser.

1263. Corneil, P. H., Pimentel, G. C., *J. Chem. Phys.*, 49, 1379 (1968), Hydrogen-chloride explosion laser II DCl.

1264. Henry, A., Bourcin, F., Arditi, I., Charneau, R., Menard, J., *C. R. Acad. Sci. Ser. B*, 267, 616 (1968), Effect laser par reaction chimique de l'hydrogene sur du chlore ou du chlorure de nitrosyle.

1265. Pickworth, J., Thompson, H. W., *Proc. R. Soc. London, Ser. A*, 218, 37 (1953), The fundamental vibration-rotation band of deuterium chloride.

1266. Webb, D. U., Rao, K. N., *J. Mol. Spectrosc.*, 28, 121 (1968), Vibration rotation bands of heated hydrogen halides.

1267. Skribanowitz, N., Herman, I. P., Osgood, R. M., Jr., Feld, M. S., Javan, A., *Appl. Phys. Lett.*, 20, 428 (1972), Anisotropic ultrahigh gain emission observed in rotational transitions in optically pumped HF gas.

1268. Skribanowitz, N., Herman, I. P., Feld, M. S., *Appl. Phys. Lett.*, 21, 466 (1972), Laser oscillation and anisotropic gain in the $1 \rightarrow 0$ vibrational band of optically pumped HF gas.

1269. Ultee, C. J., *Rev. Sci. Instrum.*, 42, 1174 (1971), Compact pulsed HF lasers.

1270. Suchard, S. N., Gross, R. W. F., Whittier, J. S., *Appl. Phys. Lett.*, 19, 411 (1971), Time-resolved spectroscopy of a flash-imitated H_2-F_2 laser.

1271. Kwok, M. A., Giedt, R. R., Gross, R. W. F., *Appl. Phys. Lett.*, 16, 386 (1970), Comparison of HF and DF continuous chemical lasers. II. Spectroscopy.

1272. Goldsmith, J. E. M., *J. Opt. Soc. Am. B* 6, 1979 (1989), Two photon excited stimulated emission from atomic hydrogen in flames.

1273. Kwok, M. A., Giedt, R. R., Varwig, R. L., *AIAA J.*, 14, 1318 (1976), Medium diagnostics for a 10-kW cw HF chemical laser.

1274. Deutsch, T. F., *Appl. Phys. Lett.*, 10, 234 (1967), Molecular laser action in hydrogen and deuterium halides.

1275. Gerber, R. A., Patterson, E. L., *J. Appl. Phys.*, 47, 3524 (1976), Studies of a high-energy HF laser using an electron-beam-excited mixture of high-pressure F_2 and H_2.

1276. Eng, R. S., Spears, D. L., *Appl. Phys. Lett.*, 27, 650 (1975), Frequency stabilization and absolute frequency measurements of a cw HF/DF laser.

1277. Pummer, H., Proch, D., Schmailzl, U., Kompa, K. L., *Opt. Commun.*, 19, 273 (1976), The generation of partial and total vibrational inversion in colliding molecular systems initiated by ir-laser absorption.

1278. Gensel, P., Kompa, K. L., Wanner, J., *Chem. Phys. Lett.*, 5, 179 (1970), IF_5-H_2 hydrogen fluoride chemical laser involving a chain reaction.

1279. Dolgov-Savel'ev, G. G., Zharov, V. F., Neganov, Yu. S., Chumak, G. M., *Sov. Phys. JETP*, 34, 34 (1972), Vibrational-rotational transitions in an $H_2 + F_2$ chemical laser.

1280. Mayer, S. W., Taylor, D., Kwok, M. A., *Appl. Phys. Lett.*, 23, 434 (1973), HF chemical lasing at higher vibrational levels.

1281. Suchard, S. N., Pimentel, G. C., *Appl. Phys. Lett.*, 18, 530 (1971), Deuterium fluoride vibrational overtone chemical laser.

1282. Ultee, C. J., *IEEE J. Quantum Electron.*, QE-8, 820 (1972), Compact pulsed deuterium fluoride laser.

1283. Spencer, D. J., Mirels, H., Jacobs, T. A., *Appl. Phys. Lett.*, 16, 384 (1970), Comparison of HF and DF continuous chemical lasers. I. Power.

1284. Petersen, A. B., Braverman, L. W., Wittig, C., *J. Appl. Phys.*, 48, 230 (1977), H_2O, NO, and N_2O infrared lasers pumped directly and indirectly by electronic-vibrational energy transfer.

1285. Deutsch, T. F., *Appl. Phys. Lett.*, 9, 295 (1966), NO molecular laser.

1286. Pollack, M. A., *Appl. Phys. Lett.*, 9, 94 (1966), Molecular laser action in nitric oxide by photodissociation of NOCl.

1287. Giuliano, C. R., Hess, L. D., *J. Appl. Phys.*, 38, 4451 (1967), Chemical reversibility and solar excitation rates of the nitrosyl chloride photodissociative laser.

1288. Callear, A. B., Van den Bergh, H. E., *Chem. Phys. Lett.*, 8, 17 (1971), An hydroxyl radical infrared laser.

1289. Wanchop, T. S., Schiff, H. I., Welge, K. H., *Rev. Sci. Instrum.*, 45, 653 (1974), Pulsed-discharge infrared OH laser.

1290. Rice, W. W., Jensen, R. J., *Appl. Phys. Lett.*, 22, 67 (1973), Aluminum fluoride exploding-wire laser.

1291. Jensen, R. J., *Handbook of Chemical Lasers* (1976), 703, John Wiley & Sons, New York, Metal-atom oxidation laser.

1292. Rice, W. W., Beattie, W. H., Oldenborg, R. C., Johnson, S. E., Scott, P., *Appl. Phys. Lett.*, 28, 444 (1976), Boron fluoride and aluminum fluoride infrared lasers from quasicontinuous supersonic mixing flames.

1293. Bergman, R. C., Rich, J. W., *Appl. Phys. Lett.*, 31, 597 (1977), Overtone bands lasing at 2.7-3 μm in electrically excited CO.

1294. Znotis, T. A., Reid, J., Garside, B. K., Ballik, E. A., *Opt. Lett.*, 4, 253 (1979), 4.3-μm TE CO_2 lasers.

1295. Guelachvili, G., *J. Mol. Spectrosc.*, 79, 72 (1980), High-resolution Fourier spectra of carbon dioxide and three of its isotopic species near 4.3 μm.

1296. Petersen, A. B., Wittig, C., *J. Appl. Phys.*, 48, 3665 (1977), Line-tunable CO_2 laser operating in the region 2280-2360 cm^{-1} pumped by energy transfer from Br ($4^2P_{1/2}$).

1297. Rao, D. R., Hocker, L. O., Javan, A., Knable, K., *J. Mol. Spectrosc.*, 25, 410 (1968), Spectroscopic studies of 4.3 μm transient laser oscillation in CO_2.

1298. Freed, C., Bradley, L. C., O'Donnell, R. G., *IEEE J. Quantum Electron.*, QE-16, 1195 (1980), Absolute frequencies of lasing transitions in seven CO_2 isotopic species.

1299. Ernst, G. J., Witteman, W. J., *IEEE J. Quantum Electron.*, QE-7, 484 (1971), Transition selection with adjustable outcoupling for a laser device applied to CO_2.

1300. Patel, C. K. N., *Phys. Rev.*, 136, 1187 (1964), Continuous-wave laser action on vibrational-rotational transitions of CO_2.

1301. Brown, C. O., Davis, J. W., *Appl. Phys. Lett.*, 21, 480 (1972), Closed-cycle performance of a high-power electric discharge laser.

1302. Gerry, E. T., *Laser Focus*, 27, (1970), The gas dynamic laser.

1303. Richardson, M. C., Alcock, A. J., Leopold, K., Burtyn, P., *IEEE J. Quantum Electron.*, QE-9, 236 (1973), A 300-J multigigawatt CO_2 laser.

1304. Kildal, H., Eng, R. S., Ross, A. H. M., *J. Mol. Spectrosc.*, 53, 479 (1974), Heterodyne measurements of $^{12}C^{16}O$ laser frequencies and improved Dunham coefficients.

1305. Schappert, G. T., Singer, S., Ladish, J., Montgomery, M. D., *J. Opt. Soc. Am.*, 68, 668 (1978), Comparison of theory and experiment on the performance of the Los Alamos Scientific Laboratory eight-beam 10 kJ CO_2 laser.

1306. Siemsen, K. J., Whitford, B. G., *Opt. Commun.*, 22, 11 (1977), Heterodyne frequency measurements of CO_2 laser sequence-band transitions.

1307. Whitford, B. G., Siemsen, K. J., Reid, J., *Opt. Commun.*, 22, 261 (1977), Heterodyne frequency measurements of CO_2 laser hot-band transitions.

1308. Hartmann, B., Kleman, B., *Can. J. Phys.*, 44, 1609 (1966), Laser lines from CO_2 in the 11 to 18 micron region.

1309. Kasner, W. H., Pleasance, L. D., *Appl. Phys. Lett.*, 31, 82 (1977), Laser emission from the 13.9-μm $10^0 \to 01^10$ CO_2 transition in pulsed electrical discharges.

1310. Paso, R., Kauppinen, J., Anttila, R., *J. Mol. Spectrosc.*, 79, 236 (1980), Infrared spectrum of CO_2 in the region of the bending fundamental n_2.

1311. Osgood, R. M., Jr., *Appl. Phys. Lett.*, 32, 564 (1978), 1-mJ line-tunable optically pumped 16-μm laser.

1312. Buchwald, M. I., Jones, C. R., Fetterman, H. R., Schlossberg, H. R., *Appl. Phys. Lett.*, 29, 300 (1976), Direct optically pumped multi-wavelength CO_2 laser.

1313. Freed, C., Ross, A. H. M., O'Donnell, R. G., *J. Mol. Spectrosc.*, 49, 439 (1974), Determination of laser line frequencies and vibrational-rotational constants of the $^{12}C^{18}O_2$, $^{13}C^{16}O_2$, and $^{13}C^{18}O_2$.

1314. Siddoway, J. C., *J. Appl. Phys.*, 39, 4854 (1968), Calculated and observed laser transitions using $C^{14}C^{16}O_2$.

1315. Sadie, F. G., Buger, P. A., Malan, O. G., *J. Appl. Phys.*, 43, 2906 (1972), Continuous-wave overtone bands in a CS_2-O_2 chemical laser.

1316. DePoorter, G. L., Balog, G., *IEEE J. Quantum Electron.*, QE-8, 917 (1972), New infrared laser lines in OCS and new method for C atom lasing.

1317. Kildal, H., Deutsch, T. F., *Tunable Lasers and Applications* (1976), 367, Springer-Verlag, New York, Optically pumped gas lasers.

1318. Maki, A. G., Plyler, E. K., Tidwell, E. D., *J. Res. Natl. Bur. Stand.* Sect. A, 66, 163 (1962), Vibration-rotation bands of carbonyl sulfide.

1319. Deutsch, T. F., *Appl. Phys. Lett.*, 8, 334 (1966), OCS molecular laser.

1320. Schlossberg, H. R., Fetterman, H. R., *Appl. Phys. Lett.*, 26, 316 (1975), Optically pumped vibrational transition laser in OCS.

1321. Deutsch, T. F., Kildal, H., *Chem. Phys. Lett.*, 40, 484 (1976), A reexamination of the CS_2 laser.

1322. Patel, C. K. N., *Appl. Phys. Lett.*, 7, 273 (1965), CW laser oscillation in an N_2-CS_2 system.

1323. Nelson, L. Y., Fisher, C. H., Byron, S. R., *Appl. Phys. Lett.*, 25, 517 (1974), unspecified.

1324. Basov, N. G., Kazakevich, V. S., Kovsh, I. B., *Sov. J. Quantum Electron.*, 10, 1131 (1980), Electron-beam-controlled laser utilizing the first overtones of the vibrational-rotational transitions in the CO molecule, I and II.

1325. Bushnell, A. H., Jones, C. R., Buchwald, M. I., Gundersen, M., *IEEE J. Quantum Electron.*, QE-15, 208 (1979), New HF laser pumped molecular lasers in the middle infrared.

1326. Jones, C. R., *Laser Focus*, 68, (1978), Optically pumped mid-ir lasers.

1327. Petersen, A. B., Wittig, C., Leone, S. R., *Appl. Phys. Lett.*, 27, 305 (1975), Infrared molecular lasers pumped by electronic-vibrational energy transfer from $Br(4^2P_{1/2})$: CO_2, N_2O, HCN, and C_2H_2.

1328. Turner, R., Poehler, T. O., *Phys. Lett.*, A, 27, 479 (1968), Emission from HCN and H_2O lasers in the 4- to 13-μm region.

1329. Benedict, W. S., Pollack, M. A., Tomlinson, W. J., *IEEE J. Quantum Electron.*, QE-5, 108 (1969), The water-vapor laser.

1330. Hartmann, B., Kleman, B., Spangstedt, G., *IEEE J. Quantum Electron.*, QE-4, 296 (1968), Water vapor laser lines in the 7-μm region.

1331. Wood, O. R., Burkhardt, E. G., Pollack, M. A., Bridges, T. J., *Appl. Phys. Lett.*, 18, 112 (1971), High-pressure laser action in 13 gases with transverse excitation.

1332. Djeu, N., Wolga, G. J., *IEEE J. Quantum Electron.*, QE-5, 50 (1969), Observation of new laser transitions in N_2O.

1333. Howe, J. A., *Phys. Lett.*, 17, 252 (1965), R-branch laser action in N_2O.

1334. Moeller, G., Rigden, J. D., *Appl. Phys. Lett.*, 8, 69 (1966), Observation of laser action in the R-branch of CO_2 and N_2O vibrational spectra.

1335. Whitford, B. G., Siemsen, K. J., Riccius, H. D., Hanes, G. R., *Opt. Commun.*, 14, 70 (1975), Absolute frequency measurements of N_2O laser transitions.

1336. Auyeung, R. C. Y., Cooper, D. G., Kim, S., Feldman, B. J., *Opt. Commun.*, 79, 207 (1990), Stimulated emission in atomic hydrogen at 656 nm.

1337. Jones, C. R., *Laser Focus*, 70, (1978), Optically pumped mid-ir lasers.

1338. Cord, M. S., Peterson, J. D., Lojko, M. S., Haas, R. H., Monograph 70, Vol. 4 (1968), U. S. National Bureau of Standards, Microwave spectral tables.

1339. DeTemple, T. A., *Infrared and Millimeter Waves Sources of Radiation*, Vol. 1 (1979), Academic Press, London, Pulsed optically pumped FIR lasers.

1340. Henningsen, J. O., Jensen, H. G., *IEEE J. Quantum Electron.*, QE-11, 248 (1975), The optically pumped FIR laser.

1341. DeTemple, T. A., Danielewicz, E. J., *IEEE J. Quantum Electron.*, QE-12, 40 (1976), CW CH_3F waveguide laser at 496 μm: theory and experiment.

1342. Temkin, R. J., *IEEE J. Quantum Electron.*, QE-13, 450 (1977), Theory of optically pumped submillimeter lasers.

1343. Kim, K. J., Coleman, P. D., *IEEE J. Quantum Electron.*, QE-16, 1341 (1980), Calculated-experimental evaluation of the gain/absorption spectra of several optically pumped NH_3 systems.

1344. Curtis, J., Ph.D. Thesis (1974), Ohio State University, Vibration-Rotation Bands of NH_3 in the Region 670-1860 cm^{-1}.

1345. Henningsen, J. O., *Infrared and Millimeter Waves, Sources of Radiation* (1981), Academic Press, London, Spectroscopy of molecules by far IR laser emission.

1346. Woods, D. R., Ph.D. Thesis(1970), University of Michigan, The High Resolution Infrared Spectra of Normal and Deuterated Methanol Between 400 and 1300 cm^{-1}.

1347. Berdnikov, A. A., Derzhiev, V. I., Murav'ev, I. I., Yakovlenko, S. I., Yancharina, A. M., *Sov. J. Quantum Electron.*, 17, 1400 (1987), Penning plasma laser utilizing new transitions in the helium atom resulting in the emission of visible light.

1348. Zolotarev, V. A., Kryukov, P. G., Podmar'kov, Yu. P., Frolov, M. P., Shcheglov, V. A., *Sov. J. Quantum Electron.*, 18, 643 (1988), Optically pumped pulsed IF (B → X) laser utilizing a CF_3I-NF_2-He mixture.

1349. Hocker, L. O., Javan, A., Rao, D. R., Frenkel, L., Sullivan, T., *Appl. Phys. Lett.*, 10, 147 (1967), Absolute frequency measurement and spectroscopy of gas laser transitions in the far IR.

1350. Frenkel, L., Sullivan, T., Pollack, M. A., Bridges, T. J., *Appl. Phys. Lett.*, 11, 344 (1967), Absolute frequency measurement of the 118.6 μm water vapor laser transition.

1351. Evenson, K. M., *Appl. Phys. Lett.*, 20, 133 (1972), Extension of absolute frequency measurements to the CW He-Ne laser at 88 THz (3.39 μm).

1352. Hodges, D. T., Hartwick, T. S., *Appl. Phys. Lett.*, 23, 252 (1973), Waveguide lasers for the far infrared pumped by a CO_2 laser.

1353. Danielewicz, E. J., Coleman, P. D., *Appl. Opt.*, 15, 761 (1976), Hybrid metal mesh-dielectric mirrors for optically pumped far IR lasers.

1354. Fetterman, H. R., Tannenwald, P. E., Parker, C. D., Melngailas, J. et al., *Appl. Phys. Lett.*, 34, 123 (1979), Real-time spectral analysis of far IR laser pulses using a SAW dispersive delay line.

1355. Mathis, L. E. S., Parker, J. T., *Appl. Phys. Lett.*, 3, 16 (1963), Stimulated emission in band spectrum of nitrogen.

1356. Lide, D. R., *Appl. Phys. Lett.*, 11, 62 (1967), On the explanation of the so-called CN laser.

1357. Chang, T. Y., Budgee, T., *Opt. Commun.*, 1, 423 (1970), Laser action at 452, 496, and 541 μm in optically pumped CH_3F.

1358. Hodges, D. T., *Infrared Phys.*, 18, 375 (1978), A review of advances in optically pumped far IR lasers.

1359. Tucker, J. R., *IEEE Conference Digest 17, Int. Conf. on Submillimeter Waves and Their Applications*, Theory of a FIR gas laser.

1360. Chang, T. Y., *IEEE Trans. Microwave Theory Tech.*, MTT-22, 983 (1974), Optically pumped submillimeter-wave sources.

1361. Coleman, P. D., *J. Opt. Soc. Am.*, 67, 894 (1977), Present and future problems concerning lasers in the far IR spectra range.

1362. Amos, K. B., Davis, J. A., *IEEE J. Quantum Electron.*, QE-16, No. 5, 574-575 (1980), Additional CW far-infrared laser lines from CO_2 laser pumped cis 12 $C_2H_2F_2$.

1363. Arimondo, E., Inguscio, M., *J. Mol. Spectrosc.*, 75, 81-86 (1979), The rotation-vibration constants of the $^{12}CH_3F$ n_3 band.

1364. Arimondo, E., Inguscio, A., *Digest for the 4th Ann. Conf. on Infrared and Millimeter Waves*, IEEE, Piscataway, NJ, Assignments of laser lines in optically pumped CH_3CN.

1365. Arimondo, E., Inguscio, A., *Int. J. Infrared Millimeter Waves*, 1, No. 3, 437-458 (1980), Assignments of laser lines in optically pumped CH_3CN.

1366. Baskakov, O. I., Dyubko, S. F., Moskienko, M. V., Fesenko, L. D., *Sov. J. Quantum Electron.*, 7, No. 4, 445-449 (1977), Identification of active transitions in a formic acid vapor laser.

1367. Bean, B. L., Perkowitz, S., *Opt. Lett.*, 1, No. 6, 202-204 (1977), Complete frequency coverage for submillimetre laser spectroscopy with optically pumped CH_3OH, CH_3OD, CD_3OD and CH_2CF_2.

1368. Benedict, W. S., Pollack, M. A., Tomlinson, W. J., III, *IEEE J. Quantum Electron.*, QE-5, No. 2, 108-124 (1969), The water-vapor laser.

1369. Blaney, T. G., Bradley, C. C., Edwards, G. J., Knight, D. J. E., *Phys. Lett.*, A, 43, No. 5, 471-472 (1973), Absolute frequency measurement of a Lamb-dip stabilised water vapour laser oscillating at 10.7 THz (28 mm).

1370. Blaney, T. G., Knight, D. J. E., Murray-Lloyd, E., *Opt. Commun.*, 25, No. 2, 176-178 (1978), Frequency measurements of some optically-pumped laser lines in CH_3OD.

1371. Blaney, T. G., Cross, N. R., Knight, D. J. E., Edwards, G. J., Pearce, P., *J. Phys. D.*, 13, No. 8, 1365-1370 (1980), Frequency measurement at 4.25 THz (70.5 μm) using a Josephson harmonic mixer and phase-lock techniques.

1372. Belland, P., Veron, D., Whitbourn, B., *Appl. Opt.*, 15, No. 12, 3047-3053 (1976), Scaling laws for cw 337-μm HCN waveguide lasers.

1373. Belland, P., Veron, D., *IEEE J. Quantum Electron.*, QE-16, No. 8, 885-889 (1980), Amplifying medium characteristics in optimised 190 μm/195 μm DCN waveguide lasers.

1374. Chang, T. Y., Bridges, T. J., *Opt. Commun.*, 1, No. 9, 423-426 (1970), Laser action at 452, 496 and 541 μm in optically pumped CH_3F.

1375. Chang, T. Y., Bridges, T. J., Burkhardt, E. J., *Appl. Phys. Lett.*, 17, No. 6, 249-251 (1970), cw submillimeter laser action in optically pumped methyl fluoride, methyl alcohol and vinyl chloride gases.

1376. Chang, T. Y., Bridges, T. J., Burkhardt, E. J., *Appl. Phys. Lett.*, 17, No. 9, 357-358 (1970), cw laser action at 81.5 and 263.4 μm in optically pumped ammonia gas.

1377. Chang, T. Y., McGee, J. D., *Appl. Phys. Lett.*, 19, No. 4, 103-105 (1971), Millimeter and submillimeter wave laser action in symmetric top molecules optically pumped via parallel absorption bands.

1378. Chang, T. Y., McGee, J. D., *IEEE J. Quantum Electron.*, QE-12, No. 1, 62-65 (1976), Millimeter and submillimeter wave laser action in symmetric top molecules optically pumped via perpendicular absorption bands.

1379. Dangoisse, D., Deldalle, A., Splingard, J-P., Bellet, J., *C. R. Acad. Sci. Ser.* B, 283, 115-118 (1976), Mesure precise des emissions continues du laser submillimetrique a acide formique.

1380. Dangoisse, D., Deldalle, A., Splingard, J-P., Bellet, J., *IEEE J. Quantum Electron.*, QE-13, No. 9, 730-731 (1977), CW optically pumped laser action in D_2CO, HDCO and $(H_2CO)_3$.

1381. Dangoisse, D., Willemot, E., Deldalle, A., Bellet, J., *Opt. Commun.*, 28, No. 1, 111-116 (1979), Assignment of the HCOOH cw-submillimeter laser.

1382. Dangoisse, D., Duterage, B., Glorieux, P., *IEEE J. Quantum Electron.*, QE-16, No. 3, 296-300 (1980), Assignment of laser lines in optically pumped submillimetre lasers: HDCO, D_2CO.

1383. Daneu, V., Hocker, L. O., Javan, A., Rao, D. R., Szoke, A., Zernike, F., *Phys. Lett.*, A, 29, No. 6, 319-320 (1969), Accurate laser wavelength measurements in the infrared and far infrared using a Michelson interferometer.

1384. Deldalle, A., Dangoisse, D., Splingard, J-P., Bellet, J., *Opt. Commun.*, 22, No. 3, 333-336 (1977), Accurate measurements of cw optically pumped FIR laser lines of formic acid molecule and its isotopic species $H^{13}COOH$, HCOOD and DCOOD.

1385. Deroche, J-C., *J. Mol. Spectrosc.*, 69, 19-24 (1978), Assignment of submillimeter laser lines in methyl chloride.

1386. Danielewicz, E. J., Galantowicz, T. A., Foote, F. B., Reel, R. D., Hodges, D. T., *Opt. Lett.*, 4, No. 9, 280-282 (1979), High performance at new FIR wavelengths from optically pumped CH_2F_2.

1387. Duxbury, G., Gamble, T. J., Herman, H., *IEEE Trans. Microwave Theory Tech.*, MTT-22, No. 12, 1108-1109 (1974), Assignments of optically pumped laser lines of 11 difluoroethylene.

1388. Duxbury, G., Herman, H., *J. Phys. B*, 11, No. 5, 935-949 (1978), Optically pumped millimetre lasers.

1389. Danielewicz, E. J., Reel, R. D., Hodges, D. T., *IEEE J. Quantum Electron.*, QE-16, No. 4, 402-405 (1980), New far-infrared CW optically pumped cis-$C_2H_2F_2$ laser.

1390. Danielewicz, E. J., Weiss, C. O., *IEEE J. Quantum Electron.*, QE-14, No. 4, 222-223 (1978), Far infrared emission from $^{15}NH_3$ optically pumped by a CW sequence band CO_2 laser.

1391. Danielewicz, E. J., Weiss, C. O., *IEEE J. Quantum Electron.*, QE-14, No. 7, 458-459 (1978), New CW far-infrared laser lines from CO_2 laser-pumped CD_3OH.

1392. Danielewicz, E. J., Weiss, C. O., *IEEE J. Quantum Electron.*, QE-14, No. 10, 705-707 (1978), New efficient CW far-infrared optically pumped CH_2F_2 laser.

1393. Danielewicz, E. J., Weiss, C. O., *Opt. Commun.*, 27, No. 1, 98-100 (1978), New cw far-infrared D_2O, $^{12}CH_3F$ and $^{14}NH_3$ laser lines.

1394. Danielewicz, E. J., *Digest for the 4th Ann. Conf. on Infrared and Millimeter Waves*, IEEE, Piscataway, NJ, Molecular parameters determining the performance of CW optically pumped, FIR lasers.

1395. Dyubko, S. F., Svich, V. A., Valitov, R. A., *JETP Lett.*, 7, No. 11, 320 (1968), SO_2 submillimetre laser generating at wavelengths 0.141 and 0.193 μm.

1396. Dyubko, S. F., Svich, V. A., Fesenko, L. D., *JETP Lett.*, 16, No. 11, 418-419 (1972), Submillimeter-band gas laser pumped by a CO_2 laser.

1397. Dyubko, S. F., Svich, V. A., Fesenko, L. D., *Sov. Phys. Tech. Phys.*, 18, No. 8, 1121 (1974), Submillimeter CH_3OH and CH_3OD lasers with optical pumping.

1398. Dyubko, S. F., Svich, V. A., Fesenko, L. D., *Zh. Prikl. Spectrosk. (USSR)*, 20, No. 4, 718-719 (1974), Stimulated emission of submillimeter waves by hydrazine excited by a CO_2 laser.

1399. Dyubko, S. F., Svich, V. A., Fesenko, L. D., *Opt. Spectrosc.*, 37, No. 1, 118 (1974), Submillimeter laser emission of CH_3I molecules excited by CO_2.

1400. Dyubko, S. F., Svich, V. A., Fesenko, L. D., *Sov. J. Quantum Electron.*, 3, No. 5, 446 (1974), Submillimeter laser using formic acid vapor pumped with carbon dioxide laser radiation.

1401. Dyubko, S. F., Fesenko, L. D., Baskakov, O. I., Svich, V. A., *Zh. Prikl. Spectrosk. (USSR)*, 23, No. 2, 317-320 (1975), Use of CD_3I, CH_3I and CD_3Cl molecules as active substances for submillimeter lasers with optical pumping.

1402. Dyubko, S. F., Svich, V. A., Fesenko, L. D., *Izv. Vuz. Radiofiz. (USSR)*, 18, No. 10, 1434-1437 (1975), An experimental study of the radiation spectrum of submillimeter laser on CD_3OH molecules.

1403. Dyubko, S. F., Svich, V. A., Fesenko, L. D., *Sov. Phys. Tech. Phys.*, 20, No. 11, 1536-1538 (1975), Submillimeter HCOOH, DCOOH, HCOOD, and DCOOD laser.

1404. Dyubko, S. F., Efimenko, M. N., Svich, V. A., Fesenko, L. D., *Sov. J. Quantum Electron.*, 6, No. 5, 600-601 (1976), Stimulated emission of radiation from optically pumped vinyl bromide molecules.

1405. Edwards, G. J., unpublished (1978), private communication, National Physical Laboratory.

1406. Epton, P. J., Wilson, W. L., Jr., Tittel, F. K., Rabson, R. A., *Appl. Opt.*, 18, No. 11, 1704-1705 (1979), Frequency measurement of the formic acid laser 311-mm line.

1407. Evenson, K. M., Wells, J. S., Matarrese, L. M., Elwell, J. B., *Appl. Phys. Lett.*, 16, No. 4, 159-162 (1970), Absolute frequency measurements of the 28- and 78-μm CW water vapor laser lines.

1408. Fetterman, H. R., Schlossberg, H. R., Parker, C. D., *Appl. Phys. Lett.*, 23, No. 12, 684-686 (1973), cw submillimeter laser generation in optically pumped Stark-tuned NH_3.

1409. Graner, G., *Opt. Commun.*, 14, No. 1, 67-69 (1975), Assignment of submillimeter laser lines in CH_3I.

1410. Grinda, M., Weiss, C. O., *Opt. Commun.*, 26, No. 1, 91 (1978), New far infrared laser lines from CD_3OH.

1411. Hard, T. M., *Appl. Phys. Lett.*, 14, No. 4, 130 (1969), Sulfur dioxide submillimeter laser.

1412. Heppner, J., Weiss, C. O., Plainchamp, P., *Opt. Commun.*, 23, No. 3, 381-384 (1977), Far infrared gain measurements in optically pumped CH_3OH.

1413. Henningsen, J. O., *IEEE J. Quantum Electron.*, QE-13, No. 6, 435-441 (1977), Assignment of laser lines in optically pumped CH_3OH.

1414. Henningsen, J. O., *IEEE J. Quantum Electron.*, QE-14, No. 12, 958-962 (1978), New FIR laser lines from optically pumped CH_3OH measurements and assignments.

1415. Henningsen, J. O., Petersen, J. C., *Infrared Phys.*, 18, No. 5, 6, 475-479 (1978), Observation and assignment of far-infrared laser lines from optically pumped $^{13}CH_3OH$.

1416. Henningsen, J. O., *Digest for the 4th Ann. Conf. on Infrared and Millimeter Waves*, IEEE, Piscataway, NJ, Spectroscopy of the lasing CH_3OH molecule.

1417. Henningsen, J. O., Petersen, J. C., Petersen, F. R., Jennings, D. A., Evenson, K. M., *J. Mol. Spectrosc.*, 77, No. 2, 298-309 (1979), High resolution spectroscopy of vibrationally excited $^{13}CH_3OH$ by frequency measurement of FIR laser emission.

1418. Henningsen, J. O., *J. Mol. Spectrosc.*, 83, No. 1, 70-93 (1980), Stark effect in the CO_2 laser pumped CH_3OH far infrared laser as a technique for high resolution infrared spectroscopy.

1419. Henningsen, J. O., *J. Mol. Spectrosc.*, 85, No. 2, 282-300 (1981), Improved molecular constants and empirical corrections for the torsional ground state of the C-O stretch fundamental of CH_3OH.

1420. Hodges, D. T., Reel, R. D., Barker, D. H., *IEEE J. Quantum Electron.*, QE-9, No. 12, 1159-1160 (1973), Low-threshold CW submillimeter- and millimeter-wave laser action in CO_2-laser-pumped $C_2H_4F_2$, $C_2H_2F_2$, and CH_3OH.

1421. Hocker, L. O., Javan, A., Ramachandra-Rao, D., Frenkel, L., Sullivan, T., *Appl. Phys. Lett.*, 10, No. 5, 147-149 (1967), Absolute frequency measurement and spectroscopy of gas laser transitions in the far infrared.

1422. Hocker, L. O., Javan, A., *Phys. Lett.*, A, 25, No. 7, 489-490 (1967), Absolute frequency measurements on new cw HCN submillimeter laser lines.

1423. Hocker, L. O., Javan, A., *Appl. Phys. Lett.*, 12, No. 4, 124-125 (1968), Absolute frequency measurements of new cw DCN submillimeter laser lines.

1424. Hocker, L. O., Ramachandra-Rao, D., Javan, A., *Phys. Lett.*, A, 24, No. 12, 690-691 (1967), Absolute frequency measurement of the 190 μm and 194 μm gas laser transitions.

1425. Herman, H., Prewer, B. E., *Appl. Phys.*, 19, 241-242 (1979), New FIR laser lines from optically pumped methanol analogues.

1426. Hubner, G., Hassler, J. C., Coleman, P. D., Steenbeckeliers, G., *Appl. Phys. Lett.*, 18, No. 11, 511-513 (1971), Assignments of the far-infrared SO_2 laser lines.

1427. Inguscio, M., Moretti, A., Strumia, F., *Opt. Commun.*, 32, No. 1, 87-90 (1980), New laser lines from optically-pumped CH_3OH measurements and assignments.

1428. Inguscio, M., Moretti, A., Strumia, F., *Digest for the 4th Ann. Conf. on Infrared and Millimeter Waves*, IEEE , Piscataway, NJ, IR-FIR Stark spectroscopy of CH_3OH around the 9-P(34) CO_2 laser line.

1429. Bionducci, G., Inguscio, M., Moretti, A., Strumia, F., *Infrared Phys.*, 19, 297-308 (1979), Design of FIR molecular lasers frequency tunable by Stark effect: electric breakdown of CH_3OH, CH_3F, CH_3I, and CH_3CN.

1430. Jennings, D. A., Evenson, K. M., Jimenez, J. J., *IEEE J. Quantum Electron.*, QE-11, No. 8, 637 (1975), New CO_2 pumped CW far infrared laser lines.

1431. Karlov, N. V., Petrov, Yu. B., Prokhorov, A. M., Stel'makh, O. M., *JETP Lett.*, 8, 12-14 (1968), Laser based on boron trichloride.

1432. Kon, S., Hagiwara, E., Yano, T., Hirose, H., *Jpn. J. Appl. Phys.*, 14, No. 5, 731-732 (1975), Far infrared laser action in optically pumped CD_3OD.

1433. Kon, S., Yano, T., Hagiwara, E., Hirose, H., *Jpn. J. Appl. Phys.*, 14, No. 11, 1861-1862 (1975), Far infrared laser action in optically pumped CH_3OD.

1434. Kramer, G., Weiss, C. O., *Appl. Phys.*, 10, 187-188 (1976), Frequencies of some optically pumped submillimetre laser lines.

1435. Landsberg, B. M., *IEEE J. Quantum Electron.*, QE-16, No. 6, 684-685 (1980), Optically pumped CW submillimetre emission lines from methyl mercaptan CH_3SH.

1436. Landsberg, B. M., *IEEE J. Quantum Electron.*, QE-16, No. 7, 704-706 (1980), New optically pumped CW submillimetre emission lines from OCS, CH_3OH and CH_3OD.

1437. Landsberg, B. M., *Appl. Phys.*, 23, 127-130 (1980), New CW FIR laser lines from optically pumped ammonia analogues.

1438. Landsberg, B. M., unpublished (June 1980), See Reference 1439.

1439. Landsberg, B. M., *Appl. Phys.*, 23, 345-348 (1980), New CW optically pumped FIR emissions in HCOOH, D_2CO and CD_3Br.

1440. Lund, M. W., Davis, J. A., *IEEE J. Quantum Electron.*, QE-15, No. 7, 537-538 (1979), New CW far-infrared laser lines from CO_2 laser-pumped CH_3OD.

1441. Lide, D. R., Maki, A. G., *Appl. Phys. Lett.*, 11, No. 2, 62-64 (1967), On the explanation of the so-called CN laser.

1442. Lourtioz, J-M., Pontnau, J., Morillon-Chapey, M., Deroche, J-C., *Int. J. Infrared Millimeter Waves*, 2, No. 1, 49-63 (1981), Submillimetre laser action of cw optically pumped CF_2Cl_2 (fluorocarbon 12).

1443. Lourtioz, J-M., unpublished (July 1980), private communication.

1444. Lourtioz, J-M., Pontnau, J., Meyer, C., *Int. J. Infrared Millimeter Waves*, 2, No. 3, 525-532 (1981), Optically pumped CW CF_3Br FIR laser. New emission lines and tentative assignments.

1445. Mathias, L. E. S., Crocker, A., Wills, M. S., *IEEE J. Quantum Electron.*, QE-4, No. 4, 205-208 (1968), Spectroscopic measurements of the laser emission from discharges in compounds of hydrogen, carbon and nitrogen.

1446. Muller, W. M., Flesher, G. T., *Appl. Phys. Lett.*, 10, No. 3, 93-94 (1967), Continuous-wave submillimeter oscillation in discharges containing C, N and H or D.

1447. Maki, A. G., *Appl. Phys. Lett.*, 12, No. 4, 122-124 (1968), Assignment of some DCN and HCN laser lines.

1448. Maki, A. G., *J. Appl. Phys.*, 49, No. 1, 7-11 (1978), Further assignments for the far-infrared laser transitions of HCN and HC^{15}N.

1449. Maki, A. G., Olson, W. B., Sams, R. L., *J. Mol. Spectrosc.*, 36, No. 3, 433-447 (1970), HCN rotational-vibrational energy levels and intensity anomalies determined from infrared measurements.

1450. Moskienko, M. V., Dyubko, S. F., *Izv. Vuz. Radiofiz. (USSR)*, 21, No. 7, 951-960 (1978), Identification of generation lines of submillimeter methyl bromide and acetonitrile molecule laser.

1451. Ni, Y. C., Heppner, J., *Opt. Commun.*, 32, No. 3, 459-460 (1980), New cw laser lines from CO_2 laser pumped CH_3OD.

1452. Patel, C. K. N., Gas Lasers in *Lasers: A Series of Advances*, Vol. 2 (1968), Marcel Dekker, New York.

1453. Petersen, F. R., U. S. National Bureau of Standards, unpublished (Nov. 1978), private communications.

1454. Petersen, F. R., Scalabrin, A., Evenson, K. M., *Int. J. Infrared Millimeter Waves*, 1, No. 1, 111-115 (1980), Frequencies of cw FIR laser lines from optically pumped CH_2F_2.

1455. Petersen, F. R., Evenson, K. M., Jennings, D. A., Wells, J. S., Goto, K., *IEEE J. Quantum Electron.*, QE-11, No. 10, 838-843 (1975), Far infrared frequency synthesis with stabilized CO_2 lasers: accurate measurements of the water vapor and methyl alcohol laser frequencies.

1456. Pontnau, J., Lourtioz, J-M., Meyer, C., *IEEE J. Quantum Electron.*, QE-15, No. 10, 1088-1090 (1979), Submillimetre laser action of CW optically pumped CF_3Br.

1457. Pollack, M. A., Frenkel, L., Sullivan, T., *Phys. Lett.*, A, 26, No. 8, 381-382 (1968), Absolute frequency measurement of the 220 µm water vapor laser transition.

1458. Pollack, M. A., Bridges, T. J., Tomlinson, W. J., *Appl. Phys. Lett.*, 10, No. 9, 253-256 (1967), Competitive and cascade coupling between transitions in the cw water vapor laser.

1459. Radford, H. E., *IEEE J. Quantum Electron.*, QE-11, No. 5, 213-214 (1975), New CW lines from a submillimeter waveguide laser.

1460. Reid, J., Oka, T., *Phys. Rev. Lett.*, 38, No. 2, 67-70 (1977), Direct observation of velocity-tuned multiphonon processes in the laser cavity.

1461. Redon, M., Gastaud, C., Fourrier, M., *IEEE J. Quantum Electron.*, QE-15, No. 6, 412-414 (1979), New CW far-infrared lasing in $^{14}NH_3$ using Stark tuning.

1462. Redon, M., Gastaud, C., Fourrier, M., *Opt. Commun.*, 30, No. 1, 95-98 (1979), Far-infrared emission in NH_3 using "forbidden" transitions pumped by a CO_2 laser.

1463. Redon, M., Gastaud, C., Fourrier, M., *Infrared Phys.*, 20, 93-98 (1980), Far-infrared emissions in ammonia by infrared pumping using a N_2O laser.

1464. Redon, M., Gastaud, C., Fourrier, M., *Int. J. Infrared Millimeter Waves*, 1, No. 1, 95-100 (1980), New CW FIR laser lines obtained in ammonia pumped by a CO_2 laser using the Stark tuning method.

1465. Sattler, J. P., Worchesky, T. L., Tobin, M. S., Ritter, K. J., Daley, T., *Int. J. Infrared Millimeter Waves*, 1, No. 1, 127-138 (1980), Submillimeter-wave emission assignments for 11 difluoroethylene.

1466. Scalabrin, A., Evenson, K. M., *Opt. Lett.*, 4, No. 9, 277-279 (1979), Additional cw FIR laser lines from optically pumped CH_2F_2.

1467. Scalabrin, A., Petersen, F. R., Evenson, K. M., Jennings, D. A., *Int. J. Infrared Millimeter*, 1, No. 1, 117-126 (1980), Optically pumped cw CH_2DOH FIR laser: new lines and frequency measurements.

1468. Steenbeckeliers, G., Bellet, J., *J. Appl. Phys.*, 46, No. 6, 2620-2626 (1975), New interpretation of the far-infrared SO_2 laser spectrum.

1469.   Tanaka, A., Tanimoto, A., Murata, N., Yamanaka, M., Yoshinaga, H., *Opt. Commun.*, 22, No. 1, 1721 (1977), CW efficient optically pumped far-infrared waveguide NH_3 lasers.

1470.   Tanaka, A., Tanimoto, A., Murata, N., Yamanaka, M., Yoshinaga, H., *Jpn. J. Appl. Phys.*, 13, No. 9, 1491-1492 (1974), Optically pumped far infrared and millimeter wave waveguide lasers.

1471.   Tanaka, A., Yamanaka, M., Yoshinaga, H., *IEEE J. Quantum Electron.*, QE-11, No. 10, 853-854 (1975), New far-infrared laser lines from CO_2 laser pumped CH_3OH gas by using a copper waveguide cavity.

1472.   Tobin, M. S., Sattler, J. P., Wood, G. L., *Digest for the 4th Ann. Conf. on Infrared and Millimeter Waves*, IEEE, Piscataway, NJ, CD_3F optically-pumped near millimeter laser.

1473.   Worchesky, T. L., *Opt. Lett.*, 3, No. 12, 232-234 (1978), Assignments of methyl alcohol submillimeter laser transitions.

1474.   Wagner, R. J., Zelano, A. J., Ngai, L. H., *Opt. Commun.*, 8, No. 1, 46-47 (1973), New submillimeter laser lines in optically pumped gas molecules.

1475.   Weiss, C. O., Grinda, M., Siemsen, K., *IEEE J. Quantum Electron.*, QE-13, No. 11, 892 (1977), FIR laser lines of CH_3OH pumped by CO_2 laser sequence lines.

1476.   Wood, R. A., Davis, B. W., Vass, A., Pidgeon, C. R., *Opt. Lett.*, 5, No. 4, 153-154 (1980), Application of an isotopically enriched $^{13}C^{16}O_2$ laser to an optically-pumped FIR laser.

1477.   Wood, R. A., Vass, A., Pidgeon, C. R., Colles, M. J., Norris, B., *Opt. Commun.*, 33, No. 1, 89-90 (1980), High power FIR lasing in $^{15}NH_3$ optically pumped with an isotopically enriched, CO_2 laser.

1478.   Ziegler, G., Durr, U., *IEEE J. Quantum Electron.*, QE-14, No. 10, 708 (1978), Submillimeter laser action of CW optically pumped CD_2Cl_2, CH_2DOH and CHD_2OH.

1479.   Redon, M., Universite Pierre et Marie Curie, unpublished (May, 1980), private communications.

1480.   Radford, H. E., unpublished (August 1976), private communication.

1481.   Petersen, F. R., Evenson, K. M., Jennings, D. A., Scalabrin, A., *IEEE J. Quantum Electron.*, QE-16, No. 3, 319-323 (1980), New frequency measurements and laser lines of optically pumped $^{12}CH_3OH$.

1482.   Dinev, S. G., Hadjichristov, G. B., Stefanov, I. L., *Opt. Commun.*, 75, 273 (1990), Stimulated emission by hybrid transitions via a heteronuclear molecule.

1483.   Radford, H. E. Petersen, K. M., Jennings, D. A., Mucha, J. A., *IEEE J. Quantum Electron.*, QE-13, 92-94 (1977), Heterodyne measurements of submillimeter laser spectrometer frequencies.

1484.   Redon, M., Gastaud, C., Fourrier, M., *Infrared Phys.*, 20, 93-98 (1980), Far-infrared emissions in ammonia by infrared pumping using a N_2O laser.

1485.   Tobin, M. S., Sattler, J. P., Wood, G. L., *Opt. Lett.*, 4, No. 11, 384-386 (1979), Optically pumped CD_3F submillimeter wave laser.

1486.   Powell, H. T., Murray, J. R., Rhodes, C. K., unpublished (Oct. 1974), Laser oscillation on the green bands of xenon oxide.

1487.   Humphreys, C. J., unpublished, Line list for ionized xenon.

1488.   Eckardt, R., Telle, J., Haynes, L., *Conference of Laser and Electro-Optical Systems*, Tech. Digest 5 (1980), Isotopically pumped CF_4, IEEE, Piscataway, NJ.

1489.   Cooper, H. G., Cheo, P. K., *Physics of Quantum ElectronicsConference*, San Juan, P. R., 690-697 (1966), McGraw-Hill, New York, Ion laser oscillations in sulfur.

1490.   Johnson, D. E., Eden, J. G., *IEEE J. Quantum Electron.*, QE-19, 1462 (1983), Lasing on the lithium resonance line at 670.7 nm.

1491.   Hemmati, H., Collins, G. J., *IEEE J. Quantum Electron.*, QE-16, 1014-1016 (1980), Atomic Sn, Sb and Ge photodissociation lasers.

1492.   Xu-hui, Zh., Jian-bang, L., *Appl. Phys. B.*, 29, 291-292 (1982), A hollow cathode bismuth ion laser.

1493.   White, J. C., Boker, J., Henderson, D., *IEEE J. Quantum Electron.*, QE-18, 320-322 (1982), Optically pumped atomic thulium lasers.

1494.   Gevasimor, V. A., Prokopev, V. E., Sokovikov, V. G., Soldatov, A. N., *Sov. J. Quantum Electron.*, 14, 426-427 (1984), New laser lines in the visible and infrared parts of the spectrum of a thulium vapor laser.

1495.   Basov, N. G., Baranov, V. V., Danilychev, V. A. et al., *Sov. J. Quantum Electron.*, 15, 1004-1006 (1985), High-pressure power laser utilizing 3p-3s transitions in generating radiation of wavelengths 703 and 725 nm.

1496.   Basov, N. G., Alexsandrov, A. Yu., Danilychev, V. A. et al., *JETP Lett.*, 41, 191-194 (1985), Intense quasi-cw lasing in the visible region in a high pressure mixture of inert gases.

1497.   Bunkin, F. V., Derzhiev, V. I., Mesyats, G. A. et al., *Sov. J. Quantum Electron.*, 15, 159-160 (1985), Plasma laser emitting at the wavelength of 585 nm with penning clearing of the lower level in a dense mixtures with neon excited by an electron.

1498.   Brown, E. L., Gundersen, M. A., Williams, P. F., *IEEE J. Quantum Electron.*, QE-16, 683-685 (1980), New infrared laser lines in argon, krypton and xenon.

1499.   Macklin, J. J., Wood, O. R., Silfvast, W. T., *IEEE J. Quantum Electron.*, QE-18, 1832-1835 (1982), New recombination lasers in Li, Al, Ca and Cu in a segmented plasma device employing foil electrodes.

1500.   Muller, W., McClelland, J. J., Hertel, I. V., *Appl. Phys. B.*, 31, 131-134 (1983), Infrared laser emission study in a resonantly excited sodium vapor.

1501.   Sharma, A., Bhaskar, N. D., Lu, Y. Q., Happer, W., *Appl. Phys. Lett.*, 39, 209-211 (1981), Continuous-wave mirrorless lasing in optically pumped atomic Cs and Rb vapors.

1502.   Madigan, M., Hocker, L. O., Flint, J. H., Dewey, C. F., *IEEE J. Quantum Electron.*, QE-16, 1294-1296 (1980), Pressure dependence of the infrared laser lines in barium vapor.

1503.   Gerck, E., Fill, E., *IEEE J. Quantum Electron.*, QE-17, 2140-2146 (1981), Blue-green atomic photodissociation lasers in Group IIb: Zn, Cd and Hg.

1504.   Shigenari, T., Vesugi, F., Takuna, H., *Opt. Lett.*, 7, 362-364 (1982), Superfluorescent laser action at 404.7 nm from the transition of mercury atom by photodissociation of $HgBr_2$ by an ArF laser.

1505.   Xu-hui, Zh., Jian-bang, L., Fu-Cheng, L., *Appl. Phys. B.*, 29, 111-115 (1982), Laser action of C, N, O, F, Cl and Br in hollow cathode discharges.

1506.   Gnadig, K., Fu-Cheng, L., *Appl. Phys.*, 25, 273-274 (1981), A hollow cathode carbon atom laser.

1507.   Ozcomert, J. S., Jones, P. L., *Chem. Phys. Lett.*, 169, 1 (1990), An optically pumped K_2 supersonic beam laser.

1508.   Petersen, A. B., *New Developments and Applications in Gas Lasers*, .SPIE 106(1987), Enhanced CW ion laser operation in the range 270 nm to 380 nm.

1509.   Petersen, A. B., *Conference Digest, Annual Meeting (IEEE/LEOS)*(1989), IEEE/LEOS, Piscataway, NJ, CW ion laser operation in krypton, argon and xenon at wavelengths down to 232 nm.

1510.   Latush, E. L., Sém, M. F., Chebotarev, G. D., *Sov. J. Quantum. Electron.*, 19, 1537 (1989), Recombination gas-discharge lasers utilizing transitions in multiply charged O III and Xe IV ions.

1511.   Hirata, K., Yoshida, H., Ninomiya, H., *Appl. Phys. Lett.*, 57, 1709 (1990), Amplification by stimulated emission from optically pumped nickel vapor in an ultraviolet region.

1512.   Hooker, S. M., Webb, C. E., *IEEE J. Quantum Electron.*, 26, 1529 (1990), F_2 pumped NO: Laser oscillation at 218 nm and prospects for new laser transitions in the 160-250 nm region.

1513.   Hooker, S. M., Webb, C. E., *Opt. Lett.*, 15, 437 (1990), Observation of laser oscillation in nitric oxide at 218 nm.

1514. Dinev, S. G., Hadjichristov, G. B., Marazov, O., *Appl. Phys. B* 52, 290 (1991), Amplified stimulated emission in the NaK (D → X) band by high power copper vapor laser pumping.

1515. Hamada, N., Sauerbrey, R., Wilson, W. L., Tittel, F. K., Nighan, W. L., *IEEE J. Quantum Electron.*, QE-24, 1571-1578 (1988), Performance characteristics of an injection-controlled electron-beam pumped XeF (C → A) laser system.

1516. Mandl, A., Litzenberger, L. N., *Appl. Phys. Lett.*, 53, 1690-1692 (1988), Efficient, long pulse XeF (C → A) laser at moderate electron beam pump rate.

1517. Zuev, V. S., Kashnikov, G. N., Kozlov, N. P., Mamaev, S. B., Orlov, V. K., *Sov. J. Quantum Electron.*, 16, 1665-1667 (1986), Characteristics of an XeF (C-A) laser emitting visible light as a result of optical pumping by surface-discharge radiation.

1518. Chaltakov, I. V., Minkovsky, N. I., Tomov, I. V., *Opt. Commun.*, 65, 33-36 (1988), Transverse discharge pumped triatomic Xe_2F excimer laser.

1519. Loree, T. R., Showalter, R. R., Johnson, T. M., Birmingham, B. S., Hughe, *Opt. Lett.*, 11, 510-512 (1986), Lasing XeO in liquid argon.

1520. McCown, A. W., Eden, J. G., *Appl. Phys. Lett.*, 39, 371-373 (1981), ZnI (B → X) laser: 600-604 nm.

1521. McCown, A. W., Ediger, M. N., Eden, J. G., *Opt. Commun.*, 40, 190-194 (1982), Quenching kinetics and small signal gain spectrum of the ZnI photodissociation laser.

1522. Greene, D. P., Eden, J. G., *Appl. Phys. Lett.*, 42, 20-22 (1983), Discharge pumped ZnI (599-606 nm) and CdI (653-662 nm) amplifiers.

1523. Uehara, Y., Sasaki, W., Saito, S., Fujiwara, E., Kato, Y., Yamanaka, M., *Opt. Lett.*, 9, 539-541 (1984), High power argon excimer laser at 126 nm pumped by an electron beam.

1524. Wellegehausen, B., *IEEE J. Quantum Electron.*, QE-15, 1108-1130 (1979), Optically pumped CW dimer lasers.

1525. Drosch, S., Gerber, G., *J. Chem. Phys.*, 77, 123-130 (1982), Optically pumped cw molecular bismuth laser.

1526. Wodarczyk, F. J., Schlossberg, H. R., *J. Chem. Phys.*, 67, 4476-4482 (1977), An optically pumped molecular bromine laser.

1527. Diegelmann, M., Grieneisen, H. P., Hohla, K., Hu, X.-J., Krasinski, J., *Appl. Phys.*, 23, 283-287 (1980), New TEA lasers based on D' → A' transitions in halogen monofluoride compounds: ClF (284.4 nm) BrF (354.5 nm), IF (490.8 nm).

1528. Bondbey, V. E., *J. Chem. Phys.*, 65, 2296-2304 (1976), Sequential two photon excitation of the C_2 Swan transitions and C_2 relaxation dynamics in rare gas solids.

1529. Ediger, M. N., McCown, A. W., Eden, J. G., *Appl. Phys. Lett.*, 40, 99-101 (1982), CdI and CdBr photodissociation lasers at 655 and 811 nm: CdI spectrum identification and enhanced laser output with $^{114}CdI_2$.

1530. Dinev, S. G., Daniel, H.-U., Walther, H., *Opt. Commun.*, 41, 117-120 (1982), New atomic and molecular laser transitions based on photodissociation of CdI_2.

1531. Greene, D. P., Eden, J. G., *Appl. Phys. Lett.*, 43, 418-420 (1983), Lasing on the B → X band of cadmium monoiodide (CdI) and ^{114}CdI in a UV-preionized transverse discharge.

1532. Zhang, J., Cheng, B., Zhang, D., Zhao, L., Zhao, Y., Wang, T., *Opt. Commun.*, 68, 220 (1988), Atomic lead photodissociation laser at 722.9 nm.

1533. Greene, D. P., Eden, J. G., *Opt. Lett.*, 10, 59-61 (1985), Injection locking and saturation intensity of a cadmium iodide laser.

1534. Diegelmann, M., Hohla, K., Kompa, K. L., *Opt. Commun.*, 29, 334-338 (1979), Interhalogen UV laser on the 285 nm band of ClF*.

1535. Pummer, H., Egger, H., Luk, T. S., Srinivasan, T., Rhodes, C. K., *Phys. Rev. A*, 28, 795-801 (1983), Vacuum-ultraviolet stimulated emission from two photon excited molecular hydrogen.

1536. Cefalas, A. C., Skordoulis, C., Nicolaides, C. A., *Opt. Commun.*, 60, 49-54 (1986), Superfluorescent laser action around 495 nm in the blue-green band of the mercury trimer Hg_3.

1537. Bazhulin, S. P., Basov, N. G., Zuev, V. S., Leonov, Yu. S., Stoilov, Yu., *Sov. J. Quantum Electron.*, 8, 402-403 (1978), Stimulated emission at $\lambda=502$ nm as a result of prolonged optical pumping of $HgBr_2$ vapour.

1538. Bazhulin, S. P., Basov, N. G., Bugrimov, S. N., Zuev, V. S., Kamrukov, A., *Zh. TF Pis'ma Red.*, 12, 1423-1429 (1986), Photodissociation molecular laser emitting in blue-green region with energy of 3J per pulse.

1539. Greene, D. P., Killeen, K. P., Eden, J. G., *Appl. Phys. Lett.*, 48, 1175-1177 (1986), $X^2 S \rightarrow B^2 S$ absorption band of HgBr: optically pumped 502 nm laser.

1540. Greene, D. P., Killeen, K. P., Eden, J. G., *J. Opt. Soc. Am.* B, 3, 1282-1287 (1986), Excitation of the HgBr $B^2S +1/2 \rightarrow X^2S +1/2$ band in the ultraviolet.

1541. Ediger, M. N., McCown, A. W., Eden, J. G., *IEEE J. Quantum Electron.*, QE-19, 263-266 (1983), Spectroscopy and efficiency of the $^{200}Hg^{81}Br$ and ^{64}ZnI photodissociation lasers.

1542. Hanson, F. E., Rieger, H., Cavanaugh, D. B., *Appl. Phys. Lett.*, 43, 622-623 (1983), Relative efficiency of $^{200}Hg^{79}Br$, $Hg^{79}Br$ and HgBr electric discharge lasers.

1543. Sugii, M., Sasaki, K., *Appl. Phys. Lett.*, 48, 1633-1635 (1986), Improved performance of the discharge pumped HgBr and Hg cell lasers by adding SF_6.

1544. Bazhulin, S. P., Basov, N. G., Bugrimov, S. N., Zuev, V. S., Kamrukov, A., *Sov. J. Quantum Electron.*, 16, 990-993 (1986), Mercury-halide vapor molecular laser pumped by a wide-band optical radiation and emitting three-color visible radiation.

1545. Berry, A. J., Whitehurst, C., King, T. A., *J. Phys. D.*, 21, 855-858 (1988), The tunable operation of mixed mercury halide lasers.

1546. Bazhulin, S. P., Basov, N. G., Bugrimov, S. N., Zuev, V. S., Kamrukov, A., *Sov. J. Quantum Electron.*, 16, 836-838 (1986), Green-emitting mercury chloride laser pumped by wide band optical radiation.

1547. Bazhulin, S. P., Basov, N. G., Bugrimov, S. N., Zuev, V. S., Kamrukov, A., *Sov. J. Quantum Electron.*, 16, 663-665 (1986), Blue-violet HgI/HgI_2 laser with wide-band optical pumping by a linearly stabilized surface discharge.

1548. Ewing, J. J., Brau, C. A., *Appl. Phys. Lett.*, 27, 557-559 (1975), Laser action on the 342 nm molecular iodine band.

1549. Shaw, M. J., Edwards, C. B., O'Neill, F., Fotakis, C., Donovan, R. J., *Appl. Phys. Lett.*, 37, 346-348 (1980), Efficient laser action on the 342 nm band of molecular iodine using ArF laser pumping.

1550. Zuev, V. S., Mikheev, L. D., Shirokikh, A. P., *Sov. J. Quantum Electron.*, 12, 342-348 (1982), Investigation of an $I_2(D'- A')$ laser pumped by wide-band radiation.

1551. Zuev, V. S., Mikheev, L. D., Shirokikh, A. P., *Sov. J. Quantum Electron.*, 13, 567-568 (1983), Permissible heating of a medium and specific ultraviolet output energy of an optically pumped I_2 laser.

1552. Sauerbrey, R., Langhoff, H., *IEEE J. Quantum Electron.*, QE-21, 179-181 (1985), Excimer ions as possible candidates for VUV and XUV lasers.

1553. Zuev, V. S., Mikheev, L. D., Startsev, A. V., Shirokikh, A. P., *Sov. J. Quantum Electron.*, 9, 1195-1196 (1979), Pulse periodic operation of an iodine ultraviolet laser pumped by radiation from quartz flashlamps.

1554. Killeen, K. P., Eden, J. G., *Appl. Phys. Lett.*, 43, 539-541 (1983), Gain on the green (504 nm) excimer band of I_2.

1555. Eden, J. G., Killeen, K. P., *Proc. SPIE*, 476, 34,35 (1984), I_2 amplifier in the green.

1556. Kaslin, V. M., Petrash, G. G., Yakushev, O. F., *Sov. J. Quantum Electron.*, 9, 639 (1979), Pulsed stimulated emission due to electronic transitions in the I_2 molecule pumped optically by a copper vapor laser.

1557. Dlabal, M. L., Eden, J. G., *Appl. Phys. Lett.*, 38, 487-491 (1981), Gain and transient absorption profiles for the iodine monofluoride 490 nm and iodine.

1558. Dlabal, M. L., Eden, J. G., *J. Appl. Phys.*, 53, 4503-4505 (1982), On the upper state formation kinetics and line tunability of the iodine monofluoride discharge laser.

1559. Dlabal, M. L., Hutchison, S. B., Eden, J. G., Verdeyen, J. T., *Appl. Phys. Lett.*, 37, 873-875 (1980), Multiline (480-496 nm) discharge-pumped iodine monofluoride laser.

1560. DeYoung, R. J., *Appl. Phys. Lett.*, 37, 690-692 (1980), Lasing characteristics of iodine-monofluoride.

1561. Dlabal, M. L., Hutchison, S. B., Eden, J. G., Verdeyen, J. T., *Opt. Lett.*, 6, 70-72 (1981), Iodine monofluoride 140 kW laser: small signal gain and operating parameters.

1562. Steigerwald, F., Emmert, F., Langhoff, H., Hammer, W., Griegel, T., *Opt. Commun.*, 56, 240-242 (1985), Observation of an ionic excimer state in CsF^+.

1563. Kubodera, S., Frey, L., Wisoff, P. J., Sauerbrey, R., *Opt. Lett.*, 13, 446-448 (1988), Emission from ionic cesium fluoride excimers excited by a laser-produced plasma.

1564. Basov, N. G., Voitik, M. G., Zuev, V. S., Kutakhov, V. P., *Sov. J. Quantum Electron.*, 15, 1455-1460 (1985), Feasibility of stimulated emission of radiation from ionic heteronuclear molecules. I. Spectroscopy.

1565. Basov, N. G., Zuev, V. S., Mikheev, L. D., Yalovoi, V. I., *Sov. J. Quantum Electron.*, 12, 674 (1982), Blue-green laser emission from IF subjected to wide-band optical pumping.

1566. Mikheev, L. D., *Izv. Akad. Nauk SSSR*, Ser. F, 1377-1386 (1987), Visible and UV photochemical lasers.

1567. Zolotarev, V. A., Kryukov, P. G., Podmar'kov, Yu. P., Frolov, M. P. et al., *Sov. J. Quantum Electron.*, 18, 643-646 (1988), Optically pumped pulsed IF (B $\rightarrow$ X) laser utilizing a CF_3I-N, F_2-He mixture.

1568. Davis, S. J., Hanko, L., Wolf, P. J., *J. Chem. Phys.*, 82, 4831-4837 (1985), Continuous wave optically pumped iodine monofluoride B^3P (O^+) $\rightarrow$ X^1S^+ laser.

1569. Davis, S. J., Hanko, L., Shea, R. F., *J. Chem. Phys.*, 78, 172-182 (1983), Iodine monofluoride B^3P (O^+) $\rightarrow$ X^1S^+ lasing from collisionally pumped states.

1570. Davis, S. J., Hanko, L., *Appl. Phys. Lett.*, 37, 692-694 (1980), Optically pumped iodine monofluoride B^3P (O^+) $\rightarrow$ X^1S^+ laser.

1571. Kurosawa, K., Sasaki, W., Fujiwara, E., Kato, Y., *IEEE J. Quantum Electron.*, QE-24, 1908-1914 (1988), High-power narrow-band operation and Raman frequency conversion of an electron beam pumped krypton excimer laser.

1572. Basov, N. G., Zuev, V. S., Kanaev, A. V., Mikheev, L. D., *Sov. J. Quantum Electron.*, 15, 1449, 1450 (1985), Lasing in optically excited Kr cell.

1573. Tittel, F. K., Smayling, M., Wilson, W. L., Marowsky, G., *Appl. Phys. Lett.*, 37, 862-864 (1980), Blue laser action by the rare gas halide trimer Kr_2F.

1574. Basov, N. G., Zuev, V. S., Kanaev, A. V., Mikheev, L. D., Stavrovskii, D., *Sov. J. Quantum Electron.*, 10, 1561-1562 (1980), Stimulated emission from the triatomic excimer Kr_2F subjected to optical pumping.

1575. Basov, N. G., Voitik, M. G., Zuev, V. S., Kutakhov, V. P., *Sov. J. Quantum Electron.*, 15, 1461-1469 (1985), Feasibility of stimulated emission of radiation from ionic heteronuclear molecules, II. Kinetics.

1576. Rajaei-Rizi, A., Bahns, J. T., Verma, K. K., Stwalley, W. C., *Appl. Phys. Lett.*, 40, 869-871 (1982), Optically pumped ring laser oscillation in the $^6Li_2$ molecule.

1577. Kaslin, V. M., Yakushev, O. F., *Sov. J. Quantum Electron.*, 12, 201-203 (1982), Optically pumped pulsed Li_2 laser.

1578. Eden, J. G., *Appl. Phys. Lett.*, 36, 393-395 (1980), 406 nm laser on the C → B band of N$_2$.

1579. Sauerbrey, R., Langhoff, H., *Appl. Phys.*, 22, 399-402 (1980), Lasing in an e-beam pumped Ar-N$_2$ mixture at 406 nm.

1580. Chou, M. S., Zawadzkas, G. A., *J. Quantum Electron.*, QE- 17, 77-81 (1981), Long-pulse N$_2$ lasers at 357.7, 380.5 and 405.9 nm in N$_2$/Ar/Ne/He mixtures.

1581. Basov, N. G., Aleksandrov, A. Y., Danilychev, V. A., Dolgikh, V. A. et al., *JETP Lett.*, 42, 47-50 (1985), Efficient high-pressure quasi CW laser using the first negative system of nitrogen.

1582. Emmert, F., Dux, R., Langhoff, H., *Appl. Phys. B.*, 47, 141-148 (1988), Improved lasing properties of the He-N$^+_2$ system at 391 nm and 428 nm by H$_2$ admixture.

1583. Burrows, M. D., Baughcum, S. L., Oldenborg, R. C., *Appl. Phys. Lett.*, 46, 22-24 (1985), Optically pumped NO (A^{2}S $^+$ → X^{2}P) ultraviolet laser.

1584. Wang, Z. G., Ma, L. A., Xia, H. R., Zhang, K. C., Cheng, I. S., *Opt. Commun.*, 58, 315-318 (1986), The generation of UV and violet diffuse band stimulated radiation in a sodium dimer.

1585. Basov, N. G., Voitik, M. G., Zuev, V. S., Klementov, A. D., Kutakhov, V., *Sov. J. Quantum Electron.*, 17, 106-107 (1987), Efficiency of rare gas-alkali ionic molecules in stimulated emission of ultraviolet and far ultraviolet radiation.

1586. Wu, C. Y., Chen, J. K., Judge, D. L., Kim, C. C., *Opt. Commun.*, 48, 28-32 (1983), Spontaneous amplified emission of a Na$_2$ diffuse violet band produced through two photon excitation of sodium vapor.

1587. Bahns, J. T., Stwalley, W. C., *Appl. Phys. Lett.*, 44, 826-828 (1984), Observation of gain in the violet bands of sodium vapor.

1588. Jones, P. L., Gaubatz, U., Hefter, U., Bergmann, K., Wellegehausen, B., *Appl. Phys. Lett.*, 42, 222-224 (1983), Optically pumped sodium dimer supersonic beam laser.

1589. Bahns, J. T., Verma, K. K., Rajaei-Rizi, A. R., Stwalley, W. C., *Appl. Phys. Lett.*, 42, 336-338 (1983), Optically pumped ring laser oscillation to vibrational levels near dissociation and to the continuum in Na$_2$.

1590. Kanorskii, S. I., Kaslin, V. M., Yakushev, O. F., *Sov. J. Quantum Electron.*, 10, 1275-1276 (1980), Optically pumped Na$_2$ laser.

1591. Dinev, S. G., Koprinkov, I. G., Stefanov, I. L., *Opt. Commun.*, 52, 199-203 (1984), Na$_2$ b^{3}S$_g$- ^{3}S$_u$ excimer laser emission in the ir.

1592. Wang, Q., Wu, T., Lu, Z., Xing, D., Liu, W., Ma, Z., *Bull. Am. Phys. Soc.*, 33, 1673 (1988), Na$_2$ 1^3S$_g$-1^3S$_u$ lasing with peak around 892.0 nm.

1593. Wellegehausen, B., Luhs, W., Topouzkhanian, A., d'Incan, J., *Appl. Phys. Lett.*, 43, 912-914 (1983), Cascade laser emission of optically pumped Na$_2$ molecules.

1594. Bernage, P., Niay, P., Bocquet, H., *J. Mol. Spectrosc.*, 98, 304-314 (1983), Laser transitions among the C^{1}P$_u$, ^{1}S $+_g$, A^{1}S$_u$ and X^{1}S$_g$ states of Na$_2$.

1595. Herman, P. R., Madej, A. A., Stoicheff, B. P., *Chem. Phys. Lett.*, 134, 209-213 (1987), Rovibronic spectra of Ar$_2$ and coupling of rotation and electronic motion.

1596. Shahdin, S., Wellegehausen, B., Ma, Z. G., *Appl. Phys. B.*, 29, 195-200 (1982), Ultraviolet excited laser emission in Na$_2$.

1597. Kaslin, V. M., Yakushev, O. F., *Sov. J. Quantum Electron.*, 13, 1575-1579 (1983), Optically pumped gas laser using electronic transitions in the NaRb molecule.

1598. Leone, S. R., Kosnik, K. G., *Appl. Phys. Lett.*, 30, 346-348 (1977), A tunable visible and ultraviolet laser on S$_2$.

1599. Girardeau-Montaut, J. P., Moreau, G., *Appl. Phys. Lett.*, 36, 509-511 (1980), Optically pumped superfluorescence S$_2$ molecular laser.

1600. Epler, J. E., Verdeyen, J. T., *IEEE J. Quantum Electron.*, QE-19, 1686-1691 (1983), Broad-band gain in optically pumped S$_2$.

1601.   Wellegehausen, B., Topouzkhanian, A., Effantin, C., d'Incan, J., *Opt. Commun.*, 41, 437-442 (1982), Optically pumped continuous multiline Se_2 laser.

1602.   Topouzkhanian, A., Wellegehausen, B., Effantin, C., d'Incan, J. et al., *Laser Chem.*, 1, 195-209 (1983), New continuous laser emissions in Te_2.

1603.   Topouzkhanian, A., Babaky, O., Verges, J., Willers, R., Wellegehausen, B, *J. Mol. Spectrosc.*, 113, 39-46 (1985), Fourier spectroscopic investigations of $^{130}Te_2$ infrared fluorescence and new optically pumped continuous laser lines.

1604.   Basov, N. G., Zuev, V. S., Kanaev, A. V., Mikheev, L. D., *Sov. J. Quantum Electron.*, 15, 1289-1290 (1985), Stimulated emission from an optically pumped Xe_2 cell laser.

1605.   Sauerbrey, R., Tittel, F. K., Wilson, W. L., Nighan, W. L., *IEEE J. Quantum Electron.*, QE-18, 1336-1340 (1982), Effect of nitrogen on XeF (C $\rightarrow$ A) and Xe_2 cell laser performance.

1606.   Nahme, H., Kessler, T., Markus, R., Chergui, M., Schwentner, N., *J. Lumin.*, 40-41, 821-822 (1988), High density excitation of rare gas crystals for stimulated emission.

1607.   Gross, R. W. F., Schneider, L. E., Amimoto, S. T., *Appl. Phys. Lett.*, 53, 2365-2367 (1988), XeF laser pumped by high power sliding discharges.

1608.   Zuev, V. S., Isaev, I. F., Kanaev, A. V., Mikheev, L. D., Stavrovskii, D., *Sov. J. Quantum Electron.*, 11, 221-222 (1981), Lasing as a result of a B-X transition in the excimer XeF formed as a reult of photodissociation of KrF_2 in mixtures with Xe.

1609.   Zuev, V. S., Mikheev, L. D., Stavrovskii, D. B., *Sov. J. Quantum Electron.*, 14, 1174-1178 (1984), Efficiency of an optically pumped XeF laser.

1610.   Shahidi, M., Jara, H., Pummer, H., Egger, H., Rhodes, C. K., *Opt. Lett.*, 10, 448-450 (1985), Optically excited XeF* excimer laser in liquid argon.

1611.   Fisher, C. H., Center, R. E., Mullaney, G. J., McDaniel, J. P., *Appl. Phys. Lett.*, 35, 26-28 (1979), A 490 nm XeF electric discharge laser.

1612.   Burnham, R., *Appl. Phys. Lett.*, 35, 48, 49 (1979), A discharge pumped laser on the C $\rightarrow$ A transition of XeF.

1613.   Campbell, J. D., Fisher, C. H., Center, R. E., *Appl. Phys. Lett.*, 37, 348-350 (1980), Observations of gain and laser oscillation in the blue-green during direct pumping of XeF by microsecond electron beam pulses.

1614.   Nighan, W. L., Nachson, Y., Tittel, F. K., Wilson, W. L., Jr., *Appl. Phys. Lett.*, 42, 1006-1008 (1983), Optimization of electrically excited XeF (C $\rightarrow$ A) laser performance.

1615.   Nighan, W. L., Tittel, F. K., Wilson, W. L., Nishida, N., Zhu, Y., Sauer, *Appl. Phys. Lett.*, 45, 947-949 (1984), Synthesis of rare gas-halide mixtures resulting in efficient XeF (C $\rightarrow$ A) laser oscillation.

1616.   Tittel, F. K., Marowsky, G., Nighan, W. L., Zhu, Y. et al., *IEEE J. Quantum Electron.*, QE-22, 2168-2173 (1986), Injection-controlled tuning of an electron beam excited XeF (C $\rightarrow$ A) laser.

1617.   Huestis, D. L., Hill, R. M., Eckstrom, D. J. et al., SRI International Report No. MP78-07 (May, 1978), New electronic transition laser systems.

1618.   Deutsch, T. F., *Appl. Phys. Lett.*, 11, 18 (1967), Laser emission from HF rotational transitions.

1619.   Welling, H., Wellegehausen, B., *Laser Spectroscopy III* (1977), 365-369, Springer-Verlag, New York, Optically pumped continuous alkali dimer lasers.

1620.   Niay, P., Bernage, P., *C. R. Acad. Sci.*, Ser. II, 294, 627-632 (1982), Spectroscopie de polarisation du sodium moléculaire sur des transitions $^1S_g \rightarrow A^1S_u$ situees dans l'infraroug.

1621.   Miller, H. C., Yamasaki, K., Smedley J. E., Leone, S. R., *Chem. Phys. Lett.*, 181, 250 (1991), An optically pumped laser on SO (B $^3\Sigma^-$ - X $^3\Sigma^-$).

1622.   Sasaki, W., Kurosawa, K., Herman, P. R., Yoshida, K., Kato, Y., *Proc. Conf. on Short Wavelength Coherent Radiation*, Generation and applications of intense coherent radiation in the VUV and XUV region with electron beam pumped rare gas excimer lasers.

1623. Eden, J. G., *Proc. Int. Conf. on Lasers '82*, 346-356(1982), Metal halide dissociation lasers.

1624. Jeffers, W. Q., *AIAA J.*, (1988), Short wavelength chemical lasers.

1625. Petersen, F. R., Wells, J. S., Maki, A. G., Siemsen, K. J., *Appl. Opt.*, 20, 3635 (1981), Heterodyne frequency measurements of $^{14}CO_2$ laser hot band transitions.

1626. Drozdowicz, Z., Rudko, R. I., Kinhares, S. J., Lax, B., *IEEE J. Quantum Electron.*, QE-17, 1574 (1981), High gain 4.3-4.5 μm optically pumped CO_2 laser.

1627. Rutt, H. N., *Infrared Phys.*, 24, 535 (1984), Optically-pumped laser action in monodeuteroacetylene.

1628. Fischer, T. A., Wittig, C., *Appl. Phys. Lett.*, 41, 107 (1982), Rotationally relaxed, grating tuned laser oscillations in optically pumped C_2D_2.

1629. Kroeker, D. F., Reid, J., *Appl. Opt.*, 25, 86 (1986), Line-tunable cw ortho- and para-NH_3 lasers operating at wavelengths of 11 to 14 mm.

1630. Rolland, C., Reid, J., Gardise, B. K., *Appl. Phys. Lett.*, 44, 380 (1984), Line-tunable oscillation of a cw NH_3 laser from 10.7 to 13.3 μm.

1631. Pinson, P., Delage, A., Girard, G., Michon, M., *J. Appl. Phys.*, 52, 2634 (1981), Characteristics of two-step and two-photon-excited emissions in $^{14}NH_3$.

1632. Morrison, H. D., Reid, J., Garside, B. K., *Appl. Phys. Lett.*, 45, 321 (1984), 16-21 μm line-tunable NH_3 laser produced by two-step optical pumping.

1633. Wessel, R., Theiler, T., Keilmann, F., *IEEE J. Quantum Electron.*, QE-23, 385 (1987), Pulsed high-power mid-infrared gas lasers.

1634. Bobrovskii, A. N., Kiselev, V. P., Kozhevnikov, A. V., Kokhanskii, V. V., *Sov. J. Quantum Electron.*, 13, 1521 (1983), Two-photon optical pumping of NH_3 in a multipass cell.

1635. Akhrarov, M., Vasilev, B. I., Grasyuk, A. Z., Soskov, V. I., *Sov. J. Quantum Electron.*, 14, 572 (1984), Middle-infrared laser utilizing isotopically substituted $^{15}NH_3$ ammonia molecules.

1636. Akhrarov, M., Vasilev, B. I., Grasyuk, A. Z., Soskov, V. I. et al., *Sov. J. Quantum Electron.*, 16, 1016 (1986), $^{15}NH_3$ laser with two-photon optical pumping.

1637. Rutt, H. N., Travis, D. N., Hawkins, K. C., *Int. J. Infrared Millimeter Waves*, 5, 1201 (1984), Mid infrared and far infrared studies of the propyne optically pumped laser.

1638. Hon, J. F., Novak, J. R., *IEEE J. Quantum Electron.*, QE-11, 698 (1975), Chemically pumped hydrogen fluoride overtone laser.

1639. Bashkin, A. S., Igoshin, V. I., Leonov, Yu. S., Oraevskii, A. N. et al., *Sov. J. Quantum Electron.*, 7, 626 (1977), An investigation of a chemical laser emitting due to an overtone of the HF molecule.

1640. Bradley, L. C., Soohoo, K. L., Freed, C., *IEEE J. Quantum Electron.*, QE-22, 234 (1986), Absolute frequencies of lasing transitions in nine CO_2 isotopic species.

1641. Vedeneev, A. A., Volkov, A. Yu., Demin, A. I., Kudryavtsev, E. M., *Sov. Tech. Phys. Lett.*, 8, 110 (1982), 18.4 μm CO_2-Ne gas dynamic laser.

1642. Akimov, V. A., Volkov, A. Yu., Demin, A. I., Kudryavtsev, E. M. et al., *Sov. J. Quantum Electron.*, 13, 556 (1983), Continuous-wave CO_2-Ar gasdynamic laser emitting at 18.4 μm.

1643. Johnson, R. P., Cornelison, D., Duzcek, C. J., *Appl. Phys. Lett.*, 44, 162 (1984), 4.4 μm cascade $^{13}C^{16}O_2$ laser.

1644. Johnson, R. P., *Appl. Phys. Lett.*, 44, 1119 (1984), Further characterization of the 4.4 μm $^{13}C^{16}O_2$ cascade laser.

1645. Jeffers, W. Q., unpublished (1988), Private communication.

1646. Knight, D. J. E., *Handbook of Laser Science and Technology*, Volume II, Gas Lasers, CRC Press, Boca Raton, FL, Vol. II, Tables of far-infrared cw gas laser lines.

1647. Whitford, B. G., *IEEE Trans. Instr. Meas.*, IM-29, 168 (1980), Measurement of the absolute frequencies of CO_2 laser transitions by multiplication of CO_2 laser difference frequencies.

1648. Saxon, G., *Review of Laser Engineering*, 13, 398 (1985), Free electron lasers - a survey.

1649. Davies, P. B., Jones, H., *Appl. Phys.*, 22, 53-55 (1980), New cw far-infrared molecular lasers from ClO_2, HCCF, FCN, CH_3NC, CH_3F and propynal.

1650. Taubmann, G., Jones, H., Davies, P. B., *Appl. Phys. B.*, 41, 179 (1986), New cw optically pumped FIR laser lines.

1651. Dyubko, S. F., Svich, V. A., Fesenko, L. D., *Sov. Tech. Phys. Lett.*, 1, 192-193 (1975), Magnetically tuned submillimeter laser based on paramagnetic ClO_2.

1652. Bugaev, V. A., Shliteris, E. P., *Sov. J. Quantum Electron.*, 14, 1331 (1984), submillimeter lasing transitions in isotopic modifications of SO_2.

1653. Calloway, A. R., Danielewicz, E. J., *Int. J. Infrared Millimeter Waves*, 2, 933 (1981), Predicted new optically pumped FIR molecular lasers.

1654. Yoshida, H., Ninomiya, H., Takashima, N., *Appl. Phys. Lett.*, 59, 1290 (1991), Laser action in a KrF laser pumped Ta vapor.

1655. Shafik, S., Crocker, D., Landsberg, B. M., Butcher, R. J., *IEEE J. Quantum Electron.*, QE-17, 115 (1981), Phosphine far-infrared CW laser transitions: optical pumping at more than 100 MHz from resonance.

1656. Jones, H., Davies, P. B., *IEEE J. Quantum Electron.*, QE-17, 13 (1981), A powerful new optically pumped FIR laser - formylfluoride.

1657. Jones, H., Davies, P. B., Lewis-Bevan, W., *Appl. Phys. B.*, 30, 1 (1983), New FIR laser lines from optically pumped DCOF.

1658. Temps, F., Wagner, H. G., *Appl. Phys. B.*, 29, 13 (1982), Strong far-infrared laser action in carbonyl fluoride and vinyl fluoride.

1659. Tobin, M. S., *Opt. Lett.*, 7, 322 (1982), Carbonyl fluoride (COF_2): strong new continuous-submillimeter-wave laser source, Erratum: *Opt. Lett.*, 8, 509 (1983).

1660. Tobin, M. S., Daley, T. W., *Int. J. Infrared Millimeter Waves*, 7, 1649 (1986), Heterodyne frequency measurements of COF_2, CD_3F, $^{13}CD_3F$ and C_2H_3I lasers.

1661. Davis, I. H., Pharaoh, K. I., Knight, D. J. E., *Int. J. Infrared Millimeter Waves*, 8, 765 (1987), Frequency measurements on far-infrared laser emissions of carbonyl fluoride (COF_2).

1662. Tobin, M. S., *IEEE J. Quantum Electron.*, QE-20, 5 and 985 (1984), SMMW laser emission and frequency measurements in doubly deuterated methyl fluoride (CHD_2F).

1663. Tobin, M. S., Felock, R. D., *IEEE J. Quantum Electron.*, QE-17, 825 (1981), Submillimeter wave laser emission in optically pumped methyl fluoride - $^{13}CD_3F$.

1664. Davies, P. B., Stern, D. P., *Int. J. Infrared Millimeter Waves*, 3, 909 (1982), New cw FIR laser lines from optically pumped silyl fluoride (SiH_3F).

1665. Inguscio, M., Moruzzi, G., Evenson, K. M., Jennings, D. A., *J. Appl. Phys.*, 60, R161 (1986), A review of frequency measurements of optically pumped lasers from 0.1 to 8 THz.

1666. Scalabrin, A., Tomaselli, J., Pereira, D., Vasconcellos, E. C. C., Evenson, K. M., *Int. J. Infrared Millimeter Waves*, 6, 973 (1985), Optically pumped $^{13}CH_2F_2$ laser: wavelength and frequency measurements.

1667. Vasconcellos, E. C. C., Petersen, F. R., Evenson, K. M., *Int. J. Infrared Millimeter Waves*, 2, 705 (1981), Frequencies and wavelengths from a new, efficient FIR lasing gas: CD_2F_2.

1668. Tobin, M. S., Sattler, J. P., Daley, T. W., *IEEE J. Quantum Electron.*, QE-18, 79 (1982), New SMMW laser transitions optically pumped by a tunable CO_2 waveguide laser.

1669. Anacona, J. R., Davies, P. B., Ferguson, A. H., *IEEE J. Quantum Electron.*, QE-20, 829 (1984), Optically pumped FIR laser action in chlorofluoromethane (CH_2ClF).

1670. Tobin, M. S., Felock, R. D., *Opt. Lett.*, 5, 430 (1980), Fluoroform-d optically pumped submillimeter-wave laser.

1671. Tobin, M. S., Leavitt, R. P., Daley, T. W., *J. Mol. Spectrosc.*, 101, 212 (1983), Spectroscopy of the v5 band of CDF_3 by heterodyne measurement of SMMW laser emission frequencies.

1672. Tobin, M. S., Daley, T. W., *IEEE J. Quantum Electron.*, QE-16, 592 (1980), Optically pumped $CHClF_2$ and C_2H_3I submillimeter wave lasers.

1673. Telle, J., *IEEE J. Quantum Electron.*, QE-19, 1469 (1983), Continuous wave 16 μm CF_4 laser.

1674. Vasconcellos, E. C. C., Wyss, J. C., Petersen, F. R., Evenson, K. M., *Int. J. Infrared Millimeter Waves*, 4, 401 (1983), Frequency measurements of far infrared cw lasing lines in optically pumped $CHCl_2F$.

1675. Gastaud, G., Redon, M., Fourrier, M., *Int. J. Infrared Millimeter Waves*, 8, 1069 (1987), Further new cw laser investigations in CH_3I, $^{13}CH_3I$ and CD_3I: new lines and assignments.

1676. Inguscio, M., Evenson, K. M., Peterson, F. R., Strumia, F., Vasconcellos, E. C. C., *Int. J. Infrared Millimeter Waves*, 5, 1289 (1984), A new efficient far infrared lasing molecule: $^{13}CD_3OH$.

1677. Ioli, N., Moretti, A., Strumia, F., D'Amato, F., *Int. J. Infrared Millimeter Waves*, 7, 459 (1986), $^{13}CH_3OH$ and $^{13}CD_3OH$ optically pumped FIR laser: new large offset emission and optoacoustic spectroscopy.

1678. Vasconcellos, E. C. C., Evenson, K. M., *Int. J. Infrared Millimeter Waves*, 6, 1157 (1985), New far infrared laser lines obtained by optically pumping $^{13}CD_3OD$.

1679. Landsberg, B. M., Shafik, M. S., Butcher, R. J., *IEEE J. Quantum Electron.*, QE-17, 828 (1981), CW optically pumped far-infrared emissions from acetaldehyde, vinyl chloride, and methyl isocyanide.

1680. Gilbert, B., Butcher, R. J., *IEEE J. Quantum Electron.*, QE-17, 827 (1981), New optically pumped millimeter wave cw laser lines from methyl isocyanide.

1681. Duxbury, G., Petersen, J. C., *Appl. Phys. B*, 35, 127 (1984), Optically pumped submillimetre laser action in formaldoxime and ammonia.

1682. Dyubko, S. F., Fesenko, L. D., Shevyrev, A. S., Yartsev, V. I., *Sov. J. Quantum Electron.*, 11, 1248 (1981), New emission lines of methylamine and methyl alcohol molecules in optically pumped laser.

1683. Dyubko, S. F., Fesenko, L. D., Shevyrev, A. S., Yartsev, V. I., *Sov. J. Quantum Electron.*, 11, 1247 (1981), Optically pumped CH_3NO_2 and CH_3 COOD submillimeter lasers.

1684. Fourrier, M., Redon, M., *Opt. Commun.*, 64, 534 (1987), A new cw FIR lasing medium: methyl fluoroform.

1685. Bugaev, V. A., Shliteris, E. P., *Sov. J. Quantum Electron.*, 13, 150 (1983), Optically pumped molecular laser utilising C_2H_5Br and C_2H_3I halogen derivatives of ethane.

1686. Bugaev, V. A., Shliteris, E. P., Klement'ev, Yu. F., Kudryashova, V. A., *Sov. J. Quantum Electron.*, 12, 304 (1982), Laser spectroscopy, submillimeter lasing, and passive Q switching when dimethyl ether is pumped with CO_2 laser radiation.

1687. Clairon, A., Dahmani, B., Filimon, A., Rutman, J., *IEEE Trans. Instr. Meas.*, IM-34, 265 (1985), Precise frequency measurements of CO_2/OsO_4 and He-Ne/CH_4 stabilized lasers.

1688. Kantowicz, P., Palluel, P., Pontvianne, J., *Microwave J.*, 22, 57 (1979), New developments in submillimeter-wave BWOs.

1689. Evenson, K. M., Jennings, D. A., Petersen, F. R., *Appl. Phys. Lett.*, 44, 576 (1984), Tunable far-infrared spectroscopy.

1690. unknown, *Reviews of Infrared and Millimeter Waves*, Vol. 2 (1984), Optically pumped far-infrared lasers, Plenum Press, New York.

1691. Ninomiya, H., Hirata, K., *J. Appl. Phys.*, 66, 2219 (1989), Laser action of optically pumped atomic titanium vapor.

1692. Knight, D. J. E., NPL Report QU 45(Feb. 1981), Ordered list of far-infrared laser lines (continuous, $\lambda > 12$ mm), National Physical Laboratory, Teddington, Middlesex, England.

1693. Ninomiya, H., Hirata, K., Yoshino, S., *J. Appl. Phys.*, 66, 3961 (1989), Optically pumped titanium vapor laser at 625.8 nm.

1694. Dyubko, S. F., Fesenko, L. D., *3rd Int. Conf. on Submillimetre Waves and Their Applications*, 70, Frequencies of optically pumped submillimeter lasers.

1695. Davies, P. B., Ferguson, A. H., Stern, D. P., *Proc. 3rd Int. Conf. on Infrared Physics* (July 1984), New optically pumped lasers containing silicon.

1696. Dyubko, S. F., Fesenko, L. D., Polevoy, B. I., *VIII Conf. on Quantum Electronics and Nonlinear Optics*, Poznan, Poland, Spectrum investigation of the submillimeter CF_2HCl laser radiation.

1697. Fourrier, M., Belland, P., Gastaud, C., Redon, M., *Proc. 3rd Int. Conf. on Infrared Physics*, 803 (1984), New cw FIR lasing lines in optically pumped vinyl halides: CH_2CHCl, CH_2CHF, CH_2CHBr.

1698. Flynn, G. W., Feld, M. S., Feldman, B. J., *Bull. Am. Phys. Soc.*, 12, 15 (1967), New infrared-laser transition and g-values in atomic oxygen.

1699. Tobin, M. S., *IEEE Quantum Electron.*, QE-17, 825 (1981), Submillimeter wave laser emission of optically pumped methyl fluoride - $^{13}CD_3F$.

1700. Tobin, M. S., Daley, T. W., *Int. J. Infrared Millimeter Waves*, 7, 1649 (1986), Heterodyne frequency measurements of COF_2, CD_3F, $^{13}CD_3F$ and C_2H_2I lasers.

1701. McCombie, J., Petersen, J. C., Duxbury G., *Electronic & Electr. Opt.*, Knight, P. L., Ed. (1981).

1702. Vasconcellos, E. C. C., Jennings, D. A., *Int. J. Infrared Millimeter Waves*, 6, 157 (1985), New far infrared laser lines obtained by optically pumping $^{13}CD_3OD$.

1703. Inguscio, M., Evenson, K. M., Petersen, F. R. Strumia, F., Vasconcellos, *Int. J. Infrared Millimeter Waves*, 5, 1289 (1984) E., A new efficient far infrared lasing molecule: $^{13}CD_3OH$.

1704. Scalabrin, A., Tomaselli, J., Pereira, D. et al., *Int. J. Infrared Millimeter Waves*, 6, 973 (1985), Optically pumped $^{13}CH_2F_2$ laser: wavelength and frequency measurements.

1705. Gastaud, C., Redon, M., Fourrier, M., *Int. J. Infrared Millimeter Waves*, 8 1069 (1987), Further new cw laser investigations in CH_3I, $^{13}CH_3I$ and CD_3I: new lines and assignments.

1706. Pereira, D., Scalabrin, A., *Appl. Phys. B*, 44, 67 (1987), Measurement and assignment of new FIR laser lines in $^{12}CH_3OH$ and $^{13}CH_3OH$.

1707. Ioli, N., Moretti, A., Strumia, F., D'Amato, F., *Int. J. Infrared Millimeter Waves*, 7, 459 (1985), $^{13}CH_3OH$ and $^{13}CD_3OH$ optically pumped FIR laser: new large offset emission and optoacoustic spectroscopy.

1708. Belland, P., Fourrier, M., *Int. J. Infrared Millimeter Waves*, 7, 1251 (1986), Submillimeter emission lines from CD_2Cl_2 optically pumped lasers.

1709. Shevirev, A. S., Dyubko, S. F., Fesenko, L. D.and Yartsev, V. I., *Sov. J. Quantum Electron.*, 16, 568 (1987), New emission lines of a submillimeter molecular CD_2Cl_2 laser.

1710. Vasconcellos, E. C.C., Petersen, F. R., Evenson, K. M., *Int. J. Infrared Millimeter Waves*, 2, 705 (1981), Frequencies and wavelengths from a new, efficient FIR lasing gas: CD_2F_2.

1711. Dyubko, S. F., Fesenko, L. D., Ferguson, A. H., *3rd Int. Conf. Submm. Waves and Appl.*, Univ. Surrey (1978).

1712. Landsberg, B. M., *Appl. Phys. B*, 23, 345 (1980), New cw optically pumped FIR emission in HCOOH, D_2CO and CD_3Br.

1713. Mathias, L. E. S., Crocker, A., Phys. *Lett.*, 13, 35 (1964), Stimulated emission in the far-infrared from water vapour and deuterium oxide discharges.

1714. Vasconcellos, E. C. C., Scalabrin, A., Petersen, F. R., Evenson, K. M., *Int. J. Infrared Millimeter Waves*, 2, 533 (1981), New FIR laser lines and frequency meaurements in CD_3OD.

1715. Pereira, D., Vasconcellos, E. C. C., Scalabrin, A., Evenson. K. M., Petersen, F. R., Jennings, D. A., *Int. J. Infrared Millimeter Waves*, 6, 877 (1985), Measurement of new FIR laser lines in CD_3OD.

1716. Fourrier, M., Kreisler, A., *Appl. Phys. B*, 41, 57 (1986), Further investigation on ir pumping of CH_3OD and CD_3OD by a cw CO_2 laser.

1717. Petersen, J. C., Duxbury, G., *Appl. Phys. B*, 37, 209 (1985), New submillimetre laser lines from CH_3OD and CD_3OD.

1718. Garelli, G., Ioli, N., Moretti, A., Pereira, D., Strumia, F., *Appl. Phys. B*, *44*, 111 (1987), New large effect far-infrared laser lines from CD_3OH.

1719. Pereira, D., Ferrari, C. A., Scalabrin, A., *Int. J. Infrared Millimeter Waves*, 7, 1241 (1986), New optically pumped FIR laser lines in CD_3OD.

1720. Sigg, H., Bluyssen, H. J. A., Wyder, P., *IEEE Quantum Electron.*, QE-20, 616 (1984), New laser lines with wavelengths from $\lambda = 61.7$ μm down to $\lambda = 27.2$ μm in optically pumped CH_3OH and CD_3OH.

1721. Tobin, M. S., Leavitt, R. P., Daley, T. W., *J. Mol.. Spectrosc.*, 101, 212 (1983), Spectroscopy of the n5 band of CDF_3 by heterodyne measurement of SMMW laser emission frequencies.

1722. Tobin, M. S., Sattler, J. P., Daley, T. W., *IEEE Quantum Electron.*, QE-18, 79 (1982), New SMMW laser transitions optically pumped by a tunable CO_2 waveguide laser.

1723. Lourtioz., J.-M., Ponnau, J., Morillon-Chapey, M., Deroche, J.-C., *Int. J. Infrared Millimeter Waves*, 2, 49 (1981), Submillimeter laser action of CW optically pumped CF_2Cl_2 (fluorocarbon 12.

1724. Lourtioz., J.-M., Ponnau, J., Meyer, C., *Int. J. Infrared Millimeter Waves*, 2, 525 (1981), Optically pumped CF_3Br FIR laser. New emission lines and tentative assignments.

1725. Herman. H., Wiggleworth, M. J., *Int. J. Infrared Millimeter Waves*, 5, 29 (1984), New FIR lasers from CO_2 laser-excited 1,1-difluoethene.

1726. Fourrier, M., Gastaud, C., Redon, M., Deroche, J. C., *Opt. Commun.*, 48, 347 (1984), New CW far infrared lasing lines in optically pumped 1,1-difluoroethylene.

1727. Belland, P., Gastaud, C., Redon, M., Fourrier, M., *Appl. Phys. B*, 34, 175 (1984), New cw fir laser emission from CO_2 laser-pumped vinyl bromide,

1728. Fourrier, M., Belland, P., Mangili, D., *IEEE Quantum Electron.*, QE-20, 85 (1984), New CW FIR laser action in optically pumped vinyl chloride.

1729. Taubmann. G., Jones, H., Davies, P. B., *Appl. Phys. B*, 41 179 (1986), New cw optically pumped FIR laser lines.

1730. Gastaud, C., Redon, M., Belland, P., Fourrier, M., *Int. J. Infrared Millimeter Waves*, 5, 875 (1984), Far-infrared laser action in vinyl chloride, vinyl bromide, and vinyl fluoride optically pumped by a cw N_2O laser.

1731. Gastaud, C., Redon, M., Belland, P., Kreisler. A., Fourrier, M., *Int. J. Infrared Millimeter Waves*, 6, 63 (1985), Optical pumping by CO_2 and N_2O lasers: new CW FIR emission in vinyl cyanide.

1732. Redon, M., Gastaud, C., Fourrier, M., *Opt. Lett.*, 9, 71 (1984), Far infrared laser lines in vinyl fluoride optically pumped by a CO_2 laser near 10 μm.

1733. Temps, F., Wagner, H. Gg., *Appl. Phys. B*, 29, 13 (1982), Strong far-infrared laser action in carbonyl fluoride.

1734.   Horiuchi, Y., Murai, A., *IEEE Quantum Electron.*, QE-12, 547 (1976), Far-infrared laser oscillations from H_2CO.

1735.   Akitt, D. P., Yardley, J. T., *IEEE Quantum Electron.*, QE-6, 113 (1970), Far-infrared laser emission in gas discharges containing boron trihalides.

1736.   Herman. H., Wiggleworth, M. J., *Int. J. Infrared Millimeter Waves*, 3, 395 (1982), Far infrared stimulated emissions from CO_2 laser excited dichloromethane.

1737.   Anacona. J. R., Davies, P. B., Ferguson, A. H., *IEEE Quantum Electron.*, QE-20, 829 (1984), Optically pumped FIR laser action in chlorofluoromethane (CH_2ClF).

1738.   Petersen, F. R., Scalabrin, A., Evenson, K. M., *Int. J. Infrared Millimeter Waves*, 1, 111 (1980), Frequencies of cw FIR laser lines from optically pumped CH_2F_2.

1739.   Golby, J. A., Cross, N. R., Knight, D. J. E., *Int. J. Infrared Millimeter Waves*, 7, 1309 (1986), Frequency; measurements on far-infrared emission from $^{12}C^{16}O$-pumped methyl chloride and from $^{12}C^{16}O$-pumped difluoromethane.

1740.   Duxbury, G., Petersen. J. C., *Appl. Phys. B*, 35, 127 (1984), Optically pumped sub-millimetre laser action in formaldoxime and ammonia.

1741.   Ioli, N., Moretti, A., Pereira, D., Strumia, F., Garelli, G., *Appl. Phys. B*, 48, 299 (1987), A new efficient far-infrared optically pumped laser gas: $CH_3^{18}OH$.

1742.   Bugaev, V. A., Shliteris, E. P., *Sov. J. Quantum Electron.*, 15, 547 (1985), Use of ethyl alcohol and its deuteroderivatives CH_3CHDOH and in CH_3CD_2OH generation of submillimeter radiation.

1743.   Fourrier, M., Redon, M., *Opt. Commun.*, 64, 534 (1987), A new cw FIR lasing medium: methyl fluoroform.

1744.   Bugaev, V. A., Shliteris, E. P., *Sov. J. Quantum Electron.*, 13, 150 (1983), Optically pumped molecular laser utilising C_2H_2Br and C_2H_2Cl halogen derivatives of ethane.

1745.   Douglas, N. G., Krug, P. A., *IEEE Quantum Electron.*, QE-18, 1409 (1982), CW laser action in ethyl chloride.

1746.   Müller, W. M., Flesher, G. T., *Appl. Phys. Lett.*, 8, 217 (1966), Continuous wave submillimeter oscillation in H_2O, D_2O and CH_3CN.

1747.   Fourrier, M., Belland, P., Redon, M., Gastaud, C., *Proceedings Third Conference on Infrared Physics (CIRP3)*, New cw FIR lasing lines in optically pumped vinyl halides: CH_2CHCl, CH_2CHF,CH_2CHBr. See alsoReferences 1727, 1728 and 1730.

1748.   Landsberg, B. M., Shafki, M. S., Butcher, R. J., *Int. J. Infrared Millimeter Wave*, 2, 49 (1981), Submillimeter laser action of cw optically pumped CH_2Cl_2 (fluorocarbon 12).

1749.   Dyubkko, S. F., Fesenko, L. D., Shevyrev, A. S., Yartsev, V. I., *Sov. J. Quantum Electron.*, 11, 1247 (1981), Optically pumped CH_3NO_2 and CH_3COOD submillimeter lasers.

1750.   Gilbert, B., Butcher, R. J., *IEEE Quantum Electron.*, QE-17, 827 (1982), New optically pumped millimeter wave cw laser lines from methyl isocyanide.

1751.   Landsberg, B. M., *Appl. Phys. B*, 23, 127 (1980), New cw FIR lines from optically pumped ammonia analogues.

1752.   Dyubko, S. F., Fesenko, L. D., Shevyrev, A. S., Yartsev, V. I., *Sov. J. Quantum Electron.*, 11, 1248 (1981), New emission lines of methylamine and methyl alcohol molecules in optically pumped lasers.

1753.   Hassler, J. C., Coleman., P. D., *Appl. Phys. Lett.*, 14, 135 (1969), Far infrared lasing in H_2S, OCS, and SO_2.

1754.   Bugaev, V. A., Shliteris, E. P., Klement'ev, Yu. F., Kudryashova, V. A., *Sov. J. Quantum Electron.*, 12, 304 (1982), Laser spectroscopy, submillimeter lasing, and passive Q switching when dimethyl ether is pumped with CO_2 laser radiation.

1755.   Deutsch, T. F., *IEEE Quantum Electron.*, QE-3, 419 (1967), New infrared laser transitions HCl, HBr, DCl, and DBr.

1756. Yin, G. Y., Barty, C. P. J., King, D. A., Walker, D. J., Harris, S. E., Young, J. F., *Opt. Lett.*, 12, 331 (1987), Low energy pumping of a 108.9-nm xenon Auger laser.

1757. Moore, C. E., *Atomic Energy Levels*, Circular 467, Vols. 1, 2, and 3, U.S. National Bureau of Standards, Washington, DC (1949, 1952, and 1958).

1758. Beck, R., English, W., Gurs, K., *Tables of Laser Lines in Gases and Vapors*, 3rd. edition, Springler-Verlag, Berlin (1980).

1759. Striganov, A. R., Sventitskii, N. S., *Tables of Spectral Lines of Neutral and Ionized Atoms*, IFI/Plenum Press, New York (1968).

1760. Wernsman, B., Prabhuram, T., Lewis, K., Gonzalez, F., Villagran, M., Rocca, J. J., *IEEE J. Quantum Electron.*, 24, 1554 (1988), CW silver ion laser with electron beam excitation.

1761. Kielkopf, J., *J. Opt. Soc. Am. B* 8, 212 (1991), Lasing in the aluminum and indium resonance lines following photoionization and recombination in the presence of H_2.

1762. Lomaev, M. I., Nagornyi, D. Yu., Tarasenko, V. F., Fedenev,A. V., Kirillin, G. V., *Sov. J. Quantum Electron.*, 19, 1321 (1989), Lasing due to atomic transitions of rare gases in mixtures with NF_3.

1763. Qi, N., Krishnan, M., Phys. *Rev. Lett.*, 59, 2051 (1987), Photopumping of a C III ultraviolet laser by Mn VI line radiation.

1764. Diemer, U., Demtröder, W., *Chem. Phys. Lett.*, 175, 135 (1991), Infrared atomic Cs laser based on optical pumping of Cs_2 molecules.

1765. Goldsmith, J. E. M., *J. Opt. Soc. Am. B* 6, 1979 (1989), Two photon excited stimulated emission from atomic hydrogen in flames.

1766. Auyeung, R. C. Y., Cooper, D. G.,Kim, S., Feldman, B. J., *Opt. Commun.*, 79, 207 (1990), Stimulated emission in atomic hydrogen at 656 nm.

1767. Berdnikov, A. A., Derzhiev, V. I., Murav'ev, I. I., Yakovlenko, S. I., Yancharina, A. M., *Sov. J. Quantum Electron.*, 17, 1400 (1987), Penning plasma laser utilizing new transitions in the helium atom resulting in the emission of visible light.

1768. Zolotarev, V. A., Kryukov, P. G., Podmar'kov, Yu. P., Frolov, M. P., Shcheglov, V. A., *Sov. J. Quantum Electron.*, 18, 643 (1988), Optically pumped pulsed IF (B→X) laser utilizing a CF_3I-NF_2-He mixture.

1769. Dinev, S. G., Hadjichristov, G. B., Stefanov, I. L., *Opt. Commun.*, 75, 273 (March 1990), Stimulated emission by hybrid transitions via a heteronuclear molecule.

1770. Ozcomert, J. S., Jones, P. L., *Chem. Phys. Lett.*, 169, 1 (1990), An optically pumped K_2 supersonic beam laser.

1771. Dinev, S. G., Hadjichristov, G. B., Marazov, O., *Appl. Phys. B* 52, 290 (1991), Amplified stimulated emission in the NaK (D→X) band by high power copper vapor laser pumping.

1772. Hooker,S. M., Webb, C. E., *Opt. Lett.*,15, 437 (1990), Observation of laser oscillation in nitric oxide at 218 nm.

1773. Hooker,S. M., Webb, C. E., *IEEE J. Quantum Electron.*, 26, 1529-1535 (September 1990), F_2 pumped NO: Laser oscillation at 218 nm and prospects for new laser transitions in the 160-250 nm region.

1774. Hirata, K., Yoshino, S., Ninomiya, H., *Appl. Phys. Lett.*, 57, 1709 (1990), Amplification by stimulated emission from optically pumped nickel vapor in an ultraviolet region.

1775. Latush, E. L., Sém, M. F., Chebotarev, G. D., *Sov. J. Quantum. Electron.*, 19, 1537 (1989), Recombination gas-discharge lasers utilizing transitions in multiply charged O III and Xe IV ions.

1776. Zhang, J., Cheng, B., Zhang, D., Zhao, L., Zhao, Y., Wang, T., *Opt. Commun.*, 68, 220 (1988), Atomic lead photodissociation laser at 722.9 nm.

1777. Miller, K., H. C. Yamasaki, H. C., Smedley, J. E., Leone, S. R., *Chem. Phys. Lett.*, 181, 250 (1991), An optically pumped laser on SO (B $^3\Sigma^-$ - X $^3\Sigma^-$).

1778. Yoshida, H., Ninomiya, H., Takashima, N., *Appl. Phys. Lett.,* 59, 1290 (1991), Laser action in a KrF laser pumped Ta vapor.

1779. Ninomiya, H., Hirata, K., *J. Appl. Phys.,* 66, 2219 (1989), Laser action of optically pumped atomic titanium vapor,

1780. Ninomiya, H., Hirata, K., Yoshino, S., *J. Appl. Phys.,* 66, 3961 (1989), Optically pumped titanium vapor laser at 625.8 nm.

1781. Hirata, K., Yoshino, S., Ninomiya, H., *J. Appl. Phys.,* 67, 45 (1990), Optically pumped titanium laser at 551.4 nm.

1782. Hirata, K., Yoshino, S., Ninomiya, H., *J. Appl. Phys.,* 68, 1460 (1990), Characteristics of an optically pumped titanium vapor laser.

1783. Ninomiya, H., Hirata, K., *J. Appl. Phys.,* 68, 5378 (1990), Visible laser action in N_2 laser pumped Ti vapor.

1784. Ninomiya, H., Takashima, N., Hirata, K., *J. Appl. Phys.,* 69, 67 (1991), Measurement of a population inversion on the 551.4 nm transition in an optically pumped Ti vapor laser.

1785. Gerasimov, V. A., Yunzhakov, B. P., *Sov. J. Quantum. Electron.,* 19, 1532 (December 1989), Investigation of a thulium vapor laser.

1786. Gerasimov, V. A., Yunzhakov, B. P. *Sov. J. Quantum. Electron.,* 19, 1323 (1989), Stimulated emission from thulium bromide vapor.

1787. Ninomiya, H., Abe, M., Takashima, N. *Appl. Phys. Lett.,* 58, 1819 (1991), Laser action of optically pumped atomic vanadium vapor.

1788. Basov, Yu. G., Denisov, L. K., Zuev, V. S. et al., *Sov. J. Quantum Electron.,* 11, 779 (1981), Coumarins in the gaseous phase. I. Stimulated emission of coumarin vapors in the range 485–520 nm.

1789. Logunov, O. A., Startsev, A. V., Stoilov, Yu. Yu., *Sov. J. Quantum Electron.,* 11, 780 (1981), .Coumarins in the gaseous phase. II. Investigation of stimulated emission from coumarin vapors in the range 470–540 nm.

1790. Wernsman, B., Rocca, J. J., Mancini, H. L., *IEEE Phot. Technol. Lett.,* 2, 12 (1990), CW ultravioltet and visible laser action from ionized silver in an electron beam generated plasma.

1791. Hocker, L. O., Javan, A., *Phys. Lett.,* 26A, 255 (1968), Laser harmonic frequency mixing of two different far infrared laser lines to 118 μ.

1792. Hocker, L. O., Javan, A., *Appl Phys. Lett.,* 12, 124 (1968), Absolute frequency measurement of new CW DCN submillimeter laser lines.

1793. Hocker, L. O., Small, J. G., Javan, A., *Phys. Lett.,* 29A, 321 (1969), Extension of absolute frequency measurements to the 84 μ range.

1794. Bennedict, W. S., Pollack, M. A., Tomlinson, W. J., *IEEE J. Quantum Electron.,* QE-5, 108 (1969), The water-vapor laser.

1795. Kneubühl, F. K. and Sturzenegger, Ch., *Infrared and Millimeter Waves* 3, chapter 5 (1980).

1796. Hocker, L. O., Javan, A., *Phys. Lett.,* 25A, 489 (1967), Absolute measurements on new CW HCN submillimeter laser lines.

1797. Hocker, L. O., Javan, A., Rao, R. D., Frenkel, L., Sullivan, T., *Appl Phys. Lett.,* 10, 147 (1967), Absolute frequency measurement and spectroscopy of gas laser transitions in the far infrared.

1798. Dangoisse, D., Depannemaecker, J. C., Monnanteuil, N., *Int. J. Infrared Millimeter Waves* 4, 913 (1983), The optically pumped ozone laser.

1799. Plant, T. K., Coleman, P. D., and DeTemple, *IEEE J. Quantum Electron.,* QE-9, 962 (1973), New optically pumped far-infrared lasers.

1800. McCombie, J., Petersen, J. C., Duxbury G., *Electronic & Electr. Opt.,* Proc. 5th National Quantum Electronic Conference, Knight, P. L., Ed. (1981).

1801. Danielewicz, E. J., Reel, R. D., Hodges, D. T., *IEEE J. Quantum Electron.*, QE-16, 402 (1980), New far infrared CW optically pumped cis-$C_3H_2F_2$ laser.

1802. Vasconcellos, E. C. C., Jennings, D. A., Evenson, K. M., *Int. J. Infrared Millimeter Waves*, 7, 291 (1986), Frequency measurements of the solitary ethyl alcohol laser line.

1803. Carelli, G., Ioli, N., Moretti, A., Pereira, D. Strumia, F., Densing, *Appl. Phys.* B45, 97 (1988), Frequency and efficiency measurements of FIR laser lines close to the N^+ $^3P_2$ – $^3P_1$ transition..

1804. Ioli, N., Moretti, A., Pereira, D. ,Strumia, F., Carelli, G., *Appl. Phys. B*, 48, 299 (1989), A new efficient far infrared optically pumped laser gas: $CH_3^{18}OH$.

1805. Dyubko, S. F., Svich, V. A., Fesenko, L. D., *JETP Lett.*, 16, 418 (1972), Submillimeter-band gas laser pumped by a CO_2 laser.

1806. Carelli, G., Ioli, N., Moretti, A., Pereira, D., Strumia, F. , *Appl. Phys. B*, 44, 111 (1987), New large offset far-infrared laser lines.

1807. Yoshida, T., Yamabayashi, N., Sakai, K., Fujisawa, K., *Infrared Phys.*, 22, 293 (1982), The Stark effect on optically pumped CH_3OD and CD_3OH lasers.

1808. Gastaud, G., Redon, M., Fourrier, M., *Optics Lett.*, 9, 71 (1984), Far-infrared laser lines in vinyl fluoride optically pumped by a CO_2 laser near 10 μm.

1809. Ioli, N., Moretti, A., Strumia, F., *Appl. Phys. B*, 48, 305 (1989), High efficiency cw far-infrared laser at 119 μm and 127 μm.

1810. Davies, P. B., Vass, A., Pidgeon, C.R, Allan, G. R., *Opt. Commun.* 37, 303 (1981), New FIR laser lines from an optically pumped far-infrared laser with isotopic $^{13}C^{16}O_2$.

1811. Kachi, T., Kon, S., *Int. J. Infrared Millimeter Waves*, 4, 767 (1982), Assignment of optically pumped CD_3OH laser lines.

1812. Benard, D. J., Winker, B. K., *J. Appl. Phys.*, 69, 2805 (1991), Chemical generation of optical gain at 471 nm.

1813. Benard, D. J., Boehmer, E., *Appl. Phys. Lett.*, 65, 1340 (1994), Chemically pumped visible-wavelength laser with high optical gain.

1814. Laymon, A. J., Silfvast, W. T., Erisman, T. L., U.S. Patent #5,257,278.

1815. White, J. D., Reid, J., *Appl. Opt.* 32, 2053 (1993), Efficient NH_3 laser operation in the 16- to 21-micrometer region.

1816. Maki, A. G., Chou, Che-Chung, Evenson, K. M., Zink, L. R., Shy, Jow-Tsong, *J. Mol. Spectrosc.*, 167, 211 (1994), Improved molecular constants and frequencies for the CO_2 laser from new high-J regular and hot-band frequency measurements,

1817. Chou, Che-Chung, Evenson, K. M., Zink, K. R., Maki, A. G., and Shy, Jow-Tsong, *IEEE J. Quantum Electron.*, 31, 343 (1995), New CO_2 laser lines in the 11-micrometer wavelength region: new hot bands.

1818. Gromoll-Bohle, M., Bohle, W., and Urban, W., *Optics Commun.*, 69, 409 (1989), Broadband CO laser emission on overtone transitions $\Delta v = 2$.

1819. Schneider, M., Evenson, K. M., Vanek, M. D., Jennings, D. A., Wells, J. S., Stahn, A., Urban, W., *J. Mol. Spectrosc.*, 135, 197 (1989), Heterodyne frequency measurements of $^{12}C^{16}O$ laser transitions.

1820. Maki, T., Evenson, K. M., Zink, L. R., Maki, A. G., *IEEE J. Quantum Electron.*, 32, 1732 (1996), 9- and 10-micrometer N_2O laser transitions.

1821. Lang, P. T., Kelly, M., Renk, K. F., Halonen, L., *IEEE J. Quantum Electron.*, 27, 480 (1991), Observation of a large number of new laser lines from $^{12}CD_3F$ gas optically pumped with a continuously tunable high-pressure pulsed CO2 laser.

1822. Telles E. M., Moraes J. C. S., Scalabrin, A., Pereira, D., Moretti, A., Strumia, F. *Appl. Phys.*, B 52, 36 (1991), New FIR laser lines from CD_3OD optically pumped by a CO2 waveguide laser.

1823. Telles E. M., Moraes J. C. S., Scalabrin, A., Pereira, D., Carelli, G., Ioli, N., Moretti, A., Strumia, F., *IEEE J. Quantum Electron.*, 30, 2946 (1994), Optically pumped $^{13}CD_3OH$: far infrared laser lines and assignment.

1824. Telles E. M., Scalabrin, A., Pereira, D., *IEEE J. Quantum Electron.*, 31, 754 (1995), Frequency measurements of optically pumped far-infrared laser lines.

1825. Bertolini, A., Carelli, G.,Moretti, A., Moruzzi, G., *IEEE J. Quantum Electron.*, 35, 124 (1999), Laser action in hydrazine: observation and characterization on new large offset FIR laser lines.

1826. Lang, P. T., Leipold, I., Knott, W. J., Semenov, A. D., Gol'tsman, G. N., Renk, K. F., *Appl. Phys. B*, 53, 207 (1991), New far-infrared laser lines from CH_3Cl and CH_3Br optically pumped with a continuous tunable high pressure CO_2 laser.

1827. Moraes J. C. S., Scalabrin, A., Pereira, D., Carelli, G., Ioli, N., Moretti, A., Strumia, F., *Appl. Phys. B*, 54, 24 (1992), IR and FIR spectroscopy of $^{13}CD_3OH$ around the 10P(22) and 10P(24) CO_2 laser lines: frequency measurements and assignments.

1828. Moraes J. C. S., Bertolini, A., Carelli, G., Ioli, N., Moretti, A., Strumia, F., *IEEE J. Quantum Electron.*, 32, 1737(1996), The $^{13}CD_3OH$ optcally pumped far-infrared laser: new lnes and frequency measurements.

1829. Bertolini, A., Carelli, G., Massa, C. A., Moretti, A., Strumia, F., Qui, M. X., Münddlein, U., *IEEE J. Quantum Electron.*, 34, 238(1998), CW submillimeter laser ction in ethyl chloride.

1830. Luis, G. M. R. S., Telles E. M., Scalabrin, A., Pereira, D. *IEEE J. Quantum Electron.*, 34, 767 (1998), Observation and characterization of new FIR laser lines from formic acid.

1831. Zerbetto, S. C., Vasconcellos, E. C. C., Zink, L. R., Evenson, K. M., *J. Opt. Soc. Am. B*, 15, 1839 (1998), Hydrazine—an excellent optically pumped far-infrared lasing gas: review.

1832. Viscovini,R. C., Telles E. M., Scalabrin, A., Pereira, D., *Appl. Phys. B*, 65, 33 (1997), Large offset far infrared laser lines generation from the C-O stretching Q-branch of $^{13}CD_3OD$.

1833. Telles E. M., Odashima, H., Zink, L. R., Evenson, K. M., *J. Mol. Spectrosc.*, 195, 360 (1999), Optically pumped FIR laser lines from CH_3OH: new laser lines, frequency measurements, and assignments.

1834. Bertolini, A., Carelli, G., Moretti, A., Strumia, F., *Infrared Phys. Technol.*, 38, 437 (1997), Qui, M. X., Corbalan, R., Cojocaru, C. M., FIR spectroscopy of ethyl bromide and trioxane: new laser transitions and assignments.

1835. Carelli, G., Moretti, A., Pereira, D., Strumia, F., *Infrared Phys. Technol.*, 35, 855 (1994), Characterization and frequency measurement of a new far infrared laser line from CH_2F_2.

1836. Carelli, G., Ioli, N., Messina, A., Moretti, A., Strumia, F., Telles E. M., Moraes J. C. S., Scalabrin, A., Pereira, D., *Infrared Phys.*, 31, 323 (1991), CD_3OD optically pumped by a wave guide CO_2 laser: new FIR laser lines and frequency measurements.

1837. Telles E. M., Zink, L. R., Evenson, K. M., *Int. J. Infrared Millimeter Waves*, 19, 1627 (1998), New FIR laser lines from CD_3OH.

1838. Vasconcellos, E. C. C., Zerbetto, S. C., Zink, L. R., Evenson, K. M. *Int. J. Infrared Millimeter Waves*, 19, 465 (1998), Laser lines and frequency measurements of fully deuterated isotopomers of methanol.

1839. Davies, P. B., Liu, Y., Liu, Z., *Int. J. Infrared Millimeter Waves*, 14, 2395 (1993), New FIR laser lines from optically pumped C_2H_3F, C_2H_3Cl, C_2H_3Br, C_2H_3CN, C_2H_5F, CH_2CF_2, HCOOH and CH_3Br.

1840. Bertolini, A., Carelli, G., Ioli, N., Massa, C. A., Moraes J. C. S., Moretti, A., Strumia, F., *Int. J. Infrared Millimeter Waves*, 18, 779 (1997) Optically pumped submillimeter laser lines from CD_2Cl_2 using a large tunability CW CO_2 laser.

1841. Bertolini, A., Carelli, Massa, C. A., Moretti, A., Strumia, F., *Infrared Phys. Technol.*, 40, 33 (1999), The H^{13}COOH optically pumped laser: a new large offset FIR laser emission and assignments,

1842. Bertolini, A., Carelli, G., Ioli, N., Massa, C. A., Moretti, A., Strumia, F., *Int. J. Infrared Millimeter Waves*, 18, 1281 (1997), Optically pumped CW FIR laser: new submillimeter laser emissions from CH$_2$DOH, CH$_3$I, CD$_3$I and trioxymethylene.

1843. Moraes J. C. S., Telles E. M., Cruz, F C., Scalabrin, A., Pereira, Garelli, C., Ioli, N., Moretti, A., Strumia, F., *Int. J. Infrared Millimeter Waves*, 12, 1475 (1991), New FIR lasers lines and frequency measurements from optically pumped ^{13}CD$_3$OH.

1844. Telles E. M., Zink, L. R., Evenson, K. M., *Int. J. Infrared Millimeter Waves*, 12, 1475 (1991), New optically pumped FIR laser lines and frequency measurements from CH$_3$OD.

1845. Zerbetto, S. C., Vasconcellos, E. C. C., Zink, L. R., Evenson, K. M., *Int. J. Infrared Millimeter Waves*, 18, 2301 (1997), ^{12}CH$_2$F$_2$ and ^{13}CH$_2$F$_2$ far-infrared lasers: new lines and frequency measurements.

1846. Vasconcellos, E. C. C., Zerbetto, S. C., Xu, L.-H., Lees, R M., Zink, L. R., Evenson, K. M., *Int. J. Infrared Millimeter Waves*, 19, 719 (1998), New far infrared laser lines and frequency measurements of acetaldehyde and vinyl fluoride.

1847. Vasconcellos, E. C. C., Tachikawa, M., Zink, L. R., Evenson, K. M., *Int. J. Infrared Millimeter Waves*, 18, 2295 (1997), Far-infrared hydrazine laser pumped by an N$_2$O laser.

1848. Sarkkinen, H., Paso, R., Anttila, R., *Infrared Phys. Technol.*, 37, 643 (1996), Assignment of methyl cyanide FIR laser lines.

1849. Bennett, W. R., Jr., *Atomic Gas Laser Transition Data, A Critical Evaluation*, Plenum, New York (1979).

1850. Carelli, G., Ioli, N., Moretti, A. et al., *Infrared Phys. Technol.*, 35, 743 (1994), CH$_3$^{18}OH FIR laser line frequency measurements and assignments.

1851. Moruzzi, G., Moraes J. C. S., Strumia, F., *Int. J. Infrared Millimeter Waves*, 13, 1269 (1992), Far infrared laser lines and assignments of CH3OH - a review.

1852. Telles E. M., Zink, L. R., Evenson, K. M., *J. Mol. Spectrosc.*, 191, 206 (1998), Frequencies of optically pumped sub-Doppler far-infrared laser lines of methanol.

1853. Pereira, D., Moraes J. C. S., Telles E. M., Scalabrin, A. et al. *Int. J. Infrared Millimeter Waves,* 15, 1 (1994), A review of optically pumped far-infrared laser lines from methanol isotopes.

1854. Patel, C. K. N., Faust, W. L., McFarlane, R. A., Garrett, C. G.B., *Proc. IEEE*, 52, 713 (1964), Laser action up to 57.335 µm in gaseous discharges.

1855. Xu, L.-H., Lees, R. M., Elza, C. C. et al., *IEEE J. Quantum Electron.*, 32, 392 (1996), Methanol and the optically pumped far-infrared laser.

1856. Zerbetto, S. C., Vasconcellos, E. C. C., Zink, L. R., Evenson, K. M., Lees, R. M., Xu, L.-H., *J. Mol. Spectrosc.*, 188, 102 (1998), C-13 methanol far-infrared laser: newly discovered lines, predictions, and assignments.

Section 4: Other Lasers

Section 4.1
EXTREME ULTRAVIOLET AND SOFT X-RAY LASERS

4.1.1 Introduction

Lasing in the extreme ultraviolet (EUV) and soft x-ray (SXR) regions (from 87.4 to 3.56 nm) has been achieved in highly ionized plasmas, generally produced by pulses from high-power lasers incident on solid targets. This may involve a nanosecond prepulse to form and ionize the plasma followed by a more intense heating pulse; the latter may vary from a much higher energy pulse of similar time duration to a pulse of similar energy but of much shorter time duration (e.g., picoseconds). Extreme ultraviolet lasing has also been obtained by electric excitation of the plasma confined in a narrow capillary channel, the discharge plasma being formed by a low-induction, high-voltage current pulse. Population inversion results from transient electron collisional excitation from an ionic ground state into the upper laser level or from rapid recombination as a collisionally stripped plasma cools adiabatically after the pulse.

The highly ionized lasing media are specified by their oxidation state. The electron configurations of the highly ionized states are similar to those of neutral atoms with the same number of electrons. Thus Al^{10+}, for example, which has only three electrons is described as "Li-like" and Se^{24+} (with ten electrons) is described as "Ne-like". As shown in Figure 4.1.1, lasing has been reported for one-half of the elements in the periodic table. The wavelength ranges of laser transitions for the various types of ion species are shown in Figure 4.1.2.

In the tables to follow, extreme ultraviolet and soft x-ray lasers are grouped by their oxidation state. The lasing ion, transition, lasing wavelength, and reported gain are given together with the primary references. The error in the gain numbers is generally ±10%. Representative energy level diagrams and lasing transitions for H-like, Li-like, Ne-like, and Ni-like lasers are shown in Figures 4.1.3–4.1.6.

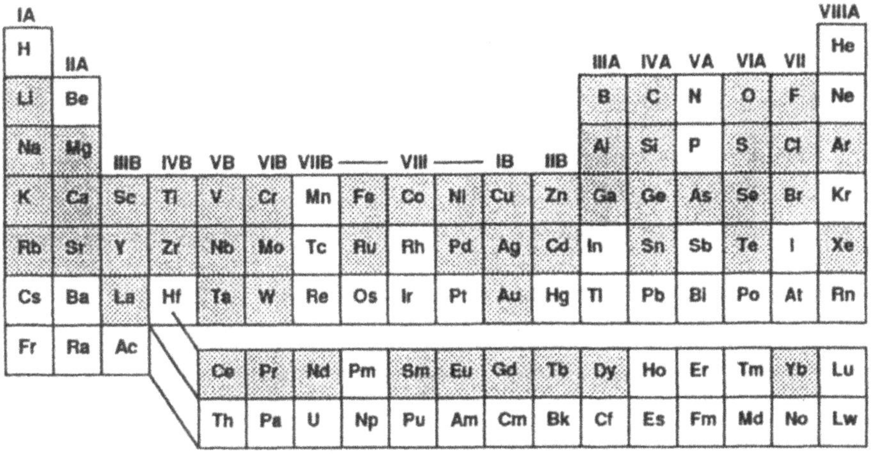

Figure 4.1.1   Periodic table of the elements showing the elements (shaded) that have been reported to exhibit laser action in the extreme ultraviolet and soft x-ray regions.

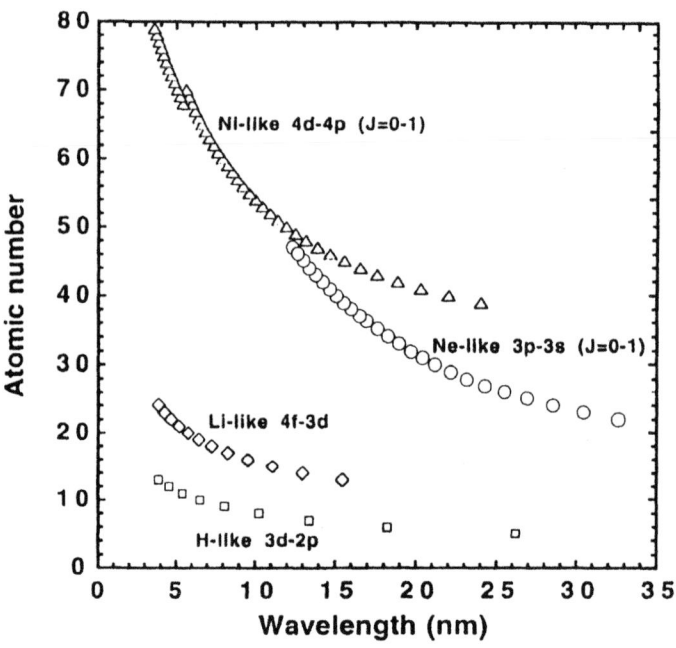

Figure 4.1.2   Soft x-ray laser wavelengths plotted versus atomic number for various ion species (courtesy of J. Nilsen).

Extreme ultraviolet and soft x-ray lasers included in this section are presented in the following order:

H-like Ions	Li^{2+}, B^{4+}, C^{5+}, O^{6+}, F^{7+}, Na^{10+}, Mg^{11+}, Al^{12+}	Table 4.1.1
Li-like Ions	Al^{10+}, Si^{11+}, Cl^{14+}, Ca^{17+}, Cr^{21+}	Table 4.1.2
Be-like Ions	Al^{9+}	Table 4.1.3
Ne-like Ions	Si^{4+}, S^{6+}, Cl^{7+}, K^{9+}, Ca^{10+}, Sc^{11+}, Ti^{12+}, V^{13+}, Cr^{14+}, Fe^{16+}, Co^{17+}, Ni^{18+}, Cu^{19+}, Zn^{20+}, Ga^{21+}, Ge^{22+}, As^{23+}, Se^{24+}, Br^{25+}, Rb^{27+}, Sr^{28+}, Y^{29+}, Zr^{30+}, Nb^{31+}, Mo^{32+}, Ru^{34+}, Ag^{37+}	Table 4.1.4
Co-like Ions	Yb^{43+}, Ta^{46+}	Table 4.1.5
Ni-like Ions	Y^{11+}, Zr^{12+}, Nb^{13+}, Mo^{14+}, Pd^{18+}, Ag^{19+}, Cd^{20+}, Sn^{22+}, Te^{24+}, Xe^{26+}, La^{29+}, Ce^{30+}, Pr^{31+}, Nd^{32+}, Sm^{34+}, Eu^{35+}, Gd^{36+}, Tb^{37+}, Dy^{38+}, Yb^{42+}, Ta^{45+}, W^{46+}, Au^{51+}	Table 4.1.6
Pd-like Ions	Xe^{8+}	Table 4.1.7

Further Reading

Attwood, D., *X-ray and Extreme Ultraviolet Radiation*, Cambridge University Press, Cambridge (1999).

Avrorin, E. N., Lykov, V. A., Loboda, P. A. and Politov, V. Yu., Review of theoretical works on X-ray laser research performed at RFNC-VNIITF, *Laser and Particle Beams* 15, 3 (1997).

Elton, R. C., *X-Ray Lasers*, Academic Press, San Diego (1990).

Falcone, R. W. and Kirz, J., Eds., *Short Wavelength Coherent Radiation: Generation and Applications*, Optical Society of America, Washington, DC (1989).

Kapteyn, H. C., Da Silva, L. B., and Falcone, R. W., Short-wavelength lasers, *Proc. IEEE* 80, 342 (1992).

Matthews, D. and Freeman, R., Eds., *The Generation of Coherent XUV and Soft X-Ray Radiation*, Special Dedicated Volume, *J. Opt. Soc. Am. B* 4 (1987).

Matthews, D. L., X-Ray Lasers, in *Handbook of Laser Science and Technology, Suppl. 1: Lasers*, CRC Press, Boca Raton, FL (1991), p. 559.

Suckewer, S. and Skinner, C. H., Soft x-ray lasers and their applications, *Science* 247, 1553 (1990).

Waynant, R. W. and Ediger, M. N., Eds., *Selected Papers on UV, VUV, and X-Ray Lasers*, SPIE Milestone Series, Vol. MS71, SPIE Optical Engineering Press, Bellingham, WA (1993).

See, also, proceedings of the x-ray laser conferences held every two years:

X-ray Lasers 1990, Proceedings of the 2nd International Colloquium on X-ray Lasers, Tallents, G.J., Ed., IOP Conf. Series 116, IOP Publishing, Bristol, U.K. (1990).

X-ray Lasers 1992, Proceedings of the 3rd International Colloquium on X-ray Lasers, Fill, E. E., Ed., IOP Conf. Series 125, IOP Publishing, Bristol, U.K. (1992).

X-ray Lasers 1994, Proceedings of the 4th International Colloquium on X-ray Lasers, Eder, D. C. and Matthews, D. L., Eds., AIP Conf. Proc. 332, American Institute of Physics, New York (1994).

X-ray Lasers 1996, Proceedings of the 5th International Colloquium on X-ray Lasers, Svanberg, S. and Wahlström, C-G., Eds., IOP Conf. Series 151, IOP Publishing, Bristol, U.K. (1996).

X-ray Lasers 1998, Proceedings of the 6th International Colloquium on X-ray Lasers, Kato, Y., Ed., IOP Conf. Series 159, IOP Publishing, Bristol, U.K. (1999).

4.1.2 Lasing Transitions of H-Like Ions

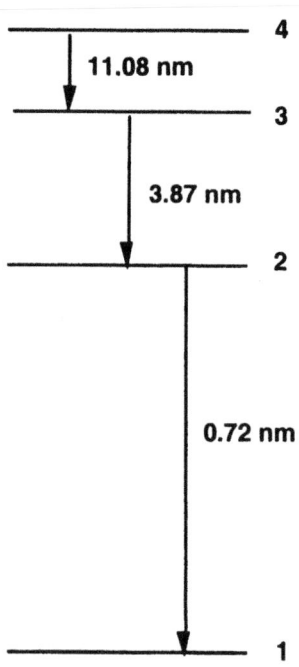

Figure 4.1.3 Simplified energy level diagram and H-like n=3 to n=2 and n=2 to n=1 recombination laser transitions. Transition wavelengths are shown for the element Al.

Table 4.1.1
Lasing Transitions of H-Like Ions

Ion	Transition	Wavelength (nm)	Gain (cm⁻¹)/ gain-length	Reference
Li^{2+}	$2 \rightarrow 1$	13.5	$\approx 11/\approx 5.5$	57,62
B^{4+}	$3 \rightarrow 2$	26.2	$\approx 14-19/\geq 5$	58
C^{5+}	$3 \rightarrow 2$	18.210	$4.1 \pm 0.6, 6.5$	31–33
O^{7+}	$3 \rightarrow 2$	10.243	~3	16
F^{8+}	$3 \rightarrow 2$	8.091	1.4	7,12
Na^{10+}	$3 \rightarrow 2$	5.419	$\approx 5 \pm 1.0$	4
Mg^{11+}	$3 \rightarrow 2$	4.553		4
Al^{12+}	$3 \rightarrow 2$	3.879		4

4.1.3 Lasing Transitions of Li-Like Ions

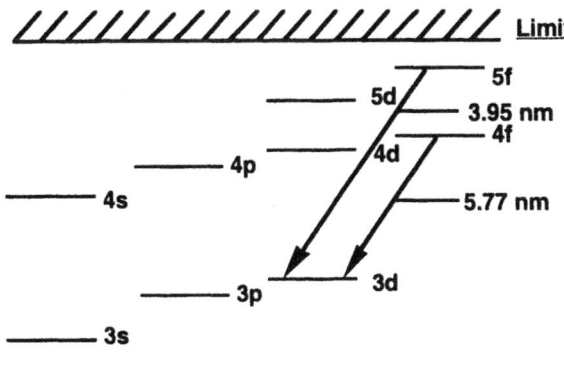

Figure 4.1.4   Simplified energy level diagram and Li-like 4f-to-3d and 5f-to-3d recombination laser transitions. Transition wavelengths are shown for the element Ca.

Table 4.1.2
Lasing Transitions of Li-Like Ions

Ion	Transition	Wavelength (nm)	Gain (cm⁻¹)/ gain-length	Reference
Al^{10+}	4f→3d	15.466	4.5±0.5/3–4	12,17,18
Al^{10+}	5f→3d	10.57	3.4±0.4	12,17,18
Si^{11+}	4f→3d	12.989	—/≈1–2	17, 23
Si^{11+}	5f→3d	8.73		14
Si^{11+}	5f→3d	8.89		14
S^{13+}	5g→4f	20.65	2.5±0.7	39
Cl^{14+}	4f→3d	8.32	~2.5	12
Ca^{17+}	4f→3d	5.77	4.3±0.9	7, 63
Ca^{17+}	5f→3d	3.95		63
Cr^{21+}	4f→3d	3.86		63
Cr^{21+}	5f→3d	2.64		63

4.1.4 Lasing Transitions of Be-Like Ions

The energy level diagram for Be-like ions is similar to that in Figure 4.1.4.

Table 4.1.3
Lasing Transitions of Be-Like Ions

Ion	Transition	Wavelength (nm)	Gain (cm^{-1})/ gain-length	Reference
Al^{9+}	4f→3d	17.78	3.5±0.5	18
Al^{9+}	5d→3p	12.35	3.4±0.5	18

4.1.5 Lasing Transitions of Ne-Like Ions

Lasing of neon-like ions involves 2p^{5}3p - 2p^{5}3s transitions as shown in Figure 4.1.5. In Table 4.1.4 below, the term 2p^5 is omitted to simplify the notation. The jj coupling designation gives the j of the L shell hole, the j of the n=3 electron, and the total J. Experimental wavelengths of neon-like 3p→3s laser lines are compared with relativistic multi-configurational Hartree-Fock calculations in Reference 55.

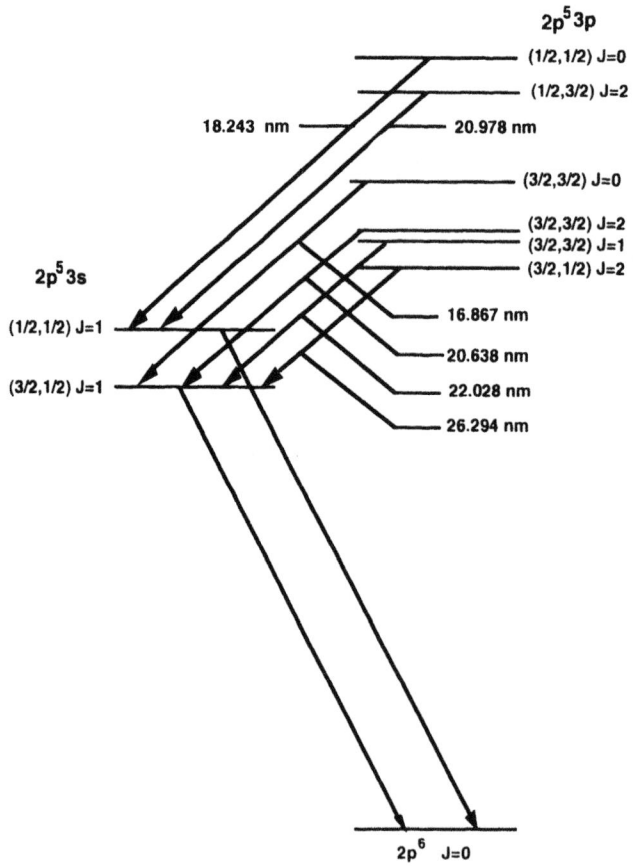

Figure 4.1.5   Simplified energy level diagram and laser transitions for Ne-like 3p-to-3s collisional excitation lasers. Lasing wavelengths are for the element Se.

Table 4.1.4
Lasing Transitions of Ne-Like Ions

Ion	Transition	Wavelength (nm)	Gain (cm^{-1})/ gain-length	Reference
Si^{4+}	3p→3s (J=0→1)	87.4		50
S^{6+}	3d→3p (J=1→1)	60.1		50
S^{6+}	3p→3s (J=0→1)	60.8		50
Cl^{7+}	3p→3s (J=0→1)	52.9		46
Ar^{8+}	3d→3p (J=1→1)	45.1		48
Ar^{8+}	3p→3s (J=0→1)	46.9		15,48
Ar^{8+}	3p→3s (J=0→1)	46.9	>25	49
K^{9+}	3p→3s (J=0→1)	42.1	2.2	46
Ca^{10+}	3p→3s (J=0→1)	38.3		46
Sc^{11+}	3p→3s (J=0→1)	31.2		46
Sc^{11+}	3p→3s (J=0→1)	35.2		46
Ti^{12+}	3d→3p (J=1→1)	30.15		65
Ti^{12+}	3p→3s (J=0→1)	32.63	19/~9.5	37,43,47
Ti^{12+}	3p→3s (J=2→1)	47.22		43
Ti^{12+}	3p→3s (J=2→1)	50.76		43
V^{13+}	3p→3s (J=0→1)	26.1		45
V^{13+}	3p→3s (J=0→1)	30.4		45
Cr^{14+}	3p→3s (J=0→1)	24.02		37,43
Cr^{14+}	3p→3s (J=0→1)	28.55		37,43
Cr^{14+}	3p→3s (J=2→1)	40.22		37,43
Cr^{14+}	3p→3s (J=2→1)	44.07		37,43
Fe^{16+}	3p→3s (J=0→1)	20.42		37,43
Fe^{16+}	3p→3s (J=0→1)	25.49		37,43
Fe^{16+}	3p→3s (J=0→1)	25.5	9.2/≈16	59
Fe^{16+}	3p→3s (J=2→1)	34.76		37,43
Fe^{16+}	3p→3s (J=2→1)	38.89		37,43
Co^{17+}	3p→3s (J=0→1)	24.24		43
Co^{17+}	3p→3s (J=2→1)	31.80		43
Co^{17+}	3p→3s (J=2→1)	32.45		43
Co^{17+}	3p→3s (J=2→1)	36.73		43
Ni^{18+}	3p→3s (J=0→1)	23.11		42,43
Ni^{18+}	3p→3s (J=2→1)	29.77		42,43

Lasing Transitions of Ne-Like Ions—*continued*

Ion	Transition	Wavelength (nm)	Gain (cm⁻¹)/ gain-length	Reference
Ni^{18+}	$3p \rightarrow 3s$ (J=2→1)	30.36		42,43
Ni^{18+}	$3p \rightarrow 3s$ (J=1→1)	31.48		42,43
Ni^{18+}	$3p \rightarrow 3s$ (J=2→1)	34.75		42,43
Cu^{19+}	$3p \rightarrow 3s$ (J=0→1)	22.111	2.0	35,42
Cu^{19+}	$3p \rightarrow 3s$ (J=2→1)	27.931	1.7	35,42,43
Cu^{19+}	$3p \rightarrow 3s$ (J=2→1)	28.467	1.7	35,42,43
Cu^{19+}	$3p \rightarrow 3s$ (J=1→1)	29.62		35,42,43
Cu^{19+}	$3p \rightarrow 3s$ (J=2→1)	33.15		35,42,43
Zn^{20+}	$3p \rightarrow 3s$ (J=0→1)	21.217	2.3,	40,41
			4.9±0.2	39
Zn^{20+}	$3p \rightarrow 3s$ (J=2→1)	26.232	2.0	40,41
			2.3±0.6	39
Zn^{20+}	$3p \rightarrow 3s$ (J=2→1)	26.723	2.0	40,41
			2.6±0.4	39
Ga^{21+}	$3p \rightarrow 3s$ (J=2→1)	24.670		40
Ga^{21+}	$3p \rightarrow 3s$ (J=2→1)	25.111		40
Ge^{22+}	$3p \rightarrow 3s$ (J=0→1)	19.606	3.1	35
Ge^{22+}	$3p \rightarrow 3s$ (J=2→1)	23.224	4.1	35
Ge^{22+}	$3p \rightarrow 3s$ (J=2→1)	23.626	4.1	35,44
Ge^{22+}	$3p \rightarrow 3s$ (J=1→1)	24.732		35
Ge^{22+}	$3p \rightarrow 3s$ (J=2→1)	28.646		35
As^{23+}	$3p \rightarrow 3s$ (J=2→1)	21.884		40
As^{23+}	$3p \rightarrow 3s$ (J=2→1)	22.256	5.4	40
Se^{24+}	$3p \rightarrow 3s$ (J=0→1)	16.867	—	29
Se^{24+}	$3p \rightarrow 3s$ (J=0→1)	18.243	2.6	24,25,29,34
Se^{24+}	$3p \rightarrow 3s$ (J=2→1)	20.638	4.9	24,25,38
Se^{24+}	$3p \rightarrow 3s$ (J=2→1)	20.978	4.9	24,25
Se^{24+}	$3p \rightarrow 3s$ (J=1→1)	22.028		24,29
Se^{24+}	$3p \rightarrow 3s$ (J=2→1)	26.294		24,29
Br^{25+}	$3p \rightarrow 3s$ (J=0→1)	17.63		20
Br^{25+}	$3p \rightarrow 3s$ (J=2→1)	19.47		20
Br^{25+}	$3p \rightarrow 3s$ (J=2→1)	19.78		20
Br^{25+}	$3p \rightarrow 3s$ (J=1→1)	20.79		20
Br^{25+}	$3p \rightarrow 3s$ (J=2→1)	25.24		20
Rb^{27+}	$3p \rightarrow 3s$ (J=0→1)	16.50	very weak	30
Rb^{27+}	$3p \rightarrow 3s$ (J=2→1)	17.35	1.9	30

Lasing Transitions of Ne-Like Ions—*continued*

Ion	Transition	Wavelength (nm)	Gain (cm^{-1})/ gain-length	Reference
Rb^{27+}	3p→3s (J=2→1)	17.61	1.4	30
Rb^{27+}	3p→3s (J=1→1)	18.52	weak	30
Rb^{27+}	3p→3s (J=2→1)	23.35	weak	30
Sr^{28+}	3p→3s (J=0→1)	15.98		28
Sr^{28+}	3p→3s (J=2→1)	16.41	4.4	28
Sr^{28+}	3p→3s (J=2→1)	16.65	4.0	28
Sr^{28+}	3p→3s (J=1→1)	17.51		28
Sr^{28+}	3p→3s (J=2→1)	22.49		28
Y^{29+}	3p→3s (J=2→1)	15.4985	~10–15/20±2	24–27,61
Y^{29+}	3p→3s (J=2→1)	15.71	~4.0	24,25
Y^{29+}	3p→3s (J=1→1)	16.490		25
Zr^{30+}	3p→3s (J=0→1)	11.89		20
Zr^{30+}	3p→3s (J=2→1)	14.66		20
Zr^{30+}	3p→3s (J=2→1)	14.86		20
Zr^{30+}	3p→3s (J=0→1)	15.040		20
Zr^{30+}	3p→3s (J=1→1)	15.63		21
Zr^{30+}	3p→3s (J=2→1)	20.96		20
Nb^{31+}	3p→3s (J=0→1)	11.25		20
Nb^{31+}	3p→3s (J=2→1)	13.86		20
Nb^{31+}	3p→3s (J=2→1)	14.04		20
Nb^{31+}	3p→3s (J=0→1)	14.590		20
Nb^{31+}	3p→3s (J=1→1)	14.76		20
Nb^{31+}	3p→3s (J=2→1)	20.25		20
Mo^{32+}	3p→3s (J=0→1)	10.64	4.1	19
Mo^{32+}	3p→3s (J=2→1)	13.10	4.2	19
Mo^{32+}	3p→3s (J=2→1)	13.27	2.9	19
Mo^{32+}	3p→3s (J=1→1)	13.94		19
Mo^{32+}	3p→3s (J=0→1)	14.16		19
Ru^{34+}	3p→3s (J=2→1)	11.7		22
Ru^{34+}	3p→3s (J=2→1)	11.8		22
Ag^{37+}	3p→3s (J=0→1)	8.156		13
Ag^{37+}	3p→3s (J=2→1)	9.9365	9.4±1.5/8.5±1.4	13
Ag^{37+}	3p→3s (J=2→1)	10.0377	6.4±2.0/8.5±1.8	13
Ag^{37+}	3p→3s (J=1→1)	10.508		13
Ag^{37+}	3p→3s (J=0→1)	12.300		13

4.1.6 Lasing Transitions of Co-Like Ions

Lasing of cobalt-like ions involve $3d^8 4d \rightarrow 3d^8 4p$ transitions similar to those in Figure 4.1.6. In the table below, the term $3d^8$ and the LSJ state designation are omitted to simplify the notation.

Table 4.1.5
Lasing Transitions of Co-Like Ions

Ion	Transition	Wavelength (nm)	Gain (cm^{-1})/ gain-length	Reference
Yb^{43+}	$4d \rightarrow 4p$	5.176	0.7±0.3	5
Ta^{46+}	$4d \rightarrow 4p$	4.607	2.2±0.4	5

4.1.7 Lasing Transitions of Ni-Like Ions

Lasing of nickel-like ions involve $3d^9 4d\ ^1S_0 \rightarrow 3d^9 4p\ ^1P_1$ and $3d^9 4f\ ^1P_1 \rightarrow 3d^9 4d$ $^1P_1$ transitions. In the table below, the notation, the term $3d^9$ and the LSJ state designation are omitted to simplify the notation. The jj coupling designation gives the j of the M shell hole, the j of the n=4 electron, and the total J. Summaries of experimental wavelengths and multi-configurational calculations of Ni-like $4d \rightarrow 4p$ and $4f \rightarrow 4d$ laser lines are given in References 63 and 64.

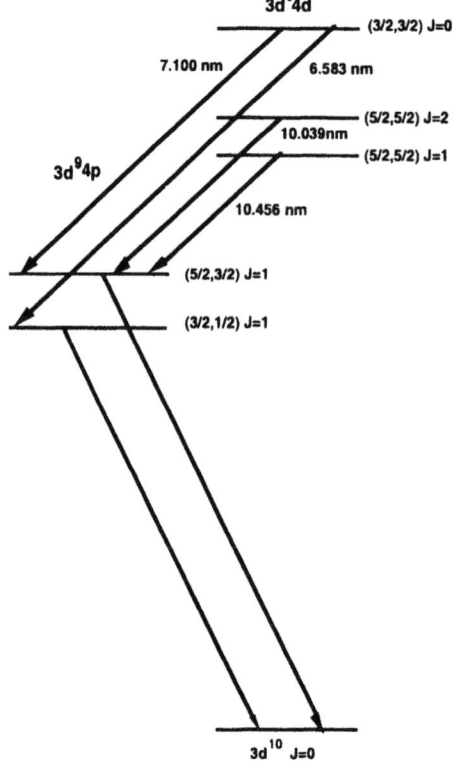

Figure 4.1.6 Simplified energy level diagram and laser transitions for Ni-like 4d-to-4p collisional excitation lasers. Lasing wavelengths are for the element Eu. Lasing has also been observed on 4f-to-4d transitions (not shown) for Zr^{12+} and Nb^{13+}.

Table 4.1.6
Lasing Transitions of Ni-Like Ions

Ion	Transition	Wavelength (nm)	Gain (cm^{-1})/ gain-length	Reference
Y^{11+}	4d→4p (J=0→1)	24.011		56
Zr^{12+}	4d→4p (J=0→1)	22.020	17	56,60
Zr^{12+}	4f→4d (J=1→1)	27.10		56,64
Nb^{13+}	4d→4p (J=0→1)	20.334	26±2/10.7±0.3	56,60
Nb^{13+}	4f→4d (J=1→1)	24.64		64
Mo^{14+}	4d→4p (J=0→1)	18.895	21/11.6±0.3	56,60
Mo^{14+}	4f→4d (J=1→1)	22.59		60,64
Pd^{18+}	4d→4p (J=0→1)	14.679	35/~12.5	54,56,66
Ag^{19+}	4d→4p (J=0→1)	13.892		51,56
Cd^{20+}	4d→4p (J=0→1)	13.166		56
Sn^{22+}	4d→4p (J=0→1)	11.473		21
Sn^{22+}	4d→4p (J=0→1)	11.910		21
Te^{24+}	4d→4p (J=0→1)	11.1		51
Xe^{26+}	4d→4p (J=0→1)	10.0		15
La^{29+}	4d→4p (J=0→1)	8.9		51
Ce^{30+}	4d→4p (J=0→1)	8.6		51
Pr^{31+}	4d→4p (J=0→1)	8.2		51
Nd^{32+}	4d→4p (J=0→1)	7.906	3.1±0.1/7.8	8,11,51
Sm^{34+}	4d→4p (J=0→1)	6.832		10
Sm^{34+}	4d→4p (J=0→1)	7.331	2.6±0.1	8,10,53
Eu^{35+}	4d→4p (J=0→1)	6.583	0.61±0.14	9
Eu^{35+}	4d→4p (J=0→1)	7.100	1.1±0.12	9
Eu^{35+}	4d→4p (J=0→1)	8.483		9
Eu^{35+}	4d→4p (J=2→1)	10.039	0.08±0.14	9
Eu^{35+}	4d→4p (J=1→1)	10.456	~0.07±0.19	9
Gd^{36+}	4d→4p (J=0→1)	6.39		8
Gd^{36+}	4d→4p (J=0→1)	6.92	2.8±0.2	8
Tb^{37+}	4d→4p (J=0→1)	6.11		8
Tb^{37+}	4d→4p (J=0→1)	6.67	4.0±0.8	8

Table 4.1.6—*continued*
Lasing Transitions of Ni-Like Ions

Ion	Transition	Wavelength (nm)	Gain (cm^{-1})/ gain-length	Reference
Dy^{38+}	4d→4p (J=0→1)	5.88		8
Dy^{38+}	4d→4p (J=0→1)	6.37		8
Yb^{42+}	4d→4p (J=0→1)	5.023	2.2±0.2	1,6
Yb^{42+}	4d→4p (J=0→1)	5.611	—	1,6
Yb^{42+}	4d→4p (J=2→1)	8.107		1
Yb^{42+}	4d→4p (J=1→1)	8.441		1,6
Ta^{45+}	4d→4p (J=0→1)	4.483	2.3±0.2	1,3
Ta^{45+}	4d→4p (J=0→1)	5.097		1,6
Ta^{45+}	4d→4p (J=2→1)	7.442		1,3
Ta^{45+}	4d→4p (J=1→1)	7.747		1,3
W^{46+}	4d→4p (J=0→1)	4.318		1,3
W^{46+}	4d→4p (J=2→1)	7.240		1,3
W^{46+}	4d→4p (J=1→1)	7.535		1,3
Au^{51+}	4d→4p (J=0→1)	3.560		1,2

4.1.8 Lasing Transitions of Pd-Like Ions

Lasing of palladium-like ions has involved a 4d^{9}5d ^{1}S$_0$ → 4d^{9}5p ^{1}P$_1$ transition shown in Figure 4.1.7. In the table below, the term 4d^9 and the LSJ state designation are omitted to simplify the notation.

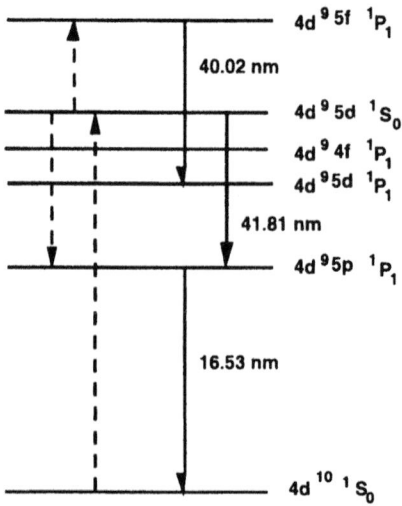

Figure 4.1.7 Partial energy level diagram for palladium-like Xe^{8+} and the laser transition at 41.81 nm. Dash lines indicate the dominant collisional excitation and deexcitation channels of the upper laser level (adapted from Reference 52).

Table 4.1.7
Lasing Transitions of Pd-Like Ions

Ion	Transition	Wavelength (nm)	Gain (cm^{-1})/ gain-length	Reference
Xe^{8+}	$5d \rightarrow 5p$	41.8	— / ~11	52

4.1.9 References

1. MacGowan, B. J., DaSilva, L. B., Fields, D. J., Fry, A.R., Keane, C. J., Koch, J. A., Matthews, D. L., Maxon, S., Mrowka, S., Osterheld, A. L., Scofield, J. H., and Shimkaveg, G., Short wavelength nickel-like x-ray laser development, *Proceedings of the 2nd International Colloquium on X-ray Lasers*, York, U.K., Sept. 1990, Inst. Phys. Conf. Series 116, 221 (1991), and Energies of nickel-like 4d to 4p lasing lines, Scofield, J. H., and MacGowan, B. J., *Physica Scripta* 46, 361 (1992).

2. MacGowan, B. J., DaSilva, L. B., Fields, D. J., Keane, C. J., Koch, J. A., London, R. A., Matthews, D. L., Maxon, S., Mrowka, S., Osterheld, A. L., Scofield, J. H., Shimkaveg, G., Trebes, J. E., and Walling, R. S., Short wavelength x-ray laser research at the Lawrence Livermore National Laboratory, *Phys. Fluids B* 4, 2326 (1992).

3. MacGowan, B. J., Maxon, S., DaSilva, L. B., Fields, D. J., Keane, C. J., Matthews, D. L., Osterheld, A. L., Scofield, J. H., Shimkaveg, G., and Stone, G. F., Demonstration of x-ray amplifiers near the carbon K edge, *Phys. Rev. Lett.* 65, 420 (1990).

4. Kato, Y., Miura, E., Tachi, T., Shiraga, H., Nishimura, H., Daido, H., Yamanaka, M., Jitsuno, T., Takagi, M., Herman, P.R., Takabe, H., Nakai, S., Yamanaka, C., Key, M.H., Tallents, G.J., Rose, S.J., and Rumsby, P.T., Observation of gain at 54.2 Å on the Balmer-alpha transition of hydrogenic sodium, *Appl. Phys. B* 50, 247 (1990).

5. MacGowan, B. J., DaSilva, L. B., Fields, D. J., Keane, C. J., Maxon, S., Osterheld, A. L., Scofield, J. H., and Shimkaveg, G., Observation of $3d^8 4d$ - $3d^8 4p$ soft x-ray laser transitions in high-Z ions isoelectronic to Co I, *Phys. Rev. Lett.* 65, 2374 (1990).

6. MacGowan, B. J., Maxon, S., Keane, C. J., London, R. A., Matthews, D. L., and Whelan, D. A., Soft X-ray amplification at 50.3 Å in nickel-like ytterbium, *J. Opt. Soc. Am. B* 5, 1858 (1988).

7. Xu, Z., Fan, P., Lin, L., Li, Y., Wang, X., Lu, P., Li, R., Han, S., Sun, L., Qian, A., Shen, B., Jiang, Z., Zhang, Z., and Zhou, J. Space- and time-resolved investigation of short wavelength x-ray laser in Li-like Ca ions, *Appl. Phys. Lett.* 63, 1023 (1993).

8. Daido, H., Kato, Y., Murai, K., Ninomiya, S., Kodama, R., Yuan, G., Oshikane, Y., Takagi, M., Takabe, H., and Koibe, F., Efficient soft x-ray lasing at 6 to 8 nm with nickel-like lanthanide ions, *Phys. Rev. Lett.* 75, 1074 (1995).

9. MacGowan, B. J., Maxon, S., Hagelstein, P. L., Keane, C. J., London, R.A., Matthews, D. L., Rosen, M. D., Scofield, J. H., and Whelan, D. A., Demonstration of soft x-ray amplification in nickel-like ions, *Phys. Rev. Lett.* 59, 2157 (1987).

10. Lewis, C. L. S., O'Neill, D. M., Neely, D., Uhomoibhi, J. O., Burge, R., Slark, G., Brown, M., Michette, A., Jaegle, P., Klisnick, A., Carillon, A., Dhez, P., Jamelot, A., Raucourt, J.P., Tallents, G.J., Krishnan, J., Dwivedi, L., Chen, H. Z., Key, M. H., Kodama, R., Norreys, P., Rose, S. J., Zhang, J., Pert, G. J., and Ramsden, S. A., Collisionally excited x-ray laser schemes: progress at Rutherford Appleton Laboratory, Proceedings of SPIE's 1991 International Symposium on Optical and Optoelectronic Applied Science and Engineering, San Diego, CA, July 1991, *SPIE Proceedings* 1551, 49 (1992).

11. Nilsen, J., and Moreno, J. C., Lasing at 7.9 nm in nickel-like neodymium, *Optics Lett.* 20, 1386 (1995).

12. Lewis, C. L. S., Corbett, R., O'Neill, D., Regan, C., Saadat, S., Chenais-Popovics, C., Tomie, T., Edwards, J., Kiehn, G. P., Smith, R., Willi, O., Carillon, A., Guennou, H., Jaeglé, P., Jamelot, G., Klisnick, A., Sureau, A., Grande, M., Hooker, C., Key, M. H., Rose, S. J., Ross, I. N., Rumsby, P. T., Pert, G. J., and Ramsden, S. A., Status of soft x-ray laser research at the Rutherford-Appleton Laboratory, *Plasma Phys. Controlled Fusion* 30, 35 (1988).

13. Desenne, D., Berthet, L., Bourgade, J-L., Bruneau J., Carillon, A., Decoster, A., Dulieu, A., Dumont, H., Jacquemot, S., Jaeglé, P., Jamelot, G., Louis-Jacquet, M., Raucourt, J-P., Reverdin, C., Thébault, J-P., and Thiell, G., X-ray amplification in Ne-like silver: Gain determination and time-resolved beam divergence measurement, in X-ray Lasers 1990, Proceedings of the 2nd International Colloquium on X-ray Lasers, York, England, edited by G.J. Tallents, *IOP Conf. Series* 116, 351 (1991); also, Fields, D. J., Walling, R. S., Shimkaveg, G., MacGowan, B. J., DaSilva, L. B., Scofield, J. H., Osterheld, A. L., Phillips, T. W., Rosen, M. D., Matthews, D. L., Goldstein W. H., and Stewart, R. E., Observation of high gain in Ne-like Ag lasers, *Phys. Rev. A* 46, 1606 (1992).

14. Xu, Z., Fan, P., Zhang, Z., Chen, S., Lin, L., Lu, P., Wang, X., Qian, A., Yu, J., Sun, L., and Wu, M., Soft x-ray lasing and its spatial characteristics in a lithium-like silicon plasma, *Appl. Phys. Lett.* 56, 2370 (1990).

15. Fiedorowicz, H., Bartnik, A., Li, Y., Lu, P., and Fill, E. E., Demonstration of soft X-ray lasing with neon-like argon and nickel-like xenon ions using a laser-irradiated gas puff target, *Phys. Rev. Lett.* 76, 415 (1996).

16. Matthews, D. L., Campbell, E. M., Estabrook, K., Hatcher, W., Kauffman, R. L., Lee, R. W., and Wang, C. L., Observation of enhanced emission of the O VIII H_α line in a recombining laser-produced plasma, *Appl. Phys. Lett.* 45, 2226 (1984).

17. Jaeglé, P., Jamelot, G., Carillon, A., Klisnick, A., Sureau, A., and Guennou, H., Soft x-ray amplification by lithium-like ions in recombining hot plasmas, *J. Opt. Soc. Am. B* 4, 563 (1987).

18. Hara, T., Kozo, A., Kusakabe, N., Yashiro, H., and Aoyagi, Y., Soft x-ray lasing in an Al plasma produced by a 6 J laser, *Jap. J. Appl. Phys.* 28, L1010 (1989).

19. MacGowan, B. J., Rosen, M. D., Eckart, M. J., Hagelstein, P. L., Matthews, D. L., Nilson, D. G., Phillips, T. W., Scofield, J. H., Shimkaveg, G., Trebes, J. E., Walling, R. S., Whitten, B. L., and Woodworth, J. G., Observation of soft X-ray amplification in neon-like molybdenum, *J. Appl. Phys.* 61, 5243 (1987).

20. Nilsen, J., Porter, J. L., MacGowan, B., Da Silva, L. B., and Moreno, J. C., Neon-like x-ray lasers of zirconium, niobium, and bromine, *J. Phys. B* 26, L243 (1993).

21. Enright, G. D., Dunn, J., Villeneuve, D. M., Maxon, S., Baldis, H. A., Osterheld, A. L., La Fontaine, B., Kieffer, J. C., Nantel, M., and Pépin, H., A search for gain in an Ni-like tin plasma, Proceedings of the 3rd International Colloquium on X-Ray Lasers, Schliersee, Germany, May 1992, X-ray lasers 1992, Fill, E. E., Ed., *Inst. Phys. Conf. Series* 125, 45 (1992).

22. Nilsen, J., Moreno, J. C., Koch, J. A., Scofield, J. H., MacGowan, B. J., and Da Silva, L. B., Hyperfine splittings, prepulse technique, and other new results for collisional excitation neon-like x-ray lasers, *AIP Conference Proceedings 332 - X-ray Lasers 1994*, Eder, D. C., and Matthews, D. L., Eds., American Institute of Physics, New York (1994), p. 271.

23. Kim, D., Skinner, C.H., Wouters, A., Valeo, E., Voorhees, D., and Suckewer, S., Soft x-ray amplification in lithium-like Al XI (154Å) and Si XII (129Å), *J. Opt. Soc. Am. B* 6, 115 (1989).

24. Matthews, D. L., Hagelstein, P. L., Rosen, M. D., Eckart, M. J., Ceglio, N. M., Hazi, A. U., Medecki, H., MacGowan, B. J., Trebes, J. E., Whitten, B. L., Campbell, E. M., Hatcher, C. W., Hawryluk, A. M., Kauffman, R. L., Pleasance, L. D., Rambach, G., Scofield, J. H., Stone, G., and Weaver, T. A., Demonstration of a soft x-ray amplifier, *Phys. Rev. Lett.* 54, 110 (1985).

25. Matthews, D., Brown, S., Eckart, M., MacGowan, B., Nilson, D., Rosen, M., Shimkaveg, G., Stewart, R., Trebes, J., and Woodworth, J., Status of the Nova x-ray laser experiments, in Proceedings of the O.S.A. Topical Meeting on Short Wavelength Coherent Radiation: Generation and Applications, Monterey, California, Attwood, D., and Bokor, J., Eds., *AIP Conference Proceedings* 147, 117 (1986).

26. Da Silva, L. B., MacGowan, B. J., Mrowka, S., Koch, J. A, London, R. A., Matthews, D. L., and Underwood, J. H., Power measurements of a saturated yttrium x-ray laser, *Optics Lett.* 18, 1174 (1993).

27. Koch, J. A., Lee, R.W., Nilsen, J., Moreno, J. C., MacGowan, B. J., and Da Silva, L. B., X-ray lasers as sources for resonance-fluorescence experiments, *Appl. Phys. B* 58, 7 (1994).

28. Keane, C. J., Matthews, D. L., Rosen, M. D., Phillips, T. W., Whitten, B. L., MacGowan, B. J., Louis-Jacquet, M., Bourgade, J. L., DeCoster, A., Jacquemot, S., Naccache, D., and Thiell, G., Study of soft x-ray amplification in laser produced strontium plasma, *Phys. Rev. A* 42, 2327 (1990).

29. Eckart, M. J., Scofield, J. H., and Hazi, A. U., XUV emission features from the Livermore soft x-ray laser experiments, Proceedings of the International Colloquium on UV and X-ray Spectroscopy, Beaulieu-sur-Mer, France, September 1987, *J. Phys. Paris* 49, C1-361 (1988).

30. Nilsen, J., Porter, J. L., Da Silva, L. B., and MacGowan, B. J. 17-nm rubidium-ion x-ray laser, *Optics Lett.*. 17, 1518 (1992).

31. Jacoby, D., Pert, G., Shorrock, L., and Tallents, G. J., Observations of gains in the extreme ultraviolet, *Phys. Rev. B* 15, 3557 (1982).

32. Suckewer, S., Skinner, C. H., Milchberg, H., Keane, C., and Voorhees, D., Amplification of stimulated soft x-ray emission in a confined plasma column, *Phys. Rev. Lett.* 55, 1004 (1986).

33. Chenais-Popovics, C., Corbett, R., Hooker, C. J., Key, M. H., Kiehn, G. P., Lewis, C. L. S., Pert, G. J., Regan, C., Rose, S. J., Sadaat, S., Smith, R., Tomie, T., and Willi, O., Laser amplification at 18.2 nm in recombining plasma from a laser-irradiated carbon fiber, *Phys. Rev. Lett.* 59, 2161 (1987).

34. Nilsen, J., and Moreno, J.C., Nearly monochromatic lasing at 182 angstroms in neon-like selenium, *Phys. Rev. Lett.* 74, 3376 (1995).

35. Lee, T. N., McLean, E. A., and Elton, R. C., Soft X-ray lasing in neon-like germanium and copper plasmas, *Phys. Rev. Lett.* 59, 1185 (1987).

36. Basu, S., Hagelstein, P. L., Goodberlet, J. G., Muendel, M. H., and Kaushik, S., Amplication in Ni-like Nb at 204.2 Å pumped by a table-top laser, *Appl. Phys. B*, 57, 303 (1993).

37. Nilsen, J., MacGowan, B. J., Da Silva, L. B., and Moreno, J. C., Prepulse technique for producing low-Z Ne-like XUV lasers, *Phys. Rev. A* 48, 4682 (1993).

38. Koch, J. A, MacGowan, B. J., Da Silva, L. B., Matthews, D. L., Underwood, J. H., Batson, P. J., and Mrowka, S., Observation of gain-narrowing and saturation behavior in Se x-ray laser line profile, *Phys. Rev. Lett.* 68, 3291 (1992).

39. Jaeglé, P., Carillon, A., Gauthe, B., Goedtkindt, P., Guennou, H., Jamelot, G., Klisnick, A., Moller, C., Rus, B., Sureau, A., and Zeitonn, P., Lasing near 200Å with neon-like zinc and lithium-like sulfur, *Appl. Phys. B* 57, 313 (1993).

40. Lee, T. N., McLean, E. A., Stamper, J. A., Griem, H. R., and Manka, C. K., Laser driven soft x-ray laser experiments at NRL, *Bull. Am. Phys. Soc.* 33, 1920 (1988).

41. Fill, E. E., Li, Y., Schloëgl, D., Steingruber, J., and Nilsen, J., Sensitivity of lasing in neon-like zinc at 21.2 nm to the use of the prepulse technique, *Optics Lett.* 20, 374 (1995).

42. Nilsen, J., Moreno, J. C., MacGowan, B. J., and Koch, J. A., First observation lasing at 231Å in neon-like nickel using the prepulse technique, *Appl. Phys. B* 57, 309 (1993).

43. Nilsen, J., MacGowan, B. J., Da Silva, L. B. Moreno, J.C, Koch, J. A., and Scofield, J. H., Reinterpretation of the neon-like titanium laser experiments, *Opt. Eng.* 33, 2687 (1994).

44. Carillon, A., Chen, H. Z., Dhez, P., Dwivedi, L., Jacoby, J., Jaeglé, P., Jamelot, G., Zhang, J., Key, M. H., Kidd, A., Klisnick, A., Kodama, R., Krishnan, J., Lewis, C.L.S., Neely, D., Norreys, P., O'Neill, D., Pert, G.J., Ramsden, S. A., Raucourt, J. P., Tallents, G. J., and Uhomoibhi, J., Saturation and near-diffraction-limited operation of an XUV laser at 23.6 nm, *Phys. Rev. Lett.* 68, 2917 (1992).

45. Li, Y., Pretzler, G., and Fill, E. E., Observation of lasing on the two J = 0-1, 3p-3s transitions at 26.1 and 30.4 nm in neonlike vanadium, *Optics Lett.* 20, 1026 (1995).

46. Li, Y., Pretzler, G., and Fill, E.E., Neon-like ion lasers in the extreme ultraviolet region, *Phys. Rev. A* 52, R3433 (1995).

47. Boehly, T., Russotto, M., Craxton, R. S., Epstein, R., Yaakobi, B., Da Silva, L. B., Nilsen, J., Chandler, E. A., Fields, D. J., MacGowan, B. J., Matthews, D. L., Scofield, J. H., and Shimkaveg, G., Demonstration of a narrow divergence x-ray laser in neon-like titanium, *Phys. Rev. A* 42, 6962 (1990).

48. Nilsen, J., Fiedorowicz, H., Bartnik, A., Li, Y., Lu, P., and Fill, E. E., Self photopumped neonlike X-ray laser, *Optics Lett.* 21, 408 (1996).

49. Rocca, J. J., Shlyaptsev, V., Tomasel, F. G., Cortazar, O.D., Hartshorn, D., and Chilla, J. L. A., Demonstration of a discharge pumped table top soft x-ray laser, *Phys. Rev. Lett.* 73, 2192 (1994); Rocca, J. J., Tomasel, F. G., Marconi, M. C., Shlyaptsev, V. N., Chilla, J. L. A., Szapiro, S. T., and Guidice, G., Discharge-pumped soft x-ray laser in neon-like argon, *Phys. Plasmas* 2, 2547 (1995).

50. Li, Y., Lu, P., Pretzler, G., and Fill, E. E., Lasing in neonlike sulfur and silicon, *Opt. Commun.* 133, 196 (1997).

51. Daido H., Ninomiya S., Imani, T., Kodama, R., Takagi, M., Kato Y., Murai K., Zhang, J., You, Y., and Gu, Y., Nickellike soft-x-ray lasing at the wavelengths between 14 and 7.9 nm, *Optics Lett.* 21, 958 (1996).

52. Lemoff, B. E., Yin, G. Y., Gordon III, C. L., Barty, C. P. J., and Harris, S. E., Demonstration of a 10-Hz femtosecond-pulse-driven XUV laser at 41.8 nm in Xe IX, *Phys. Rev. Lett.* 74, 1574 (1995).

53. Zhang, J., MacPhee, A. G., Lin, J. et al., A saturated x-ray laser beam at 7 nanometers, *Science* 276, 1097 (1997).

54. Dunn, J., Osterheld, A. L., Shepard, R., White, W. E., Shlyaptsev, V. N., and Steward, R. E., Demonstration of x-ray amplification in transient gain nickel-like palladium scheme, *Phys. Rev. Lett.* 80, 2825 (1998).

55. Nilsen, J. and Scofield, J. H., Wavelengths of neon-like $3p \rightarrow 3s$ x-ray laser transitions, *Physica Scripta* 49, 588 (1994).

56. Li, Y., Nilsen. J., Dunn, J., and Osterheld, A. L., Wavelengths of the Ni-like 4d $^1S_0$–4p $^1P_1$ x-ray laser line, *Phys. Rev. A* 58, R2668 (1998).

57. Korobkin, D. V., Nam, C. H., and Suckewer, S., Demonstration of soft x-ray lasing to ground state in Li III, *Phys. Rev. Lett.* 77, 5206 (1996).

58. Korobkin, D., Goltsov, A., Morosov, A., and Suckewer, S., Soft x-ray amplification at 26.2 nm with 1-Hz repetition rate in a table-top system, *Phys. Rev. Lett.* 81, 1607 (1998).

59. Loewenthal, F., Tommasini, R., and Balmer, J. E., Observation of saturated lasing on the 3p–3s, J=0–1 transition at 25.5 nm in neonlike iron using a double-prepulse technique, *Optics Commun.* 154, 325 (1998).

60. Dunn, J., Nilsen, J., Osterheld, A. L., Li, Y., and Shlyaptsev, V. N., Demonstration of transient gain x-ray lasers near 20 nm for nickellike yttrium, zirconium, niobium, and molybdenum, *Optics Lett.* 24, 101 (1999).

61. Cauble, R., Da Silva, L. B., Barbee, Jr., T. W., Celliers, P., Decker, C., London, R. A., Moreno, J. C., Trebes, J. E., Wan, A. S., and Weber, F., Simultaneous measurement of local gain and electron density in x-ray lasers, *Science* 273, 1093 (1996).

62. Nagata Y., Midorikawa K., Kubodera S., Obara M. et al., Soft-x-ray amplification of the Lyman-alpha transition by optical-field-induced ionization, *Phys. Rev. Lett.* 71, 3774 (1993).

63. Scofield, J. H., and MacGowan. B. J., Energies of nickel-like 4d and 4p laser lines, *Physica Scripta* 46, 361 (1992).

64. Nilsen J. Li, Y., Dunn, J., Osterheld, Ryabtsev, A., and Churilov, S., Measuring the wavelengths of the Ni-like 4d $^1S_0 \rightarrow 4p$ $^1P_1$ and 4f $^1P_1 \rightarrow 4d$ $^1P_1$ x-ray laser lines, in *X-ray Lasers 1998*, IOP Conf. Series 159, IOP Publishing, Bristol, U.K. (1999).

65. Kalachnikov, M. P., Nickles, P. V., Schnürer, M. et al., Saturated operation of a transient collisional x-ray laser, *Phys. Rev. A* 57, 4778 (1998).

66. Tommasini, R.., Lowenthal, F., and Balmer, J. E., Saturation in a Ni-like Pd soft-x-ray laser at 14.7 nm, *Phys. Rev. A* 59, 1577 (1999).

Section 4.2
FREE ELECTRON LASERS

4.2.1 Introduction

Free electron lasers (FEL) provide radiation over a large wavelength range spanning six orders of magnitude—from the ultraviolet to millimeter waves (~200 nm to ~100 mm). They are based on a beam of high energy electrons traversing a spatially varying magnetic field (wigglers, undulators) which cause the electrons to oscillate and emit radiation. Free electron laser configurations include (1) oscillators with reflectors at the ends of the magnet array, thus providing multi-pass, low-gain operation, (2) amplifiers in which electrons are injected into an undulator in synchronism with a signal derived from a conventional laser source, and (3) self-amplified spontaneous emission amplified by a single pass through a wiggler.

Various types of electron accelerators are used for FELs: storage rings, rf and induction linacs, electrostatic and pulse-line accelerators (operating in single-shot mode), power supplies, microtrons, modulators, and ignition coils. The wavelength ranges of FELs based on different types of accelerators are shown in Figure 4.2.1.

Free electron lasers are divided into short and long (> 0.5 mm) wavelengths because different physical processes are involved. For the latter, higher currents and lower energy beams are typical, space-charge effects are more important, and the dominant interaction mechanism is often coherent Raman scattering.

Short wavelength FELs are grouped in Table 4.2.1 by accelerator and in order of increasing wavelength. The type of FEL (amplifier or oscillator), operating wavelength or range, electron pulse length divided by c, electron beam energy E, peak current I, number of undulator periods N, undulator wavelength λ_0, and undulator parameter parameter K = $eB\lambda_0/2\pi mc^2$, where B is the rms undulator field strength, are included in the table. Much of this data is from tabulations of W. B. Colson (see Further Reading); this reference also includes FELs in design studies or under construction.

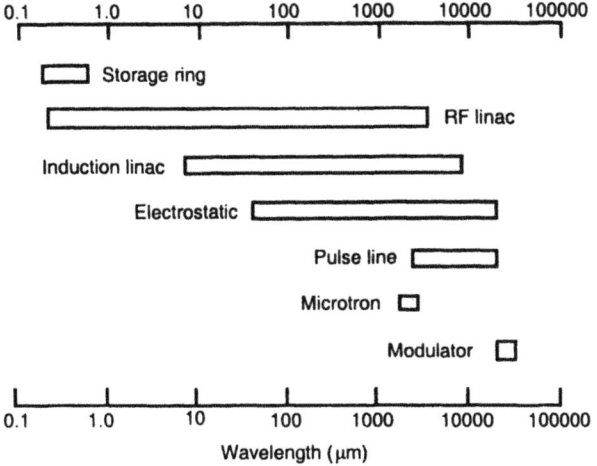

Figure 4.2.1   Reported ranges of output wavelengths of free electron lasers for various types of electron accelerators.

Long wavelength FELs are also grouped by accelerator and in order of increasing wavelength in Table 4.2.2. The type of FEL (amplifier, oscillator, or self-amplified spontaneous emission), operating wavelength and frequency, peak power P, pulse time t_p, electron voltage V and current I, number of undulator periods N, wiggler period λ_w, and wiggler parameter K = $0.0934B_w\lambda_w$, where B is the wiggler amplitude in kG, and type of wiggler (C – circular, H – helical, P – planar, T – tapered). Much of this data is from tabulations of H.P. Freund and V. L. Granatstein (see Further Reading); this reference also includes FELs in design studies or under construction.

Further Reading

Brau, C. A., *Free-Electron Lasers*, Academic Press, Boston (1990).

Colson, W. B., Short wavelength free electron lasers in 1998, *Nucl. . and Meth.* A 429, 37 (1999).

Colson, W. B., Pellegrini, C., and Remeri, A., Eds., Free Electron Lasers, in *Laser Handbook*, Vol. 6, North Holland, Amsterdam (1990).

Colson, W. B., and Prosnitz, D., Free Electron Lasers, in *Handbook of Laser Science and Technology, Suppl. 1: Lasers*, CRC Press, Boca Raton, FL (1991), p. 515.

Couprie, M. E., Storage rings FELs, *Nucl.. Instr. and Meth.* A 393, 13 (1997).

Freund, H. P., and Autonsen, T. M., Jr., *Principles of Free Electron Lasers*, 2nd edition, Chapman and Hall, London (1996).

Freund, H. P., and Granatstein, V. L., Long wavelength free-electron lasers in 1998, *Nucl. Instr. and Meth.* A 429, 33 (1999).

Freund, H. P., and Parker, R. K., Free-Electron Lasers, in *Encyclopedia of Physical Science and Technology*, Academic Press, San Diego (1991), p. 49.

Gallerano, G. P., Doria, A., Giovenale, E., and Renieri, A, Compact free electron lasers: From Cerenkov to waveguide free electron lasers, *Infrared Phys. & Tech.* 40, 161 (1999).

Granatstein, V. L., Parker, R. K., and Spangle, P. A., Millimeter and Submillimeter Lasers, in *Handbook of Laser Science and Technology, Vol. I: Lasers and Masers* , CRC Press, Boca Raton (1981), p. 441.

Luchini, P., and Motz, H., *Undulators and Free-Electron Lasers*, Oxford University Press, Oxford (1990).

Poole, M. W., FEL Sources: Present and future prospects, *Rev. Sci. Instrum.* 63, 1528 (1992).

Prosnitz, D., Free Electron Lasers, in *Handbook of Laser Science and Technology, Vol. I: Lasers and Masers*, CRC Press, Boca Raton, FL (1981), p. 425.

See, also, Proceedings of the International Free Electron Laser Conferences in *Nucl. Instr. and Meth.* A 272 (1988), A285 (1989), A 296 (1990), A 304 (1991), A 318 (1992), A 331 (1993), A 341 (1994), A 358 (1995), A 375 (1996), A 393 (1997), A 429 (1999).

4.2.2 Short Wavelength Free Electron Lasers

Table 4.2.1
Short Wavelength Free Electron Lasers

Type	λ (μm)	Pulse length	E (MeV)	I (A)	N	λ_u (cm)	K (rms)	Ref.
Radio Frequency Linac								
oscillator	0.25	6 ps	70	100	70	0.88	0.4	1
oscillator	0.3–0.7	5 ps	155	60	67	4	0.7–1.4	41
oscillator	0.37–0.38	8–15 ps	46	≤200	73	1.38	—	2
oscillator	0.5	6 ps	50	100	70	0.88	0.4	1
oscillator	0.525	—	115	2.4	131	3.6	—	3
oscillator	0.63 (TH)	10 ps	68	42	78	3.8	1	4
oscillator	0.662	—	109	270	115	2.2	—	5
oscillator	1.57	—	66	2.5	131	3.6	—	6
oscillator	1.8—17.5	1–10 ps	32–58	50–80	48	4	—	7
oscillator	1.88	10 ps	68	42	78	3.8	1	4
oscillator	1.9–8.1	—	25–45	40	47	2.3	—	8
oscillator	2–6	2 ps	170	100	50	6	1.3	9
oscillator	2.2–9.6	2 ps	47	50	52	2.3	1	10
oscillator	3	3 ps	44	20	47	2.3	1	11
amplifier	3	—	38	35	80	2.5	—	12
oscillator	3–12	0.7–3 ps	22–45	10	72	3.1	0.8	13
oscillator	3–53	0.2–0.4 ps	21°50	80	38	5	1.4	14
oscillator	3.04	—	38	15	80	2.5	—	15
oscillator	4–6	10 ps	15	200	24	1	0.3	16
oscillator	4–40	—	21	500	37	2.7	—	17
oscillator	4–200	1 ps	50	50	38	6.5	1.8	42
oscillator	4.2	—	38	60	47	2.3	—	18
oscillator	4.65[1]	—	33	—	34	3.4	0.5–1.5	43
oscillator	4.8	0.4 ps	48	60	40.5	2.7	1.0	52
oscillator	4.8–9.7	10 ps	31	42	58	3.4	0.43–1.6	4
oscillator	5	2 ps	40	2.7	80	3.2	1	19
oscillator	5–35	5 ps	25	50	38	6.5	1.2	20, 21
oscillator	5.5	10 ps	33.2	42	58	3.4	1	4
oscillator	6.6–7.8	2 ps	30	2.7	80	3.2	1	53
oscillator	10	4 ps	30	14	50	3	1	22
amplifier	12	5 ps	18	170	100	2	0.7	44
oscillator	12–21	5 ps	9–14	100	73	1.4	0.2	23
oscillator	15	7 ps	25	5	50	4.4	1	24
oscillator	15–65	1–5 ps	15–32	14	25	6	1	25
oscillator	15.5	15 ps	17	300	200	2	0.9	45
amplifier	16	3 ps	13.5	80	40	1.5	1	46
amplifier	16.3	16 ps	17	300	200	2	0.9	55
oscillator	18–40	10 ps	33	40	30	8	1.3–1.7	47

Table 4.2.1—*continued*
Short Wavelength Free Electron Lasers

Type	λ (μm)	Pulse length	E (MeV)	I (A)	N	λ_u (cm)	K (rms)	Ref.
oscillator	19.4[1]	—	33	—	30	8.0	1.3–3.4	43
oscillator	20	30 ps	18	100	30	3	0.8	26
oscillator	20–110	3–5 ps	45	50	38	6.5	1.8	20, 21
oscillator	21–126	30 ps	10–19	50	32	3–12	0.01–1.5	56
oscillator	24	—	16	—	53	3.3	0.7	54
oscillator	40	30 ps	17	50	32	6	1	27
oscillator	43	10 ps	13	20	40	4	0.7	28
oscillator	47	3 ps	8	50	50	0.66	0.5	29
oscillator	80–200	15 ps	4	8	50	1	0.7	31
Electron Storage Ring								
oscillator	0.2–0.4	2.5 ps	300–750	12	2x33	10	2.4	40, 51
oscillator	0.24	6 ps	500	5	2x8	11	2	30
oscillator	0.24–0.69	35 ps	350	6	2x33	10	1.6	32
oscillator	0.3	6 ps	500	5	2x8	11	2	33
oscillator	0.3–0.6	15 ps	800	0.1	2x10	13	4.5	49
oscillator	0.23–0.35	160–280 ps	300	5	2x42	7.2	2	34, 35, 57
oscillator	0.35	20 ps	20	0.1	2x10	13	4	36
oscillator	0.46–0.68	—	160–230	4	2x17	7.8	—	37
amplifier	0.5145	—	625	0.018	20	11.6	—	38
Electrostatic Accelerator								
oscillator	60	25 μs	6	2		150/2	0.1	39
oscillator	340	25 μs	6	2		42/7.1/2	0.1	39
Smith-Purcell								
oscillator	200	cw	0.04	0.001	50	300	–	50

(1) Simultaneous two-color lasing.

References—Table 4.2.1

1. Batchelor, K., Ben-Zvi, I., Fernow, R. C. et al., Status of the visible free-electron laser at the Brookhaven Accelerator Test Facility, *Nucl. Instr. and Meth. A* 318, 159 (1992).
2. O'Shea, P. G., Bender, S. C., Byrd, D. A. et al., Demonstration of ultraviolet lasing with a low energy electron laser, *Nucl. Instr. and Meth. A* 341, 7 (1994).
3. Edighoffer, J. A., Neil, G. R., Fornaca, S., Thompson, H. R., Smith, T. I., Schwettman, H. A., Hess, C. E., Frisch, J., and Rohtagi, R., Visible free-electron-laser oscillator (constant and tapered wiggler), *Appl. Phys. Lett.* 52, 1569 (1988).
4. Kobayashi, A., Saeki, K., Oshita, E. et al., Optical properties of infrared-FELs at the FELI, *Nucl. Instr. and Meth. A* 375, 317 (1996).
5. Shoffstall, D. et al., 1989 U. S. Particle Accelerator Conference, Chicago, IL (1989).
6. Edighoffer, J. A., Neil, G., R., Hess, C. E., Smith, T. I., Fornaca, S. W., and Schwettman, H. A., Variable-wiggler free-electron laser oscillation. *Phys. Rev. Lett.* 52, 344 (1984).

7. Ortega, J. M., Operation of the CLIO infrared laser facility, *Nucl. Instr. and Meth. A* 341, 138 (1994); Prazeres, R., Glotin, F., and Ortega, J. M., Optical mode analysis on the CLIO infrared FEL, *Nucl. Instr. and Meth. A* 341, 54 (1994).

8. Benson, S. et al. 1989 U. S. Particle Accelerator Conference, Chicago, IL (1989).

9. Asakawa, M. et al., *Proc. 15th Annual Meeting of the Laser Society of Japan*, Osaka, Japan (1995).

10. Brau, C., The Vanderbilt University Free Electron Laser Center, *Nucl. Instr. and Meth. A* 318, 38 (1992).

11. Bhowmik, A., Curtin, M. S., McMillin, W. A., Benson, S. V., Madey, J. M. J., Richardson, B. A., and Vintro, L., First operation of the Rocketdyne/Stanford free electron laser, *Nucl. Instr. and Meth. A* 272 (1988).

12. Benson, S. V. et al., The Stanford Mark III infrared free electron laser, *Nucl. Instr. and Meth. A* 250, 39 (1986).

13. Schwettman, H. A., Smith, T. I., and Swent, R. L., *Nucl. Inst., and Meth. A* 375, 662 (1996); Smith, T., and Marziali, A., Feedback stabilization of the SCA/FEL wavelength, *Nucl. Instr. and Meth. A* 331, 59 (1993); Smith, T. et al., *Proc. SPIE* 1854, 23 (1993).

14. Ortega, J. M., Berset, J. M., Chaput, R. et al., Activities of the CLIO infrared facility, *Nucl. Instr. and Meth. A* 375, 618 (1996).

15. Bhowmik, A., Curtin, M. A., McMullin, M. A., Benson, S., Richman, B. A., and Vintro, L., First operation of the Rocketdyne/Stanford free-electron laser. *Nucl. Inst., and Meth. A* 272, 10 (1988).

16. Nguyen, D. C. et al., Initial performance of Los Alamos advanced free electron laser, *Nucl. Instr. and Meth. A* 341, 29 (1994).

17. Warren., R., Sollid, J. E., and Feldman, D. W., Near-ideal lasing with a uniform wiggler, *Nucl. Instr. and Meth. A* 285, 1 (1989).

18. Feinstein, J., Fisher, A. S., Reid, M. B., Ho, A., Ozcan, M., Dulman, H. D., and Pantell, R. H., Experimental results on a gas-loaded free-electron laser, *Phys. Rev. Lett.* 60, 18 (1988).

19. Auerhammer, J. et al., First observation of amplification of spontaneous emission achieved with the Darmstadt IR-FEL, *Nucl. Instr. and Meth. A* 341, 63 (1994).

20. Bakker, R. J., van der Geer, A. J., Jarosznski, D. A., van der Meer, A. F. G., Oepts, D., and van Amersfoort, P. W., Broadband tunability of a far-infrared free-electron laser, *J. Appl. Phys.* 74, 1501 (1993).

21. Jaroszynski, D. A., Bakker, R. J., Oepts, D. et al., Limit cycle behavior in FELIX, *Nucl. Instr. and Meth. A* 331, 52 (1993).

22. Xie, J., Zhuang, J., Huang, Y. et al., Status of the Beijing FEL project, *Nucl. Instr. and Meth. A* 331, 204 (1993: First lasing of the Beijing FEL, *Nucl. Instr. and Meth. A* 341, 34 (1994).

23. Lehrman, I. S.., Krishnaswamy, J., Hartley, R. A., Austin, R. H. et al., First lasing of the compact infrared free-electron laser, *Nucl. Inst., and Meth. A* 393, 178 (1997).

24. Castellano, M. et al., Status report of the IR FEL project on the superconducting linac LISA at LNF-Frascati, *Nucl. Instr. and Meth. A* 304, 204 (1991).

25. Berryman, K. W., and Smith, T. I., First lasing, capabilities and flexibility of FIREFLY, *Nucl. Instr. and Meth. A* 375, 6 (1996).

26. Guimbal, P. et al., First results in the saturation regime with the ELSA FEL, *Nucl. Instr. and Meth. A* 341, 43 (1994).

27. Okuda, S., Ohkama, J., Suemue, S. et al., Amplification of spontaneous emission with two high brightness electron bunches of the ISIR linac, *Nucl. Instr. and Meth. A* 341, 59 (1994).

28. Nishimura, E., Saeki, K., Abe, S. et al., Optical performance of the UT-FEL at first lasing, *Nucl. Instr. and Meth. A* 341, 39 (1994).

29. Asakawa, M. et al., *Proc. 15th Annual Meeting of the Laser Society of Japan*, Osaka, Japan, January 19-20, 1995.

30. Hama, H., Yamozaki, J., and Isoyama, G., FEL experiment on the UVSOR storage ring, *Nucl. Instr. and Meth. A* 341, 12 (1994).

31. Lewellen, J. W., Schmerge, J. F., Huang, Y. C., Feinstein, J., and Pantell, R. H., Preliminary emission characteristic measurements for a $300k FIR FEL, *Nucl. Instr. and Meth. A* 358, 24 (1995).

32. Drobyazko, I. B., Kulipanov, G.N., Litvinenko, V. N., Pinoyev, I. V., Popik, V. M., Silvestrov, I. G., Skrinsky, A. N., Sokolov, A. S., and Vinokurov, N. A., Lasing in visible and ultraviolet regions in optical kylstron installed on the VEPP-3 storage ring, *Nucl. Instr. and Meth. A* 282, 424 (1989).

33. Takano, S., Hama, H., and Isoyama, G., Lasing of a free electron laser in the visible on the UVSOR storage ring, *Nucl. Instr. and Meth. A*331, 20 (1993.

34. Hara, T., Hama, H., and Isoyama, G., Lasing of a free electron laser in the visible on the UVSOR storage ring, *Nucl. Instr. and Meth. A* 341, 21 (1994).

35. Yamazaki, T., Yamada, K., Sugiyama, S., Ohgaki, N., Tomimasu, T., Noguchi, T., Mikado, T., Chiwakei, M., and Suzuki, R., Lasing in visible of a storage-ring free electron laser at ETL, *Nucl. Instr. and Meth. A* 309, 343 (1991).

36. Yamazaki, T. Yamada, K., Sugiyama, S. et al., First lasing of the NIJI-IV storage-ring free-electron laser, *Nucl. Instr. and Meth. A* 331, 27 (1993), Yamazaki, T. et al., Present status of the NIJI-IV free-electron laser, *Nucl. Instr. and Meth. A* 341, ABS 3 (1994).

37. Billardon, M., Elleaume, P., Lapierre, Y., Ortego, J.M., Bazin, C., Bergher, M., Marilleau, J., and Petroff, Y., The Orsay storage ring free-electron laser - new results, *Nucl. Instr. and Meth.* 250, 26 (1986).

38. Barbini, B. R., Vignola, G., Trillo, S., Boni, R., DeSimone, S., Faini, S., Guiducci, S., Preger, M., Serio, M., Spataro, B., Tazzari, S., Tazzioli, F., Vescovi, M., Cattoni, A., Sanelli, C., Castellano, M., Cavallo, N., Cevenini, F., Masullo, M. R., Patteri, P., Rinzivillo, R., Solimeno, S., and Cutolo, A., Preliminary results of the Adone storage ring FEL experiment, LELA, *J. Phys.* (Paris), 44, C1-1 (1983).

39. Ramian, G., The new UCSB free-electron lasers, *Nucl. Instr. and Meth. A* 318, 225 (1992).

40. Litvinenko, V. N. et al., First UV/visible lasing with the OK-4/Duke storage ring FEL, *Nucl. Inst., and Meth. A* 407, 8 (1997).

41. Tomimasu, T., Oshita, E., Okuma, S., Wakita, K. et al., First lasings at 0.27-0.37 µm and 0.51-0.63 µm using an S-band electron linac with a thermionic gun, *Nucl. Inst., and Meth. A* 393, 188 (1997) .

42. Oepts, D., van der Meer, A. F. G., and van Amersfoort, P. W., The free electron laser facility FELIX, *Infrared Phys. Technol.* 36, 297 (1995).

43. Zako, A., Kanazawa, Y., Konishi, Y., Yamaguchi, S., Nagai, A., and Tomimasu, T., Simultaneous two-color lasing in the mid-IR and far-IR region with two undulators and on RF linac at the FELI, *Nucl. Inst., and Meth. A* 429, 136 (1999).

44. Hogan, M., Pellegrini, C., Rosenzweig, J.,Travish, G. et al., Measurements of gain larger than 10^5 at 12 µm in a self-amplified spontaneous-emission free-electron laser, *Phys. Rev. Lett.* 81, 4867 (1998).

45. Sheffield, R. L. et al., *Proc. SPIE* 2988, 28 (1997).

46. Hogan, M., Pellegrini, C., Rosenzweig, J.,Travish, G. et al., Measurements of high gain and intensity fluctuations in a self-amplified, spontaneous-emission free-electron laser, *Phys. Rev. Lett.* 80, 289 (1998).

47. Takii, T., Oshita, E., Okuma, S., Wakita, K. et al., First lasings at IR and FIR range using the FELI FEL facility 4, *Nucl. Inst., and Meth. A* 407, 21 (1998).

48. Tecimer, M., and Elias, L. R., Hybrid microundulator designs for the creol compact cw-FEL., *Nucl. Inst., and Meth. A* 341, B126 (1994).

49. Couprie, M. E., Nutarelli, D., Roux, R., Nahon, L. et al., The Super-ACO FEL operation with shorter positron bunches, *Nucl. Inst., and Meth. A* 407, 215 (1998).

50. Urata, J., Goldstein, M., Kimmitt, M. F., Naumov, A. et al., Superradiant Smith-Purcell emission., *Phys. Rev. Lett.* 80, 516 (1998).

51. Litvinenko, V. N., Park, S. H., Pinayev, I. V. et al., OK-4/Duke storage ring FEL lasing in the deep-UV, *Nucl. Instr. and Meth. A* 429, 151 (1999).

52. Benson, S., Biallas, G., Bohn, C. et al., First lasing of the Jefferson Lab IR demo FEL, *Nucl. Instr. and Meth. A* 429, 27 (1999).

53. Brunken, M., Döbert, S., Eichhorn, R. et al., First lasing of the Darmstadt cw free electron laser, *Nucl. Instr. and Meth. A* 429, 21 (1999).

54. Minehara, E. J., Sugimoto, M., Sawamura, M., Nagai, R., Kikuzawa, N., Yamanouchi, T., and Nishimori, N., A 0.1 kW operation of the JAERI superconducting RF linac-based FEL, *Nucl. Instr. and Meth. A* 429, 9 (1999).
55. Nguyen, D. C., Sheffield, R. L., Clifford, M. F. et al., First lasing of the regenerative amplifer FEL, *Nucl. Instr. and Meth. A* 429, 125 (1999).
56. Kato, R., Okuda, S., Kondo, G. et al., Upgrade of the ISIR-FEL at Osaka University and oscillation experiments in the sub-millimeter wavelength region, *Nucl. Instr. and Meth. A* 429, 146 (1999).
57. Yamada, K., Sei, N., Yamazaki, T. et al., Lasing down to the deep UV in the NIJI-IV FEL, *Nucl. Instr. and Meth. A* 429, 159 (1999).

4.2.3 Long Wavelength Free Electron Lasers

Table 4.2.2
Long Wavelength Free Electron Lasers

Type	λ(mm)/ f (GHz)	P_{peak} (MW)	τ_p (ms)	V_b(MV)	I_b(A)	λ_w(cm)	K(type)	Ref.
Electrostatic Accelerator								
oscillator	1.5/206	0.73	12	1.77	7.2	2	0.67 (P)	40
oscillator	2/130– 1/260	0.29	5	2	6.25	2	0.67	12
oscillator	≤2.5/120	≤0.015	6	6	2	7.14	7/1.0 (P)	2
oscillator	3/100	0.012	70	1.4	1.4	4.4	0.82 (P)	42
oscillator	12/26	0.001	10-30	0.4	2	3.2	0.39 (H)	1
oscillator	68/4.4	0.0035	5	0.07	0.8	4.4	0.12 (P)	3, 34
Pulse Line Accelerator								
oscillator	3/100	10	0.25	2.5	100	10	1.9 (H)	8
SASE	3/100	1	0.02	0.3	400	1	0.14 (H)	14
SASE	3/100	1	1.5×10^{-2}	0.5	100	2	0.28 (H)	15
amplifier	3.5/85	0.25	0.02	0.45	17	0.96	0.34 (P)	5
oscillator	4/75	200	1	1	2000	4	0.3 (P)	9,14
oscillator	6.7/45	7	0.025	0.5	120	2.4	0.07 (P)	10
amplifier	8/35	60	0.025	0.75	300	3.1	0.4 (H)	6
amplifier	8/35	50	0.03	1.8	400	8	2.24 (H)	7
SASE	8/35	2.3	0.1	0.5	750	3	0.53 (H)	16
oscillator	9.4/32	0.75	0.1	0.3	50	2.3	0.64	13
oscillator	10.3/29	7	0.0025	0.5	120	2.4	0.07 (P)	14
amplifier	12.5/24	0.2	0.15	0.6	100	1.85	0.2 (H)	4
Cerenkov	16/19	$\sim 10^{-4}$	0.01	—	~0.1	—	—	35,36
oscillator	26/12	3.6	0.4	0.43	190	3.27	0.27 (C)	11
Radio Frequency Linac								
oscillator	0.4/650	—	10	40	0.2	3	1.4 (P)	19
oscillator	~2.3/~130	1.6×10^{-5}	3×10^{-6}	6.85	1.5	6	4 (P)	18
oscillator	2.7/109	1	4×10^{-6}	9	50	6	4 (P)	17

Table 4.2.2—*continued*
Long Wavelength Free Electron Lasers

Type	λ(mm)/ f (GHz)	P_{peak} (MW)	τ_p (ms)	V_b(MV)	I_b(A)	λ_w(cm)	K(type)	Ref.
Induction Linac								
amplifier	2.0/150	100	—	3.5	950	9.8	—	20
amplifier	3/9.4	150	0.005	2	170	16	3.3 (P)	26
amplifier	6/45	10	0.1	1	3000	4.5	1.3 (P)	21
oscillator	8/31	48	0.2	0.8	170	6	0.84	43
amplifier	8/35	140	0.05	3.4	800	11	3.5 (P)	22
oscillator	8/35	7	0.0025	0.5	120	2.4	0.07 (P)	14
amplifier	8/35	90	0.03	2.2	800	12	1.2 (H)	38
amplifier	8/35	80	0.03	2.2	800	12	1.2 (H)	39
SASE	8/35	40	0.03	2.2	800	12	1.2 (H)	38
amplifier	8.7/34	1000	0.01-0.02	3.5	850	9.8	— (T)	23,24
oscillator	9.7/31	31	0.2	0.8	150	6	0.84 (H)	27
amplifier	32/9.4	100	cw	1.5	450	16	1.5 (P)	25
Modulator								
amplifier	8/35	0.027	1	0.1	10	0.64	0.2 (CH)	41
oscillator	30/10	10^{-5}	1000	0.01	0.2	2	0.02 (P)	28
oscillator	35/8.2	10^{-6}	cw	0.13	0.0018	3.8	0.16 (P)	29
Microtron								
oscillator	2–3.5/ 85–150	0.0015	5.5	2.3	0.35	2.5	1.4 (P)	30
Power Supply								
amplifier	30/9.9	10^{-6}	cw	0.05	0.01	1.9	0.03 (P)	31
oscillator	25–37/ 8–12.4	2×10^{-5}	cw	0.12	0.18	3	28(P)	37
SASE	32–37/ 8–9.3	10^{-5}	cw	0.04-0.08	0.01	1.9	0.03 (P)	32
Electron Gun								
SASE	~300/~1		~50	~0.001	<0.5	2.0	— (FF)	33

C – circular, CH – CHI wiggler, FF – folded-foil, H – helical wiggler, P – planar wiggler, T – tapered wiggler

References—Table 4.2.2

1. Lee, B. C., Cho, S. O., Kim, S. K., Jeong, Y. U., Cha, B. H., and Lee, J., First lasing of the KAERI millimeter-wave free electron laser, *Nucl. Instr. and Meth.* A 375, 28 (1996).
2. Ramian, G., The new UCSB free-electron lasers, *Nucl. Instr. and Meth.* A 318, 225 (1992).
3. Cohen, M., Eichenbaum, A., Kleinman, H., Arbel, M., Yakover, I.M., and Gover, A., Report of first masing and single-mode locking in a prebunched beam FEM oscillator, *Nucl. Instr. and Meth.* A 375, 17(1996).
4. Liu, Y. H., and Marshall, T. C., Harmonic millimeter radiation for a microwave free-electron-laser amplifier, *Phys. Rev. E* 56, 2161 (1997).

5. Cheng, S., Granatstein, V. L., Destler, W. W., Levush, B., Rodgers, J., and Antonsen, Jr., T. M., FEL with applications to magnetic fusion research, *Nucl. Instr. and Meth.* A 375, 160 (1996).

6. Conde, M. E., and Bekefi, G., Experimental study of a 33.3-GHz free-electron-laser amplifier with a reversed axial guide magnetic field, *Phys. Rev. Lett.* 67, 3082 (1991).'

7. Rullier, J. L., Devin, A., Gardelle, J., Labrouche, J., and Le Taillandier, P., Strong coupling operation of a FEL amplifer with an axial magnetic field, *Nucl. Instr. and Meth.*, A 358, 118 (1995).

8. Pasour, J. (personal communication–H. P. Freund).

9. Arzhannikov, A. V., Agafonov, M. A., Ginzburg, N. S. et al., presented at the 18th FEL Conf., Rome, Italy (1996).

10. Peskov, N. Yu., Bratman, V. L., Ginzburg, G. G., Denisov, N. S., Kol'chugin, B. D., Samsonov, S. V., and Volkov, A. B., Experimental study of a high-current FEM with a broadband microwave system, *Nucl. Instr. and Meth.* A 375, 377 (1996).

11. Mizuno, T., Ohtsuki, T., Ohshima, T., and Saito, H., Experimental mode analysis of a circular free electron laser, *Nucl. Instr. and Meth. A* 358, 131 (1995).

12. Valentini, M., Caplan, M., van der Geer, C. A. J. et al., Commissioning of the 1 MW, 130-260 GHz fusion-FEM and simulations on beam transport and radiation gain, *Nucl. Instr. and Meth. A* 393, II-55 (1997).

13. Phelps, A. D. R., Cross, A. W., Jaroszgneski, D. A., Whyte, C. G., He, W., Ginzburg, N. S., and Yu., N., paper presented at the 22nd International Conference on Infrared and Millimeter Waves, Wintergreen, VA (1997).

14. Peskov, Yu. N. (personal communication–H. P. Freund).

15. Renz, G. (personal communication–H. P. Freund).

16. van der Slot, P. J. M., and Witteman, W. J., Energy and frequency measurements on the Twente Raman free-electron laser, *Nucl. Instr. and Meth.* A331, 140 (1993).

17. Asakawa, M., Sakamoto, N., Inoue, N., Yamamoto, T., Mima, K., Nakai, S., Chen, J., Fujita, M., Imasaki, K., Yamanaka, C., Agari, T., Asakuma, T., Ohigashi, N., and Tsunawaki, Y., Experimental study of a waveguide free-electron laser using the coherent synchrotron radiation emitted from electron bunches, *Appl. Phys. Lett.* 64, 1601 (1994).

18. Asakawa, M., Sakamoto, N., Inoue, N. et al., Experimental study of a wave guide free-electron laser using the coherent synchrotron radiation emitted from electron bunches, *Appl. Phys. Lett.* 64, 1601 (1994).

19. Ponds, M. L., Feng, Y., Madey, J. M. J., and O'Shea, P. G., Non-destructive diagnosis of relativistic electron beams using a short undulator, *Nucl. Instrum. Meth. A*, 375, 136 (1996).

20. Throop, A. L., Orzechowski, T. J., Anderson, B. R., Chambers, F. W., Clark, J. C., Fawley, W. M., Jong, R. A., Paul, A. C., Prosnitz, D., Scharlemann, E. T., Steve, R. D., Westenskow, G. A., and Yarema, S. M., Experimental characteristics of high-gain free-electron laser amplifier operating at 8-mm and 2-mm wavelengths, presented at AIAA 19th Fluid Dynamics & Lasers Conf., June 8, 1987, American Institute of Aeronautics and Astronautics, Honolulu, Hawaii.

21. Maebara, S. (personal communication–H. P. Freund).

22. Deng, J. J., Ding, B. N., Hu, S. Z. et al., *High Power Particle Beams* (China) 6(4) (1995).

23. Orzechowski, T. J., Anderson, B. R., Fawley, W. M., Prosnitz, D., Scharlemann, E. T., Varema, S. M., Hopkins, D. B., Paul, A. C., Sessler, A. M., and Wurtele, J. S., Microwave radiation from a high-gain free-electron laser amplifier, *Phys. Rev. Lett.* 54, 889 (1985).

24. Orzechowski, T. J., Anderson, B. R., Clark, J. C., Fawley, W. M., Paul, A. C., Prosnitz, D., Scharlemann, E. T., Yarema, S. M., Hopkins, D. B., Sessler, A. M., and Wurtele, J. S., High-efficiency extraction of microwave radiation from a tapered wiggler free-electron laser, *Phys. Rev. Lett.* 57, 2172 (1986).

25. Saito, K., Takeyama, K., Ozaki, T., Kishiro, J., Ebihara, K., and Hiramatsu, S., X-band prebunched FEL amplifier, *Nucl. Instr. and Meth.* A 375, 237 (1996).

26. Takayama, K., Kishiro, J., Hiramatsu, S., and Katoh, H., 1.5 MeV ion channel guided X-band free-electron laser amplifier, *J. Appl. Phys.* 77, 5467 (1995).

27. Kaminsky, A. K., Bogachenkov, V. A., Ginzburg, N. S. et al., presented at the 17th Int. Free Electron Laser Conf., New York, *Nucl. Instr. and Meth.* A 375, (1996), and Kaminsky, A. K., Kaminsky, A. A., Sedykh, S. N. et al., in *Free Electron Lasers*, Dattoli, G., and Renieri, A., Eds. North-Holland, Amsterdam (1996).

28. Einat, M., Jerby, E., and Shahadi, A., Dielectric-loaded free-electron maser in a stripline structure, *Nucl. Instr. and Meth.* A 375, 21 (1996).

29. Al'Shamma'a, A., Stuart, R. A., and Lucas, J., A wiggler magnet for a CW-FEM, *Nucl. Instr. and Meth.* A 375, 424(1996).

30. Doria, A., Gallerano, G. P., Giovenale, E., Kimmitt, M. F., and Messina, G., The ENEA F-CUBE facility: trends in RF driven compact FELs and related diagnostics, *Nucl. Instr. and Meth.* A 375, ABS 11 (1996).

31. Benson, S. V. et al., The Stanford Mark III infrared free electron laser, *Nucl. Instr. and Meth. A* 250, 39 (1986).

32. Bhowmik, A., Curtin, M. S., McMillin, W. A., Benson, S. V., Madey, J. M. J., Richardson, B. A., and Vintro, L., First operation of the Rocketdyne/Stanford free electron laser, *Nucl. Instr. and Meth. A* 272 (1988).

33. Drori, R., Jerby, E., Shahadi, A., Einat, M., and Sheinin, M., Free-electron maser operation at the 1 GHz/1 keV regime, *Nucl. Instr. and Meth.* A 375, 186 (1996).

34. Abramovich, A., Pinhasi, Y., Arbel, M., Gilutin, L. et al., Experimental investigation of mode build-up and mode competition process in a prebunched free-electron maser oscillator, *Nucl. Inst., and Meth. A* 407, 87 (1998).

35. van der Slot, P. J. M., Couperus, J., Witteman, W. J., et al., A Cherenkov free electron laser with high peak power, *Nucl. Instr. and Meth. A* 385, 100 (1995).

36. Wieland, J., Couperus, J., van der Slot, P. J. M., and Witteman, W. J., First lasing of a Cherenkov free-electron laser, *Nucl. Instr. and Meth. A* 429, 17 (1999).

37. Al'Shamma'a, A., Shaw, A., Stuart, R. A., and Lucas, J., BNFL Contract FEL3, Report 101A, Department of Electrical Engineering and Electronics, University of Liverpool (1996).

38. Donohue, J. T., Gardelle, J., Lefevre, T., Marchese, G., Padois, M., and Rullier, J. L., Comparison of beam bunching in amplifier and SASE modes at the CEA-CESTA free-electron laser, *Nucl. Instr. and Meth. A* 429, 202 (1999).

39. Gardelle, J., Lefevre, T., Marchese, G., Padois, M. Rullier, J. L., and Donohue, J. T., Production of high power microwaves for particle acceleration with an FEL bunched electron beam, *Nucl. Instr. and Meth. A* 429, 111 (1999).

40. Urbanus, W. H., Bongers, W. A., van der Geer, C. A. J. et al., First lasing of the Dutch Fusion-FEM: 730 kW, 200 GHz, *Nucl. Instr. and Meth. A* 429, 12 (1999).

41. Taccetti, J. M., Jackson, R. H., Freund, H. P., Pershin, D. E., and Granatstein, V. L., A Ka-band CHI-wiggler free-electron maser: experimental results, *Nucl. Instr. and Meth. A* 429, 116 (1999).

42. Abramovich, A., Arensburg, A., Chairman, D. et al., Anticipated performance parameters of the TAU tandem FEL, *Nucl. Inst., and Meth. A* 393, II-41 (1997).

43. Ginzburg, N. S. (personal communication–H. P. Freund).

Section 4.3
NUCLEAR PUMPED LASERS

4.3.1 Introduction

Nuclear pumped lasers (NPL) are gas lasers excited directly or indirectly by high energy particles or gamma rays resulting from nuclear reactions (fission, fusion, radioisotope). This may occur in either a reactor or a nuclear explosion, thus NPLs can be grouped into two broad categories: reactor pumped lasers and nuclear device pumped lasers. Both types provide direct conversion of nuclear energy into directed optical energy. Nuclear pumped lasers have been demonstrated to operate pulsed or steady-state over a wavelength range extending from the vacuum ultraviolet (0.17 μm) to the infrared (5.6 μm) using a variety of gases and molecules. An example of the excitation pathways for a ^{3}He–Ar nuclear-pumped laser system is shown in Figure 4.3.1.

Tables 4.3.1 and 4.3.2 list performance data for reactor pumped lasers and nuclear device-pumped lasers (published data for the latter are limited because much of the research is classified). The lasing medium (gas or gas mixture) is given in the first column followed by the nuclear reaction giving rise to excitation of the lasing medium and then the lasing wavelength, threshold, and efficiency. Primary references are given in the final column. Table 4.3.1 was adapted from a table provided by G. H. Miley and E. G. Batyrbekov (private communication).

Further Reading

Karelin, A. V., Sinyanskii, A. A., and Yakovlenko, S. I., Nuclear-pumped lasers and physical problems in constructing a reactor-laser, *Quantum Electronics* 27, 375 (1997).

Lipinski, R. J. et al., Survey and comparison of mission for a nuclear-reactor-pumped laser, *Proc. Nuclear Technologies for Space Exploration* (NTSE-92), 564, Jackson Hole, WY (1992).

Magda, E. P., Analysis of experimental and theoretical research of nuclear-pumped lasers at the Institute of Technical Physics, *Laser and Particle Beams* 11, 469 (1993).

McArthur, D. A., Nuclear Pumped Lasers, in *Encyclopedia of Lasers and Optical Technology*, Meyers, R. A., Ed., Academic Press, San Diego (1991).

Miley, G. H., Overview of nuclear-pumped lasers, *Laser and Particle Beams* 11, 575 (1993).

Miley, G. H., DeYoung, R., McArthur, D., and Prelas, M., Fission reactor pumped lasers: history and prospects, in *50 Years With Nuclear Fission*, American Nuclear Society, New York (1989), p. 333.

Mis'kevich, A.I., Visible and near-infrared direct nuclear-pumped lasers, *Laser Physics* 1, 445 (1991).

Schneider, R. T., and Hohl, F., Nuclear-pumped lasers, *Adv. Nucl. Sci. Technol.* 16, 123 (1984).

Shaban, Y. R., and Miley, G. H., Practical, visible wavelength nuclear-pumped laser, *Laser and Particle Beams* 11, 559 (1993).

Thom, K., and Schneider, R. T., Nuclear pumped gas lasers, *AIAA Journal* 10, 400 (1972).

Voinov, A, M., Pulsed Nuclear Reactors in Nuclear-pumped laser research, *Laser and Particles Beams* 11, 635 (1993).

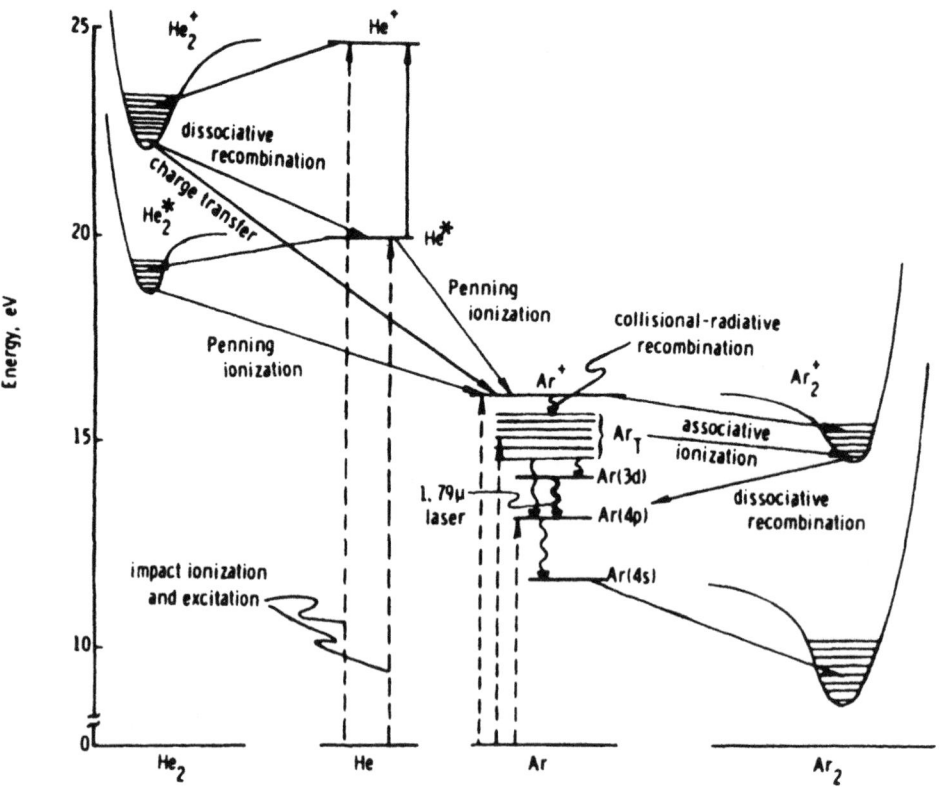

Figure 4.3.1   Energy level diagram and major energy pathways of the ^{3}He–Ar nuclear-pumped laser system [from Wilson, J. W., DeYoung, R. J., and Harries, W. L., *J. Appl. Phys.* 50, 1226 (1979), with permission].

4.3.2 Reactor Pumped Lasers

Table 4.3.1
Reactor Pumped Lasers

Active medium	Pumping reaction	Wavelength (μm)	Threshold (n/cm^2·s)	Eff. (%)	Ref.
Ar–Xe	$B^{10}(n,\alpha)Li^7$	1.73	1×10^{14}	0.5-1	22
Ar–Xe	$U^{235}(n,f)F$	1.732	2×10^{12}	1	24
Ar–Xe	$U^{235}(n,f)F$	2.026	2×10^{12}	1	15
Ar–Xe	$U^{235}(n,f)F$	2.03	—	3.0	23
Ar–Xe	$U^{235}(n,f)F$	2.48	3×10^7	0.01	36
Ar–Xe	$U^{235}(n,f)F$	2.482	2×10^{12}	1	25
Ar–Xe	$U^{235}(n,f)F$	2.6	3.5×10^{16}	—	1
Ar–Xe	$U^{235}(n,f)F$	2.627	2×10^{12}	1	41
CO	$U^{235}(n,f)F$	5.1–5.6	5×10^{16}	≥ 1	33,34
He–Ar	$U^{235}(n,f)F$	1.149	10^{16}	0.1	15
		1.190			
		2.397			
He–Ar–Xe	$U^{235}(n,f)F$	1.73	—	—	23
He–Ar–Xe	$U^{235}(n,f)F$	1.732	5×10^{13}	2.2	25,42
		2.03			
		2.60			
He–Cd	$U^{235}(n,f)F$	0.4416	7×10^{14}	0.52	42
		0.5337			
		0.5378	6×10^{14}	0.36	
		0.8066			
		0.8531	3.2×10^{15}	0.02	
		1.4300			
		1.6500	6×10^{14}		
He–CO, He–CO$_2$, Ar–CO$_2$	$B^{10}(n,\alpha)Li^7$	1.4550	1×10^{15}	10^{-2}	18–20
He–Hg	$B^{10}(n,\alpha)Li^7$	0.615	1×10^{16}	10^{-6}	9,10
He–Kr	$U^{235}(n,f)F$	1.78	6.2×10^{13}	0.6	15,26
		2.52			
		3.07			
He–Ne–Ar	$U^{235}(n,f)F$	0.5852	1×10^{16}	<0.01	
He–Ne–Ar	$U^{235}(n,f)F$	0.5852	4×10^{15}	0.16	1,6,7
He–Ne–Ar	$U^{235}(n,f)F$	0.5852	4×10^{17}	0.1	42
He–Ne–H$_2$	$B^{10}(n,\alpha)Li^7$	0.5852	1.5×10^{13}	0.01	8
He–Ne–Kr	$U^{235}(n,f)F$	0.7245	1.2×10^{16}	0.02	42
He–Xe	$U^{235}(n,f)F$	2.627	6×10^{13}	1.2	15,31

Table 4.3.1—*continued*
Reactor Pumped Lasers

Active medium	Pumping reaction	Wavelength (μm)	Threshold (n/cm^2-s)	Eff. (%)	Ref.
He–Xe	$U^{235}(n,f)F$	3.5080	3×10^{15}	0.01	32
He–Xe–Hg–H$_2$	$U^{235}(n,f)F$	0.5461	10^{15}	0.4	4
He–Zn	$U^{235}(n,f)F$	0.7479	1×10^{17}	0.06	42
^{3}He–Ar	^{3}He(n,p)T	1.27	1.4×10^{16}	0.1	16,17,37
		1.79		1.4	3
^{3}He–Ar	^{3}He(n,p)T	1.27	9×10^{13}	~0.6	35
		1.69			
		1.79			
^{3}He–Ar	^{3}He(n,p)T	2.19, 2.52	5×10^{16}	≤0.1	17,30
^{3}He–Cd	^{3}He(n,p)T	0.4416		0.3	2,3,41
		0.5337	2×10^{14}		
		0.5378			
^{3}He–Cd	^{3}He(n,p)T	1.587	7×10^{15}	<0.1	17
^{3}He–Kr	^{3}He(n,p)T	2.19	5×10^{16}	<0.1	17
^{3}He–Kr	^{3}He(n,p)T	2.5	5×10^{16}		30
^{3}He–Ne	^{3}He(n,p)T	0.6328	$2 \times 10^7 - 1 \times 10^{14}$	0.03	38
^{3}He–Ne–Ar	^{3}He(n,p)T	0.5852	1.4×10^{15}	0.1	5
^{3}He–Xe	^{3}He(n,p)T	2.026		<0.1	17,28–30
		3.508	~4×10^{15}	0.01	
		3.652			
^{3}He–Zn	^{3}He(n,p)T	0.7479	2.1×10^{15}	—	11
Ne–CO, Ne–CO$_2$	$B^{10}(n,\alpha)Li^7$	1.4550	1×10^{15}	10^{-4}	18,21
Ne–Kr(Ar)	$U^{235}(n,f)F$	0.7032	3×10^{14}	0.02	5, 7
		0.7245			
Ne–N$_2$	$B^{10}(n,\alpha)Li^7$	0.8629	1×10^{15}	3×10^{-5}	12–14
		0.9393			

f and F denote light and heavy fission fragments emitted in the reaction of neutrons (n) with ^{235}U; T – triton.

4.3.3 Nuclear Device Pumped Lasers

Table 4.3.2
Nuclear Device Pumped Lasers

Active medium	Pumping reaction	Wavelength (μm)	Reference
Xe_2	gamma rays	0.17	39
Ne-Xe-NF_3	gamma rays	0.353	40
HF	gamma rays	2.7	39
SF_6-H_2	gamma rays	3	40
unknown	x-rays	0.0014	43

4.3.4 References

1. DeYoung, R. J., Multiple-path fission-foil nuclear lasing of Ar-Xe, *Appl. Phys. Lett.* 50, 39, 585 (1981).
2. Dmitirev, A. B., Il'ishenko, V. S., Miskevich, A. I. et al., *Zh. Tekh. Fiz.* 52, 2235 (1982) (in Russian).
3. Miskevich, A. I., Dmi'triev, A. B., Il'ishenko, V. S. et al., *Pis'ma Zh. Tekh. Fiz.* 6, 818 (1980), *Sov. Phys. Tech. Phys.* 27, 260 (1982).
4. Bochkov, A. V., Kryzhanovskii, V. A., Magda, E. P. et al., Quasi-cw lasing on the 7^3S_1-6^3P_2 atomic mercury transition, *Sov. Tech. Phys. Lett.* 18, 241 (1992).
5. Copai-Gora, A. P., Miskevich, A. I., and Salamadia, B. S., *Pis'ma Zh. Tekh. Fiz.* 16, 23 (1990) (in Russian).
6. Hebner G. A., and Hays, G. N., Fission-fragment-excited lasing at 585.3 nm in He/Ne/Ar gas mixtures. *Appl. Phys. Lett.* 57, 2175 (1990).
7. Voinov, A. V., Krivonosov, V. N., Mel'nikov, S. P. et al., Quasicontinuous lasing on the 3p–3s transitions of a xenon atom in mixtures of inert gases excited by uranium fission fragments, *Sov. Phys. Dokl.* 35, 568 (1990).
8. Shaban, Y., and Miley, G. H., A practical visible wavelength nuclear-pumped laser, *Proceeding of Specialist Conference on Physics of Nuclear Induced Plasma and Problems of Nuclear Pumped Lasers*, Obninsk, Russia, Vol. 2, 241 (1993).
9. Akerman, M. A., and Miley, G. H., A helium-mercury direct nuclear pumped laser, *Appl. Phys. Lett.* 30, 409 (1977).
10. Akerman, M.A., Demonstration of the first visible wavelength DNPL, Ph.D. Thesis, Department of Nuclear Engineering, U. of Illinois at Urbana-Champaign (1976).
11. Miskevich, A. I., Copai-Gora, A. P., and Salamadia, B. S., *Pis'ma Zh. Tekh. Fiz.* 16, 62 (1990) (in Russian).
12. DeYoung, R., A direct nuclear pumped neon-nitrogen laser, Ph.D. Thesis, Department of Nuclear Engineering, U. of Illinois at Urbana-Champaign (1976).
13. DeYoung, R., Wells, W. E., Miley, G. H., and Verdeyen, J. T., Direct nuclear pumped Ne-N_2 laser, *Appl. Phys. Lett.* 28, 519 (1976).
14. Cooper, G., Verdeyen, J. T., Wells, W., and Miley, G. H., The pumping mechanism for the neon-nitrogen nuclear-excited laser, *Proceedings, 3rd Conf. Uranium Plasmas and Applications*, Princeton, NJ (June 1976).
15. Voinov, A. M., Dovbych, L. E., Krivonosov, V. N. et al., *Sov. Tech. Phys. Lett.* 5, 171 (1979).
16. DeYoung, R. J., Jalufka, N. W., Hohl, F., and Williams, M. D., Direct nuclear pumped lasers using the volumetric ^{3}He reaction, *Conf. on Partially Ionized and Uranium Plasmas*, 96, Princeton, NJ (1976).

17. DeYoung, R. J., Jalufka, N. W., and Hohl, F., Direct nuclear-pumped lasers using $He^3(n,p)T$ reaction, *AIAA Journal* 16, 991 (1978).

18. Prelas, M. A., Anderson, J. H., Boody, F. P. et al., Nuclear pumping of a neutral carbon laser, *Progress in Astronautics and Aeronautics, Radiation Energy Conversion in Space* 61, 411 (1978).

19. Prelas, M. A., Akerman, M. A., Boody, F. P., and Miley, G. H., A direct nuclear pumped 1.45μ atomic carbon laser in mixtures of He-CO and He-CO_2, *Appl. Phys. Lett.* 31, 428 (1977).

20. Prelas, M. A., Akerman, M. A., Boody, F. P., and Miley, G. H., A direct nuclear pumped 1.45μ atomic carbon laser in mixtures of He-CO and He-CO_2, *4th Workshop on Laser Interaction and Related Plasma Phenomena*, Plenum Press, New York (1976), p. 249.

21. Prelas, M. A., Anderson, J. H., Boody, F. P. et al., A nuclear pumped laser using Ne-CO and Ne-CO_2 mixtures, *30th Annual Electronics Conference* (Oct. 1977).

22. Batyrbekov, E. G., Poletaev, E. D., Suzuki, E., and Miley, G. H., $B^{10}(n,a)Li^7$ pumped Ar-Xe laser *Transactions of 11th International Conference on Laser Interactions and Related Plasma Phenomena* (Oct. 1993), p. 152.

23. Alford, W. J., and Hays, G. H., Measured laser parameters for reactor-pumped He-Ar-Xe and Ar-Xe lasers, *J. Appl. Phys.* 65, 3760 (1990).

24. Voinov, A. M., Dovbych, L. E., Krivonosov, V. N. et al., *Sov. Tech. Phys. Lett.* 27, 819 (1982).

25. Voinov, A. M., Zobnin, V. G., Konak, A. I. et al., *Pis'ma v ZTF* 16, 34 (1990) (in Russian).

26. Voinov, A. M., Dovbych, L. E., Krivonosov, V. N. et al., *Pis'ma v ZTF* 52, 1346 (1982).

27. Jalufka, N. W., DeYoung, R. J., Hohl, F., and Williams, M. D., Nuclear pumped He^3-Ar laser excited by $He^3(n,p)T$ reaction, *Appl. Phys. Lett.* 29, 188 (1976).

28. Mansfield, C. R., Bird, P. F., Davis, J. F. et al., Direct nuclear pumping of a He^3-Xe laser, *Appl. Phys. Lett.* 30, 640 (1977).

29. Jalufka, N. W., Nuclear pumped lasing of He^3-Xe at 2.63 μ, *Appl. Phys. Lett.* 39, 535 (1981).

30. DeYoung, R. J., Jalufka, N. W., and Hohl, F., Nuclear-pumped lasing of He^3-Xe and He^3-Kr, *Appl. Phys. Lett.* 30, 19 (1977).

31. Voinov, A. M., Dovbych, L. E., and Krivonosov, V. N. et al., Low-threshold nuclear-pumped lasers using transitions of the atomic xenon, *Sov. Phys. Dokl.* 24, 189 (1979).

32. Helmick, H. H., Fuller, J. I., and Schneider, R. T., Direct nuclear pumping of helium-xenon laser, *Appl. Phys. Lett.* 26, 327 (1975).

33. McArthur, D. A., and Tollefsrud, P. B., Observation of laser action in CO gas excited only by fission fragments, *Appl. Phys. Lett.* 26, 187 (1975).

34. McArthur, D. A., Schmidt, T. R., Tollefsrud, P. B., and Walker, J. V., Preliminary designs for large (1-MJ) reactor-driven laser systems, *IEEE International Conf. on Plasma Science*, Ann Arbor, MI (May 1975).

35. Voinov, A. V., Krivonosov, V. N., Mel'nikov, S. P., Mochkaev, I. N., and Sinyanskii, A. A., Quasi-cw nuclear-pumped laser utilizing atomic transitions in argon, *Sov. J. Quantum Electron.* 21, 157 (1991).

36. Bochkov, A. V., Kryzhanovskii, V. A., Lyubimov, O., Magda, E. P., and Mukhin, S., Nuclear reactor-pumped laser atomic xenon operated at 2.48 μm, *Laser and Particle Beams* 11, 491 (1993).

37. DeYoung, R. J., and Harries, W. L., Nuclear-pumped ^{3}He-Ar laser modelling, *J. Appl. Phys.* 50, 1226 (1979).

38. Carter, B. D., Rowe, M. J., and Schneider, R. T., Nuclear-pumped CW lasing of the ^{3}He-Ne system, *Appl. Phys. Lett.* 36, 115 (1980).

39. McArthur, D. A., Nuclear pumped lasers, in *Encyclopedia of Lasers and Optical Technology*, Meyers, R. A., Ed., Academic Press, San Diego (1991), p. 385.

40. Bonyushkin, E. K., Varaksin, V. V., Lazhintsev, B. V., Lakhtihov, A. E., Morovov, A. P., Nor-Arevyan, V. A., and Pavlovskii, A. I., On the investigation of thermonuclear fusion at targets of "cannonball" type with high-power pulsed lasers pumped by the nuclear explosing gamma-radiation, *Transactions of Second Specialist International Conference*

on Physics of Nuclear Induced Plasmas and Problems of Nuclear Pumped Lasers, Arzamas, Russia, (1994), p. 167.

41. Dmi'triev, A. B., Il'ishenko, V. S., Miskevich, A. I., and Salamakha, B. S., Observation of laser transitions in parametallic gas mixtures using neutron products from nuclear reactions, *Sov. Phys. Tech. Phys.* 27, 1373 (1982).

42. Magda, E. P., Grebyonkin, K. F., and Kryzhanovsky, V. A., Nuclear pumped lasers at the Institute of Technical Physics, *Transactions, Lasers '90*, 827, San Diego, CA (1991).

43. Avrorin, E. N., Lykov, V. A., Loboda, P. A., and Politov, V. Yu., Review of theoretical works on X-ray laser research performed at RFNC-VNIITF, *Laser and Particle Beams* 15, 3 (1997).

Section 4.4
NATURAL LASERS

Introduction

Naturally occurring maser action is frequently found in clouds of molecular gases in our galaxy where water or other molecules amplify radiation from stars. Whereas natural masers operating at microwave and millimeter wavelengths have been known for several decades and now number in the hundreds (see tabulations by Elitzur and by Moran in the references listed below under Further Reading), shorter wavelength natural lasers are much rarer. Those discovered thus far operate at infrared wavelengths of CO_2 in the mesosphere and thermosphere of Mars and Venus and, more recently, at the submillimeter wavelengths of hydrogen in interstellar and circumstellar sources.

Natural lasers are listed in Table 4.4.1 by the active gas molecule and the object from which the radiation was observed. The lasing transition and wavelength (in micrometers) and the primary reference(s) are given in the subsequent columns.

Further Reading

Clegg, A. W., and Nedoluha, G. E., Eds. *Astrophysical Masers*, Springer-Verlag, Berlin (1993).

Deming, D., Espenak, F., Jennings, D., Kostiuk, T., F. Mumma, M. J., and Zipoy, D., Modeling of the 10-μm natural laser emission from the mesospheres of Mars and Venus, *Icarus* 55, 356 (1983).

Elitzur, M., *Astronomical Masers*, Kluwer, New York (1992).

Moran, J. M., Maser action in nature, in *Handbook of Laser Science and Technology, Suppl. 1: Lasers*, CRC Press, Boca Raton, FL (1991), p. 579.

Moran, J. M., Maser action in nature, in *Handbook of Laser Science and Technology, Vol. I: Lasers and Masers*, CRC Press, Boca Raton, FL (1982), p. 483.

Mumma, M. J., Natural lasers and masers in the solar system, in *Astrophysical Masers*, Clegg, A. W., and Nedoluha, G. E., Eds., Springer-Verlag, Berlin (1993), p. 455.

Table 4.4.1
Natural Lasers

Gas – Object	Transition	Wavelength (μm)	Reference
CO_2 – Venus	v_3-2v_2	10.4[a]	1, 2
CO_2 – Mars	v_3-2v_2	10.4[a]	1–3
H – Cygnus MWC349	H10α	52.5	4
H – Cygnus MWC349	H12α	88.8	4
H – Cygnus MWC349	H15α	169.4	4
H – Cygnus MWC349	H21α	453	5
H – Cygnus MWC349	H26α	850	9
H_2O – stars[b]	$v_2=1$, $1_{10}-1_{01}$	456	6
H_2O – [c]	$6_{42}-5_{51}$	636.6	7
H_2O – [c]	$6_{43}-5_{50}$	682.7	7
H_2O – VY CMa	$17_{4,13}-16_{7,10}$	844.9	8
H_2O – S269	7_3-6_3 E_1 [d]	885.4	8
	$7_{-3}-6_{-3}$ E_2	885.5	
H_2O – S252	7_3-6_3	885.5	8
H_2O – VY CMa	$5_{2,3}-6_{1,6}$	891.6	8
para-H_2O – SFR[e]	$5_{15}-4_{22}$	922.0	10
ortho-H_2O – SFR[e]	$10_{29}-9_{36}$	933.4	11
H_2O[f]	$3_{13}-2_{20}$	1635	12
CH_3OH – S231	$7_{-4}-6_{-4}$	885.4	8
CH_3OH – S252	7_3-6_3	885.6	8

(a) Predicted amplifications are very small (≤ 1.1)
(b) VY CMa, R Leo, R Crt, RT Vir, W Hya, RX Boo, S CrB, U Her, VX Sgr, NML Cyg
(c) Interstellar and circumstellar water
(d) Ambiquity of the line identification
(e) SFR – star forming regions
(f) Various molecular clouds, star forming regions, and evolving stars

References

1. Johnson, M. A., Betz, A. L., McLaren, R. A., Sutton, E. C., and Townes, C. H., Nonthermal 10 micron CO_2 emission lines in the atmospheres of Mars and Venus, *Astrophys. J.* 208, L145 (1976).

2. Deming, D., Espenak, F., Jennings, D., Kostiuk, T. F., Mumma, M. J., and Zipoy, D., Observations of the 10-μm natural emission from the mesospheres of Mars and Venus, *Icarus* 55, 347 (1983).

3. Mumma, M. J., Buhl, D., Chin, G., Deming, D., Espenak, F., and Kostiuk, T., Discovery of natural gain amplification in the 10-micrometer carbon dioxide laser bands on Mars: a natural laser, *Science* 212, 45 (1981).

4. Strelnitski, V., Haas, M. R., Smith, H. A., Erickson, E. F., Colgan, S. W. J., and Hollenbach, D. J., Far-infrared hydrogen lasers in the peculiar star MWC 349A, *Science* 272, 1459 (1996).

5. Thum, C., Matthews, H. E., Harris, A. I., Tacconi, L. J., Schuster, K. F., and Martin-Pintado, Detection of H21α maser emission at 662 GHz in MWC349, *Astron. Astrophys.* 288, L25 (1994).

6. Menten, K. M., and Young, K., Discovery of strong vibrationally excited water masers at 658 GHz toward evolved stars, *Astrophys. J.* 450, L70 (1995).

7. Melnick, G. J., Submillimeter water masers, in *Astrophysical Masers*, Clegg, A. W., and Nedoluha, G. E., Eds., Springer-Verlag, Berlin (1993), p. 41.

8. Feldman, P. A., Matthews, H. E., Amano, T., Scappini, F., and Lees, R. M., Observations of new submillimeter maser lines of water and methanol, in *Astrophysical Masers*, Clegg, A. W., and Nedoluha, G. E., Eds., Springer-Verlag, Berlin (1993), p. 65.

9. Thum, C., Matthews, H. E., Martin-Pintado, J., Serabyn, E., Planesas, P., and Bachiller, R. A., Submillimeter recombination line maser in MWC 349, *Astron. Astrophys.* 283, 582 (1994).

10. Menten, K. M., Melnick, G. J., Phillips, T. G., and Neufeld, D. A., A new submillimeter water maser transition at 325 GHz, *Astrophy. J.* 363, L27 (1990).

11. Menten, K. M., Melnick, G. J., and Phillips, T. G., Submillimeter water masers, *Astrophy. J.* 350, L41 (1990).

12. Cernicharo, J., Thum, C., Hein, H., John, D., Garcia, P., and Mattioco, F., Detection of 183 GHz water vapor maser emisssion from interstellar and circumstellar sources, *Astron. Astrophys.* 231, L15 (1990).

Section 4.5
INVERSIONLESS LASERS

It is possible to extract energy from a medium even if there are more atoms in the lower level than the upper level. Two schemes, Λ and V, have been used to obtain lasing without inversion (LWI). In the first scheme, the optical fields have a common upper level and LWI is achieved via a coherence between the two lower levels. In the V scheme, the fields have a common lower level and LWI is achieved via a coherence between two upper levels. Gain and cw laser oscillation have been observed in metal vapors using both Λ and V schemes at wavelengths in the visible-near visible region.

Table 4.5.1 presents experiments involving either lasing or amplification in inversionless systems arranged in order of increasing wavelength. The lasers have been operated in a continuous wave mode; the amplifier experiments have measured transient gain or gain on a probe laser pulse. For each experiment, the lasing atomic species, LWI scheme, operative transition and wavelength, and pump transition and/or wavelength are given. Primary references to the experiments are listed in the final column.

Further Reading

Alam, S., *Lasers without Inversion and Electromagnetically Induced Transparency*, SPIE Press, Bellingham, WA (1999).

Harris, S. E., Lasers without inversion: interference of lifetime-broadened resonances, *Phys. Rev. Lett.* 62, 1033 (1989).

Khurgin. J. B., and Rosencher, E., Practical aspects of lasing without inversion in various media, *IEEE J. Quantum Electron.* 32, 1882 (1996), and Practical aspects of optically coupled inversionless lasers, *J. Opt. Soc. Am. B* 14, 1249 (1997).

Kocharovskaya, O., Amplification and lasing without inversion, *Phys. Rep.* 219, 175 (1992).

Kocharovskaya. O., and Mandel, P., Basic models of lasing without inversion: general form of amplification condition and problem of self-consistency, *Quantum Optics* 6, 217 (1994).

Kocharovskaya. O., Kolesov, R., and Rostovtsev, Y., Lasing without inversion: a new path to gamma-ray laser, *Laser Physics* 9, 745 (1999).

Scully, M. O., Zhu, S.-Y., and Gavrielides, A., Degenerate quantum-beat laser: lasing without inversion and inversion without lasing, *Phys. Rev. Lett.* 62, 2813 (1989).

See, also, Papers on Atomic Coherence and Interference, Crested Butte Workshop 1993, in *Quantum Optics* 6 (1994).

Table 4.5.1
Inversionless Lasers and Amplifiers

Medium	Scheme	Laser/Amplifier Transition	Wavelength (nm)	Pump	Reference
Lasers:					
^{23}Na atomic beam	Λ	D_1:$3^2S_{1/2}(F=1) - 3^2P_{1/2}(F=1)$	589.76	D_2:$^2S_{1/2}(F=2 - F=1)$	1
^{87}Rb vapor, magnetic field	V	D_1:$5^2S_{1/2} - 5^2P_{1/2}$	794	D_2:$(F=1 - F=2)$	2
Amplifiers:					
^{112}Cd vapor, magnetic field	Λ	$^3S_1 - {}^3P_1$	479	$^1S_0 - {}^3P_1$ (326 nm), $^3P_1 - {}^3S_1$ (479 nm)	3
Sm vapor, magnetic field	Λ		571	531 nm	4, 9
^{23}Na vapor, magnetic field	Λ	D_1:$3^2S_{1/2}(F=2) - 3^2P_{1/2}(F=2)$	589	D_1:$^2S_{1/2}(F=1) - {}^2P_{1/2}(F=2)$	7
^{23}Na vapor, magnetic field	V	D_1:$3^2S_{1/2} - 3^2P_{1/2}$	589.0/589.6	589.6/589 nm	5, 6
K vapor, He buffer gas	*	D_1:$4S(F=2) - 4P_{1/2}(J=1/2)$	770	D_2:$4S - 4P(J=3/2)$	8
Ba, atomic beam	**	$^1D_2 - {}^1P_1$	821	554 nm	10
Cs vapor, magnetic field	V	D_1:$(F=3,4) - (F''=3,4)$	894	D_2:$(F=3,4) - (F'=4,5)$, 852 nm	11

*   Four-level Raman driven amplification

**   Three-level cascade system

References

1. Padmabandu, G. G., Welch, G. R., Shubin, I. N., Fry, E. S., Nikonov, D. E., Lukin, M. D., and Scully, M. O., Laser oscillation without population inversion in a sodium atomic bean, *Phys. Rev. Lett.* 76, 2053 (1996).

2. Zibrov, A. S., Lukin, M. D., Nikonov, D. E., Hollberg, L., Scully, M. O., Velichansky, V. L., and Robinson, H. G., Experimental demonstration of laser oscillation without population inversion via quantum interference in Rb, *Phys. Rev. Lett.* 75, 1499 (1995).

3. van der Veer, W. E., van Diest, R. J. J., Dönszelmann, A. and van Linden van den Heuvell, H. B., Experimental demonstration of light amplification without population inversion, *Phys. Rev. Lett.* 70, 3243 (1993).

4. Nottelmann, A., Peters, C., and Lange, W., Inversionless amplification of picosecond pulses due to Zeeman coherence, *Phys. Rev. Lett.* 70, 1783 (1993).

5. Gao, J.-Y., Zhang, H.-Z., Cui, H.-F., Guo, X.-Z., Jiang, Y. Wang, Q.-W., Jin, G.-X., and Li, J.-S., Inversionless light amplification in sodium, *Opt. Commun.* 110, 590 (1995).

6. Gao, J., Guo, C., Guo, X., Jin, G., Wang, P., Zhao, J., Zhang, H., Jiang, Y., Wang, D., and Jiang, D., Observation of light amplification without population inversion in sodium, *Opt. Commun.* 93, 323 (1995).

7. Fry, E. S., Li, X., Nikonov, D. et al., Atomic coherence effects within the sodium D_1 line: lasing without inversion via population trapping, *Phys. Rev. Lett.* 70, 3235 (1993).

8. Kleinfeld, J. A., and Streater, A. D., Observation of gain due to coherence effects in a potassium-helium mixture, *Phys. Rev. A*, 49, R4301 (1994).

9. Lange, W., Nottelman. A., and Peters, C., Observation of inversionless amplification in Sm vapour and related experiments, *Quantum Optics* 6, 273 (1994).

10. Sellin, P. B., Wilson, G. A., Meduri, K. K., and Mossberg, T. W., Observation of inversionless gain and field-assisted lasing in a nearly ideal three-level cascade-type atomic system, *Phys. Rev. A* 54, 2402 (1996).

11. Fort, C., Cataliotti, F. S., Hänsch, T. W., Inguscio, M., and Prevedelli, M., Gain without inversion on the cesium D_1 line, *Optics Commun.* 139, 31 (1997).

Section 4.6

AMPLIFICATION OF CORE-VALENCE LUMINESCENCE

In materials where the energy difference between the uppermost core level and the conduction band is less than twice the fundamental band gap ($E_c < 2E_g$), a core hole may decay by radiative recombination with a valence band electron because Auger processes are energetically forbidden. This results in core-valence luminescence (the emission has also been called Auger-free luminescence and cross luminescence). Core-valence luminescence has been fobserved from several wide-band-gap alkali and alkaline earth halide crystals of the heavier alkali (K, Rb, Cs) and alkaline earth (Sr, Ba) ions and involves an np(anion) – n'p(cation) transition.

Excitation and emission transitions associated with core-valence luminescence are shown schematically in the energy band diagram Figure 4.6.1. Excitation resulting in a core level hole creates an inverted electron population with respect to the filled valence band and hence laser action is possible. Table 4.6.1 presents results of experiments involving amplification of core-valence luminescence following core level excitation. Experiments were performed at room temperature using radiation from an undulator beamline at an electron storage ring. Evidence of lasing has been based on the enhancement of the peak intensity, the sharpening of the emission spectrum, and the shortening of the luminescence decay time.

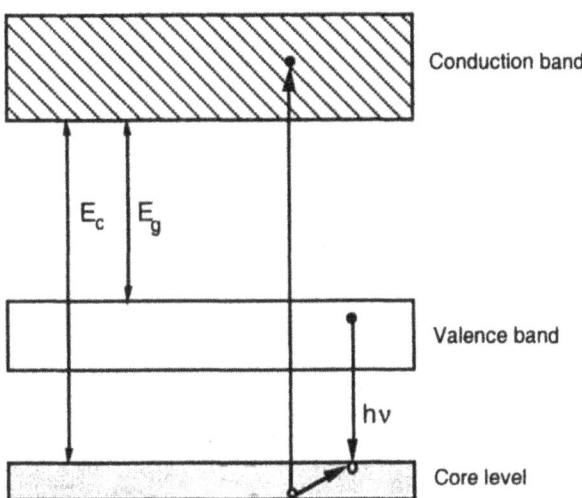

Figure 4.6.1 Energy band diagram for materials where $E_c < 2 \; Eg$ showing core level excitation and core-valence luminescence.

Further Reading

Rodnyi, P. A., Core-valence band transitions in wide-gap ionic crystals, *Sov. Phys. Solid State* 34, 1053 (1992).

van Eijk, C. W. E., Cross-luminescence, *J. Lumins.* 60&61, 936 (1994).

Table 4.6.1
Amplified Core-Valence Luminescence

Crystal	Excitation transition	Excitation energy (eV)	Emission transition	Emission wavelength (nm)	Ref.
BaF_2	Ba(5p) – CB	36.0	F(2p)–Ba(5p)	219	1
$Rb_{0.82}Cs_{0.18}Cl$	Cl(3p) – CB	15	Cl(3p)–Cs(5p)	275	2
$Rb_{0.82}Cs_{0.18}Cl$	Rb(4p) – CB	36	Cl(3p)–Cs(5p)	275	2

CB – conduction band

References

1. Itoh, M., and Itoh, H., Stimulated ultraviolet emission from BaF_2 under core-level excitation with undulator radiation, *Phys. Rev. B* 46, 15509 (1992).
2. Mikhailik, V. B., Itoh, M., Asaka, A., Bokumoto, Y., Murakami, J., and Kamada, M., Amplification of impurity-associated Auger-free luminescence in mixed rubidium-caesium chloride crystals under core-level excitation with undulator radiation, *Optics Commun.* 171, 71 (1999).

Appendices

APPENDIX I
LASER SAFETY

Laser products are grouped into one of four general hazard classes as shown in Table 1. Safety measures become more restrictive with increasing hazard class.

Table 1
Laser Hazard Classification

Class 1	Very low power, eye-safe, output beam below maximum permissible exposure; safe to view.
Class 2	Low-power visible lasers only; safe for brief (<0.25 s) viewing; intentional extended viewing is considered hazardous.
Class 3A	Medium power; safe for brief (<0.25 s) viewing; direct beam should not be viewed with magnifying optics such as a microscope, binoculars, or telescope.
Class 3B	Medium power; not safe for brief viewing of direct beam or specular reflections; control measures should eliminate this possibility.
Class 4	High power; not safe for momentary viewing; potential for skin, fire, or diffuse reflection hazard.

Potential hazards to the eye and skin from laser radiation depend on the wavelength, exposure duration, and viewing conditions. Safety procedures are published in the American National Standards Institute ANZI Z136 series of standards and are followed by the U.S. Occupational Safety and Health Administration (OSHA). Information about safety standards and procedures can be obtained from:

American National Standards Institute (ANSI)
11 West 42nd Street
New York, NY 10036
Safe Use of Lasers (Z136.1)
Safe Use of Lasers in Health Care Facilities (Z136.3)

Laser Institute of America
12424 Research Parkway, Suite 125
Orlando, FL 32826
www.laserinstitute.org/publications
LIA Laser Safety Guide (corresponds with the latest revisions of Z136.1)
LIA Guide to Medical Laser Safety

IEC
3 rue de Varembe
P.O. Box 131, CH-1211
Geneva 20, Switzerland
www.iec.ch

FDA, U.S. Department of Health and Human Services
CDRH, Office of Compliance
2098 Gaither Road
Rockville, MD 20850

Exposure Limits

Occupational exposure limits (ELs) are referred to as maximum permissible exposures (MPEs) by ANSI, as ELs by the World Health Organization and the International Radiation Protection Association, and as threshold limit values (TLVs) by the American Conference of Governmental Industrial Hygienists. All organizations recommend virtually identical exposure limits. Exposure limits are provided for exposure durations from one nanosecond to 8 hours and for wavelengths between 180 nm in the ultraviolet (at the extreme end of the vacuum ultraviolet band in the UV-C) to 1 mm in the extreme infrared IR-C band (at the edge of the microwave spectrum). Selected exposure limits for representative common lasers are given in Tables 2 to 4. Biological effects and limits for different spectral regions are summarized in Table 5.

Table 2
Selected Occupational Exposure Limits for Representative Lasers

Laser	Wavelengths	Maximum permissible exposure (MPE)
Gas Lasers		
Argon fluoride	193 nm	3.0 mJ/cm^2 over 8 h
Xenon chloride	308 nm	40 mJ/cm^2 over 8 h
Nitrogen	337 nm	1.0 J/cm^2 over 8 h
Argon ion	488, 514.5 nm	3.2 mW/cm^2 for 0.1 s; 2.5 mW/cm^2 for 0.25 s 1.8 mW/cm^2 for 1.0 s; 1.0 mW/cm^2 for 10 s varies for t > 10 s
Copper vapor	510, 578 nm	3.2 mW/cm^2 for 0.1 s; 2.5 mW/cm^2 for 0.25 s 1.8 mW/cm^2 for 1.0 s; 1.0 mW/cm^2 for 10 s varies for t > 10 s
Krypton ion	568, 647 nm	3.2 mW/cm^2 for 0.1 s; 2.5 mW/cm^2 for 0.25 s 1.8 mW/cm^2 for 1.0 s; 1.0 mW/cm^2 for 10 s varies for t > 10 s
Gold vapor	628 nm	3.2 mW/cm^2 for 0.1 s; 2.5 mW/cm^2 for 0.25 s 1.8 mW/cm^2 for 1.0 s; 1.0 mW/cm^2 for 10 s varies for t > 10 s
Helium-neon	632.8 nm	3.2 mW/cm^2 for 0.1 s; 2.5 mW/cm^2 for 0.25 s 1.8 mW/cm^2 for 1.0 s; 1.0 mW/cm^2 for 10 s varies for t > 10 s
Carbon monoxide	~5 mm	100 mW/cm^2 for 10 s to 8 h, limited area 10 mW/cm^2 for >10 s for whole-body exposure
Carbon dioxide	10.6 mm	100 mW/cm^2 for 10 s to 8 h, limited area 10 mW/cm^2 for >10 s for whole-body exposure
Solid State Lasers		
Doubled Nd:YAG	532 nm	0.5 mJ/cm^2 for 1 ns to 18 ms
Ruby	694.3 nm	0.5 mJ/cm^2 for 1 ns to 18 ms

Table 2—*continued*
Selected Occupational Exposure Limits for Representative Lasers

Laser	Wavelengths	Maximum permissible exposure (MPE)
Diodes	850 nm (typical)	1.0 mJ/cm^2 for 1 ns-18 ms 2 mW/cm^2 for 10 s
Nd:YAG	1.064 mm	5.0 mJ/cm^2 for 1 ns to 50 ms No MPE for t < 1 ns 5 mW/cm^2 for 10 s
Nd:YAG	1.334 mm	5.0 mJ/cm^2 for 1 ns to 50 ms 40 mW/cm^2 for 10 s
Nd:YAG	1.44 mm	0.1 J/cm^2 for 1 ns to 1 ms
Holmium (pulsed)	2.1 mm	0.1 J/cm^2 for 1 ns to 1 ms
Holmium/thulium	1.9–2.2 mm	100 mW/cm^2 for 10 s to 8 h, limited area 10 mW/cm^2 for >10 s for most of body

Not all standards/guidelines have MPEs below 200 nm.
Sources: ANSI standard Z-136.1-1993, ACGIH TLVs (1993), and IRPA (1988).
Note: To convert MPEs in mW/cm^2 to mJ/cm^2, multiply by exposure time t in seconds; for example, the HeNe or argon MPE at 0.1 s is 0.32 mJ/cm^2.

Table 3
Intrabeam Exposure Limits for Common cw Lasers

Laser type	Wavelength(s) (nm)	Exposure limit value — Eye	Exposure limit value — Skin
Helium-Cadmium	325	1 J/cm^2 for 10 – 1000 s; 1 mW/cm^2 for t > 1000 s	1 J/cm^2 for 10 – 10^3 s; 1 mW/cm^2 for t > 1000 s
Nitrogen	337	1 J/cm^2 for 10 – 1000 s; 1 mW/cm^2 for t > 1000 s	1 J/cm^2 for 10 – 10^3 s; 1 mW/cm^2 for t > 1000 s
Helium-Cadmium	441.6	2.5 mW/cm^2 for 0.25 s; 10 mJ/cm^2 for 10-10^4 s; W ccm^{-2} for > 10^4 s	0.2 W/cm^2
Argon	488 514	2.5 mW/cm^2 for 0.25 s; 10 mJ/cm^2 for 10-10^4 s; W ccm^{-2} for > 10^4 s	0.2 W/cm^2

Table 3—*continued*
Intrabeam Exposure Limits for Common cw Lasers

Laser type	Wavelength(s) (nm)	Exposure limit value Eye	Skin
Frequency doubled Nd:YAG	532	2.5 mW/cm^2 for 0.25 s; 10 mJ/cm^2 for 10-10^4 s; W ccm^{-2} for > 10^4 s	0.2 W/cm^2
Helium-Neon	632.8	2.5 mW/cm^2 for 0.25 s; 10 mJ/cm^2 for 10-10^4 s; W/cm^2 for > 10^4 s	0.2 W/cm^2
Krypton	647.1	2.5 mW/cm^2 for 0.25 s; 10 mJ/cm^2 for 10-10^4 s; W ccm^{-2} for > 10^4 s	0.2 W/cm^2
Gallium arsenide (300 K)	905	0.8 mW/cm^2 for t > 100s	0.5 W/cm^2
Neodymium:YAG	1064	1.6 mW/cm^2 for t >100s	1.0 W/cm^2
Carbon dioxide (and other lasers)	10.6 μm 1.4–1000 μm	0.1 W/cm^2 for t > 10 s	0.1 W/cm^2 for t > 10 s

Table 4
Intrabeam Exposure Limits for Common Pulsed Lasers

Laser type	Wavelength(s) (nm)	Pulse duration	Exposure limit value Eye	Skin
Normal-pulsed ruby laser	694.3	~ 1 ms	10^{-5} J/cm^2	0.2 J/cm^2
Q-switched ruby	694.3	5-100 ns	5x10^{-7} J/cm^2	0.02 J/cm^2
Rhodamine 6G dye laser	~500-700	0.5-20 ms	5 x 10^{-7} J/cm^2	0.03-0.01 J/cm^2
Normal pulsed neodymium laser	1064	~1 ms	5x 10^{-8} J/cm^2	1.0 J/cm^2
Q-switched neodymium laser	1064	5-100 ns	5 x 10^{-6} J/cm^2	0.1 J/cm^2

Table 5
Summary of Biological Effects and Limits for Different Spectral Regions

Wavelength region	Representative lasers	Adverse effects	Symptoms	Representative maximal levels of exposure for a collimated laser beam
190 nm to 315 nm	ArF, KrF, XeF (repetitively pulsed)	(A) Photokeratoconjunctivitis, normally reversible within 48 hr. (B) Erythema (sunburn), normally reversible within 1 week. (C) Skin cancer. (D) Cataract of lens (300-315 nm).	(A) Uncontrolled blinking, painful sensation of "sand" in eyes, inflammation of the cornea and conjunctiva, delayed in onset 6-12 hr. (B) Reddening of skin (after delay). (C,D) Delayed in onset by many years.	3 mJ/cm^2 for 200 nm to 302 nm for single or multiple pulse exposure; 0.56 $t^{1/4}$ J/cm^2 for 1 ns to 10 s; 1 J/cm^2 for single pulse or pulse train at 315 nm.
315 nm to 380-400 nm	He-Cd, Argon	Same as above, except that threshold doses are orders of magnitude greater.	Same as above. Some accelerated tanning of skin. Very low risk skin cancer.	0.56 $t^{1/4}$ J/cm^2 for 1 ns to 10 s; 1 J/cm^2 for 10 sec to 1000 sec ; 1 mW/cm^2 for 1000-s exposure or greater.
400 nm to 550 nm	He-Cd, Argon	Retinal photochemical injury ("Blue-Light Injury"). Permanent scotoma in severe cases; some recovery after two to six weeks.	Delayed onset of 24-48 hr. Retinal lesion visible by ophthalmoscope; blind-spot (scoloma) also develops in 24-48 hr.	10 mJ/cm^2 for t = 10 s to 10,000 s; 1 μW/cm^2 for t = 10^4 s (2.8 hr).

Note: The expressions for exposure limits with time t raised to a fractional exponent of 1/4 or 3/4 require the use of seconds for t.

Ultrashort Pulses: No exposure limits for subnanosecond pulses have been set; however, a conservative approach recommended by the IEC would require the exposure in terms of irradiance (W/cm^2) not to exceed the irradiance limit for a l-ns pulse. There is some limited experimental evidence to support that approach.

Extended Sources: The limits given above for wavelengths between 400 and 1400 nm apply strictly only low intrabeam (direct) exposure for the eye. More lenient limits not shown apply to laser radiation exposure of the eye from extended source, scattered radiation.

Table 5—*continued*

Summary of Biological Effects and Limits for Different Spectral Regions

Wavelength region	Representative lasers	Adverse effects	Symptoms	Representative maximal levels of exposure for a collimated laser beam
400 nm to 1400 nm	Argon, Ruby, He-Ne, Nd:YAG, GaAs	(A) Retinal thermal injury; permanent scotoma.	Rapid appearance of ophthalmoscopically visible retinal lesion with associated blind spot that does not improve.	5 $\mu J/cm^2$ at 1064 nm or 0.5 $\mu J/cm^2$ for t = 1 ns to 18 μs at 400 nm to 700 nm; 1.8 $t^{3/4}$ C_A mJ/cm^2 for 100 μs to 10 s; 320 C_A 1 $\mu W/cm^2$ for 700 − 1400 nm at t = 1000 s.
		(B) Thermal burns of the skin.	Rapid appearance of red spot (erythema) at site of exposure. Generally healing (sometimes with scar) within 4-6 weeks.	2 C_A x 10^{-2} J/cm^2 for 1 ns to 100 ns; 1.1 C_A $t^{1/4}$; J/cm^2 for 100 ns to 10 s; 200 C_A mW/cm^2 for t = 10 s (small spots).
1400 nm to 1 mm	HF, DF, Er, Ho, CO, CO_2	Corneal thermal burns, skin burns.	White flaky patch from pulsed laser near threshold; erythema from CW exposure to deep burn at much higher levels. Corneal vacuolization.	10 mJ/cm^2 for 1 ns to 100 ns (exception: @ 1.54 μm, 1 J/cm^2, 1-100 ns); 0.56 $t^{1/4}$ J/cm^2 for 100 ns to 10 s; 100 mW/cm^2 for t > 10 s.

Protective Eyewear

Laser eye protection is necessary for all class 3B and 4 laser systems where exposure above the MPE is possible. ANSI Z136.1 provides general guidance for the use of eyewear. Figure 1 indicates the required optical density of eyewear at the laser wavelength for non-Q-switched (curve 1) and Q-switched (curve 3) maximum output energies measured in mJ, and for continuous wave intrabeam viewing (curve 2) and diffuse viewing (curve 4) for maximum output powers measured in W.

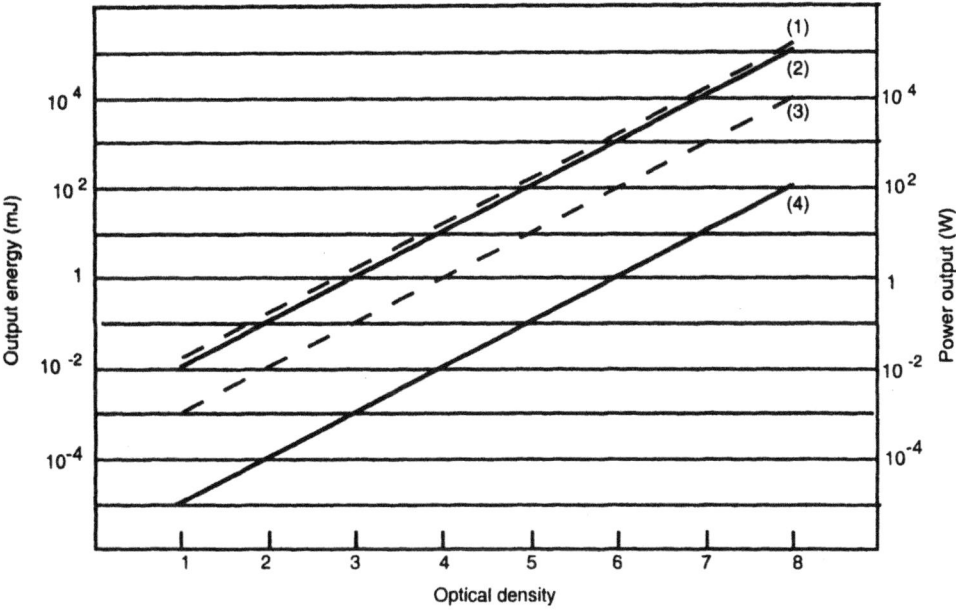

Figure 1. Optical density of eyewear required for pulsed and cw laser operation.

Further Reading

American Conference of Governmental Industrial Hygienists (ACGIH) (1996), *TLV's, Threshold Limit Values and Biological Exposure Indices for 1996*, American Conference of Governmental Industrial Hygienists, Cincinnati, OH.

Cain, C. P. et al., Retinal damage and laser-induced breakdown produced by ultrashort-pulse lasers, *Graefe's Arch. Clin. Exp. Ophthalmol.* 234: suppl. 1, S28 (1996); Cain, C.P. et al., Visible retinal lesions from ultrashort laser pulses in the primate eye, *Invest. Ophthalmol.* 36, 879 (1995).

Duchene A. S., Lakey, J. R. A., and Repacholi, M. H., (Eds.), *IRPA Guidelines on Protection Against Non-Ionizing Radiation*, New York, MacMillan (1991).

International Commission on Non-Ionizing Radiation Protection (ICNIRP), 1996, Guidelines on Limits for Laser Radiation of Wavelengths between 180 nm and 1000 mm, *Health Physics* 71, 804 (1996).

Lund, D. J. et al., Ocular Hazards of Tunable Continuous Wave Near-Infrared Laser Sources, *SPIE Proceedings Volume 2674*, SPIE Press, Bellingham, WA (1996), p. 53.

Sliney, D. H., Ed., *Selected Papers on Laser Safety*, Volume MS117, SPIE Press, Bellingham, WA (1995).

Sliney, D. H., Laser safety, *Lasers in Surgery and Medicine* 16, 215 (1995).

Sliney, D. H., Laser effects on vision and ocular exposure limits, *Appl. Occup. Environ. Hyg.* 11, 313 (1996).

Sliney, D. H., Laser and LED eye hazards: safety standards, *Optics and Photonics News* (September 1997), p. 31.

Sliney, D. H., and LeBodo, H., Laser eye protectors, *J. Laser Appl.* 2, 9 (1990).

Sliney, D. H., and Trokel, S. L., *Medical Lasers and Their Safe Use*, New York, Springer Verlag (1992).

Wolbarsht, M. L., Ed., *Laser Applications in Medicine and Biology* (Plenum Publishing Corp., New York (1989).

APPENDIX II

Acronyms, Abbreviations, Initialisms, and Common Names for Types of Lasers, Laser Materials, Laser Structures and Operating Configurations, and Systems Involving Lasers

A-FPSA	— anti-resonant Fabry-Perot saturable absorber
ADM	— add-drop multiplexer
AGRIN	— axial gradient index (lens)
ANSI	— American National Standards Institute (laser safety standards)
AO	— acousto-optical
AON	— all optical network
AOS	— acoustic-optic switch
AOTF	— acousto-optic tunable filter
APM	— additive pulse mode-locked (laser)
AR	— antireflection (coating)
ASE	— amplified spontaneous emission
ASRL	— anti-Stokes Raman laser
ATM	— asynchronous transfer mode
ATTA	— atom trap trace analysis (laser manipulation)
AWI	— amplification without inversion
BB	— broadband
BC	— buried crescent (laser)
BEL	— bound-electron laser
BEL	— lanthanum beryllate (laser crystal)
BER	— bit error rate
BFA	— Brillouin fiber amplifier
BFAP	— barium fluoroapatitie (laser crystal)
BGSL	— broken-gap superlattice (laser)
BH	— buried heterostructure (laser)
BIG	— bundle-integrated-guide (laser)
BOG	— buried optical guide (laser)
BRL	— Brillouin ring laser
BRS	— buried ridge structure (laser)
BS	— beamsplitter
BTRS	— buried twin-ridge structure (laser)
BVSIS	— buried V-groove substrate inner stripe
BWDM	— bandpass wavelength-division multiplexing
BYF	— barium yttrium fluoride (laser crystal)
C^3	— cleaved coupled cavity (laser)
CARS	— anti-Stokes Raman spectroscopy

Acronyms, Abbreviations, Initialisms, and Common Names for Types of Lasers, Laser Materials, Laser Structures and Operating Configurations, and Systems Involving Lasers

CBH	—	circular buried heterostructure (laser)
CC-CDH	—	current confined constricted double heterostructure (laser)
CCGSE	—	concentric-circle grating surface emitting (laser)
CCL	—	color center laser
CDH	—	constricted double heterostructure (laser)
CEL	—	correlated emission laser
CLR	—	coherence laser radar
CLSM	—	confocal laser scanning microscopy
CLUE	—	compensated laser ultrasonic evaluation
CMBH	—	capped-mesa buried-heterostructure (laser)
CMC	—	coupled microcavity (laser structure)
COD	—	catastrophic optical damage
COIL	—	chemical oxygen-iodine laser
CPA	—	chirped pulse amplification
CPM	—	colliding-pulse mode-locked (laser)
CPM	—	corrugation-pitch-modulated (laser)
CPM	—	critical phase matching
CPS	—	coherent photon seeding
CRL	—	compact, rugged laser
CRLAS	—	cavity ringdown laser absorption spectroscopy
CSM	—	cascaded second-order-nonlinearity modelocking
CSN	—	cascade second-order nonlinear-optical (process)
CSP	—	channeled-substrate planar (laser)
CTH:YAG	—	Cr,Tm,Ho-doped yttrium aluminum garnet (laser crystal)
CTLM	—	computer tomography laser mammography
CVL	—	copper vapor laser
CW	—	continuous wave
D-SAM	—	dispersive saturable absorber mirror
D^3	—	directly doubled diode (laser system)
DASAR	—	darkness amplification by stimulated absorption of radiation
DBR	—	distributed Bragg reflection (laser)
DC	—	direct current (continuous output)
DCFL	—	double clad fiber laser
DCPBH	—	double channel planar buried heterostructure (laser)
DD-WGM	—	dye-doped whispering galley mode (laser)
DDS	—	deep-diffuse stripe (laser)
DFB	—	distributed feedback (laser)

Acronyms, Abbreviations, Initialisms, and Common Names
for Types of Lasers, Laser Materials, Laser Structures and Operating
Configurations, and Systems Involving Lasers

DFC	—	distributed forward coupled (laser)
DFG	—	difference-frequency generation
DFWM	—	degenerate four-wave mixing
DH	—	double heterostructure (laser)
DIAL	—	differential absorption lidar
DLM	—	diffraction limited (beam)
DOE	—	diffractive optical element
DOVE	—	doubly vibrationally enhanced (four-wave mixing)
DPL	—	diode-pumped laser
DPSSL	—	diode-pumped solid-state laser
DR	—	distributed reflector (laser)
DRSFM	—	double resonant sum-frequency mixing
DS	—	diffused stripe (laser)
DSF	—	dispersive shifter fiber
DSM	—	dynamic-single-mode (laser)
DUV	—	deep ultraviolet
DWDM	—	dense wavelength-division multiplexing
DWELL	—	dots in a well (laser)
EAM	—	electro-absorptive-modulated (laser)
EARS	—	excited absorption reflection switch
ECDL	—	external cavity diode laser
EDFA	—	erbium-doped fiber amplifier
EEDL	—	edge-emitting diode laser
EEL	—	edge-emitting laser
EFA	—	erbium fiber amplifier
EFISH	—	electric-field-induced second harmonic (generation)
EML	—	electroabsorption modulated laser
ErF	—	erbium fiber (laser)
ESA	—	excited state absorption
ESA-FEL	—	electrostatic-accelerator based free-electron laser
ESF	—	effective saturation fluence
ETDL	—	energy transfer dye laser
ETU	—	energy transfer upconversion (laser)
EUV	—	extreme ultraviolet
excimer	—	excited dimer (laser)
F-(center)	—	Farbe, German word for color (laser)
FAP	—	calcium fluoroapatite (laser crystal)

Acronyms, Abbreviations, Initialisms, and Common Names for Types of Lasers, Laser Materials, Laser Structures and Operating Configurations, and Systems Involving Lasers

FBG	—	fiber Bragg grating
FCSEL	—	folded cavity surface emitting laser
FD	—	frequency doubled (laser)
FEDL	—	flashlamp-excited dye laser
FEL	—	free-electron laser
FEM	—	free-electron maser
FFH	—	fifth harmonic
FG-ECL	—	fiber-grating external-cavity laser
FH	—	fourth harmonic
FHG	—	fourth harmonic generation
FIR	—	far infrared
FITL	—	fiber in the loop
FL	—	flashlamp
FM	—	frequency modulation
FOG	—	fiber optic gyroscope
FP	—	Fabry-Perot (resonator)
FR	—	Faraday rotator
FRL	—	fiber Raman laser
FROG	—	frequency-resolved optical grating
FRSL	—	fiber Raman soliton laser
FSFL	—	frequency-shifted feedback laser
FTIR	—	frustrated total internal reflection (shutter)
FTTC	—	fiber to the curb
FTTH	—	fiber to the home
FWHM	—	full-width half-maximum
FWM	—	four-wave mixing
FWPO	—	four-wave parametric oscillator
GC	—	grating coupled (laser)
GDD	—	group delay dispersion
GDL	—	gas dynamic laser
GGG	—	gadolinium gallium garnet (laser crystal)
GI-POF	—	graded-index plastic optical fiber
GRENOUILLE	—	grating-eliminated, no-nonsense observation of ultrafast incident laser light e-field
GRIN	—	gradient (graded) index (lens)
GRINSCH	—	graded-index separate confinement heterostructure (laser)
GRM	—	gradient reflectivity mirror

Acronyms, Abbreviations, Initialisms, and Common Names
for Types of Lasers, Laser Materials, Laser Structures and Operating
Configurations, and Systems Involving Lasers

GSA	—	ground state absorption
GSE	—	grating surface emitting (laser)
GSGG	—	gadolinium scandium gallium garnet (laser crystal)
GVD	—	group velocity dispersion
GVL	—	gold vapor laser
HAP	—	high average power (laser)
HAP	—	high-average-power (laser glass)
HDL	—	homodyne laser
HEL	—	high enery laser
HENE	—	He-Ne (gas laser)
HHG	—	high-harmonic generation
HMPGI	—	hybrid multiple-prism grazing-incidence (cavity)
HOE	—	holographic optical element
HP	—	high power
HPP	—	high peak power (laser)
HR	—	high reflectivity (coating)
HRO	—	heteroepitaxial ridge overgrown (laser)
ICF	—	inertial confinement fusion (laser)
ICL	—	interband cascade laser
IFE	—	inertial fusion energy
IFRA	—	ignition feedback regenerative amplifier
IM	—	intensity modulation
IO	—	integrated optical (device)
IR	—	infrared
IRED	—	infrared emitting diode
ISDN	—	integrated services digital network
IVB	—	intervalence band (laser)
KLM	—	Kerr-lens mode locking
LADAR	—	laser detection and ranging
LAN	—	local area network
LASD	—	laser airborne depth sounder
LASER	—	light amplification by stimulated emission of radiation
LASIK	—	laser in-situ keratomilensis
LBU	—	laser-based ultrasound
LC	—	liquid crystal
LCLV	—	liquid crystal light valve
LCM	—	laser capture microdissection

Acronyms, Abbreviations, Initialisms, and Common Names for Types of Lasers, Laser Materials, Laser Structures and Operating Configurations, and Systems Involving Lasers

LD	—	laser diode
LDMS	—	laser desorption mass spectrometry
LDV	—	laser doppler velocimeter
LEB	—	laser enhanced bonding
LEC	—	long external cavity (laser)
LED	—	light emitting diode
LiCAF	—	lithium calcium aluminum fluoride (laser crystal)
LIDAR	—	light detection and ranging (light radar system)
LIF	—	laser-induced fluorescence
LiSAF	—	lithium scandium aluminum fluoride (laser crystal)
LISER	—	laser-induced stimulated emission of radiation
LIVAR	—	laser-illuminated viewing and ranging (system)
LMA	—	lanthanum magnesium hexaluminate (laser crystal)
LOC	—	large optical cavity (laser)
LOFT	—	laser optical feedback tomography
LOLA	—	laser-offset locking accessary
LSA	—	laser subassembly
LTK	—	laser thermal keratoplasty
LWI	—	lasing without inversion
M^2	—	far field beam divergence angle
M^2	—	times-diffraction-limit number
MASELA	—	matrix-addressable surface-emitting-laser array
MASER	—	microwave amplification by stimulated emission of radiation
MCA	—	monolithic crystal array
MCS	—	modified channeled substrate (laser)
MDC	—	mirror dispersion controlled (oscillator)
MDR	—	morphology-dependent resonance (laser)
MFA	—	monolithically integrated flared amplifier
MIFROG	—	multipulse interferometric frequency-resolved optical grating
MIH	—	monolithically integrated hybrid (laser)
MIR	—	mid-infrared
ML	—	multilayer (coating)
ML	—	multiline
MLQD	—	multiple layer quantum dot
MMI	—	multi-mode interference
MO	—	master oscillator
MOC	—	mirco-optical component

Acronyms, Abbreviations, Initialisms, and Common Names
for Types of Lasers, Laser Materials, Laser Structures and Operating
Configurations, and Systems Involving Lasers

MOPA	—	master oscillator-power amplifier
MOPFA	—	master oscillator-fiber power amplifier
MOPO	—	master oscillator-power oscillator
MPL	—	microgun-pumped laser
MQB	—	multi-quantum barrier (laser)
MQW	—	multiple quantum well (laser)
MSA	—	multisegment amplifier
MSL	—	microgun-pumped semiconductor laser
MUG (Russian)	—	molekulyarnyy usilitel i generator (molecular amplifier and oscillator)
MVL	—	metal vapor laser
MWIR	—	mid-wave infrared
NA	—	numerical aperture (optical fiber)
NALM	—	nonlinear amplifying loop mirror
NAM	—	nonabsorbing mirror
NCPM	—	non-critical phase matching
NDFA	—	neodymium fiber amplifier
NDFWM	—	nondegenerate four-wave mixing
NDPL	—	nuclear device pumped laser
NDTWM	—	nondegenerate two-wave mixing
NIR	—	near infrared
NOLM	—	nonlinear optical loop mirror
NOPC	—	nonlinear optical phase conjugation
NPL	—	nuclear-pumped laser
NPRO	—	nonplanar ring oscillator
NYAB	—	neodymiun yttrium aluminum borate (laser crystal)
OCL	—	optical confinement layer
OCT	—	optical coherence tomography
ODT	—	optical diffraction tomography
OEIG	—	optoelectronic integrated circuit
OFDM	—	optical frequency division multiplexing
OKG	—	optical Kerr gate
OLA	—	optical power-limiting amplifier
OLED	—	organic light-emitting diode
OPA	—	optical parametric amplifier
OPAL	—	optical parametric amplifier laser
OPC	—	optical phase conjugation
OPG	—	optical parametric generator

Acronyms, Abbreviations, Initialisms, and Common Names for Types of Lasers, Laser Materials, Laser Structures and Operating Configurations, and Systems Involving Lasers

OPL	—	optical power limiter
OPO	—	optical parametric oscillator
OPOL	—	optical parametric oscillator laser
OPPO	—	optical parametric power oscillator
OPS	—	optically pumped semiconductor (laser)
OSL	—	organic semiconductor laser
OTDR	—	optical time-domain reflectrometry
OVCSEL	—	organic vertical cavity surface-emitting laser
PASER	—	pion amplification by stimulated emission of radiation
PBC	—	p-type buried crescent (laser)
PBC	—	planar buried crescent (laser)
PBH	—	planar buried heterostructure (laser)
PC	—	phase conjugation
PC	—	Pockels cell
PCM	—	phase conjugation mirror
PCPA	—	parametric chirped pulse amplification
PCR	—	phase conjugate resonator
PCR	—	polarization-coupled resonator
PCSEL	—	planar cavity surface-emitting laser
PDFFA	—	praseodymium-doped fluoride fiber amplifier
PDL	—	photonic delay line
PDT	—	photodynamic therapy
PELA	—	peripheral excimer-laser angioplasty
PESO	—	piezoelectrically induced strain-optic (modulator)
PIC	—	photonic intregated circuit
PIFI	—	polarization insensitive fiber isolator
PIL	—	photolylic iodine laser
PINSCH	—	periodic-index separate-confinement heterostructure (laser)
PIV	—	particle image velocimetry
PLC	—	planar lightwave circuit
PM	—	phase modulator
PM	—	polarization maintaining (optical fiber)
PMD	—	polarization mode dispersion
POF	—	polymer optical fiber
POFA	—	polymer optical fiber amplifier
PON	—	passive optical network (system)
POWA	—	planar optical waveguide amplifier

Acronyms, Abbreviations, Initialisms, and Common Names for Types of Lasers, Laser Materials, Laser Structures and Operating Configurations, and Systems Involving Lasers

PRF	—	pulse repetition frequency
PRFA	—	praseodymium fiber amplifier
PRK	—	photorefractive keratectomy
PTK	—	phototherapeutic keratectomy (laser process)
Q	—	quality factor (laser cavity)
QB	—	quantum box (laser)
QC	—	quantum cascade (laser)
QCL	—	quantum cascade laser
QCW	—	quasi-continuous wave
QD	—	quantum dot (laser)
QF	—	quantum film (laser)
QMAD	—	quiet multiaxial mode doubling
QPL	—	quantum parallel laser
QPM	—	quasi-phase match
QW	—	quantum well (laser)
QW	—	quantum wire (laser)
QWH	—	quantum well heterostructure (laser)
QWL	—	quantum well laser
QWR	—	quantum well ridge (laser)
QWS	—	quantum well structure (laser)
R-SEED	—	resistor-biased self-electro-optic effect device
RCLED	—	resonant cavity light-emitting diode
REDF	—	rare-earth doped fiber
RFA	—	Raman fiber amplifier
RGH	—	rare gas halide (laser)
RLG	—	ring laser gyroscope
RMI	—	resonant holographic interferometry
ROW	—	resonant optical waveguide (diode laser array)
RPL	—	reactor pumped laser
RS	—	Raman-shifted (laser)
RSA	—	reverse saturable absorption (molecules, material)
RW	—	ridge waveguide (laser structure)
S-SEED	—	symmetric self-electro-optic effect device
SA	—	saturable absorber
SAR	—	synthetic aperture radar (optical processing)
SASE	—	self-amplified spontaneous emission
SBL	—	space based laser

Acronyms, Abbreviations, Initialisms, and Common Names
for Types of Lasers, Laser Materials, Laser Structures and Operating
Configurations, and Systems Involving Lasers

SBR	—	saturable Bragg reflector
SBR	—	selective buried ridge (laser structure)
SBS	—	stimulated Brillouin scattering (amplifier)
SCH	—	separate carrier heterostructure (laser)
SCH	—	separate confinement heterostructure (laser)
SCL	—	semiconductor laser
SCLA	—	semiconductor laser amplifier
SD-FROG	—	self-diffraction frequency-resolved optical grating
SDL	—	semiconductor diode laser
SDWL	—	simultaneous dual wavelength lasing
SE	—	stimulated emission
SEA	—	surface-emitting (laser) array
SEED	—	self-electrooptic effect device
SEL	—	surface-emitting laser
SELD	—	surface-emitting laser diode
SELDA	—	surface-emitting laser diode array
SFD	—	self-frequency doubled (laser)
SFG	—	sum frequency generation
SFM	—	sum frequency mixing
SGDBR	—	sample-grating distributed-Bragg-reflector
SH	—	second harmonic
SHG	—	second harmonic generation
SLA	—	semiconductor laser amplifier
SLM	—	single longitudinal mode
SLM	—	spatial light modulator
SLQD	—	single layer quantum dot
SLS	—	strained layer superlattice (laser structure)
SM	—	submillimeter (laser)
SOA	—	semiconductor optical amplifier
SOI	—	silicon-on-insulator (waveguide structure)
SONET	—	synchronous optical network
SP	—	self phase (modulation)
SPAM	—	self phase and amplitude modulation
SP-APM	—	stretched-pulse additive pulse mode-locked (laser)
SPL	—	short pulse laser
SPM	—	self-phase modulation
SPML	—	synchronously-pumped mode-locked (laser)

Acronyms, Abbreviations, Initialisms, and Common Names for Types of Lasers, Laser Materials, Laser Structures and Operating Configurations, and Systems Involving Lasers

SPPO	—	synchronously pumped parametric oscillator
SQW	—	single quantum well (laser)
SRFEL	—	storage ring free-electron laser
SRS	—	stimulated Raman scattering (amplifier)
SSD	—	smoothing by spectral dispersive
SSL	—	serpentine superlattice (laser structure)
SSM	—	single spatial mode
STF	—	soliton transmission fiber
SXPL	—	soft x-ray projection lithography
SXR	—	soft x-ray
S–FAP	—	strontium fluoroapatite (laser crystal)
S–VAP	—	strontium vanadium fluoroapatite (laser crystal)
T-ray	—	tetrahertz ray (0.1–3 THz)
T2QWL	—	type II quantum well laser
T^3	—	table-top-terawatt (laser)
TADPOLE	—	temporal analysis by dispersing a pair of light E-fields
TAL	—	thin active layer (laser)
TAPS	—	tapered stripe (laser structure)
TCL	—	taper coupled laser
TCL	—	twin-channel laser
TCSM	—	twin-channel substrate mesa (laser)
TDL	—	tunable diode laser
TDM	—	time-division multiplexing
TEA	—	tranverse excited atmospheric pressure (laser)
TEM_{00}	—	lowest order or fundamental tranverse mode
TFR	—	tightly folded resonator (laser)
TG-FROG	—	transient-grating frequency-resolved optical grating
TH	—	third harmonic
THG	—	third harmonic generation
TIE	—	tunable interdigital electrode (DBR laser)
TII	—	time-integrated interferometry
TILDAS	—	tunable infrared laser differential absorption spectroscopy
TJS	—	transverse junction stripe (laser structure)
TMI	—	two-mode interference
TOAD	—	terahertz optical asymmetric demultiplexer
TOPAS	—	traveling-wave optical parametric amplifier of superfluorescence
TOPG	—	traveling-wave optical parametric generator

Acronyms, Abbreviations, Initialisms, and Common Names for Types of Lasers, Laser Materials, Laser Structures and Operating Configurations, and Systems Involving Lasers

TPFM	—	two-photon fluorescence microscopy
TREEFROG	—	twin-recovery of electric-field components by use of FROG
TRI	—	time-resolved interferometry
TRS	—	twin-ridge structure (laser)
TSL	—	tilted superlattice
TTT	—	transpupillary thermotherapy
TTW-SLA	—	traveling-wave semiconductor laser amplifier
T W	—	traveling wave
TWM	—	three-wave mixing
TWM	—	two-wave mixing
UCL	—	upconversion laser
UFL	—	upconversion fiber laser
UV	—	ultraviolet
UVA	—	ultraviolet (320—400 nm; CIE 315—400 nm)
UVB	—	ultraviolet (280—320 nm; CIE 280—315 nm)
UVC	—	ultraviolet (200—280 nm; CIE 100—280 nm)
VCSEL	—	vertical cavity surface-emitting laser
VECOD	—	vertical-coupled quantum dot (laser)
VECSEL	—	vertical-external-cavity surface-emitting laser
VLAP	—	visual laser prostatectomy
VLD	—	visible laser diode
VPBS	—	variable polarization beamsplitter
VSAL	—	very-small-aperture laser
VSIS	—	V-channeled substrate inner stripe (laser)
VSQW	—	variable-strained quantum well (laser)
VUV	—	vacuum ultraviolet
WAN	—	wide area network
WDM	—	wavelength-division multiplexing
WDM-OS	—	wavelength-division multiplexing optical switch
WG	—	waveguide (optical)
WGM	—	whispering galley mode (laser)
WGR	—	waveguide grating router
WSOL	—	wavelength selective optical logic
XDL	—	times diffraction limit
XFEL	—	x-ray free-electron laser
XM	—	cross modulation
XPM	—	cross phase modulation

Acronyms, Abbreviations, Initialisms, and Common Names for Types of Lasers, Laser Materials, Laser Structures and Operating Configurations, and Systems Involving Lasers

XRL	—	x-ray laser
XUV	—	extreme ultraviolet
YAB	—	yttrium aluminum borate (laser crystal)
YAG	—	yttrium aluminum garnet (laser crystal)
YALO	—	yttrium aluminate (laser crystal)
YAP	—	yttrium aluminum perovskite (laser crystal))
YBF	—	yttrium barium fluoride (laser crystal)
YDFA	—	ytterbium-doped fiber amplifier
YGG	—	yttrium gallium garnet (laser crystal)
YLF	—	lithium yttrium fluoride (laser crystal)
YSAG	—	yttrium scandium aluminum (laser crystal)
YSGG	—	yttrium scandium gallium (laser crystal)
YSO	—	yttrium silicon oxide, yttrium orthosilicate (laser crystal)
YVO	—	yttrium vanadate (laser crystal)
Z-laser	—	zone laser (self-focusing)
ZBLAN	—	Zr-Ba-La-Al-Na fluorozirconate (laser glass)

*For a more complete list of abbreviations, acronyms, initialisms, and minerological or common names for laser materials, see Appendix 2 of the *Handbook of Laser Science and Technology: Supplement 2 – Optical Materials*, CRC Press, Boca Raton, FL (1995).

APPENDIX III

Electron Configurations of Neutral Atoms in the Ground State

Atomic number	Element	K n=1 s	L n=2 s p	M n=3 s p d	N n=4 s p d f	O n=5 s p d f	P n=6 s p d	Q n=7 s
1	H	1						
2	He	2						
3	Li	2	1					
4	Be	2	2					
5	B	2	2 1					
6	C	2	2 2					
7	N	2	2 3					
8	O	2	2 4					
9	F	2	2 5					
10	Ne	2	2 6					
11	Na	2	2 6	1				
12	Mg	2	2 6	2				
13	Al	2	2 6	2 1				
14	Si	2	2 6	2 2				
15	P	2	2 6	2 3				
16	S	2	2 6	2 4				
17	Cl	2	2 6	2 5				
18	Ar	2	2 6	2 6				
19	K	2	2 6	2 6	1			
20	Ca	2	2 6	2 6	2			
21	Sc	2	2 6	2 6 1	2			
22	Ti	2	2 6	2 6 2	2			
23	V	2	2 6	2 6 3	2			
24	Cr	2	2 6	2 6 4	1			
25	Mn	2	2 6	2 6 5	2			
26	Fe	2	2 6	2 6 6	2			
27	Co	2	2 6	2 6 7	2			
28	Ni	2	2 6	2 6 8	2			
29	Cu	2	2 6	2 6 10	1			
30	Zn	2	2 6	2 6 10	2			
31	Ga	2	2 6	2 6 10	2 1			
32	Ge	2	2 6	2 6 10	2 2			
33	As	2	2 6	2 6 10	2 3			
34	Se	2	2 6	2 6 10	2 4			
35	Br	2	2 6	2 6 10	2 5			
36	Kr	2	2 6	2 6 10	2 6			
37	Rb	2	2 6	2 6 10	2 6	1		
38	Sr	2	2 6	2 6 10	2 6	2		
39	Y	2	2 6	2 6 10	2 6 1	2		
40	Zr	2	2 6	2 6 10	2 6 2	2		
41	Nb	2	2 6	2 6 10	2 6 4	1		
42	Mo	2	2 6	2 6 10	2 6 5	1		
43	Tc	2	2 6	2 6 10	2 6 5	2		

Electron Configurations of Neutral Atoms in the Ground State

Atomic number	Element	K n=1 s	L n=2 s p	M n=3 s p d	N n=4 s p d f	O n=5 s p d f	P n=6 s p d	Q n=7 s
44	Ru	2	2 6	2 6 10	2 6 7	1		
45	Rh	2	2 6	2 6 10	2 6 8	1		
46	Pd	2	2 6	2 6 10	2 6 10	1		
47	Ag	2	2 6	2 6 10	2 6 10	1		
48	Cd	2	2 6	2 6 10	2 6 10	2		
49	In	2	2 6	2 6 10	2 6 10	2 1		
50	Sn	2	2 6	2 6 10	2 6 10	2 2		
51	Sb	2	2 6	2 6 10	2 6 10	2 3		
52	Te	2	2 6	2 6 10	2 6 10	2 4		
53	I	2	2 6	2 6 10	2 6 10	2 5		
54	Xe	2	2 6	2 6 10	2 6 10	2 6		
55	Cs	2	2 6	2 6 10	2 6 10	2 6	1	
56	Ba	2	2 6	2 6 10	2 6 10	2 6	2	
57	La	2	2 6	2 6 10	2 6 10	2 6 1	2	
58	Ce	2	2 6	2 6 10	2 6 10 1	2 6 1	2	
59	Pr	2	2 6	2 6 10	2 6 10 3	2 6	2	
60	Nd	2	2 6	2 6 10	2 6 10 4	2 6	2	
61	Pm	2	2 6	2 6 10	2 6 10 5	2 6	2	
62	Sm	2	2 6	2 6 10	2 6 10 6	2 6	2	
63	Eu	2	2 6	2 6 10	2 6 10 7	2 6	2	
64	Gd	2	2 6	2 6 10	2 6 10 7	2 6 1	2	
65	Tb	2	2 6	2 6 10	2 6 10 9	2 6	2	
66	Dy	2	2 6	2 6 10	2 6 10 10	2 6	2	
67	Ho	2	2 6	2 6 10	2 6 10 11	2 6	2	
68	Er	2	2 6	2 6 10	2 6 10 12	2 6	2	
69	Tm	2	2 6	2 6 10	2 6 10 13	2 6	2	
70	Yb	2	2 6	2 6 10	2 6 10 14	2 6	2	
71	Lu	2	2 6	2 6 10	2 6 10 14	2 6 1	2	
72	Hf	2	2 6	2 6 10	2 6 10 14	2 6 2	2	
73	Ta	2	2 6	2 6 10	2 6 10 14	2 6 3	2	
74	W	2	2 6	2 6 10	2 6 10 14	2 6 4	2	
75	Re	2	2 6	2 6 10	2 6 10 14	2 6 5	2	
76	Os	2	2 6	2 6 10	2 6 10 14	2 6 6	2	
77	Ir	2	2 6	2 6 10	2 6 10 14	2 6 7	2	
78	Pt	2	2 6	2 6 10	2 6 10 14	2 6 9	1	
79	Au	2	2 6	2 6 10	2 6 10 14	2 6 10	1	
80	Hg	2	2 6	2 6 10	2 6 10 14	2 6 10	2	
81	Tl	2	2 6	2 6 10	2 6 10 14	2 6 10	2 1	
82	Pb	2	2 6	2 6 10	2 6 10 14	2 6 10	2 2	
83	Bi	2	2 6	2 6 10	2 6 10 14	2 6 10	2 3	
84	Po	2	2 6	2 6 10	2 6 10 14	2 6 10	2 4	
85	At	2	2 6	2 6 10	2 6 10 14	2 6 10	2 5	
86	Rn	2	2 6	2 6 10	2 6 10 14	2 6 10	2 6	
87	Fr	2	2 6	2 6 10	2 6 10 14	2 6 10	2 6	1
88	Ra		2 6	2 6 10	2 6 10 14	2 6 10	2 6	2

Electron Configurations of Neutral Atoms in the Ground State

Atomic number	Element	K n = 1 s	L n = 2 s p	M n = 3 s p d	N n = 4 s p d f	O n = 5 s p d f	P n = 6 s p d	Q n = 7 s
89	Ac	2	2 6	2 6 10	2 6 10 14	2 6 10	2 6 1	2
90	Th	2	2 6	2 6 10	2 6 10 14	2 6 10	2 6 2	2
91	Pa	2	2 6	2 6 10	2 6 10 14	2 6 10 2	2 6 1	2
92	U	2	2 6	2 6 10	2 6 10 14	2 6 10 3	2 6 1	2
93	Np	2	2 6	2 6 10	2 6 10 14	2 6 10 4	2 6 1	2
94	Pu	2	2 6	2 6 10	2 6 10 14	2 6 10 6	2 6	2
95	Am	2	2 6	2 6 10	2 6 10 14	2 6 10 7	2 6	2
96	Cm	2	2 6	2 6 10	2 6 10 14	2 6 10 7	2 6 1	2
97	Bk	2	2 6	2 6 10	2 6 10 14	2 6 10 9	2 6	2
98	Cf	2	2 6	2 6 10	2 6 10 14	2 6 10 10	2 6	2
99	Es	2	2 6	2 6 10	2 6 10 14	2 6 10 11	2 6	2
100	Fm	2	2 6	2 6 10	2 6 10 14	2 6 10 12	2 6	2
101	Md	2	2 6	2 6 10	2 6 10 14	2 6 10 13	2 6	2
102	No	2	2 6	2 6 10	2 6 10 14	2 6 10 14	2 6	2
103	Lr	2	2 6	2 6 10	2 6 10 14	2 6 10 14	2 6 1	2
104	Rf	2	2 6	2 6 10	2 6 10 14	2 6 10 14	2 6 2	2

Reference: *CRC Handbook of Chemistry and Physics*, Lide, D., Ed., CRC Press, Boca Raton, FL. (1999).

APPENDIX IV

Fundamental Physical Constants

Quantity	Symbol	Value
speed of light in vacuum	c	299 792 458 m/s
permeability of vacuum, $4\pi \times 10^{-7}$	μ_0	1.256 637 061 4$\times 10^{-6}$ N/A^2
permittivity of vacuum, $1/\mu_0 c^2$	ε_0	8.854 187 817$\times 10^{-12}$ F/m
Planck constant	h	6.626 075 5$\times 10^{-34}$ J s
elementary charge	e	1.602 177 33$\times 10^{-19}$ C
magnetic flux quantum, $h/2e$	Φ_0	2.067 834 61$\times 10^{-15}$ Wb
electron mass	m_e	9.109 389 7$\times 10^{-31}$ kg
proton mass	m_p	1.672 623 1$\times 10^{-27}$ kg
fine structure constant, $\mu_0 c e^2/2h$	α	7.297 353 08$\times 10^{-3}$
inverse fine-structure constant	$1/\alpha$	137.035 989 5
Rydberg constant, $m_e c \alpha^2/2h$	R_y, R_∞	10 973 731.534 m^{-1}
Bohr radius, $\alpha/4\pi R_\infty$	a_0	0.529 177 249$\times 10^{-10}$ m
Hartree energy, $e^2/4\pi\varepsilon_0 a_0 = 2R_\infty hc$	E_h	4.359 748 2$\times 10^{-18}$ J
in eV, E_h/e		27.211 396 1 eV
Compton wavelength, $h/m_e c$	λ_C	2.426 310 58$\times 10^{-12}$ m
classical electron radius, $\alpha^2 a_0$	r_e	2.817 940 92$\times 10^{-15}$ m
Bohr magneton, $eh/4\pi m_e$	μ_B	9.274 015 4$\times 10^{-24}$ J/T
nuclear magneton, $eh/4\pi m_p$	μ_N	5.050 786 6$\times 10^{-27}$ J/T
electron magnetic moment	μ_e	9.284 770 1$\times 10^{-24}$ J/T
magnetic moment anomaly, $\mu_e/\mu_B - 1$	a_e	1.159 653 193$\times 10^{-3}$
electron g factor, $2(1 + a_e)$	g_μ	2.002 319 304 386
proton gyromagnetic ratio	γ_p	2.675 221 28$\times 10^8$ s^{-1}T^{-1}
Avogadro constant	N_A	6.022 136 7$\times 10^{23}$ mol^{-1}
Boltzmann constant, R/N_A	k	1.380 658$\times 10^{-23}$ J/K
Faraday constant, $N_A e$	F	96 485.309 C/mol
molar gas constant	R	8.314 510 J/mol K
Stefan-Boltzmann constant	s	5.670 51$\times 10^{-8}$ W/m^2 K^4

References:

Cohen, E. R., and Taylor, B. N., The 1986 adjustment of the fundamental physical constants, *Rev. Mod. Phys.* 59, 1121 (1987).

Taylor, B. N., and Cohen, E. R., Recommended values of the fundamental physical constants: a status report, *J. Res. Natl. Inst. Stand. Technol.* 95, 497 (1990).

See, also, NIST Web site: physics.nist.gov/constants.

Index

INDEX

N

O

W

X

Y

Z

CONVERSION FACTORS

Energy E (joule)	Frequency ν (hertz) $E = h\nu$	Wavelength λ (nanometer) $E = hc/\lambda$	Wavenumber σ (centimeter)$^{-1}$ $E = hc\sigma$	Electron volt eV (volt) $E = 10^8\, eV/c$	Temperature T (kelvin) $E = kT$
E	$E = h\nu$	$E = hc/\lambda$	$E = hc\sigma$	$E = 10^8\, eV/c$	$E = kT$
1 J	1.5092×10^{33}	1.9864×10^{-16}	5.0341×10^{22}	6.2415×10^{18}	7.2429×10^{22}
6.6261×10^{-34}	1 Hz	2.9979×10^{17}	3.3356×10^{-11}	4.1357×10^{-15}	4.7992×10^{-11}
1.9864×10^{-16}	$2{,}9979 \times 10^{17}$	1 nm	10^7	1239.8	1.4388×10^7
1.9864×10^{-23}	2.9979×10^{10}	10^7	1 cm^{-1}	1.2398×10^{-4}	1.4388
1.6022×10^{-19}	2.4180×10^{14}	1239.8	8065.5	1 eV	11,604
1.3807×10^{-23}	2.0837×10^{10}	1.4388×10^{y}	0.69504	8.6174×10^{-5}	1 K

NOTE: The unit at the top of a column applies to all values below; all values on the same line of the table are equal.

Other conversions:

1 cm^{-1} = 1 kayser = 10^4 angstrom (Å) = 1.331×10^{-13} atomic mass unit = 2.426×10^{-10} electron mass unit

1 cm^{-1} ≈ 1.44 K ≈ 30 GHz ≈ 1.24×10^{-4} eV ≈ 2×10^{-16} erg ≈ 10^4 tesla (g = 2)

1 hartree = 27.2116 eV = 4.35975×10^{-18} joule

1 bohr = 0.52918 angstrom (Å)